W9-BTC-818

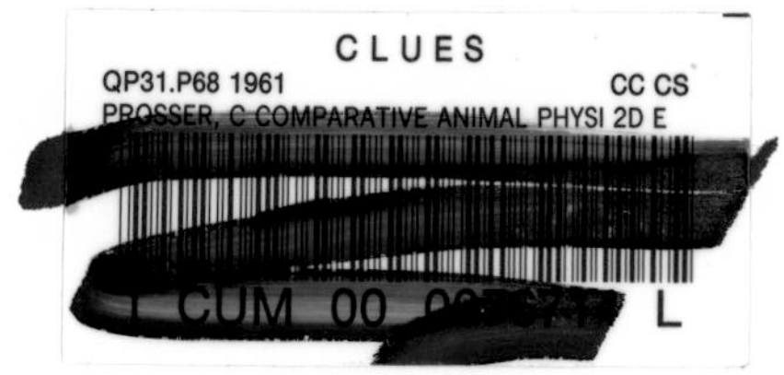
CLUES
QP31.P68 1961 CC CS
PROSSER, C COMPARATIVE ANIMAL PHYSI 2D E

WITHDRAWN

80-302

QP 31 P68 1961
Prosser, Clifford Ladd, 1907-.
Comparative animal physiology

JUL 2000
DATE
JUN 2004
JUL 09
JUL X X 2015

DUE DATE
05/17/05

© THE BAKER & TAYLOR CO.

CONTRIBUTORS

Frank A. Brown, Northwestern University
John D. Anderson, University of Illinois
Eric A. Barnard, State University of New York at Buffalo
James W. Campbell, Rice University
Lawrence I. Gilbert, Northwestern University
Timothy H. Goldsmith, Yale University
Peter W. Hochachka, University of British Columbia
Ronald R. Novales, Northwestern University
C. Ladd Prosser, University of Illinois

COMPARATIVE ANIMAL PHYSIOLOGY

Edited by

C. Ladd Prosser
Professor of Physiology and Zoology
Department of Physiology and Biophysics
University of Illinois at Urbana–Champaign

Third Edition

Cumberland County College
Library
P.O. Box 517
Vineland, NJ 08360

SAUNDERS COLLEGE PHILADELPHIA

80-302

Saunders College Publishing
West Washington Square
Philadelphia, PA 19105

QP
31
P68

Listed here is the latest translated edition of this book together with the language of the translation and the publisher. 2nd Ed. Vol. I & II. 7/21/77 Arabic Franklin Book Program New York, New York

Listed here is the latest translated edition of this book together with the language of the translation and the publisher. 3rd Ed. 1/12/78 Russian MIR Publishers Moscow, USSR

Listed here is the latest translated edition of this book together with the language of the translation and the publisher. 3rd Ed. 4/14/78 Russian MIR Publishers Moscow, Russia

Comparative Animal Physiology ISBN 0-7216-7381-3

© 1973 by Saunders College Publishing / Holt, Rinehart and Winston. Copyright 1950 and 1961 by W. B. Saunders Company. Copyright under the International Copyright Union. All rights reserved. This book is protected by copyright. No part of it may be reproduced, stored in a retrieval system, or transmitted in any form or by any means, electronic, mechanical, photocopying, recording, or otherwise, without written permission from the publisher. Made in the United States of America. Press of W. B. Saunders Company. Library of Congress Catalog card number 72-80793.

0123 147 9876

PREFACE

This is an entirely new edition of the survey *Comparative Animal Physiology*. This book is intended to serve (1) as a text in courses at the advanced undergraduate and beginning graduate level, (2) as a reference source for beginning investigators, and (3) as a review of selected topics for workers in other fields. This edition has more contributors than previous editions (1950 and 1961) and it is broader in coverage. The field of Comparative Physiology has expanded greatly during the past decade, and physiological phenomena are increasingly explained in biochemical and biophysical terms.

At the request of many readers, tables of biological data are more extensive than in the Second Edition. The coverage of literature is both extensive and selective. Many detailed examples of physiological phenomena are given, and these are interpreted by frequent summaries. In the reference lists, a few older classical papers are retained, but emphasis is placed on papers published since 1960. Some data are given without reference when citations appeared in previous editions of this book or when the facts are commonly given in elementary physiology texts. The titles are abbreviated.

Certain factual background of the reader is assumed. Acquaintance with the principal phyla and classes of animals is assumed, but common names are used extensively. A knowledge of elementary cellular physiology is assumed, and cellular phenomena are used to arrive at comparative physiological generalizations. Knowledge of elementary biochemistry is essential for understanding the chapters on digestion, nitrogen metabolism, respiration, and intermediary metabolism. Some knowledge of organ function as usually presented in mammalian and human physiology texts is assumed. Mammals are freely used in this edition to illustrate comparative principles.

Comparative Animal Physiology has been expanded in an effort to keep abreast of the field. The first half of the book deals with adaptational and environmental physiology; it should be useful in courses in ecological physiology. This half also contains a considerable amount of comparative biochemistry. The second half of the book deals with sensory, effector, and integrative (neural and endocrine) physiology; it should be useful in courses in behavioral physiology. A chapter on comparative physiology of excitable membranes has been added.

The organization, as in previous editions, is by function systems. The arrangement for each system and the index should permit readers to find material concerning specific environments or specific animal groups or enzyme systems.

Each chapter has been read by critical specialists; the authors have endeavored to balance depth of coverage with breadth. We are indebted to many colleagues for useful comments. Some of these are:

Introduction – M. V. Greenberg
Chs. 2, 3 – L. B. Kirschner, J. Willis
Ch. 8 – A. Riggs, K. VanHolde

Ch. 9	—J. E. Heath, J. Hazel
Ch. 11	—J. McReynolds, I. Parnas, D. Kennedy, D. Hirst
Ch. 12	—L. Aitken, T. Trahiotis, G. Offutt, B. A. Budelman
Ch. 15	—D. Kennedy, N. Kotchabhakdi, R. Josephson
Ch. 16	—L. Barr, M. Holman, G. Campbell, R. Robson
Ch. 17	—M. V. Bennett
Ch. 19	—M. Child
Ch. 21	—K. R. Rao
Ch. 23	—K. R. Rao, M. Fingerman
Ch. 24	—J. W. Hastings, J. F. Case, J. B. Buck

The editor thanks Mrs. M. Walden, who has typed parts of all three editions of this book. He expresses special thanks to his wife, H. B. Prosser, who contributed with unflagging zeal to bibliographic search, to clarification and condensation of the text, and to manuscript preparation.

CONTENTS

II: SENSORY, EFFECTOR, AND NEUROENDOCRINE PHYSIOLOGY

INTRODUCTION

By C. Ladd Prosser

The science of physiology is the analysis of function in living organisms. Physiology is a synthesizing science which applies physical and chemical methods to biology. An understanding of comparative animal physiology requires some background in general zoology, animal morphology, biochemistry, and cellular physiology.

THE FIELDS OF PHYSIOLOGY

For practical purposes, physiology can be divided into three categories, as follows.

Cellular Physiology. Cellular or general physiology treats of those basic characteristics common to most living organisms. A vast amount of biochemical evolution occurred in protoplasm before multicellular organisms appeared, and cells are exceedingly complicated in their functional organization. In any cell—yeast, muscle fiber, or leaf parenchyma—the fundamental properties of differential permeability, oxidative enzyme activity, role of nucleotides, nuclear-cytoplasmic interaction, bioelectrics of excitable membranes, and many other properties are much the same. At the cellular level all organisms are more alike than they are different, and this basic similarity forms the starting point for evolutionary theory. Cellular specialization has led to some diversity of cell types and has often brought with it the loss of one function with enhancement of another. In this sense, one may speak of a comparative physiology of the cells in one organism. The characters treated in cellular physiology are nearly universal and are extremely stable with respect to the environment; these characters will not be discussed in this book, except to provide necessary explanations of organ functions.

Physiology of Special Groups. The physiology of special groups of organisms treats of functional characteristics in particular kinds of plants and animals. Different kinds of animals carry out similar functions in different ways. The physiology of some animal groups has been examined in great detail. Traditionally, the basic animal physiology is human and mammalian physiology, and this science provides the rational basis for much of medicine and animal husbandry. The physiology of higher plants is important as a basis for agriculture. Insect physiology, fish physiology, and the physiology of parasites are other specialties.

Comparative Physiology. Comparative physiology treats of organ function in a wide range of groups of organisms. Comparative animal physiology integrates and coordinates functional relationships which occur in more than one group of animals. It is concerned with the ways in which diverse organisms perform similar functions. Genetically dissimilar organisms may show striking similarities in functional characteristics and in responses to the same environmental stimulus. Conversely, closely related animals frequently react differently to their surroundings. While other branches of physiology use such variables as light, temperature, oxygen tension, and hormone balance, comparative physiology uses, in addition, species or animal type as a variable for each function. Use of kind of animal as an experimental variable leads to unique kinds of biological generalizations.

Comparative environmental and behavioral physiology constitutes a bridge between molecular and organismic biology, between the reductionistic and holistic philosophies.

An important function of comparative physiology is to put man into perspective in biological history and phylogenetic relationships. Medical physiology is necessarily anthropocentric. Biological man is the result of a long history of natural selection of physiological processes. The physiology of man can sometimes be elucidated by the study of an animal in which a given function is more patent than it is in man.

PHYLOGENY

The physiology of any animal group reflects the evolutionary history of that group. The phylogenist uses the data of paleontology when they are available, but relies also on taxonomy and comparative morphology. Since we cannot carry out physiological experiments on fossils, it is important to learn how far functional analogy and homology agree with morphological evidence for animal relationships.

Homology assumes a common ancestral gene pool; hence, it is based on structure—at one level, proteins and lipids, and at another level, whole organs or their parts. Physiological homology refers to function of a genetically related system (or enzyme) for a similar purpose, e.g., use of a sodium pump in dissimilar organs, or use of rhodopsin in eyes of diverse animals. Physiological analogy refers to evolutionary convergence or similar solution of a given life problem by unrelated and different means, e.g., the use of various metalloproteins for O_2 transport.

Since comparative physiology uses taxonomic types—from sub-species to phyla—as experimental variables, it is important to know something of the relationships among organisms and how these evolved. In a survey of some physiological process, e.g., excitation-contraction coupling in muscle, a knowledge of phylogeny helps in selecting appropriate animals for study and in answering questions such as whether the process arose once or repeatedly. Function in a given group of animals is clearly limited by the ancestry of the group. Physiological analyses are useful in elucidating evolutionary relationships.

The principal phyla evolved more or less simultaneously during the Cambrian; hence, currently accepted evolutionary trees are more polyphyletic and less monophyletic than former ones. There are many phyla which are not fitted with assurance into a given position, and numerous "missing links" between phyla are postulated. A useful phyletic chart is given in Figure 1.

Flagellates are the most primitive of protozoans, and other modern protozoans, as well as sponges, are offshoots from the direct phylogenetic line. At a primitive level, the acoelomate animals such as Nemathelminthes emerged. Near this level, a branching resulted in two important parallel lines. On one side are the annelids, arthropods, and molluscs; on the other side, the echinoderms and chordates. The cephalopods show the greatest specialization among the molluscs, the insects among the arthropods, and the birds and mammals among the chordates. In general, embryological differences have been used to distinguish different lines. The annelids-arthropods-molluscs show determinate or spiral cleavage; i.e., the blastomeres arrange themselves in a stereotyped pattern and there is little cellular equipotentiality, each cell having a fixed prospective role, whereas in the echinoderms and chordates cleavage is indeterminate. In the former group, mesoderm formation begins with a particular cell in the blastula which starts two lateral mesodermal bands; in the echinoderms-chordates, mesoderm arises as an outpouching from the archenteron, i.e., from endoderm. In the annelid-arthropod line the coelom is hollowed out from mesodermal bands, while in echinoderms and chordates the coelom pushes out from the archenteron. Several small phyla are pseudocoelomate; i.e., the body cavity is not a true coelom. In the annelid-arthropod-mollusc line the blastopore gives rise to the mouth (protostomes), and the anus is opened secondarily at the opposite end, whereas in the echinoderms-chordates the blastopore becomes the anus and the mouth opens secondarily (deuterostomes).

SPECIATION

There are several definitions of species, and no one of them can apply to all organisms, largely because of their different types of reproduction. The typological definition of species makes use of stable distinctive characters—usually morphological or protein isozymal—that are not necessarily adaptive

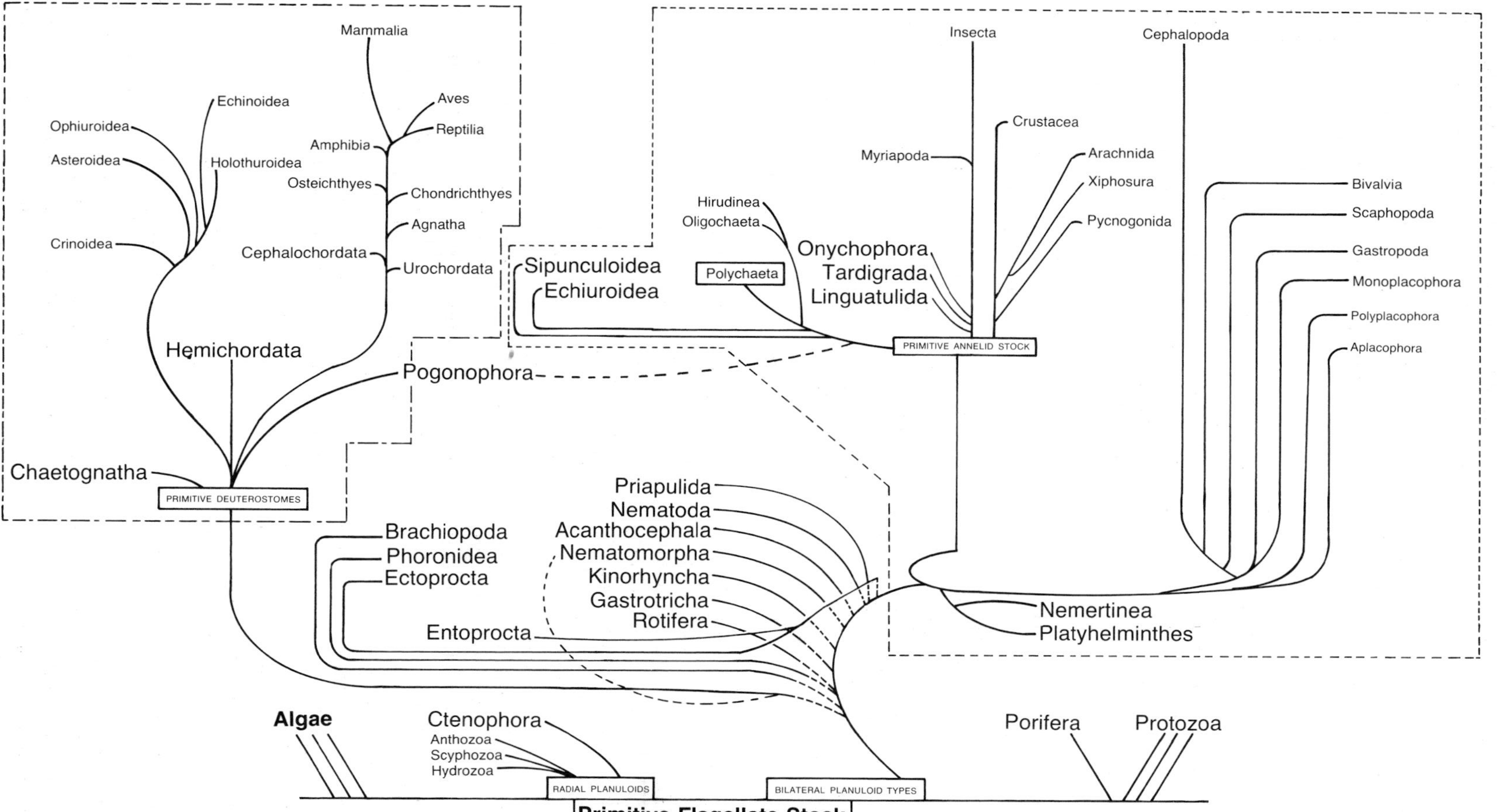

Figure 1. Phylogenetic tree of the animal kingdom. Current views favor polyphyletic origins. The coelenterates—Ctenophora and Cnidaria (Anthozoa, Scyphozoa, and Hydrozoa)—are related to one another, but are not considered as being the direct ancestors of higher animals; sponges (Porifera) also are not in a direct line. The protostome radiation arose from an extinct flatworm-like stock and consists of three branches: (i) the modern flatworms and nemertines, (ii) the Mollusca (classes shown on the right), and (iii) the annelid-arthropod lineage including various annelid-like worms. The origins and interrelationships of the several pseudocoelomate phyla, such as Rotifera and Nematoda, are uncertain. The position of some coelomate animals, such as the brachiopods, ectoprocts, and phoronids, is similarly obscure. Primitive deuterostomes arose independently, giving rise to the echinoderms and chordates. (Courtesy of M. J. Greenberg; *Syst. Zool. 22,* 1973.)

characters, and it is most useful in classification. Classification hopefully corresponds to phylogeny, but it need not do so to be useful in its own right. Taxonomy is thus the *result* of phylogenetic speculation, not the *evidence* for phylogeny.

The characters of cellular physiology are too universal to be used in detailed classification, although some special cellular characters may distinguish large taxonomic groups. Many physiological characters are highly sensitive to the environment, and hence these are unstable taxonomically. Some of the physiological characters which have been analyzed genetically are based on multiple factors, and frequently adaptive characters have large safety factors and parallel or alternate functional pathways. However, some highly similar species (sibling species), morphologically similar, are physiologically isolated.

The second and most useful definition of a species is the biological species, the population of like animals within which there can be flow of genes. Populations which remain distinct when living together (i.e., are sympatric) are clearly different species. When two populations are spatially separated (i.e., are allopatric) there may develop isolating mechanisms which keep them reproductively separated if they come together later in geological time. The biological definition breaks down for series of populations—in clines, circles, etc.—in which the two terminal populations are incapable of cross-breeding although each population in the series can breed with adjacent ones. In nature, many breeding populations are spatially restricted, yet hybridization between taxonomic species is more common than is sometimes recognized. Reproductive isolation may result from morphological differences, seasonal or diurnal variance in habits, chromosomal, hormonal, or behavioral differences, physiological and psychological incompatibility, and ecological separation.

The third species concept, the physiological species, is based on two generalizations—(1) that no two species can occupy the same ecological niche at the same time throughout their life cycles, and (2) that no two species from similar niches can simultaneously occupy the same geographic range throughout corresponding stages of their life cycles. This implies that every species is uniquely adapted to its ecological niche and geographic range, and that if the functional adaptations were fully understood, an evolutionarily meaningful description of the species would be possible. *Adaptation* is a general term referring to any alteration or response of an organism which favors survival in a changed environment. The physiological description of species requires both field and laboratory observations, which are possible with only a few organisms. First there must be a description of physiological variation in natural populations in terms of critical characters, a statistical analysis of a capacity deemed adaptive. Second, such variation as is found must be analyzed for the genetic component and for that which is environmentally induced; this requires cross-acclimatization, transplantation, and breeding experiments. Finally, an analysis of physiological mechanisms underlying the varying characters is needed.

Irrespective of which definition of species is used, the sequence in speciation involves several physiological steps. This sequence is:

(1) Environmentally induced variation of labile characters.

(2) Behavioral extension of range to exploit new niches and range limits.

(3) Random genetic change establishing varieties, races, subspecies, etc.

(4) Selection of genetic variants adapted to environmental niches or ranges.

(5) Isolation—ecologic or geographic—of the selected populations.

(6) Establishment of secondary characters which ensure reproductive isolation.

Physiological characters come into play in steps (1), (2), (4), and (6).

In nature, genetic variation may be greater than phenotypic variation, but natural selection acts on phenotypes. Thus, a functional character located far from its genetic code may be either selected or eliminated. For example, selection acts on the handling of ions by an excretory organ rather than on the DNA which codes the proteins of the transport enzymes.

PHYSIOLOGICAL VARIATION

In any population analysis it is essential to separate variations which are genetically fixed from those which are environmentally induced. Every population contains a variety of genotypes, and these fall on a distribution curve. An adaptive genetic change occurring in a population near the limit of the range is more likely to be retained than if it had occurred in the less stressful mid-range. In general, the amount of genetic variation in populations (e.g., chromosome band patterns of *Drosophila* species) is greater in the middle than at the boundaries of the range. Environmentally induced variation can take place only within the limits set for an individual animal by its genotype. Genetic and environmentally based variation can be distinguished by acclimation or acclimatization and ultimately by cross-breeding. *Acclimation* refers usually to the compensatory change in an organism under maintained deviation of a single environmental factor (usually in the laboratory). If acclimation is complete, a measured rate function is the same under one environmental condition as under another. *Acclimatization* refers to those compensatory changes in an organism undergoing multiple natural deviations of milieu—climatic, physical, and biotic.

Criteria of physiological variation which can be used in evaluating adaptiveness of genotypes take account of all function systems and all levels of animal organization. Most of these concern environment-organism interactions. Among the most useful are measurements of functional differences between animals under environmental stress. Some of these are the following:

(1) Survival tests at environmental limits, for example median lethal values for heat, cold, salinity, or oxygen supply. Survival may be measured for intact animals, for tissues, or for isolated enzymes (inactivation). In general, survival limits are widest for enzymes, narrower for tissues, and most narrow for whole organisms. Not all parts of an animal are equally subject to functional failure, and in metazoans the nervous system is usually more sensitive than other systems. Survival limits of organisms or parts of an organism can be modified within limits by prior experience with respect to the environmental factor.

(2) Environmental limits for reproduction. Embryos are often more sensitive to stresses than adults, yet completion of full life cycles is essential. Too few physiological studies have been made on early stages in life histories.

(3) Internal state as a function of the environment (Fig. 2). Some animals change internally to conform to the environment, for example, poikilotherms in varying temperature. Other animals maintain relative internal constancy in a changing environment;

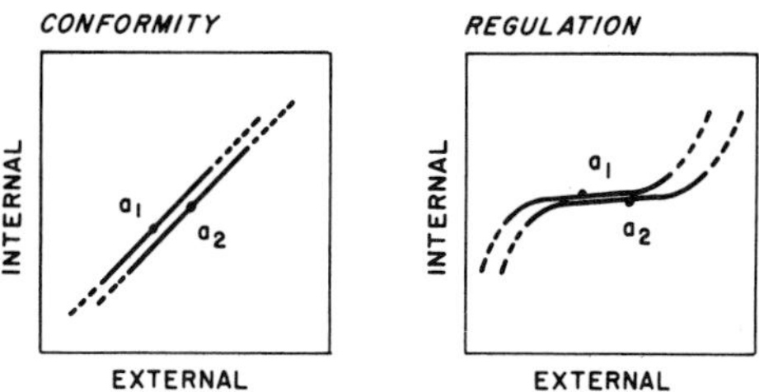

Figure 2. Diagrams representing internal state as function of external condition for a given parameter. Two degrees of acclimation indicated as a_1 and a_2. Solid lines lie within the range of normal tolerance; broken lines lie within the range tolerated for brief periods only. In the pattern of conformity, the variations of the internal state correspond with the external variations. In the pattern of regulation, the internal state is somewhat independent of the external.

that is, they regulate (e.g., homeotherms). Measurements of conformity and regulation can be made for all the physical factors of the environment. In general, conformers tolerate wide internal variation but narrow environmental limits, whereas regulators tolerate only narrow internal variation but a wider environmental range. Acclimation can shift the tolerated internal limits for a conformer; in a regulator it can change the critical limits for either activation or failure of homeostatic controls. Both patterns, conformity and regulation, are homeostatic in the sense of permitting survival in a changing environment, and most animals show elements of both patterns.

(4) Recovery from a deviated state. Animals tend toward certain norms, and when deviated, as by excess hydration or dehydration, heating or cooling, they may compensate, or on removal of the stress they may return in a definite pattern to the original norm. The rate of return is characteristic for a particular kind of animal. Too little attention has been given to what establishes the norm, and to what sensing devices detect deviations.

(5) Rate functions. Rates can be measured for movements, for metabolism, and for enzymatic reactions, both *in vivo* and *in vitro*. Rates are critically measured as a function of environmental or of internal condition. For enzymatic rates, two methods are useful: (1) measurement of maximum velocity where enzyme activity is rate limiting, and (2) measurement of Michaelis constants (K_m's), which give rates in the more physiological range of substrate. Variations in reaction rates, for example, as a function of tem-

perature, depend on genetic and environmental determinants.

(6) Macromolecular diversity. Similar proteins can be separated by electrophoresis, amino acid analysis, or immunological reactions. Similar nucleic acids can be characterized by base ratios. Isozymes are variant forms of the same protein, each determined by a different gene or combination. Allozymes represent multiple alleles of the same gene and occur in balanced polymorphism within a population. Biochemical diversity is meaningful physiologically when adaptive value is discovered for given chemical configurations.

(7) Behavior. This is shown in taxic responses, selection of "preferred" environments in gradients, and in complex behavior, such as courtship, mating, and rearing of young.

COMPARATIVE ENVIRONMENTAL PHYSIOLOGY

Where an animal lives is related to where its ancestors lived. There is no simple way to summarize all of the environmental factors influencing an organism, but we may distinguish four main habitats—marine, freshwater, terrestrial, and endoparasitic. Each of these has been extensively subdivided by marine biologists, zoogeographers, and ecologists. The animals in different environments are confronted with different problems; conversely, in the same environment different animals react to stress in different ways.

The comparative physiologist considers the organism as played upon by a variety of environmental factors—water, inorganic ions, organic food, oxygen and carbon dioxide, light, high and low frequency mechanical waves, pressure, gravity, ionizing and other radiation, and temperature. Comparative physiologists are concerned with the adaptive responses to these environmental factors. In addition, an animal is influenced in its environment by other organisms. Understanding of the biotic environment requires study of animal behavior.

The range of a species is determined through natural selection by its limits of tolerance. One environmental factor, such as salinity, may limit the distribution of one group; another factor, such as temperature, may limit another group. Over an ecological range, individuals may vary within limits set by the genotype; a phenotype results from the balance of genetic and environmental factors.

The use of physiological measurements on an animal under various stresses or environmental conditions was mentioned previously. The net effect of adaptive changes in an altered environment is homeokinesis. *Homeostasis* refers to constancy of internal state and refers usually to physiological characters which are regulated. *Homeokinesis* refers to constancy of life functions, such as locomotor activity and energy liberation, all of which permit survival in an altered environment, sometimes even when internal state varies. Thus, homeokinesis for a given function can occur in the absence of homeostasis.

Three general time scales can be distinguished in response to an environmental alteration. First, there are direct reactions to the environmental change. These may be mediated by stimulation of sense organs which initiate chains of responses, or they may be direct effects on metabolic reactions. Changes in ions, temperature, oxygen, or foodstuffs may lead to alterations in enzymatic reaction rates. Frequently if the environmental change is abrupt or great, as with cooling or warming, the animal shows an initial overshoot or series of oscillatory metabolic reactions before settling down to a new stabilized rate. The time for the direct response or new rate is usually minutes or hours.

The second time period in the response of an animal to an environmental alteration may require days or weeks. This is the period of acclimation or compensation, and the magnitude of the physiological adjustment varies with the amount of environmental change and with the genotype of the animal. In seasonal changes the compensatory alterations may develop gradually. Changes associated with wet and dry seasons, slow salinity shifts in estuaries, and seasonal changes in temperature and available food may involve migrations or changes in hormone balance and in reproductive activity. In the laboratory, compensatory acclimations take place via a variety of biochemical alterations, the net effect of which is homeokinesis or maintenance of some constancy of energy output. Another type of response to environmental change is for the animal not to compensate but to alter its physiological state, as in hibernation, estivation, or premature aging. In these cases animals turn their physiological activity on or off without true acclimatization.

A third type of response to environmental change is that of regulators of internal state with respect to environmental parameters. In these organisms, secondary changes, often morphological (e.g., insulation of birds and mammals in cold), reduce the energy cost of the altered state and thus result in metabolic and behavioral acclimation.

The third temporal scale of biological response occurs over many generations, the time for selection of genetic variants. The time may be many years, depending on the kind of animal and the environmental pressure. To a large extent, evolutionary changes are influenced by biotic factors of competition, predation, and social interaction. Identification of genetically fixed physiological differences between populations and between related species living in slightly different niches or ranges permits ecological and phylogenetic correlations. These are the variations that make for speciation.

Physiological variation in each of the three time periods can be examined for genetic and environmental components. The cellular mechanisms are usually different for the direct responses, the compensatory acclimations, and the selected long-term variations. In the first period there are direct changes in chemical reactions within the animal with alteration in such factors as ionic strength, oxygen, and temperature. In the second there are feedback actions to protein synthesis, to modification of cofactors of metabolism, and to membrane structure. In the third period, genetic changes are manifested by differences in critical proteins and in neural function. In all three, the interaction between environment and organism is complex and subtle. These interactions exemplify biological balance, a series of involvements from macromolecules to whole organisms and populations.

It is evident that the identification of the physiological variations in environment-organism interaction, and of the molecular mechanisms underlying these variations, is a challenge for environmental physiologists. An understanding of the interactions is important for theory of the history and distribution of living things. Environmental physiology also provides a rational basis for approaching many practical problems. Man is faced with the need for more food resources and for protection of the physical environment against deterioration. Granted the delicate physiological balance between organisms and their environment, practical action must be based on an understanding of these interactions. Efforts to regulate the physical environment, to alter land and water, to cause geographic change as by dams and canals, to dispose of the wastes of civilization, should be preceded by critical studies of the biological effects. Similarly, introduction of new food sources for man and of means of biological control of disease and pests must be based on knowledge of physiological adaptedness of the particular organisms. Comparative environmental physiology has, therefore, much to contribute to both biological theory and human welfare.

COMPARATIVE BEHAVIORAL PHYSIOLOGY

Behavioral extension of range is important in the exploitation of environmental limits, new foods, and new niches; also, differences in reproductive behavior are important in the isolation of true or incipient species. The initial responses to environmental changes are usually behavioral, and many sense organs are exquisitely sensitive to specific stimuli. Much of behavior is programmed in genetically fixed patterns of neuronal networks. Yet these patterns are modifiable within limits which vary according to type of nervous system. Effector systems permit wide ranges of response to environmental stimuli—fast or slow movements, color changes, luminescence, or electric discharge. Behavioral mechanisms—sense organs, effectors, integrating nervous and endocrine systems—show greater variations from one animal group to another than do metabolic mechanisms.

The cellular bases of behavior are general—polarized cell membranes, electrical and chemical interaction between cells, use of contractile proteins for movement, and transduction from one form of energy to another. Yet in detail, the methods which have been employed by cells are extremely diverse, and in no other area of physiology has study of specialized structures been so rewarding.

For example, understanding of the nature of nerve impulses has come in most detail from study of giant nerve fibers. The goal of comparative behavioral physiology is not only to throw light on animal evolution and distribution but to understand diversity in coping with commonly encountered stresses. The behavioral physiologist seeks the variations on common themes: for example, light sensitivity of different types of photoreceptors; movement by fast and slow muscles, by cilia and by protoplasmic flow; and chemical signalling by neurosecretory, synaptic, and endocrine cells.

One of the biological challenges for the future is to learn the basis for central nervous patterns. How much and in what way is behavior programmed genetically and developmentally? What is the basis for central modification, for the many types of "conditioning"? One approach to nervous function is observation of the behavior of whole animals. Another approach is the study of cellular and molecular mechanisms in neurons. Comparative neurophysiology bridges these two approaches, for example, in neural analysis in *Aplysia*. The implications for human societies, medical applications, and the future of civilization are evident.

PLAN FOR STUDY OF COMPARATIVE PHYSIOLOGY

An animal does not react to a complex environmental situation with a single organ or organ system. The parts of an organism interact with one another, and in combination their degrees of freedom are less than if separated. An organism constantly interacts with its micro-environment; hence, it cannot be described without considering its range of environmental interaction. For these reasons the properties of a whole organism can be either greater or less than those of the sum of its parts, depending on the interactions. Out of the whole organism emerge unique characteristics not present in any of the isolated parts. It is important, therefore, to examine the relation between components of the environment and the whole organism and to analyze the interactions in terms of organ and cell physiology. The plan of this book is to consider the reactions of animals to transitions in environmental factors and then to consider sensory, effector, and coordinating mechanisms. An attempt is made to arrive at general biological principles which can be reached only with *kind* of organism as an experimental variable.

CHAPTER 1

WATER: OSMOTIC BALANCE; HORMONAL REGULATION

by C. Ladd Prosser

INTRODUCTION

Water is an essential constituent of all living things; it is the universal biological solvent, the continuous phase in which most of the cellular reactions of metabolism occur, and an environmental constituent most necessary to life. Life undoubtedly began in a watery medium; numerous exits from water to land have been made in the course of evolution, but only a few groups of animals have been successful in maintaining themselves out of water. Each group which has made the exit from water has used its own set of adaptations to life in air, some being more successful than others. Insects have made the exit more completely than have other animals, yet a few insects return to water for part of their life cycle. All other animals, including birds and mammals, return to a watery medium, at least for embryonic life.

One problem of animal life is to maintain inside the organism just the proper amount of water—not too much, not too little. Water content *per se* is less important than water concentration (chemical activity), and this limits the dissolved ions which are critical for life. Terrestrial animals must retain and use what water is available; freshwater animals must exclude water to prevent self-dilution; some marine and parasitic animals are in osmotic equilibrium with their medium, whereas others are more dilute than the medium and have the problem of taking in enough water while living in a plenitude of it. The environmental range with respect to water, from fresh water through the sea to salt lakes, from humid swamps to dry deserts, is far greater than the tolerated range of concentrations of body fluids; hence animals must have a variety of mechanisms for regulating osmotic balance.

PHYSICAL CONSIDERATIONS

Properties of Water. The uniqueness of water as the solvent within which biological reactions occur has been frequently emphasized, classically by L. J. Henderson in his "Fitness of the Environment." It may be argued that the unique properties of water permitted the chemical evolution which led to life as we know it. Water is remarkable in being a liquid over the range from 0°C to 100°C. This spans the temperatures of most parts of the earth where life can occur. Water has a very high specific heat; that is, much heat must be added or withdrawn to alter its temperature. Water has a high latent heat of vaporization; hence it evaporates slowly from ponds and lakes, and when it evaporates from a body surface, heat is removed. Water in its solid state is less dense than in its liquid state and, in fact, its density is maximum at 3.98°C (D_2O maximum at 11.23°C) above its freezing point.[193] This property has permitted life to develop in temperate and subpolar regions where ice, lower in density than water, forms in winter only on the surfaces of bodies of water.

Water is remarkable as a solvent for electrolytes and for most organic non-electrolytes, and also for oxygen and carbon dioxide. Only non-polar compounds such as some fats do not dissolve in water. The dissolving of inorganic salts depends on the hydrogen bonding of water to the ions, and this liberates energy which is used to dissociate the salt into ions. Also, some molecules become ionized as they become hydrated; for example, hydrogen ions in water are largely present as hydronium (H_3O^+).

Many of the remarkable properties of water are due to the orientation of its atoms. The two hydrogens make an angle of about 105° with respect to the oxygen, and the O—H distances are about 0.95 Å. The hydrogens provide positive charges and the oxygen provides two negative charges; hence a water molecule is an electric dipole. The dipolar structure favors the orientation toward an ion or a charged group on a protein molecule, with resulting hydration. Water forms hydrogen bonds with many molecules which have negative charges; and in the solid state (ice), water molecules are hydrogen bonded so that the oxygen atoms lie in tetrahedra with separations of 2.76 Å. The energy of a hydrogen bond is low, only 5 cal/mole as compared to 50 to 100 cal/mole for covalent bonds; as a result, the hydrogen bond is a relatively weak one. In liquid water a considerable amount of hydrogen bonding persists, particularly in the biological temperature range, and the ordering this causes in the liquid probably accounts for many of the remarkable physical properties of water. In steam all of the hydrogen bonds are broken. The statistical uncertainty and unpredictability of the degree of H-bonding and "crystallization" in solutions leads to uncertainty concerning the state and amount of free water in protoplasm. Nuclear magnetic resonance measurements show that more of the water in tissue is in a "crystalline" structure than is that in liquid water (27% muscle).[71] Free or solvent water as opposed to bound water (or water of hydration) varies according to protein content; the amount of solvent water varies according to whether inorganic ions or non-electrolytes serve as solutes. Similarly, the freezing points of biological systems that are rich in organic solutes (which form multiple hydrogen bonds) may not agree with osmotic data. These properties permit some animals to become supercooled (see pp. 364–367).

Colligative Properties of Solutions. The state of water in protoplasm is not well understood, but a first approximation may be made by considering biological fluids as dilute solutions. The water content of a solution, either within an animal or outside it, gives little indication of the actual water activity (in a thermodynamic sense, the effective water concentration). The total effective concentration of all solutes present, or the *osmotic concentration*, is often expressed in osmoles, that is, the total number of moles of solute per liter of solvent; the osmolal concentration is determined by the colligative properties of the solution. Any one of the colligative properties can be calculated from any of the others; the higher the concentration of solute, the greater are the osmotic pressure, the lowering of the vapor pressure, the elevation of the boiling point, and the depression of the freezing point of the solution. The osmotic pressure is that pressure necessary to prevent entry of water across a semipermeable membrane (one which permits only solvent to pass into a solution). Strictly semipermeable membranes rarely if ever exist in living organisms, for if there were, there would be no exchange of solutes; and cells are rarely if ever bathed by pure water. For these reasons, osmotic pressure is a less useful concept *per se* than osmotic concentration.

The history of the theory of solutions started with observations by biologists. The Dutch plant physiologist Pfeffer in 1877 observed volume changes in stamen hairs in different sugar solutions, and from these observations van't Hoff showed that dilute solutions behave like gases. By analogy, the osmotic pressure (π) equals the osmolal concentration (C) multiplied by the gas constant ($R = 0.082$ liter atm/°C/mole) and by the absolute temperature (T).

$$\text{solutions:} \quad \pi = CRT$$

$$\text{gases:} \quad P = \frac{N}{V} RT$$

Another botanist, DeVries, found that equimolal solutions of salts exert more osmotic pressure than do sugar solutions, and from this the Danish physical chemist Arrhenius formulated the theory of electrolytes. The osmotic concentration of an electrolyte is given by the product of an isotonic coefficient (i) and the molal concentration; hence for electrolytes,

$$\pi = iCRT$$

Despite the fact that weak electrolytes in dilute solutions are completely dissociated (as shown by conductivity measurements), i is less than 2 for univalent salts and less than 3 for salts which dissociate into three ions. This is the effect of interionic interactions; hence, values of i must be determined empirically.

A one osmolal aqueous solution freezes at −1.86°C; hence, the lowering of the freezing point is

$$\Delta_{fp} = -1.86\ iC$$

Since an osmolal solution has an osmotic pressure (π) of 22.4 atmospheres,

$$\pi = \Delta \frac{22.4}{1.86} = 12.06\ \Delta$$

For practical purposes osmotic concentrations are usually given as osmolal (or milliosmolal), as lowering of the freezing point, or in equivalent NaCl concentration (millimolal). A few representative values from the Bureau of Standards table for NaCl follow:

%NaCl	*Molality*	*i*	Δ_{fp}(°C)
4.08	0.7	1.806	−2.38
2.92	0.5	1.81	−1.69
1.75	0.3	1.83	−1.02
1.17	0.2	1.84	−0.68
0.58	0.1	1.87	−0.34

PHYSICAL PRINCIPLES OF FLUID AND SOLUTE EXCHANGE

The cellular physiology of passive and active fluxes of ions and of water is treated in numerous books on general physiology. However, a few of the commonly used equations are presented here in order that the principles of water and ion movements across gradients may be understood for animals in different environments.[188] The symbol J will be used for flux or rate of transfer from one point in space to another.

Principles of Fluid and Solute Movement. A solute may move in a solution under the influence of several forces; two of the most important are gradients of chemical potential (concentration) and electrical potential. For diffusion in one dimension, as through a membrane, the solute flux (J) can be given by

$$J_{net} = -AD\left(\frac{dC}{dx} + \frac{zF}{RT} \cdot C\,\frac{dE}{dx}\right)$$

↗ concentration gradient $\left(\frac{\text{moles/cm}^3}{\text{cm}}\right)$ ↖ electrical gradient (volts/cm)

where A = area (cm²)
D = diffusion coefficient (cm²/sec)
z = charge on solute
F = the Faraday (96,500 coulombs/mole)
R = gas constant (8.314 volt-coulombs/degree/mole)
T = absolute temperature (degrees Kelvin)
x = distance (cm)

The equation in this form is never useful in biological work, but it can often be simplified. For example, a non-electrolyte has no charge ($z = 0$), and the entire second term in the bracket becomes zero, giving one form of Fick's Law:

$$J_{net} = -AD\frac{dC}{dx}$$

If, further, the gradient is linear through the membrane $\left(\text{i.e., } \frac{dC}{dx} = \frac{\Delta C}{\Delta x} = \frac{C_2 - C_1}{x}\right)$, where x is the membrane thickness, then

$$J_{net} = -A\frac{D}{x}(C_2 - C_1) = -AP(C_2 - C_1)$$

which permits one to assess the permeability (P) of the membrane to the test compounds.

But many solutes of biological interest are ions (i.e., are charged so that $z \neq 0$) and the voltage term (E) cannot be neglected. Further, we seldom know the form of the concentration gradient, and the linear case must be rare, especially for ions moving through membranes containing charges. Other restricted solutions of the diffusion equation have been used for these less simple situations. One, the Goldman Equation (p. 464), has been used to estimate membrane permeability to ions. Another, the Ussing Equation, provides a criterion for active transport.[360] In this case, it was shown that if, instead of considering the net flux of an ion, one considers two ion streams (unidirectional fluxes), one passing from compartment 1 to compartment 2 and the other from 2 to 1:

$$\frac{J_{1\to 2}}{J_{2\to 1}} = \frac{C_1}{C_2} \cdot e^{\frac{zF}{RT}(E_1 - E_2)} = \frac{C_1}{C_2} \cdot e^{\frac{zF}{RT}E_m}$$

where E_m is the potential difference across the membrane. That is, if an ion conforms to the diffusion equation, the ratio of unidirectional fluxes can be calculated from the concentrations and the potential across the membrane. If the actual ratio (measured with tracers) does not agree with the calculated value, active transport is assumed. Two simplifications are also useful. For nonelectrolytes $z = 0$ and the electrical term $= 1$; therefore, the *expected* flux ratio equals the ratio of concentrations. If the system does not behave this way, active transport is assumed. If the system is in a steady state ($J_{1\to 2} = J_{2\to 1}$), the flux ratio is unity and

$$(E_1 - E_2) = E_m = \frac{RT}{zF} \ln \frac{C_2}{C_1}$$

This is the Nernst Equation, and it is used, just as is the more general Ussing Equation, as a criterion for active ion transport, but only when there is no net movement of the ion from one compartment to the other.

Consideration of the Nernst Equation leads to the question of what might cause such an asymmetric distribution of diffusible ions at equilibrium. One case occurs when there exists on one side of a membrane an ion species that is not freely diffusible (e.g., proteins to which membranes are impermeable). If Na^+ and Cl^- are diffusible and A^- is not, then the distribution would appear as:

inside (i)	outside (o)
Na_i^+	Na_o^+
Cl_i^-	Cl_o^-
A_i^-	

To preserve electrostatic neutrality, $Na_o^+ = Cl_o^-$; but $Na_i^+ = Cl_i^- + A_i^-$, and hence $Na_i^+ \neq Cl_i^-$. It can be shown that when such a system is at equilibrium $[Na_i^+] \cdot [Cl_i^-] = [Na_o^+] \cdot [Cl_o^-]$ or

$$\frac{Na_i^+}{Na_o^+} = \frac{Cl_o^-}{Cl_i^-}$$

and the ratios cannot equal 1. Thus, asymmetric distribution of diffusible ions can be caused by the presence of an indiffusible ion in one compartment. This is called a Donnan Equilibrium, and the ratios are called Donnan ratios or distributions. From the Nernst Equation it follows that a potential difference must exist at equilibrium. It can also be shown that the total solute concentration inside ($\Sigma(Na_i^+ + Cl_i^- + A_i^-)$) is greater than that outside ($\Sigma(Na_o^+ + Cl_o^-)$). This has important consequences for osmotic equilibrium, as will be seen later.

It was mentioned on page 2 that conventional notation (molarity, molality) is a poor index of the chemical potential of water. Yet a difference in chemical potential will cause water to diffuse just as surely as solute, and it is important to be able to express the force quantitatively. The appropriate measure of the chemical potential of water in a solution is the osmotic pressure, π, of a difference in chemical potentials between different solutions $\pi_1 - \pi_2 = \Delta\pi$. The water flux (osmosis) is proportional to the $\Delta\pi$:

$$J_{osm} = AL\Delta\pi$$

where L is the osmotic permeability coefficient. Since $\pi = cRT$, we have $\Delta\pi = (c_1 - c_2)RT$ where c_1 and c_2 are total solute concentrations in compartments 1 and 2. For a cell at or near Donnan equilibrium, total intracellular solute concentration exceeds extracellular concentration; in the example above,

$$\Sigma(Na_i^+ + Cl_i^- + A_i^-) > \Sigma(Na_o^+ + Cl_o^-),$$

so there should be osmotic inflow with cell swelling. Plant cells have rigid cellulose walls which resist the tendency to swell; enormous internal ("turgor") pressures develop, and when the turgor pressure = $\Delta\pi$, the cell is at water equilibrium.

If a flux is greater than can be accounted for by the physical forces (i.e., if it is not passive), then active transport is assumed. Additional evidences for active transport are: requirement of extra energy, blocking by specific inhibitors, and competitive inhibition by related substances. A useful way to compare effectiveness of active transport systems is to measure the amount transported for different concentrations of the substance. The concentration for half-maximum transport corresponds to a K_s for an enzyme system, and represents half-saturation of the transport system. Another useful method is to measure the threshold concentration for activation or turning on of a transport mechanism.

The preceding paragraphs considered driving forces; we now consider the mode of transport through a membrane. Movement across (or within) a membrane may be by the following means:

(1) Passive flux, as through pores under concentration and electrical gradients. This is calculated from concentrations and potentials according to preceding equations.

(2) Exchange diffusion (passive), whereby a carrier molecule in the membrane passively loads and unloads at different positions so that individual molecules of a given kind exchange between the two sides. This is estimated from tracer efflux into media with and without the ion in question.

(3) Active transport, whereby energy is expended by a carrier system and the transport may (but need not) occur against a concentration gradient. This is measured as the difference between total flux and that which is computed as passive.

(4) Facilitated diffusion, whereby a carrier combines with a substance and speeds its movement from the side of a higher to that of a lower concentration (as transfer of oxygen by myoglobin).

The flux of any material, such as an ion or water, between the body fluids of an animal and its environment is usually measured with a labeled isotope. An extended treatment of tracer kinetics is beyond the scope of this chapter, but one simple case is worth developing here because it is often used, and because it shows how flux parameters are developed. Suppose we are interested in the movement of an ion between the body fluids of an aquatic animal and its environment. A known amount of isotope can be added to the external medium at the beginning of an experiment, and some of it will disappear into the animal. If the amount of labeled isotope in the medium is denoted Q^*_{out}, the rate of its disappearance can be written

$$-\frac{dQ^*_{out}}{dt} = J_{in}X_{out} - J_{out}X_{in} \qquad (1)$$

where J_{in} and J_{out} are in the inward and outward fluxes of the ion, and X_{out} and X_{in} are specific activities (proportion of labeled to total ions of the test species; i.e., Q^*/Q_{total}). When the animal is in a steady state, $J_{in} = J_{out}$, and if the measurements are made before much tracer enters the animal, $X_{in} = 0$. Equation 1 can be rewritten as

$$-\frac{dQ^*_{out}}{dt} = J_{in}X_{out} = \frac{J_{in}}{Q_{out}} \cdot Q^*_{out} \qquad (2)$$

which can be integrated simply to give

$$\frac{\ln Q^*_{out(t)}}{\ln Q^*_{out(0)}} = -\left[\frac{J_{in}}{Q_{out}}\right] \cdot t = -kt \qquad (3)$$

where $Q^*_{out(t)}$ is the quantity of radioactivity at the time of a measurement, and $Q^*_{out(0)}$ is the initial quantity added at zero time. As long as the animal is in a steady state J_{in} and Q_{out} are both constant, and their ratio (J_{in}/Q_{out}) is also constant. The latter, denoted k, is called the rate constant for movement of the test ion.

If the constraints of this analysis are observed (the animal must be in a steady state and the measurements taken before much isotope enters), evaluation of the kinetics is fairly straightforward. A quantity of isotope is added at zero time, and a series of samples are taken and analyzed for the amount re-

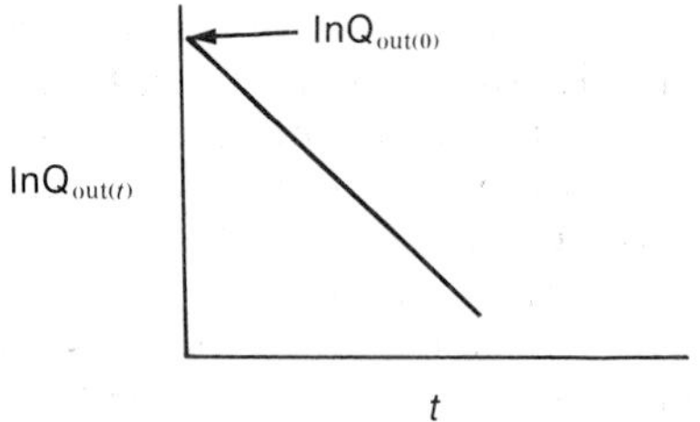

maining. Logarithms of these values are plotted against time, giving a straight line as shown in the figure above. The ordinate intercept is ln $Q^*_{out(0)}$ and the slope equals the rate constant k.

If Q_{out} (total amount of the ion in the medium) is known, the influx (J_{in}) can be calculated. Rates of movement are usually expressed in terms of J or k. Sometimes the time for Q^*_{out} to fall to one-half the initial value (the "half-time", $t_{1/2}$) is used.

Variations of this approach involve injection of tracer into the animal, or analyzing the animal instead of the external medium by measuring the isotope in the animal with a whole body counter or in blood samples. The equations pertaining differ slightly in form from equation 3, but permit graphical evaluation of k as long as the boundary conditions described above hold. For example, if the tracer is injected into the animal, and the tracer's appearance in the external medium followed,

$$Q^*_{out(t)} = Q^*_{out(eq)}(1 - e^{-kt}) \qquad (4)$$

where $Q^*_{out(eq)}$ is the amount of tracer in the medium at equilibrium.If a series of measurements are made and $\ln(Q^*_{out(eq)} - Q^*_{out(t)})$ is plotted against t, a straight line with slope k is obtained.

More complicated situations occur when the compound is distributed among three or more compartments (e.g., intracellular, extracellular, and external medium). They are amenable to analysis, but the analyses are more complicated. The subject has been reviewed many times (Solomon, A. K., Compartmental methods of kinetic analysis, *in* C. F. Comar and F. Bronner (eds.), *Mineral Metabolism,* Vol. 1, Part A, Academic Press, New York, 1960; Sheppard, C. W., *Basic Principles of the Tracer Method,* John Wiley & Sons, New York, 1962; Robertson, J. S., Theory and use of tracers in determining transfer rates in biological systems, Physiol. Rev. *37*:133–154, 1957).

PATTERNS OF BIOLOGICAL RESPONSE TO OSMOTIC CONDITIONS IN THE ENVIRONMENT

One solution is said to be *isosmotic* with another if the two are equal in osmotic concentration. An *isotonic* solution is one in which a cell (or organism) does not change its volume. An isosmotic solution of a substance to which a cell is permeable, but which is initially absent from the cell, is not isotonic. For example, red blood cells show no volume change in 0.3 M sucrose, to which they are not permeable, whereas in urea of isosmotic concentration the cells swell because the urea enters and water follows. The term "isosmotic" is preferable to "isotonic" in referring to equal osmotic concentrations. Similarly, a *hyposmotic* solution is more dilute and a *hyperosmotic* solution is more concentrated than the comparison solution.

Nearly all freshwater and terrestrial plants, by virtue of their cellulose walls and active plasma membranes, maintain their cell constituents (particularly their vacuolar sap) at concentrations higher than those of the fluids which bathe their tissues. The cells are continually more concentrated than extracellular fluids, and hence are turgid because they are prevented from swelling by rigid cell walls. In desert plants, osmotic pressures of some 50 atmospheres may be attained between cell and extracellular water; the osmotic pressure of cells of xerophytes exceeds that of mesophytes grown under similar conditions, so the differences are genetic. In animals, inelastic cell walls are absent and, although the outer gel cytoplasm maintains some turgidity, the intracellular concentration is approximately the same as that in the extracellular fluid. Excised tissues placed in isosmotic saline may swell, particularly at low temperature and in anoxia; water is gained by the tissue and is lost on return to higher temperature and adequate oxygen. This is because under the Donnan conditions (p. 4) an energy-requiring mechanism is needed to maintain osmotic equilibrium. This is done by (1) a Na-K pump which keeps intracellular ion concentrations slightly lower than the equilibrium value to compensate for protein (cooling or anoxia stops the pump, permitting ions to diffuse toward equilibrium and allowing the cells to gain ions and then swell), and (2) variation of the intracellular concentration of free amino acids (especially in marine invertebrates) to provide osmotic equilibrium.

The *water content* of tissues differs greatly with type of tissue, age, metabolic activity, and environmental conditions. Water content of muscle is typically 75 to 82 per cent, bone 35 per cent, and fat 10 per cent. A jellyfish, with a water content of 95 per cent or more, has a higher osmotic concentration than does a fish, which is only 70 per cent water but which contains relatively more organic solutes and less electrolytes. Similarly, the density or *specific gravity* of a solution is not a direct function of osmotic concentration, but depends on the nature of the solute, its concentration, and the temperature and barometric pressure.

The range of environmental osmotic conditions tolerated by animals is great; the tolerated range of internal osmoconcentration is much narrower. Table 1–1 shows that environments form a continuum from very dilute fresh water to concentrated saline lakes, and from water-saturated air to dry deserts; biological fluids show only a 20-fold range of osmoconcentration.

The ocean is the ancestral animal home. The osmotically effective concentration of the ocean has increased only slightly since the earliest appearance of life. The water of mid-ocean is now equivalent to a 3½ per cent salt solution; smaller seas and bays are diluted by inflow of fresh water, and in estuaries and river mouths brackish water merges with fresh. In seas such as the Mediterranean, where evaporation is high, the salinity exceeds that of the ocean. Fresh-water dilutions of ocean water are often expressed as per cent sea water, and *salinity* or total salt content is given in parts per thousand (0/00). An average sea water is as follows: 100% S.W. = salinity 34.5 0/00 = Δ_{fp} − 1.86° C = specific gravity at 10° of 1.0215 = chlorinity 18.8 0/00 = osmolality 1.01.[151]

Any water with a salinity less than 0.5 0/00 ($\Delta_{fp} < \sim 0.05°$) may be considered as fresh. We shall refer to anything between fresh water and salinity 30 0/00 as brackish; ecologists recognize at least four intermediate ranges, each with its own plant and animal communities.

Animals and plants which are restricted to a narrow range of salinity, usually to full sea water, are called *stenohaline;* those that are tolerant of a range of salinities are *euryhaline.* The limits of tolerance of salinity change by individual animals at various stages in the

TABLE 1–1. Ranges of Variation in Respect to Osmoconcentration in Body Fluids, in Aquatic Environments, and in Respect to Water Concentration of Air.

Medium	Δ_o, °C	*Animal*	Δ_i, °C
Fresh water <5 0/00	−0.01	Freshwater mussel	−0.08
		Pelomyxa	−0.14
		Freshwater fish	−0.50 to −0.55
		Frog	−0.45
		Crayfish	−0.82
Brackish water 5 to 35 0/00	−0.2 to −0.5	Euryhaline invertebrates	−0.5 to −1.8
Sea water 35 0/00	−1.85	Marine fish	−0.65 to −0.7
		Marine invertebrates	−1.8 to −1.85
		Marine elasmobranchs	−1.85 to −1.92
Salt lakes 50 to 250 0/00	−13.5 to −15	Brine shrimp (*Artemia*)	−1.2 to −1.6
Humid niches (80 to 95% Rel. humidity)		Earthworms	−0.3 to −0.4
		Some insect larvae	−0.5 to −1.0
		Mammals	−0.5 to −0.58
		Insect adults	−0.8 to −1.2
Dry air (less than 10% Rel. humidity)		Reptiles	−0.6 to −0.7
		Insects	−0.8 to −1.2

life cycle are determined, within genetic limits, by their acclimatization experience.

Many exits from the ocean to fresh water, and from water to land, have been made in the course of evolution, as well as many separate reinvasions of water. All phyla and most classes of animals have representatives in the marine habitat; some animals have remained in the ancestral home, and some have returned to it. Members of fewer phyla have taken to living in brackish water, and comparatively few meet the demands imposed by fresh water. Some animals have invaded land directly from the ocean, others through the avenues of estuarine and fresh water. The parasitic habit has been assumed by marine, freshwater, and soil-dwelling groups. Evidence regarding the osmotic limitations to distribution of a group of animals can be obtained by observing the responses to *osmotic stress.*

There are two extreme patterns of response to osmotic stress, Animals may be osmotically labile (dependent) and their body-fluid concentrations may change with the medium; these are *osmoconformers.* Other animals are osmotically stable (independent), and when the medium changes, their internal concentration remains relatively constant; these are *osmoregulators.* The terms *poikilosmotic* and *homoiosmotic* are sometimes applied respectively to conforming and regulating animals. Many gradations exist between the extremes of lability and constancy; an animal may conform to the medium osmotically in one concentration range and regulate in another. Osmotic changes may bring about gain or loss of water and thus volume changes; if appropriate solute transfer also occurs, the body volume is kept constant when the animal's concentration changes with the medium. Osmoconformers tolerate wider variation in internal osmoconcentration, while osmoregulators can withstand a wider environmental range. Both patterns are homeostatic in that they permit continued life functions in an altered environment.

Animals tend to maintain an "optimum" osmotic concentration for a given environment. When, after a period of dehydration, they are returned to a normal environment, water is taken up, and when after a period of overhydration animals are returned to normal water access, water is lost until the internal osmoconcentration reaches the "optimum" for the particular animal.[2] The time course of recovery of water balance from the deviated state may be characteristic of a given species. The mechanisms of homeostatic rehydration or dehydration include sensing devices, neural and hormonal coordination, and behavioral effectors.

OSMOTIC CONFORMITY IN CELLS

Cellular Homeostasis in Respect to Water Activity and Cell Volume. Most marine invertebrates are isosmotic with the environment, and most cells of higher multicellular animals are likewise isosmotic to extracellular fluid. The limitation is probably the dilution at which protoplasmic organization can still permit basal metabolic functions. An organism or cell which is bounded by a membrane permeable to water only (semipermeable) acts as a simple osmometer and swells or shrinks in proportion to its solute concentration gradient; there is no volume regulation. When the volume of a marine invertebrate egg or of a red blood cell is measured after reaching equilibrium in different dilutions of sea water or of physiological saline, these cells, as osmometers, should approach the relation:

$$\text{pressure } (\pi) \times \text{volume } (v) = \text{a constant } (K).$$

However, the volume change is less than that predicted by this simple relation.

A certain proportion of the cell is apparently osmotically inactive. If a correction is made for this osmotically inactive volume (b), then

$$\pi(v-b) = K$$

Some values of b are: unfertilized *Arbacia* eggs, 7.3% of initial volume; fertilized eggs, 27.4%; human erythrocytes, 11%.

One explanation of the deviation from calculated osmotic behavior is that not all of the water in the cell is available as solvent. Other evidence, such as the percentage of intracellular water that can be frozen without irreversible damage (p. 365) agrees with the osmotic evidence that a fraction (some 15 to 25 per cent) of cell water is bound as water of hydration. However, for red blood cells the volume of water that can serve as a solvent for chloride ions[70] and for glucose[289] approaches total cell water; also, the osmotically inactive volume (*b*) increases with decreasing cell volume. The solvent fraction for Cl varies with pH; hence, chloride may substitute in varying proportions for hemoglobin as an anion in maintaining Donnan equilibrium.[70, 121, 289, 311, 312]

In striated muscle of frog, volume is linearly related to osmotic concentration (O.C.) from 0.77 to 1150 mOsm.[37] However, approximately 33% of the volume is osmotically inactive, or 13% of the cell water. The tubular system (T and S tubules, p. 720) accounts for 11 to 18 per cent of cell volume; hence, possibly some of the osmotically inactive volume is in the tubules.[37] Crustacean muscle has a very extensive tubular system into which water and ions can be sequestered so that the entire cell does not change in volume during osmotic stresses. The percentage of muscle water that is osmotically active is for frog, 85; crab, 87; crayfish, 78; barnacle, 66.[157] In giant nerve fibers of squid and lobster, the axon proper swells and shrinks more than the whole fiber; the sheath constitutes a dead space of 20 to 40 per cent of the total volume.[117]

In conclusion, the value of *b*, which is a measure of the deviation of measured swelling from theoretical osmotic swelling, varies greatly from species to species for corresponding cell types. Values of *b* range from 25 to 50% of cell volume in sea urchin eggs, to 60 to 80% in amphibian eggs, to 90% in eggs of sturgeon and axolotl.[388] Bound water is an insufficient explanation for all of the deviation of measured swelling from the theoretical value, and apparently some of the water bound to protein can serve as solvent for small molecules and ions.

Water Permeability. From rates of volume change, one can calculate the osmotic permeability to water in terms of volume of water entering per unit surface area per unit time for a given osmotic gradient. From rates of entrance and exit of tracer water (D_2O or THO), rate constants of diffusive permeability are obtained. Table 1–2 gives data where it has been possible to convert the two kinds of measurements to the same units. It is found, for a wide variety of tissues, that osmotic permeability to water significantly exceeds diffusive permeability. The differences in free diffusion rates between D_2O or THO and H_2O are insufficient to account for the permeability differences.[310, 311]

One explanation proposed for red blood cells is that water-filled channels exist through which free diffusion is limited.[311, 312] Another explanation, based on water permeability of frog skin at different rates of stirring of the solutions, is that an unstirred layer of water impedes free diffusion close to the membrane.[82] This explanation has been rejected for red cells.[312] In artificial membranes made from brain lipids, no difference between diffusive and osmotic permeabilities was

TABLE 1–2. Comparison of Permeability Coefficients Determined by Diffusion (P_d) and by Osmotic Pressure (P_F).*

	P_d (cm/sec)	P_f (cm/sec)	P_f/P_d
human red cells	3×10^{-3}	7×10^{-3}	2.3
beef red cells	2.7×10^{-3}	8×10^{-3}	3.0
dog red cells	2.4×10^{-3}	10.5×10^{-3}	4.4
Amoeba	0.023×10^{-3}	0.037×10^{-3}	1.6
frog, body cavity egg	0.075×10^{-3}	0.13×10^{-3}	1.7
Xenopus egg	0.09×10^{-3}	0.159×10^{-3}	1.8
zebra fish, ovarian egg	0.068×10^{-3}	2.93×10^{-3}	43
zebra fish shed, nondeveloping egg	0.036×10^{-3}	0.045×10^{-3}	1.3
frog, ovarian egg	0.128×10^{-4}	8.9×10^{-3}	69
human amnion	3.62×10^{-4}	3.7×10^{-2}	102
frog skin (poorly stirred)	6.5×10^{-5}	5.4×10^{-4}	12
(well stirred)	11.1×10^{-5}	1.2×10^{-5}	0.108

*Calculations made from data in:
Paganelli, C. V., and A. K. Solomon, J. Gen. Physiol. *41*:259–277, 1957.
Prescott, D. M., and E. Zeuthen, Acta Physiol. Scand. *28*:77–94, 1953.
Rich, G. T., et al., J. Gen. Physiol. *50*:2391–2405, 1967.
Sha'afi, R. I., et al., J. Gen. Physiol. *50*:1377–1399, 1967.
Villegas, R., et al., J. Gen. Physiol. *42*:355–369, 1958.
and references 25, 82, and 311.

noted provided stirring was adequate; however, the H_2O permeability (measured osmotically) was higher in artificial membranes having a high ratio of cholesterol to phospholipid than when the ratio was low.[111]

Another complication is that exosmosis may not equal endosmosis. In sea urchin eggs the rate of swelling is less than the rate of shrinking. In red cells, however, the two rates are the same in the near isotonic range.

In some osmoconforming marine animals, the principal gain or loss of osmotically active solute in hemolymph and muscle is in small organic molecules, while the ionic concentrations remain relatively constant (p. 82). When a European flounder migrates from sea water to fresh water, its plasma is reduced in osmotic concentration by 20%. If red cells are removed from a flounder and placed in a medium diluted by 20%, they initially swell and then shrink to their original volume; analysis shows that the volume regulation is due to the loss of a ninhydrin positive substance.[119] Tissues such as nerves may show volume regulation by virtue of changes in free amino acids. A crab nerve initially swells when put into a dilute solution; then its volume decreases due to loss of amino acids.[204] A nerve from a freshwater crab, *Eriocheir*, in saline isosmotic with the blood, loses free amino acids; in more concentrated saline the concentration of free amino acids in the nerve increases.[126]

Further evidence for some volume regulation in isosmotic cells is found in the varied shapes of cells in different tissues and the rarity of spherical cells. The occurrence of relatively inelastic membranes and of cortical layers of high viscosity, and of contractile proteins, is widespread. Many cells of specific nonspherical shape become spherical when they swell in a hypotonic medium.

Osmotic measurements with different solutes permit estimates of the diameter of the pores through which non-electrolytes (and water) may pass. The Staverman coefficient, σ, is a measure of deviation from complete permeability for a molecule. If the membrane is as freely permeable for a non-electrolyte as for water, σ is 0; if the membrane is semipermeable with respect to the non-electrolyte, σ is 1. By comparing instantaneous volume changes in solutions of various compounds, one can calculate the diameter of the smallest molecule that cannot pass readily through the membrane and, hence, the "pore" size for complete semipermeability. Estimated pore radii in red blood cells are:[130]

human	3.5 Å
beef	4.1 Å
dog	7.4 Å

It has been suggested that cells of some tissues may be hypertonic (i.e., turgid) with respect to extracellular fluid. However, corrected measurements of osmoconcentration of intracellular and extracellular fluids show that most animal cells are isotonic. In some instances intracellular hydrostatic pressure may permit some hypertonicity, but for practical purposes we may consider the osmotic concentrations of cells to equal those of the extracellular fluid.

It is concluded that volume and shape regulation of cells is not fully understood, and that most cells do not behave as simple osmometers; they do, however, approach the behavior of ideal solutions. Permeability of cell membranes is less than in free diffusion in bulk solution, and varies greatly among cells. The difference between diffusive and osmotic permeability is not fully understood.

OSMOCONFORMERS

Multicellular Animals

In most marine invertebrates, the osmotic concentration of body fluids (extracellular and intracellular) is in passive equilibrium with that of the ocean although, because of Donnan effects, the total osmoconcentration of blood may be slightly higher than that of the medium. Echinoderms, sipunculids, cephalopods, ascidians, and most coelenterates are limited in habitat to full-strength sea water (i.e., they are highly stenohaline), and they conform to the medium over narrow osmotic variations. These animals show varying degrees of volume regulation and of ionic regulation; all of them regulate ionically at the cellular level, and to a lesser extent in extracellular fluids. One way to achieve ionic regulation and osmotic conformity is to vary the nitrogenous organic molecules while inorganic ions are held more constant.

The osmotic concentration of marine jelly-

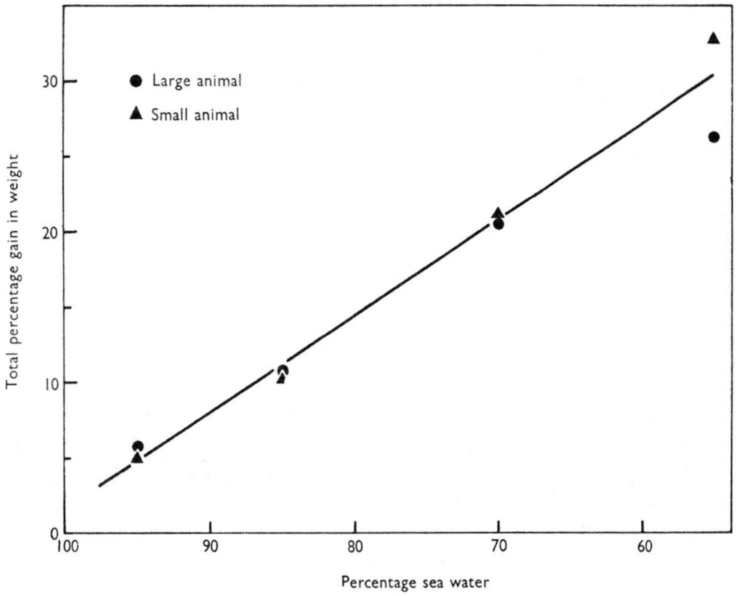

Figure 1-1. Volume of starfish *Asterias* in different percentages of sea water. (From Binyon, J., J. Marine Biol. Assoc. U.K. *41*:161–174, 1961. Reprinted by permission of Cambridge University Press.)

fish is essentially the same as that of the ocean where they occur. They are relatively intolerant of dilution; corals and ctenophores are said to tolerate 20% reduction in salinity. The blood of *Ascidia* and *Molgula* is approximately isosmotic to the medium for a limited range of dilution or concentration of sea water.

When echinoderms (holothurians, echinoids,[124] or starfish[35, 36]) are transferred to dilute (80%) sea water they gain weight due to water entry; when they are transferred to concentrated (110%) sea water they lose weight and little or no recovery of volume occurs, at least during several days (Fig. 1–1). The salinity limit for survival of starfish from the North Sea (salinity 35 0/00) was 23 0/00, while for starfish from the Baltic (salinity 15 0/00) the limit was 8 0/00.[35, 36]

The biological cost of osmoconformity by internal dilution may be considerable. *Asterias* collected in the Baltic have a soft integument, increased water content, lessened tolerance of heat, and reduced metabolism as compared to North Sea starfish. *Asterias* does not breed in the Baltic, and populations there represent a surplus from *Asterias* populations which breed in the North Sea. In *Asterias rubens*, approximately half of the osmotic concentration in tissues is due to free amino acids and the nitrogenous base taurine; these compounds change in concentration more than do tissue salts as the animal goes from one salinity to another:[173, 206] (See table below.)

Sipunculid worms swell and shrink as osmometers, with an osmotically inactive volume of about 25 per cent. In *Golfingia*, no volume regulation occurs,[2, 364] but in *Dendrostomum* salt is lost by the gut and especially by the nephridia, and some volume recovery occurs after 4 to 6 days in dilute medium.[177] *Golfingia* contains free amino acids to the amount of 550 mM in body wall cells and 80 mM in coelomic fluid. In reduced salinity, the amino acid concentrations decrease sharply, thus preventing excessive hydration and swelling without loss of ions.[364] The sipunculid *Themiste* is an osmoconformer, with chloride concentration in coelomic fluid

	Sea Water	*60% Sea Water*
Δ_{fp} medium	−2.09°	−1.23°
Δ_{fp} body fluid	−2.12°	−1.24°
Δ_{fp} in gastric caeca due to amino acids and taurine	−0.6°	−0.32°
Δ_{fp} in gastric caeca due to other nitrogen compounds	−0.35°	−0.30°
concentration in gastric caeca, taurine	3.9 mM/100 ml	2.2 mM/100 ml
glycine	17.3 mM/100 ml	9.5 mM/100 ml

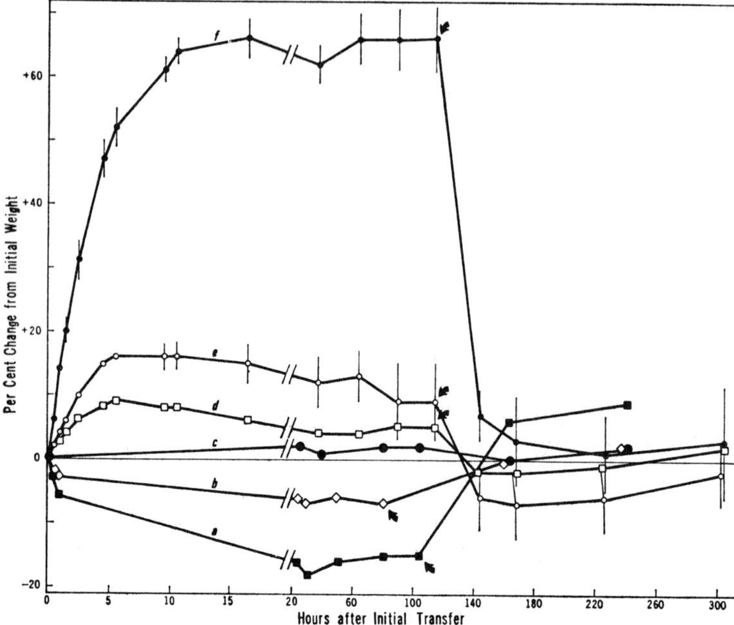

Figure 1–2. Time course of weight changes in sipunculid *Themiste* after transfer at zero time to the following: lowest curve, *a,* 139% S.W.; *b,* 111% S.W.; *c,* 98% S.W.; *d,* 89% S.W.; *e,* 80% S.W.; *f,* 49% S.W. At points indicated by arrows worms were returned to 98% S.W. No volume regulation. (From Oglesby, L. C., Comp. Biochem. Physiol. *26*:155–177, 1968.)

equalling that in concentrated media and diminishing in dilute media. It swells freely in dilute media and shrinks somewhat in concentrated media[258] (Fig. 1–2).

The flatworm *Gunda* (now *Procerodes*) *ulvae* lives in intertidal zones, where it may be exposed alternately for several hours twice each day to fresh water and then to undiluted sea water. In soft tap water *Procerodes* swells rapidly and dies within 48 hours, but in diluted sea water or stream water containing significant amounts of calcium, swelling is retarded. Substances other than calcium salts fail to retard swelling.[373] In dilute sea water, water passes to the parenchyma, which swells; then it collects in vacuoles in endodermal cells lining the gut. Thus, body volume is kept relatively constant and other body cells are kept from becoming diluted. The vacuoles remain while the worm is in dilute salt water, but some active process prevents further volume increase.

Marine molluscs are osmoconformers with varying degrees of stenohalinity. Cephalopods are strictly limited to full salinity. Some of the shell-less gastropods (nudibranchs and prosobranchs such as *Doris* and *Onchidium*) swell or shrink in dilute or concentrated sea water, with little volume regulation for many hours. Others (*Aplysia*) show some gain or loss of salt in limited volume regulation.

In the limpet *Acmaea* a linear relation of water content of the whole animal and of muscle to salinity was observed; there was no volume regulation during a week of altered salinity. The extravisceral space provides a sort of osmotic buffer for rapid changes; it is large in concentrated medium and small in dilute medium.[372]

Behavioral responses are made to salinity change. Reattachment by limpets to a substratum fails beyond a limited dilution. Snails such as the brackish-water *Hydrobia* alter blood concentration rapidly if they are active, and more slowly when withdrawn into the shell; but they remain isosmotic at all concentrations of the medium.[10] *Littorina* survives in diluted water or in air in proportion to the tightness of fit of the operculum. Spontaneous electrical activity in a particular neuron of the visceral ganglion of *Aplysia* decreases when dilute sea water is applied to specific osmoreceptors on the osphradium, presumably triggering avoidance behavior.[340]

When such bivalve molluscs as *Crassostrea, Modiolus,* and *Mytilus* are transferred to diluted sea water, the valves remain closed for many days and the animals survive, with their hemolymph becoming diluted very little if at all. However, if the valves are forced open, there is weight gain and rapid dilution; there is volume regulation for small dilutions, but no volume regulation for two days in

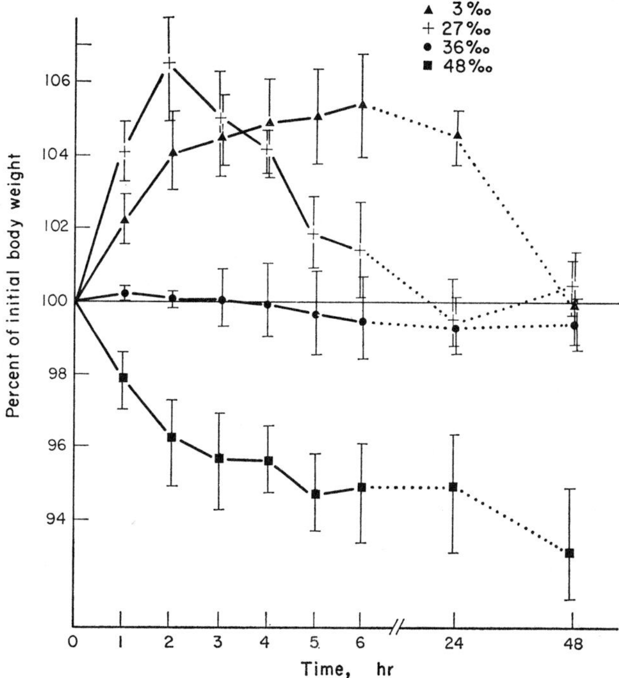

Figure 1-3. Time course of weight change in *Modiolus* with valves propped open. Animals transferred from normal habitat (36 0/00 salinity) at time zero to concentrations shown. Some recovery of volume shown in 27 0/00 salinity. (From Pierce, S. K., Comp. Biochem. Physiol. *39A*:103–117, 1971.)

greater dilution. Only such species as *Mytilus* can live in dilutions of 50 per cent[268] (Fig. 1-3). Differences in salinity tolerance among genera are correlated with natural habitat.[362]

Careful measurements of fluids in several molluscs show slight gradients in osmoconcentration corresponding to Donnan effects in the following series: S.W. = mantle fluid < pericardial fluid < hemolymph; each body fluid falls or rises in concentration with the medium, but hemolymph concentration is higher than that in sea water by a constant amount.[267] In *Modiolus,* with valves propped open, volume is regulated with a decrease in internal osmoconcentration in dilute saline by loss of amino acids, whereas in concentrated sea water there is little volume regulation; i.e., water is lost. Blood remains about 20 mOsm more concentrated than S.W. at different concentrations (i.e., it conforms but

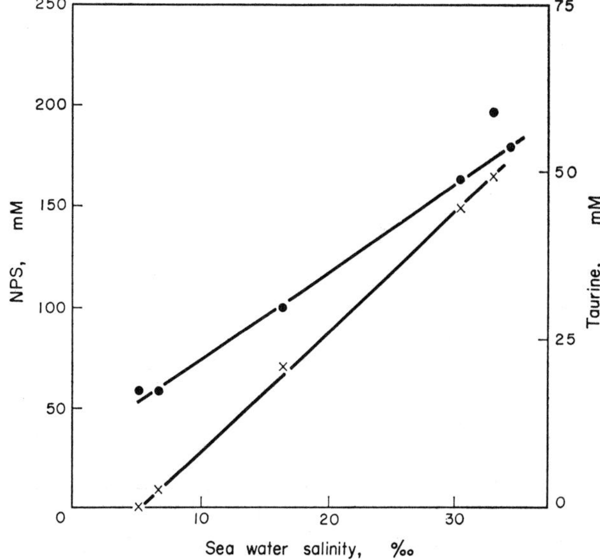

Figure 1-4. Concentrations of non-protein nitrogenous compounds (ninhydrin positive substances) (●) and taurine (×) in body fluid of *Mytilus* in different salinities (0/00). (From Lange, R., Comp. Biochem. Physiol. *10*:173–179, 1963.)

is in Donnan equilibrium).[268] When *Mytilus* conforms to different salinities, most of the change in intracellular concentration, and part of that in the hemolymph, is due to small organic molecules rather than to salts. Figure 1–4 shows that the increases or decreases in osmoconcentration are principally in ninhydrin positive substances (amino acids and especially taurine) in muscle.[205] In several bivalves (*Rangia*[5] and *Mya*) the free amino acids vary in proportion to the total osmoconcentration; taurine concentration may vary later. In adductor muscle of *Mya,* glycine and alanine concentrations are proportional to the salinity of the medium.[365] Thus, osmoconformity is permitted with maintenance of relative constancy of ionic concentrations within the animal.

In brackish water, as in the Baltic, oxygen consumption by Mytilus is higher than in normal sea water (North Sea); the heart rate, ciliary activity, and heat resistance all become reduced in dilute sea water.[299] The rate of filtering is reduced at both high and low salinities, and the limits are different for Baltic and North Sea populations; however, after 7 to 10 days of cross-acclimation, the original North Sea animals show maximum filtering at a lower salinity and Baltic *Mytilus* at a higher concentration than originally[349] (Fig. 1–5). It appears that the population differences in salinity optima are non-genetic. There are species differences in the dilution limit for gill cilia; *Aquipecten* gills stop activity in sea water at 12 to 15 0/00, and *Modiolus* and *Crassostrea* stop at 3 to 4 0/00.[362]

Some polychaetes, such as *Arenicola marina, Nereis pelagica, and Perinereis cultrifera,* show osmotic conformity with the medium and have little or no volume regulation. The intracellular concentration in *Arenicola* muscles is adjusted by variations in free amino acids, much as in *Mytilus.*[98]

Marine crustaceans and Arachnoidea over-

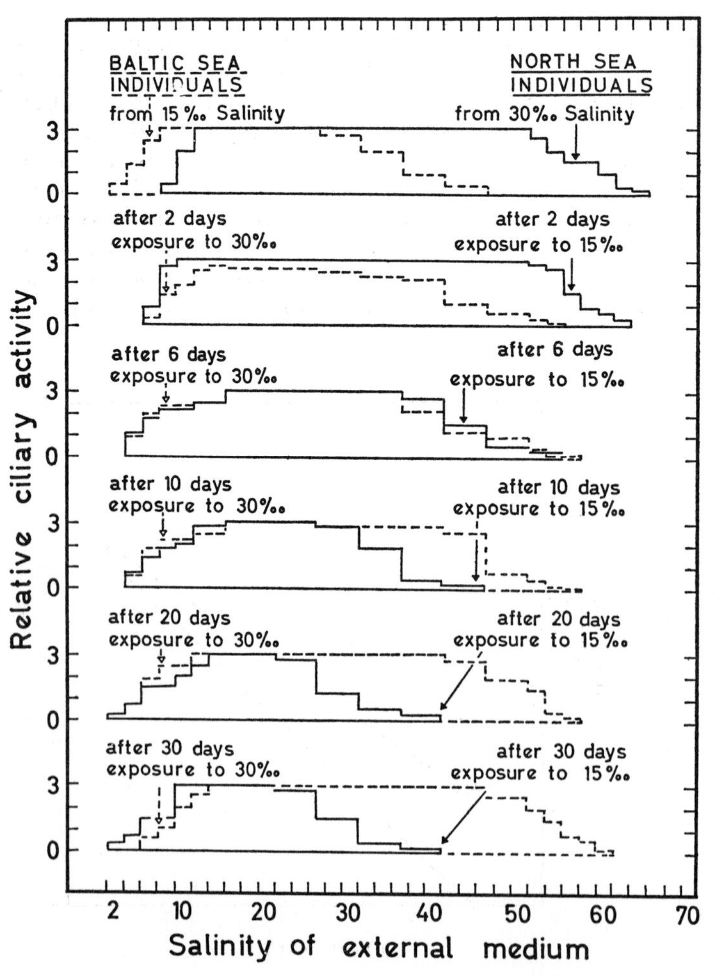

Figure 1–5. Relative ciliary activity in gills of *Mytilus* from Baltic and from North Sea adapted to either 15 0/00 or 30 0/00 and measured over a wide range of salinities. Initial differences between the two populations are largely reversed by cross-acclimation. (Data from Theede, H., *in* Kinne, O., Neth. J. Sea Res. *3*:222–224, 1966.)

lap some brackish water species in osmotic properties. Deep water crabs such as *Maja* survive only a few hours in sea water diluted by more than 20%; in this dilution the crabs swell initially, but within three hours their weight decreases and the blood is isosmotic with the medium. Numerous other marine crustaceans behave similarly—*Palinurus, Portunus, Cancer antennarius, Hyas,* and *Pagurus. Libinia* fails to survive in medium more dilute than 90 to 95% sea water. *Homarus* is isosmotic with the medium, and produces urine which is isosmotic but different in composition from the hemolymph.[57] A barnacle, *Balanus balanoides,* remains closed in air or in dilute (50%) sea water; it opens when bathed with sea water.[13] A planktonic copepod, *Acartia tonsa,* is an osmoconformer but remains slightly hyperosmotic at all external concentrations, probably due to a Donnan effect.[203] Similarly, the king crab *Paralithoides* maintains hemolymph osmoconcentration about 15 mOsm above the medium at all concentrations.[225] In *Limulus* the OC_i/OC_o curve is slightly above the isosmotic line at concentrations below normal sea water.[294] In both *Paralithoides* and *Limulus,* the variation in blood osmoconcentration is produced by salts, but the osmotic concentration variation of *Limulus* muscle is produced by amino acids, homarine, and other nitrogenous compounds.[210]

In summary, the osmoconforming marine invertebrates are not completely at the mercy of their environment. Their responses to osmotic changes are homeokinetic, in that biological activity can continue within a limited range of internal osmotic concentration. Many of these animals show some volume regulation by salt excretion or uptake. In many conformers the changes in nonprotein nitrogenous compounds, especially amino acids, are larger than the changes in salts, especially in muscle and nerve; hence, osmotic concentration can vary more than can ion concentrations. A few euryhaline species maintain fluid concentration parallel to but greater than the medium; they do not regulate to a constant internal concentration but to a constant difference from the environment. A few osmoconformers have regions of fluid storage as buffers to tissue cells; some show behavioral responses for avoiding dilution. Regulation of ions is apparently more important than regulation of volume, and volume control in turn is more important than regulation of osmotic concentration.

LIMITED HYPEROSMOTIC REGULATION

Adaptations to Brackish Water

Some phyla and classes that are predominantly marine have certain genera which live in brackish or fresh water. The nereid polychaetes show increases in ability to regulate volume and in tolerance of dilute media as follows: *Perinereis cultrifera, Nereis vexillosa* < *Nereis pelagica* < *Neanthes virens* < *Nereis diversicolor, Nereis succinea, Laeonereis culveri* < *Nereis limnicola, Neanthes lighti.*[332] The first three species are stenohaline marine; *N. virens* can regulate its volume in a medium whose salinity is as low as 3 to 8 0/00, and can live in the laboratory with salinities below 1 0/00 (i.e., 1.4% S.W. or 2 to 4 mM Cl). *Laeonereis culveri* can live in salinities below 10 mM Cl, and *Nereis limnicola* can live and reproduce in fresh water[256, 257, 328, 329] (see Fig. 1–6). Distribution of *Nereis diversicolor* in Finland appears to be limited by its low tolerance of diluted sea water at low temperatures; larvae are more sensitive than adult worms.

In dilute sea water *N. diversicolor* remains hyperosmotic to the medium, but is not able to maintain the level of blood concentration that it had in sea water; oxygen consumption increases initially on transfer to dilute sea water, partly because of muscular resistance to swelling. In dilute sea water, osmotic and volume regulation fail if cyanide is added, if O_2 is much reduced, or if the medium lacks calcium.

Tracer studies show a greater chloride exchange and water permeability in *N. virens* than in *N. diversicolor;* also, *N. cultrifera* has a much greater Na^+ permeability than *N. diversicolor.* When *N. diversicolor* is in very dilute sea water, its permeability to ions and water is less than when it is in sea water,[175] and there is active absorption through the body surface of chloride from the dilute medium. In sea water *N. diversicolor* normally exchanges some 8 per cent of its body Cl^- per hour.[175] In distilled water the rate of salt loss is 2.5 times greater and the rate of swelling (osmotic uptake) is 4 times greater in *N. succinea* than in *N. limnicola.* Water permeability, as measured by influx of D_2O, is similar in *N. succinea* and *N. limnicola* and is independent of osmotic gradient.[328] Diffusional influx is greater than that predicted from osmotic swelling. Specimens of *N.*

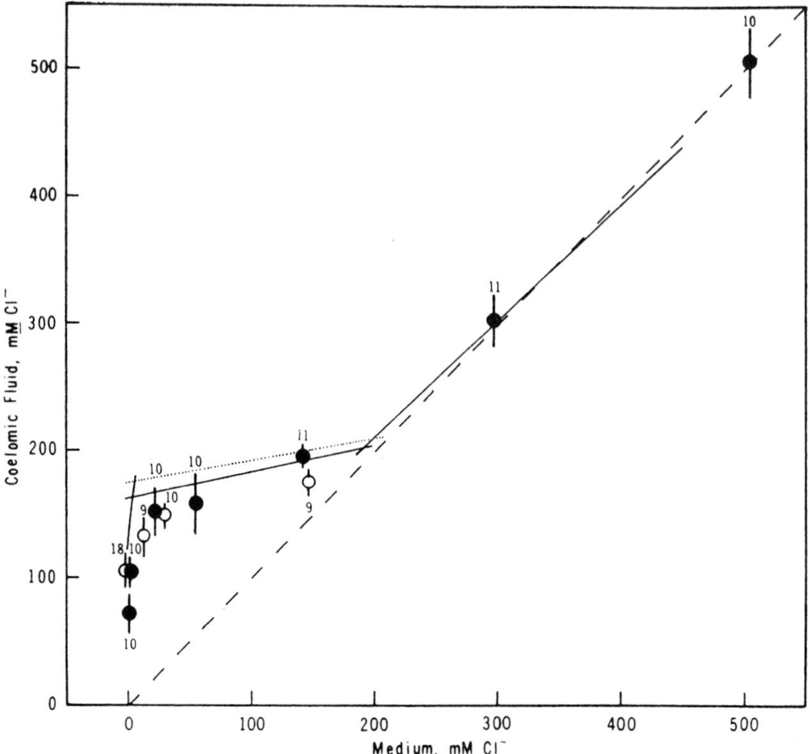

Figure 1–6. Chloride concentration in coelomic fluid of *Nereis limnicola* (two different populations) acclimated to media of different salinities. Regulation over the range from 20 to 180 mM Cl^-. (From Oglesby, L. C., Biol. Bull. *134*:118–138, 1968.)

limnicola collected in the upper, fresh portions of a river show nephridial canal diameters of 20.8 μ, while the worms from the brackish regions near the mouth of the river have nephridial canal diameters of 8.5 μ.[174] *N. limnicola* is reproductively adapted to dilute water in that it reproduces viviparously, rather than by shedding eggs and sperm as do marine and brackish worms.[329, 332] Salt loss and water gain are given for four species in distilled water:[328]

	salt loss, mM/hr	*vol. gain, %/hr*	*location*
marine *N. succinea*	0.068	55	Kiel
marine *N. succinea*	0.039	57	Berkeley
brackish *N. diversicolor*	0.0129	24	Kiel
freshwater *N. limnicola*	0.0086	10	San Francisco

N. limnicola from the San Francisco Bay area undergo seasonal fluctuations from nearly fresh water to 85% S.W.; volume regulation is good and salt fluxes are slower than water fluxes. When specimens which had been adapted to 49% S.W. were compared to those which had been adapted to 3 to 5% S.W. and then transferred to higher concentration, those previously adapted to the lower salinity showed a much lower salt permeability; hence, cellular changes had been induced by acclimation.[256, 257] One of the initial adjustments shown by numerous polychaetes on transfer to reduced salinity is the movement of amino acids out of tissues into coelomic fluid.[64]

Nereis diversicolor is a brackish water polychaete which tolerates extreme dilutions. It actively absorbs chloride against a concentration gradient from a medium of 20 mM Cl^-, particularly if it has previously been exposed to fresh water (less than 10 mM Cl^-). Passive permeability to water and to chloride decreases, while active uptake of Cl^- increases. The potential across the body wall was −17.3 mV in pond water, −3.7 mV in 10% S.W., and zero in 50% S.W. Water exchange across the body wall (measured by D_2O) dropped from 12.53% of body water per hour in S.W. to between 5.9 and 6.9% in pond water. However, the relation between chlorinity of the medium and loss of Cl and water by urine is not linear; rather, urine loss is maximal at 10% S.W. (62 mM Cl^-). This

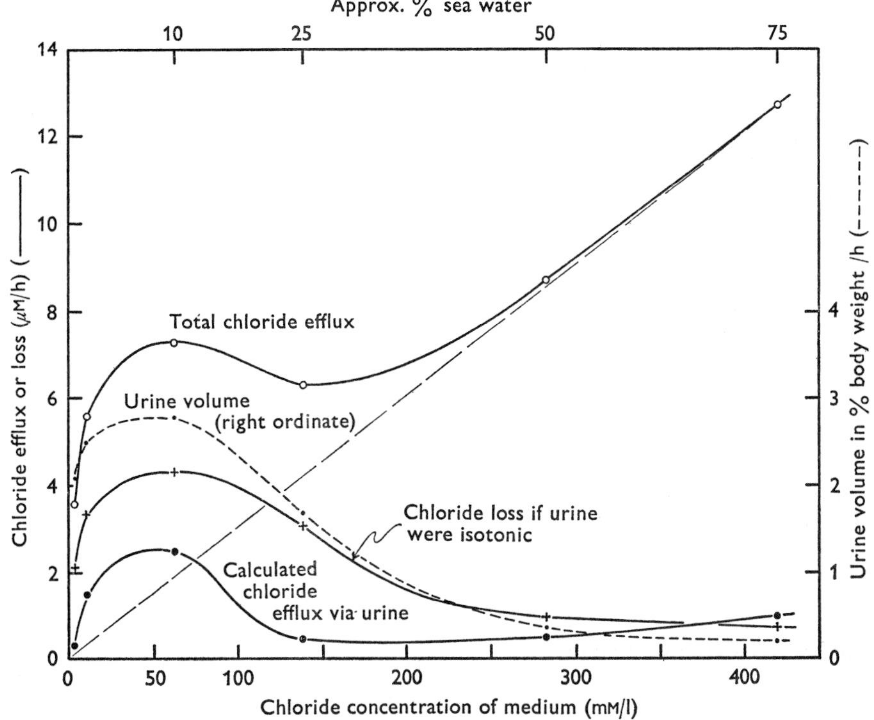

Figure 1–7. Diagram of chloride concentrations and urine volume of *Nereis diversicolor* in media of salinities indicated by chloride concentrations. Urine volume and chloride loss are maximal in low range of external chloride. (From Smith, R. I., J. Exp. Biol. *53*:93–100, 1970.)

may be related to a very steep drop in water permeability below that concentration and maximal Cl uptake in 10% S.W.[332] (Fig. 1–7).

Both Ca^{++} and Mg^{++} concentrations in coelomic fluid of *Nereis diversicolor* are less than those predicted by the electrochemical gradient. Calcium flux is proportional to salinity, but not to external calcium concentration. Influx of calcium as measured by tracer is accomplished in two steps, as follows:

	fast exchanging	*slow exchanging*
in 100% S.W.	$J = 4.3$ mM/kg $T_{1/2} = 8$ hr.	0.8 mM/kg 50 hr.
in 1% S.W.	$J = 0.62$ mM/kg $T_{1/2} = 22$ hr.	0.62 mM/kg 300 hr.

Hence, the fast flux decreases on dilution as do the chloride and water fluxes.

In summary, nereid worms show adaptive differences by species in the following characteristics: larval tolerance of dilution, permeability to salts and water, active absorption of ions, nephridial excretion, and mode of reproduction. Euryhaline species such as *N. diversicolor*, according to salinity of habitat, adjust permeability to water and ions, and the active uptake of ions.

CRUSTACEANS – MARINE, BRACKISH-WATER, FRESHWATER, AND TERRESTRIAL

Crustaceans show a greater variety of osmotic behavior than does any other animal group. Many are stenohaline osmoconformers restricted to full-strength sea water. A number (especially shore crabs) are able to regulate hyperosmotically in dilute sea water, although they are isosmotic or even hyposmotic in higher salinities. As shown in Figures 1–8 through 1–11, there is much variation in the capacity of adults to maintain the hemolymph at an osmoconcentration higher or lower than that of the medium. Some crustaceans live most of their lives on land, and a few will drown in sea water. Brine shrimps live in hypersaline water and are remarkable for their hyposmotic regulation. A few crabs and all crayfish live complete life cycles in fresh water.

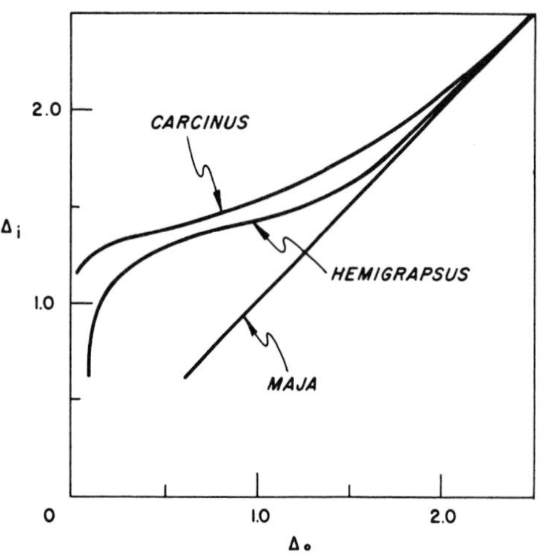

Figure 1-8. Blood osmoconcentration (Δ_i) of several marine Crustacea as a function of environmental osmoconcentration (Δ_o).

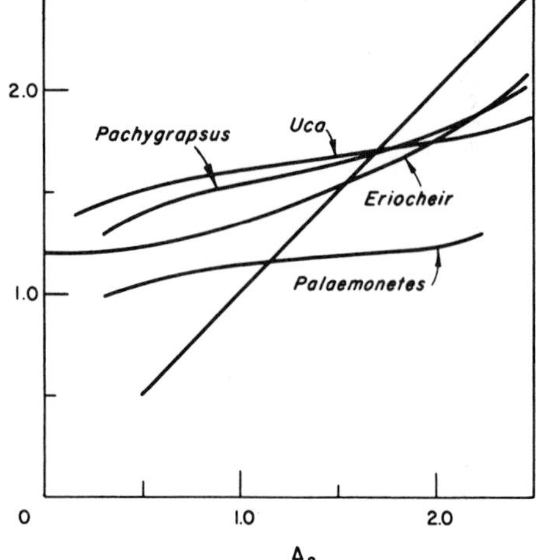

Figure 1-9. Blood osmoconcentration (Δ_i) of several hyperosmotically and hyposmotically regulating crustaceans as a function of environmental osmoconcentration (Δ_o).

Hyperosmotic Regulation; Entry into Estuaries; Shore Crabs

Carcinus from the Baltic are more hyperosmotic to a medium of 5 to 10 0/00 than are *Carcinus* from the North Sea; in higher salinities they are similar. Cross-acclimation reduces the differences between the two populations, which are partly genetic and partly environmentally induced.[350]

Blue crabs, *Callinectes,* can live as adults in potable water, particularly if it is high in calcium, as in the upper St. Johns River in Florida. However, they must return to the river mouth to breed, and some of the upriver population may not reproduce. The zoeal larvae take up water at the molt stages and may be very sensitive to dilution of the medium. *Rithropanopeus* regulates hyperosmotically from the time of hatching, but in the fifth day as a zoea it osmoconforms; then, before molting to a megalops, it again becomes a hyperosmotic regulator. *Callinectes* hyper-regulates except for a 48-hour period in larval life.[76] Thus, these species can spend their entire lives in dilute water except for a brief larval period when they must be in sea water. *Libinia* is isosmotic at hatching, develops ability to hyper-regulate, and then

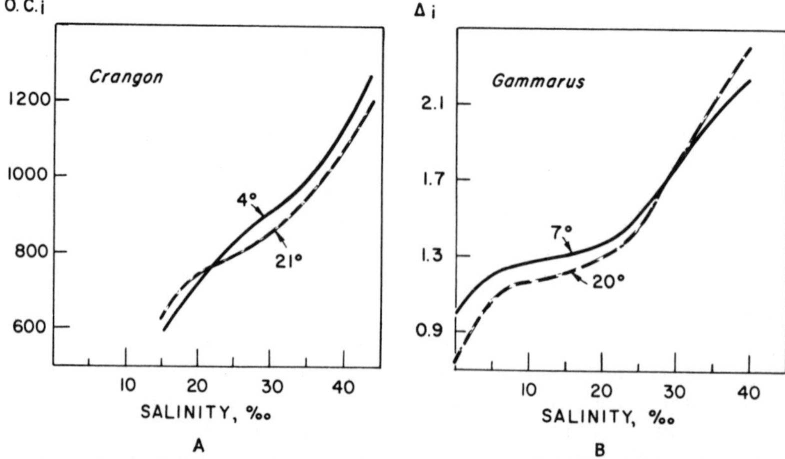

Figure 1-10. *A,* blood concentration in relative conductivity units as a function of salinity of medium for *Crangon crangon,* adapted to 4° C and 21° C. *B,* blood osmoconcentration, Δ_i as a function of salinity of medium for *Gammarus duebeni,* adapted to 7° C and 20° C.

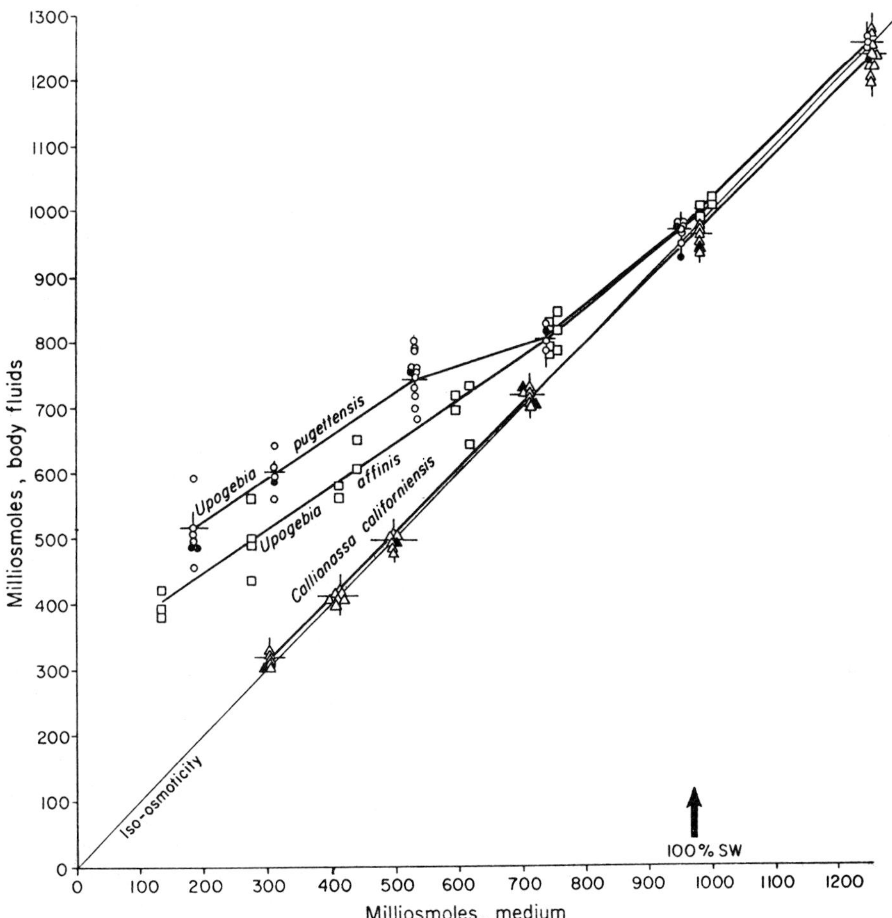

Figure 1-11. Osmotic concentration of blood (open symbols) and urine (closed symbols) of an osmoconformer (*Callianassa*) and two limited regulators (two species of *Upogebia*). (From Thompson, L. C., and A. W. Pritchard, Biol. Bull. *136*:114–129, 1969.

loses it; hence, it must remain in the sea.[176] A few decapod crabs have made the complete transition to fresh water, and breed as well as live their entire life cycles there.

Temperature interacts with salinity to modify regulatory capacity in some crabs, and to control migrations of some shrimps and prawns. *Callinectes* is capable of hyposmotic regulation only at high temperature; it can live in hypersaline lagoons in the summer. Its hemolymph concentration is higher at low than at high temperature, and it moves out of both low and high salinity waters in the winter.[12] *Hemigrapsus oregonensis* also has higher blood concentration in winter than in summer; *Hemigrapsus nudus* shows no such difference.[90] Survival of larvae to first crab instar in *Halicarcinus australis* was zero at 30°C at all salinities; it was nearly 100% at 25°C at chlorinity 6.4 to 12.1 0/00, at 19°C at chlorinity 6.4 to 19.3 0/00, and at 13°C at chlorinity 12.4 0/00 only.[220] It is evident that under natural conditions, the osmoconcentrations of many estuarine crustaceans depend on temperature as well as on concentration of the medium.

Permeability to Water and Ions. One of the most important mechanisms permitting euryhalinity is reduced permeability of the body surface to water and salts; shore crabs are less permeable than stenohaline marine crabs. For example, the pelagic *Maja* recovers its normal volume after initial swelling in dilute sea water much faster than does the shore crab *Cancer*. Rate constants for flux of tracer water, given in hr^{-1}, show decreasing permeabilities in the following series:[296]

Macropipus (Portunas) in S.W.	2.39
Carcinus in S.W. (or in 70% S.W.)	0.79
Palaemonetes in 120% S.W. and in 70% S.W.	0.64
in 10% S.W.	0.55
Astacus in F.W.	0.20

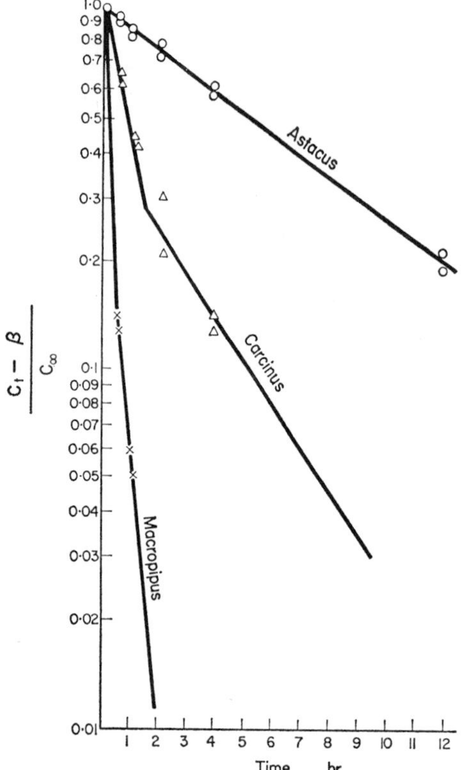

Figure 1–12. Water permeability in an osmoconformer *(Macropipus),* a moderate regulator *(Carcinus),* and a freshwater crustacean *(Astacus),* showing decreasing permeabilities in the series. Amount of tritiated water remaining in crab at different times after removal from labeled medium. (From Rudy, P. P., Comp. Biochem. Physiol. *22*:581–589, 1967.)

Similar evidence for decreased permeability in dilute media is given by water turnover in *Carcinus*:[330, 331]

	Medium		
	30% S.W.	*75% S.W.*	*100% S.W.*
daily influx of H_2O (% body wt.)	17.2	11.5	nil
daily urine volume (% body wt.)	31.3	11.1	3.6
% body water exchanged per hour	1.76	2.73	2.36

At steady state in a given medium, influx equals outflux. Osmotic fluxes of water are estimated from the urine volumes observed in different media; in general, the osmotic flux is larger than that measured with labeled water. In marine crab species the water permeability is greater than that in brackish or freshwater species, and the marine animals cannot alter their permeabilities as can the freshwater forms. The internal water of *Carcinus* and *Macropipus* exchanges as a two-pool system; in *Macropipus* the fast pool constitutes 70 per cent of body water[296] (Fig. 1–12). In *Rhithropanopeus* the passive influx of D_2O is less in dilute than in normal sea water.[331]

Not only water permeability but also salt permeability is decreased in dilute media, as shown in the following data in μEq Na/20 g crab/hr for *Pachygrapsus*:[296]

	Medium		
	160% S.W.	*100% S.W.*	*60% S.W.*
passive gain	814	495	234
active gain	–	–	158
urine loss	6	14	91
passive loss	749	489	91
active excretion	59	–	–

In *Hemigrapsus nudus* the passive influx of sodium is 20.4 μM/g/hr in full sea water and only 3.9 μM/g/hr in 35% S.W. *Pachygrapsus* in intermolt in sea water exchanges sodium at the rate of 20 μEq/g/hr; the exchangeable Na is in two pools, and about 15 per cent of body Na is not exchangeable. In the soft-shelled stage the exchange is four times faster than in intermolt. Most of the exchange is across the gills; virtually none is across the carapace. Thus, the species that can live in brackish water have lower permeabilities for water and ions than do strictly marine species; as the former go from concentrated to dilute medium, their permeabilities decrease. The membrane properties responsible for such permeability changes are not known.

Excretion. A second capacity which permits invasion of dilute media is the ability to increase volume and dilution of urine. The urine of *Carcinus* in sea water can be isosmotic or slightly hyperosmotic or hyposmotic to the hemolymph; it serves mainly in ionic rather than in osmotic regulation. In dilute sea water the urine is hyposmotic to the hemolymph and hence is important in osmoregulation; however, it is more concentrated than the medium. In *Carcinus* the volume of urine flow increased from 3.6 per cent of body weight per day in sea water to 33 per cent of body weight per day in 40% S.W.[315] The sodium loss by urine was 2.7 per cent of the total efflux in sea water and 21 per cent of total efflux in 40% S.W. Each antennary gland in *Carcinus* is an excretory organ consisting of a coelomosac across which filtration can occur, a tubular labyrinth, and a bladder. The ratio of urine to blood (u/b) concentration of non-reabsorbed substances such as inulin and sorbitol is 1.0 in sea water, indicating that no water is reabsorbed after filtration.

Glucose is actively reabsorbed with a threshold of 150 mg per cent in the blood, and when reabsorption is blocked by phlorizin, the u/b ratio for glucose approaches 1.0. In 50% S.W., urine flow increases four times and glucose reabsorption increases only two times.[34]

Homarus is an osmoconformer with OC_i higher by a small constant amount than OC_o; urine is isosmotic with blood, and rectal fluid is slightly hyperosmotic.[84] In *Hemigrapsus nudus* sodium loss in urine was 0.95 μM/g/hr in 100% S.W. and 4.49 μM/g/hr in 35% S.W.[282] The u/b ratio for chloride was 1.26 in isosmotic media and 1.63 in hyperosmotic media, showing that water was reabsorbed faster than Cl from tubular fluid. In hypotonic media the u/b ratio for Cl was 0.64, indicating faster reabsorption of Cl than of water.

In general, the urine flow is less in marine than in brackish water animals and in euryhaline genera; strictly freshwater forms tend to have lower urine flow than do shore crabs placed in fresh water. *Eriocheir,* the wool-handed crab of western Europe, is extremely euryhaline and shows little variation in urine volume with change in medium concentration; it relies on low permeability and active uptake of salt by the gills for osmoregulation. A comparison of normal loss of Na by urine in various species follows:[317]

	Na loss by urine *μM/100 g/hr*
Carcinus	1782 in S.W.
Eriocheir	20 in Br.W.
Potamon	80 in F.W.
Astacus	15 in F.W.

Active Uptake of Ions. A third and very important mechanism of regulation in dilute media is active uptake, largely by gill membranes, of sodium and chloride ions. When *Carcinus* is placed in a dilute medium, salt is lost; and at a critical blood concentration of sodium, mobilization of amino acids from tissues and active absorption of sodium from the medium are activated.[33] The active absorption mechanism is half saturated at a concentration of 20 mM Na in the medium. The activation is a function of internal sodium concentration;[315,316] absorption starts when blood sodium falls to about 400 mM and is saturated at about 300 mM. *Carcinus* increases Na uptake by 13 times for a 10 to 30 mM decrease in blood concentration.

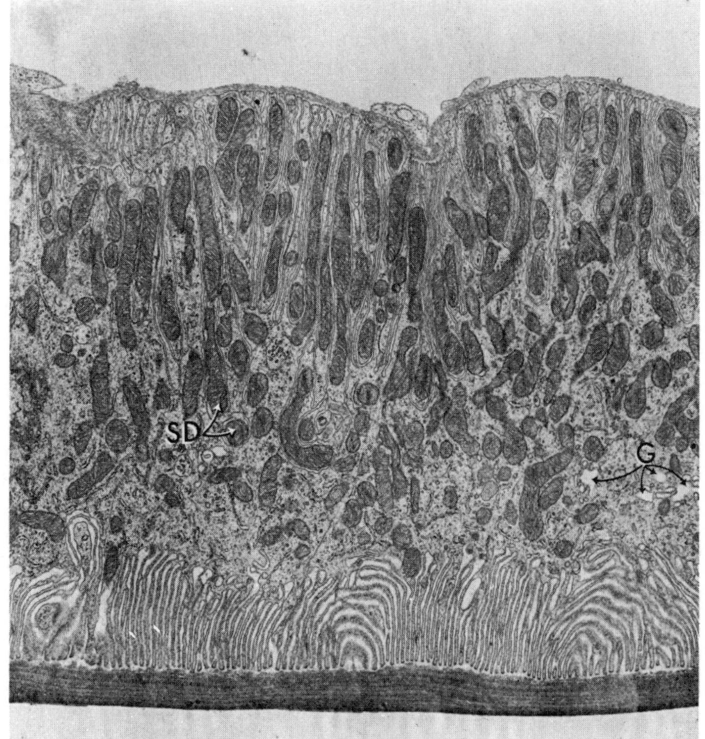

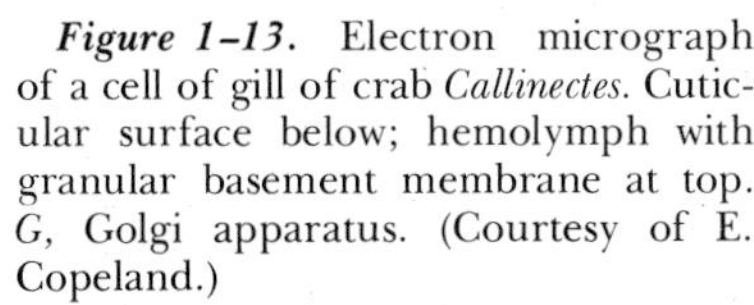
Figure 1–13. Electron micrograph of a cell of gill of crab *Callinectes.* Cuticular surface below; hemolymph with granular basement membrane at top. *G,* Golgi apparatus. (Courtesy of E. Copeland.)

Active uptake by *Hemigrapsus* becomes stimulated in 50% S.W.

Callinectes actively absorbs salt from a dilute medium, and the gills have areas that are readily stained with silver. The gill patches are much more extensive in animals from dilute brackish or fresh water; the cells have long, flat mitochondria between folds or interdigitations on the hemolymph side (Fig. 1-13). Strictly marine species lack the capacity for active salt absorption; when the concentrations of external sodium for half-saturation of the uptake process are compared for different genera, it is seen that the brackish forms such as *Carcinus* saturate in a medium of higher concentration than do the freshwater species.

A correlation of transport of sodium with ecology follows:[316]

Environment, Animal	*blood Na (mM)*	*Na loss (μM/hr/ 50 g)*	*external concentration for half saturation of uptake (mM)*
marine *Carcinus*	300	891	20
brackish *Eriocheir*	280	102	1.0
brackish *Gammarus duebeni*			1.5
freshwater *Astacus*	184	7.5	0.2 to 0.3
freshwater *Gammarus pulex*			0.15

Potentials Across Body Wall. Related to permeability and to active transport by gills are electric potentials which can be measured between blood and medium in intact animals or across isolated gills. The blood in intact *Callinectes* is slightly negative with respect to normal sea water, and more so in 50% S.W.; it is positive with respect to 150% S.W. Replacement of chloride by acetate or sulfate in the medium reverses the negativity, suggesting an active chloride absorption in the dilute medium.[234] Uptake of ions by gills of *Callinectes* is diminished by ouabain or by substituting Li for some of the Na. Similarly, *Pachygrapsus* hemolymph is negative by 3 mV in 80% S.W., by 2 mV in 100% S.W., and by 1 mV in 140% S.W.[296] In a stenohaline marine crab, *Libinia,* no potentials were recorded, even in 90% S.W., the lowest dilution the crab could tolerate; presumably this crab lacks active transport mechanisms. In the prawn *Palaemonetes* the potential is minimal in a medium isosmotic to the blood (50 to 65% S.W.) and is more negative if the medium is more or less concentrated than the blood. This is interpreted to mean that chloride is actively absorbed in a dilute medium and that sodium is actively extruded in a concentrated medium.[275] Presumably both chloride and sodium contribute to the potential in varying proportions in different species. No potential was recorded across the gill of *Maja,* in which ion flux is passive; but the inner surface of a *Carcinus* gill was slightly positive when $Na_o = Na_i$ and was negative when $Na_o < Na_i$.[184]

In order to ascertain whether sodium uptake was active, the flux ratio analysis (p. 4) was applied. The potentials across the gills of *Maja* and *Carcinus* were shown to be nearly zero through a range of sea water dilutions. Therefore, if an ion is diffusing passively, J_i/J_o must equal C_o/C_i. The following data indicate active uptake by *Carcinus* but not by *Maja* in dilute media, where the ratio C_o/C_i is reduced:[184]

Na_o/Na_i	1.0	0.76	0.4	0.2
Maja, J_i/J_o	1.0	0.76	0.38	0.19
Carcinus, J_i/J_o	0.95	0.90	0.70	0.60

Metabolism. The oxygen consumption by a euryhaline crustacean (*Carcinus, Gammarus, Locusta,* or *Uca*) is higher in a dilute medium than it is in full strength sea water. Part of the extra energy requirement must be for active absorption of ions and production of dilute urine. Other crustaceans, such as *Ocypode* and *Palaemonetes,* show increased oxygen consumption in either dilute or concentrated sea water, but the metabolism of *Eriocheir* is independent of salinity.[185] The oxygen consumption by mitochondria isolated from crab gills increases in a dilute medium, and the increase is greater in mitochondria from brackish-water crabs than in those from marine crabs; this may be a result of mitochondrial swelling which could also occur *in vivo.*[183] Uptake of ions by gills of *Callinectes* is diminished by ouabain and by Na-Li medium; this fraction of metabolism is related to Na transport.[234]

Changes in Amino Acids. It was mentioned earlier (p. 15) that in many osmoconformers osmotic changes, especially in tissues, are due more to changes in amino acid concentrations than to inorganic ion movements. When *Carcinus* is transferred to a dilute medium, muscle amino acids decrease and blood amino acids increase; the u/b ratio for amino acids then rises:[33]

	Medium	
	SEA WATER	FRESH WATER
free amino acids, muscle	271 mM	160 mM
free amino acids, blood	5.6 mM	2.4 mM

When prawns (*Leander*) from 30% S.W. and from 100% S.W. were compared, the blood and tissues were higher in osmotic concentration in the S.W. animals by 30 per cent; amino acids in muscle of the S.W. animals were higher by 23% than those in the dilute animals, as compared with a difference of 300% O.C. in the medium.[173] In *Eriocheir*, the muscle concentrations of alanine, glutamic acid, proline, and glycine (but not of arginine) were higher when the animals were in sea water than when they were in fresh water. Also, in sea water the excretion of ammonia was less than in fresh water, indicating amino acid synthesis in sea water.[113] Muscle amino acid concentration decreased by 41 per cent after the crabs had remained in fresh water for 16 days following a prior sojourn in sea water.[363]

Behavior. Adaptive behavior may be important in salinity selection by some estuarine and intertidal crustaceans. *Pachygrapsus*, after adaptation to high salinity, tends to select lower concentrations when placed in a gradient; the opposite behavior occurs in individuals adapted to dilute medium. An African marine crab, *Jasus*, has receptors on the antennules which respond to dilute sea water. Postlarvae of the pink shrimp *Penaeus*, which hatch in the sea and enter estuaries as larvae, tend to go shoreward toward low salinity, and can perceive a gradient of 1 0/00; juveniles and subadults, as they later return to the sea, are positively rheotactic at high salinities but not at low.[169]

Gammarids – Euryhaline Animals

The interaction of mechanisms of osmoregulation in various proportions is illustrated by the genus *Gammarus*, which has marine, brackish, and freshwater species. Some species are restricted in salinity range, while others are extremely euryhaline. Marine species of *Gammarus* show some hyperosmotic regulation in dilute medium; but regulation fails at higher concentrations in the marine species than in brackish or freshwater species (Fig. 1–14). The marine species *G. locusta* and the brackish species *G. duebeni* can be contrasted with the freshwater species *G. pulex* and *G. lacustris*. *Gammarus duebeni* is tolerant of a wide range of salinities and temperatures; it occurs in such places as the Black Sea, freshwater ponds in England, and warm springs in Iceland. Temperature and salinity tolerances are reciprocally related in that only optimal salinities (8 to 20 0/00) can be tolerated at extreme temperatures, and extreme salinities (high or low) can be

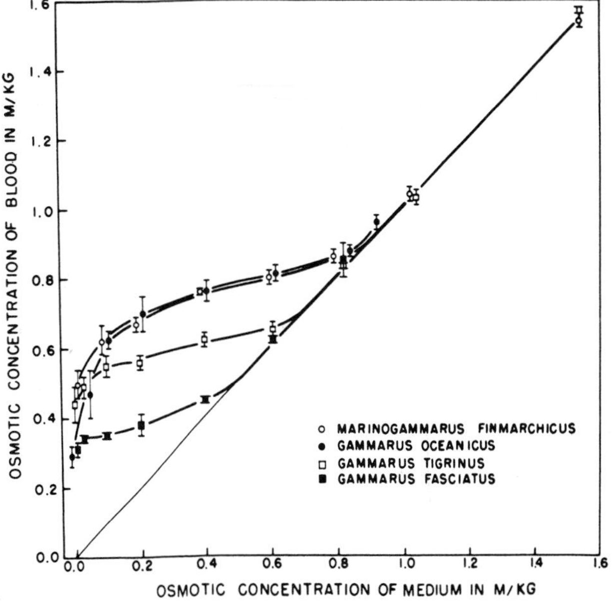

Figure 1–14. Osmoconcentration of blood as a function of the osmoconcentration of sea water in four species of gammarids. (From Werntz, H. O., Biol. Bull. *124*:225–239, 1963.)

tolerated only at optimal temperatures (4°C to 16°C). Also, hyperosmotic regulation is better at low temperatures than at high ones.[184a] In *G. duebeni* the hemolymph sodium decreased from 230 mM to 175 mM on going from medium containing 6 mM Na to that containing 0.5 mM. Production of hyposmotic urine began within two hours after transfer to the dilute medium even though the blood was not much diluted; hence, stimulation cannot be by blood concentration alone.[213, 214, 215, 216] In sea water, the urine is isosmotic to the blood, and 80 per cent of sodium efflux is by urine; in dilute medium the urine volume increases and its concentration decreases, so the excretory organs conserve sodium. In the freshwater *G. pulex* the urine was more dilute and less variable with external concentration than in *G. duebeni.* The excretory tubules of *G. duebeni* are longer than those of the marine *G. locusta.* In *G. pulex* at concentrations of Na below 0.2 mM, 82 per cent of Na loss is by diffusion and 18 per cent is by urine.[344]

As with crabs, the marine species *Marinogammarus finmarchicus* is more permeable to water and salts than are the brackish species *G. zaddachi* and *G. tigrinus.*[344] Another euryhaline amphipod produces hyposmotic urine in media of concentrations below 20 0/00 and isosmotic urine in media above this. The ions Na, K, Ca, and Cl are more concentrated, but Mg is consistently less concentrated, in the blood than in the medium.[242]

An important difference between the species of *Gammarus* is in their sodium uptakes. The uptake saturates at higher external concentrations of sodium in euryhaline *G. duebeni* than in freshwater *G. pulex.* In *G. lacustris* the uptake mechanism is activated to half maximum (K_m) at 0.14 mM external sodium, but in *G. duebeni,* K_m is at 1.5 to 2.5 mM Na.[344] Active absorption from the medium and reabsorption by the excretory organ are increased as the external concentration is reduced. The sensing mechanism and the nature of the membrane regulation of active uptake are unknown, but must vary with species and with acclimation of individuals.

In summary, numerous adaptations permit marine crustaceans to invade estuaries and to enter fresh water in varying degrees. The adaptations are partly genetic and partly environmentally modifiable. Larvae are more sensitive to dilution than are adults, and the sensitivity is increased at temperature extremes. The adaptations for hyperosmoticity are: limited and modified permeability to water and salts; excretion of copious, dilute urine; active absorption of ions; potential gradients across gills; use of amino acids to buffer osmotic changes; and adaptive behavior.

Hyposmotic Regulation

Invasion of Land and of Hypersaline Waters. A few of the crustaceans that regulate hyperosmotically also regulate hyposmotically; that is, they maintain their blood concentrations lower than that of the medium. Their blood-medium curves show varying degrees of flatness out to limiting concentrations at each end (Fig. 1–15). Some of the hyporegulating crustaceans are normally more dilute in body fluids than sea water, some live in hypersaline lakes, and many of them spend much time on land. Brackish-water calanoid copepods show both hyperosmotic and hyposmotic regulation; freshwater species show only hyperosmotic regulation.[51]

Intertidal and Land Crabs. Many crabs that spend time on land are capable of hyposmotic regulation. *Pachygrapsus crassipes* lives among rocks above low tide, but shows weak hyposmotic regulation in concentrated sea water. The fiddler crab, *Uca,* spends long periods on sand and mud flats; it is a good hyposmotic regulator and is protected against desiccation. Its blood concentration increases less during prolonged exposure to air than does that of rock crabs such as *Hemigrapsus.* The coconut crab of the South Pacific, *Birgus,* is a hyposmotic regulator under experimental conditions; it is normally terrestrial and can be drowned by immersion in sea water. *Birgus* drinks fresh water and places water in its gill chamber without becoming immersed in it.[144] In *Uca* living in either 100% S.W. or 175% S.W., the urine may be slightly hyperosmotic to the blood.

A desiccated *Pachygrapsus* put into 140% S.W. shows net uptake of water, and in air it may secrete water outward into the branchial chamber. In both *Pachygrapsus* and *Uca* in 175% S.W., the urine concentration of magnesium is higher and that of sodium is lower than when they are in 100% S.W.; extrarenal excretion of sodium is indicated. Urine/serum ratios of calcium, magnesium, and total

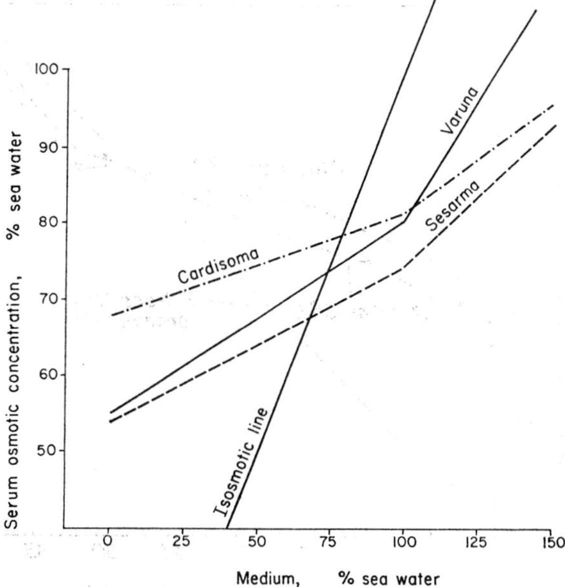

Figure 1-15. Serum osmotic concentrations of three species of land crab as a function of salinity of the medium. (From Gross, W. J., et al., Comp. Biochem. Physiol. *17*:641–660, 1966.)

osmoconcentration in the land crab *Cardisoma* are as follows:[145]

	Medium		
	TAPWATER	SEA WATER	150% SEA WATER
Ca	0.42	0.47	0.42
Mg	1.93	4.4	15.7
O.C.	0.99	0.99	0.99

Urine Mg increases as its osmotic concentration in hyperosmotic medium in *Cardisoma, Sesarma, Pachygrapsus,* and *Uca,* but not in the highly terrestrial *Gecarcinus.* The integument is the main route of water loss in land crabs; the crab *Potamon* in air loses less water than do insects, but less than do crayfish. The meaning of the capacity for hyposmotic regulation in land crabs is not clear; it has been suggested that such regulation is an adaptation to evaporation from the branchial chamber. However, the branchial fluid constitutes less than 3 per cent of body volume, and the salts contained in it would not put much stress on the blood. The degree of hypertonicity of the urine is small, and it may be related to the active secretion of magnesium by the antennary glands. In the entire series – subtidal *Cancer;* brackish to hypersaline *Hemigrapsus;* semiterrestrial *Pachygrapsus;* nocturnal, high-on-sand *Ocypode;* diurnal, muddy sand dweller *Uca;* and highly terrestrial *Gecarcinus* – the osmotic urine/blood ratio is close to unity.[144] Capacity to regulate blood sodium and osmoconcentration is lost at high temperatures; *Uca* and *Ocypode* die at 30°C in concentrated sea water.[125]

Land crabs use a variety of adaptations of respiration, water retention, and behavior. The anomuran *Coenobita* has blood isosmotic to 80% S.W.; its urine is isosmotic and of low volume. This crab normally enters the sea briefly at night to moisten its gills.[144] *Gecarcinus* lives well on moist sand, from which it can absorb water. It dies if the sand is moistened with sea water. *Gecarcinus* has large membranous pericardial sacs which help to maintain high humidity in the gill chamber, and which swell with water in proecdysis, providing water for the succeeding soft-shelled stage.[39] A molt-inhibiting hormone from the central nervous system at other times decreases or limits the swelling which prevails at the molt.[39] The foregut of *Gecarcinus* in intermolt is normally permeable to water and salts; if the eyestalk has been removed, the foregut is impermeable, but a thoracic ganglionic extract restores permeability.[233] *Gecarcinus* will drink water if any is available; water from the sand is conducted by capillarity by tufts of hairs between the second and third walking legs to the pericardial sacs, where it may be stored or conducted to the branchial chamber. From there it may be absorbed through the gills. Thus, the pericardial sacs both store water and provide a

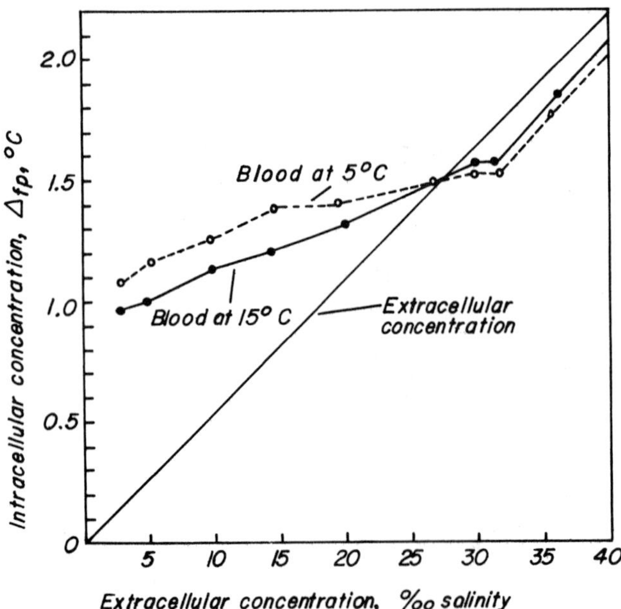

Figure 1-16. Osmotic concentration of blood of prawn *Crangon* as a function of salinity of medium at 5° C and 15° C. Hyperosmotic regulation better in cold than in warm environment. (After Flügel, H., Kieler Meersforsch. *16*:186–200, 1960.)

hydraulic assist in the molt. The gills are small and the branchial chambers may have an air-filled branchial space.[38, 40]

Land crabs rely extensively on visual and acoustic cues; at high temperatures, often in midday, they retreat into burrows. Most evaporative loss is through the integument, and a series in order of increasing cuticular permeability is: *Gecarcinus* < *Cardisoma* < *Uca* < *Sesarma* < *Callinectes* < *Menippe*.[156]

Shrimps, Brine Shrimps, Some Isopods. Capacity for hyposmotic regulation is found in many shrimps and prawns, some of which migrate between sea and brackish water. The European prawn *Palaemonetes varians* is normally isosmotic with 65% S.W., and is thus hyposmotic in normal sea water. Its urinary output increases in either dilute or concentrated sea water. In *Crangon* from the Baltic Sea, the resistance to both low and high salinity is maximum at 5°C; it cannot withstand brackish water at the freezing point. The blood-medium osmotic curves in both *Crangon* and *Gammarus duebeni* cross, so that at low salinity the blood is more concentrated at 5°C than at 15°C, and at high salinity the converse is true[114] (Fig. 1-16). Of two caridean shrimps, *Syncaris* produces hyposmotic urine in dilute medium while in *Palaemon* the urine is blood-isosmotic in both dilute and concentrated sea water.[47]

Palaemonetes varians is hyposmotic in normal sea water; its urine is isosmotic to the blood. The following data show the importance of extra-renal regulation:[276]

	100% S.W.	*65% S.W.*	*2% S.W.*
Medium Na	460 mEq		
Blood Na	328 mEq		
Medium Cl	540 mEq		
Blood Cl	342 mEq		
rate constant, Na influx	1.25 hr^{-1}		
rate constant, Na efflux	1.09 hr^{-1}		
Na flux, mM/kg/hr	150	73.5	18
Cl flux, mM/kg/hr	50	20.7	28.2
potential (negative)	13–40 mV	7–10 mV (50% S.W.)	11–32 mV

In 100% S.W. sodium is actively excreted extrarenally, making the blood negative and keeping the chloride in equilibrium. In 2% S.W. chloride is actively absorbed, again causing electrical negativity; some sodium is also actively absorbed. Regulation fails below 2% S.W.

Artemia, the brine shrimp, thrives in Great Salt Lake at 220 0/00, and in salterns where sodium chloride is precipitating. It lives in sea water as dilute as 2.6 0/00, and requires that the principal salt be NaCl.[78] In a salinity of 58 0/00 the blood was equivalent osmotically to 13 0/00; in a medium three times more concentrated, the blood concentration increased less than two-fold. The blood is

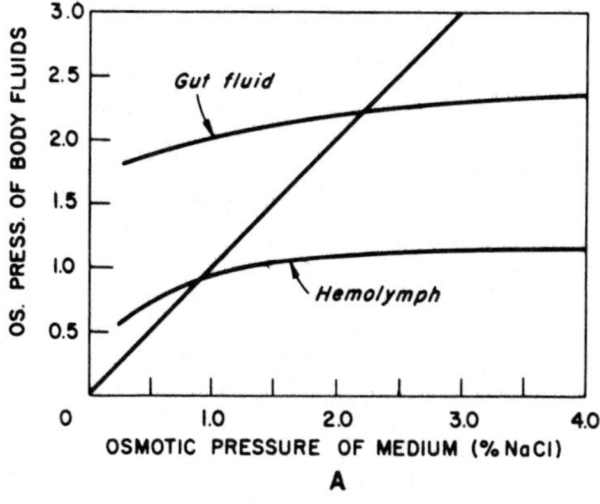

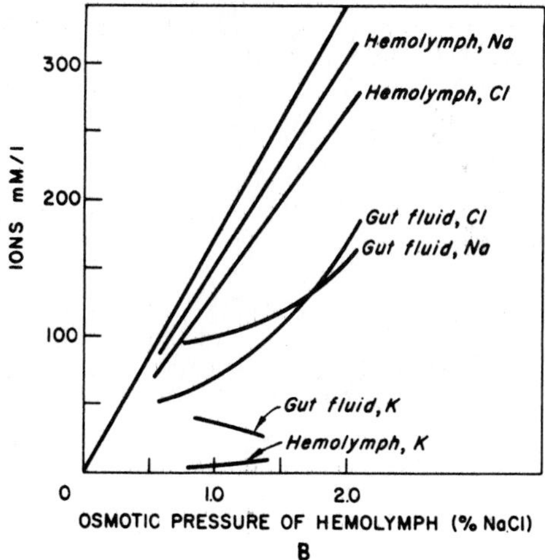

Figure 1–17. *A, Artemia:* osmotic concentration of gut fluid and hemolymph in different salinities, indicated in per cent NaCl. *B, Artemia:* Na^+, Cl^-, and K^+ in hemolymph and gut fluid at different osmotic concentrations of hemolymph. (From Croghan, P. C., J. Exp. Biol. *35*:219–233, 1958.)

hyposmotic to all media more concentrated than 25% S.W. The permeability to water and salts is low as compared with that of most other microcrustaceans; *Artemia* obtains water by swallowing the medium, absorbing the water, and leaving the salts to be defecated. *Artemia* may drink the equivalent of 3 per cent of its body weight per hour. Figures 1–17 and 1–18 show that the osmoconcentration of the gut fluid is greater than that of the hemolymph at all sea water concentrations, but that the gut fluid concentration is lower than that of a concentrated medium. The sodium and chloride concentrations in gut fluid are lower than in hemolymph; hence, NaCl and water are absorbed from the gut, leaving other salts to be voided from the intestine. The epithelia of the first 10 gills

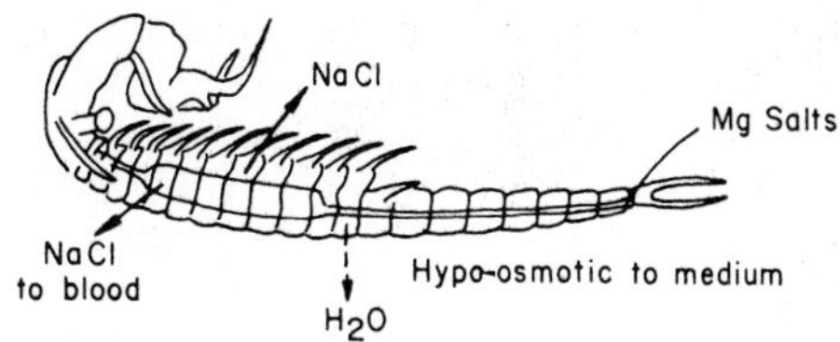

Figure 1–18. *Artemia:* main paths of ion and water movement. Solid arrows indicate active, and broken arrows passive, movement.

actively excrete NaCl into a hyperosmotic medium, and may absorb it from a dilute medium. Thus, the principal organ of water balance in *Artemia* is the gut, while that of NaCl regulation is the gill.[78]

The branchiae of *Artemia* have patches of silver-staining epithelium, and the ion transporting cells contain stacks of mitochondria lying between flat sinusoids on the hemolymph side[73] (Fig. 1–19). An ouabain-sensitive ATPase extracted from *Artemia* from 400% S.W. is five times more active than the same ATPase extracted from *Artemia* in 50% S.W.; ouabain, which blocks sodium-transporting ATPase, causes the animal to die in high salinities.[9]

The total sodium turnover is very high, as shown by tracer fluxes, and is greater in concentrated than in dilute media:

	Na turnover % per hour	*Eq/hr/g*
in S.W.	42	42.5
in 50% S.W.	35	31.7

The influx by the gut is one-third of the total; much of the remainder is by exchange diffusion or by passive diffusion across the gills, and this diminishes on transfer to a dilute medium (see p. 99). The reduction is a response to osmotic concentration, not to Na or Cl. However, Na^+ and Cl^- exchange independently.[354] There is a striking resemblance between *Artemia* and euryhaline marine fish in general mechanisms of hyposmotic control.

The dormant cysts of *Artemia* are remarkably resistant to desiccation, dehydration, and cold. In normal nymphs, trehalose is the principal blood sugar, and this may be converted to glycogen. Prior to encystment (which can be caused by a marked increase in

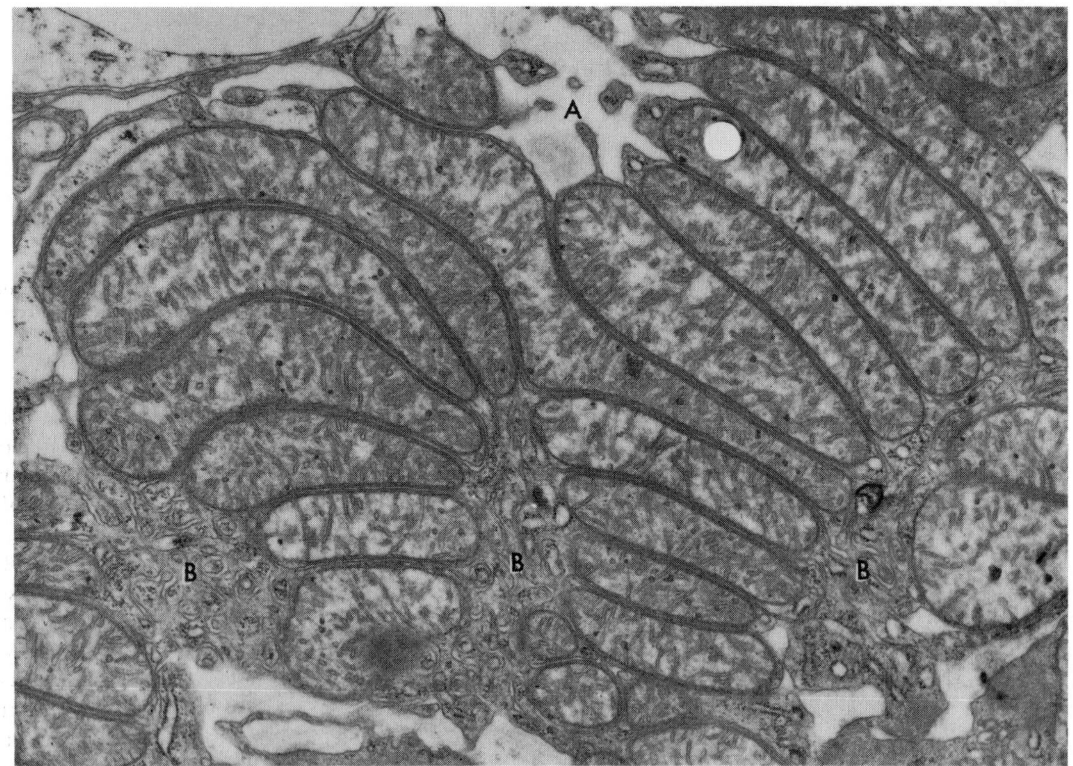

Figure 1–19. Electron micrograph of secretory cell of gill of *Artemia*, showing arrangement of mitochondria. Hemolymph out of figure at top. Sinusoids of dark cell, *A*, with canaliculi intertwined with endoplasmic reticulum, *B*. (From Copeland, D. E., Protoplasma *63*:363–384, 1967.)

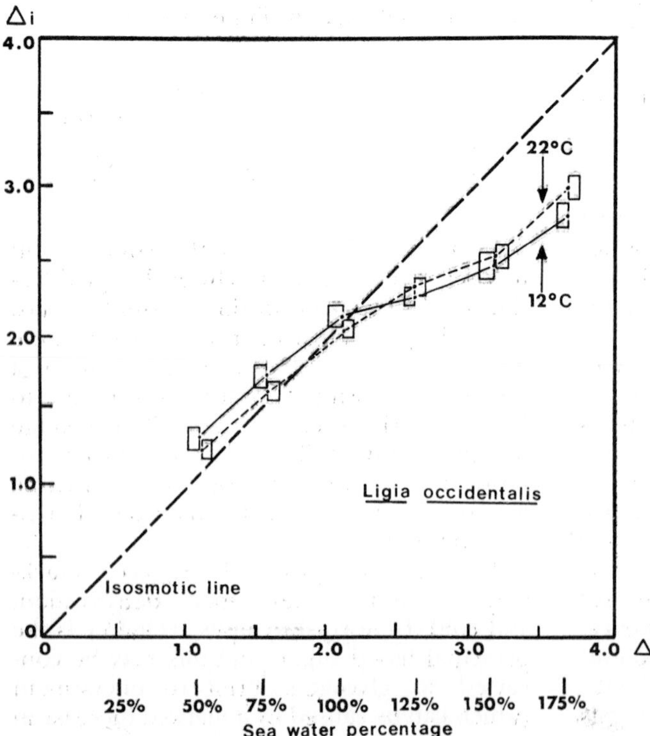

Figure 1–20. Osmotic concentration of hemolymph of isopod *Ligia* as a function of salinity of medium. Hyposmotic but not hyperosmotic regulation. (From Wilson, W., Biol. Bull. *138*:96–108, 1970.)

osmotic concentration of the medium), the oxygen consumption decreases, as does trehalose concentration, while glycerol concentration increases to high levels. Glycerol accumulates between the embryo and the shell, and is released into the medium upon emergence.[67]

Isopods occur in sea water, fresh water, and a variety of terrestrial environments. *Ligia oceanica* lives between rocks at high tide and above; it is hyperosmotic in sea water (Δ_i of −2.15 in Δ_o of −1.98) and shows good hyperosmotic regulation, at least down to 50% S.W. In moist air the hemolymph concentration falls to Δ_i of −1.44, and in dry air Δ_i is −3.48 (Fig. 1–20). Thus, *Ligia* tolerates a wide range of osmotic concentrations. In contrast, another marine isopod, *Idotea,* occurs in the intertidal weed zone and is isosmotic.[355] Baltic and Scandinavian lake populations of *Mesidotea* were compared. The Baltic population fails at a dilution below 5.5 mM Na/liter, while the lake population reproduces in fresh water. The body surface of the isopod lake population is less permeable, and the isopods' active uptake of sodium is half-saturated at 2.4 to 5 mM, in contrast to 10 to 14 mM for the Baltic population.[80] Thus, evolution toward fresh water is being brought about by decreased permeability to NaCl and increased affinity for Na of the active absorption system.[80] Another isopod, *Halmiscus,* lives in hypersaline lakes in Australia; it is isosmotic at 19 0/00 and lives over the range from 1 to 160 0/00.[15]

In summary, capacity for hyposmotic regulation is found in crustaceans which spend much time on land, in inhabitants of hypersaline lakes, and in some marine shrimps and prawns. Most of these animals show very low permeability of body surfaces, which protects against desiccation in land crabs and against osmotic loss in brine crustaceans. Land crabs have elaborate structural and behavioral adaptations for water absorption. Aquatic forms show active secretion of salts, mostly by gills. In none is production of hyperosmotic urine significant.

Good Hyperosmotic Regulation; Freshwater Crustaceans

Freshwater crustaceans have relatively higher hemolymph osmoconcentrations than other freshwater animals. The Italian crab *Telphusa fluviatilis* has a Δ_i of −1.17, while the crayfish *Astacus* and *Orconectes* have Δ_1 of −0.80 to −0.85. Freshwater crabs have not been much studied for osmoregulation. When crayfish are put into 50% S.W. the hemolymph concentration rises and hypertonicity is maintained, but in higher concentrations (60% S.W.) they are isosmotic. In 50% S.W. they may survive for several months, but crayfish are not normally found in very brackish water. The following data indicate that when *Astacus* from fresh water is acclimated for one month to 50% S.W., the muscle amino acids increase more than do the salts:[99]

	F.W.	*50% S.W.*
Δ_o	−0.02°	−0.95°
Δ_i	−0.8°	−1.04°
muscle amino acids	139.9 mM	204 mM
muscle K	95 mEq	105 mEq
blood K	4.6 mEq	6 mEq

Another crayfish, *Pacificastacus,* when adapted to 60% (or 70%) S.W. for 48 hours, showed relatively more increase in blood ions. Urine output decreased; hence, filtration decreased, although u/b inulin was unchanged.[280a] These data are given in Table 1–3.

TABLE 1–3. Ions in Crayfish *Pacificastacus* Before and After Acclimation to 60% Sea Water.

	Fresh Water		*60% Sea Water*	
	VALUE	U/B	VALUE	U/B
blood osmoconcentration	421 mOsm	0.095	500 mOsm	0.38
blood Na	195 mM	0.07	232 mM	0.266
blood Cl	180 mM	0.016	213 mM	0.278
blood K	4.9 mM		5.5 mM	
muscle K	111.9 mM		122.4 mM	
inulin		2.11		2.13
inulin clearance	2.45 ml/kg/hr		0.03 ml/kg/hr	

If crayfish excretory pores are plugged, the animals gain weight in fresh water and lose weight in hyperosmotic media. The urine output in fresh water is slightly higher than that in many marine crustaceans (4.37% B.W./day[178] or 8.27% B.W./day[54]). *Astacus* in fresh water produces urine equivalent to 4.6% B.W./day.[25] This indicates a low but not negligible degree of water permeability. Isolated gills are permeable to water to the extent of 19 liters/g/day; the permeability constant is calculated to be 0.12 sec^{-1}.[25] Crayfish urine is very dilute, as demonstrated by the following data:

	medium	*hemolymph*	*urine*
O.C., Δ_{fp}	0.018	0.81	0.09
sodium, mM	0.4	203	6.0
chloride, mM		195	10

In no condition has a crayfish been observed to excrete a salt-free urine. A 50 g crayfish loses 600 mM of chloride daily. Some salt is normally obtained from food, but crayfish can remain in salt balance without food for weeks in tapwater. Krogh washed out crayfish with distilled water for three days; when they were then put into 0.02 Ringer solution, they took up chloride initially at the rate of 2.3 mM/hr even though the external concentration was 2 mM and the internal concentration was 100 mM. Crayfish actively absorb Br^-; they take up Cl^- from KCl, Na^+ and Cl^- from NaCl, and Na^+ from Na_2SO_4.[196] When a crayfish was transferred from tapwater to distilled water, the urine Na decreased from 8 mM to 1 mM, and the Na outflux in distilled water was 64.5 per cent of that in tapwater.[54] After some sodium depletion during a sojourn in distilled water and subsequent return to tapwater, the net uptake increased by 6.8 times, and sodium loss was greater than normally occurs in tapwater; hence, there was increased sodium exchange but net uptake.

If a crayfish with normal blood sodium is placed in a series of increasing concentrations, the blood sodium shows little or no increase until the medium concentration exceeds the blood concentration, 200 mM Na. At an external concentration exceeding 300 mM, the blood and urine concentrations rise to a similar level. By use of labeled sodium, this element is seen to exchange as if it were in a single compartment equivalent to 46.9% of the body volume.[54] The permeability constant in fresh water for sodium outflux at a blood concentration of 310 mM was 0.0076 hr^{-1}; at a blood concentration of 186 mM, the constant was 0.0047 hr^{-1}. This corresponds to a sodium outflux of 0.87 μEq/g/hr; the sodium lost in the urine was 0.049 μEq/ml blood/hr or 5.6 per cent of the total Na outflux. The relations between loss by urine, by exchange, and by diffusion are shown for various concentrations of the medium in Figure 1-21.[54] A balance for a normal crayfish is as follows: at equilibrium influx equals outflux, when $Na_i = 200$ mM and $Na_o = 2$ mM. Outflux is 6% by urine and 94% by surface (probably gills); of the latter, 2/3 is by passive diffusion and 1/3 is by exchange. In intact crayfish the blood is positive by 6.6 mV with respect to tapwater.[79]

The difference in chloride content between urine and blood suggests that salt is reabsorbed into the hemolymph by the kidney. The crayfish kidney (antennary gland) consists proximally of a coelomic sac penetrated by blood vessels and sinuses lined by a single epithelial layer. Then follows the green tubu-

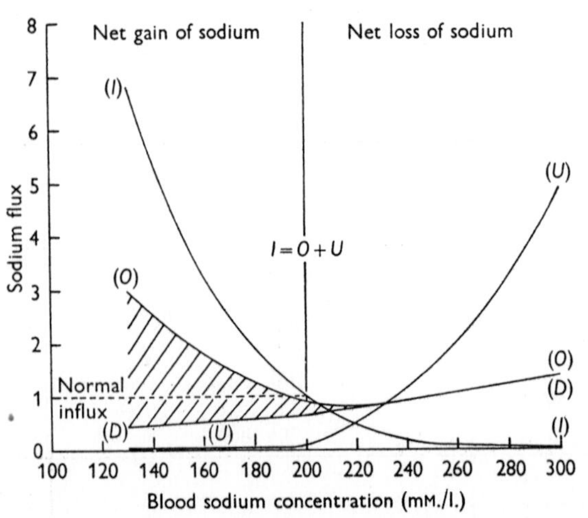

Figure 1-21. Diagram of sodium fluxes in crayfish as a function of blood sodium concentration. Net gain or loss. *U*, urine loss; *O*, total outflux; *D*, diffusion loss; *I*, influx of ^{22}Na. Cross-hatched area indicates active component. (From Bryan, G. W., J. Exp. Biol. *37*:113–128, 1960.)

lar labyrinth, a sponge-like structure, and a long (3 cm) nephridial canal. The canal empties into the urinary bladder. The blood supply, opening from vessels into hemocoel spaces, is plentiful to all parts except the bladder. In the lobster, which excretes isosmotic urine, the labyrinth is larger but the nephridial tubule is not well differentiated. *Carcinus* has no tubule. The nephridial tubule in the freshwater *Gammarus pulex* is longer than in the marine *Gammarus locusta.* The entire crustacean kidney is, in principle, like one unit of a vertebrate kidney. Epithelial cells of the coelomosac have extending processes with fenestrations and wide intercellular gaps, much as in the podocytes of the mammalian glomerulus.[198] Evidently, the long nephridial tubule is adaptive in freshwater animals for excreting dilute urine.

Fluid from the various regions of the crustacean kidney has the compositions shown in the table below.[291]

It is postulated that the fluid in the coelomic sac and labyrinth may be a blood filtrate, essentially isosmotic with the blood, and that in the nephridial canal chloride and other solutes may be reabsorbed, leaving a dilute urine to enter the bladder. Evidence for filtration comes from measurements of the concentration of inulin (a substance that is filtered but not reabsorbed). At a urine flow of 3% of body weight per 24 hours, inulin clearance was 4.4% of body weight, and the ratio of inulin concentration in urine to that in blood was 1.8 in the distal tubule, indicating water reabsorption.[291] Apparently, the filter begins to retain carbohydrate polymers of molecular weights 50,000 to 100,000.[186] In the lobster, where urine is isosmotic, the u/b ratio for inulin was 1.0. In the lobster[57] and crayfish[186] glucose in the blood is high, but is essentially zero in the urine; however, after phlorizin (which poisons the glucose-absorbing mechanism in mammals) is administered, the glucose levels in lobster and crayfish urine are the same as those in blood. With increased salinity of the medium, inulin clearance diminishes and urine volume decreases; hence, the proportion of water reabsorbed remains constant. Decreased formation of primary urine is the main response.

Cells of the distal tubule of the crayfish kidney show extensive folding of the basal membrane and a smooth distal (lumen) surface; hence, active transport must occur on the blood side. Labyrinth cells have a brush border of microvilli and large protruding vacuoles. The epithelial cells of the coelomosac and tubules of kidneys in crayfish in fresh water contain large vacuoles. Formed bodies (vesicles) are produced by tubule cells and are released into the lumen.[292]

Hormonal control of urine formation is indicated for *Procambarus.* After the eyestalks are removed, urine flow increases and the blood becomes diluted. Injection of an extract of brain or eyestalk restores normal urine flow and blood concentration; blood chloride was 195 mM in normal crayfish, 175 mM in eyestalk-less animals, and 197 mM in eyestalk-less animals injected with brain homogenate.[178]

In *Eriocheir sinensis,* the wool-handed crab of Northern Europe, hyperosmotic regulation is so good that the crab matures in fresh water; however, it does return to the sea in autumn to breed. Its permeability to water and salts, as measured by tracers, is very low. The blood concentration in fresh water is only slightly less ($\Delta_i = -1.22$) than in the North Sea ($\Delta_i = -1.66$). The urine output is low (only 3 to 5 ml/day from a 60 g crab) and urine may be slightly hyperosmotic to the blood irrespective of the medium concentration. Chloride and ammonia losses are the same with excretory pores closed as with them open. *Eriocheir* absorbs salt from very dilute medium. Isolated gills filled with blood having a sodium concentration of 300 mM absorb sodium from a medium of 8 mM Na. Thus, the kidneys are increasing the loss of ions in fresh water and putting out relatively little water, and the crab can live in fresh water because of its low salt and water permeability and the ability of its gills to absorb salt actively.

In summary, most freshwater crustaceans show extreme elaboration of the hyperosmotic mechanisms of brackish water crabs: extremely low permeability, excretion of dilute urine, and active uptake of sodium which is turned on at low external concentrations.

	blood	*coelomosac*	*labyrinth*	*proximal tubule*	*distal tubule*	*bladder*
Cl, mM	196	198	209	212	90	10
Na, mM	242	207	168	168	182	12.4
K, mM	7	8.5	6.8	13.3	12	0.9

EGGS OF MARINE FISH AND FRESHWATER ANIMALS

The eggs of marine invertebrates are isosmotic to the environment and swell and shrink to correspond with changes in the medium. The eggs of marine teleosts, on the other hand, are hyposmotic to sea water, whereas all freshwater eggs are hyperosmotic to their medium. Some means must be provided to maintain the osmotic gradients prior to the development of excretory organs.

The fertilized egg of the marine killifish, *Fundulus,* has an internal concentration corresponding to $\Delta_i = -0.76°C$, less than half that of the surrounding sea water; when placed in various dilutions of sea water it shows little change in freezing point. The chloride content of embryos of several marine teleosts is one-fiftieth to one-half that of sea water. After fertilization, a *Fundulus* egg develops hydrostatic pressure as high as 150 mm Hg, and the electrical resistance of the egg membrane rises from 3450 to 13,290 ohms/cm^2.[181] The outer chorion is inelastic and prevents swelling. Between the chorion and the egg is a perivitelline space occupied by a clear colloid, which swells by imbibition and causes the egg proper to shrink and to come under hydrostatic pressure. Transfer experiments into different media indicate that the chorion is permeable to water and salts but that the vitelline membrane has very low permeability.

Eggs of freshwater animals must likewise be protected against dilution. In the trout *Salmo,* the embryo has a Δ_i of $-43°$ to $-49°$ (220 mOsm); the perivitelline fluid is 31 mOsm in river water of 2 mOsm. Freshly shed trout eggs are relatively permeable, and if shed into salt solution they remain permeable; but when hardened by shedding into fresh water, the permeability decreases rapidly during three hours. The osmotic permeability constants are:

	P in μ^3/μ^2/sec
salmon eggs before shedding	0.06
salmon eggs water hardened	less than 0.004
Amoeba	0.35
Sialis larvae	0.04 to 0.05
frog ovarian eggs	0.89
frog shed eggs	1.2
red blood cells	125

Other data show the membrane permeability to water in frog eggs to be be 5 to 10 times greater than in salmon eggs. It is calculated that the salmon egg may take up water at 1/300 of its volume per day;[217] the water permeability, while not zero, is exceedingly low. Similarly, sodium exchange is confined to the perivitelline fluid, and sodium accumulation by the embryo does not begin until the eyed state is reached.[277] Permeability of shed fertilized eggs which was lower than that of the ovarian eggs has also been seen in zebra fish and brook lamprey. In frog eggs the osmotic concentration falls from the equivalent of 120 mM NaCl at the time of laying to 80 mM at the time the blastopore closes, because of initial swelling followed by reduced permeability.

It has been assumed that eggs of freshwater invertebrates have relatively impermeable membranes. However, the embryos of planorbid snails lose the vitelline membrane at gastrulation, and the capsular membrane (chorion) is no barrier to water and ions. Embryos can be removed from their capsules and can then regulate salt and water content. If metabolism is depressed, they swell and fluid accumulates extracellularly, particularly in the cleavage cavity. It appears that, normally, fluid is actively removed from the cleavage cavity, that embryonic cells can actively absorb sodium, and thus that there may be cellular regulation at very early stages.[16, 17]

It may be concluded that in embryos before development of excretory systems, permeability to water and ions is very low; however, active transport mechanisms have not been much considered in early aquatic embryos.

ADAPTATION TO SEA WATER, FRESH WATER, SOIL, AND THE ENDOPARASITIC ENVIRONMENT

PROTOZOA

Protozoa comprise a great variety of distantly related forms which are found in an extreme diversity of environments. Marine and parasitic Protozoa appear to have no osmotic problem; they are isosmotic. However, they do regulate ionically and do maintain cell volume and shape; some of them are euryhaline and can shift from being isosmotic to hyperosmotic. Freshwater Protozoa are hyperosmotic to their medium, and the contractile vacuole eliminates excess water load. A survey of many ciliates shows that few freshwater species can survive in more than 5 0/00 salinity, that brackish species are extremely euryhaline, and that a few marine species can live at as low a concentration as 3 0/00.[11]

The water balance of Protozoa has been examined for volume changes, vacuolar output in different media, and analyses of cytoplasm and vacuolar contents. Volume regulation is usually more rapid in marine species than in freshwater forms.

The osmotic concentration of the large rhizopod *Pelomyxa carolinensis*, as estimated by vapor pressure measurement, is 103 mOsm, and for *Spirostomum* it is 89 mOsm. Volume measurements gave a value of 94 mOsm for *Pelomyxa*. The osmoconcentration of *Paramecium* is largely due to small molecules: O.C. = 111 mOsm; Na 6.5 mM; K 28.8 mM; Cl 9.2 mM; amino acids 56.5 mM. When the osmoconcentration of the medium is increased by addition of sucrose, the internal concentration rises because of an increase in the amount of amino acids; hence, intracellular ions remain relatively constant[342] (Fig. 1–22).

Water can enter protozoans by exchange diffusion, by osmotic movement, and by food vacuoles. Rhizopods have a lower water permeability than ciliates, but all values for Protozoa are low compared with those for aquatic eggs. On the basis of observed permeability to heavy water and the computed surface area, it is calculated that in *Pelomyxa* some thirteen times the cell volume passes in and out by diffusion exchange per hour but that only the equivalent of 2 per cent of the cell volume moves in osmotically; this is the amount which must be eliminated per hour by the contractile vacuole.[219] The observed value, 3.8 per cent of the volume evacuated per hour, is in agreement with this estimate.

Quantities of water enter with the food, and in ciliates the cytopharynx membrane appears to be more permeable than the rest of the body surface. In freshwater peritrichs, the water taken up by food vacuoles may be 8 to 20 per cent of the output of the contractile vacuole (c.v.), and in *Paramecium* this amount is 30 per cent. An increase in vacuolar output during feeding has been observed in many species. In a marine *Amoeba mira* there is virtually no vacuolar output in the nonfeeding state, but large

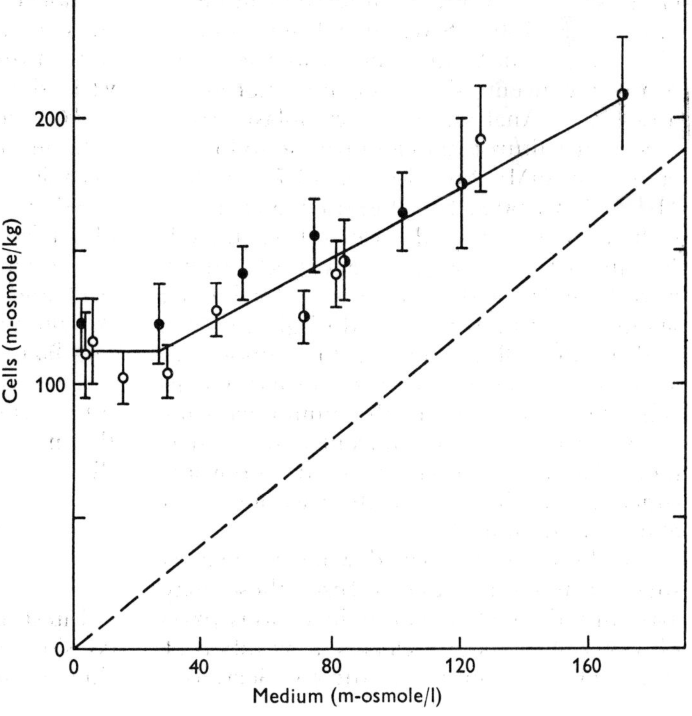

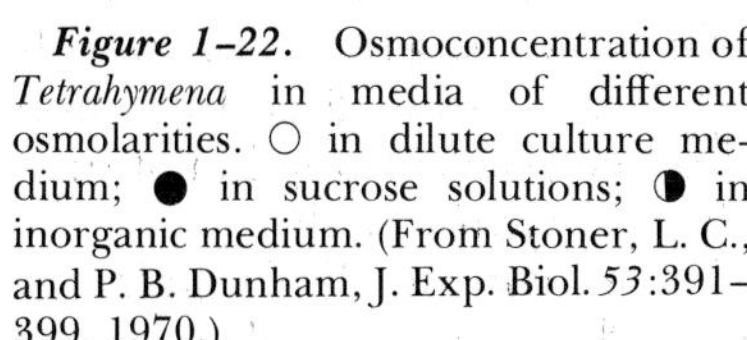
Figure 1–22. Osmoconcentration of *Tetrahymena* in media of different osmolarities. ○ in dilute culture medium; ● in sucrose solutions; ◑ in inorganic medium. (From Stoner, L. C., and P. B. Dunham, J. Exp. Biol. *53*:391–399, 1970.)

vacuoles appear when the animal begins to feed. In *Paramecium* the c.v. activity increases not only during feeding but also when water is merely pumped toward the mouth. Some ciliates, such as the parasitic Ophryoscolecidae, can close the oral passage, and then contractile vacuole activity is slowed.

Contractile vacuoles are present in all fresh-water Protozoa, in many marine ones, and in a few endoparasitic species. From volume and rate of pulsation measurements, the osmotic function is indicated. Output is greater in freshwater than in marine species. Vacuoles may appear in euryhaline species only when in a dilute medium. A marine ciliate, *Amphileptus,* showed a 21 per cent increase in vacuolar rate when transferred from sea water to 70% S.W. When the fresh-water *Amoeba verrucosa* was cultured in increasing concentrations of sea water, the vacuolar pulsations slowed and no vacuole was seen when the animals were in 50% S.W. When *Paramecium* is first put into a hyperosmotic solution, the body becomes flattened but later recovers its normal shape, and vacuolar output is markedly reduced. In one saline-tolerant species of *Paramecium* the average intervals between vacuolar pulsations increased from 13 seconds in fresh water to 32 seconds in 50% S.W. and to 65 seconds in 100% S.W.[118]

The marine ciliate *Miamiensis* in 25% S.W. eliminates the equivalent of its cell volume in 0.5 hr., in 100% S.W. in 1.1 hr., and in 175% S.W. in 2 hr.; the response is to osmotic concentration and not merely to ions *per se.* Analyses of the cytoplasm from sea-water-cultured animals give the following values in mM: Na 87.9, K 73.7, Ca 3.7, Mg 28.5, Cl 60.8. In different media intracellular K, Ca, and Mg are well regulated; Na and Cl are less well regulated. Amino acids total 317 mM/kg in cells in 100% S.W.; alanine constitutes most of this, glycine less, and proline still less, but the three amount to 75% of the total amino acids. In varying concentrations of sea water the amino acids increase with osmotic concentration much more than do the electrolytes. An important function of the contractile vacuole is to eliminate sodium.[180]

Usually a c.v. is formed when numerous small vacuoles or vesicles fuse; these may arise in a defined region or in various parts of the cell. In some ciliates a number of canals fill and empty into the c.v. Secretory work is done in c.v. production. When a respiratory inhibitor, such as cyanide, dinitrophenol, or amytal, is added to a dilute medium containing ciliates, the vacuolar output declines and cell volume increases. Vacuolar output is highly temperature dependent. Mitochondria and other granules are abundant in the region of vacuole formation, and fine tubules are noted between canals and sarcoplasmic reticulum. How the small vacuoles or canals fill is not known. Analyses of cytoplasm and vacuole contents indicate that the small vacuoles are isosmotic, that they lose solute by active absorption, and that they fuse when they become hypotonic:

	O.C., mOsm			*Na, mM*		*K, mM*	
	CYTOPLASM	C.V.	MEDIUM	CYTOPLASM	C.V.	CYTOPLASM	C.V.
Pelomyxa carolinensis[290]	117	51	2	0.57	19.9	31	4.6
Amoeba proteus[304]	101	32	7				

These results indicate that potassium is reabsorbed into cytoplasm, while sodium is excreted, and the vacuole formation provides both ionic and osmotic regulation.

What determines the terminal size or the time of emptying of a contractile vacuole is unknown. In some ciliates the vacuoles empty at a fixed point or pore. Fibrillar bundles (which may be contractile) extend from the vacuole of *Paramecium* to the pellicle. When a suctorian, *Discophrya,* was subjected to hydrostatic pressures of about 2500 lb/in^2 there was visible creasing of the pellicle, an increase in vacuole frequency, and a decrease in its volume; at pressures known to cause protoplasmic solation, the vacuole output diminished and the cell rounded up.[190] The surface of an isolated c.v. can be seen to contract, like some muscle proteins, when treated with ATP and Mg.[281] It is possible, therefore, that the emptying results from contraction of protein filaments in the vacuole wall.

The functions of contractile vacuoles in ionic, osmotic, and volume regulation and the mechanisms of filling and emptying may differ in different kinds of Protozoa.

Parasitic Helminths

Intestinal parasites may be subjected to considerable variations in the medium from time to time. A few measurements of intestinal cestodes indicate that they are normally isosmotic to the medium but that they can show limited regulation in different media.

Nematodes have nephridia-like excretory organs which may function in ionic and osmotic regulation. The excretory ampullae of third stage larvae of *Nippostrongylus* and *Ancylostoma* are observed to pulsate more frequently in dilute than in concentrated media.

Analyses of coelomic fluid of the roundworm *Ascaris* indicate that it may be hyposmotic to the intestinal fluid in which it occurs. Specimens gain weight in dilute medium, less so if the gut is closed by ligation; hence, the body wall must be somewhat permeable.[159] However, in saline media there is no osmotic regulation. *Ascaris* in an isotonic salt medium excretes a urine which is isosmotic with the coelomic fluid:

	mOsm	*Na*	*K*
medium	310	140	3
coelomic fluid	333	107	8.5
urine	329	110	40

It appears that the excretory organs function more for ionic than for osmotic regulation. When inulin was injected, it was not excreted; hence, a filtration mechanism is unlikely. The nephridia have a secretory structure.[199] A cestode, *Hymenolepis,* is isosmotic with its medium.

More accurate and extensive data regarding salt and water regulation of various parasitic worms with and without excretory organs might reveal a variety of degrees of regulation.

Regulation by Invertebrates in Fresh Water

Freshwater sponges, coelenterates, and bivalve molluscs are the most dilute of animals. By maintaining a low gradient to the medium, they reduce their energy requirements, but they are restricted in locomotor and metabolic activity. The osmotic concentration of cells in freshwater *Spongilla* is equivalent to 25 to 30 mM NaCl, and rises several times higher at the time of gemmulation. Contractile vacuoles have been seen in amoebocytes and choanocytes of freshwater but not marine sponges.

Cells of isolated tentacles of *Hydra viridis* and *Pelmatohydra oligactis* decreased in volume in sucrose more concentrated than 40 mM, and digestive cells of *Chlorohydra viridissima* were isotonic in 50 mM sucrose. Osmotic permeability corresponds to movement of 0.3 to 0.6 μ^3 H_2O/μ^2/min. atm.[211] The volume of *Chlorohydra* cells decreased in proportion to the concentration of the medium above 200mM, thus acting as osmometers (with osmotically inactive volume of 70 per cent); in a solution below 200 mM concentration, hydrostatic pressure developed to a maximum (i.e., volume remained nearly constant down to 50 mM, below which the cells again acted as osmometers).[192] *Hydra* accumulates potassium and extrudes sodium; the enteric cavity is positive by 30 mV with respect to the outside of the body, probably because of net transport of sodium inward. A freshwater medusa, *Craspedacusta,* maintained tissue osmolality of 27 mOsm and water content of 99% in a medium of 7 mOsm, by active absorption of Na and K. Hyperosmotic regulation was observed in solutions up to 69 mOsm, and above that the animal decreased in volume and was isosmotic.[154] These jellyfish had an Na content of 15 mM and a K content of 0.6 mM in a medium containing 2 mM Na and 0.2 mM K. It is postulated that these dilute animals actively absorb Na and then excrete it in hyperosmotic solution into the gut, thus drawing water from the body. Permeability to water is low but not insignificant.[154]

Rotifers are slightly more concentrated than freshwater coelenterates:[52]

	mOsm	*Na, mEq*	*K, mEq*
pond water	17.7	4.2	1.7
body fluid	81	21	7
urine	42	13.4	2.7

Rotifers have a syncytial protonephridium consisting of terminal flame bulbs, tubules, and vesicle (bladder). Frequency of flame bulb activity is maximal in a dilute medium.[271] In pond water, urine is produced (as judged from bladder volume) at 47×10^{-9} ml/min, and sodium is excreted at 0.63×10^{-9} mEq/min; a ratio of urine to body fluid concentration for inulin was 1.42, and total filtration rate was 33.1×10^{-9} ml/min.[52]

Freshwater bivalve molluscs such as *Anodonta* have a very low blood concentration ($\Delta_i = -0.08°C$ or approximately 42 mOsm). In solutions of sea water more concentrated than $\Delta_o = -0.1$, *Anodonta* loses weight. Anesthetized or chilled animals gain weight in pond water. Hence, *Anodonta* is permeable to water and excretes a dilute urine which may amount to 50 per cent of body volume per day. Fluid from the pericardial cavity passes through the nephrostome into the kidney. The blood is under a hydrostatic pressure of about 6 cm H_2O and the colloid osmotic pressure of blood is some 3.8 mm H_2O. Hence, there may be filtration through the wall of the heart into the pericardial cavity. From inulin clearance measurements, it is estimated that a filtration of 1 to 1.3 ml/hr/100 g wet weight takes place; urine concentration is about 23.6 mOsm, about half that of blood.[272]

A freshwater snail, *Lymnaea*, accumulates Na and Ca from pond water.[141] In tapwater the snail's blood was electrically negative by 16.6 mV, and this was independent of external Na concentration. This potential would support a blood concentration of 0.67 mM, but the concentration was actually 57 mM. Active uptake of tracer Na was found to vary with external concentration, with a value of 0.25 mM for half-saturation of the uptake mechanism. The influx of Na into snails adapted to 0.35 mM Na was found to be 0.132 μM/g/hr, of which 0.053 was estimated to be an exchange component. Snails partially depleted of Na showed a doubled uptake. Calcium influx into *Lymnaea* also increased with increase in Ca in the medium, and a K_m of 0.3 mM Ca was found. The negative blood potential increased as Ca decreased, and could account for the Ca uptake except at low concentrations where uptake may be active. Evidently, freshwater molluscs, despite low osmotic concentrations of body fluids, make use of a dilute urine and of active absorption of Na^+, and possibly also of Ca^{++}.

Earthworms and Leeches

There are many freshwater annelids—Oligochaeta and Hirudinea. Earthworms are adapted to life in moist soil, where osmotic stress is intermediate between that of fresh water and that of air. When an earthworm is transferred from soil to tapwater, it absorbs water equivalent to as much as 15 per cent of its initial weight in about five hours. Water-adapted worms removed to soil or air lose water. Earthworms can tolerate loss of 50 to 80 per cent of body water, *Allolobophora* more so than *Pheretima* or *Eisenia*.[140] Water passes out of a worm after excess hydration more quickly than it is taken in after dehydration. In a gradient, earthworms orient toward a moist surface and away from a dry one. In air, the body surface is kept moist by excreted water; upon irritation, coelomic fluid is extruded through dorsal pores.

Representative values of coelomic fluid and urine concentration follow:[283]

	coelomic fluid		*urine*	
	Δ_{fp}	Cl (mEq)	Δ_{fp}	Cl (mEq)
Lumbricus from tapwater	−0.31	46.5	−0.10	3.4
Lumbricus from Ringer = 65% NaCl	−0.49	74	−0.23	17.

When placed in several dilutions of sea water, the earthworm maintained coelomic fluid hyperosmotic to the medium; chloride, however, could be kept lower than that in the medium. At all concentrations the urine was blood-hyposmotic.[283]

In some oligochaetes, nephridia are large and occur as one pair per segment; in others they are small and very numerous in each segment. The best known nephridia open to the coelom by ciliated funnels, the nephrostomes, but others are closed internally. Some worms (e.g., *Pheretima posthuma* from dry soils) have some nephridia which empty into the intestine and thus conserve water, whereas in others (e.g., *Lumbricus*) they open externally. Coelomic fluid enters the nephrostomes under the force of ciliary beat. Dyes and breakdown products from hemoglobin may accumulate in the tubular epithelium. Nephridial excretion of urine by earthworms in tapwater is 2 to 2.5 per cent of body weight per hour. The urine chloride concentration was 7.7 per cent of the coelomic fluid Cl^- in *Lumbricus* and 4.6 per cent in *Pheretima*. In *Pheretima* the blood glucose was 100 mg/100 ml, but no glucose was found in coelomic fluid or urine.

Samples removed from nephridia isolated in Ringer solution showed little or no reduction in Cl in the proximal and middle tubes but a marked reduction in the distal tube and bladder.[283] Recent measurements yielded the following values as per cent of the Cl^- in the bathing Ringer solution: ciliated canal, 72%; proximal canal, 61%; distal canal, 31%; bladder, 26%. The lumen of the canal is electrically negative with respect to the outside by 14 mV, which suggests that Na^+ is actively reabsorbed.[43, 46] Sodium efflux from the nephridial tubule was calculated and compared with other absorbing tubes as follows:[43, 46]

Lumbricus nephridium proximal tubule	μEq/mm²/sec 0.91×10^{-5}
dog ileum	2.01×10^{-5}
rat proximal tubule	8.5×10^{-5}
frog bladder	0.44×10^{-5}
Necturus proximal tubule	0.062×10^{-5}

The ratio of urine to coelomic fluid concentration of inulin is greater than unity; hence, water is probably reabsorbed.

As in most euryhaline animals, passive fluxes (indicative of permeability) decrease and active uptake of ions increases when the worms are adapted to dilute pond water. Passive fluxes for Na and Cl were calculated from flux equations (p. 3), using measured concentrations and the potential of −20 mV found for coelomic fluid when the animal is in pond water. Observed flux ratios for Na and Cl are 20 to 50 times greater than those calculated; thus influx is active for both ions. Active uptake of Na is six times greater, and that of Cl is three times greater, in worms transferred directly from soil than in worms previously adapted to pond water. Measurements of uptake at different concentrations show that it increases to a maximum at 0.94 μEq/10 g/hr, compared with 1.5 for crayfish and 3.0 for salamander; the concentration for half-saturation of the uptake mechanism (K_m) is 1.3 mM NaCl, compared with 0.2 for crayfish and 20 for *Carcinus*.[182a]

A freshwater leech has osmotic concentration of its body fluids of 202 mOsm; oxygen consumption is minimal in 25% S.W. Urine is formed at the rate of 3 to 6 liters/cm² body surface/hr. Blood is hyperosmotic to urine over a wide range of external concentrations[45] (Fig. 1–23). Data for leeches in fresh water follow:

	Na, mM	*Cl, mM*	*O.C., mOsm*	*urine flow*
Hirudo blood	125	36	202	
urine	6	6	15	46.5 μl/g/hr
Lumbricus blood			167	25 μl/g/hr

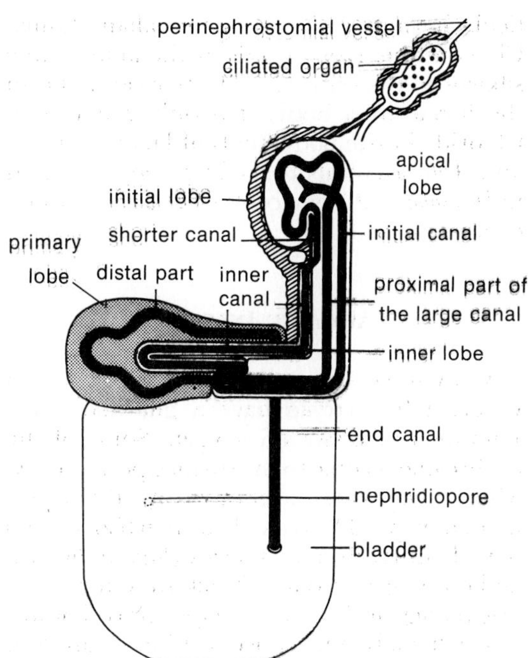

Figure 1–23. Diagram of the nephridium of a leech. (From Boroffka, I., H. Altner, and J. Haupt, Z. vergl. Physiol. *66*:421–438, 1970.)

The leech excretory organ is a nephridium, with a ciliated organ surrounded by the perinephridial blood vessel. Filtration occurs from blood into a system of canaliculi; fluid then passes through tubules to a bladder.[44, 45] Leeches show very active absorption of salt from dilute Ringer solution when they are placed in it after living in distilled water for some days.

The resemblance of function in annelid nephridia to that of vertebrate nephrons is striking.

INSECTS

Insects are mostly terrestrial, and their osmotic problem is to obtain enough water from food and air to resist dehydration. A few insects live in fresh water, at least during the larval period, and a few dipteran larvae live in salt lakes, brackish water, or sea water.

Osmotic concentrations of insect hemolymph have been recorded in the range of Δ_{fp} −0.4 to −1.3°C, which is equivalent to 120 to 376 mM NaCl/liter. Insects, in general, differ from other animals in the small fraction of the osmotic pressure of the body fluid that is due to inorganic chlorides and the large fraction due to organic substances, particularly amino acids. In mammalian plasma, Cl^- constitutes two-thirds of the anions; in a silkworm it is only 12 to 18 per cent, and in the larvae of a botfly it is only 7 per cent. Chloride in the body fluids of larvae of *Culex* and *Aedes* accounts for only 25 per cent, and in *Drosophila* 20 per cent, of the total osmoconcentration.

AQUATIC INSECTS

Many insects are air-breathers, yet live in water; a few larvae have a gill system for utilization of dissolved oxygen. Some of the air-breathers come to the surface periodically, while others carry a bubble with them permanently (p. 180). Water boatmen (Corixidae) have both freshwater species (*Sigura distincta* and *Corixa dentipes*) and brackish water forms (*Sigura lugubris*). *Corixa dentipes* in fresh water has a hemolymph concentration equivalent to 149 mM NaCl and excretes a rectal fluid equivalent to 57 mM NaCl; its daily water output is 49 per cent of its body weight.[338]

Aquatic insects vary considerably in the permeability of the cuticle to water. Gills are more permeable than body surfaces, and larvae and nymphs tend to be more permeable than adults. The evaporation in air shows a critical temperature above which water loss is markedly increased, and this transition temperature is lower for terrestrial species.[18] In aquatic *Collembola,* the cuticle is relatively non-wettable; on abrasion of its waxy coat, it becomes wettable. The hydrofuge properties of aquatic insect cuticle may serve to ensure non-wettability rather than to prevent desiccation.[253]

Chironomid (midge) larvae live in a wide variety of aquatic environments. An African species breeds in small temporary pools, survives drought, and revives after a rain. In dry air these larvae may contain only 3 per cent water, which in six hours in humid air rises to 33 per cent; they may tolerate water temperatures of 41°C for short periods.[158] A freshwater species (*Chironomus thummi*) lives in water no more concentrated than 0.5 0/00, and *C. salinarius* lives in concentrations of 1 to 37 0/00.[252a]

In several dipteran larvae, anal papillae become very large in a dilute medium but are small when the larvae are reared in saline solution. In *Chironomus thummi* the papillae hypertrophy in the absence of Cl^-, but not if Ca^{++} is absent. In larvae of the euryhaline mosquito *Aedes aegypti,* the anal papillae are responsible for 90 per cent of Cl^- exchange and can actively absorb Cl^- from dilute solutions of NaCl, less from KCl, $CaCl_2$, and NH_4Cl. Part of the Cl^- uptake may be in exchange for OH^- or HCO_3^- from the hemolymph, since Cl^- influx is reduced by raising the concentrations of these ions in the medium. Sodium is actively absorbed from NaCl, less from $NaNO_3$, $NaHCO_3$, or Na_2SO_4. Half-saturation of the absorbing mechanism is at 0.2 to 0.5 mM for Cl and at 0.55 mM for Na.[340a, 341]

The permeability of anal papillae to water is shown by ligation experiments in media of different osmoconcentrations. Larvae of *Aedes argenteus* normally swallow little water; water enters osmotically through anal papillae and is excreted by malpighian tubules. Hemolymph is maintained relatively constant in various salinities up to the equivalent of 0.65% NaCl, but it conforms osmotically at higher salinities. However, after gradual acclimation larvae can live in sea water diluted to the equivalent of 1.4% NaCl. The anal papillae of *Culex pipiens* larvae reared in 0.65% NaCl are very small; in 0.006% NaCl they are larger, and in distilled water they become very large and the epithelial cells may become vacuolated. Anal papillae of *Aedes detritus* larvae in Algerian ponds of salinity equivalent to 1.2 to 10% NaCl are smaller than those of larvae of *Aedes aegypti* from fresh water.[377] Endothelial cells of the anal papillae of *A. aegypti* show deep basal (cuticular) infoldings with long flat mitochondria between them. The distal (hemolymph) border has vesicular canaliculi and scattered mitochondria.[72] It is probable that these endothelial cells actively absorb Na and Cl, but the fluxes have not been measured.

Excretion is by the malpighian tubules, with reabsorption in the rectum. In larvae of the trichopteran *Limnophilus* reared in a salinity equivalent to 0.01% NaCl, the osmotic concentration of body fluid was equivalent to 0.031 to 0.048% NaCl; that of the malpighian tubule fluid was similar, but the urine was equivalent to less than 0.009% NaCl. In freshwater mosquito larvae (*Aedes aegypti*) the intestinal fluid (equal to malpighian tubule fluid) is isosmotic to hemolymph, but rectal urine is very dilute (Fig. 1–24). In a larva of *A. detritus* living in dilute sea water, the blood-isosmotic intestinal fluid is made hyperosmotic in the rectum:[282a]

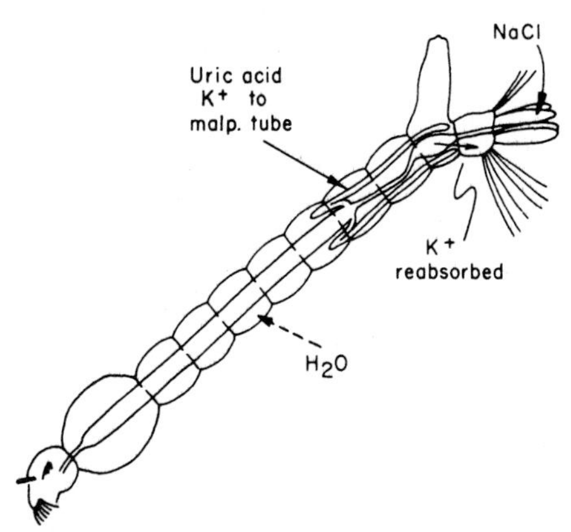

Figure 1-24. Diagram representing main paths of movement of ions and water in osmoregulation of mosquito larva. Passive entrance of water, active uptake of NaCl from medium, active reabsorption of potassium from hindgut.

	Equivalent % NaCl		
	RECTAL FLUID	BODY FLUID	INTESTINAL FLUID
A. aegypti (in distilled H_2O)	0.07	0.63	0.57
A. detritus (in sea water)	3.5	0.93	1.2

Evidently, in mosquito larvae the malpighian tubules form a fluid which is not very different osmotically from hemolymph, and the rectal epithelium absorbs water and salts. There is evidence for control by a hormone from thoracic ganglia and retrocerebral complex.[341]

Larvae of *Sialis* (freshwater hellgrammites) are unlike those of mosquitoes in that there is no special surface for ion absorption, and there is high water permeability. The rectal excretory fluid has 65% as much electrolytes as the hemolymph, mainly NH_4HCO_3, but rectal fluid is much lower than hemolymph in Cl^-, Na^+, and K^+. The lumen of the rectum is 24 mV positive with respect to the hemolymph; this may indicate active reabsorption of anions. Data follow:[313a]

	Na	*K*	*Cl*	*NH_4*	*HCO_3*
hemolymph, mEq	109	5	31		
excreted fluid, mEq	12	4	0	100	91

Brackish and freshwater species of caddis larvae (*Limnophilus*) are compared as follows:[343]

L. affinis	*L. stigma*
brackish water	fresh water
hemolymph can be hyperosmotic to medium	die in brackish water
urine can be hyperosmotic to hemolymph	rectal fluid hypo- or isosmotic to hemolymph
time constant of Cl exchange $t_{1/2} = 400$ hr	$t_{1/2} = 40$ hr
low permeability	permeable to ions
drinks little	drinks in salt medium and regurgitates
can tolerate 75% S.W. or 410 mOsm	dies at more than 60 mOsm

Larvae of the dipteran genus *Ephydra* can live in saline lakes. *Ephydra riparia,* a brackish water form, is hyposmotic to its medium. *E. cinerea* occurs in Great Salt Lake in a medium of 250 atm of osmotic pressure and has a hemolymph of 20.5 atm ($\Delta_{fp} - 1.89°$); in distilled water the hemolymph has an osmotic pressure of 20.1 atm or $\Delta_{fp} - 1.67°$. These larvae must be very impermeable to salts.[252]

Two sibling species of *Anopheles* appear identical although their eggs are distinguishable and the hybrids are sterile; one species, *A. gambiae,* develops in fresh water while the other, *A. melas,* normally breeds in brackish water but can develop in concentrations as high as 150% S.W.

Terrestrial Insects, Diplopods, and Arachnids

Terrestrial insects have very successfully solved the problems of life in air, and they vary in their humidity requirements. Their osmoconcentration tends to be higher than that of aquatic insects:

		Δ_i
arachnids	scorpion	−1.125
	spider	−0.894
terrestrial insects	*Carabus*	−0.94
	mantis	−0.885
	Ephestia larva	−1.12
	Acheta	−0.72
	Drosophila larva	−0.70
aquatic insects	mosquito larva	−0.65

The principal loss of water from terrestrial insects is by evaporation; a mammal may have a surface-to-volume ratio of 0.5, whereas in an insect this ratio may be 50.[18] An important avenue of evaporative loss is through the spiracles; most insects keep these closed between respiratory movements. Species which lack spiracular control have high rates of water loss. When the spiracles are forcibly kept open, as by exposure to CO_2 or in activity, the water loss is increased many times. In *Locusta* in dry air, ventilation decreases and water loss diminishes in proportion to saturation deficit; hyperventilation at high temperature may double water loss. Ventilation varies according to water reserves and relative humidity.[218] In several desert beetles respiratory water loss is much less than cuticular loss, and is much less increased by rising temperature up to 40°C (at which point respiratory loss increases steeply).[3] Similarly, in scorpions, cuticular loss of water is relatively greater at low temperatures ($<38°$) and respiratory loss is greater at high temperature ($>40°$).[148]

In dehydration, a cockroach, *Periplaneta,* may show a reduction of hemolymph volume from 17.5 to 13.5 per cent and a decrease of hemolymph water content from 16.4 to 11.0 mg/100 mg; there is an increase in O.C. from 410 to 467 mOsm and a decrease in Cl from 132 to 127 mM. Thus, chloride is regulated and organic solutes (probably amino acids) may increase.[104, 105]

Properties of the Integument in Relation to Water Balance

The insect integument is a complex structure of chitin, tanned lipoprotein, waxes, polyphenols, and cements. In general, the low evaporative loss depends on cuticular wax, a layer approximately 0.2 μ thick. The permeability constant of wax for water is some two-thousand-fold less than that of red blood cells. Aquatic insects in air tend to lose more water by evaporation than do terrestrial insects: active plant feeders more than pupae, *Limnophilus* larvae 11, *Corix* 2, *Hydrobius* adult 0.3, *Tenebrio* pupae 0.02 mg $H_2O/cm^2/hr$ at 20°C.[161] Water loss is increased after abrasion of the cuticle, after removal of wax by appropriate lipid solvents, or by changing the physical properties of the wax by raising its temperature. As the temperature rises, water loss increases gradually over a moderate range, then more steeply above a transition point[19] (Fig. 1–25). At high temperatures, death may be due to desiccation rather than to temperature *per se.* High humidity retards the evaporative loss at elevated temperatures. In a desert scorpion the permeability of the cuticle shows two steps as a function of temperature at 35° and 60°; for desert arthropods, water loss in % B.W./hr at 30°C in dry air is less than for those accustomed to more water: scorpion 0.028 to 0.091, tarantula 0.147, *Locusta* 0.32, tenebrionids 0.12 to 0.245.[148]

A unique property of the cuticle of insects and ticks is the capacity to absorb water from an atmosphere which is humid but less than saturated. Water absorption from the atmosphere has been demonstrated in ticks, mites, fleas, *Tenebrio* larvae, and lepidoptera, but not in *Periplaneta* or *Blatta.*[104] Uptake is enhanced after dehydration and diminished by hydration of the animal. The relative humidity (in per cent saturation) above which water is absorbed from the atmosphere is for *Ixodes* 94, *Tenebrio* 88, *Xenopsylla* 65, *Thermobia* 50.[21] Third stage larvae of the flea *Xenopsylla* absorb 2 μg/hr/mm² from 93% R.H. whereas at 0% R.H. they lose 0.9 μg/hr/mm²; that is, their rate of uptake in moist air is twice that of loss in dry air.[191] The firebrat *Thermobia* loses water by evaporation in 43% R.H.; it gains water in 63% R.H.

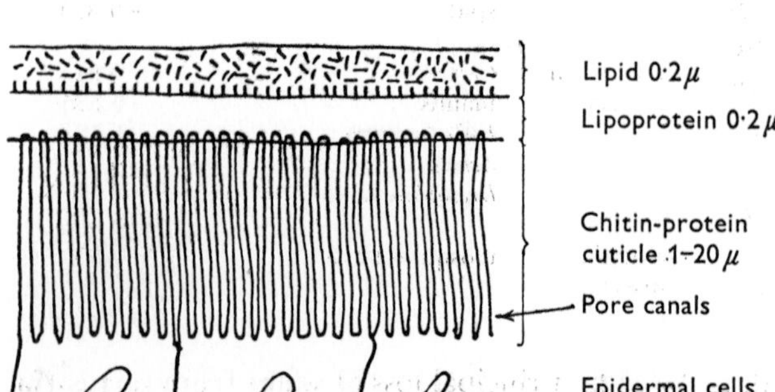

Figure 1–25. Diagram of insect cuticle. (From Beament, J. W. L., Soc. Exp. Biol. Symp. *19*:273–298, 1965.)

The mechanism of absorption of atmospheric water is unclear. The cuticle is penetrated by pore canals, 0.1 μ diameter, and water moves inward three times more readily than it moves outward, even in roaches which do not absorb water. Inward movement from a drop is independent of the osmoconcentration of the fluid. It is postulated that a water droplet may change the angle of orientation of monolayers of lipid in the cuticular pore. Also, tanned protein with wax on one side is asymmetric with respect to water movement. The absorption occurs only in living insects; hence, water which enters through the cuticle must be distributed by hemolymph to tissues.[253] In *Thermobia* ligation experiments indicate that water uptake is entirely by way of the rectum.[253]

In summary, insect evaporative loss of water is restricted by spiracles, and is restricted by waxy cuticle up to a critical temperature; water can be absorbed from moist air by virtue of the asymmetrical properties of the cuticular pores.

BEHAVIOR

Another adaptive mechanism is behavioral selection of a "preferred" humidity. Slightly desiccated animals tend to go to higher humidities and hydrated ones to drier air. *Tribolium* at 27°C can distinguish between 95% and 100% R.H., but cannot distinguish so small a difference in the range 40 to 53% R.H. A freshly collected clover mite, *Bryobia praetiosa*, chooses low humidity in a gradient but, if desiccated, may go to the moist end.[381] Adult *Drosophila* from a normal culture prefer drier air, but they reverse this preference if their water content is reduced by 12 to 16 per cent; hydroreceptors occur on the funiculus segment of antennae, on palps, and on the proboscis. *Tenebrio confusum* distinguishes differences in humidity only if peg organs on antennal segments 7 to 11 are intact. Coeliconic sensillae on antennae of *Locusta* are of three types: responding to moisture; responding to dry air; and increasing discharge in warm air, decreasing it in cool air.[366] In some lepidopteran caterpillars, the third antennal segment has three receptor neurons which increase their discharge with a rise in temperature and three others which increase spontaneous firing in moist air and decrease it in dry air. They can sense the changes due to evaporation from leaves; hence, by the information from temperature and humidity receptors they can tell fresh from wilted leaves.[92] One bipolar neuron in the taste sensilla of *Phormia* (blowfly) is stimulated by water; the other is stimulated by salt or sugar solutions.[107]

Humidity preferences may be important in the distribution of many insects. The tropical forest mosquitoes, *Anopheles bellator* and *A. homunculus*, overlap for part of their ranges in the forest canopy, but *A. bellator* rises higher in the evening and descends less far in the day because of its greater tolerance of low humidity. Of two *Drosophila* species, one, *D. persimilis*, occupies a colder, wetter range than *D. pseudobscura*. Mosquitoes select different salt solutions for oviposition, probably by salt taste receptors on tarsi and tibiae. Water taste and vapor-sensitive receptors are widely distributed.

EXCRETION IN RELATION TO WATER AND SALT BALANCE

The excretory system of insects consists of Malpighian tubules, midgut, and rectum. The Malpighian tubules usually end blindly in the body cavity and empty at the junction of midgut with hindgut. In a hydrated *Periplaneta* or *Calliphora* the rectal fluid (urine) is hyposmotic to hemolymph; in dehydrated *Periplaneta* urine O.C. may be twice that of hemolymph.[370] In *Schistocerca* the rectal fluid is always hyperosmotic and only a small part of the solute is inorganic ions:[262]

	Δ_{fp} hemolymph	Δ_{fp} rectal fluid
fed tapwater	−0.74°C	−1.55°C
fed hypertonic saline	−0.96	−3.4

Rhodnius, a blood-sucking bug, has a diuresis for 2 to 3 hours after a meal. A high rate of formation of tubular fluid is noted immediately after isolation of tubules in an appropriate saline; a later decline in fluid formation can be reversed by adding hemolymph to the medium. A factor extracted from neurosecretory cells of the mesothoracic ganglion potentiates fluid formation by isolated tubules.[227]

Isolated tubules from the stick insect (*Carausius*) produce a slightly hyposmotic fluid, higher in K and lower in Na than the

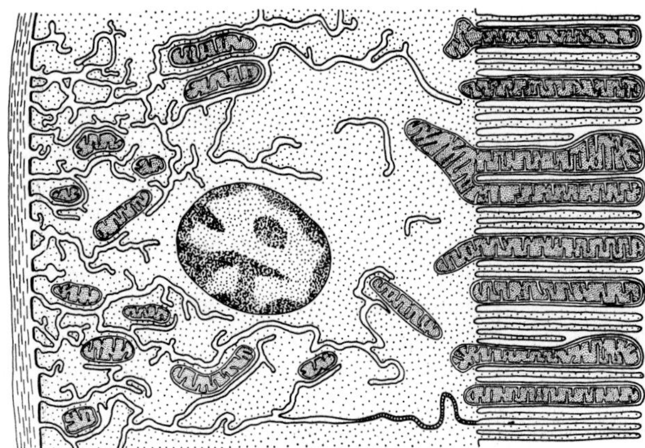

Figure 1–26. Secretory cell of malpighian tubule of cockroach. Mitochondria between indentations on hemolymph side; branched intercellular tubules in lumen side. (From Oschman, J. L., and M. J. Berridge, Fed. Proc. *30*:49–56, 1971.)

hemolymph or equivalent medium. The tubule lumen is electrically positive by 15 to 30 mV.[284] Secretion by the tubules is stopped if the hydrostatic pressure in the lumen is raised above 20 cm H_2O. Isolated tubules from *Calliphora* are stimulated to form urine by cations in the series K > Rb > Ca > Na. The urine may be slightly hyperosmotic to the medium, but its formation is independent of external osmotic concentration. *Calliphora* tubule cells are of two types; the primary cells show a basal (outer) zone with extensive infolding of the membrane and an apical (luminal) zone with microvilli richly interspersed with mitochondria[27] (Fig. 1–26). Stellate cells lie between the primary cells; they lack mitochondria in their apical microvillae. It is postulated that sodium diffuses into the primary cells at the basal end, and that this ion is pumped back in exchange for potassium; also, stellate cells may return sodium to the hemolymph. At the apical end an electrogenic potassium pump is indicated.[29, 31] Standing concentration gradients may exist in the spaces of infoldings on the basal side, where fluid becomes increasingly hyposmotic toward the closed end of the space. Water and chloride appear to move passively down their gradients into the lumen (Fig. 1–27). The pump is not affected by ouabain; it is inhibited by Cu^{++} in the presence of Cl^- but not of $HPO_4^=$. Numerous small anions support the urine formation, but larger anions do not; a pore size of 3.6 Å

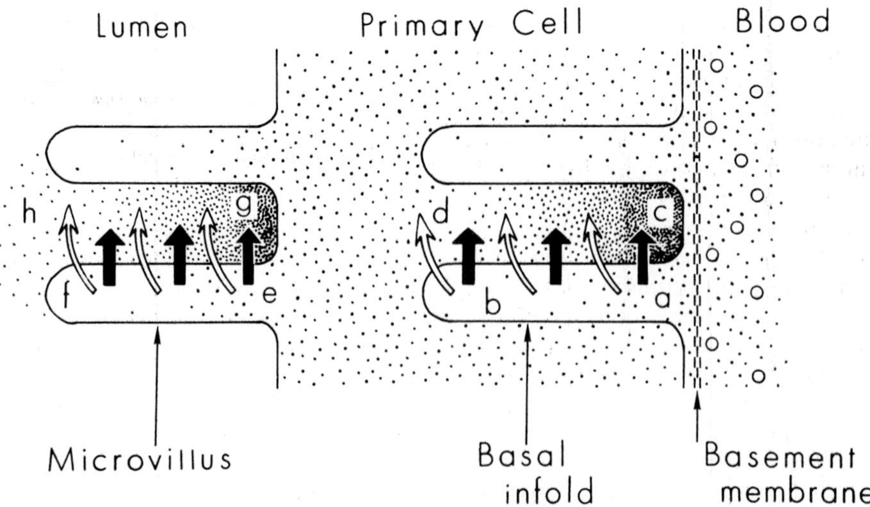

Figure 1–27. Diagram of movement of water and solute in malpighian tubule. Solid arrows indicate transport of solute; open arrows indicate passive flow of water. (From Berridge, M. J., and J. L. Oschman, Tissue & Cell *1*:247–272, 1969.)

is calculated.[30, 31] Similarly, in *Carausius* the Malpighian tubules transport K from hemolymph (18 mM) to lumen urine (145 mM), a transport unaffected by ouabain but stopped by DNP.[270] Isolated Malpighian tubules from a lepidopteran larva form tubular fluid of 13 mM K in a medium of less than 0.45 mM, but can secrete Na if K_o is low and Na_o is high; K is reabsorbed by the lower part of this tubule.[171]

In *Rhodnius* the fluid in the upper part of the Malpighian tubules is clear, isosmotic, and alkaline; in the lower part of the tubules uric acid crystals are noted and the fluid is acidic, suggesting some water reabsorption. In maximum diuresis, the tubules of *Rhodnius* can secrete *in vitro* up to 3.3 $\mu l/cm^2/min$, and less if there is no sodium outside; but there is no reduction by low Na if there is no K. In a choline medium, secretion is proportional to sodium or to potassium or to their sum. Secretion is isosmotic and not affected by ouabain. Secretion can go against a gradient, especially of potassium (from 14 to 85 mM). Separate pumps are postulated for Cl, Na, and K.[227]

In most insects after the fluid enters the rectum, both potassium and water are reabsorbed into the hemolymph. In *Aedes* larvae the rectal fluid becomes markedly hyposmotic; the intestinal fluid is nearly isosmotic with the hemolymph.

Isolated recta of the locust *Schistocerca* can transport water from the lumen even when the solute in the lumen is a non-electrolyte such as trehalose; the water transport is independent of osmotic concentration[261] (Fig. 1–28). Transport of water is, however, more effective if the rectum contains electrolyte solution. Recta of saline-fed locusts transported water 2.5 times better than those of tapwater-fed animals. The lumen of the rectum of *Schistocerca* is positive to the hemolymph by 2 to 30 mV. When saline of composition resembling that of the hemolymph, and made hyperosmotic by addition of 1 M xylose, is placed in the rectum, there is absorption of 31% of the Na, 75% of the K, and 44% of the Cl against concentration gradients, so that the rectal fluid (Na + Cl) is three to five times that in hemolymph. Water movement out of the rectum is 4 to 12 times faster with K than with Na. Calculations from permeability to different molecules indicate a pore size of radius 6.5 Å in the pads along the rectum.[262] Formation of hyperosmotic

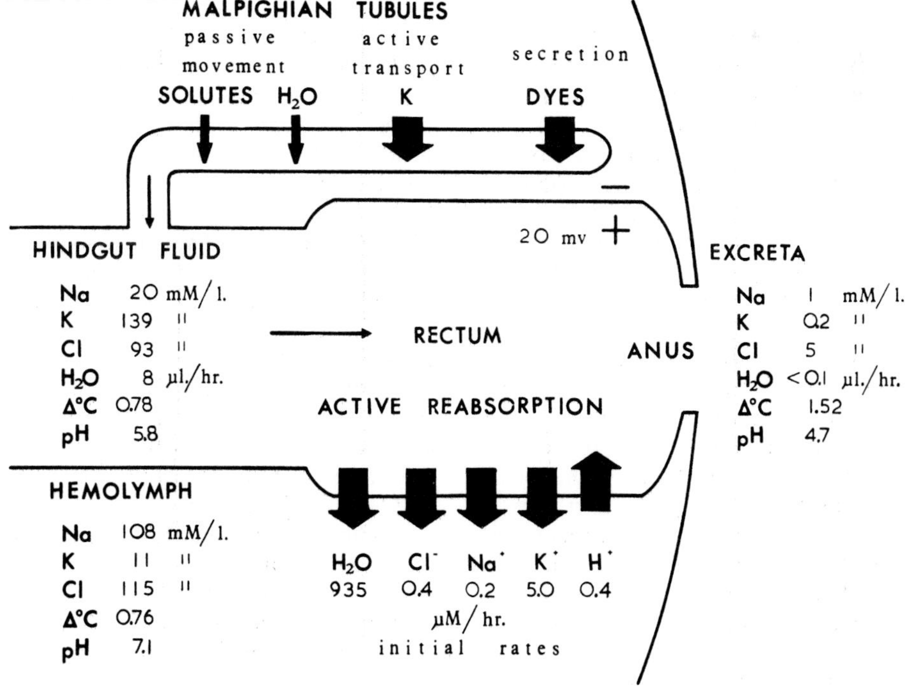

Figure 1–28. Summary of concentrations and ion and water movements in excretory system of locust. (From Phillips, J. E., Amer. Zool. *10*:413–436, 1970.)

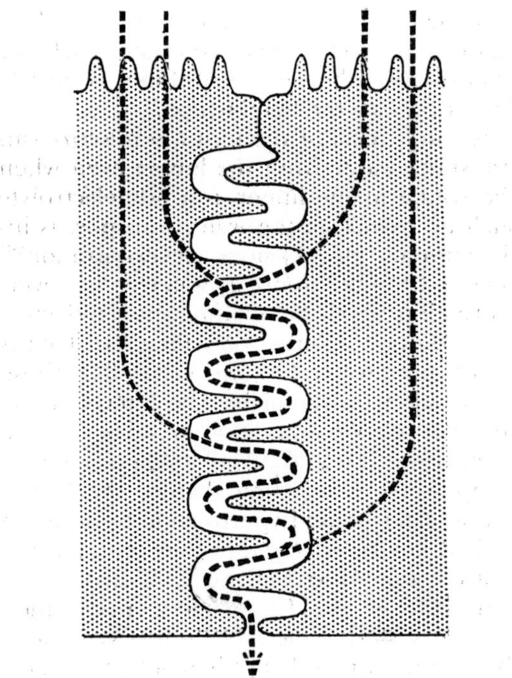

Figure 1–29. Schema of water movement in standing gradient pump by which solute is secreted into intercellular spaces, creating an osmotic concentration into which water flows passively. Intercellular space is open to lumen, closed at other end. (From Tormey, J. M., and J. M. Diamond, J. Gen. Physiol. *50*:2031–2060, 1967.)

fluid, even from non-electrolytes, led to the postulate that there exists some form of active transport of water coupled with solute.

From certain vertebrate epithelia, a model to account for rectal transport in insects has been based on a scheme developed by Diamond[93, 94, 356] (Fig. 1–29). Two cells are separated by a space which is closed at one end, and narrowly open in the direction of secretion. Solute (ion) is secreted at the apical end, making the fluid in the intercellular space hyperosmotic; water then moves out of the cells into the space so long as the permeability for water exceeds that for solute. A concentration gradient (also probably a hydrostatic gradient) exists down the channel, and the secreted fluid may be isosmotic or hypertonic. The channels enlarge when fluid is actively secreted. The application of this model to secretion of bile in the gall bladders of mammals has been demonstrated.

To this model, add (1) a membrane (possibly cuticle) which is selectively permeable to water, as on the lumen side, and (2) a mechanism for reabsorbing the solute (usually Na) by a membrane beyond the secretion channel. Thus, sodium would be cycled out and back into the cells, and water would be transported against an overall gradient as from lumen to hemolymph, leaving hyperosmotic urine[369] (Fig. 1–30).

Studies of ultrastructure of the rectum of a cockroach, *Calliphora,* and of *Locusta* corroborate the above model. In *Calliphora,* cortical cells occur in cones which project into the lumen of the rectum. Large intercellular sinuses formed by infolding of lateral membranes are distended in flies injected into the rectum with hyposmotic media, and are collapsed in starved flies. The sinuses open into the hemocoel, and along their surfaces Mg-ATPase can be visualized histochemically. It is postulated that cations are pumped into the intercellular spaces, making them hyperosmotic, and that water is then drawn from the lumen by the osmotic gradient and thus absorbed. Such a mechanism could function without ions in the lumen, and thus could account for water absorption when non-electrolyte solution is in the rectum.[27]

The walls of the rectum of *Periplaneta* have pads of columnar cells with septate desmosomes at the apical and basal ends; between the cells are dilations and apical sinuses. Water can be withdrawn from the lumen to the medium (hemolymph). When normally hydrated, the lumen contents (including fluid from the malpighian tubules) may be at 275 mOsm and the hemolymph at 379 mOsm, whereas in a dehydrated animal the lumen contents may reach 1000 mOsm. It is postulated that solute (probably an organic molecule) is pumped to the intercellular dilations and that water from the lumen enters across the lining cuticle osmotically; the solute then diffuses back via hemolymph and epithelial cells. Potassium ions are cycled, and are more concentrated in the posterior lumen than in hemolymph. However, Na and K in intercellular fluid are insufficient to account for the amount of water reabsorbed; some other solute must be active.

Amino acids are also reabsorbed by the rectum. The absorbing system is inhibited by DNP, and the rate of water absorption is increased by extracts of corpora allata and of thoracic and abdominal ganglia.[370]

In the mealworm *Tenebrio molitor,* feces enter the rectum as liquid and leave the anus

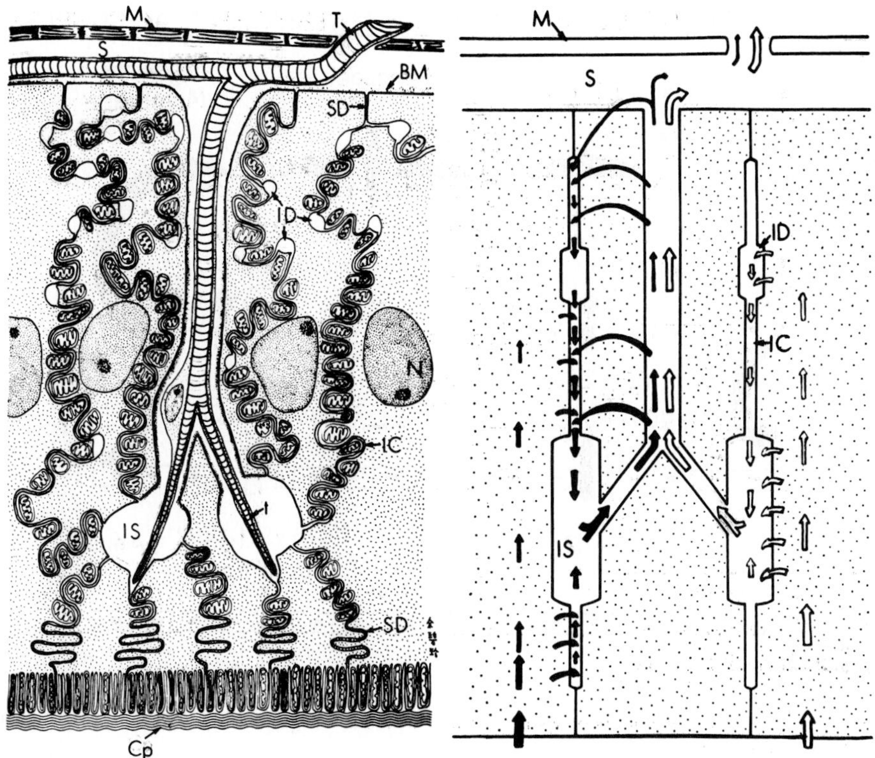

Figure 1–30. Drawing and schema of rectal pad of *Periplaneta.* M is muscle layer facing hemolymph and Cp is cuticle facing rectal lumen. Tracheae penetrate through subepithelial sinus, S, and into intercellular sinuses, IS. Cells with nuclei, N, are separated by intercellular channels, IC, and larger dilations, ID. In schema, water flow is represented by open arrows and solute by solid arrows. (From Wall, B. J., Fed. Proc. *30*:42–48, 1971.)

as dried pellets. On a moist diet, the hemolymph osmoconcentration corresponds to a freezing point depression of −0.75°C; on dry food it is −1.5°C. The Malpighian tubules are not free in the body cavity but are applied close to the rectum (perirectal tubes), the whole cryptonephric rectal complex being surrounded by a perinephric membrane[143,286] (Fig. 1–31). The contents of the upper hindgut are isosmotic with hemolymph; the lower gut is hyperosmotic. Potassium is absorbed from the rectum into the perirectal space, from which it passes into the tubules; also, K is taken from the hemolymph by the perirectal tubules. The lumen of the tubule is 50 mV positive with respect to the hemolymph. Water then leaves the rectum by the osmotic gradient and moves into the tubules, which return it to the hemocoel. Thus, flow in these modified Malpighian tubules in the op-

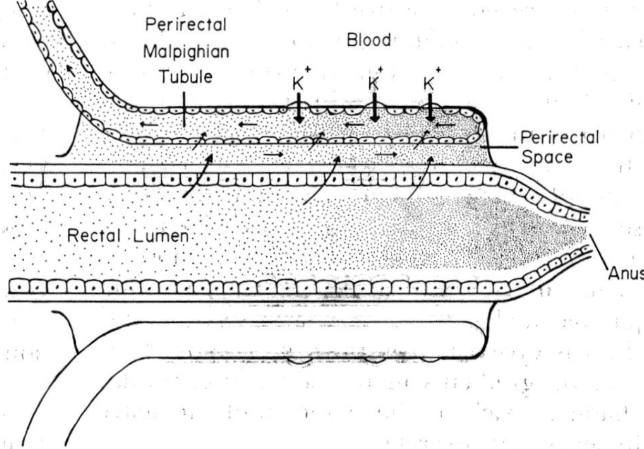

Figure 1–31. Cryptonephric complex of *Tenebrio.* (From Wall, B. J., Fed. Proc. *30*:42–48, 1971.)

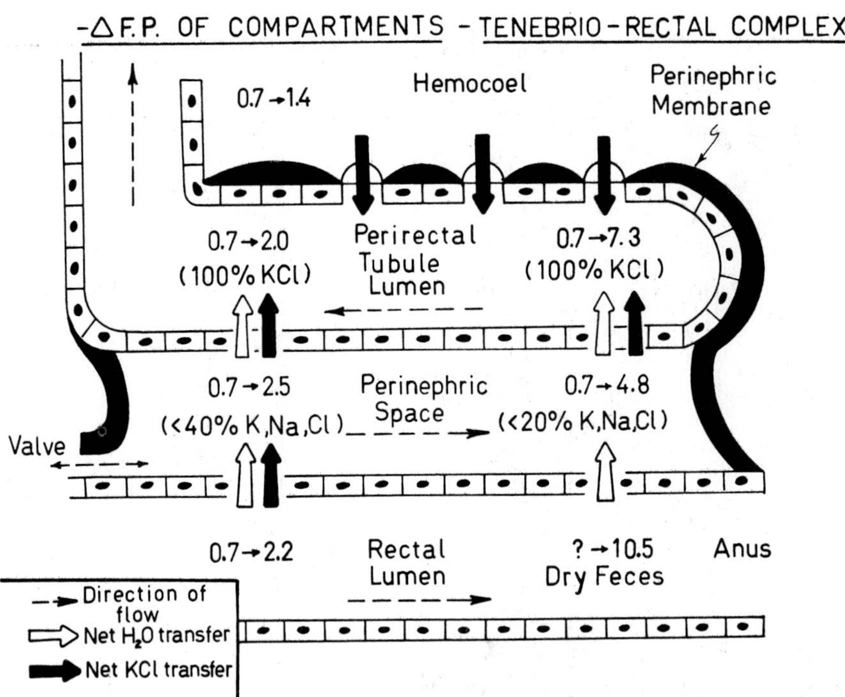

Figure 1–32. Summary of movements of KCl and water in rectal complex of *Tenebrio* larvae. (From Phillips, J. E., Amer. Zool. *10*:413–436, 1970.)

posite direction to that in most insects. The inner sheath of the perinephric membrane is multilaminate, and the cells of the tubule walls have infoldings on the sides facing the perirectal space. Thus, by a combination of K uptake from hemolymph and transport out of the rectum, an osmotic gradient is created and the perirectal (Malpighian) tubules transport water to the hemolymph[261] (Fig. 1–32).

Certain common features are evident in the Malpighian tubules and the rectum. Each of them actively transports potassium, and each can create an osmotic gradient. The pumps are not sensitive to ouabain, and are clearly not sodium transporting ATPase systems. In each there are deep mitochondria-lined infoldings on the side of the cell toward which transport occurs. Insects appear to be the only invertebrates that can produce markedly hyperosmotic urine and thus conserve water. The mechanisms by which they accomplish this are different from those in mammals, the only other animals with comparable ability. In terrestrial insects, potassium is cycled and water can be moved via osmotic gradients in restricted intercellular channels back to the hemolymph, to leave hyperosmotic excreta.

FISHES

The origin of vertebrates is obscure. All prochordates are marine, and the phylum Chordata doubtless arose from marine ancestors. Freshwater origin of fish is suggested by the following facts: blood of all modern bony fish is dilute, and blood of cartilaginous fish is low in salt content; kidneys are glomerular, adapted for excreting a hyposmotic urine. However, filtration with reabsorption of solutes does occur in marine crustaceans and molluscs; thus it is not proof of a freshwater origin. The ancient coelacanth *Latimeria* is essentially isosmotic with sea water, having an osmoconcentration of 1181 mOsm.[266]

Whatever their origin, Osteichthyes and Chondrichthyes evidently migrated from fresh water to the sea in the Cretaceous. Many modern fish (teleosts) have reinvaded fresh water, whereas a few others (holosteans) have lived throughout vertebrate history continuously in fresh water. A few marine fish are not now independent of fresh water; some, like salmon (anadromous), return to the fresh water to spawn, while some, like eels (catadromous), breed in the ocean and reach maturity as adults in fresh water. In river mouths some fish, such as *Fundulus*, occur in either fresh or brackish water both as young and adults; others, such as menhaden and gray squeteague, move to fresh water when young.

AGNATHA (CYCLOSTOMES)

The myxinoids (hagfishes) are exclusively marine; they (*Myxine* and *Polistotrema*) are essentially isosmotic with the medium, over a range of 600 to 1540 mOsm. They show considerable ionic regulation, and vary the total osmoconcentration largely by changes in amino acids. For a 2½-fold change of osmotic concentration in a hagfish, 80 per cent of the variation in muscle concentration was due to changes in amino acids, particularly proline, alanine, and leucine.[62]

Petromyzonts (lampreys) are either anadromous (spawning in fresh water) or live exclusively in fresh water. Blood from *Petromyzon* has a Δ_{fp} of $-0.48°C$ in fresh water and $-0.59°C$ in sea water. When marine lampreys enter fresh water they stop drinking, but if returned to brackish water they drink and extrarenal excretion of Na and Cl is activated. In fresh water they can actively absorb chloride. Lake Michigan specimens had urine/plasma (u/p) ratios for Na^+ of 13.5/124 and for K^+ of 2.0/4.1.[232]

In Britain *Lampetra planeri* is restricted to fresh water; *Lampetra fluviatilis* is anadromous, and the larvae migrate to brackish water where they mature and then ascend the river to spawn. The young *L. fluviatilis* show hyposmotic regulation:

medium	Δ_{fp} *medium*	Δ_{fp} *plasma*
F.W.		0.46
33% S.W.	0.66	0.54
50% S.W.	0.97	0.57

L. fluviatilis maintains hypotonicity by swallowing brackish water and by excreting Cl^- extrarenally. As the lampreys mature they lose the capacity for surviving in dilute sea water. *L. planeri* is several times more permeable to water than *L. fluviatilis*.[149,245]

Lampetra adults in fresh water resemble fishes; their blood concentration is about 255 mOsm and urine volume is 200 ml/kg/day.[55] The larval forms are slightly more dilute (215 mOsm). They actively absorb Na by a transport system which has a K_m (half-saturation) at 0.13 mM and which can transport a maximum of 1.08 μM/g/hr.[246]

In summary, hagfishes and lampreys are very different in osmotic capacities—hagfishes are isosmotic, but lampreys show both hyposmotic and hyperosmotic regulation.

CHONDRICHTHYES (ELASMOBRANCHS)

Sharks, skates, and rays are largely marine; a few are found in fresh water. They are unique among marine fishes in that the blood osmoconcentration is always higher than that of the medium (see Table 1–4).

In marine elasmobranchs the plasma salt concentration is of the same order as in teleost fishes, but the osmotic fraction due to NaCl is only about 45 per cent of the total. Most of the osmotic concentration is made up by retention of quantities of urea and some trimethylamine oxide (TMO) in all body tissues and fluids. Thus, an elasmobranch can maintain the same osmotic gradient in sea water as in fresh water and can continue to excrete a blood-hyposmotic urine. Reabsorption of urea in the dogfish kidney varies inversely in accordance with blood urea concentration. Skin and gills of elasmobranchs are relatively impermeable to urea; the young are provided with urea, either by viviparity or by placement in an impermeable egg case containing urea which lasts until they can develop mechanisms for retaining it.

There is much species variation in the urea concentration. *Potamotrigon* from the upper Amazon has only 1.07 mM urea, whereas *Carcharhinus* from Lake Nicaragua has 180 mM urea (Table 1–4).

Marine elasmobranch species differ in tolerance of slightly brackish water. The dogfish *Scyliorhinus* failed to survive in 60% S.W. but survived if sucrose or urea were added to raise the osmoconcentration to that of sea water.[4] A skate, *Raja eglanteria*, tolerated acclimation over a salinity range from 16.2 0/00 to 31 0/00, and the increases in blood urea (35%) were slightly greater than those in chloride (31%).[280] Lemon sharks (*Negaprion*) were acclimated to 50% S.W. for a week; the blood osmotic concentration declined by 40%, plasma urea by 55%, trimethylamine oxide by 60%, and chloride by 20%; urea clearance in the kidney increased threefold.[132] When a skate (*Raja erinacea*) was

TABLE 1–4. Osmotic Concentrations in Members of Chondrichthyes.

Genus	Δ_o	Δ_{serum}	Δ_{urine}	Blood Cl mEq	Blood Urea mM
Marine					
Raja[327]	−1.85 (1000 mOsm)	−1.94 (1041 mOsm)	−1.68 (950 mOsm)	230	357 (2140 mg %)
Freshwater					
Pristis[327]	0	−1.0 (538 mOsm)	−0.1 (54 mOsm)	170	107 (642 mg%)
Carcharhinus[353]				219	180 (504 mg urea N/100)
Potamotrigon[353]		308 mOsm			1.07 (3.0 mg urea N/100)
Raja eglanteria[280]	−1.52 −1.13	−1.57 −1.37		222 195	368 320

transferred from S.W. to 50% S.W., its urine flow increased six times, its glomerular filtration ratio four times, its urea clearance 22 times, its Cl^- clearance four times, and its TMO clearance 13 times.[131]

As in many marine bony fish, the kidneys are the primary route for excretion of magnesium, phosphate, and sulfate; urine is blood-hyposmotic, and an extrarenal route of NaCl excretion has evolved. This is by rectal glands which, in dogfish, produce a secretion which may be isosmotic with blood but is much more concentrated in NaCl and contains virtually no Ca^{++}, Mg^{++}, or urea. In sea water, a spiny dogfish excretes 0.47 ml/kg/hr of fluid containing 495 mM Cl/liter. In an unfed dogfish, the inactive rectal gland secretion has a chloride ratio to plasma of 1.24, but when the fish has been injected with NaCl, an increase in secretion occurs. When sucrose or urea was injected, less stimulation was noted; secretion by rectal glands is stimulated by osmotic and volume receptors, and specifically by NaCl.[56] The rectal gland (9 species) is rich in ATPases requiring Mg, Na, and K, and its activity is inhibited by ouabain; hence, the transport resembles other membrane sodium pumps.[41] The epithelial cells of the gland show lateral interdigitations and apical vesicles.[96] A dogfish, *Squalus*, has plasma Na^+ of 250 mM in a medium of 440 mM Na; tracer measurements show a net influx of 0.52 mM/kg/hr, mostly across the gills. This is balanced by the following effluxes: renal, 0.15 mM/kg/hr; rectal gland, 0.24 mM/kg/hr; and head (gill), 0.16 mM/kg/hr.[164] Whether there is active uptake of Na by elasmobranchs in fresh water is unknown.

Osteichthyes (Teleosts)

Ecological Correlates. Freshwater bony fish have plasma osmoconcentrations of 130 to 170 mOsm, and the urine is copious but dilute, as in the garpike:

Δ_o	Δ_i	Δ_u	*urine volume ml/kg/day*
−0.03	−0.57	−0.08	200 to 400

The skin is relatively impermeable and little water is taken in by the mouth, but some water enters osmotically through the gills and oral membrane. The inward stream provides a water load, and the kidneys have well developed glomeruli that filter a considerable volume. As the filtrate passes down the tubules, most of the solutes are reabsorbed, leaving a dilute urine. However, the urine is not as dilute as the fresh water medium, and some salt is lost by diffusion and a small amount by feces (Fig. 1–33). The salt loss is compensated partly by food and partly by active absorption from the medium by special cells of the gills. Chloride is absorbed against a gradient from very dilute media. Some species depend mainly on food for salts (*Acerina, Perca*), while others have more active salt-absorbing systems (*Leuciscus, Carassius*).

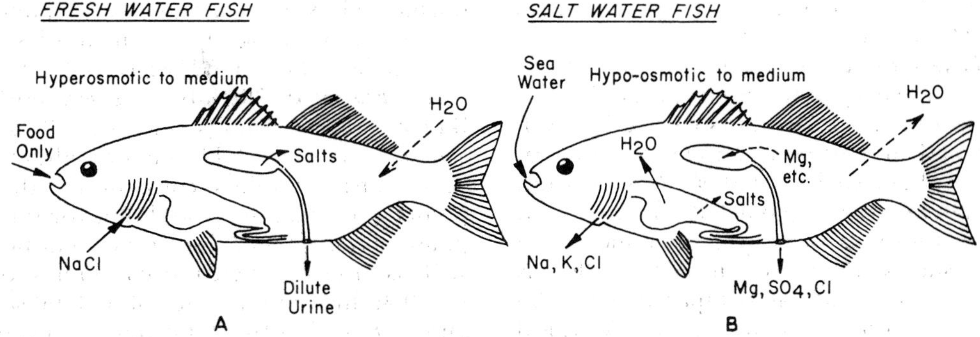

Figure 1-33. Schematic representation of main paths of ion and water movement in osmoregulation of freshwater bony fish, *A*, and marine bony fish, *B*. Solid arrows indicate active transport; broken arrows show passive transport.

Catadromous fish breed in the sea and mature in fresh water. Both American and European eels (*Anguilla vulgaris*) breed in the Sargasso Sea of the mid-Atlantic; then the young elvings of the two subspecies migrate, each to the appropriate continent. Osmoconcentration of blood of eels in sea water is Δ_{fp} −0.69° to −0.73°; in fresh water it is −0.61° to −0.62°. The kidney of *Anguilla* has glomeruli and a distal segment. The skin and gills have low permeability to ions and water. In fresh water, urine flow is 84 ml/kg/day; in sea water it is only 15.1 ml/kg/day. In sea water, eels actively excrete Cl^- through the gills. In fresh water, the gills absorb some Cl^-, and some ions are obtained in food.

Anadromous fish such as salmon spawn in fresh water, and the young then migrate to the sea as smolt. The blood of salmon freezes at −0.76°C when at sea and at −0.67°C in the spawning grounds. Thus, king salmon blood is diluted only 12% in going from the sea to spawning lakes. Coho salmon change more; their blood osmolality in sea water is 400 mOsm, while in fresh water it is only 295 mOsm. *Oncorhynchus kisutch* in streams develop adaptations for survival in sea water during the spring, and large individuals migrate to the sea; small fish retain the adaptations and migrate the following year. Steelhead salmon (*Salmo gairdneri*) de-adapt if they fail to migrate.[68, 136] Red salmon (*Oncorhynchus nerva*) adapt before migration and have a urine concentration of Δ_u −0.44°C; if held in fresh water for a few days they de-adapt and produce urine which is more dilute (Δ_u −0.23°C). Blood Na is 172 mM in S.W. and 150 mM in F.W.[386]

The osmotic problem of marine teleosts is to conserve water and exclude salt; in this the water-excreting glomerular kidney is a liability (cf. Fig. 1-33). The blood of marine teleosts is not much more concentrated than that of freshwater fish (Δ_i −0.7°C); hence, a mechanism for maintaining a high degree of hyposmoticity is necessary. The urine is hyposmotic to the blood and is scanty. Marine fish, unlike freshwater ones, swallow some water, although less than was formerly thought. As the sea water passes down the intestine, water is absorbed, as well as some salts (more Na^+ and Cl^- than Mg^{++} or $SO_4^=$). Hence, the concentration of Na^+ and Cl^- decreases while that of K^+, Ca^{++}, Mg^{++}, and $SO_4^=$ increases down the intestine.[327a] Some of the sea salts, particularly Mg^{++} and $SO_4^=$, are lost in the feces; the urine salts are largely Mg^{++}, Ca^{++}, $SO_4^=$, and phosphate, with some K^+, while most of the Na^+ and Cl^- are excreted extrarenally by way of the gills.

By perfusion of a heart-gill preparation from an eel, the gills were clearly shown to be the major route of NaCl excretion.[196] Chloride is secreted outward into sea water which contains three times as much Cl^- as does the eel's blood. The secretion varies with the Cl^- concentration in the solution perfusing the gill vessels. When eels were transferred from F.W. to S.W., the plasma Na increased from 108 to 153 mEq; Na outflux went from 8.3 to 700 μEq/100 g/hr. Infusion of NaCl also increased Na outflux; any change in Na_i is a stimulus for a change in gill exchange. The internal receptor for Na^+ is unknown.[236]

Normally such fish as sculpin, goosefish, flounder, and haddock show zero or low Cl^- in their urine, but with handling they become diuretic and Cl^- in urine is increased. Nitrogenous wastes are largely ammonia, and are excreted by diffusion from the gills. A flounder drinks sea water amounting to 0.5 to 1 per cent of body weight per hour; *Serranus* drinks 0.5% B.W./hr, and a blenny drinks only 0.03% B.W./hr.[249, 250]

Aglomerular Kidneys. In view of the reduced urine output, it is appropriate that the kidneys of many marine fishes are less well developed than those of freshwater ones. In many of them, such as toadfish (*Opsanus*) and goosefish (*Lophius*), glomeruli are absent in adults; there may be pseudoglomeruli in young stages. In the toadfish, urine flow is only 2.5 ml/kg/day as compared with 300 ml/kg/day in a freshwater catfish. A typical marine teleostean kidney also lacks the distal convoluted tubule. Even in those marine teleosts having glomeruli, the filtration rate is lower than in freshwater fish. Urine flow in per cent of body weight per day is: eel, 1.5; flounder, 1.4; *Fundulus,* 1; and *Xiphister,* 0.34. Some reabsorption in the bladder is indicated for flounder.[200]

Excretion by an aglomerular kidney is essentially tubular secretion. The kidney can excrete Mg^{++}, $HPO_4^=$, $SO_4^=$, Cl^-, and creatinine, and can concentrate small dye molecules such as phenol red. It does not excrete inulin or glucose, even after phlorizin (a drug which blocks glucose reabsorption in glomerular kidneys) has been administered. In *Lophius,* with $\Delta_p - 0.67$ and $\Delta_u - 0.57$, the urine is essentially free of Cl^-, but it may contain a hundred-fold concentration of Mg^{++} and $SO_4^=$, and a lesser concentration of Ca^{++}.[115]

Some aglomerular fish can be adapted to fresh water; their urine volume increases and their concentration decreases, as shown in the following data for the toadfish *Opsanus tau*:[201]

		Medium		
		Sea Water	*50% S.W.*	*Fresh Water*
plasma	Na(mEq)	159.8	153	107
	K(mEq)	5.2	4.5	7.1
	mOsm	392	385	250
ureteral urine	Na(mEq)	72.7	71.5	87
	K(mEq)	9.0	7.8	6.6
	mOsm	356.3	331	213
	flow(ml)	0.018	0.019	0.087
bladder urine	Na(mEq)		62.4	107
	K(mEq)		8.5	8.5
	mOsm	348.5	300	231

The most spectacular change is in extrarenal efflux of sodium which, in sea water, is 840 μEq/100 g/hr; after one or two days in fresh water it drops to 19.4 μEq/100 g/hr, and after three to four days it is 5.7 μEq/100 g/hr. The normal efflux in sea water corresponds to 16% of body Na per hour, which is less than for *Xiphister* (23%) and *Tilapia* (66%).[200]

Ion Fluxes in Teleosts in Fresh and Sea Water. Measurements of Na efflux for *Fundulus heteroclitus* indicate two sodium compartments; the efflux constant for the fast phase (k_1) is 0.55 hr^{-1}, and that for the slow phase (k_2) is 0.16 hr^{-1}. In other marine fish, such as *Pelates,* a single rate constant is found; it is 0.49 hr^{-1} in sea water of 35 0/00 salinity and 0.18 hr^{-1} in 10 0/00 salinity.[83] In general, both efflux and influx constants decrease in reduced salinities, as follows for *Fundulus*:[273]

Medium	k_e (Na)	
S.W.	0.464 hr^{-1}	
40% S.W.	0.135	
F.W.	0.017	
	k_i (Na)	
S.W.	0.261	or 20.5 mM Na/kg/hr
40% S.W.	0.154	
F.W.	0.012	0.58 mM Na/kg/hr

The results show that the exchange of ions is much reduced when a fish goes from sea water to fresh water.

The role of the intestine is indicated by measurements of ion and water movements across the wall of isolated intestine. With Ringer solution on both sides of sections of intestine from the marine *Cottus scorpius,* there was net absorption of 11.5 $\mu Eq/cm^2/hr$, accompanied by net uptake of H_2O. There is essentially no electrical potential across the intestinal wall; hence, an electroneutral pump is indicated.[166] Similarly, in an eel intestine perfused *in vivo,* absorption of sodium is active, and water absorption is linked to sodium transport so that it can go against an osmotic gradient. Sodium transport, and with it water absorption, from near-isosmotic solution is greater in marine eels than in eels adapted to fresh water; it is also greater than in frog or dog intestine:[325]

	eel in S.W.	*eel in F.W.*	*frog colon*	*dog ileum*
net Na transport 10^{-5} $mEq/mm^2/sec$	6.03	2.77	1.03	2.01
water transport 10^{-4} $\mu l/mm^2/sec$	4.5	1.83		0.5–1

Balance studies for sodium and water have been done for a number of fish in different salinities. An intertidal *Blennius* has a blood sodium concentration of 170 mM/liter; in

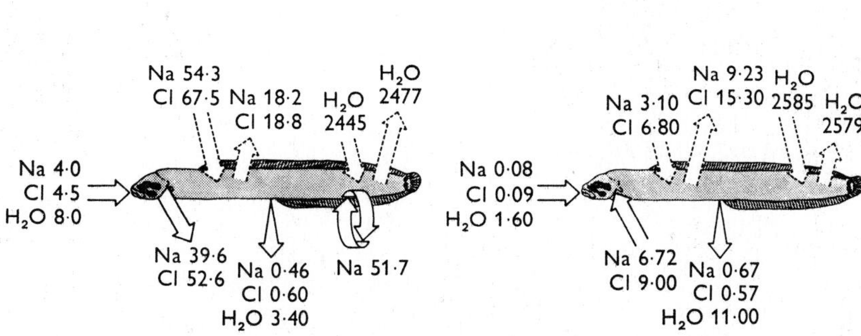

Figure 1–34. Flux balance in *Xiphister:* movement of water, Na, and Cl is by different routes when fish is in S.W. and in 10% S.W. (From Evans, D. H., J. Exp. Biol. *47*:525–534, 1967.)

S.W., a Na exchange flux of 100 mM Na/liter blood/hr was observed, while in 10% S.W. this exchange was reduced to 20 mM Na/liter blood/hr. The blood is normally electrically positive by 23 mV to 100% S.W., but it is negative by 3 mV to 10% S.W. Hence, in sea water Cl^- is actively excreted, probably by the gills, and in 10% S.W. Cl^- is taken up.[165] A similar summary for a blenny, *Xiphister,* is given in Figure 1–34.[108] From Na losses in isosmotic mannitol as compared with losses in sea water, it is estimated that 47% of Na loss in sea water is by exchange diffusion; this type of loss is negligible in 10% S.W.

A balance sheet for an intertidal fish, *Pholis gunnelus,* is given in Table 1–5.[109]

A marine species, *Cottus bubalis,* survives only a few hours in fresh water; the fresh-water species, *C. morio,* cannot live in sea water. Sodium fluxes, in per cent of body Na exchanged per hour, are as follows:[116]

	In Fresh Water	*In Sea Water*
Cottus bubalis		16
Cottus morio	1.1	
Cottus scorpius		12.5
Fundulus heteroclitus	1.4	29
Tilapia	4.9	55.5

TABLE 1–5. Balance Sheet for *Pholis gunnelus.*

Influx mM/kg/day			*Efflux mM/kg/day*	
Na	Cl		Na	Cl
Sea Water				
5.1	5.9	drinking	–	–
–	–	urine	0.5	0.5
47.6	36.4	diffusion & potential	37.5	5.6
64.3	25.1	diffusion only	27.7	8.1
–	–	active transport	14.8	36.2
		potential, +18 mV		
Fresh Water				
0.14	0.16	drinking	–	–
–	–	urine	0.5	0.5
7.2	3.7	diffusion & potential	7.6	5.3
6.6	4.1	diffusion only	8.3	5.8
0.69	1.87	active transport	–	–
		potential, −6 mV		

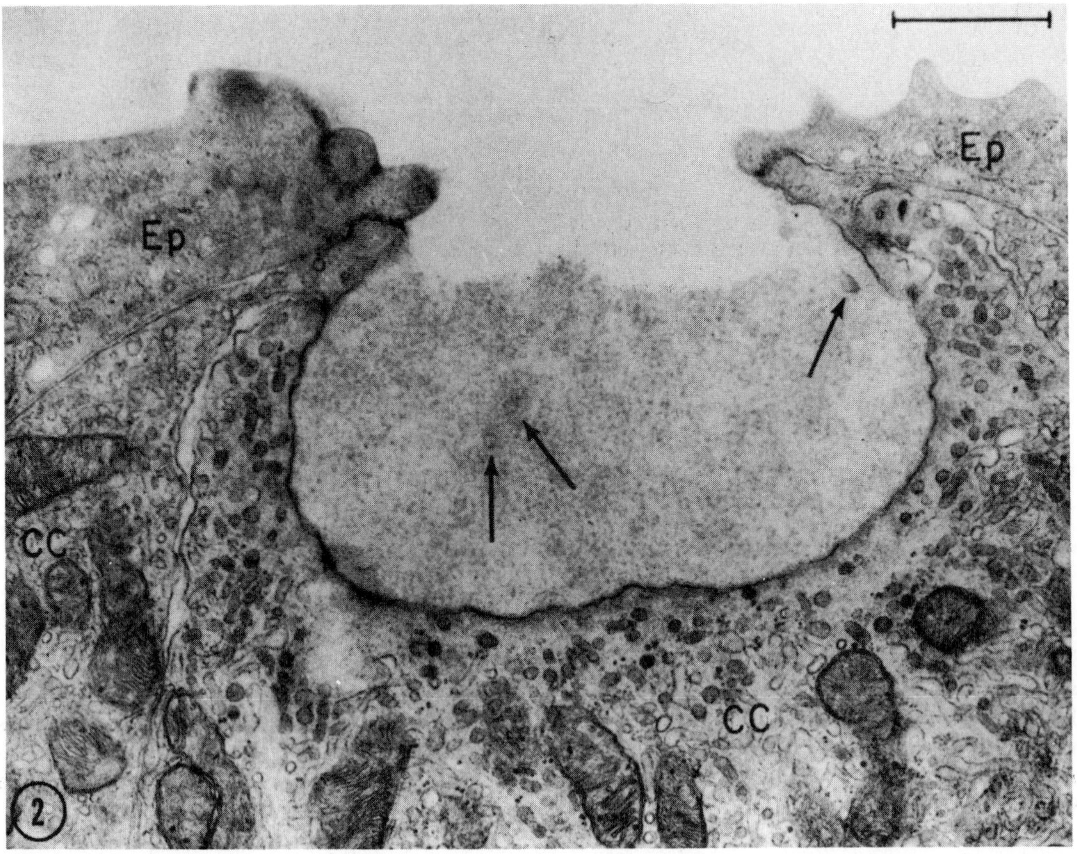

Figure 1–35. Chloride cells in gill of *Fundulus* adapted to sea water. Three cells share a single apical cavity; amorphous material in the cavity is being released. (From Philpott, C. W., and D. E. Copeland, J. Cell Biol. *18*:389–404, 1963.)

Apparently, *C. scorpius* and *C. bubalis* cannot reduce Na loss enough to live in fresh water, and *C. morio* cannot get rid of sufficient salt to tolerate sea water.

Fundulus heteroclitus, which is normally marine, lives well in either fresh or sea water. On direct transfer it acclimates in body density by changes in gas volume in the swim bladder —during six hours on going from F.W. to S.W., and during 24 hours on going from S.W. to F.W. In sea water there is a high influx of Na (20 mM/kg/hr), of which approximately half is by diffusion across the gills and half by drinking. In fresh water the influx of Na is 0.6 mM/kg/hr, all of it by active uptake. Reduced permeability to Na appears after a few minutes in fresh water; increased permeability in sea water develops during several hours.[273] The gills of *Fundulus* contain secretory cells, rich in mitochondria and endoplasmic tubules which, in sea water, have prominent distal vesicles whose bases stain for chloride (Figs. 1–35, 1–36). Gills of freshwater *Fundulus* lack the distal vesicles.[263] It

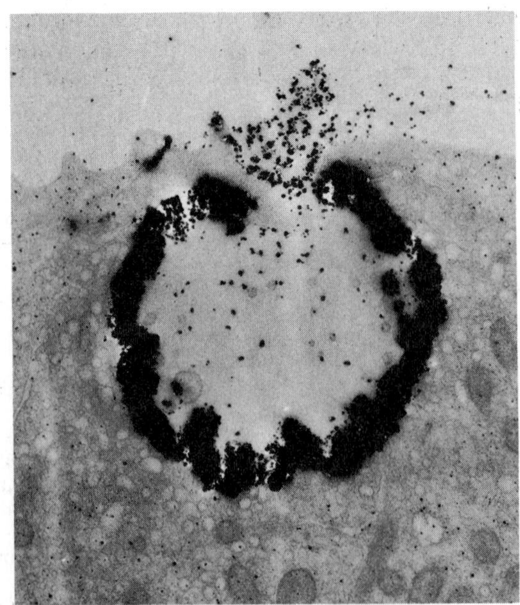

Figure 1–36. Chloride cells of *Fundulus* adapted to sea water. Cl precipitated by silver stain. (From Philpott, C. W., Protoplasma *60*:7–23, 1965.)

	filtration ml/kg/hr	u/p ratio inulin	mEq/liter Na	K	Cl	mOsm
in S.W.						
urine	0.24	4	59.7	2.5	145	275
plasma			142	3.4	168	297
in F.W.						
urine	0.42	2.3	30.4	1.2	36	90
plasma			124	3	132	240

is suggested that these cells actively secrete Cl^- (and with it Na^+) in sea water; the cells which absorb NaCl from fresh water are not identified.

Another species of *Fundulus (F. kansae)* is found in hypersaline lakes; this fish survives in solutions from 0.2 mM Na plus 0.1 mM Ca to 250% S.W. It is much less permeable to Na, with influx of 1.7 mM/kg/hr in S.W. as compared to 20 mM/kg/hr in *F. heteroclitus*.[274]

The alterations in Na fluxes in different salinities have been studied in the euryhaline flatfish *Platichthys*.[248] In sea water there is a sodium exchange of 2.6 mEq/100 g/hr; in fresh water it is 0.01 to 0.2 μEq/100 g/hr. The flux across the gills has two components: in S.W., 85% of the transfer is by exchange diffusion and 15% is by net outward excretion. Possible mechanisms of the change in permeability of gills to sodium are discussed on page 98.

In general, in stenohaline species which cannot tolerate transfer to fresh water, the Na efflux is high: *Serranus* 58%, *Uranoscopus* 51%, as compared with euryhaline *Platichthys* 15.9% and *Anguilla* 19.2% of the total Na in the fish per 100 g per hour.[247]

There are also changes in renal function—more water reabsorption in S.W. and more salt reabsorption in F.W.—as shown for *Platichthys*[200] (see table above).

Some reabsorption of solutes and water in the bladder is indicated. Also, clearance differences suggest that glomerular recruitment (variation in number of functional glomeruli) may take part in regulation.

Not only are there changes in gill permeability to Na and in kidney function when fish go from sea water to brackish water, but there are also changes in distribution of water and ions between cells and extracellular fluid. When the flounder *Pleuronectes* was transferred from S.W. to 75% S.W., the volume of extracellular fluid initially rose; then as adjustments in muscle solutes occurred, organic nitrogenous compounds changed so that at equilibrium, ninhydrin positive material had decreased from 71 to 44 mM/kg and tri-

TABLE 1–6. Permeability Constants for Fish and Amphibian Membranes.

	Diffusional Water Flow μl/cm²/hr	Intake by Drinking μl/100 g/hr	Net Osmotic Flow μl/mm²/hr	P_{osm} cm/sec × 10^4	P_{diff} cm/sec × 10^4
Sea water adapted					
Anguilla	64	325	−294	0.19	0.18
Platichthys	61	192	−145	0.14	0.17
Serranus	33	277	−207	0.10	0.10
Fresh water adapted					
Anguilla	91	135	+403	0.79	0.25
Platichthys	97	37	+250	0.70	0.27
Carassius	121	51	+1394	2.08	0.34
Amphibians					
Xenopus skin				2.8	0.3
Rana esculenta skin				5.6	0.55
bladder				7.5	1.0
Rana temporaria skin				3.3	0.65
Bufo bufo skin				23.6	1.48
Bufo marinus bladder				6.9	0.95

methylamine oxide from 30 to 12 mM/kg, while Na had decreased only from 14.7 to 10 mEq and Cl from 41 to 30 mEq.[207]

Water Permeability. Another difference between freshwater and marine fish is gill permeability to water. Diffusional permeability, as measured with THO, is higher in fresh water than in sea water; also, osmotic flow, measured as the difference between urine output and intake by drinking, is higher in fresh water. Permeability constants (diffusional and osmotic) are calculated on the basis of gill area; a number of constants for fish are compared with those for some amphibian membranes in Table 1–6.[250]

Much species variation occurs in the effect of salinity on water permeability. The yellow *Anguilla* shows a much higher water flux in fresh water than in sea water, whereas in silver eels the fluxes are similar in both environments;[110] in *Pholis* the water permeability is similar in 100% S.W. and in 20% S.W.[109] The lower permeability to water in high salinity may reflect a Ca-Mg effect on gill membranes (p. 100).

In several invertebrates (*Carcinus, Rhithropanopeus,*[331] *Nereis*[332]) the permeability to water is either little affected in going from sea water to fresh water or decreases like the sodium permeability. Euryhaline fishes show greater differences; when in S.W. the water permeability is less than when in F.W. Regulation of diffusive loss may be more important in fishes, since the kidney is adapted for producing dilute urine and in sea water a steep outward gradient for water must be maintained.

Effects of Temperature. Some marine fish which are seasonally subjected to cold show changes in osmotic and ionic properties at low temperatures. Some which live near the surface where ice is forming have blood which freezes at −1.5° to −1.8°C in winter and at −0.7° to −0.8°C in summer; the lowered freezing point is due to organic solutes in the blood (see p. 367). Other fish, particularly deep-water species, show little increase in osmotic concentration but survive by supercooling. In general, marine fish have higher concentrations of blood ions at low than at high temperatures. Freshwater fish generally have lower blood ion concentrations in winter or when cold-acclimated than in summer. Rainbow trout have higher Cl and osmotic concentrations at 5°C than at 15°C. The net effect for both marine and freshwater teleosts is to reduce slightly the osmotic work necessary in low temperatures (see Chapter 10). The Antarctic *Trematomus* excretes more Cl and Na in its urine in cold temperatures, and more Mg when it is warmer.

Enzymes and Hormones. Sodium transport is accomplished in many animals by ouabain-sensitive Na-K ATPase. The activity of this enzyme in gills of marine teleosts is high; in gills of elasmobranchs and freshwater teleosts it is low. The activity in eels doubles when the animals go from fresh water to sea water. (See p. 98.)

Several hormones may function in ionic and osmotic regulation in fishes. These are discussed on pp. 69–70.

In summary, fishes, both cartilaginous and bony, have wide-ranging osmotic capacities. Elasmobranchs are essentially marine, but some live in fresh water; they are consistently hyperosmotic and maintain high blood concentration by urea retention. Some have very active salt-secreting rectal glands. Marine bony fish are hyposmotic, and rely especially on salt secretion by the gills. Freshwater fish produce copious urine and actively absorb salts through the gills. Euryhaline fish, which can migrate between fresh and salt water, change their permeability and transport mechanisms to maintain relative constancy of internal concentration.

AMPHIBIA

Ecological Correlates with Osmoregulation

Some amphibians are permanently aquatic; others spend at least part of their lives on land. They normally lay eggs in water or in a very moist environment, such as leaf mold, or between leaves which form a cup to collect moisture. The rate of water uptake by tadpoles just after hatching is several times greater than that by adults. Some desert amphibians develop very rapidly, and the adults burrow underground, to emerge when rains come. A few frogs are known to breed in brackish water, but none are truly marine.

Water content is similar in terrestrial and aquatic anurans (toads and frogs), but the tolerable water loss on desiccation decreases as shown in the following series:[300]

Species	*Habitat*	*% Weight Loss Tolerated*
Scaphiopus hammondi	fossorial	50, 60
Pseudacris sequita	arboreal	58
Rana pipiens	semiaquatic	46, 52
Rana clamitans	aquatic	39
Rana septentrionalis	aquatic	35

After dehydration and return to water, the frog *Rana pipiens* rehydrates rapidly, by as much as 12 per cent of body weight per hour. After excessive hydration, as by injection of water, the rate of loss of water on return to normal water load is slower than the rate of rehydration, and is a linear function of the excess water load. An aquatic salamander rehydrates more slowly than a terrestrial species; salamanders can withdraw water from the soil in proportion to their hydration deficit.[337] A desert toad, *Bufo cognatus,* can store up to 30% of the equivalent of its gross body weight as water in its urinary bladder. In another desert toad, *Bufo punctatus,* the ventral pelvic integument presses against a moist substratum and can take up 423 mg $H_2O/cm^2/hr$; the bladder can also store water, which is then used during a drought.[238, 239]

Evaporative loss through the skin from a frog or salamander is 1/30 that from a free water surface, but more than that from a millipore filter.[155] Efflux of water to air is the same as that to a sucrose solution when osmotic gradient is converted to an equivalent vapor pressure gradient.[224] When the brackish water toad *Bufo marinus* is dehydrated by 80%, muscle sodium concentration increases by 43% and plasma sodium by 33%; if NaCl is injected into a dehydrated *B. marinus,* cutaneous uptake of water increases and urine water loss decreases.[319]

The permeability of the skin to water is less in terrestrial than in aquatic species, and can be modified as follows:[61]

		Osmotic Water Flux $\mu l/cm^2/hr$
semiaquatic *Lepidobactracus*	in water ad lib	502
	in air 3 days	44
terrestrial *Bufo*	in water ad lib	441
	in air 3 days	372

The blood of a frog is hyperosmotic to pond water; the urine is hyposmotic to the blood and is more copious in a dilute than in a concentrated medium:

Δ_o	Δ_i	Δ_u
−0.07°C	−0.44° to −0.56°C	−0.17°C

A desert spadefoot toad, *Scaphiopus,* remained burrowed in sand for 10 months of drought. On emergence its blood concentration was 456 mOsm; Cl was 126 mEq and urea was 148 mM, compared with values at normal hydration of 326 mOsm, 118 mEq Cl, and 69 mM urea.[239]

A brackish water toad, *Bufo marinus,* when dehydrated uses water from the bladder, and the ratio of osmotic concentration of bladder urine to that of plasma may approach unity. When the toad has adapted to brackish medium, urine production is reduced; conversely, when it is adapted to fresh water, sodium excretion by the kidney is reduced. In this way, when the toad is dehydrated, both water and sodium are reabsorbed from the bladder. A reduced blood volume also stimulates water reabsorption from the bladder.[244] The aquatic toad *Xenopus laevis* is normally ammonotelic (see Chapter 7), but after living in saline for 2 to 3 weeks its blood urea increases 15-fold, urine flow is reduced, and it excretes urea instead of ammonia. The hormone arginine vasotocin, which is an antidiuretic in terrestrial amphibians and a diuretic in fish, has no effect on *Xenopus.*[237]

Salt and Water Movement Across Skin

In frogs, the capacity for selective salt absorption by the skin is well developed. When frogs that have been kept in distilled water for several days, and have lost some body salts, are returned to tapwater or dilute solutions, active uptake of salt occurs. Chloride is absorbed with Na from dilute NaCl, but without the cation from KCl, NH_4Cl, and $CaCl_2$; Br^- is absorbed but I^- is not.[196] Active ion pumps persist in skin isolated from a frog. Sodium can be absorbed from a medium as dilute as 10^{-5} M, and the transport against a steep gradient uses energy and establishes a potential. Frogs in pond water (1.4 mM Na) absorb 4.5 μEq/g/day, which balances Na loss.[142] The efflux is 10% by skin and 90% by kidneys. In a tadpole before stage 22, body fluids are negative with respect to the external bath; later the body fluid becomes positive, and at this

TABLE 1-7. Blood Concentrations in Anurans.

	Medium	*mOsm*	*Na mEq*	*Urea mM*	*Cl mEq*
B. viridus[137]	F.W.	279	113	35.8	99
S.E. Europe	40% S.W.	441	173	71.6	99
B. viridus[308a]	144 mOsm	484	145	89	118
	618 mOsm	544	230	101	179
R. cancrivora[138]	F.W.	290	122	40	113
Vietnam	60% S.W.	830 (80% S.W.)	181	300	131
B. boreas[137]	F.W.	156	64	23.3	61
America	40% S.W.	366	183	31.6	172
B. bufo[308a]	F.W.	213	145	12.5	
	410 mOsm	318	179	10.5	
R. temporaria[308a]	F.W.		104	5.9–11.7	75.4
	407 mOsm	510	172	4.4	134

time isolated skin shows positivity on the inside when both sides are bathed in Ringer solution. In early developmental stages, ion absorption is by gills; in adults it is by skin.[6]

A brackish water frog, *Rana cancrivora,* from South Vietnam is compared with a brackish water toad, *Bufo viridis,* of southeastern Europe and a euryhaline American toad, *B. boreas,* in Table 1-7.

The increase in plasma osmoconcentration due to urea is greatest in *R. cancrivora,* less in *B. viridus,* and least in *B. boreas,* which increases Cl_p to maintain hyperosmoticity. *R. cancrivora* maintains blood hypertonicity and urine hypotonicity to the blood by retaining urea, reducing renal filtration, and increasing water reabsorption in the kidney tubules (Fig. 1-37). In going from fresh water to 40% S.W., the muscle K increased only 23%, while the urea increased 770%; thus, urea compensates osmotically while ionic changes are minimal.[137, 138] In contrast to *R. cancrivora* and *B. viridis,* species such as *R. temporaria* and *B. bufo* have limited survival ability in brackish water; in these species the increase in blood osmoconcentration is

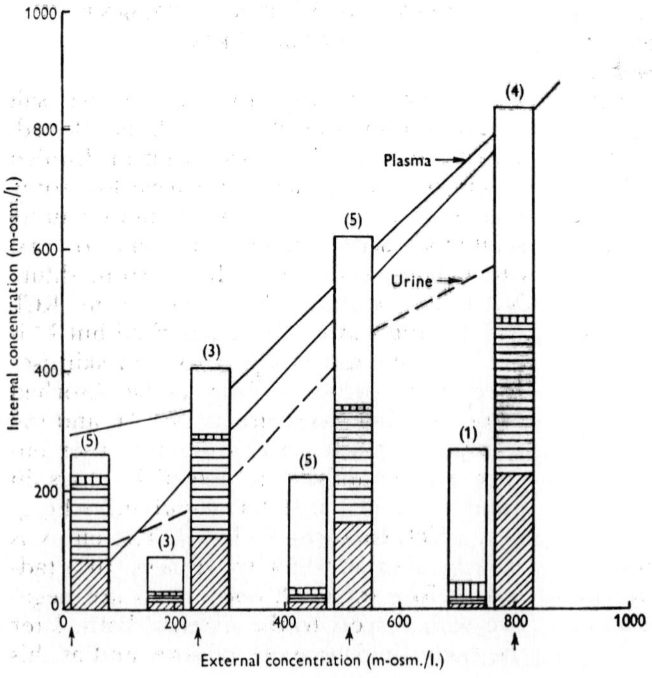

Figure 1-37. Plasma and urine concentrations: mOsm, Cl, Na, K, and urea in *Rana cancrivora* acclimated to salinities of indicated osmolalities. ▨ Cl^-, ▤ Na^+, ▥ K^+, □ urea. (From Gordon, M. S., et al., J. Exp. Biol. *38*:659–678, 1961.)

due entirely to ions rather than to urea. No amphibians have the capacity to maintain blood hyposmotic by extrarenal salt excretion as do marine teleosts.

Amphibian tadpoles are usually ammonotelic, and they shift to ureogenesis as they approach metamorphosis (Chapter 7). Tadpoles of *R. cancrivora* in brackish water have high salt concentrations; they shift to the ureogenesis of adults at developmental stage 25.[138] When adult *B. viridis* is transferred from fresh water to 30% S.W., the arginase activity of liver increases as urea is made and retained.[150]

Kidney Function; Bladder

The capacity to retain salts by action of the kidneys is well developed in amphibians; the kidneys of *Rana* and *Necturus* have been favorable objects of study because of the small number of relatively large nephrons. The glomeruli lie in a layer close to the mesial border of the kidney. Beyond each glomerulus is a ciliated neck, then the thick-walled proximal tubule, then a narrow intermediate tubule, and finally the distal tubule, which empties into a collecting tubule. Blood supplies to glomeruli and tubules are separate.

The amount of glomerular filtration is ascertained from the "clearance" of some substance which is neither reabsorbed, stored, nor secreted by the tubules. Clearance of a substance is given by: $C = \frac{UV}{P}$, where C = ml of protein-free plasma filtered per unit of time, U = urine concentration of the substance being measured, V = volume of urine formed in the given time, and P = concentration of the substance in plasma water. For substances which are filtered in the glomeruli but are neither reabsorbed nor secreted by the renal tubules, such as inulin or creatinine, the clearance equals the filtration rate; for substances which are reabsorbed to some extent (Na or Cl or glucose) the clearance is less than the filtration rate, whereas substances secreted by the tubules (*p*-aminohippuric acid, PAH) have clearance ratios greater than unity. In frogs, filtration rates of about 30 ml/kg/hr have been observed under normal conditions.

Fluid can be collected from various regions of a nephron and analyzed for solutes. In frog and *Necturus* the glomerular fluid has essentially the same chloride and sugar concentrations as the blood plasma; it is free of blood proteins. The filtration pressure at the glomerulus is the difference between the blood pressure (29 mm Hg) and the colloid (protein) osmotic pressure of the blood (7.7 mm Hg). Since the chloride content decreases along the distal tubule, chloride reabsorption must occur in that segment. Glucose diminishes in concentration early in the proximal tubule; after injection of the drug phlorizin, the glucose is not reabsorbed and actually increases in concentration because water is reabsorbed. Water and salt may be absorbed in both proximal and distal portions of the tubule. Acidification of the urine occurs in the distal tubules. The distal tubule removes salt but has low water permeability; hence, the urine is hyposmotic. Antidiuretic hormone (known as arginine vasotocin or ADH) increases water permeability of the distal tubule and thus increases urine concentration and decreases its volume. In the bullfrog, *Rana catesbiana,* water reabsorption is insufficient to account for the concentration of urea; clearance of urea can be 7 to 10 times higher than the filtration rate. Thus, some urea must be secreted by the tubules. Urea can be secreted by the tubules when the renal arteries are ligated and only the renal portal circulation remains. Tubules can also secrete injected dyes such as phenol red.

In *Necturus,* some 30% of filtered NaCl is reabsorbed in the proximal tubules. Studies on isolated tubules show that Na can be absorbed against electrical and concentration gradients.[123] Short-circuiting of a tubule requires current equivalent to the transported Na, and the absorption can be blocked by ouabain or DNP. Water is also absorbed, as indicated by increases in the concentration of inulin in the tubules. There is normally insignificant net influx of K—an influx of 52 pM/cm²/sec and an efflux of 49 pM/cm²/sec. When account is taken of water movement, the net absorption of potassium and of chloride appears to be passive, and that of sodium is active. By varying the osmotic concentration in the tubule, a water flux of 0.15×10^{-8} ml/cm²/sec/cm H_2O was measured (see p. 9). Passive water movement is in proportion to Na transport. However, in unoperated freshwater amphibia and fish, inulin u/b ratios are 1.1 to 1.3; this suggests that there is little water reabsorption and that the *in vitro* evidence may reflect an antidiuretic response.[189]

There is active transport not only by the skin and kidney tubules, but also by the amphibian bladder. In desert anurans the bladder capacity may equal 50% of the body weight; in tropical *B. marinus* it is 25%, and in aquatic *Xenopus* only 1%. When *B. marinus* is adapted to a brackish medium, urine production is reduced; cloacal urine may have a sodium concentration 10 to 12 times that in ureteral urine (i.e., in dehydration, water is absorbed from the bladder); and in a hydrated state sodium is reabsorbed. Reduced blood volume also stimulates reabsorption of water from the bladder.[244]

REPTILES

Reptiles are a heterogeneous group with representatives in sea water, fresh water, dry air, and moist air. The horny scales have been thought to limit evaporative loss. However, recent measurements indicate nearly as great evaporation rates from reptiles as from some amphibians, and when reptiles are submerged the cutaneous uptake of water may be considerable. Tracer measurements with tritiated water also indicate considerable skin permeability.[23]

Evaporative loss from reptiles correlates well with their ecology; cutaneous loss in mg $H_2O/cm^2/hr$ is as follows: subtropical *Sphenomorphus,* 0.25; *Anolis,* 0.19; *Gehyra,* 0.21; semi-arid *Uta,* 0.10; *Sauromalus,* 0.05.[66] Similarly, snakes from the desert lose very little water, while species from tropical rain forests evaporate much. When in air, the cutaneous route accounts for more of the water loss in aquatic reptiles than in desert lizards; 87% of total loss is cutaneous in *Caiman,* 78% in *Pseudemys,* 72% in *Iguana,* 66% in desert *Sauromalus,*[22,23] and 76% in *Anolis* and *Uta.*[66] Most of the remainder is respiratory water loss. The evaporative loss in air is proportional to the saturation deficit. The soft-shelled turtle *Trionyx* is four times more permeable to water than is the horny *Pseudemys.*[24] Pulmonary loss is half as much (expressed as a percentage of the total loss) in two desert lizards as it is in a skink.[89] Evaporative water loss is more temperature-dependent in lizards that have high temperature preferenda.[348] Evaporation through pieces of skin to dry air was 0.5 ml/cm²/hr for *Lacerta* as compared with 8.3 ml/cm²/hr in *Rana.* The flux of water through the skin is increased by ADH in the reptile *Uromastrix,* as it is in the amphibians *Bufo* and *Rana.* Terrestrial reptiles show behavioral preferences for temperature and humidity which tend to minimize water loss.

The osmotic concentration of reptile blood tends to be higher than that in amphibians but similar to birds and mammals. Values of the freezing point are: marine turtle *Caretta,* −0.66°; land turtle *Testudo,* −0.6°; grass snake, −0.68°; desert lizard *Trachysaurus,* −0.65°.

Freshwater reptiles may actively absorb ions. A soft-shelled turtle, *Trionyx,* actively absorbs sodium from a dilute (5 μM) medium, probably by its pharynx;[100,101] it shows no active uptake of K or Cl.

The amount of urine formed is small, and in some snakes and lizards it may be solid or semisolid. In the alligator, the urine flow is 0.4 to 1.2 ml/kg/hr, as compared with 1.5 to 20 ml/kg/hr in *Rana*; in the alligator the glomerular filtration rate is 1.5 to 3.4 ml/kg/hr as compared with 2.8 to 40 ml/kg/hr in a frog. A lizard, *Trachysaurus,* when hydrated formed 5.7 ml/kg/hr of urine; when dehydrated, it formed only 0.24 ml/kg/hr. An aquatic turtle, *Pseudemys,* is ureotelic, and dehydration reduces the glomerular filtration, increases water reabsorption, and reduces urine flow; in contrast, a land tortoise, *Gopherus,* is uricotelic and shows no anuria on dehydration until blood osmoconcentration is raised at least 100 mOsm.[86] In *Gopherus* the tubular urine is hyposmotic, and bladder urine may be isosmotic; hence, there is water reabsorption in the bladder. A brackish water terrapin, *Malaclemys,* on transfer from 3.3% salinity to fresh water showed an increase in urine volume from 0.44 to 2.4 ml/g/day and a decrease in urine osmotic concentration from 207 to 57 mOsm. Reabsorption of ions may also occur in the cloaca. In a crocodile in fresh water, hyposmotic urine is produced in the kidney at the rate of 3.5 ml/kg/hr; in the cloaca sodium is reabsorbed, so that during a three-hour period the concentration can be reduced by half.[22]

In a crocodile under normal water load, the glomerular filtration rate is 9 ml/kg/hr, and about half of the fluid and more than half of the sodium are reabsorbed in the proximal tubule. Comparative data for glomerular filtration rates and urine/plasma ratios (osmotic) follow for reptiles and a frog under normal states of hydration:

		u/p_{osm}	GFR ml/kg/hr
Rana	frog	0.2	34.2
Pseudemys	turtle	0.62	4.7
Gopherus	desert tortoise	0.34	4.7
Hemidactilus	gecko	0.64	10.4
Phrynosoma	horned toad	0.93	3.5
Tropidurus	lizard	0.91	3.4
Crocodylus	crocodile	0.82	9.6

In a crocodile there is reabsorption of water and sodium in the proximal tubules, but in the cloaca most of the Na^+ and Cl^- and some of the water (but not NH_4^+) are reabsorbed:

	mOsm	Na	Cl	NH_4
ureteral urine, mEq	234	61	45	53
cloacal urine, mEq	251	5.9	16.5	118

Cloacal reabsorption of water may be by the high colloid osmotic pressure in blood.[305]

In the freshwater snake *Natrix* in water diuresis some 10% of filtered Na and 60% of K may be excreted. The inulin u/p ratio increases while the sodium u/p ratio decreases; hence, the distal tubules reabsorb water and sodium.[85] Under varying water loads the excretion is controlled by varying the number of functional tubules and by a 10-fold change in the filtration rate.[208] Reabsorption from the distal tubules is influenced by ADH. The glomerular filtration rate is much less in the bull snake (*Pituophis*) than in the water snake (*Natrix*); injection of hypertonic saline increases filtration in *Pituphis* but has no effect or decreases filtration in *Natrix*.[194]

Horned toads and Galapagos lizards reabsorb some 55% of glomerular filtrate; geckos reabsorb 75%. In the toad and lizard, tubular cells are not folded and are poor in mitochondria, while in the gecko the cells have deep infoldings and many mitochondria.[293] Water is reabsorbed from the cloaca of *Iguana* and of the lizard *Varanus;* arginine vasotocin increases the absorption by increasing both permeability to water and transport of Na.[53]

A diamondback terrapin, *Malaclemys centrata,* lives in salt marshes and estuaries. Blood Na and Cl increase on going from fresh to brackish water; urea increases more after going from brackish to sea water:[127]

		Na mEq	Cl mEq	K mEq	urea mM	mOsm
F.W.	blood	129	88	3.1	21.5	309
	urine	4.4		16.8	22	107
50% S.W.	blood	156	113	4.1	30.1	355.5
S.W.	blood	163.4	136.6	3.8	115	458.8
	urine	7.9		59.7	107.4	372

An extreme example of resistance is the Australian desert lizard, *Amphibolorus,* which lives in dry hot regions and eats sodium-rich ants. Sodium is retained in extracellular fluid, which shifts water out of cells; the extracellular fluid volume decreases from 33% of body volume to 25.4% during the desert summer, and is restored when rains come and sodium is excreted. Thus, sodium retention protects against fluid loss, although osmotic concentrations must vary considerably.[49, 50]

Extrarenal excretion of salt occurs by nasal glands, which are found in loggerhead turtles, crocodiles, marine snakes, and lizards. In the iguana *Ctenosaura* the nasal gland secretion may contain Na at 190 times the concentration in plasma; secretion may be produced at the rate of 190 μl/hr. Injection of NaCl increases potassium excretion by these glands. In *Sauromalus* the ratio K_{sec}/K_{pl} is 77.2 and for sodium the ratio is 0.67; in *Ctenosaura* the secretion/plasma ratio for K is 69 and for Na is 0.43.[347] After sodium chloride is injected, sodium is secreted, more in the herbivorous *Dipsosaurus* than in *Sceloporus.* In sea snakes (*Laticauda*), stimulation of salt gland secretion can be caused by injected NaCl or sucrose; hence, osmoreceptors are involved. The maximum excretion of Na is 72 μM/100 g/hr; for K it is 3 μM/100 g/hr.[102] A marine turtle excretes more sodium and potassium by the salt glands than by the kidney; it could not maintain water balance by renal excretion alone. The secretion from the salt gland of a large turtle contained 685 mEq Na and 21 mEq K per liter.[163] A comparison of several reptiles on the basis of salt gland secretion concentrations follows:[101, 102]

	Na mM	Cl mM	K mM	Na/K ratio
marine iguana *Amblyrhynchus*	1434	1256	235	6.7
land iguana *Conolophus*	692	486	214	3.4
sea snake *Palamis*	607	627	28	21.7
sea turtle *Lepidochelys*	713	486	29	24.5

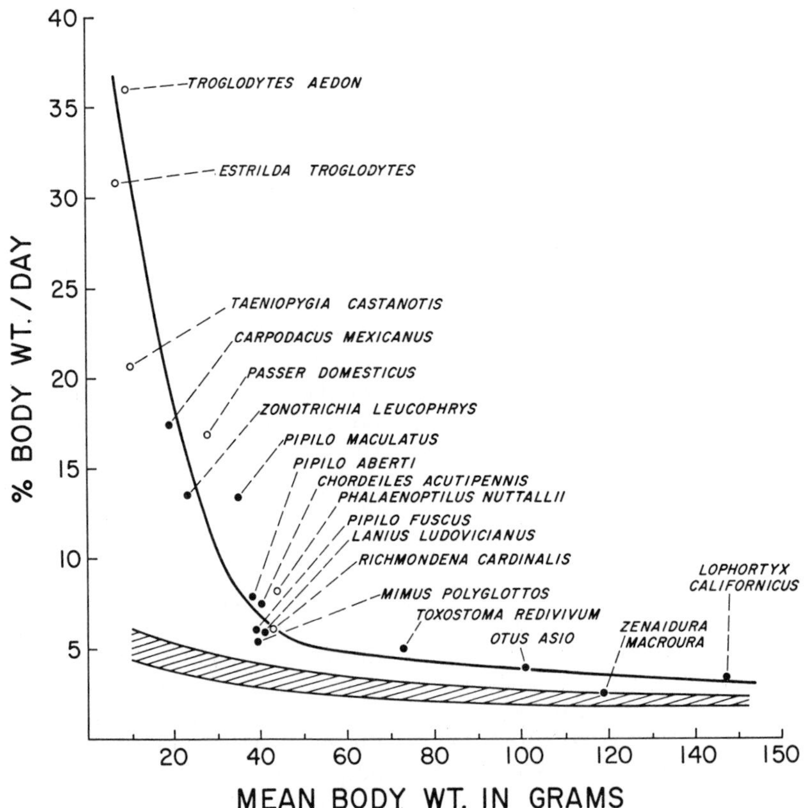

Figure 1–38. Daily water loss from birds as a function of body weight at ambient temperature of 25° C. (From Bartholomew, G. A., and T. C. Cade, Auk *80*:504–539, 1963.)

In general, land reptiles have high potassium; marine reptiles have high sodium secretion as in birds. This situation is probably related to the high K content of plants. The marine iguana feeds on algae.

BIRDS

Most birds have access to fresh water. However, some live on dry seeds in the desert and must rely on metabolic water; others, which live in salt marshes or at sea, have severe problems of water supply. Metabolism of birds is high, and respiratory loss of water is relatively large. Cloacal urine may be a viscous paste of uric acid crystals, and water reabsorption may occur in kidney, cloaca, and lower intestine.

The percentage of body weight lost as water each day on a seed diet is higher in small birds than in large ones, and increases more than does the production of metabolic water[14] (Fig. 1–38). A series of birds that normally have access to fresh water, when offered salt solution for drinking, increase the volume of fluid ingested and may take salt water equal to body weight daily; maximum urine/serum ratio for Cl^- was 1.9 in mourning dove and 2.4 in house finch[278] (Fig. 1–39). California quail drank a constant amount irrespective of salinity. A Savannah sparrow subspecies *beldingii* from salt marshes and a budgerigar decreased fluid intake as salinity increased; these birds can produce a concentrated urine. Of two Savannah sparrows, *beldingii* can maintain its weight on 0.6 to 0.7 M NaCl, whereas *brooksi* loses weight on 0.3 M NaCl.[279]

A red crossbill consumes water ad lib equivalent to 22% of its body weight per day. When offered only saline solution, its intake increases with concentration up to 0.25 M NaCl; then intake is reduced. It shows no preference between 0.1 M NaCl and H_2O, but does distinguish 0.2 M NaCl.[88]

Some desert birds require very little water and must rely on metabolic water. An Aus-

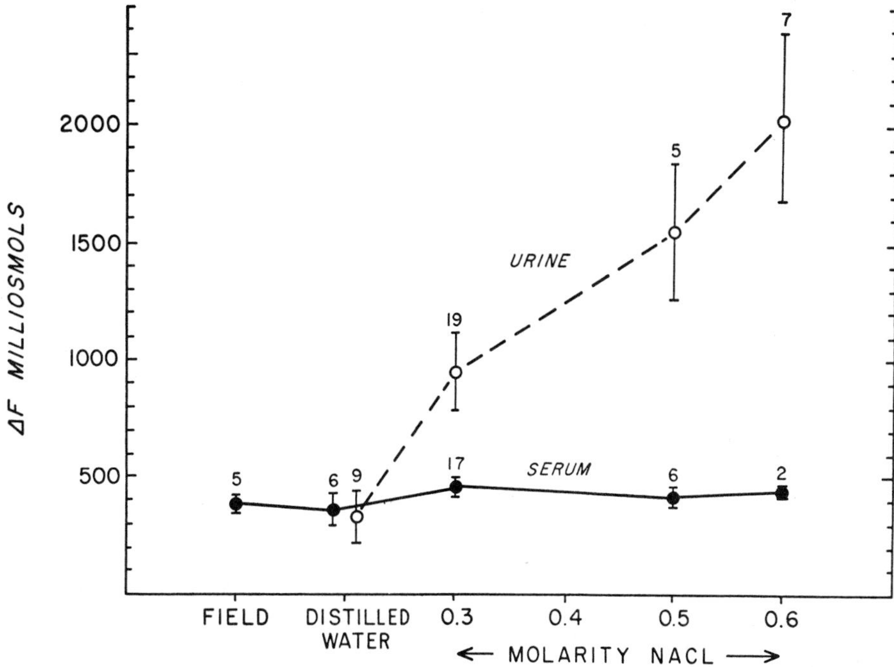

Figure 1–39. Osmoconcentrations of urine and serum in Savannah sparrow allowed to drink various solutions ad libitum as indicated. (From Poulson, T. L., and G. A. Bartholomew, Physiol. Zool. *35*:109–119, 1962.)

tralian zebra finch lives well on dry seeds, but when offered water it consumes 24 ml/100 g/day; in contrast, an African waxbill dies on dry seeds alone, and ad libitum it consumes 42 ml H_2O/100 g/day.[60] Total water balance for a 16 gram lark from a southwestern African desert showed daily intake as follows: seed consumption, 1.9 g, of which 0.17 g was water; metabolic water, 0.93 to 1.1 g. Daily output was 1.38 g by evaporative loss (mainly respiratory) and 0.37 to 1.75 g in excreta. Therefore, only 0.65 g of water need be replaced per day from green plants and insects.[379] A zebra finch, fed dry seeds, will drink 0.2 M NaCl but not 0.3 M; its urine/plasma ratio for Cl^- may reach 2. Cutaneous evaporation loss decreases in high humidity, but the finch has an obligatory respiratory loss of 0.54 mg H_2O/ml O_2 consumed.[209] Excreta of a sparrow normally contained 81% H_2O; if the bird was kept on an H_2O-free diet, the excreta content was 57%.[333]

The kidney of a chicken can vary the osmoconcentration of urine according to the state of hydration of the bird and, unlike all lower vertebrates, can produce a urine which may be hyperosmotic to the blood. On hydration, the glomerular filtration rate and urine flow increase over those in dehydration. Osmotic u/p ratios were 0.37 in water-loaded roosters, 1.1 in salt-loaded ones, and 1.6 to 2.0 in dehydrated ones. A gradient of increasing osmolarity exists in the kidneys of dehydrated chickens and turkeys from 447 mOsm in the cortex to 463 mOsm in the medulla and 522 mOsm in the urine.[326] Infusion into the cloaca with regurgitation into the large intestine showed some absorption of sodium and water; Na absorption is independent of concentration above 80 mEq, and water absorption follows a gradient. In a dehydrated state, 50% of the sodium and 15% of the water in ureteral urine may be absorbed in the cloaca and large intestine:[324]

	Na *mEq/kg/min*	*K* *mEq/kg/min*	*Cl* *mEq/kg/min*	*H_2O* *μl/kg/min*
ureter urine	3.58	1.21	1.66	17.94
voided urine	1.26	1.43	0.92	7.39

However, in normal hydration, cloacal absorption of water is small.

By a combination of renal and cloacal (plus intestinal) reabsorption, the voided urine can be hyperosmotic in a chicken. In going from the dehydrated to the hydrated state, the glomerular filtration rate increased by 23%, urine flow increased from 20 μl/kg/min to 300 μl/kg/min, and the u/p ratio dropped from 1.58 to 0.3.[323] In chickens, injection of arginine vasotocin decreased urine flow by 23%, increased u/p_{osm} from 0.33 to 1.11, and decreased NaCl excretion by 20% as compared with 50% in frog and 25% in lizard. Maximum u/p_{osm} in dehydration for a chicken was 1.58.[7]

Marine birds—sea gulls, cormorants, and pelicans—and some desert birds and ducks have nasal glands which provide a route for extrarenal excretion of salt. The balance for a cormorant and a sea gull is as follows:[240]

	Cl *mEq*	*Na* *mEq*	*K* *mEq*	H_2O *ml*
Cormorant fed 60 g fish and 3 g NaCl				
intake	54	54	4	50+
cloacal elimination	27.5	25.6	2.7	108.9
nasal gland loss	26.1	23.8	0.3	51.4
Sea gull given 0.5 g NaCl				
plasma,				
gland inactive	124	169		
gland active	154	190	4.5	
nasal fluid	690	783	37.8	
cloacal fluid	191	184	18.2	

Gulls excrete much Na^+ and K^+ extrarenally even in the absence of osmotic stress.[170]

In a duck, the nasal gland Na secretion may be seven times more concentrated than urine, and nasal K secretion may be 3.0 times as concentrated as urine. Secretion is stimulated by a salt load or by hyperosmotic sucrose, both of which cause an increase in plasma volume by withdrawal of tissue water. The secretory response (in duck *Anas*) is abolished by adrenalectomy, and is stimulated by injection of a glucocorticoid extracted from the adrenal cortex.[162] Secretion is also triggered by cholinergic agents such as methacholine.[351] The lumen of the gland is electrically positive to the blood; ouabain inhibits both the Na transport and the positive potential.[351] The sizes of the salt glands decrease in the series: marine > brackish > freshwater. Data on secretion follow:[385]

	Δ_{fp} *plasma*	Δ_{fp} *secretion*	*secretion conc. in mEq* Na	K
herring gull (*Larus*)	−0.77	−3.46	775	34
Arctic tern	−0.68	−2.96	765	33
guillemot	−0.73	−3.37		
oyster catcher	−0.70	−3.72	836	90

Some desert-living birds, such as the ostrich and desert partridge, secrete nasal gland fluid that is rich in potassium rather than sodium.

The cells of nasal glands show deep infolding of membranes from base to apex, thus providing channels through the cells. Between the infoldings are many slender mitochondria.[95, 195]

Secretion by the salt glands is normally initiated by cholinergic nerves, and injected acetylcholine is an effective agent. In a homogenate of salt gland, ACh causes the formation of a phosphatidic acid fraction. ACh-stimulated slices of the gland accumulate sodium. Extracts of salt glands have an abundance of Na-K ATPase, which probably functions in the sodium transport.[160] The postulated first stage is Na accumulation; the second is transport into the lumen, probably with the aid of Na-K ATPase.

MAMMALS

ROUTES OF WATER LOSS

Regulation of water balance in mammals permits them to live in moist or dry air, in fresh or sea water, and over a wide range of environmental temperatures. Mammals subjected to osmotic stress and dehydration maintain water balance by varying their intake of water and by controlling the avenues of water loss. Mammals have more capacity than birds for producing a blood-hyperosmotic urine, and they have no major extrarenal routes of salt excretion other than sweat.

In man, for example, the daily water loss for a 70 kg individual is 600 to 2000 ml by urine, 50 to 200 ml by feces, 350 to 700 ml by insensible evaporation from the skin, 50 to 400 ml by sweat, and 350 to 400 ml from the lungs. A lactating mother may lose an additional 900 ml as milk. Thus, the normal daily loss may range from one to more than nine liters (up to 12 liters/day in the tropics), depending on temperature, exercise, state of hydration, and other factors. This loss must be made up by water drunk, water in food, and metabolic water.

Most mammals have a plasma concentration of about 0.30 osmolar (equivalent to 0.95% NaCl or $\Delta_i = -0.8°C$). In man, the urine concentration is usually about 0.65 Osm, and the maximum urine concentration in thirsting man is 1.4 Osm ($\Delta_u = -2.6°C$).[382] The permeability of the skin of mammals is extremely low, which is important in reducing evaporative loss in air and excessive hydration in water. Some mammals use water evaporation, either by sweat glands or by panting, as a means of cooling; this water loss may become critical. Evaporative (insensible) loss from human skin decreases non-linearly with vapor pressure of water in the air above the skin.[133] Small mammals, such as rodents, have little or no evaporative cooling. For man, loss of 10% of body water causes serious illness; for mice, camels, and sheep, loss of 30% of body water results in death. Man in water may take up a small amount of water through the skin, mostly by imbibition by the stratum corneum. Lactation may double the water turnover of the camel and provide a serious drain on the water reserve.

A beaver, *Aplodontia*, normally consumes water to the extent of 32.7% of its body weight per day.[254, 303] A desert heteromyid, rodent, *Liomys salvani*, consumes water ad libitum up to 5.7% of its body weight per day.[168] To maintain balance in a hot environment, the waterbuck must consume 12% of its body weight in water per day. Rectal temperature in well hydrated ungulates in tropical Africa was below air temperature, but when gazelle and oryx were dehydrated the rectal temperature conformed to air temperature. Dehydration lowers the rates of sweating and panting and raises the critical temperature at which panting starts.[345] In Somali donkeys, food consumption ceased after 20 to 22% dehydration; the donkey could survive 30% weight loss as water, and could drink in 2 to 5 minutes some 24 to 30 liters to restore a water deficit. Under comparable summer conditions a camel loses water by evaporation equivalent to 1% of body weight daily; in the donkey this figure is 4.5%, and in man it is 7%. Pulmocutaneous losses in neutral conditions are as follows:

	mg H_2O/ml O_2 consumed
Liomys (rodent)	0.9–1.0
albino rat	0.94
man	0.84
Dipodomys (rodent)	0.54
Perognathus (rodent)	0.50

A reciprocal relation holds between sweat and urine concentrations of Na, K, and H_2O; i.e., when sweat is voluminous it is more dilute, and under this condition urine volume decreases and is more concentrated.

A porpoise, *Tursiops*, loses only 30% as much by evaporation as do comparable terrestrial mammals. If the exhaled air had the same water content as the alveolar air sacs, a 140 kg porpoise would lose 290 g/day, whereas the actual evaporative loss is 60 g/day. The ventilation rate is low and O_2 withdrawal is high. In a porpoise, the temperature in the upper respiratory tract is 9°C lower than that in the lungs; the condensation due to this drop permits reclamation of 80% of the water prior to exhalation.[77]

In each of several terrestrial environments, series of mammals are found that have decreasing water turnover as measured under comparable conditions. In each series, the highest turnover is in animals with high metabolism.[22] Several such series are listed in Table 1–8.

The half-time of tritiated water in a camel in the desert is 12.2 days in winter and 6.4 days in summer. In each climate, evidently,

TABLE 1–8. Water Turnover Series in Terrestrial Environments.

Climate	*Animal*	*Water Turnover ml/kg/day*
tropical, moist	*Bos bubalus* (buffalo)	200
	Bos taurus (shorthorn)	161
	Bos indicus	123
desert, dry	boran cattle	135
	sheep	107
	goat	96
	camel	61
Arctic (Alaska)	reindeer	128
	moose	111
	sheep	62
	musk ox	35
desert marsupials	*Sminthopsis crassicaudata*	461
	Dasycercus cristicauda	134
	Dasycercus byrnei	125

some species are better able than others to meet the demands of a given niche.

A harbor seal, *Phoca,* has a total water flux of 0.5 to 1.8 liters/day; metabolic water exceeds that from food at low food intake rates, but fish food provides more water at high intake levels.[91] A possum, *Trichosurus,* licks its fur for evaporative cooling; if licking is forcibly prevented, panting increases and total evaporative cooling remains approximately constant.[87] A fish-eating bat, *Pizonyx,* evaporates 12 to 14 mg H_2O/g/hr or 1.2 to 3.7 ml daily; it loses 1.3 ml H_2O by feces and some 6.35 ml by urine. A diet of 13 g of shrimp provides 11.3 ml of water, and thus balance is maintained.

Water loss by feces varies greatly in mammals. A camel without access to drinking water excretes feces containing 76 g water per 100 g dry weight; with access to water, the feces contain 109 g water per 100 g dry weight. Comparable values of water content per 100 g dry feces are: white rat, 225 g; grazing cow, more than 566 g. A 400 to 500 kg camel with or without access to drinking water may put out 1.5 liters of urine per day; grazing animals excrete 0.5 to 8 liters per day. When deprived of drinking water and fed dry dates and hay, a camel lost in 8 days some 17 per cent of its initial weight or 30% of its body water; this represented a loss of 38% of the interstitial water and 24% of the intracellular water.[306] The camel uses rumen water for cooling.[221] A donkey may lose 1300 g H_2O by feces and 1 to 1.2 liters by urine—some 2.5% of body weight per day.

Water turnover, as measured by THO, in ruminants at 41°C (desert) in ml/kg/day was: camel, 61; merino sheep, 110; shorthorn cattle, 148. A camel cools by sweat at the rate of 300 ml/m^2/hr, but produces only 2 to 3 ml urine/kg/day on any feed. Urine concentration in a camel was 3.17 mOsm, compared with 3.8 mOsm in a dehydrated sheep.[221]

After dehydration, recovery by drinking follows different patterns in various mammals: a camel may make up a large deficit in 10 minutes and may drink as much as 25% of its body weight in one draught; a rat may drink enough to overshoot its normal state of hydration; man is usually slow to replace water by drinking.[2] The time required to excrete comparable water loads also varies greatly: to excrete 40% of an ingested water load of 5 ml/100 g, *Dipodomys* (desert kangaroo rat) required 367 minutes, compared to 84 minutes for a white rat.[63]

Some desert rodents and marsupials (as well as sheep and camels in winter) do not require drinking water, but may get enough from metabolic water. Rodents that remain in burrows by day may reduce respiratory loss by 25%. In the kangaroo rat (*Dipodomys*) breathing dry air, the evaporative loss from the lungs was 0.54 mg H_2O/ml O_2 consumed, as compared with 0.84 for man and 0.94 for a laboratory rat. When fed dry grain at humidities above 10% R.H., the kangaroo rats showed no negative water balance; they used water of oxidation, which amounts to the following:[306]

	grams H_2O/gram food	*grams H_2O/kcal*
fat	1.07	0.113
carbohydrate	0.556	0.133
protein	0.396	0.092

The relative efficiencies of the kangaroo rat and the laboratory white rat are as follows:

	loss in g H_2O/100 Cal food metabolized at 22 mg H_2O/liter air				
	feces	*urine*	*evaporation*	*total loss*	*oxidative intake*
white rat	4	5	14	23	13
kangaroo rat	0.63	3.4	3.4	8	13

Many mammals drink salt solutions up to a limit. A cat or a camel can maintain its water balance on sea water. A kangaroo rat survives well when it drinks sea water. In man the limitation seems not to be in the kidneys but in the gastrointestinal and vascular effects of the sea salts.

Kidney Function

Mammalian kidneys are highly adaptive in their capacities to produce blood-hyperosmotic urine. Reptiles cannot make a hyperosmotic urine; birds may concentrate urine to twice the plasma concentration, and some mammals by as much as 20 times. In a canyon mouse, *Peromyscus crinitus,* the maximum urine osmolality is 3430 mOsm; maximum u/p_{osm} is 10.4, and maximum u/p_{urea} is 127.[1]

The limit of urine dilution in man is 50 mOsm; the limit of concentration is 1400 mOsm. A desert rat can produce urine of more than 5000 mOsm. Data on mechanisms of control of urine concentration have been

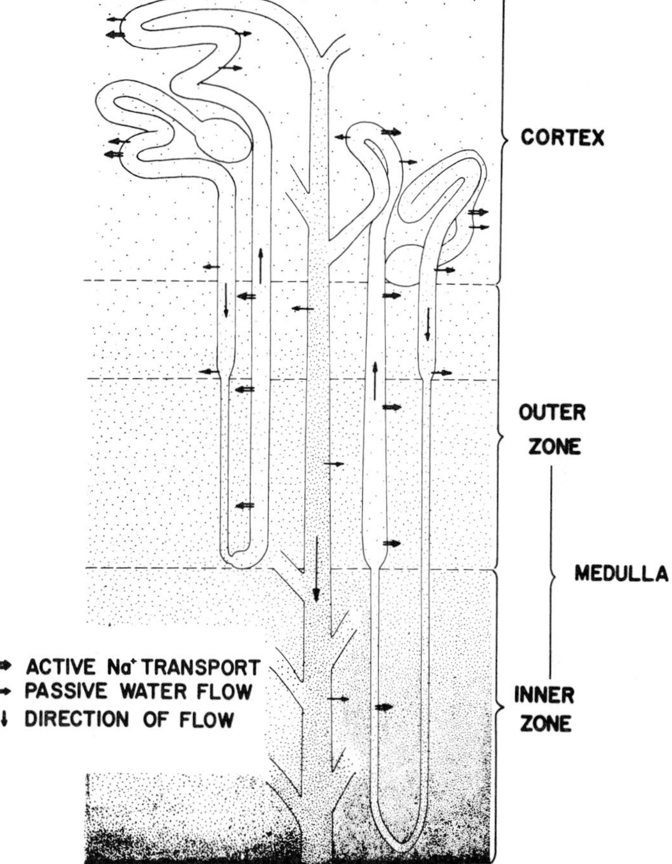

Figure 1-40. Diagram of two nephrons of mammalian kidney, one with long loop and the other with short loop of Henle. Osmoconcentration in regions of kidney indicated by density of dots. (From B. Schmidt-Nielsen.)

obtained by micropuncture analysis of tubular fluid, by stopping flow by back-pressure in the ureter for a time, and by analysis of slices of kidney. Those animals which can produce the most concentrated urine have the longest loops of Henle (long papillae) (Fig. 1-40).

Glomerular filtration in a dog is about 4.3 ml/min/kg body weight; in man it is 1.73. Thus, the equivalent of the total extracellular fluid of the body is reworked by the kidney about 16 times each day. In man, 1100 liters of plasma circulated per day in the glomeruli yield 180 liters of filtrate; of this, 178.5 liters are reabsorbed by the kidney tubules, leaving 1.5 liters per day of urine. The filtrate is protein-free but contains all smaller plasma solutes determined by ionic balancing of plasma proteins. Filtration is largely determined by the difference between hydrostatic (blood) pressure and the colloid osmotic pressure of the plasma proteins. The size of protein molecule that is retained is set by a complicated system of slit pores. The fluid passes through the pores in the endothelial cells, and then through a permeable basement membrane; it emerges between processes of the covering cells (podocytes) on the lumen side of the capsule.

Of the filtered solutes, glucose, amino acids, vitamins, and electrolytes are reabsorbed in the proximal tubule; paraminohippuric acid (PAH) is actively secreted in this region. The fluid in the proximal tubule is isosmotic with plasma. In the loop of Henle or thin segment, the tonicity increases steadily toward the renal papilla, and decreases again in the ascending limb toward the cortex. In the early distal convolutions the fluid is hyposmotic to plasma. The loops of Henle act as countercurrent multiplier systems in which diffusion, combined with active transport of sodium, results in a higher concentration in the papilla than in the cortex. Flow is in opposite directions in the ascending and descending arms of the loop; sodium is moved actively out of the ascending arm and passively into the descending arm (that is, it can be cycled within the loop). Absorption of sodium creates an osmotic gradient between the fluid in the ascending limb and the collecting duct. The osmotic concentration in the descending limb rises because of diffusion of water out to the hyperosmotic interstitium and because of diffusion of Na into the descending limb. The osmotic gradient between the two limbs of the loop of Henle and the resulting exchange (countercurrent multiplier) create an increasing osmotic concentration from the outer to the inner zone of the medulla. Fluid in the early distal tubule is hyposmotic and water may diffuse out, so that the fluid is isosmotic by the time it enters the collecting duct. As the collecting duct passes through the increasingly hyperosmotic environment in the medulla, water diffuses out, making the final urine nearly isosmotic with the papilla and hyperosmotic to the blood. In water diuresis the permeability of the distal tubule and collecting duct is low and the urine remains hyposmotic; in antidiuresis, when the antidiuretic hormone vasopressin is present in the blood, the permeability of the distal tubule and collecting duct increases and water leaves, resulting in hyperosmotic urine.

Regulation of urine concentration in *Macaca* is by tubular reabsorption, and maximum u/p_{osm} may be 22.[81] In dehydration in chinchilla, the urine osmolality increases from 2350 mOsm to 7599 mOsm; when the animal is hydrated it drops to 130 mOsm. Under normal conditions 32% of the water entering the loop of Henle and 38% of the Na are reabsorbed in the distal tubule.[375]

Micropuncture data from rats show a ratio of Na in proximal tubule fluid to Na in plasma to be 1.0 in antidiuresis and less than 1 in diuresis; the ratio of Na in distal tubule fluid to Na in plasma is 0.6 in antidiuresis and 0.2 in diuresis. Similarly, the urea concentration in the proximal tubule may be 15 mM, in the early distal tubule 65 mM, and in the late distal tubule 190 mM, while in the medulla of the kidney, urea concentration is 135 mM[358] (Fig. 1-41). Thus, both sodium and water leave the proximal tubule and the collecting duct and re-enter in the loop of Henle. Na is actively transported out of the proximal and distal tubules, and Cl follows. Both NaCl and urea enhance the reabsorption of water by making the interstitium hyperosmotic. Transport from the distal tubule requires energy from glycolysis, while the transport from the proximal tubule depends more on the electron transport system. Vasopressin makes the collecting tube more permeable to both water and urea.[120]

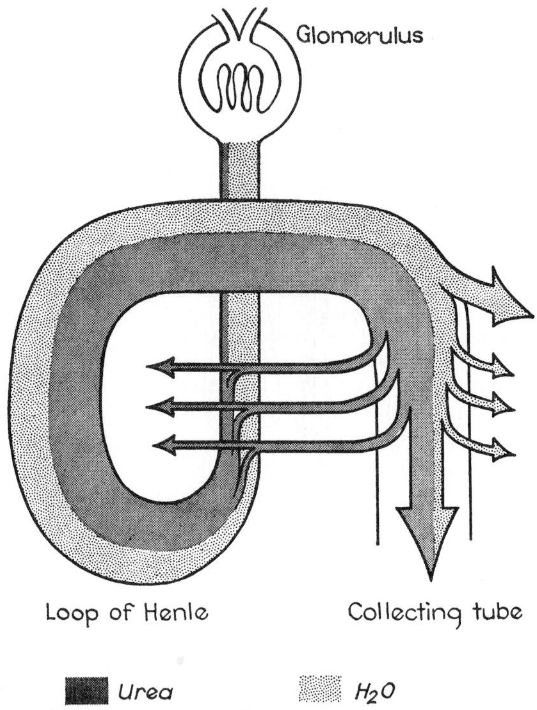

Figure 1-41. Cycle of movements of urea and H_2O during antidiuresis (water retention). (From Gardner, K. D., and R. H. Maffly, J. Clin. Invest. *43*:1968–1975, 1964.)

Maximum renal concentrating capacities of mammals are summarized in Table 1-9.

The kangaroo rat can concentrate to a u/p ratio of 14, as compared with 4 in man and 9 in the white rat. The urea u/p ratio may be 400 in the kangaroo rat, as compared with 170 in man and 200 in the white rat. In a desert rodent, *Psammomys*, 9% of the filtered water reaches the tip of the loop and 0.9% is in the final urine; in the loop, 64% of the solute is Na and 19% is urea, whereas in the collecting duct the effective solute is mostly urea.[139] A vampire bat consumed blood equal to 37% of its body weight in 7 hours; initially it voided urine of 475 mOsm and later 4656 mOsm, and 73% of the osmolality was due to urea.

The effectiveness of the kidney in concentrating urine solutes is related to the relative percentages of long and short loops present. Desert animals have predominantly long loops, while mammals with little concentrating ability have only short loops:[302, 303]

TABLE 1-9. Maximum Concentrating Capacities of Mammals. Data from References 128, 147, 167, 223, 254, 255, 287.

	urine mOsm	*u/p$_{osm}$*
desert mouse *Notomys*	5540–6400	
desert mouse *Psammomys*	5700–6000	17
desert rat *Dipodomys*	5500–6000	14
heteromyid *Liomys*	3580–4000	
ground squirrel *Citellus*	3900	9.5
camel	2500–3170	8
laboratory rat	2560–3000	8.9
laboratory cat	2100–3250	
laboratory hamster	3000	
laboratory dog	2006–2660	
seal	2150–2420	
chinchilla	2006	
fish-eating bat *Pizonyx*	2006	
porpoise	1833	
rabbit	1502–1910	
man	1400–1480	4.2
Macaca	1191	
laboratory pig	1100	
pollock whale	900–1340	
marsupial *Trichosurus*	1066	4.8
mountain beaver *Aplodontia*	495–770	2.4
sloth *Choloepus*	460–907	

animal	*relative medulla thickness*	*percentage long loops*	*urine maximum* Δ_{fp}
beaver	1.3	0	0.96
pig	1.6	3	2.0
man	3.0	14	2.6
cat	4.8	100	5.8
jerboa	9.3	33	12.0
desert mouse	10.7	100	9.2

An exception is the chinchilla, which has a long papilla but is less effective than a laboratory rat in concentrating urine.[147]

In animals like the rat, the osmotic ceiling of urine concentration increases with high protein diet; the kidney has a deep medulla with pelvic fornices, and the maximum u/p ratio is greater than 2. In other species, such as the beaver (*Aplodontia*), the maximum osmotic concentration of urine does not increase with high protein diet; the pelvic fornices are rudimentary, and the maximum u/p ratio is low. In muskrat, the u/p ratio is 3.3 on high protein diet and 2.5 on low protein diet:[384]

	urine osmolality, mOsm ON HIGH PROTEIN DIET	ON LOW PROTEIN DIET
white rat	2834	2078
beaver	640	672
muskrat	1063	934

It is only in mammals that glomerular filtration increases in response to added salt load. In all other vertebrates, especially in amphibians, glomerular filtration is reduced when salt is injected. The control in mammals is by means of hormonal effects on blood flow and on permeability of renal tubules.

The blood level of vasopressin in mammals is controlled mainly by reflex, by osmoreceptors located anteromedially in the supraoptic hypothalamus. These receptors respond to small (1 to 2 per cent) changes in plasma osmoconcentration. Intracarotid infusion of a small amount of hyperosmotic NaCl reflexly increases blood vasopressin, and this reduces urine flow. Injection of a small amount of hyperosmotic saline, such as 0.005 to 0.01 ml of 2 to 3 per cent NaCl, into the paraventricular nucleus of a goat elicits polydipsia (excessive drinking) followed by polyuria. After intracarotid injection of hyperosmotic solutions, increased electrical activity is recorded from the supraoptic nucleus.

Other receptors in the walls of arterioles, particularly in the liver and kidneys, are sensitive to plasma volume and hence to plasma concentration. They reflexly activate release of aldosterone from the adrenal cortex, which causes active absorption of sodium in the kidney and sweat glands and thus favors water retention and increased plasma volume. Osmoreceptors in the liver send afferent impulses by the vagus to the brain.

Water taste receptors on the tongue have been identified in the dog, pig, and cat; laboratory rats and man lack a water taste. Thirst is a complex sensation which involves stimulation of brain osmoreceptors, general level of tissue hydration, and mouth dryness.

Marine mammals have blood only a little more concentrated than that of land mammals. The urine of marine mammals can be more concentrated than sea water. In a mammal such as a seal (*Phoca vitulina*) no water is lost for temperature regulation; lungs account for about 10% of water loss, feces 20%, and urine 70%. It has been calculated that food (1250 g of herring) would yield urea and salt sufficient to give 800 ml of urine a freezing point of −2.7°C, which is within the normal range for feeding seals. Marine mammals living on fish, which are hyposmotic to sea water, appear to get ample water from their food to keep the blood more dilute than sea water. They do not drink sea water. However, walrus and whalebone whales feed on invertebrates that are more concentrated than the blood of the mammals themselves. Less water is lost from the lungs by whales than by terrestrial mammals; evidently, the kidney can excrete the salt and urea with water from the food and can produce a hyperosmotic urine which is sufficient to keep the blood hyposmotic.

In summary, a critical problem faced by terrestrial vertebrates—most reptiles, birds and mammals—is to reduce evaporative loss of water. Adaptations for low surface evaporation are mostly structural—horny scales, feathers, and hair. Respiratory water loss is reduced by lessening locomotor activity at high temperatures, and by condensation of water in cool nasal passages. The use of respiratory evaporation, sweat secretion, and salivation for cooling at high ambient temperatures increases water turnover and requires central nervous balancing of the stress of dehydration against overheating. Osmoreceptors in the brain, water taste, and fluid volume receptors provide signals for behavioral and hormonal responses. Water is conserved by low urine output, and in some birds and reptiles by use of nasal glands for extrarenal excretion of salts. Mammals have developed the highest capacity for producing urine which has high osmoticity relative to blood. The hormonal control of renal blood flow, of tubular permeability, and of the cycling of sodium within a nephron permit a wide range of urine concentration according to hydration. Renal adaptations serve equally well for desert mammals that survive long periods without drinking and for marine mammals that consume food of high osmoconcentration.

HORMONAL CONTROL OF OSMOTIC AND IONIC BALANCE

(See also Chapter 21.)

Several classes of hormones function in water and salt balance of vertebrates. The best known from a comparative viewpoint are the neurohypophyseal polypeptides.[33, 34] Those with a basic amino acid such as arginine or lysine in the 8-position are called vasopressins and have marked antidiuretic (ADH) or water-conserving activity. Those with neutral 8-amino acids are classed as oxytocins and have weak ADH activity. The function of the neutral principle in non-mammals is unknown. These polypeptides are produced by the supraoptic (paraventricular) nucleus of the hypothalamus; they pass along the infundibular stalk to the neurohypophysis, where they are stored and released on demand.

All mammals except some Suiformes have arginine vasopressin (AVP); some Suiformes have both AVP and lysine vasopressin (LVP), and others (wild pig and hippopotamus) have only LVP.[13a] (Fig. 1–42). The finback whale has five times more oxytocin relative to vasopressin than other mammals.[1] In birds, amphibia, reptiles, and most fishes the active polypeptides are oxytocin and arginine vasotocin, the latter being the principal ADH (Fig. 1–43). Cyclostomes (lamprey) have only one of the polypeptides, arginine vasotocin, which is the most widespread and probably the primitive agent. The holocephalan ratfish *Hydrolagus* has arginine vasotocin plus oxytocin. The actinopterygian *Polypterus* and the dipnoan *Protopterus* both have arginine vasotocin plus a neutral principle; in *Polypterus* the neutral principle is isotocin (4-serine, 8-isoleucine oxytocin) as in holosteans and teleosts, whereas in *Protopterus* it is mesotocin (8-isoleucine oxytocin) as in amphibians and birds.[2] Elasmobranchs have two oxytocins which differ from those in other vertebrates; one of these (from *Raja*) appears to be 4-serine, 8-glutamine oxytocin (glumitocin), and the other may possibly be 3-serine,

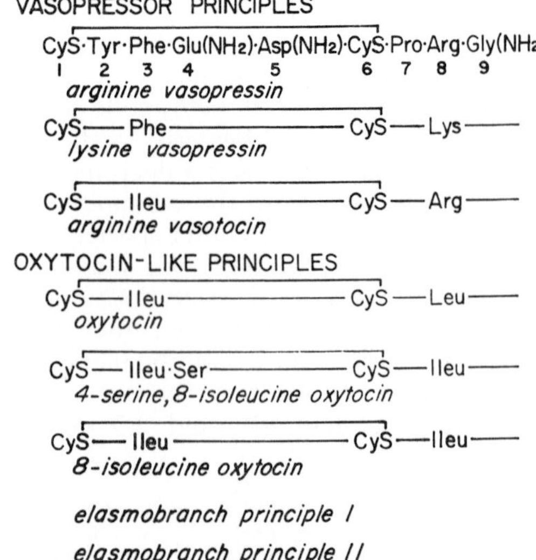

Figure 1–42. Formulae of neurohypophyseal principles of importance in water balance in various classes of vertebrates. (From Sawyer, W. H., Arch. d'Anat. Micr. Morph. Exper. *54*:295–312, 1965.)

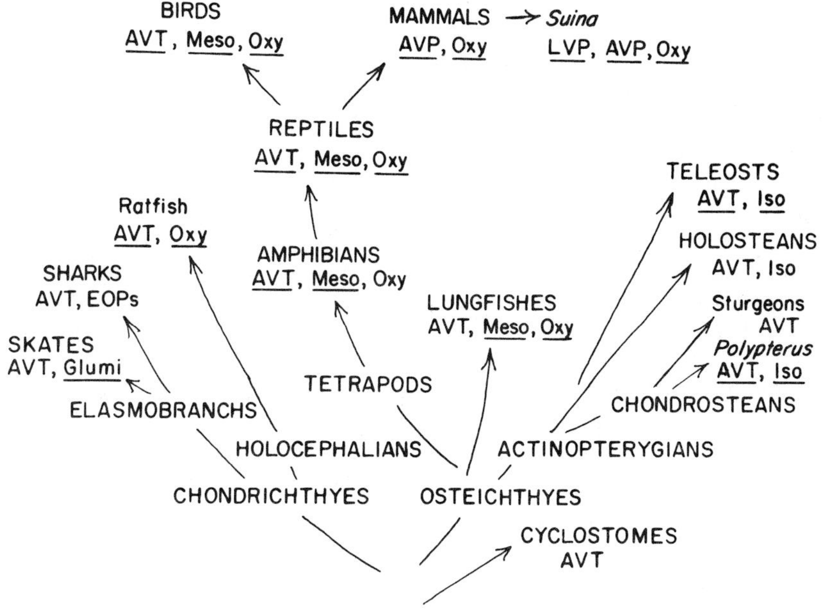

Figure 1–43. Scheme showing the distribution of neurohypophyseal peptides among the vertebrates in which they have been studied. (From Sawyer, W. H., personal communication.) When the abbreviation for a peptide is underlined, this indicates that its identification is based on chemical analysis in addition to pharmacological and chromatographic characteristics. The abbreviations are:

AVP = arginine vasopressin
LVP = lysine vasopressin
AVT = arginine vasotocin
Oxy = oxytocin
Meso = mesotocin (8-isoleucine oxytocin)
Iso = isotocin (4-serine, 8-isoleucine oxytocin)
Glumi = glumitocin (4-serine, 8-glutamine oxytocin)
EOPs = at least two unknown elasmobranch oxytocin-like principles. These are not glumitocin or any of the other known principles found among the vertebrates.

8-isoleucine oxytocin.[37, 38] Hypothetical schemes have been proposed to show how the various basic and neutral peptides could have arisen by a series of point mutations.[39]

The mode of action of vasopressins is different in the various vertebrate classes. In mammals and birds the principal action is to increase the permeability of the renal tubules, particularly the distal tubules and collecting ducts, to water and thus to permit more water reabsorption and to reduce the production of urine. In reptiles and amphibians, AVT is also antidiuretic, but in addition to a small tubular effect the hormone reduces the glomerular filtration rate, probably by constriction of the glomerular arterioles.[31, 39] An additional action in anurans is to increase permeability of the skin to sodium and, in variable degrees, to water; this (in combination with a sodium pump) favors salt uptake.

The action of AVT on frog skin is to increase the permeability of the outer surface to Na; the effect occurs only if AVT is applied to the inner side.[10] Oxytocin and norepinephrine also enhance uptake of Na; their effect is mimicked by 3,5 cyclic AMP, and the action of norepinephrine is via α-receptors.[4]

Another function of AVT is to increase water reabsorption from the bladder. The water permeability of toad bladder is sensitive to AVT in concentrations as dilute as 10^{-11} M.[17] In the urodele *Triturus*, regulation is by the kidney and the bladder.[5]

In freshwater fishes (eel, goldfish), AVT increases urine flow and glomerular filtration rate.[18] Thus, the action on kidneys of freshwater fish is opposite to that in tetrapods, namely to elicit diuresis. Sodium uptake by the gills is increased. Hypophysectomy in the eel decreases urine production and sodium uptake; in goldfish, plasma sodium is lowered.[8] In *Opsanus*, an aglomerular marine fish, AVT increases blood pressure but does not cause diuresis as in goldfish.[21] In the lungfish *Protopterus*, AVT increases glomerular filtration rate as in freshwater fish, but in addition it enhances sodium excretion; i.e., it is natriuretic.[39] Possibly the hormone is important as a diuretic and natriuretic when the lungfish emerges from estivation. In lamprey, AVT also promotes sodium excretion.[35] The function of the unique oxytocins in elasmobranchs is not known.

In summary, arginine vasotocin is the primitive ADH, vasopressin is restricted to mammals, and elasmobranchs have two unique oxytocins. The site of action differs in various vertebrates according to their water requirements, water being conserved in terrestrial animals and excretion of water being promoted in hyperosmotic aquatic forms.

In bony fish, prolactin from the anterior pituitary enhances production of a dilute urine. A hypophysectomized euryhaline fish, *Fundulus heteroclitus,* fails to survive well when transferred from sea water to fresh water; survival is furthered by injection of prolactin, but not by arginine vasotocin.[29] Hypophysectomized freshwater *Fundulus* (also stickleback *Gasterosteus* and *Xiphophorus*) loses Na rapidly, mostly by the gills, but injection of prolactin reduces the rate of Na loss to one-fourth and the fish survive.[23,30] Thus, prolactin is critical for euryhaline fish in migration between fresh and salt water.

Sticklebacks (*Gasterosteus*) move to the sea in autumn and up freshwater streams in the spring. In winter in a salinity gradient they select sea water; in the late spring they select fresh water. In winter they show a more marked drop in osmoconcentration when transferred from sea water to fresh water than they do in late spring. Prolactin administered to winter specimens diminishes the drop in plasma concentration and favors production of dilute urine. Thus, prolactin enhances freshwater osmoregulation by both gills and kidney.[23]

In bony fish a secretory structure at the base of the spinal cord, the urohypophysis, may have some function in water balance.[45] Action potential spikes have been recorded from the neurosecretory cells when the fish had been injected with salt solution. An extract of urohypophysis of fishes increases water movement across the bladder of toad and stimulates contraction of the bladder of trout.[16]

In *Fundulus,* the activity of the thyroid gland (as judged by the height of epithelial cells, by density of staining of colloids, and by loss of incorporated ^{131}I) is less in fresh water than in sea water and less at low than at high temperatures.[25]

Several principles from the adrenal cortex also function in maintaining salt balance. Aldosterone increases sodium reabsorption by the bladder of a toad, *Bufo marinus.*[42] In mammals, aldosterone favors sodium retention and potassium excretion; Na reabsorption in the distal tubules is enhanced. In a sodium-depleted dog, low plasma sodium increases secretion of aldosterone and corticosterone.[6] Corticosterone enhances potassium excretion and decreases sodium excretion, in salivary glands as well as in kidney. Aldosterone appears to act on toad bladder by inducing the synthesis of some mediator or transport protein, since its effect is blocked by inhibitors of protein synthesis. The action of vasopressin on Na transport by bladder is not so blocked.[11,13] In birds, glucocorticoids from the adrenal cortex stimulate salt output by the salt glands.[14,15] In lip-shark, deoxycorticosterone and cortisol decrease secretion by the rectal gland, but hypertonic saline injected with cortisol increases rectal gland output of Na. In the cyclostome *Lampetra,* aldosterone and ACTH decrease sodium loss.[5] The extent to which adrenocortical hormones are normal regulators of sodium and potassium in fishes is uncertain.

In mammals, elevated blood sodium causes the release from juxtaglomerular cells of the kidney a hormone, renin, which enzymatically converts angiotensin precursor in the blood to angiotensin I; this goes to angiotensin II, which then acts on the adrenal cortex to cause the release of aldosterone. Aldosterone stimulates reabsorption of sodium in the kidney, which in turn favors water retention. Renin is found in fish, more in freshwater than in marine species, and it seems to be absent from elasmobranchs and cyclostomes.[27,28] When an eel is transferred from fresh water to sea water, the renin level in the kidney decreases over a three-week period.[43] Assays of fish kidneys for renin (in rat units) follow:

aglomerular toadfish in S.W.	3.7
aglomerular toadfish in dilute S.W.	47
freshwater goldfish	49
anadromous shad in 2.7% NaCl	0.6
anadromous shad in F.W.	5.6
alewife in 2.7% NaCl	0
alewife in F.W.	2.7

Another hormone active in ion regulation is thyrocalcitonin, which is released from the thyroid in the presence of high serum calcium. It lowers plasma calcium by inhibiting bone resorption. In fish, cells of the ultimobranchial bodies release a polypeptide of 32 amino acids which lowers blood Ca level of eel and catfish, but not of *Fundulus*.[9]

Neuroendocrine regulation of salt and water balance has also been observed in crustaceans and insects. A grapsoid crab which regulates hyperosmotically and hyposmotically has increased Na-K ATPase activity in gills when placed in dilute sea water; if the eyestalks are ligated, the ATPase decreases and blood and urine osmoconcentrations fall in the dilute medium. An extract from the ventral nerve cord restores the ATPase level.[20] In crayfish, ligation of the eyestalks increases water influx and urine production when in fresh water, and increases blood osmoconcentration when in 50% S.W.; a brain extract counteracts these effects.[19]

In the blood-sucking bug *Rhodnius*, a hemolymph-borne factor is needed for producing the diuresis after a blood meal, and this factor can stimulate formation of fluid in isolated Malpighian tubules (see p. 43). The source of the hormone is the neurosecretory cells of the mesothoracic ganglion[24] and of the cerebral ganglion.[3] In cockroaches the source is the last abdominal ganglion.[26]

Endocrine secretions modulate osmotic and ionic balance in both vertebrates and invertebrates by action on permeable membranes (gills and skin) and on excretory organs, and on tissues in relation to extracellular ions. The complete control cycle, particularly the sensing mechanisms, are known only for mammalian neurohypophyseal principles.

SUMMARY

Regulation of osmotic concentration or water activity (in a thermodynamic sense) is intimately bound to regulation of ionic concentration and of cell and body volume. Organisms have exploited a wide range of salinities—from hypersaline lakes to rain water—and of humidities—from rain forests to dry deserts—and relatively similar adaptive mechanisms are used, with variations, for each environment. Life originated in the sea, and a most universal property of organisms is to separate the interior from the environment by a selective boundary layer. All living cells are bounded by a membrane which is restrictive in its permeability to water, ions, and non-electrolytes. Unrestricted permeation is an early sign of damage leading to cell death.

Most cells, whether free-living or in multicellular organisms, are in osmotic equilibrium with their medium. A few free-living unicellular organisms may osmoregulate for a period in the life cycle by initial impermeability to water (freshwater eggs) or by developing complex organelles for water excretion (freshwater protozoans). However, all cells differ ionically from their medium, and it is most important for intracellular ionic composition to be regulated. Frequently, when a cell is subject to osmotic change, this is brought about mainly by alteration in organic solutes, particularly amino acids, while inorganic ions are held relatively constant. In addition to selective permeability, Donnan effects, and other passive properties, most cells have ion pumps by which some metabolic energy is used to hold specific ions at concentrations different from those reached in passive equilibrium.

When cellular isotonicity is disturbed or when membranes become leaky, water moves passively into or out of cells, and their volume may change. It is important for "optimum" metabolism that cell volume, like ionic composition, be maintained relatively constant. Volume regulation is dependent on water movement together with resistance to deformation, the control of which in animal cells is not understood.

The amount of solvent water (i.e., water activity) in a cell is less than the total water. Yet water bound electrostatically to a protein molecule can apparently still serve as a solvent for small ions. Thus, estimates of "free" and "bound" water vary with the method used. Also, water permeability as estimated by diffusion of tracer is consistently less than that obtained by osmotic flux. Water, like most cell constituents, is compartmentalized; also, unstirred layers of molecular thickness may prevent uniform distribution. In addition, it is increasingly evident that most water in protoplasm is hydrogen-bonded into "liquid ice." It must be concluded that any present attempts to define active water in and around a cell are rough approximations.

We may start with the approximations for cell water and note that the basic mechanisms for control of water, ions, and volume evolved very early in the course of chemical evolution. The modifications in multicellular organisms are more quantitative than qualitative, although some new regulatory mechanisms appeared with complex biological system organization.

The basic and probably most primitive osmotic state is conformity, which is equal or steady state concentration of body fluids with respect to the environment. Marine invertebrate eggs and some marine invertebrate animals are osmoconformers, nearly isosmotic with the sea; they need not regulate volume, and hence they shrink or swell within tolerable ranges. A large number of osmoconformers have means of excreting or gaining ions, so that over some time in an altered salinity they can regulate their volume. It is interesting, however, that some of the relatives of osmoconformers, when they live in fresh water, have good mechanisms for restricting swelling or for pumping out water (e.g., *Hydra,* probably freshwater sponges, freshwater jellyfish, and annelids); yet some others—echinoderms, cephalopods—have no representatives that can regulate osmotically and are limited to full strength sea water.

Very many exits from sea water to fresh water have been made, and many animals have some limited osmoregulatory capacity. A few, such as tide-flat bivalves, merely "clam up" and exclude fresh water for brief periods. A few may store water in special vacuolated structures. Most estuarine invertebrates are less permeable to water and ions in dilute than in full sea water. The ability to change permeability, especially of gills, with external concentration is widespread, but its mechanism has scarcely been investigated. In addition, many estuarine animals bring excretory organs from the sea, where they were used mainly in ionic regulation, and modify them for eliminating osmotic water. The most successful invaders of dilute media have special cells (usually on gills) for actively absorbing ions from the environment. The appearance of active absorbing mechanisms represents a modification of the widespread ion pumps which extrude some substances, particularly Na, from many cells irrespective of osmotic condition. Similarly, the excretory organs conserve solutes by pumping ions back into blood. A few estuarine animals do well in near-fresh water as adults, and are bound to the sea only for brief larval periods.

Some estuarine invertebrates or their relatives, especially specific orders of crustaceans, annelids, and molluscs, have made the transition to fresh water for their full life cycles. Usually they have very "tight" or impermeable gills and body surfaces. They have well developed kidneys which produce copious dilute urine, and they actively absorb ions to supplement those in food.

Some marine invertebrates, mainly crabs, leave the sea for very long periods, even for full life cycles, and live in moist sand. These, together with some intertidal species, are able to regulate hyposmotically; i.e., they can maintain their blood more dilute than the medium. This ability is also found in a few marine crustaceans, and it is most highly developed in inhabitants of saline lakes (e.g., *Artemia*). Hyposmotic regulation carries to the extreme the extrusion of ions which diffuse in. This is done partly by kidneys and gut, but mostly by the actively transporting cells of the gills. Here the direction of transport is the reverse of that in freshwater species. Differences between the pumps in the two directions are not understood. The kidneys function mainly for ionic regulation in these species.

The most successful transition to land was made early in the evolution of arthropods by insects. Terrestrial insects have a number of adaptations not found elsewhere. Their waxy cuticle, and a respiratory system that can be opened and closed, minimize evaporative loss. Many insects can absorb water through the cuticle from moist air, a capacity not found in other animals. They also have external hygroreceptors which trigger precise behavioral orientation. They have a very complicated excretory system which permits production of hyperosmotic excreta under conditions of extreme water deprivation. The excretion system of malpighian tubules and hindgut cycles solutes (K plus uric acid and some unidentified organic substances) into narrow intercellular channels, where osmotic gradients can draw water from the rectum to tissue spaces and hemolymph.

Chordates are marine in origin, but vertebrate lines evolved early and prospered in fresh water. Some vertebrates—lampreys and various bony fish—can migrate between sea water and fresh water. Others—the hagfishes—are marine osmoconformers; the elasmobranchs are unique in maintaining hypertonicity to sea water by retaining urea. Many of them excrete salt via a rectal gland. Both the euryhaline bony fish and those restricted to either fresh or salt water are hyperosmotic in fresh water and hyposmotic in sea water. The hyposmoticity depends largely on active salt excretion via the gills, just as in hyposmotic brine shrimps. The gills of marine fish are rich in Na-K ATPase and have special secretory (chloride) cells. In fresh water the kidneys excrete a copious dilute urine, and the gills actively absorb salt against a gradient (probably by different cells from those which excrete it in sea water). Amphibians absorb salt through gills as tadpoles; they shift to skin and bladder absorption as adults. Euryhaline fish can decrease their permeability to Na within a few minutes on transfer to fresh water; they are often less permeable to water when in sea water than when in fresh water, the reverse of the changes in some invertebrates.

Other examples of extrarenal salt excretion are the salt glands of marine and desert birds and reptiles. The principal cation of the nasal gland excreta is either Na or K. Both birds and mammals use evaporative cooling in varying degrees, and this aggravates water loss in the desert. Mammals, like insects, have the capacity to produce markedly blood-hyperosmotic excreta, but the mechanisms are different. There is much variation among mammals in the ability to concentrate urine; this ability is determined partly by the length of the loops of Henle and their countercurrent multiplier system.

Endocrine control of skin or gill permeability, of glomerular filtration, and of tubular reabsorption is widely present among vertebrates, and frequently several hormones are involved. Hormonal release, particularly from the hypophysis, is triggered by neural reflexes, which are stimulated by volume- and osmo-receptors.

The evolution of osmoregulation permitted utilization of a wide variety of habitats. On close examination, one finds many exceptions to the general trends for specific animal groups, as well as marked differences within groups in capacity to tolerate wide environmental osmotic ranges. Certain properties of cell membranes—selective and adjustable permeabilities, active transport of one or more ions across the face of a cell—evolved early and have been modified according to different osmotic situations. The integration of water, ionic, and volume regulation has persisted at cellular levels, and has been expanded to include numerous organ systems in complex animals. Water is an essential commodity of life, and many ways have evolved to provide it to cells in proper quantities.

REFERENCES

Water: Osmotic Balance

For references to hormonal control, see page 78.

1. Abbott, K. D., Comp. Biochem. Physiol. *38A*: 37–52, 1971. Water economy in *Peromyscus.*
2. Adolph, E. F., Physiological Regulations. Jacques Cattell Press, 1943. 502 pp.
3. Ahearn, G. A., J. Exp. Biol. *53*:573–595, 1970. Water loss in desert beetles.
4. Alexander, M. D., E. S. Haslewood, G. A. D. Haslewood, D. C. Watts, and R. L. Watts, Comp. Biochem. Physiol. *26*:971–978, 1968. Osmotic control and urea synthesis in elasmobranchs.
5. Allen, K., Biol. Bull. *121*:419–424, 1961. Effect of salinity on amino acids in Pelecypoda.
6. Alvarado, R. H., and L. B. Kirschner, Comp. Biochem. Physiol. *10*:55–67, 1963. Ionic regulation in *Ambystoma.*
7. Ames, E., K. Steven, and E. Skadhauge, Amer. J. Physiol. *221*:1223–1228, 1971. Arginine vasotocin effects on renal function in chicken.
8. Anderson, E., and W. R. Harvey, J. Cell Biol. *31*:107–134, 1966. Active transport in midgut of *Samia cecropia.*
9. Augenfeld, J. M., Life Sci. *8*:973–978, 1969. Na-K activated ATPase in *Artemia,* different salinities.
10. Avens, A. C., Comp. Biochem. Physiol. *16*:143–153, 1965. Osmotic balance in brackish water gastropod.
11. Ax, P., and R. Ax, Biol. Zentralbl. *79*:7–31, 1960. Salinity tolerance, brackish and freshwater ciliates.
12. Ballard, B. S., and W. Abbott, Comp. Biochem. Physiol. *29*:671–687, 1969. Osmotic regulation in *Callinectes.*
13. Barnes, H., D. M. Finlayson, and J. Piatigorsky, J. Animal Ecol. *32*:233–252, 1963. Behavior of barnacles out of water.
14. Bartholomew, G. A., and T. C. Cade, Auk *80*:504–539, 1963. Water economy of land birds.
15. Bayley, I. A. E., and P. Ellis, Comp. Biochem. Physiol. *31*:523–528, 1969. Hyposmotic regulation, aquatic isopod.
16. Beadle, L. C., J. Exp. Biol. *50*:473–479, 1969. Osmotic and ionic regulation, embryos of aquatic pulmonates.
17. Beadle, L. C., J. Exp. Biol. *50*:491–499, 1969. Regulation of water during development, pulmonate embryo.
18. Beament, J. W. L., Biol Rev. *36*:281–320, 1961; J. Exp. Biol. *38*:277–290, 1961. Water relations of insect cuticle.
19. Beament, J. W. L., Nature *191*:217–221, 1961. Electrical properties of oriented lipid membranes.
20. Beament, J. W. L., Adv. Insect Physiol. *2*:67–129, 1964. Water movement in insect cuticle.
21. Beament, J. W. L., Soc. Exp. Biol. Symp. *19*:273–298, 1965. Active uptake of water by insects; models and mechanisms.
21a. Beament, J. W. L., and J. E. Treherne, editors, Insects and Physiology. Elsevier, New York, 1968. 378 pp.
22. Bentley, P. J., and K. Schmidt-Nielsen, J. Cell. Comp. Physiol. *66*:303–310, 1965. Permeability to water and sodium, crocodilian.
23. Bentley, P. J., and K. Schmidt-Nielsen, Science *151*:1547–1549, 1966. Cutaneous water loss, reptiles.
24. Bentley, P. J., and K. Schmidt-Nielsen, Comp. Biochem. Physiol. *32*:363–365, 1970. Water exchange, aquatic turtles.
25. Bergmiller, E., and J. Bielawski, Comp. Biochem. Physiol. *37*:85–91, 1970. Water permeability of gills, crayfish.
26. Berridge, M. J., J. Exp. Biol. *43*:511–521, 523–533, 1965. Excretion in cotton stainer *Dysdercus.*
27. Berridge, M. J., and B. L. Gupta, J. Cell. Sci. *2*:89–112, 1967. Ultrastructure and ion and water transport in renal papillae of *Calliphora.*
28. Berridge, M. J., and B. L. Gupta, J. Cell. Sci. *3*:17–32, 1968. Localization of ATPase in rectum of *Calliphora.*
29. Berridge, M. J., J. Exp. Biol. *48*:159–174, 1968; *50*:15–28, 1969. Urine formation by malpighian tubules, *Calliphora.*
30. Berridge, M. J., pp. 329–347 *in* Insect Physiology, edited by J. E. Beament and J. E. T. Trehern, American Elsevier Publishing Co., 1968. Ion and water transport across insect epithelia.
31. Berridge, M. J., and J. L. Oschman, Tissue & Cell *1*:247–272, 1969. Structural basis for fluid secretion, malpighian tubules.

32. Biber, T. U. L., and P. F. Curran, J. Gen. Physiol. *56*:83–99, 1970. Uptake of Na by frog skin.
33. Binns, R., J. Exp. Biol. *51*:1–10, 11–16, 1969. Urine formation by antennal gland of *Carcinus*.
34. Binns, R., J. Exp. Biol. *51*:17–27, 29–39, 1969. Glucose and amino acid reabsorption in antennal gland of *Carcinus*.
35. Binyon, J., J. Marine Biol. Assoc. U. K. *41*:161–174, 1961. Salinity tolerance and water permeability, *Asterias*.
36. Binyon, J., J. Marine Biol. Assoc. U. K. *42*:49–64, 1962. Ionic regulation in low salinity, *Asterias*.
37. Blinks, J. R., J. Physiol. *177*:42–57, 1965. Osmotic properties of striated muscle fibers.
38. Bliss, D. E., Phylogeny and Evolution of Crustaceans. Harvard University Press, 1963, pp. 59–78. Pericardial sacs of terrestrial crustaceans.
39. Bliss, D. E., Amer. Zool. *6*:197–212, 1966. Water balance in land crab, *Gecarcinus*.
40. Bliss, D. E., Amer. Zool. *8*:355–392, 1968. Transition from water to land in decapod crustaceans.
41. Bonting, S. L., Comp. Biochem. Physiol. *17*:953–966, 1966. Na-K ATPase in rectal glands of elasmobranchs.
42. Boonkoom, V., and R. H. Alvarado, Amer. J. Physiol. *220*:1820–1824, 1971. ATPase in gills of *Rana* tadpoles.
43. Boroffka, I., Z. vergl. Physiol. *51*:25–48, 1965. Ion transport in nephridium of *Lumbricus*.
44. Boroffka, I., Z. vergl. Physiol. *57*:348–375, 1968. Osmotic and volume regulation in *Hirudo*.
45. Boroffka, I., H. Altner, and J. Haupt, Z. vergl. Physiol. *66*:421–438, 1970. Function and structure of nephridium of *Hirudo*.
46. Boroffka, I., and A. Frick, Pflüger. Arch. *281*:20 (abstr.), 1964. Ion excretion in nephridium of *Lumbricus*.
47. Born, J. W., Biol. Bull. *134*:235–244, 1968. Osmoregulation in caridean shrimps.
48. Bourguet, J., B. Lahlou, and J. Maetz, Gen. Comp. Endocr. *4*:563–576, 1964. Regulation of ion and water balance in goldfish.
49. Bradshaw, S. D., Comp. Biochem. Physiol. *36*:689–719, 1970. Seasonal changes in water and ion balance, lizards.
50. Bradshaw, S. D., and V. H. Shoemaker, Comp. Biochem. Physiol. *20*:855–865, 1967. Water and electrolyte balance in desert lizards.
51. Brand, G. W., and I. A. E. Bagly, Comp. Biochem. Physiol. *38B*:361–371, 1971. Osmoregulation in four species of copepod.
52. Braun, G., G. Kummel, and J. A. Mangos, Pflüger. Arch. *289*:141–154, 1966. Ultrastructure and excretory function in protonephridium of a rotifer.
53. Braysher, M., and B. Green, Comp. Biochem. Physiol. *35*:607–614, 1970. Water and ion balance in Australian lizard.
54. Bryan, G. W., J. Exp. Biol. *37*:83–99, 100–112, 113–128, 1960. Sodium regulation in crayfish.
55. Bull, J. M., and R. Morris, J. Exp. Biol. *47*:485–494, 1967. Osmoregulation in larva of *Lampetra*.
56. Burger, J. W., Physiol. Zool. *35*:205–217, 1962. Function of rectal gland, dogfish.
57. Burger, J. W., Biol. Bull. *113*:207–223, 1965. Excretion, lobster.
58. Butler, D. G., Comp. Biochem. Physiol. *18*:773–781, 1966. Hormonal regulation of water balance in *Anguilla*.
59. Cade, T. J., pp. 237–256. *In* Thirst: Proceedings. International Symposium on Thirst in the Regulation of Body Water, edited by M. J. Wayner. Pergamon Press, 1963. Water and salt balance in birds.
60. Cade, T. J., C. A. Tobin, and A. Gold, Physiol. Zool. *38*:9–33, 1965. Water economy in finches.
61. Carlisky, N. J., et al., Comp. Biochem. Physiol. *36*:321–337, 1970. Water balance in excretion of urea in amphibians.
61a. Carpenter, R. E., Comp. Biochem. Physiol. *24*:951–964, 1968. Salt and water metabolism in marine fish-eating bat, *Pizonyx vivesi*.
62. Cholette, C., A. Gagnon, and P. Germain, Comp. Biochem. Physiol. *33*:333–346, 1970. Isosmotic adaptation in *Myxine*.
63. Chew, R. M., Biol. Rev. *36*:1–31, 1961. Water metabolism of desert vertebrates.
64. Clark, M. E., Biol. Bull. *134*:252–260, 1968. Effect of osmotic dilution on amino acids in polychaetes.
65. Clarke, M. R., E. J. Denton, and J. B. Gilpin-Brown, J. Physiol. *203*: 49P–50P, 1969. Buoyancy mechanisms in squid.
66. Claussen, D. L., Comp. Biochem. Physiol. *20*:115–130, 1967. Water loss in lizards.
67. Clegg, J. S., J. Exp. Biol. *41*:879–892, 1964. Osmotic function of glycerol in cysts of *Artemia*.
68. Conte, F. P., H. H. Wagner, F. Fessler, and C. Gnose, Comp. Biochem. Physiol. *18*:1–15, 1966. Development of osmotic and ionic regulation in young salmon.
69. Conway, E. J., Biol. Rev. *20*:56–72, 1945. Electrolytes in muscle and plasma.
70. Cook, J. S., J. Gen. Physiol. *50*:1311–1325, 1967. Nonsolvent water in human erythrocytes.
71. Cope, F. W., Biophys. J. *9*:303–319, 1969. Structure of water in tissues.
72. Copeland, D. E., J. Cell Biol. *23*:253–264, 1964. Mitochondrial osmotic pump in anal papillae, mosquito larvae.
73. Copeland, D. E., Science *151*:470–471, 1966; Protoplasma *63*:363–384, 1967. Salt secreting cells in *Artemia*.
74. Copeland, D. E., Amer. Zool. *8*:417–432, 1968. Ultrastructure of salt transport cells in *Gecarcinus*.
75. Copeland, D. E., Z. Zellforsch. *92*:1–22, 1968. Ultrastructure of salt absorbing cells in gills of *Callinectes*.
76. Costlow, J. D., C. G. Bookhout, and R. Monroe, Physiol. Zool. *39*:81–100, 1966. Effect of salinity on larval development of crab *Rhithropanopeus*.
77. Coulombe, H. N., S. H. Ridgway, and W. E. Evans, Science *149*:86–88, 1965. Respiratory water exchange in porpoises.
78. Croghan, P. C., J. Exp. Biol. *35*:219–233, 243–249, 1958. Osmotic and ionic regulation in *Artemia*.
79. Croghan, P. C., R. A. Carra, and A. P. M. Lockwood, J. Exp. Biol. *42*:463–474, 1965. Electrical potentials across gills of crayfish.
80. Croghan, P. C., and A. P. M. Lockwood, J. Exp. Biol. *48*:141–158, 1968. Ionic regulation in aquatic isopods.
81. Cronin, R. J., S. Solomon, and E. L. Klingler, Comp. Biochem. Physiol. *37*:511–516, 1970. Renal function in *Macaca*.
82. Dainty, J., and R. House, J. Physiol. *185*:172–184, 1966. Permeability to water of frog skin.
83. Dall, W., and N. E. Milward, Comp. Biochem. Physiol. *30*:247–260, 1969. Water and ion balance in amphibious and aquatic fishes.
84. Dall, W., J. Fish. Res. Bd. Canad. *27*:1123–1130, 1970. Osmoregulation in lobster.
85. Dantzler, W. H., Comp. Biochem. Physiol. *22*:131–140, 1967. Renal function in water snake, *Natrix*.
86. Dantzler, W. H., and B. Schmidt-Nielsen, Amer. J. Physiol. *210*:198–210, 1966. Excretion by *Pseudemys*.
87. Dawson, T. J., Comp. Biochem. Physiol. *28*:401–407, 1969. Evaporative water loss in possum, *Trichosurus*.
88. Dawson, W. R., V. H. Shoemaker, H. B. Trodoff, and A. Borut, Auk *82*:606–623, 1965. Salt balance in red crossbill.
89. Dawson, W. R., V. H. Shoemaker, and P. Licht, Ecology *47*:589–594, 1966. Evaporative water losses in Australian lizards.
90. Dehnel, P. A., Physiol. Zool. *39*:259–265, 1966. Chloride regulation in crab, *Hemigrapsus*.
91. Depocas, F., and J. S. Hart, Canad. J. Physiol. Pharmacol. *49*:53–62, 1971. Water fluxes in seal, *Phoca*.
92. Dethier, V. G., and L. M. Schoonhoven, J. Insect Physiol. *14*:1049–1054, 1968. Humidity receptors on antennae of caterpillars.
93. Diamond, J. M., Soc. Exp. Biol. Symp. *19*:329–347, 1965; Fed. Proc. *30*:6–13, 1971. Mechanism of isotonic water absorption.
94. Diamond, J. M., and W. H. Bossert, J. Gen. Physiol. *50*:2061–2083, 1967. Standing-gradient osmotic flow.
95. Doyle, W. L., Exp. Cell Res. *21*:386–393, 1960. Ultrastructure of salt gland in marine birds.
96. Doyle, W. L., Amer. J. Anat. *111*:223–229, 1962. Tubule cells of rectal gland of sting ray.
97. Duchateau-Bosson, G., and M. Florkin, Arch. Int. Physiol. Biochem. *70*:345–355, 1962. Osmotic regulation in crab, *Eriocheir*.
98. Duchateau-Bosson, G., and M. Florkin, Arch. Int. Physiol. Biochem. *70*:393–396, 1962. Intracellular amino acids in polychaetes.
99. Duchateau-Bosson, G., and M. Florkin, Comp. Biochem. Physiol. *3*:245–247, 1963. Intracellular amino acids in crayfish in different salinities.
100. Dunson, W. A., and R. D. Weymouth, Science *149*:67–69, 1965. Absorption of sodium by soft-shelled turtle, *Trionyx*.
101. Dunson, W. A., J. Exp. Zool. *165*:171–182, 1967. Sodium fluxes in freshwater turtles.
102. Dunson, W. A., and A. M. Taub, Amer. J. Physiol. *213*:975–982, 1967; *216*:995–1002, 1969. Extrarenal salt excretion in sea snake.
103. Dunson, W. A., Comp. Biochem. Physiol. *32*:161–174, 1970. Salt and water balance in estuarine reptiles.
104. Edney, E. B., Comp. Biochem. Physiol. *19*:387–408, 1966. Absorption of water vapor by insects.
105. Edney, E. B., Water Relations of Terrestrial Arthropods. Cambridge University Press, 1957. 108 pp.
106. Epstein, F. H., A. J. Manitius, T. Weinstein, A. I. Katz, and G. E. Pickford, Yale J. Biol. Med. *41*:388–393, 1969. Na-K ATPase in kidney of euryhaline *Fundulus*.
107. Evans, D. R., and D. Mellon, Jr., J. Gen. Physiol. *45*:487–513, 1962. Water receptor in taste sensilla of blowfly.
108. Evans, D. H., J. Exp. Biol. *47*:513–517, 525–534, 1967. Chloride and water balance of intertidal teleost, *Xiphister*.
109. Evans, D. H., J. Exp. Biol. *50*:179–190, 1969. Water and salt balance in teleost, *Pholis*.
110. Evans, D. H., J. Exp. Biol. *50*:689–703, 1969. Water permeability of marine, freshwater, and euryhaline teleosts.
111. Finkelstein, A., and A. Cass, Nature *216*:717–718, 1967. Cholesterol effects on water permeability of lipid membranes.
112. Fleming, W. R., and D. H. Hazelwood, Comp. Biochem. Physiol. *23*: 911–915, 1967. Ionic and osmotic regulation in freshwater medusa.
113. Florkin, M., G. Duchateau-Bosson, C. Jeuniaux, and E. Schoffeniels, Arch. Int. Physiol. Biochem. *72*:892–906, 1964. Osmotic and ionic regulation in *Eriocheir*.
114. Flügel, H., Kieler Meeresforsch. *16*:186–200, 1960. Osmotic resistance and regulation in prawn, *Crangon*.
115. Forster, R. P., F. Berglund, and B. R. Rennick, J. Gen. Physiol. *42*:319–327, 1958. Tubular secretion of organic bases in aglomerular kidney of *Lophius*.
116. Foster, M. A., Comp. Biochem. Physiol. *30*:751–759, 1969. Osmotic and ionic regulation in *Cottus*.
117. Freeman, A. A., J. P. Reuben, P. W. Brandt, and H. Grundfest, J.

Gen. Physiol. *50*:423–445, 1966. Osmotic properties of nerve membranes.
118. Frisch, J. A., Anat. Rec. *89*:571, 1944. Contractile vacuole functions in paramecia.
119. Fugelli, K., Comp. Biochem. Physiol. *22*:253–260, 1967. Regulation of red cell volume in flounder.
120. Gardner, K. D., and R. H. Maffly, J. Clin. Invest. *43*:1968–1975, 1964. Urea absorption in kidney.
121. Gary-Bobo, C. M., and A. K. Solomon, J. Gen. Physiol. *52*:825–853, 1968. Solvent water in red blood cells.
122. Giebisch, G., J. Gen. Physiol. *44*:659–678, 1961. Transtubular potentials in *Necturus* kidney.
123. Giebisch, G., and E. E. Windhager, Amer. J. Physiol. *204*:387–391, 1963. Chloride movement in proximal tubules of *Necturus*.
124. Giese, A. C., and A. Farmanfarmaian, Biol. Bull. *124*:182–192, 1963. Osmotic resistance of sea urchins.
125. Gifford, C. A., Publ. Inst. Marine Sci. U. Texas *8*:97–125, 1962. Osmotic regulation, crabs.
126. Gilles, R., and E. Schoffeniels, Comp. Biochem. Physiol. *31*:927–939, 1969. Osmotic regulation in crab nerves.
127. Gilles-Baillien, M., J. Exp. Biol. *52*:691–697, 1970. Urea osmoregulation in terrapin.
128. Goffart, M., Function and Form in the Sloth. Pergamon Press, 1971. 225 pp.
129. Goffart, M., and J. Nys, Arch. Int. Physiol. Biochem. *73*:166–168, 1964. Water balance in sloths.
130. Goldstein, D., and A. K. Solomon, J. Gen. Physiol. *44*:1–17, 1960. Pore sizes in red cells.
131. Goldstein, L., and R. P. Forster, Amer. J. Physiol. *220*:742–746, 1971. Osmoregulation and urea metabolism in skate, *Raja*.
132. Goldstein, L., W. W. Oppelt, and T. H. Maren, Amer, J. Physiol. *215*:1493–1497, 1968. Osmotic and urea metabolism in shark, *Negaprion*.
133. Goodman, A. B., and A. V. Wolf, J. Appl. Physiol. *26*:203–207, 1969. Water loss from human skin.
134. Gordon, M. S., K. Schmidt-Nielsen, and H. M. Kelly, J. Exp. Biol. *38*:659–678, 1961. Osmotic regulation in brackish water frog.
135. Gordon, M. S., J. Exp. Biol. *39*:261–270, 1962. Osmotic regulation in green toad.
136. Gordon, M. S., Biol. Bull. *124*:45–54, 1963. Chloride exchange in rainbow trout.
137. Gordon, M. S., Biol. Bull. *128*:218–229, 1965. Intracellular osmoregulation in toad muscle.
138. Gordon, M. S., and V. A. Tucker, J. Exp. Biol. *49*:185–193, 1968. Salinity adaptation in crab-eating frog.
139. Gottschalk, C. W., Amer. J. Med. *36*:670–685, 1968. Concentration of urine by mammalian kidney.
140. Grant, W. C., Ecology *36*:400–407, 1955. Tolerance of water loss in earthworms.
141. Greenaway, P., J. Exp. Biol. *53*:147–163, 1970; *54*:199–214, 1971. Sodium and calcium fluxes in snail, *Limnaea*.
142. Greenwald, L., Physiol. Zool. *44*:149–161, 1971. Sodium balance in frog.
143. Grimstone, A. V., A. M. Mullinger, and J. A. Ramsay, Phil. Trans. Roy. Soc. Lond. B *253*:343–382, 1968. Water and ion movements in rectal complex of *Tenebrio*.
144. Gross, W. J., Biol. Bull. *126*:54–68, 1964. Water balance in land crabs.
145. Gross, W. J., et al., Comp. Biochem. Physiol. *17*:641–660, 1966. Salt and water balance in Madagascar crabs.
146. Gupta, B. L., and M. J. Berridge, J. Morph. *120*:23–82, 1966. Fine structure of rectum of *Calliphora*.
147. Gutman, Y., and Y. Beyth, Life Sci. *9*:37–42, 1970. Urine concentration by kidney of chinchilla.
148. Hadley, N. F., J. Exp. Biol. *53*:547–558, 1970. Water relations of desert scorpion.
149. Hardisty, M. W., J. Exp. Biol. *34*:237–251, 1957. Osmotic regulation in larval and adult lampreys.
150. Harpur, R. P., Canad. J. Zool. *46*:295–301, 1963. Osmoregulation and urea metabolism in toads and frogs.
151. Harvey, H. W., The Chemistry and Fertility of Sea Waters. Cambridge University Press, 1955. 224 pp.
152. Harvey, W. R., and K. Zerahn, J. Exp. Biol. *50*:297–306, 1969. Potassium transport in *Samia cecropia* midgut.
153. Harvey, W. R., and S. Nedergaard, Proc. Nat. Acad. Sci. *51*:757–765, 1964. Active transport of potassium in midgut of *Samia cecropia*.
154. Hazelwood, D. H., W. T. W. Potts, and R. R. Fleming, Z. vergl. Physiol. *67*:186–191, 1970. Sodium and water balance in freshwater medusa.
155. Heatwole, H., F. Torres, S. B. de Austin, and A. Heatwole, Comp. Biochem. Physiol. *28*:245–269, 1969. Evaporative water loss in frog.
156. Herreid, C. F., Comp. Biochem. Physiol. *29*:423–429, 1969. Permeability of crabs to water.
157. Hinke, J. A. M., J. Gen. Physiol. *56*:521–541, 1970. Solvent water in muscle fibers of barnacle.
158. Hinton, H. E., Nature *188*:336–337, 1960. Tolerance of desiccation in chironomid larvae.
159. Hobson, A. D., et al., J. Exp. Biol. *29*:1–17, 22–29, 1952. Osmotic concentration in intestinal parasites.
160. Hokin, M. R., and L. E. Hokin, J. Gen. Physiol. *50*:793–811, 1967. Salt secretion in bird salt gland.
161. Holdgate, M. W., and M. Seale, J. Exp. Biol. *33*:82–118, 1956. Water loss through insect cuticle.
162. Holmes, W. N., J. G. Phillips, and D. G. Butler, Endocrinology *69*: 483–495, 1961. Effects of adrenal hormones on renal function in duck.
163. Holmes, W. N., and R. L. McBean, J. Exp. Biol. *41*:81–90, 1964. Ion excretion in turtle, *Chelonia*.
164. Horowicz, P., and J. W. Burger, Amer. J. Physiol. *214*:635–642, 1968. Sodium balance in dogfish, *Squalus acanthias*.
165. House, C. R., J. Exp. Biol. *40*:87–104, 1963. Osmotic regulation in brackish water fish, *Blennius*.
166. House, C. R., and K. Green, J. Exp. Biol. *42*:177–189, 1965. Ion and water transport by intestine of teleost, *Cottus*.
167. Hudson, J. W., pp. 211–236. *In* Thirst: Proceedings. International Symposium on Thirst in the Regulation of Body Water, edited by M. J. Wayner. Pergamon Press, 1963. Water metabolism in desert mammals.
168. Hudson, J. W., and J. A. Rummel, Ecology *47*:345–354, 1966. Water metabolism and temperature regulation in heteromyids.
169. Hughes, D. A., Biol. Bull. *136*:43–53, 1969. Osmotic responses of shrimp, *Penaeus*.
170. Hughes, M. R., Comp. Biochem. Physiol. *32*:315–325, 807–812, 1970. Cloacal and salt gland excretion in seagull.
171. Irvine, H. B., Amer. J. Physiol. *217*:1520–1527, 1969. Sodium and potassium secretion by insect malpighian tubules.
172. Jampol, L. M., and F. H. Epstein, Amer. J. Physiol. *218*:607–611, 1970. Na-K ATPase and osmotic regulation in fishes.
173. Jeuniaux, C., S. Bricteux-Gregoire, and M. Florkin, Cahiers de Biol. Marine *2*:373–379, 1961; *3*:107–114, 1962. Role of amino acids in osmoregulation of invertebrates.
174. Jones, M. L., Biol. Bull. *132*:362–380, 1967. Morphology of nephridia of *Nereis*.
175. Jorgensen, C. B., and R. P. Dales, Physiol. Comp. Oecol. *4*:357–374, 1957. Volume and osmotic regulation in nereid polychaetes.
176. Kalber, F. A., and J. D. Costlow, Amer. Zool. *8*:411–416, 1968. Osmoregulation in larvae of land crab *Cardisoma*.
177. Kamemoto, F. I., and E. J. Larson, Comp. Biochem. Physiol. *13*:477–480, 1964. Chloride concentration in coelomic and nephridial fluids of sipunculids.
178. Kamemoto, F. I., and J. K. Ono, Comp. Biochem. Physiol. *27*:851–857, 1968. Hormone control of urine flow in crayfish.
179. Kamiya, M., and S. Utida, Comp. Biochem. Physiol. *31*:671–674, 1969. Na-K ATPase in gills of fish.
180. Kaneshiro, E. S., P. B. Dunham, and G. G. Holz, Biol. Bull. *136*:63–75, 1969; *137*:161–169, 1969. Osmoregulation in a marine ciliate.
181. Kao, C. Y., J. Gen. Physiol. *40*:91–105, 107–119, 1956. Membrane properties in *Fundulus* eggs.
182. Kerley, D. E., and A. W. Pritchard, Comp. Biochem. Physiol. *20*:101–113, 1967. Osmoregulation in crayfish.
182a. Kerstetter, T. H., et al., J. Gen. Physiol. *56*:342–359, 1970. Sodium uptake by earthworms.
183. King, E. N., Comp. Biochem. Physiol. *17*:245–258, 1966. Oxidative activity of crab gill mitochondria.
184. King, E. N., and E. Schoffeniels, Arch. Int. Physiol. Biochem. *77*:105–111, 1969. Gill potentials in crab.
184a. Kinne, O., Z. wiss. Zool. *157*:427–491, 1953. Gammarids from different salinities.
185. Kinne, O., Neth. J. Sea Res. *3*:222–224, 1966. Metabolism of invertebrates in different salinities.
186. Kirschner, L. B., and S. Wagner, J. Exp. Biol. *43*:385–390, 1965. Filtration and reabsorption in crayfish antennal gland.
187. Kirschner, L. B., Amer. J. Physiol. *217*:596–604, 1969. Sodium fluxes in suffused eel gills.
188. Kirschner, L. B., Amer. Zool. *10*:365–376, 1970. Salt transport in aquatic animals.
189. Kirschner, L. B., T. H. Kerstetter, D. Porter, and R. H. Alvarado, Amer. J. Physiol. *220*:1814–1819, 1971. Adaptation of larval *Ambystoma* to concentrated solutions.
190. Kitching, J. A., J. Exp. Biol. *31*:56–67, 68–73, 76–83, 1954. Contractile vacuoles.
191. Knulle, W., J. Insect Physiol. *13*:333–357, 1967. Absorption of water vapor by small insects.
192. Koblick, D. C., and L. Yu-Tu, J. Exp. Zool. *166*:325–330, 1967. Osmotic properties of digestive cells of *Chlorohydra*.
193. Kohn, P. G., Soc. Exp. Biol. Symp. *19*:3–16, 1965. Physical and chemical properties of water.
194. Komadina, S., and S. Solomon, Comp. Biochem. Physiol. *32*:333–343, 1970. Renal function in snakes.
195. Komnick, H., Protoplasma *56*:274–314, 385–419, 605–636, 1963. Ultrastructure of salt glands of birds.
196. Krogh, A., Osmotic Regulation in Aquatic Animals. Cambridge University Press, 1939. 242 pp.
197. Kummel, G., Z. Zellforsch. *57*:172–201, 1962; Z. Naturforsch. *16*b: 692–697, 1961. Evolutionary cytology of primitive excretory organs.
198. Kummel, G., Zool. Beitr. *10*:227–252, 1964. Structure of antennal glands of crayfish.
199. Kummel, G., L. Dankworth, G. Braun-Schubert, and K. H. Gertz, Z. vergl. Physiol. *64*:118–134, 1969. Structure and function of nephridia of *Ascaris*.
200. Lahlou, B., Comp. Biochem. Physiol. *20*:925–938, 1967. Renal function in flounders.

201. Lahlou, B., J. W. Henderson, and W. H. Sawyer, Amer. J. Physiol. *216*:1266–1272, 1273–1278, 1969. Renal adaptations of toadfish, *Opsanus*.
202. Lahlou, B., J. W. Henderson, and W. H. Sawyer, Comp. Biochem. Physiol. *28*:1427–1433, 1969. Sodium fluxes in goldfish.
203. Lance, J., Comp. Biochem. Physiol. *14*:155–165, 1965. Respiration and osmotic behavior of copepod *Acartia*.
204. Lang, M. A., and H. Gainer, Comp. Biochem. Physiol. *30*:445–456, 1969. Osmotic regulation and volume control in crab muscle fibers.
205. Lange, R., Comp. Biochem. Physiol. *10*:173–179, 1963. Osmotic function of amino acid and taurine in *Mytilus*.
206. Lange, R., Comp. Biochem. Physiol. *13*:205–216, 1964. Osmotic adjustment in echinoderms and molluscs.
207. Lange, R., and K. Fugelli, Comp. Biochem. Physiol. *15*:283–292, 1965. Osmotic adjustments in euryhaline teleosts.
208. LeBrie, S. J., and I. D. W. Sutherland, Amer. J. Physiol. *203*:995–1000, 1962. Renal function in water snake.
209. Lee, P., and K. Schmidt-Nielsen, Amer. J. Physiol. *220*:1598–1605, 1971. Water balance in zebra finch.
210. Levy, R. A., Comp. Biochem. Physiol. *23*:631–644, 1967. Homarine in osmotic regulation of *Limulus*.
211. Lilly, S. J., J. Exp. Biol. *32*:423–439, 1955. Water osmoregulation in *Hydra*.
212. Lloyd, S. J., K. D. Gerlid, R. C. Reba, and A. E. Seeds, Amer. J. Physiol. *26*:274–276, 1969. Water permeability of human placenta.
213. Lockwood, A. P. M., J. Exp. Biol. *38*:647–658, 1961. Urine in two species of *Gammarus*.
214. Lockwood, A. P. M., Biol. Rev. *37*:257–305, 1962. Osmoregulation of crustacea.
215. Lockwood, A. P. M., Animal Body Fluids and their Regulation. Harvard University press, 1964. 177 pp.
216. Lockwood, A. P. M., J. Exp. Biol. *42*:59–69, 1965; *53*:737–751, 1970. Sodium balance in amphipod *Gammarus*.
217. Loeffler, C. A., and S. Lovtrup, J. Exp. Biol. *52*:291–298, 1970. Water balance in salmon egg.
218. Loveridge, J. P., J. Exp. Biol. *49*:15–29, 1968. Water loss from spiracles of locust.
219. Lovtrup, S., and A. Pigon, C. R. Lab. Carlsberg Sci. Chim. *28*:1–36, 1951. Water permeability of cells.
220. Lucas, J. S., and E. P. Hodgkin, Austral. J. Marine and Freshwater Res. *21*:163–173, 1970. Growth of *Halicarcinus* in different salinities and temperatures.
221. MacFarlane, W. V., R. J. H. Morris, and B. Howard, Nature *197*:270–271, 1963. Water turnover in desert mammals.
222. MacFarlane, W. V. et al., Nature *234*:483–484, 1971. Series of water requirements in mammals from different habitats.
223. MacMillen, R. E., and A. K. Lee, Comp. Biochem. Physiol. *28*:493–514, 1969; *35*:355–369, 1970. Water metabolism of Australian mice.
224. Machin, J., Amer. J. Physiol. *216*:1562–1568, 1969. Water permeability of skin of *Bufo*.
225. Mackay, W. C., and C. L. Prosser, Comp. Biochem. Physiol. *34*:273–280, 1970. King crab, ions in blood.
226. Macklin, M., J. Cell. Physiol. *70*:191–196, 1967. Osmotic regulation in *Hydra*.
227. Maddrell, S. H. P., J. Exp. Biol. *45*:499–508, 1966; *51*:71–97, 1969. Hormonal regulation of function in malpighian tubules of *Rhodnius*.
228. Maetz, J., and Lahlou, B., J. Physiol. Paris *58*:249, 1966. Na and Cl exchange in elasmobranch.
229. Maetz, J., Science *166*:613–615, 1967, Na-K exchange in branchiopump of flounder.
230. Maetz, J., W. H. Sawyer, G. E. Pickford, and N. Mayer, Gen. Comp. Endocr. *8*:163–176, 1967. Ionic balance in *Fundulus*.
231. Maloiy, G. M. O., Amer. J. Physiol. *219*:1522–1527, 1970. Water economy of desert donkey.
232. Malvin, R. L., E. Carlson, S. Legan, and P. Churchill, Amer. J. Physiol. *218*:1506–1509, 1970. Renal function of freshwater lamprey.
233. Mantel, L. H., Amer. Zool. *8*:433–442, 1968. Role of foregut in ion and water balance in crab, *Gecarcinus*.
234. Mantel, L. H., Comp. Biochem. Physiol. *20*:743–753, 1967. Gill potentials and ion regulation in *Callinectes*.
235. Martin, D. W., and J. M. Diamond, J. Gen. Physiol. *50*:295–315, 1966. Na and Cl transport in gall bladder.
236. Mayer, N., and J. Nibelle, Comp. Biochem. Physiol. *35*:553–556, 1970. Salt balance in *Anguilla*.
237. McBean, R. L., and L. Goldstein, Amer. J. Physiol. *219*:1115–1123, 1124–1130, 1970. Renal function in *Xenopus*.
238. McClanahan, L., Comp. Biochem. Physiol. *20*:73–99, 1967. Adaptations of desert toad.
239. McClanahan, L., and R. Baldwin, Comp. Biochem. Physiol. *28*:381–389, 1969. Water uptake through skin of desert toad.
240. McFarland, L. Z., Nature *204*:1202–1203, 1964. Salt excretion by seagulls.
241. McFarland, W. N., and W. A. Wimsatt, Comp. Biochem. Physiol. *28*:985–1006, 1969. Renal function of vampire bat.
242. McLusky, D. S., J. Marine Biol. Assoc. *48*:769–780, 1968. Osmotic and ionic regulation in amphipod *Corphium*.
243. Marshall, E. K., Physiol. Rev. *14*:133–159, 1934. Vertebrate kidneys.
244. Middler, S. A., C. R. Kleeman, and E. Edwards, Comp. Biochem. Physiol. *25*:335–341, 1968; *26*:57–68, 1968. Water reabsorption in bladder of toad.
245. Morris, R., J. Exp. *33*:235–248, 1956; *35*:649–665, 1958. Osmoregulation of *Lampetra* during spawning.
246. Morris, R., and J. M. Bull, J. Exp. Biol. *52*:275–290, 1970. Osmoregulation of larvae of *Lampetra*.
247. Motais, R., Ann. Inst. Oceanogr. Monaco *45*:1–83, 1967. Ionic and osmotic regulation, water and salt fluxes in marine and freshwater fishes.
248. Motais, R., C. R. Acad. Sci. Paris *253*:724–726, 1961. Sodium exchange in *Platichthys* in sea and fresh water.
249. Motais, R., and J. Maetz, Gen. Comp. Endocr. *4*:210–224, 1964. Pituitary hormones, sodium exchange in teleosts.
250. Motais, R., I. Rankin, and J. Maetz, J. Exp. Biol. *51*:529–546, 1969. Water fluxes in marine fishes.
251. Motais, R., F. G. Romeu, and J. Maetz, J. Gen. Physiol. *50*:391–422, 1966. Exchange diffusion in teleosts.
252. Nemenz, H., J. Insect Physiol. *4*:38–44, 1960; Sitzungsb. Öster. Akad. Wiss., Mat.-Nat. Kl. *169*:17–41, 1960. Osmotic and ionic regulation, *Ephydra* larvae.
252a. Neumann, D., Kieler Meeresforsch. *18*:38–54, 1962. Ions in brackish-water animals.
253. Noble-Nesbitt, J., J. Exp. Biol. *50*:745–769, 1969; *52*:193–200, 1970. Water balance in the firebrat, *Thermobia*.
254. Nungesser, W. C., and E. W. Pfeiffer, Comp. Biochem. Physiol. *14*:289–297, 1965. Kidney function in moutain beaver, *Aplodontia*.
255. O'Dell, R. M., H. M. Radwin, L. H. Bernstein, and J. W. Schlegel, J. Appl. Physiol. *24*:366–368, 1968. Renal function in *Macaca*.
256. Oglesby, L. C., Comp. Biochem. Physiol. *14*:621–640, 1965; *16*:437–455, 1965. Water and chloride regulation in nereids.
257. Oglesby, L. C., Biol. Bull. *134*:118–138, 1968. Responses of estuarine polychaetes to osmotic stress.
258. Oglesby, L. C., Comp. Biochem. Physiol. *26*:155–177, 1968. Osmotic responses of sipunculid worms.
259. Oschman, J. L., and M. J. Berridge, Fed. Proc. *30*:49–56, 1971. Structural basis of fluid secretion.
260. Oschman, J. L., and B. J. Wall, J. Morph. *127*:475–510, 1969. Fluid transport in rectal pads of *Periplaneta*.
261. Phillips, J. E., Amer. Zool. *10*:413–436, 1970. Review of excretion in insects.
262. Phillips, J. E., J. Exp. Biol. *41*:15–38, 39–67, 69–80, 1964; *48*:521–532, 1968. Rectal absorption and excretion in the desert locust.
263. Philpott, C. W., Protoplasma *60*:7–23, 1965; Philpott, C. W., and D. E. Copeland, J. Cell Biol. *18*:389–404, 1963. Chloride cells in gills of *Fundulus*.
264. Pickering, A. D., and R. Morris, J. Exp. Biol. *53*:231–243, 1970. Osmoregulation in cyclostomes.
265. Pickford, G., et al., Comp. Biochem. Physiol. *18*:503–509, 1966. Effect of calcium on survival of hypophysectomized *Fundulus* in fresh water.
266. Pickford, G. E., and F. B. Grant, Science *155*:568–570, 1967. Osmotic concentration in coelacanth *Latimeria*.
267. Pierce, S. K., Comp. Biochem. Physiol. *36*:521–533, 1970. Water balance of mollusc *Modiolus*.
268. Pierce, S. K., Comp. Biochem. Physiol. *38A*:619–635, 1971. Volume regulation in marine molluscs.
269. Pierce, S. K., Comp. Biochem. Physiol. *39A*:103–117, 1971. Volume regulation by marine mussels.
270. Pilcher, D. E. M., J. Exp. Biol. *52*:653–655; *53*:465–484, 1970. Hormone effects on urine secretion in *Carausius*.
271. Pontin, R. M., Comp. Biochem. Physiol. *17*:1111–1126, 1966. Osmotic function of secretory vesicles in a rotifer.
272. Potts, W. T. W., J. Exp. Biol. *31*:614–617, 618–630, 1954. Urine production in *Anodonta*.
273. Potts, W. T. W., and D. H. Evans, Biol. Bull. *133*:411–425, 1967. Ion fluxes in *Fundulus*.
274. Potts, W. T. W., and W. R. Fleming, J. Exp. Biol. *53*:317–327, 1970. Hormone effects on ion and water regulation in *Fundulus*.
274a. Potts, W. T. W., M. A. Foster, P. P. Rudy, and G. P. Howells, J. Exp. Biol. *47*:461–470, 1967. Sodium and water balance in *Tilapia*.
275. Potts, W. T. W., and G. Parry, J. Exp. Biol. *41*:591–601, 1964. Na and Cl balance in prawn, *Palaemonetes*.
276. Potts, W. T. W., and G. Parry, Osmotic and Ionic Regulations in Animals. Pergamon Press, 1964. 423 pp.
277. Potts, W. T. W., and P. P. Rudy, J. Exp. Biol. *50*:223–237, 1969. Water balance in eggs of salmon.
278. Poulson, T. L., and G. A. Bartholomew, Condor *64*:245–252, 1962. Salt utilization in house finch.
279. Poulson, T. L., and G. A. Bartholomew, Physiol. Zool. *35*:109–119, 1962. Salt balance in Savannah sparrow.
280. Price, K. S., and E. P. Creaser, Comp. Biochem. Physiol. *23*:65–76, 77–82, 1967. Water, urea, and sodium regulation in skate, *Raja*.
280a. Pritchard, A. W., and D. E. Kerley, Comp. Biochem. Physiol. *35*:427–437, 1970. Kidney function in crayfish.
281. Prusch, R. D., and P. B. Dunham, J. Cell Biol. *49*:431–434, 1970. Contractions of contractile vacuole of amoeba.
282. Ramamurthi, R., and B. T. Scheer, Personal communication. Sodium balance in crab, *Hemigrapsus*.
282a. Ramsay, J. A., J. Exp. Biol. *27*:145–157, 1950; *28*:62–73, 1951. Osmoregulation in mosquito larvae.
283. Ramsay, J. A., J. Exp. Biol. *26*:46–56, 65–75, 1949. Osmotic relations and urine production in *Lumbricus*.

284. Ramsay, J. A., J. Exp. Biol. *32*:200–216, 1955. Excretion by malpighian tubules, *Dixippus*.
285. Ramsay, J. A., J. Exp. Biol. *35*:871–891, 1958. Excretion in *Tenebrio*.
286. Ramsay, J. A., Phil. Trans. Roy. Soc. Lond. B *248*:279–314, 1964. Rectal complex of meal worm, *Tenebrio*.
287. Reid, I. A., and I. R. McDonald, Comp. Biochem. Physiol. *25*:1071–1079, 1968. Renal function in marsupial, *Trichosurus*.
288. Rich, G. T., R. I. Sha'afi, T. C. Barton, and A. K. Solomon, J. Gen. Physiol. *50*:2391–2405, 1967. Permeability of red cells.
289. Rich, G. T., R. I. Sha'afi, A. Romualdez, and A. K. Solomon, J. Gen. Physiol. *52*:941–954, 1969. Osmotic permeability of red cells.
290. Riddick, D. H., Amer. J. Physiol. *215*:736–740, 1968. Contractile vacuoles in amoeba.
291. Riegel, J. A., J. Exp. Biol. *38*:291–299, 1961. Function of crayfish antennal gland.
292. Riegel, J. A., Comp. Biochem. Physiol. *36*:403–410, 1970. Fluid movement in crayfish antennal gland.
293. Roberts, J. S., and B. Schmidt-Nielsen, Amer. J. Physiol. *211*:476–486, 1966. Excretion in terrestrial lizards.
294. Robertson, J. D., Biol. Bull. *138*:157–183, 1970. Osmotic and ionic regulation in *Limulus*.
295. Romeu, F. G., A. Salibean, and S. Pezzani-Hernandez, J. Gen. Physiol. *53*:816–835, 1969. Na and Cl uptake through skin of frog.
296. Rudy, P. P., Comp. Biochem. Physiol. *22*:581–589, 1967. Water permeability in decapod crustaceans.
297. Ruibal, R., Physiol. Zool. *35*:133–147, 218–223, 1962. Osmotic capacity of toads.
298. Sawyer, W. H., Amer. J. Physiol. *218*:1789–1794, 1970. Hormone effect on lungfish kidney.
299. Schlieper, C., Helg. wiss. Meeresunters. *14*:482–502, 1966. Osmotic resistance of gills of *Mytilus*.
300. Schmid, W. D., Ecology *46*:261–269, 1965. Water economy of amphibians.
301. Schmidt-Nielsen, B., and D. F. Laws, Ann. Rev. Physiol. *25*:631–658, 1963. Survey of comparative aspects of excretion in invertebrates.
302. Schmidt-Nielsen, B., and R. O'Dell, Amer. J. Physiol. *200*:1119–1124, 1961. Structure and function in mammalian kidney.
303. Schmidt-Nielsen, B., and E. W. Pfeiffer, Amer. J. Physiol. *218*:1370–1375, 1970. Kidney function in moutain beaver, *Aplodontia*.
304. Schmidt-Nielsen, B., and C. R. Schrauger, Science *139*:606–607, 1963. Contractile vacuoles of amoeba.
305. Schmidt-Nielsen, B., and E. Skadhauge, Amer. J. Physiol. *212*:973–980, 1967. Renal function in crocodile.
306. Schmidt-Nielsen, K., Desert Animals. Oxford University Press, 1964. 277 pp.
307. Schmidt-Nielsen, K., and P. Lee, J. Exp. Biol. *39*:167–177, 1962. Kidney function in crab-eating frog.
308. Schmidt-Nielsen, K., and P. J. Bentley, Science *154*:911, 1966. Cutaneous water loss in tortoise.
308a. Schoffeniels, E., and R. R. Tercafs, Ann. Soc. Roy. Zool. Belgique *96*:23–29, 1965. Osmotic and ionic balance in brackish water amphibians.
309. Schreibman, M. P., and K. D. Kallman, Gen. Comp. Endocr. *6*:144–155, 1966. Endocrine control of fresh water tolerance in teleosts.
310. Seeds, A. E., Amer. J. Physiol. *219*:551–554, 1970. Water movement across human placenta.
311. Sha'afi, R. I., G. T. Rich, D. C. Mikulecky, and A. K. Solomon, J. Gen. Physiol. *55*:427–450, 1970. Urea entrance into red cells.
312. Sha'afi, R. I., G. T. Rich, V. W. Sidel, W. Bosset, and A. K. Solomon, J. Gen. Physiol. *50*:1377–1399, 1967. Red blood cell permeability.
313. Sharrat, B. M., I. C. Jones, and D. Bellamy, Comp. Biochem. Physiol. *11*:9–18, 1964. Water and salt composition and renal function in eel.
313a. Shaw, J., J. Exp. Biol. *32*:353–382, 1955. Water relations of aquatic insect *Sialus*.
314. Shaw, J., J. Exp. Biol. *36*:126–144, 1959. Sodium balance in freshwater crayfish.
315. Shaw, J., J. Exp. Biol. *38*:135–152, 1960. Sodium fluxes in crayfish.
316. Shaw, J., J. Exp. Biol. *38*:153–162, 1961. Sodium balance in *Eriocheir*.
317. Shaw, J., Soc. Exp. Biol. Symp. *18*:237–254, 1964. Salt balance in crustaceans.
318. Shaw, J., and R. H. Stobbart, Adv. Insect Physiol. *1*:315–399, 1963. Osmotic and ionic regulation in insects.
319. Shoemaker, V. H., Comp. Biochem. Physiol. *15*:81–88, 1965. Responses to dehydration in toads.
320. Shoemaker, V. H., P. Licht, and D. R. Dawson, Comp. Biochem. Physiol. *23*:255–262, 1967. Water and salt excretion in lizards.
321. Siebert, B. D., and W. W. MacFarlane, Austral. J. Agric. Res. *20*:613–622, 1969. Water balance in ungulates.
322. Siebert, B. D., and W. W. MacFarlane, Physiol. Zool. *44*:225–240, 1971. Water balance in camels in the desert.
323. Skadhauge, E., Comp. Biochem. Physiol. *23*:483–501, 1967. Cloacal water and salt reabsorption in chicken.
324. Skadhauge, E., Comp. Biochem. Physiol. *24*:7–18, 1968. Cloacal function in chicken.
325. Skadhauge, E., J. Physiol. *204*:135–158, 1967. Salt and water fluxes in *Anguilla*.
326. Skadhauge, E., and B. Schmidt-Nielsen, Amer. J. Physiol. *212*:793–798, 973–980, 1313–1318, 1967. Renal function in domestic fowl.
327. Smith, H. W., Biol. Rev. *11*:49–82, 1936. Urea in elasmobranchs.
327a. Smith, H. W., Quart. Rev. Biol. *7*:1–26, 1932. Evolution of excretion in fishes.
328. Smith, R. I., Biol. Bull. *125*:322–343, 1963. Salt loss in nereids.
329. Smith, R. I., Biol. Bull. *126*:142–149, 1964. Water uptake in two nereids.
330. Smith, R. I., Biol. Bull. *133*:643–658, 1967. Osmoregulation and water permeability in estuarine crab.
331. Smith, R. I., Biol. Bull. *139*:351–362, 1970. Water permeability of *Carcinus*.
332. Smith, R. I., J. Exp. Biol. *53*:75–92, 93–100, 1970. Ionic regulation in nereids.
333. Smyth, M., and G. A. Bartholomew, Condor *68*:447–458, 1966. Water economy of desert birds.
334. Smyth, M., and G. A. Bartholomew, Auk *83*:597–602, 1966. Water and salt balance in mourning doves.
335. Sohal, R. S., and E. Copeland, J. Insect Physiol. *12*:429–439, 1966. Ultrastructure of anal papillae of *Aedes*.
336. Solomon, A. K., J. Gen. Physiol. *43*:1–15, 1960. Red cell membrane structure and ion transport.
337. Spight, T. M., Comp. Biochem. Physiol. *20*:767–771, 1967; Biol. Bull. *132*:126–132, 1967. Water economy of salamanders.
338. Staddon, B. W., J. Exp. Biol. *41*:609–619, 1964. Water balance in hemipteran, *Corixa*.
339. Stanley, J. G., and W. R. Fleming, Science *144*:63–64, 1964. Excretion of hypertonic urine by teleosts.
340. Stinnakre, J., and L. Tauc, J. Exp. Biol. *51*:347–361, 1969. Central nervous responses to osmoreceptors of *Aplysia*.
340a. Stobbart, R. H., J. Exp. Biol. *42*:29–43, 1965; *47*:35–37, 1967. Ion fluxes in larvae of *Aedes*.
341. Stobbart, R. H., J. Exp. Biol. *54*:29–66, 1971. Sodium uptake by mosquito larvae.
342. Stoner, L. C., and P. B. Dunham, J. Exp. Biol. *53*:391–399, 1970. Osmotic and volume regulation of *Tetrahymena*.
343. Sutcliffe, D. W., J. Exp. Biol. *38*:521–530, 1961; *39*:141–160, 1962. Osmotic and ionic regulation in aquatic insects.
344. Sutcliffe, D. W., and J. Shaw, J. Exp. Biol. *46*:499–518, 519–528, 529–550, 1967. Sodium balance in freshwater amphipod, *Gammarus*.
345. Taylor, C. R., et al., Amer. J. Physiol. *220*:823–827, 1971. Water balance in African ungulates.
346. Taylor, C. R., C. A. Spinage, and C. P. Lyman, Amer. J. Physiol. *217*:630–634, 1969. Water relations of African antelope.
347. Templeton, J. R., Comp. Biochem. Physiol. *11*:223–229, 1964. Nasal salt secretions in lizards.
348. Tercafs, R. R., Arch. Int. Physiol. Biochem. *71*:318–320, 1963. Permeability of skin of reptiles.
349. Theede, H., Kieler Meeresforsch. *19*:20–41, 1963; *21*:153–166, 1965. Effect of salinity on cilia of two populations of *Mytilus*.
350. Theede, H., Marine Biol. *2*:114–120, 1969. Osmoregulation in *Carcinus*.
351. Thesleff, S., and K. Schmidt-Nielsen, Amer. J. Physiol. *202*:597–600, 1962. Electrophysiology of salt glands of gulls.
352. Thompson, L. C., and A. W. Pritchard, Biol. Bull. *136*:114–129, 1969. Osmoregulation of crustaceans.
353. Thorson, T. W., pp. 265–270. *In* Sharks, Skates, and Rays, edited by P. W. Gilbert, R. F. Mathewson, and D. P. Rall, Johns Hopkins Press, 1967. Osmoregulation in freshwater elasmobranchs.
354. Thuet, P., R. Motais, and J. Maetz, Comp. Biochem. Physiol. *26*:793–819, 1968. Ionic regulation in *Artemia*.
355. Todd, M. E., J. Exp. Biol. *40*:381–392, 1963. Osmoregulation in marine isopods.
356. Tormey, J. M., and J. M. Diamond, J. Gen. Physiol. *50*:2031–2060, 1967. Fluid transfer in gall bladder.
357. Tosteson, D., and J. Hoffman, J. Gen. Physiol. *44*:169–194, 1960. Criteria of active transport in red blood cells.
358. Ullrich, K. J., B. Schmidt-Nielsen, R. O'Dell, G. Pehlling, G. W. Gottschalk, W. E. Lassiter, and M. Mylle, Amer. J. Physiol. *204*:527–531, 1963. Micropuncture studies of rat kidney.
359. Ullrich, K. J., G. Rumrich, and B. Schmidt-Nielsen, Pflüger. Arch. *295*:147–156, 1967. Urea transport in collecting duct of rat kidney.
360. Ussing, H., Acta Physiol. Scand. *19*:43–56, 1949. Theory of active and passive fluxes across membranes.
361. Utida, S., M. Kamiya, and N. Shirai, Comp. Biochem. Physiol. *38A*:443–447, 1971. ATPase in eel gill.
362. Vernberg, F. J., C. Schlieper, and D. E. Schneider, Comp. Biochem. Physiol. *8*:271–285, 1963. Effects of salinity on cilia of molluscan gills.
363. Vincent-Marique, C., and R. Gilles, Life Sci. *9*:509–512, 1970. Amino acids in blood and muscle of *Eriocheir*.
364. Virkar, R. A., Comp. Biochem. Physiol. *18*:617–625, 1966. Amino acids in salinity adaptation of *Golfingia*.
365. Virkar, R. A., and K. L. Webb, Comp. Biochem. Physiol. *32*:775–783, 1970. Amino acids in clam, *Mya*.
366. Waldron, W., Z. vergl. Physiol. *69*:249–283, 1970. Humidity receptors in locusts.
367. Wall, B. J., J. Insect Physiol. *13*:565–578, 1967. Hormonal control of rectal water absorption in cockroach.
368. Wall, B. J., J. Insect Physiol. *16*:1027–1042, 1970. Effects of hydration and dehydration on *Periplaneta*.
369. Wall, B. J., Fed. Proc. *30*:42–48, 1971. Osmotic gradients in rectal pads of an insect.
370. Wall, B. J., and J. L. Oschman, Amer. J. Physiol. *218*:1208–1215, 1970; Science *167*:1497–1498, 1970. Water and salt uptake by rectal pad of *Periplaneta*.

371. Webb, D. A., Proc. Roy. Soc. Lond. B *129*:107–136, 1940. Osmotic and ionic regulation in *Carcinus.*
372. Webber, H., and P. A. Dehnel, Comp. Biochem. Physiol. *25*:49–64, 1968. Water and ion regulation in limpet.
373. Weil, E., and C. F. A. Pantin, J. Exp. Biol. *8*:73–81, 1931. Adaptations of *Gunda* to salinity.
374. Weinstein, S. W., and R. M. Klose, Amer. J. Physiol. *217*:498–504, 1969. Micropuncture studies on mammalian kidney.
375. Weisser, F., F. B. Lacey, H. Weber, and R. L. Jamison, Amer. J. Physiol. *219*:1706–1713, 1970. Renal function in chinchilla.
376. Werntz, H. O., Biol. Bull. *124*:225–239, 1963. Osmoregulation in gammarids.
377. Wigglesworth, V. B., Principles of Insect Physiology, 6th ed. Methuen, 1965.
378. Wigglesworth, V. B., and M. M. Salpeter, J. Insect Physiol. *8*:299–307, 1962. Histology of malpighian tubules in *Rhodnius.*
379. Willoughby, E. J., Comp. Biochem. Physiol. *27*:723–745, 1968. Water economy of African birds.
380. Wilson, W., Biol. Bull. *138*:96–108, 1970. Osmoregulation in isopods.
381. Winston, P. W., and V. E. Nelson, J. Exp. Biol. *43*:257–269, 1965. Regulation of transpiration in mites.
382. Wolf, A. V., Thirst. Charles C Thomas, Springfield, 1958. pp. 340–372.
383. Wood, J. L., P. S. Farrand, and W. R. Harvey, J. Exp. Biol. *50*:169–178, 1969. Active transport of potassium by midgut of *Samia cecropia.*
384. Zahn, T. J., Comp. Biochem. Physiol. *25*:1021–1033, 1968. Protein effects on secretion in muskrats.
385. Zaks, M. G., and M. M. Sokolova, Sechenov Physiol. J. (U.S.S.R.) *47*:120–127, 1961. Salt glands of marine birds.
386. Zaks, M. G., and M. M. Sokolova, Fed. Proc. Transl. Suppl. *25*:T531–534, 1966. Osmotic regulation in migrant salmon.
387. Zaugg, W. S., and L. R. McLain, Comp. Biochem. Physiol. *35*:587–596, 1970. ATPase in gills of salmon.
388. Zotin, A. I., Soc. Exp. Biol. Symp. *19*:365–384, Uptake of water by embryos.

REFERENCES

Hormonal Control

For references to water and osmotic balance, see page 73.

1. Archer, R., J. Chauvet, and M. T. Chauvet, Nature *201*:191–192, 1964. Neurohypophyseal hormones in finback whale.
2. Archer, R., and J. Chauvet, Europ. J. Biochem. *17*:509–513, 1970. Mesotocin in birds.
3. Baehr, J. C., and N. Bandry, C. R. Acad. Sci. Paris *270*:3134–3136, 1970. Hormonal control of water balance in *Rhodnius.*
4. Bastide, F., and S. Jard, Biochim. Biophys. Acta *150*:113–123, 1968. Action of hypophyseal polypeptides.
5. Bentley, P. J., and H. Heller, J. Physiol. *109*:124–129, 1965. Hormonal control in *Salamandra.*
6. Brown, T. C., J. O. Davis, and C. I. Johnston, Amer. J. Physiol. *211*: 437–441, 1966. Responses of renin and aldosterone secretion in dogs.
7. Capelli, J. P., L. G. Wesson, and G. E. Aponte, Amer. J. Physiol. *218*: 1171–1178, 1970. Phylogenetic study of the renin-angiotensin system.
8. Chan, D. K., I. C. Jones, and W. Mosley, J. Endocr. *42*:91–98, 1968. Hormonal control of water and ions in *Anguilla.*
9. Copp, D. H., pp. 377–398. *In* Fish Physiology, Vol. 2, edited by W. S. Hoar and D. J. Randall, Academic Press, 1969. Calcitonin in fish.
10. Cuthbert, A. W., and E. Painter, J. Physiol. *199*:593–612, 1968. Hormonal effects on frog skin.
11. Edelman, I. S., and G. M. Timognori, Rec. Prog. Horm. Res. *24*:1–34, 1968. Action of vasopressin on Na transport in bladder.
12. Eggena, P., O. L. Schwatz, and R. Walter, J. Gen. Physiol. *52*:465–481, 1968. Effects of vasotocin on toad bladder.
13. Fanestil, D. D., and I. S. Edelman, Fed. Proc. *25*:912–916, 1966. Effects of vasopressin on toad bladder.
13a. Ferguson, D. R., and H. Heller, J. Physiol. *180*:846–863, 1965. Neurohypophyseal hormones in mammals.
14. Holmes, W. N., D. G. Butler, and J. G. Phillips, J. Endocr. *23*:53–61, 1961. Saline stimulation of salt gland in gull.
15. Holmes, W. N., J. D. Phillips, and I. Chester Jones, Rec. Prog. Horm. Res. *10*:619–666, 1963. Glucocorticoids in ion balance of birds.
16. Ireland, M. P., J. Endocr. *43*:133–138, 1969. Effect of urophysectomy on survival of *Gasterosteus* in osmotic stress.
17. Jard, S., J. Physiol. Paris *58*; Supp. *15*:1–124, 1966. Hormonal effects on water balance in frog.
18. Jones, I. C., et al., J. Endocr. *43*:21–31, 1968. Effects of AVT in freshwater fishes.
19. Kamemoto, F. I., and J. K. Ono, Comp. Biochem. Physiol. *29*:393–401, 1969. Endocrine control of osmoregulation, crustacea.
20. Kato, K. N., and F. I. Kamemoto, Comp. Biochem. Physiol. *28*:665–674, 1969. Endocrine control of osmoregulation, crab.
21. Lahlou, B., and W. H. Sawyer, Gen. Comp. Endocr. *12*:370–377, 1969. Ion balance in hypophysectomized goldfish.
22. Lam, T. J., Comp. Biochem. Physiol. *31*:909–913, 1969. Effect of prolactin on stickleback fish.
23. Lam, T. J., and J. F. Leatherland, Comp. Biochem. Physiol. *33*:295–302, 1970. Prolactin in osmotic and ionic balance, *Fundulus.*
24. Maddrell, S. H. P., J. Exp. Biol. *45*:479–508, 1966; *51*:71–97, 1969. Hormonal regulation of function in malpighian tubules of *Rhodnius.*
25. McNabb, R. A., and G. E. Pickford, Comp. Biochem. Physiol. *33*:783–792, 1970. Thyroid function in euryhaline *Fundulus.*
26. Mills, R. R., J. Exp. Biol. *46*:35–41, 1967. Hormonal control of excretion in cockroach.
27. Mizogami, S., M. Oguri, H. Sokabe, and H. Nishimura, Amer. J. Physiol. *215*:991–994, 1968. Renin in teleost kidneys.
28. Nishimura, H., M. Oguri, M. Ogawa, H. Sokabe, and M. Imai, Amer. J. Physiol. *218*:911–915, 1970. Absence of renin in elasmobranchs and cyclostomes.
29. Pickford, G. E., Comp. Biochem. Physiol. *18*:503–509, 1966. Effect of prolactin on *Fundulus* in various salinities.
30. Potts, W. T. W., and D. H. Evans, Biol. Bull. *131*:362–368, 1966. Prolactin in fish.
31. Sawyer, W. H., Amer. J. Physiol. *164*:457–466, 1951. Glomerular antidiuretic effects in amphibians.
32. Sawyer, W. H., Endocrinology *75*:981–990, 1964. Vertebrate neurohypophyseal principles.
33. Sawyer, W. H., Arch. d'Anat. Micro. Morph. Exper. *54*:295–312, 1965. Review of action of pituitary polypeptides.
34. Sawyer, W. H., Gen. Comp. Endocr. *5*:427–439, 1965. Neurohypophyseal principles from lamprey and two Chondrichthyes.
35. Sawyer, W. H., Amer. J. Physiol. *210*:191–197, 1966. Hormonal effects on ion balance in lamprey.
36. Sawyer, W. H., Amer. J. Med. *42*:678–686, 1967. Evolution of antidiuretic hormones.
37. Sawyer, W. H., J. Endocr. *44*:421–435, 1969. Hormones in elasmobranchs.
38. Sawyer, W. H., M. Manning, E. Heinicke, and A. M. Parks, Gen. Comp. Endocr. *12*:387–390, 1969. Elasmobranch hormones of water balance.
39. Sawyer, W. H., pp. 257–269. *In* Miami Winter Symposia. Vol. 1. Homologies in Enzymes and Metabolic Pathways, edited by W. J. Whelan and J. Schultz, Elsevier, New York, 1970. Homologies among neurohypophyseal peptides.
40. Sawyer, W. H., Amer. J. Physiol. *218*:1789–1795, 1970. Hormones of water balance in lungfish, *Protopterus.*
41. Sawyer, W. H., R. J. Freer, and T. Tseng, Gen. Comp. Endocr. *9*:31–37, 1967.
42. Sharp, G. W. G., and A. Leaf, Rec. Prog. Horm. Res. *22*:431–466, 1966. Action of aldosterone on toad bladder.
43. Sokabe, H., S. Mizogami, and A. Sato, Japan. J. Pharmacol. *18*:332–343, 1968. Role of renin in adaptation to sea water in euryhaline fishes.
44. Stanley, J. G., and W. R. Fleming, Biol. Bull. *131*:155–165, 1966. Hypophyseal effects on Na transport in *Fundulus kansae.*
45. Takasugi, N., and H. S. Bern, Comp. Biochem. Physiol. *6*:289–303, 1962. Caudal neurosecretory organ in *Tilapia.*

CHAPTER 2

INORGANIC IONS

*by C. Ladd Prosser**

The osmotic concentration of a solution depends on the total number of solute particles, irrespective of kind. Many animals conform osmotically to their environment but regulate specific ion concentrations in their body fluids. In no organism are the intracellular and extracellular concentrations of ions identical. One of the most universal qualities of living things is the selection at the cell surface of specific ions and the exclusion of others. At both the cellular and organismic levels, ionic regulation is a general and primitive capacity.

Life originated in the sea, and geochemical evidence indicates that the ionic composition of sea water has not changed greatly since the early Cambrian, although total salinity may have increased slightly. A list of elements in decreasing abundance follows:[140, 172]

Earth's crust:
- per cent atoms
  - O 49, Si 26, Al 7.5, Fe 4.7, Ca, Na, K, Mg, H

Oceans:
- per cent atoms of salt
  - O, H, Cl 55.2, Na 30.6, Mg 3.7, Ca 1.6, K 1.1, S, P

Protoplasm
- per cent by weight
  - O 76, C 10, H 10, N 2, K, Ca=Mg, Na, Cl, Fe=S, P

In igneous rocks, potassium approximately equals sodium; but upon leaching out, the potassium is more strongly adsorbed to clays so that soils contain more potassium than sodium. Consequently, rivers and the sea contain much less potassium than sodium. Sodium is the principal cation of extracellular fluids of most animals, and it is needed by marine plants but not by most terrestrial plants. However, potassium is the principal intracellular cation of all plants and animals. Thus, there is little relation between abundance of elements and their availability and biological utilization.

In large part, the chemical properties of an element determine its biological usefulness. Metals have special uses, particularly in association with specific proteins, and some of the most abundant elements in the earth's crust, such as silicon and aluminum, have little biological usefulness. Marine animals accumulate a higher proportion of potassium than sodium from sea water, and exclude magnesium and sulfate. The concentration of hydrogen ions tends to be higher in body fluids than in surrounding aqueous media. The amounts of particular ions bound to proteins rather than free in the organism are not readily ascertained and may change with time. The mechanisms of ionic regulation are extremely complex and incompletely known.

INTRACELLULAR REGULATION OF IONS

All cells have an inorganic composition different from that of the fluids that bathe them; cell membranes selectively regulate

*With contributions by L. B. Kirschner.

TABLE 2–1. Total Ionic Composition of Tissues and Plasma.
Values in mM/liter or mM/kg. pl = plasma; m = muscle; RBC = red blood cells; bl = blood.

Animal	*Tissue*	*Na*	*K*	*Ca*	*Mg*	*Cl*
Aplysia[95]	bl	492	9.7	13.3	49	543
	m	325	47.8	14.7	95.5	378
Buccinium[95]	bl	413	7.7	10.6	42	
	m	62.4	82.1	7.8	35.0	121
Acmaea (snail)[246]	bl	432	11.8	10	48.9	497
	m	45.7	162	39.8	22.6	28.8
Caudina[124]	pl	460	11.8	10.7	50.5	523
	m	191	138	89.0	39.0	122
Mytilus[19]	bl	490	12.5			573
	m	73	158			56
Ostrea[19]	white m	274	61.5	12.5		290
	yellow m	182	102	6.75		165
Sepia[196]	pl	465	21.9	11.6	57.7	591
	m	67.7	174.7	2.7	22.3	91.4
Loligo[196]	pl	419	20.6	11.3	51.6	522
	m	78	152	3.0	15.2	91.3
Eledone[196]	pl	432	14.4	11.2	54.2	516
	m	91.9	144.5	4.3	22.8	123
Paralithodes[142]	bl	461	13.5	13	36.9	
	m	62	105	9	8	
Squilla[129]	pl	442	15	9.6	21	431
	m	163	87	8.5	21.2	176.7
Carcinus[205]	pl	468	12.1	35	47.2	524
	m	54	120	5.5	17.5	54
Nephrops	pl	517	8.6	16.2	10.4	527
	m	83.2	166.6	5.2	19.1	109.9
Limulus[197]	pl	445	11.8	9.7	46	514
	m	126	99	3.7	20	159
Carausius[233]	pl	20	33.7	6.4	61.8	
	nerve cord	86.3	556	61.8	10.7	
Anodonta[95]	pl	15.4	0.38	5.3	0.35	10.5
	m	5.2	10.5	5.4	2.5	10.6
Limnaea[85]	bl	57	1.8	4.9		43.9
	m	37.2	14.7	26.6		
Myxine[195]	serum	560	9.5	6.1	18.4	570
	m	122	117	1.8	12.9	107
Anguilla (eel)[204] in F.W.	pl	155	2.7			106
	m	30.4	110.1			
Muraena (fish)[194]	pl	21.2	1.95	3.9	2.4	188
	m	25	165	9.3	7.4	23.7
frog[45a]	pl	104	2.5	2.0	1.2	74.3
	m	11.4	126			
		(23.9)	(84.6)	(2.5)	(11.3)	(10.5)
Dipsosaurus (lizard)[151]	pl	168	3.4			120
	m	44.9	122			30
Anas (duck)[102]	pl	138	3.1	2.3		103
	m	21	83.3			13.1
rat[31, 39]	pl	145	9.6	3.6	1.9	116
			(6.2)	(3.1)	(1.6)	
	RBC	34	105	1.0	5.4	
	stomach m			11.0	13.6	
	m	8.4	185	1.5	11.0	
		(26.6)	(101.4)			
dog[31]	pl	159	4	3.7	1.7	
	RBC	93	5.8	0.14	8.4	
	stomach m			7.0	19.6	
man[221]	pl		151	4.9	1.9	
	RBC			0.12	4.7	
	m		150			
cat[45a]	pl	143	3.2			
	RBC	104	5.9			

cytoplasmic ion composition. Table 2–1 gives examples of the compositions of various cells and tissues in relation to their media. Unfertilized sea-urchin eggs contain some twenty times more potassium and nine times less sodium than sea water. The sum of the cations is 292 mEq/kg_{H_2O}, while that of inorganic anions is less than 100 mEq; the necessary balance is given by phosphates and organic anions.[201] Mammalian red blood cells usually contain higher concentrations of potassium than does the plasma; however, carnivores tend to have higher Na relative to K in red cells. In some strains of sheep the red cells change in content from potassium to sodium predominance early in development; this character is inherited in Mendelian ratios:[64]

Sheep Blood (mM):

	K_o	Na_o	K_i	Na_i	Cl_i/Cl_o	vol_{H_2O}
high K strain	4.9	139	121	36.1	0.67	0.69
low K strain	5.0	139	17.3	139	0.70	0.69

Chloride is distributed passively in these cells, while Na and K are pumped using energy from glycolysis. The half-time of permeability for ^3H_2O is a few milliseconds, for Cl several hundred milliseconds, and for Na and K several hours.[231]

Nuclei of amphibian oocytes have higher water content, lower Na and slightly higher K than the cytoplasm; exchange data indicate two cytoplasmic pools of Na.[32] It is evident that *Tetrahymena* retains K and extrudes Na; a culture was adapted to a high Na medium (220 mM), and soon after transfer the cells had 105 mEq Na^+, after two weeks 43 mEq, and after 22 months 21 mEq. Evidently, selection for sodium extrusion had taken place.[58]

Ionic regulation in Protozoa is shown by *Tetrahymena*:[58, 59]

	Na	*K*	*Cl*
medium, mM	36.5	4.7	28.7
ciliate cells, mM	12.7	31.7	6.4

High concentrations of K and Mg in amoebae are indicated as follows:[72]

μEq/mg dry weight	*K*	*Na*	*Mg*	*Ca*	*Fe*
Amoeba proteus	0.376	0.026	0.434	0.127	0.102
Pelomyxa carolinensis	0.554	0.021	0.413	0.124	0.031

In multicellular tissues, intracellular concentrations are calculated after measuring extracellular space by means of some substance, such as inulin, which does not readily penetrate cell membranes. Extracellular spaces (as determined by inulin) comprise between 18 and 30 per cent of tissue volume. In some tissues, such as muscle, chloride is low inside cells and chloride (and sometimes also sodium) washout curves show initial replacement of the ion in extracellular space. From the analysis of a tissue and the measurement or calculation of extracellular space, the intracellular concentrations can be obtained (Table 2–1).

Analyses of muscles (and of some other tissues) of a wide variety of animals reveal a striking constancy of intracellular potassium concentration—from 110 to 160 mM, with a mode at 150 mM (Table 2–2). The range for sodium is very much greater (Fig. 2–1). Only in freshwater molluscs and in Protozoa is intracellular potassium significantly lower than in most animals, from 15 to 32 mM. Potassium activates some enzymes (e.g., transferases) and is critical for maintenance of normal membrane excitability; the constancy of intracellular potassium, with varying total osmotic concentration or habitat, may represent a very old cellular character.[220, 221] In general, whole plants contain more K than Na, while in animals the total amounts of Na and K are nearly equal. Herbivorous animals have a high K diet, while carnivores eat more Na. It is concluded

TABLE 2–2. Ratios of Intracellular to Extracellular Concentrations of Na and K in Muscles of Various Animals. (Modified from References 221 and 196.)

Animal	Na_i/Na_o	K_i/K_o
Strongylocentrotus	0.72	14.3
Golfingia	0.17	14.0
Sepia	0.006	8.6
Eledone	0.077	11.6
Loligo	0.03	11.0
Carcinus	0.11	9.0
Eriocheir	0.07	18.0
Nephrops	0.16	19.3
Limulus	0.28	8.4
Limnaea	0.65	8.2
Carausius (nerve cord)	4.3	7.4
Myxine	0.22	12.3
Muraena	1.2	8.5
Rana	0.15	49.
Gallus	0.05	20
Rattus	0.02	38

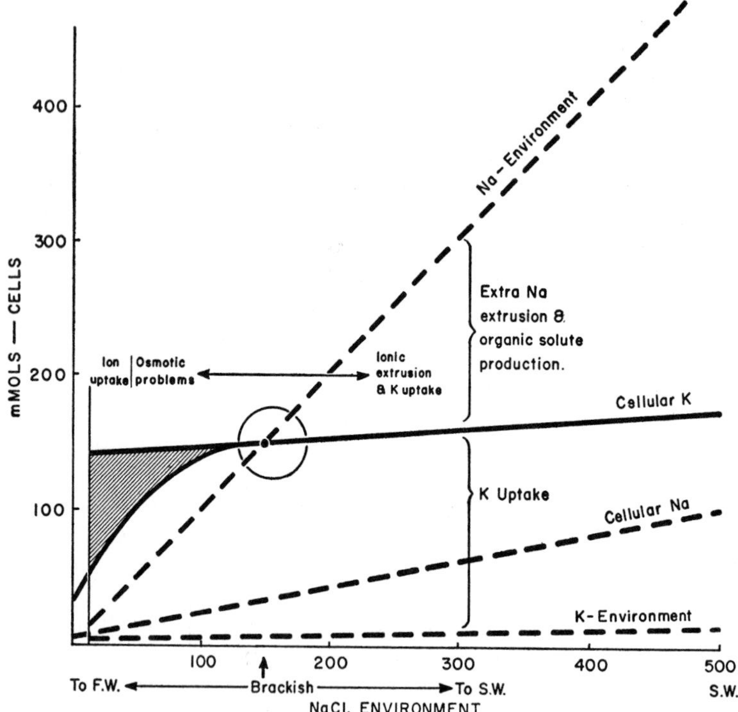

Figure 2–1. Diagram relating intracellular Na and K to environmental salinity, expressed as NaCl equivalents. Shaded area represents extremes of variation of cell potassium concentrations of animals in dilute environments. (From Steinbach, H. B., Perspectives in Biology and Medicine 5:338–355, 1962.)

that the sodium content of tissues is variable, that the ratio of intracellular potassium to sodium is more constant, and that intracellular potassium concentration is one of the most constant quantities in animals.

Lowered content of intracellular sodium and chloride in various tissues occurs in invertebrates which show little ionic regulation of their body fluids with respect to the sea. The sea urchin *Strongylocentrotus* is an osmoconformer that varies the osmoconcentration of its intestinal cells in different salinities, but it keeps the cellular ionic composition relatively constant while varying the nitrogenous compounds. Ninhydrin positive substances constitute as much as 50 per cent of the intestinal solutes; this quantity is much greater when the animal is in concentrated than when in dilute sea water.[127]

Within cells of several molluscs—marine *Acmaea, Ostrea,* and *Mytilus,* terrestrial *Strophocheilus,* and freshwater *Anodonta*—the concentrations of K and Ca are higher than in blood, and Na and Mg concentrations are lower than in blood. The energy required to maintain the cell-blood gradient in *Mytilus* is some six times greater than that in *Anodonta;* however, *Anodonta* must maintain a blood-medium gradient.[174] Sodium efflux curves for nervous tissue of *Anodonta* indicate two sodium compartments. The marine molluscs use organic ions—amino acids and neutral nitrogenous compounds—as cellular solutes more than do freshwater molluscs.[19]

Cephalopod tissues show very diverse composition, yet are iososmotic with sea water. Ratios of intracellular (muscle) to plasma ions are as follows:[196]

	Na	*K*	*Ca*	*Mg*	*Cl*	SO_4	*Phosphates*
Sepia	.006	8.6	.16	.33	.076	.34	115
Eledone	.077	11.6	.28	.32	.106	.22	322

Of the total osmotic concentration in cephalopod muscle, some 10 per cent is due to phosphates, 40 per cent to amino acids, and 15 to 20 per cent to nitrogenous bases:[196]

mM/kg_{H_2O} muscle	*amino acids*	*trimethylamine oxide*	*betaine*	Cl^-
Sepia	442	78.8	99	89.3
Loligo	388	117	71	
Eledone	279	33.7	99.7	128

In the giant nerve fibers of squid, glycerols and myoinositol constitute 13 to 18 per cent of the dry weight. A balance sheet of cations, anions, and neutral solutes in dry axoplasm is as follows:[47]

μEq/100 g dry weight

cations		*anions*		*neutral*	
K	344	Cl	140	glycine	10
Na	65	HPO_4	26.6	alanine	8
Ca	7	isethionate	152	serine	4
Mg	20	aspartate	73	leucine	2
arginine	3.2	glutamic	19.6	valine	2
lysine	2.4			threonine	2
ornithine	1.8			proline	1
				tyrosine	0.7
				homarine	20.4
				glycerol	82
				betaine	68
				taurocholine	98
				myoinositol	7

The high concentrations of isethionate, homarine, and betaine are of particular interest. Whether these have functions other than to provide nonreactive solutes to balance the inorganic cations and to provide for isotonicity is unknown.

In marine crustaceans also a large proportion of intracellular solutes are organic molecules. In the lobster *Nephrops,* muscle concentrations in mEq are: Na 83, K 176, Cl 110; amino acid concentration in the muscle is 405 mM, trimethylamine oxide 59 mM, and betaine 66 mM. Thus, amino acids constitute half of the osmotic concentration.[196]

Myxinoid and elasmobranch fish have blood and tissues at higher osmoconcentrations than do higher vertebrates (Table 2–1). The organic constituents, particularly anionic ones, constitute a high proportion of solute in *Myxine* muscle. Calcium is remarkably low in hagfish blood and muscle.[196]

Muscle	*inorganic ions + organophosphates mM (% solutes)*	*organic constituents mM (% solutes)*
Cyclostome (*Myxine*)	491 (42%)	690 (58%)
Teleost (*Muraena*)	308 (74%)	111 (28%)
Mammal (Rat)	273 (87%)	41 (13%)

The presence of urea and, to a lesser extent, amino acids which provide muscle solutes in high osmoconcentrations has been mentioned for elasmobranchs (p. 47) and brackish-water frogs (p. 55). The holocephalan *Hydrolagus* is isosmotic with sea water, but much of the serum solute is NaCl, rather than urea as in elasmobranchs.[185]

Many phytophagous insects have nerves in which, as in other animals, the resting potential is determined by high K_i/K_o, and spikes by a steep inward gradient of Na; but the hemolymph may be low in sodium and high in potassium, and the nerves function because of selective and/or actively ion-transporting sheaths (see pp. 479–480, Chap. 11).

In higher vertebrates the central nervous system is buffered against changes in the ionic composition of the blood. The cerebrospinal fluid is much lower in protein than the blood and is formed by both active and passive ion movements. Ratios of plasma to cerebrospinal fluid ions in man are:[45] Na 1.04, K 1.8, Cl 0.86, Ca 2.1, Mg 0.84, protein 300. In *Necturus* in which plasma K has been increased five-fold, the cerebrospinal fluid K increases two-fold; glial and nerve cells adjust their membrane potentials to the cerebrospinal fluid and not to plasma, whereas muscle cells adjust membrane potentials to plasma concentrations.[38] In the leech nervous system, glia may constitute half of the nervous system; K_i for glia is 108 mEq, and is 138 mEq for neurons.[161]

COMPOSITION OF BODY FLUIDS

Since the sea is the ancestral animal home and since sponges, coelenterates, and echinoderms use sea water to some extent as a body fluid, one might expect a correspondence between sea salts and the constituents of body fluids. It was formerly hypothesized that specialized animals, especially those with closed circulation, might resemble in body fluid ionic proportions the oceans at the time that their ancestors' circulation became closed. This hypothesis is contraindicated by the ionic variation in closely related animals, by ionic regulation when the environment changes, and by geochemical evidence for relative constancy in proportions of sea ions since the Cambrian. However, it is useful to compare sea water with body fluids of marine animals.

The composition of "average" sea water is given in Table 2–3. Tables for computing concentrations of common ions at different salinities are available.[4, 93] The pH of sea water is determined largely by the ratio of H_2CO_3 to HCO_3^-, and may be taken as 8.16 in equilibrium with air at 20°C.

(Text continued on page 88.)

TABLE 2–3. Composition of "Average" Sea Water.[4]

	$mM/l_{S.W.}$	gm/kg_{H_2O}
Salinity 34.33‰ } 20°C		
Chlorinity 19‰ } 20°C		
Sodium	470.2	10.933
Magnesium	53.57	1.317
Calcium	10.23	0.414
Potassium	9.96	0.394
Strontium	0.156	0.014
Chloride	548.3	19.657
Sulfate	28.25	2.744
Bromide	0.828	0.067
Borate as H_3BO_3	0.431	0.027
Bicarbonate	2.344	0.145
Osmotic equivalence {0.557 M NaCl, 0.949 M sucrose}		

TABLE 2–4. Concentrations of Ions in Body Fluids. (Values in mM/L except where stated otherwise.)
hl = hemolymph
SW = sea water
FW = fresh water
coel. fl. = coelomic fluid
vertebrate blood values are for plasma

	Na	*K*	*Ca*	*Mg*	*Cl*	SO_4	*Protein (mg/ml)*
Marine Invertebrates							
Coelenterates							
Aurelia mM/L H_2O	454	10.2	9.7	51.0	554	14.58	0.4
SW	459	9.8	10.1	52.5	538	26.55	
Echinoderms							
Strongylocentrotus[127]							
coel. fl.	452	14	10.1	50.9	511		
SW	458	9.5	10	51.1	519		
Echinus[192]	444	9.6	9.9	50.2	522	34.0	
Asterias[9]							
perivascular fl.	428	9.5	11.7	49.2	487	26.7	
ambulacral fl.	418	15.1	9.7	50.3	481	25.5	
SW	429	9.5	10.8	49	494	25.4	
Holothuria mM/L H_2O	489	10.7	11.0	58.5	573	28.4	0.7
Marthasterias[192]	459	10.8	10.1	51.2	540	26.5	0.6
Sipunculids							
Golfingia[192] mM/L H_2O	508	11.5	11.2	38.8	561	26.8	0.9
Annelids							
Aphrodite[192]	456	12.3	10.1	51.7	538	26.5	0.9
Arenicola[192]	459	10.1	10.0	52.4	537	24.4	0.2
Molluscs							
Mytilus[174] mM/L H_2O	502	12.5	12.5	55.6	585	29.4	1.5
Ostrea	544	14.7	10.9				
Strombus[133] hl (prosobranch)	496	10.9	11.0	58	558	20.5	
SW	457	10.1	10.3	54	554	28.6	
Scutus[235] hl	491	11.3	11.9	49		17.4	0.98
SW	487	10.9	11.6	48		17.8	
Aplysia[95] hl	492	9.7	13.3	49	543	28.2	
Sepia[192] mM/L H_2O	460	23.7	10.7	56.9	589	4.7	109.0
Sepia[196] hl	465	21.9	11.6	57.7	591	6.3	
SW	492	10.5	10.8	56.1	575	29.6	
Eledone[196] hl	432	14.4	11.2	54.2	516	20.6	
SW	438	9.3	9.6	50	512	26.4	
Loligo[196]	419	20.6	11.3	51.6	522	6.9	
Octopus[178]	371	10.3	8.2		447	18.0	
SW	407	9.1	8.9		475	24.5	

(Table 2–4 continued on opposite page.)

TABLE 2-4. Concentrations of Ions in Body Fluids. (Values in mM/L except where stated otherwise.)
hl = hemolymph
SW = sea water
FW = fresh water
coel. fl. = coelomic fluid
vertebrate blood values are for plasma
(Continued)

	Na	*K*	*Ca*	*Mg*	*Cl*	SO_4	*Protein (mg/ml)*
Crustaceans							
Palinurus[193] mM/L H_2O	545	10.3	13.4	16.6	557	20.5	
Palaemon[166]	394	7.7	12.6	12.6	430	2.6	
Squilla[129] hl	442	15	19.2	42	431		
Nephrops[193] hl	517	8.6	16.2	10.4	527	18.7	
SW	457	9.8	10.1	52.2	535	27.5	
Lithodes[193]	476	12.8	12.3	52.2	536	24.7	
Homarus	472	10	15.6	6.8	470		
Maja[192] mM/L H_2O	500	12.7	13.9	45.2	569	14.3	
Carcinus[205]	468	12.1	17.5	23.6	524		
Pachygrapsus[179]	465	12.1	11.4	29.2			
Carcinus[190]	525	12.7	14.3	21.2	502		
Gecarcinus[89] } on	459	9.3	20.9	13.5			
Coenobita[89] } sand	465	10.5	14.7	30.9			
Ocypode[77]	449	7.0	15	28.6	475	24.1	56
Eriocheir[50] serum	407	8.8	12.2	12.1	398	8.2	
Limulus[197] serum	445	11.8	9.9	46	514	14.1	
SW	440	9.4	9.7	50.2	514	26.5	
Paralithodes[142] hl	461	13.5	13	36.9	577		
SW	452	12.1		52	494		
Pandalus[142] hl	395	7.4	12.3	5.8	466		
SW	380	7.8	9.3	52	493		
Chionoectes[142] hl	403	11.1	8.1		495		
SW	420	11.4	10.4		482		
Fresh Water and Humid Air Invertebrates							
Molluscs							
Anodonta	13.9	0.28	11.0	0.31	12	(pH 7.5)	3.0
FW	0.48	0.059	2.7	0.375			
Anodonta[174]	15.55	0.487	8.4	0.19	11.7	0.73	
Viviparus[132]	34	1.2	5.7	1	31		
FW medium	2.5	0.2	3.0		8		
Otala[26]	70.7	3.4	7.1	5.7			
Helix pomatia[25]	5.1	3.7	10.0	12.6			
Helix aspersa[28]	67.9	2.86	6.2	3.6			
Arion[191]	61.5	2.7	2.3	5.7	52.5		
Arion[28]	60	2.7	6.9	5.4			
Limnaea[85]	57	1.8	4.9		43.9		
Annelids							
Lumbricus[14]							
coel. fl.	75.6	4.0	2.9		42.8		
blood	85.7	5.5	8.3		39.0		
Lumbricus[183]							
coel. fl.	83.2	12.4			46		
blood	105	8.9			43		
Hirudo[15] blood	125	4			36		
Heliodrilus[114]							
coel. fl.	78.6	5.8	3.6		35.6		
blood	88.5	7.9	13.4		35.9		
Pheretima[49] blood	42.8	6.7	3.6		54.3		0.13
Glossoscolex[49] blood	74.5	8.4	2.3		78.7		0.14
Nematodes							
Ascaris[96]	129	24.6	5.9	4.9	52.7		

(Table 2-4 continued on following page.)

TABLE 2–4. Concentrations of Ions in Body Fluids. (Values in mM/L except where stated otherwise.)
hl = hemolymph
SW = sea water
FW = fresh water
coel. fl. = coelomic fluid
vertebrate blood values are for plasma
(Continued)

	Na	*K*	*Ca*	*Mg*	*Cl*	SO_4	*Protein (mg/ml)*
Crustaceans							
Cambarus	146	3.9	8.1	4.3	139		
FW	0.65	0.01	2.0	0.21	0.48		
Asellus[134]	137	7.4			125		
Triops[104]	74	4.5	1.7	0.9	56.2		
Pacifastacus[117]	195	4.9	12.6		180		
Paratelphusa[181]	330	6.8	7.8	7.8	255		
Insects							
Aeschna[224] larva	145	9	3.8	3.6	110		
Aedes[222] larva	100	4.2			51.3		
Chironomus larva	104	2.1	10	15			
Dytiscus[224]	115	20			52		
Terrestrial Invertebrates							
Insects							
Diapherma	4.8	15.5	7.5	88.8	69.2		
Carausius[248]	13.3	15.4	8.6	86.1	68.6		
Carausius[233]	20.1	33.4	6.4	61.8			
Carausius[260]	15	18	7.5	53	101		
Carausius[10]	21	25					
Dixippus[182]	8.7	27.5	8.1	72.5	93		
Locusta[56a]	60	12	17	25			
Periplaneta	161	7.9	4.0	5.6	144		
Calliphora[56a]	140	26	21	34			
Gastrophilus larva	175	11.5	2.8	15.9	14.9		
Ephestia[56a]	33	33	41	51			
Samia pupa	2.6	42	9.4	32.6			
Bombyx[56a]	14	46	24	81			
Apis[56a]	11	31	18	21			
Antherea[247] *larva*	12	30	7	49			
Leptinotarsa[56a]	3.5	55	47	188			
Arachnids							
scorpion[164]	210	1.4	6.1		266		
scorpion[18]	262	7.9	2.4		270		
Capiennis[137a]	258	6.8	8.0		258		
Marine Fishes							
Cyclostomes							
Myxine[194]	558	9.6	6.25	19.4	576	6.6	
Myxine[75]	533	8.3			567		
Polistotrema[194]	450				500		
Elasmobranchs							
Narcine[202a]	134	7	12	3	159		
Rhinobatus[202a]	143	12.8	7.3	2	144		
Dasyatis[5]	238	17	8	3	238	Urea 1064	41.2
Coelacanths							
Latimeria[171]	181	51.3	6.9	28.7	199	Urea 355	

(Table 2–4 continued on opposite page.)

TABLE 2–4. Concentrations of Ions in Body Fluids. (Values in mM/L except where stated otherwise.)
hl = hemolymph
SW = sea water
FW = fresh water
coel. fl. = coelomic fluid
vertebrate blood values are for plasma
(Continued)

	Na	*K*	*Ca*	*Mg*	*Cl*	SO_4	*Protein (mg/ml)*
Teleosts							
Lophius[22]	185–200	5	6	5	153		
Corregonus[194]	141	2.8	2.7	1.7	117	2.29	41.9
mackerel	188	9.8			167		
barracuda	215	6.4			189		
Hemilepidotus[179a]	184	5.6	2.8	2.5	170		
Myoxocephalus[179a]	194	4.3	3.0	2.4	177		
Freshwater Fishes							
Cyclostomes							
Lampetra	116.4	7.5			95.1	2.7	36
Petromyzon[23]							
larva	103	3.4	2.4	1.6	91		
adult	139	6.2	2.4	1.9	113		
Elasmobranchs							
Pristis[227]	217	6.5	8.3	1.7	193		37
Carcharhinus[228]	246	6.4	8.9	3.2	219		30
Potamotrygon[227]	150	5.9	3.6	1.7	149		18
Teleosts							
Salmo	149	5.1			140.5		
Salmo[73]	144	6.0	5.3		151		
Oncorhynchus[150]	147	8.6	3.4	0.6	117		
Oncorhynchus[167]	176	3.0	3.4		172		
Cyprinus[106]	130	2.9	2.1		125		
Carassius[179a]	163	3.1			122		
Amphibians							
Necturus	99	2					
frog	104	2.5	2.0	1.2	74.3	1.9	
Reptiles and Birds							
Varanus[91]							
summer	181	3.4	3.2	1.1	148		
winter	131	4.3	3.6	1.8	135		
alligator[40]	141	3.8	2.1	1.4	112		51.7
Dipsosaurus[151]	168	3.4			120		
Anas[102]		3.1	2.3		103		
Gallus	154	6.0	5.6	2.3	122		
Mammals							
Tursiops[149]	166	4.2	2.2	0.92	116		75
horse[149]	150	3.0	5.5	0.9	100		75
dog[149]	147	4.5	10	1.8	108		70
dog[45a]	150	4.4	5.3	1.8	2.0		58
rat[16]	152	3.7	2.6	1.1	114		
rat	145	6.2	3.1	1.6	116		59
man	147	6.3	2.6	1.1	109		75

The ionic composition of extracellular body fluids varies with the animal's environmental conditions, nutrition, and stage in the life cycle. Selected values of ion concentrations in blood (plasma) and coelomic fluid are given in Table 2–4.

If for marine animals the differences between ionic concentrations of body fluid and sea water were due to protein binding, the values before and after dialysis against sea water should be the same. Protein binding of calcium is important in fluids high in protein content. However, in all marine animals the differences remaining after dialysis are so small compared with the ionic differences between normal blood and sea water that protein binding must be considered insignificant as a regulating mechanism.

Marine Invertebrates. Less regulation of inorganic ions in body fluids is found in echinoderms than in any other marine animals. Their coelomic fluid has slightly higher K^+ and much higher H^+ concentrations than sea water. In *Asterias,* potassium is at a higher concentration in the ambulacral system than it is in the perivisceral fluid; in the latter, calcium is more concentrated than in sea water.[9] The jelly of medusae (coelenterates) is essentially an extracellular fluid and has higher K^+ and lower $SO_4^=$ than sea water. The low content of sulfate, 40 to 85 per cent of SO_4 in sea water, imparts a low specific gravity so that the jellyfish can float. SO_4 extrusion may occur in cells lining the gastrovascular cavity.[141] The sipunculid worms maintain higher K^+ and lower Mg^{++} and $SO_4^=$ in body fluid than occur in sea water (Table 2–4). This is significant because some sipunculids have such low salt permeability as to appear semipermeable for many hours (p. 11). Prosobranch gastropods are poikilosmotic and cannot survive sea water at dilutions below 85 per cent. They have higher K^+ and Mg^{++} than sea water.[235] *Nereis diversicolor* is an osmoconformer, but at all salinities has the same ratio of Na and Cl in coelomic fluid as in the medium.[161a]

In marine pelecypods, the blood potassium and calcium are high relative to sea water; gastropods have higher K^+ and lower $SO_4^=$ concentrations. Cephalopods have low $SO_4^=$, high K^+, and unusually high NH_4^+ concentrations in body fluids.[36] Polychaetes tend to have elevated K^+ concentrations. Ascidians have low blood sulfate. The sipunculids, molluscs, and most polychaetes are osmoconformers, yet they have well developed excretory organs which function primarily in their limited ionic regulation (p. 14). This suggests that excretory organs originated as a means of ionic control.

Marine arthropods present a variety of degrees of ionic regulation (Table 2–4). Most of them accumulate potassium, while in a few, such as *Homarus,* potassium is as low as in sea water. Magnesium and sulfate are kept lower than in the medium. Highly active crustaceans—*Carcinus, Portunas, Pachygrapsus, Nephrops,* and *Palaemon*—have low Mg (14 to 18 per cent of that in sea water), whereas more sluggish crustaceans—*Maja, Hyas*—have higher magnesium concentrations.[193] Most marine crustaceans have higher Ca^{++} concentrations than does sea water, and *Homarus* is remarkable in having more Ca^{++} than K^+ in its blood. In brackish water, crustaceans maintain about the same proportion of ions as in sea water; calcium is high but varies with the molt cycle.

Freshwater Invertebrates. The ion concentrations in all freshwater animals are necessarily different from those in their media. Freshwater crustaceans have relatively high ion concentrations, and freshwater molluscs and coelenterates are the most dilute of multicellular animals. However, calcium is maintained at similar levels in both groups.

Insects. No group of animals presents such diversity in composition of blood ions, particularly with respect to sodium and potassium. The protection of the nervous system against the high K and low Na of hemolymph is discussed on page 479. The magnesium content is relatively low in some of those insects with high hemolymph potassium, and the ratio Ca/Mg increases in the order: herbivorous $<$ omnivorous $<$ carnivorous $<$ blood-feeding. The pH of the blood of some insects is between 6.4 and 6.8. Insect blood tends to have high amino acid and low Cl concentrations.

Vertebrates. Vertebrates show less variation in ionic ratios of blood than do many invertebrates (Table 2–4). The magnesium level is low, and the ratio (Na + K)/(Ca + Mg) is higher in vertebrate bloods than in body fluids of invertebrates. In marine vertebrates every ion is regulated. The hagfishes are isosmotic with sea water, and NaCl accounts for 88 per cent of osmotic solutes, but they show regulation of all ions. The freshwater lampreys resemble freshwater teleosts in ionic regulation. Marine teleosts tend to have higher Mg^{++} and Ca^{++} concentrations than do freshwater fish. Elasmobranchs resemble teleosts in ionic composition, but in addition they maintain high urea concentrations (p. 47). The coelacanth *Latimeria* is high in Mg and urea.[171]

REGULATION OF IONS IN RELATION TO ECOLOGY AND LIFE CYCLE

In crustaceans, calcium concentration in hemolymph is related to stage in the molt cycle. In premolt stages, calcium is stored in the hepatopancreas, and blood levels of calcium are high. In *Homarus,* for example, plasma calcium was 40 mM in the premolt stage and 13 mM postmolt; hepatopancreas Ca was in mg/g_{dw}: 85 in early premolt, 285 in late premolt, 326 in newly molted and 65 to 95 in intermolt. Mg and phosphate, like Ca, are stored in the hepatopancreas in the premolt stage.[80] In *Carcinus* as molt is approached the blood Ca increases from 126% to 164% of the level in blood in sea water; Mg increases from 37.1% to 55.5%. After molt, water equal to 66 per cent of the premolt weight is absorbed.[193] A softshelled *Pachygrapsus* shows Na exchange four times greater than the exchange in intermolt.[202]

As mentioned on page 30, when euryhaline crustaceans enter dilute water, they begin active absorption of ions in relation to blood sodium concentration. In *Carcinus,* active uptake of sodium increases abruptly when Na_i falls to 400 mM; a 13-fold increase in uptake occurs for a 10 to 30 mM decrease in Na_i.[205] In the euryhaline *Gammarus duebeni,* Na uptake is saturated at 10 mM Na_o, whereas the external concentration of Na for 50% saturation of the Na transport in *Eriocheir* is 1 mM, for *Astacus* it is 0.25 mM, and for freshwater *Gammarus pulex* it is 0.15 mM.[159, 225] Uptake in fresh water is decreased by addition of 2 to 10 μM Ca.

When a freshwater flatworm (*Dugesia*) was compared with a marine flatworm (*Bdelloura*), the exchange of Na and of K was faster for *Bdelloura*; Na and K effluxes in pM/cm²/sec were:[219]

	Na	*K*
Bdelloura	14.5	1.26
Dugesia	0.006	0.015

Disproportionate changes in concentrations of various ions occur when marine animals are placed in different dilutions of sea water or in artificial sea water in which not all ions are altered in proportion. For example, when the holothurian *Caudina chilensis* was placed in artificial sea water with ions changed in disproportionate amounts, within five days the coelomic fluid conformed in all respects to the new medium.[124] K in the animal increased by three times, and Ca by 2.5 times; Cl, Na, and SO_4 decreased. In *Golfingia* the content of Cl in the muscles increased more than did the K content when these elements had been increased proportionately in the medium. In *Carcinus* transferred to 2/3 sea water, all ions decreased; Mg and SO_4 decreased more than did Na, K, and Cl, so that the ratio (In/Out) for Mg and SO_4 decreased while this ratio increased for the other ions.[243] In *Eriocheir* transferred from fresh water to sea water, body fluid Na and Cl increased more than did K and Ca. *Pachygrapsus* and *Uca* in hyperosmotic sea water increased serum Mg more than serum Na and K. It is concluded that in different salinities, blood Mg and SO_4 are more variable and Na, K, and Cl are better regulated.

In vertebrates the ionic ratios tend to be more fixed than in animals with open circulatory systems. Euryhaline fish show some changes in ionic regulation associated with transitions between fresh water and sea water:

	Na in plasma	*Na space*
Eel		
in S.W.	139 mEq	32 ml/100 g
in F.W.	122	24
Flounder		
in S.W.		34
in F.W.		32

Coho salmon (*Oncorhynchus kisutch*) in downstream migration to the sea show a transient increase in plasma sodium and magnesium. Drinking begins after 12 hours in sea water, enhanced renal excretion of divalent ions begins after 24 hours, and by 36 hours all blood ions are at steady state at slightly higher levels than in fresh water.[150] However, in sea water the increase in Cl^- is proportionately less than the increase in Na^+; hence organic ions are important.[167] Similarly, in eels transferred from F.W. to S.W. plasma sodium and chloride increase in concentration, while potassium increases hardly at all.

Rana esculenta	*In Fresh Water*			*In 0.8% NaCl*			*In Ringer*		
	Na	K	mOsm	Na	K	mOsm	Na	K	mOsm
plasma mM	92	2.7	210	119	2.9	260	105	2.8	249
urine mM	4.6	1.6	25.9	56	1.2	168	39	2.0	113

Eels' gills secrete NaCl in S.W. and absorb it in F.W. The gills have a dual circulation—one path traverses lamellae which are opened by epinephrine and high O_2 demand, whereas the other path or shunt is opened by acetylcholine; the latter path bypasses respiratory demand, and functions primarily for ion exchange.[121a]

At low temperatures freshwater fish show a small decrease in serum ions; under the same conditions marine fish show an increase. For example, the figures for *Fundulus* in fresh water are:[237]

Temperature (°C)	*Na (mM)*	*Cl (mM)*
0.1	114	76
4.0	127	85.5
11.0	162	130

In the dogfish *Squalus* in sea water, the ionic gradient is inward and virtually the entire influx is across the gills, being 0.9 to 1 mM Na/kg/hr.[24, 105]

In amphibians in a dilute salt solution, the blood sodium increases disproportionately to the potassium concentration, as shown for *Rana esculenta* (see table above).[146]

Changes in concentrations of blood ions occur in hibernation. A lizard *Varanus* in winter hibernation has in its blood 55 per cent more magnesium than during its active phase.[91] Hibernating mammals (hedgehogs, marmots, bats) also have elevated blood magnesium levels.[188]

Mammals and birds tend by behavior to regulate salt intake according to blood levels. Tropical birds can distinguish 0.05 M NaCl, but not 0.03 M NaCl, from water.[29] Sodium deficient animals show an increased appetite for salt. Thirsty rats drink more of water than of salt solution, but normally hydrated rats select isosmotic saline. Sheep depleted of $NaHCO_3$ by parotid fistula have increased appetite for sodium salts; central receptors together with taste receptors coordinate intake so that blood levels are regulated.[51]

Some insects show a relation of blood ions to the life cycle and diet. Blood of some Lepidopteran larvae contains more Mg and Cl, and less K and phosphate, than blood of pupae. Cockroach hemolymph shows higher Ca and lower K on a milk powder diet than on wheat flour or lettuce:[170]

diet	*Na*	*K*	*Ca*
lettuce	119	28	6.2
wheat flour	183	30	5.9
milk powder	122	21	8.7

On the contrary, in a series of aquatic insects no correlation was found between diet and hemolymph ratio of Na/K.[224]

Among insects the Na/K ratio in hemolymph is highest in blood-feeding species (11 to 27), high (1 to 18) in carnivorous species, and low (<1) in herbivorous insects. A silkworm has in its body fluids about the same sodium concentration as the mulberry leaves on which it feeds, but on pupation it loses practically all of its sodium; the pupal tissues contain much potassium and negligible sodium. In insects, both genetic and dietary factors control the blood ion concentrations.

EXCRETION

It is less wasteful of energy for an excretory organ to reabsorb a substance which has been filtered to the same concentration as in the blood than it is to absorb this substance from a more dilute medium. Non-utilizable ions which had entered the animal by diffusion or feeding may be concentrated in excretory fluid. Kidneys evolved in marine animals which are isosmotic with their medium but which regulate specific ions. Coelenterates and echinoderms lack special excretory organs, and they show the least ionic regulation. It is reasonable to assume that in annelids, molluscs, and arthropods kidneys arose and still function in ionic regulation; on migration to fresh water, kidneys took over osmotic regulation secondarily. In some crustaceans (e.g., *Eriocheir* in fresh water) the kidneys put out isosmotic urine, and hence excrete salt along with water which enters osmotically. In fishes, kidneys were early used in water regulation; on migration to the sea, the vertebrate kidneys were inadequate for ionic regulation and extrarenal routes were developed for Na^+ and Cl^-, leaving the renal route for Mg^{++} and $SO_4^{=}$. The function of excretory organs in eliminating nitrogenous wastes was a very late development in excretory function. No direct evolutionary series exists from contractile vacuoles, nephridia, antennary glands, and flame cells to glomerular nephric units. However, the basic principles of filtration, reabsorption, and secretion have been used by quite different sorts of kidney, and the presence of filtration-reabsorption is so widespread that it gives no clue to ecological origin of a group of animals.

Evidence for active excretion of selected ions is given by the ratios of urine to plasma

TABLE 2–5. Excretion Data: Urine/Plasma (u/p) Ratios of Ions.

Animal	*Na*	*K*	*Ca*	*Mg*	*Cl*	SO_4
MARINE						
Invertebrates						
Osmoconformers						
Strombus[133]	0.96	1.22	1.4	1.0	1.0	1.06
Octopus[178]	0.98	1.3	0.53		0.9	1.8
Sepia[192]	0.79	0.5	0.7	0.68	1.0	2.15
Homarus[192]	0.99	0.91	0.64	1.8	1.0	1.59
Nephrops[192]	0.98	0.83	0.81	1.3	1.0	1.1
Maja[192]	0.99	0.98	0.99	1.1	0.98	2.14
Hyperosmotic regulators						
Carcinus[243]	0.95	0.78	0.94	3.9	0.98	2.24
Carcinus[190]	0.92	0.98	1.2	4.2	1.02	
Pachygrapsus[90]	0.78	0.82	1.17	13.6		
Uca[88]	0.65	1.87	0.64	5.32		
Hemigrapsus[88]	1.14	1.69	1.31	2.43		
Gecarcinus[89] S.W.	1.08	1.19	0.35	3.08		
F.W.	1.06	1.03	1.01	2.43		
Ocypode[77]	0.85	1.4	0.8	1.13	0.92	0.79
Eriocheir[50] S.W.		1.15	0.93	2.17	1.14	
S.W.	0.95	1.3	0.70	3.9	0.97	5.6
Fish						
Cyclostome						
Myxine[152]	0.95	0.89	1.6	1.13	1.02	2.2
Teleosts						
Hydrolagus[185]	0.54	2.3	3.9	8.7	0.87	50.6
Sygnathus	0.41	0.24	1.85	10.75	1.70	
Lophius[22]	0.94	0.4	1.0–3.3	62.	1.1	
Anguilla (eel)[204]						
S.W.	0.11	0.75			0.77	
Muraena[204]	0.58	0.12	3.6	26.3	1.75	
FRESH WATER						
Invertebrates						
Eriocheir[50] F.W.		1.39	0.47	0.22	0.95	
F.W.	1.07	1.14	0.42	0.49	1.05	0.97
crayfish[189] F.W.	0.092	0.176				
F.W.	0.46	0.14	0.09	0.45	0.082	
leech[15]	0.04					
Fish						
Anguilla (eel)[204]						
F.W.	0.042	0.24				
Salmo[73]	0.065	0.2	0.25		0.067	
Salmo[103]						
pre-smolt	0.62	1.49			0.06	
smolt	0.09	1.45			0.087	
FRESH WATER AND TERRESTRIAL						
Rana[146]	0.05	0.6				
Ambystoma[52]	0.055	0.11			0.018	
Alligator[40]	0.0004	0.1	0.006	0.01	0.0005	
Anas[102]	0.062	12.	0.52		0.10	
man	1.16	7.5	1.87	2.4	1.63	

concentrations (u/p ratios), and a few selected values are given in Table 2–5. The fact that ratios are different for various ions indicates that each ion is processed separately. A u/p ratio less than unity indicates reabsorption of this ion from filtrate; a u/p ratio greater than unity indicates either water reabsorption or secretion of the ion outward. The ratios are subject to variation with diet and life cycle, but they do indicate the efficiency of the kidneys. Inulin is a substance which is commonly filtered and not reabsorbed; hence, an increase in its concentration may be taken as a measure of water reabsorption.

In marine crustaceans the kidney (antennary gland) is important for retention of potassium and calcium and for elimination of magnesium and sulfate. In *Pachygrapsus* and *Uca* kept in hyperosmotic sea water the urine concentrations of Mg and SO_4 increase much more than do those of K, Ca, and Cl; Na output actually decreases. Extrarenal routes of excretion—gills and gut—are probably important, especially for sodium. Some dyes injected into a lobster appear in the stomach, from which they are regurgitated; the role of the crustacean digestive tract in excretion is poorly known.[89, 179] When a series of crabs that are increasingly adapted to a terrestrial environment is exposed to water of increasing salinity, they show a declining u/p ratio for Na through the series and an increasing ratio for Mg. In *Pachygrapsus,* for example, u/p for inulin was 1.11 in S.W. and 1.16 in 158% S.W.; u/p for Mg increased from 8.7 in S.W. to 13.3 in 158% S.W.[90] It appears that in the higher salinity, some calcium and sodium are reabsorbed, but magnesium may be secreted into the urine. The response seems to be to salinity and not specifically to the Mg in the medium. In *Carcinus,* u/p for Mg increased from 4.5 to 10 after four days in moist air.[190] In *Hemigrapsus* the unique processing of Mg supports the possibility of secretion.[48] The largest proportionate increase in urine Mg with increasing osmotic concentration seems to be in grapsoid and similar crabs which spend much time in air. However, in the land crab *Gecarcinus* in S.W., the u/p for Mg is only 1.74, and in hypersaline medium it fails to increase; in *Ocypode* also u/p for Mg is low.[77] Hence hyposmotic regulation, Mg retention, and habitation of land are separate functions; there are marked species differences.

Evidence for filtration and reabsorption of ions in the tubule of the freshwater crayfish has been presented (p. 31). Inulin u/p increased, in going from coelomosac to distal tubule, from 1.2 to 1.8; potassium increased from 7.4 to 11.2 mM. These shifts indicate water reabsorption. Sodium decreased from 242 to 183 mM; hence Na (and Cl) is reabsorbed. Sodium and potassium are absorbed in the bladder.[189]

In molluscs, urine formation may be by filtration across the heart wall into the pericardium, from which fluid enters the kidney where reabsorption and secretion may occur. In a marine snail, *Strombus,* urine is produced at a rate of 3 ml/kg/hr, and pericardial fluid is lower in protein and Ca and higher in K than hemolymph; extrarenal excretion of SO_4 is indicated.[133] In a freshwater snail, *Vivaparus,* urine is formed at 15 to 50 ml/kg/hr, more when blood pressure is raised; final urine is more dilute than pericardial fluid, indicating salt reabsorption.[132] In the land snail *Achatina* the inulin u/p is 2.8; after poisoning with dinitrophenol the ratio becomes 1.0, and hence water is normally reabsorbed. Glucose and chloride are also reabsorbed; phenol red and p-aminohippuric acid (PAH) are secreted.[144b] Similarly, in abalone (*Haliotis*) inulin is filtered and α-amino hippurate is secreted.[92] In *Octopus,* filtration occurs in the branchial hearts, but the kidney secretes K, SO_4, Cu, Zn, NH_4, PAH, and dyes. Filtration from the branchial hearts is proportional to the difference between systolic and diastolic pressures down to 4 cm H_2O, which balances the blood colloid osmotic pressure of 4.3 to 5 cm H_2O.[178]

Annelids have nephridia which receive coelomic fluid at the ciliated nephrostome. The function of the nephridia in earthworm and leech is discussed on p. 37. Some representative values in two annelids are as follows:[14, 15]

Lumbricus	*coelomic fluid*	*urine*
O.C.	182 mOsm	34 mOsm
Cl	46 mM	3.5 mM
NH_4	0.0012 mM	0.0019 mM
urea	0.00099 mM	0.0012 mM
Hirudo	*blood*	*urine*
O.C.	202 mOsm	15 mOsm
Na	125 mM	6 mM
K	4 mM	
Cl	36 mM	6 mM

Marine annelids have well developed nephridia and are osmoconformers with limited ionic regulation. Ionic regulation by the nephridia has not been much investigated.

A generalized nephron of a vertebrate is described on page 57; its function in producing hyposmotic urine is discussed on page 57, and its function in producing hyperosmotic urine is discussed on page 65. The role of the vertebrate kidney in ion regulation is shown in the marine hagfish *Myxine*. Urine Cl approximates plasma Cl; Na is reabsorbed, but urine Ca and Mg are slightly higher than in plasma; and urine SO_4 is much higher than in plasma.[152] Similarly, a freshwater cyclostome *Lampetra* retains potassium relatively more than sodium. Migratory lampreys in 50% S.W. drink up to 99 ml/kg/day; they absorb Na, K, and Cl for mainly extrarenal excretion; and they eliminate much Ca, Mg, and SO_4 by both gut and kidney. The u/p ratios for Mg and SO_4 are 11 and 12.9 respectively, compared to 0.42 for Na.[23, 153] Rectal glands excrete much NaCl in the dogfish *Squalus* as follows: total efflux of Na 0.87 to 0.92 mM/kg/hr, renal loss 0.1 to 0.2; hence only 10 to 30 per cent of efflux is by kidney.[24]

In the glomerular kidney of *Salmo* there is much more absorption of sodium and chloride than of potassium and calcium; chloride probably moves passively.[73] Aglomerular kidneys of marine fish form urine entirely by secretion. As plasma levels of various ions are raised, transport maxima are reached for excretion of Mg, thiosulfate, SO_4, etc. Elevated Mg in plasma depresses Ca excretion, and high thiosulfate in plasma diminishes SO_4 excretion; hence cations and anions are excreted by separate transport systems. Neither is influenced by the drug benemid, which diminishes excretion of PAH. The aglomerular fish *Opsanus tau* survives in dilute sea water and even in fresh water. Urine flow in fresh water is five times that in S.W. or in 50% S.W.; some inulin is excreted despite absence of a filtration mechanism. Chloride is absorbed from the bladder as shown in the following data:[126]

	medium	*F.W.*	*½ S.W.*	*S.W.*
Cl (mEq)	plasma	94.2	159	159
	ureteral urine	38	104	169
	bladder urine	0	7	40

ACTIVE TRANSPORT OF IONS

Ions may move in and out of cells passively according to the permeability, as determined by the size and charge of pores in the plasma membrane. Passive movement may be by kinetic exchange between ions of the same kind in and out ("exchange diffusion"). This cannot cause concentration changes, and hence plays no role in ionic regulation. Passive movement may also be by diffusion or leakage from higher to lower concentration or through an electrical potential gradient (cf. p. 3, Chap. 1). The maintenance of an ionic steady state in the presence of diffusive leaks requires an equivalent active transport against chemical and electrical gradients. Active movement is indicated if the flux in or out is greater than can be accounted for by the electrochemical gradient, i.e., if the ions are not in Donnan equilibrium. Such transfer requires metabolic energy (see p. 5).

There are two situations in which active ion transport is required: regulation of intracellular ionic concentrations when the extracellular fluids have a very different ionic composition, and regulation of stable body fluid concentrations when the external environment has a different pattern. Some aspects of the second problem were discussed earlier in connection with osmotic regulation. Thus, one requirement for invasion of fresh water was development of active ion uptake mechanisms in the body surface. Conversely, maintenance of blood hypotonicity in concentrated media and conservation of water on land require renal and extrarenal mechanisms for active extrusion of ions. It is beyond the scope of this chapter to discuss the detailed physical biochemistry of ion transport, but a comparative survey suggests variations on a common theme.

Cellular Ionic Regulation. Most cells retain high concentrations of potassium and exclude sodium, as noted earlier. Tracer measurements show that concentration differences are not due to membrane impermeability, and since the gradients of the two cations are in opposite directions the distributions cannot be potential-dependent; hence active transport must be involved. The characteristics of the membrane ion pump can be shown in the following experiments. Muscle cells, normally high in potassium and low in sodium, lose K and gain Na (by diffusion) when stored in the cold; when they are rewarmed, they restore the normal gradients by extruding Na and pumping K in. Coupling of the two fluxes is obligatory. If K is omitted from the bathing solution, Na extrusion from the muscle stops. Exactly the same thing happens in most mammalian red blood cells. Sodium extrusion (and K influx) is associated with ATP splitting and is inhibited by cardiac glycosides (digitalis, ouabain). The pump is saturatable; K_m for external K is 1 to 2 mM, and for intracellular Na it is 10 to 20 mM. In stored human red cells the exchange is 3 Na for 2 K ions,[173] and in the squid giant axon a similar ratio is seen.[97]

The energy requirement for transport varies with the tissue and conditions. In crab nerve, 1.9 to 3.4 Na ions are extruded per ATP split. In red cells the ratio is about 3 ions per split.[74, 173, 250, 251] In ground squirrel and hamster kidney slices, the ratio of K uptake to oxygen consumed varies from 17 to 66 (K/ATP = 3 to 12). Ouabain reduces O_2 consumption from 16 to 10 ml/mg dry wt./hr. Removal of Na reduces it by a similar amount; hence respiration for active transport is 35 to 45 per cent of the total.[257] Although O_2 consumption is commonly used as an index of the energetic requirement, Na-K transport is clearly dependent on intracellular ATP, as shown for red cells[98, 99] and squid axons.[97]

The enzyme responsible for transport-related ATP hydrolysis was first isolated by Skou[211, 212] for crab nerves; it has since been isolated from nearly every animal tissue examined, as well as from a few plant cells.[11] Its characteristics parallel almost exactly those of the ion pump. It is found in the membrane fraction of cell homogenates, requires Mg plus both Na and K for activity, and is inhibited by cardiac glycosides, oligomysin, and fluoride (as is coupled Na extrusion and K accumulation). In many cell systems the ratio active Na flux/total ATPase activity is 2/3, similar to the ratio of transport to ATP split.[12] The enzyme is clearly implicated in the energetic demands of transport; whether it plays a more direct role in translocating the ions is still speculative.[212]

If Na extrusion is exactly balanced by K influx, no charge transfer occurs, but if the movements are unequal the pump may be electrogenic (i.e., contribute directly to the membrane potential). The 3 Na-2 K coupling in human erythrocytes may represent such a case. In a snail ganglion cell, injection of Na at the rate of 4.4 mM/min stimulated the Na pump and increased the resting potential by 30 mV in 10 minutes. The hyperpolarization was blocked by ouabain or by removing K from the external bathing solution.[116]

Epithelia: Inward-Directed Transport. A number of complex multicellular membranes —skins, gills, kidney tubules, urinary bladders —actively transport ions from one face to the other, thereby playing a role in the regulation of body fluid composition. This process involves moving an ion not simply between the cell interior and the extracellular fluid, but across the cell, from a fluid bathing one surface (e.g., the external environment) to one bathing the other surface (e.g., blood). The membranes of these epithelial cells must have very different characteristics on the two sides. But transport shows some characteristics similar to those of the Na-K pump described for symmetrical cells.

Freshwater animals have the capacity of moving NaCl from a dilute environment into a more concentrated blood. Movements of Na and Cl are independent,[125] and hence must occur by ion exchange to preserve electrical neutrality in the bathing solutions. Excretion of ammonia in the goldfish *Carassius* in tap water is roughly equal to Na absorption;[143] the same is true in *Ambystoma tigrinum* larvae.[55] Injection of NH_4 into the fish *Carassius* increases Na influx; addition of NH_4 to the medium depresses it. Injection of HCO_3 stimulates Cl intake, while HCO_3 in the medium depresses it. Inhibition of carbonic anhydrase inhibits both Na and Cl movements. Similar results are obtained with ammonium ion in crayfish.[206] It was proposed that NH_4 exchanges for Na and HCO_3 for Cl as in Figure 2–2. However, in the trout *Salmo gairdneri*, NH_4 excretion is not always equivalent to Na uptake and does not vary when Na influx changes. Instead, H^+ appears to be the cation to which sodium influx is coupled. This was also clearly shown in the frog *Calytocephalella,* which is ureotelic and excretes little NH_3. Na uptake is exactly equivalent to H^+ excretion across the skin. Procaine blocks Na transport, and also H^+ output. Cl is pumped independently. If a frog is preadapted to NaCl, the Na and Cl uptakes are equal; if the frog is selectively depleted of Cl, then in NaCl solution the Cl pump is more active, while if the frog has been Na depleted the Na pump predominates. Cl uptake is by exchange with HCO_3 and can be inhibited by pentobarbiturate.[199] Larvae of the mosquito *Aedes aegypti* also have

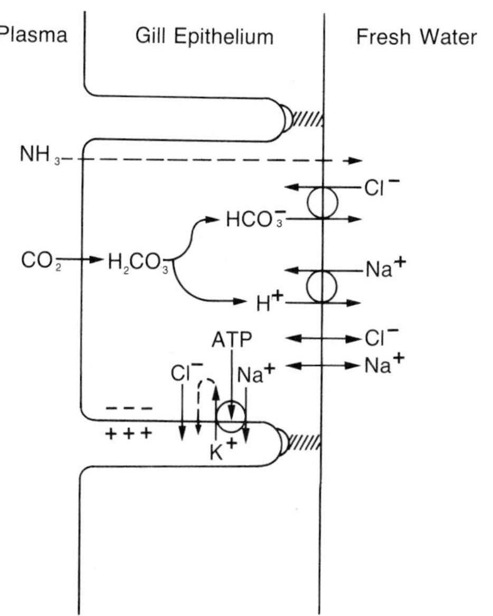

Figure 2–2. Schema of active absorption of sodium and chloride by gill of freshwater fish. Uptake of Na^+ in exchange for H^+, and of Cl^- in exchange for HCO_3^-, are indicated. Passive inward leak of Na and Cl may be important in high environmental concentrations of NaCl. In addition, a Na–K pump is indicated as maintaining appropriate Na and K gradients between cell and plasma.

independent ion transport systems. In Na_2SO_4 solution, uptake of Na is balanced 33% by loss of K, 49% by H^+ excretion, and 18% by SO_4 uptake. In KCl solution the larvae absorb Cl, which is balanced 36% by exchange for an unknown anion, about 41% by HCO_3 (and possibly by OH), and 23% by K influx.[223]

The transport systems are saturatable, with K_m values for Na^+ as given in Table 2–6.[86a] Ion transfer is often accompanied by the generation of potential differences across the transport epithelium. The latter are dependent on sodium concentration in the external medium in frogs[21] and in salamanders,[55] but no clear relation to Cl transport has been observed in vivo. The colon of *Bufo marinus* transports Na from mucosa to serosa; the permeability on the mucosal side limits the Na available to the serosal pump; aldosterone increases the permeability to Na, and hence increases the intracellular Na pool.[37]

Several in vitro epithelial preparations have been used to investigate transport mechanisms. In isolated frog skin, with Ringer solution on each side, the inner face is electrically positive with respect to the

TABLE 2–6. The Affinities for Sodium (K_m) of the Sodium Uptake Systems of Different Animals.

Animal	*Class*	*Affinity mM Na*	*Habitat Preference*	*Reference*
Carcinus maenas	Crustacea	20	marine to brackish estuaries	205
Marinogammarus finmarchius	Crustacea	6–10	marine to brackish estuaries	225
Fundulus heteroclitus	Osteichthyes	2.0	brackish, does not survive well in fresh water	176
Gammarus duebeni (brackish race)	Crustacea	1.5–2.0	brackish water	207
Gammarus tigrinus	Crustacea	1–1.5	brackish water	225
Gammarus zaddachi	Crustacea	1–1.5	brackish water	225
Lumbricus terrestris	Oligochaeta	1.3	moist earth	54
Bufo americanus	Amphibia	>1	terrestrial	86
Eriochier sinensis	Crustacea	1.0	breeds in sea water; matures in fresh water	205
Salmo gairdneri (gills)	Osteichthyes	0.5	fresh water, can live in sea water	118
Gammarus duebeni (freshwater race)	Crustacea	0.4–0.5	fresh water	225
Rana cancrivora	Amphibia	0.4	fresh water to sea water	86
Ambystoma gracile	Amphibia	0.3–0.5	freshwater larvae of terrestrial form	2
Limnea stagnalis	Gastropoda	0.25	fresh water	85
Astacus pallipes	Crustacea	0.25	fresh water	206
Rana pipiens	Amphibia	0.2	aquatic to semi-terrestrial	86
Amphiuma	Amphibia	~0.2	fresh water	86
Gammarus pulex	Crustacea	0.15	fresh water	207
Gammarus lacustris	Crustacea	0.1–0.15	fresh water	207
Lampetra planeri	Agnatha	0.13–0.26	fresh water; after metamorphosis will migrate to the ocean	153
Carassius auratus	Osteichthyes	<0.05	fresh water	143
Xenopus laevis	Amphibia	<0.05	fresh water	86

outer face. The potential can be opposed by an external battery, and the short-circuit current (SCC) needed to balance out the skin potential can be measured. The SCC is equal to net inward flux of Na^+ as measured with tracers, and hence is a simple electrical measure of sodium transport.[17] Transport is saturable[120, 121, 215] with a K_m of about 10 mM, and can be stopped by metabolic inhibitors. High calcium concentrations decrease the SCC by reducing permeability of the outer face to Na,[41, 42, 187] and the renal diuretic amiloride also stops sodium entry at the outer membranes. The SCC requires K in the inner bathing solution and is abolished by ouabain, suggesting that the transport ATPase (see p. 94) is involved in the inner membranes. A membrane-bound ATPase has been demonstrated on the lateral and inward-facing membranes of skin cell in all layers of the epidermis.[33] Tadpole skin has much less ATPase than does adult skin, and it generates neither a potential difference nor SCC; instead, the transport system, including ATPase, is in the gills.[13] About 18 Na ions are transported per O_2 molecule consumed, or roughly 3 Na/ATP.[264] A bullfrog tadpole absorbs Na and Cl independently; the body fluid potential reverses from − to + at the time of appearance of the skin potential late in larval development.[2]

If chloride concentration is high on the outside of isolated skin, as in Ringer, influx of Cl is passive. In contrast, chloride uptake is definitely active in intact frogs, and if dilute KCl, instead of Ringer, bathes the outside of isolated skin, active Cl transport can be demonstrated.[145] The isolated skin of the South American frog *Leptodactylus* transports Cl even when bathed on both sides by Ringer solution.[262]

The toad urinary bladder behaves in vitro like the frog skin. When bathed on both sides with Ringer it has a potential of some 50 mV, serosa positive. Ion substitution shows that the asymmetry potential across the single layer of epithelial cells requires external Na, and net Na movement from mucosa to serosa equals SCC.[65, 70, 71, 203] About 16 Na ions are transported per O_2 consumed. The transport of Na stops if K is omitted from the serosal side, and the potential reverses. It was postulated that K acts by increasing P_{Na} on the mucosal side. However, amphotericin B increases mucosal P_{Na} in K-free medium without enhancing Na transport; addition of K then increases Na transport. Hence K is required for operation of the serosal pump.[65] As in the frog skin, SCC is inhibited by ouabain and by metabolic inhibitors.

Chloride transport is easier to demonstrate in turtle bladder. The net Na transfer exceeds SCC by 44 μa; the difference represents an anion current due to Cl and HCO_3.[83] Thus, Na and Cl are actively transported from the mucosal to the serosal side, and mucosal acidification may be due to HCO_3 absorption.[20] Na transport depends on oxidative metabolism; about 20 Na ions are transported per O_2 consumed.[123, 130]

Isolated tubules of the kidney of *Necturus* reabsorb salts from the lumen, and transport can be studied by analysis of injected drops sealed at the tubular ends with oil. A potential of 20 mV, lumen negative, exists across the tubule when lumenal and serosal fluids are both Ringer (Fig. 2–3). The interior of the tubular cells is negative with respect to both bathing solutions, by 73 mV to the lumen and by 53 mV to the interstitium; the transtubular potential is the sum of these membrane potentials. Electrical asymmetry is lost if the lumenal Na is replaced by K or choline, or if ouabain is added to the preparation. A sodium-potassium-coupled pump is postulated at the peritubular (serosal) face. Short circuiting gives a net ion current of 4.8×10^{-6}

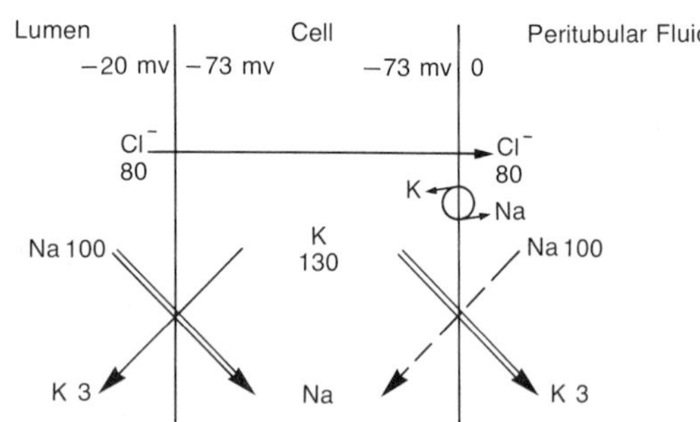

Fluxes (pM/cm² sec)

	lumen → cell	cell → peritubular fluid
Na	308	132
K	57	378
	lumen ← cell	**cell ← peritubular fluid**
Na	246	89
K	57	378

Figure 2–3. Schema of potentials, ionic gradients, concentrations, and fluxes across proximal tubule of *Necturus* kidney. (Modified from Whittembury, G., N. Sugino, and A. K. Solomon, J. Gen. Physiol. *44*:689–712, 1961; and from Solomon, A. K., Amer. J. Physiol. *204*:381–386, 1963.)

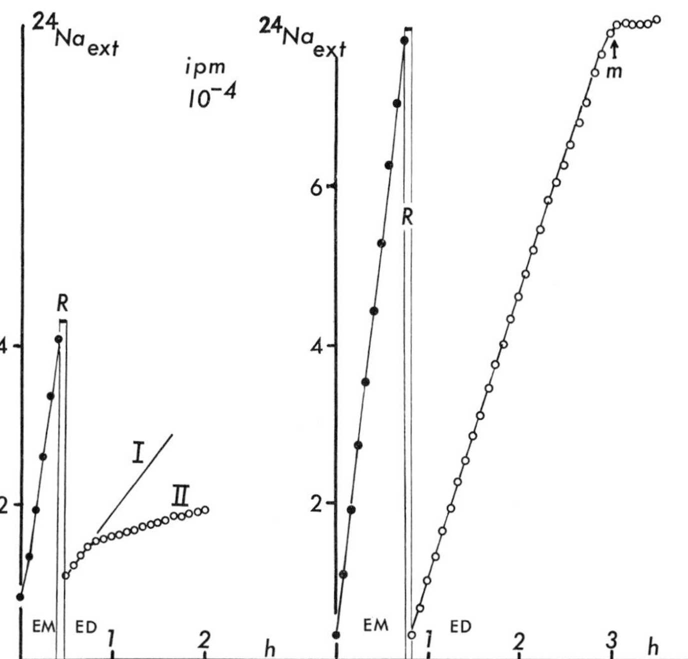

Figure 2–4. Efflux of ^{24}Na from euryhaline flounder *Platichthys* (on left) and stenohaline *Serranus* (on right) in sea water (EM) and after direct transfer to fresh water (ED). Time in hours. Slopes give rates of efflux. Flounder shows a two-step drop in Na efflux; *Serranus* shows a one-step drop. (From Motais, R., F. G. Romeu, and J. Maetz, C. R. Acad. Sci. Paris *261*:801–804, 1965.)

amp/cm^2 or 5.0×10^{-11} equiv/cm^2 sec. Fluxes across mucosal and serosal surfaces are shown in Figure 2–3.

In surface epithelia such as skins, gills, and urinary bladders the combined data suggest a common transport system, and the model shown in Figure 2–4 would encompass many of the in vivo and in vitro observations. The external membrane has a very low K permeability. Entry of Na and Cl may be by diffusive leak or by exchange; in dilute media the latter predominates, and in concentrated solutions the former is more important. Carbonic anhydrase may (in fish gills) or may not (in frog skin) be necessary to ensure adequate rates of CO_2 hydration for generation of H^+ and HCO_3^-. Sodium that has entered the cell is extruded by the coupled Na-K pump, perhaps into lateral intercellular channels; anion movement into the channel may be passive, and water must move into the channel by osmosis. Thus, a salt solution continually enters the channel and moves through into the inside bathing medium. Ouabain sensitivity of transport and the K requirement in the inner solution suggest that the inner step in Na movement depends on the same coupled Na-K transport system described for cells in tissues like muscle and nerve. The exact manner in which these ion movements generate a potential across the epithelium is not well understood; there is not even agreement on whether, in the frog skin, it rises in two[61] or possibly more[7] discrete steps, or whether it rises gradually on going from outside to inside.[35]

Preparations from invertebrates have been much less well investigated; in vivo studies indicate that transport mechanisms in surface epithelia (e.g., crayfish gills) and excretory tubules resemble those in vertebrates. Isolated gills of crayfish are negative inside, and active inward transport of Cl is indicated. The Cl transport system is saturated in 0.2% NaCl.[8] There are several reports of inhibition by anticholinesterases of Na uptake by freshwater crustacean gills, and also from the urinary bladder. These inhibitors have been shown to stop sodium transport across frog skin also.

Gills of the land crab *Cardisoma* have an active Na-K ATPase which is inhibited by ouabain; O_2 consumption by excised gill is maximal in 75% S.W. In diluted S.W. these crabs regulate hyperosmotically, although in higher salinities they conform.[180]

In the midgut of caterpillars of the moth *Hyalophora cecropia* potassium is transported from blood to gut lumen.[3, 94] The intestinal lumen is positive by 84 mV, and under quasi-physiological conditions all of the current (SCC) can be accounted for by K transport; the potential is insensitive to Na, and Rb can be transported as well as K. In a medium low or lacking in K, Na can be transported; when concentrations are equal only K is transported. The system is indifferent to Ca and Mg, and ouabain has no effect on transport of either Na or K. However, transport is inhibited by anoxia or DNP. The gut epithelium is columnar with basal infoldings, between which are long mitochondria and apical microvilli projecting into the lumen in the direction of K transport.

Epithelia: Outward-Directed Transport. Many animals face the problem of salt loading; this is true for aquatic hypotonic regulators, for elasmobranchs which are isosmotic or hyperosmotic but must cope with a steep inward salt gradient in sea water, and for birds and reptiles whose diet may include marine invertebrates. Such animals must extrude salt to maintain steady state, and the organs involved include the gills of teleosts and crustacea, the salt glands of birds and reptiles, and the rectal glands of elasmobranchs. Only the last has been worked on in vitro, but information from studies on intact animals gives some clues about mechanisms.

Fish gills transport salt outward in sea water and actively absorb it in fresh water.[142a] Thus, in a euryhaline fish the transport vector must reverse on going from one medium to another. Potentials across the gills of fish in S.W. are about 20 mV (eels) or 30 mV (blenny eel), body fluids positive, and the concentration gradient is oriented inward; Cl extrusion across the gill is clearly active, but the situation for Na is more ambiguous because the electrical gradient is about enough to balance the concentration gradient, and hence to maintain the ion in Donnan equilibrium. Tracer fluxes have been studied in several fish, and are extremely rapid in most of them. In the euryhaline flounder, eel, and toadfish and in the stenohaline sea trout, these fluxes are between 1 and 3 mEq/100 gm/hr. The net amount of salt that must be extruded can be calculated from the amount drunk (see p. 49); in the eel, for example, this is only 0.17 mEq/100 gm/hr.[209] This suggests that most of the tracer flux may be by exchange diffusion (Na-Na exchange between blood and S.W. which is physiologically meaningless). This was shown to be the case by abruptly transferring the fish from S.W. to F.W. In the flounder there is an initial rapid decrease in Na efflux on transfer to F.W. from 2.6 to 0.4 mEq/100 gm/hr, followed in 30 minutes by a further drop to 0.1 mEq/100 gm/hr (Fig. 2–4). The rapid change in efflux is sensitive to Na concentration; the slow phase is sensitive only to osmotic concentration (as in a non-electrolyte). Hence, the fast phase is attributed to exchange diffusion, while the slow phase is related to some unidentified property of the gill with respect to osmotic permeability. In the stenohaline *Serranus* (also *Gobius* and *Uranoscopus*) only the fast change was noted, and the animals continued to lose salt rapidly in F.W.[154] The fast change was seen in *Fundulus* as well as in *Platichthys*.[156] The rainbow trout shows much less exchange diffusion, and the total flux in S.W. is only 0.1 mEq/100 gm/hr,[87] which is close to the salt ingested.[204]

No useful isolated gill preparation has been described since the initial work on the perfused eel gill,[119a] but some inferences about mechanisms can be drawn. As noted above, Cl is actively transported, and Na less certainly so. Marine teleosts have higher gill Na-K ATPase activities than do freshwater fish.[110, 111, 115] Elasmobranchs, which have no salt extrusion mechanism in the gills, also have low levels of ATPase in the gills; however, they have an effective salt transport system with high ATPase in the rectal glands.[111]

In eel, goby, and trout the Na-K ATPase of gill is higher in sea water than in fresh water; ouabain decreases activity of the enzyme and diminishes excretion of Na by the gill of the eel.[115] In gills of the euryhaline Japanese eel, the activity of Na-K ATPase is low in F.W. and increases to a maximum after 30 days in S.W. In eels transferred from fresh water to sea water for two weeks, the number of Cl cells and the Na-K ATPase activity increased 5- to 6-fold.[240] In *Fundulus* the activity of gill Na-K ATPase from S.W. fish is six times greater than in gills of F.W. fish; the increase is less in hypoxia. If *Fundulus* is transferred from S.W. to F.W., gill Na-K ATPase decreases while the enzyme in the kidney increases.[62]

A comparison of freshwater and marine fish with regard to ATPase activity of gills follows:[110, 111]

	Na-K ATPase activity of gills
Freshwater teleosts	
Netropis	1.0
Anguilla	6.0
Marine teleosts	
Myoxocephalus	9.4
Fundulus	11.0
Pseudopleuronectes	16.0
Lophius	21.3
Anguilla	11.4
Marine elasmobranch	
Squalus	2.7

The same increase on S.W. adaptation is seen in gills of the salmon *Oncorhynchus*.[263] The enzyme is inhibited by injection of ouabain into the ventral aorta.

In order to extrude Na from the gill by a coupled Na-K pump, the ATPase should be located in a membrane bathed by the external environment, which is the source of K. In confirmation, Na extrusion is inhibited when the eel or trout is put into F.W., but normal extrusion resumes if 10 mM KCl is added; no other constituent of sea water is necessary.[87, 143]

The brine shrimp *Artemia* maintains relatively constant blood concentration over a wide range of salinities (p. 27). In sea water it drinks the equivalent of 2 per cent of its body weight per hour, and excretes most of the ingested salt via the gills. When the shrimp is transferred to a dilute medium, active outward transport by the gills is stopped. In Na-free sea water, Na efflux decreases by 70 per cent. Efflux decreases in dilute sea water or in mannitol of O.C. equal to that of S.W.; hence, the response is to NaCl and not to osmoconcentration. Cl flux is also rapid, and the Na and Cl transport mechanisms are independent.[229] *Artemia* blood is 23.4 mV positive to sea water, and by ion substitution it was concluded that this potential is largely due to active chloride extrusion. As in fish, the blood Na concentration is near to that predicted from a Nernst relationship, whereas chloride is far from equilibrium. About 70 per cent of the Cl efflux in S.W. is by exchange diffusion, the other 30 per cent being by active transport; Na fluxes are passive.[142a, 213]

The salt glands of reptiles and birds and the rectal glands of elasmobranchs extrude salt from a less concentrated blood to a more concentrated excretory fluid. Although not much is known about mechanisms of active transport, some observations indicate some parallels with the systems in gills. Thus, high concentrations of Na-K ATPase are present in functional salt glands, less when the glands involute during salt deprivation.[68, 100] Salt secretion is inhibited by injecting ouabain into the blood, but not by retrograde injection into the glandular lumen. The rectal gland also has very high Na-K ATPase, and ouabain inhibits secretion. Both of these organs develop potential differences during activity, but in contrast to marine fish gill, the blood is negative with respect to the glandular lumen.[226] This means that outward Na transport is active, while the situation for Cl is equivocal.

Other active transport systems are described in other chapters. For example, absorption of sugar and amino acids by intestine is coupled with active transport of sodium (p. 157). Evidence for active transport of Mg^{++} by crustacean kidneys was mentioned earlier (p. 92; see also references[90, 137, 179]); mechanisms are unknown.

Active transport of calcium occurs in certain tissues of animals which accumulate it, as in molluscs for shell formation.[112] Active uptake of Ca^{++} by an aquatic snail *Lymnaea* may occur at low external concentrations.[85] In muscle, granules of the sarcoplasmic reticulum contain an enzyme, Mg ATPase, which causes uptake of Ca^{++} by the particles (p. 103). The mucosa of the shell glands of birds secretes Ca^{++} and generates a counter-current due to $CO_3^{=}$; the current may be independent of Ca^{++} transport.[60] Calcium transport may occur more widely than is now known. From flux measurements and analyses, it appears that a jellyfish actively extrudes $SO_4^{=}$, and *Aequorea* can maintain a gradient of 9 mM between its jelly and the sea water.[141]

Modification of Transport and of Permeability to Ions. Calcium is a general regulator of permeability of cell membranes to water and ions; high calcium generally decreases permeability and low calcium increases it. In frog skin, Ca^{++} on the outside decreases influxes of Na^+ and Cl^-.[41, 42] Addition of calcium to fresh water reverses the negative blood potential of a trout to positive by 5 to 6 mV, but has little effect on Na or Cl fluxes.[121] In *Fundulus* there is a higher rate of exchange of THO in sea water lacking Ca^{++} than in its presence:[177]

THO rate constant (hr^{-1}), Fundulus

in fresh water	2.02
F.W. + 5 mM Ca^{++}	1.36
F.W. + 10 mM Ca^{++} + Mg^{++}	0.93
100% S.W.	0.87
Ca-free S.W.	1.75

Nereis diversicolor is unable to osmoregulate in Ca-free brackish water.[214] Calcium reduces Na fluxes in *Platichthys*.[155] It has been suggested that calcium reduces the hydration of polar organic molecules by reducing the repulsion between fixed anions; thus, closer packing of the organic molecules occurs and permeability of membranes containing such molecules is increased.

The decrease in sodium flux on going from sea water to fresh water is greater for euryhaline than for stenohaline species of fish.[155] This is shown for a euryhaline flounder and a stenohaline sea perch in Figure 2–4 and as follows:

Flux of Na in mEq/100 gm/hr

	Outflux		*Influx*		
	S.W.	F.W.	S.W.	F.W.	*Reference*
Platichthys	2.6	0.022	2.2	0.014	154
Anguilla	1.45	0.048	–	0.021	154
Opsanus	0.80	0.013	–	0.003	126
Tilapia	2.0	0.015	2.4	0.011	177a
*Salmo**	0.17	0.014	–	0.019	118, 87
Serranus†	2.6	0.55	2.6	0.54	154

*Flux across the gills only. All others are total fluxes and include efflux via the urine and influx by drinking.

†Stenohaline marine. All other fish shown are euryhaline.

A similar decrease in Na efflux occurs in *Salmo gairdneri* on transfer from S.W. to F.W.; the flux is then increased if potassium is added to the fresh water, suggesting Na-K exchange.[121] When a flounder is transferred from sea water to a dilute medium, there is a two-step decrease in Na efflux. On return to S.W., the efflux also rises by two steps, but adaptation is slower than on transfer from S.W. to F.W. The Na efflux after adaptation to F.W. is higher in *Serranus* and *Uranoscopus* (58 and 51% of that in S.W., respectively) than in *Platichthys* and *Anguilla* (15.9% and 19.2%).[156] The adaptive value is that euryhaline fish are able to conserve Na in fresh water better than stenohaline fish. However, the molecular mechanisms in gill membranes remain obscure.

It may be concluded from the preceding examples that gill permeability to Na is reduced in many animals on going from sea water to fresh water. The extent to which this is due to changes in permeability (diffusion + exchange), to transepithelial potentials, and to active transport carriers is not clear. Some of the effects are responses to NaCl, while others are to osmotic concentration per se. The extent to which these are direct effects on epithelial cells or indirect effects via sense cells and hormonal or nervous influences is also unknown.

SECONDARY AND TRACE ELEMENTS

Of 48 elements in sea water, animals are known to make use of approximately half. Some inorganic elements become accumulated in tissues by storage excretion. Some of the minor and trace elements are widely distributed in living matter; others are variable in occurrence. Some elements which are necessary in trace amounts are toxic in high concentrations. Sporadic distribution of certain elements in tissues is not readily understood, and in analyses it is important to distinguish elements concentrated by the animal from those in associated organisms. Activation analyses are revealing traces of heavy elements that had formerly escaped chemical detection. Extensive tables of the elementary composition of biological material have been published.[242]

Halogens. Chlorine is the common halogen of animals (but not of plants) and is the most widely dispersed and innocuous anion. Bromine can sometimes be substituted for chlorine, but it is not a normal constituent of most animals. *Aplysia* concentrates bromine in its digestive gland to some 10 per cent of its dry weight; the Br may come from red algae on which it feeds.[258] Sea water contains several thousand times more bromine than iodine, yet iodine is used more by animals. A series of keratinous sponges have an I/Br ratio of 3.9; nonkeratinous ones have a lower content and a ratio of 2.1.

Iodine-containing proteins are widely distributed, as spongin in sponges and as scleroproteins in polychaetes. *Nereis diversicolor* binds iodine where sclerotization occurs.[66, 67] Diiodotyrosine and iodotyrosine are widely distributed, and in the vertebrates the derivative thyroxine is synthesized in the thyroid. In tunicates such as *Ciona,* iodine is bound in the endostyle (in larvae) and in the test (in the adult). In *Branchiostoma* (amphioxus) the endostyle and mucous glands of the gut accumulate iodine, and the endostyle has thyroxine activity. The ammocoete larva of the lamprey has an endostyle which concentrates iodine into thyroglobulin and hydrolyzes this into thyroxine.

Fluorine is present in the mantle of certain gastropods (*Archidoris*) and as fluorite in the shell. It occurs concentrated in some sponges. Fluorine is found in traces in vertebrate bone and aids the hardening of dental enamel, but in excess it causes abnormal bone structure.

Metals. The sporadic concentration of heavy metals has long been noted. Many are essential factors for specific enzymes. Iron is present in heme proteins, such as the cytochromes, catalase and peroxidase in aerobic cells. Heme pigments, hemoglobins, have evolved several times as oxygen carriers. The iron content of some snails is high, and the radular teeth of *Patella* contain Fe_2O_3 to the extent of 50 per cent of the ash weight. Freshwater mussels accumulate iron and manganese in inactive form in storage sites in gills, mantle, and digestive glands.[95a]

Copper occurs in sea water, in which its concentration is about 1 mg/100 liters. Some marine molluscs and arthropods concentrate copper to 1 mg/100 ml blood, where it functions in the respiratory pigment hemocyanin (p. 327). Copper is necessary for the manufacture of some of the ferroproteins. Ceruloplasmin is a copper-containing protein in blood plasma; it has ferroxidative activity and is essential for the transfer of iron between plasma and cells.[198] Cu is essential for polyphenol oxidase, cytochrome oxidase, uricase, laccase, ascorbic acid oxidase, and tyrosinase. In crabs (*Maja*) the Hcy content of blood is reduced at the time of molt, and part of the Cu lost from the blood is stored in the hepatopancreas to be reused in synthesis of new Hcy. In starvation Hcy may be metabolized, and the hepatopancreas accumulates the released copper (*Crangon*).[56] In *Helix* the concentration of Cu in the hepatopancreas is high in the summer; it occurs as Hcy in hemolymph in the winter.[249] The Cu content of land isopods is high, especially in the cells of the pancreas, which lose Cu during the molt. Cu concentration in the hepatopancreas, in $\mu g/mg_{ww}$, is in *Porcellio* 0.21, in marine *Ligia* 0.037, in *Gammarus* 0.009.[255] *Octopus* accumulates Cu in hemocyanin and Zn to a lesser amount in the blood:

	Concentration, $\mu M/liter\ H_2O$		
	Br	*Zn*	*Cu*
Sea water	720	0.1–0.3	0.02–0.04
Octopus blood	1010	6.1	870
Octopus urine	600	13.8	14.9

Chickens with copper deficiency die from aortic rupture associated with increased mucopolysaccharide in the aorta and reduced cytochrome oxidase level in liver ahd heart.[109] In the red pigment, turacin, of touraco feathers, copper may reach a concentration of 7 per cent. A man needs 2 mg Cu per day, more while growing, and blood concentration is 0.5 to 1 ppm.[44]

Zinc is found widely in higher concentrations than copper in animal tissues. It is present in large amounts in bivalve molluscs (0.04 per cent of dry weight in *Pecten* and *Ostrea*). *Octopus* urine may contain zinc 170 times more concentrated than it is in sea water,[178] Zinc is concentrated in the prostate of mammals and is present in ejaculate. On a low-zinc diet, female rats show a reduced number of pregnancies and increased resorption of fetuses.[3a] Deficiency of zinc leads to defects in feather and skin development in birds. Zinc is an essential component of carbonic anhydrase (p. 354). It is an essential activator for pancreatic carboxypeptidase and occurs in glutamic and lactic dehydrogenases, alcohol dehydrogenase, and alkaline phosphatases.

Molybdenum is found in trace levels in marine animals (0.0002 to 0.002 per cent of dry weight). In mammals a trace of molybdenum is needed, but an excess is toxic; it is required for proper utilization of copper. High molybdenum causes diarrhea in cattle; it leads to accumulation of Cu in the kidneys. Mo may be important for several flavoproteins, such as xanthine oxidase. High levels of Mo reduce the utilization of copper. Synthesis of xanthine oxidase in rats is induced by Mo; this is followed by elevated uric acid, which induces Cu-containing urate oxidase.[124a]

Cobalt is a constituent of vitamin B_{12}, and microorganisms use Co in synthesis of this vitamin. Ruminants require cobalt for the rumen organisms; this requirement is some 0.05 mg/day in sheep. Cobalt is one of the few metals required and accumulated by *Tetrahymena.*

Vanadium is present in tissues of many animals in small amounts (some 0.1 ppm). It may function in some cellular oxidations. When rats were reared on a trace-element-controlled diet, a requirement of 10 μg of vanadium per 100 grams of diet was demonstrated for best growth.[202c] Vanadium may be concentrated by 5×10^5 times above the sea water level by certain ascidians.[6] It is present in plasma of two families of ascidians, and in special vanadocytes in two others. Sea water contains vanadium at 2 ppm dry weight, and some ascidians contain 6500 ppm. *Styela* and *Molgula* concentrate niobium rather than vanadium.[113] It has been calculated from the concentrations of vanadium in sea water and in *Ciona* and from the animals' pumping rate that they could not extract more than 45 per cent of their vanadium content in a year; hence they must obtain the element from particulate matter. The vanadium concentration is higher in tunicates from warm waters than in those from cold waters.[244] Why vanadium should be concentrated by these few animals, and niobium by a few of their relatives, is a mystery.

Mercury is accumulated by many animals, and at levels above 24 ppm it is toxic. Methyl-Hg compounds are particularly toxic and cause central nervous system damage in vertebrates.

Manganese occurs at levels of 5 to 10 mg/100 gm ash weight in many animals. Oysters have high Mn content, particularly in the gills. In mammals Mn is essential and is stored in the liver. It functions as a cofactor in oxidative phosphorylation, in isocitric dehydrogenase and L-malic dehydrogenase, and in liver arginase. Ducklings require 15 ppm of Mn in the diet. Mn can substitute for Mg in some enzymes.

Magnesium is an essential element for all animals. *Tetrahymena* requires Mg rather than Ca. Magnesium is the active metal in chlorophyll, and is found in all green plants. It is a cofactor for some ATPases, and for hexokinase, enolase, pyruvate kinase, and muscle myosin ATPase.

Aluminum may occur in traces, and at higher levels in some gastropods. No enzymatic function has been assigned to Al. Nickel occurs in keratinous tissues, especially feathers; it is present in liver and thymus. Tin, silver, and titanium are elements of sea water that have rarely been reported in animals. Lead is widely distributed, more in molluscs (0.05% of ash weight in some) than in other animals. Boron occurs widely, and in *Helix* may be 0.25% of ash weight.[245] Selenium is required in traces; in chickens and lambs muscle dystrophy develops in selenium deficiency. Chickens show poor feathering and pancreatic defects.

Toxicity of heavy metals, particularly uranium, lead, and mercury, constitutes an industrial hazard. Lead from gasoline fumes is concentrated by plants, and some 90% of the lead accumulated in mammals is deposited in bone, where it may cause structural defects. Lead, like UO_2^{++}, complexes with phosphoryl groups, particularly of proteins at cell surfaces; at 0.06 mM/liter, lead inhibits the ATPase of red blood cells. Mercury reacts strongly with sulfhydryl groups and is a good diuretic; it inhibits NaCl reabsorption by the kidney and is accumulated in the renal cortex. Heavy metals, by reacting with $PO_4^{=}$, SH^-, and COO^- groups on proteins, can inhibit many enzymes. Lead is toxic for rats at 25 ppm of diet, and selenium at 3 ppm.[144a, 167a, 202b]

SKELETAL ELEMENTS

Animals use both organic compounds and inorganic salts to provide skeletal support. Cellulose, the general plant skeletal carbohydrate, is also used in tunicates as an exoskeleton; some chordate connective tissue has a cellulose-like material.[232] Chitin, a glucosamine with 6% nitrogen, is widely used, particularly in arthropods, annelids, and some molluscs; it may be infiltrated by calcium compounds for strength, as in many crabs. Protein materials used in skeletons are numerous: spongin in *Porifera,* conchiolin in certain molluscs, keratin and collagen in vertebrates. Collagen may be replaced by bone, but in some vertebrates (e.g., elasmobranchs and cyclostomes) the skeleton is cartilaginous, sometimes with apatite microcrystals; sharks give no response to vitamin D or to parathormone.[238]

Two general classes of inorganic skeletons occur—the oxides of silicon and the carbonates of calcium. Silicon is widely used for support in plants and diatoms; *Equisetales* and grasses are examples. Among protozoa the *Heliozoa* have spicules of SiO_2; radiolarians have more silicon dioxide than oxides of calcium and magnesium. Two groups of sponges, Demospongiae and Hyalospongiae, or glass sponges, have spicules of $SiO_2 \cdot H_2O$; siliceous sponge fossils are found in the Precambrian. Silicon occurs in a few bryozoan skeletons.

Calcareous skeletons are of several sorts. Calcium carbonate is the most extensively used skeletal material; it may be amorphous, or it may be as calcite, often combined with $MgCO_3$ when it consists of hexagonal crystals. It may occur as aragonite in rhombic crystals, sometimes with $SrCO_3$. Calcium also occurs in various phosphates such as the apatites ($Ca_3P_2O_8 \cdot CaCO_3$) of vertebrate bone. The shells of doriid gastropods contain fluoride.

Foraminiferan shells of $CaCO_3$ constitute

vast deposits, some of them associated with petroleum. Coral deposits are primarily $CaCO_3$, but some have 1% $CaSO_4$ and others 10% $MgCO_3$. Corals actively concentrate calcium from sea water, adsorb it onto mucopolysaccharides, and then combine it with bicarbonate formed from CO_2 by the aid of carbonic anhydrase to form the calcium carbonate of the corallum skeleton.[84] The CO_2 comes mainly from symbiotic algae, the zooxanthellae. Calcification can be reduced by inhibitors of carbonic anhydrase, or by removal of the zooxanthellae. Medusae (*Aurelia*) have statoliths consisting of $CaSO_4$; they develop no statoliths in SO_4-free sea water.[217]

Brachipod shells may contain 50% $Ca_3(PO_4)_2$ in addition to the carbonate; some of them also have CaF_2. Serpulid worms and echinoderm skeletons are largely $CaCO_3$, as calcite, with small amounts of phosphate and sulfate and some $MgCO_3$. Worm tubes formed in summer have more aragonite than do those formed in the winter. Gastropods have more aragonite, while pelecypods have varying amounts of calcite and aragonite.[138] Barnacle shells are calcite. The radular teeth of chiton are capped with more than 65% magnetite (80% Fe_2O_3 and 20% FeO).[139]

Molluscan shells are formed by the activity of the epithelium at the edge of the mantle. A birefringent protein matrix, the periostracum, is first formed, and $CaCO_3$ is deposited on this. The calcium may be distributed from the gut to the mantle by means of blood; it can also be taken directly from the water by the mantle cells and deposited as shell.[112] During long periods out of water pelecypods produce metabolic acids, which may be buffered by calcium eroded from the shell so that the blood pH remains constant.[57] The larval shell of *Crassostrea* is aragonite, but the adult shell is mostly calcite; molluscs with very hard shells have more aragonite than calcite.[232]

Some arthropods calcify the chitinous exoskeleton. Crabs may store calcium in the hepatopancreas before molt; crayfish may store it in gastroliths.Gastrolith deposition and molt are under the control of an eyestalk hormone. When eyestalks are removed, Ca deposition begins in gastroliths and molt occurs.[148] The isopod *Limnoria* molts in two steps, posterior half first; then a large amount of calcium is shifted from anterior to posterior half before the anterior half molts. Thus, crustaceans have several means of limiting loss of calcium at molt.

In earthworms the esophageal epithelium of the calciferous glands secretes spheres of $CaCO_3$ crystals, apparently as a means of calcium storage.

In vertebrate bone the $CaCO_3$ is as aragonite, and the ratio of $Ca_3P_2O_8$ to $CaCO_3$ is about 7 in mammals and 11 in fish (approaching apatite). Teeth and tusks have 2% Mg salts; bone contains less than 0.3%. Vitamin D regulates the Ca in blood and the mineralizing of bone by aiding absorption from the gut, and with parathormone and calcitonin, by mobilizing Ca from bone.

The ratio of Sr to Ca varies from 0.1 to 0.8; it tends to be higher in marine than in freshwater animals, but it is relatively constant within a class.[200] The ratio Sr/Ca is slightly higher than that in sea water for the coelenterates, bryozoans, polychaetes, and cirripedes; it is lower for gastropods, pelecypods, and cephalopods. Fossil molluscs have higher Sr/Ca ratios than do modern shells, although the ratio in the oceans has remained constant for 10^8 years.[138, 139] It appears that genetic factors are more important than environmental ones in determining the Sr/Ca ratio.

BALANCED SALT MEDIA

Numerous media have been devised for the culture of aquatic animals. Tissue culture media are complex, and provide organic nutrients as well as inorganic requirements. Frequently a mixture of trace elements is added. An element which is not necessary for survival may nevertheless accelerate growth. Sterile conditions are essential, and antibiotics are needed. A culture medium for tissues of ticks includes 14 amino acids in addition to salts, sugar, and serum albumin.[186] Discussion of culture media is beyond the scope of this chapter. However, a few typical ones are listed in Table 2-7.[168, 259]

The maintenance of excised tissues is essential for physiological experiments, particularly those with excitable tissues. The design of a physiological saline is usually empirical, and its composition is not necessarily the same as that of blood plasma or even interstitial fluid. Usually Na and Cl are similar to their concentrations in blood; K may be slightly higher and Ca similar to or

TABLE 2–7. Physiological Salines. (Values in mM except where designated otherwise.)

	NaCl	*KCl*	*$CaCl_2$*	*$MgCl_2$*	*$NaHCO_3$*	*Other Constituents*
Artificial sea water	493	10.95	10.65	28.0		$MgSO_4$ 31
Sea water[93]	405	8.9	10.0	53.0	2.25	Na_2SO_4 27.6 KBr 0.8 H_3BO_3 0.42 $SrCl_2$ 0.15 NaF 0.07
Homarus[261]	455	13.5	16.5	4.0	4.0	H_3BO_3 + NaOH buffer
Crab	466	10	10.2	24.8	28.2	
Gecarcinus[210]	430	5	9	7		tris buffer 10
Carcinus[165]	585	14	12.7	24.4		$NaHCO_3$ to pH 7.0
Strombus[133]	430	11	10	58	10	Na_2SO_4 20
Squid	475	10.3	10.6	53.1	25.7	phosphate to pH 7.6
Squid[119]	436	10.0	9.9	24.3		$MgSO_4$ 2.72 pH buffer to 7.6
Dogfish[108]	224	1.35	1.8	1.1	2.38	glucose 15 urea 333
Elasmobranch	134	6.8	6.1	1.1		glucose 5 urea 208
Artificial pond water[54]	0.5	0.05	0.4		0.2	
Opalina[157]	60	5	0.1	1.0		tris buffer to pH 7.8 glucose 0.2 casein
Ascaris	129	19.2	4.1	4.43		phosphate to pH 6.7
Leech[160]	115	4	1.8			glucose 10 tris buffer 10
F.W. snail[34]	110	4.9	6.4		18.3	$MgSO_4$ 3.5
Arion[191]	43.7	3.5	3.0			$MgSO_4$ 4.1 Na_2HPO_4 0.25
Crayfish[241]	207	5.4	13.6	2.64	2.4	
Tinca[162]	121.5	2.7	1.8	1.07	11.9	glucose 5.5 NaH_2PO_4 0.42
Cottus[107]	163.9	2.6	2.7	1.0	18	glucose 7 NaH_2PO_4 3.2
Gasterosteus[126a]	170	6	1.6	1.0	2.3	glucose 15 KH_2PO_4 0.5
F.W. fish	101.8	3.38	1.36		2.5	$MgSO_4$ 1.19 KH_2PO_4 1.2
Frog	112	1.9	1.1		2.4	glucose 11.1 NaH_2PO_4 0.7
Toad[30]	124	3.8	1.1		2.4	glucose 1.1 NaH_2PO_4 0.85
Amphiuma (kidney perfusion)	75	3.0	1.8	1.0	20	Na_2HPO_4 0.56 NaH_2PO_4 0.14 glucose 12 glycine 250 mg PVP 15 g
Locusta[168]	120	2.7	1.8	1.05	6	NaH_2PO_4 0.17 glucose 4.5
Carausius[260]	15	18	7.5	50		$H_2PO_4^-$ 6 $HPO_4^=$ 4.5
Cockroach[169]	210	3.1	1.8			phosphate buffer to pH 7.2
Cockroach[158]	188	21.2	7.7	1.8		
Antheraea[247] larva	12	30	3	18		
pupa	4	40	3	18		
Locust	130.5	10.1	2.0	1.98	4	KH_2PO_4 5.5
Lizard	130	3.0	3.0	2.0	2.0	
Lizard[30]	116	3.2	1.2		2.0	NaH_2PO_4 0.3 glucose 1.7 $MgSO_4$ 1.4
Chicken[78]	150	5	5	2	20	glucose 12
Bird Ringer	117	2.33	5.8		28	$MgSO_4$ 2.12
Mammal Tyrode	138	2.7	1.84	1.06	11.9	NaH_2PO_4 0.5
Krebs	118.9	4.8	2.56		2.5	$MgSO_4$ 1.2 KH_2PO_4 1.3 glucose 20.4

(Table 2–7 continued on opposite page.)

TABLE 2–7. Physiological Salines. (Values in mM except where designated otherwise.) *(Continued)*

	NaCl	*KCl*	*$CaCl_2$*	*$MgCl_2$*	*$NaHCO_3$*	*Other Constituents*
Krebs (modified)	119	5.0	2.5	1.0	2.5	NaH_2PO_4 0.5 glucose 11.0
Krebs Bohr	119	4.7	1.6		14.9	$MgSO_4$ 1.17 KH_2PO_4 1.18
Rat[16]	108.5	3.5	1.5		26.2	$MgSO_4$ 0.69 NaH_2PO_4 1.67 glucose 5.55 Na gluconate 9.64 sucrose 7.6
			CULTURE MEDIA			
Mammal Krebs culture medium	95.4	4.8	2.54		2.5	$MgSO_4$ 1.2 KH_2PO_4 1.25 glucose 20.4 Na pyruvate 0.542 g Na fumarate 0.745 g Na l-glutamate 0.813 g
Frog tissue culture[131]	114	4.8	1.8	1.44		100 μ penicillin 7% fetal calf serum glucose 10
Fish tissue culture[259]	125	5.1	1.9		1.2	NaH_2PO_4 3.0 $MgSO_4$ 0.94 glucose 5.6 serum 7%
Mammalian tissue culture, Hanks[259]	138	5.4	1.3	0.49		Na_2HPO_4 0.25 $MgSO_4$ 0.41 serum 7% glucose 5.6

raised slightly above the free ionic concentrations in blood plasma; and Mg may be at blood levels or slightly lower. A balance of elements, especially Na + K/Ca + Mg, is needed.[136] Appropriate buffering is necessary, and metabolic substrates are added, especially for prolonged experiments. The chief defect of most physiological salines is the absence of the variety of organic substances which normally bathes the tissues. Recently, there has been a tendency to add a number of amino acids, usually rather empirically. Another defect of most salines is the lack of large molecules to provide colloid osmotic pressure, and for perfusion experiments the addition of polymers is usual.

A few solutions for use in maintaining tissues of some relatively common experimental animals are listed in Table 2–7.

SUMMARY

The specificity of certain inorganic ions for life processes was probably established before much biological differentiation began. This specificity is not related to abundance of particular elements in the earth's crust or in the sea, but must be determined by charge, shape, size, reactivity (especially with water), and by bonding capacities. Specificity is often evident in activation of an enzyme, and in determining the properties of excitable membranes. Constancy of intracellular ion concentrations is more important than osmotic constancy. The most highly regulated of cell ions is potassium; sodium is more variable, and is usually much lower in concentration intracellularly than extracellularly. Organic anions constitute a large proportion of intracellular solutes; inorganic anions serve as buffers and to provide negative charges to some organic bases. Intracellular calcium is largely bound, and hence the ion concentration is very low. Particular levels of ion concentrations in cells are genetically fixed, and variants occur among species—such as high and low potassium red cells, plants that can dispense with Na, or variable-potassium bacteria. Within one animal, concentrations of ions with specific functions vary from tissue to tissue. Some tissues store particular elements as reserves or as a way of eliminating them.

Cell regulation preceded total organismic regulation; changes in ion composition of extracellular fluids are reflected in abnormal functioning of tissues. Regulation of ions in plasma is by restricted permeabilities, binding to proteins, selective excretion (renal and extrarenal), and active absorption. These processes are often under hormonal control. Excretory organs occur in osmoconformers and presumably originated for ionic regulation. The use of excretory organs for osmoregulation came only with invasion of fresh water and land, and their use for eliminating nitrogenous wastes came as a late specialization. Every excretory system requires transport mechanisms for specific ions. Some of these serve for selective reabsorption from filtered fluid, some for active secretion into a urine, and others for active transport by nonrenal excretory organs. Some of these systems are very complex and involve active secretion into intercellular channels to provide local osmotic gradients, followed by active reabsorption, with the net effect of a cycling of a given ion, as in insect Malpighian tubules and rectal complex.

No transport system for a particular ion is really understood. Most widespread is a sodium "pump" which uses a specific ATPase, activated by external K and intracellular Na in the presence of Mg; the sodium pump can often be blocked by the glycoside ouabain. Energy from ATP is required, and usually more than ten sodium ions are transported per oxygen molecule consumed. This type of Na transport has been well studied in red blood cells, nerve and muscle fibers, and kidney cells.

Some epithelial systems actively transport ions outward, some inward. Examples of outward transport are excretory organs, gills of marine fish, and salt glands of birds. Transport is inward in gills of freshwater fish and the skin of frogs. Some transepithelial Na pumps use the same mechanisms of K-coupling as does cellular extrusion. In some epithelia there may be uptake of Na^+ in exchange for H^+, or of NH_4^+ and Cl^- in exchange for HCO_3^-. The molecular mechanisms of spatial translocation of transported ions remain to be determined.

Other pumps are less well studied—chloride transport as in frog skin and bladder, and potassium transport by a system that can handle other univalent cations in the absence of potassium as in *Cecropia* midgut. Pumps for magnesium and calcium exist but have not been characterized.

Modulation and control of ion permeability and active transport are of adaptive importance. The gills of many euryhaline aquatic animals become less permeable to water and sodium when they are in fresh water than when they are in sea water. Part of the control is in permeability, and part is carrier mediated (as in certain fish). Calcium ions are necessary for regulating permeability of many membranes. The sensing mechanisms, the neural and hormonal integration, as well as the direct cellular responses to changed ionic environment are poorly known. Pituitary polypeptides, varying among vertebrate classes, modulate permeabilities of skin, gills, and kidneys to water and sodium.

In addition to the ions of general biological function, there are others which are used for special purposes. Many of these are needed in near-trace amounts, as for specific enzymes. Others (e.g., Fe and Cu) have a role in pigments, and Ca, Mg, Sr, and Si function in different skeletal complexes. Nutritional and endocrine factors presumably control tissue levels of trace and skeletal elements. Scarcity of some elements in certain habitats limits the distribution of some animals.

Inorganic ions have set limits for determining many biological functions. Yet organisms have been highly selective and have exerted more regulation of ionic composition than of most other physical parameters. The reasons, at the atomic level, why certain elements are essential for various processes remain a chapter for the future.

REFERENCES

1. Alvarado, R. H., and L. B. Kirschner, Comp. Biochem. Physiol. *10*:55–67, 1963. Ionic regulation in *Ambystoma*.
2. Alvarado, R. H., and A. Moody, Amer. J. Physiol. *218*:1510–1516, 1970. Na and Cl transport in tadpoles.
3. Anderson, E., and W. R. Harvey, J. Cell Biol. *31*:107–134, 1966. Active transport in midgut of *Samia cecropia*.
3a. Apgar, J., J. Nutr. *100*:470–476, 1970. Requirement of female rats for zinc.
4. Barnes, H. J., J. Exp. Biol. *31*:582–588, 1954. Ionic composition of sea water.
5. Bernard, G. R., R. A. Wynn, and G. G. Wynn, Biol. Bull. *130*:18–27, 1966. Ions in body fluids of sting ray.
6. Bertrand, D., Bull. Amer. Mus. Nat. Hist. *94*:407–455, 1950. Biogeochemistry of vanadium.
7. Biber, T. U. L., and P. F. Curran, J. Gen. Physiol. *56*:83–99, 1970. Sodium uptake by frog skin.
8. Bielawski, J., Comp. Biochem. Physiol. *13*:423–432, 1964. Chloride transport into gill of crayfish.

9. Binyon, J., J. Marine Biol. Assoc. *42*:49–64, 1962. Ions in body fluids of *Asterias*.
10. Boné, G. J., Ann. Soc. Roy. Zool. Belg. *75*:123–132, 1944. Sodium-potassium ratio in insect hemolymph.
11. Bonting, S. L., Chapter 8, pp. 257–363. *In* Membranes and Ion Transport, edited by E. E. Bittar, Wiley-Interscience, 1970.
12. Bonting, S. L., and L. L. Caravaggio, Arch. Biochim. Biophys. *101*: 36–46, 1963. NaK-ATPase in various tissues.
13. Boonkoom, V., and R. H. Alvarado, Amer. J. Physiol. *220*:1820–1824, 1971. ATPase in gills of *Rana* tadpoles.
14. Boroffka, I., Z. vergl. Physiol. *51*:25–48, 1965. Ion transport in nephridium of *Lumbricus*.
15. Boroffka, I., Z. vergl. Physiol. *57*:348–375, 1968. Osmo- and volume regulation in *Hirudo*.
16. Bretag, A. H., Life Sci. *8*:319–329, 1969. Synthetic interstitial fluid for mammalian tissue.
17. Bricker, N. S., T. Biber, and H. H. Ussing, J. Clin. Invest. *42*:88–99, 1963. Transport of ions by frog skin.
18. Bricteux-Gregoire, S., G. Duchateau-Bosson, C. Jeuniaux, E. Schoffeniels, and M. Florkin, Arch. Int. Physiol. Biochem. *71*:393–400, 1963. Ions in body fluids of scorpions.
19. Bricteux-Gregoire, S., G. Duchateau-Bosson, C. Jeuniaux, and M. Florkin, Arch. Int. Physiol. Biochem. *72*:267–275, 1964. Ions in muscle of oyster.
20. Brodsky, W. A., and T. P. Schilb, Fed. Proc. *26*:1314–1321, 1967. Mechanism of acidification in turtle bladder.
21. Brown, A. C., J. Cell. Comp. Physiol. *60*:263–269, 1962. Na absorption by frogs.
22. Brull, L., and Y. Cuypers, J. Marine Biol. Assoc. U. K. *34*:637–642, 1952. Kidney function in *Lophius*.
23. Bull, J. M., and R. Morris, J. Exp. Biol. *47*:485–494, 1967. Ions in tissues of larva of *Lampetra*.
24. Burger, J. W., and D. C. Tosteson, Comp. Biochem. Physiol. *19*:649–653, 1966. Ion fluxes in *Squalus*.
25. Burton, R. F., Comp. Biochem. Physiol. *25*:501–508, 1968. Ion regulation in *Helix*.
26. Burton, R. F., Comp. Biochem. Physiol. *25*:509–516, 1968. Ion balance in pulmonates.
27. Burton, R. F., Comp. Biochem. Physiol. *27*:763–773, 1968. Plasma and muscle ions, invertebrates and vertebrates.
28. Burton, R. F., Comp. Biochem. Physiol. *39A*:267–275, 1971. Cations in land snails.
29. Cade, T. J., C. A. Tobin, and A. Gold, Physiol. Zool. *38*:9–33, 1965. Water economy in finches.
30. Campbell, G., G. Burnstock, and M. Wood, Quart. J. Exp. Physiol. *49*:268–276, 1964. Saline for lizard.
31. Cassidy, M. M., and C. S. Tidball, Amer. J. Physiol. *217*:674–679, 1969. Ions in mammalian tissues.
32. Century, T. J., I. R. Fenichel, and S. B. Horowitz, J. Cell. Sci. 7:5–14, 1970. Concentrations of ions in amphibian oocytes.
33. Cereijido, M., and C. A. Rotunno, J. Physiol. *190*:481–497, 1967. Sodium transport across frog skin.
34. Chiarandini, D. J., Life Sci. *3*:1513–1518, 1964. Saline for pulmonates.
35. Chowdhury, T. K., and F. M. Snell, Biochim. Biophys. Acta *94*:461–471, 1965. Microelectrode study of asymmetry potentials, frog skin, toad bladder.
36. Clark, M. R., E. J. Denton, and J. B. Gilpin-Brown, J. Physiol. *203*: 49P–50P, 1969. Cephalopod buoyancy.
37. Cofre, G., and J. Crabbe, J. Physiol. *188*:177–190, 1967. Sodium transport by colon of *Bufo*.
38. Cohen, M. W., H. M. Gerschenfeld, and S. W. Kuffler, J. Physiol. *197*:363–380, 1968. Ionic environment of neurones and glia in *Necturus*.
39. Coldman, M. F., and W. Good, Comp. Biochem. Physiol. *21*:201–206, 1967. Ions in blood of mammals.
40. Coulson, R. A., and T. Hernandez, Biochemistry of the Alligator. Louisiana State University Press, 1964. 133 pp.
41. Curran, P. F., and M. Cereijido, J. Gen. Physiol. *48*:1011–1033, 1965. Na and K fluxes, frog skin.
42. Curran, P. F., and J. R. Gill, J. Gen. Physiol. *45*:625–641, 1962. Effect of calcium on sodium transport, frog skin.
43. Daniel, E. E., and K. Robinson, J. Physiol. *154*:421–444, 1960. Ion transport in muscle of uterus.
44. Davis, G. K., *in* Copper Metabolism, edited by W. McElroy and B. Glass. New York, Academic Press, 1950, pp. 216–229. Requirement of cattle for copper.
45. Davson, H., Physiology of the Cerebrospinal Fluid. London, Churchill and Boston, Little, Brown & Co., 1967. 445 pp.
45a. Davson, H., A Textbook of General Physiology, 4th ed., Vols. I and II. London, Churchill, 1970. 1694 pp.
46. Deffner, G. G., and R. S. Hafter, Biochim. Biophys. Acta *32*:362–374, 1959. Ion analysis of squid axon.
47. Deffner, G. G., Biochim. Biophys. Acta *47*:378–388, 1961. Amino acids in squid blood and axon.
48. Dehnel, P. A., and T. H. Carefoot, Comp. Biochem. Physiol. *15*:377–397, 1965. Ion regulation, intertidal crabs.
49. de Jorge, F. B., et al., Comp. Biochem. Physiol. *16*:491–496, 1965. Ions in giant earthworm.
50. de Leersnyder, M., Etude Exper. Cahiers de Biol. Marine *8*:295–321, 1967. Ion composition of blood and urine in *Eriocheir*.
51. Denton, D. A., and J. R. Sabine, J. Physiol. *157*:97–116, 1961. Appetite for salt in sheep.
52. Deyrup, I. J., *in* Physiology of Amphibia, edited by J. A. Moore. New York, Academic Press, 1964, pp. 251–328. Water balance in kidney.
53. Diecke, F. P., and D. J. Weidler, Z. vergl. Physiol. *64*:372–399, 1969. Effect of cations on conduction in nervous system of insect *Carausius*.
54. Dietz, T. H., and R. H. Alvarado, Biol. Bull. *138*:247–261, 1970. Ionic regulation in *Lumbricus*.
55. Dietz, T. H., L. B. Kirschner, and D. Porter, J. Exp. Biol. *46*:85–96, 1967. Transepithelial P.D.'s in larval salamanders.
56. Djangmat, J. S., Comp. Biochem. Physiol. *32*:709–731, 1970. Copper in tissues and blood in *Crangon*.
56a. Duchateau, G., M. Florkin, and J. Leclercq, Arch. Int. Physiol. Biochem. *61*:518–549, 1953. Ions in insect hemolymph.
57. Dugal, L. P., J. Cell. Comp. Physiol. *13*:235–251, 1939. Calcium in shell as buffer in clams.
58. Dunham, P. B., Biol. Bull. *126*:373–390, 1964. Adaptation of *Tetrahymena* to NaCl.
59. Dunham, P. B., and F. M. Child, Biol. Bull. *121*:129–140, 1961. Ion regulation and *Tetrahymena*.
60. Ehrenspeck, G., H. Schraer, and R. Schraer, Amer. J. Physiol. *220*: 967–972, 1971. Calcium transfer by avian shell gland.
61. Engbaack, L., and T. Hoshiko, Acta Physiol. Scand. *39*:349–355, 1957. Electrical potential gradients through frog skin.
62. Epstein, F. H., A. I. Katz, and G. E. Pickford, Science *156*:1245–1247, 1967. Na-K ATPase in gills of *Fundulus*.
63. Evans, D. H., J. Exp. Biol. *50*:179–190, 1969. Water and salt balance in teleost *Pholis*.
64. Evans, J. V., et al., Proc. Roy. Soc. Lond. B *148*:249–262, 1958. Genetic differences in K^+ and Na^+ in red cells, sheep.
65. Finn, A. L., Amer. J. Physiol. *218*:463–469, 1970. Ion transport in toad bladder.
66. Fletcher, C. R., J. Exp. Biol. *53*:425–443, 1970. Regulation of Ca and Mg in *Nereis*.
67. Fletcher, C. R., Comp. Biochem. Physiol. *35*:105–123, 1970. Metabolism of iodine by polychaete.
68. Fletcher, G. L., I. M. Stainer, and W. N. Holmes, J. Exp. Biol. *47*:375–391, 1967. ATPase in salt glands of duck.
69. Florkin, M., G. Duchateau-Bosson, C. Jeuniaux, and E. Schoffeniels, Arch. Int. Physiol. Biochem. 72:892–906, 1964. Osmotic and ionic regulation in *Eriocheir*.
70. Frazier, H. S., E. F. Dempsey, and A. Leaf, J. Gen. Physiol. *45*:529–543, 1962. (See reference 71.)
71. Frazier, H. S., and A. Leaf, J. Gen. Physiol. *46*:491–515, 1963. Active sodium transport in toad bladder.
72. Friz, C. T., Comp. Biochem. Physiol. *38A*:477–482, 1971. Cations in amoebae.
73. Fromm, P. O., Comp. Biochem. Physiol. *10*:121–128, 1963. Ion excretion in *Salmo*.
74. Garrahan, P. J., and I. M. Glynn, Nature *207*:1098–1099, 1965. Sodium pump in red blood cells.
75. Germain, P., and A. Gagnon, Comp. Biochem. Physiol. *26*:371–375, 1968. Ions in blood of hagfish.
76. Giebisch, G., J. Gen. Physiol. *44*:659–678, 1961. Transport by nephrons of *Necturus*.
77. Gifford, C. A., Publ. Inst. Marine Sci. U. Texas *8*:97–125, 1962. Osmotic and ionic regulation, *Callinectes* and *Ocypode*.
78. Ginsberg, B. L., J. Physiol. *150*:707–717, 1960. Chicken Ringer solution.
79. Glynn, I. M., Brit. Med. Bull. *24*:165–169, 1968. Membrane ATPase and cation transport.
80. Glynn, J. P., Comp. Biochem. Physiol. *26*:937–946, 1968. Ions in relation to moult cycle of *Homarus*.
81. Goffart, M., Function and Form in the Sloth. Pergamon Press, 1971. 225 pp.
82. Goffart, M., and J. Nys, Arch. Int. Physiol. Biochem. *73*:166–168, 1964. Water balance in sloths.
83. Gonzalez, C. F., Y. E. Shamoo, H. R. Wyssbrod, R. E. Solinger, and W. A. Brodsky, Amer. J. Physiol. *213*:333–340, 1967. Transport across turtle bladder.
84. Goreau, T., Endeavour *20*:32–39, 1961. Calcium deposition in reef corals.
85. Greenaway, P., J. Exp. Biol. *54*:199–214, 1971. Calcium regulation in snail *Limnaea*.
86. Greenwald, L., Physiol. Zool. *44*:149–161, 1971. Sodium balance in frog.
86a. Greenwald, L., Physiol. Zool. *44*:149–161, 1972. K_m values.
87. Greenwald, L., and L. B. Kirschner, Amer. Zool. *11*:664– , 1971.
88. Gross, W. J., Biol. Bull. *127*:447–466, 1964. Water and salt regulation in crabs.
89. Gross, W. J., Physiol. Zool. *36*:312–324, 1963. Ion balance in terrestrial crabs.
90. Gross, W. J., and R. L. Capen, Biol. Bull. *131*:272–291, 1966. Excretion of ions in crab *Pachygrapsus*.
91. Haggag, G., K. A. Raheem, and F. Khalil, Comp. Biochem. Physiol. *16*:457–465, 1965. Ions in blood of reptiles.

92. Harrison, F. M., J. Exp. Biol. *39*:179–192, 1962. Excretion in *Haliotis*.
93. Harvey, H. W., The Chemistry and Fertility of Sea Waters. Cambridge University Press, 1955. 224 pp.
94. Harvey, W. R., J. A. Haskell, and K. Zerahn, J. Exp. Biol. *46*:235–248, 1967; *54*:269–274, 1971. Active transport by midgut of moth larva.
95. Hayes, F. R., et al., J. Marine Biol. Assoc. U. K. *26*:580–589, 1947. Inorganic constituents of molluscs.
95a. Hobden, D. J., Canad. J. Zool. *48*:83–86, 1970. Storage excretion of heavy metals, freshwater mussels.
96. Hobson, A. D., et al., J. Exp. Biol. *29*:22–29, 1952. Ions in *Ascaris*.
97. Hodgkin, A. L., and R. D. Keynes, J. Physiol. *128*:28–60, 1955. Active transport of cations in giant axons from *Sepia* and *Loligo*.
98. Hoffman, J. F., pp. 13–17 *in* Biophysics of Physiological and Pharmacological Action, edited by A. M. Shanes. AAAS meeting, 1960. Cation transport.
99. Hoffman, J. F., Circulation *26*:1201–1213, 1962. Cation transport and structure of the red cell plasma membrane.
100. Hokin, M. R., and L. E. Hokin, J. Gen. Physiol. *50*:793–811, 1967. Salt secretion in bird salt gland.
101. Holley, A., and P. Régondaud, Compt. Rend. Soc. Biol. *157*:1100–1102, 1963. Ionic composition of isopod *Porcellio*.
102. Holmes, W. N., G. L. Fletcher, and D. J. Stewart, J. Exp. Biol. *48*: 487–508; 509–520, 1968. Salt excretion by duck.
103. Holmes, W. N., and I. M. Stainer, J. Exp. Biol. *44*:33–46, 1966. Renal excretion of salts by trout.
104. Horne, F. R., Comp. Biochem. Physiol. *19*:313–316, 1966. Ionic regulation in tadpole shrimp.
105. Horowicz, P., and J. W. Burger, Amer. J. Physiol. *214*:635–642, 1968. Sodium transport in dogfish *Squalus*.
106. Houston, A. H., and I. A. Madden, Nature *217*:969–970, 1968. Plasma electrolytes in carp.
107. Hudson, R. C. L., Comp. Biochem. Physiol. *25*:719–725, 1968. Physiological saline for fish.
108. Hugnel, H., Z. vergl. Physiol. *42*:63–102, 1959. Saline for fish heart.
109. Hunt, C. E., J. Landesman, and D. M. Newberne, Brit. J. Nutr. *24*: 607–614, 1970. Copper deficiency in chickens.
110. Jampol, L. M., and F. H. Epstein, Amer. J. Physiol. *218*:607–611, 1970. Na-K ATPase in gills of marine and freshwater fish.
111. Jampol, L. M., and F. H. Epstein, Amer. J. Physiol. *218*:607–611, 1970. Salt glands of elasmobranchs.
112. Jodrey, L. H., and Wilbur, K. M., Biol. Bull. *108*:346–358, 1955. Calcium deposition in oyster shells.
113. Kakubu, N., and T. Hidaka, Nature *205*:1028–1029, 1965. Vanadium and niobium in ascidians.
114. Kamemoto, F. I., A. E. Spalding, and S. M. Keister, Biol. Bull. *122*: 228–231, 1962. Ion balance in earthworms.
115. Kamiya, M., and S. Utida, Comp. Biochem. Physiol. *26*:675–685, 1968. ATPase in gills of eel.
116. Kerkut, G. A., and R. C. Thomas, Comp. Biochem. Physiol. *14*:167–183, 1965. Sodium pump in nerve cells.
117. Kerley, D. E., and A. W. Pritchard, Comp. Biochem. Physiol. *20*:101–113, 1967. Osmoregulation in crayfish.
118. Kerstetter, T. H., L. B. Kirschner, and D. D. Rafuse, J. Gen. Physiol. *56*:342–359, 1970. Na uptake by rainbow trout.
119. Keynes, R. D., J. Physiol. *117*:119–150, 1952. Squid saline.
120. Kidder, C. W., M. Cereijido, and P. Curran, Amer. J. Physiol. *207*: 935–940, 1964. Potential differences across frog skin.
121. Kirschner, L. B., J. Cell. Comp. Physiol. *45*:61–87, 1955.
121a. Kirschner, L. B., Amer. J. Physiol. *217*:596–604, 1969. Sodium fluxes in perfused eel gills.
121b. Kirschner, L. B., Amer. Zool. *10*:365–376, 1970. NaCl transport in aquatic animals.
122. Kirschner, L. B., T. Kerstetter, D. Porter, and R. H. Alvarado, Amer. J. Physiol. *220*:1814–1819, 1971. Adaptation of *Ambystoma* larvae to saline medium.
123. Klahr, S., and N. S. Bricker, J. Gen. Physiol. *48*:571–580, 1965. Sodium transport by turtle bladder.
124. Koizumi, T., Sci. Rep. Tohoku Univ., ser. IV, *10*:269–275, 277–286, 1935. Inorganic composition of tissues in holothurian *Caudina*, and effect of changes in the medium.
124a. Koval'skiĭ, V., and I. E. Vorotaitskaya, Dokl. Akad. Nauk SSSR *187*:1422–1424, 1969. Interaction of Mo and Cu.
125. Krogh, A., Osmotic Regulation in Aquatic Animals. Cambridge University Press, 1939. 242 pp.
126. Lahlou, B., J. W. Henderson, and W. H. Sawyer, Amer. J. Physiol. *216*:1266–1278, 1969. Renal adaptation in toadfish *Opsanus*.
126a. Lam, T. J., Comp. Biochem. Physiol. *31*:909–913, 1969. Ringer solution for stickleback fish.
127. Lange, R., Comp. Biochem. Physiol. *13*:205–216, 1964. Osmotic adjustments, echinoderms and molluscs.
128. Leaf, A., Ann. Rev. Physiol. *22*:111–168, 1960. Ion transport, toad bladder.
129. Lee, B. D., and W. N. McFarland, Publ. Inst. Marine Sci. U. Texas *8*:126–142, 1962. Osmotic and ionic regulation in *Squilla*.
130. LeFevre, M. E., J. F. Gennaro, and W. A. Brodsky, Amer. J. Physiol. *219*:716–723, 1970. Transport by isolated turtle bladder.
131. Liebowitz, A., Amer. J. Hygiene *78*:173–180, 1963. Frog tissue culture medium.
132. Little, C., J. Exp. Biol. *43*:23–37, 1965. Osmotic and ionic regulation in snail *Viviparus*.
133. Little, C., J. Exp. Biol. *46*:459–474, 1967. Ionic regulation in conch *Strombus*.
134. Lockwood, A. P. M., J. Exp. Biol. *36*:546–555, 556–565, 1959. Ion regulation in fresh water isopod *Asellus*.
135. Lockwood, A. P. M., Animal Body Fluids and their Regulation. London, Heinemann, 1963. 177 pp.
136. Lockwood, A. P. M., Comp. Biochem. Physiol. *2*:241–289, 1969. Catalog of physiological salines.
137. Lockwood, A. P. M., and J. A. Riegel, J. Exp. Biol. *51*:575–587, 1969. Excretion of magnesium by *Carcinus*.
137a. Loewe, R., et al., Z. vergl. Physiol. *66*:27–34, 1970. Spider *Cupiennis* hemolymph.
138. Lowenstam, H. A., J. Geol. *69*:241–260, 1961. Mg and Sr in fossil brachiopods.
139. Lowenstam, H. A., pp. 114–132 *in* Isotopic and Cosmic Chemistry, edited by H. Craig. New York, Humanities Press, 1963. Sr/Ca ratios in fossil gastropods.
140. Lowenstam, H. A., pp. 137–195 *in* The Earth Sciences, edited by T. W. Donnelly. Chicago, University of Chicago Press, 1963. Composition of earth sediments.
141. Mackay, W. C., Comp. Biochem. Physiol. *30*:481–488, 1969. Sulfate regulation in jellyfish.
142. Mackay, W. C., and C. L. Prosser, Comp. Biochem. Physiol. *34*:273–280, 1970. King crab concentrations.
142a. Maetz, J., Phil. Trans. Roy. Soc. Lond. B *262*:209–249, 1971. Review; ion transport by fish gills.
143. Maetz, J., and F. G. Romeu, J. Gen. Physiol. *47*:1195–1207, 1964. Na and Cl uptake by gills of goldfish.
144. Mandel, L. J., Nature *225*:450–451, 1970. Ionic transport across biological membranes.
144a. Maniloff, J., et al., Effects of Metals on Cells, Subcellular Elements and Macromolecules. Charles C Thomas, Springfield, 1970. 397 pp.
144b. Martin, A. W., D. M. Stewart, and F. M. Harrison, J. Exp. Biol. *42*:99–123, 1965. Urine formation in pulmonate land snail *Achatina*.
145. Martin, D. W., and P. F. Curran, J. Cell. Physiol. *67*:367–374, 1966. Reversed potentials in isolated frog skin.
146. Mayer, N., Comp. Biochem. Physiol. *29*:27–50, 1969. Ion regulation in *Rana*.
147. Mayer, N., and J. Nibelle, Comp. Biochem. Physiol. *31*:589–597, 1969. Sodium space in eels.
148. McWhinnie, M. A., Comp. Biochem. Physiol. 7:1–14, 1962. Gastroliths in crayfish.
149. Medway, W., and J. Gerach, Amer. J. Physiol. *209*:169–172, 1965. Blood composition in dolphin *Tursiops*.
150. Miles, H. M., and L. S. Smith, Comp. Biochem. Physiol. *26*:381–398, 1968. Ionic regulation in migrating salmon.
151. Minnich, J. E., Comp. Biochem. Physiol. *35*:921–933, 1970. Ion balance in desert *Iguana*.
152. Morris, R., J. Exp. Biol. *42*:359–371, 1965. Salt balance in *Myxine*.
153. Morris, R., and J. M. Bull, J. Exp. Biol. *52*:275–290, 1970. Osmoregulation of larva of *Lampetra*.
154. Motais, R., Ann. Inst. Oceanogr. Monaco *45*:1–84, 1967. Mechanisms of osmotic and ionic regulation in fish.
155. Motais, R., F. G. Romeu, and J. Maetz, C. R. Acad. Sci. Paris *261*:801–804, 1965. Responses of flounder and serran on transfer to fresh water.
156. Motais, R., F. G. Romeu, and J. Maetz, J. Gen. Physiol. *50*:391–422, 1966. Exchange diffusion in teleosts.
157. Naitoh, Y., Zool. Mag. *73*:267–274, 1964. Physiological saline for *Opalina*.
158. Narahashi, T., Adv. Insect Physiol. *1*:175–256, 1963. Physiological salines for insects.
159. Neumann, D., Kieler Meeresforsch. *18*:38–54, 1962. Ions in brackish water animals.
160. Nicholls, J. G., and D. A. Baylor, J. Neurophysiol. *31*:740–756, 1968. Physiological saline for a leech.
161. Nicholls, J. G., and S. W. Kuffler, J. Neurophysiol. *27*:645–671, 1964. Ion distribution in nervous system of leech.
161a. Oglesby, L. C., Comp. Biochem. Physiol. *36*:449–466, 1970. Salt and water balance in *Nereis*.
162. Ohnesorge, F. K., Z. vergl. Physiol. *58*:153–170, 1968. Saline for fish *Tinca*.
163. Okin, D. E., G. Whittenburg, E. E. Windhage, and S. K. Solomon, Amer. J. Physiol. *204*:372–376, 1963. Na fluxes in tubules of *Necturus*.
164. Padmanabhanaidu, B., Comp. Biochem. Physiol. *17*:157–166, 1966. Ions in blood of scorpion.
165. Pantin, C. F. A., J. Exp. Biol. *11*:11–27, 1934. Saline for *Carcinus*.
166. Parry, G., J. Exp. Biol. *31*:601–613, 1954. Ionic regulation in prawn *Palaemon*.
167. Parry, G., J. Exp. Biol. *38*:411–427, 1961. Osmotic and ionic changes in migrating salmon.
167a. Passoro, H., et al., Pharmacol. Rev. *13*:185–224, 1961. Pharmacology of heavy metals.
168. Paul, J., Cell and Tissue Culture, 4th ed. Baltimore, Williams & Wilkins, 1970.
169. Pichon, Y., J. Exp. Biol. *53*:195–209, 1970. Ionic content of hemolymph, cockroach.

170. Pichon, Y., and J. Boistel, J. Insect Physiol. *9*:887–891, 1963; J. Exp. Biol. *47*:357–373, 1967; *49*:31–38, 1968.
171. Pickford, G. E., and F. B. Grant, Science *155*:568–570, 1967. Osmotic concentration in coelacanth *Latimeria.*
172. Poldervaart, A., pp. 119–144 *in* Geol. Soc. Amer. Sp. Paper 62, edited by A. Poldervaart, New York, 1955. Chemistry of the earth's crust.
173. Post, R. L., pp. 19–30 *in* Biophysics of Physiological and Pharmacological Action, edited by A. M. Shanes. AAAS meeting, 1960. Sodium and potassium transport in human red cells.
174. Potts, W. T. W., J. Exp. Biol. *31*:376–385, 1954. Composition of blood, *Mytilus* and *Anodonta.*
175. Potts, W. T. W., Ann. Rev. Physiol. *30*:73–104, 1968. Osmotic and ionic regulation.
176. Potts, W. T. W., and D. H. Evans, Biol. Bull. *133*:411–425, 1967. Sodium balance in *Fundulus.*
177. Potts, W. T. W., and W. R. Fleming, J. Exp. Biol. *53*:317–327, 1970. Hormone effects on ion and water regulation in *Fundulus.*
177a. Potts, W. T. W., M. A. Foster, P. P. Rudy, and G. P. Howells, J. Exp. Biol. *47*:461–470, 1967.
178. Potts, W. T. W., and M. Todd, Comp. Biochem. Physiol. *16*:479–489, 1965. Kidney function in *Octopus.*
179. Prosser, C. L., J. W. Green, and T. J. Chow, Biol. Bull. *109*:99–107, 1955. Ionic balance in *Pachygrapsus.*
179a. Prosser, C. L., W. Mackay, and K. Kato, Physiol. Zool. *43*:81–89, 1970. Ions in marine and fresh-water fish.
180. Quinn, D. J., and C. E. Lane, Biol. Bull. *133*:245–254, 1967; Comp. Biochem. Physiol. *19*:533–543, 1966. Ion transport by gill of land crab.
181. Ramamurthi, R., Comp. Biochem. Physiol. *23*:599–605, 1969. Metabolic responses to salinity stress in fresh water crab.
182. Ramsay, J. A., J. Exp. Biol. *31*:104–113, 1954; *32*:183–199, 1955; *33*:697–708, 1956. Excretion by stick insect *Dixippus.*
183. Ramsay, J. A., J. Exp. Biol. *26*:46–56, 65–75, 1949. Chloride excretion by earthworms.
184.*Ramsay, J. A., J. Exp. Biol. *30*:358–369, 1953; Symp. Soc. Exp. Biol. *8*:1–15, 1954. Movements of electrolytes in insects and other invertebrates.
185. Read, L. J., Comp. Biochem. Physiol. *39A*:185–192, 1971. Body fluids and urine of holocephalan fish.
186. Rehacek, J., and H. W. Brzostowski, J. Insect Physiol. *15*:1431–1436, 1969. Physiological saline for tissues of ticks.
187. Rider, J., and S. Thomas, J. Physiol. *203*:72P–73P, 1969. Vasopressin influence on Na transport by frog skin.
188. Riedesel, M. L., and Volk, G. E., Nature *177*:668, 1956. Serum magnesium in hibernants.
189. Riegel, J. A., J. Exp. Biol. *43*:379–384, 1965; *48*:587–596, 1968. Renal functions in crayfish.
190. Riegel, J. A., and A. P. M. Lockwood, J. Exp. Biol. *38*:491–499, 1961. Excretion by *Carcinus.*
191. Roach, D. K., J. Exp. Biol. *40*:613–623, 1963. Ion analysis, blood of snail.
192. Robertson, J. D., J. Exp. Biol. *26*:182–200, 1949; *30*:277–296, 1953. Ionic regulation in marine invertebrates.
193.*Robertson, J. D., pp. 229–246 *in* Physiology of Invertebrate Animals, edited by B. T. Scheer. University of Oregon Press, 1957. Ionic regulation by invertebrates.
194. Robertson, J. D., J. Exp. Biol. *31*:424–442, 1954. Composition of blood of lower chordates.
195. Robertson, J. D., J. Exp. Biol. *37*:879–888, 1960. Composition of muscle of hagfish and eel.
196. Robertson, J. D., J. Exp. Biol. *38*:707–728, 1961; *42*:153–175, 1965. Composition of muscle in lobster and cephalopods.
197. Robertson, J. D., Biol. Bull. *138*:157–183, 1970. Osmotic and ionic regulation in *Limulus.*
198. Roeser, H. P., G. R. Lee, S. Nacht, and G. E. Cartwright, J. Clin. Invest. *49*:2408–2417, 1970. Ceruloplasmin in iron metabolism.
199. Romeu, F. G., A. Salibean, and S. Pezzani-Hernandez, J. Gen. Physiol. *53*:816–835, 1969. Na and Cl uptake through skin of frog.
200. Rosenthal, H. L., M. M. Eves, and O. A. Cochran, Comp. Biochem. Physiol. *32*:445–450, 1970. Strontium and calcium in skeletons.
201. Rothschild, L., and H. Barnes, J. Exp. Biol. *30*:534–544, 1953. Ions in sea-urchin eggs.
202. Rudy, P. P., Comp. Biochem. Physiol. *18*:881–907, 1966. Sodium balance in *Pachygrapsus.*
202a. Salome Pereira, R., and Sawaya, P., Bol. Fac. Fil. Cien. Univ. Sao Paulo Zool. no. 21, 85–92, 1957. Ions in blood of elasmobranchs.
202b. Schroeder, H. A., et al., J. Nutr. *100*:59–68, 1970. Metal toxicity in rats.
202c. Schwarz, K., and D. B. Milne, Science *174*:426–428, 1971. Growth effects of vanadium in rat.
203. Sharp, G. W. G., C. H. Coggins, N. S. Lichtenstein, and A. Leaf, J. Clin. Ivest. *45*:1640–1647, 1966. Permeability of toad bladder.
204. Sharrat, B. M., I. C. Jones, and D. Bellamy, Comp. Biochem. Physiol. *11*:9–18, 1964. Water and salt composition and renal function in eel.
205. Shaw, J., J. Exp. Biol. *32*:383–396, 1955. Ionic regulation in muscle of *Carcinus.*
206. Shaw, J., J. Exp. Biol. *36*:126–144, 1959; *38*:135–152, 1960. Sodium fluxes in crayfish.
207. Shaw, J., and D. W. Sutcliffe, J. Exp. Biol. *38*:1–15, 1961. Sodium balance, two species of *Gammarus.*
208. Shehadeh, Z. H., and M. S. Gordon, Comp. Biochem. Physiol. *30*:397–418, 1969. Osmotic and ionic regulation in rainbow trout.
209. Skadhauge, E., and J. Maetz, C. R. Acad. Sci. Paris *265*:347–350, 1967. Water permeability of teleost gill.
210. Skinner, D. M., D. J. Marsh, and J. S. Cook, Biol. Bull. *129*:355–365, 1965. Physiological saline for crab *Gecarcinus.*
211. Skou, J. C., Biochim. Biophys. Acta *42*:6–23, 1960. Na-K ATPase in nerve membranes.
212.* Skou, J. C., Prog. Biophys. Molec. Biol. *14*:133–166, 1964; Physiol. Rev. *45*:596–617, 1965. Active transport by cell membranes.
213. Smith, P. G., J. Exp. Biol. *51*:727–738, 739–757, 1969. Ionic relations of *Artemia.*
214. Smith, R. I., J. Exp. Biol. *53*:75–92, 93–100, 1970. Chloride regulation at low salinities by *Nereis.*
215. Snell, F. M., and C. P. Leeman, Biochim. Biophys. Acta *25*:311–320, 1957. Na transport by frog skin.
216. Solomon, A. K., Amer. J. Physiol. *204*:381–386, 1963. Fluxes across prixImal tubules, *Necturus* kidney.
217. Spangenberg, D. B., J. Exp. Zool. *169*:487–500, 1969. Statoliths in *Aurelia.*
218. Steinbach, H. B., J. Gen. Physiol. *44*:1131–1142, 1961. Sodium extrusion by frog sartorius.
219. Steinbach, H. B., Biol. Bull. *127*:310–319, 1962. Ionic balance in planarians.
220. Steinbach, H. B., pp. 677–720 *in* Comparative Biochemistry, vol. 4, edited by M. Florkin and H. S. Mason. New York, Academic Press, 1962. Comparative biochemistry of alkali metals.
221. Steinbach, H. B., Perspectives in Biology and Medicine *5*:338–355, 1962. The prevalence of potassium.
222. Stobbart, R. H., J. Exp. Biol. *36*:641–653, 1959. Sodium regulation in larvae of *Aedes.*
223. Stobbart, R. H., J. Exp. Biol. *54*:19–27, 1971. Na and Cl exchanges during salt uptake, *Aedes* larva.
224. Sutcliffe, D. W., J. Exp. Biol. *39*:325–343, 1962. Composition of hemolymph, aquatic insects.
225. Sutcliffe, D. W., J. Exp. Biol. *55*:325–370, 1971. Ionic balance in gammarids.
226. Thesleff, S., and K. Schmidt-Nielsen, Amer. J. Physiol. *202*:597–600, 1962. Electrophysiology of salt glands of gulls.
227.* Thorson, T. B., Life Sci. *9*:893–900, 1970; pp. 265–270 *in* Sharks, Skates and Rays, edited by P. Gilbert, R. Mathewson, and D. P. Rall, Johns Hopkins University Press, 1967. Osmoregulation and urea content in elasmobranchs from sea, brackish and fresh waters.
228. Thorson, T. B., C. M. Cowan, and D. E. Watson, Science *158*:375–377, 1967.
229. Thuet, P., R. Motais, and J. Maetz, Comp. Biochem. Physiol. *26*:793–819, 1968. Ionic regulation in *Artemia.*
230.* Tosteson, D. C., Fed. Proc. *22*:19–26, 1963. Genetics of active transport, red cells.
231. Tosteson, D. C., Fed. Proc. *26*:1805, 1967. Ion composition in red cells.
232.* Tracey, M. V., Adv. Comp. Physiol. Biochem. *3*:233–270, 1968. Biochemistry of skeletons.
233. Treherne, J. E., J. Exp. Biol. *42*:7–27, 1965. Ion distribution and axon function in stick insect *Carausius.*
234. Treherne, J. E., N. J. Lane, R. B. Moreton, and Y. Pichon, J. Exp. Biol. *53*:109–136, 1970. Potassium fluxes in *Periplaneta* nervous system.
235. Tucker, L. E., Comp. Biochem. Physiol. *36*:301–319, 1970. Ionic regulation in gastropod *Scutus.*
236. Turekian, K. K., and R. L. Armstrong, J. Marine Res. *18*:133–151, 1960. Sr, Ba, and calcite-aragonite in shells.
237. Umminger, B. L., Nature *225*:294–295, 1970. Ions in blood of *Fundulus.*
238. Urist, M. R., pp. 151–179 *in* Bone Biodynamics, edited by H. M. Frost. Boston, Little, Brown & Co., 1964. Composition of skeleton in fishes.
239. Ussing, H. H., and E. E. Windhagen, Acta Physiol. Scand. *61*:484–504, 1964. Paths of sodium transport in frog skin.
240. Utida, S., et al., Comp. Biochem. Physiol. *38A*:443–447, 1971. ATPase activity in eel gill.
241. Van Harreveld, A., Proc. Soc. Exp. Biol. Med. *34*:428–432, 1936. Physiological saline for crayfish.
242.* Vinogradov, A. P., The Elementary Chemical Composition of Marine Organisms. Memoir Sears Found. Mar. Res. II. Yale University Press, 1953. 647 pp.
243. Webb, D. A., Proc. Roy. Soc. Lond. B *129*:107–136, 1940. Osmotic and ionic regulation in *Carcinus.*
244. Webb, D. A., Publ. Staz. Zool. Napoli *28*:273–288, 1956. Vanadium in marine invertebrates.
245.*Webb, D. A., and W. R. Fearon, Sci. Proc. Roy. Dublin Soc. *21*:487–503, 505–539, 1937. Chemical elements in marine animals.

*Reviews.

246. Webber, H. H., and P. A. Daniel, Comp. Biochem. Physiol. *25*:49–64, 1968. Ion balance in gastropod *Acmaea*.
247. Weevers, R. de G., J. Exp. Biol. *44*:163–175, 1966. Physiological saline for moths.
248. Weidler, D. J., and F. P. J. Diecke, Z. vergl. Physiol. *69*:311–325, 1970. Regulation of Na/K in relation to nervous functions in *Carausius*.
249. Weischer, M. L., Zool. Beitr. *11*:517–540, 1965. Copper metabolism in *Helix*.
250. Whittam, R., pp. 313–325 *in* The Neurosciences, edited by F. O. Schmitt, Rockefeller Press, New York, 1967. Molecular mechanisms of active transport.
251. Whittam, R., and M. E. Ager, Biochem. J. *97*:214–227, 1965. The connection between active cation transport and metabolism in erythrocytes.
252. Whittam, R., and J. S. Willis, J. Physiol. *168*:158–177, 1963. Ion accumulation by kidney slices.
253. Whittembury, G., N. Sugino, and A. K. Solomon, J. Gen. Physiol. *44*:689–712, 1961. (See reference 253a.)
253a. Whittembury, G., and E. R. Windhagen, J. Gen. Physiol. *44*:679–687, 1961. Ion permeability and electrical potentials in *Necturus* kidney tubules.
254. Wiederholt, M., et al., J. Gen. Physiol. *57*:495–525, 1971. Saline for *Amphiuma*.
255. Wieser, W., J. Marine Biol. Assoc. U. K. *45*:507–523, 1965. Copper metabolism in isopods and amphipods.
256.*Williams, R. J. P., Biol. Rev. *28*:381–415, 1953. Metal ions in biological systems.
257. Willis, J. S., Biochim. Biophys. Acta *163*:516–530, 1968. Cation transport and respiration in kidney slices.
258. Winkler, L. R., Veliger *11*:268–271, 1969. Bromine in *Aplysia*.
259. Wolf, K., and M. C. Quimby, pp. 253–305 *in* Fish Physiology, vol. III, edited by W. Hoar and D. J. Randall. New York, Academic Press, 1969. Culture media for fish tissues.
260. Wood, D. W., J. Physiol. *138*:119–139, 1957. Physiological saline for *Carausius*.
261. Wright, E. B., and T. Tomita, Amer. J. Physiol. *202*:856–864, 1962. Na and K carriers in crustacean axons.
262. Zadunaisky, J. A., and O. A. Canolia, Nature *195*:1004, 1962. Active transport of Na and Cl by the isolated skin of frog *Leptodactylus*.
263. Zaugg, W. S., and L. R. McLain, Comp. Biochem. Physiol. *38*B:501–506, 1971. Increased gill Na-K ATPase in salmon in sea water.
264. Zerahn, K., Acta Physiol. Scand. *36*:300–318, 1956. O_2 consumption and active Na transport in isolated and short circuited frog skin.

Students are referred to a symposium on active transport of salts in Phil. Trans. Roy. Soc. Lond. B *262*:83–342, 1971.

CHAPTER 3

NUTRITION

By C. L. Prosser

Nutritive patterns of different animals show wide variation in what is required in the diet and what can be synthesized. Organic nutrients are essential in three different orders of magnitude: (1) Energy-yielding compounds must be available in sufficient quantity to permit growth and work, usually in quantities of several grams per kilogram of body weight daily. (2) Substances such as amino acids, purines, and some lipids are needed in milligram amounts in diet or by synthesis to provide carbon skeletons for more complex organic molecules. (3) Specific factors—vitamins, coenzymes—are needed in quantities of micrograms per kilogram body weight daily. Animals differ greatly in their capacity to use different energy-yielding compounds, to convert compounds of one class of foodstuff to another, and to synthesize the specific dietary requirements.

What is a vitamin for one animal may not be a dietary requirement for another. The amino acids which constitute proteins are similar in all organisms, but the proportions vary from one protein to another; the coenzymes of cellular metabolism, such as the phosphopyridine nucleotides, are similar in all cells. Animals differ as to which amino acids and coenzymes they can synthesize and which they must take as food. In many animals, certain essential substances are synthesized in insufficient amounts, so that they are required in the diet. Many animals obtain essential nutrients from symbiotic bacteria.

The nutrition of a given animal reflects its microhabitat, but food selection is often more influenced by taste preferences than by nutritive values (Chapter 13). Herbivores, carnivores, omnivores, and dietary specialists usually differ in digestive capacities (Chapter 4). The basic aspects of nutrition were established before the appearance of animals as we know them, and the evolution of animal nutrition has many examples of loss of ability for specific synthesis and subsequent dependence on external sources.

ORIGIN OF NUTRITIVE TYPES

Organisms fall into three classes with respect to sources of energy: (1) Chemotrophic organisms (largely bacteria) obtain energy from inorganic reactions such as oxidation of iron or sulfur. (2) Phototrophic cells use sunlight in photosynthesis. (3) Heterotrophic organisms oxidize preexisting organic compounds of varying degrees of complexity; some organisms can use lower fatty acids such as acetate, but most need sugars or higher carbohydrates. With respect to nitrogen sources, the following classes occur: (1) autotrophs (most photosynthesizing algae and higher plants) use inorganic nitrogen, as nitrate or ammonia. (2) Mesotrophs can rely on a single amino acid, or NH_4 plus fatty acid. (3) Metatrophs require many amino acids, either in compounds or in mixtures. All animals except green flagellate protozoans are ultimately dependent on photoautotrophic plants for their carbon sources of energy. Flagellates present the widest range of nutritional patterns: some can use nitrate, some use ammonia, and a few are mesotrophic or metatrophic; many of them require a few vitamins, particularly cobalamin. Parasitic flagellates have highly specialized requirements. Nutritionally, flagellates have both plant-like and animal-like aspects.

Modern ideas of the origin of life stem from the work of Oparin, and it is postulated that the first organisms were heterotrophic forms which used preexisting compounds; then phototrophs appeared, to be followed by the heterotrophic forms we now know as animals.

The age of the earth is placed by geochemists at between 4.5 and 5×10^9 years. Release of gas from the alkaline crust and upwelling of volcanic gases resulted in an atmosphere containing N_2, H_2, CO, H_2O, and some CO_2, and fixed the pH of the oceans at between 8 and 9. Action of ultraviolet radiation from the sun caused formation of HCN. It was formerly thought that methane and ammonia were the first compounds of C and N for biosynthesis, but now HCN appears to be a more likely starting substance. Irradiation of HCN in solution by either ultraviolet light or electric discharge yields NH_3 plus amino acids, particularly serine and glycine, as well as pyruvate and some methane.[1] A mixture of HCN, NH_3, and H_2O leads to synthesis of adenine and amides.[98] Also, irradiation of solutions of methane, ammonia, and hydrogen results in synthesis of a variety of amino acids, lower fatty acids, and purines.[91] Similar small organic molecules have been found in meteorites, and laboratory experiments show that organic compounds can be formed by the action of interstellar energy even in the absence of water. An additional source of energy in the primitive earth was volcanic heat. With mild heating, amino acids have been found to polymerize in certain preferred sequences, and some of the resulting protenoids have the catalytic activities of hydrolysis, decarboxylation, amination, deamination, and oxidoreduction.[45, 46, 56] Artificial peptides tend to coil in an alpha helix as soon as eight to ten members are formed, and polymerization accelerates as coiling proceeds.[50] Under primitive earth conditions, the amino acids which became incorporated into proteins were L-isomers. Metallo-organic compounds probably appeared early; iron porphyrins would have been effective catalysts of energy-yielding oxidations and could have developed to photochemical reactions. Thus, synthesis of complex proteins apparently went to considerable complexity before mechanisms of translation appeared. Protein "protocells" may have been formed by the establishment of surfaces with selective properties.[46]

Purines and pyrimidines, as well as riboses, are among the small organic molecules formed in primitive earth conditions.[13] Heating of nucleosides together with inorganic phosphates leads to formation of nucleotides – such compounds as cyanogen are condensing agents for formation of uridine phosphate and for cyclization of this compound.[85] Nucleic acids were probably formed independently of proteins, and complexing with proteins came later.

Chains of amino acids complex with nucleotides in specific ways. Positive amino acid residues bind to negative nucleotides, and the latter stabilize polypeptide chains.[82] Polylysine chains complex with adenine-thymidine chains, and polyarginine chains with guanisyl-cytosyl nucleotides.[44] If, in a polylysine chain, alternate lysines were replaced by other amino acids, the chain would bind solely to one chain of a double strand. Various models for establishment of codons and translation of a nucleic acid segment to an amino acid sequence have been proposed.[121]

The aggregates of nucleoproteins having capacity for duplication, and associated with other proteins capable of enzymatic degradation of organic substances for energy, were the first organisms.[50] In these heterotrophic pro-organisms, those properties basic to all life evolved – use of high energy phosphate bonds for transferring energy for biological work, stepwise electron transfer in intermediary metabolism, use of metallo-protein catalysts, use of L-amino acids in natural proteins, active transport of ions by cell surfaces, the use of certain ions intracellularly and rejection of others, selective permeability of membranes, and ultimately the control of protein synthesis by nucleic acids, and other universal properties of living things. Biochemically and biophysically, far more evolution took place before there were living organisms as we know them than has occurred since. The time during which chemical evolution occurred was longer than the time for organic evolution. The similarity of living organisms is more impressive than their diversity.

As the supply of small organic compounds became depleted, mechanisms for synthesizing them had selective value, and chemoautotrophs (followed by photoautotrophs) evolved from the heterotrophic pro-organisms. The early heterotrophs and the chemo-

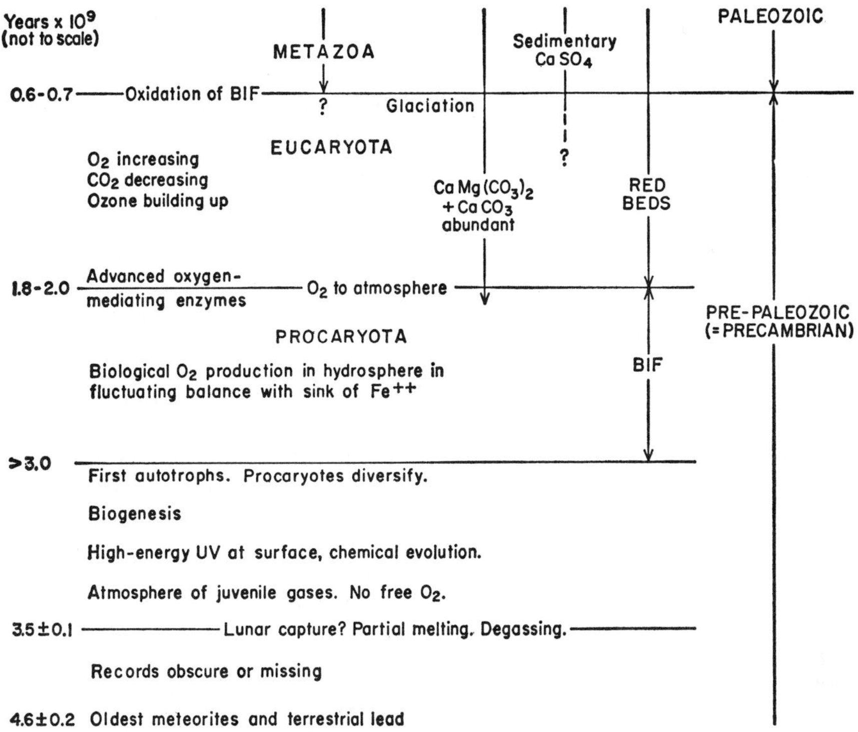

Figure 3–1. Postulated main features of interacting biospheric, lithospheric, and atmospheric evolution on the primitive earth. (From P. Cloud, Science *160*:729–736, 1968.)

autotrophs were presumably prokaryotic bacteria, and the first photoautotrophs were prokaryotic blue-gree algae.[111b] The first photoautotrophs appeared about 1.8×10^9 years ago (2×10^9 according to Barghoorn and Tyler[7]) (Fig. 3–1). This is the age of some deposits of banded iron oxides. Possibly the iron removed the O_2 which was formed in photosynthesis, and which was toxic to the existing anaerobic prokaryotes. The oxygen formed by the photoautotrophs caused a shift from an anaerobic (reducing) to an aerobic (oxidizing) atmosphere, which blocked the growth of anaerobes and screened out solar ultraviolet radiation.

The differences between prokaryotes and eukaryotes are much more than nutritional: a condensed nucleus, mitochondria as sites for energy reactions, the presence of endoplasmic reticulum, and a pattern of 9+2 microtubules in each flagellum all characterize eukaryotes; absence of these structures characterizes prokaryotes.[89a] Early fossils of eukaryotic algae are considered to be 1.6×10^9 years old.[7] In general, the ability to use light or inorganic reactions to provide energy for CO_2 reduction accompanies the ability to use nitrate or ammonia.[17]

With the change in the atmosphere and the accumulation of organic compounds formed in photosynthesis, heterotrophy was again possible and the evolution of animals began. The break between phototrophy and heterotrophy is not necessarily sharp; some photosynthesizing flagellates are phagocytic, and many non-green flagellates are predatory. Heterotrophy evolved in eukaryotes both with and without loss of photosynthesizing capacity. All of the O_2 in the atmosphere comes from photosynthesis, and it is renewed every 2000 years; the CO_2 turnover time is 300 years.[8]

Metazoa evolved rapidly and polyphyletically early in the Cambrian period, which extended from 520 to 440×10^6 years ago, because virtually all ecological niches were unoccupied. Beginning with the generalized patterns of nutrition in the flagellates, animals showed increasing specialization and dependence on environmental sources for nutrients. The type of basic foodstuff used by an animal depends partly on its position in a food chain. Many animals use symbiotic microorganisms not only to synthesize vitamins but to digest basic foodstuffs. A few animals contain photosynthetic zooxanthellae within their tissues and use products from these organisms as food.

CARBON AND NITROGEN REQUIREMENTS

Phototrophic green flagellates (chloroflagellates) reduce CO_2 directly in light, and they use the resulting organic acids and sugars for energy in the dark. Modern colorless flagellates (leucophytes) were derived from green forms, and some species can change color and form according to the amount of ambient light. Leucophytes have been cultured extensively on acetate and other organic acids, and there are species differences as to which acids can be used. Numerous flagellates use lower fatty acids and alcohols but cannot use sugars. *Prototheca* uses saturated fatty acids up to palmitic, and some alcohols, but not formic acid or dicarboxylic acids. *Polytoma caeca* can use acids of up to five carbons (valerate) but not caproate, yet it can use hexyl alcohol.[120] *Chlamydomonas* uses only the even-numbered fatty acids, the even-numbered monohydric alcohols, and lactic and tricarboxylic acids, while the odd-numbered alcohols above C_1 inhibit growth. *Chilomonas* uses only even-numbered carbon fatty acids, alcohols, lactic acid, and the acids of the TCA cycle; odd-numbered acids above C_1 inhibit growth, and there is no use of sugars.[18] Some flagellates can alter their ability to use specific substrates by a process of enzyme induction.

A number of coelenterates contain symbiotic algae which, by photosynthesis, provide fatty acids to the host. In a green hydra, some 10% of the carbon fixed by symbiotic algae is assimilated by the animal, mostly as glucose or maltose.[93] In reef corals, 35 to 50% of total fixed carbon appears in animal tissue, much of it as glycerol.[94] Similar use of intracellular algae occurs in a few protozoans, sponges, free-living flatworms, and molluscs. Another example of dependence of metazoa on microorganisms to provide organic compounds is in ruminants. The rumen bacteria (and to a lesser extent rumen Protozoa) provide lower fatty acids, particularly acetate, which are absorbed by the host (see p. 147). Termites appear to be nourished mainly by lower fatty acids formed by symbiotic flagellates from cellulose in the digestive tract.

Many parasites are able to use only foods which have been partly degraded. Several cestodes, for example, can use only a few monosaccharides and no disaccharides. *Lacistorphychus,* which is parasitic in a dogfish, uses glucose and galactose but not mannose, fructose, maltose, sucrose, or lactose.[104]

The ciliate *Tetrahymena* can make carbohydrates and fatty acids from acetate. It lives well with a combination of glycerol, acetate, and Krebs cycle acids, and is commonly cultured with glucose as the principal energy source. It can use also levulose, mannose, and maltose but cannot use some thirteen other sugars, one of them being sucrose.[68] *Tetrahymena,* like plants and many bacteria, uses the glyoxalate cycle, which permits the use of glycine as an energy source; metazoa lack two enzymes of the glyoxylate cycle—isocitrase and malate synthetase.

Some adult insects normally live exclusively on a diet of sugars. Adult honeybees can survive on any of seven sugars which are sweet to them and on six which are tasteless, but they cannot use five others which are tasteless. Adult blowflies can use all pentoses except fucose and all hexoses except sorbose; they can use maltose, sucrose, trehalose, and melezitose well, but use lactose and cellobiose poorly. There is a poor correlation between utilization and taste stimulation of the tarsi.[57] Blowflies can live on α-glucosides and α-galactosides but not on other glycosides.[47] If sucrose, glucose, or fructose was omitted from cultures of a wood borer, *Ips cembrae,* they did not thrive.[6]

The interconversion of carbohydrates, fats, and proteins has been studied more in mammals than in other organisms. Certain animal tissues, as well as bacteria, can fix some CO_2 in the reaction from pyruvic to oxaloacetic acids; it is probable that such CO_2 fixation can occur in most cells. When mice breathe $C^{14}O_2$, some of the C^{14} appears in liver glycogen within a few minutes. The normal equilibrium in intermediary metabolism is such that CO_2 fixation is negligible in the over-all carbon economy. Acetate can be substituted for part of the carbohydrate in a rat's diet. Some protein and fat are needed to provide essential amino acids and fatty acids, but it is possible for mammals to dispense with dietary carbohydrate. The total requirement for carbon compounds is largely determined by the caloric needs. For an average man these needs range from 65 kcal/hr when asleep to 500 kcal/hr at strenuous work. The energy yield of foodstuffs in kcal per gram is approximately 4.1 for protein and carbohydrate and 9.3 for fat.

Relatively few organisms—nitrogen-fixing bacteria and fungi—can utilize atmospheric nitrogen. Most photosynthetic plants can utilize nitrates. Blue-green algae can use nitrate or ammonia but not free nitrogen. Some colorless flagellates (*Astasia* and *Chilomonas*) require ammonia as a nitrogen source.[70]

All animals other than flagellates require mixtures of amino acids. Ruminant mammals harbor symbiotic bacteria, which build essential amino acids from ammonia and urea. Purines are formed in the rumen from NH_3 and amino acids in such amounts as to permit high purine N excretion.[39] Other mammals have been found to utilize small amounts of ammonia in combination with a mixture of essential amino acids. Many animals—Protozoa, insects, and vertebrates—have been cultured or maintained on mixtures of amino acids as the sole nitrogen source. Some need exogenous purine-pyrimidine, while others can make these necessary compounds. Cestode worms appear to be unable to utilize whole protein and to use only amino acids; they normally live in a medium rich in amino acids, the intestine of a host. Some ciliates (*Tetrahymena*) do well on amino acids as nitrogen source, while some carnivorous ciliates (*Stylonychia, Euplotes*) have not been cultured without whole protein or polypeptides.[77] *Paramecium aurelia* does not need a supplementary carbon source if adequate amino acids are supplied.[115]

In conclusion, intermediate degradation products of the three classes of basic foodstuffs enter common enzymatic pathways; hence, there is some common usage and interconversion. Among heterotrophs, dietary differences exist in requirements for carbon and nitrogen compounds, resulting mainly from differences in digestive enzymes and in enzymes for the initial stages of degradation.

IDENTIFICATION OF SPECIFIC FACTORS IN NUTRITION

In biochemical evolution, certain types of organic compounds came to be essential for protoplasmic structure and function. Synthesis of proteins, carbohydrates, fats, and nucleic acids requires certain structural components. In addition, energy-yielding and synthetic reactions came to be catalyzed by enzymes consisting of proteins plus specific cofactors. Later, specialized functions (often in single or several tissues of certain groups of animals) developed, with unique requirements for specific carbon compounds.

The evolution of the nutrition of special requirements took place partially by the loss of capacity for synthesis, so that what is a dietary essential for one animal is not required by another. Frequently, symbiotic microorganisms synthesize the essential nutrient, so that a host lacking the microorganisms needs the "vitamin" but one with normal symbionts does not need it. A parasite may live in a medium rich in essential substances and use exogenous supply. As an example of need for dietary balance, larvae of the fly *Agria* grow best on a ratio of 2.25 parts amino acid to 0.5 parts glucose.[67]

Many dietary essentials are needed in larger amounts for growth than for adult maintenance; hence, they are called growth factors. Need for an essential nutrient is recognized by retarded growth or by some physiological defect in adults after they have been fed a diet deficient in a given substance. Experimentally, nutritionists aim first at axenic culture, i.e., culture in the absence of other living organisms, particularly bacteria, yeasts, and protozoans. This usually means supplying a source of protein and carbohydrate plus a sterile extract of yeast, liver, or other mixture. The next steps are to fractionate the protein and ultimately culture the organism on a mixture of amino acids, and to substitute various known vitamins for the organic extract.

Many kinds of animals have been cultured in media containing the following ingredients: a single protein such as casein as the source of amino acids, sugar (or lower fatty acid), yeast or liver extract, plus some vitamins. Tables 3–1a and 3–1b give the culture media for a few selected species which have been grown successfully with known dietary components. These mixtures are arrived at empirically and are not minimal.

The quantitative approach to nutrition is made difficult by the fact that a given nutrient is sometimes essential only if some other nutrient is lacking. Precursors may be substituted for an essential nutrient, as may related substances. A compound may obviate the need for a more usual nutrient, and may or may not completely replace it. Another difficulty is that nutrients must be ingested

TABLE 3–1a. Axenic Culture Media for Three Protozoans. Values are in mg/100 ml of Culture Medium.

	Tetra-hymena[68]	*Paramecium aurelia*[115]	*Crithidia*[79]
glycine	–	40	–
valine	15	80	66
leucine	50	60	97
serine	15	80	–
lysine	30	50	76
isoleucine	30	40	63
histidine	250	20	21
arginine	40	60	43
phenylalanine	10	40	50
tyrosine	5	20	20
tryptophan	12	15	12
proline	–	40	–
glutamic acid	250	–	–
threonine	30	80	44
methionine	25	25	34
thiamine	0.5	1.5	0.2
biotin	0.0001	0.0001	0.02
riboflavin	0.3	0.5	0.2
niacin	2.0	0.5	0.5
pyridoxine	0.2	0.5	0.4
folic acid	0.05	0.5	0.2
Ca pantothenate	3	1.0	0.8
sterols	–	stigmasterol 0.2	–
choline	–	–	0.3
purines	guanylate 12	adenosine 22.5 guanosine 37.5	adenine 5
pyrimidines	uracil 6	cytidine 22.5 thymidine 20.0 uridine 22.5	–
thioctic acid	0.01	0.01	–
glucose (equiv.)	400	50	100

TABLE 3–1b. Axenic Culture Media for Four Insects. Values in mg/100 ml of Culture Medium, Except *Tribolium* Values in g/100 g Dry Weight.

	Tri-bolium[81]	*Onco-peltus*[116]	*Hymen-olepis*[10]	*Hy-lemya*[49]
glycine	120	80	15	175
alanine	400	100	17.8	110
valine	2440	80	23.4	136
leucine	1470	80	–	235
serine	240	80	21	88
lysine	1800	120	36.4	134
isoleucine	1950	80	26.2	126
histidine	1050	80	31	48
arginine	600	270	42	80
phenylalanine	1470	40	33	101
tyrosine	730	40	36.2	124
tryptophan	490	80	40.8	175
proline	240	80	23	168
hydroxyproline	120	–	–	38
glutamic acid	2440	140	29.4	440
aspartic acid	500	–	26.6	122
threonine	1700	140	21.3	38
methionine	980	40	29.8	34
cysteine	–	40	–	48
cystine	240	–	48	–
asparagine	–	550	26.4	–
glutamine	–	150	29.2	–
thiamine	0.1	2.5	–	0.15
biotin	0.005	0.1	–	0.002
riboflavin	0.2	0.5	–	0.24
niacin	0.8	10	–	1.0
pyridoxine	0.1	2.5	–	3.0
folic acid	0.002	0.5	–	0.6
Ca pantothenate	0.4	5.0	–	0.6
sterols/cholesterol	1000	–	–	10
choline	150	50	–	2
carnitine	0.001	–	–	–
purine	–	–	40	RNA 100
inositol	–	50	–	33
pyrimidine	–	–	30	thymine 0.4
sugar	69,300	50,000	200	1500

before they can be utilized, and certain diets may be adequate nutritionally but are not accepted by an animal. Requirement for a nutrient may first appear in a second generation on a diet lacking it; hence, small but sufficient amounts may be carried in the egg for one generation, particularly in insects.

Some nutrients are required by classes or possibly by phyla of animals, while others are needed only by certain genera or species. Most of the so-called essentials are used in cellular metabolism by all animals but are synthesized to a varying extent. Some synthesizing abilities have been retained by all animals; some must have been lost in animal evolution, and others at the establishment of certain classes or phyla; still others have been genetic losses in single species or strains.

There is no satisfactory classification of all dietary essentials; they can, however, be grouped according to the way they are used: (1) Some organic compounds are needed because they provide carbon skeletons or essential side-groups of organic molecules. These substances are needed in relatively large amounts (0.5 to 2 g/day by man). Usually, organic homologues or substances which are similar in some key structure can be substituted for the normal nutrient. The best known of these are the essential amino acids. (2) Similar to the preceding category in amount required are some small organic molecules which are incorporated in large ones. The requirements for these compounds are extremely variable in different animals. These substances include purines and pyrimidines, essential fatty acids, and some sterols. (3) Other dietary essentials are used as coenzymes in specific reactions of intermediary metabolism. They are needed in small amounts (0.11 to 1 mg/day in man), and they may be grouped together as water-soluble vitamins. (4) Still other substances are less universally needed and may function only in specialized tissues. Examples are some fat-soluble vitamins. These resemble hormones and may have specific target organs. The need for coenzymes evolved very early and is most widespread; that for specific amino acids appeared early in animal evolution. The specialized needs are often for substances in a class of compounds used previously for other functions.

AMINO ACID REQUIREMENTS

During growth, new protein is continually being synthesized. In adult animals, the body proteins are being replaced at variable rates; the half-time for replacement of total protein in a rat is 17 days, while that in man is 80 days.[111a] For the synthesis of protein, animals need varying proportions of the twenty common amino acids from whatever source. In addition, specific amino acids are needed for the synthesis of various nitrogen-containing compounds other than proteins. The need for certain amino acids as carbon skeletons can be satisfied by analogues; e.g., phenylpyruvic acid can be aminated to form phenylalanine, and this can then go to alanine.

Photoautotrophic organisms can synthesize each of the twenty normally occurring amino acids, but animals have lost the ability to synthesize some of them. Some proteins are inadequate nutritionally in that they are deficient in essential amino acids. Zein is deficient for mammals in its low content of tryptophan, lysine, cysteine, and hydroxyproline; human and beef hemoglobins are low in isoleucine and rich in methionine, whereas dog hemoglobin is rich in isoleucine and lacking in methionine. Egg albumin is low in leucine and glutamic acid; casein is low in glycine. The cockroach *Blatella* grows well on fibrin or casein, less well on egg albumin or wheat glutin, and very poorly on zein, gelatin, or hemoglobin.[83] *Tetrahymena* uses proteins of fish meal or egg albumin twice as well as those of soybeans or linseed, and uses gelatin hardly at all. Lactalbumin is a poorer protein than casein for the nutrition of *Tenebrio,* but it can be improved in nutrient quality if arginine or tyrosine, or mixtures of these with other amino acids, are added.[28] In human and domestic animal nutrition, supplementation of grain proteins by synthetic amino acids is now providing a balanced diet, where previously amino acid deficiencies have been common. Also, genetic strains of grains rich in certain amino acids are being selected, such as high-lysine corn. Supplementation of wheat flour, where it is the sole protein source, by lysine at 0.75 to 1 g/kg protein is recommended. Lysine is the first limiting amino acid in corn, and methionine is second, except that on a high-zein diet tryptophan becomes limiting.

The essential amino acids have been identified by (1) systematic omission of each from the diet and (2) measurement of capacity, or lack of it, to incorporate C^{14} from fatty acids or sugars into amino acids. A few of the amino acids are dispensable for adults but stimulate growth and are needed as dietary supplements at certain developmental stages. The required amount of amino acid often depends on the balance of other amino acids. Interconversions among amino acids are numerous, and some may be used as precursors for others. An increase of one may cause a deficiency of the next most limiting amino acid. For example, in rats, if threonine is added alone to a suboptimal diet, growth is retarded unless tryptophan is also added in proportion. If leucine is added to a 9% casein diet, growth is less than on the casein diet alone, but if isoleucine also is added, normal growth resumes.[40]

Toxicity of single amino acids may be antagonized not only by other amino acids but sometimes by B vitamins. In rats, high methionine in a low casein diet is a depressant that is counteracted by pyridoxine.[40] Natural amino acids are the L-acids, and these are better used than D-amino acids. However, there is variation as to whether a D-acid can be used, is indifferent, or is toxic. Statements of essentiality of amino acids must, therefore, be qualified by reference to other dietary components.

With recognition of the preceding restrictions, Table 3–2 summarizes selected data on essential and dispensable amino acids. The most striking feature is the similarity in re-

TABLE 3–2. Requirements of Animals for Amino Acids.

Animal	*glycine*	*alanine*	*valine*	*leucine*	*serine*	*lysine*	*isoleucine*	*histidine*	*arginine*	*phenylalanine*	*tyrosine*	*tryptophan*	*proline*	*hydroxyproline*	*glutamic acid*	*aspartic acid*	*threonine*	*methionine*	*cysteine*	*cystine*
Protozoa																				
Tetrahymena pyriformis[38, 77]	–	–	+	+	–,+	+	+	+	+	+	–	+	–	–	–	–	+	+	–	–
Paramecium aurelia[115]	+	–	+	+		+	+	+	+	+	+	+	+		–	–	+	+	+	
Paramecium multimicronucleatum[71]	+	stim	+	+	or gly	+	+	+	+	+	+	+	+	–	stim	stim	+	+		
Crithidia[79]	–	–	+	+	–	+	+	+	or cit	+	+	+	–	–	–	–	stim	or cys	or met	
Roundworms																				
Caenorhabditis[33]	–	–	+	+	–	+	+	+	+	+	+	+	–		–		+	+	–	
Insects																				
Phormia[60]	–	–	+	+	–	+	+	+	+	+	–	–	+	–	or asp	or glu	+	or cys	or met	
Phormia[74]	–	–	+	+	–	+	+	+	+	or tyr	or φ ala	+	–		–	–	+	+		+
Bombyx[69]	–	–	+	+	–	+	+	+	+	+	–	+	+		+ or asp	+ or glu	+	+		–
Musca[14]	–	–	+	+	–	+	+	+	+	+	–	+	–	–	–	–	+	+	–	
Calliphora[112]	–	–	+	+	–	+	+	+	+	+	–	–	–	–	–	–	+	+	–	+pupa
Drosophila[59]	stim	–	+	+	–	+	+	+	or cit	+	–	+	–	–	–	–	+	+		
Aedes larva[114]	– (+pupa)	–	+	+	stim	+	+	+	+	+	–	+	stim	stim	–	–	+	+	–	
Agrotis[49]		–		+	–	+	+	+	+	+	or φ ala		–		–	–		stim	–	+
Apis[86]	–	–	+	+	–	+	+	+	+	+	+	+	–	–	–	+	+	+		+
Apis[29]	–	–	+	+	–	+	+	+	+	+	–	+	stim	–	–	–	+	stim		–
Argyrotaenia[107]			+	+		+	+	+	+	+		+					+	+		
Crustacea																				
Astacus[122]	–	–	+	+	–	+	+	+	–	+	–		–	–	–	–	+			
Vertebrates																				
salmon[55]	–	–	+	+	–	+	+	+	+	+	–	+	–	–	–	–	+	+	–	
chicken[4]	stim	–	+	+	–	+	+	+	+	+	–	+	–	–	stim	–	+	+		
rat[109]	–	–	+	+	–	+	+	+	+ growth	+	–	+	–	–	–	–	+	+		
man[109]	–	–	+	+	–	+	+	+ growth	+ growth	+	–	+	–	–	–	–	+	+		

Key: stim = stimulates growth, but not required
– = not required
+ = required
+ growth = needed for growth, not maintenance
+ pupa = needed for that stage of life

quirements by all animals. Apparently, the general pattern of lack of synthesizing capacity of some nine (or ten) amino acids was established very early in animal evolution. Superimposed on this general pattern of similar amino acid requirements, there are specific requirements for certain animals. Some amino acids are indicated as stimulating growth but not necessary for adult maintenance; pairs of others are indicated as interchangeable.

Tetrahymena gelei, strain W, was the first ciliate to be cultured on a synthetic medium of known chemicals, and ten amino acids are required (Table 3–2). Methionine can be replaced by homocysteine as a sulfur-containing amino acid[77] and growth is stimulated by serine or glycine. Of sixteen strains of *Tetrahymena gelei,* only one needs exogenous serine; this strain lacks an aldolase for synthesis of serine from threonine. Fourteen strains can synthesize serine from glycine or threonine provided the vitamin folic acid is present.[30] In *Tetrahymena,* tyrosine spares but cannot replace phenylalanine, and both citrulline and ornithine can spare arginine.[76] Histidine is toxic; in high concentration, the D-isomers of threonine, isoleucine, and tryptophan are not utilized but D-arginine is used.[77] Cysteine and homocysteine can spare methionine for *Tetrahymena;* quantitative needs depend on the balance among amino acids. In *Paramecium aurelia* culture, glycine can be replaced by serine if thymidine is present.

For a trypanosomid flagellate from mosquito, *Crithidia,* phenylpyruvic acid can replace phenylalanine; cysteine, cystathionine, or homocysteine can replace methionine; and citrulline but not ornithine can replace arginine.[79] Tapeworms use amino acids from the host, and *Hymenolepis* supplied with α-ketoglutaric, pyruvic, or oxaloacetic acids plus ammonium ions can synthesize L-amino acids.[25]

Several insects have been grown on mixtures of amino acids as the sole nitrogen source (Table 3–1b). Larvae of the blowfly *Calliphora* do not require proline. Larvae of *Phormia* can use either methionine or cysteine and either glutamic or aspartic acid.[60] The grain beetle *Oryzaephilus* needs arginine, and this can be replaced by citrulline but not by ornithine; arginine can be synthesized from alanine and proline, and these two amino acids are not needed if arginine is available in quantity. Cysteine can replace cystine and is toxic in high concentrations, but the toxicity is reversed by increased methionine.[27] A cutworm, *Agrotis,* cannot incorporate carbon from glucose to form tyrosine, but can synthesize it from phenylalanine.[73] The roach *Blatella* can mature but fails to produce viable eggs in the absence of leucine, isoleucine, and valine.[54] A diet was prepared for the leafroller *Argyrotaenia* based on carcass analysis. Unbalanced dietary mixtures could be used if amino acids were at high levels or if indispensable amino acids were supplemented by dispensable ones; cysteine is sparing for methionine, and tyrosine for phenylalanine.[107] *Phormia* can dispense with tyrosine if alanine is present in the diet.[74] For the silkworm *Bombyx,* histidine is not indispensable but if it is added, larval growth improves, especially if aspartate is not present; cysteine spares methionine, and tyrosine spares phenylalanine.[69]

Dispensable and indispensable amino acids for the insect *Heliothis zea* were determined by (1) dietary omission and (2) incorporation of C^{14} from glucose into amino acids. The two methods agreed in showing that the indispensable amino acids are leucine, lysine, histidine, arginine, valine, methionine, and isoleucine; the dispensable amino acids are aspartic, glutamic, glycine, serine, alanine, and cysteine. Tryptophan is indispensable by feeding, but it is hydrolyzed in the isolation procedure; hence, it cannot be tested by C^{14}. Phenylalanine is indispensable and tyrosine is dispensable by feeding, but they overlap radiometrically and tyrosine is synthesized from phenylalanine. Threonine and methionine are labelled but apparently are not synthesized in sufficient amounts; hence, they are indispensable by feeding. Thus, the two methods agree except for special cases.[106]

Quantitative differences exist in the reserves of some amino acids for different insects. In onion maggots (*Hylemya antiqua*), lack of several amino acids caused death in the first instar; lack of others caused death in the third instar, and lack of still other amino acids resulted in death just before pupation.[49] Cysteine is necessary for emergence of adult mosquitoes, *Aedes aegypti,* and for growth of their larvae either phenylalanine or tyrosine (or both) is needed.[53] Strain differences exist in amino acid requirements for *Drosophila melanogaster.*

The amino acid requirements of mammals have been investigated by omitting various amino acids from a food mixture and observing weight and nitrogen balance, as well as by observing incorporation of C^{14} from other sources into amino acids. Arginine is synthesized by rats, but not at a rate sufficient for normal growth. Cysteine is also needed and can be synthesized if adequate methionine is present. Adult humans require eight amino acids, whereas growing rats need ten (Table 3–2). Human infants need histidine and arginine in addition to the eight other amino acids required by adults.[2a] Those amino acids required in the diet by man have six or more enzymes in their synthetic path, whereas the amino acids not required have three or fewer steps; when the exogenous supply of an amino acid is high, feedback inhibition of synthesis of that amino acid occurs.[26] Man requires most of the essential amino acids in amounts varying from 0.25 to 1.0 g per day. In general, D-acids cannot be used well, but phenylalanine and methionine can be used in the D-form.[109] Tyrosine is needed by rat and man if the ration is low in phenylalanine. When cysteine and tyrosine supply are adequate or high, the amounts of methionine and phenylalanine can be reduced. When lambs are fed urea as the sole nitrogen source to rumen organisms, nitrogen balance is more positive if supplementary methionine is supplied; lysine is less effective, as is threonine, and either of these favors N retention only if given with methionine.[95] Chickens require glycine and arginine; citrulline but not ornithine can be substituted for arginine, and tyrosine is needed if the ration is low in phenylalanine.[4] Salmon require the same nine amino acids as do mature rats.[117]

In conclusion, the basic pattern of requirement of essential amino acids is similar in all animals. Some specific differences occur, even for genetic strains; these represent lack of one or more synthetic enzymes. Quantitative requirements depend on the amounts of precursors or on replacement by substances with comparable carbon skeletons.

SPECIAL DIETARY REQUIREMENTS OF SOME ANIMALS

Choline. Choline, sometimes classed with the water-soluble vitamins, provides essential structural components for several syntheses. It is a constituent of lecithin (phosphatidyl choline) and has a quaternary nitrogen; hence, it is important in such compounds as acetylcholine, and it is an important source of labile methyl groups needed in transmethylation processes.

$$HO—CH_2—CH_2—\overset{+}{N}{\equiv}(CH_3)_3$$

Choline

Choline is synthesized from methionine and ethanolamine in mammals, and as a methyl donor it acts via betaine and methionine. On a diet that is low in protein (particularly the amino acid methionine) and high in fat, mammals show a requirement for choline; in choline deficiency, rats have local hemorrhages, fatty liver, and ultimate renal damage. Normally, when there is adequate methionine, mammals do not need exogenous choline. For many insects, choline is an essential needed in relatively large amounts. A cockroach does not make choline from ethanolamine and methionine, although it can use betaine (trimethyl glycine) in place of choline.[96] In *Drosophila*, choline is needed for pupation and can be replaced by lecithin;[110, 111] it is an absolute requirement for motile sperm, and promotes optimal growth.[51] In synthetic diets for most insects, choline is supplied in relatively large amounts (100 mg/g). In axenic culture, larvae of *Aedes aegypti* need 100 parts choline to 10 parts niacin, 2 parts thiamin, and 0.1 part biotin.[2]

Purines and Pyrimidines. Purines and pyrimidines are essential for the synthesis of nucleic acids (RNA and DNA). Most animals appear to synthesize sufficient amounts of them, and purine components have been traced from glycine, CO_2, NH_3, and acetate. Purines and pyrimidines are not vitamins for vertebrates and insects, but growth of some flies is increased when RNA is supplied. In *Drosophila*, adenine and thymine are as effective as RNA but one strain of *Drosophila* requires either RNA or the pyrimidine cytidine.[111] Several Protozoa, however, need purines and pyrimidines in the diet. One species of the flagellate *Chlamydomonas* needs the pyrimidine uracil. The trypanosomiid *Crithidia* needs a purine and a pyrimidine; if folic acid in the diet is high, there is no need for the pyrimidine, but otherwise thymine and methionine are needed. *Tetrahymena* lacks orotic acid decarboxylase and must be supplied with exogenous purines and pyrimidines, usually as adenosine, guanosine, cytidine, thymidine, and uridylic acid. Adenine and hypoxanthine can spare but not replace guanine, and xanthine and uric acid are ineffective; hence, *Tetrahymena* is unable to close the imidazole ring.[76] Of the pyrimidines, *Tetrahymena* needs uracil, although cytidine and cytidylic acid can be substituted whereas cytosine, thymine, and orotic acid cannot. Thymine cannot spare uracil but can spare the vitamin folic acid. For *Paramecium multimicronucleatum*, the purine need can be supplied by guanine, guanosine, or guanylic acid but not by adenine; the pyrimidine can be supplied by cytidine or uracil, but not by thymidine.[71] *Glaucoma* needs guanine and uracil (or equivalents).[78] In conclusion, the requirement of higher Protozoa for purines and pyrimidines represents a dietary specialization.

Hematin. All aerobic organisms have oxidative enzymes and electron carriers (cytochromes) which contain iron in a porphyrin ring, in some cases similar to and in others identical with hematin (p. 319). The heme of cytochrome is synthesized by many plants and animals which do not make hemoglobin. Various flagellates that are parasitic in the blood of birds and mammals—*Leishmania* and *Trypanosoma*—require hematin or its equivalent when cultured in vitro.[87] A parasitic trypanosome from the digestive tract of a mosquito requires hematin, while trypanosomes from hemipterans and flies do not require it.[88] This is another instance of loss of a synthetic function by a parasite, with increasing dependence on the host. A blood-sucking hemipteran, *Triatoma*, also has a nutritional requirement for hematin.[88] Hematin is, therefore, a vitamin for certain blood-feeding animals.

Carnitine (β-Hydroxybutyrobetaine). Muscles of all animals appear to contain nitrogenous acids such as carnosine, anserine, and carnitine. These acids probably function as anions to balance the excess of inorganic cations present in muscle (p. 83). Carnitine concentration in mammalian muscle is 1000 μg/g, and in brain it is 87 μg/g.[48] Carnitine is synthesized by virtually all animals, and the only ones known to require it in their diet are grain-eating beetles of the family Tenebrionidae.[48] Larvae of the fly *Phormia* convert choline, but not betaine, to carnitine. Substitution of different compounds shows the essentiality of the quaternary ammonium group separated from the carboxyl group by three carbon atoms.

$$(CH_3)_3\equiv\overset{+}{N}—CH_2—CHOH—CH_2—COO^-$$

Carnitine

Carnitine is a vitamin for the flour-eating *Tenebrio molitor*, *T. obscurus*, *T. confusum*, *T. castaneum*, and *Palorus*; it must be supplied at about one part carnitine to seven parts choline. Carnitine is not required by *Gnathocerus*.[80] *Dermestes* and *Phormia* synthesize much of their carnitine requirement.[80] *Drosophila melanogaster* requires choline for growth, but if it is given carnitine, this can be decarboxylated and methylcholine can be synthesized; in the presence of choline, larval growth is improved by addition of carnitine.[52]

Essential Fatty Acids. Most animals grow and live well with little or no dietary fat, but fat can be interconverted to and from carbohydrate and sometimes to and from protein; fat may be deposited in an animal after high caloric intake even when no fat is fed. In many animals, after ingesting fat, the deposited fat may resemble but differ in detail from the food fat in melting point or degree of saturation. Rats on a low fat diet develop hematuria, loss in weight, and scaly feet and tail, all of which are cured by feeding linoleic acid as sodium linoleate, at about 30 mg/day. Arachidonic acid, but not linolenic acid, can also be used by rats. Essential fatty acids are needed for uptake of P^{32} into brain microsomal phosphatidic acids. Of a series of precursors of phospholipids, only 4 out of 26 supported emergence of adult *Phormia* from puparia.[61]

Man on a low fat diet shows a decline in blood concentration of linoleic acid; apparently man can synthesize more of this acid than can a rat.

Several lepidopterans (e.g., the pink bollworm *Ephestia* and the rice moth *Corcyra* and webworm *Loxostege*) require linoleic acid for successful development, emergence, and normal appearance of the scales.[83] Linoleic acid deficiency in cockroaches is shown by aborted egg-cases and death of second-generation nymphs; linolenic acid can substitute in *Ephestia* but not in cockroach.[54] *Drosophila* can synthesize (from acetate) a series of saturated and mono-unsaturated fatty acids but not polyunsaturated acids.[75] A lepidopteran, *Argyrotaenia*, needs linoleic or linolenic acid for normal development of wing and body scales.[108] A boll weevil, *Anthonomus*, needs fatty acids for reproduction and does best on trilinolein or trilinolenin.[36] The waxmoth *Galleria* normally feeds on beeswax, the essential component of which is myricin, a long-chain ester of palmitic acid and myricyl alcohol; simple fatty acids from C_{14} to C_{30} cannot be utilized by *Galleria*.[21]

Of four species of *Tetrahymena*, three need an unsaturated fatty acid or phospholipid, but one (*T. pyriformis*) has no lipid requirement.[43] All *Tetrahymena* contain linoleic acid.[62] The trypanosomiid *Crithidia* requires an unconjugated pteridine, and this need can be spared by a mix of fatty acids; unsaturated fatty acids are made from saturated ones.[31]

Unsaturated fatty acids function in various ways—structurally and in energy metabolism. Specific ones are needed by most if not all animals, and some animals are unable to synthesize enough saturated fatty acids for their needs.

Sterols. Sterols constitute a large class of fused carbon-ring compounds, and the most common of animal sterols is cholesterol. Apparently, all animals use cholesterol in lipid metabolism but there are variations in their ability to synthesize it. Most animals, including vertebrates, synthesize sufficient cholesterol from acetate. However, many insects and the cephalopod *Sepia* are unable to synthesize sufficient cholesterol.[122]

Most insects seem to require a dietary sterol, usually cholesterol. Other sterols can be used by some insects, especially the herbivorous ones. *Bombyx* normally eats plant sitosterol, yet 85% of its body sterol is cholesterol. A cockroach does best on cholesterol but can use several plant sterols, and can convert ergosterol to 22-dehydrocholestrol, possibly by symbiotic microorganisms.[16] Larvae of houseflies, *Musca*, do well on cholesterol or cholesteryl acetate, and half as well on 7-dehydrocholesterol or ergosterol, but they cannot use 7-hydroxy- or 7-keto-cholesterol.[84] The carnivorous *Dermestes* can use only animal sterols—cholesterol or 7-dehydrocholesterol; usual precursors of these are not used.[16] A boll weevil needs 20 mg of cholesterol per 100 g of ration, and the cholesterol turnover time is 15 days.[37] A scotylid beetle, *Xyleborus*, can use cholesterol or lanosterol for larval development, but requires ergosterol or 7-dehydrocholesterol for pupation.[15] Plant-sucking aphids can be maintained on a liquid diet (sucked from parafilm sacs) which contains sucrose, amino acids, vitamins, and minerals but which has no sterols or lipids; possibly, symbiotes supply sterols.[5]

Some parasites require a sterol. A parasitic nematode, *Caenorhabditis*, can be grown on *E. coli* if sterols are added.[58] A facultative parasitic species of *Tetrahymena* requires cholesterol or a precursor of it.[63] *Trichomonas columbae*, an intestinal parasite of pigeons, requires cholesterol or a related sterol.[76] Evidently, the ability to synthesize cholesterol requires specific configurations of the dietary precursors; herbivorous insects have wider synthetic capacities than the carnivorous ones, and omnivorous species such as roaches are intermediate.

The principal sterol in insects and vertebrates, where the unsaponifiable portion is less than 7% of total fats, is the 27-carbon cholesterol. In most invertebrates the unsaponifiable fraction is greater. Sponges contain a variety of sterols, including C_{27}, C_{28}, and C_{29} sterols, some of which are unique. Gastropod and pelecypod molluscs contain mainly cholesterol, but pelecypods also have higher melting-point sterols, mostly with 28 carbons. The snail *Helix* needs a dietary sterol and normally uses sitosterol. Among echinoderms, the crinoids, ophiuroids, and echinoids have cholesterol, while the asteroids and holothurians have different com-

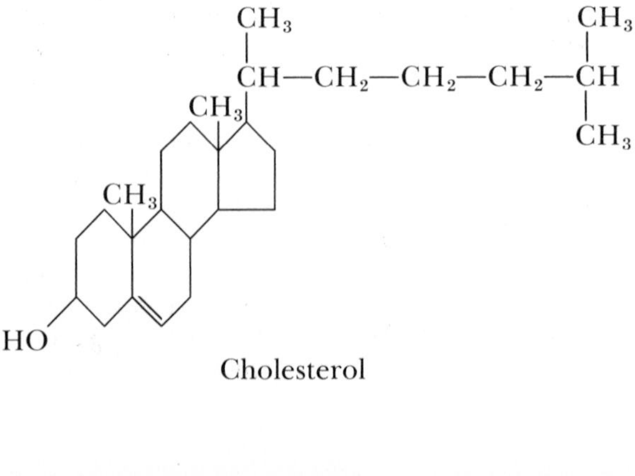

CH_3 CH_3 | | CH—CH_2—CH_2—CH_2—CH (CH_3) | CH_3 CH_3 HO

Cholesterol

TABLE 3–3. Requirements of Animals for Water-Soluble Vitamins. sl = slight

	thiamine	*cobala-min*	*biotin*	*ribo-flavin*	*niacin*	*pyridox-ine*	*folic acid*	*α-lipoic acid*	*Ca pan-tothenate*	*choline*	*choles-terol or other sterol*	*purine*	*pyrimi-dine*	*inositol*	*carnitine*	*ascorbic acid*
Tetrahymena[77]	+	−?	−	+	+	+	+	+	+	−		+	+		−	
Paramecium[71]	+	−	−	+	+		+	−	+		+	+	+			
Caenorhabditis[33]	+	?	?	+	+	+	+	?	+							
Pseudosarcophaga[65]	+	stim	+	+	+	−	−		+	+		stim		−		
Tenebrio[47]	+		+	+	+	+	+		+	+					+	
Calliphora[112]	+	−?	+	+	+	+	+		+	+	+			−		
Musca[11]	+	−	+	+	+	+	+	−	+	stim	+	+			−	
Phormia[12]	+		stim	+	+	+	growth stim		+	stim	+	−			−	
Hylemya[49]	+	stim	+	+	+	+	+			+	+	+				
Tribolium[89]	+	stim	+	+	+	+	+	stim	+	+	+				+	
Palorus[23]	+		−	+	+	+	−		+	+	+			−		
Neomyzus[41]	+		+	+	+	+	+		+	+				+		+
Locusta[20]	+		+	+	+	+	+		+	+				+		
Acheta[90]	+		+	+	+	+	+		+	+						
Xenopsylla[99]				+	+				+		+					
Artemia[103]	+		+	+	+	+	+		+	sl +				sl +		
salmon[55]	+	+?	+	+	+	+	+		+	+				+		−
guinea pig[105]	+	−	−	+	+?	+	+		+	+				−		+
rat[3]	+	+	+	+	+	+	+		+	or met	−	−				

pounds, the stellasterols.[9] Apparently, the biochemistry of steroid metabolism displays extreme diversity.

Water-Soluble Vitamins

A number of dietary essentials, the water-soluble vitamins, function as coenzymes in specific metabolic reactions that are common to most if not all animal cells. Certain of the water-soluble vitamins are widely required; others can be synthesized (in part or in whole); and still others are required by only a few organisms that have lost the capacity for synthesizing them. Some vitamins are needed in such small amounts that need has not been demonstrated for all animals and synthesis by them has been unjustifiably postulated. The B vitamins will be considered in order of decreasing requirement, from the most general to the more specific (Table 3–3).

Thiamine. The first B vitamin to be discovered was thiamine, which is active in preventing beriberi in man and polyneuritis in birds. Thiamine plus adenosine triphosphate forms cocarboxylase (diphosphothiamine), an important coenzyme in oxidative decarboxylation of pyruvic acid (p. 213). Thiamine also functions in other decarboxylations and in interactions between the pentose shunt and glucose. Thiamine is widely distributed in plant and animal tissues, and probably all cells, aerobic or anaerobic, require it. The thiamine molecule consists of two parts, a pyrimidine and a thiazole moiety.

Thiamine occurs in natural waters, both fresh and marine, particularly in regions near where there is bacterial growth; it also occurs in soils. Many bacteria, molds, and some flagellates can synthesize both the pyrimidine and thiazole rings, and from these the complete thiamine molecule. Some microorganisms require either the pyrimidine or the thiazole as nutrient; others need both.[87] Many Protozoa and some slime molds need

Thiazole

N, NH_2, S

H_3C—C, C, HC, C—CH_2—CH_2OH

N, C, N—C—CH_3, +

Pyrimidine, C H, C H_2

Thiamine

the complete thiamine molecule, and all multicellular animals need it (Table 3–3). Flagellates, including green ones, vary considerably in their requirement of thiamine. *Chilomonas* can synthesize some thiamine if supplied with the pyrimidine and thiazole, but *Chilomonas* multiplication is doubled by addition of thiamine, and increased fivefold by thiamine plus iron. All animal flagellates—*Trypanosoma, Leishmania,* and *Strigomonas*—need thiamine.[76]

Insects need thiamine in small amounts—*Phormia* needs 3 μg/g of diet, and *Tenebrio* needs 1 μg/g.[47] Trout need it in amounts of 0.16 mg/kg body weight/day.[100] For man, the recommended amount is 0.5 mg/1000 calories of diet. Many animals, especially insects and mammals, have microorganisms in the digestive tract, and these animals do not need thiamine in the diet. In deer, synthesis by symbionts may exceed intake by 5 to 12 times.[118]

Cobalamin (Vitamin B_{12}). Cobalamin is a cobalt-containing compound that occurs in natural waters, often in concentrations of 0.1 to 1 μg/ml. The molecule consists of two portions, a porphyrinlike ring with cobalt and a nucleotide, ribazole phosphate. It is usually isolated as the cyanide derivative, cyanocobalamin.

Vitamin B_{12} functions in the metabolism of 1-carbon fragments. It is a cofactor for several isomerases in methyl transferases, and may function in synthesis of proteins.

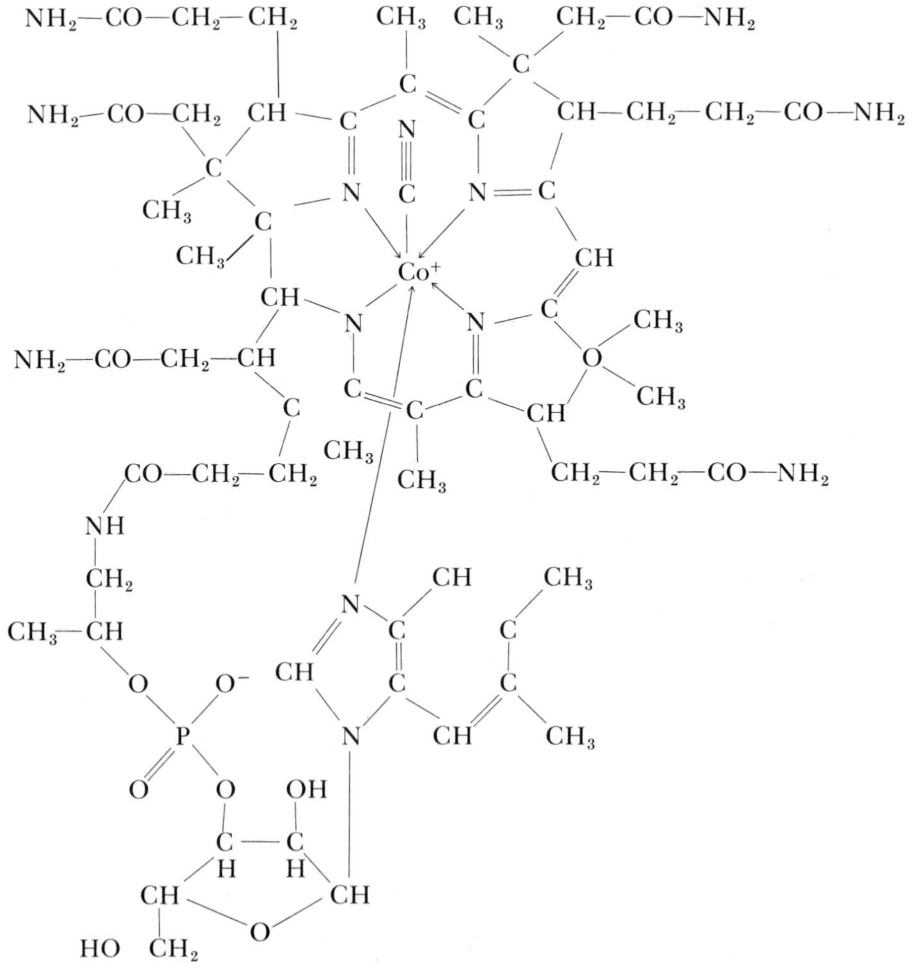

Cyanocobalamin

Cobalamin is required by many flagellates, both green and colorless, and *Euglena gracilis* is a sensitive organism for bioassay. The B_{12} required by algae and protozoans is normally supplied by bacteria, and the availability of this vitamin may be a factor limiting gross productivity of some natural waters. Of a large number of algae, 80% needed B_{12}, 53% needed thiamine also, and 10% needed biotin also.[34]

Daphnia reared on specific flagellates needs B_{12} and thiamine, which apparently act via the flagellate.[24] Cobalamin is synthesized by symbiotic microorganisms in many animals. A cockroach grows well but does not produce viable eggs without cobalamin; need for the vitamin has not been demonstrated in other insects. In mammals, lack of B_{12} results in anemia due to failure of maturation of red blood cells. The vitamin is an antipernicious anemia factor of which about 1 mg/day is needed by man. Weanling rats on a cobalamin-deficient diet showed intensified depression of growth if protein content of the diet was increased; hence, cobalamin is concerned with metabolic disposal of the carbon skeletons of amino acids.[35]

Biotin. Biotin functions in CO_2 fixation in bacteria and mammals, and is probably involved in oxaloacetate synthesis. Egg yolk is rich in biotin, and egg white contains an antibiotin factor. Biotin is present in natural waters and is required by a few flagellates, some of them green (Table 3–3). Biotin stimulates growth in flour beetles and metamorphosis in mosquitoes. It is needed in very small amounts by most insects which have been carefully examined. *Drosophila* species differ in requirements for B vitamins; one (*D. simulans*) has no need for biotin, whereas *D. melanogaster* requires it.[42] In silkworm, a deficiency in biotin results in a decrease in content of several fatty acids; the optimal level is 0.2 μg/g of dry diet.[64]

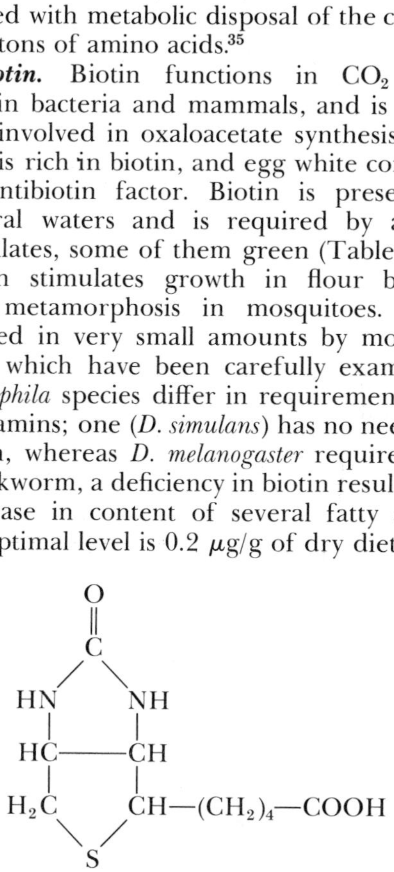

Biotin

Biotin is synthesized by symbiotic bacteria in the intestines of many animals. Not enough is synthesized to satisfy the needs of the hamster, but probably enough is produced by symbionts in rabbits. Deficiency of biotin in chickens results in high embryonic mortality and skeletal deformities.[19] Biotin deficiency can be produced in monkeys, calves, and other experimental mammals by sterilization of the intestinal tract or by feeding the antibiotin factor of egg white. The recommended amount for man is less than 10 μg/day.

Riboflavin. Riboflavin, an alloxazine derivative, functions in flavin mononucleotide (FMN) or flavin adenine dinucleotide (FAD) as the prosthetic group of flavoprotein enzymes which act in cellular respiration (Chap. 6). It is a cofactor for xanthine oxidase, D-amino acid oxidase, and cytochrome reductase.

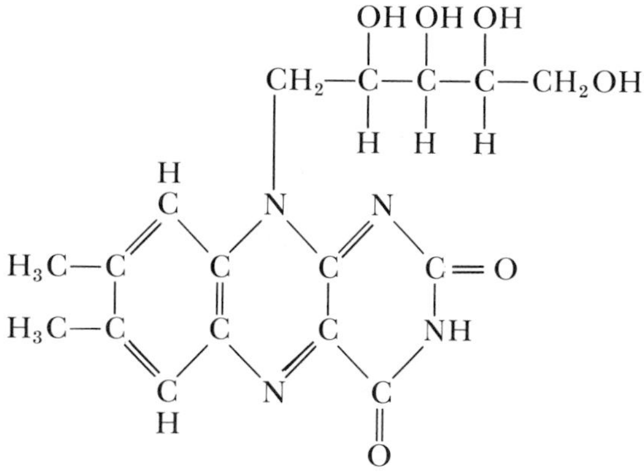

Riboflavin

Riboflavin is synthesized by many microorganisms—yeasts, bacteria, and probably flagellates. It is required by *Tetrahymena* and by numerous insects (Table 3–3). Some insects (*Blatella*) appear to get sufficient riboflavin from intestinal symbionts but others (*Tenebrio, Drosophila*) require it in the diet. Trout need some 0.5 mg of riboflavin per kg of body weight per day. Rats deficient in riboflavin develop skin lesions, shedding of hair, and ocular lesions. Because it is synthesized in many animals, the true requirement is difficult to establish. Rabbits may excrete in feces 10 to 15 times their intake, and they grow well without added dietary riboflavin, either because of cecal absorption or from eating feces.[97] In cattle, the rumen may contain riboflavin concentrated a hundredfold, and the milk may contain ten times as much as the dietary intake. Calves lacking rumen organisms need riboflavin in the diet. The recommended intake for man is 1.0 to 1.5 mg/day.

Niacin (Nicotinic Acid). Nicotinic acid amide (niacinamide) combines with adenine, a pentose, and a phosphate to form DPNH (diphosphopyridine nucleotide or coenzyme I) and TPN (triphosphopyridine nucleotide or coenzyme II). Niacin is synthesized from tryptophan by many animals and by symbiotic bacteria. This synthesis has been observed in the flagellate *Chilomonas*, but the ciliates *Tetrahymena* and *Colpoda* are incapable of it.

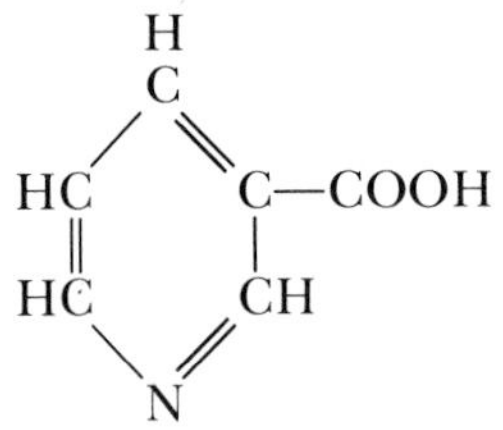

```
        H
        C
      /   \\
   HC       C—COOH
   ||       |
   HC       CH
      \   //
        N
```

Niacin

Niacin is required in the diet of a variety of insects (Table 3–3) but is produced by symbiotic microorganisms in some of them; hence, a nutritional need is not always demonstrated. *Phormia* cannot substitute tryptophan for niacin. In rats, mice, and men, tryptophan is converted to niacin, partly by digestive microorganisms but largely in the liver; more niacin than is ingested may be excreted even after the intestinal organisms have been poisoned, and especially after injection of tryptophan.[72] Calves lacking rumen organisms can synthesize sufficient niacin, but dogs and foxes deficient in niacin develop "black tongue," which is cured by feeding nicotinic acid. In man, pellagra results in part from a deficiency of niacin. Pellagra is a complex condition which is common on a diet of corn (maize), the protein of which, zein, is deficient in tryptophan; the corn usually also contains an antiniacin factor. Recommended intake of niacin by man is 10 to 15 mg/day.

Pyridoxine (Pyridoxal, Pyridoxamine). Vitamin B_6 is a pyridine derivative and can be provided as pyridoxine, pyridoxal, or pyridoxamine.

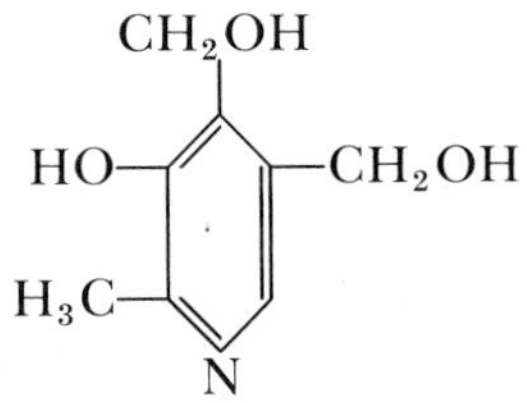

```
          CH2OH
            |
   HO—    /   \    —CH2OH
         |     ||
  H3C—    \\   /
            N
```

Pyridoxine

Pyridoxal phosphate is the coenzyme for transamination of alpha-amino acids; hence, it is important in protein metabolism. Deficiency in mammals results in a decrease in phosphorylases. Pyridoxine stimulates growth but is not essential for *Chilomonas*. It must be supplied in large amounts for optimal growth of *Tetrahymena* and *Colpoda*. Pyridoxine is a general dietary requirement for insects except when it is synthesized by their symbiotic flora (Table 3–3). A cestode, *Hymenolepis*, shows B-vitamin deficiency if its host rats are prevented from coprophagy, but relief is provided by administration of pyridoxine, which appears to be critical for the cestode.[101]

The rumen contents of sheep and the milk of cows may contain ten times as much pyridoxine as the food; synthesis by rumen organisms is indicated. In man, pyridoxine is synthesized by intestinal organisms. Convulsions appear in extremely pyridoxine-deficient rats and pigs, dermatitis in rats, and anemia in dogs, foxes, and pigs.

Pantothenic Acid (Pantoyl-Beta-Alanine). Pantothenic acid (P.A.) is combined with adenylic acid, 2-mercaptoethylamine, and

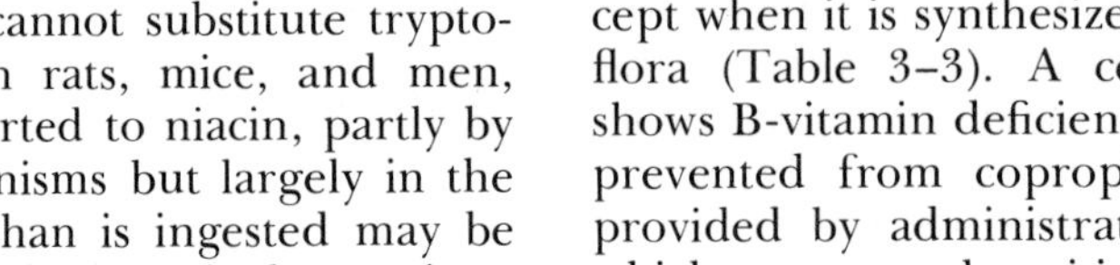

```
          CH3  OH   O
           |   |    ||
HO—CH2—C—CH—C—N—CH2—CH2—COOH
           |            |
          CH3           H
```

Pantothenic acid

three phosphates to form coenzyme A, which functions in the transfer of acyl groups. It is used in transfer of acetyl groups from pyruvic acid to the tricarboxylic cycle, and in acetylation of choline to form acetylcholine. It is usually supplied as calcium pantothenate.

Colpidium requires pantothenic acid in concentrations 20 times greater than does *Tetrahymena*.[77] Insects generally need it even when not depleted of their microorganisms (Table 3–3). Ruminants show marked synthesis by rumen organisms, 6 to 25 times above the concentration in food; P.A. is synthesized by cecal bacteria in rabbits and in the colon of man. An indication of pantothenate deficiency in rats is hemorrhagic necrosis of the adrenal cortex and reduced adrenocortical function.

Folic Acid (Pteroylglutamic Acid). Folic acid consists of glutamic acid, *p*-aminobenzoic acid, and a substituted pterin. In some cells (yeast and chicken liver), folinic acid (citrovorum factor) occurs and this substance may be more active as the vitamin than is folic acid. Folic acid functions in intermediary metabolic reactions involving 1-carbon fragments. Folic acid is required for purine and pyrimidine synthesis, and in *Tetrahymena* (but not in *Glaucoma*) thymine can spare the folic acid need.[78] Folic acid is antagonized by aminopterin, which in low concentrations can stop hematopoiesis and growth in mice. Folic acid is partly replaced in *Drosophila* diet by thymine, a purine, or high serine.[110] Folic acid is widely required in small amounts, but is commonly synthesized by symbiotic microorganisms. In man, there is evidence that intestinal bacteria supply much of this vitamin, but deficiency may lead to anemia.

α-Lipoic Acid (Thioctic Acid). This is a sulfur-containing acid (6,8-dithio-*n*-octanoic acid) which condenses with thiamine to form lipothiamide, which receives the acyl group from pyruvate in the oxidative decarboxylation of this intermediate. Then the acyl

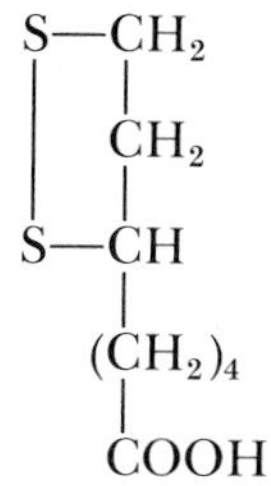

$$\begin{array}{l} S-CH_2 \\ |\qquad | \\ |\quad\ CH_2 \\ |\qquad | \\ S-CH \\ \qquad | \\ \quad (CH_2)_4 \\ \qquad | \\ \quad COOH \end{array}$$

α-Lipoic acid

is transferred to coenzyme A, with subsequent reactions with DPN. It is of interest that four vitamins—thiamine, α-lipoic acid, pantothenic acid, and niacin—are required for this important part of cellular metabolism.

α-Lipoic acid is synthesized by most animals and can be obtained from mammalian liver, chick embryos, and other active tissues. The only animal in which a dietary need for α-lipoic acid has been demonstrated is *Tetrahymena.*

Ascorbic Acid (Vitamin C). Asorbic acid, like glutathione, is reversibly oxidized or reduced by many oxidants and reductants; it is not assigned to any known coenzyme. It is utilized in the animal in oxidation of tyrosine.

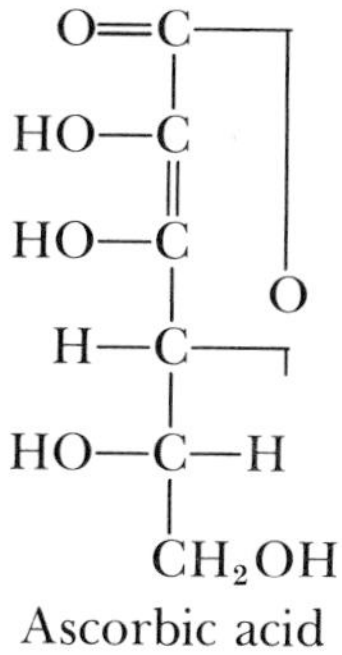

$$\begin{array}{l} O{=}C{-}\!\!-\!\!\urcorner \\ \quad\ \ |\qquad | \\ HO{-}C\qquad | \\ \quad\ \ \|\qquad | \\ HO{-}C\qquad | \\ \quad\ \ |\qquad O \\ \ \ H{-}C{-}\!\!-\!\!\urcorner \\ \quad\ \ | \\ HO{-}C{-}H \\ \quad\ \ | \\ \quad CH_2OH \end{array}$$

Ascorbic acid

Ascorbic acid is widely found in plant and animal tissues; it occurs in high concentration in the adrenal cortex, and under body stress the cortical ascorbic acid decreases rapidly.

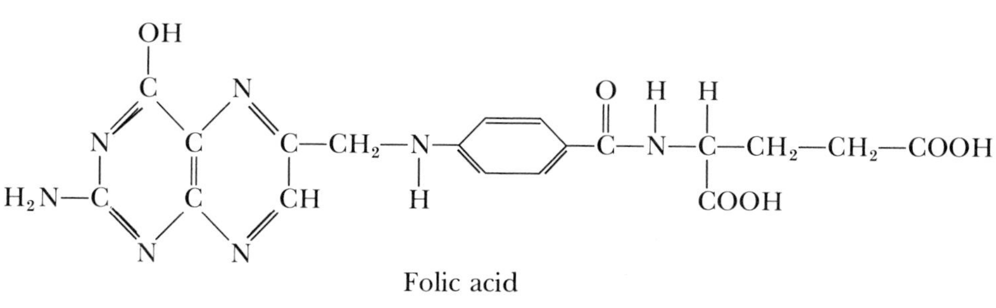

$$\text{(pterin: } C(OH),\ N,\ C,\ N,\ H_2N{-}C,\ N,\ C,\ C,\ N,\ CH) {-}C{-}CH_2{-}\underset{H}{\underset{|}{N}}{-}C_6H_4{-}\overset{O}{\overset{\|}{C}}{-}\overset{H}{\overset{|}{N}}{-}\overset{H}{\overset{|}{\underset{COOH}{\underset{|}{C}}}}{-}CH_2{-}CH_2{-}COOH$$

Folic acid

Ascorbic acid was early discovered as the antiscurvy vitamin in man, and lack of it results in scattered hemorrhages, particularly of the gums, and in swollen joints. Ascorbic acid is synthesized in sufficient amounts by most vertebrates. It is not needed in the diet of rat and hamster, but it must be fed to guinea pigs, monkeys, and men. The human requirement is 1.6 mg/kg/day.[113] Ascorbic acid is synthesized by insects such as honeybee larvae, and is not needed in the diet of most insects or of ciliate protozoans. However, many of the trypanosomes which are mammalian blood parasites, such as *Leishmania* and *Trypanosoma*, require ascorbic acid, while free-living trypanosomes, *Strigomonas* and *Leptomonas*, do not.[88] Three cotton-feeding insects – boll weevil *Anthonomus*, bollworm *Heliothis*, and march caterpillar *Estigmene* – all need 0.4 g of ascorbic acid per 100 g of diet.[119]

Fat-Soluble Vitamins

The fat-soluble vitamins are more restricted in their loci of action than the water-soluble vitamins. Requirement for vitamins A, D, and K appears to be restricted to vertebrates; need for vitamin E may be more general.

Vitamin A. This is known in two forms, vitamins A_1 and A_2. Two molecules of vitamin A are formed from one molecule of β-carotene, which occurs widely in plants. The conversion of β-carotene or a provitamin occurs partly in the liver but mainly in the intestine; the vitamins are stored in the liver and in the retina, in both receptor and pigment cells. The function of vitamin A_1 and A_2 is discussed in Chapter 14. Vitamin A_1 occurs in higher vertebrates and marine fishes; A_2 is predominant in freshwater fishes. Deficiency in vitamin A in man results first in night blindness and later in xerophthalmia; in extreme deficiency there may be retarded skeletal growth and lesions of the skin. An excess of dietary vitamin A may be toxic. The dietary requirement depends on the animal's synthetic capacity and the amount of dietary carotenes. The concentration in liver in different animals varies by some thousandfold, and is lowest in guinea pig and highest in polar bear and seal.[92] The

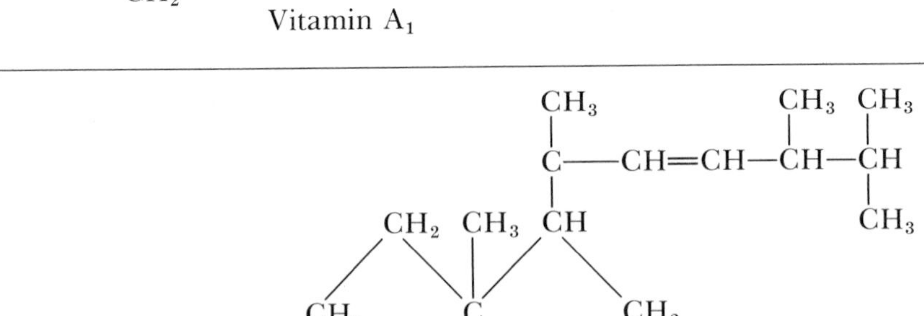

```
        H3C     CH3
           \   /
             C              CH3                 CH3
           /   \             |                   |
     H2—C       C—CH=CH—C=CH—CH=CH—C=CH—CH2OH
        |       ||
     H2—C       C—CH3
           \   /
            CH2
```

Vitamin A_1

```
                                   CH3                 CH3  CH3
                                    |                   |    |
                                    C——CH=CH—CH—CH
                                    |                        |
                    CH2  CH3  CH                      CH3
                   /    \   |  /    \
              CH2        C          CH2
               |         |           |
              CH2        CH————CH2
       CH2        \      /
      /    \  CH2   \   /
     /      \ ||      C
  CH2        C        ||
   |         |        CH
HO—CH        C       /
     \     /   \\   /
      CH2        CH
```

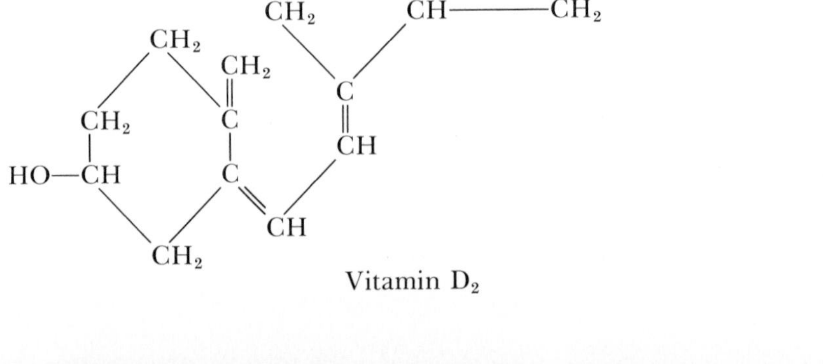

Vitamin D_2

polar bear stores so much vitamin A that its liver is toxic to man.

No invertebrates have been found to need dietary vitamin A, but many of them contain it. Squid liver contains much vitamin A, which is required for vision in squid. In a variety of marine crustaceans—euphausids, amphipods, copepods—90% of the vitamin A is found in the eyes. Carotenoids are made only by plants, but many animals modify those they consume to form vitamin A.

Vitamin D. This is a class of substances, the most important of which are D_2 or calciferol, formed by ultraviolet irradiation of ergosterol, and D_3, formed by activation by irradiation of 7-dehydrocholesterol. Vitamin D_3 is present in fish liver and is synthesized in the skin of higher animals under the action of sunlight. Birds secrete the precursor in the preen gland, and place it on the feathers, where it is activated; then they either eat it or absorb the vitamin directly.[92] In man, vitamin D_3 is formed in the skin by irradiation of 7-dehydrocholesterol; it is then transformed in the liver to 25-hydroxychole-calciferol, which in the kidney is then converted to 1, 25-dihydrovitamin D. In that it is synthesized at one site, altered at other sites, and transported to target organs by the blood, vitamin D_3 shows more characteristics of a hormone than of a vitamin.

Vitamin D aids calcium absorption from the digestive tract and, with parathormone, regulates the blood level of calcium and the mobilization of calcium for mineralization of bone. Deficiency of vitamin D results in rickets in man; in a rat, no rickets appears unless there is also an imbalance of calcium and phosphorus in the diet. No form of vitamin D has been found essential for any invertebrate, but sterols are widely utilized.

Vitamin K. Vitamin K occurs in two forms, K_1 and K_2. Vitamin K_1 is 2-methyl-3-phytyl-1,4-naphthoquinone. Vitamin K_1 is commonly obtained from alfalfa, and K_2 from fish meal. Vitamin K has a role in photosynthesis, and hence it is widely distributed. It is also similar to the pigment echinochrome in echinoderms. Vitamin K is made by symbiotic microorganisms, and man seems to absorb enough from intestinal bacteria, although newborn infants may show a deficiency. Chickens are very sensitive to a deficiency and are used for assay. Vitamin K stimulates the formation of prothrombin and is necessary for normal clotting of blood. Deficiency leads to hemorrhages.

Vitamin E (α-Tocopherol). Three tocopherols differ in number and position of substituent methyl groups. α-Tocopherol contains a phytol chain, as do chlorophyll and vitamin K.

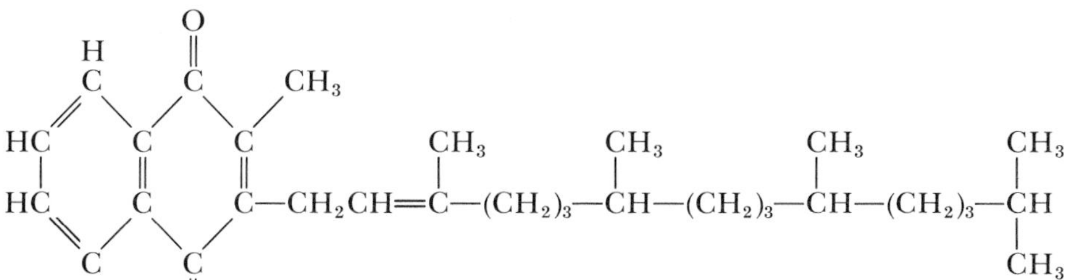

```
          O
    H     ‖
    C     C      CH3
  //  \  /  \   /
HC     C     C              CH3          CH3            CH3            CH3
 |     ‖     ‖               |            |              |              |
HC     C     C—CH2CH═C—(CH2)3—CH—(CH2)3—CH—(CH2)3—CH
  \\  /  \  /                                                            |
    C     C                                                             CH3
    H     ‖
          O
```

Vitamin K_1

```
      CH3
       |
       C      CH2
     //  \   /   \
HO—C      C       CH2                                                              CH3
   |      ‖        |                                                                |
H3C—C     C       C—CH2—CH2—CH2—CH—CH2—CH2—CH2—CH—(CH2)3—CH
     \\  /  \   / |                 |                   |                |
       C      O   CH3              CH3                 CH3              CH3
       |
      CH3
```

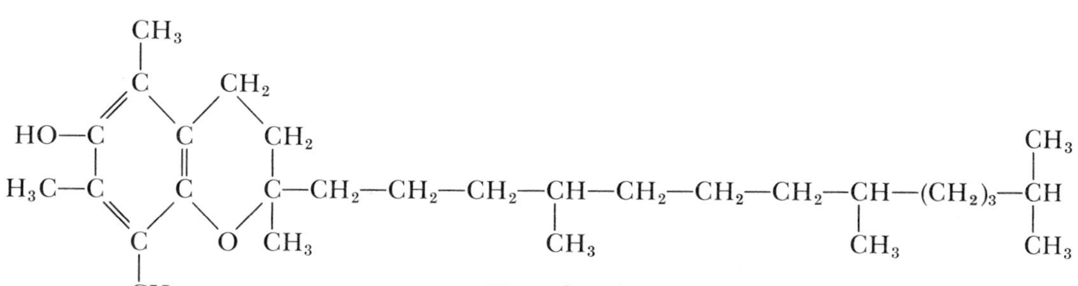

α-Tocopherol

Deficiency in vitamin E in mammals results in degeneration of epithelium, production of immobile sperm, and resorption of embryos. Hepatic necrosis, local hemorrhages, and lesions in the testis have been observed in vitamin E deficient chickens, tadpoles, and fishes. The rabbit and guinea pig are more sensitive than the rat. It is likely that vitamin E is a dietary requirement of all vertebrates. Extracts of heart and skeletal muscle of rats contain α-tocopherol, and the cytochrome reductase of these tissues shows enhanced activity for oxidation of DPNH and succinate when tocopherols are added. In addition, vitamin E may have a general antioxidant action. A laying hen requires 1.3 mg of α-tocopherol per day.[32]

Vitamin E may be required by some invertebrates. Crickets reproduce only if males are provided with vitamin E in the last nymphal stage; it has no effect on the females.[90] Need for α-tocopherol is also reported for the beetle *Oryzaephilus*[26] and for maturation of females of the fly *Agria*.[66]

CONCLUSIONS

Nutritional requirements of animals are related to patterns of intermediary metabolism, many of which were established during the period of chemical evolution before there were organisms that we would recognize as such. The syntheses of organic molecules from simple nitrogen and carbon sources in a reducing atmosphere occupied a long period, but could not be repeated once photosynthesizing organisms and an oxidizing atmosphere existed. Exobiologists have clearly demonstrated the synthesis of amino acids from cyanide and water, and the synthesis of various fatty acids, sugars, purines, and pyrimidines from mixtures of CN, NH_3, CO, H_2O, CH_4, and other small molecules under the influence of ultraviolet or electrical sources of energy. Organic compounds can be observed to be formed in absence of water, thus making possible the discovery of steps in chemical evolution on extraterrestrial bodies.

During the long period of chemical evolution much selection occurred, often for unknown reasons. The general use of amino acids and sugars, the cyclization of purines and pyrimidines and their combination with sugar and phosphates in nucleic acids, the limited number of amino acids and their polymerization in patterns called proteins, the catalytic functions of proteins, usually with smaller cofactors, the sequence and branching of energy-yielding pathways—all these reactions and many more established the kinds of molecules available for utilization by metazoa, once they appeared. Chemo- and photo-autotrophs elaborated new combinations of enzymes, and many unique features of eukaryotes appeared. The transition from prokaryotes to eukaryotes was accompanied by marked nutritional changes. Many transitional eukaryotes are photosynthetic but are capable of heterotrophic metabolism. A few heterotrophic eukaryotes carry over primitive pathways, e.g., the use of the glyoxylate cycle by the ciliate *Tetrahymena*.

Heterotrophs evolved metabolically in the Precambrian, and metazoa radiated rapidly. They have common nutritional needs—for example, need for similar amino acids and vitamin cofactors. They retained, in varying degrees, the ability to interconvert sugars, lipids, and amino acids and to use these as energy sources. Some (e.g., *Tetrahymena*) can rely exclusively on amino acids, but most animals need carbohydrates for energy.

Further evolution in animals took place mainly as loss or reduction of synthetic pathways. The amino acids which are required in the diet of most animals are synthesized by more complex paths than are those amino acids which are dispensable. Many animals can synthesize essential amino acids but in insufficient amounts. Some animals have lost the ability to synthesize essential purines and pyrimidines. Many insects require animal sterols, while vertebrates can synthesize them. The requirements for specific higher fatty acids occur sporadically among different animal groups.

The physiology of nutrition clearly illustrates the interdependence of organisms. Photosynthetic flagellates need cobalamin and sometimes thiamin, formed by bacteria. Symbionts may contribute basic foodstuffs as in corals, termites, and ruminants, or "vitamins" as in many insects and vertebrates. It is difficult to rid many animals of symbiotic microorganisms in the digestive tract without damage. In nature, the dependence on symbionts is such that nutritional requirements, especially for B vitamins, are much less than after sterilization of the digestive tract. Reverse dependence is found for parasitic helminths, some of which cannot use protein but must have the amino acid products of the host's digestion. Blood feeders need heme for cytochrome.

Some nutritional needs are highly specialized, and some represent changes in function of a type of compound with animal evolution. For example, the rhodopsin system in vision requires vitamin A, and calcification in vertebrate bone uses vitamin D.

Many animals do better on a synthetic medium if extracts of yeast, liver, or other organic materials are supplied; they are cultured axenically (i.e., in absence of other organisms), but unidentified substances are needed. Many associations of animals with microorganisms may be based on unknown substances. For example, bark beetles appear to obtain some unknown factor from fungus, since they fail to pupate when fed sterile bark.

Nutrients cannot be used unless they are ingested and digested. It has been difficult to culture some forms without proper feeding conditions. *Artemia* needs particulate matter. Aphids can use a liquid diet if they suck it from a sac resembling a stem or leaf. Food must be palatable. Feeding often depends on chemical attraction or repellance, as in phytophagous insects. It is possible to control some pests by making use of chemosensitive behavior.

Axenic culture is increasingly possible. It is easier to maintain an animal in a simple medium than it is to obtain reproduction: growth requires more factors than adult maintenance.

Dietary needs are influenced by the interdependence among nutritional substances. Some substances within a class are converted to others; e.g., supply of one amino acid may protect against need for another. Similar sparing actions exist between some water-soluble vitamins and specific amino acids. Within each class of dietary requirements a balance is needed. Dietary requirements may change within limits according to nutritional experience. Enzymes for some syntheses are inducible.

Nutritional needs for man and for agricultural animals can be best appreciated by an understanding of nutritional history, and of the enzymatic function of many specific substances. Most mammalian vitamins are probably known. An urgent contemporary problem is to provide balanced protein sources for man, especially on exclusively plant protein diets. Supplementation by synthetic amino acids is very important. Human diet is already in part synthetic, and for the world as a whole is largely based on plant sources. Increased use of microorganisms and of synthetic components is likely.

REFERENCES

1. Abelson, P. H., Proc. Nat. Acad. Sci. *55*:1365–1372, 1966. Chemical events on primitive earth.
2. Akov, S., J. Insect Physiol. *8*:319–335, 1962. Axenic culture of *Aedes*.

2a. Albanese, Z., J. Clin. Nutr. *1*:44–51, 1952. Effects of amino acid deficiencies, rat and man.

3. Albritton, E. C., editor, Standard Values in Nutrition and Metabolism. W. B. Saunders Co., Philadelphia, 1954.
4. Almquist, H. J., and E. Mecchi, J. Biol. Chem. *135*:355–356, 1940; *also* J. Nutr. *28*:325–331, 1944. Amino acid requirements of chickens.
5. Auclair, J. L., Ent. Exp. Appl. *12*:623–641, 1969. Nutrition of plant-sucking insects.
6. Balogun, R. R., J. Insect Physiol. *15*:141–148, 1969. Essential amino acids and sugars for *Ips* (= *Scotylus*) *cembrae*.
7. Barghoorn, E. S., and S. A. Tyler, Science *147*:563–575, 1965. Precambrian fossils.
8. Beerstecher, E., pp. 119–220 *in* Comparative Biochemistry, Vol. 6, edited by M. Florkin and H. Mason. Academic Press, New York, 1964. Chemical history of the earth.
9. Bergman, W. J., Org. Chem. *12*:67–75, 1947; *also* J. Marine Res. *8*:137–176, 1949; *also* pp. 435–444 *in* Cholesterol, edited by R. S. Cook, Academic Press, New York, 1958. Evolutionary aspects and comparative biochemistry of animal sterols.
10. Berntzen, A. K., Ann. N. Y. Acad. Sci. *139*:176–189, 1966. Axenic cultures of parasitic worms.
11. Brookes, V. J., and G. Fraenkel, Physiol. Zool. *31*:208–223, 1958. Nutrition of fly *Musca*.
12. Brust, M., and G. Fraenkel, Physiol. Zool. *28*:186–204, 1955. Nutrition of larvae of blowfly *Phormia*.
13. Calvin, M., Proc. Roy. Soc. Lond. A *288*:441–466, 1965. Chemical evolution, primeval molecules. Dehydration condensation of amino acids.
14. Chang, J. T., and M. Y. Wang, Nature *181*:566, 1958. Nutrition of *Musca*.
15. Chu, H. M., D. M. Norris, and L. T. Kok, J. Insect Physiol. *16*:1379–1387, 1970. Sterol requirements of bark beetles.
16. Clark, A. J., and K. Black, J. Biol. Chem. *234*:2578–2582, 2583–2588, 2589–2594, 1959. Sterol requirements and sterol synthesis in insects.
17. Cloud, P., Science *160*:729–736, 1968. Atmospheric and hydrospheric evolution in primitive earth.
18. Cosgrove, W. B., and B. K. Swanson, Physiol. Zool. *25*:287–292, 1952. Culture of *Chilomonas*.
19. Couch, J. R., J. Nutr. *35*:57–72, 1948; *also* Arch. Biochem. *21*:77–86, 1949. Biotin requirements of chicken.
20. Dadd, R. H., J. Insect Physiol. *6*:1–12, 1961. B-vitamin requirements of locusts.
21. Dadd, R. H., J. Insect Physiol. *12*:1479–1492, 1966. Nutrition of wax-moth *Galleria*.
22. Dadd, R. H., and D. L. Krieger, J. Insect Physiol. *14*:741–763, 1968. Amino acid requirements of aphid *Myzus*.
23. Dadd, R. H., et al., J. Insect Physiol. *13*:249–272, 1967. Requirements of aphid for water-soluble vitamins.
24. D'Agostino, A. S., and L. Provasoli, Biol. Bull. *139*:485–494, 1970. Axenic culture of *Daphnia*.
25. Daugherty, J. W., Proc. Soc. Exp. Biol. Med. *85*:288–291, 1954. Use of amino acids and ammonia by tapeworm *Hymenolepsis*.
26. Davis, B. D., Cold Spring Harbor Symp. Quant. Biol. *26*:1–10, 1961. Enzymes for synthesis of amino acids.
27. Davis, G., J. Insect Physiol. *14*:1247–1250, 1968. Amino acid requirements of beetle *Oryzaephilus*.
28. Davis, G. R. F., Arch. Int. Physiol. Biochem. 77:741–748, 1969. Protein nutrition of *Tenebrio*.
29. De Groot, A. P., Experientia *8*:192–194, 1952; *also* Physiol. Comp. Oecol. *3*:1–90, 1953. Amino acid requirements of honeybee.
30. Dewey, V. C., and G. W. Kidder, J. Gen. Microbiol. *22*:72–78, 1960. Nutrition of strains of *Tetrahymena*.
31. Dewey, V. C., and G. W. Kidder, Arch. Biochem. Biophys. *115*:401–406, 1966. Lipid nutrition of *Crithidia*.
32. Dju, M. Y., et al., Amer. J. Physiol. *160*:259–263, 1950. Use of tocopherols in laying hens.
33. Dougherty, E. C., et al., Ann. N. Y. Acad. Sci. *77*:176–217, 1959. Axenic culture of nematode *Caenorhabditis*.
34. Droop, M. R., pp. 141–159 *in* Physiology and Biochemistry of Algae. edited by R. A. Lewin. Academic Press, New York, 1962. Organic nutrients.
35. Dryden, L. P., and A. M. Hartman, J. Nutr. *101*:579–588, 1971. Vitamin B_{12} deficiency in rat.
36. Earle, N. W., B. Slatten, and M. L. Burks, J. Insect Physiol. *13*:187–200, 1967. Fatty acid requirements in boll weevil.
37. Earle, N. W., et al., Comp. Biochem. Physiol. *16*:277–288, 1963. Steroid metabolism of adult boll weevil.
38. Elliott, A. M., Physiol. Zool. *22*:337–345, 1949; *23*:85–91, 1950. Amino acid and growth factor requirements, *Tetrahymena*.

39. Ellis, W. C., and W. H. Pfander, Nature *205*:974–975, 1965. Role of rumen organisms in nutrition of host.
40. Elvehjem, C. A., Fed. Proc. *15*:965–970, 1956. Amino acid imbalance.
41. Erhardt, P., Z. vergl. Physiol. *60*:416–426, 1968. Vitamin requirements of squid.
42. Erk, F. C., and F. H. Sang, J. Insect Physiol. *12*:43–52, 1966. Nutrition of two *Drosophila* species.
43. Erwin, J., and K. Block, J. Biol. Chem. *238*:1618–1624, 1963. Lipid metabolism of ciliates.
44. Felsenfeld, G., Biochim. Biophys. Acta *29*:133–144, 1958. Properties of synthetic polyribonucleotides.
45. Fox, S. W., Bioscience *14*:13–21, 1964. Polymerization of peptides.
46. Fox, S. W., pp. 361–382 *in* Origin of Prebiological Systems, edited by S. W. Fox. Academic Press, New York, 1965. Simulation of prebiotic synthesis of proteins.
47. Fraenkel, G., Ann. N. Y. Acad. Sci. 77:267–274, 1959. Dietary requirements of insects.
48. Fraenkel, G., et al., Physiol. Zool. *27*:40–56, 1954; *also* Arch. Int. Physiol. Biochem. *64*:601–622, 1956; *also* J. Nutr. *65*:361–395, 1958. Carnitine, chemistry and nutritional properties.
49. Friend, W. G., et al., Canad. J. Zool. *34*:152–162, 1956; *35*:535–543, 1957; *also* Ann. N. Y. Acad. Sci. 77:384–393, 1959. Nutrition of onion maggot *Hylemya*.
50. Gaffron, H., pp. 39–84 *in* Evolution After Darwin. Vol. 1, edited by S. Tax. University of Chicago Press, Chicago, 1960. The origin of life.
51. Geer, B. W., Biol. Bull. *133*:548–566, 1967. Choline requirements of *Drosophila*.
52. Geer, B. W., et al., J. Exp. Zool. *176*:445–460, 1971. Metabolism of carnitine in *Drosophila*.
53. Goldberg, L., and B. DeMeillon, Biol. Chem. *43*:379–387, 1948. Protein and amino acid requirements of *Aedes* larvae.
54. Gordon, H. T., Ann. N. Y. Acad. Sci. 77:290–338, 1959. Nutrition of roach *Blatella*.
55. Halver, J. E., J. Nutr. *62*:225–243, 1957; *63*:95–105, 1957. Nutrition of chinook salmon.
56. Harada, K., and S. W. Fox, Nature *201*:335–336, 1964. Thermal synthesis of amino acids.
57. Hassett, C. C., et al., Biol. Bull. *99*:446–453, 1950. Nutritive value of carbohydrates for blowfly.
58. Hieb, W. F., and M. Rothstein, Science *160*:778–779, 1968. Sterol requirements of a nematode.
59. Hinton, T., Ann. N. Y. Acad. Sci. 77:366–372, 1959. Amino acid and growth factor requirements of *Drosophila*.
60. Hodgson, E., et al., Canad. J. Zool. *34*:527–531, 1956. Choline substitutes and amino acid requirements of *Phormia*.
61. Hodgson, E., et al., Comp. Biochem. Physiol. *29*:343–359, 1969. Choline and phospholipid requirements of *Phormia*.
62. Holz, G., J. Protozool. *13*:1–4, 1966. Unique features of *Tetrahymena*.
63. Holz, G., B. Wagner, and J. Erwin, Comp. Biochem. Physiol. *2*:202–217, 1961. Sterol requirements of *Tetrahymena*.
64. Horie, Y., and S. Nakasone, J. Insect Physiol. *14*:1381–1387, 1968. Nutrition of silkworm *Bombyx*.
65. House, H. L., Canad. J. Zool. *32*:331–365, 1954; *also* Ann. N. Y. Acad. Sci. 77:394–405, 1959. Nutrition of *Pseudosarcophaga*, dipteran parasite of spruce budworm.
66. House, H. L., J. Insect Physiol. *12*:409–417, 1966. Vitamin E requirement for fly *Agria*.
67. House, H. L., J. Insect Physiol. *17*:1225–1238, 1971. Dietary requirements of fly larva *Agria*.
68. Hutner, S. H., et al., pp. 85–177 *in* Biology of Nutrition, edited by R. M. Fiennes. Pergamon Press, New York, 1972. Nutrition and metabolism in Protozoa.
69. Ito, T., and N. Arai, J. Insect Physiol. *12*:861–869, 1966; *13*:1813–1824, 1967. Amino acid requirements of silkworm *Bombyx*.
70. Jahn, T. L., J. Protozool. *2*:1–5, 1955. Flagellate nutrition.
71. Johnson, W. H., Physiol. Zool. *25*:10–19, 1952; *30*:106–113, 1957; *also* Ann. Rev. Microbiol. *10*:193–212, 1956. Amino acid requirements of *Paramecium multinucleatum*.
72. Jungulira, P. B., and B. S. Schweigert, J. Biol. Chem. *175*:535–546, 1948. Synthesis of nicotinic acid by rats.
73. Kasting, R., and A. J. McGinnis, J. Insect. Physiol. *8*:97–103, 1962. Nutrition of cutworm *Agrotis*.
74. Kasting, R., and A. J. McGinnis, Ann. N. Y. Acad. Sci. *139*:98–110, 1966. Amino acid requirements of *Phormia*.
75. Keith, A. D. Comp. Biochem. Physiol. *21*:587–600, 1967. Fatty acid metabolism of *Drosophila*.
76. Kidder, G. W., pp. 162–196 *in* Biochemistry and Physiology of Nutrition, Vol. 2, edited by G. H. Bourne and G. W. Kidder. Academic Press, New York, 1953. Nutrition of invertebrate animals.
77. Kidder, G. W., and V. C. Dewey, Arch. Biochem. *8*:293–301, 1945; *20*:433–443, 1949; *21*:58–65, 66–73, 1949; *also* Proc. Nat. Acad. Sci. *33*:347–356, 1947; *34*:81–88, 566–574, 1948. Amino acid and growth factor requirements of *Tetrahymena*.
78. Kidder, G. W., et al., Arch. Biochem. Biophys. *55*:126–129, 1955. Nitrogen, purine, and pyrimidine requirements of ciliate *Glaucoma*.
79. Kidder, G. W., and B. N. Dutta, J. Gen. Microbiol. *18*:621–638, 1958. Nutrition of trypanosomiid flagellate *Crithidia*.
80. Leclerq, J., Arch. Int. Physiol. Biochem. *62*:101–108, 1954; *65*:337–346, 1957. Carnitine in nutrition of *Tenebrio* and *Gnathocerus*.
81. Lemonde, P., and R. Bernard, Canad. J. Zool. *29*:80–83, 1951. Nutrition of *Tribolium*.
82. Leng, M., and G. Felsenfeld, Proc. Nat. Acad. Sci. *56*:1325–1332, 1966. Binding of polypeptides to nucleic acid chains.
83. Levinson, Z. H., Riv. Parassitol. *16*:113–138, 189–204, 1955. Nutritional requirements of insects.
84. Levinson, Z. H., and E. D. Bergman, Biochem. J. *65*:254–260, 1957. Steroid use and fatty acid synthesis by *Musca* larvae.
85. Lohrmann, R., and L. E. Orgel, Science *161*:64–66, 1968. Phosphorylation under prebiotic conditions.
86. Lue, P. T., and S. E. Dixon, Canad. J. Zool. *45*:595–599, 1967. Nutritional value of royal jelly of honeybees.
87. Lwoff, M., L'évolution physiologique. Hermann, Paris, 1944.
88. Lwoff, M., pp. 129–177 *in* Biochemistry and Physiology of Protozoa, Vol. 1, edited by S. H. Hutner and A. Lwoff. Academic Press, New York, 1951. Nutrition of parasitic flagellates.
89. Magis, N., Bull. Ann. Soc. Entomol. Belg. *90*:49–58, 1954. Nutritional requirements of *Tribolium*.
89a. Margulis, L., pp. 342–368 *in* Exobiology, edited by C. Ponnamperuma. North Holland Publ. Co., Amsterdam, 1972. Properties of prokaryotes.
90. Meikle, J. E. S., and J. E. McFarlane, Canad. J. Zool. *43*:87–98, 1965. Lipids in nutrition of cricket *Acheta*.
91. Miller, S. L., J. Amer. Chem. Soc. 77:2351–2361, 1955. Production of organic substances under possible primitive earth conditions.
92. Moore, T., pp. 265–290 *in* Biochemistry and Physiology of Nutrition, Vol. 2, edited by G. H. Bourne and G. W. Kidder. Academic Press, New York, 1953. The fat-soluble vitamins.
93. Muscatine, L., Comp. Biochem. Physiol. *16*:77–92, 1965. Symbionts of *Hydra*.
94. Muscatine, L., and E. Cernichiari, Biol. Bull. *137*:506–523, 1969. Utilization of products of zooxanthellae by corals.
95. Nimrock, K., et al., J. Nutr. *100*:1273–1300, 1970. Amino acid supplementation of urea in diet of lambs.
96. Noland, J. L., and C. A. Baumann, Ann. Entomol. Soc. Amer. *44*:184–188, 1951. Choline and amino acid requirements of *Blatella*.
97. Olcese, O., et al., J. Nutr. *35*:577–590, 1948. Synthesis of B vitamins in rabbits.
98. Oro, J., Nature *190*:387–390, 1961; *also* Arch. Biochem. Biophys. *93*:166–171, 1961. Origin of life.
99. Pausch, R. D., and G. Fraenkel, Physiol. Zool. *39*:202–222, 1966. Nutrition of rat flea *Xenopsylla*.
100. Phillips, A. M., Trans. Amer. Fish. Soc. *74*:81–87, 1944; *76*:34–45, 1946. Vitamin requirements of brook trout.
101. Platzer, E. G., and L. S. Roberts, J. Parasit. *55*:1143–1152, 1969. Developmental nutrition of cestodes.
102. Provasoli, L., and R. S. D'Agostino, Amer. Zool. *2*:439, 1962. Vitamin requirements of *Artemia*.
103. Provasoli, L., and A. D'Agostino, Biol. Bull. *136*:434–453, 1969. Synthetic medium for *Artemia*.
104. Read, C. P., Exp. Parasit. *6*:288–293, 1957. Carbohydrate utilization by cestodes.
105. Reid, M. E., Proc. Soc. Exp. Biol. Med. *85*:547–550, 1954. B vitamin requirements of guinea pig.
106. Rock, G. C., and E. Hodgson, J. Insect Physiol. *17*:1087–1098, 1971. Amino acid requirements of *Heliothis*.
107. Rock, G. C., and K. W. King, J. Insect Physiol. *13*:175–186, 1967. Amino acid requirements of leaf-roller *Argyrotaenia*.
108. Rock, G. C., R. L. Patton, and C. H. Glass, J. Insect Physiol. *11*:91–101, 1965. Fatty acid requirements of *Argyrotaenia*.
109. Rose, W. C., Physiol. Rev. *18*:100–136, 1938; *also* Fed. Proc. *8*:546–552, 1949. Amino acid requirements in rat and man.
110. Sang, J. H., J. Exp. Biol. *33*:45–72, 1956; *also* Ann. N. Y. Acad. Sci. 77:352–365, 1959. Nutritional requirements of *Drosophila*.
111. Sang, J. H., and R. C. King, J. Exp. Biol. *38*:793–809, 1961. Axenic culture of *Drosophila*.
111a. Schoenheimer, R., The Dynamic State of Body Constituents. Harvard University Press, Cambridge, 1942.
111b. Schopf, J. W., pp. 12–61 *in* Exobiology, edited by C. Ponnamperuma. North Holland Publ. Co., Amsterdam, 1972. Precambrian fossils.
112. Sedee, D. J. W., Experientia *9*:142–143, 1953; *also* Acta Physiol. Pharm. Neerl. *3*:262–269, 1954. Nutrition of blowfly larvae.
113. Sherman, H. C., The Chemistry of Food and Nutrition. Macmillan Co., New York, 1946.
114. Singh, K. R. P., and A. W. A. Brown, J. Insect Physiol. *1*:199–220, 1957. Nutritional requirements of *Aedes aegypti*.
115. Soldo, A. T., and W. J. van Wagtendonk, J. Protozool. *16*:500–506, 1969. Culture medium for *Paramecium aurelia*.
116. Srivastava, P. N., and L. Auclair, Ent. Exp. Appl. *13*:208–216, 1970. Dietary responses of milkweed bug *Oncopeltus*.
117. Tarr, H. L., Ann. Rev. Biochem. *37*:223–244, 1958. Fish nutrition.
118. Teeri, A. E., et al. J. Mammal. *36*:553–557, 1955. Vitamin excretion, deer.
119. Vanderzant, G. S., et al., J. Insect. Physiol. *8*:287–297, 1962. Ascorbic acid in nutrition of cotton insects.
120. Wise, D. L., J. Protozool. *2*:156–158, 1955; *6*:19–23, 1959. Carbon nutrition, *Polytomella*.
121. Woese, C., *in* Exobiology, edited by C. Ponnamperuma. North Holland Press, Amsterdam, 1972. Origin of nucleic acid coding of protein structure.
122. Zandee, D. I., Arch. Int. Physiol. Biochem. *74*:35–44, 1966. Biosynthesis of amino acids by crayfish.
123. Zandee, D. I., Arch. Int. Physiol. Biochem. *75*:487–491, 1967. Cholesterol requirement of *Sepia*.

CHAPTER 4

COMPARATIVE BIOCHEMISTRY AND PHYSIOLOGY OF DIGESTION

I. *DIGESTION*

By E. A. Barnard

The Digestive Apparatus

All animals (with the exception of certain endoparasites, to be discussed later in this chapter) utilize as their main source of exogenous nutrients a set of macromolecular or otherwise complex components; for the cleavage of these to simple metabolizable forms, specific enzymes are universally required. It seems best to confine the term "digestion" to the processes involved in this cleavage. The enzymes engaged in the digestion of food are entirely hydrolytic, other possible processes (e.g., phosphorolysis or oxidation) never being known to occur at these stages. It is easy to see that a requirement for a co-substrate other than water in a reaction occurring in the alimentary tract would confer disadvantage in natural selection.

The macro-nutrients comprise carbohydrates, proteins, lipids, and nucleic acids. Sets of digestive enzymes have evolved that are capable of degrading each of these to the level of simple units that are absorbed and metabolized. These sets will be dealt with in turn in this section. Not all potential nutrients are in fact digested in all species; the β-glycosidic polysaccharides and the nucleic acids are important examples of those discarded in most vertebrates. These will be discussed separately, in the light of special cases outlined in Section III. The complement of digestive enzymes varies significantly in detail among animal groups, reflecting in a complex way the influences of adaptation to diet and of evolutionary descent.

Intracellular and Extracellular Digestion. The simplest digestive system is intracellular; food vacuoles are generally formed in the Protozoa and in the sponges. The secretion of digestive enzymes extracellularly into a cavity commences at the level of the Coelenterata and the Ctenophora. The coelenterates retain vacuole digestion for secondary stages.[206] Vacuolar digestion is often found, too, in the Platyhelminthes.[169] In the latter phylum, purely extracellular digestion is also known, as in *Fasciola hepatica*[79, 169] and the polyclad *Cycloporus*,[169] as well as purely intracellular digestion, as in the acoel *Convoluta convoluta*.[169] Supplementary intracellular digestion occurs[24] also in some higher invertebrates, despite their general trend to development and specialization of the alimentary tract; it has been found in nemerteans, echinoderms, some annelids, and many (but not all) of the molluscan groups. In most of these animals the intracellular component is provided by wandering amoebocytes in the wall of the alimentary tract, which are phagocytic and endowed with a concentration of digestive enzymes. This amoebocytic digestion usually supplements extracellular digestion, but various other patterns occur, strongly linked to the feeding habit of the species.[24, 74, 280] Thus, among the polychaetes, *Arenicola,* which removes organic matter ingested with sand, both secretes enzymes into its tract and employs phagocytic epithelial cells and amoebocytes for maximum extraction,[74, 186] whereas *Clymenella,* which feeds on diatoms, and the Terebellidae,[74] which feed on particles by means of ciliated tentacles, exhibit entirely extracellular digestion. In the cephalopod molluscs, crustaceans, insects, and nematodes, intracellular digestion is absent or rare.[24] Even in the chordates secondary intracellular digestion occasionally occurs, as in *Amphioxus*.[24]

Current thinking on intracellular digestion of food particles emphasizes the role of the lysosome. This is a vesicular organelle in the cytoplasm which contains an extensive set of hydrolases (phosphatase, petidases, glycosidases, and so on) of acidic pH optimum, and which has a well established role in the removal of foreign bodies.[81] It is believed that as a general mechanism (including the case of protozoan digestion[244]), the vacuole containing captured food material fuses with lysosomes, so that the digestion is by acid hydrolases.[167, 257] The lysosomal enzymes, which are essentially one distinctive set wherever they occur,[81] are quite dissimilar to the typical enzymes of extracellular digestion, and probably reflect an extremely remote evolutionary divergence.

The Sites of Extracellular Digestion. The digestive tract is always basically a tube, but displays innumerable and complex variations in adaptation to the food type and mode of feeding of the organism. There is, however, a general pattern of functional sequence:

(a) Reception of food. The mouth and its appendages assume a variety of forms in different animal groups, related to specializations in the intake of food (see Section IV). When the mouth is well developed, oral glands are common. These secrete a watery or mucous saliva; the main function of the saliva in relation to digestion is lubrication, both for mastication and for passage through the canal. In ruminant mammals, the saliva has additional alimentary roles (see Section III). Mucous lubrication is important when dentition is used effectively, and all vertebrates produce a saliva rich in the glycoproteins of mucus. The total composition of saliva in man and a few other mammals has been tabulated.[8]

The role of the saliva in enzymatic digestion is very restricted. Evidence for such a function is found in certain insects, in some gastropods, and in those predatory cephalopods, centipedes, and snakes that secrete paralyzing toxins together with digestive enzymes. In the insects, it is known[71, 95, 366] that the saliva of the silkworm, the desert locust, some Hemiptera, and cockroaches contains strong activities of carbohydrases, proteases, and lipase that are significant in digestion; but this is not general in the Insecta, and many flies and beetles are known not to possess salivary enzymes.[366] They are also absent, so far as is known, in the Crustacea.[353] Among the molluscs, some gastropods[126] (e.g., *Helix*) and chitons[232] have salivary glands whose secretions are important in carbohydrate digestion, while in many carnivorous snails these glands secrete aspartic acid or sulfuric acid up to 1N concentration to dissolve shells of the prey.[356] In the mammals, salivary amylase is known[8, 56, 356] to be present in significant amounts only in man and other primates, rodents, and the rabbit, being minor or absent in the others tested. Certain carnivorous vertebrates such as frogs[258, 356] may secrete salivary amylase, although the substrate present in the prey (glycogen) is unavailable until intestinal digestion has occurred. Little happens to ingested starch in the mouth or stomach of vertebrates, and their salivary amylase may be vestigial.

(b) Conduction and storage. This occurs in the esophagus and crop, which are usually muscular. The crop (ingluvies) is a food storage chamber, found in most birds;[42, 328] analogous structures occur in some invertebrate groups (e.g., those insects that are infrequent feeders). Softening and much carbohydrate digestion can occur there, usually from salivary amylase and disaccharidases.[42, 71, 328, 366]

Storage of food to await time for digestion also occurs in the stomach of some vertebrates, in sacs or non-glandular regions, as in some fishes (e.g., eels) and some fish-eating birds.

(c) Early digestion. Structures for internal grinding or stirring (styles, gastric mills, gizzards) occur at this stage in rotifers and many of the annelids, crustaceans, molluscs, and insects. In vertebrates, the highly contractile gizzard (muscular stomach) of most birds, which contains grit and has a non-glandular, keratinized lining, is in this category. Similar divisions of the stomach specialized thus for mechanical function are found in the crocodiles and in certain fish (e.g., *Mugil* and *Mormyrus*). When these grinding structures are well developed, oral mastication is not important. The rotating crystalline style found in many herbivorous molluscs (bivalves other than Septibranchia, and some gastropods) is also a major source of enzymes; the style is a mucoprotein rod seated in a pouch of the anterior stomach, and carries amylase and lipase[24, 126] and other enzymes.[235] It probably represents an insoluble enzyme source, needed because of the lack of isolation of the gut of these animals from the environment.

The midgut or stomach region of the higher invertebrates is generally a major site of digestion. Its digestive enzymes are secreted either by unicellular glands in the wall, or (in mollusca and many arthropods) by glandular diverticula, termed the midgut gland (sometimes called the hepatopancreas). This gland is sometimes purely secretory (e.g., in prochordates and some molluscs[24]), but entry into, and digestion of food material within, its ceca occur in other molluscs, in crustaceans, and in arachnids. It is often, also, a site of absorption. This gland in most cases simultaneously performs storage (of glycogen and fat) and metabolic functions, analogous to those of the vertebrate liver.

The separation of a purely pancreatic type of gland, containing protease zymogens, occurs in the Cephalopoda. In the most primitive vertebrates (cyclostomes) the liver is already of the mammalian type, with a distinct gall bladder;[1] there is no exocrine pancreas, but the intestinal epithelium contains isolated cells apparently of the pancreatic zymogen type.[252] The exocrine pancreas has become the acinar type, functionally fully differentiated, in the elasmobranch fishes, as in all higher vertebrates. The highly developed system of digestive enzyme secretion in the vertebrates is further discussed below. It should be noted, however, that even in the more primitive groups of the Metazoa where extracellular digestion occurs (e.g., in the gastric filaments of coelenterates[121, 169] and the gastrodermis of some platyhelminths[169] and nematodes[205]), secretory cells have been found containing dense granules which are similar to pancreatic zymogen granules and are packed with the proteases and other enzymes to be discharged into the lumen. These proteases can be present in the granules as precursor, activatable molecules,[121] as in the vertebrates. In the epithelium of the digestive gland of bivalve molluscs, certain basophil cells have been shown to have secretory granules and a fine structure resembling the vertebrate pancreatic acinar cells,[262] and are presumed to be the source of the extracellular proteases[280] found there. The basic mechanisms of zymogen granule production probably arose, therefore, very early in metazoan life.

(d) Final digestion and absorption of nutrients. This occurs in most of the intestine or the posterior midgut. In the vertebrates the intestine is complex in structure, the surface being highly folded in several ways. In man, for example, the absorptive surface area of the small intestine is increased two- to three-fold by circular folds, and a further eight- to ten-fold by the projecting finger-like villi, which are about 1 mm in length. Villi are not present in the small intestine of all vertebrates,[228] being pronounced in man, the hamster, and the pigeon but absent and replaced by long ridges of the surface in the rat, the guinea pig, and the crane (*Grus grus*),[382] giving a smaller surface area increase and apparently less absorption.[228] The characteristic "brush border" is always present in the absorptive cells of the vertebrate intestinal epithelium, being formed of microvilli (Fig. 4–1), which further increase the area of the surface of the cells of the villi[49] by a factor of about 40. Microvilli also occur in invertebrates with well developed epithelia in the gut (e.g., the insect midgut). The experimental separation by Crane and co-workers[236, 260] of the layer of these microvilli in mammals and the determination of its functional components has emphasized its central importance in terminal digestion and absorption. It was formerly thought that a digestive secretion, the "succus entericus," was liberated in the small intestine, but it is now firmly established that in various mammals,[105, 236, 260, 307] birds,[312, 382] reptiles,[382] amphibia,[264] and probably in all vertebrate species, there is no such enzymatic liberation, the digestive enzymes produced by the intestine being firmly bound in the brush border of the epithelium (see Fig. 4–1). These include the intestinal carbohydrases,

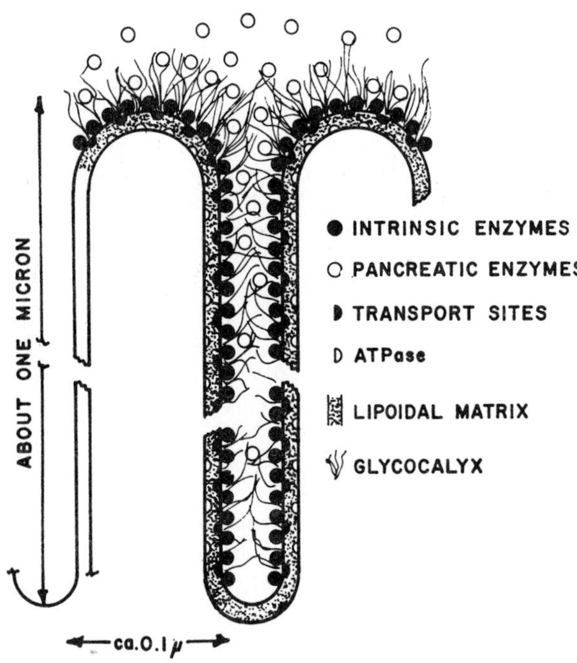

Figure 4–1. Schematic representation of the intestinal brush-border functional system. The open circles represent the pancreatic hydrolases, which are likely[347] in part loosely to adsorb to the glycocalyx. The filled circles represent the hydrolases (Table 4–3) intrinsic to the membrane of the microvillus. Carrier molecules for sugars, and also for amino acids, are shown in close orientation to the membrane hydrolases. (From Crane, R. K., *in* Code, C. F. (ed.), Handbook of Physiology, Sec. 6, Vol. 5, pp. 2535–2542, Amer. Physiol. Soc., Washington, D.C., 1968.)

lipases, enterokinase,[255] certain peptidases,[156, 283] and alkaline phosphatase.[156] The terminal stages of digestion are, therefore, closely linked to the absorption of products by the mucosal cells, and a cooperative organization of enzymes and carriers is believed to be present in the brush border (Fig. 4–1). It has been shown[236] that absorption of the newly-cleaved sugars there is faster than that of the same products if added to the lumen, suggesting that a "kinetic advantage" is provided by this intimate association. The only significant secretion known with reasonable assurance to occur in the vertebrate intestine (apart from mucin from the goblet cells) is that of water and electrolytes, needed in a balanced state for intestinal digestion and absorption to proceed efficiently.[149]

(e) Elimination of undigested residues. The terminal region of the digestive tract is conspicuous in many terrestrial animals in which an important function is the absorption of water from the fecal material, as in insect hindgut and vertebrate colon. The whole intestine of crustaceans and of most molluscs comprises this region, as does the mid- and hind-intestine of arachnids. In bivalves and many gastropods the undigested material is twisted together with mucus into a fecal string, which is moved by ciliary action and peristalsis down the intestine; covering the feces by a mucous coat in this manner prevents reingestion by these filter feeders.

In stages (d) and (e) a significant role is played by the bacterial inhabitants of the intestine in many animals (see Section III).

Protection of Digestive Epithelia. In all animals employing extracellular digestion, the surfaces involved must be protected against attack by the secreted enzymes, especially the proteases. Further, where the food after passing through the mouth is coarse, protection against abrasion is needed. The commonest device, serving both purposes, is the secretion of mucus. This is found in all vertebrates, and also in many others (e.g., gastropods, crustaceans, and holothurians). The digestive epithelia are probably universally impermeable to the extracellular enzyme proteins. The mucosa of the vertebrate stomach is also protected by the "tight junctions" between the tops of its epithelial cells, which help maintain the impermeability of that surface to hydrogen ions.[76] For the intestinal epithelium of the vertebrates, it is known that a protective outer coat of mucopolysaccharide, termed the glycocalyx, is attached.[164] A reversible adsorption to the mucosal surface of some of the luminal enzymes, without penetration, has been reported to occur;[7, 346] this presumably is onto the glycocalyx.

To protect against mechanical erosion, a toughened lining is produced for regions of the tract where coarse, abrasive food is held. In crustaceans and insects eating hard food, the foregut and the hindgut have a heavy chitinous lining. The midgut epithelium is unlined, but in most insects the food mass there is enveloped in a protective "peritrophic membrane," composed of protein and chitin and less than 0.5 μm thick. Digestive secretions initially mixed with the food perform their actions within this membrane, which must be specialized to permit the exit of end-products for their absorption in the gut; indeed, in Diptera there is evidence[383] that it has selective permeability. Analogous membranes enveloping digesta are found also in some polychaetes and in a few vertebrates; thus, some of the fish-eating birds secrete an elastic covering for digesta in the gizzard,[349] as do the cyclostomes in the intestine.[1] Keratinized, non-glandular surfaces are commonly developed in vertebrates in regions of the gut where special mechanical protection is needed, as in the gizzard of birds and the complex stomach of ruminants (see page 147).

Secretion of Digestive Enzymes in the Vertebrates

The vehicles of extracellular digestive enzymes of the vertebrates are the gastric secretion and the pancreatic juice. The bile has also been reported to carry alkaline phosphatase[8] and amylase[37,57] in vertebrates of various classes, the amylase being significant in omnivores and herbivores. While the various glands of the stomach almost always secrete HCl and acid-requiring proteases in mammals, reptiles,[316,372] birds,[99] and amphibians,[256,279] this is not universally true in vertebrates, nor even within mammals. In the few monotremes that have been investigated

TABLE 4–1. Enzymes Secreted by the Pancreas of Vertebrates

*Enzyme or Precursor**	*Substrates*	*References (Occurrence)*
A. PROTEASES		
Chymotrypsinogen	Mainly on C-terminal side of Tyr, Phe, and Trp bonds	100, 184, 237, 252, 277, 293, 295, 300, 330, 343, 367, 381
Trypsinogen	On C-terminal side of Lys and Arg bonds	45, 100, 184, 192, 252, 278, 293, 295, 300, 367, 381
Pro-elastase	Wide specificity for neutral amino acid bonds, especially in elastin	120, 184, 199, 200, 223, 252, 278, 310, 381
Pro-carboxypeptidase A	From C-terminus, at all except Lys, Arg, or Pro; fastest at a C-terminal aromatic or non-polar amino acid	107, 184, 198, 249, 252, 278, 381
Pro-carboxypeptidase B	From C-terminus, at Lys and Arg	184, 249, 252
B. OTHERS		
Amylase	$\alpha(1 \rightarrow 4)$-glucosides	37, 100, 184, 220, 230, 265, 278, 288, 370
Lipase†	Triglycerides, to form 2-monoglycerides and fatty acids (see Fig. 4–3)	46, 57, 99, 100, 130, 184, 220, 226, 240, 265, 298, 354
Carboxylesterase and cholesterol esterase‡	Water-soluble or bile-salt-dispersed mono-esters	61, 93, 226, 240, 298
Pro-phospholipase A	3-*Sn*-phosphoglycerides, at C-2 (Forms, e.g., lysolecithin)	9, 31, 82, 217, 365
Ribonuclease	RNA, forming Cp, Up, and oligonucleotides	21, 130, 184, 278, 281
Deoxyribonuclease I	DNA, forming oligonucleotides	184, 278, 297

*Some of these enzymes (or zymogens) exist as multiple forms, even within one animal. For earlier references, see the second edition of this book.

†Whether a separate pancreatic lecithinase acts on the ester bonds in phospholipids such as lecithin is uncertain.

‡It has been deduced[240] that these two activities are in one enzyme, and depend on whether the substrate requires bile salts for its solubility.

and the grasshopper mouse (*Onychomys toridus*), the stomach is non-glandular;[135] the contents are at neutral pH, permitting the abundant salivary amylase to act. Some fishes lack a stomach altogether, notably the chimaeras, lungfish, and some genera of certain families of teleosts (e.g., Cyprinidae and Ladridae),[23] as well as the cyclostomes.[1] So far as has been investigated, neither HCl nor pepsins are produced in any of those species.[23, 252] In the stomachless fish the food enters directly into the neutral or alkaline intestine, the first part of which is expanded but is not structurally differentiated from the rest.[4, 23] The great majority of fish species, on the other hand, possess a true stomach, but only one type of secretory cell has been identified in their gastric glands, apparently producing both HCl and pepsinogen.[4, 23, 318, 363] In most of the higher vertebrates these functions are differentiated into separated parietal (oxyntic) and chief (peptic) cells. In many of the Amphibia, the peptic glands are placed separately in the esophagus[256, 279] for unknown reasons, while the acid-producing glands are in the epithelium of the stomach. In all of the classes of the vertebrates, strong HCl secretion occurs and gastric pH values of 1 to 2 have been recorded.[318] This initial acidic phase of digestion permits a wide spectrum of proteolytic activity, kills bacteria and the cells of food if alive when swallowed, may dissolve calcified layers, and probably facilitates iron absorption.[299]

The exocrine pancreas secretes a remarkable array of digestive enzymes (Table 4–1) through all the major vertebrate classes. The enzymes are stored in zymogen granules, which can reach a content of 40 per cent of the total protein of the gland. In the elasmobranch fishes,[23] including the primitive chimaeras,[110] the exocrine pancreas is a distinct, acinar organ of the mammalian type, with a pancreatic duct discharging into the intestine and with the characteristic pancreatic enzymes.[252, 381] In most, though not all, teleost fishes[4, 23, 57] the exocrine pancreas is diffuse, often in thin threads, and it may penetrate the liver; in many cases no glandular structure is evident in gross anatomy, and clumps of pancreatic tissue are widely dispersed along the intestine. In these teleosts, the intestine usually carries a number of appendices (the pyloric ceca or appendices pyloricae), which are folded, narrow extensions of the intestine,[4, 23, 356] increasing the surface area for digestion and absorption. The disseminated pancreatic cells usually discharge into these ceca so that separate analysis of the pancreatic enzymes is then very difficult. The proteases, for example, can be isolated from these teleosts only in the activated form, although they probably occur in the pancreatic cells as zymogens.[381]

pH Values in Extracellular Digestion

In general, the later stages, at least, of digestion are maintained at an alkaline or neutral pH value. The usual occurrence of an acidic gastric phase in the vertebrates has already been reviewed. In contrast, a preliminary phase of distinctly acidic secretion has been found in very few invertebrates (e.g., in certain echinoderms which need to dissolve calcareous material,[208] and in the carnivorous snails noted above). A few dipteran insects have this specialization for protein digestion, since their midgut is at pH 2 to 3 and secretes proteases with an optimum in that range.[158] Such an acid midgut has been found in the larvae of *Calliphora, Musca,* and *Stomoxys.*[129] The pH values found in insect alimentary tracts have been extensively tabulated;[158] they largely fall in the range pH 6 to 8, which is characteristic of invertebrates in other than special cases. In the known cases of acid secretion in insects, phosphoric acid appears to be produced.[158] The gastric juice pH and its proteolytic optimum are at pH 7 to 8 in 20 species of amphipod, isopod, and decapod Crustacea,[85] but the barnacle *Balanus nubilus*[85] and the shrimp *Penaeus setiferus*[118] have acidic values (pH 4 to 5). Rhynchocoelan flatworms have in their foregut glandular, carbonic anhydrase-containing cells similar to vertebrate oxyntic cells, producing an acid secretion that is used both in killing the prey and for proteolysis by an acid protease which is secreted by the gastrodermis.[168, 169] Similar exceptions due to dietary adaptation probably occur in other invertebrate groups.

The pancreatic juice of vertebrates is at a pH of 7.5 to 8.5, and contains, when secretion is stimulated, a high concentration of sodium bicarbonate. The intestinal mucosa actively secretes or absorbs sodium, bicarbonate, chloride, and water,[149] so that an appropriate ionic balance is maintained for the neutralization of the acid from the stomach and the buffering at pH 7 to 8.5, which is maintained[8] in the intestine of virtually all vertebrates.

SUMMARY

Although it is a simplification of a great variety of cases, the statement is generally true that the digestive system shows evidence of increasing complexity for nutrient gain throughout the evolution of successive invertebrate and chordate lines. The higher vertebrates have in their tract and associated glands a greater variety of cells, specialized anatomically and biochemically for the secretion of a wider range of enzymes (and hormones; see Section V) than in any of the other groups.

II. *COMPARATIVE BIOCHEMISTRY OF DIGESTIVE ENZYMES*

By E. A. Barnard

DIGESTION OF CARBOHYDRATES

The great bulk of the carbohydrate encountered in the diet of animals is of plant origin. Carnivores, therefore, deal with smaller proportions of carbohydrate. In man, about 50 per cent of the caloric intake is accounted for by carbohydrates, this being almost entirely starch, sucrose, lactose, and fructose.[155, 380] This 50 per cent level is greatly exceeded in the human diet in many underdeveloped regions of the world. Through food chains, ultimately all food of animals is derived, of course, from plants or bacteria.

Of the polysaccharides in the diet of animals in general, two classes are significant: (a) structural polysaccharides which are indigestible by vertebrates—cellulose, lignin, dextrans, mannans, inulin, pentosans, pectic acids, alginic acids, agar, and chitin; and (b) universally digestible polysaccharides—principally starch and glycogen. Starch is the most important food storage product of plants, and consists of amylose and amylopectin (Table 4–2), the latter being more external in the starch grain. Glycogen is the corresponding animal storage polysaccharide. The structures and cleavages of the carbohydrates of dietary significance are summarized in Table 4–2.

The devices used by herbivores to utilize the structural polysaccharides are considered separately (Section III).

Digestive Carbohydrases of Vertebrates. Pancreatic α-amylase hydrolyzes amylose (Table 4–2), *not* at random so as to yield a significant amount of glucose, but rather by a multiple attack on two to six adjacent bonds in the polymer at a time.[132, 287] Maltose and maltotriose are essentially resistant to α-amylase under physiological conditions, and these and a smaller quantity of larger oligosaccharides are the sole products of the digestion of amylose by α-amylase.[132, 287] Amylopectin (and, similarly, glycogen) also does not yield glucose, but rather maltose, maltotriose, and a considerable range of branched oligosaccharides,[148, 284] in which the $\alpha(1 \rightarrow 6)$-glucosidic bonds prevent further cleavage. Hence, intestinal enzymes acting on oligoglucosides and maltose are essential, since only free monosaccharides can be absorbed.

The digestion of the oligosaccharides and disaccharides (released by amylase or ingested directly) in the vertebrate intestine is accomplished by specific enzymes (Table 4–3), which are firmly membrane-bound in the microvilli. The absorption of the monosaccharides across the epithelial cell membrane occurs as they are formed there by the enzymatic cleavages.[67, 236] Part of this final digestion is accomplished by an intestinal γ-amylase, which releases glucose stepwise from the ends of oligoglucosides and maltose (as well as from starch and glycogen).[7, 73, 89] γ-Amylase does not show the activation by chloride that is a general characteristic of animal α-amylases, and it is much more heat-stable than the latter.[7, 73, 89] Various α-glucosidases have been termed maltases; the number of such enzymes present in human intestinal brush border is still uncertain. A current classification is shown in Table 4–3. These enzymes in mammals can be distinguished by their differing rates of heat inactivation[72, 193] and sugar affinities,[89, 193] and can be separated chromatographically.[72, 193] In the vertebrates other than mammals, the same pattern of glycosidases occurs, but one or another of these types may be missing.[187, 307, 382] One of the maltases occurs in various mammals as the "sucrase-isomaltase complex," having equal parts of these two activities.[193, 307, 382] These appear to be caused by two independent active centers[193] that be-

TABLE 4–2. Principal Carbohydrates Digested by Animals

Carbohydrate	*Structure**	*Enzyme*	*Action*
POLYSACCHARIDES			
Starch: Amylose	(Gluc-α(1 → 4) Gluc)$_n$	α-Amylase *or*	Endo-; cleaves α(1 → 4) bonds
Starch: Amylopectin	(Gluc-α(1 → 4) Gluc)$_m$ branched through α(1 → 6) Gluc bonds; $m = 6$ to 12	β-Amylase† *or*	Exo-; removes maltose units stepwise
Glycogen	As for amylopectin, with $m = 3$ to 6	γ-Amylase (Glucoamylase)	Exo-; removes glucose units stepwise
Cellulose	(Gluc-β(1 → 4) Gluc)$_n$	Cellulase†	Endo-
Chitin	(AcGluc-β(1 → 4) AcGluc)$_n$	Chitinase†	Endo-
OLIGOSACCHARIDES			
α-Glucosides: Maltose	Gluc-α(1 → 4) Gluc	Maltase and γ-amylase	
Isomaltose	Gluc-α(1 → 6) Gluc	Isomaltase	
Sucrose	Gluc-α(1 → 2)β-Fruc	Sucrase	
α,α-Trehalose	Gluc-α(1 → 1)α-Gluc	Trehalase	
β-Glucosides: Cellobiose	Gluc-β(1 → 4) Gluc	β-Glucosidase	
β-Galactosides: Lactose	Gal-β(1 → 4) Gluc	Lactase	
α-Galactosides: Raffinose	Gal-α(1 → 4) Gluc-α(3 → 2)β-Fruc	α-Galactosidase	Releases α-galactose
β-Fructosides: Sucrose	Fruc-β(2 – 1)α-Gluc	Invertase	

*Gluc = D-glucopyranosyl; Fruc = D-fructofuranosyl; Gal = D-galactopyranosyl; AcGluc = 2-N-acetyl-D-glucosaminyl. The linkage between the successive monosaccharide units is from carbon-1 to carbon-4 (1 → 4), etc. n is a large and indefinite number.

†Secretion by animals is rare (see text).

have as though they were under a single genetic control.[308] In the intestinal mucosa of birds[312,382] and of a turtle,[312] both activities are also present, but not in the same ratios.

TABLE 4–3. Enzymes Found in the Brush Border of the Intestinal Mucosa of Vertebrates

Enzyme	*References (Occurrence)*
γ-Amylase (=Maltase 1)*	73, 89, 7
Maltase 2	72, 193
Maltase-sucrase (=Maltases 3 & 4)	72, 193, 307, 312, 382
Maltase-isomaltase (=Maltase 5)	72, 193, 307
Trehalase	187, 307, 312
Lactase	6, 187, 307, 336, 382
β-Glucosidase	219, 307
Alkaline phosphatase	156
Aminopeptidase	109, 156, 283
Oligopeptidases	88, 182, 266, 286
Enterokinase	140, 255
Carboxylesterases†	77, 218

*Numbering of maltases is according to the classification of Semenza.[307] An alternative classification has been made by Dahlqvist.[72]

†Lipolytic activities due to the brush-border enzymes, rather than intracellular or luminal enzymes, have not been well defined.

Another intestinal α-glucosidase of widespread occurrence from humans to fishes (but missing in cats, pinnipeds, the koala, and birds)[187,307,312] is trehalase. Its only known substrate is α, α-trehalose (see Table 4–2). This disaccharide occurs in algae, mushrooms, and insects; although it is negligible in the diet of Western man, it is significant for some Oriental populations and was so, apparently,[307] for primeval man.

Lactase occurs only in the mammals,[307,382] in which it is bound in the brush border. Other β-galactosidases also occur in the mammalian intestinal mucosa, but they are readily soluble and their lactase activity is lower and has an acid pH optimum; this latter activity is lysosomal and is not involved in extracellular digestion in the brush border.[6,15] Only a single lactase seems to be present at the latter site, from studies made in man,[6,15] a monkey,[336] and the rat,[6] and has an optimum at pH 6.0. A β-glucosidase has been found in the mammalian brush border.[219,307]

The occurrence of intestinal carbohydrases shows an overall correlation with the diets of vertebrates. Lactose is synthesized as the principal carbohydrate nutrient (3 to 7 g per

100 g)[8] in the milk of all land mammals, and lactase is always abundant in the intestinal brush border of these animals at or before birth.[307] It is absent there in the non-mammalian vertebrates,[307,312] whose young are not suckled. Some marine mammals, notably Pinnipedia (seals, sea lions, and walruses), do not secrete lactose (or any carbohydrate) in their milk, and their young do not possess intestinal lactase.[307,335] Marsupials resemble the placental land mammals in being positive in both of these characters.[187] Lactase in the intestine of mammals usually decreases after weaning, when lactose disappears from the diet. In man, this decrease in lactase is much less than in other mammals.[307]

Sucrase, also, is absent in the Pinnipedia,[307,335] and is very low in certain fish-eating birds,[187] again corresponding to the diet. Mammals (except man) possessing intestinal sucrase and maltases develop them only after weaning,[307] when their substrates begin to be encountered. In a number of ruminants, which metabolize sugars in the rumen, these intestinal enzymes are correspondingly low or absent at all stages.[72,311,357]

A deficiency of one of the enzymes for digesting disaccharides in an animal leads to intolerance of that disaccharide, which will produce acid diarrhea and, eventually, death.[29,335] These symptoms follow from the normal lack of absorption of any of the disaccharides; if they are not hydrolyzed, they create an excessive osmotic load in the intestine. This observation illustrates the selection pressure that operates to ensure that an animal has sufficient intestinal enzyme to deal with a disaccharide occurring significantly in its diet. This factor is important in human medicine also, since a few children and a considerable global proportion of adults (especially in Asiatic and African races) are deficient in the brush-border lactase.[127,128] A combined sucrase-isomaltase congenital deficiency is also recognized.[127,308]

Invertebrates. The carbohydrate digestive enzymes are less fully investigated in invertebrates than in vertebrates, although amylases and disaccharidases have been shown to be extremely widely distributed. In the Crustacea, these enzymes are principally secreted by the midgut gland; their known occurrence in this class has been listed by van Weel.[353] So far as has been tested, all crustaceans secrete an α-amylase; where it has been studied in detail, as in *Carcinus maenas*,[39] this enzyme appears to be similar to vertebrate α-amylase (e.g., in activation by chloride and by calcium). The purified *Carcinus* enzyme[39] has a molecular weight about one-quarter that of the mammalian[101,348] pancreatic α-amylases. In insects, also, a chloride-activated α-amylase is widely secreted by the midgut.[20,60,71,95,157] It is interesting that the pH optimum of this enzyme varies very widely according to the midgut pH range; the midguts of Lepidoptera are very alkaline and secrete an amylase with an optimum near pH 10, whereas some Heteroptera have corresponding values near pH 4. Salivary α-amylase in insects occurs only in certain groups.[20,71,366] This supplementary polysaccharide digestion is generally associated with storage of plant food in the crop. β-Amylase is very rare in animals, although it is common in plants.

Lactase is generally absent in invertebrates, but a few exceptions are known (e.g., in the intestine of some oligochaetes,[209,251] *Helix*,[247] and a few insects such as honeybee larvae, cockroaches,[104,335] locusts,[78,95] and dermestid beetles[60]). In this and in other cases of exotic enzyme activities in invertebrate digestive systems, the absence of a contribution by microbial or protozoan symbionts should be (and usually has not been) thoroughly excluded. A careful inventory of the digestive carbohydrases of a locust[78] and a beetle,[60] including lactase, has been provided. The occurrence of significant lactase in the Crustacea has not been established.[353]

Activity on sucrose occurs very widely in the gut of invertebrates, including all feeding insects examined, and is generally due to an α-glucosidase that is also active on maltose (see Table 4–2), unlike the plant and fungal invertases. Exceptions occur, since it is known that some insects,[24] including silkworms[242] (which use inulin) and some beetles[60] and roaches,[20,104] possess β-fructosidase activity. Ingested β-glucosidic disaccharides are split by many invertebrates, mostly herbivorous, including certain insects, gastropods, and crustaceans.[20,24,95,242,353] Larvae of moths and butterflies have sucrase in addition to lipase and protease in the midgut, whereas those adults that suck nectar have sucrase but no other digestive enzymes, and non-feeding moths have no digestive enzymes whatever.

Sugar digestion is usually weak in carnivorous invertebrates. For example, the blood-sucking tsetse fly *Chrysops* has no salivary carbohydrases and has amylase and sucrase only in the posterior midgut, whereas the closely related non-bloodsucking *Calliphora* has amylase in the salivary glands and amylase, maltase, and sucrase in both anterior and posterior midgut.[366] Similarly, plankton-feeding and herbivorous gastropods have a high activity of extracellular amylase both in the crystalline style (see Section I) and secreted from the hepatopancreas, as well as a wide spectrum of other carbohydrase activities.[24,261] In the carnivorous gastropods, on the other hand, the style is absent and a weak amylase is produced in the hepatopancreas.[261]

Trehalase is found in the gut of many insects[20,24,95,373] and in various invertebrates feeding on fungi in which trehalose occurs (e.g., earthworms[209] and some other oligochaetes,[251] whose diets include fungal spores). It may also be significant in insectivorous animals, since trehalose is probably used as a storage compound in all adult insects,[345,373] reaching a concentration of up to 7 per cent in the hemolymph (see page 222). One role of trehalose in insects is thought to be to facilitate intestinal glucose absorption by removing glucose from the hemolymph.[345] The trehalase in the insect gut may be needed to deal with back-diffusing trehalose.[345]

The correlation of carbohydrase activity of invertebrates with diet is not always clear, however. Some essentially protein-feeding insects secrete a wide range of carbohydrases.[60] The β-galactosides, if any, in the diet of those invertebrates having notable intestinal lactase activity are unknown. These and similar characters must at present be explained as vestigial.

Digestion of Proteins

The digestive endopeptidases attack particular internal peptide bonds in proteins and in peptides; the exopeptidases attack, instead, stepwise from either the amino or carboxyl terminus. Some of the latter enzymes attack both proteins and peptides, while some, also of digestive significance, attack only peptides (e.g., the mammalian brush-border tri- or dipeptidases).[283] The enzymes employed in the digestion of proteins of the food must be distinguished from the proteases present within cells for other purposes. Thus, "cathepsins" are a group of intracellular proteinases of wide specificity, mostly acting at fairly acidic pH and being lysosomal in location; they may be confused experimentally with extracellular digestive acid proteases if extracts of whole tissues or organisms are used.

Acid Proteases. Extracellular acid proteases are rare, but not unknown, in invertebrates (see Section I); little is known about them, but they appear to be cathepsin-like.[169] On the other hand, all vertebrates tested (except stomachless fish,[23] cylostomes,[252] and a few species with a non-glandular stomach[135]) secrete a pepsinogen, of molecular weight about 39,000.[338] This is activated autocatalytically in an acidic medium, to yield the protease pepsin, having an optimum activity at pH 1.8 to 2.0. A number of variant forms of pepsin occur in a single species, as described from man and the pig.[296,306] In addition, at least one other chemically distinct protease, termed gastricsin,[160,338] has been characterized from human gastric juice. This is indicated as occurring also in other vertebrates, being recognized by its pH optimum of 3.3 to 4.0. Rennin is a distinct, milk-clotting protease, with pH optimum 3.7, occurring only in suckling ruminants. Each of these must come from a different zymogen.[160]

Parts of the amino acid sequences of the chains of pepsins, a gastricsin, and a rennin are known, and are closely related. From these comparisons, it has been proposed[160] that they arose by evolution from a common ancestral protease (1), by gene doubling, via a hypothetical, now extinct protease (2):

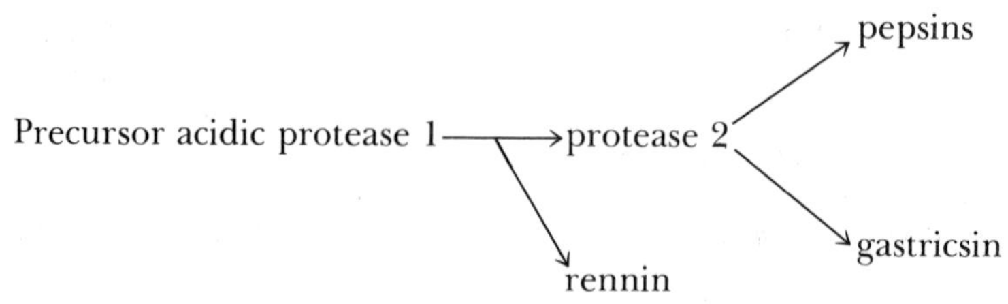

The pH of food just after entering the vertebrate stomach is usually higher than pH 2 until sufficient acid is secreted, and may still remain higher if within a food mass,[318] so that the occurrence of gastric enzymes with a higher pH optimum than that of pepsin can be of advantage.

The Serine-Protease Series. Chymotrypsinogens and trypsinogens are universally secreted by the vertebrate pancreas so far as is known[45, 143, 367, 381] (Table 4–1). They have molecular weights close to 25,000. Amino acid sequence comparisons in several vertebrate species show strong homology between the chymotrypsins, trypsins, and elastase,[45, 143, 310, 367] indicating that they arose from a common ancestral neutral protease, in a series similar to that proposed for the pepsins. All of these enzymes have a common active center structure, involving a serine and a histidine residue.[143, 381] The full molecular structures of bovine chymotrypsin A and porcine elastase are known.[38, 359]

Trypsinogen is activated by enterokinase in the vertebrate intestine. Trypsin then activates more trypsinogen, and also is responsible for the activation of all of the other protease zymogens of Table 4–1. The activation process is always initiated by the cleavage of one or more peptide bonds in the zymogen molecule. Two types of pancreatic proteinaceous, specific inhibitors of trypsin are known, the Kunitz (basic) inhibitor[183] which has been found in the pancreas and other tissues of ruminants, and the Kazal (acidic) inhibitor.[52] Only the latter inhibitor is secreted in the pancreatic juice, in all mammals investigated. Presumably it controls the autoactivation process until enterokinase is encountered in the intestine, when a cascade of protease activation occurs.

At least three alternate forms of the chymotrypsinogen molecules, A, B, and C, are known in vertebrates, and can occur together in the same species. These differ in structure and (for C) in secondary specificity.[237, 343] They should not be confused with the variants α, γ, δ, and so on of one given form (e.g., chymotrypsin A), which arise from various secondary cleavages of peptide bonds in the activation of its zymogen.

Elastase appears to be produced by all mammals investigated.[310] It is secreted in the pancreatic juice as the zymogen, proelastase,[120, 199] which loses a small peptide in forming the enzyme (in the pig) of molecular weight 26,000. Elastase attacks many proteins, but is unique among pancreatic enzymes in its ability to digest elastin (a fibrous cross-linked protein in the vascular walls and connective tissue of vertebrates). Proelastase would be expected to be much more abundant in the pancreas of carnivorous vertebrates than in that of herbivores, but this has not yet been established.[310] Elastase was found to be absent in the intestine of a cyclostome.[252] It has been found in the pancreas of numerous mammals,[184, 310] the chicken,[223] and several fishes.[200, 252, 278, 381] The pancreas also secretes enzymes with mucolytic and esteratic activities that have been described as associated with elastase ("elastomucase" and "elastolipoproteinase"), and act with it in attacking carbohydrate and lipid bridges in elastic tissue.[142, 212]

In intestinal secretions of invertebrates, a trypsin-like protease has very generally been found;* it is recognized by its activity on N-acyl-L-arginine (or α-N-acetyl-L-lysine) esters or amides at alkaline pH values. This is the minimum requirement for such an identification; almost all of the earlier literature (prior to about 1960) on invertebrate digestive proteases is vague owing to lack of specificity definitions or even, in some cases, of pH control (as Baldwin[18] pointed out), or failure to distinguish extracellular from intracellular proteases (mixed tissues or even the whole animal, in many cases, being extracted). The use of these small substrates, more or less specific for mammalian trypsin, is not in itself a completely rigorous criterion. In a number of cases, further evidence for a relationship to vertebrate trypsin has been obtained by showing specific tryptic cleavages of protein substrates, or (a valuable tool[381]) by inhibition by the active-site-directed trypsin alkylator[302] TLCK (1-chloro-3-tosylamido-7-amino-2-heptanone), or by immunological cross-reactivity.[14]

Chymotrypsins, too, have been identified by parallel methods in some invertebrates, although this activity has been found to be absent in certain invertebrates in which a trypsin is present.[36, 118] However, a chymotrypsin has been found (along with a trypsin) in the digestive tract of several coelenterates[62, 121] and echinoderms,[75] so that the divergence of the two types of protease presumably occurred at an early stage of metazoan phylogeny. These invertebrate chymotrypsins and trypsins have the active site chemistry of the mammalian serine

*See references 36, 46, 54, 62, 75, 85, 267, 367, and 384.

proteases, wherever this has been investigated.

Evidence for zymogen forms of the chymotrypsin-like and trypsin-like secreted proteases of invertebrates has been obtained in a few cases,[34,54,62,75,121] and can be expected to be found widely. In at least one case,[34] the release from the precursor is known to be by a limited proteolytic (tryptic) cleavage as in the vertebrates. The invertebrate trypsins and chymotrypsins have been found to have molecular weights fairly close to that of the mammalian forms (about 25,000), but a much smaller protein (12,500) with the properties of chymotrypsin has been isolated from the midgut of larvae of the hornet *Vespa orientalis*.[330] Proteases of similar low molecular weight, which resemble in specificity neither trypsin nor chymotrypsin but have a similar pH optimum, have also been found in the digestive fluid of crustaceans.[47,267]

A collagenase that degrades undenatured collagen has been found in the hepatopancreas of the crab *Uca pugilator*; it is believed that this enzyme is used by such crustaceans for scavenging animal material.[90] This enzyme, unlike the known collagenases unrelated to the alimentary tract, is a serine proteinase.[90] Involvement of such an enzyme in vertebrate digestive systems has not been established. It occurs, however, in the venom of various snakes.[314]

Enterokinase. This enzyme cleaves an N-terminal peptide[278,375] highly selectively from the chain of trypsinogen to produce trypsin (Fig. 4–2). So far, it has been identified in the intestine of all vertebrates examined; it can be suggested, but not yet proven, that it also occurs in many invertebrates. Enterokinase recognizes a sequence of at least three acidic residues just before the lysyl bond cleaved, in the trypsinogens so far known (Fig. 4–2), and does not attack other lysyl or arginyl bonds.[222,278] It has been suggested[222] that such a code is likely to occur in all vertebrate trypsinogens, to protect them, and hence all the pancreatic zymogens, from premature activation by other enzymes. Enterokinase has been shown to be located in the brush border of the intestinal villous epithelial cells of mammals.[140,255] However, bile salts appear to release the enzyme into the lumen, so that trypsinogen activation would occur both at the surface and in the lumen.[339] Enterokinase deficiency has been found as a congenital defect in humans,[269,339] and is alleviated by feeding of pancreatic extract.

Exopeptidases. Carboxypeptidases A and B (Table 4–4) of vertebrates are formed from the specific zymogens, pro-carboxypeptidase A and B. In the cow, the activated procarboxypeptidase A also generates a chymotrypsin as a separate subunit,[30] but this is not the case in various other vertebrates.[107,198,277] Carboxypeptidase B acts in concert with trypsin, since it attacks the peptides terminating in a basic residue produced by the latter enzyme. Carboxypeptidase A similarly acts in concert with chymotrypsin, so that an efficient stepwise system has evolved.

The final stage of complete degradation to amino acids is accomplished by aminopeptidases and tri- and dipeptidases. In the vertebrate intestinal villi, aminopeptidases are bound in the brush border membrane.[109,156,283] The enzymes specifically hydrolyzing various di- and tripeptides are also located in these mucosal cells,[88,266,286] but most of this activity

	↓
Cow	Val-Asp-Asp-Asp-Asp-Lys-
Sheep, goat, deer	Val-Asp-Asp-Asp-Asp-Lys-
	Phe-Pro-Val-Asp-Asp-Asp-Asp-Lys-
Pig	Phe-Pro-Thr-Asp-Asp-Asp-Asp-Lys-
Horse	Ser-Ser-Thr-Asp-Asp-Asp-Asp-Lys-
Dogfish	Ala-Pro-Asp-Asp-Asp-Asp-Lys-
Lungfish	Phe-Pro-Ile-Glu-Glu-Asp-Lys-

Figure 4–2. Activation peptides of vertebrate trypsinogens. Each of these is liberated by enterokinase (or by auto-activation) from the corresponding trypsinogen by an attack at the lysyl bond marked by the arrow; these sequences, therefore, represent the N-terminal region of the trypsinogen. In sheep, goat, red deer (*Cervus elaphus*), and roe deer (*Capreolus capreolus*), two chains are cleaved, giving two peptides. Note that there is in common a sequence of at least three acidic residues (penultimate to the bond cleaved). These acidic groups bind calcium, to promote the activation process. For references to the sequences, see Reeck and Neurath.[278]

TABLE 4–4. Distribution of RNase in Pancreas of Vertebrates[21a]

Group A: High Content (200 to 1200 μg per gram pancreas)
Cow, bison, sheep, goat, elk, kangaroos (red and grey), wallaby, mouse, iguana, Uganda kob, rat, hamster, guinea pig.

Group B: Moderate-to-low Content (20 to 100 μg per gram)
Pig, turtles (*Chelydra serpentina, Podocnemis unifillis, Pseudemys elegans*), hippopotamus, alligator, armadillo, horse, turkey, chicken.

Group C: Very Low Content (0 to 20 μg per gram)
Opossum, whale, dogfish, bullfrog, snakes (two species), turtle (*Chrysemis picta*), Necturus, frog (*Rana pipiens*), grouper (fish), lungfish, dolphin, sting ray, toads (two species) pigeon, shark, tuna, barracuda, monkeys (4 species), man, *Amphiuma,* elephant, rabbit, dog, cat.

is soluble within the cells.[266,286] It has been suggested that the small peptides are mainly absorbed as such and hydrolyzed within the absorptive cells,[225,250,266] although this is still uncertain.[347]

Carboxypeptidases and other peptidases have also been found in invertebrate intestine.[24,169,261,353] In animals showing a definite phase of secondary intracellular digestion, the peptidases have been detected in the cells responsible and not in the gut lumen or secretory cells (e.g., in various turbellarian flatworms[168,169]).

Nucleic acid digestion will be considered in Section III.

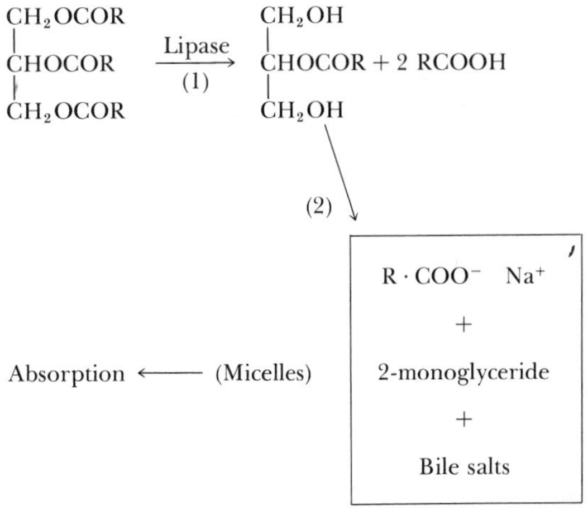

Digestion of Lipids

The animal and plant triglycerides are of major dietary importance, and lipases, which hydrolyze the longer chain esters (Fig. 4–3) for digestion of these components, occur universally. Carboxylic esterases hydrolyze shorter chain esters, but the distinction between the two types is not sharp.

Vertebrates. Pancreatic lipase is the major enzyme involved, although certain cleavages of lipids are accomplished by other pancreatic enzymes (see Table 4–1). This lipase is present in the zymogen granules as the active enzyme, and is secreted in all vertebrates examined.[46,57,100,130,346,354] It acts in

CH_2OCOR | $CHOCOR$ | CH_2OCOR $\xrightarrow[(1)]{\text{Lipase}}$ CH_2OH | $CHOCOR + 2\ RCOOH$ | CH_2OH

(2) ↓

Absorption ⟵ (Micelles) [$R \cdot COO^-$ Na^+ + 2-monoglyceride + Bile salts]

Figure 4–3. Cleavage of triglyceride by pancreatic lipase. The hydrocarbon chains R may be different at each position in dietary fats. Reaction (1) is the main one brought about by lipase in the vertebrate intestinal lumen, and involves assistance by bile salts (see text). Reaction (2) is the association of products to micelles (represented by rectangle), which act at the mucosal cell surface in absorption (see Figure 4–8).

the intestinal lumen, and its substrates are emulsified by bile salts. It has been proposed[46] that the bile salts (Fig. 4–4) protect lipase from unfolding at the interface; co-lipase, a small protein in the pancreatic juice,[227] probably in turn protects the lipase from inhibition by bile salts.[44] Fatty acids and monoglycerides are released from the sub-

The Bile Salts of Man

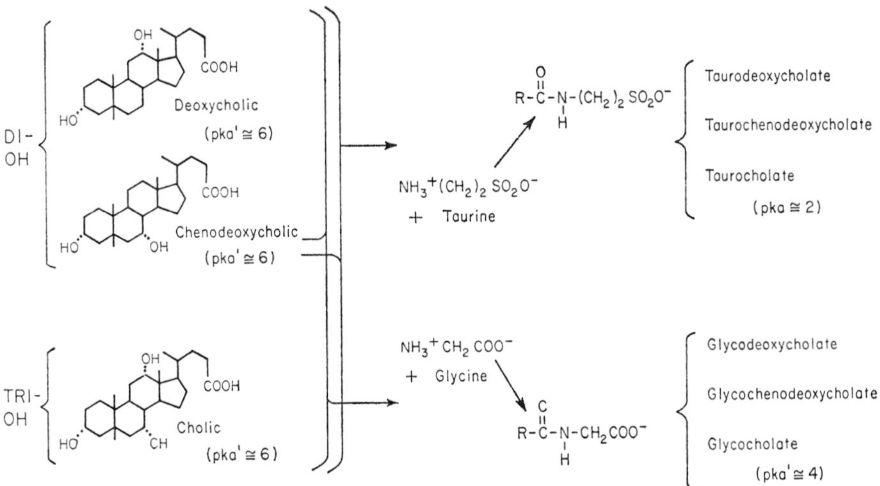

Figure 4–4. Predominant bile salts of human bile. The bile acids—cholic, chenodeoxycholic, and deoxycholic—are conjugated with glycine or taurine in a peptide bond that is resistant to all digestive enzymes. Conjugation increases the solubility range; the unconjugated bile salts precipitate from solution below pH 7. (From Hofmann, A. F., and Small, D. A., Ann. Rev. Med. *18*:333, 1967.)

strates and absorbed by the epithelial cells. This absorption also requires conjugated bile salts, which form mixed micelles with these products.[153]

Dietary cholesterol esters and lecithin[303] are rapidly hydrolyzed by esterases, presumed to be pancreatic, in the intestinal lumen, and the cholesterol and lysolecithin formed are absorbed; again, these processes require bile salt assistance. The same or associated enzymes (Table 4–1) also act on truly dissolved ester substrate molecules in the absence of bile salts, as well as on monoesters in bile-salt-dispersed micelles.

A stomach lipase has also been reported to be widely distributed;[165] this enzyme can act only on previously emulsified fats (e.g., those in milk).

Invertebrates. Simple esterases seem to be predominant in lipid digestion in invertebrates. In crustaceans, the midgut gland has been found to be rich in esterase and, in some cases, lipase activity.[2,189,224,352] In insects, digestive esterases have been widely found.[19,71,123,158] In bivalve molluscs, both intracellular and extracellular digestive esterases occur,[261,281] and both types are probably also involved in gastropods.[248,290] Digestive esterases have been detected in the gut of polychaetes,[175] echinoderms[75,208] (although much of the lipid digestion in that phylum is probably intracellular), and nematodes,[96] but are absent in many leeches.[170] Histochemical evidence indicates that the digestive esterase of *Ascaris* is stored in granules of secretory cells of the gut, which are discharged into the lumen.[205] A variety of carboxylic esterases, with evidence for lipases, too, has been found in protozoans.[243] These enzymes can be associated by histochemical means with the food vacuoles.

Emulsifiers. Although they are not enzymes, these substances play an important role in increasing the efficiency (see above) of both the digestion and the absorption of fats. The bile salts perform this function in all vertebrates[153] (where a separate bile is secreted into the intestine). Bile salts do not affect the non-lipolytic digestive enzymes,[210] but they can protect chymotrypsin and trypsin from inhibition by fatty acids.[289] The bile salts are based upon cholesterol (e.g., cholic acid, I (Fig. 4–4), is the commonest form in the higher vertebrates). They have an acidic group in the side-chain so that the molecule is a detergent. In most vertebrates this acidic group is provided by taurine or glycine conjugation through a terminal carboxyl, but in some of the fishes and amphibians it is an alcohol sulfate, probably a more primitive and less efficient structure.[144] For example, the principal bile salt in teleost fishes of the family Cyprinidae and some allied forms is 5α-cyprinol sulfate (II).

In general, the bile acids or alcohols vary in detailed structure among the various vertebrate classes and orders in an evolutionary direction.[144]

There is a correlation with diet in the mammalian bile acids, herbivores generally making dihydroxy or monohydroxymonoketo bile acids and carnivores making trihydroxy bile acids.[144] Carnivores (and ruminants) have both types well represented (Fig. 4–4).

Emulsifiers of some type are also secreted for lipid digestion and absorption in at least some of the invertebrates with more specialized digestive tracts.[356,374] In some crustaceans (e.g., *Cancer pagurus*) acylsarcosyltaurines (III) play this role.[350] Evolution to the steroid bile acid structure has not, so far, been shown to have occurred in invertebrates.

III. $CH_3(CH_2)_nCO\ N(CH_3)\ CH_2CO\ NH(CH_3)_2SO_3^-$

II.

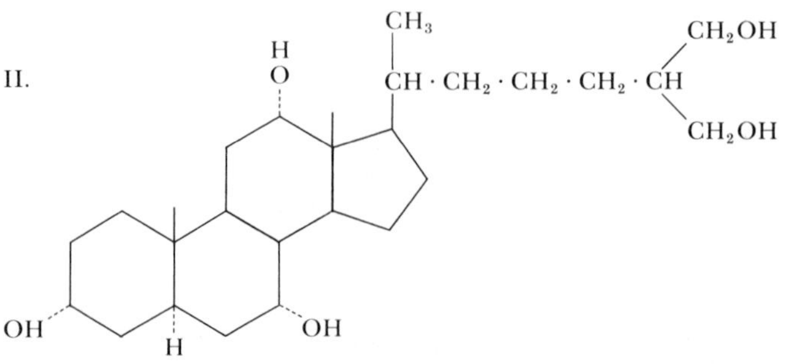

5α-Cyprinol

III. *BIOCHEMICAL ADAPTATIONS TO DIET*

By E. A. Barnard

While all animals show some form of anatomical and physiological adaptation to their diet, as noted in Section I, some major types of adaptation bring important changes at the biochemical level. The most prominent among these are the specializations of herbivory and of endo-parasitism.

The Devices for Herbivorous Specialization

The colonization in the Tertiary period of a great variety of environments by green plants offered the possibility of widespread development of herbivorous fauna, based on the vast energy source of plant polysaccharides. However, the enzymes necessary to cleave those β-glycosidic structural polymers have rarely been developed by animals. The herbivorous mode has developed in certain groups only, based on one of several specializations: (1) secreted cellulase and related hydrolases, found so far only in a very few animal groups (see below) and never in the vertebrates and tunicates;[174,378] (2) adoption of cellulolytic micro-organisms as gastric symbionts, fermenting food prior to final digestion (i.e., the ruminant mode); (3) use of those symbionts for a secondary fermentation in the hind-gut after the main digestion has occurred conventionally, as in the horse and elephant; (4) coprophagy (i.e., reingestion of excreta) so that secondary bacterial attack can be exploited, as in the rabbit; or (5) use of micro-organisms for secretion of ancillary enzymes in the gut for hydrolysis without fermentation, as in many of the phytophagous insects. The structural polysaccharides, generally indigestible by other groups of animals, are degraded to varying degrees in these classes of herbivores.

Ruminants

The true ruminants have a multi-compartmental stomach. The largest compartment, the rumen, and the associated compartment, the reticulum, maintain a continuous anaerobic culture of up to 10^{11} bacteria and 10^{6} protozoa per milliliter.[161] The total contents can reach a mass of one-seventh that of the entire animal. Cellulose, xylans, pectin, and lignin are partially digested by the micro-organisms, and the products and other digestible carbohydrates are fermented, producing volatile fatty acids, H_2CO_3, and CH_4. The acids are absorbed directly across the non-glandular epithelium of the rumen; acetate and butyrate are oxidized directly for energy, while the proprionate is used for hexose and lipid synthesis.[35,239] The blood sugar level of adult ruminants is only about one-half that of other mammals,[239] since little hexose is available for intestinal absorption. Ruminants show a low sensitivity to insulin.[11] Dietary protein is hydrolyzed, and mostly also deaminated and fermented, in the rumen, and its nitrogen is largely (usually 70 to 90 per cent) incorporated into microbial protein and nucleic acid.[91,145,324,360] Fiber digestion is facilitated by the rumination (regurgitation) process.

The micro-organisms continually pass into the abomasum (true stomach), which is acidic. They are digested, initially there and terminally in the intestine. Hence, most of the true food of the ruminant is its own micro-organisms. The ruminant secretes a copious saliva (e.g., on the order of 100 liters/day in the cow[16]), and this conveys much $NaHCO_3$ and urea to the rumen.[16,161] The $NaHCO_3$ neutralizes excess fermentation acids. CO_2 from this source and from the fermentation itself is partly fixed by other ruminal bacteria, partly absorbed, and partly (along with CH_4) discharged by belching. The ammonia released by protein fermentation is either utilized by the ruminal bacteria[51] or absorbed through the rumen wall, transported to the liver, and there converted to urea. Urea back-diffuses continuously from the blood to the rumen,[146] and the excess is returned to the rumen via the salivary stream, which contains about 70% of its N in this form.[329] Hence, a specific "ruminant nitrogen cycle" for digestion has been deduced[12,21] (Fig. 4–5). The proteases and peptidases of the true stomach, pancreatic juice, and intestine, which are similar to those of other mammals, complete this cycle by releasing amino acids from the microbial food. Unlike other vertebrates, ruminants utilize urea efficiently as a nitrogen source,[145,146,229] because of the activities of their bacteria. This recycled urea is increased in amount under adverse dietary conditions

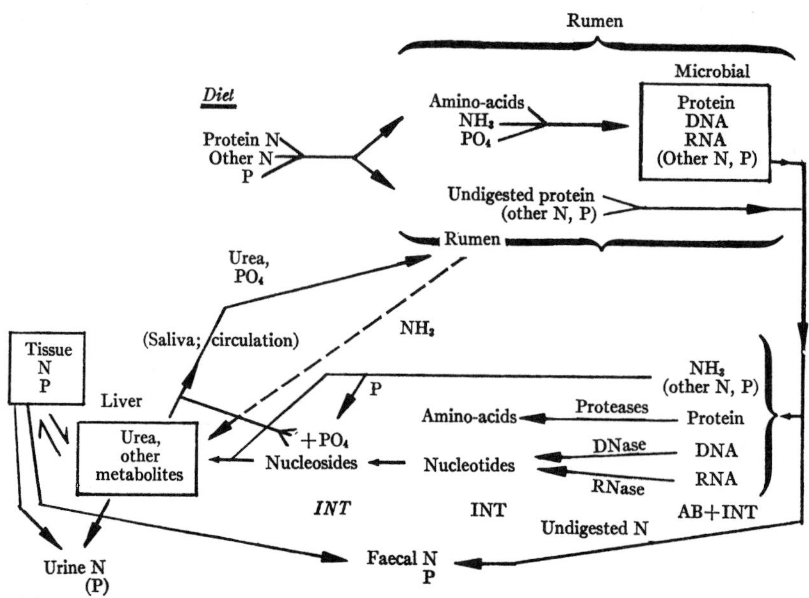

Figure 4–5. Nitrogen and phosphorus cycle for the ruminant. Rectangles indicate processes occurring within cells. The broken line represents the exit of some ammonia via the portal circulation when dietary N is in abundance. In nutritional insufficiency, the excretory loss of N is much reduced, and there is a much greater return of N and phosphate to the rumen via the saliva and the circulation. Note the role of the pancreatic nucleases (RNase and DNase) in releasing both N and phosphate for this cycle from the bacterial nucleic acids. AB = abomasum (true stomach); INT = intestine. (From Barnard, E. A., Nature *221*:340–344, 1969.)

(Fig. 4–5), providing an important adaptive mechanism for foraging animals. Addition of urea to cattle fodder, as a cheap nitrogen source, is important, therefore, in agricultural practice.

Nucleic acid digestion is important in the ruminant economy. Of the total microbial N entering the stomach, roughly 20% is in the form of polynucleotide N.[325] Hence, the excretion of purine products (allantoin and uric acid) is unusually high.[40, 344] The pancreatic nucleases are, therefore, produced in much larger amounts than in other animals (Table 4–4), to salvage polynucleotide N. An efficient break-down (75 to 85 per cent) of the microbial RNA and DNA has been shown to occur in the duodenum of ruminants.[325] Part of the N released enters the urea cycle of Figure 4–5, but an important advantage is believed to be the release of the relatively large amount of phosphorus locked up in the microbial nucleic acids.[21] This is required to meet the phosphorus needs[229] of the host and, after recycling, of the ruminal micro-organisms. The P cycle (Fig. 4–5) is completed by both the saliva[320] and the blood flow to the rumen.[273]

Unsaturated fatty acids from plants are hydrogenated to saturated *trans* isomers by the ruminal microflora,[309] and enter the host depot fat in that form. Lipolysis also occurs in the rumen, with fermentation there of most of the glycerol produced.[371]

Ruminant-like Vertebrates. The true ruminants are confined to the large sub-order Ruminantia of the mammalian order Artiodactyla. The camels, alpaca, and similar animals, which are in the allied sub-order Tylopoda, have a different anatomy and histology in their multi-compartmented stomachs,[41] but their rumination, fermentation,[368] and urea cycling[301] show that their digestive system is essentially the same as that of the true ruminants.

Some much more distantly related mammals show a ruminant-like convergence. The macropod marsupials have a sacculated, disproportionately large stomach, which contains bacteria and protozoa at the same density as in the ruminants.[358] That the same type of economy prevails is shown by the evidence of the similar fermentation products, methane and CO_2 production, frequently low blood glucose, low N excretion, and urea recycling.[239, 358] These marsupial herbivores also have the high level of pancreatic ribonuclease characteristic of the ruminants (Table 4–4), suggesting that microbial nucleic acid components are recycled as in the latter. The ruminant-like economy equips these marsupials to survive under seasonal highly adverse conditions.[50,334] Certain primates—the colobid and langur monkeys—are leaf-eaters, and also have a multi-compartmented stomach with a rich bacterial population which readily digests cellulose.[27,239] Convergence with the ruminants is evidenced by the fermentation, CH_4, CO_2 and NH_4^+ production, and volatile fatty acid release and absorption that occur in the stomach.[27,197] The leaf-eating tree sloths (Bradypodidae) also have very large and complex stomachs[238] which contain cellulolytic bacteria.[172] The rumination movements, however, do not appear to have developed in these ruminant-like groups, other than in the Tylopoda,[197,239] probably either caused by or determining a lesser amount of roughage in their diet.

A lesser degree of convergence to the ruminant system is shown by certain herbivorous rodents. Many genera among the Myomorpha have a complex stomach with a microbial population,[239] which can utilize both cellulose and urea. This is found in the golden hamster, the pack rat (*Neotoma*), and the Gambian rat, but not in the laboratory rat. The gastric fermentation is supplementary here to a conventional digestive system.

Non-ruminant Herbivores

Many herbivorous mammals with a simple stomach[33] have a greatly enlarged cecum or colon (or both) with a rich microflora. The orders Perissodactyla (e.g., the horse) and Lagomorpha (the rabbits and hares), as well as the elephant and the hyrax, follow this pattern.[3] Their digestive system is conventional except that structural polysaccharides are digested and fermented microbially beyond the small intestine. The sugars and fatty acids released are absorbed there, and urea can be recycled,[146] but the microbial bodies are not available for subsequent digestion as in the ruminant-like species. There is, therefore, no advantage to break-down of microbial nucleic acids, and pancreatic nucleases are absent in these species (see Table 4–4). Fermentation of sugars to volatile fatty acids in the tract of the horse, occurring in its greatly enlarged colon and cecum, is more important nutritionally than in the rabbit,[233] but also occurs in the latter, the dietary cellulose being bacterially hydrolyzed in its cecum.[64] Among the Suiformes, both types of solution can be found: the pigs employ a cecal fermentation, whereas the hippopotamuses are ruminant-like,[239] and there is some evidence that some of the peccaries are also ruminant-like.[239] Ruminants are more efficient digestively than non-ruminant herbivores, and have spread at the expense of the latter since the Oligocene.[313]

A device employed by the lagomorphs and most rodents to increase the efficiency of cecal digestion of cellulose is coprophagy,[340] that is, fecal reingestion. This tends to recover at least the nitrogen of the cecal microflora, as well as their vitamin content.[22] The reingested soft fecal pellets[340] may also persist for some hours in the fundus of the conventional stomach, buffered by their phosphate content, to provide an additional digestion and fermentation of polysaccharide.[136]

In many other vertebrates, microflora are also present in the alimentary tract.[321] Bacteria often proliferate in the anterior stomach and small intestine after feeding, but are subsequently largely depleted because of the stomach acid; in the large intestine, however, large populations of certain bacteria are permanent. This has been found in 15 species of mammals and birds examined, but in cold-blooded vertebrates the numbers seem to be much lower.[321] These bacteria serve secondary nutritional roles (e.g. in synthesis of biotin, folic acid, vitamin K[139] and other co-factors) in many species including man, as well as assisting in the deconjugation and hydroxylation of bile salts.[185] Such activities, long inferred, have been proven by the use of germ-free animals.[63,139,185] Thus, rats maintained with an absolutely sterile gastrointestinal tract require dietary co-factors of those types,[63] and also fail to modify their bile salts.[153] In some fruit-eating vertebrates (e.g., bats) a significant activity of the intestinal microflora is the digestion of pectins (galacturonic acid polymers).

Devices (1) and (5) (as listed above) of the herbivores are those that are well developed in the herbivorous invertebrates. Cellulase activity in invertebrate gut is usually due to symbiotic bacteria, protozoa, or fungi, but their presence is easy to overlook and has often been disputed. Endogenous cellulase and cellobiase (persisting after antibiotic sterilization) are reported to be secreted by herbivorous gastropods.[111,248,263,378] Bacterial cellulase is often also present, however.[106] The wood-boring bivalves (e.g., *Teredo*)[131] and certain crustaceans[353] may secrete endogenous cellulase. Cellulases are important in the digestive equipment of the annelids and echiurids examined, and have been suggested to be endogenous.[378] The agar of algae is digested by sea urchins, by means of an alginase;[201] there is some doubt as to whether this is bacterial,[97] although echinoderms also harbor a rich microflora and microfauna in their gut. Gastropods, too, often secrete alginase, this being correlated primarily with an herbivorous habit.[108,261] β-Glucuronidase can also be secreted by the digestive gland of such molluscs, probably for digestion of sulfated polysaccharides in seaweeds.[65,87,261]

Among the wood-eating insects, there are four groups. (1) Most appear to use bacterial or protozoal cellulases.[71,158,292,364] (2) Some, such as the silverfish *Ctenolepisma*,[201] secrete a cellulase in the gut,[95] or in the saliva (as in the cockroach *Periplaneta*[364]). (3) Strict bark-feeders have not been found to have any cellulase activity, but have a range of glycosidases (including pectinase and xylanase) appropriate for the carbohydrates of the cambium.[19] (4) Some can use only the stored starch and sugars of the wood.[158]

The termites[71,254] are a major group of cellulose consumers. They digest cellulose in wood or, in some species, leaves or humus. In the lower termites this is accomplished by symbiotic flagellates in the hind intestine. In the higher termites (Termitidae) these are replaced by cellulolytic bacteria.[254] It has been suggested that some endogenous cellulase and cellobiase can also be produced.[378] The pentosans of hemicellulose are also hydrolyzed by the symbionts. Anaerobic fermentation occurs in each case, and volatile fatty acids are produced and metabolized, as in the ruminants.[254] In the Macrotermitinae, the lignin (excreted after the cellulose of wood is digested) is used to grow a symbiotic fungus, *Termitomyces*, which is then also consumed by the termites, providing them with organic nitrogen and other nutrients.[297a]

Digestion of Nucleic Acids

As described above, digestion of microbial nucleic acids by pancreatic nuclease is important in the intestine of ruminants, including the ruminant marsupials. The amounts of nucleases available from the pancreas of other herbivorous vertebrates are generally lower (Table 4–4). In some cases where a cecal microflora is abundant, as in the pig and the horse, sufficient pancreatic ribonuclease occurs to suggest that there is opportunity for digestion of some of the cecal microbial nucleic acids. Other non-ruminant herbivores, such as the rabbit, have an extremely low amount of pancreatic ribonuclease (as do carnivorous and omnivorous vertebrates) (Table 4–4). Yet administered RNA, at least at moderate levels, is digested by such species,[21] including man.[44a] This paradox has not been resolved; possibly, low levels of ribonuclease from the microflora are sufficient.

The occurrence and use of pancreatic deoxyribonuclease has not been explored. No information is available on the extent of digestion of nucleic acids in invertebrates.

Chitinolysis, Endoparasitism and Other Adaptations

Chitin is, after cellulose, the most abundant polysaccharide, occurring in fungal cell walls and also extensively in the cuticle of arthropods and in molluscs, annelids, and coelenterates; it is not normally digestible, but chitinolytic activity has evolved in many predators of those animals. In earthworms[194,378] (which consume fungi) and in some crustaceans,[173,194] chitinase that is at least in part endogenous seems to be secreted. In seven genera of gastropods, including nine species of pulmonates, Jeuniaux[174] has found that, contrary to earlier reports, an endogenous chitinase is abundantly secreted in the digestive gland, being unaffected by antibiotic sterilization. Exceptions can occur, as in *Aplysia depilans*, which has only bacterial chitinase. In many other predaceous invertebrates, a bacterial origin of digestive chitinase is evident or not excluded.[174,356] Endogenous chitinase and the associated chitobiase (β-acetylglucosaminidase, needed to complete the digestion) appear to be secreted by many vertebrates that eat chitin-covered prey (e.g., insectivorous bats, birds, and reptiles, or fishes that eat chitinous molluscs or crustaceans). These enzymes are not found in herbivores, nor in carnivores preying only on other vertebrates.[171,174]

Keratin of feathers and wool, which is strongly disulfide-bridged and usually indigestible, is hydrolyzed in the midgut of certain insect larvae—some tineids, dermestids, and Mallophaga.[158,366] These larvae are unusual in having a very low redox potential in the midgut, permitting reduction of the disulfides. They may also have a protease adapted to digest reduced keratin.[272]

Certain groups of animals have, in adapting to a highly specialized feeding habit, abandoned the biosynthesis of some digestive hydrolases. Thus, aphids feed on plants in the phloem sap only, using its free amino acids and sugars, and they lack proteases and polysaccharidases.[71] Leeches appear to have lost the capacity to secrete endopeptidases in the gut, using instead the gut flora and exopeptidases, and having very long digestive periods.[170]

Endoparasites often represent the extreme of such adaptations to minimize digestive requirements. The Cestoda and Acanthocephala have no digestive tract, and no evidence for their use of digestive enzymes has been found.[274] They actively transport amino acids,[274,294,370] glucose,[103] and purines and pyrimidines[215] from the host through their body surface, and absorb lipids[17] (see Section VI). The vertebrate intestine tends to maintain a homeostatic state[275] for amino acids (but not for carbohydrates) by means of the exocrinic secretions and epithelial cell shedding, and this appears to be utilized by parasites such as cestodes.[275] In digenetic trematodes, digestive proteases and peptidases are secreted for the digestion of the host tissue proteins imbibed, except that the strigeid trematodes use these secretions extracorporeally[92,326] and some forms such as *Schistosoma mansoni* combine digestion of host blood in their gut with direct absorption of simple nutrients through their body surface. Parasitic nematodes possess protease,[96,288] lipase, and amylase activities in their gut; these are probably secreted in the gut. Malaria parasites take in protein and sugars from the host; the protein (especially hemoglobin) is digested in the vacuoles or vesicles by proteases, but most other digestive enzymes appear to be absent in these protozoans.[265]

Symbiosis with unicellular algae (zooxanthellae and zoochlorellae) occurs in many invertebrate groups, and in some cases is linked to digestive adaptation. In some sea anemones, corals, and hydra, sugars and glycerol are provided to the host by the intracellular photosynthesizing algae,[206,246] improving growth on limited exogenous food. This adaptation can proceed further, as in the turbellarian *Convoluta roscoffcensis*, whose digestive tract disappears in the adult as zoochlorellae multiply in its cells.[361] In certain molluscs, such as the Tridacnidae, the superficial tissues are packed with photosynthesizing algae, which are apparently digested by amoebocytes to provide nutrients directly to the tissues.[379]

Whether the complement of hydrolases present in many snake venoms represents, in part, a specialization for promoting digestion of the injected prey has not been established.[191] An extensive review of the enzymes found in these venoms is available.[176]

Adaptive phenotypic differences in digestive enzyme production are well established. Intake of a given substrate induces the increased secretion of its digestive hydrolase in some, but certainly not in all, cases. In the various mammals that have been studied, pancreatic amylase and protease outputs increase respectively with the starch or the protein content of the diet,[32,159,327] but lipase does not follow its substrates,[276] nor does ribonuclease.[21] These effects have been correlated with an increased biosynthesis of the respective proteins,[276] although the overall protein synthesis in the exocrine pancreas appears to be unaffected by feeding.[271] The products of the digestion appear to provide the stimuli for these adaptive effects on the pancreas,[32,216,271,276,327] but the intertwined factors have not yet been unravelled. In the frog, different effects occur, in that the total protein synthesis in the pancreas increases greatly upon feeding.[270] Intestinal disaccharidases of mammals, also, can be elevated after some days of feeding of their substrates.[291]

In invertebrates, only a few phenotypic adaptations have been studied biochemically. Snails produce more amylase on a starch diet, but adaptation of proteases to different protein intakes was not found.[351] In snails that change with maturity from herbivorous to omnivorous habits, the digestive amylase/protease ratio declines upon maturation.[319] Enzymic changes are frequently correlated with dietary change during development. The gastric milk-clotting protease of ruminants, rennin,[102] and mammalian lactase[307] (other than in the exceptional case of humans[117,127,307]) are produced abundantly by suckling infants only.

SUMMARY

In general, we can conclude that the digestive enzymes of animals present a basic uniformity that suggests that they have evolved from only a few general types. Overlying this uniformity is a great diversity, in which we can often, but not invariably, see adaptive specialization to exploit all possible sources of nutrition.

IV. *MECHANISMS OF FOOD INTAKE*

By C. L. Prosser

Methods of feeding differ according to types of food: (1) organic molecules dissolved in natural waters (or tissues); (2) particulate matter, most of which is non-living detritus but part is living phytoplankton; and (3) large particles, food masses, or captured prey.

Soluble Food. The concentration of dissolved organic material in natural waters is of the order of 2 to 20 mg/liter. The amount in oceanic water varies much with location and season. In the Atlantic at 30°N, values of 4.7 mg/liter are given for subsurface waters; at 38°N the amount is 0.6 to 1.7 mg/liter. In some Scandinavian lakes the total organic matter was 14.3 mg/liter, of which phytoplankton constituted 1.7 mg, colloidal detritus was 1.0 mg, and dissolved organic matter was 11.6 mg.[196] Similar quantities (about 15 mg/liter) were found in Wisconsin lakes in summer. The organic material was estimated as 2 times the measured carbon, and protein was estimated as 6.25 times the measured nitrogen. Determinations for the Mediterranean were 1.4 mg/liter, of which 0.81 mg was protein. The mid-Pacific showed 3.2 mg/liter of organic solutes, with a minimum concentration at depths of 500 to 1000 meters.[180] In general, the carbohydrate content as sugar equivalents is 0.5 to 3.5 mg/liter; protein is 0.5 to 0.6 mg/liter, and dissolved lipid is 40 to 50 μg/liter.[180]

Measurements of uptake of dissolved organic molecules indicate active absorption by some, but not by all, aquatic animals. The amount absorbed is probably not sufficient for maintenance of many animals, but may be a significant contribution to nutrition.

Many endoparasites, such as cestodes and parasitic protozoa, live in a rich organic soup and can absorb enough so that they do not need specialized digestive systems (see Section III). Some marine animals, also, lack digestive tracts. For example, the Pogonophora have no gut and have been shown to incorporate amino acids into tissue cells from concentrations as low as 10^{-7} M.[211] Considerable variation in ability to absorb amino acids actively from sea water has been noted. The solitary coral *Fungia* removed, in one hour, some 25% of labeled glucose or tyrosine from concentrations as low as 1 mg/liter.[331] The clam *Spisula* showed a reduction of 14% in concentration of glycine between inflow and outflow water when inflow contained glycine at 1.5 mg/liter. *Littorina* removed virtually all the glycine from its container. Reduction in glycine in a low concentration medium in 24 hours by other animals were: polychaete *Chaetopterus,* 50%; ascidian *Molgula,* 90%; sponge *Microciona,* 90%; and starfish *Asterias,* 97%. Crustaceans generally fail to absorb amino acids from solution; *Artemia* showed insignificant uptake, and *Balanus* removed none of the glycine from solution.[332,333] Some animals lose amino acids to the medium, and loss may exceed uptake in the turbellarian *Bdelloura.*[177] However, incorporation into tissues, and active uptake kinetics and absorption by animals without digestive tracts, demonstrate use of dissolved organic substances; the relative importance for energy varies greatly.

A few insects, such as aphids, suck in large quantities of plant juices and extract the nutrients. Many kinds of animals suck blood and make use of plasma solutes as well as of cellular components.

Particulate Food. Particulate food consists of detritus particles, organic substances adsorbed on inert particles, and dead plankton; it also comprises living phytoplankton (mainly diatoms) and some flagellates and bacteria. In surface sea water, particulate organic matter may be present at 0.2 to 1.7 mg/liter, of which some 10 per cent is described as protein. The amount of particulate organic matter decreases rapidly with depth in the ocean and is usually less in open ocean than near the continents. Phytoplankton varies with season, latitude, and depth; in inshore water it may range in amount from zero to 700 μg/liter. In fresh water lakes the phytoplankton may reach 2 mg/liter near the surface in summer and may be absent in winter. In the North Atlantic the total particulate material is 500 to 1500 μg/liter, of which phytoplankton constitute 14 to 60 per cent.[180]

Many kinds of animals live exclusively by

suspension feeding, and they pump large quantities of water through filtering devices. There are two general methods of filter feeding—by the use of mucoid nets, and by cilia or flagella. Mucoid-secreting animals form sheets that filter out particulates, which are then eaten. The tube-dwelling polychaete *Chaetopterus* and the echiuroid *Urechis* secrete a mucous net, pump water through it by means of posterior appendages, and swallow the sacful of trapped food about every 15 minutes. The nets of *Chaetopterus* retain not only small particles but also protein molecules as large as hemocyanin (but not as small as ovalbumin); hence, they have a pore size of about 40 Å.[214] Some molluscs and ascidians have both mucoid and ciliary filters. The gastropod *Crepidula* has one mucous filter at the entrance of the mantle cavity and another over the frontal surface of the gill; these collect food and are periodically rolled up and eaten.[362]

Ciliary mechanisms for fine-particle feeding are found in the ciliate protozoans, sponges, tubicolous annelids, echiurids, rotifers, bryozoans, brachiopods, many gastropod and lamellibranch molluscs, and many tunicates and cephalochordates. In sponges, a water current is generated by flagellated choanocytes; 1 μm particles are retained by the freshwater sponge *Spongilla. Ephydatia* pumps 1200 times its body volume daily.[188]

Lamellibranchs have three mechanisms of filtering: sets of cilia on the gill filaments sort particles by size and can transport food to the palps; straining occurs between the rows of latero-frontal cilia of the gill filaments; and mucous sheets may be secreted.[214] In mussels, particles are caught by adhesion to the long latero-frontal cilia, are wiped off by the frontal ones, and are transported to the palps.[337] The labial palps can discriminate between food and inedible particles; the latter are discarded as pseudofeces. The ciliary filter of *Mytilus* keeps back 30 μm particles, can retain or pass 7 μm particles, and cannot retain 1.5 μm particles.[180] Adult oysters pump about 10 liters/hr/animal.[195] In *Crassostrea* the retention of particles between 1 and 3 μm in size is 33 per cent; the distance between latero-frontal cilia is 2 to 3 μm.[147] The rate of pumping increases after a period of starvation or when there are nutrients in the water.[341] The filtering rate of *Ostrea* is six times greater than that of *Mya* and *Venus*.[5]

Copepods retain particles larger than 5 to 10 μm. The amount of water cleared of particulates such as unicellular algae decreases with increased concentration. The copepod *Calanus* may clear 29 to 150 ml/day/animal.[180] The amount of water that must be pumped for respiratory needs is usually greater than that needed for food, and bivalves and ascidians commonly pump 10 to 20 liters of water per ml O_2 consumed; each liter of water must usually contain 0.05 mg utilizable food.[179]

Daphnia regulates its feeding by its pumping rate and by rejection of excess food; when fed on suspensions of *Chlorella* or yeast, the intake is proportional to the concentration of the suspension up to 10^5 cells/ml and rejection occurs at 10^6 cells/ml.[231]

Phagocytosis, the engulfing of particles by cytoplasmic processes, is employed by the protozoa, leading to digestion in food vacuoles. This mechanism also occurs in specialized cells in many metazoa (see Section II).

Mechanisms for Consuming Food Masses. Discussion of the variety of methods of obtaining food in various habitats is beyond the scope of this chapter. Many burrowing animals swallow quantities of mud or sand and extract food from it, expelling the residue. Some animals have mechanisms for scraping rocks, others for boring into wood. Many predatory animals have mechanisms for poisoning prey (e.g., the nematocysts of coelenterate tentacles, or venom glands). Structures for seizing or masticating prey are jaws and radulas of various sorts. In other cases, seizure is linked to external digestion. Some animals have structures for piercing and sucking fluid, as in blood-suckers and in insects that suck sap.

Nutrition by Symbiosis. Many animals make use of symbiotic organisms to provide nutrients. The ruminant state, and polysaccharide digestion by symbionts, are discussed in Section III. Some invertebrates have algae associated with them (see above), and these may supplement the food that is taken (e.g., in hydroids and corals) by tentacles and digested in the coelenteron. In some coral heads the contained algal plant biomass is three times that of the host animal, and much photosynthesis can occur. When sea water contains $C^{14}O_2$ and the corals are exposed to light, organic products of photosynthesis can be detected in coral cells within a few hours.[245]

Food Selection. Food selection is based on

many types of chemical stimulation. Some animals are strict carnivores, some are strict herbivores, and many are omnivorous to various degrees. Specialization with respect to diet is highly developed in insects, both carnivorous and herbivorous. Choice of food plants is determined by taste preference (see Chapter 21), and each plant presents a unique profile of chemical stimulation.

Ingestion by *Hydra* can be induced by specific products of tissue break-down, especially glutathione, which activates the feeding reaction at a concentration as low as 5×10^{-6} M. This feeding reaction is blocked by N-methyl maleimide or by heavy metals; the SH group of glutathione is evidently not needed, since the reaction can be obtained with some tripeptides.[206] Similarly, in some corals, feeding responses are induced by amino acids such as proline; analogs of proline and glutathione are also effective at higher concentrations.[206,221]

Blood-sucking animals can be stimulated to feed from artificial diets (solutions) by various components of blood. For example, feeding by a rat flea, *Xenopsylla*, can be initiated by salt or sugar solutions of osmotic concentration similar to that of plasma, whereas mosquitos suck fluid only if adenine is present.[112] The tsetse fly *Glossina* is stimulated to artificial feeding by ATP.[115] A tick, *Ornithodoros*, is stimulated to suck fluid via one set of receptors by glutathione and ATP, and by another set of receptors by amino acids.[112] The leech *Hirudo* is attracted by various sugars and by NaCl; NaCl apparently induces the sucking reflex, and intake increases in the presence of sugar.[113]

In the blowfly *Phormia*, food intake is regulated reflexly by distension of the foregut irrespective of the solute, whereas emptying of the crop into the midgut is controlled by blood sugar concentration by a mechanism that does not involve the nervous system.[119] Intake of fluid by a blowfly depends upon its state of hydration, and injection experiments show that volume, rather than osmotic or dilution factors, is important.[83]

Food intake by mammals is regulated in varying amounts by caloric need and by taste. In rats, the caloric properties of food are critical when metabolism is in balance or surfeit, whereas chemosensory properties are more important when the rats are food-deprived. The state of energy balance biases the way in which the central nervous system handles information from taste and caloric receptors.[166]

V. *STIMULATION OF SECRETION OF DIGESTIVE FLUIDS*

By C. L. Prosser

In most animals, feeding is periodic and secretion is elicited to correspond with the presence of food; in starvation there is scanty, dilute, but continuous secretion. Gland cells can be stimulated directly by food in the digestive lumen, by chemical agents via blood (hormonally), by nerve impulses, or by all three methods.

Vertebrates. Control of salivary secretion in mammals is entirely nervous, and all three pairs of glands receive parasympathetic innervation; they also have a less important sympathetic innervation. Both parasympathetic and sympathetic nerves stimulate secretion of saliva, but the nature of the fluid secreted differs according to the nerves stimulated and the glands activated. Some salivary secretion is continuous, and food in the mouth can stimulate it reflexly; conditioned reflexes by other stimuli associated with food also elicit salivary secretion. The stimulating action of the parasympathetics is prevented by atropine, probably by blocking the action of acetylcholine (ACh) at the nerve endings. The gland cells themselves also contain much ACh, the function of which is not known. Some gland (acinar) cells show increased electrical negativity on neural activation, whereas others are depolarized; the gland cells are not electrically excitable.[53] The secreted fluid is initially hyperosmotic to plasma, and in the main ducts Na^+ is reabsorbed.

Secretion of gastric juice in vertebrates is stimulated by the vagus nerve, and the sympathetic elicits some mucous secretion. The basal gastric secretion is reduced after vagotomy. Control of gastric secretion can be conditioned reflexly and is related to emotional experience. Local stretch receptors send afferent fibers to the brain via the vagus. Food or mechanical distension, or ACh, in the stomach elicit the liberation from cells of the antral mucosa of a substance which passes into the blood and stimulates parietal cells to secrete HCl and, to a lesser degree, the chief cells to secrete pepsinogen. This hormone is gastrin,[134] which has been isolated in two very similar forms, gastrin I and II. The gastrins and various homologs of them have been synthesized. Gastrin is a polypeptide having a molecular weight of about 2000, containing 17 amino acid residues.[138] The N-terminal group is pyroglutamyl and the C-terminal group is phenylalanine; in the chain of gastrin II there is a sulfated tyrosine. Histamine also stimulates gastric secretion, but this probably has negligible physiological importance.

The vagus has a weak stimulating action on the pancreas, particularly on enzyme liberation (in the dog). When the acid chyme from the stomach enters the duodenum, a hormone is liberated from the intestinal mucosa; this substance, secretin, is carried in the blood and strongly stimulates secretion of pancreatic juice, enzymes, and especially HCO_3^-. Secretin[315] is a linear polypeptide of 27 amino acid residues, having a molecular weight of 3200. Secretin also stimulates pepsin secretion.

A second hormone also is liberated from intestinal mucosa; initially this was called pancreozymin, from its stimulation of secretion of pancreatic enzymes. Also, contraction of the gallbladder to liberate bile is caused by an intestinal hormone, cholecystokinin. It is now considered that these activities are from the same substance, for which the abbreviation is CCK-PZ. This hormone is a polypeptide of 33 amino acid residues.[315] It is similar to a decapeptide, caerulin, from the skin of an Australian frog.[94]

Glucagon is a substance that is liberated from intestinal and stomach mucosa and from alpha cells of the pancreas; it stimulates insulin secretion, relaxes gastrointestinal smooth muscle, and inhibits gastric acid secretion. Glucagon has 29 amino acids, of which 14 are in the same positions as in secretin.[134] Insulin also decreases gastric secretion.

Nervous influences on intestinal gland secretion are relatively slight. The glands respond to a substance from the mucosa, and this has been called enterocrinin. CCK-PZ also stimulates the intestinal glands,[315] and whether enterocrinin is a different substance is not established. The mucus-secreting cells of the intestine are continually replaced by new cells from the bases of the crypts.

While each of the hormones—gastrin, secretin, and cholecystokinin-pancreozymin—has its own primary action, there is good evidence that all can have the same action in varying amounts. Target effects are listed by Grossman[137] as follows: stomach—secretion of acid and of pepsin, and increase of motility; pancreas—acinar cells, secretion of HCO_3^- and of enzymes; gallbladder—contraction; small intestine—secretion by Brunner's glands and stimulation of motility; pancreas islet cells—release of insulin.

In summary, a progressive series from nervous control of salivary glands, to mixed nervous and hormonal control in stomach, to predominantly hormonal control in pancreas and intestine is evident. The gastrointestinal hormones are linear polypeptides, and each of them can act in varying degrees on all of the appropriate targets. The cellular mechanisms of action are not understood.

In birds, secretion of gastric juice by the proventriculus is stimulated by the vagus nerves, and also by gastrin. Conditioned reflex secretion has been established.[318]

In frogs, mechanical distension of the stomach reflexly stimulates secretion, and this is not abolished by extrinsic denervation. Vagotomy does not reduce acid secretion, but does diminish pepsin secretion by some 40 per cent. Acetylcholine is an effective stimulant, and it is postulated that local reflexes involve cholinergic neurones in the wall of the stomach. Active blood circulation to the stomach is necessary for the reflex responses to occur.[190, 317, 318] Low doses of gastrin stimulate acid secretion in frogs; higher doses are needed to stimulate pepsin secretion.[241]

In teleosts, as in other vertebrates, food in the stomach stimulates gastric secretion. There is evidence for both vagal and hormonal phases of the secretion. The holocephalan fish *Chimaera* lacks a stomach, yet its intestinal mucosa produces a substance that stimulates its pancreas to secrete the digestive enzymes. Also, an intestinal mucosal extract from *Chimaera* can elicit contraction of mammalian gall bladder and HCO_3^- production by pancreas; hence, secretin and CCK-PZ are probably produced by this fish.[252] They must be triggered by some agent other than acid, since there is no stomach. In elasmobranchs, food in the stomach stimulates acid secretion; both histamine and acetylcholine are stimulatory, and there is a gastric nerve plexus. Secretin has been obtained from the intestinal mucosa.[28]

In summary, it is probable that the basic pattern of polypeptide gastrointestinal hormones is similar in all vertebrates. Intrinsic reflexes in the enteric plexuses may differ in relative importance.

Invertebrates. Stimulation of secretion of digestive fluids has not been much studied in invertebrates. In cephalopods, the salivary glands are multi-functional; they secrete poisons that paralyze prey, they secrete proteases, and they probably have endocrine function. Esophagus and intestine of cephalopods have a rich nerve plexus. In snails and crayfish, secretory activity as seen cytologically is markedly increased after feeding. In the earthworm *Lumbricus*, stimulation of the ventral nerve cord elicits secretion of digestive enzymes in crop and intestine.[150] Activity of the secretory cells of *Dytiscus* is maximal 45 minutes after feeding.[80] A blood-borne substance from fed *Tenebrio* stimulates mitoses in the midgut of starved individuals; during starvation, proteases accumulate in midgut cells, but on feeding the enzymes are liberated into the crop.[70] The production of intestinal protease in the fly *Calliphora* is influenced by diet, and is stimulated very little if there is no meat in the diet. If neurosecretory cells have been removed from *Calliphora* brain, intestinal protease is decreased to one-third. Injection of corpora cardiaca restores the protease production; hence, a neural hormone is indicated.[342]

In invertebrates, therefore, very little clear distinction between direct stimulation by food, by hormones, and by the nervous system has been established. Food is the ultimate stimulus for secretion of digestive fluids in all animals, but there is variation in its mode of action.

VI. *ABSORPTION*

By C. L. Prosser

Absorption of the products of digestion is partly by diffusion and partly by active transport.[116] The relative importance of the two processes is not fully understood, and it is probable that active carriers for many substances remain to be discovered. Absorption of sugars and of amino acids, whether in mammalian intestine or in parasitic worms, is coupled to active transport of sodium, and there are probably several carriers that are more or less alike. Absorption has been studied by transfer *in vivo* from intestine to blood, by transfer into inverted intestinal sacs *in vitro,* and by uptake into epithelial cells from mucosal solutions. The properties of the mucosal and serosal boundaries of the cell are different, and in all cases there is active sodium transport across the serosal membrane, making the gut lumen (mucosal side) electrically negative.[117]

Sugars. Glucose can be absorbed against a concentration gradient into epithelial cells; aerobic energy is required, and phlorizin competitively inhibits the glucose uptake at the mucosal surface. Sugar accumulation and transfer does not occur in the absence of Na^+, and in active absorption a potential is developed across the epithelium (serosa positive) that is proportional to the Na^+ transport.[68] Ouabain, added on the serosal side, reduces the potential and also reduces sugar absorption; as was stated previously (page 94), ouabain inhibits the active transport of sodium in many tissues.[67]

A generalized model (Fig. 4–6) is that a carrier molecule has sites for binding of glucose and of Na^+, with some cooperative interaction; sodium which enters the cell with glucose is released from the carrier and is then actively pumped out, mainly on the serosal side, by a transport system that is sensitive to ouabain. The transport is also inhibited by metabolic inhibitors such as DNP. The potential is determined by the Na^+ extrusion and is proportional to Na^+ concentration in the cell. Glucose, by coupling with a carrier, diffuses across the serosal membrane; glucose also stimulates the Na^+ pump. Absorption of sugar is inhibited if K^+ or Li^+ is substituted for Na^+. Transfer of glucose, too, is reduced by Li^+ and K^+; they decrease the potential more than Tris does. If epithelial cells are permitted to accumulate glucose and Na^+ in the presence of DNP, which limits the pump, and the cells are then placed in a Na^+-free but glucose-rich medium, glucose is transferred outward against a gradient (i.e., the transport is reversed).[67, 68]

Xylose and 6-deoxyglucose can apparently be bound by the same carrier as glucose, but the concentration for half-saturation of the carrier is about 2 mM for glucose and 100 mM for xylose; i.e., the affinity is much greater for glucose. D-glucose can be absorbed against a 20-fold gradient, whereas L-glucose enters only passively and reaches a 1:1 distribution. Fructose may be accumulated by a different process, which is non-concentrating.[67] D-glucosamine is not transported, but it inhibits glucose absorption. Galactose is transported, and increases the mucosal potential, but its transport does not alter accumulation of glucose, and galactose is not metabolized by

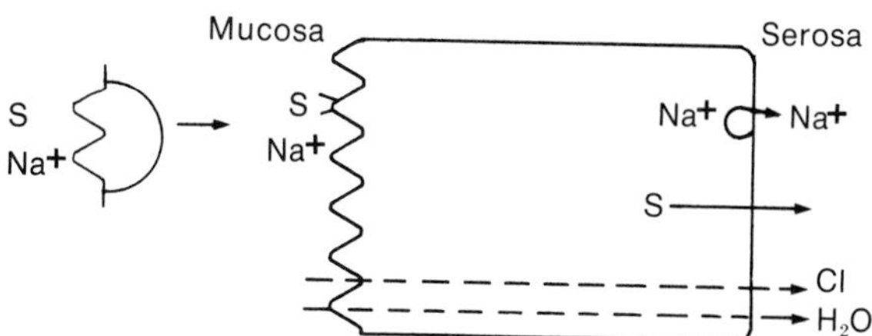

Figure 4–6. Diagram to illustrate transport of a sugar. The carrier molecule combines with sodium (Na) and the sugar (S) at the mucosal border, and dissociates after crossing the cell; the sugar then diffuses across the serosal border, and sodium is actively extruded. (Modified from Crane, R. K., *in* Code, C. F. (ed.), Handbook of Physiology, Sec. 6, Vol. 3, pp. 1323–1351, Amer. Physiol. Soc., Washington, D.C., 1968.)

the cells; mannose is metabolized but is not actively transported and has no effect on the potential.[25]

In the intestine of lower vertebrates, the mechanisms of sugar transport are basically similar to those in mammals but may differ in details. In the bullfrog, *Rana catesbiana*, D-glucose and D-galactose are absorbed against concentration gradients, while D-xylose and D-arabinose are not.[203] In the goldfish, glucose and water are transported to the serosal side; the serosa is positive to the mucosa in a steady-state potential and in a transient glucose-stimulated potential. The steady-state potential occurs in the absence of glucose, but is absent if the mucosal medium is free of Na^+. The response to glucose is rapid; phlorizin inhibits the glucose-evoked potential, and ouabain inhibits the steady-state potential.[323]

Cestodes,[103, 274] and an acanthocephalan from duck intestine,[69] absorb glucose against a concentration gradient; the absorption system is half-saturated at 0.25 mg/ml. Another acanthocephalan, from insects, can use trehalose, unlike species from other hosts.[69] A chiton (*Cryptochiton*) transports D-glucose, methylglucose, and D-galactose, but not mannose or fructose, across its intestine.[204] The intestine of the holothurian *Stichopus* shows no active transport of glucose, methyl-D-glucoside, or galactose.[204]

Amino Acids. Absorption of the amino acids[369] formed by protein digestion is selective and may be against a gradient; it is largely by active transport. As with sugars, there is isomeric specificity and sodium dependence. The L-amino acids are absorbed much more than D-acids, and the L-D ratio in absorption by rat intestine is 6 for histidine, and 1.6 for methionine. However, some competition occurs between D and L forms of the same amino acid. The attainable concentration gradient varies with amino acid; e.g., isolated hamster intestine with 20 mM amino acid (a.a.) concentration on both sides reaches a serosa-mucosa gradient of 2 for proline, 1.4 for valine, and 1.2 for methionine and leucine. Competition between various amino acids indicates separate carriers for the following classes: neutral, acidic, and basic amino acids, and imino acids.[151, 369]

The dicarboxylic (anionic) acids, aspartic and glutamic, appeared not to be transported, but recent evidence suggests that they are transferred into epithelial cells and are rapidly metabolized (transaminated) so that their components appear as alanine on the serosal side.[304] Transfer of these acids into the cells, like that of neutral acids, requires Na^+ on the mucosal side. The differences between the postulated carriers for the different a.a. groups may be slight, and some acids may use more than one carrier molecule. Also, the transport of glucose by the epithelium may be inhibited by a.a. transport; hence, there may be some competition between sugars and a.a.'s for carriers. Active absorption of L-phenylalanine by sections of rat intestine requires Na^+ in the medium, and aerobic conditions; serosal exit of labeled phenylalanine from loaded cells is stimulated by the presence of the unlabeled acid in the medium.[285]

In rabbit ileum, mounted to measure the short-circuit current due to sodium transport as well as movement of labeled a.a.'s and Na^+ into epithelial cells and into serosal fluid, net transport of alanine and Na^+ is abolished by cyanide, dinitrophenol, iodoacetamide, and ouabain; but these agents do not stop uptake across the brush border into the cells.[58] No drug prevents the complexing of a.a. with carrier on the mucosal side, as phlorizin prevents glucose combination with its carrier. The serosa is electrically positive to the mucosa, and ouabain is effective in blocking Na^+ transport out of the serosal surface. Alanine uptake by epithelial cells is inhibited in Na^+-free medium, but is not affected by changes in intracellular concentration of Na^+. In 140 mM Na^+ and 5 mM alanine on both sides, a mucosa-to-serosa flux of alanine was measured as 1.4 μM/hr/

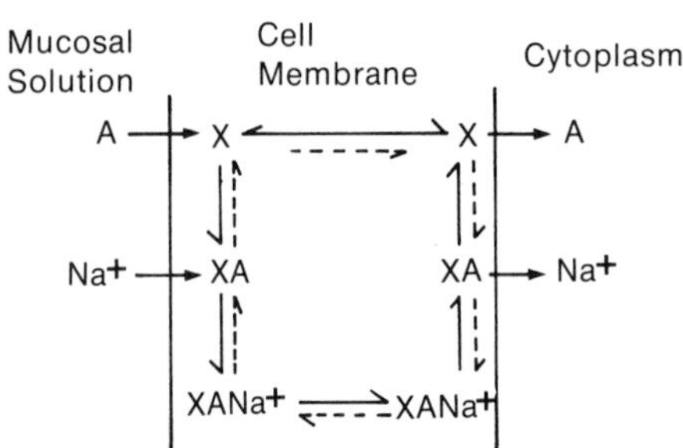

Figure 4–7. Diagram of transport of an amino acid across the lumen membrane of a mucosal cell by a carrier molecule (X), which complexes with Na^+ and the amino acid (A). (Modified from Schultz, S. G., Yu-tu, L., Alvarez, O. O., and Curran, P. F., J. Gen. Physiol. *56*: 621–638, 1970.)

cm^2, whereas in zero Na^+ (choline, Tris, Li^+, or K^+ substitution) and 5 mM alanine, there was no net movement of the amino acid. The saturation level of alanine leaving the cells is proportional to the amount of sodium being pumped. The net transport of lysine is less than that of alanine, yet uptake by the epithelial cells is the same for the two acids; hence, the efflux of lysine across the serosal surface must be less.[58]

On the basis of the preceding evidence plus extensive kinetic measurements, the following model has been proposed.[58, 304] At the mucosal surface a binary complex is first formed between a.a. and carrier (X), and then a tertiary complex with Na^+ is formed (Fig. 4–7). Cationic acids form more stable binary complexes than anionic ones; possibly the cationic acids can mimic Na^+. Several univalent cations inhibit the Na^+-dependent influx in the affinity series H > K > Rb > Li > Cs, Na. It is postulated that the tertiary complex moves across the mucosal membrane, and first the a.a. and then the Na^+ are released inside the cell, leaving the carrier to accept more a.a. and Na^+ at the surface. Na^+ is then actively pumped out, and a.a. diffuses out across the serosal membrane. Transport of a.a. across the serosal border is independent of Na^+ in the cells, but is proportional to the concentration of a.a. in the cells. Na^+ extrusion is blocked by inhibitors, whereas the effect of Na^+ on uptake of a.a. at the mucosal side is not affected by inhibitors. The receptor sites on the carrier have not yet been well characterized; removal of the amino group from phenylalanine reduces the affinity for the receptor 50-fold, and removal of the COOH reduces the affinity 12-fold. Chain length is also critical. Binding with hydrogen ions indicates that the carrier has anionic groups with a pK of 4.[141, 369]

Absorption of amino acids by non-mammalian animals has not been much studied by modern methods. Goldfish intestine actively transfers thr, ala, ser, his, val, met, phe, and leu, and this transport is determined by the rate of transport of sodium from the mucosal to the serosal side.[323] The marine teleost *Maemalon* shows active transport of glycine, which is Na^+-dependent and requires aerobic energy; threonine competitively reduces transport of glycine.[322] Active transport of several amino acids has been demonstrated in tortoise intestine. In the holothurian *Stichopus*, active absorption of L-alanine and glycine has been noted.[204]

Tapeworms actively absorb amino acids from the medium; absorption requires sodium. Sugar in the medium interferes with a.a. absorption, and competition between leucine and glucose has been noted.[163, 274] Similarly, in mammalian intestine, L-glucose transport is inhibited by his, met and ala, but not by those a.a.'s which are transaminated (glu and asp); this is taken as evidence for competition between amino acids and hexoses for carriers.[152]

Fatty Acids and Glycerides. Absorption of the products of fat digestion is very different from absorption of the products of carbohydrate and protein digestion. It has long been known that most fat is absorbed in mammals into the lymphatics as triglyceride, and is thus transported as neutral fat to the circulation. Pancreatic lipase acts on emulsified fats in the intestine (see Figure 4–3, page 145), hydrolyzing them in the 1 and 3 positions to fatty acids and monoglycerides. Evidence from electron microscopy of intestinal mucosa from mammals at different times after a meal of fat indicates that the free fatty acids (mainly long-chain acids) and monoglycerides aggregate as micelles of 40–50 Å diameter,[43] with the aid of bile salts. One emulsion particle can make 10^6 micelles. These micelles enter the epithelial cells, probably by diffusion, since the uptake of monoglycerides occurs at 0°C and is independent of energy.[43, 178] Whether any membrane carrier is involved is unknown.[125, 178] Short-chain fatty acids can be transferred against a concentration gradient and may be handled differently from long-chain acid.[26] Inside the mucosal cells, the free fatty acids in the presence of the appropriate enzymes, ATP, Mg^{++}, and coen-

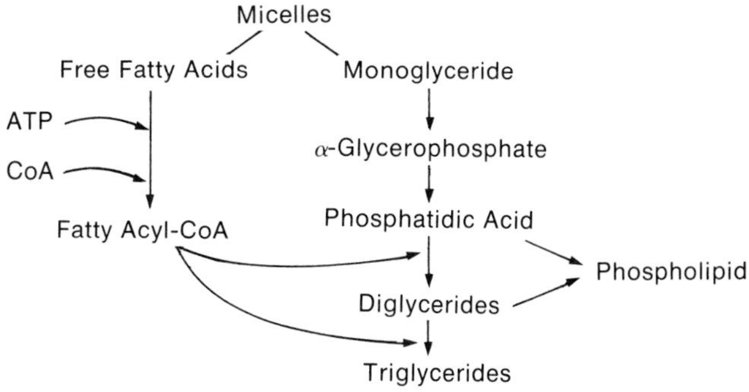

Figure 4–8. Diagram of reactions in a mucosal cell after uptake of lipid micelles. (Modified from Isselbacher, K. J., Gastroenterology *50*:78–82, 1966.)

zyme A form a CoA-fatty acid complex (Fig. 4–8). Monoglycerides are converted both to phosphatidic acids and again to triglycerides. The triglyceride or neutral fat appears as droplets in endoplasmic reticulum. Parallel to triglyceride biosynthesis, phospholipids plus proteins and cholesterol interact to form β-lipoproteins, which coat or stabilize the neutral fat; the resulting droplet is a chylomicron. These chylomicrons are extruded by a negative pinocytosis from the serosal surface into the lymphatics.[162, 178]

Cestodes contain much lipid, particularly as triglycerides and unsaturated fatty acids; they are unable to synthesize the fatty acids, and hence they must absorb monoglycerides and fatty acids from micelles of the host. Saturated fatty acids such as monostearic and palmitic are absorbed, at half the rate of oleic and linoleic.[17] There is evidence in the tapeworm *Hymenolepis* for one absorptive transport system for short-chain fatty acids (acetate) and a second system for higher fatty acids (longer than 12 carbons).[13, 55]

In conclusion, only a very few species have been studied so far with respect to the mechanism of absorption of nutrients. In view of the extensive variety of such systems in the animal kingdom, general comparative principles of absorption are as yet far from established.

SUMMARY

Nutrition, feeding, and digestion of animals include many adaptations which permit full utilization of food plants. All animals, whether herbivores, carnivores, or omnivores, are ultimately dependent on plants for organic compounds. Some animals maintain symbiotic algae within themselves—e.g., corals and giant bivalves with zooxanthellae. Feeding methods permit exploitation by different animals of a wide range of possible sources of nutrients. Organic molecules dissolved in natural waters constitute a supplement to the diet of particulate substances for many aquatic invertebrates. Filter feeders use the same water currents for supplying nutritional and respiratory needs. Many endoparasites obtain nutrients entirely by absorption of small molecules through their body surface. A wide variety of anatomical structures have evolved for taking in, dividing, and digesting particulate foods.

Digestion is a sequential process, and all digestive systems (other than intracellular, used in whole or in part in the simpler invertebrates) have regions for an initial phase of lubrication and sometimes preliminary hydrolysis; a second phase, acid in pH in almost all vertebrates but only in special cases in invertebrates, and where preliminary stages of protein hydrolysis occur; and later phases of hydrolysis of all nutrient macromolecules to small molecules and of absorption and elimination. Enzymes that perform these hydrolyses are secreted by glands, which are suitably located at each stage, and which become more specialized in higher animals. Dietary generalists and specialists are adapted in both feeding mechanisms and digestive enzymes.

Certain digestive enzymes are very widespread, others are restricted in animal distribution, and still other hydrolyses are carried out for animals by microorganisms. In the relatively universal category are α-amylases and α-glucosidases, tryptic and chymotryptic proteases, and several peptidases as well as lipases or related esterases. More restricted or scattered in distribution are β-glucosidases, lactase, trehalase, and pepsins. Similarly, few animals digest nucleic acids. Proteases of single classes often occur in multiple forms, as in chymotrypsinogens. A few animals have cellulase, chitinase, lignase, keratinase, and collagenase; but where cellulose is commonly used, reliance is upon symbiotic bacteria, used in various fashions by the different types of herbivores. The ruminants have perfected an entire digestive economy based upon this symbiosis.

Lipid digestion is related to mechanisms of absorption, since the fatty acids and monoglycerides formed are generally absorbed as micelles, and they may then be further hydrolyzed and reassembled inside mucosal cells. Absorption of amino acids and of sugars, whether by aquatic animals from the medium or, as better known, from a digestive lumen, is coupled with transport of sodium, and the cation pumps are a basic component of the absorbing system.

Interaction between nutrients and the pattern of feeding and digestion permits some environmentally induced variation. Many animals vary their intake according to nutritive value as well as taste. Production of the enzymes for hydrolysis can increase or decrease within an individual according to diet. The nature of such feedback controls is not known. Liberation of enzymes for extracellular digestion is often initiated by local hormones, and sometimes by nervous reflexes.

Specialization in dietary selection and in digestive capacity, whether by enzymes or by microorganisms, permits a few animals to use nutrients not available to most. Few physiological systems so well illustrate common patterns with superimposed specializations for particular ecological situations as do digestive systems.

REFERENCES

1. Adam, H., pp. 256–288. *In* Biology of *Myxine*, edited by A. Brodahl and R. Fänge, Universitets Forlaget, Oslo, Norway, 1963. Physiology of digestion in cyclostomes.
2. Agrawal, V. P., J. Zool. *143*:133, 1964. Digestion in *Orchestia*.
3. Alexander, F., Brit. Vet. J. *110*:146, 1954. Digestion in horse.
4. Al-Hussaini, A. H., J. Morph *80*:251, 1947; Quart. J. Micr. Sci. *90*:109, 1949. Functional morphology of teleost alimentary tracts.
5. Allen, J. A., J. Marine Biol. Assoc. U. K. *42*:609–623, 1962. Filter feeding in molluscs.
6. Alpers, D. H., J. Biol. Chem. *244*:1238–1246, 1969. β-Galactosidases: rat, human.
7. Alpers, D. H., and M. Solin, Gastroenterology *58*:833, 1970. Intestinal amylase.
8.*Altman, P. L., and D. S. Dittmer (editors, for the Committee on Biological Handbooks), Metabolism. Fed. Amer. Soc. Exp. Biol., Bethesda, Maryland, 1968. Tables of biological values.
9. Amesjo, B., J. Barrowman, and B. Borgström, Acta Chem. Scand. *21*:2897, 1967. Pancreatic pro-phospholipase A.
10. Anderson, J. A., and S. C. Stephens, Marine Biol. *9*:242–249, 1969. Absence of uptake of dissolved organic molecules by crustaceans.
11. Annison, E. F., and R. R. White, Biochem. J. *80*:162–169, 1961. Glucose utilization in sheep.
12.*Annison, E. F., and D. Lewis, Metabolism in the Rumen. Methuen, London, 1959.

*Review.

13. Arme, C., and C. P. Read, Biol. Bull. *135*:80–91, 1968. Absorption of acetate and butyrate by a cestode.
14. Arnon, R., and H. Neurath, Proc. Nat. Acad. Sci. *64*:1323, 1969. Immunochemical relations of trypsins.
15. Asp, N.-G., Biochem. J. *121*:299–308, 1971. Human intestinal β-galactosidase.
16. Bailey, C. B., Brit. J. Nutr. *15*:443–451, 1961. Saliva secretion in cattle.
17. Bailey, H. H., and D. Fairbairn, Comp. Biochem. Physiol. *26*:819–836, 1968. Absorption of fatty acids and monoglycerides by a cestode.
18. Baldwin, E., An Introduction to Comparative Biochemistry. Cambridge Univ. Press, London and New York, 1948.
19. Balogun, R. A., Comp. Biochem. Physiol. *29*:1267–1270, 1969. Digestive enzymes of larch bark beetle.
20. Banks, W. M., Science *141*:1191–1192, 1963. Carbohydrate digestion in the cockroach.
21. Barnard, E. A., Nature *221*:340–344, 1969. Function of pancreatic ribonuclease.
21a.*Barnard, E. A., Ann. Rev. Biochem. *38*:677–732, 1969. Ribonucleases.
22. Barnes, R. H., G. Fiala, and E. Kwong, Fed. Proc. *22*:125, 1963. Effects of coprophagy.
23.*Barrington, E. J. W., pp. 109–161. *In* The Physiology of Fishes, Vol. 1, edited by M. E. Brown, Academic Press, New York, 1957. Digestion in fishes.
24.*Barrington, E. J. W., Adv. Comp. Biochem. Physiol. *1*:1–65, 1962. Digestive enzymes.
25. Barry, R. J., C. J. Eggenton, and D. H. Smyth, J. Physiol. *204*:299–310, 1969. Relation of Na^+ pump to sugar absorption.
26. Barry, R. J., and D. H. Smyth, J. Physiol. *152*:48–66, 1960. Absorption of fatty acids.
27. Bauchop, T., and R. W. Martucci, Science *161*:698–699, 1968. Ruminant-like digestion of langur monkey.
28. Bayliss, L. E., and E. H. Starling, J. Physiol. *29*:174, 1903. Secretin.
29. Becker, D. E., D. E. Ullrey, S. W. Terrill, and R. A. Notzold, Science *120*:345–346, 1954. Failure of newborn pig to digest sucrose.
30. Behnke, W. D., R. D. Wade, and H. Neurath, Biochemistry *9*:4179–4188, 1970. Bovine pro-carboxypeptidase A.
31. Belleville, J., and J. Clément, Bull. Soc. Chim. Biol. (Paris) *50*:1419–1424, 1968. Pancreatic phospholipase (man, rat).
32. Ben Abdeljlil, A., and P. Desnuelle, Biochim. Biophys. Acta *81*:136–149, 1964. Adaptation of pancreatic enzymes to diet.
33. Bensley, R. R., Amer. J. Anat. *2*:105, 1902. Stomach morphology of mammals.
34. Berger, E., F. C. Kafatos, R. L. Felsted, and J. H. Law, J. Biol. Chem. *246*:4131–4137, 1971. Cocoonase.
35. Bergman, E. G., R. S. Reid, M. G. Murray, J. M. Brockman, and F. G. Whitelaw, Biochem. J. *97*:53–66, 1965. Volatile fatty acids of sheep rumen.
36. Bewley, G. C., and E. J. DeVillez, Comp. Biochem. Physiol. *25*:1061–1069, 1968. Digestive proteases of earthworm.
37. Bhattacharya, S., and K. C. Ghose, Comp. Biochem. Physiol. *37*:581–587, 1970. Amylase in bile.
38. Birktoft, J. J., D. M. Blow, R. Henderson, and T. A. Steitz, Phil. Trans. Roy. Soc. Lond. B *257*:67–76, 1970. Chymotrypsin structure.
39. Blandamer, A., and R. B. Beechey, Biochim. Biophys. Acta *118*:204–206, 1966. Amylase of crab hepatopancreas.
40. Blaxter, K. L., and A. K. Martin, Brit. J. Nutr. *16*:397–407, 1962. Nitrogen metabolism of sheep.
41. Bohlken, H., Proc. Zool. Soc. London *134*:207, 1960. Stomach of camels.
42. Bolton, W., Brit. Poultry Sci. *6*:97, 1965. Crop, in fowl.
43. Borgstrom, B., Biochim. Biophys. Acta *106*:171–183, 1965; *also* Johnston, J. M., and B. Borgstrom, Biochim. Biophys. Acta *84*:412–423, 1964. Bile salt micelles; intestinal absorption of fats.
44. Borgstrom, B., and C. Erlanson, Biochim. Biophys. Acta *242*:509–513, 1971. Co-lipase.
44a. Bowering, J., D. H. Calloway, S. Mangen, and N. A. Kaufman, J. Nutr. *100*:249–261, 1970. RNA digestion in man.
45. Bradshaw, R. A., H. Neurath, R. W. Tye, K. A. Walsh, and W. P. Winter, Nature *226*:237–239, 1970. Dogfish trypsinogen.
46. Brockerhoff, H., J. Fish. Res. Bd. Canad. *23*:1835, 1966; J. Biol. Chem. *246*:5828, 1971. Intestinal digestion of fats.
47. Brockerhoff, H., R. J. Hoyle, and P. C. Hwang, J. Fish. Res. Bd. Canad. *27*:1357, 1970. Digestive enzymes of lobster.
48.*Brooks, F. P., Control of Gastrointestinal Function. Macmillan Co., New York, 1970. Uptake of fatty acids.
49. Brown, A. L., Jr., J. Cell Biol. *12*:623–627, 1962. Microvilli of human intestinal mucosa.
50. Brown, G. D., and A. R. Main, Austral. J. Zool. *15*:7, 1967. Marsupial nutrition.
51. Bryant, M. P., and I. M. Robinson, J. Dairy Sci. *46*:150, 1963. Ammonia utilization in the rumen.
52. Burck, P. J., Methods Enzymol. *19*:906, 1970. Pancreatic inhibitors of trypsin.
53.*Burgen, A. S. V., pp. 561–580. *In* Handbook of Physiology, Vol. 2, Sec. 6, edited by C. F. Code. Amer. Physiol. Soc., Washington, D.C., 1968. Secretion by salivary glands.

54. Camacho, Z., J. R. Brown, and G. B. Kitto, J. Biol. Chem. *245*:3968–3972, 1970. Starfish trypsin.
55. Chappell, L. H., C. Arme, and C. P. Read, Biol. Bull. *136*:313–326, 1969. Absorption of fatty acids by cestodes.
56. Chauncey, H. H., B. L. Henriques, and J. M. Tanzer, Arch. Oral Biol. *8*:615, 1963. Salivary amylase: occurrence.
57. Chesley, L. C., Biol. Bull. *66*:133–144, 1934. Amylase in bile of fishes.
58. Chez, R. A., R. R. Palmer, S. G. Schultz, and P. F. Curran, J. Gen. Physiol. *50*:2357–2375, 1967. Effects of inhibitors in amino acid absorption.
59. Cheung, A. C., and R. H. Gooding, Comp. Biochem. Physiol. *37*:331–338, 1970. Insect trypsin.
60. Chinnery, J. A. B., J. Insect Physiol. *17*:47, 1971. Carbohydrases in midgut of beetle.
61. Clément, J., and B. Rigollot, Nutr. Dieta *6*:61, 1964. Lipases and phosphatidase.
62. Coan, M. H., and J. Travis, Comp. Biochem. Physiol. *32*:127–139, 1970. Coelenterate proteases.
63.*Coates, M. E. (editor), The Germ-free Animal in Research. Academic Press, 1968.
64. Cools, A., and C. Jeuniaux, Arch. Int. Physiol. Biochem. *69*:1, 1961. Digestion of cellulose in rabbit cecum.
65. Corner, E. D. S., Y. A. Leon, and R. D. Bulbrook, J. Marine Biol. Assoc. U. K. *39*:51, 1960. Sulphatases and β-glucuronidase in marine invertebrates.
66. Crane, R. K., pp. 2535–2542. *In* Handbook of Physiology, Vol. 5, Sec. 6, edited by C. F. Code. Amer. Physiol. Soc., Washington, D.C., 1968. Intestinal brush border function.
67.*Crane, R. K., pp. 1323–1351. *In* Handbook of Physiology, Vol. 3, Sec. 6, edited by C. F. Code. Amer. Physiol. Soc., Washington, D.C., 1968. Intestinal absorption of sugars.
68. Crane, R. K., G. Forstner, and A. Eichholz, Biochim. Biophys. Acta *109*:467–477, 1965. Relation of Na^+ to sugar transport.
69. Crompton, D. W. T., and A. P. M. Lockwood, J. Exp. Biol. *48*:411–425, 1968. Absorption of glucose by acanthocephalans.
70. Dadd, R. H., J. Exp. Biol. *33*:311–324, 1956. Control of proteolysis in beetles *Tenebrio* and *Dytiscus.*
71.*Dadd, R. H., pp. 117–142. *In* Chemical Zoology, Vol. V, Part A, edited by M. Florkin and B. T. Scheer, Academic Press, New York, 1970. Digestion in insects.
72. Dahlqvist, A. J. Clin. Invest. *41*:463, 1962; *also* Dahlqvist, A., and U. Telenius, Biochem. J. *111*:139, 1969. Intestinal disaccharidases.
73. Dahlqvist, A., and D. L. Thomson, Biochem. J. *89*:272–277, 1963. Rat intestinal amylases.
74. Dales, R. P., J. Marine Biol. Assoc. U. K. *34*:55, 1955. Digestion in polychaetes.
75. Das, P. K., R. A. Watts, D. C. Watts, and E. J. Dimelow, Comp. Biochem. Physiol. *39B*:979–997, 1971. Distribution of some proteases and esterases in starfish.
76. Davenport, H. W., Sci. Amer. *226*:86–93, 1972. Protection of the stomach against digestion.
77. David, J. S. K., P. Malathi, and J. Ganguly, Biochem. J. *98*:662, 1966. Absorption of cholesterol, rat.
78. Davis, G. R. F., Arch. Int. Physiol. Biochem. *71*:166, 1963. Carbohydrases of grasshoppers.
79. Dawes, B., Parasitology *52*:483, 1962. Digestion in *Fasciola.*
80.*Day, M. F., and D. F. Waterhouse, pp. 273–349. *In* Insect Physiology, edited by K. D. Roeder, John Wiley and Sons, New York, 1953. Feeding and digestion in insects.
81.*DeDuve, C., and R. Watteaux, Ann. Rev. Physiol. *28*:135, 1966. Lysosomes.
82. DeHaas, G. H., N. M. Posterna, W. Niewenhuizen, and L. L. M. Van Deenen, Biochim. Biophys. Acta *159*:118–129, 1968. Pancreatic prophospholipase A.
83.*Dethier, V. G., Ch. 6, pp. 79–96. *In* Handbook of Physiology of the Alimentary Canal, edited by C. F. Code. Amer. Physiol. Soc., Washington, D.C., 1967. Feeding and drinking behavior of invertebrates.
84. Devigne, J., and C. Jeuniaux, Arch. Int. Physiol. Biochem. *69*:223, 1961. Chitinase of earthworms.
85. DeVillez, E., and K. Buschlen, Comp. Biochem. Physiol. *21*:541–546, 1967. Trypsin in crustaceans.
86. DeVillez, E., and R. M. Reid, Comp. Biochem. Physiol. *38B*:235–238, 1971. Polychaete trypsin.
87. Dodgson, K. S., J. M. Lewis, and B. Spencer, Biochem. J. *55*:253–259, 1953. Sulphatases.
88. Dolly, J. O., and P. F. Fottrell, Clin. Chim. Acta *26*:555, 1969. Dipeptidases of human intestinal mucosa.
89. Eggermont, E., Europ. J. Biochem. *9*:483, 1969. Glucosidases of human intestinal mucosa.
90. Eisen, A. Z., and J. J. Jeffrey, Biochim. Biophys. Acta *191*:517–526, 1969. Collagenase of crustacean hepatopancreas.
91. Ellis, W. C., and W. H. Pfander, Nature *205*:974–975, 1965. Rumenal nucleic acids.
92. Erasmus, D. A., and C. Ohman, Ann. N.Y. Acad. Sci. *113*:7, 1963. Adhesive organ of trematodes.
93. Erlanson, C., and B. Borgström, Biochim. Biophys. Acta *167*:629–631, 1968. Pancreatic esterase.
94. Erspamer, V., Gut *11*:79–87, 1970. Caerulein.
95. Evans, W. A. L., and D. W. Payne, J. Insect Physiol. *10*:657, 1964. Digestive carbohydrases of a locust.
96.*Fairbairn, D., pp. 361–378. *In* Chemical Zoology, Vol. III, edited by M. Florkin and B. T. Scheer. Academic Press, New York, 1969. Lipid metabolism in nematodes.
97. Farmanfarmaian, A., and J. H. Phillips, Biol. Bull. *123*:105–120, 1962. Digestion of algae in sea urchin.
98. Farner, D. S., Poultry Sci. *22*:245, 1943; Biol. Bull. *84*:240–243, 1943. Amylase in bile of fowl.
99.*Farner, D. S., pp. 411–467. *In* Biology and Comparative Physiology of Birds, Vol. I, edited by A. J. Marshall. Academic Press, New York, 1960.
100. Figarella, C., Bull. Soc. Chim. Biol. *48*:97, 1966. Enzymes of mammalian pancreatic juice.
101. Fischer, E. H., and E. A. Stein, pp. 313–343. *In* The Enzymes, Vol. 4, edited by P. D. Boyer, H. Lardy, and L. Myrback. Academic Press, New York, 1960. Mammalian α-amylase.
102. Fish, J. C., Nature *180*:345, 1957. Rennin.
103. Fisher, F. M., and C. P. Read, Biol. Bull. *140*:46–62, 1971. Sugar transport in tapeworm.
104. Fisk, F. W., and B. R. Rao, Ann. Entom. Soc. Amer. *57*:40, 1964. Digestive carbohydrases of the cockroach.
105. Flock, A. H., S. Van Norden, and H. N. Spiro, Gastroenterology *52*: 230, 1967. Gastric and intestinal mucosal enzymes, primates.
106. Florkin, M., and F. Lozet, Arch. Int. Physiol. Biochem. *57*:201–207, 1949. Chitinase.
107. Folk, J. E., and E. W. Schirmer, J. Biol. Chem. *240*:181–192, 1965. Chymotrypsin C (pig).
108. Franssen, J., and C. Jeuniaux, Arch. Int. Physiol. Biochem. *71*:301, 1963. Alginase in molluscs.
109. Friedrich, M., R. Novack, and G. Schenk, Biochem. Zeitschr. *343*:346, 1965. Peptidases of rat intestinal mucosa.
110. Fujita, T., Zellforsch. Mikr. Anat. *57*:487, 1962. Pancreas of *Chimaera.*
111. Galli, D. R., and A. C. Giese, J. Exp. Zool. *140*:415, 1959. Carbohydrate digestion in herbivorous snail.
112. Galun, R., Life Sci. *5*:1335–1342, 1966. Stimulation of feeding in rat flea.
113. Galun, R., and S. H. Kindler, Comp. Biochem. Physiol. *17*:69–73, 1966. Stimulation of feeding in leech.
114. Galun, R., and S. H. Kindler, J. Insect Physiol. *14*:1409–1421, 1968. Stimulation of feeding in tick *Ornithodoros.*
115. Galun, R., and J. Margalib, *in* First Int. Symp. Tsetse Fly Breeding, edited by J. de Azevedo, Lisbon, 1970. Stimulation of feeding in tsetse fly.
116. Gardner, J. D., M. S. Brown, and L. Laster, New Eng. J. Med. *283*: 1196, 1970. Transport in mammalian intestinal epithelium.
117. Gardner, J. D., M. S. Brown, and L. Laster, New Eng. J. Med. *283*: 1264–1271, 1317–1324, 1970. Digestion in mammalian intestinal epithelium.
118. Gates, B. J., and J. Travis, Biochemistry *8*:4483, 1969. Shrimp trypsin.
119. Gelperin, A., J. Insect Physiol. *12*:331–345, 829–841, 1966. Control of crop emptying and food intake in blowfly.
120. Gertler, A., and Y. Birk, Europ. J. Biochem. *12*:170, 1970. Pig pancreatic pro-elastase.
121. Gibson, D., and G. H. Dixon, Nature *222*:753–756, 1969. Proteases of sea anemone.
122. Giebel, W., R. Zwilling, and G. Pfleiderer, Comp. Biochem. Physiol. *38B*:197–210, 1971. Proteases of honeybee.
123.*Gilbert, L. I., Adv. Insect. Physiol. *4*:69, 1967. Lipid metabolism in insects.
124. Gooding, R. H., and L. T. Huang, J. Insect Physiol. *15*:325, 1969. Beetle trypsin and chymotrypsin.
125. Gordon, S. G., and F. Kern, Biochim. Biophys. Acta *152*:372–378, 1968. Absorption of fatty acids; functions of bile.
126. Graham, A., Proc. Zool. Soc. Lond. *122*:543, 1939. Amylases of lamellibranch style.
127. Gray, G. M., Ann. Rev. Med. *22*:391, 1971. Intestinal malabsorption of carbohydrates.
128. Gray, G. M., and N. A. Santiago, J. Clin. Invest. *48*:716, 1969. Human intestinal carbohydrases.
129. Greenberg, B., Ann. Entomol. Soc. Amer. *61*:365, 1968. pH of maggot digestive tracts.
130. Greene, L. H., C. H. W. Hirs, and G. E. Palade, J. Biol. Chem. *238*: 2054–2070, 1963. Enzymes of pancreatic zymogen granules (cow).
131. Greenfield, L. J., and C. E. Lane, J. Biol. Chem. *204*:669–672, 1953. Utilization of cellulose by *Teredo.*
132.*Greenwood, C. T., and E. A. Milne, Adv. Carbohydrate Chem. *23*:281, 1968. Starch-degrading enzymes.
133. Gregory, H., P. M. Hardy, D. S. Jones, G. W. Kemmer, and R. C. Sheppard, Nature *204*:931–932, 1964. Gastrin.
134.*Gregory, R. A., Gastroenterology *51*:953–959, 1966; *in* Handbook of Physiology, Vol. 2, Sec. 6, pp. 827–834, edited by C. F. Code, Amer. Physiol. Soc., Washington, D.C., 1967. Gastrin.
135. Griffiths, M., Comp. Biochem. Physiol. *14*:357–375, 1965. Digestion in echidna.
136. Griffiths, M., and D. Davies, J. Nutr. *80*:171–180, 1963. Soft pellets in rabbit stomach.
137.*Grossman, M. I., pp. 835–864. *In* Handbook of Physiology, Vol. 2, Sec. 6, edited by C. F. Code. Amer. Physiol. Soc., Washington, D.C., 1967. Stimulation of gastric secretion.
138. Grossman, M. I., Lancet *1*:1088–1089, 1970. Gastrin, cholecystokinin, and secretin.

139. Gustaffson, B. E., F. S. Paft, E. G. McDaniel, J. C. Smith, and R. J. Fitzgerald, J. Nutr. *78*:461–468, 1962. Vitamin K-deficient germfree rats.
140. Hadorn, B., N. Steiner, C. Sumida, and T. J. Peters, Lancet *1*:165, 1971. Intestinal enterokinase.
141. Hajjar, J. J., and P. J. Curran, J. Gen. Physiol. *56*:673–691, 1970. Amino acid transport in intestine.
142. Hall, D. A., Arch. Biochim. Biophys. *98*(Suppl.) *1*:239–246, 1962. Elastase complex.
143. Hartley, B. S., Phil. Trans. Roy. Soc. Lond. B *257*:77–87, 1970. Pancreatic proteases.
144.*Haslewood, G. A. D., J. Lipid Res. *8*:535, 1967. Bile salt evolution.
145. Hatfield, E. E., Fed. Proc. *29*:44, 1970. Ruminant nitrogen metabolism.
146. Haupt, T. R., Amer. J. Physiol. *197*:115–120, 1959; *205*:1144–1150, 1963. Utilization of urea; ruminants, rabbits.
147. Haven, D. S., and R. Morales-Alamo, Biol. Bull. *139*:248–264, 1970. Filter feeding in oyster.
148. Heller, J., and M. Schramm, Biochim. Biophys. Acta *81*:96–100, 1964. α-Amylase action.
149.*Hendrix, T. R., and T. M. Bayless, Ann. Rev. Physiol. *32*:139, 1970. Secretion of electrolytes and water in mammalian intestine.
150. Heran, H., Z. vergl. Physiol. *39*:44–62, 1956. Digestion in earthworms.
151. Hillman, R. E., and L. E. Rosenberg, Biochim. Biophys. Acta *211*: 318–326, 1970. Amino acid transport in renal tubules.
152. Hindmarsh, J. T., D. Kilby, and G. Wiseman, J. Physiol. *186*:166–174, 1966. Effect of amino acids on sugar absorption.
153.*Hoffmann, A. F., and D. A. Small, Ann. Rev. Med. *18*:333, 1967. Function of bile salts.
154. Hogben, C. M., Gastroenterology *50*:78–82, 1966. Fat absorption.
155. Hollingsworth, D. F., and J. P. Greaves, Amer. J. Clin. Nutr. *20*:65–72, 1967. Human carbohydrate consumption.
156. Holt, J. H., and D. Miller, Biochim. Biophys. Acta *58*:239, 1962. Phosphatases and peptidases of intestinal brush border.
157. Hori, K., J. Insect Physiol. *16*:373, 1970. Salivary amylase, hemipterans.
158.*House, H. L., pp. 815–858. *In* The Physiology of Insects, Vol. 2, edited by M. Rockstein. Academic Press, New York, 1965. Digestion in insects.
159. Howard, F., and J. Yudkin, Brit. J. Nutr. *17*:281–294, 1963. Pancreatic amylase and trypsin; responses to feeding.
160. Huang, W.-Y., and J. Tang, J. Biol. Chem. *245*:2189–2193, 1970. Human gastrin and pepsin.
161.*Hungate, R. E., The Rumen and its Microbes. Academic Press, New York, 1966.
162. Isselbacher, K. J., Gastroenterology *50*:78–82, 1966. Biochemistry of fat absorption.
163. Isseroff, H., and C. P. Read, Comp. Biochem. Physiol. *30*:1153–1159, 1969. Amino acid absorption by trematodes.
164. Ito, S., J. Cell Biol. *27*:475, 1965. Surface coat on intestinal microvilli.
165. Itoh, R., J. Biochem. (Tokyo) *33*:269, 1941. Gastric lipase.
166. Jacobs, H. L., and K. N. Sharma, Ann. N.Y. Acad. Sci. *157*:1084–1121, 1969. Sensory and metabolic control of food intake in mammals.
167. Jacques, P. J., pp. 395–420. *In* Lysosomes in Biology and Pathology, Vol. 2, edited by J. T. Dingle and H. B. Fell. North Holland Publishing Co., Amsterdam, 1969. Digestion by lysosomes.
168. Jennings, J. B., Biol. Bull. *122*:63–72, 1962; *123*:571–581, 1962. Digestive enzymes: turbellarian, rhynchocoelan.
169.*Jennings, J. B., pp. 305–323. *In* Chemical Zoology, Vol. 2, edited by M. Florkin and B. T. Scheer. Academic Press, New York, 1968. Digestion in Platyhelminthes.
170. Jennings, J. B., and V. M. Van Der Lande, Biol. Bull. *133*:166–183, 1967. Digestion in leeches.
171. Jeuniaux, C., Nature *192*:135–136, 1961. Chitinase in digestive tract of vertebrates.
172. Jeuniaux, C., Arch. Int. Physiol. Biochem. *70*:407, 1962. Polysaccharidases of stomach in sloth.
173. Jeuniaux, C., Arch. Int. Physiol. Biochem. *71*:307, 1963. Chitinases of vertebrates and invertebrates.
174.*Jeuniaux, C., Chitine et chitinolyse. Masson, Paris, 1963. 181 pp.
175. Jeuniaux, C., pp. 64–91. *In* Chemical Zoology, Vol. 4, edited by M. Florkin and B. T. Scheer. Academic Press, 1969. Digestion in annelids.
176. Jimenez-Poraz, J. M., Clin. Toxicol. *3*:389, 1970. Enzymes of snake venoms.
177. Johannes, R. E., S. J. Conrad, and H. L. Webb, Comp. Biochem. Physiol. *29*:283–288, 1969. Uptake of dissolved amino acids by marine invertebrates.
178.*Johnston, J. M., pp. 1353–1375. *In* Handbook of Physiology, Vol. 3, Sec. 6, edited by C. F. Code. Amer. Physiol. Soc., Washington, D.C., 1968. Lipid absorption.
179. Jorgensen, C. B., and E. D. Goldberg, Biol. Bull. *105*:477–489, 1953. Filtration feeding in molluscs and ascidians.
180.*Jorgensen, C. B., Biology of Suspension Feeding. Pergamon Press, Oxford, 1967. 357 pp.
181. Jorpes, J. E., Gastroenterology *55*:157–164, 1968. Secretin and cholecystokinin.
182. Joseffson, L., T. Lindberg, and L. Ojesj, Scand. J. Gastroent. *3*:207, 1968. Dipeptidases in human intestine.
183. Kassell, B., Methods Enzymol. *19*:844, 1970. Pancreatic trypsin inhibitor.
184. Keller, P. J., and B. J. Allan, J. Biol. Chem. *242*:281–287, 1967. Pancreatic enzymes in human.
185. Kellogg, T. F., and B. S. Wastmann, J. Lipid Res. *10*:495, 1969. Bile acids from germfree rats.
186. Kermack, D. M., Proc. Zool. Soc. Lond. *125*:347, 1955. Digestion in *Arenicola*.
187. Kerry, K. R., Comp. Biochem. Physiol. *29*:1015–1022, 1969. Intestinal disaccharidases: birds, marsupials, monotremes.
188. Kilian, E. F., Z. vergl. Physiol. *34*:407–447, 1952. Ingestion and water currents in sponges.
189. Kleine, R., Z. vergl. Physiol. *55*:51, 1967; *56*:142, 1967. Peptidases of hepatopancreas and stomach juice of crayfish.
190. Klok, J. L., and H. Smit, Comp. Biochem. Physiol. *7*:251–254, 1962. Gastric secretion in frog.
191. Kochva, E., and C. Gans, Clin. Toxicol. *3*:363, 1970. Snake venoms.
192. Koide, A., and Y. Matsuoko, J. Biochem. (Tokyo) *66*:541, 1969. Pancreatic proteases in whale.
193. Kolinská, J., and G. Semenza, Biochim. Biophys. Acta *146*:181–195, 1967. Sucrase-isomaltase complex of rabbit intestine.
194. Kooiman, P., J. Cell Comp. Physiol. *63*:197, 1964. Carbohydrases of hepatopancreas of crayfish and lobster.
195. Koringa, P., Quart. Rev. Biol. *27*:266–308, 339–365, 1952. Filter feeding by oysters.
196. Krogh, A., Biol. Rev. *6*:412–422, 1931. Utilization of dissolved foods by aquatic animals.
197. Kuhn, H.-J., Folia Primat. *2*:193, 1964. Stomach of leaf-eating monkeys.
198. Lacko, A. G., and H. Neurath, Biochemistry *9*:4680, 1970. Fish procarboxypeptidase A.
199. Lamy, F., and S. Tauber, J. Biol. Chem. *238*:939–944, 1963. Elastase.
200. Lansing, A. I., T. B. Rosenthal, and M. Alex, Proc. Soc. Exp. Biol. Med. *84*:689, 1953. Elastase of fish pancreas.
201. Lasker, R., and R. A. Boolootian, Nature *188*:1130, 1960. Digestion of algae by echinoderms.
202. Lasker, R., and A.C. Giese, J. Exp. Biol. *33*:542, 1956. Cellulose digestion in insects.
203. Lawrence, A. L., Comp. Biochem. Physiol. *9*:69–73, 1963. Sugar transport, bullfrog intestine.
204. Lawrence, A. L., and D. C. Lawrence, Comp. Biochem. Physiol. *22*:341–357, 1967. Sugar absorption, *Cryptochiton*.
205. Lee, D. L., Parasitology *52*:241, 1962. Esterases in *Ascaris*.
206.*Lenhoff, H. M., pp. 158–222. *In* Chemical Zoology, Vol. 2, edited by M. Florkin and B. T. Scheer. Academic Press, New York, 1967. Digestion in coelenterates.
207. Lenhoff, H. M., pp. 203–229. *In* Biology of *Hydra*, edited by H. M. Lenhoff and M. F. Loomis. University of Miami Press, Coral Gables, Fla., 1961. Activation of feeding reflex in *Hydra*.
208. Lewis, J. B., Canad. J. Zool. *42*:549, 1964. Digestion in sea urchin.
209. Li, Y.-T., and M. R. Shetlar, Comp. Biochem. Physiol. *14*:275–279, 1965. Glycosidases in earthworm.
210. Lippel, K., and J. A. Olson, Biochim. Biophys. Acta *127*:243–245, 1966. Bile salts on non-lipolytic enzymes of pancreas.
211. Little, C., and B. L. Gupta, Nature *218*:873–874, 1963. Uptake of dissolved nutrients by Pogonophora.
212. Loeven, W. A., Europ. J. Biochem. *12*:170, 1970; Clin. Chim. Acta *30*:165, 1970. Elastase complex.
213. Luppa, H., Acta Histochem. *12*:137, 1961. Stomach glands, Amphibia.
214. MacGinitie, C. E., Biol. Bull. *77*:115–118, 1939; *88*:107–111, 1945. Feeding in *Chaetopterus* and *Urechis*.
215. MacInnis, A. J., F. M. Fisher, Jr., and C. P. Read, J. Parasitol. *51*:260–267, 1965. Purine and pyrimidine transport in cestodes.
216. Magee, D. F., and E. G. Anderson, Amer. J. Physiol. *181*:79–82, 1955. Induction of pancreatic enzymes by dietary protein.
217. Magee, W. L., J. Gallai-Hatchard, H. Snados, and R. H. S. Thompson, Biochem. J. *83*:17–25, 1962. Pancreatic phospholipase A in human.
218. Malathi, P., Gastroenterology *52*:1106, 1967. Cholesterol esterase in intestinal mucosa.
219. Malathi, P., and R. K. Crane, Biochim. Biophys. Acta *173*:245–256, 1969. Intestinal β-glucosidase in hamster.
220. Marchis-Mouren, G., Bull. Soc. Chim. Biol. *47*:146, 1965. Pancreatic juice enzymes in mammals.
221. Marisal, R. N., and H. M. Lenhoff, J. Exp. Biol. *49*:689–699, 1968. Stimulation of feeding in corals.
222. Maroux, S., J. Baratti, and P. Desneulle, J. Biol. Chem. *246*:5031–5039, 1971. Enterokinase activation of trypsinogen.
223. Marrama, P., C. Ferrari, R. Lapiccirella, and U. Parisoli, Ital. J. Biochem. *8*:280, 1959. Elastase of chicken pancreas.
224. Martin, A. L., J. Zool. *148*:515, 1966. Digestion in crustaceans.
225. Matthews, D. M., M. T. Lis, B. Cheng, and R. F. Crampton, Clin. Sci. *37*:751, 1969. Peptide absorption in rat intestine.
226. Mattson, F. H., and R. A. Volpenstein, J. Lipid Res. *7*:536, 1966. Esterase of pancreatic juice.
227. Maylie, M. F., M. Charles, C. Gache, and P. Desnuelle, Biochim. Biophys. Acta *229*:286–289, 1971. Pancreatic co-lipase.
228. McCarthy, C. F., and M. P. Tyor, Gastroenterology *46*:891, 1964. Morphology of intestinal surface in vertebrates.

229.*McDonald, I. W., Nutr. Abstr. Rev. *38*:381–400, 1968. Ruminant digestion.
230. McGeachin, R. L., and W. P. Welbourne, Comp. Biochem. Physiol. *38A*:457–460, 1970. Amylase in frog tissues.
231. McMahon, J. U., and J. H. Ryler, Canad. J. Zool. *41*:321–332, 1963. Regulation of feeding in *Daphnia*.
232. Meeuse, B. J. D., and W. Fleugel, Arch. Neerl. Zool. *13*(Suppl. 1):301, 1958. Amylase of lamellibranch style.
233. Mehring, J. S., and W. J. Tyznik, J. Animal Sci. *30*:764, 1970. Glucose tolerance in horse.
234. Michel, C., Ann. Histochem. *15*:19, 1970. Intestinal proteases in annelids.
235. Michelson, E. H., and L. Dubois, Comp. Biochem. Physiol. *38B*:263–268, 1971. Enzymes of crystalline style of *Oncomelania* (mollusc).
236. Miller, D., and R. T. Crane, Biochim. Biophys. Acta *52*:293–298, 1961. Enzymes of brush border of mammalian intestine.
237. Mockel, W., and E. A. Barnard, Biochim. Biophys. Acta *178*:354–363, 1969. Reptile chymotrypsins.
238.*Moir, R. J., pp. 1–14. *In* Physiology of Digestion in the Ruminant, edited by A. W. Dougherty. Butterworths, London, 1963. Ruminants and ruminant-like mammals.
239.*Moir, R. J., pp. 2673–2694. *In* Handbook of Physiology, Vol. 5, Sec. 6, edited by C. F. Code. Amer. Physiol. Soc., Washington, D.C., 1968. Digestion in ruminants.
240. Morgan, R. G. H., J. Borrowman, J. Filipek-Wender, and B. Borgström, Biochim. Biophys. Acta *167*:355–366, 1968. Pancreatic lipases in rat.
241. Morrissen, S. M., and C. S. Yuk, Comp. Biochem. Physiol. *34*:521–533, 1970. Gastrin effects on secretion in *Rana*.
242. Mukaiyama, F., Y. Horie, and T. Ito, J. Insect Physiol. *10*:247–254, 1964. Silkworm disaccharidases.
243.*Müller, M., pp. 351–380. *In* Chemical Zoology, Vol. 1, edited by M. Florkin and B. T. Scheer. Academic Press, New York, 1967. Digestion in protozoans.
244. Müller, M., and I. Toro, J. Protozool. *9*:98, 1962. Lysosomes and food vacuoles in Protozoa.
245. Muscatine, L., and C. Hand, Proc. Nat. Acad. Sci. *44*:1259–1263, 1958. Symbiotic algae in corals.
246. Muscatine, L., Science *156*:516–518, 1967. Symbiotic algae in corals.
247. Myers, F. L., and D. H. Northcote, J. Exp. Biol. *35*:639, 1958. Digestive enzymes of *Helix*.
248. Meyers, F. L., and D. H. Northcote, Biochem. J. *71*:749–755, 1959. Cellulase from *Helix*.
249.*Neurath, H., R. A. Bradshaw, and R. Arnon, *in* Proceedings of the International Symposium on Structure-Function Relationships of Proteolytic Enzymes, edited by M. Otteson. Munksgaard, Copenhagen, 1969. Evolution of proteases.
250. Newey, H., and D. H. Smyth, J. Physiol. *164*:527–551, 1962. Intestinal amino acid transport.
251. Nielson, C. V., Oikos *13*:200, 1962. Carbohydrases in annelids.
252. Nilsson, A., and R. Fänge, Comp. Biochem. Physiol. *31*:147–165, 1969; *32*:237–250, 1970. Digestive proteases of cyclostome and *Chimaera*.
253. Nilsson, A., Comp. Biochem. Physiol. *32*:387–390, 1970. Gastrointestinal hormones, *Chimaera*.
254. Noirot, C., and C. Noirot-Timothee, pp. 49–88. *In* Biology of Termites, Vol. 1, edited by K. Krishna and F. W. Weesner. Academic Press, New York, 1969. Digestion of wood by termites.
255. Nordstrom, C., and A. Dahlqvist, Biochim. Biophys. Acta *242*:209–225, 1971. Enterokinase in brush border.
256. Norris, J. L., J. Exp. Zool. *141*:155, 1959. Esophageal and gastric glands of frog.
257.*Novikoff, A. B., p. 36. *In* Ciba Foundation Symposium on Lysosomes, edited by A. V. De Reuck and M. P. Cameron. Churchill, London, 1963. Enzymes of lysosomes.
258. Okamoto, A., J. Biochem. (Tokyo) 37:269, 1950. Salivary amylase of frog.
259. Ondetti, M. A., et al., Amer. J. Digest. Diseases *15*:149–156, 1970. Cholecystokinin and pancreomyzin.
260. Overton, J., A. Eichholz, and R. K. Crane, J. Cell Biol. *26*:693, 1965. Enzymes of intestinal brush border.
261.*Owen, G., pp. 53–96. *In*: Physiology of Mollusca, Vol. 2, edited by K. M. Wilbur and C. M. Yonge. Academic Press, New York, 1966. Amylase in gastropods.
262. Owen, G., Phil. Trans. Roy. Soc. Lond. B *258*:245, 1970. Secretory cells of mollusc digestive gland.
263. Parnas, I., J. Cell. Comp. Physiol. *58*:195, 1961. Snail cellulase.
264. Parsons, D. S., and J. S. Prichard, Nature *208*:1097–1098, 1965. Disaccharide hydrolysis in amphibian intestine.
265.*Peters, W., Trop. Diseases Bull. *66*:1–29, 1969. Biochemistry of plasmodia.
266. Peters, T. J., Biochem. J. *120*:195, 1970. Intestinal proteases in guinea pig.
267. Pfleiderer, G., R. Zwilling, and H. H. Sonneborn, Z. Physiol. Chem. *348*:1319, 1967. Crayfish trypsin.
268. Pilgrim, M., Proc. Zool. Soc. Lond. *147*:387, 1966. Alimentary tract of maldonid polychaetes.
269. Polonovski, C., and H. Bier, Acta Paed. Scand. *59*:458, 1970. Enterokinase deficiency in man.
270. Poort, C., and J. J. Geuze, Z. Zellforsch. *98*:7, 1969. Feeding effects on frog pancreas.
271. Poort, C., and M. F. Kramer, Gastroenterology *57*:689, 1969. Feeding effects on mammalian pancreas.
272. Powning, R. F., and H. Irzykiewicz, J. Insect Physiol. *8*:275–286, 1962. Digestive protease of clothes moth larvae.
273. Preston, R. L., and W. H. Pfander, J. Nutr. *83*:369–378, 1964. Phosphorus metabolism in sheep.
274.*Read, C. P., Amer. Zool. *8*:139, 1968. Nutrition in parasites.
275.*Read, C. P., and J. R. Simmons, Physiol. Rev. *43*:263–305, 1963. Biochemistry of tapeworms.
276. Reboud, J. P., A. Ben Abdeljlil, and P. Desnuelle, Biochim. Biophys. Acta *58*:326, 1962; *also* Reboud, J. P., et al., Biochim. Biophys. Acta *117*:351–367, 1966. Feeding effects on pancreatic enzymes.
277. Reeck, G. R., W. P. Winter, and H. Neurath, Biochemistry *9*:1398, 1970. Pancreatic enzymes of lungfish.
278. Reeck, G. R., and H. Neurath, Biochemistry *11*:503, 1972. Trypsinogen activation peptides: fishes, mammals.
279.*Reeder, W. G., pp. 99–149. *In* The Physiology of Amphibia, edited by J. A. Moore. Academic Press, New York, 1964. Digestion in amphibians.
280. Reid, R. G. B., Comp. Biochem. Physiol. *17*:417–433, 1966. Digestive proteases and esterases in bivalves.
281. Reid, R. G. B., Comp. Biochem. Physiol. *24*:727–744, 1968. Digestive enzymes of lamellibranchs.
282. Reid, R. G. B., and K. Rauchert, Comp. Biochem. Physiol. *35*:689–695, 1970. Proteases of bivalve molluscs.
283. Rhodes, J. B., A. Eichholz, and R. K. Crane, Biochim. Biophys. Acta *135*:959–965, 1967. Intestinal brush border enzymes.
284. Roberts, P. J. P., and W. J. Whelan, Biochem. J. *76*:246–253, 1960. α-Amylase action.
285. Robinson, L. A., C. L. Churchill, and T. T. White, Biochim. Biophys. Acta *222*:390–395, 1970. Human pancreatic juice content.
286. Robinson, G. B., Biochem. J. *88*:162–168, 1963. Intestinal mucosal peptidases.
287. Robyt, J. F., and D. French, Arch. Biochem. Biophys. *122*:8, 1967; J. Biol. Chem. *245*:3917–3927, 1970. α-Amylase action.
288.*Rogers, W. P., pp. 179–428. *In* Chemical Zoology, Vol. 3, edited by M. Florkin and B. T. Scheer. Academic Press, New York, 1969. Nitrogen metabolism in Acanthocephala and Nematoda.
289. Rogers, A. I., and P. S. Buchorik, Biochim. Biophys. Acta *159*:200–202, 1968. Bile salt protection of chymotrypsin.
290. Rosenbaum, R. M., and B. Ditzion, Biol. Bull. *124*:211–224, 1963. Digestive gland enzymes in *Helix*.
291. Rosenweig, N. S., R. H. Herman, and F. B. Stifel, Amer. J. Clin. Nutr. *24*:65–69, 1971. Feeding effect, intestinal saccharidases (man).
292. Rossler, M. E., J. Insect Physiol. *6*:62, 1968. Digestion in scarabalid larvae.
293. Rothman, S. S., Amer. J. Physiol. *218*:372–376, 1970. Zymogens of rabbit pancreas.
294. Rothman, A. M., and F. M. Fisher, J. Parasitol. *50*:410, 1964. Uptake of amino acids by parasitic worms.
295. Ryan, C. A., Arch. Biochim. Biophys. *110*:169–183, 1965. Chymotrypsin and trypsin in bird pancreas.
296. Ryle, A. P., and M. P. Hamilton, Biochem. J. *101*:176–183, 1966. Pepsinogen.
297. Salnikow, J., S. Moore, and W. H. Stein, J. Biol. Chem. *245*:5685–5690, 1970. Bovine pancreatic deoxyribonuclease.
297a. Sands, W. A., pp. 495–524. *In* Biology of Termites, Vol. 1, edited by K. Krishna and F. W. Weesner. Academic Press, New York, 1969. Fungal symbiont of termite.
298. Sarda, L., M. F. Maylie, J. Roger, and P. Desnuelle, Biochim. Biophys. Acta *89*:183–185, 1964. Pancreatic lipase.
299. Schade, S. G., R. J. Cohen, and M. E. Conrad, New Eng. J. Med. *279*:672, 1968. Effect of stomach acid on iron absorption.
300. Schingoethe, D. J., A. D. L. Gorrill, J. W. Thomas, and M. G. Yang, Canad. J. Physiol. Pharmacol. *48*:43, 1970. Pancreatic protease output of mammals.
301. Schmidt-Nielsen, B., K. Schmidt-Nielsen, T. P. Haupt, and S. A. Jamieson, Amer. J. Physiol. *188*:477–484, 1957. Urea excretion in camel.
302. Schoellmann, G., and E. Shaw, Biochemistry *2*:252, 1963. Active reaction center of chymotrypsin.
303. Scow, R. O., Y. Stein, and O. Stein, J. Biol. Chem. *242*:4919–4924, 1967. Lipids in chylomicrons.
304. Schultz, S. G., L. Yu-tu, O. O. Alvarez, and P. F. Curran, J. Gen. Physiol. *56*:621–638, 1970. Amino acid transport in rabbit ileum.
305. Schultz, S. G., P. F. Curran, R. A. Chez, and R. E. Fuisz, J. Gen. Physiol. *58*:1261–1286, 1971. Amino acid transport in intestine.
306. Seijffers, M. J., H. L. Segal, and L. L. Miller. Amer. J. Physiol. *205*:1099–1105, 1963. Human pepsins.
307.*Semenza, G., pp. 2543–2566. *In* Handbook of Physiology, Vol. 5, Sec. 6, edited by C. F. Code. Amer. Physiol. Soc., Washington, D.C., 1968. Intestinal carbohydrases.
308. Semenza, G., S. Auricchio, A. Rubino, A. Prader, and J. D. Welsh, Biochim. Biophys. Acta *105*:386–389, 1965. Genetic deficiency of intestinal maltases (man).
309. Shorland, F. B., R. O. Weenink, and A. T. Johns, Nature *175*:1129–1130, 1955. Effects of rumen on dietary fat.

310. Shotton, D. M., Methods Enzymol. *19*:113, 1970. Pancreatic elastase.
311. Siddons, R. C., Biochem. J. *108*:839–844, 1968. Digestive carbohydrases, cow.
312. Siddons, R. C., Biochem. J. *112*:51–59, 1969. Intestinal carbohydrases, chicken.
313. Simpson, G. G., Bull. Amer. Mus. Nat. Hist. *85*:1, 1945. Radiation of ungulates.
314. Simpson, J. W., A. C. Taylor, and B. M. Levy, Comp. Biochem. Physiol. *39B*:963–967, 1971. Collagenase in snake venoms.
315.*Singleton, J. W., Gastroenterology *56*:342–362, 1969. Humoral effects of pancreas on gastrointestinal tract.
316. Skoczylas, R., Comp. Biochem. Physiol. *35*:885–903, 1970. Digestion in snakes.
317. Smit, H., Comp. Biochem. Physiol. *13*:129–141, 1964. Pepsin secretion in frog.
318.*Smit, H., pp. 2791–2806. *In* Handbook of Physiology, Vol. 5, Sec. 6, edited by C. F. Code. Amer. Physiol. Soc., Washington, D.C., 1968. Non-mammalian gastric secretion.
319. Smith, A. C., and P. B. Van Weel, Experientia *16*:60, 1960. Midgut gland enzymes in developing snails.
320. Smith, A. H., M. Kleiber, A. L. Black, and G. P. Lofgreen, J. Nutr. *58*:95–111, 1956. Transfer of phosphate in digestive tract.
321. Smith, H. W., J. Path. Bact. *89*:95, 1965. Microflora of alimentary tract of vertebrates.
322. Smith, M. W., J. Physiol. *182*:559–590, 1966. Sodium-glucose interactions in goldfish intestine.
323. Smith, M. W., Comp. Biochem. Physiol. *35*:387–401, 1970. Active transport of amino acids by fish intestine.
324.*Smith, R. H., J. Dairy Res. *36*:313, 1969. Nitrogen metabolism and the rumen.
325. Smith, R. H., and A. B. McAllan, Brit. J. Nutr. *25*:181–190, 1971. Nucleic acid metabolism in ruminants.
326.*Smyth, J. D., The Physiology of Trematodes. Oliver and Boyd, London, 1966.
327. Snook, J. T., J. Nutr. *87*:297–305, 1965. Adaptation of pancreatic enzymes to diet.
328. Soldarno, D., M. R. Kare, and R. H. Wasserman, Poultry Sci. *20*:123, 1961. Crop in birds.
329. Somers, M., Austral. J. Exp. Biol. *39*:145, 1961. Nitrogen cycling via saliva, sheep.
330. Sonneborn, H. H., G. Pfleiderer, and J. Ishay, Z. Physiol. Chem. *350*:389, 1969. Hornet larva chymotrypsin.
331. Stephens, G. C., Science *131*:1532, 1960. Uptake of glucose by coral.
332. Stephens, G. C., Amer. Zool. *8*:95, 1968. Dissolved organic matter as nutrient.
333. Stephens, G. C., and R. A. Schinske, Limnol. Oceanogr. *6*:175–181, 1961. Dissolved organic matter as nutrient.
334. Storr, G. M., Austral. J. Biol. Sci. *17*:469, 1964. Marsupial nutrition.
335. Sunshine, P., and N. Kretchmer, Science *144*:850–851, 1964. Absence of intestinal disaccharidases in sea lions.
336. Swaminathan, N., and A. N. Rhadhakrishnan, Arch. Biochem. Biophys. *135*:288, 1969. Intestinal disaccharidases.
337. Tammes, P. M. L., and A. D. G. Dral, Arch. Neerl. Zool. *11*:87–112, 1955. Food intake in mussels.
338. Tang, J., J. Mills, L. Chiang, and L. DeChiang, Ann. N.Y. Acad. Sci. *140*:688, 1967. Pepsin.
339. Tarlow, M. J., B. Hadorn, M. W. Artherton, and J. K. Lloyd, Arch. Dis. Child. *45*:651, 1970. Enterokinase deficiency.
340. Thacker, E. J., and C. S. Brandt, J. Nutr. *55*:375–385, 1955. Coprophagy in rabbit.
341. Theede, H., and J. Lassig, Helg. wiss. Meeresunters. *16*:119–129, 1967. Ciliary pumping in molluscs.
342. Thomsen, E., and I. Moller, J. Exp. Biol. *40*:301–321, 1963. Hormonal control of protease in blowfly.
343. Tobita, T., and J. E. Folk, Biochim. Biophys. Acta *147*:15–25, 1967. Chymotrypsin C, pig.
344. Topps, J. H., and R. C. Elliott, Nature *205*:498–499, 1965. Nucleic acid digestion, ruminants.
345.*Treherne, J. E., Ann. Rev. Entomol. *12*:43, 1967. Intestinal absorption, insects.
346. Ugolev, A. M., Physiol. Rev. *45*:555–595, 1965. Contact (membrane) digestion.
347. Ugolev, A. M., and R. F. Kooshuck, Nature *212*:859–860, 1966. Intestinal absorption of peptides.
348. Vandermeers, A., and J. Christophe, Biochim. Biophys. Acta *154*: 110–129, 1968. Amylase and lipase in rat pancreas.
349. Van Dobben, W. H., Ardea *40*:1, 1952. Protection in cormorant gizzard.
350. Van Den Oord, A., Comp. Biochem. Physiol. *17*:715–718, 1966; *also* Van Den Oord, A., H. Danielsson, and R. Ryhage, J. Biol. Chem. *240*:2242–2247, 1965. Emulsifiers in crustaceans.
351. Van Weel, P. B., Z. vergl. Physiol. *42*:433–448, 1959. Adaptation of amylase in snails.
352. Van Weel, P. B., Z. vergl. Physiol. *43*:567–575, 1960. Crustacean midgut gland enzymes.
353.*Van Weel, P. B., pp. 97–115. *In* Chemical Zoology, Vol. 5, Part A, edited by M. Florkin and B. T. Scheer. Academic Press, New York, 1970. Digestion in Crustacea.
354. Verger, R., G. H. DeHaas, L. Sarda, and P. Desnuelle, Biochim. Biophys. Acta *188*:272, 1969. Pancreatic lipase, pig.
355. Vonk, H. J., Arch. Int. Physiol. Biochem. *70*:67, 1962. Emulsifiers in invertebrate digestive fluids.
356.*Vonk, H. J., pp. 347–401. *In* Comparative Biochemistry, Vol. 6, edited by M. Florkin and H. S. Mason. Academic Press, New York, 1964. Comparative aspects of digestion.
357. Walker, D. M., J. Agric. Sci. *53*:375, 1959. Carbohydrases of lamb.
358.*Waring, H., R. J. Moir, and C. H. Tyndale-Biscoe, Adv. Comp. Physiol. Biochem. *2*:238, 1966. Comparative physiology of marsupials.
359. Watson, H. C., D. M. Shotton, J. M. Cox, and H. Muirhead, Nature *225*:806–811, 1970. Elastase structure.
360. Weller, R. A., A. F. Pilgrim, and F. V. Gray, Brit. J. Nutr. *16*:83–90, 1962. Digestion in rumen of sheep.
361. Welsh, M. F., Biol. Bull. *70*:282, 1936. Zooxanthellae in turbellarian.
362. Werner, B., Zool. Anz. *146*:97–113, 1951. Filter feeding in *Crepidula*.
363. Western, J. R. H., and J. B. Jennings, Comp. Biochem. Physiol. *35*: 879–884, 1970. Secretion of HCl and pepsinogen in teleost stomach.
364. Wharton, D. R., M. L. Wharton, and J. E. Lola, J. Insect Physiol. *6*:62, 1964. Digestion in scarabalid larvae.
365. White, D. A., D. J. Pounder, and J. N. Hawthorne, Biochim. Biophys. Acta *242*:99–107, 1971. Phospholipase A of guinea pig pancreas.
366.*Wigglesworth, V. B., The Principles of Insect Physiology. Methuen, London, 1965.
367. Wilcox, P. C., Methods Enzymol. *19*:64, 1970. Chymotrypsins.
368. Williams, V. H., Nature *197*:1221, 1963. Fermentation in camel stomach.
369.*Wiseman, G., pp. 1277–1307. *In* Handbook of Physiology, Vol. 3, Sec. 6, edited by C. F. Code. Amer. Physiol. Soc., Washington, D.C., 1968. Absorption of amino acids.
370. Woodward, C. K., and C. P. Read, Comp. Biochem. Physiol. *30*:1161–1177, 1969. Transport of amino acids in tapeworm.
371. Wright, D. E., New Zealand J. Agr. Res. *4*:216, 1961. Lipid metabolism in rumen.
372. Wright, R. D., H. W. Florey, and A. B. Sanders, Quart. J. Exp. Physiol. *42*:1, 1957. Gastric mucosa of reptiles.
373.*Wyatt, G. R., Adv. Insect. Physiol. *5*:287, 1967. Trehalose in insects.
374. Yamasaki, K., T. Usui, T. Iwata, S. Nakasome, M. Hozumi, and S. Takatsuki, Nature *205*:1326–1327, 1965. Absence of bile acids in crayfish.
375. Yamashina, I., Acta Chem. Scand. *10*:739, 1956. Trypsinogen activation.
376. Yamashita, K., and K. Ashida, J. Nutr. *99*:267–273, 1969. Lysine metabolism in rats fed lysine-free diet.
377. Yang, Y. J., and D. M. Davies, J. Insect Physiol. *14*:205, 1968. Trypsin in dipteran insects.
378. Yokoe, Y., and I. Yasumasu, Comp. Biochem. Physiol. *13*:323–338, 1964. Cellulase in invertebrates.
379. Yonge, C. M., pp. 86–88. *In* Physiology of Mollusca, Vol. 2, edited by K. M. Wilbur and C. M. Yonge. Academic Press, New York, 1966. Algal symbionts in molluscs.
380. Yudkin, J., Amer. J. Clin. Nutr. *20*:108–115, 1967. Dietary carbohydrates in man.
381. Zendzian, E., and E. A. Barnard, Arch. Biochem. Biophys. *122*:699, 1967. Proteases and ribonuclease in vertebrate pancreas.
382. Zoppi, G., and D. H. Shmerling, Comp. Biochem. Physiol. *29*:289–294, 1969. Intestinal disaccharidase, vertebrates.
383. Zhuzhikov, D. P., J. Insect Physiol. *10*:273, 1964. Permeability of insect peritrophic membrane.
384. Zwilling, R., G. Pfleiderer, H. H. Sonneborn, V. Kraft, and I. Studky, Comp. Biochem. Physiol. *28*:1275–1287, 1969. Crayfish trypsin.

CHAPTER 5

OXYGEN: RESPIRATION AND METABOLISM

By C. Ladd Prosser

Chemical evolution began when the earth had a reducing atmosphere, and the early synthesis of organic compounds occurred under anaerobic conditions. It is probable that both the transfer of energy from organic substrates in a stepwise fashion by such compounds as adenosine triphosphate (ATP) and the enzymes catalyzing anaerobic metabolism, as in glycolysis and fermentation, evolved very early. Some oxygen may have been released from water by photochemical reactions catalyzed by metalloporphyrins, but all of the oxygen in the earth's present atmosphere has come from photosynthesis. The rate of photosynthesis is such that the atmospheric oxygen could all have been formed in 2000 years. Aerobic metabolism followed anaerobic and is less universal but more efficient; both must have appeared before there was differentiation of organisms. A few microorganisms obtain energy exclusively by anaerobic means; all plants and animals retain some of the enzymes of anaerobic pathways, but obtain most energy oxidatively. A few animals, such as intestinal parasites, which can live for long periods without oxygen have apparently acquired the anaerobic capacity secondarily (see Chap. 6).

This chapter considers: (1) environmental levels of oxygen and carbon dioxide; (2) methods of uptake of oxygen and their regulation; and (3) oxygen consumption by intact animals and the factors that modify it. The next chapter considers the enzyme pathways of metabolism.

AVAILABILITY OF OXYGEN

Oxygen which enters an animal across a respiratory surface or into cells must diffuse via an aqueous solution; the availability of oxygen depends on its concentration immediately outside the organism. The maximum oxygen concentration is normally limited by air, of which oxygen comprises 20.95% (dry air); the partial pressure of oxygen in air is 159 mm Hg at sea level. In water-saturated air, the partial pressure of oxygen decreases with rising temperature by about 10 mm Hg from 0°C to 37°C. At higher altitudes, the partial pressure of oxygen decreases in proportion to the decrease in barometric pressure (approximately 11.3 per cent per kilometer). At 18,000 feet elevation the barometric pressure is one-half that at sea level. Carbon dioxide is normally present in dry air as 0.03% of the total (under STP conditions, this is a partial pressure of 0.228 mm Hg); the proportion of CO_2 is considerably higher in industrial areas. There is evidence for a gradual increase in CO_2 in the earth's atmosphere with time, caused by combustion of fossil fuels by man.

In fresh water the amount of a gas dissolved (ml/liter) is given by the product of the solubility coefficient (α) and the percentage of an atmosphere of the gas in equilibrium with the water. The amount of O_2 or CO_2 in water may be expressed as a concentration (ml O_2/liter, mg O_2/liter, or mM O_2/liter), or as the partial pressure (mm Hg) of the gas

Temperature	α_{H_2O}	O_2 *Content (at equilibrium with air)* WATER (pure)	SEA WATER (salinity 34.96 ‰)
5°C	0.44	9.22 ml/li (13.2 mg)	6.89 ml/li (9.8 mg)
20°C	0.031	6.51 ml/li (9.4 mg)	5.05 ml/li (7.8 mg)

phase in equilibrium with the water.* The solubility coefficient decreases with rising temperature and increasing salinity as shown above.

Figure 5–1 presents partial pressures in air and water of O_2 and CO_2 for air-breathers and water-breathers when there is an equal exchange of O_2 and CO_2 (respiratory quotient = 1). For an air-breather the slope is 1 and P_{CO_2} may reach high values, whereas for water-breathers the slope corresponds to the ratio of the solubilities of O_2 and CO_2 in water. The environmental P_{CO_2} for a water-breather cannot, therefore, exceed 5 mm Hg, whereas for an air-breather it could theoretically reach 150 mm Hg. Values of $\dot{V}$ give the number of milliliters of air or water which must pass over the respiratory surface per minute if 1 ml of O_2 is to be withdrawn. Thus, at P_{O_2} of 100 mm Hg, 480 ml/min of water are required or only 17 ml/min of air; at this P_{O_2} the P_{CO_2} is 50 mm Hg in air and less than 2 mm Hg in water. In natural fresh waters the oxygen concentration may vary considerably according to temperature, degree of equilibration with air, photosynthesis, and oxygen depletion by organisms. Below the thermocline there is little mixing, and the oxygen concentration may be much lower than at the surface. In warm surface waters, both in fresh water and in the sea, where there is much photosynthesis, the oxygen concentration may exceed air saturation. In the ocean there is, in general, sufficient mixing that the concentrations of oxygen and CO_2 are relatively constant. Concentrations at the surface are high when there is much algal activity, and oxygen tends to become reduced in the mid-depths by non-photosynthesizing organisms; here an oxygen minimum layer occurs (Fig. 5–2), but in the deep regions the oxygen level tends to correspond to air saturation. The *hydrostatic pressure* increases by about one atmosphere per 10 meters depth, but the dissolved gases remain at the same percentage of an atmosphere as at the surface; hence their concentrations (mg/liter) remain constant. This is in contrast to high altitudes, where the *barometric pressure* is due to the gases themselves and the concentrations change with altitude. Very low oxygen concentrations may be encountered in muds where non-biological oxidations occur; also, some parasites living in mammalian intestines are subjected to very low P_{O_2}.

*1 liter O_2 = 1429 mg O_2
1 mg O_2 = 0.031 mM O_2

The diffusion coefficient D (diffusivity) for a gas is given by the diffusion equation:

$$\frac{dv}{dt} = -AD\frac{dc}{dx}$$

which states that the amount diffusing in a given time is equal to the product of the diffusion coefficient, the area, and the concentration gradient over the distance x. The

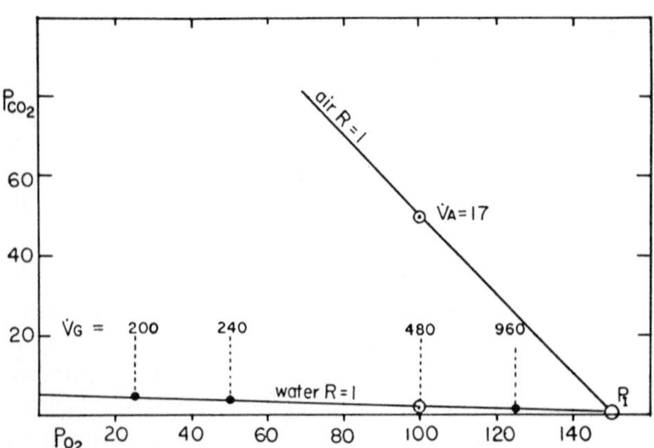

Figure 5–1. Diagram showing partial pressures of O_2 and CO_2 in air and water when there is equal respiratory exchange of CO_2 and O_2. (From Rahn, H., *in* Oxygen in the Animal Organism, edited by F. Dickens and E. Neil, Pergamon Press, 1965.)

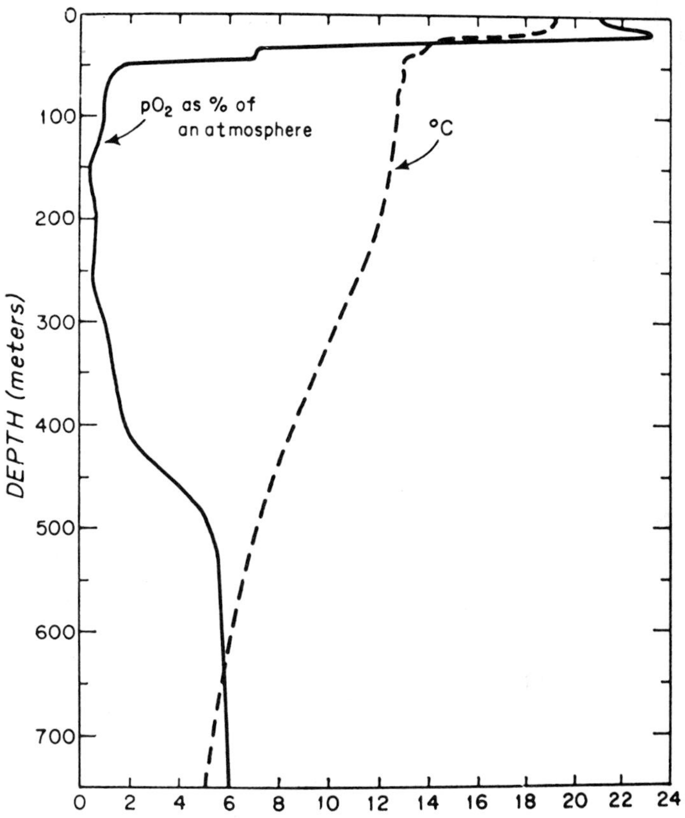

Figure 5–2. Profiles of temperature and oxygen at different depths in southern Pacific Ocean. (From Teal, J. M., and F. G. Carey, Limnol. Oceanogr. *12*:548–550, 1967.)

units of D reduce to $cm^2 time^{-1}$, and D increases by about 3 per cent per centigrade degree rise in temperature. A biological diffusion coefficient, K, was introduced by Krogh as the product of D and the solubility coefficient α, and this reduces in units to $cm^2 time^{-1} atm^{-1}$. Since the solubility of oxygen decreases by about 1.6% per °C rise in temperature, the net increase in K is 1.4% per °C. Values of K in $cm^2 min^{-1} atm$ given by Krogh for oxygen at 20°C are as follows:[121]

air	11.0	muscle	0.000014
water	0.000034	chitin	0.000013

Thus, diffusion of oxygen in air is some 3 million times greater than in water. Carbon dioxide is 28 times more soluble in water at 20°C than is oxygen, and its measured diffusion coefficient in water is about 25 times greater than that of oxygen. In addition, CO_2 forms a hydrate (H_2CO_3), which dissociates as a weak acid; hence, an equilibrium exists among CO_2, H_2CO_3, and HCO_3^-.

In summary, animals are rarely subjected to environmental Po_2 higher than 155 mm Hg, but they may be subject to reduced oxygen approaching zero levels.

UPTAKE OF OXYGEN; EXTERNAL RESPIRATION

Direct Diffusion. All cell surfaces are permeable to oxygen, even those that are relatively impermeable to water. In animals which have oxygen transport systems in body fluids, certain epithelia have become specialized for external respiration. Where no transport system exists, oxygen enters directly by diffusion from the medium into the cells where it is used. For a metabolizing organism, availability is limited by the length of the diffusion path. It is calculated that for a spherical organism with a given Vo_2 (oxygen consumption) in ml/g/min and C_0 (the O_2 concentration in the medium), the limiting size (or thickness) of the organism is given by:

$$\text{limiting thickness} = \sqrt{8\ C_0\ \frac{K}{Vo_2}}$$

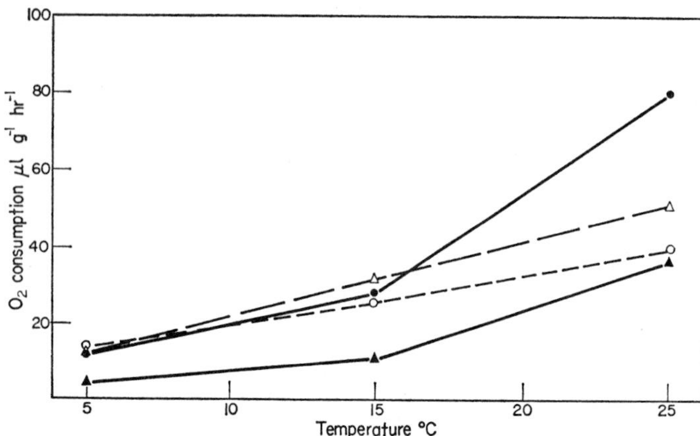

Figure 5–3. Pulmonary and cutaneous exchange in frogs acclimated at the temperatures indicated. Solid circles, pulmonary O_2; solid triangles, pulmonary CO_2; open circles, cutaneous O_2; open triangles, cutaneous CO_2. (From Guimond, R. W., and V. H. Hutchinson, Comp. Biochem. Physiol. *27*:177–195, 1968.)

where K is the diffusion coefficient as before.[64] For small aquatic organisms, diffusion suffices if the organism is no more than 1 mm in thickness. However, in animals with tissues that are more than 0.5 mm distant from an O_2 supply, some supplementary transport system is necessary unless metabolism is very low. Intracellular values of Po_2 are discussed on page 196.

In Protozoa, Porifera, Coelenterata, many Platyhelminthes, Nematoda, Rotifera, a few Annelida, and Bryozoa, oxygen is obtained by diffusion directly from the surrounding water. There may be channels through which water passes; oxygen may diffuse through several cell layers; and transport may be facilitated by cytoplasmic streaming. In animals with circulatory systems, the tolerable distance from a blood capillary or sinus carrying O_2 to a metabolizing cell is limited by diffusion, and in mammals capillaries are rarely more than 0.1 to 0.3 mm apart.

Integument as Respiratory Organ. Since all cells are permeable to oxygen, all multicellular animals have some cutaneous respiration. In many free-living embryos the circulatory system develops before special organs of respiration, and the blood is oxygenated at the body surface.

In many annelids (oligochaetes, hirudineans, and some polychaetes) the integument is highly vascular and provides the principal route of exchange. Respiration is integumentary in sipunculids, echiuroids, and some aquatic insects. In eels, oxygen uptake by skin is 60% of the total and that by gills is only 40%; in pulmonate snails, half is by lung and half by integument.[121] Blowfly larvae obtain 2.5% to 10% of their oxygen cutaneously.[31]

In frogs, the oxygen uptake through the skin is some 50 ml/kg/hr throughout the year, while the uptake by the lungs is negligible in the winter but reaches 130 ml/kg/hr in the spring. Carbon dioxide elimination via skin in frogs exceeds that by the lungs at all times of the year, even in summer when it may be 2.2 times the lung output[48] (Fig. 5–3). In salamanders, 80% of the CO_2 release and 65% to 74% of the O_2 uptake is by the skin.[215] In the lizard *Sauromalus*, about 4 per cent of the total CO_2 output and less than 2 per cent of the O_2 uptake is via the skin; the percentage of CO_2 lost through the skin increases as blood levels of CO_2 go up.[38] In a bat, *Eptesicus*, the CO_2 lost by the skin (mainly by the wings) may be 11.5% of the total[78a] as compared with 1.4% of total CO_2 loss via skin in man. Oxygen uptake through skin of mammals may be used mostly for skin metabolism, and oxygen can pass into the skin in man either from the atmosphere or from the blood.[54] Regulation of integumentary breathing is principally vasomotor.

Gills. Several types of well-vascularized respiratory appendages fall into the category of gills. They may be ciliated; they may move by muscles, or may be located in regions of water currents generated by auxiliary structures. In fish, blood in gill capillaries flows in a direction opposite to water flow; this countercurrent flow provides for almost complete oxygen saturation of blood. The efficiency of gills is given by the percentage withdrawal of oxygen from water flowing through the gill chamber; regulation can be by changes in frequency and amplitude of gill movement and by changes in blood flow.

Polychaetes have various gill-like structures such as parapodia (nereids), gills (*Arenicola*), and branchial tufts (Terebellidae and most Sabellidae) which supplement integumentary uptake of oxygen. *Arenicola* lives in a U-shaped tube, and it withdraws 30 to 35% of the oxygen at several Po_2's. Intermittent ventilatory movements pump water through the tube, and the pattern varies according to dissolved gases.[213] At low oxygen concentration the pumping may change from intermittent to continuous. Utilization by terebellid worms is higher (50 to 60 per cent) when pumping is only for respiration than when feeding is also occurring (5 to 10 per cent O_2 utilization). An active tube-dwelling polychaete can remove 60% of the dissolved O_2 in the tube; an inactive one can only remove 14%. It would require one hour to reduce the burrow O_2 to 10% of the initial concentration; after this the worm would switch to anaerobiosis.[144, 145]

The oligochaetes *Tubifex* and *Limnodrilus* occur in mud of stagnant pools where oxygen may be scarce. In ample oxygen no ventilatory movements are necessary; when oxygen is reduced slightly below air saturation, the tail waves slowly back and forth. As oxygen is depleted the body is extended, and may reach 10 to 12 times its usual length; at very low oxygen concentrations, corkscrew motions tend to pull upper layers of water toward the worm, and in complete anoxia the worm collapses. CO_2 may inhibit the rhythm initiated by oxygen lack. A swamp oligochaete, *Alma*, living in hypoxic mud carries down air bubbles with the upfolding edges of its vascular hind end.[17]

In most echinoids and asteroids the tube feet constitute the main route of O_2 uptake; some of them have dermal papillae which bring the slowly moving coelomic fluid into close association with the cilia-stirred external water. Some holothurians, such as *Cucumaria*, have a coelomic respiratory tree, which opens at the cloaca and accounts for 60 per cent of total O_2 consumption. In *Holothuria*, the rate of cloacal pumping first increases as dissolved oxygen diminishes, but below 60 to 70 per cent air saturation of the water pumping stops and the animal becomes inactive.[158]

Most molluscs have extensive gills which are well supplied with cilia, and water flow is controlled by the cilia in response to oxygen in the water. Bivalve molluscs periodically or reflexly close the valves forcibly, thus clearing the mantle cavity. A medium-sized *Mytilus* (75 to 166 g) pumps 2.2 to 2.9 liters of water per hour. Oxygen withdrawal from the water by *Mytilus* is 3 to 5 per cent at 13°C, and more at higher temperatures. *Mercenaria* pumps 2.9 liters of water per ml O_2 consumed;[140] oxygen withdrawal is constant but ventilation increases as Po_2 is reduced. After a period of anaerobiosis when pumping had virtually stopped, a return to aerated water yielded oxygen withdrawal initially greater than 20%, and 5 to 10 per cent after several hours. Many bivalves show periodicity of opening and closing their shells, and in anoxia the valves remain shut (in *Mytilus, Mercenaria,* and *Ostrea* but not in *Cardium* or *Pecten*). The withdrawal of oxygen by several marine gastropods is higher (40 to 80 per cent) than by pelecypods. In cephalopods (*Octopus*), as in pelecypods, exposure to low O_2 concentration increases pumping, and *Octopus* is also sensitive to increased CO_2 levels. Oxygen withdrawal by *Octopus* may be as high as 80 per cent.

Aquatic Crustacea have gills which are usually ventilated by the paddle-like movements of special appendages such as the scaphognathites. Crabs that live in the low intertidal zone have more gills than do beach crabs. Active aquatic crabs have larger total gill area (1367 mm^2/g in *Callinectes*) than intertidal crabs (624 mm^2/g in *Uca*), and these have larger area than land crabs (325 mm^2/g in *Ocypode*). In some grapsoid land crabs, not only is there a reduction in number and size of gills, but vascular membranous tufts project into the branchial cavity and these structures need be moistened only occasionally to permit oxygen uptake on land. In *Carcinus,* water in the gills flows in a direction opposite that of blood flow and can be shunted along the edges of the gill or through the lamellae; scaphognathite rhythmic activity superimposes pressures of 3 to 12 mm Hg on the small negative pressure in the branchial space[99] (Fig. 5–4). *Carcinus* (and other crustaceans) periodically reverses the current, and can reflexly reverse it to clear the gills of particulates in a kind of "cough." In low O_2 the volume of water pumped increases and utilization may rise to 40 per cent. In the lobster *Homarus,* as the oxygen in the water decreased from 5.8 to 2.4 ml/liter, the rate of pumping remained constant (9.5 liters/hr) but withdrawal increased from 31 to 55 per cent.[196] Similarly, *Carcinus* and *Balanus* showed no increase of external respiration at reduced O_2 or elevated CO_2 levels.

Stimulation of ventilation by reduced O_2 was observed in *Squilla* and *Pandalus*, and when O_2 dropped from 6.6 to 2.1 ml/liter

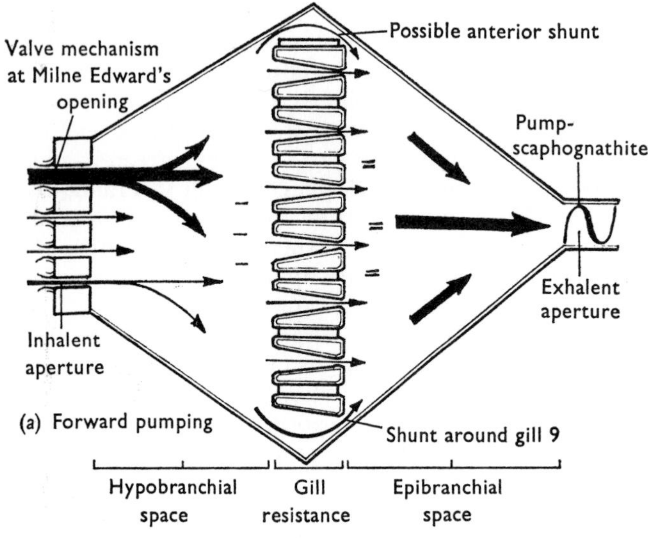

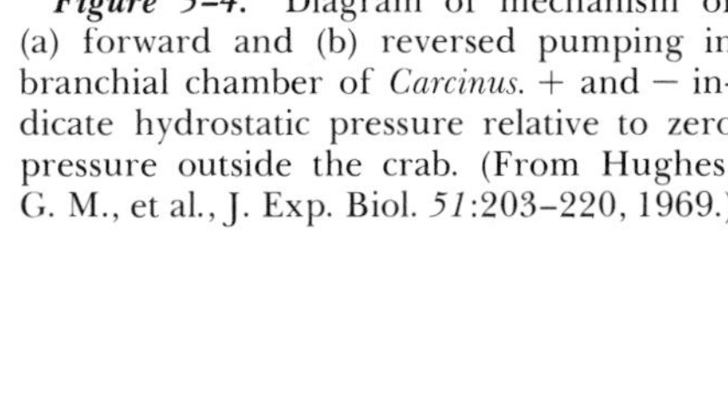

Figure 5-4. Diagram of mechanism of (a) forward and (b) reversed pumping in branchial chamber of *Carcinus*. + and − indicate hydrostatic pressure relative to zero pressure outside the crab. (From Hughes, G. M., et al., J. Exp. Biol. *51*:203–220, 1969.)

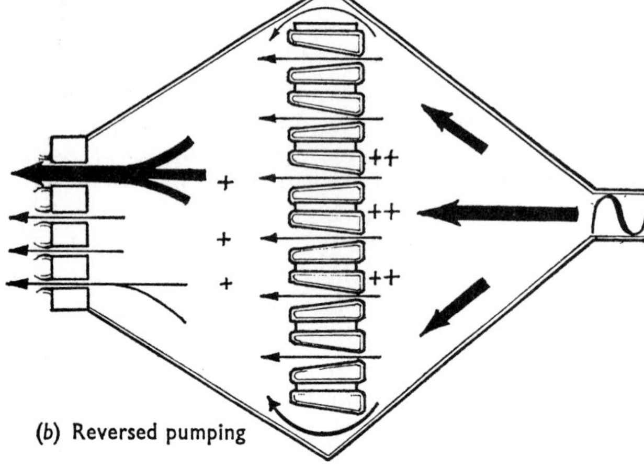

the crayfish *Astacus* tripled the breathing rate and doubled the ventilation volume. In crayfish, as in lobster, consumption decreases linearly with Po_2 but percentage withdrawal remains constant (Fig. 5–5). Pumping in crayfish increases in low O_2 and decreases in high CO_2; heart rate decreases in low O_2. Hypoxia stimulates receptors somewhere in the circulatory system to bring about bradycardia and increased ventilation; CO_2 applied to the base of periopods or at the gill aperture elicits bradycardia.[6,7] CO_2 had little effect until after the carapace was coated with collodion, after which breathing frequency increased upon a four-fold increase in CO_2. Brackish-water isopods and gammarids showed respiratory stimulation in decreased oxygen, but elevated CO_2 had no effect. In *Limulus*, breathing frequency is proportional to O_2 concentration, and in anoxia or excess CO_2 breathing stops.

Numerous aquatic insects have gills through which oxygen passes directly into the profusely branched tracheae. Tracheal gills occur in nymphs of Odonata, Plecoptera, Trichoptera, and Ephemeridae. Dependence on tracheal gills in aquatic Hemiptera is shown as follows:

	O_2 Uptake (ml O_2/kg/hr) with Gills	*O_2 Uptake (ml O_2/kg/hr) without Gills*
Naucoris	241	24
Notonecta	225	29
Corixa	305	47

In fishes, both Chondrichthyes and Osteichthyes, variations on a basic pattern of respiration are adaptive for different levels of activity and responsiveness to environ-

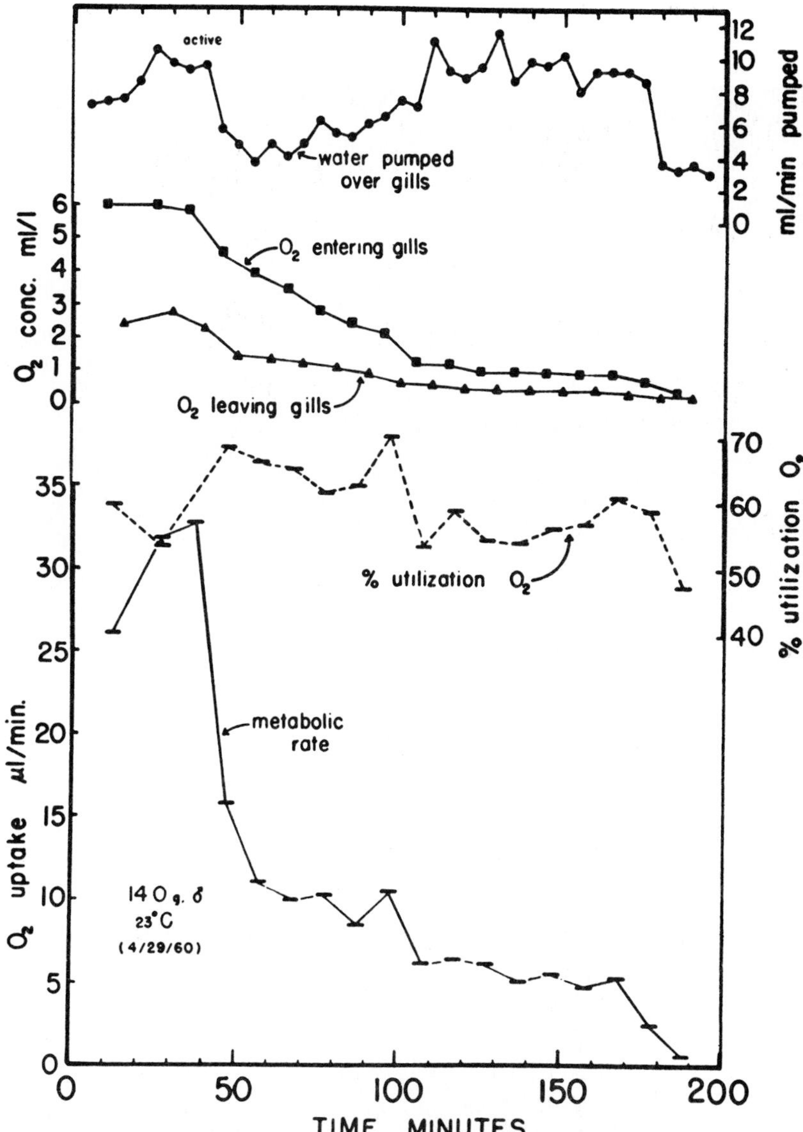

Figure 5–5. Changes in respiration and oxygen uptake by a crayfish during 200 minutes in a closed bottle as oxygen concentration (ml/liter) entering gills declined. (From Larimer, J. L., and A. H. Gold, Physiol. Zool. *34*:167–176, 1961.)

mental factors. In very active fish the total gross gill area is larger than in more sluggish fish:[93]

menhaden	1241 mm²/g
mackerel	1040 mm²/g
sea trout	275 to 432 mm²/g
toadfish	151 to 189 mm²/g

Scanning electron micrographs of the lamellae on gill filaments show extensive surface area due to micro-ridges (Fig. 5–6). This micro-surface has not yet been correlated with O_2 transfer.

Fast-swimming fish are said to keep the mouth open continually and thus provide maximum water flow over the gills. In general, however, the gills lie between two pumps, the buccal pump powered by a set of muscles in the oral floor and the branchial pump powered by opercular movements (Figs. 5–7, 5–8). A cycle starts with the mouth open and a small negative pressure in both cavities; as the mouth closes, positive pressure builds up in the buccal cavity, water flows through the gills, and slight positive pressure is found in the branchial cavity. The opercular valve opens while both pressures are positive,

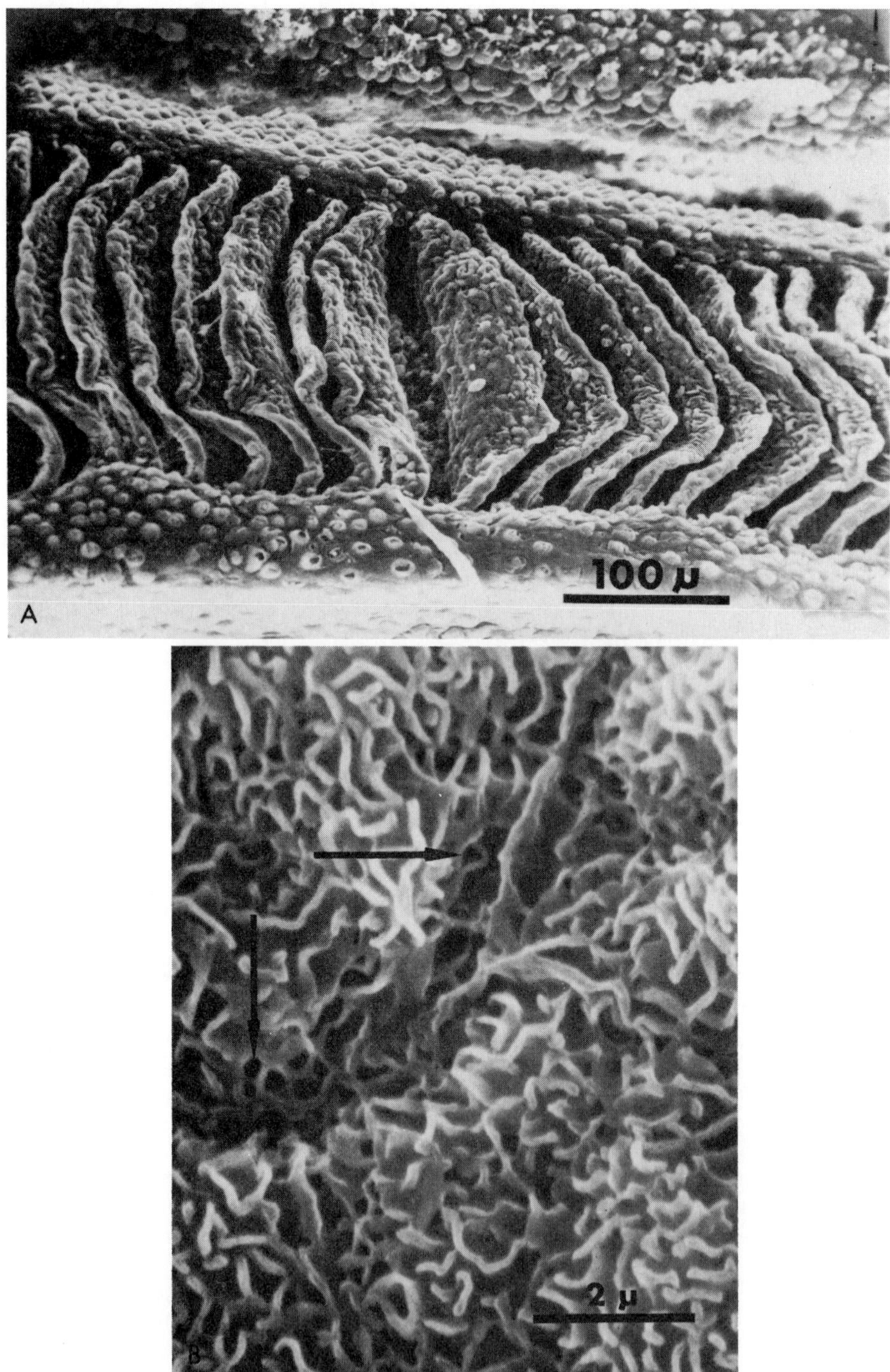

Figure 5–6. Scanning electron micrographs of gill of trout *Salmo gairdneri.* (A) Several secondary lamellae and body of a filament which is covered with mucus secreting cells. (B) Section of surface of a secondary lamella. Epithelial cell surface consists of extensions of the cell as micro-ridges. Arrows indicate regions which appear as pores. (Courtesy of K. R. Olson and P. O. Fromm.)

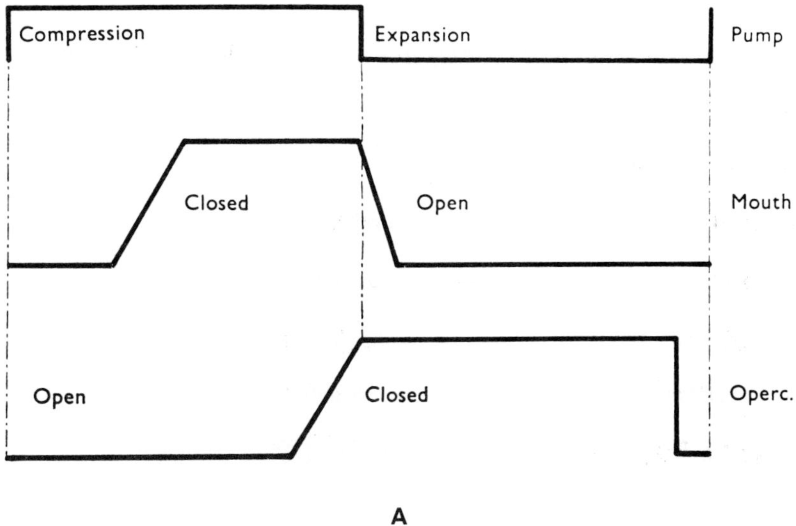

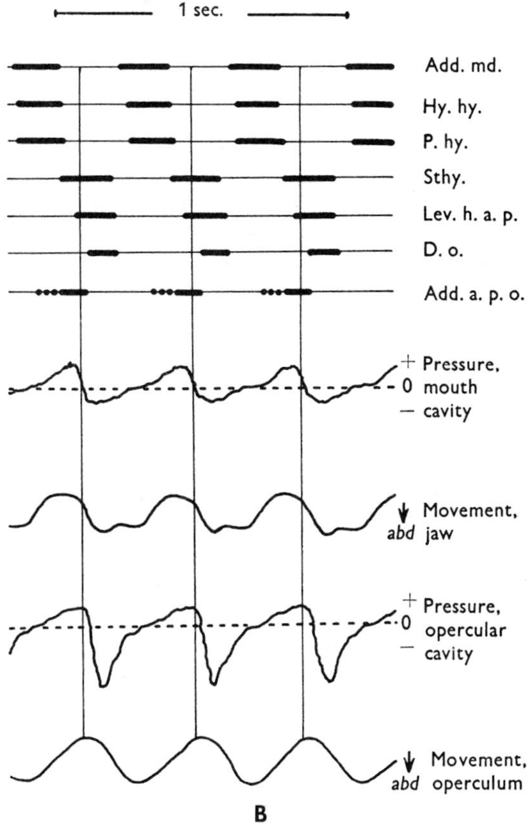

Figure 5–7. A, Diagram of one respiratory cycle in carp. Top trace, movement of buccal and branchial cavities; middle trace, size of mouth opening; lower trace, size of opercular opening. (From Ballintijn, C. M., J. Exp. Biol. *50*:569–591, 1969.) B, Relation between activity in several muscles and the movements and pressures of the respiratory pumps in trout. (From Ballintijn, C. M., and G. M. Hughes, J. Exp. Biol. *43*:349–362, 1965.)

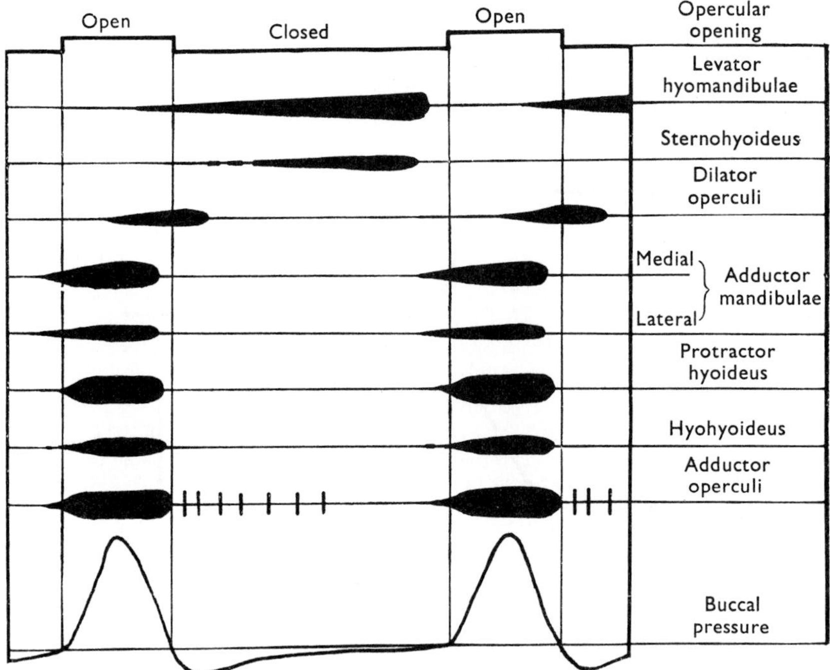

Figure 5–8. Diagram to show phase during which various muscles are active during the respiratory cycle of dragonet fish. (From Hughes, G. M., and C. M. Ballintijn, J. Exp. Biol. *49*:583–602, 1968.)

the buccal being higher than the branchial. Both pressures then drop, the buccal leading the branchial, and the opercula close and the mouth opens again. In some fish, such as bullhead and carp, flow over the gills is nearly continuous; in others, such as trout, dragonet, and roach, flow is periodic. In a dogfish at rest, ventilation is by rhythmic contraction of the branchial muscles and elastic recoil of the skeleton. In an active dogfish, swimming movements are coupled with breathing.[177] In trout, shallow breathing is mainly by buccal muscles; in deeper ventilation the branchial muscles are active.[96] When a pressure gradient was maintained experimentally between the two pumps, flow rate could be varied; percentage of oxygen withdrawn decreased at high flow rates and, as the pressure gradient increased, respiratory frequency was constant but ventilatory volume increased (Fig. 5–9). In a dragonet, O_2 consumption was maximal at zero pressure difference,[94] while in a dogfish it increased with positive buccal pressure.[94] In many fishes, the normal breathing rhythm is interrupted periodically by a closure of the opercula and a reversal of flow, a cough that clears the gill cavity. Stimulation of the gills in a shark by particles in the water can cause a reflex parabronchial cough, and chemicals may cause an orobronchial cough.

Removal of oxygen from the water during a respiratory cycle may be given as percentage withdrawal of O_2; it may also be given as the volume of water that is pumped to provide a given amount of oxygen (e.g., liters of water per ml O_2 removed). Respiratory effectiveness takes account of the oxygenation of blood entering the gills as follows:[90]

$$\text{effectiveness of transfer} = \frac{\text{amount of } O_2 \text{ passing from water to blood}}{\text{amount of } O_2 \text{ that could be taken up by blood}}$$

or, more quantitatively,

$$\text{effectiveness} = \frac{V_{O_2}}{Q\alpha_{b_{O_2}}[P_{i_{O_2}} - P_{v_{O_2}}]} \times 100$$

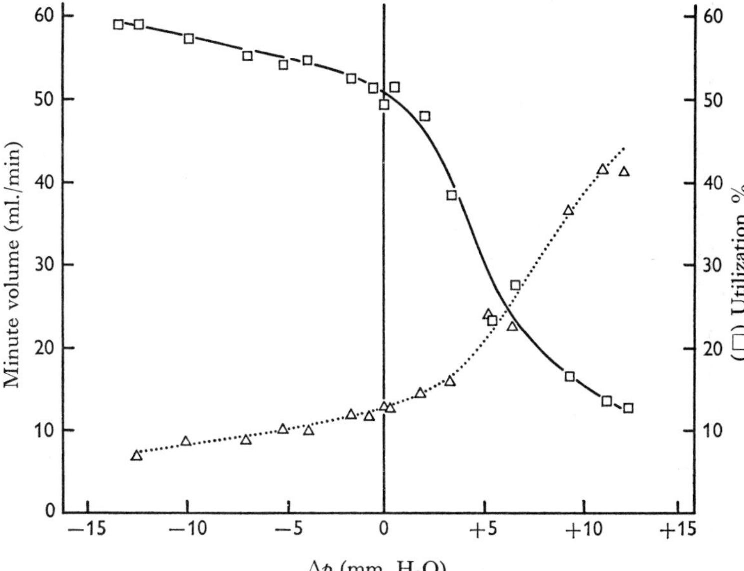

Figure 5–9. Effect of changing the pressure gradient between the buccal and opercular chambers on respiratory minute volume (triangles) and on percentage utilization of oxygen (squares). (From Hughes, G. M., and C. M. Ballintijn, J. Exp. Biol. *49*:583–602, 1968.)

where Vo_2 = oxygen uptake, Q = volume of blood passing through gills in a given time, $\alpha_{b_{O_2}}$ = solubility of oxygen in blood, $P_{i_{O_2}}$ = pressure of O_2 in incurrent water, and $P_{v_{O_2}}$ = pressure of O_2 in venous blood entering gills. Maximum uptake would occur if blood leaving the gills has the same Po_2 as the inspired water. The effectiveness of transfer by a trout approaches 100%, but removal of oxygen from the water is 11 to 30 per cent.[171] This high effectiveness is an indication that blood flow is countercurrent to water flow.

Water flow is laminar over the gill surface, and 80 to 90 per cent of the resistance to O_2 transfer from water to blood is caused by slow transfer of O_2 in the thin unstirred layer surrounding the gill lamellae.[81]

Countercurrent flow of water and blood is also indicated by dye injection experiments in a shark. The arterial oxygen, $P_{a_{O_2}}$, is less than the mean of the O_2 in inflowing and outflowing water, but is greater than the outflow content.[72] In trout, the Po_2 of dorsal aorta blood is higher than that of expired water, and in an anesthetized dogfish Po_2 is the same in efferent blood and exiting water, thus proving a countercurrent arrangement of blood and water flow in the gills.[81] There are vascular shunts in the margins and middle of the gill lamellae in *Salmo gairdneri* which can permit changes in flow pattern with need.[173a]

Nomograms for trout, carp, and dogfish show the relation between arterial oxygen ($P_{a_{O_2}}$), oxygen consumption (Vo_2), and both water and blood flow; the $P_{a_{O_2}}$ increases as the ratio of ventilation volume to cardiac output increases (Fig. 5–10).[111]

When the Po_2 in water decreases, the ventilatory volume increases markedly, breathing rate increases slightly, stroke volume of the heart increases very much, and heart rate is reduced. When Pco_2 increases, respiratory amplitude increases, frequency decreases, and heart rate increases.[90] The control by Po_2 is much more important than that by Pco_2. In trout, ventilation increased sevenfold following a brief hypoxia.[41] Respiratory and circulatory responses combine to maintain relatively constant oxygen consumption over a wide range of Po_2 in the water:[171, 172]

	Aerated Water (155 mm Po_2)	*Hypoxic Water* (30 mm Po_2)
breathing rate	85 min^{-1}	120 min^{-1}
ventilation volume	500 ml	3500 ml
O_2 consumption	100 ml/kg/hr	100 ml/kg/hr
heart rate	75 min^{-1}	22–28 min^{-1}
blood pressure (ventral aorta)	60 mm Hg	80 mm Hg
blood lactate	12.77 mg/100 ml	34.86 mg/100 ml
blood Po_2 (dorsal aorta)	120 mm Hg	20 mm Hg
cardiac output	80 ml/min	80 ml/min

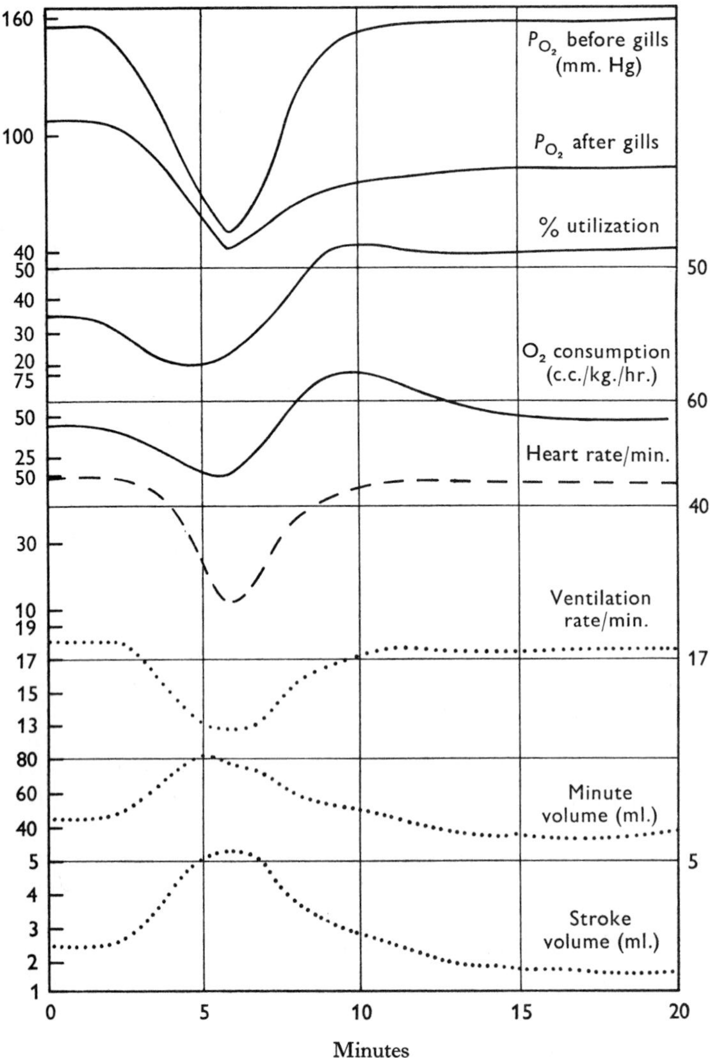

Figure 5-10. Summary of effect of lowering P_{O_2} of water for six minutes, as shown by values of P_{O_2} before gills. (From Hughes, G. M., and S. I. Umezawa, J. Exp. Biol. *49*:565–582, 1968.)

In hypoxia the effectiveness of O_2 uptake is reduced. In many fish, O_2 consumption increases as P_{O_2} is reduced, because of the greater ventilatory effort; hence, a double hazard is created. If temperature is raised, ventilation volume increases first and rate later; but at a critical temperature, heart rate slows with inadequate oxygenation and the two respiratory pumps may become uncoupled.[95]

Rhythmic respiratory movements are triggered by a spontaneously active center in the central medulla oblongata. Rhythmic discharges of nerve impulses have been recorded from this region after immobilization by tubocurarine or succinylcholine,[177] and also after cutting all sensory nerves to this part of the brain.[100] The location of O_2 and CO_2 receptors is uncertain, since responses were present but reduced after sectioning of the ninth and tenth cranial nerves, which serve the gills. Probably there are both peripheral and central receptors. Oxygen lack increases both frequency and depth of breathing in *Tinca tinca* so that ventilation volume increases in low oxygen; CO_2 excess has little effect but may bring about a slight decrease in breathing rate and a small increase in depth.[63] Eel (*Anguilla*) and trout remove some 80 per cent of oxygen at air saturation, and a smaller percentage at low O_2 concentrations. They may increase ventilation four-fold when the O_2 content of the water falls below 4 ml/liter.[121]

In a choice gradient, salmonid and centrarchid fishes avoid low O_2 concentrations;

chinook salmon avoided 4.5 mg/liter, coho salmon 6 mg/liter, and bluegill 1.5 mg/liter. These behavioral regulations suggest a sensitivity greater than is needed for respiratory control and avoidance of concentrations well above hypoxic levels.[216]

Lungs. All air-lungs present moist surfaces across which oxygen and carbon dioxide diffuse and from which water can evaporate; hence, all are protected against evaporation.

DIFFUSION LUNGS. Many arachnids—spiders and scorpions—have book lungs which open to the exterior by a spiracle and contain tubes that serve to aerate the blood. Diffusion lungs are best known in pulmonate snails, in which the mantle cavity is modified as a lung that opens to the exterior by a pneumostome, which can be opened or closed according to respiratory conditions. Diffusion is adequate, even with 1 to 2 mm Hg gradients, because the surface area is large and metabolism is low. The opening of the pneumostome is stimulated by CO_2 in snails such as *Limax, Helix,* and *Arion*; in 3 to 5 per cent CO_2 the pneumostome remains open.

Planorbis rises to the surface and fills the lung when lung oxygen content falls to between 1 and 4 per cent; *Lymnaea* does so at between 6 and 13 per cent. During a dive the lung oxygen decreases much more than the CO_2 increases; hence, CO_2 must be lost cutaneously. The duration of a dive is longer at high than at low O_2 levels.[110]

GAS BLADDERS OF TELEOSTS AND LUNGS OF DIPNOANS. There are two types of gas bladder, the open type that occurs in physostome teleosts, and the closed type that occurs in physoclistous fish. The physostomes can gulp air into the bladder, and both types can secrete gas into it. The gas bladder is primarily a hydrostatic organ, but in some fish it contains much oxygen that can be used in hypoxic conditions. The mechanism of gas secretion is discussed on page 205. In lungfish (dipnoans) the lung arises from the floor of the pharynx, and the fish use the lung during long periods out of water. Increase in P_{CO_2} in the water causes cessation of gill breathing, a response that is abolished by atropine. CO_2 in the lung has no such effect; hence, CO_2 receptors in the branchial region are associated with responses to external CO_2.[137] At rest in aerated water, *Protopterus* uses both branchial and pulmonary breathing; skin and gills account for most of the loss of CO_2. Increase in dissolved CO_2 decreases branchial and favors lung breathing.[137]

VENTILATION LUNGS. It is probable that lungs evolved in primitive fish and were replaced by swim-bladders in more recent fish; but as lungs they persisted in the side-branch Dipnoi and were present in the crossopterygian ancestors of the Amphibia. Dipnoans have a buccal force pump, and exhalation is not due to rib movement. It appears that air-breathing in terrestrial vertebrates was derived from the buccal pump of aquatic forms; the buccal force pump was earliest used to power gill irrigation, then respiratory pouches, and then lungs. Active lung ventilation is a more recent development. In some primitive urodeles the lungs are smooth; in most amphibians and reptiles they have partitioning septa; and in birds and mammals they are alveolar. In frogs pulmonary ventilation is by nostrils, glottis, and buccal floor. Pulmonary pressure is constant. Buccal ventilation is independent but is reflected in pulmonary volume. The walls of the lungs of many amphibians and reptiles contain smooth muscle, contraction of which sets a tonic level on which thoracic and buccal pumps act. In a frog the lung smooth muscle may aid in emptying the lungs, and in a Japanese toad isolated lungs show spontaneous rhythmicity at 10 to 20 contractions per hour.[119] Toad lung contracted upon stimulation of sympathetic nerves, and relaxed upon vagus stimulation.[186]

The two types of ventilation are: (1) by positive pressure, in which air is forced into the lung by swallowing or buccal movements (as in a frog), and (2) by negative pressure, in which air is drawn in by increasing the space around the lungs (as in mammals).

During eupnea (normal breathing) in mammals, inspiration is active and expiration is largely passive except when forced; but some animals—birds and turtles—effect both movements by active muscular contraction. In lizards there is both buccal and thoracic movement, and oxygen consumption (and respiration) may be periodic. In lizards, pulmonary ventilation is constant down to 10% O_2; it then decreases. CO_2 stimulates breathing, especially in hypoxia.[158a] Water turtles supplement lung and pharyngeal respiration with dermal and cloacal aquatic respiration. Buccal ventilation is mainly for olfactory sampling. Pulmonary ventilation is brought about by movements of abdominal, thoracic, and leg muscles; both inspiration and expiration are active.[66] When a turtle breathes gas containing less than 2 per cent O_2, ven-

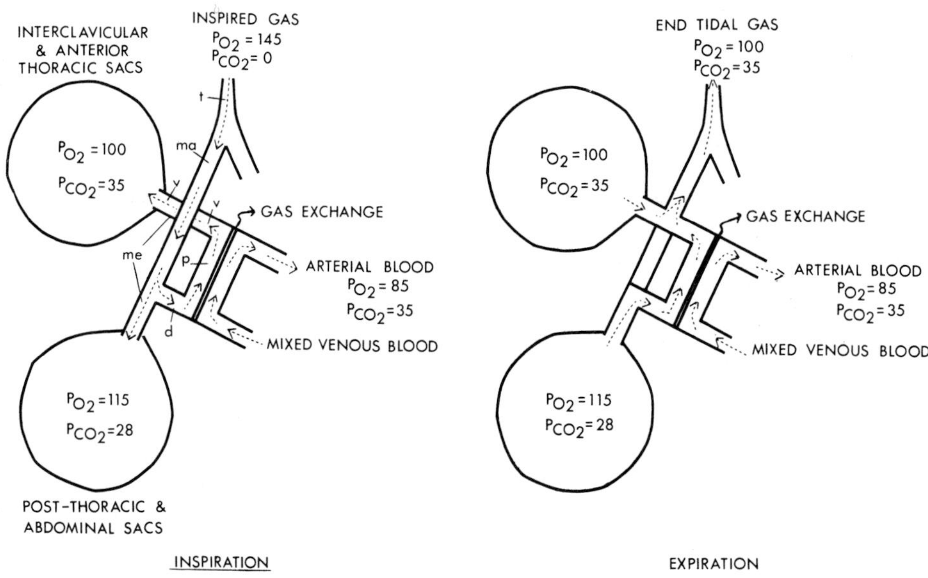

Figure 5–11. Model of respiratory pathways in goose to illustrate the probable directions of gas flow in inspiration and expiration. *ma,* Main bronchus; *me,* mesobronchus; *d,* dorsal bronchi; *p,* parabronchi; *v,* ventral bronchi. (From Cohn, J. E., and R. Shannon, Respir. Physiol. 5:259–268, 1968.)

tilation increases, probably by a vagal reflex.[62]

Reptiles (*Caiman*) and amphibians (*Xenopus*) show periodic bursts of breathing. In *Caiman* there is much variation in duration of ventilatory periods, in intervals between such periods, and in number of breaths per burst.[156] In *Xenopus* during a ventilatory burst, blood flow in the pulmocutaneous artery increases to a maximum while systemic flow is relatively constant and blood pressure falls; hence, there must be respiratory vasodilation. Blood-flow adjustments such as this may tend to stabilize O_2 absorption during intermittent breathing.[185]

Some birds use the respiratory system in body cooling. In cool dry air, most tidal air passes through the lungs and heat loss is low, whereas in hot humid air, when the major function of respiration is cooling, only a small fraction of inspired air traverses the lung passages. Panting may lead to hypocapnia and resulting alkalosis. Birds have extensive air sacs that are devoid of respiratory epithelium but may have three or four times the volume of the lungs; these permit two-way flushing of the lungs so that there is gas exchange during both inspiration and expiration (Fig. 5–11). Analysis of O_2 and CO_2 in lungs and air sacs of geese showed that thoracic and abdominal sacs have higher O_2 and CO_2 than the exhaled gas:[36]

		P_{O_2}, mm Hg	P_{CO_2}, mm Hg
inspired air		145	0
end tidal air		100	35
air sacs	post-thoracic, abdominal	115	28
	interclavicular	100	35

Breathing may be increased during flight by synchronization of wing and respiratory movements. In birds of the size of a pigeon or larger (but not in smaller birds), wingbeat and respiration are synchronized; heat loss may be increased five-fold in flight.[200, 201] In a flying duck, telemetered recordings showed one respiratory cycle per two wing beats and an increase in respiration of seven-fold during flight.[141]

Diving mammals respire rapidly while at the surface, and dive with the lung partially collapsed. A pilot whale had only 5.6 liters of gas remaining in the lungs between blows; its tidal volume was 39.5 liters and its lung capacity was 45.1 liters.[160] Exhalation in whales is very rapid (0.5 sec) and is due to elastic recoil, but at inspiration the esophageal pressure was 20 cm Hg below atmospheric pressure.

The mammalian respiratory center consists of a network of self-exciting neurones, the activity of which is modulated by sensory inputs; it is located between the mesencephalon and the hindbrain ventrally at the level of the pons. Inspiratory and expiratory neurons are intermingled in the respiratory center of the cat, and sectioning of the midline leaves cell activity but stops breathing; neuronal activity persists after elimination of vagal and proprioceptive afferents.[176] The most sensitive control of the respiratory center is by CO_2 concentration in the blood that reaches the center. In addition, chemoreceptors in the carotid sinus are sensitive to oxygen lack (P_{O_2} below 100 mm Hg for cat and man, 60 mm Hg for dog), and are also sensitive to relatively high CO_2. Carotid and aortic bodies are stimulated by hypoxia, as in breath holding, and they are sensitive to cyanide.[47] Carotid body chemoreceptors are readily stimulated by acetylcholine, and the nervous responses are blocked by atropine, hexamethonium, and tubocurarine. Acetylcholine is released from the carotid body during hypoxic stimulation, and glomus cells make synaptic contact with sensory nerve endings. It appears that low P_{O_2}, high P_{CO_2}, and acid stimulate the glomus cells, which probably stimulate the nerve endings by release of acetylcholine.[55]

A chicken has CO_2 receptors in the lungs, which send afferent fibers to the vagi; respiration is stimulated within half a second by 15% CO_2 in the lung.[163] In some birds—English sparrow, starling, mallard duck—carbon dioxide stimulates breathing, while in others—muscovy and pekin ducks—it causes apnea, probably by inhibitory reflexes from chemoreceptors in the nasopharynx. In diving mammals and birds the sensitivity of the respiratory center to CO_2 is less than in non-divers.

As air is inhaled by birds, mammals, or warm reptiles, it is warmed and water is added; as the air is exhaled it is cooled and some water is condensed onto air passages. In animals in which conservation of water is important, there is exchange of water such that when the temperature gradient between ambient air and the body is high, much of the respiratory water is recovered. A cactus wren recovers 75%, and a kangaroo rat 88%, of the water added in the lung and passages to its inhaled air; an iguana in $T_A = 30°C$, with $T_B = 42°C$, exhales air at 35°C and recovers 31% of respiratory water. Respiratory passages in all of these animals are multiple, narrow, and have much surface area for condensation.[155, 178]

Tracheae. Tracheal respiration carries air directly to the metabolizing cells without intervention of blood. Tracheae are found in insects, onychophorans, some spiders, diplopods, and chilopods. Each trachea has an outer epithelium continuous with the hypodermis, and an intima continuous with the cuticle which contains beadlike taenidia for support. Tracheae usually open by spiracles; some, however, receive oxygen from gills. Single spiracles may serve for both inspiration and expiration, but more often there are numerous spiracles, some (commonly anterior) for inflow and others for outflow of air. Tracheae branch extensively and open into small tracheoles which may be fluid-filled; each tracheole is surrounded by a tracheal cell at its origin as a sort of valve. Tracheoles may penetrate into cells such as muscle fibers and thus bring oxygen close to mitochondria. Some insects have air sacs which can build up tracheal pressure.

In a few insects, such as fleas, the spiracles remain open. Usually, however, there are ventilatory movements, which are controlled by segmental ganglia; there are several degrees of ganglionic interaction. In *Periplaneta*, the isolated abdominal ganglia can maintain a breathing rhythm faster than that which occurs when it is under thoracic control; treatment of thoracic ganglia with CO_2 speeds the rhythm. In dragonfly larvae, expiratory nerve impulses activate dorsoventral muscles; other nerves activate the transverse muscles of inspiration. Spike frequency rises to a maximum during an expiratory burst.[92]

Most insects can open the spiracles either independently or in coordination with breathing. Adults of the fly *Phormia* have valved spiracles; those of the larvae are fixed.[30, 31] Low concentrations of CO_2 stimulate the spiracles to remain open; higher concentrations may cause more active ventilation, and very high (25%) concentrations may anesthetize. When O_2 is reduced, ventilatory movements are enhanced, probably by central nervous stimulation. In adult dragonflies the spiracle rhythm is not usually synchronized with ventilatory movements; however, in three species, CO_2 caused synchronization. Hypoxia increases the spiracular sensitivity to CO_2, and high O_2 raises the CO_2 threshold. In addition to CO_2 sensitivity in spiracles, and sometimes in central ganglia, some insects (e.g., honeybees) have antennal CO_2 recep-

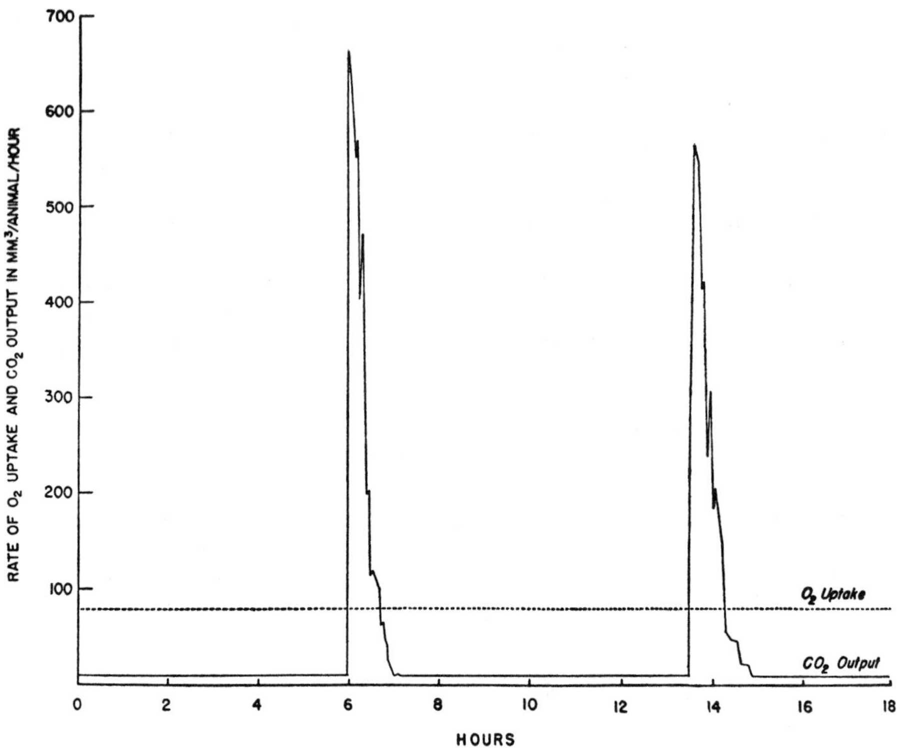

Figure 5–12. Rate of oxygen consumption and carbon dioxide production by cecropia pupae. (From Schneiderman, H. A., and C. M. Williams, Biol. Bull. *109*:123–143, 1955.)

tors.[123] Hive bees are stimulated by low O_2 or high CO_2 to ventilate the hive.

Continuous records of respiration of some insects, such as *Phormia* larvae and *Hyalophora* (moth) pupae, show periodic bursts of CO_2 output with more or less continuous O_2 consumption. An adult carabid shows a 30-second rhythm, and a *Sphinx* moth pupa shows a 24-hour period; at 25°C a cecropia pupa shows a CO_2 burst every 7.3 hours, at 1°C every 3 to 4 days. At low oxygen or high CO_2 the bursts disappear and spiracles remain open. If spiracles are forced open by small tubes, the CO_2 bursts disappear. The bursts are not due merely to the opening of spiracles in response to CO_2, since movement of gaseous CO_2 from inside to outside would not be recorded manometrically. It appears that during an interburst, O_2 enters faster than CO_2 leaves and that the O_2 is absorbed by tissues, thus creating a slight negative pressure in the tracheae. When a spiracle opens, the P_{CO_2} falls and bound CO_2 is released from tissue fluids.[31, 179, 180] Cecropia pupae burst intervals are 4 to 17 hours at 20°C, and are relatively constant for an individual; the function of the bursts is to reduce water loss[112] (Fig. 5–12).

Aquatic insects show a variety of respiratory adaptations. Some larvae pierce weed stems and use the contained gases. Others, such as *Notonecta* and *Dytiscus*, hold a bubble of air by hydrofuge hairs, and the spiracle opens into it. As oxygen is used from the bubble, more diffuses in from the water; nitrogen diffuses out when its partial pressure becomes greater than that in the water. Hence, the bubble can transfer many times its volume of oxygen and can last for many hours without renewal. Other aquatic Hemiptera and Coleoptera have a dense hair mat with some 2.5×10^8 hairs/cm^2. This surface is nonwettable and holds a thin gas layer, or plastron, which serves as a permanent physical gill through which oxygen diffuses. In some aquatic larvae, a plastron is formed by spiracles modified as gills. The plastron insects, although air-breathers, never need to come to the surface.[199] In eggs of some aquatic flies, air is held between vertical columns in the inner part of the chorion, providing a large air-water interface when the egg is immersed; this is a plastron that resists pressure and permits O_2 diffusion.[83]

Auxiliary External Respiration. Examples have been mentioned earlier of animals that shift from one respiratory mode to another under hypoxic stress. It has been suggested that such supplementary modes of respiration were critical in the evolution of vertebrates as they moved from aquatic to terrestrial habitats. An eel removed from water first uses oxygen from its swim-bladder; it then increasingly absorbs oxygen through the skin, and then breathes air over the gills. In air, the gills account for a third of respiration, while in water they account for 85 to 90 per cent of it. Metabolism in air drops to half that in water, and heart rate is reduced. After some hours in air, an O_2 debt is accumulated and blood lactic acid may go from 10 to 120 mg/100 ml. The active breathing of air is greatly increased after O_2 is reduced; it is not stimulated by CO_2, but may be inhibited if this gas is applied to the gills. All of these adaptations permit the eel to survive up to a day out of water.[22]

A mudsucker, *Gillichthys*, has a vascular buccopharynx; it gulps air whenever O_2 in the water is reduced, and surfaces more into a N_2-CO_2 mixture than into pure N_2.[199a] Similarly, an electric eel, *Electrophorus*, takes air into the mouth at a rate which increases at high CO_2 or at reduced O_2.[104] Another swamp fish, *Symbranchus*, in air periodically inflates the gill chamber and saturates its blood with O_2; arterial CO_2 increases in air breathing and decreases in water breathing.[104] *Amia* increases breathing into an air bladder as temperature rises or as it becomes more active; at 30°C it obtains three times as much O_2 from air as from water. Most CO_2 output is by gills; reduced P_{O_2} in water stimulates air breathing and reduces use of the gills.[106]

The African lungfish, *Neoceratodus*, in hypoxic water increases both branchial and aerial breathing; elevated CO_2 decreases branchial and increases aerial breathing. Chemoreceptors are located on the external gills or efferent branchial vessels. Hypoxia stimulates both gill and lung ventilation; hypercapnia increases gill ventilation more than lung ventilation, and the combination of hypoxia and hypercapnia increases lung ventilation and decreases gill ventilation.[103a] A South American lungfish, *Lepidosiren*, is stimulated to air breathing by hypoxia and hypercarbic water.[105] Several African catfishes have air-breathing chambers as diverticula of opercular and branchial cavities. *Sacobranchus* normally obtains 41% of its O_2 from air, and in hypoxic water it relies increasingly on air breathing and shows reduced metabolism.[97a] *Clarias* normally obtains 58% of its O_2 from air, and when removed from water its V_{O_2} is reduced from 93 to 65 ml O_2/kg/hr and 17% of its O_2 is obtained via skin.[186a]

Amphibians have various combinations of dependence on lungs, skin, and gills. Pulmonary respiration becomes increasingly important in frogs at higher temperatures, proportionately more for O_2 uptake than for CO_2 output.[73] *Ambystoma opacum* takes in 34% of its O_2 by lungs and buccopharyngeal mucosa and 66% by skin; 80% of the CO_2 output is through the skin.[215] Tadpoles reared in low oxygen have large, branched gills; in high oxygen the gills are small and thick-walled. However, the oxygen consumption of *Ambystoma* was the same with or without gills from 21 per cent O_2 down to 10 per cent; it was slightly lower at 5 per cent O_2. The hemipteran *Rhodnius* increases growth of tracheae to body regions that are low in O_2, and aquatic dipteran larvae develop highly branched peripheral tracheoles in low O_2. Lung size, hemoglobin content, vital capacity, and pulmonary ventilation increase in men acclimatized to high altitudes. Many animals, when kept in reduced oxygen, increase the synthesis of transport hemoglobin (Chap. 8).

OXYGEN CONSUMPTION BY INTACT ANIMALS; MODIFYING AGENTS

Biological literature records many values of oxygen consumption by various animals, yet metabolic values are meaningful only for the particular conditions of measurement, and tabulated values are subject to necessary qualifications. Rate of oxygen consumption is influenced by activity, temperature, body size, stage in life cycle, season, and time of day, as well by previous oxygen experience and genetic background.

Activity. One of the intrinsic modifiers of O_2 consumption most difficult to control is muscular and other activity. By definition, "basal" metabolism represents the oxygen consumption for maintenance only. In man this is measured in a postabsorptive state, in a relaxed but awake subject, usually in mid-morning. For measurements on animals, movements are minimized by darkness, quiet, and habituation to the metabolism chamber. Measurements may be made on curarized or anesthetized animals, but mus-

(Text continued on page 186.)

TABLE 5-1. Standard Metabolism

Animal	*Body Weight grams*	Vo_2 *ml O_2/g/hr*	T_A, °C
Small mammals			
shrew *Sorex*	3.4–3.6	7–10.6	
house mouse *Mus*	17	1.7	
red-backed mouse	21.7	2.27	
Marmosa	13	1.44	
Microdipodops[13]	15.2	1.3	
kangaroo mouse *Microdipodops*[29]	10–14	2–3 (active)	5
		0.035 (torpid)	5
Perognathus longimembris[33]	11.5	2.06	
Perognathus crinitus[149a]	23.1	0.92	
Perognathus hispidus[210]	39.5	1.25	
Peromyscus ermicus[149a]	21.5	1.48	
Peromyscus californicus[149a]	45.5	1.03	
heteromyid *Liomys salvani*[89]	43.8	1.07	
heteromyid *Liomys irroratus*[89]	48.1	1.12	
Heterocephalus[149]	32	0.26	
Microtus	31	2.65	
Baiomys[89]		1.95	
chipmunk *Tamias*[210]		1.03	
Gerbillus	72–145	0.80	
Cricetus	435	0.87	
Rattus	280	0.88	
Cavia	500	0.76	
Citellus	225	0.6–0.9	
Large mammals			
Lepus	1581	0.96	
rabbit	2700–3700	0.55	
cat[20]	3000	0.45	
dog	7000–20000	0.36	
baboon *Papia*	17000	0.49–0.6	
seal[3]	26000	0.59	
porpoise[3]	170000	0.22–0.34	
pig	48000	0.35	
sheep	46800	0.25	
man	60000–71500	0.21–0.24	
cow	500000	0.124	
camel	407000	0.099	
horse	170000–330000	0.03–0.04	
elephant	3700000	0.07–0.11	
bottlenose whale[3]		0.144	
Bats			
Megachiroptera			
Synconyctris[11]	17.5	1.93	
Nyctimene[11]	28.2	1.43	
Eumops[134]	57	0.51	
Macroderma[134]	148	0.94	
Paranyctimene[11]	21	1.38	
Dobsonia[11]	87	1.26	
Pteropus scapulatus[11]	362	0.67	
Pteropus poliocephalus[11]	598	0.53	
Rousettus	130	0.48	
Microchiroptera			
Myotis[159]		1.74	
Miniopterus[153]		0.74	(neutral temperature)
		2.5	0
Myotis	25	2.2	
Noctilio	27	1.2	
Desmodus	29.4	1.2	

TABLE 5-1. Standard Metabolism *(Continued)*

Animal	*Body Weight grams*	Vo_2 *ml O_2/g/hr*	T_A, °C
Miscellaneous and primitive mammals			
armadillo *Dasypus*	3700	0.27	
		0.97	
Heterohyrax	1292	0.52	
edentate *Bradypus*[68a]	3400	0.13	
Dasypus[68a]	3700	0.2–0.27	
marsupial *Dasyuroides*[141]	89	0.87	
Dasycercus[141]	88	0.52	
	910	0.45	
Antechinus[141]	8.5	1.26	
	22.1	1.53	
fat tail mouse *Sminthopsis*[43]	14.1	1.32	
brown marsupial mouse *Antechinus*[43]	36.5	1.0	
bandicoot *Isodon*[43]	880	0.445	
Trichosurus[43]	1982	0.315	
kangaroo *Megaleia*[42]	32490	0.177	
wallaby *Macropus*[42]	4960	0.29	
Sarcophilus[141]	5050	0.28	
monotreme *Tachyglossus*	2500–3500	0.2–0.25	
echidna	3800	0.22	
Birds			
Hummingbirds			
Selasphorus[127]	3	4.0	
Stellula[127]	3	3.9	
Calypte anna[127]	5.4	3.8	
Calypte costae[127]	3.2	2.9	
Archilochus[127]	3.3	3.6–4.3	
Eugenes[132]	6.6	2.7 (at rest)	20
		0.1 (torpid)	15
		1.18 (torpid)	27
Lampornis[132]	7.9	2.3	
Patagona[131]	19	2.7 (at rest)	
		0.25 (torpid)	16
Sparrows			
zebra finch	11.7	3.28	
house sparrow *Passer domesticus*[88]	25.5	3.53 (day)	
		2.34 (night)	
Junco	18	2.94	
Fringilla	24.8	3.3	
Zonotrichia	26.4	2.8	
Harris sparrow[46]	33	4.1	
Other birds			
red crossbill *Loxia*[46]	29.4	3.1	
white crossbill *Loxia*[46]	29.8	2.8	
oriole[175]		2.4	
budgerigar[202]	30–40	4.5 (rest)	
		21.9 (active)	
waxbill *Estrilda*[130]		3.7	
grosbeak *Hesperiphona*	60	2.5	
arctic gray jay *Perisoreus*	71.2	1.75	
night hawk	75	1.1	
Inca dove *Scardafella*[142]	42	1.11	
ptarmigan	32	1.3	
poorwill *Phalaenoptilus*[128]	49.5	0.72	
nightjar *Eurostopodus*[44]	88	0.83	
quail *Lophortyx*	150 (male)	0.96	
	127 (female)	1.05	
bob-white *Coturnix*	97	2.06	
pigeon *Columba*	150	0.98	
chicken	2000	0.43	
emu[37]	38300	0.023	
rhea[37]	21700	0.031	

(Table 5-1 continued on following page.)

TABLE 5-1. Standard Metabolism *(Continued)*

Animal	*Body Weight grams*	V_{O_2} *ml O_2/g/hr*	T_A, °C
Reptiles			
skink *Lygosoma*[87]	1.5	0.295	30
		0.392	40
	30	0.5	40
Anolis	3–6	0.117	20
Lacerta	6.3	0.245	19
Scelophorus	0.5–5	0.12	25
Gerrhonotus	30	0.55	20
		0.298	35
Crotaphytus[45]	30	0.1	20
		0.2	30
Dipsosaurus[45]	34–94	0.05	20
Iguana[152]	369–1200	0.06 (rest)	20
		0.25 (active)	
agamid *Amphibolurus*[14]	373	0.14	28
Alligator	49000	0.079	28
turtle *Chrysemys*	100–300	0.031	18
turtle *Pseudemys*	150–350	0.08	24
South American boa[65]	3270	0.018	20
Natrix	84	0.07	16
Python	5000	0.007	18
Eunectes	11300	0.021	20
Crotalus	2000–5000	0.0078	17
Amphibians			
salamander *Ambystoma*	13.4	0.075	14
frog *Acris*[49]		0.035 (April)	15
		0.100 (May)	15
Rana	32	0.055	15
Bufo	61	0.055	15
Necturus	129	0.012	15
Amphiuma	213	0.007	15
Rana tadpoles		0.065	20
Fishes			
mudskipper *Periophthalmus*[69]	10–15	0.084 (in S.W.)	
		0.065 (in F.W.)	
goldfish[18]	90	0.042	20
brook trout[18]	112	0.205	20
	64	0.150	15
white sucker[18]	70	0.109	15
brown bullhead[18]	147	0.092	20
carp[18]	146	0.067	20
bullhead *Ictalurus*	43–127	0.07	
Antarctic *Nothenia*[170]	200	0.056	0
Chaenocephalus[170]		0.063	0
Trematomus[170]		0.059	0
Lampetra	37	0.098	16
ammocoetes *Ichthyomyzon*[80]		0.09	22
		0.022	9.5
Gadus	1000	0.03	3
Anguilla	40	0.088	17
Carassius	33	0.077	15
Cyprinus	74	0.09	14
Tinca	30	0.067	18
Squalus	440	0.038	15
Scyliohinus	149	0.060	14
Squalus		0.042	
Mugil		0.065–0.1	
Lepomis	30	0.06	20
Girella	70	0.131	20

TABLE 5-1. Standard Metabolism (*Continued*)

Animal	*Body Weight grams*	$\dot{V}_{O_2}$ *ml O_2/g/hr*	T_A, °C
Insects and related arthropods			
Chironomus[166] larva	0.38×10^{-3}	4.75	30
propupa	0.44×10^{-3}	4.7	30
pupa	0.34×10^{-3}	5.5	30
Collembola[225] (series)	2.14 to 7.45×10^{-3}	0.096	18
Tetrodentophora[225]	85 to 140×10^{-6}	0.681	18
thysanuran *Lepisma*[225]	0.01 to 0.013	0.141	18
oribatid *Eugetes*[57]	0.32×10^{-3}	0.161	18
Tenebrio larva		0.182	12
Spiders			
Achaearanea[5]	0.073	0.356	
Phidiphor[5]	0.337	0.147	
Phidippus[5]	0.568	0.093	
Filistata[5]	0.571	0.05	
Lycosa[5]	0.970	0.097	
Isopods			
Porcellio[218]	0.082	0.096	
Armadillidium[218]	0.085	0.076	
Syspastus[218]	0.120	0.012	
Crustaceans			
gammarid *Orcomella*		0.176	17
Uca pugilator[195]	2.39	0.08	28
Uca pugilator and *U. pugnax*[195]		0.04 (in air)	
Uca pugilator[195]		0.06 (in water)	
Uca pugnax[195]		0.10 (in water)	
crayfish *Orconectes*[126]	14	0.126–0.15	
Callinectes[115]		0.097	
Carcinus[115]		0.094	
Libinia[115]		0.037	
Maja[115]		0.054	
Homarus americanus[196]	189	0.025	10
Homarus vulgaris[196]	352	0.04	15
Pachygrapsus	10	0.035	16
Balanus	15	0.11	12
Emerita		0.11	20
Talorchestia[225]	0.27	0.176	
Molluscs			
Ancylus[21]	0.02	0.177	16
Lymnaea[21]		0.24	
Mytilus		0.076	15
Octopus		0.28	25
Pecten		0.67	
Anodonta		0.002	10
Echinoderms			
Asterias	10	0.4	
Parastichopus	50	0.006	
Pteraster[108]		0.0105	11
Pisaster[108]		0.0105	11
Annelids			
Clymenella torquata[144]	0.05	0.294	
Clymenella mucosa[144]	0.109	0.265	
Clymenella zonalis[144]	0.023	0.324	
terebellid *Eupolymnia*[39]	3	0.09	
terebellid *Thelepus*[39]	3	0.06	
oligochaete *Lumbricus*[225]	1	0.065	
Arenicola	1	0.062	

(*Table 5-1 continued on following page.*)

TABLE 5–1. Standard Metabolism (*Continued*)

Animal	*Body Weight grams*	$\dot{V}O_2$ *ml O_2/g/hr*	T_A, °C
Other worms			
cestode *Schistocephalus*[40]	0.01	3.0	20
echiuroid *Ochestoma*		0.069	
echiuroid *Bonellia*		0.048–0.056	
Eurechis		0.012	
Ascaris		0.50	
Protozoans			
Bresslaua	0.015 μl	7.3	
Tetrahymena		2.85	
Paramecium	0.6 μl	1.3	
Pelomyxa	10 μl	0.5	

cle tone varies with level of anesthesia. "Routine" metabolism refers to oxygen consumption measured with uncontrolled but minimum motor activity. Measurements may be made during enforced activity, maximum levels being called "activity" metabolism. When different degrees of activity are used, oxygen consumption may be extrapolated to zero activity; this extrapolated value is "standard" or "rest" metabolism. Table 5–1 lists standard metabolisms for a number of animals.

When standard and activity metabolisms are measured as a function of some environmental parameter, such as temperature, the difference between the two curves is the "scope for activity" (Fig. 5–13). When the two measurements are made at different oxygen pressures, the activity metabolism remains constant down to an incipient limiting level; the standard metabolism is constant to a lower PO_2, which is the level at which there can be no excess activity and below which the needs of maintenance cannot be met. Activity and standard metabolisms involve somewhat different enzymatic pathways.[63]

In man, oxygen consumption in exercise may be fifteen to twenty times the resting value; in insects, flight may increase consumption by fifty to two hundred times. In trout, active metabolism is four times the standard at optimal temperature. Bees and butterflies have been shown to consume 90 liters O_2/kg/hr in sustained flight, or fifteen times the body volume in oxygen every minute.[121] Other data for insects follow:

	Rest	*Flight*
butterflies	0.4 to 0.7 ml O_2/g/hr	40 to 100 ml O_2/g/hr
Drosophila	1.68 ml O_2/g/min	21 ml O_2/g/min
Schistocerca	0.63 ml O_2/g/hr	10 to 30 ml O_2/g/hr

In a moth, *Hyalophora cecropia*, the O_2 consumption associated with heat production during activity increases as T_A is reduced, but the $\dot{V}O_2$ during inactivity decreases with lowered T_A.[76] Hummingbirds consume four to six times more oxygen when active (hovering) than when at rest, and six times more at rest than in torpor at the same thermoneutral temperature, as follows:[127, 132]

	O_2 consumption (ml/g/hr)		
	HOVERING	AT REST	IN TORPIDITY
Calypte	42.4	10	0.17
Selasphorus	85	14	

Similar differences between active and rest metabolism exist for larger birds as follows:

	$\dot{V}O_2$ (ml O_2/g/hr)
budgerigar	
resting	4.5–3.3
flying	21.9
pigeon	
resting	0.89
flying	11.9

The oxygen consumption of a series of mammals (body weight 21 g to 18 kg) increased linearly with running speed, but in different proportions in such a way that the minimum cost of running a given distance is greater for small species than for large ones.[193] In goldfish, the efficiency in net O_2 consumption at different swimming velocities increased with velocity according to the equation:[187]

$$E = 0.031\ V^{1.65}$$

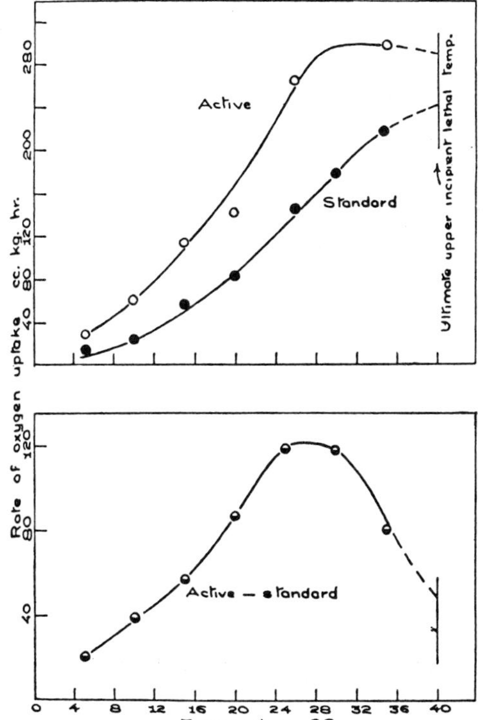

Figure 5–13. Rate of oxygen consumption of young goldfish in activity and under standard conditions at various temperatures. Lower curve gives difference between active and standard metabolism, i.e., energy available for work. (From Fry, F. E. J., Publ. Ontario Fish Res. Lab. *68*:1–52, 1947.)

Increased O_2 consumption need not be proportional to the increased activity, and animals differ significantly in their ability to use energy from anaerobic routes and to incur oxygen debts. In *Iguana*, the anaerobic production of lactic acid provides 3/4 of the energy of activity.[152] The contribution of muscle to total metabolism increases in man from 20% in basal condition to 80% in exercise.

Nonmuscular activity is also reflected in metabolic increase, as in reproductive glands and related structures during a reproductive season. *Mytilus* consumes twice as much oxygen in June and July as in winter. Excitement may markedly increase O_2 consumption without overt muscular activity in fish, and at high swimming speeds the pattern of movement may change to provide increased speed at constant O_2 consumption.[152] Evidence from metabolic inhibitors suggests different metabolic routes for rest and activity. In frog muscle and nerve, the metabolism of activity, but not that of rest, is inhibited by azide. In *Gammarus*, the Q_{10} of standard metabolism is 1.4, while that of routine metabolism is 2.4.[74] (See Chapter 9, p. 363 for the definition of Q_{10}.)

Active fish such as darters have several times higher metabolic rates than sluggish fish. This difference is reflected at the tissue level: excised brain of active menhaden and mullet had a V_{O_2} (ml O_2 /kg/min) of 11 to 14; brain from less active sea bass and croaker, 7.7 to 9.7; and sluggish flounder and toadfish, 5.6 to 6.9.[206]

Temperature and Seasons. The effects of temperature on metabolism are discussed in detail in Chapter 9. In general, homeothermic animals have a range of minimal metabolism at neutral temperatures, and their O_2 consumption rises both when body temperature rises and in the cold. In most poikilotherms the metabolism rises and falls with body temperature by about two and one-half times per 10°C in the physiological range. The Q_{10} of standard metabolism in *Iguana* was 2.3; that of post-activity metabolism was 1.8.[152] Standard metabolism of fish increases continuously with temperature up to lethal levels; active metabolism may either increase to a plateau or pass through a peak at a temperature above which the fish are incapable of increased work (Fig. 5–14a). The difference between active and standard metabolism, or the scope for activity, increases to a maximum and then decreases. Both scope and motor activity are maximal at approximately the same temperature (Fig. 5–14b).[26] Aquatic poikilotherms that remain active over a wide range of temperatures show acclimation; that is, individuals acclimated to cold tend to have higher metabolic rates at a given temperature than individuals acclimated to high temperatures, although at the temperatures of acclimation those in the cold metabolize less than do those in warm water. The net effect is to compensate for environmental changes with season and latitude (see pp. 373–376).

An activity peak occurs for brown trout at 15°C and for carp at 25°C, but no such peak occurs for bullheads.[18] Active bony fish generally show good temperature compensation; hagfish show none but become lethargic in the cold.[154] In two freshwater snails, the V_{O_2} is 40 per cent lower in August and September than in June.[21] The crab *Hemigrapsus* in low salinity has higher metabolism after summer acclimatization than in the winter, whereas in high salinity there is no acclimation or else the winter animals have slightly

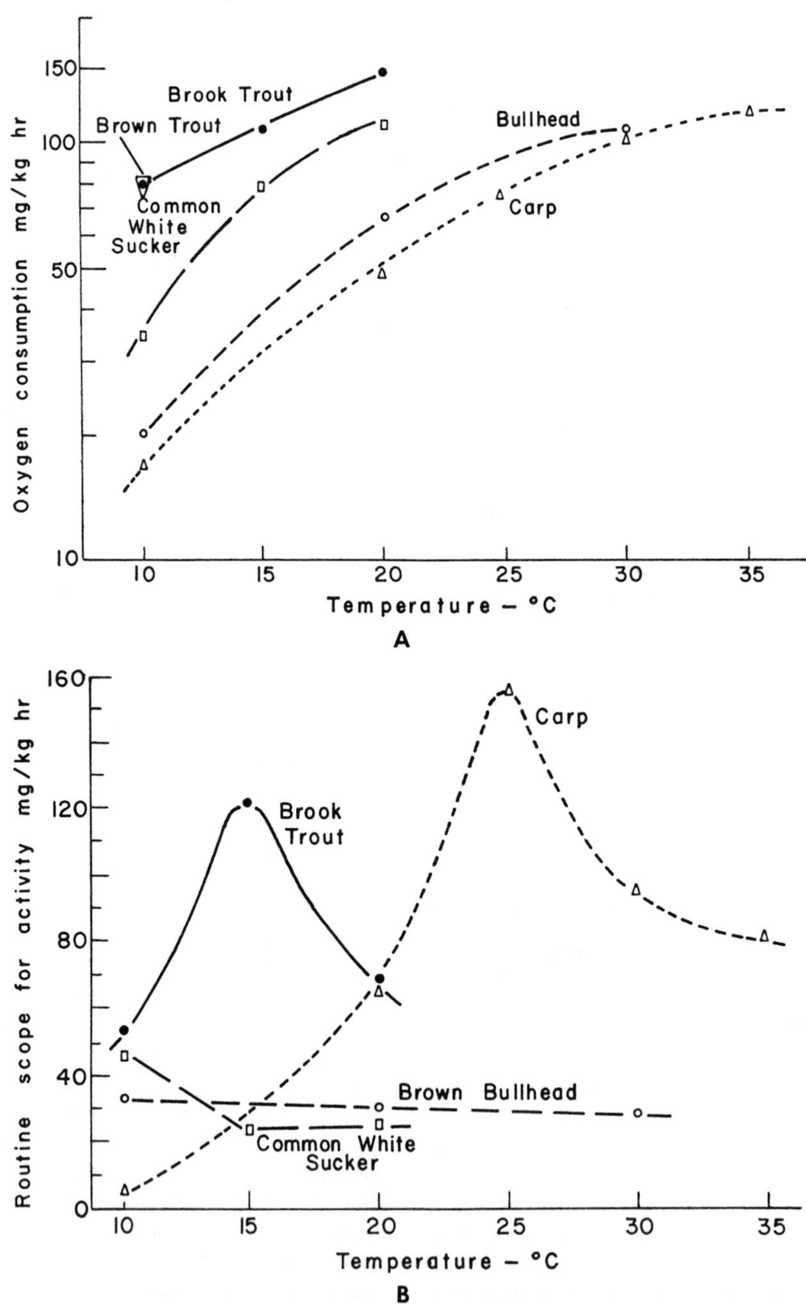

Figure 5-14. A, Standard O_2 consumption by 100 g specimens of several fish as a function of temperature. B, Influence of temperature on scope for activity (difference between active and standard metabolism) in freshwater fish. (From Beamish, F. W. H., and P. S. Mookherjii, Canad. J. Zool. *42*:177–188, 1964. Reproduced by permission of the National Research Council of Canada from the Canadian Journal of Zoology, Vol. 42, pp. 177–188, 1964.)

higher metabolism.[47] *Gammarus oceanicus* shows a Q_{10} of 2.4 for routine metabolism and 1.4 for standard metabolism; mitochondria reflect the Q_{10} level of standard metabolism.[74]

The oxygen consumption of many aquatic animals becomes reduced in high CO_2 concentrations. In speckled trout, the reduction in active metabolism at elevated CO_2 is greater at 10°C than at 20°C, but in bullheads the effect of CO_2 is similar at both temperatures.[16]

Salinity. For aquatic animals of a given kind, the V_{O_2} of freshwater species tends to be higher than that of marine species. Those species that are euryhaline (e.g., *Mytilus edulis*) consume more oxygen in dilute brackish water than in sea water; stenohaline animals (e.g., *Asterias*) consume less. Prawns from a marine environment showed maximum V_{O_2} at 25% S.W.; those from a brackish environment had a maximum metabolic rate in tap water. Metabolism of *Artemia* in sea water is 25 per cent higher than in concentrated brine.[68] V_{O_2} of *Artemia* reared in low salinity may be twice that for individuals reared in sea water.[68] When individuals were transferred from sea water to reduced salinity, V_{O_2} was enhanced in *Carcinus* (up 33% in 50% S.W.), *Callinectes* (up 53% in 50% S.W.), *Eriphia, Gammarus duebeni, Uca, Hemigrapsus,* and *Potamon*;[115] it was reduced in *Maja* and *Libinia* (down 33% in 50% S.W.).[115] V_{O_2} is minimal in brackish water for *Ocypode, Palaemonetes,* and *Metapenaeus,* whereas V_{O_2} is constant at all salinities for *Eriocheir.* Oxygen consumption by gills is higher at low salinity in *Mercenaria* and *Modiolus*; it is similar at both high and low salinity in *Crassostrea* and *Mytilus*; and it is lower at low salinity in *Dreissensia.* In general, metabolism is increased at reduced salinity in euryhaline species, but there are exceptions.

Photoperiod and Rhythms. Many animals show diurnal and lunar rhythms of oxygen consumption, which persist under laboratory conditions that are constant with respect to light, temperature, and pressure. Circadian and other rhythms are discussed in Chapter 10.

In *Uca,* oxygen consumption is 3 to 50 per cent higher at 0600 than at 1800, and in a lunar cycle V_{O_2} is 30 to 50 per cent higher when the moon is at either its highest or lowest transit than when it is at the horizon.[211] An intertidal isopod shows higher V_{O_2} at times corresponding to high tide than at low tide.[218] *Diadema* from moderate depths showed no diurnal fluctuations.[138]

Photoperiod can also affect metabolism. *Pachygrapsus* on an 8-hour photoperiod consumed 55 per cent more oxygen than when it was on a 16-hour day.[47] Oxygen consumption in sunfish was higher at all temperatures above 10°C in individuals on a 9-hour photoperiod than in those on a 15-hour photoperiod. In a nocturnal mammal, *Rattus,* succinic oxidase of liver is 40% more active when the tissue is taken at night than when it is taken during the day, and P/O ratios are higher.[68a]

Hormones. Metabolic differences between the sexes are widespread, and are not always in the same direction. Human males at all ages have higher metabolism than females, but the difference is less on a fat-free basis. The male housefly, *Musca,* has a higher V_{O_2} than the female, but the difference diminishes in the reproductive phase. In male adult silk moths, V_{O_2} is three times higher than in the females at high but not at low temperatures. A male waxmoth, *Galleria,* consumes more oxygen than the female during early pupal development; in later development the situation is reversed. In *Tribolium* and in some intertidal isopods, the metabolism of females is greater than that of males. Excised muscles from male cockroaches consume five times more O_2 than muscles from females.[15]

Many of the effects of physical factors on metabolism are mediated by hormones. The sinus gland of crustaceans may regulate metabolism indirectly; removal of eyestalks from crabs increases their oxygen consumption, and injection of eyestalk extract into eyestalk-less crabs decreases their V_{O_2}. Crustacean metabolism increases prior to molting. In adult blowflies, *Calliphora,* removal of the corpora allata decreases V_{O_2}, and implantation of the gland increases it.[198]

Thyroid extracts increase metabolism in amphibians, but the thyroid seems not to affect metabolism in fish or lampreys. Gonadal steroids may increase fish metabolism. When lizards (*Eumeces*) were injected repeatedly with thyroxine over a period of three weeks, the nonmuscular tissues (especially liver and brain) showed increased V_{O_2}.

Development. Stage in development or life cycle may affect metabolism independently of body size. In general, old animals have lower metabolism than young ones, but embryos may respire less and a maximum may occur late in development. A mammalian fetus has a lower V_{O_2} than a newborn:[79]

O_2 Consumption by Human Babies	
30–36 hours old	0.316 ml O_2/g/hr
640 hours old	0.282 ml O_2/g/hr
adults	0.21 to 0.24 ml O_2/g/hr

The metabolic peak occurs in man at about the time of weaning, in cattle between weaning and puberty, in swine at puberty, and in horses when mature.[28] In chickens a peak occurs four weeks after hatching.

In *Rana*, V_{O_2} increases abruptly at the end of gastrulation; in *Ambystoma* this occurs just before establishment of circulation.[24] The V_{O_2} of sea urchin eggs (*Arbacia*) increases seven-fold on fertilization.

Holometabolous insects show a gradual increase in metabolic rate during larval development and a drop early in pupal life, followed by a rise prior to emergence as an adult; this U-shaped metabolic curve is found in many insects (*Tenebrio, Drosophila, Phormia*) (Fig. 5–15). A chironomid pupa shows as high metabolism (5.5 mm³ O_2/mg/hr) as does the larva (4.75 mm³ O_2/mg/hr).[166] Hemimetabolous insects show a rise in V_{O_2} just after each molt and a decline during the intermolt; genera showing this behavior include the milkweed bug, *Oncopeltus*, and the blood-sucking bug, *Rhodnius*.

In Protozoa, the metabolism tends to decline as cultures mature and age. On encystment, the V_{O_2} of a ciliate may drop by 98 per cent.

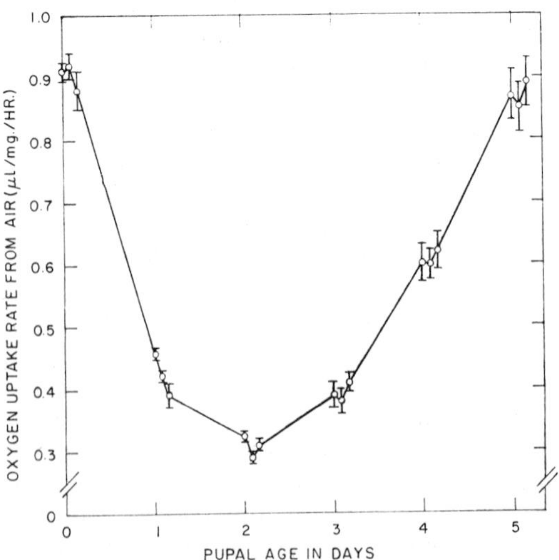

Figure 5–15. Oxygen consumption by pupae of blowflies at different pupal ages. (From Park, H. D., and J. Buck, J. Insect. Physiol. *4*:220–228, 1960.)

Body Size. The rate of oxygen consumption is sometimes given in volume of O_2 per animal, more often volume of O_2 per gram or kilogram of body weight (wet or dry), rarely per unit of nitrogen content, and often as some exponential function of body weight. The symbol Q_{O_2} refers to ml O_2 (at STP) consumed per gram dry weight per hour; V_{O_2} refers to the same quantity per gram wet weight (or volume) of tissue.

When large and small adults of a species, or differently sized species of the same general type of animal, are compared, it is found that the total metabolism (ml O_2 per animal) is higher in the larger animals but that the metabolic rate (Q_{O_2} or V_{O_2}) is higher in the smaller animals. In general, metabolism is more uniform when expressed as a power function of body size, and many papers have discussed what measure of body size gives most consistency and meaning.[20,116,117] If M = total O_2 consumed per unit time and W = body weight, then the power function is expressed as

$$M = K W^b$$

and

$$\log M = b \log W + \log K.$$

The constant b is obtained from the slope of a plot of the logarithm of oxygen consumption against the logarithm of body weight; K is obtained from the intercept. If oxygen consumed per unit time per unit weight is desired, the function is

$$\log \frac{M}{W} = (b - 1) \log W + \log K.$$

K determines the position of the metabolism-weight curve on the metabolism axis (i.e., when $W = 1$), and is sometimes called the weight-independent metabolism. The constant b gives the rate at which O_2 metabolism changes with size; if metabolism is directly proportional to weight, then $b = 1$. Actually, b is usually less than 1.0. This means that in similar species or during growth, metabolism increases less than does body mass. Similar exponential allometry applies to appendage lengths and to body surface area as functions of body weight.

Extensive studies have related metabolism

as heat production (M) or as oxygen consumption (Vo_2) to body size as follows (W in kg):[117]

$$M(\text{kcal/day}) = K\,W^{0.74}.$$

For some 36 kinds of eutherian mammals, $K = 70$ and

$$M = 70\,W^{0.74}$$

and

$$\log M = 0.74 \log W + 1.845$$

For an animal weighing 1 kg, metabolism is $M = 70$ kcal/day. To convert from heat equivalent to O_2 consumption, the following relations are used:[153]

Taking 4.8 as the kcal equivalent of 1 liter of O_2, converting to ml O_2 and from 24 hours to one hour, for $W = 1000$ g:

$$Vo_2 = \frac{70\ \text{kcal/day}}{4.8 \times 24} \times 1000 = 604\ \text{ml}\ O_2\,/\text{g/hr}$$

To obtain K, we substitute into the oxygen-weight relation:

$$604 = K(1000)^{0.74} = K \times 158$$

from which $K = 3.8$ and

$$Vo_2\ (\text{ml/hr}) = 3.8\,W^{0.74}.$$

In most of the published data for birds and mammals, O_2 consumption is converted to kcal/day. The K values set the level of the metabolism-weight (MW) curve, but K varies with a number of conditions and is necessarily an extrapolated value. The exponent b is important as a means of relating metabolism to body size.

Table 5–2 presents selected examples of metabolism as a function of body weight. Other data are found in the Biological Handbook.[2]

In mammals, surface area is given by a constant times $W^{0.66}$, whereas oxygen consumption for different animals is nearly constant when expressed per $W^{0.75}$. This is illustrated by the famous mouse-to-elephant curve (Fig. 5–16). It has been argued that O_2 consumption is closely correlated with surface area because of the greater heat dissipation in small than in large mammals; if a whale produced as much heat per gram as a mouse, its interior temperature would approach the boiling point. This is an oversimplification; heat is lost by lungs and avenues other than the body surface; some mammals function with much of their body surface at low temperature (approaching ambient). Body temperature is nonadaptive in that it does not vary with climate, and metabolism follows the same function at different environmental temperatures. The rate of O_2 consumption rises so steeply that it would be virtually impossible for a mammal weighing less than 3.5 g to obtain sufficient food when active. Many mammals and birds show a b value of 0.75 but have a low K; hence, their M vs. W curve lies below the theoretical. For a few, K is high and the curve lies above the theoretical one. The value of K, and hence the standard O_2 consumption, is some 25 to 30 times higher for average homeotherms than for poikilotherms of equal size[78] (Fig. 5–17).

In hibernating mammals, the O_2 correlates with body weight better than with body surface area. In newborn human infants, metabolism is proportional to weight ($Vo_2 = 7.2\,W$), but the exponential relation develops so slowly[79] that at 10 to 12 kg (about 18 months) $Vo_2 = 20\,W^{0.6}$. Dasyurid marsupials show a b value of 0.74 but a low K value, so that their metabolism-weight line is 32.1% below that of placental mammals.[141] A large number of passerine birds show a b value of 0.724 and a K of 129, whereas in non-passerines b is 0.723 but $K = \log 78.3$.[131] Hence, passerines have a higher weight-specific metabolism than non-passerines. A canyon mouse (*Peromyscus crinitus*) had a weight-specific metabolism of only 59% of the predicted value, and a mole rat (*Heterocephalus glaber*) had a metabolism that was 18% of what would be expected for a mammal of its size.[149] In contrast, *Microtus montanus* has a metabolism 75% higher than predicted. In mammals, a high metabolism (and heat production) for a constant body temperature and given size implies either high capillary density, especially in the periphery, or high unloading Po_2 for hemoglobin. The correlation of weight-specific metabolism with P_{50} for hemoglobin is good (see Chapter 8).

That more than heat relations is relevant to the exponential function of metabolism is shown by the good logarithmic relation between O_2 consumption and weight in poikilotherms. In an extensive survey, Hem-

TABLE 5-2. Effect of Body Size on Metabolism

Animals	*Size Range*	*b*	*K*
Mammals			
series of 36 species[116,190]		0.75	70 kcal/kg/day 3.4 ml O_2/g/hr 105 ml O_2/m^2/hr
8 Australian marsupials[43]	16–28700 g	0.74	2.56 ml O_2/g/hr 70 kcal/kg/day
dasyurid marsupials[141]	7.2–5050 g	0.74	2.45 ml O_2/g/hr
megachiropteran bats[11]	18–600 g	0.73	3.8 ml O_2/g/hr
Heterocephalus (rodents)[149]		0.75	3.4 ml O_2/g/hr
Birds			
non-passerines[131]		0.72	78.3 kcal/kg/day 4.3 ml O_2/g/hr
passerines[131]		0.724	129 kcal/kg/day 7.0 ml O_2/g/hr
hummingbirds in torpor[132]		0.848	
Reptiles (for poikilotherms, *K* is in ml O_2/g/hr)			
series of lizards[14]	16–11000 g	0.62	0.234 (30°C)
series of skinks[87]	1–1.3 g	0.63	0.3 (30°C)
tropical snakes[65]		0.86	
tropical boids[65]		1.09	
hylid frog *Acris* (May)[49]		1.0	0.088
Testudo		0.71	0.215
Testudo[90a]	118–35000 g	0.97 (active) 0.82 (rest)	140 (active) 45 (rest)
Fish			
salmon *Oncorhyncus*[26]	3–1400 g	0.97 (active) 0.78 (standard)	
goldfish[187]	10–400 g	0.85	
goldfish[18]		0.86	
Gadus[18]		0.70	
starry flounder[18]		0.86	
Squalus[18]		0.74	
lingcod[18]		0.78	
Antarctic *Notothenia*[170]	30–2000 g	0.785	0.235
ammocoetes of lamprey[80]	0.14–3.49 g	0.72	0.054
Insects			
Hymenoptera series[113]	12–1200 mg	0.92	
Coleoptera series[113]	100–1000 mg	0.77	
moth larvae[182]	0.5–29 g	0.87	3.2 ml O_2/g dry wt/hr
Collembola[225]	0.1–7 mg	1.0	
Chironomus larvae[50]	0.89–12.8 g	0.70	
Crustaceans			
Ligia oceanica	0.04–1.03 g	0.73	
Antarctic gammarids[5a]		0.42 (2°C)	
Balanus	1.2–80 g	0.66	
Artemia[68]		0.66 (female in S.W.) 0.88 (male in S.W.)	
Homarus[147]		0.88	
Uca pugnax[195]		0.68	
Uca pugilator[195]		0.77	

TABLE 5–2. Effect of Body Size on Metabolism (*Continued*)

Animals	*Size Range*	*b*	*K*
Molluscs			
Patella[39a]	2–15 g	0.696	
Lymnaea[48a] (lab reared)	0.014–3.6 g	1.0	
shedding parasites		0.77	
Mercenaria[140]		0.66	0.11 ml O_2/g/hr
F. W. snail *Theodopus*		0.87	
prosobranch snails		0.67	
Brachidontes[172a]		0.7	
Mytilus	0.43–3 g dry wt	0.65	
F.W. limpet[21]		0.73	
Echinoid			
sea urchin[152a]		0.7 (winter)	
		0.65 (summer)	
Worms			
Schistocephalus[40]		0.52	
Clymenella[144]	7.6–22.3 mg	0.48	
Miscellaneous			
soil nematode[78]		1.0	
small poikilotherms; larvae[78]		1.0	
eggs, Protozoa, bacteria[78]		0.756	

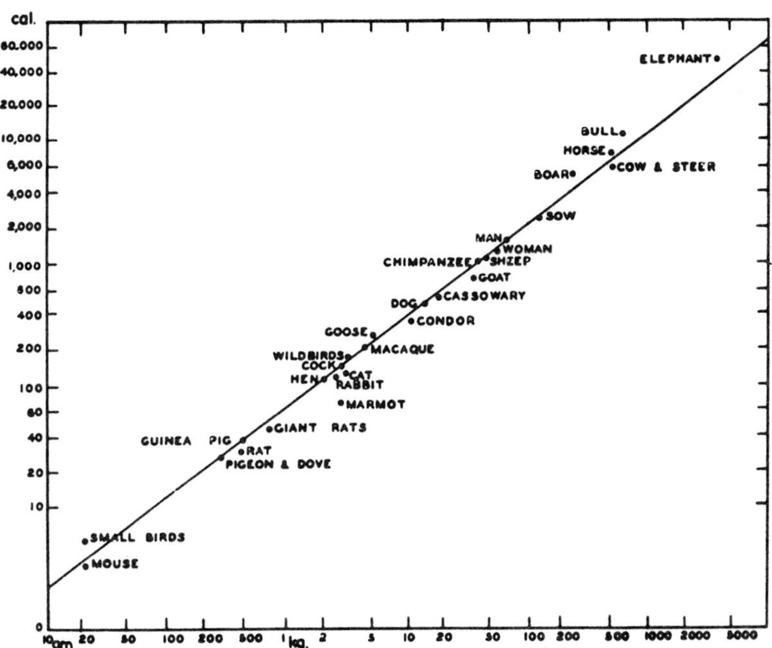

Figure 5–16. Double logarithmic plot of average total heat production and average body weight of birds and mammals. (From Benedict, F. G., Carnegie Inst. Washington Publ. *503*:1–215, 1938.)

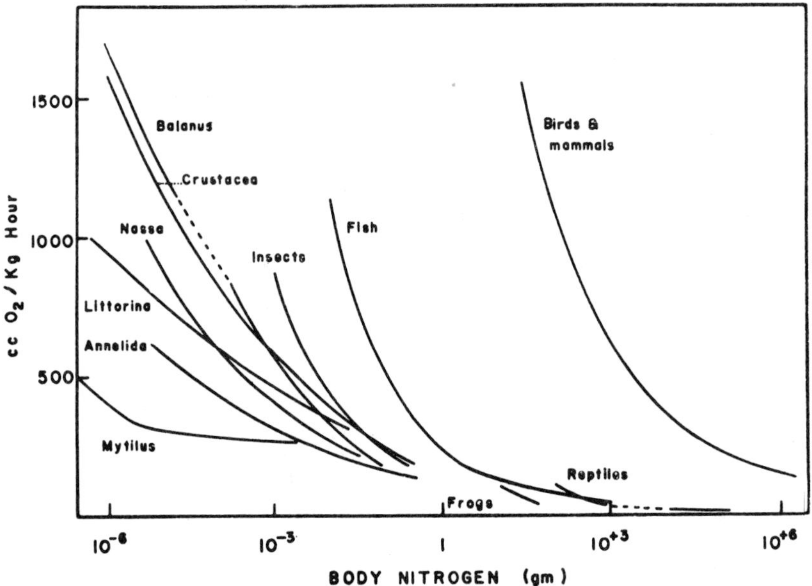

Figure 5–17. Comparison of metabolic rates of various animals as functions of body nitrogen. (From Zeuthen, E., C. R. Lab. Carlsberg, Ser. Chim. *26*:17–161, 1947.)

mingsen[78] concluded that $M = K\ W^b$ with b not far from 0.73 in poikilotherms, homeotherms, and beech trees! Table 5–2 includes b values for many invertebrates. Among unicellular organisms, bacteria have a very high metabolism, some as high as 100 to 500 liters O_2/kg/hr. Protozoa consume less than this; e.g., *Pelomyxa* is 10^7 to 10^8 times larger than the bacteria *Salmonella* and *Escherichia*, and consumes only 0.2 liter O_2/kg/hr.[224]

Small multicellular poikilotherms—microcrustaceans, lower worms and molluscs—have steeper log-log plots (Fig. 5–18). Soil nematodes give a b value of 0.9, and a number of invertebrates weighing about 40 mg have an average b of 0.95.[222, 223] In larger poikilotherms, the slope is less steep: in small crustaceans b was 0.95, and at about 40 mg body weight and larger it was 0.80. *Pachygrapsus* of a wide size range gave a b of 0.66; *Mytilus*, 0.67; *Artemia* from sea water, 0.88; *Artemia* from brine, 0.62. Cockroaches, walking sticks, and tarantulas had a b of 0.8.[52] In some holometabolous insects b approaches 1, while in hemimetabolous insects the values are lower.[50] Values of b may vary with temperature; for the tropical fish *Etroplus*, b is 1 at 0°C and drops to 0.67 at 35°C. For several gobiids, b was higher in the range 10 to 17°C than in the range 21 to 31°C.[10a] In a series of snakes, heat production is not strictly proportional to surface area; in 50 species C = 0.34 $W^{0.86}$.[65] In the skink *Lygosoma*, b was 0.63 at 30°C and was 0.4 at 18.5°C.[87]

Intraspecific measurements indicate that in very early development (embryos less than 40 mg net weight) b values are 0.7 to 0.8; later in development, values of 0.9 to 1.0 are found, and in larger mature animals the values decline again. These three-step curves have been observed for *Mytilus, Artemia, Asterias,* and amphibians. In the salmon *Oncorhyncus*, b for rest metabolism is 0.78 and for activity it is 0.97.[27] For rest metabolism, K is 0.37; for activity K is 4.29. The b value in *Mercenaria* is the same for metabolism as for rate of water pumping.[140] In the sea urchin, *Eucioraris*, b is 0.70 in winter and 0.65 in summer. In the frog, *Acris*, values of b are high in early spring and lower after the breeding season.[49]

In summary, in unicellular organisms, oxygen consumption is nearly proportional to cell surface and size may be limited by the surface-to-volume ratio. Multicellular animals, even of the size of Protozoa, have higher metabolism, and O_2 consumption is no longer limited by surface area; enzymes increase with body weight. At larger sizes (above 40 mg) enzymes increase less in proportion than body weight, and may even decrease as maximum size is reached. Zeuthen points out that animals can either be small, grow rapidly, and live intensely, or be

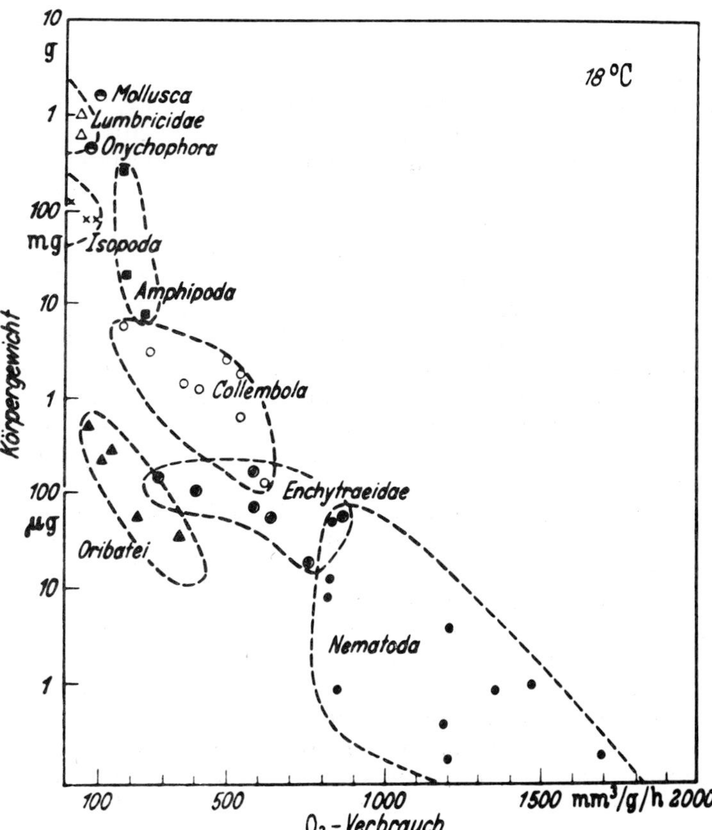

Figure 5-18. Rate of oxygen consumption (mm^3/g/hr) as a function of the logarithm of body weight for numerous species of small invertebrates. (From Zinkler, D., Z. vergl. Physiol. *52*:99–144, 1966.)

large, grow slowly, and metabolize at a low rate. To what extent these relations can be explained by cell surface areas is uncertain, but the metabolism-size relation may be better related to respiratory surface and oxygen transport.

Mechanisms of Size-Metabolism Regulation. One factor in the decreased metabolic rate with increased body size is the disproportionate increase of tissues of low metabolic rate—skeleton, fat, and connective tissue. Because of different relative weights of various organs, the rate of oxygen consumption of tissues need not be expected to change with the same exponent of body weight as the total metabolism.

The metabolism-size relation implies higher activity of oxidative enzymes in corresponding tissues of small animals. Tissues from nine mammalian species showed a limited correlation in the proper direction: metabolic rate of an intact horse was 11% of that of a mouse, yet the horse brain and kidney rates were 47% of those in the mouse. The liver Q_{O_2} of the horse was 13 to 23 per cent of that of the mouse, yet liver Q_{O_2} was similar in guinea pig, cat, and dog.[120] In *Salmo clarki*, values of b of 0.85 for gill and kidney were obtained; these are higher than the total metabolism. For the crab, *Hemigrapsus*, the O_2 consumption by isolated gills was 170 µl/g for gills from a crab weighing 2.4 g and 110 µl/g for gills from one weighing 12.5 g.[47]

In several mammalian species, a good correlation was found between abundance of mitochondria and total metabolism; the total mitochondrial mass is less in a steer than in a rat, in proportion to $W^{0.77}$.[188] Cytochrome oxidase and malic dehydrogenase increase in a series from large to small mammals, but not in proportion to the increase in metabolic rate.

The relation between animal size and metabolic rate is a complex allometric adaptation. In virtually no organism is metabolism proportional to weight, nor is it proportional to surface area. According to Hemmingsen,[78] most of the deviant values of b can be brought close to 0.74 if corrected for temperature and other factors. However, Hemmingsen does agree with Zeuthen[224] that unicellular organisms have a b close to 0.74, small multicellular poikilotherms have a higher *b* of 0.95, and the value for larger poikilotherms, plants, and homeotherms is 0.74. The second group, small metazoans (marine larvae and some soil nematodes), overlaps the other two but covers the size range of 1 μg to 1 mg.

It was indicated on page 168 that the size of an organism that depends on diffusion for oxygen is limited; for a single cell this means that there is a limit on the total surface area, and hence multicellularity provides for more surface area within an organism of given dimensions. However, attempts to relate the metabolism-size curves to total number of cells or to cell surface area have not been successful, largely because of the great diversity in shapes and sizes of cells.

A second suggestion for animals with gills or lungs is that the proportionality may be with respiratory surfaces. This may be important in some animal groups, but not for all. In homeotherms, the critical factors are rate of heat transfer and circulation of blood to tissues.

The fact that many tissues in vitro show relations similar to the total metabolism of the animal indicates a remarkable cellular adaptation. Per unit of tissue weight (or nitrogen), the activity of oxidative enzymes is higher in tissues from small animals than in those from large animals. However, this applies only in a general sense because of large variations in basal levels between tissues from the same animal. There must be some subtle control system relating the metabolic need of the whole organism to enzyme activity in critical cell masses.

METABOLISM AS RELATED TO ENVIRONMENTAL OXYGEN

Conformity and Regulation. The principles of regulation and conformity with respect to oxygen can be tested by measuring the amount of oxygen made available in the blood at different levels of environmental oxygen. This is done by measuring the percentage saturation of respiratory pigment. Ideally, one needs to know the Po_2 inside cells, at mitochondrial cristae, where oxidative enzymes function. One method has been to measure oxygen concentration by means of an O_2 electrode either at the surface of tissue (for example, on cerebral cortex) or intracellularly (as in muscle). A promising method is the use of fluorescence to estimate the amounts of reduced and oxidized nucleotides (DPN $\rightleftharpoons$ DPNH); this method is limited to favorable tissue preparations. Variation in O_2 supply and utilization in different guinea pig muscles is shown by the following values of intracellular Po_2: gracilis, 6 mm Hg; soleus, 19.7; gastrocnemius, >0. For the cat, Po_2 in the outer layers of the ventricle was 10 mm Hg, and in the inner layers it was 5 mm Hg. Venous Po_2 frequently exceeds that in the tissues; hence, many cells function normally at very low Po_2.[214]

For intact animals, limits of survival in hypoxia may be ascertained and the oxygen consumption measured over a range of environmental Po_2. Many animals regulate their metabolism; i.e., they maintain constant consumption as oxygen is reduced down to some critical pressure (P_c) below which their O_2 consumption declines rapidly. Such animals show a wide range of independence of oxygen. In other animals (conformers), the O_2 consumption is proportional to the environmental O_2 and increases when the oxygen level rises above air saturation toward an atmosphere of pure O_2; they are O_2 dependent. In metabolic regulators the P_c is below air saturation; in conformers the potential use of oxygen is not reached at air saturation, although there may be some leveling off at Po_2 above 155 mm Hg. Some animals show a decreasing slope (i.e., reduced rate of metabolic increase) at Po_2 above air saturation; these are intermediate types. The position of the metabolism-oxygen curve on the Vo_2 axis and the value of the critical pressure are influenced by internal and external factors, and are often correlated with the animal's ecology (Fig. 5–19). Below P_c, oxygen availability limits metabolism; above P_c, the limit is imposed by substrates, enzymes, or cofactors. Values of P_c are not necessarily sharp, but cover a range of Po_2 for the specific conditions.

A list of metabolic regulators includes Protozoa, freshwater and terrestrial annelids, echinoderm eggs, many molluscs and crustaceans, most aquatic vertebrates, some aquatic insects, and probably all terrestrial insects and vertebrates. Oxygen-conforming animals include a few Protozoa, numerous coelenterates, free-living annelids and marine worms, most parasitic worms, some molluscs and crustaceans, adult echinoderms, and a few aquatic insects and vertebrates. Metabolic conformity is particularly common in marine invertebrates, especially large sluggish ones

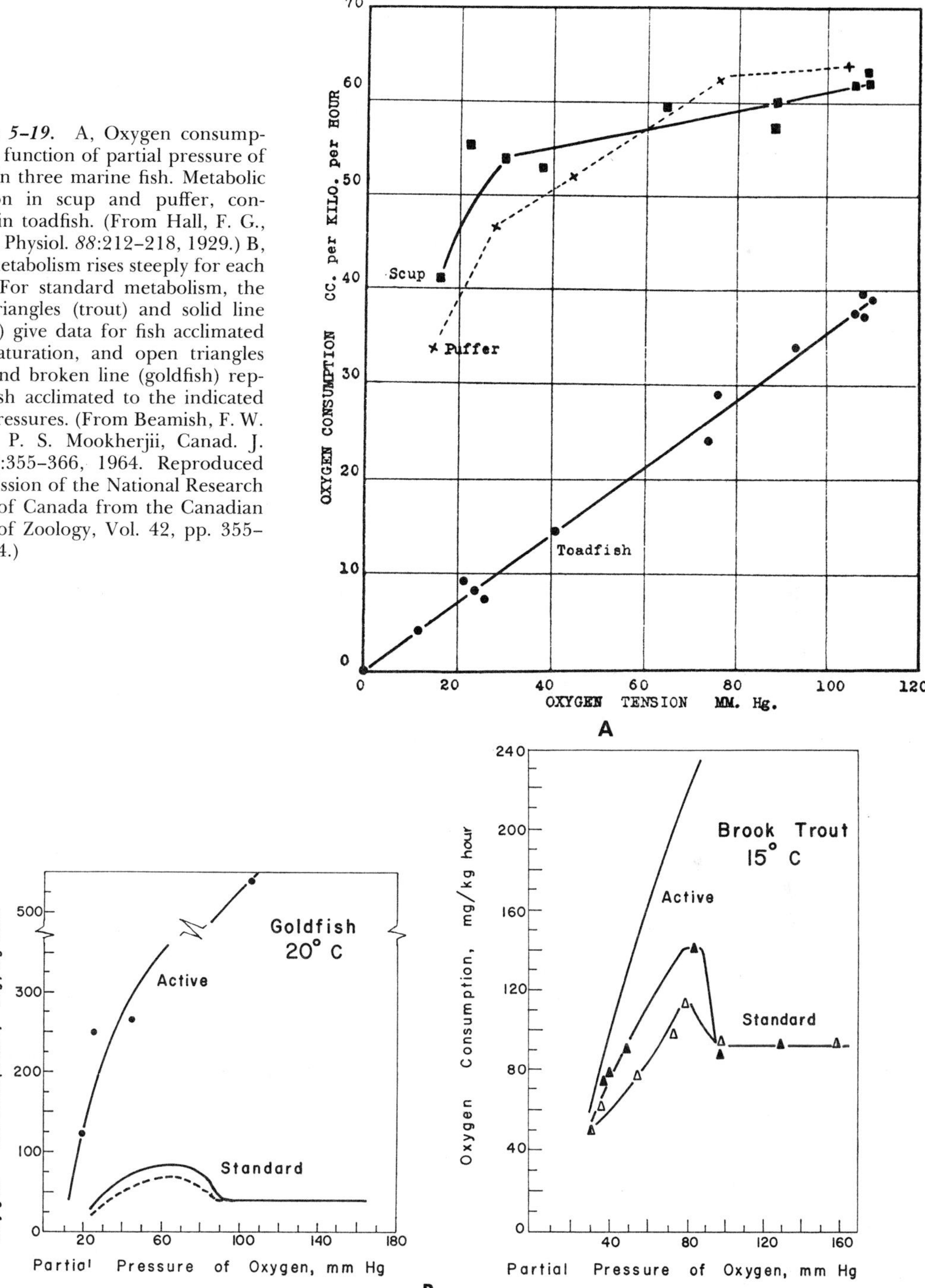

Figure 5-19. A, Oxygen consumption as a function of partial pressure of oxygen in three marine fish. Metabolic regulation in scup and puffer, conformity in toadfish. (From Hall, F. G., Amer. J. Physiol. *88*:212–218, 1929.) B, Active metabolism rises steeply for each species. For standard metabolism, the closed triangles (trout) and solid line (goldfish) give data for fish acclimated to air saturation, and open triangles (trout) and broken line (goldfish) represent fish acclimated to the indicated partial pressures. (From Beamish, F. W. H., and P. S. Mookherjii, Canad. J. Zool. *42*:355–366, 1964. Reproduced by permission of the National Research Council of Canada from the Canadian Journal of Zoology, Vol. 42, pp. 355–366, 1964.)

where diffusion may limit the oxygen supply to tissues. Here the V_{O_2} of tissue slices may be higher than that of the intact organism because of reduced diffusion path length. P_c may be very low where the diffusion path is short (e.g., 2.5 mm Hg in *Tetrahymena*[8]). In some species with circulation, such as crayfish, the larger individuals tend to have a higher P_c than small individuals.

Modifiers of Oxygen Dependence and Mechanisms of Regulation. Activity metabolism is more O_2 dependent than standard metabo-

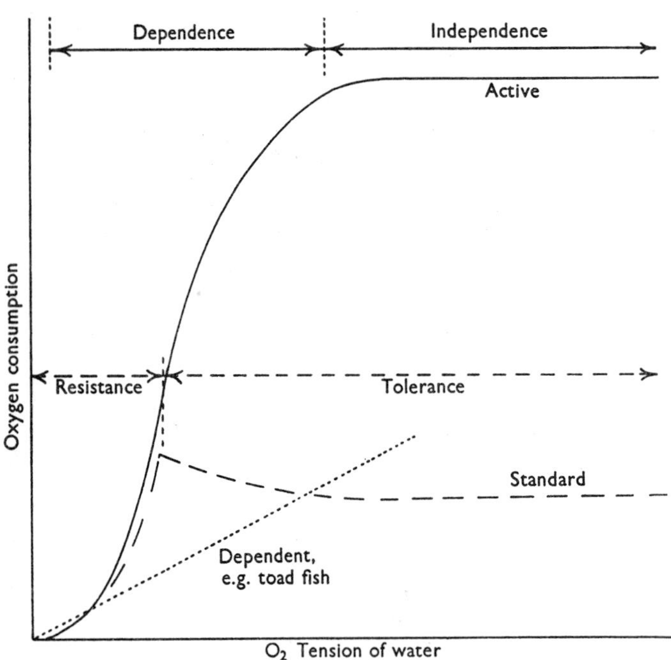

Figure 5-20. Diagram of the relation between oxygen consumption of a fish and oxygen in the water. Solid line, active metabolism; dashed line, resting metabolism, which is divided into zones of tolerance (indefinite survival) and resistance (limited survival). (From Hughes, G. M., Soc. Exp. Biol. Symp. *18*:81–107, 1964.)

lism, and P_c for each is higher at elevated temperatures, as shown in the trout *Salvelinus fontinalis*.[70] Maximum swimming speed of trout parallels oxygen consumption as a function of environmental oxygen (Fig. 5–20). P_c is higher in bluegills in high than in low temperatures, and is also higher when Po_2 is reduced abruptly than when it is changed slowly. Active fish have much higher values of P_c than do sluggish fish (mackerel, 70 mm Hg; tautog, 1.6 mm Hg; goldfish, 80 mm Hg; bullhead, 50 mm Hg). The sluggish toadfish is metabolically dependent and shows no regulation down to zero oxygen.

Animals from high oxygen environments have higher O_2 consumption rates at air levels and higher values of P_c than do animals from low oxygen environments; these properties correlate with survival under anaerobic conditions and represent genotypic differences. In certain aquatic insect larvae, increasing dependence on oxygen is found in a series from stagnant ditches, ponds, and fast streams. The stream species are less resistant to hypoxia, they have lower thermal tolerance, and their O_2 consumption declines more with low environmental oxygen.[209] The marine mysid crustacean *Gnathophausia* lives at oxygen concentrations ranging from 0.2 to 1.25 ml/liter, and regulates O_2 consumption in a milieu of down to 0.26 ml O_2/liter.[34]

Acclimation to low oxygen can change the P_c for an animal down to lower O_2 concentrations. Chironomid larvae become more independent of environmental oxygen after some hours of adaptation to low O_2.[209] Goldfish acclimated for several days to low oxygen (1/6 air saturation) show a lower P_c and lower standard metabolism than before acclimation.[167] The brook trout, *Salvelinus*, showed no change in Vo_2 when tested at acclimation pressures from 160 to 88 mm Hg Po_2; below 80 mm Hg there was a slight increase and then a sharp drop in oxygen consumption. If the trout were acclimated to low Po_2, the O_2 consumption was less than if they were acclimated to air saturation. Similar results were obtained with carp.[18] Apparently, some sort of induction phenomenon occurs.

Mechanical factors modify the oxygen responses of some aquatic larvae. Caddis fly (Trichoptera) larvae may require fast streams, not so much because of O_2 concentration as because of agitation of the water. Ephemerid larvae show higher O_2 consumption and are metabolic conformers when they are free-swimming in a bottle; but in the presence of small pebbles, they consume less O_2 and regulate metabolism down to 1 ml O_2/liter.[53] Pelagic euphausids regulate their metabolism with P_c corresponding to about 5% of an atmosphere, but deep-sea species respire at a relatively constant rate as oxygen is reduced nearly to zero.[194]

An important mechanism of metabolic

regulation and determinant of P_c is external respiration ventilation, and ability to withdraw oxygen from the air or water supplied to the respiratory epithelium. Examples of increased ventilation were given on pages 169–171. The dragonet fish, *Callionymus*, is a metabolic conformer; as Po_2 is reduced, respiratory frequency decreases, heart rate slows, and O_2 consumption is reduced. Oxygen-sensitive receptors are located in the gill region.[94]

Some invertebrates without an active blood pigment are metabolic conformers, while those with blood pigments and insects with tracheae are regulators. A good transport system does not, however, guarantee metabolic regulation. *Carcinus* regulates but *Homarus* and *Limulus* are conformers; metabolic conformity may be related to the fact that the hemocyanin is normally never fully saturated (see page 352). Most transport pigments become saturated at Po_2 levels well below air saturation. After the pigment (e.g., hemoglobin) is inactivated, the P_c may become higher. In an earthworm, the O_2 consumption is normally independent of environmental oxygen down to 3% (22 mm Hg), but when the hemoglobin has been inactivated by CO the metabolism is regulated only down to 8% (60 mm Hg).

Animals that live in low oxygen and can rely on glycolysis for energy tend to be oxygen conformers. In aquatic annelids and insects, life in low O_2 favors regulation and low values of P_c, but parasites, which make more use of glycolysis, tend to be O_2 conforming. Among the metabolic regulating parasites are some blood-dwelling protozoans, *Trypanosoma cruzi, Plasmodium,* and the nematode *Trichinella*, which normally lives in well oxygenated muscle. Examples of metabolic conformers are the trematode *Fasciola*, the cestode *Diphyllobothrium*, and the nematode *Ascaris*. It is doubtful that diffusion is critical, because tissue brei as well as intact *Ascaris* show O_2 conformity, and other nematodes weighing less than 1 mg are O_2 dependent.[25]

Freshwater turtles survive forced submergence for several hours; heart rate is slowed, blood oxygen drops to low levels, and some lactic acid accumulates. If the turtles have been injected with the glycolytic inhibitor iodoacetate (IAA), survival is reduced to less than one hour; normally in submergence the metabolism shifts predominantly to glycolytic.[19, 101, 102]

In general, oxidative enzymes become limiting at low levels of O_2. Measurements of DPN reduction fluorometrically or of O_2 levels by bacterial luminescence are much more sensitive than measurement by O_2 electrodes. Half-maximal reduction of heart cytochrome-C is at an O_2 concentration of 7×10^{-8} M (~0.015 mm Hg); similar values occur for cytochrome reduction in intact yeast cells.[16a] When DPN reduction was followed *in vivo* in rat brain and breathing was shifted from air to nitrogen, cessation of brain activity occurred at 80% reduction of DPN. Half-values for O_2 affinity of the cytochrome chain are 0.05 to 0.5 μM O_2.[32a] As O_2 in extracellular fluid is diminished, a pressure is reached intracellularly at which cytochrome oxidase becomes reduced and glycolysis is stimulated.

Metabolic Correlates with Genetically Determined Life Habits. Most animals, when available oxygen is reduced, make direct responses which permit relative constancy of energy; these are usually short-term responses. Some examples are known of acclimatization to low oxygen—reduced metabolism, and increase in transport pigment. Genetically determined correlates with oxygen availability in different habitats have been mentioned—adaptations to fast or sluggish water and to high altitude. Other genetic correlations relate to locomotor activity and general behavior.

Behavioral correlates are well illustrated by mammals and birds. Primitive mammals—monotremes and edentates—have low metabolic rates and marsupials are intermediate, as compared with Eutheria; yet the size relation (*b* value) is similar for all of them. Metabolic rate of desert rodents tends to be below that of mesic rodents; metabolism of a mole rat, *Heterocephalus*, is 18% of that predicted from the eutherian curve (0.26 ml O_2/g/hr), while *M* of a canyon mouse (*Peromyscus crinitus*) is somewhat higher (0.92 ml O_2/g/hr) but is still 41% below prediction.[149] Yet *Microtus*, a very lively mouse, has an *M* that is 75% higher than predicted. Among birds, a large series of non-passerines showed a weight-independent O_2 consumption (*K* in the *M-W* curve) of 78.3 cal/day, whereas in passerines the corresponding value is 129.[129, 131] A relatively sluggish bird, the poorwill, has O_2 consumption that is only 46% of the predicted value.[128] The large bats, members of Megachiroptera, have metabolic rates similar to those of higher placental

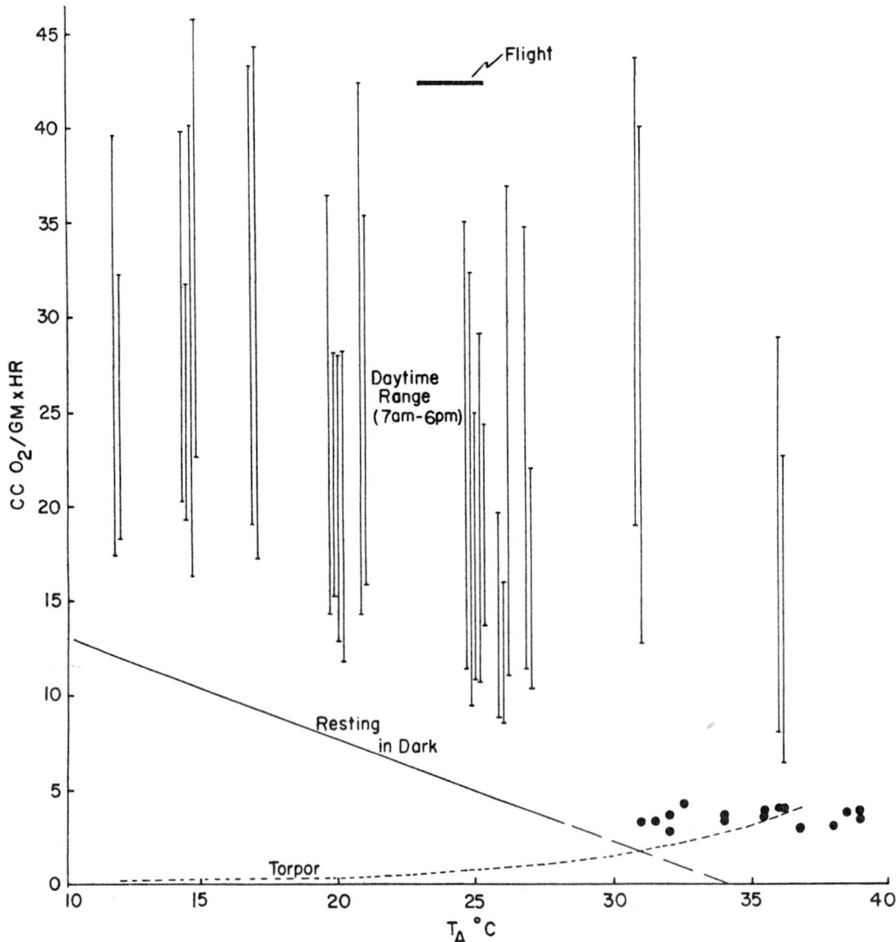

Figure 5-21. Oxygen consumption in hummingbirds at different temperatures in torpid state; resting awake in dark (solid circles give O_2 consumption at T_n); in flight at $T_A = 24°C$. Vertical lines are daytime ranges of metabolism. (From Lasiewski, R. C., Physiol. Zool. *36*:122–140, 1963.)

mammals (Table 5–2), yet the Microchiroptera have lower *K* values.

In general, homeotherms that can undergo torpidity have more variable and somewhat lower *K* values than those unable to become torpid. The torpid state permits survival without motor activity and thus conserves fuel reserves. This is illustrated for a hummingbird in Figure 5–21. The bat *Nictimene albiventer* when awake at 25°C uses 2.59 ml O_2/g/hr; in torpor at the same temperature, it uses 0.67 ml O_2/g/hr. Changes in *M* during exercise were mentioned on page 186. A pigeon, *Columba,* in flight has an *M* value of 22 kcal/hr; at rest, it is 2.69 kcal/hr.[133]

Among fishes the active species have higher standard metabolism than sluggish species, as follows for fish of similar body size at 15°C: trout, 100 mg O_2/kg/hr; white sucker, 10 mg O_2/kg/hr; bullhead, 40 mg O_2/kg/hr; carp, 30 mg O_2/kg/hr.[18] Fishes from cold water tend to have high metabolic rates. The high amount of oxygen dissolved in blood at low temperatures is sufficient for some Antarctic fishes, and at 0°C those without hemoglobin have O_2 consumption (0.012 to 0.022 ml O_2/g/hr) similar to and only slightly lower than those that have hemoglobin (0.022 to 0.037 ml O_2/g/hr).[170]

The cost of locomotion, calculated on a body weight basis, is higher for small animals than for large ones, whether for flying or for running (Fig. 5–22). Biological flight is more efficient than running for animals of the same weight. Great variation in efficiency of swimming among fishes is related to body form.

Examples suggest that populations of species living in warm climates may have lower metabolism than populations of the

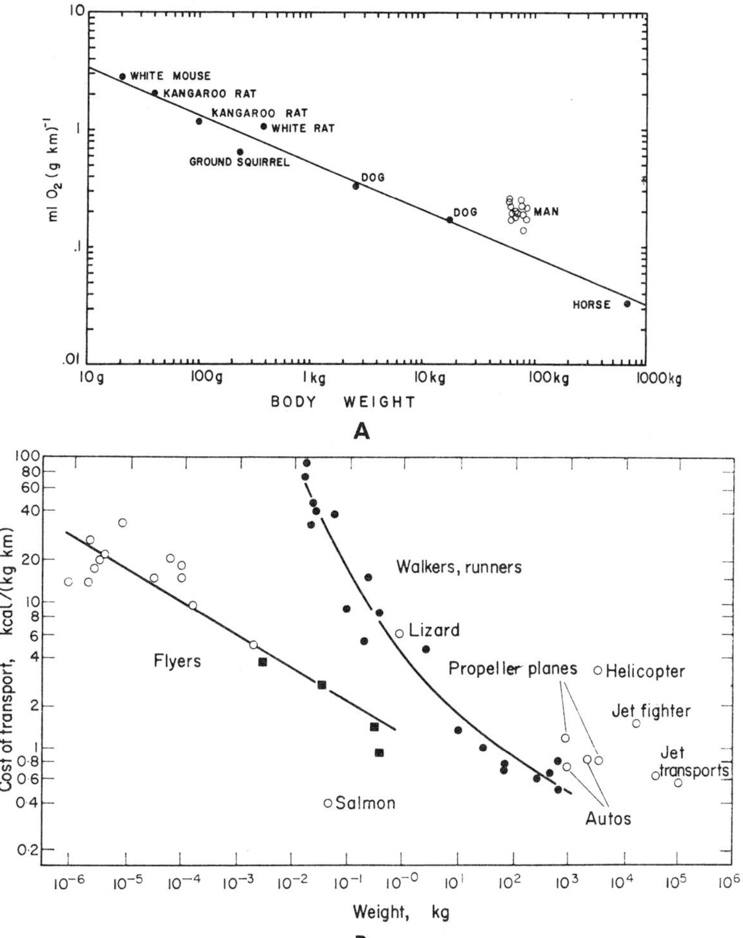

Figure 5–22. A, Minimum cost of running, as slope of curve relating V_{O_2} to speed for animals of different sizes. (From Taylor, C. R., K. Schmidt-Nielsen, and J. L. Raab, Amer. J. Physiol. *219*:1104–1107, 1970.) B, Minimum cost of transport for insects (open circles), birds (solid squares), other animals, and man-made vehicles as indicated. (From Tucker, V. A., Comp. Biochem. Physiol. *34*:841–846, 1970.)

same species from cooler environments (see page 375). A Connecticut population of the polychaete *Clymenella torquator* has higher V_{O_2} than a North Carolina population, and in *C. mucosa* the North Carolina population showed a higher V_{O_2} than a Puerto Rican population.[144] A southern England population of terrestrial isopods has a higher metabolic rate than a population from Austria.[218, 219]

Oxygen dependence or metabolic conformity to tolerable O_2 concentrations in the environment is often found in less active animals, such as intertidal or bottom-dwelling invertebrates and sluggish fish. Metabolic regulation is the rule in highly active animals. Similarly, the ability to shift from aerobic to anaerobic metabolism for long periods is common in prolonged divers, such as turtles. Anaerobic metabolism without payment of oxygen debt is common among parasitic worms.

ADAPTATIONS FOR DIVING

Many air-breathing animals survive long periods of submergence by virtue of auxiliary respiratory mechanisms, reduced oxidative requirements, and changes in metabolic pattern. Pulmonate snails absorb significant amounts of oxygen by the body surface; diving insects may have plastron respiration by a maintained bubble or may have tracheal gills. Anurans (frogs) under water can obtain about half as much oxygen through the skin as they obtain in air by the lungs; when submerged, their O_2 consumption drops, the heart rate is slowed (bradycardia), and in

	Sternothaerus	*Pseudemys*
V_{O_2} in air	26.6	38.5
V_{O_2} submerged at air equilibrium	3.4	4.1
V_{O_2} submerged, O_2 saturated	22.2	1.6
P_c in air	8	8
P_c in water		425
Survival time in water at air saturation	59.7 hours	26.5 hours
Survival time in water, O_2 saturated	more than 4300 hours	28.9 hours
Survival time in anoxic water	12.7 hours	24.7 hours
Survival time in anoxic water after IAA treatment	0.32 hour	

some seasons they accumulate some oxygen debt. When they emerge, the heart rate increases faster than the renewal of O_2 and the elimination of CO_2, and in fact recovers even if they emerge into nitrogen. The initial rise in heart rate is prevented by bilateral vagotomy, yet the bradycardia does not occur on submergence in water at 760 mm Hg of O_2; hence, the heart response is partly a submergence reflex and partly an anoxic response.[109] Data for two species of turtles are given above. *Sternothaerus,* but not *Pseudemys,* is stimulated to buccopharyngeal ventilation on submersion. *Pseudemys* shows a delayed bradycardia. *Sternothaerus* relies on auxiliary respiration and glycolysis to provide for much longer submergence than *Pseudemys* can survive.[19]

Many fish, such as trout, show bradycardia when removed from water; this is not caused by reduced O_2 *per se,* because bradycardia does not follow transfer to hypoxic water. A mudskipper, *Periophthalmus,* survives up to 27 hours out of water and then dies of dehydration; this fish fails to show the emergence bradycardia so common in less amphibious fish.[69]

Among birds and mammals, some relax and slow the heart on submergence, while others become excited, struggle, and die. A Pekin duck can dive for 15 minutes, and a duck in a prolonged dive may reduce metabolism by 90% and heart rate to between 5 and 8 per cent of the pre-dive rate; the body temperature may drop. The period of apnea may be prolonged if pure O_2 has been breathed prior to the dive.[164] The submersion bradycardia can be elicited by stimulation of a region in the mesencephalon.[56]

Maximum durations of dives for a few mammals are: harbor seal (*Phoca vitulina*), 15 minutes; sperm whale (*Physeter macrocephalus*), 1½ hours; finback whale (*Balaenoptera physalus*), ½ hour; bottlenose whale, 2 hours; man, 2½ minutes. Prolonged diving is made possible by interacting respiratory, circulatory, and behavioral adaptations. Divers' rate of oxygen consumption in air does not differ greatly from that of land mammals – 546 ml O_2/kg/hr in seal, 180 in manatee, and 250 in man. Tidal air in divers is significantly large – 80% of lung volume in porpoise as compared with 20% in man. The oxygen capacity of the hemoglobin of a seal is 1.78 ml O_2/g Hb, compared with 1.23 for hemoglobin in man; the blood volume in the seal is somewhat larger. Myoglobin is high in diving mammals; seal muscle yields 7715 mg Mb/100 g, compared with 1084 in beef muscle. Scholander calculated the total oxygen stores – in lungs, blood, muscle, and body fluid – of a 29 kg bladdernose seal as 1520 ml. During a 25 minute period at rest, this seal consumes 6250 ml of O_2, which is four to five times the stores available for a dive. Muscle oxygen is depleted in 5 to 10 minutes in a dive, after which the muscles rely on glycolysis. Oxygen consumption during the dive may be only 20 to 25 per cent of that at rest, yet the seal is swimming actively. In whales, the muscle store of oxygen may be 40% of the total.

Oxygen stores vary most for myoglobin and lung, as follows:[136]

Oxygen Stores in ml/kg

		BLOOD		TISSUES		
	LUNG	arterial	venous	myoglobin	dissolved	TOTAL
ribbon seal	12.6	14.3	22.6	27.2	2	78.7
sea otter	51.2	6.1	9.4	8.8	2	77.5
harbor seal	13.6	12.2	18.8	18.6	2	65.2
fur seal	21.8	6.7	9.9	11.7	2	52.1
sea lion	16.5	5.0	7.2	8.1	2	38.8

Lactic acid does not accumulate in the blood to any great amount during the dive, but on emergence the level may rise (e.g., from 70 to 175 mg/100 ml in the seal *Cystophora*). During the dive lactic acid accumulates in the muscle, from which the circulation is virtually excluded. All pinnipeds have a striated sphincter around the posterior vena cava just ahead of the diaphragm; thus, the blood supply to the head may be maintained while the supply to the rest of the body is occluded. Another important circulatory adaptation is the bradycardia that occurs in many good divers (including trained humans). Korean and Japanese Ama women may dive to 18 or 20 meters for 60 to 80 seconds for pearl oysters or abalone. They show marked bradycardia, which starts within a few seconds after the dive begins. The stimulus for bradycardia is not hypoxia, since it does not occur with breath-holding but only if the face is immersed, especially in cold water. These divers have elevated basal metabolic rate (BMR) during the winter.[169]

In a seal, the heart slows from a resting rate of 80 to 10 per minute during a dive. Bradycardia is a vagal reflex; the venous sphincter of seals is innervated by the phrenic nerve.

Sensitivity of the respiratory center to CO_2 is lower in divers than in non-diving animals; the threshold is high in beaver, seal, muskrat, and porpoise. Brain activity, too, as indicated by alpha waves, persists to lower P_{O_2} than in terrestrial mammals.[51] Utilization of inspired oxygen is high and tidal air volume is large; a porpoise has a respiratory capacity of 7 liters and a tidal volume of 5 to 6 liters, and oxygen utilization is some 10 per cent, two or three times more efficient than in man. Whales take several rapid breaths at the surface and may exhale just before submergence; porpoises dive on inspired air, while Weddell seals, elephant seals, and harbor seals dive on expiration. At depth, the lungs and bronchi are collapsed; thoracic collapse was seen to begin in a porpoise at 50 meters, with the total lung volume reduced to 200 to 260 ml at 300 meters depth. Because of this lung compression, probable thickening of the alveolar walls, and reduction in pulmonary circulation, conditions are not suitable for nitrogen invasion of the blood and, hence, bubbles do not appear in the blood when the animals surface.

The ability of diving mammals to remain submerged is, then, based on several mechanisms: large tidal volume, limited oxygen storage, especially in muscle myoglobin, ability to tolerate oxygen debt in muscles, relative insensitivity of the respiratory center to CO_2 and lactic acid, circulatory shunts restricting posterior blood flow but permitting it to heart and brain, heart slowing, and locomotor efficiency.

ADAPTATIONS TO HIGH ALTITUDE

Mountain sickness is characterized by reduction in respiratory and cardiovascular function brought on by the reduced partial pressure of oxygen at high altitudes. At 10,000 feet the barometric pressure is 523 mm Hg and the P_{O_2} is 110 mm Hg, 71 per cent of the sea level pressure. On going from sea level to 17,000 feet, alveolar P_{O_2} in man is reduced from 105 to 45 mm Hg and P_{CO_2} is reduced from 40 to 25 mm Hg. When the oxygen saturation of blood in man is reduced to 85% of normal, appreciable mental impairment occurs; at 70 to 80 per cent of normal saturation, there is serious confusion. The initial response of man on going to high altitude is hyperventilation, which increases alveolar P_{O_2} but decreases alveolar P_{CO_2}, causing alkalosis and upsetting the acid-base balance. Alkalosis due to low CO_2 in the presence of hypoxia results in cerebral vasoconstriction, whereas in the absence of hypocapnia a comparable hypoxia would cause vasodilation. Thus, dizziness and reduced brain function are early signs of altitude sickness. An initial vascular response is increased cardiac output.

After a few days or weeks at altitude, man acclimatizes. The increase in hemoglobin content of blood during acclimatization and the high Hb content in animals living at altitude are discussed on pages 348–349. The cardio-respiratory changes consist of increased alveolar ventilation, adjustment of acid-base balance, and adjustment of the elevated cardiac output nearly to normal. The curve relating ventilatory response to CO_2 is shifted to the left and the slope of the curve is steeper. However, the respiratory response to low O_2 is less than at sea level, and highlanders living at sea level retain low sensitivity to hypoxia. Their ventilation increases only if $P_{a_{O_2}}$ falls to 50 to 60 mm Hg, and they show less increase in ventilation during exercise than persons acclimated to sea level.[124] High altitude natives have high lung capacity and high normal ventilation volume, as well as some hypertrophy of the right ventricle. In tissues of animals living at altitude, high levels of some oxidative enzymes have been reported. Hemoglobin synthesis is stimulated (see page 348). Some persons and animals show better acclimatization to altitude than others, and natives in the Andes and Himalayas may represent selected genetic variants.

OXYGEN TOXICITY

Animals have elaborate mechanisms for preventing cellular O_2 from becoming too low. Conversely, cellular damage can result from an excess of oxygen. Some bacteria are obligate anaerobes, and a few animals do best with an oxygen supply less than air saturation. Termites can be rid of intestinal flagellates by putting them into pure oxygen, and 1 atm of O_2 is more toxic at low than at high temperatures.[35] The parasitic protozoan *Trichomonas* is a facultative aerobe. Some parasitic helminths, such as *Ascaris*, survive for less than an hour in pure O_2. Besides the parasites, some free-living animals from regions normally low in oxygen are sensitive to atmospheric O_2. Chironomid larvae, *Tubifex*, and an ostracod, all of which normally live in the mud at the bottoms of sluggish streams, live best at 4% O_2; they survive a short time at air saturation and only 2 to 4 days in 100% O_2. *Tubifex* cannot regenerate lost segments at very low Po_2, yet oxygen above air saturation blocks the differentiation and causes the worms to break apart in a few days; cyanide relieves the O_2 toxicity.[4] Thus, mud dwellers and parasites that normally live in low oxygen levels are more readily damaged by high oxygen concentration than are animals that usually live at air saturation. The effects of pure oxygen, as in insects, are not caused by high pressure *per se.* The hymenopteran *Habrobracon*, when exposed to 1 atm of O_2 for an hour or to 2 atm for 5 minutes, shows cytological damage; life span is shortened in 1 to 7 atm O_2. In *Drosophila*, a given dose of high pressure of oxygen is less toxic if given intermittently; recovery is about 30% in 3 to 7 days.[58] Eight hours daily at 1 atm O_2 or 1/2 hour daily at 2 to 5 atm O_2 reduces the longevity of *Drosophila.*

The time to death of 50% of a group of animals in 1 atm O_2 is: rats, 3 days; mice, 5½ days; chickens, 9½ days; quail, 14 days. Hamsters kept in 0.7 atm O_2 showed retinal degeneration and damage to germinal epithelium of the testes. Cell divisions of Protozoa can stop in pure O_2. Human divers breathing pure O_2 may show signs of toxicity. The effects of pure oxygen resemble those of ionizing radiation, and a nematode, *Caenorhabditis*, is highly resistant to both radiation and pure O_2. The action of oxygen is caused by the production of free radicals such as [OH] and H_2O_2. The cellular effects are oxidation of sulphydryl coenzymes such as CoA, inactivation of SH-enzymes, peroxidation of lipids, and oxidation of various SH-containing intermediates. Normally, cells have a variety of antioxidants—glutathione, catalase, vitamin E—that tend to protect SH-compounds from auto-oxidation.[77] The nucleus is relatively anaerobic, the mitochondrial cytochrome system needs some 4 μM of O_2, and the maximum that most cells can tolerate is 65 μM.[67]

GAS SECRETION: SWIM-BLADDERS

The transfer of oxygen across the respiratory membranes into blood is caused by diffusion gradients, even at high altitudes where external O_2 pressure is low. However, a number of unrelated aquatic animals concentrate gases in bubbles or gas chambers against considerable pressures. The fishes best studied are those in which the swim-bladders serve functions of resonance in hearing, sound production (see Chapter 12), auxiliary supply of respiratory oxygen, and, above all, regulation of hydrostatic pressure to provide buoyancy.[1] Bladder volume of 3 to 5 per cent of body volume is needed to provide neutral buoyancy in sea water, and somewhat more in fresh water. Physoclist fishes secrete all of the gas in the bladder; physostomes may swallow some gas, but if they are kept submerged they can secrete it.

In most fishes with swim-bladders, there are mechanisms for both gas secretion and absorption; there are both glandular and muscular mechanisms. Inflatory muscles have cholinergic nerve endings, and deflatory reflexes involve adrenergic nerves. These transmitters act on both muscle and glands.[13]

The gas that is secreted into swim-bladders is very different from air. Hydrostatic pressure in deep bodies of water increases about 1 atm per 10 meters depth; thus, gas is secreted against a pressure of 10 atm at a depth of 100 meters, and the total pressure equals the hydrostatic pressure. When the fish rises rapidly, the gas expands to the point of rupture of the swim-bladder. The partial pressure of N_2 at the surface is 0.8 atm, and that of O_2 is 0.2 atm (150 mm Hg). At depth, under hydrostatic pressure, the amount of gas dissolved increases by 14% per 100 atm, so that at 1000 m the partial pressure of O_2 is 0.228 atm. The gas in deep-sea fish bladders, however, may be 65 to 95 per cent O_2.[19] Thus, the total O_2 pressure in a swim-bladder is about 90 atm at a depth of 1000 m; hence, the fish is transporting oxygen against a 394-fold gradient. Many fish live at greater depths.[8, 10, 11]

The gas in the swim-bladder of bluefish is 65 to 85% O_2, 4 to 7% N_2, and 17 to 37% CO_2.[25] In some freshwater fish (e.g., white-fish), the gas may be 95% nitrogen. The argon-to-nitrogen ratio is similar to that in air, but in different mixtures of inert gases the soluble argon is high compared to the less soluble neon and helium; hence, even inert gases may be moved against a hydrostatic pressure gradient. Oxygen isotopes are maintained in the same ratio as in air (or a mixture), and hence the oxygen that is secreted is molecular oxygen.[23]

When a swim-bladder is artificially deflated, the time needed to replace the gas volume is 16 to 18 hours for eel, 18 to 24 hours for toadfish, 24 hours for tautog, 48 hours for sea robin, 48 hours for *Fundulus*, and 4 hours for bluefish.[25]

A gas gland lies along one side of the swim-bladder. Secretory cells in the gland have paravascular folds, between which is the Golgi apparatus; gas bubbles may appear between the folds.[4] In addition, there is an extensive rete of parallel capillaries, which may occur central to the gland or at the level of the gland. Fänge[12] estimates that the rete of an eel has 200,000 capillaries having a total length of 800 meters. The rete provides a morphological basis for countercurrent exchange of oxygen.

One hypothesis concerning secretion is that the gland produces lactic acid and CO_2, and that the acid causes release of O_2 from hemoglobin (Hb) by shifting the equilibrium curve to the right (Bohr effect) and by reducing the maximum saturation (Root effect). By the countercurrent arrangement, oxygen may cycle from the vein from the swim-bladder into the artery coming from the heart, and then from the systemic vein into the swim-bladder artery; thus, a high concentration can build up in the gland. A positive or "Root-on" effect occurs in the bladder artery with a half-time of 10 to 20 seconds, whereas the normal or "Root-off" effect occurs in the bladder epithelial drainage with a half-time of 50 milliseconds.[2]

Preparations of gas glands perfused under a microscope can be shown to liberate gas bubbles.[20] The gas gland produces lactic acid when under as much as 51 atm O_2. It is low in the TCA cycle enzymes and high in those of anaerobic glycolysis, and its metabolism is relatively insensitive to O_2 poisoning.[6, 7] Lactic acid in blood from the gland increases during gas secretion. An eel gas gland secretes 0.37 ml O_2 and 0.02 ml N_2 per hour. The secretion is reduced by diamox, and inhibitor of carbonic anhydrase.[13] In barracuda, the post-rete blood may contain as much as 40 mg of lactic acid per 100 ml of blood.[10]

Analyses of blood at the proximal and distal sides of the rete in eel are given by Steen[20, 21] (Fig. 5–23). They show that, while acid from the gland favors liberation of O_2 from Hb, the O_2 in the afferent vessels diffuses by countercurrent flow into the efferent capillaries, thus cycling O_2 in the rete. This permits higher pressures of O_2 to build up in the gas bladder. The Po_2 in venous blood at the exit from the rete decreases during secretion.[20]

The swim-bladders of some deep-sea fish have a large central mass of lipid, which has a ratio of phospholipid to cholesterol of 1:1; the gas glands of deep-sea fishes synthesize cholesterol. Oxygen may be soluble in this cholesterol-rich lipid so that there may be little free gas in the swim-bladders of deep-sea fishes. Shallow-water fishes, in contrast to deep-sea fishes, have a thin lipid lining in the swim-bladder.[17]

Another postulated mechanism is that of salting out.[14, 15, 24] Addition of lactate to the blood in the gland reduces the solubility of dissolved gases. This mechanism is particularly important for nitrogen and argon. In salting out, the percentage of gases present in the swim-bladder is related to solubility rather than to partial pressure.

The Root effect is slow, and the swim-bladder gas pressures in deep-sea fishes may exceed those in the blood. The combination of the CO_2 effect on the Hb equilibrium curve and the countercurrent exchange in the rete, together with the low permeability of the bladder wall, is marginal for deep-sea

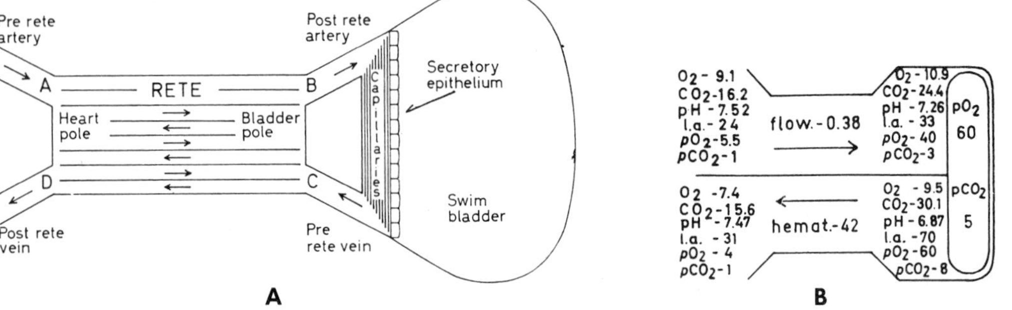

Figure 5-23. A, Diagram of gas gland with rete and bladder. *B*, Composition of inflowing and outflowing blood at the two ends of the rete, and of gas in bladder. O_2 and CO_2 in ml/100 ml blood; lactic acid (l.a.) in mg/100 ml blood; flow in ml/min; Po_2 and Pco_2 in percent atm. Samples were analyzed after 3 hours of secretion. (From Steen, J. B., Acta Physiol. Scand. *59*:221–241, 1963.)

fishes and cannot explain the secretion of nitrogen. However, it is calculated that, given enough time, oxygen accumulation would occur at high pressures.

Siphonophores (Portuguese men-of-war) secrete gas into a large float or gas bladder. They secrete gas on ascent and release it on descent. The composition of the float gas is 74.4% N_2, 8.9% CO, 14.4% O_2, 1.1% argon, and 0.4% CO_2.[18, 22] At the base of the float is a gas gland, which in ultrastructure resembles the gland in fishes.[5] CO secretion occurs in vivo at the rate of 7.5 to 120 μl/hr/animal and in vitro at 0.06 mm^3/mg gland/hr. The CO is derived from serine.[22] The floats of some giant kelps also contain much CO (5 to 10%).

In cuttlefish (*Sepia*), the cuttlebone occupies 9 per cent of the body volume and has a density of 0.6 g/cm^3; if the bone is removed, the animal is 4 per cent denser than sea water. The pores in the cuttlebone contain much gas, together with a fluid that is hyposmotic to blood in the deep sea and isosmotic to blood in shallow water. Na and Cl are actively withdrawn from the cuttlebone fluid, thus creating an osmotic gradient that draws water out of the bone and balances the hydrostatic pressure, which would otherwise force water into the air space. By means of this osmotic pump, the cuttlefish maintains buoyancy.[9]

SUMMARY

Anaerobic (glycolytic and fermentative) metabolism is more widespread among living organisms than is aerobic metabolism. This fact, together with geochemical evidence, indicates that in chemical evolution anaerobiosis preceded oxidative metabolism. All animals are essentially aerobic, but they can use anaerobic routes of energy production in varying degrees and at different times.

Ecologically, oxygen is limiting mostly when it is in low supply. High pressures of oxygen are toxic, but in nature only a few animals are poisoned by oxygen at air concentrations (155 mm Hg of Po_2); these are animals that live normally in restricted O_2—some mud-dwelling worms and intestinal parasites. For air-breathing animals, environmental oxygen is limiting at high altitudes and in diving. For water breathers, the levels of oxygen are restrictive in stagnant pools, in warm waters, and in some polluted waters. For each of these environments there are adaptations which enhance O_2 uptake or which provide for better distribution and utilization of oxygen by some animals; some adaptations relate to the permanent life habit of the species, while others are effective only for brief exposures to reduced O_2 supply, as in exercise.

Oxygen must enter all cells by diffusion, and O_2 diffusivity may determine cell size in both free-living cells and multicellular organisms. Oxygen limits the attainable size of an animal that does not have a circulatory system, and it limits the distance between capillaries in those that do. For most metabolic rates and Po_2 levels, a diffusion distance of 0.5 to 1 mm is maximal. Intracellular Po_2 levels are probably normally one to two orders of magnitude lower than extracellular Po_2.

Several modes of getting oxygen into multicellular animals have evolved. In only a few animals is O_2 carried directly to tissues—in sponges and coelenterates and in tracheates (mostly insects). Control of ventilation in insects is by breathing movements and spiracles. In all other animals some sort of fluid—coelomic fluid or blood, separated from the water or air by a thin epithelial membrane—is interposed between the O_2 source and the tissues. Ventilation of membranes where exchange occurs may be accomplished by elaborate muscular mechanisms that are reflexly controlled. These reflexes are initiated mainly either by reduced O_2 or by elevated CO_2, the former more commonly in aquatic animals and the latter more in terrestrial animals. Receptors sensitive to low O_2, and others sensitive to high CO_2, may be external, as in tetrapod vertebrates. The reflex responses are usually increased ventilation (rate or amplitude of breathing or both), or they may be changed blood flow. The net effect is to tend to maintain a constant supply of O_2 to the organism. Sometimes increased ventilation has secondary effects, such as the alkalosis caused by blowing off CO_2, that complicate the attempted increase in O_2 uptake. Also, the extra energy consumed in increased ventilation may make the increased supply of O_2 less useful, as in some fish.

There are many auxiliary respiratory organs, such as the oral membranes of turtles, the swim-bladders of physostome fish, and the skin of amphibians. CO_2 output and O_2 uptake do not take place in the same proportions in the auxiliary and main respiratory organs in a given animal.

Aerobic metabolism is measured by oxygen consumption, less accurately (because of recycling) by CO_2 production, and by calorimetry, which is calculated as heat equivalents. Metabolic rate varies considerably according to kind of animal, body size, environmental conditions, and locomotor activity. There is incomplete evidence that activity metabolism may proceed by enzyme pathways slightly different from those of rest metabolism. One useful way to separate rest or standard metabolism from activity metabolism is to drive the animal to different speeds of locomotion and then extrapolate to zero activity; for some animals it is necessary to measure as standard metabolism the O_2 consumption during periods when there is no perceptible movement. The difference between rest and activity metabolism may be one to three orders of magnitude; discordance in published metabolic data is often caused by absence of activity control. Effects of environmental and body temperature, hormonal state, nutrition, and nature of the foodstuffs being oxidized are considerable.

For a given kind of animal, small individuals and species have a higher metabolic rate than large individuals. Metabolism is proportional to body weight in a few small animals, but in unicellular and large animals it is proportional to body weight to a power (often 0.75), which is close to but not the same as the power function for body surface. There is very incomplete evidence that oxidative activity, on a protein or weight basis, is higher at the cellular (perhaps mitochondrial) level in small animals. How this is controlled is unknown. Apart from similar metabolism-size relations, the basal metabolism (for unit body weight) shows genetically determined variations related to life habits—sluggish versus highly active, and capable versus not capable of torpor or hibernation.

An important modifier of metabolism is the amount of environmental oxygen. Many animals, and probably most cells, are metabolic conformers; they increase oxygen utilization in proportion to available oxygen, and at normal atmospheric levels they are not metabolizing to capacity. Other animals are metabolic regulators and have constant metabolism down to critical levels of P_{O_2}, which are related to their life habits. Most of the regulation in animals is by variation in external respiration (ventilation, and circulation to the respiratory organ).

Some animals shift from aerobic to anaerobic metabolism for brief periods (as air breathers during diving, or as in severe exercise or in periods of anoxia). Some animals, after a period of hypoxia, show increased O_2 consumption and pay back the metabolic debt. Others live for long times on glycolytic energy and excrete the resultant acids (e.g., intestinal parasites).

The balance between aerobic and anaerobic pathways and their coupling represent a complex and beautifully integrated system. Regulation of O_2 supply by CO_2 and O_2 sensors and central nervous reflexes are on the level of control by the whole organism. How enzyme levels are set at the cellular level to reflect body size, environmental stresses and available oxygen is only partly known. Some of the control mechanisms, particularly hormonal control, are discussed in the following chapter.

REFERENCES

Oxygen: Respiration

For references to swim-bladders, see page 211.

1. Adolph, E. A., Respir. Physiol. 7:356–368, 1969. Survival in anoxia, infant mammals.
2. Altman, P. L., and D. S. Dittmer, eds., Biological Handbook, Respiration and Circulation. Fed. Assoc. Soc. Exp. Biol., 1971. 930 pp.
3. Andersen, H. T., Physiol. Rev. *46*:213–243, 1966. Physiological adaptations of diving vertebrates.
4. Anderson, J. C., Biol. Bull. *111*:179–189, 1956. Relation of metabolism to regeneration in *Tubifex.*
5. Anderson, J. F., Comp. Biochem. Physiol. *33*:51–72, 1970. Metabolic rates of spiders.

5a. Armitage, K. B., Biol. Bull. *123*:225–232, 1962. Temperature and oxygen consumption, gammarid.

6. Ashby, E. A., and J. L. Larimer, Physiol. Zool. *37*:21–32, 1964. Cardiac responses of crayfish to carbon dioxide stress.
7. Ashby, E. A., and J. L. Larimer, J. Cell. Comp. Physiol. *65*:373–379, 1965. Cardiac and respiratory responses to carbohydrate chemoreception, crayfish.
8. Baker, E. G. S., and J. P. Baumberger, J. Cell. Comp. Physiol. *17*:285–304, 1941. Critical O_2, *Tetrahymena.*
9. Ballintijn, C. M., J. Exp. Biol. *50*:569–591, 1969. Muscle coordination of respiratory pump in carp.
10. Ballintijn, C. M., and G. M. Hughes, J. Exp. Biol. *43*:349–362, 1965. Respiratory pumps in trout.

10a. Barlow, G. W., Biol. Bull. *121*:209–229, 1961. Metabolism of gobiid fishes.

11. Bartholomew, G. A., W. R. Dawson, and R. C. Lasiewski, Z. vergl. Physiol. *70*:196–209, 1970. Thermal regulation and oxygen consumption in flying foxes.
12. Bartholomew, G. A., et al., Physiol. Zool. *37*:179–198, 1964. Metabolism of bats.
13. Bartholomew, G. A., and R. E. MacMillen, Physiol. Zool. *34*:177–483, 1961. O_2 consumption of kangaroo mouse.
14. Bartholomew, G. A., and V. A. Tucker, Physiol. Zool. *37*:341–354, 1964. Oxygen consumption in relation to temperature and size in varanid lizards.
15. Barron, E. S. G., and T. N. Tamisian, J. Cell. Comp. Physiol. *32*:57–70, 1948. Metabolism of cockroach muscle.
16. Basu, S. P., J. Fish. Res. Bd. Canad. *16*:175–212, 1959. Active respiration of fish in relation to ambient concentrations of oxygen and carbon dioxide.
17. Beadle, L. C., J. Exp. Biol. *34*:1–10, 1957. Respiration of swamp oligochaete, *Alma.*
18. Beamish, F. W. H., and P. S. Mookherjii, Canad. J. Zool. *42*:161–175, 177–188, 355–366, 847–856, 1964. Standard and activity metabolism of brook trout and goldfish in relation to temperature and body size.
19. Belkin, D. A., Fed. Proc. *22*:634, 1963; Respir. Physiol. *4*:1–14, 1968. Aquatic respiration and diving ability of turtles.
20. Benedict, F. G., Carnegie Inst. Washington Publ. *503*:1–215, 1938. Vital energetics.
21. Berg, K., P. M. Jonasson, and K. W. Ockelmann, Hydrobiologia *19*: 1–39, 1962. Respiration of freshwater invertebrates.
22. Berg, T., and J. B. Steen, Comp. Biochem. Physiol. *15*:469–484, 1965. Aerial respiration in eel.
23. Berkson, H., Comp. Biochem. Physiol. *18*:101–119, 1966. Diving in Pacific green turtle.
24. Boell, E. J., Ann. N.Y. Acad. Sci. *49*:773–800, 1948. Metabolism of amphibian embryos.
25. von Brand, T., pp. 177–234, 434. *In* Biochemistry and Physiology of Protozoa, Vol. 1, edited by S. H. Hutner and A. Lwoff. Academic Press, 1951. J. Cell. Comp. Physiol. *29*:33–48, 1947; *45*:421–434, 1955; Zool. Anz. *157*:119–123, 1956. Intermediary metabolism of trypanosomes.
26. Brett, J. R., J. Fish. Res. Bd. Canad. *21*:1183–1226, 1964; *22*:1491–1501, 1965. Respiration in relation to swimming speed in sockeye salmon.
27. Brett, J. R., Amer. Zool. *11*:99–113, 1971. Energetics of salmon.
28. Brody, S., Bioenergetics and Growth. Reinhold Publishing Co., 1945. 1023 pp. Chaps. 13–15. Metabolism in relation to body size.
29. Brown, J. H., and G. A. Bartholomew, Ecology *50*:705–709, 1969. Energetics of kangaroo mouse.
30. Buck, J., Ann. Rev. Entomol. 7:27–56, 1962. Physical aspects of insect respiration.
31. Buck, J. B., and M. L. Keister, Biol. Bull. *105*:402–411, 1953; *119*: 144–163, 1955; J. Exp. Biol. *32*:681–691, 1955; Physiol. Zool. *29*: 137–146, 1956. Cyclic CO_2 release by saturnid moth pupae; cutaneous and tracheal respiration; tracheal filling in fly larvae.
32. Cade, T. J., et al., Physiol. Zool. *38*:9–33, 1965. Water economy and metabolism of finches.

32a. Chance, B., J. Gen. Physiol. *49*:163–188, 1965. Respiratory controls in cells.

33. Chew, R. M., R. G. Lindberg, and P. Hayden, Comp. Biochem. Physiol. *21*:487–505, 1967. Effects of temperature on metabolism in pocket mouse.
34. Childers, J. J., Science *160*:1242–1243, 1964. Distribution of a mysid crustacean as limited by oxygen minimum layer.
35. Cleveland, L. R., and A. U. Burke, J. Protozool. *3*:74–77, 1956. Oxygen toxicity in symbiotic protozoa.
36. Cohn, J. E., and R. Shannon, Respir. Physiol. *5*:259–268, 1968. Respiration in geese.
37. Crawford, E. C., and R. C. Lasiewski, Condor *70*:333–339, 1968. Oxygen consumption and respiratory evaporation of emu and rhea.
38. Crawford, E. C., and R. Schultetus. Copeia 179–180, 1970. Cutaneous respiration in lizard, *Sauromalus.*
Cutaneous respiration in lizard, *Sauromalus.*
39. Dales, R. P., Physiol. Zool. *34*:306–311, 1961. Oxygen uptake and water irrigation by terebellid polychaetes.

39a. Davies, P. S., J. Marine Biol. Assoc. U.K. *46*:47–61, 1967. Metabolism of *Patella.*

40. Davies, S. P., and M. Walkey, Comp. Biochem. Physiol. *18*:415–425, 1966. Metabolism of cestodes.
41. Davis, J. C., and J. N. Cameron, J. Exp. Biol. *54*:1–18, 1971. Water flow and gas exchange in rainbow trout.
42. Dawson, T. J., et al., Comp. Biochem. Physiol. *31*:645–653, 1969. Metabolism of marsupial, *Macropus.*
43. Dawson, T. J., and A. J. Hulbert, Amer. J. Physiol. *218*:1233–1238, 1970. Standard metabolism of Australian marsupials.
44. Dawson, W. R., and C. D. Fisher, Condor *71*:49–53, 1969. Metabolism of nightjar.
45. Dawson, W. R., and J. R. Templeton, Ecology *47*:759–765, 1966. Metabolism of lizards.
46. Dawson, W. R., and H. B. Tordoff, Condor *61*:388–395, 1959. Metabolism of birds.
47. Dehnel, P. A., Nature *181*:1415–1417, 1958; Biol. Bull. *118*:215–249, 1960. Photoperiod, temperature, and salinity effects on oxygen consumption of crabs.
48. Dolk, H. E., and N. Postma, Z. vergl. Physiol. *5*:417–444, 1927. Frog respiration.

48a. Duerr, F. G., Comp. Biochem. Physiol. *20*:391–398, 1967. Size-metabolism relation in *Lymnaea.*

49. Dunlap, D. G., Comp. Biochem. Physiol. *31*:555–570, 1969. Modification of metabolism in hylid frog, *Acris.*
50. Edwards, G. A., Physiol. Comp. Oecol. *2*:34–50, 1950; *in* Insect Physiology, edited by K. D. Roeder, John Wiley and Sons, 1953, pp. 55–95. Hormone metabolism of *Uca*; insect respiration.
51. Elsner, R., et al., Respir. Physiol. *9*:287–297, 1970. Cerebral tolerance to hypoxemia in Weddell seals.
52. Enger, P. S., and P. Savalov, J. Insect Physiol. *2*:232–233, 1958. Metabolism of tropical arthropods.
53. Eriksen, C. H., Verh. Int. Verein Limnol. *15*:903–911, 1964. Effect of substrate on respiration of burrowing mayfly naiads.
54. Evans, N. T. S., Respir. Physiol. *3*:21–37, 1967. Systemic oxygen supply to human skin.
55. Eyzaguirre, C., and P. Zapata, J. Physiol. (London) *195*:557–588, 589–607, 1968. Acetylcholine and stimulation of carotid sinus.
56. Feigl, E., and B. Folkow, Acta Physiol. Scand. *57*:99–110, 1963. Diving in ducks.
57. Feldmeth, C. R., Comp. Biochem. Physiol. *32*:193–202, 1970. Respiratory energetics of caddis fly larvae.
58. Fenn, W. O., M. Philpott, C. Meehan, and M. Henning, Amer. J. Physiol. *213*:663–670, 1967; *also* Fenn, W. O., M. Henning, and M. Philpott, J. Gen. Physiol. *50*:1693–1708, 1967. Oxygen toxicity in *Drosophila.*
59. Ferrante, F. L., Amer. J. Physiol. *218*:363–371, 1970. Diving in nutria.
60. Fishman, A. P., pp. 215–219. *In* Sharks, Skates and Rays, edited by P. W. Gilbert, R. F. Mathewson, and D. P. Rall. Johns Hopkins Press, 1967. Respiration and circulation in dogfish, *Squalus.*
61. Forster, R. E., pp. 393–407. *In* Oxygen in the Animal Organism, edited by F. Dickens and E. Neil. Pergamon Press, 1965. Exchange of oxygen between blood and tissue.
62. Frankel, H. M., et al., Comp. Biochem. Physiol. *31*:535–546, 1969. Respiratory responses of turtles to changes in blood gases.
63. Fry, F. E. J., Publ. Ontario Fish Res. Lab. *68*:1–52, 1947. Effects of temperature and oxygen on animal activity.
64. Fuhrman, F. A., and D. A. Farr, Helg. wiss. Meeresunters. *9*:324–329, 1964. Oxygen diffusion in rat diaphragm.
65. Galvao, P. E., J. Tarasantchi, and P. Guertzenstein, Amer. J. Physiol. *209*:501–506, 1965. Heat production of tropical snakes.
66. Gans, C., and G. M. Hughes, J. Exp. Biol. *47*:1–20, 1967. Mechanics of lung ventilation in tortoise.
67. Gerschman, R., pp. 475–492. *In* Oxygen in the Animal Organism, edited by F. Dickens and E. Neil. Pergamon Press, 1965. Oxygen toxicity.

68. Gilchrist, B. M., Proc. Roy. Soc. Lond. B *143*:136–146, 1954; Hydrobiologia *8*:54–65, 1956; *12*:27–37, 1958. Hemoglobin, oxygen consumption by *Artemia* at different salinities.

68a. Glick, J. L., and W. D. Cohen, Science *143*:1184–1185, 1965. Nocturnal changes in enzyme activity of liver mitochondria.

68b. Goffart, M., Function and Form in the Sloth. Pergamon Press, 1971. 225 pp.

69. Gordon, M. S., J. Bretinus, D. H. Evans, R. McCarthy, and L. C. Oglesby, J. Exp. Biol. *50*:141–149, 1969. Respiration of amphibious fishes.

70. Graham, J. M., Canad. J. Res. D *27*:270–288, 1949. Effects of temperature, oxygen pressure, and activity on metabolism of trout.

71. Grigg, G. C., Comp. Biochem. Physiol. *29*:1253–1257, 1969. Critical O_2 pressures for bullhead fish.

72. Grigg, G. C., J. Exp. Biol. *52*:565–568, 1970; Z. vergl. Physiol. *73*: 439–451, 1971. Countercurrent flow of blood and water in shark gill.

73. Guimond, R. W., and V. H. Hutchinson, Comp. Biochem. Physiol. *27*:177–195, 1968. Temperature and photoperiod effects on gas exchange in frog.

73a. Hainsworth, F. R., and L. L. Wolf, Science *168*:368–369, 1970. Metabolism of hummingbirds.

74. Halcrow, K., and C. M. Boyd, Comp. Biochem. Physiol. *23*:233–242, 1967. Oxygen consumption and swimming activity of *Gammarus*.

75. Hall, F. G., Amer. J. Physiol. *88*:212–218, 1929. O_2 withdrawal and consumption by marine fish.

76. Hanegan, J. L., and J. E. Heath, J. Exp. Biol. *53*:611–627, 629–639, 1970. Activity patterns and energaetics of Cecropia moth.

77. Haugaard, N., Physiol. Rev. *48*:311–373, 1968. Cellular oxygen toxicity.

78. Hemmingsen, A. M., Rep. Steno Mem. Hosp. Nordisk Insulin Lab. *9*:1–110, 1960. Metabolism in relation to body size.

78a. Herreid, C. F., W. L. Bretz, and K. Schmidt-Nielsen, Amer. J. Physiol. *215*:506–508, 1968. Cutaneous gas exchange in bats.

79. Hill, J. R., and K. A. Rahimtulla, J. Physiol. (London) *180*:239–265, 1965. Metabolism of human infants in relation to temperature.

80. Hill, B. J., and I. C. Potter, J. Exp. Biol. *53*:47–57, 1970. O_2 consumption in lamprey ammocoetes.

81. Hills, B. A., and G. Hughes, Respir. Physiol. *9*:126–140, 1970. Theoretical analysis of countercurrent water and blood flow in fish gills.

82. Hinton, H. E., J. Insect Physiol. *4*:176–183, 1960; J. Marine Biol. Assoc. *47*:319–327, 1967. Plastron respiration in eggs of blowfly and of a marine fly.

83. Hinton, H. E., Adv. Insect Physiol. *5*:65–162, 1968. Spiracular gills.

84. Hoar, W. S., Canad. J. Zool. *36*:113–121, 1958. Thyroxin and gonadal steroids, effects on metabolism of goldfish.

85. Holeton, G. F., and D. J. Randall, J. Exp. Biol. *46*:297–305, 317–327, 1967. Effect of hypoxia on ventilation and metabolism of rainbow trout.

86. Houlihan, D. F., J. Insect Physiol. *16*:1607–1622, 1970. Respiration of insects that respire from aquatic plants.

87. Hudson, J. W., and F. W. Bertram, Physiol. Zool. *39*:21–29, 1966. Metabolism of skink, *Lygosoma*.

88. Hudson, J. W., and S. L. Kimzey, Comp. Biochem. Physiol. *17*:203–217, 1966. Metabolism of house sparrow.

89. Hudson, J. W., and J. A. Rummel, Ecology *47*:345–354, 1966. Metabolism of primitive rodents.

90. Hughes, G. M., Soc. Exp. Biol. Symp. *18*:81–107, 1964. Fish respiratory homeostasis.

90a. Hughes, G. M., et al., J. Exp. Biol. *55*:651–665, 1971. Metabolism of giant tortoise.

91. Hughes, G. M., and C. M. Ballintijn, J. Exp. Biol. *43*:363–383, 1965. Respiratory pumps in dogfish.

92. Hughes, G. M., and L. Mill, J. Exp. Biol. *44*:317–333, 1966. Patterns of ventilation in dragonfly larvae.

93. Hughes, G. M., J. Exp. Biol. *45*:177–195, 1966. Gill areas in relation to respiration of fish.

94. Hughes, G. M., and S. I. Umezawa, J. Exp. Biol. *49*:557–564, 565–582, 1968. Respiration of the dragonet and dogfish.

95. Hughes, G. M., and J. L. Roberts, J. Exp. Biol. *52*:177–192, 1970. Respiration in rainbow trout.

96. Hughes, G. M., and R. L. Saunders, J. Exp. Biol. *53*:529–545, 1970. Respiratory control in trout.

97. Hughes, G. M., and B. N. Singh, J. Exp. Biol. *53*:281–298, 1970. Respiration in air-breathing fish, *Anabas*.

97a. Hughes, G. M., and B. N. Singh, J. Exp. Biol. *55*:667–682, 1971. Gas exchange in catfish *Sacobranchus*.

98. Hughes, G. M., and G. Shelton, pp. 275–364. *In* Advantages in Comparative Physiology and Biochemistry, Vol. 1, edited by O. Lowenstein, Academic Press, 1962. Respiratory mechanisms and their nervous control in fish.

99. Hughes, G. M., et al., J. Exp. Biol. *51*:203–220, 1969. Control of gill ventilation in crab, *Carcinus*.

100. Hukuhara, T., and H. Okada, Japan. J. Physiol. *6*:313–320, 1956. Respiratory center in fishes.

101. Jackson, D. C., J. Appl. Physiol. *24*:503–509, 1968. Metabolic depression during diving in turtles.

102. Jackson, D. C., and K. Schmidt-Nielsen, J. Cell. Comp. Physiol. *67*: 225–232, 1961. Heat production in diving turtles.

103. Jackson, D. C., and K. Schmidt-Nielsen, Proc. Nat. Acad. Sci. *51*: 1192–1197, 1964. Countercurrent heat exchange in respiratory passages.

103a. Jesse, M. J., C. Shub, and A. P. Fishman, Respir. Physiol. *3*:267–287, 1967. Control of respiration in lungfish.

104. Johansen, K., Comp. Biochem. Physiol. *18*:383–395, 1966. Air breathing in fish, *Symbranchus*.

105. Johansen, K., Fed. Proc. *25*:389, 1966. Chemoreception in respiration of lungfish.

106. Johansen, K., et al., Respir. Physiol. *9*:162–174, 1970. Stimulation of air breathing in fish, *Amia*.

107. Johansen, K., and R. L. Vadas, Biol. Bull. *132*:16–22, 1967. Critical oxygen pressure in sea urchins.

108. Johansen, K., and J. A. Petersen, Z. vergl. Physiol. *71*:365–381, 1971. Metabolism of starfish.

109. Jones, D. R., J. Exp. Biol. *44*:387–411, 1966; Comp. Biochem. Physiol. *20*:691–707, 1967. Recovery from diving in amphibia.

110. Jones, D. R., Comp. Biochem. Physiol. *12*:297–310, 1964. Metabolism in swamp pulmonates.

111. Jones, D. R., D. J. Randall, and G. M. Jarman, Respir. Physiol. *10*: 285–298, 1970. Analysis of oxygen transfer in fish.

112. Kanwisher, J. W., Biol. Bull. *130*:96–105, 1966. Tracheal gas dynamics of cecropia silkworm.

113. Kayser, C., and A. Heusner, J. Physiol. Paris *56*:489–524, 1964. Review of metabolism in relation to size and hibernation.

114. Kempner, W., Cold Spring Harbor Symp. *7*:269–289, 1939. Role of oxygen tension in biological oxidations.

115. King, E. N., Comp. Biochem. Physiol. *15*:93–102, 1965. Oxygen consumption of intact crabs and excised gills.

115a. Kinne, O., Neth. J. Sea Res. *3*:222–244, 1966. Respiratory adaptations of marine invertebrates in different salinities.

116. Kleiber, M., and H. H. Cole, Amer. J. Physiol. *161*:294–299, 1950. Body size and metabolic rate in rats.

117. Kleiber, M., The Fire of Life. John Wiley and Sons, 1961. 453 pp.

118. Knight, A. W., and A. R. Gaufin, J. Insect Physiol. *12*:347–355, 1966. Oxygen consumption by stoneflies.

119. Kobayasi, S., S. Tsuchiya, and K. Takahashi, Japan. J. Physiol. *11*: 194–204, 1961. Rhythmic contractions in lung of toad.

120. Krebs, H. A., Biochim. Biophys. Acta *4*:249–269, 1950. Body size and tissue respiration.

121. Krogh, A., Comparative Physiology of Respiratory Mechanisms. University of Pennsylvania Press, 1941. 172 pp.

122. Kutty, M. N., Marine Biol. *4*:239–242, 1969. Oxygen consumption in mullet, *Liza*, at different swimming velocities.

123. Lacher, V., Z. vergl. Physiol. *54*:75–84, 1967. Reactions of honeybees to carbon dioxide.

124. Lahiri, S., and N. H. Edelman, Respir. Physiol. *6*:375–385, 1969. Respiratory sensitivity to hypoxia in mountain natives.

125. Lahiri, S., J. P. Szidon, and A. P. Fishman, Fed. Proc. *29*:1141–1148, 1970. Respiratory and circulatory adjustments to hypoxia in lungfish.

126. Larimer, J. L., and A. H. Gold, Physiol. Zool. *34*:167–176, 1961. Response of crayfish to respiratory stress.

127. Lasiewski, R. C., Physiol. Zool. *36*:122–140, 1963; Condor *64*:324, 1962. Metabolism of hummingbirds.

128. Lasiewski, R. C., Amer. J. Physiol. *217*:1504–1509, 1969. Metabolism of poorwill.

129. Lasiewski, R. C., and W. A. Calder, Respir. Physiol. *11*:152–166, 1971. Allometric analysis of respiratory variables in birds.

130. Lasiewski, R. C., and W. R. Dawson, Condor *66*:477–490, 1964; *72*:332–338, 1970. Metabolism of night hawk and of frog-mouth owl.

131. Lasiewski, R. C., and W. R. Dawson, Condor *69*:13–23, 1967. Relation between standard metabolism and body weight in birds.

132. Lasiewski, R. C., et al., Comp. Biochem. Physiol. *23*:797–813, 1967; Auk *84*:34–48, 1967. Metabolism of hummingbirds.

133. LeFebvre, E. A., Auk *81*:403–416, 1964. Energy metabolism of pigeon at rest and in flight.

134. Leitner, P., Comp. Biochem. Physiol. *19*:431–443, 1966. Metabolism of bats.

135. Leitner, P., and J. E. Nelson, Comp. Biochem. Physiol. *21*:65–74, 1967. Metabolism of vampire bat.

136. Lenfant, C., et al., Respir. Physiol. *9*:277–286, 1970. Oxygen stores in diving mammals.

137. Lenfant, C., and K. Johansen, J. Exp. Biol. *49*:437–452, 453–468, 1968. Respiration in lungfish, *Protopterus*.

138. Lewis, J. B., Physiol. Zool. *41*:476–480, 1968. Respiration of sea urchin, *Diadema*.

139. Lindroth, A., Arkiv Zool. *30B*:1–7, 1938. Respiratory regulation in crayfish.

140. Loveland, R. E., and D. S. K. Chu, Comp. Biochem. Physiol. *29*:173–184, 1969. Respiratory movements and pumping in *Mercenaria*.

141. Lord, D. R., I. C. Bellrose, and W. W. Cochran, Science *137*:39–40, 1962. Telemetered flight in ducks.

142. MacMillen, R. E., and J. E. Nelson, Amer. J. Physiol. *217*:1246–1251, 1969. Metabolism of dasyurid marsupials.

143. MacMillen, R. E., and C. H. Trost, Comp. Biochem. Physiol. *23*:243–253, 1967. Metabolism of Inca dove.

144. Mangum, C. P., Comp. Biochem. Physiol. *10*:335–349, 1963. Oxygen consumption in different species of polychaete worms.

145. Mangum, C. P., Amer. Sci. *58*:641–647, 1970. Respiratory physiology in annelids.

146. Marvin, D. E., and A. G. Heath, Comp. Biochem. Physiol. *27*:349–355, 1968. Cardiac and respiratory responses to hypoxia in three species of fish.

147. McLeese, D. W., Helg. wiss. Meeresunters. *10*:7–18, 1964. Oxygen consumption of lobster.
148. McMahon, B. R., J. Exp. Biol. *52*:1–16, 1970. Gas exchange across lungs and gills in African lungfish.
149. McNab, B. K., Comp. Biochem. Physiol. *26*:337–343, 1968. Effect of fat on basal metabolism of desert rodents.
149a. McNab, B. K., and P. Morrison, Ecol. Monogr. *33*:63–82, 1963. Metabolism of *Peromyscus*.
150. Miller, P. L., J. Exp. Biol. *39*:513–535, 1962. Control of spiracles in dragonflies.
151. Mines, A. H., and S. C. Sorenson, J. Appl. Physiol. *28*:826–831, 1970. Ventilatory responses of goats to hypoxia.
152. Moberly, W. R., Comp. Biochem. Physiol. *27*:1–20, 1968. Oxygen consumption of iguana.
152a. Moore, H. B., and B. F. McPherson, Bull. Marine Sci. *15*:855–871, 1965. Productivity, sea urchins.
153. Morrison, P., and P. Ryser, Physiol. Zool. *37*:90–103, 1959. Metabolism of small mammals.
154. Munz, F. W., and R. W. Morris, Comp. Biochem. Physiol. *16*:1–6, 1965. Oxygen consumption of hagfish, *Eptatretus*.
155. Murrish, D. E., and K. Schmidt-Nielsen, Respir. Physiol. *10*:151–158, 1970. Water conservation in respiratory passages of lizards.
156. Naifeh, K. H., et al., Respir. Physiol. *9*:31–42, 349–357, 1970. Respiratory patterns in crocodilian reptiles.
157. National Academy of Sciences, Handbook of Respiration. W. B. Saunders Co., 1958. 403 pp.
158. Newell, R. C., and W. A. M. Courtney, J. Exp. Biol. *42*:45–57, 1965. Respiratory movements in *Holothuria*.
158a. Nielsen, B., J. Exp. Biol. *39*:107–117, 1962. Regulation of respiration in lizards.
159. O'Farrell, M. J., and E. H. Studier, Comp. Biochem. Physiol. *35*:697–703, 1970. Metabolism of *Myotis*.
160. Olsen, C. R., R. Elsner, and F. C. Hale, Science *63*:953–955, 1969. The "blow" of the pilot whale.
160a. Oshino, R., et al., Biochem. Biophys. Acta *273*:5–17, 1972. Bacterial luminescence probe for O_2.
161. Otis, A. B., pp. 315–321. *In* Oxygen in the Animal Organism, edited by F. Dickens and E. Neil. Pergamon Press, 1965. Response of man to altitude.
162. Pearson, O. P., Condor *52*:145–152, 1950. Metabolism of hummingbirds.
163. Peterson, D. F., and M. R. Fedde, Science *162*:1499–1501, 1969. CO_2 receptors in lungs of chickens.
163a. Phleger, C. F., and A. A. Benson, Nature *230*:122, 1971. Hyperbaric oxygen in some deep sea fish.
164. Pickwell, G. V., Comp. Biochem. Physiol. *27*:455–485, 1968. Metabolism of ducks during diving.
165. Piiper, J., D. Baumgarten, and M. Meyer, Comp. Biochem. Physiol. *36*:513–520, 1970. Effect of hypoxia on respiration and circulation in dogfish.
166. Platzer-Schultz, I., Z. vergl. Physiol. *67*:179–185, 1970. Oxygen consumption, life stages of *Chironomus*.
167. Prosser, C. L., et al., Physiol. Zool. *30*:137–141, 1957. Acclimation to reduced oxygen in goldfish.
168. Rahn, H., pp. 3–23. *In* Ciba Foundation Symp. on Devel. of Lung, edited by A. V. S. de Reuck and R. Porter, 1966. Gas transport from environment to cell.
169. Rahn, H., and J. B. West, The Physiologist *6*:259, 1963. Comparison of air and water breathing.
169a. Rahn, H., and T. Yokoyama, eds., Physiology of Breath-Holding; the Ama of Japan. Publ. 1341, Nat. Acad. Sci. Washington, 1967. 369 pp.
170. Ralph, R., and I. Everson, Comp. Biochem. Physiol. *27*:299–307, 1968. Respiration of Antarctic fish.
171. Randall, D. J., G. F. Holeton, and E. D. Stevens, J. Exp. Biol. *46*:339–348, 1967. Respiratory efficiency in trout.
172. Randall, D. J., and G. Shelton, Comp. Biochem. Physiol. *9*:229–339, 1963. Respiratory responses to O_2 lack and CO_2 excess in fish, *Tinca*.
172a. Reid, K. R. H., Comp. Biochem. Physiol. 7:89–101, 1962. Relation between body size and O_2 consumption, brachiodonts.
173. Reynafarje, C., et al., J. Appl. Physiol. *24*:93–97, 1968. Erythropoiesis at high altitudes.
173a. Richards, E. D., and P. Fromm, Comp. Biochem. Physiol. *29*:1063–1070, 1968. Blood flow in gills of trout.
174. Ridgway, W. H., and D. G. Johnston, Science *151*:456–457, 1966. Blood oxygen and ecology of porpoises.
175. Rising, J. D., Comp. Biochem. Physiol. *25*:327–333, 1969; *31*:915–925, 1969. Metabolism of orioles and sparrows.
176. Salmoiraghi, G. C., and R. von Baumgarten, J. Neurophysiol. *24*:203–218, 1961. Electrical responses from respiratory neurones in brain of cat.
176a. Salt, W. G., Biol. Rev. *39*:113–136, 1964. Respiratory evaporation in birds.
177. Satchell, G. H., Comp. Biochem. Physiol. *27*:835–841, 1968. Neurological basis for swimming and respiration in dogfish.
178. Schmidt-Nielsen, K., et al., Respir. Physiol. *9*:263–276, 1970. Countercurrent heat exchange in respiratory passages.
178a. Schmidt-Nielsen, K., et al., J. Cell. Physiol. *67*:63–72, 1966. Oxygen consumption in echidna.
179. Schneiderman, H. A., and A. N. Schechter, J. Insect Physiol. *12*:1143–1170, 1966. Discontinuous respiration in insects.
180. Schneiderman, H. A., and C. M. Williams, Biol. Bull. *109*:123–143, 1955. Discontinuous respiration in silkworm.
181. Schreuder, G. R., Exp. Parasitol. *2*:236–241, 1953; *5*:138–148, 1956. Intermediary metabolism of hemoflagellates.
182. Schroeder, L., and D. G. Dunlap, Comp. Biochem. Physiol. *35*:953–957, 1970. Respiration in cecropia moths.
183. Schwartz, E., Z. vergl. Physiol. *65*:324–339, 1969. Respiration in air-breathing fish.
184. Severinghaus, J. W., C. R. Brainton, and A. Carcelen, Respir. Physiol. *1*:308–334, 1966. Respiratory insensitivity to hypoxia in chronically hypoxic man.
185. Shelton, G., Respir. Physiol. *9*:183–196, 1970. Circulatory adjustments to periodic respiration in *Xenopus*.
186. Shimada, K., and S. Kobayasi, Acta Med. Biol. *13*:297–303, 1965. Neural control of pulmonary smooth muscle in toad.
186a. Singh, B. N., and G. M. Hughes, J. Exp. Biol. *55*:421–434, 1971. Respiration in air-breathing catfish.
187. Smit, H., et al., Comp. Biochem. Physiol. *39A*:1–28, 1971. Oxygen consumption of goldfish.
188. Smith, R. E., Ann. N.Y. Acad. Sci. *62*:403–422, 1956. Relation between mitochondrial properties and body size in mammals.
189. Spitzer, K. W., D. E. Marvin, and A. G. Heath, Comp. Biochem. Physiol. *30*:83–90, 1969. Effects of temperature on respiration and cardiac response of bluegill sunfish to hypoxia.
190. Stahl, W. R., J. Appl. Physiol. *22*:453–460, 1967. Scaling of respiratory variables in mammals.
191. Steen, J. B., and A. Kruysse, Comp. Biochem. Physiol. *12*:127–142, 1964. Respiratory function of fish gills.
192. Stevens, E. D., and D. J. Randall, J. Exp. Biol. *46*:307–315, 329–337, 1967. Effects of exercise on respiration in rainbow trout.
193. Taylor, C. R., K. Schmidt-Nielsen, and J. L. Raab, Amer. J. Physiol. *219*:1104–1107, 1970. Oxygen consumption during running, mammals of different sizes.
194. Teal, J. M., and F. G. Carey, Deep-Sea Res. *14*:725–733, 1967; Limnol. Oceanogr. *12*:548–550, 1967. Effect of pressure and temperature on respiration of euphausiids.
195. Teal, J. M., and F. G. Carey, Physiol. Zool. *40*:83–91, 1967. Metabolism of marsh crabs at reduced oxygen levels.
196. Thomas, H. J., J. Exp. Biol. *31*:228–251, 1954. Oxygen uptake of *Homarus*.
197. Thompson, R. K., and A. W. Pritchard, Biol. Bull. *136*:274–287, 1969. Respiratory adaptation of burrowing crustaceans.
198. Thomsen, E., J. Exp. Biol. *26*:137–149, 1949. Hormone effects on metabolism in flies.
199. Thorpe, W. H., Biol. Rev. *25*:344–390, 1950. Plastron respiration in aquatic insects.
199a. Todd, E. S., Biol. Bull. *130*:265–288, 1966. Air breathing by mudsucker.
200. Tucker, V. A., Science *154*:150–151, 1966. Oxygen consumption of a flying bird.
201. Tucker, V. A., J. Exp. Biol. *48*:55–66, 1968. Respiration of sparrows in high altitude flight.
202. Tucker, V. A., J. Exp. Biol. *48*:67–87, 1968. Respiration and evaporative water loss in flying budgerigar.
203. Tucker, V. A., Comp. Biochem. Physiol. *34*:841–846, 1970. Energetics of locomotion.
204. Tucker, V. A., Amer. Zool. *11*:115–124, 1971. Energetics of flight and walking.
205. Veghte, J. H., Physiol. Zool. *37*:316–328, 1964. Metabolism of gray jay.
206. Vernberg, F. J., and I. E. Gray, Biol. Bull. *104*:445–449, 1953. Oxygen consumption by tissues of teleosts of different sizes.
207. Vlasblom, A. G., Comp. Biochem. Physiol. *36*:377–385, 1970. Oxygen transfer by physical gill in aquatic insects.
208. Walker, J. G., Biol. Bull. *138*:235–244, 1970. Oxygen poisoning in annelid, *Tubifex*.
209. Wälshe, B. M., J. Exp. Biol. *25*:35–44, 1948. Oxygen requirements of chironomid larvae.
210. Wang, L. C., and J. W. Hudson, Comp. Biochem. Physiol. *32*:275–293, 1970; *38A*:59–90, 1971. Metabolism of chipmunk and *Perognathus*.
211. Webb, H. M., and F. A. Brown, Physiol. Rev. *39*:127–161, 1959. Biological rhythms.
212. Weis-Fogh, T., J. Exp. Biol. *47*:561–587, 1967. Respiration and tracheal ventilation in flying insects.
213. Wells, G. P., Proc. Roy. Soc. Lond. B *140*:70–82, 1932; J. Marine Biol. Assoc. U.K. *28*:447–464, 1949. Respiratory movements in polychaetes, especially *Arenicola*.
214. Whalen, W. J., The Physiologist *14*:69–82, 1971. Intracellular oxygen concentrations in muscle.
215. Whitford, W. G., and V. H. Hutchison, Copeia 573–577, 1966. Pulmonary and cutaneous gas exchange in salamanders.
216. Whitmore, C. M., C. E. Warden, and P. Dondaroff, Trans. Amer. Fish Soc. *89*:17–26, 1960. Behavior of fish in oxygen gradient.
217. Wiens, A. W., and K. B. Armitage, Physiol. Zool. *34*:39–54, 1961. Oxygen consumption of crayfish as a function of temperature and O_2 saturation.
218. Wieser, W., Z. vergl. Physiol. *45*:247–271, 1962. Oxygen consumption of land isopods.
219. Wieser, W., J. Marine Biol. Assoc. U.K. *42*:665–682, 1962. Adaptations of intertidal isopods.

220. Wohlschlag, D. E., Biol. of Antarctic Seas *1*:33–62, 1964. Metabolism and ecology of Antarctic fishes.
221. Yousef, M. K., R. R. J. Chaffee, and H. D. Johnson, Comp. Biochem. Physiol. *38A*:709–712, 1971. Effect of temperature on oxygen consumption of tree shrews.
222. Zeuthen, E., Ann. Rev. Physiol. *17*:459–482, 1955. Comparative physiology of respiration.
223. Zeuthen, E., C. R. Lab. Carlsberg, Ser. Chim. *26*:17–161, 1947. Oxygen consumption of developing animals.
224. Zeuthen, E., Quart. Rev. Biol. *28*:1–12, 1953. Oxygen uptake as related to size of organism.
225. Zinkler, D., Z. vergl. Physiol. *52*:99–144, 1966. Comparative metabolism of invertebrates.

REFERENCES

Swim-Bladders

For references to oxygen and respiration, see page 208.

1. Alexander, R. M., Biol. Rev. *41*:141–176, 1966. Review of function of swim-bladders.
2. Berg, T., and J. B. Steen, J. Physiol. *195*:631–638, 1968. Time constants of Root effects.
3. Coburn, R. F., and L. B. Mayers, Amer. J. Physiol. *220*:66–74, 1971.
4. Copeland, D. E., Z. Zellforsch. *93*:305–331, 1969. Ultrastructure of gas gland in *Fundulus*.
5. Copeland, D. E., Biol. Bull. *135*:486–500, 1968. Structure of gas gland in siphonophore float.
6. D'Aoust, B. G., Science *163*:576–578, 1969. Oxygen toxicity in fish.
7. D'Aoust, B. G., Comp. Biochem. Physiol. *32*:637–668, 1970. Insensitivity of gas gland to hyperbaric O_2.
8. Denton, E. J., Prog. Biophys. and Biophys. Chem. *11*:178–234, 1961. Po_2 in oceans.
9. Denton, E. J., J. B. Gilpin-Brown, and J. V. Howarth, J. Marine Biol. Assoc. U.K. *41*:351–364, 1964. Osmotic flotation mechanisms of cuttlefish.
10. Enns, T., E. Douglas, and P. F. Scholander, Adv. Biol. Med. Phys. *11*: 231–244, 1967. Function of rete in gas secretion.
11. Enns, T., P. F. Scholander, and E. D. Bradstreet, J. Phys. Chem. *69*: 389–391, 1965. Gas pressures at high hydrostatic pressures.
12. Fänge, R., Acta Physiol. Scand. *30* (Suppl. 110):1–133, 1953. Rete of an eel.
13. Fänge, R., Physiol. Rev. *46*:299–322, 1966. Function of gas gland, secretion and resorption.
14. Kuhn, H. G., P. Moser, and W. Kuhn, Pflüger. Arch. *275*:231–237, 1962. Modification of gas solubility by solutes from gas gland.
15. Kuhn, W., A. Ramel, H. G. Kuhn, and E. Marti, Experientia *19*:497–511, 1963. Salting out as a mechanism of gas concentration.
16. Larimer, J. L., and E. A. Ashby, J. Cell. Comp. Physiol. *60*:41–48, 1962.
17. Phleger, C. F., and A. A. Benson, Nature *230*:122, 1971. Lipid concentration in swim-bladder of deep-sea fish.
18. Pickwell, G. V., Navy Elect. Lab. Rep. 1369, pp. 1–47, 1966. Composition of float in siphonophores.
19. Scholander, P. F., and L. van Dam, Biol. Bull. *104*:75–86, 1953; *107*: 247–259, 1954; J. Cell. Comp. Physiol. *48*:517–522, 1956. Oxygen in swim-bladders of deep-sea fish.
20. Steen, J. B., Acta Physiol. Scand. *58*:124–137, 138–149, 1963; *59*: 221–241, 1963. Direct measurements of blood gases in secreting gland and rete of eel.
21. Steen, J. B., pp. 413–443. *In* Fish Physiology, Vol. IV, edited by W. S. Hoar and D. J. Randall. Academic Press, 1970. Review of gas secretion in swim-bladder.
22. Wittenberg, J. G., J. Exp. Biol. *37*:698–705, 1960; *also* Wittenberg, J. G., et al., Biochem. J. *85*:9–15, 1962. Origin of gases in siphonophore float.
23. Wittenberg, J. B., J. Gen. Physiol. *44*:521–526, 527–542, 1961. Proof that secreted gas is molecular O_2.
24. Wittenberg, J. B., Physiol. Rev. *50*:559–636, 1970. Theory of transport of gases; salting out effect.
25. Wittenberg, J. B., M. J. Schwerd, and B. A. Wittenberg, J. Gen. Physiol. *48*:337–355, 1964. Gas in bladder of bluefish; inert gases.

CHAPTER 6

COMPARATIVE INTERMEDIARY METABOLISM

*by P. W. Hochachka**

METABOLIC PATHWAYS

INTRODUCTION

Classically, metabolic pathways have been divided into two types—catabolic and anabolic. The former defines essentially degradative processes in which large organic molecules are broken down to simple cellular constituents with the release of free energy. The latter defines synthetic processes that produce complex organic cellular components from simpler precursors and frequently involve reductive energy-requiring reactions. The remarkable diversity of metabolites handled in the metabolism of different organisms is balanced by an equally remarkable order and simplicity in metabolic patterns. Of these patterns, none is more significant than the central area of metabolism that provides a link between catabolic and anabolic routes.

*The bulk of this chapter was written while I was a visiting investigator at the Oceanic Institute, Waimanalo, Hawaii. Especial thanks are due to the Oceanic Institute for providing facilities and working space. B. Clayton-Hochachka assisted in much of the library search.

GLOSSARY OF ABBREVIATIONS

INTERMEDIATES

AMP, ADP, ATP—the adenylates, adenosine mono-, di-, and triphosphate
CoA—Coenzyme A
DHAP—dihydroxyacetone phosphate
1,3 DPG—1,3 diphosphoglycerate
FAD, $FADH_2$—oxidized and reduced flavin adenine dinucleotide
FDP—fructose diphosphate
F1P—fructose-1-phosphate
F6P—fructose-6-phosphate
FFA—free fatty acid
GAP—glyceraldehyde-3-phosphate
G1P—glucose-1-phosphate
G6P—glucose-6-phosphate
αGP—α-glycerophosphate
αKGA—α-ketoglutarate
NAD, NADH—oxidized and reduced nicotinamide adenine dinucleotide
NADP, NADPH—oxidized and reduced nicotinamide adenine dinucleotide phosphate
OXA—oxaloacetate
PEP—phosphoenolpyruvate
6PG—6-phosphogluconate
PGA—phosphoglycerate
UDPG—uridine diphosphate glucose
UDPAG—uridine diphosphate acetylglucosamine

ENZYMES

FDPase—fructose diphosphatase
G6Pase—glucose-6-phosphatase
G6PDH—glucose-6-phosphate dehydrogenase
HK—hexokinase
LDH—lactate dehydrogenase
MDH—malate dehydrogenase
PC—pyruvate carboxylase
PEPCK—phosphoenolpyruvate carboxykinase
PFK—phosphofructokinase
PK—pyruvate kinase
TDH—triose phosphate dehydrogenase (glyceraldehyde-3-P dehydrogenase)

In the initial anaerobic phases of catabolism, large molecules are broken down to yield, apart from CO_2 and H_2O, a quite restricted group of small organic molecules, liberating about *one-third of the available free energy in the process.* For carbohydrates, these product molecules are triose phosphates and/or pyruvate; for fats, they are acetyl CoA, propionyl CoA, and glycerol; for proteins, they are acetyl CoA, oxalacetate, α-ketoglutarate, fumarate, and succinate.

A major and unifying theme in biochemistry is that the same set of reactions is involved in three crucial phases of metabolism: (1) the interconversion of various products of catabolism mentioned above; (2) their complete combustion to CO_2 and H_2O, which releases to the organism the remaining two-thirds of the available free energy supply; and (3) the supply of crucial intermediates for biosynthetic, anabolic processes. These central pathways are composed of relatively few reactions. In essence, they consist of these steps: triose phosphate $\rightleftharpoons$ pyruvate; pyruvate $\rightarrow$ acetyl CoA; oxalacetate $\rightleftharpoons$ aspartate; α-ketoglutarate $\rightleftharpoons$ glutamate; and the citric acid cycle reactions designed to catalyze the complete combustion of acetyl-CoA to CO_2 and H_2O.

A fundamental distinction between anabolic and catabolic routes is that they rarely follow the same enzyme pathways in detail. This is apparent when the product of catabolism is not identical with the carbon source in anabolism, as is the case with many amino acids. With fatty acids, catabolism leads to acetyl CoA as the end product, and biosynthesis commences with this intermediate. However, the enzyme reactions of fatty acid synthesis and degradation are different, and also are located in different cellular compartments. Even in the biosynthesis of glucose, which proceeds in large part by a reversal of a number of glycolytic reactions, biosynthesis and degradation differ at several critical points in the enzyme sequence. Thus, glucose is converted to G6P during catabolism in a reaction requiring ATP, but it is formed in anabolism by simple hydrolysis of the phosphate ester. Pyruvate is produced in catabolism from PEP by a transphosphorylation to ADP; it is utilized in gluconeogenesis in most organisms by two linked reactions, which first carboxylate the pyruvate to oxalacetate and then transform the latter to PEP. Two important implications arise: (1) the two pathways cannot be maximally active simultaneously, for this would lead to a kind of short circuit in metabolism, and (2) controls on the two pathways must be integrated to some extent. The latter requirement leads to a very close connection between related anabolic and catabolic pathways. The following discussion is organized in this framework.

The Pathways of Glycolysis and Gluconeogenesis

The major route of catabolism of glucose and glycogen in most cells is encompassed by the series of reactions that convert glucose to pyruvate. Such diverse physiological work functions as skeletal muscle contraction (at times in a relatively anaerobic environment), heart muscle contraction (in a relatively aerobic environment), ion transport functions in the central nervous system, in the nephric tubules, and in the red blood cell, and many of the "fermentations" of glucose by invertebrate facultative anaerobes, all make use of this pathway. Because the pathway does not require oxygen, it is accepted as being the primeval mechanism for providing requisite cellular energy supplies under completely anaerobic conditions.[154] The first end product of the pathway, pyruvate, is reduced to lactate by an NADH-dependent lactate dehydrogenase in most vertebrate tissues. In the invertebrates, other metabolic fates of pyruvate are common and will be discussed in more detail below.

It is generally accepted that the major route of glucose synthesis *from* pyruvate involves carboxylation of pyruvate to OXA and the conversion of OXA to PEP, followed by a series of reactions leading ultimately to glucose. In the case of four-carbon precursors, the pathway leads directly to OXA, because the pyruvate carboxylation step is of course not required.

The situation found in complex tissues, such as liver, is shown in Figure 6–1. The diagram indicates several pivotal reactions in the two pathways (located at the points of metabolism of glycogen, G6P, F6P, and PEP). At each of these points the catabolic and anabolic reactions are catalyzed by enzymes which are kinetically and structurally distinct. In the synthesis and breakdown of carbohydrate, however, part of the sequence appears to be shared—from G1P to F6P and from FDP to PEP. The evidence supporting this view is equivocal and does *not* exclude the

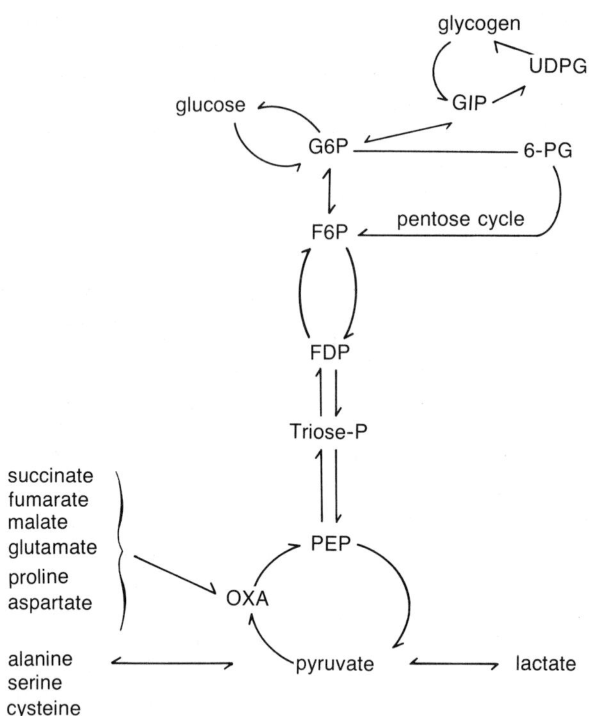

Figure 6–1. The glycolytic and gluconeogenic pathways and some pathways of G6P metabolism.

existence of unique gluconeogenic enzyme steps in this sequence. It is known, for example, that triose phosphate dehydrogenase (TDH) occurs in at least two kinetically distinguishable forms; the form in muscle favors glycolysis, while liver contains a second form which is kinetically better adapted for gluconeogenesis.

Central to an understanding of the operation of these pathways of catabolism and anabolism are the factors which control or limit the flow of carbon through them. The first and simplest approach to this problem is based on the relative maximal catalytic activities of component enzymes under optimal conditions of substrate, modulators, pH, and so forth. The results, if valid, define the upper limit of the capacity of a given tissue to catalyze the metabolic pathway under consideration, and may help to identify those individual reactions which have low catalytic capacities.

In the pathway from glucose to pyruvate, three enzymes (HK, PFK, and PK) are unique to the glycolytic sequence (Table 6–1). Since PFK may be the rate limiting enzyme under

TABLE 6–1. Maximum Catalytic Activities of Enzymes Unique to Glycolysis in Rat Tissues, Expressed in μM/min/g net weight at 37°C. (Modified from reference 127.)

Enzyme or Process	*Liver*	*Kidney*	*Muscle*	*Heart*	*Brain*
HK (low K_m)	0.7	7.9	1.8	10	17
HK (high K_m)	4.3	—	—	—	—
PFK	3.3	5.7	80	20	33
PK	50	50	780	167	200
glycolysis (μM glucose utilized)	0.2	0.17	0.03 (resting) 1.7 (tetanized)	0.0 (aerobic) 2.8 (anaerobic)	—

TABLE 6–2. Maximum Catalytic Activities of Enzymes Unique to Gluconeogenesis in Rat Tissues, Expressed in μM/min/g net weight at 37°C. (Modified from reference 127.)

Enzyme or Process	*Liver*	*Kidney*	*Muscle*	*Heart*	*Brain*
PC	6.7	5.8	0	0	0.5
PEPCK	6.7	6.7	<0.1	0.5	0.3
FDPase	15	17	0.7	0.7	2.3
G6Pase	17	13	–	2	0
glucogenesis (μM pyruvate converted to glucose)	1.7	1.4			

most conditions, the extent of control of glycolysis may be evaluated by comparison of the capacity of this enzyme with observed rates of glycolysis (Table 6–1). Such comparison suggests that this pathway operates from one to two orders of magnitude *below* its maximal capacity under most conditions.

High maximal capacities of the enzymes unique to the pathway from pyruvate to glucose (PC, PEPCK, FDPase, and G6Pase) are present only in those tissues, such as liver and kidney, that catalyze significant rates of glucose synthesis (Table 6–2). Comparison of PC and PEPCK with measured rates of glucose synthesis from 3- and 4-carbon precursors indicates that this pathway operates at about 1/3 to 1/2 its maximal capacity. This relationship is in contrast to that noted above for the glycolytic pathway.

The enzymes of gluconeogenesis may also be present in other tissues, such as brain, heart, and muscle, where they may be integrated into (and subserve) different physiological functions.[97]

Typically, the maximal capacities of enzymes that are common to glycolysis and glucogenesis are very high, and large differences between strictly glycolytic or glucogenic tissues are not observed (Table 6–3).

The capacities of the enzymes in glycogen metabolism and the pentose phosphate pathway are shown in Table 6–4. If phosphorylase is assumed to function only in glycogen breakdown *in vivo*, the glycogenolytic activity of most tissues appears to exceed glycogen synthesis by at least one order of magnitude. Comparisons among different mammalian tissues indicate that kidney cortex and brain have a lesser capacity for overall glycogen metabolism than liver, heart, or muscle, in accord with the absence of glycogen from the normal kidney and brain.

The metabolic roles of the pentose cycle usually are assumed to be the provision of NADPH for biosynthetic processes (particularly fatty acid synthesis) and pentoses for nucleotide and nucleic acid synthesis. The contribution of the pentose cycle to G6P metabolism has been estimated *in vivo* as being less than 5% in all tissues except adipose. In rat liver, the maximal capacities for the first two enzymes which characterize this pathway (Table 6–4) are greater than the capacities observed for PFK and glycogen synthetase (Tables 6–1 and 6–4). Hence, the small contribution of the pentose phosphate pathway to overall G6P metabolism cannot be explained on the basis of maximal capacities of enzymes competing for G6P.

The qualitative differences among various selected tissues are summarized in Figure 6–2. Neither the quantitative contribution of

TABLE 6–3. Maximum Catalytic Activities of Enzymes Common to Glycolysis and Gluconeogenesis in Rat Tissues, Expressed in μM/min/g net weight at 37°C. (Modified from reference 127.)

Enzyme	*Liver*	*Kidney*	*Muscle*	*Heart*	*Brain*
FDP-aldolase	10	10	77	24	17
triose phosphate dehydrogenase	170	200	590	180	160
3-phosphoglycerate kinase	150	170	340	150	150
enolase	17	53	210	20	30
LDH	230	170	490	450	130

TABLE 6–4. Maximum Catalytic Activities of Enzymes of Glycogen Metabolism and the Pentose Phosphate Pathways in Rat Tissues, Expressed in μM/min/g net weight at 37°C. (Modified from reference 127.)

Enzyme	*Liver*	*Kidney*	*Muscle*	*Heart*	*Brain*
PGM	43	7	100	20	7
UDPG pyrophosphorylase	3	–	3	–	–
glycogen synthetase*	5	0.5	4	4	1.2
phosphorylase*	37	–	100	36	–

*Phosphorylase and glycogen synthetase activities are total capacities of different forms of these enzymes.

metabolites from each tissue to the blood nor the uptake of metabolites by each tissue from the blood is well worked out, but in a qualitative sense it is clear that brain, muscle, and possibly heart tissues depend upon glucose uptake from the blood, while liver and kidney can release it into the blood for further metabolism elsewhere. Similar differences among the tissues are evident as regards the overall metabolism of lactate and fatty acids (Figure 6–2). More detailed descriptions of these pathways are available in all modern biochemistry textbooks and need not be repeated here.

The glycolytic and glucogenic pathways appear to be ubiquitous in the animal kingdom; evidence for and documentation of the phylogenetic distribution of the pathways are adequately covered elsewhere. We will stress, for a small number of systems, the roles of these pathways in adaptation of organisms to their conditions of existence.

Phylogenetic Distribution of the Glycolytic and Gluconeogenic Pathways

Glycolysis and Gluconeogenesis in Parasitic Helminths: Metabolic Adaptation to Low Oxygen Tensions. One of the most fascinating biochemical adaptations is evident in the aerobic-anaerobic transitions found in parasitic metazoans. Through the *deletion* of certain enzymic reactions and the *alteration* of the kinetic properties of persisting enzymes,

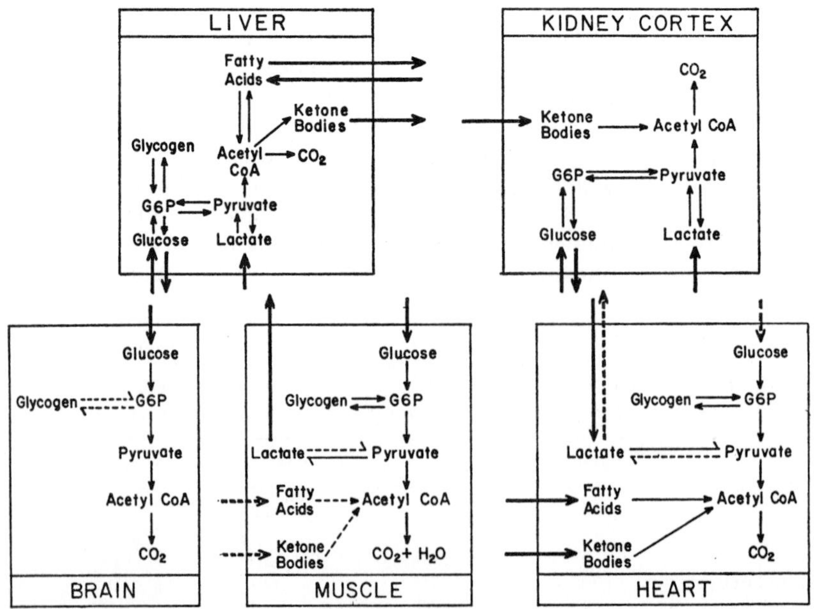

Figure 6–2. Qualitative summary of tissue capacities for reaction pathways of carbohydrate metabolism and other related pathways. (From Scrutton, M. C., and M. F. Utter, Ann. Rev. Biochem. *37*:249–302, 1968.)

many parasites display an ability to exploit the unique intraorganism environment in which they must exist.

Since environmental factors have been shown to be critically important in patterns of metabolism in parasitic helminths, it is not safe to generalize from the parasitic to the free-living forms. Nevertheless, three different kinds of evidence (presence of glycolytic enzymes, presence of glycolytic intermediates, and predictable labelling patterns of end-products following administration of specifically labelled ^{14}C-glucose) indicate that the first phases of the glycolytic pathway occur commonly among the helminths. The terminal reaction sequences of the glycolytic pathway, however, are often strikingly modified in accord with the primary special condition (anaerobiosis) which is often encountered in the normal life cycle of helminths and other parasites. This field has been reviewed by von Brand,[153] Read,[111] and others. We will focus our attention on the intestinal parasite *Ascaris lumbricoides,* since, within the helminths, its metabolism is perhaps most completely understood.

The problems likely to arise for an intestinal parasite, such as *Ascaris lumbricoides,* can be best appreciated if we consider the outcome of strict anaerobiosis in a typical aerobic metazoan. The catabolism of glucose, for example, could proceed to the level of lactate, but no further. For the aerobic organism, a momentary truncating of glucose catabolism would ordinarily be relieved, following resupply of oxygen, and activation of gluconeogenic reactions takes place in the liver. However, for the parasite, no relief from the lack of O_2 occurs. Hence, exploitation of this environment requires the development of metabolic reactions to circumvent such metabolic "dead-ends." This is indeed what occurs in the parasitic helminths (Fig. 6–3).

In initial phases, the carbohydrate catabolizing reactions of the parasite are comparable to the glycolytic pathway in aerobic cells. Important differences occur at the level of PEP. Phosphoenolypyruvate (PEP), which normally would be converted to pyruvate (and then reduced to lactate) in a typical aerobic metazoan, is shunted away from the normal glycolytic reactions towards oxaloacetate (OXA) synthesis. OXA then follows a series of transformations which, in effect, is a reversed and abbreviated segment of the Krebs citric acid cycle (see p. 241). The ultimate product of this fermentative sequence is succinate. Succinate is far from being a metabolic "dead-end" like lactate. Instead, succinate can be drawn into propionate biosynthesis and, therefore, this

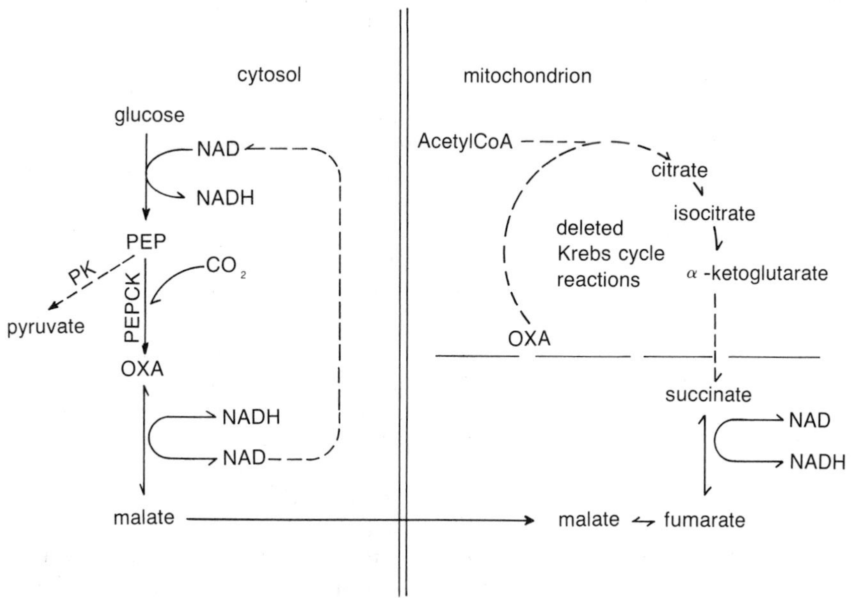

Figure 6–3. Proposed pathway for utilization of glucose in *Ascaris* muscle. (After Saz, H. J., and O. L. Lescure, Comp. Biochem. Physiol. *30*:49–60, 1969; Saz, H. J., Amer. Zool. *11*:125–135, 1971.)

fermentative pathway ultimately can lead to lipid synthesis rather than to the build-up of an acidic end product such as lactate.

Several features in this arrangement indicate specific adaptations for maintaining proper *vectors* and *rates* of carbon flow under anaerobic conditions:

(1) The production of lactate, which is a metabolic *cul de sac,* is avoided by the presence of only low activities of pyruvate kinase. This *deletion* prevents a large accumulation of pyruvate from PEP. Moreover, lactate dehydrogenase (LDH) activities in *Ascaris* tissues are low; hence, any pyruvate which is formed is not strongly "drained" by this reaction.

(2) In *Ascaris,* the PEPCK reaction appears to favor OXA synthesis rather than PEP synthesis (as occurs in mammals and birds). The kinetic basis for this difference is an *increased* affinity for PEP and a *decreased* affinity for OXA. In addition, under physiological conditions, OXA formed by PEPCK is quickly removed by MDH conversion to malate. OXA concentrations are thereby maintained so low that significant reversal of the PEPCK reaction to PEP does not occur.

(3) The "usual" function of the LDH reaction in most organisms is the regeneration of NAD for further glycolysis. In *Ascaris,* this function is taken by the active malate dehydrogenase (MDH) reaction. Unlike the LDH reaction, *the MDH reaction is very much in the mainstream of carbon flow at this point.* Its product, malate, is quickly and reversibly converted to fumarate by the enzyme fumarase.

(4) The final, specific adaptation of helminth metabolism at this level is the succinoxidase catalyzed *reduction* of fumarate to succinate. In most organisms, this enzyme serves to oxidize succinate, and indeed it is one of the most characteristic steps in the Krebs cycle. In *Ascaris,* however, *the system appears to be better adapted for fumarate conversion to succinate.* The NAD formed by this reaction is probably made available for further glycolysis.[123] The succinate formed can be converted to propionate, which, as the CoA derivative, is an important precursor for fatty acid synthesis.

It is evident that the *Ascaris* kind of anaerobic metabolism provides greater flexibility than is typically found in the vertebrates. Its development clearly has been the outcome of selective forces acting upon the helminth parasite. In *Ascaris,* natural selection has molded energy metabolism so that the species is entirely committed to a metabolism which is largely anaerobic, for *its development led to the complete deletion of a functional Krebs cycle.* The sites of the deletion represent the first span of the Krebs cycle: *citrate synthase, aconitase, isocitrate dehydrogenase, and α-ketoglutarate dehydrogenase.* This deletion is as indicative of adaptation to anaerobiosis as are the specific "modifications" of the end reactions of glycolysis.

However, in parasitic helminths which routinely encounter *anaerobic* conditions alternating with *aerobic* ones, the maintenance of enzymes of aerobic metabolism would be favored. One would expect that in the various groups of helminths, spanning the spectrum from fully free-living forms to highly parasitic forms, there would occur a spectrum in the *degree of dependence* upon aerobic versus anaerobic metabolism. The *Ascaris* example is in fact an extreme one of an *obligate anaerobe.* In other parasitic helminths (such as the cestode *Echinococcus granulosus*), the Krebs cycle enzymes which are deleted in *Ascaris* are present.[153] Whether or not they are functional under normal parasitic conditions is unknown, but most likely they are not. By the same token, they undoubtedly *are* functional during aerobic exposures. These organisms, then, are *facultative anaerobes.* Mechanisms involved in the switch-over from anaerobic to aerobic metabolism, however, are not known in any detail. Further comparative studies of parasitic and free living helminths are needed to fill this gap in our knowledge.

The CO_2 fixing reaction catalyzed by PEPCK is also important in glucose synthesis from noncarbohydrate sources, but control of glucogenesis has not been fully examined in the helminths.[124]

Glycolysis and Gluconeogenesis in the Molluscs. By virtue of their mode of existence, many molluscs, and particularly bivalves, appear to be facultative anaerobes. The degree of anaerobiosis which these organisms can tolerate has been an issue of some dispute. Organisms such as the oyster are capable of living in the complete absence of oxygen. One important component of this adaptation has been the evolutionary development of enzyme systems which do not lead to lactic acid as the major end product of anaerobic glucose dissimilation. Rather, in these organisms as in the helminths, succinic acid is accumulated as the primary end product.[51] The precise metabolic situation apparently depends upon the tissue under consideration.

In muscle, a highly glycolytic tissue, glycogen is the major source of energy. It is degraded to PEP by the usual glycolytic enzymes. The fate of PEP depends upon the environment. In the absence of oxygen, PEP is converted to OXA (via PEPCK); OXA is then converted via malate and fumarate to succinate. Thus, significant amounts of lactate do not accumulate. In the presence of oxygen, PEP is converted to pyruvate (via PK). Pyruvate is then oxidized by the usual Krebs cycle reactions. At the PEP branching point, then, essentially two alternate routes are available, one favored by the presence of oxygen, the other favored by its absence:

(1) PEP $\rightharpoonup$ OXA $\rightharpoonup \rightharpoonup$ succinate

(2) PEP $\rightharpoonup$ pyruvate $\rightharpoonup \rightharpoonup$ $CO_2 + H_2O$

In the mantle, an important gluconeogenic tissue in molluscs, the situation is more complex. During net glycolytic flow, PEP is converted to pyruvate under aerobic conditions. During anaerobiosis, PEP carboxylation to OXA occurs in the usual fashion. To this extent the system is similar to that in muscle. However, during gluconeogenesis from pyruvate or 4-carbon precursors, OXA must be converted to PEP. That is, in this tissue, depending upon metabolic requirements at any given time, PEPCK function must be allowed in both directions. Furthermore, under conditions favoring OXA decarboxylation to PEP, there is a strict requirement for PK to be held in a shut-off conformation, for any appreciable PK activity would lead to a wasteful cycling of carbon at this site.

From these considerations, it is clear that the control requirements for both PK and PEPCK in mantle differ in very fundamental ways from the requirements in muscle. In evolutionary terms, one obvious way of meeting these distinct control necessities is the elaboration of tissue-specific forms of these two enzymes, each displaying different regulatory properties. That this is indeed the case for PK in oysters has recently been established;[96] however, comparable data are not yet available for the PEP carboxykinases in oyster muscle and mantle.

In the *obligate* anaerobe, such as adult *Ascaris,* the metabolic reactions of the organism are at once simplified, through deletion of unnecessary enzymes, and modified, through changes in enzyme-substrate affinities and enzyme quantities, in such a manner that *only* an anaerobic mode of life remains possible. In contrast, in those parasitic species which encounter oxygen, the capacity for aerobic metabolism remains an important feature of the organism's biological organization. A third pattern of anaerobiosis is found in the case of intertidal molluscs.

Here, the crux of the control problem is that the intertidal bivalve must maintain the potential for directing PEP flux into two different directions: towards the Krebs cycle under aerobic conditions, and towards OXA synthesis during anaerobiosis. Since the time-course of PEP redirection is short (possibly as short as a single tidal cycle), the bivalve must maintain both pyruvate kinase and PEP carboxykinase in its tissues simultaneously. The *de novo* synthesis of one of the two enzymes to greet the incoming tide and the degradation of this enzyme and concomitant synthesis of the other enzyme to greet the falling tide is an approach which is precluded simply by the short time interval. Hence, solutions to this complex regulatory problem must involve control of carbon flow through the PEP metabolic "crossroads." The salient features of this control circuitry as they are currently documented are as follows:

First, pyruvate kinase is under tight regulation by FDP feedforward activation, ATP feedback inhibition, and alanine feedback inhibition. The K_m for PEP is strongly pH dependent and is about seven times greater at pH 6.5 than it is at pH 8.5. Consequently, at low substrate levels the reaction velocity is strikingly inhibited by H^+. The maximum velocity also shows a distinct alkaline pH optimum (five times greater activity at pH 8.5 than at pH 6.5). FDP abolishes these effects of pH.

Secondly, PEPCK displays a distinct acid pH optimum, and at this pH it has a 10-fold greater affinity for PEP than than does PK. Thus, in the absence of any other regulatory factor, H^+ production in effect activates PEPCK and inhibits PK. In addition, alanine increases the K_m for PK but decreases the K_m for PEPCK. This also would favor PEPCK activity whenever alanine accumulated.

From these data, Mustafa and Hochachka[96] have postulated that H^+ is pivotal in the channelling of carbon through this branch point. When there is an absence of O_2, the pH of the oyster tissues would be expected to drop because of the build-up of various acid products. This would lead to an effective activation of PEPCK (because of an increase in affinity

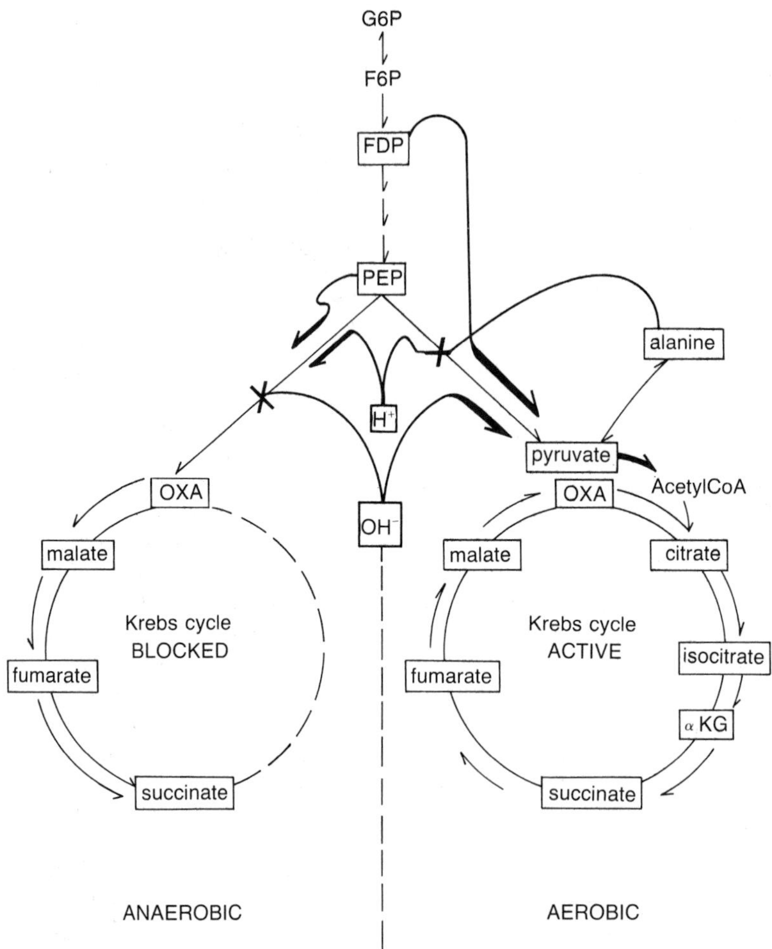

Figure 6–4. Postulated switch-over mechanisms in oyster tissue metabolism during aerobic and anaerobic conditions.

for PEP and an increase in V_{max}) and an inhibition of PK (by a decrease in affinity for PEP and a decrease in V_{max}). Any accumulation of alanine at this time would potentiate these effects. The more acidic the anaerobic system became, the more avidly would PEPCK channel PEP towards OXA (Fig. 6–4).

Upon return to aerobic conditions, the pH would be expected to rise again, and the above events would be reversed. As H^+ concentration fell, the PEPCK activity would fall concomitantly with activation of PK. Since PK and PFK activities are often integrated, one might also expect transient increases in FDP concentration; FDP overcomes H^+, alanine, and ATP inhibition of PK and would therefore dramatically potentiate the initial OH^- activation.

The most important feature of this regulatory system is that it is a kind of autocatalytic cascade. Once either PK activation or PEPCK activation is initiated, all the various regulatory interactions potentiate one another. Consequently, the system would switch from the "PEPCK open, PK closed" position to the "PK open, PEPCK closed" position with extreme efficiency. Such "on-off" behavior indeed appears to typify the action of many enzymes that are positioned at strategic branch points in metabolism. There is little doubt that the specific control components at this point in the metabolism of molluscan facultative anaerobes are the outcome of selective tailoring of the two enzymes functioning at this point. Many of the control properties of these enzymes in molluscs are shared by other organisms. Evolution seems to have arranged these characteristics in a functional manner in the molluscs.

Carboydrate Metabolism in Insects. In their carbohydrate metabolism, insects make use of substrates, enzymes, and operational principles that are common to other groups. However, these fit together to form a distinctive system, the main features of which are only now being fully discerned. Outstanding among the unique aspects of insect carbohydrate metabolism are the following, each of which we shall discuss in turn: (a) Trehalose is the predominant blood sugar and occupies a central position in the metabolism of carbohydrate, serving as a major carbon source for glycolysis. (b) The end products of the glycolytic pathway are pyruvate and α-glycerophosphate (αGP), and their further metabolism follows patterns which appear to play important functional roles in flight metabolism. (c) Characteristic of insects, and other arthropods, is the presence of chitin as a structural component of the exoskeleton, which in its deposition and resorption interacts closely with the metabolism of other carbohydrates. Finally, (d) the bulk conversion of glycogen to glycerol and sorbitol appears to be largely restricted to insects, and plays an important role in the well developed cold hardiness of this group.

TREHALOSE. The most characteristic sugar of insect hemolymph is trehalose, a symmetrical disaccharide of glucose (α-D-glucopyranosyl-α-D-glucopyranoside) which is nonreducing because the anomeric carbon atoms of both glucose moieties are bound in a glycosidic linkage.

Although trehalose was first discovered and named some centuries ago, its association with insects was generally forgotten and it was long regarded as a sugar characteristic of lower plants. In 1956 trehalose was rediscovered in this animal group almost simultaneously in the work of three different laboratories.[169] In some insects, the hemolymph contains so much trehalose that the sugar can easily be obtained from it in crystalline form.

REGULATION OF BLOOD SUGAR. Earlier workers generally assumed that the blood sugar levels of insects were determined largely by the nutritional state of the organism. However, evidence of three kinds suggests that blood trehalose concentrations are regulated by the organism: (1) When the trehalose level is lowered experimentally or by starvation, the initial level is restored at the expense of fat body glycogen. (2) On the other hand, when blood trehalose levels are experimentally elevated by injection of the sugar, it soon declines, establishing a new steady state somewhat above the normal. (3)

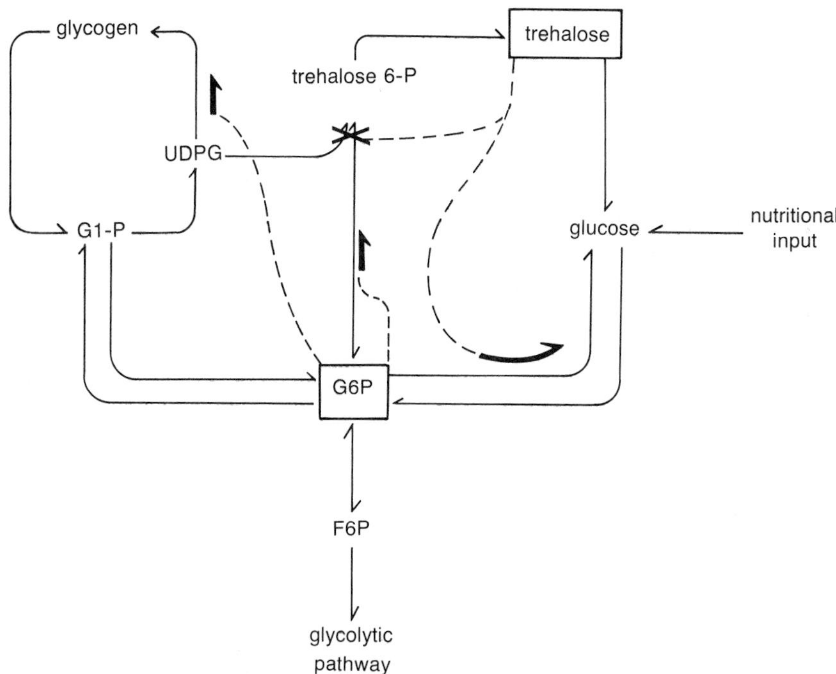

Figure 6–5. Metabolic pathways linking glucose, trehalose, and glycogen in insects, with a number of regulatory interactions indicated.

Although trehalose levels vary widely, there appears to be a characteristic level for any given species and stage, which tends to be maintained despite changes in nutrition. Thus, in the *Hyalophora cecropia* pupa during 6 months of diapause and subsequent adult development, while fat body glycogen and blood glycerol undergo extreme changes, blood trehalose is rather constant, first at a level characteristic of diapause, and then a new, higher level established early in development.[169] Experiments with ^{14}C-glucose have shown that trehalose is not inert in these systems, but is undergoing constant turnover. It is evident that the synthesis and catabolism of trehalose is a process of some importance in the economy of insects and that this process is intimately linked to the metabolism of glucose and glycogen (as indicated in Figure 6–5).

SYNTHESIS OF TREHALOSE. The activity of the fat body in trehalose synthesis has been unequivocally shown by incubating fat body tissue *in vitro* with ^{14}C-glucose and then isolating radioactive trehalose. Fat body probably is not the sole site of trehalose synthesis. Conversion of ^{14}C-labelled glucose to trehalose has been observed in insect muscle, mid-gut, and head tissues. Considering the bulk of the fat body in most insects, however, its contribution to trehalose production must be by far predominant. The enzyme system specific for trehalose synthesis in insect fat body was first examined by Candy and Kilby,[24] and evidence was obtained in support of the reaction sequence previously discovered in yeast. In this scheme, UDPG serves as a donor of one of the glucose moieties, while the other hexose moiety derives from G6P, which yields trehalose-6-phosphate upon condensation with the nucleotide. This is followed by dephosphorylation of trehalose-6-phosphate by a specific phosphatase:

(1) Trehalose phosphate synthetase:
UDP-glucose + glucose-6-P→trehalose-6-P + UDP

(2) Trehalose-6-phosphatase:
trehalose-6-P + H_2O→trehalose + P_i

Trehalose phosphate synthetase from larval fat body of *H. cecropia* can be obtained free of glycogen synthetase by high-speed centrifugation of fat body homogenates, during which the latter enzyme forms a sediment with the particulate glycogen.[95]

Control of the activities of these enzymes may be basic to the regulation of blood levels of trehalose.[119] Thus, any elevation in glucose (by ingestion or other process) causes a rise in G6P, and this will activate both trehalose-6-phosphate synthetase and glycogen synthetase. Trehalose-6-phosphate synthetase has a higher affinity for UDPG than does glycogen synthetase; hence, this will allow preferential synthesis of trehalose levels when UDPG levels are low. When trehalose accumulates sufficiently to inhibit the trehalose-6-phosphate synthetase, the UDPG level may rise and thus allow increased synthesis of glycogen. Such a control circuitry can provide for rapid production of both trehalose and glycogen when glucose input is high. In addition to inhibiting the trehalose-6-phosphate synthetase, G6Pase is specifically and greatly increased by trehalose. This mechanism may result in an indirect kind of negative feedback control, which removes a necessary substrate for the synthetase reaction (Fig. 6–5).

CLEAVAGE AND USE OF TREHALOSE. The metabolic use of trehalose depends upon its hydrolysis to glucose by the enzyme trehalase:

$$\text{trehalose} + H_2O \rightarrow 2\ \text{glucose}$$

The enzyme catalyzing this reaction has been substantially purified from whole insects of several species, and its kinetic properties are fairly well documented. The enzyme occurs in at least two forms (termed muscle and gut trehalases after the tissues from which they are characteristically extracted). Both enzyme variants are specific for trehalose and are not general α-glucosidases. In insect muscle, the bulk of the enzyme activity is membrane bound, but its precise intracellular positioning is still unsettled.[119, 169]

It has been pointed out by Sacktor and Wormser-Shavit[120] that a marked decrease occurs in the concentration of trehalose at the onset of flight in the blowfly, which is coincident with a rapid, marked increase in glucose levels. These opposite changes, occurring when there has been a considerable increase in glycolytic flux, clearly indicate that the cleavage of trehalose to glucose is greatly facilitated at flight initiation. Such trehalase activation is undoubtedly under strict regulation, but mechanisms of regulation of this membrane-bound enzyme in insect muscle remain unknown.[119]

END PRODUCTS OF GLYCOLYSIS IN INSECT TISSUES. Glucose, produced from trehalose

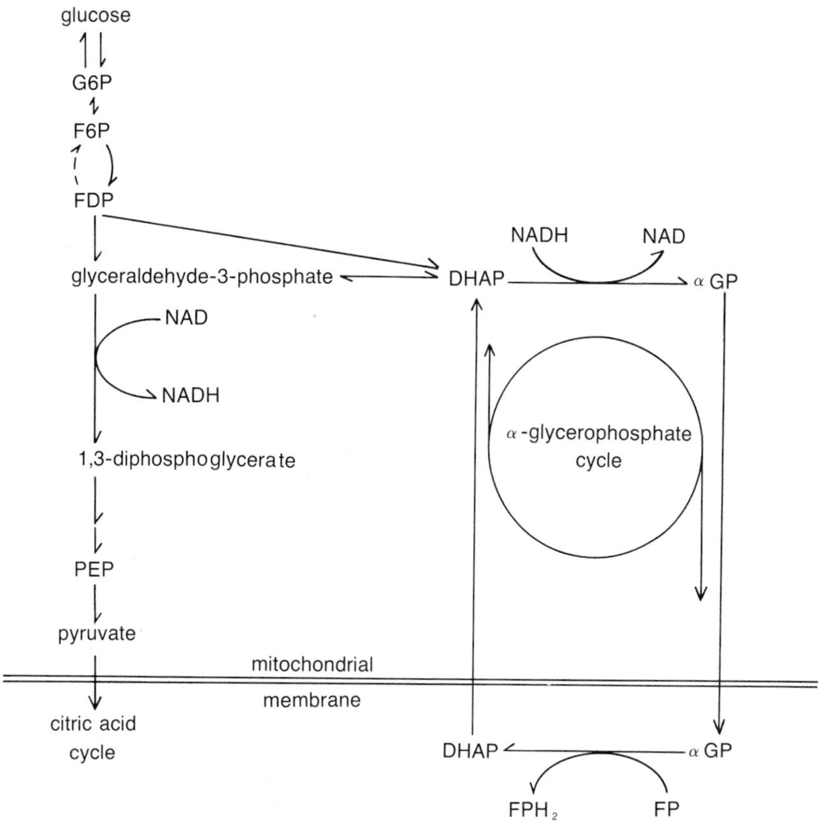

Figure 6–6. Glycolysis and the αGP cycle in insect flight muscle.

or from glycogen, is metabolized in insect tissues by the enzymes of the glycolytic pathway. In insects, this pathway differs operationally from that in vertebrates at the level of fructose diphosphate (FDP) cleavage.

The cleavage products of FDP, namely glyceraldehyde-3-phosphate (GAP) and dihydroxyacetone phosphate (DHAP), can be isomerized and subsequently handled in either of two ways (Fig. 6–6): (1) The DHAP can be reduced to αGP with the concomitant oxidation of NADH, or (2) the GAP can be oxidized in five enzymatic steps to pyruvic acid with the concomitant reduction of NAD^+.

The significance of the formation of pyruvate and αGP in glycolysis became evident with the discovery of the αGP cycle.[119] As shown in Figure 6–6, the αGP cycle provides a mechanism whereby extramitochondrially formed NADH becomes oxidized. In other organisms, glycolytically formed NADH is oxidized by the lactate dehydrogenase (LDH) reaction; but in insects (with a virtual absence of LDH), NADH reoxidation occurs concomitantly with the reduction of dihydroxyacetone phosphate by αGP dehydrogenase. The αGP formed in the cytosol diffuses freely into the mitochondria, where it is in turn reoxidized to DHAP by the mitochondrial αGP dehydrogenase, a flavoprotein (FP). This reaction thus regenerates dihydrooxyacetone phosphate, which can diffuse freely out of the mitochondria where it is again available for NADH oxidation. Accordingly, the cycle is a shuttle system in which NAD-linked substrates, in reduced and oxidized states respectively, enter and leave the mitochondria. In this way, hydrogen or reducing equivalents from the extramitochondrial pool of NADH pass the cytosol-mitochondrial barrier and are oxidized by the mitochondrial respiratory chain. Furthermore, the cycle is self-generating in that only a catalytic quantity of DHAP is needed to oxidize the NADH being formed continuously by glycolysis. This suggests that most of the triose phosphate can be isomerized to glyceraldehyde-3-phosphate and that essentially all of the carbon of the carbohydrate metabolized in prolonged work is convertible to pyruvate. Thus, the system is remarkably efficient, in that end products of glycolysis such as lactate need not accumulate wastefully as they do in exercising vertebrate muscle.

GLYCOGEN IN INSECT FLIGHT. Although changes in the concentration of glycogen are not great during the initial few seconds following flight initiation, after about 2 minutes of flight of the blowfly the glycogen levels decrease dramatically in the flight muscle, indicating that the polysaccharide does serve as a major energy reserve.[120] The rapid utilization of muscle glycogen shortly after commencement of muscle work is achieved through coordinated facilitation of glycogen phosphorylase and inhibition of glycogen synthetase.[26] The catalytic and regulatory mechanisms involved in mobilization and synthesis of glycogen are discussed at length below (pp. 248–250).

CHITIN AND OTHER INSECT GLYCOPROTEINS. Chitin is a long-chain polymer consisting of β-1,4-linked N-acetylglucosamine (together, apparently, with a certain variable percentage of glucosamine). In the insect cuticle, it occurs bound to protein, and the complex is thus a glycoprotein. Extensive studies on its structure by physical and chemical methods have been discussed elsewhere.[50]

Other glycoproteins in insects have been virtually ignored until recently. The insect peritrophic membrane contains protein, chitin, and other mucopolysaccharides, including hyaluronic acid.[169] In studies of glycogen metabolism, a relation occurs between this polysaccharide and chitin as reserve nutrients, with quantitative changes in the two being to some extent reciprocal under certain conditions. The metabolic mobility of chitin is supported by several kinds of observations: (1) In *Hyalophora cecropia*, some 80 to 85% of the integumentary chitin is digested and resorbed during the pupal moult; (2) in *Rhodnius* and some other insects, resorption of the endocuticle occurs during prolonged starvation, and deposition of endocuticle follows feeding; and (3) in *Periplaneta americana*, the synthesis and degradation of cuticle polysaccharide is continuous throughout the moulting cycle.[169]

Some evidence concerning precursors and reserves for chitin synthesis has been obtained with radioisotopes. Bade and Wyatt injected ^{14}C-glucose into *H. cecropia* silkworms at the end of the fourth instar and showed that by the end of the feeding period in the fifth instar the retained radioactivity was very largely in the cuticle. The subsequent new pupal cuticle, taken two days after pupation, was again highly radioactive, the specific activity of its chitin being close to that of the chitin of mature larvae and much greater than that of their glycogen, free sugars, or lipids. This appears to indicate specific transfer of material from the larval to the pupal cuticle. Such transfer might occur by products of digestion of the old cuticle being directly incorporated into the new as they are resorbed.[169]

Despite the dominant place of chitin in arthropod economy, very little has been established concerning the enzymes that make and degrade it.

In *Schistocerca gregaria*, Candy and Kilby[25] demonstrated by direct analysis and by *in vivo* isotope incorporation that the legs and wings are active in chitin synthesis for approximately seven days after the adult moult. Wings from two to three day old adults were therefore used as an enzyme source, and all of the enzymic activities required to convert glucose to UDP-acetylglucosamine were demonstrated. Attempts to detect activity of chitin synthetase itself, however, which would incorporate the latter compound into chitin, were unsuccessful, although the extracts were fortified and conditions were varied in a number of ways.

The digestion of the chitin (along with the proteins) of the endocuticle is a central feature of the arthropod moult. The process has not been extensively studied. In the studies of Jeuniaux,[68] evidence is presented for two steps in the digestion of chitin:

$$\text{chitin} \xrightarrow{\text{chitinase}} \text{chitobiose} \xrightarrow{\text{chitobiase}} \text{N-acetylglucosamine}$$
$$(+\text{chitotriose} \; +\text{acetylglucosamine})$$

Chitinase is assayed by loss of turbidity in a suspension of chitin, and chitobiase by colorimetric estimation of acetylglucosamine.

This area clearly is in need of further and detailed study.

GLYCEROL AND SORBITOL PRODUCTION FROM GLYCOGEN. A specialized conversion that occurs in the overwintering stage of a number of insects is that of glycogen to polyhydric alcohols. This was discovered independently in the diapause embryo of *Bombyx mori,* which produces sorbitol and glycerol,[27,28] and in the cecropia silkmoth pupa, which produces glycerol.[170] Although glycerol production in these species coincides with the onset of diapause, glycerol can be produced in non-diapausing insects, as illustrated by its appearance in adult carpenter ants upon exposure to cold.[40,136] The remarkably high level of 5 molal (about 25% of body weight) has been recorded for diapause prepupae of the wheat-stem sawfly parasite, *Bracon cephi.*[121] The production of glycerol under various conditions has been carefully studied in the slug caterpillar, *Monema flavescens.*

Ecologically, the accumulation of glycerol and sorbitol appears to be connected with resistance to cold, and these compounds probably act by lowering the supercooling point rather than by conferring the ability to survive freezing.[1,2,121] This is in contrast to the situation in polar fishes, in which resistance to freezing is achieved by production of true anti-freeze compounds.[36] This topic is discussed further in Chapter 9.

The biochemical pathway of glycerol production is presumably via reduction from the triose level of glycolysis, and appropriate enzymes have been demonstrated in insects. It is believed that glycerol production depends on reduction of dihydroxyacetone phosphate to α-glycerophosphate (for glycerophosphate dehydrogenase is highly active in insect tissues) with subsequent hydrolysis of αGP by α-glycerophosphatase.

The Glycolytic and Gluconeogenic Pathways in Fishes. Many of the features of carbohydrate metabolism in fishes, insofar as these have been studied, appear to be qualitatively similarly to those in other vertebrates, particularly the mammals.[58,137] However, some of the more intriguing aspects of metabolic adaptation typical of the lower vertebrates as a whole are carried to the extreme in this group of organisms; among these are the phenomena of (1) temperature adaptation of metabolism, (2) metabolic adaptation to low oxygen tensions, and (3) aestivation of the lungfishes during periods of drought.

TEMPERATURE ADAPTATION. Temperature adaptation is discussed in greater detail in Chapter 9. At this point, we wish mainly to stress that the organization of cellular metabolism in fishes seems to be strongly affected by the state of thermal acclimation. During cold acclimation in the salmonid fishes, new steady state levels of glycolysis, gluconeogenesis, and the pentose phosphate pathway are established[47] (see Chapter 9). A variety of molecular mechanisms are potentially available to achieve this new steady state, the alteration in the relative activities of different isozymic components in these pathways being one of the more basic processes involved. According to recent evidence, different isozymes within a given series are adapted for optimally controlled catalysis over different portions of the biological temperature range. Hence, changes in relative activities of these isozymes could lead to changes in steady state activities of pathways such as glycolysis and gluconeogenesis.

ADAPTATIONS TO LOW OXYGEN TENSIONS. It is not widely appreciated that some of the lower vertebrates become facultative anaerobes to survive certain environmental circumstances. For example, during winter conditions European carp often become "ice-locked" in small ponds which gradually grow anaerobic and remain O_2-free for two to three months until the spring thaw. The carp show no apparent ill effects after this extreme exposure to anoxic conditions. Blazka[15] was probably the first to recognize the fundamental consequences of this habit in carp. Unlike fishes such as the salmonids, which depend upon an aerobic metabolism, the carp do not accumulate an O_2 debt during anaerobiosis, and as in the intertidal molluscs, the usual end products of anaerobic breakdown of carbohydrates do not accumulate.

A similar situation probably occurs in closely related goldfish species. When ambient O_2 tensions become low, goldfish derive considerable energy for active and basal metabolism from anaerobic reactions. This partial anaerobiosis can be sustained for a long period. At oxygen concentrations near 15% of air saturation the goldfish sustains a respiratory quotient (CO_2/O_2) of about 2 for week-long periods,[78] and it is clear that metabolic CO_2 is produced even in the total absence of O_2.

Little or nothing is known of the metabolic pathways of anaerobic metabolism in fishes, and this is an area that is clearly in need of much further research. Reaction pathways of anaerobic metabolism are much better understood in various invertebrate facultative anaerobes and in tissues such as the kidney in higher vertebrates which must supplant their aerobic metabolism with important anaerobic decarboxylations. Chief among these is α-ketoglutarate decarboxylation, which is coupled to oxaloacetate reduction (see diagram below).

Cohen[31] has shown that in mammalian kidney cortex about 9 kcal/mole of α-ketoglutarate are "trapped" as ATP by these reactions (compared to 18 kcal/mole of glucose during anaerobic glycolysis). On a weight basis, these α-ketoglutarate decarboxylation reaction rates are approximately equal to the glycolytic rates and hence contribute significantly to energy required for various transport functions of the kidney tubule cells. Similar anaerobic decarboxylations may account for the high CO_2/O_2 ratio observed in goldfish during anoxic stresses and in other fishes during excursions into anaerobic waters.

In addition to CO_2, large amounts of long-chain fatty acids are accumulated during anaerobiosis in the carp.[15] Again, exact metabolic pathways are unknown. Nevertheless, as carbohydrate is almost certainly the major source of acetyl-CoA for lipogenesis in this species, it is of interest to consider the overall process of fatty acid synthesis from glucose. The process of glycolysis and pyruvate decarboxylation results in the formation of two moles of CO_2 and two moles of acetyl-CoA for each mole of glucose. Thus, four of every six carbons of glucose are available for fatty acid synthesis under ideal (energy-saturating) conditions. For palmitic acid synthesis, for example, the process may be represented stoichiometrically as

$$\underset{\text{Glucose}}{4C_6H_{12}O_6} \rightarrow \underset{\text{Palmitic Acid}}{C_{16}H_{32}O_2} + 8CO_2 + 8H_2O$$

The theoretical energetic efficiency of this process is relatively high. Four moles of glucose approximate 2744 kcal, whereas one mole of palmitate approximates 2400 kcal when completely oxidized. Incorporation of glucose carbon into lipid has been observed in fishes, and, at low temperatures, the amount of energy which can be obtained from the conversion apparently is adequate to meet both maintenance and active metabolic requirements.

Adaptations to high oxygen tensions. A primary function of a countercurrent "exchanger" (or rete) arrangement of blood vessels at the swim-bladder in those fishes which use this organ in hydrostatic function appears to be the secretion of O_2 from the blood into the swim-bladder, sometimes against exceedingly high concentration gradients (p. 205). The mechanism of gas deposition seems to depend upon acidification of the blood. Thus, the pH of the blood *leaving* the rete is about 1 pH unit lower than the pH of the blood *entering* the rete system. This pH change is brought about largely (if not solely) by lactic acid, which is presumably produced as an end product of glycolysis in the swim-bladder

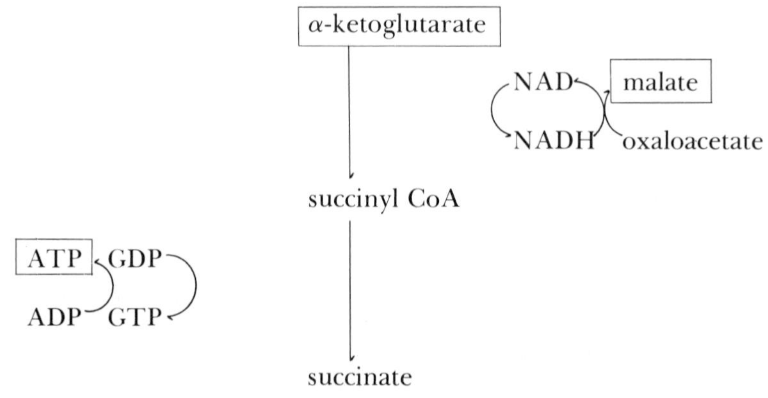

epithelium.[34] These conditions raise two important problems: (1) high glycolytic rates do not normally occur in the presence of high tensions of O_2 because of the Pasteur effect (inhibition of glycolysis by high O_2 tensions), and (2) variability in intracellular pH may be expected to be high, being a function of the rate of O_2 secretion.

The Pasteur effect is brought about by the development of a high energy charge in the cell and the subsequent inhibition of phosphofructokinase by high ATP concentrations. In the swim-bladder, the Pasteur effect is absent,[5] either (1) because mitochondrial metabolism is reduced or (2) because swim-bladder epithelium possesses forms of phosphofructokinase which are not sensitive to inhibition by high ATP.

We have little information on the pH responses of enzymes of the bladder epithelium. Swim-bladder LDH, particularly at high substrate values, appears to be less sensitive to pH change than do other LDH's that have been examined. In this way, this particular enzyme appears to be well adapted for function in the microenvironment of the swim-bladder epithelium. However, we do not know whether the same is true for other enzymes.

METABOLIC ADAPTATION TO AESTIVATION IN THE LUNGFISH. Available information on this problem is remarkably sparse.[58] Aestivating lungfish are capable of surviving for several years housed in a mud cocoon and in a state of suspended animation. Metabolic rates are as low as 1/100th of normal *basal* values. During this time enzyme activities remain potentially quite high, for the lungfish can arouse from aestivation within about 15 minutes (a time much too short for major resynthesis of enzyme systems), but the mechanisms by which metabolic processes can be shifted down by such a large factor are entirely unknown. Also, during aestivation, carbohydrate reserves are maintained over exceedingly prolonged periods of starvation. Again, we are only beginning to examine the various control mechanisms which are available to these fascinating organisms.[11]

Role of Glycolysis in Resistance to Anoxia in the Reptiles and Amphibians. Certain fishes seem to be the only vertebrates capable of surviving for long periods in the complete absence of O_2. In the evolution of the vertebrates, they would appear to represent the last examples of facultative anaerobes. However, many other vertebrates have an impressive capacity for tolerating anaerobic conditions. Of a wide array of such species which have been examined, the Chelonia are clearly the most superior in tolerance of anoxic conditions. The turtles display a remarkable adaptability to a surprising diversity of environments; they are successful in the deserts, in various fresh and brackish water areas, and even in open-ocean, pelagic conditions. Of the various chelonian groups, the aquatic forms probably are most tolerant to anoxia and certainly depend upon it most frequently because of, and during, prolonged diving. A characteristic physiological response to diving in all vertebrates is a lowered metabolism. In the most skilled of mammalian and avian divers (the cetaceans and the marine and freshwater birds), metabolic adaptations during the dive involve a peripheral vasoconstriction, forcing many peripheral tissues to rely more and more upon anaerobic glycolysis for energy requirements. This mechanism "spares" the O_2 in the blood for the brain and the heart. The *performance limits* which are set upon these divers are probably determined by the production of energy in the peripheral tissues; the *time limits* of the dive are set by the central nervous system and the heart. The work functions of these tissues are linked in an obligatory way to aerobic metabolism; hence, in mammals and birds, when the O_2 delivery to these tissues can no longer be sustained, the animal must surface. Such is not the case in the diving turtles; they can survive many hours—at times even many days!—after the blood O_2 tensions have dropped to zero. Indeed, the tolerance of turtles to anoxia is so great that whether or not indefinite *survival at low temperatures can occur under totally anoxic conditions* is still an open question. Whatever the outcome of the discussions over this question, it is nonetheless clear that the strategy of adaptation to low or zero O_2 levels in the turtles does *not* involve the elaboration of distinct metabolic steps designed to avoid accumulation of compounds such as lactate. As far as is known, the organization of cellular metabolism in turtles is not unlike that of typical, more aerobic, vertebrates. Their tolerance to anoxia depends upon (1) a high glycolytic capacity leading to very high lactate accumulation, and (2) the elaboration of mechanisms in the central nervous system and other tissues which allow the tissue to carry out its specific work functions even if O_2 is absent.[13]

THE DEPENDENCE ON ANAEROBIC GLYCOLYSIS DURING PROLONGED DIVING IN TURTLES. On the basis of direct calorimetric measurements of metabolism, Jackson[67] has divided the diving period of turtles into three metabolic phases:

(1) During phase I, lasting about a half hour at 24°C, the metabolic rate persists at the same rate as in the prediving condition, but O_2 tensions in the blood and the lungs are rapidly falling.

(2) During phase II, which also lasts about 30 minutes, the metabolic rate falls to 40 per cent of the initial rate and remaining O_2 reserves are exhausted.

(3) The remainder of the dive (phase III), which can last for many hours (and in some species at low temperature, for many days), is *totally anoxic.* All maintenance and work functions in phase III are sustained by anaerobic metabolism. On the basis of heat measurements, Jackson has estimated that this metabolism can yield 15 to 20 per cent of the total energy which is available in the prediving state. Apparently, anaerobic glycolysis accounts for all of the energy generated during this time. The unusual activity of this pathway in these species leads to unusually large lactate accumulations in the blood and the tissues, and accounts for the extreme insensitivity of turtles to metabolic poisons such as cyanide, which specifically blocks oxidative metabolism but does not affect glycolysis.

Belkin[12] provided an unequivocal demonstration of the critical importance of anaerobic glycolysis to the anoxic turtle. He selected a metabolic poison, iodoacetate, whose chief locus of action is known to be the triose phosphate dehydrogenase step in glycolysis. If anoxic survival depended upon glycolysis, inhibition of this enzyme by injection of iodoacetate into the turtle should lead to a greatly reduced tolerance to anoxia. This prediction was verified. Furthermore, Jackson then observed by direct calorimetry that the injection of iodoacetate leads to a predicted drop in the remaining energy metabolism of the anoxic turtle. These experiments emphatically underline the pivotal role of glycolysis in the adaptation to anoxia, but they do not explain the tolerance of the central nervous system to it.

THE NATURE OF ANOXIA TOLERANCE OF THE CHELONIAN CENTRAL NERVOUS SYSTEM. To gain an insight into the anoxia tolerance of the CNS of turtles, it is instructive to consider the mechanisms underlying anoxia sensitivity of other nervous systems, such as the mammalian brain. At the outset, it should be emphasized that these mechanisms are not yet known unequivocally. However, it is clear that the mammalian brain, although absolutely dependent upon oxidative metabolism, does possess a functional glycolytic pathway. Paradoxically, this tissue also displays an absolute dependence upon glycolysis because glucose is the only significant carbon and energy source for its metabolism. The cellular organization of the glycolytic enzymes is unique in that they are all attached to—or are a part of—a membrane structure intimately associated with the mitochondria. Although control sites reside at the hexokinase, PFK, and PK reactions here as in other tissues, it is clear that this glycolytic system is very intimately coupled with aerobic mitochondrial metabolism. Presumably, this structural arrangement of the glycolytic enzymes serves to channel glucose carbon as pyruvate into the Krebs cycle with high efficiency. But although there is a tight coupling between the aerobic and anaerobic phases of glucose catabolism in the mammalian brain, its dependence upon oxygen does not lie in this coupling but rather in the coupling between aerobic ATP generation and the Na^+-K^+ ATPase function of the CNS.

The basic function of the nerve cell is propagation of action potentials. To sustain these, specific Na^+ and K^+ concentration gradients must be and are maintained across the neuronal membrane by the action of the Na^+-K^+ ATPase-linked ion pump. In the mammalian brain, this Na^+-K^+ ATPase is apparently tightly linked to one or more of the ATP-generating steps in the electron transfer system. A comparable situation prevails in the well studied squid giant axon. Here, as in the mammalian brain, axonal function depends critically upon ATP generated by aerobic metabolism and therefore is sensitive to metabolic poisons which interfere with oxidative metabolism.[37] *Extreme anoxia sensitivity* arises from this coupling of the Na^+-K^+ ATPase to ATP generating site(s) in the electron transfer system. In contrast, we postulate that the *anoxia insensitivity* of the turtle CNS stems from Na^+-K^+ ATPase function which is not linked to aerobic energy metabolism. There is some precedent for this view.

Although Na^+-K^+ ATPase is linked to aerobic metabolism in other tissues as well as the brain (the gill of crustaceans and fishes, the mammalian kidney tubules, and the avian salt gland), in a number of tissues it is *linked in absolute manner to glycolysis*. The mammalian red blood cell is an excellent case in point for which there is, additionally, a large backlog of information.[103] Here, Na^+-K^+ ATPase function *must* be tied to glycolytically generated ATP, since the mature red blood cell does not possess mitochondria. Recent evidence indicates that Na^+-K^+ ATPase function is very intimately associated with the phosphoglycerate kinase step (1,3 DPG + ADP $\rightharpoonup$ 3PG + ATP) in glycolysis. The precise nature of this coupling is unknown. Nevertheless, it is clear that the ATP *product* of the kinase reaction serves as the *substrate* for Na^+-K^+ ATPase. An increase in the kinase activity leads to increased Na^+-K^+ ATPase function; conversely, an inhibition of Na^+ and K^+ pumping leads to an inhibition of phosphoglycerate kinase activity. Both functions are probably components of the red blood cell membrane.

Other comparable examples are known. The electric organ of the electric eel is also largely glycolytic, and its normal function also displays a dependence upon glycolysis. Again, the site of the interaction probably involves phosphoglycerate kinase.[89] A third example of this interaction is to be found in the toad bladder.[52] The evidence here does not identify a single locus coupling Na^+-K^+ ATPase function to glycolysis, but it is unequivocal in demonstrating an absolute dependence of Na^+ and K^+ pumping on glycolysis. Thus, experimental activation of glycolysis leads to an activation of Na^+ pumping; conversely, hormonally induced alterations in the Na^+ pumping rate are sustained by alterations in glycolytic rate. Several control sites are involved in the glycolytic adjustments, including PFK, phosphoglycerate kinase, and PK. However, exclusive dependence of the pump upon only one of these sites either does not occur or has not been demonstrated.

In none of these tissues is there an absolute Na^+-K^+ ATPase dependence upon O_2-supported metabolism. Additionally, these tissues are quite insensitive to anoxia. Since the CNS of turtles can function in the total absence of O_2, we postulate that this insensitivity to anoxia also stems from the elaboration of Na^+-K^+ ATPase function which is *dependent* upon glycolytically generated ATP, but is *independent* of oxidative metabolism. Direct testing of this postulate must await further research. However, the postulate is consistent with the observation that Na^+-K^+ ATPase in other tissues of aquatic turtles *is* sustained by glycolytically generated ATP.[115]

Fat Metabolism

Introduction. Since the classical studies of Zuntz, it has been evident that fat must by necessity be the principal source of energy in working animals, a result adequately corroborated by Krogh and Lindhard in 1920 and by many others since then. But the elucidation of the glycolytic pathway and the Krebs cycle was of such strong inherent interest and put so much emphasis on carbohydrate metabolism that biochemistry tended to neglect fat metabolism or to assume that fat was used only indirectly. All this has changed, and the mobilization, combustion, and synthesis of lipids now represent areas of very active research.

The lipids are a heterogeneous class of substrates characterized by a variable solubility in organic solvents and an insolubility for the most part in water. The principal classes of complex lipids are the neutral fats and oils, the phospholipids, and the steroids. The essential features of the chemistry of lipids have been summarized many times and will not be our main concern. Rather, we will focus our attention on what is known of catabolic and anabolic mechanisms.

Degradation and Synthesis of Triglycerides. Triesters of glycerol and fatty acids, the triglycerides, compose two of several important classes of complex lipids, the fats and the oils. The distinction between these classes of lipids is based on the physical state of the triglyceride at room temperature, the oils being liquid and the fats being solid. The melting point of a triglyceride is determined by its fatty acid composition. In general, the higher the proportion of short-chain acids and unsaturated acids, the lower the melting point.

The hydrolysis of ingested triglycerides in mammals occurs largely in the small intestine and is catalyzed by lipolytic enzymes elaborated by the pancreas. The pancreatic lipases (glycerol ester hydrolases) appear to be of two types, one specific for ester linkages at the α-position of triglycerides and the second for ester linkages at the β-position. The complete hydrolysis of triglycerides proceeds in a stepwise fashion, with rapid hydrolysis of the α and α' linkages followed by slow hydrolysis of the β-monoglyceride. Little information on the catalytic and regulatory properties of these enzymes is available.

The catecholamines are known to activate the mobilization of triglycerides in both brown fat cells and white fat cells. This hormonal activation of lipolysis in adipose tissue involves a catecholamine activation of adenyl cyclase; the subsequent increase in cyclic AMP production leads to an activation of triglyceride lipase, which hydrolyzes triglycerides to free fatty acids and glycerol.[22, 30, 112] This hormonal control of mobilization of storage fat has many features which are analogous to the cyclic AMP control of mobilization of storage carbohydrate[106] (see p. 266). Drummond and his coworkers have recently found that the catecholamines activate the maximal capacity of the adenyl cyclase, with no major changes in other kinetic properties of the enzyme.[38, 106]

Although the hydrolysis of triglycerides is in principle reversible, thus providing one possible pathway for the biosynthesis of these compounds, this pathway apparently is of little functional significance *in vivo*. The biosynthetic pathway which is used was first formulated by Weiss and Kennedy in 1956. Free glycerol is phosphorylated at the α-position in the presence of glycerokinase and ATP, yielding L-α-GP; alternatively, the latter may arise from reduction of DHAP. One mole of αGP is subsequently acylated with two moles of fatty acyl-SCoA, yielding an α,β-diglyceride phosphate termed a phosphatidic acid. The synthesis of triglyceride is completed by dephosphorylation, yielding an α,β-diglyceride, followed by esterification with a third mole of fatty acyl-SCoA. In adipose tissue and intestinal mucosa, glycerophosphate arises largely from the NADH-dependent reduction of dihydroxyacetone phosphate, glycerokinase being virtually absent from these tissues. With this exception, the biosynthetic pathway leading to triglycerides is similar to that previously established in liver.

The catabolic and anabolic pathways are summarized below. The principle of separation of related catabolic and anabolic processes is again preserved. Because triglycerides probably represent the major storage form of energy in most animals, further study of regulation of triglyceride mobilization and triglyceride deposition is a matter of some importance.

The Degradation and Synthesis of Phospholipids. Several very important classes of complex lipids are derived from phosphatidic acid. These include phosphatidylcholines, also termed lecithins, and phosphatidlyethanolamines. The phosphatidylcholines and phosphatidylethanolamines are major phospholipid constituents, and they play important structural and functional roles in membranes of cell walls and organelles. Most phosphatidlycholines contain one saturated fatty acid, esterified at the α-position, and one unsaturated fatty acid, esterified at the β-position. Some members of this class contain two saturated or two unsaturated fatty acids.

Free phosphatidic acid has been demonstrated to occur naturally, but it does not appear to account for an appreciable fraction of total phospholipid. The phosphatidylcholines and phosphatidylethanolamines are very widely distributed in nature and are frequently the major lipid constituent of tissues.

The enzymatic degradation of the phospholipids is incompletely understood. At

Triglyceride
↓
diglyceride + FFA_1 (or FFA_3)
↓
monoglyceride + FFA_3 (or FFA_1)
↓
glycerol + FFA_2

Triglyceride
↑
α,β-diglyceride + fatty acyl-SCoA
↑
α,β-diglyceride phosphate
↑
α-glycerophosphate + fatty acyl-SCoA
↑
glycerol + ATP

least five enzymes that cleave one or more of the ester linkages of phosphatides are known: phosphatidases (or phospholipases) A, B, C, and D, and lysophosphatidases. These enzymes, acting in concert, can lead to the complete removal of fatty acids from phosphatides, but the stepwise fashion in which these and related enzymes degrade phospholipids to their components has not been established.

The biosynthetic pathways leading to the formation of phosphatidylcholine, phosphatidylethanolamine, and phosphatidylserine have been summarized by Kennedy.[75] Phosphatidylcholine and phosphatidylethanolamine arise from the reaction of D-α, β-diglyceride with CDP-choline and CDP-ethanolamine, respectively. Phosphatidylethanolamine may also arise from the decarboxylation of phosphatidylserine, and may give rise to phosphatidylcholine via a methylation reaction.

As noted above, phosphatides are of near universal occurrence. Within the cell, phosphatides are found concentrated in the nuclear, mitochondrial, and microsomal fractions. These structures are, of course, membranous and the phosphatides appear to be found predominantly, if not solely, in the membranes themselves. It is generally held that the phosphatides perform several important functions, including those of structural maintenance, ion transport, permeability regulations, and electron transfer within the mitochondrial oxidative phosphorylation system. Wilson and his coworkers[168] have demonstrated that the phospholipids strongly determine the properties of various transport enzymes which are themselves thought to be structural components of the membrane.

Fatty Acid Oxidation. The oxidative degradation of fatty acids appears to be a universal biochemical capacity among living organisms. In mammals, such oxidation occurs in a variety of tissues, including liver, kidney, and heart, but the quantitative contribution of fatty acids to overall metabolism varies from tissue to tissue. Within the cell, fatty acid oxidation occurs principally within the mitochondria. In contrast, the bulk of fatty acid biosynthesis is catalyzed by enzyme machinery located in the cytosol. The capacity for fatty acid biosynthesis also varies from tissue to tissue.

The fundamental features of the pathway of fatty acid oxidation were clearly detailed by Knoop in 1904 and shortly thereafter by Embden and Dakin. Knoop arrived at the theory of "β-oxidation" through feeding experiments employing fatty acids labeled in the terminal position with phenyl groups—a tracer technique nearly half a century ahead of its time! These tracer compounds permitted the identification of the end products in the urine of experimental animals. Knoop's results were consistent with a successive removal of two-carbon fragments from the parent fatty acid molecule. Nearly fifty years were required for full experimental confirmation of Knoop's brilliant conclusions.

The key milestones in the understanding of the pathway of fatty acid oxidation, following the work of Knoop, were (1) the observation of fatty acid oxidation in cell-free preparations from the guinea pig liver by Leloir and Munoz in 1939, and (2) the elucidation of the structure of "activated" fatty acids as the S-acyl derivatives of coenzyme A by Lynen and Reichert in 1951. The use of cell-free preparations permitted the identification of the individual enzymes involved in fatty acid oxidation and the determination of the properties of the catalytic reactions. The discovery of the acyl derivatives of CoASH, which followed the discovery of coenzyme A by Lipmann, provided a solution to the most perplexing problem of fatty acid oxidation—the failure to detect any of the postulated intermediates or any free short-chain fatty acids in the course of oxidation of longer molecules. It is now clear that all the intermediates, including the short-chain fatty acids formed in the course of oxidation, occur only as the acyl-SCoA derivatives.

In the pathway for the oxidation of fatty acids, the parent fatty acid is activated by conversion to the fatty acyl-SCoA, oxidized to the α,β-unsaturated compound, hydrated, oxidized to the β-keto derivative, and finally subjected to a thiolytic cleavage yielding acetyl-SCoA, and the fatty acyl-SCoA containing two less carbon atoms, which, in turn, undergoes the same series of reactions. Discussions of each of the individual enzyme steps are available in all modern biochemistry texts.

Little information is available on molecular control mechanisms in fatty acid oxidation, partly because until recently it has been difficult to study the direct combustion of free fatty acids in mitochondrial suspensions. Now it is known that carnitine plays a major role in the transport of free fatty acids into the mitochondria, where β-oxidation occurs.

Activated fatty acids are probably carried through the mitochondrial membranes as acylcarnitines. The acylcarnitines are formed according to the reaction, acyl-CoA + carnitine ⇌ acylcarnitine + CoA, which is catalyzed by two different enzyme systems. Carnitine acetyltransferase is active in the transfer of short-chain fatty acids, whereas carnitine palmityltransferase catalyzes the transfer of long-chain fatty acids.[100] Norum and Bremer[101] examined the distribution of the latter enzyme in a variety of organisms and tissues. In cod, for example, the specific activity of the enzyme is highest in liver, heart, and kidney and is least in muscle, spleen, gonad, and intestinal tissue. These activities seem to correlate with the tissue ability to oxidize fatty acids in fishes[70] as well as in other animals.[101]

The central role of carnitine palmityltransferase in the transfer of activated fatty acids to the site of β-oxidation in the mitochondria suggests that it may play a key role in the regulation of fatty acid oxidation. However, studies by Bremer and his coworkers[16] indicate that this transferase does not limit fatty acid oxidation rates under most circumstances. However, availability of other oxidizable substrates appears to be of importance. Succinate, in particular, inhibits acylcarnitine oxidation in coupled mitochondria, possibly by competing for CoA required for succinyl-CoA formation. Further work in this area is required before control of fatty acid oxidation is fully understood.

Fatty Acid Biosynthesis. Each step in the oxidation of fatty acids is reversible. Hence, it was proposed that fatty acid synthesis might occur by a simple reversal of the oxidative pathway, but it is now apparent that the major pathway of fatty acid biosynthesis consists of extramitochondrial components.

One of the first suggestions that fatty acid synthesis may occur via a pathway distinct from that of oxidation was the demonstration by Van Baalen and Gurin[150] and by Tietz and Popjak[142] that fatty acids could be synthesized by soluble extracts of avian liver and mammary gland. Cofactor requirements for fatty acid synthesis by soluble cell extracts include an absolute dependence upon bicarbonate, ATP, Mn^{++}, and NADPH. These cofactor requirements clearly distinguish the synthetic and oxidative pathways, since (1) bicarbonate has no effect on fatty acid oxidation and (2) NAD, rather than NADP, is the oxidation-reduction coenzyme in the oxidative pathway.

The synthesis of long-chain fatty acids is a multistep process in which at least six different enzymes participate. In bacteria and plants, all the enzymes involved occur in the soluble fraction of the cell and are readily separable from each other. However, in yeast, birds, and mammals, the enzymes are contained in macromolecular multienzyme complexes, referred to as fatty acid synthetases. The synthetase from yeast, the most adequately studied of these complexes, has a molecular weight of about 2.3 million, and it displays a high degree of substructural order under the electron microscope. The structure can be dissociated to units having molecular weights of about 100,000. From these facts, and the further demonstration of seven terminal amino acids in the complex, Lynen postulated that the complex is made up of six or seven enzymes, each catalyzing a specific step of fatty acid synthesis.

With the discovery of malonyl-CoA as an intermediate of fatty acid biosynthesis, the investigation of the process entered an advanced phase. Using different enzyme systems, derived from animal tissues, yeast cells, and bacteria as well as from plants, it was demonstrated that synthesis of fatty acid from malonyl-CoA also requires small amounts of acetyl-CoA besides NADPH as a reducing agent. The two carbon atoms from acetyl-CoA are incorporated only into the methyl ends of the fatty acids synthesized, whereas all other carbon atoms derive from malonyl-CoA. Thus, it was obvious that acetyl-CoA serves as some kind of "primer" in the synthetic process. The formation of fatty acids may be described by the following equation:

$$\text{acetyl-CoA} + n\ \text{malonyl-CoA} + 2n\ \text{NADPH} + 2n\ \text{H}^+ \rightleftharpoons$$
$$\text{CH}_3(\text{CH}_2\text{CH}_2)_n\text{CO-CoA} + n\ \text{CO}_2 + n\ \text{CoA} + 2n\ \text{NADP}^+ + n\ \text{H}_2\text{O}$$

The enzyme systems from animals, bacteria, and plants produce free fatty acids, whereas the enzyme system from yeast produces the CoA derivatives.

The first insight into the chemical details of the synthetic process came from Lynen's experiments with the enzyme system of yeast. These experiments demonstrated that *the same carboxylic acid intermediates were formed during fatty acid synthesis as during fatty acid degradation.* But an important difference between degradation and synthesis can be seen in the fact that during the oxidative process the intermediates are bound to coenzyme A, whereas during synthesis *protein-bound carboxylic acid intermediates are formed, with sulfur forming a bridge between the carboxylic acid intermediates and the protein.* Two different types of sulfhydryl groups are involved in the synthetic process, denoted as "central" and "peripheral" SH-groups for purposes of differentiation. In the scheme of fatty acid synthesis shown in Figure 6–7 (first presented in 1961 by Lynen), the two SH-groups are distinguished by bold-face and normal type.

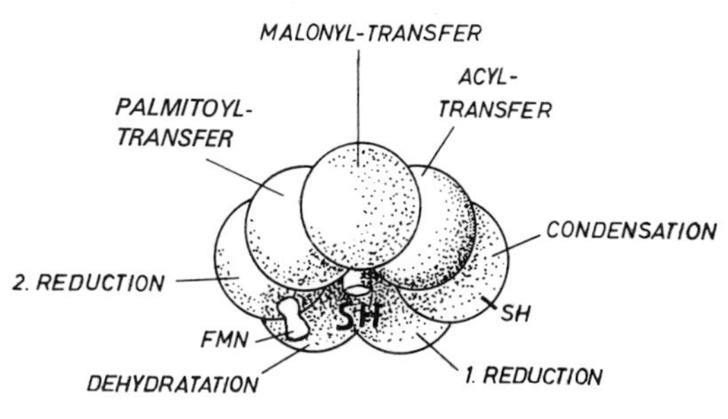

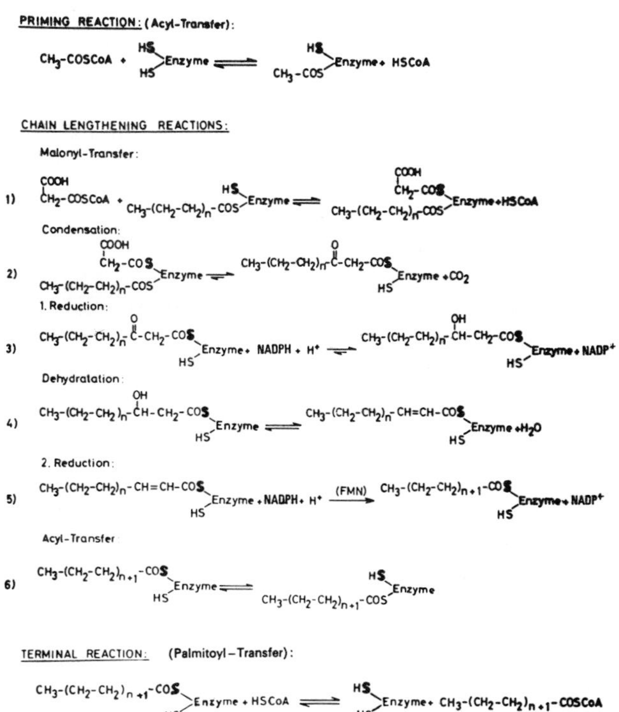

Figure 6–7. Mechanism and reactions of fatty acid biosynthesis and the hypothetical structure of the multienzyme complex of fatty acid synthetase. (From Lynen, F., D. Oesterhelt, E. Schweizer, and K. Willeckek, *in* Cellular Compartmentalization and Control of Fatty Acid Metabolism, edited by F. Gran. Universitetsforlaget, Oslo, 1968.)

In the "priming reaction" of synthesis, an acetyl residue from acetyl-CoA is transferred to the "peripheral" SH-group. It is followed by the transfer of a malonyl residue from malonyl-CoA to the "central" SH-group. The acetyl-malonyl-enzyme thus produced is transformed into acetoacetyl-enzyme with the concomitant release of CO_2. In this reaction the sulfur-bound acetyl group attacks the nucleophilic methylene group of the malonyl thioester. This release of CO_2, coupled to the formation of acetoacetyl-enzyme, is important from the thermodynamic point of view, for the decarboxylation drives the reaction to the right side, that is, to the side of condensation. Following the formation of acetoacetyl-enzyme, the β-ketoacyl protein derivative undergoes stepwise reduction, dehydration, and reduction to form the butyryl-enzyme derivative. The butyryl group is finally transferred to the peripheral sulfhydryl group, leaving the central sulfhydryl free for attachment of a second malonyl residue. Thus, the whole reaction sequence is initiated again and is repeated until long-chain saturated fatty acids with 16 to 18 carbons are formed. Little is known about why chain lengthening stops here.

All efforts to dissociate the complex into individually active enzymes have thus far been unsuccessful. Evidence for the postulated scheme is derived from three kinds of data: (a) studies on the roles of the two kinds of sulfhydryls; (b) data showing that acetyl-CoA and malonyl-CoA react to form a protein-bound acetoacetyl derivative that is covalently bound in thiolester linkage to the central sulfhydryl residue; and (c) studies with substrate analogs which are converted by the fatty acid synthetase complex to predicted products. Strong supporting evidence also comes from studies of fatty acid synthesis in *E. coli*. In this organism, all the enzymes of the fatty acid synthesis are found in the soluble fraction of cell-free preparations; these have been separated and studied separately by Vagelos and his co-workers and by Wakil's group. Basic mechanisms are entirely analogous to the mechanism proposed by Lynen for the synthetase complex.[49,87]

What advantages can a cell obtain from the specific ordering of enzymes into functionally important aggregates? It has been suggested that the fatty acid synthetases of *E. coli* and yeast (as well as animal tissues) might represent two stages in the evolution of the enzyme system. According to this concept, the synthesis of fatty acids from malonyl-CoA, NADPH, and acetyl-CoA may originally have been carried out by separate enzymes and an easily dissociable "acyl carrier protein" as in the case of *E. coli* system. In the course of evolution the various component enzymes and the "acyl carrier protein" came to be associated into a form of fatty acid synthetase in which the components were firmly bound to each other, as in the yeast multienzyme complex.

The highly organized structure of the yeast fatty acid synthetase may be more efficient catalytically because of two effects:

(a) From the kinetic point of view, the transfer of intermediates from one enzyme to the next, without the need of free diffusion and dilution in the cytoplasm, would seem to offer much. If all intermediate products were covalently bound to the complex, the encounter of an intermediate produced by one component with the next catalytic site in the sequence would seem to be very much more probable in an organized complex than in a mixed solution of separate enzymes. This advantage may, on the other hand, be partly neutralized by the fact that only some of the catalytic sites may be occupied at any given instant. The net effect, however, is a complex process of greater efficiency, judging by the high specific activity. In the case of the freshly prepared yeast enzyme, one enzyme complex with a molecular weight of 2.3 million catalyzes the conversion of up to 6,000 moles of malonyl-CoA to fatty acids per minute (25°C; pH 6.5). Of equally great value is the fact that the covalently bound intermediates are not accessible to competing processes. In the case of fatty acid synthesis, interference due to enzymes of fatty acid degradation is avoided. In this sense, the multienzyme complex may be compared to a strict compartmentation in the smallest possible space.

(b) Studies have indicated that a second advantage in assembling the enzymes of fatty acid synthesis into a single complex may be related to the maintenance of active conformations of the component polypeptide chains. Thus, dissociation of the yeast complex leads to a loss of certain catalytic activities which are regained on reassociation, suggesting that catalytically active conformations depend upon specific enzyme-enzyme interactions within the fatty acid synthetase complex. Similar dissociation-reassociation experiments are being carried out on fatty acid synthetase from rat and pigeon liver. In these studies, too, dissociation leads to the complete loss of some activities and the reduction in catalytic capacities of others, but the reassociated complex shows the same specific activity for fatty acid synthesis as the untreated complex.[77]

Fatty acid synthesis in the liver and lactating mammary gland of mammals is predominantly an extramitochondrial process.[132] Yet acetyl-CoA, formed from pyruvate and fatty acid oxidation, is produced within the mitochondria; hence, provision must be made for diverting it from the mitochondria to the extramitochondrial space of the cell. A number of possible mechanisms have been considered.[85] The pathway considered most likely involves citrate or a near biochemical relative of citrate. Citrate is formed in the mitochondria by the condensing enzyme reaction of the Krebs cycle. On diffusion out of the mitochondria, citrate can be cleaved by the citrate cleavage enzyme to produce acetyl-CoA and oxaloacetate. Acetyl-CoA produced in this way can then be used as a precursor for fatty acid synthesis. That this pathway can operate is established,[85] but whether it is the sole or even major source of acetyl-CoA for fatty acid synthesis is still an open question.

The pathway of fatty acid synthesis, starting with citrate, can be considered to involve the reactions shown below. Reaction (1) is catalyzed by the citrate cleavage enzyme; reaction (2), by acetyl-CoA carboxylase; and reaction (3), by the fatty acid synthetase complex.

When the citrate cleavage enzyme is saturated with respect to its substrates, the reaction catalyzed by acetyl-CoA carboxylase is the rate limiting step of fatty acid synthesis. This enzyme in mammals is activated by carboxylic acids; of these, citrate is by far the most effective.[85] This short-term regulatory effect by citrate is complicated by the action of Mg^{++}, which also activates the enzyme, and by the cosubstrate, ATP, which at high concentrations prevents the citrate activation.

Palmityl-CoA also inhibits fatty acid synthesis from malonyl-CoA. Since palmityl-CoA may be regarded as an "end product" of fatty acid synthesis, it is probable that fatty acid synthesis is subject to feedback inhibition by end product at multiple points. Further studies will probably uncover regulatory metabolites as well as mechanisms which integrate control of fatty acid synthesis and oxidation.

In addition to short-term regulation of fatty acid synthesis, which depends upon metabolite modulation of specific regulatory enzymes such as citrate cleavage enzyme, much information is available on long-term regulation of this pathway. This kind of control mechanism requires up to several days (in mammals) to be maximally effective. Prolonged starvation, diabetes, or a diet high in fat all lead in time to significant decreases in fatty acid synthesis rates; diets high in carbohydrate or insulin treatments lead to increased fatty acid synthesis. These kinds of long-term effects have been found to correlate with adaptive changes in the level of enzymes involved in fatty acid synthesis, presumably brought about by a change in synthesis or degradation rates of component enzymes of the pathway.[85]

Comparative Aspects of Lipid Metabolism. FLYING INSECTS. Running or swimming animals may stop and be at least partly supported at any moment, but the flapping flight of insects (and, indeed, of small birds) requires sustained muscular work. This means that wing muscles cannot become fatigued, for even a few seconds of anaerobiosis will flood the body with lactate or similar intermediates. Hence, it is not surprising that active wing muscle is an almost exclusively aerobic tissue with exceptionally favorable conditions for gas exchange by diffusion, particularly in flying insects, and probably also in birds.[160] In birds, the only fuel of any quantitative significance during prolonged flight is fat. In insects, on the other hand, two extreme types are found. Most dipterans and hymenopterans are able to utilize only carbohydrate for flight, although fat is also metabolized when the animals are not flying.

---

(1) citrate + ATP + CoA ⟶ acetyl-CoA + ADP + P_i + oxalo-acetate
(2) acetyl-CoA + ATP + CO_2 ⟶ malonyl-CoA + ADP + P_i
(3) malonyl-CoA + NADPH ⟶ fatty acyl-CoA + NADP + CO_2

All lepidopterans, which display a remarkably high metabolic rate, depend exclusively on fat metabolism both during flight and during rest. Many other insects, such as the Orthoptera, Homoptera, and Coleoptera, supplement fat with carbohydrate during initial flight; but during sustained flight, fat is the only fuel of importance, the store of trehalose in the blood and glycogen in the fat body being depleted during the first hour or so of flight.

During the past decade, many biochemical studies have devoted themselves to the synthesis, transport, and degradation of fat and fatty acids in flying insects. From these studies, it has become evident that carbohydrates and fatty acids in the diet are incorporated in the fat body mainly as triglycerides, and are released when required as diglycerides loosely bound to protein in the hemolymph. The fatty acids are then split off and activated in the flight muscle fibers, where they are catabolized by the usual β-oxidation "spiral" to acetyl-CoA. These tissues of flying insects possess the most active fat catabolism known in nature, and it is worth discussing in some detail certain specializations that have evolved.

Hexoses, acetate, and free palmitate are readily incorporated in the fat body as neutral triglycerides (mainly palmitic and oleic acid) both *in vitro* and *in vivo*.[29,140,141] During flight of the migratory locust, the triglycerides and free fatty acids of the blood increase by a factor of about three-to-four-fold; a smaller but significant increase in fatty acids occurs in the muscle, presumably due to uptake from the blood. At the same time, the concentrations of total carbohydrate in the fat body and of blood trehalose drop to essentially zero.[141] The main lipid component released into the blood from storage depots is not free fatty acids (bound to albumin) as in mammals, but is diglyceride bound to protein, and this liberation itself seems to demand metabolic energy.[29] Release of the diglyceride appears to be under hormonal control.

In earlier studies, many attempts to measure the direct combustion of free fatty acids by flight muscle homogenates ended in failure, probably because the wing muscles of locusts require carnitine as a cofactor for the transport of fatty acids into the mitochondria. Beenakkers and Klingenberg[10] found a carnitine-CoA transacetylase in the mitochondrial fraction of locust flight muscle which was as active as in pigeon heart and two to three times more active than in breast muscle. It was also noted that the activity was absent in the flight muscle of the bee, which only utilizes carbohydrate for flight. Carnitine is a vitamin for certain bettles, but other insects which require it seem to be able to synthesize it.

Inside the muscle mitochondria, the breakdown of fatty acids follows the usual pathway, as revealed by studies of component enzymes.[10] In a comparison of the locust, which relies solely on fat as a fuel source, and the bee, which relies solely on sugar, Beenakkers found that the activities of enzymes in the fatty acid oxidation pathway were as much as 100 times greater in the locust flight muscle. A similar analysis from the flight muscles of moths and butterflies would be particularly interesting because of their high metabolic rate.

Swimming; mostly fishes. Although swimming at high speeds requires considerable power, both the speed and the necessary muscular work have been somewhat exaggerated in the past; the best recent estimates, for example, show that a Pacific dolphin (*Tursiops gilli*) has a maximum speed of about 30 km/hour and does not require more power per unit weight of muscle than is known for human athletes.[79] Comparable figures are not readily available for fast swimming fishes, but it is a fair guess that they fall into the same range as the dolphin; indeed, the efficiency of locomotion of the salmon is one of the highest known.[146] Although the power costs are not too unusual and the swimming efficiency is high, certain species of fishes, such as the salmon, are capable of performing quite remarkable feats. Thus, the Pacific salmon species, in their spawning migration, undergo journeys many hundreds, and even thousands, of miles long, during which time the fish does not eat to replenish energy stores. In contrast to birds and mammals, fish do not have a specific adipose tissue, but triglycerides are stored throughout the body, particularly about or in muscle fibers.[138] There is no doubt that these triglycerides constitute the main source of energy for the salmon migration, since about 90% of the energy expenditure derives from fat;[65] slices of salmonid muscle, particularly the red (fat loaded) lateral muscle, can oxidize fatty acids.[14] In these features, as in the enzymic mechanisms for fatty acid oxidation and fatty acid synthesis, lipid metabolism in fishes appears to be rather similar to that in mammals.[58] The fishes do, however, present special problems in lipid metabolism during thermal adaptation. These effects may not be unique to fishes, but most available experimental data are to be found in studies of fishes, so we shall consider the problem in this context.

A basic experimental observation is that cold acclimation in fishes leads to rather marked increases *both* in fatty acid oxidation rates[35] and in rates of fatty acid synthesis.[57] Several studies[116,117] indicate that during the acclimation process, changes in the kind of membrane-based phospholipids being synthesized are more important than mere changes in concentrations of any given components. That is, in thermal acclimation the *kind* of fatty acid being incorporated into phospholipids may be more important than the *amount* of any given component being stored. As far as is known, these events occur only in organisms which are capable of thermal acclimation. This includes many fishes and probably many invertebrates. The basic observation raises intriguing questions: What control mechanisms are operative which lead to the preferential synthesis of certain fatty acids and of certain phospholipids? How is temperature involved in triggering that control? These changes are very dramatic in the nervous system. What are the functional consequences?

Wilson and his co-workers[168] have obtained data which may have an important bearing on the above questions. Working with an *E. coli* mutant deficient in its ability to synthesize unsaturated fatty acids, these workers were able to experimentally modify the fatty acid composition of membrane-based phospholipids. They discovered that the properties of β-galactoside and β-glucoside transport systems of the cell membrane depend upon the kind of fatty acid which is incorporated into the membrane phospholipid. Arrhenius plots of activity of these enzymes versus 1/T show distinct breaks; the temperature at which these breaks occur also depends upon the fatty acid incorporated into the membrane. These experiments suggest two fundamental conclusions: (1) the membrane may exist in two physical states, one at temperatures above a critical point and the other at temperatures below a transition point determined by the composition of the lipid phase (see Chapter 9), and (2) the properties of enzymes located in the membrane (or which are a part of the structure of the membrane) may be determined, to some extent at least, by the phospholipid of that membrane. These conclusions may have long range implications to further developments in the area of temperature adaptation.

CETACEANS. Aside from phenomenal migrations spanning the world's oceans, during which time these animals rely exclusively on fat as a fuel, the cetaceans have developed a special modification of fat metabolism for the purpose of laying down thermo-insulating layers of blubber. The blubber depots are apparently metabolically inert, as they do not contribute to the fat reserve for energy production and are not used by cetaceans even under conditions of extreme starvation. The mechanisms involved in regulating the flow of carbon between these different fat reserves are entirely unknown.

UNGULATES. The metabolism of wild species of ungulates is also quite unexplored. It is mentioned here only because some unique specializations of fat metabolism may have evolved in this group. Thus, the mule deer of North America are known to be able to withstand extreme and extended starvation; under comparable conditions, domestic ungulates develop extreme ketosis of the liver due to overproduction of acetoacetate and β-hydroxybutyrate. This characteristic is presumably common in related species such as the arctic caribou, and would be of adaptive significance during their annual and very distant migrations when food supplies may be reduced. Molecular mechanisms underlying this metabolic peculiarity are also entirely unknown.

BROWN ADIPOSE TISSUE IN MAMMALS. Brown adipose tissue is the only organ known in homeotherms whose primary physiological function is the production of heat. Numerous studies indicate that it is a specialized site of thermogenesis in mammals during arousal from hibernation, during adaptation to cold, and in the newborn of certain species.[8,54,129] For this purpose, the distribution of brown adipose around vital organs and the special arrangement of the vascular network which allows warmed blood to return to the thorax are of particular significance.[130] The molecular basis for heat production is not entirely clear. Ultrastructurally, the brown adipose cells are tightly packed with large circular mitochondria, each of which is in turn heavily loaded with cristae.[45] During early development or during cold acclimation, large fat droplets accumulate around the mitochondria,[8] suggesting a high rate of fat metabolism. Because heat production is probably determined by the degree of respiratory activity of the mitochondria, it is evident that the control of thermogenesis and the control of catabolism are closely related problems.

In tissues such as skeletal muscle, heart, and liver, there is good evidence that cellular respiration can be controlled by the availability of ADP as the work load (metabolic or contractile) is increased. However, brown adipose tissue seems to be poorly equipped with mechanisms for the transduction of metabolic energy into physiologically useful forms of work; rather, the energy from the catabolism of fats is apparently released in the form of heat. The situation, as far as it is understood today, can be summarized as follows:

(1) Both *in vitro* and *in vivo* very large (up to 30-fold) increases in oxygen consumption by brown adipose cells occur in response to the catecholamines, and presumably these hormones are involved in the *in vivo* control of this organ[44, 55, 69] (Fig. 6–8).

(2) These very high rates of respiration are achieved by catecholamine activation of the adenyl cyclase system; subsequent lipase activation leads to the mobilization of triglycerides to fatty acids, which are the immediate substrate of energy metabolism in this tissue.

(3) The high rates of respiration also imply either a very high rate of ATP hydrolysis (for example, in lipogenesis) or a lack of respiratory control by phosphate acceptor

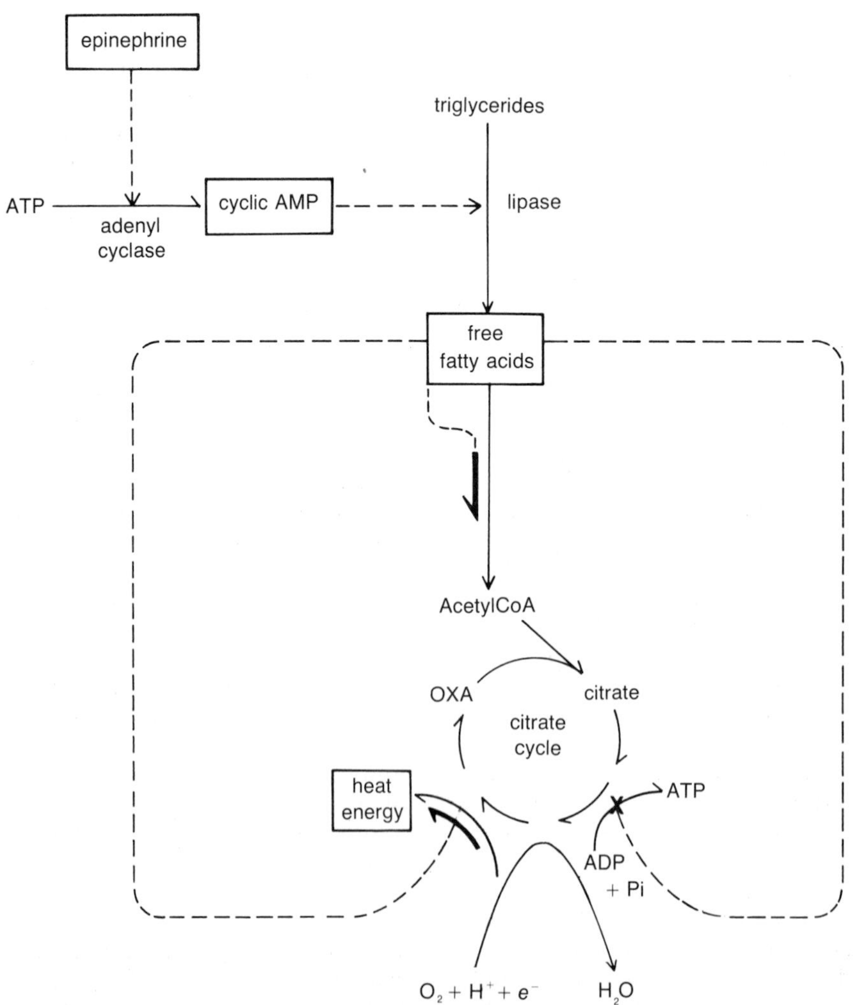

Figure 6–8. Control of ATP synthesis and heat production in mitochondrial metabolism of brown adipose tissue.

(i.e., by ADP). In the latter case, the mitochondria would be uncoupled. The degree to which they are uncoupled and the control of uncoupling have been controversial questions.

(4) Some studies[76] indicate that the rate of triglyceride synthesis is too low for fatty acid esterification to account for the utilization of the ATP which could be formed, assuming normal phosphorylation efficiency.

(5) Most authors therefore agree that a certain amount of uncoupling must occur during active thermogenesis by this tissue.

(6) Williamson[164] has presented evidence that the fatty acids which are released upon norepinephrine stimulation provide *both the substrate for the high respiration and the messenger for uncoupling.* Whereas there is some evidence that fatty acids at high concentration themselves could cause uncoupling, there is also strong evidence against its operation under *in vivo* conditions.[164] Thus, the details of the uncoupling mechanism remain to be resolved (see Chapter 9 for further discussion of this interesting problem).

THE PROSTAGLANDINS. A striking development in biology over the last several years has been the emergence of a group of highly potent compounds termed the prostaglandins. These arise from a variety of long-chain polyunsaturated fatty acids (such as, for example, arachidonic acid), which undergo cyclization reactions catalyzed by microsomally bound enzymes. The prostaglandins have been detected in all tissues so far examined, and a basal rate of release can be detected *in vitro* or during neural and/or hormonal stimulation.

Upon release into the blood stream, these compounds are efficiently and rapidly metabolized by the liver, and possibly by the lung. They exhibit a broad spectrum of pharmacological activity; their physiological activities are not known in any definitive manner, although a direct involvement or an adenyl cyclase mediated involvement with hormone action is considered as one possible major function for prostaglandins. This is an area which will be attended in greater depth in the near future.[109]

THE CITRATE CYCLE

The citrate cycle—frequently also called the Krebs cycle or the tricarboxylic acid cycle—constitutes the hub of metabolism in most cells. Originally proposed to account for the complete combustion of pyruvate and of the 2- and 3-carbon end products of fatty acid oxidation, it also provided one of the earliest recognized examples of an integrated process in which all intermediates are involved as catalysts in the operation of the overall cycle. The importance of the citrate cycle transcends these and other purely catabolic functions, for as the hub of central pathways of cellular metabolism, it supplies crucial reactions and substrates for the biosynthesis (anabolism) of a host of important metabolites, ranging from amino acids, purines, and pyrimidines to long-chain fatty acids and porphyrins.

Studies of the oxidation of mono-, di-, and tricarboxylic acids by preparations from animal tissues were initiated over a half century ago. Early investigators found that of many compounds tested, only succinate, fumarate, malate, and citrate were oxidized at rapid enough rates to be of any possible significance as intermediates in the catabolism of fats and carbohydrates. In 1936, Martius and Knoop suggested that the following set of reactions intervened between citrate and succinate: citrate → isocitrate → oxalosuccinate → α-ketoglutarate → succinate. The next year, Krebs pointed out, on the basis of his demonstration of citrate formation from pyruvate and oxaloacetate in pigeon breast muscle, that these reactions could be used in a cyclical fashion to account for the complete oxidation of pyruvate to three molecules of CO_2 as follows: pyruvate + oxaloacetate → citrate → cis-aconitate → isocitrate → α-ketoglutarate → succinate → fumarate → malate → oxaloacetate.

The essential features of this pathway proposed by Krebs have been well established. The evidence that this sequence constitutes the major pathway for the catabolism of pyruvate and acetate (as acetyl-CoA) in most animal tissues can be briefly summarized as follows:

(1) The various intermediates postulated by Krebs are present in most tissues and are oxidized at rapid rates.

(2) Addition of the various metabolites of the cycle stimulates endogenous respiration but also leads to an uptake of oxygen far above that accountable for by the addition of the intermediates themselves. The latter data forcefully underline the cyclical nature of the oxidation process: *the added substrates act catalytically rather than stoichiometrically, because they are being (at least in part) regenerated in the course of the oxidative reaction.*

(3) Malonate at low concentrations is a highly specific inhibitor of the succinate → fumarate conversion, and predictably leads to an inhibition of respiration and an accumulation of succinate.

(4) The distribution of ^{14}C in the various carbon atoms of the Krebs cycle intermediates during oxidation of specifically labelled substrates (such as glucose or pyruvate) is in complete agreement with the scheme.

(5) The individual reactions postulated can be demonstrated to occur in crude homogenates as well as with highly purified enzymes.

(6) In most organisms, all the individual reactions of the Krebs cycle are localized in the mitochondria, and although enzyme activities vary from tissue to tissue, they occur in nearly a constant proportion to one another.

Contemporary versions of the citrate cycle, basic descriptions of the reactions involved, and the nature of the enzymes catalyzing each individual step in the cycle are available in all modern biochemistry tests and will not be repeated here.

Catabolic Functions of the Citrate Cycle. There exist three great classes of organic compounds in a relatively reduced state that are used as common metabolic fuels: carbohydrates, fats, and proteins. The citrate cycle can serve as a catalytic mechanism for the complete oxidation of each of these. Any metabolite capable of being catabolized to any of the intermediates of the Krebs cycle can then serve as a substrate for cellular respiration. In the case of carbohydrates, the combustion occurs through the production of pyruvate and acetyl-CoA. The fatty acids, which form the bulk of the energy supply available in lipids, also produce acetyl-CoA as a consequence of the β-oxidation reactions. Acetyl-CoA is the sole product of β-oxidation, if, as is the usual case, the fatty acid under consideration contains an even number of carbon atoms in an unbranched chain. Other fatty acids (those with an uneven number of carbon atoms or certain branched fatty acids) lead to propionyl-CoA as well as to acetyl-CoA. The complete oxidation of propionyl-CoA involves a 2-step enzyme catalyzed conversion to succinyl-CoA. While acetyl-CoA can undergo a variety of metabolic reactions, the metabolic fate of succinyl-CoA is mostly limited to conversion to succinate, with subsequent substrate-level phosphorylation of GDP to GTP. As is evident in Figure 6–9, acetyl-CoA and succinyl-CoA are located at critical points in the cycle and together account for the bulk of carbon flowing into the cycle under most circumstances.

In certain species and tissues, amino acid catabolism via the citrate cycle is quantitatively significant. The catabolic pathways for the various amino acids will not be discussed in detail. However, in Figure 6–9, the sites at which various amino acids can "feed into" the Krebs cycle are clearly indicated. In general, there exist two main pathways for the conversion of α-amino acids to the corresponding α-keto acids—the first step in the catabolism of most of the compounds of this class: (1) dehydrogenation by flavoprotein amino acid oxidases and (2) transamination with α-ketoglutarate by a series of pyridoxal-phosphate-linked transaminases. In mammals, transaminations predominate for the naturally occurring L-amino acids, except L-proline. Such metabolism of amino acids by transamination constitutes a dead end as far as the $-NH_2$ group is concerned unless some provision is available for its conversion to a "free" form such as NH_3 or NH_4^+. This consideration ascribes a special role to L-glutamate, which is converted to α-ketoglutarate with the release of NH_4^+. The enzyme catalyzing this reaction, glutamate dehydrogenase, is a widely distributed activity in nature and attests to the importance of this reaction in cellular metabolism.

It is evident that the citrate cycle as a metabolic device for the complete oxidation of pyruvate or acetyl-CoA requires the reoxidation of the reduced coenzymes (NADH and $FADH_2$) produced during its operation. This is achieved through a sequence of electron carriers (the electron transfer system) which link the various dehydrogenases to molecular oxygen. During the period from 1937 to 1941, it was recognized by Belitzer, Tsibabova, and Kalckar that the large amounts of free energy liberated during the complete oxidation of metabolites via the citrate cycle could be utilized to drive the synthesis of ATP, a process which came to be known as oxidative phosphorylation. The earlier studies were greatly expanded in the subsequent decade, so that by 1950 it was clear (1) that oxidative phosphorylation is an exclusive property of mitochondria, (2) that respiration (i.e., oxidation) and phosphorylation, normally "tightly coupled," can be dissociated or "uncoupled" from each other by

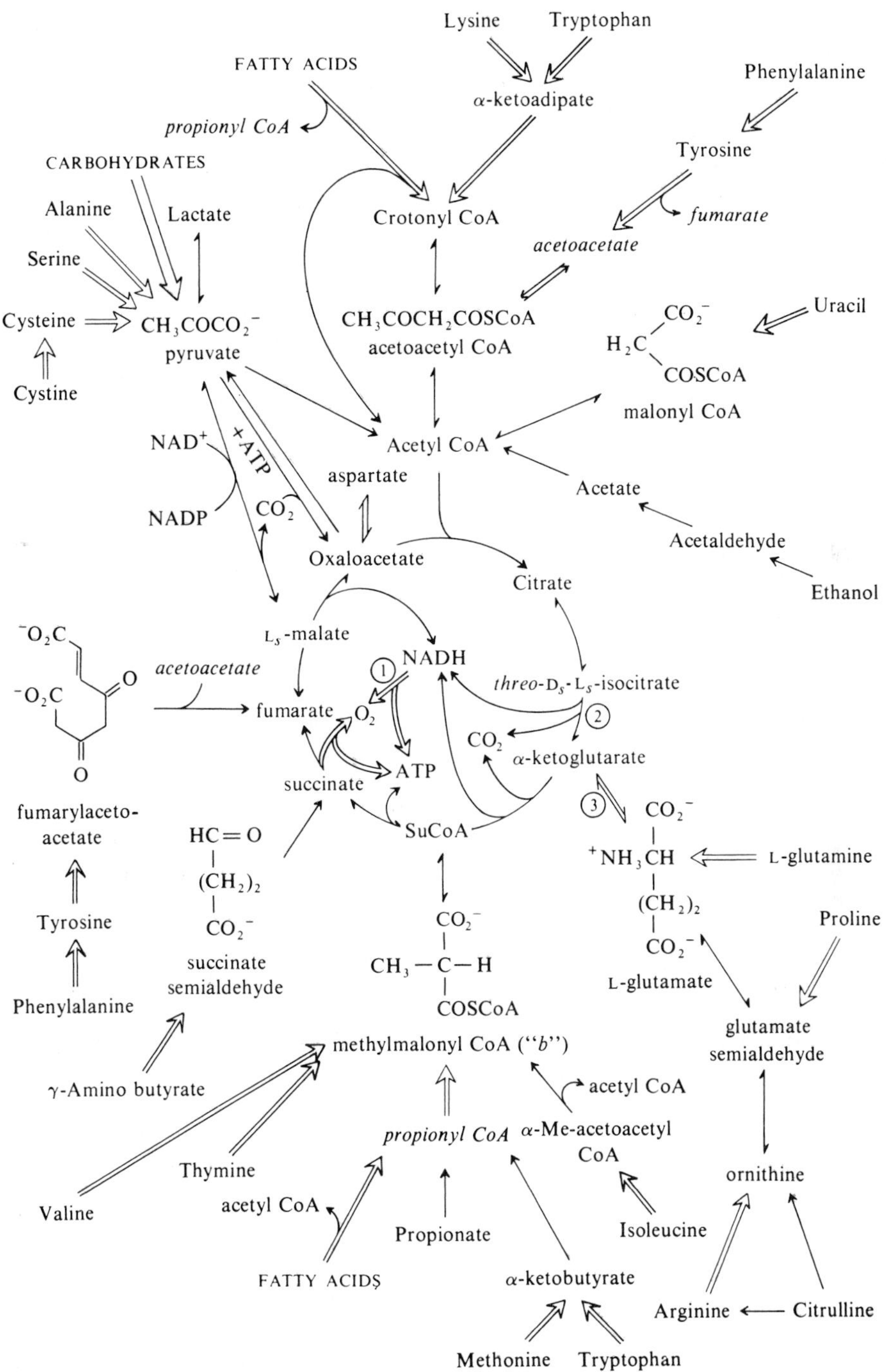

Figure 6–9. The citric acid cycle in catabolism. Multistep reactions indicated by ⇒. The terms in italics are joint intermediates. (From Mahler, H. R., and E. H. Cordes, Biological Chemistry. Harper & Row, New York, 1966.)

appropriate agents, (3) that phosphorylation is almost exclusively linked to the electron transfer system rather than to substrate level oxidations (specifically, 11 out of 12 moles of ATP from the oxidation of one mole of acetyl-CoA, or 14 out of 15 moles of ATP from the oxidation of one mole of pyruvate, are generated as a consequence of mitochondrial oxidative phosphorylation), and (4) that the number of moles of ATP formed relative to the gram atoms of oxygen consumed (i.e., the P/O ratio) approaches integral values for different substrates undergoing one-step oxidations. P/O ratios approaching 4 are obtained for α-ketoglutarate going to succinate; values near 3 are obtained for malate conversion to oxaloacetate, for glutamate conversion to α-ketoglutarate, or for β-hydroxybutyrate conversion to acetoacetate; and P/O ratios of 2 are observed for succinate dehydrogenation to fumarate.

The essential components of the electron transfer system, the reactions catalyzed, and their kinetics all appear to be quite similar in many cells and can be summarized as shown below.

Green and his collaborators have utilized heart mitochondria for the preparation of even smaller submitochondrial electron transfer particles, leading finally to four complexes, each capable of catalyzing one of the component reactions of the electron transport. Recombinations have yielded more complicated arrays capable, in turn, of catalyzing two or more of these steps. These kinds of study, along with sophisticated spectroscopic measurements of reduction and re-oxidation of various respiratory particles and studies with specific respiratory poisons, have identified the major phosphorylation sites: NADH → NADH dehydrogenase, cyt b → cyt c_1, and cyt a → a_3. ATP synthesis is thought to occur coincidentally with these reactions, but the mechanism of phosphorylation is still an open question. This area of cellular bioenergetics is a vast and expanding one which we shall not deal with in any further detail here; the interested student should consult recent texts and annual reviews.

Mitochondria contain relatively large amounts of lipid – more than 90% of which is phospholipid – as an integral part of the membrane system. Isolated submitochondrial particles still retain this lipid component. These lipids, obtained from beef heart muscle, consist of about equal amounts of phosphatidylcholine and phosphatidylethanolamine, together accounting for about 70% of the total lipid. Mitochondrial phospholipids exhibit a particularly high degree of unsaturation, which appears to be of functional importance. The catalytic activity of any of the four complexes can be reduced upon extraction with lipid solvents, but it is reactivated by unsaturated phospholipids. Usually, the greater the degree of unsaturation, the greater the reactivation. The function of the phospholipid appears to be to stabilize the active conformations of the various proteins of the respiratory chain both individually and collectively, and to allow requisite interactions between them.

Control of the Catabolic Activities of the Citrate Cycle. ENZYME LEVELS. The maximum catalytic activities of mitochondrial enzymes, including dehydrogenases characteristic of the citrate cycle and the carriers of the electron transfer system, tend to occur in constant relative proportion. This appears to be true both of insect (locust) flight muscle and of various rat tissues, and has led to the suggestion that the synthesis and turnover rates of key mitochondrial enzymes are integrated, probably at a genetic level.

SUBSTRATE LEVELS. An important controlling feature for any reaction sequence is the availability of initiating substrates. A good indication of the metabolic activity of any given metabolite is its half-life. In rat liver and kidney, half-lives of all of the citrate cycle intermediates are in the order of seconds; oxaloacetate is an exception, for the half-life of this intermediate is in the order of tenths of seconds. These data suggest an important role for oxaloacetate in controlling mitochondrial metabolism; recent studies strongly confirm this observation.[110, 167]

RESPIRATORY CONTROL. The rate of mitochondrial respiration also depends in a most fundamental way upon the coupling between respiration and phosphorylation. Intact mitochondria are usually "tightly" coupled (when ADP and P_i are not limiting), and their rate

substrate → NAD → NADH dehydrogenase → cyt b → cyt c_1 → cyt c → cyt (a + a_3) → O_2
succinate → succinate dehydrogenase ——————↑

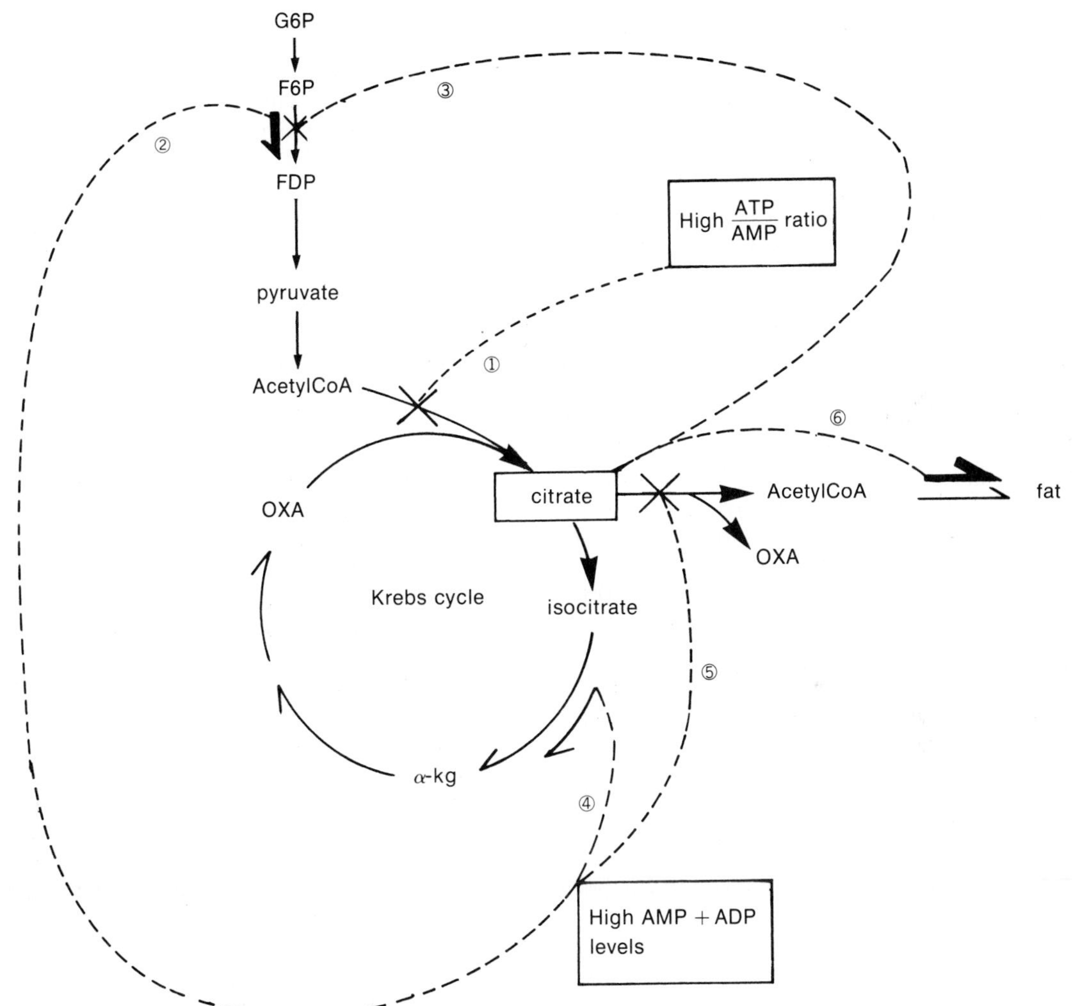

Figure 6–10. Six regulatory interactions affecting the production and utilization of citrate. Broken lines link modifiers to the reactions they control. Negative modifier action is indicated by a cross, and positive modifier action by an arrow. (After Atkinson, D. E., *in* Metabolic Roles of Citrate, edited by T. W. Goodwin. Academic Press, London, 1968.)

of respiration is strongly affected by the ratio of ADP/ATP. When this ratio is high ("state 3"), most of the intramitochondrial adenine nucleotide is in the form of ADP and oxygen uptake is high. In contrast, when ATP synthesis from ADP + P_i is high (that is, when ATP utilization cannot keep pace with ATP formation), respiration wanes ("state 4"). These phenomena are rather loosely termed "respiratory control" in the literature. In passing, mention should be made of "states 1 and 2." In "state 1," the mitochondria are lacking both ADP and substrate, while in "state 2" ADP is present but substrate is still lacking. Oxygen uptake is low in both.

REGULATION AT THE CITRATE BRANCHING POINT AND INTEGRATION WITH OTHER PATHWAYS. The best available evidence on control of the catabolic functions of the citrate cycle suggests that the cycle operates as two reaction spans: (a) acetyl-CoA → α-ketoglutarate, controlled by the citrate synthase and isocitrate dehydrogenase reactions, and (b) α-ketoglutarate → oxaloacetate, controlled in part at least by the α-ketoglutarate dehydrogenase reaction.[4, 110, 167] Central to control of the first span of the citrate cycle is control of the production and utilization of citrate.

At least six enzyme-modifier interactions are known that seem to be directly involved in regulation of the production and utilization of citrate (Fig. 6–10). Three of these interactions affect citrate production: (1) Citrate synthase itself is controlled by the adenylates. ATP inhibition is competitive with respect to both acetyl-CoA and oxaloacetate at least in the rat liver.[48] Thus, conditions of high ATP would tend to reduce the apparent affinity of the enzyme for both of its substrates. Under most conditions, the activity of citrate synthase appears to be most dependent upon the apparent affinity for oxaloacetate.[110, 167] It is probable that these controls assure that the entry of acetyl-CoA (produced from carbohydrate or fatty acid metabolism) into the citrate cycle is kept in step with the energy needs of the cell. (2) In addition, the production rate of pyruvate and acetyl-CoA from hexose precursors is regulated at the PFK step of glycolysis, where AMP is a positive modulator. Thus, when ATP levels are high (AMP levels low) pyruvate production from hexose precursor would be reduced by PFK inhibition. (3) Citrate, which itself is a negative modifier of the PFK catalyzed step, may accumulate under these conditions and contribute to maintaining PFK in a "shut-off" conformation. On the other hand, this effect may insure that glycolysis is accelerated when concentrations of citrate fall below usual levels.

The other three interactions shown in Figure 6–10 affect the utilization of citrate: (4) AMP is a positive modulator of NAD-linked IDH, and this effect is in the right direction to increase ATP generation (by channelling citrate into the citrate cycle) if its concentration is low. This regulatory effect may couple with control at the citrate synthase step. (5) ADP is a negative modifier of the citrate cleavage enzyme; thus, citrate and ADP will be consumed in this reaction only when the ATP/ADP ratio is high. These mechanisms, taken together, appear to be responsible for the channelling of citrate between oxidation by the citrate cycle (with concomitant ATP regeneration) and cleavage for the production of acetyl-CoA for fatty acid biosynthesis (with concomitant ATP utilization). (6) Finally, citrate itself is a positive modifier for acetyl-CoA carboxylase, as we have already pointed out. We can consider this effect as a kind of feed-forward activation of a pathway leading from citrate to fatty acids. This regulatory effect would be favored under energy-saturating conditions when citrate and ATP levels are high.

Regulatory mechanisms in the second span (α-ketoglutarate → oxaloacetate) are not as well understood. Most studies of respiratory control have involved substrates such as succinate, α-ketoglutarate, and malate; under these conditions, control of respiration by availability of ADP must somehow be integrated with control of specific enzyme reactions in the citrate cycle,[48] but we must await further work before this integration is better understood.

Anabolic Functions of the Citrate Cycle. A large number of biosynthetic pathways (Fig. 6–11) emanate from the citrate cycle and its metabolite derivatives, and the cycle therefore plays a pivotal role in anabolism as well as in catabolism. The problem of integration of anabolic and catabolic functions arises again, for any intermediate which is bled off from the cycle must be replenished at the same or a different locus in the cycle in order for steady-state operation to be sustained.

In most tissues which are capable of *de novo* synthesis of carbohydrates, the synthesis

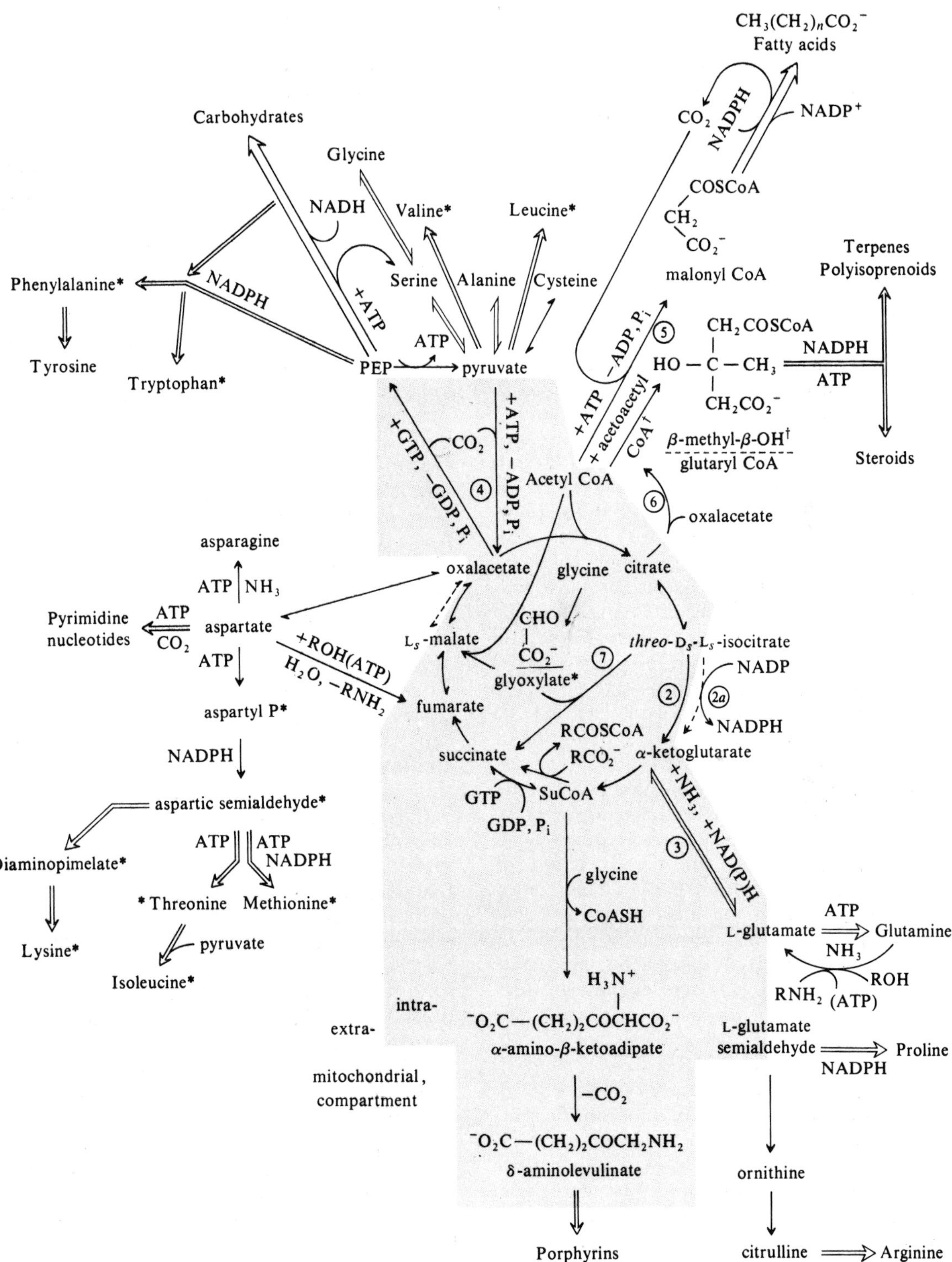

Figure 6–11. Biosynthetic functions of the citric acid cycle. (From Mahler, H. R., and E. H. Cordes, Biological Chemistry. Harper & Row, New York, 1966.)

requires the presence of stoichiometric amounts of one of the dicarboxylic or tricarboxylic acids of the citrate cycle or of compounds that can give rise to one of these intermediates. This requirement is met by a large number of organic intermediates containing a minimum of three carbon atoms (since oxaloacetate can be formed from pyruvate by CO_2 fixation reactions discussed above). Those cells containing the two enzymes peculiar to the glyoxylate bypass (i.e., isocitrate lyase and malate synthetase) are capable of generating 4-carbon dicarboxylic acids and hence carbohydrate from 2-carbon precursors. The biosynthesis of carbohydrate from acetyl-CoA (derived from fatty acids) can take place only in cells containing both enzymes. Most animals cells therefore are quite incapable of affecting this conversion.

The integration of control mechanisms of the citrate cycle and gluconeogenesis are not well worked out. Concentration ratios of ATP/AMP are certainly involved in the regulation of the PEP $\leftrightarrows$ pyruvate and the F6P $\rightleftharpoons$ FDP control sites in the pathway.

We have already discussed the basic features of the biosynthesis of lipids. Here we wish to reemphasize that acetyl-CoA, the key intermediate involved, is generated by two principal pathways, both localized in the mitochondria—by the thiolytic cleavage of acetoacetyl-CoA (generated by β-oxidation of the fatty acids) and by the oxidative decarboxylation of pyruvate (catalyzed by pyruvate dehydrogenase). Yet, as we have pointed out, the biosynthesis of fatty acids requires as an obligatory first step the carboxylation of acetyl-CoA to malonyl-CoA. This latter reaction, as well as the subsequent fatty acid synthetase reactions, are catalyzed by an extramitochondrial complex of enzymes. The role of the citrate cleavage enzyme in generating extramitochondrial acetyl-CoA for fatty acid and other biosynthesis has already been stressed above.

Protein synthesis requires metabolic energy (as ATP) and a supply of monomeric precursors (about 20 of the common L-amino acids), both of which are at least in part provided for by the operation of the citrate cycle. Of the 20 required amino acid precursors, most higher animals are able to synthesize about half. The metabolic pathways for formation and degradation will not be discussed at this time; however, several aspects are worth mentioning now. There exist three groups of amino acids that originate from the three keto acids (pyruvate, oxaloacetate, and α-ketoglutarate) and the corresponding α-amino acids (alanine, aspartate, and glutamate). The group related to α-ketoglutarate and glutamate is apparently found throughout nature. Only some of the amino acids arising from pyruvate are synthesized in higher organisms. Those for which the carbon skeleton arises from oxaloacetate and aspartate are not produced at all by mammals and probably not by other animals, either.

The synthesis of purines and pyrimidines, fundamental constituents of coenzymes and of nucleic acids, is also intimately related to the citrate cycle. Thus, the carbon skeleton of pyrimidines is provided by aspartate, whereas aspartate and glutamine are involved in the biosynthesis of the ureido nitrogens of purines, as well as the amino nitrogen for both the amino purines and the amino pyrimidines. Recent information on these important metabolic pathways is available in modern biochemistry texts and annual reviews.

Finally, brief mention should be made of the role of the citrate cycle in the biosynthesis of the porphyrins. These are vital components of respiratory pigments (such as hemoglobin) and of the electron transfer system. The biosynthetic pathway is initiated in the citrate cycle at the level of succinyl-CoA and thus constitutes yet another significant and continuous drain on the carbon flowing through the cycle.

Comparative Aspects of the Citrate Cycle. RELATIONSHIP OF STRUCTURE OF THE MITOCHONDRION TO ITS FUNCTIONS. For those tissues which are well examined, such as rat liver, it is known that the mitochondria contain all the enzymes and cofactors required for the operation of the citrate cycle, respiratory chain and substrate level phosphorylations, fatty acid and pyruvate oxidations, and acetoacetate synthesis. For continued function in the isolated state, the only further requirements are a suitable osmotic support medium, oxygen, substrate, phosphate, and a phosphate acceptor (such as ADP). In special cases, further additions are required to activate transport systems. For example, carnitine must be added for the oxidation of long chain fatty acids, and malate must be present before exogenous isocitrate can be oxidized.[48]

There is widespread agreement that these mitochondrial functions either are associated with the inner membrane and cristae (as in the case of the respiratory chain) or are located in the matrix of the cell (as in the case of enzymes such as citrate synthase and isocitrate dehydrogenase). The inner membrane is not freely permeable to proteins, nucleotides, the great majority of non-lipid soluble cations and anions, and uncharged but hydrophylic compounds such as sucrose. These permeability properties of the inner membrane, in association with various transport systems, provide the structural basis for concentration gradients between the mitochondrial and extramitochondrial compartments.

Also, it is probable that the inner membrane is at least in part composed of elements (protein and phospholipid) which are involved directly in such functions as electron transfer. Indeed, Green and his collaborators have shown that the overall ultrastructure of the mitochondrion depends upon its metabolic state. Their observations have led them to postulate the existence of energized and de-energized states of the mitochondrion; the mechanical events involved in going from one to the other are thought to be closely associated with phosphorylation. We have very little information on mechanisms of control at this higher organizational level. That it is an important level of control of mitochondrial function is evident by the extreme specializations of mitochondrial ultrastructure in different tissue types. We will discuss a number of these separately.

TISSUE SPECIFICITIES IN MITOCHONDRIAL ULTRASTRUCTURE. The basic problem here is the relationship between the various exergonic reactions and the peculiar work functions which mitochondrial metabolism must support in any given tissue. At the outset, we would stress that although this problem of coupling between energy production and work function is well recognized, very little information is available. Intuitively, it does seem clear that this coupling depends upon the tissue (i.e., upon the type of work function the tissue performs, chemical, osmotic, or mechanical), and it appears to involve important control components at the organizational level of mitochondrial structure. Thus, certain tissues possess *highly characteristic mitochondrial ultrastructure as well as highly characteristic organization of mitochondria within the cell.* Some of the more notable examples are listed below:

(1) In all "fast" muscles, particularly in hummingbird flight muscle and in the flight muscle of insects, the mitochondria are arranged in a very regular manner along the myofibrils. The cristae are numerous and the shape of the mitochondria is typically fairly long and narrow. Often, a fat droplet is formed adjacent to or contiguous with each mitochondrion, presumably supplying each mitochondrion with its own fuel depot. The coupling mechanism accounting for delivery of ATP to the contractile elements is not known.

(2) In the proximal tubule and the ascending limb of the loop of Henle in the kidney, the mitochondria are so large they are referred to as "giant mitochondria" in the literature. Their arrangement in the cell is such as to polarize it: the lumen side of the cell is essentially barren of mitochondria, while the blood side is packed with them. Here they come to lie in special infoldings of the cell plasma membrane, each giant mitochondrion often being "housed" by the membrane on essentially all but one side. The work function of these mitochondria clearly is to support the active transport of cations (mainly Na^+). A comparable ultrastructural organization of the mitochondria is found in the absorptive cells of the small intestine, and in this tissue we again see the mitochondria being localized to one region of the cell.

(3) A major work function of the gills in fishes and crustaceans is cation transport. In these tissues, too, there typically occurs a highly specialized ultrastructure and a localization of most of the large mitochondria at the basal membrane of the cell.[34, 64] A comparable ultrastructure occurs in the cells of the salt gland in sea birds. This ultrastructure is considered an integral part of this tissue's Na^+ transport abilities.[41]

(4) Brown adipose tissue of mammals, as we have already stressed, possesses a distinct thermogenic function. In our context, this is not to be regarded as a work function *per se*. In a sense, one might suppose that mitochondria in this tissue have been to some extent "released" from work—energy normally formed as ATP, then either stored in some other metabolite or utilized in support of some work process, here is largely dissipated as heat. The ultrastructure of these mitochondria appears to be quite unique to this tissue, again attesting to an important relationship between mitochondrial structure and function.

These observations suggest that the milieu

in which mitochondria are formed determines to some extent their ultrastructure. It is evident that giant mitochondria from nephric tubules or gills would not be functional in the milieu of muscle or of brown adipose tissue, but we are only beginning to examine the reasons why.

METABOLIC REGULATION

Introduction

The metabolic properties which an organism is capable of expressing, like any other phenotypic characteristics, must be shaped by evolutionary processes. But no reaction or pathway can be considered adaptive in the absence of regulation. Thus, we expect that the evolution of every metabolic sequence has been accompanied by the development of control mechanisms which can allow the sequence to respond automatically and adaptively to the needs of the organism under a wide range of internal and external parameters. Some of these mechanisms we have already briefly described. However, the area of regulation of metabolic processes has undergone such an extensive flowering in the last decade that a separate discussion of some of the major principles and findings seems necessary.

During this last decade, four major mechanisms for regulating the participation of given enzymic reactions and reaction pathways have been widely recognized: (1) *regulation of the steady-state concentration of enzymes,* (2) *regulation of the enzyme type (or isozyme) responsible for the catalysis of a given reaction,* (3) *regulation of the catalytic activities of strategically placed enzymes by low-molecular weight intermediates, which themselves are often involved in the same or related metabolic pathways,* and (4) *regulation through the proper positioning of enzymes within the cell.* The first two mechanisms presumably involve changes in rates of enzyme synthesis or degradation or both, and provide for only a rough level of control; response times here are in the order of hours to days. Fine control, usually with response times in seconds to milliseconds, is achieved through modulation of enzyme activity. We shall now discuss specific examples of each of these general mechanisms in turn.

Regulation of Enzyme Concentrations

Insulin and Glucocorticoid Regulation of Metabolism. It is well established that insulin and the glucocorticoids act in an antagonistic manner in the regulation of glycolysis and gluconeogenesis. Known sites of action of insulin and glucocorticoid hormones are shown in Figure 6–12. One of the first events in glucocorticoid activation of glucose synthesis is the mobilization of amino acid precursors from peripheral tissues, mainly muscle. These are carried to the liver, where they are transaminated to appropriate keto acids and are then incorporated into the gluconeogenic pathway. A part of the hormonal control in liver involves the induction of key glucogenic enzymes (pyruvate carboxylase, PEPCK, FDPase, G6Pase and important transaminases). These (200 to 300 per cent) changes in enzyme levels are due to changes in rates of enzyme synthesis and contribute to increased glucose synthesis.[43, 158] Comparable effects of the glucocorticoids are known to occur in the kidney cortex, another major glucogenic tissue in the vertebrate body.[128]

Of the various amino acid precursors to glucose, alanine is quantitatively the most important, and under conditions of starvation in man, when glucogenic events are activated, control of gluconeogenesis can occur at the level of alanine mobilization and/or delivery to the glucogenic tissues.[43, 92] Krebs has argued that most of this alanine cannot be derived from protein breakdown and therefore must be derived metabolically. Under conditions of starvation, when urea synthesis slows down because of lack of ATP, alanine may serve as an ammonia-binding agent; the mobilization of alanine in peripheral tissues[92] may therefore be the expression of a carrier function of alanine, concerned with transporting nitrogen from peripheral tissues to the liver. Krebs also suggests that alanine may serve as the carrier of carbon atoms of glucogenic amino acids. Thus, glutamate or aspartate may be partially degraded in muscle and other tissues to the level of pyruvate (thus yielding a certain amount of energy); the three carbon atoms might then be transported to the liver, together with the nitrogen of these amino acids, in the form of alanine. Whatever the secondary functions, alanine is therefore a major amino acid precursor of glucose, and

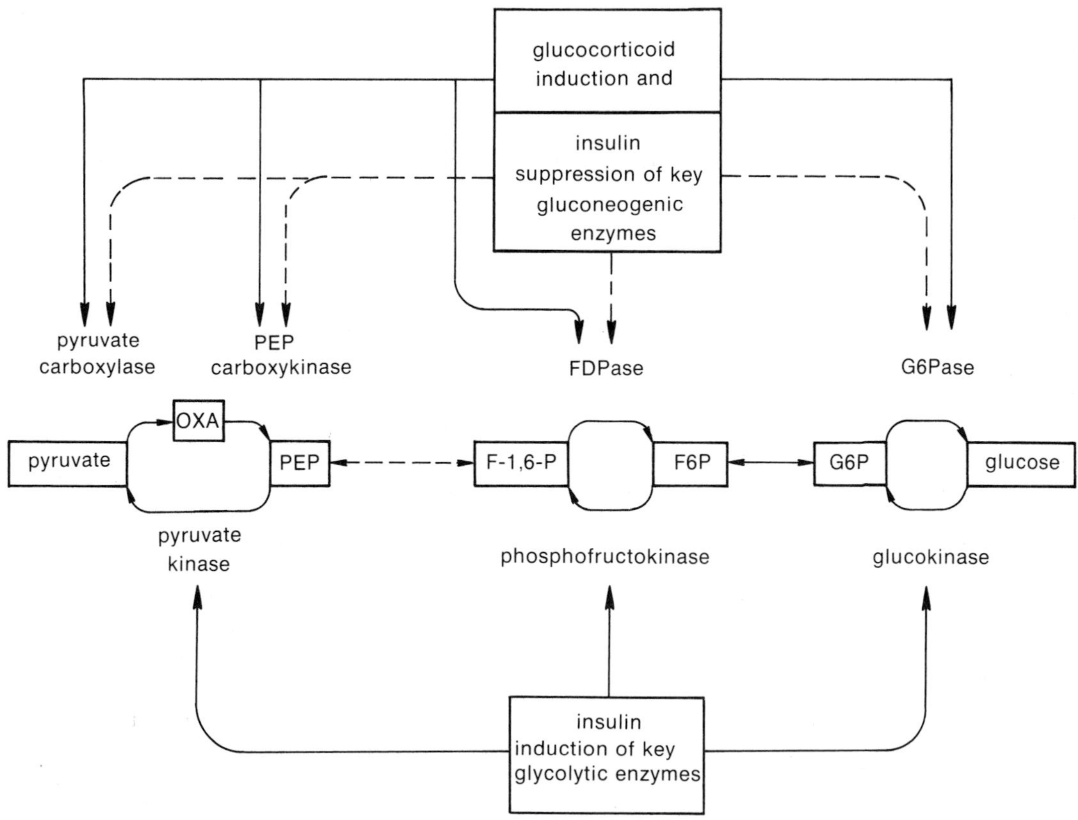

Figure 6–12. Insulin and glucocorticoid control of glycolysis and gluconeogenesis through the regulation of enzyme concentrations. (After Weber, G., R. L. Singhal, and S. K. Srivastava, *in* Advances in Enzyme Regulation, Vol. 3, edited by G. Weber. Pergamon Press, Oxford, 1965.)

its transamination to pyruvate in the liver is a hormonally controlled process.

When glucose levels rise, glucose does not feedback and inhibit its further production, as might be expected in a simple system of negative feedback control. Rather, it serves as a signal at the pancreas for the release of insulin. The basic effects of insulin are antagonistic to those of the glucocorticoids (Fig. 6–12). Thus, key glycolytic enzymes (Hk, PFK, and PyK) are induced, while glucogenic ones are repressed, when control is transferred to insulin.[158] One can then visualize a kind of oscillating hormonal control of glucose metabolism, the glucocorticoids favoring carbon flow in the direction of glucose and insulin favoring carbon flow in the direction of glycolysis. But this simple scheme is complicated by important interactions with other metabolic pathways, some of which are not yet well understood.

Thus, both insulin and the glucocorticoids stimulate glycogen synthesis by comparable activation of glycogen synthetase.[56] The significance of the insulin effect is that it presumably represents activation of a pathway which uses glucose as a precursor and thus will lead to reduced glucose levels in the blood. But the significance of the glucocorticoid effect is unclear.

Secondly, there are important interactions between fatty acid metabolism and the pathways of glucose metabolism. Thus, perfusion of rat liver with a fatty acid such as oleic acid leads to a doubling of rates of glucose synthesis from a variety of precursors (lactate, pyruvate, and alanine). This effect is abolished by inhibiting palmitylcarnitine transferase (which blocks fatty acid transfer into the mitochondria) *without altering control rates of glucose synthesis.* These experiments prove conclusively that gluconeogenesis can be controlled by the rate of fatty acid oxidation. Changes in liver levels of intermediates in this pathway indicate that pyruvate carboxylase is the major control site and that the activity of this enzyme is being modulated by acetyl-CoA produced by fatty acid oxidation.[165] It is important to remember that fatty acids are not being used as carbon precursors of glucose, for as we have pointed out above, no net conversion of fatty acids to carbohydrate can occur in organisms which do not possess the glyoxylate by-pass for priming the citrate cycle. The importance of these interactions between fatty acid oxidation and glucose synthesis, however, should not be minimized; the integration is also observed at the hormone level of control, for insulin is known to favor glycolysis and fatty acid synthesis, while the activation of glucose synthesis is correlated with increased fatty acid oxidation.[43, 99, 158]

In this connection, it has been reported[81, 157] that fatty acids inhibit several glycolytic enzymes selectively without altering the activities of either glucogenic enzymes or bifunctional enzymes of glucose metabolism. It is therefore suggested that free fatty acids may regulate glucogenesis directly by suppressing flux through glycolysis. More recent studies have found that these may be nonspecific effects, as free fatty acids and palmityl-CoA also strongly inhibit gluconeogenic enzymes (PEPCK, PC, and G6Pase).[104]

The mechanisms by which insulin and glucocorticoids alter enzyme synthesis rates, that is, the general mechanisms of action of these hormones *per se*, are still not entirely clear, and this represents an area of very active current research. Kenney and his collaborators have approached this problem by examining the induction of specific enzymes in cultured hepatoma cells. In the case of tyrosine α-ketoglutarate transaminase (a liver enzyme induced by both glucocorticoids and insulin), enzyme induction by the two hormones differs kinetically, and a synergistic response is obtained when both are added together. This result would be expected if the mechanisms of induction were different for the two hormones. The best current evidence suggests that the steroid hormone (hydrocortisone) induces transaminase synthesis by acting on genetic transcription, while the polypeptide hormone (insulin) acts at some post-transcriptional or translational event in the synthesis of the enzyme.[83, 113]

Glucagon and Catecholamine Effects on Liver Glucose Metabolism. Glucagon also plays a significant role in the regulation of glucose output by the liver. A part of this effect is mediated through glycogen mobilization by activation of the adenyl cyclase system, but a part of the effect depends upon an activation of gluconeogenesis. Thus, glucagon strongly activates the incorporation of ^{14}C-labelled carbon from lactate, pyruvate, or alanine into glucose. The response time in perfused rat liver is in the order of 40 seconds, and therefore the mechanism of glucagon action is probably quite different from that of insulin and glucocorticoids. It has been suggested that the glucagon activation depends upon the adenyl cyclase system.[43] The reactions affected by cyclic AMP have not been identified but are located between pyruvate and PEP in the gluconeogenic pathway. The effects of the catecholamines are comparable to that of glucagon, and the mode of action is thought also to be similar.[43]

Thyroid Hormone Effects on Metabolism. The basic action of thyroid hormone in higher vertebrates involves a stimulation of cellular metabolism, but the mechanism by which thyroxine accelerates cellular reactions is not yet clear. Injection of the hormone into animals stimulates most enzymic systems studied, such as glucose oxidation, amino acid incorporation into protein, liver G6Pase, and G6PDHs. When added to mitochondrial preparations, thyroxine uncouples a part of the oxidative phosphorylation occurring during the oxidation of β-hydroxybutyrate. Concomitant changes in the distribution of water lead to mitochondrial swelling. This kind of uncoupling of respiration and phosphorylation could channel oxidative energy into heat rather than into ATP synthesis, as appears to be the case in brown fat mitochondria. But the quantitative importance of these effects *in vivo* is not established, and indeed the thyroxine-mediated uncoupling may be pharmacological, particularly in mammals, since thyroxine concentrations required have been well beyond the physiological range. In fishes, prolonged treatment with low doses of thyroxine leads to decreased phosphorylative activity in mitochondria.[93] Upon addition of thyroxine to mitochondrial preparations *in vitro*, uncoupling occurs at thyroxine levels between 5×10^{-7} and 5×10^{-8}M. Comparable figures for mammalian systems are about 5×10^{-5}M. In addition, Massey and Smith[93] observed increased specific activities of oxidative enzyme systems in fish following prolonged low-dosage treatment with thyroxine, suggesting an effect of the hormone on enzyme synthesis rates. Control of the synthesis of enzymes of urea metabolism is the established mechanism for thyroid hormone action during metamorphosis in amphibians, when the organism switches from NH_4^+ to urea excretion.

Comparative Aspects of Regulation of Enzyme Concentrations. HORMONAL REGULATION OF INTERMEDIARY METABOLISM IN LOWER VERTEBRATES AND INVERTEBRATES. At the outset, it should be stressed that molecular mechanisms of hormone action in the lower vertebrates and the invertebrates are not understood mainly because they have not been studied extensively. Insect endocrinology is an important exception to this statement, and we have already cited the example of hormonal regulation of blood trehalose in insects. Advances in our understanding of these problems in insects can be expected with confidence. The lack of developments in the area of control of metabolism in such organisms as the Crustacea, the molluscs, and the fishes is most unfortunate, and hopefully the situation will change in the future.

A large literature has developed on the significance of regulating enzyme concentrations during thermal adaptation of metabolism in poikilotherms, particularly fishes.[47, 57, 61, 108, 131] The general observation is that enzyme concentrations tend to be increased during cold adaptation, but nothing is known of the signals leading to this response. The ultimate signal is probably a physical parameter (such as temperature) or a combination of physical parameters (such as temperature and photoperiod). If hormone signals are important, nothing definitive is known of which ones are involved.

Conte and his co-workers[32] have described an entirely analogous situation in salinity adaptation of the salmon gill. These workers have established that fundamental changes in the synthesis rates of specific proteins occur in response to a salinity change in the outer environment. In nature, this event presumably occurs during the parr-smolt transformation. The proteins involved probably are part of the Na-K ATPase complex of the gill cell membrane, but again the signals initiating these events are not known.

CONTROL OF GLUCONEOGENESIS AND LIPOGENESIS IN RUMINANTS. In non-ruminant mammals, glucose derived from the diet is the major precursor for the synthesis of lipids, but the situation is less clear in ruminants. Here, dietary carbohydrate is converted into various short-chain intermediates, such as the trioses, acetate, and butyrate, by rumen metabolism. Two adaptations arise which are peculiar to the ruminants: (1) Since adult ruminants do not have a dietary source of glucose, these animals are entirely dependent upon gluconeogenesis in the liver and kidney for the production of all required glucose (plus derivative carbohydrates), and (2) the obvious premium which is placed on this source of carbohydrate suggests that acetate, the product of rumen metabolism, and not glucose, is the major carbon precursor for lipogenesis.

This unusual dependence on glucogenesis as the sole source of glucose in ruminants has led to higher steady state concentrations of key gluconeogenic enzymes, such as FDPase. Also, during lactation, when gluconeogenesis in the liver of these animals is very rapid, very high activities of pyruvate carboxylase and PEPCK occur.[6] As other enzyme pathways which compete for oxaloacetate (such as the citrate cleavage pathway) are reduced in these organisms, the high activities of pyruvate carboxylase and PEPCK would direct most of the oxaloacetate formed from amino acids in the direction of PEP and hence glucose.[53]

The situation as regards lipogenesis in the ruminant is more complex. We have already indicated that fatty acid synthesis in nonruminants from glucose in adipose tissue and liver involved the breakdown of glucose to pyruvate in the cytosol, the decarboxylation of pyruvate to yield acetyl-CoA within the mitochondria, the condensation with oxaloacetate to form citrate, and the transport of acetyl-CoA carbons from the mitochondria as citrate. The subsequent cleavage of citrate in the cytosol to yield extramitochondrial acetyl-CoA and oxaloacetate is now accepted as an important control site in fatty acid synthesis.[155] Lardy and his coworkers[171] suggested that the oxaloacetate produced is converted to malate (by NAD-linked MDH) and that the malate is oxidatively decarboxylated (by the NADP-linked malic enzyme) to pyruvate. The NADPH generated in this latter reaction supplies reducing power for fatty acid biosynthesis (Fig. 6–13). These reactions constitute the

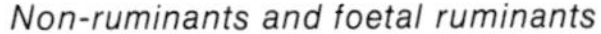

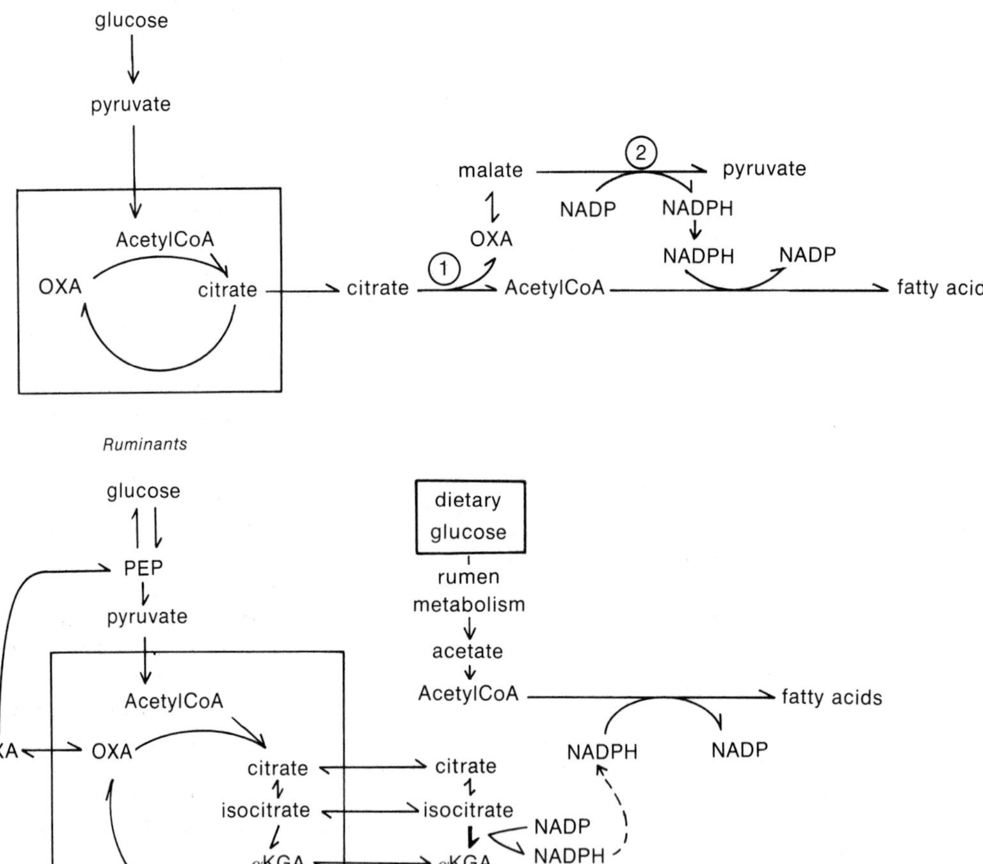

Figure 6–13. Pathways of lipid synthesis in liver of ruminants and non-ruminants.

so-called citrate-cleavage pathway, and *in the ruminant mammals this pathway is largely inoperative.* The two major "molecular lesions" occur at the citrate cleavage and malic enzymes, the activities of both enzymes being unusually low in adipose, liver,[53] and mammary tissues of domestic ruminants.[9]

In place of the citrate-cleavage pathway, ruminant animals utilize dietary acetate as the main carbon source of lipid synthesis. In the cytosol of liver and adipose tissue, acetate is converted to acetyl-CoA by extramitochondrial acetate thiokinase, and the acetyl-CoA can then feed directly into the pathway of fatty acid synthesis (Fig. 6–13).[7] Bauman also pointed out that in the cow extramitochondrial NADP-linked IDH specific activities are about 20 times higher than in homologous rat tissue, and these reactions therefore probably generate all or most of the requisite NADP for fatty acid synthesis.[9]

A fascinating "control" to these experiments is the ruminant fetus, which, unlike its mother, has an excellent and continuous "dietary" source of glucose; namely, the *glucose delivered by the maternal blood to the placenta.* Hence, during fetal life, glucose is an abundant lipogenic precursor and the citrate cleavage pathway is operative. The specific activities of the citrate cleavage and malic enzymes are about 20 times greater in fetal liver than in the liver of the adult ruminant.[9, 53] Although these specific metabolic patterns depend upon changes in enzyme levels (presumably by adjustments in enzyme synthesis rates), hormonal or other triggers setting off these molecular adaptations are not yet identified.

CONTROL OF LIPOGENESIS DURING LACTATION IN MAMMALS. In the mammary gland of the rat, large increases in citrate cleavage enzyme activity occur within two days of the onset of lactation. Weaning results in a rapid decline of citrate cleavage activity within about a day. These effects are presumably general in mammary gland tissue, and also suggest that lipogenesis here is controlled by modulating the level of citrate cleavage enzyme.[85] Reducing power, in the form of NADPH, for lipogenesis is apparently supplied by the G6P dehydrogenase reaction catalyzed by a specific isozyme which is induced at lactation.[114] The activity of this form of G6P dehydrogenase increases 20-fold at this time.

HORMONAL CONTROL OF LACTOSE SYNTHESIS IN THE MAMMARY GLAND. The disaccharide lactose, like casein, is made only by the mammary gland. Certain characteristic features of the development of the lactose synthetase system have led to an enhanced appreciation of the interplay between enzymes and hormones in the mammary gland during pregnancy.

Lactose synthetase catalyzes the formation of lactose according to the reaction:

UDPgalactose + glucose ⇀ lactose + UDP

The transfer of galactose to glucose is mediated by two proteins, referred to as A and B.[18] The A enzyme apparently is associated with the microsomes[19] and is a galactosyl transferase which catalyzes the reaction:[17]

UDPgalactose + *N*-acetylglucosamine →
N-acetyllactosamine + UDP

The B-protein is identical with one of the common milk proteins, α-lactalbumin.[20] Unlike the A-enzyme, it has no known enzymatic activity by itself. In the presence of the B-protein, however, the A-enzyme is able to utilize glucose as an acceptor of the galactosyl portion of the UDPgalactose.[17] In essence, then, the *B-protein directs the A-enzyme to catalyze lactose rather than N-acetyllactosamine formation.*

Maximal stimulation of the synthesis of the A- or B-proteins by mammary gland explants requires the presence of three hormones—insulin, hydrocortisone (or some other glucocorticoid) and prolactin, and under these *in vitro* conditions the A- and B-proteins develop at about the same time. However, in intact mice the A-enzyme activity increases rapidly at about the middle of pregnancy and reaches a maximum shortly before parturition; the B-protein does not increase markedly until after parturition[147] (Fig. 6–14). The difference in the *in vivo* versus the *in vitro* response can be ascribed to progesterone, for this steroid has been shown to exert a selective inhibitory effect on B-protein synthesis by mammary explants.[148] During pregnancy, then, the synthesis of B-protein is presumably depressed by high circulating levels of progesterone, but the synthesis of the A-enzyme is unaffected by this hormone (Fig. 6–14). Under these circumstances, N-acetyllactosamine should accumulate.

These biosynthetic and metabolic events in the mammary gland are closely associated with ultrastructural changes in the alveolar cells, and indeed, with other specific biosyn-

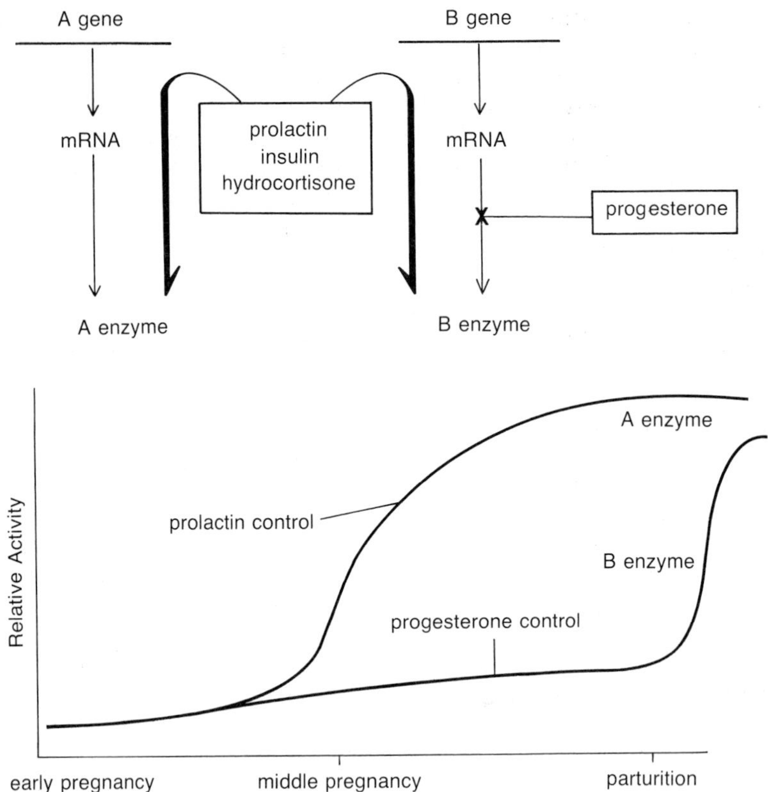

Figure 6–14. Diagrammatic representation of hormonal control of A and B enzymes of lactose synthesis.

theses (such as casein biosynthesis) which occur concomitantly.[143]

Role of Isozymes in the Regulation of Metabolism

Earlier studies of hormonal regulation of metabolism at the enzyme level assumed implicitly or explicitly that the enzymes being induced were of the same type as those already present in the tissue. In many if not most cases, this assumption is incorrect. In the regulation of glucose metabolism, for example, insulin and presumably also the glucocorticoids appear to *regulate the synthesis of specific isozymic forms of enzymes, rather than the total activity of the enzyme reaction under consideration. In such cases, it appears that the enzyme forms being induced are in some way particularly well "adapted" for function in the new organizational pattern of metabolism.* This is true both in hormonal adaptation of enzymes in mammals[72, 135] and in various enzyme adaptations to the environment in lower vertebrates and invertebrates.[61] Thus, during insulin induction of glycolysis in the liver, a special isozyme of pyruvate kinase (PK) is induced which displays catalytic and regulatory properties "better suited" for net gluconeogenesis than is the muscle form of the enzyme. The latter form, however, is present in liver and contributes to PK activity under condition of glycolysis.[135] A similar situation is found in the case of insulin induction of hexokinase activity in various tissues.[71] Similarly, thermal acclimation in fishes leads to the appearance of new isozymic forms of enzymes which are better suited for controlled catalytic function in the new thermal regime.[61] There are a number of other examples of isozyme modulation during metabolic regulation; we shall, however, restrict ourselves to discussing only a few of these.

Regulation of PK Isozymes. Gluconeogenesis in vertebrates is a function that not only serves the requirements of the liver cell or the kidney cell *per se,* but also subserves basic energy and carbon requirements of other tissues, particularly the central nervous system. That is, gluconeogenesis is an intertissue process. Hence, control of the process is unusually complex, involving integration with the metabolism of other tissues. Within the cells of glucogenic tissues *per se*, the control requirements during periods of net glycolysis clearly are distinct from the requirements during periods of net glucose synthesis. An obvious way of meeting such distinct needs is the elaboration through evolution of specific isozyme forms of any given enzyme with regulatory properties which fulfill the special metabolic demands of the moment. Tanaka et al.[135] were the first to stress such a role for the two kinds of pyruvate kinases which are present in mammalian liver cells. The L-PK is responsive to induction by factors favoring gluconeogenesis, while the M-PK seems to be kinetically better adapted for glycolytic function. These first studies have been greatly expanded, so that it is possible to separate two basic kinds of control of the L-PK isozyme (Fig. 6–15): control of its synthesis, by insulin or by dietary substrates such as sucrose and fructose,[135,156] and control of its catalytic activity by various allosteric modulators.[84]

Sucrose or fructose induction of L-PK appears to be a process that is initiated at the level of the gene (since it is blocked by inhibitors of RNA synthesis), and in this way the induction is similar to insulin regulation of this isozyme. However, the two kinds of inducers behave in essentially an additive way; thus, when fructose and insulin are given together to diabetic rats, L-PK activity increases to 560% of control values, which is the most marked induction reported for this enzyme.[156] It is therefore likely that the substrate induction occurs by a mechanism different from the hormone induction. Further evidence suggests the same conclusion. Thus, insulin appears to induce glycolytic enzymes belonging to the same functional genetic unit (and include Hk, PFK, and PK), leading to activation of the entire pathway, while

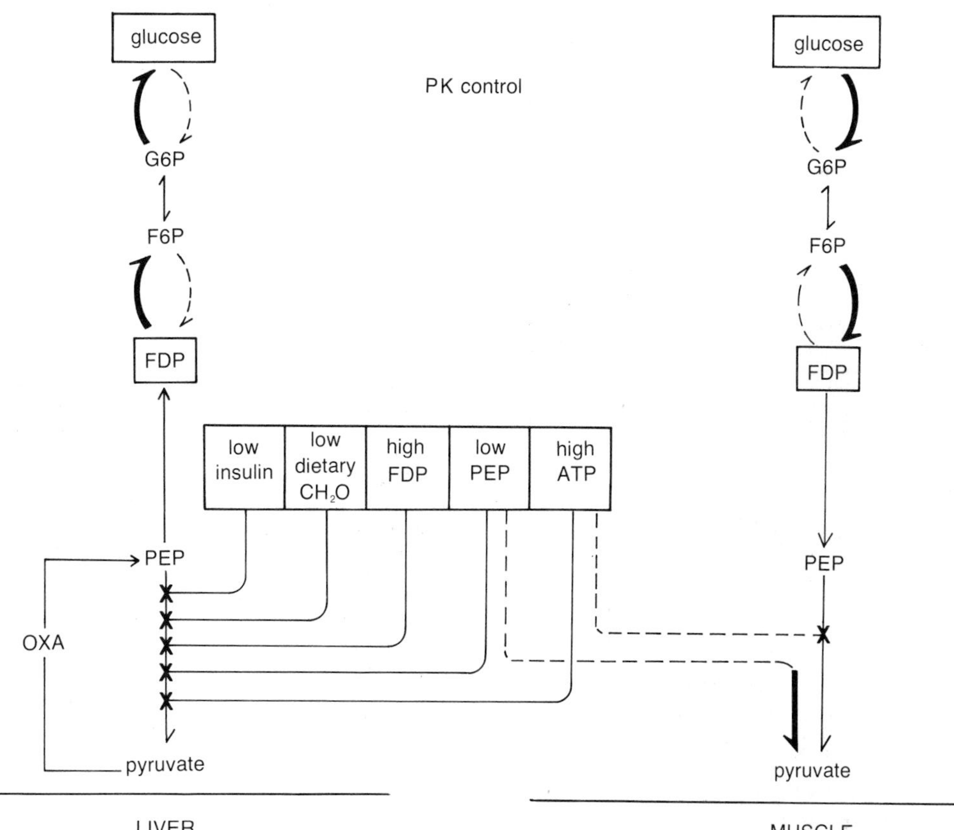

Figure 6–15. Control of PK isozymes in liver and in muscle. M-PK has a low K_m of PEP and a high K_i of ATP, and is insensitive to FDP, hormones, and dietary substrates.

sucrose or fructose induction involves an increased functioning of only a part of the pathway.

A great deal of information is available on the catalytic and regulatory properties of pyruvate kinases. In general, the kinetic properties fall into two categories: those typical of L-PK, and those typical of the muscle isozyme, M-PK. In fresh extracts of liver or kidney (the two major glucogenic tissues in mammals), PK activity appears to be modulated (1) by PEP, which serves both as a substrate and as an allosteric activator, (2) by alanine and ATP, each of which decreases enzyme-substrate affinity, and (3) by FDP, which greatly increases the apparent enzyme-substrate affinity.[84] Thus, under conditions of high energy (ATP) and high amino acid precursor (alanine) supplies, these mechanisms would prevent the diversion of PEP from its way to glucose; decreased FDP levels occurring as a result of activation of gluconeogenesis (and deinhibition of FDPase in particular) would also sharply reduce the potential drain of PEP from glucose synthesis by the pyruvate kinase reaction.

The isozymes of PK which are found in heart and muscle do not show any of these regulatory effects. Some interesting regulatory properties of PK in adipose tissue have been described,[107] but these do not seem to be present in freshly prepared extracts. Hence, Llorente et al. conclude that, kinetically, PK in adipose tissue does not differ significantly from the PK forms prevailing in other nonglucogenic tissues.[84]

We have already mentioned the role of PK isozymes in the regulation of carbon flow through the PEP crossroads in molluscan tissue metabolism.

Hexokinases. Dating back to the early reports from the Cori laboratory during the period from 1945 to 1947, a possible direct action of insulin on the hexokinase reaction has been the subject of a large number of studies. But the nature and magnitude of the effect of insulin on hexokinases have not been resolved, and the field even today remains rather controversial. A part of the difficulty arises from what must be described as a poor understanding of the nature of this enzyme system. Thus, it was only in the mid-1960's that multiple molecular forms of this enzyme were discovered and it became clear (1) that liver hexokinase activity was not the sum of a "low K_m" plus a "high K_m" enzyme, but rather consisted of at least four distinct isozymes and (2) that at least three hexokinases are typically found in all other

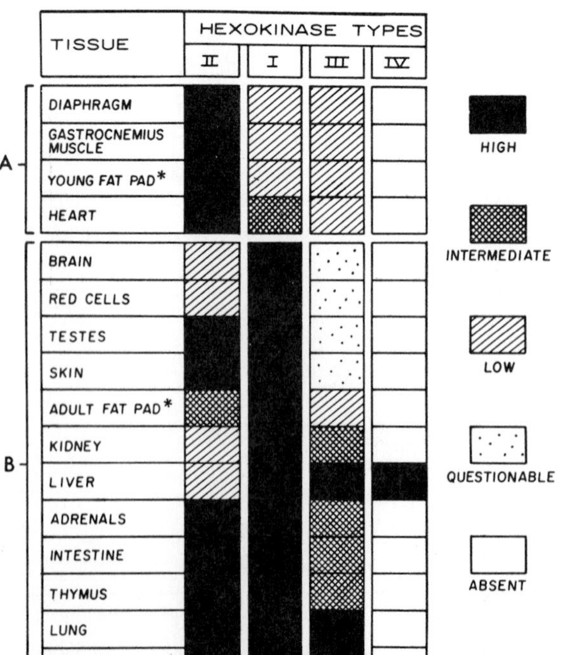

Figure 6–16. Summary of tissue distribution of hexokinases in the rat. Quantities of hexokinase types determined by a starch gel electrophoretic procedure. Tissues are arranged from top to bottom in order of their decreasing sensitivity to insulin. "A" refers to highly insulin-sensitive (dependent on insulin for glucose uptake) tissues, and "B" refers to tissues relatively insensitive to insulin (non-dependent on insulin). Asterisks(*) designate fat pads taken from 125 gm rats (young fat pad) and from 250 gm rats (adult fat pad). Hexokinase types arranged (except for Type IV) from left to right in order of decreasing K_m for glucose. (From Katzen, H. M., *in* Advances in Enzyme Regulation, Vol. 5, edited by G. Weber. Pergamon Press, Oxord, 1967.)

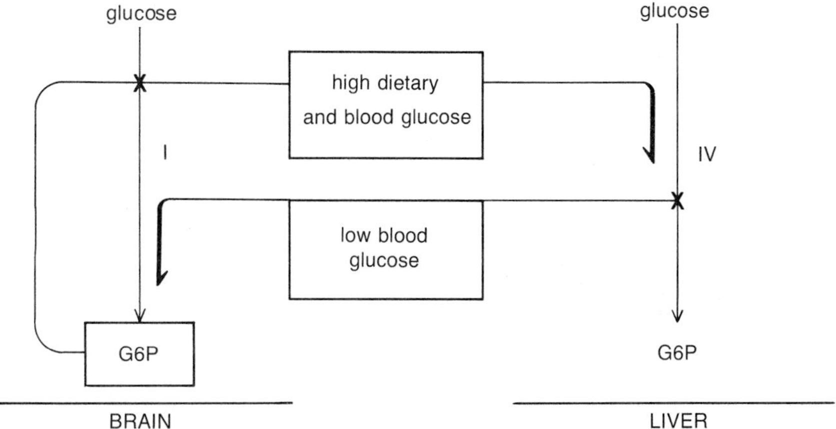

Figure 6–17. The effects of different levels of blood and dietary glucose on Type I and Type IV hexokinase function in brain, which depends upon blood glucose as a sole carbon and energy source, and in liver, which can either take up or release glucose.

tissues.[71, 72] Recent studies have clarified the situation substantially.

Based on their degree of mobility (towards the anode) during starch gel electrophoresis, most tissues contain in varying proportions isozymes I, II, and III. Types I and II are predominant in heart, type II in skeletal muscle, and type I in brain, but for the most part at least small amounts of each isozyme are found in most tissues (Fig. 6–16). Liver tissue is unique in possessing a fourth hexokinase, termed type IV. Type IV hexokinase has been shown to be the "high K_m glucokinase" of earlier studies on two criteria: (1) on kinetic criteria, since it displays a very high K_m (10^{-2} M) for glucose, and (2) because of its disappearance in fasted or diabetic animals and its reappearance under the influence of insulin.[71] The catalytic properties of these different isozymes differ from each other (for example, the K_m of glucose is about 0.02, 0.2, and 0.005 mM for types I, II, and III respectively), but each individual form of the enzyme is uniform from tissue to tissue.[71]

As pointed out above, hexokinase activity in the liver is subject to induction by insulin, and similar effects of insulin on hexokinase activities in all insulin-sensitive tissues probably occur. In liver, the response to insulin is due primarily to changes in activities of isozyme types II and IV, and particularly the latter, which is more abundant. In other tissues which lack type IV, insulin induces the type II isozyme; conversely, this isozyme disappears during diabetes.[71, 72]

The existence and particular kinetic properties of the liver type IV isozyme fit well with the particular metabolic requirements of the liver for glucose phosphorylation (Fig. 6–17). The liver must maintain a constant supply of blood glucose for use by other tissues; it deposits glucose (as glycogen) only when dietary, and hence blood, glucose levels are high but it must release glucose when none is available in the diet. The K_m of glucose is consistent with type IV isozyme function only when glucose is plentiful from the alimentary system. If it functioned at physiological blood glucose concentrations (about 5 mM), *the liver would withdraw glucose from, rather than add glucose to, the circulation.* This may indeed account for type IV being the predominant isozyme in liver and suggests that the activities of liver hexokinases I, II, and III are usually maintained in an inactive state (probably by G6P product inhibition or by inhibition by high substrate concentrations). Kinetic information currently available supports this supposition: although G6P inhibits all hexokinases, the concentration required to inhibit type IV hexokinase (the "high K_m glucokinase") is about 100 times as great as for types I, II, and III. Thus, type IV hexokinase will continue to convert glucose to G6P even at very high glucose and G6P concentrations.[126]

The physiological basis for differences in the distribution and properties of hexokinases I, II, and III, on the other hand, is not clear at present. Lowry and Passonneau[86] have shown that brain hexokinase, predominantly type I, is one of at least three important control sites in glycolysis, and this is

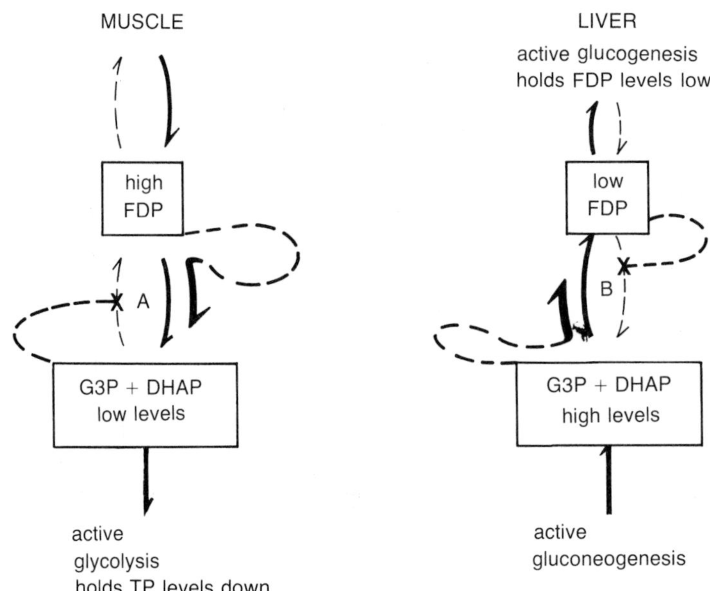

Figure 6–18. Regulation of activities of aldolase A in muscle and aldolase B in liver.

undoubtedly true in other tissues which rely on blood glucose as an important carbon and energy source. Since the different hexokinase isozymes have different kinetic properties, it is apparent that control requirements would potentially be somewhat different for each isozyme and thus for each tissue. Different tissue distributions of type I and II are particularly noteworthy. Thus, type I is the predominant enzyme in brain and kidney, two tissues whose metabolic activity is relatively constant. On the other hand, type II hexokinase is present in tissues whose metabolic activity can be varied greatly, such as skeletal muscle. But the full significance of these and a similar observation[72, 126] must await further study.

Aldolases. In homologous tissues of various vertebrate species, three unique forms of aldolases have been identified: aldolase A (the classical muscle isozyme), aldolase B (which along with A is found in liver) and aldolase C (which along with A is found in brain). These isozymes have similar physical properties (four subunits; similar molecular weight) but are immunologically distinct. The A and B isozymes have significantly different amino acid compositions and fingerprint patterns, but similar data are not yet available for the C isozyme. The catalytic properties of the three isozymes, so far as these are currently documented, are summarized by Rutter et al.[118] The general conclusion allowed is that aldolase A appears to be well adapted for glycolytic function, while aldolase B is better suited than is the A isozyme for gluconeogenic function (Fig. 6–18).

The A isozyme has a high affinity for FDP (about 0.6 mM) relative to the triose phosphates (about 1 to 2 mM), with a ratio of maximum FDP synthesis to FDP cleavage of about 2.0. All of these catalytic parameters favor aldolase A catalysis of FDP cleavage; i.e., operation in the glycolytic direction.

Aldolase B, on the other hand, displays kinetic properties more favorable for FDP synthesis. The apparent affinity of the B isozyme for the triose phosphates is about 10 times greater than in the case of the muscle type aldolase. The ratio of maximum FDP synthesis to maximum FDP cleavage rates for the liver adolase B is about 10.0. It is therefore evident that aldolase B is better adapted for gluconeogenic function than is aldolase A, and this presumably is the basis for its presence in liver.

As yet there is no catalytic attribute of aldolase C which is suggestive of a specialized physiological function. This isozyme type has a fairly low K_m for FDP, similar to that of aldolase B, and this may be of metabolic significance. Other possibilities, such as unique regulatory properties or enzyme-enzyme interactions to form complexes with brain mitochondria, are yet to be examined. Also, it should be stressed that the regulatory properties of none of the aldolase isozymes have been well worked out.

Isozymes of Glyceraldehyde-3-phosphate Dehydrogenase. In mammalian muscle, triose phosphate dehydrogenase (TDH) occurs as a single protein species, which kinetically appears to be well adapted for operation in a glycolytic direction. Since the reaction catalyzed by TDH can become rate limiting under certain glycolytic and gluconeogenic conditions,[165] its regulatory characteristics are of some interest.

The TDH isozyme typical of skeletal muscle appears to be quite distinct from a liver isozyme, which also occurs as a single protein species. The kinetic constants of the liver TDH (apparent affinities for substrates and coenzymes) are such as to favor catalysis in the direction of glucose synthesis.[102]

In fishes, TDH occurs in many (over 10) distinct molecular forms. These are tissue specific and probably are generated by the random assembly of two kinds of subunits into tetramer holoenzymes.[82] Nothing is known of their functions.

Regulation of the Synthesis of Serine Dehydratase Isozymes. Inuoe and Pitot[66] have identified two isozymic forms of the enzyme serine dehydratase. The amounts of the two forms of the enzyme vary considerably under different dietary, hormonal, and developmental conditions. Enzyme 2 appears to be under the control of glucagon, while the synthesis of enzyme 1 appears to be controlled by dietary substrates and by corticosteroids. At this time, nothing is known of the possible physiological significance of these results.

Isozymes of Lactate Dehydrogenase. The glycolytic enzyme, lactate dehydrogenase (LDH), catalyzes the reversible interconversion of pyruvate to lactate. The enzyme occurs in highest activities in tissues such as white muscle, which may have to function at times in essentially anaerobic environments. In vertebrate systems, it occurs in concentrations much higher than those of other glycolytic enzymes, and is "rate limiting" only when its generation of nicotinamide adenine dinucleotide (NAD) for the triose phosphate dehydrogenase step cannot meet demands of the pathway.

In most vertebrates, lactate dehydrogenase occurs as five isozymic types derived from self-assembly of A and B subunit polypeptide chains to form the active tetramers, A_4, A_3B_1, A_2B_2, A_1B_3, and B_4.[91] In teleost fishes, isozyme patterns are complicated by the presence of more than two kinds of subunits. In tetraploid fishes, such as the salmonids, at least eight subunit types are involved in generating multiple isozymic LDH tetramers. These are all thought to be derived by duplication of the genes specifying the A and B subunits.[62] The functions of these various LDH isozymes in different tissues has been the subject of lively research.

One popular theory of their function suggests that A-type LDH's are predominant in tissues such as skeletal muscle, in which an active glycolysis is often the sole source of energy. These LDH's are insensitive to pyruvate inhibition and hence allow large accumulations of lactic acid. B-type LDH's are thought to function in more aerobic tissues, where sensitivity to pyruvate inhibition prevents lactate accumulation and thus favors channelling of pyruvate into the Krebs cycle. This theory has been severely challenged by Vesell and his co-workers.[152]

It is possible that the major function of these different LDH isozymes relates to proper localization in specific cells, rather than to important kinetic differences.[161, 162]

Role of Isozymes in Metabolic Regulation in Bacteria. Studies of the regulation of various metabolic activities in bacteria have led to the discovery of isozymes whose functions also are clearly concerned with regulatory processes. In general, it appears that *isozymes with control functions are elaborated when a given enzymic reaction is required for more than one metabolic process.* The existence of these isozymic types in bacteria was first recognized by H. E. Umbarger and co-workers in their studies on feedback regulation of isoleucine and valine biosynthesis. In both of these biosynthetic pathways the first step under feedback control is a reaction that, under certain conditions, is also involved in catabolism.

Under these circumstances, *if only a single enzyme catalyzed the critical common step, its feedback control by excesses of the final end product could lead to a deficiency in energy metabolism.* In such instances, when the critical step is needed for energy metabolism, the organisms elaborate another, inducible isozyme that catalyzes the same reaction, but which is insensitive to feedback control by the biosynthetic end product. These findings led

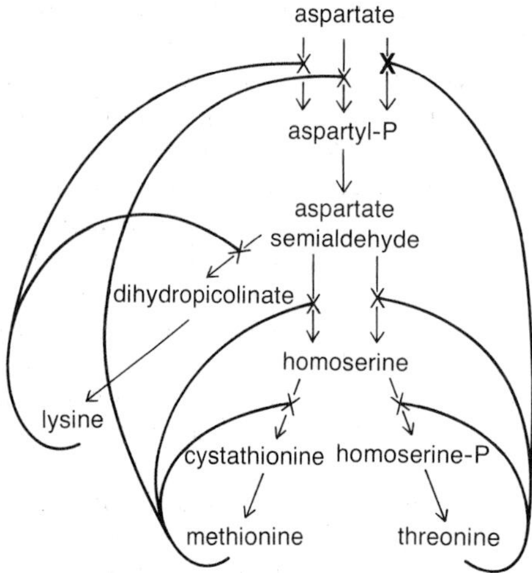

Figure 6–19. Feedback regulation of lysine, threonine, and methionine biosynthesis in *Escherichia coli.* Number of isozymes of aspartokinase and of enzymes at each branch in metabolic map is indicated by the number of arrows. Crosses indicate steps subject to feedback control. (Modified after Stadtman, E. R., Ann. N. Y. Acad. Sci. *151*:516–530, 1968.)

Umbarger to propose that, whenever an enzyme is under rigid end-product inhibition and repression, another isozyme (or isozymes) will be needed by the cell if there is some other essential role for that enzyme. This principle has found support from numerous studies.

Stadtman and his co-workers demonstrated the first branched biosynthetic pathway shown to be under isozymic control. The pathway concerned the biosynthesis of lysine, methionine, and threonine. They showed that these amino acids are all derived from aspartate by a branched pathway in which the phosphorylation of aspartate (catalyzed by aspartokinase) is the first reaction step. As shown in Figure 6–19, there are three independent aspartokinases in *Escherichia coli,* and they are differentially inhibited by the three end products. Moreover, there are only two separate homoserine dehydrogenases catalyzing the first step that are uniquely involved in the formation of methionine and threonine, and these two dehydrogenases are differentially controlled by methionine and threonine. In addition, the first reaction step of each branch of the pathway is under specific end-product control by the ultimate product of the divergent pathway. A consequence of this kind of control circuitry is that *all aspartokinase activity is inhibited when all three end products are in excess.* In essence, the production of multiple enzymes catalyzing the first common step in a branched pathway provides for flexibility in the control process: by avoiding the drastic consequences that would arise from unrestrained end-product control of a single enzyme, such isozymes confer an obvious selective advantage on the organism.[133]

Role of FDPase Isozymes in Channelling Carbon Towards Chitin and Glycogen. Isozyme functions comparable to those of bacterial aspartate kinases have recently been shown for FDPases in crustacean hypodermis and gill. The major reaction pathways are shown in Figure 6–20, and indicate that two chief functions of the hypodermis are the biosyntheses of large quantities of chitin and of glycogen. The first "committed" metabolite in the pathway to chitin is uridine diphosphate acetyl glucosamine (UDPAG); the first "committed" metabolite in the pathway to glycogen is UDPglucose (UDPG). During the molt cycle in these organisms, a reciprocal relationship is maintained in the relative activities of these two pathways—during intermolt and early premolt, glycogen synthesis is high and chitin synthesis does not occur, while chitin synthesis is high during later premolt and early post-molt, when glycogen breakdown is activated.[105] Situated strategically at the primary branching point between these major biosynthetic pathways is the FDPase conversion of FDP to F6P + P_i (Fig. 6–20). It is evident that efficient channelling of carbon through this "bottleneck" should favor the glycogen pathway (in intermolt) or the chitin pathway (during molt) *but not both simultaneously.* This regulatory requirement has been met in part through the elaboration of at least two kinds of FDPases in the hypodermis. FDPase I is largely insensitive to UDPAG and UDPG. FDPase II is under stringent feedback regulation by both of these metabolites: at low concentrations, they activate the reaction by reducing the K_m for FDP and increasing the maximum velocity, while at high concentrations, they potently inhibit this isozyme. Both FDPase I and II are subject to the usual control by AMP (see p. 266). Through an interplay between these various metabolites, it is postulated that an effective channelling of carbon is achieved through this "bottleneck" towards either glycogen or chitin, but not both at the same time.[59]

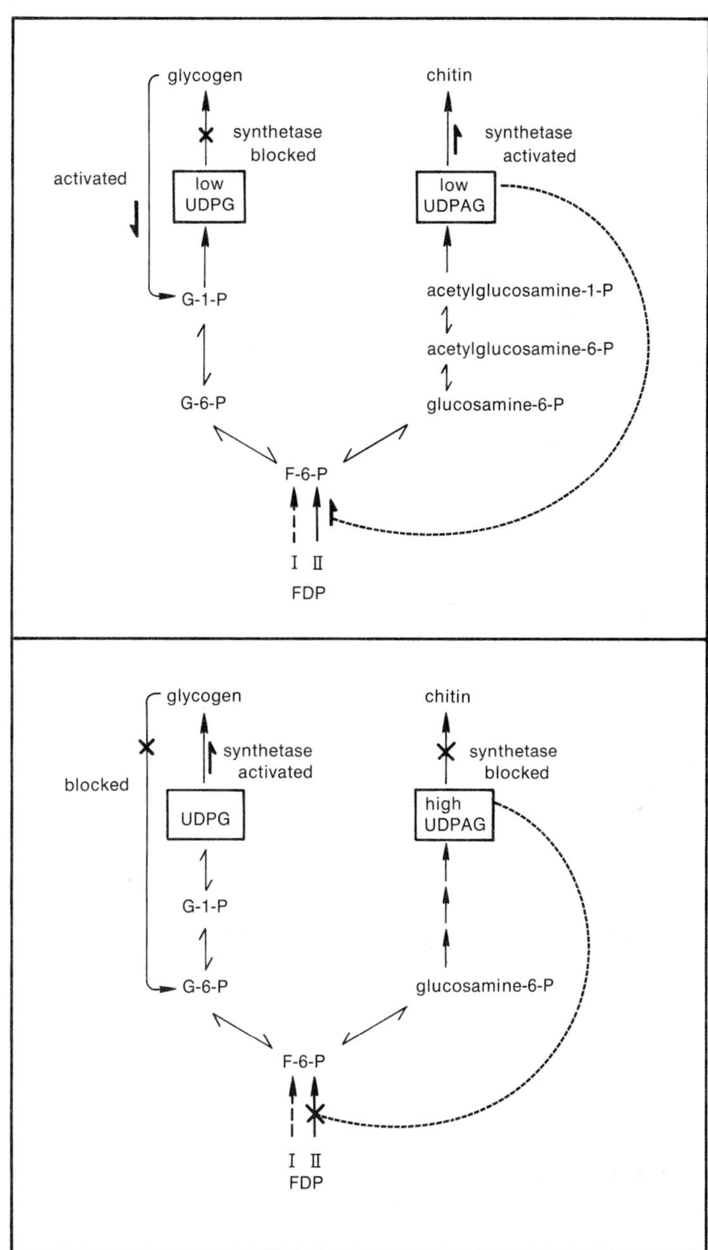

Figure 6–20. Postulated feedback regulation of FDPase II activity in the gill and hypodermis during periods of chitin synthesis (top) and in the hypodermis during periods of glycogen synthesis (bottom). In the gill, FDPase II accounts for about 90% of total FDPase activity; in the hypodermis of the intermolt animal, FDPase II accounts for only about 30% of total FDPase activity. Potential control interactions are described in the text. (From Hochachka, P. W., Biochem. J. *127*:781–793, 1972.)

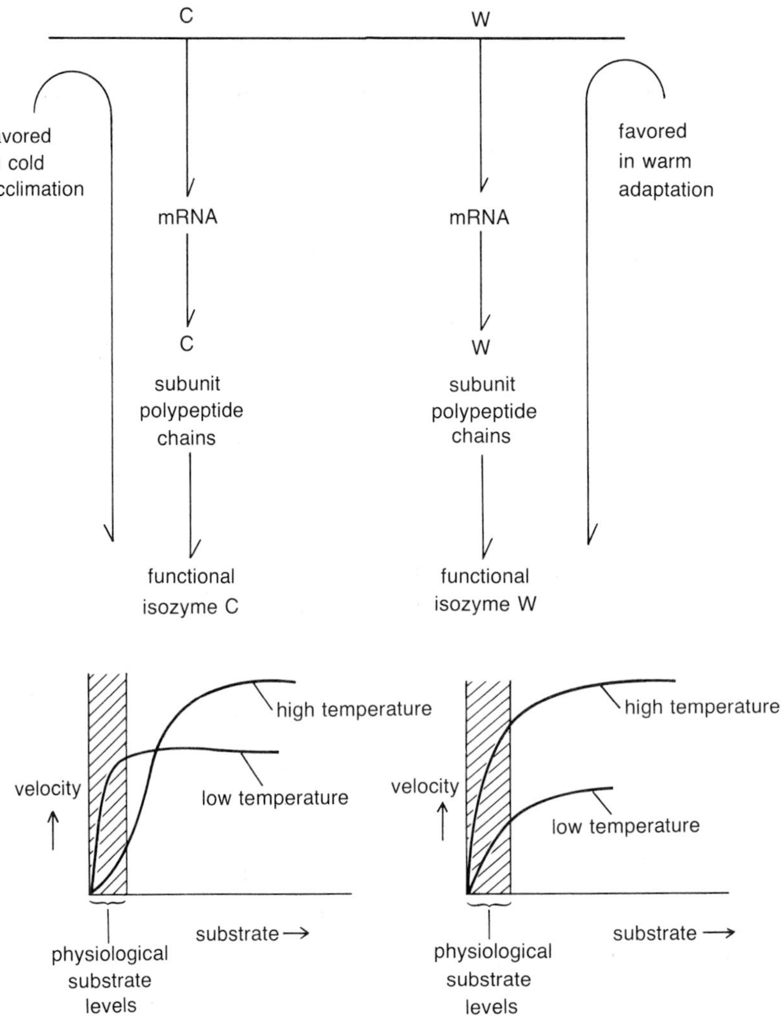

Figure 6–21. Summary model of events occurring during thermal acclimation.

Role of Isozymes in Thermal Acclimation of Salmonid Fishes. Isozymes play analogous regulatory functions during thermal acclimation in the salmonid fishes and possibly in eurythermal poikilotherms in general. In these fishes, the rates of enzymic reactions may be highly independent of temperature. In many poikilothermic enzymes, active sites are tailored so that a decrease in thermal energy is compensated for by an increase in enzyme-substrate affinity. However, at temperatures above or below a critical range, which depends upon species and isozymes considered, enzyme-substrate affinity drops: that is, the Michaelis-Menten constant rises dramatically. This means that the K_m values of substrate may be many times higher than substrate concentrations in the animal. Under such conditions these enzymes become highly inefficient and possibly entirely inactive. To circumvent this problem, new isozymes are elaborated during low-temperature acclimation which differ kinetically from those of warm-acclimated animals in having higher absolute affinities for substrate or in having maximal affinities for substrate at lower temperatures, or both.[61]

The primary functional and selective advantage of employing "better" isozymes in thermal acclimation, as opposed to produc-

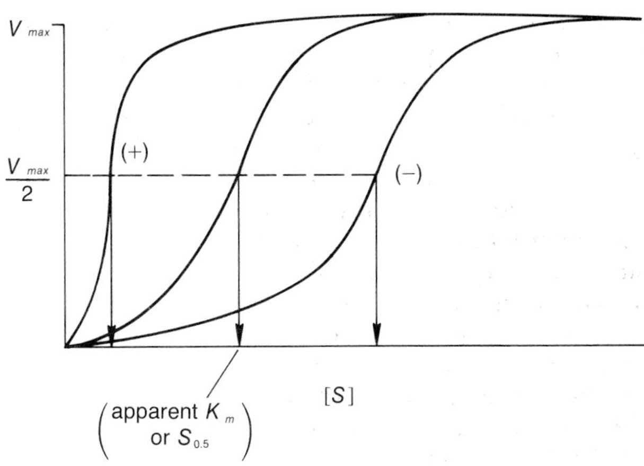

Figure 6–22. Relation between substrate concentration and velocity of reaction catalyzed by a regulatory enzyme in the presence of positive (+) and negative (−) modulators. The apparent K_m or $S_{0.5}$ values are shown by arrows.

ing altered quantities of single enzyme species, is the production of enzymes with K_m values in a range likely to be optimal for regulation of catalytic activity (Fig. 6–21). Thus, at low temperatures, small changes in substrate concentration or small changes in enzyme-substrate affinity can lead to large changes in the activities of "cold" forms of these enzymes, a condition which is admirably suited to controlling reaction rates. In the case of "cold" variants of these enzymes at high temperatures, very large changes in substrate concentration or in enzyme-substrate affinities are required to yield small changes in reaction rates. This condition is clearly not one which allows efficient control of reaction rates. In evolutionary terms it appears that, during thermal acclimation, there is a strong selection for the biosynthesis of new isozymes which are optimally suited for catalysis and control of catalysis at the given acclimation temperature. This is reflected in the patterns of new isozymes of lactate dehydrogenases, pyruvate kinases, aldolases, phosphofructokinases, citrate synthases, acetylcholinesterases, isocitrate dehydrogenases, and presumably many other enzymes during acclimation and during evolutionary adaptation to temperature.

Control of Catalysis

Introduction (General Principles). Current attempts to relate the regulation of metabolic processes to the known properties of enzyme systems are primarily centered on the properties of allosteric or "regulatory" enzymes. Thus, as we have already briefly pointed out, preferential use of particular metabolic pathways is correlated with particular regulatory characteristics of enzymes which are components of those pathways. Two characteristics are of particular importance: (a) the ability of specific metabolites to alter enzymic activity by binding to specific enzyme sites distant from the catalytic site, and (b) the unusual (i.e., nonhyperbolic) dependence of the initial velocity of the reaction on substrate or ligand concentration.[3, 127, 149] The basic catalytic response to positive and negative modulators is diagramed in Figure 6–22, which indicates that (a) because of the sigmoidal saturation of the enzyme with substrate, small changes in substrate concentration can lead to large changes in the catalytic activity of the en-

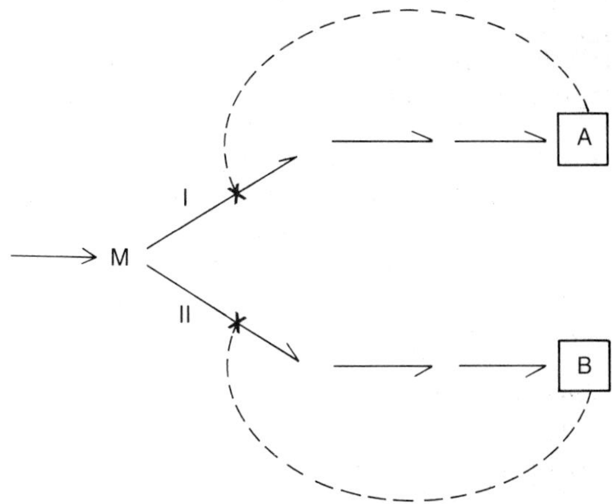

Figure 6–23. Generalized pattern of feedback regulation at a metabolic branch point. (After Atkinson, D. E., *in* Metabolic Roles of Citrate, edited by T. W. Goodwin. Academic Press, London, 1968.)

zyme, and (b) modulators lead to an alteration in the apparent enzyme-substrate affinity with no necessary effects on the maximum velocity (V_{max}) of the reaction. Thus, positive modulators lead to a decrease in the apparent enzyme-substrate affinity, while negative modulators lead to an increase in the apparent $S_{0.5}$ value (the concentration of substrate required for 50% of maximum activity of the enzyme). This means that such metabolite regulation of enzyme activities occurs only at low substrate concentrations (in the range of the K_m or the $S_{0.5}$ values of the substrate). As we have pointed out above (p. 215), pathways such as glycolysis seem to function *in vivo* at rates far lower than those which would be expected if the component enzymes were operating under saturating conditions (i.e., at maximum velocities). The generalization that most enzymes *in vivo* function at low substrate levels and never "see" saturating concentrations of their substrates is fundamental to all current schemes of metabolic control.

Another important generalization, arising initially from studies of control of biosynthetic pathways in microorganisms, is that regulation occurs at metabolic branch points; the usual modulator is a metabolite that is competed for by two or more enzymes, at least one of which is kinetically regulated by one or more additional modulators. Product feedback inhibition is but a special case of this situation.

Thus, in a simple generalized case (Fig. 6–23), enzymes I and II compete for the branch point metabolite M. The outcome of this competition depends on the relative enzyme-substrate affinities for M, and these affinities are regulated by the concentrations of the end products A and B. The pool level of A is in large part controlled by the affinity for A of the regulatory site on enzyme I; and similarly, the concentration of B depends upon its binding affinity for the regulatory site on enzyme II. Thus, the concentration of M will be a complex function of A-enzyme I and B-enzyme II interactions and will depend indirectly on the cell's momentary needs for A and B.

A final generalization that has arisen in these studies is that particular metabolic control networks (i.e., specific enzyme-modulator interactions) are determined on grounds of physiological usefulness and not on grounds of chemical similarity. Adenylate control of many enzymes in intermediary metabolism is an excellent case in point. Thus, as far back as 1938, Cori and his coworkers discovered that AMP strongly enhances the activity of glycogen phosphorylase. Atkinson and others found that AMP activates NAD-linked IDH at low isocitrate concentrations, because the AMP causes an increase in the affinity of the enzyme for isocitrate. At about the same time, several workers noted that AMP similarly increases the affinity of PFK for its substrate, F6P. These effects posed a basic problem, for the functions of these three enzymes are quite distinct: phosphorylase catalyzes the phosphorylytic cleavage of an acetal bond; PFK catalyzes the transfer of a phosphate group; and the reaction catalyzed by IDH is an NAD-linked electron transfer. The reactions have nothing in common chemically; yet they are all affected in a similar manner by AMP.

It is evident that the explanation must be sought in terms of functional metabolic interactions rather than chemical similarity. Thus, all three reactions participate in the overall oxidation of polysaccharide, which furnishes electrons for the electron transfer system. Indirectly, these enzymes are involved in the generation of ATP. By analogy with product feedback inhibition in biosyntheses, it appears that glycolysis and the citrate cycle are inhibited by ATP, the "end product" of energy metabolism. Since concentrations of AMP and ATP vary in an opposite manner in the cell, stimulation by AMP is metabolically equivalent to inhibition by ATP.

The three reactions are also strategically well placed for efficient regulation. Phosphorylase catalyzes the first step in the mobilization of glycogen; PFK catalyzes the first unique and physiologically irreversible step in glycolysis and one that has been implicated as a major control point in glycolysis on the basis of various kinds of evidence; and NAD-linked IDH catalyzes the first oxidative step of the citrate cycle. These, and a host of similar considerations, have led Atkinson and his co-workers to propose that the relative concentrations of the adenylates play a most critical role in partitioning metabolic intermediates between oxidative pathways (leading to ATP production) and biosynthetic pathways (leading to the production of storage compounds). Their adenylate control concept is discussed more fully elsewhere.[4] For our purposes, it will be sufficient to examine in detail the kinetic properties of only a few enzymes involved in metabolic control. In particular, we will examine regulatory enzymes which are *oppositely poised in related catabolic and anabolic pathways*: glycogen phosphorylase and glycogen synthetase, and PFK and FDPase. By a consideration of control at these particular sites, the student may gain insight into the kinds of molecular mechanisms which are available for maintaining "order" in the face of hundreds of chemical reactions which a cell is potentially capable of activating at any given time.

Control of Glycogen Phosphorylase and Glycogen Synthetase in Liver. In the vertebrate liver, more than in any other tissue, the concentration of glycogen can vary to a large extent and at a rapid rate. Extreme values, for example, can be as low as 0.01% wet weight (after fasting) and as high as 12% of wet liver weight (obtained by refeeding the animal). The rate of synthesis under these conditions is in the order of 1% of the wet weight per hour! In the normally fed mammal, however, both the synthesis and the degradation are slow and the half-life of liver glycogen has been estimated at about 24 to 36 hours. These simple calculations indicate that the synthesis and degradation of liver glycogen are both closely regulated and integrated processes.[56]

It is now well known that this regulation occurs by the interconversion of two forms of the key enzymes that catalyze the rate limiting steps in glycogen metabolism; one of these forms, *a*, is active and the other, *b*, is inactive under probable ionic and metabolite concentrations to be found in the liver cell *in vivo*. The interconversion of the key enzymes (glycogen phosphorylase in glycogen breakdown to G1P; glycogen synthetase in glycogen synthesis from UDPG) is achieved by phosphorylation and dephosphorylation of the two enzyme forms. The major difference between the two systems is that phosphorylation (itself an enzyme-catalyzed process under regulation) causes *activation of the phosphorylase* (conversion of $b \rightarrow a$), while it causes *inactivation of the synthetase* (conversion of $a \rightarrow b$). The reverse is the case when the two enzymes are dephosphorylated by specific protein phosphatases. This inverse

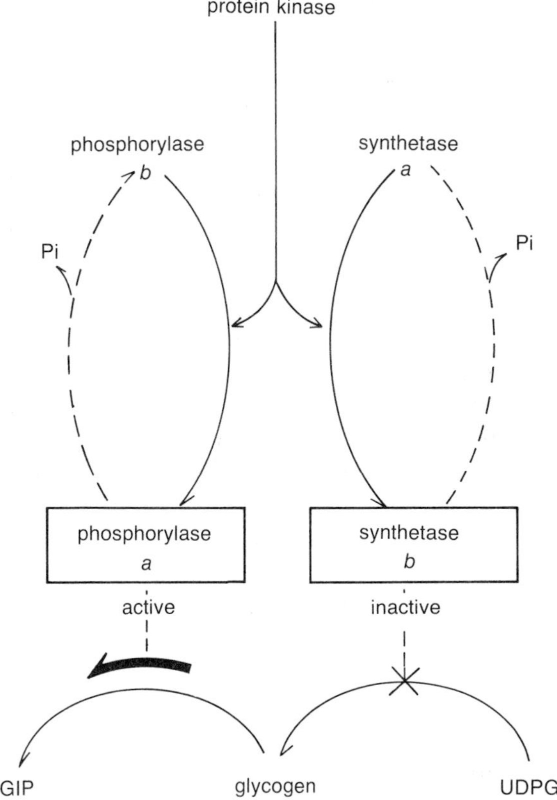

Figure 6–24. Generalized scheme of control of glycogen breakdown and glycogen synthesis.

polarity is a key feature in a biological sense, for it is clear that synthesis and degradation of glycogen will not occur simultaneously at any significant rate (Fig. 6–24). Although comparable control mechanisms in other tissues, such as heart and muscle, are better understood and were worked out at an earlier date, regulation in the liver appears at this time to be somewhat less complex. Hence, our discussion will begin with a consideration of these mechanisms.

The control of liver glycogen metabolism can occur by several different mechanisms. Of these, none is more important than the activation of glycogen breakdown by epinephrine in particular. This *hormonal* control of glycogen metabolism is discussed at this point rather than in an earlier section (pp. 248–250) because *it does not depend for its action upon changes in synthesis rates of the enzymes being modulated.* Unlike the hormonal controls already described, the response times here are in the order of seconds to minutes. In other words, we are dealing here with a true molecular modulation of enzymic activities, and not with effects on enzyme concentrations.

The basic observations on the effects of various hormones on cyclic AMP production were first made about 10 years ago by Sutherland and Rall. The earlier work and the much expanded studies of the following decade led to the hypothesis that many hormones act by a two-messenger system. According to this concept, the hormones themselves are the first messengers, whose basic effect at their target tissues is to trigger the release of the second messenger, which then does the real work of the hormone. Cyclic AMP is considered the first of such intracellular messenger carriers.[22, 106] Most of the detailed definitive work on the mechanism of action of cyclic AMP derives from studies by Krebs and Fisher on phosphorylase action. Their work has led to the hypothesis that hormones whose actions are mediated by cyclic AMP bring about their physiological effect by promoting the phosphorylation of specific protiens as shown below.

In the case of glycogen breakdown and synthesis in the liver,[56] epinephrine appears to act by activating the adenyl cyclase system, a membrane bound enzyme which catalyzes the synthesis of cyclic 3,5-AMP from ATP. At concentrations of between 10^{-8} and 10^{-7} M, cyclic AMP leads to a strong activation of phosphorylase kinase; this conversion of an inactive to an active phosphorylase kinase involves an autocatalytic phosphorylation of the enzyme and is catalyzed by a specific cyclic AMP-stimulated protein kinase. The activated kinase catalyzes the further phosphorylation and activation of phosphorylase *b* to phosphorylase *a.* In muscle tissues, this activation involves a polymerization of the phosphorylase (from a dimer to a tetramer), but such a change in polymerization state does not occur in the liver system. ATP serves as the phosphate donor in both systems. Phosphorylase *a* apparently accounts for essentially all of liver glycogen breakdown, because the *b* form in liver is essentially inactive even in the presence of AMP. As we shall discuss below, the *b* form in muscle is active in the presence of AMP. By means of this cascade of activation reactions (Fig. 6–25), the original cyclic AMP signal is amplified many-fold by the time its effect impinges on glycogen breakdown to G1P. Despite such efficiency, the effect of mobilizing these reactions could be nullified (in a wasteful cyclic way) if glycogen synthetase was not inhibited concomitantly.

The essential features of cyclic AMP inhibition of glycogen synthetase are formally analogous to the activation of glycogen phosphorylase. Thus, cyclic AMP activates a protein kinase which phosphorylates a non-activated synthetase kinase into an activated form. The activated synthetase kinase in

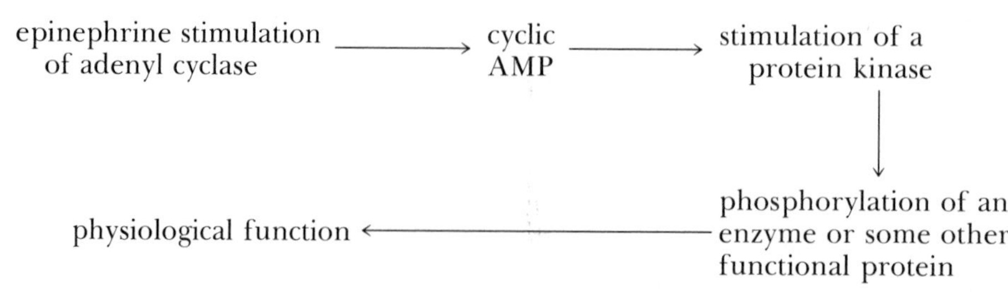

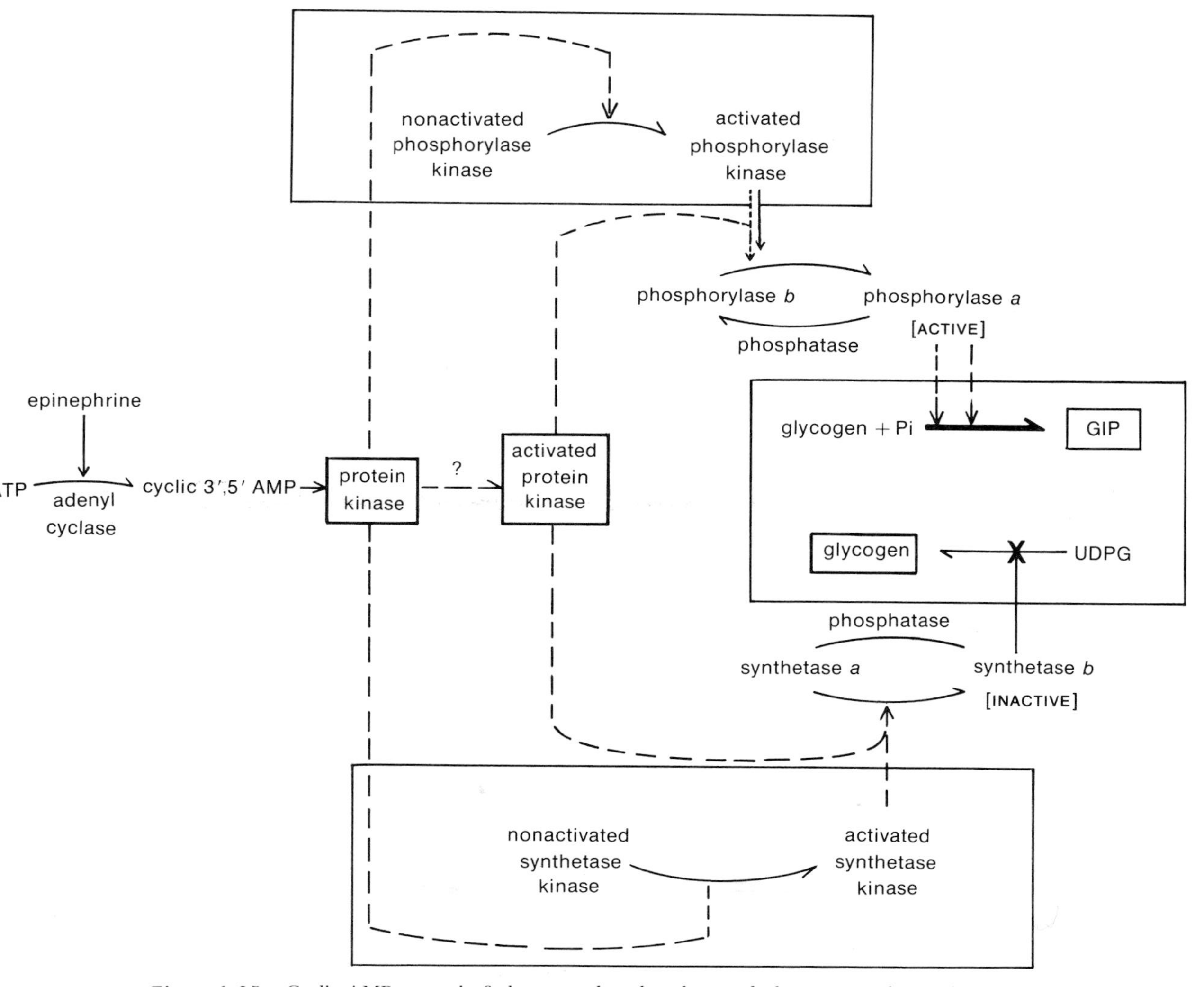

Figure 6–25. Cyclic AMP control of glycogen phosphorylase and glycogen synthetase in liver.

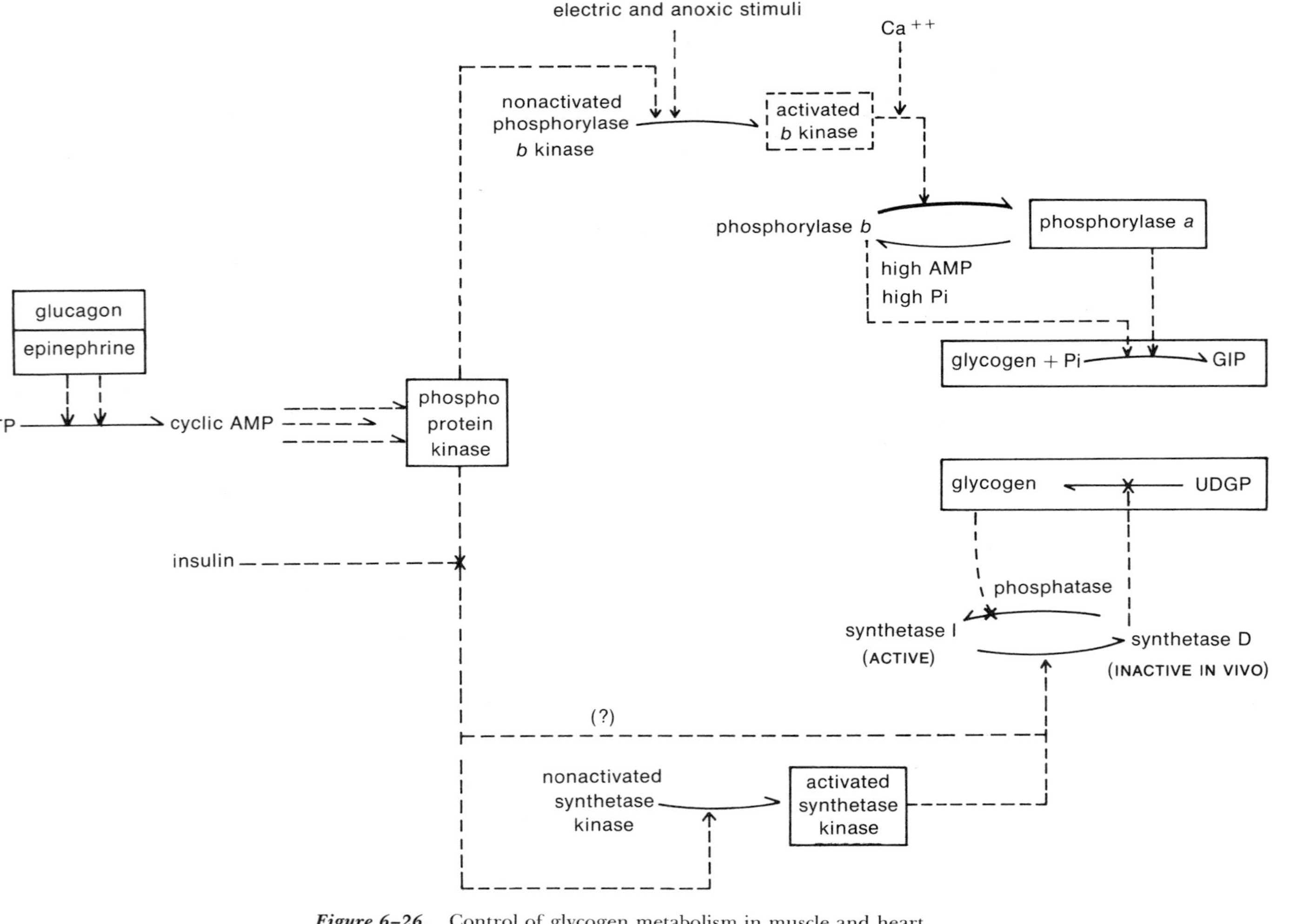

Figure 6–26. Control of glycogen metabolism in muscle and heart.

turn catalyzes the phosphorylation and conversion of synthetase *a* to the *b (inactive) form* of the enzyme. Thus, at the time when the liver glycogen phosphorylase system is being activated, by comparable mechanisms liver glycogen synthetase is being sharply inhibited. Integration of the control of these two enzymes prevents their simultaneous function and thus prevents carbon cycling at this point.

The reverse of this sequence of events occurs when the phosphatase activities exceed the kinase activities. Control of the entire cascade could theoretically lie at this point, and indeed this appears to be the case in the adrenal cortex.[22] But in most tissues thus far examined, cyclic AMP effects ultimately seem to focus upon the phosphorylase kinase catalyzed $b \rightarrow a$ conversion, rather than upon the phosphatase catalyzed $a \rightarrow b$ inactivation. Also, currently available evidence suggests that the cyclic AMP-activated protein kinase is the same in both the glycogen phosphorylase and the glycogen synthetase schemes,[21] and this is the way it is drawn in Figures 6–24 and 6–25.

Control of Glycogen Phosphorylase in Skeletal Muscle and Heart. In heart and muscle, as in liver, the glycogenolytic response to epinephrine in the blood or to stimulation of sympathetic nerves (i.e., norepinephrine) is mediated by cyclic AMP. In these tissues, the activation pathway has been worked out mainly by Edwin Krebs and his co-workers. The basic sequence can be schematized as in Figure 6–26: Epinephrine or glucagon activates adenyl cyclase so as to produce an increase in cyclic AMP. Cyclic AMP is required for the activity of a phosphoprotein kinase (probably the same kinase as is used in glycogen synthetase control),[21] which catalyzes the phosphorylation and activation of *b* kinase. Activated *b* kinase, in the presence of ATP, Mg, and Ca, catalyzes the phosphorylation of phosphorylase *b* and its concomitant polymerization to the *a* form. The latter form of the enzyme is more active, but in muscle and heart, unlike in liver, phosphorylase *a* is not the exclusive catalyst of glycogen cleavage. Under conditions of high AMP and P_i all glycogen cleavage can be accounted for on the basis of phosphorylase *b* function. The phosphorylation of $b \rightarrow a$ does not occur in an "all or none" fashion as originally proposed, but rather in a stepwise fashion in which partially phosphorylated intermediates are produced. These hybrids show catalytic activity and physicochemical properties distinct from those of either the *b* or *a* forms. This cascade system in heart and muscle is necessary for the effect of the catecholamines and glucagon on glycogen breakdown. But it soon became evident that the scheme is subject to other mechanisms of regulation at several sites. Some of these have been at least partially charted and can be summarized as follows.[94]

In heart there are at least two receptor sites for adenyl cyclase activation: one for epinephrine, which is blocked by β-adrenergic blocking agents, and a second site sensitive to glucagon. In this tissue, the transformation of *b* to *a* may be neither sufficient nor necessary to mobilize glycogen, for phosphorylase *b* may be a potent physiological catalyst under conditions in which ATP concentrations in the heart are reduced while AMP and P_i levels are much increased (for example, during extreme work loads or partial anoxia). On the other hand, maximal formation of phosphorylase *a* by epinephrine will cause no net mobilization of glycogen in the heart as long as the concentrations of high energy phosphate compounds are maintained. Thus, it is clear that under certain conditions glycogen phosphorylase is regulated by alterations in heart muscle metabolites, and not by the cascade control system of Figure 6–26. Similarly, after removal of Ca^{++} from the perfusion medium of rat hearts, epinephrine causes cyclic AMP formation, as one would predict, but not *b* to *a* transformation. Excess Ca^{++}, in contrast, produces *b* to *a* conversion in the absence of epinephrine. Regulation of phosphorylase $b \rightarrow a$ conversion and activation by Ca^{++} is also suggested by the experiments of Drummond et al.,[39] in which electrical stimulation of skeletal muscle produced rapid formation of phosphorylase *a*, *without either an increase in the concentration of cyclic AMP or activation of phosphorylase kinase.* Thus, the release of free intracellular Ca^{++} by hormones or by muscle depolarization may provide a second mechanism of regulation of the phosphorylase system.

Control of Glycogen Synthetase in Skeletal Muscle and Heart. The two forms of glycogen synthetase were first discovered by Larner and his co-workers in studies of insulin effects on glycogen synthesis in 1960. In muscle and heart (see Figure 6–26), glycogen synthetase occurs in two distinct forms: an I (or active) form is converted upon phosphorylation to a D (or inactive) form. The degree of catalytic activity in this case is not of an "all or none" nature. The I form is active in the absence of G6P, but the D form is only active when G6P is bound to an allosteric site. In this sense, G6P appears to fulfill a role analogous to that of AMP for phosphorylase, except that here G6P is effective with the fully phosphorylated D form and G6P is a direct, stoichiometric precursor of the eventual polysaccharide product.[80] The conversion of the D form to the I form is a dephosphorylation catalyzed by a specific phosphatase which is inhibited by glycogen. Other controls (insulin and epinephrine) on this system ultimately affect the activity of the kinase catalyzing I → D conversion, and are comparable to those already described above for the liver glycogen synthetase system. In these tissues, as in liver, it is clear that conditions favoring glycogen degradation also favor inhibition of glycogen synthesis and vice versa.

Catalytic and Regulatory Properties of PFK in the Conversion of F6P to FDP. The phosphorylation of F6P to form FDP is catalyzed by a specific enzyme, PFK, which on the basis of a variety of evidence is considered to constitute a major control site in the glycolytic pathway. PFK shares a number of properties of other kinases (requirements for ATP and Mg^{++}, and a negative ΔG^0) and catalyzes the first reaction characteristic of the glycolytic sequence. Lardy and Parks first recorded in the mid-1950's that ATP in concentrations above the molar equivalent of Mg^{++} inhibited the activity from skeletal muscle, but the first detailed kinetic studies were carried out by Mansour and his coworkers on the enzyme from the liver fluke. These workers found that the enzyme may occur *in vivo* in active and subactive forms, conversion to the active form being favored by anoxic conditions. At non-inhibitory concentrations of ATP, the F6P saturation curve did not have the Michaelis-Menten form, but rather was sigmoidal in nature. Hill plots for this saturation curve $\left(\log \frac{v}{V_{max}-v}\right.$ versus s, where v is velocity at any given substrate concentration, V_{max} is maximal velocity, and s is substrate concentration) yielded Hill coefficients (the slopes of the plots) greater than one. The Hill coefficient, n, is not an elementary kinetic parameter, but rather is a measure of both the number of substrate binding sites and their strength of interaction. Values of n greater than 1 indicate that the binding of the first F6P molecules facilitates binding of subsequent F6P molecules; that is, F6P binding exhibits positive cooperativity. In a sense, one can then look upon F6P both as a substrate molecule and as a positive modulator of PFK. Addition of cyclic 3,5-AMP or AMP increases the affinity of PFK for F6P, converts the sigmoidal saturation curve toward a more normal hyperbolic curve, and therefore decreases the Hill coefficient (the value of n approaches 1 as the sigmoidal curve approaches the classical hyperbolic Michaelis saturation curve). On the other hand, increasing the ATP concentrations is tantamount to inhibiting the enzyme. Also, increasing the ATP concentrations decreases the affinity of the enzyme for cyclic AMP. In all these experiments, AMP has effects analogous to those of cyclic AMP; *in vivo*, AMP is probably the physiological activator of the enzyme. The interactions between ATP and AMP suggest that when ATP is high, AMP will not be as effective an activator of the enzyme.[90]

Subsequent to the kinetic studies mentioned above, PFK enzymes isolated from several sources were examined, with remarkably similar results. A number of additional modulators were discovered: citrate as a negative modulator, P_i as an activator, and FDP and ADP as product activators. Kinetic studies with mammalian heart PFK revealed that at alkaline pH (8.2) the enzyme loses its regulatory properties: F6P saturation curves exhibit normal Michaelis properties rather than the positive site-site interactions that occur at lower pH values. Under the latter conditions (pH 6.9) and with low levels of F6P, the heart enzyme is inhibited by ATP, citrate, and creatine phosphate. Once inhibited, the enzyme can be deinhibited (or activated) by F6P, FDP, AMP, cyclic AMP, and P_i. The kinetic data suggest that the effects of these activators occur at an ATP site separate from the catalytic site (which binds both ATP and F6P). Thus, while the apparent K_m for ATP is not changed by these activators, the apparent K_i is increased. As in the case of the fluke enzyme, the activators affect the heart enzyme by decreasing the apparent K_m for F6P. Further evidence that the binding sites of substrates

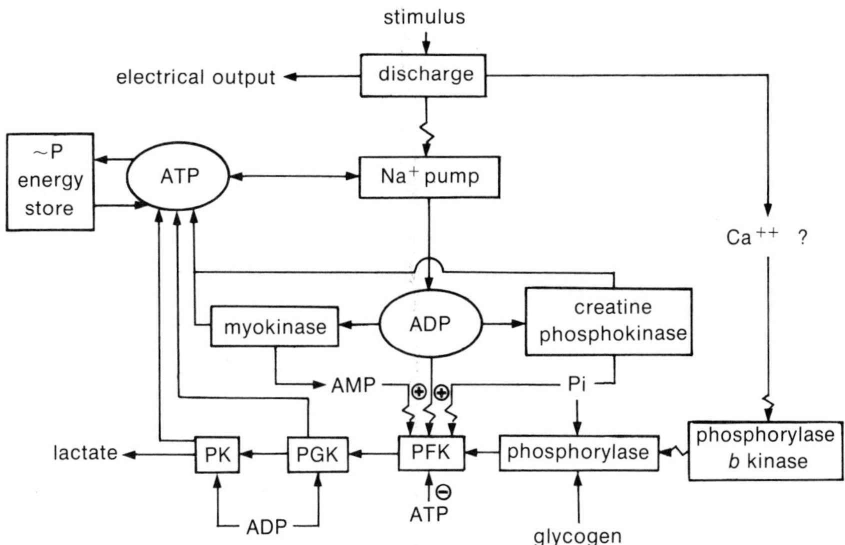

Figure 6–27. Scheme indicating the main feedback pathways between electrical activity and control of glycogenolysis in the main electric organ of *Electrophorus*. (From Williamson, J. R., W. Y. Cheung, H. S. Coles, and B. E. Herczeg, J. Biol. Chem. *242*:5112–5118, 1967.)

and modulators are quite distinct comes from experimental modification of the enzyme and by direct binding studies.[90]

The effects of these various metabolites are in the appropriate direction and are of appropriate magnitude to account for PFK regulation of glycolysis in many tissues, such as liver, heart, muscle, and nerve. Evidence that this occurs is indirect and is based largely on measurements of transient changes in metabolites during transition from one level of glycolytic flux to another (higher or lower) level. Most critical work here has been carried out by Williamson and his co-workers. In their studies of the *Electrophorus* electric organ, for example, discharge of which is almost totally supported by energy produced by glycolysis, Williamson et al. identified the principle sites of glycolytic control at the phosphorylase and PFK reactions.[166] The phosphorylase activation apparently involves Ca^{++} (it definitely does *not* depend upon cyclic AMP) in a manner similar to that in skeletal muscle. The increased input of hexose phosphates to the glycolytic pathway are accommodated by concurrent activation of PFK; hence the levels of G6P and F6P drop while FDP levels increase greatly. Kinetic measurements of the adenylates, creatine phosphate, and P_i show that a disappearance of creatine phosphate occurs within seconds of the onset of discharge. The levels of ATP are maintained by the high activity of creatine phosphokinase, and marked increases in ADP and AMP are not observed until after the end of a 60-second discharge, when the creatine phosphate reserves are depleted. PFK activity, as depicted by changes in the PFK mass action ratio, [FDP] × [ADP]/[F6P] × [ATP], increases during discharge and shows a further large but transient increase during recovery. Judging from the transient changes in metabolites, Williamson et al. point out that the initial activation of PFK is caused by the early rise of P_i; thereafter, the change of the PFK mass action ratio closely follows the changes in ADP and AMP, both of which are strong activators of PFK, and the changes in FDP product activation.

The probable events occurring when the electric organ is stimulated to discharge are summarized briefly in Figure 6–27. The peculiar properties of PFK are ideally suited to provide for a rapid "switch on" of glycolysis in response to increased metabolic demands of the electric organ during discharge, and thus make this enzyme a key control site. Thus, once activation is initiated by an increase of P_i, further activation is rapidly and efficiently induced by positive feedback from both reaction products, FDP and ADP, and later by AMP. Activity of the enzyme is curtailed either by depletion of substrate or by removal of ADP and FDP. Rapid removal of these products is ensured by high activities of aldolase and triose phosphate dehydrogenase, and by phosphorylation of the ADP at subsequent glycolytic steps.

Comparable data on transient changes in the concentration of glycolytic intermediates and the adenylates are available from studies of the mammalian heart, liver, skeletal muscle, and brain,[86,163] and from studies of insect flight muscle,[119] crustacean leg muscle,[60] and epaxial muscle of fish.[46] In all cases, a pivotal role for PFK is indicated, and in all cases, one or several of the above modulators are involved in regulation of PFK activity.

In insects such as the blowfly and other Diptera, flight initiation occurs with a concomitant 100-fold increase in glycolytic flux. This probably exceeds any comparable process in nature. Examination of transient changes in metabolites during the first few seconds of flight indicates that, in spite of the high flux rates involved, the control mechanisms are so efficient that steady state concentrations of most of the component intermediates are reached within only 15 to 20 seconds of flight.[120]

The F6P ⇌ FDP Interconversion: Integration of PFK and FDPase Activities. The control of PFK is complicated in tissues such as the liver and kidney in vertebrates or the mantle in molluscs, where significant flow of carbon can occur in either gluconeogenic or glycolytic directions. Although the PFK reaction is theoretically reversible, large kinetic and thermodynamic barriers prevent any extensive use of this reaction during net gluconeogenesis. Through evolution, organisms have surmounted this difficulty by the elaboration of a second enzyme, FDPase, which hydrolyzes FDP to F6P and P_i. It is important to stress that physiologically this enzyme achieves the reversal of the PFK reaction, but of course chemically the reactions are entirely different: PFK catalyzes a transphosphorylation while FDPase catalyzes a simple hydrolysis. The occurrence of both of these enzymes in a single cell compartment, however, raises the possibility of a short-circuit in both carbon and energy metabolism at this point, for it is clear that simultaneous function of both enzymes would lead to a futile carbon cycling with a net hydrolysis of ATP. It is therefore evident that in tissues such as the liver and kidney, possessing both PFK and FDPase, regulation of the two enzymes must be tightly integrated. An entirely analogous problem of cycling arises wherever two oppositely directed enzymes occur within a single cell. The glucose ⇌ G6P interconversions (catalyzed by hexokinases and G6Pases) and the PEP ⇌ pyruvate interconversions (catalyzed by PK and PC + PEPCK) constitute further examples of this problem of carbon and energy cycling within tissues such as liver and kidney. In principle, all of these problems have been solved in the same manner: *intracellular conditions which favor catalysis in the catabolic direction are highly unfavorable for catalysis in the anabolic direction, and vice versa.* Hence, conditions that favor hexokinase function do not favor G6Pase function; conditions which favor PFK activity do not favor FDPase catalysis; conditions which favor PK do not favor PC or PEPCK functions, and so forth. Since in principle these are all similar, we shall focus our attention on integrative control of FDPase and PFK.

As is often the case in science, the discovery of special regulatory functions of FDPase were made essentially simultaneously and independently in at least three different laboratories—by Krebs and his co-workers, by Horecker, Pontremoli, and their co-workers, and by Pogell and his laboratory.[63,98] In all vertebrates thus far examined, liver FDPase displays an absolute requirement for a divalent cation (Mg^{++} or Mn^{++}), shows a complex pH profile (usually with an alkaline pH optimum), and is specifically inhibited by AMP. The AMP inhibition kinetics are sigmoidal with n values approaching 4, suggesting fairly strong positive cooperativity in the binding of AMP. In this case, the n value and the true number of binding sites correspond. Both the degree and the nature of the AMP inhibition can be altered by H^+ and the divalent cation, and thus intracellular Mg^{++}/Mn^{++} ratios and pH may supply mechanisms for a finer modulation of the FDPase reaction. FDP binding may show positive cooperativity in some systems, but in the case of most mammalian liver FDPases the FDP saturation curve is hyperbolic. In lower vertebrates, the FDP curve may be sigmoidal.[11]

It is widely accepted that AMP serves as the major control signal for integrating the activity of FDPase with that of PFK. Thus, under gluconeogenic conditions, ATP/AMP ratios are high; the FDPase pathway is maintained open by the low AMP levels, while the PFK reaction is closed by the high ATP con-

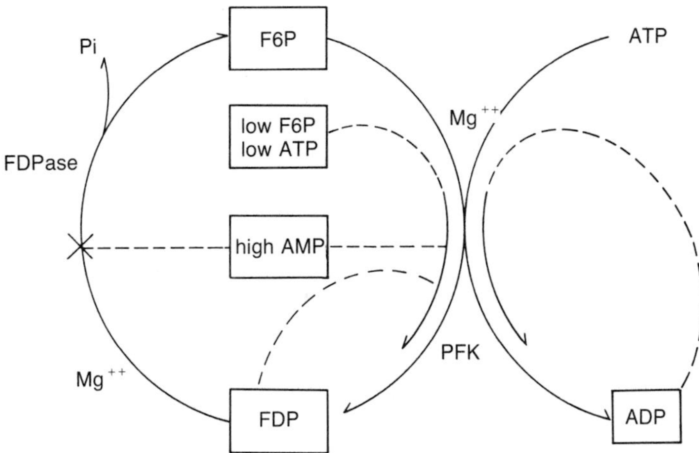

Figure 6–28. Integration of PFK and FDPase activities within a single cell compartment.

centrations. When the system swings over to net glycolysis, which would be favored under energy-depleting conditions, the raised AMP levels dampen FDPase activity at the same time as they activate PFK. The efficiency of the system is further improved by FDP and ADP product activation of PFK: this product activation leads to a "flare-up" of PFK activity initiated by AMP (Fig. 6–28).

Finally, an overlooked control component at this site may be H^+ *per se*. H^+ is one of the products of the PFK reaction and, as we have indicated, has profound effects on the properties of both enzymes.[11, 46, 144] Hence, it may also contribute to establishing oppositely directed control characteristics of these two enzymes.

The Role of "Proper Positioning" in Regulation of Enzyme Activity

The potential problem of futile carbon cycling through simultaneous function of oppositely poised, unidirectional enzymes can also be solved by their physical separation in space and/or time within a single cell. The glucose $\rightleftharpoons$ G6P interconversion is an excellent case in point. In most tissues hexokinase is thought to occur at or near the cell membrane; in the central nervous system, it is thought to be plated out on cellular membrane intrusions which come to lie adjacent to the mitochondria. In contrast, the enzyme G6Pase, wherever it is found, appears to be a phospholipid protein complex and *in vivo* occurs bound quite tightly to the endoplasmic reticulum. Thus, although the two enzymes catalyze oppositely directed reactions, their spatial separation probably precludes major "futile" cycling of carbon and energy.[127] This specific separation of enzymes is probably fairly common in living systems. But how have mechanisms evolved to allow the structure of each enzyme to "recognize" its proper location within the cell?

Given the coexistence in the cell of thousands of different macromolecular polyelectrolytes and many more kinds of smaller molecules, it is evident that the solvent capacity of the cell water can be conserved only by maintaining the concentration of constituent solutes at low levels. Hence, the biochemical complexity which we see in the living cell could have evolved only with parallel development of effective mechanisms for limiting concentrations. In the case of proteins, two opposing needs (the need for each enzyme at concentrations allowing for adequate catalytic activity *plus* the need for maintenance of low concentrations within the cytoplasm) may have added selective pressures for mechanisms incorporating enzymes into the membrane structures of the cell matrix, the mitochondria, the nucleus, and the cell plasma membrane itself. Thus, it has become increasingly clear that most intracellular enzyme systems are to some extent associated with particulate portions of the cell. This concept was elegantly and directly verified by Kempner and Miller,[73, 74] who used a centrifugal stratification of the intracellular contents of living

Euglena, a protozoan about 30 μ in length, to demonstrate that there is no free or unbound protein in the cell. Of the wide spectrum of enzymes assayed histochemically, all appear to be associated with particulate components of the cell; *no enzymes occur in the cytosol, which indeed appears to be free of macromolecules in general.* The full implications of this striking observation to our current theories of enzyme regulation are not yet fully recognized.

SUMMARY

Despite the apparent complexity of intermediary metabolism, its basic functions are at once few in number and readily comprehensible. In essence, cellular metabolism in all organisms must at all times perform the following "tasks":

(1) High energy compounds such as ATP must be generated in sufficient quantities to supply the cell with the "energy currency" needed for such vital work functions as ionic regulation, contractility, biosynthesis, and so forth.

(2) The intermediates needed for biosynthesis must be formed in adequate quantities to meet the demands for the synthesis of large molecules, such as nucleic acids, proteins, lipids, and carbohydrates.

(3) Biological reducing power such as NADPH must be generated to support the reductions which occur in biosynthetic pathways.

(4) Lastly, through a union of the above three processes, the organism must synthesize the large molecules which serve as the basis for the distinct "biological" properties of living systems.

Superimposed on all of these essential functions, or more accurately, upon the reaction sequences which conduct these functions, is an intricate and efficient control system which ensures that the participation of each metabolic process is consistent with the needs of the organism as a whole. No metabolic pathway is a free-wheeling process. The activity of each pathway, and in fact of each regulatory enzyme, is subject to an array of controls which links its function to the local chemistry of the cell, and ultimately, into the basic requirements of the whole organism. These control functions represent important sites of interaction with the environment, are highly sensitive to natural selection, and consequently display interesting species specificities.

A key fact emerging from studies of the last decade is that regulation is achieved through a hierarchy of control mechanisms. Since essentially all cellular functions are enzyme-catalyzed, control of metabolism reduces itself to control of the *rates, types,* and *directions* of enzyme function. It is now clear that there are only two ways in which this can be achieved: either (1) *the amounts of enzymes can be varied,* or (2) *the activities of enzymes—the extent to which catalytic potentials are actually utilized—can be regulated.* The hierarchical nature of control arises from the simple fact that regulation of enzyme function can be achieved by placing "on-off" signals at either of these two levels—either at the level of enzyme synthesis or at the level of enzyme activity. It will be evident that the properties of these two control processes differ in terms of:

(a) Speed of response: the first is slow, usually requiring at least hours; the second is so fast that for practical purposes it is considered instantaneous.

(b) Sensitivity of response: the first allows for only a "coarse" level of control; the second allows for a much finer "tuning" of enzyme function.

(c) Versatility of response: the first is the more versatile in that it allows the organism to modulate the relative *amounts* and the *kinds* of enzymes functioning at any given locus in metabolism; in the second strategy of control, the qualitative composition of the enzyme battery cannot be changed, and only its relative activity can be modulated.

At any given time, enzyme activity is largely determined by (a) the availability of substrates and cofactors, (b) the influence of physical environmental parameters such as temperature and pressure, and (c) the interaction with a class of metabolites termed "enzyme modulators." These latter compounds are instrumental in providing metabolism with its most *rapid, accurate, and sensitive* mechanism of regulation.

Enzyme modulators fall into two categories: positive modulators increase enzymic activity, whereas negative modulators have the opposite (inhibitory) effect. Enzymes which are regulated by modulators are termed regulatory enzymes. The effect of modulators on most regulatory enzymes is to alter the enzyme-substrate affinity rather than the maximum catalytic potential. Since physiological substrate concentrations almost always are too low to saturate enzymes, it is clear that the catalytic rate of an enzyme is highly sensitive to changes in enzyme-substrate affinity. For most regulatory enzymes, *slight changes in enzyme-substrate affinity can lead to large changes in catalytic rate,* a property that is the key to much of enzyme regulation.

The metabolites which serve as enzyme modulators frequently bear little chemical similarity to the substrates, products, or cofactors of the enzyme activity being regulated. This is only to be expected, for what an enzyme needs to "sense," if it is to gear its activity according to the demands placed by the cell for products of the *pathway* of which it is an integral part, is the concentration of the final, key products of that pathway. For example, many of the mainline catabolic pathways have as their primary function the generation of ATP. Enzymes acting as "valves" in these pathways should be responsive to changes in the ATP concentration of the cell—even though ATP may be very dissimilar, structurally, from the substrates of the enzyme. That is, regulatory enzymes must exhibit affinity for modulators which, for steric reasons, cannot bind at the substrate-binding site. We thus find that most regulatory enzymes have distinct regions for binding substrates, on the one hand, and for binding modulators, on the other. The modulator-binding sites have been termed "allosteric" sites (meaning sterically different) to stress their steric and spatial separation from the site where catalysis *per se* occurs.

The mechanism of allosteric regulation of catalysis reveals most elegantly the level of sophistication which exists in protein "design," for it involves modulator-induced changes in the conformation of the enzyme protein, such that the binding of the modulator will determine whether the substrate-binding site will have the proper geometry and charge configuration to permit the ready binding of substrate. We see, then, an important evolutionary "rationale" for the "macro" dimensions of biological catalysts: not only must enzymes supply complex sites for substrate attachment, but they must also possess the proper degree of structure to allow for the allosteric regulation of catalysis.

It is common to refer to the vast interlinked network of metabolic reactions as a "metabolic map." This is an apt metaphor and yields an important insight into the basic design of metabolic pathways and metabolic control circuitry. Thus, in addition to a complex series of "routes" through which metabolites are channelled, there also exist highly coordinated "traffic signalling" systems which keep the rates and directions of metabolic flow consistent with the overall interests of the cellular economy. Without these controls and their proper positioning, the vast catalytic potential of the "metabolic map" would be nonfunctional, much as a modern freeway system would be chaotic without traffic signals, offramps, and so forth.

The overall design of metabolic control circuitry is remarkably simple, and is based on two fundamental characteristics:

(1) The regulatory enzymes which "direct" metabolic flow are usually *strategically positioned* either (a) at the beginning of metabolic pathways, or (b) at vital metabolic branchpoints, where two or more pathways diverge or converge.

(2) Regulatory enzymes, as already noted, are sensitive to the cell's needs for the product of the pathway as a whole. This sensitivity is of two distinct kinds: (a) All regulatory enzymes, whether linked to biosynthetic or to catabolic pathways, are "locked" into the adenylate charge of the cell. The importance of adenylate sensitivity is readily appreciated, for the adenylates (AMP, ADP, ATP) are the most important metabolic coupling agents between energy-yielding and energy-requiring processes. (b) In addition, regulatory enzymes are sensitive to the levels of certain metabolites which are characteristic intermediates or products of the particular pathway. An enzyme "pacing" the overall synthesis rate of a particular amino acid, for example, typically is feed-back inhibited by this amino acid, the terminal product of the pathway.

These two types of regulatory sensitivities can be seen to offer "coarse" and "fine" control potentials for regulatory enzymes. Energy charge modulation is a relatively coarse control, whereas modulation by specific pathway products permits a fine-tuning of enzyme activity. A high energy charge, for example, generally stimulates ATP-utilizing pathways, but specific metabolites supply the cell with the additional information required to determine which energy-requiring processes are of particular need at the moment.

The complex, hierarchical nature of metabolic control has several important implications concerning the interactions between the organism's biochemistry and the organism's environment. Firstly, an environmental change may affect one to several events in the metabolic control hierarchy. Thus, the metabolic machinery of the cell is "vulnerable" at a number of different sites. Secondly, if we examine the other side of the above argument, biochemical adaptations to different environments can at least in theory be achieved in a number of different ways: the organism has available to it a number of different strategies of biochemical adaptations and survival. Thirdly, the interaction between environmental parameters and metabolic activities may be sudden or relatively slow, depending upon where in the control hierarchy the environmental factor impinges. Altering rates of enzyme synthesis, for example, would exert a slower effect on metabolism than would a direct influence of the environment on the activity of pre-existing enzymes, and the organism's response would be adjusted accordingly. Thus, it is not surprising to find a great deal of species-specificity in regard to (a) the presence or absence of specific metabolic pathways, (b) their "physiological" functions, and (c) their overall integration with the rest of cellular metabolism. In a sense, each particular "metabolic map" and each "control circuit" represents the successful outcome of one kind of adaptational strategy. Hence, each can be fully appreciated only in the context of the organism's life style and in the context of its interactions with its environment. By this comparative approach to intermediary metabolism, we may hope gradually to develop more profound insights into the biochemical bases for the spectacular diversity of living things.

REFERENCES

1. Asahina, E., Fed. Proc. Suppl. *24*:183–187, 1965. Freezing and injury in cells.
2. Asahina, E., pp. 1–49. *In* Advances in Insect Physiology, Vol. 6, edited by J. W. L. Beament, J. E. Treherne, and V. B. Wigglesworth. Academic Press, London, 1969. Frost resistance in insects.
3. Atkinson, D. E., Ann. Rev. Biochem. *35*:85–124, 1966. Regulation of enzyme activity.
4. Atkinson, D. E., pp. 23–40. *In* Metabolic Roles of Citrate, edited by T. W. Goodwin. Academic Press, London, 1968. Citrate in the regulation of energy metabolism.
5. Ball, E. Q., C. S. Strittmatter, and O. Cooper, Biol. Bull. *108*:1–17, 1955. Gas gland of the swimbladder.
6. Ballard, F. J., R. W. Hanson, and D. S. Kronfield, Biochem. Biophys. Res. Comm. *30*:100–104, 1968. Mitochondrial oxaloacetate in liver during spontaneous bovine ketosis.
7. Ballard, F. J., R. W. Hanson, and D. S. Kronfield, Fed. Proc. *28*:218–231, 1969. Gluconeogenesis and lipogenesis in ruminant and nonruminant animals.
8. Barnard, T., J. Skala, and O. Lindberg, Comp. Biochem. Physiol. *33*:499–508, 1970. Changes in interscapular brown adipose tissue of the rat.
9. Bauman, D. E., R. E. Brown, and C. L. Davis, Arch. Biochem. Biophys. *140*:237–244, 1970. Fatty acid synthesis in mammary gland of rat, cow, and sow.
10. Beenakkers, A. M. Th., and M. Klingenberg, Biochim. Biophys. Acta *84*:205–207, 1964. Carnitine co-enzyme A transacetylase in mitochondria.
11. Behrisch, H. W., and P. W. Hochachka, Biochem. J. *112*:601–607, 1969. Temperature and the regulation of enzyme activity in poikilotherms.
12. Belkin, D. A., Physiologist *5*:105, 1962. Anaerobiosis in diving turtles.
13. Belkin, D. A., Resp. Physiol. *4*:1–14, 1968. Underwater survival of two freshwater turtle species.
14. Bilinski, E., Canad. J. Biochem. Physiol. *41*:107–112, 1963. Utilization of lipids by fish.
15. Blazka, P., Physiol. Zool. *31*:117–128, 1958. The anaerobic metabolism of fish.
16. Bremer, J., pp. 65–88. *In* Cellular Compartmentalization and Control of Fatty Acid Metabolism, edited by F. C. Gran. Fed. Europ. Biochem. Soc. (Fourth Meeting), Universitetsforlaget, Oslo, 1968. Carnitine-dependent oxidation of fatty acids.
17. Brew, K., T. C. Vanaman, and R. L. Hill, Proc. Nat. Acad. Sci. *59*: 491–497, 1968. α-Lactalbumin and the A protein in lactose synthetase.
18. Brodbeck, U., and K. E. Ebner, J. Biol. Chem. *241*:762–764, 1966. Resolution of a soluble lactose synthetase into components.
19. Brodbeck, U., and K. E. Ebner, J. Biol. Chem. *241*:5526–5532, 1966. Distribution of the proteins of lactose synthetase in bovine and rat mammary tissue.
20. Brodbeck, U. W. L. Denton, N. Tanahashi, and K. E. Ebner, J. Biol. Chem. *242*:1391–1397, 1967. Identification of the B protein of lactose synthetase.
21. Brostrom, M. A., E. M. Reimann, D. A. Walsh, and E. G. Krebs, pp. 191–203. *In* Advances in Enzyme Regulation, Vol. 8, edited by G. Weber. Pergamon Press, Oxford, 1970. A cyclic 3′, 5′-AMP-stimulated protein kinase from cardiac muscle.
22. Butcher, R. W., G. A. Robinson, J. G. Hardman, and E. W. Sutherland, pp. 357–389. *In* Advances in Enzyme Regulation, Vol. 6, edited by G. Weber. Pergamon Press, Oxford, 1968. Cyclic AMP in hormone actions.
23. Arion, W. J., and R. C. Nordlie, J. Biol. Chem. *239*:2752–2757, 1964. Liver microsomal glucose-6-phosphatase, inorganic pyrophosphatase, and pyrophosphate-glucose phosphotransferase.
24. Candy, D. J., and B. A. Kilby, Biochem. J. *78*:531–536, 1961. The biosynthesis of trehalose.
25. Candy, D. J., and B. A. Kilby, J. Exp. Biol. *39*:129–140, 1962. Studies on chitin synthesis.
26. Childress, C. C., and B. Sacktor, J. Biol. Chem. *245*:2927–2936, 1970. Regulation of glycogen metabolism in insect flight muscle.
27. Chino, H., Nature *180*:606–607, 1957. Conversion of glycogen in the diapause egg of the bombyx silkworm.
28. Chino, H., J. Insect Physiol. *2*:1–12, 1958. Carbohydrate metabolism in the diapause egg of the silkworm *Bombyx mori.*
29. Chino, H., and L. I. Gilbert, Biochim. Biophys. Acta *98*:94–110, 1965. Lipid release and transport in insects.
30. Christian, D. R., G. S. Kilsheimer, G. Pettett, R. Paradise, and J. Ashmore, pp. 71–82. *In* Advances in Enzyme Regulation, Vol. 7, edited by G. Weber. Pergamon Press, Oxford, 1969. Regulation of lipolysis in cardiac muscle.
31. Cohen, J. J., Proc. Int. Union Physiol. Sci. *VI*:233–234, 1968. Renal gaseous and substrate metabolism *in vivo.*
32. Conte, F. P., H. H. Wagner, J. Fessler, et al., Comp. Biochem. Physiol. *18*:1–15, 1966. Development of osmotic and ionic regulation in juvenile coho salmon.
33. Conte, F. P., and T. N. Morita, Comp. Biochem. Physiol. *24*:445–454, 1968. Cell differentiation in gill epithelium of euryhaline *Oncorhynchus.*
34. Copeland, D. E., Amer. Zool. *8*:417–432, 1968. Fine structure of salt and water uptake in the land crab.
35. Dean, J. M., Comp. Biochem. Physiol. *29*:185–196, 1969. Metabolism of tissues of thermally acclimated trout.
36. DeVries, A. L., *in* Fish Physiology, Vol. 6, edited by W. S. Hoar and D. J. Randall. Academic Press, New York, in press. Freezing resistance in fishes.
37. De Weer, P., J. Gen. Physiol. *56*:583–620, 1970. Sensitivity of sodium efflux from squid axon to external sodium and potassium.
38. Drummond, G. I., and L. Duncan, J. Biol. Chem. *245*:976–983, 1970. Adenyl cyclase in cardiac tissue.
39. Drummond, G. I., J. P. Harwood, and C. A. Powell, J. Biol. Chem. *244*:4235–4240, 1969. Activation of phosphorylase in skeletal muscle.
40. Dubach, P., F. Smith, D. Pratt, and C. M. Stewart, Nature *184*:288–289, 1959. Glycerol and the winter-hardiness of insects.
41. Ernst, S. A., and R. A. Ellis, J. Cell Biol. *40*:305–321, 1969. Surface specialization in the secretory epithelium of the avian salt gland.
42. Exton, J. H., and C. R. Park, *in* Advances in Enzyme Regulation, Vol. 6, edited by G. Weber. Pergamon Press, Oxford, 1968. Cyclic AMP in liver metabolism.
43. Krebs, E. G., and D. A. Walsh, Fed. Proc. *31*:xiv, 1972. The mechanism of action of cyclic AMP in mammalian systems.
44. Fain, J. N., N. Reed, and R. Saperstein, J. Biol. Chem. *242*:1887–1894, 1967. The isolation and metabolism of brown fat cells.
45. Fawcett, D. W., The Cell. W. B. Saunders Company, Philadelphia, 1966.
46. Freed, J. M., Comp. Biochem. Physiol. *39*:765–774, 1971. Temperature effects on muscle phosphofructokinase of the Alaskan king crab.
47. Fry, F. E. J., and P. W. Hochachka, pp. 79–134. *In* Comparative Physiology of Thermoregulation, edited by C. Whittow. Academic Press, New York, 1970. Fish.
48. Garland, P. B., pp. 41–60. *In* Metabolic Roles of Citrate, edited by T. W. Goodwin. Academic Press, New York, 1968. Control of citrate synthesis in mitochondria.

49. Ginsburg, A., and E. R. Stadtman, Ann. Rev. Biochem. *39*:429–472, 1970. Multienzyme systems.
50. Hackman, R. H., Chap. 8. *In* The Physiology of Insecta, Vol. 3, edited by M. Rockstein. Academic Press, New York, 1964. Chemistry of the insect cuticle.
51. Hammen, C. S., Amer. Zool. *9*:309–318, 1969. Metabolism of the oyster.
52. Handler, J. S., A. S. Preston, and J. Orloff, J. Biol. Chem. *244*:3194–3199, 1969. Aldosterone and glycolysis in the toad urinary bladder.
53. Hanson, R. W., and F. J. Ballard, Biochem. J. *108*:705–713, 1968. The metabolic fate of the products of citrate cleavage in ruminants and nonruminants.
54. Hayward, J. S., and E. G. Ball, Biol. Bull. *131*:94–103, 1966. Brown adipose tissue thermogenesis during arousal from hibernation.
55. Heim, T., and D. Hull, J. Physiol. *186*:42–55, 1966. Blood flow and oxygen consumption of brown adipose tissue in newborn rabbit.
56. Hers, H. G., H. De Wulf, W. Stalmans, and G. Van der Berghe, pp. 171–190. *In* Advances in Enzyme Regulation, Vol. 8, edited by G. Weber. Pergamon Press, Oxford, 1970. The control of glycogen synthesis in the liver.
57. Hochachka, P. W., pp. 177–203. *In* Molecular Mechanisms of Temperature Adaptation. Amer. Assoc. Adv. Sci., Washington, D.C., 1967, Publ. No. 84. Metabolism during temperature compensation.
58. Hochachka, P. W., pp. 351–389. *In* Fish Physiology, Vol. 1, edited by W. S. Hoar and D. J. Randall. Academic Press, New York, 1970. Intermediary metabolism in fishes.
59. Hochachka, P. W., Variants of fructose-1,6-diphosphatase in a marine crustacean. Biochem. J. *127*:781–793, 1972.
60. Hochachka, P. W., J. M. Freed, G. N. Somero, and C. L. Prosser, Int. J. Biochem. *2*:125–130, 1971. Control sites in glycolysis of crustacean muscle.
61. Hochachka, P. W., and G. N. Somero, *in* Fish Physiology, Vol. 6, edited by W. S. Hoar and D. J. Randall. Academic Press, New York, 1971. Biochemical adaptation to the environment.
62. Holmes, R. S., and C. L. Markert, Proc. Nat. Acad. Sci. *64*:205–211, 1969. Immunochemical homologies in trout lactate dehydrogenase isozymes.
63. Horecker, B. L., S. Pontremoli, O. Rosen, and S. Rosen, Fed. Proc. *25*(5):1521–1528, 1966. Structure and function in fructose diphosphatase.
64. Ritch, R., and C. W. Philpott, Exp. Cell. Res. *55*:17–24, 1969. Repeating particles associated with an electrolyte-transport membrane.
65. Idler, D. R., and I. Bitners, Canad. J. Biochem. Physiol. *36*:739–798, 1958. Cholesterol, fat, protein, and water in the flesh of sockeye salmon.
66. Inoue, H., and H. C. Pitot, pp. 289–296. *In* Advances in Enzyme Regulation, Vol. 8, edited by G. Weber. Pergamon Press, Oxford, 1970. Regulation of synthesis of serine dehydratase isoenzymes.
67. Jackson, D. C., J. Appl. Physiol. *24*:503–509, 1968. Metabolic depression and O_2 depletion in the diving turtle.
68. Jeuniaux, C., Chitine et chitinolyse. Masson et Cie, Paris, 1963.
69. Joel, C. D., J. Biol. Chem. *241*:814–821, 1966. Stimulation of metabolism of rat brown adipose tissue.
70. Jonas, R. E. E., and E. Bilinski, J. Fish. Res. Bd. Canad. *21*:653–656, 1964. Fatty acid oxidation by various tissues from sockeye salmon.
71. Katzen, H. M., pp. 335–356. *In* Advances in Enzyme Regulation, Vol. 5, edited by G. Weber. Pergamon Press, Oxford, 1967. Multiple forms of mammalian hexokinase and their significance to the action of insulin.
72. Katzen, H. M., D. D. Soderman, and V. J. Cirillo, Ann. N.Y. Acad. Sci. *151*:351–358, 1968. Tissue distribution and physiological significance of multiple forms of hexokinase.
73. Kempner, E. S., and J. H. Miller, Exp. Cell. Res. *51*:141–149, 1968. Cellular stratification by centrifuging, *Euglena*.
74. Kempner, E. S., and J. H. Miller, Exp. Cell. Res. *51*:150–156, 1968. Enzyme localization in *Euglena*.
75. Kennedy, E. P., Fed. Proc. *20*:934, 1961. Biosynthesis of complex lipids.
76. Kornacker, M. S., and E. G. Ball, J. Biol. Chem. *243*:1638–1644, 1968. Respiratory processes in brown adipose tissue.
77. Kumar, S., J. A. Dorsey, R. A. Muesing, and J. W. Porter, J. Biol. Chem. *245*:4732–4744, 1970. Pigeon liver fatty acid synthetase complex and its subunits.
78. Kutty, M. N., J. Fish. Res. Bd. Canad. *25*:1689–1728, 1968. Respiratory quotients in goldfish and rainbow trout.
79. Lang, T. G., and K. S. Norris, Science N.Y. *151*:588–590, 1965. Swimming speed of a Pacific bottlenose porpoise.
80. Larner, J., C. Villar-Palasi, N. D. Goldberg, J. S. Bishop, F. Huijing, J. I. Wenger, H. Sasko, and N. B. Brown, pp. 409–423. *In* Advances in Enzyme Regulation, Vol. 6, edited by G. Weber. Pergamon Press, Oxford, 1968. Control of transferase phosphatase and transferase I kinase.
81. Lea, R. J., and G. Weber, J. Biol. Chem. *243*:1096–1102, 1968. Inhibition of glycolytic enzymes by free fatty acids.
82. Legherz, H. G., and W. J. Rutter, Science *157*:1198–1200, 1967. Glyceraldehyde-3-phosphate dehydrogenase variants in phyletically diverse organisms.
83. Lee, K. L., J. R. Reel, and F. T. Kenney, J. Biol. Chem. *245*:5806–5812, 1970. Regulation of tyrosine α-ketoglutarate transaminase in rat liver.
84. Llorente, P., R. Marco, and A. Sols, Europ. J. Biochem. *13*:45–54, 1970. Regulation of liver pyruvate kinase.
85. Lowenstein, J. M., pp. 61–86. *In* Metabolic Roles of Citrate, edited by T. W. Goodwin. Academic Press, New York, 1968. Citrate and the conversion of carbohydrate into fat.
86. Lowry, O. H., and J. V. Passonneau, J. Biol. Chem. *239*:31–42, 1964. The relationships between substrates and enzymes of glycolysis in brain.
87. Lynen, F., D. Oesterhelt, E. Schweizer, and K. Willeckek, pp. 1–24. *In* Cellular Compartmentalization and Control of Fatty Acid Metabolism, edited by F. C. Gran. Fed. Europ. Biochem. Soc. (Fourth Meeting), Universitetsforlaget, Oslo, 1968. The biosynthesis of fatty acids.
88. Mahler, H. R., and E. H. Cordes, Biological Chemistry. Harper & Row, New York, 1966.
89. Maitra, P. K., A. Ghosh, B. Schoener, and B. Chance, Biochim. Biophys. Acta *88*:112–119, 1964. Glycolytic metabolism following electrical activity in *Electrophorus*.
90. Mansour, T. E., pp. 37–51. *In* Advances in Enzyme Regulation, Vol. 8, edited by G. Weber. Pergamon Press, Oxford, 1970. Properties of phosphofructokinase.
91. Markert, C. L., Ann N.Y. Acad. Sci. *151*:14–40, 1968. The molecular basis for isozymes.
92. Marliss, E., T. T. Aoki, P. Felig, T. Pozefsky, and G. F. Cahill, Jr., pp. 3–11. *In* Advances in Enzyme Regulation, Vol. 8, edited by G. Weber. Pergamon Press, Oxford, 1970. The regulation of gluconeogenesis in fasting man.
93. Massey, B. D., and C. L. Smith, Comp. Biochem. Physiol. *25*:241–255, 1968. Thyroxine and mitochondrial respiration and phosphorylation in trout.
94. Mayer, S. E., D. H. Namm, and J. P. Hickenbottom, pp. 205–216. *In* Advances in Enzyme Regulation, Vol. 8, edited by G. Weber. Pergamon Press, Oxford, 1970. The phosphorylase activating pathway in intact cardiac and skeletal muscle.
95. Murphy, T. A., and G. R. Wyatt, J. Biol. Chem. *240*:1500–1508, 1965. Glycogen and trehalose synthesis in silk moth fat body.
96. Mustafa, T., and P. W. Hochachka, J. Biol. Chem. *246*:3196–3203, 1971. Properties of pyruvate kinases in tissues of a marine bivalve.
97. Newsholme, E. A., B. Crabtree, S. G. Higgins, S. D. Thornton, and C. Start, Biochem. J. *128*:89–97, 1972. Activity of FDPase in flight muscles from bumblebee and its role in heat generation.
98. Newsholme, E. A., and W. Gevers, Vitamins and Hormones *25*:1–87, 1967. Glycolysis and gluconeogenesis in liver and kidney cortex.
99. Nikkilä, E. A., Adv. Lipid Res. *7*:63–134, 1970. Control of plasma and liver triglyceride kinetics.
100. Norum, K. R., Biochim. Biophys. Acta *89*:95–108, 1964. Palmityl-CoA: carnitine palmityltransferase. Purification and some properties.
101. Norum, K. R., and J. Bremer, Comp. Biochem. Physiol. *19*:483–487, 1966. The distribution of palmityl CoA: carnitine palmityltransferase in the animal kingdom.
102. Papadopoulos, C. S., and S. F. Velick, Fed. Proc. *26*:557, 1967. An isozyme of glyceraldehyde-3-phosphate dehydrogenase in rabbit liver.
103. Parker, J. C., and J. F. Hoffman, J. Gen. Physiol. *50*:893–916, 1967. Membrane phosphoglycerate kinase in the control of glycolytic rate in human red blood cells.
104. Parvin, R., and K. Dakshinamurti, J. Biol. Chem. *245*:5773–5778, 1970. Inhibition of gluconeogenetic enzymes.
105. Passano, L. M. pp. 473–536. *In* Physiology of Crustacea, Vol. I, edited by T. Waterman. Academic Press, New York, 1960.
106. Pastan, I., and R. L. Perlman, Nature *229*:25–32, 1971. Cyclic AMP in metabolism.
107. Pogson, C. I., Biochem. J. *110*:67–77, 1968. Adipose-tissue pyruvate kinase.
108. Prosser, C. L., pp. 351–376. *In* Molecular Mechanisms of Temperature Adaptation. Amer. Assoc. Adv. Sci., Washington, D.C., 1967, Publ. No. 84. Temperature adaptation in relation to speciation.
109. Ramwell, P. W., and J. E. Shaw, pp. 139–187. *In* Recent Progress in Hormone Research, Vol. 26, edited by E. B. Astwood. Academic Press, New York, 1970. Biological significance of the prostaglandins.
110. Randle, P. J., P. J. England, and R. M. Denton, Biochem. J. *117*:677–695, 1970. The tricarboxylate cycle during acetate utilization in rat heart.
111. Read, C. P., pp. 327–357. *In* Chemical Zoology, Vol. 2, edited by M. Florkin and B. T. Scheer. Academic Press, New York, 1968. Intermediary metabolism of flatworms.
112. Reed, N., and J. N. Fain, pp. 207–224. *In* Brown Adipose Tissue, edited by O. Lindberg. American Elsevier Co., New York, 1970.
113. Reel, J. R., K. L. Lee, and F. T. Kenney, J. Biol. Chem. *245*:5800–5805, 1970. Regulation of tyrosine α-ketoglutarate transaminase in rat liver.
114. Richards, A. H., and R. Hilf, Fed. Proc. *30*:698, 1971. Glucose-6-phosphate and lactate dehydrogenase isoenzymes in the rat mammary gland.
115. Robin, E. D., J. W. Vester, H. V. Murdaugh, Jr., and J. E. Millen, J. Cell. Comp. Physiol. *63*:287–297, 1964. Anaerobic metabolism in the freshwater turtle.
116. Roots, B. I., Comp. Biochem. Physiol. *25*:457–466, 1968. Temperature and phospholipids of goldfish brain.
117. Roots, B. I., and P. V. Johnston, Comp. Biochem. Physiol. *26*:553–560, 1968. Plasmalogens of the nervous systems and environmental temperature.

118. Rutter, W. J., T. Rajkumar, E. Penhoet, M. Kochman, and R. Valentine, Ann. N.Y. Acad. Sci. *151*:102–117, 1968. Aldolase variants.
119. Sacktor, B., pp. 267–347. *In* Advances in Insect Physiology, Vol. 7, edited by J. W. L. Beament, J. E. Treherne, and V. B. Wigglesworth. Academic Press, New York, 1970. Control mechanisms in insect flight muscle.
120. Sacktor, B., and E. Wormser-Shavit, J. Biol. Chem. *241*:624–631, 1966. Concentrations of some intermediates in insect flight muscle during flight.
121. Salt, R. W., Canad. J. Zool. *37*:59–69, 1959. Role of glycerol in the cold-hardening of *Bracon cephi.*
122. Salt, R. W., Ann. Rev. Entomol. *6*:55–74, 1961. Principles of insect cold-hardiness.
123. Saz, H. J., Amer. Zool. *11*:125–135, 1971. Facultative anaerobiosis in invertebrates: pathways and control systems.
124. Saz, H. J., and O. L. Lescure, Comp. Biochem. Physiol. *22*:15–28, 1967. Gluconeogenesis, fructose-1,6-diphosphatase and phosphoenolpyruvate carboxykinase activities of *Ascaris lumbricoides* adult muscle and larvae.
125. Saz, H. J., and O. L. Lescure, Comp. Biochem. Physiol. *30*:49–60, 1969. Functions of phosphoenolpyruvate carboxykinase and malic enzyme in anaerobic formation of succinate in *Ascaris.*
126. Schimke, R. T., and L. Grossbard, Ann. N.Y. Acad. Sci. *151*:332–350, 1968. Studies on isozymes of hexokinase in animal tissues.
127. Scrutton, M. C., and M. F. Utter, Ann. Rev. Biochem. *37*:249–302, 1968. Regulation of glycolysis and gluconeogenesis in animal tissues.
128. Seubert, W., H. V. Henning, W. Schoner, et al., Adv. Enz. Reg. *6*:153–187, 1968. Effects of cortisol on levels of metabolites and enzymes controlling glucose production from pyruvate.
129. Smith, R. E., and B. A. Horowitz, Physiol. Rev. *49*:330–425, 1969. Brown fat and thermogenesis.
130. Smith, R. E., and J. C. Roberts, Amer. J. Physiol. *206*:143–148, 1964. Thermogenesis of brown adipose tissue in cold acclimated rats.
131. Somero, G. N., Biochem. J. *114*:237–241, 1969. Pyruvate kinase variants of Alaskan king crab.
132. Spencer, A., L. Corman, and J. M. Lowenstein, Biochem. J. *93*:378–388, 1964. Citrate and the conversion of carbohydrate to fat.
133. Stadtman, E. R., Ann. N.Y. Acad. Sci. *151*:516–530, 1968. Multiple enzymes in branched metabolic pathways.
134. Steen, J. B., Acta Physiol. Scand. *59*:221–241, 1963. The physiology of the swimbladder of the eel.
135. Tanaka, T., Y. Harano, F. Sue, and H. Morimura, J. Biochem. (Japan) *62*:71–87, 1967. Two types of pyruvate kinase from rat tissues.
136. Tanno, K., Low Temp. Sci. Ser. B *20*:25–34, 1962. Glycerol and frost resistance in carpenter ant.
137. Tarr, H. L. A., *in* Fish in Research, edited by O. W. Neuhaus and J. E. Halver. Academic Press, New York and London, 1969.
138. Tashima, L., and G. F. Cahill, pp. 55–58. *In* Handbook of Physiology, Sec. 5. Amer. Physiol. Soc., Washington, D.C., 1965. Fat metabolism in fish.
139. Threadgold, L. T., and A. H. Houston, Exp. Cell Res. *34*:1–23, 1964. The chloride cell of *Salmo salar* L.
140. Tietz, A., J. Lipid Res. *3*:421–426, 1962. Fat transport in the locust.
141. Tietz, A., pp. 45–54. *In* Handbook of Physiology, Sec. 5. Amer. Physiol. Soc., Washington, D.C., 1965. Metabolic pathways in the insect fat body.
142. Tietz, A., and G. Popjack, Biochem. J. *60*:155–165, 1955. Coenzyme A dependent reactions in mammary gland.
143. Topper, Y. J., pp. 287–308. *In* Recent Progress in Hormone Research, Vol. 26, edited by E. B. Astwood. Academic Press, New York, 1970. Multiple hormone interactions in the development of mammary gland *in vitro.*
144. Trivedi, B., and W. H. Danforth, J. Biol. Chem. *241*:4110–4114, 1966. pH and frog muscle phosphofructokinase.
145. Trump, B. F., S. M. Duttera, W. L. Byrne, and A. U. Arstila, Proc. Nat. Acad. Sci. *66*:433–440, 1970. Lipid-protein interactions in microsomal membranes.
146. Tucker, V. A., Amer. Zool. *11*:115–124, 1971. Flight energetics in birds.
147. Turkington, R. W., K. Brew, T. C. Vanaman, and R. L. Hill, J. Biol. Chem. *243*:3382–3387, 1968. Control of lactose synthetase in the developing mouse mammary gland.
148. Turkington, R. W., and R. L. Hill, Science *163*:1458–1460, 1969. Lactose synthetase.
149. Umbarger, H. E., Ann. Rev. Biochem. *38*:323–370, 1969. Regulation of amino acid metabolism.
150. Van Baalen, J., and S. Gurin, J. Biol. Chem. *205*:303–308, 1953. Cofactor requirements for lipogenesis.
151. Vaughan, H., and E. A. Newsholme, Biochem. J. *114*:81–82, 1969. Calcium ions and the activities of hexokinase phosphofructokinase and fructose-1,6-diphosphatase.
152. Vesell, E. S., and P. E. Pool, Proc. Nat. Acad. Sci. *55*:756–762, 1966. Lactate and pyruvate concentrations in exercised ischemic canine muscle.
153. Von Brand, T., Biochemistry of Parasites. Academic Press, New York, 1966.
154. Wald, G., Proc. Nat. Acad. Sci. *52*:595–611, 1964. The origins of life.
155. Watson, J. A., and J. M. Lowenstein, J. Biol. Chem. *245*:5993–6002, 1970. Citrate and the conversion of carbohydrate into fat.
156. Weber, G., pp. 15–40. *In* Advances in Enzyme Regulation, Vol. 7, edited by G. Weber. Pergamon Press, Oxford, 1969. Regulation of pyruvate kinase.
157. Weber, G., M. A. Lea, H. J. Convery, et al., pp. 257–300. *In* Advances in Enzyme Regulation, Vol. 5, edited by G. Weber. Pergamon Press, Oxford, 1967. Regulation of gluconeogenesis and glycolysis.
158. Weber, G., R. L. Singhal, and S. K. Srivastava, pp. 43–75. *In* Advances in Enzyme Regulation, Vol. 3, edited by G. Weber. Pergamon Press, Oxford, 1965. Regulation of biosynthesis of hepatic gluconeogenetic enzymes.
159. Weber, G., R. L. Singhal, N. B. Stamm, M. A. Lea, and E. A. Fisher, pp. 59–81. *In* Advances in Enzyme Regulation, Vol. 4, edited by G. Weber. Pergamon Press, Oxford, 1966. Synchronous behavior pattern of key glycolytic enzymes.
160. Weiss-Fogh, T., pp. 143–159. *In* Insects and Physiology, edited by J. W. L. Beament and J. E. Treherne. Academic Press, London, 1968. Metabolism and weight economy in migrating animals.
161. Whitt, G. S., Science *166*:1156–1158, 1969. Homology of lactate dehydrogenase genes.
162. Whitt, G. S., Arch. Biochem. Biophys. *138*:352–354, 1970. Directed assembly of polypeptides of the isozymes of lactate dehydrogenase.
163. Williamson, J. R., J. Biol. Chem. *240*:2308–2321, 1965. Inhibition of glycolysis in the isolated, perfused rat heart.
164. Williamson, J. R., J. Biol. Chem. *245*:2043–2050, 1970. Control of energy metabolism in hamster brown adipose tissue.
165. Williamson, J. R., E. T. Browning, and M. S. Olson, pp. 67–100. *In* Advances in Enzyme Regulation, Vol. 6, edited by G. Weber. Pergamon Press, Oxford, 1968. Fatty acid oxidation and gluconeogenesis in perfused rat liver.
166. Williamson, J. R., W. Y. Cheung, H. S. Coles, and B. E. Herczeg, J. Biol. Chem. *242*:5112–5118, 1967. Glycolytic intermediate changes in the main organ of *Electrophorus electricus.*
167. Williamson, J. R., R. Scholz, and E. T. Browning, J. Biol. Chem. *244*: 4617–4627, 1969. Interactions between fatty acid oxidation and the citric acid cycle in perfused rat liver.
168. Wilson, G., S. P. Rose, and C. F. Fox, Biochem. Biophys. Res. Comm. *38*:617–623, 1970. Membrane lipid unsaturation and glycoside transport.
169. Wyatt, G. R., pp. 287–360. *In* Advances in Insect Physiology, Vol 4, edited by J. W. L. Beament, J. E. Treherne, and V. B. Wigglesworth. Academic Press, New York, 1967. The biochemistry of sugars and polysaccharides in insects.
170. Wyatt, G. R., and W. L. Meyer, J. Gen. Physiol. *42*:1005–1011, 1959. The chemistry of insect hemolymph: glycerol.
171. Young, J. W., E. Shrago, and H. A. Lardy, Biochemistry *3*:1687–1692, 1964. Enzymes involved in lipogenesis and gluconeogenesis.

CHAPTER 7

NITROGEN EXCRETION

By James W. Campbell

INTRODUCTION

The classical concepts of excretory nitrogen metabolism were established by the early 1930's. The comparative implications of these concepts were popularized by Ernest Baldwin in his book, *An Introduction to Comparative Biochemistry,* first published in 1937. Briefly, these included the observation that the normal dietary intake of protein by animals provides amino acids in excess of the amounts required for the synthesis of new protein to sustain protein turnover. In order to oxidize the carbon skeleton of these amino acids (*carbon catabolism*), the α-amino group is first removed as ammonia (*nitrogen catabolism*). Because ammonia is generally toxic to animals, it must be eliminated from the body as an end-product. The three major end-products of nitrogen catabolism in animals are ammonia itself, usually as the ammonium ion, urea, and uric acid. Animals can thus be classified on the basis of which of these three compounds predominates in their excreta: those excreting mainly ammonium ion are *ammonotelic*, those excreting mainly urea are *ureotelic*, and those excreting mainly uric acid are *uricotelic.** All three end-products, as well as other nitrogenous compounds, are

**Purinotelism* is a generic term that includes those species excreting mainly uric acid as well as those species that excrete another purine such as guanine, i.e., the *guanotelic* arachnids.

TABLE 7–1. General Classification of Animals Based on Their Major Nitrogenous End-Product

Mode of Nitrogen Excretion	*Major Nitrogenous End-Product*	*Representatives*
Ammonotelism (Aquatic)	NH_4^+	Freshwater and marine invertebrates; freshwater and marine teleosts; larval and permanently aquatic amphibians
Ammonotelism (Terrestrial)	NH_3	Terrestrial isopods
Mixed Ammonotelism–Ureotelism	NH_4^+–urea	Earthworms; metamorphosing amphibians
Ureotelism	Urea	Land planaria; adult amphibians; mammals
Mixed Ureotelism–Uricotelism	Urea–uric acid	Chelonid and rhynchocephalid reptiles
Uricotelism	Uric acid	Terrestrial gastropods; terrestrial insects; squamate reptiles; birds
Mixed Ammonotelism–Uricotelism	NH_4^+–uric acid	Crocodilid reptiles
Guanotelism	Guanine	Scorpions; spiders

normally present in animal excreta; this classification is based on which end-product accounts for 50 per cent or more of the total excretory nitrogen. As shown in Table 7–1, not all animals fall neatly into one category or another since some may exhibit mixed patterns of nitrogen excretion, the predominant type depending upon specific physiological conditions.

An additional concept of excretory nitrogen metabolism is that the nature of the major nitrogenous end-product of a species is correlated with that species' environment. As can be seen in Table 7–1, aquatic species are generally ammonotelic, whereas terrestrial species are either ureotelic or purinotelic. The basis for this correlation is the availability of water. Aquatic organisms, especially those in fresh water, have access to large quantities of water and eliminate ammonia, as the ammonium ion, by exchange with the environment. Terrestrial animals, on the other hand, are more restricted in their water supply and, to prevent the build-up of ammonia in tissues and body fluids, must therefore detoxify this compound by converting it to a non-toxic end-product, urea or a purine.

Of primary importance in the correlation of end-product and environment are the conditions under which the embryo develops.[187, 195] Ammonia can serve as the end-product of aquatic species because there is relatively free direct exchange between the environment and the embryo across the egg membranes. When the exchange between the embryo and the environment is mediated via a maternal circulation, such as occurs in ovoviviparous and viviparous species, ammonia is converted to urea for elimination. Development within a cleidoic egg, an egg that is relatively independent of the environment save for gaseous exchange, requires the conversion of ammonia to a purine, usually uric acid, since the accumulation of urea in a cleidoic egg would create unfavorable osmotic conditions for the embryo. This osmotic problem is avoided with uric acid, whose limited solubility results in precipitation which renders it osmotically inactive. The elimination of urea as an end-product by adult animals requires considerable amounts of water because of its high solubility, whereas the elimination of a purine does not. Purinotelism is therefore generally found in those species best adapted to arid environments.

This, then, constitutes a brief outline of the classical concepts in excretory nitrogen metabolism. These concepts will be examined in detail in this chapter and interpreted, whenever possible, in terms of our present knowledge of cellular metabolism, its physical organization, and metabolic control.

THE FORMATION OF AMMONIA: NITROGEN CATABOLISM OF THE AMINO ACIDS

Proteins ingested in the diet are hydrolyzed in the digestive tract to their constitutive amino acids by proteolytic enzymes (see Chapter 4). These amino acids are actively concentrated by the gut epithelium[11] and enter the hepatic portal circulation to be carried to the liver and other tissues. Amino acids are actively accumulated in mammalian liver cells[50] for carbon[109, 231] and nitrogen catabolism. The main enzymes involved in nitrogen catabolism of the amino acids are the transaminases, glutamate dehydrogenase, specific amino acid deaminases, and amino acid oxidases.

Enzymes of Amino Acid Catabolism

Transamination. Most of the amino acids normally found in protein undergo transamination reactions. Reaction 1, which is

(1) $$\underset{\text{L-Alanine}}{CH_3{-}\underset{\displaystyle NH_3^+}{\underset{|}{CH}}{-}COO^-} + \underset{\alpha\text{-Ketoglutarate}}{^-OOC{-}CH_2CH_2{-}\underset{\displaystyle O}{\underset{\|}{C}}{-}COO^-} \rightleftarrows \underset{\text{L-Glutamate}}{^-OOC{-}CH_2CH_2{-}\underset{\displaystyle NH_3^+}{\underset{|}{CH}}{-}COO^-} + \underset{\text{Pyruvate}}{CH_3{-}\underset{\displaystyle O}{\underset{\|}{C}}{-}COO^-}$$

catalyzed by glutamate-pyruvate transaminase (EC 2.6.1.2, L-alanine: 2-oxoglutarate aminotransferase),* illustrates the transfer of an α-amino group in a typical transamination reaction.

Derivatives of vitamin B_6 (pyridoxal phosphate or pyridoxamine phosphate) serve as co-factors for the transaminases. In general, the highly purified enzymes are relatively specific for their substrates. However, in most crude tissue preparations, all of the major protein amino acids with the exception of serine and threonine undergo transamination reactions.[239] In mammalian cells, transaminases are localized in both the soluble cellular fraction (*cytosol*) and mitochondria.[302] During gluconeogenesis (the formation of glucose from non-carbohydrate precursors) induced by a high protein diet, experimental diabetes, or hormone treatment, the cytosol isozymes of glutamate-pyruvate and glutamate-oxaloacetate transaminases show a marked increase in activity[136, 284] indicating that they are the main ones involved in amino acid catabolism. The mitochondrial isozymes function in aspartate[196] or alanine[30] formation.

As can be seen in Reaction 1, transamination does not result in ammonia release but only in the transfer of an amino group to an α-keto acid. α-Ketoglutarate acts as the major acceptor keto acid, and thus the transamination of most amino acids results in L-glutamate formation. The removal of the α-amino group of glutamate as ammonia involves the oxidative deamination of this amino acid by glutamate dehydrogenase.

Glutamate Dehydrogenase. Glutamate dehydrogenase catalyzes Reaction 2.

*Number and systematic name of the enzyme as established by the Commission on Enzymes of the International Union of Biochemistry. A complete listing of the systematic nomenclature for all enzymes discussed in this chapter is given in Volume 13 of *Comprehensive Biochemistry*, edited by M. Florkin and E. H. Stotz and published by Elsevier, Amsterdam, 1964.

This enzyme has been purified from a number of animals, and its molecular and kinetic properties have been examined in detail.[92] Except for rat and dogfish,[144] the vertebrate liver enzymes undergo association reactions to form high molecular weight polymers and are affected by purine nucleotides.[93] In the presence of NADH, adenine nucleotides (ADP, ATP, etc.) promote association and activate enzyme activity. Guanine nucleotides (GDP, GTP, etc.), on the other hand, promote dissociation of the high molecular weight complex and inhibit enzyme activity. Either NAD or NADP may be used as co-factors of the vertebrate enzymes, but their effectiveness may vary depending upon the direction of the reaction being measured and upon the experimental conditions used. The overall equilibrium constant for the glutamate dehydrogenase reaction measured *in vitro* favors glutamate synthesis and not glutamate deamination.[92] However, as discussed below, it is clear that the enzyme functions *in vivo* for glutamate deamination due to its unique association with other enzymes.

Transdeamination. Braunstein and Bychkov[26] first suggested that amino acids are oxidized to ammonia by the combined action of the transaminases and glutamate dehydrogenase. This mechanism is outlined in Figure 7-1. The experimental evidence for *transdeamination* includes the observation that α-keto acids are required for amino acid oxidation by cell-free preparations of ureotelic and uricotelic vertebrate liver and kidney tissues.[8] In addition, the rates of transamination of most amino acids with α-ketoglutarate by mammalian liver and kidney tissue slices also correspond to their rate of deamination.[236] Certain amino acids may first undergo transamination with pyruvate to form alanine, which then transaminates with α-ketoglutarate to form glutamate for subsequent deamination.[237] Rather convincing evidence for the transdeamination of alanine has been obtained using the perfused mammalian liver system.[46]

(2)

$$\underset{\text{L-Glutamate}}{{}^{-}OOC{-}CH_2CH_2{-}\underset{\substack{|\\ NH_3^+}}{CH}{-}COO^-} + \underset{\text{Oxidized co-factor}}{NAD^+\ (NADP^+)} \rightleftharpoons$$

$$\underset{\alpha\text{-Ketoglutarate}}{{}^{-}OOC{-}CH_2CH_2{-}\underset{\substack{\|\\ O}}{C}{-}COO^-} + NH_4^+ + \underset{\text{Reduced co-factor}}{NADH\ (NADPH) + H^+}$$

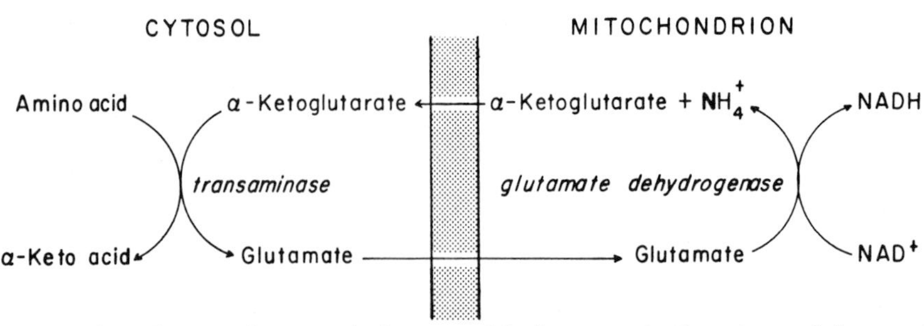

Figure 7–1. Function of transaminases and glutamate dehydrogenase in the release of the α-amino group of amino acids via transamination.

Although only transamination in the cytosol is depicted in Figure 7–1, the mitochondrial transaminase isozymes are also active.[236] However, glutamate dehydrogenase is a mitochondrial enzyme,[10] and the release of ammonia during amino acid catabolism via the transdeamination pathway is an intramitochondrial process.

Specific Deamination Reactions. Serine and threonine generally are not transanimated in animal tissues. These two hydroxy amino acids are deaminated by a specific enzyme or enzymes which are referred to as *dehydratases* because of their initial action in removing water, as shown in Reaction 3.

Threonine undergoes a similar deamination reaction. Both enzyme activities appear to be localized in the cytosol of mammalian liver cells,[205] and their release of ammonia is an extramitochondrial process. They, like the transaminases, are vitamin B_6 enzymes.

Amino Acid Oxidases. The L-amino acid oxidases catalyze the direct formation of ammonia from several amino acids, as shown in Reaction 4. In the absence of catalase, hydrogen peroxide may be formed in this reaction. Catalase is normally present in liver and should rapidly decompose any hydrogen peroxide formed during L-amino acid oxidation. The role of L-amino acid oxidases in mammalian liver amino acid catabolism is questionable. These flavoproteins may, however, function in kidney and in nonmammalian liver metabolism. An L-amino acid oxidase has been crystallized from mammalian kidney mitochondria,[183] and L-amino acid oxidases are present in reptilian tissues[325] and avian liver.[24,279] In the latter tissue, L-amino acid oxidase activity is associated with the microsomal fraction.[279] D-Amino acid oxidases and other peroxide-forming enzymes are now thought to be localized in peroxisomes along with catalase in most tissues. This may also be true for certain L-amino acid oxidase activities in mammalian kidney.[67a]

Comparative Aspects of Amino Acid Catabolism

Although transaminases are ubiquitous among animal species and tissues, their presence does not necessarily indicate that they function with glutamate dehydrogenase for the release of ammonia during amino acid catabolism and, indeed, in some cases, evidence for this is lacking. The equilibrium constants for most enzyme-catalyzed transamination reactions is near unity so the en-

---

(3) $$\underset{\text{L-Serine}}{\mathrm{H_2\underset{\displaystyle HO}{\underset{|}{C}}{-}\underset{\displaystyle NH_3^+}{\underset{|}{C}H}{-}COO^-}} \rightleftarrows \mathrm{H_2O} + \underset{\text{(intermediate)}}{\mathrm{HC{=}\underset{\displaystyle NH_3^+}{\underset{|}{C}}{-}COO^-}} \rightarrow \underset{\text{Pyruvate}}{\mathrm{H_3C{-}\underset{\displaystyle O}{\underset{\|}{C}}{-}COO^-}} + \mathrm{NH_4^+}$$

---

(4) $$\underset{\text{L-Amino acid}}{\mathrm{R{-}\underset{\displaystyle NH_3^+}{\underset{|}{C}H}{-}COO^-}} + \underset{\text{Oxidized flavoprotein}}{\mathrm{FAD}} + \tfrac{1}{2}\,\mathrm{O_2} \rightarrow \underset{\text{α-Keto acid}}{\mathrm{R{-}\underset{\displaystyle O}{\underset{\|}{C}}{-}COO^-}} + \underset{\text{Reduced flavoprotein}}{\mathrm{FADH_2}} + \mathrm{NH_4^+}$$

zymes function equally well for the synthesis of one amino acid from another, depending upon the cellular concentrations of the reactants. As shown in the perfused mammalian liver,[141] amino acids synthesized via transamination reactions in the liver may enter the blood for utilization by extrahepatic tissues. Glutamate dehydrogenase functions in transdeamination to remove one of the reaction products of transamination, and this enzyme is therefore critical for the release of ammonia via this pathway.

Transdeamination in Vertebrates and Invertebrates. This is the major route for amino acid catabolism in liver tissue of most vertebrate animals with the possible exception of certain reptiles.[60, 61] Glutamate dehydrogenase has been studied in fish,[58, 93, 168, 303] amphibians,[68, 85, 93, 316] and birds.[91, 93] Although these enzymes differ somewhat in their properties from each other and from the mammalian enzyme, they are all capable of functioning in the direction of glutamate deamination.

The process of transdeamination has not been extensively studied in invertebrate animals. The major transaminases have been demonstrated in representatives of most major invertebrate phyla and have been shown to operate *in vivo* in the interconversion of amino acids.[38] That is, following the incubation of whole animals or tissues with substrates such as ^{14}C-carbon dioxide and ^{14}C-glucose, ^{14}C is incorporated into alanine, aspartate, and glutamate. On the other hand, glutamate dehydrogenase has been studied in only a few invertebrate species, mainly arthropods. The enzyme is present in insects,[64, 166] and the best characterized insect glutamate dehydrogenase is that from larval *Drosophila melanogaster*.[22] This enzyme is similar to the vertebrate enzymes in showing activation by adenine nucleotides and inhibition by guanine nucleotides. The extremely high K_m for ammonia of *Drosophila* glutamate dehydrogenase (0.1 to 0.6 M) indicates that it functions mainly in the direction of glutamate deamination. In this direction, NAD is the more effective co-factor, whereas in the direction of glutamate synthesis, NADP is the more effective co-factor. These properties plus the mitochondrial localization of the enzyme in insects suggest that it functions in these species in conjunction with the transaminases[59] in the release of ammonia via transdeamination.

Crustacean glutamate dehydrogenase has received considerable attention with respect to its role in osmoregulation. The concentration of free amino acids in euryhaline invertebrates is at a maximum in a hypertonic environment. Upon transfer to a hypotonic environment, the concentration of free amino acids decreases concomitant with an increase in ammonia excretion.[83] It has been suggested that glutamate dehydrogenase plays a key role in this adjustment.[88, 251] Oxygen consumption and ammonia excretion are low in a hypertonic environment due, according to the hypothesis, to the utilization of ammonia, NADH, and α-ketoglutarate for glutamate synthesis. Under these conditions, higher concentrations of glutamate are thus available for the synthesis of other amino acids by transamination. In a hypotonic environment, the converse is true and glutamate is deaminated, causing an increased excretion of ammonia. The withdrawal of glutamate under the latter condition serves to decrease the concentration of other amino acids due to the action of the transaminases. In a hypotonic environment, oxygen consumption is also increased due to the availability of NADH and α-ketoglutarate. Consistent with this hypothesis is the reported activation of crustacean glutamate dehydrogenase in the direction of glutamate synthesis by cations (Na^+).[249, 252] Anions (Cl^-) may also be involved in the regulation of glutamate dehydrogenase under these circumstances.[47] In the absence of an activating adenine nucleotide, gill glutamate dehydrogenase of the spiny lobster *Palinurus vulgaris* has been reported not to deaminate glutamate with either NAD or NADP as co-factors.[250] However, glutamate dehydrogenase from muscle of the crab *Carcinus maenas* and the lobster *Homarus vulgaris* is a completely reversible enzyme utilizing NAD in the direction of deamination.[47] This is also true of the enzyme in the terrestrial crustacean *Oniscus asellus*.[115] Transdeamination thus appears to take place in the crustacean arthropods and may possibly be a key pathway in the metabolic adjustment shown by euryhaline species.

The mitochondrial glutamate dehydrogenase of gut tissue of the earthworm *Lumbricus terrestris* also acts to deaminate glutamate and could thus operate in conjunction with the transaminases present in this tissue for transdeamination.[38] Polychaete annelids, on the other hand, do not appear to possess glutamate dehydrogenase activity.[16]

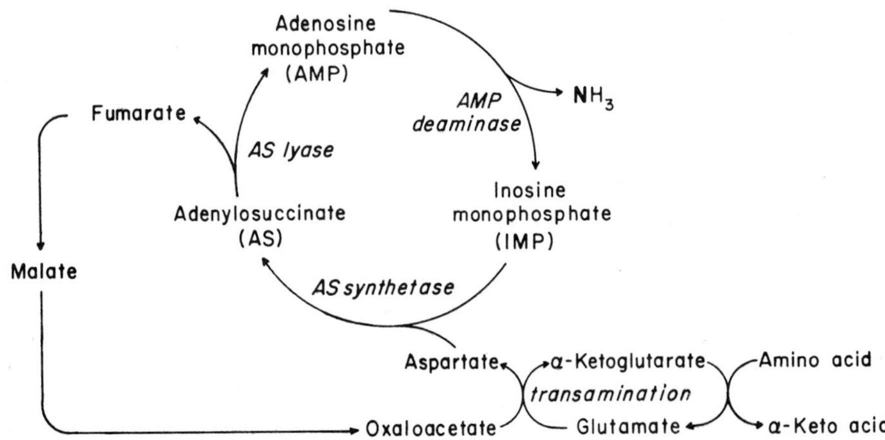

Figure 7–2. Release of the α-amino group of amino acids by the combined action of the transaminases and the purine nucleotide cycle.

In contrast to arthropods and annelids, there is no good evidence that molluscs catabolize amino acids via the transdeamination pathway. Glutamate dehydrogenase, if present in these species, appears not to operate in the direction of glutamate deamination,[38, 43] nor do preparations of molluscan tissues oxidize glutamate or proline.[39] Proline oxidation is usually dependent upon its conversion to glutamate, which then undergoes oxidative deamination via glutamate dehydrogenase. Also, in contrast to the broad substrate specificity shown by crude preparations of vertebrate and arthropod transaminases, the substrate specificity of the transaminases in the bivalve molluscs *Mytilus edulis* and *Modiolus modiolus* is limited to L-alanine, L-aspartate and L-glutamate.[218]

Molluscan L-Amino Acid Oxidase. L-Amino acid oxidase activity has been reported in representatives of several invertebrate phyla, but is especially common in molluscan tissues.[39] The molluscan L-amino acid oxidases are in general most active with the basic amino acids and, in this respect, are similar to the soluble enzyme in turkey liver.[24] They are inactive with several common amino acids including glutamate, alanine, aspartate, proline, isoleucine, glycine, serine, and threonine, so their function as an alternative to transdeamination for amino acid catabolism in molluscs would appear to be limited. The comparative role of L-amino acid oxidase in molluscan amino acid catabolism has not, however, been critically evaluated. The molluscan enzymes are usually associated with a particulate cellular fraction, although they may also occur in the soluble fraction in some species.[100]

Alternate Routes of Amino Acid Deamination. Adenylate (AMP) deaminase is present in many tissues, and Braunstein[25] suggested that this enzyme might participate indirectly in the deamination of amino acids. A modified outline of the scheme originally proposed is shown in Figure 7–2. Transamination of glutamate with oxaloacetate to form aspartate without the release of ammonia normally occurs in mammalian mitochondria,[196] and the purine nucleotide cycle as outlined in Figure 7–2 has been shown to be the major route for ammonia production by working muscle.[155, 155a] This pathway has been proposed as a means of ammonia release in fish[163] and in polychaete annelids that lack glutamate dehydrogenase.[16]

AMMONIA TOXICITY

The concept of ammonia toxicity is a fundamental one in animal nitrogen metabolism. It is, however, by no means a universal concept. Ammonia tolerance is fairly widespread among the procaryotes and lower eucaryotes. An organism similar to the Precambrian microfossil *Kakabekia umbellata* requires culture under ammonia gas,[264] and a mutant of *Penicillium notatum* shows some metabolic activity when cultured in liquid ammonia.[263] The extreme ammonia tolerance shown by the *Kakabekia*-like organism is of interest since life may have arisen in an ammonia-rich environment.[194] The unprotonated form of ammonia (NH_3) is toxic to higher fungi.[149] Although most higher plants can assimilate the ammonium ion (NH_4^+), an excess is also toxic to them.[211a] In this section, certain hypotheses concerning the mechanism of ammonia toxicity in animal cells will be considered along with the results of comparative studies on ammonia toxicity in animals.

Mechanism of Ammonia Toxicity

General pH Effects. Because of the effects of pH on enzyme catalysis, the uncontrolled release of ammonia in the unprotonated form would be expected to alter cellular metabolism by the increased alkalinity in the immediate area of its release. The K_a for the reaction $NH_3 + H^+ \rightleftarrows NH_4^+$ is 5.5×10^{-10} at 25°C. At physiological pH's (pH 7 to 7.4), approximately 99% of the ammonia molecules exist in the protonated form as NH_4^+. Several enzymes release ammonia in the unprotonated form, which then takes up a proton from the medium resulting in a local alkalinization. In the glutamate dehydrogenase reaction (Reaction 2), the proton apparently comes from the reactants and there is no local alkalinization. Ammonia toxicity thus cannot be treated in all cases as a simple effect of pH on enzyme-catalyzed reactions.

In addition to the general effect of pH on enzyme activity, this parameter also affects membrane structure. Alkaline pH's, which would result from the uncontrolled release of ammonia, appear to decrease membrane stability.[9] Calcium uptake and release by the sarcoplasmic reticulum is markedly affected by pH,[182] so there is also a general pH effect on certain membrane transport systems.

Substrate Depletion Effects. One of the first hypotheses for ammonia toxicity was that any intramitochondrial increase in this compound would cause a reversal of the glutamate dehydrogenase reaction, withdrawing α-ketoglutarate from the tricarboxylic acid cycle as well as decreasing the amount of NADH available for oxidation.[13] Under these conditions, the increased glutamate concentration would then serve to lower the cellular concentration of ATP due to an increased conversion of glutamate to glutamine, a conversion mediated by glutamine synthetase (Reaction 5). Mammalian brain tissue has been the main tissue used to test this hypothesis, but the experimental results are equivocal.[119, 262] The amount of α-ketoglutarate that can be measured in brain of rats during experimental ammonia intoxication is actually greater than normal, rather than less as predicted, and there is no significant change in the levels of ATP. During ammonia intoxication, there is, however, an increase in the cytosol lactate/pyruvate ratio, indicating that oxidative metabolism is impaired. This latter result is in keeping with the known effects of ammonium salts on brain oxidative metabolism.[171]

Membrane Effects. Ammonium salts are known to inhibit the active transport of Na^+ in a number of tissues including fish gill,[162] crayfish,[258] toad bladder,[110] and avian kidney.[193] In the latter tissue, ammonium salts also inhibit Cl^-, HCO_3^-, and water reabsorption. In toad bladder, the inhibition of Na^+ transport by ammonium salts also serves to inhibit K^+ accumulation.[110] It is thus clear that NH_4^+ can inhibit the transmembrane movement of several biologically important ions. This does not, however, appear to be the primary effect on nerve tissue since there is no change in either Na^+ or K^+ in mammalian brain during experimental ammonia intoxication.[241] In addition to interfering with membrane transport, it also now seems clear that NH_4^+ itself can traverse biological membranes.[110] It was previously thought that membranes were permeable only to the unprotonated molecule (NH_3).[175]

---

(5) $$^-OOC{-}CH_2CH_2{-}\underset{\displaystyle NH_3^+}{\underset{|}{CH}}{-}COO^- + ATP + NH_4^+ \rightarrow$$

L-Glutamate

$$H_2N{-}\underset{\displaystyle O}{\underset{\|}{C}}{-}CH_2CH_2{-}\underset{\displaystyle NH_3^+}{\underset{|}{CH}}{-}COO^- + ADP + P_i$$

L-Glutamine

The active transport of ions, especially K^+, by the mitochondrial membrane is a critical process in mitochondrial energy metabolism.[210] Any interference with these transport systems by NH_4^+ might thus be expected to interfere with oxidative metabolism. In addition to the metallic cations, there is also the very critical role of the transmembrane movement of H^+ in mitochondria. According to the Mitchell hypothesis (see ref. 212), the driving force for phosphorylation is the H^+ gradient across the mitochondrial (or chloroplastal) membrane. During oxidative phosphorylation, protons pass out of the mitochondrion and, during photosynthetic phosphorylation, they pass into the chloroplast. In chloroplasts, ammonium salts inhibit phosphorylation. According to one mechanism that has been proposed for this inhibition, NH_3 penetrates the chloroplast and abolishes the proton gradient by binding H^+ taken up during the light-dependent process.[66] Beef heart mitochondria swell in the presence of ammonium chloride and the ionophore valinomycin,* indicating that NH_4^+ penetrates under these circumstances. Inside the mitochondrion, where the reaction is alkaline, NH_4^+ is converted to NH_3 which then effluxes, thereby serving to equilibrate protons across the mitochondrial membrane. This abolishes the H^+ gradient necessary for phosphorylation.[28] The effect of NH_4^+ in this system is overcome by K^+ due to the greater affinity of the ionophore for the latter cation. Thus, the displacement of NH_4^+ by K^+ allows for the reestablishment of the H^+ gradient. The isolation of a naturally occurring ionophore in beef heart mitochondria has been reported,[20] indicating that the results obtained *in vitro* could possibly relate to an actual situation *in vivo*. In any event, the NH_4^+/NH_3 system can affect cation transport by mitochondria, and this may be an important consideration in formulating a universal mechanism of ammonia toxicity in animal cells. At present, it is not possible to outline such a mechanism, since ammonia toxicity could involve one or a combination of the effects of the NH_4^+/NH_3 system discussed above. That ammonia is toxic to higher animals has been repeatedly demonstrated in studies on whole animals.

*One of several cyclic compounds that facilitate membrane transport of cations.[210]

Comparative Aspects of Ammonia Toxicity

Environmental Ammonia Tolerance. The ability of mammals to survive different atmospheric contents of ammonia is illustrated in Table 7–2. The initial symptoms of exposure to toxic levels of atmospheric ammonia in mammals are usually those associated with damage to the respiratory membranes. The ability of the guano bat *Tadarida brasiliensis* to withstand high atmospheric ammonia concentrations appears to be based on its capacity to neutralize ammonia with mucus and thereby prevent damage to these membranes.[281] This is an adaptive mechanism since *Tadarida* frequents caves showing high ammonia contents.[176, 280] This bat can also tolerate high blood levels of ammonia that cause little or no change in blood pH due, apparently, to its ability to neutralize excess ammonia by carbon dioxide retention.[282] On return to a normal atmosphere, the guano bat "blows off" excess ammonia at the lungs along with carbon dioxide.[280] Most mammals

TABLE 7–2. Ammonia Tolerance of Mammals*

	Maximum Time Until Death (min)					
	ATMOSPHERIC NH_3 CONTENT (ppm)					
Species	500	1000	3000	5000	7000	10,000
Man	60	—	—	–	—	–
Laboratory mouse	–	960	180	20	—	–
Laboratory rat	–	960	—	40	—	–
Bats:						
Myotis californicus	–	—	540	60	—	–
M. lucifugus	–	—	—	–	45	–
Eptesicus fuscus	–	—	—	–	120	20
Tadarida brasiliensis	–	—	—	>5760	180	20

*Compiled by Studier, Beck and Lindeborg.[283]

TABLE 7–3. Experimental Ammonia Toxicity in Vertebrates

Species	Ammonium Salt	Route of Administration	mMole/kg Body Wt.	
			LD^{*}_{50}	$LD^{*}_{99.9}$
Mouse[306]	Chloride		7.0	—
	Acetate		6.9	—
	Bicarbonate	Intravenous	5.3	—
	Carbonate		4.6	—
	Hydroxide		2.6	
Mouse[308]	Acetate	Intravenous	6.8 (in 21% O_2)	—
			2.9 (in 9% O_2)	—
Mouse[318]	Acetate	Intravenous	5.6	7.7
		Intraperitoneal	10.8	18.0
Chicken[318]	Acetate	Intravenous	2.7	4.9
		Intraperitoneal	10.4	26.2
Trout[317]	Acetate	Intraperitoneal	17.7	40.7
Catfish[317]	Acetate	Intraperitoneal	25.7	41.0
Goldfish[317]	Acetate	Intraperitoneal	29.5	76.0

*Dose at which 50% or 99.9% of the animals die during the conditions described in the individual references.

are far less tolerant than the guano bat of atmospheric ammonia, succumbing in some instances to low concentrations in only minutes.

Fish are especially intolerant of dissolved ammonia. According to Spotte,[272] the culture water for these and other aquatic animals should contain no more than 0.1 ppm ammonia. As with atmospheric ammonia, the initial damage caused by dissolved ammonia is to the respiratory epithelium. The toxicity of ammonia to fish is determined by the pH of the water; the greater the alkalinity, the more its toxicity.[74] This may be interpreted on the basis of the greater permeability of the unprotonated form of ammonia, the concentration of which is increased with increasing alkalinity.[175] In keeping with the general effect of ammonia on oxidative metabolism, hypoxia also increases the toxicity of ammonia to fish.[74]

Experimental Ammonia Toxicity: Mammals. Experimental studies on ammonia toxicity in mammals involve the direct injection of ammonium salts. The toxicity of these salts is directly related to their effect on blood pH; the more alkaline the salt, the greater its toxicity[306] (Table 7–3). Again, this is presumed to be due to the greater passage of NH_3 into the brain through the blood-brain barrier as well as into other tissues at more alkaline pH's.[273, 307] Lowering the blood pH of mice by the infusion of fixed acid, but not carbon dioxide, protects against ammonia toxicity because it increases the amount of ammonia present as NH_4^+, which is impermeable to the blood-brain barrier.[309] A major symptom of ammonia intoxication in mammals is the eventual loss of consciousness, indicating that brain is a major tissue affected. The ability of the tissues, especially the liver, to detoxify ammonia is exceeded in experimental ammonia intoxication. This also occurs clinically in conditions of liver malfunction such as cirrhosis, and the resulting loss of consciousness is referred to as "hepatic coma."[327] As in fish, hypoxia potentiates the toxicity of ammonium salts in mammals[308] (see Table 7–3).

Comparative Studies on Ammonia Toxicity in Vertebrates. As shown in Table 7–3, birds (uricotelic) are similar to mammals (ureotelic) in their tolerance of exogenous ammonium salts. In both birds and mammals, larger doses of ammonium salts are tolerated when they are injected intraperitoneally rather than intravenously because in the former route, ammonia is absorbed into the portal circulation and detoxified on first passage through the liver. Death due to ammonia intoxication in higher vertebrates has been attributed to cardiac malfunction. Abnormal electrocardiograms are obtained immediately after injecting a toxic dose of ammonium salts into experimental animals,[318, 319] and these are also observed clinically prior to hepatic coma.[327] Cardiac malfunction could possibly be associated with the observed effect of pH on calcium metabolism in cardiac muscle.[182]

Freshwater fish show a greater tolerance to exogenous ammonia than do either birds or mammals (Table 7–3), and this may be related to their ammonotelic mode of nitrogen metabolism.

Blood Ammonia Levels. In most mammals, the levels of ammonia in blood are low. In man and dogs, they range from 0.03 to 0.08 μmole per ml.[29, 184, 238] These levels are normally closely regulated but may increase during certain pathological conditions including hepatic coma. When ammonia is generated endogenously in the rabbit by the action of urease on blood urea, death occurs at around 4 μmoles per ml.[145] However, under other circumstances, higher levels are tolerated. During ammonia inhalation, blood ammonia normally increases to around 14 μmoles per ml in the guano bat.[282] The LD_{50} of ammonia acetate for the mouse (Table 7–3) corresponds to well over 100 μmole per ml assuming a plasma volume of around 50 ml per kg for this species. In the chicken, the plasma ammonia concentration averages around 0.13 μmole per ml but may increase to more than twice this value during egg laying.[223]

Normal blood ammonia levels in aquatic vertebrates appear to average only slightly higher than those in terrestrial species. The concentration reported in fish is around 0.2 to 0.3 μmole per ml[71, 103] and may be increased by the oral administration of ammonium salts.[71] The concentration in the aquatic toad *Xenopus laevis* ranges from 0.2 to 0.5 μmole per ml.[295]

Ammonia Toxicity in Invertebrates. There appear to have been no systematic studies on ammonia tolerance in invertebrates and there are only a few reports of the effect of ammonia on invertebrate tissues. Although invertebrates are generally thought to be more tolerant of ammonia than vertebrates, many of the reported values for ammonia in the body fluids of invertebrates fall within the range tolerated by at least some vertebrates, including mammals. Some of these values are, in μmoles per ml: 7.1 for blowfly *Lucilia cuprina* larvae;[19] 2 to 6.9 for the land crab *Cardisoma guanhumi*;[98, 121] 1 for the marine crab *Carcinus maenas*;[14] 0.1 to 0.2 for terrestrial snails;[289] 0.2 to 0.7 for *Octopus dofleini*;[209] and 0.5 to 4.4 for several annelids.[186]

Ammonium chloride has been observed to decrease oxygen consumption by crayfish ventral nerve chord.[88] This effect is similar to that discussed above for mammalian nerve tissue. It is not, however, universal to crustacean nerve tissue since ammonium salts do not inhibit oxygen consumption by nerve tissue from the lobster *Homarus vulgaris*. During the ammonium ion-induced decrease in oxygen uptake by crayfish nerve chord, there is a two-fold increase in glutamate synthesis indicating a utilization of α-ketoglutarate and NADH. An increased formation of glutamate in the presence of NH_4^+ is not observed with lobster nerve tissue. Nor would ammonia be expected to affect molluscan nerve tissue by the withdrawal of α-ketoglutarate and NADH because of the negligible glutamate dehydrogenase activity in this tissue.[228]

During insect development, diapause is a condition of dormancy characterized by low oxidative metabolism. Ammonium salts have the curious effect of terminating diapause, and this effect has been shown to be specific for the ammonium ion. During normal diapause the NADH/NAD ratio should be high, and the injection of ammonium salts might be expected to lower this ratio by causing NADH to be utilized for glutamate synthesis via the glutamate dehydrogenase reaction. An alternate explanation for the experimental termination of diapause by ammonium ion is that high concentrations of this ion cause an increased synthesis of glutamine by glutamine synthetase (Reaction 5). Free glutamine increases during normal diapause development and also following the injection of ammonium salts. L-Glutamine may therefore be critical to processes involved in the termination of diapause.[160]

DETOXICATION PATHWAYS

There are two major metabolic pathways for the detoxication of ammonia released during amino acid catabolism: the urea pathway, first elucidated in mammalian liver, and the uric acid pathway, first elucidated in avian liver. In addition to these two pathways, glutamine synthesis may also serve as a general detoxication mechanism in tissues that possess neither the urea nor the purine pathways. Glutamine synthesis could also be important in ammonotelic species for which a cellular ammonia detoxication mechanism has not been described.

Cellular Aspects

The Mammalian Liver System. The mechanism of ammonia detoxication in mam-

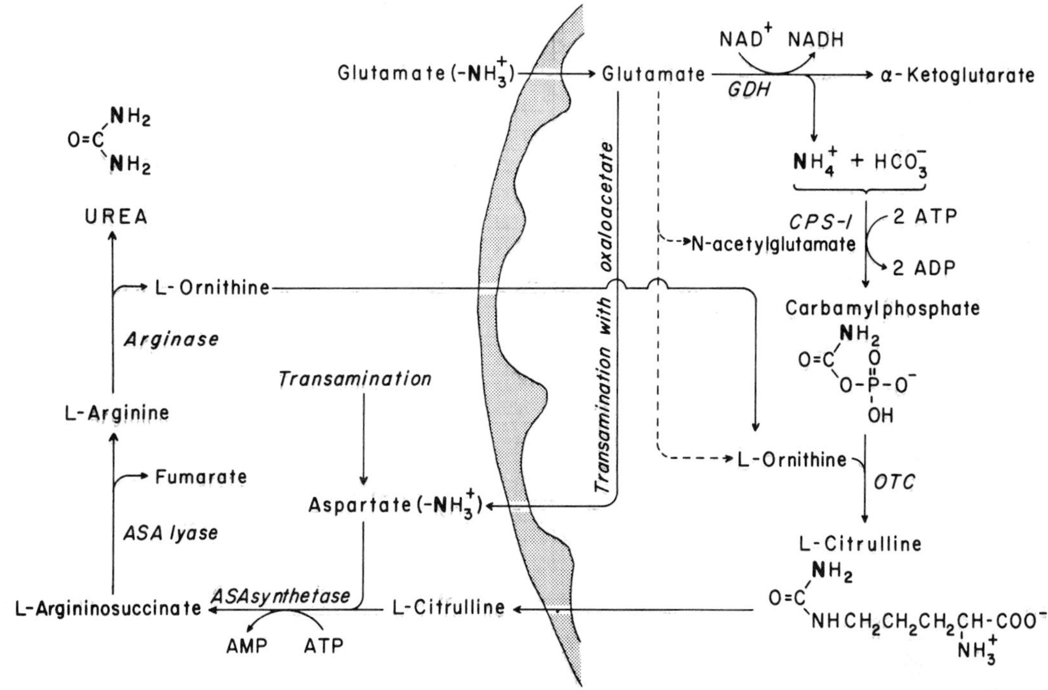

Figure 7–3. Detoxication of ammonia released via transdeamination by conversion to urea. The abbreviations are: *GDH*, glutamate dehydrogenase; *CPS-I*, carbamylphosphate synthetase-I; *OTC*, ornithine transcarbamylase; and *ASA*, argininosuccinate.

malian liver is outlined in Figure 7–3. Ammonia released by the action of glutamate dehydrogenase is converted to carbamylphosphate by carbamylphosphate synthetase-I, a mitochondrial enzyme with a high affinity for ammonia.[53] It is the withdrawal of ammonia by this reaction that appears to be responsible for the deamination of glutamate by glutamate dehydrogenase *in vivo*, a reaction that is kinetically unfavorable *in vitro*. The intramitochondrial formation of carbamylphosphate requires ATP generated by oxidative metabolism.[108] Ornithine transcarbamylase, also an intramitochondrial enzyme, serves to convert carbamylphosphate to citrulline in the presence of L-ornithine. There is some question as to the main source of ornithine used in this reaction. Mitochondria utilize exogenous ornithine for citrulline synthesis[48] and take up ornithine by an energy-dependent process.[96] On the other hand, there is also some evidence that ornithine may be synthesized intramitochondrially from glutamate by the action of ornithine δ-transaminase.[299] It has also been suggested that the activity of carbamylphosphate synthetase-I, and therefore the rate of urea biosynthesis, is regulated *in vivo* by *N*-acetyl-L-glutamate.[259, 260] *N*-Acetylglutamate is an intermediate in the pathway for ornithine biosynthesis that occurs in microorganisms so this pathway, and not the ornithine δ-transaminase pathway, could be the major route of ornithine synthesis in mitochondria. If this were the case, it would serve to tie in the regulatory mechanism for carbamylphosphate synthesis with the synthesis of ornithine. Citrulline, synthesized intramitochondrially, exits the mitochondrion and is converted to arginine in the cytosol by the combined actions of argininosuccinate synthetase and argininosuccinate lyase. In this conversion occurring in the cytosol, an α-amino group of aspartate is added. This aspartate could be formed either in the mitochondrion or in the cytosol by transamination reactions in which oxaloacetate serves as the acceptor keto acid. The carbon skeleton of oxaloacetate is released as fumarate during the formation of arginine from citrulline. Fumarate can be recycled to oxaloacetate by reactions of the tricarboxylic acid cycle present in mitochondria. Arginine thus formed in the cytosol is converted to urea by arginase, thereby regenerating ornithine to complete the "urea cycle."

Greenstein and co-workers[320] and others[114,313] have shown that L-arginine, when injected into mammals, protects against experimental ammonia intoxication caused either by high levels of amino acids or by injected ammonium salts. Protection against the latter is also afforded by a mixture of L-ornithine and L-aspartic acid.[240] The protective action of these intermediates is interpreted as being due to their enhancement of the activity of the urea cycle in detoxifying ammonia.

The Avian Liver System. The biosynthesis of the purine ring was elucidated by Buchanan and co-workers[35] in the 1950's using pigeon liver preparations. A review of the more recent literature on this pathway has been made by Hartman.[116] Although the detoxication of ammonia via uric acid formation is an old and well established concept in comparative biochemistry, far less is known of the cellular details of this mechanism than is known of those involved in urea formation. The most direct entry of ammonia into the purine pathway is via L-glutamine, the *amide-N* of which contributes N-3 and N-9 to the purine ring. The other nitrogens of the ring are contributed by glycine (N-7) and aspartate (N-1) (Fig. 7–4). The soluble cellular fraction of avian liver is capable of incorporating all of these nitrogen precursors into the purine ring,[267] indicating that uric acid synthesis occurs mainly in the cytosol. It has been found recently that in birds and other uricotelic vertebrates, glutamine synthetase occurs mainly in mitochondria.[301] This is in contrast to the enzyme's localization in the cytosol of mammalian liver cells.[323] The mitochondrial localization of glutamine synthetase in birds therefore suggests that the function of this enzyme in these species is analogous to the function of carbamylphosphate synthetase-I in mammals. A tentative cellular mechanism for ammonia detoxication via uric acid synthesis incorporating this concept is shown in Figure 7–4. Certain details of this mechanism are remarkably similar to those of urea synthesis. Ammonia generated intramitochondrially by the action of glutamate dehydrogenase exits the mitochondrion as the *amide-N* of glutamine in uric acid formation and as a *ureido-N* of citrulline in urea formation. In both mechanisms, glutamate is a key compound in that it may also provide the "carrier skeleton" for ammonia: this is ornithine in citrulline formation and

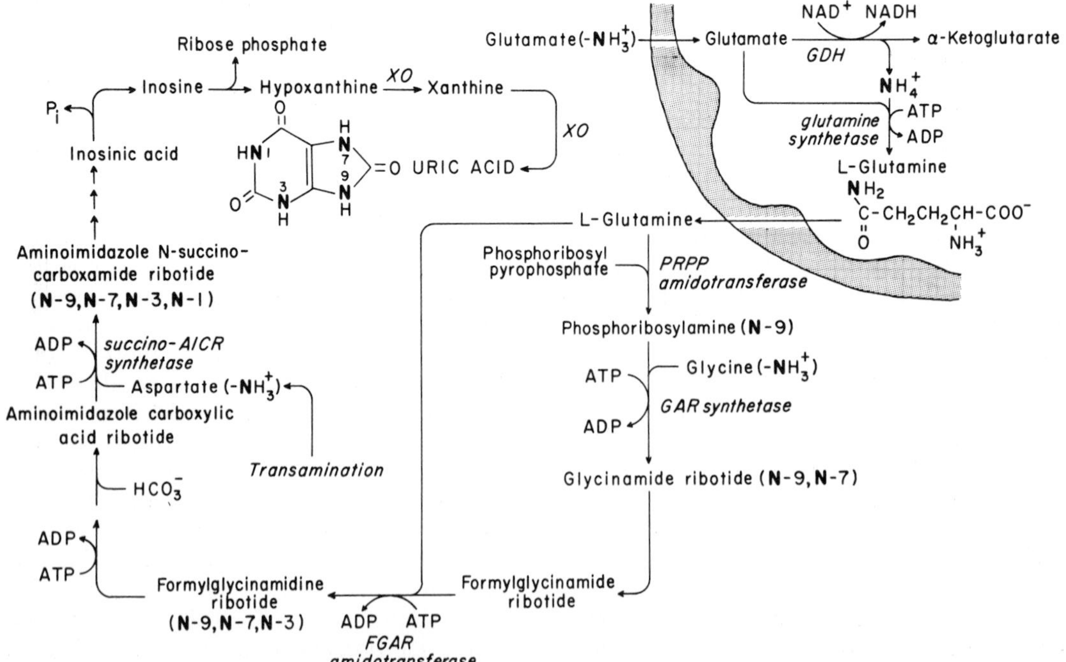

Figure 7–4. Detoxication of ammonia released via transdeamination by conversion to uric acid. The abbreviations are: *PRPP*, phosphoribosyl pyrophosphate; *GAR*, glucinamide ribotide; *FGAR*, formylglycinamide ribotide; *succino-AICR*, aminoimidazole N-succinocarboxamide ribotide; and *XO*, xanthine oxidase.

glutamate itself in glutamine formation. Also, in both mechanisms the α-amino group of aspartate is added as a nitrogen of the end-product in the cytosol. As discussed above, this aspartate could originate either in the mitochondrion (not shown in Figure 7–4) or in the cytosol via transamination of glutamate or other amino acids with oxaloacetate.

In the uric acid pathway, the formation of inosine monophosphate (inosinic acid) represents the formation of the purine ring. Inosine monophosphate is then converted to the nucleoside inosine by the action of 5′-nucleotidase. Inosine is in turn converted to hypoxanthine by nucleoside phosphorylase. Xanthine oxidase acts to convert hypoxanthine to xanthine and xanthine to uric acid. There are two forms of xanthine oxidase: one, the *oxidase*, utilizes oxygen as an electron acceptor and the other, the *dehydrogenase*, utilizes an acceptor such as NAD. In mammalian liver, these two forms may be interconvertible.[277] In avian liver, the dehydrogenase predominates.[253] When xanthine oxidase activity is inhibited in birds, hypoxanthine and xanthine, rather than uric acid, are excreted.[311]

Just as intermediates of the urea cycle have been shown to have a protective effect in experimental ammonia intoxication in mammals, glycine has been shown to exert a similar protective effect in birds,[21] presumably because of its enhancement of uric acid biosynthesis.

General Cellular Mechanisms. In mammals, brain and intestine appear to be the only extrahepatic tissues in which citrulline synthesis may possibly take place for the mitochondrial detoxication of ammonia.[112, 133] The rapid incorporation of exogenous ammonia into glutamine in mammals[78] suggests that glutamine synthetase acts in many tissues as a general cellular mechanism for ammonia detoxication. In keeping with this, the enzyme is widely distributed among the extrahepatic tissues of mammals and also birds (Table 7–4). Glutamine synthetase is an extramitochondrial enzyme in mammalian brain[255] and, in the extrahepatic tissues, it may act mainly for the detoxication of exogenous ammonia. Its cellular localization in tissues other than liver of birds is not known. Glutamine is a major blood amino acid in mammals, and glutamine synthesized by the extrahepatic tissues is transported to the liver where the *amide-N* is ultimately converted to urea. For this, the enzyme glutaminase (Reaction 6) releases the *amide-N* intramitochondrially[147] as ammonia, which is then converted to citrulline.[46, 48] The synthesis of glutamine and its hepatic utilization for urea formation is a process that is normally quantitatively important in mammals because of the high percentage of the daily production of urea that is recycled through the gut as ammonia. This is due to urea breakdown in the lumen by bacterial urease.[305, 322] Ammonia arising in this manner is presumably converted to glutamine prior to its hepatic conversion to urea.[157]

The possible interactions of glutamate dehydrogenase, glutamine synthetase, and glutaminase in the cellular regulation of ammonia levels[134] are shown in Figure 7–5.

TABLE 7–4. Tissue Distribution of Glutamine Synthetase in Rat and Pigeon

Tissue	Enzyme Activity (μmole glutamine/min/g tissue at 37°) RAT[158]	PIGEON[300]
Liver	8.8	28.1
Heart	< 0.5	13.6
Brain:		
Cortex	2.3	–
Cerebrum	–	11.4
Optic lobe	–	10.4
Cerebellum	–	7.5
Kidney	2.3	1.3
Spleen	2.5	5.0
Testis	2.4	1.9
Retina	6.9	–
Muscle	< 0.5	1.5
Gastrointestinal tract:		
Stomach	–	1.5
Gizzard	–	1.3
Small intestine	< 0.5	7.4
Caecum	–	1.1
Large intestine	–	2.3
Pancreas	< 0.5	8.9
Adrenal	< 0.5	–
Lung	< 0.5	1.4
Erythrocytes	< 0.5	–

$$\text{(6)}\quad H_2N{-}\underset{\underset{O}{\|}}{C}{-}CH_2CH_2\underset{\underset{NH_3^+}{|}}{CH}{-}COO^- \rightarrow {}^-OOC{-}CH_2CH_2\underset{\underset{NH_3^+}{|}}{CH}{-}COO^- + NH_4^+$$

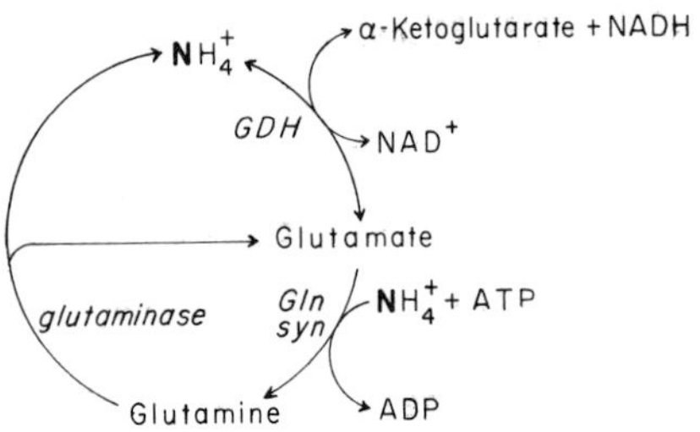

Figure 7-5. General roles of glutamate dehydrogenase (*GDH*), glutamine synthetase (*Gln syn*), and glutaminase in regulating cellular levels of ammonia.

Cellular Mechanisms in Ammonotelic Vertebrates. The major source of ammonia excreted by freshwater and marine fish is blood ammonia[71,103,199] that originates, at least in part, in the liver[201,202] via transdeamination.[127,168] Blood amino acids contribute to excretory ammonia, but to a lesser extent than ammonia itself.[103] Blood ammonia may also be the major source of ammonia excreted by aquatic amphibians.[86,296] The idea that excess ammonia formed in mitochondria of liver cells of aquatic vertebrates by transdeamination simply diffuses into the blood for peripheral elimination poses some basic questions concerning the universality of the cellular basis for ammonia toxicity and of cellular detoxication mechanisms. There is, of course, a quantitative consideration in that the rates of amino acid catabolism in poikilothermic vertebrates, and therefore the demands for ammonia detoxication, are less than in homoiothermic vertebrates. This is certainly true in the alligator, where the clearance of injected amino acids from the blood requires some fourteen times longer than in the rat.[61]

The general rationale for ammonotelism in aquatic vertebrates is that their water flux serves to keep ammonia concentrations low. While this may be true for freshwater species, it certainly does not apply to marine species that are actually faced with the problem of dehydration in a hypertonic environment. It is also an organismal concept in which the cellular mechanisms for ammonia release and detoxication are not considered. Ammonia is generally eliminated extrarenally by aquatic organisms through the gills or, in some cases, the skin. This is true, for example, of freshwater crustaceans,[197] freshwater fish,[94,103,220] some freshwater amphibians,[2,72,86] marine crustaceans,[14,15] cephalopods,[209] and marine fish.[321] Of the vertebrates studied, only in the secondarily aquatic toad *Xenopus laevis* does the kidney function as the main route of ammonia excretion.[5]

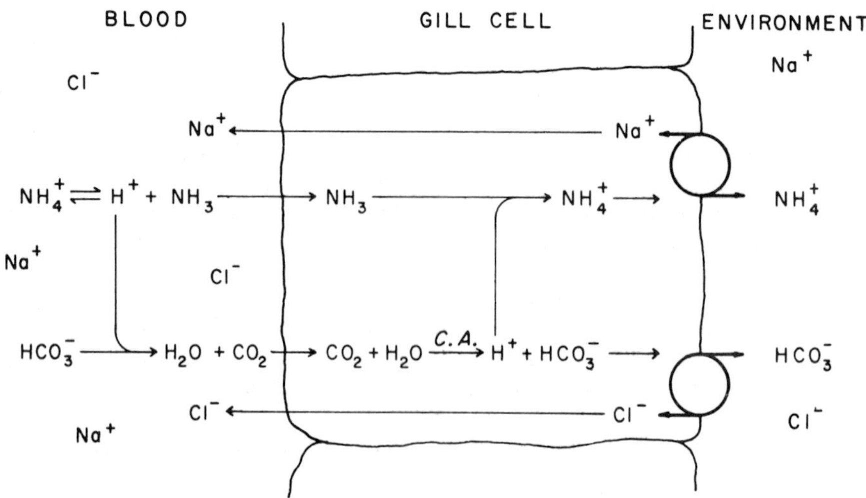

Figure 7-6. Ionic exchange mechanism for NH_4^+ and HCO_3^- excretion by freshwater animals. *C.A.* is the abbreviation for carbonic anhydrase, which catalyzes the reaction $H_2O + CO_2 \rightleftharpoons H^+ + HCO_3^-$. (After Maetz and García-Romeu.[162])

Because the organs through which ammonia is excreted are those concerned with osmotic regulation, Smith[265] suggested that ammonia excretion is part of a regulatory process primarily involved in acid-base and cation balance, a view that has recently received considerable support.[161] In freshwater species, there are two mechanisms involved in the elimination of ammonia. One is by simple diffusion[95] and the other, by the ionic exchange mechanism outline in Figure 7–6. This exchange mechanism could function in freshwater species for the absorption of Na^+ and Cl^- from their environment, a process required to maintain osmotic balance. Sodium ion uptake is not necessarily accompanied by NH_4^+ excretion[257,258] and, in the absence of NH_4^+, H^+ can presumably exchange for Na^+, thereby contributing to acid-base balance.

Even though most aquatic amphibians possess all of the hepatic urea cycle enzymes including the mitochondrial enzymes for citrulline synthesis, carbamylphosphate synthetase-I and ornithine transcarbamylase,[4] ammonia, not urea, is the major excretory product. This is also true of marine and freshwater teleosts[125] which, for some time, were thought not to possess the urea cycle enzymes.[56,57] Because certain aquatic vertebrates are capable of synthesizing urea but do not excrete it as a major end-product, the term *ureogenic* has been proposed[125] to distinguish them from *ureotelic* species that excrete urea as a major end-product. The possible role of urea synthesis in the hepatic detoxication of ammonia in ureogenic vertebrates is not known.[191]

Glutamine synthetase is present in liver and other tissues of marine and freshwater vertebrates[159] and might therefore serve as a mechanism for hepatic ammonia detoxication. Glutamine is added to the blood by the liver, at least in some species.[200,202] Glutaminase, which would be required for the peripheral release of ammonia from glutamine (Fig. 7–5), is present in fish gill[163] but not in the skin of aquatic amphibians, in which this organ serves as the major route for ammonia excretion.[86] Glutamine is a precursor of urinary ammonia in the aquatic toad *Xenopus* but is not as effective a precursor as ammonia itself.[296] There is as yet no good evidence for a central role of glutamine synthesis in ammonia detoxication in ammonotelic species.

Physiological Responses of the Mammalian and Avian Pathways

Gluconeogenesis. An increase in nitrogen catabolism of the amino acids is elicited by starvation, high protein diets, hormones such as glucagon of the pancreas and the glucocorticoids of the adrenal cortex, and diseased states such as diabetes mellitus. The general response of the enzymes associated with the detoxication pathways to this condition of gluconeogenesis is consistent with their adaptation to the higher levels of ammonia generated by transdeamination. During gluconeogenesis, especially that induced by high protein diets, the transaminases,[90,266] glutamate dehydrogenase,[312] glutaminase,[135] and the enzymes of the urea cycle[242,243,244] all show marked increases in mammalian liver. That the increased levels of the urea cycle enzymes confer a greater capacity for detoxication *in vivo* is shown by the increased tolerance of animals adapted to high protein diets to experimental ammonia intoxication.[313] Physiological states in which gluconeogenesis is decreased, such as those following adrenalectomy and growth hormone treatment, result in a decrease in the activities of the urea cycle enzymes.[172]

During gluconeogenesis in birds, induced by either high protein diets or starvation, there is a marked increase in both hepatic glutamine synthetase and glutamine phosphoribosylpyrophosphate amidotransferase activity,[135] the latter being the first enzyme unique to purine biosynthesis. In the rat, these two enzymes remain the same during gluconeogenesis, and their increase in the bird is consistent with their function in hepatic ammonia detoxication as outlined in Figure 7–4. Xanthine oxidase, the terminal enzyme in uric acid biosynthesis, is decreased in birds[150] as well as in mammals[234] fed protein-deficient diets. Its activity increases following re-feeding of protein. During starvation in birds, xanthine oxidase is conserved but does not show a marked increase in activity.[150] The response of xanthine oxidase in birds is thus somewhat comparable to that of arginase, the terminal enzyme in urea biosynthesis, in mammals.[243]

Metabolic Acidosis. The overall reactions for either urea or uric acid synthesis result in a net formation of H^+ since, during these processes, either NH_4^+ or $—NH_3^+$ are converted to $—NH_2$ or $=NH$ groups in the final

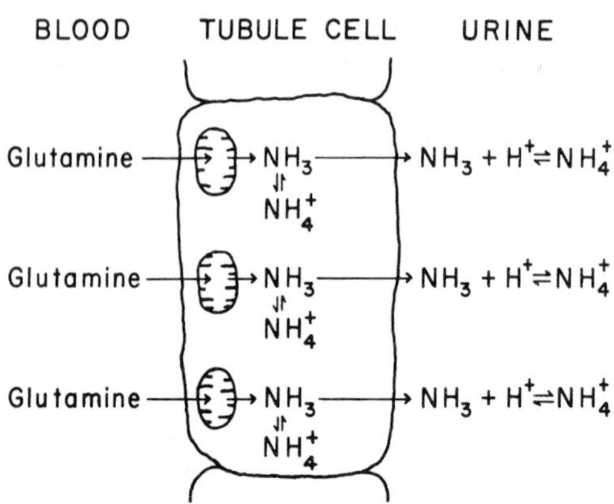

Figure 7–7. Excretion of ammonium ion by the mammalian kidney during metabolic acidosis.

end-products. During metabolic acidosis, the synthesis of urea in mammals, and presumably of uric acid in birds, is decreased. The decreased synthesis of urea by mammalian liver is due to a decreased utilization of glutamine by this organ.[157] There is an increased renal excretion of ammonium ion during metabolic acidosis, the major portion of which is formed in the kidney by the action of mitochondrial glutaminase on glutamine extracted from the blood (Fig. 7–7). Ammonia formed in this manner in the kidney neutralizes the excess protons of acidosis either directly by a "diffusion trapping" mechanism[206, 207] or indirectly through its effect as the ammonium ion on renal cation exchange systems.[170] The decreased utilization of glutamine by the liver in metabolic acidosis therefore serves to divert this compound to the kidney, where it is utilized in renal ammoniagenesis.[106]

Nutritional Function of the Detoxication Pathways

Arginine Biosynthesis. In addition to its function in ammonia detoxication in ureotelic vertebrates, the urea cycle (hereafter referred to as the arginine-urea pathway) also serves as a source of arginine which is required for protein synthesis[75, 76] and in other reactions such as those of phosphagen biosynthesis.[186, 304] Arginine formation is the main function of the pathway in microorganisms, as indicated by its regulation through the availability of arginine.[298] In animal species lacking a functional arginine-urea pathway, such as birds[285, 286] and insects,[224, 225] arginine is required in the diet. Although mammals can synthesize arginine, some species, including the rat[23] and rabbit,[174] nevertheless show a dietary requirement for this amino acid for optimal growth. During postnatal growth of the rat, the greater reliance by the liver on dietary (exogenous) arginine as opposed to biosynthetic (endogenous) arginine is reflected in the pattern of incorporation of arginine from these two sources into protein. The ratio of incorporation of exogenous/endogenous arginine is about three in the neonate and declines to one in the adult animal, where dietary arginine is not required for nitrogen balance.[75] In the frog *Rana catesbiana*, there is some question as to whether biosynthetic arginine ever meets the complete metabolic requirement of the liver for this amino acid, since the incorporation ratio of exogenous/endogenous arginine is greater than one throughout its life cycle.[76] The arginine-urea pathway in vertebrate liver thus appears to have become so specialized for ammonia detoxication during the evolution of some species that there is little reserve capacity for arginine biosynthesis. This is true, for example, of the rat but perhaps not of man.[139]

The liver of mammals is the main site of urea and therefore of arginine biosynthesis. Other tissues, such as kidney[27] and brain,[216] can, however, convert citrulline to arginine. This conversion also takes place in birds and insects since their dietary requirement for arginine can be replaced by L-citrulline. Recent studies with the perfused mammalian liver indicate that this organ may supply citrulline to the extrahepatic tissues, especially kidney, for conversion to arginine for use in protein synthesis.[77] This organ interaction is of interest because of a metabolic

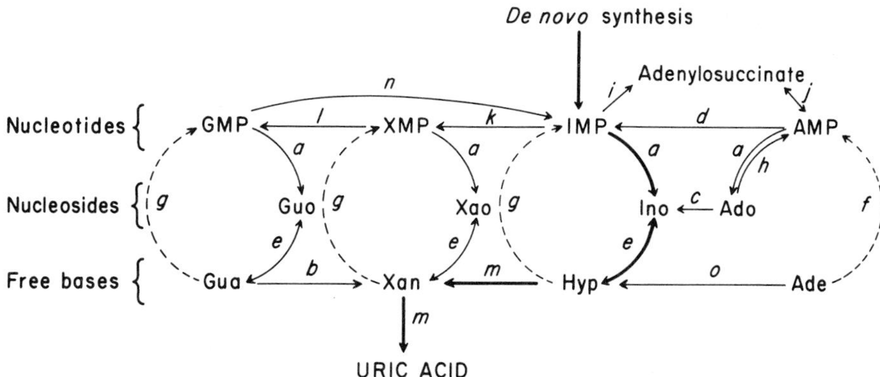

Figure 7–8. Interconversions of the nucleotides, nucleosides, and free bases in animals. The "shunt pathway" is indicated by heavy arrows and the "salvage pathway," by dashed arrows. The enzymes mediating the interconversions are: *a*, 5′-nucleotidases; *b*, guanine deaminase (guanase); *c*, adenosine deaminase; *d*, AMP deaminase; *e*, nucleoside phosphorylase; *f*, adenine phosphoribosyltransferase; *g*, hypoxanthine-guanine phosphoribosyltransferase; *h*, adenosine kinase; *i*, adenylosuccinate synthetase; *j*, adenylosuccinate lyase; *k*, IMP dehydrogenase; *l*, XMP aminase; *m*, xanthine oxidase; *n*, GMP reductase; and *o*, adenine deaminase (adenase).

defect in man, citrullinaemia, in which there is an increased urinary excretion of citrulline and its *N*-acetyl derivative.[278] The excretion of these compounds suggests that there is a decreased utilization of citrulline by the kidney for arginine synthesis. Argininosuccinuria is another metabolic defect in the arginine-urea pathway in man, in which the ability to synthesize urea appears not to be impaired.[314] In both conditions the extrahepatic tissues and especially the brain, since mental retardation is a symptom, may be deprived of arginine.

Nucleotide Biosynthesis. Just as the arginine-urea pathway serves both an excretory and nutritional function in ureotelic vertebrates, so does the purine pathway in uricotelic vertebrates. The purine pathway is also present in ureotelic vertebrates and functions for the synthesis of adenine and guanine nucleotides required in nucleic acid turnover. The synthesis of the nucleotides from inosine monophosphate and their subsequent interconversions are shown in Figure 7–8. In bacteria[188] and mammals,[45] where the purine pathway serves mainly a nutritional function, adenosine and guanosine monophosphates are potent inhibitors of the first enzyme of the pathway, glutamine phosphoribosylpyrophosphate amidotransferase (Fig. 7–4). Because the pathway is directed toward the synthesis of these nucleotides, this represents end-product inhibition. The avian enzyme is less sensitive to adenosine monophosphate inhibition than is the mammalian enzyme,[135] indicating that a modification of the regulation of the purine pathway has taken place so the pathway may serve its excretory function in uricotelic species. In the latter, far more inosine monophosphate is synthesized than is required for conversion to adenosine and guanosine monophosphates,[81] and the excess represents that amount formed during ammonia detoxication. This excess is converted to uric acid by the "shunt pathway," which consists of 5′-nucleotidase, nucleoside phosphorylase and xanthine oxidase (Fig. 7–8). This shunt pathway is present in man and is responsible for the excretion of some uric acid as a minor excretory product. Its function in man has been considered vestigial and the cause of excessive uric acid production in gouty conditions.[137, 276]

In mammals, the liver is also the main site of purine biosynthesis, although other tissues such as brain may have a limited capacity for this.[124] The extrahepatic tissues are supplied purines by the liver in the form of the free bases.[211] This supply is critical to those tissues such as blood whose cells (erythrocytes) cannot synthesize purines *de novo*.[156] Adenine and hypoxanthine-guanine phosphoribosyltransferases are required for the utilization of the free bases[181] and constitute the "salvage pathway" in purine metabolism. These enzymes are found in most extrahepatic tissues.[180] In man, the Lesch-Nyhan syndrome is a metabolic defect associated with the salvage pathway,[254] specifically the hypoxanthine-guanine enzyme. This syndrome is characterized by an overproduction of uric acid which may be due to defective regulation of the purine pathway.[138, 268]

EVOLUTION OF THE DETOXICATION PATHWAYS

ORIGIN

Establishment of Nutritional Pathways. According to current hypotheses for the origin of life on Earth, the first metabolizing systems arose in a nutritionally complete medium whose components were products of a prior chemical evolution.[192] As outlined by Horowitz,[122] metabolic pathways became established as specific components of this nutritionally complete medium—the "primitive soup"—were systematically depleted by biological activity. The evolution of complete metabolic sequences proceeded from what are now the terminal enzymes of a pathway to its initial enzymes. This sequence may not apply to all metabolic pathways, but it is especially attractive as an explanation for the origin of those nutritional pathways supplying small molecule precursors of proteins and nucleic acids which the arginine-urea and purine pathways represent. One can, for example, envision the origin of the pathway for arginine synthesis as follows: when arginine was depleted from the environment in which life arose, those metabolizing systems possessing even limited catalytic activity for the conversion of citrulline to arginine would have a selective advantage over those systems lacking this activity. Similarly, when citrulline was then depleted, selection would favor those systems capable of citrulline formation from ornithine. One can also envision a similar sequence for establishing the interconversions of the nucleotides and eventually the purine pathway itself.

The Detoxication Function. There is, at present, no documented case among animals in which the arginine portion of the arginine-urea pathway functions solely for the synthesis of arginine. As we shall see in a later discussion of the lower vertebrates, the synthesis of urea itself may have originally been selected for because of the function of this compound as an osmotic component. Nevertheless, there are no real clues to the mechanism by which the original nutritional pathway was usurped for ammonia detoxication. One of the first requirements to be met for ammonia detoxication was the acquisition of a carbamylphosphate synthetase with a high affinity for ammonia. In vertebrates, there are two carbamylphosphate synthetases: Enzyme-I, which is specific for the arginine-urea pathway, and Enzyme-II, which is specific for pyrimidine biosynthesis.[132] The former is a mitochondrial enzyme that utilizes low concentrations of ammonia and requires *N*-acetyl-L-glutamate as a co-factor; the latter is an extramitochondrial enzyme that utilizes L-glutamine and does not require a co-factor. In some procaryotic cells, a single glutamine-utilizing enzyme is present and functions in both arginine and pyrimidine biosynthesis. Because of the subunit structure of this microbial enzyme, it has been suggested that carbamylphosphate synthetase-II arose during evolution by the association of an ammonia-utilizing carbamylphosphate synthetase and glutaminase,[294] both of which are mitochondrial in higher animals.

Some invertebrate animals contain, in addition to a carbamylphosphate synthetase-I-like enzyme, an enzyme that is distinct from either vertebrate carbamylphosphate synthetase. This mitochondrial enzyme, first characterized in land snails[290] and later shown also to be present in a land planarian and an earthworm,[291] utilizes L-glutamine but, unlike carbamylphosphate synthetase-II, shows an absolute requirement for *N*-acetyl-glutamate. There is indirect evidence that an enzyme very similar to the invertebrate enzyme may be present in elasmobranchs.[310] Because of its unique combination of properties, the invertebrate enzyme might possibly represent a primitive animal type that could have given rise to vertebrate Enzyme-II by losing its co-factor requirement during evolution, or to Enzyme-I by modifying its substrate affinity so that low concentrations of ammonia would substitute for glutamine.

In the evolution of the arginine-urea pathway, the citrulline-forming system, of which carbamylphosphate synthetase is a part, appears to be the more labile. This system is absent from cells and organisms that do not have the ability to synthesize arginine *de novo* but can utilize citrulline. This is true, for example, in some protozoa,[143] insects,[126,224,225] birds,[285,286] and mammalian cells in tissue culture that have undergone dedifferentiation.[245] The limited occurrence of the citrulline-forming system in the extrahepatic tissues of mammals also suggests that it is the more labile during tissue differentiation. Cells and tissues lacking the ability to synthesize citrulline can nevertheless provide carbamylphosphate for pyrimidine formation,[132] so the major defect is associated with the absence of carbamylphosphate synthetase-I. Birds also lack the ability to synthesize ornithine.[286] The other component of the citrulline-forming system, ornithine transcarbamylase, also exists in at least two forms. One, a catabolic enzyme, functions in conjunction with carbamate kinase in microorganisms for citrulline catabolism[214] and the other, an anabolic enzyme, for citrulline synthesis. The latter appears to be the only form of the enzyme found in animal tissues. Because the citrulline-forming system is mitochondrial in higher animals, the origin and stability of this system during evolution is of interest relative to the proposed symbiotic origin of mitochondria.[165]

A second requirement for the detoxication of ammonia via the arginine-urea pathway is that the arginine formed be converted to urea by arginase. The occurrence of arginase in animal tissues does not, however, necessarily indicate the capacity for urea synthesis *de novo*, since this enzyme also functions during gluconeogenesis in the conversion of arginine to glutamate or proline, both of which are oxidizable substrates. The separateness of this catabolic function is indicated by the retention and control of arginase by mammalian liver cells in tissue culture that lose the ability to synthesize arginine *de novo*.[82] Arginase also occurs in many species incapable of arginine biosynthesis. These include uricotelic vertebrates[31,33,178] and several invertebrate species.[115,224] The function of the enzyme in insects in the conversion of arginine to proline, a substrate for flight muscle, has been demonstrated.[225] It is thus probable that the urea appearing as a minor component in the excreta of insects[217] and of other species incapable of arginine synthesis arises in part from the action of arginase on exogenous (dietary) arginine or arginine formed during protein catabolism. Arginase functions primarily as a catabolic enzyme in procaryotic cells,[213] so it seems likely that the enzyme occurred in cells and tissues prior to its integration into the arginine-urea pathway. Although there are several different molecular forms of arginase in animals,[7,120,179,208,226] thus far there has not been an association of a specific form with a specific physiological function.

Even fewer clues are available as to the evolutionary mechanism whereby the purine pathway was usurped for a detoxication function. As outlined in Figure 7–4, the function of glutamine synthetase in uricotelic vertebrates is analogous to the function of carbamylphosphate synthetase-I in ureotelic vertebrates, in that both enzymes initially bind the ammonia released during transdeamination. An ammonia-utilizing system for phosphoribosylamine formation has been described in avian liver, but the physiological role of this enzyme is not known.[116,227] Glutamine synthetase is present in procaryotic cells, and functions to assimilate ammonia.[274] As discussed earlier, this enzyme is also present in most tissues of higher vertebrates and has a general function in controlling ammonia levels, as outlined in Figure 7–5. Glutamine synthetase was thus predisposed for its proposed role in ammonia detoxication in uricotelic species and required only association with the mitochondrion to function as shown in Figure 7–4. L-Glutamine, the product of glutamine synthetase, is required for protein synthesis, which again illustrates the close relationship between nutrition and excretion in animals.

Because of the ubiquitous distribution of the enzymes converting inosine monophosphate directly to uric acid—the shunt pathway (Fig. 7–8)—these too provide few clues as to the origin of the excretory function of the purine pathway, and it seems that the major modifications in the basic pathway have been regulatory ones. Glutamine synthetase responds to gluconeogenesis in birds but not mammals,[135] and the sensitivity of avian glutamine phosphoribosylpyrophosphate amidotransferase to feedback inhibition by the nucleoside monophosphates, a labile property during purification of the enzyme,[235] is less than that of the mammalian enzyme. It has been suggested that the salvage pathway for the purines in uricotelic species is quantitatively different from that in mammals,[81] thereby contributing to the "over-production" of uric acid such as is seen in the Lesch-Nyhan syndrome in man. Some differences may also exist in the nature and regulation of the xanthine oxidases of avian and mammalian liver.[253]

One evolutionary event considered critical for excreting uric acid as an end-product was the loss of the enzyme uricase.[137] This enzyme catalyzes the first step in uric acid degradation by animals, as outlined in Figure 7–9.

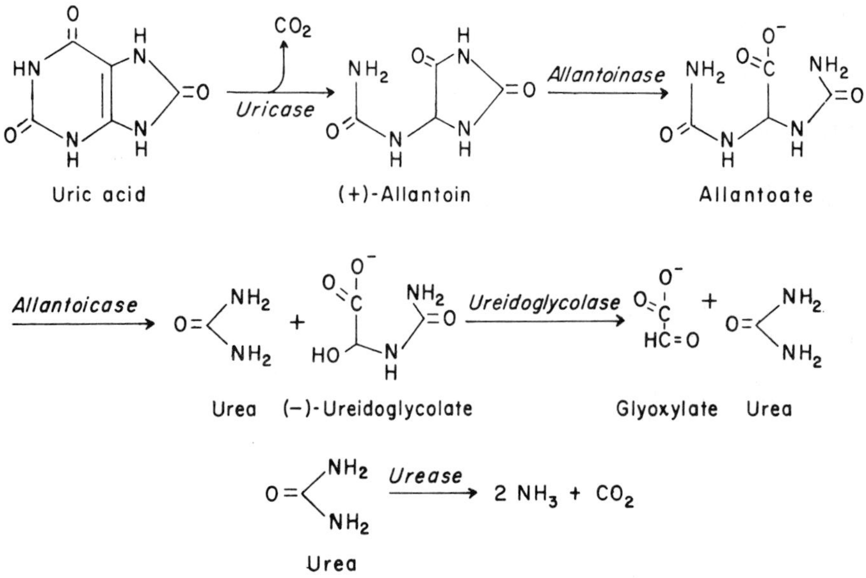

Uric acid —Uricase (CO_2)→ (+)-Allantoin —Allantoinase→ Allantoate

—Allantoicase→ Urea + (−)-Ureidoglycolate —Ureidoglycolase→ Glyoxylate + Urea

Urea —Urease→ $2\ NH_3 + CO_2$

Figure 7–9. Degradation of uric acid by animals.

The enzymes of the degradative sequence are widely distributed in invertebrates[38] and lower vertebrates.[102, 293] According to the scheme of Florkin and Duchâteau,[87] there has been a systematic evolutionary deletion of the enzymes of uric acid degradation starting with urease and culminating in the loss of all enzymes, including uricase, in certain uricotelic species and also in man and the higher primates. Because only a few species were examined in the original survey of the uricolytic enzymes in animals,[87] the distribution of these enzymes, especially in invertebrate species, is now known to differ from that originally proposed.[38] However, some uric acid is presumably formed in all animals due to nucleic acid turnover and to the catabolism of dietary nucleic acids. Depending upon which of the enzymes of the uricolytic sequence are present in a given species, either ammonia, urea, allantoin, or uric acid itself appears as a minor component in the excreta and represents the nitrogenous end-product of nucleic acid catabolism. The loss of uricase in the evolutionary line leading to man and the higher apes has been considered by some to be fortuitous, since the resulting hyperuricemia could have stimulated brain development; the limitations of this hypothesis have been discussed.[137] In any event, the absence of uricase from man is a contributing factor to gouty conditions, since in those animals possessing uricase, inhibition of the enzyme *in vivo* leads to hyperuricemia as well as an increased excretion of uric acid.[131] Because of the presence of the shunt pathway in man and the absence of uricase, not all the oxypurines excreted are derived from nucleic acid catabolism. Some fraction of these appear to arise by the same mechanism as in birds.[81]

As with the arginine-urea pathway, cells that lose the capacity for purine synthesis *de novo* nevertheless in some cases retain a functional portion of the purine pathway. Mature mammalian erythrocytes, for example, that can no longer synthesize purines, can convert the intermediate 5-amino-4-carboxamide ribotide to purines,[156] and this also appears to be true in the tunicate *Molgula manhattensis*.[189] The biochemical lesion in the purine pathway in ciliate protozoans that require purines for growth involves their inability to synthesize glycinamide ribotide.[297]

Evolution of the Pathways in Invertebrates

The major excretory products of invertebrate animals are ammonia and purines. Aquatic species are, in general, ammonotelic and terrestrial species are purinotelic. Uric acid is usually the major purine excreted although other purines, such as xanthine and guanine, may account for a high percentage of the total excretory nitrogen.[270] In some arachnids, guanine accounts for more than 90 per cent of the excretory nitrogen. Degradation products of purines such as allantoin may also account for a high percentage of the excretory nitrogen of some insects[217] but not others.[288] The excretion of urea as a major end-product, although rare in invertebrate animals, does occur in the land planarian *Bipalium kewense*[37] and the earthworm *Lumbricus terrestris* during fasting.[18]

Certain terrestrial invertebrates exhibit a unique type of ammonotelism in excreting gaseous ammonia rather than the ammonium ion. The amount of NH_3 excreted by the terrestrial isopods *Oniscus asellus* and *Porcellio scaber* can account for the total nitrogen ingested by these species[315] and is comparable to the amount of nitrogen excreted as NH_4^+ by freshwater and marine crustaceans.[15] Ammonia gas release also occurs in terrestrial gastropods[269] but does not constitute their major end-product.[270] The direct release of NH_3 could very well represent the most primitive method for excreting nitrogen in the terrestrial environment. When coupled with the release of CO_2, this does not require a drastic alteration in acid-base balance, as can be illustrated by an inspection of the exchange of NH_3 and CO_2 between the blood and gill cells of aquatic organisms outlined in Figure 7–6. The conversion of HCO_3^- to CO_2 in the blood requires an uptake of a proton, causing a local alkalinization which serves to dissociate NH_4^+ to form NH_3. As outlined in this figure, CO_2 and NH_3 penetrate the gill cell to be reconverted to HCO_3^- and NH_4^+. In terrestrial organisms excreting NH_3, both would be lost to the atmosphere. The only difference between the excretion of NH_3 and CO_2 in the terrestrial environment and the excretion of NH_4^+ and HCO_3^- in the aquatic environment is the appearance of the components of water in the latter end-products.

The Arginine-Urea Pathway. Most protozoans do not have the capacity for arginine synthesis *de novo*, as evidenced by their nutritional requirement for L-arginine.[297] The capacity for arginine biosynthesis did, however, continue into the lower metazoa, as evidenced by the synthesis *de novo* of both arginine and urea by the land planarian *Bipalium kewense*.[36] Enzymes of the arginine-urea pathway also occur in other flatworms, and there is evidence that these may function *in vivo* in some species for arginine and/or urea synthesis.[38]

The main phylogenetic line of the invertebrates, from which the flatworms branch, diverges into two main branches. One branch leads via the echinoderms and protochordates to the vertebrates, and the other leads to the arthropods. Since the echinoderm-protochordate line gave rise to the vertebrates, the arginine-urea pathway must have continued into and through this line. No representative taxons along this phylogenetic route have, however, been investigated for their ability to synthesize and hydrolyze arginine. The arginine-urea pathway also persisted into the line leading to the arthropods and appears in terrestrial gastropods, where it functions *inter alia* for the synthesis of protein arginine.[40, 292] The nutritional requirements for amino acids by these organisms have not been established, but their rate of arginine synthesis relative to that of alanine, aspartate, and glutamate (amino acids not normally required by animals) indicates that the pathway is nutritionally significant. The land snails *Otala lactea, Helix aspersa*, and *Strophocheilus oblongus* are the only invertebrates in which the tissue distribution of the arginine-urea pathway enzymes have thus far been examined.[41, 292] Unlike vertebrates, where a high degree of tissue differentiation serves to locate the pathway mainly in liver, thereby creating a dependence of the extrahepatic tissues on this tissue for arginine,[77] there is little tissue differentiation with respect to the arginine-urea pathway in the snails. This greater tissue independence is perhaps indicative of the relative inefficiency of operation of their primitive circulatory system in tissue interactions.

Although the arginase present in gastropod tissues[97] is capable of hydrolyzing biosynthetic arginine[271] to urea, it seems unlikely that the main function of the arginine-urea pathway in these organisms is an excretory one. The mitochondrial release of α-amino-N by transdeamination is questionable, and their major mitochondrial carbamylphosphate synthetase utilizes L-glutamine as a substrate but does not utilize low concentrations of ammonia. Urea formed by the action of arginase in these snails is hydrolyzed to ammonia and carbon dioxide by urease, an enzyme also present in most tissues of some species[269] but apparently not others.[292] Urea could not be considered an end-product of the pathway under these circumstances. The turnover of urea by the snails, which accounts for some of the NH_3 released by certain species, may be related to their unique acid-base balance prerequisite for the deposition and maintenance of a calcium carbonate shell.[38,42] There is also some evidence that urea turnover in certain species of terrestrial (and possibly amphibious) gastropods is so regulated as to allow this compound to be accumulated in the blood for purposes of osmotic water retention during periods of desiccation.[43] Irrespective of the possible alternate functions of the arginine-urea pathway in gastropod molluscs, the occurrence of the pathway in these organisms is, in itself, a unique situation since they are thus far the only purinotelic species[270] examined that have retained the capacity for arginine synthesis *de novo.*

The arginine-urea pathway continued along the invertebrate line of evolution and is present in oligochaete annelids where, in addition to its nutritional function, it also appears to function in ammonia detoxication, at least during certain physiological states.[18] The cellular mechanism for ammonia detoxication in the earthworm *Lumbricus terrestris* is somewhat similar to that described in vertebrates, since a mitochondrial glutamate dehydrogenase is present in this species[43] as is a carbamylphosphate synthetase-I-like enzyme.[17,291] Both are necessary components for the release of ammonia from amino acids by transdeamination and its subsequent detoxication through urea formation. The detoxication function of the arginine-urea pathway described in oligochaete annelids, specifically *Lumbricus terrestris*, may not apply to marine polychetes since, as previously discussed, amino acid catabolism in the latter may involve the purine nucleotide cycle[16] (Fig. 7–2).

The arginine-urea pathway appears to have been lost in the evolutionary line leading from the Annelida to the Arthropoda, since neither crustaceans[62,115,324] nor insect arthropods[123] have the ability to synthesize arginine *de novo* (even though some of the latter species have retained a portion of the pathway). The onychophorans, a transitional group between the annelids and arthropods, have not been investigated in this respect, although it has been suggested that they are uricotelic like the majority of terrestrial insects.[164]

During early studies on the comparative aspects of animal nitrogen metabolism, the occurrence of the arginine-urea pathway in invertebrates was somewhat questionable,[3] but it is now clear that the pathway does occur in representatives of at least three phyla. This gives an evolutionary continuity to the pathway from its basic nutritional function in procaryotic cells to its more specialized functions in animals. The persistence of the pathway through the evolution of the invertebrates is not surprising since, in addition to their requirement for arginine for protein synthesis, this amino acid, as phosphoarginine, is their major muscle phosphagen[84] and is therefore required for movement in feeding, reproduction, and other activities. Considerable attention has been given the distribution of phosphoarginine relative to that of phosphocreatine in the echinoderm-protochordate line of evolution[230,275] but, unfortunately, no information about arginine synthesis *de novo* in this line is available.

The synthesis of excretory urea in certain land planaria and earthworms via the arginine-urea pathway indicates that the usurpation of the basic nutritional pathway for ammonia detoxication first took place in invertebrate animals. The details of the operation of the pathway in its excretory capacity in these or other invertebrates is not, however, sufficiently well-known to formulate a mechanism for this basic metabolic adaptation. Certain associations of the pathway with specific physiological responses in invertebrates are nevertheless indicated. The possible function of the pathway in the acid-base balance of molluscs that form a calcium carbonate shell has been referred to, and Needham[185] has also suggested that urea formation as well as calcium carbonate are involved in the acid-base balance of earthworms. In this context, it should be recalled that one of the physiological conditions to which the pathway responds in mammals is an abnormal acid-base balance. Osmotic water retention may have been a primary factor in the selection for a high rate of urea synthesis in primitive vertebrates, and we have seen that this may also occur in certain terrestrial gastropods. The adaptations of the arginine-urea pathway seen in extant vertebrates may thus all have evolutionary precedence in the Invertebrata.

The Purine Pathway. The synthesis *de novo* of the purine ring has been documented in only a few species of invertebrates. The small molecule precursors of uric acid in the land snails *Helix* and *Otala* are the same as those in birds[107, 130, 151] and some of these precursors are utilized for uric acid synthesis in insects.[70, 117, 169] No enzymes mediating the incorporation of these precursors have been characterized, but the pattern of their incorporation indicates that the pathway is the same in these two groups of purinotelic invertebrates as in other species capable of synthesizing the purine ring. Little is known, however, of the pathway's function in either a nutritional or excretory capacity. Purines or their derivatives are common nutritional requirements of protozoa,[297] and in some the specific defect responsible for this has been identified. The failure of the tunicate *Molgula* to incorporate precursors of the purine ring indicates a purine requirement in this species. Even though the purine pathway is present in insects, some, such as the housefly, show specific dietary requirements for adenine and guanine[1] or other nucleic acid components.[123, 190] Insects can synthesize the major small molecule precursors of the purine ring (serine, glycine, and glutamine[59]) and possess such ancillary enzymes as those for formate activation,[326] so these purine requirements must reflect a failure of the pathway to provide adequate amounts of guanine and adenine nucleotides.

Glutamine synthetase, which functions specifically in ammonia detoxication via the purine pathway as well as more generally in the regulation of tissue levels of ammonia, is present in insects and is similar to the vertebrate enzyme in several of its properties.[153] Glutamine synthesis also takes place *in vivo* in terrestrial snails.[40, 140] The cellular localization of glutamine synthetase in these purinotelic invertebrates relative to its function in ammonia detoxication has not been determined. The high concentrations of free glutamine in insects and snails[40, 49, 152] may be related to its role as a purine precursor. The high concentration of glutamine in non-purinotelic invertebrates[146] may, on the other hand, be related to its more general role in tissue ammonia metabolism.

Adenase, guanase, xanthine oxidase, and specific purine nucleoside and nucleotide deaminases occur in invertebrates and presumably function in the interconversions of the purines and their ribotides as outlined in Figure 7–8. Hypoxanthine-guanine phosphoribosyltransferase has also been detected in several tissues of the snail *Otala*.[38] The xanthine oxidase present in insects is the dehydrogenase type[198] as it is in birds, and the different isozymes of insect xanthine oxidase may be involved in the regulation of the purine pathway.[261]

A unique aspect of invertebrate purine metabolism is the excretion of large amounts of guanine by arachnids.[111, 215] As we have seen, uric acid or its degradation products appear as minor components of the excreta in most animals because they represent end-products of nucleic acid catabolism. The flow of nitrogen in this case is through guanine and adenine nucleotides (Fig. 7–8). In uricotelic species, the excess inosine monophosphate formed during ammonia detoxication is converted directly to uric acid via the shunt pathway. The excretion of guanine as an end-product by arachnids thus represents a different kind of purinotelism than is found in vertebrates, assuming that the basic interconversions are the same in both groups. The routing of inosine monophosphate through guanosine monophosphate to form guanine as an end-product could have an adaptive value, since guanine contains the highest percentage of nitrogen of all the purines. The conversion of xanthosine monophosphate to guanosine monophosphate could serve to "detoxify," directly or indirectly, an extra ammonia. Adenine is not known to be formed as a major nitrogenous end-product in animals.

Storage "Excretion" of Purines. Many invertebrates, including gastropod molluscs,[79, 80, 154] crustaceans,[98] insects,[12, 113, 177] and urochordates,[105, 189] accumulate high concentrations of purines in their tissues. This process appears to be directed toward the storage of purines for subsequent metabolic utilization since, in some species, there is a temporal variation in the amounts of tissue purines present. Just as with gout in man, the accumulation of purines in invertebrate tissues may reach pathological proportions. The symbionts present in fat body tissue of the cockroach *Periplaneta americana* degrade uric acid, thereby preventing an excessive accumulation.[73] In the wax moth *Galleria mellonella*, uric acid is converted to its ribotide to facilitate its removal from the tissues.[148]

Evolution of the Excretory Function in Vertebrates

Unlike their invertebrate ancestors, the vertebrates have widely exploited the

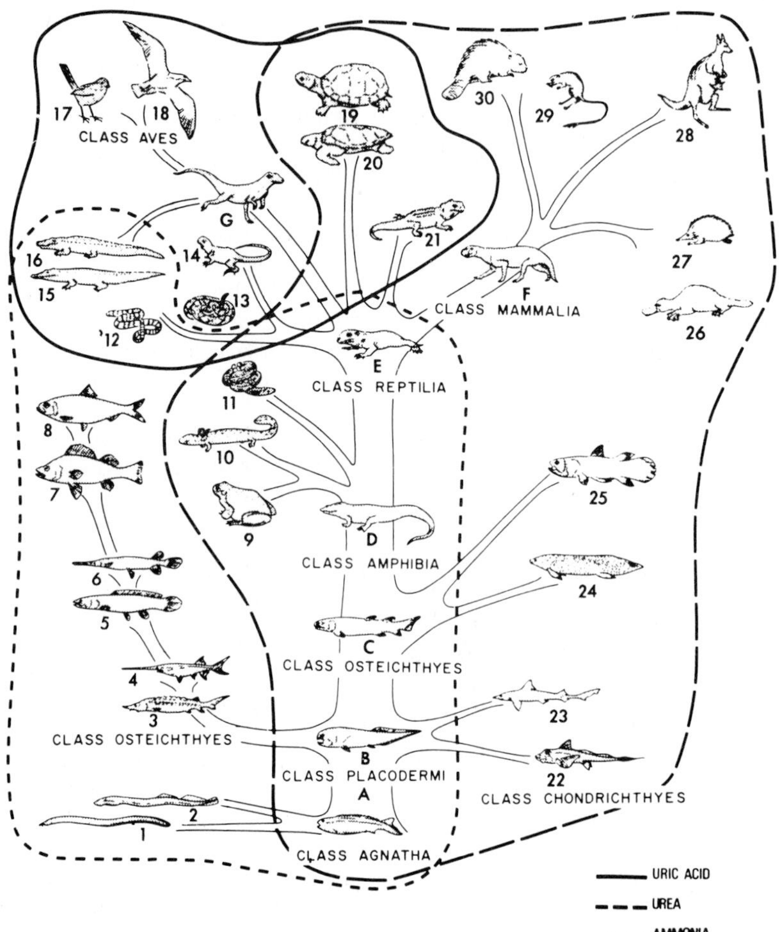

Figure 7-10. Evolution of nitrogen excretion in the Vertebrata. Capital letters denote extinct transitional forms, and numbers denote representative extant species. A, ostracoderm—oldest fossil vertebrate; B, arthrodire—early jawed fish; C, coelacanth—crossopterygian ancestor of lungfish and amphibians; D, labyrinthodont amphibian; E, cotylosaurian—stem reptile; F, therapsid—early mammalian relative of the reptiles; G, archosaurian—ruling reptiles; 1, 2, cyclostomes—the lamprey *Petromyzon* and hagfish *Myxine*; 3, 4, chondrosteans—the sturgeon *Scaphyrhynchus* and paddlefish *Polydon*; 5, 6, holosteans—the bowfin *Amia* and gar *Lepidosteus;* 7, 8, teleosts—perch and herring; 9, anuran—the frog *Rana*; 10, urodele—the mudpuppy *Necturus*; 11, apodan—the legless *Chtonerpton*; 12, 13, snakes—representative serpentine reptiles; 14, lizard—representative lacertilid reptile; 15, 16, crocodilid reptiles—the crocodile *Crocodylus* and *Alligator*; 17, 18, passeriform and charadriiform birds; 19, 20, chelonid reptiles; 21, rhynchocephalid reptile—the tuatara *Sphenodon*; 22, holocephalan—the ratfish *Chimaera*; 23, elasmobranch—the shark *Squalus*; 24, dipnoan—the lungfish *Neoceratodus*; 25, coelacanth—*Latimeria*; 26, 27, protherian mammals—the duckbill platypus *Ornithorhynchus* and anteater *Echnida*; 28, marsupial mammal—the kangaroo *Macropus*; 29, 30, true mammals—the kangaroo rat *Dipodomys* and beaver *Castor*. (Courtesy of B. Schmidt-Nielsen and W. C. Mackey.[247])

TABLE 7–5. Enzymes of the Arginine-Urea Pathway in Representative Animal Species

Species	*Enzyme Activity (μMole product/g tissue/hr)**				
	CARBAMYL-PHOSPHATE SYNTHETASE	ORNITHINE TRANS-CARBAMYLASE	ARGININO-SUCCINATE SYNTHETASE	ARGININO-SUCCINATE LYASE	ARGINASE
Invertebrates					
Mollusca:					
Helix aspersa (T)†	<0.1	80	0.3	7	2200–14400
Otala lactea (T)	0.1	200	0.1	10	2000–45000
Strophocheilus oblongus (T)	0.5	190	0.2	17	30
Annelids:					
Lumbricus terrestris (T, fasted)	5	2300	6	50	2000
Vertebrates					
Elasmobranchs:					
Scylliorhinus canicula (M)	1	600	8	25	1800
Dasyatis americana (M)	7	14400	20‡		34900
Urolophus jamaicensis (M)	5	8500	17		14000
Potamotrygon sp. (FW)	0.4	1600	10		4300
Teleosts:					
Anguilla anguilli (FW → M)	0.2	6	1.6	1.8	4600
Salmo salar (M → FW)	<0.1	4	0.6	0.7	1100
Cyprinus carpio (FW)	0.4	1	0.5	0.4	60
Gadus callarias (M)	0.1	1	4	1	250
Mullus barbatus (M)	0.2	180	3.4	1.7	700
Tinca vulgaris (FW)	0.2	2.4	3.6	1.3	370
Opsanus tau (M)	10	10200	15	180	31800
Lungfish:					
Protopterus aethiopicus (A)	30	1700	7	60	35000
Neoceratodus forsteri (A)	0.5	150	4.3	13	1500
Amphibians:					
Necturus maculosus (A)	16	2100	9	56	20000
Amphiuma means (A)	206	10200	30	370	18700
Rana catesbiana tadpole (A)	55	2300	2	25	1050
Rana catesbiana adult (ST)	980	9700	46	500	24600
Rana temporaria (ST)	365	9500	51	120	19800
Xenopus laevis (A)	24	250	22	40	10000–180000
Chtonerpeton indistinctum (ST)	97	12780	21		10800
Chelonid reptiles:					
Chrysemys picta (SA)	644	12800	45		6030
Pseudemys scriptae (SA)	140	8000	45		45000–55000
Kinosternon hirtipes (SA)	125–300	7000–9000	40		38000–60000
Gopherus flavomarginatus (T)	9	275	10		28000
Gopherus berlandieri (T)	27	900	18		29600
Rhynchocephalid reptiles:					
Sphenodon punctatus (T)	– –	7800	45		35200
Mammals:					
Rattus norvegicus	600	22800	85	150	30000
Myotis lucifugus	2800	53400	>90		52000
Macca rhesus	420	14000	40		11200
Man	240	8500	50	175	7800

*Approximate values for gastropod hepatopancreas,[41,97,291,292] earthworm gut,[18] and vertebrate liver.[6,33,34,44,101,104,118,125,178]

†Habitat: M, marine; FW, freshwater; FW → M, catadromous; M → FW, anadromous; A, aquatic; ST, semiterrestrial; SA, semi-aquatic; T, terrestrial.

‡Activity of overall arginine synthetase system (argininosuccinate synthetase plus argininosuccinate lyase).

arginine-urea pathway for ammonia detoxication. The occurrence of the arginine-urea pathway in primitive fishes suggests that the genetic potential for this pathway had an evolutionary continuity from the echinoderm-protochordate line into the vertebrates (Fig. 7-10). A complete complement of enzymes has not yet been detected in cyclostomes,[220] the only surviving members of the Agnatha, but all enzymes are present and functional in holocephalans[219] and elasmobranchs (Table 7-5). From these primitive fishes, the pathway continued into the teleost line[125, 222] and along the main line of vertebrate evolution leading to the mammals. It is present in extant lungfish, all classes of amphibians, the chelonid and rhynchocephalid reptiles, and probably also the coelacanth *Latimeria*.[32] The pathway became dysfunctional in the reptilian line leading to the birds, and this may also have occurred in the reptilian lines giving rise to the Squamata and Crocodilia.[33] In birds, the enzymes for arginine synthesis are below limits of detection in liver but not kidney,[285] and they also appear to be absent from liver of the rattlesnake *Crotalus*.[178] All enzymes of the arginine-urea pathway except for a carbamylphosphate synthetase are, however, present in liver of the lizard *Ctenosaura*.[178] Studies *in vivo* with the alligator suggest that at least some reactions of the pathway may also take place in these organisms.[60, 61] The evolution of the pathway in the Reptilia is thus not clear.

Selection for Urea Synthesis: Osmotic Water Retention. Irrespective of whether their habitat was a freshwater or a marine one,[229] the earliest vertebrates are presumed to have possessed the components of the arginine-urea pathway; these were at least functional in the metabolism of arginine during protein turnover. The possible function of the arginine-urea pathway in supplying arginine in extant fish, especially the ureogenic teleosts,[125] has not been investigated. The teleost species thus far examined all show a dietary requirement for arginine, indicating they are incapable of synthesizing this amino acid.[63] It seems probable that the selective pressures for the integration of the components of the arginine-urea pathway for an increased rate of urea synthesis first came about in the marine environment. Whether this was originally directed toward ammonia detoxication, the synthesis of urea as an osmotic component, or both is a moot point. Differences in water and ionic regulation in the marine environment as opposed to the freshwater environment may have made the excretion of ammonia as a major end-product of nitrogen catabolism disadvantageous. On the other hand, the accumulation of urea as an osmotic component was certainly advantageous, and the function of the arginine-urea pathway in this capacity is seen today in elasmobranchs, holocephalans,[221] and probably coelacanths.[204] Amino acids are major osmotic components in many marine invertebrates.[88] The utilization of an *end-product* of nitrogen catabolism rather than amino acids themselves for osmotic purposes has an energetic advantage, in that the energy derived from carbon catabolism of the amino acids is not lost. The adaptation of the arginine-urea pathway for ammonia detoxication was not new to the vertebrates, having been used to this end by terrestrial invertebrates such as the earthworm, nor apparently was the retention of urea for osmotic purposes new (since this may possibly occur in terrestrial gastropods and also in crustaceans adapted to a hypertonic environment[256]). The retention of urea by early vertebrates in the marine environment required, in addition to the necessary biochemical adaptations in the arginine-urea pathway, a major physiological adaptation in the kidney, since this organ is mainly responsible for regulating plasma urea levels.[246] This latter adaptation is seen today in elasmobranchs in a hypotonic medium and amphibians in a hypertonic medium. The freshwater elasmobranch *Potamotrygon*, unlike its marine relatives, is ammonotelic[101] and also has less than 1/300 the amount of plasma urea[287] present in marine elasmobranchs. The adaptation of this species to freshwater involves (1) a lower rate of urea biosynthesis, as reflected by the lower activities of the hepatic arginine-urea pathway enzymes (Table 7-5), and (2) an increased urinary excretion of urea.[101] During the transfer of euryhaline elasmobranchs into a more dilute environment, reduction of plasma urea levels is brought about mainly by an increased urinary excretion of urea.[101] In amphibians living in or adapted to a hypertonic environment, there is a reduction in the glomerular filtration rate resulting in a decreased rate of urea excretion.[246] There is also an increased rate of hepatic urea synthesis under these conditions.[129, 167] Thus, both biochemical and physiological adaptations were required of the first vertebrates in the sea: the former involved an increased rate of urea synthesis, and the latter involved an increased tubular reabsorption of urea.

Transition to the Land. The capacity to synthesize urea during periods of restricted water availability preadapted the early vertebrates for their transition to the land. This is seen most clearly in extant dipnoans which include the African (*Protopterus*), South American (*Lepidosiren*), and Australian (*Neoceratodus*) lungfish. *Protopterus* and *Lepidosiren* can survive desiccation and estivate out of water during droughts, whereas *Neoceratodus* is restricted to an aquatic existence. *Protopterus* and *Neoceratodus* are ammonotelic when free-swiming, although both are ureogenic in that they are capable of hepatic urea synthesis. The levels of the arginine-urea pathway enzymes are higher in *Protopterus* (Table 7–5), as is its capacity for hepatic urea synthesis;[104] during estivation out of water, this species accumulates urea.[128] This accumulation does not involve an increased rate of urea synthesis[89, 101] even though the animals appear to be in continuous gluconeogenesis throughout estivation.[128] The details of metabolic adaptation in estivating lungfish are not known. Their overall metabolism is presumably lowered so that the normal level of urea synthesis may be sufficient to detoxify the amount of ammonia formed due to endogenous protein catabolism, thereby preventing it from reaching toxic levels. The failure to excrete urea allows this compound to accumulate and therefore function in osmotic water retention.

Aquatic amphibians, like lungfish, are ammonotelic and ureogenic. The ammonotelic-ureotelic transition made by metamorphosing anurans is assumed to be somewhat similar to that made by the first vertebrates—possibly labyrinthodont relatives of primitive dipnoans—to invade the land. Unlike estivating *Protopterus*, anuran amphibians show a marked increase in their capacity for hepatic urea synthesis during metamorphosis.[54] An increase in carbamylphosphate synthetase activity is especially critical in this transition to ureotelism and is due to the synthesis *de novo* of new enzyme protein, a situation that can be duplicated *in vivo* and *in vitro* with the hormone thyroxine.[55] Also, when different species of amphibians are compared with respect to their adaptation for a terrestrial existence, there is a good correlation between the degree of this adaptation and their hepatic levels of carbamylphosphate synthetase-I.[54] These observations indicate that carbamylphosphate synthetase plays a key role in the transition to a terrestrial existence.

Once established in semi-terrestrial and terrestrial amphibians, the detoxication of ammonia via the arginine-urea pathway continued through the cotylosaurians to the most primitive mammals—the echnida and platypus[69]—to become the predominant mechanism in nitrogen catabolism of the amino acids in the Mammalia.

Transition to Uricotelism. The Reptilia are the most interesting of the vertebrates with respect to the evolution of excretory function, since they exhibit all three major types of nitrogen metabolism and undergo transitions between ammonotelism and ureotelism, ammonotelism and uricotelism, and ureotelism and uricotelism. The latter transition presumably occurred in reptilian ancestors of birds prior to their loss of the ability to synthesize arginine and urea, although the arboreal amphibian *Phyllomedusa sauvagii* is known to excrete large amounts of uric acid.[261a] Again, the transition to uricotelism appears to have been influenced mainly by changing water relationships. This is illustrated today in adult chelonid reptiles. Terrestrial species excrete less urea than do aquatic or semiaquatic species, and this is correlated with the hepatic levels of the arginine-urea pathway enzymes (Table 7–5). Terrestrial tortoises normally excrete both urea and uric acid but, during dehydration, shift mainly to uric acid.[142] This shift appears to be regulated primarily by the kidney[67] and results in the retention of urea as a plasma osmotic component.[99] During dehydration of tortoises, uric acid secreted by kidney tubule cells precipitates as urate salts in the bladder, whereas urea appearing in the urine re-equilibrates with the plasma, thereby contributing to the increased plasma osmolality. Dehydration also causes an increased excretion of uric acid by aquatic turtles, although urea remains the predominant excretory product.[232] Moderate dehydration causes a decreased glomerular filtration rate in the latter species, and extreme dehydration causes a complete cessation of urine formation.[67] The details of the biochemical adaptations taking place during water restriction in these chelonid reptiles are not known.

Both urea and uric acid are also excreted by *Sphenodon punctatus*, the only surviving member of the Rhynchocephalia.[118] Although uric acid is the major excretory product of this species, the percentage urea excreted is increased with high protein diets, indicating a significant role of the arginine-urea pathway in detoxifying ammonia during gluconeogenesis.

The arginine-urea pathway may also be present in some squamate reptiles, although this has not been critically documented. Under certain circumstances, urea is the main excretory product of the lizard *Lacerta*[203] and, as indicated above, all enzymes of the pathway except carbamylphosphate synthetase are detectable in liver of the lizard *Ctenosaura*.[178] Urea also appears to be the major end-product of certain snake embryos.[51,52] Glutamine synthetase is a mitochondrial enzyme in the Western gartersnake *Thamnophis elegans*,[301] indicating that adults of this species detoxify ammonia via the purine pathway in the same manner as do birds. As first suggested by Needham,[187] it is the water relationships of the embryo that are the most critical in transitions to different modes of nitrogen excretion, and the occurrence of urea as an end-product of serpentine embryos is an important factor in considering the transition from ureotelism to uricotelism that occurred in early reptiles. Early reptilian eggs may have been relatively permeable, and ammonia generated during embryonic metabolism could have been eliminated by exchange with the environment, an ammonotelic condition inherited from their amphibian ancestors.[195] These eggs therefore required an aquatic or very moist environment for development. In order for eggs to develop under conditions of more restricted water availability, a transition to ureotelism had to be made early in embryonic development. These conditions may possibly be recapitulated today during development of serpentine reptiles.[51,52] The development of a semipermeable eggshell allowed eggs to be laid in still drier environments and urea, accumulated as an end-product, helped conserve water by decreasing evaporative water loss. With the development of the cleidoic egg—an egg with a relatively impermeable shell[233]—water retention was less of a problem and, under these circumstances, high concentrations of urea may have begun to create osmotic or other problems for the embryo. There was therefore selection against the synthesis of urea and consequently for the detoxication of ammonia via the purine pathway. According to the mechanisms outlined earlier, this selection must have been directed toward a replacement of the mitochondrial function of carbamylphosphate synthetase-I with that of glutamine synthetase, as well as a modification in the regulation of the purine pathway in such a manner that excess inosine monophosphate formed during ammonia detoxication was efficiently converted to uric acid. The loss of carbamylphosphate synthetase-I resulted in the loss of the ability to synthesize arginine, and this amino acid became an absolute dietary requirement, as is seen today in birds.

The transition to uricotelism allowed for the radiation of ancestral reptiles into arid environments, a major habitat of extant species. In addition to the basic biochemical adaptations, the adoption of uricotelism also required major physiological adaptations, especially in the kidney. In mammals, urea plays a major osmotic role in the concentration of urine, whereas in birds it does not.[246] Only minimal amounts of water are required for the excretion of uric acid, since this compound contributes little to urine osmolality.[173] Uricotelism therefore represents a complete adaptation to the terrestrial environment.

The crocodilid reptiles represent a secondary return to an aquatic existence[248] and are primarily ammonotelic.[65] Unlike mammals, in which NH_4^+ is excreted during metabolic acidosis and HCO_3^- is excreted during alkalosis, the alligator excretes NH_4^+ and HCO_3^- simultaneously. The latter anion arises in the kidney by the action of carbonic anhydrase. The renal excretion of NH_4^+ and HCO_3^- by the alligator thus appears directed toward the conservation of Na^+ and Cl^-, as is the extrarenal elimination of these ions in fresh-water invertebrates, fish, and amphibians.[61]

REGULATION OF NITROGEN EXCRETION

The synthesis and excretion of nitrogenous end-products by animals is influenced both quantitatively and qualitatively by physiological, environmental, developmental, and phylogenetic parameters. In general, physiological regulation involves a greater or lesser synthesis of end-product and, therefore, an adjustment of levels of enzyme activity. During gluconeogenesis induced by high protein diets, starvation, or glucocorticoid administration, there is an increase in the activities of the arginine-urea pathway enzymes in mammalian liver[244] and in at least some of the purine pathway enzymes in avian liver.[135] These increases in enzyme activity confer a greater capacity for synthesis of end-product and have, in many cases, been shown to be due to an increase in enzyme protein.[245a] Starvation-induced gluconeogenesis in the earthworm also elicits an increase in activity of the arginine-urea pathway enzymes; but here, the increased capacity for urea synthesis results in a change in the nature of the major end-product, a change from ammonotelism to ureotelism.[18] Anabolic states, such as that elicited by growth hormone (somatotrophin), result in decreased levels of the arginine-urea pathway enzymes in mammalian liver.[172] Environmental factors may also affect levels of enzyme activity, and this is especially noted in vertebrates that utilize urea for osmotic water retention. The arginine-urea pathway enzyme activities increase in both *Xenopus laevis*[167] and *Rana cancrivora*[57a] when these amphibians are subjected to hypertonic environments. In *Xenopus*, the increase in carbamylphosphate synthetase-I activity is due to synthesis *de novo* of new enzyme protein.[167] On the other hand, transfer of the skate *Raja erinacea* to a hypotonic environment results in a decreased production of urea, in addition to other physiological adjustments,[102a] indicating a decreased activity of the enzymes for urea biosynthesis. The marked increase in hepatic levels of the arginine-urea pathway enzymes during amphibian metamorphosis[54, 55, 93b] is an example of the influence of developmental factors upon nitrogen excretion. Here, as in the earthworm during fasting, the increased capacity for urea synthesis results in a qualitative change from ammonotelism to ureotelism. There is also an environmental factor implicated in this developmental change in amphibians, since metamorphosis normally represents a transition from an aquatic to a semi-terrestrial life. In *Xenopus*, a permanently aquatic species, there is little change in the amount of urea excreted during metamorphosis.[4] In *Rana catesbiana*, the activities of all five enzymes involved in urea biosynthesis show a concerted increase during metamorphosis[56, 320a] and, again in the case of carbamylphosphate synthetase-I, the increased activity is due to the synthesis of new enzyme protein. This response is elicited by thyroxine both *in vivo* and by intact liver tissue *in vitro*.[55]

In mammals, the enzymes for urea biosynthesis become localized primarily in liver during development.[133] Some enzymes, especially those converting citrulline to arginine (argininosuccinate synthetase and argininosuccinate lyase), are also produced in other tissues such as kidney and brain.[216] Thus, during tissue differentiation, the genetic potential for these enzymes is expressed in certain tissues and not in others. This is also seen phylogenetically. Although the enzymes for both citrulline and arginine biosynthesis are absent from avian liver, the enzymes for the conversion of citrulline to arginine are nevertheless present in kidney.[285] It seems that in some species the genetic potential for the synthesis of both urea and uric acid as end-products is present, and the utilization of one or the other pathway is dependent upon environmental or other factors. This is especially indicated in certain reptiles where the percentage urea or uric acid excreted changes with states of hydration.[142, 232] The excretion of large quantities of uric acid by amphibians adapted to arid conditions[261a] suggests that this situation may also exist in this class of vertebrates. In man, some uric acid may be formed *de novo* by a mechanism similar to that in birds and reptiles.[81] There is thus a phylogenetic implication in metabolic defects in man which result in excessive uric acid production.

The above represent but a few examples of modifications of nitrogen excretion brought about by basic cellular regulatory mechanisms. In general, physiological and environmental factors affect levels of enzyme activity, whereas developmental and phylogenetic factors affect the expression of basic genetic information for the synthesis of enzymes involved in ammonia detoxication. The fundamental cellular regulatory mechanisms involved in these adjustments are those concerned with gene transcription and translation and with modifications in enzyme activity.

Regulation of Gene Activity

In both procaryotic and eucaryotic cells the flow of genetic information is much the same.

This information, as specific nucleotide sequences of DNA, is transcribed into specific nucleotide sequences of messenger RNA (mRNA) that are, in turn, translated by the protein-synthesizing mechanism into specific amino acid sequences in the final product. For our purposes, this product is a specific enzyme. The molecular details of regulation of these processes, although well known in procaryotic cells, are only now being elucidated in eucaryotic cells.

Procaryotic cells. The operon of procaryotic cells consists of the structural genes for specific enzymes of a given metabolic pathway plus regulatory genes. The best known system is the *lac* operon of *Escherichia coli.*[328] The *structural genes* of this operon code for enzyme proteins involved in the utilization of galactosides such as lactose. These are β-galactosidase, β-galactoside permease and thiogalactoside transacetylase. The regulatory genes consist of a *promoter gene,* which is the site of initiation of mRNA transcription from the structural genes, an *operator gene* where the repressor protein binds to inhibit this transcription, and a *regulator gene* which encodes the repressor protein. Transfer of *E. coli* cells to a galactoside-containing medium results in induction of the enzymes for galactoside utilization. Induction occurs because of the interaction of the galactoside with the repressor protein, thereby preventing its binding to the operator gene. Under these conditions, mRNA for the enzymes of the pathway can be transcribed from the structural genes and subsequently translated into specific enzyme protein. This results in a marked increase in the amount of enzymes involved in galactoside utilization. In the absence of galactoside, the repressor protein binds to the operator locus, thereby preventing transcription. Cyclic AMP (adenosine-cyclic 3′,5′-monophosphate) also affects transcription of the *lac* operon, and is associated with repression of enzyme synthesis. In the presence of glucose, for example, intracellular cyclic AMP is reduced, inhibiting mRNA transcription.

Eucaryotic cells. Animal cells differ fundamentally from procaryotic cells in having the transcriptional and translational mechanisms compartmentalized in the nucleus and cytosol, respectively. In contrast to the single molecules of DNA in procaryotic cells, eucaryotic chromosomes are complex structures containing (in addition to DNA) RNA and specialized proteins. A high percentage of the DNA consists of repeated nucleotide sequences due, apparently, to gene duplication. Although eucaryotic cells contain much more DNA than do procaryotic cells, and there is a positive correlation between cellular DNA content and evolutionary complexity,[28a] only a small fraction of the genome is active at a given time. There is also, as yet, no evidence that the structural genes for the enzymes of metabolic pathways are physically organized into operons in animal cells. In addition to these differences, there is a more rapid turnover of RNA, especially intranuclear RNA, and protein in eucaryotic cells.

One model that has been proposed for gene regulation in eucaryotic cells is that of Britten and Davidson.[28a] This model consists of the following elements: producer gene, receptor gene, activator RNA, integrator gene, and sensor gene. The *producer gene* is that portion of the genome that is transcribed to yield specific mRNA, and is thus equivalent to the structural gene of procaryotic cells. The *receptor gene* is linked to the producer gene and causes its transcription when the receptor is complexed with activator RNA. *Activator RNA* forms a sequence-specific complex with the receptor gene. The *integrator gene* encodes the activator RNA and is responsible for its synthesis. The *sensor gene* is a DNA sequence linked to the integrator gene or genes, which acts as a binding site for agents that regulate genome activity. These agents include steroid and polypeptide hormones as well as embryonic inducers and products of producer genes themselves. The binding of these agents by the sensor gene results in the activation of an integrator gene or genes which, by the synthesis of activator RNA, may in turn activate a battery of producer genes. The binding of agents to the sensor gene may be mediated by specific proteins, such as steroid-binding proteins.[271a] The excess DNA in eucaryotic cells over that required for producer or structural genes is concerned with regulation in this model: the increasing cellular DNA content correlated with increasing evolutionary complexity of organisms thus reflects a demand for more complex regulatory mechanisms.

The Britten-Davidson model is especially attractive in considering developmental and phylogenetic aspects of nitrogen excretion, since these parameters affect the expression of the genetic potential of a given tissue or species for the synthesis of a functional complement of detoxication enzymes. One characteristic of the model is that the producer or structural genes are inactive unless specifically activated via sensor genes. Thus, during mammalian tissue differentiation, the genetic potential for ammonia detoxication via urea biosynthesis is presumably present in all tissues but is only finally expressed in its full functional capacity in liver (along with other enzymes of amino acid catabolism via transdeamination). During amphibian metamorphosis, where there is a concerted response of all the enzymes involved in urea biosynthesis, a response involving both mitochondrial and cytosol enzymes, there is suggestive evidence that thyroxine initiates transcription.[109a] From an evolutionary standpoint, the model also allows for the continuity of genetic potential without this potential necessarily being expressed. Expression results only after activation by specific environmental or other factors. The potential for synthesizing uric acid as a major nitrogenous end-product must, for example, be common in animals. As we have seen, uric acid or other purines are commonly accumulated in several invertebrates irrespective of their habitat, and some amphibians adapted to arid conditions appear to utilize this end-product in spite of the general utilization of ureotelism by other semiterrestrial and terrestrial amphibians. The overproduction of uric acid in man in gouty conditions may thus be due to mutations resulting in the activation of whatever mechanisms are involved in uricotelic species that allow the purine pathway to function in its excretory capacity. Because both the arginine-urea and purine pathways are fundamentally nutritional pathways, the evolutionary changes that allowed them to be utilized for excretion were most likely mainly changes involved in their regulation with only minor changes in structural genes.

Regulation of Enzyme Activity

The levels of enzyme activity in animal tissues are controlled either by modifications in the amount of enzyme protein present or by modifications in the enzyme itself that affect its activity.

Enzyme Turnover. Transcriptional control, as discussed above, is one means of regulating amounts of enzyme. In addition, there is also regulation at the translational level, possibly involving cyclic AMP.[97a, 205a] Regulation at this level is equally complex, in that several classes of compounds are involved, including transfer RNA (tRNA), amino acids, amino acyl tRNA synthetases, ribosomes, and polyribosomes. Regulation at either level, however, involves rates of protein synthesis, and this is only part of the mechanism of control of enzyme levels in animal tissues where, in addition to enzyme synthesis, there is continual enzyme degradation.[245a] A change in either process thus affects the level of enzyme activity. The following equation describes a change in enzyme content of a given tissue:

$$\frac{dE}{dt} = k_s - k_d E$$

where E is the enzyme content of the tissue, k_s is the rate constant for enzyme synthesis, and k_d is the constant for enzyme degradation. Under steady-state conditions where $dE/dt = 0$, the rate of synthesis is equal to the rate of degradation: $k_s = k_d E$ and $E = k_s/k_d$. There is thus a constant turnover of enzymes in animal cells, the turnover time or half-life ($t_{1/2}$) being unique to each enzyme that has been studied.[222a] A change in enzyme content may occur when there is an increase or decrease in the rate of enzyme synthesis due to transcriptional or translational control or when there is an increase or decrease in the rate of enzyme degradation. The mechanism of enzyme degradation in animal cells is not known, although the lysosomes have been implicated.[111a] For the enzymes that have been studied, changes in activity are more commonly due to changes in rates of synthesis although, in some cases such as the increase in mammalian liver arginase due to increased dietary protein intake, the change has been shown to be due to a change in rate of degradation.[245a]

Enzyme Modification. The best known non-covalent modifications of enzyme activity are those involving allosteric regulation. Allosteric enzymes have two or more binding sites: the catalytic site, to which substrate binds, and the regulatory site, to which low molecular weight compounds, termed effector molecules, bind. Binding of the effector at the regulatory site induces conformational changes in the enzyme protein involving subunit interactions,[93a] which affect substrate binding and, consequently, catalytic activity. Carbamylphosphate synthetase-I appears to be an allosteric enzyme, the effector molecule being *N*-acetylglutamate,[110a] and it has been suggested that the synthesis of this effector molecule is involved in the regulation of urea synthesis in mammals.[259,260] The first enzyme of purine biosynthesis, glutamine phosphoribosylpyrophosphate amidotransferase, may also be an allosteric enzyme.[107a] In mammalian liver, where the purine pathway is acting in a nutritional capacity, the amidotransferase is inhibited by most purine nucleotides, as it is in procaryotic cells.[188] The regulation of this enzyme in avian liver, where the pathway also serves an excretory function, is not clear. The avian enzyme has been reported to be sensitive to nucleotide inhibition, although less sensitive to AMP than the mammalian enzyme,[135] but to lose this sensitivity during purification.[45]

Enzyme activity is also modified by covalent changes in enzyme structure, the most familiar example being the enzymatic conversions of inactive chymotrypsinogen and trypsinogen to the active proteases, chromotrypsin and trypsin. In mammalian liver, the interconversion of xanthine oxidase types D and O may involve an enzymatic oxidation of a thiol group of the molecule.[120a]

Regulatory Mechanisms and the Excretory Pathways. The regulation of the enzymes of the excretory pathways in animal cells is complex because of the dual function of these pathways in nutrition and excretion. In procaryotic cells, where the main function of the pathways is to supply arginine for protein synthesis and nucleotides for nucleic acid synthesis, the arginine-urea and purine pathways are regulated by the availability of their end-products. Short-term regulation can be affected by simple feedback inhibition, and long-term regulation can be affected by enzyme induction or repression; both mechanisms are based on the presence and concentration of end-product. In mammals, there are indications that regulation of the arginine-urea pathway based on arginine availability also occurs. For example, under most physiological conditions in which amino acid catabolism is increased and there is therefore a greater demand for ammonia detoxication, all five enzymes of urea biosynthesis increase. On the other hand, only the four enzymes for arginine synthesis increase in animals on an arginine-free diet.[244] Also, in liver cells in tissue culture that have lost the ability to synthesize citrulline but can utilize this substrate to satisfy their arginine requirement, arginine causes a decreased activity of argininosuccinate synthetase and argininosuccinate lyase.[245] In the excretory capacity of the pathways, it is obvious that negative end-product regulation, either by enzyme repression or by feedback inhibition, could not occur: during demands for increased ammonia detoxication, urea or uric acid levels normally increase. It would thus seem more logical that the detoxication pathways be subject to positive, "feedforward" regulation in order to function in excretion. Such a regulation is the suggested allosteric activation of carbamylphosphate synthetase-I by *N*-acetylglutamate. Arginine, which is known to protect against experimental ammonia intoxication *in vivo*,[114,313,320] stimulates mitochondrial *N*-acetylglutamate synthesis,[260] so this mechanism would presumably respond primarily to increased cellular levels of arginine. An increased activity of the pathway in response to increased arginine levels is, however, opposite to the expected regulation of the pathway in its nutritional role. This illustrates the complexity of regulating the detoxication pathways in both their functional capacities. In addition to regulating the initial enzymes of the detoxication pathways, it also seems clear that arginase in the arginine-urea pathway and the enzymes involved in IMP metabolism in the purine pathway (Fig. 7–8) are critical enzymes, in that they represent branch-point enzymes between nutrition and excretion and also regulate cellular levels of arginine and IMP.

I wish to thank Dr. Bonnalie O. Campbell for her help in the preparation and editing of this chapter.

REFERENCES

1. Altman, P. L., and D. S. Dittmer, eds., Metabolism, Fed. Amer. Soc. Exp. Biol., Bethesda, 1968, pp. 95–187.
2. Alvarado, R. H., and T. H. Dietz, Comp. Biochem. Physiol. *33*:93–110, 1970. Effect of salt depletion on hydromineral balance in larval *Ambystoma gracile.*
3. Baldwin, E., An Introduction to Comparative Biochemistry, 4th Ed., Cambridge University Press, 1964.
4. Balinsky, J. B., pp. 519–637. In Comparative Biochemistry of Nitrogen Metabolism, edited by J. W. Campbell. London, Academic Press, 1970. Nitrogen metabolism in amphibians.
5. Balinsky, J. B., and E. Baldwin, J. Exp. Biol. *38*:695–705, 1961. The mode of excretion of ammonia and urea in *Xenopus laevis.*
6. Balinsky, J. B., E. L. Choritz, C. G. L. Coe, and G. S. Van Der Schans, Comp. Biochem. Physiol. *22*:53–57, 1967. Urea cycle enzymes and urea excretion during the development and metamorphosis of *Xenopus laevis.*
7. Bascur, L., J. Cabello, M. Veliz, and A. Gonzalez, Biochim. Biophys. Acta *128*:149–154, 1966. Molecular forms of human-liver arginase.
8. Bässler, K. H., and C.-H. Hammar, Biochem Z. *330*:555–564, 1958. Amino acid metabolism in cell fractions of rat liver: transamination and oxidation of L-amino acids.

9. Bell, M. L., H. M. Lazarus, A. H. Herman, R. H. Egdahl, and A. M. Rutenburg, Proc. Soc. Exp. Biol. Med. *136*:298–299, 1971. pH dependent changes in cell membrane stability.
10. Bendall, D. S., and C. De Duve, Biochem. J. *74*:444–450, 1960. Activation of latent dehydrogenases in mitochondria from rat liver.
11. Benson, J. A., and A. J. Rampone, Ann. Rev. Physiol. *28*:201–226, 1966. Gastrointestinal absorption.
12. Berridge, M. J., J. Exp. Biol. *43*:535–552, 1965. Nitrogen excretion and excretory metabolism in the cotton strainer, *Dysdercus fasciatus* Signoret.
13. Bessman, S. P., and A. N. Bessman, J. Clin. Invest. *34*:622–628, 1955. The cerebral and peripheral uptake of ammonia in liver disease with an hypothesis for the mechanism of hepatic coma.
14. Binns, R., J. Exp. Biol. *51*:41–45, 1969. The physiology of the antennal gland of *Carcinus maenas* (L.): Some nitrogenous constituents in the blood and urine.
15. Binns, R., and A. J. Peterson, Biol. Bull. *136*:147–153, 1969. Nitrogen excretion by the spiny lobster *Jasus edwardsi* (Hutton): the role of the antennal gland.
16. Bishop, S. H., and L. B. Barnes, Comp. Biochem. Physiol. *40B*:407–422, 1971. Ammonia forming mechanisms: deamination of 5′-adenylic acid (AMP) by some polychaete annelids.
17. Bishop, S. H., and J. W. Campbell, Science *142*:1583–1585, 1963. Carbamyl phosphate synthesis in the earthworm *Lumbricus terrestris*.
18. Bishop, S. H., and J. W. Campbell, Comp. Biochem. Physiol. *15*:51–71, 1965. Arginine and urea biosynthesis in the earthworm *Lumbricus terrestris*.
19. Blight, M. M., J. Insect Physiol. *15*:259–272, 1969. Volatile nitrogenous bases emanating from laboratory-reared colonies of the desert locust, *Schistocerca gregaria*.
20. Blondin, G. A., A. F. DeCastro, and A. E. Senior, Biochem. Biophys. Res. Commun. *43*:28–35, 1971. The isolation and properties of a peptide ionophore from beef heart mitochondria.
21. Bloomfield, R. A., A. A. Letter, and R. P. Wilson, Arch. Biochem. Biophys. *129*:196–201, 1969. The effect of glycine on ammonia intoxication and uric acid biosynthesis in the avian species.
22. Bond, P. A., and P. A. Sang, J. Insect Physiol. *14*:341–359, 1968. Glutamate dehydrogenase of *Drosophila* larvae.
23. Borman, A., T. Wood, H. Black, E. Anderson, M. Oesterling, M. Womack, and W. Rose, J. Biol. Chem. *166*:585–594, 1946. The role of arginine in growth with some observations on the effects of arginic acid.
24. Boulanger, P., and R. Osteux, Biochim. Biophys. Acta *21*:552–561, 1956. Action of L-amino acid dehydrogenase in the liver of the turkey (*Meleagris galloparo* L.) on the basic amino acids.
25. Braunstein, A. E., pp. 335–389. *In* Advances in Enzymology, edited by F. F. Nord. New York, Interscience, 1957. The main pathways of assimilation and dissimilation of nitrogen in animals.
26. Braunstein, A. E., and S. M. Bychkov, Nature (London) *144*:751–752, 1939. A cell-free enzymatic model of *l*-amino-acid dehydrogenase ('*l*-deaminase').
27. Bray, R. C., and S. Ratner, Arch. Biochem. Biophys. *146*:531–541, 1971. Argininosuccinase from bovine kidney: comparison of catalytic, physical, and chemical properties with the enzyme from bovine liver.
28. Brierley, G. P., and C. D. Stoner, Biochemistry (NY) *9*:708–713, 1970. Swelling and contraction of heart mitochondria suspended in ammonium chloride.
28a. Britten, R. J., and E. H. Davidson, Science *165*:349–357, 1969. Gene regulation for higher cells.
29. Bromberg, P. A., E. D. Robin, and C. E. Forkner, Jr., J. Clin. Invest. *39*:332–341, 1960. The existence of ammonia in blood *in vivo* with observations on the significance of the NH_4^+—NH_3 system.
30. Brosnan, J. T., H. A. Krebs, and D. H. Williamson, Biochem. J. *117*: 91–96, 1970. Effects of ischemia on metabolite concentrations in rat liver.
31. Brown, G. W., Jr., Arch. Biochem. Biophys. *114*:184–194, 1966. Studies in comparative biochemistry and evolution. 1. Avian liver arginase.
32. Brown, G. W., Jr., and S. G. Brown, Science *155*:570–573, 1967. Urea and its formation in coelacanth liver.
33. Brown, G. W., Jr., and P. P. Cohen, Biochem. J. *75*:82–91, 1960. Comparative biochemistry of urea synthesis. 3. Activities of urea-cycle enzymes in various higher and lower vertebrates.
34. Brown, G. W., Jr., W. R. Brown, and P. P. Cohen, J. Biol. Chem. *234*:1775–1780, 1959. Levels of urea cycle enzymes in metamorphosing *Rana catesbiana* tadpoles.
35. Buchanan, J. M., J. G. Flaks, S. C. Hartman, B. Levenberg, L. N. Lukens, and L. Warren, pp. 233–255. *In* Chemistry and Biology of Purines, edited by G. E. W. Wolstenholme and C. M. O'Conner. Boston, Little, Brown and Co., 1957. The enzymatic synthesis of inosinic acid *de novo*.
36. Campbell, J. W., Nature (London) *208*:1299–1301, 1965. Arginine and urea biosynthesis in the land planarian: its significance in biochemical evolution.
37. Campbell, J. W., pp. 48–68. *In* Urea and the Kidney, edited by B. Schmidt-Nielsen and D. W. S. Kerr. Amsterdam, Excerpta Medical Foundation, 1970. Comparative biochemistry of arginine and urea metabolism in invertebrates.
38. Campbell, J. W. *In* Biochemical Adaptation, edited by F. P. Conte. Chicago, University of Chicago Press, 1973, in press. Animal adaptations in nitrogen metabolism.
39. Campbell, J. W., and S. H. Bishop, pp. 103–206. *In* Comparative Biochemistry of Nitrogen Metabolism, edited by J. W. Campbell. London, Academic Press, 1970. Nitrogen metabolism in molluscs.
40. Campbell, J. W., and K. V. Speeg, Jr., Comp. Biochem. Physiol. *25*:3–32, 1968. Arginine biosynthesis and metabolism in terrestrial snails.
41. Campbell, J. W., and K. V. Speeg, Jr., Z. Vergl. Physiol. *61*:164–175, 1968. Tissue distribution of enzymes of arginine biosynthesis in terrestrial snails.
42. Campbell, J. W., and K. V. Speeg, Jr., Nature (London) *224*:725–726, 1969. Ammonia and the biological deposition of calcium carbonate.
43. Campbell, J. W., R. B. Drotman, J. A. McDonald, and P. R. Tramell, pp. 1–54. *In* Nitrogen Metabolism and the Environment, edited by J. W. Campbell and L. Goldstein. London, Academic Press, 1972. Nitrogen metabolism in terrestrial invertebrates.
44. Carlisky, N. J., A. Barrio, and L. I. Sadnik, Comp. Biochem. Physiol. *29*:1259–1262, 1969. Urea biosynthesis and excretion in the legless amphibian *Chtonerpeton indistinctum* (Apoda).
45. Caskey, C. T., D. M. Ashton, and J. B. Wyngaarden, J. Biol. Chem. *239*:2570–2579, 1964. The enzymology of feedback inhibition of glutamine phosphoribosylpyrophosphate amidotransferase by purine nucleotides.
46. Chamalaun, R. A. F. M., and J. M. Tager, Biochim. Biophys. Acta *222*:119–134, 1970. Nitrogen metabolism in the perfused rat liver.
47. Chaplin, A. E., A. K. Huggins, and K. A. Munday, Comp. Biochem. Physiol. *16*:49–62, 1965. Ionic effects on glutamate dehydrogenase activity from beef liver, lobster muscle and crab muscle.
48. Charles, R., J. M. Tager, and E. C. Slater, Biochim. Biophys. Acta *131*:29–41, 1967. Citrulline synthesis in rat-liver mitochondria.
49. Chen, P. S., pp. 115–135. *In* Amino Acid Pools, edited by J. T. Holden. Amsterdam, Elsevier, 1962. Free amino acids in insects.
50. Christensen, H. N., and J. B. Clifford, J. Biol. Chem. *238*:1743–1745, 1963. Early postnatal intensification of hepatic accumulation of amino acids.
51. Clark, H., J. Exp. Biol. *30*:492–501, 1953. Metabolism of the black snake embryo: nitrogen excretion.
52. Clark, H., and B. F. Sisken, J. Exp. Biol. *33*:384–393, 1956. Nitrogenous excretion by embryos of the viviparous snake *Thamnophis s. sirtalis* (L.).
53. Cohen, P. P., pp. 477–494. *In* The Enzymes, 2nd Ed., Vol. 6, edited by P. D. Boyer, H. Lardy, and K. Myrbäck. New York, Academic Press, 1962. Carbamyl group synthesis.
54. Cohen, P. P., Harvey Lect. Ser. *60*:119–154, 1966. Biochemical aspects of metamorphosis: transition from ammonotelism to ureotelism.
55. Cohen, P. P., Science *168*:533–543, 1970. Biochemical differentiation during emphibian metamorphosis.
56. Cohen, P. P., and G. W. Brown, Jr., pp. 161–244. *In* Comparative Biochemistry, Vol. 2, edited by M. Florkin and H. S. Mason. New York, Academic Press, 1960. Ammonia metabolism and urea biosynthesis.
57. Cohen, P. P., and G. W. Brown, Jr., pp. 129–138. *In* Proceedings of the Fifth International Congress of Biochemistry, Vol. 3. London, Pergamon Press, 1963. Evolution of nitrogen metabolism.
57a. Colley, L., W. C. Rowe, A. K. Huggins, A. B. Elliott, and S. E. Dicker, Comp. Biochem. Physiol. *41*B: 307–322, 1972. Effect of short-term external salinity changes on ornithine-urea cycle enzymes in *Rana cancrivora*.
58. Corman, L., L. M. Prescott, and N. O. Kaplan, J. Biol. Chem. *242*: 1383–1390, 1967. Purification and kinetic characteristics of dogfish liver glutamate dehydrogenase.
59. Corrigan, J. J., pp. 387–488. *In* Comparative Biochemistry of Nitrogen Metabolism, edited by J. W. Campbell. London, Academic Press, 1970. Nitrogen metabolism in insects.
60. Coulson, R. A., and T. Hernandez, Biochemistry of the Alligator. Baton Rouge, Louisiana State University Press, 1964.
61. Coulson, R. A., T. Hernandez, pp. 639–710. *In* Comparative Biochemistry of Nitrogen Metabolism, edited by J. W. Campbell. London, Academic Press, 1970. Nitrogen metabolism in the living reptile.
62. Cowey, C. B., and J. R. M. Forster, Marine Biology *10*:77–81, 1971. The essential amino acid requirements of the prawn *Palaemon serratus*.
63. Cowey, C. B., J. Adron, and A. Blair, J. Mar. Biol. Assoc. U.K. *50*: 87–95, 1970. Essential amino acid requirements of plaice and sole.
64. Crabtree, B., and E. Z. Newsholme, Biochem. J. *117*:1019–1021, 1970. The activities of proline dehydrogenase, glutamate dehydrogenase, aspartate-oxoglutarate aminotransferase and alanine-oxoglutarate aminotransferase in some insect flight muscles.
65. Cragg, M. M., J. B. Balinsky, and E. Baldwin, Comp. Biochem. Physiol. *3*:227–235, 1961. A comparative study of nitrogen excretion in some amphibia and reptiles.
66. Crofts, A. R., J. Biol. Chem. *242*:3352–3359, 1967. Amine uncoupling of energy transfer in chloroplasts.
67. Dantzler, W. H., and B. Schmidt-Nielsen, Amer. J. Physiol. *210*:198–210, 1966. Excretion in fresh-water turtle (*Pseudemys scripta*) and desert tortoise (*Gopherus agassizii*).
67a. DeDuve, C., Ann. N.Y. Acad. Sci. *168*:369–381, 1969. Evolution of the peroxisome.

68. DeGroot, N., and P. P. Cohen, Biochim. Biophys. Acta *59*:588–594, 1962. Studies on dehydrogenase activities during amphibian metamorphosis.
69. Denton, D. A., M. Reich, and F. J. R. Hird, Science *139*:1225, 1963. Ureotelism of echidna and platypus.
70. Desai, R. M., and B. A. Kilby, Arch. Int. Physiol. Biochim. *66*:282–286, 1958. Experiments on uric acid synthesis by insect fat body.
71. De Vooys, C. G. N., Arch. Int. Physiol. Biochim. 77:112–118, 1969. Formation and excretion of ammonia in Teleostei.
72. Dietz, T. H., L. B. Kirschner, and D. Porter, J. Exp. Biol. *46*:85–96, 1967. The roles of sodium transport and anion permeability in generating transepithelial potential differences in larval salamanders.
73. Donnellan, J. F., and B. A. Kilby, Comp. Biochem. Physiol. *22*:235–252, 1967. Uric acid metabolism by symbiotic bacteria from the fat body of *Periplaneta americana.*
74. Downing, K. M., and J. C. Merkens, Ann. Appl. Biol. *43*:243–246, 1955. The influence of dissolved oxygen concentration on the toxicity of unionized ammonia to rainbow trout (*Salmo gairdneri* Richardson).
75. Drotman, R. B., and J. W. Campbell, Amer. J. Physiol., *222*:1204–1212, 1972. Protein arginine biosynthesis by mammalian liver tissue during postnatal development.
76. Drotman, R. B., and J. W. Campbell, Amer. J. Physiol., *222*:1213–1217, 1972. Protein arginine biosynthesis by amphibian liver tissue during metamorphosis.
77. Drotman, R. B., and R. A. Freedland, Amer. J. Physiol., *222*:973–975, 1972. Citrulline metabolism in the perfused rat liver.
78. Duda, G. D., and P. Handler, J. Biol. Chem. *232*:303–314, 1958. Kinetics of ammonia metabolism *in vivo.*
79. Duerr, F. G., Comp. Biochem. Physiol. *22*:333–340, 1967. The uric acid content of several species of prosobranch and pulmonate snails as related to nitrogen excretion.
80. Duerr, F. G., Comp. Biochem. Physiol. *26*:1051–1059, 1968. Excretion of ammonia and urea in seven species of marine prosobranch snails.
81. Duggan, D. E., K. H. Pua, and G. Elfenbein, Molec. Pharmacol. *4*:52–60, 1968. Purine metabolism in the chick embryo: effect of uricogenesis and xanthine oxidase inhibition.
82. Eliasson, E. E., and H. J. Strecker, J. Biol. Chem. *241*:5757–5763, 1966. Arginase activity during the growth cycle of Chang's liver cells.
83. Emerson, D. N., Comp. Biochem. Physiol. *29*:1115–1133, 1969. Influence of salinity on ammonia excretion rates and tissue constituents of euryhaline invertebrates.
84. Ennor, A. H., and J. F. Morrison, Physiol. Rev. *38*:631–674, 1958. Biochemistry of the phosphagens and related quanidines.
85. Fahien, L. A., B. O. Wiggert, and P. P. Cohen, J. Biol. Chem. *240*: 1083–1090, 1965. Crystallization and kinetic properties of glutamate dehydrogenase from frog liver.
86. Fannelli, G. M., and L. Goldstein, Comp. Biochem. Physiol. *13*:193–204, 1964. Ammonia excretion in the neotenous newt, *Necturus maculosus* (Rafinesque).
87. Florkin, M., and G. Duchâteau, Arch. Int. Physiol. *53*:267–306, 1943. The configurations of the enzymatic system of uricolysis and the evolution of purine catabolism in animals.
88. Florkin, M., and E. Schoffeniels, Molecular Approaches to Ecology. New York, Academic Press, 1969.
89. Forster, R. P., and L. Goldstein, Science *153*:1650–1652, 1966. Urea synthesis in the lungfish: relative importance of purine and ornithine cycle pathways.
90. Freedland, R. A., and B. Szepesi, pp. 103–104. *In* Enzyme Synthesis and Degradation in Mammalian Systems, edited by M. Rechcigl, Jr. Basel, Karger, 1971. Control of enzyme activity: nutritional factors.
91. Freedland, R. A., K. D. Martin, and L. Z. McFarland, Biochem. J. *103*:6P, 1967. Properties of glutamic dehydrogenase from several tissues of the Japanese quail.
92. Frieden, C., pp. 3–24. *In* The Enzymes, 2nd Ed., Vol. 7, Pt. A, edited by P. D. Boyer, H. Lardy, and K. Myrbäck. New York, Academic Press, 1963. L-Glutamate dehydrogenase.
93. Frieden, C., J. Biol. Chem. *240*:2028–2035, 1965. Glutamate dehydrogenase.
93a. Frieden, C., Ann. Rev. Biochem. *40*:653–696, 1971. Protein-protein interaction and enzymatic activity.
93b. Frieden, E., pp. 349–398. *In* Metamorphosis, edited by W. Etkin and L. I. Gilbert. Appleton-Century-Crofts, New York, 1968. Biochemistry of amphibian metamorphosis.
94. Fromm, P. O., Comp. Biochem. Physiol. *10*:121–128, 1963. Studies on renal and extra-renal excretion in *Salmo gairdneri.*
95. Fromm, P. O., and J. R. Gillette, Comp. Biochem. Physiol. *26*:887–896, 1968. Effect of ambient ammonia on blood ammonia and nitrogen excretion of rainbow trout (*Salmo gairdneri*).
96. Gamble, J. G., and A. L. Lehninger, pp. 611–622. *In* Biochemistry and Biophysics of Mitochondrial Membranes, edited by G. F. Azzore, E. Carafoli, A. L. Lehninger, E. Quagliarello, and N. Siliprandi. New York, Academic Press, 1972. Transport of ornithine and citrulline across mitochondrial membranes during urea synthesis in the liver.
97. Gaston, S., and J. W. Campbell, Comp. Biochem. Physiol. *17*:259–270, 1966. Distribution of arginase activity in molluscs.
97a. Gelehrter, T. D., pp. 165–199. *In* Enzyme Synthesis and Degradation in Mammalian Systems, edited by M. Rechcigl. Karger, Basel, 1971. Regulatory mechanisms of enzyme synthesis: enzyme induction.
98. Gifford, C. A., Amer. Zool. *8*:521–528, 1968. Accumulation of uric acid in the land crab, *Cardisoma guanhumi.*
99. Gilles-Baillien, M., and E. Schoffeniels, Ann. Soc. Roy. Zool. Belg. *95*:75–79, 1966. Seasonal variations of the composition of the blood of the Greek tortoise *Testudo hermanni* J. F. Gmelin.
100. Glahn, P. E., P. Manchon, and J. Roche, C. R. Seances Soc. Biol. *149*:509–513, 1955. L-Amino acid oxidases of the hepatopancreas of lamellibranchs.
101. Goldstein, L., pp. 55–77. *In* Nitrogen Metabolism and the Environment, edited by J. W. Campbell and L. Goldstein. London, Academic Press, 1972. Adaptation of urea metabolism in aquatic vertebrates.
102. Goldstein, L., and R. P. Forster, Comp. Biochem. Physiol. *14*:567–576, 1965. The role of uricolysis in the production of urea by fishes and other aquatic vertebrates.
102a. Goldstein, L., and R. P. Forster, Amer. J. Physiol. *220*:742–746, 1971. Osmoregulation and urea metabolism in *Raja erinacea.*
103. Goldstein, L., R. P. Forster, and G. M. Fanelli, Jr., Comp. Biochem. Physiol. *12*:489–499, 1964. Gill blood flow and ammonia excretion in *Myxocephalus scorpius.*
104. Goldstein, L., P. A. Janssens, and R. P. Forster, Science *157*:316–317, 1967. Lungfish *Neoceratodus forsteri*: activities of ornithine-urea cycle and enzymes.
105. Goodbody, I., J. Exp. Biol. *42*:299–305, 1965. Nitrogen excretion in Ascidiacea: storage excretion and the uricolytic enzyme system.
106. Goodman, A. D., pp. 297–318. *In* Nitrogen Metabolism and the Environment, edited by J. W. Campbell and L. Goldstein. London, Academic Press, 1972. Relation of the carbohydrate metabolism and ammonia production in the kidney.
107. Gorzkowski, B., Acta Biochim. Polonica *16*:193–200, 1969. Utilization of ^{14}C-labelled purine precursors for uric acid synthesis in *Helix pomatia.*
107a. Gots, J. S., pp. 225–255. *In* Metabolic Pathways, 3rd Ed., Vol. 5, edited by H. J. Vogel. Academic Press, New York, 1971. Regulation of purine and pyrimidine metabolism.
108. Graafmans, W. D. J., R. Charles, and J. M. Tager, Biochim. Biophys. Acta *153*:916–919, 1968. Mitochondrial citrulline synthesis with exogenous ATP.
109. Greenberg, D. M., pp. 95–190. *In* Metabolic Pathways, 3rd Ed., Vol. 3, edited by D. M. Greenberg. New York, Academic Press, 1969. Carbon catabolism of amino acids.
109a. Griswold, M. D., and P. P. Cohen, J. Biol. Chem. *247*:353–359, 1972. Alteration of deoxyribonucleic acid-dependent ribonucleic acid polymerase activities in amphibian nuclei during thyroxine-induced metamorphosis.
110. Guggenheim, S. J., J. Bourgoignie, and S. Klahr, Amer. J. Physiol. *220*:1651–1659, 1971. Inhibition by ammonium of sodium transport across isolated toad bladder.
110a. Guthöhrlein, G., and J. Knappe, Eur. J. Biochem. 7:119–127, 1968. Structure and function of carbamyl phosphate synthetase. I. Transitions between two catalytically inactive forms and the active form.
111. Haggag, G., and Y. Fouad, Nature (London) *207*:1003–1004, 1965. Nitrogenous excretion in arachnids.
111a. Haider, M., and H. L. Segal, Arch. Biochem. Biophys. *148*:228–237, 1972. Some characteristics of the alanine aminotransferase- and arginase-inactivating system of lysosomes.
112. Hall, L. M., R. C. Johnson, and P. P. Cohen, Biochim. Biophys. Acta *37*:144–145, 1960. The presence of carbamyl phosphate synthetase in intestinal mucosa.
113. Harmsen, R., J. Exp. Biol. *45*:1–13, 1966. The excretory role of pteridines in insects.
114. Harper, A. E., N. J. Benevenga, and R. M. Wohlhueter, Physiol. Rev. *50*:428–558, 1970. Effects of ingestion of disproportionate amounts of amino acids.
115. Hartenstein, R., Amer. Zool. *8*:507–519, 1968. Nitrogen metabolism in the terrestrial isopod, *Oniscus asellus.*
116. Hartman, S. C., pp. 1–68. *In* Metabolic Pathways, 3rd Ed., Vol. 4, edited by D. M. Greenberg. New York, Academic Press, 1970. Purines and pyrimidines.
117. Heller, J., and M. M. Jeżewska, Bull. Acad. Pol. Sci. Ser. Biol. 7:1–4, 1959. The synthesis of uric acid in the Chinese tussur moth (*Antheraea pernyi*).
118. Hill, L., and W. H. Dawbin, Comp. Biochem. Physiol. *31*:453–468, 1969. Nitrogen excretion in the tuatara, *Sphenodon punctatus.*
119. Hindfelt, B., and B. K. Siesjö, Life Sci. *9* (Pt. II):1021–1028, 1970. The effect of ammonia on the energy metabolism of the rat brain.
120. Hirsch-Kolb, H., J. P. Heine, H. J. Kolb, and D. M. Greenberg, Comp. Biochem. Physiol. *37*:345–359, 1970. Comparative physical-chemical studies of mammalian arginase.
120a. Holzer, H., and W. Duntze, Ann. Rev. Biochem. *40*:345–374, 1971. Metabolic regulation by chemical modification of enzymes.
121. Horne, F. R., Comp. Biochem. Physiol. *26*:687–695, 1968. Nitrogen excretion in the herbivorous land crab *Cardisoma guanhumi* Latreille.
122. Horowitz, N. H., Proc. Nat. Acad. Sci. *31*:153–157, 1945. Evolution of biochemical syntheses.

123. House, H. L., pp. 769–813. *In* The Physiology of Insecta, Vol. 2, edited by M. Rockstein. New York, Academic Press, 1965. Insect nutrition.
124. Howard, W. J., L. A. Kerson, and S. H. Appel, J. Neurochem. *17*:121–123, 1970. Synthesis *de novo* of purines in slices of rat brain and liver.
125. Huggins, A. K., G. Skutsch, and E. Baldwin, Comp. Biochem. Physiol. *28*:587–602, 1969. Ornithine-urea cycle enzymes in teleostean fish.
126. Inokuchi, T., Y. Horie, and T. Ito, Biochem. Biophys. Res. Commun. *35*:783–787, 1969. Urea cycle in the silkworm, *Bombyx mori*.
127. Janicki, R., and J. Lingis, Comp. Biochem. Physiol. *37*:101–105, 1970. Mechanism of ammonia production from aspartate in teleost liver.
128. Janssens, P. A., Comp. Biochem. Physiol. *11*:105–117, 1964. The metabolism of the aestivating African lungfish.
129. Janssens, P. A., and P. P. Cohen, Comp. Biochem. Physiol. *24*:887–898, 1968. Biosynthesis of urea in the estivating African lungfish and in *Xenopus laevis* under conditions of water-shortage.
130. Jezewska, M. M., B. Gorzkowski, and J. Heller, Acta Biochim. Polon. *11*:135–138, 1964. Utilization of [1-^{14}C]glycine in purine biosynthesis in *Helix pomatia*.
131. Johnson, W. J., B. Stavric, and A. Chartrand, Proc. Soc. Exp. Biol. *131*:8–12, 1969. Uricase inhibition in the rat by *s*-triazines: an animal model for hyperuricemia and hyperuricosuria.
132. Jones, M. E., pp. 35–45. *In* Urea and the Kidney, edited by B. Schmidt-Nielsen and D. W. S. Kerr. Amsterdam, Excerpta Medica Foundation, 1970. Vertebrate carbamyl phosphate synthetase I and II. Separation of the arginine-urea and pyrimidine pathways.
133. Jones, M. E., A. D. Anderson, C. Anderson, and S. Hodes, Arch. Biochem. Biophys. *95*:499–507, 1961. Citrulline synthesis in rat tissues.
134. Kamin, H., and P. Handler, Ann. Rev. Biochem. *26*:419–490, 1957. Amino acid and protein metabolism.
135. Katunuma, N., Y. Matsuda, and Y. Kuroda, pp. 73–81. *In* Advances in Enzyme Regulation, Vol. 8, edited by G. Weber. Oxford, Pergamon Press, 1970. Phylogenic aspects of different regulatory mechanisms of glutamine metabolism.
136. Katunuma, N., M. Okada, and Y. Mishii, pp. 317–335. *In* Advances in Enzyme Regulation, Vol. 4, edited by G. Weber. Oxford, Pergamon Press, 1966. Regulation of the urea cycle and TCA cycle by ammonia.
137. Keilin, J., Biol. Rev. (Cambridge) *34*:265–296, 1959. The biological significance of uric acid and guanine excretion.
138. Kelley, W. N., Fed. Proc. 27:1017–1112, 1968. Hypoxanthine-guanine phosphoribosyltransferase deficiency in the Lesch-Nyhan syndrome and gout.
139. Kennan, A. L., and P. P. Cohen, Proc. Soc. Exp. Biol. Med. *106*:170–173, 1961. Ammonia detoxification in liver from humans.
140. Kerkut, G. A., J. T. Rick, and A. K. Huggins, Comp. Biochem. Physiol. *28*:765–770, 1969. The intermediary metabolism *in vitro* of glucose and acetate by ganglia from *Helix aspersa* and the effects of amphetamine.
141. Kerly, M., and J. E. L. Spruyt, J. Physiol. (London) *216*:11–20, 1971. Amino acid metabolism in the isolated perfused rat liver.
142. Khalil, F., and G. Haggag, J. Exp. Zool. *130*:423–432, 1955. Ureotelism and uricotelism in tortoises.
143. Kidder, G. W., J. S. Davis, and K. Cousens, Biochem. Biophys. Res. Commun. *24*:365–369, 1966. Citrulline utilization in *Crithidia*.
144. King, K. S., and C. Frieden, J. Biol. Chem. *245*:4391–4396, 1970. The purification and physical properties of glutamate dehydrogenase from rat liver.
145. Kirk, J. S., and J. B. Sumner, J. Biol. Chem. *94*:21–28, 1931. Anti-urease.
146. Kittredge, J. S., D. G. Simonsen, E. Roberts, and B. Jelinek, pp. 176–186. Amino Acid Pools, edited by J. T. Holden. Amsterdam, Elsevier, 1962. Free amino acids of marine invertebrates.
147. Kovacević, Z., J. D. McGivan, and J. B. Chappell, Biochem. J. *118*: 265–274, 1970. Conditions for activity of glutaminase in kidney mitochondria.
148. Krzyzanowska, M., and W. Niemierko, Bull. Acad. Polon. Sci. Ser. Biol. *17*:673–676, 1970. Uric riboside as one of the end products of mitrogen metabolism in the wax moth larvae (*Galleria mellonella* L.) during starvation.
149. Leal, J. A., V. G. Lilly, and M. E. Gallegly, Mycologia *62*:1041–1056, 1970. Some effects of ammonia on species of *Phytophthora*.
150. Lee, P. C., and J. R. Fisher, Biochim. Biophys. Acta *237*:14–20, 1971. Regulation of xanthine dehydrogenase levels in liver and pancreas of the chick.
151. Lee, T. W., and J. W. Campbell, Comp. Biochem. Physiol. *15*:457–468, 1965. Uric acid synthesis in the terrestrial snail, *Otola lactea*.
152. Levenbook, L., J. Insect Physiol. *8*:559–567, 1962. The distribution of free amino acids, glutamine and glutamate, in the Southern army-worm, *Prodenia eridania*.
153. Levenbook, L., and J. Kuhn, Biochim. Biophys. Acta *65*:219–232, 1962. Properties and distribution of glutamine synthetase in the Southern armyworm, *Prodenia eridania*.
154. Little, C., J. Exp. Biol. *48*:569–585, 1968. Aestivation and ionic regulation in two species of *Pomacea* (Gastropoda, Prosobranchia).
155. Lowenstein, J., and K. Tornheim, Science *171*:397–400, 1971. Ammonia production in muscle: the purine nucleotide cycle.
155a. Lowenstein, J. M., Physiol. Rev. *52*:382–414, 1972. Ammonia production in muscle and other tissues: the purine nucleotide cycle.
156. Lowry, B. A., M. K. Williams, and I. M. London, J. Biol. Chem. *237*: 1622–1625, 1962. Enzymatic deficiencies of purine nucleotide synthesis in the human erythrocyte.
157. Lueck, J. D., and L. L. Miller, J. Biol. Chem. *245*:5491–5497, 1970. The effect of perfusate pH on glutamine metabolism in the isolated perfused rat liver.
158. Lund, P., Biochem. J. *118*:35–39, 1970. A radiochemical assay for glutamine synthetase, and activity of the enzyme in rat tissues.
159. Lund, P., and L. Goldstein, Comp. Biochem. Physiol. *31*:205–210, 1969. Glutamine synthetase activity in tissues of lower vertebrates.
160. MacFarlane, J. R., and T. W. Hogan, J. Insect Physiol. *12*:1265–1278, 1966- Free amino acid changes associated with diapause in the egg of the field cricket. *Teleogryllus commodus*.
161. Maetz, J., pp. 105–154. *In* Nitrogen Metabolism and the Environment, edited by J. W. Campbell and L. Goldstein. London, Academic Press, 1972. Interaction of salt and ammonia transport in aquatic organisms.
162. Maetz, J., and F. García-Romeu, J. Gen. Physiol. *47*:1209–1227, 1964. The mechanism of sodium and chloride uptake by the gills of a freshwater fish, *Carassius auratus*.
163. Makarewicz, W., and M. Żydowo, Comp. Biochem. Physiol. *6*:269–275, 1962. Comparative studies on some ammonia-producing enzymes in the excretory organs of vertebrates.
164. Manton, S. M., Phil. Trans. Roy. Soc. London B *227*:411–464, 1937. The feeding, digestion, excretion and food storage of *Peripatopsis*.
165. Margulis, L., Science *161*:1020–1022, 1968. Evolutionary criteria in thallophytes: a radical alternative.
166. McAllan, J. W., and W. Chefurka, Comp. Biochem. Physiol. *3*:1–19, 1961. Properties of transaminases and glutamic dehydrogenase in the cockroach *Periplaneta americana*.
167. McBean, R. L., and L. Goldstein, Amer. J. Physiol. *219*:1124–1130, 1970. Accelerated synthesis of urea in *Xenopus laevis* during osmotic stress.
168. McBean, R. L., M. J. Neppel, and L. Goldstein, Comp. Biochem. Physiol. *18*:909–920, 1966. Glutamate dehydrogenase and ammonia production in the eel (*Anguilla rostrata*).
169. McEnroe, W., and A. Forgash, Ann. Entomol. Soc. Amer. *50*:429–431, 1957. The *in vivo* incorporation of C^{14}formate in the ureide groups of uric acid by *Periplaneta americana* (L.).
170. McGilvery, R. W., Biochemistry: A Functional Approach. Philadelphia, W. B. Saunders Co., 1970, pp. 619–620.
171. McKhann, G. M., and D. B. Tower, Amer. J. Physiol. *200*:420–424, 1961. Ammonia toxicity and cerebral oxidative metabolism.
172. McLean, P., and M. W. Gurney, Biochem. J. *87*:96–104, 1963. Effect of adrenalectomy and growth hormone on enzymes concerned with urea synthesis in rat liver.
173. McNabb, F. M. A., and T. L. Poulson, Comp. Biochem. Physiol. *33*:933–939, 1970. Uric acid excretion in pigeons, *Columbia livia*.
174. McWard, G. W., L. B. Nicholson, and B. R. Poulton, J. Nutr. *92*:118–120, 1967. Arginine requirement of the young rabbit.
175. Milne, M. D., B. H. Scribner, and M. A. Crawford, Amer. J. Med. *24*:709–729, 1958. Non-ionic diffusion and the excretion of weak acids and bases.
176. Mitchell, H. A., J. Mammal. *45*:568–577, 1964. Investigations of the cave atmosphere of a Mexican bat colony.
177. Mitlin, N., and J. K. Maudlin, Ann. Entom. Soc. Amer. *59*:651–653, 1966. Uric acid in the nitrogen metabolism of the boll weevil.
178. Mora, J., J. Martuscelli, J. Ortiz-Pineda, and G. Soberón, Biochem. J. *96*:28–35, 1965. The regulation of urea-biosynthesis enzymes in vertebrates.
179. Mora, J., R. Tarrab, J. Martuscelli, and G. Soberón, Biochem. J. *96*:588–594, 1965. Characteristics of arginases from ureotelic and non-ureotelic animals.
180. Murray, A. W., Biochem. J. *100*:664–670, 1966. Purine-phosphoribosyltransferase activities in rat and mouse tissues and in Ehrlich ascites tumor cells.
181. Murray, A. W., D. C. Elliott, and M. R. Atkinson, pp. 87–119. *In* Progress in Nucleic Acid Research and Molecular Biology, edited by J. N. Davidson and W. E. Cohn. New York, Academic Press, 1970. Nucleotide biosynthesis from preformed purines in mammalian cells: regulatory mechanisms and biological significance.
182. Nakamaru, Y., and A. Schwartz, Biochem. Biophys. Res. Commun. *40*:830–836, 1970. Possible control of intracellular calcium metabolism by [H$^+$]: sarcoplasmic reticulum of skeletal and cardiac muscle.
183. Nakano, M., and T. S. Danowski, J. Biol. Chem. *241*:2075–2083, 1966. Crystalline mammalian L-amino acid oxidase from rat kidney mitochondria.
184. Nathan, D. G., and F. L. Rodkey, J. Lab. Clin. Med. *49*:779–785, 1957. A colorimetric procedure for the determination of blood ammonia.
185. Needham, A. E., J. Exp. Biol. *34*:425–446, 1957. Components of nitrogenous excreta in the earthworms *Lumbricus terrestris* L. and *Eisenia foetida* (Savigny).
186. Needham, A. E., pp. 207–297. *In* Comparative Biochemistry of Nitrogen Metabolism, edited by J. W. Campbell. London, Academic Press, 1970. Nitrogen metabolism in Annelida.
187. Needham, J., Biol. Rev. (Cambridge) *13*:225–251, 1938. Contributions of chemical physiology to the problem of reversibility in evolution.
188. Nierlich, D. P., B. Magasanik, J. Biol. Chem. *240*:358–365, 1965. Regulation of purine ribonucleotide synthesis by end product inhibi-

tion. The effect of adenine and guanine ribonucleotides on the 5'-phosphoribosylpyrophosphate amidotransferase of *Aerobacter aerogenes.*

189. Nolfi, J. R., Comp. Biochem. Physiol. *35*:827–842, 1970. Biosynthesis of uric acid in the tunicate, *Molgula manhattensis.*
190. Nørby, S., Hereditas *66*:204–214, 1970. A specific nutritional requirement for pyrimidines in rudimentary mutants of *Drosophila melanogaster.*
191. Olson, K. R., and P. O. Fromm, Comp. Biochem. Physiol. *40A*:999–1007, 1971. Excretion of urea by two teleosts exposed to different concentrations of ambient ammonia.
192. Oparin, A. I., Genesis and Evolutionary Development of Life. New York, Academic Press, 1968.
193. Orloff, J., M. Kahn, and L. Brenes, Amer. J. Physiol. *201*:747–753, 1961. Renal tubular effects of ammonium salts on electrolyte transport.
194. Oró, J., pp. 137–171. *In* The Origins of Prebiological Systems, edited by S. W. Fox. New York, Academic Press, 1965. Stages and mechanisms of prebiological organic synthesis.
195. Packard, G. C., Amer. Naturalist *100*:667–682, 1966. The influence of ambient temperature and aridity on modes of reproduction and excretion in amniote vertebrates.
196. Papa, S., F. Palmieri, and E. Quagliariello, pp. 153–165. *In* Regulation of Metabolic Processes in Mitochondria, edited by J. M. Tager, S. Papa, E. Quagliariello, and E. C. Slater. Amsterdam, Elsevier, 1966. Control mechanisms of glutamate oxidation in liver mitochondria.
197. Parry, G., pp. 341–366. *In* The Physiology of Crustacea, Vol. 1, edited by T. H. Waterman. New York, Academic Press, 1960. Excretion.
198. Parzen, S. D., and A. S. Fox, Biochim. Biophys. Acta *92*:465–471, 1964. Purification of xanthine dehydrogenase from *Drosophila melanogaster.*
199. Pequin, L., C. R. Acad. Sci. Paris *225*:1795–1797, 1962. The amounts of ammonia nitrogen in the blood of the carp (*Cyprinus carpio* L.).
200. Pequin, L., Arch. Sci. Physiol. *21*:193–203, 1967. Degradation and synthesis of glutamine by the carp (*Cyprinus carpio* L.).
201. Pequin, L., and A. Serfaty, Comp. Biochem. Physiol. *10*:315–324, 1963. Ammonia excretion by a freshwater teleost: *Cyprinus carpio* L.
202. Pequin, L., and F. Serfaty, Arch. Sci. Physiol. *22*:449–459, 1968. Hepatic and intestinal regulation of blood ammonia by the carp.
203. Perschmann, C., Zool. Beitr. Ber. *2*:17–80, 1956. On the significance of the hepatic portal vein in particular for the excretion of urea and uric acid by *Testudo hermanni* and *Lacerta viridus.*
204. Pickford, G. E., and F. B. Grant, Science *155*:568–570, 1967. Serum osmolality in the coelacanth, *Latimeria chalumnae*: urea retention and ion regulation.
205. Pitot, H. C., and C. Peraino, J. Biol. Chem. *239*:1783–1788, 1964. Induction of threonine dehydrase and ornithine-δ-transaminase by oral intubation of casein hydrolysate.
205a. Pitot, H. C., J. Kaplan, and A. Čihák, pp. 216–235. *In* Enzyme Synthesis and Degradation in Mammalian Systems, edited by M. Rechcigl. Karger, Basel, 1971. Translational regulation of enzyme levels in liver.
206. Pitts, R. F., Physiology of the Kidney and Body Fluids, 2nd Ed. Chicago, Year Book Medical Publishers, 1968. pp. 195–211.
207. Pitts, R. F., pp. 277–296. *In* Nitrogen Metabolism and the Environment, edited by J. W. Campbell and L. Goldstein. London, Academic Press, 1972. Control of renal ammonia metabolism.
208. Porembska, Z., A. Baranczyk, and J. Jachimowicz, Acta Biochim. Pol. *18*:77–85, 1971. Arginase isoenzymes in liver and kidney of some mammals.
209. Potts, W. T. W., Comp. Biochem. Physiol. *14*:339–355, 1965. Ammonia excretion in *Octopus dofleini.*
210. Pressman, B. C., pp. 213–250. *In* Membranes of Mitochondria and Chloroplasts, edited by E. Racker. New York, Van Nostrand Reinhold, 1970. Energy-linked transport in mitochondria.
211. Pritchard, J. B., F. Chavez-Peon, and R. D. Berlin, Amer. J. Physiol. *219*:1263–1267, 1970. Purines: supply by liver to tissues.
211a. Puritch, G. S., and A. V. Barker, Plant Physiol. *42*:1229–1238, 1967. Structure and function of tomato leaf chloroplasts during ammonium toxicity.
212. Racker, E., pp. 127–172. *In* Membranes of Mitochondria and Chloroplasts, edited by E. Racker. New York, Van Nostrand Reinhold, 1970. Function and structure of the inner membrane of mitochondria and chloroplasts.
213. Ramaley, R. F., and R. W. Bernlohr, Arch. Biochem. Biophys. *117*: 34–43, 1966. Postlogarithmic phase metabolism of sporulating microorganisms: breakdown of arginine to glutamic acid.
214. Ramos, F., V. Stalon, A. Pierard, and J. M. Wiame, Biochim. Biophys. Acta *139*:98–106, 1967. The specialization of the two ornithine carbamoyltransferases of *Pseudomonas.*
215. Rao, K. P., and T. Gopalakrishnareddy, Comp. Biochem. Physiol. 7:175–178, 1962. Nitrogen excretion in arachnids.
216. Ratner, S., H. Morell, and E. Carvalho, Arch. Biochem. Biophys. *91*:280–289, 1960. Enzymes of arginine metabolism in brain.
217. Razet, P., L'Annee Biol. *70*:42–73, 1966. The terminal elements of nitrogen catabolism of insects.
218. Read, K. R. H., Comp. Biochem. Physiol. 7:15–22, 1962. Transamination in certain tissue homogenates of the bivalved molluscs *Mytilus edulis* L. and *Modiolus modiolus* L.
219. Read, L. J., Nature (London) *215*:1412–1413, 1967. Enzymes of the ornithine-urea cycle in the chimaera *Hydrolagus colliei.*
220. Read, L. J., Comp. Biochem. Physiol. *26*:455–466, 1968. Ammonia and urea production and excretion in the fresh-water-adapted form of the Pacific lamprey, *Entosphenus tridentatus.*
221. Read, L. J., Comp. Biochem. Physiol. *39A*:185–192, 1971. Chemical constituents of body fluids and urine of the holocephalan *Hydrolagus colliei.*
222. Read, L. J., Comp. Biochem. Physiol. *39B*:409–413, 1971. The presence of high ornithine-urea cycle enzyme activity in the teleost *Opsanus tau.*
222a. Rechcigl, M., pp. 236–310. *In* Enzyme Synthesis and Degradation in Mammalian Systems, edited by M. Rechcigl. Karger, Basel, 1971. Intracellular protein turnover and the roles of synthesis and degradation in regulation of enzyme levels.
223. Reddy, G., and J. W. Campbell, Experientia *28*:530–532, 1972. Correlation of ammonia liberation and calcium deposition by the avian egg and blood ammonia levels in the laying hen.
224. Reddy, S. R. R., and J. W. Campbell, Comp. Biochem. Physiol. *28*: 515–534, 1969. Arginine metabolism in insects: properties of insect fat body arginase.
225. Reddy, S. R. R., and J. W. Campbell, Biochem. J. *115*:495–503, 1969. Role of arginase in proline formation during silkmoth development.
226. Reddy, S. R. R., and J. W. Campbell, Comp. Biochem. Physiol. *32*: 499–509, 1970. Molecular weights of arginase from different species.
227. Reem, G. H., J. Biol. Chem. *243*:5695–5701, 1968. Enzymatic synthesis of 5'-phosphoribosylamine from ribose 5-phosphate and ammonia, an alternate first step in purine biosynthesis.
228. Roberts, N. R., R. R. Coelho, O. H. Lowry, and E. J. Crawford, J. Neurochem. *3*:109–115, 1958. Enzyme activities of giant squid axoplasm and axon sheath.
229. Robertson, J. D., Biol. Rev. (Cambridge) *32*:156–187, 1957. The habitat of the early vertebrates.
230. Rockstein, M., Biol. Bull. *141*:167–175, 1971. The distribution of phosphoarginine and phosphocreatin in marine invertebrates.
231. Rodwell, V. W., pp. 191–235. *In* Metabolic Pathways, 3rd Ed., Vol. 3, edited by D. M. Greenberg. New York, Academic Press, 1969. Carbon catabolism of amino acids.
232. Rogers, L. J., Comp. Biochem. Physiol. *18*:249–260, 1966. The nitrogen excretion of *Chelodina longicollis* under conditions of hydration and dehydration.
233. Romanoff, A. L., Biochemistry of the Avian Embryo. New York, Interscience (Wiley), 1967.
234. Rowe, P. B., and J. B. Wyngaarden, J. Biol. Chem. *241*:5571–5576, 1966. The mechanism of dietary alterations in rat hepatic xanthine oxidase levels.
235. Rowe, P. B., and J. B. Wyngaarden, J. Biol. Chem. *243*:6373–6383, 1968. Glutamine phosphoribosylpyrophosphate amidotransferase: purification, substructure, amino acid composition, and absorption spectra.
236. Rowsell, E. V., Biochem. J. *64*:235–245, 1956. Transaminations with L-glutamate and α-oxoglutarate in fresh extracts of animal tissues.
237. Rowsell, E. V., Biochem. J. *64*:246–252, 1956. Transaminations with pyruvate and other α-keto acids.
238. Rubin, M., and L. Knott, Clin. Chim. Acta *18*:409–415, 1967. An enzymatic fluorometric method for ammonia determination.
239. Sallach, H. J., and L. A. Fahien, pp. 1–94. *In* Metabolic Pathways, 3rd Ed., Vol. 2, edited by D. M. Greenberg. New York and London, Academic Press, 1969. Nitrogen metabolism of amino acids.
240. Salvatore, F., and V. Bocchini, Nature (London) *191*:705–706, 1961. Prevention of ammonia toxicity by amino acids concerned in the biosynthesis of urea.
241. Schenker, S., D. W. McCandless, E. Brophy, and M. S. Lewis, J. Clin. Invest. *46*:838–848, 1967. Studies on the intracerebral toxicity of ammonia.
242. Schimke, R. T., J. Biol. Chem. *237*:459–468, 1962. Adaptive characteristics of urea cycle enzymes in the rat.
243. Schimke, R. T., J. Biol. Chem. *237*:1921–1924, 1962. Differential effects of fasting and protein-free diets on levels of urea cycle enzymes in rat liver.
244. Schimke, R. T., J. Biol. Chem. *238*:1012–1018, 1963. Studies on factors affecting the levels of urea cycle enzymes in rat liver.
245. Schimke, R. T., J. Biol. Chem. *239*:136–145, 1964. Enzymes of arginine metabolism in mammalian cell culture.
245a. Schimke, R. T., and D. Doyle, Ann. Rev. Biochem. *39*:929–976, 1970. Control of enzyme levels in animal tissues.
246. Schmidt-Nielsen, B., pp. 79–103. *In* Nitrogen Metabolism and the Environment, edited by J. W. Campbell and L. Goldstein. London, Academic Press, 1972. Mechanism of urea excretion by the vertebrate kidney.
247. Schmidt-Nielsen, B., and W. C. Mackey. pp. 45–93. *In* Clinical Disorders of Fluid and Electrolyte Metabolism, 2nd ed., edited by M. H. Maxwell and C. R. Kleeman. New York, McGraw-Hill, 1972. Comparative physiology of electrolyte and water regulation with emphasis on sodium, potassium, chloride, urea, and osmotic pressure regulation.
248. Schmidt-Nielsen, B., and E. Skadhauge, Amer. J. Physiol. *212*:973–980, 1967. Function of the excretory system of the crocodile.
249. Schoffeniels, E., Life Sci. *3*:845–850, 1964. Effect of inorganic ions on the activity of L-glutamic acid dehydrogenase.
250. Schoffeniels, E., Arch. Int. Physiol. Biochim. *73*:73–80, 1965. L-Glutamic acid dehydrogenase activity in the gills of *Palinurus vulgaris* Latr.
251. Schoffeniels, E., Arch. Int. Physiol. Biochim. *76*:319–343, 1968. The control of intracellular hydrogen transport by inorganic ions.

252. Schoffeniels, E., and R. Gilles, Life Sci. *2*:834–839, 1963. Effect of cations on the activity of L-glutamic acid dehydrogenase.
253. Scholz, R. W., Comp. Biochem. Physiol. *36*:503–512, 1970. Comparative studies on liver and kidney xanthine dehydrogenase in two breeds of domestic chicks (*Gallus domesticus*) during prolonged starvation.
254. Seegmiller, J. E., F. M. Rosenbloom, and W. N. Kelley, Science *155*: 1682–1684, 1967. Enzyme defect associated with a sex-linked human neurological disorder and excessive purine synthesis.
255. Sellinger, O. Z., and F. de Balbian Verster, J. Biol. Chem. *237*:2836–2844, 1962. Glutamine synthetase of rat cerebral cortex: intracellular distribution and structural latency.
256. Sharma, M. L., Comp. Biochem. Physiol. *30*:309–321, 1969. Trigger mechanism of increased urea production by the crayfish *Orconectes rusticus* under osmotic stress.
257. Shaw, J., J. Exp. Biol. *37*:534–547, 1960. The absorption of sodium ions by the crayfish *Astacus pallipes* Lereboullet.
258. Shaw, J., J. Exp. Biol. *37*:548–556, 1960. The absorption of sodium ions by the crayfish *Astacus pallipes* Lereboullet.
259. Shigesada, K., and M. Tatibana, J. Biol. Chem. *246*:5588–5595, 1971. Role of acetylglutamate in ureotelism.
260. Shigesada, K., and M. Tatibana, Biochem. Biophys. Res. Commun. *44*:1117–1124, 1971. Enzymatic synthesis of acetylglutamate by mammalian liver preparations and its stimulation by arginine.
261. Shinoda, T., and E. Glassman, Biochim. Biophys. Acta *160*:178–187, 1968. Multiple molecular forms of xanthine dehydrogenase and related enzymes.
261a. Shoemaker, V. H., D. Balding, R. Ruibal, and L. L. McClanahan, Jr., Science *175*:1018–1020, 1972. Uricotelism and low evaporative water loss in a South American frog.
262. Shorey, J., D. W. McCandless, and S. Schenker, Gastroenterology *53*: 706–711, 1967. Cerebral α-ketoglutarate in ammonia intoxication.
263. Siegel, S. M., H. C. Nathan, and K. Roberts, Proc. Nat. Acad. Sci. *60*:505–508, 1968. Experimental biology of ammonia-rich environments: optical and isotopic evidence for vital activity in *Penicillium* in liquid ammonia-glycerol media at −40°.
264. Siegel, S. M., K. Roberts, H. Nathan, and O. Daly, Science *156*:1231–1234, 1967. Living relative of the microfossil *Kakabekia*.
265. Smith, H. W., From Fish to Philosopher. New York, Oxford, 1956.
266. Snell, K., Biochem. J. *123*:657–659, 1971. The regulation of rat liver L-alanine-glyoxylate aminotransferase by glucagon *in vivo*.
267. Sonne, J. C., I. Lin, and J. M. Buchanan, J. Biol. Chem. *220*:369–378, 1956. Biosynthesis of the purines: precursors of the nitrogen atoms of the purine ring.
268. Sorenson, L. B., J. Clin. Invest. *49*:968–978, 1970. Mechanism of excessive purine biosynthesis in hypoxanthine-guanine phosphoribosyltransferase deficiency.
269. Speeg, K. V., Jr., and J. W. Campbell, Amer. J. Physiol. *214*:1392–1402, 1968. Formation and volatilization of ammonia gas by terrestrial snails.
270. Speeg, K. V., Jr., and J. W. Campbell, Comp. Biochem. Physiol. *26*:579–595, 1968. Purine biosynthesis and excretion in *Otala* (=*Helix*) *lactea*: an evaluation of the nitrogen excretory potential.
271. Speeg, K. V., Jr., and J. W. Campbell, Amer. J. Physiol. *216*:1003–1012, 1969. Arginine and urea metabolism in terrestrial snails.
271a. Spelsberg, T. C., A. W. Steggles, F. Chytil, and B. W. O'Malley, J. Biol. Chem. *247*:1368–1374, 1972. Progesterone-binding components of chick oviduct. Exchange of progesterone-binding capacity from target to nontarget tissue chromatin.
272. Spotte, S. H., Fish and Invertebrate Culture. New York, John Wiley and Sons, 1970.
273. Stabenau, J. R., K. S. Warren, and D. P. Rall, J. Clin. Invest. *38*:373–383, 1959. The role of pH gradient in the distribution of ammonia between blood and cerebrospinal fluid, brain and muscle.
274. Stadtman, E. R., A. Ginsburg, J. E. Ciardi, J. Yeh, S. B. Hennig, and B. M. Shapiro, pp. 99–118. *In* Advances in Enzyme Regulation, edited by G. Weber. Oxford, Pergamon, 1970. Multiple molecular forms of glutamine synthetase produced by enzyme catalyzed adenylylation and deadenylylation reactions.
275. Stephens, G. C., J. F. van Pilsum, and D. Taylor, Biol. Bull. *129*:573–581, 1965. Phylogeny and the distribution of creatine in invertebrates.
276. Stetten, D. W., Bull. N.Y. Acad. Med. *28*:664–672, 1952. On the metabolic defect in gout.
277. Stirpe, F., and E. Della Corte, J. Biol. Chem. *244*:3855–3863, 1969. The regulation of rat liver xanthine oxidase. Conversion *in vitro* of the enzyme activity from dehydrogenase (type D) to oxidase (type O).
278. Strandholm, J. J., N. R. M. Buist, N. G. Kennaway, and H. T. Curtis, Biochim. Biophys. Acta *244*:214–216, 1971. Excretion of α-*N*-acetylcitrulline in citrullinaemia.
279. Struck, J., and I. W. Sizer, Arch Biochem. Biophys. *90*:22–30, 1960. Oxidation of L-α-amino acids by chicken liver microsomes.
280. Studier, E. H., J. Exp. Zool. *163*:79–85, 1966. Studies on the mechanisms of ammonia tolerance of the guano bat.
281. Studier, E. H., J. Exp. Zool. *170*:253–258, 1969. Respiratory ammonia filtration, mucous composition and ammonia tolerance in bats.
282. Studier, E. H., and A. A. Fresquez, Ecology *50*:492–494, 1969. Carbon dioxide retention: a mechanism of ammonia tolerance in mammals.
283. Studier, E. H., L. R. Beck, and R. G. Lindeborg, J. Mammal. *48*:564–572, 1967. Tolerance and initial metabolic response to ammonia intoxication in selected bats and rodents.
284. Swick, R. W., P. L. Barnstein, and J. L. Strange, J. Biol. Chem. *240*: 3341–3345, 1965. The response of the isozymes of alanine aminotransferase to diet and hormones.
285. Tamir, H., and S. Ratner, Arch. Biochem. Biophys. *102*:249–258, 1963. Enzymes of arginine metabolism in chicks.
286. Tamir, H., and S. Ratner, Arch. Biochem. Biophys. *102*:259–269, 1963. A study of ornithine, citrulline and arginine synthesis in growing chicks.
287. Thorson, T. B., C. M. Cowan, and D. E. Watson, Science *158*:375–377, 1967. *Potamotrygon* spp.: elasmobranchs with low urea content.
288. Tojo, S., Insect Biochem. *1*:249–263, 1971. Uric acid production in relation to protein metabolism in the silkworm, *Bombyx mori*, during pupal-adult development.
289. Tramell, P. R., and J. W. Campbell, Comp. Biochem. Physiol. *32*:569–571, 1970. Nitrogenous excretory products of the giant South American land snail, *Strophocheilus oblongus*.
290. Tramell, P. R., and J. W. Campbell, J. Biol. Chem. *245*:6634–6641, 1970. Carbamyl phosphate synthesis in a land snail, *Strophocheilus oblongus*.
291. Tramell, P. R., and J. W. Campbell, Comp. Biochem. Physiol. *40B*: 395–406, 1971. Carbamyl phosphate synthesis in invertebrates.
292. Tramell, P. R., and J. W. Campbell, Comp. Biochem. Physiol. *42B*: 439–449, 1972. Arginine and urea metabolism in the South American land snail, *Strophocheilus oblongus*.
293. Trijbels, F., and G. D. Vogels, Comp. Biochem. Physiol. *30*:359–365, 1969. Catabolism of allantoate and ureidoglycolate in *Rana esculenta*.
294. Trotta, P. P., M. E. Burt, R. H. Haschenmeyer, and A. Meister, Proc. Nat. Acad. Sci. *68*:2599–2603, 1971. Reversible dissociation of carbamyl phosphate synthetase into a regulated synthesis subunit and a subunit required for glutamine utilization.
295. Unsworth, B. R., and E. M. Crook, Comp. Biochem. Physiol. *23*:831–845, 1967. The effect of water shortage on the nitrogen metabolism of *Xenopus laevis*.
296. Unsworth, B. R., J. B. Balinsky, and E. M. Crook, Comp. Biochem. Physiol. *31*:373–377, 1969. Evidence for direct excretion of blood ammonia by an ammoniotelic amphibian.
297. Van Wagtendonk, W. J., and A. T. Soldo, pp. 1–56. *In* Comparative Biochemistry of Nitrogen Metabolism, edited by J. W. Campbell. London, Academic Press, 1970. Nitrogen metabolism in protozoa.
298. Vogel, R. H., W. L. McLellan, A. P. Hirvonen, and H. J. Vogel, pp. 463–488. *In* Metabolic Pathways, 3rd Ed., Vol. V, edited by H. J. Vogel. New York, Academic Press, 1971. The arginine biosynthetic system and its regulation.
299. Volpe, P., R. Sawamura, and H. J. Strecker, J. Biol. Chem. *244*:719–726, 1969. Control of ornithine δ-transaminase in rat liver and kidney.
300. Vorhaben, J. E., and J. W. Campbell, unpublished observations, 1971.
301. Vorhaben, J. E., and J. W. Campbell, J. Biol. Chem. *247*:2763–2767, 1972. Glutamine synthetase: a mitochondrial enzyme in uricotelic species.
302. Wada, H., and Y. Morino, pp. 411–444. *In* Vitamins and Hormones, edited by R. S. Harris, I. G. Wool, and J. A. Loraine. New York, Academic Press, 1964. Comparative studies on glutamic-oxalacetic transaminases from the mitochondrial and soluble fractions of mammalian tissues.
303. Wainwright, S. D., P. Bright-Asare, and J. C. Campbell, Canad. J. Biochem. *45*:614–618, 1967. Exploratory studies of the liver glutamic dehydrogenase of the hagfish *Myxine glutinosa*. Lack of regulation of activity by ADP and diethylstilbestrol in physiological saline.
304. Walker, J. B., pp. 43–55. *In* The Comparative Biochemistry of Arginine and Derivatives, edited by G. E. W. Wolstenholme and M. P. Cameron. Boston, Little, Brown and Co., 1965. Transamidination and the biogenesis of guanidine derivatives.
305. Walser, M., pp. 421–429. *In* Urea and the Kidney, edited by B. Schmidt-Nielsen and D. W. S. Kerr. Amsterdam, Excerpta Medica, 1970. Use of isotopic urea to study the distribution and degradation of urea in man.
306. Warren, K. S., J. Clin. Invest. *37*:497–501, 1958. The differential toxicity of ammonium salts.
307. Warren, K. S., and D. G. Nathan, J. Clin. Invest. *37*:1724–1728, 1958. The passage of ammonia across the blood-brain barrier and its relation to blood pH.
308. Warren, K. S., and S. Schenker, Amer. J. Physiol. *199*:1105–1108, 1960. Hypoxia and ammonia toxicity.
309. Warren, K. S., and S. Schenker, Amer. J. Physiol. *203*:903–906, 1962. Differential effect of fixed acid and carbon dioxide on ammonia toxicity.
310. Watts, D. C., and R. L. Watts, Comp. Biochem. Physiol. *17*:785–798, 1966. Carbamoyl phosphate synthetase in the Elasmobranchii: osmoregulatory function and evolutionary implications.

311. Weir, E., and J. R. Fisher, Biochim. Biophys. Acta *222*:556–557, 1970. The effect of allopurinol on the excretion of oxypurines by the chick.
312. Wergedal, J. E., and A. E. Harper, Proc. Soc. Exp. Biol. Med. *166*: 600–607, 1964. Glutamic dehydrogenase activity of rats consuming high-protein diets.
313. Wergedal, J. E., and A. E. Harper, J. Biol. Chem. *239*:1156–1163, 1964. Effect of high protein. intake on amino nitrogen catabolism *in vivo*.
314. Westall, R. G., Biochem. J. 77:135–144, 1960. Argininosuccinic aciduria.
315. Wieser, W., and G. Schweizer, J. Exp. Biol. *52*:267–274, 1970. A reexamination of the excretion of nitrogen by terrestrial isopods.
316. Wiggert, B. A., and P. P. Cohen, J. Biol. Chem. *241*:210–216, 1966. Comparative study of tadpole and frog glutamate dehydrogenase.
317. Wilson, R. P., R. P. Anderson, and R. A. Bloomfield, Comp. Biochem. Physiol. *28*:107–118, 1969. Ammonia toxicity in selected fishes.
318. Wilson, R. P., M. E. Muhrer, and R. A. Bloomfield, Comp. Biochem. Physiol. *25*:295–301, 1968. Comparative ammonia toxicity.
319. Wilson, R. P. L. E. Davis, M. E. Muhrer, and R. A. Bloomfield, J. Animal Sci. *23*:1221, 1964. Toxicity of ammonium carbamate.
320. Winitz, M., J. P. de Ruisseau, M. C. Otey, S. M. Birnbaum, and J. P. Greenstein, Arch. Biochem. Biophys. *64*:368–373, 1956. Effect of combined administration of nonprotective compounds and subprotective levels of L-arginine · HCl on ammonia toxicity in rats.
320a. Wixom, R. L., M. K. Reddy, and P. P. Cohen, J. Biol. Chem. *247*: 3684–3692, 1972. A concerted response of the enzymes of urea biosynthesis during thyroxine-induced metamorphosis of *Rana catesbiana*.
321. Wood, J. D., Canad. J. Biochem. Physiol. *36*:1237–1242, 1958. Nitrogen excretion in some marine teleosts.
322. Wrong, O., B. J. Houghton, P. Richards, and D. R. Wilson, pp. 461–470. *In* Urea and the Kidney, edited by B. Schmidt-Nielsen and D. W. S. Kerr. Amsterdam, Excerpta Medica, 1970. The fate of intestinal urea in normal subjects and patients with uremia.
323. Wu, C., Biochim. Biophys. Acta 77:482–493, 1963. Glutamine synthetase, intracellular localization in rat liver.
324. Zandee, D. I., Arch. Int. Physiol. Biochim. *74*:35–44, 1966. Metabolism in the crayfish *Astacus astacus* (L.): biosynthesis of amino acids.
325. Zeller, E. A., G. Ramachander, G. A. Fleisher, T. Ishimaru, and V. Zeller, Biochem. J. *95*:262–269, 1965. Ophidian L-amino acid oxidase. The nature of the enzyme-substrate complexes.
326. Zielinska, Z. M., and B. Grzelakowska-Sztabert, Acta Biochim. Polon. *15*:1–13, 1968. Formyltetrahydrofolate synthetase from a uricotelic insect, *Galleria mellonella* L. (Lepidoptera).
327. Zieve, L., Arch. Int. Med. *118*:211–223, 1966. Pathogenesis of hepatic coma.
328. Zubay, G., and D. A. Chambers, pp. 297–347. *In* Metabolic Pathways, 3rd Ed., Vol. 5, edited by H. J. Vogel. Academic Press, New York, 1971. Regulating the *lac* operon.

CHAPTER 8

RESPIRATORY FUNCTIONS OF BLOOD

By C. Ladd Prosser

Whenever a circulatory system transports oxygen from a respiratory surface to body tissues, there is usually a transport pigment in the blood. In only a few animals can the blood carry enough oxygen in solution without a pigment. In man, for example, oxygen is *dissolved* in the plasma to the extent of 0.3 volumes (vol) per cent; in whole blood this figure is 0.24 vol per cent. Actually, arterial blood contains over 19 vol per cent of oxygen; 98 per cent of the oxygen in the blood is combined with hemoglobin. In most animals the pigment functions in transport continuously; in some it functions only at low oxygen pressures, and in others it provides reserve oxygen above that in solution. Blood pigments serve additional functions as buffers in the transport of carbon dioxide and as protein for maintaining colloid osmotic pressure of the blood.

All transport pigments contain a metal in an organic complex. Most blood pigments have iron, and a few have copper; other blood pigments are known, but their respiratory function is doubtful.

DISTRIBUTION OF PIGMENTS

Hemoglobins. Hemoglobins consist of an iron-porphyrin (heme) coupled to a protein (globin). Porphyrins are widely distributed in nature. Chlorophyll is a magnesium-porphyrin. The iron-porphyrins, cytochromes, are found in all aerobic cells. The iron-protoporphyrin heme coupled with specific proteins is the best known oxygen-carrying pigment. The protein moiety or globin varies considerably in size, amino acid composition, charge, solubility, and other physical properties from animal to animal. Hemoglobins may exist extracellularly in body fluids, or they may be intracellular in special corpuscles or in tissue cells, particularly in muscle and nerve. Hemoglobin may occur in only a few genera of a phylum, and it occurs in members of unrelated phyla. There are many different hemoglobins with different proteins and similar heme.

The distribution of hemoglobins and related pigments, primarily identified spectroscopically, is as follows:

Chordates:

Vertebrates—All classes have hemoglobin (Hb) in blood corpuscles; red muscles contain myoglobin (muscle hemoglobin or Mb). Pigment is lacking in a few fish—leptocephalan eel larvae, and in three genera of Antarctic fish.

Mammalian erythrocytes are non-nucleated, circular (except in Camellidae), biconcave; erythrocytes of most other vertebrates are elliptical, nucleated, double convex. In general, the non-nucleated cells are smaller and more numerous than the nucleated ones; the largest red cells occur in amphibians. The hemoglobin content is 12 to 18 gm/100 ml blood in most mammals and birds, 6 to 10 in amphibians and reptiles, 6 to 11 in fish. The average life of a circulating red cell is in days: man 113 to 118, dog 90 to 135, rat 50 to 60, rabbit 50 to 70, chicken 28 to 38, turtle more than 11 months.

Prochordates—no Hb in *Branchiostoma* or most prochordates.
Echinoderms:
Holothurians—*Thyone, Cucumaria, Molpadia, Caudina*—have hemoglobin in corpuscles.
Annelids:
Oligochaetes—*Lumbricus, Tubifex*—and hirudineans—*Hirudo, Analastoma*—have Hb dissolved in blood plasma; *Lumbricus* has myoglobin in muscle.
Polychaetes show a variety of pigments:
A. Those with a closed circulatory system may have
1. Hemoglobin in both coelomic fluid cells and in blood plasma—*Terebella, Travisia.*
2. Hemoglobin only in blood plasma—Nereidae, *Arenicola, Amphitrite,* Cirratulidae, Eunicidae, and others.
3. Chlorocruorin in plasma—Sabellidae, Serpulidae, Chlorhaemidae, Ampharetidae. Some serpulids have both hemoglobin and chlorocruorin together in the plasma.
4. Hemerythrin in blood cells—*Magelona.*
5. No pigment in either blood or coelomic fluid—Syllidae, *Phyllodice,* Aphroditidae, Chaetopteridae, *Lepidonotus.*
B. Polychaetes lacking a functional circulatory system may have
1. Hemoglobin in cells in coelomic fluid—Capitellidae, Glyceridae, *Polycirrus hematodes, P. aurantiacus.*
2. No pigment in coelomic fluid—*Polycirrus tenuisetis, P. arenivorus.*
Echiuroids: Hemoglobin occurs in coelomic corpuscles in *Urechis, Thalassema,* and also in body wall muscles of *Urechis* and *Arhynchite.*
Phoronids: Hb in corpuscles in *Phoronis* and *Phoronopsis.*
Arthropods:
Crustacea—Hemoglobin unknown in Malacostraca, common among Entomostraca.
Anostraca—*Artemia,* Cladocera—*Daphnia,*
Notostraca—*Triops, Apus.*
Conchostraca, a parasitic copepod, an ostracod, a parasitic cirripede.
Insects—Chironomid larvae, *Gastrophilus* (a dipteran parasite, Hb within tracheal and some other cells).
Molluscs: in corpuscles in a few pelecypods—*Solen, Arca, Pectunculus,* in plasma of one gastropod, *Planorbis.* Myoglobin in radular muscle of many prosobranch gastropods and chitons.
Nemerteans: In plasma in some, in erythrocytes in others; also in some ganglion cells of *Polia.*
Platyhelminthes: A few parasitic trematodes and rhabdocoeles, *Derostoma, Syndesmis, Telorchis.*
Nemathelminthes: Hb in pseudocoelic fluid and in hypodermal cells of body wall—*Ascaris, Nippostrongylus, Eustrongylides, Camallanus.*
Protozoa: *Paramecium* and *Tetrahymena* (certain strains).
Plants: Root nodules of some legumes contain Hb, which is formed when plant and bacteria are symbiotic, not when separate.

Chlorocruorins. This pigment, green in dilute solution, contains iron in a different porphyrin from hemoglobin. It occurs in the plasma of at least four families of polychaete worms, particularly the Sabellidae and Serpulidae. Some twenty-one species of Serpulimorpha have been shown to have chlorocruorin, but *Potamilla* has muscle hemoglobin and blood chlorocruorin; in the genus *Spirorbis* the species *borealis* has chlorocruorin, *corrugatus* has blood hemoglobin, and *militaris* neither. The genus *Serpula* has both hemoglobin and chlorocruorin in the blood.[50, 53]

Hemerythrins. A third iron-containing pigment, hemerythrin, occurs in the polychaete worm *Magelona,* in the sipunculid worms Sipunculus, *Dendrostomum, Golfingia (Phascolosoma),* in the brachiopod *Lingula,* and in some priapulids. Hemerythrin is found in corpuscles, is violet in color, and the iron is not contained in a porphyrin.

Hemocyanins. The pigment which is next in importance to hemoglobin, as judged by its occurrence, is the copper-containing hemocyanin. It is found among molluscs in amphineurans, cephalopods, and some gastropods, among arthropods in many malacostracans and crustaceans; in *Limulus* and a few arachnids. Hemocyanin is a copper-protein without any porphyrin group: it always occurs dissolved in the plasma. Many molluscs, e.g., the snail *Busycon* and the amphineuran *Cryptochiton,* have hemocyanin in the blood and myoglobin in some muscles.

CHEMISTRY OF TRANSPORT PIGMENTS

HEMOGLOBINS

General Structure. Hemoglobins consist of variable numbers of unit molecules, each unit containing one heme and its associated protein. The unit molecular weight (MW) for vertebrates is approximately 16×10^3; the blood hemoglobin (Hb) of higher vertebrates has four units (i.e., MW 64 to 65×10^3). The ratio of heme to protein is less in invertebrate hemoglobins. Heme is a protoporphyrin consisting of four pyrroles with an iron at the center (Fig. 8–1).

The iron content of mammalian hemoglobin is 0.336% and the heme content is 4%; the iron is normally in the ferrous state. The iron content of human myoglobin is 0.318%. The function of hemoglobin depends on the reversible combination of the ferrous iron with oxygen (oxygenation) in proportion to the partial pressure of oxygen (Po_2). The iron can be oxidized to the ferric state by strong oxidants, resulting in methemoglobin, which is no longer capable of com-

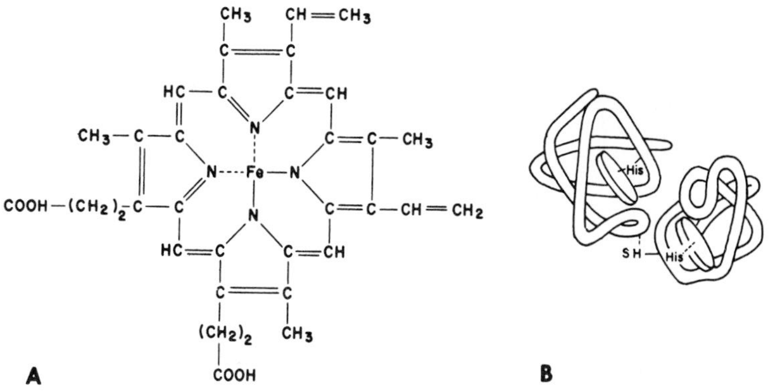

Figure 8–1. *A*, Structural formula of heme. *B*, Configuration of two of the subunit chains of hemoglobin as deduced from x-ray diffraction data. Hemes, represented as discs, connected via histidine to protein. Sulfhydryl (SH) groups are in different planes from the histidine (His). (Modified from Perutz, M. F., et al., Nature *185*:416–422, 1960.)

bining reversibly with oxygen. Oxygen combines with each atom of iron as described by the equilibrium

$$K = \frac{[FeO_2]}{[Fe][O_2]}$$

Carbon monoxide can also combine reversibly, and the affinity of hemoglobin for CO is usually much greater than that for O_2.

X-ray analysis reveals the heme at the surface of the globular protein, although it is in a crevice (Fig. 8–2). Most of the bonding points of heme to the globin are hydrophobic residues. In the β-chain one propionic acid COO^- makes contact with a serine and one with a lysine; in the α-chain a propionic acid residue contacts a histidine.[146]

Deoxygenated (reduced) hemoglobin has a positive magnetic moment (i.e., it is paramagnetic). Combination with oxygen or carbon monoxide results in loss of magnetic moment (i.e., it becomes diamagnetic, or weakly repulsed in a magnetic field), which indicates an absence of unpaired electrons. Changes in magnetic moment indicate changes in electron pairing relative to the Fe. Optic spectra and magnetic measurements indicate the state of the iron and the nature of the ligand in oxyhemoglobin.[206] Ferric and ferrous iron can exist in two spin states, which reflect the distribution of electrons in the d-orbital. By comparison of a series of iron porphyrin compounds, it is concluded that on oxygenation there is a partial transfer of an electron from ferrous iron to O_2; the iron becomes ferric and the O_2 becomes a superoxide:

$$\text{Hb (heme } d_2^6) + O_2 \rightleftharpoons \text{Hb (heme } d_2^5) \cdot OO^-$$

Thus, ferrous iron (d^5) goes to formal ferric (d^6), and O_2 goes to OO^-.

Myoglobin and hemoglobin molecules are spheroid, with a variable number of unit chains. There is one chain in myoglobin (Mb), and there are four in most vertebrate blood hemoglobins. The four-heme human hemoglobin molecule is made up of two equal halves, each with an α and a β chain fitted together. Each β chain consists of eight segments of right-handed α-helix, and each α chain has seven helical segments. Myoglobin consists of eight helical segments, with a total of approximately 150 residues. In HbA, the α chain has 141 residues and the β chain has 146 residues. Models of these molecules are discussed by Perutz et al.[144,145,146]

After cleavage of only one or two terminal amino acids by carboxypeptidase, myoglobin shows a marked decrease in Bohr effect (see page 332) and O_2 affinity. Also, these enzymes are more active on HbO_2 than on Hb. Thus, we can say that even very small conformational changes modify respiratory function of these pigments. Antonini[7] has shown that carboxypeptidase A cleaves at the COOH-terminal residues of the β chain on human Hb and that carboxypeptidase B cleaves at the same residues on the α chain.

The structure of hemoglobin (sometimes called erythrocruorin) in several invertebrates has recently been elucidated. Hb of *Paramecium aurelia* contains 100 amino acid residues.[190] The α-helix structure as shown by x-ray diffraction is similar for Hb of the insect *Chironomus*,[85] the annelid *Glycera*,[139] and the horse.[146] Hb of the polychaete *Arenicola*, unlike vertebrate Hb, has one heme per two protein chains, each chain having about 13,000 MW.[201a] The hemoglobin of perienteric fluid of *Ascaris* contains eight hemes per molecule of total MW 328,000, and the Hb of the body wall has one heme per molecule, MW 37,000.[138a]

Most animals have several different hemo-

A

B

Figure 8–2. Photographs of model of reconstruction of hemoglobin molecule. *A* view is along axis, showing contact of $\alpha_1\beta_1$: *B* view is along axis perpendicular to *A* to show contact of $\alpha_1\beta_2$. (From Perutz, M. F., et al., Proc. Roy. Soc. Lond. B *173*:113–140, 1969.)

TABLE 8–1. Maxima of Absorption Bands of Oxygenated, Carbon Monoxide, Deoxygenated, and Oxidized (met) Forms of Pigment in Nanometers

	Oxygenated		*Carbon Monoxide*			
	α	β	α	β	*deoxy*	*met*
Hemoglobins						
vertebrates	576.4 to 576.9	540 to 544	571	535	544	630
annelids						
Arenicola	574.6	540	569.8		563	
Lumbricus	575.5	538 to 544	570		563	
Hirudo	576.5	538				
Haemopis[136]	576.5	538				
Marphysa[29a]	576.	540	570	537	553	
echinoderms						
Cucumaria sp.	579	543	573	535	538	
C. miniata			570	539		
arthropods						
Chironomus[202]	578	541.8	573	539		
Artemia	577	542	569	537		
Triops[78]	574	541	569	538	566	
Moina[82]	578	542	571	540	559	
Cyzicus[10]	573	538	568	536		
molluscs						
Arca	578	541.5				
Planorbis	574.6		570.8			
nematodes						
Ascaris[189]						
perienteric fluid	578	540		540	550	630, 500
body wall	579	541				
Strongylus	578.1	540			555	637
rhabdocoels						
Phaenocora[34]	580	540			560	
protozoans						
Paramecium	581	545	570	542	553	
Myoglobins	580	541	579		555	630
Chlorocruorin	604	559	601.8	574	525	
Hemerythrin	408	510			390 to 410	
Hemocyanin						
Octopus	570					
Limulus	575 to 580	580				
Homarus	570					
Palinurus	558					
Helix[107]	575					

globins, in which a different chain may occur in substitution for one pair. For example, fetal Hb in man (HbF) differs from adult Hb in the substitution of α chains (of different amino acid sequences) for the β chains. More than 90 abnormal human hemoglobins have been described; with few exceptions, each of these differs by one amino acid residue on one of the chains. For example, substitution of one valine for a glutamic acid residue among the approximately 290 amino acid residues per half molecule decreases the solubility of the deoxygenated form of hemoglobin by a factor of 100; this is the condition of sickle-cell anemia.[90, 91]

Absorption Bands. All heme compounds show characteristic absorption spectra; the midpoints of some of these bands for a few animals of the vast number that have been examined are listed in Table 8–1. Oxygenated hemoglobin has two principal bands, an alpha band in the yellow and a beta band in the green. Carbon monoxide-Hb also has two bands at shorter wavelengths than oxy-

Hb, and the distance or span in nm or Å between the α bands of oxy-Hb and carboxy-Hb is characteristic of a given hemoglobin. Deoxygenated hemoglobin has a single broad band in the yellow-green. Methemoglobin shows four absorption bands depending on pH; two of them are most prominent. In addition to the bands in the visible spectrum, there are strong bands in the violet (Soret bands) as follows:

Soret absorption bands in mμ

	HbO_2	HbCO	Met Hb	Hb
Hemoglobin	414.5	420	406	425
Myoglobin	418	424	407	435

Different hemoglobins show (qualitatively) the same absorption bands, since the absorption is by heme. However, differences in proteins can be seen as shifts of several Å units within an animal class or as differences of several mμ from one class or phylum to another. Usually the α peak is higher than the β peak, but in some (e.g., *Ascaris* and the legume nodule Hb) the β is higher.[102] The span between the α bands of oxy-Hb and carboxy-Hb is, for most Hb's, 43 to 56 Å units; for myoglobin it is 31 to 36 Å, for the parasitic fly *Gastrophilus* it is 95 Å, and for the root nodule it is 100 Å. In most parasites the hemoglobin has slightly different absorption bands from those of the host's Hb (e.g., *Ascaris* in pig and *Gastrophilus* in horse), but *Nematodirus* has bands similar to those of its sheep host. The six Hb's of sea lamprey are of the same size but differ slightly in Soret peaks.[180]

Solubility, Isoelectric Point, and Alkaline Denaturation. In neutral phosphate buffer, horse Hb is more soluble than that of man; cow fetal Hb is six times (and sheep 20 times) more soluble than those of the adult, whereas human adult Hb is more soluble than is fetal Hb.

In general, vertebrate hemoglobins are isoelectric at pH 6.8 to 7.0, while most invertebrate Hb's have isoelectric points between pH 5.0 and 6.0. However, some fish have low isoelectric points—carp, 6.5; toadfish, 5.7 to 6.2.[173] Hb of a cyclostome (*Petromyzon*) is isoelectric at pH 5.6.[201] In the turtle *Pseudemys*, one Hb is isoelectric at pH 5.7 and another at 7.2. Isoelectric points of invertebrate Hb's differ; for example, *Gastrophilus* 6.2, *Arenicola* 5.1,[201a] and *Marphysa* 4.6.[29a] The Hb of root nodule bacteria is isoelectric at pH 4.5. *Ascaris* has body wall Hb with an isoelectric point at pH 5, and perienteric fluid 6.7.[189] An eel, *Anguilla japonica*, has two hemoglobins: E_1 has isoelectric point at pH 8.08, while E_2 is isoelectric at pH 5.96. E_1 is more resistant to alkaline denaturation.[68]

Rate of denaturation at high pH is used to characterize hemoglobins. Alkaline denaturation is much slower in human fetal hemoglobin (HbF) than in adult hemoglobin (HbA), but in many animals—sheep, tadpoles, certain fishes—the embryonic Hb denatures faster than the adult.[101] The hemoglobin of *Lumbricus* is heterogeneous; it denatures in alkali in three steps, and its organization is not known.

Molecular Size. Estimates of molecular size have been made for many hemoglobins by sedimentation in an ultracentrifuge. The sedimentation coefficient, $s_{20,w}$, is the sedimentation velocity in cm/sec in a unit centrifugal field at 20°C in a medium reduced to the viscosity of water.[193] Representative values of s_{20}, together with some calculated molecular weights, are given in Table 8–2. In general, sedimentation coefficients for vertebrate hemoglobins are 4.3 to 4.7×10^{-13} cm/sec, and those for invertebrate hemoglobins are higher. In the eel *Anguilla japonica*, one form of Hb, E_1, has a molecular weight of 65,200 and consists of four units of MW 16,300, while E_2 has MW 68,800 and consists of four units of MW 17,200.[212] Hemoglobin molecules of amphibians and reptiles tend to be slightly larger than those of mammals, birds, and fishes. In the toad *Bufo valliceps*, three components with s_{20}'s of 4.8, 7.7, and 12.5 suggest aggregation of units.[193] In a survey of 54 species of turtles, some had 4s and 7s components, and others had more than 8s.[192] Frog and turtle Hb's tend to polymerize on hemolysis.[165] In *Ascaris* perienteric fluid, the sedimentation coefficient of the major component of the Hb is 11s; some molecules occur at 8s and at 2s; and body wall Hb has a value of 1.5s.[189, 191] These correspond to eight hemes per molecule in the perienteric fluid, or a molecular weight for each heme of 40,600. *Paramecium* hemoglobin has a molecular weight of 13,000 and an s_{20} of 1.5, compared to 1.84 for Mb.[190]

Myoglobin (human) has an s_{20} of 1.81 and a molecular weight of 17,450 (150 aa residues); it contains one heme per molecule. Blood Hb has a molecular weight of 64,500 and has four hemes per molecule. The circulating Hb of cyclostomes has a MW of about 17,800 with a single heme per molecule and s_{20} of 1.9. The Hb of *Petromyzon marinus* is a monomer when oxygenated and may be a dimer or trimer when deoxygenated; hence, loading is in steps.[25] The hemoglobins of three amphibians differ in capacity for polymerization: in axolotl, deoxy-Hb is an octomer at pH below 6, but oxy-Hb is tetrameric over the entire pH range; in *Rana esculenta,* oxy-Hb does not polymerize, but deoxy-Hb may be octomeric at pH 5 to 8; and in *Triturus,* both oxy-Hb and deoxy-Hb remain as tetramers over the entire pH range.[44] Mammalian myoglobin can polymerize, especially at pH between 3.5 and 5.5.

In those invertebrates in which the Hb is in solution in the plasma, the MW is usually greater than 1,000,000. The large size tends to confine the molecules to the circulatory system. Those invertebrates with Hb in corpuscles have smaller molecules. In *Thyone* and *Gastrophilus,* with Hb in cells, the MW is about 34,000—probably in two units per molecule. However, in chironomid larvae a hemoglobin of low molecular weight is free in the plasma. In *Chironomus thummi,* there are two classes of Hb with molecular weights 16,000 and 22,000, probably monomer and dimer.[24] In the annelid *Glycera,* the Hb in the coelom has MW 18,200; the number of residues per helix is the same as in myoglobin.[139] The Hb of *Arenicola cristata* has a MW of 2.85×10^6. The intact molecule consists of two six-membered rings, 12 subunits of MW 230,000, with eight hemes each, or 96 hemes per total Hb molecule.[201a] In the clam shrimp *Cyzicus,* the blood Hb has an s_{20} of 11.4, corresponding to a MW of 2.2×10^5; the molecule consists of 12 or 13 subunits.[10]

When some mammalian Hb's are treated with agents such as urea, they split into molecules of size corresponding to MW 34,000; with difficulty, these halves can be split again.[211] X-ray diffraction data indicate that a molecule of horse Hb is a spheroid measuring 64 Å by 55 Å by 50 Å; myoglobin has dimensions of 40 Å by 35 Å by 23 Å.[103, 104] The crystals of hemoglobin from different species have characteristic forms. Electron micrographs of the large hemoglobin molecules of several annelids, especially *Lumbricus,* show that they consist of two hexagonal discs opposed to each other, with a total diameter of 230 to 265 Å; each disc consists of six particles of three 70 Å subunits each.[173a, 178a]

Electrophoretic Mobility. Electrophoretic analysis shows species differences in mobility, and usually reveals several different Hb's in the same animal. Horse myoglobin has two electrophoretic components, hemoglobin of pig has one, cow two, and sheep and buffalo two.[200] Of the two components of sheep, one is more abundant at sea level and the other at high altitudes; different breeds of sheep can be distinguished by their proportions of the two hemoglobins.[45] In adult human hemoglobin (HbA), 2.5% is a second form called HbA_2; in those having thalassemia, this fraction is 5.1%.[108] Human fetal hemoglobin (HbF) is found in umbilical blood; 20-week fetuses contain a Hb of lower electrophoretic speed—an early or primitive hemoglobin (HbP). There is evidence that HbP is made in mesoblastic cells, while HbF is made in the liver and HbA in marrow.[109] An early embryonic Hb precedes the fetal Hb in a number of other vertebrates. Hemoglobins from 112 species in 73 genera of carnivores show Hb with a mobility 0.85 times that of human HbA.[185]

Among Hb's of birds, three components were found in duck and chicken, two in each of 20 species of wild birds, and one in pigeon and penguin. Hybrids of quail and jungle fowl show a mixture of Hb's from the parental stocks.[130] Electrophoretic analysis of hemoglobins from a number of reptiles and amphibians shows many with several components. The frog *Rana catesbiana* has multiple Hb's, and none of the bands is common to tadpole and adult.[2]

The newt *Triturus* has a slow-moving hemoglobin, HbA, and fast-moving HbB; the two forms have one chain in common.[182]

The cyclostome *Petromyzon marinus* has six types of hemoglobin, all with the same molecular weight.[180] *Lampetra fluviatilis* Hb has two major components. Most fish have several circulating Hb's, and the nature of these pigments can supply taxonomic information. When different species of sunfish are hybridized (e.g., bluegill × crappie), the F_1 hybrid usually shows a mixture of the Hb's of the parents. However, the bluegill × warmouth cross yielded 25% hybrid molecules in F_1, while the green sunfish × warmouth cross gave 40% hybrid Hb's in F_1.[131]

In herring (*Clupea harengus*), eight forms of Hb have been identified; each persists throughout life, but the different forms appear successively during four years of development as if different genes were turned on at different times[207] (Fig. 8–3). Coho salmon have 10 anodally migrating Hb's and 12 cathodally migrating ones; adults at sea or mature individuals in fresh water have much more of some other cathodal components.[199]

A sea cucumber occurs in two genetic variants; one of these, "stout," has two types of Hb chain, while the other, "thin," has five.[129]

Evidently, in most animals with hemoglobin several forms are in the blood simultaneously.

Amino Acid Structure of Chains. Different hemoglobins can be split into their component chains, and each chain can be subjected to analysis for amino acids. Each different type of chain is the product of one cistron, and a point mutation can cause a change in one amino acid residue of that chain. Human HbA has two pairs of chains, $\alpha_2\beta_2$; human fetal hemoglobin, HbF, is $\alpha_2\gamma_2$. Genetic analysis shows that the α and β

TABLE 8–2. Sedimentation Coefficients and Computed Molecular Weights of Blood Pigments (Sedimentation Constants in Parentheses are Secondary Values)

	Sedimentation Coefficient ($s_{20,w}$) *cm* × 10^{-13}/*sec dyne at 20°C*	*Molecular Weight*	*Location*
horse Hb	4.4	68,000	
myoglobin		17,000	
man	4.4		
rabbit	4.4		
pigeon	4.4		
duck	4.4		
salamander	4.8		
frog *Rana*	4.5		
toad *Bufo*	4.8		
turtle *Chrysemys*	4.5		
lizard *Lacerta*	4.6		
snake *Coluber*	4.4		
lungfish *Protopterus*	4.3		
goldfish *Cyprinus*	4.4		
toadfish *Opsanus*	4.3		
skate *Raja*	4.3		
eel *Anguilla*[68]		E_2 48,800 E_1 45,200	
lamprey *Lampetra*	1.87	19,100	
Polistostrema	1.9		
Myxine	2.3	23,100	
leech *Hirudo*	58		plasma
earthworm *Lumbricus*	61	2,946,000	plasma
Marphysa[29a]	58.4		plasma
Nereis[183]	58.6	2,400,000	
Arenicola[201a]	58.7	2,850,000	
Arenicola[203]	54	(180 heme units)	
Glycera	3.5		corpuscles
Notomastus	2.1	36,000	corpuscles
holothurian *Thyone*	2.4	23,600	corpuscles
clam *Arca*	3.5	33,600	corpuscles
snail *Planorbis*	33.7	1,539,000	plasma
bivalve *Phacoides*[155]	1.8 to 2	15,000	
crustacean *Daphnia*	14.3		plasma
Moina	17.8	670,000	
Cyzicus[10]		220,000	
insect *Chironomus* sp.	2.0	31,400	plasma
Chironomus plumosus		16,000	plasma
Chironomus thummi[24]		16,000 and 32,000	

(*Table 8–2 continued on opposite page.*)

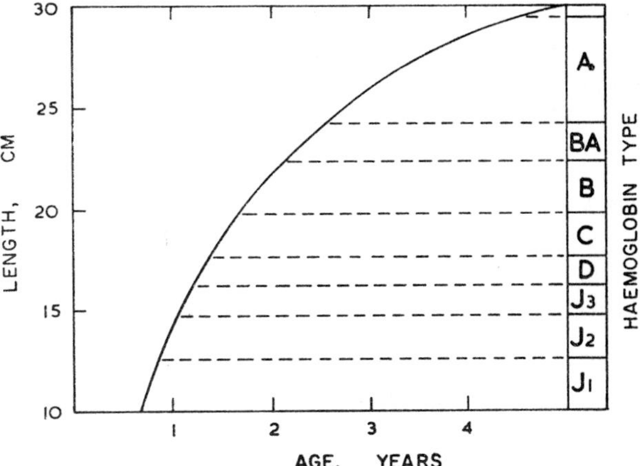

Figure 8–3. Diagram indicating the change in abundance of different hemoglobins according to age and size in herring from the North Sea. (From Wilkins, N. P., and T. D. Iles, Comp. Biochem. Physiol. *17*:1141–1158, 1966.)

TABLE 8–2. Sedimentation Coefficients and Computed Molecular Weights of Blood Pigments (Sedimentation Constants in Parentheses are Secondary Values) (*Continued*)

	Sedimentation Coefficient ($s_{20,w}$) *cm* × 10^{-13}/*sec dyne at 20°C*	*Molecular Weight*	*Location*
Ascaris[138a]			
body wall	3.1	37,000	
perienteric fluid	11.8	328,000	
Ascaris			
perienteric fluid	11.8	328,000	
flatworm *Phaenocora*[34]	2.3		
Chlorocruorin			
Serpula	59	3,000,000	plasma
Sabella	53		plasma
Spirographis[8a, 64a]	57.5	2,750,000	
Hemerythrin			
Sipunculus		66,000	corpuscles
Golfingia		108,000	corpuscles
Hemocyanin			
gastropods			
Buccinum	102 (132)		
Littorina	99.7 (132)		
Busycon[179a]	60	4,400,000	
	100	8,800,000	
	130	13,000,000	
Helix[107]	99	6,680,000	
Helix		9,000,000	
cephalopods			
Loligo[39]	59 (19, 11)	3,800,000	
Rossia	56.2	3,316,000	
Octopus	49.3	2,785,000	
Eledone	49.1		
arthropods			
Cancer[43]	25.4	940,000	
	18	480,000	
Callianassa[179a]	35	1,800,000	
Eriphia[40]	24 (16, 5)	950,000	
Homarus[148]	16	825,000	
Limulus	34.6 (56.6, 16.1, 5.9)	1,800,000	
Pandalus	17.4	397,000	
Palinurus	16.4		
Pagurus	16		
Astacus	23.3		
Carcinus	23.3 (16.7)		

genes are on different chromosomes, but the β and γ are on the same chromosome. The β chain is more subject to mutation than is the α, and the N-terminal sequence in the β chains of four human mutants is as follows:[90]

HbA:	val-	hist-	leuc-	thr-	pro-	glut-	glut-	lys
HbS:	val-	hist-	leuc-	thr-	pro-	val-	glut-	lys
HbC:	val-	hist-	leuc-	thr-	pro-	lys-	glut-	lys
HbG:	val-	hist-	leuc-	thr-	pro-	leu-	glyc-	lys

Hemoglobin is relatively insensitive to amino acid replacements on the surface, but is very sensitive to changes of non-polar contacts, especially near the hemes. Also, replacements at contacts between α and β subunits affect their function. Some amino acid replacements show no alteration in function, while others alter O_2 affinity, the heme-heme interactions, or the equilibrium between the Hb subunits.[146]

Tertiary structure of α and β chains is similar in oxy-Hb and deoxy-Hb, but quaternary structure may be different. In general, α chains are relatively constant, and Ingram postulates an evolutionary sequence as follows: the first vertebrate Hb consisted of a single chain like myoglobin; it became Hb by virtue of α-chain duplication; the differences in amino acids of α chains increase in correspondence with known phyletic separation. It is further postulated that γ and β chains diverged most recently.[91]

In man, α and β differ in respect to 85 out of 141 amino acid sequences; β and γ chains differ in only seven or eight amino acids. The γ chain (HbF) has higher glutamic acid and methionine content and lower valine and proline content than the β chain (HbA).

Human and anthropoid ape Hb's are so similar as to indicate that stable genes for α and β chains may have formed before separation of the two lines. Comparison of human and gorilla HbA shows two aa differences in the α chain, and one in the β chain; between human and horse there are 15 differences.[214] Among lower primates (prosimians) the β chains differ more than do the α chains.[71]

The 153 amino acid residues in each of the four polypeptide chains of mammalian myoglobins show 25 differences between horse and kangaroo, and 31 between whale and kangaroo.[3] For the 141 residue sequences in the α chain of hemoglobin, the number of differences for several species is shown below.[17] Thus, number of differences in aa residues agrees with other data on taxonomic relationships.

Sheep are polymorphic in having two principal hemoglobins, A and B, which differ in proportions in different strains. HbA has faster electrophoretic mobility than HbB. In anemia a third hemoglobin, HbC, replaces A. The three hemoglobins differ in their β chains, C from A by 12 aa's and B from C by 21 aa's.[22] A weak linkage of Hb's A and B with high-K and low-K strains of sheep has been suggested; the three forms differ in respiratory properties (see page 340).

In chickens, adult Hb has two bands; one, HbI, constitutes 75 to 80 per cent of the total and contains more lysine, histidine, and arginine, and less asparagine and glutamine, than the other (HbII). The chick embryo has a third type (HbIII).[88] In a series of passeriform birds, some with two and some with three Hb's as identified electrophoretically, each consists of four chains; the α_2 and β_2 of HbI are more alike than they are in HbII.

The two hemoglobins of eel consist of four polypeptide chains as in most vertebrate Hb's; each chain has a valyl residue as the terminal amino acid, but the carboxyl terminals and sequences are different in the α and β chains.[212] Hb of the cyclostome *Lampetra* is a single chain in which fewer than 30% of the sequences are the same as in the human chain.[170]

The invertebrate hemoglobins appear to have relatively less histidine and lysine, more arginine, and much more cystine than vertebrate Hb's. Lamprey Hb resembles mammalian Hb in low arginine and high lysine,

	man	*rhesus monkey*	*horse*	*cow*	*chicken*	*gray kangaroo*	*carp*
man	0	4	18	17	35	27	71
rhesus monkey	4	0	16	16	35	26	71
horse	18	16	0	18	40	29	70
cow	17	16	18	0	38	26	68
chicken	35	35	40	38	0	41	75
gray kangaroo	27	26	29	26	41	0	74
carp	71	71	70	68	75	74	0

but resembles some invertebrate Hb's in low histidine and high cystine.[173] *Arenicola* Hb yields two types of polypeptide chain, the longer one with aspartic as the terminal NH_2 and alanine as the terminal COOH.[201a] *Ascaris* Hb yields 332 amino acid residues and is rich in isoleu, glut, and arg.[138a]

Sickle cell anemia is a condition in man in which a fraction of the red cells take on the characteristic sickle or "crescent" form and do not transport oxygen in normal amounts. Sickling is promoted by low O_2 and low pH. In deer, four hemoglobins are seen on electrophoresis; one of these is responsible for sickling, which in this animal is promoted by high O_2 and alkaline pH.[204]

The extent to which different hemoglobins are present in the same red cells is undetermined. In frogs, where tadpole cells are made in liver and adult cells are made in bone marrow, immunofluorescence staining of metamorphosing tadpoles shows the embryonic and adult hemoglobins to be in different cells.[121] In mammals, it is assumed that more than one type (mutant) of molecule can occur in the same cell. Staining cells of adult or fetal humans with fluorescent antibodies for fetal and adult Hb shows that erythroblasts make more or less of HbA and HbF according to developmental stage.[34a, 59a]

Chlorocruorin

Chlorocruorin, the green iron-containing blood pigment of sabellid and serpulid worms, has a porphyrin which differs from heme in that on one pyrrole ring a vinyl chain is replaced by a formyl as follows:

$$CH_3-C=C-CHO$$

N

The amino acids of chlorocruorin resemble in proportion those of invertebrate hemoglobins. The close similarity to hemoglobin may explain the fact that some worms have both pigments, and that in the same genus one species has chlorocruorin and another has hemoglobin (p. 349). Chlorocruorin, oxychlorocruorin, and carbon monoxide-chlorocruorin have absorption bands about 20 to 25 $m\mu$ toward the red compared with the analogous hemoglobin bands (Table 8–1). *Spirorbis* chlorocruorin has a molecular weight of about 3,000,000 and that of *Spirographis* is 2,750,000.[64a] Its isoelectric point is 4.2; this pigment is always in solution in the plasma (Table 8–2). As with hemoglobin, two atoms of oxygen combined with one atom of iron.[50] *Spirographis* Chl has 80 hemes per molecule.[80] Electron micrographs of *Spirographis* chlorocruorin show the molecules to resemble some annelid hemoglobins in consisting of two identical units each of six 70 Å subunits at the corners of a hexagon[64a] (see Fig. 8–4c).

Hemocyanin

In hemocyanin, the copper is bound to the protein and does not occur in a prosthetic group. Some hemocyanins appear to be attached to carbohydrate moieties; *Helix* Hcy has some 9% carbohydrate. In bloods with hemocyanin, the pigment usually constitutes more than 90% of the soluble proteins. In *Loligo* blood the concentration of Hcy is 80 mg/ml, and in *Busycon* it is 40 mg/ml. The copper content of Hcy is 0.17% in arthropods and 0.25% in molluscs. In a number of crustaceans and *Limulus*, blood copper is 4 to 9 mg/100 ml; in cephalopod molluscs it is 25 mg/100 ml. Copper is present in deoxy-Hcy in the cuprous form; its occurrence in cupric form is now questioned.[74]

Hemocyanins appear to be colorless when deoxygenated, but like all proteins they absorb in the ultraviolet, with a peak at 280 nm. Oxygenated Hcy's are blue, with a broad peak at about 570 nm and a strong absorption band at 340 to 350 nm.

Hemocyanins are large molecules, those of molluscs having sedimentation coefficients corresponding to molecular weights of several million and those of arthropods corresponding to MW's of several hundred thousand. Each Hcy can be dissociated into smaller components. The larger aggregates of arthropod hemocyanins correspond to molecular weights of several hundred thousand. Arthropod Hcy's may have s_{20}'s of 35s (*Limulus*, *Callianassa*) and 60s (*Limulus*). At high pH, deoxy-Hcy tends to dissociate to smaller components; reassociation can occur on oxygenation. Lobster Hcy has a molecular weight of 825,000; in alkaline solution, removal of Ca^{++} favors dissociation to 12 subunits of ~69,000 each. On further dissociation, the O_2 binding sites are lost.[148] Arthropod Hcy's usually consist of either two

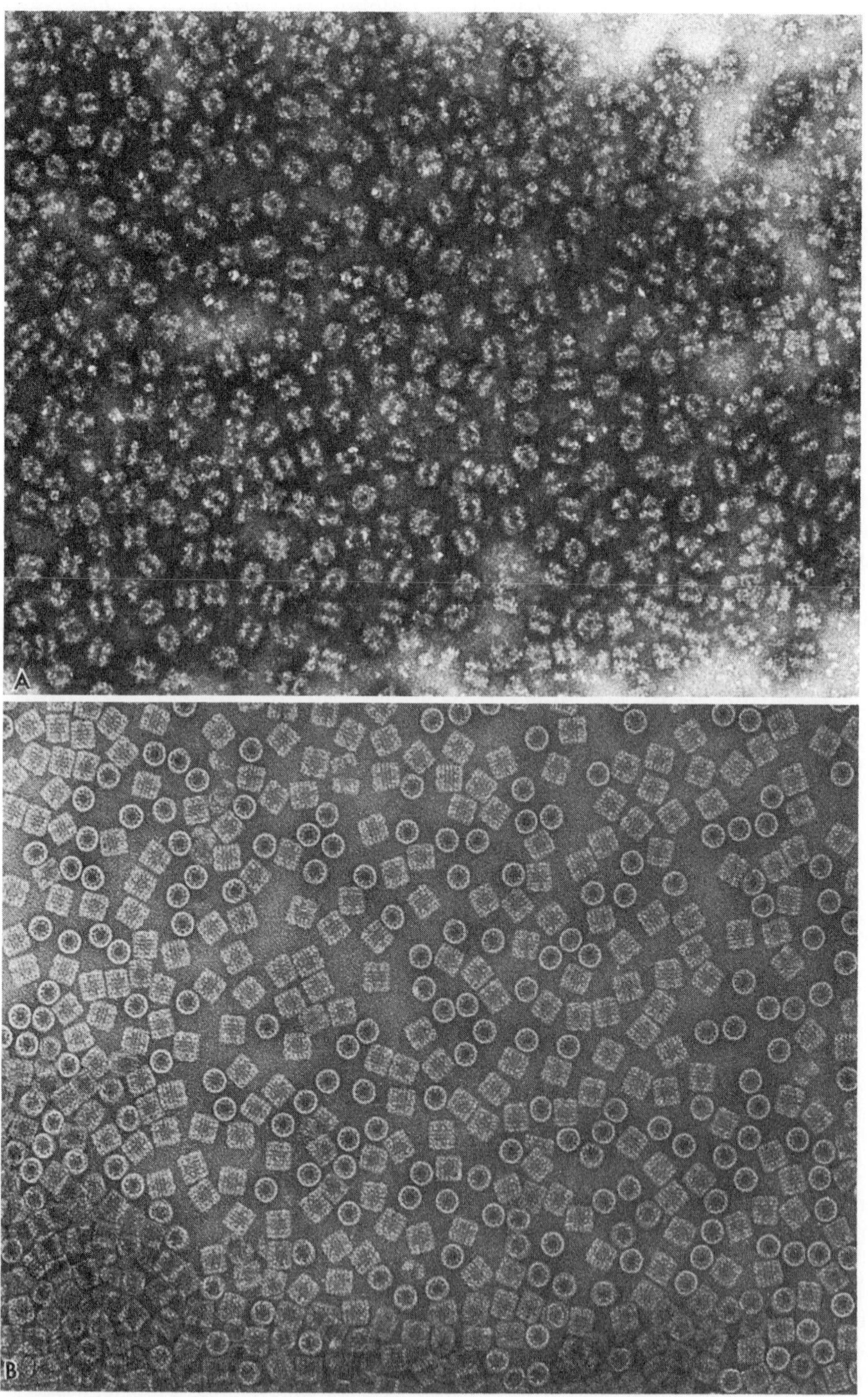

Figure 8–4. Electron micrographs of hemocyanin molecules. *A*, *Limulus* hemocyanin, × 290,000. *B*, *Helix pomatia* hemocyanin, × 184,000. (Courtesy of Dr. E. F. J. van Bruggen.)

(*Figure 8–4 continued on opposite page.*)

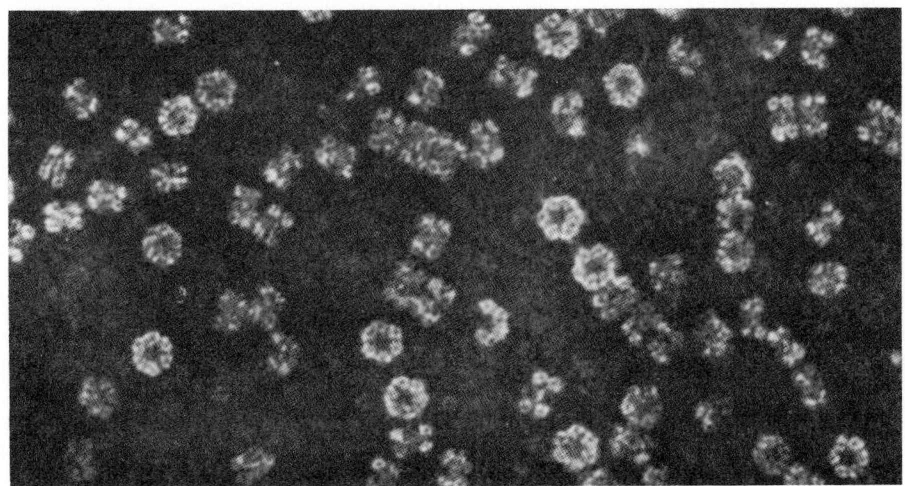

Figure 8–4 Continued. C, *Spirographis* deoxy chlorocruorin at pH 7. ×200,000. (From Guerritore, D., et al., J. Molec. Biol. *13*:234–237, 1965.)

chains of MW 37,000, each with one Cu, or three chains of MW 25,000, one of which lacks Cu.[74] Hcy of *Cancer* shows no subunit smaller than 75,000 MW.[179a]

Electron microscopy of Hcy from *Homarus*, having a sedimentation coefficient of 24s, shows a double structure of 16s components with both rectangular and hexagonal projections, with a maximum dimension of 100 to 125 Å. Scorpion Hcy shows a large rectangle as a dimer; *Limulus* shows hexagonally packed cylinders[39, 74] (Fig. 8–4).

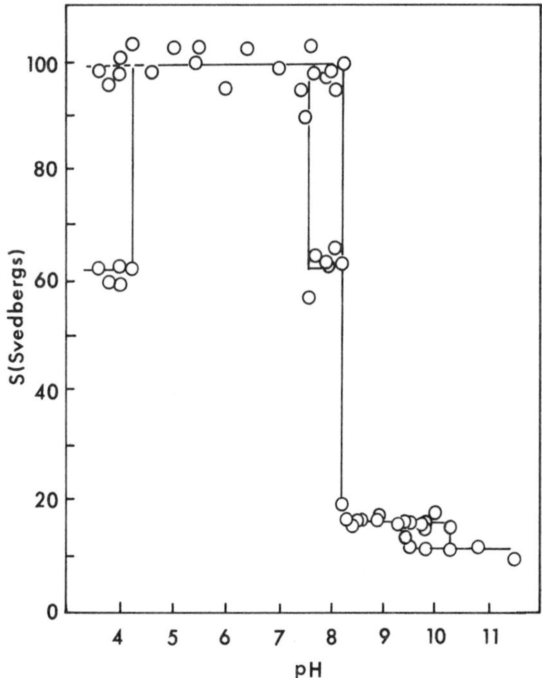

Figure 8–5. Sedimentation constants of *Helix* hemocyanin at different pH's: one at pH 5 to 7, two at pH 7.5, and one above pH 8. (From Eriksson-Quensel, I., and T. Svedberg, Biol. Bull. *71*:498–547, 1936.)

Molluscan Hcy's are more heterogeneous than those of arthropods. In gastropods, the molecules show an s_{20} of 100, or a MW of 8.7 to 9 million (Fig. 8–5). In *Busycon*, each 100s molecule can be seen in the electron micrograph as a hollow cylinder of 350 Å diameter and 380 Å height; 60s molecules are formed by a transverse division, and some smaller (11s, 15s, and 30s) particles are also formed. Oxygenation favors reaggregation, at pH 8.28, as follows:[39]

fully oxygenated	66% are 100s
deoxygenated	17% are 100s
reoxygenated	69% are 100s

In a number of gastropods, such as *Helix pomatia*, two forms of Hcy are found; in 1 M NaCl the α-Hcy dissociates into halves but the β-Hcy does not. Normally, the α and β forms occur in the proportion 3:1, each with an s_{20} of 100 and a MW of about 9 million. In *Helix aspersa*, the ratio α:β is 1:3. High pH favors dissociation to 1/2, 1/10, and 1/20 the initial size. Calcium stabilizes the polymer, and dissociation is favored by removal of divalent cations.

Squid deoxy-Hcy is predominantly 59s, or

MW 3.8×10^6, over the pH range from 6.2 to 8.4 in the presence of Mg. In the absence of Mg at pH 7.2, a 19s unit predominates; at pH 9.5, an 11s unit predominates.[38] At 80% oxygenation, only subunits are found, and at 100% oxygenation reassociation occurs.[39]

HEMERYTHRIN

This pigment contains approximately three times as much iron as hemoglobin (0.9 to 1.0%); the iron is directly attached to the protein and there is no porphyrin. Hemerythrins (Hr) occur in special cells, the blood corpuscles. Most information concerning Hr has been obtained from the sipunculid *Golfingia*.[106] Deoxyhemerythrin is colorless in the visible range but has the protein absorption peak in the ultraviolet. Oxy-Hr has a good peak at 330 nm and a low, broad peak at 500 nm. Met-Hr absorbs at 330 and 380 nm (Fig. 8–6). The isoelectric point of *Sipunculus* Hr is at pH 5.8.

Hemerythrins have molecular weight of 108,000 with an s_{20} of 7s. They can be dissociated chemically to 2s units of MW 13,500. Each subunit contains 2 Fe, whereas the normal Hr has 16 Fe. There is evidence that an equilibrium normally exists between the monomer and octomer, that there are essentially no dimers or tetramers, and that at the normal high concentration in corpuscles most of the Hr is octameric.[56]

Peptide analysis shows the monomer to be a chain of some 113 amino acids with a single cystine at position 50. The secondary structure of the chain is 75% α-helix. The two Fe atoms have a valence state of II and may occur near the middle of the folded chain.[56] In met-Hr they can be held by various ligands (L)$-O_2$, Cl, cyanate, N_3^-, or OH.

Recent evidence from chemical, magnetic, Mössbauer, and spectroscopic measurements indicates that oxygenation involves considerable conformational rearrangement. The following structures have been postulated:[56]

deoxyhemerythrin $Fe^{II} \langle \begin{matrix} H_2O & H_2O \\ H_2O & H_2O \end{matrix} \rangle Fe^{II}$

methemerythrin $Fe^{III} \langle \begin{matrix} O_2 \\ L \end{matrix} \rangle Fe^{III}$

oxyhemerythrin $Fe^{III} \langle \begin{matrix} O_2^- \\ O_2^- \end{matrix} \rangle Fe^{III}$ $O_2^- - O$----H—

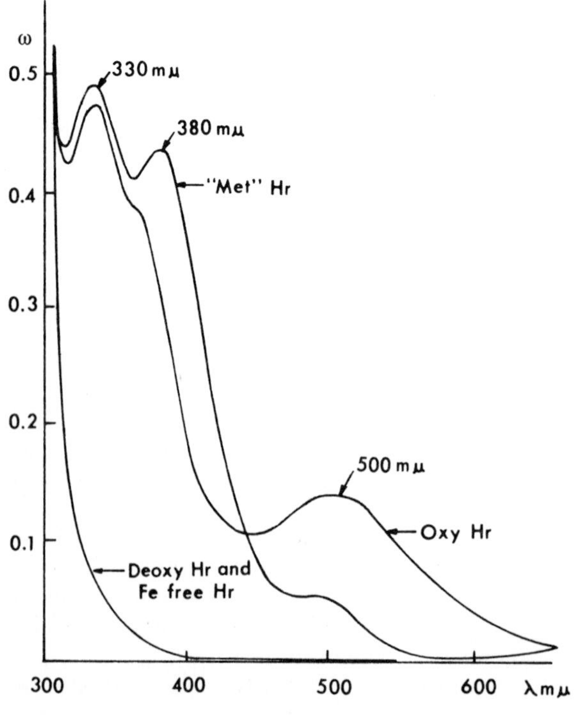

Figure 8–6. Absorption patterns of oxyhemerythrin, methemerythrin, and deoxyhemerythrin. (From Garbett, K., D. W. Darnall, and I. M. Klotz, Arch. Biochim. Biophys. *142*:455–470, 1971.)

COMBINATION OF HEMOGLOBIN WITH OXYGEN

Rates of Association and Dissociation

The rates of loading and unloading of pigments have been measured with rapidly recording spectrophotometers. Binding and release of O_2 and CO by Hb is a four-step process, and the kinetics are very complex.[179] Combination with CO is an order of magnitude slower than with O_2. Overall association velocity constants for hemoglobin in $mol^{-1}sec^{-1}$ are as follows:[58]

	oxygen	*carbon monoxide*
sheep	2.6×10^6	1.8×10^5
Lumbricus	2.3×10^6	2.2×10^5
Arenicola	2.3×10^6	2.9×10^5

Association is faster than dissociation. The reactions of mammalian myoglobin are faster than those of hemoglobin; the half-time for myoglobin association with oxygen is 0.0004 sec compared with 0.038 sec for Hb. Hemocyanins are similar in speed to hemoglobins.

When intact cells are used rather than Hb in solution, the reaction is slower. Animals with large red blood cells react more slowly than those with small cells, and hence cell size, permeability, and diffusion may limit the rate of loading and unloading:[77]

	Velocity Constant for O_2 Uptake in $mM^{-1}sec^{-1}$		*Red Cell Volume*
	CELLS	HB	
sheep	137	–	
horse	–	2.4×10^3	
goat	133	2.7×10^3	20 μm^3
man	80	2.4×10^3	90 μm^3
dog	65	–	
frog	19	2.7×10^3	680 μm^3

Structural Correlates with O_2 Loading

Several lines of evidence indicate conformational changes associated with loading and unloading of O_2. The distance between two reactive cysteinyl residues in the two β chains is 37.6 Å in deoxygenated Hb and 30 Å in oxygenated Hb; this is associated with a change in positions of the β chains relative to the α chains.[135a] Oxygenation releases protons from histidines, an average of 3.8 protons in mouse and 1.5 in elephant HbA. This is reflected by the fact that HbO_2 is a stronger acid than Hb (i.e., it is more dissociated). A shift of an electron from the ferrous iron and a change of the oxygen to a superoxide state on oxygenation was mentioned on page 319.

The affinity of various Hb's for oxygen differs. For example, in *Rana catesbiana*, HbB is half-saturated at 15.8 mm Hg Po_2 while HbC is half-saturated at 7.9 mm Hg.[2] Lamprey Hb consists of a single chain, but these chains tend to polymerize to tetramers when deoxygenated and to separate when oxygenated. In lamprey, loading of oxygen is viewed as consisting of three steps—deaggregation, oxygenation, and proton loss.[25] In the snake *Natrix*, oxygenation leads to dissociation of components.[192] Cat hemoglobin shows two electrophoretic peaks; the P_{50}'s for these components are similar.[194]

Interaction Between Chains in Relation to Oxygenation

An oxygen equilibrium curve is obtained by plotting the percentage saturation of Hb at different partial pressures of oxygen; a measure of affinity is the pressure for half-saturation (P_{50}). Such a plot is shown in Figure 8–7. With myoglobin, a single heme unit, the equilibrium curve is hyperbolic; with hemoglobin, however, the curve is sigmoid, which indicates that cooperativity in oxygen binding by the subunits is present. That is, the oxygenation of one subunit increases the affinity of the other subunits for oxygen.

The equilibrium curve is described by the approximate equation:

$$Y = 100 \times \frac{(P/P_{50})^n}{1 + (P/P_{50})^n}$$

where Y is the percentage of hemoglobin combined with O_2 or $\frac{100(HbO_2)}{(Hb + HbO_2)}$, and P is the pressure of O_2 in mm Hg. When $\frac{Y}{100-Y}$ is plotted against log P, the slope at the point of half-saturation (where Y = 50%) gives n, which is taken as a measure of interaction between heme units (Fig. 8–7). The slope may be less than n at the two ends of the curve. For myoglobin, $n = 1$, and for mammalian hemoglobin it is between 2.4 and 2.9

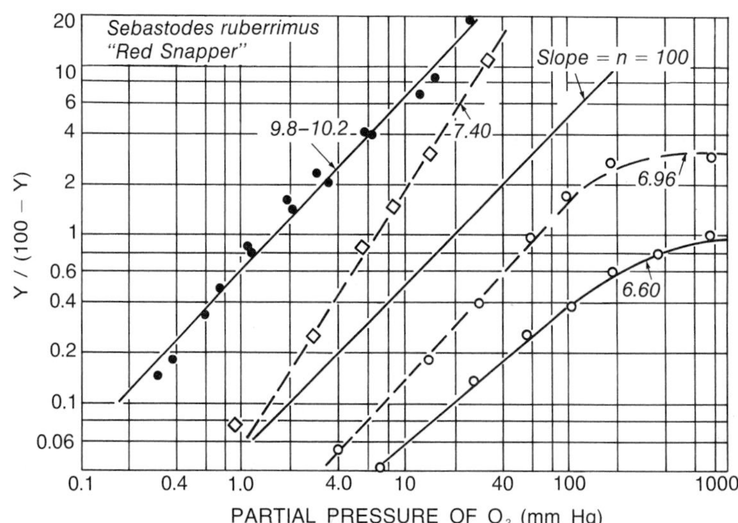

Figure 8–7. Log-log plot of oxygen equilibrium curves of hemoglobin in erythrocytes of red snapper, *Sebastodes ruberrimus.* Measurements at pH 9.8 to 10.2, 7.4, 6.96, and 6.60, showing positive Bohr effect and reduction of saturation at high Po_2 (Root effect). Also shown is decrease in heme-heme interactions or *n* value with acidification. (From Manwell, C., personal communication.)

in the middle of the curve; for large hemoglobins of invertebrates, *n* is greater than 3. For example, *Arenicola* Hb has *n* of 5.0 to 5.7.[201a] Hb of *Nereis* has two *n* values.[203] For chlorocruorin of *Spirographis*, $n = 3.2$.

The oxygen affinity of a particular heme on the Hb molecule is influenced by the state of the other hemes (i.e., by whether or not they are combined with oxygen). Such heme-heme interactions are usually facilitating, and account for the sigmoid equilibrium curve and for the value of *n*, which is greater than one but less than the number of hemes. In cyclostomes, *n* for the hagfish *Eptatretus* is 1.0; for *Ichthyomyzon* adult it is 1, and for its ammocoete larva it is 1.2;[127] for *Lampetra* larvae it is 1.25, and for adults it is 1.5; and for different components of *Petryomyzon* Hb it is 1.1 and 1.55.[9] However, at high Hb concentrations and when 90% oxygenation is approached, *n* may increase to 3, indicating polymerization.[170]

Of 54 kinds of turtles, the Hb in blood from some families polymerizes greatly, while that from others may polymerize little or not at all; *n* is pH-dependent in the turtle, but not in man. The turtle *Gopherus* has Hb with $n = 1.3$ at pH 5, $n = 2.5$ to 2.7 at pH 7.[192]

Further evidence for structural involvement in oxygenation comes from measurements on hybrid Hb molecules. The separated chains of mouse (m) and donkey (d) Hb were assembled in vitro; the P_{50} values for the combinations were as follows:[171]

α	β	
m_2	m_2	12.3
d_2	d_2	4.7
d_2	m_2	14.4
m_2	d_2	4.7

In these combinations, the affinity is largely determined by the source of the β chain.[171]

Other evidence for conformational changes on oxygenation is the change in spacing between the β chains (p. 319), as well as differences in the rates of splitting by carboxypeptidase. Oxy-Hb has two titrable SH-groups, while deoxy-Hb has none. Also, *n* varies with salt concentration for Hb in solution, and diphosphoglycerate (DPG) can reduce the value of *n*.[145] It is concluded that the sigmoid equilibrium curve facilitates loading and unloading, particularly when affinity for O_2 is high.

Effects of Acidic Groups on O_2 Binding

The Bohr Effect. When carbon dioxide enters vertebrate blood, as it does in the tissues, the affinity of the Hb for O_2 is reduced and the equilibrium curve moves to the right (normal Bohr effect). At high CO_2 pressures, or at low pH, the oxygen pressure required to load Hb is higher than under normal conditions. In the tissues, addition of CO_2 (or lactic acid) facilitates unloading of

O_2; in the lungs or gills, as CO_2 is given off, the uptake of O_2 is enhanced (Fig. 8–8). The Bohr effect represents an interaction between oxygenation equilibrium and proton dissociation (acid strength) of hemoglobin. At low pH (usually well below the physiological range), the Bohr effect may reverse and affinity increases; thus, P_{50} is maximal at a given pH. The magnitude of the Bohr effect is given by the change in P_{50} per pH unit, or $\frac{\Delta \log P_{50}}{\Delta pH}$ = number of protons released per O_2 per mole of heme; for human Hb the change in P_{50} is 1.1 per pH unit from pH 6.5 to 9.5.[7] For representative values of the Bohr effect, see Table 8–3. A normal Bohr effect is shown by chlorocruorin from polychaetes.

The protons that are released on oxygenation come mainly from imidazole groups of C-terminal histidines of β chains, and also from α amino groups of α chains. When the α amino groups of the α chain were blocked, the Bohr effect decreased but was not lost.[144] Hybridization experiments indicate a strong contribution of the β chains to the Bohr effect.[171] The CO_2 effect may be observed on loading with CO as well as with O_2.[7] The magnitude of the Bohr effect decreases with dilution of the pigment, with increasing temperature, and with increasing ionic strength due to salts. The magnitude of the Bohr effect is greater in small mammals than in large ones; $\Delta \log P_{50}/\Delta pH$ is 0.9 in mouse, 0.65 in man, and 0.45 in elephant.[195]

The Haldane Effect. The converse of the Bohr effect is the facilitation of CO_2 loss from the blood on oxygenation and facilitation of CO_2 uptake in tissues on deoxygenation (Haldane effect). An acid group present on each subunit lowers its pK upon oxidation; in other words, oxy-Hb is a stronger acid than deoxy-Hb. Thus, as O_2 is given off in the tissues, the deoxygenated Hb becomes a better buffer (see page 354), and in the lungs, as O_2 is taken up, there is an increase of negative charge and equivalent displacement of HCO_3^-, thus facilitating CO_2 loss.

The Root Effect. Not only is the O_2 affinity, as measured by the P_{50}, reduced by acidification, but the maximum amount of saturation

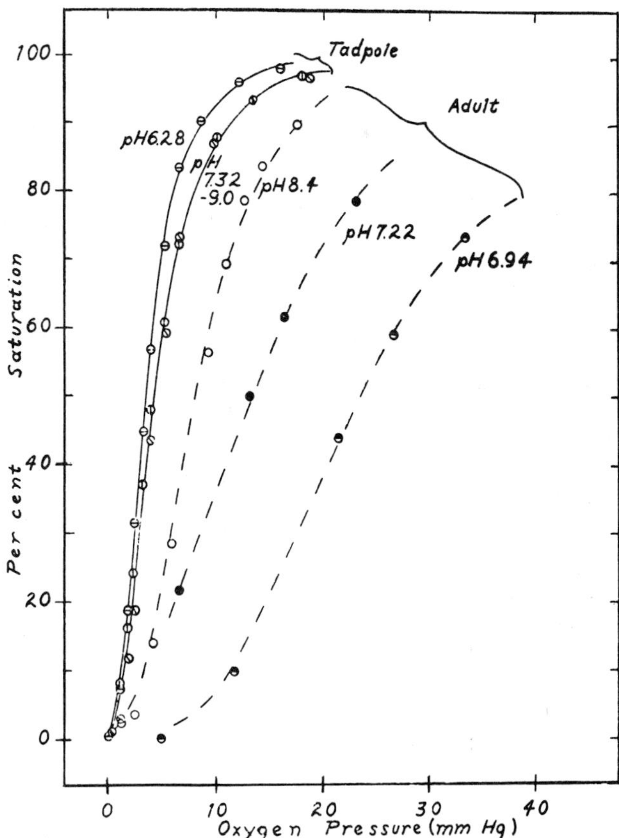

Figure 8–8. Oxygen equilibrium curves of hemoglobin of tadpole and adult bullfrogs. (From Riggs, A., J. Gen. Physiol. 35:23–40, 1951.)

TABLE 8–3. Magnitude of Bohr Effect in Blood of Various Animals

	$\Delta\ log\ P_{50}/\Delta pH$
Mammals	
man (whole blood)[194]	−0.62
(Hb solution)[164]	−0.48
gorilla[163]	−0.49
chimpanzee[163]	−0.59, −0.46
baboon[112]	−0.55
lion[142]	−0.54
cat[194]	−0.49
dog[194]	−0.65
guinea pig[167]	−0.79
mouse[167]	−0.96
rabbit[167]	−0.75
hamster[198]	−0.41
shrew[198]	−0.63
chinchilla[66]	−0.53
nutria[66]	−0.56
woodchuck[66]	−0.63
prairie dog[66]	−0.51
hedgehog[32a]	−0.62
vole	−0.54
horse[167]	−0.68
yak[12]	−0.68
llama[12]	−0.74
camel[12]	−0.84
deer[12]	−0.78
African elephant[12]	−0.58
hog[167]	−0.57
Weddell seal[113] adult	−0.613
fetus	−0.671
porpoise[79]	−0.55
beaver[32a]	−0.56
seal *Cystophora*[32a]	−0.66
fur seal[118]	−0.475
harbor seal[118]	−0.534
ribbon seal[118]	−0.48
walrus[118]	−0.525
sea lion[118]	−0.696
sea otter[118]	−0.46
kangaroo[164]	−0.54
Tasmanian devil[164]	−0.47
armadillo[38]	−0.43
echidna[196]	−0.54
Birds	
Adelie penguin[120]	−0.5
duck	−0.40
Reptiles and Amphibians	
Iguana[210]	−0.52
Necturus[117]	−0.131
Amphiuma[117]	−0.205
Rana catesbiana[117]	−0.288
Fishes	
Protopterus[116]	−0.47
Neoceratodus[116]	−0.42
Ictalurus[62]	−0.31
Salmo gairdneri[42a]	−0.54
Invertebrates	
Lumbricus[33]	−0.40
Gastrophilus	0.0
Spirographis (chlorocruorin)	−0.66
Chironomus[202]	−0.56 to −1.3
Cancer[96] (hemocyanin)	−0.27

attained at high pressures may be reduced. This reduction in the upper saturation level is particularly marked in the blood of some fishes. The Root effect, or reduced saturation, is brought about more rapidly by CO_2 than by equivalent amounts of lactic acid.

The preceding three interactions between CO_2 and oxygenation reflect communication from one part of the molecule to another.

EFFECTS OF DILUTION, IONS, AND PHOSPHATES ON O_2 AFFINITY

The P_{50} for hemoglobin in intact corpuscles is usually higher than for the same pigment extracted by hemolysis. Also, as a solution of hemoglobin is diluted, the equilibrium curve may move toward lower pressures of O_2. When a solution of Hb is dialyzed with loss of salts, the curve moves to the left. Lamprey Hb in dilute solution, at neutral pH, is monomeric when oxygenated and partly dimeric when deoxygenated; in high concentrations and at low pH, the deoxy-Hb may be tetrameric. An increase in Hb concentration of 63-fold corresponds to a 10-fold decrease in O_2 affinity.[170] The P_{50} for whole sheep blood at 6°C was 7.2 mm Hg, while for a hemoglobin solution of similar concentration it was 5 mm Hg; dilution by ten times lowered the P_{50} to about 2.5 mm Hg.[143] For a hagfish, *Polistotrema*, P_{50} was 3.5 mm Hg for cells and 1.8 mm Hg for solution; corresponding P_{50}'s for a holothurian, *Cucumaria*, were 12.5 and 3.84.[123, 124] Thus, the microenvironment in the red cell influences the binding of O_2 by Hb, and both dilution and the presence of salts lower the P_{50}. Observations made on whole blood are not to be compared quantitatively with those made on purified protein.

It has long been known that the equilibrium curves for Hb in solution lie to the left of the curves for whole blood, and that whole blood is more sensitive to pH changes. An important component of red blood cells (RBC's) that modifies the Hb affinity for oxygen is diphosphoglycerate (DPG). Human red cells contain 5×10^{-3} molar DPG, or approximately one mole of DPG per mole of Hb. DPG shifts the hemoglobin equilibrium curve to the right; it binds to the four β chains of tetrameric Hb (more to deoxy-Hb than to oxy-Hb). DPG acts like NaCl in increasing the P_{50}, and accounts for many of the differences in properties between isolated hemoglobin and RBC's.[19] In man

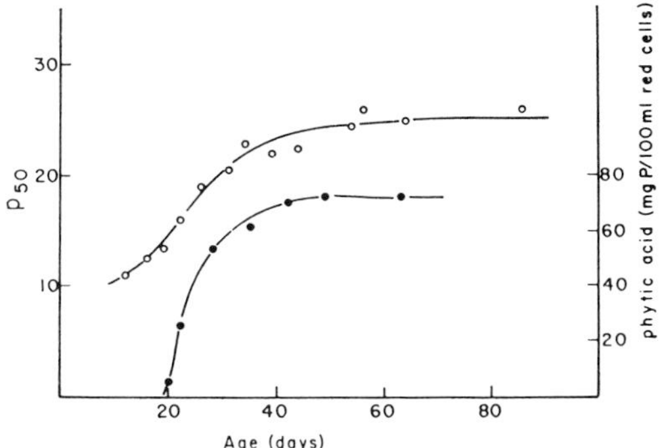

Figure 8–9. Correlation of IHP (inositol hexaphosphate) in chicken red blood cells with O_2 affinity, according to age. IHP concentration, OP_{50}. (From Benesch, R., R. E. Benesch, and C. I. Yu, Proc. Nat. Acad. Sci. *59*:526–532, 1968.)

adapted to high altitudes, the RBC's contain more DPG than in individuals adapted to sea level.[111]

In man, the DPG content is similar in fetal and adult cells, but DPG is more effective on HbA than on HbF.[197] During early postnatal life of lambs, DPG increases from 8.2 μmoles PO_4/g Hb in the fetus to 29.9 μmoles PO_4/g Hb, and the O_2 affinity and intracellular pH decreases; later, in the adult, the DPG content diminishes to 1.1 μmoles PO_4/g Hb as HbA becomes predominant.[15]

In birds and turtles, inositol hexaphosphate (IHP) occurs rather than DPG (Fig. 8–9). In fishes, the same function is served by ATP, which may be present at 1 to 2×10^{-3} M. These phosphorylated compounds represent 45% of the phosphorus in red blood cells; they decrease the Hb affinity for O_2 by 30-fold (i.e., they facilitate unloading):[19]

	log P_{50} of Hb	
	MAN	CHICK
0.01 M NaCl	−0.64	−0.59
0.01 M NaCl + DPG	+0.87	+1.11
0.01 M NaCl + IHP	+1.35	+1.35

In freshwater cichlid fish, ATP is present at 0.76 mole per mole of Hb; the Bohr effect increases with increased ATP.[59] Cyclostomes have no organophosphates in the red cells, and their Hb is unaffected by these compounds.[170]

The Bohr effect is increased by addition of DPG to mammalian Hb or of IHP to avian Hb; similarly, in some fish Hb's, the number

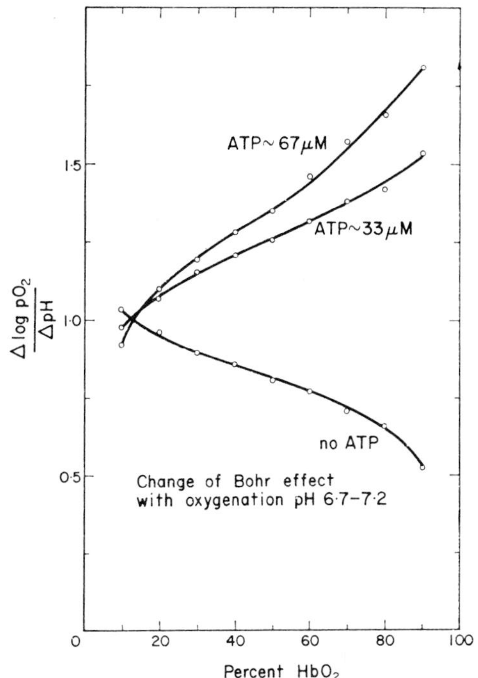

Figure 8–10. Dependence of the Bohr effect as given by H^+ discharged per oxygen, based on the presence of ATP and on level of oxygenation. Data for Hb from cichlid fish. (From Gillen, R. G., and A. Riggs, Comp. Biochem. Physiol. *38*B:585–595, 1971.)

of protons released on oxygenation increases by 60% on addition of ATP (Fig. 8–10). Diphosphoglycerate lowers the O_2 affinity of Hb; it binds more strongly to deoxy-Hb than to oxy-Hb, and more at low pH and low ionic strength. It appears that DPG binds between the terminal NH_2 groups of the β chains; CO_2 competes for this binding site, and partial release of the DPG upon oxygenation is accompanied by release of protons.[169]

SPECIES SPECIFICITY AND ECOLOGICAL CORRELATES OF O_2 AFFINITY

OXYGEN CAPACITY

The prime requisite of an oxygen-transport pigment is ability to combine reversibly with enough oxygen to supply the needs of the animal. The oxygen capacity is the amount of oxygen carried in saturated blood. Oxygen content is usually expressed as volumes per cent or moles per liter of oxygen in whole blood or cells. Equilibration with air may not

(*Text continued on page 339.*)

TABLE 8–4. Respiratory Characteristics of Blood. Data for Mammals and Birds at Physiological pH or P_{CO_2} and Temperature; Conditions Specified for Others. Data for Whole Blood Except Where Indicated as Pigment in Solution.

Animal	P_{50} *P_{O_2} in mm Hg*	*O_2 capacity ml O_2/100 ml blood*	*n*
Mammals			
man[186]	27.8		
man[112]	29.6–29.7		
man[134]	27	2.9	
man[134] adult	30	14.4	
fetus	20	8.4	
baboon	33.2		
baboon[112]	37.2		
rhesus monkey[112]	35.2		
chimpanzee[163]	26.4	17.4	
gorilla[163]	25	16.7	
macaque[134] adult	32	15.5	
fetus	16	18.8	
macaque[18] adult	32.9		
fetus	17.4		
deer[14]	22.2		
camel[14]	20.9		
sheep[133] adult	32–37		2.9
fetus	16		
sheep[37] HbA	20		
HbB	30		
goat[134] adult	32.2	14.2	
fetus	19.0	12.2	
alpaca[30]	18.4	18	
horse[134]	26	16.7, 14	2.9
African elephant[164] adult	22.8		2.65
fetus	17.2		
tiger[142]	42.1	15.8	
lion[142]	42	19	
cat[137] adult	36.2	15	
fetus	35.6		
dog[167]	29		2.8
hamster[65]	22, 27.8		
prairie dog[65]	22		
rat[65]	38		
kangaroo rat[65]	53		
mouse	41.5		2.8
shrew[198]	36.4		
chinchilla[66]	26.8		
rabbit adult	31.6	15.6	
fetus	28		
woodchuck[65]	25		
hedgehog *Erinaceus*[32a]	34 (38°C) 8.9 (5°C)	15.3	
beaver *Castor*[32a]	23.3	16.1	
water vole *Arvicola*[32a]	38.4	19.4	
seal *Cystophora*[32a]	24	36	
Tursiops[79]	24.6	18.20	
Phocoenoides[79]	19.1	25.27	
sea lion *Zaophus*[118]	32	17.5	2.6
whale *Orcinus*[118]	30.7, 37.5	21.5	
harbor seal[118]	31	29.3	2.6
walrus[118]	34.2	23.4	2.4
fur seal[118]	29.7	19.6	2.5
sea otter[118]	30.8	21.3	2.6
Weddell seal[113] adult	28.5	31.6	
fetus	22.1	27.7	
armadillo[38]	10.1		
echidna[196]	19.5	19	
platypus[97]	10 (4 mm CO_2) 36 (30 mm CO_2)	23.5	

(*Table 8–4 continued on opposite page.*)

TABLE 8–4. Respiratory Characteristics of Blood. Data for Mammals and Birds at Physiological pH or P_{CO_2} and Temperature; Conditions Specified for Others. Data for Whole Blood Except Where Indicated as Pigment in Solution. (*Continued*)

Animal	P_{50} P_{O_2} *in mm Hg*	O_2 *capacity* *ml* O_2*/100 ml blood*	*n*
Birds			
chicken[13] adult	58		
10 day	54.8	10	
1 day	48.4	10.7	
embryo 17 day	31.3	10.3	
Adelie penguin[120]	34.4	22.4	
goose[120]	45		
pigeon[120]	42		
duck[120]	54		
Reptiles			
Pseudemys	19.5	6.6–10.8	
Chrysemys	15		
Crocodylus	38	8–10	
Alligator	28	6.7	
Heloderma[152]	32 (18–20°C)	10	
Eumeces	19	12.5	
Iguana[210]	51	10.5	
Agama[152]	34 (37 mm CO_2)		
Sauromalus[152]	24		
Amphibians			
Necturus[117] (gill breather)	14.5	6.3	
Amphiuma[117] (gill and lung)	27	7.6	
Rana catesbiana[117] (lung and skin)	39	8.1	
Rana esculenta[168] adult	13.2	9.8	
tadpole	4.6 (7.2, 20°C)	7.8	
Rana esculenta[55] 5° adapted	42.5	13.3	
20°C adapted	39.1	10.4	
Bufo marinus	42 (25°C)		
	38 (15°C)		
Fishes			
pike		9.0	
goldfish[6]		10.7	
rainbow trout[29]	30 (7–8 mm CO_2, 15°C)		
salmon	19 (1–2 mm CO_2, 15°C)		
brook trout[20]	7 (1 mm CO_2, O°C)		
	21 (1 mm CO_2, 25°C)		
electric eel	12 (0 CO_2, 18°C)	19.7	
	14 (7.4 mm CO_2)		
Japanese eel[212] Hb E_1	2.1 (7.0 mm CO_2, 20°C)	2.4	
Hb E_2	14 (7.0 mm CO_2, 20°C)	1	
Trematomus	21.5 (0.1 mm CO_2, −1.5°C)	5.3–7.7	
mackerel	16 (1 mm CO_2, 20°C)	15.7	
carp	5 (1–2 mm CO_2, 15°C)	12.5	
catfish	1.4 (0–1 mm CO_2, 15°C)	13.3	
Bagrus (deep FW)	1.5 (0 mm CO_2)		
Lates (high O_2 FW)	17 (0 mm CO_2)		
Neoceratodus[116]	11 (3.5 mm CO_2)	7.7	
Protopterus[116]	10 (6 mm CO_2, 25°C)	6.8	
Lepidosiren[95]	10.5 (6 mm CO_2, 23°C)	4.9–6.8	
Symbranchus[95]	5–6	14.7	
Scyliorhynus[150]	12 (7.0 mm CO_2, 17°C)	4.5	
Squalus[115]	17 (0.5 mm CO_2, 11°C)	4.35	
Squalus adult	16.4 (7.3 mm CO_2, 12°C)		
fetus	10.6		
Lampetra planeri[9] adult	0.77 (Hb 5 mg/ml)		
larva	0.37		
Eptatretus[123] Hb	2–4	1	
Myxine[123] Hb	8 (7.5 mm CO_2, 25°C)	1	
Petromyzon[123]	14–20	1.2	
Ichthyomyzon[123] adult	17–19 (25°C)	1.0	
ammocoete	16	1.2	

(*Table 8–4 continued on following page.*)

TABLE 8–4. Respiratory Characteristics of Blood. Data for Mammals and Birds at Physiological pH or P_{CO_2} and Temperature; Conditions Specified for Others. Data for Whole Blood Except Where Indicated as Pigment in Solution. (*Continued*)

Animal	P_{50} P_{O_2} *in mm Hg*	O_2 *capacity* *ml* O_2 */100 ml blood*	*n*
Arthropods			
Chironomus plumosus[202]	0.39 (12 mm CO_2) 0.21 (0.6 mm CO_2)	5.4–11.6	1.1–1.2
Moira[10]	3.5 (7.2 mm CO_2, 28°C)		
Daphnia[10]	3.1 (7.2 mm CO_2, 28°C)		
Cyzicus[10]	0.035 (7.2 mm CO_2, 28°C)	2.3	
Triops[78]	6.6 (7.4 mm CO_2)	3.2	
Worms			
Urechis	12.3 (8.6 mm CO_2, 19°C)	2.2–6.7	
Glycera[72]	7.0 (7.4 mm CO_2)	4.1	
Glossoscolex[98]	7.0 (7.5 mm CO_2 20°C)	14	
Lumbricus[33]	6.8 (7.4 mm CO_2, 25°C)	5.1–5.4	
Tubifex	0.6 (0 mm CO_2, 17°C)		
Arenicola	1.8–2 (7.4 mm CO_2, 18°C)		
Arenicola[201a]	4.0 (pH 7.0)	5.0	
Nephthys vascular fluid	6 (7.4 mm CO_2, 20°C)		
coelomic fluid	7.4 (7.4 mm CO_2, 20°C)		
Echinoderms			
Thyonella[129]	8		
Cucumaria	12.5 (7.4 mm CO_2, 10°C)		
Roundworms			
Ascaris[191] perienteric fluid	0.05	3	
body wall	0.2		
Ascaris[138a] perienteric fluid	0.001–0.004	1.2	
Nematodirus	0.05		
Nippostrongylus	0.2		
Molluscs			
Phacoides[155] (gill Hb)	0.19 (7.4 mm CO_2)		
Cryptochiton	18 (7.2 mm CO_2, 10°C)		
Planorbis	7 (0 mm CO_2, 20°C)		
Protozoans			
Paramecium[190]	0.6		
Chlorocruorin			
Spirographis[8]	1.65		
Sabella	27 (7.7 mm CO_2, 20°C)		
Hemerythrin			
Sipunculus	8		
Golfingia	6.9 (6.3 mm CO_2, 20°C)		
Hemocyanin			
Cancer[96]	19.6 (7.7 mm CO_2)	3.44	
crayfish[110]	1.6	3.1	
Gecarcinus[160]	17 (7.4 mm CO_2, 27°C)		
Cardisoma[157]	3.5 (25°C)	2.9	2.6
scorpion *Heterometrus*[140]	16.5	1.8	
Limulus	11 (0 mm CO_2, 23°C)	0.7–2.7	
Panulirus	6.5 (7.5 mm CO_2, 15°C)	2.0	
Homarus	14 (0 mm CO_2, 25°C)	1.3	
Helix[107]	8, 10	1.1–2.2	
Cardita[126]	11 (7.5 mm CO_2)		
Neotia[126]	13.5		
Fusitriton	7 (7.8 mm CO_2, 25°C)		
Busycon	15 (0 mm CO_2, 23°C)	2.1–3.3	
Cryptochiton	18		
Chiton[156]	20–26	1.3	2.2–3.0
Diodora[156]	5 (10°C)	0.86	1.7
Octopus[114]	15 (0.4 mm CO_2, 11°C)	3.1	
Sepia	4 (0.7 mm CO_2, 14°C)		
Loligo	36 (0 mm CO_2, 23°C)	3.8–4.5	
Mytilus		0.32	
Asterias		0.46	

permit saturation if the pigment requires a high pressure for saturation, particularly in the presence of CO_2 and of high temperatures; hence, equilibration with pure oxygen may be preferable to equilibration with air for accurate determination of O_2 capacity. Also, some blood cells, particularly nucleated erythrocytes, consume considerable amounts of oxygen which must be taken into account. The oxygen capacity is proportional to the amount of hemoglobin or other pigment in the blood after the amount of O_2 physically dissolved is subtracted.

Table 8–4 gives values of oxygen capacity of the blood of a number of animals. The O_2 capacity of the blood of mammals and birds is usually between 15 and 20 vols/100 ml. Some diving mammals have high capacities and the llama, even at sea level, has a high capacity. Young mammals tend to have lower capacities than adults. The O_2 capacities of cold-blooded vertebrates are usually between 5 and 12 vol per cent. In a few fish which can resort to air breathing, the O_2 capacity is high. Active fish like the mackerel tend to have higher O_2 capacities than sluggish ones like the toadfish. Among invertebrates, individual variation is very great. Where there is significant amount of pigment, as in *Arenicola*, *Urechis*, and some others, the O_2 capacity may be 10 times greater than it would be without the pigment.

Oxygen Affinity as Measured by Equilibrium Curves

Most respiratory transport pigments become saturated at lower partial pressures of oxygen than is present in air at sea level (155 mm Hg). The most adaptive differences among the hemoglobins of different animals are the partial pressures at which they load and unload oxygen. These pressures determine the range of usefulness of a particular pigment. In man, for example, the blood is exposed in the lungs to oxygen at a partial pressure of approximately 100 mm Hg; when the blood leaves the lungs it carries 19 vol per cent of O_2 at 80 mm Hg, and 98 per cent of the hemoglobin is saturated. In the capillaries, the blood passes through tissues where the oxygen pressure is low (5 to 30 mm Hg), and here 25 to 30 per cent of the oxygen is unloaded; venous blood carries 14 vol per cent of O_2 at about 40 mm Hg.

From oxygen equilibrium curves, the affinity can best be expressed as the P_{50} (see page 331 and Figure 8–11). A value of P_{50} is meaningful only if the state of the pigment, the P_{CO_2} (or pH), and the temperature are stated. Table 8–4 gives selected data from the very extensive literature. The data in this table refer mainly to hemoglobin contained in red cells and measured under nearly physiological conditions. They represent the

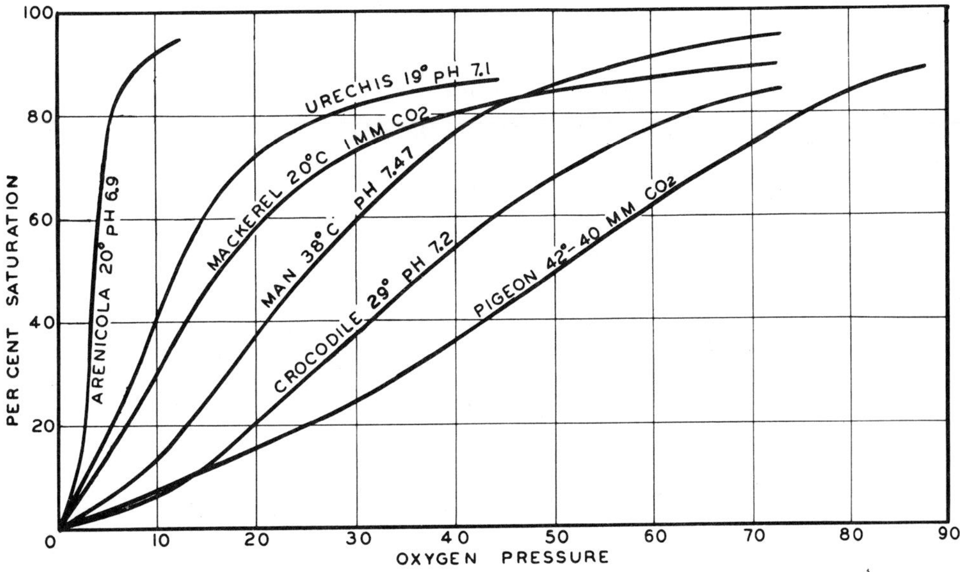

Figure 8–11. Oxygen equilibrium curves in per cent saturation of hemoglobin as a function of oxygen pressure in mm Hg in a variety of animals.

average if the hemoglobins are polymorphic. The respiratory properties of the several different molecular species are known in only a few animals.

Hemoglobins differ greatly in their oxygen affinities, and bloods differ in the amounts of oxygen they carry when saturated. Unloading pressures set the upper limit of tissue oxygen pressure and the lower limit of environmental oxygen for function of hemoglobin. The Bohr effect facilitates loading in the respiratory organs and unloading in the tissues; hence, the physiological equilibrium curve lies between the arterial and venous curves.

Muscle hemoglobin (myoglobin) has a greater affinity for O_2 than blood Hb, and it can transfer oxygen from blood to cell enzymes. When hemoglobin (or myoglobin) is added to a fine filter separating two solutions of different concentrations of oxygen, the rate of diffusion of O_2 is several hundred times greater than in the absence of the pigment. This is an example of facilitated diffusion. Dog myoglobin becomes only 40% deoxygenated at 5 mm Hg pressure of O_2; its blood Hb is 95% deoxygenated at this pressure. Myoglobin is high in concentration in A bands of some striated muscles, and Mb may be ten times more concentrated in striated muscle than in smooth muscle.

Sheep have a hemoglobin (HbC) that becomes predominant in the bloodstream (replacing normal Hb's A and B) during severe anemia. HbC has greater O_2 affinity and a larger Bohr effect than the other two, and hence appears to be adapted to anemic stress.[87]

Mammals that make quick movements (mouse, cat) tend to have higher P_{50} values than animals that are slow and steady (dog). The P_{50} of a mouse is 72 mm Hg, compared with 24 mm Hg for a woodchuck. P_{50} decreases with increasing body size according to the relation $P_{50} = 50.34\ W^{-0.054}$ (Fig. 8–12). Large animals also tend to have a smaller Bohr effect than do small animals.

Burrowing mammals (prairie dog) have lower P_{50} values than do surface or arboreal mammals. Both deep divers and high altitude mammals show high affinities.[65] The snowshoe hare, *Lepus*, nearly doubles its myoglobin concentration in winter as compared with summer.[176] In sheep, HbA has higher affinity for O_2 than HbB, and breeds with predominant HbA may suffer chronic hypoxia at sea level because of high venous P_{O_2}.[37] The Weddell seal can dive to 400 m for up to 43 minutes; its Hb shows large Bohr and Haldane effects, and it may accumulate a high HCO_3^- concentration.[113] A monotreme, *Platypus*, also shows large Bohr and Haldane effects.[97] The blood of a bladdernose seal *Cystophora* has a high O_2 capacity as a consequence of high Hb content, 26.4 g/100 ml.[32a] In a hibernator, the hedgehog, the sensitivity of Hb loading to temperature is much less than in non-hibernators;

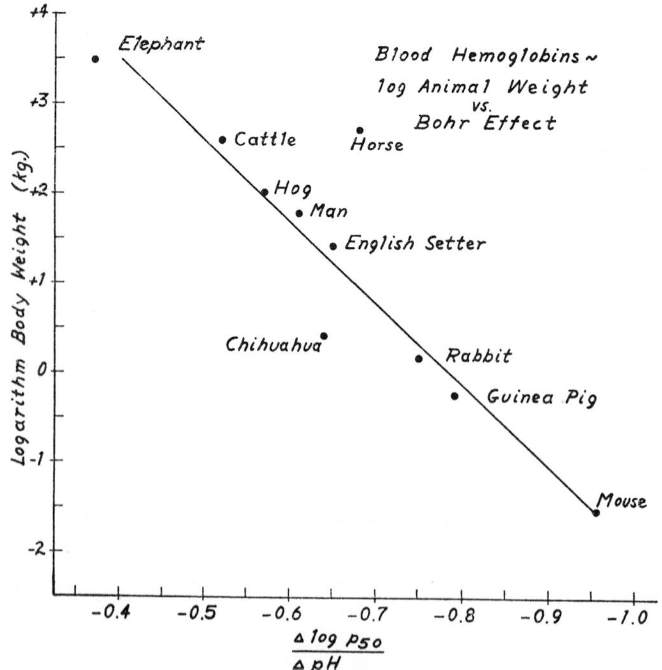

Figure 8–12. Relation between body size and Bohr effect in mammals. (From Riggs, A., J. Gen. Physiol. *43*:737–752, 1960.)

$\Delta P_{50}/\Delta T$ is 0.017 in hedgehog and 0.023 in man.[32a] The Bohr effect also is low in the hibernating hedgehog.

In general, avian hemoglobins require higher oxygen pressures than do mammalian Hb's to reach saturation. In ducks and pigeons, the difference between oxygen in arterial and venous blood shows utilization of 60 per cent compared with 27% in man, 44% in turtle, and 66% in a skate (*Raja*). A diving bird, the Adelie penguin, requires O_2 at 34.4 mm Hg for half-saturation.[13]

Those amphibians that spend much time on land have oxygen equilibrium curves to the right of those for aquatic species.[117] Adult bullfrogs show an increase in P_{50} as pH decreases to 6.2, and then it reverses; tadpole P_{50} is independent of pH from 9.0 to 6.2.[168] In a series of amphibians, the Bohr effect was least in aquatic salamander: *Rana* > *Amphiuma* > *Necturus*. An opposite series was found for turtle blood, the aquatic species having a greater Bohr effect than the terrestrial. Similarly, the terrestrial box turtle has a P_{50} of 28.5 mm Hg, and that of the aquatic loggerhead is 12 mm Hg. In the sluggish *Terrapene* the P_{50} is 11 mm Hg, and in the active swimmer *Chelydra* it is 32 mm Hg. The Hb content of blood can be reduced by treatment with phenylhydrazine to less than 1% of normal in some tadpoles and bullfrogs[47] and in turtles, and the animals may survive if kept in water well equilibrated with air; the oxygen dissolved in blood is adequate for basal activity.

In general, those fish that live in relatively stagnant water have a low P_{50}, and the Bohr effect (while it may be considerable as a percentage) does not put the dissociation curve out of the useful range. In those fish with a high P_{50}, on the other hand, an increase in CO_2 from 2 to 10 mm Hg may move the equilibrium curve so far to the right that the fish suffocate even in ample oxygen. Thus CO_2, which favors unloading in the tissues, prevents loading in the gills. For example, when the Po_2 was kept at 160 mm Hg, a shiner (*Notropos*) died at 80 mm Hg of CO_2, while a bullhead survived until the CO_2 reached 338 mm Hg.[20] Addition of CO_2 to a closed vessel containing trout blood may cause the appearance of a bubble of oxygen because of the large Bohr effect. A small increase in CO_2 may force a fish with low affinity to swim into water of high oxygen, but in a fish with high O_2 affinity the CO_2 effect is less significant. Among freshwater fish, the P_{50} of catfish, carp, and bowfin is not raised above 10 mm Hg of O_2 by 10 mm of CO_2, whereas in a sucker and three species of trout the P_{50} is raised above 35 mm Hg of O_2. Fish such as trout with low affinity and large Bohr effect show greater increases in blood lactic acid on exercise than do catfish and carp, which tolerate low O_2 and have a smaller Bohr effect. CO_2-sensitive marine fish are toadfish, mackerel, and sea robin; elasmobranchs are less affected by CO_2.

Two hemoglobins have been separated from the blood of the Japanese eel; E_1 has a P_{50} of 2.1 mm Hg and shows interaction ($n = 2.4$), while E_2 has a P_{50} of 14 mm Hg and n of 1.0. E_2 has a Bohr effect, but E_1 does not. The ecological significance of the two forms, which occur normally in the proportion $E_1:E_2$ of 3:7, is not known.[212] Two hemoglobins, one fast and one slow electrophoretically, from larvae of *Lampetra planeri* have identical O_2 affinities; of six Hb's from adult *Petromyzon marinus*, one has a high O_2 affinity and the others have similar low P_{50}'s.[9]

The Root effect, or lowering of saturation with acidification, is seen in a variety of fishes. In *Tautoga* and the red snapper *Sebastodes*, the Root effect requires more acidification after hemolysis than in intact blood; also, the O_2 equilibrium curve is more hyperbolic for red cells, and the n value decreases with acidification.[128] The lake trout, *Salvelinus fontinalis*, shows a marked Root effect[20] (Fig. 8–13). The clingfish, *Gobiesox*, shows no Root effect; in *Scorpaenichthys*, the presence of Hb in erythrocytes suppresses heme-heme interactions and reduces the Bohr and Root effects. The lungfish *Neoceratodus* shows a large Bohr effect but no Root effect.[116] When the lungfish is in air, metabolic alkalosis compensates for respiratory acidosis. In a series of African fishes, the lungfish were the only ones to live where oxygen was low and CO_2 was high. For a South American fish living near waterfalls, addition of 25 mm Hg of CO_2 reduced oxygenation by 25%, compared with a reduction of 7 to 13% in a fish inhabiting marshy ponds.

The blood of some fish is very sensitive to temperature. In three species of trout, the P_{50} rises about 1 mm Hg of O_2 with each rise of 1°C. At higher temperatures the amount of oxygen dissolved in water is also diminished, so that the combined effect on the equilibrium curve and on dissolved oxygen forces the fish to seek cool water. Sensitivity of a trout, *Salmo gairdneri*, to both high temperature and CO_2 is shown by the following values of P_{50} at different CO_2 levels:[29]

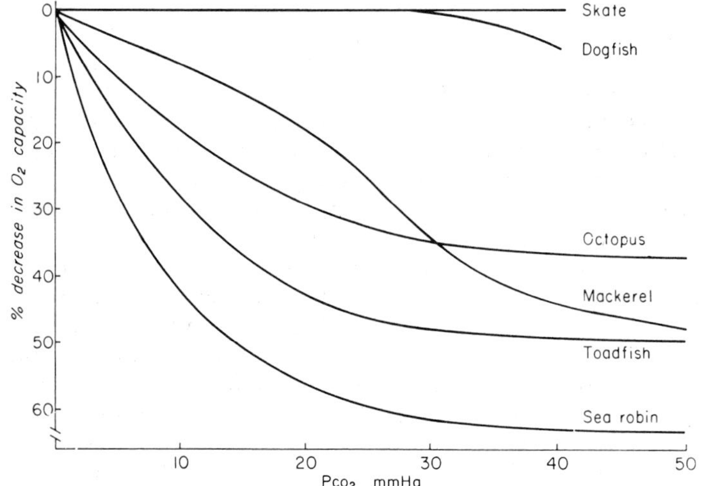

Figure 8–13. Root effect. Decrease in O_2 capacity with addition of CO_2 to blood of various fishes and *Octopus*. (From Lenfant, C., and K. Johansen, Respir. Physiol. *1*:13–29, 1966. With permission of North-Holland Publishing Co.)

Temperature	*no* CO_2	*3 mm* CO_3	*7 to 8 mm* CO_2
10°C	9	18.5	
15°C	14	20	38
20°C	18.5	27	38

In contrast to many hemoglobins, the reaction of tuna hemoglobin with O_2 is virtually insensitive to temperature changes.[178]

The dogfish *Squalus* shows no Bohr, Root, or Haldane effects, and the blood is poorly buffered.[115] The hagfish *Myxine* shows no Bohr effect, but the lamprey *Petromyzon* has a large Bohr effect.[127]

A simple method for testing the importance of Hb for survival is to poison it with CO. A goldfish can carry out routine activity for long periods with its Hb inactive,[6] but a rainbow trout can survive only 30 minutes (at 10 to 15°C).[76] For man, 1.28% CO is toxic within 3 minutes. The relative affinity of Hb for CO and O_2 is given by M in the relation

$$\frac{\text{CO Hb}}{\text{O}_2\text{ Hb}} = \text{M}\,\frac{\text{Pco}}{\text{Po}_2}$$

or

$$\text{M} = \frac{(\text{CO Hb})(\text{Po}_2)}{(\text{O}_2\text{ Hb})(\text{Pco})}$$

Differences between hemoglobins in relative affinities of blood for CO and O_2 are as follows:[6, 76, 138a]

	M
horse	550
man	300
roach (fish)	210
pike	100
eel	99–114
trout	66
carp	60.5
goldfish	63
mammalian muscle	21–51
root nodule bacteria	37
parasitic fly *Gastrophilus*	0.67
Ascaris	0.036

It is evident from the difference between trout and goldfish that high Hb affinity for CO is only one factor correlated with survival after inactivation of Hb.

The Antarctic Chaenichthyidae lack red blood cells, yet their oxygen consumption is nearly as high (45 ml/kg/hr) as in species having Hb. Their skin is highly vascularized.[154] When the blood of a pinfish, *Lagodon*, is reduced to 2.9% of normal by phenylhydrazine treatment, there is little change in oxygen consumption; hence, the Hb must normally provide reserve transport capability.[28] In *Chaenocephalus*, $\dot{V}o_2$ at 1°C is 0.02 ml O_2/g/hr, the critical Po_2 is 50 mm Hg, blood volume is high, and cutaneous respiration is high; 8% of the respiration is through the tail.[70]

In an Antarctic fish, *Trematomus*, high temperature (4.5°C compared with the normal 1°C) acts as the Root effect in reducing saturation; the P_{50} is more sensitive to temperature than in other fish, thus restricting *Trematomus* to cold water.[62]

It is apparent that fishes show a wide variety of effects of CO_2 and temperature on O_2 affinity. Much more information is needed about the detailed ecology of many species before general conclusions can be drawn.

TRANSPORT CHANGES DURING EMBRYONIC DEVELOPMENT

Structural differences between fetal and adult hemoglobins were mentioned on page 323. In human HbF, γ chain replaces each β chain. The P_{50} of adult blood is 28 mm Hg, and that of fetal blood is 20 mm Hg;[183] but in solution the difference is lost, with P_{50} values of 14.3 for fetal and 14.7 for adult hemoglobin at pH from 6.5 to 7.5. However, below pH 6.5, the P_{50} of HbF is higher than for HbA.[9] The equilibrium curves for fetal blood lie well to the left of that for adult blood. In goats, the P_{50} of whole blood increases and the O_2 capacity decreases with age.[12,14] In goats and sheep, unlike man, HbF in solution has a higher O_2 affinity than HbA (Fig. 8–14), although the effect is not as great as with whole blood.

In primates the placental circulation has multivillous streams, and there is a normal gradient of Po_2 from uterine artery to fetal vein. In sheep and goats, the fetal blood is exposed in capillaries to "pools." In rabbits, the fetal arteries parallel maternal arteries, but with the blood streams flowing in opposite directions, so that by countercurrent exchange the oxygen diffusion results in higher Po_2 in fetal veins than in maternal veins. Thus, in rabbit, countercurrent flow is more important, while in man, sheep, and goat, differences in O_2 affinity are more important[14,134] (Fig. 8–15). This is shown by the data below.[134]

In camel, the P_{50} of fetal blood is lower than that of the adult and the O_2 capacity is higher; in elephant, the shift from HbF to HbA occurs during the first third of pregnancy.[163,164] No difference exists between fetal and adult Hb in cats. In man, HbF (compared to HbA) has higher O_2 affinity and is more resistant to alkaline denaturation; n is the same for both.[197] In lambs, the decrease in O_2 affinity of red cells is due in part to high DPG; however, in adult sheep DPG is lower in concentration than in either lamb or fetus.[15] Chickens have five forms of Hb as determined electrophoretically. Hb_4 is a major component for seven days of incubation. Hb_2 is minor in the embryo but is major at hatching; Hb_3 is minor in both bloods. Bone marrow synthesis begins on day 14, but if eggs are incubated in reduced O_2 the changes are delayed.[187] The adult Hb gene is activated on the sixth day of incubation in domestic fowl, and two or three days later in turkey.[130] In a chick, the chorio-allantois becomes less efficient in providing oxygen as development proceeds, as shown by the following composition of blood from the chorio-allantoic artery and vein:[54]

	$P_{v_{O_2}}$	$P_{a_{O_2}}$	$P_{v_{CO_2}}$	$P_{a_{CO_2}}$
14 days incubation	70	29	33	46
18 days incubation	50	20	46	58
air space breathing	68	37	35	47
1 day hatch	34	109	43	28

In the viviparous garter snake *Thamnophis*, a fetal-maternal shift occurs for Hb in corpuscles; in solution the two curves are superimposable.[125] The blood of the bullfrog tadpole becomes saturated with oxygen at lower pressures than does the blood of the adult frog; tadpole P_{50} is 4.6 mm Hg compared with 13.2 mm Hg in the adult, and n is 2.8 for each. Frog blood shows a normal Bohr effect down to pH 6.2, while tadpole blood shows no Bohr effect.

In the ovoviviparous spiny dogfish *Squalus*, the fetus during its 22 month gestation has an Hb with higher affinity for O_2 than the adult Hb. An oviparous ray has a transient embryonic hemoglobin of high O_2 affinity during its first few months of development. Even in an oviparous teleost, *Scorpaenichthys*, the postlarvae show a change in hemoglobin.[122]

The fetal-maternal shift in O_2 affinity in mammals, some snakes, and elasmobranchs is an example of facilitated transport by interposing a pigment of intermediary affinity

	Maternal						*Fetal*					
	$P_{a_{O_2}}$	pH (arterial)	$P_{v_{O_2}}$	pH (venous)	P_{50}	O_2 capacity	$P_{v_{O_2}}$	pH (venous)	$P_{a_{O_2}}$	pH (arterial)	P_{50}	O_2 capacity
man	100	7.4	33	7.36	27	15.5%	29	7.32	17	7.24	19	
goat	84		46				33		14			
rabbit	80		25				46		17	(counter		
Macaca			30		32	15.5%	16		12	current)	19.2	17.8%

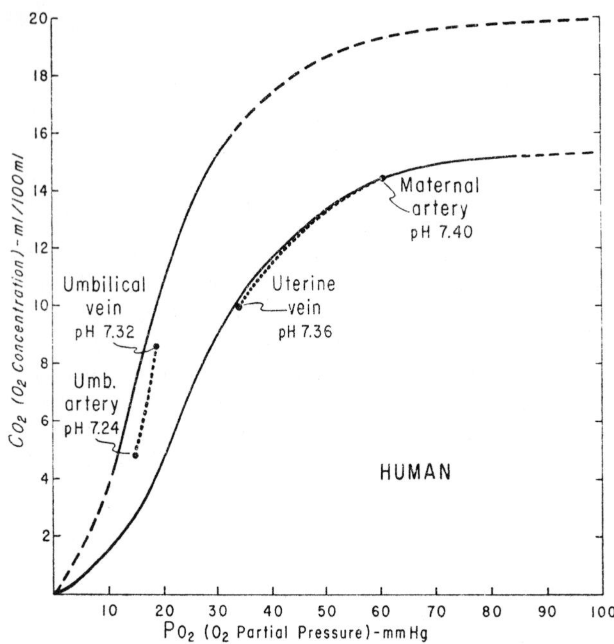

Figure 8–14. Differences between fetal and adult bloods in O_2 affinity. *A,* O_2 concentration as a function of Po_2 in human fetal and maternal blood. (From Metcalfe, J., H. Bartels, and W. Moll, Physiol. Rev. *47*:782–838, 1967.)

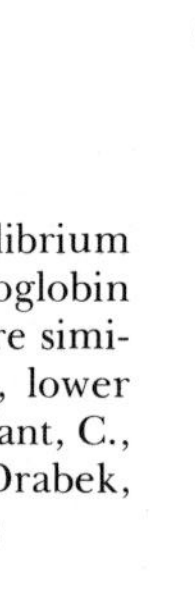

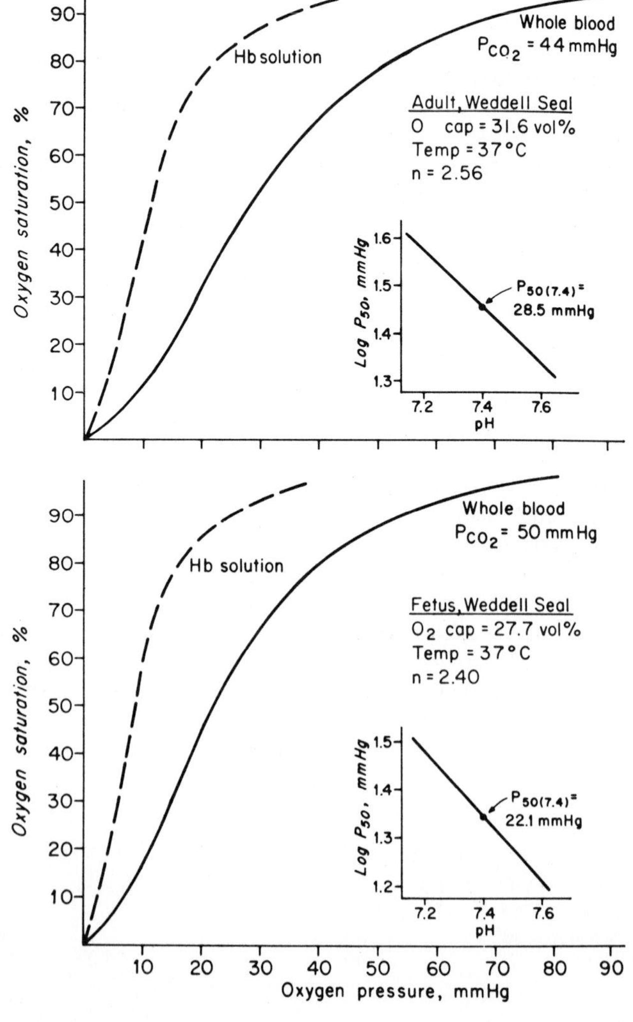

Figure 8–14 Continued. *B,* O_2 equilibrium curves for adult and fetal blood and hemoglobin in Weddell seal; hemoglobin solutions are similar, but fetal blood has lower capacity, lower P_{50}, and greater Bohr effect. (From Lenfant, C., R. Elsner, G. L. Kooyman, and C. M. Drabek, Amer. J. Physiol. *216*:1595–1597, 1969.)

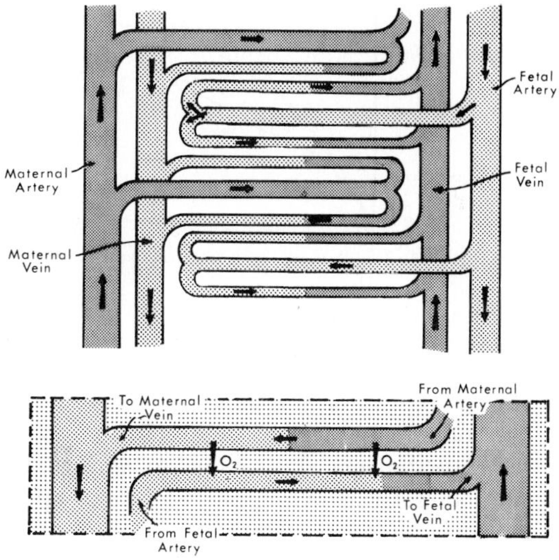

Figure 8–15. Diagram showing countercurrent arrangement of vessels in human placenta, which permits oxygen to pass from maternal artery to fetal vein, and from maternal vein to fetal artery. (From Metcalfe, J., H. Bartels, and W. Moll, Physiol. Rev. *47*:782–838, 1967.)

between primary blood at high O_2 partial pressure and embryonic tissue of low pressure. The meaning of the shift in birds and frogs may be associated with the transition from an aquatic embryonic life to air breathing. The meaning in oviparous fish is unclear.

HEMOGLOBIN FUNCTION IN INVERTEBRATES

In some invertebrates, hemoglobin functions in oxygen transport at atmospheric pressures; in others it functions only at low pressures. In some it may provide an oxygen store for times of hypoxia, and in a few it may be a store of hematin for cellular uses. The following criteria are useful in judging whether a respiratory pigment functions in O_2 transport: (1) the position of the oxygen equilibrium curve (i.e., whether the pigment loads and unloads at pressures corresponding to those at the respiratory surface and in tissues); (2) the Bohr effect (i.e., whether addition of CO_2 facilitates unloading of O_2 at tissue pressures and, conversely, aids loading at respiratory pressures); (3) the oxygen capacity of the blood and the amount of oxygen combined or in solution; (4) differences between arterial and venous blood in O_2 content; and (5) effect on oxygen consumption of CO as a result of inactivating the pigment at different O_2 pressures. In experimental carbon monoxide poisoning, it is important that maintained combination of the hemoglobin and CO be checked spectroscopically and that the amount of CO be small enough not to interfere with tissue respiration.

Some annelid worms live in near-anaerobic conditions and have functional hemoglobin. An African oligochaete, *Alma emini*, lives in swamps where the mud more than a few millimeters deep is anaerobic; the Hb is saturated at less than 2 mm Hg and is unaffected by CO_2.[16] The hemoglobin of *Tubifex*, which inhabits stagnant drainage ditches, is half-saturated at 0.6 mm Hg, but in vivo the bands of oxy-Hb disappear when the O_2 in the external water reaches about 10 mm Hg.[53] In the presence of CO, respiration is reduced, more at high than at low pressures. These worms are poisoned by hyperbaric oxygen, and there must normally exist an extremely steep gradient of oxygen from water to tissues.

Most annelids live in soil or water that is moderately aerated. The oxygen consumption of an earthworm, *Lumbricus*, is reduced by CO poisoning at partial pressures of 40 mm Hg and higher (Fig. 8–16). A rise in temperature from 15° to 25°C raised the P_{50} from 3.9 to 6.8 mm Hg; a normal Bohr effect is found (Table 8–3). Oxygen consumption is relatively constant, and the percentage of oxygen carried by Hb is about 40 from 75 to 152 mm Hg of O_2.[33] In a giant earthworm, *Glossoscolex*, the oxygen capacity

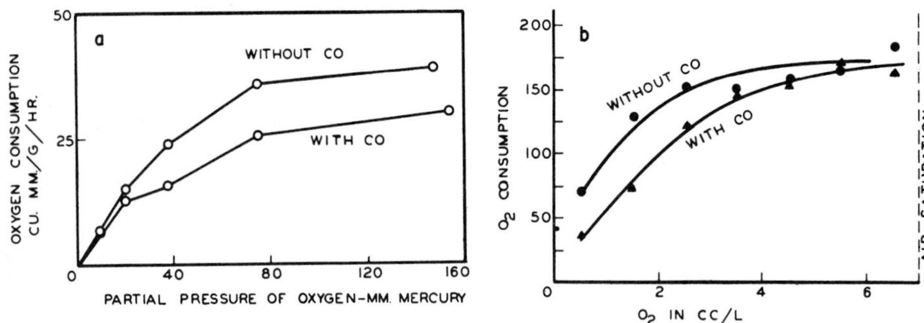

Figure 8–16. *A,* Oxygen consumption by *Lumbricus* as a function of partial pressure of oxygen with and without carbon monoxide. (From Johnson, M. L., J. Exp. Biol. *18*:266–277, 1941.) *B,* Oxygen consumption by *Chironomus* as a function of Po_2 with and without carbon monoxide. (From Ewer, R. F., J. Exp. Biol. *18*:197–205, 1941.)

of blood is 14 vol per cent; Hb in blood in the dorsal artery was 41% saturated, and the Bohr effect was negligible.[98] In three species of leeches with Hb, the O_2 consumption was reduced by CO at 10 to 20% O_2 but not at 3% O_2, while two species lacking Hb were unaffected by CO.[184]

Most hemoglobin-containing annelids are more sensitive to carbon monoxide at high Po_2 than at low; thus, at high pressure their metabolism depends on hemoglobin, but they probably become inactive and shift to glycolysis at low O_2. *Nereis diversicolor* differs from most annelids in that its respiration is completely blocked by CO at 3.3 ml O_2/liter, and is reduced by 50% at higher O_2 concentrations. Several species of *Arenicola* show moderately high O_2 affinities—P_{50} of 1.8 mm Hg and saturation at 5 to 10 mm Hg of O_2.[209] *Arenicola* hemoglobin could provide an O_2 store which would last not more than 21 minutes, much less than the low-tide period.[99] The hemoglobin of *Arenicola* shows a very large interaction ($n > 5$) and very small Bohr effect.[201a] In *Nephthys*, the P_{50} is 5.5 mm Hg for blood Hb and 7.5 mm Hg for coelomic Hb; the interstitial water in sand outside these worms averaged 5.7 mm Hg of O_2 (low tide), and in freshly filled burrows it was 13.7 mm Hg of O_2.

Another polychaete, *Glycera*, has Hb in coelomic cells; much variation in P_{50} was noted, but an average value was 7 mm Hg of O_2. Vo_2 was reduced by CO by nearly 20% at high O_2, but not at all at low O_2. The hemoglobin from the coelomocytes was separated as two fractions: a light fraction of high O_2 affinity and a heavy fraction with low affinity.[72]

Many gastropod and pelecypod molluscs have myoglobin in some tissues (e.g., the red radular muscles). In an amphineuran, the radular myoglobin has a molecular weight of 33,800 and has two hemes per molecule. In the radular muscle of an amphineuran, the myoglobin is in two forms, a dimer and a monomer. In the gills of a clam, *Phacoides*, a myoglobin exists which is probably tetrameric.[155] Myoglobin is widely found but hemoglobin is sporadic in molluscs. *Cardita* has a blood Hb in its plasma that has a P_{50} of 11 mm Hg; the myoglobin in *Mercenaria* muscle has a P_{50} of 0.55 mm Hg.[126]

In the freshwater and mud-dwelling snail *Planorbis*, the Hb is well suited for transport; P_{50} is 7.4 mm Hg at 20°C, there is a positive CO_2 effect, and in vivo the oxy-Hb bands disappear when the oxygen in the water outside reaches 25 mm Hg. High concentrations of CO_2 are probably never encountered. At Po_2 down to 54 mm Hg (7.7% saturation), skin breathing is sufficient; at O_2 between 7.2 and 3%, the snails were at the surface part of the time for lung breathing, and below 3% O_2 they were at the surface continually. From measurements of oxygen in lungs and arterial blood of *Planorbis*, it appears that the Hb is useful between 60 and 20 mm Hg of O_2. At higher pressures the venous O_2 does not fall low enough for unloading, below 40 mm arterial blood is not fully saturated, and below 20 mm both transport and storage functions are lost. The lower limit for cutaneous respiration is higher than that for lung respiration, with the minimum for full saturation being 39 mm Hg at 3.1 Pco_2 and 61.5 mm Hg at 1.9 Pco_2.[100] Another snail, *Lymnaea*, browses at the surface; it lacks Hb and does not reduce pulmonary O_2 as low as does *Planorbis*.[100]

A number of small crustaceans have Hb in the blood, particularly when in water that is low in oxygen. A freshwater cladoceran, *Moina*, synthesizes Hb during several days

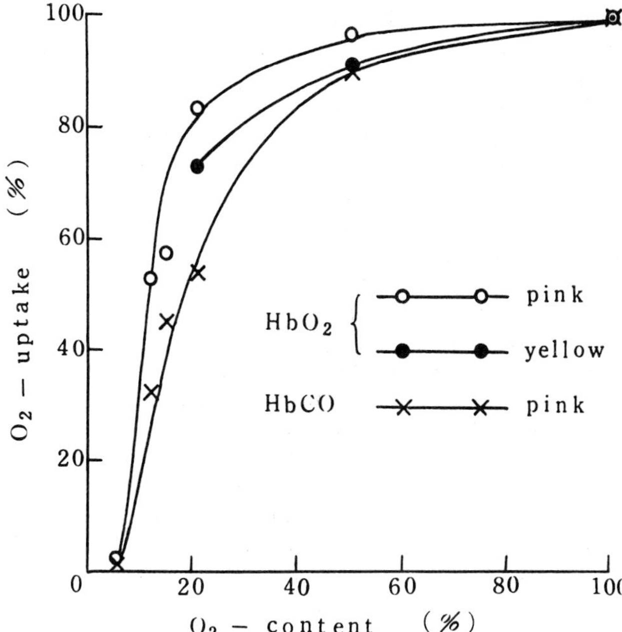

Figure 8–17. Oxygen consumption as a percentage of maximum at different O_2 concentrations by a cladoceran, *Moina*, when low in Hb (yellow), rich in Hb (pink), and pink but poisoned with CO. (From Hoshi, T., and T. Shimada, Sci. Rep. Niigata Univ. D *2*:1–12, 1965.)

in low O_2. Its P_{50} is 3.5 mm Hg at 25°C and 1.1 mm Hg at 15°C; it shows no Bohr effect (Fig. 8–17). Pink (hemoglobin-containing) specimens consume more oxygen than pale ones, and CO poisoning reduces V_{O_2}, particularly in moderately low oxygen.[81] Thus, the Hb is synthesized when the need for oxygen increases, and it has a transport function. In the brachiopod crustacean *Triops*, the oxygen content in vivo was 0.93 vol per cent, but the capacity when saturated was 3.2 ml O_2 per cent. P_{50} in *Triops* is 6.8 mm Hg; the Bohr effect was normal, but respiration is not reduced by treatment with CO.[78] In a red daphnid *Simocephalus*, CO reduced the O_2 consumption at below 2.5 ml O_2/liter, but not above that concentration.[80] In *Artemia*, the Hb content can be increased 20-fold in low oxygen.

The aquatic larvae of dipterans of the genus *Chironomus* have hemoglobin. The P_{50} is very low as measured in vitro (0.17 to 0.6 mm Hg), but in vivo the Hb is deoxygenated when the O_2 in the outside water corresponds to 13 mm Hg. The extra O_2 consumption after a period of anaerobiosis is abolished by CO. V_{O_2} is unaffected by CO down to 1/2 saturation. The Hb may provide a store of O_2 during periods of interruption of respiration for feeding. Feeding normally stops in pink species below 10% of air saturation, and at 26% in the presence of CO; however, pale species continue to feed actively in low O_2, and even in the presence of cyanide. No difference was noted in payment of O_2 debt after anaerobiosis in pale and pink species. Larvae of *Chironomus plumosus* show O_2 capacity of blood from 5.4 to 11.6 vol per cent according to the amount of Hb; P_{50} is 0.2 mm Hg at pH 8 and 0.6 mm Hg at pH 7.4.[202] V_{O_2} is 250 ml O_2/kg/hr, and it is calculated that O_2 bound to Hb would last for only nine minutes, which is close to the normal interval between ventilation periods. High temperature decreases O_2 affinity and increases the CO_2 effect. Another species, *C. thummi*, is more independent of P_{O_2} than is *C. plumosus*; it maintains constant V_{O_2} down to 18 mm Hg of O_2.[202] It seems likely that the enzymatic patterns of various crustaceans and chironomids may vary according to their dependence on Hb; other functions than O_2 consumption should be studied to learn the meaning of variation in Hb content.

Some aquatic insects (e.g., the Notonectidae) have tracheal cells that are rich in hemoglobin. They surface to breath every 2 to 5 minutes, and under CO the duration of the dives is shortened.[134a] However, some species lack the Hb.

In endoparasites, the function of hemoglobin is not clear. In a number of nematodes the P_{50} of Hb is less than 0.05 mm Hg for dilute solutions.[174] However, in vivo the Hb of *Nippostrongylus* is deoxygenated when the worms are in a medium containing 13 mm Hg of O_2. The normal oxygen concentration in the intestine where they live (4 to 9 mm Hg) is well below the critical O_2 for their aerobic metabolism. The Vo_2 was not reduced by CO in the range from 30 down to 5 mm Hg of O_2. Hence, these worms normally are not exposed to oxygen at levels that would permit maximal metabolism and saturation of their hemoglobin.

Ascaris has two very different hemoglobins—one in the perienteric fluid and another in the body wall. When in nitrogen, the Hb of the body wall becomes visibly deoxygenated but that in the perienteric fluid does not. Deoxygenation time by a reducing agent is 150 seconds for perienteric Hb, as compared with 0.008 second for sheep Hb.[35] Deoxygenation of body wall Hb has a Q_{10} of 3 and is very sensitive to pH; Q_{10} for perienteric Hb is 5, and deoxygenation is insensitive to pH. (See page 363 for definition of Q_{10}.) It is evident that the perienteric fluid Hb does not function in O_2 transport; Smith[189] has suggested that it may serve as a store of hematin, which is needed in egg production.

A freshwater nemertean, *Prostoma*, contains Hb, and Vo_2 is reduced at low Po_2 if CO is present.[151] It has long been known that *Paramecium* contains hemoglobin. That this functions as a carrier is indicated by the reduction in respiration in the presence of CO.[190]

In summary, hemoglobin occurs sporadically in several phyla of invertebrates; some species that lack it seem to survive as well as those that have it. In some invertebrates, Hb is essential for O_2 transport; in some, the O_2 gradient between environment and tissues must be very steep. In a few Hb-containing animals, no function for the Hb has been discovered.

EFFECTS OF REDUCED ENVIRONMENTAL OXYGEN; SYNTHESIS OF HEMOGLOBIN

The Hb content of the blood is normally high in animals that live at high altitudes, and when animals go from sea level to high altitudes the Hb content of their blood increases; with it, an increase in O_2 capacity occurs. When animals are first subjected to hypoxia, as at altitude, red cells are released from body stores such as spleen; later, hematopoiesis is increased under stimulation of a hormone, erythropoietin. Increased synthesis is noted in man under severe hypoxia within 12 hours, and a maximum rate is reached in three days.[188] There is an increase in blood volume, due largely to an increase in the number of red cells. Concomitant with an increased O_2 capacity, the O_2 equilibrium curve in man may shift to the right, but no change in *n* value or Bohr effect occurs:[111]

	O_2 *capacity (vol %)*	P_{50} *(mm Hg)*	CO_2 *capacity*
at sea level	18.9	26.8	43.6
at altitude, natives	26.4	30.7	36.4
at altitude, sojourners	20.7	30.4	36.6

The red cell count and hematocrit of acclimated sheep, rabbits, dogs, and men are elevated; but in native llama and vicuña, blood counts are not much different at sea level from in the mountains.[65] Oxygen equilibrium curves of llama and vicuña, as well as mountain ostrich and huallata, lie to the left of those of related species at sea level. Dogs reared at 14,890 feet had 40% more Hb in blood and 65.7% more myoglobin in muscle than dogs at sea level.[200] In highland species of Peruvian rodents, the Mb in diaphragm and leg muscles was twice that in sea-level species.[161, 162]

The increase in Hb on acclimation to altitude may not be sufficient to maintain constant oxygen supply to the tissues. Lambs showed the following differences at two altitudes:[64]

	5200 feet	*12,700 feet*
mean % saturation	84	59
$P_{a_{O_2}}$, mm Hg	70	40
$av_{P_{O_2}}$ gradient, mm Hg	28	11

The life span of circulating red blood cells in man is the same (111 to 121 days) at altitude as at sea level. Rats kept in 10% O_2 for six months increased the Hb concentration from 14.8 to 22.3 g/100 ml.

Red blood cells are made in the liver of fetal animals and in the bone marrow of adults. Hemoglobin is synthesized by the nucleated pro-erythrocytes or reticulocytes, and the RNA control system for this synthesis has been isolated from circulating reticulocytes. Stimulation of production of reticulocytes can be caused by hemorrhage as well as by reduced O_2. The hormone erythropoietin (ESF) is tested on polycythemic mice in terms of reticulocyte production. Hormone levels in blood are increased by hypoxia (as in hemorrhage) and by treatment with cobalt. Erythropoietin is normally produced in the kidney, probably by juxtaglomerular cells; the kidney is possibly the sensing organ for low O_2.[93] The action of erythropoietin is increased by testosterone and prolactin, and is decreased by estrogen. Erythropoietin has been demonstrated in mammals, birds, frogs, and fishes.[177] It acts mainly on stem cells in the marrow, causing them to take up iron and produce reticulocytes. The hormone can also stimulate synthesis of Hb by liver in tissue culture from 13-day mouse fetus. When applied to these cultures, erythropoietin increases DNA synthesis in 20 minutes, and RNA and Hb synthesis in two hours.[89] Synthesis of a 150s particle (RNA-protein) can be detected in a few minutes. Mammalian erythropoietin is ineffective on lower vertebrates, but serum from a bled fish increases erythropoiesis when injected into another fish; hence, production of a hormone in fish may be stimulated by hypoxia as in mammals.[213]

In several crustaceans, hemoglobin synthesis is markedly stimulated by low oxygen. *Daphnia* collected or reared in water that is low in oxygen are red, while those from high oxygen are pale; genetic differences exist in ability to form Hb. *Daphnia* gains or loses Hb in about 10 days; it also gains cytochrome and myoglobin in its muscles in low oxygen. *Artemia* may be brilliant red with Hb when living in hypersaline pools that are low in oxygen. Blood Hb concentration increases in reduced oxygen not only in some crustaceans (mostly cladocerans and phyllopods), but also in dipteran larvae (*Chironomus* and *Anatopynia*) and in snails (*Planorbis*). However, no such stimulation of Hb synthesis was found in a variety of other invertebrates, such as annelids and molluscs.[49, 51] Whether Hb synthesis is hormonally triggered as in vertebrates is unknown.

FUNCTION OF CHLOROCRUORIN

Despite its much larger size, chlorocruorin resembles hemoglobin functionally. The O_2 equilibrium curve is sigmoid, and a Hill plot gives n of 4.5 in the mid-saturation region; the P_{50} is sufficiently high (30 to 40 mm Hg) that the pigment is probably deoxygenated in the tissues and oxygenated in the respiratory tentacles. The O_2 capacity of the blood is high (9.1 ml O_2/100 ml).[50] Chlorocruorin has a higher affinity for CO than any hemoglobin, and the oxygen consumption of the sabellid worms that contain it is reduced by CO at all concentrations of oxygen. The blood of *Serpula* has both hemoglobin and chlorocruorin; both pigments are deoxygenated together.[52] *Potamilla* has chlorocruorin in blood and hemoglobin in muscles; probably the hemoglobin facilitates diffusion from the low-affinity chlorocruorin. It may be concluded that chlorocruorin is an oxygen carrier functioning normally at high oxygen pressures.

FUNCTION OF HEMOCYANIN

It was shown above that the hemocyanins of arthropods and of molluscs differ in physical properties; however, all hemocyanins bind oxygen reversibly in the proportion of one oxygen molecule per two copper atoms. A series of arthropods have one binding site per molecular weight of 75,800; in molluscs there is one site per 50,800.[74] Hcy is blue, with strong absorption bands in the oxy-Hcy, but not the deoxy-Hcy, form.

Cephalopod molluscs—*Loligo* and *Octopus*—transport virtually all of their oxygen by Hcy. The arterial blood of the squid is saturated and contains 4.27 vol per cent of O_2 and 3.82 vol per cent of CO_2; venous blood has 0.37 vol per cent of O_2 and 8.27 vol per cent of CO_2. Thus, approximately 92% of the oxygen is removed in the course of circulation, three times the percentage in man. The half-saturation value (36 mm Hg at 0 mm CO_2 and 23°C) is high enough to make the squid sensitive to oxygen concentration in the external water. A strong Bohr effect shifts the loading to high O_2 values as CO_2 is added; also, the amount of interaction between subunits is sensitive to pH and to Mg. Interaction is low both at high and at low pH, but $n = 3.9$ at pH 7.4 (Fig. 8–18). Deoxy-Hcy remains associated as a large molecule (59s) from pH 6 to 10; but on oxygenation, dissociation into subunits of 19s and 11s is complete at 80% oxygenation. At 100% oxygenation, reassociation occurs.[39]

In *Octopus*, the O_2 capacity is 3.0 to 5.0

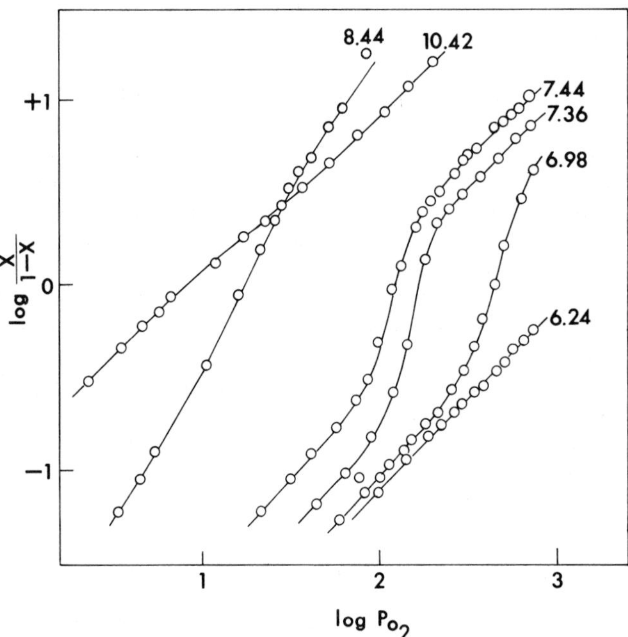

Figure 8–18. Log plots of oxygen equilibrium curves of hemocyanin of *Loligo*; determinations at the indicated pH's. (From DePhillips, H. A., et al., Biochemistry *8*: 3665–3672, 1969.)

vol per cent.[208] Addition of CO_2 increases the Po_2 for unloading (Bohr effect) and lowers the amount of oxygen bound at loading Po_2 (Root effect); high temperature also decreases the amount of O_2 bound.[114] Arterial blood was 82% saturated at a Po_2 of 77.5 mm Hg; venous blood was 9.6% saturated at a Po_2 of 9.7 mm Hg. More than 90% of the oxygen in arterial blood is bound to Hcy, and *Octopus* extracts 27% of the oxygen from respiratory water.[94] It appears that cephalopods are highly dependent on Hcy for oxygen transport, that they have little oxygen reserve in the blood, and that the dependence of association of subunits on Po_2, pH, and divalent cations is very complex.

In the snail *Helix pomatia*, two molecular types of Hcy are identified by their dissociation in 1 M NaCl, but their respiratory properties are similar. *Busycon* oxy-Hcy in dilute solution is an aggregate with an s_{20} of 100, whereas in the deoxygenated form only 17% of the molecules are 100s and the remainder are 60s and 19s (Fig. 8–19).[39] The O_2 equilibrium curve is hyperbolic in the absence of Ca and Mg when n values of 1.1 are obtained; in the presence of Ca and Mg, the equilibrium curve is sigmoid, and n is 2.3 for the α form and 1.4 to 4.6 for the β form of Hcy. The α form shows a small positive Bohr effect and the β form shows a negative effect in the physiological range of pH. Possibly the β form functions best at low CO_2 and the α form at high CO_2.[107] In *Busycon*, P_{50} is lower than in *Helix* but, as in *Helix*, n is higher in the presence of divalent cations; the Bohr effect is negative or questionable. The arterial blood has a Po_2 of 36 mm Hg, and that of venous blood is 6 mm Hg. In another marine snail, *Fusitriton*, the Bohr effect is negative in the normal pH range; it is suggested that, since the circulation is sluggish, production of CO_2 in one region would cause O_2 to move toward that region.[158, 159] In a limpet, *Diodora*, and in

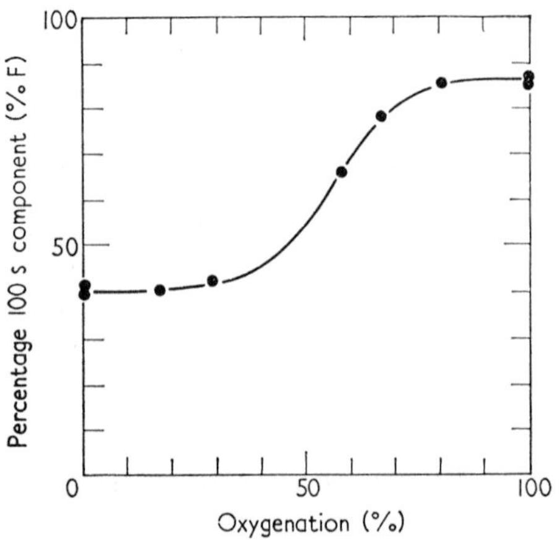

Figure 8–19. Percentage of 100s component in hemocyanin from *Busycon* as a function of percentage oxygenation at pH 8.2. (From DePhillips, H. A., et al., J. Molec. Biol. *50*:471–479, 1970.)

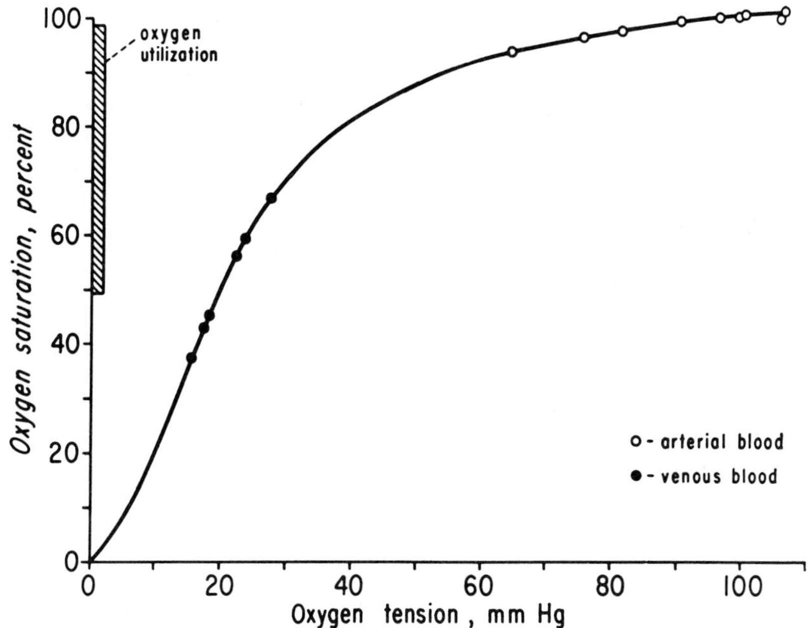

Figure 8–20. Oxygen equilibrium curve and measured saturation of arterial and venous blood of crab *Cancer*. (From Johansen, K., and T. Mecklenburg, Z. vergl. Physiol. *70*:1–19, 1970.)

Chiton, more than half of the O_2 is transported by Hcy and there is little effect of pH on the P_{50}.[156, 158] Some gastropods (*Busycon*, *Cryptochiton*) use Hcy for O_2 transport in blood but have myoglobin in some muscles, particularly the radular muscle; the O_2 affinity of the myglobin is greater than that of Hcy, and hence the muscle pigment appears to facilitate diffusion as in the vertebrates.

Among arthropods, transport of O_2 by Hcy has been examined in crustaceans, in *Limulus*, and in a few scorpions (Fig. 8–20). In the lobster *Homarus*, Hcy constitutes 88.2% of blood protein; P_{50} is 6 mm Hg at 0.5 mm CO_2 and is 25 mm Hg at 9.9 mm CO_2. The aggregation of subunits is higher at physiological pH than in alkaline or acid media, and calcium promotes a high *n* value. Removal of calcium dissociates the large molecule and shifts the equilibrium curve to the right.[148] Postbranchial blood had Hcy that was 49% saturated, while prebranchial blood Hcy was 20% saturated. In crayfish, the P_{50} rises from 1.57 to 2.79 mm Hg and *n*

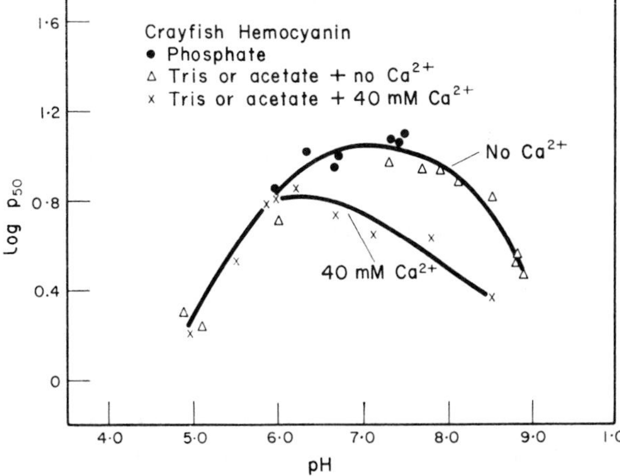

Figure 8–21. Effect of pH on P_{50} of hemocyanin from crayfish, showing reversal in O_2 binding and effect of Ca^{++} on the pH for maximum P_{50}. (From Larimer, J. L., and A. F. Riggs, Comp. Biochem. Physiol. *13*:35–46, 1964.)

decreases from 3.1 to 2.9 when temperature is raised from 15° to 25°C. P_{50} is maximal at pH 7 and declines on both sides of this value; there is less decline on the alkaline side if Ca is absent[110] (Fig. 8–21).

Respiratory characteristics of blood obtained by indwelling catheters clearly show the function of Hcy in O_2 transport in the crab *Cancer magister.* The oxygen capacity of the blood is 3.44 vol per cent, and oxygen extraction from the ventilatory stream is 16%. $P_{a_{O_2}}$ is 91 mm Hg, corresponding to complete saturation, and $P_{v_{O_2}}$ is 21 mm Hg, or 50% saturation. After five minutes of exercise, the O_2 saturation drops to 10%, and a rise in temperature shifts the saturation curve toward higher Po_2's.[96]

In two land crabs, Hcy transports most of the O_2: in *Cardisoma,* arterial blood contained 1.6 to 1.7 vol% O_2 and venous blood had 0.85 to 0.99 vol%, while in *Gecarcinus* the corresponding values were 1.45 and 0.61 vol%.[157, 160] These bloods showed a normal Bohr effect; saturation in the gills is some 85%, whereas venous blood is 20 to 40% saturated.

In a scorpion, *Heterometrus,* the O_2 capacity of the blood is 1.82 vol% and P_{50} is 16.5 mm Hg at 30°C. The Bohr effect is normal from pH 10 to 8 and reverses from pH 8 to 6; *n* increases with oxygenation.[140] In *Limulus* also, the Bohr effect is normal from pH 9.1 to 8.2 and then reverses; i.e., P_{50} is maximum at a pH above the physiological range. The meaning of the negative Bohr effect, which is common for Hcy, is not clear; it could permit the animal to obtain oxygen better from water that is rich in CO_2. In many crustaceans, pericardial blood is far from saturated, yet the oxygen bound to Hcy is much more than that in solution, and the best controlled measurements show the transport function of the pigment. Yet, in sluggish life, many animals (especially gastropods and *Limulus*) that contain Hcy could probably get along without it.

FUNCTION OF HEMERYTHRIN

In *Golfingia* (also in *Sipunculus*) the O_2 equilibrium curve of hemerythrin (Hr) shows a P_{50} of 8 mm Hg. A rise in temperature shifts the curve to the right; heats of oxygenation for hemerythrin are much higher than those for hemoglobins. The *n* value is close to (or slightly greater than) 1.0; hence, there is little interaction between monomers. In *Golfingia,* pH has little effect on the O_2 equilibrium curve. Likewise, the absence of a Bohr effect has been noted in Hr from *Dendrostomum* and from *Sipunculus.* However, hemerythrin from brachiopods (*Lingula*) shows a normal Bohr effect; brachiopod Hr has not been studied chemically as extensively as that from sipunculids. Hemerythrin is not poisoned by CO. The ratio of iron atoms to oxygen is 2 in hemerythrin and 1 in Hb, and the copper to oxygen ratio in hemocyanin is 2.

Hemerythrin from the vascular system (tentacular circulation and main contractile vessels) of the sipunculid *Dendrostomum zosteriocolum* is electrophoretically different from the Hr in the coelom; the cells containing vascular Hr have a P_{50} of 40 to 50 mm Hg, while that of coelomic Hr is 4.5 mm Hg. Apparently, the vascular Hr transports O_2 from the high pressure of sea water to the coelom, and the coelomic Hr transfers it to the tissues. In a burrowing sipunculid (*Siphonosoma ingers*) that does not use its tentacles for respiration, the vascular pigment has a higher O_2 affinity than the coelomic Hr (as with Hb in *Nephthys*); here, the transfer is through the body wall to the coelom and then to the high-affinity vascular pigment.[125]

CARBON DIOXIDE TRANSPORT

The amount of carbon dioxide in blood, like the amount of oxygen, greatly exceeds the amount in solution. Carbon dioxide in solution represents an equilibrium among dissolved CO_2, H_2CO_3, HCO_3^-, and $CO_3^=$. The equilibrium between CO_2 and H_2O is strongly toward H_2CO_3, and this reaction is accelerated by carbonic anhydrase; hence, the amount of dissolved CO_2 is given by the concentration of H_2CO_3. The pH of dissociation of HCO_3^- is sufficiently high that at physiological pH's very little $CO_3^=$ is present. Hence, the total CO_2 content is effectively the sum of H_2CO_3 and HCO_3^-.

The solubility of CO_2 in human blood is 48 vol% at 760 mm Hg of CO_2 at 37.5°C; the CO_2 pressure in alveolar air is 40 mm Hg. Hence, the amount of CO_2 which might be dissolved in blood is 2.5 vol% (the solubility coefficient multiplied by the partial pressure).

Actually, arterial blood contains 40 to 50

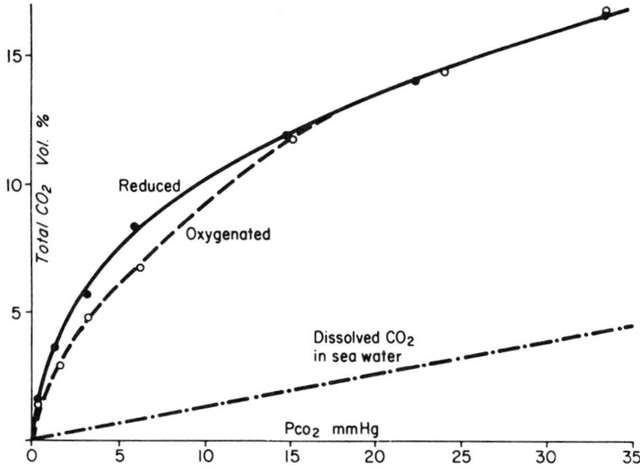

Figure 8–22. CO_2 equilibrium curves for deoxygenated and oxygenated blood of *Octopus*. (From Lenfant, C., and K. Johansen, Amer. J. Physiol. *209*:991–998, 1965.)

vol% CO_2, and venous blood contains 55 to 60 vol% CO_2. Similarly, sea water has a solubility coefficient at 24°C of 0.71; in equilibrium with air in which the CO_2 partial pressure is 0.23 mm Hg, the sea water would dissolve 0.0215 vol% CO_2. Sea water normally contains 4.8 vol% CO_2. The difference between the CO_2 dissolved and the CO_2 actually contained is due to combination as HCO_3^- with cations from various buffers. For each fluid containing buffers, a CO_2 equilibrium curve can be constructed by equilibrating with a known CO_2 pressure and measuring the volume per cent taken up. Thus, in the two examples above, 40 to 50 vol% in blood corresponds to 40 mm Hg of CO_2, and 4.8 vol% in sea water corresponds to 0.23 mm Hg of CO_2.

The relation between the amount of HCO_3^-

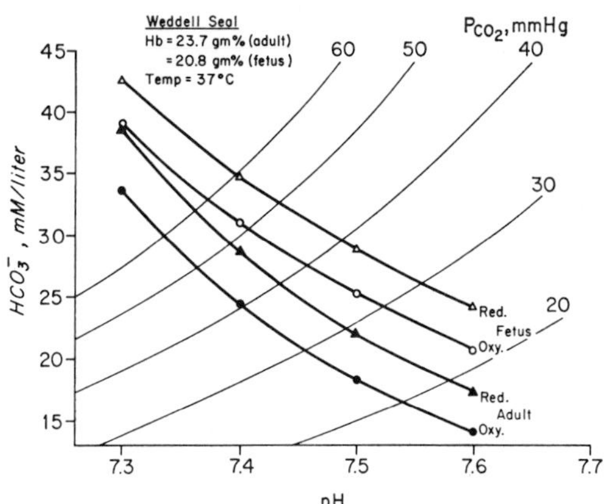

Figure 8–23. Buffer curves relating HCO_3^- to pH for deoxygenated and oxygenated blood of adult and fetal Weddell seal; CO_2 partial pressures corresponding to HCO_3^--pH combinations are also shown. (From Lenfant, C., et al., Amer. J. Physiol. *216*:1595–1597, 1969.)

and H_2CO_3 and the pH is given by the Henderson-Hasselbalch equation:

$$pH = pK + \log \frac{HCO_3^-}{H_2CO_3}$$

where pK is 6.1 for this reaction. The relation between P_{CO_2} and HCO_3^- (or H_2CO_3) is curvilinear and difficult to quantitate (Fig. 8–22). However, the relation between either of these quantities (usually HCO_3^-) and pH is linear over a midrange, and the buffer capacity is expressed as the slope of this line (that is, HCO_3^- in mmole/liter per pH unit). Three-dimensional plots relate P_{CO_2} to the buffer curves (Fig. 8–23).

The total CO_2 content of circulating blood depends on the buffering capacity of blood, the partial pressures of tissue CO_2, and the control of acid-base balance by the kidneys and respiratory organs. In terrestrial vertebrates, the CO_2 pressure to which blood is exposed in the lungs is high (40 mm Hg in many mammals), whereas in the water to which the gills of aquatic animals are exposed, the P_{CO_2} is low (0.23 mm Hg). The partial pressure of CO_2 in the blood of aquatic vertebrates is much lower than in terrestrial vertebrates, so that the gradient across the respiratory surface is similar in both, being about 6 mm Hg. In mammals, the difference between arterial and venous CO_2 indicates an unloading of about 10%; since the total in blood is lower, the percentage gained in tissues or lost in gills (or skin) is greater in aquatic vertebrates.

The CO_2 that diffuses from tissues into blood mainly passes into the red blood cells, where its hydration is catalyzed by carbonic anhydrase:

$$CO_2 + H_2O \xrightleftharpoons{\text{carbonic anhydrase}} H_2CO_3$$

Dissociation then occurs:

$$H_2CO_3 \rightleftharpoons HCO_3^- + H^+$$

and most of the HCO_3^- leaves the cells in exchange for Cl^- from the plasma. The Cl^- shifts from plasma to red cells in the tissues and back out of the red cells in the lungs. The total reaction is as follows:

Tissue	*Plasma*	*RBC*
CO_2 →	CO_2 →	CO_2 ⎤ C.A.
	H_2O →	H_2O ⎦
		H_2CO_3 ↙↘
	HCO_3^- ←	HCO_3^- H^+
	Cl^- →	Cl^-

RBC	*Plasma*	*Lung Alveolus*
CO_2 → (C.A. ↑)	CO_2 →	CO_2
→ H_2O →	H_2O	
H_2CO_3 ↑		
H^+ — HCO_3^- ←	HCO_3^-	
Cl^- →	Cl^-	

In arterial blood about 67% of the HCO_3^- is in plasma; in venous blood this figure is about 65.5%. In plasma, Na^+ is the principal cation for balancing HCO_3^-. In the red cells, most of the buffering is provided by hemoglobin:

$$BHb + H^+ \rightleftharpoons HHb + B^+$$

where B^+ refers to cation, mostly K^+.

Deoxygenated Hb is a weaker acid (pK = 7.95 in horse) than is oxygenated Hb (pK = 6.68); therefore, as O_2 is given off in the tissues, more cations (B^+) are freed, and as CO_2 is given off in the lungs the stronger oxy-Hb attracts more base, thus freezing more CO_2. For each O_2 taken up there is an increase of negative charge per Hb of about 0.6 equivalent and an equal displacement of HCO_3^-. The reciprocal effect of oxygenation on the acid strength of Hb accounts for most of the CO_2 exchange.

In addition to the CO_2 transported by the buffers noted above, some 15 per cent of the CO_2 in arterial blood cells and 20% of that in venous blood cells is present in combination with hemoglobin as a carbamino compound, coupled to an —NH group of Hb. The percentage of carbamino CO_2 is low, but it has been calculated to account for some 20 to 30% of the total change in blood CO_2 in the respiratory cycle and for much more of the CO_2 exchange in the red cells. The carbamino CO_2 is responsible for more than half of the Haldane effect.[179]

Disturbances in acid-base balance are of two sorts—respiratory and metabolic. In respiratory acidosis, as after exercise, the blood pH falls and HCO_3^- rises, while in respiratory alkalosis, as in hyperventilation, the blood pH rises and HCO_3^- falls. In metabolic acidosis, as in lactic acid accumulation or in ketosis, the blood HCO_3^- declines; in metabolic alkalosis, as in renal loss of fixed acids (e.g., in vomiting), the blood HCO_3^- rises.

In a series of birds, panting as a mechanism of cooling may reduce P_{aCO_2} values by as much as 50%, with a moderate rise in blood pH[27] (Fig. 8–24).

Table 8–5 gives selected values of buffer constants. Most of these were obtained by equilibrating whole blood with varying percentages of CO_2, and then analyzing the plasma for HCO_3^-. In man, the total buffer capacity is 29 mM HCO_3^-/pH unit; of this, 8.2 mEq is due to plasma buffers and 20.7 mEq to hemoglobin.[36] Hemoglobins that are low in histidine are poorer buffers than those

TABLE 8–5. Selected Buffer Constants. Data for Whole Blood Except Where Indicated

	$\frac{\Delta HCO_3^- (mM/liter)}{\Delta pH}$
man	30.8
plasma	6.5
dogfish[115]	9
plasma	6.5
dogfish[5]	10
plasma	9.8
lungfish[115]	13
horse	25.3
beaver[32a]	27
water vole[32a]	24
alligator	22.6
crocodile	18.2
Necturus[117]	8.0
Amphiuma[117]	9.2
Rana catesbiana[117]	16.4
Urechis	4.9
Sipunculus	3.5

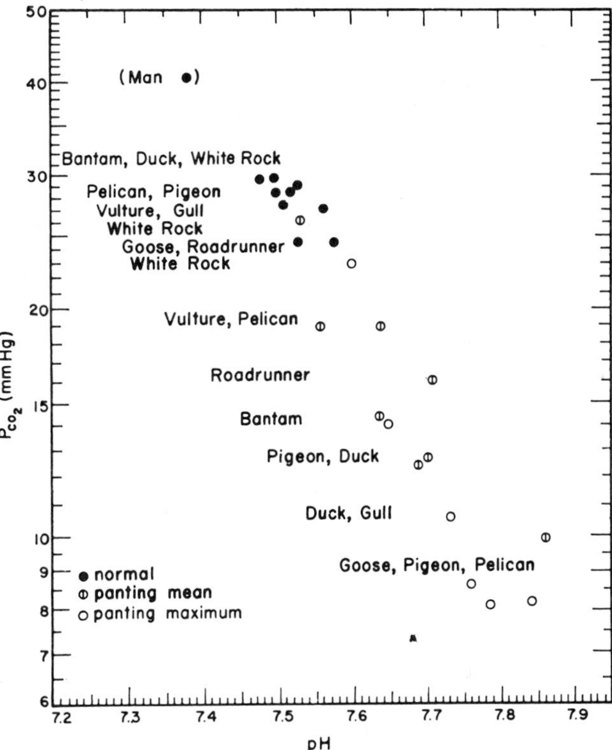

Figure 8–24. Relation between P_{CO_2} and pH in blood of various birds at rest ●, panting ⦶, and in maximum panting ○. (From Calder, W. A., and K. Schmidt-Nielsen, Amer. J. Physiol. *215*:477–482, 1968.)

that are rich in histidine. The buffering capacity of trout and mackerel is high, whereas that of carp, toadfish, and skate is low. The buffer value of the serum proteins is greater per gram of protein in skate and crocodile than in mammals. Similarly, in the dogfish, *Squalus*, whole blood is a poor buffer, but buffering in plasma is equal to that in man.[5, 115] Hypercapnia in dogfish raises the $P_{a_{CO_2}}$ from 5 mm Hg to 20 mm Hg, compared with an increase from 30 to 100 mm Hg in a mammal.

Terrestrial vertebrates have higher plasma P_{CO_2} values and higher buffer capacities than do aquatic species (e.g., in amphibians).[117] CO_2 diffuses rapidly in water, and a fish may pump, on a weight basis, 30 times more water than the air moved by lung breathers. Excretory control of acid-base balance is more important in air breathers, and malfunction of kidneys may cause pH changes in the blood. In a lungfish moved from water to air, the arterial P_{O_2} declines and P_{CO_2} rises; the fish develops metabolic alkalosis to compensate for respiratory acidosis.[116]

Water is neutral at pH 7.0 only at 22°C; its pH of neutrality decreases at high temperatures and increases at low temperatures. Water pH of neutrality at 35°C is 6.84; at 5°C it is 7.365. The blood removed from frogs, toads, and turtles acclimated to different temperatures changes pH with temperature in the ratio $\Delta pH/\Delta °C = -0.016$; the buffers of the blood are adjusted to the same slope as water, so that at different pH's the ratio of OH^- to H^+ is kept constant[83, 84] (Fig. 8–25).

In invertebrates, most of the buffering action is in blood proteins, and the principal proteins are respiratory pigments. In *Urechis*, for example, the coelomic fluid has practically no buffering capacity, but the corpuscles buffer to about the same degree as do corpuscles of a vertebrate of similar oxygen capacity. The coelomic fluid of a sea urchin contains little protein and is poorly buffered. The titration curves of hemocyanin of *Limulus* and of *Helix* show several acid- and base-binding groups. Deoxygenated Hcy of *Maja*, *Octopus*, and *Loligo* is a weaker acid, and hence binds more CO_2, than oxy-Hcy. The CO_2 equilibrium curves (CO_2 content vs. P_{CO_2}) are initially steep, and then gradually flatten. Many invertebrates, with low protein concentrations, operate on the steep part of these curves, where CO_2 content is low. In animals with calcium-containing shells (some molluscs and crustaceans) an important source

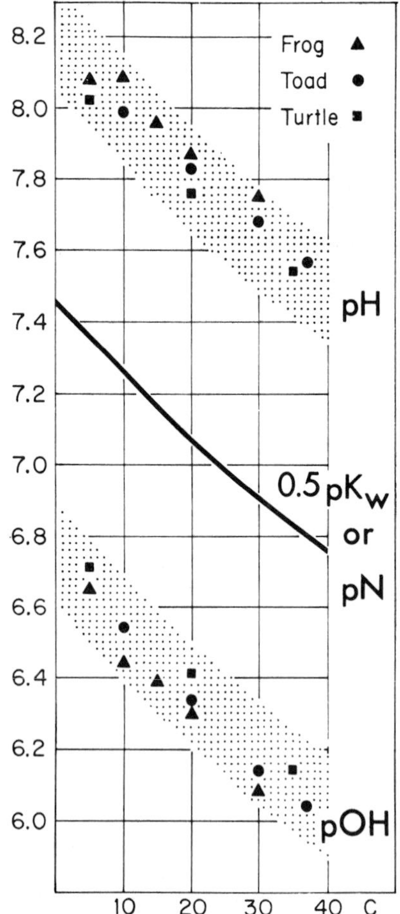

Figure 8-25. pH and pOH of blood of toads, frogs, and turtles acclimated and measured at several temperatures. Middle line gives curve of neutrality or 0.5 pK of water. (From Howell, B. J., et al., Amer. J. Physiol. *218*:600–606, 1970.)

of base for buffering blood acid is the shell; as lactic acid is produced by a clam in air, the shell may become eroded at the mantle edge.

In air-breathing insects, CO_2 is lost directly from tissues to the tracheae. The blood is buffered, partly by amino acids. In the parasitic *Gastrophilus*, serum proteins (10.75% by weight) account for about 62% of the total buffering; the P_{CO_2} is high (300 to 500 mm Hg) and the blood pH is low (6.64).

Carbonic anhydrase is widely distributed in the animal kingdom. It is essential in acidification of the urine in the vertebrate kidney and in acid secretion in the stomach. Carbonic anhydrase (CA) is present in the eye of all vertebrates, as well as in the choroid plexus and pancreas; it is present in the avian salt gland and the elasmobranch rectal gland. It is present in large concentrations in fish gills and in kidneys of freshwater (but not marine) fish.[132] The CA activity is greater in red cells of small mammals than in those of large species; this is presumably related to the higher metabolic rate in small mammals. CA may facilitate diffusion of CO_2; for example, much more CO_2 diffuses across a millipore filter containing CA for a given gradient than across one that does not contain CA.

CA is present in coelenterates (e.g., in the tentacles of Anthozoa). It occurs in the respiratory trees and gonads of echinoderms. CA is highly active in blood of *Arenicola;*[204a] it is present in the blood of earthworms and nereids. Gills of squid and gills and mantle of bivalve molluscs contain much CA. In arthropods, the gills of *Limulus* have much CA, while those of *Homarus* and *Libinia* have less.

CA from human red blood cells occurs in three isozymal forms, which are separable by electrophoresis; CA of *Sepia* gills and muscles occurs in two isozymal forms.[1] Each of the three human forms contains 0.2% zinc; the zinc content of the enzyme of duck is 0.14%, that of turtle is 0.2%, and that of tuna is 0.09%.[132] Inhibitors of CA are unsaturated aromatic sulfonamides.

Carbonic anhydrase facilitates the hydration of respiratory CO_2 in blood cells of terrestrial vertebrates and in gills of many aquatic animals; it may facilitate diffusion of CO_2 much as myoglobin facilitates diffusion of O_2. CA is important in other tissues where transfer of CO_2 and formation of HCO_3^- have functions other than in respiration.

SUMMARY

Respiratory transport pigments are of interest to workers in many areas—physiological ecology, physical biochemistry, genetics, and evolution.

All respiratory pigments are metalloproteins, and metallo-enzymes probably were well established early in the aerobic phase of chemical evolution. Three basic types of pigments are used for O_2 transport in animals: (1) hemoglobin, and chlorocruorin, which is similar; (2) the iron-protein hemerythrin in sipunculids and brachiopods; and (3) the copper-protein hemocyanin in molluscs and arthropods. It can be argued from the sporadic occurrence of Hb in different phyla that this pigment evolved numerous times. The recently established multiple differences between annelid (especially *Arenicola*) Hb and vertebrate Hb indicate independent evolution. Opposed to such a view are the widespread occurrence of myoglobin, the many forms of hemoglobin in vertebrates, and the ease with which synthesis of hemoglobin can be initiated in many animals.

Many specific properties of blood pigments can be explained as adaptive for given ecological conditions. Aquatic animals that are highly active and live in fast-moving water have pigment with half-saturation level at high Po_2, as compared with sluggish animals that often live in stagnant water (e.g., Hcy in squid compared with some crabs, hemoglobin in trout compared with carp and catfish). The O_2 equilibrium curve for blood of most vertebrate embryos lies to the left of that of maternal blood. Hemoglobin synthesis is stimulated by life in hypoxic environment, as in man at high altitude, and in *Artemia* in saline pools. Antarctic fish species and some invertebrates carry enough O_2 dissolved in plasma for life activities; other related species need pigment, especially under stress. Animals that live in regions of low oxygen—such as *Tubifex* in mud—have very low tissue Po_2's and a steep inward gradient of O_2.

Addition of CO_2 (or other acids) to blood moves the equilibrium curve to the right in animals that have high active metabolism, but in sluggish species it has little effect on O_2 affinity. The CO_2 effect facilitates O_2 unloading in the tissues, and loss of CO_2 favors O_2 loading at respiratory surfaces. The converse effect—that of deoxygenation to increase the buffering capacity of a pigment—is especially important in animals in which most of the blood buffering is by respiratory pigments. The reduction in O_2 capacity by addition of CO_2 may be meaningful in O_2 secretion by fish into swim-bladders. For a number of blood pigments (e.g., hemocyanin in *Limulus*) the effect of acid on O_2 affinity is to decrease affinity in pH ranges slightly below normal; the significance of this is not known. The presence of Hb in the body wall of the parasite *Ascaris* at Po_2's that are never low enough for unloading is not understood.

The physical measurements of blood pigments have added considerably to the knowledge of protein structure. Hemoglobins of vertebrates (except cyclostomes) consist of four chains, usually two of a kind. These have been isolated, and the amino acid sequences have been ascertained for many of them. X-ray diffraction analysis, with other physical measurements, has provided a view of the steric arrangement of the four chains in relation to the heme. Hemocyanins are not amenable to such analysis with present techniques. The charge of the metal ions and the nature of linkages of the metal to its ligand are not fully known. It has long been recognized that the copper in hemocyanin and the iron in hemerythrin go to a higher valence on oxygenation; recent evidence suggests that in hemoglobin also an electron may be transferred from the iron to oxygen on loading of the molecule. Changes in conformation are also indicated in oxygenation. Some hemocyanins depolymerize as they become oxygenated, and at least one of them reaggregates on complete oxygenation. Some polymerization is also indicated for cyclostome hemoglobin. The hemocyanins are very large molecules that often consist of multiples of units which aggregate over certain ranges of pH and of Ca^{++} or Mg^{++} concentration. Snail Hcy has two chains analogous to the two kinds of chain in Hb.

The effect of pH on O_2 affinity mentioned above results from liberation of one proton per chain on oxygenation. This also means that Hb is a stronger acid when it is oxygenated and is a better buffer when it is reduced.

Another structural puzzle in oxygenation is the cooperativity between chains in Hb or between units in Hcy. Loading of O_2 is a stepwise procedure, and a given Hb loads on the average more than one but fewer than four O_2 molecules. The pH has important effects on aggregation in Hcy, and also affects the O_2 affinity.

It has long been known that hemoglobin may have less affinity for oxygen when it is in red cells than when it is in solution. Recent evidence suggests that the increase in P_{50} and part of the difference in the Bohr effect in mammals are caused largely by diphosphoglycerate; in birds they are caused by inositol hexosephosphate, and in fishes by adenosine triphosphate. The speed of reaction of blood pigments with oxygen is high; the half-time is a few milliseconds. The reaction of Hb with carbon monoxide is slower, but the CO affinity is higher than the O_2 affinity. Many details remain to be learned, such as how so many changes as are known in pigment molecules can occur reversibly and quickly in loading and unloading oxygen.

Respiratory pigment proteins provide useful materials for genetic and developmental studies. The principal chain types of mammalian hemoglobins (alpha and beta in adult, alpha and gamma in fetal Hb) are species specific, but hybrids can be made. Each chain type differs in amino acid sequences from those in other species, and speculations have been made on evolutionary separation according to the number of differences in sequence—hardly justifiable speculations in view of all the assumptions. Not only are the chains different for different species, but many mutants for given chains have been discovered, and most vertebrates are polymorphic. In some animals (e.g., most mammals) there are one or possibly two dominant proteins, as identified electrophoretically, in the blood of adults. In other animals (e.g., most fishes) there are multiple bands—in salmon there are about a dozen anodal and a dozen cathodal bands. The proportion of the different hemoglobins present in the blood varies with development, and some types of hemoglobin differ in O_2 affinity; the meaning of this polymorphism is far from clear. Linkages between physiological and morphological characters are known, but such genetic linkages do not prove a function. Hemocyanin is known to occur in two primary forms in only one animal (the snail *Helix*); however, differences in polymerization provide a kind of polymorphism of hemocyanin that is not known for hemoglobin (although cyclostomes are possibly an exception).

Since the amino acid sequences of many Hb chains are known, the genetic coding for them can be specified. Synthesis occurs continually at a low replacement rate, but it is stimulated in prolonged hypoxia or anemia. The synthesis, at least in vertebrates, is regulated by a hormone, erythropoietin. Cells that form reticulocytes, and ultimately red blood cells, are being cultured, and hormonal effects on the protein synthesis can be studied. During development, a sequence of hemoglobins must mean that different genes are activated at different stages.

Blood pigments provide a prime example of a link between molecular and whole-organ biology and an example of environmental control of adaptive functions of specific molecules.

REFERENCES

1. Addink, A. D. F., Dissertation, Utrecht University, 1968. 161 pp. Carbonic anhydrase of *Sepia*.
2. Aggarwal, S. J., and A. Riggs, J. Biol. Chem. *244*:2372–2383, 1969. Hemoglobin in bullfrog.
3. Air, G. M., and E. O. P. Thompson, Austral. J. Biol. Sci. *24*:75–95, 1971. Amino acid sequence, myoglobin of red kangaroo.
4. Albers, C., pp. 173–208. *In* Fish Physiology, Vol. 4, edited by W. S. Hoar and D. J. Randall. Academic Press, New York, 1970. Acid-base balance.
5. Albers, C., and K. Pleschka, Respir. Physiol. *2*:261–273, 1967. Buffering by elasmobranch blood.
6. Anthony, F. H., J. Exp. Biol. *38*:109–125, 1961. Survival of goldfish in carbon monoxide.
7. Antonini, E., pp. 121–137. *In* Oxygen in the Animal Organism, edited by F. Dickens and E. Neil. Pergamon Press, New York, 1965. *Also* Physiol. Rev. *45*:123–170, 1965. Structure and function of hemoglobin and myoglobin.
8. Antonini, E., et al., Arch. Biochem. Biophys. *97*:336–342, 343–350, 1962. Studies on chlorocurin.

8a. Antonini, E., A. Rossi-Fanelli, and A. Caputo, Arch. Biochem. Biophys. *97*:343–350, 1962. Size and structure of chlorocruorin.

9. Antonini, E., J. Wyman, L. Bellelli, N. Rumen, and M. Siniscalco, Arch. Biochem. Biophys. *105*:404–408, 1964. Oxygen equilibrium of lamprey hemoglobins.
10. Ar, A., and A. Schejter, Comp. Biochem. Physiol. *33*:481–490, 1970. Hemoglobin in clam shrimp *Cyzicus*.
11. Atassi, M. Z., J. Theor. Biol. *11*:227–241, 1966. Amino acid composition, hemoglobin.
12. Bartels, H., Lancet *2*:601–604, 1964. Oxygen transport in mammals.
13. Bartels, H., G. Hiller, and W. Reinhardt, Respir. Physiol. *1*:345–356, 1966. Oxygen affinity of chicken blood.
14. Bartels, H., et al., Amer. J. Physiol. *205*:331–336, 1963; *also* Bartels, H., D. E. Yassin, and W. Reinhardt, Respir. Physiol. *2*:149–162, 1967. O_2 and CO_2 transport, adult and fetal mammals.
15. Battaglia, F. C., H. McGaughey, E. L. Makowski, and G. Meschia, Amer. J. Physiol. *219*:219–221, 1970. Effect of diphosphoglyceric acid on O_2 affinity of red cells.
16. Beadle, L. C., J. Exp. Biol. *34*:1–10, 1957. Respiration in oligochaete *Alma*.
17. Beard, J. M., and E. O. P. Thompson, Austral. J. Biol. Sci. *24*:765–786, 1971. Amino acid sequences in kangaroo hemoglobin.
18. Behrman, R. E., J. Appl. Physiol. *25*:224–229, 1968. O_2 dissociation curve, rhesus monkeys.
19. Benesch, R., R. E. Benesch, and C. I. Yu, Proc. Nat. Acad. Sci. *59*: 526–532, 1968; *also* Benesch, R. E., and R. Benesch, Fed. Proc. *29*:1101–1104, 1970. Reaction between hemoglobin and diphosphoglycerate and inositol hexaphosphate.
20. Black, E. C., D. Kirkpatrick, and H. H. Tucker, J. Fish. Res. Bd. Canad. *23*:1–13, 1966. O_2 dissociation curves of brook trout.
21. Bowen, S. T., H. G. Lebbenz, M. Poon, V. H. S. Chow, and T. A. Grigliatti, Comp. Biochem. Physiol. *31*:733–747, 1969. Hemoglobins of *Artemia*.
22. Boyer, S. H., et al., Science *153*:1539–1543, 1966. Hemoglobins in sheep.
23. Bradshaw, R. A., L. A. Rogers, R. L. Hill, and J. Buettner-Janusch, Arch. Biochem. Biophys. *109*:571–578, 1965. Amino acid composition of primate hemoglobins.
24. Braun, V., R. R. Crichton, and G. Braunitzer, Z. physiol. Chem. *349*:197–210, 1968. Polymers of hemoglobin in *Chironomus*.
25. Briehl, R. W., J. Biol. Chem. *238*:2361–2366, 1963. O_2 equilibrium and aggregation of subunits, lamprey hemoglobin.
26. Brunoni, M., E. Antonini, J. Wyman, L. Tentori, G. Vivaldi, and S. Carta, Comp. Biochem. Physiol. *24*:519–524, 1968; *also* Biochim. Biophys. Acta *133*:177–180, 1967. Equilibrium kinetics of frog hemoglobin.
27. Calder, W. A., and K. Schmidt-Nielsen, Amer. J. Physiol. *213*:883–889, 1967; *215*:477–482, 1968. Acid-base balance in birds.
28. Cameron, J. N., Comp. Biochem. Physiol. *32*:175–192, 1970. Hematology, pinfish *Lagodon* and striped mullet.
29. Cameron, J. N., Comp. Biochem. Physiol. *38A*:699–704, 1971. O_2 equilibrium curve of rainbow trout.

29a. Chew, M. Y., et al., Biochem. J. *94*:378–383, 1965. Size of Hb molecule in polychaete.

30. Chiodi, H., Acta Physiol. Lat. Amer. *12*:208–209, 1962. O_2 affinity in high altitude animals.
31. Chiodi, H., and J. W. Terman, Amer. J. Physiol. *208*:798–800, 1965. Blood gases in hen.
32. Chumley, J. H., and R. A. B. Holland, Respir. Physiol. *7*:287–294, 1969. Respiratory properties of sheep hemoglobin.

32a. Clausen, G., and A. Ersland, Resp. Physiol. *5*:221–233, 350–359, 1968; *7*:1–6, 1969. Respiratory properties of blood of hedgehog, beaver, water vole, and bladdernose seal.

33. Cosgrove, W. B., and J. B. Schwartz, Physiol. Zool. *38*:206–212, 1965. Properties of earthworm hemoglobin.

34. Crompton, D. W., and M. H. Smith, Nature *197*:118–119, 1963. Flatworm hemoglobin.
34a. Dan, M., and A. Hagiwara, Exp. Cell. Res. *46*:596–598, 1967. HbA and HbF in same red cell, human.
35. Davenport, H. E., Proc. Roy. Soc. Lond. B *136*:255–270, 271–280, 1949. Hemoglobins of nematodes.
36. Davenport, H. W., ABC of Acid-Base Chemistry. University of Chicago Press, 1969. 86 pp.
37. Dawson, T. J., and J. V. Evans, Austral. J. Biol. Sci. *15*:371–378, 1962. Relation between K levels and O_2 transport, sheep red cells.
38. DePhillips, H. A., and K. E. van Holde, Biol. Bull. *133*:462, 1967. O_2 equilibrium of hemocyanin.
39. DePhillips, H. A., K. W. Nickerson, M. Johnson, and K. E. van Holde, Biochemistry *8*:3665–3672, 1960; *also* J. Molec. Biol. *50*:471–479, 1970. O_2 dissociation of hemocyanin, subunit structure, *Busycon*, squid.
40. Di Giamberardino, L., Arch. Biochem. Biophys. *118*:273–278, 1967. Dissociation of *Eriphia* hemocyanin.
41. Eaton, J. W., G. J. Brewer, and R. F. Grover, J. Lab. Clin. Med. *73*:603–609, 1969. Diphosphoglycerate in man at altitude.
42. Eddy, F. B., J. Exp. Biol. *55*:695–711, 1971. Blood gas data for *Salmo gairdneri.*
43. Ellerton, H. D., D. E. Carpenter, and K. E. van Holde, Biochemistry *9*:2225–2232, 1970. Subunits of hemocyanin of *Cancer.*
44. Ellis, R., A. Giuliani, A. L. Tentori, E. Chiancone, and E. Antonini, Comp. Biochem. Physiol. *36*:163–171, 1970. Hemoglobin of amphibians.
45. Evans, J. V., et al., Nature *178*:849–850, 1956. Genetics of hemoglobin differences in sheep.
46. Faura, J., et al., Blood *33*:668–676, 1969. Effect of altitude on erythropoiesis.
47. Flores, G., and E. Frieden, Science *159*:101–103, 1967. Bullfrogs lacking hemoglobin.
48. Forster, R. E., and J. B. Steen, J. Physiol. *204*:259–282, 1969. The Root shift in eel blood.
49. Fox, H. M., Proc. Roy. Soc. Lond. B *135*:192–212, 1948; *136*:388–399, 1949; *138*:514–528, 1951; *141*:179–189, 1953; *also* Nature *166*:609–610, 1950; *also* Bull. Soc. Zool. France *80*:288–298, 1955. Hemoglobin synthesis and function in *Daphnia.*
50. Fox, H. M., Proc. Roy. Soc. Lond. B *136*:378–388, 1949. Comparison of chlorocruorin and hemoglobin.
51. Fox, H. M., Proc. Roy. Soc. Lond. B *143*:203–214, 214–225, 1955. Effect of O_2 on hemoglobin synthesis and function, invertebrates.
52. Fox, H. M., Nature *168*:112, 1951. Blood pigments of *Serpula.*
53. Fox, H. M., J. Exp. Biol. *21*:161–164, 1945. O_2 affinities of invertebrate hemoglobins.
54. Freeman, M. B. M., and B. H. Misson, Comp. Biochem. Physiol. *33*:763–772, 1970. Respiratory properties of blood of chickens.
55. Gahlenbeck, H., and H. Bartels, Z. vergl. Physiol. *59*:232–240, 1968. O_2 affinity of blood of bullfrog.
56. Garbett, K., D. W. Darnall, and I. M. Klotz, Arch. Biochem. Biophys. *142*:455–470, 1971. Reactivity of residues in hemerythrin.
57. Ghosh, J., Comp. Biochem. Physiol. *16*:341–360, 1965. Chemical properties of avian hemoglobins.
58. Gibson, Q. H., Prog. Biophys. *9*:1–53, 1959. Kinetics of reactions between Hb's and gases.
59. Gillen, R. G., and A. Riggs, Comp. Biochem. Physiol. *38B*:585–595, 1971. Effect of ATP on hemoglobin properties in fish.
59a. Gitlin, D., et al., Blood *32*:796–810, 1968. Immunofluorescent detection of two Hb's in same cell.
60. Gordon, A. S., Physiol. Rev. *39*:1–40, 1959. Hemopoietine.
61. Gray, L. H., and J. M. Steadman, J. Physiol. *175*:161–171, 1964. Dissociation curves for mouse and rat blood.
62. Grigg, G. C., Comp. Biochem. Physiol. *23*:139–148, 1967. Blood of Antarctic fishes.
63. Grover, R. F., Ann. N.Y. Acad. Sci. *121*:662–673, 1965. Pulmonary tension at altitude.
64. Grover, R. F., et al., J. Appl. Physiol. *18*:560–566, 567–574, 575–579, 909–912, 1963. Oxygen transport at high altitudes.
64a. Guerritore, D., et al., J. Molec. Biol. *13*:234–237, 1965. Electron microscopy of *Spirographis* chlorocruorin.
65. Hall, F. G., J. Biol. Chem. *115*:485–490, 1946; *also* J. Mammal. *18*: 468–472, 1937. Blood of animals living at high altitudes.
66. Hall, F. G., Proc. Soc. Exp. Biol. Med. *119*:1071–1073, 1965. O_2 transport in chinchilla.
67. Hall, F. G., J. Cell. Biol. *68*:69–74, 1966. Hemoglobin properties in toad.
68. Hamada, K., et al., J. Biochem. (Tokyo) *55*:154–162, 1964. Eel hemoglobins.
69. Harkness, D. R., J. Ponce, and V. Grayson, Comp. Biochem. Physiol. *28*:129–138, 1969. Phosphoglyceric acid cycle in mammalian red cells.
70. Hemmingsen, E. A., E. L. Douglas, and G. C. Grigg, Comp. Biochem. Physiol. *29*:467–470, 1969. O_2 transport in hemoglobin-free fish.
71. Hill, R. L., J. Buettner-Janusch, and V. Buettner-Janusch, Proc. Nat. Acad. Sci. *50*:885–893, 1963. Evolution of primate hemoglobins.
72. Hoffman, R. J., and C. P. Mangum, Comp. Biochem. Physiol. *36*:211–228, 1970. Coelomic hemoglobin in *Glycera.*
73. van Holde, K. E., and L. B. Cohen, Biochemistry *3*:1803–1808, 1809–1813, 1965; *6*:93–99, 1967. Ultrastructure of hemocyanins.
74. van Holde, K. E., and van Bruggen, E. F. J., *in* Biological Macromolecules, Vol. 5, edited by G. D. Fasman and S. N. Timasheff. Marcel Dekker, Inc., New York, 1971. The hemocyanins.
75. Holeton, G. F., Comp. Biochem. Physiol. *34*:457–471, 1970. Antarctic fish without hemoglobin.
76. Holeton, G. F., J. Exp. Biol. *54*:239–254, 1971. O_2 uptake in *Salmo* exposed to CO.
77. Holland, R. A. B., and R. E. Forster, J. Gen. Physiol. *49*:140–146, 199–220, 1968. Kinetics of uptake of O_2 and CO by red blood cells.
78. Horne, F. R., and K. W. Beyenbach, Amer. J. Physiol. *220*:1875–1881, 1971. Hemoglobin in brachiopod crustaceans.
79. Horvath, S. M., H. Chiodi, S. H. Ridgway, and S. Azar, Comp. Biochem. Physiol. *24*:1027–1033, 1968. Respiratory properties of hemoglobin of porpoise and sea lion.
80. Hoshi, T., Sci. Rep. Tohoku Univ. *23*:35–58, 1957. Hemoglobin in a daphnid.
81. Hoshi, T., M. Kobayashi, and H. Sugano, Sci. Rep. Niigata Univ. D *5*:87–98, 1968; *also* Hoshi, T., and H. Sugano, Sci. Rep. Niigata Univ. D *2*:13–26, 1965. Oxygen dissociation curve of hemoglobin in cladocerans.
82. Hoshi, T., and T. Shimada, Sci. Rep. Niigata Univ. D *2*:1–12, 1965. Hemoglobin in freshwater cladocerans.
83. Howell, B. J., Fed. Proc. *29*:1130–1134, 1970. Acid-base balance in transition from water to air breathing.
84. Howell, B. J., F. W. Baumgardner, K. Bondi, and H. Rahn, Amer. J. Physiol. *218*:600–606, 1970. Acid-base balance, cold-blooded vertebrates.
85. Huber, R., H. Formanek, and O. Epp, Naturwissenschaften *55*:75–77, 1968. Crystal structure analysis of erythrocruorin.
86. Huehns, E. R., N. Dance, G. H. Beaver, J. V. Keil, F. Hecht, and A. G. Motulsky, Nature *201*:1095–1097, 1964. Human embryonic hemoglobins.
87. Huisman, T. H. J., and J. Kitchens, Amer. J. Physiol. *215*:140–146, 1968. Functional properties of Hb A, B, and C in sheep.
88. Huisman, T. H. J., and M. S. Van Veen, Biochim. Biophys. Acta *88*: 352–366, 367–374, 1964. Studies in animal hemoglobins.
89. Hunter, J. A., and J. Paul, J. Embryol. Exp. Morph. *21*:361–368, 1969. Adult and fetal hemoglobin, rat.
90. Ingram, V. M., Biochim. Biophys. Acta *28*:539–545, 1958; *36*:402–411, 1959. Normal and sickle cell hemoglobins, man.
91. Ingram, V. M., The Hemoglobins in Genetics and Evolution. Columbia University Press, New York, 1963. 165 pp.
92. Itazawa, Y., Bull. Jap. Soc. Sci. Fish. *36*:571–577, 1970. Respiratory characteristics of fish.
93. Jacobsen, L. O., and S. B. Krantz, Ann. N.Y. Acad. Sci. *149*:578–583, 1968. Summary on erythropoietin.
94. Johansen, K., and C. Lenfant, Amer. J. Physiol. *210*:910–918, 1966. Gas exchange in *Octopus.*
95. Johansen, K., and C. Lenfant, J. Exp. Biol. *46*:205–218, 1967. Respiratory function in lungfish, *Lepidosiren.*
96. Johansen, K., and T. A. Mecklenburg, Z. vergl. Physiol. *70*:1–19, 1970. O_2 transport in *Cancer.*
97. Johansen, K., C. Lenfant, and G. C. Grigg, Comp. Biochem. Physiol. *18*:597–608, 1966. Respiratory properties of blood of platypus.
98. Johansen, K., and A. W. Martin, J. Exp. Biol. *45*:165–172, 1966. O_2 transport by blood of giant earthworm.
99. Jones, J. D., J. Exp. Biol. *32*:110–125, 1955. Hemoglobin function in polychaetes *Nephthys* and *Arenicola.*
100. Jones, J. D., Comp. Biochem. Physiol. *4*:1–29, 1961; *12*:283–295, 1964. Function of hemoglobin in aquatic snails.
101. Jonxis, J. H. P., pp. 261–267. *In* Haemoglobin, edited by F. J. W. Roughton and J. C. Kendrew. Interscience Publishers, New York, 1949. Properties of fetal hemoglobin.
102. Keilin, D., and Y. L. Wang, Biochem. J. *40*:855–867, 1946. Hemoglobin of *Gastrophilus.*
103. Kendrew, J. C., et al., Nature *174*:946–949, 1954; *181*:662–666, 1958; *185*:422–427, 1960. Structure of myoglobin.
104. Kendrew, J. C., Science *139*:1259–1266, 1963. Structure of myoglobin.
105. Kleihauer, E., and G. Stöffler, Molec. Gen. Genet. *101*:59–69, 1968. Embryonic hemoglobins.
106. Klotz, I. M., et al., Arch. Biochem. Biophys. *68*:284–299, 1957. Active sites in hemerythrin.
107. Konings, W. N., R. van Driel, E. F. J. van Bruggen, and M. Gruber, Biochim. Biophys. Acta *194*:55–66, 1969. Properties of *Helix* hemocyanin.
108. Kunkel, H. G., et al., J. Clin. Invest. *36*:1615–1625, 1957. Types of human hemoglobin, especially A_2.
109. Kunzer, W., Nature *179*:477–478, 1957. Human embryo hemoglobin.
110. Larimer, J. L., and A. F. Riggs, Comp. Biochem. Physiol. *13*:35–46, 1964. Properties of crayfish hemocyanin.
111. Lenfant, C., et al., J. Clin. Invest. *47*:2652–2656, 1968. Effect of altitude on O_2 binding by hemoglobin in man.
112. Lenfant, C., and C. Aucutt, Respir. Physiol. *6*:284–291, 1969. Respiratory properties of blood of monkeys.
113. Lenfant, C., R. Elsner, G. L. Kooyman, and C. M. Drabek, Amer. J. Physiol. *216*:1595–1597, 1969. Respiratory properties of blood of seal.
114. Lenfant, C., and K. Johansen, Amer. J. Physiol. *209*:991–998, 1965. Oxygen transport by blood of *Octopus.*

115. Lenfant, C., and K. Johansen, Respir. Physiol. *1*:13–29, 1966. Respiratory function of blood in dogfish, *Squalus*.
116. Lenfant, C., K. Johansen, and G. C. Grigg, Respir. Physiol. *2*:1–21, 1966. Respiratory function of blood of lungfish.
117. Lenfant, C., and K. Johansen, Respir. Physiol. 2:247–260, 1967. Respiratory adaptations in amphibians.
118. Lenfant, C., K. Johansen, and J. D. Torrance, Respir. Physiol. *9*:277–286, 1970. Respiratory properties of Hb in sea mammals.
119. Lenfant, C., D. W. Kenney, and C. Aucutt, Amer. J. Physiol. *215*: 1506–1511, 1968. Properties of hemoglobin in whale, *Orcinus*.
120. Lenfant, C., et al., Amer. J. Physiol. *216*:1598–1600, 1969. Respiratory properties of blood of penguin.
121. Maniatis, G. M., and V. M. Ingram, J. Cell Biol. *49*:390–404, 1971. Erythropoiesis in amphibians.
122. Manwell, C., Science *126*:1175–1176, 1957; *also* Physiol. Zool. *31*:39–100, 1958. Larval and adult hemoglobins in ovoviviparous dogfish and oviparous teleost.
123. Manwell, C., Biol. Bull. *115*:227–238, 1958. Hemoglobin of hagfish.
124. Manwell, C., J. Cell. Comp. Physiol. *53*:75–84, 1959. Absence of Bohr effect in *Cucumaria* hemoglobin.
125. Manwell, C., Ann. Rev. Physiol. *22*:191–244, 1960. Respiratory pigments.
126. Manwell, C., Comp. Biochem. Physiol. *8*:209–218, 1963. Hemoglobin in marine clams.
127. Manwell, C., pp. 372–455. *In* Biology of *Myxine*, edited by A. Brodal and R. Fänge. Oslo University Press (Univertetsforlaget), Oslo, 1963. Blood proteins of cyclostomes.
128. Manwell, C., pp. 49–116. *In* Oxygen in the Animal Organism, edited by F. Dickens and E. Neil. Pergamon Press, New York, 1965. Review, comparative properties of blood pigments.
129. Manwell, C., Science *152*:1393–1395, 1966. Hemoglobins of holothurians.
130. Manwell, C., and C. M. A. Baker, Proc. Nat. Acad. Sci. *49*:496–504, 1963. Survey of hemoglobins in wild birds.
131. Manwell, C., C. M. A. Baker, and W. Childers, Comp. Biochem. Physiol. *10*:103–120, 1963. Genetics of hemoglobins in hybrid fish.
132. Maren, T. H., Fed. Proc. *26*:1097–1103, 1967. Carbonic anhydrase in the animal kingdom.
133. Meschia, G., et al., Quart. J. Exp. Physiol. *46*:156–160, 1961. Properties of fetal and adult hemoglobin in sheep at altitude.
134. Metcalfe, J., H. Bartels, and W. Moll, Physiol. Rev. *47*:782–838, 1967. Gas exchange across placenta.
134a. Miller, P. L., J. Exp. Biol. *44*:529–543, 1966. Function of hemoglobin in water bug *Anisops*.
135. Moll, W., Pflüger. Arch. *299*:247–251, 1968. Facilitated diffusion of oxygen by myoglobin.
135a. Muirhead, H., et al., J. Molec. Biol. *28*:117–156, 1957. Structure of hemoglobin.
136. Needham, A. E., Nature *221*:572, 1969. Absorption spectra of leech hemoglobin.
137. Novy, M. J., and J. T. Parer, Respir. Physiol. *6*:144–150, 1969. Oxygen affinity in blood of fetal cat.
138. Novy, M. J., J. T. Parer, and R. E. Behrman, J. Appl. Physiol. *26*:339–345, 1969. O_2 dissociation curves, adult and fetal macaque.
138a. Okazaki, T., et al., Biochim. Biophys. Acta *111*:496–502, 503–511, 1965; *140*:258–265, 1967. Properties of hemoglobin from *Ascaris* perienteric fluid and body wall.
139. Padlan, E. A., and W. E. Love, Nature *220*:376–378, 1968. Structure of hemoglobin of *Glycera*.
140. Padmanabhanaidu, B., Comp. Biochem. Physiol. *17*:167–181, 1966. Properties of hemocyanin in scorpion.
141. Parer, J. T., Respir. Physiol. *2*:168–172, 1967. O_2 dissociation curve of rhesus monkey.
142. Parer, J. T., A. S. Hoversland, and J. Metcalfe, Respir. Physiol. *10*:30–37, 1970. Respiratory properties of blood, lion and tiger.
143. Paul, W., and F. J. W. Roughton, J. Physiol. *113*:25–35, 1951. O_2 equilibrium curve of sheep hemoglobin.
144. Perutz, M. F., Proc. Roy. Soc. Lond. B *173*:113–140, 1969. The hemoglobin molecule.
145. Perutz, M. F., Nature *228*:726–734, 1970. Stereochemistry of hemoglobin.
146. Perutz, M. F., et al., Nature *219*:29–32, 131–139, 1968; *also* Perutz, M. F., and H. Lehmann, Nature *219*:902–909, 1968. Crystal analysis and molecular model of hemoglobin.
147. Perutz, M. F., et al., Nature *222*:1240–1243, 1969. Function of specific residues in hemoglobin.
148. Pickett, S. M., A. F. Riggs, and J. L. Larimer, Science *151*:1005–1007, 1966. Subunit structure of lobster hemocyanin.
149. Piiper, J., and D. Schumann, Respir. Physiol. *2*:129–134, 1967. O_2 exchange in gills of dogfish.
150. Piiper, J., and D. Schumann, Respir. Physiol. *5*:317–325, 326–337, 1968. Respiratory functions of blood of dogfish.
151. Poluhowich, J. L., Comp. Biochem. Physiol. *36*:817–821, 1970. Respiratory pigment in a nemertean.
152. Pough, F. H., Comp. Biochem. Physiol. *31*:885–901, 1969. O_2 and CO_2 transport by lizard blood.
153. Rahn, H., and T. Yokoyama, eds., Physiology of Breath-holding; the Ama of Japan. Publ. 1341, Nat. Acad. Sci. Washington, 1967. 369 pp.
154. Ralph, R., and I. Everson, Comp. Biochem. Physiol. *27*:299–307, 1968. Metabolism of Hb-less Antarctic fish *Chaenocephalus*.
155. Read, K. R. H., Biol. Bull. *122*:605–717, 1962; Comp. Biochem. Physiol. *15*:137–138, 1963. Hemoglobin in bivalve mollusc.
156. Redmond, J. R., Physiol. Zool. *35*:304–313, 1962. Respiratory function in *Chiton*.
157. Redmond, J. R., Biol. Bull. *122*:252–262, 1962. O_2 transport in crab *Cardisoma*.
158. Redmond, J. R., Science *139*:1294–1295, 1963. Molluscan hemocyanin.
159. Redmond, J. R., Helg. wiss. Meeresunters. *9*:303–311, 1964. O_2 transport by blood of various invertebrates.
160. Redmond, J. R., Amer. Zool. *8*:471–479, 1968. Transport of O_2 in crab *Carcinus*.
161. Reynafarje, C., et al., Proc. Soc. Exp. Biol. Med. *116*:649–650, 1964. Humoral control of erythropoietic activity in man at altitude.
162. Reynafarje, C., and P. R. Morrison, J. Biol. Chem. *237*:2861–2864, 1962. Myoglobin in rodents at high altitudes.
163. Riegel, K., H. Bartels, E. Kleihauer, E. M. Lang, and J. Metcalfe, Respir. Physiol. *1*:138–144, 145–150, 1966. Respiratory properties of mammalian blood.
164. Riegel, K., H. Bartels, D. E. Yassin, J. Oufi, E. Kleihauer, J. T. Parer, and J. Metcalfe, Respir. Physiol. 2:173–181, 182–195, 1967. Respiratory functions in mammalian blood.
165. Riggs, A., Canad. J. Biochem. *42*:763–775, 1964. Structure and function of hemoglobins.
166. Riggs, A., Physiol. Rev. *45*:619–673, 1965. Functional properties of hemoglobins.
167. Riggs, A., J. Gen. Physiol. *43*:737–752, 1960. Bohr effect, mammalian hemoglobins.
168. Riggs, A., J. Gen. Physiol. *35*:23–40, 1951. Frog and tadpole hemoglobin.
169. Riggs, A., Proc. Nat. Acad. Sci. *68*:2062–2065, 1971. Enhancement of Bohr effect in mammalian hemoglobins by diphosphoglycerate.
170. Riggs, A., Biology of Lampreys. Academic Press, New York, 1971. Lamprey hemoglobins.
171. Riggs, A., and A. E. Herner, Proc. Nat. Acad. Sci. *48*:1664–1670, 1962. Hybridization of hemoglobins.
172. Riggs, A., and M. Rona, Biochim. Biophys. Acta *175*:248–259, 1969. Aggregation of polymerizing mouse hemoglobins.
173. Roche, J., and M. Fontaine, Ann. Inst. Oceanogr. Monaco *20*:77–87, 1940. Amino acid composition of various blood pigments.
173a. Roche, J., et al., Biochim. Biophys. Acta *41*:182–194, 1960. Electron micrographs of Hb of annelids.
174. Rogers, W. P., Austral. J. Sci. Res. B. Biol. Sci. *2*:287–303, 399–407, 1949. Hemoglobin function in nematode parasites.
175. Rosenberg, M., Proc. Nat. Acad. Sci. *67*:32–36, 1970. Electrophoretic analysis of hemoglobin.
176. Rosenmann, M., and P. R. Morrison, J. Biol. Chem. *240*:3353–3356, 1965. Myoglobin in snowshoe hare.
177. Rosse, W. F., T. Woldmann, and E. Hull, Blood *22*:66–72, 1963; *also* Rosse, W. F., and T. Woldmann, Blood *27*:654–661, 1966. Stimulation of erythropoiesis in frogs and birds.
178. Rossi-Fanelli, A., and E. Antonini, Nature *186*:895–896, 1960; *also*, Rossi-Fanelli, A., E. Antonini, and R. Giuffré, Nature *186*:896–897, 1960. Crystalline hemoglobin from tuna fish.
178a. Rossi-Fanelli, M. R., et al., Arch. Biochem. Biophys. *141*:278–283, 1970. Size and structure of earthworm Hb.
179. Roughton, F. J. W., pp. 767–825. *In* Handbook of Physiology, Vol. 1, Sec. 3, edited by W. F. Hamilton. Amer. Physiol. Soc., Washington, D.C., 1964. Transport of O_2 and CO_2 (review).
179a. Roxby, R., and K. van Holde, 1972, personal communication. Molecular size, Hcy of *Busycon*.
180. Rumen, N. M., and W. E. Love, Arch. Biochem. Biophys. *103*:24–35, 1963. Hemoglobins of sea lamprey.
181. Saha, A., and J. Ghosh, Comp. Biochem. Physiol. *15*:217–235, 1965. Properties of avian hemoglobins.
182. Sarcini, M., M. Orlando, and L. Tentori, Comp. Biochem. Physiol. *34*:751–753, 1970. Polymorphism of hemoglobin of salamander.
183. Schruefer, J. J. P., et al., Nature *196*:550–552, 1962. Dissociation curves of hemoglobin and whole blood.
184. Schweer, M., Z. vergl. Physiol. *42*:20–42, 1959. Metabolism of leeches with and without hemoglobin.
185. Seal, U. S., Comp. Biochem. Physiol. *31*:799–811, 1969. Hemoglobins of carnivores.
185a. Serfaty, A., et al., Rev. Canad. Biol. *24*:1–5, 1965. Cardiac reaction to emergence in carp.
186. Severinghaus, J. W., J. Appl. Physiol. *21*:1108–1116, 1966. Blood gas calculator.
187. Simons, J. A., J. Exp. Zool. *162*:219–230, 1966. Embryonic development of hemoglobin in chicken.
188. Siri, W. E., et al., J. Appl. Physiol. *21*:73–80, 1966. Erythropoietic responses of man to hypoxia.
189. Smith, M. H., Biochim. Biophys. Acta *71*:370–376, 1963. Combination of *Ascaris* hemoglobin with O_2 and CO.
190. Smith, M. H., P. George, and J. R. Preer, Arch. Biochem. Biophys. *99*:313–318, 1962. Hemoglobin from paramecium.
191. Smith, M. H., and D. L. Lee, Proc. Roy. Soc. Lond. B *157*:234–257,

1963; *also* Smith, M. H., and M. Morrison, Biochim. Biophys. Acta *71*:364–370, 1963. Isolation and metabolism of hemoglobin from *Ascaris*.
192. Sullivan, B., and A. Riggs, Comp. Biochem. Physiol. *23*:437–447, 449–458, 459–474, 1967. Structure, function, and evolution of turtle hemoglobin.
193. Svedberg, T., and K. O. Pederson. The Ultracentrifuge. Oxford University Press, New York, 1940; Johnson Reprint Corp., 1959. 478 pp.
194. Taketa, F., and S. A. Morell, Biochem. Biophys. Res. Comm. *24*:705–713, 1966. Oxygen affinity of cat hemoglobin.
194a. Terwilliger, R. C., and K. R. Read, Comp. Biochem. Physiol. *29*:551-560, 1969. Myoglobin in radular muscle, mollusc.
195. Tomita, S., and A. Riggs, J. Biol. Chem. *246*:547–554, 1971. Effects of body size on Bohr effect.
196. Tucker, V. A., Comp. Biochem. Physiol. *24*:307–310, 1968. O_2 dissociation curve in Echidna.
197. Tyuma, I., and K. Shimizu, Fed. Proc. *29*:1112–1114, 1970. Effect of organic phosphates on O_2 affinity of hemoglobin.
198. Ulrich, S., P. Hilpert, and H. Bartels, Pflug. Arch. 277:150–165, 1963. Respiratory properties of rodents.
199. Vanstone, W. E., et al., Canad. J. Physiol. Pharmacol. *42*:697–703, 1964. Multiple hemoglobins in coho salmon.
200. Vaughan, B. E., and N. Pace, Amer. J. Physiol. *185*:549–556, 1956. Myglobin changes at high altitudes.
201. Wald, G., and A. Riggs, J. Gen. Physiol. *35*:45–53, 1951. Hemoglobin of lamprey, *Petromyzon*.
201a. Waxman, L., J. Biol. Chem. *246*:7318–7327, 1971. Properties of *Arenicola* Hb.
202. Weber, R. E., Konink. Nederl. Akad. van Wetenschappen *66*:284–295, 1963. Hemoglobin in larva of *Chironomus*.
203. Weber, R. E., Comp. Biochem. Physiol. *35*:179–189, 1970. Properties of annelid hemoglobins.
204. Weisberger, A. S., Proc. Soc. Exp. Biol. Med. *117*:276–280, 1964. Sickling phenomenon of deer hemoglobin.
204a. Wells, R. M. G., Comp. Biochem. Physiol. *46A*:325–331, 1973. Carbonic anhydrase in *Arenicola*.
205. Whittenberg, B. A., T. Okuzuki, and J. B. Whittenberg, Biochim. Biophys. Acta *111*:485–495, 1965. Hemoglobin of *Ascaris*.
206. Whittenberg, J. B., et al., Proc. Nat. Acad. Sci. *67*:1846–1853, 1970. State of iron in oxyhemoglobin.
207. Wilkins, N. P., and T. D. Iles, Comp. Biochem. Physiol. *17*:1141–1158, 1966. Hemoglobin polymorphism in fishes.
208. Wolvekamp, H. P., Z. vergl. Physiol. *25*:541–547, 1938. Oxygen transport and hemocyanin in *Octopus*.
209. Wolvekamp, H. P., et al., Arch. Neerl. Physiol. *25*:265–267, 1941; *28*:620–629, 1947. Respiratory function in blood of *Arenicola, Helix,* and *Homarus*.
210. Wood, S. C., and W. R. Moberly, Respir. Physiol. *10*:20–29, 1970. Respiratory properties of blood of *Iguana*.
211. Wyman, J., Adv. Protein Chem. *4*:407–531, 1948. Heme proteins.
212. Yoshioka, M., et al., J. Biochem. (Tokyo) *63*:70–76, 1968. Eel hemoglobins.
213. Zanjani, E. D., M.-L. Yu, A. Perlmutter, and A. S. Gordon, Blood *33*:573–581, 1969. Hormonal control of erythropoiesis in fish.
214. Zuckerkandl, E., R. T. Jones, and L. Pauling, Proc. Nat. Acad. Sci. *46*:1349–1360, 1960. Evolution of hemoglobins.

CHAPTER 9

TEMPERATURE

By C. Ladd Prosser

INTRODUCTION

Ecological and Physical Considerations

Temperature limits the distribution of animals, and at the same time determines their activity. The range of environmental temperatures on earth is much greater than the range permitting active life. Surface temperatures in open oceans range from −2° to +30°C; the range of air temperatures is from −70° to +85°C. In general, life activities occur only within the range from about 0° to 40°C; most animals live within much narrower limits. Some survive but are inactive below 0°C, and a few tolerate freezing. Some animals live in warm springs, and a few bacteria and algae are active in springs as hot as 70°C. Limits for reproduction are narrower than those for survival of adults, but the embryos of many homeotherms tolerate a wider range than do the adults. Since temperature is a measure of molecular agitation, it controls the rate of chemical reactions and is one factor limiting growth and metabolism. The measurement of temperature is so easy—by mercury thermometer, thermistor, or thermocouple—that the literature on temperature is vast. Heat is a form of energy, and temperature is the integral of heat, or the measure of heat content.

Many animals correspond in body temperature to the environment; these are "cold-blooded" or *poikilothermic* animals. "Poikilothermic" means shifting in temperature (i.e., labile). The body temperature (T_b) need not be equal to the ambient temperature (T_a) in poikilotherms; measurements of muscle temperature of a 40 gm fish show a temperature 0.44°C above that of the environment.[445]

Fewer kinds of animals regulate body temperature; these are "warm-blooded" or *homeothermic* animals. Between poikilotherms and homeotherms are the *heterothermic* animals, those with partial temperature regulation; that is, their regulation is limited to restricted conditions or body regions.

The temperature of any metabolizing cell is necessarily higher than that of its medium, because oxidation and glycolysis liberate heat. The temperature of an animal depends on the balance of those factors that tend to add or subtract heat. Heat may be gained by metabolic thermogenesis (endothermy), or by absorption of heat originating externally, largely from solar radiation (ectothermy). Heat is lost by radiation, convection, conduction, and vaporization of water; heat loss is favored by fluids circulating from the interior to the body surface and is restricted by insulation. Poikilothermic animals are not without some thermal control—by behavior, by entering a state of dormancy, or by metabolic or nervous compensations. Homeotherms maintain relative constancy of body temperature by producing heat, by protective behavior, and by altering insulation, circulation, and other modifiers of heat transfer. Some "warm-blooded" animals undergo a drop in body temperature during periods of lethargy, and some hibernate with the physiological thermostat set at a low temperature. Sensory mechanisms signal the changes in temperature which evoke compensatory and protective responses.

Since animals are largely composed of water, the thermal properties of water are important in determining the thermal relations of animals. The heat conductivity of water (0.0014 cal/cm/sec/°C) is low compared with those of materials such as metals, but is higher than those of many liquids (e.g., ethyl alcohol, 0.00042, and olive oil, 0.000395). In addition, the specific heat of water is high, being 1.0 cal/gm/°C as compared with 0.09 for copper and 0.535 for ethyl alcohol; animal tissues, except for compact bone, require 0.7 to 0.9 cal to raise the temperature of 1 gm of tissue 1 degree Centigrade. Thermal diffusivity is given by the heat conductivity divided by the product of density and specific heat. Low thermal diffusivity results in slow warming and cooling, and in limited conduction of heat within an animal; fat is a good heat insulator but is not as good as air. Animals with much tissue mass are slow to warm or cool; most of their transfer of heat is by circulating body fluids, and a sluggish circulation makes for slow heat transfer. A 1 gm fish cools at the rate of 1.8°C/min/°C gradient, while the rate for a 100 gm fish is only 0.4°C/min/°C gradient.[445]

Natural waters (hot springs excepted), because of their high specific heat and low heat conductivity, rarely have a temperature above the limit for most aquatic animals. The freezing point of aquatic animals (except for some such as marine bony fishes or brine shrimp) is usually similar to or lower than the freezing point of the medium (Chapter 1). Since ice has a lower specific gravity than water, aquatic animals do not freeze so long as they remain in the water below ice. Terrestrial animals are subject to much greater fluctuations in temperature than are aquatic ones, and in the former, body temperature is closely tied to water balance.

Water has a high heat of fusion (79.7 cal/gm); aqueous solutions supercool by several degrees, especially in capillary spaces, and bound water is resistant to freezing. Some animals supercool, and partially dehydrated animals (or those with some of their water replaced by polyhydric organic solvents) can withstand temperatures well below the freezing point of water without their tissues becoming frozen. Water loss by evaporation has a marked cooling effect on any moist surface (585 cal lost per gm of H_2O evaporated at 20°C), and at high air temperatures animal temperature control is limited by the heat of vaporization of water.

Temperature Characteristics (Activation Energy)

Kinetic activity, or frequency of molecular collisions, is proportional to the absolute temperature, and increases by about 3 per cent per 10°C. One method of quantification of the effect of temperature on reaction rates is the Q_{10} approximation. The Q_{10} is the factor by which a reaction velocity is increased for a rise of 10°C:

$$Q_{10} = \left(\frac{K_1}{K_2}\right)^{10/(t_1 - t_2)}$$

where K_1 and K_2 are velocity constants corresponding to temperatures t_1 and t_2. Velocity (V_1 and V_2) is normally used in place of the rate constant. Some functions are linear, but most are logarithmic, with respect to temperature. Q_{10} varies over the temperature range, and is higher in low ranges than in high ranges. Hence, the temperature range for which a Q_{10} is calculated must be specified. Physical properties of solutions are less sensitive to temperature changes than are catalyzed reactions. A Q_{10} of 2.5 means an increase in rate of 9.6% per degree; many chemical reactions have values in this vicinity.

The critical incremental energy of activation of a reaction is given by the Arrhenius μ (sometimes called E*):

$$-\ln\left(\frac{K_2}{K_1}\right) = \frac{\mu}{R}\left(\frac{1}{T_2} - \frac{1}{T_1}\right)$$

$$\mu = \frac{-2.3\ R\ (\log K_2 - \log K_1)}{\frac{1}{T_2} - \frac{1}{T_1}}$$

where K_1 and K_2 are velocity constants (proportional to measured velocities) at absolute temperatures T_1 and T_2, and R is the gas constant (1.98 cal/mole). When the logarithm of a velocity is plotted against the reciprocal of the absolute temperature, most biological reactions (and all simple chemical reactions) give a straight line, the slope of which is $-\mu/2.3R = -\mu/4.6$. This relation permits the conclusion that the number of molecules exceeding a *critical energy* (e.g.,

12,000 calories) may double for a 10° rise in temperature even though the kinetic change, proportional to the absolute temperature, is much less. Definite values of μ characterize specific catalysts irrespective of the substrate, when measured at substrate concentrations such that the catalyst is rate limiting. However, at lower, more physiological substrate concentrations, values of temperature characteristics are lower. It has been postulated that the μ of a complex biological reaction is that of its limiting or pacemaker step; however, this is an extreme oversimplification.

Many metabolic reactions show Q_{10}'s of 2 to 2.5 over the biological temperature range. However, some complex rate functions (e.g., circadian rhythms) are relatively independent of temperature, and in some temperature-conforming animals the oxygen consumption shows a Q_{10} between 1 and 2 over a limited range. By comparing oxygen consumption in an animal when it is active with that when it is at rest, active and rest metabolism can be measured at different temperatures. In the mollusc *Cardium*, Q_{10} in activity is 1.84, and that at rest is 1.20; a series of invertebrates gave low resting Q_{10}'s in the range of low normal ambient water temperatures.[344, 345] In insects (e.g., *Calliphora*), the Q_{10} of muscle and of whole fly approximates 2.[463] Enzyme-catalyzed reactions may show decreasing temperature characteristics (either μ or Q_{10}) as the substrate concentration is reduced to the normal cellular levels where substrate rather than enzyme is rate-limiting. When catalyzed reactions obey Michaelis-Menten kinetics, a measure of enzyme-substrate affinity is given by the K_m (see Chapter 6) or by the substrate concentration for half-saturation of the enzyme (K_s). For most enzymes, the apparent affinity increases (the K_m is decreased) as temperature is reduced. This counteracts the decrease in activation energy and may lower the apparent Q_{10}.[206, 436] In complex reactions, with parallel and series steps, and with opposing effects on activation Q_{10} and K_m, attempts to analyze temperature effects in simple terms are hopeless.

FREEZING; WINTER HARDENING

Protoplasm, an aqueous solution, freezes at a few degrees below zero. Slow freezing is often more deleterious than fast. If the rate of cooling through the temperature range of freezing is rapid, about 100°C/sec, ordinary ice crystals do not form but the organism becomes solid. When small nematodes, protozoans, muscle fibers, or tissue cultures are placed directly into liquid air (−197°C), they are solidified; if they are then rewarmed rapidly, they revive.

One mode of resistance to freezing is increase in solute concentration, which causes *lowering of the freezing point.* This occurs in nature to a limited extent; cold-water marine fish have slightly elevated blood concentrations of salts and some have markedly higher concentrations of specific non-electrolytes (antifreeze agents). Winter-hardened insects and some plants accumulate organic solutes which increase their osmotic concentration.

More important in winter hardening and polar latitude adaptation than a lowering of freezing point is *supercooling.* When pure water or an aqueous solution is cooled, the temperature falls below the freezing point until nucleation begins; then the temperature rises while the water crystallizes and heat of fusion is released. After freezing, the ice may be cooled further. Crystallization involves hydrogen-bonding of water molecules, and this process is delayed in supercooling by various non-polar organic solutes. Usually the accumulation of protective solutes is accompanied by some dehydration.[11]

In many animals and plants, extracellular fluid freezes before cells freeze. Cell membranes, in some unknown way, protect against protoplasmic freezing. When salt solution freezes slowly, the water crystallizes, salt is trapped in the interstices, and the osmotic concentration of the unfrozen solution rises. When this occurs in extracellular fluid, water is drawn osmotically from cells and they may suffer dehydration. Replacement of some free water by non-polar organic molecules protects against freezing by inducing lattice formation and hydrogen bonding other than between water molecules (this is reviewed by Smith[424]). The rate of ice crystal growth is inversely related to concentration of hydrogen-bonding groups on protective substances.[110] When intracellular freezing occurs, mechanical disruption of protoplasmic organization may occur; this has been seen in amoebae, plant cells, and sea urchin eggs.[10] Water of hydration or that bound to charged solutes, especially proteins, is the last to freeze, and death usually occurs if this water is frozen. Collagen may bind much water which retains its liquid state, as judged by nuclear magnetic resonance, down to −50°C.[102]

At temperatures below 0°C, in addition to osmotic dehydration, mechanical disruption, and change in distribution of bound water, there is very steep reduction in enzyme activity. Diffusion of O_2 and CO_2 is 10^4 slower in ice than in fluid water. Also, denaturation of some enzymatic proteins can occur at subfreezing temperatures.

Many cells can endure brief periods of freezing but not prolonged freezing, particularly at low temperatures. In a frog sartorius muscle, ice forms at −0.42°C, but the muscle can be kept undamaged when supercooled to −4°C for two days; it is irritable when thawed after being frozen for a week at −0.9°C, for 20 hours at −2°C, or for a half hour at −3.5°C. The principal damage may be caused by removal of water from molecular combination. The ciliate *Tetrahymena* in 10% dimethyl sulfoxide (DMSO) cooled at the rate of 4.5°C/min survived at −95°C for more than 128 days; *Paramecium* recovered motility but not cell division after extreme freezing.[483]

Marine invertebrates, especially intertidal molluscs and barnacles, are frequently frozen in winter. In such animals frozen at −15°C, only 55 to 65 per cent of body water was frozen. The percentage of water frozen increases at lower temperatures, and at a given low temperature proportionately less of the water in an animal freezes than of sea water[250] (Fig. 9–1). Intertidal *Balanus*, *Littorina*, and *Mytilus* tolerate freezing at −10 to −20°C in winter, but not in summer.[438, 439] *Mytilus* is injured at −10°C, when 64% of cell water is ice, but if the animal is previously acclimated to 150% sea water its tolerance increases to −15°C and freezing of 80% of its water. In 150% sea water, its solute nonprotein nitrogen increases disproportionately to its osmotic concentration.[491]

In winter-hardened insects supercooling is important, and some insects (e.g., *Cephus*) survive supercooling to −23 to −30°C as compared with only −5 to −8°C by other, less winter-hardy species.[396] More supercooling is possible with rapid than with slow cooling.[397, 399] In *Bracon*, a parasite on sawflies, supercooling to −47.2°C was observed; the glycerol concentration in hemolymph had increased to 2.7 M. Of ten species of overwintering insects, one had high sorbitol, two probably had mannitol, and the others had glycerol[438] (Fig. 9–2). Increase in glycerol content (to 16%) parallels capacity for supercooling; when diapause breaks in the spring, the glycerol content drops. Larvae of a flour moth, *Anagasta*, showed in cold acclimation an increase in various ninhydrin-positive substances and in glucose and glycine, which resulted in more significant increase in supercooling than in lowering of freezing point:[437]

Acclimation Temperature	*Time*	*Supercooling Minimum*	Δ_{fp} *Hemolymph*
−6°C	7 days	−21°C	−1.1°C
+6°C	7 days	−18.3°C	−1.0°C

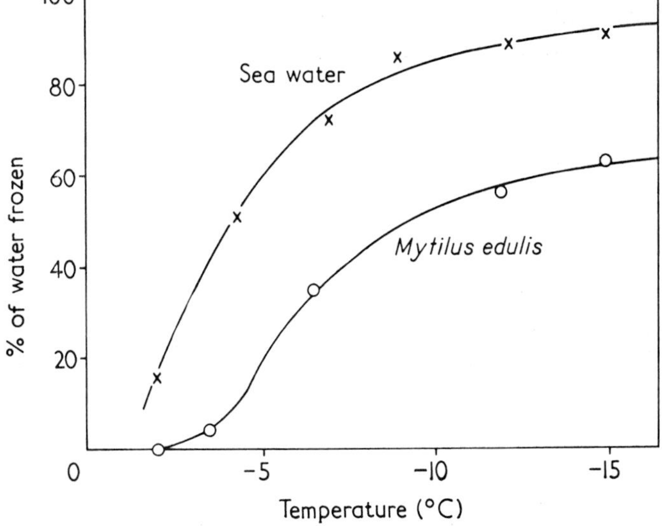

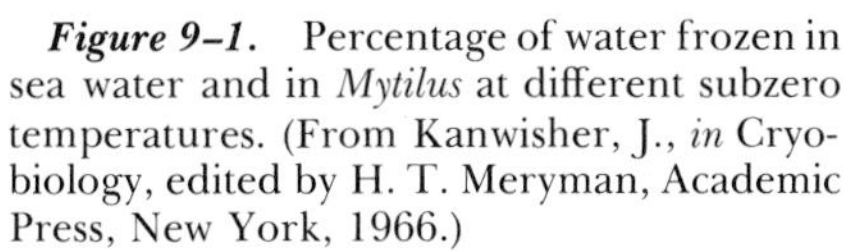
Figure 9–1. Percentage of water frozen in sea water and in *Mytilus* at different subzero temperatures. (From Kanwisher, J., *in* Cryobiology, edited by H. T. Meryman, Academic Press, New York, 1966.)

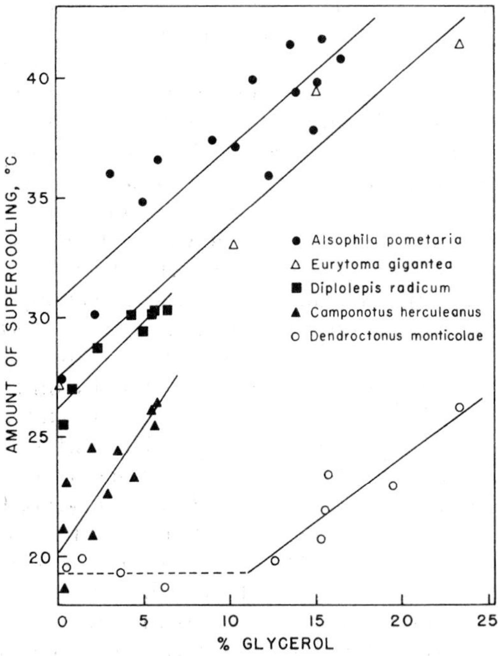

Figure 9–2. Relation between supercooling and concentrations of glycerol in various insects in winter. (From Sømme, L., Canad. J. Zool. *42*:87–101, 1964. Reproduced by permission of the National Research Council of Canada.)

Eggs of *Malacosoma* (tent caterpillar) accumulate glycerol to a concentration of 35.2% of dry weight in January, and can supercool to −40.8°C in winter as compared with −29.5°C in summer[155] (Fig. 9–3). Temperature relations of an Alaskan carabid beetle, *Pterostichus*, are as follows:[326]

	Supercooling Capacity	Δ_{fp} *(thawing)*	*Glycerol Content,* %	*Water Content,* %
Winter	−35°C	−3.5°C	25	54
Summer	− 6.6°C	−0.7°C	0	65

Red blood cells or sperm can be stored frozen without damage for many months if they are pretreated with glycerol (or propylene glycol, or DMSO), frozen rapidly, and stored at below −40°C. Other cells, such as tissue cultures of mammalian tissues, survive well if cooled at the rate of 100°C/min, and less well if cooled at 1°/min; they can be protected by such agents as PVP (polyvinyl pyrrolidine) and sucrose. Marrow has been protected well by amino acids; dextrans retard extracellular freezing in tissues.[251] Corneas have been successfully transplanted after freeze-storage. Sympathetic ganglia were electrically active after freezing and thawing from −79°C if pretreated with glycerol.[424] Nematodes of two genera survived freezing if cooling and warming rates were rapid, but below −30°C they did not survive.[343] It is argued that when cooling is rapid, more intracellular water is in the supercooled state, whereas in slow cooling water leaves the cells and damage results. Cell death in slow freezing appears to be caused more by membrane damage than by inactivation of enzymes.[309, 310, 311, 312]

Woody plants undergo remarkable cold-hardening, so that a given plant resists winter temperatures many degrees colder than those it can tolerate in summer. Metabolically active layers of bark synthesize much RNA and protein at the beginning of the winter.[419] Levels of acclimatization are recognized: the first consists of many metabolic changes that are hormonally initiated; later there may be polymerization of macromolecules with a decrease in bound water.[485] One hypothesis emphasizes the increase in intermolecular disulfide bonds on freezing, as indicated by protection by agents that retard S-S formation.[9] Another theory emphasizes reorientations within macromolecules so that they resist dehydration.[485]

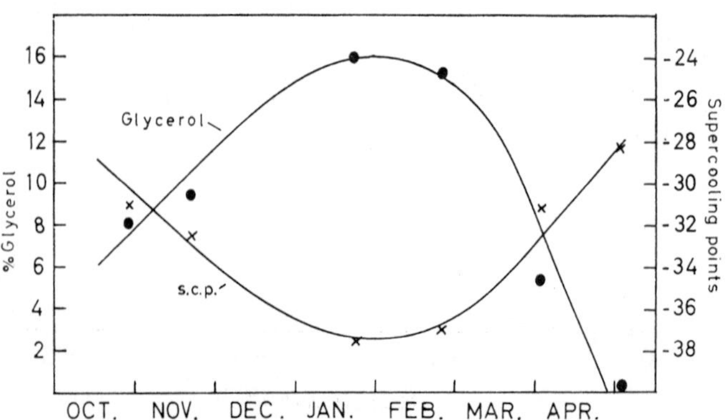

Figure 9–3. Changes in glycerol content and supercooling points (s.c.p.) in insect larvae stored outdoors in Canada in winter. (From Sømme, L., Canad. J. Zool. *43*:765–770, 1965. Reproduced by permission of the National Research Council of Canada.)

Arctic and Antarctic fish have significant resistance to freezing. *Trematomus* lives under the ice in the Antarctic, and the freezing point of the blood of two species was −1.87° to −2.07°C.[107] NaCl, urea, and amino acids constituted only half of the osmotically active solute. The blood contains much glycoprotein, which is identified by five electrophoretic bands corresponding to molecular weights of 10,500 to 78,000. The glycoproteins (as high as 1870 mg% in plasma) have remarkable antifreeze capacity, which can be inactivated by acetylation or by peptide bond cleavage.[106, 269] The glycoproteins are 16% threonine, 26% alanine, 29% N-acetylgalactosamine, and 28% galactose, in a pattern shown in Figure 9–4. These glycoproteins lower the freezing point more than would be expected solely on the basis of concentration; how this effect is achieved is unknown.

In fish living in deep fjords in Labrador, where the water temperature is −1.7°C winter and summer, the freezing point of the fish is −0.9°C; if the fish are seeded with ice crystals, they freeze at once. Surface fish that live in water at 5°C in summer and at −1.5°C in winter have a blood Δ_{fp} of −0.8°C in summer and −1.6°C in winter. The deep-water fish live in a permanent state of supercooling.[417] Winter flounder (*Pseudopleuronectes*) has blood Δ_{fp} of −0.71°C whether they are kept at −1.0° or at +15°C; yet they survive in water at −1.8°C by supercooling.[355, 469]

The New England killfish *Fundulus* changes its blood solutes seasonally as follows:[469]

	Summer	*Winter*
glucose	68	348 mg/100 ml
cholesterol	201	307 mg/100 ml
amino acids	11.9	13 mg N
protein	4	2.2 g/100 ml
sodium	183	205 mM
chloride	145	170 mM
osmolality	394	401 mOsm

Serum glucose increases and liver glycogen decreases during cold acclimation, more in winter than in summer fish. The water temperature in summer was 20°C, and that in winter was −1.5°C. The increase in lowering of freezing point is not significant, but the tolerance of supercooling increases very greatly. It is postulated that molecules that are rich in OH groups impede the hydrogen bonding of ice formation.[469]

Hamsters were anesthetized by hypercapnia, perfused with some glycerol, and chilled to a colon temperature of 15° to 20°C; they were then immersed in ice water. Respiration stopped between 2° and 6°C, and the heart beat stopped between 0.8° and 2.5°C.[424] Then the animals were put into propylene glycol at −3° to −14°C, and freezing occurred. After more than 50 minutes at −5°C they were rigid. If they were then rewarmed gradually and artificial respiration was applied, most of the hamsters revived and were normal. All of those in which less

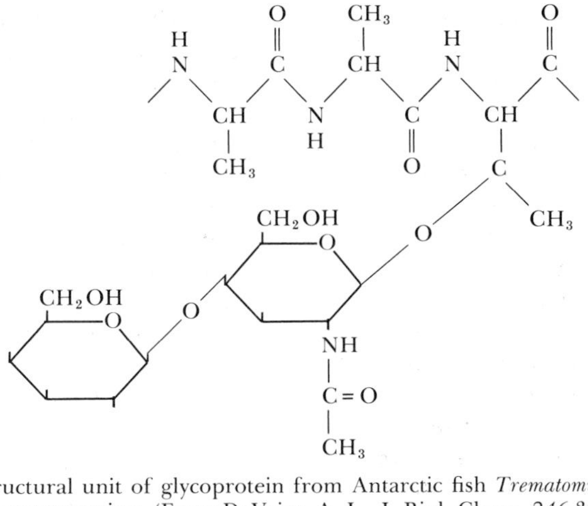

Figure 9–4. Repeating structural unit of glycoprotein from Antarctic fish *Trematomus*; this lowers freezing point more than is predicted from concentration. (From DeVries, A. L., J. Biol. Chem. *246*:305–309, 1971.)

than 15% of body water had been frozen recovered, as did one-third of those in which 45% of total body water had been frozen for one hour.[8, 424]

In summary, freezing is deleterious when extracellular fluid is concentrated as ice crystals form and cells then become dehydrated when intracellular crystals appear. Some tissues withstand freezing, and more tolerate it if the process is fast. Some animals gain limited protection in winter by increasing osmotic concentration; more animals accumulate organic substances, which permit much supercooling.

LETHAL LIMITS; RESISTANCE ADAPTATION

Causes of death at either high or low temperatures are not well understood and are certainly multiple. Death due to cold occurs at temperatures well above freezing in those animals which show no winter hardening and which cannot endure freezing. As the body temperature of either homeotherms or poikilotherms becomes lowered, cardiac and respiratory activities are slowed and hypoxia may result. Most commonly, cell membranes become permeable, ionic gradients are not maintained, and ion pumps are stopped. Energy liberation may be insufficient for maintenance of function. Components of coupled enzymatic reactions may have differences in their K_m-temperature relations. Integration by the central nervous system fails, and many animals enter a state of chill coma at temperatures well above the lethal. Mortality curves plotted as per cent of deaths as a function of time at different low temperatures show that there are multiple causes of death, each operating at a particular time-cold-intensity combination.

Heat death also occurs from multiple causes at different time-temperature combinations. When body temperatures are raised, the transport of oxygen by blood pigments is reduced, water loss by evaporation is increased in terrestrial animals, and they may become desiccated. As the temperature rises, lipids change in state and cell membranes become increasingly permeable. Enzymes pass their temperature optima at the temperature where inactivation exceeds activation. At higher temperatures (non-physiological), protein denaturation occurs and toxic substances may be released from damaged cells. The DNA double-stranded helix is not stable at very high temperatures.

Mechanisms of killing are different at the levels of the whole organism, tissue, and enzyme. The temperatures of heat death for 15-minute exposure in *Rana pipiens* were: intact tadpole, 37.5; whole frog, 38.6; gastrocnemius muscle, 40.2; heart *in situ*, 42; sciatic nerve, 43°C.[351] In general, the tolerable limits are narrowest for whole animals, somewhat wider for tissues and cells, and much wider for inactivation of isolated enzymes. However, the temperature limits within which a given enzyme functions optimally, as indicated by its point of maximum affinity for substrate, may be very narrow.[433] Both cold and heat death limits may be modified by acclimation, and both of the limits may be genetically related to the ecology of given animals. Sometimes heat and cold death points are modified in the same direction by acclimation, and in other cases they are independent; the mechanisms are different.

Without consideration of causes of death, the measurement of high and low lethal temperatures after acclimation to different temperatures characterizes species as to their "zone of tolerance." Many individuals of a given species of aquatic animal can be experimentally acclimated to several temperatures. Then groups from each acclimation temperature are placed directly in a series of baths at high temperatures for heat death determination and others in baths at low temperatures for cold death measurement. The percentage mortality (probit) is plotted as a function of time at each temperature (Fig. 9–5), and time to death of 50 per cent of each group is plotted against exposure temperature[46] (Fig. 9–6). The median tolerance temperature (LD_{50}) falls at longer times until, at a long time (about 48 hours) no further deaths occur; this is the incipient lethal temperature (high or low). Two lines are plotted, the incipient high and low lethal temperatures against acclimation temperature (Fig. 9–7); these enclose the tolerance polygon. Acclimation occurs at different rates according to species; heat acclimation is usually faster, by a few days, than cold acclimation (10 to 20 days). In sockeye salmon, the upper lethal temperature changed about 1°C for every 4°C change in acclimation temperature.[46] For goldfish, the upper lethal temperature increased 1°C for every 3°C rise

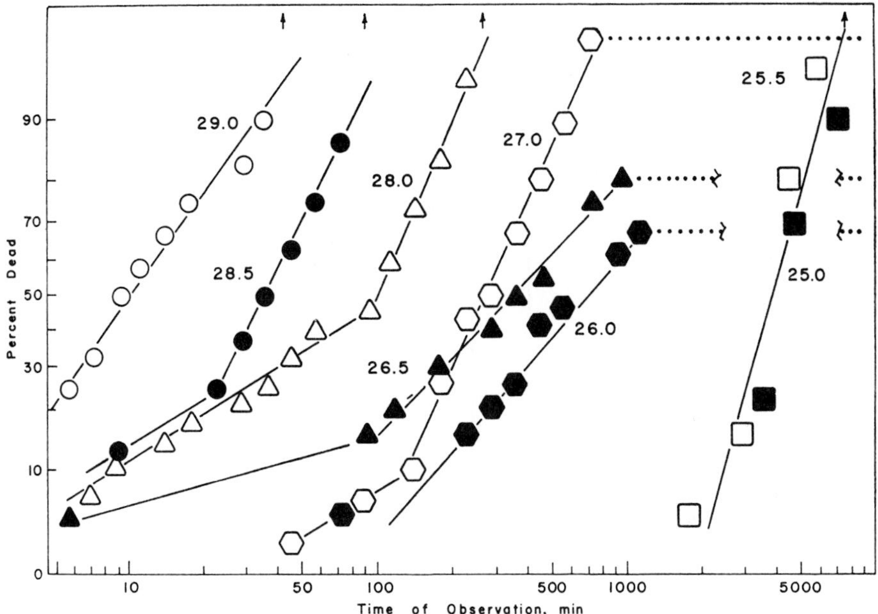

Figure 9–5. Times to death (expressed as probits) at various lethal high temperatures of minnows. Changes in slope suggest different causes of death. (From Tyler, A. V., Canad. J. Zool. *44*:349–361, 1966. Reproduced by permission of the National Research Council of Canada.)

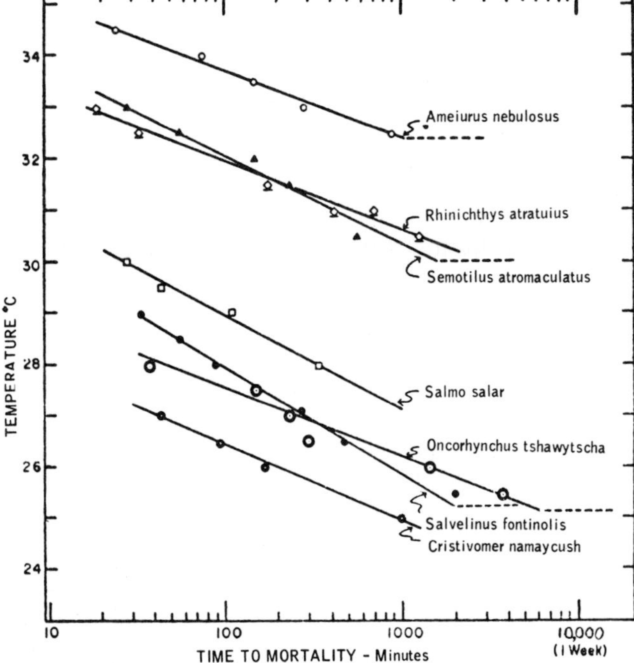

Figure 9–6. Median resistance times for seven species of fish acclimated to 20°C and tested at various temperatures. (From Brett, J. R., Quart. Rev. Biol. *31*:75–87, 1956.)

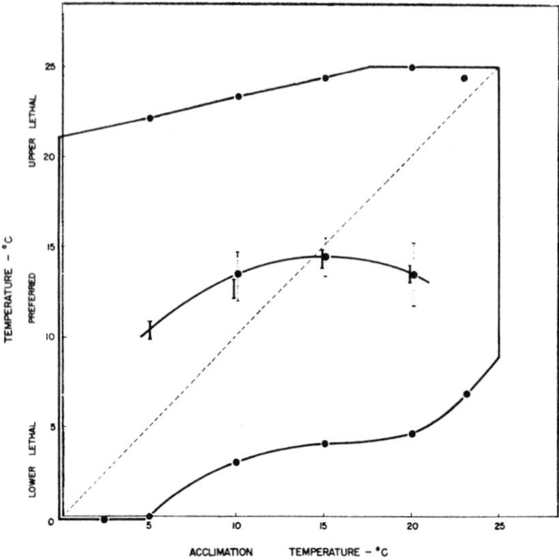

Figure 9–7. Tolerance polygon and preferred temperature as a function of acclimation temperature in sockeye salmon. (From Brett, J. R., Amer. Zool. *11*:99–113, 1971.)

in acclimation temperature up to 36.5°C; the lethal high temperature was then 41°C and could not be increased further. The low lethal temperature decreased 2°C for every 3°C fall in acclimation temperature down to 17°C, where the low lethal temperature was 0°C.[131] In earthworms (*Pheretima*), heat tolerance increases 0.3° per 1°C rise in acclimation temperature. High lethal temperature for the Mediterranean *Littorina neritoides* was 46 to 47°C, compared with 40 to 41°C for New England *L. littorina*.[126]

The tolerance polygon is fairly characteristic of a species (see Fig. 9–7). Within a tolerance polygon are smaller polygons for development and locomotor activity; these functions are more limited in temperature range than is mere survival. Fourteen species of eastern North American freshwater fish from different latitudes were examined for racial differences in tolerance polygons. Of these, three (e.g., *Notropus cornutus*) gave different polygons for populations from Ontario and from Tennessee.[164] Two geographic races of the brook trout *Salvelinus fontinalis* differ in upper lethal temperatures; a thermal race of *Daphnia atkinsoni* differs from a temperate race by 6°C in high lethal temperature.

Both thermophilic and psychrophilic strains of bacteria are known. In hot springs of Yellowstone, bacteria occur in near boiling water (90°C), and protozoans exist at 45 to 51°C; blue-green algae tolerate 73°C but show maximum growth at 55°C.[49] The melting temperature of DNA strands is characteristic of different bacteria.[255] The protein flagellin from thermophilic bacteria has more amino acids which readily form stable H-bonds; this suggests more protein stability.[267] Thermophiles have lower Q_{10} values for growth than does *E. coli*; in other words, the high-temperature strain is less sensitive to temperature changes.[49] Enzymes such as PFK are stable up to 95°C, but their K_m is temperature-dependent.[500] Many examples of genetically determined thermostability of proteins are known.[473] Thermostability of collagen, as measured by the temperature of transition (melting point, Fig. 9–8), is proportional to the proline plus hydroxyproline (pyrrolidine) content[243] (Fig. 9–9):

	Collagen From	
	Ascaris	*Lumbricus*
Number of residues		
proline	7.7	29
hydroxyproline	19	165
½ cystine	16	0
Thermal transition, °C	52	22

These amino acids are higher in amount in collagen from tropical fish than in that from cool-water fish.[144] Triplets of amino acids with neighboring pyrrolidine are thermally most stable.[243] Heat may also block specific synthesizing paths; some bacteria at 37°C need amino acids and vitamins that are not essential to them at 26°C.[279] In eleven species of related bivalves, the aspartic/glutamic transaminase of subtidal forms is more heat stable than that of upper intertidal species;[379,380] the percentage inactivation by two minutes of heating at 56°C is as follows:[378]

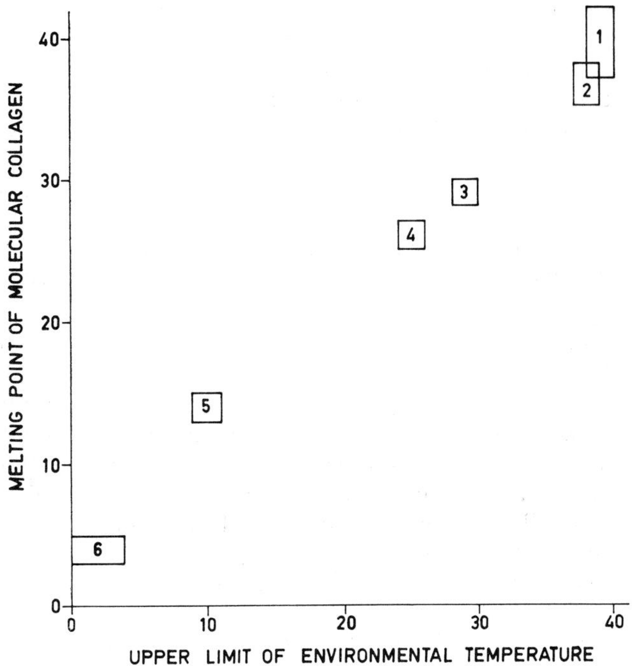

MELTING POINT OF MOLECULAR COLLAGEN

UPPER LIMIT OF ENVIRONMENTAL TEMPERATURE

Figure 9–8. Relation between melting points of collagen and approximate upper limits of environmental temperature for several species. (1) *Ascaris* and hog; (2) rat, cow; (3) snail *Helix*; (4) tuna skin; (5) cod skin; (6) Antarctic fish. (From Rigby, B. J., Biol. Bull. *135*:223–229, 1968.)

	Percentage Inactivation of asp/glut Transaminase	*Median Lethal Limit*
subtidal *Modiolus*	80 to 100	25°C
intertidal *Mytilus*	40 to 50	28°C
littoral *Brachidontes*	0 to 10	38°C

In a wide variety of aquatic animals, the thermal resistance of muscles, as measured by response to electrical stimulation after heating, is closely correlated with thermostability of many enzymes in the muscles—aldolase, cholinesterase, myosin ATPase, and succinic dehydrogenase.[473] For fishes from deep water, as compared with those from shallow water, the thermostability of muscles, heat narcosis, and 50% inactivation of muscle cholinesterase by heat parallel each other.[275, 474] In a series of reptiles, the thermostability of myosin ATPase correlates well with the temperature of maximum muscle tension and with the thermal preferences of the animals.[295]

Genetic differences in low lethal temperatures are indicated for a series of amphibians; high altitude animals from Central America resemble temperate zone (high latitude) forms, while low altitude animals resemble tropical species.[44] Australian anomurans that have a wide geographic range show more

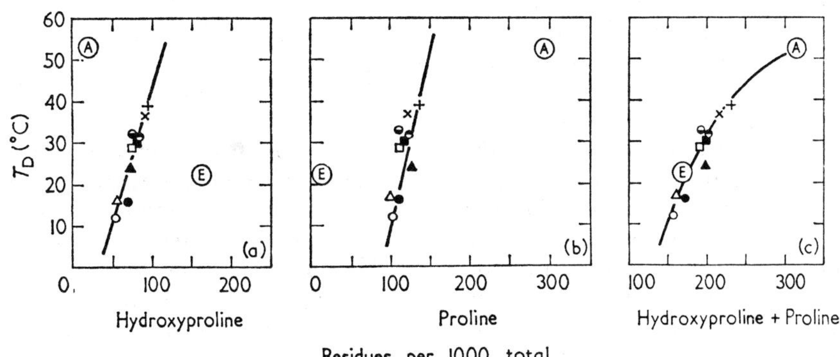

Figure 9–9. Relation between melting points of collagen and number of residues of hydroxyproline, proline, and hydroxyproline + proline per 1000 total residues for collagens from different species. (From Josse, J., and W. F. Harrington, J. Molec. Biol. *9*:269–287, 1964.)

temperature adaptability than do species with a narrow geographic range.[45]

Limits for development frequently correlate well with survival limits, as for the following two toads:[44]

	Development Limits	*Survival Limits*
Bufo cognatus (Arizona)	19.1 to 32.0°C	10 to 31°C
Bufo leutkeni (Guatemala)	32.1 to 34.8°C	22 to 39°C

Rana pipiens from northern United States have a lower temperature range for development than do southern populations[395] (possibly different species). In the milkweed bug *Oncopeltus*, the lower limit for egg survival is 5°C, while for embryonic development it is 14°C and for viable larvae (full development) it is 18°C.[382] The upper tolerance limit for *Balanus balanoides* from North Carolina is 5° higher than for the same species in England; similarly, both upper and lower temperature limits for cirral beats differ for warm-water and cold-water species of barnacles. Genetic strains of thermophilic bacteria, flagellates, and other organisms have been selected and reared for heat tolerance. A psychrophilic nematode growing at 18°C dies at 24.5°C, but by gradual warming and selection over 100 generations, nematodes were obtained that grew well at 24.5°C.[53]

Lethal temperature is influenced not only by acclimation temperature and genetic background, but also by age, size, hormonal state, diet, and environmental factors such as photoperiod, oxygen, and salinity. In lobsters, the upper lethal temperature is lowered by decreased salinity and by lower O_2 levels.[316] *Mytilus* in low salinity show decreased tolerance of high temperatures.[409] A variety of effects of dietary lipids on temperature tolerance in poikilotherms has been reported. Heat tolerance of goldfish was greater when they had been fed lard than when fed fish oil. A diet of saturated fat increased the heat tolerance of blowfly larvae.[210] *Hemigrapsus oregonensis* shows different sensitivities to temperature and salinity in winter and in summer; high environmental temperature increases heat resistance, and low salinity decreases it, as shown by the following lethal temperatures for two acclimation temperatures:[461]

	Acclimation Temperature	
Salinity	*20°C*	*5°C*
35% S.W.	33.3°	30.2°
75% S.W.	34.3°	32.6°

In *Astacus*, tissue respiration continued for an hour at high temperatures that were lethal for the whole animal; but blood Na decreased and blood K increased at the high temperature, and neuromuscular function stopped when cool crayfish were perfused with saline of similar Na and K to that found in the heated individuals. Thus, lethality may be related more to ionic changes than to enzymatic ones. The euryhaline fish *Cyprinodon* growing at 15 to 20°C do best in fresh water, whereas growth at 25 to 35°C is best at 35 to 55 0/00 salinity.[261,262] Exposure to excess oxygen (more than one atmosphere) favors survival of goldfish at high temperatures.[484]

Heat and cold coma correlate well with lethal temperatures and vary with acclimation. Coma is a measure of reversible failure of function in the central nervous system. In the guppy *Lebistes*, the temperature for heat coma increases 0.4° per 3°C rise in temperature of rearing, and the cold coma temperature decreases 1° per 3°C fall in rearing temperature.[465]

A series of intertidal snails shows heat coma which correlates with their ecology, and which is thus species specific; they also show a considerable acclimation effect:[147]

Snail Species	*Habitat*	*Acclimation Temperature*	*Mean Heat Coma Temperature*
Nerita peleronta	high intertidal Bimini	15	37
		37	45.8
Nerita versicolor	mid-intertidal Bimini	25	43
		37	44.6
Nerita tesselata	below mean water Bimini	25	41
		37	44.2
Littorina littorea	intertidal Woods Hole	5	32.5
		15	35.0
		25	38.8

The temperatures for coma in *Littorina* are essentially the same as the temperatures for heat block of spontaneous activity in the central nervous system. Snails from a hot Israel desert showed a two-hour median lethal temperature of 55°C.[415a]

In a series of insects, the temperature necessary to cause chill coma within 30 minutes correlates with the Q_{10} of myosin ATPase:[339]

	Coma Temperature	Q_{10}
Musca	6.1°C	3.9
Periplaneta	7.9	2.7
Tenebrio	7.4–10.8	2.2
Galleria	8.8	2.0

The temperature for cold coma in cockroaches can be changed by acclimation, as follows:[84]

Blatella *A.T.,°C*	*Coma T,°C*
35	7
15	−4.4
10	−4.3

Experiments in which temperature was cycled on either 24 or 48 hour schedules show very little difference in high coma temperature from continuous acclimation at the higher point of the cycle in *Salmo*[180] and conversely for cold coma in isopods.[116]

Higher nerve centers fail within narrower limits than peripheral nerve and muscle. Data for cold block in freshwater fish are shown in Figure 9–10.[371] In a skate, with heating, various reflexes disappear in a definite sequence, and muscles can be activated via nerves after spinal reflexes have been lost.[32] The temperatures for cold block in frog sartorius muscle preparations were lowered by cold acclimation,[236] while the temperature for heat block was higher after cold acclimation (a non-adaptive change).[365]

Mechanisms of cold and heat death are different and are poorly known. Secondary effects such as hypoxia may be important in complex animals; increased membrane permeability and blocking of ion pumps, the disequilibrium of coupled enzyme reactions, and the limits imposed by kinetics are all important in cells in general. Many genetic adaptations (e.g., differences in primary structure of given proteins) permit survival at temperature extremes. In general, nervous systems appear to be more sensitive and to block in function at narrower limits than other tissues. Acclimations can change lethal limits, and other environmental factors can modify the effects of extreme heat or cold.

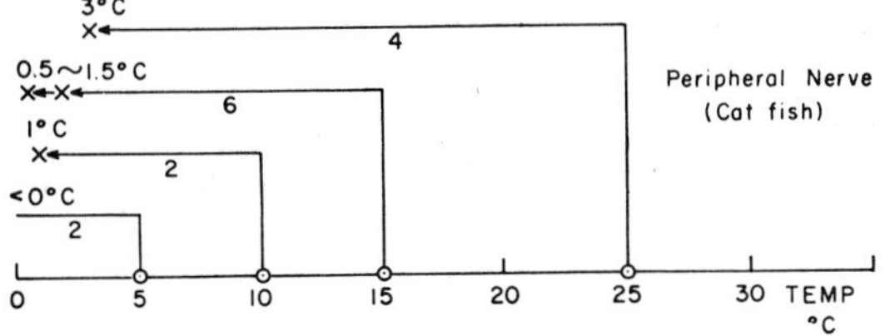

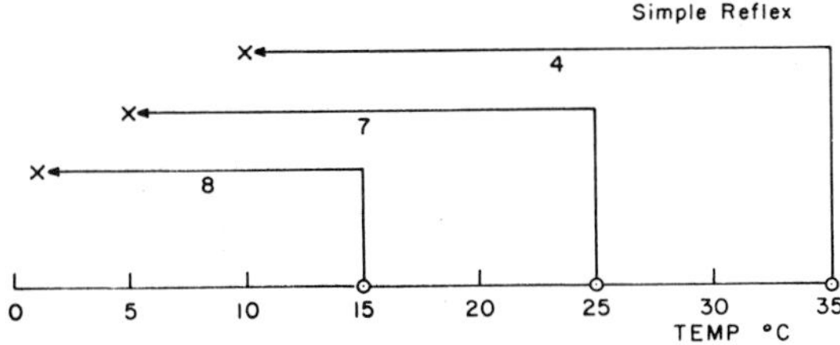

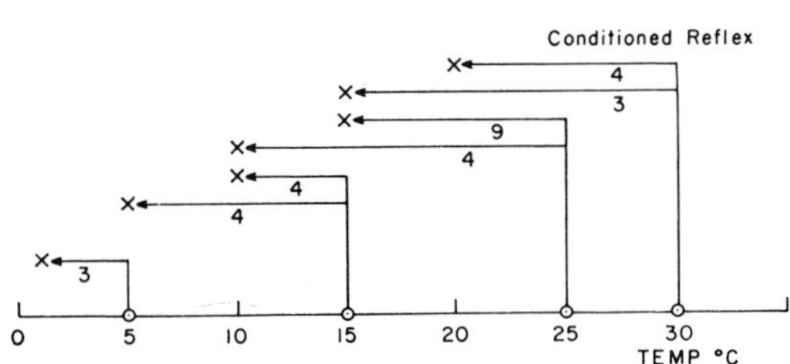

Figure 9–10. Temperature of cold block of nerve conduction, spinal reflexes, swimming, and conditioned reflexes in goldfish acclimated to specified temperatures. (From Prosser, C. L., and T. Nagai, *in* The Central Nervous System and Fish Behavior, edited by D. Ingle, University of Chicago Press, 1968.)

RATE FUNCTIONS

Capacity Adaptations to Temperature. In addition to changes in lethal temperatures, poikilotherms show alterations in various rate functions, behavior, and biochemical activity with respect to temperature. Three general time periods must be distinguished: (1) direct responses of rate functions to change in temperature persisting for hours, which are intrinsic properties of the system; (2) compensatory acclimations during days or weeks of exposure; and (3) selection over many generations of genetic variants adapted to a particular temperature regime. Natural selection may operate on those properties of direct responses which minimize changes in reaction rates; it may also act on capacity for compensatory acclimation.

When the temperature is abruptly raised or lowered, most poikilotherms show an initial overshoot or shock reaction that lasts for seconds or minutes. In *Artemia*, the O_2 consumption overshoots on warming and undershoots on cooling; in goldfish, an initial overshoot is found for either warming or cooling if the temperature change is abrupt. The magnitude of the initial effect varies with the amount of temperature change. Part of the initial metabolic response may be due

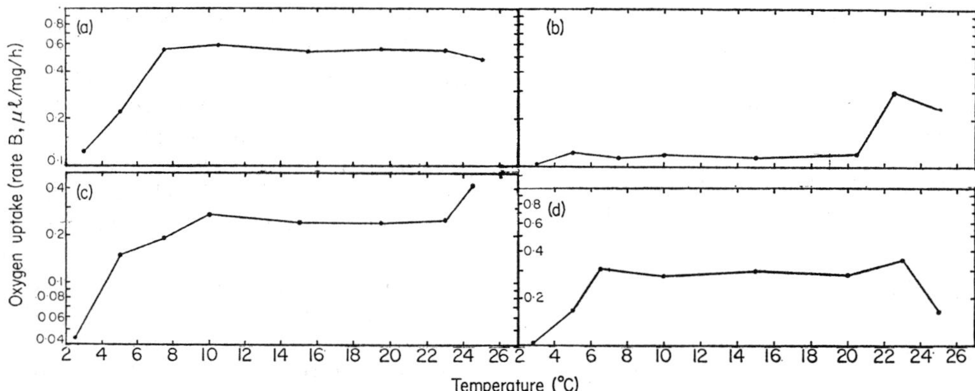

Figure 9–11. Minimal oxygen consumption at different temperatures for (a) sea anemone *Actinia*, (b) crustacean *Nephrops*, (c) snail *Littorina* and clam *Cardium*. (From Newell, R. C., and H. R. Northcraft, J. Zool. (Lond.) *151*:277–298, 1967.)

to increased motor activity resulting from sensory stimulation, but there may also be a general cellular effect, since yeast shows an overshoot.

There follows a stabilized rate which may last for many hours; this is the rate which is usually used for Q_{10} determinations. The stabilized rate depends on the acclimation state, and if the animal is returned to its original temperature during this period the rate returns to the initial level. This period of direct response is characterized by increase or decrease in energy production and in concentrations of metabolic intermediates; there may be shifts of ions, particularly sodium and potassium, between cells and extracellular fluids, and changes in pH. Spontaneous central nervous system activity may increase or decrease.

The standard metabolism of a number of marine invertebrates is relatively independent of temperature over a normal mid-range, but active metabolism shows a high Q_{10}[344, 345] (Fig. 9–11). Oxygen consumption and heart rate of a reptile, *Thamnophis*, from Manitoba (but not Florida specimens) de-

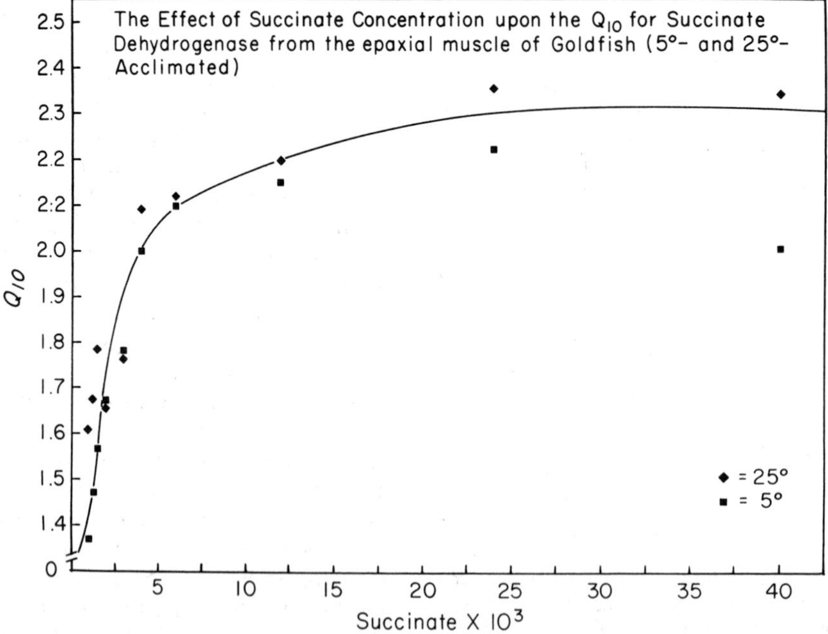

Figure 9–12. Effect of succinate concentration upon Q_{10} for succinate dehydrogenase from muscle of goldfish acclimated to 5° C and to 25° C. (From Hazel, J., Ph.D. thesis, University of Illinois, 1971.)

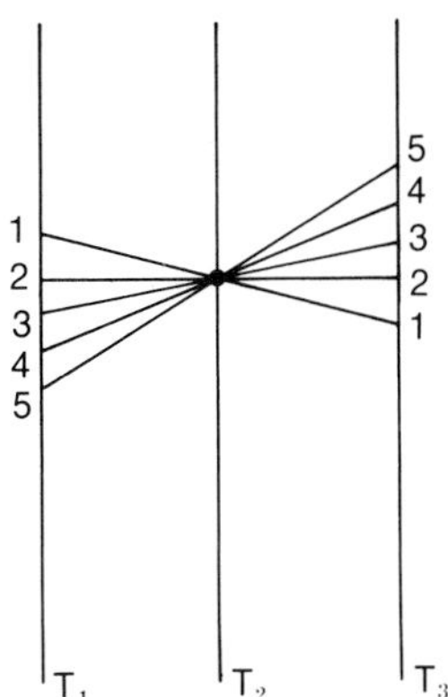

Figure 9–13. Diagram of types of temperature acclimation according to Precht. Animals acclimated at T_2 transferred to lower temperature T_1 or to higher temperature T_3. Acclimation types 1 to 5 are described in the text.

cline as temperature drops from 30° to 20°C, and then increase to a peak at 15° to 20°C, below which they fall rapidly.[4]

Q_{10} values for enzymatic reactions are commonly based on measurements made at saturating concentrations of substrates (i.e., on V_{max}'s). In vivo, substrates are rarely saturating, and therefore measurements near half-saturation are more physiological. The Q_{10} decreases as substrate concentration is reduced; hence, many in vitro measurements of Q_{10} have little physiological meaning. At physiological concentrations of substrate, Q_{10} may approach 1 (Fig. 9–12).

General Patterns of Acclimation. If the animal is kept at the altered temperature for many days, its rate functions show some compensation; i.e., it becomes acclimated. If it is now returned to the original temperature, rate function does not return directly to the original level, but rather to a higher or lower value according to the direction of acclimation. A useful classification of patterns of acclimation is that of Precht[363] (Fig. 9–13). T_2 represents an intermediate temperature from which the animal is transferred to a lower temperature (T_1) or a higher one (T_3). The rate at the new temperature falls or rises in direct response; if, after time for acclimation, no further change occurs, the pattern is Type 4. If in acclimation the rate returns to the original value, compensation is perfect (Type 2). Partial or incomplete acclimation is Type 3; overshoot or excess compensation is Type 1; and inverse or paradoxical acclimation is Type 5.

If, instead of measuring the rate function at the two temperatures of acclimation only, the measurements are made of either stabilized or acclimated rates over the entire temperature range, a more complete description is obtained (Fig. 9–14). When there is no acclimation (Pattern I), the rate-temperature curves coincide for animals from either temperature. Frequently, there is translation of the curve without change in slope, so that the cold-acclimated rate is higher than the warm-acclimated rate at all temperatures (Pattern IIA); inverse translation is Pattern IIB. There may be rotation of the rate-temperature curve about a midpoint (i.e., a change in slope or Q_{10}, Pattern III). Commonly, acclimation is a combination of translation and rotation; if the Q_{10} of the cold-acclimated rate is less than that of the warm-acclimated rate, the change is compensatory (Pattern IVA), while if the Q_{10} of the cold-acclimated rate is more than that of the warm-acclimated rate, the effect is noncompensatory (Pattern IVB). If the curves intersect at a low temperature, a high Q_{10} for cold acclimation is compensatory (Pattern IVC) and a low one is not (Pattern IVD).

A common pattern of acclimation is translation combined with clockwise rotation (Pattern IVA) so that the two rate curves intersect by extrapolation above the normal temperature range, and the Q_{10} is reduced by cold acclimation; examples of this effect are pumping of water by *Mytilus* and the heart beat of a newt.

Translation without change in slope (Pattern IIA) is shown by oxygen consumption in many animals—beetles, carp, *Planaria, Pachygrapsus,* and *Mytilus* gills are some examples. *Chironomus* cultured at 20° and at 30°C show nearly complete compensation, acclimated O_2 consumption being the same at both temperatures.[359]

When temperature is cycled diurnally, the acclimation curve is intermediate between the curves of continuous acclimation at the two temperatures (in the fish *Idus*).[36]

Precht's types of acclimation response are useful for characterizing whether or not temperature compensation occurs for a given function. Prosser's patterns add the dimension of Q_{10}. Each classification has been used for intact animals, tissues, and enzymes from differently acclimated animals, and for comparing populations or species from different latitudes or seasons. Generalizations regard-

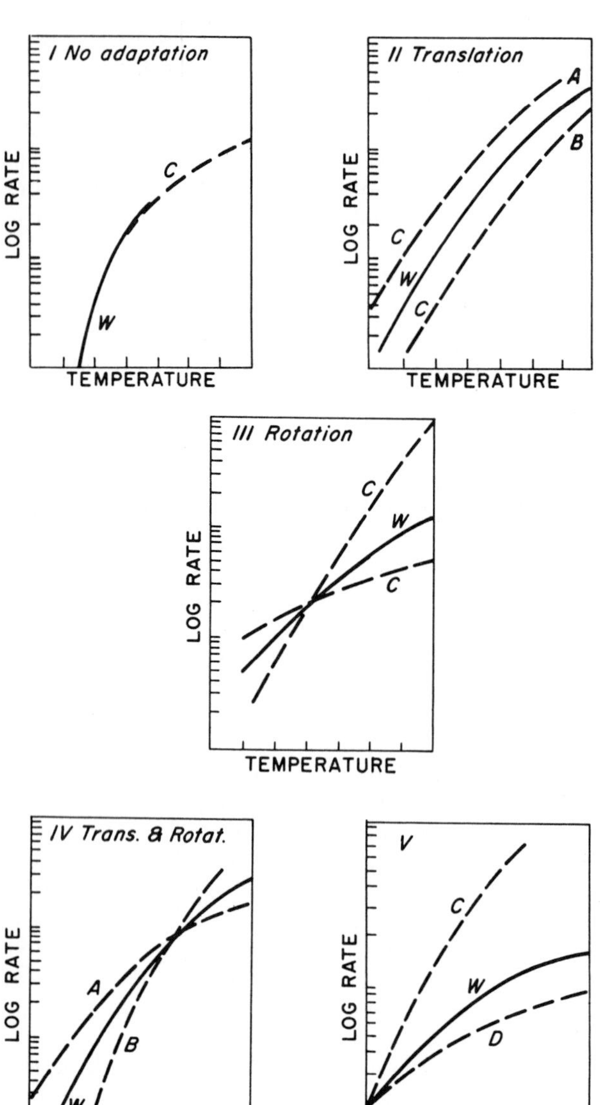

Figure 9–14. Patterns of rate functions measured at different temperatures for animals acclimated to two temperatures. In each diagram (except IV) C = cold acclimation, W = warm acclimation.

ing mechanisms of acclimation cannot be drawn from either type of description alone. However, in combination with correlated data, the patterns of acclimation help in understanding the long-term responses of animals to temperature.

Modifying Factors for Acclimation. The first qualification of an acclimation pattern is the temperature range of measurement. Recent data show that compensatory acclimations may occur over a mid-range of temperature, but not above or below it. The mosquito *Culex* compensates in O_2 consumption well between 15° and 25°C, but not between 25° and 35°C.[56] In a sunfish, the O_2 consumption measured at various temperatures of acclimation compensates (Precht Type 2) in the range from 16° to 17.5°C, but not higher or lower[388] (Fig. 9–15). Thus, for any statement concerning acclimation, the temperature range must be specified.

A related complication is the difference in Q_{10} for acute measurements under different conditions and in different temperature ranges. In enzyme reactions, the Q_{10} decreases as the substrate concentration decreases (p. 382). Also, decreased kinetic activity (collisions) in cold is counteracted by an increase in apparent enzyme-substrate affinity (p. 382); hence, the interpretation is difficult, and the value questionable, of temperature characteristics for complex systems. O_2 consumption, especially rest metabolism, may be relatively independent of temperature; i.e., it may show a low Q_{10} over a limited but normal range. In *Littorina* and some other littoral invertebrates, standard metabolism shows a Q_{10} approaching 1.0 in the mid-range whereas Q_{10} for active metabolism is higher in the same temperature range.[345] Similarly, *Balanus balanoides* has a Q_{10} approaching 1.0 in the range from 10° to 15°C except in January.[18] These results with marine invertebrates are in contrast to those in the fly *Calliphora*, in which metabolism rises in proportion to temperature with greater slope (Q_{10}) for 10°C-acclimated than for 20° or 30°C-acclimated flies.[463]

The pattern of compensatory acclimation varies with season, hormonal state, and nutrition. The standard metabolism of the trout *Salvelinus* at the time of spawning in late fall is twice the level in March and April, and this is reflected in the temperature response.[33] Metabolism of crayfish differs according to stage in the molt cycle; the hepatopancreas shows no temperature compensation in V_{O_2}, but does show increased utilization of hexose monophosphate shunt over the glycolytic pathway in cold-acclimated animals in intermolt but not in premolt.[321] Data on acclimation of digestive tract enzymes are extremely variable, probably because nutritional state has not been controlled. In a parasite, *Schistocephalus*, from stickleback fish, the V_{O_2} curves for winter and for summer intersect at about 35°C, and the Q_{10} is higher for summer animals. Thus, summer animals from all parts of the range acclimated at 15°C resemble winter ones from 10°C in V_{O_2}.[89]

Another reason for variability in patterns of acclimation is that different tissues and functions in the same animal show different patterns. Figure 9–16 shows the different relations of development and oxygen consumption in a grasshopper, *Melanoplus*, to temperature.[382] The trout, *Salmo gairdneri*,

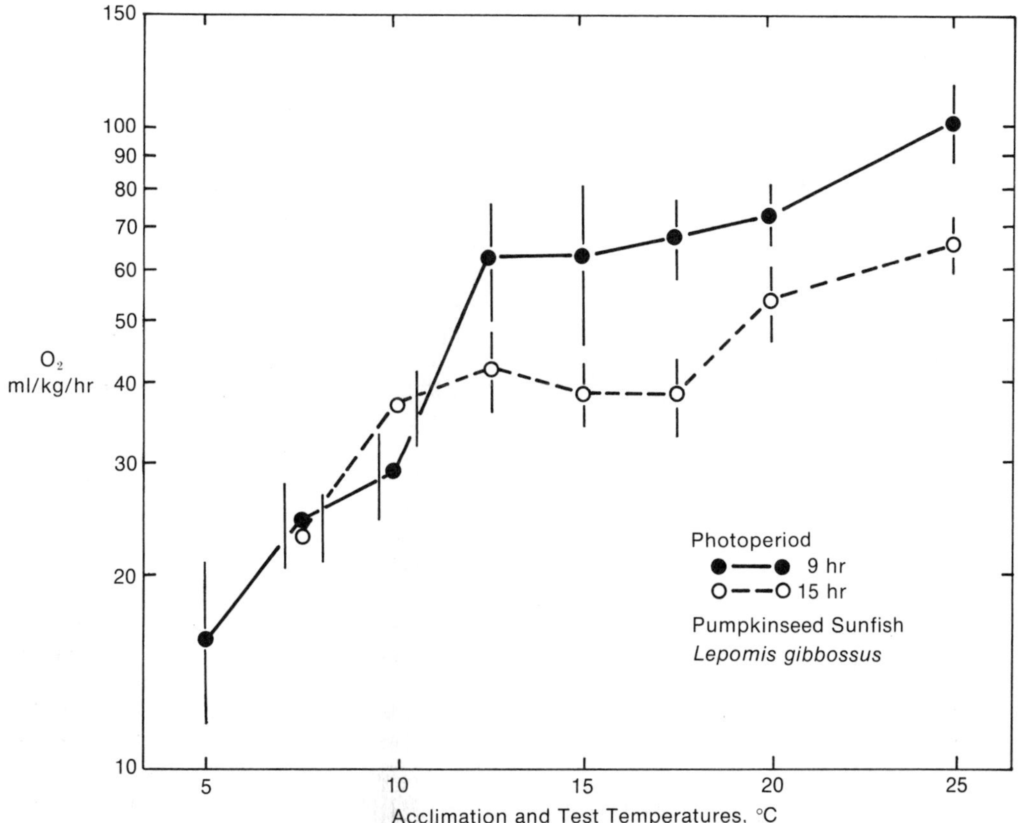

Figure 9–15. Standard metabolism of sunfish measured at temperatures of acclimation; two photoperiods indicated. Nearly complete temperature compensation in mid-range. (Courtesy of J. L. Roberts.)

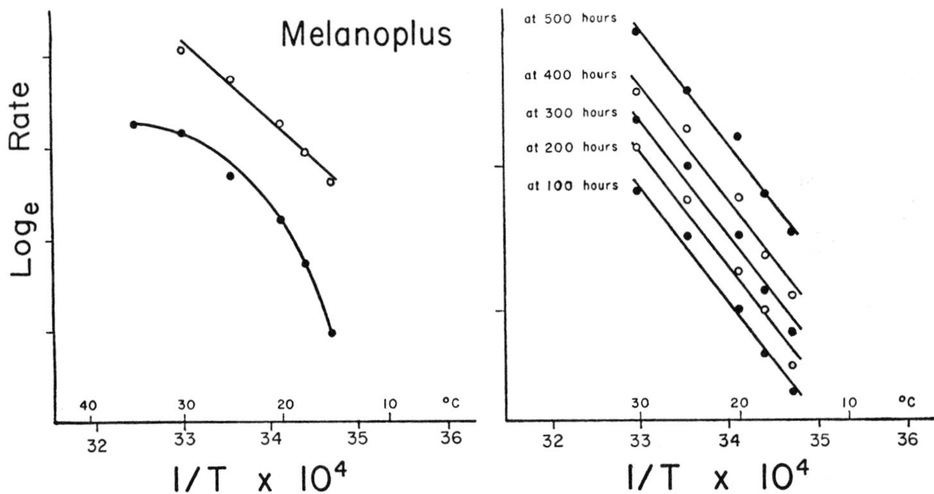

Figure 9–16. Arrhenius plots of oxygen consumption (open circles) and development rates (closed circles) of grasshopper *Melanoplus*. (From Richards, A. G., Physiol. Zool. *37*:199–211, 1964.)

acclimated and measured at 16° and 8°C show the following Precht types for endogenous respiration: whole fish, 3; gill, 4; brain, 2; liver, 1.[122] Eels were placed in tubes with inlets at the two ends and outflow in the middle, and muscle was removed from the two ends for measurement of O_2 consumption. When both head and tail were at the same temperature, the V_{O_2} of head muscle was 3/4 that of tail muscle; but when the head was cooled for a week, the consumption at both ends became equal, and if the tail was cooled the difference increased.[418] The reason for differences between head end and hind end muscle metabolism is not known.

In a lizard, *Sauromalus*, the heart rate increases over the range from 15° to 45°C but arterial pressure is constant from 25° to 45°C and pulse pressure is constant from 10° to 45°C; thus, regulation of the vascular bed is very different from the response of the heart.[455] In snails that were shifted between 20° and 30°C, hepatopancreas and gonads showed rapid and complete acclimation, whereas in foot muscle the compensation was partial.[368] *Uca pugnax* from North Carolina show Precht Type 3 acclimation for whole animal and Type 2 for brain cytochrome oxidase; *U. rapax* shows Type 4 acclimation for both whole animal and brain.[478] Gills show Type 3 acclimation for cytochrome oxidase in *Libinia*, but not in *Callinectes* or in *Ocypode*; muscles of *Libinia* show inverse or Type 5 acclimation.[479, 480] Crayfish heart and gill show some compensation, hepatopancreas shows none, and abdominal muscle shows inverse or paradoxical (Type 5) acclimation.[43, 245] It must be concluded that generalization regarding acclimation response, even of tissues of one animal, is impossible.

Most tests of acclimation have been made with oxygen consumption. Yet the acclimated curves for maximum active metabolism rise to a peak, whereas the curve for rest (standard) metabolism rises progressively[33] (Fig. 9–17), and most measurements of metabolism do not distinguish active from rest metabolism. In *Gammarus oceanicus*, acclimation V_{O_2} is relatively independent of temperature from 5° to 10°C, whereas routine (including active) metabolism rises steeply in this range.[146]

Environmental factors other than temperature may complicate the acclimation curves. Salinity and temperature interact (p. 54). A gammarid from Alaskan ponds shows higher metabolism in summer than in winter, probably as an acclimatory response to reduced O_2 in the winter. Salinity modifies temperature effects on metabolism in a complex manner in the crustaceans *Hemigrapsus*[103] and *Homarus*.[316] Also, salinity and temperature interact on rate of development in fish (gobiids[262]). The crab *Hemigrapsus* in low salinity has higher V_{O_2} after summer acclimatization than in the winter, whereas in high salinity either there is no acclimatization or the winter animals have slightly higher metabolism.[103]

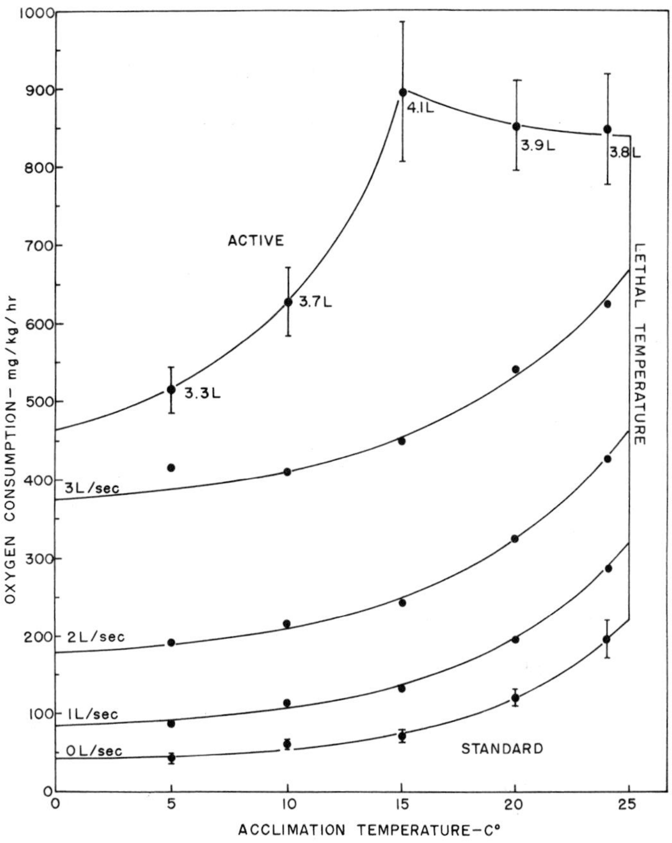

Figure 9–17. Oxygen consumption of young sockeye salmon measured at various acclimation temperatures and at a series of swimming speeds given in body lengths per second. Bottom curve (0 lengths per second), standard metabolism. Top curve (maximum swimming rates), active metabolism. (From Brett, J. R., Amer. Zool. *11*:99–113, 1971.)

In general, animals that become torpid or lethargic in the cold show much less capacity for acclimation than those that remain active over a wide temperature range. Attempts to demonstrate metabolic acclimation in crayfish, frogs, toads, and turtles have yielded equivocal results; the changes that have been reported are small compared with those in active fish. Some fish (such as cunner) show no difference between summer and winter. Cricket frogs (*Acris*) show essentially no metabolic differences at different temperatures according to acclimation.[114] Both a snail (*Doroceras*) and an insect (*Carausius*) show no acclimation in low temperatures, and both are dormant in the winter. Species that become dormant seasonally may show some temperature acclimation during their active phase; failure to take torpidity into account at low temperatures has diminished the value of many acclimation studies. Metabolic acclimation is marked in the shore crab, *Callinectes,* which is active in all but 2 to 3 cold months, whereas the land crab *Ocypode* shows little acclimation capacity and is dormant whenever the temperature drops below 15°C.[480]

Genetic capacity for temperature acclimation may be greater in animals from environments where temperature fluctuations (annual and diurnal) are considerable, as in temperate zones, than in related animals from areas where the temperature regime is relatively constant (tropics and polar regions). An actinian, *Metridium,* from Woods Hole shows Precht Type 3 acclimation of O_2 consumption, whereas the actinians *Haliplanella* and *Diadumene* from Virginia show no such acclimation.[401] Enzymes from Antarctic fish show much less capacity for acclimation than do corresponding enzymes from trout.[435] A polychaete from the North Atlantic shows a Q_{10} of about 2 over a wide temperature range; southern populations have a high Q_{10} at temperature extremes and a low value in the mid-range.[305] Populations of a gobiid fish from a low latitude show low Q_{10} over a wide range; populations from high latitudes show a low Q_{10} in high temperatures and a high Q_{10} in low temperatures.[16] *Uca* is a

TABLE 9–1. Enzymes Showing Compensation (Types 2 or 3).[171] M = Muscle; L = Liver; K = Kidney; G = Gill; B = Brain; I = Intestine

		Reference
Glycolytic enzymes		
phosphofructokinase	goldfish M	129
aldolase	crucian carp G	120
aldolase	golden orfe M,L,G	231a
lactic dehydrogenase	goldfish M,L	201
	brook trout M,L	201
	lake trout	201
Hexose monophosphate shunt enzymes		
6-phosphogluconate DH	crucian carp G	120
TCA cycle and electron transport enzymes		
succinic DH	*Rhodeus* L	273
	goldfish M	172
	golden orfe M	231a
	earthworm M	400
malic DH	golden orfe M,L,G	231a
cytochrome oxidase	goldfish M	128
	goldfish M,L,G,B	69
	fiddler crab M	479
	golden orfe M,L,G	231a
succinate-cytochrome C reductase	goldfish G	69
NAD-cytochrome C reductase	goldfish G	69
Synthetic enzymes		
aminoacyl transferase	toadfish L	167
Miscellaneous enzymes		
Na-K ATPase	goldfish I	430
	goldfish G	
protease	*Tenebrio* I	
	Helix I	

widely distributed crustacean with marked species and latitudinal differences. *Uca pugnax* from North Carolina shows Type 3 (partial) temperature compensation, while *U. rapax* shows no compensation when laboratory acclimated; populations of *U. pugnax* from Florida and Jamaica also fail to show acclimation,[478] and Brazilian specimens show very little acclimation.

Some enzymes from metabolic acclimations show good compensation, while others do not. A survey of enzymes identified as showing partial or complete acclimation (Precht Types 2 or 3) is given in Table 9–1.[171] All of these are enzymes associated with energy liberation (enzymes of glycolysis, pentose shunt, TCA cycle, electron transport, fatty acid oxidation). In addition, protein synthesis is greater in cold-acclimation than in warm-acclimation. Aminoacyl transferase of toadfish liver shows 60% more activity in 10°-acclimated than in 21°-acclimated toadfish.[167, 168, 169] The net effect of these compensations is that acclimating animals can maintain relative constancy of activity (elevated in cold and reduced in warm).

Table 9–2 lists enzymes that show either no acclimation or inverse acclimation. These are mostly enzymes that have to do with degradation of metabolic products, such as peroxidases, catalase, acid phosphatase, D-amino acid oxidase, acetylcholinesterase, allantoicases, and allantoinase.

Since no energy-yielding system gives perfect (Type 2) acclimation, the concentrations of products may be higher in warm and lower in cold; hence, it would be non-adaptive if enzymes of degradation were reduced at high temperatures or enhanced in the cold. The situation for digestive enzymes is unclear: some show compensation, and others do not. The variation probably depends on the nutritional state of the animal.

The situation for transport enzymes is mixed. Active transport of glucose by goldfish intestine is coupled with Na transport (see Chapter 2). An ouabain-sensitive Na-K ATPase of the intestine is more active in cold-acclimated than in warm-acclimated fish, and the affinity for Na is less in fish acclimated at 30°C than in those acclimated at 8°C. Mg ATPase shows an acclimation effect opposite to that of the Na-K ATPase.[427, 428, 430, 431] Transfer of amino acids,

TABLE 9–2. Enzymes Showing Reverse or No Compensation (Types 4 and 5). Abbreviations as in Table 9–1.

		Reference
Peroxisomal and lysosomal enzymes		
catalase	crayfish M	
	eel M	367
	carp M,G	120
	goldfish K	171
peroxidase	goldfish K	171
	yeast	363
acid phosphatase	goldfish K	171
D-amino acid oxidase	goldfish K	171
Miscellaneous enzymes		
Mg ATPase	goldfish G,B	
	goldfish I	431
choline acetyl transferase	goldfish B	182
acetylcholine esterase	goldfish B	171
	killifish B	171
alkaline phosphatase	lizards	296
allantoinase	carp L	476
allantoicase	carp L	476
uricase	carp L	476
amylase	*Tenebrio* I	
lipase	hibernating snail I	
malic enzyme	frog L	15
G-6-P DH	carp G	120
	frog unfed L	15

especially the non-lipophilic ones, is three times as fast in 8°C fish as in 30°C fish. Likewise, the Na-K ATPase of goldfish gills shows temperature acclimation.

In summary, acclimation patterns may vary according to temperature range in that compensation may occur in the mid-range but not at extreme temperatures; they may be modified according to season, reproductive and hormonal state, and nutrition. Metabolic responses differ according to whether resting (standard) or activity (and routine) metabolism is measured. The pattern of acclimation need not be the same for all functions or tissues of a given animal, and it may be modified by other strong environmental parameters such as oxygen availability and salinity. In general, animals that enter a torpid state in winter show much less acclimation than those that maintain activity. Some species that live in relatively unvarying environments may be less capable of compensatory acclimation than species that are subject to wide variations. Enzymes that function in energy liberation tend to show more compensation than those that degrade metabolic intermediates and products. Both compensatory acclimation (Types 1, 2, and 3) and no or inverse acclimation (Types 4 and 5) are adaptive for specific physiology or ecology.

Mechanisms of Enzyme Acclimations. Capacity adaptation is shown for a variety of enzymes removed from animals that are genetically fitted for specific temperatures, as well as for enzymes from animals that are acclimated to cold or to warm. Enzyme effects are measured in terms of activity of an enzyme extract or of a metabolic pathway, by means of a temperature coefficient and a kinetic measure such as K_m. Molecular mechanisms of compensation may consist of changes in synthesis or amount of a given enzyme, differences in kinetics, changes in the proportion of the isozymes suitable for particular temperatures, and changes in cofactors such as lipids, coenzymes, or other factors such as pH and ions.

Changes in primary structure—amino acid sequence—are unlikely during acclimation, since these are genetically coded and the processes of acclimation occur during short periods in one individual. However, conformational changes, as in quaternary structure, could result from some of the activity modifying factors mentioned below.

Protein synthesis, as measured by the incorporation of labeled amino acids, is enhanced during cold acclimation. Incorporation of radio-leucine into proteins is greater in liver, muscle, and gill of cold-acclimated goldfish.[88] When synthesis is measured at the same temperature for both cold- and warm-acclimated fish, perfused toadfish liver shows protein synthesis of 29 μg/g/min for 21°C-acclimated fish and 46 μg/g/min for 10°C-acclimated fish. Aminoacyl transferase of toadfish liver microsomes shows 60% more activity in 10°C-acclimated fish than in 21°C-acclimated fish.[167] RNA turnover is enhanced by cold in earthworms and fish.[87] It is possible that one of the first steps of biochemical adaptation is stimulation of the transfer enzymes of RNA associated with protein synthesis. Measurements by immunological methods show selective synthesis of specific proteins (cytochrome oxidase) in cold acclimation.[495a]

In multi-enzyme systems there is evidence for a shift in balance between relative pathways that are coordinated through metabolite-induced regulation of their component enzymes. The O_2 consumption by gills from cold-acclimated goldfish is greater, and shows greater sensitivity to cyanide and less sensitivity to iodoacetate, than the metabolism of warm-acclimated fish gills.[119] In *Streptococcus cremoris*, growth at high temperatures favors production of more acetoin and less butylene glycol; low-temperature cultures tested at the same temperature produce more lactic acid than do high-temperature cultures. The Antarctic fish *Trematomus* from −1.9°C water was acclimated at −2° and at +4°C; inhibition by IOA (indicating glycolysis) was greater for 4°-acclimated fish than for −2°-acclimated ones, whereas CN sensitivity was greater for the −2°-acclimated fish.[435] Trout (*Salvelinus fontinalis*) were acclimated at 15°C and at 4°C; the warm-acclimated tissues showed relatively more incorporation of $^{14}CO_2$ into positions 1, 2, 5, and 6 of glucose, and in utilization of glucose the ratio C_6/C_1 was higher in the warm. The hexose monophosphate shunt was three times more active in cold than in warm, glycolysis was reduced in cold, and fat synthesis was greater in the cold (i.e., more acetate entered fat in the cold).[203] Gluconeogenesis was favored in cold-acclimated fish.[34] Lipid metabolism is greater in cold-acclimated than in warm-acclimated *Salmo*.[101]

Enzymes that must function at low temperatures in nature are differently adapted

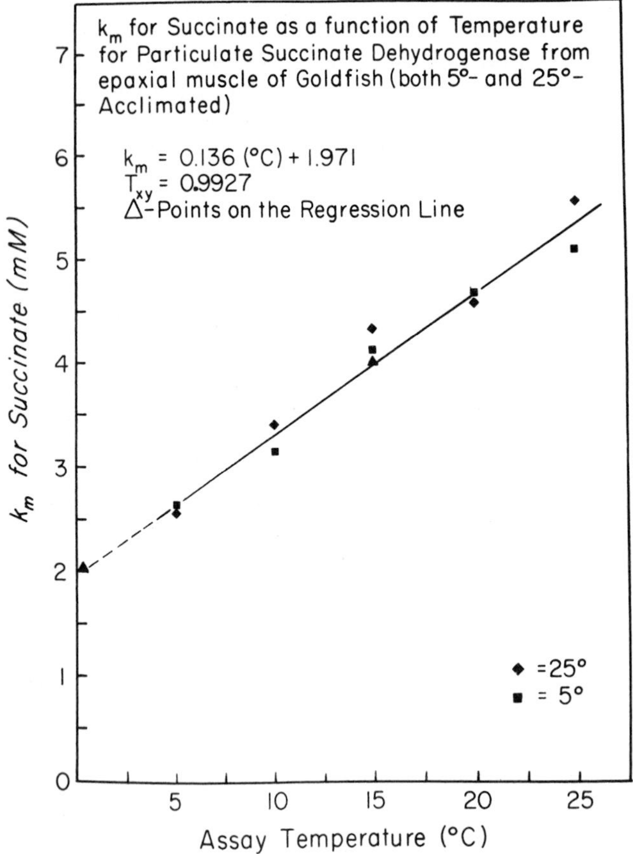

Figure 9–18. K_m of succinate dehydrogenase of muscle from goldfish acclimated to 5° C and to 25° C, and measured at various temperatures. (From Hazel, J., Ph.D. thesis, University of Illinois, 1971.)

from those that must function at high temperatures. In a few (mostly isozymal) enzymes the plot of K_m versus temperature goes through a minimum (maximum substrate affinity), and this point correlates with the temperature of acclimation[14] (see p. 375). For most enzymes, the K_m decreases continuously as temperature falls; this compensates for the normal Q_{10} effect on kinetic activity (Fig. 9–18). Thus, over certain ranges the Q_{10} may approach 1.0; this effect may explain relative independence of temperature in standard metabolism of some animals in their normal temperature ranges. Most enzymatic reactions in cells probably occur at much less than substrate saturation. Hence, shifts in the shape or position of the velocity-substrate curve with temperature acclimation may permit function, at an altered temperature, of an enzyme at cellular concentrations of substrate, whereas function at maximal concentrations might not be possible.[202, 436] In evolution, selection may be more for enzyme affinities than for molecular activity, as shown by turnover numbers.[433, 436]

One mechanism of enzyme acclimation may be selective synthesis of isozymes which function best (i.e., have lowest K_m's) in a particular temperature range.[205, 436] In brain of rainbow trout, acetylcholinesterase occurs in two isozymes, one after acclimation at 17°C and the other after acclimation at 2°C; both are present after acclimation at 12°C. The K_m values of the enzyme variants are correlated with the temperature of synthesis[14] (Fig. 9–19). Similarly, pyruvate kinase of king crab muscle shows two forms, one functioning below 10°C and having a minimum K_m at 5°C, and the other functioning above 9°C with a minimum K_m at 12°C.[434] Citrate synthase from trout liver occurs in two forms, cold and warm; the K_m's are similar below 15°C, but at higher temperatures the K_m of the cold form increases (i.e., its affinity for substrate decreases) much more than that of the warm form[205] (Fig. 9–20). In a sloth, the LDH from forearm muscle shows a 16-fold increase in K_m from 20° to 42°C, but the enzyme from heart muscle is less sensitive to temperature.[204]

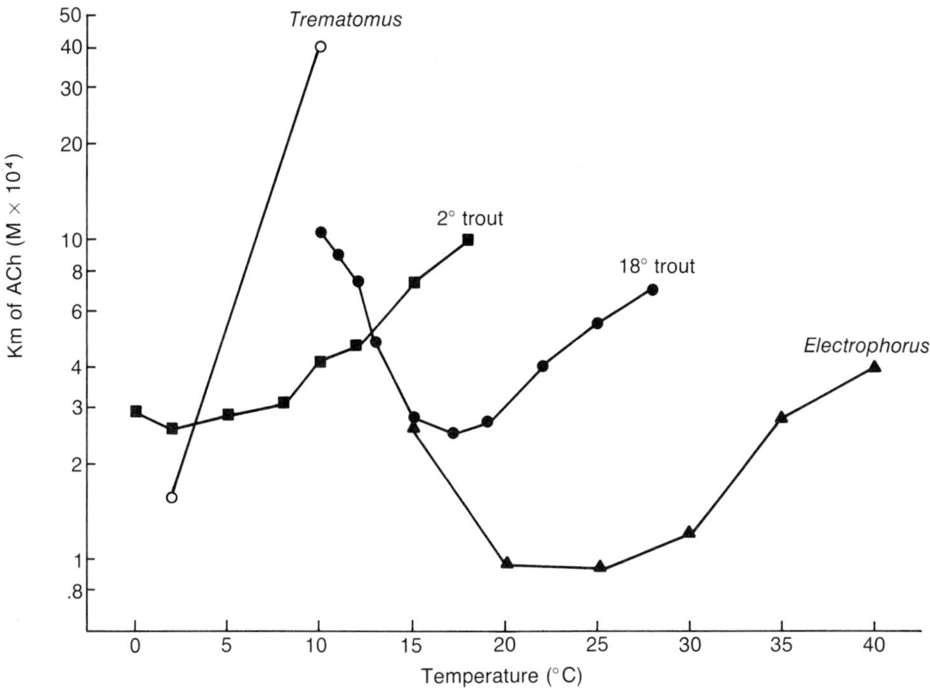

Figure 9–19. K_m as a function of temperature for brain acetylcholinesterase from Antarctic *Trematomus*, from trout acclimated to 2°C and to 18°C, and from Amazon *Electrophorus*. (From Baldwin, J., Ph.D. thesis, University of British Columbia, 1970.)

Lactate dehydrogenase (LDH) occurs in several genetically determined isozymes in fish. The proportion of these in a tissue does not change during temperature acclimation (goldfish, green sunfish[495a]) but geographic selection does occur. Populations of the killifish *Fundulus heteroclitus* show variation with latitude with respect to the two forms of liver LDH in the proportion of homozygotes to heterozygotes. Kinetic studies show that one form (B′B′) is at an advantage in warm waters and the other (B B) below 12.2°C.[362a] Thus, geographic polymorphism in respect to LDH isozymes has a kinetic interpretation.

Another important mechanism of acclimation is change in lipids. In microorganisms, plants, and animals, the lipids that are deposited at low temperatures tend to have more unsaturated fatty acids, higher iodine numbers, and hence lower melting points than those deposited at high temperatures. This has been shown for fly larvae, frog adipose and liver tissue,[15] goldfish brain[392] and other tissues,[265] and planktonic copepods. *Chlorella* reared at 38°C has 46% of its lipids composed of saturated, short-chain fatty acids; at 22°C most lipids are triunsaturated.[353]

In bacteria (*Bacterium megaterium*), transfer from 30°C to 20°C induces synthesis of an enzyme that desaturates palmitic acid; this enzyme is not present significantly at 30°C, and is synthesized in greater amounts after transfer from 30° to 20° than in cultures maintained at 20°C.[134]

Turnover of total phospholipids, as measured by incorporation of labelled acetate and phosphorus, is enhanced in cold acclimation.[6, 265] There are more unsaturated fatty acids in phosphatidyl choline and phosphatidyl ethanolamine in plasmalogens of brain in cold-acclimated than in warm-acclimated goldfish,[391, 392] and proportionately more phosphatidyl choline and phosphatidyl inositol in the phospholipids of gills from 5°-acclimated than from 25°-acclimated goldfish.[6] In goldfish phospholipids, warm acclimation halved the percentage of $C_{20:1}$, $C_{20:4}$, and $C_{22:6}$ fatty acids and nearly doubled the percentage of $C_{18:0}$ and $C_{20:3}$ fatty acids.[254] Phospholipids extracted from mitochondria of cold-acclimated goldfish activate succinic dehydrogenase from either cold- or warm-acclimated fish about 50% more than do the phospholipids from warm-acclimated fish (Fig. 9–21). The proteins appear to be the

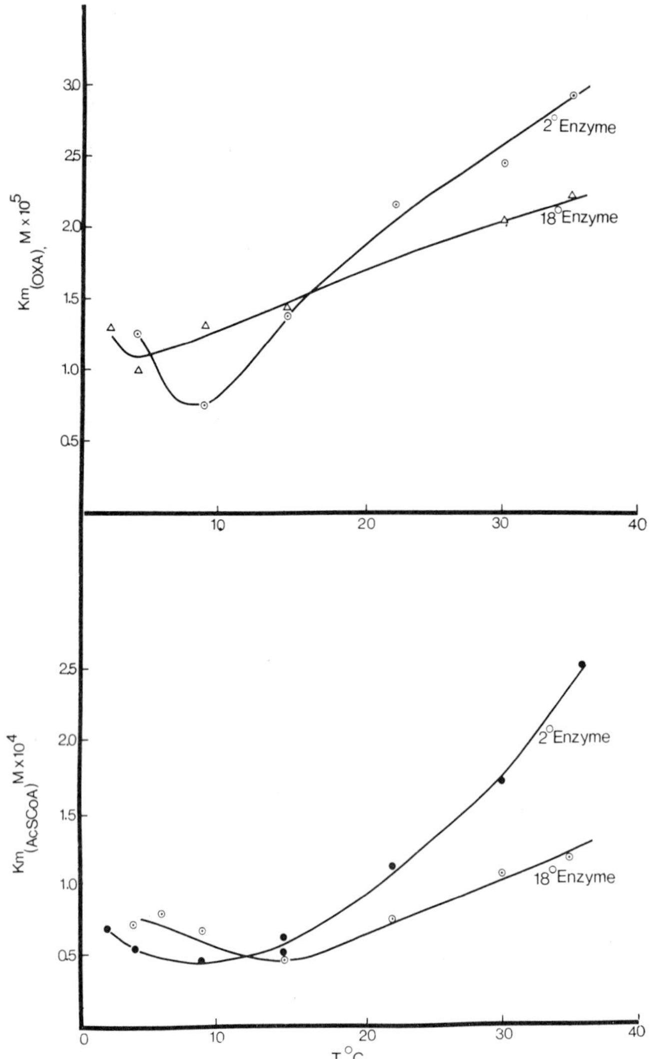

Figure 9–20. K_m of citrate synthase from 2° C-acclimated and 18° C-acclimated trout at different assay temperatures. Upper figure, enzyme assayed with oxalacetate; lower figure, enzyme assayed with acetyl CoA. (From Hochachka, P., and J. K. Lewis, J. Biol. Chem. *245*:6567–6573, 1970.)

same from the two temperatures, but the compensatory acclimation in activity can be explained by differences in the lipid cofactors.[172]

Low temperature decreases lipid fluidity, and unsaturation tends to counteract this physical change. Hence, the variation in composition of fatty acids, particularly in phospholipids, may preserve a constant physical state of both cell and mitochondrial membranes. In artificial membranes, a thermotropic (temperature induced) transition of phospholipids (e.g., lecithin) from crystalline to a liquid-crystal state, as well as a lyotropic (state-dependent) transition from lamellar to hexagonal phase, relate to percentage of unsaturation and kind of phospholipid. Lecithin forms membranes at 36° but not at 20°C.[38, 79] In *E. coli*, glucose and galactose are actively taken up, and transport of sugars measured at different temperatures shows two different slopes on an Arrhenius plot, reflecting two different physical states of the membrane. When the cells are reared with oleic acid in the medium, the Arrhenius plot of sugar transport (Fig. 9–22) shows a break at 13°C. When linoleic acid is used instead, the break comes at 7°C, thus providing evidence for membrane states that vary with fatty acid composition.[496] Thus, it is concluded that the maintenance of a fluid state of cellular and subcellular membranes by appropriate fatty acid composition is an important aspect of temperature acclimation.

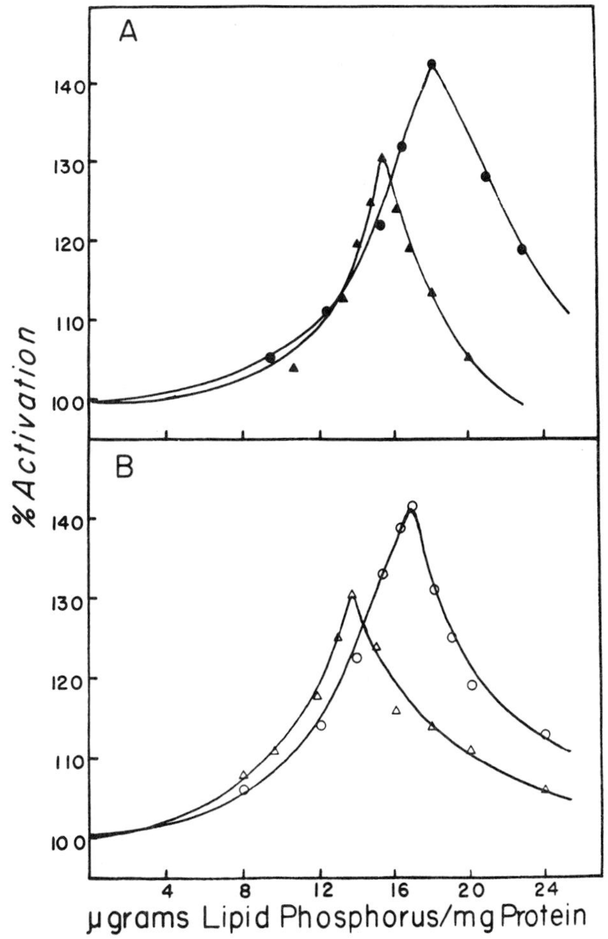

Figure 9–21. Activation of lipid-free succinate dehydrogenase from muscles of goldfish acclimated at 5°C (A) and at 25°C (B); phospholipid from mitochondria from 5°C fish shown as circles and that from 25°C fish as triangles. (From Hazel, J., Comp. Biochem. Physiol. *43*:837–882, 1972.)

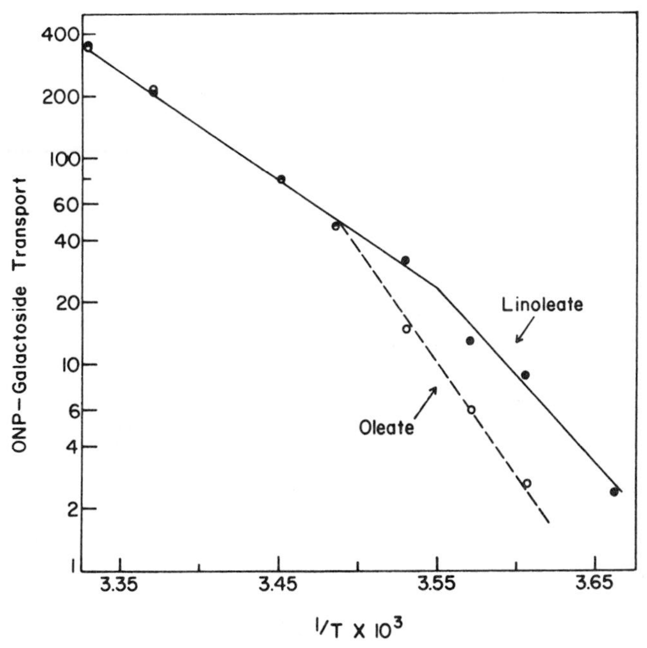

Figure 9–22. Temperature dependence of galactoside transport by *E. coli* reared with oleate or with linoleate in medium. (From Wilson, G., S. P. Rose, and C. F. Fox, Biochem. Biophys. Res. Comm. *38*:617–623, 1970.)

Another possible mechanism of temperature acclimation is related to the control of enzymes by modulating substances, which change in concentration as a part of the direct response to temperature. Also, changes in concentrations of intermediates could repress or induce synthesis of enzymes. Formation of metabolic intermediates, such as phosphorylated sugars and nucleotides, is enhanced directly in warm and diminished in cold.[129] The pH of blood, and probably of tissues, in frogs, turtles, and some fish decreases as the temperature rises, so that $\Delta pH/\Delta T = -0.016$; the net effect is to keep the ratio OH/H constant.[211] Some critical enzymes, such as PFK,[129] have narrow pH ranges. The K_m of pyruvate for LDH of trout and tuna increases sharply with temperature at pH 7.5, but hardly at all at pH 7.0.[204, 205] There are reports of small changes in concentrations of blood ions; Mg and Ca increase, and Na and Cl decrease, in warm fish.[375] However, these changes are probably not sufficient to modify tissue metabolism.

An enzyme that is sensitive to a variety of modulators—metabolic intermediates, ions, pH, and so forth—can be markedly altered in its activity. For example, phosphofructokinase of goldfish is modulated by F6P, ATP, AMP, Mg, and citrate, and is very sensitive to pH. The relation of direct changes in concentrations of cellular constituents to long-term acclimations of enzyme systems is not known.

In conclusion, the molecular mechanisms of thermal acclimation are not fully understood. They comprise stimulation of protein and phospholipid synthesis or turnover, selective synthesis of adapted isozymes, modulation by membrane lipids (which are more unsaturated in cold), possible effects of co-factors, and such ions as (H^+). The mechanisms of thermal acclimation are multiple, and this is not surprising in view of the all-pervasive character of temperature as an environmental parameter.

Direct or Indirect Cellular Action. The manner in which biochemical changes are initiated in acclimation (i.e., whether by direct action of temperature at the cellular level or by nervous or hormonal stimulation) is not clear. Changes in enzyme activity in microorganisms in the absence of genetic selection, and in isolated molluscan gills and tentacles, must be direct.[368] Photoperiod modifies temperature acclimation in crucian carp and sunfish, with better acclimation in a 9-hour photoperiod than in a 15-hour one.[388] Goldfish on a long photoperiod are more resistant to sudden heat, and on a short photoperiod they are more resistant to chilling; this suggests hormonal control. There was no effect of hypophysectomy on metabolic acclimation in a minnow,[159] and no effect of thyroidectomy or of thyroid inactivation on metabolic acclimation in goldfish.[263] Treatment with thiourea did not alter upper lethal temperatures of white chad.[81] Thyroxine increases cold resistance but does not affect heat resistance in *Xiphophorus*;[457] mammalian thyroid extract (but not goldfish thyroid) slightly increased cold resistance in goldfish.[200] It must be concluded that the evidence is very contradictory or negative for hormonal control of temperature acclimation in fish. Seasonal differences in readiness for acclimation may reflect annual hormonal cycles, and may make many hormonal studies difficult to interpret.

BEHAVIORAL AND LOCOMOTOR ADAPTATIONS

In addition to metabolic acclimations which favor survival at temperature extremes, and which permit capacity for normal functioning of poikilotherms over a wide range of temperatures, most animals show adaptive behavior. Hence, the nervous system plays a key role in both resistance and capacity adaptation.

Gradient Selection and Kineses. A simple adaptive response consists of movement in a thermal gradient toward some "selected" or "preferred" temperature. Paramecia aggregate at a mid-temperature in a chamber that is cold at one end and warm at the other; the "preferred" temperature may be that of minimal kineses.[127] Free-living nematodes move to an optimum mid-temperature, whereas nematodes parasitic on warm-blooded animals move to the warm end of the gradient even if they suffer thermal damage there.[314] Snails (*Limax*) go to a preferred temperature that is slightly higher than the acclimation temperature, but which varies with acclimation.[387]

Herring show selective aggregation in water temperature gradients of 0.5°C and may be able to detect smaller gradients. The temperature selected by the fish *Girella* rose from 18° to 24.3°C during acclimation from 10° to 30°C.[347] A similar rise in selected temperature with higher acclimation temperature was found in sockeye salmon and carp. What determines the selection temperature is not clear; it could be the temperature at which ion pumps maintain critical membrane potentials. Goldfish were trained to press a lever at one-minute intervals to get food; the time sense was not altered by a change in temperature.[394]

Insects also aggregate at some point in a thermal gradient; for bees, ants, locusts, flour beetles, and ticks, aggregation depends on acclimation temperature. Ants acclimated between 3° and 5°C aggregated at 23.5°C, while those acclimated between 25° and 27°C aggregated at 32°C. An Alaskan carabid beetle in summer aggregates at 7° to 9°C; in winter it goes to the extreme cold end (−25°C) even where it freezes.[326] The selected temperature may be relatively sharp or may cover a broad range. Selected temperature may vary according to season, diet, and stage in the life cycle. Dung maggots while actively feeding select a temperature of 30° to 37°C, which is within the range of their medium; when they are ready to enter the ground for pupation, they aggregate at 15°C in a gradient. A feeding housefly larva has a thermoselection range of 15° to 33°C; for prepupating larvae the range is 8° to 20°C. Bloodsucking insects—mosquitoes, *Rhodnius*—are attracted to their prey partly by odors but also by temperature. Selection by insects depends more on air temperature than on substrate temperature.

Related to gradient behavior are general activity, kineses, and maximum sustained swimming speed in fish. Spontaneous movements of trout show two temperatures of maximum activity: one at the temperature of selection, and the other just below the high lethal temperature. The maximum cruising speed of sockeye salmon from open lakes occurs at 15.5°C, while that of coho salmon from warmer streams occurs at 20°C. The temperature of maximum cruising speed of goldfish corresponds to the selected temperature.[133] The maximum distance moved by a salmon or trout in response to a shock increases to a peak at the selected temperature, which varies with acclimation.

When the rate of oxygen consumption at rest (standard metabolism) at different acclimation (and measurement) temperatures is subtracted from the O_2 consumption at maximum activity (active metabolism), the difference gives the cost of activity. For lake trout, the difference rises to a peak near the temperature of maximum steady swimming. Cruising speed may be limited by available energy; however, electrical stimulation can elicit higher rates of O_2 consumption than does maximal swimming. Also, swimming speed and temperature selection can be altered by brain lesions. Possibly oxygen utilization, rather than determining the speed, reflects the demand created by activity. In fingerling sockeye salmon, the rate of digestion increases with rising temperature to above 25°C, but the amount of food eaten is maximal at 15°C; hence, growth is determined more by behavior than by digestion.[48] Acclimation may not only change metabolism but also better permit neurally controlled behavior at particular temperatures. A Manitoba toad burrows to increasing depths as winter advances, always staying below the freezing level in the soil.[456]

Metabolic Warming. Another behavioral adaptation provides for heat production by active muscles and for heat absorption from sun rays or from substratum. This permits limited regulation of body temperature by ectothermy.

Absorption of heat from solar radiation is important for terrestrial insects and reptiles, and may be correlated with pigmentation. A light-colored snout beetle (*Compus niveus*) absorbs 26% and a dark carrion beetle (*Silpha obscura*) absorbs 95% of the incident infrared radiation. Desert locusts (*Schistocerca*) are very inactive below 17°C and begin to move at between 17° and 20°C; at this temperature in the morning, they take a position on an eastern slope with their bodies oriented perpendicularly to the sun's rays. At 28°C they start to migrate and at about 40°C may rest on a bush, parallel with the sun's rays, thus receiving minimal radiation. In the late afternoon, when the temperature falls, the locusts again assume a position perpendicular to the sun's rays. Certain butterflies bask in the sun, and absorb radiation by their wings; at high temperatures they seek shade and increase evaporative cooling. Other butterflies show warm-up movements before flight, especially at low ambient temperatures. *Vanessa* warmed up for more than six minutes at 11°C, for 1½ minutes at 23°C, for 18 seconds at 34°C, and not at all at 37°C. *Danaus* raises its thoracic temperature by 1.3°C/min to levels 5° to 8°C higher than the ambient temperature.[247]

Moths are nocturnal and many, such as the

sphingids, have furry insulation and are endothermic, heat-producing. The muscles that move the wings undergo shivering in an uncoordinated fashion, and flight is initiated at a thoracic temperature of 32°C or higher. If the thoracic ganglia are warmed by a thermode, flight can occur without muscle warm-up. The thorax warms at 4°C/min, and T_b is 32° to 36°C at T_a of 17° to 29°C. Preflight warm-up increases oxygen consumption by some 2.3 times, and fat is the principal fuel used.[1, 156] A hawkmoth warms at 4.2°C/min, and flight temperature for the individual varies from 34° to 45°C.[111] A desert moth *Manduca* does not initiate flight until the thorax temperature rises to 38°C; it maintains thoracic temperature at 42°C in flight at ambient temperatures from 17° to 32°C. Heat produced in the thorax is transferred by the dorsal blood vessel to the abdomen, and if this vessel is occluded the moth overheats and stops flying.[184]

A cecropia moth (*Hyalophora*) cannot fly if T_b is below 32°C, and in flight it maintains T_b at 37°C. Behavioral regulation of temperature continues when connections between the heart and thorax are severed, but interference with the connections within the thoracic ganglia disturbs thermoregulatory set points.[313] Local heating of the thoracic nerve cord causes a cecropia moth to switch from heat-generating to flight-motor patterns; hence, the temperature of the nerve cord is more important than that of the muscles, even though the latter normally produce the heat.[156] The behavior triggered in cicadas at specific temperatures is shown in the following table:[177]

	Magicicada	*Hyalophora*
Heat death	45°C	46°C
Heat torpor	43	44
Maximum continuous flight	37	38
Chorus singing	30–34	–
Minimum for flight	22	34
Cold torpor	14	15
Cold death	0	0

A beetle, *Acillius*, shows preflight warming movements, producing sounds with frequencies up to 140/sec.[294] A katydid, *Neoconocephalus*, starts singing at 33.5°C T_b by chirps at 150 to 200/sec, and may warm during the singing to maintain T_b at 34.4°C, or a gradient of some 15° between T_b and T_a.[178]

Colonial Hymenoptera and termites make use of metabolic heat. Dancing honeybees may have T_b at 1°C above T_a, with the thorax at 36°C.[121] The temperature optimum for brood development of the ant *Formica rufa* is between 23° and 29°C; for honeybees it is between 34.5° and 35°C. Brood nest temperature is fairly well regulated. In warm weather, with T_a as high as 48°C, the brood temperature in a hive is kept below 37.6°C by evaporative cooling; ventilatory activity is proportional to the nest temperature.[293] Workers transport and distribute water and increase evaporation by fanning. In winter the bees cluster, and the cluster expands or contracts according to temperature. Sugar utilization (and metabolism) increases as temperature decreases.[421] If the air tempera-

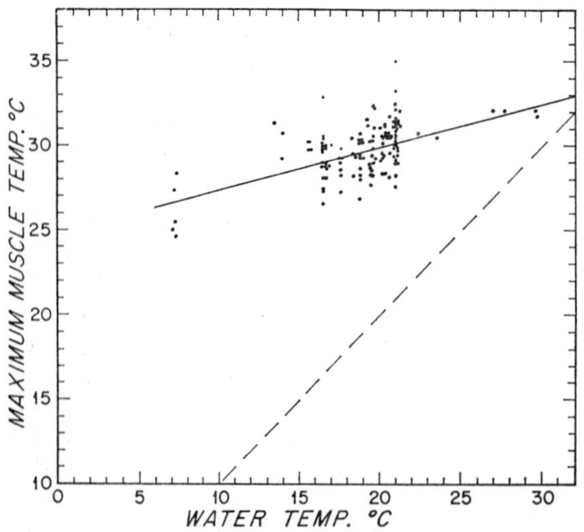

Figure 9–23. Maximal muscle temperatures of bluefin tuna caught at various water temperatures. (From Carey, F. G., and J. M. Teal, Comp. Biochem. Physiol. *28*:205–213, 1969.)

ture falls to 8° to 10°C, the bees show uneasiness; outer bees are more active than those in the interior of the cluster, and outer and inner bees often change places. Once chill coma is reached at about 7°C T_b, there is no movement, and death results if cold coma persists for more than 50 hours.[121] In an occupied termite mound the temperature may be 14° to 18°C higher than in an unoccupied mound.

A striking example of heterothermy associated with motor performance and heat production in large fish makes use of two countercurrent heat exchange systems of vascular retia. In tuna, the temperature of the swimming muscle is maintained at a mean of 30°C within a 5° range, in ambient temperatures varying from 10° to 22°C (Fig. 9–23). The tuna head and tail are more nearly at T_a (27°C in brain and 30°C in muscle in water at 21°C). In lamnid sharks, the temperature of heavy red muscle can be 7° to 10°C above T_a; most blood is carried in lateral arteries and veins with retia between them, a system that is analogous to the tuna's[71, 72, 73, 74] (Fig. 9–24).

Thermoreceptors of Poikilotherms. Thermal sensing which initiates regulatory behavior in poikilotherms is performed by peripheral receptors and by the sensitive central nervous system. These have been studied in insects, fish, and snakes.

In crickets and the centipede *Lithobius*, the antennae perceive air temperature and tarsi perceive ground temperature. From ablation experiments, it has been ascertained that *Rhodnius*, bedbugs, honeybees, and phasmids have antennal thermoreceptors, and a beetle has antennary and maxillary palp receptors. *Locusta migratoria* has paired sensory patches on the antennae, thorax, and abdomen, which are sensitive to heat; after one antennal patch is destroyed, the grasshopper no longer turns toward a heat source on that side. In *Periplaneta*, antennal receptors discharge spontaneously at a rate of 45/sec; they show a phasic increase in frequency of impulses when cooled and a decrease when warmed, with a maximum rate of 200 impulses/sec/°C and an absolute sensitivity of 0.6 impulses/sec/°C.[299] *Periplaneta* has temperature-sensitive pads on the tarsi that show increased activity when the insect is cooled below 10° or warmed above 30°C.[259] The wood-boring buprestid beetle *Melanophila* has specific receptors on the coxae of the legs that are sensitive to infrared radiation (2 to 4 μ wavelength). These receptors could sense heat at a considerable distance, as from a forest fire toward which they could then fly.[123, 124]

Spontaneous activity of the ganglia of cockroach, crayfish, or slugs is depressed transiently by a rise in temperature and stimulated temporarily by a fall in temperature; the nerve cord of a roach acclimated to 30°C is stimulated at 27°C. The temperature optimum in terms of minimum number of impulses is the same as that of behavioral aggregation and is the same as the acclimation temperature, at least in the mid-range.[257]

Fish have thermoreceptors in the skin, and the lateral line and brain are very sensitive to temperature. Aggregation in a gradient, fin reflex responses to local applications of cold or warm water, and conditioned responses to temperature are not abolished by cutting the lateral line; they are probably mediated by skin receptors. *Ictalurus* and trout can be conditioned to respond to rapid changes of less than 0.1°C. In *Leuciscus*, receptors sensitive to temperature alone were not found, but mechanoreceptors were maximally sensitive to touch at 18°C after 5°C acclimation and at 22°C after 15°C acclimation.[442]

In elasmobranchs there are large sensory bulbs on the surface of the head, called the ampullae of Lorenzini, which are sensitive to various stimuli, one of which is temperature. The nerve fibers from these ampullae show autonomous rhythmic activity that is maximal in frequency at 20°C and declines to cessation at below 2°C and above 34°C. Rapid cooling causes a sudden and transient increase in frequency; warming causes a sudden decrease. Response to as little as 0.05°C change can be detected.[189] Ampullae of Lorenzini are highly sensitive to mechanical stimulation and to small electrical fields. Whether they serve as thermoreceptors has not been demonstrated, but their activity is modulated both tonically and phasically by temperature.

Temperature sense is well developed in pit vipers (Crotalidae), which have facial temperature-sensitive pits, and in some boas with labial pits. The membrane at the base of the pit of the viper contains many free nerve endings spreading palmately over 1500 μ^2, and there is an air space behind the membrane which reduces thermal loss.[58] Fibers in the branches of the trigeminal nerve to the pit show spontaneous activity. The organ is specialized for radiant heat detection and responds to long infrared (0.5 to 15 μ)

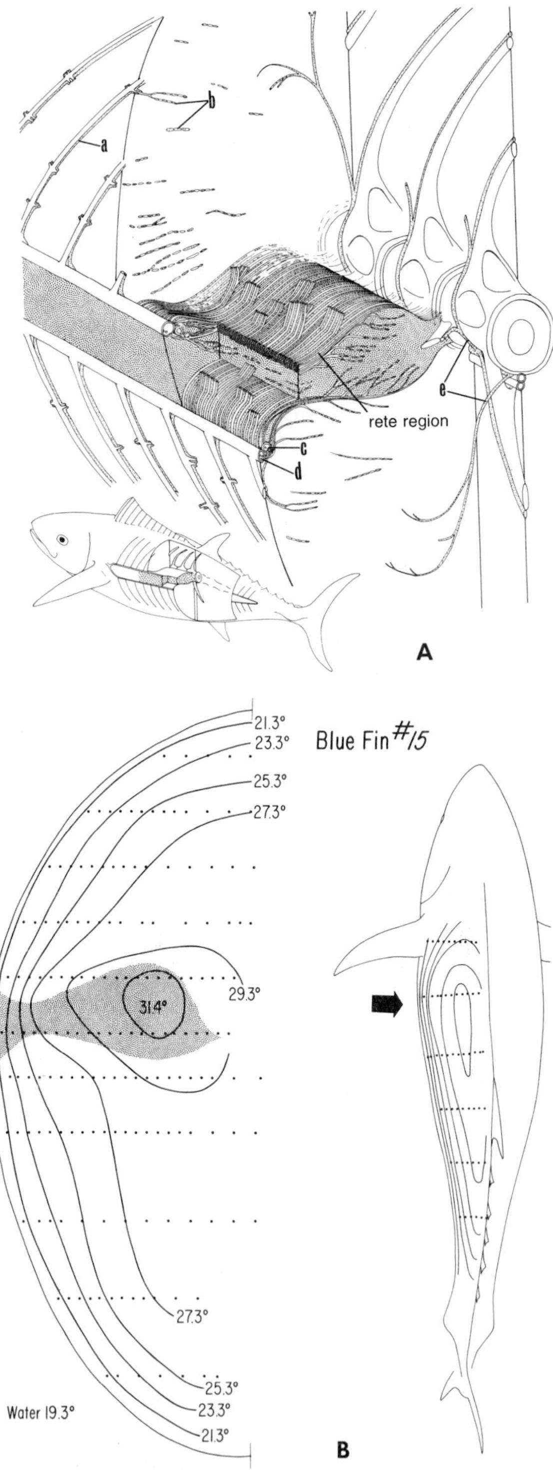

Figure 9–24. (A) Diagram of section of a bluefin tuna to show region of rete; (B) diagram of section through tuna to show temperature profiles when in water at 19°C and at 35°C. (Courtesy of F. G. Carey and J. M. Teal.)

but not to near infrared or visible rays. Temperature detection is directional, and temperature shadows passing over the pit are readily detected. From the modulation of the steady discharge, it was calculated that as small a change as 0.003°C in the pit membrane could be detected; that is, the threshold was 5×10^{-4} cal/0.1 sec. A 0.4°C temperature rise increased the frequency of nerve impulses from 18/sec to 68/sec.[58] In another pit viper sensitive down to 2.7 μ wavelength, the threshold was 0.1 milliwatt/cm^2. The response to heat is a phasic burst followed by a new tonic level of discharge, and the response is independent of body temperature.[139] The adaptive value of these highly sensitive and directional receptors is the ability to detect warm-blooded prey after dark.

TEMPERATURE EFFECTS ON EXCITABLE MEMBRANES IN GENERAL

Many receptors, as well as central nervous systems, are spontaneously active, and temperature modulates the frequency of firing. In addition, temperature affects nerve excitability and conduction. These temperature effects could be upon enzyme systems which set energy levels for membrane events, or they could be directly on the excitable membranes. Probably both actions are important for behavioral temperature adaptations. Resting potentials are relatively independent of temperature up to a high limiting value; action potentials in squid axons decrease in amplitude as temperature rises, are of maximal amplitude at low temperatures, and can be recorded at near-freezing.[208]

The rates of rise and fall of an action potential decrease on cooling; cold block occurs for cat saphenous non-myelinated axons at 2.7°C and in myelinated axons at 7.2°C. Heat and cold block of *Homarus* axons occur at 26° and at 3°C, respectively. The velocity of *Lumbricus* giant axons is greater for cold-acclimated than for warm-acclimated earthworms; the absolute refractory period is lengthened by warm acclimation as follows:[278]

	Acclimation Temperature 13°C	23°C
test temperature 6°C	6.5 msec	12 msec
13°C	3.1 msec	3.7 msec

Callinectes axons respond to cooling by spike initiation; 0.5°C cooling stimulates when the animals are in a medium containing 10 mM Ca, and 0.3°C cooling stimulates when Ca is reduced to 8 mM.[418a] In mechanoreceptors such as cat pacinian corpuscles, the generator potential increases on warming, with a Q_{10} of 2.5 for rate of rise and a Q_{10} of 2.1 for amplitude.[227] In contrast, a crayfish stretch receptor on cooling[59] shows an increase in generator potential and a decrease in threshold for firing.

In general, nerves of hibernating animals block at lower temperatures than those from non-hibernators, and non-myelinated fibers are more susceptible to cold than are myelinated ones.[328] The phrenic nerve from an active hamster blocked at 5°C, but one from a hibernating hamster did not block down to 0°C.[441] Hamster tibial nerve blocked at 2.7°C, while that of a rat blocked at 11° to 12°C.[327, 328] Blocking of caudal nerves correlates with the normal exposure of animals: beaver, −5°; muskrat, −4.4°; red squirrel (Alaska), −3.8°; coyote, +1.0°; fox, +3.1°C.[325] The velocity of conduction of nerve impulses decreases more per degree of cooling in nerves of rats kept indoors than in those of rats kept outdoors. The velocity decreases more for fox than for muskrat.

In herring gulls that had been living in cold water, the tibial nerve blocked between 11.7° and 14.4°C, and the metatarsal nerve blocked between 2.8° and 3.9°C.[80] In single fibers of mammalian nerves, the Q_{10} for velocity is 4.8 from 8° to 18°C, 2.5 from 18° to 28°C, and 1.6 from 27° to 37°C.[352] It is concluded from the effects of cold on conduction that the properties of nerve membranes are adaptively related to the function in appropriate temperature ranges, and that these properties are subject to modification.

Central nervous systems are so sensitive as to serve as temperature receptors. In *Lumbricus*, the spontaneity in some small neurones of the ventral nerve cord is maximal at 12°C.[288] Single giant neurons in ganglia of *Aplysia* differ in the effect of temperature on resting membrane potential (RP); in some cells the RP is minimal at 20° to 25°C, and in other cells it is minimal at 15°C, while still others are progressively more negative as temperature is lowered.[337] Motoneurones of toad spinal cord are hyperpolarized by cold and depolarized by warmth; spike discharge may be induced by elevated temperatures.[454] Cold block of central nervous functions occurs at higher temperatures for "higher" levels of integration than for "lower" levels in goldfish.[371, 393]

In cat spinal cord, the magnitude of reflexes is enhanced on cooling and is maximal at 20° to 30°C.[50] Similar enhancement on cooling occurs for evoked potentials in goldfish cerebellum to visual stimulation; maximum response for 25°C-acclimated fish was at 15° to 20°C.[357a]

In general, conducting membranes (such as those in axons and muscle fibers) are functional over a wider temperature range than are junctions (synapses and motor endplates). In frog sartorius muscle, the resting potential of muscle fibers is relatively independent of temperature; muscle and nerve membranes are responsive down to 0°C and −3°C, while endplates block at 1°C (for 10°C-acclimated frogs). The effect of cooling is to decrease the frequency of spontaneous (miniature) endplate potentials; however, the amplitude increases on cooling from 20° to 15°C and then decreases. Similar enhancement at 15° to 20°C was noted for response to applied acetylcholine.[236] Rat phrenic-diaphragm shows neuromuscular block at 10°C, and the nerve blocks at 5°C.

Heart contractions generally continue at lower and at higher temperatures than do skeletal muscle contractions. In heart of a skate (*Raja*), heat block is at 22° to 25°C, and nerve-muscle block is at 26° to 28°C.[32]

Several central nervous functions are blocked with less cooling or warming than are peripheral nerve and muscle. In sunfish and goldfish, induced swimming and conditioned reflexes block with less cooling than spinal reflexes (Fig. 9–10). In Purkinje cells of goldfish cerebellum, responses to antidromic stimulation block at 6.5°C, whereas synaptic responses to parallel fiber stimulation block at 8 to 9°. The tactile receptive field of central sensory neurons in the facial lobe is reduced in areas as the fish is cooled.[357a] Upper and lower limits for respiratory movements and for a jaw response to brain stem stimulation are modified by acclimation:[365]

Acclimation Temperature	*Respiratory Movements*		*Jaw Response*	
	HIGH	LOW	HIGH	LOW
31°C	40	10.6	40.8	5.6
19°C	38	8.6	39.6	3.1

At 16°C, the upper limit for pitch discrimination in *Phoxinus* is 800 to 1260 Hz; at 25°C, it is 1260 to 1420 Hz.

Temperature-sensitive neurons which resemble some of those in mammalian thermoregulatory centers have been found in brains of lower vertebrates. In a trout diencephalon, out of many units sampled, 12 responded to warming and 5 responded to cooling.[141] In a lizard (*Tiliqua*), a few units in the preoptic diencephalon increased firing with brain warming, and a few with brain cooling.[63, 152]

In a free-swimming sculpin, heating the forebrain by an implanted thermode decreased the time the fish spent at the warm end of a tank, while cooling the brain had the reverse effect.[154] In lizards, local warming and heating of the brain elicits behavioral thermoregulation.[150, 152, 173] The lizard *Tiliqua* moves to warm or cold regions about a "preferred" brain temperature of 29.3°C. Increasing the T_a lowers the T_b at which exit to the lower temperature occurs. Local warming of the preoptic region lowers the T_b for exit from a warm region, and local cooling has the opposite action. It appears that the reptile combines information of brain temperature and general body temperature in determining its thermal behavior.[341]

TEMPERATURE REGULATION BY TRANSPIRATION AND BEHAVIOR IN TERRESTRIAL POIKILOTHERMS

The temperature of a land snail or frog is below that of a dry-bulb or a black-bulb thermometer unless the air is fully saturated, but T_b is higher than a wet-bulb thermometer reading. In water, the T_b is that of the medium. In still air, the body temperature of a frog is closer to the dry-bulb reading, while in moving air it approaches the wet-bulb thermometer reading. In a toad at 27.6°C and relative humidity of 82%, the T_b (cloacal) was 26.5°C; at the same temperature with relative humidity of 27%, the deep body temperature was 17.5°C.[322] At 20°C the metabolism of a frog corresponded to a heat production of 6 cal/hr, but it lost 3.2 g of water per hour, which absorbed 1850 cal.[322] Hence, in moist air the heat loss by vaporization normally exceeds heat production, and T_b is lower than in water of the same environmental temperature.

Insects. The body temperature of insects and isopods may be different from that of the air because of heat loss by vaporization, production of heat by metabolism, and ab-

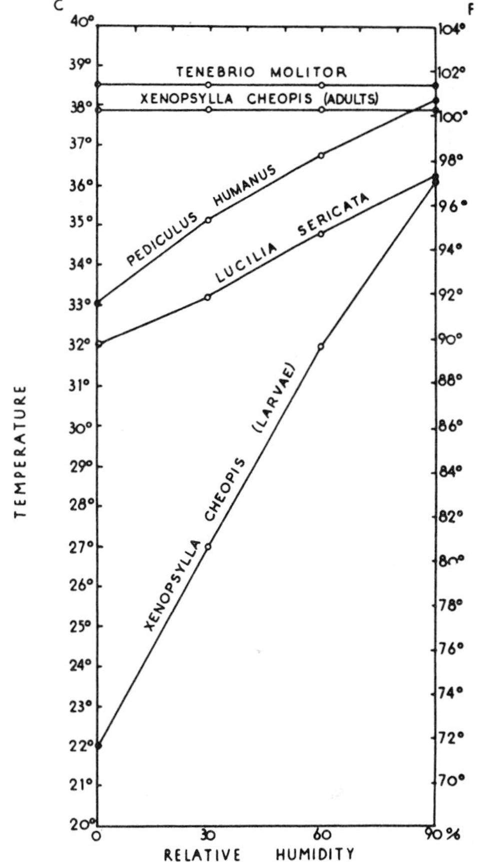

Figure 9–25. Highest temperature at which insects can survive exposure for 24 hours at different humidities. (From Mellanby, K., J. Exp. Biol. *18*:55–61, 1942.)

sorption of heat by radiation.[83] Large insects, such as *Blatta, Periplaneta, Schistocerca,* and the grasshopper *Gastrimargus,* showed body temperatures 2.6°C lower than air; they evaporated 0.06 g H_2O/hr, while the isopod *Armadillidium* was only 0.4°C below air temperature and lost 0.002 g H_2O/hr. The temperature of a live *Ligia* (high intertidal isopod) is higher than that of a dead one in moist air, but the live *Ligia* temperature is lower in dry air in the sun; heat gained by conduction is low, but gain by radiation is balanced by loss by convection and evaporation.

Most water loss by insects is from the spiracles, and death in warm dry air may result from desiccation rather than from heat *per se.* The temperature which was lethal within one hour for a series of small insects—*Xenopsylla, Pediculus, Lucilia,* and *Tenebrio*—was independent of humidity. However, for death within 24 hours, the lethal temperatures were much lower in dry air than in moist air[322] (Fig. 9–25). Apparently, the 1-hour death was thermal, while the 24-hour death was by desiccation. Chinch bugs' survival at temperatures below 50°C is favored by high humidity; at higher temperatures survival is greater at low humidities. Death at below 48° to 50° is caused by desiccation, while death at higher temperatures is caused by some direct effect of heat.

In the locust *Schistocerca gregaria,* the body temperature at about 40°C is 3 to 4° lower in dry than in moist air. The epicuticular layer of lipid becomes permeable at about 30°C.[83] *Blatta orientalis* tolerates 37° to 39°C for 24 hours in moist air, and 34° to 36°C in dry air. At high temperatures, tracheal ventilation is markedly increased in tenebrionid beetles. In the tsetse fly (*Glossina*), opening and closing of the spiracles commences at about 30°C, and they are fully open at 39° to 41°C; T_b may be 1.66°C below T_a in 45°C dry air but not in moist air.[117,196] Evidently, the water loss through the spiracles is used for evaporative cooling by some insects; water loss through the cuticle is not controlled, but increases when protection by the waxy cuticle is impaired at high temperatures, where desiccation may overbalance any cooling effect.

Reptiles. Reptiles use both behavioral and metabolic methods to function below but close to their critical thermal maximum. The thermal maximum is evidenced by disorganized locomotion and then death. Dawson arranged the families of lizards as thermophilic or heat resistant, with heat preferenda above 35°C, and as less thermophilic and more heat-sensitive, with preferred temperatures between 30° and 35°C.[95] Some snakes are thermotolerant, with resistance that is little altered by acclimation:[229]

Acclimation Temperature:	*15°C*		*30°C*	
CRITICAL THERMAL:	MAXIMUM	MINIMUM	MAXIMUM	MINIMUM
Natrix	40	6	41	8.2
Thamnophis	39	4	42	8

Numerous behavioral patterns come into play. Many snakes and lizards bask in the sun and absorb radiant heat; they also gain heat by conduction from rocks and sand. Skin reflectance of radiant heat varies with pigmentation:[414]

	Skin Reflectance	Absorbance
Iguana	6.2	74
Phrynosoma	35	65

A marine iguana on shore showed a preferred T_b of 35° to 37°C, yet in the sea they fed at a temperature of 25°C. When on rocks where the black-bulb thermometer reading exceeded 50°C, the iguana kept its T_b below 40°C by postural means.[20] Sea snakes bask in sunlight on the surface of the ocean.

A horned lizard, *Phrynosoma,* in a terrarium with light and shade normally keeps its body at 35°C, with a maximum voluntary tolerance of 39°C and minimum of 28°C. It seeks shade at 37.7°, burrows in sand at 40.5°, and pants at 43.2°C.[173] *Iguana* emerges at 38.3°, retreats at 39.2°, and becomes inactive below 36°C.[315] The desert dragon *Amphibolurus* shows coordinated movements at temperatures up to 48.5°C, but cannot survive heating to 49.3°C; it may live close to its thermal limit.[181] *Dipsosaurus* regulates by body position; its preferred T_b is 38.5°, but it may voluntarily tolerate 42°C.[108] Other behaviors include orientation toward or away from the sun, seeking shelter or shade, and running with the body elevated above the substratum. Many desert lizards and snakes enter burrows when temperature is high, and are nocturnal in activity. A female python broods its eggs and generates heat by muscle contraction. At T_a below 33°C, the T_b may be 7.3°C higher than the substratum. At T_a of 25.5°C, the O_2 consumption during brooding is 9.3 times that of a non-brooding female python.[221]

Reptiles lose less water by vaporization than do amphibians because of lower dermal permeability. When the relative humidity was decreased from 100% to 7% at 20°C, the T_b of horned toad, turtle, and alligator was lowered by only 1°C, as compared with 8°C in salamander and frog. Many reptiles (e.g., the skinks) pant when T_b is near the critical temperature maximum; others merely keep the mouth open. In an iguana, evaporative cooling dissipates half of the total metabolic heat at 32° to 40°C, and 1¼ times the metabolic heat at 44°C.[414] *Sauromalus* loses water as follows:[85]

	Acclimation Temperature	
	26°C	43.5°C
cutaneous loss, mg $H_2O/cm^2/day$	2.0	9.5
respiratory loss, mg $H_2O/cm^2/day$	1.3	8.0

The thermal conductance of its skin is three times that of a mammal.

Sphenodon is active at night; it heats faster than it cools.[497] Similarly, the lizard *Amphibolurus* cools slowly, and selection of basking position may act on the rate of change of T_b.[28] A marine iguana cools in air at 0.3°C/kg/min, or a third of the rate of cooling in water.[25] A basking *Varanus* in the morning warms at 0.14°C/min, but it cools at night at 0.02°C/min.[443] Turtles (*Pseudemys*) in water heat faster than they cool; in air the changes are reversed. Cutaneous blood flow increases in heating and decreases in cooling. These changes depend on local skin temperature rather than on body temperature. The differences in rates of warming and cooling of reptiles mainly reflect circulatory adaptations. The metabolism is lower and the thermal conductance is higher in a reptile than in a mammal of comparable size.

HOMEOTHERMS

Characterization. By definition, a poikilotherm has variable body temperature (T_b), and a homeotherm has a regulated T_b; heterotherms are incomplete regulators or have different regions of the body at very different temperatures. It was shown above that many so-called poikilotherms can, under some conditions (particularly activity), maintain some constancy of T_b; this may be by absorption of external heat (ectothermy) as in behavioral orientation, or by production of heat internally (endothermy) as in muscular warm-up. It will be shown below that many so-called homeotherms have variable T_b and can shift to virtual poikilothermy. Despite these limitations of terminology, the distinction between "warm-blooded" birds and mammals and all other animals with respect to control of body temperature is useful.

A comparison of ectotherms (lizards) and heterothermic and euthermic mammals with active and less active birds follows:[92]

	Metabolic Heat Production at T_b in kcal/kg$^{3/4}$/day	Heat Production Corrected to T_b of 38°C in kcal/kg$^{3/4}$/day	Normal T_b
lizards	7.5	19.5	30
monotremes	34	62	30
marsupials	49	62	35.5
eutherian mammals	69	62	38
non-passerine birds	83	72	39.5
passerine birds	143	114	40.5

This table shows that reptiles, which have only behavioral and vasomotor methods of temperature regulation, also have much lower heat production than relatively poor homeotherms.

Responses to Cooling. When the skin cold receptors of a mammal or bird are stimulated, a series of reflex responses tend to conserve heat. Blood vessels in the skin constrict, hairs or feathers may be erected, and an animal may reduce its exposed surfaces by appropriate posture. Vasoconstriction in man may reduce heat loss by one sixth to one third. For many animals, the layer of superficial tissues, especially fat, may serve as insulation. Hairless rats lose more heat than normal rats by radiation and convection up to a T_a of 34°C.[198] In animals with a relatively long nasal passage, water condensation and a heat exchange surface permit retention of both water and heat while gas is exchanged in the respiratory passages[228] (Fig. 9–26).

When physical cooling continues, and insulative mechanisms are inadequate to maintain body temperature, muscle activity can be detected and oxygen consumption increases. In many animals the superficial tissues, such as fat, serve as an insulating

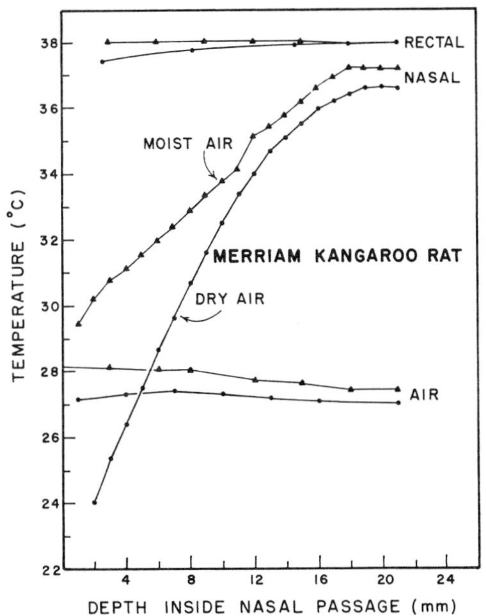

Figure 9–26. Temperature at different depths in nasal passage of a kangaroo rat breathing dry or moist air. (From Jackson, D. C., and K. Schmidt-Nielsen, Proc. Nat. Acad. Sci. *51*:1192–1197, 1964.)

"shell" and remain at a lower temperature than the "core."

The thermoregulating center in the hypothalamus coordinates the peripheral reflexes and activates hormonal stimulation of metabolism even before blood temperature is reduced. Part of the heat production is by shivering in muscle (in birds and mammals), and part is by non-shivering thermogenesis in muscle, brown fat, and other tissues (in mammals). Thermogenesis in chilled mammals is normally activated by norepinephrine (NE); the increase in O_2 consumption is proportional to the blood concentration of NE.[234] Injected NE has a calorigenic action that is greater in cold-acclimated animals than in warm-acclimated ones; this action can occur in curarized and eviscerated animals, and hence has a direct effect on muscle metabolism. The effectiveness of NE is maximal for small mammals and is negligible for mammals larger than 10 kg.[187] The general metabolic increase in cold also involves the thyroid. Thyroidectomy abolishes responsiveness to NE. Liver mitochondria decrease in unsaturated free fatty acids, but the decrease is prevented if thiouracil has been given. A normal rat regulates its body temperature in air down to −10°C, while a rat lacking thyroid or adrenal glands regulates only down to −2°C; a rat lacking both glands regulates only to +10°C. NE is the immediate trigger in mammals, but in birds the thyroid is of prime importance; there is no stimulation by injected epinephrine or norepinephrine in a variety of birds.[96,165]

Continued exposure to cold for a few hours activates the adrenal cortex, probably via the pituitary and corticotrophic hormone (ACTH); general stress signs, such as decrease in eosinophil and lymphocyte counts, may result. In man there may be hemoconcentration and decrease in blood volume. Heat conservation is favored for various animals by behavioral responses such as huddling, nest building, burrowing beneath snow, and reducing surface area by curling.

Norepinephrine brings about lipolysis and increased oxidation of fatty acids.[214] Rat serum lipids are decreased initially, and recover slowly. Cooled rats may show a twofold increase in blood NE and a corresponding increase in O_2 consumption by muscle (but not by kidney).[233] Man can produce a maximum of 370 kcal/hr and can maintain constant T_b in air from 20° to 90°F (−6.7° to 32°C).[222] An Eskimo at a T_a of 17°C increases

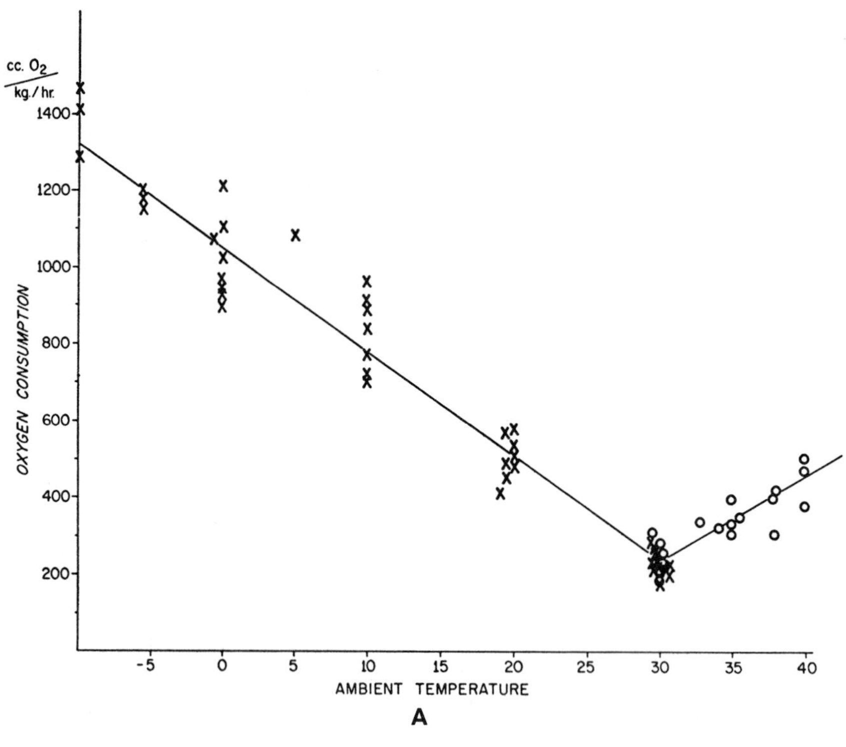

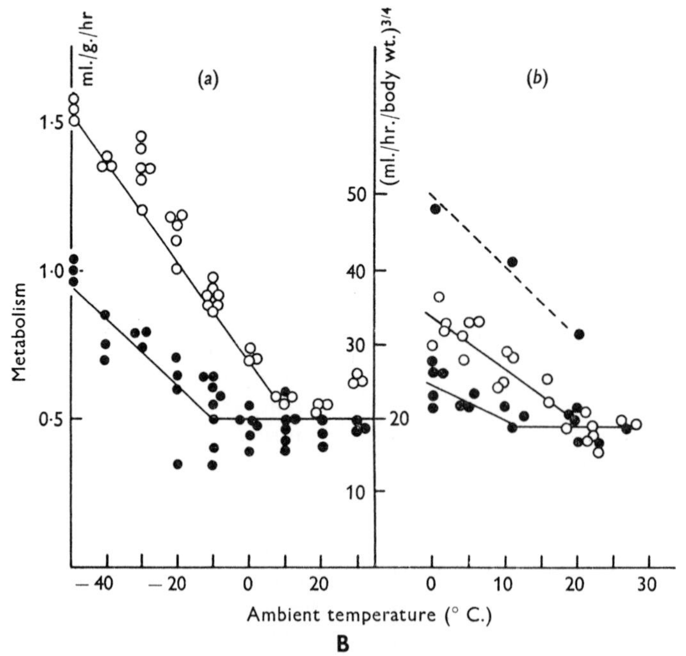

Figure 9–27. Oxygen consumption of some mammals as a function of ambient temperature. (A) Armadillo. (From Johansen, K., Physiol. Zool. *34*:126–144, 1961.) (B) Red fox (a) and harbor seal (b); open circles are summer values and closed circles are winter values. (From Hart, J. S., Symp. Soc. Exp. Biol. *18*:31–48, 1964.)

(Figure 9–27 continued on opposite page.)

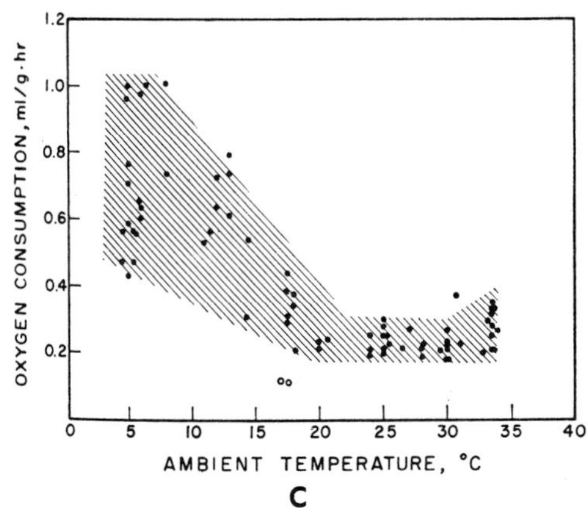

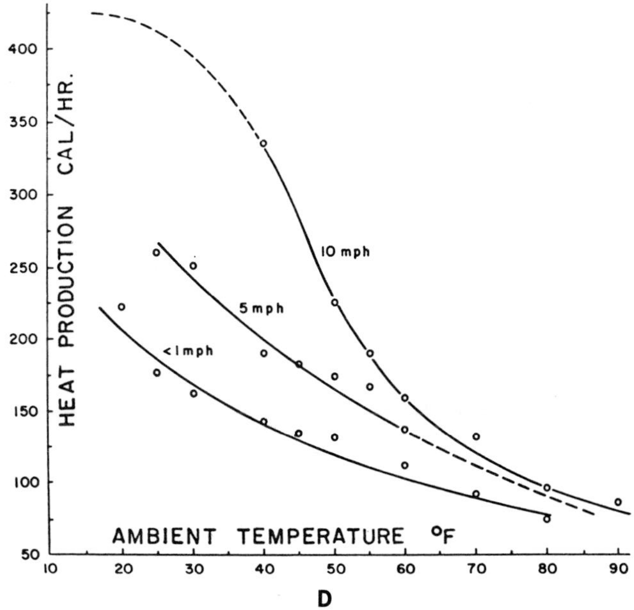

Figure 9–27 Continued. (C) Echidna. (From Schmidt-Nielsen, K., et al., J. Cell. Comp. Physiol. *67*:63–72, 1966.) (D) Nude men, heat production when exposed to different wind velocities at ambient temperatures (°F). (From Iampietro, P. F., et al., J. Appl. Physiol. *15*:632–634, 1960.)

his metabolism by 22 cal/m^2/hr above his basal 55 cal/m^2/hr.[2] Japanese quail (*Coturnix*) at 4°C decrease muscle glycogen and increase plasma fatty acids.[130] Thermogenesis in cold results from shivering and non-shivering metabolic responses in varying proportions; exercise can be substituted for the shivering thermogenesis. A cold-acclimated rat extends its tolerance range by non-shivering thermogenesis which, unlike shivering, is not elevated by exercise.[163]

When metabolism (M) is measured at different temperatures, a neutral range of minimum O_2 consumption is found, usually at an air temperature slightly below T_b. This thermoneutral zone may be broad in large mammals, or may be a minimum point on the M/T curve for small mammals and birds. In some birds acclimated to the temperature of measurement, M continues to decline to the tolerated maximum temperature.[486] For most birds and mammals, at the low end of the neutral zone there is a critical temperature (low T_c) below which physical mechanisms are inadequate and metabolism increases if T_b is to be maintained. At the other end of the curve, above the thermoneutral zone there is an upper critical temperature (high T_c) above which cooling mechanisms are ineffective and metabolism increases. In diurnally active house sparrows, T_n is a point, while in their nocturnal sleep it is a plateau[218, 283, 284] (Fig. 9–27). Table 9–3 summarizes values of T_n for a number of different species. In general, low T_n values are associated with low surface-to-weight ratios and with good insulation by feathers and hair. As insulation improves in the winter,

TABLE 9–3. Body Temperatures (T_b) During Activity, and Thermoneutral Zones.

Animal	T_b, °C	*Thermoneutral Temperatures* LOW	HIGH	*Animal*	T_b, °C	*Thermoneutral Temperatures* LOW	HIGH
Large Mammals				Large Birds			
man[499]	37	27	32	quail *Lophortyx*[217]	40.6	30	
baboon[135]	38.1			*Lophortyx*[54]	40.6	27.3	37.5
mountain sheep[41]	37.9 night			Inca dove[316a]	38.8–42.7	35	
	39.8 day			pigeon[377]		25	30
goat[451]	37–40			oriole[386]	40.0	34.5	
fur seal[225]	38	20 water		red-footed booby[21]	40.3 day		
		0–3 air			38.0 night		
camel[411]	38.1			masked booby[21]	40.7 day		
humpback whale[330]	36				38.3 night		
				night hawk[283]	36–40	35	
Rodents and Related Mammals				nightjar *Eurostopodius*[96]	39–42		
Baiomys[215]	35	29	36	chicken[130]	39.8		
Perognathus[468]	36–38	32.5		ostrich[86]	39.3		
Perognathus hispidus[482]	36.8–38.7	30	33.5	owl	38	25	37
Perognathus longimembris[82]	34.7	31		cormorant	39–40		40
Liomys[219]		31	34	wood stork[246]	40.7		
Microdipodops[51]	38.8	30	35	shearwater[212]	39.5 day		
Peromyscus eremicus[320]	36.6	30	35		37.7 night		
Peromyscus californicus[320]	36.4	27	34.5	ptarmigan[241]	39.6	4	36
Tamias[482]	36–40.3	28.5	32	evening grosbeak[486]	41 day	16	34
Citellus[300]	35.5–39.5				38.5 night		
hyrax *Procavia*[453]	37–38.5	20	30	crossbill *Loria*[98]	38.5–40	15	28.5
				gray jay *Prisoreus*[475]	42.3 day	36 summer	
Bats					41 night	7 winter	
Macroderma[292]	35–39	30	35				
Pteropus polycephalus[31]	35–39	18	35	Passeriformes			
Pteropus scaphilalus[31]		24	35	zebra finch[65]	41.5–42.2	36	42
Syconycteris[31]	36.4	33		zebra finch[67]	39.8–42.4	30	40
Dobsonia[23]	37	34		Harris sparrow[385]		35	
Eumops[291]	31.4–33.1	31		house sparrow[218]	41.1 day		
Myotis[349]		32.5	34.5		38.3 night		
Eptesicus[264]	35.6						
Miniopterus[332]	37–39.1			Hummingbirds			
				giant *Patagona*[287]		27	
Primitive Mammals				*Eugenes*[287]		31	
armadillo *Dasypus*[238]	34–36	30		*Lampornis*[285]		31	
echidna[413]	30.7	20	30	blue-throated			
marsupial *Macropus*[92]	33.9–34.5	30.5		hummingbird			
opossum *Marmosa*[333]	33.2 day	28		*Lampornis*[285]	38.5–40	31–33	
	35.7 night			*Eulampis*[145]	40	30	
phalanger *Cercaertus*[24]	32–38	31	35	*Stellula*[281]	35–40	27–30	

the T_n is lowered. There are correlations with habitat—T_n for Abert towhee from the desert is 25° to 35°C, while for a brown towhee from cool coastal slopes it is 23° to 33°C; for an ortolan bunting from southern California, T_n is 32° to 38°C, while for a yellow bunting from farther north it is 25° to 32°C.

The low T_c below which metabolism increases is +26°C for large tropical animals and for man, while for the Arctic fox it is −30°C. Animals with low T_c tend to have lower slopes of their M/T curves, and these curves frequently extrapolate to T_b. A low slope indicates a low heat loss (weight specific thermal conductance).

When all regulation fails and body temperature falls, circulation of blood is reduced and brain function may be impaired. Blood flow ceases in vessels of rat mesocecum at 20°C, and in hamster cheek pouch at 5° to 10°C. A cat loses consciousness at a T_b of 25° to 27°C, but some reflexes continue down to about 16°C. In man, some mental defect is noted at a T_b of 35°C; breathing continues down to 25°C. Breathing stops in a dog between 20° and 25°C. The critical limit for

TABLE 9–4. Thermal Conductance at Below Neutral Temperature.

Animal	*Body Weight, Grams*	*Weight Specific Conductance, ml O_2/g/hr/°C*	*Thermal Conductance, ml O_2/cm²/hr/°C*
Rodents and Related Mammals			
Perognathus sp.[468]	21	0.19	
Perognathus hispidus[482]	39.5	0.201	
Perognathus longimembris[82]	11.55	0.28	
Peromyscus maniculatus[317]	24	0.17	
Peromyscus eremicus[320]	21.5	0.18	0.22
Peromyscus californicus[320]	45.5	0.10	0.18
Microtus[335]	47	0.166	
Dipodomys[93]	35	0.174	
Sorex[333]	3.3	0.54	
Marmosa[333]	13	0.26	
Bats			
Macroderma[292]	148	0.11	
Eumops[291]	56	0.56	
Pteropus polycephalus[23]	598	0.026	
Pteropus syconycteris[23]		0.66	
Larger Mammals			
guinea pig	430	0.04	
rabbit[166]	1581	0.029	
Procavia[453]	2630	0.02	
dog[150]	6666	0.022	
wallaby *Macropus*[92]	4960	0.16	
red fox[92]		0.14	
Birds			
gray jay *Perisoreus*[475]	71.2	0.06	
nightjar[96]	88		1.7
evening grosbeak[98]	60		1.8
evening dove *Scardafella*[316a]	42	0.20 day 0.14 night	
cardinal *Richmondena*[280]	40	0.10	
Hummingbirds			
Patagona[287]	19.1	0.17	
Eugenes[287]		0.30	
Archilochus[287]	3.3	0.50	
Lampornis[285]	7.9	0.23	
Eugenes[285]	6.6	0.30	
Calypte[280]	5.4	0.44	
Stellula[280]	3	0.61	
Selasphorus[280]	3	0.57	

locomotion in a rat is 21°C, for respiration 13°C, and for heart beat 8°C.[362] Rats on a 50% fat diet showed cessation of heart beat at a T_b of 2.9°C when fed saturated fats and at 7.2°C when fed unsaturated fats.[163]

Various measures of the effectiveness of insulative mechanisms have been used. The critical temperatures (low and high) and the extent of the thermoneutral zone are related to body size and insulation.

Specific thermal conductance is heat loss per unit surface area for a given temperature gradient:

$$K = \text{cal/cm}^2\text{/hr/°C}.$$

Surface area can be calculated from the 2/3 rule (see Chapter 5); but uncertainties of measurement of area make the use of body weight more practical; thus, weight-specific thermal conductance is

$$C = \text{ml } O_2\text{/g/hr/°C gradient}$$

or

$$C = \frac{M}{T_b - T_a}$$

where M = ml O_2/g/hr. To convert this value to kcal/g/hr/°C, multiply C by 4.74, which is the caloric equivalent of one liter of O_2.

Metabolic efficiency is the energy from total metabolism divided by the heat loss, or M/C.

A significant heat loss is caused by evaporation from skin and respiratory tract; thus, the thermal conductance by non-evaporative mechanisms is given by

$$C = \frac{M - \text{evaporative heat loss}}{(\text{area})(T_b - T_a)}$$

Some values of C are given in Table 9–4.

If an animal follows Newton's law of cooling and maintains constant body temperature, the increase in heat production is linear with decreasing ambient temperature. The slope of the M/T_a curve is a measure of thermal conductance.

Table 9–4 also gives some values of thermal conductance as measured by O_2 consumption at different ambient temperatures and constant body temperature. Data obtained by measuring the rates of cooling of dead animals agree reasonably well, in that a small mammal cools faster than a large one[335] (Fig. 9–28):

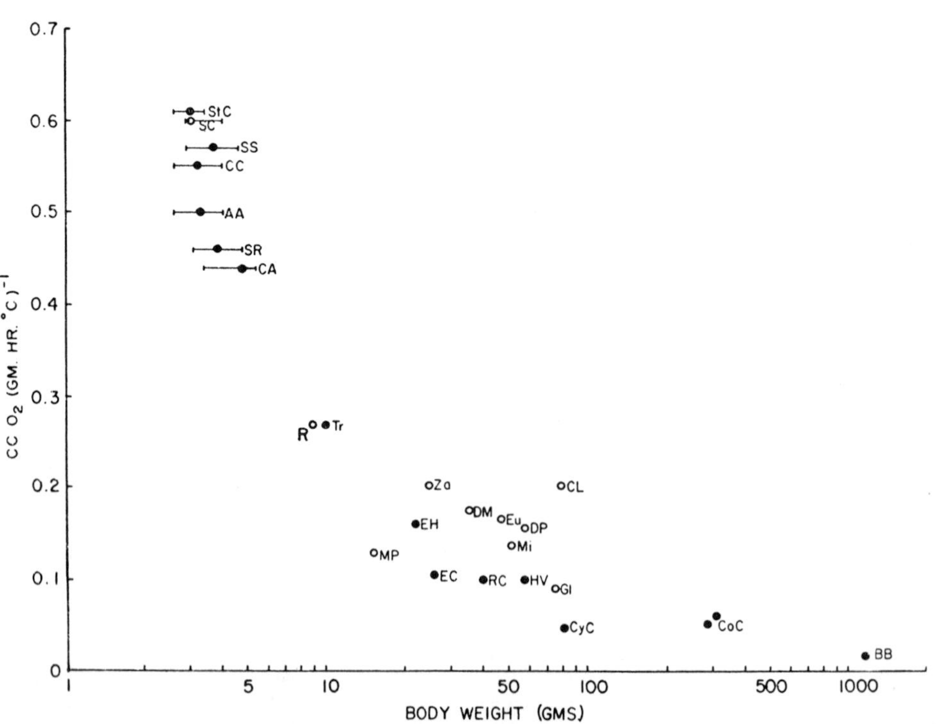

Figure 9–28. Relation between specific weight thermal conductance (C) and body weight of various small birds and mammals. Open circles, mammals; closed circles, birds. (From Lasiewski, R. C., Physiol. Zool. *36*:122–140, 1963.)

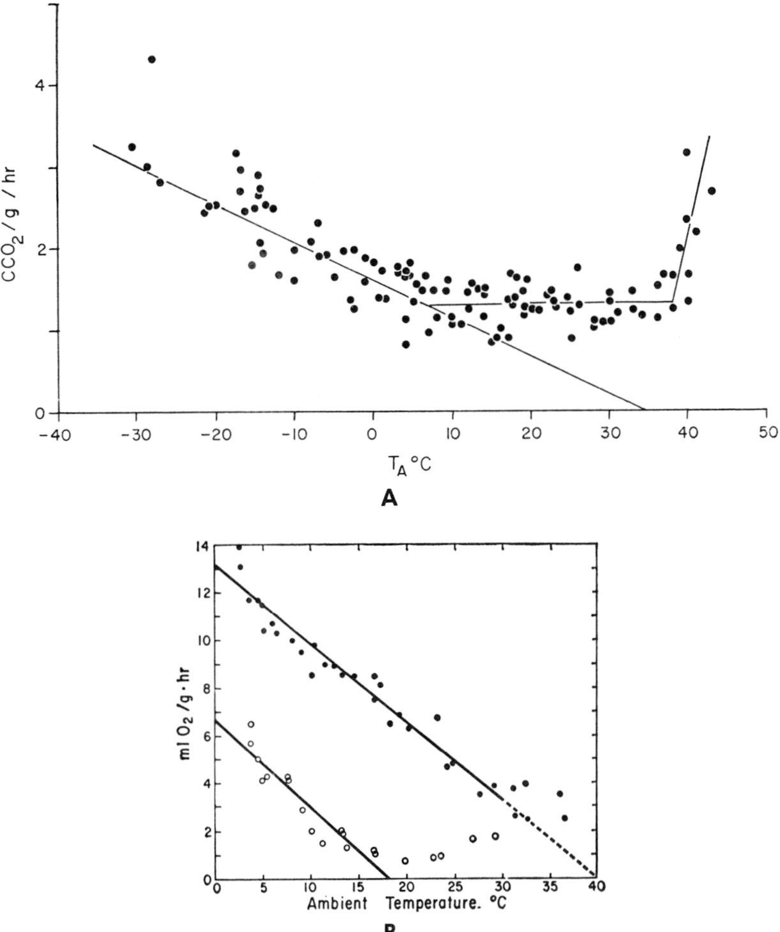

Figure 9–29. Metabolism as a function of ambient temperature in birds. (A) White-tailed ptarmigan. (From Johnson, R. E., Comp. Biochem. Physiol. *24*:1003–1014, 1968.) (B) Hummingbird *Eulampis*; solid circles are resting birds, and open circles are birds in torpor. (From Hainsworth, F. R., and L. L. Wolf, Science *168*:368–369, 1970. Copyright 1970 by the American Association for the Advancement of Science.)

	Weight	*Thermal Conductance cal/cm²/hr/°C*
Sorex	2.3 g	0.31 (limbs extended)
Clethrionomys	14.3 g	0.47
Microtus	20.4 g	0.43

In general, arctic and winter animals have lower slopes of M/T_a than do tropical and summer animals. This reflects improved insulation, and hence lower thermal conductance in cold acclimatization (Fig. 9–29; see p. 404). Other data corroborate the seasonal changes in thermal conductance:[385]

	Thermal Conductance ml O₂/g/hr/°C	
	Winter	*Summer*
cardinal	0.098	0.205
evening grosbeak	0.10	
Harris sparrow		0.23
white-crowned sparrow	0.125	

Interpretations of thermal conductance are difficult in animals such as those capable of torpidity (e.g., Microchiroptera), in that as temperature is lowered, activity may be reduced.

Acclimation to Cold. After birds and mammals have been maintained in the cold for several weeks, various signs of acclimation appear; these signs are different from those in poikilotherms. There is extension of the tolerable low temperature when measured as either survival limit or ability to maintain thermal balance. Dogs in cold water died at a T_b of 18.6°C, but if they were previously acclimated to a low T_a they survived chilling to 14.9°C. In a rat moved from 28°C to 5°C, metabolism rose in three days from 55 to 88 ml O_2/hr/$W^{0.52}$, and at three weeks the metabolism was twice that of rats kept at 28°C. If the metabolism is measured at 30°C rather than at 5°C, it is not elevated initially, but after a week in the cold the elevated metabolism is seen at all temperatures of measurement.[232] The responsiveness to NE is much greater in cold-acclimated rats than in warm-acclimated ones.

Rats were maintained at 9° and at 28°C, and the V_{O_2} of the 9°C group corresponded to 5.5% of body weight per day, while that of the 28°C group was 3.0%. Food consumption was 8.6% greater in the cold group.[258] Muscles account for 57% of total metabolism in the rat; liver accounts for 22%, and skin for 6%.[232] The thyroid is necessary for maintenance of elevated O_2 consumption, and the thyroid glands show increased activity after a week in the cold. Men kept at 8°C for nine days showed an increase in rest V_{O_2} from 0.23 to 0.46 liters/minute and an increase in active V_{O_2} from 1.50 to 1.53 liters/minute; the increase in basal metabolism is proportionally much greater than that in active metabolism.[390]

The recorded data concerning biochemical changes in cold acclimation are contradictory because of differences among tissues and species; differences between non-hibernators and hibernators active or in hiberation at low temperatures; and differences between tissue homogenates and whole mitochondria. Also, most assays are at saturating concentrations of substrates, so that data are not available for K_m's or for enzyme activities at physiological substrate concentrations.

In rats acclimated to low temperature (5°C for more than 10 days), the oxygen consumption by various tissues is elevated when measured at 37°C; the increase in muscle and liver metabolism is less if the rats have been thyroidectomized. Oxygen consumption by skeletal muscle of lemmings acclimated to 1°C was 40% greater than for lemmings from 17°C.

Homogenates of liver and muscle from cold-acclimated rats show increased activity of succinic DH, malic DH, and cytochrome oxidase, as well as oxidation of isocitrate, α-ketoglutarate, and glutamate; they show a decrease in the activities of the pentose shunt enzymes—6-phosphogluconic acid DH and glucose-6-PDH. They show a decrease or no change in lactate DH and DPNH cytochrome C reductase.[157, 158] After rats and hamsters had been at 5°C for two weeks, their kidney tissues showed enhanced capacity for glucose formation from pyruvate and from amino acids (gluconeogenesis).[62] The cytochrome C content of abdominal muscles was 33 μg/g for 28°C rats and 50.5 μg/g for 6°C rats.[105] Cold-acclimated rats have increased ability to oxidize fatty acids, and are more responsive to injected NE.[213, 214] Hamsters and rats in cold acclimation showed increased unsaturation in white fat; hamsters, but not rats, show increased unsaturation in liver fats.[490] Perirenal and subcutaneous fat of hamsters in cold acclimation have decreased amounts of linoleic and palmitic acids and increased amounts of oleic and stearic acids.[266] Rat liver shows a marked decrease in P/O ratio, especially with β-hydroxybutyrate; this indicates more active TPN-linked oxidation of fatty acids.[158] Several of the enzymatic changes can be caused at higher temperatures by injection of thyroxine; in the cold, the thyroid glands show signs of increased activity.

In contrast to the homogenates, isolated mitochondria of cold-acclimated rats show no change in oxidation of succinate, glutamate, or β-hydroxybutyrate or in cytochrome oxidase.[78, 298] Also, no change in P/O ratios was found in rat mitochondria. However, in hamsters that were cold-exposed but not hibernating, there was increased oxidation by liver mitochondria of succinate and β-hydroxybutyrate and reduced P/O ratio, especially when measured at low temperature.[298] Also, mitochondria from 5°C-acclimated rats showed a decrease in unsaturated fatty acids, an effect prevented by thiouracil or caused at high temperatures by thyroxine injections.[354] Liver mitochondria from hamsters acclimated to 35°C showed reduced utilization of succinate, glutamate, isocitrate, and β-hydroxybutyrate as compared with mitochondria from 23°C hamsters.[77]

Mitochondria from hamsters and ground squirrels in hibernation showed marked increases in oxidation of succinate and β-hydroxybutyrate, and increased cytochrome oxidase, when measurements were made at 37°C. In brown fat mitochondria from hibernants, the oxidation of α-glycerophosphate and isocitrate is increased 3 to 4 times more than in liver.[78]

Evidently, therefore, in non-hibernants such as rats, extra-mitochondrial mechanisms, probably under thyroid control, cause enhanced respiration in the cold. In hibernants, changes occur in the mitochondria as well. There is a trend in cold exposure in both

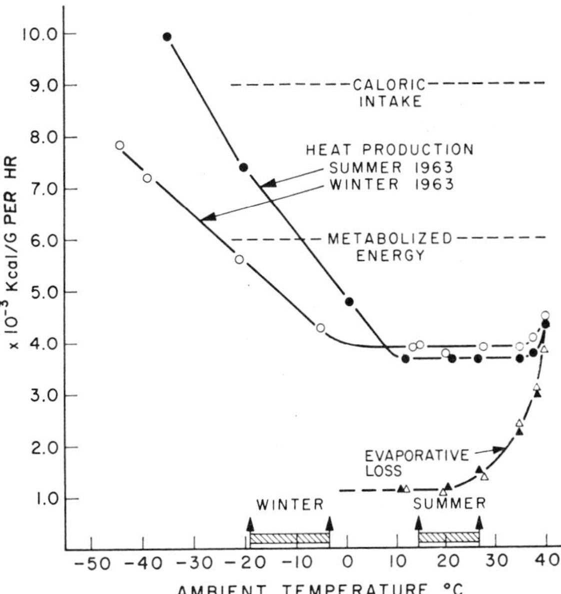

Figure 9–30. Average heat production (circles) and evaporative heat loss (triangles) in varying hare. Closed symbols, summer; open symbols, winter. (From Hart, J. S., H. Pohl, and J. S. Tener, Canad. J. Zool. *43*:731–744, 1965. Reproduced by permission of the National Research Council of Canada.)

groups toward unsaturation of deposited fatty acids, toward increased oxidation of fats, and toward reduced P/O ratios due to increased oxygen consumption. The net effect is heat production with less formation of ATP.

The lower T_c or temperature below which M increases is not lowered in rats and rabbits by acclimation to cold in the laboratory; this is in contrast to the outdoors situation, where acclimatization lowers T_c. In fact, in the laboratory cold, rats and rabbits show increased skin vascularity, which appears to be non-adaptive. Heart rate in vivo increases as T_b drops; in vitro the reverse occurs, but the heart of 30 to 33° acclimated animals beats at a lower rate in vitro than that of cold-acclimated animals when measured at the same temperature.[215]

Acclimatization in Cold Climates. In cold climates the compensations of increased heat production, as observed in laboratory cold, can be only a temporary solution to lasting cold stress. Animals in winter conditions or in the Arctic or Antarctic show different changes from those reported for acclimation for a few weeks in a cold room. The time of natural exposure is longer, seasonal cold increases gradually, and photoperiod, nutrition, and behavior interact to provide a different picture from that seen in the laboratory.

Winter mammals and birds are better able to survive extreme cold than summer ones. *Peromyscus maniculatus* showed 200 minutes survival at −35°C in winter, and at −15°C when tested in summer. Decreasing slopes of the metabolism/temperature curves indicate improved insulation (*Peromyscus, Zapus, Citellus*). Changes in CNS are probable, as indicated by improved resistance. Wild rats show less shivering in cold but more non-shivering thermogenesis in response to NE than indoor cold rats.[194] In small birds (house sparrow, grosbeak, pigeon) the M/T curves are similar in summer and winter, but food consumption increases and heat production is maintained longer under cold stress in winter than in summer[165, 166] (Fig. 9–30). Winter fat is more unsaturated than fat deposited in the summer.

Thicker and more effective winter coats have been noted in numerous birds, and in porcupines, squirrels, dogs, and rabbits. Plumage in English sparrows is 29% heavier in winter than summer, and the sparrow gains cold tolerance during the fall at a rate of 6°C per month.[19] The insulative value of the gray jay *Perioseus* increased from 0.16°C m^2hr/kcal in summer to 1.0 in winter.[475] The low critical temperature is 36°C in summer and 7°C in winter. Low critical temperatures in winter are for evening grosbeak 16°, cardinal 18°, snow bunting 10°C.[98] The grosbeak relies more on metabolism, the jay more on insulation. The lower limit for cold tolerance in the sparrow, *Passer domesticus,* was −25°C in January, and 0°C in August-September; fat content was high in winter birds, and fatty acids were more unsaturated.[19] In white rats kept outdoors in winter at average T_a of 10°C, the fur became thick, while fur of rats kept indoors at 6°C did not; both groups showed increased vascularization of skin. There were fewer frost lesions, such as to ears, in the acclimatized than in the acclimated rats.[194]

In ground squirrels (*Citellus tereticaudus*), the T_c is 3°C higher in summer than in winter.[216] In wild hares the critical T_a where regulation fails is in summer +10°C, in winter −5°C; fur is 27% heavier in winter than in summer. Shivering response is greater in summer than winter, but food intake is similar in both seasons. Rabbits were shaved and some were kept at 6°C, some at 28°C; food consumption increased 41% in the cold, and metabolism in response to a given dose of NE increased 41% in the 28°C rabbits, 113% in the 6°C ones.[194]

In Arctic mammals, not only is the fur thicker, but it is better insulation. Maximal insulation (except for a vacuum) is given by still air, and this has a value of 4.7 clo/inch. (The clo unit is one developed for clothing and is based on heat loss for a given temperature gradient, area, and time.) A plot of insulation against thickness of fur for a series of arctic animals gives a slope of 3.7 clo/inch (Fig. 9–31). On the average, the insulation of arctic mammals is nine times better than that of tropical ones. Seal blubber is such good insulation that the skin temperature is only slightly above that of surrounding water while the visceral temperature is kept constant at 37°C. The insulation increase in winter is greater for large mammals (52% for black bear and 41% for wolf) than for small mammals (21% for deermouse and 16% for hare). Insulation increases in a series from shrew to rabbit.[163, 166]

An exposed extremity may continue to function at a temperature much lower than that tolerable by the whole body. The temperature of the foot of a gull on an ice floe, the lower legs of a reindeer, the flippers of a seal, or the flukes of a whale may be

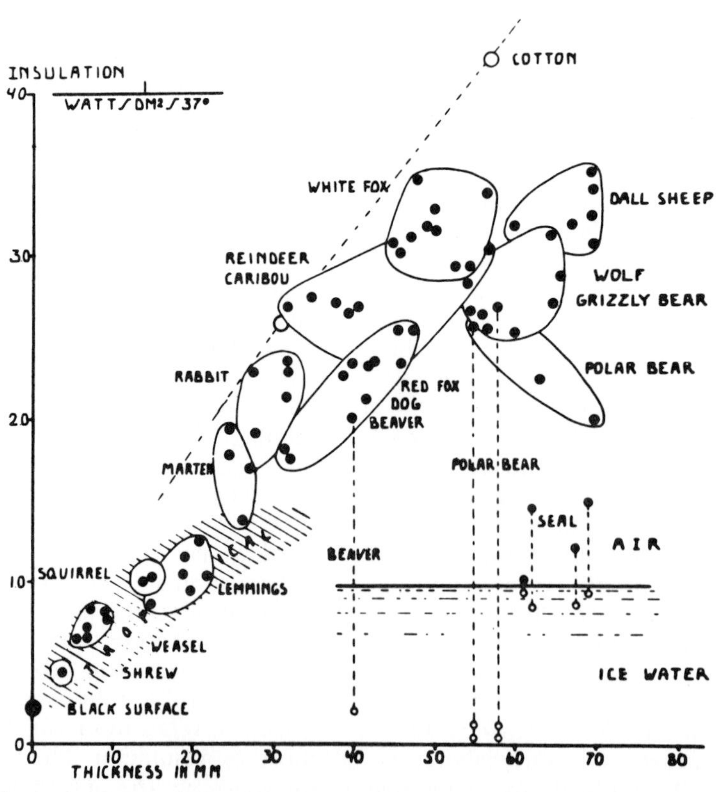

Figure 9–31. Maximum insulation of fur of arctic and tropical mammals as a function of thickness. Values for aquatic animals are given in air (●) and in ice water (○). Insulating value of medium weight cotton is given by broken line. (Courtesy of P. F. Scholander.)

below 10°C. When the domestic pig is in air at 10 to 12°C, the skin on the dorsal surface is at 10 to 12°, the nose is at 17 to 18°, the lower leg is at 9°, and the deep body temperature is 38.4°C.[224] Properties of subcutaneous fat in a pig are:[100]

	Melting Point	*Ratio of Saturated to Unsaturated Fats*
outer back	26.4°C	0.66
inner back	27.8°C	0.79
perinephric	29.6°C	

The melting point of fat declines from the thigh down the leg of a caribou; the melting point of marrow fat of phalanges of Arctic mammals (fox, caribou, eskimo dog) may be 10 to 15°C, and that from near the head of the femur 45°, while foot pad fat may remain soft down to 0°C.[226]

Enzymes of cells in the skin and appendages of many mammals are heterothermic, in that they tolerate a range of some 35°—wider than in many poikilotherms. Different regions of the same cell (e.g., a nerve fiber) differ chemically; in the metatarsal portion of the leg nerve of a cold-acclimated herring-gull, conduction stops at 2.8 to 3.9°, whereas the tibial portion of the same nerve is blocked at 11.7 to 14.4°C.[80] The tibial nerve of a non-acclimated hamster conducts at temperatures down to 3.4°; that of a rat conducts to 9°.

Behavioral adaptations are multiple. Rats kept outdoors in groups survive longer in cold and have greater capacity for heat production, less shivering, greater sensitivity to NE, higher rest metabolism, and less increase in pelt than when the rats are kept separated; the thyroid increases more in separated rats.[194] A colony of mice was bred at −3° for 10 years and, as compared with controls in a warm environment, they consumed more food, had smaller litters, and showed less exploring, better nest building, and smaller size (especially of appendages), but both rat groups had the same rectal temperatures.[19a]

Vasomotor adaptations are particularly important for exposed extremities. In some animals the arteries and veins run close to each other or are intertwined around one another in a rete. Where there is close apposition between arterial and venous blood, there is heat transfer by countercurrent exchange. Hence, an extremity may remain cool but receive adequate oxygen if heat flow is from artery to vein. In a sloth where a rete occurs in the limbs, a cooled forelimb or leg rewarms slowly; the limb of a coati or monkey, which has no rete, rewarms after cooling five times faster than the sloth's.[416] There is a rete in the axilla of penguins. Wherever a rete is present, heat dissipation is not proportional to the total body surface. Ptarmigans and pheasants have retes in the upper leg, a venous network around the artery; thigh temperature in the cold is 37.5°C, while 2 cm lower in the leg the temperature is 9.7°, and in the digit it is 2.7°C. In herons and gulls, the legs and feet provide for giving off most of the excess metabolic heat, particularly at high T_A; panting stops at once when the legs enter water.[444a]

Gaspé fishermen, accustomed to handling icy nets, show less increase in blood pressure when one hand is immersed in ice water than do controls. Temperature of the fingers is higher, indicating vasomotor conditioning.[290] Finger temperature after immersion in 0°C water for 10 minutes averaged 4.8°C for Eskimos, and 2.5°C for Caucasians; the difference persisted in Eskimos living in warm climates.[115] A group of men lived for five weeks, some in cold, some in warm, with minimal protection; some in each group exercised regularly, while others did not. The exercised group maintained higher metabolism at night in the cold, had higher heat conductivity and heart rate, and showed some shift from shivering to non-shivering thermogenesis on cooling; blood flow in the fingers increased in the exercised but not in the non-exercised group. Thus, exercise may increase tolerance of cold by increasing metabolism and blood flow.[5] Similarly, men living in Antarctica for 12 months had higher toe temperatures on immersion in 5° water, but they had the same neutral zone of metabolism as temperate zone men.[498]

Acclimatization in cold takes place by increased thickness and quality of insulation, greater sensitivity to noradrenaline, improved circulation in skin, and more tolerance of low temperature by the central nervous system.

Responses and Acclimation to Heat. When an animal is placed in a hot environment, the heat receptors of the skin are stimulated, and reflex responses favoring heat dissipation result. Cutaneous vessels dilate, there is increased blood flow in the skin, thermal conductance of the peripheral tissues increases 5 to 6 times, and there is increased heat loss. Insensible water evaporation increases. If balance is not maintained, the skin temperature rises and reflex sweating occurs in animals with sweat glands, and increased respiratory loss occurs in others.

The relative loss of water by respiratory tract, by sweat glands, and by evaporation from skin varies from species to species. Man under basal conditions loses about 20% of his heat by vaporization; in exercise this figure is 75 to 80%. Loss of heat by radiation exceeds loss by vaporization at low temperatures, but above 31°C loss by vaporization predominates in man at rest. Heat loss by radiation, convection, and conduction is reduced by insulation. Heat loss by mammals with or without hair is twice as great in water as in air. Man in cool air may lose 1 liter of water per day by evaporation; in hard work on the desert he may evaporate 1.5 liters per hour. Threshold temperature for sweating is 30° to 32°C and varies for different body regions. Output of sweat increases 20 gm/hr for each degree rise in T_a. Evaporation increases abruptly in a camel at 35°C, and in a donkey at 31°C.[414] Man loses much salt in sweat; the sweat of a burro contains much less chloride. As water and salt are lost by sweat, the urinary loss becomes reduced.

In man, the sweat glands of the palms are emotionally controlled, while those of the other regions are thermally controlled. Sweating or fluid secretion by eccrine glands is initiated by sympathetic nerve fibers which are cholinergic. Apocrine glands are destroyed as they liberate their secretion; they are activated by norepinephrine and epinephrine, and must be replaced after breakdown. All higher mammals except rodents and lagomorphs have skin glands, in which the efficiency for producing sweat varies. In sheep, dog, and pig, secretion is continuous, but expulsion is periodic and is due to contraction of myo-epithelium and not in response to heat or epinephrine.[40] Most nonprimates have apocrine glands which open into or close to hair follicles, while the eccrine glands of primates are often separate from hair follicles. A burro sweats in cycles of 2 or less per minute; the entire skin is synchronized. When the skin is cooled, the amplitude decreases but the cycle persists; both core and skin temperature determine sweating. In gazelle and oryx, sweating is periodic, while in buffalo it is sustained.[58, 389] Cows kept at 50°F and at 80°F showed percentage distribution of heat loss as follows:[241]

		Acclimation Temperature	
		50°F	80°F
Brown Swiss	skin	16	41
	respir. tract	8	12
Brahmin	skin	19	46
Shorthorn	skin	26	53

In a cow, pulmonary ventilation increases threefold on going from T_a of 68°F to 111°F. Animals which pant (e.g., dogs and cattle) evaporate from the respiratory tract, and blood chloride may increase and an alkalosis may develop. However, in a panting dog, most of the air enters the nose and leaves by the mouth, thus giving a unidirectional flow which does not greatly increase air flow through the lungs.[412] In a ram, increasing the skin temperature of the scrotum to above 35.3°C induces sweating and a fall in T_b; thus, heat receptors in the scrotum induce sweating independently of core temperature. A ewe similarly is induced to sweat if the mammary glands are locally warmed.[481] Man sweats and tolerates ambient temperatures up to 49.3°, and a dog pants and in T_a 41.1° its T_b may be 42°C; a baboon is intermediate, and its rectal temperature may rise from 38.1° to 40.6°C in air at 45°C.[135] Some ungulates are predominantly panters, and others are predominantly sweaters. Cooling by cutaneous loss of water may begin under high solar radiation before a rise in deep body temperature occurs; hence, sweating may be partly a local response.

Some mammals (e.g., opossums and some rodents) salivate profusely and lick their fur when heated. Most mammals have less control than man of evaporative heat loss at high temperatures and show a greater rise in body temperature; man's tropical origin is reflected in his temperature relations.

Many mammals seek shade, are active nocturnally in the desert, or burrow to escape heat. Pigs have been trained to push a switch to turn on heat or turn off a cooling draft.[13]

The horn of a goat has a rich blood supply; at below 15°C the horn temperature approaches the ambient, and between 30° and 39° the horn temperature exceeds ambient; if the horn is experimentally covered with insulation, brain temperature rises dangerously. At T_a of 30°C, the horns dissipate 3% of metabolic heat, and at 22°C they dissipate 2%. The horns vasodilate in warm air and constrict in cold air.[450, 451]

When cooling mechanisms fail and body temperature rises, the O_2 consumption increases because of the direct cellular effect of heat, and also because of increased ventilation; the increased metabolism may be part of the regulatory mechanism, since it is less in thyroidectomized and hypophysectomized rats. The zone of neutrality does not extend to high air temperatures; i.e., the upper T_c is usually near or just below body temperature. In small birds and mammals the upper and lower T_c may be the same. A rat transferred from 28° to 34°C increases its metabolism transiently for 48 to 72 hours. Cows transferred from 70° to 84°F (29°C) showed an initial increase in metabolism, but after 9 weeks it decreased by 15% and skin vaporization increased 85%. Cows in hot air had increased sodium excretion, decreased nitrogen retention, increased glucocorticoid excretion and decreased mineralocorticoid excretion. Plasma volume increased initially in the cold, and later red cell volume also increased.

Thermal conductance was defined on page 400, and Table 9–4 gives values of C for a number of animals. The thermal conductance of *Perognathus* decreases as T_a falls, as follows:[82]

C, ml O_2/g/hr/°C	*T_a, °C*
2.5	35
1.3	32
0.58	25
0.43	15
0.42	5

Heat shock (hyperpyrexia) results if the thermoregulatory center fails and body temperature rises critically. Heat exhaustion can occur in man without rise in body temperature and is largely due to dehydration and changes in salt balance; exhaustion may occur in ample hydration if excessive amounts of salt are lost. Heat stroke results from brain damage. Heat death is due in large part to cardiovascular failure. The tolerated temperature is less with external than with internal heating; a man loses consciousness when heated externally to 38.6°C (rectal), yet in fever that temperature may go to 42°, and in exercise the body temperature may reach 40°C without harm.

The net effect of acclimatization to heat is higher capacity for activity and less discomfort on heat exposure. In man the capacity for sweating may be doubled; sweating starts at a lower skin temperature; skin temperature is more readily reduced, and the sweat contains less salt. Cardiovascular efficiency improves; rise in heart rate with increased air temperature is less in native Nigerians than in Europeans. Bantu Africans show less rise in rectal temperature with the same T_a, less increase in sweating, lower critical temperature for increased M, and less insulation than Caucasians.[499]

Many tropical mammals are genetically adapted to heat. Indian cattle are more tolerant of heat than European cattle. At 27°C, Indian cattle gained weight as readily as at 10°C, while shorthorn cows decreased food consumption and decreased growth in the heat.[240] African ungulates use either of two strategies to survive in hot deserts. Large animals (eland) have low T_b in the morning (32.8°C) and warm slowly; small animals (gazelle) warm rapidly to temperatures above air temperature, having T_b 42° in T_a 40°; water is conserved by each method. At a critical T_b both groups increase evaporation, by sweating or increased respiration; the temperature for panting is higher if the animal is dehydrated than if it is well hydrated.[452] Fur is a barrier to heat gain; the T_b of a camel in the sun rises faster after shearing than before. The rectal temperature can vary from 34° to 40°C if the camel is without water, and it is less variable if water is supplied. When the tolerated temperature is high, there is less gradient and less heat gain from the environment.[411]

Birds use a variety of means, many of them behavioral, to survive heat in the desert.[97] They do not sweat, but have two methods of respiratory cooling—increased breathing and gular flutter. In a pigeon (Inca dove), breathing rate increased from 30/min at T_a less than 40°C to 650/min at 45°C; pulmonary ventilation (deduced from blood chemistry) increased less than total ventilation, and hence the air-sacs are important in total air flow. However, lung ventilation increases, more CO_2 is lost, and alkalosis may develop. In addition, at high temperatures a gular flutter appears that is synchronized with the fast ventilation rates.[68,316a] In a young egret in heat stress, gular flutter was at 965/min while breathing was only 44/min. California quail *Lophortyx* in heat stress shows both panting (ventilatory) and gular flutter; it may also defecate and moisten its feet to provide additional evaporative cooling.[54] Evaporative water loss by a poorwill increased from 2.9 mg/g/hr at 35° to 23 mg/g/hr at 47°C.[282] In a nighthawk with a body temperature of 36°C, evaporative loss rises steeply as T_a increases above 35°: for T_a of 1.2 , the loss is 1.1 mg/g/hr, and at T_a of 35.5°, the loss is still only 2.8 mg/g/hr; at a T_a of 44°, however, the loss is 17.9 mg/g/hr.[283] Respiratory loss from an ostrich is 0.4 g in cold to 4.5 g at T_a of 45°C.[86] An Alaskan ptarmigan, *Lagopus,* starts panting when ambient temperature is 21°C.[241] A survey of evaporative water loss from birds at moderate temperatures shows a strong weight dependence: $EWL = 0.432W^{0.585}$ is the empirical description of the weight loss function.[85a]

Several kinds of behavior retard absorption of radiant heat and enhance evaporative cooling. An ostrich fluffs its feathers at ambient temperatures above 25°C. Thermoregulation in an ostrich is so efficient that metabolism is constant over a wide environmental range.[86] An ostrich under natural conditions holds its T_b below 40°C in T_a as high as 50°C; evaporative cooling increases, as does panting, but no alkalosis develops because of a functional air shunt which bypasses the lungs.[415] Small birds dust their feathers; this may increase heat loss by conduction. Desert birds spread their wings, seek shade, or soar at considerable heights. In many of these birds the normal T_b is 41° to 42.6°C (Table 9–3, p. 398), and survival fails at 46° to 47°C.[97] A booby with a normal range of T_b from 38.3° to 40.7°C orients to the sun, shades its feet, and displays gular flutter; it broods eggs at 38 to 40°C and may sense the temperature of eggs and young with its feet.[21]

Water-birds (herons and gulls) may lose metabolic heat through the legs and feet. An overheated stork pants and voids excreta on its legs; both of these processes cease after cooling.[246] An albatross has a ventral incubation patch at a temperature lower than T_b; young are brooded at 38.8°C, and at higher temperature the young may pant and balance on their heels with their feet off the ground.[212]

Development of Homeothermy. Newborn mammals or newly hatched birds show little temperature regulation. A two-day-old laboratory mouse is essentially poikilothermic; its O_2 consumption is maximal at 32° (rectal).[358] At 10 days of age it regulates at moderate air temperatures, at 20 days at extremes. Wild mice – the smallest species an 8 g adult – show no metabolic response to cooling until 5 to 8 days old; others (larger species) respond at one day. Sustained increase in metabolism occurs at the time of first shivering and piloerection.[82] Young laboratory mice tolerate more cooling than adults; they show high levels of activity of succinate DH in their brown fat when young. They do not respond metabolically to injected epinephrine during the first four days of life, and they give adult responses only after three weeks. The thyroid becomes active during this period, and injection of thyroxine into newborn mice causes them to be responsive to NE.[276] During the first 10 postnatal days, a hamster increases V_{O_2} as T_a rises from 25° to 36°C; i.e., it is poikilothermic. For the next seven days the rate of increase of V_{O_2} with rising temperature declines to zero, and between 17 and 21 days homeothermic stability is achieved for a thermoneutral zone of 30° to 36°C, with increased metabolism at lower or higher temperatures.[384]

A 6-day-old rat has a body temperature only 1.3° to 1.8°C above T_a; adult control is attained at 25 days. Tundra voles at 10 to 12 days had a temperature of 35.4° in the nest, but outside in air of 0°C their T_b ranged from 2.5° to 16°C. A pig shows fair regulation at one day. An opossum is born in an immature state and shows its first regulation at 60 days.

Young naked birds (altricial) such as herons require much parental care and show wide temperature fluctuations, but less variation occurs after they acquire down. Precocial birds, such as gulls, have good temperature control as soon as they are hatched. House wrens show partial regulation down to 26°C T_a at 9 days, and at 15 days they maintain a body temperature of 40°C down to 10° T_a. Young altricial birds are at a disadvantage not only for lack of feathers but also because of their large surface/volume ratio. Vesper sparrows through the second day are less than 3°C warmer than the ambient air; at four days they are 10°C warmer than T_a, and at 7 to 9 days T_b is maintained above 35°. From 1 to 4 days their metabolism rises and falls with temperature; at days 5 and 6, V_{O_2} data are scattered; and at 7 days O_2 consumption increases with cooling, in the adult pattern[94] (Fig. 9–32). In quail *Excalfactoria*, cutaneous evaporation accounts for 70% of endogenous heat loss in immatures, while in adults it accounts for 13%; attainment of homeothermy is by decrease in skin evaporation and increase in metabolic heat production.[37]

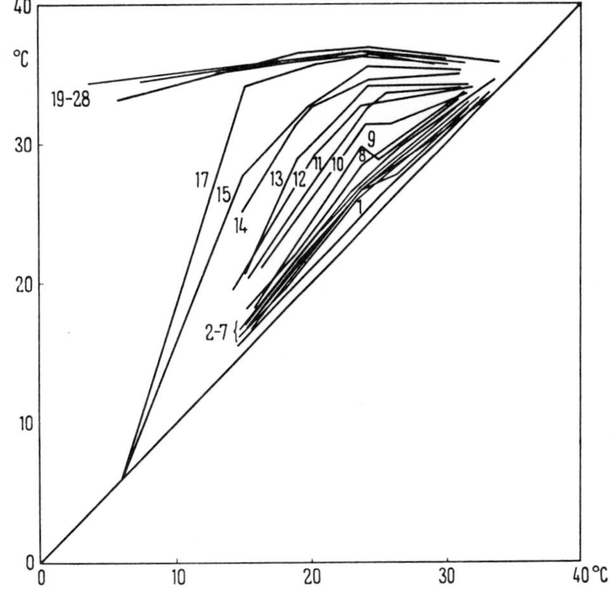

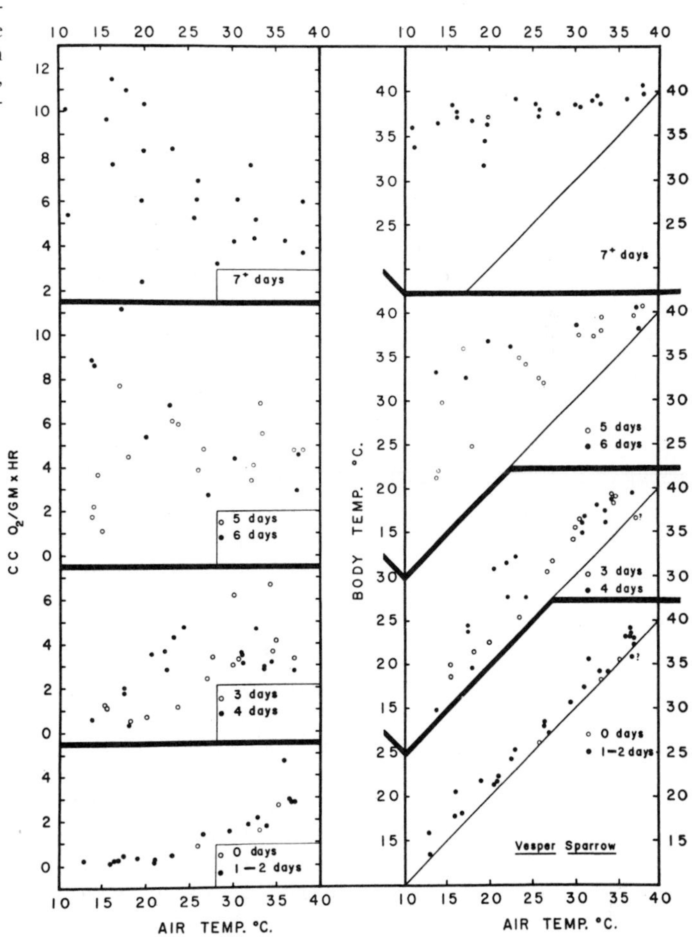

Figure 9–32. Development of homeothermy. (A) Stabilized body temperatures at different ambient temperatures for mice of ages indicated in days. (From Lagerspetz, K. Y. H., Helg. wiss. Meeresunters. *14*:559–571, 1966.) (B) Oxygen consumption figures at left and body temperature figures at right for vesper sparrows from hatching to seven days. (From Dawson, W. R., and F. C. Evans, Condor *62*:329–340, 1960.)

Lack of thermal regulation in young birds and mammals is reflected in lower cold lethal temperatures.[3] The acquisition of regulation cannot be assigned to a single factor, neural, endocrine or metabolic.

Aquatic Mammals. Aquatic mammals face thermal stresses greater than those placed on terrestrial mammals because the thermal conductance of water is greater than that of air. Many aquatic mammals also spend some time in air; hence, they may be subject to widely varying temperatures. Small animals such as water shrews, weighing 8–18 g, have an air layer trapped in their fur which reduces heat loss. The index of conductance is much higher for small than for large animals.[66]

Fur, with its trapped air, can be good insulation so long as it is not penetrated by water. A muskrat coat is 21.5% air. Young Greenland seals spend their first one and a half months in air and have fur as insulation, but adults have a layer of fat.[89] Similarly, fur seal pups tolerate hypothermia and their fur wets, unlike that of adults.[225]

A baby seal, when immersed, increases its metabolism, unlike an adult; the O_2 consumption of pups is 1048 ml/kg/hr, compared with 444 by adults. A seal in air showed a lower T_c of 0°C; in water the T_c was estimated at −20°C.[163] Thus, subcutaneous fat is sufficiently good insulation that metabolism-temperature curves fail to reveal a lower critical temperature in water. A harp seal in 10°C water has skin surface temperature of 10°C; at 22 mm depth below the skin the temperature is 22°C, and at 35 mm depth it is 39°C. The arteries and veins of fin flippers and flukes have a rete arrangement. All major arteries to fins and flippers in a seal have trabeculate veins. (See page 390.)

Insulation of a polar bear is by fur, blubber, and peripheral muscles; skin temperature was 30° to 36°C in air of −15° to −2°C.[350]

Whales have a countercurrent arrangement of vessels to tail flukes and flippers, with a large central artery surrounded by spiralling rings of small veins; the appendages function at lower temperatures than does the body.

Aquatic mammals have, at times, the converse problem of heat dissipation or keeping cool. In large animals such as cetaceans, the heat loss is twice that of a land mammal but the specific heat is so high that they could not prevent overheating if they were on land. The temperature of a dead whale dropped less than 1°C in 28 hours even when flushed with water, and the cooling half-time is about two weeks.[249] In a porpoise, a 30° gradient exists across two centimeters of blubber. The oxygen consumption by a porpoise is only 380 ml/min, or about half that of a land animal of similar size.[249] A beaver is able to regulate its T_b well with its tail in cold water and its body in air, but if the tail is also in air, the beaver overheats; thus, the tail is used for controlling heat loss.[444]

Primitive Mammals. Monotremes, some marsupials, armadillos, sloths, and anteaters tend to have low body temperatures and to show a range of T_b according to environmental temperature. Echidna in field conditions showed diurnal variation in T_b from 28.7°C to 31.5°C; in the laboratory it had a mean T_b of 30.7° in T_a from 5° to 25°C and showed a thermoneutral zone from 20° to 30°C; at 33°C it may be hyperthermic, and evaporation accounts for one third of its metabolic heat.[413] However, echidna can become torpid in the cold, especially when food is scanty; its T_b drops to about 5.7°C in air of 5°C, and torpidity may last for 5 to 10 days.[12] *Platypus* regulates its temperature well at T_b of 30° to 31° in air as cold as 0°C, but it is unable to maintain normal T_b in cold water (11°C), and hence its dives for food are of short duration; it does not show the torpid response of echidna.[62a] Among Australian marsupials, the best regulation occurs in the macropods; it is less good in phalangers, and poor in marsupial mice. A nocturnal Central American opossum, *Metachirus*, had a fairly steady T_b of 36°C at night for a T_a from 10° to 30°, but T_b was 33°C at a T_a of 10°; it was 34° at 30°C, and 40° at a T_a of 35°C.[333]

An armadillo, *Dasypus*, that is active at night shows a nocturnal temperature of 35°C down to a T_a of 0°C. In heat it vasodilates and pants, but is unable to sweat; in cold it rolls into a ball. The tenrec, a primitive insectivore, is heterothermic in showing variable, limited regulation. Its metabolism is some 50% lower than predicted for a mammal of its size, and in high T_a it can reduce its T_b by evaporative cooling.[197]

Numerous marsupials resemble higher mammals in temperature relations. The American opossum *Didelphis* increases metabolism in cold; decortication did not impair regulation, but a spinal animal could not maintain constant T_b in the cold. A marsupial dasyurid has a metabolism 32% below that of placental mammals of similar size, yet individuals larger than 100 g maintain a T_b of 37.2°C.[316a] A Tasmanian phalanger has T_b varying between 32° and 38°, averaging 34.9°C, with a thermoneutral range of 31° to 35°C.[24]

It is evident that primitive mammals use regulatory mechanisms similar to those of placentals but that their temperature is held somewhat lower.

Animals Capable of Torpor. Bats (Chiroptera) show considerable diversity in temperature regulation. Most Megachiroptera and large Microchiroptera are good regulators. They are put into alarm by cold and do not become lethargic; this is true of the fruit-eating bats. Small microchiropterans display daily torpor, and most of them can hibernate in winter. A few families—vespertilionids, rhinolophids, and desmodontids—are intermediate, and each species has its own pattern.[301]

The 80 g megachiropteran *Dobsonia* has a stable T_b of 36° to 38°C even at 40°C T_a. Its evaporative loss is 4.5 mg/g/hr over the range from 5° to 35°C, and is greater above 35°C.[23] The false vampire *Macroderma* maintains T_b at 35° to 39° in T_a from 0° to 35°C; at 38°C it pants, salivates, and licks its appendages as do many rodents.[292] Thermal conductance at 10°C is 71% of that at 30°C. In a series of megachiropterans ranging in size from 17.5 to 598 grams, metabolism at thermoneutral temperature decreased with body size according to the relation[23]

$$\text{ml } O_2\text{/g/hr} = 3.8\ W^{-0.27}.$$

A series of microchiropterans hold T_b at a level 2.5°C above T_a. At T_a of 41.5°C, the bats kept T_b below 42.5°C by evaporative cooling, but 43.5°C air temperature was lethal.[297] *Eptesicus fuscus* uses its wings for cooling. If the bat is warmed rectally, vasodilation occurs and wing temperature increases by as much as 4°C. Local heating of the preoptic hypothalamus leads to increased wing temperature, and preoptic cooling results in lowering of wing temperature. When lesions were made in the hypothalamus, this bat could still regulate but regulation began only after more rectal heating than in the intact bat. When air was blown into the mouth, hypothalamic temperature fell below rectal temperature. Low sensitivity of the hypothalamus to temperature may be adaptive for flying with the mouth open. Extreme heterothermy, possibly associated with an insensitive hypothalamus, is found in flying insect-feeding bats and birds.[264]

Myotis in summer shows diurnal fluctuations or periods of daytime torpidity. In winter it hibernates and, if T_a drops below 5°C, T_b tends to be higher than T_a and some individuals arouse.[381] *Myotis* can arouse from winter hibernation and warm from a T_b of 3°C up to 35°C within 40 to 60 minutes while the air temperature remains low, whereas if cooled to 3°C in summer torpor the bat cannot raise T_b by more than 1 or 2°C.[323] Winter bats can arouse from a lower T_b than can summer bats. Injection of norepinephrine can increase oxygen consumption in arousal 10-fold. *Myotis* makes the transition to the non-homeothermic state in late September. The mastiff bat, *Eumops*, shows daily torpor but no hibernation.[291] Small neotropical insectivorous bats readily enter torpor. The ability to undergo torpor may have originated in the tropics in small bats which needed to conserve energy reserves by low body temperatures when they were inactive.[318]

The smallest hummingbird weighs 1.7 g, and one of the largest weighs 19.1 g. Thermally, a 3 g hummingbird resembles a 3 g shrew. The metabolism of the smallest hummingbird resting in daylight is 11 to 16 ml O_2/g/hr; in flight it is 70 to 85 ml O_2/g/hr. A blue-throated hummingbird has a rest metabolism of 2.3 ml O_2/g/hr, and a giant hummingbird has a value of 2.7 ml/g/hr. These energy requirements are high relative to larger birds and mammals. Also, the index of thermal conductance is high: C = 0.17 ml O_2/g/hr/°C for large hummingbirds and 0.3 for the blue-throated variety,[145] and heart rate is very fast—7/sec to 21/sec in blue-throated hummingbird. The energy expended in flight is 6 to 12 times that at rest. Hummingbirds, therefore, have high energy requirements, and they can turn off their homeothermic mechanisms when they are not needed. In torpor a blue-throated hummingbird has a minimum O_2 consumption of 0.1 ml/g/hr at a T_a of 15°C and of 1.18 ml/g/hr at 27°C T_a. The entry into torpidity is rapid and abrupt, with a Q_{10} of 3.7 to 4.3, and without test drops such as occur in mammals entering hibernation.[145, 285] A T_b below 8°C is lethal for hummingbirds. The metabolic saving in torpidity is illustrated by the following data for O_2 consumption in torpid and in resting but awake hummingbirds at 15 to 16°C:[287]

Hummingbird	*Body Weight, Grams*	*M, ml O_2/g/hr* TORPID	RESTING
Calypte costae	3.2	0.17	10.1
Calypte anna	6.8	0.17	9.8
Eugenes	6.6	0.12	7.0
Lampornis	7.9	0.12	6.5 (2.3 at T_n of 31°C)
Phalaenoptilus	40.0	0.15	2.7
Chordeiles	75.0	0.18	2.4

Arousal from torpidity is very rapid. A 4 g hummingbird warms from 25° to 37.5°C by expending 0.057 cal, whereas a 40 g poorwill would need 0.57 cal. Similarly, cooling in a small bird can occur in seconds, compared with hours in a large bird. Hence, torpor (usually diurnal) can be used only by small birds and mammals.

A few other birds can enter the torpid state—poorwills and nightjars (*Caprimulgus*) can survive with T_b of 5 to 8°C for some hours. Lowering the body temperature to below 5°C does not cause arousal as it does in hibernators. Also, arousal occurs only after the body has become warmed to 13°C (hummingbirds) or 15°C (poor-wills)[282, 283] (Fig. 9–33).

Among mammals, some desert rodents can become torpid; their metabolism varies directly with T_a when they are torpid, and inversely when they are active. The pigmy mouse *Baiomys*, if left without food, permits its T_b to drop from a normal 32° to 36°C down to 23° to 25°C, but no further; these mice have been cooled artificially to 15°C for four days, and to 6°C for five hours[215] and survived. The pocket mouse *Perognathus* becomes torpid when T_b drops below 30°C, and its metabolism falls from 0.9 to 0.19 ml O_2/g/hr.[468] Starved *Perognathus* show torpor in T_a of 0° to 23°C, and maintain T_b well above T_a. They do not remain torpid for more than 12 hours at a time and show no test drops of T_b; the diurnal torpor lasts longer if food is restricted than if they are well fed.[306] A Brazilian murine opossum, *Marmosa,* is in torpor daily from 0900 to 1800. A kangaroo mouse, *Microdipodops,* adjusts the duration of its torpor periods from hours to days according to T_a and food supply in such a way as to conserve food.[51]

It is concluded that the ability to enter a torpid state occurs in a number of unrelated small birds and mammals, and that it has probably evolved several times. The distinction between torpor and hibernation is not sharp, and one state can merge with the other; torpidity, however, is not seasonal and is often diurnal, arousal is faster, and body temperature is not so low; some bats show both phenomena. Torpor is uniquely suited for energy conservation in animals that are small enough to warm or cool rapidly.

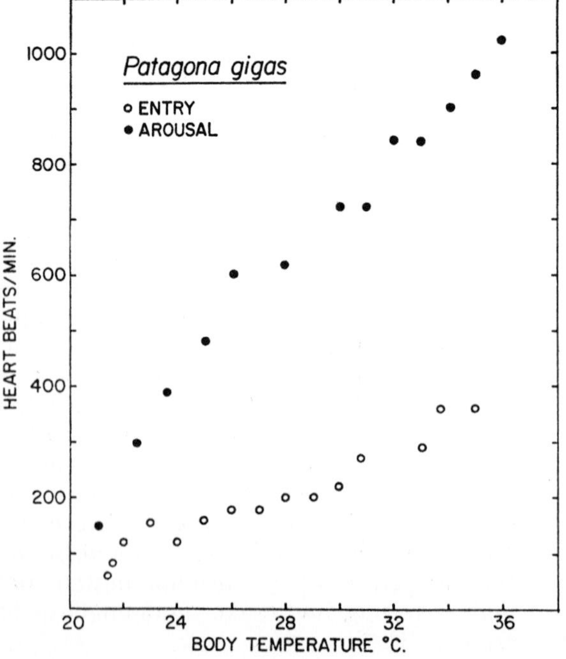

Figure 9–33. Heart rate as a function of body temperature during entry into and arousal from torpor for a hummingbird, *Patagona.* (From Lasiewski, R. C., W. W. Weathers, and M. Bernstein, Comp. Biochem. Physiol. *23*:797–813, 1967.)

Metabolism is much reduced, but nothing is known of individual enzymes or of central nervous system function in the torpid state.

Brown Fat. Most young mammals and all hibernants have an active tissue, brown fat or adipose tissue (BAT), which is histologically very different from white adipose tissue and which has a remarkable ability to generate heat. In a newborn rabbit, brown fat constitutes 4.3% of body weight, and decreases rapidly with age.[220] Brown fat diminishes as homeothermy develops. The thermogenic response to norepinephrine (NE) in young rabbits is proportional to the amount of brown fat. Normally the brown fat becomes depleted if the rabbits are placed in cold; however, if the cervical sympathetics are cut, the depletion is diminished, and if the sympathetics are stimulated, production of heat by the brown fat increases. Hence, oxidation of fat is mediated by the sympathetic nervous system.[220] In newborn and cold-acclimated adult hamsters, the cervical brown fat heats cervical vertebrae and suppresses shivering.[52] In rats, BAT accounts for all of the nonshivering thermogenesis in the first three days, and later for as much as three-fourths of it.[17] BAT increases in amount during cold acclimation in many mammals; it amounts to nearly 3% of body weight in cold-acclimated shrews and in hibernating marmots and ground squirrels. The BAT increase when animals have been kept in the cold can be matched in the warm by injections of thyroxine plus norepinephrine.[289] Cold acclimation increases α-GPDH, LDH, and CoA in brown fat.[17]

The metabolic pathways of BAT produce heat with little production of ATP; i.e., P/O ratios are low. The mitochondria normally shown an uncoupling of phosphorylation from oxidation.[143] BAT from 6°C-acclimated rats oxidizes α-ketoglutarate at twice the rate of BAT from 26° rats, but P/O ratios are similar.[432] Heat generation is mainly from oxidation of fatty acids. Triglycerides account for 99 per cent of the total lipid, and norepinephrine (NE), acting via cyclic AMP, releases fatty acids from the triglyceride. Epinephrine, norepinephrine, and cyclic AMP (also oleate) increase respiration by 30 to 40-fold; uncouplers of phosphorylation increase respiration and decrease the response to NE.[492] Brown adipose tissue from a bat shows lower activation energy (temperature coefficient) and greater oxidation of palmitic, oleic, and succinic acids than BAT from a rat.[113] The fatty acids which are released from BAT by NE not only provide oxidative heat but also regulate energy transfer, apparently by acting prior to incorporation of inorganic phosphate much like uncouplers of oxidative phosphorylation.[374]

Treatment of rats for two weeks with oxytetracycline, an inhibitor of protein synthesis, prevents the metabolic response of BAT to cold or to NE; synthesis of specific oxidative enzymes in mitochondria of BAT (but not of liver or muscle) is essential for cold acclimation.[200]

NEURAL MECHANISMS OF THERMOREGULATION

The effect of cold on nerves was discussed on page 391.

Thermoreception. Mechanisms of heat regulation are activated from thermoreceptors in the skin, in deep organs, and in various parts of the central nervous system.

Peripheral thermal sense. Heat and cold receptors are distributed in a definite pattern in the skin, where warmth receptors are usually deeper than cold receptors; cold receptors are the more abundant. Some thermoreceptors are encapsulated, but many are free nerve endings. Much sensory summation occurs, in that the sensation threshold for stimulation of large areas is less than that for stimulation of single sensory endings. Sensations can be perceived for changes in temperature of a few thousandths of a degree in a second. Heat (infrared) stimulation can also elicit pain, and both heat and pain are conducted in small fibers, but different endings are probably involved. Some cold receptors are also sensitive to pressure.

Impulses have been recorded in various mammals from nerve bundles and from single sensory fibers from cold and warmth receptors.[192] Both types of receptor show steady (spontaneous) activity; the fibers from warm receptors are active over the range from 20° to 47°C with a maximum frequency between 38° and 43°C, while the fibers from cold receptors are active between 10° and 40°C with maximum frequency between 20° and 34°C (Fig. 9–34). The maximum steady frequency from cold receptors is about 10/sec, and that from warmth receptors is 3.7/sec, but the warmth receptors discharge more irregularly.

When the skin temperature is abruptly

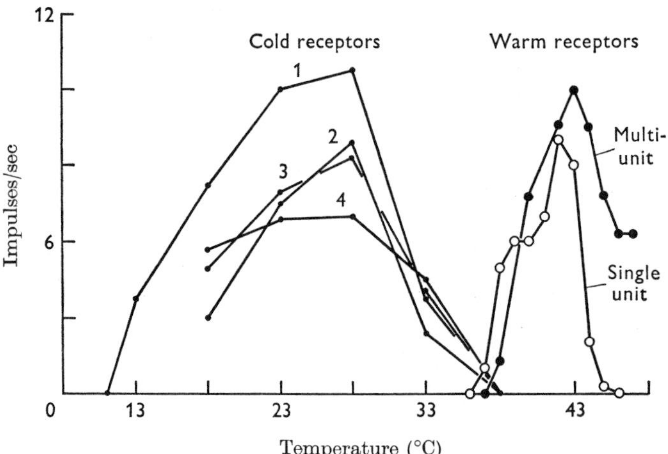

Figure 9–34. Frequency of impulses in nerve fibers from cutaneous cold and warm receptors in primates. (From Iggo, A., J. Physiol. *200*:403–430, 1969.)

raised, there is a sudden brief increase of discharge from heat receptors which lasts a second or two; then the discharge settles down to the frequency characteristic of the temperature. On removal of the heat stimulus, the warmth fiber temporarily reduces its frequency or stops firing. Similarly, when the temperature is lowered, a cold fiber gives a brief extra discharge (at a frequency as high as 140/sec) and then settles at its new frequency, which is higher than before because the temperature is lower; on removal of the cold stimulus the activity of the cold fiber is temporarily inhibited. Cold fibers are silent between about 35° and 45°C, but between 45° and 50°C they show increased discharge, a paradoxical response of cold fibers to heat. By measurement of cutaneous temperature and by applying thermal stimulation beneath the skin as well as from the outside, it was determined that stimulation depends on the absolute temperature of the endings rather than on the thermal gradient.[192]

Temperature receptors in the nasal tract of the cat send impulses in the infraorbital nerve with maximum frequency of 45 to 47/sec for warmth receptors and 27/sec for cold receptors.[190, 191] The rate of change of firing of nerve impulses per degree change in temperature varies from 20 to 80 imp/sec/°C. Similarly, in *Rhesus*, warmth receptors fired steadily at 32°; they increased frequency up to a maximum at 40° to 44° for warming at 20°/sec, and then declined in frequency at higher temperatures.[191] Some cutaneous thermoreceptors fire in bursts, and others fire continuously; in dog the sensitivity of cold fibers was maximum at 31° to 37°C, and that of warm fibers at 40°C.[223]

There are deep body receptors which initiate shivering even though skin and brain temperatures are kept constant. Some deep body receptors are located in veins; warming the femoral vein of a dog elicits a rise in blood pressure.[460]

Central thermoreceptors and the temperature regulating center. Cooling the spinal cord induces shivering and vasoconstriction, while warming it stops shivering, and brings about vasodilation and panting in dog; the effect of changing the temperature of the spinal cord is additive with the hypothalamic changes, and cord temperature can cancel temperature effects on hypothalamus.[237] Possibly cold tremor can start in spinal motoneurons. Evidence for sensitivity of spinal cord to cold and warmth has been obtained in dog,[237] cat,[270] and pigeon.[377] Some neurons in the preoperative hypothalamus change their rate of firing if either spinal cord or hypothalamus is cooled or warmed.[142]

Impulses from peripheral receptors ascend via lateral spinothalamic tracts and thalamus to the hypothalamus, from which various autonomic reflex responses are activated. In the hypothalamus there are also centers which are sensitive to temperature. Here cold and heat sensitive neurons are not spatially separated but appear to be intermingled, and there may be some mutual inhibition. Dogs with lesions in the posterior hypothalamus lost the ability to regulate in the cold, and cats with lesions in posterior lateral hypothalamus became somewhat poikilothermic. Local warming of the anterior hypothalamus of a cat caused peripheral vasodilation and increased breathing. At a T_a of 25°, local cooling of the preoptic hypothalamus leads to hyperthermia, while at a T_a of 35° similar cooling is ineffective. In bats, local warming of the preoptic area of the hypothalamus causes vasodilation in the wings and consequent heat loss; hypothalamic cooling had the opposite action. Rats, trained to press a bar for heat when cooled, pressed when only the hypothalamus was cooled by a thermode.[75]

In a baboon, local cooling of the preoptic region of the hypothalamus induces shivering and vasoconstriction, liberation of NE and epinephrine, and an increase in plasma 17-ketosteroids, while cessation of cooling leads to vasodilation; warming the preoptic hypothalamus decreases liberation of NE and epinephrine and decreases plasma 17-ketosteroids.[136, 462] In goats, cooling the anterior hypothalamus with local thermodes activates the thyroid and sympathetics, and raises the concentration of PBI within a few minutes; cutting the pituitary stalk abolishes the effect of hypothalamic cooling and blocks the NE effect.[7]

Electrical activity was recorded from each of some 1000 units in the anterior hypothalamus in cats.[162] Sixty per cent of the units were unaffected by local temperature change; of the others, 4/5 increased in frequency on heating and 1/5 on cooling. Similarly in rabbit, 1 to 2 mm from the midline of the hypothalamus, 10 per cent of the tested units responded to temperature; 2/3 of these increased frequency on warming and decreased it on cooling, 1/3 showed the reverse response, and a few cells fired at frequencies proportional to temperature from 2° above T_b to 2° below it.[64, 188] Local cooling of preoptic hypothalamus elicits shivering, or if access to a heat source is available, to turning on the heat. Rats with hypothalamic lesions showed impaired physiological warming but kept the behavioral warming response.[403, 404]

Temperature-sensitive neurons are evidently widely distributed; peripheral, deep body, spinal cord, and hypothalamic receptors may all contribute to integrated thermoregulation. The hypothalamus is the controlling center which acts via autonomic and neuroendocrine systems.

One hypothesis[125] of the action of the hypothalamus is that the responses depend on a balance between release of epinephrine and 5HT in the diencephalon. Norepinephrine infused into the hypothalamus of cat, dog, or monkey causes initial vasodilation and fall in T_b. 5HT causes shivering and rise in T_b, an effect antagonized by NE. The content of NE in the hypothalamus increases after heat exposure. In rabbits, sheep, goats, oxen, and rats opposite effects occur. NE causes a hyperthermia and 5HT a hypothermia.[125] See p. 497 for data on 5HT.

Another hypothesis is that changes in the ratio of sodium to calcium occur in critical regions of the brain; monkeys perfused in the brain ventricles showed a rise in T_b when Na was increased and a fall in T_b when Ca was increased.[340]

The set point hypothesis of Hammel is that each given regulatory response (e.g., vasoconstriction) has a set threshold temperature:

$$R - R_0 = A_R\,(T_{hypo} - T_{set\,R})$$

where R is the regulating response in cal/kg/hr, R_0 is the basal level of this responding reaction, A_R is a factor which is positive for heat loss and negative for heat production, T_{hypo} is the temperature of the hypothalamus, and $T_{set\,R}$ is the threshold or reference point.

The response is proportional to $(T_{hypo} - T_{set\,R})$. $T_{set\,R}$ can vary with both skin and core temperature so that either a high T_b or a warm T_a can lower T_{set} for vasoconstriction or low T_b or T_a can raise it; thus, evaporative heat loss is a function of both T_{hypo} and of T_a.[153] Set points are different for each regulatory response – panting, shivering, and vasomotor responses. In the cat, the thermoregulatory component of metabolism V_{O_2} at different rectal temperatures extrapolates to a preoptic temperature of 43°C. Preoptic temperature thresholds can be modified by skin temperature.[230]

Numerous block diagrams have been presented as models of thermoregulation.[153, 160, 161, 446] In each there is a controlled system (some portion of T_b) for which a disturbance in temperature is a sensory signal from the environment. A feedback signal from the controlled system goes to the controlling system in the hypothalamus. Here, comparison is made with a reference signal or set point. The controlling system then signals appropriate responses to bring the

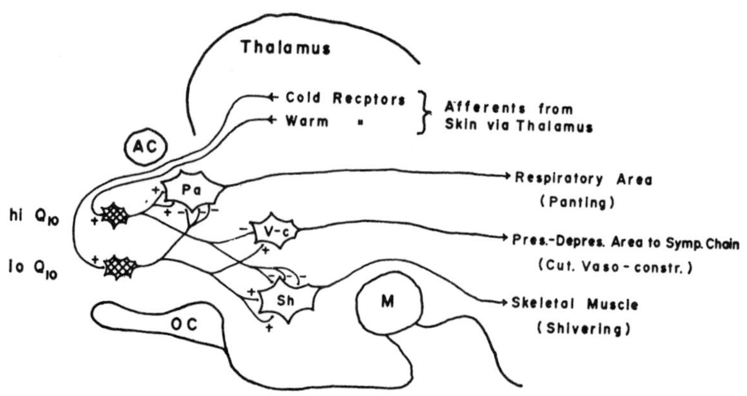

Figure 9–35. Diagram of possible arrangement of neurons in thermoregulatory center in hypothalamus. Afferent fibers from skin receptors impinge on neurons, which themselves have different thermal sensitivities. These in turn act on respiratory, vasomotor, and shivering-inducing centers, each with its own set point for balancing input from the thermoreceptors. (From Hammel, H. T., *in* International Symposium on Natural Mammalian Hibernation 1965: Mammalian Hibernation III, edited by K. C. Fisher, American Elsevier Publ. Co., New York, 1968.)

controlled system to a point of minimum feedback signal[148] (Fig. 9–35).

Temperature regulation can be modelled by comparing the output of the effector systems to the temperature change at the site of the central nervous temperature receptor. Four basic forms of controllers used industrially to control temperature, on-off, proportional, rate, and integral, have been compared to biological temperature regulation.[160, 446] The equations governing these types of controllers are:

On-off

$$T_b > T_{set};\ Y = \max$$
$$T_b \leq T_{set};\ Y = 0$$

Proportional

$$Y_1 - Y_0 = -\alpha(T_b - T_{set})$$

Rate

$$Y_1 - Y_0 = -\gamma \frac{dT_b}{dt}$$

Integral

$$Y_1 - Y_0 = -\beta \int_{t_0}^{t} (T_t - T_{set})dt$$

In these equations, the following notation applies:

T_b = regulated temperature
T_{set} = initial or preset temperature
Y = effector response
$Y_1 - Y_0$ = differential effector response
α, β, γ = proportionality constants

On-off and proportional controls are particularly useful approximations to biological temperature control.[161] On-off models describe behavioral thermoregulation and physiological mechanisms in which the effector response cannot be graded—as in cases in which a lizard goes from sun to shade.[177] Proportional control is useful for modeling heat production and evaporative heat loss.[150] Constancy of T_b may be obtained with a proportional controller if T_{set} varies with environmental temperature, deep body temperature, and state of activity.[150] The change in T_{set} is inversely proportional to body size.[179]

HIBERNATION

A spectrum of patterns of hypothermia ranges from daily torpor, usually with shallow cooling, to seasonal dormancy or hibernation, which may be shallow or deep. Hibernation differs from daily torpor in that an animal usually enters hibernation by test drops of temperature, and in that T_b may go to lower levels. Recovery from each is by endothermy. Hibernation is usually on an annual cycle with much preparation; dormancy may occur daily or with mild cooling or food deprivation. The biochemistry and neurophysiology of the two states may be different, but it is not understood for either one.

Hibernation involves a setting of the thermoregulator down to a low level (often about 2°C) and such changes in most organ systems that homeokinesis is maintained even in a relatively inactive state. Only a few mammals—some monotremes, insectivores, rodents, and bats—are capable of true hibernation. Some large mammals, such as bears, go without food for long periods but maintain body temperature at about 31°C. In terms of the theory of thermoregulation, the set points for heat conservation would be lowered by input from the reticular activating system of the brain, and mechanisms of heat conservation would be turned off. In arousal from hibernation, thermal conductance of the periphery decreases, heat is produced particularly in the thorax, and T_b rises rapidly.

There is good evidence for an annual cycle underlying hibernation. Ground squirrels (*Citellus*), maintained for two years at constant temperatures (32°, 70°, and 97°F) and 12 hour photoperiod, showed 1-year cycles in weight and food consumption; those at 32° and 70°F hibernated, and more hibernated if food was limited. Other animals, observed for four years at 12°C or 3°C, showed an annual rhythm of temperature and hibernation. The annual cycle under constant conditions was about 15% shorter than one calendar year.[356, 357]

Evidence has been presented[91] that serum from a hibernating ground squirrel can induce hibernation in a recipient, even in summer.

Preparation for Hibernation. Some hibernators, such as *Citellus*, accumulate body fat in the fall; they have a respiratory quotient greater than 1.0 and synthesize much fat from carbohydrate. Others, such as the hamster *Mesocricetus*, do not accumulate fat but in the cold they store food and build a nest. In the cold (5°C), hamsters desaturate their fat and deposit fats of lower melting point. In hedgehogs, the insulin level in blood becomes elevated. The thyroid becomes involuted, and injected thyroxin can retard hibernation. Pituitrin injections can arouse hibernating hamsters. Adrenalectomized animals do not hibernate, and in hibernation the adrenal cortex is depleted. Hibernating animals do not respond by thermogenesis to injections of norepinephrine (NE). While several endocrines are involved, there is no causal relation between any single endocrine and hibernation.

Some species like hamsters delay entry into hibernation when in cold, and then enter smoothly; others do not delay but enter directly, as *Perognathus*; still others enter by periodic shivering, rewarming, and dropping of temperature. Body temperature drops slowly, 2° to 4°C per hour; the drops are associated with peripheral vasodilation and reduction in muscle tone. At 33° to 34°C T_b there is abrupt drop in heart rate. Breathing shows stepwise slowing; the respiratory system can still be stimulated by hypoxia.

The State of Hibernation. As body temperature falls, the heart slows to a minimal rate, and at slightly below 3°C the beat is irregular in most species. The heart of an active ground squirrel beats at 200 to 400/min, while that of a dormant one beats at 7 to 10/min. The P-T interval of the electrocardiogram is lengthened and block of conduction may result in uncoordinated beats. In hibernating hamster, stimulation of the vagus is without effect on the heart. Isolated hearts of hibernators beat at lower temperatures (below 1°C for *Citellus*) than do hearts of nonhibernators (13° to 16°C for *Sciurus*);[299a] hedgehog and hamster hearts stop at 1.5° and 6.0°C, and rat and rabbit hearts at 16° to 18°C.

When atria of ground squirrels are cooled to 6°C, the resting potential shows little decline and the action potential actually increases, whereas in rabbit the RP declines below 25°C and the AP disappears below 17°C.[308] Cardiac output of a hibernating ground squirrel was 1/65 that of an active one.[361] Cardiac arrest occurs at the following temperatures: guinea pig 13.4°C, rat 6°C, non-hibernating hamster 10.2°C, hibernating hamster 1°C.[426] In a hedgehog, *Erinaceus*, below 6.4°C, the vagus has little effect on heart and blood pressure, and in rewarming the heart is unresponsive to the vagus until 20°C is reached; at the low temperature some responses to acetylcholine and epinephrine were observed.[239] Similar insensitivity of the heart to the vagus was noted in hibernating *Citellus*.[300] Neuromuscular block in a phrenic nerve-diaphragm preparation is at 10°C in rat, at 5°C in awake hamster, and below 5°C in hibernating hamster.[441]

Activity in the nervous system does not entirely stop when the body cools. Cortical activity (brain waves) is lost in a hamster at a T_b of about 19°C, although some cortical responses to peripheral nerve stimulation could be seen as low as 9.1°C. In ground squirrel, brain waves persisted to 12°C, and in woodchuck to 11°C. *Citellus* shows low-amplitude cortical waves at 5°C, and at a brain temperature of 6.1°C it can still localize sound, erect pinnae, vocalize, and move.[447]

When lesions have been made in the posterior hypothalamus of hamsters, they fail to enter hibernation; when lesions are made in the anterior hypothalamus of ground squirrels, they may enter hibernation but they fail to arouse.[404, 406] Forced breathing of CO_2 in hibernation causes an increase in breathing rate and then in heart rate. When stimulated by CO_2, a hamster remained asleep with faster breathing.

The temperature of hibernating mammals is essentially that of the surrounding air, and its fluctuations may roughly follow those of the environment. Colon and esophagus temperatures may be 2 to 3°C above air temperature. *Citellus* in air of 5.5°C had a brain temperature of 8.7°C, and in air of 2°C it had a brain temperature of 6°C. When the air temperature approaches 0°C, some animals (hamster, woodchuck, dormouse) tend to hold T_b at about 2°C and their oxygen consumption can increase. They retain some capacity for defending body temperature from cooling too low. Many hibernants awaken periodically, some of them to eat and drink. Tagged bats show much individual variation in winter movements. Ground squirrels were awake 7% of the time in hibernation. A ground squirrel during an arousal of a few hours (on an average of once every 11 days) uses as much body reserves as in 10 days of sleep. Hamsters awaken every few days.

In hibernation the metabolism is reduced by 20 to 100 times. A marmot, awake, produced 2.8 kcal/kg/hr, but in hibernation it produced only 0.09 kcal/kg/hr. The respiratory quotient corresponds to metabolism of fat. Despite the low metabolic rate, considerable weight is lost—a brown bat lost 33% of body weight in 180 days, and fat decreased from 28% to 10% of body weight.

The magnesium content in serum is high in hibernating hedgehogs and marmots, in bats hibernating at 13°C but not at 17 to 20°C, and in hibernating ground squirrels. Awakening does not require a lowering of serum Mg, and hence high Mg concentration is not causal.

Blood sugar is regulated at low levels in some species, and recovers, after an initial drop, to nearly normal values in others. Leukocyte count may be reduced and hematocrit diminished slightly. Clotting time is prolonged, apparently by a decrease in prothrombin. Hibernating ground squirrels and bats are temporarily resistant to x-irradiation; after irradiation there is little apparent effect until arousal some days later, when the usual drop in blood cells occurs.

The endogenous respiration of slices of a variety of tissues (excluding liver) of ground squirrels is higher in the fall than in the spring, and is lower if they are from hibernating animals.[158] Kidney slices from hibernating *Citellus* have enhanced capacity for gluconeogenesis and for glycolysis, properties that may be related to the tolerance of hypoxia in the hibernating animals.[60, 61] The Q_{10} of respiration of mitochondria prepared from the liver of hibernating ground squirrels was higher than from non-hibernants; i.e., their temperature sensitivity was enhanced.[298] However, the Q_{10}'s of oxidative phosphorylation by heart mitochondria and of oxygen consumption and glycolysis by brain are similar for hibernating and non-hibernating hamsters.[440] Differences in effects of extramitochondrial factors have been noted between hibernating and non-hibernating species. After five to seven weeks of exposure of rats to 5°C, liver mitochondria showed unaltered respiration, whereas O_2 consumption by homogenates (with added succinate) was increased and the P/O ratio was decreased. In contrast, the respiration of liver mitochondria from hamsters was increased and the P/O ratio decreased by cold-exposure without hibernation, while the reverse was true in hibernation.[298] In the brown fat mitochondria of cold-exposed rats, respiration (succinate) was increased and the sensitivity to stimulation by ADP at low temperatures (5° to 15°C) was reduced, whereas mitochondria from cold-exposed hamsters (brown fat) and hibernating ground squirrels (liver) were more sensitive to ADP at 5° to 15°C.[298] It must be concluded that no general pattern of metabolic properties distinguishes hibernants from non-hibernants.

The capacity for lipogenesis in liver is four times greater in a hamster than in a rat, and on arousal from hibernation hamster lipogenesis increases rapidly.[104]

It is essential that hibernants maintain normal ionic gradients at low temperatures, and evidence is accumulating for uniqueness of their cell membranes. The net effect of maintaining normal ionic distribution is prevention of conduction block and swelling in cold and continuance of protein synthesis

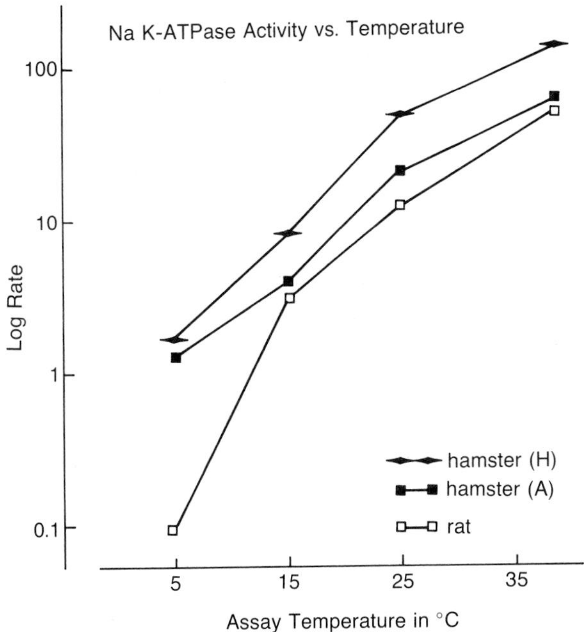

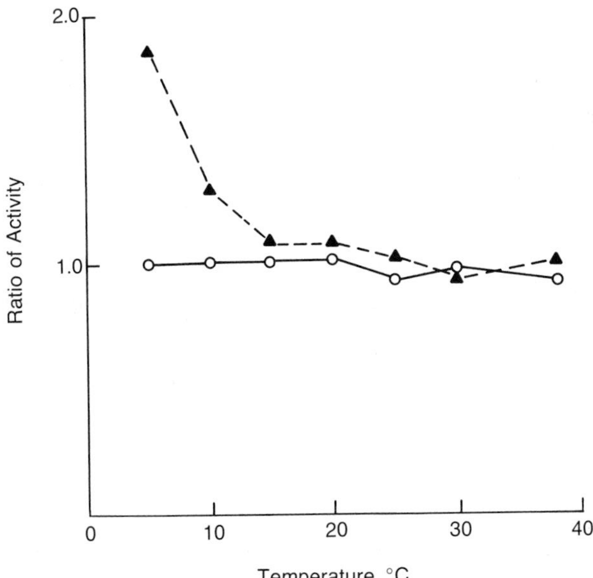

Figure 9–36. Na-K ATPase of hibernator and non-hibernator. (A) Enzyme activity of slices of kidney from hibernating and from awake hamster and from rat; measurements at indicated temperatures. (Courtesy of J. S. Willis and L. Fang.) (B) Ratios of activities of ATPases of brain from hibernating hamster to activities of brain from awake hamster, measured at different temperatures: triangles are Na-K ATPase, and circles are Mg ATPase. (Courtesy of S. Goldman and J. S. Willis.)

and absorption of sugar and of amino acids. Ionic gradients are maintained by properties of passive fluxes and of active transport, and differences in active transport between hibernators and non-hibernators are well documented. Ion transport of K is depressed more by cold in nonhibernants than in hibernants. For example, slices of ground squirrel kidneys retain potassium concentration during several days of storage at 3°C, but slices of rat kidney lose potassium.[493] Swelling occurs at 6°C in slices of heart and diaphragm of rat more than in slices of hamster or ground squirrel.[494] The portion of potassium influx into red blood cells which is active (i.e., is ouabain inhibited) is maintained at 5°C in ground squirrels but drops to zero in guinea pigs and humans.[259] The Na-K ATPase of brain of hibernating hedgehog and of kidney of hamster is cold resistant; the corresponding enzyme in non-hibernants is cold-sensitive. Evidence for compensatory acclimation such as occurs in poikilotherms (page 375) is that the Na-K ATPase of kidney from hibernating hamster is twice that from awake hamsters (Fig. 9–36). It may well be that membrane properties are much more critical than are energy liberating systems for the survival of hibernants at low temperatures.[495]

Arousal. Arousal from hibernation is a rapid awakening, with warming starting in the thorax. Bats kept in a refrigerator for 144 days without food were able to fly within 15 minutes at room temperature. A ground squirrel's temperature may rise from 4°C to 35°C in four hours; a birchmouse warms at 1° per minute. The arousal of a mammal is a process of self-warming and does not require external heating. The temperature of the chest rises first, and blood distribution brings about warming of the anterior half of the body. Vasoconstriction occurs posteriorly and flow to thoracic muscles increases 16-fold.[239] The temperature of thoracic brown fat increases rapidly, and oxidation of triglycerides is activated by adrenergic sympathetic nerves, and possibly partly indirectly via ACTH.[422] In the first hour of arousal, CO_2 production increases 15-fold, and the respiratory quotient is 0.7; evidently, fats are being used.[449] Early in arousal of ground squirrels, CO_2 production increased 15-fold and there was acclerated turnover of triglyceride.[449] In a bat, the temperature of brown fat rises faster than that of heart; thermogenesis is non-shivering, and arousal is possible but slowed if the animal is curarized. However, in ground squirrel and *Glis*, warming is greatly impeded if shivering is blocked. During arousal in *Myotis*, blood glycerol increased from 1.3 mg per cent to 25 mg per cent, and if α-adrenergic receptors were blocked with propanolol the time required for arousal was doubled; nonshivering thermogenesis accounted for 55% of the total heat in arousal.[186]

The heart accelerates rapidly; e.g., in a spermophile asleep at 5°C, the heart rate was 3/min, on awakening at 8°C it rose to 20/min, at 14°C it was 200/min, and at 20°C it was 300/min. Breathing becomes regular early in arousal, and then accelerates. As the animal emerges from hibernation, there is a metabolic overshoot; oxygen consumption by a hamster rose from 0.5 ml/kg/hr to 8000 ml/kg/hr, and then after temperature was stabilized it settled down to about 5000 ml/kg/hr. Electrical activity of the brain cortex shows a marked increase at about 20°C, varying with species. The termination of hibernation starts with the autonomic nervous system, and it would be informative to have measurements from the basal ganglia and reticular arousal system.

Hibernators differ from non-hibernators in many respects—deposition of much brown fat, capacity of cell membranes to maintain activity at low temperature, low sensitivity of nerves to cold-block, and unique properties of many enzymes and of protein synthesis. The most critical event in entering hibernation is the presumed change in setting of the temperature-regulating center, and that in arousal is the activation of the sympathetic nervous system. Hibernation cannot be ascribed to any single system; rather the hibernator retains integration of all body systems at low levels of activity.

SUMMARY

Temperature is a measure of the heat content of a physical body, and in solutions temperature reflects molecular agitation and thus determines velocity of chemical reactions. The energy of activation of a reaction can be ascertained by measuring the velocity at different temperatures and calculating the thermal increment from the different velocities. In complex systems, such as living cells, a useful measure is the Q_{10}, but the value of this varies with temperature range, substrate concentration, and modulating factors. Also, the apparent enzyme-to-substrate affinity (K_m) increases (perhaps because of persistence of the ES complex) as temperature is lowered. The K_m is genetically determined and is adaptive for particular temperature ranges corresponding to the ecology of animals.

Temperature limits for biological activity are closely related to the properties of water, to denaturation of proteins, and to liquefaction of lipids. Water as solid crystals of ice inside cells is incompatible with life; freezing of water outside cells leads to dehydration. Resistance of animals to freezing can be (1) by replacement of some water by some organic solvents, such as glycerol, (2) by lowering freezing point by an increase in solutes (some glycoproteins acting out of proportion to their colligative properties), (3) by supercooling with the aid of organic hydrogen-bonding molecules, and (4) by resistance to ice formation by unknown properties of cell membranes. Cold death occurs at temperatures well above freezing; it may result, in different kinds of animals, from the slowing of energy-yielding reactions, mismatching of components in metabolic chains, central nervous failures, or reduced supply of oxygen, and most certainly from cell membranes becoming leaky, partly because of slowing of ion pumps.

High temperatures are limiting or lethal by enzyme denaturation at high extremes, by acceleration of some reactions relative to others, by exceeding the temperature optimum of single reactions, and by the "melting" of lipids, especially in cell and mitochondrial membranes, so that they become freely permeable.

The limits for survival of intact animals are narrower than for isolated cells and tissues, and these survive within a narrower temperature range than the enzymes extracted from them. In general, central nervous systems impose close limits for activity; the causes of neural failure at low and high temperatures are not known. Limits of survival of both whole animals and some tissues can be changed with acclimation, which probably accelerates synthesis of specific lipids and proteins. Cold hardening in plants involves a series of biochemical modifications. In addition, genetic variants of many enzymes in animals and microorganisms function best within restricted temperatures.

Capacity to function well over a "normal" environmental temperature range involves many types of compensations. One group of animals—temperature conformers, poikilotherms—vary body temperature with the environment, but they compensate metabolically so that activity can remain relatively constant over a temperature range. Another group—temperature regulators, homeotherms—keep relatively constant body temperature but when cooled increase metabolism to provide body heat. A third pattern, which can be characteristic of either a poikilotherm or homeotherm, is to enter a state of torpor, diapause, or hibernation in environmental cold. Compensations for heat are more limited and more specific to individual kinds of animals than are compensations for cold.

Three periods can be distinguished in the time course of response to a change in environmental temperature. First, there are direct responses, immediate and adaptive accelerations or decelerations of reactions, both sensory and behavioral; these may last for minutes or hours. The magnitude of response depends on amount and rate of temperature change, and its nature varies with particular kind of animal. Second, there are the delayed alterations of acclimation or acclimatization which compensate for the environmental change. Finally, there may, over many generations, be genetic changes (acclimatization) which provide the material for selection of mutants adapted to particular temperature ranges.

The second category, capacity acclimation changes, in poikilotherms vary according to the temperature range, the hormonal and nutritional state, and whether the animal becomes torpid or remains active in the cold. Animals from relatively constant environments may show less acclimation than those from variable environments. Enzymes concerned with energy liberation compensate, while those concerned with degradation of metabolites often fail to compensate or their activity changes in the opposite direction. Molecular mechanisms of compensation are enhanced protein (and probably RNA) turnover, selective synthesis of isozymes with suitable kinetic properties, and changes in modulating agents, especially phospholipids (increased unsaturation in cold). It is suggested that membrane lipids may be critical for both resistance and capacity adaptation.

Acclimation in temperature regulators involves (1) changes in insulation—vasomotor, fat, hair, and feather coats, (2) changes in sensitivity to regulating hormones, especially of the adrenal and thyroid glands, (3) changes in levels of some enzyme activities, which are poorly known, and (4) changes in overall sensitivity to temperature extremes. Acclimation to cold is more effective than to heat.

The distinction between temperature conformers and regulators is not sharp. Poikilotherms are not totally dependent upon the Q_{10} relation; many of them show long-term metabolic compensation in cold and warmth, some suspend activity below certain temperatures, and some, especially insects and reptiles, show behavior which permits absorption of external heat or dissipation of excess heat. A few poikilotherms, especially moths and large fish, produce heat in body muscles and raise their temperature endothermically. Some have peripheral thermoreceptors, and some sense the temperature of the nervous system to trigger behavioral or endothermic reactions. Some brain structures are more thermosensitive than others. Similarly, many temperature regulators change the set point of their thermostats and conserve energy in the cold by becoming lethargic or going into hibernation. The receptors in homeotherms are multiple—cutaneous, deep vessel, spinal cord, and especially hypothalamus. A hierarchy of controls is evident according to temperature stress. A widespread mechanism for reducing body temperature is evaporative cooling; other animals permit body temperature to rise and limit water loss.

It should be emphasized that function systems of all animals remain coordinated over a range of temperatures. A hibernator's brain and cardiovascular system maintain control whether at 37° or at 5°C. Once one system fails, as when the brain of a non-hibernator goes below 25°C, life is in jeopardy. Similarly, in compensating poikilotherms, many enzyme systems change adaptively. Temperature is an all-pervasive physical parameter, and the responses and adjustments to change in temperature are multiple. In the evolution of species, very many mutations have been selected that permit survival over wide ambient temperature ranges and favor extension of geographic ranges. Thermophilic and psychrophilic organisms have appropriately set enzyme systems for functions at high or low temperatures.

REFERENCES

1. Adams, P. A., and J. E. Heath, Nature *201*:20–21, 1964. Temperature regulation in sphinx moth *Celerio*.
2. Adams, T., and B. G. Covino, J. Appl. Physiol. *12*:9–12, 1958. Racial variations in response to cold stress.
3. Adolph, E., Origins of Physiological Adaptations. Academic Press, New York, 1968. 144 pp. Development of thermal tolerance.
4. Aleksiuk, M., Comp. Biochem. Physiol. *39A*:495–504, 1971. Temperature-dependent metabolism of reptile *Thamnophis*.
5. Andersen, K. L., and O. Wilson, Acta Univ. Lundensis, Sec. II, 1966, Nos. 11–20. Tolerance of men to cold in field conditions.
6. Anderson, T. R., Comp. Biochem. Physiol. *33*:663–687, 1970. Temperature adaptation, membrane phospholipids in goldfish.
7. Andersson, B., et al., Acta Physiol. Scand. *59*:12–33, 1963; *61*:182–191, 1963. Control of secretion of thyrotropic hormone and of catecholamines by heat loss center.
8. Andjus R. K., Symp. Soc. Exp. Biol. *23*:351–394, 1969. Mammalian tolerance of low body temperature.
9. Andrews, S., and J. Levitt, Cryobiology *4*:85–89, 1967. Cryoprotective agents in protection against freezing.
10. Asahina, E., pp. 451–486. *In* Cryobiology, edited by H. T. Meryman, Academic Press, New York, 1966. Freezing and frost resistance in insects.
11. Ashwood-Smith, M. J., pp. 5–42. *In* Current Trends in Cryobiology, edited by A. U. Smith, Plenum Press, New York, 1970. Resistance to freezing in microorganisms.
12. Augee, M. L., et al., J. Mammal. *49*:446–454, 1968; *51*:561–570, 1970. Body temperature and torpor in echidna.
13. Baldwin, B. A., and D. L. Ingram, Physiol. & Behav. *2*:15–21, 1967. Behavioral thermal regulation in pigs.
14. Baldwin, J., and P. W. Hochachka, Biochem. J. *116*:883–887, 1970. Interpretation of isozymes in thermal acclimation.
15. Barańska, J., and P. Wlodawer, Comp. Biochem. Physiol. *28*:553–570, 1969. Effect of temperature on fatty acids and lipogenesis in frog tissues.
16. Barlow, G. W., Biol. Bull. *121*:209–229, 1961. Metabolic differences in northern and southern populations of gobiids.
17. Barnard, T., J. Skála, and O. Lindberg, Comp. Biochem. Physiol. *33*:499–508, 509–528, 1970. Brown fat in developing rats.
18. Barnes, H., and M. Barnes, J. Exp. Marine Biol. Ecol. *4*:36–50, 1969. Effects of temperature on O_2 consumption of barnacles.
19. Barnett, L. B., Comp. Biochem. Physiol. *33*:559–578, 1970. Seasonal temperature acclimatization of house sparrow.
19a. Barnett, S. A., Biol. Rev. *40*:5–51, 1965. Adaptations of mice to cold.
20. Bartholomew, G. A., Copeia 241–250, 1966. Temperature relations of marine iguanas.
21. Bartholomew, G. A., Condor *68*:523–535, 1966. Behavioral regulation of temperature in booby.
22. Bartholomew, G. A., Symp. Soc. Exp. Biol. *18*:7–29, 1964. Behavioral thermal regulation in desert animals.
23. Bartholomew, G. A., W. R. Dawson, and R. C. Lasiewski, Z. vergl. Physiol. *70*:196–209, 1970. Thermal regulation and heterothermy in small flying foxes.
24. Bartholomew, G. A., and J. W. Hudson, Physiol. Zool. *35*:94–107, 1962. Temperature relations of pygmy possums, *Cercaertus*.
25. Bartholomew, G. A., and R. C. Lasiewski, Comp. Biochem. Physiol. *16*:573–582, 1965. Heating and cooling rates in marine iguana.
26. Bartholomew, G. A., and R. E. MacMillen, Physiol. Zool. *34*:177–183, 1961. Temperature regulation in kangaroo mouse.
27. Bartholomew, G. A., and M. Rainy, J. Mammal. *52*:81–95, 1971. Regulation of body temperature in hyrax.
28. Bartholomew, G. A., and V. A. Tucker, Physiol. Zool. *36*:199–218, 1963. Temperature control in agamid lizard.
29. Bartholomew, G. A., and V. A. Tucker, Physiol. Zool. *37*:341–354, 1964. Temperature relations of varanid lizards.
30. Bartholomew, G. A., V. A. Tucker, and A. K. Lee, Copeia 169–173, 1965. Thermal relations of Australian skink *Tiliqua*.
31. Bartholomew, G. A., et al., Physiol. Zool. *37*:179–198, 1964. Temperature relations of bats.
32. Battle, H. I., Trans. Roy. Soc. Canad. *20*:127–143, 1926; *also* Contrib. Canad. Biol. & Fisheries, N. S. *4*:497–526, 1929. Lethal temperatures in relation to reflexes of skate.
33. Beamish, F. W. H., and P. S. Mookherji, Canad. J. Zool. *42*:161–175, 1964. Routine and standard metabolism in trout at different seasons.
34. Behrisch, H. W., and P. W. Hochachka, Biochem. J. *111*:287–295, 1969. Fructose diphosphatase in poikilotherms.
35. Berg, K., Hydrobiologia *5*:331–350, 1953. Temperature and metabolism of freshwater limpets.
36. Berkholz, G., Z. wiss. Zool. *174*:377–399, 1966. Temperature adaptation in fish *Idus*.
37. Bernstein, M. H., Comp. Biochem. Physiol. *38A*:611–617, 1971. Cutaneous and respiratory regulation of temperature, quail.
38. Bishop, D. G., and J. L. Still, J. Lipid Res. *4*:87–90, 1963. Temperature and fatty acid metabolism in bacteria.
39. Blazka, P., Physiol. Zool. *36*:117–128, 1958. Metabolic depression in winter, crucian carp.
40. Bligh, J., Envir. Res. *1*:28–45, 1967. Mechanisms of sweat secretion.

41. Bligh, J., et al., J. Physiol. *176*:136–144, 1965. Telemetered temperatures in ungulates.
42. Bollinger, R. E., and C. O. McKinney, J. Exp. Zool. *161*:21–28, 1966. Development of temperature tolerance in anurans.
43. Bowler, K., J. Cell. Comp. Physiol. *62*:119–132, 133–146, 1963. Mechanisms of acclimation and heat death in *Astacus.*
44. Brattstrom, B. H., Comp. Biochem. Physiol. *24*:93–111, 1968. Thermal acclimation of anurans at different latitudes and altitudes.
45. Brattstrom, B. H., Comp. Biochem. Physiol. *35*:69–103, 1970. Thermal acclimation in Australian amphibians.
46. Brett, J. R., Publ. Ontario Fish. Res. Lab. *63*:1–49, 1944; J. Fish. Res. Bd. Canad. *9*:265–323, 1952; *21*:1183–1226, 1964; *22*:1491–1501, 1965; Quart. Rev. Biol. *31*:75–87, 1956. Lethal temperatures of freshwater fish.
47. Brett, J. R., Amer. Zool. *11*:99–113, 1971. Energetic responses of salmon to temperature.
48. Brett, J. R., and D. A. Higgs, J. Fish. Res. Bd. Canad. *27*:1767–1779, 1970. Effect of temperature on gastric digestion in young salmon.
49. Brock, T. D., Science *158*:1012–1019, 1967. Life at high temperature.
50. Brooks, C. M., K. Koiaumi, and J. L. Malcolm, J. Neurophysiol. *18*: 205–216, 1955. Effects of temperature on cat spinal cord.
51. Brown, J. H., and G. A. Bartholomew, Ecology *50*:705–709, 1969. Energetics of torpor in kangaroo mouse.
52. Bruck, K., and W. Wunnenberg, Pflug. Arch. *290*:167–183, 1966. Role of brown fat in thermal regulation of guinea pigs.
53. Brun, J. L., Ann. Biol. Anim. Biochem. Biophys. *6*:127–158, 267–300, 439–466, 1966. Adaptation to high temperatures in a nematode.
54. Brush, A., Comp. Biochem. Physiol. *15*:399–421, 1965. Temperature relations of California quail.
55. Buetow, D. E., Exp. Cell Res. *27*:137–142, 1962. Effects of temperature on growth of *Euglena.*
56. Buffington, J. D., Comp. Biochem. Physiol. *30*:865–878, 1969. Temperature acclimation in *Culex.*
57. Bullard, R. W., D. B. Dill, and M. K. Yousef, J. Appl. Physiol.*29*:159–167, 1970. Responses of burro to desert heat.
58. Bullock, T. H., Fed. Proc. *12*:666–672, 1953. Pit viper perception of heat.
59. Burckhardt, D., Biol. Zentralbl. *78*:22–62, 1959. Effect of temperature on crayfish stretch receptors.
60. Burlington, R., Comp. Biochem. Physiol. *17*:1049–1052, 1966. Biochemistry of kidney from cold-exposed rats and hamsters.
61. Burlington, R., and J. Klein, Comp. Biochem. Physiol. *22*:701–708, 1967. Gluconeogenesis during hibernation and arousal.
62. Burlington, R., and J. E. Wiebers, Comp. Biochem. Physiol. *17*:183–189, 1966. Glycolysis in heart of hibernators and nonhibernators.
62a. Burwick, R. E., personal communication.
63. Cabanac, M., T. Hammel, and J. D. Hardy, Science *158*:1050–1051, 1967. Temperature-sensitive units in lizard brain.
64. Cabanac, M., J. A. Stolwijk, and J. D. Hardy, J. Appl. Physiol.*24*:645–652, 1968. Single unit responses in temperature center of rabbit brain.
65. Cade, T. J., et al., Physiol. Zool. *38*:9–33, 1965. Temperature relations of finch.
66. Calder, W. A., Comp. Biochem. Physiol. *30*:1075–1082, 1969. Temperature relations of water shrew.
67. Calder, W. A., and J. R. King, Experientia *19*:603–604, 1963. Evaporative cooling in zebra finch.
68. Calder, W. A., and K. Schmidt-Nielsen, Proc. Nat. Acad. Sci. *55*:750–756, 1966. Evaporative cooling and respiratory alkalosis in pigeon.
69. Caldwell, R. S., Comp. Biochem. Physiol. *31*:79–93, 1969. Thermal compensation of respiratory enzymes, goldfish.
70. Caldwell, R. S., and J. F. Vernberg, Comp. Biochem. Physiol. *34*:179–191, 1970. Temperature effects on lipid composition of fish mitochondria.
71. Carey, F. G., and J. M. Teal, Proc. Nat. Acad. Sci. *56*:1464–1469, 1966. Heat conservation in tuna fish muscle.
72. Carey, F. G., and J. M. Teal, Comp. Biochem. Physiol. *28*:199–204, 1969. Warm-bodied sharks.
73. Carey, F. G., and J. M. Teal, Comp. Biochem. Physiol. *28*:205–213, 1969. Regulation of body temperature in tuna.
74. Carey, F. G., J. M. Teal, J. W. Kanwisher, and K. D. Lawson, Amer. Zool. *11*:137–145, 1971. Temperature regulation in tuna.
75. Carlisle, H. J., Nature *209*:1324–1325, 1966. Hypothalamic temperature-sensitive cells in rat.
76. Carpenter, D. O., Comp. Biochem. Physiol. *35*:371–385, 1970. Temperature effects on membrane Na pump in *Aplysia* neurones.
77. Cassuto, Y., and R. R. J. Chaffee, Amer. J. Physiol. *210*:423–426, 1966. Effect of heat on cell metabolism of hamster.
78. Chaffee, R. R. J., et al., Lipids *5*:23–29, 1970. Temperature effects on enzymes of brown fat.
79. Chapman, D., pp. 123–146. *In* Thermobiology, edited by A. H. Rose, Academic Press, New York, 1967. Effect of heat on cell membranes.
80. Chatfield, P. O., C. P. Lyman, and L. Irving, Amer. J. Physiol. *172*: 639–644, 1953. Temperature adaptations of nerves in leg of gull.
81. Cheverie, J. C., and W. G. Lynn, Biol. Bull. *124*:153–162, 1963. High temperature tolerance, thyroid activity of teleosts.
82. Chew, R. M., R. G. Lindberg, and P. Hayden, Comp. Biochem. Physiol. *21*:487–505, 1967. Temperature regulation in mouse *Perognathus.*
83. Cloudsley-Thompson, J. L., Ann. Rev. Entomol. 7:199–222, 1970. Thermal relations of insects.
84. Colhoun, E. H., Entomol. Exp. Appl. *3*:27–37, 1960. Acclimation to cold in insects.
85. Crawford, E. C., and G. Kampe, Amer. J. Physiol. *220*:1256–1260, 1971. Responses of lizard to temperature changes.
85a. Crawford, E. C., and R. C. Lasiewski, Condor *70*:333–339, 1968. Water loss and body temperature of large birds.
86. Crawford, E. C., and K. Schmidt-Nielsen, Amer. J. Physiol. *212*:347–353, 1967. Temperature regulation and evaporative cooling in ostrich.
87. Das, A. B., Comp. Biochem. Physiol. *21*:469–485, 1967. RNA and protein turnover in cold acclimation.
88. Das, A. B., and C. L. Prosser, Comp. Biochem. Physiol. *21*:447–467, 1967. Protein synthesis.
89. Davies, P. S., and M. Walkey, Comp. Biochem. Physiol. *18*:415–425, 1966. Temperature effects on metabolism of cestodes.
90. Davydoff, A. F., and A. R. Makarova, Fed. Proc. Transl. Suppl. *24*:T563–T566, 1964. Newborn seals entering water.
91. Dawe, A. R., and W. A. Spurrier, Science *163*:298–299, 1969. Hibernation inducing factor in blood of ground squirrels.
92. Dawson, T. J., M. S. Denny, and A. J. Hulbert, Comp. Biochem. Physiol. *31*:645–653, 1969. Thermal balance in marsupial *Macropus.*
93. Dawson, T. J., and A. J. Hulbert, Amer. J. Physiol. *218*:1233–1238, 1970. Metabolism, body temperature, and surface area of Australian marsupials.
94. Dawson, W. R., and F. C. Evans, Condor *62*:329–340, 403–405, 1960. Development of temperature regulation in young vesper sparrows.
95. Dawson, W. R., pp. 230–257. *In* Lizard Ecology: A Symposium, edited by W. W. Milstead, University of Missouri Press, Columbia, Mo., 1967. Temperature relations of lizards.
96. Dawson, W. R., and C. D. Fisher, Condor *71*:49–53, 1969. Responses to temperature changes by nightjar *Eurostopodius.*
97. Dawson, W. R., and K. Schmidt-Nielsen, pp. 481–492. *In* Handbook of Physiology, Sec. 4, edited by D. B. Dill et al. Amer. Physiol. Soc., Washington, D.C., 1964. Terrestrial animals in dry heat.
98. Dawson, W. R., and H. B. Tordoff, Condor *61*:388–395, 1959; Auk *81*:26–35, 1964. Metabolism in relation to temperature in evening grosbeak and crossbills.
99. Dawson, W. R., and J. R. Templeton, Physiol. Zool. *36*:219–236, 1963; Ecology *47*:759–765, 1966. Physiological response to temperature in lizards.
100. Dean, H. K., and T. P. Hilditch, Biochem. J. *27*:1950–1956, 1933. Body fat in pig.
101. Dean, J. M., Comp. Biochem. Physiol. *29*:185–196, 1969. Metabolism of thermally acclimated trout.
102. Dehl, R. E., Science *170*:738–739, 1970. Water content of collagen fibers.
103. Dehnel, P. A., Biol. Bull. *118*:215–249, 1960. Temperature and salinity effects on metabolism of crabs.
104. Denyes, A., and J. Baumberg, Ann. Acad. Sci. Fenn. A. IV 71/9, 131–139, 1964. Lipogenesis of cold and hibernating hamsters.
105. Depocas, F., Canad. J. Physiol. Pharmacol. *44*:875–880, 1966. Cytochrome C in rat muscles.
106. DeVries, A. L., Science *172*:1152–1155, 1971; *also* DeVries, A. L., S. K. Komatsu, and R. E. Feeney, J. Biol. Chem. *245*:2901–2908. 1970. Chemistry of anti-freeze glycoproteins from Antarctic fishes.
106a. DeVries, A. L., pp. 157–190 *in* Fish Physiology, Vol. 6, edited by W. S. Hoar and D. J. Randall. Academic Press, New York, 1971.
107. DeVries, A. L., and D. E. Wohlschlag, Science *163*:1073–1075, 1969. Freezing resistance of Antarctic fishes.
108. DeWitt, C. B., Physiol. Zool. *40*:49–66, 1967. Thermoregulation in desert iguana *Dipsosaurus.*
109. Dingley, F., and J. M. Smith, J. Insect Physiol. *14*:1185–1194, 1968. Temperature acclimation in *Drosophila.*
110. Doebber, G. F., Cryobiology *3*:2–11, 1966. Cryoprotective compounds.
111. Dorsett, D. A., J. Exp. Biol. *39*:579–588, 1962. Preparation for flight by hawk moths.
112. Drury, D. E., and J. G. Eales, Canad. J. Zool. *46*:1–9, 1968. Temperature influence on thyroid activity in brook trout.
113. Dryner, R. L., J. R. Paulsund, D. J. Brown, and K. Mavis, Lipids *5*:15–22, 1970. Oxidation of fatty acids by brown fat.
114. Dunlap, D. G., Comp. Biochem. Physiol. *38A*:1–16, 1971. Metabolism-temperature curves in frog *Acris.*
115. Eagan, C. J., Int. J. Biometeor. *10*:293–304, 1966. Effect of cold on circulation in fingers of Eskimos and whites.
116. Edney, E. B., Physiol. Zool. *37*:364–394, 1964. Acclimation to temperature in terrestrial isopods.
117. Edney, E. B., and R. Barrass, J. Insect Physiol. *8*:469–481, 1962. Body temperature of tsetse fly.
118. Edwards, G. A., and W. L. Nutting, Psyche *57*:33–44, 1950. Metabolism of firebrat and snow cricket.
119. Ekberg, D. R., Biol. Bull. *114*:308–316, 1958. Temperature acclimation and tissue metabolism, goldfish.
120. Ekberg, D. R., Comp. Biochem. Physiol. *5*:123–128, 1962. Anaerobic and aerobic metabolism of carp gills in relation to temperature.
121. Esch, H. Z., vergl. Physiol. *43*:305–335, 1960. Temperature relations of honeybee.
122. Evans, R. M., F. C. Purdie, and C. P. Hickman, Canad. J. Zool.*40*:107–118, 1962. Temperature effects on metabolism of rainbow trout.

123. Evans, W. G., Nature *202*:211, 1964. Infrared receptors in buprestid beetles.
124. Evans, W. G., Ann. Entom. Soc. Amer. *59*:873–876, 1966. Infrared receptors in beetles.
125. Feldberg, W., et al., Nature *200*:1325, 1963; *also* J. Physiol. *191*:501–515, 1967; *197*:221–231, 1968. Regulation of body temperature by monoamines in hypothalamus.
126. Fraenkel, G., Ecology *42*:604–606, 1961. Resistance to high temperatures in snail *Littorina*.
127. Fraenkel, G., and D. L. Gunn, Orientation of Animals. Dover Press, New York, 1961.
128. Freed, J. M., Comp. Biochem. Physiol. *14*:651–659, 1965. Temperature acclimation of cytochrome oxidase in goldfish.
129. Freed, J. M., Comp. Biochem. Physiol. *39B*:747–764, 765–774, 1971. Phosphofructokinase in temperature adaptation.
130. Freeman, B. M., Comp. Biochem. Physiol. *33*:219–230, 1970; *34*:871–881, 1970. Thermoregulation in quail and other fowl.
131. Fry, F. E. J., Publ. Ontario Fish. Res. Lab. *66*:1–35, 1946; *68*:1–52, 1947. Environmental effects on activity of fish.
132. Fry, F. E. J., J. R. Brett, and G. H. Clawson, Rev. Canad. Biol. *1*:50–56, 1942. Lethal temperatures for young goldfish.
133. Fry, F. E. J., and J. S. Hart, J. Fish. Res. Bd. Canad. 7:169–175, 1949. Swimming speed of goldfish at different temperatures.
134. Fulco, A. J., Biochim. Biophys. Acta *218*:558–560, 1970. Induction of fatty acid desaturation by temperature in bacteria.
135. Funkhouser, G. E., E. A. Higgins, T. Adams, and C. C. Snow, Life Sci. *6*:1615–1620, 1967. Response of savanna baboon to heat.
136. Gale, C. C., M. Jobin, D. W. Proppe, D. Notter, and H. Fox, Amer. J. Physiol. *219*:193–201, 1970. Endocrine responses to hypothalamic cooling in baboons.
137. Galster, W. A., and P. Morrison, Amer. J. Physiol. *218*:1228–1232, 1970. Carbohydrate changes during hibernation in ground squirrel.
138. Gatt, S., Science *164*:1422–1423, 1969. Thermal lability of galactosidase from salmon liver.
139. Goris, R. C., and M. Nomoto, Comp. Biochem. Physiol. *23*:879–892, 1967. Infrared reception by facial pits of crotaline snakes.
140. Grainger, J. N. R., Comp. Biochem. Physiol. *29*:665–670, 1969. Heat death in terrestrial snail *Arianta*.
141. Greer, G. L., and D. R. Gardner, Science *169*:120–122, 1970. Temperature-sensitive neurones in brain of brook trout.
142. Guieu, J. D., and J. D. Hardy, J. Appl. Physiol. *29*:675–683, 1970. Effect of temperature change of spinal cord on hypothalamic neurones in rabbit.
143. Guillory, R. J., and E. Racker, Biochim. Biophys. Acta *153*:490–493, 1968. Brown fat oxidation.
144. Gustavson, K. H., Chemistry and Reactivity of Collagen. Academic Press, New York, 1956. 342 pp.
145. Hainsworth, F. R., and L. L. Wolf, Science *168*:368–369, 1970. Metabolism and body temperature in torpor, hummingbird.
146. Halcrow, K., and M. Boyd, Comp. Biochem. Physiol. *23*:233–242, 1967. Metabolism of *Gammarus*.
147. Hamby, R., Ph.D. Thesis, Univ. Chicago, 1969. Heat coma in intertidal snails.
148. Hammel, H. T., Spec. Rep. Univ. Missouri Agric. Exp. Station *73*:1–34, 1966. Theory of temperature regulation in mammals.
149. Hammel, H. T., pp. 86–96. *In* International Symposium on Natural Mammalian Hibernation 1965: Mammalian Hibernation III, edited by K. C. Fisher. American Elsevier Publ. Co., New York, 1968. Theory of temperature regulation.
150. Hammel, H. T., Ann. Rev. Physiol. *30*:641–710, 1968. Regulation of internal body temperature.
151. Hammel, H. T., pp. 71–97. *In* Physiological Controls and Regulations, edited by W. S. Yamamoto and J. R. Brobeck. W. B. Saunders Company, Philadelphia, 1965. Neurons and temperature regulation.
152. Hammel, H. T., F. T. Caldwell, and R. M. Abrams, Science *156*:1260–1262, 1967. Regulation of body temperature in lizard *Tiliqua*.
153. Hammel, H. T., D. C. Jackson, J. A. J. Stolwyk, J. D. Hardy, and S. B. Stromme, J. Appl. Physiol. *18*:1146–1154, 1963. Temperature regulation by hypothalamic control center.
154. Hammel, H. T., S. B. Stromme, and K. Myhre, Science *165*:83–85, 1969. Behavioral thermal regulation in Arctic sculpins.
155. Hanec, W., J. Insect Physiol. *12*:1443–1449, 1966. Cold hardiness in tent caterpillar *Malacosoma*.
156. Hanegan, J. L., and J. E. Heath, J. Exp. Biol. *53*:349–362, 1970. Control of body temperature in moth *Hyalophora*.
157. Hannon, J. P., Fed. Proc. *19*:100–105, 1960. Intermediary metabolism in cold-acclimated rat.
158. Hannon, J. P., and A. M. Larson, Amer. J. Physiol. *203*:1055–1061, 1962. Fatty acid metabolism in thermogenesis of cold-acclimated rat.
159. Hanson, R. C., and J. G. Stanley, Comp. Biochem. Physiol. *33*:871–879, 1970. Hypophysectomy and thermal acclimation in mud minnow.
160. Hardy, J. D., Physiol. Rev. *41*:521–605, 1961. Thermoregulating center in mammalian brain.
161. Hardy, J. D., pp. 98–116. *In* Physiological Controls and Regulations, edited by W. S. Yamamoto and J. R. Brobeck. W. B. Saunders Company, Philadelphia, 1965. The "set-point" concept in temperature regulaion.
162. Hardy, J. D., R. F. Hellon, and K. Sutherland, J. Physiol. *175*:242–253, 1964. Temperature-sensitive neurones in dog hypothalamus.
163. Hart, J. S., Symp. Soc. Exp. Biol. *18*:31–48, 1964. Insulative and metabolic adaptation to cold in vertebrates.
164. Hart, J. S., Publ. Ontario Fish. Res. Lab. *72*:1–79, 1952. Lethal temperatures of fish from different latitudes.
165. Hart, J. S., Physiol. Zool. *35*:224–236, 1962. Seasonal acclimation in small birds.
166. Hart, J. S., H. Pohl, and J. S. Tener, Canad. J. Zool. *43*:731–744, 1965. Seasonal acclimatization in hare.
167. Haschmeyer, A. E. V., Proc. Nat. Acad. Sci. *62*:128–135, 1969. Synthesis of polypeptide chains in temperature acclimation of *Opsanus*.
168. Haschmeyer, A. E. V., Biol. Bull. *135*:130–140, 1968; Comp. Biochem. Physiol. *28*:535–552, 1969. Protein synthesis by liver in temperature acclimation of toadfish.
169. Haschmeyer, A. E. V., Biol. Bull. *136*:28–32, 1969. Oxygen consumption in temperature acclimation of toadfish.
170. Hayward, J. S., and C. P. Lyman, pp. 346–355. *In* International Symposium on Natural Mammalian Hibernation 1965: Mammalian Hibernation III, edited by K. C. Fisher. American Elsevier Publ. Co., New York, 1968. Nonshivering heat production in arousal from hibernation.
171. Hazel, J., and C. L. Prosser, Z. vergl. Physiol. *67*:217–228, 1970. Interpretation of inverse acclimation to temperature.
172. Hazel, J., Ph.D. Thesis, University of Illinois, 1971. Effects of mitochondrial lipids on succinic dehydrogenase.
173. Heath, J. E., Univ. Calif. Publ. Zool. *64*:97–134, 1965. Temperature regulation by behavior in horned lizards.
174. Heath, J. E., pp. 259–278. *In* Evolution and Environment, edited by E. T. Drake, Yale University Press, New Haven, Conn., 1968. Evolution of thermal regulation.
175. Heath, J. E., The Physiologist *13*:399–410, 1970. Theory of behavioral regulation of temperature in poikilotherms.
176. Heath, J. E., and P. A. Adams, J. Exp. Biol. *47*:21–33, 1967. Regulation of heat production by large moths.
177. Heath, J. E., et al., Amer. Zool. *11*:147–158, 1971. Behavioral thermoregulation in insects.
178. Heath, J. E., and R. K. Josephson, Biol. Bull. *138*:272–285, 1970. Body temperature and singing in katydids.
179. Heath, J. E., A. Williams, and S. Mills, Int. J. Biometeor. *15*:254–257, 1971. Hypothalamic thermosensitivity.
180. Heath, W. G., Science *142*:486–488, 1963. Thermoperiodism in trout.
181. Heatwole, H., Ecol. Monogr. *40*:425–457, 1970. Thermal regulation in desert dragon *Amphibolurus*.
182. Hebb, C., D. Morris, and M. W. Smith, Comp. Biochem. Physiol. *28*:29–36, 1969. Acetyl transferase in brain of goldfish from different temperatures.
183. van Heel, W. H. D., Experientia *12*:75–77, 1956. Temperature effect on pitch discrimination in minnow, *Phoxinus*.
184. Heinrich, B., Science *168*:580–582, 1970. Thoracic temperature in free flying moths.
185. Heinrich, B., J. Exp. Biol. *54*:141–152, 153–166, 1971. Temperature regulation of sphinx moth *Manduca*.
186. Heldmaier, G., Z. vergl. Physiol. *63*:59–84, 1969. Heat production in bat *Myotis*.
187. Heldmaier, G., Z. vergl. Physiol. *73*:222–240, 1971. Calorigenic responses to noradrenaline.
188. Hellon, R. F., J. Physiol. *193*:381–395, 1970. Thermal stimulation of hypothalamic neurones in rabbit.
189. Hensel, H., Z. vergl. Physiol. *37*:509–526, 1955; Pflüger. Arch. *263*:48–53, 1956. Temperature sensitivity of ampullae of Lorenzini.
190. Hensel, H., Pflüger. Arch. *313*:150–152, 1969. Cutaneous heat receptors in primates.
191. Hensel, H., and D. R. Kenshalo, J. Physiol. *204*:99–112, 1969. Warm receptors in nasal region of cat.
192. Hensel, H., and Y. Zotterman, Acta Physiol. Scand. *22*:96–105, 106–113, 1951; *23*:291–319, 1951. Heat and cold receptors.
193. Henshaw, R. E., and G. E. Folk, Physiol. Zool. *39*:223–236, 1966. Thermal regulation in seasonally changing climate, bats.
194. Heroux, O., Fed. Proc. *22*:789–794, 1963. General patterns of temperature adaptation in mammals.
195. Herreid, C. F., and B. Kessel, Comp. Biochem. Physiol. *21*:405–414, 1967. Thermal conductance in birds and mammals.
196. Heusner, A., and T. Stussi, Insectes Sociaux *11*:239–265, 1964. Energy metabolism of bee.
197. Hildwein, G., Arch. Sci. Physiol. *24*:55–71, 1970. Seasonal thermoregulation in tenrec.
198. Hildwein, G., and O. Champigny, Arch. Sci. Physiol. *21*:45–58, 1967. Thermoregulation of hairless rats.
199. Himms-Hagen, J., Canad. J. Physiol. Pharmacol. *49*:545–553, 1971. Effect of tetracycline on noradrenaline action on brown fat.
200. Hoar, W. S., and J. G. Eales, Canad. J. Zool. *41*:653–669, 1963. Thyroid and temperature resistance in goldfish.
201. Hochachka, P. W., Arch. Biochim. Biophys. *111*:96–103, 1965. LDH isozymes in temperature adaptation of goldfish.
202. Hochachka, P. W., Comp. Biochem. Physiol. *25*:107–118, 1968. Glucose and acetate metabolism in temperature adaptation.
203. Hochachka, P. W., and F. R. Hayes, Canad. J. Zool. *40*:261–270, 1962. Temperature acclimation and pathways of glucose metabolism in trout.

204. Hochachka, P. W., and J. K. Lewis, J. Biol. Chem. *245*:6567–6573, 1970. Enzyme variance in thermal acclimation.
205. Hochachka, P. W., and J. K. Lewis, Comp. Biochem. Physiol. *39B*: 925–934, 1971. Effects of temperature and pH on fish LDH.
206. Hochachka, P. W., and G. N. Somero, Comp. Biochem. Physiol. *27*: 659–668, 1968. Adaptation of enzymes to temperature.
207. Hochachka, P. W., pp. 351–389. *In* Fish Physiology, Vol. 1 edited by W. S. Hoar and D. J. Randall, Academic Press, New York, 1970. Adaptation of enzymes to temperature.
208. Hodgkin, A. L., and B. Katz, J. Physiol. *109*:240–249, 1949. Effect of temperature on action potentials in squid giant axon.
209. Horowitz, B. A., and L. Nelson, Comp. Biochem. Physiol. *24*:385–394, 1968. Mitochondrial respiration in hibernator and nonhibernator.
210. House, H. L., et al., Canad. J. Zool. *36*:629–632, 1958. Effect of diet on temperature resistance, insects.
211. Howell, B. J., F. W. Baumgartner, K. Bondi, and H. Rahn, Amer. J. Physiol. *218*:600–606, 1970. Acid-base balance as a function of body temperature.
212. Howell, T. R., and G. A. Bartholomew, Auk *78*:343–354, 1961; Condor *63*:185–197, 1961. Temperature regulation in petrels and shearwaters; incubation temperature of albatross.
213. Hsieh, A. C. L., J. Physiol. *169*:851–861, 1964. Metabolism of cold-adapted rats.
214. Hsieh, A. C. L., et al., Fed. Proc. *25*:1205–1212, 1966. Metabolic effects of noradrenaline on cold-adapted rats.
215. Hudson, J. W., pp. 30–46. *In* International Symposium on Natural Mammalian Hibernation 1965: Mammalian Hibernation III, edited by K. C. Fisher. American Elsevier Publ. Co., New York, 1968. Patterns of torpidity in small mammals.
216. Hudson, J. W., Ann. Acad. Sci. Fenn. A. IV, 71/15, 217–233, 1964. Temperature regulation in ground squirrels.
217. Hudson, J. W., and A. H. Brush, Comp. Biochem. Physiol. *12*:157–170, 1964. Comparative metabolic study of quail and dove.
218. Hudson, J. W., and S. L. Kimzey, Comp. Biochem. Physiol. *17*:203–217, 1966. Metabolism and temperature regulation in house sparrow.
219. Hudson, J. W., and J. A. Rummel, Ecology *47*:345–354, 1966. Temperature regulation of primitive rodent, *Liomys*.
220. Hull, P., and M. M. Segall, J. Physiol. *181*:449–457, 468–477, 1965. Brown fat in heat production of newborn rabbit.
221. Hutchinson, V. H., H. G. Dowling, and A. Vinegar, Science *151*:694–696, 1966. Thermal regulation in brooding Indian pythons.
222. Iampietro, P. F., et al., J. Appl. Physiol. *15*:632–634, 1960. Heat production from shivering.
223. Iggo, A., J. Physiol. *200*:403–430, 1969. Cutaneous thermal receptors in mammals.
224. Irving, L., J. Appl. Appl. Physiol. *9*:414–420, 1956. Temperature of skin in pig.
225. Irving, L., L. J. Peyton, C. H. Bahn, and R. S. Peterson, Physiol. Zool. *35*:275–284, 1962. Temperature regulation in fur seal.
226. Irving, L., K. Schmidt-Nielsen, and N. S. Abrahamson, Physiol. Zool. *30*:93–105, 1957. Melting points of animal fats in cold climates.
227. Ishiko, N., and W. R. Loewenstein, J. Gen. Physiol. *45*:105–124, 1961. Effects of temperature on pacinian corpuscles.
228. Jackson, D. C., and K. Schmidt-Nielsen, Proc. Nat. Acad. Sci. *51*:1192–1197, 1964. Temperature relations of kangaroo rat.
229. Jacobson, E. R., and W. G. Whitford, Comp. Biochem. Physiol. *35*: 439–449, 1970. Effect of acclimatization on responses to temperature in snakes.
230. Jacobson, F. H., and R. D. Squires, Amer. J. Physiol. *218*:1575–1582, 1970. Thermal regulation center in brain of cat.
231. Jankowsky, H. D., Helg. wiss. Meeresunters. *13*:402–407, 1966. Metabolism and temperature adaptation in eel.
231a. Jankowsky, H. D., Personal communication. Metabolic acclimation in golden orfe.
232. Jansky, L., Canad. J. Biochem. *41*:1847–1854, 1963. Cytochrome oxidase in cold- and warm-acclimated rats.
233. Jansky, L., and J. S. Hart, Canad. J. Biochem. *41*:953–964, 1963. Nonshivering thermogenesis in cold-acclimated rats.
234. Jansky, L., et al., Physiol. Bohemoslovaca *16*:366–371, 1967. Noradrenaline thermogenesis in rats.
235. Javaid, M. Y., and J. M. Anderson, J. Fish. Res. Bd. Canad. *24*:1507–1513, 1967. Thermal acclimation and temperature selection in salmon trout.
236. Jensen, D., Comp. Biochem. Physiol. *41A*:685–695, 1972. Effect of cooling on frog endplate.
237. Jessen, C., E. Simon, and R. Kullman. Experientia *24*:694–695, 1968. Interaction of spinal and hypothalamic temperature detectors in dog.
238. Johansen, K., Physiol. Zool. *34*:126–144, 1961. Temperature regulation in armadillo, *Dasypus*.
239. Johansen, K., J. Krog, and O. Reite, Ann. Acad. Sci. Fenn. A. IV, 71/17, 245–253, 1964. Nervous control of heart in hypothermic regulator.
240. Johnson, H. D., et al., Univ. Missouri Coll. Agric. Res. Bull. *683*:1–31, 1958. Temperature tolerance of cattle.
241. Johnson, R. E., Comp. Biochem. Physiol. *24*:1003–1014, 1968. Temperature regulation in ptarmigan.
242. Johnston, P. V., and B. J. Roots, Comp. Biochem. Physiol. *11*:303–310, 1964. Brain lipid fatty acids and temperature acclimation.
243. Josse, J., and W. F. Harrington, J. Molec. Biol. *9*:269–287, 1964. Composition of collagen as a function of temperature.
244. Jungreis, A. M., Comp. Biochem. Physiol. *24*:1–6, 1968. Glycogen and long-term temperature acclimation in crayfish.
245. Jungreis, A. M., and A. B. Hooper, Comp. Biochem. Physiol. *26*:91–100, 1968. Cold resistance adaptation in crayfish.
246. Kahl, M. P., Physiol. Zool. *36*:141–151, 1963. Thermal regulation in wood stork.
247. Kammer, A. E., Z. vergl. Physiol. *68*:334–344, 1970. Thoracic temperature and flight in monarch butterfly.
248. Kanungo, M., and C. L. Prosser, J. Cell. Comp. Physiol. *54*:259–263, 265–274, 1960. Biochemical changes in temperature acclimation, goldfish.
249. Kanwisher, J., Topics in the Study of Life in the Sea, pp. 209–214. Harper & Row, New York, 1970. Temperature regulation in marine mammals.
250. Kanwisher, J. W., pp. 487–494. *In* Cryobiology, edited by H. T. Meryman, Academic Press, New York, 1966. Freezing in intertidal animals.
251. Karow, A. M., and W. A. Webb, Cryobiology *2*:99–108, 1965. Tissue freezing.
252. Kayser, C., Rev. Canad. Biol. *16*:303–389, 1957; Ann. Rev. Physiol. *19*:83–120, 1957. Hibernation.
253. Keister, M., and J. Buck, J. Insect Physiol. 7:51–72, 1961. Temperature and respiration in *Phormia*.
254. Kemp, P., and M. W. Smith, Biochem. J. *117*:9–15, 1970. Effect of temperature on fatty acid composition in goldfish.
255. Kempner, C. L., Science *142*:1318–1319, 1963. Upper temperature limits of life.
256. Kendeigh, S. C., Wilson Bull. *81*:441–449, 1969. Energy responses of birds to thermal environment.
257. Kerkut, G. A., and B. J. R. Taylor, Nature *178*:426, 1956; J. Exp. Biol. *34*:486–493, 1957; Behaviour *13*:259–279, 1958. Effects of temperature on nervous systems of cockroach, slug, and crayfish.
258. Kibler, H. H., H. D. Silsby, and H. D. Johnson, J. Geront. *18*:235–239, 1963. Cold acclimation in rats.
259. Kimzey, S. L., and J. S. Willis, J. Gen. Physiol. *58*:620–633, 634–649, 1971. Resistance of red blood cells to loss in hibernation and in cold storage.
260. King, J. R., Comp. Biochem. Physiol. *12*:13–24, 1964. Oxygen consumption and body temperature in white-crowned sparrow.
261. Kinne, O., Oceanogr. Marine Biol. Ann. Rev. *1*:301–340, 1963. Effects of temperature and salinity on invertebrates.
262. Kinne, O., and E. M. Kinne, Canad. J. Zool. *40*:231–253, 1962. Development of fish embryos at different temperatures.
263. Klicka, J. K., Physiol. Zool. *38*:177–189, 1965. Physiological effects of ACTH, TSH, and thyroid hormones in goldfish.
264. Kluger, M. J., and J. E. Heath, Comp. Biochem. Physiol. *32*:219–226, 1970. Thermal regulation by bat wing.
265. Knipprath, W. G., and J. F. Mead, Lipids *3*:121–128, 1968. Temperature effects on fatty acids in goldfish.
266. Kodama, A. M., and N. Pace, Fed. Proc. *22*:761–765, 1963. Cold-dependent changes in tissue fat.
267. Koffler, H., G. E. Mallett, and J. Adye, Proc. Nat. Acad. Sci. *43*:464–477, 1957. Molecular basis of heat stability of bacteria.
268. Kohler, H., Marine Biol. *5*:315–324, 1970. Acclimation to temperature and salinity in an oligochaete.
269. Komatsu, S. K., A. L. DeVries, and R. E. Feeney, J. Biol. Chem. *245*: 2909–2913, 1970; *also* Komatsu, S. K., H. T. Miller, A. L. DeVries, D. T. Osuga, and R. E. Feeney, Comp. Biochem. Physiol. *32*:519–527, 1970. Glycoproteins and lipoproteins in blood of Antarctic fish.
270. Kosaka, M., and E. Simon, Pflüger. Arch. *302*:357–373, 1968. Spinal receptors for cold.
271. Kosaka, M., E. Simon, R. Thauer, and O. E. Walther, Amer. J. Physiol. *217*:858–863, 1969. Thermal stimulation of spinal cord.
272. Krog, J., Biol. Bull. *107*:397–410, 1954. Seasonal effects on metabolism and lethal temperature in Alaskan gammarid.
273. Kruger, G., Z. wiss. Zool. *167*:87–104, 1962. Temperature adaptation in bitterling *Rhodeus*.
274. Kulzer, E., Z. vergl. Physiol. *50*:1–34, 1965. Temperature regulation of bats from different climates.
275. Kusakina, A. A., Tsitologiya *4*:68, 1962; transl. in Fed. Proc. Transl. Suppl. *22*:T123–T126, 1963. Thermal stability of muscle enzymes in fishes.
276. Lagerspetz, K. Y. H., Helg. wiss. Meeresunters. *14*:559–571, 1966. Development of thermal regulation in laboratory mice.
277. Lagerspetz, K. Y. H., and D. Dubitscher, Comp. Biochem. Physiol. *17*:665–671, 1966. Temperature acclimation of cilia in gill of *Anodonta*.
278. Lagerspetz, K. Y. H., and A. Talo, J. Exp. Biol. *47*:471–480, 1967. Temperature acclimation and conduction in nerve fibers, *Lumbricus*.
279. Langridge, J., Ann. Rev. Plant Physiol. *14*:441–462, 1963. Biochemical aspects of temperature response.
280. Lasiewski, R. C., Proc. XIII Int. Ornith. Cong. 1095–1103, 1963. Energetic cost of small size in hummingbirds.
281. Lasiewski, R. C., Physiol. Zool. *36*:122–140, 1963. O_2 consumption of torpid, resting, and active hummingbirds.
282. Lasiewski, R. C., Amer. J. Physiol. *217*:1504–1509, 1969. Physiological response to heat stress in the poorwill.
283. Lasiewski, R. C., and W. R. Dawson, Condor *66*:477–490, 1964; *72*: 332–338, 1970. Temperature responses in night hawk and frogmouth *Podargus*.
284. Lasiewski, R. C., Condor *66*:212–220, 1964. Energetic relationships of very small passerine birds to temperature.

285. Lasiewski, R. C., and J. J. Lasiewski, Auk *84*:34–48, 1967. Responses of hummingbirds to temperature changes.
286. Lasiewski, R. C., and G. K. Snyder, Auk *86*:529–540, 1969. Temperature in nesting cormorants.
287. Lasiewski, R. C., W. W. Weathers, and M. Bernstein, Comp. Biochem. Physiol. *23*:797–813, 1967. Metabolic responses of hummingbirds to temperature.
288. Laverack, M. S., Comp. Biochem. Physiol. *3*:136–140, 1961. Effect of temperature on activity of nerve cord of earthworm.
289. Leblanc, J., and A. Villemaire, Amer. J. Physiol. *218*:1742–1745, 1970. Thyroid and adrenalin in cold resistance, brown fat.
290. Leblanc, J., et al., J. Appl. Physiol. *15*:1031–1034, 1960; *17*:950–952, 1962; Canad. J. Physiol. Pharmacol. *44*:287–293, 1966. Habituation of Gaspe fishermen to cold extremities; laboratory acclimation.
291. Leitner, P., Comp. Biochem. Physiol. *19*:431–443, 1966. Metabolism and temperature regulation in mastiff bat *Eumops*.
292. Leitner, P., and J. E. Nelson, Comp. Biochem. Physiol. *21*:65–74, 1967. Temperature, metabolism, and heart rate of vampire bat.
293. Lensky, Y., J. Insect Physiol. *10*:1–12, 1964. Bee colony, temperature extremes.
294. Leston, D., J. W. S. Pringle, and D. C. S. White, J. Exp. Biol. *42*:409–414, 1965. Insect warmup by muscle contraction.
295. Licht, P., Comp. Biochem. Physiol. *13*:27–34, 1964. Thermal dependence of contractility in reptilian muscles.
296. Licht, P., Copeia 428–436, 1965; Amer. Midl. Nat. *79*:149–158, 1968. Preferred body temperatures in lizards.
297. Licht, P., and P. Leitner, Comp. Biochem. Physiol. *22*:371–387, 1967. Responses to high temperatures in bats.
298. Liu, C., J. L. Frehn, and A. D. LaPorta, J. Appl. Physiol. *27*:83–89, 1969. Liver and brown fat responses to cold.
299. Loftus, R., Z. vergl. Physiol. *59*:413–455, 1968; *63*:415–433, 1969. Antennal cold receptors in *Periplaneta*.
299a. Lyman, C. P., and D. C. Blinks, J. Cell. Comp. Physiol. *54*:53–63, 1959. Temperature effects on hearts of hibernators.
300. Lyman, C. P., and R. C. O'Brien, Ann. Acad. Sci. Fenn. A. IV 71/22, 213–320, 1964. Effects of autonomic drugs on hibernating ground squirrels.
301. Lyman, C. P., and W. A. Wimsatt, Physiol. Zool. *39*:101–109, 1966. Temperature regulation in vampire bat *Desmodus*.
302. MacMillen, R. E., and C. H. Trost, Comp. Biochem. Physiol. *23*:243–253, 1967. Nocturnal hypothermia in Inca dove.
303. MacMillen, R. E., and J. E. Nelson, Amer. J. Physiol. *217*:1246–1251, 1969. Bioenergetics in dasyurid marsupials.
304. Malan, A., Arch. Sci. Physiol. *23*:47–87, 1969. Hypothalamic control of thermal regulation and hibernation in hamster.
305. Mangum, C. P., and C. Sassaman, Comp. Biochem. Physiol. *30*:111–116, 1969. Temperature sensitivity and metabolism in polychaete annelid.
306. Markardt, J. E., Comp. Biochem. Physiol. *33*:423–439, 1970. Daily torpor in mouse *Peromyscus*.
307. Markert, C. L., and I. Faulhaber, J. Exp. Zool. *159*:319–332, 1965. LDH isozymes of fish.
308. Marshall, J. M., and J. S. Willis, J. Physiol. *164*:64–76, 1962. Effects of temperature on membrane potentials in isolated atria of ground squirrel *Citellus*.
309. Mazur, P., pp. 213–315. *In* Cryobiology, edited by H. T. Meryman, Academic Press, New York, 1966. Basis of cellular injury in freezing and thawing.
310. Mazur, P., Ann. Rev. Plant Physiol. *20*:419–448, 1969. Freezing injury in plants.
311. Mazur, P., Science *168*:939–949, 1970. The freezing of biological systems.
312. Mazur, P., et al., Cryobiology *6*:1–9, 1969. Survival of tissue culture cells after freezing and thawing.
313. McCrea, M., and J. E. Heath, J. Exp. Biol. *54*:415–435, 1971. Dependence of flight on temperature regulation in moth *Manduca*.
314. McCue, J. F., and R. E. Thorson, J. Parasitol. *50*:67–71, 1964. Behavior of parasitic helminth in thermal gradient.
315. McGinnis, S. M., and L. L. Dickson, Science *156*:1757–1759, 1967. Iguana temperature.
316. McLeese, D. W., J. Fish. Res. Bd. Canad. *13*:247–272, 1956. Effects of temperature, salinity, and oxygen on survival of lobster.
316a. McMillen, R. E., and C. H. Trost, Comp. Biochem. Physiol. *23*:243–253, 1967. Thermal regulation in Inca dove.
317. McNab, B. K., Ecology *44*:521–532, 1963. Energy balance in wild mice.
318. McNab, B. K., Comp. Biochem. Physiol. *31*:227–268, 1969. Economics of temperature regulation, neotropical bats.
319. McNab, B. K., J. Exp. Biol. *53*:329–348, 1970. Body weight and energetics of temperature regulation.
320. McNab, B. K., and P. Morrison, Ecol. Monogr. *33*:63–82, 1963. Body temperature and metabolism in *Peromyscus*.
321. McWhinnie, M. A., and J. D. O'Connor, Comp. Biochem. Physiol. *20*:131–145, 1967. Metabolism and temperature acclimation in crayfish.
322. Mellanby, K., J. Exp. Biol. *18*:55–61, 1942; Nature *181*:1403, 1958. Body temperature, frogs and insects.
323. Menaker, M., J. Cell. Comp. Physiol. *59*:163–174, 1962. Hibernation in bat *Myotis*.
324. Meryman, H. T., pp. 1–114. *In* Cryobiology, edited by H. T. Meryman, Academic Press, New York, 1966. Review of biological freezing.
325. Miller, L. K., Comp. Biochem. Physiol. *21*:679–686, 1967. Temperature and conduction in mammalian nerve.
326. Miller, L. K., Science *166*:105–106, 1969. Freezing tolerance in carabid beetle.
327. Miller, L. K., and P. J. Dehlinger, Comp. Biochem. Physiol. *28*:915–921, 1969. Neuromuscular function in cold.
328. Miller, L. K., and L. Irving, Amer. J. Physiol. *204*:359–362, 1963. Sensitivity of rat nerve to cold.
329. Moore, J. A., Evolution *3*:1–21, 1949. Geographic variation of temperature adaptation in *Rana pipiens*.
330. Morrison, G. R., F. E. Brock, D. T. Sobral, and R. E. Shank, Arch. Biochim. Biophys. *114*:494–501, 1966. Rat liver enzymes after cold acclimation.
331. Morrison, P., Biol. Bull. *123*:154–169, 1962. Body temperature of humpback whale.
332. Morrison, P., Biol. Bull. *116*:484–497, 1959. Heat loss in small mammals.
333. Morrison, P., and B. K. McNab, Comp. Biochem. Physiol. *6*:57–68, 1962. Daily torpor in opossum.
334. Morrison, P., and R. A. Ryser, Physiol. Zool. *37*:90–103, 1959. Body temperature in *Peromyscus leucopus*.
335. Morrison, P., and W. J. Tietz, J. Mammal. *38*:78–86, 1957. Insulation in small mammals.
336. Mount, L. E., J. Physiol. *217*:315–326, 1971. Metabolic responses of hairless mice to cold.
337. Murray, R. W., Comp. Biochem. Physiol. *18*:291–303, 1966. Temperature effects on membrane properties of neurones in *Aplysia*.
338. Murrish, D. E., Comp. Biochem. Physiol. *34*:853–858, 859–869, 1970. Responses to diving and temperature in the dipper *Cinclus*.
339. Mutchmor, J. A., and A. G. Richards, J. Insect Physiol. 7:141–158, 1961. Low temperature and muscle apyrase of insects.
340. Myers, R. D., and W. L. Veale, Science *170*:95–97, 1970; J. Physiol. *217*:381–392, 1971. Ionic mechanisms of temperature control in hypothalamus.
341. Myhre, K., and H. T. Hammel, Amer. J. Physiol. *217*:1490–1495, 1969. Behavioral regulation of temperature in lizard *Tiliqua*.
342. Nakayama, T., H. T. Hammel, J. D. Hardy, and J. S. Eisenman, Amer. J. Physiol. *204*:1122–1126, 1963. Thermal stimulation of neurones in preoptic hypothalamus.
343. Namatov, T., Dokl. Akad. Nauk S.S.S.R. *185*:351–354, 1969. Resistance of nematodes to freezing temperatures.
344. Newell, R. C., Nature *212*:426–428, 1966; J. Zool. (Lond.) *151*:299–311, 1967; *also* Newell, R. C., and H. R. Northcraft, J. Zool. (Lond.) *151*:277–298, 1967. Metabolic independence of temperature over limited ranges in poikilotherms.
345. Newell, R. C., and V. I. Pye, Comp. Biochem. Physiol. *38B*:635–650, 1971. Relation between metabolism and temperature in *Littorina*.
346. Nopp, H., and A. Z. Farahat, Z. vergl. Physiol. *55*:103–118, 1967. Temperature and metabolism in snails.
347. Norris, K. S., Ecol. Monogr. *33*:23–62, 1963. Temperature in the ecology of percoid fish *Girella*.
348. Nowell, N. W., and D. C. White, Life Sci. *8*:239–245, 1969. Noradrenaline in cold acclimation of rat heart.
349. O'Farrell, M. J., and E. H. Studier, Comp. Biochem. Physiol. *35*:697–703, 1970. Metabolism in relation to temperature of *Myotis*.
350. Oritsland, N. A., Comp. Biochem. Physiol. *37*:225–233, 1970. Temperature regulation of polar bear.
351. Orr, P. R., Physiol. Zool. *28*:290–302, 1955. Heat death, whole animal and tissues.
352. Paintal, A. S., J. Physiol. *180*:1–35, 1965. Cold block of mammalian nerve.
353. Patterson, G. W., Lipids *5*:597–600, 1970. Effect of culture temperature on fatty acids of *Chlorella*.
354. Patton, J. F., and W. S. Platner, Amer. J. Physiol. *218*:1417–1422, 1970. Cold acclimation and thyroid effects on liver fatty acids.
355. Pearcy, W. G., Science *134*:193–194, 1961. Seasonal changes in osmotic concentration of flounder.
356. Pengelley, E. T., and S. M. Asmundson, Comp. Biochem. Physiol. *30*:177–183, 1969. Circadian rhythms in ground squirrel *Citellus*.
357. Pengelley, E. T., and K. C. Fisher, Canad. J. Zool. *41*:1103–1120, 1963. Annual cycles in ground squirrels.
357a. Peterson, R. H., and C. L. Prosser, Comp. Biochem. Physiol. *42A*:1019–1038, 1972. Cold block of central nervous responses in goldfish.
358. Pichotka, J., Helg. wiss. Meeresunters. *9*:274–284, 1964. Temperature regulation in newborn mice.
359. Platzer, I., Z. vergl. Physiol. *54*:58–74, 1967. Temperature adaptations in tropical chironomids.
360. Pocrnjic, Z., Arch. Biol. Nauka *17*:139–148, 1965. Thermal adaptation of oxygen consumption in the newt.
361. Popovic, V., Amer. J. Physiol. *207*:1345–1348, 1964. Cardiac output in hibernating ground squirrels.
362. Popovic, P., et al., Biometeorology *2*:276–282, 1966. Rats, adaptation to hypothermia.
362a. Powers, D. A., Personal communication. Polymorphism of LDH in *Fundulus*.
363. Precht, H., pp. 50–78. *In* Physiological Adaptation, edited by C. L.

Prosser. Amer. Physiol. Soc., Washington, D.C., 1958. Theory of temperature adaptation in cold-blooded animals.
364. Precht, H., Z. vergl. Physiol. *42*:365–382, 1959. Resistance adaptation in tropical fish.
365. Precht, H., Z. wiss. Zool. *164*:336–354, 354–363, 1960. Thermal resistance of muscles and muscle endplates.
366. Precht, H., J. Christophersen, and H. Hensel, Temperatur und Leben. Springer Verlag, Hamburg, 1955. 514 pp.
367. Precht, H., Z. vergl. Physiol. *44*:451–462, 1961. Temperature adaptation of metabolism of eels.
368. Precht, H., and J. Christophersen, Z. wiss. Zool. *171*:197–209, 1965. Temperature adaptation of ciliated epithelium of clam gills.
369. Prosser, C. L., pp. 167–180. *In* Physiology Adaptation, edited by C. L. Prosser. Amer. Physiol. Soc., Washington, D.C., 1958.
370. Prosser, C. L., pp. 11–25. *In* Handbook of Physiology, Sec. 4, edited by D. B. Dill et al., Amer. Physiol. Soc., Washington, D.C., 1964. Theory of adaptation.
371. Prosser, C. L., and T. Nagai, pp. 171–180. *In* The Central Nervous System and Fish Behavior, edited by D. Ingle. University of Chicago Press, Chicago, 1968.
372. Prosser, C. L., and E. Farhi, Z. vergl. Physiol. *50*:91–101, 1965. Effects of temperature on conditioned reflexes.
373. Prosser, C. L., pp. 351–376. *In* Molecular Mechanisms of Temperature Adaptation, edited by C. L. Prosser. Amer. Assoc. Adv. Sci., Washington, D.C., 1967.
374. Prusiner, S., J. Biol. Chem. *245*:382–389, 1970. Oxidative phosphorylation in hamster brown fat.
375. Rao, K. P., pp. . *In* Molecular Mechanisms of Temperature Adaptation, edited by C. L. Prosser. Amer. Assoc. Adv. Sci., Washington, D.C., 1967.
376. Rapatz, G., and B. Luyet, Cryobiology *4*:215–222, 1968. Cryoprotective agents in preservation of blood.
377. Rautenberg, W., Z. vergl. Physiol. *62*:235–266, 1969. Spinal cord in relation to thermal regulation in pigeons.
378. Read, K. R. H., Comp. Biochem. Physiol. *9*:161–180, 1963. Thermal inactivation of aspartic-glutamic transaminase from molluscs.
379. Read, K. R. H., Proc. Malac. Soc. Lond. *37*:233–241, 1967. Thermal tolerance of Puerto Rican bivalves.
380. Read, K. R. H., pp. 93–106. *In* Molecular Mechanisms of Temperature Adaptation, edited by C. L. Prosser. Amer. Assoc. Adv. Sci., Washington, D.C., 1967. Thermal stability of proteins in invertebrates.
381. Reiter, O. B., and W. H. Davis, Proc. Soc. Exp. Biol. Med. *121*:1212–1215, 1966. Chill in bats.
382. Richards, A. G., Physiol. Zool. *37*:199–211, 1964. Temperature effects on development and oxygen consumption in insect eggs.
383. Rigby, B. J., Biol. Bull. *135*:223–229, 1968. Melting temperature of collagen.
384. Rink, R. D., J. Exp. Zool. *170*:117–124, 1969. Metabolism of brown fat in newborn hamsters.
385. Rising, J. D., Comp. Biochem. Physiol. *25*:327–333, 1968. Temperature effect on metabolism of a sparrow.
386. Rising, J. D., Comp. Biochem. Physiol. *31*:915–925, 1969. Metabolism and evaporative water loss in orioles.
387. Rising, T. L., and K. B. Armitage, Comp. Biochem. Physiol. *30*:1091–1114, 1969. Acclimation to temperature in snails.
388. Roberts, J. L., Helg. wiss. Meeresunters. *9*:459–473, 1964. Metabolic responses of sunfish to photoperiod and temperature.
389. Robertshaw, D., and C. R. Taylor, J. Physiol. *203*:135–143, 1969. Sweat gland activity in African ungulates.
390. Rodahl, K., et al., J. Appl. Physiol. *17*:763–767, 1962. Human metabolism, cold stress.
391. Roots, B. I., Comp. Biochem. Physiol. *25*:457–466, 1968. Phospholipids of goldfish brain.
392. Roots, B. I., and P. V. Johnston, Comp. Biochem. Physiol. *26*:553–560, 1968. Plasmalogens and environmental temperature.
393. Roots, B. I., and C. L. Prosser, J. Exp. Biol. *39*:617–629, 1962. Temperature acclimation and nervous system of fish.
394. Rozin, P. N., Science *149*:561–563, 1965; *also* Rozin, P. N., and J. Mayer, Science *134*:942–943, 1961. Temperature discrimination in goldfish.
395. Ruibal, T., Evolution *15*:98–111, 1961. Graded temperature preferences of five species of *Anolis*.
396. Salt, R. W., Canad. J. Zool. *34*:1–5, 283–294, 391–403, 1956. Ice formation and supercooling in insects.
397. Salt, R. W., Canad. J. Zool. *39*:349–357, 1961. Survival of freezing in larvae of sawfly.
398. Salt, R. W., Nature *193*:1207–1208, 1962. Intracellular freezing in insects.
399. Salt, R. W., Symp. Soc. Exp. Biol. *23*:331–350, 1969. Survival of insects at low temperatures.
400. Saroja, K., and K. P. Rao, Z. vergl. Physiol. *50*:35–54, 1965. Biochemical changes in invertebrates in temperature acclimation.
401. Sassaman, C., and C. P. Mangum, J. Marine Biol. *7*:123–130, 1970. Lethal temperatures in North Atlantic actinians.
402. Sastry, A. N., Biol. Bull. *138*:56–65, 1970. Seasonal variations in breeding of Atlantic scallops.
403. Satinoff, E., Amer. J. Physiol. *206*:1389–1394, 1964. Behavioral thermal regulation after brain cooling in rats.
404. Satinoff, E., Science *148*:399–400, 1965. Effect of brain lesions on temperature regulation in hibernators.
405. Satinoff, E., Science *155*:1031–1033, 1967; Amer. J. Physiol. *212*:1215–1220, 1967. Disruption of hibernation and thermal regulation by hypothalamic lesions in ground squirrels.
406. Satinoff, E., Prog. Physiol. Psych. *3*:201–236, 1970. Hibernation and central nervous system.
407. Satinoff, E., and J. Rutstein, J. Comp. Physiol. Psych. *71*:77–82, 1970. Behavioral thermal regulation in rats with hypothalamic lesions.
408. Scheit, H. G., Marine Biol. *6*:158–166, 1970. Effect of temperature adaptation on denervated muscle of eel.
409. Schlieper, C. R., et al., Kieler Meeresforsch. *14*:3–10, 1958. Temperature and salinity effects on bivalve molluscs.
410. Schmid, W. D., Ecology *46*:559–560, 1966. Thermal tolerance in toads.
411. Schmidt-Nielsen, K., et al., Amer. J. Physiol. *188*:103–112, 1957. Temperature relations in camels.
412. Schmidt-Nielsen, K., W. L. Bretz, and C. R. Taylor, Science *169*:1102–1104, 1970. Panting in dogs.
413. Schmidt-Nielsen, K., T. J. Dawson, and E. C. Crawford, J. Cell. Comp. Physiol. *67*:63–72, 1966. Temperature in echidna.
414. Schmidt-Nielsen, K., and W. R. Dawson, pp. 467–480. *In* Handbook of Physiology, Sec. 4, edited by D. B. Dill. Amer. Physiol. Soc., Washington, D.C., 1963. Terrestrial animals in dry heat.
415. Schmidt-Nielsen, K., J. Kanwisher, R. Lasiewski, J. E. Cohn, and W. L. Bretz, Condor *71*:341–352, 1969. Temperature regulation of ostrich.
415a. Schmidt-Nielsen, K., et al., J. Exp. Biol. *55*:385–398, 1971. Heat tolerance of desert snails.
416. Scholander, P. F., et al., J. Appl. Physiol. *8*:279–282, 1955; *10*:404–411, 1957. Countercurrent heat exchangers.
417. Scholander, P. F., et al., J. Cell. Comp. Physiol. *49*:5–24, 1957. Supercooling and osmoregulation in Arctic fish.
418. Schultze, D., Z. wiss. Zool. *172*:104–133, 1965. Temperature adaptation of muscle of eel.
418a. Shea, S., D. Sigafoos, and D. Scott, Comp. Biochem. Physiol. *28*:701–708, 1969. Thermoexcitability of *Callinectes* nerve.
419. Siminowitch, D., et al., pp. 3–40. *In* Molecular Mechanisms of Temperature Adaptation, edited by C. L. Prosser. Amer. Assoc. Adv. Sci., Washington, D.C., 1967.
420. Simmonds, M. A., and L. L. Iverson, Science *163*:473–474, 1969. Effect of temperature on hypothalamic norepinephrine.
421. Simpson, J., Science *133*:1327–1333, 1961. Honeybee colonies in cold.
422. Smalley, R. L., and R. L. Dryer, pp. 325–345. *In* International Symposium on Natural Mammalian Hibernation 1965: Mammalian Hibernation III, edited by K. C. Fisher. American Elsevier Publ. Co., New York, 1968. Brown fat in hibernation.
423. Smit, H., Comp. Biochem. Physiol. *21*:125–132, 1967. Effect of temperature on gastric secretion in bullhead *Ictalurus*.
424. Smith, A. U., Monogr. Physiol. Soc. No. 9. Williams & Wilkins, Baltimore, 1961. Biological effects of freezing and supercooling.
425. Smith, C. L., J. Exp. Biol. *28*:141–164, 1951. Temperature and pulse rate in frog.
426. Smith, D. E., and B. Katzuns, Amer. Heart J. *71*:515–521, 1966. Mechanical properties of heart from hibernating and nonhibernating mammals.
427. Smith, M. W., J. Physiol. *182*:559–590, 1966; *183*:649–657, 1966. Effects of temperature on sodium and glucose transport in goldfish intestine.
428. Smith, M. W., Biochem. J. *105*:65–72, 1967. Temperature-dependent intestinal ATPase.
429. Smith, M. W., Comp. Biochem. Physiol. *35*:387–401, 1970. Regulation of amino acid transport by intestine of goldfish.
430. Smith, M. W., V. E. Colombo, and E. A. Munn, Biochem. J. *107*:691–698, 1968. Effect of temperature on ionic activation of ATPase.
431. Smith, M. W., and J. C. Ellory, Comp. Biochem. Physiol. *39A*:209–218, 1971. Temperature-induced change in Na transport and ATPase in goldfish.
432. Smith, M. W., J. C. Roberts, and K. J. Hittelman, Science *154*:653–654, 1966. Nonphosphorylating respiration of brown fat mitochondria.
433. Somero, G. N., Amer. Nat. *103*:517–530, 1969. Enzymic mechanisms of temperature compensation.
434. Somero, G. N., Biochem. J. *114*:237–241, 1969. Pyruvate kinase of Alaskan king crab.
435. Somero, G. N., A. C. Giese, and D. E. Wohlschlag, Comp. Biochem. Physiol. *26*:223–233, 1968. Cold adaptation of Antarctic fish.
436. Somero, G. N., and P. Hochachka, Amer. Zool. *11*:159–167, 1971. Biochemical adaptations to temperature in poikilotherms.
437. Sømme, L., J. Insect Physiol. *12*:1069–1083, 1966. Blood composition and supercooling in larvae of flour moths.
438. Sømme, L., J. Insect Physiol. *13*:805–814, 1967. Effect of temperature and anoxia on hemolymph in overwintering insects.
439. Sømme, L., Nytt Mag. Zool. *13*:52–55, 1967. Seasonal changes in cold tolerance of intertidal animals.
440. South, F. E., Physiol. Zool. *31*:6–15, 1958. Respiration and glycolysis of heart and brain from hibernators and nonhibernators.
441. South, F. E., Amer. J. Physiol. *200*:565–571, 1961. Nerve-diaphragm in relation to temperature and hibernation.
442. Späth, M., Z. vergl. Physiol. *56*:431–462, 1967. Action of temperature on mechanoreceptors of fish *Leuciscus*.
443. Stebbins, R. C., and R. E. Barwick, Copeia 541–547, 1968. Telemetry of temperature from *Varanus*.

444. Steen, I., and J. B. Steen, Comp. Biochem. Physiol. *15*:267–270, 1965. Thermal regulation by beaver's tail.
444a. Steen, I., and J. B. Steen, Acta Physiol. Scand. *63*:285–291, 1965. Heat loss via legs of aquatic birds.
445. Stevens, E. D., and F. E. J. Fry, Canad. J. Zool. *48*:221–226, 1970. Cooling curves of fish.
446. Stolwiyk, J. A. J., and J. D. Hardy, Pflüger. Arch. *291*:129–162, 1966. Theory of temperature regulation in man.
447. Strumwasser, F., Amer. J. Physiol. *196*:8–30, 1959. Brain activity in hibernating *Citellus*.
448. Sumner, F. B., and U. N. Lanham, Biol. Bull. *82*:313–327, 1942. Respiration of warm and cold spring fishes.
449. Tashima, L. S., S. J. Adelstein, and C. P. Lyman, Amer. J. Physiol. *218*:303–309, 1970. Glucose utilization by active, hibernating, and arousing ground squirrels.
450. Taylor, C. R., Arctic Aeromed. Lab. Bull. AAL-TR-63-31, 1–103, 1966. Thermal regulatory functions of horns in Bovidae.
451. Taylor, C. R., Physiol. Zool. *39*:127–139, 1966. Thermal regulatory function of horns in goats.
452. Taylor, C. R., Amer. J. Physiol. *219*:1131–1135, 1136–1139, 1970. Temperature regulation in African ungulates.
453. Taylor, C. R., and K. Schmidt-Nielsen, Amer. J. Physiol. *220*:823–827, 1971. Temperature relations of African hunting dog.
454. Tebecis, A. K., and J. W. Phillis, Comp. Biochem. Physiol. *25*:1034–1047, 1968. Effect of temperature on spinal cord of toad.
455. Templeton, J. R., Physiol. Zool. *37*:300–306, 1964. Vascular responses to temperature in *Sauromalus*.
456. Tester, J. R., and W. J. Breckenridge, Ann. Acad. Sci. Fenn. A. IV *31*:423–430, 1964. Winter behavior in Manitoba toad.
457. Thiede, W., Z. wiss. Biol. *172*:305–346, 1965. Hormonal basis of adaptation in fish.
458. Thiessen, C. I., and J. A. Mutchmor, J. Insect Physiol. *13*:1837–1842, 1967. Effects of temperature on muscle apyrase in *Periplaneta* and *Musca*.
459. Thoenen, H., Nature *228*:861–863, 1970. Catecholamine metabolism in cold exposed rats.
460. Thompson, F. J., and C. D. Barnes, Life Sci. *9* (Part 1):309–312, 1970. Thermosensitive elements in femoral vein.
461. Todd, M. E., and P. A. Dehnel, Biol. Bull. *118*:150–172, 1960. Effects of temperature and salinity on heat tolerance in grapsoid crabs.
462. Toivola, P., and C. C. Gale, Neuroendocrinology *6*:210–219, 1970. Temperature effects on amines in hypothalamus of baboon.
463. Tribe, M. A., and K. Bowler, Comp. Biochem. *25*:427–436, 1968. Temperature dependence of standard metabolism in flies.
464. Troshin, A. S., ed., International Symposium on Cytoecology–Leningrad–1963. Pergamon Press, Oxford, 1967. 462 pp. The cell and environmental temperatures.
465. Tsukuda, H., Biol. J. Nara Women's Univ. *10*:11–14, 1960. Temperature tolerance in the guppy.
466. Tsukuda, H., J. Biol. (Osaka City, Japan) *12*:15–45, 1961. Temperature acclimatization in fish.
467. Tsuyuki, H., and F. Wold, Science *146*:535–537, 1964. Enolase in fish muscle.
468. Tucker, V. A., J. Cell. Comp. Physiol. *65*:405–414, 1965. Relation between torpor cycle and heat exchange in *Perognathus*.
468a. Tucker, V. A., J. Exp. Biol. *48*:55–66, 1968. High-altitude flight in sparrows.
469. Umminger, B. L., Biol. Bull. *139*:574–579, 1970. Effects of subzero temperature on blood of winter flounder.
470. Umminger, B. L., J. Exp. Zool. *172*:283–302, 409–424, 1970. Blood constituents and supercooling in killifish.
471. Umminger, B. L., J. Exp. Zool. *173*:159–174, 1970. Carbohydrate metabolism in fish at subzero temperatures.
472. Umminger, B. L., Nature *225*:294–295, 1970. Osmoregulation in fish in fresh water near freezing.
473. Ushakov, B. P., Int. J. Life in Oceans and Coastal Waters *1*:153–160, 1968; *also* pp. 107–130 *in* Molecular Mechanisms of Temperature Adaptation, edited by C. L. Prosser, Amer. Assoc. Adv. Sci., Washington, D.C., 1967. Cellular thermostability of aquatic animals.
474. Ushakov, B. P., B. Iamosova, I. Pashkova, and I. Chernokozhera, J. Exp. Zool. *167*:381–390, 1968. Individual variability in heat resistance of cells.
475. Veght, J. H., Physiol. Zool. *37*:316–328, 1964. Metabolic regulation of temperature in gray jay.
476. Vellas, F., Ann. Limnol. *1*:435–442, 1965. Temperature effects on uricolytic enzymes in carp.
477. Vernberg, F. J., C. Schlieper, and D. E. Schneider, Comp. Biochem. Physiol. *8*:271–285, 1963. Temperature and salinity effects on cilia of molluscan gills.
478. Vernberg, F. J., and R. E. Tashian, Ecology *40*:589–593, 1959; *also* Vernberg, W. B., and F. J. Vernberg, Comp. Biochem. Physiol. *17*:363–374, 1966. Physiological variation of tropical and temperate zone *Uca*.
479. Vernberg, W. B., and F. J. Vernberg, Amer. Zool. *8*:449–458, 1968. Diversity in metabolism of marine and terrestrial crustaceans.
480. Vernberg, W. B., and F. J. Vernberg, Comp. Biochem. Physiol. *26*:499–508, 1968. Effect of temperature on cytochrome C oxidase of *Uca*.
481. Waites, G. M. H., Quart. J. Exp. Physiol. *471*:314–323, 1962; Nature *190*:172–173, 1961; *also* Waites, G. M. H., and J. K. Voglmayr, Austral. J. Agric. Res. *14*:839–851, 1963. Temperature regulation by scrotum of ram.
482. Wang, L. C., and J. W. Hudson, Comp. Biochem. Physiol. *32*:275–293, 1970; *38A*:59–90, 1971. Temperature regulation in chipmunk *Tamias* and in *Perognathus*.
483. Wang, G. T., and W. C. Marquardt, J. Protozool. *13*:123–128, 199–202, 1966. Survival of *Tetrahymena* and *Paramecium* following freezing.
484. Weatherley, A. H., Biol. Bull. *139*:229–238, 1970. Effects of high oxygen on thermal tolerance of goldfish.
484a. Weathers, W. W., and F. N. White, Amer. J. Physiol. *221*:704–710, 1971. Thermoregulation in turtles.
485. Weiser, C. J., Science *169*:1269–1271, 1970. Cold resistance and injury in woody plants.
486. West, G. C., and J. S. Hart, Physiol. Zool. *39*:171–184, 1966. Metabolic response of grosbeak to temperature.
487. Wieser, W., and J. Kanwisher, Limnol. Oceanogr. *6*:262–270, 1961. Ecology of marine nematodes from salt marsh.
488. Williams, B. A., and J. E. Heath, Amer. J. Physiol. *218*:1654–1660, 1970. Preoptic heating and cooling in ground squirrel.
489. Williams, D. D., Comp. Biochem. Physiol. *27*:567–573, 1968. Effect of cold exposure on norepinephrine heat production in hamster.
490. Williams, D. D., and W. S. Platner, Arctic Aeromed. Lab. Bull. AAL-TR-67-5, 1–12. Cold-induced changes in fatty acids in rat and hamster.
491. Williams, R. J., Comp. Biochem. Physiol. *35*:145–161, 1970. Freezing tolerance in *Mytilus*.
492. Williamson, J. R., S. Prusint, M. S. Olson, and M. Fukami, Lipids *5*:1–14, 1970. Control of metabolism in brown fat.
493. Willis, J. S., J. Physiol. *164*:51–63, 64–76, 1962. Resistance of tissues of hibernators to cold swelling.
494. Willis, J. S., Amer. J. Physiol. *214*:923–928, 1968. Cold resistance of kidney of hibernators.
495. Willis, J. S., and N. M. Li, Amer. J. Physiol. *217*:321–326, 1969. Cold resistance of Na-K ATPase of kidney.
495a. Wilson, F. R., Personal communication. Specific protein synthesis in acclimation.
496. Wilson, G., S. P. Rose, and C. F. Fox, Biochem. Biophys. Res. Comm. *38*:617–623, 1970. Lipid unsaturation and sugar transport by *E. coli*.
497. Wilson, K. J., and A. K. Lee, Comp. Biochem. Physiol. *33*:311–322, 1970. Metabolism and heart rate of *Sphenodon* at various temperatures.
498. Wyndham, C. H., R. Plotkin, and A. Munro, J. Appl. Physiol. *19*:593–597, 1964. Reactions of men in Antarctic.
499. Wyndham, C. H., et al., J. Appl. Physiol. *19*:583–592, 868–876, 1964. Reactions of African natives to cold.
500. Yoshida, M., T. Oshima, and K. Imahori, Biochem. Biophys. Res. Comm. *43*:36–39, 1971. Thermo-stable enzyme PFK from thermophilic bacteria.
501. Zhirmunsky, A. V., pp. 209–218. *In*: International Symposium on Cytoecology–Leningrad–1963, edited by A. S. Troshin. Pergamon Press, Oxford, 1967. Cellular thermal stability of marine invertebrates.

CHAPTER 10

BIOLOGICAL RHYTHMS

By F. A. Brown, Jr.

Biological rhythmicity, with periods correlated with the major periods of the fluctuating physical environment, is clearly widespread; it encompasses animals, plants, and microorganisms, and is probably a universal attribute of life on the earth. These rhythms possess special properties, differing strikingly from those of rhythms without such correlates (such as heart and respiratory ones).

The general properties of the rhythms are so remarkably similar among all of the wide spectrum of living creatures in which they have been investigated that there is reason to postulate that they comprise a very fundamental and ancient characteristic. Consequently, knowledge of the nature of this phenomenon has, not surprisingly, been contributed from organisms representing the gamut of life. The periods appear to be essentially uninfluenced by temperature and by virtually all known depressants, inhibitors, or stimulants of metabolic processes. The differences among the rhythms involve chiefly the nature of the species-specific biochemical, physiological, and behavioral phenomena that comprise the rhythmic variations.

In this chapter there will be no attempt to describe all of the innumerable kinds of reported rhythmic processes that have been, and are continuing to be, detailed and which suggest that geophysically correlated periodisms are the rule. Descriptions of specific rhythmic variations can more properly be treated in relation to the physiological functions for which they have special relevance. Here we shall concern ourselves with the phenomenon of biorhythmicity *per se*, using only selected examples. A number of reviews and treatises are available which include abundant examples.*

Components of Rhythms. The recurring cycles of the rhythms of organisms in their natural environment result from two simultaneous kinds of sources: (a) a continuing direct response to the fluctuating, periodic environment, both living and non-living, and (b) an inherent tendency of an organism to repeat in successive cycles the forms of the physiological and behavioral patterns, whether genetic or environmentally imposed, of preceding cycles. As a consequence, the detailed cycle forms vary somewhat for successive cycles in response to alterations in the forms of the environmental contributions, both during the concurrent cycle and the recently preceding ones.

The most influential contributors to the rhythmic patterns from the environment are the variations in such biologically significant factors as light, temperature, humidity, tidal changes, and feeding times, to which the effected alterations appear typically to comprise adaptive responses. Of less influence, but measurable, are alterations in cyclic patterns as responses to subtle, pervasive geophysical parameters, probably principally electromagnetic, whose periodic changes are steadily distorted or otherwise modified by

*See references 18, 38, 56, 61, 62, 63, 92, 120, 147, 155, 161.

weather-system passages, periodisms of other frequencies, and other unidentified factors.

When organisms are held in unvarying conditions with respect to light, temperature, and all other obviously influential factors, there persist the continuing responses to those subtle physical variables of the environment from which the organisms still remain unshielded. There may also persist for a longer or shorter time, depending upon the organism and system, continuing recurrences of the preceding adaptive rhythmic patterns, reflecting the operation of the inherent rhythmic tendency. These latter, even in the constant conditions, continue to display periods which are highly resistant to alteration by temperature change or application of metabolic modifiers.

There is rapidly mounting evidence that the rhythmic nature comprises an important, often probably critical, temporal framework for biological integration at three levels.

1. It serves to time alterations in organismic phenomena to enable organisms to maintain, within the range of their other limitations, optimal adaptive states to meet such rhythmically varying physical environmental factors as light, temperature, humidity, and ocean tides, including effective anticipation and advance preparation for the more and less favorable times.

2. It is employed usefully to coordinate reproductive activities within species, to coordinate the activities of ecologically interdependent species, and to time the normal splendidly integrated ecosystem dynamics. In the latter, for maximal efficiency and productivity, the environment is used only part time (but at times of their optimal adaptation) by each of the innumerable species comprising the diversity that contributes to ecologic stability.

3. Within the individual organism, evidence increasingly supports the hypothesis that independent rhythmicity of individual organs, tissues, and cells, and even of components of cells, contributes a temporal orderliness to vital phenomena, which enables a degree of integration that would be impossible to achieve were events within each of the numerous interrelated and interacting processes deprived of such orderliness.

There are probably no such almost universally occurring phenomena in biology as generally accepted as the "biological clocks" and the rhythms which are dependent upon them (solar-day, tidal and monthly, and annual), for which our knowledge of their bases and mechanisms is so obscure and speculative and which have given way so little to conventional experimental approaches.

Solar-Day Rhythms. The most extensively investigated of the biological rhythms are those that are correlated with the 24-hour solar day. Under natural conditions, all the factors that contribute to the recurring patterns display in common the 24-hour period. The contribution of any one of them to the total observed recurring patterns is difficult to resolve under these conditions but may be, in some measure, differentiated by observations conducted under controlled laboratory conditions. Subjection of an organism to an experimental regime of 12L:12D (12 hours of light alternating with 12 hours of darkness) in otherwise constant conditions of all other obvious environmental factors will disclose that a normally nocturnal animal will set the time of activity in its solar-day rhythm to the time of darkness, a diurnal organism will be active during the light interval, and a crepuscular one will display maximum activity at or near the times of the transitions between light and darkness. Still other animals may display times of maximum activity restricted to very short periods at specific times of day. That light is an environmental factor to which the animal adjusts its genetically determined behavioral pattern to local time of day is evident from the fact that when the times of the two light transitions in the 12L:12D cycles are shifted to different clock hours of the day, the animal typically within a few cycles will have shifted its time of activity in the day to accord again in its characteristic manner with the new light and dark times. A major effect of the 24-hour environmental rhythm in illumination is, therefore, to relate adaptively to the day-night changes the phase of an animal's genetically determined solar-day behavioral pattern. In comparable manner, the altered times of light and darkness which result from a rapid eastward or westward translocation of an organism to a different longitude is followed by the appropriate readjustment in its activity-rest rhythm. In this capacity light is playing the role of a rhythm-phase adjusting factor, or "Zeitgeber." Twenty-four-hour rhythms in temperature, especially in the absence of light cycles, may often serve similarly as a phase-setting factor, as do also in some instances rhythms in sound, feeding, and even social interactions.

That factors other than light and other

phase-determiners are contributing to the observed 24-hour cycles is readily evident from the discovery that even with careful regulation of constancy of all ordinary factors while providing precise repetition of the light cycles, the daily patterns commonly differ significantly in their detailed form from one day to the next, and that the imposition of the 12L:12D cycle in one relationship to the local-time clock hour may not be exactly the equivalent of its imposition in another relationship. The cycle may also display semimonthly and monthly modulations. Still uncontrolled environmental parameters are, therefore, eliciting responses from the animals, parameters all of which are continuing to reflect in some degree the 24-hour and other natural periodicities of the terrestrial environment.

The precise mean 24-hour periodicity of animals in nature despite large weather-associated day-to-day mean temperature changes or other organismic stressing factors, or in the laboratory in 12L:12D over a large range of experimental temperatures, is readily accounted for in terms of the obvious 24-hour rhythmic light input from the environment.

Circadian Rhythms. When animals are deprived of variations in both of their major phase-setting periodic environmental cues, light and temperature, the rhythmic solar-day patterns may continue and very commonly cease to maintain their precise mean 24-hour periodicity. The recurring cycles may become slightly longer or shorter than 24 hours. The times of activity may systematically drift to progressively earlier local clock times of day, to generate an overt rhythmic period less than 24 hours; or they may drift to progressively later times, to yield a rhythmic period longer than 24 hours.

The discovery of such altered periodicities under these experimental conditions led to the coining of the term "circadian" (circa—about; diem—day)[87] to emphasize the fact that they only approximated a day in length. The term circadian has become widely adopted and has become synonymous with diurnal, solar-day, and 24-hour. Since these rhythmic changes no longer remained, of course, in any fixed relationship to the clock hours of the day, they were termed "free-running"[67] (Fig. 10–1).

The properties of "free-running" circadian rhythms have been extensively investigated in numerous laboratories both for populations and single individuals.[73,95] Circadian periods vary widely among individuals of the same species held under the same carefully controlled conditions of environmental constancy, and even may vary, apparently spontaneously, in a single individual with time (Fig. 10–2) with the period changes sometimes gradual and at other times abrupt. Other factors being equal, the periods appear to be a function of the level of the constant illumination. In diurnal vertebrates whose free-running periods tend to be shorter than 24 hours, and in nocturnal vertebrates with periods generally longer than 24 hours (Aschoff's rule)[4] (Fig. 10–3), the brighter the illumination the shorter and longer the cycles, respectively. Especially in poikilotherms, the period length appears to display a very slight, comparable temperature-dependence.[162] For most organisms the Q_{10} appears to range between 1.0 and 1.2, but for some species the temperature relationship has been found to be the reverse, with a Q_{10} in the range of 0.8 to 1.0. The free-running period is, other factors being equal, also determined by genetic factors.[51,52,110]

Numerous attempts to alter the "free-running" period through application of metabolic stimulants or depressants have generally led to negative or equivocal results, or to results that could be most plausibly explained in terms of the agent effecting transient phase shifts that accounted for a few apparently altered periods while the agent was thereafter without effect on the continuing periodicity. Nor have pulsed applications of chemical agents led to any clear alteration in period. Such negative results from chemical investigations have even involved interference with DNA, RNA, and protein synthesis.[104,166,167] The timing system for the relatively regular free-running periods seems, therefore, to be extraordinarily stable and highly resistant to period alteration.[96] The forms and amplitudes of the cycles are more labile.

One chemical agent, D_2O, does have an influence on the period of a free-running rhythm.[50,132,160] The effect is a small degree of lengthening of the cycles, the amount of change being proportional to the concentration of the heavy water. This has been noted for organisms as diverse as *Euglena,* birds, and mice. Period appears also influenced by very weak natural electric fields and alterations in them.[181]

Investigations of the influence of light

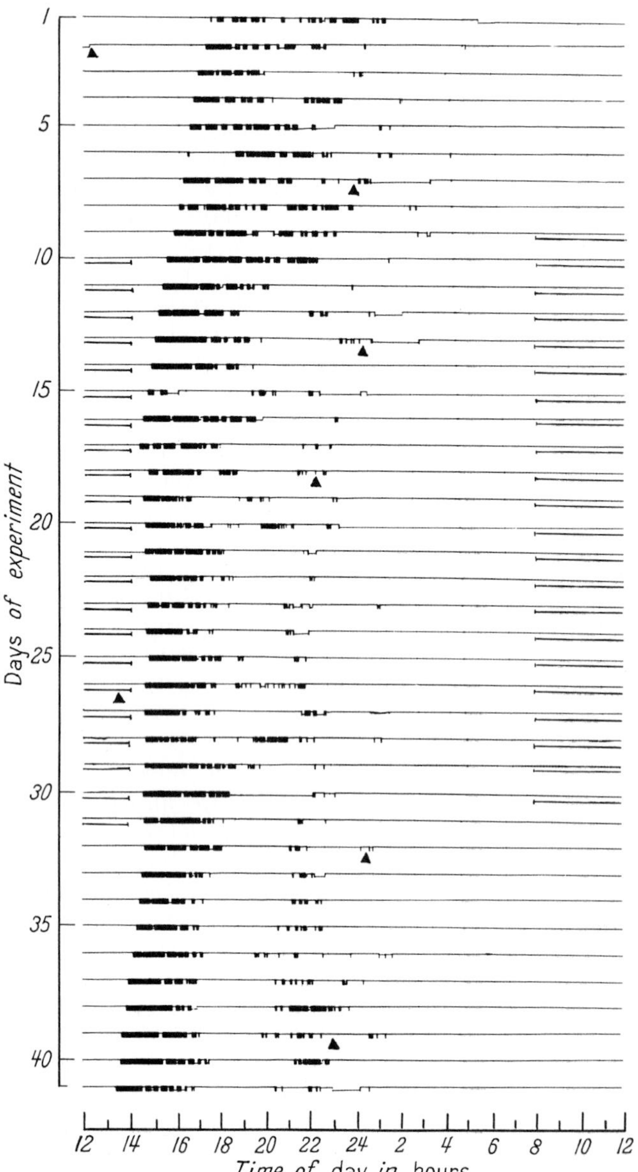

Figure 10–1. Free-running period of a flying squirrel in continuous darkness (days 1–9), with entrainment to 24-hour LD cycles (days 10–32) with bars indicating the light hours, and then resumption of a free-running period after the light cycles were discontinued (days 32 onward). (From DeCoursey, P. J.; Effect of Light on the Circadian Activity Rhythm of the Flying Squirrel, *Glaucomys volans.* Z. vergl. Physiol. *44*:331–354, 1961. Berlin-Göttingen-Heidelberg: Springer.)

changes on altering the phases of circadian rhythms to effect the species' appropriate relationship to the 24-hour LD cycles led to the discovery that L to D or D to L changes did not effect equivalent amounts of phase-shifts in the circadian rhythms at all times of day.[37, 58, 68, 99, 140, 173] Organisms exhibited a diurnal variation in their phase-shifting response to light, the changing response being termed a phase-response curve. Using, for example, the flying squirrel, a nocturnal mammal, held in constant darkness (DD) it

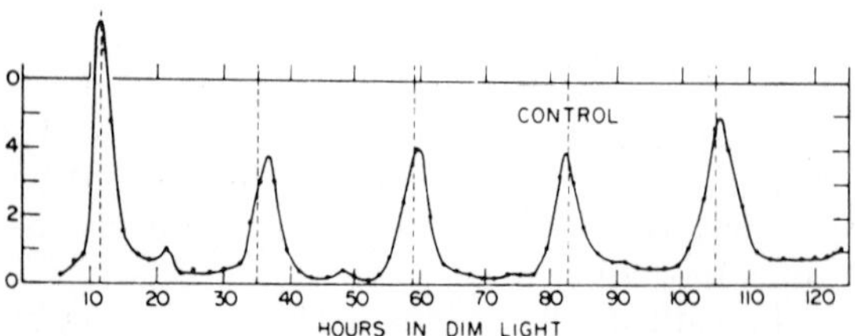

Figure 10–2. Persistent rhythm of bioluminescent glow in *Gonyaulax polyhedra.* Cells were grown in LD 12:12 and placed in constant dim light and constant temperature (21° C) at the end of a light period (zero time on the graph). Luminescence is expressed in arbitrary units. (Redrawn from Hastings, J. W., New Engl. J. Med. *282*:435–441, 1970.)

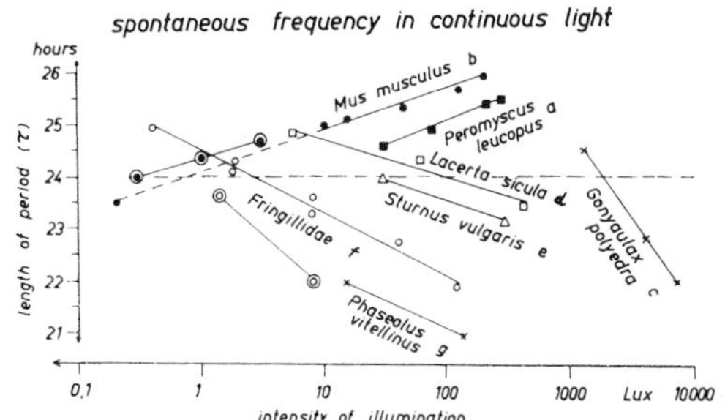

Figure 10–3. Free-running period of various organisms in constant illumination at different intensities. Encircled points indicate a repetition of the earlier study represented by unencircled similar points. Measured is activity except for *Gonyaulax* (bioluminescence) and *Phaseolus* (leaf movement). (From Aschoff, J., Cold Spring Harbor Symp. Quant. Biol. *25*:11–27, 1960.)

was learned that the phase-shifting effect of a 10-minute light pulse varied with the time within the organism's circadian cycle.[67] Since under these experimental conditions the squirrels were usually displaying frequencies deviating slightly from 24 hours, the times of activity (the subjective night) and inactivity (the subjective day) of the animals systematically drifted over the clock hours of the day. It was discovered that a light pulse over the onset of the animal's activity period resulted in maximum amount of phase delay (to a later time), while an equivalent pulse administered about the time of activity termination effected a maximum in phase advance, with intergrading through no phasing response to the light change during the intervening times within the cycles (Fig. 10–4).

Phase-response curves are presumed to describe the property of the circadian system responsible for the normal phase adjustment of the rhythm pattern to the natural LD cycles and are postulated, therefore, to be a fundamental property of the circadian rhythmic systems. The phase-response cycles, phase-locked to the activity-rest cycles, will (for the flying squirrel, for example) be expected to delay the activity to the L to D evening transition and advance it from the D to L morning one. An equilibrium phase adjustment would be expected to result when the daily advance and delay contributions from the curve become equal as portions along the curve spanned the natural, gradual LD changes.

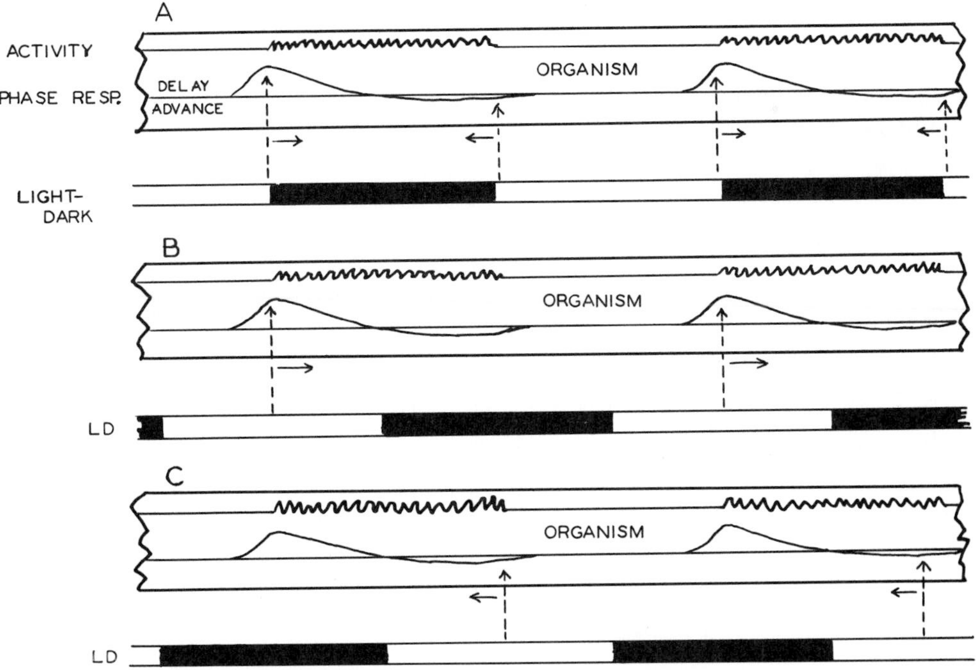

Figure 10–4. A, Diagram of the relationship between the activity-rest cycle of a nocturnal mammal and its tightly associated phase response curve, together with the relationship of these to the environmental light cycles in an animal in phase equilibrium. B, The comparable state of affairs immediately after an abrupt phase-delay in the environmental light cycle. C, The comparable state of affairs immediately after an abrupt phase-advance in the light cycle. The expected phasing influences of light are indicated by arrows. (From Brown, F. A., Jr., Amer. Scientist *60*:756–766, 1972.)

Details of forms and amplitudes of phase response curves appear to vary somewhat from one individual to another, and hence slight differences in the specific phase-adjustment relationships of individual animals to the same LD cycles would be expected, and indeed are commonly observed. It has been noted, for example, among finches in 12L: 12D that the activity of some of the birds anticipates the D to L change while onsets for others lag on the light change.[5] Those finches that led the DL transition were found to have, when examined in continuous light, free-running periods shorter than 24 hours, while those that lagged on the DL change had periods longer than 24 hours. An investigation with kangaroo rats in continuous dim light in which the free-running periods among the animals ranged from shorter than to longer than 24 hours disclosed a systematic quantitative relationship between the observed free-running period for an animal and the amount of phase delay or advance for its response curve over the time of activity onset.[124] For rats with periods shorter than 24 hours the curve displayed phase advances to the light change over onsets, and for the rats with periods longer than 24 hours, delays at this same time (Fig. 10–5).

Phase-response curves are present for temperature also.[55, 58, 157, 188]

Free-running circadian rhythms in constant illumination and temperature may persist apparently indefinitely, even increasing or fluctuating in amplitude, or may show varying rates of damping with ultimate apparent cessation. Even abrupt cessation has been reported, lasting for a few or many cycles, only to become spontaneously reestablished later in the expected phase relationship at that time had the rhythm been simply continuing uninterruptedly.[92] In instances where a rhythm has been damped to cessation, it has often been demonstrated to be possible to reinstate an ongoing circadian rhythm by an imposed LD cycle, or even by a single brief light shock which imparts no 24-hour cue. Such a stimulus not only may commence a rhythm but may determine its phase setting as well. Results such as these indicate that the circadian rhythmic patterns are coupled to some deeper rhythmic variation and may under some circumstances become uncoupled. Loss of an overt circadian rhythm in one system within an organism does not necessarily mean a loss in other systems within the same individual.

The recurring patterns of circadian rhythms may be experimentally modified. This can be accomplished, for example, by light. There may, in continuous light (LL), be a gradual reduction in amplitude over a number of successive cycles. The cycles that persist thereafter when the animals are transferred to constant darkness retain for at least many cycles the depressed amplitudes that had been attained at the time that the trans-

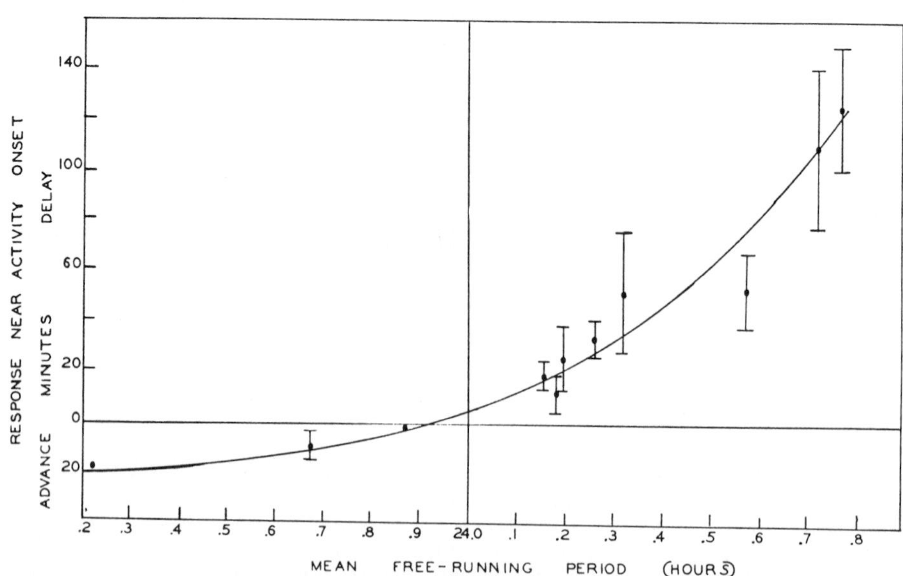

Figure 10–5. Relationship between amount of phase advance or delay in the circadian rhythm of kangaroo rats to constant light shocks imposed over the time of onset of activity. (From Natalini, J. J., Physiol. Zool., *45*:153–166, 1972.)

fer had occurred.[39] A brighter light flash at a particular point in a cycle of an ongoing rhythm in continuous dim light has been reported to induce in the bean seedling a brief behavioral response, which behavior then recurs at the corresponding points in the succeeding cycles, persisting in the absence of further stimulation.[54] In a comparable manner, time-training may be accomplished. Bees have been trained to appear at a feeding station at a specific time within the day or circadian cycle; thereafter, even in the absence of further feedings, the bees return at the same time of day for a number of days.[146] The recurring patterns of the circadian systems clearly reflect both environmentally imposed and gentically determined elements.

Studies have been made of the influences of rapid translocations by airplane of organisms from one terrestrial longitude to another.[48, 146] The usual results when the experiment is conducted with free-running rhythms is that the cycles continue quite as would be expected had the translocation not occurred. When, however, the translocation occurs under natural conditions, though again the general ongoing circadian pattern persists, the altered times of light, temperature, and other environmental "Zeitgeber" at the new longitude operate gradually to readjust the phases of the circadian rhythms over the succeeding several cycles until a new equilibrium is achieved.[99, 158]

Circadian rhythms have been shown to persist even in the Arctic during the seasons of continuous illumination[114, 117] or darkness, and observations made at the south geographic pole[89] have established that typical free-running periodicities also persist in organisms held there in carefully maintained constancy of every obvious environmental factor. The unique and fundamental nature of approximately 24 hours for circadian rhythms is emphasized by a number of kinds of observations. For organisms as different as flowering plants and mice,[109, 165] attempts to have the circadian rhythmic cycles lengthened or shortened by subjecting the organisms to LD rhythms over a range of frequencies have demonstrated that, while the organisms can often adjust to cycle as short as about 10L:10D or as long as about 14L:14D, altering the LD cycles beyond these relatively narrow limits commonly results in the organism breaking away from the imposed cycles and reverting to its characteristic circadian frequency. In one experiment with fiddler crabs[173] it was discovered that subjection of the crabs to cycles of 16L:16D led to a gradual repression of an initially clear 24-hour rhythm and the adoption of a 96-hour variation, apparently the consequence of beats between the continuously imposed 32-hour and the stubbornly retained 24-hour ones. The limits to which mice can be entrained to LD cycles can be significantly extended by using gradual light transitions instead of abrupt on-off ones.[105]

Not only does circadian rhythmicity occur in organisms as a whole, but individual organs,[3] tissues, and even single cells can independently exhibit such rhythmicity.[7, 148, 153, 163] Investigations of unicellular organisms have even shown that nucleus and cytoplasm may each separately possess circadian rhythmicity, and that if the nucleus of one cell is implanted into an enucleated cytoplasmic fragment whose circadian rhythm is not in phase synchrony with it, the nucleus will gradually shift the phase of the cytoplasmic rhythm into phase with itself.[154]

The organism appears to comprise an integrated complex of rhythmic component systems at all levels of organization. In its normal equilibrium state, in response to an ongoing 24-hour rhythmic variation in the natural phase-adjustors or Zeitgeber, the rhythmic components offer 24-hour variational patterns which are temporally interrelated in their phases for optimal facilitation of interactions. The diverse rhythmic components may pass through their maxima and minima at quite different times of day, with lags and leads on one another in good part as a consequence of the numerous sequential cause-effect relationships, together with the times required for completion of processes comprising the various links in the circadian chain of events. The description of the various temporal relationships of the rhythmic components that underlie the circadian rhythmicity of an organism as a whole constitutes a circadian "phase map."

When an organism is subjected to an abruptly altered Zeitgeber or a complex of Zeitgeber (either naturally as by rapid translocation to a different time belt, or experimentally by subjecting the organism to an equivalent abrupt shift in phase of the LD cycles), readjustment commences during the first cycle; but commonly several or many cycles have passed before the organism's total circadian system has attained a new equilibrium state, with all the component rhythms coming to possess their normal phase-map relations. Different components appear to shift at different rates toward the new equilibrium state. During this transitional period the various transient cyclic elements may be dissociated from one another relative to their condition either initially or at their eventual new equilibrium state. This transitional period appears to be correlated with lowered efficiency for the jet traveler, and presumably an equivalent transient state for other organisms as well.

A comparable dissociation has been reported for some free-running rhythms in constancy of all effective Zeitgeber. For example, in man the sleep-waking rhythm has been described in one instance to be more than 33 hours, while the body temperature one was only 24.8 hours.[6]

Transient cycles for circadian rhythms may also be noted following other types of experimental procedures. A single phasing stimulus may induce a phase shift in the immediately ensuing cycle of a free-running rhythm, a phase shift which is gradually lost over the course of a few succeeding cycles. Alternatively, a single phasing stimulus may induce little or no obvious phase shift on the next ensuing cycle but a series of phase-shifts, or transient cycles, may continue over the next few succeeding cycles until a new phase equilibrium is achieved and the frequency becomes restored essentially to its initial value. It is presumed that these two different behaviors depend upon whether or not a dominant phase-determining rhythmic component has been phase-shifted by the stimulus.

While circadian rhythms in the absence of any obvious Zeitgeber cues commonly free-run with periods deviating slightly from 24 hours, it has also been reported that free-running periods may sometimes retain a period statistically insignificantly different from 24 hours even for extended intervals.[47, 119, 164] The latter has been described particularly for mice and rats held in continuous darkness, and for populations of fiddler crabs, kangaroo rats, and gila monsters. The retention of this precision of 24 hours, it has been pointed out, occurs more commonly than would be expected from a random sample of circadian periods. Animals may retain for many days or even weeks a free-running period of exactly 24 hours and then suddenly adopt a different frequency at the later time. Conversely, an animal that has been free-running with a period deviating from 24 hours for some time may abruptly adopt a period of 24 hours without evident explanation. The 24-hour period for circadian rhythmic patterns of animals in general appears, therefore, to occupy a special position, with a greater propensity for precise 24-hour free-running than for any other single period deviating significantly from it.

Another reported property of the circadian system is its apparent capacity, at least in rare instances, to have its phase adaptively adjusted to local time even in the absence of any obvious environmental cues. Young chick embryos incubated in carefully controlled constancy of all known phasing stimuli, upon first achieving a fully differentiated sensory-neuromotor system have been reported to exhibit not only a circadian pattern of activity but one which is adaptively phase-related to local time for the genetically diurnal chick.[101] This suggests that subtle atmospheric variations correlated with the solar-day tides can in this instance serve as a Zeitgeber.

Geophysically Dependent Rhythms. Another periodic component that persists in organisms even after all environmental factors which are able to serve as phase-setters for circadian rhythms are held constant appears to be a consequence of a continuing organismic response to still uncontrolled, subtle residual environmental variables. Other factors being equal, the rate of biological activity varies with the local time of day, or hour-angle of the sun. These rate variations are of relatively small mean amplitude and hence are usually obscured by the much larger circadian and other overt rhythmic patterns of variation. In addition, the periodic environmental complex upon which evidence indicates that they depend, the atmospheric electromagnetic field, itself has its successive cycles modulated by natural geophysical rhythms of other frequencies, by the passages of weather systems, as well as by intrinsic alterations in the activities of the sun and earth. In short, the living system appears steadily to derive information about fluctuations in the terrestrial atmosphere, as evidenced by the ready demonstrability of continuing cross-correlations with subtle environmental parameters whether the organisms are subjected to the natural day-night environment, held in fields of artificial periodic Zeitgeber (as, for example, light), or are shielded from all periodic Zeitgeber.

The simplest circumstance under which

to observe and characterize the geophysically dependent mean 24-hour pattern is with an organism lacking an overt circadian pattern and held in constancy of all obvious environmental stimuli, or with populations of large numbers of individual organisms held in constancy of potential Zeitgeber synchronizers and therefore displaying a presumably random distribution of phase relations for their free-running rhythms. In instances where conspicuous free-running circadian rhythms are evident, the 24-hour pattern imposed as a response to the uncontrolled physical variables can ordinarily be extracted by determining the mean rate of activity for each hour of the day over a period of time just adequate for the circadian pattern systematically to scan across the day a simple integral number of times. The number of days must include not only exactly the number of days for the circadian pattern to scan the solar day once, or a simple integral number of times, but *simultaneously* a simple integral number of synodic months, since there exists for some animals at the same time an overt lunar tidal rhythmic component and, probably for all organisms, an atmospheric dependent lunar-day variation of the same general order of magnitude as for the solar-day one. By the foregoing procedure it is evident that what is obtained is the rate of activity for each clock hour of the day, with the phase angles of the circadian pattern (as well as of any lunar tidal or lunar-day periodicity) essentially randomized.

The most extensively investigated rate variations for the resolution of the persistent geophysically dependent cycles have been for organismic oxidative metabolism and for the rate of spontaneous motor activity, which, of course, exhibits a high correlation with standard metabolic rate of the organism. For the former, automatic recording respirometers designed to permit very long periods of monitoring in constancy of all obvious factors, including ambient pressure, have been employed. For spontaneous motor activity, actographs of various types linked with appropriate recording systems have similarly permitted the required long continuing studies.

Unlike the circadian patterns which appear to adjust each species to its solar-day environmental changes in specific and adaptive manners, the geophysically dependent variations are less specific and tend to assume one or another of a limited number of forms (Fig. 10–6). One of the most common is a trimodal variation with maxima at about 7 A.M., noon, and 6 P.M., with a daily range generally about 4 to 20 per cent of the daily mean. This general pattern has been reported for organisms as diverse as potatoes, beans, carrots, mealworms, four- and five-day chick embryos during the spring months, and the rat.[29, 44, 101] The same general 24-hour pattern has also been observed for the effects of very weak gamma radiation on mice.[43] Another reported general pattern, for fiddler crabs, is a unimodal one with maximum about 5 to 6 A.M. and minimum about 5 to 6 P.M.[174] Young chick embryos in fall and winter have been reported to display a 24-hour variation which is roughly the mirror image of that of the crab.[8]

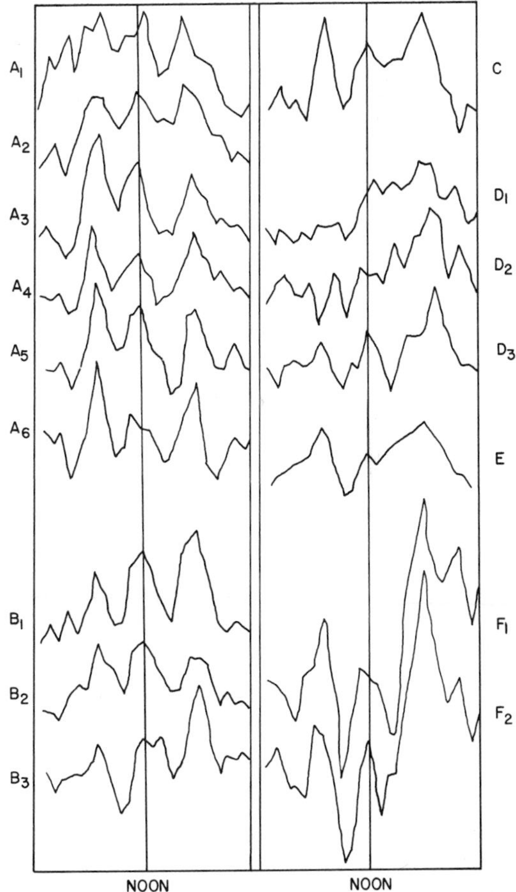

Figure 10–6. Mean patterns of solar-day variations in a variety of organisms held for extended periods in constant conditions. A, Oxygen consumption in potatoes for each of 6 years; B, in sprouting bean seeds for each of 3 years; C, in carrot slices for 8 months; D, for chick embryos of 5, 6, and 8 days of age; E, in influence of a 5-fold increase in background radiation on mouse spontaneous activity; F, of *Tenebrio* larvae for 9 months: F_1 semimonths centered over full moons and F_2 over new moon.

Attempts have been made to identify the particular atmospheric parameters to which the organisms are responsive. These efforts led first to the disclosure that the organismic rates at specific times of day exhibited a slight lag correlation with the rate of barometric pressure change. During late spring and summer, potatoes and fiddler crabs possessed mean metabolic rates for the 5–7 A.M. and 5–7 P.M. periods of the day, which, independently, correlated highly significantly with the mean 2 to 6 A.M. and 2 to 6 P.M. rates of barometric pressure change, respectively.[24, 25] The crabs appeared to derive their morning maximum and afternoon minimum by retaining the same sign of the correlation for both morning and afternoon. The potatoes, whose sign of correlation was positive for the morning relationship and negative for the afternoon, displayed a maximum in metabolic rate for each of these times of day. Chicks have been found to exhibit the corresponding afternoon correlation.[8] Since all these organisms displayed their correlations despite being maintained in constant ambient pressure, the organisms were not responding to the atmospheric pressure changes *per se*, but instead were responding at one or two times of day, 12 or 24 hours apart, to some atmospheric parameter or parameters whose stimulating influence at these sharply defined intervals was correlated with the concurrent rates of atmospheric pressure change. By this means, influences of the well-known solar-day tides of the atmosphere could be indirectly imposed upon the organism.

A suggestion of the probable nature of at least one of the mediating forces is the recent finding in a study of gerbils in 12L:12D that the day-to-day differences in the mean amount of gerbil activity between 3 and 6 P.M. are highly significantly correlated with the corresponding day-to-day differences in the rate of change in strength of the horizontal vector of geomagnetism at these same hours, and that no other correlation between these two parameters is present at any other time of day or in any other temporal relationship.[159] The horizontal vector of geomagnetism is known to display a well defined mean 24-hour variation.

The range from midnight to noon of the day-to-day 24-hour cycles of the potato hermetically sealed in constant conditions (including ambient temperature) shows a highly significant correlation with the concurrent mean daily temperatures, having a positive correlation with temperatures below, and a negative one with temperatures above, 57°F, indicating that some additional atmospheric parameter is also contributing to the observed, characteristic 24-hour geophysically dependent patterns.[24]

Since it can be assumed that fluctuations in any single atmospheric parameter exhibit at least some degree of correlation with every other atmospheric parameter, the discoveries of correlations of rate variations in organisms in constant conditions with amplitudes of daily variations in general background radiation[45] and with patterns of variation in primary cosmic radiation[49] were not surprising, despite the fact that for the latter there could be, as for the previously described atmospheric pressure and temperature correlations, no possible direct cause-effect relationship involved in view of the known screening action of the atmosphere.

Investigations to discover for what atmospheric electromagnetic parameters the organism might display adequate sensitivity have only emphasized further the complexity of the problem. Simple, direct experiments to assay a living organism's capacity to perceive very weak magnetic, electric, and radiation fields of the order of strength of the earth's have disclosed that the organism possesses the capacity to perceive all three types, and moreover appears to exhibit its maximum ability to resolve changes in strengths and vector directions of these fields at the level of the ambient intensities of their natural environment.[33, 107, 116, 122, 139, 177, 180, 184]

Lunar-Related Rhythms. Another terrestrial period, the lunar day of 24 hours and 50 minutes, is associated with rhythmic variation in the physical environment of organisms, especially for those that inhabit the littoral intertidal area. The ocean tides, whose periods are determined chiefly by the rotation of the earth relative to the moon, typically display two high and two low tides each day; the heights of the two may be nearly equal, or they may be unequal and exhibit differences from one geographic region to another (ranging from slight differences in heights of the two to having only one high and one low tide in the day). The tidal patterns of any given area typically change in amplitude through the synodic month as a consequence of the superimposed gravitational attraction of the sun to produce the semimonthly high high, or spring tides, at the times the sun and moon reinforce one another (new and full moons) and low highs, or neap tides, at times of minimum reinforcement. The tides at some locations also display semimonthly or monthly changes in their diurnal patterns.

The temporal patterns of the activities of the seashore organisms follow the tidal changes in an adaptive manner. Some, like diatoms, *Convoluta,* fiddler crabs, and birds that feed on the beaches exposed at ebb tide, display the maxima in their activity cycles during low tides; others, like snails, barnacles, and green crabs, resort to inactivity and various devices to prevent desiccation and withstand the reduced availability of oxygen when exposed by the ebbing tide, and display their activity peak during the flood tide. Organisms at the upper and lower limits of the beach are subjected to the rhythmic tidal submergence only at semimonthly or monthly intervals.

It was discovered long ago that the green flatworm, *Convoluta,* taken from its tidal environment into the laboratory would continue to follow tidal periodicity of rising to the surface of the sand over low tides, and descending over high tides.[19, 82] Other littoral organisms, crustaceans and molluscs, were reported to exhibit in a comparable manner a persistent tidal rhythmicity in their behavior. More recently, tide-pool dwelling protozoans and unicellular algae, diatoms inhabiting tidal flats, fiddler crabs, green crabs, and some fishes of the littoral area have been described as showing persistent tidal rhythms.*

In nature, the animals while subjected to the tidal variations are simultaneously subjected to the 24-hour cycles of light, and therefore the phases of the tidal cycles systematically move across the day at the mean rate of 50 minutes a day, with the peaks coming to occur at all times of day and night. Some organisms, such as fiddler crabs, appear in the field to follow the tidal cycles in their activity throughout the 24-hour day; others, like the diurnal shore-feeding birds, and diatoms which depend upon illumination for photosynthesis, have a strong diurnal component superimposed over their tidal cycles. Activity bouts follow a tidal frequency but occur only during daytime.

Just as for the solar-day rhythms, many organisms possess species-specific tidal behavioral patterns which they are able to adjust adaptively to the phases of the tidal changes of specific beaches. The times of tidal changes on beaches only a few miles apart may differ by up to 5 or 6 hours as a consequence of the influence of the local land topography on the passage of the ocean tidal waves. It is evident, however, that there exists no significant correlation between the tidal variations and the solar-day ones, and hence such well known phasing cues for circadian rhythms as the natural daily changes in light and temperature are not involved.

It has been reported that a number of littoral species, fiddler crabs[11] and mussels,[144] will retain (at least for a few cycles in the laboratory away from tidal changes) tidal rhythms in phase with the tides of the specific beach from which they have been taken (Fig. 10–7), and furthermore, when transferred to a beach with different tidal times will have readjusted within a few days or weeks to the new tidal schedule. The environmental rhythmic factors that serve for the phase-setting to the tides still remain to be firmly established.

Tidal rhythms of fiddler crabs[13] in constant darkness and temperature and away from the tides, as well as tidal rhythms of oysters, have been reported to alter their phase relations gradually over a week or two from local ocean tidal phases to display their maxima at times of upper and lower transits of the moon, times which are closely correlated with maxima in the lunar-day tides of the atmosphere. This suggests that subtle atmospheric variations correlated with the lunar-day tides may serve as phase setters for tidal rhythmic patterns in the absence of the normally dominant ones associated with ocean tidal changes. Fiddler crabs collected on beaches that chance to have their low tides (maximum crab activity) coinciding with times of upper and lower transits of the moon have continued to have accurate tidal rhythms of activity persisting for more than 45 days in the laboratory away from the ocean tidal changes.

One of the major differences between the tidal and solar-day circadian rhythmic patterns, therefore, is that the former, unlike the latter, cannot be synchronized to 24-hour light-dark cycles. Not only can the tidal rhythms persist despite the presence of the natural or artificial LD cycles, but under some

*See references 16, 36, 77, 80, 83, 102, 125, 133, 134, 144, 183.

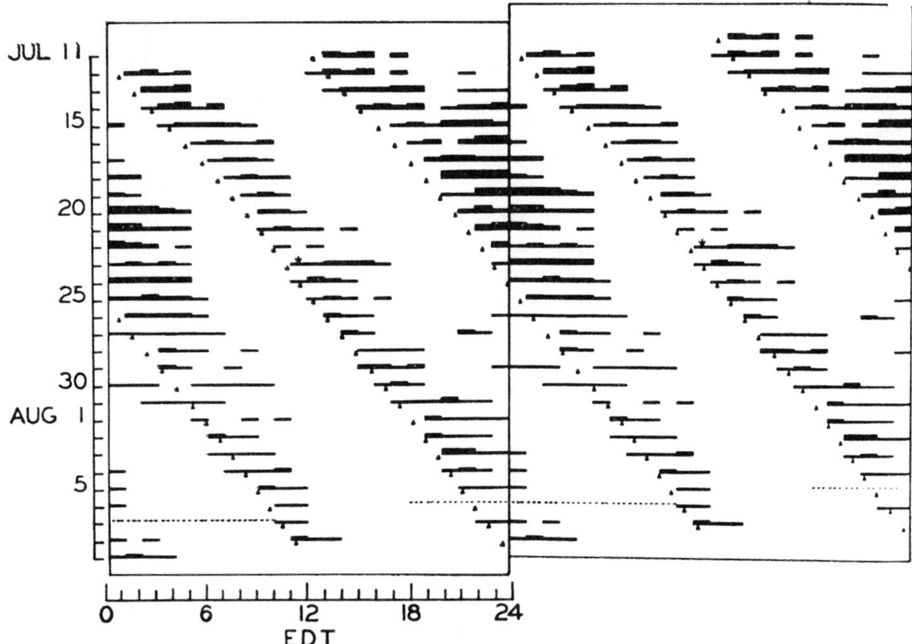

Figure 10–7. Tidal rhythm of *Uca pugnax* in the laboratory under natural conditions of illumination. Dots signal predicted time of high tides of the beach of their collection. (From Barnwell, F. H., Biol. Bull. *130*:1–17, 1966.)

circumstances they appear to display greater precision of tidal frequency with than without the daily light changes.[10, 176] Light can, however, exercise one kind of influence on the tidal rhythm. A phase shift in the circadian pattern by an altered time of light and dark changes will induce an immediate, corresponding amount of phase shift in the tidal cycle.[36] The tidal cycle behaves as if it were functionally associated with the circadian cycle only during the shift itself, thereafter resuming its original lunar-day frequency despite the continuing LD cycles. Indeed, in the fiddler crabs it has been shown that when the circadian rhythm is given a daily 50-minute phase-delay shift by subjecting the crabs to a 24.8-hour LD rhythm, the overt tidal cycles of activity reflect the daily 50-minute shift and display, in addition, the usual 50-minute delay on the circadian to generate an overt 25 hour and 40 minute activity cycle.[176]

Apparently related to the foregoing is the observation that fiddler crabs continuing in darkness (DD) and constant temperature to have an accurate mean 24-hour free-running circadian rhythm of color change have simultaneously an accurate tidal rhythm of spontaneous motor activity; but in continuous illumination, a condition that favors a free-running circadian period longer than 24 hours, the tidal cycles appear similarly to have a free-running period longer than 24.8 hours, or now exhibit a circa-tidal rhythm.[10]

Lunar tidal and solar-day circadian rhythms can clearly persist simultaneously in the same individual both in nature and in controlled constancy of all obvious factors in the laboratory (Fig. 10–8). The two frequencies may be seen to regulate two different processes as in the fiddler crab; the circadian predominates in the regulation of the color changes, while the lunar tidal dominates the spontaneous running cycles. However, for both of the foregoing a careful quantitative study usually reveals a modulating influence by the other period. In *Sesarma*[131] and *Uca*,[9, 10] both periods may be evident in records of overt spontaneous activity. In the diatom, *Hantzia*, even in constant conditions, the tidally timed vertical migrations appear to interact with a circadian component in such a manner that tidally timed activity occurs only during the daytime phase of the circadian rhythm.[134]

The concurrent presence of circadian and lunar-tidal rhythmicities yields (by periodic interference, or beats) semimonthly and monthly variations (Fig. 10–9). These, for the organisms in their natural tidal environ-

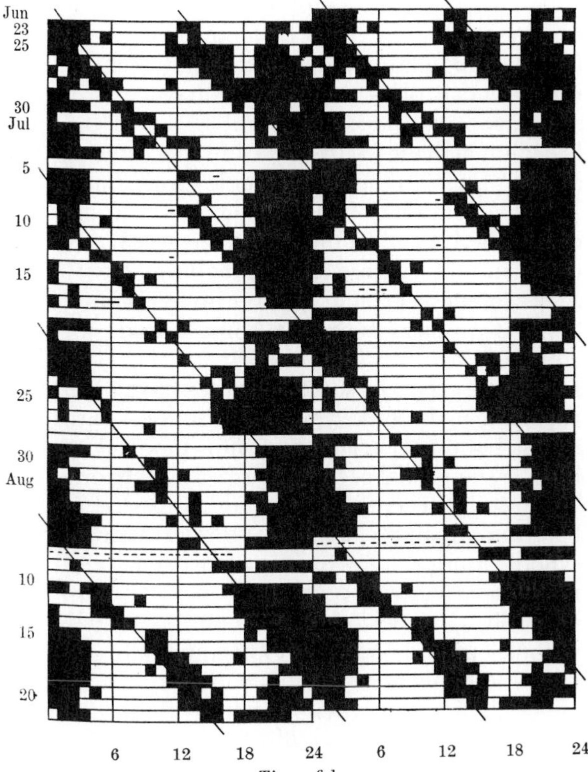

Figure 10–8. Pattern of activity in a population of the crab *Sesarma* in constant conditions. Darkened blocks indicate all hours when activity was equal to or above daily mean value. (From Palmer, J. D., Nature *215*:64–66, 1967.)

ment, can also include direct responses to the semimonthly and monthly physical environmental patterns resulting from interactions between the tides and solar days. A monthly environmental variation is also evident in the amounts, and times during the nighttime, of lunar illumination as the moon and sun systematically alter their relationships during the synodic months averaging 29.53 days.

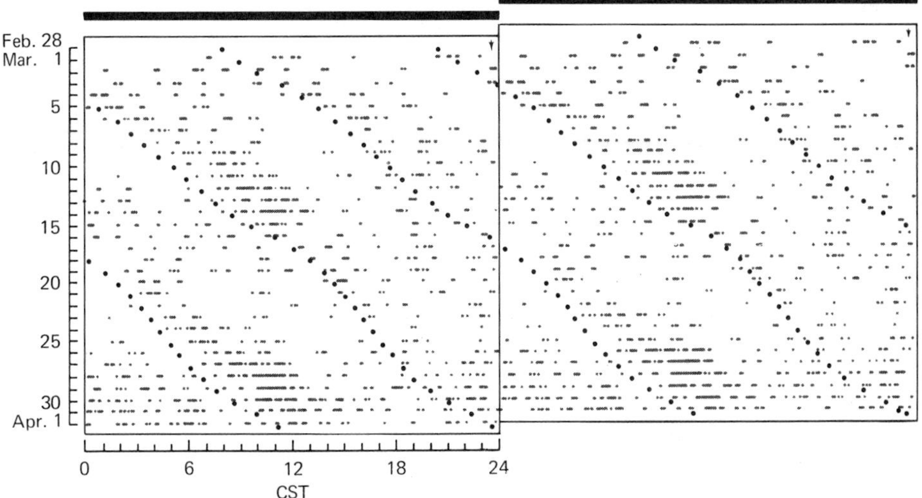

Figure 10–9. The spontaneous locomotor activity of a Costa Rican fiddler crab (*Uca princeps*) during a month's sojourn in continuous darkness. The lines of circles mark the times of high tide on the crab's home beach. Note the two bursts of greater activity which occur in late morning with a semimonthly interval. The record is repeated and displaced upward one day to facilitate the viewing of the passage of the patterns of activity from one day to the next. (After Barnwell, F. H., from Brown, F. A., Jr., J. W. Hastings, and J. D. Palmer, The Biological Clock: Two Views, Academic Press, New York, 1970.)

Perhaps the most common evident role of semimonthly and monthly periodicities is for the regulation of genetic patterns of reproductive physiology and behavior. The synchronization of period and phase of the breeding activities within marine species such as the alga, *Dictyota*,[59] the palolo worms and Atlantic fireworms, and the grunion is accomplished by this means. Also accurately timed to moon phase is the breeding activity of a Japanese crinoid.[65] Monthly swarming behavior is also seen for some species of chironomids and mayflies.[93, 126, 127, 128] The menstrual cycles of some primates have been reported to be linked to lunar phases, with menstruation occurring over new moon. The human menstrual cycle is a circa-monthly one which some recent studies have suggested to be fundamentally related to the natural synodic month.[69, 121] Human menstrual cycles which in natural conditions deviate substantially in period from the 29.5-day synodic period can be experimentally induced to converge toward a common length, about 29.5 days, when the individuals are subjected to nighttime illumination during the 14th through 17th days following the onset of menstruation.[69]

The breeding rhythm-phase relative to moon phase in some organisms, such as *Dictyota*, appears to be determined by the phase-angle between the time of day of the high and low tides on the specific beach. The day of gamete release relative to moon phase will thus be the same for all the individuals on the same beach, but different from that for the individuals of another beach where the tidal times are different. For some other species, such as *Platynereis*, some chironomids, and suggestively for some wild primates, there is evidence suggesting that variations in nighttime illumination serve as the phasing factor for the lunar-related reproductive cycles.[98]

Overt periodicities, whether correlated with earth-moon relations by themselves, or in association with the sun, are clearly able to persist as free-running ones in organisms in the absence of all obvious rhythmic variations with these frequencies.

Monthly or semimonthly variations persisting in presumed constant conditions have been reported for many organisms and phenomena. These have ranged, for example, from responses of guppies[112, 113] and planarian worms to light, responses of snails and planarians to very weak magnetic fields, responses of planarians to very weak gamma radiation fields, or spontaneous activity in mice and hamsters held in controlled 12L:12D regimes; rate of learning in planarians, and rate of activity and orientation in insects, to tendency toward hemorrhaging following surgery in man.[2, 17, 26, 28, 41] Spontaneous activity in freshwater crayfishes and in rats[44] has been described to vary with hour angle of the moon or with moon phase, even in the absence of all obvious lunar cues. Mean lunar-related variations have thus been reported for numerous organisms for which no evident role for such a periodism exists.

Lunar-day environmental variations, probably correlated with the lunar-day tides of the atmosphere, were noted above as apparently able gradually to determine phase-setting of free-running tidal rhythms in oysters and crabs. There have been comparable reports that even free-running circadian patterns in mammals may be entrained by subtle lunar-day atmospheric rhythms.[30, 46] Hamsters and rats were described to adopt lunar-day frequencies of their circadian rhythms with apparent favored phase-relations to the rhythms of the physical environment. The body temperature of a man held in constant conditions over an extended period of time was described, possibly coincidentally, to follow a 24.8-hour period.[6] Other reports on free-running human rhythms also have suggested that the lunar tidal period of 24.7 to 24.9 hours may occur more frequently than by chance.

In general, lunar-related periodicities appear to be of rather wide occurrence, having been demonstrated in most instances where they have been sought in the proper manner. As with the geophysically dependent solar-day rhythmic component, which reflects biological responses to the atmospheric rhythmic elements having that frequency, lunar-day geophysically dependent variations also appear to be widespread and possibly universal. The latter seem similarly to display a limited number of general forms or their mirror images. For essentially unimodal cycles, such as those described for rat activity, snail metabolic rate, or clam valve opening, the maximum or minimum tends to occur at upper lunar transit (Fig. 10–10). For essentially bimodal lunar-day cycles, such as those seen for the fiddler crab or for *Tenebrio* respiration, the maximum or minimum tends to occur at both upper and lower lunar transits. In a number of species, including the mouse, earthworm, and salamander, a composite pattern is evident; it resembles the crab or *Tenebrio* pattern except that the alternate components in the semi-diurnal tidal cycles are inverted. Hence, all of the several reported cycles appear reducible to a semi-diurnal lunar-tidal variation, with each of the two semi-diurnal elements capable of independent inversion.

Annual and Sidereal Rhythms. Another very conspicuous rhythmic period of the physical environment for much of the earth is the year with its seasonal variations in sun altitude and daylight lengths, together with correlated variations in such factors as illumination and temperature. In addition, there is about a 3% annual variation in the distance separating earth and sun, with minimum in January (perihelion) and maximum in July (aphelion). Annual patterns of variation of organisms are the rule and include importantly species-characteristic sequences in reproduction and growth, migrations, and physiological and morphological changes. Annual variations have been reported even for organisms inhabiting equatorial regions; these variations are often, but not always, associated with annually rhythmic environmental patterns of such factors as rainfall and drought.

A phase-determining factor for the annual genetic and adaptive patterns for numerous organisms has been demonstrated to be the annual changes in photoperiod,* with long-day species responding to the lengthening days and shortening nights of spring by being brought into the active reproductive state. Short-day species are fall breeders, being activated by the shortening daylight periods of fall. Substantial evidence supports a hypothesis that the mechanism by which organisms differentiate among the changing photoperiods involves the circadian system acting as a reference clock, the environmental changes being distinguished by the changing relations within the circadian cycles of the times of dawn and dusk.

Temperature change, a less regular and dependable annual variant, has been less investigated in this connection than light, but appears generally to be of lesser importance than light.

As with the solar-day and lunar-related rhythms, annual ones (even with relatively large amplitudes) may persist despite the

*See references 20, 53, 64, 78, 79, 88, 90, 143, 187.

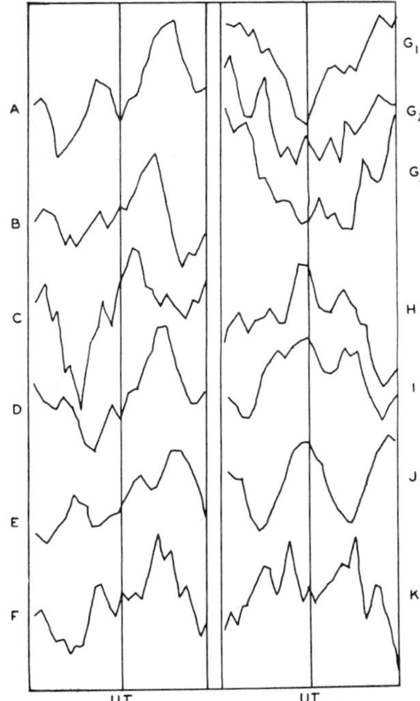

Figure 10–10. The forms of mean *lunar-day variations.* (A) *Solanum,* (B) *Daucus,* (C) *Lumbricus,* (D) *Triturus,* (E) *Mus,* (F) *Fucus,* (G_1) *Rattus* (1955–56), (G_2) and (G_3) *Rattus* (1959), (H) *Venus,* (I) *Nassarius,* (J) *Uca,* (K) *Tenebrio.* U.T. signifies upper lunar transit. Oxygen consumption was monitored factor for A, B, C, D, F, I, J, and K; spontaneous motor activity for E, G, and H. (From Brown, F. A., Jr., pp. 231–261 *in* Circadian Clocks, edited by J. Aschoff, North Holland Publ. Co., Amsterdam, 1965.)

constancy of all factors which impart obvious information as to time of year. In a constant environment, except with a steady 12L:12D regime, weaver finches have been observed to continue through two nearly normal annual reproductive cycles.[118] Some other species of birds, investigated in constant conditions, appear to display a circannual rhythm with the successive cycles recurring two or three months earlier in succeeding years.[85, 86] In still other instances, irregular variations of the general order of length of a year have been reported.

In contrast with the persistent recurrences in these species-specific behavioral patterns, which may deviate from a true annual period in length, rather precise annual variations for a number of apparently less specific rate functions have been described. Annual variation in the rate of seed germination has been reported[57] which persisted accurately whether seeds were stored at $-22°$ or $+45°C$, and whether they were in air or in N_2, rendering highly improbable any internal metabolic measurement of cycle length. Rate of nitrate reduction by an alga, response of earthworms and humans to light, response of planarian worms to a very weak gamma source, rate of O_2 consumption in sprouting beans and potatoes, and rate of food intake in woodchucks vary systematically with season of the year, even when the organisms are maintained in conditions that are constant with respect to every obvious factor.[14, 31, 40, 66, 72, 108]

The mean precision of the annual period for the foregoing rate variations argues for their direct environmental dependency through response to subtle annual physical variations. Such a derivation is also supported by (a) the fact that many of these cycles pass through maxima and minima in July-August and October-December, respectively, (b) the discovered inability to alter experimentally the phases of the woodchuck food-intake cycle by manipulations of light periods or temperature,[66] and (c) the mean diurnal cycle forms and amplitudes for respiratory rate in potatoes,[24] which reflected the day-by-day weather-correlated mean air temperatures and distortions in the form of the solar-day tide of the atmosphere, following both of these parameters as they underwent their well known, systematic annual variations in these same parameters.

Another natural period which is related to the earth's rotation is 24 hours and 56 minutes, the period of rotation of the earth relative to the celestial sphere. For all practical purposes the solar and sidereal days reproduce their relative phase relations exactly once a year. Any systematic annual variation in the solar-day cycle (and we have noted earlier that one exists) is indistinguishable from any existing modulating influence of a sidereal-day cycle, which retrogresses over the solar-day one at a rate of about four minutes a day. It is of significance, nevertheless, that an apparent mean sidereal-day variation has been discovered for a terrestrial organism held in constancy of all obvious factors.[22] This one, reported for potatoes, was confirmed in an eleven-year continuous study of O_2 consumption.[34] The mean cycle, whose range was about 1%, was essentially sinusoidal with maximum and minimum occurring about the 6th and 18th hours sidereal time, respectively. The 6th sidereal hour reaches upper transit at solar-day noon on the summer solstice; the 18th hour reaches upper transit at solar-day noon on the winter solstice. Suggestive of an influence of solar activity, the two extremes in the ranges in the sidereal variation over eleven years appeared to occur close to times of sunspot maximum and minimum.

18-Year and 18.6-Year Cycles. Some other rhythms with geophysical correlates appear to result from simultaneous possession of periodisms treated earlier. Many invertebrates with external fertilization, especially polychaetes, have limited and well defined restricted annual breeding intervals, with spawning also simultaneously clocked to first or third, or both first and third, lunar quarters. Spawning is commonly additionally clocked to a specific hour of the day. The best documented of these cases have been the reproductive cycles of the Pacific palolo worm, *Eunice viridis*,[60] and the sea lily, *Comanthus japonica*.[65] The date (time of year) on which these species shed their gametes drifts within the breeding period of the year following the moon phases. The interval between years (beat cycle) when the same phase of moon occurs on the same dates is the "metonic cycle" of 18 years. This is essentially the period reported for the two species. However, it has not been possible to distinguish whether the cycle is truly a correlate of the metonic cycle or, instead, of the cycle of "lunar nutation" or rotation of the moon's orbital plane, with its resultant 18.6-year variation in dates of maximal lunar declination from the ecliptic, times that in 1971 nearly coincided with the moon's first and third quarters.

Biological Rhythms, Homing, and Navigation. Both time (or phase angle within the biological rhythms that have been described in this chapter) and geographic space (longitude and latitude) can be precisely defined only in terms of their celestial relationships. It is, therefore, not surprising that investigations of biological rhythms on the one hand and animal homing and navigation on the other should have led into a common area with fundamental problems common to the two.[1, 129]

When it became apparent about two decades ago that birds were able to employ the sun as a geographic-compass reference, a solar-day chronometer was recognized as obviously essential.[111] It was quickly demonstrated that the circadian rhythmic system was able to provide for this need. It was shown that not only birds, but also fishes, turtles, amphibians, insects, crustaceans, and spiders were able to use the sun in association with the capacity to resolve time within the 24-hour rhythmic framework to obtain dependable geographic information, apparently correcting steadily for the relative movements of the sun and earth.* The clock system exhibited the typical circadian rhythmic property of phase lability, and could be phase-shifted experimentally by LD cycles to elicit corresponding directional misjudgements of animals orienting by the sun (Fig. 10–11).

Later studies with amphipod crustaceans, which normally migrate to and from the water's edge, demonstrated that the moon could be used as a geographic directional reference.[135, 136] Here, of course, a chronometer with a lunar-day period was demanded. Not only was the rate of the apparent movement of the moon across the sky on a single night involved, but day after day the direction of the moon for a given time of night was systematically changing with a mean 29.5-day period for a full cycle. A chronometer was also needed to measure the synodic month.

Studies with a migratory European warbler, using the facilities of a planetarium, led to the conclusion that these birds were able to obtain directional and geographic spatial information for their normal migratory flights using only the constellations as their reference.[149] The rate of apparent stellar movement in the sky (except, in the northern sky, for Polaris), and the changing directional star pattern in the heavens with time of year as well as with points in space, made it evident that the stars could be related to homing or navigation only with the aid of a sidereal chronometer and, in addition, a means for accurately distinguishing the time of year. Orientation of indigo buntings in a planetarium has also been investigated.[75, 76]

Using the celestial bodies as geographic directional and spatial references seemed clearly to demand the operation of a remarkably complex, precise, and dependable calendar-clock system, and also to require fair weather with an adequately clear view of the sky. The system was given some insurance, however, by the finding that many organisms, including insects and fishes, could dispense with direct sight of the sun and substitute any patch of blue sky.[172] From the patch, its light polarization pattern could be used by the organism to gain indirect knowledge of the hidden sun's detailed position.

However, it had been popular knowledge for a long time that organisms were not exclusively fair-weather homers and navigators, even in strange territory where landmarks could not have been learned. Recent extensive studies with migratory birds using radar tracking and radio telemetry techniques have disclosed that apparently normal migratory flight directions can be maintained in complete overcast and without sight of the ground. Furthermore, homing pigeons are able to home with nearly normal ease in overcast weather from strange territory.[106, 170] In addition, it has been discovered that, whatever the directional cues that are employed under the latter circumstances, the circadian system is not directly involved; altering the phase of the circadian rhythmic system by altered times of day of experimental LD cycles does not interfere with the determination by the bird of its home direction. The bird clearly possesses some means, by employment of subtle geophysical fields, to distinguish geographic compass directions and also to be informed of the direction of its displacement from home. Therefore, the biological capacity to orient, navigate, and recognize geographic spatial sites, like the biological rhythms with their concomitant apparent capacity to recognize points in geophysical temporal cycles, appears to depend in part on perception of subtle geophysical parameters. All such subtle parameters of the earth's atmosphere vary continuously in the four dimensions of space and time. Orientation in space is as universal an attribute of terrestrial animals as is their continuous variation in time.

Space, Time, and Responses to Geoelectromagnetic Fields. There have been a number of experimental studies over about the past decade of both the timing mechanism for the biological rhythmic periods and the means employed by homing and navigating organisms. For investigators concerned with biological rhythms, the attack on the problem was beset by the dilemma that the experimental results could invariably be interpreted in two ways and neither could be excluded. If an event recurred with a period the same as, or close to, a natural environmental one, one could never distinguish whether or not

*See references 81, 84, 94, 100, 115, 137, 138, 146, 150, 151, 171.

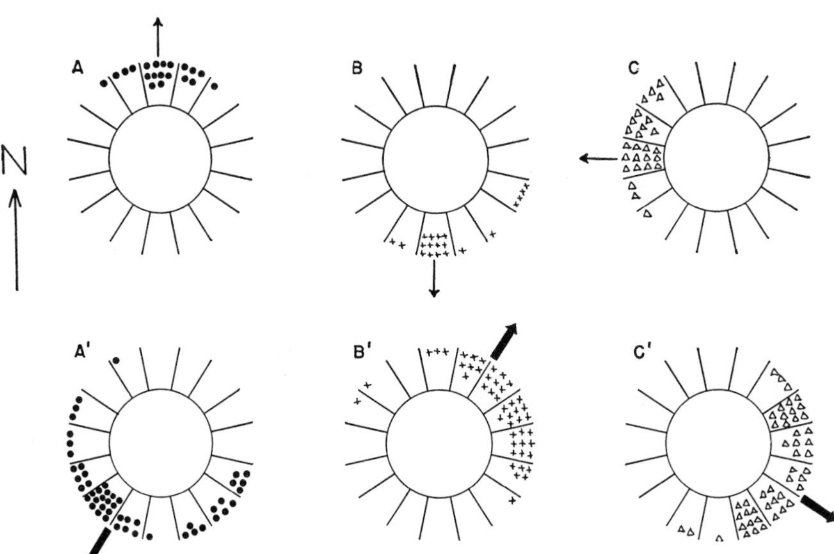

Figure 10–11. Resetting of navigational "clock." Upper row: last experiment before resetting. Fish A is trained to the north, Fish B to the south, Fish C to the west, as noted by the thin black arrows. Lower row: 7th to 13th days after the resetting. The solid black arrow indicates the compass direction in which the fish must swim, if the sun azimuth changed 15°/hour, and the fish compensated 15°/hour in the other direction. Each point is a critical choice. A 14 hour, 10 minute light shift occurs between A B C and A′ B′ C′. (From Braemer, W., Cold Spring Harbor Symp. Quant. Biol. *25*:413–427, 1960.)

the interval of the period was dependent upon the rhythmic external environment. If, in an experimental manipulation of one of the natural subtle environmental parameters, one obtained an alteration in the rhythmic phase, amplitude, or period, one could still not distinguish whether one were modifying, synchronizing, or otherwise influencing an endogenous system capable of normal rhythmic variation even without that environmental parametric field, or indeed any others. Of great importance for the problem of rhythms was the question, could an organism distinguish phase-angle points along the temporal geophysical cycles? And a second critical question was if so, did an organism possess the capacity to employ such temporally organized information to provide itself with a reference framework upon which to arrange either or both of genetically derived and environmentally imposed physiological and behavioral patterns?

It was evident that the same subtle geophysical parameters that varied with time also varied systematically in the three dimensions of space. The temporal questions could perhaps be resolved by transposing the experimental study to space, and simplified even more by restricting the spatial investigation to a single plane, the horizontal.

Using such a system, first with mudsnails and planarian worms,[33] it was demonstrated that (a) an organism, at any given point in time, could differentiate among the compass directions using only subtle geophysical cues, and (b) the response of the organism for any given geographic direction varied systematically with time and included the major geophysical frequencies of the solar day, the lunar-related periods, and the year.

Employing the foregoing system for assay, it was discovered that the organisms were extraordinarily sensitive to strength and geographic directional changes of the horizontal vector of very weak magnetic fields,[26] exhibiting their maximum capacity to perceive experimental alterations of these at strengths simulating the natural ambient ones. In the case of experimentally altered strengths, the organisms required at least half an hour for apparent accommodation to the new strength level. The horizontal vector of magnetism was associated with the rhythmic system in still another way (Fig. 10–12). It was discovered that when planarians were north-directed with a light source to their right, the strength of their apparent negative phototaxis displayed a clear overt monthly variation, with minimum response over full moon and maximum over new moon. This monthly variation could be 180° phase-shifted instantaneously[32] by either of two techniques: (1) rotating the whole experimental set-up 180° to point south with the light now from the west, or (2) performing the magnetic equivalent change by reversing the field with a weak bar magnet, even by one providing a reversed field of as little as 0.04 gauss.

Other studies with planarians disclosed an ability of the worms to distinguish the presence and geographic directions of horizontal electrostatic gradients of no more than 0.1 μv/cm.[27] The response of the worms to exactly the same imposed directional fields varied with time of day.

When gamma fields ranging from 2 to 25 times the background level were applied from horizontal sources to right and left of the assay system, the worms distinguished strengths and even vector directions of the imposed fields.[28] Again, the responsiveness varied systematically with times within the natural geophysical cycles, including even the year.

A series of investigations with the European robin[122, 123, 184, 184a] has convincingly shown that this bird, during its spring and fall migratory restlessness, is able to select the appropriate migratory direction even when detained in cages in the laboratory and deprived of all obvious directional cues (Fig. 10–13). The direction of the birds can be experimentally reversed by simulating a reversed geomagnetic field with the use of Helmholtz coils. Stronger or weaker experimental fields, even by a factor of as little as about 50%, fail to be used in the same manner by the birds until they have accommodated to the new strengths over 2 or 3 days. Other organisms also appear to display orientational responses to very weak magnetic fields.[12, 15, 116, 130, 152, 177]

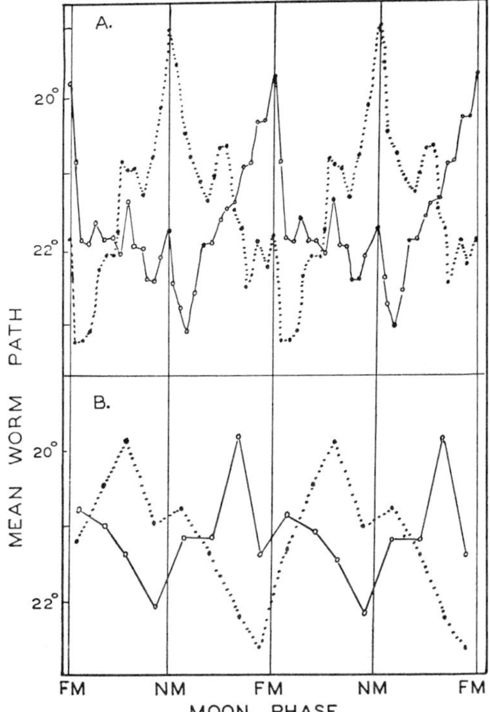

Figure 10–12. Variations with phase of moon for planarian worms initially north-directed (0°) in a two-light (from south and east) field during experiments spanning several months. The mean paths of worm samples were determined (solid-line curves) and then the mean paths for the same worms were redetermined right after the direction of the horizontal vector of magnetism was rotated 180° relative to the light field (dotted-line curves) by (A) rotating the whole apparatus, including its light field, geographically by 180°, and (B) application of a reversed magnetic vector of 0.05 gauss by a bar magnet appropriately placed beneath the apparatus. FM = full moon; NM = new moon. (Reproduced by permission of the National Research Council of Canada from the Canadian Journal of Botany, *47*, pp. 287–298 (1969).)

Relevant to the general problem of the great sensitivity of living systems to the natural electromagnetic fields and their variations are recent findings of extraordinary sensitivities of some elasmobranch fish to electric fields[70, 103] produced by action potentials within the bodies of their otherwise hidden prey, sensitivities which extend even to 0.01 μv/cm. Electric fields have also been reported to be in some manner concerned with circadian-rhythm period and phase[71, 180, 181, 182] and possibly with navigation of eels in the ocean.[148a]

Relations Between Obvious Stimuli and Subtle Electromagnetic Field Responses. The ability of a living system to relate the responses to ordinary stimuli to concurrent inflowing information concerning the subtle electromagnetic fields is implicit in the demonstration of an organism's ability to experience the existence of a home site, and then, after displacement to a new site never experienced in its lifetime and in thick overcast screening out all ordinary celestial references, return home at a rate significantly exceeding that expected from chance encountering of familiar landmarks nearer home. Such behavior has been shown, for example, for homing pigeons.

There has been a report of a successful conditioning of heart-rate response of pigeons to very weak magnetic fields.[145] Strongly suggestive of an association of genetic behavioral patterns of birds with subtle geomagnetic directional cues has been the finding of a significant directional tendency in young gulls for their normal fall migratory direction, a behavior which weakens, vanishes, and even reverses direction in a manner correlated with increasing

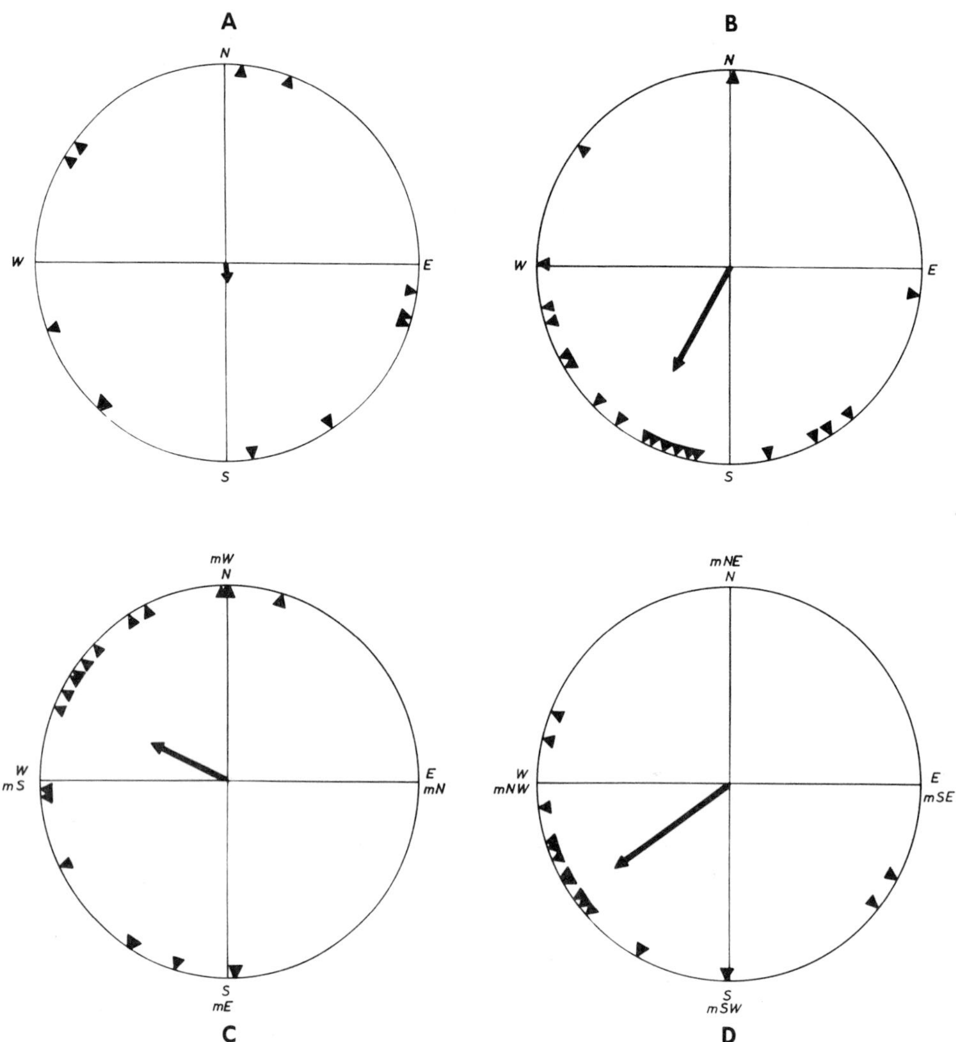

Figure 10–13. A, Mean direction of choice of European robin during time of migratory unrest (normal choice, SW) on a single night when bird has been held in a 0.41 gauss magnetic field and is tested in a 0.14 gauss field. Arrow length indicates mean directional tendency. B, When held in a 0.14 gauss field and tested in the same strength. C, When held in a 0.14 gauss field and tested in an artificial 0.30 gauss field rotated to directions indicated. D, When held in a 0.14 gauss field and tested in an artificial 0.41 gauss one rotated to directions indicated. (From Merkel, F. W., and W. Wiltschko, Die Vogelwarte *23*:71–77, 1965.)

concurrent degrees of geomagnetic disturbance (increased field strengths).[156]

An ability of an organism, the planarian, to associate a spatial light pattern with a geographic grid comprising exclusively the natural ambient geoelectromagnetic field has been described (Fig. 10–14).[42] Not only is the animal's response determined in part by the specific orientation of the imposed light pattern, but a persistent modification is produced which determines in some measure responses to other light-patterns imposed several minutes later.

Biological Clock Hypotheses. The establishment of the ability of living systems to maintain rhythmic patterns even when screened from every environmental periodism to which they were known to respond, and to do so by means providing relatively stable and dependable periods despite changes in the temperature or chemical character of the environment, suggested a clock analogy. The organisms appeared to possess the equivalent of man-made clock systems, highly specialized for time-keeping, and measuring off relatively accurately the natural geophysical periods. There developed the concept of biological clocks which timed the periods of the various observed rhythms. The biochemical and physiological changes which comprised the recurring patterns included many concurrent and consecutive reactions and processes, but none of them appeared by itself to possess the requisite clock-like properties. Collectively, however, they appeared to be temporally integrated as if they comprised the hands and face of a complex calendar-clock system.

From the first demonstrations of the phenomenon of persistent rhythmicity, the basic question has obviously been the following: Are these persistent periodicities independent of all physical environmental periodicities of the correlated frequencies and generated by the organism as an autonomously oscillating system, or are the observed free-running periodicities dependent for their clock-like properties on a continuing input by still uncontrolled environmental rhythms? The first of these two alternative possibilities has become known as the endogenous-clock hypothesis. The second is

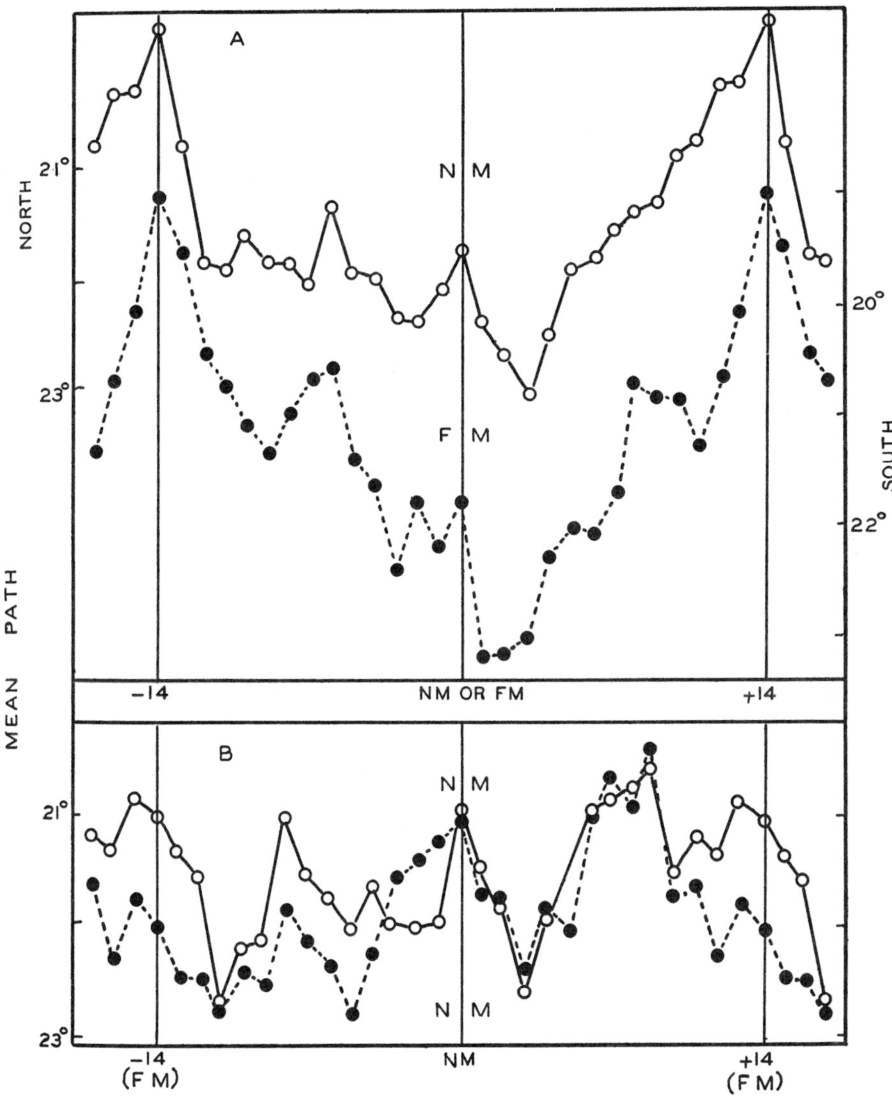

Figure 10–14. A, The mean monthly variations in strength of turning of planarians away from a light source on their right when initial direction of worm is N (open circles) and immediately afterward, following rotation of the apparatus 180° to S (dots). B, The comparable monthly variations when the initial worm direction is S (dots) and the apparatus is rotated to 180° to N (open circles). (From Brown, F. A., Jr., and Y. H. Park, Biol. Bull. *132*:311–319, 1967.)

referred to as the exogenous-clock hypothesis. It has not been possible to date to exclude either of these two hypotheses.[38, 168]

THE ENDOGENOUS-CLOCK HYPOTHESIS. This is logically the simpler of the hypotheses, but it postulates that the organism has evolved a temporal precision and stability for which no plausible biological scheme has yet been advanced. In terms of this hypothesis many of the reported properties of free-running rhythms can be interpreted, and most of the remainder may be accounted for by postulating responsiveness to uncontrolled subtle geophysical rhythms which under some circumstances may act as phasing stimuli, determining the phase adjustment of a rhythmic pattern whose periods are being generated by the endogenous clock. Offered as evidence for an endogenous clock are such observed properties as the following: (1) a rhythm persists when as many environmental parameters as possible are prevented from varying; (2) a rhythm phase can often be experimentally shifted and thereafter continue in the new phase but with the former period; (3) the observed period of a rhythm may deviate from that of the correlated geophysical one; (4) a rhythm can sometimes be initiated by a single brief stimulus; (5) a rhythm can sometimes be phase-delayed by metabolic depressants or inhibitors; (6) the period of a free-running rhythm can be a function of the temperature or illumination levels, or of D_2O concentration; and (7) a rhythm may persist unaltered during rapid eastward or westward translocations of the organisms, or even at the geographic south pole.

It is postulated that the periods of the biological clocks and the observed rhythmic periods are the same, and that any observed alteration in rhythm phase or period reflects a comparable alteration in the biological clock. In the organism there is postulated to be a multiplicity of clocks, a fact which is believed to be established in all instances where period or phase can become dissociated, or rhythms are observed to continue in isolated portions of the organism.

With the working hypothesis of endogenous rhythmicity, many attempts have been made to discover a cellular or biochemical mechanism with the requisite clock characteristics, including temperature independence. There have been few attempts to formulate a concrete scheme as to the possible nature of the clock. However, one based in molecular biology is the chronon concept (Fig. 10–15).[74] Very long DNA segments are postulated to have the rate of genetic transcription along them regulated by a complex

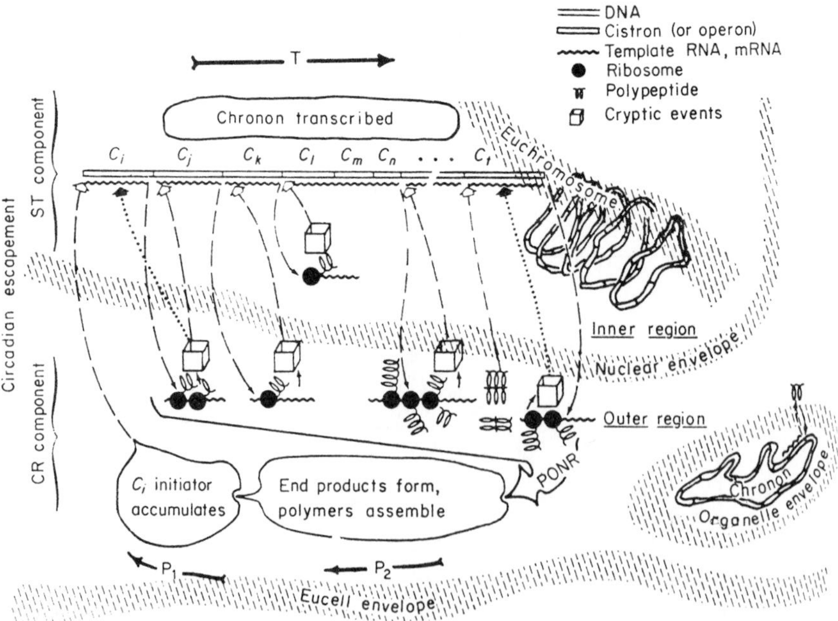

Figure 10–15. The chronon hypothesis: Synthesis of mRNA begins at one end of a postulated polycistronic DNA complex. The mRNA diffuses outward to ribosomes and protein synthesis, with back-diffusion of an initiator for actions of the next cistrons. About 24 hours is needed to complete the complex with initiation of a new cycle. (From Ehret, C. F., and E. Trucco, J. Theoret. Biol. *15*:240–262, 1967.)

of related events in the cell in such a manner that the system can serve as a clock. For the most part, however, the investigations of the nature of the postulated endogenous clock have been focussed upon the descriptive treatment of the rhythmic systems as theoretical oscillator models and examination of the results of experimental perturbations of the biorhythms for degrees of conformity with expectations from these models.[141, 142, 178, 179, 185, 186]

Interpreted in terms of the endogenous-clock hypothesis, the persisting mean patterns of variation of geophysical frequencies which have been referred to as the geophysically dependent periodisms are considered to be endogenous rhythmic components which are held in phase and period synchrony by subtle geophysical rhythms. While these rhythmic components with precise geophysical periods exist, these same subtle factors fail to be effective Zeitgeber for other adaptive rhythmic patterns, whose periods are dominated by the autonomous endogenous clock and are believed to be entrainable through day-by-day phase-corrections by much stronger Zeitgeber (such as light) to maintain in nature the usually observed natural geophysical frequencies.

THE EXOGENOUS-CLOCK HYPOTHESIS. While this hypothesis is logically less simple, it has the advantage of accounting easily for the extraordinary stability of rhythms in the face of temperature and chemical changes for the organisms. It assumes no autonomous rhythmic capacity for these clock-timed periods.[23, 24, 25, 29, 32] All the observed properties of free-running rhythms are interpreted in terms of a biological clock depending upon a continuing rhythmic input from the environment. A primary difference between this and the endogenous-clock view is that for exogenous timing of the rhythms the observed rhythms may differ slightly in period from the clock upon whose timing they depend. The clock, whose period and phase is postulated to be exogenously determined, retains accurate mean geophysical periodicity, and hence its period stability resembles that of those periodisms referred to as geophysically dependent (Fig. 10–16). The genetic and environmentally imposed rhythmic patterns (such as the circadian, circa-lunar and circannual) are postulated in the natural environment to be simultaneously period-synchronized with both the correlated biological clocks and the environmental

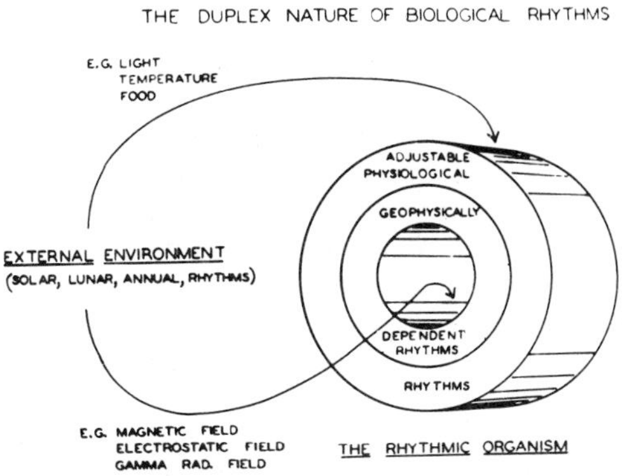

Figure 10–16. Exogenous-clock hypothesis: The organism is postulated to possess a deep-seated rhythmic complex (geophysically dependent) resulting from direct response to the geophysical rhythms of subtle parameters. This deep-seated rhythmic complex is considered to provide the major clock properties, namely the relative stability of periods of the correlated genetic and environmentally determined (adjustable physiological) cycles at or near geophysical frequencies. The observed frequencies of the adjustable cyclic patterns may, in constant conditions, deviate slightly from the geophysical ones by continuous advancing or delaying phase shifts. (From Brown, F. A., Jr., Adv. Astronaut. Sci. *17*:29–39, 1964.)

rhythms in such obvious factors as light and temperature.

For the circadian rhythm, when an organism is placed under conditions in which all forces serving as effective Zeitgeber or phase-determiners are held constant, the organism's normal phasing-adjusting system (which usually is in operation in response to the periodic Zeitgeber) is postulated to effect a slight phase shifting, autophasing, occurring cycle after cycle through an interaction between the constant Zeitgeber parameter and the organism's cyclic responsiveness to it.[23, 35] Such a periodic phase shift produces a new cycle period indistinguishable from a regular oscillation with this period. The period is postulated to differ among individuals of the same species on the basis of individual differences among response curves in forms and amplitudes, as indeed has recently been shown. These differences may be in part genetic and in part environmental modifications. At the same time, this postulated method of genesis of the periods deviating from the natural geophysical ones would be expected to vary slightly with the strength level of any constant Zeitgeber parameter with which the phase response system was interacting; in fact, we see small changes in free-running period with changes in the levels of the two best known Zeitgeber, light and temperature. The quantitative and qualitative similarities of these last two effects suggest that in its slight influence on free-running period, temperature is acting in some degree in its phase-setting capacity rather than in a nonspecific kinetic thermal one. Similarly, D_2O is postulated, for exogenous timing, to exert its influence through the phase shifting mechanism.

It is postulated that circadian patterns are essentially as freely labile in their relationship to the underlying clock upon which they depend for timing as, for example, the hands of a man-made clock in relation to its own timer, and that the phase and pattern is normally dominated by Zeitgeber and other obvious stimulus factors. In other words, the patterns are believed to have no conspicuously favored phase relationship to the underlying clock, nor usually to reflect any significant Zeitgeber influence from subtle environmental parameters.

The exogenous clock hypothesis deals easily with the problem of concurrent presence of solar- and lunar-day frequencies without mutual entrainment, and the persistence of free-running monthly and annual rhythms for which there is postulated to be a quantitatively increasing degree of direct association of the environmental factors and the organismic phenomena.

An endogenous-exogenous-clock hypothesis. The two hypotheses that have been discussed above have served very fruitfully in stimulating designs of experiments that have resulted in extension of our knowledge of diverse characteristics of the rhythmic processes and relationships of them to two domains of environmental stimuli: (a) that of the obvious ones reflecting external events which are of practical importance in the survival of the organism, and (b) that of the subtle fields of the earth, whose influences have been more difficult to discern and understand. Since the extremes of the two hypotheses, exclusively internal versus exclusively external timing, seemed wholly incompatible, there was little beyond the simple descriptive aspects of the rhythms and their behavior that was common to the two. However, the progress that has been made from both points of view has been so substantial that it is clearly evident that neither of them can be excluded in the ultimate accounting for the phenomena of biological clocks and rhythms.

It appears most probable that the long-elusive mechanism for these omnipresent rhythms resides in organismic systems which are themselves genetically, potentially rhythmic, with inherited behavioral programs adapting the organisms each to the general and specific aspects of the rhythmic characteristics of its physical and biological environment.[35] With these patterns comprising orderly series of biochemical and physiological events, the organism has intrinsically the means to cycle with any one or a combination of periods, with different patterns adapted to different cyclic time courses and to the systematically changing phase relations among such similar periods as the solar and lunar days. It seems plausible to postulate on the basis of our current knowledge of biochemistry and physiology that these could cycle autonomously, but with periods that would greatly reflect differences in temperature levels and influences of diverse chemical and other physical alterations in both the internal and external environments. With such systems alone, cycles of variable and irregular lengths could be expected. The observed clock properties of period regularity and independence of temperature and other environmental factors would be lacking either completely or to degrees inversely related to the extent to which the organism had evolved general temperature regulation, or other physiological or behavioral homeostatic mechanisms.

The remaining contributions to the observed properties of the rhythmic systems of organisms are postulated to be derived from the subtle-field rhythmic physical environment, to which it is well established that organisms display a highly specialized sensitivity. These environmental rhythms are postulated to entrain the genetically diurnal, lunar, and annual basic metabolic patterns that comprise the clocks underlying biological rhythms, and upon which the rhythms depend for their regularity and stability in the face of diverse, normally disturbing stimuli.

The exogenous factors could also simultaneously provide organisms with information concerning phase angles in both time and space, since such subtle fields as the electromagnetic family vary in four dimensions. Points and vector directions in time and space could be biologically identified and could elicit adaptive responses of survival value, through their contributions to phase-setting of some components of the organism or in recognition of homing direction (and possibly even home site). The ability of birds to home either with celestial light references with employment of biological clocks or by subtle-field means could be considered as simply utilization of different portions over the range of the space-time continuum. The organism, with all its biological "clocks," "compasses," and "sextants," would be employing environmental information from all potential sources with maximum efficiency.

General Conclusions. Periodic variations with geophysically correlated frequencies appears to characterize all living things, plant, animal, and microorganism. The variations under natural conditions are usually synchronized with the day-night, lunar tidal and monthly, or annual variations in the physical environment. These fluctuations (1) describe organismic responses to the obvious environmental stimulus changes, (2) include genetically dependent and environmentally modified recurrent behavioral patterns adjusted in an optimal manner to the environmental cycles, and (3) reflect changes in organisms dependent upon continuing responses to periodically varying ambient subtle fields, such as the electromagnetic ones.

If an organism is shielded from variations in all obvious environmental stimuli normally eliciting direct overt responses, and from all rhythmic variations operating as phase-determiners or Zeitgeber to adjust the phases of the adaptive recurring patterns to their characteristic favored relationship to the external cycles, the rhythms may then "free-run," exhibiting periods deviating slightly from the natural geophysical ones. The periodisms which are responses to the subtle geophysical fields meanwhile, however, retain their accurate mean frequencies.

Rhythmic changes within organisms apparently may serve as chronometers to measure the relative lengths of the daily light and darkness to phase-adjust annual biological cycles adaptively to time of year. They also appear sometimes to serve as clocks for use in conjunction with sun, moon, and possibly even the stars for geographic homing and navigation.

Recent investigations have disclosed not only that organisms held in presumably constant conditions continue to respond to fluctuating parameters of uncontrolled subtle geophysical fields providing information concerning phase-angle variations in all the natural geophysical cycles, but that some organisms are able to derive geographic directional information directly by means of subtle geophysical fields alone. The search for the nature of the subtle temporal and spatial cues that convey the information to the organism has disclosed that living things display a highly specialized sensitivity to the earth's very weak magnetic, electric, and electromagnetic fields, and that this capacity is linked to the mechanisms by which organisms normally orient in space and derive their geophysically dependent periodicities. There is evidence that organisms can inherit a specific manner of response to these fields comparable to inherited character of response to such a factor as light, and that, in addition, the organism can learn and retain at least for a time the relationships of obvious events to spatio-temporal relationships in the subtle fields.

The genetic and environmentally modified behavioral patterns, especially those which are normally synchronized to natural environmental periods but which may "free-run" in constancy of all Zeitgeber, have been postulated by some to be timed by wholly autonomous, temperature-compensated, organismic oscillations whose periods only approximate the geophysical ones, and must depend for their observed period in nature on continuous Zeitgeber phase correcting (endogenous hypothesis). The patterns have been postulated by others to be timed by the pervasive rhythmic geophysical variations which have remained uncontrolled (exogenous hypothesis), with the diverse "free-running" frequencies resulting from an intraorganismic phase-shifting (autophasing or self-resetting) through interaction of the organisms' natural phase-shifting mechanism with the unvarying Zeitgeber factors.

A hypothesis is suggested that both endogenous and exogenous components, collaborating with one another, provide the most probable explanation of the many ordinary and extraordinary properties of the biological rhythms. The endogenous clock is postulated to contribute the genetic and obvious-factor imposed cyclic patterns, along with the biochemical and physiological sequences of processes upon which they depend (including an inherent capacity to synchronize with the environmental physical rhythms). The exogenous clock is postulated to provide the rhythm regularities and the essential stability of the periods despite temperature changes or most kinds of metabolic disturbances. The exogenous clock also provides a continuing highly dependable, celestial-related reference for all the diverse aspects of organismic orientation in time and in space.

REFERENCES

1. Adler, H. E., Animal Behavior *11*:566–577, 1963. General problems in animal migration.
2. Andrews, E. J., J. Florida Med. Assoc. *46*:1362–1366, 1960. Monthly variations in postoperative hemorrhaging.
3. Andrews, R. V., and G. E. Folk, Comp. Biochem. Physiol. *11*:393–490, 1964. Circadian rhythm in cultured hamster adrenals.
4. Aschoff, J., Cold Spring Harbor Symp. Quant. Biol. *25*:11–27, 1960. Circadian period and light intensity.
5. Aschoff, J., Comp. Biochem. Physiol. *18*:397–404, 1966. Phase-angle differences and circadian periods.
6. Aschoff, J., pp. 160–173. *In* Life Sciences and Space Research, edited by A. H. Brown and F. G. Favorite. North-Holland Publishing Co., Amsterdam, 1967. Circadian rhythms.
7. Barnett, A., J. Cell. Physiol. *67*:239–270, 1966. Circadian rhythm of *Paramecium* mating type.
8. Barnwell, F. H., Proc. Soc. Exp. Biol. Med. *105*:312–315, 1960. Geophysically dependent periodisms in chick.
9. Barnwell, F. H., Biol. Bull. *125*:399–415, 1963. Tidal rhythms in *Uca.*
10. Barnwell, F. H., Biol. Bull. *130*:1–7, 1966. *Uca* from Woods Hole region.
11. Barnwell, F. H., Amer. Zool. *8*:569–583, 1968. Tidal rhythms in *Uca.*
12. Becker, G., Naturwissenschaften *50*:664, 1963. Insect orientation to magnetic fields.
13. Bennett, M. F., Z. vergl. Physiol. *47*:431–437, 1963. Rephasing of *Uca* tidal cycles.
14. Bennett, M. F., Z. vergl. Physiol. *60*:34–40, 1968. Annual rhythm in earthworm light response.
15. Bennett, M. F., and J. Huguenin, Z. vergl. Physiol. 440–445, 1969. Earthworm response to geomagnetism.
16. Bennett, M. F., J. Shriner, and R. A. Brown, Biol. Bull. *112*:267–275, 1957. Tidal rhythm in *Uca.*
17. Best, J. B., Animal Behavior *14* (Suppl. 1):69–75, 1966. Monthly rhythm in planarian "learning."
18. Biological Clocks, edited by A. Chovnick. Cold Spring Harbor Symp. Quant. Biol. *25*, 524 pp., 1960. General account of rhythms.
19. Bohn, G., C. R. Acad. Sci. Paris *137*:576–578, 1903. Tidal rhythm in *Convoluta.*
20. Borthwick, H. A., and S. B. Hendricks, Science *132*:1223–1228, 1960. Photoperiodism and plants.
21. Braemer, W., Cold Spring Harbor Symp. Quant. Biol. *25*:413–427, 1960. Sun orientation by fish.
22. Brown, F. A., Jr., Biol. Bull. *115*:81–100, 1958. Mean sidereal-day cycle.
23. Brown, F. A., Jr., Science *130*:1534–1544, 1959. Autophasing, genesis of circadian period.
24. Brown, F. A., Jr., *in* Cold Spring Harbor Symp. Quant. Biol., edited by A. Chovnick, *25*:57–71, 1960. Geophysically dependent rhythms.
25. Brown, F. A., Jr., N.Y. Acad. Sci. *98*:775–787, 1962. Geophysically dependent rhythms.
26. Brown, F. A., Jr., Biol. Bull. *123*:264–281, 1962. Magnetism and *Dugesia* and *Paramecium.*
27. Brown, F. A., Jr., Biol. Bull. *123*:282–294, 1962. Planarian response to electrostatic fields.
28. Brown, F. A., Jr., Biol. Bull. *125*:206–225, 1963. *Dugesia* response to weak gamma radiation.
29. Brown, F. A., Jr., pp. 231–261. *In* Circadian Clocks, edited by J. Aschoff. North Holland Publishing Co., Amsterdam, 1965. Geophysically dependent periodisms.
30. Brown, F. A., Jr., Proc. Soc. Exp. Biol. Med. *120*:792–797, 1965. Lunar-day propensity in hamsters.
31. Brown, F. A., Jr., Scientia, 1968. Geophysically-dependent rhythms.
32. Brown, F. A., Jr., Canad. J. Bot. *47*:287–298, 1969. Extrinsic timing of circadian rhythms.
33. Brown, F. A., Jr., Ann. N.Y. Acad. Sci. *188*:224–241, 1971. Responses to weak electromagnetic fields.
34. Brown, F. A., Jr., Sidereal Variation in Potato Metabolism. (Unpublished manuscript.)
35. Brown, F. A., Jr., Amer. Sci. *60*:756–766, 1972. The "clocks" timing biological rhythms.
36. Brown, F. A., Jr., M. Fingerman, M. I. Sandeen, and H. M. Webb, J. Exp. Zool. *123*:29–60, 1953. Tidal rhythm in *Uca.*
37. Brown, F. A., Jr., M. Fingerman, and M. N. Hines, Biol. Bull. *106*: 308–317, 1954. Diurnal variation in phase shifting response in *Uca.*
38. Brown, F. A., Jr., J. W. Hastings, and J. D. Palmer. The Biological Clock: Two Views. Academic Press, New York, 1970, 94 pp. Review of clock theories.
39. Brown, F. A., Jr., and M. N. Hines, Physiol. Zool. *25*:56–70, 1952. Persistence of experimentally modified circadian cycle forms.
40. Brown, F. A., Jr., and Y. H. Park, Nature *202*:469–471, 1964. Gammataxis, seasonal variations, planarians.
41. Brown, F. A., Jr., and Y. H. Park, Proc. Soc. Exp. Biol. Med. *125*:712–715, 1967. Monthly rhythm in hamsters.
42. Brown, F. A., Jr., and Y. H. Park, Biol. Bull. *132*:311–319, 1967. Association formation between light and subtle geophysical fields.
43. Brown, F. A., Jr., Y. H. Park, and J. R. Zeno, Nature *211*:830–833, 1966. Geophysically dependent periodism in mice.
44. Brown, F. A., Jr., J. Shriner, and C. L. Ralph, Amer. J. Physiol. *184*: 491–496, 1956. Geophysically dependent periodisms in rat.
45. Brown, F. A., Jr., J. Shriner, and H. M. Webb, Biol. Bull. *113*:103–111, 1957. Organismic activity and background radiation.
46. Brown, F. A., Jr., and E. D. Terracini, Proc. Soc. Exp. Biol. Med. *101*:457–460, 1959. Lunar-day propensity in rat.
47. Brown, F. A., Jr., and H. M. Webb, Physiol. Zool. *21*:371–381, 1948. 24-hour precision in constant conditions in *Uca.*
48. Brown, F. A., Jr., M. F. Bennett, and H. M. Webb, Proc. Nat. Acad. Sci. *41*:93, 1955. Geographic translocation experiment.
49. Brown, F. A., Jr., H. M. Webb, and M. F. Bennett, Amer. J. Physiol. *195*:237–243, 1958. Organismic correlations with cosmic radiation.
50. Bruce, V. G., and C. S. Pittendrigh, J. Cell. Comp. Physiol. *56*:25–31, 1960. D_2O and *Euglena* circadian period.
51. Bünning, E., Jahrb. Bot. *77*:283–320, 1932. Inheritance of free-running period in *Phaseolus.*
52. Bünning, E., Jahrb. Bot. *81*:411–418, 1935. Inheritance of free-running period in *Phaseolus.*
53. Bünning, E., Ber. Deutsch. Bot. Ges. *54*:590–607, 1937. Circadian rhythms in photoperiodism.
54. Bünning, E., Naturwiss. Rundschau. *9*:351–357, 1956. Persistence of experimentally modified circadian cycle forms.
55. Bünning, E., Cold Spring Harbor Symp. Quant. Biol. *25*:1–9, 1960. Phase response curve for temperature in *Phaseolus.*
56. Bünning, E., The Physiological Clock. Springer-Verlag, New York, 1967, 167 pp. General account of rhythms.
57. Bünning, E., and E. W. Bauer, Zeitschr. Bot. *40*:67–76, 1952. Annual rhythm in seed germination.
58. Bünning, E., and I. Moser, Planta *69*:101–110, 1966. Phase response curve in *Phaseolus.*
59. Bünning, E., and D. Müller, Z. Naturforsch. *16b*:391–395, 1962. Monthly rhythm in *Dictyota.*
60. Caspers, H., Int. Rev. Ges. Hydrobiol. *46*:175–183, 1961. Swarming rhythm in palolo worm.
61. Circadian Clocks, edited by J. Aschoff. North Holland Publishing Co., Amsterdam, 1965, 479 pp. International symposium.
62. Cloudsley-Thompson, J. L., Rhythmic Activity in Animal Physiology and Behavior. Academic Press, New York, 1961, 236 pp. General account of rhythms.
63. Conroy, R. T. W. L., and J. N. Mills, Human Circadian Rhythms. J. and A. Churchill, London, 1970, 236 pp.
64. Cumming, B. G., Canad. J. Bot. *47*:309–324, 1969. Circadian rhythm in photoperiodism.
65. Dan, K., and H. Kubota, Embryologia *5*:21–37, 1960. Monthly, annual, and 18-year rhythms in *Comanthus* spawning.
66. Davis, D. E., Physiol. Zool. *40*:391–402, 1967. Annual rhythm in woodchuck.
67. DeCoursey, P., Z. vergl. Physiol. *44*:331–354, 1961. Phase-response curves of *Glaucomys.*
68. DeCoursey, P., J. Cell. Comp. Physiol. *63*:189–196, 1964. Phase-response curve of hamsters.
69. Dewan, E. M., Science and Technology, Jan. 20–28, 1969. Monthly rhythm in man.
70. Dijkgraaf, S., and A. J. Kalmijn, Z. vergl. Physiol. *53*:187–194, 1966. Electroperception.
71. Dowse, H. B., and J. D. Palmer, Nature *222*:564–566, 1969. Circadian entrainment by electrostatic fields.
72. Dresler, A., Licht *10*:79–82, 1940. Annual variations of man to spectral colors.
73. Edmunds, L. N., Jr., and R. R. Funch, Science *165*:500–503, 1969. Free-running rhythm in *Euglena* population.
74. Ehret, C. F., and E. Trucco, J. Theoret. Biol. *15*:240–262, 1967. Chronon concept for circadian clock.
75. Emlen, S. T., Auk *84*:309–342, 1967. Star navigation in birds.
76. Emlen, S. T., Auk *84*:463–489, 1967. Star navigation in birds.
77. Enright, J. T., Z. vergl. Physiol. *46*:276–313, 1962. Tidal rhythm in Amphipods.
78. Farner, D. S., Proc. XIV International Ornithological Congress 107–133, 1967. Photoperiodism and bird reproductive cycles.
79. Farner, D. S., Envir. Res. *3*:119–131, 1970. Photoperiodic control of annual cycles.
80. Fauré-Fremiet, E., Biol. Bull. *100*:173–177, 1951. Tidal rhythms in microorganisms.

81. Ferguson, D. E., H. F. Landreth, and McKeown, Animal Behavior *15*:45–53, 1967. Sun navigation in frogs.
82. Gamble, F. W., and F. Keeble, Quart. J. Micr. Sci. *47*:363–431, 1904. Tidal rhythm in *Convoluta.*
83. Gompel, M., C. R. Acad. Sci. Paris *205*:816–818, 1937. Tidal rhythms in fish.
84. Gould, E., Biol. Bull. *112*:336–348, 1957. Sun orientation in turtles.
85. Gwinner, E., Naturwissenschaften *54*:447, 1967. Circannual rhythms in mice.
86. Gwinner, E., J. Ornithol. *109*:70–95, 1968. Circannual rhythms in migratory birds.
87. Halberg, F., Z. Vitamin-, Hormon und Fermentforschung *10*:225–296, 1959. Coining of term "circadian."
88. Hamner, K. C., Cold Spring Harbor Symp. Quant. Biol. *25*:269–277, 1960. Circadian rhythms in photoperiodism.
89. Hamner, K. C., J. C. Finn, G. S. Siroli, T. Hoshizaki, and B. H. Carpenter, Nature *195*:476–480, 1962. Circadian rhythms at south pole.
90. Hamner, W. M., Science *142*:1294–1295, 1963. Circadian rhythms in photoperiodism.
91. Harker, J. E., J. Exp. Biol. *35*:251, 1958. Phase dissociation of tissues and tumor production.
92. Harker, J. E., The Physiology of Diurnal Rhythms. Cambridge University Press, 1964, 114 pp. General account of rhythms.
93. Hartland-Rowe, R., Rev. Zool. Bot. Africaine *58*:185–202, 1958. Monthly rhythm in mayfly emergence.
94. Hasler, A. D., R. M. Horrell, W. J. Wisby, and W. Braemer, Limnol. Oceanogr. *3*:353–361, 1958. Sun and fish homing.
95. Hastings, J. W., New Engl. J. Med. *282*:435–441, 1970. Review.
96. Hastings, J. W., and A. Keynan, pp. 167–182. *In* Circadian Clocks, edited by J. Aschoff, North Holland Publishing Co., Amsterdam, 1965. Metabolic rhythms in *Gonyaulax.*
97. Hastings, J. W., and B. M. Sweeney, Biol. Bull. *115*:440–458, 1958. Phase response curve in *Gonyaulax.*
98. Hauenschild, C., Cold Spring Harb. Symp. Quant. Biol. *25*:491–497, 1960. Lunar periodicity in *Platynereis.*
99. Hauty, G. T., and T. Adams, Aerospace Med. *37*:1257–1262, 1966. Human circadian rhythms during geographic translocations.
100. Hoffman, K., Z. Tierpsychol. *11*:453–475, 1954. Sun navigation and "clock"-shifting.
101. Johnson, L. G., Biol. Bull. *131*:308–322, 1966. Circadian-rhythm self setting to local time.
102. Jones, D. A., and E. Naylor, J. Exp. Mar. Biol. Ecol. *4*:188–199, 1970. Tidal rhythms in isopods.
103. Kalmijn, A. J., Nature *212*:1232, 1966. Electroperception in sharks and rays.
104. Karakashian, M. W., and J. W. Hastings, J. Gen. Physiol. *47*:1–12, 1963. Inhibitors of macromolecule synthesis and rhythms.
105. Kavanau, J. L., Nature *194*:1293–1295, 1962. LD transitions and extended limits for circadian entrainment.
106. Keeton, W. T., Science *165*:922–928, 1969. Pigeon homing without sun.
107. Keeton, W. T., Proc. Nat. Acad. Sci. *68*:102–106, 1971. Magnets and pigeon homing.
108. Kessler, E., and F. C. Czygan, Experientia *19*:89, 1962. Annual variations in *Ankistrodesmus.*
109. Kleinhoonte, A., Arch. Néerl. Sci. Exp. Nat. *5*:1–110, 1929. Limits of circadian entrainment in *Phaseolus.*
110. Konopka, R. J., and S. Benzer, Proc. Nat. Acad. Sci. *68*:2112–2116, 1971. Inheritance of free-running period in *Drosophila.*
111. Kramer, G., Ibis *101*:399–416, 1959. Sun navigation in birds.
112. Lang, H. J., Z. vergl. Physiol. *56*:296–340, 1967. Monthly rhythms in guppy light response.
113. Lang, H. J., Verhandl. Deutsch. Zool. Gesellsch. Innsbruck (1968): 291–298, 1968. Monthly rhythm in guppy light response.
114. Lewis, P. R., and M. C. Lobban, J. Physiol. *133*:670–680, 1956. Human circadian rhythms in Arctic.
115. Lindauer, M., Cold Spring Harbor Symp. Quant. Biol. *25*:371–377, 1960. Clocks and bee sun-compass.
116. Lindauer, M., and H. Martin, Z. vergl. Physiol. *60*:219–243, 1968. Magnetic responsiveness of bees.
117. Lobban, M. C., Quart. J. Exp. Physiol. *52*:401–410, 1967. Human circadian rhythms in Arctic.
118. Lofts, B., Nature *201*:523–524, 1964. Annual rhythm in *Quelea.*
119. Lowe, C. H., D. S. Hinds, P. J. Lardner, and K. E. Justice, Science *156*:531–534, 1967. 24-hour precision in circadian rhythms in kangaroo rats and gila monsters.
120. Luce, G. G., Biological Rhythms in Psychiatry and Medicine. Public Health Service Publ. #2088, 183 pp., 1970. General review.
121. Menaker, W., and A. Menaker, Amer. J. Obstet. Gynecol. *77*:905–914, 1959. Monthly period in human reproduction.
122. Merkel, F. W., Ann. N.Y. Acad. Sci. *188*:283–294, 1971. Magnetic fields in bird navigation.
123. Merkel, F. W., and W. Wiltschko, Die Vogelwarte *23*:71–77, 1965. Magnetic fields and bird orientation.
124. Natalini, J. J., Physiol. Zool. *45*:153–166, 1972. Phase response curve of kangaroo rat.
125. Naylor, E., J. Exp. Biol. *35*:602–610, 1958. Tidal rhythm in *Carcinus.*
126. Neumann, D., Z. vergl. Physiol. *53*:1–61, 1966. Monthly rhythms in *Clunio.*
127. Neumann, D., Z. vergl. Physiol. *60*:63–78, 1968. Tidal-cycle entrainment of semimonthly rhythm in *Clunio.*
128. Neumann, D., Oecologia *3*:166–183, 1969. Timing by combinations of rhythms periods.
129. Orientation: Sensory Bases, edited by H. E. Adler. Ann. N.Y. Acad. Sci. *188*:1–408, 1971.
130. Palmer, J. D., Nature *198*:1061–1062, 1963. Responses of *Volvox* to weak magnetic fields.
131. Palmer, J. D., Nature *215*:64–66, 1967. Diurnal and tidal rhythms in *Sesarma.*
132. Palmer, J. D., and H. B. Dowse, Biol. Bull. *137*:388, 1969. D_2O and mouse and bird circadian period.
133. Palmer, J. D., and F. E. Round, J. Marine Biol. Assoc. U.K. *45*:567–582, 1965. Tidal rhythm in *Euglena.*
134. Palmer, J. E., F. E. Round, Biol. Bull. *132*:44–55, 1967. Tidal rhythm in diatoms.
135. Papi, F., Cold Spring Harbor Symp. Quant. Biol. *25*:475–480, 1960. Amphipod orientation by moon.
136. Papi, E., and L. Pardi, Biol. Bull. *124*:97–105, 1963. Lunar orientation in amphipods.
137. Papi, F., and P. Tongiorgi, Ergeb. Biol. *26*:259–280, 1963. Celestial orientation by wolf-spiders.
138. Pardi, L., Cold Spring Harbor Symp. Quant. Biol. *25*:395–401, 1960. Sun orientation by amphipods.
139. Picton, H. D., Nature *211*:303–304, 1966. *Drosophila* response to weak electromagnetic fields.
140. Pittendrigh, C. S., Cold Spring Harbor Symp. Quant. Biol. *25*:159–182, 1960. Phase response curve in *Drosophila.*
141. Pittendrigh, C. S., Harvey Lectures, Ser. *56*:93–125, 1961. Academic Press, New York. Circadian rhythms as endogenous complex oscillations.
142. Pittendrigh, C. S., and V. G. Bruce, Photoperiodism and Related Phenomena in Plants and Animals. Amer. Assoc. Adv. Sci., Washington, D.C., 1959, pp. 475–505. Daily rhythms as complex oscillator systems.
143. Pittendrigh, C. S., and D. H. Minis, Amer. Nat. *98*:261–294, 1964. Circadian rhythms in photoperiodism.
144. Rao, K. P., Biol. Bull. *106*:353–359, 1954. Tidal rhythm in *Mytilus.*
145. Reille, A., J. Physiol. Paris *30*:85–92, 1968. Pigeon conditioning to magnetic fields.
146. Renner, M., Naturwiss. Rundschau *14*:296–305, 1961. Persistence of experimentally modified circadian cycle forms; geographic translocation experiment.
147. Rhythmic Functions in the Living System, edited by W. Wolf. Ann. N.Y. Acad. Sci. *98*:753–1326, 1962. General account.
148. Richter, G. Z., Z. Naturforsch. *18B*:1085, 1963. Circadian rhythms in enucleated *Acetabularia.*
148a. Rommel, S. A., Jr., and J. D. McCleave, Science *176*:1233–1235, 1972. Electric field perception by eels.
149. Sauer, F., Z. Tierpsychol. *14*:29–70, 1957. Star navigation by birds.
150. Schmidt-Koenig, K., Z. Tierpsychol. *15*:301–331, 1958. Sun orientation and "clock" shifting.
151. Schmidt-Koenig, K., Z. Tierpsychol. *18*:221–224, 1961. Sun orientation and "clock" shifting.
152. Schneider, F., Ergebn. Biol. *26*:147, 1963. Insect orientation to magnetic fields.
153. Schweiger, E., H. G. Walraff, and H. G. Schweiger, Z. Naturforsch. *19B*:499–505, 1964. Circadian rhythms in enucleated *Acetabularia.*
154. Schweiger, E., H. G. Walraff, and H. G. Schweiger, Science *146*:658, 1964. Nuclear determination of circadian phase in *Acetabularia.*
155. Sollberger, A., Biological Rhythm Research. Elsevier Publ. Co., Amsterdam, 1965, 461 pp. General review.
156. Southern, W. E., Ann. N.Y. Acad. Sci. *188*:295–311, 1971. Geomagnetic strength and gull orientation.
157. Stephens, G. C., Physiol. Zool. *30*:55–69, 1957. Phase response for temperature in *Uca.*
158. Strughold, H., Ann. N.Y. Acad. Sci. *134*:413–422, 1965. Human circadian rhythms during space flights.
159. Stutz, A., Ann. N.Y. Acad. Sci. *188*:312–323, 1971. Gerbil response to geomagnetism.
160. Suter, R. B., and K. S. Rawson, Science *160*:1011–1015, 1968. D_2O and mouse circadian period.
161. Sweeney, B. M., Rhythmic Phenomena in Plants. Academic Press, New York, 1969, 147 pp. General review.
162. Sweeney, B. M., and J. W. Hastings, Cold Spring Harbor Symp. Quant. Biol. *25*:87–104, 1960. Temperature and circadian rhythms.
163. Sweeney, B. M., and F. T. Haxo, Science *134*:1361, 1961. Circadian rhythms in enucleated *Acetabularia.*
164. Terracini, E. D., and F. A. Brown, Jr., Physiol. Zool. *35*:27, 1962. Propensity for 24-hour periods in mice.
165. Tribukait, B., Z. vergl. Physiol. *38*:479–490, 1956. Limits of circadian entrainment in mice.
166. Vanden Driessche, T., Biochim. Biophys. Acta *126*:456–470, 1966. Nucleus and circadian rhythm in *Acetabularia.*
167. Vanden Driessche, T., and S. Bonotto, Biochim. Biophys. Acta *179*: 58–66, 1969. Circadian rhythm in RNA synthesis in *Acetabularia.*
168. Van Laar, W., Acta Biotheoretica *19*:95–139, 1970. The concept of "Biological Clock."
169. Vielhaben, V., Zeitschr. Bot. *51*:156–173, 1963. Monthly rhythm in *Dictyota.*

170. Walcott, C., and M. C. Michener, J. Exp. Biol. *54*:291–316, 1971. Sun navigation in pigeons.
171. Wallraff, H. G., Proc. XIV International Ornithological Congress, 1967, pp. 331, 358. Review of bird homing problems.
172. Waterman, T. H., Nature *228*:85–87, 1970. Polarized light and fish orientation.
173. Webb, H. M., Physiol. Zool. *23*:316–337, 1950. Diurnal variations in phasing shifting response in *Uca.*
174. Webb, H. M., and F. A. Brown, Jr., Biol. Bull. *115*:303–318, 1958. Metabolic periodisms in fiddler crabs.
175. Webb, H. M., and F. A. Brown, Jr., Physiol. Rev. *39*:127–161, 1959. RMI hypothesis for biological rhythms.
176. Webb, H. M., and F A. Brown, Jr., Biol. Bull. *129*:582–591, 1965. Daily and tidal rhythms in fiddler crabs.
177. Wehner, R., and T. Labhart, Experientia *26*:967–968, 1970. *Drosophila* response to geomagnetism.
178. Wever, R., Kybernetik *2*:127–144, 1964. Model of a 24-hour clock.
179. Wever, R., Z. angewandte Math. Mech. *46*:148–157, 1966. Mathematical model for biological clock.
180. Wever, R., Z. vergl. Physiol. *56*:111–128, 1967. Electromagnetic fields and human circadian rhythms.
181. Wever, R., Naturwissenschaften *55*:29–32, 1968. Influence of weak electromagnetic fields on circadian period.
182. Wever, R., Life Sciences and Space Research VIII, North Holland Publ. Co., Amsterdam, 1970, pp. 177–187. Electric fields and human circadian rhythms.
183. Williams, B. G., and E. Naylor, J. Exp. Biol. *51*:715–725, 1969. Tidal rhythm in *Carcinus.*
184. Wiltschko, W., Z. Tierpsychol. *25*:537, 1968. Magnetic-field orientation in birds.
184a. Wiltschko, W., and R. Wiltschko, Science *176*:62–65, 1972. Magnetic compass of European robins.
185. Winfree, A. T., J. Theoret. Biol. *16*:15–42, 1967. Biological rhythms and populations of coupled oscillators.
186. Winfree, A. T., J. Theoret. Biol. *28*:327–374, 1970. Integrated view of resetting a circadian clock.
187. Wolfson, A., pp. 1–49. *In* Photophysiology, Vol. 2, edited by A. C. Giese, 1965. Photoperiodism.
188. Zimmerman, W. F., C. S. Pittendrigh, and T. Pavlidis, Insect Physiol. *14*:669–684, 1968. Phase response curve for temperature in *Drosophila.*

CHAPTER 11

EXCITABLE MEMBRANES

By C. Ladd Prosser

The cellular basis of animal behavior resides largely in the conduction of signals by cell membranes and in transmission from cell to cell. Sensory structures provide information about the environment; central nervous systems integrate sensory information with neural patterns already present; muscles and other effectors bring about behavioral reactions. These three components of animal behavior have common cellular features. Membrane function in neuronal and muscular tissues has its basis in electrical properties of excitable cells in general and in the functional coupling between cells of non-excitable tissues.

A distinction must be made, for all nervous and many non-nervous tissues, between conduction within a cell and transmission from cell to cell. Electrical conduction (1) may be graded and decremental, or (2) may be regenerative, self-propagating, and all-or-none. Intercellular transmission may be electrical or chemical. Electrical transmission may be: (a) graded, electrotonic, and bidirectional; (b) unidirectional, or rectifying in one direction; or rarely, (c) inhibitory by exerting external negative fields. Chemical transmission is based on liberation of specific agents from presynaptic vesicles, which requires presynaptic membrane depolarization and is blocked by low Ca^{++} and/or high Mg^{++} in the medium; the specific agents chemically excite postsynaptic membranes.

NON-NERVOUS CONDUCTION; RESPONSES OF SINGLE CELLS

All cells are bounded by a surface layer, the plasma membrane, which lies beneath the limiting membrane and is electrically polarized, normally negative on the protoplasmic interior. Excitation usually consists of reduction or reversal of the normal polarization. The depolarization may be local or propagated; conduction is a wave of depolarization passing along the plasma membrane. It was formerly thought for ciliate Protozoa that intracellular fibrils could serve in conduction from one part of the organism to another and could coordinate ciliary beat. Recent observations on *Euplotes* show coordinated changes in direction and frequency of beat of the three types of cilia after "neuromotor" fibers have been cut; also, cell fragments show ciliary coordination.[178] Evidently, conduction in Protozoa is by the plasma membrane, as it is in other excitable cells.

The ciliate *Opalina* has a membrane potential of -27 mv, and the intracellular negativity becomes reduced when ciliary beat is reversed in response to electrical stimulation.[233] In *Paramecium*, intracellular electrodes record a resting potential of -30 to -40 mv, and stimulation—tactile, chemical, or electrical—evokes an active electrical response, reversing the ciliary beat. Tactile stimulation at the anterior end elicits a fast depolarizing response, which is conducted posteriorly and is dependent in amplitude on external calcium concentration. Stimulation at the posterior end elicits a localized hyperpolarizing response, which decreases when external potassium is increased, and hence is due to efflux of K from the cytoplasm. The membrane transient potential associated with reversal of ciliary beat represents an increase in Ca conductance, and the $[Ca^{++}]$ gradient is inward. The membrane shows repetitive firing of regenerative action potentials in response to electrical or chemical $[Ba^{++}]$ stimulation, and the action potentials may overshoot zero potential. It is suggested that ciliates give both local and regenerative membrane potentials, and that the primitive action potential may depend on Ca^{++} rather than on Na^{+} conductance change.[57–60, 164, 165]

In sponges, spindle-shaped cells around the oscula are contractile. Contractions in response to mechanical stimulation are local, and no propagation from cell to cell occurs; there are no nerve elements. The contractile responses to touch require both univalent and divalent cations but can occur when all sodium in the medium has been replaced by potassium (500 mM). The contractile system of sponges is probably stimulated directly (p. 773). Some potassium-depolarized vertebrate smooth muscles also can be stimulated to contract by neurohumoral agents such as acetylcholine.

TRANSMISSION FROM CELL TO CELL IN NON-NERVOUS TISSUES

Transmission in various tissues, such as sheets of epithelium or early embryos, has

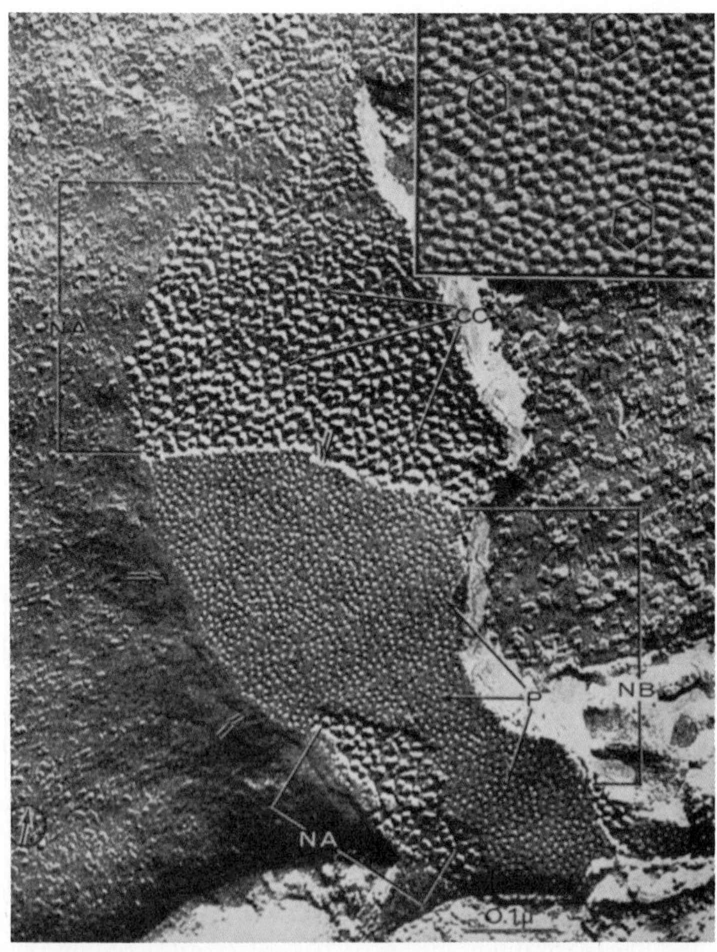

Figure 11–1. A, Face view of a nexus in freeze-cleaved and -etched, platinum-shadowed heart muscle. Nexus face NB contains regular pits (P) 35 to 50Å in diameter. Protruding inward from nexus face NA are many contact cylinders (CC), which are shown enlarged in upper insert. (From NcNutt, N. S., and R. S. Weinstein, J. Cell Biol. *47*:666–688, 1970.)

(Figure 11–1 continued on opposite page.)

been shown to occur independently of nerves. Such tissues show electrical coupling between cells, in that current pulses injected into one cell are recorded with little reduction in nearby cells. The coupling coefficient is given by the percentage of a potential applied in one cell (V_2) that is recorded in an adjacent one (V_1): coefficient $= \frac{V_1}{V_2}$.

In the classification of junctions between cells, several terms have been used for a given structure, and in one intercellular junction there may be several types of contact. Two general categories are: (a) separated membranes (100 to 150 Å apart), and (b) direct contact or fusion of outer membrane layers. The separated membranes include: (1) desmosomes or dense bars on one or both sides of a region of aligned but separated membranes; (2) chemical synapses with synaptic vesicles near the presynaptic membrane, and sometimes desmosome-like structures in the postsynaptic membrane; and (3) septate junctions such as occur between some invertebrate epithelial cells.

Electron microscopically, the direct contact junctions show 3, 5 or 7 alternating light and dense bands according to staining methods. Some of these have been called tight junctions, which occur as a "zonula occludens" or ring, forming a seal around one end of a cell; "facia occludens" as a strip; and "macula occludens" as a patch of tight junction. In face view, each of these has a fibrillar structure in the plane of the membrane. Tight junctions occur mainly in vertebrate epithelia. Another type of direct contact is the nexus or gap junction. It has 60 to 70 Å subunit projections forming contacts between the outer membrane laminae,

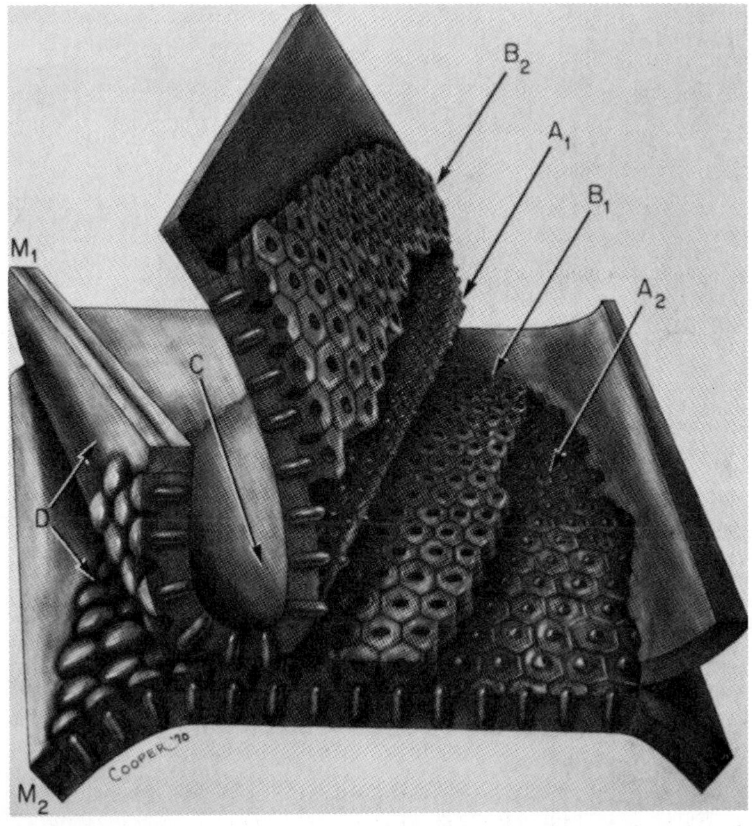

Figure 11-1 (Continued). B, Drawing of model of nexus. M_1 and M_2 are the two plasma membranes. Each membrane is cleaved into two lamellae with faces A and B. (From McNutt, N. S., and R. S. Weinstein. J. Cell Biol. *47*: 666–688, 1970.)

separated by 20 Å areas which seem continuous with extracellular space; the projections are seen in face view as hexagons with 90 to 100 Å distance between centers[151, 200, 238] (Fig. 11–1). Nexuses occur wherever there is low-resistance electrical coupling between cells—in septate giant nerve fibers, between smooth and cardiac muscle fibers, between electrically coupled neurons, and between some epithelia in vertebrates and in invertebrates[23, 49] (Fig. 11–2). Nexus regions often occur interspersed among either septate or tight (occludens) areas.[18] Electrical coupling may occur via nexuses with or without zonula occludens or septate junctions.

The salivary glands of dipteran larvae have

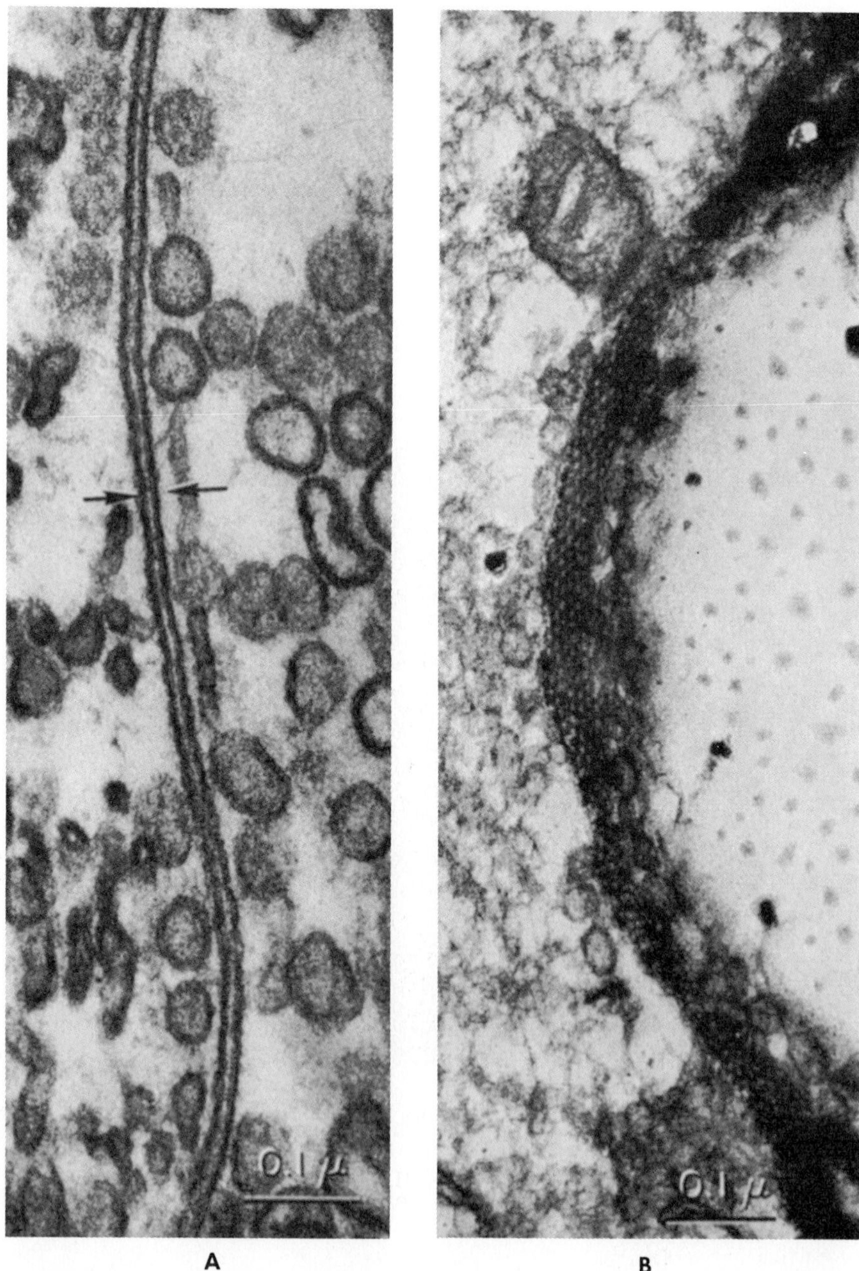

Figure 11–2. High magnifications of junctional regions of septa of lateral giant axons of crayfish. A, Thin section; overall thickness of junction 180 Å. B, An approximately tangential section through junction after fixation in presence of lanthanum, which outlines the hexagonal array. (From Pappas, G. D., Y. Asada, and M. V. L. Bennett, J. Cell Biol. *49*:173–188, 1971.)

rows of large epithelial cells. Electrical coupling ratios of 0.8 to 0.9 are normally found between adjacent cells, and the electrical resistance between adjacent cells may be only 1/100 of what it is between a cell and the outer medium.[136, 137, 138, 139] When dyes (fluorescein of MW 330 or procion of MW 550) are injected into one cell, they soon appear in adjacent cells; but polypeptides, which are larger, fail to pass. In dipteran salivary gland epithelium, septate and gap (nexus) regions are intermingled[18] (Fig. 11-3). The coupling is related to the amount of calcium ions bound to the cell membrane. If calcium in the medium is high (10^{-4} M for *Chironomus*, 10^{-3} M for mouse liver) the intercellular junctions become impermeable; also, if calcium is removed, the cells become uncoupled. Intercellular permeability is high when intracellular Ca^{++} is low ($<10^{-6}$ M). Injection of Ca^{++}, treatment with DNP, substitution of Li_o^+ for Na_o^+, or washing with Ca^{++}-free and Mg^{++}-free medium all decrease junctional permeability. Electrical uncoupling is accompanied by cell depolarization, and recoupling can be induced by repolarization.[166, 204a]

Electrical coupling has been noted between cells of toad bladder, mouse liver, dipteran salivary glands, and malpighian tubules. When dissociated sponge cells fuse in a culture they form contacts of low electrical resistance.[137, 138]

In squid embryos, most growing cells are electrically coupled and dye can pass from one cell to another; the embryo appears to be an electrical syncytium.[196] Similarly, in embryos of the fish *Fundulus* during cleavage stages, there is nearly complete electrical connection and both gap and tight junctions occur; coupling is less in gastrulae.[16] Starfish embryos show electrical coupling in the 2-cell stage, and none after the second, third, or fourth cleavages; but low resistance connections are again established in the 32-cell stage.[232]

Conduction through non-nervous tissue has been established in a number of coelenterates. In the swimming bells of siphonophores, nerves and muscle fibers are absent

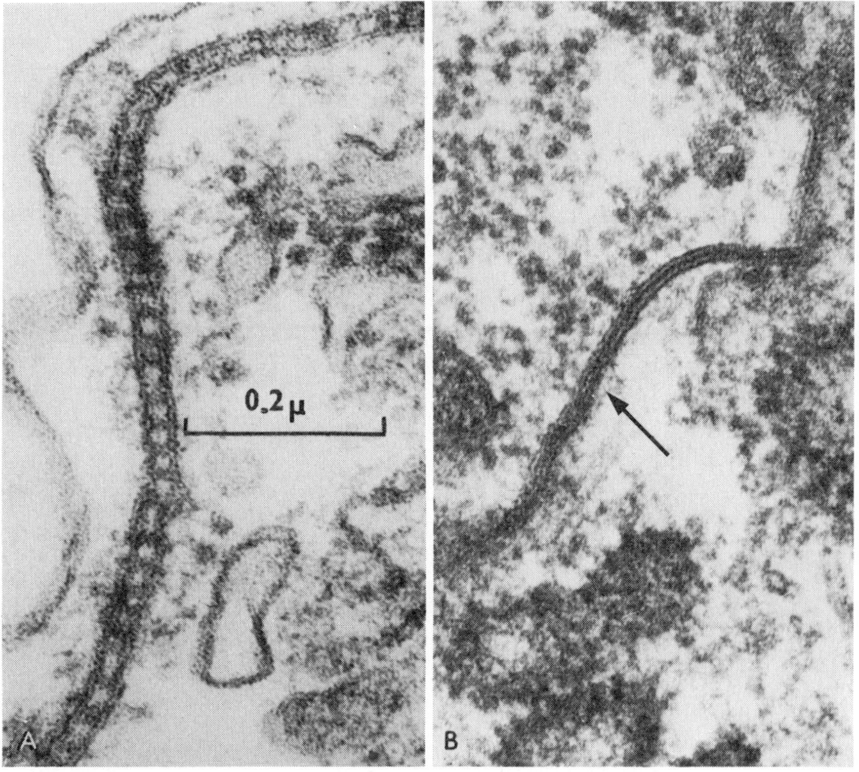

Figure 11-3. Electron micrographs of sections of junctional areas between two cells of *Chironomus* salivary gland. A, septate junction; B, gap junction (nexus). (From Berger, W. K., and B. Uhrik: Membrane Junctions between Salivary Gland Cells of *Chironomus thummi*. Z. Zellforsch. *127*:116–126, 1972. Berlin-Heidelberg-New York: Springer.)

from stretches of the epithelium; yet non-decrementing, non-polarized conduction of electrical potentials can be observed in the epithelium at 20 to 25 cm/sec with a refractory period of 2 to 3 msec.[141,142,143] In colonial hydroids, conduction occurs from polyp to polyp, . partly by non-nervous tissue. In hydromedusae, biphasic pulses occur in epithelial cells; these are 1 to 15 msec in duration, are conducted at 15 to 35 cm/sec, and are not blocked by magnesium ions, which do block the nerve net.[144] It is likely that conduction within a slow system in hydroids—6 cm/sec with high threshold and no direct effect on behavior—is non-nervous.[98,99] *Hydra* shows several types of electrical activity—rhythmic spontaneous spikes originating in pacemaker regions, small potentials associated with tentacle movement, and large conducted potentials travelling at 5 cm/sec. The body cavity is 15 to 40 mv positive to the outside due to asymmetry potentials, and the large conducted action potentials represent reversals of potential across the epithelial layer. Nexuses permit lateral flow of current from cell to cell and prevent current flow out of the cells through high resistance external membranes.[100] In *Hydra,* septate junctions join the outer surfaces of epidermal cells, and below them are gap junctions of plaques containing many hexagons, 95 to 110 Å center to center.[87]

In late embryos and young tadpoles of the toad *Xenopus* before nerves develop, cells of the outer skin show both tight and gap junction; all-or-none impulses, generated by stimulation of superficial cells, are conducted through the skin sheet at 7.7 cm/sec. If sodium is replaced by ions such as Tris in the inside solution, or if the skin is poisoned by TTX (see p. 468), conduction fails. The skin cells have resting potentials of −75 mv and give 100 msec depolarizing spikes of 110 mv amplitude. In later embryos, such skin potentials can excite nerve endings.[201]

Evidently, a primitive type of signal transmission is by electrical coupling between cells which are closely joined by gap (nexus) or tight junctions. In embryos, such coupling permits functional continuity; in some epithelia, as in coelenterates and amphibian embryos, it may mediate primitive types of behavior. Examples of similar electrical coupling between nerve cells are presented below (pp. 481–483).

CONDUCTION IN EXCITABLE TISSUES; ORGANIZATION OF MEMBRANES, ESPECIALLY IN NEURONS

Every excitable cell is bounded by a plasma membrane some 50 to 100 Å thick, outside which there may be various investing sheaths. Some non-myelinated nerve fibers (if they are large, as some crustacean and cephalopod axons) may be surrounded by a glial sheath. Other kinds of non-myelinated nerve bundles may have one multinucleate sheath cell surrounding a group of small fibers. Myelinated nerve fibers of vertebrates have a spiral of many layers of myelin, with the sheath cell outside. In myelinated nerve fibers of some invertebrates (e.g., prawns) the nucleus of the sheath cell lies inside the single loose layer of lipid or myelin; in others, myelin occurs in several spiral layers. Vertebrate myelinated fibers have the sheath interrupted at nodes of Ranvier; in fibers of some invertebrates, frequent side branches function much as nodes. In general, sheaths become thin and myelin may be absent as axons approach their terminals or cell bodies.

The excitable cell membrane consists of lipid and protein and has high resistance and capacitance (Fig. 11–4). The core-conductor properties of a cylindrical cell such as an axon or muscle fiber provide the basis for

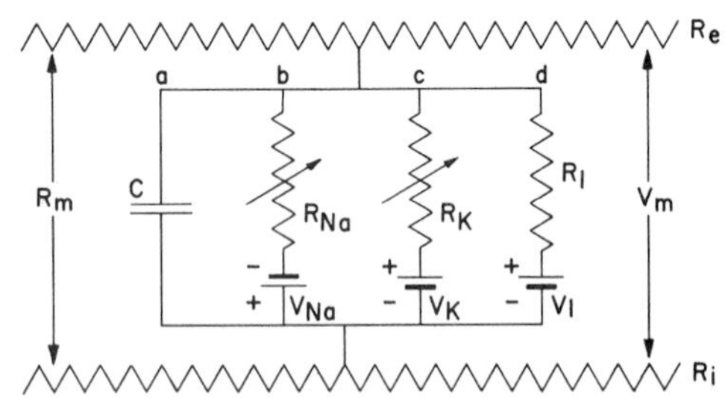

Figure 11–4. Diagram of an element of an excitable membrane of a nerve fiber. *C,* constant membrane capacity; R_{Na}, variable permeability channel for Na^+; R_K, variable permeability channel for K^+; R_l, channel for other ions, especially Cl^-; V_m, membrane potential; R_e, external resistance; R_i, internal resistance.

Average values for squid giant axon: R_m = 400 to 1100 ohms cm^2; R_e = 20 ohms cm; R_i = 30 ohms cm; C_m = 1.1 $\mu F/cm^2$; V_K = −85 to −90 mv; V_m = −60 to −70 mv; V_l = −55 to −60 mv; V_{Na} = +45 to +50 mv. (Modified from Hodgkin, A. L., Proc. Roy. Soc. Lond. B *148*:1–37, 1958.)

the time constant τ and space constant λ, which express respectively the time of decline and the spatial spread of a transient electric pulse to $1/e$ (approximately 37%) of the applied potential (Table 11–1).

The time constant in seconds is given by $C_m R_m$ where C_m is membrane capacity in farads/cm² and R_m is membrane-specific resistance in ohms cm².* The passive spread of potential is given by

$$V_x = V_o e^{(-x/\lambda)}$$

where V_o is applied voltage and V_x is measured voltage at distance x. Then λ is given by the slope when log V_o/V_x is plotted against x in mm. Also, λ is determined by the ratio of membrane resistance to internal and external resistivity,

$$\lambda = \sqrt{\frac{r_m}{r_i + r_o}}$$

where r_m is transverse resistance times unit length of membrane in ohms cm, r_i is intracellular resistance per unit length in ohms/cm, and r_o is extracellular resistance per unit length in ohms/cm.

The resistance of the cell membrane exists across only 100 Å; hence, the specific resistivity of the membrane is high—some 10^9

*Specific resistance (r) refers to resistance of a cubic centimeter of material and has units of ohm cm²/cm = ohm cm. For resistance across a cell membrane, the thickness (length) is unknown; hence, the specific membrane resistance R_m is the resistance of 1 cm² of surface irrespective of thickness, ℓ, or $R_m = r\ell$. Measured resistance is $R = \frac{r\ell}{A}$; hence, $R_m = r\ell = RA$ or ohms cm². Capacitance is proportional to dielectric constant D and area A for a given thickness ℓ, or $C = \frac{DA}{\ell}$; hence, specific membrane capacity $C_m = \frac{D}{\ell} = C/A$ or farads/cm².

TABLE 11–1. Membrane Constants of Selected Neurons
(For similar data for muscle membranes, see Table 16–4, p. 748.)

Animal	*Membrane Resistance (input resistance in ohms, or specific resistance in ohms cm²)*	*Membrane Capacitance (input capacitance in F, or specific capacitance in F/cm²)*	*Time Constant*	*Velocity*
Myxicola[19]				
giant fiber	1.2 × 10³ Ω cm²			
Eudistyla[85a]				
giant fiber	1.5–4.5 × 10³ Ω cm²	2–3 μF/cm²		
Leech[128]				
large neurons	1.7 × 10³ Ω cm²			
Cockroach[246]				
giant fibers	0.8 × 10³ Ω cm²	6.3 μF/cm²	4.2 msec	1.5–3.5 m/sec
Lobster[130a]				
giant fibers				8 m/sec
Shrimp[130a]				
giant fibers				90 m/sec
Goldfish[67a]				
Mauthner cells	65 Ω cm²	6.5 μF/cm²	0.39 msec	
motor cells	600 Ω cm²	5.0 μF/cm²	3.1 msec	
Onchidium[178a]				
giant neurons	6.8 MΩ	0.27 μF		
Helix[146]				
neurons	0.98 × 10³ Ω cm²	29 μF/cm²	30 msec	
Aplysia[93a]				
H cells	4–10 × 10³ Ω cm²		100–200 msec	
Cockroach[32]				
sensory				3.3 m/sec
Frog[90]				
node	10–20 Ω cm²	3–7 μF/cm² or 0.6–1.5 pF		
Lamprey[150]				
touch cells	5.4 MΩ	271 pF	1.2 msec	
pressure cells	6.3 MΩ	375 pF	2.3 msec	
nociceptive cells	8.3 MΩ	479 pF	3.9 msec	
Mauthner cell[204b]	2 MΩ (16,000 Ω cm²)	2.8 μF/cm²		

ohms cm. The resistivity of the axoplasm is low, approaching that of plasma or sea water ($r_i = 200$ ohm cm). Conductance (g) is the reciprocal of resistance, $1/R$. Specific resistance of many excitable cell membranes is about 1000 ohms cm^2. A useful quantity for cells of indeterminate shape or surface area is the input resistance, or the resistance measured by a microelectrode in reference to the medium.

Capacity in many excitable cells is about 1 $\mu F/cm^2$ (Table 11–1). Capacity of striated muscle fibers and cardiac Purkinje fibers is high and can be resolved into two components, one in the outer membrane and the other in the transverse tubules; in frog sartorius fibers, the two values are 2.6 $\mu F/cm^2$ and 4.1 $\mu F/cm^2$, and in crayfish muscle fibers the corresponding values are 3.9 and 17 $\mu F/cm^2$ (reference 64); for Purkinje fibers the two values are 2.4 and 7 $\mu F/cm^2$ (reference 67).

Resting Potentials. Across plasma membranes of unstimulated cells, there is a resting potential (RP). Most cells are symmetrical, but some flattened cells are asymmetrical in that the two faces have different electrical properties.

Symmetrical resting potentials. The resting potential can be measured by intracellular microelectrodes in large fibers or cell bodies, or by electrodes at nodes, well insulated as by an air gap on each side. The resting potential of many neurons is about −70 mv, that of many skeletal muscle fibers is −90 mv, and that of many smooth muscle fibers is −40 mv. Values from various cell types range from −4 mv to −100 mv. Cytoplasm contains much more potassium, less chloride, less sodium, and much less calcium ion than does extracellular fluid (p. 81). The resting potential can be formally described by the Goldman constant field equation, in which the concentration of each ion is multiplied by a permeability constant, P, which takes into account the apparent ionic mobilities in the membrane:

$$E = \frac{RT}{nF} \ln \left(\frac{P_K[K_i] + P_{Na}[Na_i] + P_{Cl}[Cl_o]}{P_K[K_o] + P_{Na}[Na_o] + P_{Cl}[Cl_i]} \right)$$

where E = equilibrium potential, $\frac{RT}{nF}$ = a conversion factor consisting of gas constant R, absolute temperature T, valence n, and Faraday constant F; and o and i refer to outside and inside the neuron. Much of our knowledge of nerve impulses has come from observations on squid giant axons, which are large enough (500 to 1000 μ) for insertion of two or more intracellular electrodes. For squid giant axons, $P_{Na} = 0.008\ P_K$ at rest and $P_{Na} = 30\ P_K$ during the early phase of activity.

Relative permeabilities for *Aplysia* neurons are: for K, 1; Na, 0.01; Cl, 0.45.[207] P_{Cl} is higher for crab nerve than for vertebrate and squid nerve. In frog twitch-type muscle, P_{Cl} is relatively high, $P_K:P_{Na}:P_{Cl} = 1: 0.9: 1.90$, and RP is close to E_{Cl}. P_{Na} in cardiac muscle is also high. Because of the low P_{Na} the resting potential in most excitable tissues is sensitive to K_o and not to Na_o; i.e., the resting membrane behaves as a potassium electrode. However, in the range of physiological concentrations of potassium, the effect of Na on RP becomes significant, so that RP deviates from the linear relation to log K_o.[1]

As previously described (p. 83), non-penetrating anions make up the bulk of intracellular negative charges. In squid axons, the principal anion is isethionate; in crustacean axons they are amino acids, and in some mammalian neurons it is *N*-acetyl aspartate. The observed potential in some neurons is close to or slightly positive to the Cl^- equilibrium potential. However, in squid axons $[Cl_i]$ is 2 to 3 times that expected in equilibrium; i.e., RP is more negative than E_{Cl}. The resting potential is far from the sodium equilibrium potential, which is 20 to 50 mv positive. The steep sodium gradient is maintained by an active sodium pump by which extrusion of the sodium, which enters in small amounts at rest, is coupled with uptake of potassium (p. 94). Crustacean nerves provided some of the earliest evidence for association of membrane ATPase with a Na-K pump. The concentration of calcium ions in squid axoplasm is only 0.0001% of that outside, so that E_{Ca} approaches +100 mv; the low concentration of Ca^{++} results partly from internal binding and partly from active extrusion, which depends on $[Na_o^+]$. Mg^{++} also is extruded by a Na-coupled mechanism.[90] In a K-free medium, Na transport is decreased; high Na_o reduces Na transport, and high Na_i increases it.

The net effects of the following factors account for the resting potential: the presence inside the fiber of non-diffusible organic anions which permit small anions to approach Donnan equilibrium, differential resting permeabilities; active mechanisms for Na extrusion coupled with K retention; and a Na-coupled extrusion of Ca. Excitability is determined by the level of resting potential; if RP declines (i.e., the membrane is depolarized, as in anoxia), excitability can be restored by cathodal repolarization.

As seen from Figure 11–4, responses of a cell membrane could result from changes in the transmembrane resistances for specific ions (conductance or 1/R), or from changes in the ionic batteries; in general, capacity changes are small. The resistance changes may be asymmetrical so that current may flow more readily in one direction than the other; i.e., the membrane may act as a rectifier.

ASYMMETRY POTENTIALS. Asymmetry potentials develop across flattened cells where the outer and inner faces differ in electrical properties. In general, the permeabilities of the two cell surfaces differ, and across one of them active transport, usually of sodium, occurs. Asymmetry potentials are best known for epithelia, and descriptions were given previously for such potentials across frogskin (p. 95), toad bladder (p. 96), gills of crustaceans and fishes, and gastrointestinal epithelia; under physiological conditions the fluids bathing the two cell faces are often very different in composition. Thus, asymmetric cell potentials result from the active pumping of ions from one fluid compartment to another and are important in maintenance of ionic regulation. Examples of asymmetric action potentials are given for some electric organs in Chapter 17.

Activity Potentials. Many types of electrical activity are known in excitable cells, and the various categories are not entirely mutually exclusive. For practical purposes the following categories will be considered: (1) local or non-propagating potentials within single cells which may be passive (charging the membrane capacitance) or active (due to membrane conductance changes); (2) propagating, regenerative, all-or-none action potentials; and (3) junctional potentials, excitatory or inhibitory postsynaptic potentials (epsp's or ipsp's). The basis for a fourth category—changes in electrogenic ion pumps—was discussed previously (p. 94). In this chapter most of the examples of action potentials will be drawn from nerve tissues; muscle and electric organs are considered later.

Local Potentials and Electrotonic Conduction. An axon or striated muscle fiber resembles a cable with a core of low resistance, separated by a membrane of high resistance from a medium of low resistance. Local potentials, graded in amplitude and from whatever source, can be conducted passively and decrementally through the RC circuit of the cable. When a subthreshold current is passed through a nerve or muscle fiber, there is current spread according to the time and space constants of the membrane. This polarization or electrotonic potential is symmetrical in size but in reverse polarities at the cathode and anode. As the stimulus strength increases to about half rheobase (threshold for very long pulses), a local response or prepotential appears at the cathode. The prepotential is nonpropagated and graded, but it increases out of proportion to the stimulus strength; hence, it represents an active decrease in the resistance to inward current. At a critical amplitude, the local potential may trigger a conducted spike.

When a nerve impulse is blocked locally (as by cold or anesthetic), an electrotonic potential can be recorded just beyond the block. The currents which propagate an impulse will flow passively in a local circuit from depolarized to non-depolarized regions of a fiber. The pathway of current flow is forward through the cytoplasmic core, out through the non-depolarized membrane, and back through the external medium. The outward current depolarizes the membrane through which it flows. Increasing the resistance of either the core or the external medium (for example, by surrounding a fiber by oil or air) lowers the conduction velocity.

Many examples of graded or local potentials are given in following chapters. Each represents an increase in permeability for one or more kinds of ions. Pacemaker potentials represent graded rhythmic changes in resting potential, or they represent specific time- and space-dependent conductance changes; for example, heart pacemakers (p. 840). Sensory potentials in mechano- or chemo-sensitive nerve endings and generator potentials in sensitive epithelial cells trigger nerve impulses; examples for chemoreceptors are given on p. 560, for stretch receptors on p. 510, for phonoreceptors on p. 529, and for photoreceptors on p. 597. Junctional or synaptic potentials result from graded depolarizations or hyperpolarizations of postsynaptic membranes; synaptic potentials are discussed on p. 472, and neuromuscular junction potentials on p. 742.

The electrotonic spread of a local potential

carries charge according to the space constant of the membrane and, in short neurons, spikes are not necessary for information transfer to adjacent neurons. Electrotonic spread without spikes may transmit information from receptors to central nervous systems, e.g., in the short proprioceptive neurons of small crustaceans[27, 185] (Fig. 11–5) and in the second-order optic fibers of the eye of flies.[248] It is probable that in each of these tissues the electrotonic potential initiates the liberation of a chemical transmitter. In the vertebrate retina, several cells function in series, transmitting information only by graded slow potentials, which may be either hyperpolarizing or depolarizing (p. 626). This type of information transfer probably occurs in many situations where a neuron is activated at one end and its axon is not much longer than its space constant.

Electrotonic spread may be important for axonless neurons and for transmission from dendrites of one neuron to dendrites of another (e.g., in crustacean cardiac ganglia). The dendrites of some neurons appear not to be capable of generating spikes, yet they can transfer information, for instance between granule and mitral cells of olfactory bulb (p. 671),[198] and between highly branched dendrites of ventral thalamus.[199] Dendrites of Purkinje cells of the cerebellum of some species may not show spikes but may transmit synaptic signals electrotonically to the soma.[56] In pyramidal cells of the cerebral cortex, a graded depolarization in a dendrite may be attenuated in the cell soma yet be sufficient to modulate the firing rate of the cell.[96]

It is concluded that electrotonic spread of graded depolarization (or hyperpolarization) is widely used for information transfer in the absence of unitary impulses; furthermore, the sum of all positive and negative electrotonic potentials arriving at a spike-generating site determines the firing rate of the cell.

Conducted Action Potentials (Regenerative Spikes). When the stimulus to an electrically excitable cell is of threshold size, the local potential reaches a critical amplitude and out of it arises a propagated impulse (Fig. 11–6). The critical height of the local potential is about 20% of spike height in crustacean nerve fibers. The ratio of spike height to threshold depolarization level is called the safety factor. The spike (nerve or muscle impulse) is unlike current in a passive conductor in that it is self-propagating (regenerative), shows no decrement from its source (is all-or-none), and is followed by a refractory period. Gradation of information

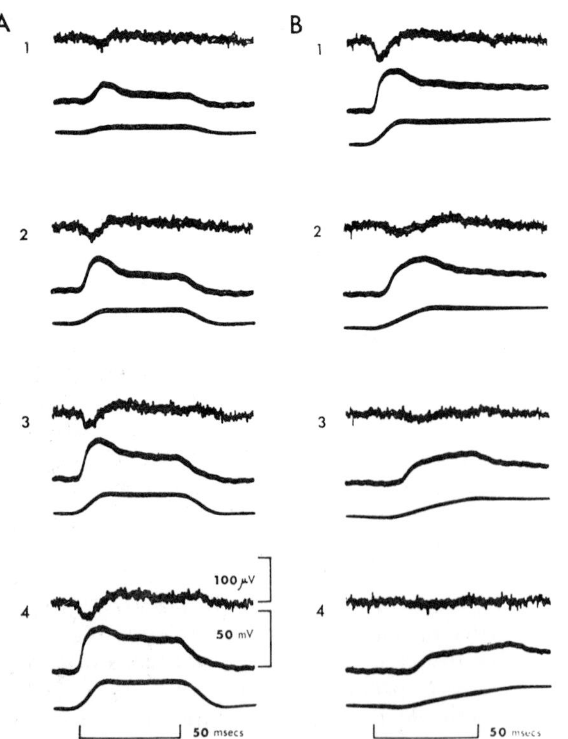

Figure 11–5. Responses in sensory fibers recorded intracellularly (middle traces) and extracellularly (upper traces) to mechanical stimulation (lower traces) of different amplitudes, A, and rates of stretch, B, of a muscle receptor of a crab. Responses are generator potentials which spread electrotonically to activate interneurons in central ganglion. (From Ripley, S. M., M. M. Bush, and A. Roberts, Nature *218*:1170–1171, 1968.)

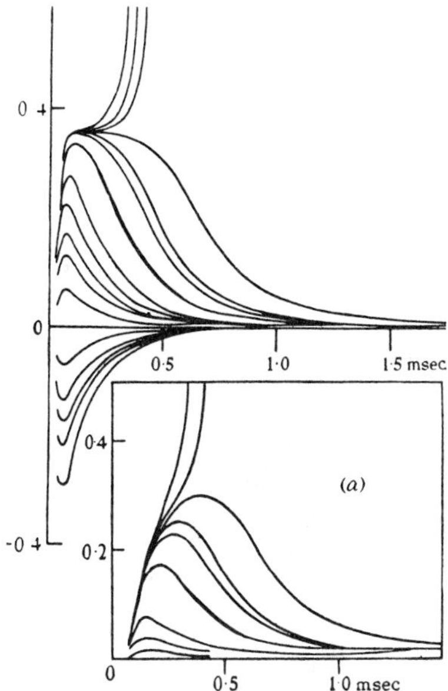

Figure 11–6. Electrical changes in crab nerve at the stimulating electrodes to shocks of increasing intensity. Electrotonic potential at anode (below zero) and in first five records at cathode (above zero); local response in upper records 6 to 9 and beginning of spike in upper three responses. Inset *a*, another series at cathode. (From Hodgkin, A. L., Proc. Roy. Soc. Lond. B *126*:87–121, 1938.)

by spikes in single cells is not by amplitude (as in electrotonic conduction) but by pulse frequency and pattern. Excitation normally consists of an increase in ion permeability expressed as increased conductance (g). During the rising phase of a nerve impulse, the membrane resistance drops by some forty times while the capacity is unchanged. Conduction occurs in the absence of oxygen, and the enzymes of the axoplasm are not essential for conduction. When the axoplasm is squeezed out of a squid axon and the fiber is perfused with an appropriate concentration of potassium and a non-penetrating anion such as sulfate or isethionate, normal impulses are conducted by the membrane.[10] However, in such a fiber the ion pumps are not functional; these require the oxidative machinery of the axoplasm.

A useful technique for identifying the ionic currents across an electrogenic membrane is voltage clamping. By means of two electrodes inserted into a large cell (e.g., a squid giant axon), one for passing current and the other for recording voltage, and with a feedback amplifier, the current required to "clamp" membrane potential at various voltage levels (depolarization or hyperpolarization) can be measured. The total membrane current I_m consists of two parts—that needed to charge the membrane capacitance (C) and that due to ionic movement (I_i).

$$I_m = C\frac{dV}{dt} + I_i$$

In clamp conditions $\frac{dV}{dt} = 0$ and $I_m = I_{\text{ionic}}$. The ionic current in nerve is the sum of sodium, potassium, and leak (mostly chloride) currents; each is determined by the conductance for the specific ion and the potential, which is given by the reversal potential gradient across the membrane. The reversal potential for each ion is the equilibrium potential given by the concentration gradient of the ion according to the Nernst equation (see p. 4).

The sodium conductance is $g_{Na} = \frac{I_{Na}}{V - V_{Na}}$

The potassium conductance is $g_K = \frac{I_K}{V - V_K}$

The concentration gradient for Na is inward and that for K is outward. Current measurements in clamp conditions, together with ion substitutions, indicate that in axons during an action potential the membrane permeabilities for Na and K both increase, but at very different rates (Fig. 11–7). During the rising phase of an action potential, the membrane permeability to Na is increased (rise in g_{Na}) and Na^+ moves into the axon since $[Na_o] > [Na_i]$ and $I_{Na_{in}} > I_{K_{out}}$; the overshoot represents an approach to the sodium equilibrium potential. The amplitude of the action potential is reduced when sodium in the medium is replaced by a non-penetrating cation, such as choline or Tris; in some nerves, Li, amines and quaternary nitrogen cations can substitute for Na^+. During the descending phase of the action potential, sodium current inactivation occurs; i.e., Na conductance falls and $I_{K_{out}} > I_{Na_{in}}$. Potassium conductance increases at a much slower rate than g_{Na} so that when sodium inactivation begins, potassium conductance is high and outward current of the falling phase of the spike is carried by K^+.[29] The essence of a nerve impulse resides in changes in Na conductance (g_{Na}) and K conductance (g_K), which are voltage-dependent and time-dependent. Tracer measurements for squid axons show an inward transfer of 4 pM

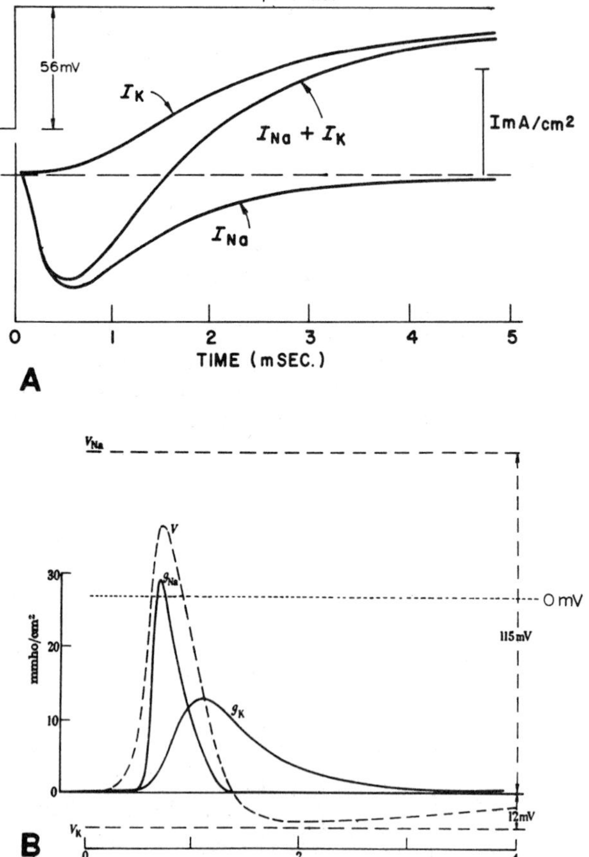

Figure 11–7. *A*, current due to movement of sodium and potassium in squid giant axon when membrane voltage is displaced (depolarized) by 56 mv by "clamping" electrode. $I_{Na} + I_K$ = current in balanced ionic medium. I_K = potassium current when Na replaced by choline, I_{Na} = obtained by difference. Inward current downward, outward current upward.

B, spike (*V*) in squid giant axon; Na^+ conductance (g_{Na}) and K^+ conductance (g_K) calculated from current during depolarization and repolarization. V_{Na} and V_K are Na^+ and K^+ equilibrium potentials. (From Hodgkin, A. L., Proc. Roy. Soc. Lond. B *148*:1–37, 1958.)

Na^+/cm^2 per impulse and an equivalent outward flux of K^+; also, Ca^{++} enters at the rate of 0.006 pM/cm^2 per impulse.[116, 117] The permeability for Na^+ increases a thousandfold, and the peak Na conductance is 0.53 mho/cm^2 in lobster giant axons and 0.12 in squid axon.[158, 159] After the peak, the conductance for outward (i.e., the late or K^+) current increases, while that for inward (i.e., early or Na^+) current decreases.

The electrically excitable channels of inward Na current are blocked specifically by poison from the puffer fish, tetrodotoxin (TTX), and from the amount required to block an impulse and the assumption of one molecule per Na channel it is estimated that there are 13 TTX-sensitive patches per μ^2 of membrane. The toxin, saxitoxin, from certain clams also blocks the conducting channels for Na; its action is irreversible, whereas that of TTX is reversible.[104, 159, 167] Tetraethylammonium (TEA) specifically blocks outward K current and thus prolongs the action potential; with this treatment an inward Cl current can permit some repolarization. The poison insecticide DDT keeps Na channels open (i.e., it delays Na inactivation) and also delays the increase in g_K; hence, it keeps the membrane depolarized.[89] A variety of evidence, particularly the effects of TTX and TEA, indicates that Na^+ and K^+ move in different channels. Conductances for various univalent cations at the peak inward current of squid axon show the following relative permeabilities:[225]

Li	Na	K	Rb	Cs
1.1	1	1/12	1/40	1/60

After passage of an action potential, the resting ionic gradients are reestablished by the action of the Na-K pump. The rate of Na influx during a spike in an axon in which axoplasm has been replaced by K_2SO_4 is 10^6 times greater than is that of extrusion of Na by the pump in a resting axon.[89]

Calcium conductance also increases during a spike in an axon, but by only 1/600 of the increase of Na conductance. A resting squid axon normally contains 0.3 μM Ca *ions* and total Ca concentration of 400 μM; it shows a 500-fold increase in intracellular Ca^{++} after 50 to 100 min in cyanide. When the axon is stimulated at 50 to 200/sec there is a slow increase in Ca_i; recovery after cessation of stimulation has a time constant of 10 to 30 sec.[7] The increase in Ca^{++} conductance is in two phases; the first is sensitive to TTX and may be via the Na channels. The late phase of Ca^{++} entry is at the time of maximum K^+ current but is not affected by TEA; it is reduced by Mn^{++}.[7]

In some cells, the Ca^{++} current is more than that carried by Na ions. For example, in giant neurons of the mollusc *Aplysia,* replacement of either Na or Ca in the medium with an ion such as Tris causes partial reduction of the overshoot; replacement of both Na and Ca abolishes the spike. The Na^+ component is blocked by TTX, and the Ca^{++} component is blocked by cobalt; in a Ca-free medium the response is determined by Na^{++}, and in an Na-free medium it is determined by Ca^{++}.[73] In *Helix,* the neurons of one class depend on Na for the spike overshoot, other neurons depend on both Na and Ca for inward current, and still others depend on Ca only.[108, 110]

In some muscle fibers (vertebrate striated), spikes result from sodium currents. In others (vertebrate intestinal smooth muscle, crustacean striated), spikes are due predominantly to calcium conductance changes. In some muscles (e.g., *Branchiostoma,* p. 744), elimination of sodium current by TTX or by replacement of external Na, and elimination of potassium current by procaine or by TEA, reveals a significant calcium current.[85b] In some vertebrate cardiac muscles, an inward calcium current occurs during depolarization.[12] In frog atrial muscle, epinephrine increases both a fast inward Na current and a slow inward Ca current.[235] It is probable that most conducting membranes have several ionic channels. The normal sodium and potassium currents are in different membrane channels, and calcium channels are probably present in most cells.

Calcium has other effects on excitable membranes. It maintains the normal membrane permeability or resistance (i.e., it stabilizes membranes), and changes in $[Ca_o]$ can shift the curves of Na-K conductance along the voltage axis.[118]

Another method of generating a depolarizing response is to increase the outward I_{Cl}. In some plant cells, such as filamentous large algae, electrical stimulation elicits an action potential which is due to increased g_{Cl}. Some crustacean muscles have an extensive system of intracellular tubules, and chloride efflux into the infolded extracellular space results in depolarization.[81] Similarly, one surface of an electrocyte of a skate, upon synaptic stimulation, shows increased outflux of Cl^-. If E_{Cl} is positive to RP, an increase in g_{Cl} causes efflux of Cl^- and depolarization, even though the chemical concentration gradient of Cl is inward. Repolarization in heart muscle is partly due to inward I_{Cl}.[67] In the smooth muscle of guinea pig taenia coli, epinephrine causes hyperpolarization by opening channels for both Cl^- and K^+; 36% of the increased conductance is for Cl^-.[177]

Late Events and Generator Potentials. Many nerve fibers show an undershoot or hyperpolarization following the spike (Fig. 11–8). This is prominent (e.g., in squid giant axons), and persists for a tenth of a second or longer in vertebrate non-myelineated fibers after tetanic stimulation. Undershoot was early observed as a positive after-potential recorded by extracellular electrodes. Tracer and other experiments indicate that the after-hyperpolarization results from continued high potassium conductance. Reduction of the potassium gradient K_i/K_o in motorneurons abolishes their after-hyperpolarization.

In other excitable fibers, following the spike and delaying its downward deflection,

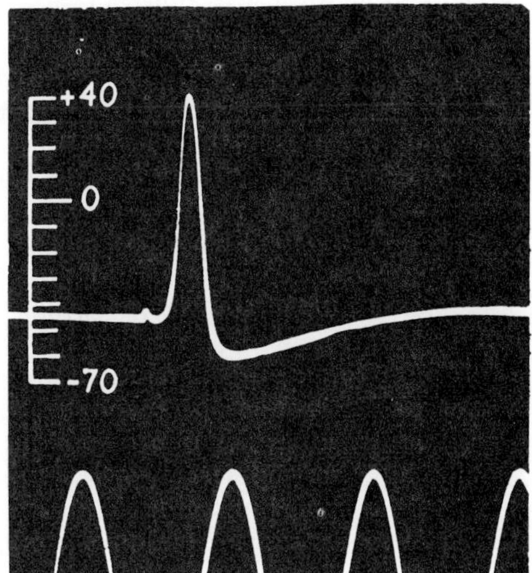

Figure 11–8. Action potential spike recorded between inside and outside of giant axon of squid showing resting potential level, overshoot of spike, and positive undershoot; time calibration 1000/sec. (From Hodgkin, A. L., and Huxley, A. F., J. Physiol. *104*:176–195, 1945.)

is an after-depolarization, observed by extracellular electrodes in some nerves as a negative after-potential. The prolonged depolarization may result in a plateau, as in heart muscle (p. 841). It is prominent in large myelinated fibers of mammals and may last for seconds in crab fibers. Repolarization is delayed in myelinated fibers by veratrine, and in many fibers by TEA. These agents cause a delay in the potassium current.

Another type of after-hyperpolarization results from sodium pump action. Components of an action potential which are due to changes in Na pump activity can be identified by lack of change in membrane conductance, by enhancement due to injection of sodium, by dependence on external potassium, and by blocking with ouabain. For example, some leech neurons show a prolonged negativity following a train of spikes, and the slow negativity is abolished by ouabain.[168] Similar hyperpolarization due to active sodium transport occurs in stretch receptors (p. 511).

There is evidence that besides sodium and calcium activation during depolarization, there are in some tissues early conductance changes for K^+. When some neurons of molluscs (*Anisodoris*) are hyperpolarized and then given a step of depolarization, an early outward potassium current (I_A) is activated; this is distinct from the initial spike current (I_{Na}) or the recovery current (I_K). The early K current delays or slows the effect of continuing applied current, but ultimately the inward Na current exceeds I_A and a spike is initiated. It is postulated that in neurons which discharge repetitively, the hyperpolarization following each spike permits activation of the early K current, which then determines the time before the next spike. The early K current is by a different channel from the K current of spike repolarization.[37]

Pacemaker cells of vertebrate heart have a high resting Na conductance relative to K conductance, which provides a persistent inward current. These conductances are time- and voltage-dependent. By appropriate coupling of these two conductances together with the threshold for spiking, repetitive firing can be obtained.[226] Some neurons of *Aplysia* show a high resting P_{Na}/P_K and their repetitive firing may be determined much as vertebrate cardiac pacemakers; however, other neurons are repetitive in the absence of Na^+ in the medium.[28]

Many mechanoreceptors give graded responses which are due to increases in sodium conductance, even though the receptor membrane does not give spikes. In some invertebrate photoreceptors (*Limulus*, barnacle) illumination triggers an increase in g_{Na} (see p. 602). Vertebrate photoreceptors in the dark have a high P_{Na}/P_K; they show a sodium current generated at the distal end of the receptor, which can be recorded extracellularly. Illumination decreases the dark current and hyperpolarizes the receptor (see p. 597). Some scallop photoreceptors have a RP lower than E_K, and a high P_{Na}/P_K. Illumination causes a hyperpolarizing response which is due to an increase in g_K.[151a]

JUNCTIONAL POTENTIALS AT CHEMICAL SYNAPSES

Postsynaptic membranes differ from those of conducting axons or muscle fibers and from pacemaker or sensory membranes. At syanaptic junctions, graded potentials precede the initiation of impulses; these are either (1) excitatory (usually depolarizing) junctional or postsynaptic potentials (epsp's) or (2) inhibitory (usually hyperpolarizing) postsynaptic potentials (ipsp's). Junctional potentials can be recorded in the absence of postsynaptic spikes after impulse initiation has been blocked by drugs (TTX or procaine) or by fatigue (as after presynaptic tetanization in squid stellate ganglion) (Fig. 11–9). The

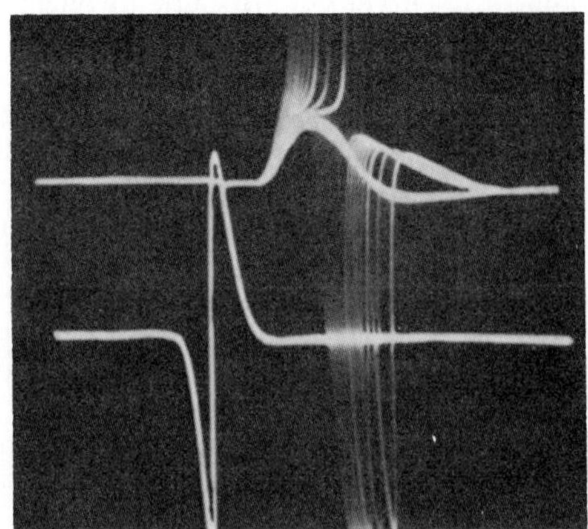

Figure 11–9. Synaptic potentials and action potential spikes from stellate ganglion of squid. *Lower record*: impulse in preganglionic fiber. *Upper record*: postganglionic spike developing later and later out of synaptic potential as fatigue occurs, finally leaving only synaptic potential. (From Bullock, T. H., Nature *158*:555–556, 1946.)

epsp lasts longer than the refractory period of the presynaptic impulse; hence, repeated impulses can cause temporal summation. A postsynaptic impulse appears earlier as the epsp increases with summation, or later as it decreases with fatigue. Synaptic potentials are graded according to the amount of transmitter liberated. At many unstimulated junctions (e.g., squid stellate ganglion, skeletal muscle endplates) the postsynaptic membrane shows spontaneous miniature junction potentials (mepsp's) of a unit amplitude. These are taken as indicating liberation of packets or quanta of transmitter molecules;[153, 154, 155] each epsp results from a number (usually several hundred) of quanta of transmitter.

In all known examples, transmitter release requires (a) some extracellular calcium and (b) depolarization of presynaptic endings; high magnesium and zero calcium diminish transmitter liberation. For example, in the giant fiber synapse of the squid stellate ganglion, an increase in calcium in the medium increases synaptic current and an increase of Mg decreases it. Presynaptic depolarization of 25 mv elicits an epsp, and mepsp's appear even in the presence of TTX, which blocks presynaptic impulses. Ca^{++} is required for the coupling between depolarization and secretion of transmitter.[129, 155] In endplates of frog sartorius, strontium can substitute for but is much less effective than calcium.[152] Measurements of quanta released by nerve endings in frog sartorius in different calcium concentrations indicate that $4Ca^{++}$ must be present at a critical position, and the number of quanta of ACh released is proportional to $(CaX)^4$, where X is an activated molecule. The number of quanta per epsp gives the fraction of X complexed with Ca^{++}.[51] In rat diaphragm, epsp's average 2.7 Ca^{++} per quantum, and miniature psp's average one Ca^{++} per quantum. In mouse diaphragm, there is evidence for cooperativity, so that the liberation of quanta accelerates with increasing Ca^{++} and increases more in high than in low K. Electrical recordings from presynaptic terminals in squid stellate ganglion indicate a regenerative electrical response at the terminal which results from an influx of Ca^{++}. This regenerative terminal action potential is normally repolarized by K current, but in the presence of TEA the Ca^{++} potential is a plateau which varies with Ca_o, is reduced by Mn^{++}, and is accompanied by a reduction in R_m.[105, 106] It is concluded that depolarization at a presynaptic terminal increases Ca conductance, and that inward Ca^{++} current elicits release of transmitter from vesicles.

Synaptic delay comprises the time for liberation and diffusion of transmitter and for its reaction with the receptor in the postsynaptic membrane. In the stellate ganglion of the squid, the epsp begins after a delay of 0.5 msec, reaches a peak at 1 msec, and has a decay constant of 1.2 msec.[24] In large ganglion cells of the sea slug *Aplysia*, the rise time of the epsp may be 100 msec and the decay time may be 500 msec.

Postsynaptic membranes are inexcitable electrically; i.e., the membrane current depends on the amount of chemical transmitter. Transmitter reacts with receptor membrane to increase the conductance for certain ions. If the postsynaptic membrane is made more negative than the equilibrium potential for these ions, an epsp is increased; or, if it is made more positive, the epsp is reversed. Converse effects occur for an ipsp, which is hyperpolarizing at membrane potentials that are positive to equilibrium.

A voltage-current curve for a postsynaptic (chemically excitable) membrane shows a linear relation between current and voltage; i.e., there is no rectification, and membrane resistance is ohmic or independent of membrane potential.

Excitatory Junctional Potentials. A nerve spike in electrically excitable membrane represents a sequential increase in conductance for Na (or Ca) and K, whereas epsp's usually result from a simultaneous conductance increase for cations—both Na and K. In a spinal motorneuron, the epsp decreases in amplitude as the membrane is clamped in a depolarized direction, and the equilibrium potential is close to zero.[44] In muscle endplates of vertebrates, the equilibrium potential is −15 mv and the epsp depends on simultaneously increased permeability for both Na and K. In a squid giant synapse, the epsp approaches the Na equilibrium (+45 mv) and there is no potassium current.[70] In frog sympathetic ganglion the reversal potential is −8 to −20 mv, much as in a motor endplate.[172, 173]

Excitatory and inhibitory equilibrium or reversal potentials measured with an electrode in a neuron soma may give an inaccurate estimate of the true synaptic reversal potential, especially when synapses are at a distance, as on dendrites or lateral axons. At all junctions, the portion of membrane which changes resistance is in parallel with some membrane which is not altered by transmitter.[80]

Several postsynaptic impulses may be initiated by a prolonged epsp; for example, in certain interneurons (Renshaw cells) of cat spinal cord, the epsp normally lasts 50 msec, but after treatment with eserine, which blocks acetylcholinesterase, the epsp may last 300 msec, during which time the cells discharge at high frequency. At other junctions, single impulses are initiated in postsynaptic cells by synchronous compound epsp's. In the two-neuron reflex arc of the cat spinal cord, the epsp in a motorneuron begins within 0.5 msec of the entry into the cord of a sensory volley; the peak is reached in 1 to 1.5 msec and the potential declines with a time constant of 4.7 msec.[38] The current rises rapidly and peaks before the voltage does (Fig. 11–10A). At 10 mv depolarization of a motorneuron with RP of −70 mv, a postsynaptic spike is generated in the initial segment of the neuron. The epsp elicited in the soma-dendrite provides the depolarization which starts a spike in the initial segment of the axon; this spike may later invade the soma-dendrite, where the threshold for spiking is 30 mv depolarization.[54, 55] In contrast to motorneurons, in Purkinje cells of the cerebellum of some species and in some pyramidal cells of cerebral cortex, spikes may be initiated synaptically in the dendrites. Junction potentials may spread extensively over the branches of a dendritic tree, as in crustacean monopolar neurons.

In cervical sympathetic ganglia of cat, rabbit, frog, and turtle, three distinct junction potentials occur in proportions varying with species. The excitatory transmitter is acetylcholine. An initial excitatory psp (cat) has latency of 4 to 9 msec and decay time of 6 to 12 msec, and appears to be referable to nicotinic (tubocurarine-sensitive) receptors. In rabbit, a late excitatory psp has latency of 200 to 300 msec and is blocked by muscarinic blocking agents (atropine). An inhibitory psp with latency of 35 msec is blocked by the alpha adrenergic blocking agent dibenamine, and appears to be the result of cholinergic activation of chromaffin cells, which in turn liberate norepinephrine on sympathetic ganglion cells. All three synaptic potentials are blocked by botulinum toxin and by lowered Ca/Mg, which depress ACh liberation from preganglionic fibers.[133, 134]

In frog (probably also in turtle) there are few or no chromaffin cells, and recordings from single neurons show a different mechanism for the slow psp's from those postulated for cat and rabbit on the basis of extracellular recordings. The fast responses of frog are nicotinic as in mammals. One cell type shows a slow epsp which appears to be due to inactivation of a resting potassium conductance, i.e., an increase in R_m and decrease in g_K. Another cell type shows a hyperpolarizing response to ACh, which is due to a decrease in Na conductance, i.e., a shift away from E_{Na}.[237]

It is evident that depolarizing excitatory postsynaptic potentials in general are variants on increases in Na and K conductance changes.

Inhibitory Postsynaptic Potentials. An inhibitory transmitter usually causes a conductance change in receptor membranes, which results in an inhibitory junctional potential, and which reduces responses to excitatory psp's; usually the latency is longer, and persistence of an ipsp is longer than for an epsp. An ipsp is often hyperpolarizing, but it may be depolarizing if the resting membrane potential is made negative to the ipsp equilibrium potential. The reversal from hyperpolarizing to depolarizing ipsp is comparable to the reversal of an epsp at a resting potential more positive than the equilibrium potential for the cations causing the epsp.

Crustacean muscles receive both excitatory and inhibitory innervation. In some crustacean muscles the inhibitory endings are postsynaptic, and inhibitory impulses increase muscle membrane conductance for anions; in other muscles inhibition is presynaptic, and reduces epsp's without effect on the muscle membrane.

The various equilibrium potentials as calculated from ion concentrations and from psp reversal voltages for cat spinal motorneurons follow:[3, 56]

E_m (RP)	−70 mv	E_{Cl}	−80 mv
E_{Na}	+40 mv	E_{epsp}	−10 to 0 mv
E_K	−90 mv	E_{ipsp}	−80 mv

In cat motorneurons and goldfish Mauthner cells the principal ion for an ipsp is Cl^-; a small amount of the current is carried by K^+.[56] Raising the intracellular concentration

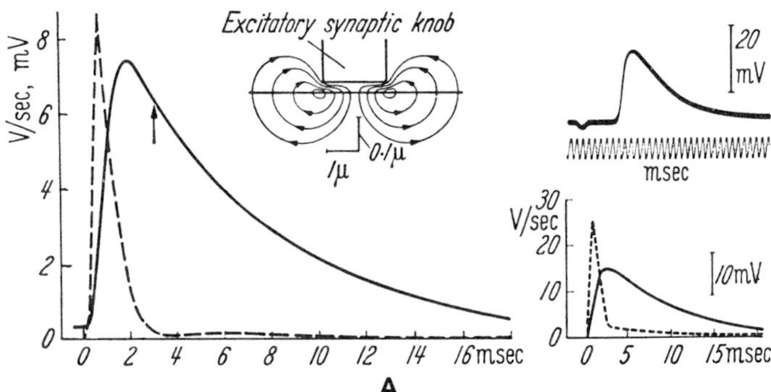

Figure 11–10. A, Excitatory postsynaptic potential of a motorneuron (graph at left, solid line) and the postsynaptic current (dotted line) that generated it. Inset shows current relative to activated synapse; vertical scale is exaggerated tenfold. Upper right, recorded epsp in frog sympathetic ganglion cell; lower right, analysis of epsp and postsynaptic current as at left. B, Inhibitory postsynaptic potential of a motorneuron (solid line) and postsynaptic current (dotted line). (From Eccles, J. C., The Physiology of Synapses, Springer Verlag. Berlin, 1964.) C, Reversal of ipsp and epsp in motorneurons of cat trochlear nucleus induced by ipsilateral vestibular nerve stimulation. At left, reversal of hyperpolarizing response by infusion of Cl^-; at right, epsp reduced and then reversed when increasing depolarizing currents were applied intracellularly. (From R. Llinas & R. Baker, J. Neurophys. *35*:484–492, 1972.)

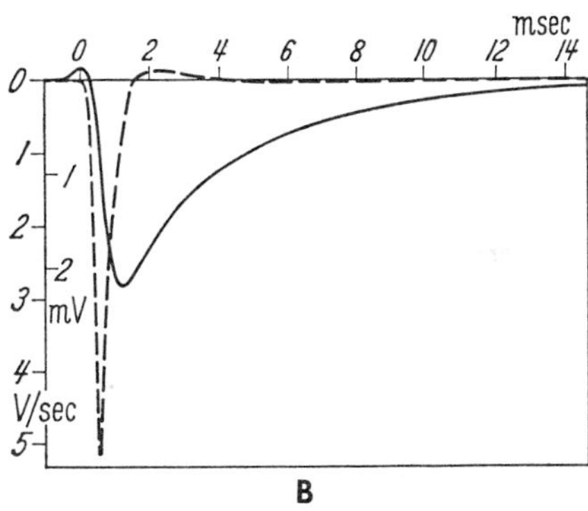

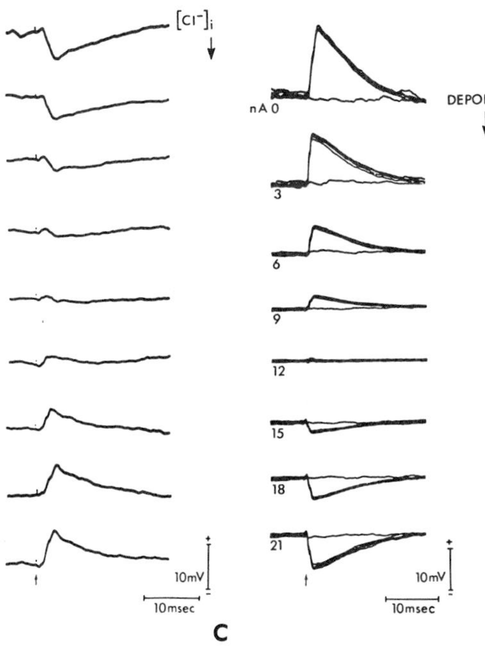

of Cl^- in a spinal motorneuron by iontophoretic methods using an intracellular electrode reduces the inward gradient, and the ipsp can be reversed (Fig. 11–10C). If a series of anions is tested on spinal motorneurons, it is found that anions smaller than ClO_4^- can enter and thus cause a hyperpolarizing ipsp; larger anions cannot do this.[56] Thus, the inhibitory transmitter opens conductance channels for small anions. Similar results were obtained with Mauthner cells of fishes.

Mollusc psp's. In molluscan ganglia, specific neural inputs elicit hyperpolarizing ipsp's, and other inputs can elicit depolarizing epsp's in the same neuron (Fig. 11–11).

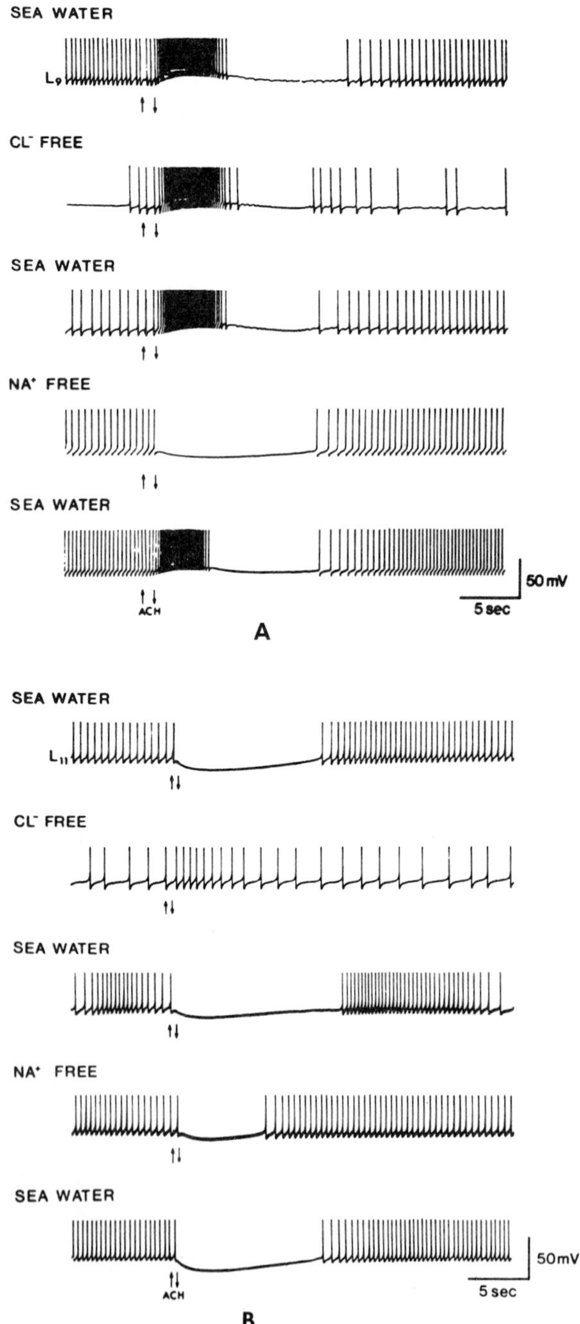

Figure 11–11. Ionic mechanisms of synaptic excitation (A) and inhibition (B) on D and H cells of abdominal ganglion of *Aplysia.* Effect of excitation input abolished in Na-free medium, and effect of inhibitory input abolished in Cl-free medium; in D cell, removal of Na causes reversal, and in B, removal of Cl causes reversal of responses. (From Kandel, E. R., pp. 385–398 *in* Electrophysiological Studies in Neuropharmacology, edited by W. Koella. Charles C Thomas, Springfield, Ill., 1968.)

A given transmitter (acetylcholine) causes epsp's in some cells (D cells) and ipsp's in others (H cells).[101a] Non-cholinergic agents may be effective at other synapses; e.g., dopamine, which at as low a concentration as 10^{-11} M depolarizes some neurons, and 5HT (serotonin), which depolarizes other neurons in a land snail, and an unknown transmitter which hyperpolarizes for many seconds (inhibitor of long duration).[33, 74–79] A variety of ionic conductance changes, indicating the diversity of membrane receptors and cell types, are shown in Table 11–2.

In some D cells the action of acetylcholine is to depolarize by increasing the Na conductance, and the equilibrium potential is close to zero.[114] However, if Na is reduced, the depolarization is sensitive to Mg^{++}, which may carry inward current.[207] In H cells of the same ganglia, acetylcholine hyperpolarizes by increasing g_{Cl}, and the response is eliminated if Cl^- is replaced by $SO_4^=$. When various anions were injected into these *Helix* neurons by iontophoresis, it appeared that those anions smaller than BrO_3^-, corresponding to a pore size of 3 Å, could carry the inward hyperpolarizing current. Apparently, 90% of this current is normally carried by Cl^-, and 10% by efflux of K^+.[114, 115] In a DINH neuron of another snail, *Cryptomphallus*, hyperpolarization is insensitive to Cl^-; but when potassium is removed from the medium the ipsp doubles, so it is due to K^+ efflux.[76]

The chloride content in H cells is lower than that in D cells (in *Helix*[110] and in *Aplysia*[207]). This favors an influx of Cl^- in hyperpolarization of the H cells. The RP of D cells in *Cryptomphallus* is more negative than E_{Cl}, and apparently the high chloride concentration in D cells is maintained by an inward Cl^- pump; depolarization can result from Cl^- efflux if the transmitter changes g_{Cl} slightly,[33] even against a concentration gradient, if the electrical gradient favors efflux[101a] (see p. 469).

In a still different cell type in *Helix*, hyperpolarization is caused by dopamine and is insensitive to K^+ and Cl^-, but is abolished by ouabain (which blocks Na^+ pumps).[110] Effects of inhibitory impulses on Na^+ pumps have been suggested but have not been confirmed.

It was mentioned previously (p. 469) that molluscan neurons differ as to whether the spike current is carried by Na^+, Ca^{++}, or both. It may now be added that postsynaptic

TABLE 11–2. Summary of Some Postsynaptic Actions in Molluscan Neurons

Animal and Cell Type	*Equilibrium Potentials for ACh and for Postsynaptic Responses, mv*	*[Cl_i], mEq*	*RP, mv*	*Mechanisms*
Aplysia[207]				
D cell	E_{ACh} −14	12	−36	↑ $g_{Na\ in}$
H cell	E_{ACh} −60	43	−36 to −48	↑ $g_{Cl\ in}$
Aplysia[71]				
D cell	E_{epsp} −14			↑ $g_{Na\ in}$
H cell	E_{ipsp} −74		−60	↑ $g_{Cl\ in}$
diphasic cells	depolarize to −14, then hyperpolarize to −63			↑ g_{Na}, then ↑ g_{Cl}
Helix[109, 110, 114, 115]				
D cell	E_{ACh} −32 E_{epsp} −34	24.7	−50	↑ $g_{Na\ in}$
H cell	E_{ACh} −72	11.2	−55	↑ $g_{Cl\ in}$
ILD cell (inhibitor of long duration) dopamine hyperpolarizes to −80; E_{ipsp} = −50				↑ $g_{K\ out}$
Cryptomphallus				
D cell	E_{ACh} −25.1		−47	↑ $g_{Cl\ out}$
H cell	E_{ACh} −56.5 E_{ipsp} −55.8			↑ $g_{Cl\ in}$
Long depolarized cell	E_{ACh} ~0			↑ $g_{Na\ in}$
Long inhibited cell, non-cholinergic	E_{ipsp} −80			↑ $g_{K\ out}$

membranes are selectively sensitive to several transmitters and that some of the ionic responses consist of increases in conductance for Na, K, Cl, possibly Mg, and possibly activation of an Na pump.

SUMMARY OF CONDUCTANCE CHANGES IN MEMBRANES

Electrically inexcitable membranes are those which respond only passively to electrical stimuli; normally such membranes are excited chemically by synaptic transmitters, or (in sensory membranes) by specific chemicals, mechanical deformation, or illumination.

This class of membrane acts as an ohmic resistor; that is, a plot of voltage (E) versus current (I) is linear (Fig. 11–12).[83, 84, 85] However, the slope of such a plot can change during activation, and the membrane potential response reverses when E_m is greater than the reversal potential specified by the intersection of the E/I lines for passive and active states (Fig. 11–12). Thus, synaptic potentials reverse for either depolarizing or hyperpolarizing responses.

Electrically excitable membranes show regenerative responses, and their E/I curves show nonlinear regions on either depolarization or hyperpolarization (Fig. 11–12). When the current increases disproportionately to voltage, the membrane is said to be activated; conductance increases and the response is regenerative. Most electrically excitable membranes show depolarizing electrogenesis, which can be either graded or all-or-none. Disproportionate increase in current on hyperpolarization, known as hyperpolarizing activation, is less common than depolarizing activation. If the nonlinearity is a deviation toward disproportionate increase in voltage, the response is said to be membrane inactivation. Most depolarizing spikes involve conductance changes in Na or Ca, for which the equilibrium potentials are highly positive. Inactivation is a decrease in conductance, and may occur with either depolarizing or hyperpolarizing currents (Fig. 11–12).

The non-linear regions of an E/I curve, or the conductance increase or decrease, correspond to a decrease or increase in membrane resistance in one direction; in engineering terms, this is rectification. The decrease in resistance or increase in conductance during the depolarization is sometimes called normal rectification; it occurs in axons. A delay in repolarization after an action potential results in a plateau, and when membrane resistance increases during the plateau it is called delayed rectification. After certain ionic treatments, some cells (muscle, electrocytes) may show an increase in resistance during depolarization or a decrease during hyperpolarization; this is anomalous rectification. It has been described for frog muscle in K_2SO_4; it occurs in the soma of a molluscan neuron which is excited at some distance on a process. Anomalous rectification also refers to rectification during hyperpolarizing currents. Physiologically, it is more useful to refer to

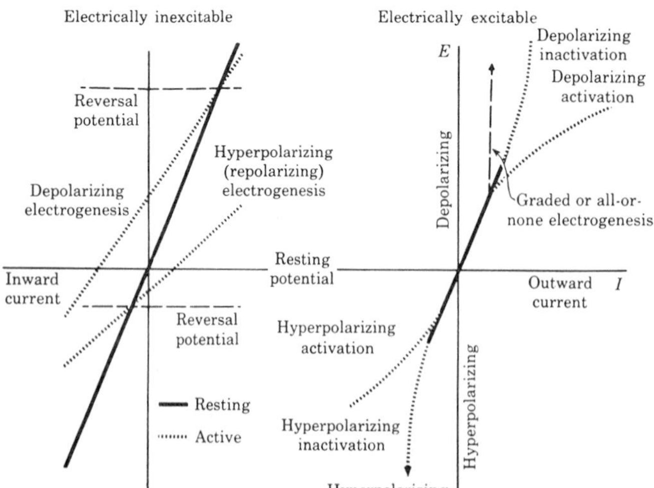

Figure 11–12. Voltage (E)/current (I) relations of electrically inexcitable and electrically excitable membranes. (From Grundfest, H., pp. 353–371 *in* The Neurosciences; A Study Program, edited by G. C. Quarton et al., Rockefeller University Press, N.Y., 1967.)

depolarizing and hyperpolarizing responses as conductance changes than as types of rectification.

A comparative survey of electrogenesis shows that, in different tissues of one animal, several mechanisms may be used and that different animals use a wide range of ionic types of electrogenesis. The following summary lists ionic mechanisms in some familiar examples.

(1) Sodium conductance increase during a depolarizing response probably occurs in all nerves and in fast muscles of vertebrates. Tetrodotoxin blocks the channel for inward fast ionic current carried by Na^+.

(2) Potassium conductance increases more slowly and peaks during the falling phase of an action potential in nerve and in many muscles. Tetraethylammonium delays the K activation and results in a plateau-type potential. Delayed K activation is normal in earthworm giant fibers, vertebrate cardiac muscle, and ureter. Continued K activation after repolarization may result in an undershoot or hyperpolarization, e.g., in squid axon.

(3) In some repetitively firing molluscan neurons, an early K current (by a different channel from that of spike repolarization) opposes a continuing inward Na current and sets the interval between spikes.

(4) Calcium conductance increase during depolarization occurs widely in spike electrogenesis of crustacean muscle and of some vertebrate smooth muscles. Significant increase in Ca conductance occurs in parallel with the Na conductance changes in some molluscan neurons,[72] in amphioxus muscle,[85b] and in some mammalian smooth muscles such as ureter. Significant inward Ca current occurs in frog heart.

(5) High P_{Na}/P_K ratio, and time- and voltage-dependent changes in Na and K conductance, form the basis for some oscillatory pacemaker potentials, as in vertebrate heart.

(6) Electrically excitable membranes of the internal tubule system of crayfish and lobster muscle, and one surface of a skate electrocyte and probably some mammalian smooth muscles, show increased outflux of Cl, and hence a depolarizing spike due to increased g_{Cl}. These responses are large in a medium low in Cl^-.[81] In the plant *Chara*, spikes are normally caused by increased Cl^- efflux.

(7) In at least one insect muscle (*Tenebrio*) the RP is relatively insensitive to K_o, and in a high K medium, activation causes increased g_K and a depolarizing response due to K influx.[13] Similarly, for a chemically excited membrane (frog sympathetic ganglion), a slow epsp results from decrease in g_K.[237]

(8) Many chemically excited postsynaptic membranes show depolarizing responses of simultaneous increase in Na and K conductance, as in the motor endplates of vertebrate striated muscle.

(9) Some postsynaptic responses to specific transmitters consist of increased conductance and influx of Cl^-; since E_{Cl} is often more negative than RP, the response is hyperpolarizing. This is the usual response at inhibitory synapses.

(10) In some molluscan neurons where E_{Cl} is less negative than RP, a depolarizing epsp may result from increased Cl^- conductance.

(11) A hyperpolarizing response due to increased g_K occurs in certain photoreceptors of a scallop, where the RP is less negative than E_K and P_{Na}/P_K is high at rest, low on stimulation.

(12) High resting sodium current may be turned off by stimulation, as in vertebrate retinal rods in the transition from darkness to illumination.

(13) In flat asymmetric cells, one surface may have different electrical properties from the opposite surface, and asymmetrical action potentials, which can add in series, are produced. Several different patterns of electrogenesis by asymmetric electrocytes will be described in Chapter 17.

(14) Electrogenic ion pumps generate currents in the direction of ion transport. Potentials are developed if the pump is not symmetrical in ion exchange. Na pumps contribute to some resting potentials, and tissues which have gained Na and lost K during storage in cold show marked negativity when restored to high temperature as the Na pump restores normal ionic gradients. Some central neurons (e.g., in leech ganglia) and some receptors (e.g., crayfish stretch receptors) show hyperpolarization after a response because of active Na extrusion.

(15) Oscillating sodium pumps are indicated as causing slow rhythms in some visceral smooth muscles and as setting the rhythm of bursts of spikes in some *Aplysia* neurons.

NON-ELECTRICAL PHYSICAL EVENTS ASSOCIATED WITH IMPULSE CONDUCTION

A number of non-electrical events occur during a nerve impulse. Heat is liberated during the upstroke of a spike (24.5 μcal/g in rabbit vagal fibers) and heat is recovered during the downstroke.[93]

Birefringence of non-myelinated crab or squid axons, as measured at 45° to the plane of polarization, decreases with a time course similar to that of an action potential.[118] Light scattering, measured at a low angle to the fiber axis, decreases; that at 90° to the fiber axis increases. The latter effect is less voltage-dependent than the low-angle effect and may represent swelling of periaxonal space.[35, 36] Some fluorescent dyes bind to the axon membrane, and changes in fluorescence during electrical activity may represent conformational changes in proteins, which allow polar compounds to pass through the non-polar portions of the membrane.[221]

It is hoped that various physical measurements may provide evidence concerning the lipoprotein alterations associated with ionic conductance changes.

ADAPTIVE FEATURES OF CONDUCTION IN NERVES

Adaptations Favoring Speed of Conduction. Speed of conduction varies in different fibers and nerve tracts, and fast conduction in reflex pathways may give an animal an advantage in escape from predators, in prey capture, and in responses to stress. Time constants of excitation parallel speed of conduction. Several commonly used excitation constants are derived from strength-duration or voltage-time curves; fast conducting fibers tend to have shorter excitation times than slow fibers. Another excitation parameter is *accommodation*, or the rise in threshold during passage of a stimulating current; the threshold for a slowly rising pulse is higher than that for a rapidly rising pulse, so the threshold can change in the absence of an all-or-none response.

Physical Factors. Speed of conduction is proportional to the depolarization-induced increase in sodium permeability; a faster rate of rise of action potential results in larger local currents and faster excitation of adjacent regions in the axon. Speed is also directly proportional to the ratio of threshold to resting potential, in that a lower threshold permits a shorter time for excitation. Speed is inversely related to membrane capacity, in that the time needed to charge a large capacity is long, while the time needed to charge a small membrane capacity is short. Speed is inversely proportional to the square root of intracellular specific resistance, in that low intracellular resistance means more mobile ions and thus greater longitudinal current for a given voltage. Speed of conduction increases, up to a limit, for a given fiber with increasing temperature.

Morphological Factors. Long nerve processes favor speed in a nervous system, in that interposed synapses slow conduction in a central pathway. In many coelenterates, the nerve nets consist of short neurons, and conduction is slow. In *Calliactis*, conduction in the net is at 0.04 to 0.15 m/sec, where conduction in through-tracts in the same animal is at 1.2 m/sec. In the ganglionic cord of annelids, conduction in the neuropil, where there are segmental synapses, is at less than 0.05 m/sec, while in the through-conducting giant fibers it is at 10 to 25 m/sec.[224]

Speed of conduction is related by some proportionality constant to fiber diameter. In large myelinated fibers of vertebrates, velocity in meters per second numerically equals the diameter in microns times a factor of 6 to 8; in small non-myelinated fibers (sympathetic fibers, olfactory nerve fibers) velocity is proportional to diameter. In giant axons of squid, the velocity is an exponential function ($V = D^{0.61}$);[247] in the giant fiber of *Myxicola*, the function is similar ($V = KD^{0.5}$).[9] During development of an animal, fiber diameters increase and, concomitantly, speed of conduction increases. Thus, the relation of speed to fiber diameter varies for different fiber types; it has not been adequately studied in particular neurons as they grow during development.

Nerve fibers specialized for extremely large size are called giant fibers. There are two general types of giant fiber, each of which has evolved independently several times. One type consists of a single large neuron which gives rise to a large axon, as the Mauthner cell of fishes. The second type of giant fiber is syncytial, in that (a) processes from a number of nerve cells fuse to form a single axon, as in giant axons of the mantle of squid, or (b) segmental neurons send processes to long fibers which have segmental septa and provide continuous conduction through many segments, as in lateral and median giant fibers of earthworms. In crustaceans (such as crayfish) the median giants are unineuronal with the soma in the brain, while the lateral giant fibers have segmental cell bodies.

The functions of giant fiber systems are discussed on p. 637. In general, they are relay rather than integrative systems, and they conduct faster than small-fiber systems of the same animal. Unicellular giant fibers are found in nemerteans, cestodes, some polychaetes, insects, and balanoglossids. In the basal portion of the medulla of bony fishes, urodeles, and anuran tadpoles, there are large Mauthner cells of the acoustic-vestibular system. Their fibers (20 to 45 μm in diameter) conduct at 50 to 80 m/sec at 15°C. Velocity for Mauthner fibers of carp is 38 to 63 m/sec. The multicellular or functional-syncytium type of giant fiber system occurs in annelids, crustaceans, and cephalopod molluscs. In each of these, the giant fibers mediate rapid movements and synchronized contractions of many muscles. In *Lumbricus*, conduction in the septate lateral giant fiber is at 7 to 17 m/sec, while in the median one it is 17 to 45 m/sec. In a squid *Loligo*, each third order giant axon (0.1 to 0.8 mm diameter) arises by fusion of processes from 300 to 1500 cells in the stellate ganglion, and each serves a large area of the mantle. Contraction forces a jet of water out of the mantle cavity through the funnel, driving the animal forward or back according to funnel tilt. Velocities in the giant fibers range from 3 to 25 m/sec, and Young has calculated that the escape or feeding reaction is twice as fast by virtue of the giant fiber system as it would be by small axons only.[247]

A myelin sheath results in faster conduction because it lowers C_m, increases R_m, and increases the space constant. The fastest vertebrate fibers are of medium size (4 to 15 μm) but have thick sheaths (30 to 50% of total fiber diameter), whereas the fastest fibers of invertebrates are large (50 to several hundred μm) but have thin sheaths (less than 1% of diameter). A 4 μm fiber of a cat saphenous nerve at 38°C conducts at about the same rate (25 m/sec) as a 650 μm squid giant axon at 20°C.

Conduction in vertebrate myelinated axons is saltatory; that is, the impulse jumps from node to node. A local circuit for conduction involves inflowing and outflowing current only at the nodes; an internode has high resistance and low capacitance. External current flows outside the myelin. The optimum length of the internode is related to the safety factor and to core resistance. Giant fibers of shrimp have no nodes, yet where each gives off a branch to a motor nerve or to the neuropil, diphasicity of spike form can be recorded and blocking of conduction can occur; thus, the branch points make for saltation. Conduction in the shrimp fiber is 90 m/sec, whereas in a lobster axon of the same size (100 to 200 μm), which is not saltatory, velocity is only 8 m/sec.[130a]

In conclusion, a variety of adaptations have been used in different groups of animals to provide for different speeds of nerve conduction. All of them are based on common physical principles of rate of current rise, membrane capacity and resistance, core conductivity, and resulting safety factor. Some animal groups have myelinated nerves with saltatory conduction, and others have axons of very large diameter with little sheath, to provide for high speed of conduction. Some animals have many short neurons in a pathway; others use fibers of very small diameter, without myelination, which give slow conduction. The range of speeds of conduction of nerve signals is from 0.1 to 100 m/sec, or a thousandfold.

CONDUCTION IN ANIMALS WITH LOW-SODIUM BLOOD

INSECTS

In some insects, the hemolymph potassium concentration is so high and sodium so low that no known nerves could conduct in such a medium (see p. 81 for concentrations). In a cockroach, the nervous system is surrounded by a perineural sheath across which there is a potential of 10 to 15 mv (inside positive). In normally sheathed cockroach giant axons *in situ*, the resting potential is −58 mv, and the action potential is 105 mv; in desheathed nerve cords in blood the RP is −67 and the AP is 86 mv. Sodium exchanges readily with glial and perineural elements, and efflux measurements show three Na compartments.[231] Degree of accessibility of K to the axon surfaces is associated with the extraneuronal or sheath potential.[193, 194] A given concentration of potassium depolarizes more if the nerve is desheathed, and nerve conduction is then more readily blocked by potassium.[227, 228] The time constant for depolarization in high K^+ is, for an intact stretched cockroach giant nerve fiber, 173 sec; for a desheathed one it is 24 sec. Dried or stretched nerve cords show lessened restriction of potassium entry.

In the stick insect *Carausius morosus,* hemolymph Na^+ is 20 meq/liter, compared with 102 meq/liter of H_2O in the nerve cord. The central nervous system (but not peripheral nerves) is surrounded by neural lamella and, outside this, by a fat-body sheath. It is postulated that cells of the fat-body sheath, or the glia in perineural elements,[227] actively transport sodium and maintain a high concentration around the axons even though blood sodium is low. In a silkmoth, the blood Na is only 4 mM; yet nerve conduction is blocked by TTX, so it depends on Na.

Evidently, phytophagous insects have nerves in which, as in other animals, the resting potential is determined by the normal high ratio of K_i/K_o, and spikes are determined by the steep inward gradient of Na. The hemolymph may be low in sodium and high in potassium, and the nerves function because of selective and/or actively ion-transporting sheaths. The means by which the sheath maintains the extra-neuronal fluid different in composition from hemolymph are incompletely known.

Molluscs

In the freshwater clam, *Anodonta,* blood Na is only 14 mM, but exchange with nervous tissue is relatively rapid; time constants of efflux indicate two sodium compartments and intracellular Na of 8.6 mM. Thus, the action potentials depend on an inward gradient of Na^+ even though the absolute concentrations inside and out are very low.[229]

GLIAL FUNCTION

Many large axons are surrounded by sheath cells; many small axons occur in bundles with a surrounding sheath of glia. Several types of glial cells are recognized between and around neurons in central nervous systems. The properties of glial sheaths have been well studied in the *Necturus* optic nerve and in the leech central nervous system, where the glia occupy most of the extraneuronal space. A leech ganglion has about 10 large glial cells which surround nerve cells and fiber bundles, and which occupy half of the extraneuronal space.[128] The glial cells have resting potentials which follow the Nernst relation for $[K_o]$ down to 20 mM, and from the E_m vs $[K_o]$ plot the $[K_i]$ is calculated to be 108 mM for glia and 138 mM for neurons.[126, 169] The *Necturus* optic nerve is surrounded by glial cells which are connected by low resistance bridges. Ions and sugar can pass through intercellular clefts, and the glia become depolarized by the K liberated by optic nerve impulses. The depolarization declines over a period of several seconds as K reenters the axons.

Leech neurons may, following a spike, show an undershoot or after-hyperpolarization which decays slowly with a time constant of 100 msec. The glial cells are depolarized slowly after a train of spikes in the neurons, and it is calculated that potassium accumulates in the 150 Å clefts outside the axons in the amount of $1.1 \times 10^{-12} M/cm^2$ (0.75 mM/liter). After a train of impulses, the neuron resting potential may be increased for several minutes. This potential is abolished by ouabain, and is due to a Na-K pump that is activated by the K which has accumulated outside the axons. The glia serve as a partial barrier which prevents loss of K from the nervous system to the blood.[11, 127, 168] Depolarization of glial cells in the retina by K released during activity of neural elements may be the basis for the b-wave in electroretinograms of vertebrates.[52]

The functions of glia are incompletely known; in insects they may provide a pump to maintain ions in the extraneuronal space at a concentration different from that in hemolymph. In other nervous systems, as in vertebrates, glia occupy much of the extracellular space and provide an ionic buffer.

EFFECTS OF SIZE DIFFERENCES IN REGIONS OF ONE NEURON

Because of the effects of fiber size on velocity, space constant, and threshold, and the differences in size of parts of one neuron, one process of a neuron may differ functionally from another, and branch-points may serve as filters of frequency. Both λ and C_m are greater for large than for small fibers, but R_m is less. When a small fiber conducts into a large diameter region, the effective current density is reduced, the current required to reach threshold is higher and spike rise-time may be prolonged since more time is required to charge the larger capacity. Conversely, at a narrowing, as from the large soma of a motorneuron to the initial segment, current density is high and a spike may be started there. Cockroach giant axons show regions of segmental narrowing where conduction may be slowed; as an impulse goes from a region of small diameter to one of widening, the safety factor is low and block may occur on repetition.[184]

At branch points, there may be differential block according to frequency. For example, one motor axon serves medial and lateral bundles of deep abdominal extensor muscles of crayfish and lobster; the branch to the medial bundle has a lower safety factor and blocks at 40 Hz, whereas the branch to the lateral bundle follows stimuli up to 80 Hz.[183] In the central nervous systems of many animals, branch points may serve as low-pass filters. High frequencies of spikes may elevate external potassium concentrations, and the resulting net depolarization may reduce the safety factor by raising the threshold.

TRANSMISSION, ELECTRICAL AND CHEMICAL, BETWEEN CELLS

Generally, neurons receive input in one set of processes (dendrites) or soma and transmit by another process (axon); i.e., they are polarized. This applies whether conduction within a neuron is graded or all-or-none, and whether transmission is electrical or chemical. However, examples of non-polarized electrical junctions are known. In many neurons, a strict distinction between dendrites and axons is not possible (amacrine cells, various stellate cells). In some of these, communication between processes (often dendritic) may occur without involving soma or all the processes of a neuron. Reciprocal dendro-dendritic synapses, as seen electron microscopically, suggest that a single process may both receive and send. Examples of apparent dendro-dendritic communication are: crustacean cardiac ganglia (p. 849), mitral-granule cell glomeruli in olfactory bulbs of vertebrates,[198] amacrines in retina,[52] in mammalian superior colliculus and thalamus,[199] and probably in frog optic tectum.

ELECTRICAL TRANSMISSION

Electrical transmission between neurons is of two sorts—bidirectional and polarized.

Electrotonic Connections Between Neurons. Electrical coupling between epithelial and embryonic cells was mentioned previously (pp. 458–462). Many examples of similar low-resistance connections between central neurons are now known. In these junctions, electrical coupling is symmetrical in the two directions, and electrical activity in one cell is reflected in corresponding activity in the second cell, of amplitude in proportion to the coupling coefficient.[181, 186, 187] Each ganglion of a leech nervous system has a pair of giant neurons; a pulse injected by a microelectrode into one cell is reflected in the other giant neuron by an electrical deflection attenuated to 1/2 to 1/5 the size of the pulse in the first cell. If one cell is depolarized to the point of spiking, small electrotonically conducted spikes in the second cell correspond to those in the first cell (Fig. 11–13).[58] If the ganglion is bathed by a Ca-free solution, the input or cellular resistance rises from 24 to 70 ΩM and functional coupling ceases.[188, 189]

Similar low-resistance connections occur between cross-branches of the lateral giant axons of crayfish and earthworm, and across the septa from segment to segment in a earthworm giant fiber. Attenuation between the two lateral fibers is to 1/3 of the applied pulse, and crossover points can shift from one segment to another.[243]

In crayfish, the medial giant axons are nonseptate and do not interact, but the lateral axons are electrically coupled from segment to segment, both across septa and from one lateral giant fiber to the other.[103] The septal resistance between segments is 2 to 4×10^5 Ω which is 1.2 to 3 Ω cm^2; a septal delay in spike conduction is 50 to 200 μsec. The dye fluorescein can pass through the septa in crayfish, but not in earthworm. A pulse of 80 mv in the preseptal segment of crayfish lateral giant fiber elicits a 30 mv deflection in the post-septal segment.[236] Septal junctions are nonpolarized and normally conduct 1:1, but when fatigued by repeated stimulation

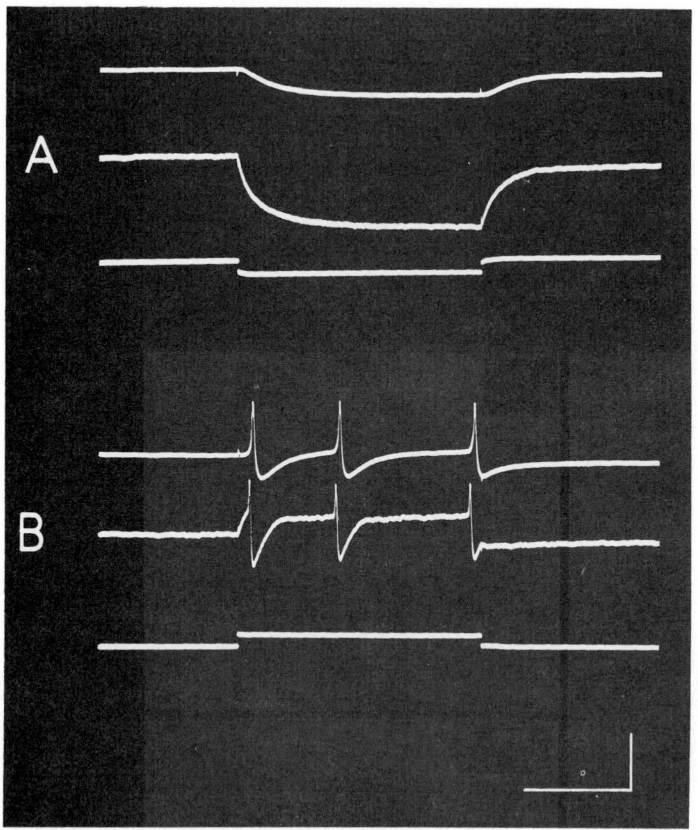

Figure 11-13. Electrical coupling between neurons in leech ganglion. Middle traces from one giant cell of leech ganglion; upper traces from another giant cell. A, hyperpolarization of first cell (middle trace) reflected in second cell (upper trace). B, When depolarization fired first cell, second cell gave synchronous spikes. (From Eckert, R., J. Gen. Physiol. *46*:573–587, 1963.)

they may require facilitation. A ladder of electrotonic junctions gives a series of delay lines, and circus activity can occur via the crossings between the two lateral giant axons.[130] Trans-septal reverberation is normally prevented because axonal conduction is by longitudinal current, which is large relative to the current across connections between lateral fibers.[130] Crayfish septa show gap junctions, electron microscopic evidence for electrotonic contacts.[86]

In the buccal ganglion of the mollusc *Navanax,* ten neurons in a cluster show weak electrical coupling, more to direct current pulses than to spikes; hence, there is attenuation of high frequencies and the coupling may not be functionally significant (Fig. 11-14).[214] In cardiac ganglia of crustaceans,

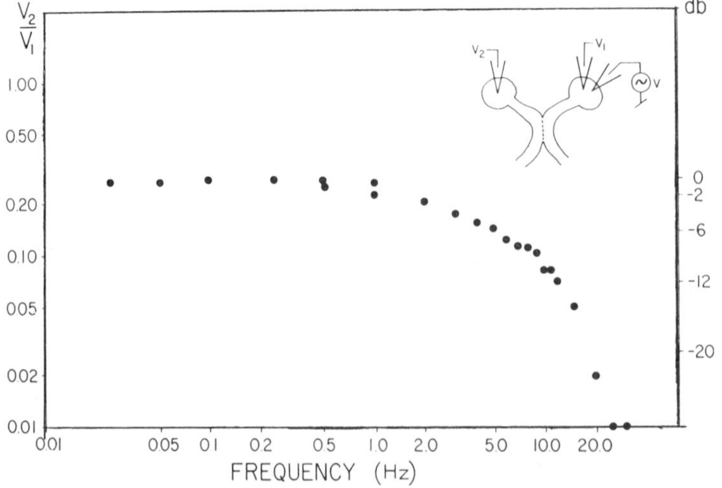

Figure 11-14. Coupling ratios (V_2/V_1) between two neurons of ganglion of mollusc *Navanax* for different frequencies of polarization of V_1. (From Levitan, H., L. Tauc, and J. P. Segundo, J. Gen. Physiol. *55*:484–496, 1970.)

high frequency signals may pass between pacemaker neurons by dendritic processes without entering the somata.

In the corpora pedunculata of the brain of a wood ant, holes in the glial sheaths allow soma-somatic connections between neurons. The junctions are of the gap type with 50 to 70 Å period; chemical synapses with 300 to 600 Å vesicles also occur.[131]

Numerous electrotonic junctions have been found in the brain of fishes.[14,15] In a puffer, two large (250 μm) neurons on the dorsal surface of the medulla activate motorneurons to the swimbladder muscle (p. 204). Afferent input from the spinal cord elicits synchronous discharge in both large neurons, and an electric pulse injected into one appears with slight attenuation in the other. The cells are coupled by electrotonic junctions between their axons. In several kinds of electric fishes, neurons which command the electric discharge are electrotonically coupled (Chapter 17). In mormyrids, the relay neurons in the medulla are activated by pacemakers higher in the brain; each medullary neuron gives two spikes per single stimulus, and the medullary neurons are electrotonically coupled by dendro-dendritic bridges. The junctions are gap junctions, with a hexagonal array having an 80 Å period between centers. The medullary neurons activate spinal motorneurons by chemical synapses, but a cluster of spinal neurons shows electrotonic coupling; the latter fire three impulses per electric organ discharge.[14] In gymnotids there is electrotonic coupling at the pacemaker level by dendro-dendritic junctions and also at medullary and spinal levels by axosomatic junctions; transmission from one level to the next is by chemical synapses. In the electric catfish *Malapterurus*, two giant electromotor neurons in the medulla are coupled electrotonically by prejunctional fibers. Each of these electrotonic junctions in electric fishes conducts bidirectionally, but polarity in the system is imposed by chemical synapses between levels; the electrotonic coupling at each level permits virtual synchrony between the coupled cells.

Oculomotor neurons of some teleosts are electrotonically coupled by axosomatic junctions. In addition, synaptic endings with vesicles are seen. Stimulation of the ophthalamic nerve initiates spikes close to the cell bodies, and these are synchronized by the electrotonic coupling.[124]

Unidirectional Electrical Synapses. A few instances of polarized electrical transmission have been described. In the synapse between lateral giant fibers and motorneurons of a crayfish, the action potential of the pre-fiber is an adequate stimulus to the motorneuron and there is no delay between pre-spike and post-response. Intracellular recordings show large transsynaptic currents in the motorneuron of orthodromic (normal direction) impulses, but not in the reverse (antidromic) direction (Fig. 11–15). The synaptic membrane is a rectifier in that it permits current to flow through a low-resistance junction orthodromically, but not in the reverse direction. A depolarizing pulse in the pre-fiber passes into the post-fiber, but a hyperpolarizing pulse does not; conversely, a hyperpolarizing pulse, but not a depolarizing pulse, passes from post- to pre-fiber. There is 50:1 rectification. Smaller third-root axons

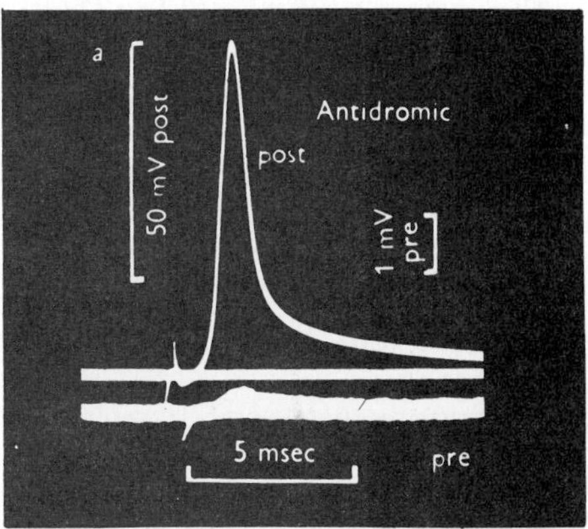

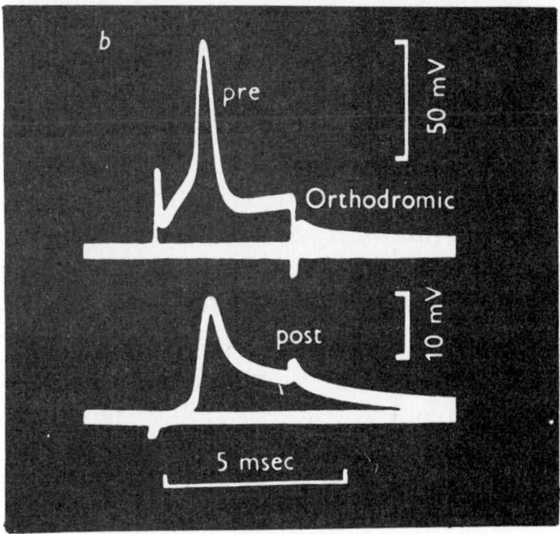

Figure 11–15. Responses at electrical synapse of crayfish. Responses in prefiber and postfiber in response to (*a*) antidromic and (*b*) orthodromic impulses. Note evidence for rectification in that transmembrane current passes orthodromically but not antidromically. (From Furshpan, E. J., and Potter, D., J. Physiol. *145*:289–325, 1959.)

also end on the motorneurons; they elicit slow junctional potentials, which reverse at −7 mv, so they are chemical synapses.[68, 69]

Another polarized electrical synapse is in the hatchet fish *Gasteropelecus* (p. 675), where giant fibers activate motorneurons with a delay of 0.05 msec. A depolarizing pulse in the giant fiber is reflected in the large motorneuron current, and a hyperpolarizing pulse does not pass orthodromically. The coupling (synaptic) resistance decreases when the giant fiber is depolarized and is increased when the motorneuron is depolarized.[5] The effect of the electrical synapse in the hatchet fish is to shorten delay and to synchronize contractions of muscles of the pectoral fins.

The physiology of Mauthner neurons is described on p. 641. They receive several inputs, including one by an excitatory chemical synapse, from the contralateral eighth nerve and another by an excitatory electrical synapse from the ipsilateral eighth nerve. Also, a recurrent process from the Mauthner axon activates an interneuron, the axon of which coils around the axon hillock of the Mauthner cell. Impulses in the interneuron set up an external field which hyperpolarizes the Mauthner axon and thus inhibits spike initiation. By this recurrent inhibition, repetitive firing after one Mauthner discharge is prevented.[69a] This example shows that electrical synapses can be either excitatory (depolarizing) or inhibitory (hyperpolarizing field).

The ciliary ganglion of a 3 to 5 day chick has both electrical and chemical transmission. The response of the postganglionic nerve to a preganglionic stimulus consists of two components separated by 1.5 to 5 msec. The second component, but not the first, is abolished by tubocurarine, and the first component is independent of membrane polarization. Antidromic stimulation elicits a small preganglionic response; hence, there is some bidirectional conduction through the path of electrical coupling.[148, 149] Numerous examples of mixed electrical and chemical synapses are known. For example, in a sea lamprey, *Petromyzon,* at the junction of axons from large Muller cells of the medulla with giant interneurons in the spinal cord, the epsp has two components, one of which is chemical (eliminated by high Mg^{++}) and the other electrical (Fig. 11–16).[150, 204b]

Neurons in each segmental ganglion of a leech give evidence for three types of electrical synapses: (1) Groups of some 14 motorneurons on each side are electrically coupled by non-rectifying junctions so that a pulse in one spreads electrotonically to the others. (2) Six sensory neurons in each hemiganglion respond to light touch on the skin, and an action potential in any one of these neurons gives rise to a short-latency potential in the others of the same segment and in three neurons in each adjacent ganglion. Between these sensory cells is an unusual double rectification, in that a depolarizing pulse can pass in *either* direction between touch cells but a hyperpolarizing pulse does not pass.[11] (3) Transmission between the touch sensory cells and motorneurons is of short latency and is rectifying, in that a depolarizing pulse passes from sensory neuron to motorneuron and a hyperpolarizing pulse passes in reverse direction.[170]

In summary, electrical transmission occurs between neurons in situations where speed and/or synchrony are important. The coupling is bidirectional in some fast systems, but transmission is unidirectional in others.

Chemical Synapses

The following characteristics of chemical synapses distinguish them from electrical synapses:[192] (1) synaptic delay, usually at least 0.5 msec; (2) lack of significant current from presynaptic impulse in postsynaptic element; (3) postsynaptic potential response (psp) (junctional potential) in excitation is usually depolarizing (epsp) and in inhibition may be hyperpolarizing (ipsp); (4) postsynaptic membrane showing conductance increase as evidenced by change in size and polarity of psp according to membrane potential, and reversal of psp at some equilibrium potential; (5) presence of synaptic vesicles in presynaptic endings, and generally specific staining of postsynaptic membrane; (6) blocking of transmitter liberation by high Mg^{++} and low Ca^{++}.

Synaptic Structure. Synaptic junctions may be at axon terminals (boutons) or in swellings along an axon (en passage); they may be from axons to dendrites, from axons to cell soma, or at an axon. Dendro-dendritic junctions may occur, usually in axonless neurons. Sheath cells lack regions of synaptic apposition, and frequently there are regions of thickening of postsynaptic membrane opposite aggregations of presynaptic vesicles. Transmitter is made mainly in the terminal, but it can be synthesized in other parts of a neuron. Presynaptic endings characteris-

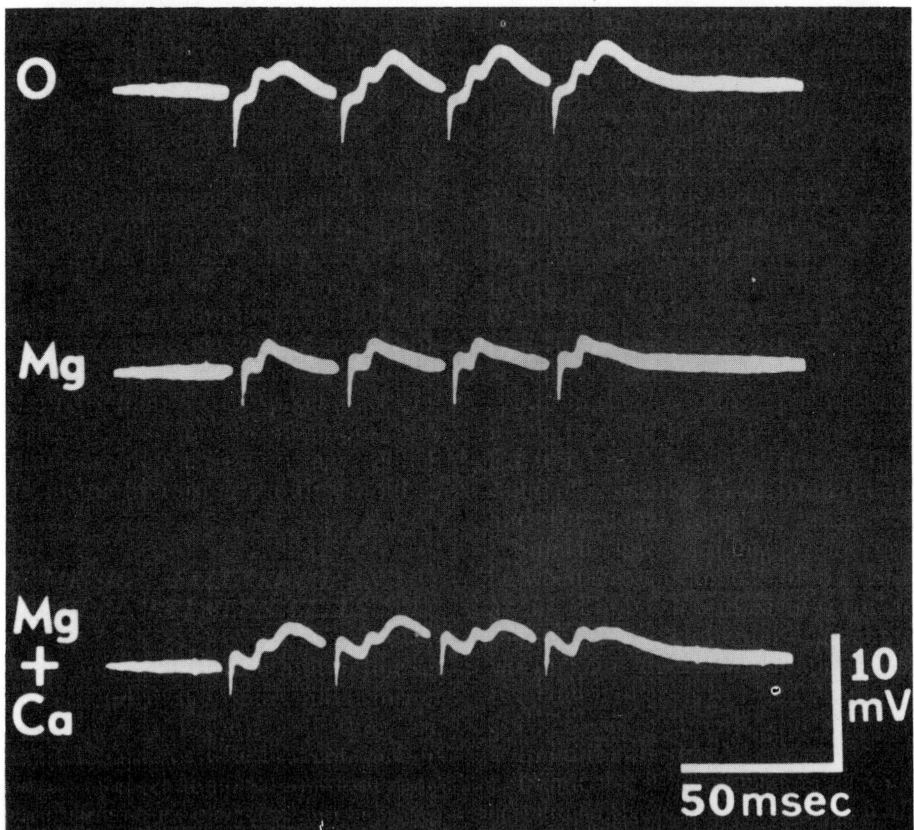

Figure 11-16. Epsp's of giant interneurons in response to stimulation of a bulbar axon in lamprey. O, series of shocks eliciting epsp's, second component facilitating. Mg, 15 minutes after Mg (20 mM) second component reduced or eliminated. Mg + Ca, calcium concentration in bath increased to 22 mM; second epsp chemical synapse, first electrical. (From Rovainen, C. M., J. Neurophysiol. *30*:1024–1042, 1967.)

tically contain mitochondria, and many vesicles of one or more size and density as seen by electron microscopy (Figs. 11–17, 11–18, 11–19). Fractionation of nervous systems yields synaptosomes, which are the membranous nerve endings containing vesicles; isolated synaptosomes contain the enzymes appropriate for synthesis of transmitter. Vesicles can be seen to become depleted of transmitter on repeated stimulation. For example, the vesicles in frog motor nerve endings are reduced in number and density after prolonged stimulation.[88] Also, catecholamine vesicles become empty after treatment with reserpine, which blocks storage of these transmitters. Transmitters such as acetylcholine (ACh) in vesicles are in equilibrium with ACh in the cytoplasm of the endings, and there is uncertainty as to exchange between the two pools; ACh released at the presynaptic membrane is probably entirely from the vesicles.[242] At a synaptic ending of frog neuromuscular junction there may be 600,000 vesicles per synapse, and after one minute of stimulation the number has been observed to drop to 400,000. Vesicles release their contents by exocytosis, and vesicle membrane appears to join plasma membrane; then, as rough-coated vesicles, they form cisternae which later provide vesicle coats for refilling with transmitter (Fig. 11–20). Thus, the vesicle protein is reused.[88a] Vesicles have been prepared from mammalian brain, and ACh vesicles in pure suspension from *Torpedo* electric organ. Each vesicle is about 84 nm in diameter and is estimated to contain about 70,000 molecules of ACh.[242]

Identification of specific vesicles with known transmitters has been attempted by comparing abundance of a type of vesicle with concentration of a given transmitter in the neural structure and, for monoamines, by fluorescence microscopy. Small agranular

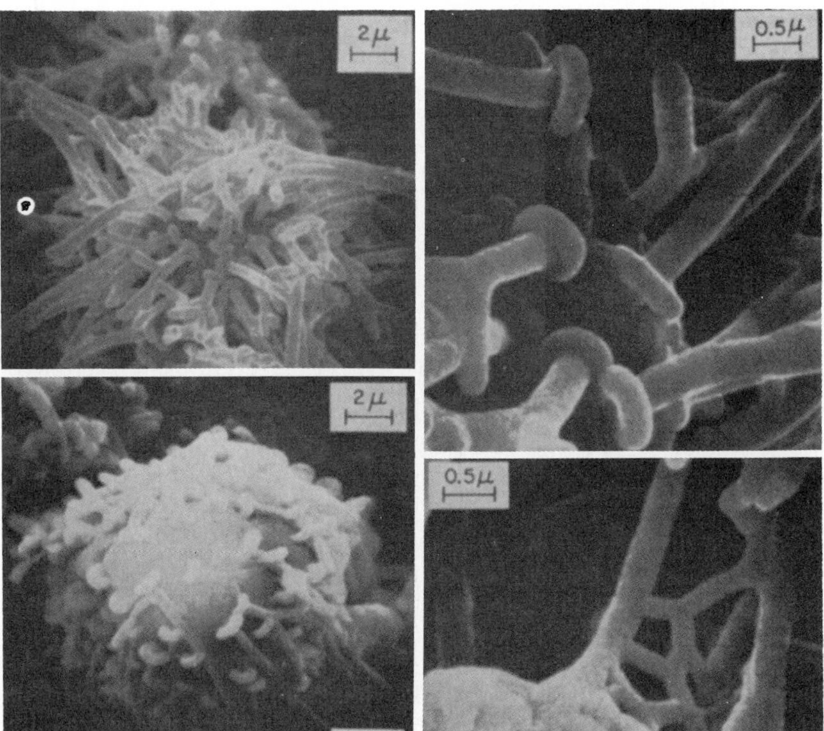

Figure 11–17. Scanning electron micrographs of synaptic regions in abdominal ganglion of *Aplysia.* (From Lewis, E. R., T. E. Everhart, and Y. Y. Zeevi, Science *165*:1140–1143, 1969. Copyright 1969 by the American Association for the Advancement of Science.)

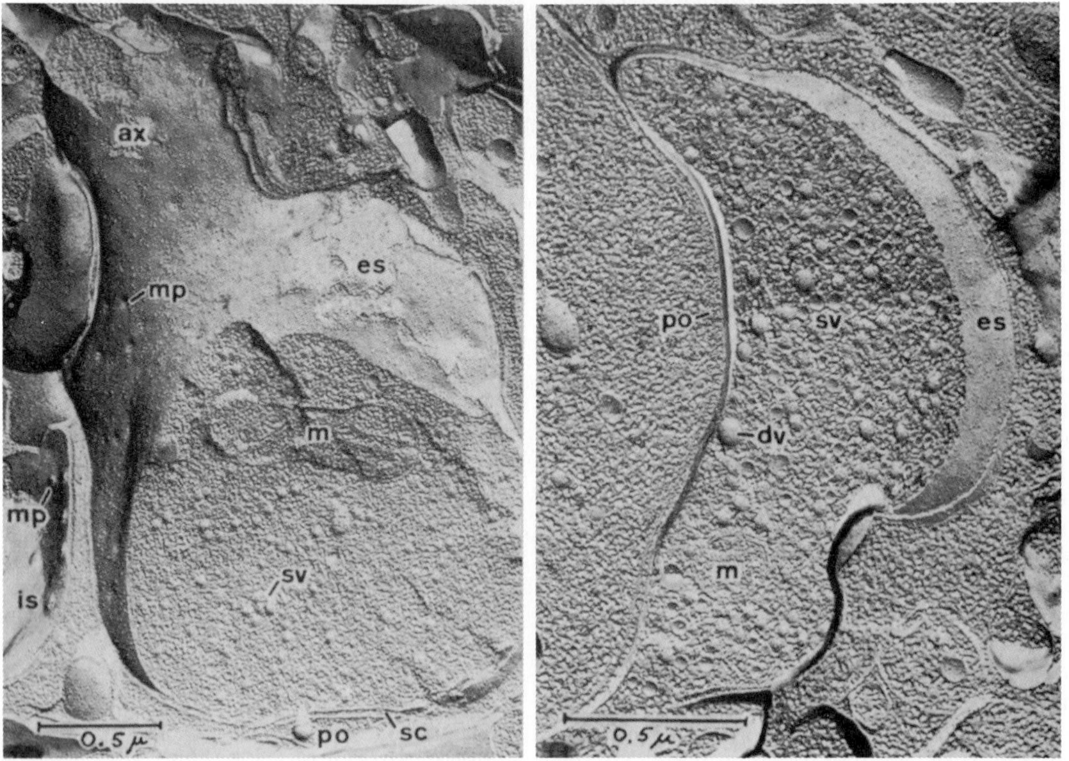

Figure 11–18. Face view of freeze-etch presynaptic terminal from subfornix of cat. Synaptic vesicles (sv), mitochondrion (m), dense-core vesicles (dv), postsynaptic dense region (po), and outer surface of presynaptic membrane (es) are shown. (From Akert, K., H. Moor, K. Pfenninger, and C. Sandri, "Contributions of new impregnation methods and freeze etching to the problems of synaptic fine structure," *in* K. Akert and P. G. Waser, eds., *Mechanisms of Synaptic Transmission,* Progress in Brain Research *31*:223–245, 1969. Elsevier, Amsterdam.)

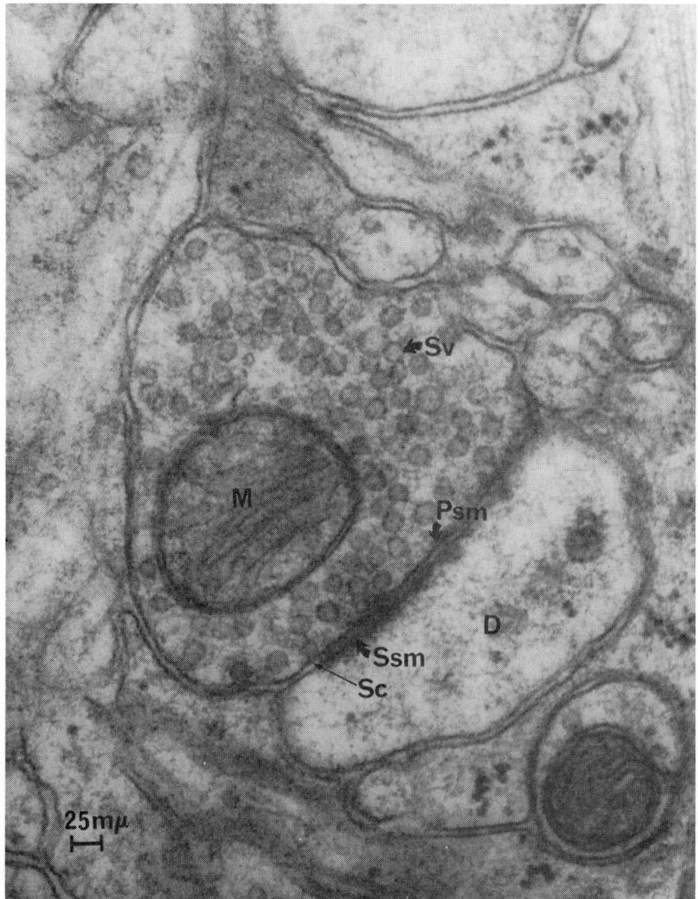

Figure 11–19. Climbing fiber synapse on dendrite (D) of Purkinje cell of cerebellum of goldfish. Sv, synaptic vesicle; m, mitochondrion; Psm, presynaptic membrane; Ssm, subsynaptic membrane; Sc, synaptic cleft. (Courtesy N. Kotchabhakdi.)

vesicles, 300 to 600 Å in diameter (average 500 Å) with clear cores, are probably cholinergic and occur at all known cholinergic endings. Small granular vesicles, 300 to 600 Å in diameter with dense cores, probably contain norepinephrine (NE); large vesicles, 900 to 1200 Å in diameter with peripheral halos, probably contain other catecholamines.[20, 21] In visceral synapses, large vesicles (1000 to 2000 Å, without halos) may store ATP.[25, 26] There is evidence from both mammalian cerebellum and crustacean neuromuscular endings that small flattened vesicles are more abundant at inhibitory endings and large spherical ones at excitatory endings.[176] In other systems this correlation seems not to hold.

Some presynaptic endings have densely staining bars perpendicular to the terminal membrane and with vesicles on each side of the bar. They occur frequently where three processes synapse and are called ribbon synapses. For example, in vertebrate retinas, ribbon synapses occur opposite two or three postsynaptic receptors, e.g., between receptor cells and bipolar and horizontal cells or between bipolar and amacrine and ganglion cells (Fig. 11–21).[52, 53]

Postsynaptic membranes differ considerably from one another, and some have desmosome-like dense regions which may, but need not, lie directly opposite a presynaptic ending. Other postsynaptic membranes have dense aggregations of granules beneath the membrane. In some cholinergic synapses (e.g., neuromuscular junctions) the postsynaptic membrane stains for acetylcholinesterase (AChE). In other cholinergic synapses there is AChE in both presynaptic and postsynaptic membranes.

Transmitter action can be stopped by: (1) breakdown of transmitter by enzymes in the postsynaptic membrane, (2) diffusion followed by transport away in the circulation,

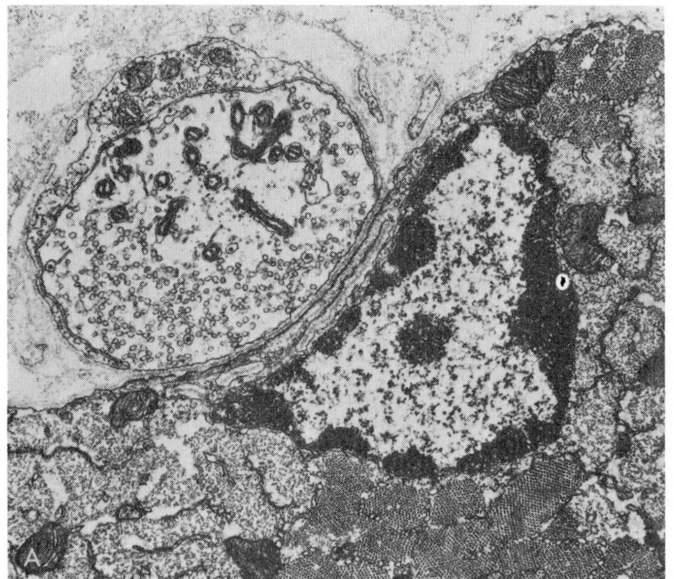

Figure 11–20. Ultrastructure of neuromuscular synapse of frog sartorius. A, Cross section; muscle shows fibrils cut at different levels of thick and thin filaments, and large nucleus; nerve ending surrounded by thin Schwann sheath except at interface with muscle; terminal contains synaptic vesicles, dense mitochondria. B, Longitudinal section of unstimulated terminal, many vesicles, dense mitochondria, presynaptic densities (arrow) opposite folds (f) in muscle surface, Schwann processes (s) between nerve and muscle. C, Longitudinal section of terminal which had been stimulated electrically for 15 minutes; many cisternae (c), coated vesicles (arrows), enlarged Schwann processes (s), mitochondria (m) swollen and pale. (From Heuser, J. E., and T. S. Reese, J. Cell Biol., 1972 [in press]).

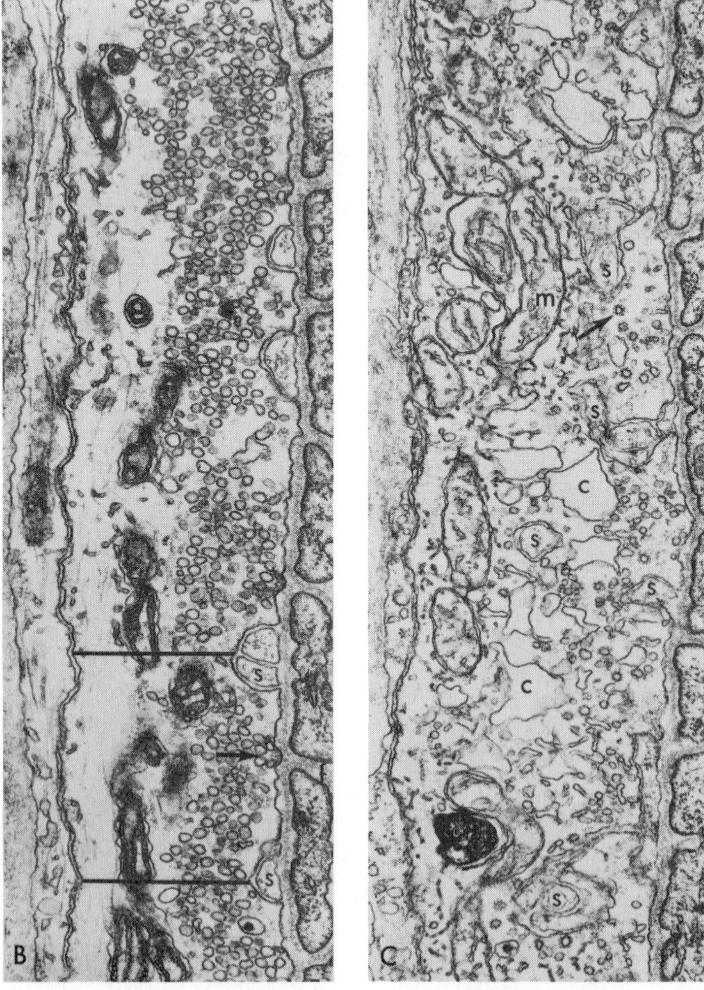

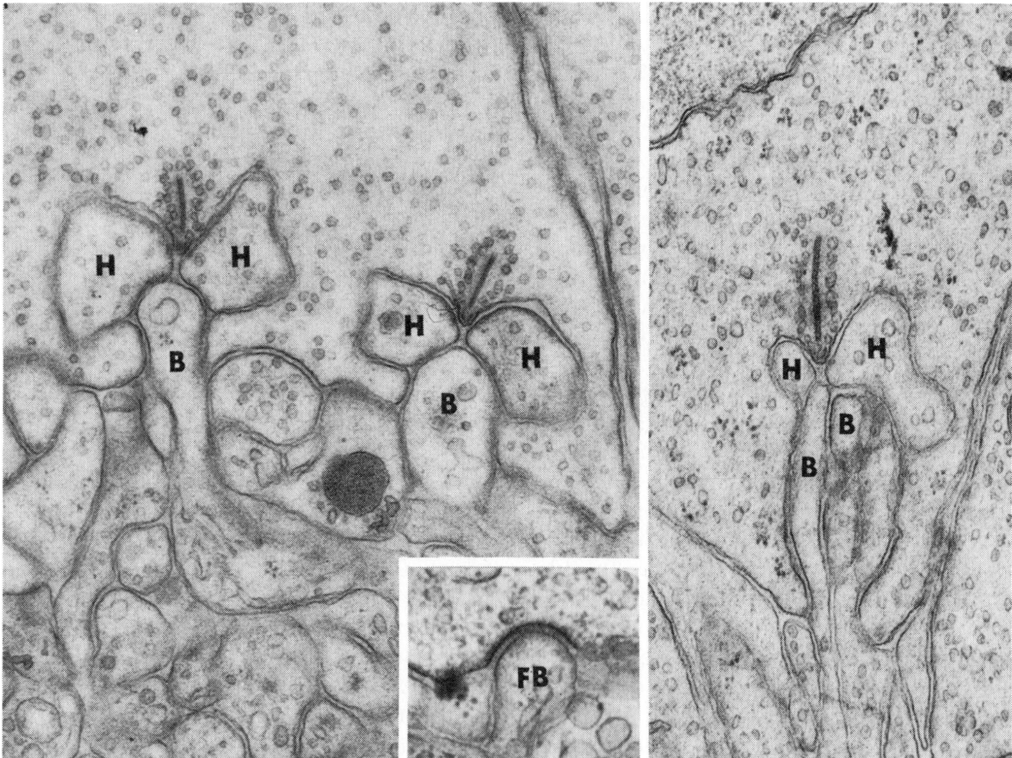

Figure 11–21. Electron micrographs of ribbon synapses of receptor terminals in retina of cat. In cone terminals (at left) three processes penetrate into each invagination of receptor base. In rod terminals (at right) four or more processes enter the invagination. H, Horizontal cell processes; B, bipolar cell processes. (From Dowling, J. E., Jour. Invest. Ophthalm. *9*:655–680, 1970.)

(3) re-uptake by presynaptic endings and repeated use of transmitter, (4) uptake followed by destruction at specific loci in presynaptic endings.

Chemical Transmitters. A transmitter is identified by the following criteria: (1) capacity of the putative transmitter, in small amounts, to mimic the effect of presynaptic impulses; (2) presence in presynaptic neurons of the transmitter and of precursors and enzymes for its synthesis; (3) release of the "transmitter" into perfusate during presynaptic nerve stimulation; (4) presence of a mechanism of inactivation, either by specific enzymatic destruction or by active re-uptake into presynaptic fibers; (5) potentiation or blocking of the responses to presynaptic impulses and to applied putative transmitter by specific drugs.[65, 240]

The number of putative neurotransmitters is small, and each of them is of molecular weight 100 to 300. The responses are more diverse than are the suspected transmitters, and it is probable that a greater diversity of receptor molecules exists than of transmitters.[95]

Subsequent chapters will consider the following examples of chemical transmission:

vertebrate skeletal muscle and electrocytes –	acetylcholine
crustacean and insect muscle: excitation –	glutamate
crustacean and insect muscle: inhibition –	gamma-amino butyric acid
vertebrate heart: excitation –	norepinephrine
vertebrate heart: inhibition –	acetylcholine
vertebrate gastrointestinal muscle: excitation –	acetylcholine
vertebrate gastrointestinal muscle: relaxation –	norepinephrine; adenosine triphosphate
molluscan heart: excitation –	5-hydroxytryptamine
molluscan heart: inhibition –	acetylcholine
bivalve catch muscle: excitation –	acetylcholine
bivalve catch muscle: relaxation –	5-hydroxytryptamine

In nervous systems, evidence exists for the presence of each of the above (except ATP) plus other catecholamines (particularly dopamine) and of at least one additional amino acid—glycine—as transmitters.

ACETYLCHOLINE. Acetylcholine has, at the choline end of the molecule, a quaternary ammonium group which bears a positive charge; at the acetyl end is a carbonyl which can be negatively charged. Thus, the molecule is an electric dipole. Acetylcholine (ACh) is synthesized from acetyl-CoA and choline (see below).

The system for synthesizing acetylcholine is present in many neural tissues, and ACh is stored in clear-cored vesicles (300 to 800 Å diameter) in cholinergic nerve endings. The hydrolytic enzyme acetylcholinesterase (AChE) can be localized by histochemical methods, and is present in postsynaptic membranes and on the surface of some presynaptic endings (Fig. 11–22). The turnover number for AChE from an electric organ is 300,000 molecules per second, or one ACh hydrolyzed in 3 to 4 μsec per enzyme site.[163] Synthesis of ACh in an electric organ can occur at the rate of 5 to 8 mg ACh/g tissue/hr.[160] The action of ACh is usually to increase the permeability of postsynaptic membranes to Na^+ and K^+ in varying proportions; its action is on Cl^- permeability in some molluscan neurons (p. 475). There may be spontaneous or resting release of packets or quanta of ACh, as judged by miniature junction potentials of constant size or of a multiple of this size; at a motor endplate, a nerve-induced epsp is produced by the release of some 300 quanta (p. 471).[121] Each quantum contains 10^3 to 10^4 molecules of ACh.

Acetylcholinesterase (AChE) is by definition a receptor, since it must complex with the molecule before hydrolysis can occur. Physiological action of ACh is potentiated by blocking AChE with any of a variety of drugs, the best known of which are physostigmine (eserine) and DFP (diisopropylfluorophosphate). AChE, isolated from electric organ as a pure protein, binds receptor inhibitors at a site distant from the catalytic site, and the dose-response curve of the AChE is shifted by both receptor activators and inhibitors. These facts argue for allosteric properties of AChE.[30] However, it is possible to block the physiological receptor, leaving the esterase action unchanged. Suspensions formed from membranes of electric organs of *Torpedo* complex with analogs of ACh (decamethonium, muscarone) which are themselves not hydrolyzed; when the AChE is inhibited, ACh can still be bound, and this binding is blocked by *d*-tubocurarine.[175] A specific blocking agent (α-toxin from a snake) combines irreversibly with the ACh receptor; when this complex was covalently bound with sepharose and the resulting particles were centrifuged out, some 75 to 100% of the receptor was removed from the original extract but only 12% of the AChE and 6% of the protein were removed.[29a, 30]

By equilibrium dialysis of solubilized *Torpedo* electric organ, the ACh receptor is shown to be a protein or proteolipid, and possibly to have several binding sites. Estimates of molecular weight range from 50,000 to 250,000. Artificial membranes prepared from proteolipids from *Electrophorus* electric organ show ion conductance increases upon addition of ACh, an effect that is blocked by *d*-tubocurarine.[182] In rat diaphragm, each endplate may contain 13,000 receptor molecules per square micron, or at a separation of 100 Å. Receptor distribution in skeletal muscle can be localized by binding of labelled bungerotoxin; the sites correspond to points of maximum sensitivity to ACh, and the spread of ACh sensitivity over the muscle membrane after nerve degeneration is due to synthesis of receptor at new sites. It is

$$\underset{\text{acetyl-CoA}}{CH_3{-}\overset{\displaystyle O}{\overset{\|}{C}}{\sim}SCoA} + \underset{\text{choline}}{HO{-}CH_2{-}CH_2{-}N^+(CH_3)_3} \xrightarrow[\text{acetylase}]{\text{choline}} \underset{\text{acetylcholine}}{CH_3{-}\overset{\displaystyle O}{\overset{\|}{C}}{-}O{-}CH_2{-}CH_2{-}N^+(CH_3)_3} + \underset{\text{coenzyme A}}{CoA{-}SH}$$

Acetylcholine is hydrolyzed as follows:

$$CH_3{-}\overset{\displaystyle O}{\overset{\|}{C}}{-}O{-}CH_2{-}CH_2{-}N^+(CH_3)_3 + H_2O \xrightarrow{\text{AChE}} \underset{\text{choline}}{HO{-}CH_2{-}CH_2{-}N^+(CH_3)_3} + \underset{\text{acetic acid}}{CH_3COOH}$$

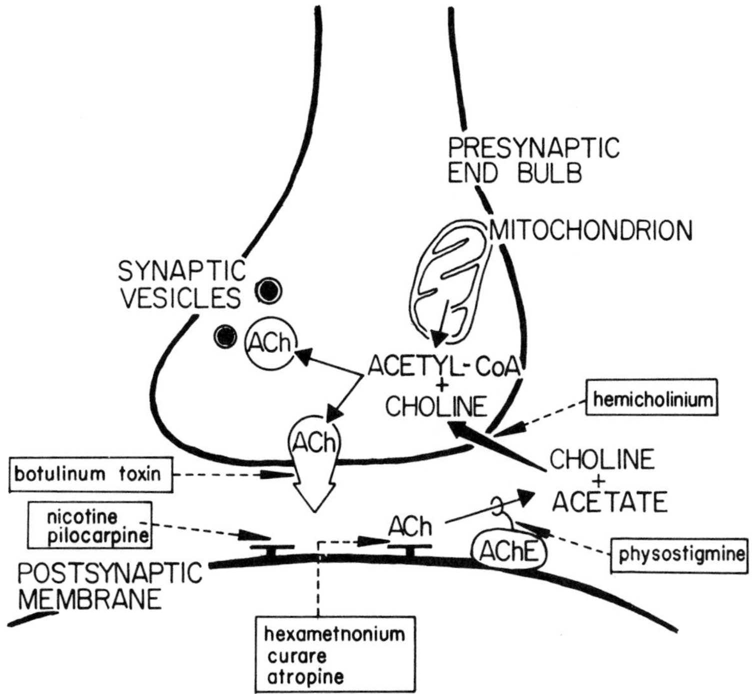

Figure 11–22. Diagram of a cholinergic synapse indicating presynaptic synthesis of acetylcholine, ACh; nicotinic and muscarinic receptor sites, postsynaptic hydrolysis by acetylcholinesterase, AChE. (From Rech, R. H., and K. E. Moore, Introduction to Psychopharmacology. Raven Press, New York, 1971.)

concluded that the receptor responsible for ion permeability changes is a protein different from AChE.

Two general types of acetylcholine receptor are suggested by the action of mimicking and blocking drugs: (1) nicotinic receptors, where the action is mimicked by low concentrations of nicotine (as in vertebrate neuromuscular junctions or sympathetic ganglia), and (2) muscarinic receptors, where the action is mimicked by muscarine and is blocked by atropine (as in vertebrate heart). A broad survey of blockers of nicotinic receptors shows two modes of action: those which depolarize, such as hexamethonium (in sympathetic ganglia), and those which block ACh by competitive inhibition, such as *d*-tubocurarine (in vertebrate neuromuscular junctions). The molecular dimensions of compounds that block nicotinic and muscarinic junctions have been compared with the structure of acetylcholine.[62] It appears that binding of nicotinic agents depends on a hydrogen-bond acceptor some 5.9 Å from the center of positive charge, and that for muscarinic agents the separation of the H-bond site from the positive charge on the receptor is 4.4 Å.[13a]

The first interneuronal junction in which ACh was clearly implicated as transmitter was in vertebrate sympathetic ganglia. In a bullfrog sympathetic ganglion the response to a preganglionic volley is a fast nicotinic epsp (which can be blocked by D-tubocurarine), followed in some 100 msec by a second epsp which is blocked by atropine; hence, both nicotinic and muscarinic types of receptor occur in the same neuron.[120] When the superior cervical ganglion of a cat is perfused via a branch of the carotid artery, and the preganglionic nerve trunk is stimulated and the ganglion is treated with eserine, ACh is liberated into the perfusate in proportion to the number of impulses entering the ganglion. Injected ACh stimulates the ganglion cells to discharge. ACh is not liberated in appreciable amounts by antidromic impulses, but it is still liberated on preganglionic stimulation after postganglionic block by tubocurarine. Cat superior sympathetic and stellate ganglia synthesize ACh at 3 mg/g_{dw}/hour and, after preganglionic fiber section, this synthesis declines by 80% in 90 hours. Synthesis can occur in isolated nerve terminals. The cat superior cervical ganglion releases, in 60 minutes of stimulation at 20/sec, five times the content of ACh in the ganglion; hence, ACh must be replaced rapidly during activity. In addition to the two cholinergic depolarizing responses, postganglionic neurons also show a hyperpolarizing response to adrenergic impulses in presynaptic fibers.[134] It is concluded that these sympathetic neurons have at least three types of receptor.

Another example of cholinergic synapse is the junction of the collateral from a spinal motorneuron onto a Renshaw interneuron, which provides feedback inhibition to the motorneuron. The responses of Renshaw cells to motor nerve stimulation (antidromic impulses) are potentiated by eserine and blocked by *d*-tubocurarine; the cells are stimulated by ACh and by nicotine. The motorneurons are also cholinergic at their endings on muscle. This is an example of the so-called Dale's principle that a neuron liberates the same transmitter at all its terminals.

In other parts of vertebrate nervous systems, the function of ACh as a transmitter is suggested but not so well established. In mammalian cerebral cortex, fewer than 30% of neurons are sensitive to ACh; some are excited and others are inhibited by it. ACh is liberated from human cortex during deep (rapid eye movement) sleep (see p. 692). The synapses between parallel fibers and Purkinje cells in the cerebellum are probably cholinergic.[42]

In the pons and medulla, 35% of the cells are excited and 22% are inhibited by ACh.[22] Responses of the visual cortex to transcallosal fiber input are enhanced by ACh and by the anticholinesterase DFP, and they are blocked by *d*-tubocurarine. Nerve terminals of the efferent olivo-cochlear bundle in the cochlea stain for AChE, and administration of ACh reduces the amplitude of eighth nerve response, as does stimulation of the efferent bundle. Possibly ACh is the transmitter from the efferent fibers to hair cells of the organ of Corti.[1b]

Acetylcholinesterase is high in concentration in spinal cord, sympathetic ganglia, thalamus, and caudate nucleus. Nonspecific cholinesterase occurs also in some nonnervous tissues—red blood cells, submaxillary gland, glial cells, and placenta. Ventral roots contain some 300 to 500 times more ACh than dorsal roots; ACh is concentrated in spinal motorneurons. Acetylcholine is found not only in excitable tissues but in spleen and placenta. Choline acetylase is low in concentration in most sensory nerves—spinal dorsal roots and optic nerve; it is high in second order sensory neurons and very high in caudate nucleus; and it is intermediate in concentration in the motor cortex, and low in the pyramidal tract.[145] The preceding examples suggest that acetylcholine may have other functions besides that of a neural transmitter; its role in the central nervous system is far from being completely understood.

Nonmammalian vertebrates resemble the mammals in distribution of cholinergic enzymes. AChE activity in fish brain is very high—193 in trout and 123 in tench (mm^3 $CO_2/100$ $mg_{ww}/30$ min from hydrolysis of 0.027 M ACh) as compared with 67 for rat and 59 for dog.[34]

The distribution of AChE and associated enzymes among organisms is very wide. In protozoans, AChE is present in ciliates and trypanosomes but absent from sporozoan plasmodia. *Tetrahymena* has significant amounts of AChE, and *Trypanosoma* contains choline acetylase. Flatworms (planaria) contain AChE, and histochemical tests show much of it to be in the nervous system.[132] Muscles of some sipunculid worms, annelids, and echinoderms have high sensitivity to ACh. In a starfish the radial nerve cord has much ACh (60 $\mu g/g_{ww}$).[190] Motor nerves in some annelids, sipunculids, and echinoderms are probably cholinergic. In earthworm muscle, epsp's are blocked by tubocurarine, and Ach increases g_{Ca}.[94]

Inhibitory nerves to the hearts of some pelecypods and gastropods are cholinergic (p. 846). ACh is released on stimulation of the nerve to the perfused salivary glands of *Octopus*.[6] The brain ganglia of cephalopods contain much ACh, and the brain of a squid has very high activity of choline acetylase and of AChE:

	ACh content
squid optic lobe	2090 nM/g
guinea pig cortex	13
dog caudate nucleus	17

The following data illustrate the flux of ACh in cephalopod nervous tissue:[135]

	ACh content mg/g$_{ww}$	*ACh synthesis μM ACh/ mg$_{pr}$/hr*	*ACh hydrolysis mg ACh/100 mg$_{ww}$/hr*
squid brain		40–80	200–400
octopus ganglia			
optic	236		97.9
stellate	33		6.2
supraesophageal	225		20
octopus mantle nerve	17		3

Although ACh is present in these ganglia at high levels, its function in the nervous system of cephalopods is not established, and the transmitter at giant synapses in the stellate ganglion is certainly not acetylcholine. In the *Aplysia* abdominal ganglion, three identified neurons contain ACh transferase, and ACh can be formed in the soma.

The ganglion of a tunicate, *Ciona*, contains 120 μg ACh/g and innervation of longitudinal muscles is probably cholinergic.[66]

In arthropods there is evidence both for and against acetylcholine as a neurotransmitter. The central nervous systems of crayfish, lobsters, and crabs contain significant amounts of ACh, AChE, and choline acetylase. However, acetylcholine is ineffective on synapses of crustacean ganglia up to very high concentrations (10^{-3}/ml), even after AChE has been blocked.[230] In the peripheral nervous system of crustaceans the usual excitatory agent appears to be glutamate rather than ACh (p. 499).

Some insects have high levels of ACh and AChE in the nervous system[36a] (see below). It is estimated that if all the ACh in the head of a blowfly were in the brain, the concentration would be 500 μg/g; in *Musca* it would be 170 μg/g. After treatment with some insecticides (DDT or tetraethylpyrophosphate), the amount of free ACh in the nervous system increases, indicating that the increased spiking activity which is observed liberates acetylcholine. A silkmoth larva contains much AChE in the neuropil, which increases some six-fold during growth and morphogenesis.[208] Cockroach ganglia are normally insensitive to applied ACh, but after they have been desheathed, ACh may stimulate. In cockroach ganglia, iontophoresis of ACh directly on some neurons at 1.3×10^{-13} M is excitatory by increasing Na conductance;[111] the reversal potential for ACh is −40.3 mv, and that for an epsp is −45 mv.[195a] Thus, insect ganglia have enzymes for making and degrading ACh, they contain ACh stores, and some neurons are locally very sensitive to ACh.[195]

It may be concluded that ACh has been demonstrated as a synaptic transmitter at neuromuscular junctions of vertebrates, at parasympathetic endings in heart and viscera of vertebrates, in mammalian sympathetic and parasympathetic ganglia, and in motorneuron synapses on Renshaw cells, and it may function in some tracts of the brain. It appears to be the transmitter in peripheral nerves of some molluscs, sipunculids, and probably annelids. It is the inhibitory transmitter to most molluscan hearts. It is probably an excitatory transmitter at some insect central synapses. The lack of correlation between distribution of ACh, AChE, and choline acetylase and the action of applied ACh and protective or blocking agents raises the probability that the acetylcholine system may have other functions, possibly "trophic" effects, besides serving as a synaptic transmitter.

Monoamines. A second class of neurotransmitters are the monoamines, which include: (1) the catecholamines, which are made up of a catechol (1,2-dihydroxybenzene) and an aliphatic chain, and (2) indolealkylamines. Of importance in the first category are dopamine, norepinephrine (NE), and epinephrine (E); most important of the second category is 5-hydroxytryptamine (5HT) or serotonin. The monoamines can be localized in small amounts in tissues by brilliant fluorescence, which for catecholamines is green and for 5HT is yellow. Both

Animal	*Region*	*ACh content, μg/g$_{ww}$*	*ACh synthesis, mg/g$_{ww}$/hr*	*ACh hydrolysis, mg/g$_{ww}$/hr*
Calliphora	head	37.2	900	2500
	brain	500 est.		
Periplaneta	brain & subes. gangl.	135	53	153
	thoracic ganglia	95	20	331
	6th abd. ganglion	63	18	314

categories of amines are found in nervous systems of animals of most phyla.

Catecholamines. Catecholamines are derived from tyrosine as follows:[157]

Tyrosine — HO–C₆H₄–CH_2–CH(COOH)–NH_2

↓ tyrosine hydroxylase

Dopa — (HO)₂C₆H₃–CH_2–CH(COOH)–NH_2

↓ aromatic acid decarboxylase

Dopamine — (HO)₂C₆H₃–CH_2–CH_2–NH_2

↓ dopamine-β-hydroxylase

Norepinephrine — (HO)₂C₆H₃–CH(OH)–CH_2–NH_2

↓ phenylethanolamine-N-methyl transferase

Epinephrine — (HO)₂C₆H₃–CH(OH)–CH_2–N(H)–CH_3

The most important of these compounds as a neurotransmitter in vertebrates is norepinephrine (NE). The NE that is liberated from adrenergic nerve endings is either (1) removed via the blood, (2) taken back into the presynaptic nerve processes, or (3) broken down by monoamine oxidase (MAO) in mitochondria or by catechol-*o*-methyl transferase (COMT) postsynaptically. Commonly used inhibitors of MAO are the drugs iproniazid and harmine. NE is stored in dense-core small (300 to 800 Å) vesicles, varicosities, and endings of adrenergic nerves (Fig. 11–23). Release is by exocytosis. Reserpine blocks the incorporation of NE into storage vesicles, whether the NE comes from synthesis or from re-uptake. The life of a vesicle is 35 to 70 days, whereas NE molecules turn over every one or two days; hence, the vesicles renew their NE many times.[48]

The adrenal medulla contains epinephrine and norepinephrine in various proportions in different vertebrates; the percentage of NE is 73% in dogfish, 83% in whale, 25% in mouse, and 2% in rabbit and guinea pig.[241] The sympathetic chain contains 7.3 μg NE/g in dog and 4 μg NE/g in cattle.[63] Postganglionic sympathetic neurons are adrenergic, and have either excitatory or inhibitory actions according to the receptor tissues. Preganglionic sympathetic fibers are cholinergic, but some interneurons in the ganglia are adrenergic, probably liberating dopamine[133, 134] (Fig. 11–24). The content of NE is high in hypothalamus, medulla, and sympathetic ganglia; it is low in cerebral cortex and basal ganglia. A rat cerebellum

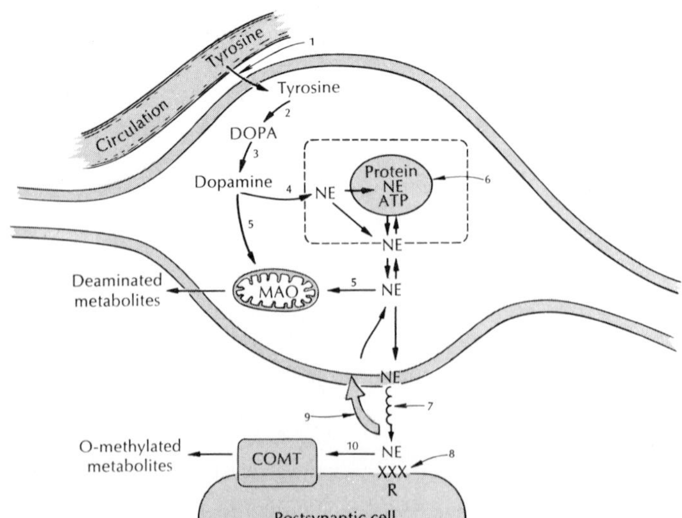

Figure 11–23. Diagram of a noradrenergic synapse at an axon swelling, showing presynaptic synthesis and packaging of NE, liberation and re-uptake of NE, and its degradation by monoamine oxidase (MAO). (From Cooper, J. R., F. E. Bloom, and R. H. Roth, Biochemical Basis of Neuropharmacology. Oxford University Press, Fair Lawn, N. J., 1970.)

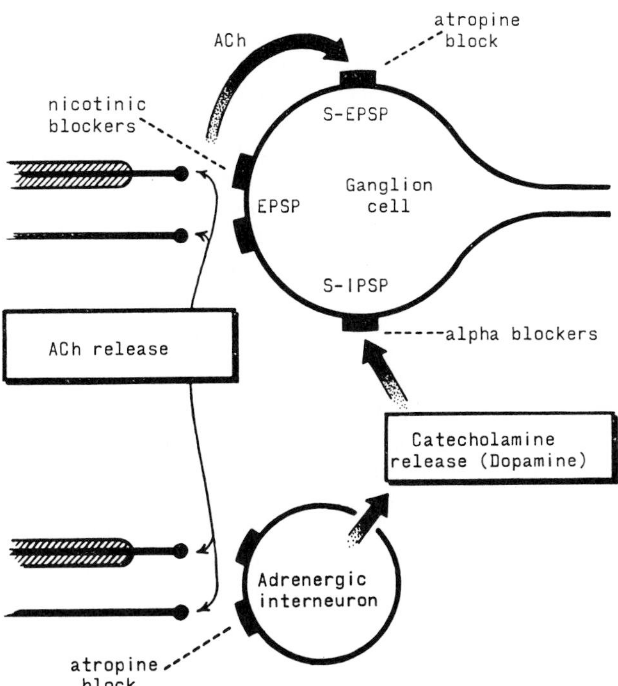

Figure 11-24. Diagram of postulated synaptic connections in rabbit sympathetic ganglion. Nicotinic and muscarinic receptor regions are shown on ganglion cell, and alpha-adrenergic receptor region. Catecholamine is postulated as coming from an interneuron which is stimulated by cholinergic presynaptic fiber. (From Libet, B., Fed. Proc. *29*:1945–1956, 1970.)

shows fluorescent axons in the molecular layer and NE-containing endings making contact with Purkinje neurons. Spontaneous activity in these neurons is reduced by NE, an effect that is enhanced by blocking the re-uptake of NE; it appears certain that some inhibitory axons in the cerebellum liberate NE, which acts via adenyl cyclase in Purkinje cells.[21,210]

The mammalian brain contains dopamine, 80% of which is in basal ganglia (caudate nucleus and putamen). Treatment of mammals with reserpine depletes the dopamine stores and the animals become lethargic; this condition is remedied by administering L-dopa.[92] Drugs such as reserpine, which depletes central synapses of catecholamines, cause psychological depression in man; conversely, MAO inhibitors increase catecholamine concentration and are antidepressants.

Postganglionic neurons of the sympathetic system of mammals are adrenergic. In lower vertebrates, however, the separation of parasympathetic as cholinergic and sympathetic as adrenergic breaks down (p. 776). A given catecholamine (e.g., NE) may be inhibitory to one effector and excitatory to another. The dual action of NE is partly explained by two classes of receptors—alpha and beta—and of subdivisions of these. Identification is by means of blocking drugs and by the relative potency of various catecholamines. Different organs in the same animal, and the same organ in different species, may have alpha receptors (blocked by phenoxybenzamine or phentolamine) or beta receptors (blocked by propranol).[1b] Norepinephrine relaxes uterus of rat, contracts uterus of rabbit and man, and has mixed effects on cat according to whether the animal is pregnant or not.[156] Stimulation of sympathetic fibers to the colon inhibits motility, and this is blocked by propranolol. Properties of alpha and beta receptors are given in Table 11-3. Release of norepinephrine is blocked by bretylium and by guanethidine, irrespective of the receptor.

The cellular action of catecholamines, as

well as that of a number of hormones, may be mediated by intracellular adenosine 3′,5′ phosphate (cyclic AMP). The enzyme adenyl cyclase occurs in the cell membrane of sensitive cells, and this catalyzes the conversion of ATP to 3′,5′-cyclic AMP, which may change the activity of various cellular enzymes leading to the specific physical effect of the catecholamine. The 3′,5′-cyclic AMP is then broken down by phosphodiesterase to 5′ AMP. Wherever catecholamines act by β receptors, there is an increase in adenyl cyclase activity and a resultant increase in the amount of cyclic AMP. For example, the adrenergic increase in amplitude of beat of the mammalian heart parallels an increase in cyclic AMP; similarly, relaxation of the uterus corresponds to increase in adenyl cyclase. Increase in cyclic AMP precedes the mechanical response to the catecholamine, and agents which block the effect of catecholamine block the increase in cyclic AMP.[203,204] Alpha effects also appear to be mediated by changes in cyclic AMP, and the difference in receptors may be due to two sites on the adenyl cyclase at the cell membrane or to two forms of the enzyme occurring in different tissues (or in the corresponding tissue of different species).

In some invertebrates, catecholamines have been identified by fluorescence of nerve endings and by analyses. The nervous system of an earthworm contains 1.4 μg E and 0.32 μg NE/g_{ww}, mostly in chromaffin cells. The nervous system of some insects (e.g., *Tenebrio* larvae) contain 1 to 2 μg NE/g; NE has been suggested as a possible transmitter to the light organ of a firefly.[211] In a sea urchin or starfish, the nervous system contains dopamine in concentrations of 3 to 8 μg/g, and NE in lower concentrations.[41]

Catecholamines, particularly dopamine, are important as transmitters in molluscs. Nervous systems of *Eledone, Octopus,* and *Sepia* have both dopamine and NE, with highest concentrations in optic lobes and superior buccal lobe.[101] Ganglia of the clam *Spisula* contain 40 to 50 μg dopamine/g and 5 to 6 μg NE/g.[39] In freshwater clams, dopamine is abundant in the visceral ganglion, and 5HT in the pedal ganglion; both are present in cerebral ganglion.[217] Three neurons in the statocyst of the clam *Anodonta* show marked catecholamine fluorescence. In *Aplysia,* the H cells (hyperpolarized by ACh) are excited (depolarized) by dopamine at 10^{-11}M.[41,78] In *Helix,* neurons which show inhibition of long duration (ILD cells) in response to pallial nerve stimulation are inhibited by dopamine.[109] Other cells in *Helix,* which are excited by 5HT, are inhibited by dopamine. There is little doubt, therefore, that dopamine is a natural inhibitory transmitter in molluscs.

TABLE 11-3. Adrenergic Receptors[1a,203,204]

	Alpha	*Beta*
relative sensitivity	epinephrine > norepinephrine > isoproterenol	isoproterenol > epinephrine > norepinephrine
blocking agents	ergot alkaloids phenoxybenzamine phentolamine dibenzyline	dichlorophenylisoproterenol (DCI) propranolol pronethalol MJ 1999 (sotalol)
sites of action	vas deferens intestinal plexus pancreatic beta cells lipolysis by fat cells constrict some vessels (skin) dilate pupil of eye contract spleen contract nictitating membrane contract uterus* chick expansor secundariorum	excitatory to heart (β_1) relax gastrointestinal muscle (β_2) dilate some blood vessels (skin) (β_2) relax bronchi and lungs relax uterus stimulate liver glycogenolysis (α and β)

*In virgin cat uterus, β receptors → relaxation; in progesterone-treated cat uterus, α receptors → increased activity.

5-Hydroxytryptamine (Serotonin). 5HT is formed as follows:

tryptophan — indole ring with side chain $-CH_2-CH(NH_2)-COOH$

↓ tryptophan hydroxylase

5-hydroxytryptophan (5HTP) — $HO-$ indole ring with side chain $-CH_2-CH(NH_2)-COOH$

↓ 5HT decarboxylase

5-hydroxytryptamine — $HO-$ indole ring with side chain $-CH_2-CH_2-NH_2$

5HT is inactivated by monoamine oxidase, which enzyme is blocked by ipromazid (Fig. 11–25). Reserpine leads to depletion and block of synthesis of 5HT, as well as of catecholamines and of enzymes for their synthesis and destruction. 5HT is widely distributed in mammalian brain, particularly in the reticular formation, especially raphe.

Serotonin content is high in thalamus, raphe, hypothalamus, limbic system, and pineal gland, and is low in cerebral cortex and cerebellum. Inhibition of synthesis of 5HT in the brain leads to insomnia, after which injection of 5HTP restores capacity for sleep.

The best case for 5HT as a transmitter is in molluscs. Many molluscan nervous systems contain 5HT, the enzymes for forming it from 5-hydroxytryptophan, and an amine oxidase for inactivating it. 5HT is released from clam hearts when excitatory nerves are stimulated, and it is probably the transmitter for relaxation of catch muscles (p. 765).

Serotonin content of tissues of the bivalve *Anodonta* follows:

Tissue	*5HT content* $\mu g/g_{ww}$
cerebral ganglion	72.5
visceral ganglion	43.5
pedal ganglion	66.5
adductor muscle	0.3
heart	0.0

The concentration in the ganglia varies according to locomotor rhythms.[206]

The innervation of the rectum of the bivalve *Tapes* is complex and acts via a peripheral plexus; apparently, both excitatory and inhibitory cholinergic actions and excitatory serotonergic effects occur.[191] 5-Hydroxytryptamine appears to be the excitatory transmitter and ACh the inhibitor to heart of the bivalve *Mercenaria.*[140]

In cephalopod nervous systems, 5HT is widespread; it is most abundant in inferior buccal and optic lobes.[101] In the clam *Sphaerium,* all fluorescent cells of the pedal ganglion contain 5HT.[217] In *Helix* and *Limax,* two symmetrically placed neurons in the cerebral ganglion are innervated by one giant

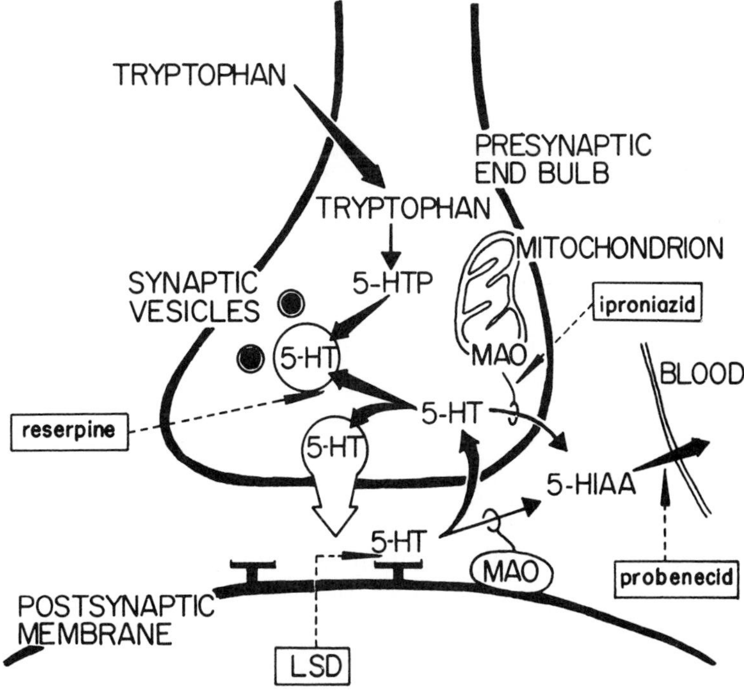

Figure 11–25. Diagram of a 5-hydroxytryptaminergic synapse to indicate pathways of synthesis and breakdown. Blocking agents given in boxes. (From Rech, R. H., and K. E. Moore, Introduction to Psychopharmacology. Raven Press, New York, 1971.)

5HT-containing neuron; the two receptor neurons are stimulated by 5HT at 10^{-7}g/ml, and their responses to nerves and to 5HT are blocked by lysergic acid diethylamide. Thus, the giant neuron probably is serotoninergic.[40,179]

In *Aplysia,* one cell type is inhibited by dopamine (10^{-10}M) and excited by 5HT (10^{-8}M).[78] The circumesophageal ganglion of *Helix* has 5.5 μg dopamine/g, and dopamine at low concentrations inhibits *Helix* ganglion cells.[112,113] In some neurons of *Helix,* 5HT increases g_{Cl}, while in other neurons 5HT increases g_K, with hyperpolarization resulting from each action.

The nervous system of a leech has 5HT-containing motor- and interneurons in the subesophageal ganglion and segmental ganglia, and has catecholamine cells mainly in supra- and sub-esophageal ganglia.[61,147] In *Lumbricus,* many neurons show 5HT fluorescence; these are probably motor cells of the nerve cord. The brain of *Limulus* has both 5HT and NE neurons, and the 5HT concentration in the nerve cord is 10 μg/g.[162]

It is concluded that there is good evidence for epinephrine as a transmitter in some autonomic nerves of vertebrates, for both NE and dopamine in specific areas of vertebrate brain, and for dopamine as a transmitter in molluscs. 5HT is widely distributed among animals, but the best evidence for its function as a transmitter is in neuropil core of vertebrates and in molluscs; it may function similarly in annelids.

Amino acids. The following three amino acids have been indicated as probable neurotransmitters:

Glutamic acid

$$HOOC—CH_2—CH_2—\overset{\displaystyle H}{\underset{\displaystyle NH_2}{C}}—COOH$$

$$\downarrow \; \to CO_2$$

Gamma-aminobutyric acid (GABA)

$$HOOC—CH_2—CH_2—CH_2—NH_2$$

Glycine

$$H—\overset{\displaystyle H}{\underset{\displaystyle NH_2}{C}}—COOH$$

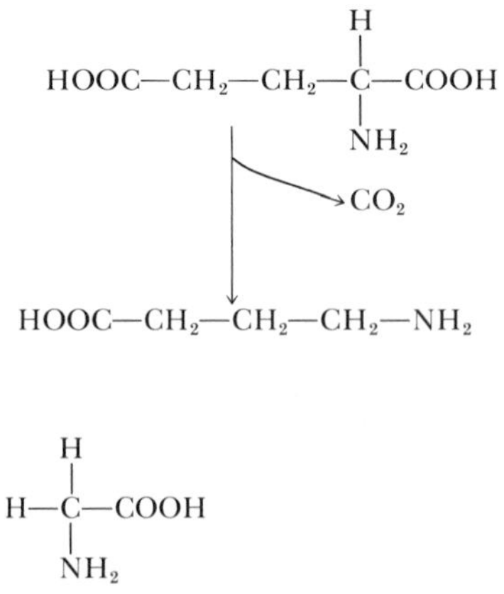

There is good evidence (p. 760) for GABA as the neuromuscular inhibitor in crustaceans and insects.[122] Picrotoxin antagonizes the inhibitory action of nerves and of GABA in these animals, and GABA is released on stimulation of inhibitory nerves.[180] There is also evidence for glutamate as an excitatory transmitter in muscles of these animals (p. 758).[121]

In mammals, GABA is taken up by specific terminals (synaptosomes) which can synthesize GABA; some 25 to 32% of the synaptosomes in the spinal cord and cortex take up GABA.[20] In mammals, the olfactory receptor neurons appear to excite granule cells by NE, and the granule cells inhibit mitral cells in the olfactory bulb by GABA.[43] GABA is inhibitory to neurons of the cerebral cortex, and it may be an inhibitory transmitter in cerebral cortex, hippocampus, cerebellum, and brainstem.[46,47] Purkinje cells of the cerebellum are inhibitory to neurons of Deiter's nucleus; GABA mimics this inhibition, and both inhibitory actions are blocked by picrotoxin.[174] GABA is probably the inhibitory transmitter from basket and stellate cells to Purkinje cells in frog cerebellum.[244]

Glycine is also inhibitory to many vertebrate central neurons, and it may well be the main inhibitory transmitter in the spinal cord. It causes hyperpolarization of motorneurons, with reversal potential of 3.8 mv negative to the RP and with a dose-response curve which is sigmoid, thus indicating the need for several molecules to initiate the reaction.[239,240] Glycine concentration in medulla and spinal cord is 2 to 5 times higher than in brain.[2] Inhibition by glycine is blocked by strychnine in low concentration; inhibition by GABA is blocked by bicuculline or picrotoxin.[43,45] In spinal cord, strychnine blocks predominantly postsynaptic inhibition, while picrotoxin blocks presynaptic inhibition.[44]

Mauthner neurons of goldfish are inhibited by GABA, with maximum sensitivity in regions of collateral endings, but with some effect in the periphery of the lateral dendrite; hence, GABA inhibition may be both presynaptic and postsynaptic.[50]

The distribution of GABA in lobster nerve cord is (in $\mu g/g_{dw}$): abdominal nerve cord, 480; sensory nerves, 9 to 14; motor nerve fibers, 6150; inhibitory nerves, 32,800.[122,123]

GABA is a very effective inhibitor of crustacean muscle; both its action and that of nerve inhibition are blocked by picrotoxin. GABA action is highly localized.[218,219] The cellular action of GABA in arthropods is to increase conductance for Cl^- in both presynaptic and postsynaptic inhibition.[220] In the central nervous system of grasshopper,

GABA inhibits the response of a large interneuron (T-cell) (see p. 663) to input from the auditory organs; after picrotoxin is applied, these cells respond to binaural stimulation, whereas normally they respond only monaurally.[216] In a cockroach ganglion, local application of GABA at 10^{-13} M stops spiking, hyperpolarizes, and increases Cl conductance and, by a lesser amount, K conductance.[111] In the sixth abdominal ganglion the reversal potential for an ipsp is −79 mv, and that for response to locally applied GABA is −75 mv; each is blocked by picrotoxin.[195a]

L-Glutamate (but not the D-isomer) is excitatory to some neurons throughout the nervous systems of some mammals, particularly in the cerebral cortex and cerebellum.[46] Glutamic acid is liberated from the cerebral cortex during electrical stimulation and during arousal, but not during sleep. It is strongly excitatory to cortex, especially to hippocampus, and to some basal ganglia.[97] Glutamate is excitatory to nerve cells in crustaceans, and the nervous system of *Carcinus* has glutamate in high concentrations (0.01 to 0.1 mg/g). Glutamate is a very effective excitatory agent on crustacean muscle. Insect muscle is also excited by glutamate, and the L-isomer is 10 times more effective than the D-isomer.[11,234] Locust muscle is stimulated by local applications of L-glutamate at 10^{-8} w/v.[17] Glutamate is excitatory at the synapses of giant axons in squid,[154] and in snail ganglia glutamate at 5×10^{-7} M excites different cells from those excited by ACh.[74,75] Glutamate is released from a *Helix* nerve-muscle preparation on nerve stimulation.[107]

A question may be raised as to how amino acids, which are normally present in plasma, can be specific neurotransmitters. One probable answer is that postsynaptic membranes are not normally bathed by plasma and that the concentration of these molecules in a synaptic cleft during transmission may be much higher than in plasma. Evidence is good for GABA as an inhibitory transmitter, particularly in crustaceans and insects and in the vertebrate cerebellum, and for glycine as an inhibitory transmitter in spinal cord. Evidence for glutamate as an excitatory transmitter is best in crustacean and insect muscle, and is less convincing for vertebrate nervous systems.

Recent evidence indicates that some autonomic neurons of vertebrates may be neither cholinergic nor adrenergic, but may use ATP as a transmitter; these are called purinergic neurons, and they have been found in gastrointestinal tract and bladder of mammal, and in innervation of the gut of fishes and amphibians (p. 776).[25]

The number of proven cases of neurotransmitters is small; however, it is unlikely that many new agents will be found. The real differences which relate to different sensitivities and different molecular actions of given transmitters reside in receptor molecules. It is probable that many specific receptors will be discovered for each transmitter.

SUMMARY

Evidence is rapidly accumulating for a comparative physiology of cell membranes. So many specific functions are localized in cell membranes that it is remarkable that all of them can coexist in the 100 Å layer of lipoproteins. Specific channels of 5 to 15 Å diameter are indicated by permeability measurements. Diversity of membranes in respect to active transport was discussed for ionic regulation in Chapter 2, and for transport of products of digestion in Chapter 4.

The first distinction for membranes of sense cells, neurons, and effectors is between those membrane areas which are normally excited electrically and those which are normally excited by chemicals or by mechanical deformation. The electrically excited membranes usually have a threshold for initiating a regenerative impulse. Structures stimulated chemically include: (1) sensory endings which are stimulated by applied chemicals or by the products of photochemical reactions, and (2) postsynaptic membranes which contain receptor molecules for specific neurotransmitters. Many neurons are excited chemically at one end, support self-propagating impulses in their axons, and then secrete a chemical at the terminal; others lack the electrically excited conducting region. Signal transmission without spikes occurs in neurons which are short relative to their space constant, as in intraretinal amacrine cells; these are axonless neurons. Electrotonic potentials may spread without spikes from a synaptic site on a dendritic tree to the soma and spiking region of the axon.

The best known electrogenesis is for all-or-none spikes in electrically excitable membranes. Most such spikes in nerves and striated muscles result from a sequential increase in conductance for sodium and potassium. Spike duration varies with time of K-activation. Post-spike events include delayed repolarization due to delayed K-activation and undershoot or hyperpolarization due to persistent K-efflux. In many muscles, particularly in arthropods and vertebrate smooth muscles, and in some molluscan neurons and in ciliate protozoans, the inward current of a spike is carried by calcium ions. Calcium spikes may be more primitive than sodium spikes, and it is probable that all electrically excitable membranes have both sodium and calcium channels but that one becomes predominant in a particular cell type or area. A few membranes depolarize by chloride efflux. Some repetitively firing neurons show two K-activations; some pacemakers oscillate because of time-dependent coupling of Na and K conductance.

Various graded conductance changes are found in chemically excited membranes. The commonest mechanism of depolarization in postsynaptic membranes shows simultaneous increase in Na- and K-conductance, while in the best known hyperpolarizing postsynaptic membranes an influx of chloride occurs. The size of the response depends on the resting potential relative to the ionic equilibrium potential. Other patterns of response involve changes in membrane resistance, e.g., the hyperpolarization of scallop photoreceptors by increasing K-conductance. Depolarization due to increase in Cl conductance can occur when the equilibrium potential is positive to RP even though the Cl concentration gradient is inward. A third type of graded electrical response, as in vertebrate photoreceptors, is the reduction, upon illumination, of an asymmetry sodium current (dark current) between the two ends of the receptor. Molluscan neurons show a wide range of conductance changes in response to neurotransmitters. In many cells an Na-pump contributes to the resting potential, and in some (e.g., crayfish stretch receptors) it provides for an after-hyperpolarization. Pump oscillations may provide for slow electrical rhythms in smooth muscle and in some neurons.

Many tissues—embryonic, epithelial, some nerve centers—show electrical coupling between cells. The degree of communication varies with condition and cell type and depends on the area of gap junctions (nexuses). Electrical transmission may be symmetrical or non-symmetrical (polarized), as in electrical synapses. In general, electrical junctions permit fast synchronized responses of groups of neurons. Some synaptic areas have both chemical and electrical junctional patches. One example of electrical inhibition by extra-axonal hyperpolarization is known.

In chemical synapses, specific transmitter substances are synthesized at the presynaptic ending or swelling, and are packaged in characteristic vesicles. Release is favored by presynaptic depolarization and influx of calcium ions. Part of the synaptic delay is due to diffusion through the synaptic cleft to receptors on the postsynaptic membrane. Transmitter inactivation occurs by postsynaptic hydrolysis, by diffusion away from the site of action, and by re-uptake into presynaptic endings. Chemical synapses occur in all nervous systems, including coelenterate nerve nets.

Known synaptic transmitters are relatively small molecules. Acetylcholine is the most widely distributed, and it may have membrane functions in many non-nervous tissues. It is established as a transmitter at synapses of vertebrate autonomic systems, some central nervous synapses, probably at some insect central synapses, at endings of regulatory nerves on vertebrate and molluscan hearts, and at neuromuscular junctions of many kinds of animals.

Catecholamines are widely distributed, as indicated by fluorescence microscopy. In vertebrate adrenergic neurons, the principal agent is norepinephrine, while in molluscs dopamine is more important. A monoamine, serotonin (5-hydroxytryptamine), is also widely distributed in nervous systems, and in molluscs it may antagonize acetylcholine and/or dopamine.

Three amino acids are transmitters—excitatory glutamic acid and inhibitory glycine and gamma-aminobutyric acid (GABA). In arthropods, glutamate and GABA are antagonists, particularly at neuromuscular junctions.

A given transmitter may have more than one action in the same animal, depending on receptor molecules of different postsynaptic cells. Acetylcholine (ACh) receptors in vertebrates are of two general classes—muscarinic and nicotinic—and in molluscan ganglia some cells are depolarized and others are hyperpolarized by ACh. The nicotinic receptor molecule has been isolated, mainly from electric organ, and it is a different protein from acetylcholinesterase. Norepinephrine acts on two classes of receptors in vertebrates, alpha and beta, and in each initiates a sequence of intracellular reactions involving cyclic AMP as an intermediate to the cellular responses. Presumably, each of the ionic conductance changes induced by transmitters depends on the nature of the receptor molecules.

It is established that the membranes of excitable cells are heterogeneous. They have patches in which permeability for specific ions may be altered. The reacting lipoprotein molecules may be stimulated by electric currents, or by specific chemicals or mechanical deformation. The responses depend in amplitude on the equilibrium potentials for the ions which move in or out across the membranes. In addition, resistance changes may alter external fields created by asymmetric cellular batteries. The mechanisms of bioelectrogenesis are very diverse, and some are more widespread than others.

REFERENCES

1. Adrian, R. H., and W. H. Freygang, J. Physiol. *163*:61–103, 1962. K and Cl conductance, frog muscle.

1a. Ahlquist, R. P., Ann. Rev. Pharmacol. *8*:259–272, 1968. Agents blocking adrenergic receptors.

1b. Amaro, J., et al., Brit. J. Pharm. Chemother. *28*:207–211, 1966. Neurotransmitter in efferent fibers of auditory system.

2. Aprison, M. H., et al., Comp. Biochem. Physiol. *28*:1345–1355, 1969. Distribution of glycine in brain and spinal cord of vertebrates.

3. Araki, T., M. Ito, and O. Oscarsson, J. Physiol. *159*:410–435, 1961. Anion permeability of motorneuron synapses.

4. Asada, Y., and M. V. L. Bennett, J. Cell Biol. *49*:159–172, 1971. Coupling resistance at septa of lateral giant axons, crayfish.

5. Auerbach, A. A., and M. V. L. Bennett, J. Gen. Physiol. *53*:183–210, 211–239, 1969. Giant fiber synapses in fish.

6. Bacq, Z. M., and F. Ghiretti, Arch. Int. Physiol. *49*:165–171, 1952; *also* Pubbl. Staz. Zool. Napoli *24*:267–277, 1953. Acetylcholine in *Octopus*.

7. Baker, P. F., Proc. XXV Internat. Physiol. Cong. *8*:17–19, 1971. Transport of Ca and Mg across nerve cell membranes.

8. Baker, P. F., et al., J. Physiol. *200*:459–496, 1969. Ouabain-sensitive fluxes of Na and K in squid giant axons.

9. Baker, P. F., A. L. Hodgkin, and E. B. Ridgway, J. Physiol. *218*:709–755, 1971. Calcium flux in squid axon.

10. Baker, P. F., et al., Nature *190*:885–887, 1961; *also* J. Physiol. *170*: 541–560, 1964. Internally perfused squid axons.

11. Baylor, D. A., and J. G. Nicholls, J. Physiol. *203*:555–569, 571–589, 1969. Ionic factors in activity of leech central nervous system.

12. Beeler, G. W., and H. Reuter, J. Physiol. *207*:165–190, 191–209, 1970. Voltage clamp experiments on mammalian ventricular fibers.

13. Belton, P., and H. Grundfest, Amer. J. Physiol. *203*:588–594, 1962. K spikes in *Tenebrio* larva muscle.

13a. Beers, W. H., and E. Reich, Nature *228*:917–922, 1970. Structure and activity of acetylcholine.

14. Bennett, M. V. L., et al., J. Neurophysiol. *30*:131–179, 180–208, 209–235, 236–300, 1967. Physiology and ultrastructure of electrotonic junctions.

15. Bennett, M. V. L., pp. 147–169 *in* Central Nervous System and Fish Behavior, edited by D. Ingle. University of Chicago Press, 1968. Electrotonic synapses.

16. Bennett, M. V. L., and J. P. Trinkaus, J. Cell Biol. *44*:592–610, 1970. Electrical coupling between cells in *Fundulus* embryos.

17. Beranek, R., and P. L. Miller, J. Exp. Biol. *49*:83–93, 1968. Glutamate stimulation of insect muscle.

18. Berger, W. K., and B. Uhrik, Z. Zellforsch. *127*:116–126, 1972. Nexuses in dipteran salivary glands.

19. Binstock, L., and L. Goldman, J. Gen. Physiol. *54*:730–740, 741–754, 755–764, 1969. Conduction in *Myxicola* giant axon.

20. Bloom, F. E., pp. 729–747 *in* The Neurosciences: Second Study Program, edited by F. O. Schmitt. Rockefeller University Press, New York, 1970. Structure-function correlation in synapses.

21. Bloom, F. E., B. J. Hoffer, and G. R. Siggins, Brain Res. *25*:501–521, 1971. Chemical transmitters in rat cerebellum.

22. Bradley, P. B., Int. Rev. Neurobiol. *11*:1–55, 1968. Chemical transmitters in brain.

23. Brightman, M. W., and T. S. Reese, J. Cell Biol. *40*:648–677, 1969. Gap junctions, tight junctions, Mauthner cells.

24. Bullock, T. H., J. Neurophysiol. *11*:343–364, 1948. Synaptic transmission, squid mantle ganglion.

25. Burnstock, G., Pharmacol. Rev. *21*:247–324, 1969; *52*:129–197, 1972. Autonomic innervation of visceral and vascular muscles in vertebrates.

26. Burnstock, G., and T. Iwayama, Prog. Brain Res. *34*:389–404, 1971. Autonomic synapses.

27. Bush, B. M. H., and A. Roberts, Nature *218*:1171–1173, 1968. Reflexes without impulses from a crab muscle receptor.

28. Carpenter, D., and R. Gunn, J. Cell. Physiol. *75*:121–127, 1970. Ionic currents in *Aplysia* neurons.

29. Chandler, W. K., and H. Meves, J. Physiol. *180*:788–820, 1965. Voltage clamp experiments on internally perfused giant axons.

29a. Changeux, J. P., J. Meunier, and M. Huchat, Molec. Pharmacol. 7:548–553, 1971. Properties of acetylcholine receptor.

30. Changeux, J. P., T. Podleski, and J. Meunier, J. Gen. Physiol. *54*: 225s–244s, 1969. Acetylcholine receptor, electric organ.

31. Chapman, R. A., J. Exp. Biol. *45*L:475–488, 1966. Repetitive firing, crab fibers.

32. Chapman, R. A., and J. H. Parkhurst, J. Exp. Biol. *46*:63–84, 1967. Conduction velocity, cockroach nerve.

33. Chiarandini, D. J., and H. M. Gerschenfeld, Science *156*:1595–1596, 1967. Ionic mechanisms, molluscan synapses.

34. Close, F., and A. Serfaty, Bull. Soc. Histoire Nat. Toulouse *92*:205–217, 1957. Cholinesterase in fish.

35. Cohen, L. B., et al., J. Physiol. *211*:495–515, 1970; *218*:205–237, 1971. Optical changes in nerve during conduction.

36. Cohen, L. B., and D. Landowne, pp. 247–263 *in* Biophysics and Physiology of Excitable Membranes, edited by W. J. Adelman. Van Nostrand Reinhold, New York, 1971. Birefringence and other optical changes in nerves.

36a. Colhoun, E. H., J. Insect Physiol. *2*:108–116, 1958; *also* Can. J. Biochem. Phys. *37*:1127–1134, 1959. Content of ACh, AChE, ACh acetylase, insect nervous systems.

37. Connor, J. A., and C. F. Stevens, J. Physiol. *213*:1–19, 21–30, 31–53, 1971. Early potassium current and repetitive firing.

38. Coombs, J. S., et al., J. Physiol. *137*:326–373, 1955; *145*:505–538, 1959. Electrical constants of motorneuron membranes.

39. Cottrell, G. A., Brit. J. Pharmacol. *29*:63–69, 1967. Adrenergic synapses.

40. Cottrell, G. A., Nature *225*:1060–1062, 1970. Serotonin synapses.

41. Cottrell, G. A., and M. S. Laverack, Ann. Rev. Pharmacol. *8*:273–298, 1968. Invertebrate pharmacology.

42. Crawford, J. M., et al., Nature *200*:579–580, 1963. Excitation of cerebellar neurons by acetylcholine.

43. Curtis, D. R., A. W. Duggan, and G. A. R. Johnston, Exp. Brain Res. *12*:547–565, 1971. Evidence for glycine as a transmitter in spinal cord.

44. Curtis, D. R., and J. Eccles, J. Physiol. *145*:529–546, 1959; *150*:374–398, 1960. Time course of synaptic actions in spinal cord.

45. Curtis, D. R., and J. C. Watkins, J. Neurochem. *6*:117–141, 1960. The excitation and depression of spinal motorneurons by amino acids.

46. Curtis, D. R., and J. C. Watkins, Pharmacol. Rev. *17*:347–391, 1965. Transmitters in central nervous system.

47. Curtis, D. R., et al., Nature *226*:1222–1224, 1971; *also* Brain Res. *32*:69–96, 1971. Evidence for GABA as transmitter in spinal cord.

48. Dahlstrom, A., and J. Haggendal, Acta Physiol. Scand. *67*:278–288, 1966. Half-life of adrenal vesicles.

49. Dewey, M. M., and L. Barr, J. Cell Biol. *23*:553–585, 1964. Structure and distribution of nexus.

50. Diamond, J., J. Physiol. *194*:669–723, 1968. GABA and glutamate sensitivity, Mauthner cells.

51. Dodge, F. A., and R. Rahamimoff, J. Physiol. *193*:419–432, 1967. Role of calcium in neuromuscular transmission.

52. Dowling, J. E., J. Invest. Ophthal. *9*:655–680, 1970. Retinal synapses.

53. Dowling, J. E., and F. S. Werblin, J. Neurophysiol. *32*:315–338, 1969. Retinal synapses in *Necturus*.

54. Eccles, J. C., The Physiology of Nerve Cells. Johns Hopkins Press, Baltimore, 1957. 270 pp.

55. Eccles, J. C., pp. 59–74 *in* Handbook of Physiology, Sec. 1, Vol. 1, edited by H. W. Magoun. Amer. Physiol. Soc., Washington, D.C., 1959. Central nervous synapses.

56. Eccles, J. C., The Physiology of Synapses. Springer-Verlag, Heidelberg, 1964. 316 pp.

57. Eckert, R., Science *176*:473–481, 1972. Bioelectric control of ciliary activity.

58. Eckert, R., J. Gen. Physiol. *46*:573–587, 1963. Electrical interaction of paired ganglion cells in the leech.

59. Eckert, R., and Y. Naitoh, J. Gen. Physiol. *55*:467–483, 1970. Bioelectrics of *Paramecium*.

60. Eckert, R., Y. Naitoh, and K. Friedman, J. Exp. Biol. *56*:683–694, 1972. Sensory mechanisms in *Paramecium*.

61. Ehinger, B., B. Falck, and H. S. Myhrberg, Histochemie *15*:140–149, 1968. *Hirudo* catecholamines and indolamines.

62. Ehrenpreis, S., J. H. Fleisch, and T. W. Mittag, Pharmacol. Rev. *21*:131–181, 1969. Molecular nature of pharmacological receptors.

63. von Euler, U. S., Noradrenaline, Chemistry, Physiology, Pharmacology, Clinical Aspects. Charles C Thomas, Springfield, Ill., 1958. 382 pp.

64. Falk, G., and P. Fatt, Proc. Roy. Soc. Lond. B *160*:69–123, 1964. Linear electrical properties of striated muscle fibers observed with intracellular electrodes.

65. Florey, E., Fed. Proc. *26*:1164–1178, 1967. Neurotransmitters and modulators in the animal kingdom.
66. Florey, E., Comp. Biochem. Physiol. *22*:617–627, 1967. Cholinergic neurons in tunicates, an appraisal of the evidence.
67. Fozzard, H., J. Physiol. *182*:255–267, 1966. Capacity of heart membranes. Personal communication, 1973.
67a. Furshpan, E. J., and T. Furukawa, J. Neurophysiol. *25*:732–771, 1962. Conduction in Mauthner cells.
68. Furshpan, E. J., and D. D. Potter, J. Physiol. *145*:289–325, 1959. Transmission at giant motor synapses of crayfish.
69. Furshpan, E. J., and D. D. Potter, J. Physiol. *145*:326–330, 1959. Slow post-synaptic potentials recorded from giant motor fiber of crayfish.
69a. Furukawa, T., Prog. Brain Res. *21*A:44–70, 1966. Electrical inhibition in the Mauthner cell.
70. Gage, P. W., and J. W. Moore, Science *166*:510–512, 1969. Synaptic current at squid giant synapse.
71. Gardner, D., and E. R. Kandel, Science *176*:675–678, 1972. Diphasic postsynaptic potential, *Aplysia*.
72. Geduldig, G., and R. Gruener, J. Physiol. *211*:217–244, 1970. *Aplysia* neurons: Na and Ca currents.
73. Geduldig, G., and D. Junge, J. Physiol. *199*:347–365, 1968. Sodium and calcium components of action potentials in *Aplysia* giant neurons.
74. Gerschenfeld, H. M., Nature *203*:415–416, 1964. Non-cholinergic synaptic inhibition in the central nervous system of a mollusc.
75. Gerschenfeld, H. M., Science *171*:1252–1254, 1971. Serotonin, snail neurons.
76. Gerschenfeld, H. M., and D. J. Chiarandini, J. Neurophysiol. *28*:710–723, 1965. Synaptic potentials and ion conductance changes, snail neurons.
77. Gerschenfeld, H. M., and A. Lasansky, Int. J. Neuropharm. *3*:301–315, 1964. Action of glutamic acid and amino acids on snail central neurons.
78. Gerschenfeld, H. M., and E. Stefani, J. Physiol. *185*:684–700, 1966. Electro-physiological study of 5HT receptors of neurons in molluscan nervous systems.
79. Gerschenfeld, H. M., and L. Tauc, J. Physiol. (Paris) *56*:360–361, 1964. Dopamine effects on *Aplysia* neurons.
80. Ginsborg, B. L., Pharmacol. Rev. *19*:289–316, 1967. Biophysics of synaptic and conducting membranes.
81. Girardier, L., et al., J. Gen. Physiol. *47*:189–214, 1963. Anion permeability of crayfish muscle fibers.
82. Grinnell, A. D., J. Physiol. *182*:612–648, 1966. Interaction between motorneurons in frog spinal cord.
83. Grundfest, H., pp. 71–99 *in* Properties of Membranes and Diseases of the Nervous System, edited by D. B. Tower. Springer-Verlag, Heidelberg, 1962. Ionic transport across neural and non-neural membranes.
84. Grundfest, H., pp. 1–116 *in* Advances in Comparative Physiology and Biochemistry, Vol. 2, edited by O. E. Lowenstein. Academic Press, New York, 1966. Comparative electrobiology of excitable membranes.
85. Grundfest, H., pp. 353–371 *in* The Neurosciences: A Study Program, edited by G. C. Quarton et al. Rockefeller University Press, New York, 1967. Synaptic and ephaptic transmission.
85a. Hagiwara, S., et al., Comp. Biochem. Physiol. *13*:453–460, 1964. Conduction in giant fibers of *Eudistyla*.
85b. Hagiwara, S., and Y. Kidokoro, J. Physiol. *219*:217–232, 1971. Ionic components of action potentials in amphioxus muscle.
86. Hama, K., Anat. Rec. *141*:275–294, 1961. Fine structure of giant fibers of crayfishes.
87. Hand, A. R., and S. Gobel, J. Cell Biol. *52*:397–408, 1972. Septate and gap junctions in *Hydra*.
88. Heuser, J. E., and R. Miledi, Proc. Roy. Soc. Lond. B *179*:247–260, 1971. Frog neuromuscular ending; La^{++} stimulation.
88a. Heuser, J. E., and T. S. Reese, J. Cell Biol. 1972 (in press). Turnover of synaptic vesicles at neuromuscular junctions.
89. Hille, B., J. Gen. Physiol. *50*:1287–1301, 1967; *51*:199–219, 1968. Pharmacological analysis of sodium and potassium channels in frog nerve.
90. Hodgkin, A. L., Conduction of the Nervous Impulse. Charles C Thomas, Springfield, Ill., 1964. 108 pp.
91. Hoffer, B. J., et al., Brain Res. *25*:523–534, 1971.
92. Hornykiewicz, O., Pharmacol. Rev. *18*:925–964, 1966. Dopamine (3-hydroxy-tryptamine) and brain function.
93. Howarth, J. V., R. D. Keynes, and J. M. Ritchie, J. Physiol. *194*: 745–793, 1968. Heat production by nerve.
93a. Hughes, G. M., J. Exp. Biol. *46*:169–193, 1967. Abdominal ganglion of *Aplysia*.
94. Ito, Y., H. Kuriyama, and N. Tashiro, J. Exp. Biol. *50*:107–118, 1969. Miniature excitatory junction potentials in somatic muscle of earthworm.
95. Iversen, L. L., pp. 768–782 *in* The Neurosciences: Second Study Program, edited by F. O. Schmitt. Rockefeller University Press, New York, 1970. Neurotransmitters.
96. Jacobson, S., and D. A. Pollen, Science *161*:1351–1353, 1968. Conduction in apical dendrites.
97. Johnson, J. L., Brain Res. *37*:1–19, 1972. Glutamic acid as a synaptic transmitter.
98. Josephson, R. K., J. Exp. Biol. *42*:139–152, 1965. Conduction systems in stalk of a hydroid.
99. Josephson, R. K., J. Exp. Biol. *47*:172–190, 1967. Conduction and contraction in the column of *Hydra*.
100. Josephson, R. K., and M. Machlin, J. Gen. Physiol. *53*:638–665, 1969. Potentials across body wall of *Hydra*.
101. Juorio, A. V., J. Physiol. *216*:213–226, 1971. Catecholamines and 5HT in nervous tissue of cephalopods.
101a. Kandel, E. R., and D. Gardner, pp. 91–144 *in* Neurotransmitters: Proceedings, Vol. 50, Association for Research in Nervous and Mental Diseases. Williams and Wilkins, Baltimore, 1972. Synaptic actions in molluscan ganglia.
102. Kanno, Y. Y., and W. R. Loewenstein, Nature *201*:194–195, 1964. Low-resistance coupling between gland cells.
103. Kao, C. Y., J. Neurophysiol. *23*:618–635, 1960. Postsynaptic electrogenesis in septate giant axons.
104. Kao, C. Y., Pharmacol. Rev. *18*:997–1049, 1966. Tetrodotoxin, saxitoxin, and the study of excitation phenomena.
105. Katz, B., and R. Miledi, J. Physiol. *203*:459–487, 1969. TTX-resistant electrical activity in presynaptic terminals.
106. Katz, B., and R. Miledi, J. Physiol. *216*:503–512, 1971. Effect of depolarization on synaptic transfer in stellate ganglion of squid.
107. Kerkut, G. A., et al., Comp. Biochem. Physiol. *15*:485–502, 1965. Glutamate as transmitter in *Helix*.
108. Kerkut, G. A., and D. R. Gardner, Comp. Biochem. Physiol. *20*:147–162, 1967. Role of calcium ions in action potentials of *Helix*.
109. Kerkut, G. A., N. Horn, and R. J. Walker, Comp. Biochem. Physiol. *30*:1061–1074, 1969. Long-lasting synaptic inhibition in the snail *Helix*.
110. Kerkut, G. A., and R. W. Meech, Comp. Biochem. Physiol. *19*:819–832, 1966. Ionic conductances in *Helix* neurons.
111. Kerkut, G. A., R. M. Pitman, and R. J. Walker, Comp. Biochem. Physiol. *31*:611–633, 1969. Iontophoretic application of acetylcholine and GABA onto insect central neurons.
112. Kerkut, G. A., C. B. Sedden, and R. J. Walker, Comp. Biochem. Physiol. *18*:921–930, 1966. Deopamine content of snail *Helix*.
113. Kerkut, G. A., C. B. Sedden, and R. J. Walker, Comp. Biochem. Physiol. *21*:687–690, 1967. Cellular localization of monoamines.
114. Kerkut, G. A., and R. C. Thomas, Comp. Biochem. Physiol. *8*:39–45, 1963. Acetylcholine potentials in snail neurons.
115. Kerkut, G. A., and R. C. Thomas, Comp. Biochem. Physiol. *11*:199–213, 1964. Effect of anion injection on several potentials of IPSP and ACh in *Helix*.
116. Keynes, R. D., J. Physiol. *114*:119–150, 152–182, 1951. Na and K content and exchange in crab and squid nerve.
117. Keynes, R. D., J. Physiol. *169*:690–705, 1963. Chloride in the squid giant axon.
118. Keynes, R. D., pp. 707–715 *in* The Neurosciences: Second Study Program, edited by F. O. Schmitt. Rockefeller University Press, New York, 1970. Optical changes during nerve conduction.
119. Koketsu, K., and S. Nishi, J. Gen. Physiol. *53*:608–623, 1969. Calcium and action potentials of bullfrog sympathetic ganglion cells.
120. Koketsu, K., S. Nishi, and H. Solda, Life Sci. *7*:955–963, 1968. Calcium and ACh potential of bullfrog sympathetic ganglion.
121. Kravitz, E. A., pp. 433–443 *in* The Neurosciences: Second Study Program, edited by F. O. Schmitt. Rockefeller University Press, New York, 1970. Gamma-aminobutyric acid and glutamic acid as neurotransmitters.
122. Kravitz, E. A., and D. D. Potter, J. Neurochem. *12*:323–328, 1965. GABA as inhibitor in lobster.
123. Kravitz, E. A., D. D. Potter, and N. M. van Gelder, Biochem. Biophys. Res. Comm. *7*:231–236, 1962. Gamma-aminobutyric acid distribution in lobster nervous system.
124. Kriebel, M., et al., Science *166*:520–524, 1969. Electrotonic coupling between neurons in fish oculomotor center.
125. Krnjevic, K., and J. W. Phillis, J. Physiol. *165*:274–304, 1963. Iontophoric studies of neurons in mammalian cerebral cortex.
126. Kuffler, S. W., Proc. Roy. Soc. Lond. B *168*:1–21, 1967. Neuroglial cells.
127. Kuffler, S. W., and J. G. Nicholls, Ergebn. Physiol. *57*:1–90, 1966. Neuroglia function in leech nervous system.
128. Kuffler, S. W., and D. D. Potter, J. Neurophysiol. *27*:290–320, 1964. Glia in leech central nervous system.
129. Kusano, K., J. Gen. Physiol. *52*:326–345, 1969. Relationship between pre- and postsynaptic potentials in squid giant synapse.
130. Kusano, K., and H. Grundfest, J. Cell. Comp. Physiol. *65*:325–336, 1965. Circus reexcitation as a cause of repetitive activity in crayfish lateral giant axons.
130a. Kusano, K., and M. M. LaVail, J. Comp. Neurol. *142*:481–494, 1971. Conduction in lobster and shrimp giant fibers.
131. Landolt, A. M., and H. Ris, J. Cell Biol. *28*:391–405, 1966; *also* Z. Zellforsch. *69*:246–259, 1966. Soma-somatic interneuronal junctions in corpus pedunculatum of wood ant.
132. Lentz, T. L., Comp. Biochem. Physiol. *27*:715–718, 1968. Acetylcholinesterase activity in a planarian.
133. Libet, B., J. Neurophysiol. *30*:494–515, 1967. Rabbit sympathetic ganglia.
134. Libet, B., and H. Kobayashi, Science *164*:1530–1532, 1969. Adrenergic and cholinergic potentials in sympathetic ganglion cells.
135. Loe, P. R., and E. Florey, Comp. Biochem. Physiol. *17*:509–522, 1966. ACh and ChEs in *Octopus*.
136 Loewenstein, W. R., Ann. N.Y. Acad. Sci. *137*:441–472, 1966. Permeability of membrane junctions.
137. Loewenstein, W. R., and Y. Kanno, J. Gen. Physiol. *6*:1123–1140, 1963. Electrical properties of a nuclear membrane.

138. Loewenstein, W. R., and Y. Kanno, J. Cell Biol. *33*:225–234, 1967. Electrotonic coupling, epithelial cells.
139. Loewenstein, W. R., M. Nakas, and S. J. Socolar, J. Gen. Physiol. *50*:1865–1891, 1967. Junctional membrane uncoupling.
140. Loveland, R. E., Comp. Biochem. Physiol. *9*:95–104, 1963. 5-Hydroxytryptamine, probable mediator of excitation in heart of *Mercenaria*.
141. Mackie, G. O., Amer. Zool. *5*:439–453, 1965. Conduction in nerve-free epithelia of siphonophores.
142. Mackie, G. O., J. Exp. Biol. *49*:387–400, 1968. Electrical activity in hydroid *Cordylophora*.
143. Mackie, G. O., Quart. Rev. Biol. *45*:319–332, 1970. Neuroid conduction and evolution of conducting tissues.
144. Mackie, G. O., and L. M. Passano, J. Gen. Physiol. *52*:600–621, 1968. Epithelial conduction in hydromedusae.
145. McLennan, H., Synaptic Transmission, 2nd edition. W. B. Saunders Co., Philadelphia, 1970. 178 pp.
146. Maiskii, V. A., Fed. Proc. Transl. Suppl. *23*:T1173–T1176, 1964. Electrical characteristics of membrane of giant nerve cells of *Helix*.
147. Marsden, C. A., and G. A. Kerkut, Comp. Biochem. Physiol. *31*:851–862, 1969. Fluorescent microscopy of central nervous system of leech.
148. Martin, A. R., and G. Pilar, J. Physiol. *168*:443–463, 1963. Dual mode of synaptic transmission in avian ciliary ganglion.
149. Martin, A. R., and G. Pilar, J. Physiol. *168*:464–475, 1963. Transmission through ciliary ganglion of chick.
150. Martin, A. R., and W. O. Wickelgren, J. Physiol. *212*:65–83, 1971. Sensory cells in spinal cord of sea lamprey.
151. McNutt, N. S., and R. S. Weinstein, J. Cell Biol. *47*:666–688, 1970. Ultrastructure of the nexus.
151a. McReynolds, J., and A. Gorman, J. Gen. Physiol. *56*:376–391, 1970; 1972 (in press). Hyperpolarizing responses of *Pecten* photoreceptors.
152. Meiri, U., and R. Rahamimoff, J. Physiol. *215*:701–726, 1971. Activation of transmitter release by strontium and calcium ions at neuromuscular junctions.
153. Miledi, R., Nature *212*:1240–1242, 1966. Miniature synaptic potentials in squid nerve cells.
154. Miledi, R., J. Physiol. *192*:379–406, 1967. Glutamate as possible transmitter in squid.
155. Miledi, R., P. Molinoff, and L. T. Potter, Nature *229*:554–557, 1971. Isolation of the cholinergic receptor protein of *Torpedo* electric tissue.
156. Miller, M. D., and J. M. Marshall, Amer. J. Physiol. *209*:859–865, 1965. Catecholamine receptors in uterus.
157. Molinoff, P. B., and J. Axelrod, Ann. Rev. Biochem. 465–500, 1971. Biochemistry of catecholamines.
158. Moore, J. W., J. Gen. Physiol. *48*: Suppl.: 11–17, 1965, Voltage clamp studies on internally perfused axons.
159. Moore, J. W., and T. Narahashi, Fed. Proc. *26*:1655–1663, 1967. Sodium channels in squid axon.
160. Morris, D., G. Bull, and C. O. Hebb, Nature *207*:1295, 1965. Acetylcholine in the electric organ of *Torpedo*.
161. Murdock, L. L., Comp. Gen. Pharmacol. *2*:254–274, 1971. Catecholamines in arthropods.
162. Myhrberg, H. E., Z. Zellforsch. *81*:311–343, 1967. Monoaminoergic neurons in *Limulus*.
163. Nachmansohn, D., Chemical and Molecular Basis of Nerve Activity. Academic Press, New York, 1959. 235 pp.
164. Naitoh, Y., and R. Eckert, Z. vergl. Physiol. *61*:427–452, 453–472, 1968. *Paramecium*, reversal of cilia, electrical responses.
165. Naitoh, Y., R. Eckert, and K. Friedman, J. Exp. Biol. *56*:667–681, 1972. Regenerative calcium spikes in *Paramecium*.
166. Nakas, M., S. Higashino, and W. R. Loewenstein, Science *151*:89–91. 1966. Uncoupling of an epithelial cell membrane junction by calcium-ion withdrawal.
167. Narahashi, T., pp. 119–131 *in* Electrical Activity of Single Cells. Igakushoin, Hongo, Tokyo, 1960. Excitation and electrical properties of giant axons of cockroaches.
168. Nicholls, J. G., and D. A. Baylor, Science *162*:279–281, 1968. Hyperpolarization after activity of neurons in leech.
169. Nicholls, J. G., and S. W. Kuffler, J. Neurophysiol. *27*:645–671, 1964. Ionic composition of glial cells and neurons of leech.
170. Nicholls, J. G., and D. Purves, J. Physiol. *209*:647–668, 1970. Monosynaptic chemical and electrical connections between sensory and motor cells in CNS of leech.
171. Nicholls, J. G., and D. E. Wolfe, J. Neurophysiol. *30*:1574–1592, 1967. Extracellular spaces in cells of leech CNS.
172. Nishi, S., and K. Koketsu, J. Cell. Comp. Physiol. *55*:15–30, 1960. Synaptic potentials in frog sympathetic ganglion.
173. Nishi, S., and K. Koketsu, Life Sci. *6*:2049–2055, 1967. Origin of ganglionic inhibitory postsynaptic potential, bullfrog sympathetic ganglion.
174. Obata, K., et al., Exp. Brain Res. *4*:43–57, 1967. Norepinephrine and cyclic AMP excitation of cerebellar nuclei.
174a. O'Brien, R. D., M. E. Eldefrawi, and A. T. Eldefrawi, Ann. Rev. Pharmacol. *12*:19–34, 1972. Isolation of acetylcholine receptors.
175. O'Brien, R. D., and L. P. Gilmour, Proc. Nat. Acad. Sci. *63*:496–503, 1969; *65*:438–445, 1970. Muscarone-binding material in electroplax.
176. Ochizono, K., Nature *214*:833–834, 1967. Structure of synaptic vesicles.
177. Ohashi, H., J. Physiol. *212*:561–575, 1971. Cl and K currents in taenia coli.
178. Okajima, A., and H. Kinosita, Comp. Biochem. Physiol. *19*:115–131, 1966. Ciliary activity and coordination in *Euplotes*.
178a. Oomura, Y., et al., Seitai no Kagaku, Japan *13*:31–38, 1962. Membrane constants.
179. Osborne, N. N., and G. A. Cottrell, Z. Zellforsch. *112*:15–30, 1971. Biogenic amines in *Limax*.
180. Otsika, M., et al., Proc. Nat. Acad. Sci. *56*:1110–1115, 1966. Lobster claw, GABA inhibition.
181. Pappas, G. D., and M. V. L. Bennett, Ann. N.Y. Acad. Sci. *137*:495–508, 1966. Electrical transmission between neurons.
182. Parisi, M., E. Rivas, and E. de Robertis, Science *172*:56–57, 1971. ACh receptors from electric organ.
183. Parnas, I., J. Neurophysiol., 1972 (in press). Differential presynaptic block at high frequency of branches of a single axon innervating two muscles.
184. Parnas, I., et al., J. Exp. Biol. *50*:635–649, 1969. Non-homogeneous conduction in giant axons of nerve cord of *Periplaneta americana*.
185. Paul, D. H., Science *176*:680–682, 1972. Decremental conduction over "giant" afferent processes in an arthropod.
186. Payton, B. W., M. V. L. Bennett, and G. D. Pappas, Science *165*:594–597, 1969. Electrotonic synapse.
187. Payton, B. W., M. V. L. Bennett, and G. D. Pappas, Science *166*: 1641–1643, 1969. Junctional membranes at an electrotonic synapse.
188. Payton, B. W., and W. R. Loewenstein, Biochim. Biophys. Acta *150*: 156–158, 1968. Electrical coupling in leech giant nerve cells.
189. Penn, R. D., and W. R. Loewenstein, Science *151*:88–89, 1966. Uncoupling of nerve cell membrane junction by calcium-ion removal.
190. Pentreath, V. W., and G. A. Cottrell, Comp. Biochem. Physiol. *27*: 775–785, 1968. Acetylcholine and cholinesterase in radial nerve of *Asterias verbens*.
191. Phillis, J. W., Comp. Biochem. Physiol. *17*:909–928, 1966. Neurotransmitters in mollusc intestine.
192. Phillis, J. W., The Pharmacology of Synapses. Pergamon Press, Oxford, 1970. 358 pp.
193. Pichon, Y., R. B. Moreton, and J. E. Treherne, J. Exp. Biol. *54*:757–798, 1971. Ionic basis of extraneuronal potential changes in central nervous system of cockroach.
194. Pichon, Y., and J. Boistel, J. Exp. Biol. *47*:343–355, 1967. Current-voltage relations in the isolated giant axon of cockroach.
195. Pitman, R. M., Comp. Gen. Pharmacol. *2*:347–371, 1971. Transmitter substances in insects: a review.
195a. Pitman, R. M., and G. A. Kerkut, Comp. Gen. Pharmacol. *1*:221–230, 1970. ACh and GABA in cockroach neurons.
196. Potter, D. D., E. J. Furshpan, and E. S. Lennox, Proc. Nat. Acad. Sci. *55*:328–336, 1966. Connections between cells of developing squid.
197. Prince, W. T., and M. J. Berridge, J. Exp. Biol. *56*:323–333, 1972. Salivary gland potentials in insects.
198. Rall, W., and G. M. Shepherd, J. Neurophysiol. *30*:1138–1168, 1967; *31*:884–915, 1968. Reconstruction of field potentials and dendrodendritic synaptic interactions in olfactory bulb.
199. Ralston, H. J. III, Nature *230*:585–587, 1971. Activity in presynaptic dendrites.
200. Revel, J. P., and M. J. Karnovsky, J. Cell Biol. *33*:C7–C12, 1967. Electrical synapses of Mauthner cells and liver.
201. Roberts, A., and C. A. Stirling, Z. vergl. Physiol. *71*:295–310, 1971. Propagation in skin of young tadpoles.
202. Robertson, J. D., pp. 715–728 *in* The Neurosciences: Second Study Program, edited by F. O. Schmitt. Rockefeller University Press, New York, 1970. Ultrastructure of synapses.
203. Robison, G. A., R. W. Butcher, and E. W. Sutherland, Ann. N.Y. Acad. Sci. *139*:703–723, 1966. Adenyl cyclase as an adrenergic receptor.
204. Robison, G. A., and E. W. Sutherland, Circ. Res. *26*: Suppl.: 147–161, 1970. Role of cyclic AMP in adrenergic responses.
204a. Rose, B., and W. R. Loewenstein, J. Membrane Biod. *5*:20–50, 1971. Effects of calcium on electrical coupling between epithelial cells.
204b. Rovainen, C. M., J. Neurophysiol. *30*:1000–1023, 1024–1042, 1967. Electrical and chemical synapses in lamprey giant neurons.
205. Rushforth, N. B., Biol. Bull *140*:255–273, 1971. Contraction pulses in *Hydra*.
206. Salanki, J., Recent Developments in Neurobiology, Hungary *3*:67–89, 1972. Serotonin in neural regulation of bivalve.
207. Sato, N., et al., J. Gen. Physiol. *51*:321–345, 1968. Ionic permeability changes during ACh-induced responses of *Aplysia* ganglion cells.
208. Shappirio, D. P., D. M. Eichenbaum, and B. R. Locke, Biol. Bull. *132*:108–124, 1967. Cholinesterase in brain of cecropia silkmoth during metamorphosis and pupal diapause.
209. Shaw, S., J. Physiol. *220*:145–175, 1972. Electrotonic spread in barnacle eye.
210. Siggins, G. R., A. P. Oliver, B. J. Hoffer, and F. E. Bloom, Science *171*:192–194, 1971. Cyclic AMP and norepinephrine effects on transmembrane properties of cerebellar Purkinje cells.
211. Smalley, K. N., Comp. Biochem. Physiol. *16*:467–477, 1971. Catecholamines in *Photinus*.
212. Smith, T. G., W. K. Stell, J. E. Brown, J. A. Freeman, and G. C. Murray, Science *162*:454–456, 457–458, 1968. Conductance changes associated with receptor potentials in *Limulus* photoreceptors.
213. Socharov, D. A., Ann. Rev. Pharmacol. *10*:335–352, 1970. Cellular aspects of invertebrate neuropharmacology.
214. Spira, M. E., and M. V. L. Bennett, Brain Res. *37*:294–300, 1972. Molluscan synaptic control of electrotonic coupling between neurons.

215. Strumwasser, F., J. Psychiat. Res. *8*:237–257, 1971. Spontaneously active neurons of *Aplysia*.
216. Suga, N., and Y. Katsuki, J. Exp. Biol. *38*:759–770, 1961. Pharmacological studies in auditory synapses in a grasshopper.
217. Sweeney, D. C., Comp. Biochem. Physiol. *25*:601–613, 1968. Distribution of monoamines in a freshwater bivalve mollusc.
218. Takeuchi, A., and N. Takeuchi, J. Gen. Physiol. *45*:1181–1193, 1962. Electrical changes in pre- and postsynaptic axons of giant synapse of *Loligo*.
219. Takeuchi, A., and N. Takeuchi, J. Physiol. *170*:296–317, 1964; *177*: 225–238, 1965. Localized action of gamma-aminobutyric acid in the crayfish muscle.
220. Takeuchi, A., and N. Takeuchi, J. Physiol. *183*:433–449, 1966. Permeability of presynaptic terminal in crayfish neuromuscular junction during synaptic inhibition by GABA.
221. Tasaki, I., A. Watanabe, and M. Hallett, J. Membrane Biol. *8*:109–132, 1972. Extrinsic fluorescence changes in squid axon.
222. Tauc, L., and H. M. Gerschenfeld, J. Neurophysiol. *25*:236–282, 1962. Cholinergic mechanism (inhibitory synaptic transmission) in a molluscan nervous system.
223. Tauc, L., Physiol. Rev. *47*:521–593, 1967. Transmission in invertebrate and vertebrate ganglia.
224. Taylor, G. W., J. Cell. Comp. Physiol. *18*:233–242, 1941; *20*:359–372, 1942. Relation between fiber size, birefringence, and conduction velocity.
225. Taylor, R. E., pp. 305–312 *in* The Neurosciences: A Study Program, edited by G. C. Quarton et al. Rockefeller University Press, New York, 1967. Initiation of nerve impulses.
226. Trautwein, W., and D. G. Kassebaum, J. Gen. Physiol. *45*:317–330, 1961. Pacemaker activity of heart.
227. Treherne, J. E., Neurochemistry of Arthropods. Cambridge University Press, 1966. 156 pp.
228. Treherne, J. E., N. J. Lane, R. B. Moreton, and Y. Pichon, J. Exp. Biol. *53*:109–136, 1970. Permeability of sheaths of *Periplaneta* nervous system.
229. Treherne, J. E., DeQ. Mellon, and A. D. Carlson, J. Exp. Biol. *50*: 711–722, 1969. Ionic basis of axonal conduction in central nervous system of *Anodonta cygnea*.
230. Treherne, J. E., and D. S. Smith, J. Exp. Biol. *43*:13–21, 1965. Penetration of acetylcholine into central nervous tissues of an insect.
231. Tucker, L. E., and Y. Pichon, J. Exp. Biol. *56*:441–457, 1972. Na efflux from CNS of cockroach.
232. Tupper, J., et al., J. Cell Biol. *46*:187–191, 1970. Electrical coupling between cells of starfish embryo.
233. Ueda, K., Ann. Zool. Japan *34*:99–110, 161–179, 1961. Electrical properties of *Opalina*.
234. Usherwood, P. N. R., P. Machili, and G. Leaf, Nature *219*:1169–1172, 1968. L-Glutamate at insect excitation nerve-muscle synapses.
235. Vassort, G., et al., Pflüg. Arch. *309*:70–81, 1969. Effects of epinephrine on inward Ca current in frog atrium.
236. Watanabe, A., and H. Grundfest, J. Gen. Physiol. *45*:267–308, 1961. Impulse propagation at the septal and commissural junctions of crayfish lateral giant axons.
237. Weight, F. F., and J. Votava, Science *170*:755–758, 1970. Late synaptic potentials in frog sympathetic ganglia.
238. Weinstein, R. S., and N. S. McNutt, New Eng. J. Med. *286*:521–524, 1972; *also* J. Cell Biol. *47*:666–688, 1970. Membrane to membrane contacts.
239. Werman, R., R. A. Davidoff, and M. H. Aprison, J. Neurophysiol. *31*:81–95, 1968. Glycine as inhibitory transmitter in spinal cord.
240. Werman, R., XXV Internat. Physiol. Cong., *8*:188–189, 1971. Spinal cord transmitters.
241. West, C. B., Quart. Rev. Biol. *30*:116–137, 1955. Comparative pharmacology of adrenal gland.
242. Whittaker, V. P., Ann. N.Y. Acad. Sci. *137*:982–998, 1966; *also* pp. 761–767 *in* The Neurosciences: Second Study Program, edited by F. O. Schmitt, Rockefeller University Press, New York, 1970. Properties of synaptosomes.
243. Wilson, D. M., Comp. Biochem. Physiol. *3*:274–284, 1961. Connections between lateral giant fibers of earthworms.
244. Woodward, D. J., et al., Brain Res. *33*:91–100, 1971. Frog cerebellum; synaptic transmission.
245. Woodward, D. J., B. J. Hoffer, G. R. Siggins, and F. E. Bloom, Brain Res. *34*:73–97, 1970. Development of synaptic activation in rat cerebellar Purkinje cells.
246. Yamasaki, T., and T. Narahashi, J. Insect Physiol. *3*:230–242, 1959. Electrical properties of cockroach giant axon.
247. Young, J. Z., Proc. Roy. Soc. Lond. B *162*:49–79, 1965. Diameters of fibers of peripheral nerves of *Octopus*.
248. Zettler, F., and M. Järvilehto, Z. vergl. Physiol. *76*:233–244, 1972. Electrotonic spread in an insect eye.

CHAPTER 12

MECHANORECEPTION, PHONORECEPTION, AND EQUILIBRIUM RECEPTION

By C. Ladd Prosser

CELLULAR BASIS OF SENSITIVITY TO MECHANICAL STIMULATION

The basic function of sense organs is to transduce physical stimuli into nerve impulses. Sense organs may be stimulated in three basic ways: chemical, mechanical, or by electric current. Gustatory and olfactory receptors are sensitive to specific chemicals, photoreceptors are stimulated by products of a photochemical reaction, and temperature receptors respond by modulation of chemically controlled spontaneity. Static mechanoreceptors are stimulated by maintained deformation of a membrane, as in some muscle tension receptors and in blood-pressure sensors; dynamic mechanoreceptors are stimulated during the process of deformation, at low frequency as in some tactile endings, or at high frequency as in sound reception. Mechanoreceptors are important for signalling posture, position of appendages, orientation with respect to gravity and acceleration, vibrations of low and high frequency, contact with surfaces, velocity of water and wind, and depth of water. Proprioceptors give information about the relative positions of parts of an animal, and exteroceptors give information about stimuli outside the body. A few kinds of fish have electroreceptors (p. 792).

In evolution, the differentiation of certain epithelial cells into receptors probably occurred before differentiation of conducting nerve cells. The simplest mechanoreceptors are free nerve endings which lie between or beneath epithelial cells. Many types of specialized terminal bulbs, discs, corpuscles, and spirals have been described. Frequently, the nerve ending is surrounded by a capsule or is connected to a projecting spine or hair. The surrounding structures serve as mechanical filters or amplifiers for applied force; they may also favor stimulation by movement in one plane or direction. Arthropods have many sensory hairs and setae. Insects have a variety of specialized sense cells associated with a semiflexible area of cuticle; these cells may be spines, campaniform sensilla, or chordotonal organs. In the campaniform sensillum, a terminal filament ends under a thin cuticular dome which magnifies surface strains. A chordotonal sensillum has a sensory filament on an elastic strand stretched between two points on the exoskeleton. Many hair-plate sensilla have a ciliary structure where the nerve ending enters the sensory hair. Muscles, tendons, and joints contain proprioceptive organs, often associated with specialized muscle fibers.

There is wide variation in the location of nuclei of sensory neurons. In an earthworm, epidermal sense cells synapse in a peripheral plexus, and there are many epidermal sense cells per afferent fiber in the segmental nerves. In many crustacean receptors, the nucleus of the neuron lies at the base of the hair or in the stretch receptor. The cell bodies of most tactile and proprioceptive endings of vertebrates are located in dorsal spinal ganglia or in the brain stem.

In some mechanoreceptors, stimulation impinges directly on nerve endings; in others, stimulation is transmitted via specialized epithelia to surrounding nerve endings. All mechanoreceptors generate graded sensory or generator potentials, which may sum and trigger conducted nerve impulses. Some mechanoreceptors are fast-adapting (i.e., give only one or a few nerve impulses per stimulus), while others are slow-adapting (give prolonged discharge during or following stimulation). Fast-adapting receptors vary greatly in the rate of stimulation which they follow. Many mechanoreceptors have their sensitivity modulated by efferent impulses from the central nervous system; some are spontaneously active, and stimulation increases or decreases the frequency of the background activity.

Many excitable nerves and muscles can be stimulated by mechanical deformation. Why some cell membranes are very sensitive to deformation and others are relatively insensitive is not understood. Some muscles are depolarized by stretch; hearts beat more strongly when distended, and many unitary smooth muscles respond actively to a quick stretch. Stretching a striated muscle can enhance neuromuscular transmission. An amoeba stops locomotion when stimulated by a mechanical tap. When nerve fibers are stimulated by a mechanical pulse, a graded increase in membrane conductance occurs and at a critical conductance a spike is initiated. In myelinated fibers of frog nerve, a 2 to 5 μm displacement triggers a spike, while in lobster giant axons a 10 to 15 μm displacement is needed. The conductance increase does not occur in the absence of sodium, but it can occur when depolarization of the membrane is prevented by voltage clamping.[112]

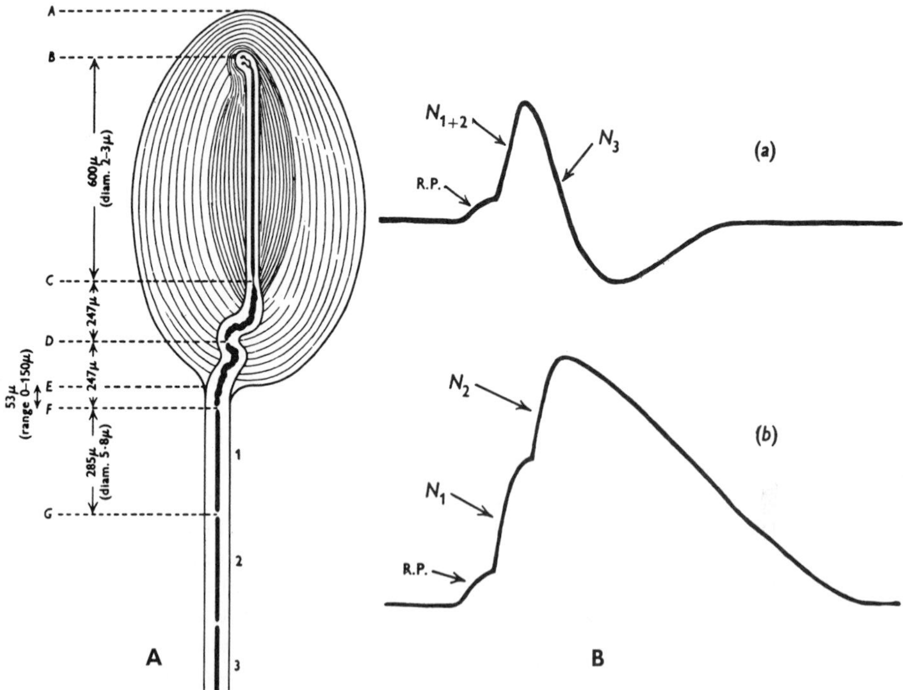

Figure 12–1. A, Diagram of a pacinian corpuscle, showing location of nodes, D, F, and G. (From Quilliam, T. A., and M. Sato, J. Physiol. *129*:167–176, 1955.) B, Electrical responses to mechanical stimuli recorded from pacinian corpuscle, (*a*) without polarizing current and (*b*) with anodal polarizing current. *R.P.*, receptor potential; N_1, N_2, N_3, and N_{1+2}, phases of response appearing at respective nodes. (From Diamond, J., J. A. B. Gray, and M. Sato, J. Physiol. *133*:54–67, 1956.)

In hair cells of vertebrate ears, stimulation may be occasioned upon movement of only 1 Å (see p. 528). Presumably, slight deformation alters the charges in cell membranes so that sodium permeability is increased.

TOUCH, MOTION, AND STRETCH RECEPTORS

Pacinian Corpuscles

Pacinian corpuscles from mammalian mesentery, tendon, and foot-pad may be taken as generalized mechano-sense organs. A large pacinian corpuscle consists of a number of concentric sheath layers around a nonmyelinated nerve terminal; the myelin sheath begins and one node occurs inside the capsule, and another node is located just outside the capsule (Fig. 12–1). An effective stimulus can be a 0.5 μm movement applied during 0.1 msec, or 0.2 to 0.4 μm over the 700 μm length of the nerve ending.[134] The electrical response from a pacinian corpuscle is a sensory potential, graded in amplitude and latency according to the strength of mechanical stimulation; at a critical amplitude, a spike is initiated at the intracapsular node. The sensory potential lasts 4 to 8 msec for a brief tap and conforms to a space constant of 0.8 mm.[39, 135] When nerve spikes are blocked by tetrodotoxin (TTX) the sensory potential continues, yet it cannot be initiated in the absence of Na. Hyperpolarization enhances, and depolarization decreases, the sensory potential. For a prolonged stimulation, the sensory potential normally shows "on" and "off" deflections; if the capsule is removed, the sensory potential is prolonged during deformation (Fig. 12–2). Hence, the capsule acts as a high-pass mechanical filter. The temperature coefficient of a sensory potential is twice that of a nerve spike. Normally, one or at most three spikes are initiated per sensory potential wave; i.e., the receptor is fast-adapting.[136] Spikes can follow repeated stimuli up to 150/sec at low strength of tap, and up to 280/sec for strong taps; the generator potential follows stimuli well at 650/sec. The rapid adaptation is partly in spike initiation and partly in the mechanical coupling of the capsule.[136]

Tactile Endings

A tactile ending in frog skin is rapidly adapting; it gives a single impulse to a brief (less than 0.1 sec) stimulation and 4 to 12 impulses to maintained deformation. On repetitive brief stimulations, spike initiation fails according to the frequency of stimulation. The latency to mechanical stimulation is 2.8 to 4.8 msec, and that to electrical stimulation is 2.3 to 2.8 msec, compared with 0.8 msec for direct nerve stimulation. The latency decreases when rate of deformation

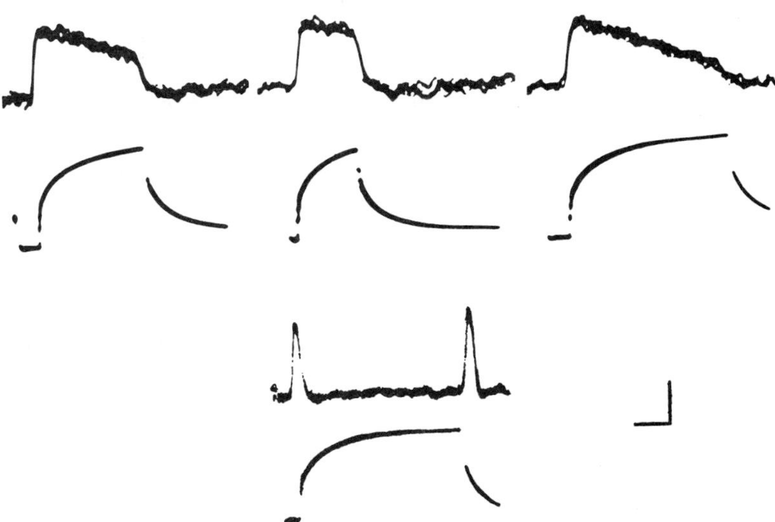

Figure 12–2. Receptor potentials from pacinian corpuscles to mechanical stimulation. Lower record of each pair indicates stimulus; upper record is receptor potential. Bottom recording from intact corpuscle; upper recordings from decapsulated endings. (From Loewenstein, W. R., and M. Mendelson, J. Physiol. *177*:377–397, 1965.)

increases, and for a given latency the effective velocity is higher after treatment of the skin with collagenase. There is a critical ramp slope for stimulation. It is concluded that the skin provides mechanical coupling to the nerve endings, that there are both elastic and viscous components, and that at low velocity of deformation the elastic component is more important than the viscous one. The threshold rises during repetitive stimulation as frequency of stimulation increases, and a subthreshold stimulus can raise the threshold for spike initiation.[25, 26, 27] The fatigue or failure to follow at high frequencies resides in the spike-initiating mechanism.

Receptors associated with tactile hairs usually adapt rapidly to stimulation. Each sensory hair along the margin of the telson and uropods of a crayfish gives only one impulse when flexed. In birds and mammals, several types of receptor associated with a feather or hair have been identified as to rates of adaptation and sensitivity.[27] In cat and rabbit, the receptors of down-hair have low thresholds and are fast-adapting, while those of guard hairs have high thresholds and adapt more slowly.[15] The frequency of sensory impulses is proportional to speed of hair movement in μm/sec raised to some exponent. Sensory endings in hairy skin show a low level of spontaneity. The response (R) is related to stimulus (S) by the power function $R = KS^n$, where n is less than 1.0 for the initial response and is about 1.0 for the later parts of the response.[256] Cat vibrissae activate several nerve endings, some of which respond only to velocity, and others to both velocity and amplitude of deflection.[83] In a duck, the down-feathers are quickly adapting and have large sensory fields and low thresholds; skin touch receptors have localized fields, a higher threshold, and a range of adaptation. Some receptors—probably Herbst corpuscles—follow stimulation at 400 to 800/sec.[46] In alligator skin, fast-adapting endings show on and off responses, and spikes follow at 100 to 200/sec; slow-adapting receptors fire at a frequency proportional to the displacement.[230] In some receptors the frequency is linearly proportional to displacement, and in others to the logarithm of displacement. The slowly adapting endings of reptilian skin fall into two categories with respect to instantaneous frequency, spontaneity, sensitivity, and speed of nerve conduction.[118] The primary sensory neurons in the spinal cord of a sea lamprey correlate in electrical properties with type of cutaneous ending:[145]

Receptor Type	*Sensory Neuron Resting Potential (mv)*	*Time Constant (msec)*
touch (fast-adapting)	70	1.4
pressure (slow-adapting)	68	2.3
(strong stimulus)	56	3.9

The frequency of impulses at various levels of sensory paths in the nervous system of monkeys was measured as a function of stimulus strength, which was indicated by skin indentation. The relation between sensory impulses and strength of tactile stimulation was non-linear; for hair skin the frequency initially increased steeply, and then much less per skin indentation; but for glabrous skin the initial frequency increase was slight, and with deeper indentation it was proportionately greater. The same relation was found in thalamic neurons and in the cerebral cortex; hence, the coding for stimulus strength is in the sense endings, and the same code continues to the highest level of the brain.[159]

Three types of mechanoreceptors are recognized in the foot-pad of the cat: (1) pacinian corpuscles, fast-adapting and most sensitive at 200 Hz; (2) receptors firing for a 500 msec burst; and (3) spontaneously discharging receptors for which the frequency is given by

$$F = K(S - S_0)^n$$

where S_0 is the threshold, S is the displacement stimulus, and n (the slope) has a value of 0.5.[106]

In the fish *Leuciscus*, tactile endings in the head give a maximal response to a tap at a particular temperature, and this changes with adaptation temperature.[234]

In the foot of a clam, phasic receptors give one spike at the beginning and one at the end of a displacement; threshold is 60 to 100 μm at a rate of 10 mm/sec, and for a sinusoidal displacement one spike occurs at the peak of each wave.[177] Some tactile receptors of crayfish are large enough that the soma or axon can be penetrated by microelectrodes; receptor potentials are small or rare, and spikes appear to originate in the dendrites.[151]

In a polychaete, *Harmothoë*, highly sensitive fast-adapting mechanoreceptors occur in widely distributed bristles; the sensory neurons appear to branch centrally so that a segmental nerve contains afferent fibers from its own segment and efferent processes of sensory cells from other segments.[98]

The campaniform sensilla of insects are intermediate in speed of adaptation to response in the cuticle. The discharge from joint receptors of insect legs starts at a high frequency (100 to 300/sec) and falls after 1 to 2 seconds to low levels.[188]

Intersegmental Stretch Receptors in Crustaceans

In crayfish, lobster, and some other decapod and stomatopod crustaceans, large stretch receptors occur; they are attached to the dorsal wall between abdominal segments and between thorax and abdomen. Each segment has two pairs of muscle receptor organs —RMO_1 (slow-adapting) and RMO_2 (fast-adapting) organs. Each receptor consists of a thin muscle, which is embraced by the finger-like dendrites of a large sensory neuron whose axon passes centrally (Fig. 12–3). The muscle fibers are innervated by some of the same motor axons that serve the extensor muscles, fibers from the innervation of the superficial extensors going to RMO_1, and fibers to deep extensors also serving RMO_2. In *Squilla*, the two receptor muscles attach to the same tendon. Each receptor cell receives inhibitory innervation from the central nervous system, a thick accessory inhibitor going to both organs of a pair (in lobster), and a thin accessory going only to RMO_1.[4] The sarcomeres are narrower in the muscle of RMO_2 than in RMO_1.[67]

Intracellular records from the sense cells show resting potentials of −70 mv. Stretch depolarizes the cell in a graded fashion. The sensory (generator) potential spreads electrotonically from the dendrites over the sense cell soma and, at a critical depolarization of 10 mv in the slow-adapting cell and 20 mv in the fast-adapting receptor, an impulse is

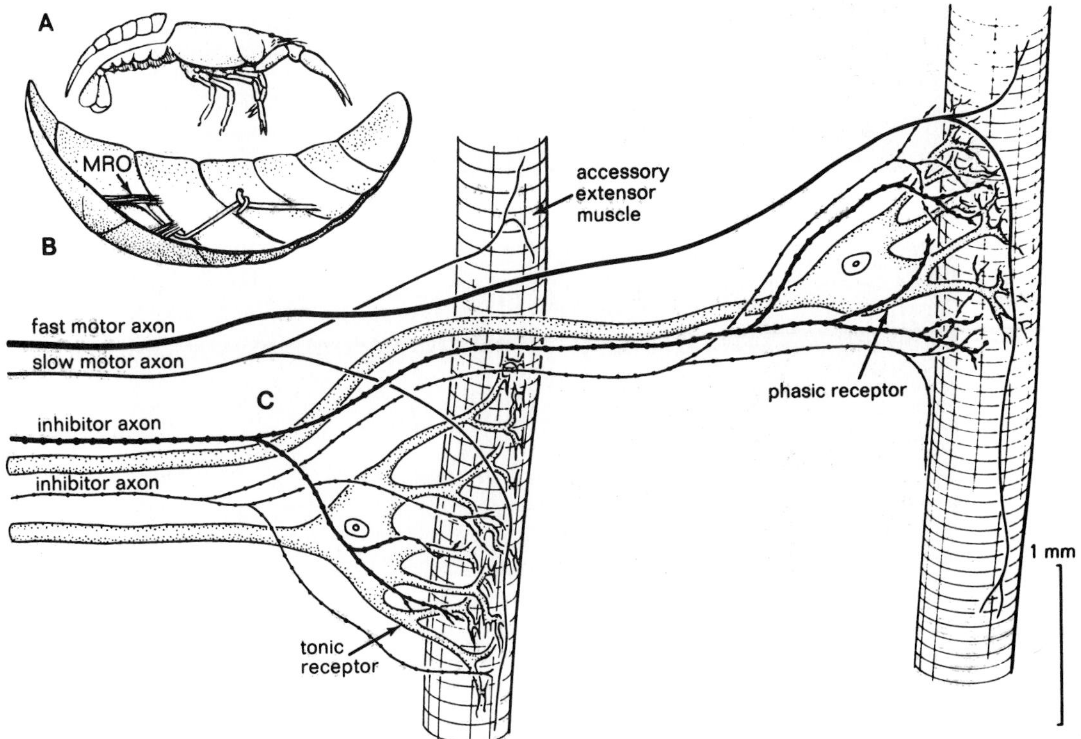

Figure 12–3. Diagrams of abdominal stretch receptor in a decapod crustacean. A and B, Location of receptor organ (MRO) between segments on dorsal abdomen. C, Phasic and tonic receptors attached to their respective muscles. Motor, inhibitor, and sensory nerve fibers are shown. (From Physiology of Invertebrate Nervous Systems, edited by C. A. Wiersma, University of Chicago Press.)

initiated. The sensory potential can be maintained and spike discharge can continue for several hours during prolonged stretch of RMO_1, but impulses stop in a few seconds in RMO_2. Receptor potentials originate in dendrites and spread over the soma; spikes originate at the base of the axon and may spread back over the soma. Combination of applied stretch with nerve-triggered contraction of the muscle is synergistic; i.e., the frequency of discharge of RMO_1 for a given stretch is increased by stimulation of motor axons, and there are 50 sensory impulses per second in RMO_1 when motor activation is at 50 to 80/sec.[16] Stimulation of the efferent (accessory) fiber or fibers inhibits the sensory discharge by preventing the sensory potential from attaining the firing level. If the receptor is under stretch and slightly depolarized, the inhibitor drives the membrane more negative, whereas in a relaxed receptor the inhibitor depolarizes. The inhibitor tends to

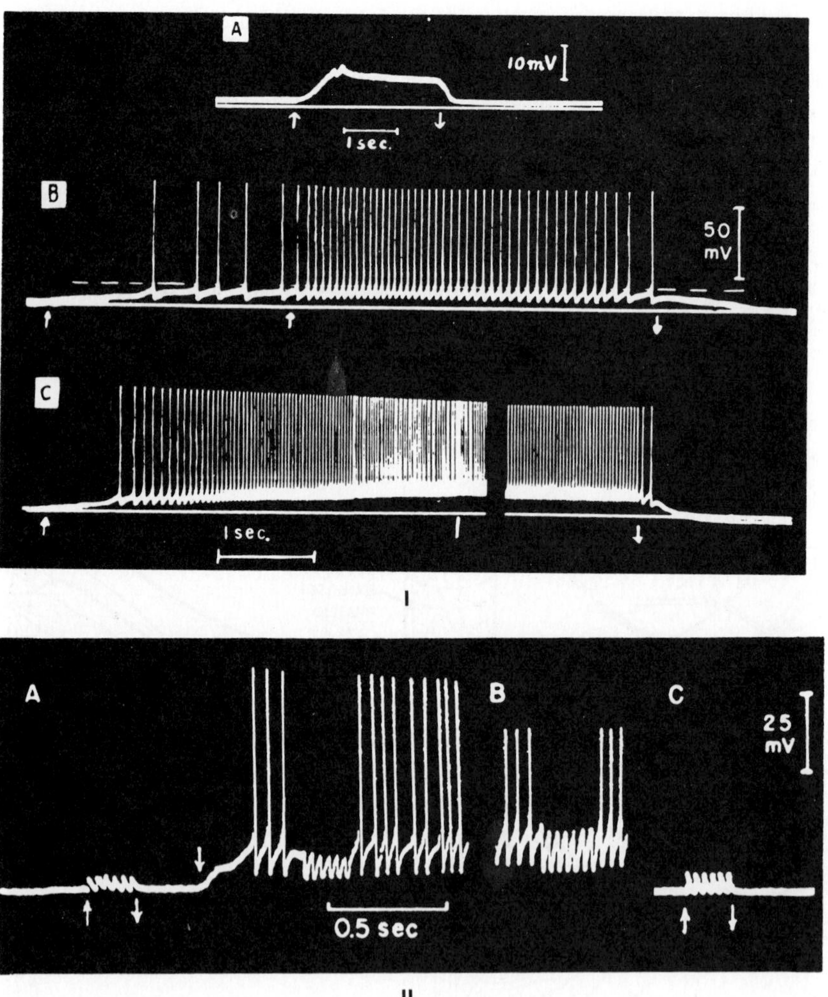

Figure 12–4. *I,* Potentials recorded intracellularly from crustacean stretch receptor. *A,* generator potential during subthreshold stretch, graded according to "deformation." Arrows show beginning and end of stretch. *B,* at critical level of depolarization, conducted impulses appear. At arrow, additional stretch increases firing frequency. *C,* stretch gradually increased between first arrow and vertical line. Relaxation followed by transient hyperpolarization. (From Kuffler, S. W., Exp. Cell. Res., suppl. *5*:495–519, 1958.) *II,* Inhibitory potentials recorded intracellularly from crustacean receptor. *A,* receptor partly relaxed; six inhibitory impulses cause depolarization potentials (arrows); at third arrow, stretch causes 20 mv depolarization and three sensory impulses, followed by inhibitory train, setting up repolarization potentials; continued stretch results in afferent impulses. *B,* inhibitory train inhibits sensory impulses and causes repolarization potentials. *C,* during complete relaxation of cell, inhibitory depolarization potentials. (From Kuffler, S. W., and C. Eyzaguirre, J. Gen. Physiol. *39*:155–184, 1955.)

place the membrane at an equilibrium about 6 mv more positive than the resting potential, and if the membrane is set at this potential by applied current the inhibitor can prevent impulse generation without altering the membrane potential[64] (Fig. 12–4). A microelectrode can record intracellular miniature inhibitory junction potentials (mipsp's) which, like the inhibitory potential, reverse if the receptor membrane is stretched. Injection of Cl^- increases the mipsp's; a KCl electrode records them as depolarizing and a potassium citrate electrode records them as hyperpolarizing; so the inhibitory transmitter (probably GABA) acts by increasing Cl^- conductance.[105]

The difference in rate of adaptation between RMO_1 and RMO_2 is seen with either stretch or intracellular current. When spikes are abolished by TTX, the generator potentials continue and are similar for both receptors. Voltage-current curves show similar membrane resistances; resting potentials are similar. It appears that about 30% of the difference in adaptation is due to spike-generating membrane and 70% to visco-elastic properties of the muscles.[162,163] If a constant current is applied by an intracellular electrode, spikes occur at decreasing frequency, much as in normal adaptation. Following a spike train, a post-tetanic hyperpolarization occurs which inhibits spike production; this hyperpolarizing component of adaptation is due to a sodium pump.[164,233] In either receptor, adaptation is faster at 24°C than at 5°C; sudden warming decreases and sudden cooling increases frequency of sensory discharge.[20]

The reflex function and control of crustacean stretch receptors are not well understood. Some evidence suggests that the receptors are reflexly inhibitory to the extensor muscles.[50] During abdominal extension the same motorneuron activates both the working fast extensor muscle and the slow RMO muscle; afferents from the slow RMO then reflexly excite a second motorneuron, which activates the tonic extensors. During flexion the extensors are centrally inhibited, and afferents from the SRMO reflexly inhibit the receptors in the next anterior segment. There is no evidence for RMO inhibition of flexor muscles and no evidence concerning the functions of the fast RMO's.[67] Intersegmental RMO's are found in swimming crustaceans but appear to be absent from brachyurans such as *Cancer*.

Vertebrate Muscle Spindles

Many skeletal muscles of vertebrates have sensory spindles and tendon organs that are sensitive to stretch. In mammals, extrafusal muscle fibers are innervated by large α motorneurons.[147] Within the muscle are spindles that consist of two to twelve intrafusal (fusimotor) muscle fibers, which receive motor innervation from small-diameter γ motorneurons; however, in some species, branches from the α motorneurons may also go to the intrafusal fibers.[9a] In the middle of the spindle are sense endings of two types: the primary annulospiral endings and the secondary flower-spray endings. The primary endings are connected to fast (70 to 120 m/sec) nerve fibers (12 to 20 μm in diameter), and the secondary endings are connected to slow (24 to 72 m/sec) nerve fibers (4 to 12 μm in diameter). Many mammalian spindles have both primary and secondary endings, while others have only primary ones. At the tendon of a muscle are Golgi endings of tension receptors connected to large afferent fibers. Muscle spindles are in parallel with extrafusal fibers, as shown by the interruption of their discharge during a muscle contraction; tendon organs are in series with extrafusal fibers, as shown by their increased frequency of firing during a muscle contraction. When the extrafusal fibers of the muscle contract, the stretch on the spindle is decreased and sensory discharge is momentarily suspended (Fig. 12–5). In isometric contraction under high tension, there may be enough pull by extrafusal fibers to stimulate the spindle.[35]

Fusimotor fibers can be subdivided into two groups according to their effects on the primary endings of muscle spindles when a muscle is stretched at constant velocity.[148] Dynamic fusimotor fibers cause the velocity response of the primary ending to be increased during the stretch. In contrast, a second group (the static fusimotor fibers) cause the velocity response to be decreased, in spite of a considerable excitatory action on the spindle with the muscle at constant length.

The central nervous system may exert independent control over static and dynamic fusimotor fibers. The role of the dynamic fibers may be to control the velocity response of the spindle and thus the sensitivity of the reflex loop connecting spindles with extrafusal fibers. Static fibers, with their powerful excitatory action on the spindle, may be important in preventing the spindle from being

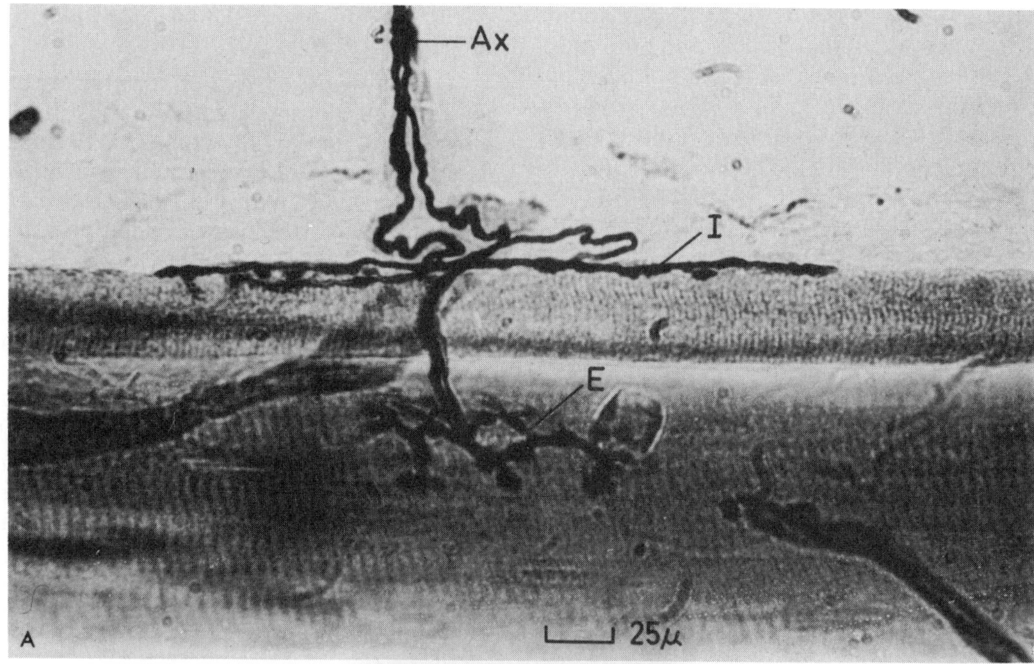

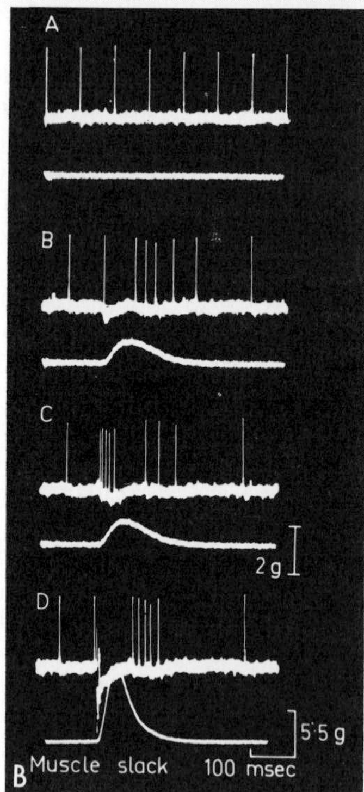

Figure 12–5. (a) A motor axon (Ax) in the semitendinosus muscle of the lizard *Tiliqua nigrolutea,* branching to supply an extrafusal (E) and an intrafusal (I) muscle fiber with "en grappe" motor endings. Silver stain. (Courtesy of U. Proske.) (b) Responses from a muscle spindle in the iliofibularis muscle of the same lizard. The action potentials were recorded from fine filaments of the dissected sciatic nerve while single electric shocks of increasing strength were applied to the motor nerve. The upper trace in each record represents the nerve discharge; muscle tension is recorded on the lower trace. Note the change in sensitivity of the myograph from C to D. The stimulus artifact in records B and C is obscured by the direct spike evoked by the stimulus. A, Subthreshold stimulus. B, Threshold for small contraction; resting spindle contraction ceased during contraction and increased during relaxation, as for parallel relation of the two kinds of fibers. C, Slightly stronger contraction; pull on spindle as in series. D, Very strong contraction of extrafusal fibers swamps pull on spindle. (Courtesy of U. Proske.)

silenced during reflex contraction of the muscle. This would permit a spindle to monitor lengths over a wide range.

The intrafusal fibers are excited tonically from higher centers, and the motorneuron discharge continues at 20 to 60/sec after spinal deafferentation.[17, 51] Intrafusal fibers contract only at their ends, and thus pull on the central region containing sense endings. Gamma fiber activity in mammals keeps the spindle tonically active, and less stretch is needed to excite a sensory spindle during tonic activity of the intrafusal fibers than in its absence.[78, 101] Tendon sense organs respond to muscle contraction and cease firing on muscle relaxation. Tendon organs have a higher threshold to passive stretch than to active contraction. Each tendon organ is activated by contraction of the small group of muscle fibers to which it is connected, but if the organ is discharging in response to passive stretch, contraction of fibers not in series with it can stop the firing by unloading the tendon receptor.[99] When a muscle is suddenly stretched, the spindles are stimulated and their afferent impulses elicit monosynaptic reflex contractions of the extrafusal fibers. Primary endings of cat are sensitive to vibration, but secondary endings are vibration-insensitive (see p. 518).[10]

Direct recording from a spindle of mammal or frog shows graded sensory potentials during stretch; these give rise to nerve afferent spikes. The receptor potential in frog spindles is sensitive to both Na^+ and Ca^{++}.[178] Early (rapid) adaptation decreases with decreasing speed of stretch and reflects changes in tonic properties of the sensory membrane in relation to length of the spindle.[179] It is estimated that 15 to 20% of adaptation is due to mechanical properties, and the rest to membrane changes.[103]

The muscle spindles of birds (duck) resemble those of mammals. When motor units contract, the firing of a spindle is interrupted; hence, the spindles are in parallel with extrafusal fibers and receive their own fusimotor supply.[46] In the lizard *Tiliqua,* the spindles interrupt their firing during a muscle twitch, but some increase their frequency of firing during the rising phase of a contraction; hence, they are in series with extrafusal fibers. By stimulation of nerve twigs and observation of muscle contraction and spindle discharge, it was concluded that some nerve fibers supply both fusimotor and extrafusal fibers.[190]

Frog muscle spindles, like those of reptiles, receive branches from the same axons that supply extrafusal fibers; i.e., the frog lacks a γ motorneuron system. However, frogs have two types of extrafusal innervation: fast muscles receive α fibers, and slow (graded) muscles receive γ innervation (p. 673).[148] Frog muscles have sensory endings of only the primary type. Fishes have no muscle spindles, but they do have proprioceptors in the fins.

In summary, stretch and tactile receptors are very diverse, particularly in respect to the auxiliary or mechano-coupling structures. Their directionality for stimulation, persistence of discharge (phasic or tonic), and adaptation all depend in part on mechanical properties and in part on receptor membranes. Probably all such receptors produce local or generator potentials at the sensing endings, and at the appropriate amplitude these potentials give rise to sensory nerve impulses.

Stretch Receptors and Stretch Reflexes in General

Stretch receptors generally adapt more slowly than tactile receptors. Impulses from single spindles in a frog toe muscle under load start at 120 to 260/sec and decline to 20/sec, a rate maintained for many minutes. Virtually no adaptation occurs in the pressure receptors of the carotid sinus of mammals. Some of the endings are tonically active and increase in frequency of discharge with each pulse; others fire four or five spikes per pulse. When the carotid artery in a cat is at steady pressure, there is little stimulation until the pressure rises above 40 to 50 mm Hg, and frequency increases continuously up to pressures of 200 mm Hg.[131]

Sense organs responding to stretch or movement of a skeletal element occur in many appendages. A dogfish lacks muscle spindles, but subcutaneous corpuscles in the fins discharge at a frequency proportional to the angular velocity and show slow adaptation to sinusoidal mechanical stimulation; discharge is maximal near the time of peak velocity, so that a corpuscle gives two bursts per locomotor cycle. In a lobster, three mouthpart receptors, each consisting of several sense cells on elastic strands, discharge phasically or tonically with mandibular and esophageal movements. The third walking

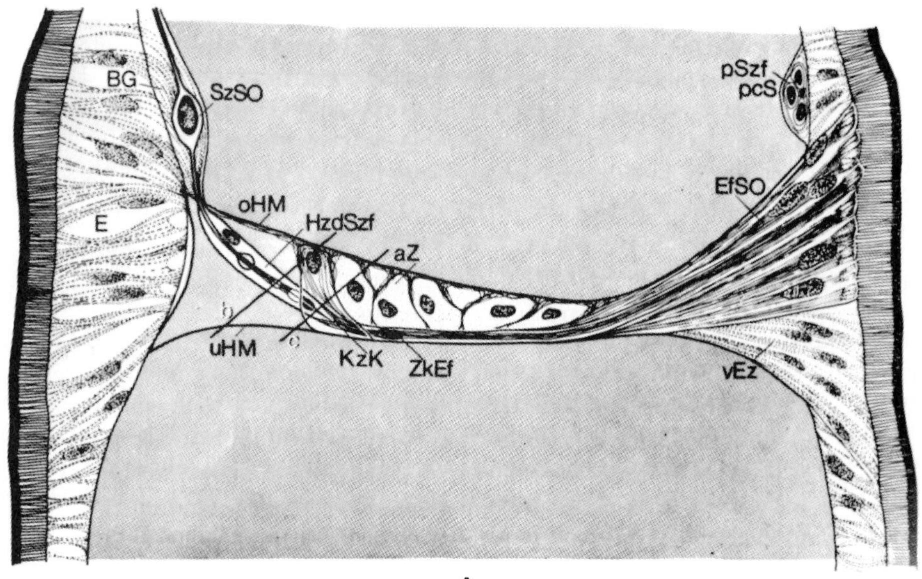

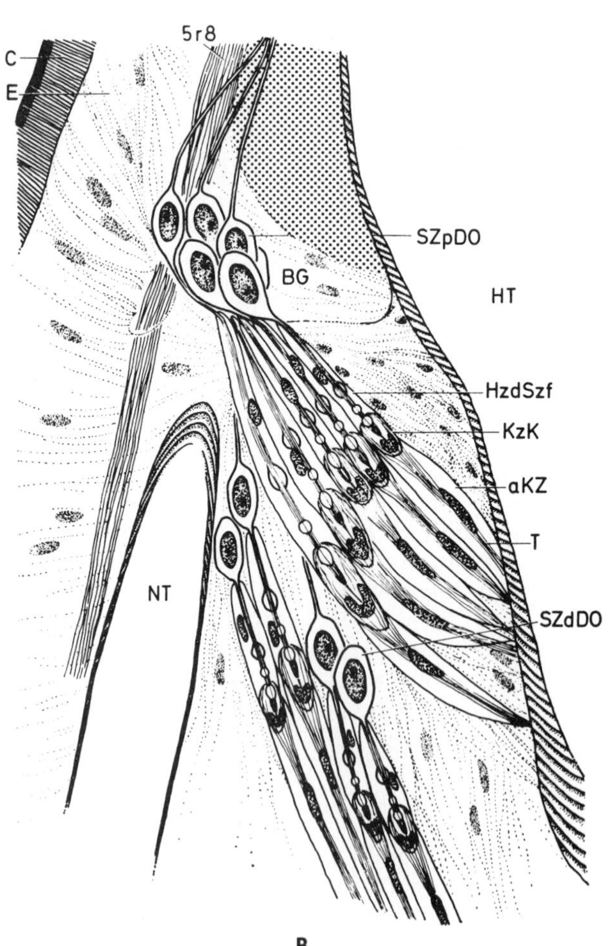

Figure 12–6. Drawings of two types of mechanoreceptors in cockroach leg. A, Subgenual organ, sensilla which respond to tension and vibration in cuticle. B, Distal organ, which responds to changes in blood pressure. SzSo, sensory neuron; EfSo, nerve endings; SZpDO, sensory neurons; HzdSzf, sensory fiber; KzK, cap over nerve endings. (From Schnorbus, H., Z. vergl. Physiol. *71*:14–48, 1971.)

leg of *Limulus* has 350 proprioceptive neurons, most of them in nine articular membrane organs.[94] Crabs have myochordotonal organs which signal both position and movement at a joint, some of them unidirectionally.[29, 91, 260] According to their orientation on a transverse ligament, some cells signal extension, and some flexion, of a segment.[23, 33] When the dactylus of *Carcinus* is passively opened, the opener and closer muscles are reflexly inhibited; passive stretching similarly inhibits the bender and stretcher muscles. Both actions result from mechanical stimulation of chordotonal organs at the joint.[23] Receptors in coxal muscles of several crabs generate large receptor potentials which are conducted passively without spike initiation to the nearby central ganglia.[197] One sensory axon receives from S (static) endings, which are in elastic strands in parallel with the muscle, and S responds in proportion to length; another axon receives from T (dynamic) endings, which are in series, adapt rapidly, and respond in proportion to velocity of stretch.[23a] In crayfish, neurons of the first and second ganglionic roots have dendrites extending to the hypodermis in association with insertion of superficial flexors and bases of swimming appendages; by axon reflexes, these neurons suppress motor discharges to flexors and swimmerets.[181]

There are two kinds of mechanoreceptors in arthropods: (1) those which lie between two exoskeletal elements and respond to vibrations in the cuticle, and (2) those which attach to tendons and are sensitive to movements of the tendon and to pressure changes (Fig. 12–6).

The foregut of the blowfly *Phormia* contains stretch receptors consisting of two bipolar neurons; when the gut is distended, stimulation of these receptors reflexly inhibits the positive response to chemostimulation and feeding.[71] In a scorpion, *Euscorpius*, each pedipalp bears 65 sensory hairs, each of which is stimulated by bending in its own plane toward the dendrite of the sense cell. Impulses are initiated by movement at 6° to 8° per second for a total arc amplitude of 2° to 3°; adaptation occurs if the hair is held toward the dendrite, and de-adaptation occurs if it is held in the opposite direction.[96]

Receptor potentials from single cells of cervical hair plates in honeybees decay with two time constants, suggesting two processes in the receptor potential.[250] The threshold deformation for stimulation of insect sensilla is calculated as compression of 30 Å or 0.5% of tubule diameter; maximum response is to movement of 0.1 μm or 15% compression.[250] Receptors in leg joints of insects serve as gravity detectors.[195] Phasic receptors at the base of the wings signal wingbeat, while tonic receptors signal wing position.[182]

The antennae of insects have a variety of mechanoreceptors, some of which are important for wind and wave detection. In a fly, the funiculus rotates around its axis and perpendicular to the arista; the resonant frequency of rotation is close to that of wingbeat, and if the funiculus is immobilized, antennal control of wing speed and position is lost.[22]

A number of beetles orient to wind or air puffs by receptors of the pedicellus joint of the antennae;[132] the antennae measure the angle between flight direction and wind. A whirligig beetle, *Gyrinus*, senses water waves by a resonant (250 Hz) vibration system of pedicellus and flagellum; motion between the flagellum and the pedicellus is detected, and slight wave motions can be signalled.[206]

Each sense hair on the cercus of *Periplaneta* contains a trichogen (supporting) cell and a sense cell; sensory potentials lead to generation of spikes.[251] The hair cell depolarizes if the hair is bent in one direction and hyperpolarizes on reverse bending, and the amplitude of the sensory potential is proportional to the extent of bending.[168] In *Locusta*, receptor potentials and spikes are proportional to bending, but no directionality such as that in *Periplaneta* was found.[232]

Behavioral Reactions to Touch and Pressure

Some reflex responses to skin and muscle receptors were mentioned above. Many other examples are known from behavioral studies. A cat, trained to lift its foot in response to tactile stimulation, can perceive (as shown behaviorally) the input from a single tactile pad.[245] Most fish, when stimulated on one side near the tail, turn the caudal fin toward the side of stimulation. Many sessile animals withdraw by body contraction from stimulation at the oral end. Many crawling animals show thigmotaxis or positive reactions to crevices and edges. For example, an earthworm crawling along one side of an object turns toward that side when it reaches the end of the stimulating surface. If, after the worm has responded positively to the edge, the stimulating object is withdrawn, the worm then straightens the anterior end to be in line with the posterior end. Contact reactions depend on unequal stimulation of the two sides, and have been demonstrated in annelids, diplopods, insect larvae, and young mammals.

Another type of contact orientation is rheotaxis, or orientation to a stream of water. Planaria react positively to a stream by receptors scattered over the body surface. Paramecia are normally positively rheotactic. Blind fish do not orient in a stream unless they touch the bottom; orientation is then controlled by the relation between the bottom and the current.

Anemotaxis, orientation to air currents, occurs in most flying animals. In flies, antennal sense organs are sensitive to wind and elicit postural changes according to direction and speed of air flow. Flight by locusts and lepidopterans can be initiated by blowing on wind-sensitive hairs on the head, and orientation in flight is maintained by wind pressure on the wings with stimulation of sensilla at the wing base.

Absence of ventral surface stimulation initiates righting or turning-over in snails, starfish, and some insects. Lifting the sucker of a leech initiates swimming, and flight by many insects is initiated by removal of contact stimulation of the tarsi. Position of a honeybee with respect to gravity is controlled by reflexes originating in sensory bristles on the episternum.

Many animals select their habitat by mechanical cues. In crayfish of the genus *Orconectes*, *O. fodiens* is restricted to mud bottoms and sluggish water and *O. propinquus* to gravel and rock bottom and running water.[13] The larvae of barnacles settle more on rough than smooth surfaces. An amphipod, *Corophium*, selects small sand grains over large ones for its tubes, with a 9 to 1 preference when in the dark, and less when in the light.[150]

Hydrostatic pressure increases approximately 1 atmosphere with each 10 meters depth of water. Two-thirds of the earth's surface is covered by water, and 90% of the oceans or 56% of the biosphere is under pressures of 100 to 1100 atm. Sea water is compressed by 4% at 10 km depth (100 atm). Pressures of a few hundred atmospheres decrease protoplasmic viscosity, block cell division, and kill many surface-adapted animals. Some barophilic bacteria from ocean depths where pressures reach 1000 atm do not grow at 1 atm. The enzymes from deep-sea fish differ in sensitivity to hydrostatic pressure from those of surface fish. Techniques are not yet available for bringing deep-sea animals alive to the surface under pressure, and little is known about biochemical and physiological adaptations to great pressures.

Many planktonic organisms tend to occupy certain levels in the water, and some of them migrate diurnally from one pressure to another.[213] Some crustaceans respond to increases or decreases in hydrostatic pressure by locomotor activity. An amphipod can react to changes as small as 0.01 atm; but no gas-containing structures are present and the sensors are unknown.[58] A polychaete worm, *Nephthys*, is stimulated to swim by an increase in hydrostatic pressure and is inhibited by a decrease; stimulation must be for at least 5 sec, and there is a latency of 10 sec.[158]

The swim-bladder of fish is more important in hydrostatic orientation than in respiration (see p. 204). A small decrease in pressure results in an increase in gas content of the swim-bladder; gas must be secreted to maintain the fish at a density which keeps it at a constant level. Gas is extruded when hydrostatic pressure increases. Physostomes swallow or belch air according to their buoyancy. *Phoxinus* responds to a change of 5 cm H_2O pressure; by recording from a swim-bladder nerve, receptors were localized in the wall of the bladder, which seem to be involved in the gas-spitting reflex.[194] Physoclyst fish from which the gas had been withdrawn and which were forced to remain near the surface regenerated only half as much gas as when kept at a depth. Abrupt changes in swim-bladder pressure caused responses of heart, breathing, and swimming—all of which ceased after nerves to the bladder were cut. Evidently, both the inner ear and bladder sense-endings participate in responses to hydrostatic pressure in fishes.

Vibration Sense

Rapidly adapting stretch receptors grade into vibration receptors, and the distinction between them is rather arbitrary. Vibration sensors may be defined as exteroceptors which respond to mechanical stimuli at low frequency, below a few hundred Hz; they must be rapidly adapting if they are to signal frequency of vibration, and like all mechanoreceptors they may be stimulated dynamically (by acceleration) or statically (by amplitude of displacement). Vibration receptors range from position detectors to audioreceptors; for terrestrial animals, vibrations are conveyed mostly by substratum and sound is mainly transmitted by air. Vibrations travel at a lower velocity than sound, and sound velocity is inversely related to frequency. For aquatic animals the distinction is less sharp; a hydroid senses ripples and bends toward a local disturbance in the water.[111]

Arthropods are extremely sensitive to vibration carried in solid substratum, in surrounding water, or in the moving air. Some receptors for vibrations are sensitive hairs, and others are campaniform sensilla. The sensilla may consist of sensitive neurons on connective tissue, or muscle strands attached to chitin, sometimes across a joint; vibration may be transmitted by the rigid exoskeleton, and movement of the sensillum may be damped by the hemolymph which surrounds it.

Vibration-sensitive hairs and joint sensilla have been described for many crustaceans and insects. For example, cercal hairs of a cockroach respond to puffs of air or to substrate movement.[168] In the cockroach tibia, distal to the femur-tibia joint, are three receptors: one responds to tension changes in cuticula with a threshold of 10^{-6} cm and a response range of 30 to 500 Hz; a second senses variations in blood pressure; and a third is a vibration receptor with a maximum sensitivity range from 1000 to 5000 Hz and a threshold of 10^{-7} to 10^{-10} cm.[219] Similarly, metathoracic femoral segments of locust and grasshopper have chordotonal organs with both tonic and phasic receptors; the tonic receptor signals the femur-tibial angle, and the phasic one signals velocity of displacement at the joint.[253] Stretch receptors at the basal joints of the wings in a locust discharge at 10 to 20 pulses per second, with frequency varying according to rate of sinusoidal wing movement.[180]

The antennae of flies (*Calliphora*) have campaniform sensilla (as Johnston's organs) at the pedicellus-funiculus joint; these show a phasic response to lateral torsion of the funiculus in air currents in flight. This response occurs for movement of 2 to 5°, is proportional to the logarithm of rate of movement, and saturates at a speed of 100°/sec; lateral movement is excitatory and medial movement is inhibitory.[212] In blowflies, these phasic receptors can follow movements up to 500 cycles/sec.[21] In mosquitoes, the same kind of organ signals direction of sound.[198] In beetles, the antennal Johnston's organs permit orientation in wind, each antenna measuring its deviation from an oriented position with or against the wind.[132] In honeybees, the corresponding receptors initiate a reflex which moves the antennae closer together as head-wind increases; motion of the flagellum induces efferent impulses from the brain. Afferent spikes are at 15 to 150 per second, two per movement wave; in a wind of 7.4 m/sec the sensory impulses indicate a vibration of the flagellum at 37 Hz. Behavioral experiments show that the sense organ at the joint of the flagellum with pedicellus signals direction and strength of wind as well as frequency of movement at the joint.[95] Sensory hairs in *Locusta* respond to displacement by wind, and may discharge initially at 245/sec, and on steady stimulation by wind at 50 to 70/sec.

The backswimmer *Notonecta* locates prey by surface waves in the water; it is most sensitive at 100 to 150 Hz, and receptors in the distal part of the first and second legs can detect water movements of 1 μ amplitude.[144] Water-striders have vibration receptors at tibial-tarsal joints.[160] A male generates water waves at 20 Hz, and females are attracted to the specific vibration pattern. Leaf-cutting ants have sensitive receptors at the joints of the forelegs, with a displacement detection threshold of 1.3×10^{-7} cm at 1 to 3 kHz and an acceleration detection threshold of 2.5 cm/sec at 0.1 to 2 kHz. Nerve spikes follow 100 Hz one-to-one, and at higher frequencies give less regular responses.[143]

Crustacean vibration receptors from antennae and walking legs have been described. A hermit crab, *Petrochirus*, has a chordotonal organ with bipolar sense cells in the basal segment of the antennal flagellum; sensory spikes synchronize with water vibrations up to 200/sec and show synchronous response to vibrations up to 1000/sec.[248] In a crayfish, an interneuron is found in each circumesophageal connective, which integrates water-borne vibrations detected by statocysts and by antennal and antennular flagellar receptors; information from these receptors is antagonized by input from movement of the legs.[249] The ghost crab, *Ocypode*, has a myochordotonal organ attached to a thin window in the exoskeleton in the merus of the walking leg, which responds to air-borne sound and substratum vibration; the maximum sensitivity for air signals is at 1.5 to 2.0 kHz, and that for substrate vibration is at 1 to 2 kHz.[97] Male fiddler crabs (*Uca*) signal to females by vibrations of legs and by tapping the substratum; some fiddlers also produce sound by stridulation. Detection is by receptors in the legs, which respond to substrate-transmitted vibrations; response to substrate displacement is maximal at low frequencies and declines sharply towards 1 kHz, whereas the threshold curve for the acceleration component for the neural response shows one peak at 600 Hz and another at 1500 Hz.[209] By conditioned cardiac response, a lobster shows maximum sensitivity at 37 Hz (on a pressure wave basis).[176]

Web-spinning spiders can discriminate live from dead prey on the web; the natural signal has an irregular form with most of its energy at 50 Hz.[184] The house spider *Achaearanea* has lyriform organs (sensilla) beneath thin chitin near the tarsal-metatarsal joints; each organ has 10 receptors, each sharply tuned (mostly in the range from 80 to 800 Hz), and they are more sensitive to air-borne than to web-transmitted vibrations.[254]

Some mechanoreceptors of vertebrates are important for detecting substrate vibrations. Mechanoreceptors in the skin of snakes respond to substrate vibrations. Cutaneous sensory fibers in a snake give one impulse per vibration up to 300 Hz, with maximum sensitivity at 150 to 200 Hz; they follow alternate vibrations at 400 to 600 Hz, fire irregularly at 600 to 800 Hz, and give no responses above 800 Hz[191] (Fig. 12–7). Thus, the frequency range of cutaneous vibration sense overlaps that of hearing. Pacinian corpuscles in a cat, particularly in the joints of the legs, follow vibrations of the substratum in a 1:1 fashion up to about 700 Hz, and give steady responses at vibration frequencies of 100 Hz.[211] Similar receptors in interosseous corpuscles of the leg or wing of a duck respond 1:1 up to 800 or 1000 Hz, and give regular responses from 1600 to 2000 Hz with threshold displacements of 0.5 μm.[45, 47] Pri-

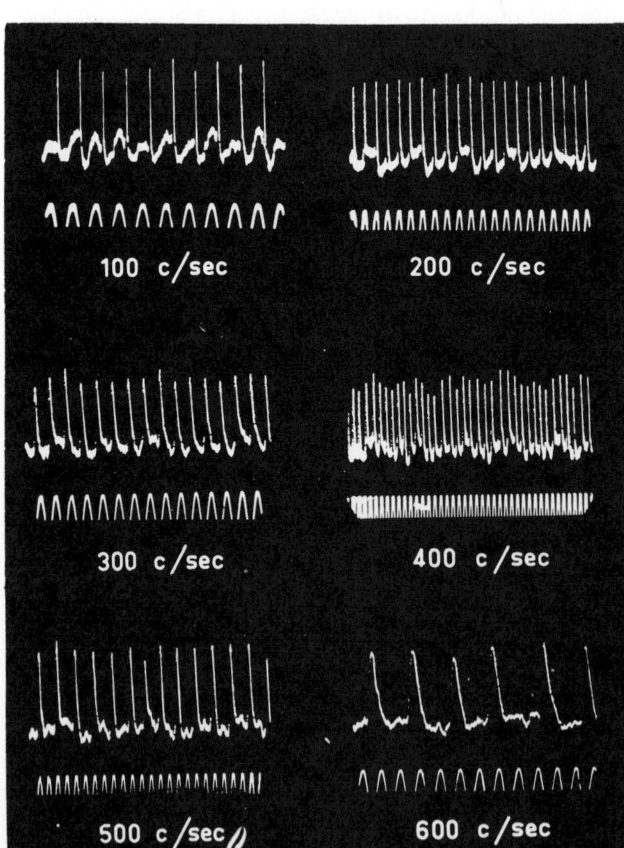

Figure 12–7. Nerve impulses initiated in a vibration receptor in the skin of a black snake. Upper traces are impulses; lower traces are the applied sinusoidal vibration. (From Proske, W., Exp. Neurol. *23*:187–194, 1969.)

mary muscle spindles and tendon organs of cats are sensitive to vibration and elicit reflex contraction of the muscle that is vibrated.[73]

In summary, vibration receptors are highly phasic and sensitive mechanoreceptors located mostly at joints or in skin, which are able to follow relatively high frequencies. They overlap in range between the low-frequency response of some tactile and stretch receptors and the high-frequency response of hearing. Vibration sense serves as "distance touch" and is important in signalling substrate disturbances.

Lateral Line Organs

Teleost and elasmobranch fish and a few amphibians have a complex system of sensory canals along the sides of the body, and often over the head, which may be open or may be closed except for occasional exits (particularly at the ends). In the wall of the canal, supported by epithelial cells, are hair cells or neuromasts. These may have many (40 to 50) small stereocilia and one or a few large kinocilia (see Ch. 19). The hair cells may be grouped in sensory hillocks with the cilia projecting into a gelatinous cupula (Fig. 12–8). The neuromasts are innervated by terminals of fibers of the lateral line nerve, which enters the brain mainly with the tenth and partly with the seventh and ninth cranial nerves. Vesicles in the bases of the hair cells indicate possible chemical transmission to the nerve endings (Fig. 12–9). The lateral line organ responds to a near-field displacement rather than to pressure, and the cupula provides for shearing over the cilia.[68] Some nerve endings on the hair cells contain vesicles; these may be efferent axons to hair cells.[84]

Electrical recording from the cupula of the

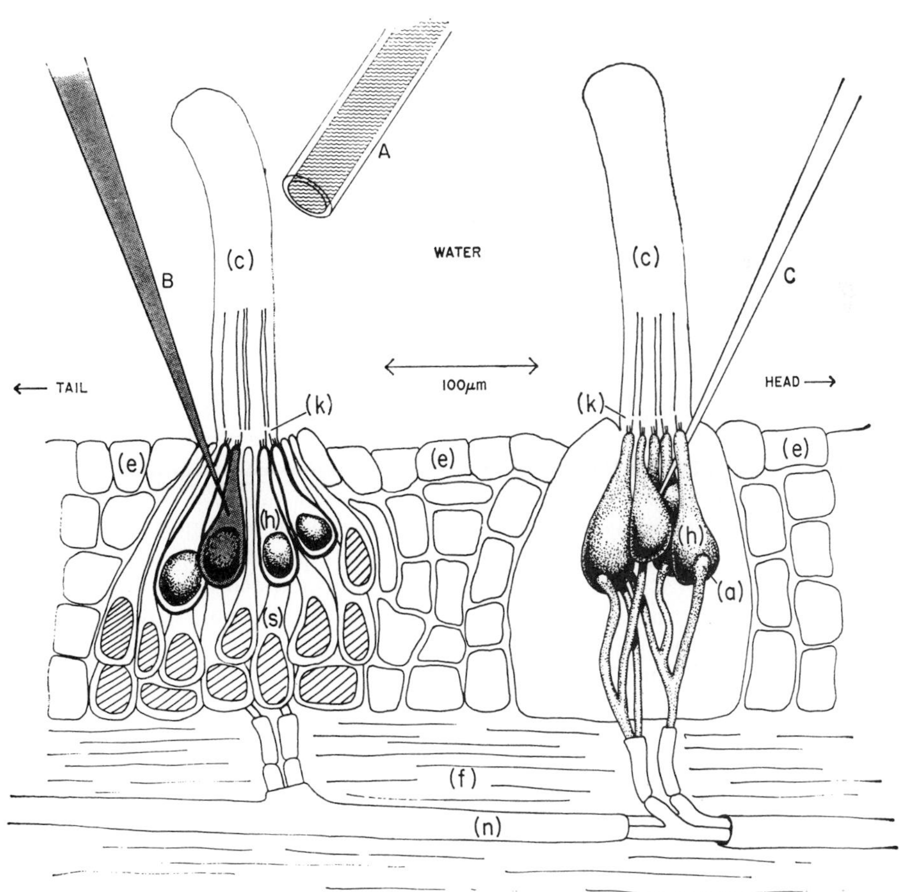

Figure 12–8. Two neuromasts in lateral line on tail of *Necturus.* (c), cupula; (k), kinocilium surrounded by stereocilia of the eight hair cells; (h), a synapse of sensory axon; (a), sensory axon; (e), epithelium; A, B, C, electrodes for stimulating and recording. (From Harris, G. G., L. S. Frishkopf, and A. Flak, Science *167*:76–79, 1970.)

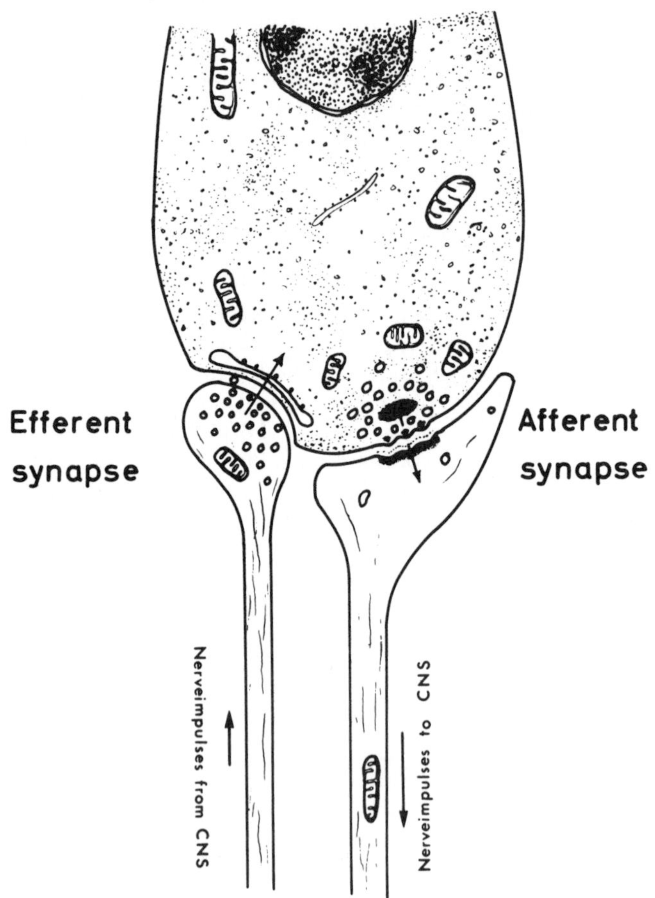

Figure 12–9. Diagram of relation between basal synapse from hair cell to afferent nerve fiber and efferent synapse from axon to hair cell of lateral line organ. (From Flock, A., pp. 163–197 *in* Lateral Line Detectors, edited by P. Cahn, Indiana University Press, 1967.)

fish *Acerina* shows a generator (microphonic) potential which varies in voltage with the amplitude of stimulating waves and may be double their frequency; sensitivity is maximal at 50 to 150 Hz.[129] Intracellular recordings from hair cells of the salamander *Necturus* give sensory potentials of 800 μv.[87, 88] Sensory fibers of the lateral line nerve show a background of spontaneous activity which originates in the neuromasts, with superimposed responses to mechanical stimulation. Single thick fibers show several spikes per cycle of low-frequency wave stimulation (20/sec), one spike at 20 to 50/sec, and irregular spiking at higher frequencies. In the fish *Ictalurus* the nerve fibers synchronize to 100/sec, and in *Fundulus* to 180/sec.[114] Modulation by mechanical stimulation of spontaneously active neuromasts has been seen also in the toad *Xenopus*.

Efferent nerve fibers regulate the sensory discharge. In an eel, *Astroconger*, stimulation of one afferent branch reflexly inhibits the sensory and spontaneous impulses in another nerve branch.[93] In *Xenopus* and *Ambystoma*, the efferent impulses inhibit afferent activity.[207]

The behavioral significance of the lateral line system has not been established. Electrical recording demonstrates the sensitivity to low-frequency waves in the water, yet conditioning experiments show that fish can detect low-frequency sound by other means (e.g., the ear and cutaneous receptors). Rheotaxis or stream orientation is mediated primarily by visual and ventral contact receptors.[41, 42] Ablation of groups of cupulae, particularly from the head of the fish *Aplocheilus*, results in behavioral evidence for loss of perception of waves from a narrow angle.[224] Probably the lateral line input is combined with that from other receptors to provide information essential for orientation, for food detection, and possibly for schooling.

SOUND RECEPTION

Phonoreception and Sound Production in Insects

It has long been known that some insects respond to sounds in the human audio range, and recently they have been shown to respond to ultra-high frequencies such as are produced by bats. The ear of a noctuid moth consists of a tympanic membrane on the side of the thorax, behind which is an air sac transected by ligaments; in one of the ligaments is a nerve containing three sensory fibers. Two of these fibers arise in the tympanic organ from A-cells or phonoreceptors; the third is from a mechanoreceptor (B-cell).[200] The B-cell is spontaneously active at 10 to 20/sec, and when the organ is stimulated by touch its frequency may increase to as much as 200/sec; it functions as a proprioceptor to measure internal displacements. The A-cells also are spontaneous; they respond to sound, one with a lower threshold than the other. The A-cells do not synchronize with sound waves, but discharge a burst of impulses of decreasing frequency as adaptation occurs during stimulation by a pure tone (Fig. 12–10). They signal changes in sound intensity by variation of the frequency of spikes, by shorter latency with high intensity, by duration of after-discharge following a sound stimulus, and by the 20 db difference in threshold of the two cells. Above some 40 db intensity, the frequency of spikes reaches a limit. The A-cells respond to sound frequencies above the human audible range. In the hawk moth *Celerio*, the phonoreceptor on the labial palps is most sensitive in the range from 15 to 170 kHz. The moth ear is a pulse detector, not a frequency discriminator; nerve spikes do not correspond to sound frequency. Sphyngid moths have an ear in the labial palps which has an air space in the first two segments; it is most sensitive to sounds of 20 to 40 kHz, and the threshold is 10 to 15 db higher than that of the noctuid tympanic organs.[199, 202, 203] The intensity of sound impinging on the tympanic organs of noctuid moths may vary by 40 db according to position of the wings relative to the direction of the sound source. When the wings are up, the ipsilateral stimulation is 20 to 40 db greater than the contralateral; when the wings are down, sound coming from below is stronger than that from above. Thus, a moth can determine the direction and, to some extent, the distance of a source by comparing the intensity reports from the two tympanic organs during a wing cycle.[187, 201]

The tympanic organ of a locust has four groups of sense cells, located at different positions relative to the tympanum and with different optimal frequency and sensitivity

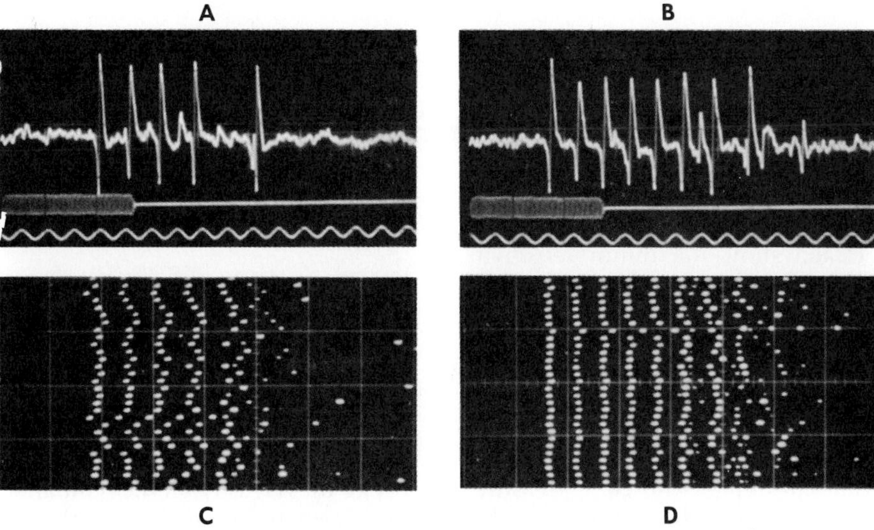

Figure 12–10. Spike responses in tympanic nerve fibers recorded in neuropile of a noctuid moth. A, single response; C, series of responses recorded in dot diagram to 5.5 msec pulse at 8 db. B, D, reponses at 18 db. Sound stimuli indicated below responses in A and B. Latency is 3 to 6 msec; spike train continues after end of sound. (From Roeder, K. D., J. Insect Physiol. *12*:1227–1244, 1966.)

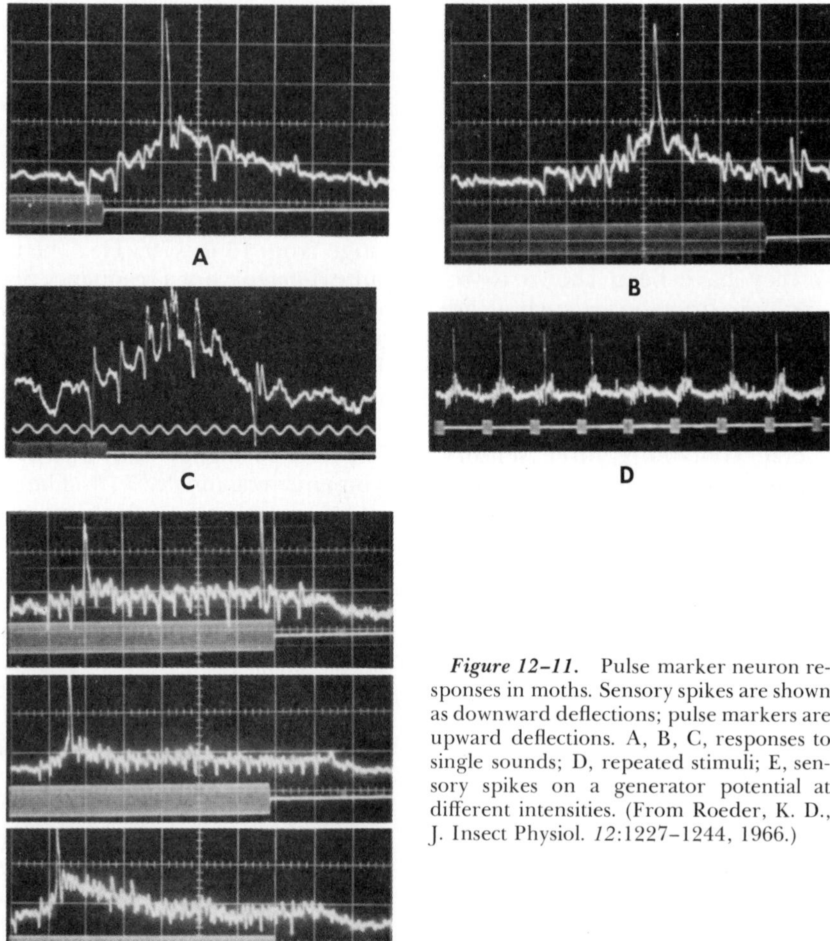

Figure 12–11. Pulse marker neuron responses in moths. Sensory spikes are shown as downward deflections; pulse markers are upward deflections. A, B, C, responses to single sounds; D, repeated stimuli; E, sensory spikes on a generator potential at different intensities. (From Roeder, K. D., J. Insect Physiol. *12*:1227–1244, 1966.)

characteristics. The tympanic membrane has two centers of vibration, one near the middle and the other at one end of the tympanum. The tissues, and particularly the fat, surrounding the intact ear serve as a low-pass filter.[153] Impulses in the tympanal nerve of a Brazilian cicada show maximum sensitivity of the ear at 6 to 9 kHz; there is no adaptation to trains of clicks, but there is fast adaptation to pure tones.[57] A lacewing, *Chrysopa,* has a tympanic organ on a vein near the wing attachment; the sensory discharge occurs with a latency of 4 to 6 msec to sound stimuli of 15 to 150 kHz, and sensory impulses follow repetition rates up to 150/sec.[154] Measurements of amplitude of vibration of a cricket tympanum by means of a Möissbauer effect show maximum vibration at 4 to 6 kHz, the same optimum as for the nerve fiber responses; thus, the tympanum is mechanically tuned to a narrow band. The threshold amplitude is estimated at $\sim 0.0005\ \mu m$.[109]

The high-frequency sensitivity of tympanic organs, as compared with chordotonal Johnston's (antennal) organs and hair cells, is shown as follows:[7]

	Optimum, kHz	
Anopheles	0.34	Johnston's organ
Calliphora	0.15– 0.25	Johnston's organ
Locusta	5 –12	Tympanal organ
Prodenia	15 –60	Tympanal organ
Parametera (spider)	0.3 – 0.7	Slit organ
Gryllus	0.5	Hair cells
Oxya	0.4 – 0.5	Hair cells

The messages from the tympanal receptors have been followed into the thoracic ganglia in noctuid moths.[199, 201] One interneuron on each side is a spike marker (i.e., it follows each spike in the train from an A-receptor) (Fig. 12–11). Another interneuron is a pulse marker, firing once (or twice) per train, thus marking the train interval. A third interneuron responds to each train with a series of spikes at intervals independent of input frequency, thus marking train duration. Neurons near the midline sum the input from the two ears, and one of these responds to the contralateral ear but is inhibited by the ipsilateral one. In hawk moths, in which the sound receptors are in the labial palps, interneurons show more convergence than in noctuids.[203]

Single neurons in the supraesophageal ganglion of *Locusta* differ in response pattern: some are high-tone, others are low-tone, and still others are broad-band interneurons; excitation may be interrupted by inhibition in a portion of the frequency range.[1a, 2] Also, in the mesothoracic ganglion of *Locusta* there are interneurons which show a very sharp rise in their frequency response curve, particularly on the low side (3 to 4 kHz); when nerves other than those to the tympanum are cut, this sharp rise is lost, so there must be inhibition of the interneuron from other inputs.[266] Four types of second-order neuron were found: those making (a) tonic responses to high (10 to 20 kHz) frequency, and "on" discharge to lower frequencies; (b) responses to high frequencies only, being tonic at threshold and "on" at higher intensity; (c) tonic responses to stimuli of 3 to 20 kHz; and (d) tonic response within a narrow belt of intensities.[113, 243]

The behavioral meaning of sound reception and its central processing is to give a moth some chance of escaping a predator bat (Fig. 12–12). The sensitivity of a moth ear in the frequency range of bat cries is 0.01 to 0.03 dyne/cm^2, which enables it to detect the cries of a bat at 30 meters. A moth can detect direction of sound with considerable accuracy, as shown by mercator contours of ear sensitivity with the wings in different positions (Fig. 12–13). A moth normally flies an irregular up-and-down path, but when a bat comes near, the moth dives steeply (probably upon activation of the A_2-receptor of the ear). A bat emits pulses of high-frequency sound at about 10 pulses per second; it can detect a moth by the echo at about three meters, and then its pulse repetition rate rises to as much as 100 per second. The bat is a faster flyer than the moth once it zeroes in on its prey. Thus, the two are well matched; sometimes the moth escapes, and sometimes it is caught.[200] Certain arctuid moths have tymbal organs in the metathorax which generate sound signals in the same frequency range as a bat's cry. One suggestion was that the moth uses its sound to jam the bat's sonar.[12] An alternative is that these particular moths appear to be distasteful to bats, and bats may disregard species which answer back![48]

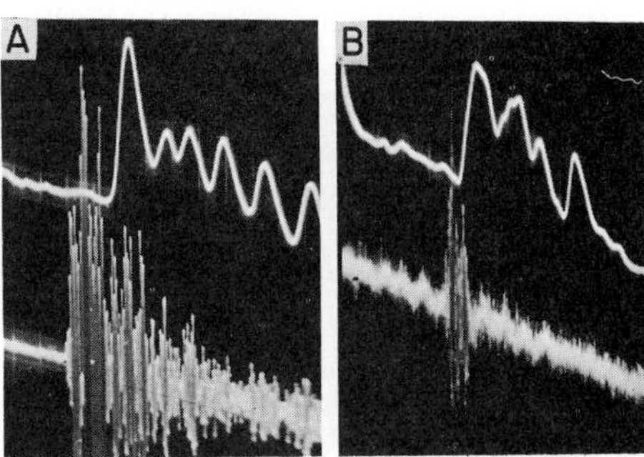

Figure 12–12. Upper records, response in tympanic nerve of moth, *Prodenia.* Lower records, cries of bat near moth. A, Bat cry in sonic range, predominant frequencies 10 to 15 kHz. B, Shorter bat cry, frequency above 20 kHz. Response of moth ear persists after stimulus has ceased. (From Roeder, K. D., and A. E. Treat, J. Exp. Zool. *134*:127–157, 1957.)

Many orthopterans produce sound by elaborate stridulating organs. A grasshopper rasps its hind femora, each of which has bead-like prominences, against the outer surface of the wings. Male crickets and tettigonids rub the two forewings together; one wing has a rasp, and the other a file region. The forewings of a tettigonid may move at 35 to 45/sec; the toothed pulses are emitted at 650 to 800 per second with vibrations of 18 to 42 kHz on opening and 40 to 66 kHz on closing. The maximum sensitivity of the tympanic organ is in the same range as the high frequency sound production.[240] Courtship and rivalry calls of male crickets can be identified as species specific, each produced by a patterned sequence of muscle contractions[130] (Fig. 12–14). Male lycosid spiders produce courtship and threat sounds by scraping the palps, and the frequency of the courtship sounds increases when a female displays leg waving.[205] A male grasshopper's song has subsequences of chirps resulting from a sequence of muscle contractions that must be highly programmed in the nervous system.[52] Comparison among many insects of the sound signals that are produced with the thresholds for tympanic organs shows close matching.[116] Cicadas sing species-specific songs, consisting of complex patterns of clicks and rasps. In *Magicicada,* each tymbal muscle twitch causes sequential

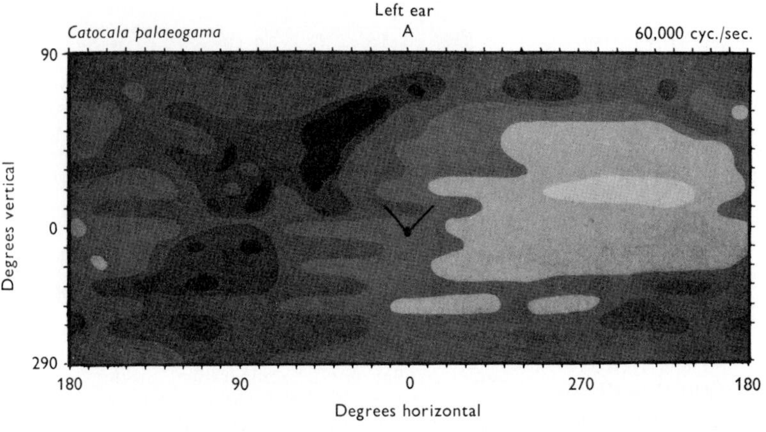

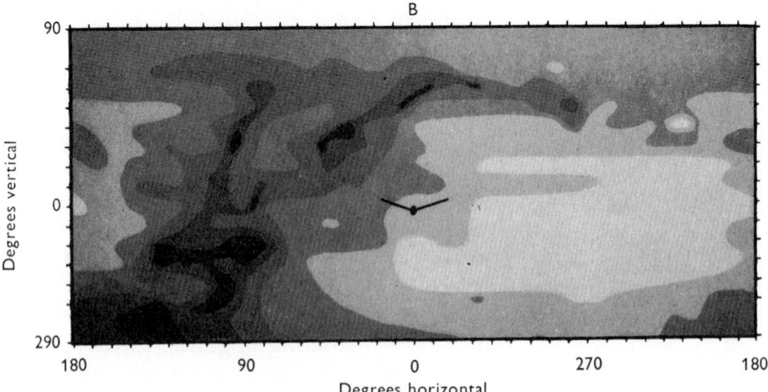

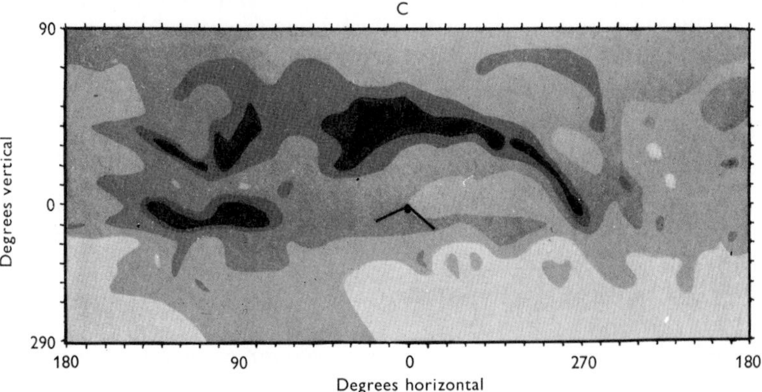

Figure 12–13. Mercator projection of sound intensity originating in different positions as judged from sensory response in left ear of moth when wings were (A) raised, (B) horizontal, or (C) lowered. (From Payne, R. S., K. D. Roeder, and J. Wallman, J. Exp. Biol. *44*:17–31, 1966.)

buckling of a series of stiff ribs on the tymbal membrane; each collapse makes a damped oscillation in an air sac, so a muscle twitching at 50/sec causes sounds resonating at 1000/sec.[196] In the locust *Schistocerca,* flight is initiated in response to sound at frequencies of lowest tympanal nerve threshold.[266a]

Some species of bees signal in the hive, without a dance, the location of feeding places, and honeybees that signal the location of food by specific dances also produce sounds as part of the communication from foragers to other workers. The sounds are produced by skeletal movements caused by contractions of wing muscles at 250 Hz, and there are high-frequency overtones, probably from other skeletal vibrations. The duration of each sound period correlates with the distance to food: 0.4 sec if food is near the hive, and 1.5 sec for food 700 meters away. Bees can be induced to leave the hive when such sounds, having been taped, are played to them; vibration reception is by Johnston's organs on the antennae (p. 517).[59, 60, 61]

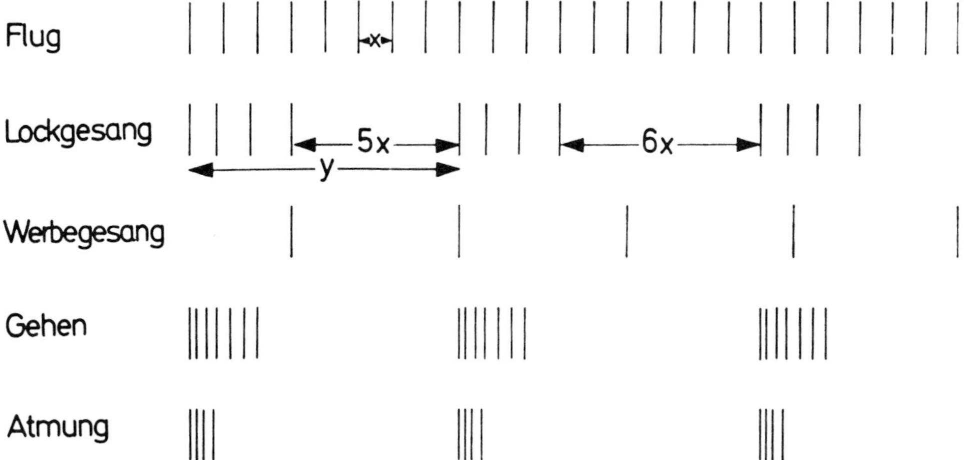

Figure 12-14. Schema of several rhythmic movements in crickets: flight, calling song, courtship song, walking, breathing. The rhythm can be explained as being based on two central oscillators; *x* gives the period of the fast one (30 Hz) and *y* gives the period of the slow one (3 to 4 Hz). (From Kutsch, W., Z. vergl. Physiol. *63*:335-378, 1969.)

Vibrations in substrate, water, and air are important to insects as warning signals, aids to locomotor or posture control, and (for a few of them) communication. Receptors at certain skeletal joints are exquisitely sensitive to vibration. Sound is perceived by moths, which then veer aside from bats that they hear getting a "fix" on them. Sound serves many orthopterans for social communication—it is rasped out by skeletal structures and is perceived by the tympani. Coding of sound in insects is mostly by temporal patterning of pulses by amplitude, and not much by frequency and harmonics as in vertebrates.

Phonoreception in Vertebrates

Properties of Sound as a Stimulus. Sound consists of regular mechanical waves which, in pure tones, can be described in terms of frequency and amplitude. As sound waves travel through a medium, particles move back and forth with oscillating pressure waves. The speed of transmission and attenuation with distance vary with the medium. Sound intensity is expressed as sound pressure level in dynes/cm^2. The faintest sound heard by man is about 0.0001 dyne/cm^2 (at a frequency of 1 kHz); conversation level is about 1 dyne/cm^2, and painfully loud sound is about 300 dynes/cm^2. While the sound pressure is given in the absolute units of dynes/cm^2, the standard practice in audiology is to express intensity in the relative unit, the decibel, which gives the logarithm of the ratio of two energies. In terms of intensities, the decibel is defined as

$$1 \text{ db} = 10 \log \frac{I}{I_r}$$

where I_r is a reference intensity (usually 10^{-16} watt/cm^2). In terms of sound pressure, since pressure varies as the square of intensity, the defining equation is

$$1 \text{ db} = 20 \log \frac{P}{P_r}.$$

The reference pressure, P_r, is often taken as an approximation to the human threshold, 2×10^{-4} dyne/cm^2 (particularly for air-borne sound). It may alternatively (for sound in other media, especially water) be taken as 1 dyne/cm^2. It should be noted for reference that 1 dyne/cm^2 is 74 db above 2×10^{-4} dyne/cm^2.[119]

Intensity is given by $I = P^2/\rho C$, where I is intensity in ergs/cm^2, P is pressure in dynes/cm^2, ρ is density in g/cm^3, and C is velocity of propagation in cm/sec. The term ρC is called the acoustic resistance, and values are approximately as follows, at biological temperatures:

	ρ	C
air	1.2 kg/m³	344 m/sec
water	1000.0 kg/m³	1481 m/sec
sea water	1026.0 kg/m³	1500 m/sec

Thus, the acoustic resistance is much higher in water than in air, and sound velocity is also higher; the intensity for a given sound pressure is lower in water than in air.

Phonoreceptors are sense organs in which sound waves in air or water are transmitted through tissues and are transduced into nerve impulses. At the low-frequency end of an audiogram (~50 Hz), phonoreception overlaps vibration sense. Sound above the 20,000 Hz limit perceived by man is sometimes referred to as ultrasonic or supersonic. The properties for frequency discrimination by a given receptor do not necessarily guarantee that the central nervous system will "hear" to a corresponding degree; hence, behavioral measurements and brain recording must accompany the study of receptor properties.

Morphology and Stimulation of the Vertebrate Ear. The external ear (pinna) aids in localization of sound and tends to concentrate high-frequency sound by reflection. The auditory canal (meatus) is separated from the middle ear by the tympanic membrane. The middle ear contains the ossicles (malleus, incus, and stapes), which rotate in a complex lever motion. The foot-plate of the stapes fills the oval window to the inner ear. The tensor tympani muscle is attached to the malleus, and the stapedius muscle is attached to the stapes; these muscles limit the vibration of the bones in response to sounds with relatively high intensities. The eustachian tube connects the middle ear with the pharynx and equalizes the pressure in the middle ear with environmental pressure. Below the oval window is the round window, which is closed by an elastic membrane that allows for fluid movement in the inner ear. The function of the ossicles is to couple sound waves in air to the fluid of the inner ear with minimum loss by reflection (i.e., to match the impedances of the two media). In man, the pressure is increased from tympanic membrane to oval window by about 13 db. In an iguana, the middle ear gives a 35 db transformer action.[258]

In cetaceans, the external meatus is reduced, and sound is apparently transmitted through tissues, particularly fat and bone, to the middle ear.[149, 169, 170] In non-mammals, the middle ear has a single bone, the columella. Ostariophysid fishes have a set of bones, the Weberian ossicles, connecting the swim-bladder with the inner ear, and sound thresholds are much lower than in fishes that lack the swim-bladder-ossicle coupling.

Measurements of amplitude and velocity of motion of the ossicles and basilar membrane have been made by using a tiny gamma-ray source and measuring the Mössbauer effect.[3] Figure 12–15 shows the stapes move-

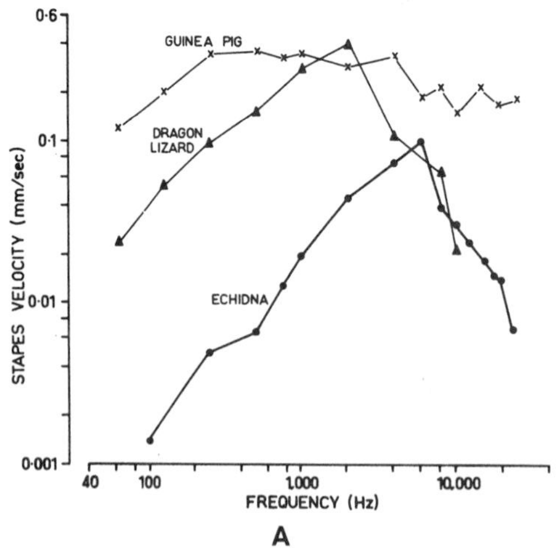

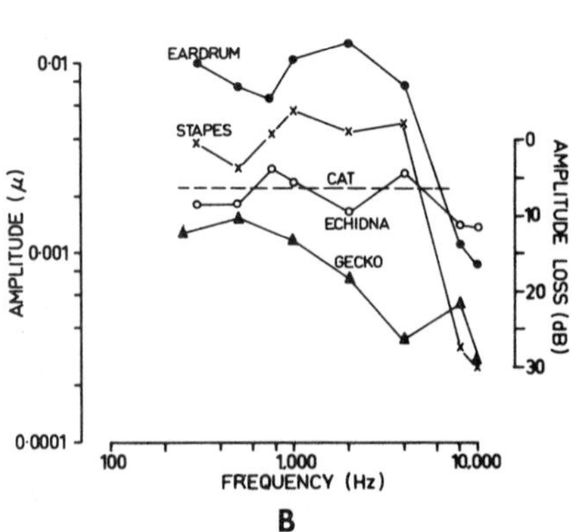

Figure 12–15. A, Comparison of velocity of movement of stapes in guinea pig and echidna and of columella in lizard *Amphibolurus* at constant intensity and various frequencies. Measurements by Mössbauer effect. B, Comparison of amplitude of movement of tympanic membrane and stapes of echidna, and ratio of stapes amplitude to tympanic membrane amplitude as loss in db in echidna, cat, and gecko. (From Aitkin, L. M., and B. M. Johnstone, J. Exp. Zool., *180*:245–250, 1972.)

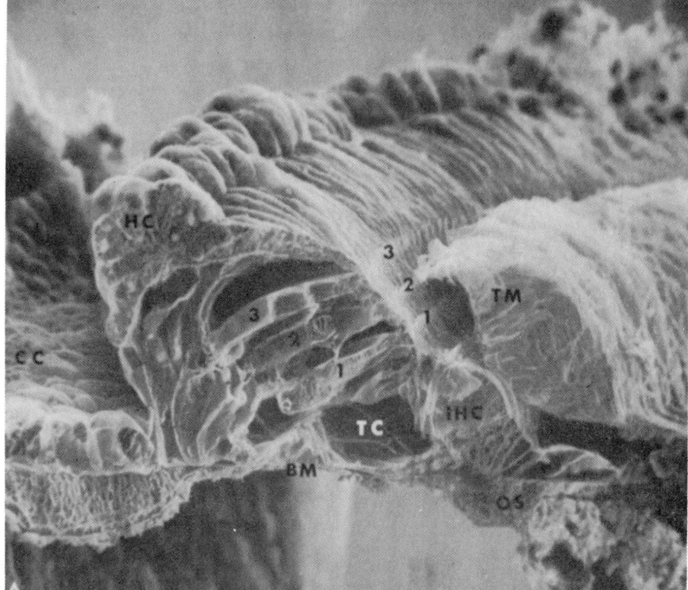

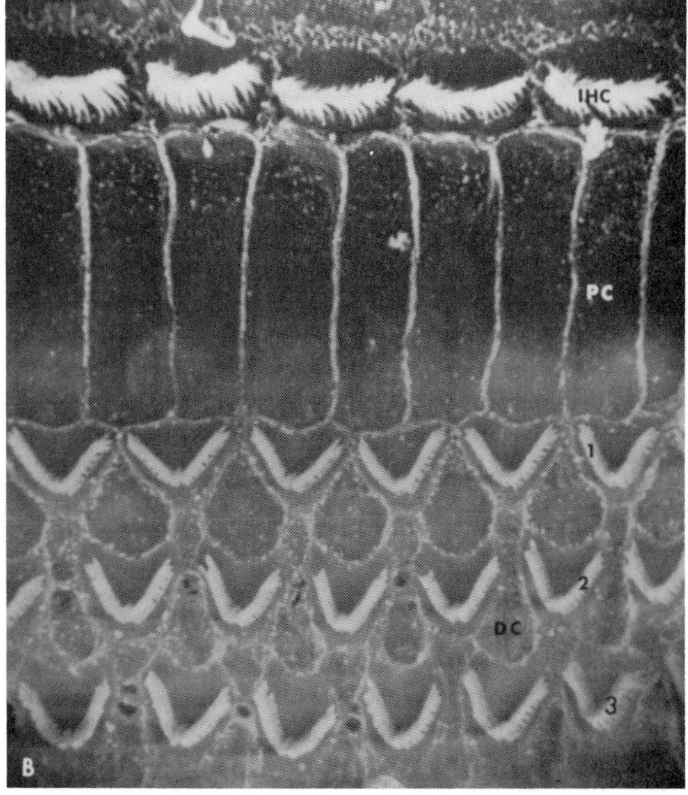

Figure 12–16. A, Scanning electron micrograph of cochlea of guinea pig. Organ of Corti shows radial section and continuation around an eighth of a turn, showing: inner hair cell (IHC), outer hair cells (1, 2, 3), basilar membrane (BM), tectorial membrane (TM), osseous spiral lamina (OS), Hensen's cells (HC), Claudius' cells (CC), and tunnel of Corti (TC). (From Engstrom, H., H. W. Ades, and G. Bredberg, pp. 127–156 *in* Ciba Foundation Symposium on Sensineural Hearing Loss, edited by G. Wolstenholme and J. Knight. Churchill, London, 1970.) B, Scanning electron micrograph of surface of organ of Corti from directly above, showing: inner hair cell (IHC), pillar cells (PC), outer hair cells (1, 2, 3), and phalangeal process of Deiters' cell (DC). (Courtesy of H. W. Ades.)

ment for guinea pig, echidna, and lizard ears over a wide range of impinging sound frequencies. In a bat, the velocity of eardrum movement is constant from 10 to 70 kHz; in reptiles and amphibians, stapes movement declines at frequencies above 1 to 3 kHz. The mechanical properties of the middle ear correspond to the range of that animal's optimal hearing.

The inner ear is a fluid-filled system of cavities; one portion constitutes the labyrinth organs of equilibrium, and the other is the cochlea. The auditory portion of the inner ear appears first as the lagena in fish and amphibians; this lengthens in mammals, where it is coiled in a spiral. The distance from the oval window to the helicotrema at the apex is 55 mm in the elephant and 7 mm in the mouse. The spiral cochlear tube is divided longitudinally by Reissner's membrane and the basilar membrane to form three canals. The two outer canals, the scala vestibuli and the scala tympani (as well as the tunnel of Corti) are filled with perilymph. The scala media, between Reissner's membrane and the basilar membrane, contains endolymph. This fluid contains thirty times as much potassium and one tenth as much sodium as does the perilymph. In the chick, the endolymph has three times more K^+ than Na^+.[156] The endolymph is probably secreted by the stria vascularis,[36, 37] and a potassium pump is postulated. Sound stimulation increases the Na concentration and decreases the K concentration in the endolymph.[244]

The organ of Corti, located on the basilar membrane, contains sensory cells: three rows of external hair cells and one row of internal ones (Fig. 12–16). The hair processes are capped by the fibrillar (but acellular) tectorial membrane. The hair cells are innervated by dendrites of neurons whose cell bodies are in the cochlear ganglion and whose axons compose much of the eighth cranial nerve; each neuron innervates one or more hair cells. Electron micrographs show vesicles at the bases of the hair cells; there are also nerve endings with synaptic vesicles innervating the hair cells and probably serving as efferent axons from the brain.

Wave propagation along the basilar membrane has been elucidated by von Bekesy[8, 9] by direct microscopic observations of vibrations in different regions and by Mössbauer effect measurements when the ear was excited by sinusoidal displacement of the stapes. Both methods led to the conclusion that movement of the outer membrane of a hair cell may be as small as 0.005 μm at the frequency of maximum sensitivity.[108] The maximum amplitude of the displacement varies in its position along the cochlea according to frequency. The normal human ear is sensitive to frequencies from between 15 and 20 Hz to between 16,000 and 20,000 Hz; its sensitivity to the high frequencies diminishes with age. The threshold is lowest at 1500 to 2000 Hz. Fibers arising in the basal turn respond to tones of any audible frequency, but those in the upper turn respond only to low frequencies. Local lesions in the organ of Corti produced by prolonged exposure to loud tones are seen in degeneration of patches of hair cells and loss of sensitivity in the corresponding range; low frequencies cause more damage to the distal than to the proximal row of hair cells.[236] Recordings from single auditory fibers of the guinea pig show fairly sharp limits of cut-off toward high frequencies.

Several electrical phenomena in the cochlea can be measured. The endolymph of the scala media is electrically positive with respect to surrounding tissue and perilymph (Fig. 12–17). This dc potential is called the endocochlear potential (EP). An electrode in the organ of Corti (probably extracellular) measures 80 to 90 mv negative to the scala tympani

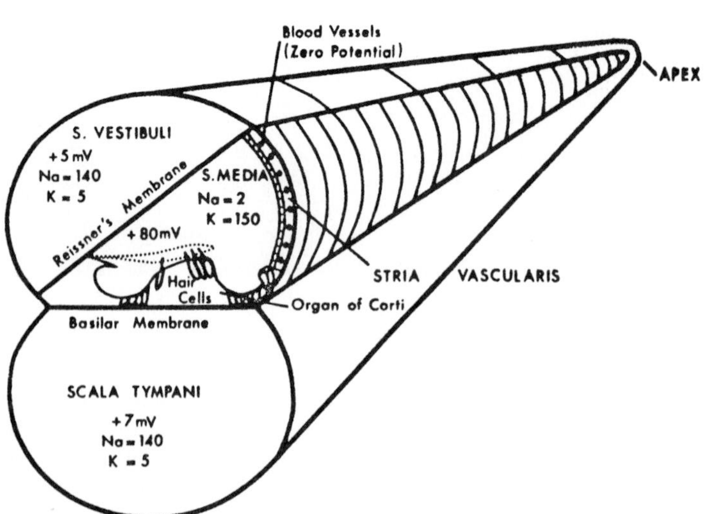

Figure 12–17. Electrochemistry of the cochlea: ionic composition and potentials in scala vestibuli (perilymph), scala media (endolymph) and scala tympani (perilymph). (From Johnstone, B. M., pp. 167–184 *in* Membranes and Ion Transport, Vol. 1, edited by E. E. Bittar, John Wiley and Sons, 1970.)

(perilymph). Thus, a potential of 160 to 180 mv exists between the fluid above the hair cells and the fluid surrounding them. The EP is maximal near the stria vascularis; it is sensitive to anoxia and presumably results from active ion transport. Marginal cells of the stria vascularis have extensive lateral and basal infoldings, between which are elongated mitochondria and microvilli at the lumen surface; such a structure is typical of cells that serve in ion transport (p. 247).[49] The endocochlear potential is increased by injection of Na into the scala media; it is slightly reduced by the Na-K ATPase inhibitor ouabain,[128] and it is very sensitive to oxygen and DNP. Apparently, the dc potential is generated by an active pump which takes K into the endolymph in exchange for Na. Chloride is at the same concentration in both endolymph and perilymph; thus, it is not in electrochemical equilibrium, and a Cl pump is postulated.[107] In birds, the EP is less (about 20 mv) and is anoxia-sensitive. In a series of reptiles, the dc potential between endolymph and perilymph is only 2 to 7 mv and is insensitive to anoxia.[214] Not all auditory chambers with high K^+ and low Na^+ have large positive potentials:

	K^+, mM	*Na^+, mM*	*Potential, mv*
turtle cochlea	114	2.7	+6
lizard cochlea	140	10	+9

Thus, permeability differences supplement active transport in determining endolymph potential.[107]

When the ear is stimulated by sound, an alternating potential (the cochlear microphonic, or CM) is generated in the cochlea. It can easily be recorded from the round window (Fig. 12–18). The CM varies in amplitude according to sound intensity (i.e., it shows no threshold); it is an alternating current response of the hair cells, which follows the waveform of impinging sound faithfully over the audible range. The CM is a generator potential, a reduction in EP, apparently elicited in the hair cells by shearing, since the tectorial membrane over the hair cells and the basilar membrane are "hinged" at different places. The CM is sensitive to Po_2; it is reduced by cyanide, and it is temperature-sensitive.[165, 246]

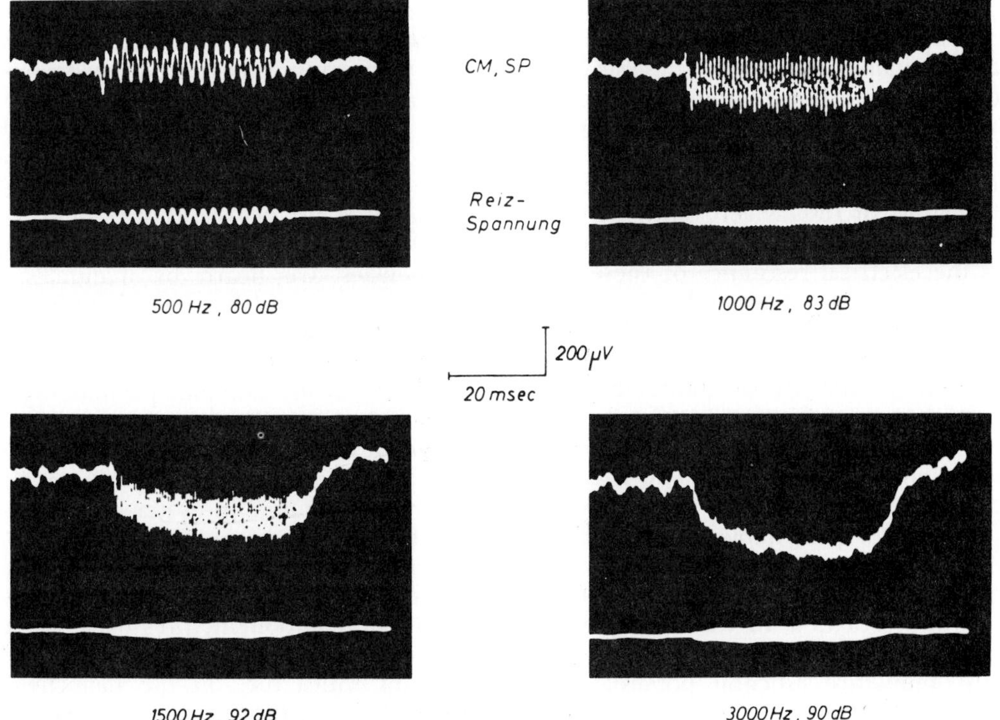

Figure 12–18. Cochlear potentials (CM) and summing potentials (SP) in responses of different frequencies in a pigeon. (From Necker, R., Z. vergl. Physiol. *69*:367–425, 1970.)

An additional response to a tone is a dc response called the summating potential (SP).[37] This also arises in the hair cells; usually the scala media becomes more negative to the scala tympani. Bending the hair cells by means of a microneedle can elicit potentials in either direction much like the CM and SP.

Auditory nerve impulses or action potentials (AP) are all-or-none spikes with refractoriness of about 1 msec. Single fibers show relatively narrow response areas or frequencies of maximum sensitivity, presumably correlated with the point of origin in the organ of Corti. A fiber may give one to three impulses per wave at low frequencies. At low frequencies and moderate intensities, many fibers fire in synchrony, even those which originate in the basal turn.[204] In the middle range, spikes in one fiber occur in some multiple of the stimulus cycles, not in a regular pattern but at random.[246] At high frequencies, the primary fibers fire in characteristic temporal patterns. At a given sound frequency, the spike frequency increases with increasing sound intensity. Spontaneous activity at less than 100/sec is observed in many auditory fibers. This may be due to hair cell movement or to low ambient noise. When a neuron is excited by tone bursts of optimum frequency, a second tone properly timed can inhibit the response on both sides of the excitation area; this suggests some peripheral interaction in the cochlea.[5]

One theory of the mode of stimulation of the auditory neuron endings is that it is electrical. The EP plus the cell potentials provide a battery of 160 mv across the hair cells. It is postulated that shearing of the hair cells alters the electrical resistance of these cells, thus producing the CM and SP. Current flows due to the IR drop across the membrane, and this current may stimulate the nerve endings directly or indirectly.[257] An alternative hypothesis is that a chemical transmitter is liberated from the hair cells to stimulate the nerve endings. The hair cells contain many vesicles at the basal end, and thus resemble presynaptic endings.[57a, 115] Recordings from auditory nerve fibers show junction potentials with a delay after the CM, with fluctuation in size indicative of quantal release of transmitter, and spontaneous miniature junction potentials.[70, 104] Since, in well known synapses (e.g., motor endplates), depolarization is needed for transmitter release, possibly the cochlear microphonic serves that function for the release from the hair cells of a transmitter that stimulates the nerve endings.

Birds. The lagena of the inner ear of birds is elongated and has a central basilar membrane. The ranges of sensitivity and pitch discrimination are greater than in lower vertebrates. The cochlear microphonic of a pigeon follows sound up to 25,000 Hz, with the lowest threshold at 3200 Hz; after the lagena was removed, good microphonic potentials were obtained in the range of 100 to 3000 Hz from ampullae of the semicircular canals.[226]

Reptiles. In reptiles, the floor of the lagena is similar to the basilar membrane of mammals. Reptiles other than snakes have a middle ear with bony columella; snakes have a columella attached to the quadrate bone. Cochlear microphonics have been recorded from turtles; thresholds are high and the frequency maximum is 5000 Hz.[259] In a skink, the CM is maximal in the frequency range from 750 to 3000 Hz.[110]

Amphibians. In amphibians, the lagena is slightly longer than in fishes, and anurans (but not urodeles) have a middle ear with a bony columella which connects the drum with the inner ear. Microphonic potentials have been recorded from frog inner ears at frequencies from 50 to 3500 Hz, with maximum sensitivity between 400 and 1500 Hz.[237]

Comparisons of maximum sensitivities and frequency ranges for many species of reptiles, birds, and mammals show that birds have much greater sensitivity than reptiles, with the mammals' range overlapping those of the other two classes. Best frequencies for reptiles are in the low range, while the highest frequencies are heard by mammals. One evolutionary trend has been lengthening of the auditory membrane, and with it development of some place analysis of frequency; the cochlear coil in mammals permits high frequency detection.[140]

Aquatic Vertebrates; Cochlear, Central Nervous, and Behavioral Responses to Sound. Sound stimulation in water has near-field and far-field components. In the near field the vector components—particle velocity and particle displacement—are large relative to the scalar component, pressure; in the far field, the vector and scalar components have constant ratios (i.e., in the near field the phase relationship between pressure and particle displacement is variable, whereas in the far field it is constant). In the near field, velocity decreases as the inverse square of the distance from the sound source; in the far field, velocity and pressure decrease in direct proportion to the distance. Therefore, the far-field pressure wave is 90° out of phase with the near-field wave.[86] An arbitrary definition is that the near field is less than $\lambda/2\pi$ from the sound source (where λ is the wavelength of the sound) and the far field is farther away than that.[228]

In water, as in air, the intensity is given by $I = P^2/\rho C$ (see p. 525). Sound intensity is relatively independent of frequency for a near source, but falls off at low frequencies for a distant source.[54, 247]

Elasmobranchs lack swim-bladders, yet they respond to low-frequency sounds even after the lateral line nerves are cut (but not after the eighth nerve is cut). A shark is attracted to sound sources of frequencies between 55 and about 500 Hz.[161] In the skate *Raja*, impulses are recorded in response to low-frequency sound (below 120 Hz) from nerve twigs to two sensory maculae in the sacculus and one in the utriculus, but no response occurs in the nerve from the lagena.[137]

Fish have a variety of sound receptors—cutaneous, lateral line, and three labyrinthine chambers—and they appear to use them according to the type of stimulation; the first two are used to detect displacement and low frequencies. In training experiments, some fish appear to use their less sensitive receptors first, and then to switch to the more sensitive ones.[247] Representative audiograms obtained by behavior observation are given in Figure 12–19. In fish, near-field stimulation is by lateral line and by bone conduction to the ear. A sculpin lacks a swim-bladder; no microphonics were recorded from the sacculus for sound sources farther away than one meter, and there was no response to sounds above 400 Hz. A cod has a swim-bladder; good responses were recorded from far sound sources and at frequencies below 1000 Hz (Fig. 12–20A). A clupeid, the herring, has a double swim-bladder with the anterior chamber directed to the sensory epithelium of the utricle; nervous responses were obtained at low threshold at frequencies from 13 to 1200 Hz, and at higher intensities from 1200 to 4000 Hz.[53] Maximum sensitivity in most fish is at 500 to 800 Hz; fish having Weberian ossicles coupled to the swim-bladder have upper limits much higher (6000 to 8000 Hz) than the upper limits for fish without Weberian apparatus (800 to 3000 Hz).[120, 247] Thus, high audiosensitivity is well correlated with swim-bladder presence and complexity.[55, 56] In *Ameiurus*, deflation of the swim-bladder had little effect on sensitivity at low frequencies (200 to 300 Hz) but it decreased sensitivity by 30 db at 1500 Hz; cutting the lateral line nerves virtually abolished responses at low frequencies[122] (Fig. 12–20B). In the midshipman *Porichthys*, microphonic potentials from the ear were

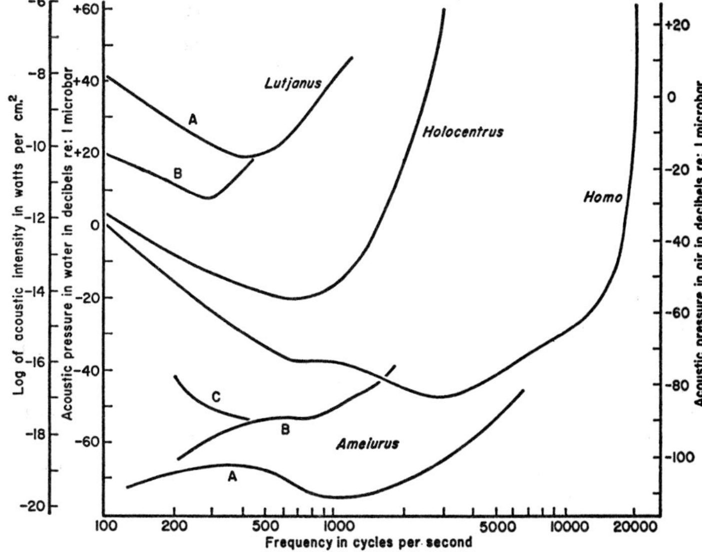

Figure 12–19. Audiograms as measured behaviorally in three fish, *Lutjanus*, *Holocentrus*, and *Ameiurus*, compared with that of man. (From Tavolga, W. N., and J. Wodinsky, Bull. Amer. Mus. Nat. Hist. *126*:179–239, 1963.)

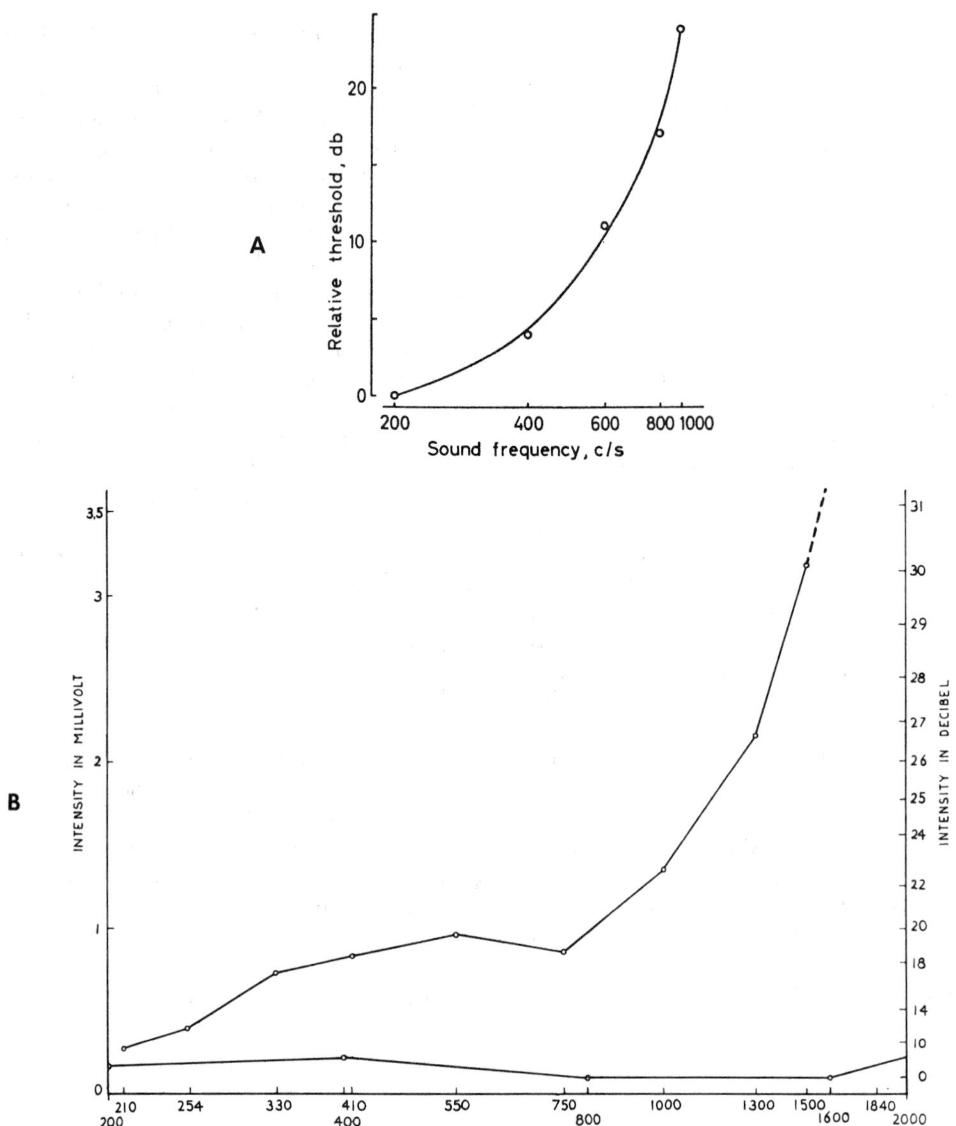

Figure 12–20. A, Audiogram as measured by cochlear microphonics in codfish. (From Enger, P. S., Comp. Biochem. Physiol. *22*:517–525, 1967.) B, Audiogram of catfish *Ameiurus* as measured by conditioned responses. Lower curve, normal fish; upper curve, fish with mutilated swim-bladder. (From Kleerekoper, H., and P. A. Roggenkamp, Canad. J. Zool. *37*:1–8, 1959.)

little affected by removing the lagena or one edge of the sacculus, but were eliminated by complete cautery of the saccular macula.[34]

Many fish produce sounds that are used in communication. The sounds are mostly produced by contraction of muscles along the swim-bladder or between the pectoral girdle and the swim-bladder; they are amplified by the resonance of the swim-bladder. *Porichthys* produces grunts of maximum amplitude at 85 to 600 Hz, and buzzes at 85 to 1000 Hz.[34] A tiger-fish, *Therapon,* makes sounds for threats and fighting. A sea robin has a grunt for alarm and a staccato breeding call. The blenny's grunt has maximum energy in the 80 to 180 Hz range; the call of a toadfish is maximum at 250 Hz. Local electrical shocks to the midbrain either facilitate or suppress the response to clicks, recorded in the auditory nucleus, according to time intervals between shocks and clicks.[187a]

In general, the thresholds for behavioral responses are lower than those for neural responses. Records from single fibers in the acoustic region of the medulla of the herring *Clupea* show much variation in highest frequency detected and in the frequency range for best response; greatest sensitivity is in the range from 100 to 1200 Hz.[55] In goldfish brain, the spread of an audiogram for units in the auditory nucleus is wider than for units in the torus semi-circularis, and much wider than for units in the tegmentum; in the latter region, some neurons respond to both sound and visual input.[183] Pulsed acoustic signals, but not pure tones, are effective and have been used in attracting predatory fishes (e.g., sharks) away from human swimming regions.

A goldfish can hear tones to 3500 Hz; little loss resulted from injury to the utriculus, but when the sacculus and lagena were also damaged, reactions to low tones (250 Hz) only remained. A minnow, *Phoxinus,* hears up to 6000 Hz; its tonal discrimination is lost on removal of the lagena.

Terrestrial Vertebrates; Central Nervous and Behavioral Responses to Sound. AMPHIBIANS. Frogs emit mating calls which are species-specific. Phonograms of songs from various cricket frogs show spectral peaks of energy and patterned clicks which differ not only with species but with locality, and the female responds only to the call of a male of her own species and locality[24] (Fig. 12-21). Recordings from neurons of the auditory nucleus show sensitivities which match the peak sensitivities of the basilar membranes and which correspond to the range of natural calls for mating, territoriality, and distress.[69, 217] The lowest threshold for evoked potential responses in the midbrain of six species of anurans corresponds to the dominant frequency of the call. In the complex known as *Rana pipiens,* four populations are so distinct in their calls that they are reproductively isolated species.[133]

REPTILES. Alligators and crocodiles give cochlear potentials to stimuli in the range from 300 to 3000 Hz, with relatively high thresholds.[258, 259] Records from single neurons of the acoustic nucleus in the medulla of a gecko, *Coleonyx,* show maximum sensitivity from 800 to 2000 Hz, but there are responses over the range from 100 to 17,000 Hz; individual neurons vary greatly and may show inhibition of response within narrow frequency ranges.[242]

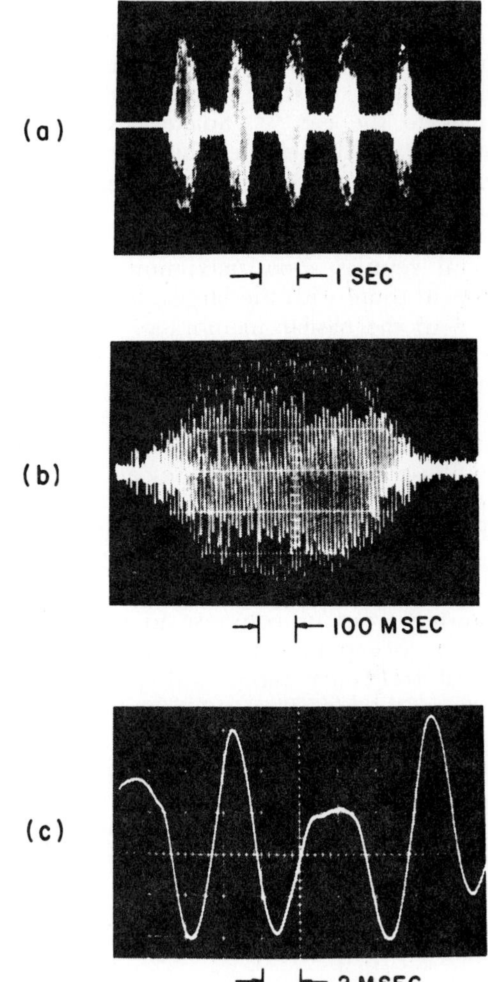

Figure 12-21. Bullfrog mating call: (a) mating call in its entirety, with five croaks; (b) third croak on faster sweep; (c) portion of third croak on very fast sweep. (From Capranica, R. R., Vocal Response of the Bullfrog. MIT Press, 1965.)

Some primary auditory fibers of a skink, *Trachysaurus,* are spontaneously active; these respond to a tone by an initial increase in firing rate, an after-inhibition, and a rebound. The optimal frequency varies from fiber to fiber, and peaks were noted at 700 Hz and 3 kHz.[110]

Both slow (evoked) responses and spikes were recorded from the tectum of different species of snakes. Air-borne tones of from 50 to 1000 Hz evoked responses of relatively high threshold; when sound was directed locally, response occurred to stimulation over much of the body surface. This sensitivity was unaltered by spinal section, but was eliminated by inner ear ablation; hence, airborne sounds must be transmitted through tissues to the ear.[89, 90] Snakes also have a somatic (spinal) system which senses substrate vibration over a broad range of frequencies (see p. 519).

Recordings from the cochlear nucleus of several reptiles show maximum frequency range in those with the largest length/width ratios of the basilar membrane. The range for a caiman is 70 to 2900 Hz. In the torus semicircularis of the caiman, units are grouped in columns; those with the best responses at low frequencies occur dorsally, and those with best responses at high frequencies occur at a depth.[141]

BIRDS. Neurons of the lateral mesencephalic nucleus (inferior colliculus) of a dove are spontaneously active and show the following types of response to sounds: inhibition of spontaneity, excitation in the central frequency range with inhibition at higher and lower frequencies, and inhibition together with narrow excitatory regions.[11] The thresholds of neurons in the cochlear nucleus correspond to best behavioral sensitivity: canary, 2.5 kHz; house sparrow, 6.5 kHz; junco, 8.6 kHz (Fig. 12–22).[124, 125] Songbirds do not produce sounds below 1 kHz but can hear them. In the starling, some neurons of the acoustic nucleus show broad response curves and some show narrow ones; most of them peak between 250 and 600 Hz.[226]

Conditioning experiments show that parrots and crossbills discriminate tones over the range from 40 to 14,000 Hz; pheasants discriminate up to 10,500 Hz,[235] and a number of songbirds were conditioned best at about 3200 Hz. Song patterns have species characteristics that are genetically determined and, in addition, local variations that are imprinted on young birds before they start to sing; this may be an important isolating mechanism leading to speciation.[126, 171] A general function of calls in songbirds is the establishment of territory; songs also function in attraction of a mate. Crows and gulls have a variety of calls: alarm, distress, assembly, chorusing, and others.

Birds that normally live in caves—the oilbird *Steatornis* and the swiftlet *Collocalia*—use echo-orientation when flying in the dark. *Steatornis* emits brief clicks, each within the frequency range from 6100 to 8570 Hz and without an ultrasonic component; *Collocalia* emits 5 to 10 clicks per second with a 4500

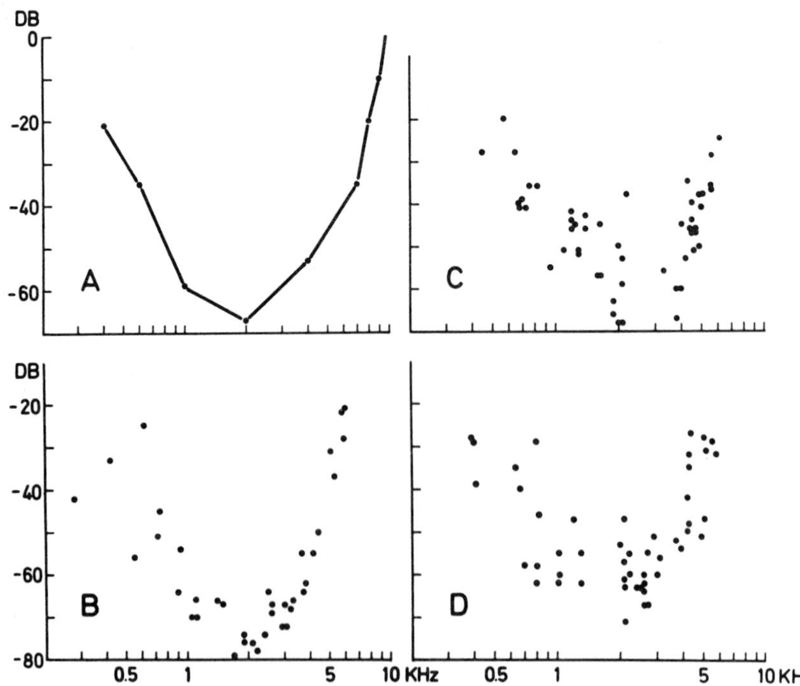

Figure 12–22. A, Behavioral audiogram; B, C, D, audiograms of single neurons in cochlear nucleus of starling. (From Konishi, M., Z. vergl. Physiol. *66*:257–272, 1970.)

to 7500 Hz component. These birds can use visual cues in light, or sound in the dark, to avoid objects such as rods and wires.[79, 173] Owls (*Tyto*) are very sensitive to sounds caused by movement of prey; the owl ear is highly directional for sounds above 8.5 kHz.[186]

MAMMALS (EXCEPT BATS). In all mammals, the cochlea is coiled: in the platypus, a quarter turn; whale, 1.5 turns; horse, 2 turns; man, 2.75 turns; cat, 3 turns; pig and guinea pig, nearly four turns. The audiogram of man shows maximum sensitivity in the range from 1000 to 4000 Hz with an upper limit of about 18 kHz. Some dogs hear up to 35 kHz, and rats and guinea pigs up to 40 kHz. A deer mouse, *Peromyscus*, can be conditioned over the range from 16 to 65 kHz. A dormouse has a minimum threshold (about 0.01 dyne/cm^2) at 20 kHz.

In general, primates have a lower cut-off for high frequency discrimination than do other mammals – 22 kHz for chimpanzee, compared with 120 kHz for bat and dolphin. Primates tend to hear better at low frequencies than mammals such as opossum and hedgehog.[146]

The neural pathways of hearing in mammals are shown in Figure 12–23. From the cochlear nucleus, fibers go to the superior olive complex, and then via the lateral lemniscus to the inferior colliculus, the medial geniculate, and the auditory cortex; crossings from one side of the brain to the other occur at the cochlear nucleus, at the olive, between the inferior colliculi, and in the cortex. In no class of vertebrates other than the mammals is the forebrain necessary for hearing; interactions from both ears occur first in the superior olive, but localization of sound source is achieved in the cortex.[166, 261] Olivo-cochlear efferent fibers which cross medially innervate the outer hair cells of the organ of Corti. Electrical stimulation of the crossed olivo-cochlear bundle reduces the afferent responses in single auditory nerve fibers to a tone; the CM, however, can

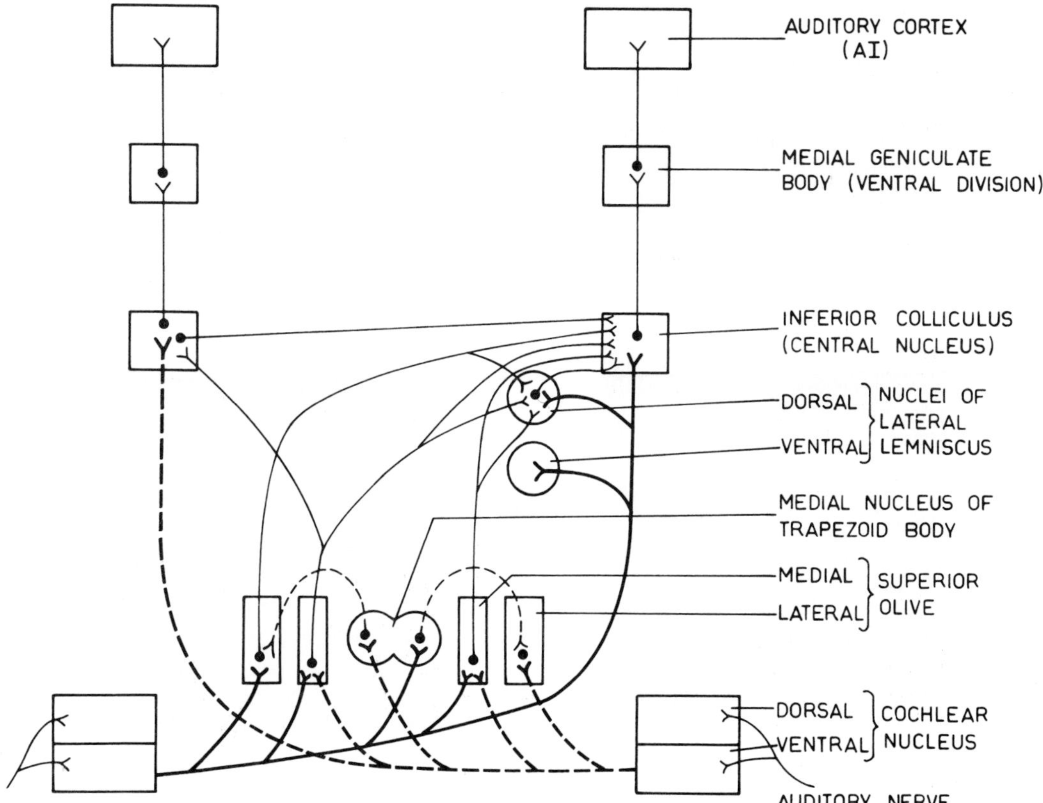

Figure 12–23. Principal ascending connections of the mammalian brain auditory system. Emphasis is given to the large bilateral component in the projection to superior olivary nucleus and higher structures. Several pathways, including those originating in the dorsal cochlear nucleus, are not shown. (Courtesy of L. M. Aitkin.)

be enhanced, and spontaneous activity may be unaffected.[262] The efferent fibers elicit inhibitory potentials and appear to act presynaptically in the basal cochlear turn, and possibly postsynaptically in the upper cochlea.[127] The olivo-cochlear efferents are activated reflexly by sound; they may sharpen discrimination and reduce the masking effect of background noise.[65, 66] Descending fibers also terminate on cells of the cochlear nucleus, where they may be inhibitory (especially at high frequencies)[117] or facilitatory.[261]

Cortical stimulation inhibits some geniculate cells and facilitates others.[255] The inferior colliculi are centers for auditory reflexes; they serve also as relays to the cortex, mostly via the medial geniculate. Tonotopic localization is maintained throughout the auditory pathways. For example, the dorsal and ventral lateral lemniscal nuclei provide one cell station between the cochlear nucleus and the inferior colliculus. Sampling of neurons in a vertical track in the inferior colliculus of cat shows a regular sequence of best frequencies for sound stimulation, with low frequencies (420 Hz) in the dorsal region and high frequencies (2180 Hz) in the ventral region.[204] A similar tonotopic arrangement occurs in both lemniscal nuclei; the dorsal receives binaural input, while the ventral receives contralateral input.

Single neurons of the auditory cortex show considerable tone discrimination; some respond poorly to a pure tone, but respond well to a swept tone (frequency modulation or FM). Tonotopic organization is weak, with low frequencies better represented posteriorly and high frequencies anteriorly. Neurons with similar best frequencies are not found in cortical columns.[1, 75, 76] Most cortical

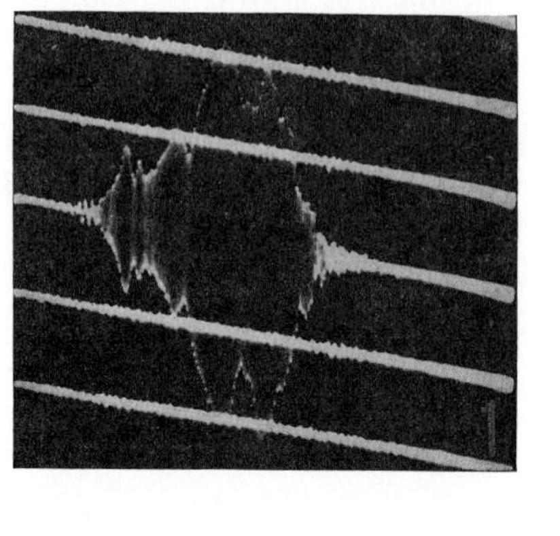

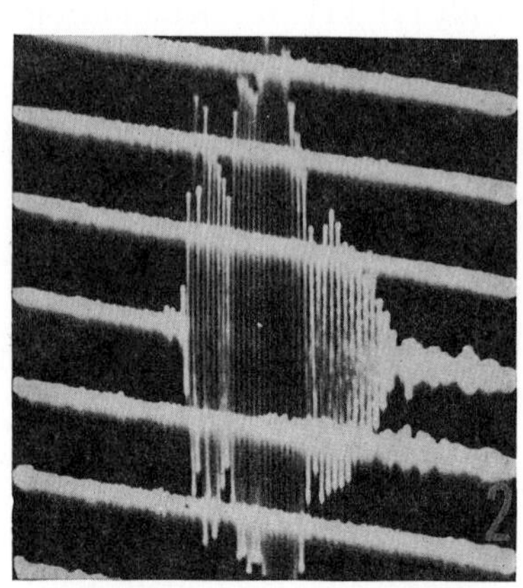

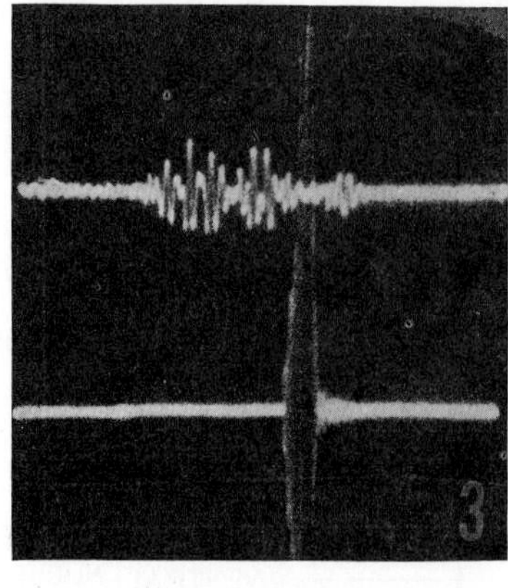

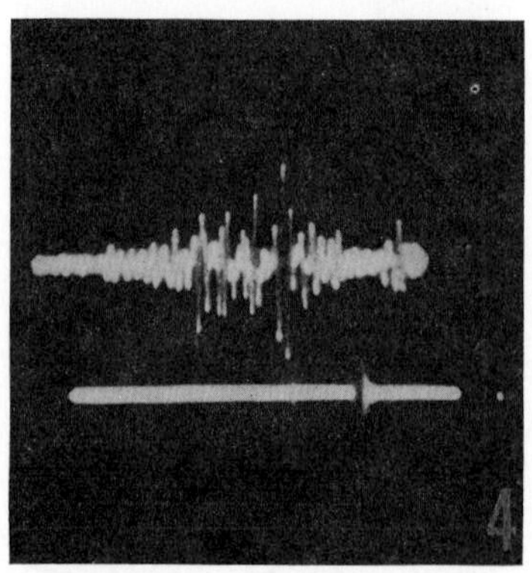

A

Figure 12–24. A, Sound pulses emitted by bat *Myotis.* (1) Ultrasonic orientation pulse, sweep duration 5 msec. (2) Orientation pulse after cutting muscularis branch of superior laryngeal nerve. Note lower frequency. (3) Upper trace, electrical recording from cricothyroid muscle during orientation pulse shown in lower trace. (4) Recording from right thyrohyoid muscle. (From Novick, A., and D. R. Griffin, J. Exp. Zool. *148*:125–141, 1961.)

(Figure 12–24 continued on opposite page.)

cells are influenced by signals from both ears; however, responses are often stronger to the contralateral ear.[75]

Cetaceans can detect the direction of sound under water. A porpoise, *Tursiops*, with its eyes covered, emits sound pulses of 1 to 1.5 msec duration, repeated at about 16/sec when cruising and accelerating to 190/sec when near a reflecting object such as a fragment of fish.[170] Evoked responses have been recorded from porpoise inferior colliculus, medulla, and medial geniculate as a result of very brief (0.1 msec) sounds; maximum sensitivity is at 60 kHz, the threshold rises steeply below 20 kHz and above 70 kHz, and the highest detected frequency is between 120 and 140 kHz. Masking can be by a narrow frequency band, and frequency modulated pulses give large responses; resolution of repeated clicks is good up to 2000/sec. The cone of best reception is forward of the head and downward, and sound probably reaches the cochlea via fat masses and the lower jaw. As a porpoise approaches a target it moves its head up and down and back and forth, presumably to note the head position at which the echo return is loudest at each ear. The emitted sound pulses are loudest at 30 to 40° to either side of the rostrum and a little above the horizontal.[19] The means of sound production by cetaceans are multiple and are poorly understood.

Recording of sound production by humpback whales shows that individuals emit sequences of sound, which may constitute songs lasting several minutes and which are repeated many times. An individual whale may be distinguished by its song; the intensities are high, and in the layer of water frequented by the whales the songs may carry for many miles.[185]

Echolocation is also used by some shrews (*Blarina*), which emit sounds of 4 to 33 msec duration and 30 to 60 kHz intrapulse frequency.[77]

HEARING AND ECHOLOCATION IN BATS. Among bats, the microchiropterans produce sounds by the larynx and the megachiropterans produce it by tongue clicking. The Microchiroptera feed largely on flying insects, and in the laboratory they can avoid small objects such as suspended wires in a dark room. As the vespertilionid *Myotis* flies, it continuously emits vocal cries; these consist of periodic clicks, each lasting about 2.3 msec. Each sound pulse starts at about 70 to 80 kHz and ends at 30 to 45 kHz, with maximum energy at 50 kHz[192] (Fig. 12–24). The

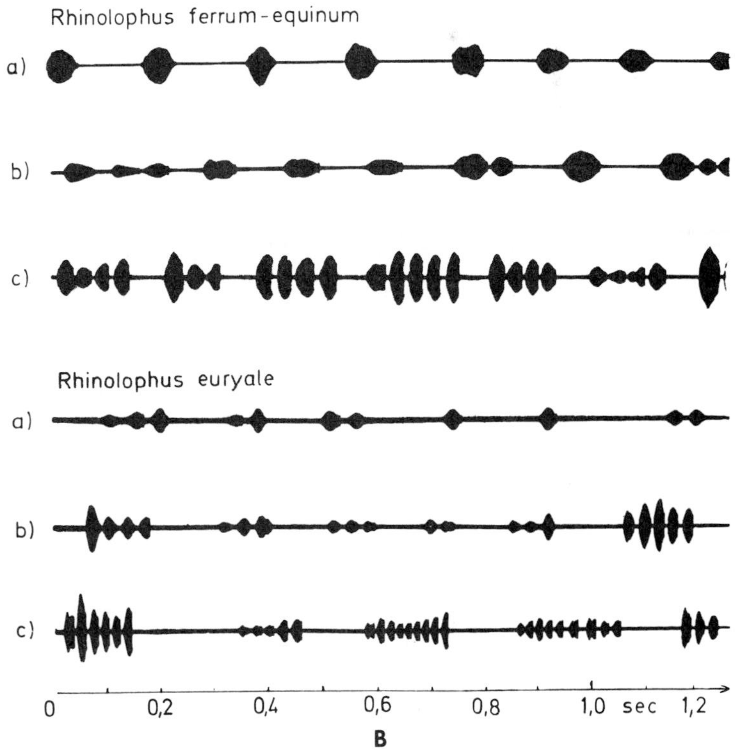

Figure 12–24 (Continued). B, Calls of two species of bats: (a) exploring without target; (b) taking bearings on target; (c) orienting toward food (mealworm) in air. (From Schnitzler, H., Z. vergl. Physiol. 57:376–408, 1968.)

pulses are repeated in free flight at about 30/sec; as an object is approached, the repetition rate rises to 50/sec or, sometimes, to 150/sec. *Myotis* also emits sounds at about 7 kHz and clicks audible to man. Sounds are produced by tensing the laryngeal membranes by means of the cricothyroid muscles, which are under vagal control.[174, 175] In *Chilonycteris*, search pulses are of 4 msec duration, repeated at 18/sec; on detection of an echo at 4 to 7 meters distance, the pulse duration increases to give pulse-echo overlap of 18 to 21 msec. On approach the pulse is shortened, and terminally the pulse duration is only 0.8 msec; during the approach, a constant pulse-echo overlap of about 1.2 msec is maintained, and the overlap could be used for estimating distance.

Rhinolophus flying at 4 m/sec and emitting sounds at 81.1 kHz receives echoes from directly ahead at 83.3 kHz (due to the Doppler effect) and from a 90° angle at 81.4 kHz; recording of evoked potentials from the colliculus shows a sharp lowering of the threshold at 83.3 kHz (Fig. 12–25A), which also increases the signal-to-noise ratio. Maximum sensitivity is for 30° above the horizon, and by alternating back-and-forth movements of the two ears the directionality of the response is enhanced by some 80%. On approach, the sound pulses are shortened to 10 msec and the frequency drops by 13 to 16 kHz, thus compensating for the Doppler shift. By the amount of this compensation and by contralateral and ipsilateral contrast, the bat can fly with sufficient accuracy to avoid wires of 0.08 mm diameter.[167, 218]

Megaderma emits sounds with most of their energy in the range from 58.5 to 78.5 kHz, with pulse durations in flight of 1.2 msec and on approach of 0.7 msec; it can perceive wires 60 μ in diameter in darkness.[155] *Vampyrum* was trained to detect targets of different shapes by echolocation; the target material was then changed to give the same amplitude but different frequency cues, and only one of the bats appeared to have relied on frequency.[14]

Myotis hears best (as noted by conditioning) at 40 kHz, in the range from 10 to 120 kHz[44]

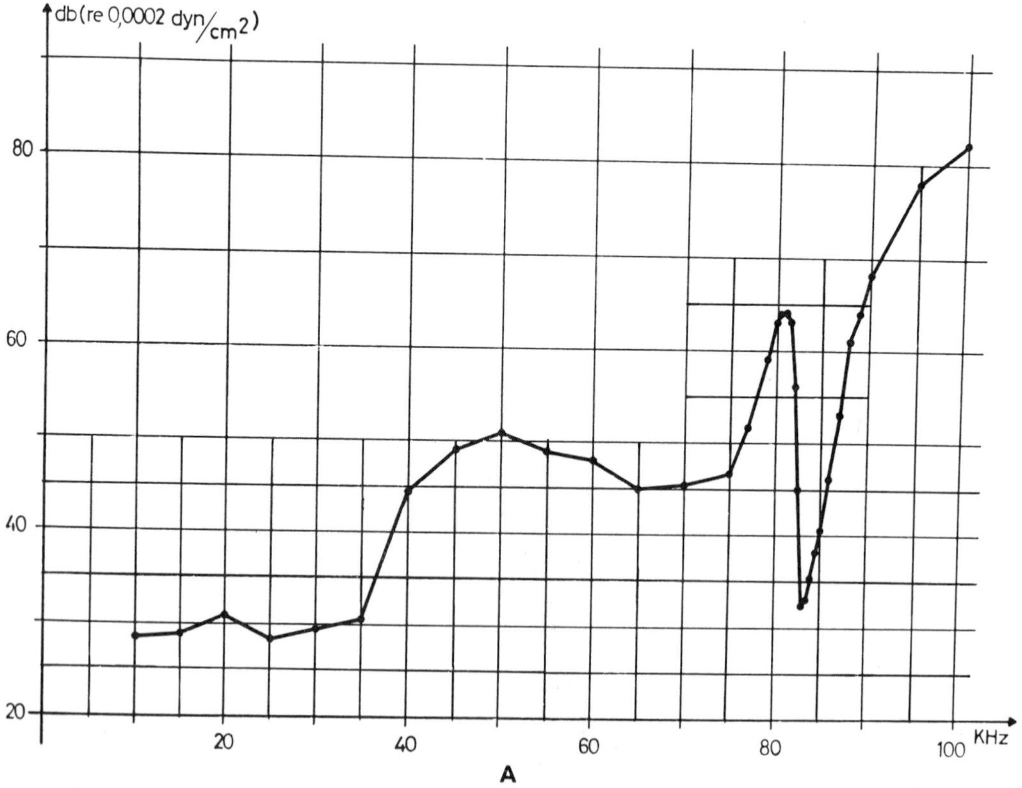

Figure 12–25. A, Audiogram as measured by collicular response in bat *Rhinolophus*. Note sharp rise in sensitivity at 83 kHz. (From Neuweiler, G., Z. vergl. Physiol. *67*:273–306, 1970.)

(Figure 12–25 continued on opposite page.)

(Fig. 12–25B). The basal turn of the cochlea is usually large and the brain is built around the central acoustic system. The inferior colliculus is hypertrophied in bats, but the geniculate and auditory cortex are not. Recordings of evoked response and of single unit response from the inferior colliculus show maximum sensitivity in the range of the emitted pulses for a given species, from 42 to 80 kHz.[81] Collicular responses show directional localization, binaural interaction, and maximum response to a source 15 to 45° contralateral and minimum response to a source 60 to 90° ipsilateral; the contralateral response may inhibit the ipsilateral one by 20 db, and there is some ipsilateral inhibition of contralateral response. One ear may reduce the masking effect of noise by orienting to the angle of maximum sensitivity. The collicular units have narrow response patterns, and sensitivity decreases by as much as 35 db/kHz at either side of the maximum.[80, 81] Responses of single neurons can repeat at 4 msec intervals and can occur to a 1 msec sound pulse. The response areas of units in the cochlear nucleus are wider than for most collicular units. In wide collicular units, frequency modulation is the same for either direction of frequency change, while for narrow units the FM response varies with direction and range. A tone of frequency outside the response area of a cell may inhibit; some units are inhibited in the middle of their response area, and most are inhibited at the ends. Collicular neurons do not respond with a fixed impulse frequency to a pure tone, and thus not only the carrier frequency but also the amplitude are important. When the rise time of a tone stimulus is lengthened, the threshold may go up, and some collicular neurons may show changes of excitatory and inhibitory ranges.[239, 241] Inhibition of some frequency ranges also occurs in units of the cochlear nucleus[238] (Fig. 12–26).

Bilateral ablation of the dorso-medial region of the inferior colliculus had no effect on obstacle avoidance, but removal of the ventral half interfered with obstacle avoidance in proportion to the amount of ablation.[241] The presence of the auditory cortex in cats is necessary for localization of a sound source. However, bilateral removal of the auditory cortex in bats led to a defect in one-third, but no defect in avoidance behavior in half, of the bats on which the operation was performed; hence, the auditory cortex is not required for echolocation but

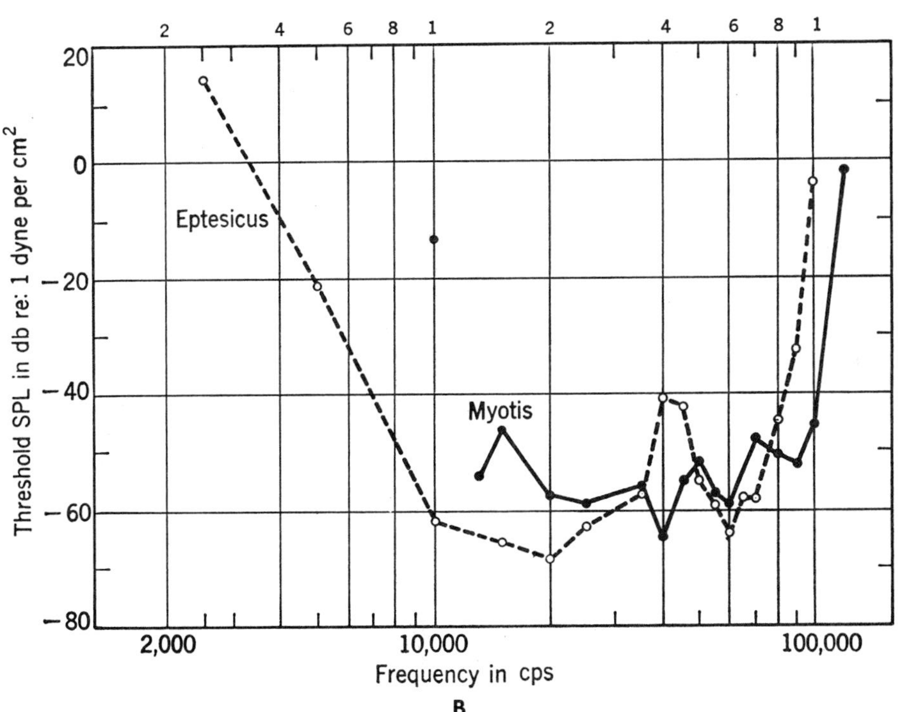

Figure 12–25 (Continued). B, Audiogram of two bats, *Eptesicus* and *Myotis*, measured by conditioned responses. (From Dollard, J. I., Science *150*:1185–1186, 1965.)

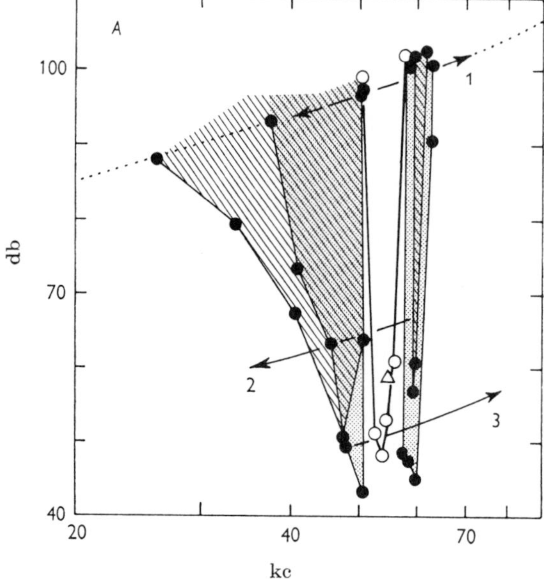

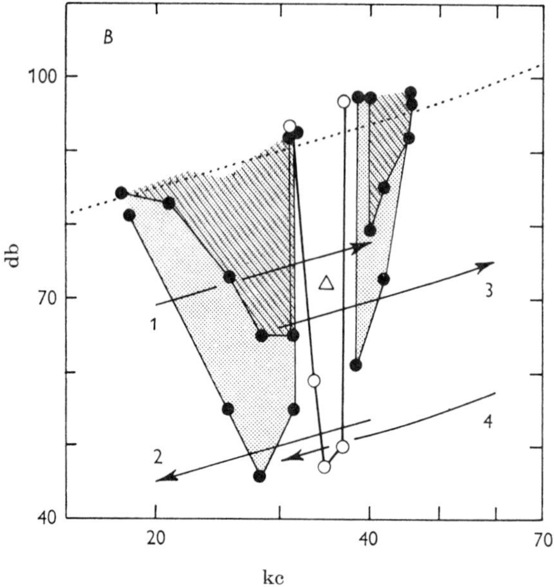

Figure 12–26. Audiograms of two neural units in inferior colliculus of a bat. Open circles give excitation areas; shaded or stippled regions give inhibitory areas. Neuron A responded to FM tone pulses. B, a neuron which showed lower threshold for downward sweep of FM tone pulse. Arrows indicate direction and range of frequency sweeps. (From Suga, N., J. Physiol. *179*:26–53, 1965.)

the ventral inferior colliculus is required.[241] Neurons of the auditory cortex have narrow frequency response areas; some show potentiation or inhibition by a tone outside the response area. More cortical units than collicular cells are sensitive to FM than to pure tones; also, more tonic units occur in the cortex. High-frequency units occur more anteriorly, and low frequency units occur posteriorly.

The activation of the stapedius muscle of the middle ear is linked in bats to sound production; it contracts just before the CM response, and hence it attenuates response to the call and restores sensitivity to an echo. The stapedius remains contracted as the pulse rate increases just prior to a catch.[193]

Megachiroptera (old-world fruit and vampire bats) rely mostly on vision for orientation; one (*Rousettus*) orients either visually or acoustically according to the illumination.[172] Its sounds are 4 msec pulses of decreasing amplitude and of 10 to 18 kHz frequency.

In summary: in vertebrate ears and lateral line organs, hair cells manifest changes in membrane resistance in response to slight displacement and then stimulate, probably by chemical synapses, the primary nerve fibers at their bases. Various coupling devices match the impedance to sound in air or water to that in the fluid which bathes the hair cells. In fishes, the auditory receptor is in one of the labyrinth chambers, and sound is transmitted best in some fishes via swimbladder and special ossicles; fishes discriminate tones, but rarely above a few kHz. In terrestrial vertebrates, the receptor chamber lengthens, and becomes a long coil in mammals; this permits some frequency localization along the basilar membrane. Coupling is by a middle ear—a single bone or three ossicles. Some mammals can hear up to 80 kHz or more. Sounds are produced by a few kinds of fish by resonating the swim-bladder. Calls of frogs and birds are important in species isolation; echolocation is used in prey capture by a few birds, most bats, and some cetaceans. The central nervous system codes frequency and intensity, with single higher neurons responding to narrow ranges, often with excitatory and inhibitory areas in the sound spectrum. Efferent impulses descend to the receptor cells and modulate the auditory input.

EQUILIBRIUM RECEPTION

Every animal maintains an equilibrium position with respect to gravity, arrived at from sensory signals—visual, proprioceptive, tactile, and special equilibrium sense.

Orientation to Gravity; Geotaxis. Many organisms orient to gravity by unknown means. In growing seedlings, the shoots show a negative and the roots a positive geotropism. When *Avena* seedlings are rotated so as to equalize the time in any one position, growth is disorganized; by appropriate deviations in rotation, a seedling can be shown to be sensitive to unidirectional forces of 10^{-3} to 10^{-2} g for the shoot and 10^{-4} to 10^{-3} g for the root. The receptor system appears to be the Golgi apparatus in cells of growing tips.[227]

Protozoans such as *Paramecium* show negative geotaxis. Many burrowing invertebrates —polychaetes, snails, and beetles—show a positive geotaxis.

Many other terrestrial animals—pulmonate snails, crabs, some beetles, young rats—show negative geotaxis mediated by static stretch receptors. When these animals are placed on an inclined plane they choose a direction of ascent which deviates from the horizontal by an angle θ which is related to the angle of inclination of the plane, α. The pull of gravity on an inclined plane is sin α, and the angle of orientation up the plane often varies as the logarithm of sin α. Addition of a slight weight to the abdomen of a beetle increases θ and adding weight at the anterior end decreases θ. Male fiddler crabs show a greater θ when the larger of the two claws is on the downward side of the animal. A young rat tends to roll when the orientation angle θ is low or the inclination α is high; an increase of α is compensated by an increase of θ. The geotaxis of these animals is dependent on a series of postural reflexes elicited via statocysts, proprioceptors, and cutaneous sense organs.

Equilibrium Receptors of Invertebrates. Equilibrium receptors are of two kinds: (1) Static or fixed-position receptors convey information about gravity, and (2) dynamic or accelerational receptors respond to either linear or angular acceleration. Accelerational receptors can be influenced by gravity, but they can respond to acceleration irrespective of gravity stimulation. Many kinds of animals have separate organs for the two functions; some have static or gravity receptors only. A common principle in all specialized equilibrium receptors is shearing stimulation of hairs; at least three types of hair structures are known.

Many molluscs have statocysts in which primary neurons bear cilia (usually long kinocilia) which project into a fluid-filled cavity containing a calcareous statolith. A few species of cephalopods have, besides some six gravity-sensitive statocysts, paired cristae or acceleration receptors.[6] Scyphomedusan and hydromedusan coelenterates have statocysts around the margin of the bell; each contains statoliths of calcium sulfate, and processes of primary sensory neurons project into the statocyst cavity. Ctenophore statocysts have ciliary processes on sensory neurons. In decapod crustaceans, the statocysts are spherical sacs at the base of the antennules, containing statoliths replaced at each molt by new inclusions such as sand grains. The hairs are cuticular and vary in length and form. Several hairs form a chorda which pushes on a process of a sensory neuron, and the nerve process has a ciliary structure.[221]

Functions of Statocysts. In scyphozoan jellyfish, removal of one or more statocysts disturbs the equilibrium during swimming; hence, the statocysts are gravity receptors. Many molluscs have statocysts, often close to pedal ganglia. In *Aplysia*, each of two statocysts is a vesicle of 200 to 250 μm diameter with 13 ciliated sensory neurons and a statolith of chalk particles. Rotation of the snail to the left causes a reflex turning of the head to the right, and vice versa; these responses are lost if statocyst nerves are cut.[43] In several land pulmonates, the tentacles move in compensation for rotation; afferent impulses in the nerve from the statocyst are most frequent when the snail is inverted. Efferent impulses in a nerve cut between the ganglion and one statocyst correlate with the position of the other statocyst.[263, 264]

An octopus has paired statocysts, each consisting of three gravity receptors in different planes (each with macula and statolith), and angular acceleration receptors (cristae). Removal of one statocyst has no obvious effect, but removal of both statocysts causes disorientation in swimming, which is greater if the eyes are also blinded.[267] Eye movements compensate for rotational stimulation of the statocysts. The effects of the six statocysts are additive.[18] Compensatory eye movements are a function of the direction rather than of the magnitude of the shearing force of the statolith on the sensory endings.[18a] Angular rotation on a turntable stimulates the cristae and leads to compensatory reflexes.[40] Records of nerve impulses originating in angular acceleration receptors show responses from the longitudinal crista system during rotation in the transverse axis, and from the vertical crista system during rotation in head-up or head-down position around the vertical axis.

In some water bugs, a thin plastron of air is held to the body surface by fine hairs (p. 180). If the body is turned sideways, the air bubble is displaced and adjacent sensory hairs are stimulated. The bubble is a "gaseous statolith."

Crustacean (e.g., lobster) statocysts are spherical sacs at the base of the antennules, which contain statoliths replaced at each molt from foreign bodies such as sand grains (Fig. 12–27). When iron particles were substituted for statoliths, a magnet held over a lobster caused the animal to turn over on its

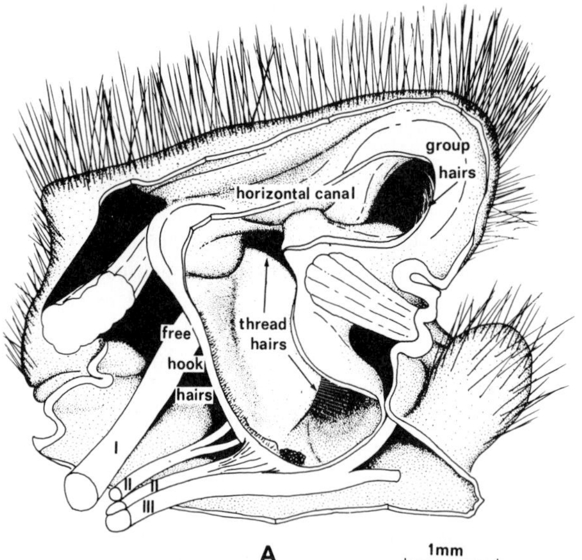

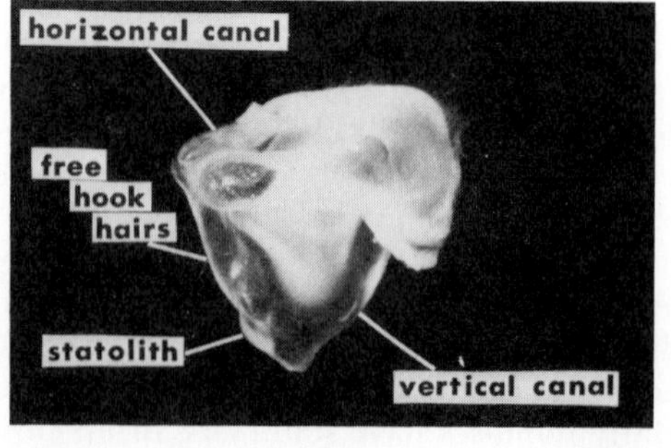

Figure 12–27 A, Dissection of basal joint of antennule of a crab to show statocyst. Posterior walls of horizontal canals have been removed to show group hairs, thread hairs, free hook hairs, and statolith. B, Stereo photograph of a statocyst viewed from the posterior end. (From Sandeman, D. C., and A. Okajima, J. Exp. Biol. 57:187–204, 1972.)

back. The walls of the statocyst of a crab are indented, forming two hollow toroids, the vertical and the horizontal canal. Rotation of the body can be imitated by irrigating the canals. Free hook hairs are located in the posterior vertical canal, "group hairs" in the lateral part of the horizontal canal, thread hairs over a central sensory cushion, and statolith hairs in a patch at the base of the vertical canal. The thread hairs are sensitive to rotation about the vertical axis; the free hook hairs respond to rotation about the horizontal axis. The statolith hairs are static receptors, and their neurons show no adaptation when in a stimulating position.

Threshold displacement of a hair terminal is 0.5 μm, and different hairs have "optimal" positions for firing. The statocysts are maximally stimulated by elevation of the rostrum, and they reflexly cause compensatory movement of the eyestalks. The thread hairs are acceleration receptors, increasing or decreasing firing rate according to direction of movement; their neuron firing adapts when the hair is held in a given position.[32] Gross frequency of sensory impulses from the statocyst is maximum when a lobster is rotated at an angle of 96 to 120° about the transverse axis.[30, 31, 32] Stretch receptors at the bases of antennules act antagonistically to the statocysts and control antennular movement.[223]

Vertebrate Labyrinths. The vestibular labyrinth of the inner ear of all vertebrates consists of two portions: the lower portion comprises the sacculus and lagena or cochlea, and the upper portion consists of the utriculus and the semicircular canals. There are three semicircular canals in all vertebrates (except certain cyclostomes). In the sacculus, utriculus, and fish lagena, the sensory hair cells occur in a cluster on a plate or macula, and there are sometimes two or more maculae per chamber. The labyrinth tubes are filled with the fluid endolymph. Each sacculus and utriculus contains one or more statoliths. Fish statoliths are composed of aragonite having a specific gravity of 2.9; those of mammals are calcite with a specific gravity of 2.7.[252] Hair cells of the semicircular canals are contained in ampullae at the junction with the utriculus; they are grouped in cristae, above which rests a gelatinous cupula (Fig. 12–28). The hair cells transmit signals synaptically to branched processes from neurons of the vestibular ganglion.

Stimulation of the hair cells is by a shearing or tangential displacement parallel to the sensory epithelium. Each cell has many (50 to 60) small stereocilia and a single large kinocilium at one side of the stereocilia (Fig. 12–29). Movement in the direction of the kinocilium is excitatory; movement away from it is inhibitory.[138] In goldfish saccular macula, the kinocilium is below the stereocilia in hair cells of the ventral half and is above the stereocilia in the dorsal half of the macula.[85] Two types of sensory nerve endings have been noted in birds and mammals; some have a chalice form, and others are nerve bulbs at the base of the hair cell (Fig. 12–30). Fish and frogs have only the sensory knobs. Vesicles in the hair cells near the nerve endings suggest chemical transmission of excitation. In addition, nerve endings with vesicles end on hair cells and on receptor-neuron endings; these are presumably efferent fibers. As with other hair cell stimulation (neuromasts, organ of Corti), the macular hair cell layers generate a dc potential—depolarizing and hyperpolarizing on displacement.[252] Single fibers of the vestibular nerve of the skate *Raja clavata* show a background of spontaneous activity. Fibers from the utriculus show a decrease in spontaneous frequency when the rostrum of the fish is up, an increase when the side from which records are being taken is up, and a decrease with that side down. Fibers from the lagena are maximally active in the normal position and decrease in frequency with either side down. Discharge from the utriculus has both static and dynamic components. In this skate, all three otoliths participate in the response to gravity and the response is proportional to the sine of the shearing angle.[137, 138] Similar results have been obtained in recordings from the vestibular nucleus of the medulla of bony fish (*Tinca, Esox, Ameiurus*).[220]

Each semicircular canal is stimulated by angular acceleration in the plane of that canal, which acts as a torsion pendulum in that the endolymph lags behind the canal wall in acceleration. In fish, a dc potential of 80 mv is recorded in the endolymph; the endolymph contains 44.4 mM potassium and 15.8 mM sodium, and the perilymph contains 4.8 mM potassium. Reduction in the dc potential is proportional to the stimulation at the cupula.[252] In *Raja*, the discharge in fibers from the horizontal canal increased on rotation ipsilaterally at 3°/sec and decreased on contralateral rotation. The nerve fiber firing frequency varies with rate of linear or rota-

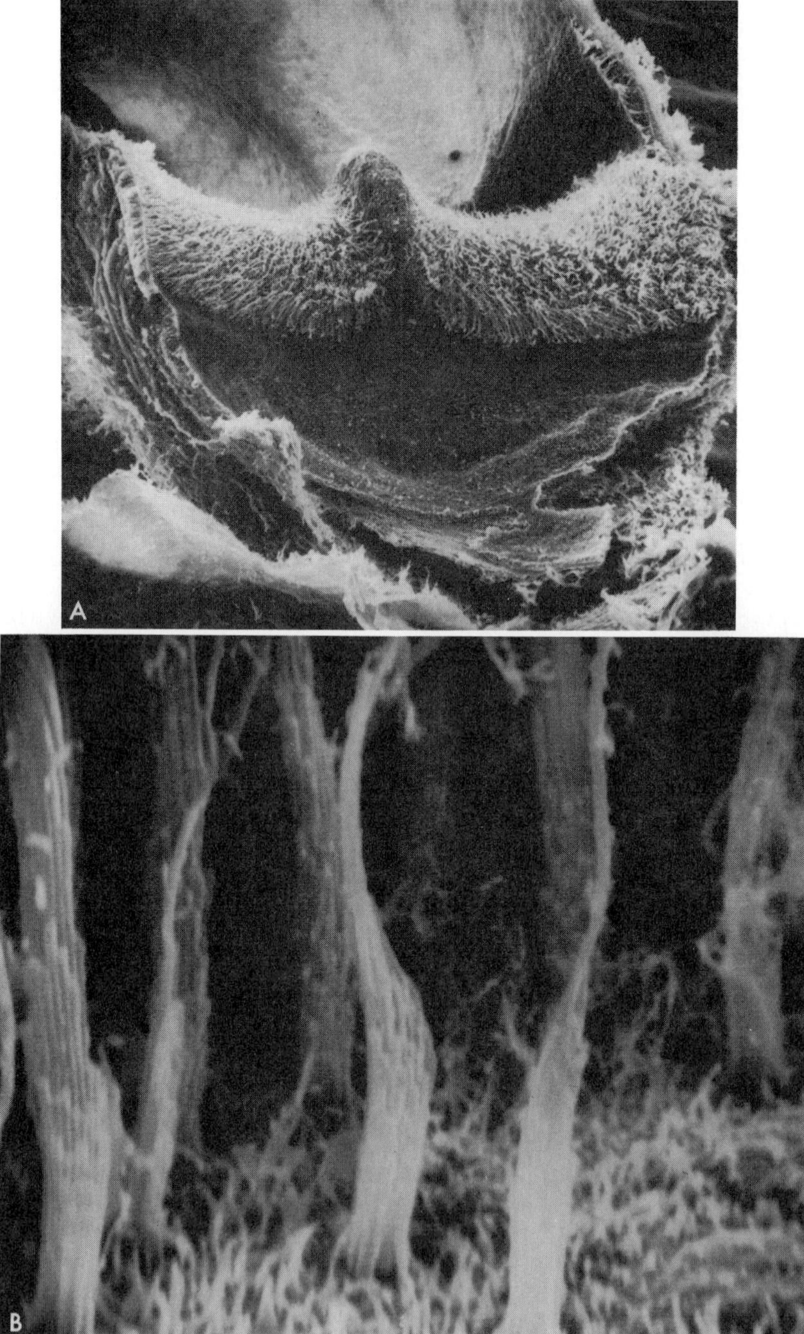

Figure 12–28. A, Crista of semicircular canal of guinea pig. Low-power electron scanning micrograph. B, High-power electron scanning micrograph of crista. Clumps of cilia rising from each cell. Longest in kinocilium. Note microvilli on supporting cells. (Courtesy of H. W. Ades.)

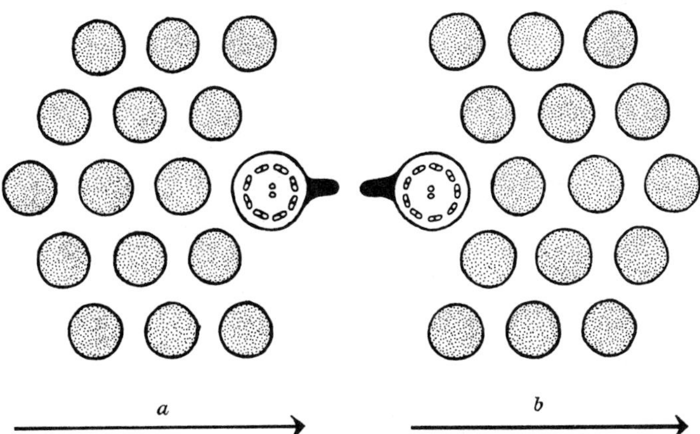

Figure 12–29. Schema of two sensory hair bundles in cross section in labyrinth of elasmobranch *Raja*. Kinocilium indicated at one side of each hair; arrows show direction of displacement for (a) excitation and (b) inhibition. (From Lowenstein, O., M. D. Osborne, and J. Wersall, Proc. Roy. Soc. Lond. B *160*:1–12, 1964.)

tional acceleration, not with static position. The vertical canals respond to rotation about all three axes, each canal responding in one axis. Post-rotatory effects include after-depression or augmentation of discharge. The frequency of impulses in nerves from ampullae of the ray, in response to sinusoidal swing of the fish, leads or lags the reversals of direction. The decline in frequency with time, after a change in velocity of rotation, corresponds to a ratio of damping (moment of friction) to the cupula-restoring couple of about 40 seconds.[82]

The labyrinth of the cyclostome *Myxine* has no well defined sacculus and gives no responses to vibrations; however, impulses in vestibular nerve fibers show clear responses to position and acceleration (better

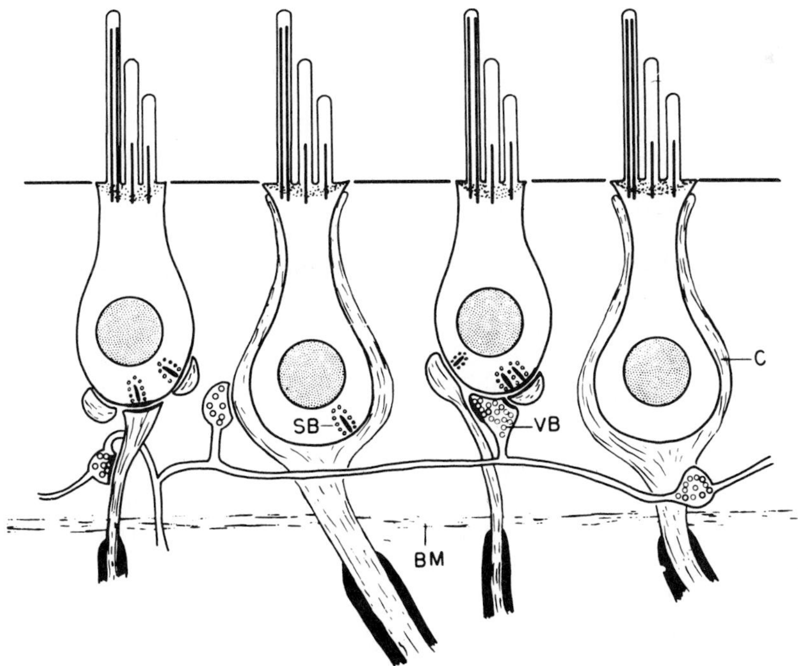

Figure 12–30. Diagram of four hair cells and two types of sensory endings in macula of chinchilla. Axon receptors are either endings opposite bar synapses (SB) or chalice terminals (C) surrounding base of hair cell. VB, efferent nerve ending. (From Smith, C. A., and G. L. Rasmussen, *in* Third Symposium on the Role of Vestibular Organs in Space Exploration [NASA SP-1521], 1968.)

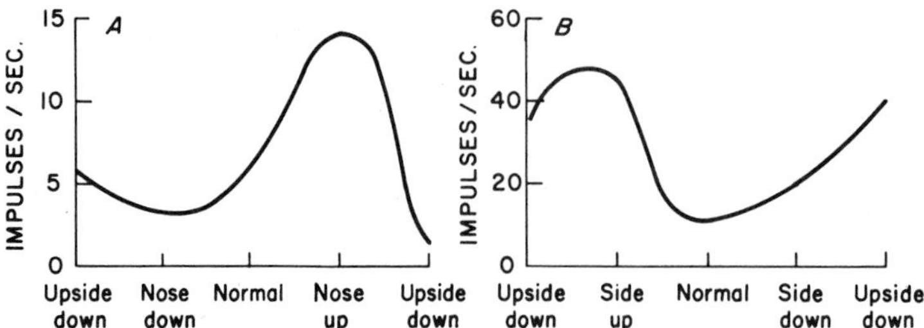

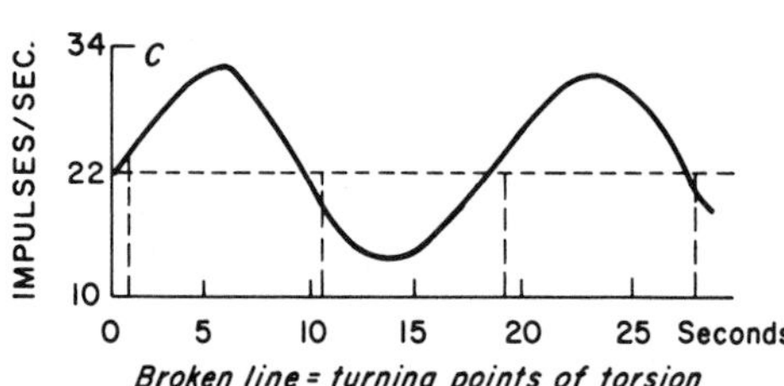

Figure 12–31. Impulses in vestibular nerves of ray (*Raja clavata*). A, Frequency of impulses coming from utriculus as head was tilted with nose down or nose up. B, Response from utriculus when side of head was tilted up or down. C, Frequency of impulses in fiber from semicircular canal during rotation at 18.5°/sec in the plane of the canal. (A and B modified from Lowenstein, O., and T. D. M. Roberts, J. Physiol. *110*: 392–415, 1950; C modified from Groen, J. J., et al., J. Physiol. *117*:329–346, 1952.)

to nose-up and nose-down movements than to lateral ones)[139] (Fig. 12–31).

Recordings from single vestibular fibers in frogs showed several patterns: response to rotation with excitation in one direction, and inhibition of spontaneous impulses on rotation in the other direction; response to tilt only; response to both rotation and tilt; and no responses to acceleration.[100] Inertia of the endolymph causes displacement of the cupula on acceleration, and frictional resistance of the canal damps the response. In frogs, efferent impulses were recorded in ampullar nerves. Fibers receiving from the horizontal canal increased their frequency on ipsilateral rotation, and stopped firing on contralateral rotation. More fibers showed non-linear frequency increase than linear with rotation. Threshold was 0.3 to 2.5°/sec^2.[187b] Efferent discharge partially inhibits afferent responses.[62, 74] In man, the caloric test is to put warm or cold water into the external ear with the head in various positions; the canal that is vertical in any given position is stimulated by the convection currents set up by the thermal gradient.[72]

Primary sensory neurons from semicircular canals of a monkey show spontaneous activity, and the units with lowest threshold to rotation have the highest spontaneous frequencies. Responses of all neurons from one canal are similar, and on excitation of the horizontal canal the cupula is deflected toward the utriculus while that of the vertical canal is deflected away from the utriculus.[74a]

Orientation in Space. Maintenance of orientation with respect to gravity results from integration of labyrinthine inputs with visual, proprioceptive, and tactile ones. In vertebrates, the utriculus is most important for static responses, while the canals are more important for rotation. In man, visual pattern stimulation can compensate for absence of labyrinthine stimulation. Motion sickness in man results when vestibular and visual inputs regarding position are dissociated, and persons lacking normal labyrinths are unaffected; there is evidence that, in weightlessness, the sensitivity to head movement may be enhanced. Compensatory eye movements are maximum at a roll of 60°;[222, 231] perceived inclination of the head is greater during acceleration than at 1 gravity. Man's macula inclines backward at 30°; hence, if the head is held forward by 30° the utricular macula is horizontal, and increased gravity stimulation is ineffective. Chickens and mice compensate so that they grow normally when maintained in a rotating (high g) chamber.[265] Removal of one utriculus from a fish results in persistent deviations of eyes and fins; damage to the sacculus and lagena has less effect.[221] Destruction of utriculi of frogs abolished responses to tilting.

In crustaceans, the angle of the eyestalk with respect to the body axis, and extension of the legs, can be used as indicators of statocyst stimulation. Crabs normally show compensatory eye movements during rotation, and after-responses (nystagmus) when turning is stopped. Blinding does not eliminate these reflexes, but removal of the statocysts eliminates the nystagmus and disturbs equilibrium. The hook hairs that bear the statoliths seem to be position detectors, and the thread hairs respond to angular accelerations or displacements about all three body axes. Eyestalk position can be altered in response to movement of a striped visual field (p. 659). Input from tactile receptors on the carapace, particularly near the eyes, triggers eye blinking. Tracking movements by eyestalks can be elicited by either statocyst or visual stimulation.[210]

Some insects, such as bees, can transpose an angle observed in relation to a light source when on a horizontal surface into a position with respect to gravity when on a vertical plane (as on the honeycomb).

Flight in insects can be initiated by wind on the head, by removing tarsal contact with the substratum, or by strong nonspecific stimuli. Maintenance of flight requires stimulation by wind on head or wing bases. Balance depends on input from eyes, wing sense organs, neck sensory structures, and, in Diptera, from halteres (modified hind wings). The halteres oscillate at about 100 to 200 Hz during flight, in synchrony with the wings. Sensory impulses from the halteres show a burst for each half cycle of oscillation; during rotation about the vertical axis, the discharge is greater than during rolling about the horizontal axis. Removal of one halter has little effect on flight, but if the eyes are blinded in addition, orientation is erratic; if both halteres are removed, the fly is unable to maintain its normal flight posture.[189]

If a frog, rabbit, or monkey is rolled around the long axis while suspended, the leading limb flexes while the trailing limb extends. If a quadrupedal animal is rotated downward, the limbs on the side toward which it is rotated extend while those on the opposite side flex. If the head is raised, extensor tone in the forelegs increases while that of the hind limbs decreases; lowering the head decreases tone in the forelimbs. A normal cat lands on its feet when it falls, but a bilaterally labyrinthectomized cat fails to do so. Visual cues are very important in righting in dogs, cats, and man, and less so in rabbits and guinea pigs. If a labyrinthectomized rabbit is placed on its side on a table, the head is righted; however, if a board is placed on the upper side of the animal, it does not right the head, so asymmetrical cutaneous cues are important for righting.

The general model for spatial orientation is as follows: the central nervous system, as a result of various sensory signals, initiates motor responses determining a given course of locomotion. Statocyst excitation is then compared in the central nervous system with a "reference level" and turning movements are initiated, which result in correspondence between the statocyst input and the reference level.

SUMMARY

The stimuli which affect equilibrium, mechano-, and phono-receptors constitute a continuum from steady pressure to rapid deformation, or from low frequency vibration to ultrasonic. Mechanoreceptors are mainly tactile or proprioceptive; equilibrium receptors are stimulated by gravity and by acceleration and deceleration; sound and vibration receptors are stimulated by regular, usually sinusoidal, vibrations of various frequencies.

Most excitable cells are sensitive to surface deformation. Why some cell membranes are depolarized by deformation and others are relatively insensitive is not known. Sensitive mechanoreceptors show generator potentials that are graded in amplitude according to amount of displacement, indicating increased permeability to ions, particularly sodium. When a generator potential reaches a critical size, an impulse is triggered, as shown in pacinian corpuscles, muscle spindles, and crustacean stretch receptors.

Two types of mechanoreceptor are: (1) those in which part of the sensory neuron is stimulated by deformation, and (2) those in which epithelial cells are stimulated and then transmit excitation to embracing nerve endings. Examples of the two types for equilibrium receptors are: for the first, filamentous nerve endings of mollusc statocysts, and for the second, the hair cells of vertebrate vestibular organs. In statocysts of decapod crustaceans, cuticular hairs connect with ciliary processes of sensory neurons. Three types of "hair" structure have evolved in specialized equilibrium receptors: processes on primary neurons (in molluscs and coelenterates), cuticular hairs (in decapod crustaceans), and epithelial hair cells which excite nerve endings (in vertebrates). These constitute an excellent example of convergent evolution.

In all mechanoreceptors there are sheaths around nerve endings or elaborate coupling devices. Some types of coupling are: thin exoskeletal membranes, as at appendage joints in arthropods; capsules around sensory endings, as in pacinian corpuscles; devices for fluid pressure gradients, as in some insect joint receptors and carotid sinuses; and bones for impedance matching, as in ossicles from the swim-bladder of fish and the middle ear of mammals, and particulate statoliths. Intersegmental stretch receptors of crustaceans and muscle spindles have contractile muscle fibers in series or in parallel.

Mechanical components of end-organs may provide directional sensitivity, matching of mechanical impedance of stimulus to tissue, regulation of sensitivity, and speed of sensory adaptation; they may thereby control frequency of response. Adaptation, or decline in response with continued mechanostimulation, results in part from changes in mechanical coupling and in part from accommodation of the excitable membrane.

Some receptors show spontaneous activity, and in these the responses represent an increase or decrease of the pre-existing level of activity. The origin of spontaneity as it occurs in lateral line nerves and auditory fibers is obscure. Many receptors receive efferent modulation. This may be inhibitory on the receptor dendritic membrane (crustacean stretch receptor) or in some undetermined way on synaptic transmission (as in lateral line or cochlear hair cells). Efferents to muscle fibers determine responses of some stretch receptors.

The stimulation of free or encapsulated nerve endings is by direct membrane deformation and resulting depolarization. Where hair cells are stimulated and excitation is by transmission to nerve endings, there are synaptic vesicles and probably chemical synapses, as in lateral line and vestibular and cochlear receptors. The significance of the very large potential difference between the endolymph and the base of hair cells in mammalian cochlea is not clear.

Receptors for sound have evolved along different lines in arthropods and vertebrates. Crustaceans may detect sound by statocysts, and insects by resonant tympanal organs. Vertebrates have built upon the basic neuromast structure of the lateral line to form sensing elements of equilibrium and acceleration (labyrinth) and high-frequency vibration (cochlea). Sound is detected in various parts of the labyrinth in fishes (utriculus in some, sacculus-lagena in many), and in the lagena of amphibians and reptiles. With lengthening of the lagena into a cochlea (in mammals), different regions of the basilar membrane came to sense different frequencies.

Different parameters of sound stimulation are important to various animals. In orthopteran insects, central neurons do not code for frequency *per se* but for pattern of sound pulses. In most vertebrates, individual neurons have optimal input frequencies, and these overlap to provide for appropriate audiograms. In bats, which hear at high frequencies, some central neurons show frequency modulation. Sound is a basis for species recognition in orthopterans, frogs, and birds, for predator detection by many animals, and for echolocation from prey for bats.

Orientation in space and posture at rest or in locomotion depend on equilibrium receptors, proprioceptors, and visual input. These are highly integrated for reflex control of body position; there is some redundancy of sensory information, and much central compensation for deviation from normally balanced inputs.

Mechanoreceptors in general occur in wider variety than other types of sense organ. The basic cellular mechanism of membrane depolarization with displacement is common, and in this respect mechanoreception may be the most primitive of sensations. The specialization comes in mechanocoupling devices, in sensitivity, and in range of frequency response.

REFERENCES

1. Abeles, M., and M. H. Goldstein, J. Neurophysiol. *33*:172–197, 1970. Functional architecture of auditory cortex.

1a. Adam, L.-J., Z. vergl. Physiol. *63*:227–289, 1969. Bioacoustics in *Locusta*.

2. Adam, L.-J., and J. Schwartzkopff, Z. vergl. Physiol. *54*:246–255, 1967. Central nervous responses to sound in *Locusta*.

2a. Aitkin, L. M., D. J. Anderson, and J. F. Brugge, J. Neurophysiol. *33*:421–440, 1970. Tonotopic organization of neurons in inferior colliculus and lateral lemniscus.

3. Aitkin, L. M., and B. M. Johnstone, J. Exp. Zool. *180*:245–250, 1972. Properties of middle ear of echidna.

4. Alexandrowicz, J. S., Biol. Rev. *42*:288–326, 1967. Stretch receptor organ in Crustacea.

5. Arthur, R. N., R. R. Pfeiffer, and N. Suga, J. Physiol. *212*:593–610, 1971. Central nervous responses to sound.

6. Barber, V. C., Symp. Zool. Soc. Lond. *23*:37–62, 1968. Structure of statocysts in cephalopods.

7. Barth, F. G., Z. vergl. Physiol. *55*:407–449, 1967. Auditory responses of spiders.

8. Bekesy, G. von, Nature *169*:241–242, 1952. Mechanics of hearing.
9. Bekesy, G. von, and W. A. Rosenblith, pp. 1075–1115. *In* Handbook of Experimental Psychology, edited by S. Stevens. John Wiley and Sons, New York, 1951. Mechanical properties of the ear.
9a. Berrou, P., et al., J. Physiol. *180*:644–672, 1965. Innervation patterns of motor muscle fibers.
10. Bianconi, R., and J. P. van der Meulen, J. Neurophysiol. *26*:177–190, 1963. Responses to vibration in mammalian muscle spindles.
11. Biederman-Thorson, M., J. Physiol. *193*:695–705, 1967. Auditory responses of neurons in midbrain of dove.
12. Blest, A. D., T. S. Collett, and J. D. Pye, Proc. Roy. Soc. Lond. B *158*: 191–207, 1963. Auditory behavior and sound production by moths.
13. Bovbjerg, R. V., Physiol. Zool. *25*:34–56, 1952. Substrate selection by crayfish.
14. Bradbury, J. W., J. Exp. Zool. *173*:23–46, 1970. Echolocation by bat *Vampyrum*.
15. Brown, A. G., and A. Iggo, J. Physiol. *193*:707–733, 1967. Responses of cutaneous receptors to touch in cat and rabbit.
16. Brown, M. C., J. Exp. Biol. *46*:445–458, 1967. Tonic responses of stretch receptors in crayfish.
17. Brown, M. C., I. Engberg, and P. B. Matthews, J. Physiol. *192*:773–800, 1967. Sensitivity to vibration in muscle receptors of cat.
18. Budelman, B.-U., Z. vergl. Physiol. *70*:278–312, 1970. Statolith organs of octopus.
18a. Budelman, B.-U. Personal communication. Nerve impulses from octopus statocysts.
19. Bullock, T. H., et al., Z. vergl. Physiol. *59*:117–156, 1968. Electrical responses in central auditory centers of cetaceans.
20. Burkhardt, D., Biol. Zentralbl. *78*:22–62, 1959. Effects of temperature on stretch receptors of crayfish.
21. Burkhardt, D., J. Insect Physiol. *4*:138–145, 1960. Responses to mechanical stimulation in antenna of *Calliphora*.
22. Burkhardt, D., and M. Gewecke, Cold Spring Harbor Symp. Quant. Biol. *30*:601–614, 1965. Mechanoreception in arthropods.
23. Bush, B. M., J. Exp. Biol. *42*:285–297, 1965. Proprioreception by chordotonal organs in leg of *Carcinus*.
23a. Bush, M. B. H., and A. Roberts, J. Exp. Biol. *55*:813–832, 1971. Coxal stretch receptors of crab.
24. Capranica, R. R., Vocal Response of the Bullfrog: A Study of Communication by Sound. MIT Press, Cambridge, Mass., 1965. 110 pp.
25. Catton, W. T., J. Physiol. *158*:333–365, 1961. Threshold recovery and fatigue of tactile receptors in frog.
26. Catton, W. T., J. Physiol. *187*:23–33, 1966. Responses of skin receptors to mechanical and electrical stimulation.
27. Catton, W. T., Physiol. Rev. *50*:297–318, 1970. Mechanoreceptor function.
28. Chadwick, L. E., pp. 577–655. *In* Insect Physiology, edited by K. D. Roeder. John Wiley and Sons, New York, 1953. Aerodynamics of flight.
29. Clarac, F., Z. vergl. Physiol. *61*:203–223, 224–245, 1968. Proprioreception in legs of *Carcinus*.
30. Clarac, F., Z. vergl. Physiol. *68*:1–24, 1970. Proprioreceptive function in legs of crayfish.
31. Cohen, M. J., J. Physiol. *130*:9–34, 1955; *also* pp. 65–108 *in* Physiology of Crustacea, Vol. II, edited by T. Waterman, Academic Press, New York, 1961. Statocysts.
32. Cohen, M. J., Proc. Roy. Soc. Lond. B *152*:30–49, 1960. Neural responses of crustacean statocysts.
33. Cohen, M. J., Comp. Biochem. Physiol. *8*:233–243, 1963. Crustacean myochordotonal organs.
34. Cohen, M. J., and H. E. Winn, J. Exp. Zool. *165*:355–370, 1967. Sound production and reception in fish *Porichthys*.
35. Crowe, A., and P. B. Matthews, J. Physiol. *174*:109–131, 1964. Muscle spindle sensitivity.
36. Davis, H., Physiol. Rev. *37*:1–47, 1957; *41*:391–416, 1961. Mechanisms of cochlear excitation.
37. Davis, H., et al., Amer. J. Physiol. *195*:251–261, 1958. Summating potential of the cochlea.
38. Desmedt, J. E., and P. J. Delwaide, Exp. Neurol. *11*:1–26, 1965. Function of the efferent cochlear bundle of pigeon.
39. Diamond, J., J. A. B. Gray, and M. Sato, J. Physiol. *133*:54–67, 1956. Site of initiation of impulses in pacinian corpuscle.
40. Dijkgraaf, S., Publ. Statz. Zool. Napoli *32*:64–86, 1961. Statocysts of octopus as rotation receptor.
41. Dijkgraaf, S., Biol. Rev. *38*:51–105, 1963. Function of lateral line detectors.
42. Dijkgraaf, S., pp. 83–95. *In* Lateral Line Detectors, edited by P. Cahn. Indiana University Press, 1967. Significance of lateral line organs.
43. Dijkgraaf, S., and H. G. A. Hessels, Z. vergl. Physiol. *62*:38–60, 1969. Structure and function of statocysts in *Aplysia*.
44. Dollard, J. I., Science *150*:1185–1186, 1965. Hearing thresholds of bats.
45. Dorward, P. K., J. Physiol. *211*:1–17, 1970. Responses of muscle receptors in ducks.
46. Dorward, P. K., Comp. Biochem. Physiol. *35*:729–735, 1970. Responses of cutaneous mechanoreceptors in ducks.
47. Dorward, P. K., and A. K. McIntyre, J. Physiol. *219*:77–87, 1971. Vibration-sensitive receptors in duck's leg.
48. Dunning, D. C., and K. D. Roeder, Science *147*:173–174, 1965. Moth sounds and insect capture by bats.
49. Echandia, E. L. R., and M. H. Burgos, Z. Zellforsch. *67*:600–619, 1965. Fine structure of stria vascularis, guinea pig.
50. Eckert, R. O., J. Cell. Comp. Physiol. *57*:149–174, 1961. Reflex function of abdominal stretch receptors of crayfish.
51. Eldred, E., et al., J. Physiol. *122*:498–523, 1953. Supraspinal control of muscle spindles.
52. Elsner, N., Z. vergl. Physiol. *60*:308–350, 1968. Pattern of muscle activity in sound production in crickets.
53. Enger, P. S., Acta Physiol. Scand. (Suppl.) *210*:1–48, 1963. Neural responses in auditory system of fishes.
54. Enger, P. S., Comp. Biochem. Physiol. *18*:859–868, 1966. Acoustic thresholds in goldfish.
55. Enger, P. S., Comp. Biochem. Physiol. *22*:517–525, 527–538, 1967. Hearing in herring, cod, and sculpin.
56. Enger, P. S., pp. 4–11. *In* Ciba Foundation Symposium on Hearing Mechanisms in Vertebrates, edited by A. De Reuck and J. Knight. Williams and Wilkins, Baltimore, 1968. Hearing in fish.
57. Enger, P. S., D. J. Aidley, and T. Szabo, J. Exp. Biol. *51*:339–345, 1969. Sound reception in cicada.
57a. Engström, H., H. W. Ades, and G. Bredberg, pp. 127–156. *In* Ciba Foundation Symposium on Hearing Loss, edited by G. Wolstenholme and J. Knight. Churchill, London, 1970. Ultra-structure of cochlea.
58. Enright, J. T., Science *133*:758–760, 1961. Pressure sensitivity of an amphipod.
59. Esch, H., Z. vergl. Physiol. *56*:199–220, 1967. Importance of sound for communication in bees.
60. Esch, H., I. Esch, and W. E. Kerr, Science *147*:320–321, 1965. Sound in communication of bees.
61. Esch, H., and D. Wilson, Z. vergl. Physiol. *54*:256–267, 1967. Sounds produced by flies and bees.
62. Eyck, M. van, Arch. Int. Physiol. *57*:102–105, 231–236, 1949. Electrical responses from semicircular canals in pigeon.
63. Eyck, M. van, Arch. Int. Physiol. *59*:236–238, 1951. Microphonic potentials.
64. Eyzaguirre, C., and S. W. Kuffler, J. Gen. Physiol. *39*:87–119, 121–153, 155–184, 1955. Excitation, inhibition, and impulse initiation in stretch receptor of lobster and crayfish.
65. Fex, J., J. Acoustical Soc. Amer. *41*:666–675, 1967. Efferent inhibition in cochlea.
66. Fex, J., pp. 169–180. *In* Ciba Foundation Symposium on Hearing Mechanisms in Vertebrates, edited by A. De Reuck and J. Knight. Williams and Wilkins, Baltimore, 1968. Efferent inhibition in cochlea.
67. Fields, H. L., J. Exp. Biol. *44*:455–468, 1966. Proprioceptive control of posture in crayfish abdomen.
68. Flock, A., pp. 163–197. *In* Lateral Line Detectors, edited by P. Cahn. Indiana University Press, 1967. Ultrastructure of lateral line organ.
69. Frishkopf, L. S., and R. R. Capranica, J. Acoustical Soc. Amer. *35*: 1219–1228, 1963; *40*:1262–1263, 1966. Auditory responses in brain of bullfrog.
70. Furukawa, T., and Y. Ishii, J. Neurophysiol. *30*:1377–1403, 1967. Auditory responses in goldfish.
71. Gelperin, A., Science *157*:208–210, 1967. A stretch receptor in foregut of blowfly.
72. Gernandt, B., J. Neurophysiol. *12*:173–184, 1949. Vestibular responses to rotation and thermal stimulation.
73. Gillies, J. D., D. J. Burke, and J. W. Lance, J. Neurophysiol. *34*:252–262, 1971. Tonic vibration reflex in cat.
74. Gleisner, L., and N. I. Henriksson, Acta Otolaryngol. (Suppl.) *192*: 90–103, 1963. Effect of efferent impulses on responses in vestibular nerve of frog.
74a. Goldberg, J. M., and C. Fernandez, J. Neurophysiol. *34*:635–684, 1971. Impulses in neurons of semicircular canals of monkey.
75. Goldstein, M. H., pp. 131–151. *In* Physiological and Biochemical Aspects of Nervous Integration, edited by F. D. Carlson. Prentice-Hall, Englewood Cliffs, N.J., 1968. Cortical responses to acoustic stimuli.
76. Goldstein, M. H., pp. 1465–1498. *In* Medical Physiology, edited by V. B. Mountcastle. C. V. Mosby, St. Louis, 1968. Auditory periphery.
77. Gould, E., N. C. Negus, and A. Novich, J. Exp. Zool. *156*:19–38, 1964. Echolocation in shrews.
78. Granit, R., Receptors and Sensory Perception. Yale University Press, New Haven, Conn., 1955. 369 pp.
79. Griffin, D. R., and R. A. Suthers, Biol. Bull. *139*:495–501, 1970. Echolocation in cave swiftlets.
80. Grinnell, A. D., Z. vergl. Physiol. *68*:117–153, 1970. Central auditory response in bats.
81. Grinnell, A. D., J. Physiol. *167*:38–127, 1963. Neurophysiology of audition in bats.
82. Groen, J. J., et al., J. Physiol. *117*:329–346, 1952. Response of horizontal semicircular canal in elasmobranch labyrinth.
83. Hahn, J. F., J. Physiol. *213*:215–226, 1971. Sensory responses from cat vibrissae.
84. Hama, K., J. Cell Biol. *24*:193–210, 1965. Ultrastructure of lateral line organ of eel.
85. Hama, K., Zellforsch. Mikr. Anat. *94*:155–171, 1969. Fine structure of saccular macula of goldfish.
86. Harris, G. G., pp. 233–247. *In* Symposium on Marine Bio-Acoustics,

American Museum of Natural History, edited by W. N. Tavolga. Pergamon Press, New York, 1964. Considerations in the physics of sound production by fishes.
87. Harris, G. G., and A. Flak, pp. 135–161. *In* Lateral Line Detectors, edited by P. Cahn. Indiana University Press, 1967. Electrical activity of lateral line in *Xenopus*.
88. Harris, G. G., L. S. Frishkopf, and A. Flak, Science *167*:76–79, 1970. Receptor potentials, lateral line.
89. Hartline, P. H., and H. W. Campbell, Science *163*:1221–1223, 1969. Auditory and vibration responses in midbrain of snakes.
90. Hartline, P. H., J. Exp. Biol. *54*:349–371, 1971. Sound detection by snakes.
91. Hartman, H. B., and E. G. Boettiger, Comp. Biochem. Physiol. *22*: 651–663, 1967. Proprioceptive organs in crab *Cancer*.
92. Harvey, R. J., and P. B. Matthews, J. Physiol. *157*:370–381, 1961. Muscle spindles in mammals.
93. Hashimoto, F., N. Katsuki, and K. Yanagisawa, Comp. Biochem. Physiol. *33*:405–421, 1970. Efferent control of lateral line of fish.
94. Hayes, W. J., and S. B. Barber, J. Exp. Zool. *165*:195–210, 1967. Proprioceptors in *Limulus* leg.
95. Heran, H., Z. vergl. Physiol. *42*:103–163, 1959. Reflex control of flight in honeybees.
96. Hoffman, C., Z. vergl. Physiol. *54*:290–352, 1967. Structure and function of mechanoreceptors of scorpion.
97. Horch, K. W., Z. vergl. Physiol. *73*:1–21, 1971. Hearing and vibration sense in *Ocypode*.
98. Horridge, G. A., Proc. Roy. Soc. Lond. B *157*:199–222, 1963. Proprioceptors and tactile receptors in polychaete *Harmothoë*.
99. Houk, J., and E. Henneman, J. Neurophysiol. *30*:466–481, 1967. Responses of Golgi tendon organs.
100. Huertas, J., and R. S. Carpenter, pp. 137–144. *In* Fourth Symposium on the Role of Vestibular Organs in Space Exploration (NASA SP-187), 1970. Single unit responses in vestibular nerve of frog.
101. Hunt, C. C., and S. W. Kuffler, J. Physiol. *113*:298–315, 1951. Muscle stretch receptors and their reflex function.
102. Hunt, C. C., J. Physiol. *155*:175–186, 1961. Vibration receptors in cat leg.
103. Husmark, I., and D. Ottoson, J. Physiol. *212*:577–592, 1971. Adaptation of muscle spindles.
104. Ishii, Y., S. Matsaura, and T. Furukawa, Japan. J. Physiol. *21*:79–98, 1971. Post-synaptic responses in auditory nerve fibers of goldfish.
105. Iwasaki, S., and E. Florey, J. Gen. Physiol. *53*:666–682, 1969. Inhibitory potentials in stretch receptors of crayfish.
106. Janig, W., R. F. Schmidt, and M. Zimmerman, Exp. Brain Res. *6*:100–115, 1968. Single unit responses and tactile efferent outflow, tactile endings of cat.
107. Johnstone, B. M., pp. 167–184. *In* Membranes and Ion Transport, Vol. 1, edited by E. E. Bittar. John Wiley and Sons, New York, 1970. Ion fluxes in the cochlea.
108. Johnstone, B. M., J. Acoustical Soc. Amer. *47*:504–509, 1970. Movements in middle ear of guinea pig and bat.
109. Johnstone, B. M., et al., Nature *227*:625–626, 1970. Mechanical responses in cricket ear.
110. Johnstone, J. R., and B. M. Johnstone, Exp. Neurol. *24*:528–537, 1969. Responses in primary auditory fibers of skink.
111. Josephson, R. K., J. Exp. Biol. *38*:17–27, 1961. Response of hydroid to water disturbance.
112. Julian, F. J., and D. E. Goldman, J. Gen. Physiol. *46*:297–313, 1962. Response of nerves to mechanical stimulation.
113. Kalmring, K., Z. vergl. Physiol. *72*:95–110, 1971. Acoustic responses of nervous system, *Locusta*.
114. Katsuki, Y., Japan. J. Physiol. *1*:87–99, 264–268, 1951. Electrical responses from lateral line organ of fish.
115. Katsuki, Y., Prog. Brain Res. *21*:71–97, 1966. Neural mechanisms appearing in cats and monkeys.
116. Katsuki, Y., and N. Suga, J. Exp. Biol. *37*:279–290, 1960. Neural mechanisms of hearing in insects.
117. Katsuki, Y., T. Watanabe, and N. Maruyama, J. Neurophysiol. *22*: 343–359, 1959. Single unit response from auditory tract of cat.
118. Kenton, B., L. Kruger, and M. Woo, J. Physiol. *212*:21–44, 1971. Adaptation in mechanoreceptors of reptiles.
119. Kinsler, L. E., and A. R. Frey, Fundamentals of Acoustics, 2nd ed. John Wiley and Sons, New York, 1962. 524 pp.
120. Kleerekoper, H., and E. C. Chagnon, J. Fish. Res. Bd. Canad. *11*:130–152, 1954 Hearing in fish.
121. Kleerekoper, H., and T. Malar, pp. 188–206. *In* Ciba Foundation Symposium on Hearing Mechanisms in Vertebrates, edited by A. De Reuck and J. Knight. Williams and Wilkins, Baltimore, 1968. Orientation through sound in fishes.
122. Kleerekoper, H., and P. A. Roggenkamp, Canad. J. Zool. *37*:1–8, 1959. Effect of swim-bladder on hearing in *Ameiurus*.
123. Kleerekoper, H., and K. Sibabin, Z. vergl. Physiol. *41*:490–499, 1959. Hearing in frogs.
124. Konishi, M., Science *166*:1178–1181, 1969. Single unit analysis of hearing in songbirds.
125. Konishi, M., Z. vergl. Physiol. *66*:257–272, 1970. Hearing and vocalization in songbirds.
126. Konishi, M., and F. Nottebohm, pp. 29–48. *In* Bird Vocalization, edited by R. A. Hinde. Cambridge University Press, 1969. Development of bird songs.
127. Konishi, M., and J. Z. Slepian, J. Acoustical Soc. Amer. *49*:1762–1769, 1971. Efferent control of cochlear potentials.
128. Kuijpers, W., N. M. Houben, and S. L. Bonting, Comp. Biochem. Physiol. *36*:669–676, 1970. ATPase activity in cochlea of chicken.
129. Kuiper, J. W., pp. 105–121. *In* Lateral Line Detectors, edited by P. Cahn. Indiana University Press, 1967. Frequency characteristics of lateral line organs.
130. Kutsch, W., Z. vergl. Physiol. *63*:335–378, 1969. Species-specific songs of crickets.
131. Landgren, S., Acta Physiol. Scand. *26*:1–56, 1952. Excitation of carotid pressure receptors.
132. Linsenmair, K. E., Z. vergl. Physiol. *64*:154–211, 1969; *70*:247–277, 1970. Detection of air currents by beetles.
133. Littlejohn, M. J., and R. S. Oldham, Science *162*:1003–1005, 1968. Mating calls in *Rana pipiens* species complex.
134. Loewenstein, W. R., J. Gen. Physiol. *41*:847–856, 1958. Facilitation and impulse origin in pacinian corpuscles.
135. Loewenstein, W. R., Ann. N.Y. Acad. Sci. *81*:367–387, 1959. Receptor potentials and spike initiation in pacinian corpuscles.
136. Loewenstein, W. R., and M. Mendelson, J. Physiol. *177*:377–397, 1965. Receptor adaptation in pacinian corpuscles.
137. Lowenstein, O., Proc. Roy. Soc. Med. *45*:133–134, 1952; *also* Lowenstein, O., and A. Sand, Proc. Roy. Soc. Lond. B *129*:256–275, 1940. Equilibrium function of otolith organs and semicircular canals.
138. Lowenstein, O., M. D. Osborne, and J. Wersall, Proc. Roy. Soc. Lond. B *160*:1–12, 1964. Innervation of sensory epithelium of labyrinth in ray.
139. Lowenstein, O., and R. A. Thornhill, Proc. Roy. Soc. Lond. B *174*: 419–434, 1970; *176*:21–42, 1970. Responses of labyrinth of hagfish.
140. Manley, G. A., Z. vergl. Physiol. *69*:363–383, 1970; *also* Nature *230*: 506–509, 1971. Evolution of hearing in vertebrates.
141. Manley, J. A., Z. vergl. Physiol. *71*:255–261, 1971. Responses in auditory units in brain of caiman.
142. Markl, H., Z. vergl. Physiol. *45*:475–569, 1962. Proprioception and tactile reception in hymenopterans.
143. Markl, H., Z. vergl. Physiol. *69*:6–37, 1970. Stridulation, leaf-cutting ants.
144. Markl, H., and K. Wiese, Z. vergl. Physiol. *62*:413–420, 1969. Wave detection by water beetles.
145. Martin, A. R., and W. O. Wickelgren, J. Physiol. *212*:65–83, 1971. Sensory fibers in spinal cord of sea lamprey.
146. Masterton, B., H. Heffner, and R. Ravizzo, J. Acoustical Soc. Amer. *45*:966–985, 1969. Evolution of mammalian hearing.
147. Matthews, B. H. C., J. Physiol. *71*:64–110, 1931; *72*:153–174, 1931; *78*:1–53, 1933. Receptors in mammalian and frog muscle.
148. Matthews, P. B., Quart. J. Exp. Physiol. *47*:324–333, 1962; *also* J. Physiol. *168*:660–677, 1963. Response of muscle spindle receptors to stretching at different velocities.
149. McCormick, J. G., E. G. Wever, J. Palin, and S. H. Ridgway, J. Acoustical Soc. Amer. *48*:1418–1428, 1970. Sound conduction in dolphin ear.
150. Meadows, P. S., J. Exp. Biol. *47*:553–559, 1967. Discrimination of substrate by amphipod *Corophium*.
151. Mellon, D., Jr., and D. Kennedy, J. Gen. Physiol. *47*:487–499, 1964. Impulse origin and propagation in crayfish tactile receptors.
152. Mendelson, M., J. Exp. Biol. *45*:411–420, 1966. Origin of impulses in stretch receptors of *Callinectes*.
153. Michelsen, A., Z. vergl. Physiol. *71*:49–62, 63–101, 1971. Physiology of locust ear.
154. Miller, L. E., and E. E. MacLeod, Science *134*:891–893, 1966. Ultrasonic sensitivity in lacewing *Chrysopa*.
155. Möhres, F. P., and G. Neuweiler, Z. vergl. Physiol. *53*:195–227, 1966. Detection of ultrasound by bats.
156. Money, K. E., et al., Amer. J. Physiol. *220*:140–147, 1971. Physical properties of fluids and membranes in vestibular apparatus, pigeon.
157. Money, K. E., M. Sokoloff, and R. S. Weaver, NASA SP-115, pp. 91–97, 1966. Physical properties of endolymph and perilymph.
158. Morgan, E., J. Exp. Biol. *50*:501–513, 1969; *51*:171–179, 1969. Responses of polychaete to hydrostatic pressure.
159. Mountcastle, V. B., pp. 393–408. *In* The Neurosciences, edited by F. O. Schmitt. Rockefeller Press, New York, 1967. Neural coding of sensory events.
160. Murphey, R. K., Z. vergl. Physiol. *72*:150–167, 168–185, 1971. Vibration sense in water-striders.
161. Myrberg, A. A., A. Banner, and J. D. Richard, Marine Biol. *2*:264–276, 1969. Acoustic stimulation of sharks.
162. Nakajima, S., Science *146*:1168–1170, 1970. Adaptation of stretch receptors in crayfish.
163. Nakajima, S., and K. Onodera, J. Physiol. *200*:161–185, 1969. Membrane properties of crayfish stretch receptors.
164. Nakajima, S., and K. Takahashi, J. Physiol. *187*:105–127, 1966. Electrogenic sodium pump in stretch receptors of crayfish.
165. Necker, R., Z. vergl. Physiol. *69*:367–425, 1970. Cochlear potentials in birds.
166. Neff, W. D., pp. 207–233. *In* Ciba Foundation Symposium on Hearing Mechanisms in Vertebrates, edited by A. De Reuck and J. Knight.

Williams and Wilkins, Baltimore, 1968. Localization and lateralization of sound in space.
167. Neuweiler, G., Z. vergl. Physiol. *67*:273–306, 1970. Electrical responses in auditory system of bat.
168. Nicklaus, R., Z. vergl. Physiol. *50*:331–362, 1965. Generator potentials and spikes from cercal nerves of cockroach.
169. Norris, K. S., pp. 297–324. *In* Evolution and Environment, edited by E. T. Drake. Yale University Press, New Haven, Conn., 1968. Evolution of acoustic mechanisms in cetaceans.
170. Norris, K. S., et al., Biol. Bull. *120*:163–176, 1961. Echolocation in porpoises.
171. Nottebohm, F., Science *167*:950–956, 1970. Ontogeny of bird songs.
172. Novick, A., J. Exp. Zool. *137*:443–461, 1958. Orientation in paleotropical bats.
173. Novick, A., Biol. Bull. *117*:497–503, 1959. Acoustic orientation in cave swiftlets.
174. Novick, A., Biol. Bull. *128*:297–314, 1965; *also* Novick, A., and J. R. Vaisnys, Biol. Bull. *127*:478–488, 1964. Echolocation of insects by bat *Chilonycteris*.
175. Novick, A., and D. R. Griffin, J. Exp. Zool. *148*:125–141, 1961. Sound production by bats.
176. Offutt, G. C., Experientia *26*:1276–1278, 1971. Acoustic stimulation of lobster.
177. Olivo, R. F., Comp. Biochem. Physiol. *35*:761–786, 1970. Mechanoreceptors in razor clam, foot withdrawal.
178. Ottoson, D., J. Physiol. *171*:109–118, 1964; *178*:68–79, 1965. Effects of ions on isolated muscle spindles.
179. Ottoson, D., J. S. McReynolds, and G. M. Shepherd, J. Neurophysiol. *32*:24–34, 1969. Sensitivity of isolated muscle spindle during and after stretch.
180. Pabst, H., Z. vergl. Physiol. *50*:498–591, 1965. Electrophysiology of stretch receptors in *Locusta*.
181. Pabst, H., and D. Kennedy, Z. vergl. Physiol. *57*:190–208, 1967. Cutaneous mechanoreceptors influencing motor output in crayfish.
182. Pabst, H., and J. Schwartzkopff, Z. vergl. Physiol. *45*:396–404, 1962. Phasic and tonic responses of joint receptors in locusts.
183. Page, C., J. Neurophysiol. *33*:116–128, 1970; *also* Page, C., and A. M. Sutterlin, J. Neurophysiol. *33*:129–136, 1970. Unit responses in central auditory system of goldfish.
184. Parry, D. A., J. Exp. Biol. *43*:185–192, 1965. Vibration signals in spider web.
185. Payne, R. S., Science *173*:585–597, 1971. Humpback whales' songs.
186. Payne, R. S., J. Exp. Biol. *54*:535–574, 1971. Directional hearing, barn owls.
187. Payne, R. S., K. D. Roeder, and J. Wallman, J. Exp. Biol. *44*:17–31, 1966. Directional sensitivity, ears of noctuid moths.
187a. Piddington, R. W., J. Exp. Biol. *55*:569–610, 1971. Central control of auditory responses, goldfish.
187b. Precht, W., R. Llinas, and M. Clarke, Exp. Brain Res. *13*:387–407, 1971. Responses of frog vestibular nerve to rotation.
188. Pringle, J. W. S., J. Exp. Biol. *15*:101–131, 1938. Proprioception in insects.
189. Pringle, J. W. S., Phil. Trans. Roy. Soc. Lond. B *233*:347–384, 1948; *also* Insect Flight, Cambridge University Press, 1957, 133 pp. Gyroscopic mechanism of halteres of Diptera.
190. Proske, U., J. Physiol. *205*:289–304, 1969. Responses of muscle spindles in lizard.
191. Proske, U., Exp. Neurol. *23*:187–194, 1969. Vibration receptors in skin of snake.
192. Pye, J. D., Ergebn. Biol. *26*:12–20, 1963. Echolocation in bats.
193. Pye, J. D., pp. 66–84. *In* Ciba Foundation Symposium on Hearing Mechanisms in Vertebrates, edited by A. De Reuck and J. Knight. Williams and Wilkins, Baltimore, 1968. Hearing in bats.
194. Qutob, Z., Arch. Neerl. Zool. *15*:1–67, 1962. Swim-bladders of fish as pressure receptors.
195. Rathmayer, W., Z. vergl. Physiol. *54*:438–454, 1967. Electrophysiology of proprioceptors in legs of spiders.
196. Reid, K. H., Science *172*:949–951, 1971. Sound production in cicadas.
197. Ripley, S. H., B. M. Bush, and A. Roberts, Nature *218*:1170–1171, 1968. Muscle receptor without nerve impulses.
198. Risler, H., and K. Schmidt, Z. Naturforsch. 22b:759–762, 1967. Structure of mechanoreceptors in antenna of *Aedes*.
199. Roeder, K. D., Science *154*:1515–1521, 1966; *also* J. Insect Physiol. *12*:1227–1244, 1966. Auditory system of noctuid moths.
200. Roeder, K. D., Nerve Cells and Insect Behavior. Harvard University Press, Cambridge, Mass., 1967. 238 pp.
201. Roeder, K. D., J. Insect Physiol. *13*:873–888, 1967. Orientation of moths in an ultrasound field.
202. Roeder, K. D., and A. E. Treat, J. Insect Physiol. *16*:1069–1086, 1970. Acoustic sense in hawkmoths.
203. Roeder, K. D., A. E. Treat, and J. S. Vandeberg, Science *159*:331–333, 1968. Auditory sense in sphingid moths.
204. Rose, J. F., et al., J. Neurophysiol. *26*:294–341, 1963; *29*:288–314, 1966; *30*:769–793, 1967. Frequency response and synchrony in auditory fibers.
205. Rovner, J. S., Animal Behavior *15*:273–281, 1967. Acoustic communication in lycosid spider.
206. Rudolph, P., Z. vergl. Physiol. *56*:341–375, 1967. Vibration sense in aquatic beetles.
207. Russell, I. J., J. Exp. Biol. *54*:621–658, 1971. Function of lateral line system in *Xenopus*.
208. Salmon, M., Animal Behavior *15*:449–459, 1967. Visual display and sound production in sex behavior, fiddler crab *Uca*.
209. Salmon, M., and K. W. Horch, pp. 60–96 *in* Behavior of Marine Animals. I, Invertebrates, edited by H. E. Wynn and B. Olla. Plenum Press, New York, 1972. Acoustic signalling in land crabs.
210. Sandeman, D. C., and A. Okajima, J. Exp. Biol. *57*:187–204, 1972. Structure and function, crustacean statocyst.
211. Sato, M., J. Physiol. *159*:391–409, 1961. Response of pacinian corpuscles to vibration.
212. Schlegel, P., Z. vergl. Physiol. *66*:45–77, 1970. Spikes and receptor potentials from joint receptors in *Calliphora*.
213. Schlieper, C., Marine Biol. *2*:5–12, 1968. Effects of hydrostatic pressure on marine animals.
214. Schmidt, R. S., Comp. Biochem. Physiol. *10*:83–87, 1963. Types of endolymphatic potentials.
215. Schneider, G., Z. vergl. Physiol. *35*:416–458, 1953. Halteres and equilibrium in blowflies.
216. Schneider, H., Z. vergl. Physiol. *47*:493–558, 1964. Bioacoustics in tiger fish *Therapon*.
217. Schneider, H., Z. vergl. Physiol. *61*:369–385, 1968. Bioacoustic studies on tree frogs.
218. Schnitzler, H., Z. vergl. Physiol. *57*:376–408, 1968. Ultrasonic detection by horseshoe bats.
219. Schnorbus, H., Z. vergl. Physiol. *71*:14–48, 1971. Detectors of pressure and vibrations.
220. Schoen, L., Z. vergl. Physiol. *39*:399–417, 1957. Reflexes and electrical responses of vestibular system, fish.
221. Schöne, H., Ergebn. Biol. *21*:163–209, 1959. Role of statolith organs and eyes in space orientation.
222. Schöne, H., Z. vergl. Physiol. *46*:57–87, 1962. Responses to gravity and rotation in man.
223. Schöne, H., pp. 223–235. *In* Gravity and the Organism, edited by S. A. Gordon and M. J. Cohen. University of Chicago Press, Chicago, 1971. Equilibrium reception in crustaceans.
224. Schwartz, E., Z. vergl. Physiol. *50*:55–87, 1965. Structure and function of lateral line.
225. Schwartz, E., and A. D. Hasler, Z. vergl. Physiol. *53*:317–327, 1966. Superficial lateral line of mud minnow *Umbra*.
226. Schwartzkopff, J., Experientia *5*:159–161, 1949; *also* Z. vergl. Physiol. *31*:527–608, 1949; *34*:46–68, 1952; *also* Proc. XIII Int. Ornith. Cong. 1059–1068, 1963. Hearing in various birds; cochlear potentials; conditioned responses.
227. Shen-Miller, J., R. Hinchman, and S. A. Gordon, Plant Physiol. *43*: 338–344, 1968; *also* Shen-Miller, J., Planta *92*:152–163, 1970. Geotropic responses in *Avena* seedlings.
228. Siler, W., J. Acoustical Soc. Amer. *46*:483–484, 1969. Near and far fields in a marine environment.
229. Siminoff, R., Exp. Neurol. *21*:290–306, 1968. Slowly adapting mechanoreceptors in alligator skin.
230. Siminoff, R., and L. Kruger, Exp. Neurol. *20*:403–414, 1968. Cutaneous and stretch receptors of reptiles.
231. Sjöberg, A., pp. 7–28. *In* Fourth Symposium on the Role of Vestibular Organs in Space Exploration (NASA SP-187), 1970. Experimental studies of motion sickness.
231a. Smith, C. A., and G. L. Rasmussen, p. 193. *In* Third Symposium on the Role of Vestibular Organs in Space Exploration (NASA SP-152), 1968. Labyrinth hair cells.
232. Smola, U., Z. vergl. Physiol. *70*:335–348, 1970. Electrical responses from hair sensilla of locusts.
233. Sokolove, P. G., and I. M. Cooke, J. Gen. Physiol. *57*:125–163, 1971. Inhibition in a sensory neuron by an electrogenic pump.
234. Spath, M., Z. vergl. Physiol. *56*:431–462, 1967. Effect of temperature on mechanoreceptors in bony fish *Leuciscus*.
235. Stewart, P. A., Ohio J. Sci. *55*:122–125, 1955. Audiogram of pheasant.
236. Stockwell, C. W., H. W. Ades, and H. Engström, Ann. Otol. Rhin. Laryngol. *78*:1144–1169, 1969. Damage of hair cells in cochlea due to intense sound.
237. Strother, W. F., J. Comp. Physiol. Psychol. *52*:157–162, 1959. Electrical responses from auditory system of frog.
238. Suga, N., J. Physiol. *179*:26–53, 1965. Echolocation by bats.
239. Suga, N., J. Physiol. *172*:449–474, 1964; *181*:671–700, 1966. Single unit activity in cochlea nucleus, inferior colliculus and auditory cortex of echolocating bats.
240. Suga, N., J. Insect Physiol. *12*:1039–1050, 1966. Ultrasonic production and reception in neotropical Tettigoniidae.
241. Suga, N., J. Physiol. *203*:707–728, 729–739, 1969; *217*:159–177, 1971. Evoked potentials in response to sound in bats.
242. Suga, N., and H. W. Campbell, Science *157*:88–90, 1967. Auditory responses in single neurons in gecko.
243. Suga, N., and Y. Katsuki, J. Exp. Biol. *38*:545–558, 1961. Central mechanisms of hearing in insects.
244. Suga, N., T. Nakashima, and J. B. Snow, Life Sci. *9*:163–168, 1970. Na and K in cochlear endolymph of guinea pig.
244a. Sviderskii, V. L., Dokl. Akad. Nauk. S.S.S.R. *172*:124–127, 1967. Nerve activity in flight control of locusts.
245. Tapper, D. N., Exp. Neurol. *26*:447–459, 1970. Behavioral properties of tactile pad receptors in cat.

246. Tasaki, I., and C. Fernandez, J. Neurophysiol. *15*:497–512, 1952. Cochlear microphonics, impulses in cochlear nucleus, guinea pig.
247. Tavolga, W. N., and J. Wodinsky, Bull. Amer. Mus. Nat. Hist. *126*: 179–239, 1963. Auditory capacity in fishes.
248. Taylor, R. C., Comp. Biochem. Physiol. *20*:709–717, 1967. Anatomy and stimulation of chordotonal organs in antenna of hermit crab.
249. Taylor, R. C., Comp. Biochem. Physiol. *27*:795–805, 1968. Water vibration reception in unrestrained crayfish.
250. Thurm, U., Science *145*:1063–1065, 1964; *also* Z. vergl. Physiol. *48*:131–156, 1964. Mechanoreceptors in cuticle of honeybee.
251. Thurm, U., Cold Spring Harbor Symp. Quant. Biol. *30*:75–94, 1965. Theory of mechanoreception.
252. Trincker, D., Symp. Soc. Exp. Biol. *16*:289–317, 1962. Transduction of mechanical stimulus to nerve impulses by labyrinth receptors.
253. Usherwood, P. N., H. I. Runion, and J. I. Campbell, J. Exp. Biol. *48*:305–323, 1968. Structure and physiology of chordotonal organ in locust leg.
254. Walcott, C., Amer. Zool. *9*:133–144, 1969; *also* J. Exp. Biol. *40*:595–611, 1963. Vibration reception by spiders.
255. Watanabe, T., K. Yanagisawa, J. Kanzaki, and Y. Katsuki, Exp. Brain Res. *2*:302–317, 1966. Cortical influence on unit responses of medial geniculate to sound.
256. Werner, G., and V. B. Mountcastle, J. Neurophysiol. *28*:359–397, 1965. Activity in cutaneous afferent nerves.
257. Wever, E. G., Physiol. Rev. *46*:102–127, 1966. Electrical potentials of the cochlea.
258. Wever, E. G., J. Exp. Zool. *175*:327–342, 1970; *also* Proc. Nat. Acad. Sci. *68*:1498–1500, 1971. Function of middle ear, reptiles.
259. Wever, E. G., and J. A. Vernon, Proc. Nat. Acad. Sci. *42*:213–220, 1956. Audiograms of turtles.
260. Whitear, M., Phil. Trans. Roy. Soc. Lond. B *245*:291–325, 1962. Structure of chordotonal organs in leg of *Carcinus*.
261. Whitfield, I. C., pp. 246–254. *In* Ciba Foundation Symposium on Hearing Mechanisms in Vertebrates, edited by A. De Reuck and J. Knight. Williams and Wilkins, Baltimore, 1968. Centrifugal control of auditory pathway.
262. Wiederhold, M. L., J. Acoustical Soc. Amer. *48*:966–977, 1970; *also* Wiederhold, M. L., and N. Y. S. Kiang, J. Acoustical Soc. Amer. *48*: 950–965, 1970. Function of efferent fibers in mammalian auditory system.
263. Wolff, H. G., Z. vergl. Physiol. *69*:326–366, 1970. Statocyst function, pulmonates.
264. Wolff, H. G., Z. vergl. Physiol. *70*:401–409, 1970. Efferent control in statocyst nerves.
265. Wunder, C. C., pp. 389–410. *In* Gravity and the Organism, edited by S. A. Gordon and M. J. Cohen. University of Chicago Press, Chicago, 1971. Effects of chronic acceleration of animals.
266. Yanagisawa, K., T. Hashimoto, and Y. Katsuki, J. Insect Physiol. *13*: 635–643, 1967. Frequency discrimination in central nervous system of locust.
266a. Yinon, U., A. Shulov, and A. T. Svilich, J. Exp. Biol. *55*:713–725, 1971. Hearing in desert locust.
267. Young, J. Z., Proc. Roy. Soc. Lond. B *152*:3–29, 78–87, 1960. Statocysts of octopus.

CHAPTER 13

CHEMORECEPTION

By C. Ladd Prosser

Chemoreception is important to animals for locating and testing food, for initiating escape from noxious agents or predators, for finding mates and conspecifics, and for identifying hosts and sites for oviposition. Animals respond to "attractants" by positive chemotaxes, and to "repellants" by negative taxes or rejection responses. In general, those receptor organs which have very high sensitivity and specificity and which are "distance chemical receptors" are called olfactory; the receptors of moderate sensitivity, usually associated with feeding, which are stimulated by dilute solutions are called taste or "contact chemical receptors," and those receptor endings which are relatively insensitive and nondiscriminating and which lead to protective responses are "general chemical sensors." The mechanisms of specificity require a molecular organization in receptor membranes, the nature of which is as yet unknown.

CHEMORECEPTION IN INVERTEBRATES (EXCEPT INSECTS)

Flagellated bacteria such as *Escherichia coli* provide the possibility of identification of specialized chemoreceptor molecules. Mutants are known for enzymes which metabolize (or fail to metabolize) specific substrates; other mutants have permeases (uptake carriers) for specific substances, and still others have sensory detection at low concentrations of substrates (10^{-7}M), as shown by aggregation toward specific substances. Among the known chemotactic mutants are the following: a serine receptor which is also very sensitive to cysteine, alanine, and glycine; an aspartate receptor which is also very sensitive to glutamate; a D-galactose receptor sensitive also to D-glucose and D-fructose. Possibly, extraction of cell membranes may permit identification of the receptor molecules.[1]

Protozoa show a general chemical sense insofar as they avoid alkali, acid, and salt. Ciliary reversal is brought about by a variety of cations in the series K > Li > Na > NH_4. Chemotaxis permits aggregation of myxamoebae; leukocytes approach food sources by positive chemotaxis.

Coelenterates show remarkable chemical discriminations. Sea anemones transport bits of food toward the mouth and inert particles away from it. *Hydra* is stimulated to a feeding response by very low concentrates of glutathione. The maximum response is pH-dependent, and a "titration" curve indicates four ionic groups in a receptor complex.[87] Other cnidarians give feeding responses to proline, tyrosine, and glutamine.

Often symbiosis, commensalism, and parasitism involve chemoreception. The polychaete *Arctonoë* goes to its commensal host, a starfish, by chemotaxis. When *Arctonoë* in a Y-tube is presented with the choice between plain sea water and starfish-inhabited sea water, the worm selects the latter, but water from an injured starfish repels *Arctonoë*.[30] A polychaete, *Podarke*, commensal with a starfish, shows positive chemotaxis, orthokinesis, and klinokinesis which lead to physiological trapping near the starfish.[30] The miracidia of digenetic trematodes attach only to their host species of snail even in the presence of other kinds of snails. A sessile rotifer, *Collotheca,* attaches only to fresh young leaves of the plant *Utricularia* even in the presence of other plants and of older leaves. Freshwater mites (*Unionicola*) live normally as parasites on the gills of a mussel, *Anodonta*; the mites, when in the water away from the host, are photopositive, but after a little water from the mantle cavity of the host species is added they reverse to negative phototaxis.

The clown fish *Amphiprion* commonly associates with a sea anemone, swimming unharmed among the tentacles; other species of fish receive nematocyst discharges from the anemone. *Amphiprion* after removal of the mucus from its skin becomes a target for the nematocysts. A rod coated with mucus from most fish stimulates nematocyst discharge, but mucus from *Amphiprion* inhibits the anemone's response to mechanical stimulation by the rod.[31]

Earthworms show a general sensitivity to acids over the body surface. Sensory impulses were recorded in nerves of *Lumbricus* on local application of solutions of pH 4.2 or lower.[85] Darwin found earthworms to distinguish between green and red cabbage, and between carrot and celery leaves, by taste.

Snails use chemical sense in locating food, and bivalves use it in regulating flow of water through the mantle cavity. *Pecten* is particularly sensitive to some chemical from its natural enemy, the starfish, and shows escape swimming responses to water from a tank containing starfish. In the snail *Buccinium,* as well as in *Aplysia,* an osphradium in the mantle cavity is a taste receptor, and nerve impulses are recorded on stimulation by dilute extracts of seaweed, oyster, shrimp, or glutamic acid.[7] One central neuron in *Aplysia* is inhibited in its spontaneous firing when solutions of seaweed or dilute sea water are applied to the osphradium.[70] A snail, *Nassarius,* is stimulated strongly by glycine and lactate, weakly by betaine, and very strongly by a heat-stable, non-volatile substance from dead shrimp.[24] An octopus can distinguish, by receptors on its suckers, quinine at a concentration 1/100 of that perceptible to human taste.[146]

Crustaceans are well equipped with taste endings—on mouthparts, antennules, and the inner surface of claws (dactyls). Receptors on the dactyl of the crab *Carcinides* are sensitive to glutamic acid at 5×10^{-5}M and can distinguish the two isomers of glutamate, aspartate, and leucine.[26] *Cancer* responds in a series of decreasing effectiveness: D, L-aminobutyric acid > taurine > L-glutamic acid > serine; L-aspartic acid is more effective as the undissociated molecule, but L-arginine is equally effective in alkaline and neutral solutions.[25] Barnacles show responses of their cirri to dilute solutions of amino acids, particularly L-glutamic and L-proline, and also to amines such as betaine and taurine.[29] A lobster, *Homarus,* detects extremely low concentrations of organic acids—alanine, glutamic, proline, succinic, malic; some synergisms and antagonisms were observed.[91] It is concluded that crustaceans have many receptors for amino acids and amines; no resolution by single endings for specific chemicals has been described.

CHEMORECEPTION IN INSECTS

Olfaction and taste are important sensory modalities for insects, particularly in mating, oviposition, and food selection.

Numerous male insects, particularly moths and cockroaches, are attracted by species-specific compounds which act as chemical stimuli at a distance; these are called pheromones.[21,86] Male silk moths and gypsy moths may be attracted from a distance of a mile or two by an odor from the scent glands of the females. Males deprived of their antennae do not orient toward the female; paper previously touched to the female scent gland will arouse normal males to attempt copulation. The following sex attractants have been identified chemically:

silk moth *Bombyx*: 10-*trans*-acetoxy-1-hydroxy-*cis*-7-hexadecene (also called bombykol)[23]

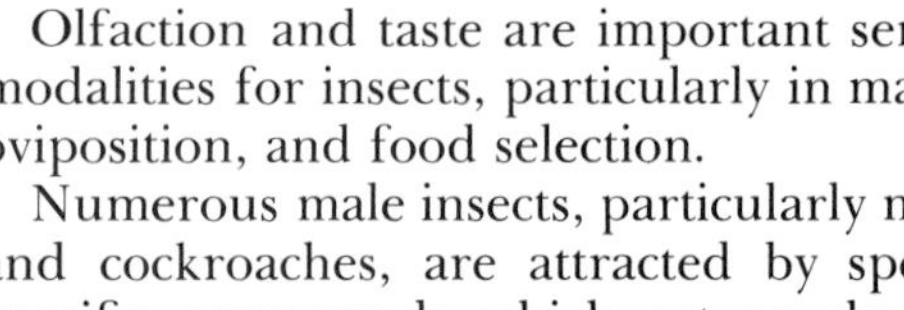

$$CH_3(CH_2)_2—\overset{H}{\overset{|}{C}}=\overset{H}{\overset{|}{C}}—\overset{H}{\overset{|}{C}}=\underset{H}{\underset{|}{C}}—(CH_2)_8—CH_2OH$$

gypsy moth *Porthetria*: *cis*-7,8-epoxy-2-methyloctadecane[14]

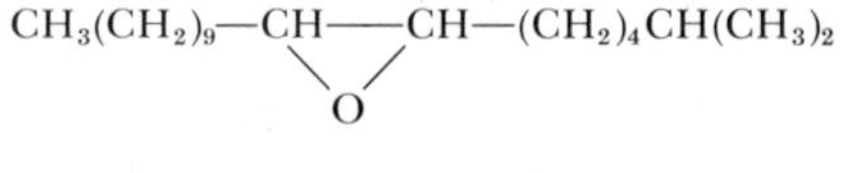

$$CH_3(CH_2)_9—\underbrace{CH—CH}_{O}—(CH_2)_4CH(CH_3)_2$$

pink bollworm moth *Pectinophora*: 10-propyl-*trans*-5,9-tridecadienyl acetate[72]

$$
\begin{array}{l}
CH_3CH_2CH_2 \\
\qquad\quad | \\
\qquad\quad C{=}CH(CH_2)_2CH{=}CH(CH_2)_4OCCH_3 \\
\qquad\quad | \qquad\qquad\qquad\qquad\qquad\quad \| \\
CH_3CH_2CH_2 \qquad\qquad\qquad\qquad\quad O
\end{array}
$$

American cockroach *Periplaneta*: 2,2-dimethyl-3-isopropylidenecyclopropyl propionate[67]

$$
\begin{array}{l}
\qquad\qquad\qquad CH_3 \\
H_3C \qquad\quad C \\
\qquad\quad C{=}C \quad | \quad CH_3 \\
H_3C \qquad\quad C \\
\qquad\qquad\quad H \quad O \\
\qquad\qquad\qquad\quad C{-}CH_2CH_3 \\
\qquad\qquad\qquad\quad \| \\
\qquad\qquad\qquad\quad O
\end{array}
$$

armyworm moth *Prodenia*: *cis*-9-tetradecen-1-ol acetate[69]

$$
\begin{array}{l}
\qquad\qquad\quad H\ \ H \qquad\qquad\qquad\quad O \\
\qquad\qquad\quad |\ \ \ | \qquad\qquad\qquad\quad \| \\
CH_3(CH_2)_3{-}C{=}C{-}(CH_2)_8{-}O{-}C{-}CH_3
\end{array}
$$

as well as *cis*-9,*trans*-12-tetradecadien-1-ol acetate

$$
\begin{array}{l}
\qquad\quad H \qquad\qquad\quad H\ \ H \qquad\qquad\qquad\quad O \\
\qquad\quad | \qquad\qquad\quad |\ \ \ | \qquad\qquad\qquad\quad \| \\
CH_3{-}C{=}C{-}CH_2{-}C{=}C{-}(CH_2)_8{-}O{-}C{-}CH_3 \\
\qquad\qquad\ \ | \\
\qquad\qquad\ \ H
\end{array}
$$

A male adult of the sugar beet wireworm is lured from a distance of 12 meters by dilute valeric acid.[68] It is evident that sex attractants do not belong to any single type of organic molecule.

The synthetic pheromones could be of practical use in attracting males of serious pests. Behavioral responses have been obtained from male *Bombyx* with dilutions of bombykol as extreme as 10^{-12} g/ml or 2×10^2 molecules/ml in an air stream.[123] It is calculated that a male moth can respond when only 40 of its 40,000 receptors receive one molecule of bombykol per second.[121]

Another use of specific chemicals by insects and diplopods is to initiate defensive reactions. Many insects (e.g., mosquitoes, bees, wasps) produce toxic materials or venoms. Some emit a toxin into the air. For example, some carabids eject formic acid; polydermids produce cyanide, and some diplopods produce a *p*-benzoquinone. A millipede *Amplelorin*, makes a toxic cyanogen, and a bombardier beetle forms a hot quinone by combining H_2O_2 with a phenol in an exergonic reaction.[44]

Mosquitoes are attracted chemically to warm-blooded animals and are sensitive to several chemicals. Carbon dioxide attracts them, and they go toward high concentrations in a gradient; they also react positively to amino acids (lysine at 0.1 ppm[22]) and to various mammalian body products; *Aedes aegypti* reacts positively to dabs of mammalian sex hormones on filter paper.[113]

Bees transmit some information concerning food sources by odors. Ants give off alarm pheromones (octanone and nonanone) from mandibular glands.[28] Army ants deposit pheromones on trails to food sources or to nest sites. Chemical communication in social insects is used for alarm, attraction, recruitment, exchange of oral and anal substances, and recognition of nest mates and of castes. Queen bee substances (e.g., 9-OH-decenoic acid) lead to bee clustering and swarm stabilization; 9-ketodecenoic acid suppresses queen-rearing; and *trans*-9-keto-2-decenoic acid is a sex attractant in nuptial flight.[147]

Many phytophagous insects feed only on a few kinds of plant. Probably none are restricted to a single species. Oligophagous insects are restricted to closely related plants—genus, family, or order—e.g., those insects that feed only on cruciferous plants. Polyphagous insects feed on many kinds of plants, but reject those which contain repellent chemicals. Specificity of selection of food plants depends not on nutrients but on secondary organic compounds which make given plants highly acceptable or make them unacceptable (repellent) for all but certain species of insects. Potato beetles and tobacco hornworms are restricted to Solanaceae. The alkaloids of tobacco are strongly repellent to the potato beetles, while those of tomato are moderately repellent and those of potato are neutral; potato beetles will eat tomato but do not develop on it. The hornworm grows on any of the Solanaceae, including tobacco.[144a]

Mexican bean beetles feed on plants of the genus *Phaseolus* and on some soybean varieties, but not on most other Leguminosae; they are attracted by phaseolunatin or related cyanogenetic glycosides.[98, 99] Of two species groups of the butterfly genus *Papilio*, one feeds only on plants of the family

Rutaceae, and the other feeds on Umbelliferae and sometimes on Rutaceae. Caterpillars conditioned to carrot are attracted to methyl chavicol; those conditioned to rue are attracted to methylnonyl ketone.[33,34] Silkworms feed mainly on Moraceae, not because of any single specific attractant but because of several nonspecific attractants.[50]

The active agents attracting cabbage butterflies (*Pieris*) to cruciferous plants are the mustard glucosides, sinigrin and sinalbin;[140] the *Pieris* caterpillar will feed on other plants that have been painted with mustard oil glucosides. A cabbage aphid can be made to feed on *Vicia* (vetch) leaf if it contains sinigrin absorbed from solution via the petiole.

Olfactory substances are attractive from a distance. Scolytid beetles are attracted to favorable breeding sites in bark by various terpenes.[114] A bark beetle, *Blastophagus,* is attracted by α-terpineol, but not by pinene or limonene.

Larval conditioning can alter chemobehavior in the adult. *Drosophila* larvae were fed on a medium containing peppermint; on emergence, the flies showed positive chemotaxis for peppermint.[139] Lepidopteran larvae (sphinx moth and *Heliothis*) were reared on different host plants within their normal range of acceptable plants; the larvae later showed preference for the food plants on which they had been fed, and the preference persisted after they had been fed on artificial diets through at least two larval molts; hence, it represents central nervous conditioning.[71] Potato beetle larvae and hornworms, reared on a synthetic diet through several molts and then given a choice of extracts from many plants, selected solanaceous plants; hence, there must be a genetic basis for the choice of plant family. However, if the larvae were fed on only one species during development, they later preferred that one, so imprinting occurs within genetically determined limits.[65,150] The return of adults to a particular plant for oviposition is probably due largely to larval conditioning.

The evolution of flowering plants during the Cretaceous was accompanied by the development of insects. Chemical attractants in flowers for certain insects facilitate cross-pollination. Repellents in plants protect them against being eaten by insects in general, but a single substance may be neutral or repellent to some insects and an attractant to others.[50]

Attraction to specific plant substances varies with the physiological state of an insect; for example, if it is dehydrated, a nymph of *Dysdercus* is less attracted by juice from cotton plants than if the nymph is well hydrated. Stimulation to feeding may be by different means than stimulation to oviposition.[116,118] Application of an attractant from a crop plant to a small area of uncultivated plants can be used as a method of insect control.

Behavioral Identification of Chemoreceptors. Chemoreceptors of insects occur on various appendages—antennae, mouthparts, and tarsi. Contact or taste chemoreceptors are often long, basiconic peg organs, and distance or olfactory receptors are short, coeloconic organs[133] (Figs. 13–1 and 13–2). The cuticle over the olfactory receptor is open at pores which are penetrated by processes of the sensory neurons that are located at the base of the organ (Fig. 13–3). Hygroreceptors are a type of distance chemoreceptor.

Certain substances are acceptable, and stimulation of taste receptors by them causes extension of the proboscis, especially in flies and lepidopterans. Other substances are unacceptable, and stimulation by these causes withdrawal of the proboscis. By local contact stimulation, taste receptors have been localized on the labellum (in a fly such as *Tabanus*), on tarsi (in many Diptera and Lepidoptera), and on maxillary and labial palpi and haustellum of the antennae (in beetles such as *Carabus*).

The threshold of taste receptors varies with physiological state, such as after feeding. Hungry insects have lower thresholds than satiated ones. By stimulating an insect simultaneously via two sets of receptors (e.g., two tarsi), central summation and inhibition can be demonstrated. When one tarsus is stimulated by sugar (acceptable) and another by alcohol (unacceptable), the proboscis is retracted.[36] Rejection is greater when the two opposing stimuli are applied to one leg than to two. Tarsal threshold is lowest in a midrange of temperature.

In blowflies, stimulation of tarsal or labellar taste receptors by water elicits proboscis extension, a response which can be abolished by injection of water or induced by bleeding (dehydration). If the recurrent nerve from

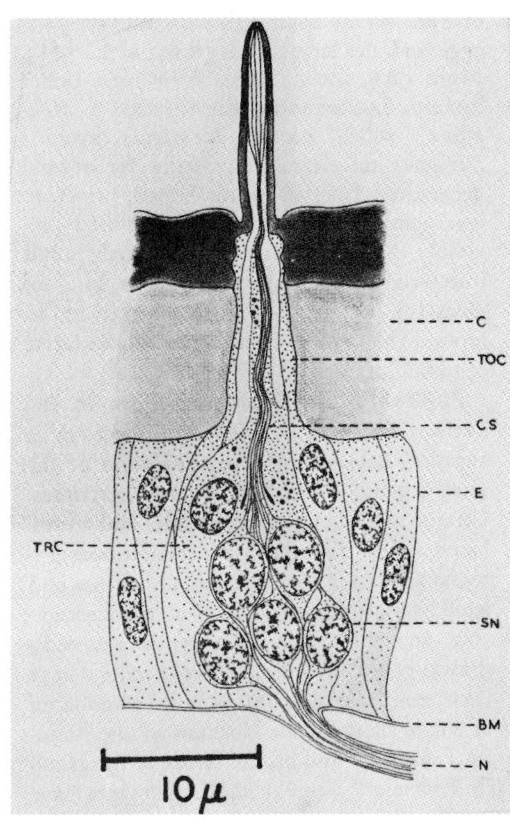

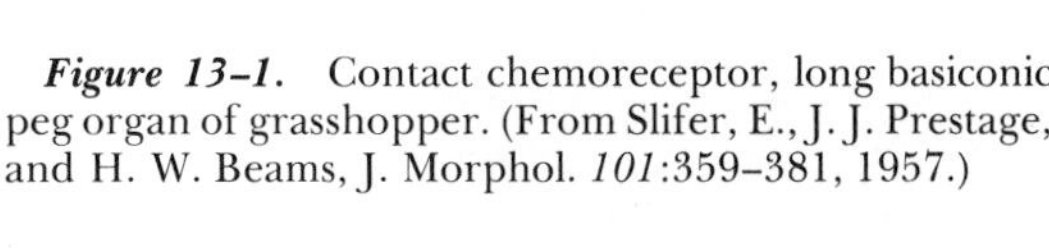
Figure 13–1. Contact chemoreceptor, long basiconic peg organ of grasshopper. (From Slifer, E., J. J. Prestage, and H. W. Beams, J. Morphol. *101*:359–381, 1957.)

the foregut is cut in a blowfly, the fly may continue drinking intermittently until it bursts.[34] Proboscis extension can be initiated by olfactory stimulation of antennae, by taste on the tarsi, and possibly also by internal factors.[33] As the mouthparts open, chemoreceptors on the interpseudotracheal papillae between the labellae must be stimulated for feeding to continue.[37] During sucking of sugar solution, sensory adaptation raises the threshold; at a certain level of sensitivity sucking stops, and then by disadaptation the threshold falls and sucking resumes. Termination of a feeding reaction is controlled by sensory impulses from the foregut.

In a mosquito, the threshold concentration in labellar lobes for sucrose to elicit proboscis extension is 0.011 M, and in tarsi it is 0.135 M, while the threshold for glucose is 10 times higher.[48] Different receptors sense water and sugar; in some caterpillars, mannose gives only the water response, and can inhibit the response to fructose (but not to glucose).[39]

When maxillae are removed from *Bombyx* larvae, the selectivity for specific plants is lost. Several sense cells in *Protoparce* maxillary sensilla discharge at higher frequency when stimulated by acceptable plant sap than when stimulated by unacceptable sap. The discharge from sense cells can be increased or decreased and each cell has its own spectrum of sensitivities.[126] Tarsal taste receptors of the potato beetle have one cell in five strongly stimulated by unacceptable alkaloids.[126] One component in larval conditioning may be development of specifically sensitive receptors; another component is probably patterning in the central nervous system.

Electrical Responses. The electrical response of a single taste receptor can be recorded by placing a glass capillary tube over a single sensillum; the capillary can contain the compound to be tested, and thus serve as both the stimulus and the recording electrode. Taste hairs of the labella in a blowfly have five sensory cells: two for salt, one for sugar, one for water, and one mechanore-

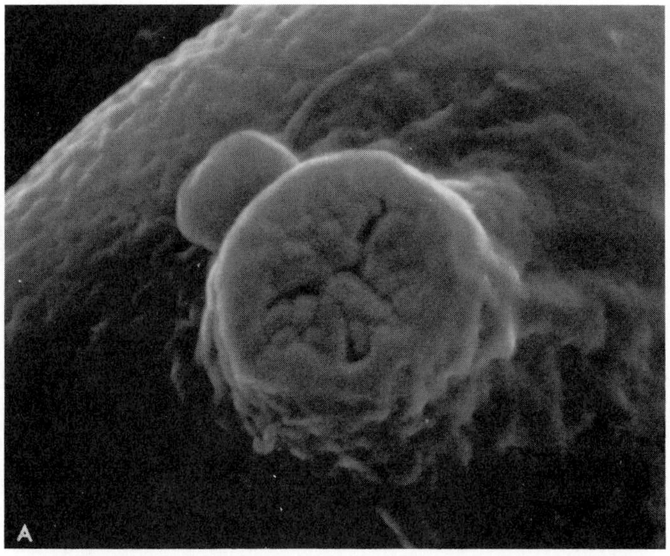

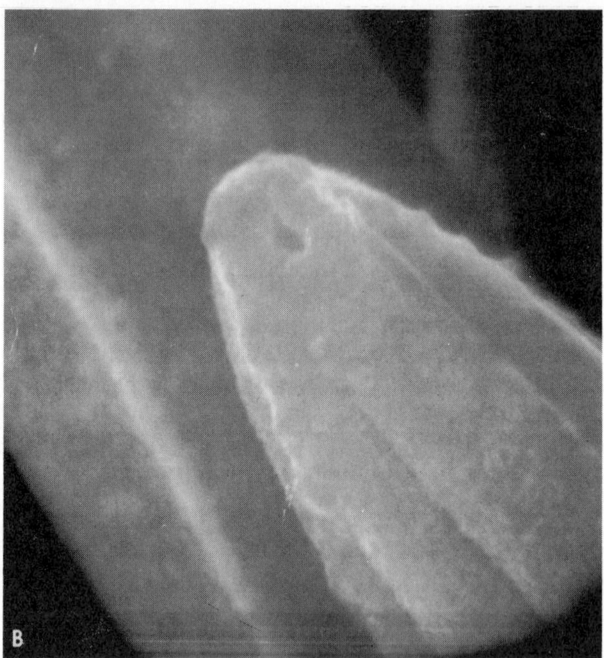

Figure 13–2. Scanning electron micrographs of taste sensilla. (A) Sensillum styloconicum of tent caterpillar *Malacosoma americana,* × 12,000. (B) Labellar hair of blowfly *Phormia regina,* × 20,000. (Courtesy of V. G. Dethier.)

ceptor. In general, the sugar fibers give smaller spikes than the salt fibers. The frequency of impulses varies with the concentration of the solution presented, and sensory adaptation occurs in from 1 to 13 seconds.[64, 94, 95] Relative sensitivities of sugar receptors in blowflies are: sucrose, 1; glucose, 0.71; fructose, 0.46.[131] Tests with various salts show that stimulation is predominantly by the cation, but the kinetics of the response indicate both cationic and anionic receptor sites.

Another type of sense cell is stimulated by water, and its response is inhibited in proportion to the osmotic concentration of either an electrolyte or a non-electrolyte solution.

Simultaneous observation of behavior and recording of impulses (Fig. 13–4) shows that

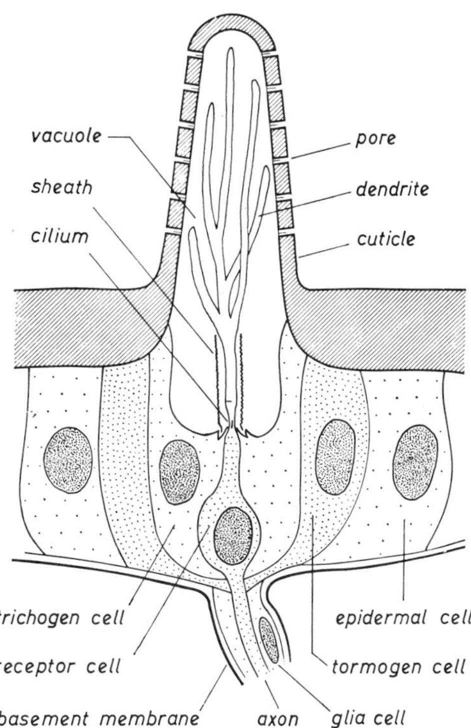

Figure 13–3. Diagram of a peg-shaped insect olfactory sensillum in longitudinal section. (From Schneider, D., and R. A. Steinbrecht, Symp. Zool. Soc. Lond. *23*:279–297, 1968.)

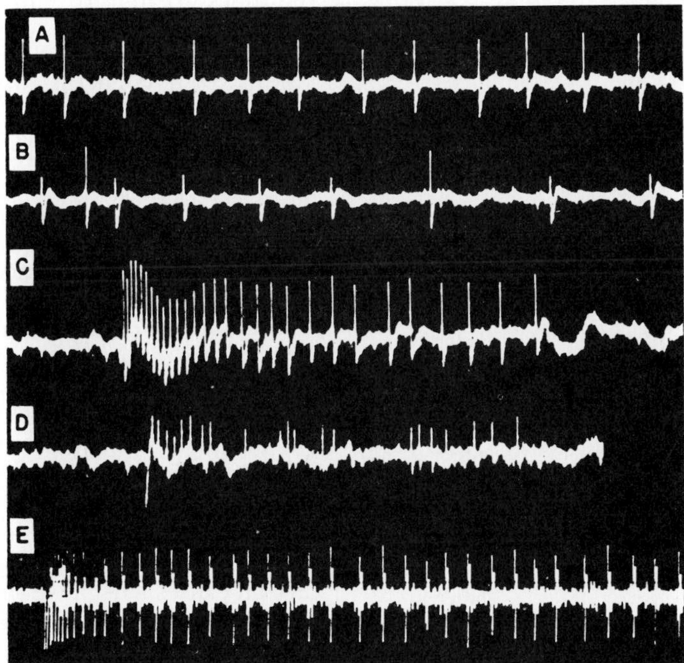

Figure 13–4. Electrical responses from contact receptor of *Phormia.* A, Maintained stimulation of large fiber by 0.5 M NaCl. B, Maintained stimulation of small fiber, two large spikes seen; 0.25 M sucrose in 0.1 M NaCl. C and D, Mechanical stimulation of large and small fibers. E, Initial stimulation of both large and small fibers by 0.5 M NaCl. (From Hodgson, E. S., and K. D. Roeder, J. Cell. Comp. Physiol. *48*:51–76, 1956.)

behavioral rejection of a high concentration of salt is mediated by the same receptor as acceptance of a low concentration, and a difference of three impulses during the first 100 msec of sensory discharge can determine behavior, as shown by the following sequence of responses to increasing concentrations of NaCl:[38]

0.05 M NaCl	only H_2O receptor active
0.1 M NaCl	salt receptor and water receptor active
0.2 M NaCl	only salt receptor active, behavioral acceptance
0.5 M NaCl	high frequency firing of salt receptor, behavioral rejection

Maxillary sensilla of ten species of lepidopteran larvae have four chemosensory neurons per peg. Many respond to dilute NaCl; usually one cell responds to both salt and some amino acids. Some cells are excited, and others are inhibited, by specific amino acids. Synergisms also occur, so that varied patterns of gustatory signals enter the central nervous system.[39]

One group of amino acids does not stimulate labellar hairs (Table 13–1); the hairs are stimulated, however, non-specifically by high concentrations of other amino acids. The salt receptors are stimulated by a few amino acids, and sugar receptors by still others.

Several morphological types of olfactory sensilla have been described. Each has a cuticular extension (hair) or a specialized thin plate or depression. No gross correlation of structure and sensitivity exists, yet molecular specialization must occur; e.g., in the receptors of males, but not females, to specific sex pheromones.[125] Each sense hair has some 3000 pores, supplied by 100 to 200 Å dendritic branches going to two or three receptor cells per hair.[15, 19] A male polyphemus moth has more than 60,000 sensilla totalling more than 150,000 receptor cells, and of these 60 to 70% are highly sensitive to bombykol. The behavioral threshold is 10^{14} molecules per ml air or 3×10^4 to 14×10^4 molecules on the antennae (or one per sensillum per 2 seconds).[122]

Sensory impulses are generated in the axon near the cell bodies at the base of the hair. A microelectrode inserted into a sensillum records a slow sensory potential, the electroolfactogram (EOG) (Fig. 13–5), followed by spikes. The EOG can be positive (hyperpolarizing) with a decrease in frequency of spontaneous spikes, or negative (depolarizing) and excitatory. The sign and rate of electrical response vary with stimulating agent. In the antennae of *Bombyx* moths, the concentration of female pheromone needed for a measurable sensory response is at least 100 times greater (10^{-8} to 10^{-10} μg/ml) than for a behavioral response (10^{-12} to 10^{-13} μg/ml).[124]

Behavioral threshold concentrations (molecules/ml air) are:[128]

	Propionic Acid	Eugenol	Phenyl Propyl Alcohol	Citral
man	4.2×10^{11}	8.5×10^{11}	6.5×10^{9}	4×10^{11}
honeybee	4.3×10^{11}	2×10^{10}	2.2×10^{9}	6×10^{10}

The antennae of carrion beetles have two types of sensilla, one long and one short, each with a single axon. They give sensory potentials and resulting spike frequencies which are proportional to the effectiveness of an olfactory stimulus. Fatty acids with chain lengths greater than six carbons are depolarizing, and shorter chain acids are hyperpolarizing.[15] In a carrion beetle, the odor of carrion causes a negative generator potential, whereas propionic acid elicits a positive response.[16, 17, 18] In *Lucilia,* the thresholds for olfactory generator potentials from sensilla basiconica are comparable to those for behavior:[76]

TABLE 13–1. Taste Responses to Amino Acids[132]

Amino Acid	*Responding Sensory Cell in Fly*	*Taste Response in Man*
glyc	no effect	sweet
ala	no effect	sweet
ser	no effect	sweet
thr	no effect	sweet
cyst	no effect	
tyr	no effect	
pro	salt	sweet
OH pro	salt	sweet
lys	non-specific	sweet
cit		sweet
glutamine		sweet
aspart	non-specific	sour
glut	non-specific	sour
hist	non-specific	sour-bitter?
arg	non-specific	bitter
val	sugar	bitter
leu	sugar	bitter
isoleu	sugar	bitter
met	sugar	bitter
ϕ ala	sugar	bitter
trypt	sugar	bitter

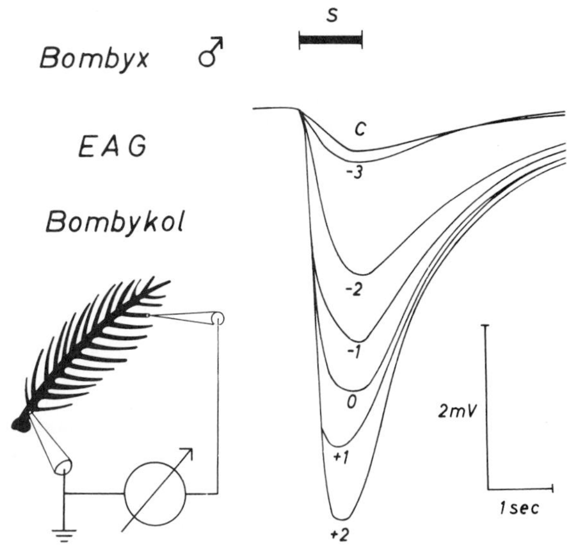

Figure 13-5. Recordings of slow electrical responses (electro-olfactograms) from isolated antenna of male silkworm *Bombyx.* Recording arrangement diagrammed at left. Responses to increasing concentrations of bombykol. (From Boeckh, J., K. E. Kaissling, and D. Schneider, Cold Spring Harbor Symp. Quant. Biol. *30*:263–280, 1965.)

	EOG (M)	*Behavior (M)*
isovaleraldehyde	10^{-7}	10^{-6}
ethanol	10^{-7}	10^{-5}
menthol	5.7×10^{-9}	
butanethiol	7.5×10^{-9}	
camphor	3.2×10^{-8}	

For other receptors of *Lucilia* the threshold is lowest for alcohols of seven carbons and for hydrocarbons of eight carbons; unit responses show a wide array of patterns[76] (Fig. 13–6).

In *Locusta,* the two or three cells in a single sensillum differ in sensitivity to various odors. One responds to odor of green grass and has thresholds (in molecules/cm^3 of air) of 10^8 for hexenal, 10^{10} for caproic acid, and 10^{15} for caprylic acid[18] (Fig. 13–7). In *Locusta,* the most effective agents are unbranched keto-acids and unsaturated fatty acids of six carbons; branched acids, halides, and unsaturated carbohydrates are relatively ineffective, and unbranched amines of four carbons are inhibitory.[74]

Fifty olfactory cells on antennae of saturnid moths were tested for 13 odors, and no two cells gave identical patterns of sensitivity[124] (Fig. 13–8). On bees, 33 compounds tested on 47 receptors showed no clear pattern of correlation of chemical properties with ability to stimulate.[84]

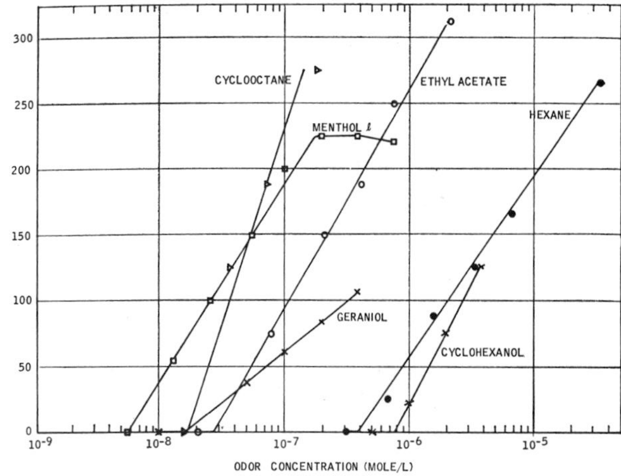

Figure 13-6. Relative amplitudes of olfactory potentials from single female fly *Lucilia* to different concentrations of odorous substances indicated. (From Kay, R. E., J. T. Eichner, and D. E. Gelvin, Amer. J. Physiol. *213*:1–10, 1967.)

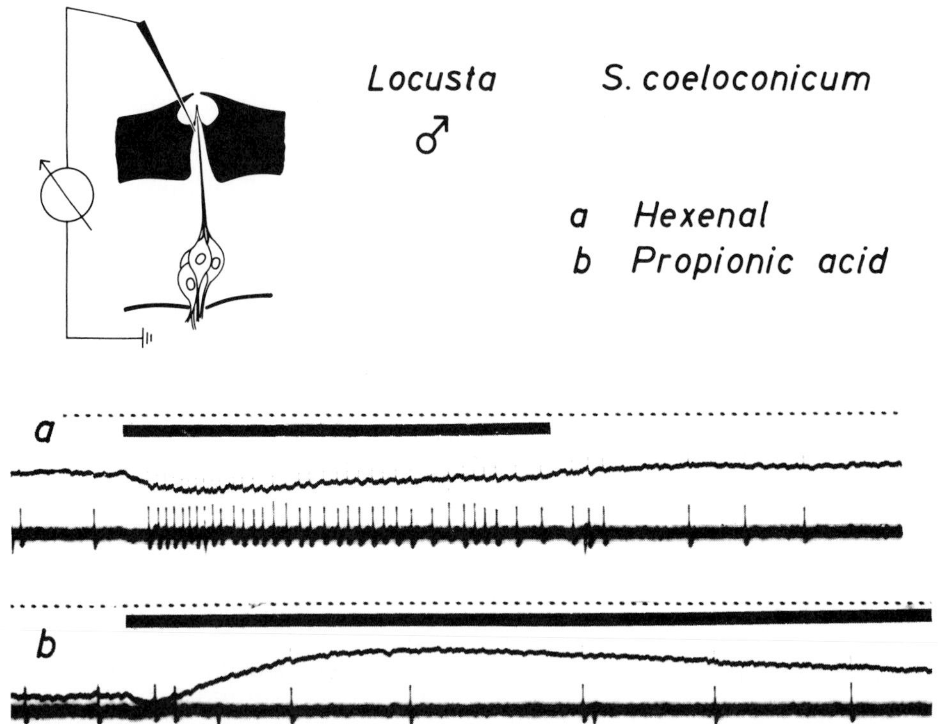

Figure 13–7. Electro-olfactograms and responses of single receptor cells from pit sensillum of *Locusta migratoria*. A, Excitatory reaction to hexenal; B, inhibitory response to propionic acid. (From Boeckh, J., K. E. Kaissling, and D. Schneider, Cold Spring Harbor Symp. Quant. Biol. *30*:263–280, 1965.)

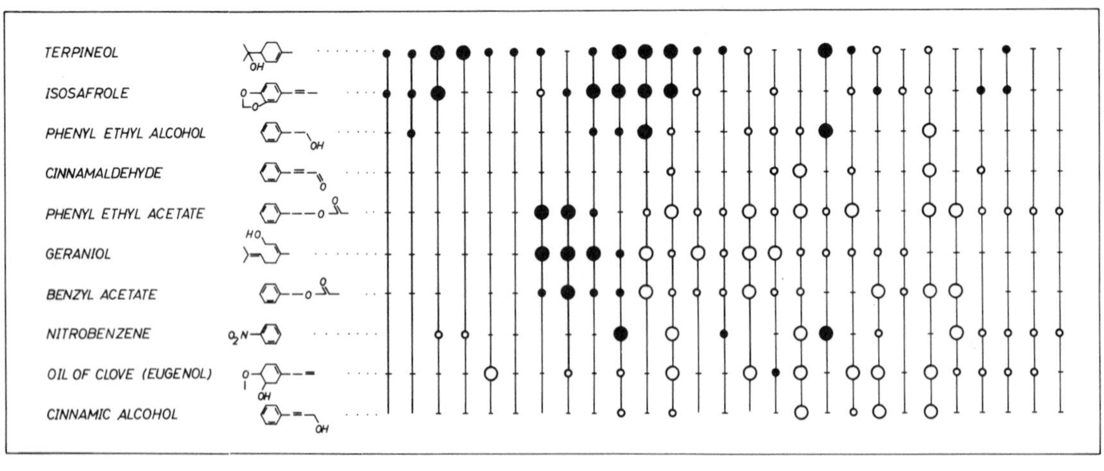

MEDIUM STRONG INCREASE OF FREQUENCY MEDIUM STRONG INHIBITION NO REACTION SUBSTANCE NOT TESTED

Figure 13–8. Reaction spectra of representative single generalized sensilla from sensory pegs of moth *Antheraea*. Each vertical line shows reaction of one cell to substances indicated at left. (From Boeckh, J., K. E. Kaissling, and D. Schneider, Cold Spring Harbor Symp. Quant. Biol. *30*:263–280, 1965.)

CHEMORECEPTION IN VERTEBRATES

In general, taste receptors of vertebrates consist of sense cells arranged in clusters or papillae in the gustatory epithelium. Nerve endings are abundant around the receptor cells. In frog and catfish,[32a] less clearly in rat, synaptic vesicles have been observed at the base of the sense cells, and stimulation of the nerve endings is probably by chemical transmitters (Figure 13–9).

Taste cells of mammals, and probably of many other vertebrates, are replaced every few days while the sensory nerve fibers remain fixed. There is developmental evidence that nerve fibers may induce epithelial cells to become specific receptor cells.

Olfactory receptors of vertebrates are primary sensory neurons, the dendrites of which extend as microvilli or cilia, usually into a mucous layer (Fig. 13–10). The receptors are surrounded by supporting cells (Fig. 13–11).[134] Axons of primary olfactory neurons go to the olfactory bulb where they synapse with secondary neurons (e.g., mitral cells), the axons of which pass in the olfactory tract to the forebrain.

FISHES AND AMPHIBIANS

Taste buds in fishes occur not only in the mouth (pharynx) but also on gills, skin, and barbels; they are innervated by the seventh, ninth, and tenth cranial nerves. Various arrangements of olfactory sacs occur in fishes. Elasmobranchs have olfactory pits, usually on the ventral side of the snout, and water enters during swimming or in respiratory movements. In most teleosts the pit is high on the head and water enters during swimming, or in movements of the mouth or by ciliary action.

Behavior. The importance of taste in fishes is shown by feeding responses toward meat juice applied on the flanks or barbels. Even the fin rays are sensitive to natural fish oils and acids, and the skin of a catfish shows general chemical sense—acid, alkaline solution, salt—as well as specific tastes. Minnows have been conditioned to discriminate sucrose by taste at 2×10^{-5} M, NaCl at 4×10^{-5} M, and quinine at 0.0025 per cent.[81]

Fish differ in apparent dependence on olfaction, and the differences are correlated with their anatomy and ecology. Both eyes and nasal sacs are important in schooling and non-predatory fish (*Phoxinus, Gobio*); olfaction is more important than vision in non-schooling predators (*Anguilla, Lota*). *Phoxinus* has been shown by conditioned reflex tests to be able to distinguish by olfaction different species of fish; this is important in discriminating their own species from other harmless ones and from predators.[56] Sex discrimination is by olfaction in *Bathygobius*.[77, 78] *Ictalurus* learns to discriminate by olfactory cues among others of the same species and to display territoriality.[72] The olfactory organ of fishes is elaborately folded, with extensive surface area on many lamellae. *Phoxinus* has 95,000 sense cells per mm^2 of the olfactory epithelium. In *Anguilla* the density of olfactory cells is $44{,}000/mm^2$ for a total of some 800,000 in a 12 cm fish. A stream of water entering the anterior nares and leaving the posterior nares is necessary for odor perception. Alarm reactions are aroused by skin extracts of the same or related species. The cyclostome *Petromyzon* has poor vision but a well developed olfactory system, and young lampreys perceive by olfactory cues the fish prey to which they attach. This response occurs even in lampreys reared in the absence of fish, but is lacking if the lampreys have been made anosmic.[78] Minnows have remarkable olfactory sensitivity for food plants and for water taken from different streams bordered by different vegetation; they can distinguish washes from related plants after extensive dilution. There is evidence that migrating fish, particularly salmon, locate their home waters by specific odors.[60, 61, 62, 63] After the olfactory sacs have been plugged, stream selection is random. Migrating elvers in Holland (*Anguilla vulgaris*) show a positive taxis to canal water by odor, and not merely to salinity dilution.[27] Experiments in multi-compartment tanks with flowing water indicate that, for several species, specific chemicals may induce a positive rheotaxis orientation to a current.

Sensitivities are high—*Phoxinus* can detect eugenol in water at 6×10^{-14} parts and phenylethyl alcohol at 4.3×10^{-14} parts.[100] Eels can detect β-phenylethyl alcohol at 2.8×10^{-18} parts, and trout can detect it at 9.9×10^{-9} parts.[138]

Newts travel overland to home territory, even if blinded, but they fail if olfactory nerves are cut; hence, they probably home in

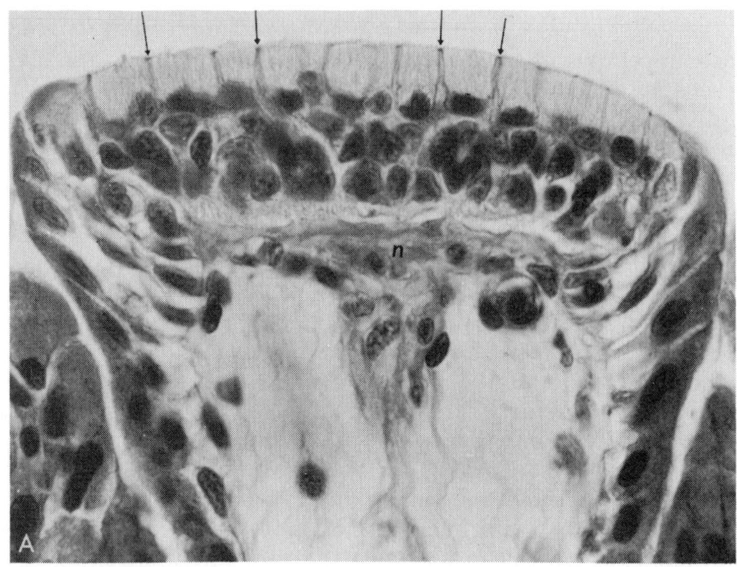

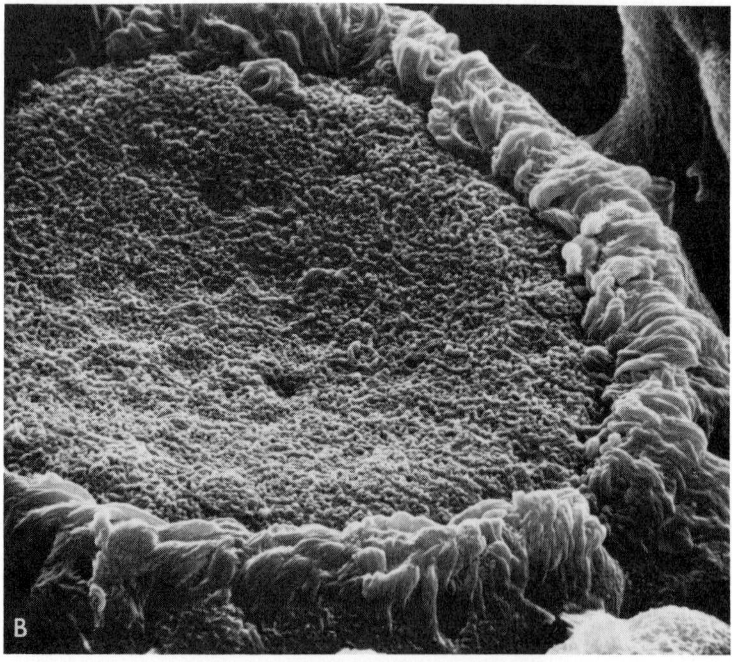

Figure 13–9. Photographs of taste receptors. *A,* Light micrograph (3600 ×) of taste organ of frog (*Rana pipiens*) showing, at arrows, the fine processes of sensory cells extending between supporting cells; a nerve bundle (n) reaches the sensory epithelium. *B,* Scanning electron micrograph (1200 ×) of taste organ of frog; a crown of cilia encircles the sensory area, made up of supporting cells among which the fine processes of secondary sense cells emerge. *C,* Transmission electron micrograph (5000 ×) of taste organ of frog. Sensory cell (SC) sends a cytoplasmic process towards the basal lamina; at arrows, this process contacts a nerve fiber (n), establishing afferent synaptic contact. Insert (35,000 ×), a synaptic contact between sense cell and nerve fiber, showing synaptic vesicles. *D,* Electron micrograph (2500 ×) of rat taste bud; stars indicate nuclei of three taste cells. Nerve profiles (n) and pore of taste bud (p) are shown. Also seen is a cell (m) in mitosis prior to maturing into a taste cell. (Courtesy of P. P. C. Graziadei.)

(Illustration continued on opposite page.)

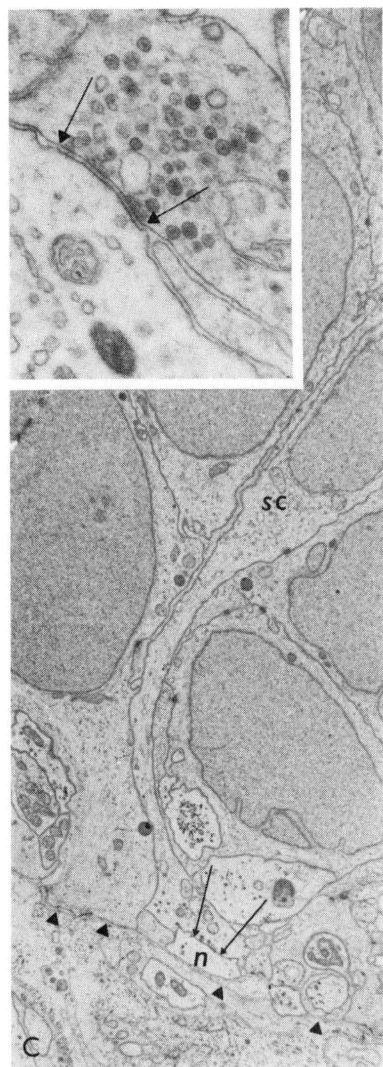

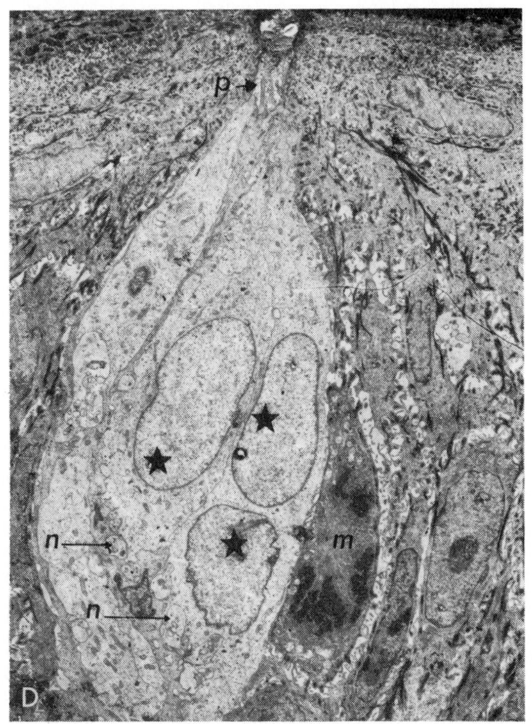

Figure 13–9 Continued.

response to combinations of odors in the air from vegetation.[57] Thresholds for butyric acid, in molecules/ml of air, follow:

Triturus	4×10^9
man	7×10^9
dog	9×10^3
bee	1.1×10^3

Electrophysiology. Recordings from sensory fibers of the palatal nerve of a carp showed several receptor types in one gustatory papilla; different fibers respond respectively to salt, acid, quinine, sugar, and human saliva.[79, 80] In *Salmo salar,* taste receptors are not sensitive to amino acids but are sensitive to sugars and to organic acids, with stimulating effectiveness increasing with chain length; high sensitivity to saliva was noted. Water sense was recorded in fibers of the palatine nerve, but not in those of the facial nerve.[135] The olfactory epithelium of *Salmo salar* is stimulated by very low concentrations of amino acids (e.g., alanine at 3×10^{-7} M); sugars, alcohols, and aliphatic acids do not stimulate olfactory receptors[135] as they do the taste receptors. Responses to amino acids are blocked by mercuric salts.

Amphibians have taste buds on the tongue, and fibers in the glossopharyngeal nerve of a frog respond to water; the response is abolished by isotonic saline but not by sugar. Other fibers in this nerve respond to hypertonic salt and to acid.[151] Frog gustatory responses show a low threshold for calcium ions (less than 0.001 M); the response is increased by prior treatment with a chelator (EDTA) and is inhibited by NaCl or KCl.[73] The threshold for NaCl is 0.02 to 0.4 M. In the

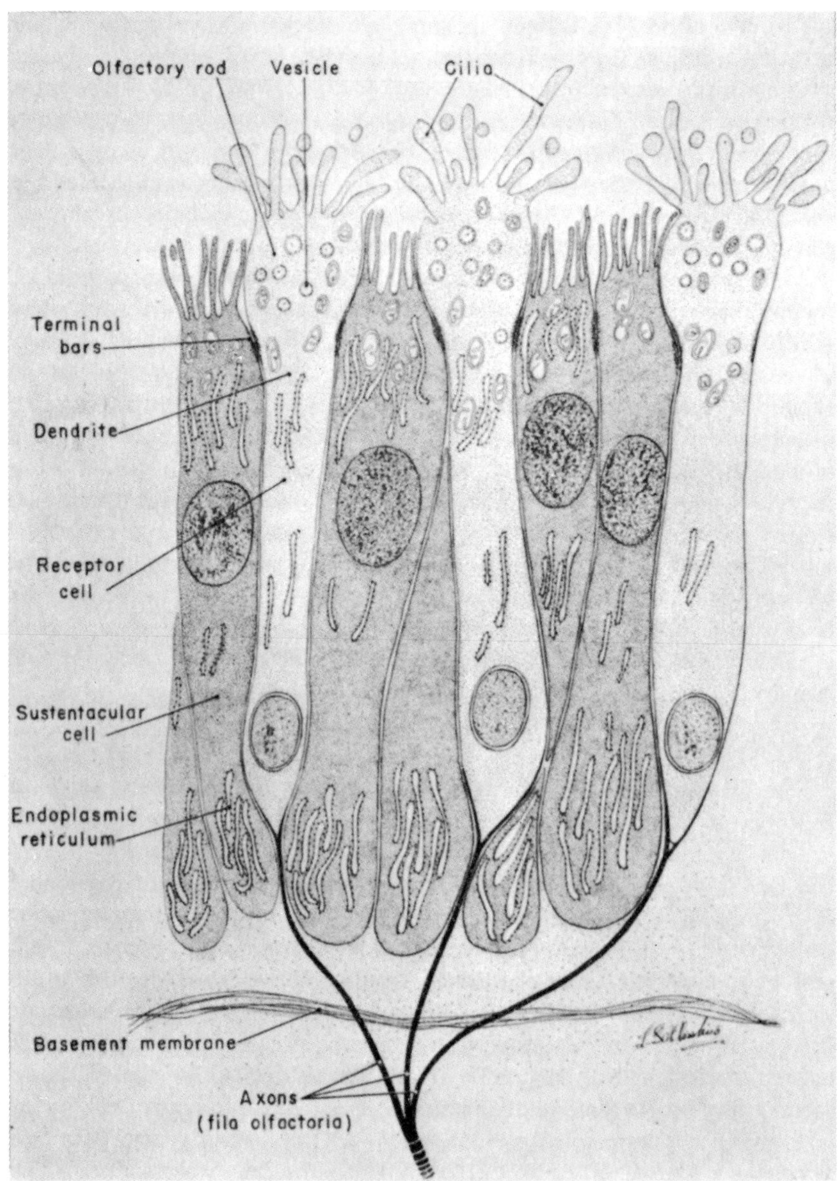

Figure 13–10. Schematic representation of olfactory mucosa of rabbit. Three receptor cells connected to axons and bounded by supporting cells. (From de Lorenzo, A. J. D., pp. 5–18, *in* Olfaction and Taste: A Symposium, edited by Y. Zotterman, Pergamon Press, New York, 1963.)

catfish *Ictalurus,* both olfactory epithelium and barbels are sensitive to amino acids.[7a]

Recordings from the olfactory epitheluim of several fishes show slow receptor potentials in response to stimulation by odors (both ON and ON-OFF responses); spontaneous spikes were recorded in the primary fibers, as well as bursts of spikes on top of the slow receptor potentials.[130] The olfactory bulb of salmon gave a response to natural waters.[58,144]

Recordings from olfactory neurons and from the olfactory bulb of frog and of the burbot fish show much spontaneous activity; some 28 substances gave responses in different patterns—some ON-OFF, some ON-maintained, and some inhibition of spontaneity.[41] Frog epithelium gives a receptor potential (electro-olfactogram or EOG) with superimposed spikes, but the amplitude of the receptor potential is proportional to the strength of the stimulus (Fig. 13–12). No two olfactory fibers gave identical responses, but there was some grouping as follows: (1) limonene, camphor, and pinene; (2) cucumarin and musk; (3) butyric, valeric, and mercaptoacetic acids; (4) benzaldehyde, nitro-

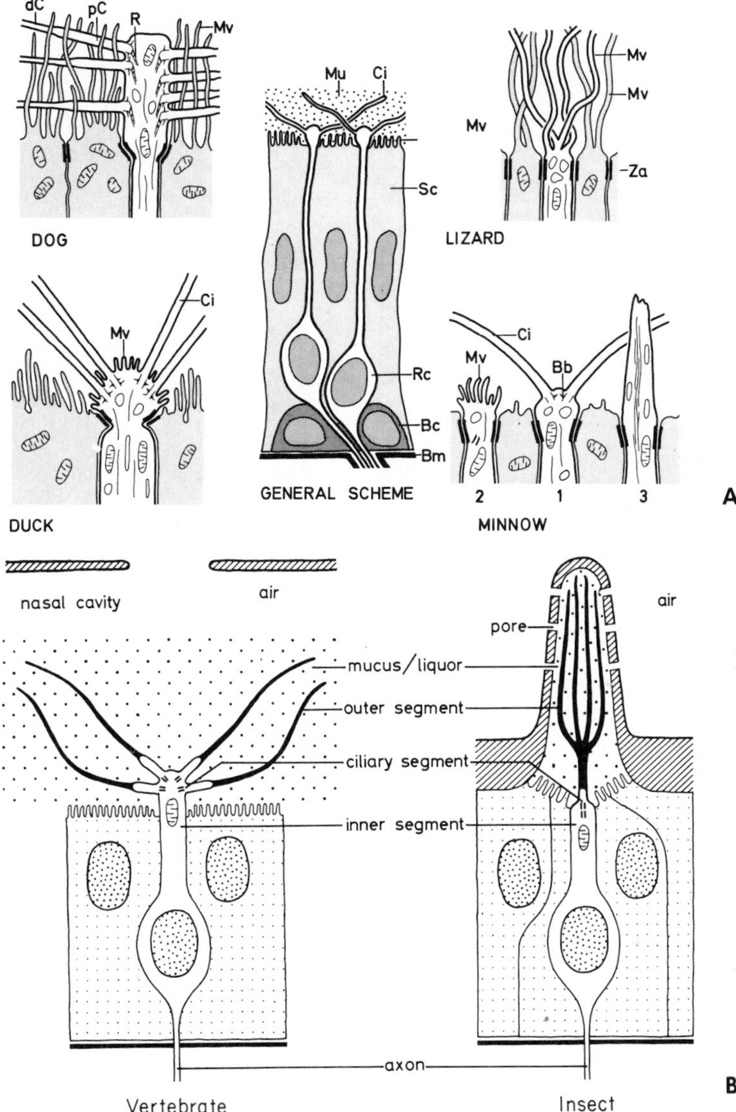

Figure 13–11. A, Diagrammatic representations of vertebrate olfactory epithelium. General scheme and modifications for animals indicated. Ci, cilium; Bc, basal cell; Mu, mucus; Rc, receptor cell; SC, supporting cell; Bm, basal membrane. B, Schematic comparison of vertebrate and insect olfactory receptors. (From Steinbrecht, R. A., pp. 3–21 *in* Olfaction and Taste, edited by C. Pfaffmann. Rockefeller Press, New York, 1969.)

benzene, and amyl alcohol; and (5) pyridine and butanol.[53, 55] The olfactory bulb potentials oscillate in correspondence to mouth and respiratory movement.[101, 103] Many odorous substances elicit a negative EOG, and some elicit a positive one. Positive responses were obtained after neural degeneration and under other circumstances when negative EOG's were absent; hence, the positive responses may arise from supporting and secretory cells, while the negative ones may be generator potentials.[136] Recordings from lateral and medial branches of the olfactory nerve of frog show differences in the proportion of response to different odors as recorded from the two nerve branches; this suggests a spatial analysis in the olfactory epithelium.[97]

In a tortoise, a variety of patterns of response was recorded from different olfactory nerve twigs; the correlation of aqueous solubility of a substance with olfaction is good.[141]

In goldfish, antidromic stimulation of the lateral respiratory tract inhibits many mitral neurons, probably by efferent neurons.[59] In the olfactory tract of a burbot (a fish), efferent impulses are partly spontaneous and partly evoked by the ipsilateral and contralateral olfactory tracts; the efferents can be inter-

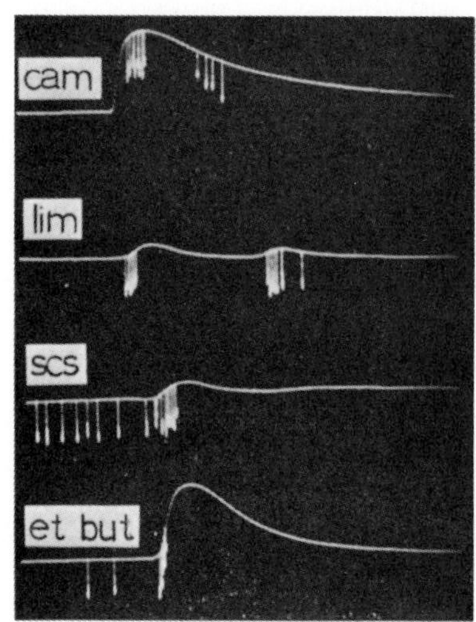

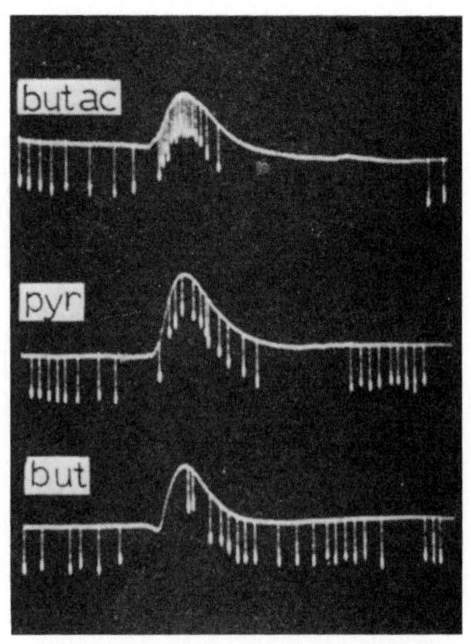

Figure 13–12. Spike responses superimposed on olfactogram (generator potential) in frog olfactory bulb. A, Responses to camphor, 2 puffs of limonene, carbon disulfide, ethyl butyrate. B, Responses to butyric acid, pyridine, and *n*-butanol. (From Gesteland, R. C., J. Y. Lettvin, W. H. Pitts, and A. Rojas, pp. 19–34, *in* Olfaction and Taste: A Symposium, edited by Y. Zotterman, Pergamon Press, New York, 1963.)

rupted by tactile stimulation of the skin.[41, 43] The function of the efferents is probably to modulate olfactory sensitivity.

Mammals and Birds

Behavior. In mammals, the recognized tastes are for sweet, bitter, salt, and acid. From behavior as well as from recordings of gustatory impulses it has been deduced that a water taste is present in rabbit and cat, but is absent or weak in rat and man. However, response to water depends on the preceding adapting solution. In man, water after salt tastes bitter, and after acid it tastes sweet. Response to water after NaCl is weak in species which are very sensitive to salt. It is now doubtful whether there are true water taste receptors. Cats are normally indifferent to sugar and have been thought to lack a sweet taste, but after NaCl they take sugar and prefer sucrose dissolved in 0.03 M NaCl to sucrose in water.[8] Sweet taste is also very weak in lambs, and pigeons fail to respond to bitter (quinine). Recordings from taste nerve fibers show that individual fibers carry responses from several tastes, but the threshold for one is much lower than for others.

In man, the lowest threshold for sweet is at the front of the tongue, that for bitter is at the back, and that for sour is at the sides, in overlapping areas. Cocaine abolishes the four taste modalities in the sequence: bitter, sweet, salt, sour. Gymnemic acid suppresses sensitivity to sweet, but not to salt or acid. An extract (protein) from the miracle fruit *Synsepalum* has no inherent taste, but suppresses the sweet taste of sucrose, cyclamate, D-amino acids, and lead acetate and makes sour food (lemon) taste sweet.[83] Sensory adaptation is faster for salt than for sweet or bitter. A given sensation (e.g., sweet) can result from stimulation by unlike compounds. Some persons are unable to taste certain bitter substances (such as *p*-ethoxyphenyl thiourea) to which other persons are very sensitive.[49]

Rats that are made salt-deficient by adrenalectomy have a greater preference for salt solutions over water than do non-deficient rats; however, taste thresholds are similar and the greater salt hunger seems to have a central nervous locus. Rats can be trained to associate a different flavor with replacement of the dietary deficiency; food and water intake can then be regulated according to the learned flavor.[89, 113a]

Many mammals depend on olfaction for much of the discrimination of food quality and for location of prey and predators; odors are also important to them in territory marking and identification, sex recognition, and reproductive behavior.[104] Glands conveying specific odors are: preputial in musk deer and beaver; anal and perineal in mustelids, civet, and ferret; perineal and sebaceous in various mammals.[112] Carnivores are more sensitive to odors than are rats, and the ability of dogs to follow trails and to locate underground objects (e.g., truffles) is well known. Mice (*Mus* and *Peromyscus*) can be trained in a Y-maze to distinguish by olfactory cues between mouse species, between sexes of the same species, and between two individual males of the same species.[20] Marsupial phalangers maintain a complex social order in darkness. They have frontal, sternal, anal, and pouch glands of scent, and mark themselves and their partners and establish territoriality and dominance in a colony without visual cues. Odors of foreign individuals evoke aggressive behavior.[127]

In most mammals, a large portion of the forebrain is olfactory; i.e., most mammals are macrosmatic. Primates and whalebone whales are microsmatic (with slight sense of smell). A rabbit has a total of 50 to 100×10^6 olfactory receptor cells.

Electrophysiology. Many fine nerve terminals serve each gustatory papilla, and probably a single neuron can be stimulated by several papillae; enhancement and depression of a surround upon one papilla are observed.[92] Each taste cell shows dc changes (receptor potentials) in response to specific chemicals.[11]

Impulses in single gustatory nerve fibers of the cat show mixed specificities. The fibers from some receptor cells show responses to acids at pH below 2.5, others to NaCl and also to choline chloride. Still others are water taste fibers, responding to the washing away of salts. Water taste endings are stimulated by solutions more dilute than 0.03 M NaCl, whereas salt receptors require NaCl more concentrated than 0.05 M. Only rare fibers in the cat respond to sugar, but in the rabbit sugar is more effective than quinine. A gustatory fiber may respond to two or more of the taste stimuli (e.g., acid and quinine), but at different thresholds.[52] In dogs, saccharin elicits either no response or impulses in smaller fibers than does sugar; gymnemic acid fails to abolish the response to saccharin but does abolish that to sugar, whereas in man it abolishes all sweet taste. Fibers of the chorda tympani of a monkey give responses to NaCl, acid, quinine, sugar, and water. Chickens respond to salt, glycerin, ethylene glycol, quinine, water, and acid but not to sucrose or saccharin. In rat chorda tympani, 48 fibers were tested for the four basic tastes; 19 of them responded to three tastes and 12 of them responded to all four tastes. In hamster, of 28 fibers tested, five of them responded to one taste, five responded to two tastes, 11 responded to three tastes, and seven responded to all four stimuli; however, the thresholds were very different.[115] Evidently, there are several kinds of taste receptors, and some of them respond with different thresholds to various compounds, giving two sensations. Behavioral tests of taste discrimination often give lower thresholds than does nerve recording. In the thalamus and somatosensory area of the cerebral cortex, receptive fields are projected spatially as they occur on the tongue.[40]

Unlike taste buds, olfactory receptor cells are primary neurons, each giving rise to an axon which passes to the olfactory bulb. Electrical responses have been recorded from the primary axons of olfactory receptors in birds and rodents.[9, 143] Bursts of impulses synchronize with respirations that draw odorous substances over the sensitive mucosa.

The primary olfactory neurons synapse in glomerular tufts in the olfactory bulb with dendrites of mitral cells; mitral axons pass in the lateral olfactory tract to the forebrain. The olfactory bulb also contains granule cells, the processes of which connect with dendrites of mitral cells, thus forming lateral interconnections within the bulb (Fig. 13–13). Impulses in mitral cells correlate with inhalations of odorous substances, and single cells differ in their responses to specific odors; hence, they must receive input from localized receptors. The olfactory bulb shows spontaneous activity, which can be replaced by lower or higher frequency rhythms during olfaction. Some mitral cells discharge continuously, others do so whenever air passes through the nose, and others respond specifically to odors.[145] Oscillatory responses (evoked potentials) of the olfactory bulb correlate well with subjective thresholds for odors in humans.[66] By recording from mitral cells during both antidromic stimulation of mitral axons and orthodromic stimulation of olfactory fibers, it was shown that spikes originate at the bases of mitral cell axons. Secondary dendrites of mitral cells (arising well below the glomerular tufts) activate granule cells, the dendrites of which in turn inhibit the same and other mitral cells. The granule cells lack axons, and their activity is probably by electrotonic spread through dendrites; thus, they resemble the amacrine cells of the retina. Reciprocal synapses with vesicles are found between mitral and granule cells, thus providing a structural basis for self- and lateral-inhibition.[108, 110, 111] In ad-

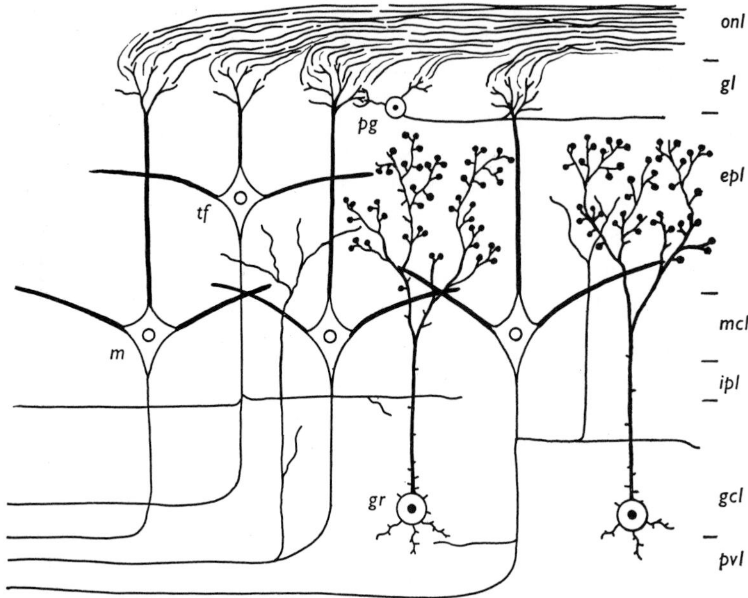

Figure 13–13. Diagram of neurons in olfactory bulb of rabbit. *Onl,* olfactory nerve layer which synapses in mitral cells (*m*) and tufted cells (*tf*); *gr,* granule cells; *pg,* periglomerular cells. (From Price, J. L., and T. P. S. Powell, J. Cell. Sci. 7:91–188, 1970.)

dition, efferent fibers from the forebrain terminate in the external plexiform layer of the olfactory bulb and presumably modulate the responses of the mitral cells.[109]

CELLULAR MECHANISMS OF CHEMOSTIMULATION

Attempts to formulate a unified theory of chemical stimulation have not been successful; rather, it appears that several reactions at cell surfaces are involved. General chemical sense represents the depolarizing effects of a variety of agents on excitable cells in general. Protozoa are stimulated by various inorganic ions and by acids in varying degrees. Similarly, free nerve endings may be stimulated to discharge after long latencies. Strong acids, particularly inorganic acids, stimulate according to hydrogen ion concentration. Weak acids, usually organic, appear to enter cells as undissociated molecules and to stimulate out of proportion to the hydrogen ion concentration. Ions such as potassium may stimulate by local depolarization. Anions such as citrate and oxalate, which form complexes with calcium, may stimulate by calcium withdrawal and the resulting membrane instability. Hypertonic solutions may change membrane properties by dehydration. The non-specificity of general chemical stimuli, the long latencies (seconds or minutes) for response, and the relatively high concentrations required distinguish general chemical sense from taste and smell.

A number of investigators of both taste and smell have adapted the equations of Michaelis-Menten enzyme kinetics to chemoreception. The response – electro-olfactogram, frequency of sensory impulses, or behavioral sensitivity – is a sigmoidal function of the logarithm of the concentration of stimulating molecules.[46, 75] The equilibrium constant K is given by

$$K = \frac{a(s-n)}{n}$$

where s = total receptor sites, n = number of occupied sites, and a = effective concentration of stimulant.[10, 11, 75]

Taste buds of various vertebrates respond to salt 20 to 25 msec after application of a stimulus; solutions from 0.0005 to 5.0 M NaCl may elicit responses, so stimulation is not purely ionic. Both cations and anions influence the response to a salt. NaCl and Na_2SO_4 taste salty, but $NaNO_3$ and $NaHCO_3$ do not; however, the chlorides vary in effectiveness according to the cation. Salt receptors on blowfly tarsi are most sensitive to KI, and less so to KCl. In a number of insects, the series for rejection reactions was, for cations,

$NH_4 > K > Na > Li$, and for anions, $I > Br > SO_4 > Cl >$ acetate.[64]

The sense of sweetness is given in man by sugars containing fructose or glucose, as well as by unrelated molecules such as saccharin, lead acetate, and some beryllium salts. Dextrorotatory asparagin tastes sweet; the levo-form is tasteless. Glycols form a series from sweet (ethylene) to bitter (hexamethylene). Of some 34 compounds tested, 30 taste sweet to man, and nine (by acceptance behavior) to a honeybee. Substances not accepted by the bee but sweet to man are mannitol, sorbitol, xylose, galactose, and melibiose.[51] Sensitivities of bee mouth parts for sugars were recorded in the following series:[36] sucrose = maltose > trehalose > fructose. Tests with flies of a number of D-glucose derivatives indicate combination of the C_3 and C_4 hydroxyl groups with the receptor site; steric hindrance of these groups prevents stimulation, and methylation makes glucose ineffective. The primary stimulating process is relatively insensitive to temperature. Relative effectiveness is: maltose, 33; fructose, 21.6; sucrose, 13.2; glucose, 1.[45] The series of decreasing effectiveness as judged by responses in the chorda tympani of cat is D-fructose > sucrose > sorbose > D-mannose > D-glucose > maltose > D-galactose > lactose. Some nerve fibers responded to all of the sugars, and some to only two of them. Some fibers respond to both sugar and salt, suggesting that one neuron may be activated from more than one taste cell. Pretreatment of a papilla with salt may depress the response to sugar.[6]

The subjective taste of bitter is given by such dissimilar substances as quinine, strychnine, and some Mg and NH_4 salts. Genetic differences in ability to taste as bitter such compounds as *n*-propylthiouracil correlate with ability to taste chlorpromazine, and they suggest chemical specificity of receptors in some taste cells. A number of human genotypes occur with specific taste deficiencies. In a series of compounds which taste bitter, a proton donor group (DH) is close enough (1 Å) to an acceptor (A) for hydrogen bonding; sweetness is inhibited by intramolecular H-bonding,[82] and sweet compounds have proton donors and their acceptors too far apart (3 Å) for H-bonding. Mutants of *E. coli* show that receptors for specific amino acids have a genetic basis.

Whether insects have a sense comparable to bitter taste is uncertain. Many compounds are rejected (proboscis withdrawn), such as alcohols, aldehydes, and ketones. The stimulating effectiveness of rejected compounds for homologous series of aldehydes, ketones, alcohols, and glycols increases (i.e., the threshold concentration falls) as the carbon chain is lengthened. Threshold molar concentrations for alcohols for blowfly rejection are: methyl, 11.3; ethyl, 3.2; *n*-butyl, 0.66; *n*-amyl, 0.1; *n*-hexyl, 0.012; secondary *n*-octyl, 0.0021.[34] Branching of a chain raises the threshold, and a second OH group decreases the stimulating effect, as does adding a halogen; *n*-ketones are more effective than iso-ketones.[34] Mono-carboxylic acids are more effective than di-carboxylic acids, and different hairs respond differently to a given acid at the same pH.[90] Homologous organic series often increase in effectiveness very slightly up to about five carbons, and then they increase in effectiveness very much more; perhaps the shorter compounds enter in an aqueous phase, while higher ones enter in a lipid phase.

Water taste receptors are stimulated by washing away salt or sugar with water; they may be viewed as tonic OFF responders to solutes. They may serve as osmoreceptors to sucrose, but are not inhibited by other solutes in very low concentration.

All contact chemical (taste) receptors, whether nerve endings or taste cells surrounding nerve endings, show graded electrical responses. It is probable that these result from the gating of cell membranes for ionic currents and that the membranes contain very specific receptors.

Olfactory or distance chemoreceptors are most sensitive and varied in specificity. Subjective measures of odor sensitivity and masking and cross-adaptation of odors have led to several classifications by psychologists—ethereal, camphoraceous, musky, minty, floral, pungent, and putrid. Evidence for multiple receptors of various proportions comes from marked differences in threshold concentrations in inhaled air for man: ethyl mercaptan, 7×10^{-13} M; amyl thioether, 5.8×10^{-8} M; ethyl ether, 7.8×10^{15} M.[105] The most complete review of the enormous literature on chemical senses is that of Moncrieff,[93] who arrived at some sixty-two general principles relating chemical structure to odor. In some series, the odor effectiveness increases with chain length to an optimum. In man, bee, and dog, the optimum for some volatile fatty acids is a four-carbon compound; for a series of volatile esters, the optimum is at 12 carbons for man and bee.[128] Possibly lipid solubility limits effectiveness on the side of smaller molecules and molecular size limits it on the higher side of the optimum. For producing an electro-olfactogram (EOG) in frog, a series of primary alcohols reached maximum effectiveness at eight carbons.[102]

Sensitive membranes apparently have highly specific receptor molecules, and different cells differ in the proportion of various receptors. This is supported by the highly specific sensitivities, as in insect olfactory responses to pheromones. Recordings from single receptor neurons in response to stimulation by many substances show individual profiles of sensitivity which differ from cell to cell; by combining the input from many such receptors, the central nervous system could distinguish an almost limitless number of odors. Antennal receptors of a male moth (*Bombyx*) can detect relatively few molecules of bombykol, but antennae of females are relatively insensitive to bombykol. Evidence for structural specificity is obtained with homologous isomers of bombykol as follows: 10 trans, 12 cis > 10 cis, 12 trans > 10 cis, 12 cis > 10 trans, 12 trans.[17] Representative reaction spectra are given in Figure 13–14. No single molecular feature—saturation, position of double bonds, chain length, functional groups, isomerization, or geometrical shape—correlates with specific thresholds.[119]

Evidence for individuality of receptor profiles in the mammalian olfactory system is based on the fact that primary fibers from regions of the olfactory epithelium are distributed spatially in different parts of the olfactory bulb. Integrated spike activity was recorded by implanted electrodes from different loci in the olfactory bulb of rabbits; six sites gave characteristic profiles of response to 12 odorants, and the same pattern was evident during several days[96] (Fig. 13–15). Similarly, recordings from two nerve branches serving the two ends of the olfac-

Tested Substances	Carrion Receptor S. basiconicum *Calliphora*	Carrion Receptor S. basiconicum *Necrophorus*	S. coeloconicum *Locusta*	Queen Subst. Receptor S. placodeum *Apis*
C_3 (Straight-Chain Fatty Acids, saturated)	0	– –	+ + +	
C_4	0	– –	+ + +	+ + +
C_5	0		+ + +	+ + +
C_6	0	+ +	+ + +	+ + +
C_7	0	+ +	+ + +	
C_8	0	+ +	+ +	+ +
C_9	0	+	0	
C_{10}	0	+ +	0	+
$C_{12,14,16,18}$			0	0
C_{18} unsaturated (Oleic ac., Linoleic ac.)	0	+	+ +	
Carrion	+ + +	+ + +		
Amines, Mercaptanes	+ + +	+ + +		
Hexenal			+ + +	+ + +
Hexanol	+ + +		0	0
Hexenol			+ +	0
Hexenylformiate			+ +	0
Queen Substance (9-Oxo-decenoic acid)			0	+ + +
9-Hydroxy-decenoic acid				+

Figure 13–14. Reaction spectra of four specialized olfactory cells in the species named. O, no response; +, excitatory; –, inhibitory effects. Number of + or – signs indicates strength of response. (From Boeckh, J., K. E. Kaissling, and D. Schneider, Cold Spring Harbor Symp. Quant. Biol. *30*: 263–280, 1965.)

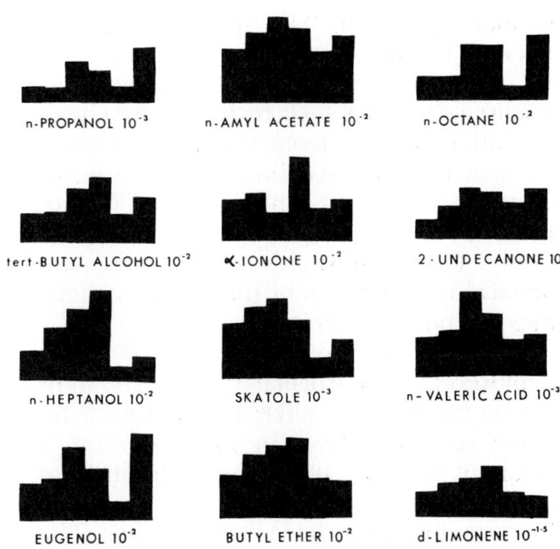

Figure 13-15. Responses to 12 odorants at six sites in the olfactory bulb. Numbers refer to concentrations expressed as exponentials of the saturated vapor at 22°C. (From Moulton, D. G., Cold Spring Harbor Symp. Quant. Biol. *30*:201–206, 1965.)

tory epithelial plate in frog showed different proportions of summed responses to different odors.[97] Just as one taste can modify the response to a succeeding substance, the olfactory response to a mixture cannot be predicted from the responses to components. In frog primary olfactory neurons, the response to a given odor varies with time and with sequence of presentation of other odors. For example, one cell gives a few spikes to butanol, but after musk (which elicits no response itself) the response to butanol is a large burst. Thus, there are not only differences from receptor cell to cell, but the same cell varies according to its immediate olfactory history.[55]

For each of four classes of odor, thresholds correlate in a general way with adsorption at an oil-water interface,[32] and with the size and shape of stimulating molecules.[2,3,4,5] Camphoraceous compounds are spherical, 7 Å in diameter; musky compounds are more disk-shaped. Recordings from single units of frog olfactory bulb support the psychophysical evidence for stereochemical correlations.[43] It is reasonable, therefore, that several molecular properties correlate with odorous stimulation, that receptor cells have heterogeneous membranes with patches of highly specific lipoprotein sites, and that the proportion of different receptor sites varies from cell to cell.

SUMMARY

For many animals the chemical senses are of critical importance in daily behavior. Striking examples of behavior based on specific chemostimulation are known—sex attraction in many insects, species and "family" recognition in some mammals, plant food identification by insects, homing by salmon and newts, defense reactions of many animals, and location of animal odors by dogs. The interest of man in wines and perfumes emphasizes the importance of the chemical senses. Specificity of plant attractants and repellants provides an approach to biological control of insect pests.

General chemical sensitivity is a universal cellular characteristic. Most cell membranes, particularly those of excitable tissues, respond to acids, alkalis, and many ions; membranes are depolarized, cells round up, cytoplasm contracts, and cilia reverse or stop. The magnitude of the cellular homeostatic response depends on the strength of the stimulus. Some ions are more effective than others, presumably because of the nature of the cell surfaces.

Much more difficult to understand, however, is the unique sensitivity to diverse organic molecules and specific inorganic ions. Merely a few molecules of some sex attractants elicit complex behavior in moths, or homing of fish. The specific sensitivity of taste receptors to certain amino acids and sugars implies many molecular receptors in sense cells. That such receptors are chemical entities follows from the fact that there are genetic mutants for sensing specific materials in bacteria, phytophagous insects, and man. The receptor sites are known mainly from thresholds of stimuli, and chemical identification of the receptor molecules has not yet been made.

Recent profiles of sensitivity of single chemoreceptors, particularly in insects, show that no two cells are alike and that each receptor cell must have in its membrane a mosaic of receptor sites. The multiplicity of receptor molecules must be very great, or else large molecules have multiple sites. The many different receptors in varying proportions permit blending of odors and flavors in varied combinations. The receptor molecules must be capable of modification, as in sensory adaptation, by changing sensitivity according to state of hydration and state of prior treatment with certain chemicals (e.g., reversal of taste sensation in man).

All chemical sense organs appear to show generator potentials in response to stimulation. In some, the dendritic endings are stimulated directly, as in most insect sensilla; in others, a receptor cell stimulates nerve endings, probably synaptically, as in vertebrate taste epithelium. In all chemosensory systems, there is considerable convergence in the central nervous system. In vertebrates, the olfactory bulb is an everted portion of the brain where an extensive neuropile "blends" the input from very many receptors.

Formerly, much "mystique" was associated with the chemical senses. Identification of pheromones, partial characterization of the nature of receptor molecules, and some understanding of central coding of chemosensory signals are beginning to make the chemical senses understandable.

REFERENCES

1. Adler, J., Science *166*:1588–1597, 1969. Chemoreceptors in bacteria.
2. Amoore, J. E., Nature *198*:271–272, 1963. Stereochemical theory of olfaction.
3. Amoore, J. E., Cold Spring Harbor Symp. Quant. Biol. *30*:623–636, 1965. Psychophysics of odor.
4. Amoore, J. E., pp. 158–171. *In* Olfaction and Taste. 3rd Internat. Symp., edited by C. Pfaffmann. Rockefeller University Press. New York, 1969. C. Pfaffmann. Rockefeller University Press. New York, 1969.
5. Amoore, J. E., Molecular Basis of Odor. Charles C Thomas, Springfield, 1970. 200 pp.
6. Andersen, H. T., M. Funakoshi, and Y. Zotterman, Acta Physiol. Scand. *56*:362–375, 1962; *also* pp. 177–192 *in* Olfaction and Taste: A Symposium, edited by Y. Zotterman, Pergamon Press, New York, 1963. Electrophysiological responses of taste receptors to sugar.
7. Bailey, D. F., and M. S. Laverack, Nature *200*:1122–1123, 1963. Central nervous responses to chemical stimulation of a gastropod receptor.

7a. Bardoch, J., et al., pp. 647–666 *in* Olfaction and Taste, edited by T. Hayashi. Pergamon Press. New York, 1967. Chemical stimulation in bullhead fish.

8. Bartoshuk, L. M., M. A. Homer, and L. H. Parks, Science *171*:699–701, 1971. Water taste in cat.
9. Beidler, L. M., J. Neurophysiol. *16*:595–607, 1953; *also* Beidler, L. M., et al., Amer. J. Physiol. *181*:235–240, 1955. Analysis of taste responses, mammals.
10. Beidler, L. M., pp. 133–145. *In* Olfaction and Taste: A Symposium, edited by Y. Zotterman. Pergamon Press, New York, 1963. Dynamics of taste cells.
11. Beidler, L. M., Cold Spring Harbor Symp. Quant. Biol. *30*:191–200, 1965. Comparison of gustatory and olfactory receptors.
12. Beidler, L. M., pp. 509–534. *In* Olfaction and Taste, Vol. 2, edited by T. H. Hayashi. Pergamon Press, New York, 1967.
13. Bell, F. R., and R. L. Kitchell, J. Physiol. *183*:145–151, 1966. Taste reception in goat, sheep, and calf.
14. Bierl. B. A., M. Beroza, and C. W. Collier, Science *170*:87–89, 1970. Sex attractants of gypsy moth.
15. Boeckh, J., Z. vergl. Physiol. *46*:212–248, 1962. Electrophysiology of odor reception in carrion beetles.
16. Boeckh, J., pp. 721–736. *In* Olfaction and Taste, Vol. 2, edited by T. H. Hayashi. Pergamon Press, New York, 1967. Single insect olfactory receptors.
17. Boeckh, J., Z. vergl. Physiol. *55*:378–406, 1967. Olfactory receptors in several insects.
18. Boeckh, J., pp. 34–51. *In* Olfaction and Taste, 3d, edited by C. Pfaffmann. Rockefeller University Press, 1969. Electrical activity in olfactory receptors.
19. Boeckh, J., K. E. Kaissling, and D. Schneider, Cold Spring Harbor Symp. Quant. Biol. *30*:263–280, 1965. Insect olfactory receptors.
20. Bowers, J. M., and B. K. Alexander, Science *158*:1208–1210, 1967. Olfactory recognition by mice.
21. Brady, U. E., J. H. Tomlinson, R. G. Brownlee, and R. M. Silverstein, Science *171*:802–804, 1971. Sex attractants in moths.
22. Brown, A. W. A., and A. G. Carmichael, Nature *189*:508–509, 1961. Lysine as a mosquito attractant.
23. Butenandt, A., Triangle (Basel) *5*:24–27, 1961. Biochemistry of ecdyson and mating scent of silk moths.
24. Carr, W. E. S., Biol. Bull. *133*:90–127, 1967. Chemoreception in snail *Nassarius*.
25. Case, J., Biol. Bull. *127*:428–446, 1964. Chemoreceptors on dactyl of *Cancer*.
26. Case, J., and G. F. Guilliam, Biol. Bull. *121*:449–455, 1961. Amino acid sensitivity of chemoreceptors of *Carcinides*.
27. Creutzberg, F., Nederl. J. Sea Res. *1*:257–338, 1961. Orientation of migrating elvers of *Anguilla*.
28. Crewe, R. M., and M. S. Blum, Z. vergl. Physiol. *70*:363–373, 1970. Alarm pheromones in ants.
29. Crisp, D. J., Biol. Bull. *133*:128–140, 1967. Chemoreception in barnacles.
30. Davenport, D., et al., Biol. Bull. *100*:71–83, 1951; Animal Behavior *8*:3–4, 1960. Commensalism between starfish and polychaetes.
31. Davenport, D., et al., Biol. Bull. *115*:397–410, 1958. Symbiosis of sea anemone and pomacentrid fish.
32. Davies, J. T., J. Theoret. Biol. *8*:1–7, 1965. Theory of odors.

32a. Desgranges, J.-C., Comp. Rend. Acad. Sci. Paris *263*:1103–1106, 1966. Ultrastructure of taste endings on barbels of catfish.

33. Dethier, V. G., Biol. Bull. *72*:7–23, 1937; *76*:325–329, 1939; *also* Amer. Naturalist *75*:61–73, 1941. Chemoreception by lepidopteral larvae in relation to food plants.
34. Dethier, V. G., Amer. J. Physiol. *165*:247–250, 1951; *also* J. Gen. Physiol. *35*:55–65, 1951; *also* Biol. Bull. *102*:111–117, 1952; *103*:178–189, 1952. Stimulation of tarsal chemoreceptors in flies by organic molecules.
35. Dethier, V. G., and L. E. Chadwick, J. Gen. Physiol. *33*:589–599, 1950. Relation between solubility and stimulating effect of organic and inorganic compounds.
36. Dethier, V. G., pp. 544–576. *In* Insect Physiology, edited by K. D. Roeder. John Wiley and Sons, New York, 1953. Review of chemoreception in insects.
37. Dethier, V. G., et al., Biol. Bull. *111*:204–222, 1956; *also* Dethier, V. G., and D. R. Evans, Biol. Bull. *120*:108–116, 1961. Relation between taste and ingestion of carbohydrates and water, blowflies.
38. Dethier, V. G., Science *161*:389–391, 1968. Chemosensory input and taste discrimination in blowfly.
39. Dethier, V. G., and J. H. Kuch, Z. vergl. Physiol. *72*:343–363, 1971. Gustatory responses of lepidopteran larvae.
40. Doetsch, G. S., J. J. Ganchron, L. M. Nelson, and R. P. Erickson, pp. 492–511. *In* Olfaction and Taste, 3d, edited by C. Pfaffmann. Rockefeller University Press, New York, 1969. Information processing in taste system.
41. Døving, K. B., Acta Physiol. Scand. *66*:290–299, 1966. Olfactory influences on secondary neurons in fish *Lota*.

42. Døving, K. B., Acta Physiol. Scand. *68*:404–418, 1966. Responses from olfactory bulb of frog.
43. Døving, K. B., and G. Genne, J. Neurophysiol. *28*:139–153, 1965; *29*:665–674, 1966. Efferent olfactory system in fish.
44. Elsner, T., and J. Meinwald, Science *153*:1341–1350, 1966. Defensive secretions of arthropods.
45. Evans, D. R., pp. 165–175. *In* Olfaction and Taste: A Symposium, edited by Y. Zotterman. Pergamon Press, New York, 1963. Chemical structure and stimulation by carbohydrates.
46. Evans, D. R., and D. Mellon, J. Gen. Physiol. *45*:487–500, 651–661, 1962. Stimulation of primary receptors by salt and water.
47. Feir, D., and S. D. Beck, Ann. Entom. Soc. Amer. *56*:224–229, 1963. Feeding, *Oncopeltus*.
48. Feir, D., J. I. Lengy, and W. B. Owen, J. Insect Physiol. *6*:13–20, 1961. Contact chemoreception in mosquito.
49. Fischer, R., and F. Griffin, Nature *200*:343–347, 1963. Quinine dimorphism.
49a. Fraenkel, G., Science *129*:1466–1470, 1959. Chemical relations of food plants and insects.
50. Fraenkel, G., Entomol. Exp. Appl. *12*:473–486, 1969. Review of function of secondary plant substances.
51. Frisch, K. von, Z. vergl. Physiol. *21*:1–156, 1934. Comparative physiology of taste, particularly in honeybees.
52. Funakoshi, M., and Y. Zotterman, Acta Physiol. Scand. *57*:193–200, 1963. Effect of salt on sugar response of taste endings.
53. Gesteland, R. C., Ann. N.Y. Acad Sci. *116*:440–447, 1964. Initial events of electro-olfactograms.
54. Gesteland, R. C., J. Y. Lettvin, W. H. Pitts, and S. H. Chung, pp. 313–322. *In* Cybernetic Problems in Bionics, edited by H. L. Oestreicher and D. R. Moore. Gordon & Breach Science Publishers, New York, 1968.
55. Gesteland, R. C., J. Y. Lettvin, W. H. Pitts, and A. Rojas, pp. 19–34. *In* Olfaction and Taste: A Symposium, edited by Y. Zotterman. Pergamon Press, New York, 1963. A code in the nose.
56. Goz, H., Z. vergl. Physiol. *29*:1–45, 1941. Olfactory discrimination by fish.
57. Grant, D., O. Anderson, and V. Twitty, Science *160*:1354–1355, 1968. Homing orientation by olfaction in newts.
58. Hara, T. J., K. Ueda, and A. Gorbman, Science *149*:884–885, 1965. Brain responses to odors in salmon.
59. Hara, T. J., Comp. Biochem. Physiol. *22*:199–208, 1967. Electrophysiology of olfactory system in goldfish.
60. Hasler, A. D., Science *132*:785–792, 1960. Guideposts of migrating fish.
61. Hasler, A. D., and W. J. Wisby, J. Amer. Fish Soc. *79*:64–70, 1950. Discrimination of stream odors and pollutants by fish.
62. Hasler, A. D., and W. J. Wisby, Amer. Nat. *85*:223–238, 1951. Discrimination of stream odors and pollutants by fish.
63. Hasler, A. D., and W. J. Wisby, Quart. Rev. Biol. *31*:200–209, 1956. Discrimination of stream odors and pollutants by fish.
64. Hodgson, E. S., and K. D. Roeder, J. Cell. Comp. Physiol. *48*:51–76, 1956. Responses of chemoreceptor hairs in Diptera.
65. Hsiao, T. H., and G. Fraenkel, Ann. Entom. Soc. Amer. *61*:44–54, 476–503, 1968. Action of secondary plant substances on Colorado potato beetle.
66. Hughes, J. R., D. E. Hendrix, H. Wetzel, and J. W. Johnston, pp. 172–192. *In* Olfaction and Taste, 3d, edited by C. Pfaffmann. Rockefeller University Press, New York, 1969. Electrophysiology of olfactory bulb and subjective response to odor in man.
67. Jacobson, M., M. Beroza, and R. T. Yamamoto, Science *139*:48–49, 1963. Sex attractants of cockroach.
68. Jacobson, M., C. E. Lilly, and C. Harding, Science *159*:208–209, 1968. Sex attractants of sugar beet wireworm.
69. Jacobson, M., et al., Science *170*:542–544, 1970. Sex pheromones of army worm moth.
70. Jahan-Parwar, B., M. Smith, and R. von Baumgarten, Amer. J. Physiol. *216*:1246–1257, 1969. Chemical stimulation in nervous system of *Aplysia*.
71. Jermy, T., F. E. Hanson, and V. G. Dethier, Entomol. Exp. Appl. *11*:211–230, 1968. Induction of food preference in lepidopteran larvae.
72. Jones, W. A., M. Jacobson, and D. F. Martin, Science *152*:1516–1517, 1966. Sex attractants of pink bollworm moth.
73. Junge, D., and M. S. Brodwick, Comp. Biochem. Physiol. *35*:623–630, 1970. Stimulation of frog gustatory units by calcium.
74. Kafka, W. A., Z. vergl. Physiol. *70*:105–143, 1970. Chemostimulation in *Locusta*.
75. Kaissling, K. E., pp. 52–70. *In* Olfaction and Taste, 3d, edited by C. Pfaffmann. Rockefeller University Press, New York, 1969. Kinetics of olfactory receptor potentials.
76. Kay, R. E., J. T. Eichner, and D. E. Gelvin, Amer. J. Physiol. *213*: 1–10, 1967; *220*:1473–1480, 1481–1487, 1971. Olfactory receptor potentials of *Lucilia*.
77. Kleerekoper, H., pp. 625–645. *In* Olfaction and Taste, 2d, edited by T. H. Hayashi. Pergamon Press, New York, 1967. Some effects of olfactory stimulation on locomotory patterns in fish.
78. Kleerekoper, H., Olfaction in Fishes. Indiana University Press, 1969. 222 pp.
79. Konishi, J., and Y. Zotterman, Acta Physiol. Scand. *52*:150–161, 1961. Taste function in carp.
80. Konishi, J., and Y. Zotterman, pp. 215–233. *In* Olfaction and Taste: A Symposium, edited by Y. Zotterman. Pergamon Press, New York, 1963. Taste functions in fish.
81. Kriner, M., Z. vergl. Physiol. *21*:317–342, 1934. Chemical discrimination by minnows.
82. Kubota, T., and I. Kubo, Nature *223*:97–99, 1969. Bitterness and chemical structure.
83. Kurihara, K., Y. Kurihara, and L. M. Beidler. *In* Olfaction and Taste: A Symposium, edited by Y. Zotterman. Pergamon Press, New York, 1963. Isolation and mechanism of taste modifiers.
84. Lacher, V., Z. vergl. Physiol. *48*:587–623, 1964. Chemical receptors of honeybee.
85. Laverack, M. S., Comp. Biochem. Physiol. *2*:22–34, 1961. Tactile and chemical perception in earthworms.
86. Law, J. H., E. O. Wilson, and J. A. McClosky, Science *149*:544–545, 1965. Pheromones in termites, bees, and ants.
87. Lenhoff, H. M., Comp. Biochem. Physiol. *28*:571–586, 1969. pH profile of a peptide receptor.
88. de Lorenzo, A. J. D., pp. 5–18. *In* Olfaction and Taste: A Symposium, edited by Y. Zotterman. Pergamon Press, New York, 1963. Ultrastructure of cell membranes and synapses in chemoreceptors.
89. Maier, S. F., D. M. Zahorik, and R. Wallin, J. Comp. Physiol. Psychol. *74*:254–262, 1971; Psychon. Sci. *17*:309–310, 1971. Specific vitamin hungers in rats.
90. McCutchan, M. C., Z. vergl. Physiol. *65*:131–152, 1969. Responses of chemoreceptors of blowfly to organic acids.
91. McLeese, D. W., J. Fish. Res. Bd. Canad. *27*:1371–1378, 1970. Chemoreception in lobsters.
92. Miller, I. J., J. Gen. Physiol. *57*:1–25, 1971. Interaction among taste papillae in rat.
93. Moncrieff, R. W., The Chemical Senses, 3rd ed. Chemical Rubber Co., Cleveland, 1967. 700 pp.
94. Morita, H., pp. 370–381. *In* Olfaction and Taste, 3d, edited by C. Pfaffmann. Rockefeller University Press, New York, 1969. Electrical signs of taste receptor activity.
95. Morita, H., and A. Shiraishi, J. Gen. Physiol. *52*:559–584, 1968. Stimulation of sugar receptors of flesh fly.
96. Moulton, D. G., Cold Spring Harbor Symp. Quant. Biol. *30*:201–206, 1965. Differential sensitivity to odors.
97. Mozell, M. M., Science *143*:1336–1337, 1964; *also* J. Gen. Physiol. *50*:25–41, 1966; *56*:43–63, 1970. Spatiotemporal analysis of odor by olfactory receptor sheet.
98. Nayar, J. K., and G. Fraenkel, J. Insect Physiol. *8*:505–525, 1962. Host plant selection in silk moth *Bombyx mori*.
99. Nayar, J. K., and G. Fraenkel, Ann. Entom. Soc. Amer. *56*:119–122, 174–178, 1963. Host selection by sphinx *Ceratomia* and Mexican bean beetle *Epilachna*.
100. Neurath, H., Z. vergl. Physiol. *31*:609–626, 1949. Olfactory sense in minnows.
101. Ottoson, D., Acta Physiol. Scand. *43*:167–181, 1958. Electro-olfactogram of frog.
102. Ottoson, D., Acta Physiol. Scand. *47*:136–172, 1959. Olfactory bulb potentials, rabbit and frog.
103. Ottoson, D., pp. 35–44. *In* Olfaction and Taste: A Symposium, edited by Y. Zotterman. Pergamon Press, New York, 1963. Generation and transmission of olfactory signals.
104. Parkes, A. S., J. Reprod. Fertil. *1*:312–314, 1960. Odors in mammalian reproduction.
105. Pfaffmann, C., pp. 257–273. *In* Olfaction and Taste: A Symposium, edited by Y. Zotterman. Pergamon Press, New York, 1963. Taste stimulation and preference behavior.
106. Pfaffmann, C., et al., pp. 361–381. *In* Olfaction and Taste, 2d, edited by T. H. Hayashi. Pergamon Press, New York, 1967. Sensory and behavioral factors in taste preferences.
107. Pfaffmann, C., editor, Olfaction and Taste, 3d, Rockefeller University Press, New York, 1969.
108. Phillips, C. G., J. P. S. Powell, and G. M. Shepherd, J. Physiol. *156*: 26*P*–27*P*, 1961; *168*:65–88, 1963. Responses of mitral cells to olfactory stimulation, rabbit.
109. Price, J. L., Brain Res. 7:483–486, 1968. Termination of centrifugal fibers in olfactory bulb.
110. Price, J. L., and T. P. S. Powell, J. Cell. Sci. 7:91–188, 1970. Granule cells of olfactory bulb.
111. Rall, W. D., and G. M. Shepherd, J. Neurophysiol. *31*:884–915, 1968. Rabbit mitral cells.
112. Ralls, K., Science *171*:443–449, 1971. Mammalian scent marking.
113. Roessler, H. P., Z. vergl. Physiol. *44*:184–231, 1961. Chemical attraction of *Aedes*.
113a. Rozin, P., pp. 411–431 *in* Handbook of Physiology, Vol. 1, edited by C. F. Code. Amer. Physiol. Soc., Washington, D.C., 1967. Thiamine specific hunger.
114. Rudinsky, J. A., Science *152*:218–219, 1966. Responses of wood beetles to terpenes.
115. Sato, M., S. Yamashita, and H. Ogawa, pp. 470–488. *In* Olfaction and Taste, 3d, edited by C. Pfaffmann. Rockefeller University Press, New York, 1969. Afferent specificity in taste.

116. Saxena, K. N., J. Insect Physiol. *9*:47–71, 1963. Ingestion by heteropterous insect *Dysdercus*.
117. Saxena, K. N., pp. 799–819. *In* Olfaction and Taste, Vol. 2, edited by T. H. Hayashi. Pergamon Press, New York, 1967. Some factors governing olfactory and gustatory responses of insects.
118. Saxena, K. N., Entomol. Exp. Appl. *12*:751–766, 1969. Changes in food attraction with hydration, *Dysdercus*.
119. Schneider, D., J. Insect Physiol. *8*:15–30, 1962. Olfactory specificity among moths.
120. Schneider, D., pp. 85–103. *In* Olfaction and Taste: A Symposium, edited by Y. Zotterman. Pergamon Press, New York, 1963. Electrophysiology of insect olfaction.
121. Schneider, D., Symp. Soc. Exp. Biol. *20*:273–297, 1965. Chemical sense communication in insects.
122. Schneider, D., Science *163*:1031–1037, 1969. Insect olfaction.
123. Schneider, D., B. C. Block, J. Boeckh, and E. Priesner, Z. vergl. Physiol. *54*:192–209, 1967. Reaction of male moths to bombykol.
124. Schneider, D., V. Lacher, and K. E. Kaissling, Z. vergl. Physiol. *48*: 632–662, 1964. Reaction spectrum of odors in saturnid moths.
125. Schneider, D., and R. A. Steinbrecht, Symp. Zool. Soc. Lond. *23*: 279–297, 1968. Check list of insect olfactory sensilla.
126. Schoonhouen, L. M., Ann. Rev. Entom. *13*:115–136, 1968; *also* Schoonhouen, L. M., and V. G. Dethier, Arch. Neerl. Zool. *16*:497–530, 1966. Sensory aspects of discrimination of host plant by insect larvae.
127. Schultze-Westrum, T., Z. vergl. Physiol. *50*:151–220, 1965. Olfactory communication in marsupial phalangers.
128. Schwarz, R., Z. vergl. Physiol. *37*:180–210, 1955. Olfactory thresholds of honeybee.
129. Scott, J. W., and C. Pfaffmann, Science *158*:1592–1594, 1967. Olfactory input to hypothalamus.
130. Shibuya, T., Japan. J. Physiol. *10*:317–326, 1960. Electrical response, olfactory epithelium of fishes.
131. Shiraishi, A., and H. Morita, J. Gen. Physiol. *53*:450–470, 1969. Labellar sugar receptors of flesh fly.
132. Shiraishi, A., and M. Kubara, J. Gen. Physiol. *56*:768–782, 1970. Responses of chemosensory hairs to amino acids.
133. Slifer, E., J. Morphol. *105*:145–191, 1959. Ultrastructure of chemoreceptors, grasshoppers.
134. Steinbrecht, R. A., pp. 3–21. *In* Olfaction and Taste, 3d, edited by C. Pfaffmann. Rockefeller University Press, New York, 1969. Comparative morphology of olfactory receptors.
135. Sutterlin, A. M., and N. Sutterlin, J. Fish. Res. Bd. Canad. *27*:1927–1942, 1970; *28*:565–572, 1971. Olfactory and taste responses of salmon.
136. Takagi, S. F., K. Aoki, M. Iiho, and T. Yajima, pp. 92–108. *In* Olfaction and Taste 3d edited by C. Pfaffmann. Rockefeller University Press, New York, 1969. Electropositive potentials in olfactory epithelia.
137. Tateda, H., and L. M. Beidler, J. Gen. Physiol. *47*:479–486, 1964. Receptor potential in taste cell of rat.
138. Teichman, H., Z. vergl. Physiol. *42*:206–254, 1959. Olfaction in fish.
139. Thorpe, W. H., Proc. Roy. Soc. Lond. B *127*:424–433, 1939. Preimaginal olfactory conditioning in insects.
140. Thorsteinson, A. J., Canad. J. Zool. *31*:52–72, 1953. Chemotactic responses of Lepidoptera.
141. Tucker, D., J. Gen. Physiol. *46*:453–489, 1963. Physical variables in olfactory stimulation.
142. Tucker, D., Nature *207*:34–36, 1965. Olfactory responses of birds.
143. Tucker, D., and T. Shibuya, Cold Spring Harbor Symp. Quant. Biol. *30*:207–215, 1965. Physiology and pharmacology of olfactory receptors, turtle.
144. Ueda, K., T. J. Hara, and A. Gorbman, Comp. Biochem. Physiol. *21*:133–143, 1967. Electro-olfactograms in salmon.
144a. Waldbauer, G. P., Entom. Exper. Appl. 7:253–269, 1964. Utilization of solanaceous plants by tobacco hornworm.
145. Walsh, R. R., Amer. J. Physiol. *186*:255–257, 1956. Single unit response, olfactory bulb of rabbit.
146. Wells, M. J., J. Exp. Biol. *40*:187–193, 1963. Taste responses in *Octopus*.
147. Wilson, E. O., Science *149*:1064–1071, 1965. Chemical communication in social insects.
148. Wolbarsht, M. L., Cold Spring Harbor Symp. Quant. Biol. *30*:281–288, 1963. Receptor sites in insect chemoreceptors.
149. Yamamoto, C., T. Yamamoto, and K. Iwama, J. Neurophysiol. *26*: 403–415, 1963. Inhibitory system in olfactory bulb.
150. Yamamoto, R. T., R. Y. Jenkins, and R. K. McClusky, Entom. Exp. Appl. *12*:504–508, 1969. Selection of plants by tobacco hornworm.
151. Zotterman, Y., Experientia *6*:57–58, 1950; *also* Acta Physiol. Scand. *37*:60–70, 1956; *also* Ann. N.Y. Acad. Sci. *81*:358–366, 1959. Species differences in water, sweet, and salt tastes.
152. Zotterman, Y., pp. 205–216. *In* Olfaction and Taste: A Symposium, edited by Y. Zotterman. Pergamon Press, New York, 1963. Studies in the neural mechanisms of taste.

CHAPTER 14

PHOTORECEPTION AND VISION

By T. H. Goldsmith

PHOTOBIOLOGY AND THE ELECTROMAGNETIC SPECTRUM

INTRODUCTION

Biologists speak frequently of adaptations of organisms to their environment while recognizing that there are two sides to the coin. As L. J. Henderson pointed out a number of years ago in his book, "The Fitness of the Environment," many of the physical features of this planet seem uniquely compatible with life. The mutual dovetailing of biological processes with their physical surroundings is seen nowhere more forcefully than in the realm of photobiology.

The electromagnetic spectrum covers a wide span, extending from γ-rays with wavelengths shorter than 10^{-3} Å (10^{-13} m) to radio waves with wavelengths measured in kilometers (10^3 m). Photobiological processes are almost exclusively confined to a narrow band of radiation at 300 to 900 nm (Fig. 14–1). There are two complementary reasons for this state of affairs.[237]

On the one hand, these are the wavelengths available near the surface of the earth. The maximum of the emission spectrum of the sun's radiant energy lies in the region of photobiological interest, and in fact the wavelength band between 300 and 1100 nm embraces three-quarters of the sun's energy. The wavelength band of solar energy reaching the biosphere is narrowed even further by absorption in the atmosphere and by water (Fig. 14–2). At the short wavelength end, a layer of ozone in the upper atmosphere prevents most of the energy at wavelengths shorter than about 290 nm from reaching the surface of the earth.[194] The long wavelength end of the spectrum is attenuated by CO_2, water vapor, and ozone. The wavelength band entering a body of clear water such as a lake or the open ocean is even more sharply filtered to a narrow band centered in the blue at 475 to 480 nm. Of course, in waters containing large amounts of silt or organic debris, the penetration of light is not as great, and the presence of additional absorbers shifts the spectrum to longer wavelengths into the yellow or even orange.[121, 224]

But availability of wavelengths is only part of the matter. Photobiological reactions are based on chemistry; energy participates in chemical reactions, and in photobiology, energy is supplied by light. The energy is absorbed by molecules in discrete packets known as *photons* or *quanta,* whose magnitude (cal) can be calculated from the expression $E = \frac{hc}{\lambda}$, where h (Planck's constant) = 1.58×10^{-34} cal · sec, c (the velocity of light) = 3×10^{10} cm · sec^{-1}, and λ is the wavelength of light in cm. Figures 14–1 and 14–2 also show the energy equivalence of various wavelengths of light in kcal Einstein^{-1} (an Einstein is 1 mole of quanta, i.e., 6.02×10^{23} photons). The activation energies of organic reactions commonly lie in the region from 15 to 65 kcal mole^{-1} (1900 to 440 nm). Activation of photochemical reactions generally

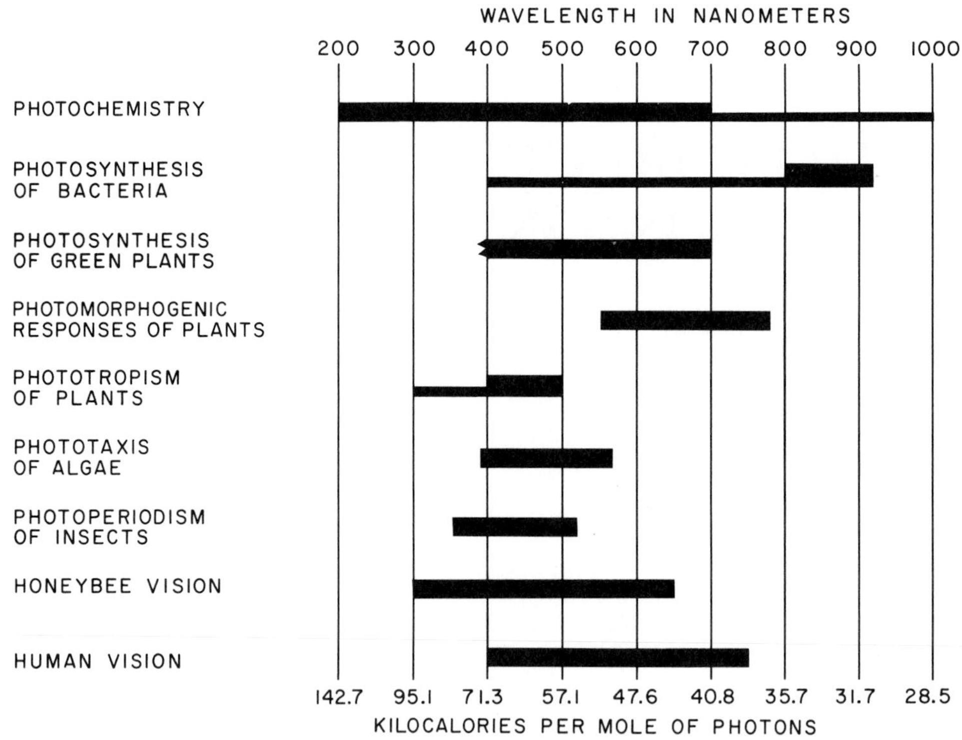

Figure 14–1. Regions of spectral response of some important photobiological processes. (For data or references to data see Wald;[237,238] Clayton;[46] Rabinowitch and Govindjee;[188] Withrow *et al.*;[266] Briggs;[28] Page;[180] Halldal;[102] Bruce and Minis;[34] Frank and Zimmerman;[83] Goldsmith;[90] Wald.[234])

involves promotion of a valence electron to a higher energy level, a process usually requiring more than 40 kcal mole^{-1} (λ shorter than about 700 nm). Thus, as one reaches farther into the infrared, the quanta become too small to provide the requisite energy for photochemical reactions.

At wavelengths shorter than about 300 nm (95 kcal mole^{-1}) there is another problem. Here the energy content of the photons becomes large enough to produce excited states and chemical alterations in proteins and nucleic acids, macromolecules whose structural integrity is crucial for life. One can consequently argue that the existence of terrestrial forms of life required as a precondition the establishment of an ozone layer in the upper atmosphere to attenuate the potentially destructive wavelengths of the solar spectrum.[238] And it is thus no accident that photobiological processes occur within the spectral region that they do.

Responses Without Eyes

What are these photobiological processes? Figure 14–1 lists the most important, and indicates the principal spectral regions involved.

The object of photosynthesis is to mobilize energy (or negative entropy) on which other living creatures can run. The other phenomena listed in Figure 14–1 are different in that they are not designed to convert efficiently the energy of sunlight into the potential energy of chemical bonds: instead they use light for triggering or informational purposes. For example, the photoperiodic responses of plants mediated by the pigment *phytochrome* are adaptive responses synchronizing aspects of the reproductive cycle with the most favorable time of the year.[214] In animals, photoperiodic responses mediated by a different pigment system regulate the time of year of reproduction[267] or, in the case of insects, the time of year or even the time of the day of various aspects of development.[34,83,137,264]

The terms *phototaxis* and *phototropism* refer to responses of organisms that are directed by light and serve to bring the organism into a more favorable relationship with its environment for more immediate purposes such as nutrition.[82] Phototropism is the bending of sessile organisms such as parts of

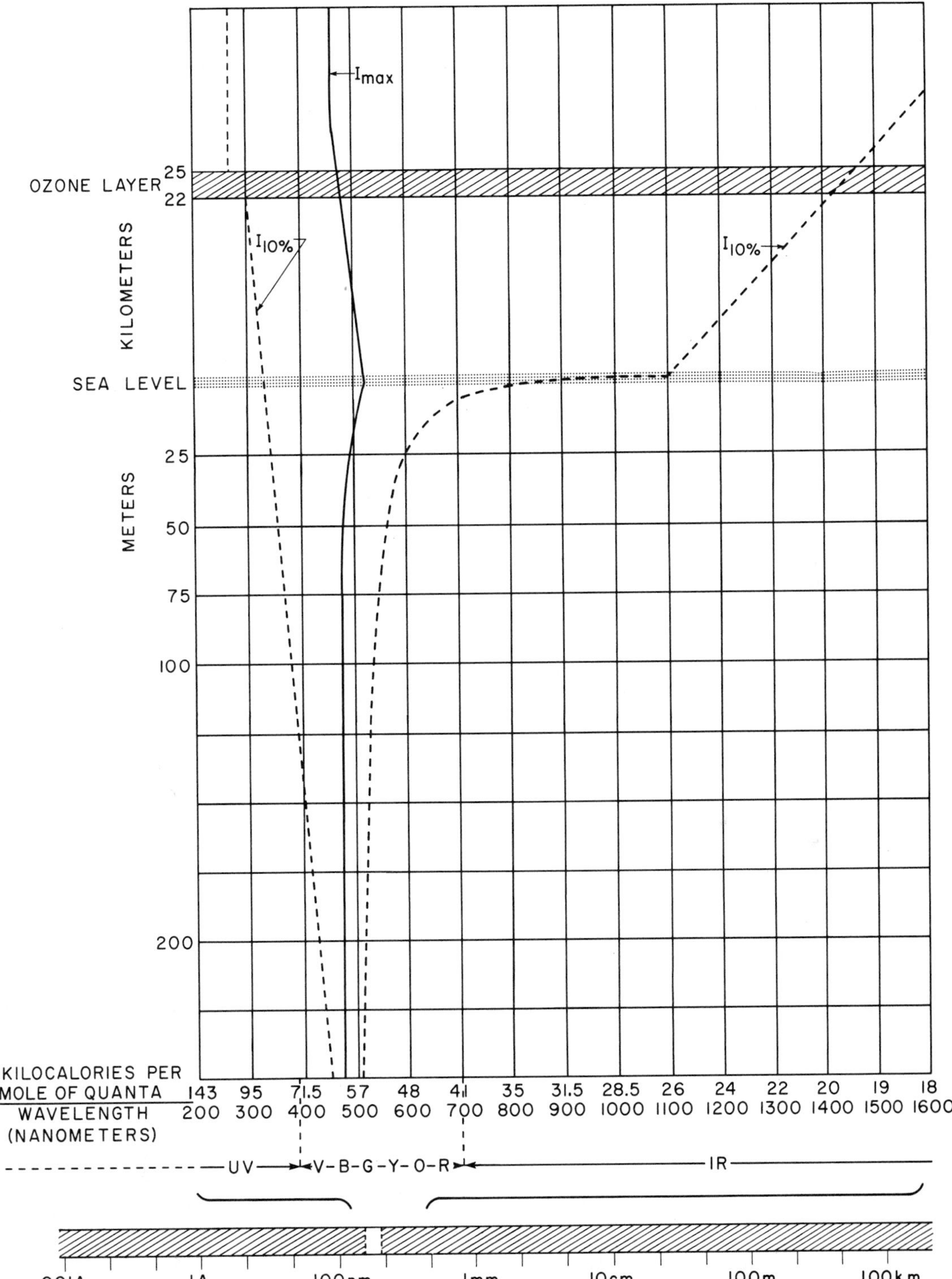

Figure 14–2. The solar spectrum is narrowed in the biosphere by absorption. Ozone in the upper atmosphere is principally responsible for removing the mid-ultraviolet, and water attenuates the longer wavelengths. The solid line (I_{max}) locates the wavelengths of maximum intensity; the broken lines ($I_{10\%}$) trace the wavelength boundaries within which 90 per cent of the solar energy is concentrated at each level in the atmosphere and ocean. The letters below the wavelength scale (UV, V, B, etc.) represent ultraviolet, violet, blue, and so forth. The small portion of the electromagnetic spectrum described in this graph is illustrated on the bottom scale. (After Wald.[237]) From Life and Light, by G. Wald. Copyright © Oct., 1959 by Scientific American, Inc. All rights reserved.

plants[28, 180] or coelenterate polyps[140] toward or away from light. Phototaxis, on the other hand, refers to the movement of unrestrained organisms.

Phototaxis and phototropism refer to the overall response of the organism rather than to the mechanism of the response to light; consequently, a number of different phenomena are embraced by these terms (Table 14–1). For example, the phototaxis (actually klino-kinesis in the terminology of Table 14–1) of photosynthetic bacteria has the same action spectrum as their photosynthesis. Falling light intensities stimulate the bacteria to reverse their direction of swimming. The result is that they congregate in regions of high intensity.[46, 47] The adaptive significance is clear.

The phototaxis of green algae (a klino-taxis) on the other hand, is different in that swimming is *directed* toward or away from a source of light. It involves a photoreceptor located near the base of the flagellum but which is not the "eyespot." As the animal swims, it spirals about its long axis, and if its trajectory is at an angle to the source of light, the photoreceptor is alternately shaded and exposed by the eyespot (or in some species perhaps the chloroplast). Presumably by altering its direction of swimming with respect to the source of light it can maximize or minimize the illumination of the photoreceptor.[47]

As a third example, in insects phototaxis is mediated by the eyes and the central nervous system.

TABLE 14–1. Simple Animal Movements. Based on Fraenkel and Gunn.[82]

kinesis	undirected; no orientation of axis of body in relation to the stimulus
ortho-kinesis	speed or frequency of locomotion dependent on intensity
klino-kinesis	frequency or amount of turning per unit time dependent on intensity
taxes	directed; axis of body oriented in relation to the stimulus
klino-taxis	attainment of orientation is indirect; receptor is subjected to successive variations in intensity by regularly alternating lateral deviations of the body
tropo-taxis	attainment of orientation is direct; bilateral pair of receptors simultaneously compare the intensities on the two sides; orientation movements equalize the intensities
telo-taxis	similar to tropo-taxis, but orientation is to a single source even in the presence of competing sources

In many animals, there are a number of photoresponses that cannot be related to a specific, multicellular receptor organ.[162] For example, locomotion of *Amoeba*,[145] movement of the spines of sea urchins (*Diadema*),[163] the firing of photoreceptor neurons in the sixth abdominal ganglion of the crayfish *Procambarus*,[126] and the effect of day length on breaking the diapause of silkworm pupae[264] are instances in which the exact location of the photoreceptor is not at all obvious, although in the latter two cases it clearly resides in the central nervous system. In earthworms (*Lumbricus*) light-sensitive cells are distributed over the body.[37, 186] There are many other instances in which small groups of specialized photosensory cells form a small eyespot or ocellus. Examples of some of these morphologically primitive photoreceptor systems are shown in Figure 14–3.

The photoreceptor pigment that absorbs light and triggers the response is unknown in many of these relatively unspecialized systems. Efforts have frequently been made to delineate the absorption spectrum of the photoreceptor pigment by measuring an *action spectrum.* To do this, the experimenter determines the relative number of photons required at each wavelength to produce some constant physiological response. An example is shown in Figure 14–35 (p. 620). It is on such data as these that the critical spectral regions in Figure 14–1 have been blocked out. Alone, this kind of experiment is rarely sufficient to establish the chemical nature of the receptor pigment.

The role of *chlorophyll* in photosynthesis[188] and of *phytochrome* in the photomorphogenic responses of plants[214, 266] is known. The receptor pigments have not been identified for photoperiodic responses such as diapause of insects,[137] the circadian rhythm of eclosion of *Drosophila*,[270] the phototaxes of most lower organisms,[47, 102] the phototropic responses of fungi[180] and higher plants,[28] as well as a variety of other responses of lower invertebrates.[93]

Most of our knowledge of photobiological processes in animals has been obtained from studies of the visual systems of animals with large eyes. The nature of the receptor pigments, the mechanisms of excitation of the receptor cells, and the initial means by which information is processed in the production of a behavioral response are the aspects of the visual system discussed in the remaining sections of this chapter.

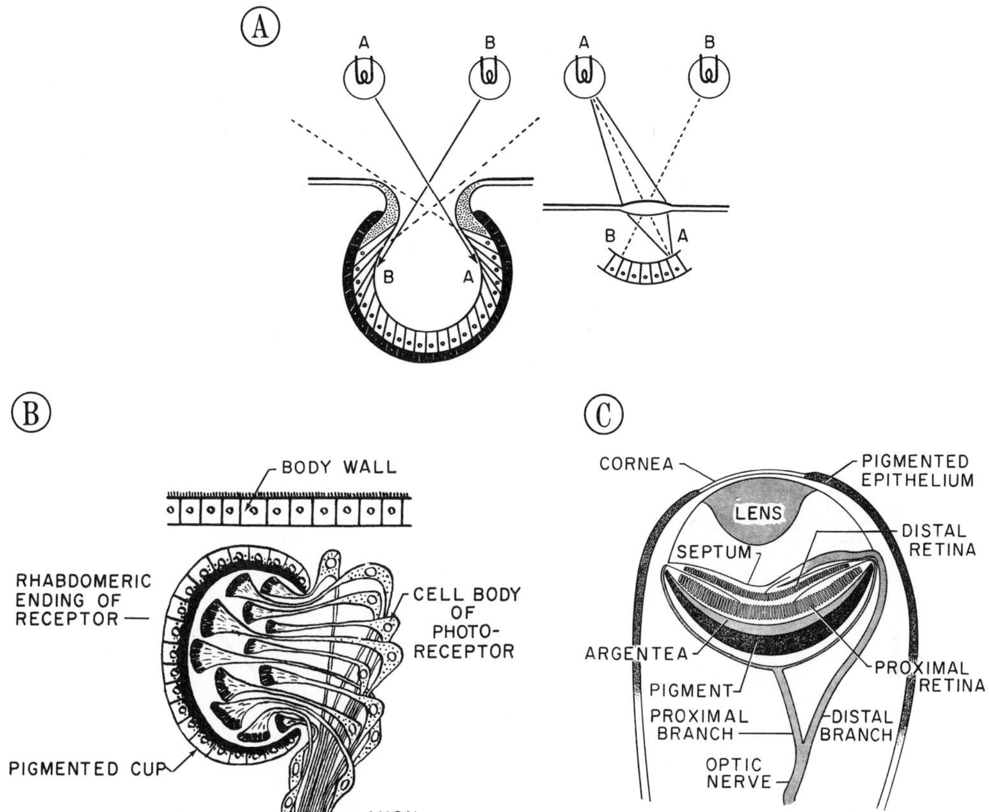

Figure 14-3. Simple eyes. *A,* The evolution of directional sensitivity and the ability to form images involves (*left*) cups lined with receptors and surrounded with dark pigment, or (*right*) a lens placed above the photoreceptors, or a combination of the two. A lens permits greater light-gathering power. (From Milne, L. J., and M. Milne, pp. 621–645 *in* Handbook of Physiology, Sec. 1, Vol. 1, American Physiological Society, 1959.) *B,* Diagrammatic section through a simple eye of the flatworm, *Planaria.* The two eyes are located on the dorsal side of the animal, and the pigmented cups open laterally. The photoreceptor neurons bear rhabdomeric endings inserted into the cavity of the pigment cup. (From Milne, L. J., and M. Milne, pp. 621–645 *in* Handbook of Physiology, Sec. 1, Vol. 1, American Physiological Society, 1959.) *C,* Double retina in the "simple" eye of the scallop *Pecten.* The distal retina contains photoreceptors derived from cilia; the proximal retina contains rhabdomeric receptors. (From McReynolds, J. S., and A. L. F. Gorman, J. Gen. Physiol. *56*:392–406, 1970.)

THE CYTOLOGICAL ORGANIZATION OF PHOTORECEPTOR CELLS

Although some cells (e.g., *Amoeba*)[145] that are responsive to light lack any apparent morphological specialization, it is more usual to possess a photoreceptor organelle formed by elaboration of the plasma membrane. These structures, which provide an increased surface area for the deposition of photopigment, fall into two general classes, depending on whether or not they form in association with the basal body of a cilium.[72, 74] The distinction is interesting, because most of the photoreceptors derived from cilia are found in vertebrates or in the Deuterostomes (those invertebrate phyla related to the vertebrate ancestral stock), whereas non-ciliary (or *rhabdomeric*) endings are characteristically found among Protostomes (annelids, molluscs, and arthropods).

Ciliary Endings in Invertebrates

The paraflagellar body of the unicellular green flagellate *Euglena* is possibly a photoreceptor in association with a flagellum that has not lost its motile function.[74, 136] The paraflagellar body is not a specialized region of the cell membrane, but rather a swelling contained completely within the flagellum. The eyespot (or *stigma*) is a cup of pigment granules (possibly carotenoid)[92] which lies outside the flagellum and is thought to be an inert shading device. The nature of the photopigment within the paraflagellar body and the means by which excitation is coupled to oriented swimming movements are unknown.

More typically, the formation of photoreceptor organelles involves an increase in the surface area of the cell membrane, rather than development of an intracellular structure analogous to the paraflagellar body.

The modification of cilia to form sensory endings is not limited to photoreceptors, for certain mechanoreceptors and olfactory endings also have ciliary origins. In general, sensory cilia have lost the two central microtubules as well as the arms on the ring of nine doublets[74] (see Chapter 12).

In the distal retina of the scallop *Pecten* (one of the places where ciliary receptors are found among Protostomes and therefore an exception to the phytogenetic generalization cited above), the photoreceptor cells bear a number of sac-like projections, each one consisting of a greatly flattened cilium.[10] In coelenterates, echinoderms, and elsewhere (Fig. 14–4) the membrane of each cilium may be thrown into finger-like folds, whereas in

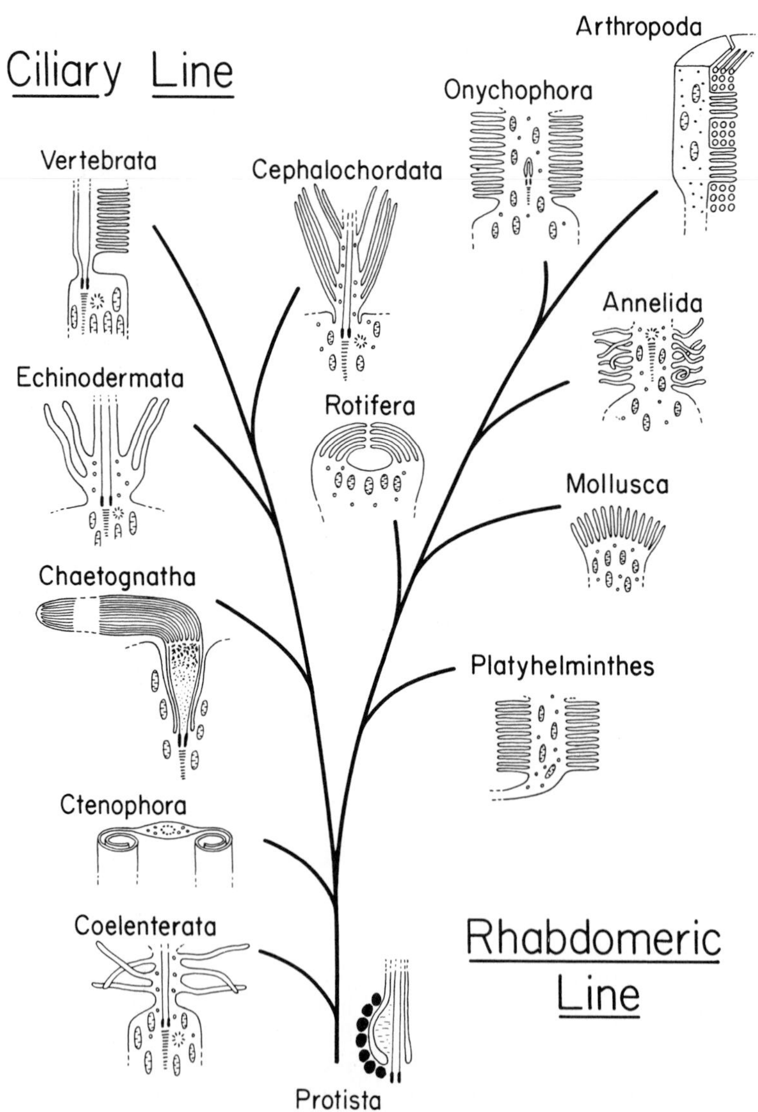

Figure 14–4. Schematic representation of photoreceptor organelles in selected examples of groups of animals along ciliary and rhabdomeric lines of evolution. (From: EVOLUTIONARY BIOLOGY, Vol. 2, edited by Th. Dobzhansky, M. K. Hecht, and Wm. C. Steere. Copyright © 1968 by Appleton-Century-Crofts, Educational Division, Meredith Corporation.)

the photoreceptor cells of larval ascidians the ciliary membrane is invaginated to form a series of parallel discs oriented roughly parallel to the long axis of the cilium.[74, 94] This latter pattern of organization is suggestive of the rods and cones of the vertebrate retina.

THE RODS AND CONES OF THE VERTEBRATE RETINA

The photoreceptor cells of the vertebrate retina are the most highly organized of the ciliary photoreceptors. An example of a *rod* cell is shown in Figure 14–5. The cell consists of an inner and an outer segment. The outer segment contains a stack of 500 to 1000 membranous discs enclosed within the plasma membrane. The outer segment is connected to the inner through a ciliary stalk, which shows the characteristic nine double microtubules when viewed in cross-section with the electron microscope. The discs arise as infoldings of the plasma membrane. In *cone* cells many of them retain continuity with the surface membrane, and their contents are therefore continuous with the extracellular space. In rods, on the other hand, the evidence indicates that the discs are completely pinched off from the plasma membrane.[48, 64] Furthermore, osmotic studies show that the disc contents behave as intracellular rather than extracellular space.[49, 64] This point takes on some importance, for as we shall see, the plasma membrane is the structure over which information is carried in the process of excitation. The apparent lack of connection between the discs of rod outer segments and the

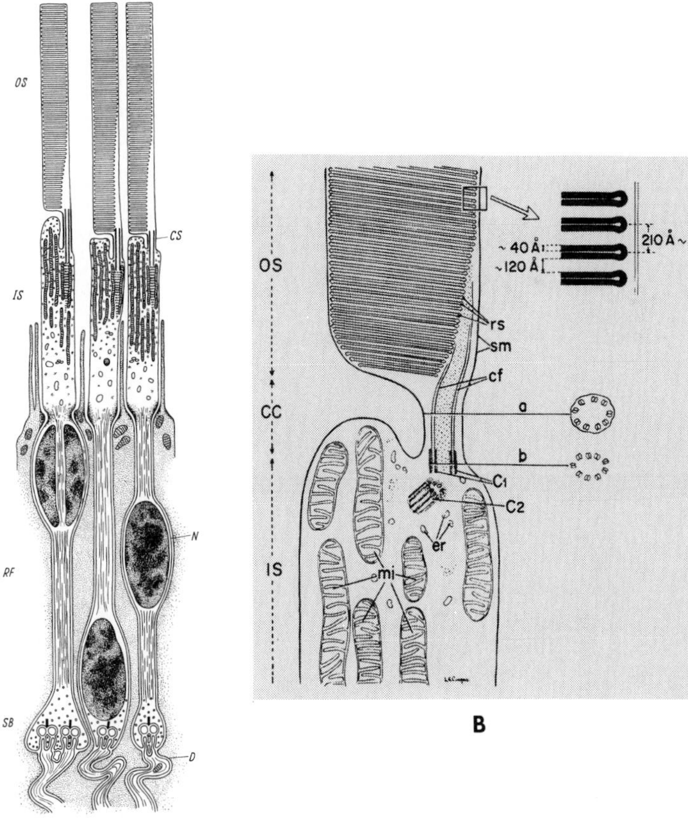

Figure 14–5. *A,* Diagram of mammalian rod cells based on electron micrographs. (From Sjöstrand, F. S., Ergebn. Biol. *21*:128–160, 1959.) *B,* Enlarged diagram of the region near the junction of the inner and outer segments. (From de Robertis, E., J. Gen. Physiol. *43* (Suppl. 2):1–6, 1960.) *OS,* outer segment; *IS,* inner segment; *CC, CS,* connecting cilium; *mi,* mitochondria (in the ellipsoid); *rs,* rod sacs or disks; *cf,* ciliary filaments (microtubules); *sm,* surface membrane; C_1, C_2, centrioles (basal body); *er,* endoplasmic reticulum; *N,* nucleus; *SB,* synaptic body (rod pedicel); *D,* dendrite of bipolar neuron.

plasma membrane of the cell raises a problem in the early events of transmission for which there is not yet a definitive explanation.

The inner segment and the remaining part of the cell is more conventional, containing the usual complement of organelles.[48, 64] There is frequently a large collection of mitochondria immediately adjacent to the outer segment, forming a compact mass which is recognized in the light microscope as the *ellipsoid.* The basal end of the rod cell is a presynaptic terminal with a synaptic ribbon and presynaptic vesicles. Further description of the relation of the receptor cells to other neurons in the retina will be found on page 623.

Retinas generally contain two kinds of photoreceptors, *rods* and *cones.* The rods mediate vision in dim light and are not capable of color vision. The cones, on the other hand, are active in brighter light and in some species subserve color vision. (Some of the evidence for this division of labor will be considered in a later section.) Morphologically, rods are frequently distinguished from cones in having longer, more cylindrical outer segments, whereas cones have shorter, more nearly conical outer segments. These differences, however, are not always evident. For example, in the retinas of primates, the foveal (central) cones are slender and not at all conical. Another difference that is present in fish, amphibians, reptiles, and birds, but not in mammals, is the presence in cones of an oil drop situated between the inner and outer segments (Fig. 14–6). The vertebrate retina is reversed, a quirk of embryology, so that the incident light passes through the inner segments prior to reaching the outer segments (Fig. 14–7). Consequently, the oil drops selectively filter light reaching the visual pigments of cones.

A feature characteristic of the retinas of all classes of vertebrates except mammals is the presence of double (amphibians, reptiles, birds) and twin (teleost fish) cones (Fig. 14–6). These are pairs of cells in close apposition over their inner segments.[48, 246] The two members of a double cone pair are not identical: the principal member tends to be

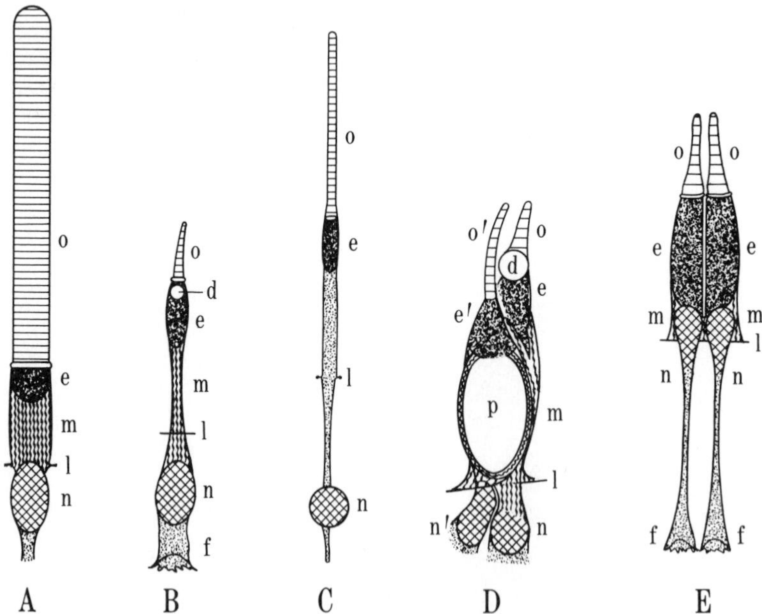

Figure 14–6. Some vertebrate rods and cones. *A,* Common or red rod of the leopard frog, *Rana pipiens*; dark-adapted, with the myoid contracted. *B,* Cone of *R. pipiens*; dark-adapted, with the myoid elongated. *C,* Human rod from near the temporal side of the macula lutea. *D,* Double cone of the painted turtle, *Chrysemys picta. E,* Twin cone of a teleost fish, the bluegill, *Lepomis macrachirus*; light-adapted, with the fused myoids contracted. *d,* oil droplet; *e,* ellipsoid (mitochondrial mass); *e',* ellipsoid of accessory member of pair; *f,* footpiece or pedicel; *ℓ*, external limiting membrane of retina; *m,* myoid; *n,* nucleus; *o,* outer segment; *o',* outer segment of accessory cone; *p,* paraboloid (glycogen). (Redrawn from Walls, G. L., The Vertebrate Eye and its Adaptive Radiation. Bulletin 19, Cranbrook Institute of Science, 1942).

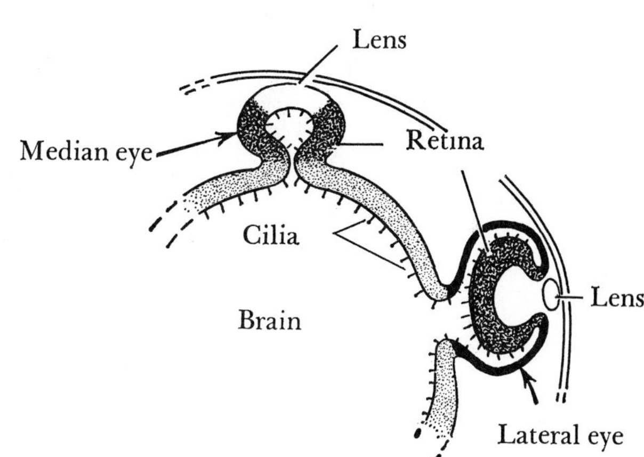

Figure 14–7. Diagram comparing the formation of vertebrate median and lateral eyes from the wall of the developing brain. The receptors (as well as other elements) arise from ciliated ependymal cells. The median eye develops from a simple outpocketing of tissue, and the photoreceptor organelles face toward the lens. In the lateral eye the outpocketing of tissue collapses back into itself, forming a double-walled cup, and the rods and cones differentiate from ciliated cells facing away from the lens. (From Eakin, R. M., Amer. Sci. *58*:73–79, 1970.)

somewhat larger with a well-developed oil droplet, whereas the accessory cone has a relatively larger inner segment but lacks the oil droplet. The two members of a pair of twin cones are morphologically similar.

The parietal and pineal eyes of fish, amphibians, and reptiles have photoreceptor cells with the same fine structure as the lateral eyes.[73] Because of differences in the development of these eyes, however, the retina is not reversed in the pineal eye (Fig. 14–7).

Rhabdomeric Endings

The second major class of photoreceptor cells is characterized by the presence of microvilli not associated with basal bodies or centrioles.[72, 74, 166] Generally one end of the cell bears numerous microvilli, whereas the other end gives rise to a neurite of variable length. In relatively unspecialized photoreceptor cells, the microvilli project from several surfaces of the soma. In the case of squid photoreceptors, the distal end of the cell extends as a long flat finger which bears microvilli on two sides.[271] In arthropods, the photoreceptor cells associate in bundles or fascicles, each cell projecting its band of microvilli towards the center to form an axial rod-like structure known as the *rhabdom.* Each small group of photoreceptor cells is surrounded by pigment-filled supporting cells. Each cluster of receptors (*retinula*) and its sleeve of pigment cells is known as an *ommatidium,* and compound eyes are constructed through the aggregation of hundreds or thousands of ommatidia. Figure 14–8 shows several ways in which rhabdoms are formed by strips of microvilli from surrounding retinular cells.

VISUAL PIGMENTS

The Effects of Light on Visual Pigments

The photoreceptor cells of eyes are sensitive to light because they contain a pigment called rhodopsin. (The term rhodopsin or visual purple also has a more restricted meaning, referring to one subclass of visual pigments found in the rod cells of vertebrates. Which meaning is intended is usually clear from the context.) The significance of rhodopsin in the visual process began to be understood almost a hundred years ago when Boll (1876)[21] and Kühne (1877)[130] showed that the pink color of a dark-adapted retina quickly faded on exposure to light. In the intervening years a great deal has been learned about the properties of rhodopsins from different sources, and all are believed to show a common response to light.[25, 168, 241]

Rhodopsin is a conjugated protein that makes up a significant part of the membranes of the photoreceptor organelle. Because of association with lipids, rhodopsin is only solubilized with the aid of detergents. The molecular weight is 28,000 to 40,000, the value depending in part on how much lipid is removed in preparing the extract.[109, 116] The rhodopsin molecules are 40 to 50 Å in diameter and are probably arranged as single layers in the membranes of the receptor organelles.[268]

The chromophore of rhodopsin is a 20-

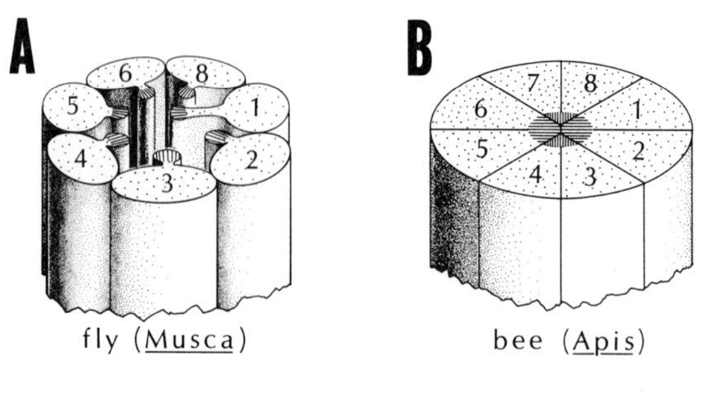

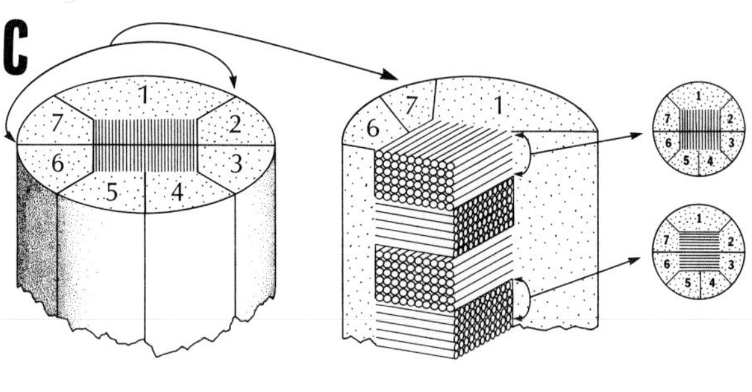

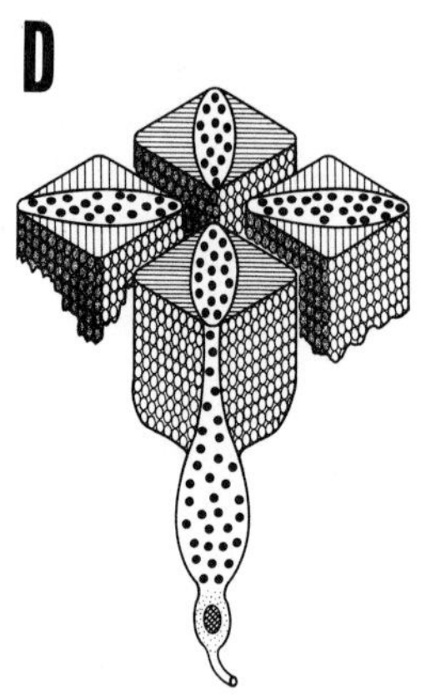

Figure 14–8. Some representative rhabdoms. *A,* A fly (Arthropoda). In Diptera and Hemiptera the rhabdomeres project into a central cavity and remain separate. Such a rhabdom is said to be "open" in contrast to the "closed" rhabdoms shown in *B* and *C.* In flies there are eight retinular cells. Six of these (numbered 2, 3, 4, 5, 6, and 8) have rhabdomeres that run the full length of the retinula. The other two (numbers 1 and 7) have shorter rhabdomeres placed end-to-end. Consequently, a cross section of the ommatidium shows only one of these rhabdomeres. In this diagram the section is from the distal half of the ommatidium and passes through cell *1,* the *superior central cell.* Cell 7, the *inferior central cell,* is restricted to the basal half of the retinula and does not show here. A more proximal section would show the central rhabdomere attached to cell *7;* cell *1* would be present but with apparently no rhabdomere. The inferior and superior central cells have different synaptic connections from the other six retinular cells. *B,* The honeybee (Arthropoda). This fused organization is more typical of insects. Eight retinular cells each contribute a wedge of microvilli along the full length of the cell. The compact mass of microvilli from all eight cells is called a rhabdom. *C,* The crayfish (Arthropoda). In decapod crustacea the microvilli project as tongues rather than continuous strips along the sides of the retinular cells. There are seven cells, and the tongues of microvilli interlock to form layers which can be up to 5 μm thick. *D,* The squid (Mollusca). The retinas of cephalopod molluscs are not organized into ommatidia. Each photoreceptor cell has an elongate "outer segment" that consists of a central strip of cytoplasm and two lateral borders of microvilli.

carbon, fat soluble molecule known as retinaldehyde (Fig. 14–9A), the aldehyde of vitamin A[9] (Fig. 14–9B). Because of its conjugated system of single and double bonds, retinaldehyde can exist in several geometrical isomers. One of these, the 11-*cis* (Fig. 14–9C), is found in rhodopsin.[178,244] The retinaldehyde is attached through Schiff's base linkage (Fig. 14–9C) to the ε-amino group of a lysyl residue of opsin,[2,23] although another school has recently argued that the Schiff's base attachment is to a phosphatidylethanolamine.[185] The protein is thought to fold around the 11-*cis* isomer in a very specific fashion.

The effect of light is to isomerize the chromophore from the 11-*cis* to the all-*trans* configuration.[118] The steric fit with the protein is no longer possible, and isomerization is followed by a sequence of conformational changes in the protein. In the case of vertebrate rhodopsin, these changes lead through a series of spectroscopically identifiable intermediates (Fig. 14–10) to exposure and hydrolysis of the chromophore.[1,146,241] Under the influence of an alcohol dehydrogenase and reduced nicotinamide adenine dinucleotide phosphate (NADPH),[88] the retinaldehyde is reversibly converted to vitamin A[120] (Fig. 14–9A, B) and (at least in the rat) is transported into the pigment epithelium.[62] The reduction is not a universal reaction, however, for in many arthropods[92] and cephalopod molluscs,[119] the reaction sequence stops at a stable intermediate analogous to metarhodopsin I (Fig. 14–10). The actual process of bleaching—that is, loss of color—which is so characteristic a property of vertebrate retinas is therefore not observed in many invertebrate visual systems. The absorption spectra of bullfrog rhodopsin and its final product of bleaching are shown in Figure 14–11A.

If 11-*cis* retinaldehyde is added to a solution of bleached rhodopsin, the chromophore condenses with the protein component (opsin), and the resynthesis of visual pigment is spontaneous[120] (Fig. 14–11B). The problem of recovery in the visual cycle therefore comes down to the mechanism by which 11-*cis* retinaldehyde is regenerated from the all-*trans* retinaldehyde (or vitamin A). The means by which this is accomplished have not been fully elucidated.[117,241]

The intermediates prelumi- and lumirhodopsin can be identified only under very

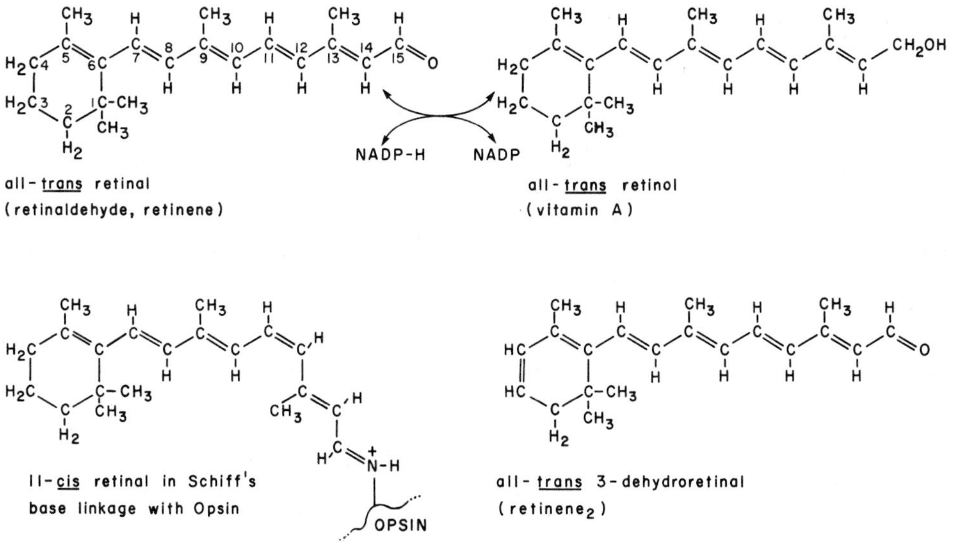

all-trans retinal
(retinaldehyde, retinene)

NADP-H NADP

all-trans retinol
(vitamin A)

11-cis retinal in Schiff's
base linkage with Opsin

OPSIN

all-trans 3-dehydroretinal
(retinene$_2$)

Figure 14–9. Retinaldehyde and related compounds. Upper left: retinaldehyde in the all-*trans* configuration, showing the system for numbering the carbon atoms. Upper right: retinol (vitamin A) which forms in the retina by reduction of the aldehyde group of retinal with reduced NADP through the mediation of an alcohol dehydrogenase. The reaction reverses during the synthesis of rhodopsin. Lower left: retinaldehyde in the 11-*cis* configuration bound to opsin through Schiff's base linkage with the ε-amino group of a lysine. The Schiff's base is shown in the protonated form. Lower right: 3-dehydroretinal in the all-*trans* form. The 11-*cis* isomer is the chromophore of the rod pigment porphyropsin. Note the extra double bond in the ring which lengthens the conjugated chain and shifts the absorption spectrum of both the chromophore and its derivative visual pigments toward the red.

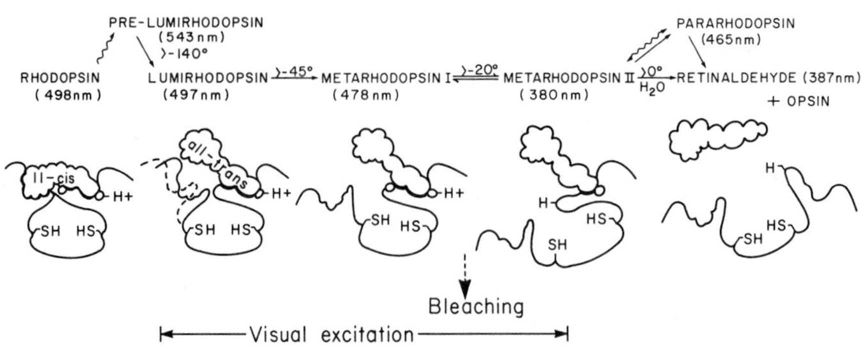

Figure 14–10. The intermediates of bleaching of cattle rhodopsin and their wavelengths of maximum absorption. The chromophore, 11-*cis* retinal, fits into a specific site on the opsin. Light isomerizes the chromophore to the all-*trans* shape, and this is followed by progressive opening of the opsin with exposure of SH groups and one acid-binding group. In vertebrates the retinaldehyde is hydrolyzed from the opsin, but this does not occur in at least some invertebrates. Bleaching, or loss of color, occurs during the formation of metarhodopsin II. The relation of pararhodopsin to the other intermediates is not yet understood. Note that cold temperatures are necessary to stabilize the early intermediates. The process of visual excitation is activated somewhere between the formation of lumi- and metarhodopsin II. (Modified from Hubbard and Kropf[118] and Wald.[241])

cold (and artificial) conditions. At the temperatures of living tissues, the life times of these substances are too short to measure. On the other hand, the later reactions in Figure 14–10 require many seconds or even minutes to occur. Consequently, attention has been focused on metarhodopsin and its formation as possible links in the process of visual excitation.[1,241] What this role might be, however, we shall consider on page 605.

The Wavelengths of Maximum Absorption of Visual Pigments

Visual pigments from different species vary considerably in their wavelengths of maximum absorption.[55] This variation has three important sources: it may depend on (1) the chromophore; (2) interspecific differences in the opsin; or (3) intraretinal differences in the opsin. We shall consider each of these in turn.

Dehydroretinaldehyde and Porphyropsin. Instead of the red rod pigment rhodopsin, the retinas of certain animals such as freshwater fish contain a more purple pigment known as porphyropsin.[27,129,233] The difference between rhodopsin and porphyropsin resides in the chromophore. Porphyropsin is not based on 11-*cis* retinal, but on the closely related 11-*cis* 3-dehydroretinal (Fig. 14–9D). Dehydroretinal has one more double bond in the conjugated chain, and consequently the absorption maximum is shifted from 387 to about 400 nm. When 3-dehydroretinaldehyde condenses with opsin, the pigment that forms also has its absorption maximum displaced towards the red (Fig. 14–12). For example, frog rhodopsin absorbs maximally at 502 nm, whereas the corresponding porphyropsin (made from the same opsin) absorbs maximally at about 522 nm.

Visual pigments based on 3-dehydroretinaldehyde are known only from vertebrates; in fact, no 3-dehydroretinaldehyde or 3-dehydroretinol has been found in any invertebrate.

The distribution and properties of porphyropsins have recently been reviewed by Bridges.[27] Classically, porphyropsins are said to be characteristic of freshwater fish, larval amphibians, and freshwater turtles. Both porphyropsin and rhodopsin are found in migratory fish that move between marine and freshwater environments. Which pigment predominates varies during the life cycle, but the shift from one pigment system to the other actually precedes migration to the environment for which the new pigment is intended. Similarly in amphibians, the retinas of larval frogs and salamanders contain porphyropsin, but this changes over

Figure 14–11. *A,* Absorption spectra of frog rhodopsin and the product of its bleaching in aqueous digitonin solution, pH 5.55. The α-band is mainly responsible for the spectral sensitivity of rod vision and depends on the chromophore. On bleaching it is replaced by the absorption of retinaldehyde, maximal at 385 nm. The γ-band is due to aromatic amino acid residues in the opsin and is not altered on bleaching. (From Wald, G., Docum. Ophthal. *3*:94–134, 1949.) *B,* Synthesis and bleaching of rhodopsin in solution (22.5°C, pH 7.0). *Left*: A mixture of 11-*cis* retinal and cattle opsin was incubated in the dark and absorption spectra recorded at (1) 0.3 min, (2) 2.5, (3) 5, (4) 10, (5) 18, (6) 30, (7) 60, (8) 120, and (9) at 180 min. The absorption band of 11-*cis* retinal (λ_{max} 380 nm) falls, while that of rhodopsin (λ_{max} 498 nm) rises. *Right*: The rhodopsin formed at the *left* was irradiated with wavelengths > 500 nm for various intervals, and the spectrum was recorded immediately after each exposure. The total exposures were (2) 5 sec, (3) 10 sec, (4) 15 sec, (5) 30 sec, (6) 120 sec. The residue was finally exposed for 45 sec longer to light of wavelengths > 440 nm (7). Note that the product of bleaching (all-*trans* retinal) has a higher molar extinction than the 11-*cis* isomer used in the synthesis (left). Note also the isosbestic point at 418 nm during synthesis and its absence during photolysis. (From Wald, G., and P. K. Brown, Nature *177*:174–176, 1956.)

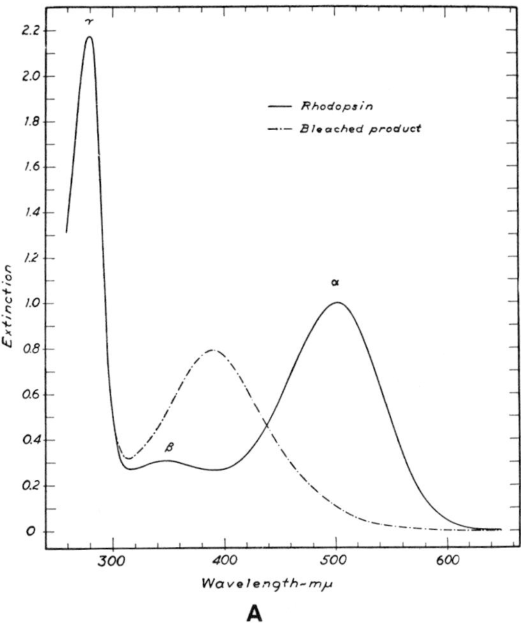

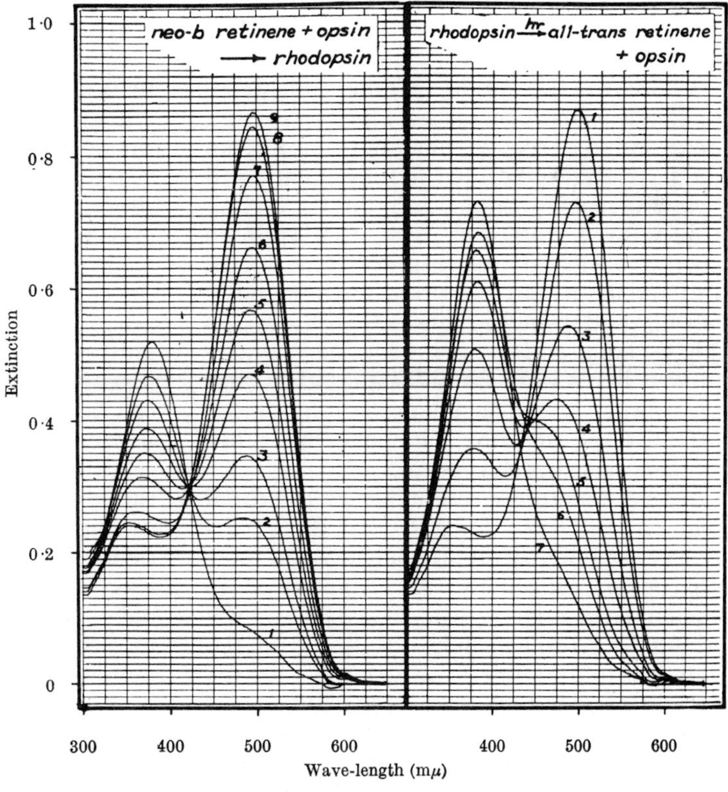

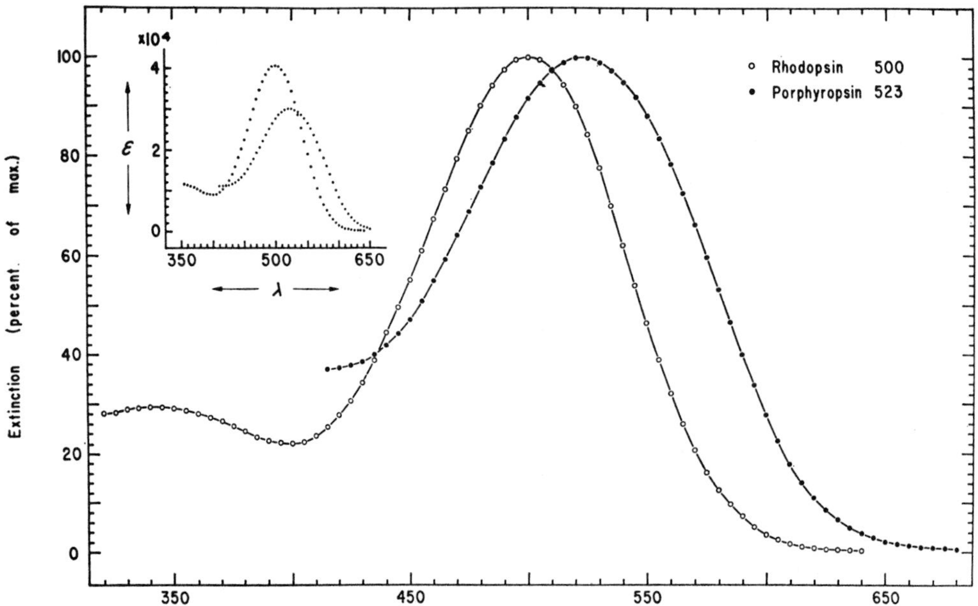

Figure 14–12. Spectroscopic properties of rhodopsin (open circles) and porphyropsin (filled circles) derived from difference spectra for total bleaches. This pigment pair (λ_{max} at 500 and 523 nm) is characteristic of many retinas, where there is good reason to believe that both are based on the same opsin. Note that the porphyropsin has a somewhat wider absorption spectrum. The inset shows the two spectra compared in terms of their relative molar extinctions. (From Bridges, C. D. B., Vision Res. 7:349–369, 1967.)

to rhodopsin during the metamorphic preparation for a more terrestrial life. The spotted newt shows a second metamorphosis from the terrestrial red eft to the fully aquatic and sexually mature adult, and this is accompanied by a shift back from rhodopsin to porphyropsin in the retina. (The toads, Bufonidae, are an exception to these rules, having rhodopsin throughout their life cycle.) These data and others have been interpreted by Wald[236] as indicating that porphyropsin is phylogenetically the more primitive of the two pigments, but other workers have been skeptical.[55]

A large amount of recent data on the visual pigments of fish[27] suggests that the occurrence of 3-dehydroretinaldehyde and porphyropsin represents an adaptation to aquatic environments where the spectral distribution of available light is shifted from the blue-green to longer wavelengths. As the sample of fish that have been studied has enlarged, it has become clear that about half of the freshwater species possess rhodopsin as well as porphyropsin, and that this includes non-migratory fish that spend their entire lives in freshwater. Moreover, a few marine forms (Labridae) have also been reported to have porphyropsin. Secondly, although there is no question that the shifts from porphyropsin to rhodopsin or *vice versa* that are associated with metamorphosis are genetically programmed and hormonally controlled and are not direct responses to the photic environment, there is now increasing evidence that in some fish the rhodopsin: porphyropsin ratio is under more immediate environmental control. For example, in juvenile salmon a shift towards rhodopsin can be achieved by raising the ambient light intensity. In non-migratory fish there are seasonal fluctuations in the proportions of these pigments, with porphyropsin maximal during the winter months; this too is dependent on the total amount of light the fish receives. Experiments on fish with individually capped eyes indicate that control is intraocular; the eyes can be influenced independently of each other and of the titer of hormones in the circulation. Similar effects of light have also been reported on the rhodopsin:porphyropsin ratio in a cyprinid fish.[4]

Microspectrophotometric measurements of single outer segments of trout rods show that rhodopsin and porphyropsin are mixed homogeneously in the outer segments of each rod cell, and that at any time all receptors have the identical mixture.[139] Other workers[172] find different proportions of the two visual pigments in different parts of the retina. More information is clearly needed on this point.

There is presumably an enzyme for interconverting retinaldehyde and 3-dehydroretinaldehyde ("terminal ring dehydrogenase"), but whether this enzyme works at the level of the aldehyde or the alcohol is not known. The mechanisms by which the synthesis and/or activity of this enzyme are controlled by both internal and external factors constitutes an experimental system for the study of differential gene action.[265]

Recently, some adult bullfrogs have been shown to have porphyropsin in the dorsal part of the retina and rhodopsin in the ventral part.[191] Which visual pigment is found in a rod is determined by the underlying pigment epithelium, which supplies the chromophore. The presence of porphyropsin is thought possibly to have adaptive value in detecting dimmer lights from the ventral part of the visual field. The environmental or other conditions that lead to the post-metamorphic retention of vitamin A-3,4 dehydrogenase in the dorsal pigment epithelium, and consequently porphyropsin in the dorsal retina, have not been elucidated.

Interspecific Variations Between Opsins. The second factor that can influence the absorption maximum of a visual pigment is the opsin. Different species possess visual pigments with different absorption maxima because (presumably) small differences in the amino acid composition of the opsin lead to proteins with slightly different abilities to interact with the chromophore. Figure 14–13 shows the spectral distribution of the wavelengths of maximum absorption of the visual pigments of a large number of fish. Clearly, the visual pigments show much interspecific variation, but the distribution of absorption maxima is not continuous. The points of maximum absorption tend to cluster around specific wavelengths, which suggests that the possible variations in amino acid sequence which lead to different functional opsins are restricted.[57] It would be most interesting to know in detail the molecular shapes of these rhodopsins and porphyropsins, as well as the different amino acid substitutions that underlie them.

An ecological generalization can also be drawn from this figure. Those species with visual pigments absorbing towards the blue tend to be found in the deeper oceanic waters, where the wavelengths of maximum transmission of sunlight are about 470 to 480 nm (cf. Fig. 14–2). On the other hand, fish with visual pigments lying to longer wavelengths tend to come from the more turbid coastal waters or from fresh waters, where the maximum available light is in the green or yellow regions of the spectrum.

The porphyropsin visual system can be viewed as an adaptation to the environment that is superimposed on interspecific variations in the opsin and that serves to shift the wavelengths of maximum sensitivity to still longer values. Note that the porphyropsin absorbing the shortest wavelength, 510 nm, is found in a marine form. Seen in this ecological context, the ability to shift seasonally between a rhodopsin and a porphyropsin system gives an animal the capacity to vary the wavelength of maximum sensitivity of its receptors between two values 20 nm or more apart. Moreover, the variation is continuous, depending on the proportion of the two pigments in the receptors. This adaptive significance remains whether the genetic program dictates the shift in pigments as a metamorphic event or enables the shift to occur as a response to changing photoperiod.

Intraretinal Differences Between Opsins: Visual Pigments of Cones. A homogeneous population of receptors, all with the same visual pigment (or mixture of pigments), cannot provide the basis for color vision. In order to distinguish differences in wavelength from differences in brightness, an animal must have two or more kinds of receptor maximally sensitive in different regions of the spectrum. In principle this could be accomplished by having two or more kinds of visual pigment segregated into different receptor cells. Alternatively, a single visual pigment could suffice if the photoreceptors were individually fitted with colored filters like the oil droplets of cones.

In 1807, Thomas Young suggested that "from three simple sensations, with their combinations, we obtain several primitive distinctions of colours; but the different proportions, in which they may be combined, afford a variety of tints beyond all calculation." The number three proved well chosen, for a century of psychophysical color matching experiments showed this to be the minimum number of spectral sensitivity functions required to account for human color vision. But

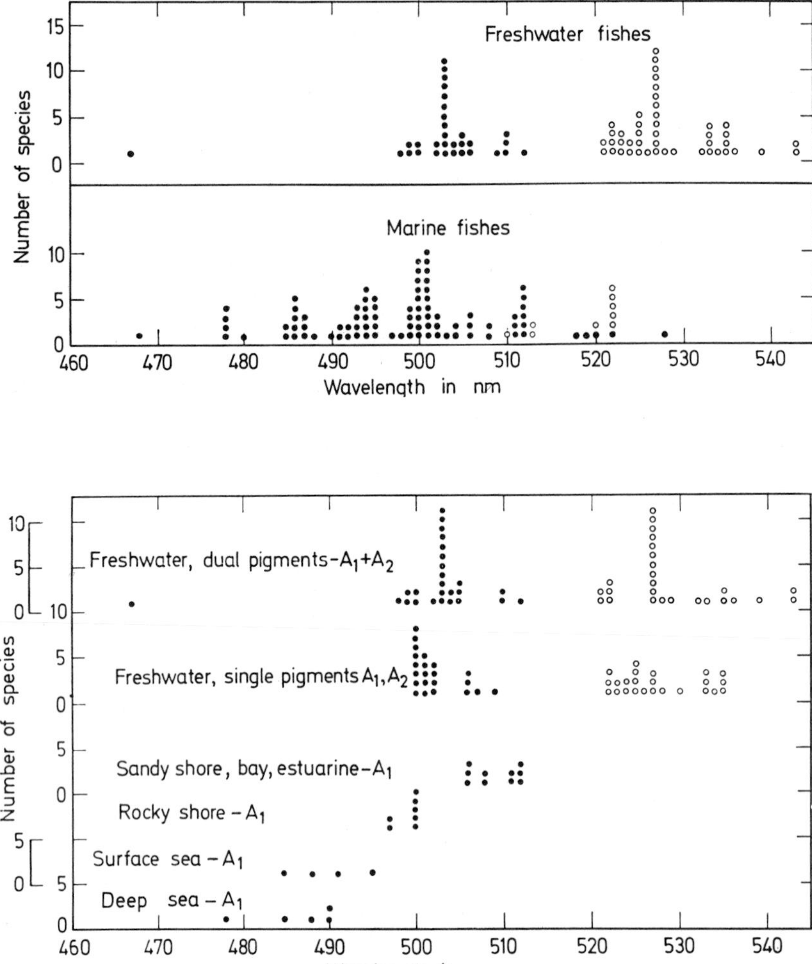

Figure 14–13. Distribution of visual pigments of teleost fishes. Each point represents the wavelength of maximal absorption of a pigment, obtained using partial bleaching analyses of extracts. The lower frame shows the data broken down by habitat. Data are from a number of sources. (From Crescitelli, F., *in* Handbook of Sensory Physiology, Vol. 7, edited by H. J. A. Dartnall. Springer-Verlag, Berlin, 1972.)

it has been only in the last decade that Young's three receptors have been demonstrated in the form of three visual pigments contained in three different cone cells (Table 14–2). The differences between these pigments seemingly reside in the opsin, rather than the chromophore.

The data in Table 14–2 are not in precise agreement. This is because they represent a relatively small number of measurements made in the face of formidable technical obstacles, some of which should be mentioned. The microspectrophotometric measurements are difficult because of the small size of the outer segments and the high photosensitivity of the visual pigments, which precludes raising the intensity of the measuring beam to improve the signal-to-noise ratio. Because of the ~ 1 μm diameter of the outer segment, spectra recorded "end on" are likely to be distorted by light that has scattered laterally into neighboring receptors.[5, 139, 192] Laterally incident beams focused on isolated outer segments could be free of this distortion, but the effective path length is very short. Use of larger fields with many receptors makes the measurement of absorption easier, but it is more difficult to analyze spectra that are a composite of several pigments, and it is impossible to tell whether the pigments are mixed in the same cone. The technique of fundal reflectometry

TABLE 14–2. The Pigments of Primate Color Vision

Blue-sensitive cones (λ_{max} in nm)	*Green-sensitive cones (λ_{max} in nm)*	*Yellow-sensitive cones (λ_{max} in nm)*	*Technique of measurement*	*Reference*
445, *455**	535	570, *570*	end-on microspectrophotometry	144
447 (primate)	540 (primate)	577 (primate)	end-on microspectrophotometry	142
450	*525*	*555*	end-on microspectrophotometry	33
440 (primate)	–	–	end-on microspectrophotometry	139
–	535 (primate)	575–580 (primate)	lateral microspectrophotometry	139
440 (?)	527, *535*	565, *565*	transmission spectrophotometry of retinal patches	32
430–440	*530–540*	*565–595*	fundal reflectometry	192, 199, 200, 257
440	*550*	*580–585*	psychophysical measurements; corrected for ocular and macular absorption	239
430	*540*	*575*		

*Figures in *italics* are from humans; others are from monkeys. Those designated "primate" are lumped data or are of unspecified origin.

has its own set of problems,[192, 200] and the definition of the "red receptor" by this means has been particularly uncertain. Table 14–2 includes data from monkey as well as human retinas, as the interspecific differences, if any, seem less than the uncertainties due to technical problems.

The last horizontal row of figures in Table 14–2 represents psychophysical measurements of spectral sensitivity made while the subject was adapted to colored light. This technique is designed to suppress selectively the output of two of the three cones, leaving the third relatively free to measure. These results indicate that the absorption spectra recorded by optical means relate to the spectral sensitivity functions of three physiological processes (but see Rushton[201a] for a discussion of the difficulties concealed in this interpretation).

Figure 14–14 summarizes the microspectrophotometric studies of Liebman[139] on retinal receptors and shows most that is known about intraretinal diversity of visual pigments in other vertebrates.

The goldfish visual system[143] is very much like that of man or monkey in that there are one rod pigment and three cone pigments. The difference is that the chromophore is dehydroretinal in the goldfish, and consequently the pigments absorb at longer wavelengths.

A digression into terminology is in order here. Wald[241] refers to four major pigments of vertebrate vision. (See below.) Most other investigators find this taxonomy unsatisfactory to varying degrees. First, it does not accommodate the pigments of invertebrates. Moreover, it does not allow for more than one kind of rod or cone pigment for each chromophore. For example, the 620 nm cone pigment of the goldfish retina is a classical cyanopsin, but what is the 535 nm pigment in the same retina? This inflexibility has led to the invention of other sets of terms, e.g., erythrolabe, chlorolabe, and cyanolabe for the human cone pigments. Note that cyanolabe, which breaks down in blue light, absorbs at the opposite end of the spectrum from cyanopsin, which is blue in color. Such proliferation of names adds little but confusion. A shorthand notation is frequently convenient, and one in common use[56, 173] is the prefix P (for pigment) followed by the λ_{max} (in nm) and a subscript 1 or 2 to designate whether the chromophore is retinal or dehydroretinal. Thus, $P620_2$ is the aforementioned pigment of the goldfish retina. One additional word indicates whether it is present in rods or cones. A slight modification of this system has been used in Figure 14–14.

Returning again to Figure 14–14, birds have but a single cone pigment and a single

---

retinal + rod opsin $\longrightarrow$ rhodopsin (λ_{max} 500 nm)
retinal + cone opsin $\longrightarrow$ iodopsin (λ_{max} 562 nm)
3-dehydroretinal + rod opsin $\longrightarrow$ porphyropsin (λ_{max} 522 nm)
3-dehydroretinal + cone opsin $\longrightarrow$ cyanopsin (λ_{max} 620 nm)

Species	Pigments (wavelength, nm: 400–600)
FROG	R432_1 R,C502_1 C575_1
SALAMANDER	R432_1 R,C502_1 C575_1
GECKO	R467_1 R518_1
TURTLE	C442_1 R,C502_1 C562_1
PIGEON	R500_1 C562_1
CHICKEN	R500_1 C562_1
GULL	R508_1 C562_1
MONKEY	C440_1 R498_1 C535_1 C575_1
MAN	C440_1 R498_1 C535_1 C575_1
GOLDFISH	C455_2 R522_2 C535_2 C620_2
MUDPUPPY	R527_2 C575_2
TADPOLE	R438_2 R,C527_2 C620_2
TURTLE	C450_2 R,C518_2 C620_2

WAVELENGTH (nm)

Figure 14–14. Summary of microspectrophotometric measurements on single outer segments showing intraretinal diversity of visual pigments. The numbers represent the wavelengths of maximum absorption of the pigments, the prefix (R, C) indicates whether present in rods or cones, and the subscript (1, 2) whether the chromophore was retinal or dehydroretinal. (After Liebman.[139])

rod pigment. The presence of red, orange, and yellow carotenoid-containing oil droplets over each outer segment, however, creates two or three kinds of receptors with λ_{max} (in the pigeon, calculated) at 625, 585, and 562 nm respectively.[139] Future work will show whether this is the sole basis for avian color vision[127a] or whether larger samples will lead to the discovery of additional cone pigments.

The adult frog has two rod pigments and two cone pigments. The most abundant pigment is the rod P502_1. The rod P432_1 is located in the so-called green rods. Single cones and the principal members of double cones contain P575_1; the accessory partners of double cones contain P502_1, which is spectroscopically indistinguishable from the predominant pigment of the rods. A corresponding dehydroretinal system is present in tadpoles.

Turtles (*Pseudemys*), too, have a "rod pigment" in their accessory cones as well as two additional cone pigments absorbing at longer and shorter wavelengths. In addition, there are four classes of oil droplets so distributed as to make (probably) five spectral classes of cone.[139]

There is ample reason to expect several visual pigments in the eyes of many invertebrates, but there are relatively fewer absorption spectra available. Spectral sensitivity measurements of receptor cells are more numerous. The interested reader can find the subject reviewed elsewhere.[92]

THE RELATION OF VISUAL PIGMENT MOLECULES TO THE RECEPTOR MEMBRANES

In the outer segments of rods and cones, the chromophores lie roughly parallel to the planes of the membrane discs, at right angles to both the long axes of the rods and the direction of propagation of the incident light (Fig. 14–15A). Because the conjugated system of single and double bonds is placed at right angles to the advancing ray, the probability of absorption should thereby be increased to 1.5 times the value observed with

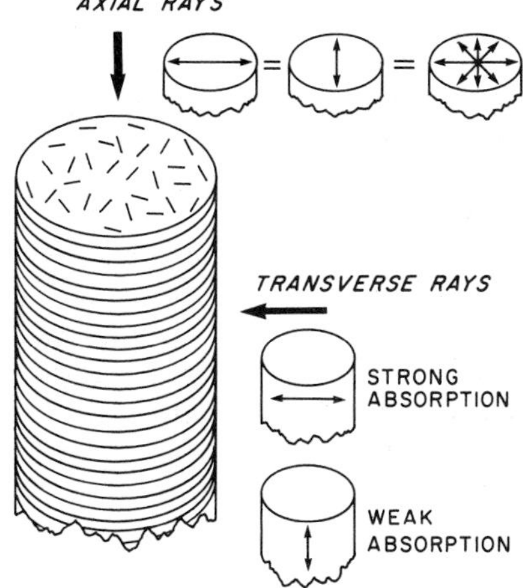

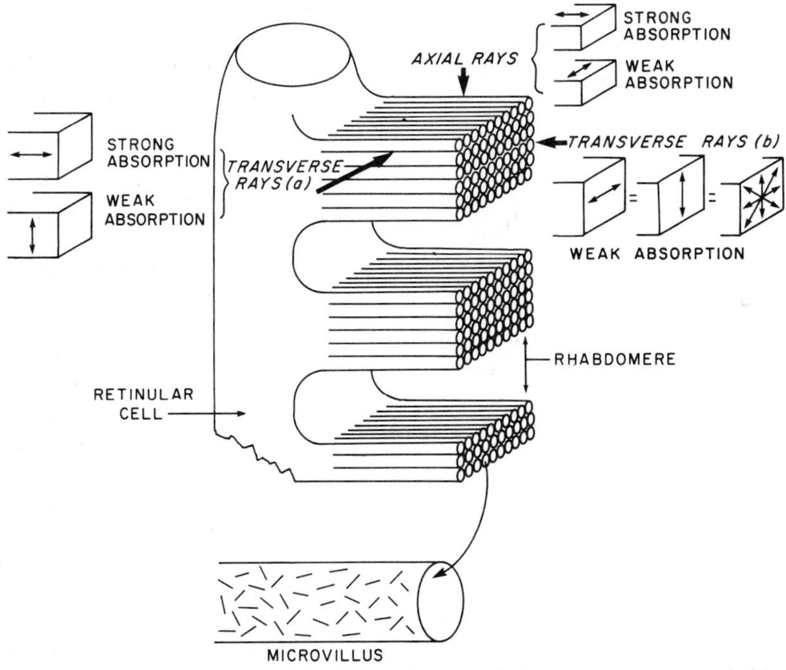

Figure 14–15. Orientation of visual pigment in the membranes of photoreceptors as revealed by dichroic absorption. *A*, In vertebrate rod and cone outer segments the chromophores lie randomly oriented in the planes of the disc membranes. Consequently, absorption of axially incident rays is independent of the plane of polarization. When illuminated from the side, however, the outer segments are dichroic: light polarized parallel to the planes of the discs (and thus the axes of many of the chromophores) is strongly absorbed, whereas light polarized at right angles to the discs is only weakly absorbed. *B*, The measurements of dichroism of decapod crustacean rhabdomeres are consistent with the view that the chromophores are randomly oriented in the planes of the photoreceptor membranes (i.e., in the tangent planes of the microvilli). Shown here is a single retinular cell and several of its tongues of microvilli. In an isolated rhabdom these tongues are interleaved with the rhabdomeres of other receptors (cf. Fig. 14–8C), but it is possible to irradiate a single band of microvilli with a laterally incident microbeam. Absorption is isotropic when the beam is incident parallel to the microvillar axes, but it is dichroic when the light is incident at right angles to the microvillar axes. This latter condition is equivalent to axial illumination, as occurs in the living eye. Absorption is strongest when the light is polarized parallel to the microvillar axes. The measured dichroic ratio of 2 is consistent with random orientation as defined above; higher dichroic ratios might be expected if there were any orientation of the chromophores parallel to the microvillar axes.[167, 253]

the same number of molecules randomly oriented in solution.* Actually, due to an average departure of several degrees from coplanarity with the discs, the measured enhancement of absorption is about 1.37.[243]

This molecular orientation is detected as a dichroic absorption of rods illuminated from the side (Fig. 14–15A).[61, 138, 243] When laterally incident light is polarized perpendicular to the rod axis (parallel to the planes of the discs), absorption is high. When the plane of polarization is rotated 90° and made parallel to the rod axis, the absorption is about 22 per cent as great, corresponding to a dichroic ratio of about 4.5. From such measurements as these, the degree of orientation of the chromophores can be estimated.

The dichroism of crustacean rhabdoms has also been measured.[107, 253] Rhabdoms detached from retinular cells were viewed from the side, and a measuring beam was placed within a single band of microvilli (cf. Figs. 14–8C and 14–15B). Rhabdoms were selected in which the microvilli of alternate layers were parallel and perpendicular to the measuring beam. When the beam was incident along the microvillar axes, absorption was independent of the plane of polarization, as one would expect from the cylindrical symmetry of the microvilli. When the beam was incident perpendicular to the microvillar axes, absorption was dichroic. The dichroic ratio was about 2, with major absorption observed when the electric vector was polarized parallel to the microvillar axes. This result is to be expected if the chromophores lie in the surfaces of the microvillar membranes, as they seemingly do in vertebrates. A higher dichroic ratio is to be expected only if there is some preferential alignment of the chromophores along the axes of the microvilli, as may be the case in squid.[97, 167]

Because of the interleaving of the bands of microvilli in the rhabdoms of decapod crustaceans (Fig. 14–8C), measurements made in single dichroic bands should provide information on the absorption properties of single rhabdomeres in axially incident light (Fig. 14–15B). Thus, the measurements of dichroism of crustacean rhabdoms are important in explaining the polarized light sensitivity of arthropods, as will be discussed on page 612.

*An extended chromophoric system (cf. Figure 14–9) absorbs most strongly when the light is polarized parallel to the axis of the conjugated chain. In solution, molecules assume all orientations. If the chromophores can be considered linear, the absorption vectors of such a random array can be resolved into three mutually perpendicular components: two at right angles and one parallel to the path of propagation of the incident beam. Absorption therefore does not vary with the plane of polarization of the incident light, and regardless of the direction of incidence, one of the absorption vectors will be oriented parallel to the axis of propagation of the light, and thus will contribute little or nothing to the absorption. When molecules are oriented, as for example in a crystal, absorption by the object is critically dependent on the plane of polarization of the incident light. This property is known as *dichroism,* and maximum absorption occurs when the plane of the E-vector is parallel to the axes of the chromophores. In the discs of rods, the molecules are partly oriented. That is, they are apparently free to rotate in the planes of the discs, but the chromophores are largely restricted to these planes. This leads to the dichroic patterns described in the text and in Figure 14–15.

THE MEMBRANE RESPONSES OF PHOTORECEPTOR CELLS TO EXCITATION

Vertebrate Rods and Cones Hyperpolarize on Illumination

As described on page 583, the photoreceptor cells of the vertebrate retina are ciliary derivatives, and in this respect they differ from most of the photoreceptor cells of invertebrates. Only recently has it been possible to record from rods and cones with intracellular microelectrodes.[226] The receptors are generally quite small, and in most species are penetrated only with difficulty. At this writing, recordings have been made from the cones of a fish (carp),[225] the cones and perhaps also the rods of the amphibian *Necturus*,[22, 227, 258] the rods of a frog,[226] rods (transmuted cones?) of a lizard (*Gekko*),[227] and the cones of a turtle (*Pseudemys*).[16] In every case unambiguous identification of the type of cell impaled was based on intracellular dye injection from the recording electrode. With *Gekko* and frog, the outer segments were impaled; in the other instances the electrode entered the inner segments.

Rod and cone cells have negative resting potentials. When the cell is illuminated, it hyperpolarizes (i.e., the inside of the cell becomes still more negative with respect to the outside). The responses are graded with intensity, and no spikes are present (Fig. 14–16). This hyperpolarization of the cell membrane is interesting, because unlike the responses of other excitable cells it is based on a

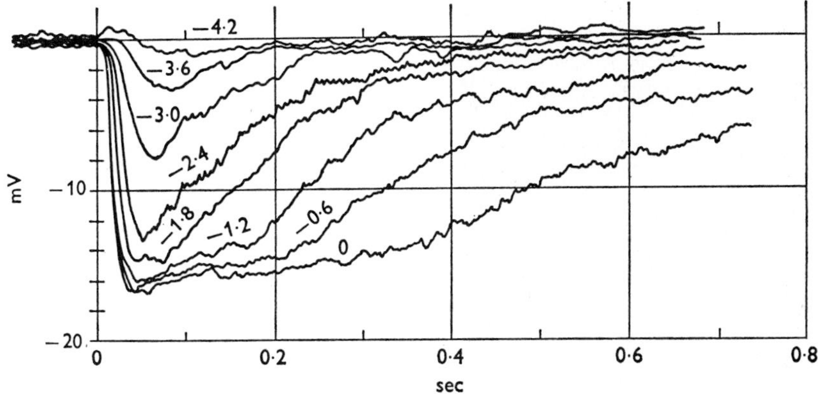

Figure 14–16. Graded, hyperpolarizing responses of a turtle cone to 10 msec flashes of increasing intensity, as indicated in relative logarithmic units for each trace. Brightest flash delivered 8.5×10^6 photons to the cone. Downward deflection indicates increasing negativity of the intracellular microelectrode. (From Baylor, D. A., and M. G. F. Fuortes, J. Physiol. *207*:77–92, 1970.)

decrease in the conductance of the membrane (i.e., on an *increase* in resistance). Figure 14–17 shows an experiment demonstrating the conductance change during illumination. The membrane responses to brief pulses of constant current injected into the cell increase from the dark value when the cell is illuminated. This is because the resistance of the membrane increases several MΩ during illumination. Control experiments (not shown in Figure 14–17) show that the increase in membrane resistance is related to the presence of light *per se* and not just hyperpolarization. That is, hyperpolarizing the cell with a steady current does not result in the increase of membrane resistance produced by light.

Sodium ions are necessary for the appearance of the retinal action potential of the whole retina,[103] and by using aspartate to isolate the receptor potential from other components of the mass response (recorded extracellularly), it can be shown that sodium ions are necessary for the response of the receptors[210, 211] (Fig. 14–18). By contrast, increasing the external potassium concentra-

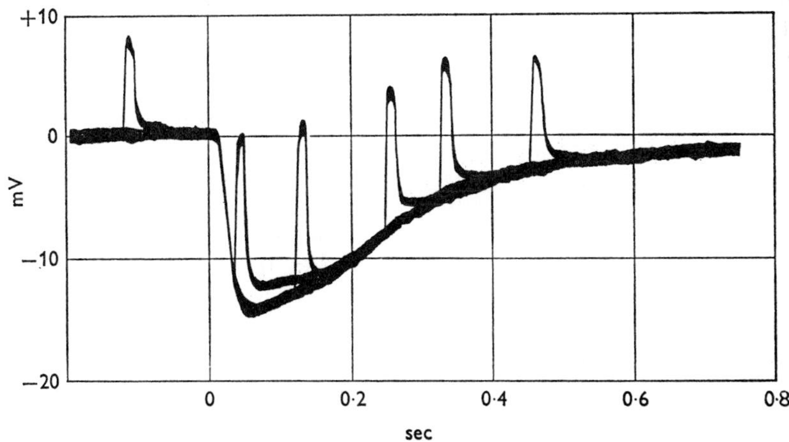

Figure 14–17. Voltage drop produced by current across a vertebrate photoreceptor membrane during response to light. Superimposed tracings of six responses to 10 msec flashes given at time 0. In each trace, a brief depolarizing current pulse of 6.1×10^{-10} A was delivered through the micro-electrode, producing the upward deflections. The size of the voltage drop produced by the pulse shows an increase during the hyperpolarizing response. The maximal change in voltage drop, occurrring at the peak of the response, is about 5 mV, corresponding to an increase in the membrane resistance of 8 MΩ. Note: the deflection produced by the current includes effects due to electrode changes, because with currents of this strength sizeable artifacts were seen with the electrode outside a cell. The *change* in voltage drop following the flash is reliable, however, and reflects a change in the cell's input resistance, since illumination of an electrode which is not within a photoreceptor produces no such effect. (From Baylor, D. A., and M. G. F. Fuortes, J. Physiol. *207*:77–92, 1970.)

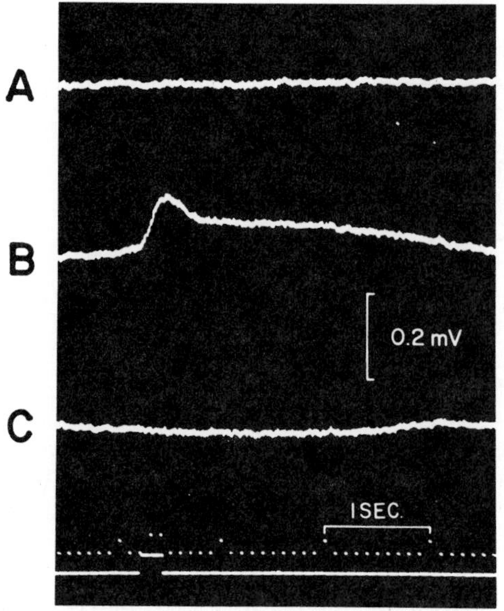

Figure 14–18. The effect of ouabain on the vertebrate photoreceptor potential. *A* shows no response after immersion of the retina for seven minutes in sodium-free solution containing 0.1 mM ouabain. *B* shows the response of the same retina after application of three drops of a solution containing the normal concentration of sodium, but also containing 0.1 mM ouabain. This experiment shows that the receptor potential requires external sodium ions but not the activity of the sodium pump. *C* shows no response after immersion of a fresh retina for seven minutes in the same sodium-ouabain solution. The response is lost because inhibition of the sodium pump in the presence of external sodium leads, after a few minutes, to an increase of internal sodium concentration and dissipation of the transmembrane sodium gradient necessary for a normal receptor potential. Extracellular recording, with other retinal potentials inhibited with aspartate. (From Sillman, A. J., et al., Vision Res. *9*:1443–1451, 1969.)

tion decreases the receptor potential. Light does not modulate the activity of an electrogenic sodium pump but directly affects the conductance of the membrane to sodium. This is strongly implied by the effects of light on membrane resistance described above, and is also shown by the fact that receptor potentials can be elicited in retinas in which the cation pump has been inhibited with the cardiac glycoside ouabain, as long as the transmembrane sodium gradient has not been allowed to dissipate[211] (Fig. 14–18).

An equivalent circuit for the membrane of the outer segment of a vertebrate photoreceptor is shown in Figure 14–19. The membrane potential is determined by the sodium and potassium gradients and the relative permeabilities of the membrane to these two ions. These factors are represented by sodium and potassium batteries in series with resistances. The sodium resistance is controlled by light, increasing during illumination and driving the membrane potential closer to the potassium equilibrium potential (E_K). This model will be compared with invertebrate photoreceptors when the latter have been considered below.

The work with intracellular electrodes meshes very well with studies of extracellular voltage fields in slices of intact retinas. From such measurements it is possible to calculate both the longitudinal extracellular currents between receptors and the membrane currents into and out of the cytoplasm.[99, 183] In the dark there is an outward current from the cytoplasm of the inner segments and an inwardly directed current through the membranes of the outer segments. On illumination, there is a photocurrent in the opposite sense, which can be thought of as a net decrease in dark current (Fig. 14–20). The photocurrent is linear with intensity, and in

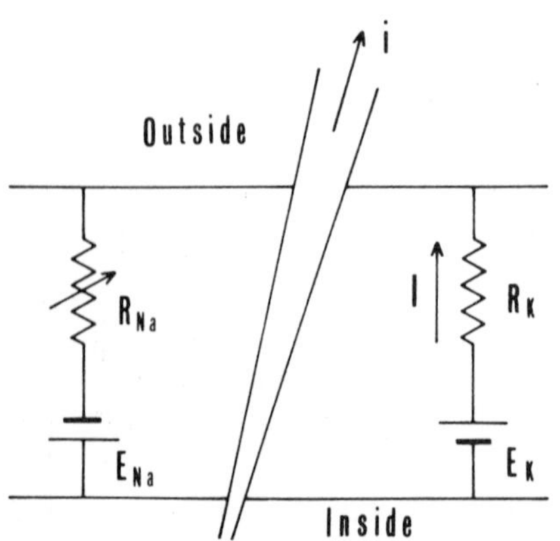

Figure 14–19. Model to account for the membrane responses of photoreceptor cells. See the text for discussion. (From Sillman, A. J., et al., Vision Res. *9*:1443–1451, 1969.)

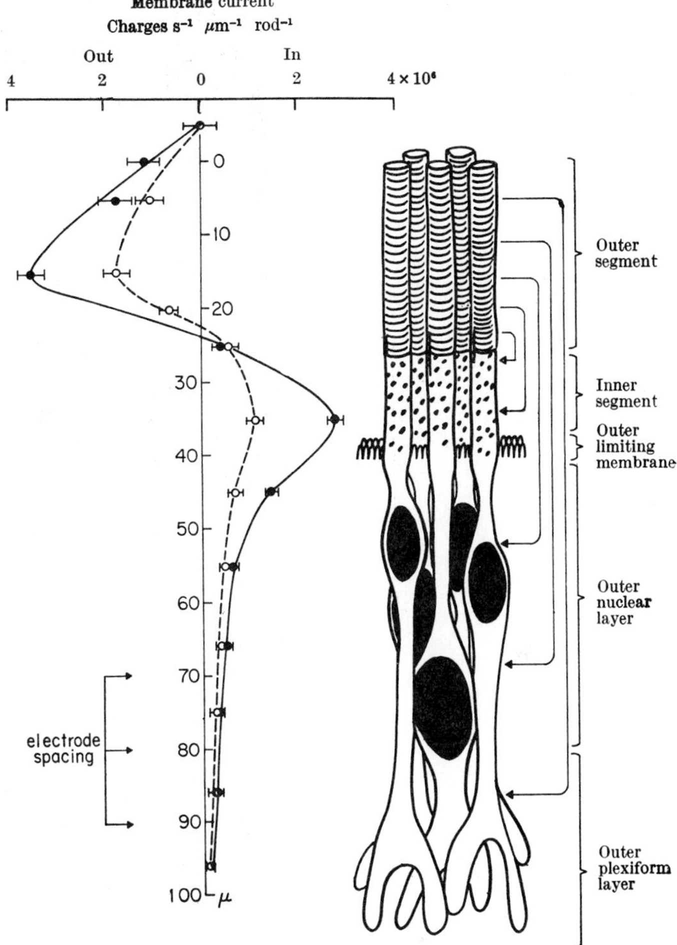

Figure 14–20. The photocurrent of vertebrate receptors. *Left:* Peak transmembrane photocurrent with depth of the center electrode of the recording triplet, calculated from extracellular measurements of longitudinal voltage fields and resistivity between the cells. Results shown are for two slices of retina stimulated with 1×10^{11} $h\nu$ cm^{-2} (open circles) and 4×10^{11} $h\nu$ cm^{-2} (filled circles) of green light at 560 nm. Error bars indicate rms noise in current tracings. *Right:* Retinal rods of the rat drawn to the same scale. Arrows indicate the flow of photocurrent. (From Penn, R. D., and W. A. Hagins, Nature *223*:201–205, 1969.)

the more recent measurements amounted to 10^6 electronic charges (ions) per photon.

Invertebrate Rhabdomeric Receptors Generally Depolarize on Illumination

A wide variety of photoreceptor cells from arthropods, molluscs, and annelids have been explored. Identification of the cell impaled has sometimes been reasonably certain without applying special techniques, but in several cases identification has been aided by intracellular dye injection.[17, 135, 169] Rhabdomeric photoreceptors respond with a depolarization which is graded with intensity and frequently shows a dynamic or transient phase followed by a static or plateau phase at higher intensities (Fig. 14–21). In the example shown, which is from the retinular cell of a fly eye, there are no spikes. In some arthropods there may be a single spike on the rising phase of the transient;[85] in still other invertebrate species there may be trains of spikes, probably generated in the axon of the cell.[152] Whether or not there are spikes depends in part on the distance from the primary photoreceptor cell to the second order unit to which it synapses.

These depolarizing receptor potentials reflect an *increase* in membrane conductance. This has been shown by the use of current pulses in a manner similar to that described for vertebrate receptors,[86] or more elegantly by the technique of voltage clamping.[30, 158]

Extracellular measurements along the slender "outer segments" of squid photoreceptors made on slices of retina cut parallel to the long axes of the receptor cells provide complementary evidence.[101] Figure 14–22 shows the voltage difference generated between two fixed microelectrodes (A, B) when a narrow slit of light is moved along the

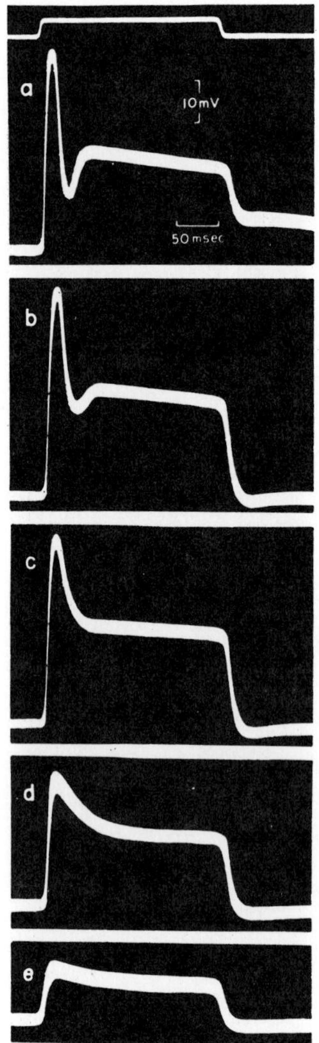

Figure 14–21. Intracellularly recorded receptor potentials from retinular cells of the blowfly *Calliphora.* Depolarizing potentials are indicated by an upward deflection of the trace. Duration of the stimulus is shown by the response of a photocell at the top of the first frame. Relative intensity decreased from *a* to *e*: 100, 25, 6, 1.5, 0.4. Ten sweeps are superimposed in each frame. (From Washizu, Y., Comp. Biochem. Physiol. *12*:369–387, 1964.)

length of the receptor. This experiment demonstrates that the region of the rhabdom directly under the spot of light becomes negative to other parts of the rhabdom, which is to be expected if there is a localized current sink under the point irradiated.

There is some evidence that the site of the conductance change is the rhabdom itself. The most direct argument for this conclusion is based on photoreceptor cells of the leech, in which the microvilli of the photoreceptive membrane project into a central "vacuole" within the cell[135] (Fig. 14–23). The vacuole is really extracellular space which is continuous through narrow channels with the fluid bathing the outside surface of the cell. A microelectrode inserted into the cytoplasm shows a positive-going (depolarizing) response when the cell is illuminated. On the other hand, a microelectrode in the vacuole (as established by dye injection) shows an increased negativity when the cell is illuminated. This is the expected result if the microvillar membranes constitute a site of in-

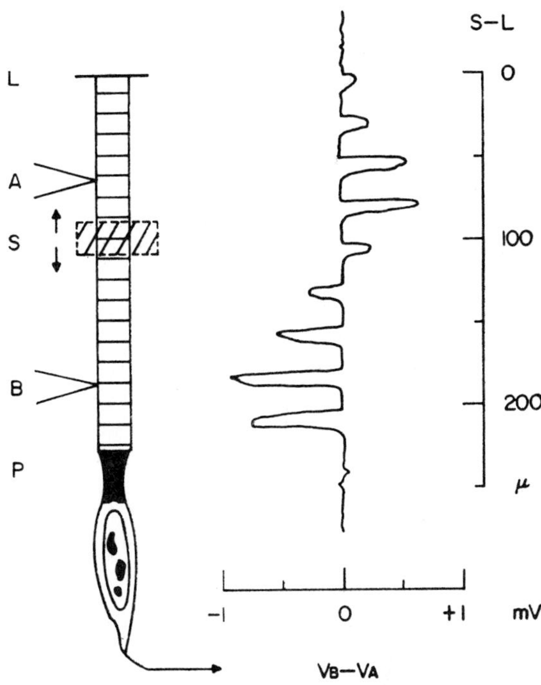

Figure 14–22. Extracellularly recorded potentials in a slice of squid retina at various depths through the layer of rhabdomeres. Ordinate, distance of the slit-shaped stimulus (*S*) from the internal limiting membrane (*L*). *A* and *B*, positions of a pair of fine electrodes. Abscissa, potential difference between the electrode tips. *P*, pigmented region of the cell. Stimulus intensity equivalent to 3×10^{11} photons cm^{-2} sec^{-1} at 500 nm. Note that local illumination produces a local current sink in the same region. (From Hagins, W. A., et al., Nature *194*:844–847, 1962.)

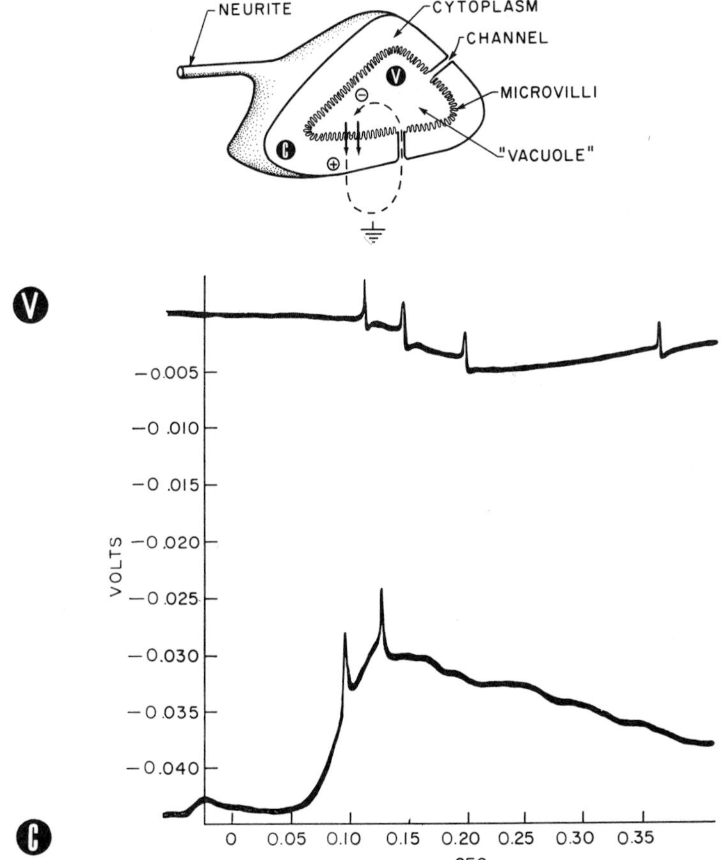

Figure 14–23. Experiment showing that the site of inward photocurrent in rhabdomeric endings is the microvillar membranes. In the leech (*Hirudo*) the microvilli line a "vacuole" which is in fact continuous through narrow channels with extracellular space. A microelectrode can be placed in either the vacuole or the cytoplasm, leading to records like those shown. In each case potential is plotted relative to a reference electrode (ground) in the bath. With an intracellular (cytoplasmic) recording site, the receptor potential is positive (depolarizing). Small, positive-going spikes also appear. They are attenuated because they arise in the neurite, at some distance from the soma. With the electrode in the vacuole, the receptor potential is negative. This is consistent with a current sink at the microvilli (solid arrows) and return current paths (broken arrows) as shown. It is left as an exercise for the reader to see that this result is not expected if the sole site of inward current is in the non-microvillar part of the soma membrane. Spikes recorded in the vacuole are positive, further evidence that their site of initiation is elsewhere. (Voltage recordings from Lasansky and Fuortes.[135])

ward current during illumination and is not anticipated if the sole site of inward photocurrent is the non-microvillar portion of the cell body.

The ionic basis for the receptor potential is not completely clear. Sodium seems to be the most important ion, with decreasing external sodium leading to a diminished receptor potential.[85, 218] In some cases calcium may also be important, but whether it functions as a charge carrier or as an agent controlling sodium conductance is not clear. Receptor potentials may persist after many hours of soaking the preparation in sodium-free solutions, a finding that has not yet been explained. In some cases the time sequence of conductance changes during the receptor potential indicates that more than one species of ion may participate in the response. Reversal potentials may lie near 0 mV,[85, 153] suggesting a general short-circuiting of the resting potential, or they may be distinctly positive and more selectively dependent on external sodium ions.[30]

Close to threshold, rhabdomeric photoreceptors show small transient depolarizations usually less than a millivolt in amplitude.[87, 203] These may reflect the opening of ionic channels in response to absorptions of single photons. As the intensity of light is increased, the frequency of occurrence of these discrete potentials increases, and with still brighter lights the responses fuse. The smooth, graded depolarization of the membrane that is observed with moderately high intensities thus results from the addition of a large number of smaller, unitary events. Tens or hundreds of thousands of ions flow through the membrane for each discrete potential.[95]

To a first approximation, the same equivalent circuit that was invoked to model the membrane responses of vertebrate rods and cones can be employed to describe invertebrate rhabdomeric receptors. The important difference is that in invertebrate systems light *decreases* the sodium resistance.

Figure 14–24 shows another way of comparing the responses of invertebrate and vertebrate photoreceptors. In invertebrate cells the photoresponse becomes larger if the membrane is hyperpolarized by extrinsic current, becomes smaller if the membrane is depolarized, and is reversed on the positive side of the equilibrium potential. This is a situation familiar to anyone who has studied the ionic basis of excitatory synapses. The effects of extrinsic current on the photoresponse of vertebrate receptors are shown in the second column of Figure 14–24. The responses are of opposite sign from invertebrate receptors. Hyperpolarization causes a larger response, depolarization causes a smaller response, and on the other side of the equilibrium potential the response reverses. To understand the reversal at the "equilibrium potential," remember that to obtain the records marked "depolarized" the microelectrode must be passing a current outward across the membrane. As the effect of light is to increase the membrane resistance, during the period of illumination this extrinsic cur-

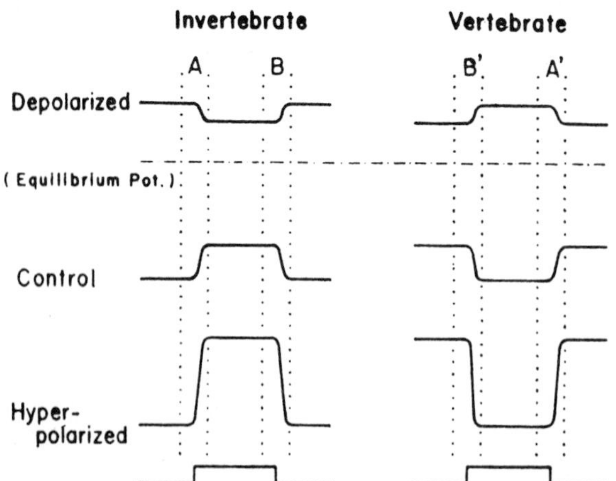

Figure 14–24. Schematic diagram comparing the membrane responses of invertebrate and vertebrate photoreceptors. Hyperpolarization and depolarization are produced by passing currents between an internal microelectrode and the external solution as shown in Figure 14–19. See the text for further description. (From Toyoda, J., et al., Vision Res. *9*:453–463, 1969.)

rent will make the inside of the cell even more positive.

Some Invertebrate Photoreceptors Give Hyperpolarizing Responses

Among the exceptions to the general rule that invertebrate photoreceptor organelles are not derived from ciliary processes are certain photoreceptor cells in molluscs. The pallial eyes of the scallop *Pecten* are the best studied example. These "simple" eyes, which lie in a row along the edges of the mantle, are remarkable in possessing a double retina consisting of two distinct layers of photoreceptor cells (Fig. 14–3C). Each layer has its own branch of the optic nerve, and peripheral synaptic interactions are believed to be absent.[152, 153]

The proximal retina holds no surprises. The photoreceptor cells have microvillar projections typical of invertebrates, and the membrane responds to light with a conventional, depolarizing receptor potential.

The photoreceptor cells of the distal retina, on the other hand, have a series of lamellar membranes arising in conjunction with ciliary microtubules. Both types of photoreceptor cells contain a visual pigment with an absorption maximum near 500 nm, but the distal cells are 2 to 3 log units less sensitive. More important, the distal cells respond to illumination with a hyperpolarizing response. This hyperpolarization is different in an important way from the responses of the ciliary photoreceptors of the vertebrate retina: it is based on an *increase* in membrane conductance rather than a decrease. Thus, if photons are likened to molecules of a synaptic transmitter, the hyperpolarization bears a formal resemblance to an inhibitory post-synaptic potential in the central nervous system. Although in the vertebrate retina the response of the receptors is hyperpolarizing, it is much less clear whether it is excitatory or inhibitory in its effects on the discharge of synaptic vesicles. Hyperpolarization of the distal photoreceptor cells of *Pecten*, however, is seemingly a case of primary inhibition. The axons of these cells discharge spikes, and the effect of light is to slow or abolish this discharge.

Behaviorally, the function of molluscan photoreceptor cells showing primary inhibition is to mediate escape responses. Thus, when a shadow falls on the eye, there is a vigorous "off" discharge as the photoreceptor axons escape from inhibition, and the shells quickly close.[133] This sensory mechanism can be contrasted with the dorsal ocelli of adult insects, which are also designed to respond to sudden cessation of illumination. In this case the receptor cells respond to light by depolarization, and they activate inhibitory synapses on the distal end of the ocellar nerve.[105] The ocellar nerve, like the axons to the distal retina of *Pecten*, therefore responds to shadows with a train of spikes.

A second example of presumed primary inhibition is found in neurons whose cell bodies lie along the pallial nerve in the surf clam, *Spisula*.[126] Isolated segments of nerve are photosensitive, the action spectrum for inhibition having a λ_{max} at about 540 nm. Although there are no intracellular recordings, this is a particularly intriguing preparation because the inhibition is antagonized by a longer latency excitatory process with maximum spectral sensitivity lying at still longer wavelengths. The available evidence suggests that both photopigments may reside in the same cell, but further analysis is required to be certain.

The possibility that a photoreceptor cell might contain two pigments, one driving excitation and the other inhibition, needs further exploration. In the vertebrate retina and perhaps commonly in arthropod eyes, individual photoreceptor cells contain a single visual pigment. When inhibitory interactions occur (as they commonly do), they are mediated by synapses between second and higher order neurons (see pages 614 and 615). Although this makes sense for an image-forming retina where color contrast can be profitably tied to spatial contours, it is no argument against an excitatory-inhibitory interaction between two photopigments in the same photoreceptor neuron in a "simple" eye. In addition to the possible example of *Spisula* cited above, there are cells in the median ocellus of *Limulus* which depolarize to near ultraviolet light and hyperpolarize to visible wavelengths (as well as other cells which show the reciprocal relationship).[176] Whether this is an example of primary inhibition superimposed on excitation or whether it is based on lateral inhibitory synapses is not yet known.

The Early Receptor Potential

The membrane responses we have just considered are sometimes referred to as late receptor potentials to distinguish them from the early receptor potential (ERP). The ERP is a low amplitude signal with no detectable latency and different origins. It was first recorded by Brown and Murakami,[31] who noticed that when the electroretinogram or ERG (the mass response of the whole retina recorded with gross extracellular contacts) was elicited by short intense flashes of light, it was preceded by a small potential that had previously not been observed. The early receptor potential was subsequently studied in several laboratories,[29, 51, 53, 98, 182] where it was shown to be a complex waveform which would survive cold and metabolic inhibitors capable of suppressing the late receptor potential. The early receptor potential is not based on changes in membrane conductance to ions, but is generated by displacements of electric charge due to conformational changes in the pigment molecules. The ERP depends on the ordered arrangement of pigment molecules in the membranes of the receptors, and is lost under conditions such as warming which lead to disorientation of the pigments in the membranes.[6, 52] The early receptor potential has been seen in both vertebrate and invertebrate receptors, and its wave form varies during the process of bleaching, depending on whether the receptor contains rhodopsin or various of the intermediates of bleaching which appear when the receptor is irradiated. Figure 14–25 shows these potentials recorded from squid receptors.

Although the magnitude of the early receptor potential is linearly related to the intensity of light, it is about a million times less sensitive than the late receptor potential. In fact, with intracellular recording, a light flash that is intense enough nearly to saturate the late receptor potential fails to evoke a detectable early receptor potential.[226] Consequently,

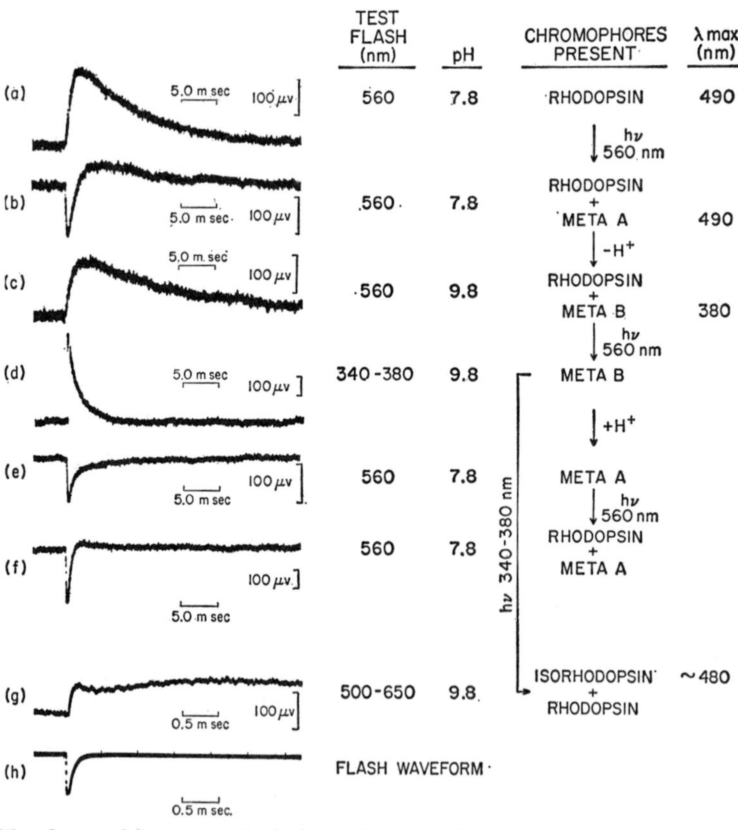

Figure 14–25. Waveforms of fast transretinal photovoltages produced by a glutaraldehyde-fixed squid retina with rhodopsin chromophores in states shown at right-hand side. Temperature, 0°C. Solution: 500 mM tris(hydroxymethyl) aminomethane-HCl titrated to pH shown. Responses are to single test flashes of the following energies (photons per square centimeter): (a) 3×10^{14}, (b and c) 2×10^{15}, (d) 6×10^{14}, (e and f) 8×10^{15}, and (g) 3×10^{16}. Adapting exposures used to convert chromophores in reaction scheme were flashes totaling at least two photons absorbed per chromophore in each case. (From Hagins, W. A., and R. E. McGaughy, Science *157*:813–816, 1967.)

the early receptor potential is not believed to be a direct, causal link in the activation of the receptor (but see Hagins and Rüppel[100]). It is nevertheless a useful experimental tool, for it provides an essentially instantaneous indication of the state of the photopigments in the receptor in a manner not possible by spectrophotometric methods.

The Link Between Photon Capture and Conductance Change

Photochemical studies of rhodopsin indicate that the most likely place to look for the key to excitation is in the conformational changes that the opsin undergoes following isomerization of the chromophore. Physiological studies on the membrane responses of single cells, on the other hand, indicate that changes in membrane conductance are the first measurable manifestation of information flow in photoreceptors. Identifying the means by which structural changes in the molecules of rhodopsin modulate the movement of sodium ions across the receptor membrane is presently the central biophysical problem in visual excitation.

As one quantum is sufficient to activate a thoroughly dark-adapted rod,[108] the idea has been advanced that the process of activation must involve considerable amplification. More explicitly, Hagins[95] has calculated that a single photon absorbed in an outer segment must lead to the movement of about 3000 electronic charges through the outer segment membrane in order to produce, through electrotonic spread, a potential at the rod pedicel that is above the noise level. More recent measurements indicate that the space constant of rod cells is only about 25μm, and several thousand ions per photon may be a significant underestimate. These calculations are attempts to express quantitatively the extent to which a single rhodopsin molecule must be able to "gate" the movement of ions. Recent experiments show that the number of ions traversing the membrane per photon absorbed is in excess of 10^6, apparently adequate for the detection of single photons.[99]

Wald[240] has discussed the problem of excitation in more biochemical terms, likening rhodopsin to a proenzyme that is converted by light into an enzyme which achieves amplification through the ability to turn over a large number of substrate molecules in a short time. Although this model originated as speculation, some recent findings of Bitensky et al.[20] indicate that rhodopsin does control enzymatic activity.

This work suggests that adenosine 3′,5′-cyclic phosphate (cyclic AMP) may be involved in visual excitation. Preparations of dark-adapted rod outer segment membranes were found to have adenyl cyclase activity in amounts ten times greater than any known tissue. On exposure of the outer segment preparation to light, adenyl cyclase activity fell to one-seventh of the dark-adapted value. Light thus serves as a switch, inactivating adenyl cyclase and decreasing the production of cyclic AMP. Here hypothesis once again succeeds experimental fact. The authors suggest that control of adenyl cyclase activity might be effected by conformational changes in contiguous rhodopsin molecules. Further, decreased production of cyclic AMP might lead (directly?) to a decrease in sodium conductance. The authors further predict that in the rhabdomeric photoreceptor cell of an invertebrate, the effect of light should be to activate adenyl cyclase. One view emerging from these experiments is that light is controlling the output of photoreceptor cells in much the same way that peptide hormones or catecholamines operate on their target cells.

At this point, recall once again that rod discs are not continuous with the surface membrane of the outer segment. As any one of the $\sim 10^9$ rhodopsin molecules is potentially able to activate a rod,[108] there remains the question of how excitation reaches the surface membrane from the center of a disc. (This is not a problem in rhabdomeric endings, where the entire surface of the microvilli is apparently in contact with extracellular space.) Most workers find little reason to appeal to various solid state processes of energy migration,[1, 96] and Yoshikami and Hagins[269] have suggested that Ca^{++} ions released from the discs diffuse to the surface of the cell and control Na^+ conductance. The absorption of one photon decreases the dark current[128a] by about 1%, or by $\sim 10^7$ ions per second, and in principle it is reasonable to suppose that an internal carrier such as Ca^{++} could control Na^+ pores to give this amplification. If a single absorption inactivates only one part in 10^9 of adenyl cyclase activity, however, it is not clear how this leads to a 1% reduction in dark current. This argument suggests that adenyl cyclase is not directly involved in modulating sodium conductance. The pace of research in this area is now so fast that doubtless by the time this account reaches print the roles of adenyl cyclase and calcium ions will be much clearer than they are now.

THE COMPOUND EYES OF ARTHROPODS

The General Plan

Compound eyes are made up of subunits called *ommatidia.* Typically a compound eye has several thousand ommatidia, but in some species with reduced eyes there may be only a few score. An ommatidium is a long, pencil-shaped structure oriented perpendicular to the corneal surface. It usually consists of eight photoreceptor cells surrounding a *rhabdom* (Figs. 14–8, 14–26). At its distal end an ommatidium has a small *cornea* and an accessory dioptric structure known as the *crystalline cone.* Viewed from the external surface, the corneal lenses give the eye its faceted appearance. Surrounding each fascicle of photoreceptor cells (*retinula* or little retina) is a sleeve of pigment cells which helps to isolate the ommatidia optically. For general accounts see Waterman,[250] Goldsmith,[91] Goldsmith and Bernard,[93] and Bullock and Horridge.[37]

Although there are usually eight retinular cells, in some species there are only seven (decapod crustacea),[75, 76, 202] and in certain forms there may be many more (*Limulus* has as many as twenty).[160] One or two of the retinular cells may be reduced in size, differ in accessory pigmentation, or occupy a characteristic position in the retinula. Examples are the basal retinular cells of moth ommatidia[91] or the superior and inferior central cells of fly eyes.[228, 229] In the compound eye of *Limulus,* each ommatidium contains one or two *eccentric cells,* which are not retinular cells but rather second order neurons that have assumed a peculiarly intimate relationship to the photoreceptors[160] (Fig. 14–26C).

At the proximal ends of the retinulae there is a *basement lamina* which is traversed by the

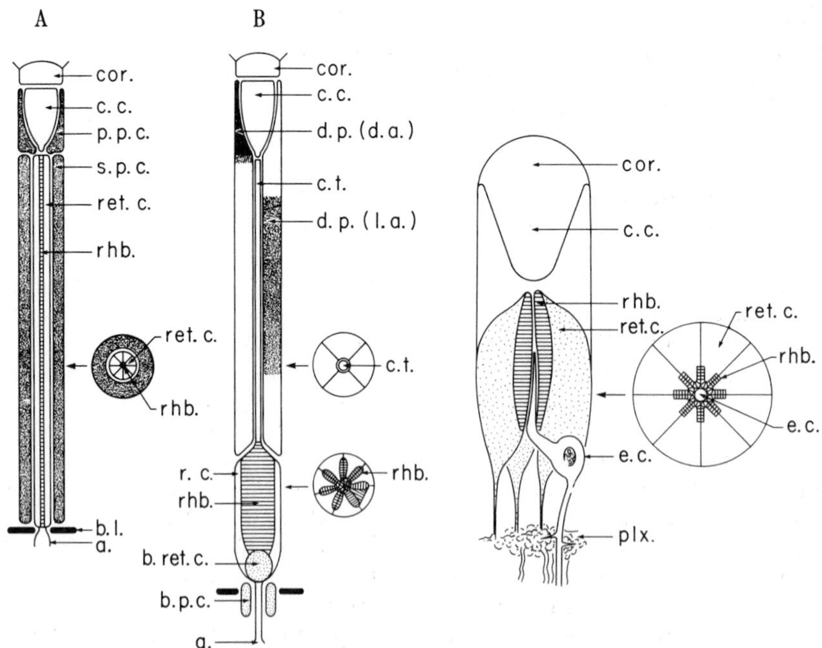

Figure 14–26. Types of ommatidia. *A,* photopic (apposition) ommatidium of an insect. *B,* scotopic (superposition) ommatidium. *C,* ommatidium of *Limulus. a,* axons of the retinular cells; *b. ℓ.,* basement lamina; *b. ret. c.,* basal retinular cell; *c.c.,* crystalline cone; *cor.,* corneal lens; *c.t.,* crystalline thread or tract; *d.p.* (*d.a.*), distal pigment in the dark-adapted position; *d.p.* (*ℓ.a.*), distal pigment in the light-adapted position; *e.c.,* eccentric cell; *ret. c.,* retinular cell; *rhb.,* rhabdom; *plx.,* plexus of retinular and eccentric cell axons and collaterals; *p.p.,* proximal pigment; *p.p.c.,* primary or iris pigment cell; *s.p.c.,* secondary pigment cell; *b.p.c.,* basal pigment cell.

axons of the retinular cells. Characteristically, the axons proceed for 50 to 100 μm or more before entering the optic lobes of the brain. The optic lobes or optic ganglia consist of three or four layers of neuropile flanked by neuron cell bodies and connected by tracts of axons.

Photopic (Apposition) and Scotopic (Superposition) Retinulae

It was Sigmund Exner[80] who in 1891 called attention to the fact that most compound eyes could be classified into two morphological types, which he called *apposition* and *superposition* eyes. His names refer to presumed modes of image formation; these are now controversial, but more will be said of this later. In apposition eyes, the rhabdoms are long, running the full length of the retinulae from the tip of the crystalline cone to the basement lamina (Fig. 14–26A). The accessory screening pigments may or may not be capable of movement, but when migrations of pigment are present they tend not to be pronounced. Exner pointed out that this kind of eye was characteristic of animals that are active during the day and under conditions of bright light; borrowing some general terminology from cone and rod vision of the vertebrate retina, it is therefore appropriate to call them *photopic* eyes.

Superposition eyes differ in having relatively short, thick rhabdoms confined to the basal ends of the retinulae. The space between the distal ends of the rhabdom and the proximal end of the crystalline cones is occupied by the *crystalline tract,* a refractile structure formed either as an extension of the cones or by the distal, non-rhabdomere bearing ends of the retinular cells.[131, 161] The accessory pigment shows pronounced migration. In the light-adapted state it is distributed along the crystalline tract. When the animal is dark-adapted, however, the pigment withdraws distally to a compact mass between the crystalline cones (some insects) or splits into two masses, a portion migrating distally and some withdrawing to the region of the basement lamina (certain crustacea) (Fig. 14–26B). This kind of ommatidium is characteristically found in animals active at night or under conditions of dim illumination. The withdrawal of pigment from around the crystalline tract is an adaptation for increasing the absolute sensitivity of the eye by about two log units.[111] The mechanism is considered in more detail in the section on image formation. In functional analogy with vertebrate rod vision, eyes with this morphology can be called *scotopic.*

Not all compound eyes fit conveniently into these two categories. The skipper butterflies are diurnal creatures that have short rhabdoms and crystalline tracts and have lost the migratory screening pigment.[161] The retina of *Dytiscus* is divided into two tiers. The distal retina is formed by a single rhabdomere from each retinula and is connected to the tip of the cone by a short crystalline tract. The proximal retina lies deeper, and the rhabdoms look like typical examples from a scotopic eye, but they are not connected to the dioptric apparatus by crystalline tracts. The proximal retina is believed to be incapable of resolving images.[115]

The Formation of Images by Compound Eyes

In 1826, Johannes Müller put forward the idea that each ommatidium of a compound eye is sensitive to a narrow pencil of light incident along the ommatidial axis.[171] Exner thought this to be true for apposition eyes, but hypothesized that in the dark-adapted superposition (scotopic) eye the entire corneal surface functions to focus light at the level of the rhabdoms. Such an image would therefore result from a superpositioning of rays that had passed through many facets. This theory has been questioned by many investigators on the grounds that superposition images are not formed where Exner supposed, nor are the concentric layers of refractive index which Exner thought to be present in the cornea and crystalline cone generally observed by modern methods.[131, 161] Nevertheless, recent work supports the superposition theory.[132] The problem is discussed further on page 611 and is critically reviewed elsewhere.[93]

Müller's view is essentially correct, at least for photopic eyes. The passage of rays through the dioptric apparatus of the honeybee has recently been examined in detail,[231] and will serve as an example. Parallel light incident on the cornea is brought to focus at a point about two-thirds of the way down the crystalline cone. As light penetrates beyond the focal point, the rays begin to diverge. Because the apical end of the crystalline cone is surrounded by cytoplasm of the primary pigment cells of a somewhat lower refractive index, the lateral rays of this cone of light are lost. The central rays proceed to the rhabdom. The rhabdom has a higher refractive index than the surrounding cytoplasm and functions as a wave guide. There are several lines of evidence for this. In slices of fresh eyes illuminated through the cornea, the rhabdoms light up as bright spots when the source is aligned with their axes. Moreover, in the bee it has been possible to photograph several of the common modes of intensity that are predicted when electromagnetic energy propagates through a wave guide. The result of this optical arrangement is that each ommatidium collects light through a narrow angle, and that energy which reaches the apex of the crystalline cone is subsequently confined to the rhabdom and a thin shell of cytoplasm around it.

The angle through which light is collected is not much larger than the inclination between adjacent ommatidia. Figure 14–27 shows the relative amount of light calculated to reach the rhabdom of a honeybee as a function of angle of incidence with the optical axis of the ommatidium. The open triangles in Figure 14–27 are actual measurements of light intensity made over the cut ends of single rhabdoms in slices of eyes illuminated with a point source which could be moved in a precise way with respect to the ommatidial axes.[131] As described in the figure caption, the most recent measurements indicate that the angular sensitivity function of the bee is actually narrower than shown in Figure 14–27. Other measurements of the angular sensitivity of single retinular cells of flies, made with intracellular microelectrodes, also give narrower functions than those shown in Figure 14–27.[204, 249]

The optimal size for corneal facets is an interesting problem which has been approached by several workers.[11, 81, 231] Larger facets will collect more light, but the price is a serious loss in visual acuity for the whole eye. The fineness of grain of the retinal image can be increased by decreasing the sizes of individual facets and thus increasing the number of retinulae, but only to a point. As facet size gets too small, diffraction limits the increase in acuity. Actual facet size seems to be determined by the best compromise—the maximum number of ommatidia that can be achieved without severe diffraction loss.

The function of the accessory pigment is to absorb oblique rays that are not refracted into the rhabdom. In a white eye mutant of flies which lacks all accessory screening pigment, the amplitude of the depolarizing response of the sense cells never becomes less than about 40 per cent of the maximum as the stimulus is moved as far as 20° off the ommatidial axis.[249]

The alternative to superposition theory is that in scotopic eyes the crystalline tract functions as a wave guide.[36] During light adaptation, when the accessory pigment is in the proximal position, the pigment granules surround the crystalline tract and increase the refractive index of the "cladding." This makes the rhabdom a less efficient wave guide, and consequently energy is lost to the surround-

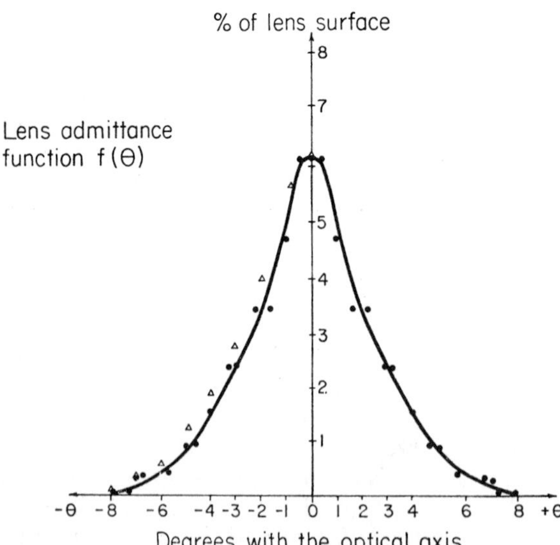

Figure 14–27. Directional sensitivity of an ommatidium of the honeybee, *Apis. Filled circles and curve*: calculated admittance function describing percentage of the total lens surface that admits rays to the rhabdom as a function of angle of incidence of rays with respect to the optic axis. (From Varela, F. G., and W. Wiitanen, J. Gen. Physiol. *55*:336–358, 1970.) *Open triangles*: measurements of light intensity in the rhabdom, scaled to fit the first curve. (Data of Kuiper,[131] from Varela, F. G., and W. Wiitanen, J. Gen. Physiol. *55*:336–358, 1970.) Recent electrophysiological results on bees indicate narrower angular acceptances, about 3° at the half-sensitivity point.[135a] Moreover, these electrophysiological data are in agreement with more recent optical measurements[77a] as well as a theoretical treatment that takes account of the angular sensitivity of the rhabdom.[215a] This curve is therefore in all likelihood too broad.

ing pigment granules. The accessory pigment thus serves as a "longitudinal pupil,"[131] decreasing the sensitivity of the photoreceptors in the light-adapted state. In the dark-adapted state, pattern vision in eyes with crystalline tracts would therefore depend on the efficiency of the dioptric structures in focusing light on the wave guide system, as well as on the absence of losses from the crystalline tract and rhabdom. Both mechanisms—superposition images and tract propagation—may function, depending on the species. Careful experiments are required to determine the relative importance of each as well as the quality of superposition images formed. In skipper butterflies the angular sensitivity of single retinular cells is narrow,[161] but this is not evidence against superposition images.[93] In the crayfish, half the light reaching a single retinular cell comes through neighboring facets,[209] and in the extreme case of the beetle *Dytiscus*,[115] there seem to be no effective wave guides between the distal and proximal retinas.

The photopic eyes of Diptera represent a special case of another kind. In species with fused rhabdoms, all of the contributing retinular cells probably view the same point in space and are equally stimulated by axial illumination. In Diptera and at least some Hemiptera, the rhabdomeres of the individual retinular cells remain separate (Fig. 14–8A). In flies, two of the eight cells, the inferior and superior central cells (cells 1 and 7 in Fig. 14–8A), have shorter rhabdomeres which lie directly over one another like two pencils placed end to end.[228] Thus, a cross-section of a fly ommatidium cut at the distal end shows seven retinular cells and seven rhabdomeres, whereas when it is cut at the proximal end it shows eight retinular cells and seven rhabdomeres. Each of the seven rhabdomeres has a slightly different optical axis; the optical axis of each of the six peripheral rhabdomeres is shared with one of the rhabdomeres in each of five neighboring ommatidia.[128] This pattern is shown in Figure 14–28. Although they lie in six different ommatidia, each of the six retinular cells that share the same optical axis sends axons to converge on the same pair of second order neurons.[24, 229] This arrangement has been called "neural superposition," and it seems to be a means of increasing the light-gathering power of the eye without increasing the sizes of the facets and decreasing visual acuity (see below).

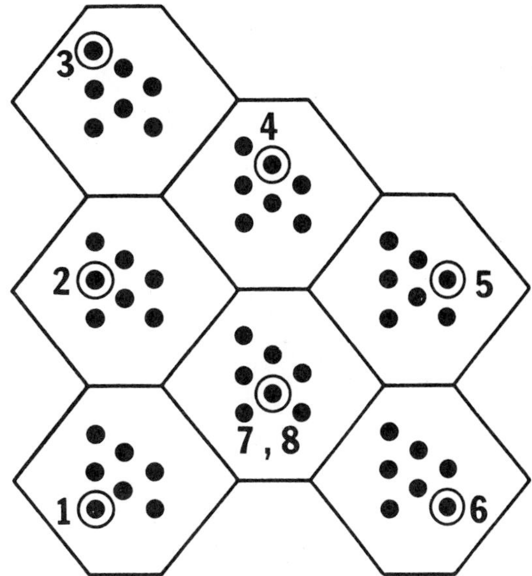

Figure 14–28. Pattern of corneal facets and underlying rhabdomeres that share the same visual field in the compound eyes of Diptera. Rhabdomeres with the same optical axis are numbered 1–6 and are circled. Axons from these six cells enter the same optical cartridge in the lamina ganglionaris, as shown in Fig. 14–29. The remaining two rhabdomeres lie end-to-end, and only one is seen in any cross section. They are the inferior and superior central cells (here labeled 7 and 8 and numbered 1 and 7 in Fig. 14–8A), and their axons are the long visual fibers that pass between cartridges and synapse in the medulla. (Modified from Kirshfeld.[128])

Synaptic Connections in the Optic Ganglia

General Arrangement of the Optic Lobe. There are usually three or four masses of neuropile in the optic lobes (Fig. 14–29A). From the periphery inward, the three regions are known in insects as the *lamina ganglionaris*, the *medulla*, and the *lobula*. In crustacea the terminology differs: *lamina ganglionaris*, *medulla externa*, *medulla interna*, and *medulla terminalis*.[37] Axons leaving the most proximal mass proceed to the brain. This discussion will be concerned with the more distal regions only.

The lamina ganglionaris contains two main classes of interneurons. The *monopolars* receive input from the sense cells. They have their cell bodies proximal to the basal ends of the ommatidia and send axons centrally to the medulla. The axons bear lateral knobs and spines as they pass through the neuropile of the lamina ganglionaris, the region where they are postsynaptic to axons of the retinular

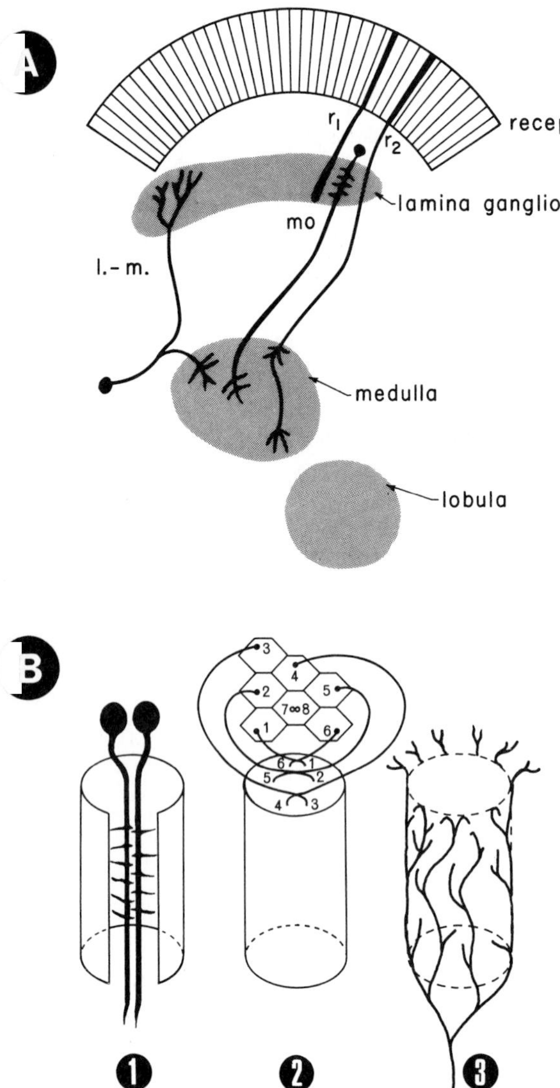

Figure 14–29. *A,* Diagram of the optic lobes of an insect showing several of the more peripheral fiber types: r_1, short receptor fiber terminating in the lamina ganglionaris; r_2, long receptor fiber terminating in the medulla; *mo,* monopolar neuron receiving input from the short retinular cell axons in the lamina ganglionaris; *ℓ.-m.,* lamina-medulla linking fiber (one class of centrifugals). *B,* Three components of Dipteran optic cartridges. The cartridges are represented by cylinders. 1, The second order monopolar neurons with synaptic spines. 2, Six retinular cell axons coming from six different ommatidia (cf. Fig. 14–28) and synapsing in the cartridge with the monopolars after a 180° reorientation. (For clarity, the fibers are not shown within the cartridge.) 3, A basket-shaped centrifugal fiber.

cells. The second type of interneurons are the *centrifugals,* whose cell bodies lie central to the lamina and whose processes project peripherally into the lamina ganglionaris (or both peripherally into the lamina ganglionaris and centrally into the medulla). The name centrifugal refers to the anatomical placement of the cell bodies with respect to the lamina ganglionaris; the function of these cells is unknown.

In many species there are clear associations of bundles of retinular cell axons with monopolar neurons in the lamina ganglionaris.[43] In those species in which such *optic cartridges* are clearly defined, there is one cartridge for each ommatidium.

Synaptic Connections in the Diptera. The details of organization of the lamina have been studied most extensively in flies. In the discussion of image formation it was pointed out that in flies six peripheral retinular cells from six different ommatidia share the same visual field. Moreover, the axons of these six retinular cells sort themselves into the same optic cartridge (Fig. 14–29 B), where they all converge on the same pair of monopolar neurons. Each of the two monopolar neurons receives input from all six presynaptic elements. An optic cartridge thus sums intensity from the same point in space viewed through six different facets.

A second kind of monopolar neuron (not

shown in Figure 14–29) sends its main fiber between cartridges, where it makes lateral connections through short collateral processes with the second order cells in three or sometimes four adjacent cartridges.[222] The axon then proceeds to the medulla. In addition, several kinds of centrifugal cell have also been described.[220] These cells branch laterally to varying extents through the lamina; some have linking fibers that connect the lamina with the medulla, whereas others (amacrine cells) have their fibers confined to the lamina ganglionaris. Very little is known of their synaptic connections.

The axons of the two central retinular cells do not terminate in the lamina ganglionaris, but traverse this region without being incorporated into cartridges. As they leave the lamina they associate with the axons of a pair of monopolars from the cartridge directly under their ommatidium of origin.[220, 228] These four fibers, all of which relate to a single point in the visual field, remain together through an anterior-posterior chiasma between the lamina and medulla, and terminate in common columns in the neuropile of the second optic ganglion. It has been reported that these "quads" of fibers are in fact accompanied by a fifth, a laminar-medulla linking fiber recognized in the lamina as one of the types of centrifugals.[113] The main point, however, is that the topographical projection of the real world that is established in the lamina ganglionaris is maintained in the medulla, although reversed along an anterior-posterior axis.

Other Arthropods. Despite the special optical features of the Dipteran retina, species with fused rhabdoms show some similarities in the organization of their optic ganglia. In a retinula with a fused rhabdom, all the rhabdomeres share the same visual field (*Limulus* may be an exception).[189] Correspondingly, Horridge and Meinertzhagen[113] report that except for the long visual fibers, all the retinular cells of ommatidia with fused rhabdoms (*Schistocerca, Apis, Notonecta,* and *Aeshna*) converge on the same pair (or group) of monopolars (i.e., the same cartridge, but cartridges are not well marked in all these species). The same conclusion was reached for *Apis* by Varela;[230] however, it has not been so obvious with other material,[221] and in fact Hámori and Horridge[104] state that in the lobster the axons from one retinula go to *different* cartridges!

In the moth *Sphinx,* which has a scotopic eye adapted for vision in dim light, the dendritic fields of the second order monopolars extend laterally for scores of microns,[221] suggesting that input funnels from many ommatidia; this is reminiscent of the high convergence ratios of rods on ganglion cells in the vertebrate retina (see p. 624). Thus, on the basis of the available evidence, the conclusion that the lamina ganglionaris invariably contains a fine-grain topographical map of the visual field seems open to question. Nevertheless, the pattern of projection of fibers from the lamina ganglionaris onto the medulla, which was described above for flies, is also found in several other insects with photopic (apposition) eyes.[113]

Some tentative conclusions about patterns of neuronal activity in the lamina ganglionaris have been inferred from the distribution of synaptic vesicles in the endings of fibers. There is agreement that receptors synapse on monopolars. In the lobster, the centrifugal fibers are thought to be presynaptic to monopolars and postsynaptic to receptor axons.[105] In the bee, the same associations are reported, as well as centrifugal-to-receptor and centrifugal-to-centrifugal contacts.[230]

There are doubtless subtleties of interaction that have been only dimly perceived through Golgi staining. That there are several different kinds of monopolars and centrifugals has already been mentioned. In addition, there are three morphological forms of long visual fibers based on the presence and distribution of lateral projections in the lamina and the type of ending in the medulla. In some species (e.g., *Pieris*) there are also three types of short retinular axons ending in the lamina ganglionaris.[221] Systematic variation of form implies specialization of function, so on anatomical grounds one might well expect some differentiation of function among the retinular cells of a single ommatidium. The following section is addressed to this point.

Differentiation of Function Within Single Retinulae

We have seen that in species with fused rhabdoms the wave guide nature of the rhabdom optically couples the retinular cells, and that even in Diptera with separate rhabdomeres, those units that have the same optical axis are physiologically coupled by convergence at the first synapse. The presence of different morphological types of retinular cell axons, however, suggests that with regard to parameters other than intensity, a single retinula may serve more than one input line.

Electrotonic Coupling Between Retinular Cells. If different retinular cells in the same ommatidium are capable of independent response, one should expect to find them electrically isolated. By putting electrodes into different cells in the same ommatidium and injecting current, it is possible to look for low resistance connections. This experiment has been performed on several species with quite different results. In *Limulus*, there is extensive coupling between all cells in the same ommatidium.[215] This is not surprising, in view of the mode of excitation of the second order (*eccentric*) cell. A dendrite of the eccentric cell lies on the axis of the ommatidium, inserted between the rhabdomeres of the retinular cells (Fig. 14–26C). Because of the presence of low resistance junctions between all cells of an ommatidium, depolarization of the rhabdomeric membranes produced by light draws current from the eccentric cell. The depolarization of the dendrite of the eccentric cell spreads electrotonically to the base of the axon, where spikes are initiated. Electrotonic coupling is not limited to the *Limulus* ommatidium, and has also been observed in retinulae of the drone honeybee.[207]

Responses to Polarized Light. Other species of arthropod with fused rhabdoms have retinular cells that are not coupled. For example, in locust it is possible to record from two retinular cells that have identical visual fields (and by inference therefore lie in the same ommatidium) yet differ in the plane of polarized light to which they are most sensitive.[205, 207, 208] In Crustacea, one can class the retinular cells into two groups, depending on whether they are most sensitive to vertically or to horizontally polarized light (Fig. 14–30). The plane of polarization to which the cells are maximally sensitive has been shown to correspond to the microvillar axes of the rhabdomeres,[77] and the ratios of sensitivity of single cells to light polarized in the optimal plane and at right angles can be as high as 10:1.[208, 252] These experiments indicate that an excitation arising in one retinular cell need not affect other units in the same ommatidium, and they strongly suggest that differential sensitivity to plane polarized light shown by the individual retinular cells in an ommatidium is based on the dichroic absorption of the microvilli (Fig. 14–15B).

The significance of these experiments extends into the realm of behavior, for many arthropods[122, 251] and cephalopod molluscs[165] show responses to the plane of polarization. Perhaps the best known example is the honeybee, which is able to analyze polarized sky light in driving its sun compass reaction.[84] Clearly, more information is needed on the synaptic fates of all of the retinular cells emanating from a single ommatidium before even the sensory side of this behavior can be understood in neurophysiological terms.

Peripheral Basis for Color Vision. The

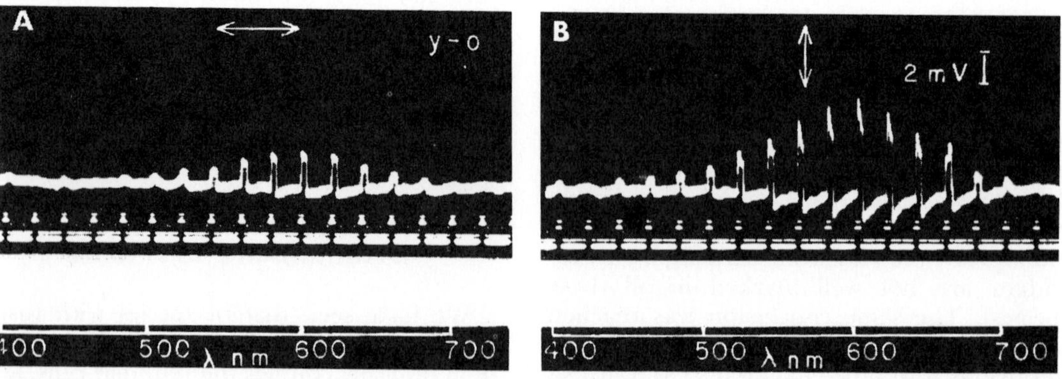

Figure 14–30. Polarized light sensitivity of an arthropod photoreceptor cell. Intracellular recordings (displayed on a slow time base) of receptor potentials evoked in a crayfish (*Procambarus*) retinular cell by 0.4 sec flashes of monochromatic light at 18 different wavelengths from 400 to 700 nm. The flashes had the same quantum content and followed each other at 2 sec intervals. Left frame: horizontally polarized light. Right frame: vertically polarized light. (From Waterman, T. H., and H. R. Fernandez, Z. vergl. Physiol. *68*:154–174, 1970. Berlin-Heidelberg-New York: Springer.)

honeybee has been shown by behavioral means involving training to be able to discriminate colors.[58,150] Spectral sensitivity functions of single receptor cells in the eyes of a number of species of arthropods have shown the presence of different kinds of cells maximally sensitive in different regions of the spectrum. Some of this information is summarized in Table 14–3 (see also Goldsmith and Bernard[93]). In most cases, little is known about the behavioral use to which this sensory information is put.

Recent work shows that different color receptors can exist in the same ommatidium. Microspectrophotometric measurements of single rhabdomeres in the eye of the fly *Calliphora* demonstrate that the six peripheral retinular cells have a visual pigment absorbing maximally at 510 nm, whereas the two central cells with long axons which bypass the lamina ganglionaris contain a different visual pigment, absorbing maximally at 470 nm.[134] The cockroach (*Periplaneta*) has a fused rhabdom and two distinct classes of receptor maximally sensitive to ultraviolet (360 nm) and green (510 nm) light.[169,247] By using two different kinds of dye-filled micropipettes[170] and by noting cell-specific movements of screening pigment following chromatic adaptation,[42] both kinds of sense cell have been identified in the same ommatidium. Other species need to be examined by equivalent techniques, and the question of regional differences between ommatidia from different parts of the eye also deserves attention.

Responses of the Second Order Cells

There is now good evidence that the monopolar neurons to which the retinular cell axons synapse in the lamina ganglionaris,[8,206] as well as the axon terminals of the ocellar nerve with which the ocellar photoreceptor cells connect,[195] respond with hyperpolarizing potentials. In the ocellus, the effect is to inhibit a spontaneous dark discharge in the ocellar nerve fibers. In the lamina ganglionaris, the monopolar cells apparently do not respond with spikes but with a slow hyperpolarizing potential that "follows" (with some filtering[122a]) the receptor potential and which presumably spreads electrotonically to deeper synaptic regions.

As was pointed out above, the anatomical organization of ommatidia in the eye of *Limulus* is different from that of insects and Crustacea. Here the second order unit, the eccentric cell, has its cell body lying within the ommatidium, is electrotonically coupled to the retinular cells, and consequently is excited when the retinular cells depolarize. The axons of the eccentric cells are believed to be the only fibers in the optic nerve that conduct spikes.[254] There are no other neuron cell bodies in the region immediately proximal to the basement lamina, although there is an extensive plexus of nerve fibers made up in part of collateral processes of the eccentric cell axons (Fig. 14–26C). This plexus mediates a series of inhibitory connections be-

TABLE 14–3. Presence of Multiple Sensitivity Maxima in the Compound Eyes of Insects.[a]

Animal	λ_{max} *(nm)*				*Method*[b]	*Reference*
Libellula	<380	420		520	selective adaptation	196
(dragonfly)		410		450–550[c]	single unit analysis	112
Periplaneta	365			510	single unit analysis	169
(cockroach)						
Locusta		430 (515)			single unit analysis	18
(locust)						
Notonecta	350	420	464?	567	single unit analysis	35
(backswimmer)						
Calliphora	345 (470)		490 (345)	520 (345)	single unit analysis	38
(fly)						
Apis ♀	345			535	selective adaptation	89
(honeybee)	340	430	460	530	single unit analysis	7
Apis ♂	345	440		535	selective adaptation	90
	340	450		530	single unit analysis	7

[a] From Goldsmith, T. H., pp. 685–719 *in* Handbook of Sensory Physiology, Vol. VII, Photochemistry of Vision, edited by H. J. A. Dartnall. Springer-Verlag, Heidelberg, 1972. Figures in parentheses refer to secondary sensitivity maxima in the same cell.

[b] Selective adaptation of the retinal action potential, recorded extracellularly, or single unit analysis of retinular cells with intracellular micropipette electrodes.

[c] Results variable. See also Autrum and Kolb, 1968.[7a]

tween eccentric cell axons of neighboring ommatidia, with the result that the presence of spike activity in one ommatidium (eccentric cell) tends to decrease the probability of firing in nearby ommatidia.[224a] In addition, there are thought to be recurrent collaterals feeding back on their own eccentric cells and underlying a system of "self-inhibition."[106, 187, 190] The function of the retinular cell axons in *Limulus* is not known.

Information Processing

Limulus. The presence of inhibitory interconnections in the plexus below the basement lamina means that an individual ommatidium responds independently of its neighbors only if it is the sole element illuminated. Quantitative studies of this system by Hartline, Ratliff, and their colleagues[106] have elucidated many of its properties, which are similar to some of the functional characteristics of the vertebrate retina where there are more different kinds of neurons.

The steady state interactions between two ommatidia can be described by a pair of simultaneous linear equations:[190]

$$r_1 = e_1 - k_{1,2}\,(r_2 - r^0_{1,2})$$

$$r_2 = e_2 - k_{2,1}\,(r_1 - r^0_{2,1})$$

The symbols r_1 and r_2 are the responses (spikes per unit time) of the two ommatidia; e_1 and e_2 are the responses of the two ommatidia when each is illuminated alone; $k_{1,2}$ is the coefficient of inhibition of unit 2 on unit 1, and $r^0_{1,2}$ is the threshold firing rate which unit 2 must have to exert an inhibitory effect on unit 1. These equations say that the response of unit 1 is determined by the amount of excitation it receives, reduced by the inhibition it gets from the other ommatidium. This inhibition is in turn directly proportional to the firing rate of the second ommatidium, once a certain threshold value is exceeded. The inhibition between ommatidia is reciprocal but it is not symmetrical; that is, the two inhibitory coefficients and the two threshold rates are in general not equal. The magnitude of the inhibition decreases as the distance between the two ommatidia increases. The effects of additional ommatidia are additive.

One of the important consequences of this system is to enhance spatial contrast at boundaries. Figure 14–31 (inset) shows a step gradient of light intensity imposed on the eye. The upper curve (open triangles) shows the responses of successive ommatidia across the eye when illumination is restricted to the unit being tested. The curve reflects the objective pattern of light intensity on the eye (inset). When the whole pattern is displayed at once, however, the outputs of the ommatidia follow the lower curve (open circles). On the light side of the boundary, the ommatidia are firing maximally because they are receiving little inhibition from units immediately across the dark zone. Conversely, the units just across the boundary on the darker side have the lowest output, because being close to the cells with the greatest firing rate, they are receiving more inhibition than any other units. The net effect is to accentuate the differences in firing rate of ommatidia on the two sides of the light-dark border.

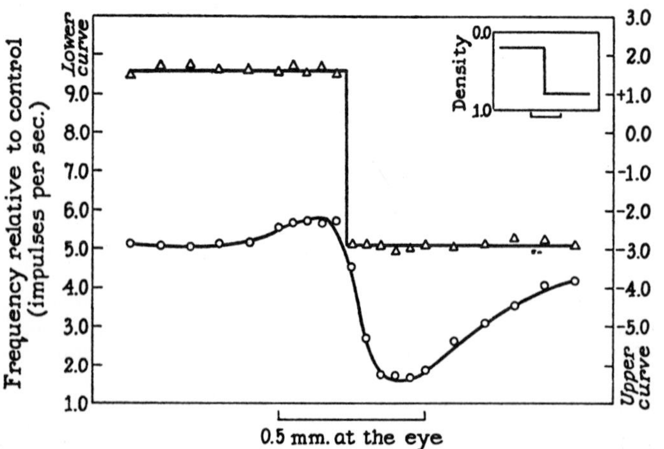

Figure 14–31. Frequency of discharge of ommatidia (eccentric cell axons) along a transect through the step gradient of intensity shown in the inset in the upper right. *Triangles*: When the ommatidia are illuminated singly, the spatial pattern of discharge follows the pattern of intensity. *Circles*: When the whole mosaic of receptors is illuminated by the pattern, the discharge of ommatidia follows the lower (curvilinear) graph. The enhancement and suppression of units on either side of the light-dark boundary is caused by lateral inhibition. (From Ratliff, F., and H. K. Hartline, J. Gen. Physiol. *42*:1241–1255, 1959.)

Other geometrical effects are possible, including disinhibition. Disinhibition can be produced in a row of three (non-adjacent) ommatidia when the two units at the ends of the row are so far apart that they exert negligible inhibition on each other. Excitation of one of the end ommatidia (the test ommatidium) causes it to fire. Simultaneous excitation of the middle ommatidium decreases the firing rate of the test unit. If the third ommatidium is now also excited, it slows the firing rate of the middle ommatidium, causing the test unit to escape from some of its previous inhibition.

The dynamic aspects of inhibition become more complicated. The presence of self-inhibition, a form of negative feedback, produces a degree of temporal sharpening of the response and may account for some but not all of the adaptation present in eccentric cell axons. The interplay of self and lateral inhibition, which operate with nonidentical time courses, can also produce transient effects. For example, the post-inhibitory rebound of a cell released from lateral inhibition may be due to a relative lag in the development of self-inhibition.[106, 187]

Crustacea and Insects. The functional properties of the optic ganglia have been approached by electrophysiological and behavioral techniques. In Crustacea, the optic ganglia are usually separated from the brain by the optic tract or optic peduncle, and single-unit activity can be sampled by thrusting needle electrodes among these fibers. In insects, equivalent records have been made by probing into the optic lobes and the brain. These experiments provide a descriptive classification of fibers in terms of their location, the kinds of effective stimuli, and their receptive fields (that region of the body surface containing receptors whose activity influences the interneuron). This type of approach has inherent limitations which it is well to recognize. Sampling of fibers is not random, and the activities of many smaller units, possibly with different functional properties, are not recorded. Secondly, as visual information reaches higher centers one may expect it to be more abstracted or "integrated." Consequently, some ingenuity is required in determining what is an adequate stimulus on the retina for a higher order visual interneuron. For example, some cells may be responsive to unidirectional movement in the visual field, but be much less responsive to the simple turning on or off of a light. Finally, particularly in regions of neuropile, it is frequently impossible with present techniques to relate the recorded response to any one morphological type of fiber.

The optic tract neurons of Crustacea have large visual fields, 30° to 180°. In *Podophthalmus,* for example, this means an input from 300 to 10,000 ommatidia.[256] Several types of visual interneurons have been described. *Sustaining fibers* are tonic (slowly adapting) fibers which signal the luminosity over wide fields. They may have weakly inhibitory surrounds, and their sensitivity is greater when the animal is in a state of arousal. *Dimming fibers* are the functional mirror images of sustaining fibers and are inhibited by light. Several kinds of *movement fibers* also are present. One is a rapidly adapting cell that is stimulated by sudden movements anywhere in the visual field. Different parts of the receptive field adapt independently, so that habituation can be overcome by displacing the test target to a previously unstimulated area of the field. These cells can be thought of as "novelty units." The more the crustacean relies on vision in directing its motor activities (crabs as opposed to crayfish, for example) the more kinds of movement fibers are present in the optic tract.[255, 259, 261, 262, 263] In *Podophthalmus* there are fibers which are sensitive to the direction of moving targets, and these may be subclassified by the velocity of movement that produces a maximal response.[256]

Another important feature of crustacean optic tracts is the large amount of efferent sensory information from mechanoreceptors, the statocyst, and the contralateral eye.[41, 260] In fact, most of the movement fibers have input from sensory hairs as well as photoreceptors. Perhaps the most fascinating of these multimodal cells are the *space constant fibers,* both movement and sustaining. In the example shown in Figure 14–32, the receptive field is a part of the upper half of the eye near the center, but there are also connections with receptors in the statocyst. When the animal is rotated about its transverse axis (pitch), the receptive field moves against the turn and the fiber continues to "see" the same part of the sky (Fig. 14–32, lower row). The ommatidia that are potentially part of the receptive field therefore occupy a large central region of the eye, but those oriented below the horizon are unable to contribute because of inhibition from the statocyst. When the animal is rotated about its long axis (roll), the size of the receptive field changes. As the eye turns downward, all of the ommatidia potentially able to contribute to excitation of the interneuron are eventually oriented below the horizon, and the fiber becomes blind. If the animal is placed on its other side, with the eye looking up, the receptive field enlarges to a maximum (Fig. 14–32, top row).

Among the insects, the locust[39, 40, 114] and grasshopper[177] have been subjected to exploratory examination. Tonic luminosity units, dimming fibers, phasic on or on-off units, and rapidly adapting non-directional movement fibers have been found. As in Crustacea, many fibers receive input from mechanoreceptors, but the relative independence of adaptation of the visual and mechanoreceptor modalities suggests that they involve different regions of the dendritic field. All of these types of neuron can be found in the medulla. More success has been reported with insects than with Crustacea in finding small-field fibers,[151] but these seem to be located in the more peripheral regions of the optic ganglia, which have not been examined in Crustacea. Because of the many studies of the optomotor response of insects, much effort has been invested in searching for directionally sensitive motion units. Such cells have been located in the lobula of bees,[123] flies,[19] and the moth *Sphinx*.[50] They have also been seen in the medulla of grasshopper and locust, species with a less well differentiated lobula. There seems to be a set of four neurons responding to movements of extended edges in the horizontal and vertical planes. In Hymenoptera and Diptera they have monocular fields; in *Sphinx*, the field is binocular. The fibers are slowly adapting and may well be involved in optomotor responses.

THE EYES OF VERTEBRATES

STRUCTURE OF THE HUMAN EYE

The human eye, shown diagrammatically in Figure 14–33, is a good example of a generalized vertebrate eye. It is roughly spherical and is surrounded by a fibrous coat of connective tissue, the *sclera*, which is modified in front as a transparent *cornea*. The lens divides the eyeball into a front compartment filled with *aqueous humor* and a larger, more posterior compartment filled with *vitreous humor*. The *iris*, which gives the eye its color, forms a diaphragm immediately in front of the lens. The aperture of this diaphragm is called the *pupil*.

Immediately internal to the sclera is a second coat called the *chorioid*, and between that and the vitreous humor is a third layer, the *retina*. The chorioid is vascularized to provide the retina with nutrients, and its internal surface is faced with a sheet of deeply pigmented epithelial cells containing melanin. The retina, which will be described in more detail below, is the light-sensitive layer and consists of receptors (rods and cones) and several kinds of neurons. The outer segments

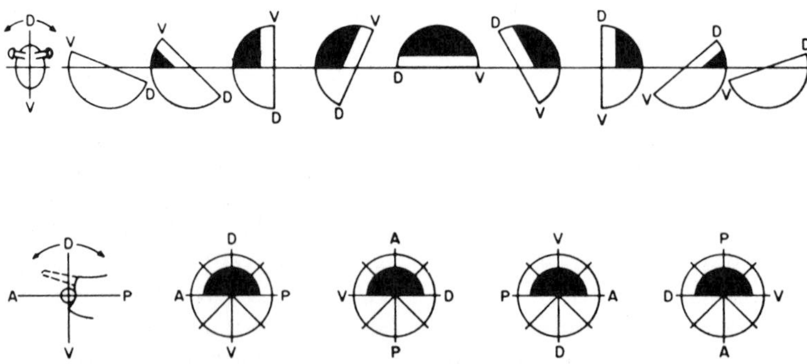

Figure 14–32. Changes in the excitatory visual field of a "space-constant" sustaining fiber in the optic tract of the crayfish *Procambarus*. *Upper row:* view of the left eye as seen from the anterior end of the animal. The normal position of the eye is shown by the third diagram from the right. The receptive field (black areas) changes as the animal is rolled about its long (anterior-posterior) axis. *Lower row:* lateral view of the left eye. The size of the visual field and its absolute position in space remain constant as the animal pitches around its transverse axis. See the text for further description. (From Wiersma, C. A. G., and T. Yamaguchi, J. Comp. Neurol. *128*:333–358, 1966.)

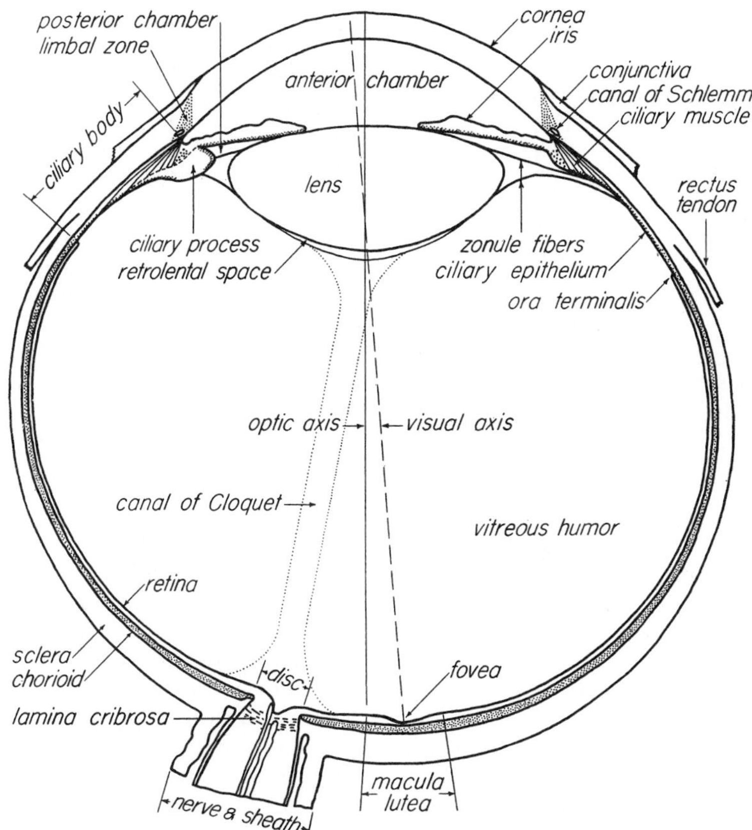

Figure 14–33. Horizontal section of the right human eye, ×4. (From Walls, G. L., The Vertebrate Eye and its Adaptive Radiation. Bulletin 19, Cranbrook Institute of Science, 1942.)

of the receptors abut against the pigment epithelium, so that light which reaches the visual pigment has passed from the vitreous humor through two layers of nerve cell bodies and the inner segments of the rods and cones (cf. Fig. 14–7). The nerve cells adjacent to the vitreous humor are called ganglion cells, and their axons pass over the surface of the retina to the *optic disc* or blind spot where they collect, turn through the sclera, and form the optic nerve.

Close to the optic axis of the eye is a small patch of retina, the *macula lutea,* which is about 1.5 mm in diameter and is yellow due to the presence of carotenoids. At the center of the macula lutea is a depressed region of the retina about 0.3 mm in diameter called the *fovea.* The fovea contains only cones; it is the region of the retina in which visual acuity is greatest and on which objects are focused when they have the attention of the observer.

The human retina contains approximately 6.5×10^6 cones and about 110 to 125×10^6 rods.[179] As Figure 14–34 shows, the mosaic of receptors is not uniform over the surface of the retina. The density of cones is very high in the fovea, but falls to very low values in the parafoveal regions. The density of rods, on the other hand, is zero in the fovea, rises to a maximum about 20° from the fovea, and then falls to lower values towards the periphery. Note that Figure 14–34 shows a transect through the optic disc, where there are no receptors.

There are only about 10^6 fibers in the optic nerve; consequently, there is much convergence of both rods and cones onto single ganglion cells. Because there are about twenty times more rods than cones, the pools of rods supplying single optic nerve fibers are larger than those of cones. In fact, in the fovea there is a 1:1 relationship between

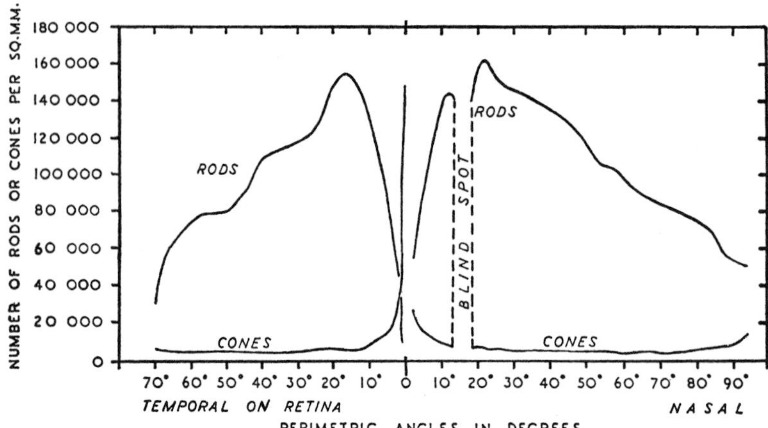

Figure 14–34. Distribution of rods and cones in the human retina. Cone density is highest in the fovea (0°), whereas rods are most abundant about 20° from the fovea. (Note that the distribution of rods and cones on the nasal side in and near the fovea is not plotted but is similar to the distribution on the temporal side.) (From Pirenne, M. H., Vision and the Eye, 2nd ed. Chapman Hall, Ltd., 1967.)

cones and ganglion cell axons. The physiological consequences of this arrangement will become clear later.

The Formation of Images

The vertebrate eye, like that of cephalopod molluscs, is frequently likened to a camera. There is a lens system, the cornea and lens, which forms an inverted image of an object on the surface of a light-sensitive layer, the retina. Stray light is suppressed by lining the chamber behind the retina with a black surface, the pigment epithelium; light entering the lens can be controlled with an adjustable diaphragm, the iris; and the lens can be altered to focus on near or distant objects, the process of accommodation.

Unlike a camera, the interior is filled with a gelatinous material with a refractive index (1.33) significantly higher than that of air. Consequently, in terrestrial animals the major refraction occurs at the air-cornea interface. The lens has a refractive index of 1.42 and serves to make fine adjustments in the focal length of the system. The lens is suspended in such a way that contraction of the ciliary muscles relieves tension on the fibers suspending the lens, which in turn permits the natural elasticity of the lens to mold it into a more convex shape. This shortens the focal length and allows the imaging of nearby objects. Conversely, when the ciliary muscles relax, the lens is pulled into a flatter shape, and the focal length increases.[246]

Like any lens system, that of the vertebrate eye is subject to various kinds of aberrations. Chromatic aberration results because red light which is refracted through the edges of the lens is brought to a focus in a plane behind the shorter wavelengths. The result is that in white light it is impossible to bring an object into sharp focus. Glass lenses can be color corrected by piecing together lens elements of different refractive index, but evolution has sought other solutions. Because the change in refractive index with wavelength becomes larger for short wavelengths, vertebrates have adopted the expedient of attenuating the short wavelength end of the spectrum. Thus the vertebrate lens is frequently yellowish, decidedly so in many diurnal species, to filter out the near ultraviolet and violet wavelengths.[127] The macula lutea is a second adaptation to attenuate blue light in the region of the fovea, and the presence of colored oil droplets in the cones of many lower vertebrates is thought by some to serve the same purpose.[12, 246]

Light which enters the edge of the pupil is several times less effective in exciting cones than rays passing through the center of the lens, even when both rays are brought to a focus on the same receptor. This directional sensitivity of cone cells is called the Stiles-Crawford effect and is based on the wave guide nature of cone outer segments.[78, 79, 219] The effect is much more pronounced for cones than for rods.

Despite the use of imperfect materials for constructing an optical device, the acuity of the human eye seems to be limited by lens aberrations only when the pupil is relatively dilated. It is possible to resolve gratings when the distance between adjacent black and white bars subtends an angle of approximately 0.5 minute of arc. Under these conditions, the image of the grating on the retina is degraded by diffraction and consists of a sinusoidal modulation of intensity about 25% between peak and trough. Experiments have been done in which sinusoidal patterns of intensity with variable period and high contrast have been produced on the retina through the interference of two coherent beams of laser light, thus bypassing the optical system. By comparing the visibility of these patterns with fine gratings viewed normally, it has been shown that with a 2 mm pupil the quality of the retinal image approaches that of a diffraction-limited system.[44, 45] Interestingly, the grain of the retinal mosaic is in balance with these optical factors. A visual angle of 0.5 minute of arc corresponds to slightly more than 2 μm of the retina, which is the approximate distance between the centers of adjacent cone outer segments.[184]

The best values of visual acuity depend on the nature of the test object. A single black line seen against an evenly illuminated bright background can be detected when it subtends in an angle of only 0.5 second of arc, a value 60 times smaller than that found with gratings. Presumably the observer detects a smeared image of the line produced by diffraction effects, and the line of cones under this shadow receives about 1% less light than the adjacent receptors.[184]

The Duplicity Theory

In 1825, J. E. Purkinje reported that as twilight falls, red flowers, which in the strong light of day seem brighter than blue flowers, appear relatively less bright, and with advancing darkness may fade from view before blue flowers. This Purkinje shift, the sliding of maximum sensitivity of the eye to shorter wavelengths with falling levels of illumination, is one manifestation of the transference of vision from cones to rods.

The different characteristics of rod and cone vision were recognized by M. Schultz in 1866 in the Duplicity Theory. Rods mediate scotopic vision or vision in low levels of light. Rod vision is also colorless and relatively blurred. Cones mediate photopic vision when the light is brighter. Cone vision is colored and sharp.[184]

Rod cells are very sensitive, and in the dark-adapted state can be excited by the absorption of a single photon.[108, 184] Because a large group of rods is connected (ultimately) to a single nerve fiber, the rod pool functions as an antenna and increases the sensitivity of that ganglion cell. The absolute threshold for vision occurs when about six or seven photons are absorbed nearly simultaneously in as many different outer segments in a limited area of the retina containing about 500 rods. Rod vision is colorless because rods contain the single visual pigment rhodopsin, and therefore differences in wavelength are perceived only as differences in brightness. Rod vision is blurred because the great convergence of rod cells on single nerve fibers makes the effective grain of the retina relatively coarse.

Cone cells take over under conditions of higher illumination. Because the cones are intrinsically less sensitive and because there is relatively less convergence of receptors on ganglion cells, the absolute sensitivity of the system is lower. But the price is paid in a good cause, for decreased convergence means improved visual acuity. Moreover, color vision is possible in a species in which there are several classes of cones containing visual pigments absorbing in different regions of the spectrum. Then, through the interaction of cone signals further along in the afferent pathway, intensity and wavelength can be distinguished as independent parameters.

The scotopic luminosity curve (the apparent brightness as a function of wavelength, measured close to threshold) is determined by the spectral absorption of rhodopsin and is maximal at about 500 nm. The photopic luminosity curve (measured at higher intensities) represents a combination of the outputs of the three kinds of cones and is maximal at about 560 nm. These two spectral sensitivity curves are shown in Figure 14–35. Only for deep red lights is the absolute sensitivity of the cones greater than that for rods. In the blue, on the other hand, rods are about 3 log units more sensitive than cones. Figure 14–35 makes clear the explanation of Purkinje's observations on the relative brightness of flowers. Note also the effect of the blue-absorbing macular pigment on the sensitivity of the foveal cones.

Figure 14–36 shows the recovery of sensitivity of the human eye following a period of exposure to bright light that has bleached the visual pigments. This recovery is known as dark adaptation, and as observed on this time scale, it clearly falls into two processes. The first or faster limb of the curve describes

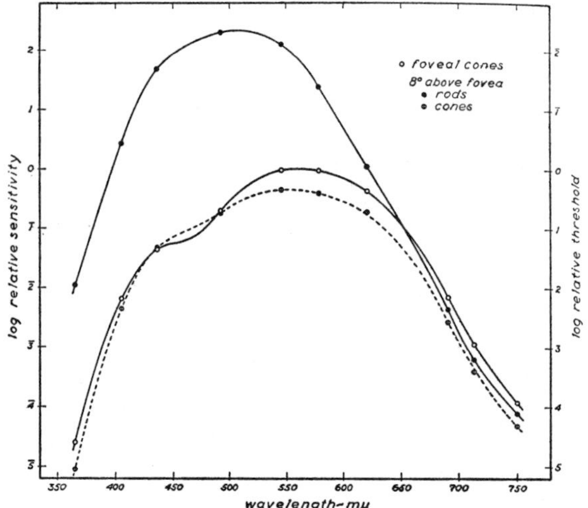

Figure 14–35. Spectral sensitivity (1/threshold energy) of rods and cones in the dark-adapted human eye. Rods, 8° above the fovea (filled circles, solid line); foveal cones (open circles, solid line); peripheral cones 8° above the fovea (broken line). Test field was 1° in diameter; test flash was 0.04 sec. All sensitivities are expressed relative to the maximum sensitivity of the fovea. Note that cones are slightly more sensitive to deep red light than are rods. The shoulder on the foveal sensitivity curve at ~450 nm is distortion caused by the carotenoids of the macula lutea. (From Wald, G., Science *101*:653–658, 1945.)

the increase of sensitivity of cones; the second, slower limb shows the recovery of rods. These curves reflect the relative rates of synthesis of cone and rod pigments. Other factors influencing dark adaptation are discussed in following sections.

Retinomotor Processes

The vertebrate eye possesses several accessory mechanisms for adjusting the amount of light reaching the receptors. These further enable animals to be active under varied conditions of illumination.[3, 246]

Pigment Migration. In many fish, anurans, and birds, and to a lesser extent in urodeles and many reptiles, the pigment granules of the pigment epithelium are capable of migrating between the outer segments of the receptors under conditions of bright illumination, and retracting towards the corioid as the ambient light level falls.

Cone Migration. In essentially the same

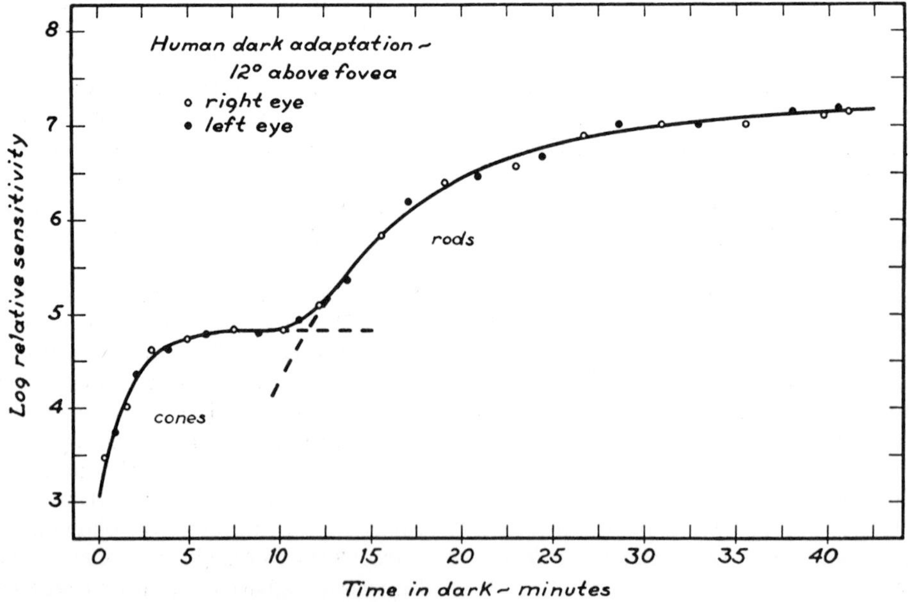

Figure 14–36. Dark adaptation of the human eye, measured in a peripheral area which contains both rods and cones. The dark adaptation of the cones is completed within about 5 minutes, that of rods within about 45 minutes. (From G. Wald, P. K. Brown, & P. H. Smith, J. Gen. Physiol. *38*:623–681, 1955.)

groups of animals that show pigment migration, the process is supplemented by movement of the cone outer segments. As the melanin advances towards the receptors, the cone outer segments retreat before it by the contraction of that part of the inner segment known as the myoid (Fig. 14–6B). Consequently, under photopic conditions the cones remain exposed.

Rod Migration. In many fish and anurans, the rod myoids move the rod outer segments in antiphase with the cones (Fig. 14–6A). Thus, when the cone outer segments retract away from the pigment epithelium, the rod outer segments advance into it, and *vice versa.*

Pupillary Adaptations. The level of light reaching the retina can also be controlled by contraction of the pupil. With the exception of a few groups such as eels and flatfish, this process is relatively poorly developed in fish. Very good pupillary control of sensitivity is achieved in some reptiles (nocturnal snakes, crocodiles), birds and mammals. Thus, with the exception of birds, strong pupillary responses are best developed in those groups of vertebrates that do not show migration of retinal pigment and cones.

LIGHT AND DARK ADAPTATION

The human eye is capable of adjusting its sensitivity over a range of intensities of about 10^{10}. Pigment migration in the corioid is absent, and the response of the pupil accounts for only about 1.2 log unit. Consequently, for an explanation one must look at processes intrinsic to the receptor cells and their associated neurons.

The recovery of rods can be conveniently studied in species or individuals that lack cones; otherwise, the first part of the recovery is not measurable because the sensitivity of the eye is determined by the photopic mechanism. In the rod retina of the rat, recovery of sensitivity after moderate adapting exposure is a relatively fast process that is completed in seconds or at most a very few minutes.[63, 65] If the adapting exposure is more intense and a significant quantity of rhodopsin is bleached, this rapid "neural" adaptation is followed by a slower phase that can be related to the amount of visual pigment in the rods. During this slower recovery, the log of the sensitivity is linearly related to the fraction of the rhodopsin that has been resynthesized. These relationships are shown

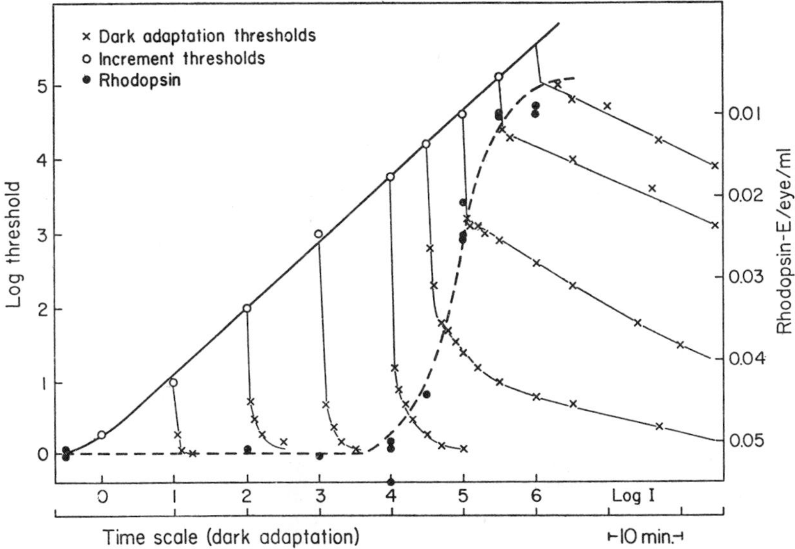

Figure 14–37. Visual adaptation in the rat eye as determined by sensitivity of the b-wave of the electroretinogram. During light adaptation (open circles, heavy line; increment thresholds) the increase in log threshold is linearly proportional to the log of the background luminance, except at the dimmest background luminances. Dark adaptation (crosses, thin lines) is rapid until the eye is adapted to background luminances bright enough to bleach significant quantities of rhodopsin (filled circles, dotted line) in the five-minute adaptation period. With bright background luminances, a slow component of dark adaptation is observed, the extent of which depends on the amount of rhodopsin bleached. (From Dowling, J. E., J. Gen. Physiol. *46*:1287–1301, 1963.)

in Figure 14–37. Rushton[197,198,199,201] has demonstrated a similar relationship for both rod and cone pigments of the living human eye by reflection densitometry—measuring the light which has been reflected out of the pupil after a double passage through the receptors; cones recover faster (Figure 14–36).

Ecological Adaptations of Vertebrate Eyes

The human eye is a generalized vertebrate eye because it possesses both rods and cones and is not heavily specialized for entirely nocturnal or diurnal life. Species whose activity has become more restricted to the day or night hours show corresponding adaptations of their eyes.[246]

Adaptations to a Diurnal Habit. Animals that are active only during the day tend to enhance their acuity at the expense of absolute sensitivity. Thus, their eyes tend to be large (within the limits imposed by the size of the head) so that more retinal cells will fall under the image. The number of cones increases at the expense of rods, and most lizards, snakes, many birds, ground squirrels, and turtles have either all-cone retinas or virtually all-cone retinas. Birds and lizards have a region of the retina (*area centralis*) in which there are no rods and the individual cones have become very long and slender. As in the human fovea, this is a region of minimal convergence on higher order neurons and of maximum visual acuity. The foveas of birds are probably the most highly developed. The retina is conspicuously thinned so that the receptors line a small depression. Walls[246] has suggested that the significance of a foveal depression lies in the fact that the refractive index of the retina is greater than that of the vitreous humor and that incident rays are refracted so as to magnify the image slightly at this point.

Adaptations for a Nocturnal Habit. Animals that have forsaken the light of day for the hours of night have sacrificed acuity for sensitivity. Thus, there is an increase in the number of rods and a loss of cones. Many species have a layer of reflecting material (*tapetum lucidum*) in the corioid so that light that traverses the retina without being absorbed is reflected back into the outer segments. This reflecting layer accounts for the retinal "glow" or eye shine of animals staring back into a source of light.

Species that are active at night but that also bask in the sun need enormous control over their pupils. A round pupil that will open wide at night cannot be closed sufficiently in the daytime by means of a sphincter muscle; this accounts for the evolutionary invention of the slit pupil, which is seen to good advantage in, for example, crotalid snakes.

Structure of the Vertebrate Retina

Five Kinds of Neurons. Being an outpost of the brain, the retina is a moderately complex bit of nervous tissue. In addition to the receptors, it contains four types of higher order nerve cells. The fibers of the optic nerve are therefore at least two synapses removed from the rods and cones.

As shown in Figure 14–38, the retina is made up of three distinct cellular layers marked by aggregations of cell nuclei. The *outer nuclear layer* is closest to the chorioid and consists of the cell bodies of the receptors.

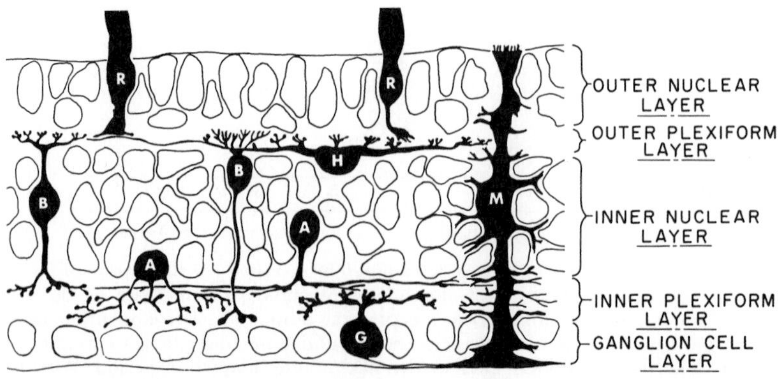

Figure 14–38. The principal types of cells and the layering of the vertebrate retina, based on mudpuppy (*Necturus*) retinas impregnated by the method of Golgi. R, receptor; H, horizontal cell; B, bipolar cells; A, amacrine cells; G, ganglion cell; and M, Müller (glial) cell. (After Dowling, Invest. Ophthal. *9*:655–680, 1970.)

The *inner nuclear layer* contains the cell bodies of three kinds of interneurons: *bipolar cells*, whose axes are vertical (radial), and the *horizontal* and *amacrine cells*, which extend horizontally (tangentially) through the retina. The *ganglion cell layer* is closest to the vitreous humor, and as described above, the axons of these cells collect at the optic disc and form the optic nerve.

Between the three layers of nuclei are two synaptic regions: in the *outer plexiform layer* the receptor cells synapse with bipolar and horizontal cell dendrites, and in the *inner plexiform layer* the bipolar cells connect to amacrine and ganglion cell dendrites.

To summarize, the simplest path of information flow through the retina is receptor → bipolar → ganglion cell. However, at the two synaptic regions there are additional interneurons, the horizontal and amacrine cells, that mediate lateral influences.

Other Cells. In addition to the types of neurons described above, in at least some species (birds) there is a small number of centrifugal fibers of unknown function that end on the processes of the amacrine cells in the inner plexiform layer.[54, 69]

Glial elements known as Müller cells run as columns connecting the cell layers of the retina (Fig. 14–38). The Müller cells bear lateral processes which invest the neural elements. The membrane potentials of these cells respond to potassium ions accumulating in the intercellular spaces of the retina as a result of neuronal activity. Interestingly, these passive glial potentials contribute a significant portion of the electroretinogram (ERG), the mass electrical response of the eye that is recorded with gross electrodes.[159]

Patterns of Synaptic Contact. In retinal tissue that has been stained by the Golgi method, it is possible to classify the four kinds of interneurons into subtypes on the basis of the shapes of their dendritic arbors. Nevertheless, electron microscopy of synaptic contacts in the retina reveals a general pattern of organization common to many species. Some striking interspecific variations on this theme occur, notably in the inner plexiform layer, and these seem to have correlates in retinal physiology (Fig. 14–39).[66, 67, 71] The original accounts should be consulted for the many details that cannot be captured in a general treatment such as this.

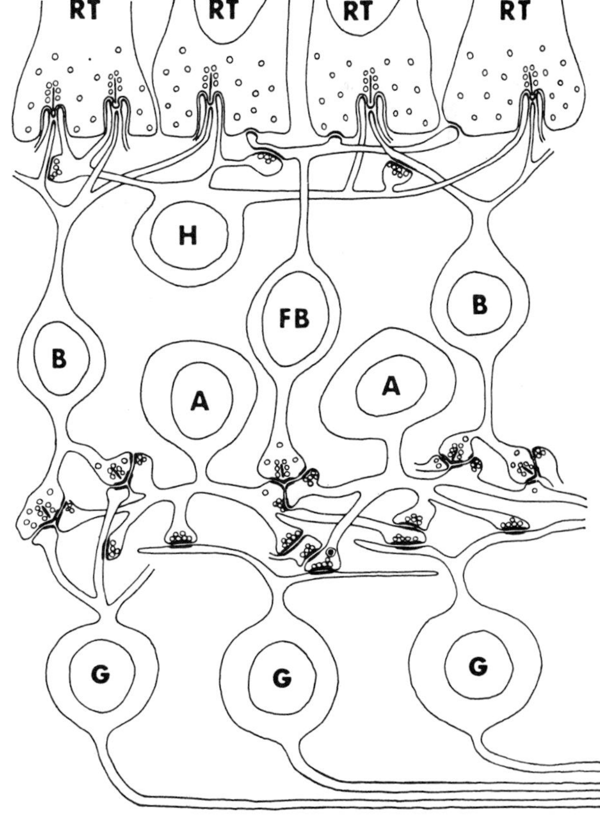

Figure 14–39. Summary diagram of the arrangements of synaptic contacts found in vertebrate retinas. In the outer plexiform layer, processes from bipolar (B) and horizontal (H) cells penetrate into invaginations in the receptor terminals (RT) and end near the synaptic ribbons of the receptor. The processes of flat bipolar cells (FB) make superficial contacts on the bases of some receptor terminals. Horizontal cells make conventional synaptic contacts onto bipolar dendrites and other horizontal cell processes (not shown). Since horizontal cells usually extend further laterally in the outer plexiform layer than do bipolar dendrites, distant receptors can presumably influence bipolar cells via the horizontal cells. In the inner plexiform layer, two basic synaptic pathways are suggested. Bipolar terminals may contact one ganglion cell dendrite and one amacrine process at ribbon synapses (left side of diagram) or two amacrine cell (A) processes (right side of diagram). When the latter arrangement predominates in a retina, numerous conventional synapses between amacrine processes (serial synapses) are observed, and the ganglion cells (G) are contacted mainly by amacrine processes (right side of diagram). Amacrine processes in all retinas make synapses of the conventional type back onto bipolar terminals (reciprocal synapses). (From Dowling, J. E., Invest. Ophthal. *9*:655–680, 1970.)

Most of the synapses that the receptors make with retinal interneurons have a characteristic morphology involving a triad of postsynaptic elements.[68, 70, 212, 217] A centrally placed bipolar terminal is flanked by two horizontal cell dendrites, and the three are inserted into a pocket in the base of a rod or cone. The receptor has both synaptic vesicles and a characteristic synaptic ribbon, an electron-dense plate oriented at right angles to the presynaptic membrane. Conventional synapses, lacking the ribbon and involving only a pair of cells, also are present between receptors and bipolars, horizontal cells and bipolars, and horizontal cells and other horizontal cells. In the outer plexiform layer the receptors are thus clearly presynaptic to both bipolar and horizontal cells, and the bipolars are postsynaptic to both receptors and horizontals. The horizontal cells, on the other hand, must be considered both pre- and postsynaptic units, even over restricted regions of their dendritic trees. These relationships are summarized in Figures 14–39 and 14–41.

In the inner plexiform layer also, both ribbon and conventional synapses are present.[67, 68, 70, 71] The bipolar terminals make ribbon synapses with pairs (dyads) of postsynaptic processes, usually either a ganglion cell and an amacrine cell dendrite or a pair of amacrine cell dendrites (Fig. 14–39), rarely a pair of ganglion cell dendrites. There are clear species differences in the identities of the pair of postsynaptic elements. In cats and primates, 80% are ganglion-amacrine pairs, whereas in frogs, 75% are amacrine-amacrine dyads. The mudpuppy *Necturus* and the rabbit are intermediate, with about equal numbers of each kind. More will be made of this comparison below.

Conventional synapses (monosynaptic, no ribbon) also are present from amacrine to bipolar cells, from amacrine to ganglion cells, and from amacrine to amacrine cells, as well as from bipolar to ganglion cells.

Just as horizontal cells make extensive lateral connections and provide an additional parallel path from receptor to bipolar, amacrine cells also spread laterally and create other transmission lines from bipolar to ganglion cells. But there are two seeming differences between horizontal and amacrine cells. The morphological evidence for horizontal cells feeding back onto receptors is not definitive. Likewise, the extent and mechanisms by which receptor output is influenced by horizontal cell activity is not yet clear. On the other hand, amacrine cells do have reciprocal synapses with bipolar terminals, and in fact an amacrine cell dendrite may be both pre- and postsynaptic to a bipolar cell within a space of a few μm.

The second point of contrast is the variation among species that amacrine synapses show. As mentioned above and diagramed in Figure 14–39, in frogs—in comparison to cats and monkeys—most ganglion cells receive input from amacrine cells rather than bipolars. Consequently, ganglion cells are fourth (or higher) rather than third order units. Along with the interposition of amacrine cells between bipolars and ganglion cells is an increase in the total number of serial junctions in which the amacrine cell is the presynaptic unit. This is shown in Table 14–4 for several species, expressed as a ratio of conventional cell synapses to ribbon synapses.

As described in the following section, these differences in the pattern of connection of retinal elements have their physiological analogs in the extent of information processing that occurs in the retina. Compared to the cat, the frog has greater numbers of synaptic pathways in the inner plexiform layer, and the responses of its ganglion cells are considerably more complicated. Conversely, the more developed are the visual centers of the brain, the less neural integration occurs in the retina.[157]

INFORMATION PROCESSING IN THE VERTEBRATE RETINA

In the past several years intracellular records have been obtained from all of the

TABLE 14–4. Relative Prominence of Amacrine Cells in the Retinas of Vertebrates. Ribbon synapses are found in the terminals of bipolar cells; in most conventional synapses, amacrine cells are the presynaptic unit. Therefore, higher ratios signify relative increases in amacrine cell synapses. (After Dubin.[71])

Animal	*Ratio of conventional:ribbon synapses in inner plexiform layer*
human (parafovea)	1.3
monkey (fovea)	2.0
(parafovea)	1.7
(periphery)	2.2
cat	2.5–2.7
rat	2.8
rabbit	3.7–3.8
ground squirrel	5.5
frog (2 species)	6.3–9.5
pigeon	10.8

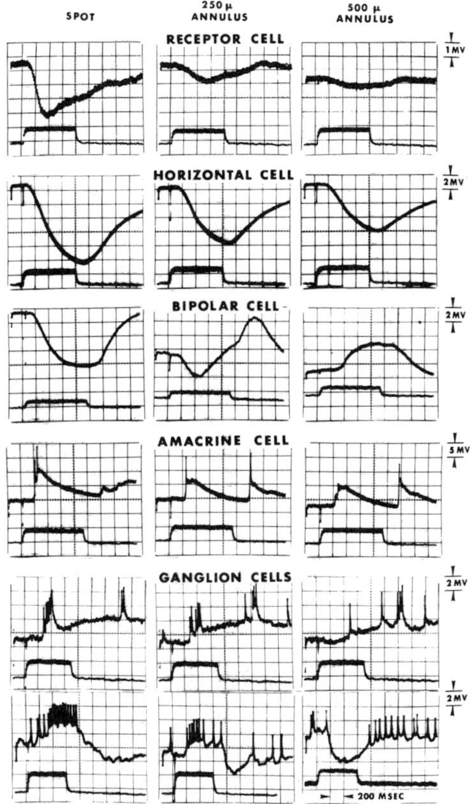

Figure 14–40. Recordings showing the major response types in the *Necturus* retina and the difference in response of a given cell type to a spot and to annuli of 250 μm and 500 μm radius. Receptors have relatively narrow receptive fields, so that annular stimulation evokes very little response. Small potentials recorded upon annular stimulation were probably due to scattered light. The horizontal cell responds over a broader region of the retina, so that annular illumination with the same total energy as the spot (left column) does not reduce the response significantly (right columns). The bipolar cell responds by hyperpolarization when the center of its receptive field is illuminated (left column). With central illumination maintained (right trace; note lowered base line of the recording and the elevated base line of the stimulus trace in the records) annular illumination antagonizes the sustained polarization elicited by central illumination, and a response of opposite polarity is observed. In the middle column the annulus was so small that it stimulated the center and periphery of the field simultaneously. The amacrine cell was stimulated under the same conditions as the bipolar cell, and gave transient responses at both the onset and cessation of illumination. Its receptive field was somewhat concentrically organized, giving a larger "on" response to spot illumination, and a larger "off" response to annular illumination of 500 μm radius. With an annulus of 250 μm radius, the cell responded with large responses at both on and off. The ganglion cell shown in the upper row was of the transient type and gave bursts of impulses at both on and off. Its receptive-field organization was similar to the amacrine cell illustrated above. The ganglion cell shown in the lower row was of the sustained type. It gave a maintained discharge of impulses with spot illumination. With central illumination maintained, large annular illumination (right column) inhibited impulse firing for the duration of the stimulus. The smaller annulus (middle column) elicited impulses at off. (From Werblin, F. S., and J. E. Dowling, J. Neurophysiol. *32*:339–355, 1969.)

types of neurons in the vertebrate retina, marking the cells unequivocally for subsequent identification by injecting dye from the micropipette.[124, 216, 258] Distally, the retina responds in a graded or tonic fashion; the proximal cells, on the other hand, introduce phasic responses as well (Figs. 14–40, 14–41). In the optic nerve, fibers are found to be sensitive not to specific patterns but to contrasts, be it color contrast, spatial contrast (borders between regions of different luminosity), or temporal contrast (abrupt changes in luminosity with time). In species such as the frog that display a high degree of intraretinal information processing, more elaborate spatio-temporal interactions are present, and optic nerve fibers are found that are sensitive to the direction of moving objects. In our present state of knowledge, we are just beginning to understand how specific responses of optic nerve fibers are generated from the physiological properties of retinal interneurons, their geometries, and their patterns of connection with other elements.

Receptors. The hyperpolarizing responses of rods and cones were discussed earlier as part of a consideration of the problem of transduction.

Bipolar Cells. The area of retinal surface that contributes to the response of a single bipolar cell—its receptive field—includes a number of receptors. The receptive fields of bipolar cells have a circular center and an annular surround with antagonistic properties (Fig. 14–40). The membrane response of a bipolar cell is graded with intensity, and there are no spikes. The response may be of either polarity, hyperpolarizing or depolarizing. If stimulation of the center of the receptive field causes hyperpolarization, light in the annular surround causes depolarization. As the receptive fields are larger than the dendritic trees of bipolar cells, influences from receptors in the periphery of the receptive field are thought to be mediated by the horizontal cells.

Horizontal Cells and S-Potentials. Identified horizontal cells have been shown to respond with slow graded hyperpolarizing responses, or with hyperpolarization to short

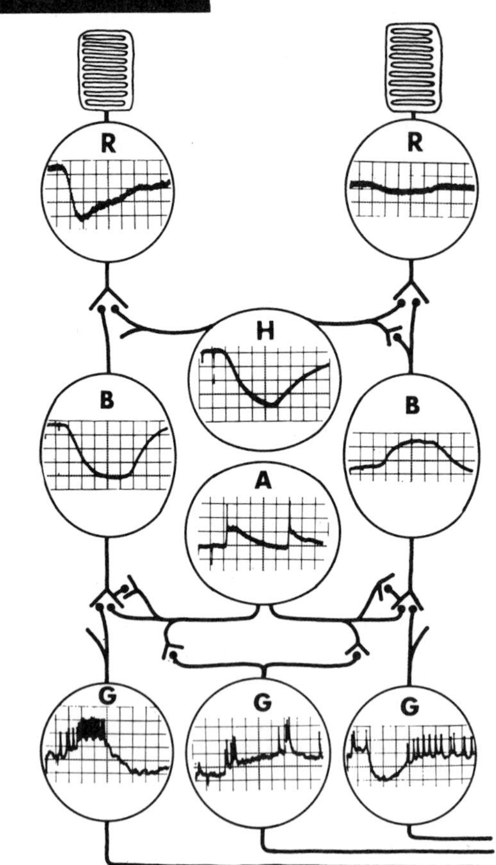

Figure 14–41. Summary diagram correlating the synaptic organization of the vertebrate retina with the intracellularly recorded responses of the neurons. R, receptors; H, horizontal cell; B, bipolar cells; A, amacrine cell; G, ganglion cells. (From Dowling, J. E., Invest. Ophthal. *9*:655–680, 1970.)

wavelengths and depolarization to long wavelengths. Thus, they correspond to a class of measurements known as S-potentials that were discovered by Svaetichin in 1953 and which have been studied most extensively in fish retinas.[141,174,175,223] S-potentials are either simple hyperpolarizations signalling luminosity (L type) (Fig. 14–40) or biphasic chromatic responses (C type). The latter are of two kinds, hyperpolarizing to green and depolarizing to red, or hyperpolarizing to blue and depolarizing to yellow. The yellow-blue units sometimes show an additional hyperpolarization in the red region of the spectrum. Selective adaptation of the retina with wavelengths at one end of the spectrum can enhance the response of C units at the other end, indicating that there are independent chromatic inputs to the mechanism of generation of these S-potentials.

There are several features of S-potentials that have made it difficult to account for them satisfactorily in terms of the properties of single neurons.[225] The amplitude of the response does not depend on the resting level of membrane potential and cannot be altered by injection of current through the micropipette. Furthermore, S-potentials are accompanied by slow conductance changes that seem to lag the potential rather than cause it. Also, C and L types of S-potential have occasionally been observed to interconvert, apparently spontaneously. And finally, the wide extent of the retina which can contribute to an S-potential, a millimeter or more, seems to be large in relation to the lateral spread of a horizontal cell's dendrites. Electrical coupling of horizontal cells has been shown in fish and may account for large receptive fields.[125]

Amacrine Cells. The first phasic responses in the afferent pathway are recorded in amacrine cells (Figs. 14–40, 14–41). Amacrine cells show transient depolarizations and an irregular number of spikes. They may signal "on," "off," or "on-off," depending on the geometrical pattern of stimulus on the retina. It has been suggested that the phasic nature of the response may be caused by local negative feedback loops based on amacrine-to-bipolar synapses in the near vicinity of bipolar-to-amacrine contacts.[67]

Ganglion Cells. Ganglion cells depolarize and give rise to bursts or trains of spikes which are propagated along the optic nerve. The depolarization is probably a generator potential in the true sense of the word, as its magnitude determines the frequency of discharge in the axon. The receptive fields of ganglion cells can be classified as either *simple* or *complex.*

The simple type of receptive field has a center-surround organization somewhat like the receptive fields described for bipolar cells. The size of the receptive field may extend beyond the dendritic arborization of the ganglion cell, and so is thought to involve lateral interactions from amacrine cells. The center of the receptive field may respond to light "on" and the surround to "off," or the ganglion cell may have an "off" center and "on" surround. The size of the receptive field should not be thought of as being static. The degree of retinal inhibition seems to increase with light adaptation, with the result that late in the process of dark adaptation the inhibitory surround of an on-center unit may drop out.[13,14,154]

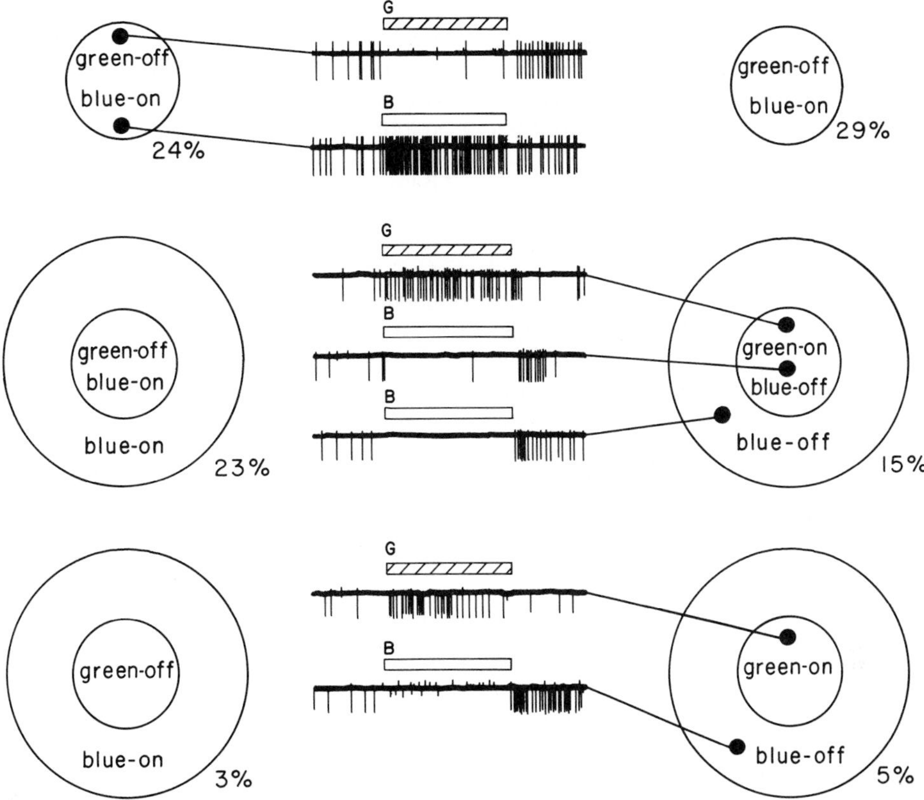

Figure 14–42. Summary diagram of the receptive fields of color-coded ganglion cells in the retina of the ground squirrel, with some sample spike trains. Two kinds of cone are present, maximally sensitive to green and blue light. Each ganglion cell receives input from both classes of cone, being inhibited by one and excited by the other. The ganglion cells can also be classified according to the degree of overlap between the excitatory and inhibitory regions of the receptive field. Percentages give relative frequency of recording of each type. (After Michael.[156, 157])

Such so-called simple ganglion cell receptive fields may be color coded if fed by cones (Fig. 14–42). In the ground squirrel, which has an all cone retina containing green-sensitive cones and blue-sensitive cones, those ganglion cells that have chromatic input may respond to green at "on" and blue at "off" or *vice versa.* For some ganglion cells, the green sensitive part of the receptive field is coincident with the blue sensitive part. For others, green sensitivity is confined to the center whereas blue sensitivity is distributed through the center and surround. For still others, there is a complete separation of the two populations of cones, green at the center and blue in the surround. Such units as these are apparently involved in the heightening of color contrast.[155]

In fish, color coded ganglion cells have been described with receptive fields up to 5 mm (40 to 60°).[60] The responses of these ganglion cells are somewhat more complicated than those of the ground squirrel. The center responds to red "on"–green "off" (or red "off"–green "on"). The retinal regions containing the two types of cone are not quite coincident, with the green-sensitive cones occupying a somewhat larger area. In fact, in the first reports of these ganglion cells this center was taken to be the whole receptive field.[232] The use of stimuli in the shape of an annulus shows that the true surround extends much further out and is truly antagonistic to the center. That is, for a red "on"–green "off" center unit, the response to light in the surround is red "off"–green "on." Such cells, unlike the green "on" center:blue "off" surround units of the squirrel's retina, will respond to heighten color contrast at a spatial border.

The complex ganglion cells are found in such animals as frogs,[148] rabbits,[14, 15] and pigeons,[147] those species with numerous amacrine cell synapses. These complex units give phasic responses to "on" and "off," but their more characteristic feature is their sensitivity to movements of edges through the receptive field. In the retina of the frog, the adequate stimulus for one group of cells is movement of a small convex edge, a feature which makes these cells admirably suited for detecting the natural prey of frogs.[148] Another characteristic of movement detectors is that they are relatively insensitive to the degree of contrast between the moving object and the background. In several species,[14,148,156] cells have been found that respond to unidirectional movement of edges across the retina; movement in the opposite direction causes a suppression of any ongoing activity. The mechanism is believed to involve the spread of a wave of inhibition for 30 microns or so in front of an object moving in the null direction, as shown in the hypothetical diagram of Figure 14–43.

VERTEBRATE AND ARTHROPOD COMPOUND EYES COMPARED

The vertebrate eye, formed by increasing the number of photosensory cells and overlaying them with a lens, has one obvious advantage over the compound eye, which can be thought of as an aggregate of simple eyes, each with slightly different directional sensitivity. The camera eye is superior in its acuity or angular resolving power. At first glance it might seem that nature has simply found two solutions to the problem of making an image-forming eye, and it is just poor luck that the compound eye is inferior. There is, however, one aspect of the vertebrate eye which would make it disadvantageous for very small animals, and that is the inability to bring into focus objects very close to the cornea. As compound eyes have no mechanism for accommodation, this is not a problem. In fact, the ability of a compound eye to resolve a pattern should increase the closer the animal approaches, and in the lives of many arthropods the ability to see objects clearly at a distance of a few millimeters may more than offset the disadvantage of relatively poor angular resolution.

For the vertebrate eye, the short wavelength limit of the spectrum is set by the need to minimize chromatic aberration. Thus, ocular filters invariably discard the near ultraviolet and frequently the violet region of the spectrum. Insects, not having a comparable lens system, are under no such constraint, and many terrestrial species have evolved visual pigments primarily sensitive in the near ultraviolet. Moreover, the ultraviolet reflectance of flowers and the pattern of polarization of sky light (primarily the near ultraviolet wavelengths) play a special role in directing the behavior of some species.[59,150]

Despite the great anatomical differences between vertebrate and arthropod eyes, one

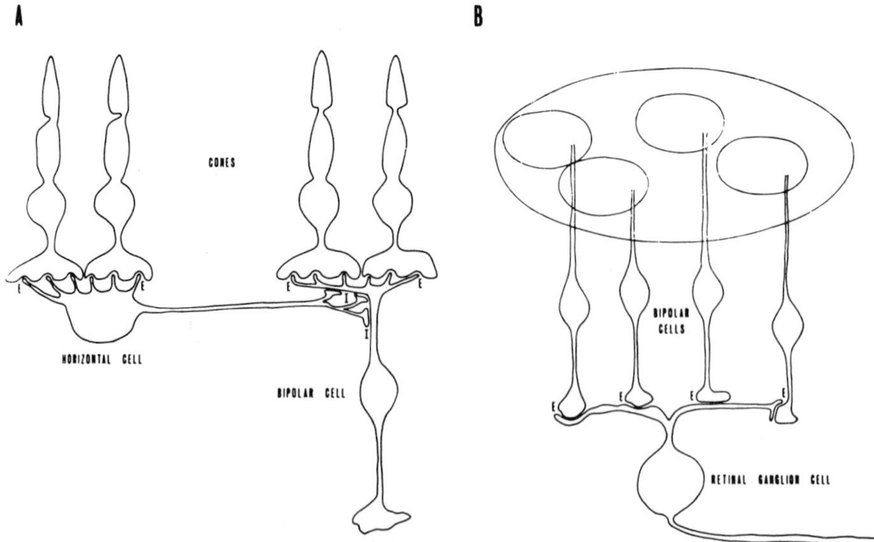

Figure 14–43. Correlation of retinal anatomy with the proposed physiological mechanism for directional selectivity. *A*, A bipolar cell receives excitatory inputs directly from one set of cones and is inhibited, via a horizontal cell, by a second set of cones, laterally displaced with respect to the first group. The sequence of illumination of the two populations of cones determines the directional selectivity of the bipolar cell. *B*, A directionally selective retinal ganglion cell receives excitatory inputs from a number of sequence-discriminating bipolars which have the same preferred-null axis. E = excitatory synapses, I = inhibitory. (From Michael, C. R., J. Neurophysiol. *31*:257–267, 1968.)

is readily struck by certain similarities. Essentially the same photochemistry is employed in both groups, although it may be coupled to somewhat different membrane responses in the two kinds of photoreceptors.

The morphology of vertebrate retinal neurons is very different from the structure of invertebrate interneurons. Correspondingly, the patterns of synaptic contact in the two groups are quite different. Nevertheless, in both vertebrates and arthropods one finds the same kinds of information being extracted by higher order visual interneurons. In neither case does one find cells responding to specific patterns, but rather the emphasis is on contrasts. The "jittery movement fibers" that are found in the optic tracts of Crustacea and the optic lobes of insects have many features in common with the "bug detector" of the frog retina. Likewise, the fibers in both systems that respond to unidirectional movement of edges in the visual field have similar properties. Although much more needs to be learned about these and other units, the suspicion is strong that both kinds of nervous system are abstracting information about the outside world in very similar fashion.

REFERENCES

1. Abrahamson, E. W., and S. E. Ostroy, Prog. Biophys. Molec. Biol. *17*:181–215, 1967. Photochemical and macromolecular aspects of vision.
2. Akhtar, M., P. T. Blosse, and P. B. Dewhurst, Biochem. J. *110*:693–702, 1968. The nature of the retinal-opsin linkage.
3. Ali, M. A., Vision Res. *11*:1225–1288, 1971. Retinomotor responses: characteristics and mechanisms.
4. Allen, D. M., Vision Res. *11*:1077–1112, 1971. Photic control of the proportions of two visual pigments in a fish.
5. Alpern, M., Ann. Rev. Physiol. *30*:279–318, 1968. Distal mechanisms of color vision.
6. Arden, G. B., and H. Ikeda, Vision Res. *6*:171–184, 1966. Effects of hereditary degeneration of the retina on early receptor potential and corneo-fundal potential of rat eye.
7. Autrum, H., *in* Ciba Foundation Symposium on Physiology and Experimental Psychology of Colour Vision. Williams & Wilkins, Baltimore, 1965. Physiological basis of color vision in honeybees.

7a. Autrum, H., and G. Kolb, Z., vergl. Physiol. *60*:450–477, 1968. Spectral sensitivity of single retinula cells in *Aeschna*.

8. Autrum, H., F. Zettler, and M. Järvilehto, Z. vergl. Physiol. *70*:414–424, 1970. Postsynaptic potentials from single monopolar neuron of ganglion opticum I of blowfly *Calliphora*.
9. Ball, S., T. W. Goodwin, and R. A. Morton, Biochem. J. *42*:516–523, 1948. Preparation of retinene$_1$ – vitamin A aldehyde.
10. Barber, V. C., E. M. Evans, and M. F. Land. Z. Zellforsch. *76*:295–312, 1967. Fine structure of the eye of mollusc *Pecten maximus*.
11. Barlow, H. B., J. Exp. Biol. *29*:667–674, 1952. Size of ommatidia in apposition eyes.
12. Barlow, H. B., pp. 163–202, *in* Photophysiology, Vol. 2, edited by A. C. Giese. Academic Press, New York, 1964. The physical limits of visual discrimination.
13. Barlow, H. B., R. FitzHugh, and S. W. Kuffler, J. Physiol. *137*:338–354, 1957. Change of organization in receptive fields of cat retina during dark adaptation.
14. Barlow, H. B., R. M. Hill, and W. R. Levick, J. Physiol. *173*:377–407, 1964. Retinal ganglion cells responding selectively to direction and speed of image motion in rabbit.
15. Barlow, H. B., and W. R. Levick, J. Physiol. *178*:477–504, 1965. Mechanism of directionally selective units in rabbit retina.
16. Baylor, D. A., and M. G. F. Fuortes, J. Physiol. *207*:77–92, 1970. Electrical responses of single cones in turtle retina.
17. Behrens, M. E., and V. J. Wulff, J. Gen. Physiol. *48*:1081–1093, 1965. Light-initiated responses of retinula and eccentric cells in *Limulus* lateral eye.
18. Bennett, R. R., J. Tunstall, and G. A. Horridge, Z. vergl. Physiol. *55*:195–206, 1967. Spectral sensitivity of single retinula cells of the locust.
19. Bishop, L. G., D. G. Keehn, and G. D. McCann, J. Neurophysiol. *31*:509–525, 1968. Motion detection by interneurons of optic lobes and brains of flies *Calliphora phaenicia* and *Musca domestica*.
20. Bitensky, M. W., R. E. Gorman, and W. H. Miller, Proc. Nat. Acad. Sci. *68*:561–562, 1971. Adenyl cyclase as a link between photon capture and changes in membrane permeability of frog photoreceptors.
21. Boll, F., Mber. Berl. Akad. Wiss. *12*:783–788, 1876. Anatomy and physiology of the retina.
22. Bortoff, A., and A. L. Norton, Vision Res. *5*:527–533, 1965. Simultaneous recording of photoreceptor potentials and PIII component of the ERG.
23. Bownds, D., Nature *216*:1178–1181, 1967. Site of attachment of retinal in rhodopsin.
24. Braitenberg, V., Exp. Brain Res. *3*:271–298, 1967. Pattern of projection in visual system of fly: retina-lamina projections.
25. Bridges, C. D. B., pp. 31–78, *in* Comprehensive Biochemistry, Vol. 27, edited by M. Florkin and E. H. Stotz. Elsevier Publishing Co., Amsterdam, 1967.
26. Bridges, C. D. B., Vision Res. 7:349–369, 1967. Spectroscopic properties of porphyropsins.
27. Bridges, C. D. B., *in* Handbook of Sensory Physiology, Vol. 7, Photochemistry of Vision, edited by H. J. A. Dartnall. Springer-Verlag, Berlin, 1972. The rhodopsin-porphyropsin visual system.
28. Briggs, W. R., pp. 223–271, *in* Photophysiology, Vol. 1, edited by A. C. Giese. Academic Press, New York, 1964. Phototropism in higher plants.
29. Brindley, G. S., and A. R. Gardner-Medwin, J. Physiol. *182*:185–194, 1966. Origin of early receptor potential of the retina.
30. Brown, H. M., S. Hagiwara, H. Koike, and R. W. Meech, J. Physiol. *208*:385–413, 1970. Membrane properties of barnacle photoreceptor examined by voltage clamp technique.
31. Brown, K. T., and M. Murakami, Nature *201*:626–628, 1964. A new receptor potential of the monkey retina with no detectable latency.
32. Brown, P. K., and G. Wald, Nature *200*:37–43, 1963. Visual pigments in human and monkey retinas.
33. Brown, P. K., and G. Wald, Science *144*:45–52, 1964. Visual pigments in single rods and cones of human retina.
34. Bruce, V. G., and D. H. Minis, Science *163*:583–585, 1969. Circadian clock action spectrum in a photoperiodic moth.
35. Bruckmoser, P., Z. vergl. Physiol. *59*:187–204, 1968. Spectral sensitivity of single sight cells of backswimmer *Notonecta glauca* L. (Heteroptera).
36. deBruin, G. H. P., and D. J. Crisp, J. Exp. Biol. *34*:447–463, 1957. Influence of pigment migration on vision of higher crustaceans.
37. Bullock, T. H., and G. A. Horridge, Structure and Function in the Nervous Systems of Invertebrates. W. H. Freeman and Co., San Francisco, 1965.
38. Burkhardt, D., Symp. Soc. Exp. Biol. *16*:86–109, 1962. Spectral sensitivity and other response characteristics of single visual cells in the arthropod eye.
39. Burtt, E. T., and W. T. Catton, J. Physiol. *133*:68–88, 1954. Electrical responses to visual stimulation in the optic lobes of the locust and certain other insects.
40. Burtt, E. T., and W. T. Catton, J. Physiol. *154*:479–490, 1960. Properties of single-unit discharges in optic lobe of locust.
41. Bush, B. M. H., C. A. G. Wiersma, and T. H. Waterman, J. Cell. Comp. Physiol. *64*:327–346, 1964. Efferent mechanoreceptive responses in optic nerve of crab *Podophthalmus*.
42. Butler, R., Z. vergl. Physiol. *72*:67–80, 1971. Identification and mapping of spectral cell types in retina of *Periplaneta americana*.
43. Cajal, S. R., and D. Sánchez, Trab. Lab. Invest. Biol. Madrid *13*:1–164. Contribución al eonocimiento de los centros nerviosos de los insectos.
44. Campbell, F. W., and D. G. Green, J. Physiol. *181*:576–593, 1965. Optical and retinal factors affecting visual resolution.
45. Campbell, F. W., and R. W. Gubisch, J. Physiol. *186*:558–578, 1966. Optical quality of human eye.
46. Clayton, R. K., Arch. Mikrobiol. *19*:107–124, 1953. Studies in phototaxis of *Rhodospirillum rubrum:* action spectrum, growth in green light, and Weber Law adherence.
47. Clayton, R. K., pp. 51–77, *in* Photophysiology, Vol. 2, edited by A. C. Giese. Academic Press, New York, 1964. Phototaxis in microorganisms.
48. Cohen, A. I., Biol. Rev. *38*:427–459, 1963. Vertebrate retinal cells and their organization.
49. Cohen, A. I., J. Cell. Biol. *37*:424–444, 1968. Linkage to extracellular space of outer segment saccules of frog cones.
50. Collett, T. S., and A. D. Blest, Nature *212*:1330–1333, 1966. Binocular directionally sensitive selective neurons possibly involved in optomotor response of insects.
51. Cone, R. A., Science *155*:1128–1131, 1967. Early receptor potential: photoreversible charge displacement in rhodopsin.
52. Cone, R. A., and P. K. Brown, Science *156*:536, 1967. Dependence of early receptor potential on orientation of rhodopsin.
53. Cone, R. A., and W. H. Cobbs III, Nature *221*:820–822, 1969. Rhodopsin cycle in living eye of rat.

54. Cowan, W. M., and T. P. S. Powell, Proc. Roy. Soc. Lond. B *158*:232–252, 1963. Centrifugal fibers in avian visual system.
55. Crescitelli, F., *in* Handbook of Sensory Physiology, Vol. 7, Photochemistry of Vision, edited by H. J. A. Dartnall. Springer-Verlag, Berlin, 1972. Visual cells and visual pigments of vertebrate eye.
56. Dartnall, H. J. A., J. Physiol. *116*:257–289, 1952. Visual pigment 467, a photosensitive pigment present in tench retinae.
57. Dartnall, H. J. A., and J. N. Lythgoe, Vision Res. *5*:81–100, 1965. Spectral clustering of visual pigments.
58. Daumer, K., Z. vergl. Physiol. *38*:413–478, 1956. Stimulation measurement investigation of color vision in bees.
59. Daumer, K., Z. vergl. Physiol. *41*:49–110, 1958. Flower colors: how bees see them.
60. Daw, N. W., J. Physiol. *197*:567–592, 1968. Color-coded ganglion cells in goldfish retina: extension of their receptive fields by means of new stimuli.
61. Denton, E. J., Proc. Roy. Soc. Lond. B *150*:78–94, 1959. Contributions of oriented photosensitive and other molecules to the absorption of the whole retina.
62. Dowling, J. E., Nature *188*:114–118, 1960. Chemistry of visual adaptation in rat.
63. Dowling, J. E., J. Gen. Physiol. *46*:1287–1301, 1963. Neural and photochemical mechanisms of visual adaptation in rat.
64. Dowling, J. E., *in* Molecular Organization and Biological Function, edited by J. M. Allen. Harper & Row, New York, 1967. The organization of vertebrate visual receptors.
65. Dowling, J. E., Science *155*:273–279, 1967. The site of visual adaptation.
66. Dowling, J. E., Proc. Roy. Soc. Lond. B *170*:205–228, 1968. Synaptic organization of frog retina: electron microscopic analysis comparing retinas of frogs and primates.
67. Dowling, J. E., Invest. Ophthal. *9*:655–680, 1970. Organization of vertebrate retinas.
68. Dowling, J. E., and B. B. Boycott, Proc. Roy. Soc. Lond. B *166*:80–111, 1966. Organization of the primate retina: electron microscopy.
69. Dowling, J. E., and W. M. Cowan, Zellforsch. Mikr. Anat. *71*:14–28, 1966. Electron microscopic study of normal and degenerating centrifugal fiber terminals in pigeon retina.
70. Dowling, J. E., and F. S. Werblin, J. Neurophysiol. *32*:315–338, 1969. Organization of retina of mudpuppy *Necturus maculosis:* synaptic structure.
71. Dubin, M. W., J. Comp. Neurol. *140*:479–505, 1970. Inner plexiform layer of the vertebrate retina.
72. Eakin, R. M., *in* Evolutionary Biology, Vol. 2, edited by T. Dobzhansky, M. K. Hecht, and W. C. Steere. Appleton-Century-Crofts, New York, 1968. Evolution of photoreceptors.
73. Eakin, R. M., Amer. Sci. *58*:73–79, 1970. A third eye.
74. Eakin, R. M., *in* Handbook of Sensory Physiology, Vol. 7, Photochemistry of Vision, edited by H. J. A. Dartnall. Springer-Verlag, Berlin, 1972. Structure of invertebrate photoreceptors.
75. Eguchi, E., J. Cell. Comp. Physiol. *66*:411–430, 1965. Rhabdom structure and receptor potentials in single crayfish retinular cells.
76. Eguchi, E., and T. H. Waterman, *in* The Functional Organization of the Compound Eye, edited by C. G. Bernhard. Pergamon Press Ltd., Oxford, 1966. Fine structure patterns in crustacean rhabdoms.
77. Eguchi, E., and T. H. Waterman, Z. Zellforsch. *84*:87–101, 1968. Cellular basis for polarized light perception in spider crab *Libinia*.
77a. Eheim, W. P., and R. Wehner, Kybernetik *10*:168–179, 1972. Visual field of central ommatidia in apposition eyes of *Apis* and *Cataglyphis* (Apidae, Formicidae; Hymenoptera).
78. Enoch, J. M., J. Opt. Soc. Amer. *50*:1025–1026, 1960. Wave guide modes: are they present and what is their role in the visual mechanism?
79. Enoch, J. M., J. Opt. Soc. Amer. *51*:1122–1126, 1961. Nature of the transmission of energy in retinal receptors.
80. Exner, S., Die Physiologie der facettieren Augen von Krebsen und Insekten [Physiology of facetted eyes of crabs and insects]. Franz Deuticke, Vienna, 1891.
81. Feynman, R. P., R. B. Leighton, and M. Sands, The Feynman Lectures on Physics, Vol. 1. Chap. 36. Addison-Wesley Publishing Co., Reading, Mass., 1964.
82. Fraenkel, G. S., and D. L. Gunn, The Orientation of Animals. Clarendon Press, Oxford, 1940.
83. Frank, K. D., and W. F. Zimmerman, Science *163*:688–689, 1969. Action spectra for phase shifts of a circadian rhythm in *Drosophila*.
84. von Frisch, K., The Dance Language and Orientation of Bees. Harvard University Press, Cambridge, 1967.
85. Fulpius, B., and F. Baumann, J. Gen. Physiol. *53*:541–561, 1969. Effects of sodium, potassium, and calcium ions on slow and spike potentials in single photoreceptors.
86. Fuortes, M. G. F., J. Physiol. *148*:14–28, 1959. Initiation of impulses in visual cells of *Limulus*.
87. Fuortes, M. G. F., and S. Yeandle, J. Gen. Physiol. *47*:443–463, 1964. Probability of occurrence of discrete potential waves in eye of *Limulus*.
88. Futterman, S., J. Biol. Chem. *238*:1145–1150, 1963. The role of reduced triphosphopyridine nucleotide in the visual cycle.
89. Goldsmith, T. H., J. Gen. Physiol. *43*:775–799, 1960. Nature of the retinal action potential, and spectral sensitivities of ultraviolet and green receptor systems in compound eye of worker honeybee.
90. Goldsmith, T. H., *in* Sensory Communication, edited by W. A. Rosenblith. MIT Press, Cambridge, 1961. Physiological basis of wavelength discrimination in eye of honeybee.
91. Goldsmith, T. H., pp. 397–462, *in* The Physiology of Insecta, edited by M. Rockstein. Academic Press, New York, 1964. The visual system of insects.
92. Goldsmith, T. H., *in* Handbook of Sensory Physiology, Vol. 7, Photochemistry of Vision, edited by H. J. A. Dartnall. Springer-Verlag, Berlin, 1972. Natural history of invertebrate visual pigments.
93. Goldsmith, T. H., and G. D. Bernard, *in* The Physiology of Insecta, 2nd edition, edited by M. Rockstein. Academic Press, New York, 1973 (in press).
94. Gorman, A. L. F., J. S. McReynolds, and S. N. Barnes, Science *172*:1052–1054, 1971. Photoreceptors in primitive chordates: fine structure, hyperpolarizing receptor potentials, and evolution.
95. Hagins, W. A., Cold Spring Harbor Symp. Quant. Biol. *30*:403–418, 1965. Electrical signs of information flow in photoreceptors.
96. Hagins, W. A., and W. H. Jennings, Disc. Faraday Soc. *27*:180–190, 1959. Radiationless migration of electronic excitation in retinal rods.
97. Hagins, W. A., and P. A. Liebman, Abstr. Biophys. Soc. 7th Annual Meeting. New York, N.Y.M.E. 6, 1963. Relationship between photochemical and electrical processes in living squid photoreceptors.
98. Hagins, W. A., and R. E. McGaughy, Science *157*:813–816, 1967. Molecular and thermal origins of fast photoelectric effects in squid retina.
99. Hagins, W. A., R. D. Penn, and S. Yoshikami, Biophys. J. *10*:380–412, 1970. Dark current and photocurrent in retinal rods.
100. Hagins, W. A., and H. Rüppel, Fed. Proc. *30*:64–68, 1971. Fast photoelectric effects and properties of vertebrate photoreceptors as cables.
101. Hagins, W. A., H. V. Zonana, and R. G. Adams, Nature *194*:844–847, 1962. Local membrane current in the outer segments of squid photoreceptors.
102. Halldal, P., Physiol. Plantarum *11*:118–153, 1958. Action spectra of phototaxis and related problems in Volvocales, *Ulva*-Gametes, and Dinophyceae.
103. Hamasaki, D. I., J. Physiol. *167*:156–168, 1963. Effect of sodium ion concentration on electroretinogram of isolated retina of frog.
104. Hámori, J., and G. A. Horridge, J. Cell. Sci. *1*:249–256, 1966. The lobster optic lamina. I. General organization.
105. Hámori, J., and G. A. Horridge, J. Cell. Sci. *1*:257–270, 1966. The lobster optic lamina. II. Types of synapse.
106. Hartline, H. K., Science *164*:270–278, 1969. Visual receptors and retinal interaction.
107. Hays, D., and T. H. Goldsmith, Z. vergl. Physiol. *65*:218–232, 1969. Microspectrophotometry of the visual pigment of the spider crab *Libinia emarginata*.
108. Hecht, S., S. Shlaer, and M. H. Pirenne, J. Gen. Physiol. *25*:819–840, 1942. Energy, quanta, and vision.
109. Heller, J., Biochemistry *8*:675–679, 1969. Comparative study of a membrane protein. Characterization of bovine, rat, and frog visual pigments.
110. Henderson, L. J., The Fitness of the Environment. Macmillan, New York, 1913.
111. Höglund, G., Acta Physiol. Scand. *69*:Suppl. 282:1–56, 1966. Pigment migration, light screening and receptor sensitivity in compound eye of nocturnal Lepidoptera.
112. Horridge, G. A., Z. vergl. Physiol. *62*:1–37, 1969. Unit studies on retina of dragonflies.
113. Horridge, G. A., and I. A. Meinertzhagen, Z. vergl. Physiol. *66*:369–378, 1970. The exact neural projection of the visual fields upon the first and second ganglia of the insect eye.
114. Horridge, G. A., J. H. Scholes, S. Shaw, and J. Tunstall, pp. 165–202 *in* The Physiology of the Insect Central Nervous System, edited by J. E. Treherne and J. W. L. Beament. Academic Press, New York, 1965. Extracellular recordings from single neurons in optic lobe and brain of locust.
115. Horridge, G. A., B. Walcott, and A. C. Ioannides, Proc. Roy. Soc. Lond. B *175*:83–94, 1970. Tiered retina of *Dytiscus*: a new type of compound eye.
116. Hubbard, R., J. Gen. Physiol. *37*:381–399, 1954. Molecular weight of rhodopsin and nature of the rhodopsin-digitonin complex.
117. Hubbard, R., J. Gen. Physiol. *39*:935–962, 1956. Retinene isomerase.
118. Hubbard, R., and A. Kropf, Proc. Nat. Acad. Sci. *44*:130–139, 1958. The action of light on rhodopsin.
119. Hubbard, R., and R. C. C. St. George, J. Gen. Physiol. *41*:501–528, 1958. The rhodopsin system of the squid.
120. Hubbard, R., and G. Wald, Proc. Nat. Acad. Sci. *37*:69–79, 1951. Mechanism of rhodopsin synthesis.
121. Hutchinson, G. E., Treatise on Limnology, Vol. 1: Geography, Physics, and Chemistry. John Wiley & Sons, New York, 1957.
122. Jander, R., and T. H. Waterman, J. Cell. Comp. Physiol. *56*:137–160, 1960. Sensory discrimination between polarized light and light intensity patterns by arthropods.
122a. Järvilehto, M., and F. Zettler, Z. vergl. Physiol. *75*:422–440, 1971. Localized potentials from pre- and postsynaptic elements in external plexiform layer of insect retina.
123. Kaiser, W. and L. G. Bishop, Z. vergl. Physiol. 7:403–413, 1970. Directionally selective motion detecting units in optic lobe of honeybee.
124. Kaneko, A., J. Physiol. *207*:623–633, 1970. Physiological and morphological identification of horizontal, dipolar, and amacrine cells in goldfish retina.
125. Kaneko, A., J. Physiol. *213*:95–105, 1971. Electrical connections between horizontal cells in dogfish retina.
126. Kennedy, D., J. Gen. Physiol. *44*:277–299, 1960. Neural photoreception in a lamellibranch mollusc.

127. Kennedy, D., and R. D. Milkman, Biol. Bull. *111*:375–386, 1956. Selective light absorption by lenses of lower vertebrates and its influence on spectral sensitivity.
127a. King-Smith, P. E., Vision Res. *9*:1391–1399, 1969. Absorption spectra and function of colored oil drops in pigeon retina.
128. Kirshfeld, K., Exp. Brain Res. *3*:248–270, 1967. Projection of the optic environment on the screen of the rhabdomere in complex eye of *Musca*.
128a. Korenbrot, J. I., and R. A. Cone, J. Gen. Physiol. *60*:20–45, 1972. Dark ionic flux and effects of light on rod outer segments.
129. Köttgen, E., and G. Abelsdorff, Z. Psychol. Physiol. Sinnesorg *12*:161–184, 1896. Absorption und Zersetzung des Sehpurpurs bei den Wirbeltieren [Absorption and decomposition of visual purples by vertebrates].
130. Kühne, W., Über den Sehpurpur Untersuchungen aus dem Physiologischen Institut der Univ. Heidelberg *1*:15–103, 1877.
131. Kuiper, J. W., Symp. Soc. Exp. Biol. *16*:58–71, 1962. Optics of the compound eye.
132. Kunze, P., Verh. Zool. Ges. Köln *64*:234–238, 1970. Behavioral physiology and optical experiments on the superposition theory of image location in compound eyes.
133. Land, M., Symp. Zool. Soc. Lond. *23*:75–96, 1968. Functional aspects of optical and retinal organization of mollusc eye.
134. Langer, H., and B. Thorell, Exp. Cell. Res. *41*:673–676, 1966. Microspectrophotometry of single rhabdomeres in insect eye.
135. Lasansky, A., and M. G. F. Fuortes, J. Cell Biol. *42*:241–252, 1969. Site of origin of electrical responses in visual cells of leech.
135a. Laughlin, S. B., and G. A. Horridge, Z. vergl. Physiol. *74*:329–339, 1971. Angular sensitivity of retinular cells of dark-adapted worker bee.
136. Leedale, G., B. J. D. Meeuse, and E. G. Pringsheim, Arch. Mikrobiol. *50*:68–102, 1965. Structure and physiology of *Euglena spirogyra*.
137. Lees, A. D., pp. 47–137 *in* Photophysiology, Vol. 4, edited by A. C. Giese. Academic Press, New York, 1968. Photoperiodism in insects.
138. Liebman, P. A., Biophys. J. *2*:161–178, 1962. *In situ* microspectrophotometric studies on pigments of single retinal rods.
139. Liebman, P. A., *in* Handbook of Sensory Physiology, Vol. 7, Photochemistry of Vision, edited by H. J. A. Dartnall, Springer-Verlag, Berlin, 1972. Microspectrophotometry of photoreceptors.
140. Loeb, J., and H. Wasteneys, Z. Exp. Zool. *19*:23–35, 1915. Relative efficiency of various parts of the spectrum for heliotropic reactions of animals and plants.
141. MacNichol, E. F., Jr., Vision Res. *4*:119–133, 1964. Retinal mechanisms of color vision.
142. MacNichol, E. F., Jr., Sci. Amer. *211*:48–56, 1964. Three pigment color vision.
143. Marks, W. B., J. Physiol. *178*:14–32, 1965. Visual pigments of single goldfish cones.
144. Marks, W. B., W. H. Dobelle, and E. F. MacNichol, Jr., Science *143*: 1181–1183, 1964. Visual pigments of single primate cones.
145. Mast, S. O., and N. Stabler, Biol. Bull. *73*:126–133, 1937. Relation between luminous intensity, adaptation to light, and rate of locomotion in *Amoeba proteus* (Leidy).
146. Mathews, R. G., R. Hubbard, P. K. Brown, and G. Wald, J. Gen. Physiol. *47*:215–240, 1963. Tautomeric forms of metarhodopsin.
147. Maturana, H. R., and S. Frenk, Science *142*:977–979, 1963. Directional movement and horizontal edge detectors in pigeon retina.
148. Maturana, H. R., J. Y. Lettvin, W. S. McCulloch, and W. H. Pitts, J. Gen. Physiol. *43*(Suppl.):129–175, 1960. Anatomy and physiology of vision in frog (*Rana pipiens*).
149. Mazokhin-Porshnyakov, G. A., *in* The Functional Organization of the Compound Eye, edited by C. G. Bernhard. Pergamon Press Ltd., Oxford, 1966. Recognition of colored objects by insects.
150. Mazokhin-Porshnyakov, G. A., Insect Vision (translated from Russian). Plenum Press, New York, 1969.
151. McCann, G. D., and J. C. Dill, J. Gen. Physiol. *53*:385–413, 1969. Properties of intensity, form, and motion perception in visual nervous systems of *Calliphora phaenicia* and *Musca domestica*.
152. McReynolds, J. S., and A. L. F. Gorman, J. Gen. Physiol. *56*:392–406, 1970. Membrane conductances and spectral sensitivities of *Pecten* photoreceptors.
153. McReynolds, J. S., and A. L. F. Gorman, J. Gen. Physiol. *56*:376–391, 1970. Photoreceptor potentials of opposite polarity in eye of scallop *Pecten irradians*.
154. Michael, C. R., J. Neurophysiol. *31*:249–256, 1968. Receptive fields of single optic nerve fibers in a mammal with an all-cone retina. I. Contrast sensitive units.
155. Michael, C. R., J. Neurophysiol. *31*:257–267, 1968. Receptive fields of single optic nerve fibers in a mammal with an all-cone retina. II. Directionally selective units.
156. Michael, C. R., J. Neurophysiol. *31*:268–282, 1968. Receptive fields of single optic nerve fibers in a mammal with an all-cone retina. III. Opponent color fibers.
157. Michael, C. R., Sci. Amer. *220*:104–114, 1969. Retinal processing of visual images.
158. Millecchia, R., and A. Mauro, J. Gen. Physiol. *54*:331–351, 1969. Ventral photoreceptor cells of *Limulus*.
159. Miller, R. F., and J. E. Dowling, J. Neurophysiol. *33*:323–341, 1970. Intracellular responses of the Müller (glial) cells of mudpuppy retina: their relation to b-wave of the electroretinogram.
160. Miller, W. H., J. Biophys. Biochem. Cytol. *3*:421–428, 1957. Morphology of ommatidia of compound eye of *Limulus*.
161. Miller, W. H., G. D. Bernard, and J. L. Allen, Science *162*:759–771, 1968. Optics of insect compound eyes.
162. Millott, N., Endeavour *16*:19–28, 1957. Animal photosensitivity, with special reference to eyeless forms.
163. Millott, N., and M. Yoshida, J. Exp. Biol. *34*:394–401, 1957. Spectral sensitivity of echinoid *Diadema antillarum* Philippi.
164. Milne, L. J., and M. Milne, pp. 621–645 *in* Handbook of Physiology, Sec. 1, Vol. 1, Neurophysiology. Amer. Physiol. Soc., Washington, D. C., 1959. Photosensitivity in invertebrates.
165. Moody, M. F., J. Exp. Biol. *39*:21–30, 1962. Evidence for intraocular discrimination of vertically and horizontally polarized light by *Octopus*.
166. Moody, M. F., Biol. Rev. *39*:43–86, 1964. Photoreceptor organelles in animals.
167. Moody, M. F., and J. R. Parriss, Z. vergl. Physiol. *44*:268–291, 1961. Discrimination of polarized light by *Octopus*.
168. Morton, R. A., and G. A. J. Pitt, pp. 97–171 *in* Advances in Enzymology, Vol. 32, edited by F. F. Nord. John Wiley & Sons, New York, 1969. Aspects of visual pigment research.
169. Mote, M. I., and T. H. Goldsmith, J. Exp. Zool. *173*:137–146, 1970. Spectral sensitivities of color receptors in compound eye of cockroach *Periplaneta*.
170. Mote, M. I., and T. H. Goldsmith, Science *171*:1254–1255, 1971. Compound eyes: localization of two color receptors in same ommatidium.
171. Müller, J., Zur vergleichenden Physiologie des Gesichtsinnes des Menschen und Tiere [On the Comparative Physiology of the Visual Senses of Men and Animals]. Cnobloch, Leipzig, 1826.
172. Muntz, W. R. A., and D. P. M. Northmore, Vision Res. *11*:551–562, 1971. Visual pigments from different parts of the retina in rudd and trout.
173. Munz, F. W., J. Physiol. *140*:220–235, 1958. The photosensitive pigments from the retinae of certain deep sea fishes.
174. Naka, K. I., and W. A. H. Rushton, J. Physiol. *185*:536–555, 1966. S-potentials from color units in the retina of fish (Cyprinidae).
175. Naka, K. I., and W. A. H. Rushton, J. Physiol. *192*:437–461, 1967. Generation and spread of S-potentials in fish (Cyprinidae).
176. Nolte, J., J. E. Brown, and T. G. Smith, Jr., Science *162*:677–679, 1968. Hyperpolarizing component of receptor potential in median ocellus of *Limulus*.
177. Northrup, R. B., and E. F. Guignon, J. Insect Physiol. *16*:691–713, 1970. Information processing in optic lobes of the lubber grasshopper.
178. Oroshnik, W., J. Amer. Chem. Soc. *78*:2651–2652, 1956. Synthesis and configuration of neo-*b* vitamin A and neoretinene *b*.
179. Østerberg, G., Acta Ophthal. Suppl. 6, 1935. Topography of the layer of rods and cones in the human retina.
180. Page, R. M., pp. 65–90 *in* Photophysiology, Vol. 3, edited by A. C. Giese. Academic Press, New York, 1968. Phototropism in fungi.
181. Pak, W. L., Cold Spring Harbor Symp. Quant. Biol. *30*:493–499, 1965. Properties of early electrical response in vertebrate retina.
182. Pak, W. L., and R. A. Cone, Nature *204*:836–838, 1964. Isolation and identification of the initial peak of the early receptor potential.
183. Penn, R. D., and W. A. Hagins, Nature *223*:201–205, 1969. Signal transmission along retinal rods and the origin of the electroretinographic a-wave.
184. Pirenne, M. H., Vision and the Eye, 2nd edition. Chapman and Hall, Ltd., London, 1967.
185. Poincelot, R. P., P. G. Millar, R. L. Kimbel, Jr., and E. W. Abrahamson, Biochemistry *9*:1809–1816, 1970. Determination of the chromophoric binding site in native bovine rhodopsin.
186. Prosser, C. L., J. Exp. Biol. *12*:95–104, 1935. Impulses in the segmental nerves of the earthworm.
187. Purple, R. L., and F. Dodge, *in* The Functional Organization of the Compound Eye, edited by C. G. Bernhard. Pergamon Press Ltd., Oxord, 1966. Self inhibition in eye of *Limulus*.
188. Rabinowitch, E., and Govindjee, Photosynthesis. John Wiley & Sons, New York, 1969.
189. Ratliff, F., *in* The Functional Organization of the Compound Eye, edited by C. G. Bernhard. Pergamon Press Ltd., Oxford, 1966. Selective adaptation of local regions in the rhabdom in an ommatidium of compound eye of *Limulus*.
190. Ratliff, F., and H. K. Hartline, J. Gen. Physiol. *42*:1241–1255, 1959. Responses of *Limulus* optic nerve fibers to patterns of illumination on receptor mosaic.
191. Reuter, T. E., R. H. White, and G. Wald, J. Gen. Physiol. *58*:351–371, 1971. Rhodopsin and porphyropsin fields in adult bullfrog retina.
192. Ripps, H., and R. A. Weale, pp. 127–168 *in* Photophysiology, Vol. 5, edited by A. C. Giese. Academic Press, New York, 1970. Photophysiology of vertebrate color vision.
193. de Robertis, E., J. Gen. Physiol. *43*(Suppl. 2):1–6, 1960. Ultrastructure and morphogenesis of photoreceptors.
194. Robinson, N., editor, Solar Radiation. Elsevier Publishing Co., Amsterdam, 1966.
195. Ruck, P., J. Gen. Physiol. *44*:605–657, 1961. Electrophysiology of the insect dorsal ocellus.
196. Ruck, P., J. Gen. Physiol. *49*:289–307, 1965. Components of visual system of dragonfly.
197. Rushton, W. A. H., J. Physiol. *156*:166–178, 1961. Dark-adaptation of regeneration of rhodopsin.
198. Rushton, W. A. H., pp. 706–723 *in* Light and Life, edited by W. D. McElroy and B. Glass. Johns Hopkins Press, Baltimore, 1961. The intensity factor in vision.

199. Rushton, W. A. H., J. Physiol. *168*:345–359, 1963. A cone pigment in the protanope.
200. Rushton, W. A. H., J. Physiol. *176*:24–37, 1965. A foveal pigment in the deuteranope.
201. Rushton, W. A. H., J. Physiol. *176*:38–45, 1965. Cone pigment kinetics in the deuteranope.
201a. Rushton, W. A. H., *in* Handbook of Sensory Physiology, vol. 7, Photochemistry of Vision, edited by H. J. A. Dartnall. Springer-Verlag, Berlin, 1972. Visual pigments in man.
202. Rutherford, D. J., and G. A. Horridge, Quart. J. Micr. Sci. *106*:119–130, 1965. Rhabdom of lobster eye.
203. Scholes, J., Cold Spring Harbor Symp. Quant. Biol. *30*:517–527, 1965. Discontinuity of excitation process in locust visual cells.
204. Scholes, J., Kybernetik *6*:149–162, 1969. Electrical responses of retinal receptors and lamina in visual system of *Musca*.
205. Shaw, S. R., Z. vergl. Physiol. *55*:183–194, 1967. Simultaneous recording from two cells in locust eye.
206. Shaw, S. R., Symp. Zool. Soc. Lond. *23*:135–163, 1968. Organization of locust retina.
207. Shaw, S. R., Vision Res. *9*:999–1029, 1969. Interreceptor coupling in ommatidia of drone honeybee and locust compound eyes.
208. Shaw, S. R., Science *164*:88–90, 1969. Optics of arthropod compound eye.
209. Shaw, S. R., Vision Res. *9*:1031–1040, 1969. Sense-cell structure and interspecies comparisons of polarized-light absorption in arthropod compound eyes.
210. Sillman, A. J., H. Ito, and T. Tomita, Vision Res. *9*:1435–1442, 1969. Studies on the mass receptor potential of the isolated frog retina. I. General properties of the response.
211. Sillman, A. J., H. Ito, and T. Tomita, Vision Res. *9*:1443–1451, 1969. Studies on the mass receptor potential of the isolated frog retina. II. On the basis of the ionic mechanism.
212. Sjöstrand, F. S., J. Ultrastruct. Res. *2*:122–170, 1958. Ultrastructure of retinal rod synapses of guinea pig eye.
213. Sjöstrand, F. S., Ergebn. Biol. *21*:128–160, 1959. Ultrastructure of retinal receptors of vertebrate eye.
214. Smith, H., Nature *227*:665–668, 1970. Phytochrome and photomorphogenesis in plants.
215. Smith, T. G., and F. Baumann, Prog. Brain Res. *31*:313–349, 1969. Functional organization within ommatidium of lateral eye of *Limulus*.
215a. Snyder, A. W., Z. vergl. Physiol. *76*:438–445, 1972. Angular sensitivity of bee ommatidium.
216. Steinberg, R. H., and R. Schmidt, Vision Res. *10*:817–820, 1970. Identification of horizontal cells as S-potential generators in cat retina.
217. Stell, W. K., Amer. J. Anat. *121*:401–424, 1967. Structure and relationship of horizontal cells and photoreceptor-bipolar synaptic complexes in goldfish retina.
218. Stieve, H., Cold Spring Harbor Symp. Quant. Biol. *30*:451–456, 1965. Interpretation of generator potential in terms of ionic processes.
219. Stiles, W. S., and B. H. Crawford, Proc. Roy. Soc. Lond. B *112*:428–450, 1933. Luminous efficiency of rays entering eye pupil at different points.
220. Strausfeld, N. J., Phil. Trans. Roy Soc. Lond. B *258*:135–223, 1970. Golgi studies on insects. II. Optic lobes of Diptera.
221. Strausfeld, N. J., and A. D. Blest, Phil. Trans. Roy. Soc. Lond. B *258*: 81–134, 1970. Golgi studies on insects. I. Optic lobes of Lepidoptera.
222. Strausfeld, N. J., and V. Braitenberg, Z. vergl. Physiol. *70*:95–104, 1970. The compound eye of the fly (*Musca domestica*): connections between the cartridges of the lamina ganglionaris.
223. Svaetichin, G., and E. F. MacNichol, Ann. N. Y. Acad. Sci. *74*:385–404, 1958. Retinal mechanisms for chromatic and achromatic vision.
224. Sverdrup, H. U., M. W. Johnson, and R. H. Fleming, The Oceans, Their Physics, Chemistry, and General Biology. Prentice-Hall, Inc., New York, 1942.
224a. Tomita, T., J. Neurophysiol. *21*:419–429, 1958. Lateral inhibition in eye of *Limulus*.
225. Tomita, T., Cold Spring Harbor Symp. Quant. Biol. *30*:559–566, 1965. Electrophysiological study of mechanisms subserving color coding in fish retina.
226. Tomita, T., Quart. Rev. Biophys. *3*:179–222, 1970. Electrical activity of vertebrate photoreceptors.
227. Toyoda, J., H. Nosaki, and T. Tomita, Vision Res. *9*:453–463, 1969. Light-induced resistance changes in single photoreceptors of *Necturus* and *Gekko*.
228. Trujillo-Cenóz, O., and J. Melamed, J. Ultrastruct. Res. *16*:395–398, 1966. Compound eye of Diptera: anatomical basis for integration.
229. Trujillo-Cenóz, O., and J. Melamed, *in* The Functional Organization of the Compound Eye, edited by C. G. Bernhard. Pergamon Press Ltd., Oxford, 1966. Electron microscope observations on the peripheral and intermediate retinas of dipterans.
230. Varela, F. G., J. Ultrastruct. Res. *31*:178–194, 1970. Fine structure of the visual system of the honeybee (*Apis mellifera*). II. The lamina.
231. Varela, F. G., and W. Wiitanen, J. Gen. Physiol. *55*:336–358, 1970. Optics of compound eye of honeybee (*Apis mellifera*).
232. Wagner, H. G., E. F. MacNichol, Jr., and M. L. Wolbarsht, J. Gen. Physiol. *43*(Suppl):45–62, 1960. Response properties of single ganglion cells in goldfish retina.
233. Wald, G., J. Gen. Physiol. *22*:775–794, 1939. Porphyropsin visual system.
234. Wald, G., Science *101*:653–658, 1945. Human vision and the spectrum.
235. Wald, G., Docum. Ophthal. *3*:94–134, 1949. Photochemistry of vision.
236. Wald, G., Science *128*:1481–1490, 1958. Significance of vertebrate metamorphosis.
237. Wald, G., Sci. Amer. *201*(4):92–108, 1959. Life and light.
238. Wald, G., Proc. Nat. Acad. Sci. *52*:595–611, 1964. Origins of life.
239. Wald, G., Science *145*:1007–1017, 1964. Receptors of human color vision.
240. Wald, G., Science *150*:1028–1030, 1965. Visual excitation and blood clotting.
241. Wald, G., Nature *219*:800–807, 1968. Molecular basis of visual excitation.
242. Wald, G., and P. K. Brown, Nature *177*:174–176, 1956. Synthesis and bleaching of rhodopsin.
243. Wald, G., P. K. Brown, and I. R. Gibbons, J. Opt. Soc. Amer. *53*:20–35, 1963. The problem of visual excitation.
244. Wald, G., P. K. Brown, R. Hubbard, and W. Oroshnik, Proc. Nat. Acad. Sci. *41*:438–451, 1955. Hindered *cis* isomers of vitamin A and retinene: the structure of the neo-*b* isomer.
245. Wald, G., P. K. Brown, and P. H. Smith. J. Gen. Physiol. *38*:623–681, 1955. Iodopsin.
246. Walls, G. L., The Vertebrate Eye and its Adaptive Radiation. Bulletin 19, Cranbrook Institute of Science, Bloomfield Hills, Michigan, 1942.
247. Walther, J. B., J. Insect Physiol. *2*:142–151, 1958. Changes induced in spectral sensitivity and form of retinal action potential of cockroach eye by selective adaptation.
248. Washizu, Y., Comp. Biochem. Physiol. *12*:369–387, 1964. Electrical activity of single retinular cells in compound eye of the blowfly *Calliphora erythrocephala*.
249. Washizu, Y., D. Burkhardt, and P. Streck, Z. vergl. Physiol. *48*:413–428, 1964. Visual field of single retinula cells and interommatidial inclination in compound eye of blowfly *Calliphora erythrocephala*.
250. Waterman, T. H., pp. 1–64 *in* The Physiology of Crustacea, Vol. 2, edited by T. H. Waterman. Academic Press, New York, 1961. Light sensitivity and vision.
251. Waterman, T. H., *in* Experimental Biology, edited by P. L. Altman and D. S. Dittmer. Fed. Amer. Soc. Exp. Biol., Bethesda, 1966. Responses to polarized light: animals.
252. Waterman, T. H., and H. R. Fernández, Z. vergl. Physiol. *68*:154–174, 1970. E-vector and wavelength discrimination by retinular cells of crayfish *Procambarus*.
253. Waterman, T. H., H. R. Fernández, and T. H. Goldsmith, J. Gen. Physiol. *54*:415–432, 1969. Dichroism of photosensitive pigment in rhabdoms of crayfish *Orconectes*.
254. Waterman, T. H., and C. A. G. Wiersma, J. Exp. Zool. *126*:59–85, 1954. Functional relation between retinular cell and optic nerve in *Limulus*.
255. Waterman, T. H., and C. A. G. Wiersma, J. Cell. Comp. Physiol. *61*:1–16, 1963. Electrical responses in decapod crustacean visual systems.
256. Waterman, T. H., C. A. G. Wiersma, and B. M. H. Bush, J. Cell. Comp. Physiol. *63*:135–155, 1964. Afferent visual responses in optic nerve of crab *Podophthalmus*.
257. Weale, R. A., Nature *218*:238–240, 1968. Photochemistry of human central fovea.
258. Werblin, F. S., and J. E. Dowling, J. Neurophysiol. *32*:339–355, 1969. Organization of retina of mudpuppy *Nectus maculosus*.
259. Wiersma, C. A. G., Proc. Konikl. Nederl. Akad. Wetenschappen (Amsterdam), series C, *73*:25–34, 1970. Neuronal components of optic nerve of crab *Carcinus maenas*.
260. Wiersma, C. A. G., B. M. H. Bush, and T. H. Waterman, J. Cell. Comp. Physiol. *64*:309–326, 1964. Efferent visual responses of contralateral origin in optic nerve of crab *Podophthalmus*.
261. Wiersma, C. A. G., and Y. Yamaguchi, J. Comp. Neurol. *128*:333–358, 1966. Neuronal components of optic nerve of crayfish as studied by single unit analysis.
262. Wiersma, C. A. G., and T. Yamaguchi, J. Exp. Biol. *47*:409–431, 1967. Integration of visual stimuli by crayfish central nervous system.
263. Wiersma, C. A. G., and T. Yamaguchi, Vision Res. 7:197–204, 1967. Integration of visual stimuli in rock lobster.
264. Williams, C. M., Symp. Soc. Exp. Biol. *23*:285–300, 1969. Photoperiodism and endocrine aspects of insect diapause.
265. Wilt, F. H., Develop. Biol. *1*:199–233, 1959. Differentiation of visual pigments in metamorphosing larvae of *Rana catesbeiana*.
266. Withrow, R. B., W. H. Klein, and V. Elstad, Plant Physiol. *32*:453–462, 1957. Action spectra of photomorphogenic induction and its photoinactivation.
267. Wolfson, A., pp. 1–49 *in* Photophysiology, Vol. 2, edited by A. C. Giese. Academic Press, New York, 1964. Animal photoperiodism.
268. Worthington, C. R., Fed. Proc. *30*:57–63, 1971. Structure of photoreceptor membranes.
269. Yoshikami, S., and W. A. Hagins, Abstr. Biophys. Soc. 15th Annual Meeting, New Orleans, TPM-E16, 1971. Light, calcium, and the photocurrent of rods and cones.
270. Zimmerman, W. F., and T. H. Goldsmith, Science *171*:1167–1169, 1971. Photosensitivity of circadian rhythm and of visual receptors in carotenoid-depleted *Drosophila*.
271. Zonana, H. V., Johns Hopkins Hosp. Bull. *109*:185–205, 1961. Fine structure of squid retina.

CHAPTER 15

CENTRAL NERVOUS SYSTEMS

*By C. Ladd Prosser**

The principal goal of neurobiology is to understand the *neural mechanisms of behavior.* Comparative neurophysiology has contributed significantly in recent years to elucidation of the nature of *nerve impulses* and *synaptic transmission* and to the analysis of *sensory reception* and *muscle contraction.* Comparative understanding of *central nervous integration* is advancing in respect to sensory and motor pathways and the functions of large interneurons.[52, 215]

METHODS OF STUDY

Neurobiologists are not in full agreement as to the most appropriate methodology. The following is a list of techniques which may provide a contemporary picture of neural function in different animals.

(1) Behavioral studies include both observations of animals under natural situations and measurements under controlled laboratory conditions. It is increasingly evident that restraint or subtle manipulations of animals bias behavioral observations. Telemetering methods provide new possibilities of observations under natural conditions.

(2) Lesions and pharmacological treatments which bring about behavioral deficits (and reinforcements) provide a first gross step toward localization of neural function. It is difficult to separate primary from secondary effects of such treatments.

(3) Physiological measurements of input-output relations, or reflex function, in anesthetized and unanesthetized animals are useful in analyzing movements such as types of locomotion, feeding, and escape responses.

(4) To understand function in nervous systems requires detailed knowledge of structure and neuroanatomy at all levels – gross connections, microscopic structure, and ultrastructure. Ultrastructural details of synapses were given on pp. 484–489.

(5) Comparison of the neurochemistry of different regions of a nervous system permits some understanding of synaptic transmitters and of metabolism. Neuropharmacology holds promise of chemical manipulation of neural function.

(6) Recording of electrical activity and measurements of membrane constants are powerful techniques; electrical properties of axons and synapses were described in Chapter 11. Two approaches are useful in analysis of central nervous integration: (a) Records can be obtained from single neurons by either extracellular or intracellular electrodes. Unit recordings are useful not only for afferent and efferent neurons but for large interneurons; however, methods are not yet available for recording from the small interneurons of the neuropil. (b) In nervous systems with a cortical or layered arrangement, as in the cerebellum and cerebrum of vertebrates, it is possible to record field potentials which represent the algebraic sum of all electrical activities of a population of neurons, particularly summed synaptic currents. Profiles of these currents

*With contributions by Donald Kennedy.

give information concerning the activation and interaction of masses of neurons.

(7) Models of neural circuits, particularly computer models of networks, have suggested hypotheses and measurements to be made with living systems, and methods for simulation of neural functions.

General Properties of Neural Integration

Most of the functional properties of nervous systems that can be measured are found in nearly all metazoans, but some kinds of animals are especially favorable for study of certain properties.

Distributing Systems; Polarity, Delay. The simplest of nervous coordinating systems is a switchboard-like distributing system. The output is some function of the input. This may be a one-to-one transfer, either with or without summation of converging inputs. A single input may be insufficient to trigger an output, and multiple inputs may be additive (*summation*) or may be multiplicative (*facilitation*). One input may reduce the response to another; i.e., it may be *inhibitory.* A single interneuron may activate several motorneurons (*divergence*) or several interneurons may converge on a single motorneuron.

In complex networks, information is coded for selective activation of specific neurons and for retention in memory patterns. Much central selectivity is based on specific interneuronal connections, wiring patterns which are often difficult to diagram. Gradation and patterning of signals may be by voltage amplitude where transmission is electrotonic, or it may be by frequency or pattern of all-or-none spikes. Possibly coding may be chemical, by areas of specific membrane lipoproteins, each determining a given pattern of ion permeability.

A few definitions applicable to fixed responses or reflexes are needed before discussion of central levels of neuronal organization in different groups of animals can be meaningful. A previous chapter (Ch. 11) considered synaptic delay and polarity. Nonpolarized electrical junctions are relatively rare; they occur in septate giant fibers and in a few central systems where there is close synchrony between a few neurons. Electrical synapses (polarized) provide for speed and synchrony but permit relatively little integration. Synaptic polarization (either chemical or electrical) guarantees one-way traffic, since it prevents antidromic impulses from entering a neural pathway. However, many neurons, particularly those with axons *leaving* a center, have recurrent branches which *feed back* via interneurons (Fig. 15–1). Usually this feedback is inhibitory and prevents repetitive discharge of the efferent neuron. Examples are the Renshaw cells of the mammalian spinal cord and recurrents from Purkinje cells in the cerebellum and from Mauthner neurons in fish. Occasionally, recurrent collaterals are excitatory, causing repetitive firing or circus conduction, as in nerve nets. A different type of control is by neurons *entering* a center and giving off inhibitory collaterals, which end on a target neuron that is later excited by a parallel pathway or by a second impulse in the primary pathway; thus, the target neuron is inhibited or has depressed excitability before the excitation arrives. This is known as *feed-forward* inhibition, and it prevents the interneuron from firing in response to a trivial input. An example is found in *Aplysia* ganglia (p. 650) and in parallel fibers of the cerebellum which activate basket cells that are inhibitory to Purkinje cells (which are themselves excited by the parallel fibers). Feedback control is shown by exiting neurons; feed-forward control is shown by entering neurons or by a branch from one interneuron in parallel with another.

Within certain neural centers, some communication may be between dendrites, as in pacemaker neurons of crustacean cardiac ganglia. In many multipolar neurons of invertebrate central nervous systems there is no distinction between dendrites and axons, and conduction in one process need not be one-way. Another mechanism of coding in branching neurons results from regions of different fiber diameter and the resulting differences in safety factor and limiting frequency at branch points (p. 479). In summary, the microcircuitry of nerve centers provides for much patterning by means of polarized interneuronic connections.

Another type of local control is that due to external fields created by synchronized impulses and junction potentials. For example, slow potentials spread out from the spinal cord in dorsal and ventral roots. In a frog, antidromic stimulation of some ventral root axons depolarizes other nearby motorneurons, and this interaction is en-

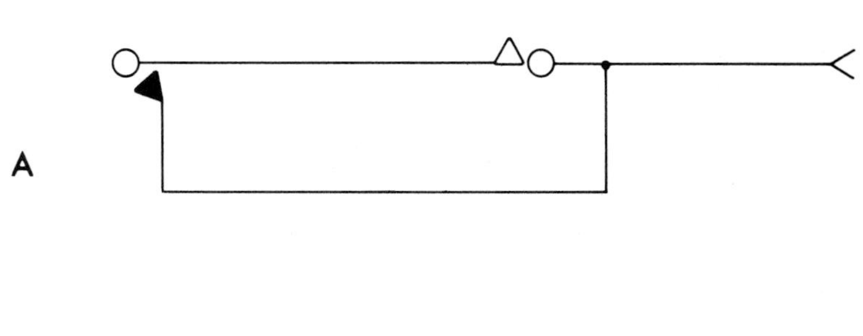

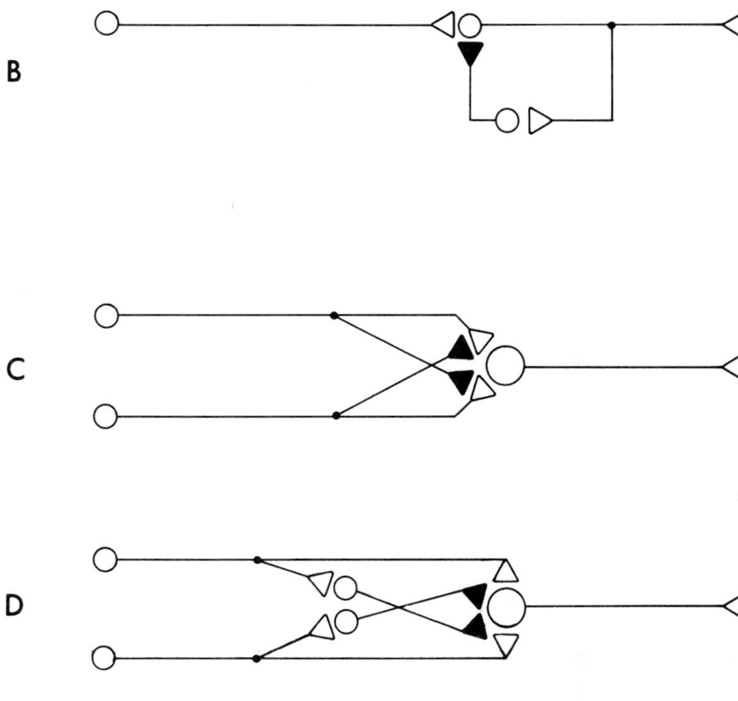

Figure 15–1. Diagrams of feedback and feed-forward inhibitory circuits. Open triangles, excitatory synapses; closed triangles, inhibitory synapses. A, Direct feedback inhibition. B, Feedback by interneuron. C, Crossed presynaptic feed-forward inhibition. D, Crossed postsynaptic feed-forward inhibition via interneurons.

hanced by orthodromic (sensory) input. Hence, there is electrical interaction between overlapping dendrites of adjacent motorneurons. Antidromic impulses in ventral root axons also initiate potentials which can be recorded externally in dorsal roots, and these, unlike the motorneuron effects, are abolished by cholinergic blocking agents; hence, they are due to synaptic connections of recurrent fibers of the motorneurons. Presumably the antidromic impulses are equivalent to activity in the recurrent collaterals of the orthodromic pathway.[173] The correlation between field potentials and probability of firing of cortical neurons is presented on p. 672.

Summation and Facilitation. Most synaptic transmission in integrative centers and in some neuromuscular junctions requires more than one incoming impulse for spike initiation. Each presynaptic impulse elicits a junctional potential, but the requirement of more than one input impulse gives a safety factor against discharge of efferent neurons to a single afferent input. In relay or distributing centers, such as some sensory nuclei, sympathetic ganglia, or squid mantle ganglion, there may be one-to-one transmission. The requirement for two input events may be spatial (two or more converging fibers) or temporal (two impulses on the same fiber within a short time). *Summation* means simple addition of junction potentials, whether initiated from the same or different sources; *facilitation* means that the two inputs give a greater effect than the algebraic sum. Thus, for two converging fibers or *spatial facilitation,* $(a + b) > a + b$; and for closely spaced inputs in one fiber or *temporal facilitation,* $(2a) > a + a$. The two kinds of facilitation can be combined, in that spatial facilitation is maximum when the two inputs are synchronous, i.e. when the two synapses are equidistant from the site of spike generation. The facilitation decays with a time-constant close to that of an epsp. In a crayfish, bending of one sensory hair on the tail sends one impulse or a brief train into the sixth abdominal ganglion, and stimulation of several (usually four) hairs is necessary to elicit one postsynaptic impulse. Single fiber inputs to crustacean muscle show much temporal facilitation (p. 757). Central nervous facilitation was originally described in terms of spike outputs; Sherrington's use of the term included *spatial summation* which leads to *facilitated spiking.* Modern usage permits distinction between summation and facilitation at the level of synaptic potentials; in preparations where junctional potentials cannot be measured, the distinction is not always clear.

A second type of facilitation, as measured by reflex output, reaches a maximum some time after an initial input. For example, in flexor reflexes of a cat a maximum response to a second afferent volley occurs 10 to 20 msec after an initial volley. *Delayed facilitation* has, in some preparations, been correlated with a by-pass or parallel circuit of several neurons such that impulses in the by-pass arrive at a postneuron synchronously with the second or facilitating input. Viewed in this way, the delayed facilitation of spike output is actually spatial summation.[110] *Disfacilitation* refers to the decline in a psp on repetition; this may provide a synaptic mechanism for behavioral habituation.

Heterosynaptic facilitation refers to prolonged potentiation of a synaptic response by pairing a weak stimulus (test input) with a strong one in another converging pathway (priming input). After a number of such pairings, the test input alone gives a large response. This phenomenon may possibly represent presynaptic facilitation.[281]

Inhibition. In most integrative centers, inhibition is as important a synaptic function as excitation. Neural output represents a balance between the two processes. Inhibition may be *presynaptic,* reducing the liberation of excitatory transmitter and preventing postneuron discharge even when no potential change occurs in the postneuron membrane. Alternatively, inhibition may be *postsynaptic,* causing a conductance change in the postsynaptic membrane and thus reducing the response to excitatory transmitter.

An example of presynaptic inhibition in the spinal cord is the following: Terminals of afferent fibers from muscle spindles receive presynaptic inhibition from afferents arising in tendons during a twitch, and thus the muscle afferents are prevented from liberating transmitter to motorneurons. An example of postsynaptic inhibition of spinal motorneurons is as follows: Collateral branches from motorneuron axons excite Renshaw interneurons, and these in turn inhibit the motorneuron and prevent repetitive firing after initial excitation.[10]

Generally, postsynaptic inhibition is mediated via interneurons, sometimes with considerable latency. Inhibition modulates spontaneous activity and sensory input. Examples of inhibition of muscle and of sense organs are given in Chapters 12 and 16. The output of neurons in a center normally depends on the algebraic sum of excitation and inhibition, and a given neuron has both excitatory and inhibitory receptor membrane patches. One method of central coding may be by variation of the relative size of the patches.

Repetitive Responses and Spontaneous Activity. Many nerve fibers discharge repetitively at frequencies higher than refractory period limitation during prolonged depolarizing stimulation. In some neural centers, certain neurons show *after-discharge* or repetitive firing as part of a reflex response. Many central neurons are repetitively active in the absence of sensory or other input; i.e., they are spontaneously firing. Incoming signals may modulate the ongoing discharge to increase or decrease its rate; this permits more gradation of information transmission than if a neuron were silent and the only effect of an input were to start it firing. Spontaneous activity is general in integrative nervous systems; it is less usual in distributive centers.

Two types of central nervous *spontaneity* are recognized. Individual cells are spontaneously active in many invertebrate ganglia and in most integrative regions of the brain of vertebrates. The frequency varies from cell to cell and there is little synchronization. One cell may fire regularly or may fire in bursts interrupted by silent periods, as in crustacean cardiac ganglia. At any one time only a few cells are spontaneously active. Spontaneity is temperature-sensitive and requires energy from oxidative metabolism, but the coupling between metabolism and rhythmic conductance changes is unknown. A second pattern of spontaneity is found mainly in cortical structures such as cerebral cortex, cerebellum, and optic tectum. Very many cells are synchronized to give large rhythms, not necessarily of impulses but rather of slow changes in membrane potential. The synchronization is due partly to reverberating circuits and partly to electrotonic coupling of neurons.

In hermit crabs and lobsters, the motor rhythm of ventilation is controlled by a pair of neurons, the membrane potentials of which oscillate without spiking but elicit spikes in motorneurons serving the muscles of ventilation.[364] In *Aplysia*, certain neurons fire uniformly during a burst, others fire at decreasing frequency, and in still others the frequency increases and then decreases (these are parabolic bursters). If spiking is prevented by use of TTX in a Ca-free medium, rhythmic slow waves corresponding to the burst frequency remain. These slow waves are abolished by ouabain or in Na-free medium, and they may represent a rhythmic sodium pump.[519] It is probable that several different mechanisms are involved in oscillating membrane potentials and repetitive spiking,[15, 63] and that spontaneous rhythmicity is characteristic of integrative neural centers.

Numerous examples will be given later for behavioral meaning of spontaneous activity and for programmed sequences in nerve centers. Observations on deafferented animals show that many coordinated movements do not require feedback input from sense organs and appendages. It is possible that sensory input may be more important in establishing rhythmic patterns during development than later in the life cycle. Once the pattern has developed, sensory input may merely trigger a programmed sequence of behavioral acts. Sensory input may also maintain a tonic level of central excitability. Numerous examples of endogenous central rhythms are known for walking and flying in insects.[589] Respiratory neurons in the brainstem of vertebrates continue to fire after deafferentation.[57, 479] Patterned vocalization in birds is unaffected by deafening, which removes the auditory feedback (p. 534). The separation of genetic from ontogenetic determination of neuronal sequences is difficult. However, it is clear that spontaneous activity occurs at the cellular level in integrative centers and that programmed sequences occur in chains and rings of neurons.

GIANT FIBER SYSTEMS

Giant fibers are examples of simple rapid distribution systems. They have evolved independently in several groups of animals and are to be considered as specialized, and certainly not primitive.

Giant fiber systems are of several types and occur in a series of increasing complexity (from through-conducting fibers, to septate neurons with electrical septal junctions, to giant fiber chains connected through chemical synapses), and they include some of the most complex single neurons known. Giant fibers were mentioned previously as adaptations providing for fast reactions (p. 478) and as examples of electrotonic junctions and electrical synapses (p. 484). The multicellular or functional syncytial type of giant fiber occurs in annelids, crustaceans, and cephalopods. Large neurons known as giants occur frequently, especially in insects and lower vertebrates.

Giant fiber systems function chiefly in rapid startle, escape, and grasping reactions, which are initiated abruptly and are not usually graded. They are efficient, since one impulse in a giant fiber activates muscles of a wide body area. Conduction in giant fibers is 2 to 10 times faster than in small fibers of the same animal.

EARTHWORMS

Earthworms have two septate lateral giant fibers and one septate median giant fiber; each giant is connected to one cell body in each segment. The two lateral fibers are cross-connected by electrical synapses, and each lateral fiber gives off five branches per segment.[384] These giant fibers mediate end-to-end twitches. Transection of individual giant fibers followed by tactile stimulation shows that the median fiber normally conducts posteriorly and the lateral fibers conduct anteriorly. In isolated cords, impulses are conducted in either direction in each fiber. The behavioral polarity results because the median fiber has sensory connections in the anterior end (first 40 segments) and the lateral complex has sensory connections in the posterior part of the worm.[50] The septa are remnants of embryological cell boundaries. Recordings of muscle contractions in response to various sequences and frequencies of giant fiber stimulation show stronger twitches when both lateral and median fibers are stimulated together than when one only is stimulated. Repeated responses also show fatigue which can be relieved by facilitating stimuli, changes in excitability lasting as long as 9 seconds. These modifications of state are localized at the junctions of giant fibers with motorneurons.[440, 441]

Polychaetes

Some polychaete worms have no giant fibers (*Aphrodite, Chaetopterus*), others have only one (*Arenicola, Pista*), and still others have several (*Neanthes, Glycera*).[52] *Myxicola* has a single 560 μm axon.[35, 395] In sabellid and serpulid worms, decussations of giant fibers occur in the midbody; on fatigue the crossing of descending impulses shifts toward more anterior decussations. In these worms, synaptic potentials can be recorded in the regions of connections between the two fibers, and hyperpolarization reduces the epsp's. Electrotonic coupling is small (30:1) and the interaction appears to be by chemical synapses. In addition, there are direct anastomoses in the head region.[180] On repeated low-frequency stimulation, the response at the junction between lateral giants and motorneurons fails (fatigues) and this can be relieved by a high-frequency stimulation (facilitation). When the junction between a lateral giant and a motorneuron has failed (in *Nereis* or *Harmothoë*), stimulation of the median giant can restore the junctional responsiveness.[208]

Squids

In squids, a pair of giant neurons occur in the posterior palliovisceral ganglion of the brain; their axons fuse and decussate, and then they synapse in the visceral part of the brain with a second set of giant neurons whose processes pass out in the mantle nerves (Fig. 15–2).[602] These second-order giants synapse in the stellate ganglion with a third set of giant neurons which serve the mantle muscle. The third-order giant axons are formed by fusion of processes from numerous (300 to 1500) neurons arising in the stellate ganglion; each third-order giant axon receives innervation from a large (200 μm) and a small (75 μm) second-order giant neuron. In addition, another group of large axons from the brain (accessory axons) pass to the stellate ganglion, some of them synapsing on the third-order giants and others passing directly through the ganglion.[59] Synaptic delay is 0.55 msec, and the two second-order fibers can summate in spike initiation. The synapse of accessory axons with the giants is more sensitive to fatigue and anoxia than is the synapse of the second-order giant fibers. One of the third-order giant axons is larger than the others—its diameter is 700 to 1000 μm. Each giant axon serves a large area of the mantle, and the largest fibers go to the posterior end so that impulses arrive nearly simultaneously in the entire mantle. Velocities are 5 to 25 m/sec at 15°C. One motor impulse in a giant fiber elicits maximum contraction of the mantle muscle that it activates. The contraction forces water out of the mantle cavity through the funnel, sending the animal rapidly forward or back according to the funnel angle.[405]

Crayfish

The lateral giant fibers of crayfish show junctional connections (a) with sensory elements and interneurons in the neuropil, (b) with the contralateral giant by electrical connection at a decussation in the cord, (c) with the lateral giant of the next posterior and anterior segment at a region of segmental overlap, and (d) with motor giant fibers and other flexor motorneurons of the next anterior segment[572] (Fig. 15–3). The overlapping junctions of the lateral giant fibers are non-polarized and normally conduct one-to-one, but after fatiguing they may respond only to several impulses (facilitation). The segmental cell bodies (somata) are contralaterally located and give off extensive dendritic arborization ipsilaterally.[435] The lateral giant fibers synapse with segmental motor giant neurons at electrical synapses; these motorneurons also receive input from the median giants through electrical synapses and from inhibitory axons of the nerve cord through chemical synapses.[572, 573] The median giant fibers are not syncytial; their cells are in the supraesophageal ganglion where the axons decussate. An impulse in a median giant leads to an inhibitory response (ipsp) in a lateral giant, and this is eliminated by picrotoxin.

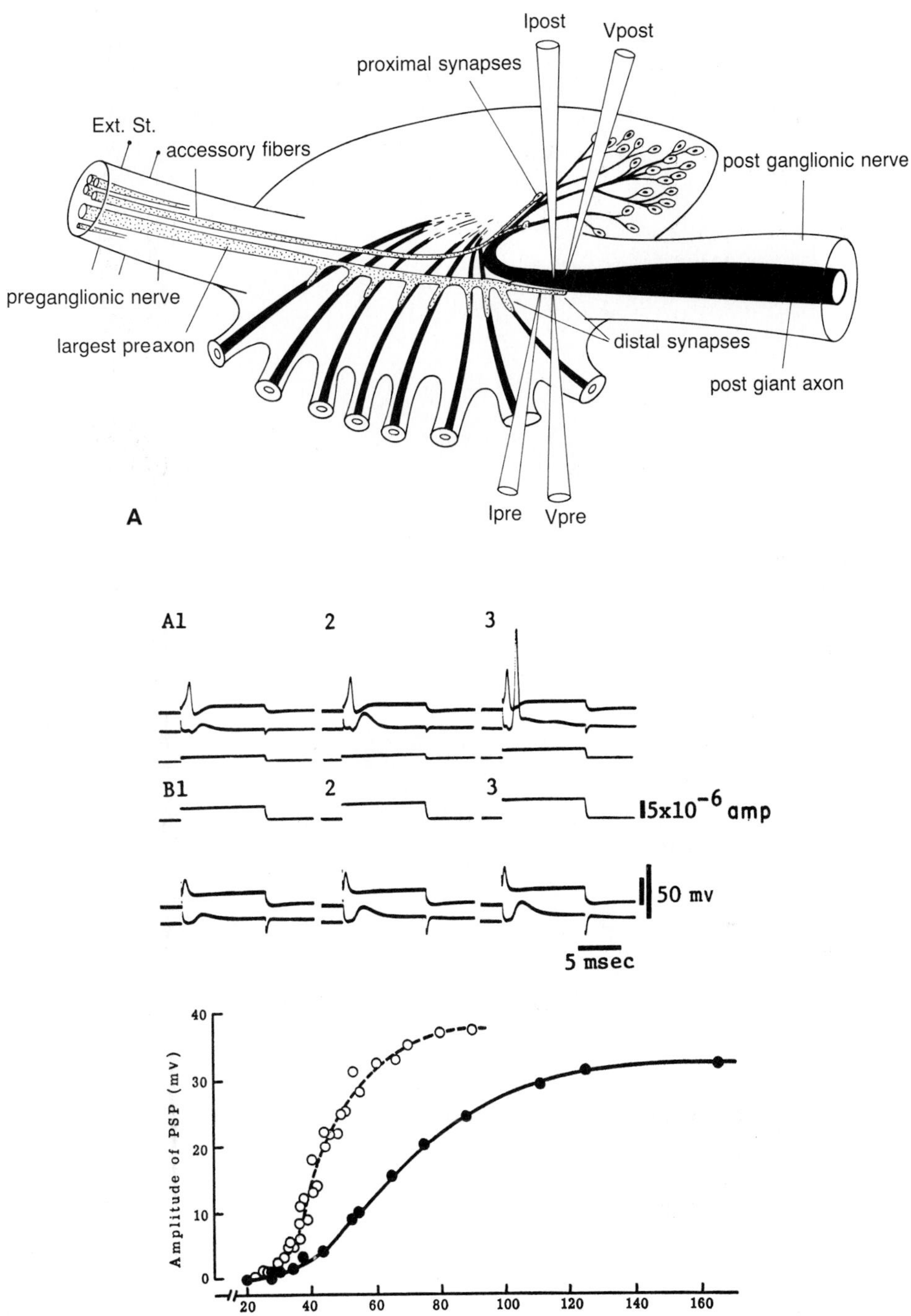

Figure 15-2. A, Diagram of giant fiber synapses in mantle ganglion of squid. Posterior giant axon formed by fusion of processes from cluster of neurons. Accessory fibers in preganglionic nerve form proximal synapses, and large pre-axon forms distal synapses across which transmission is measured by current-injecting (I) and voltage-recording (V) electrodes in pre- and post-fibers. (Courtesy of K. Kusano, modified from Bryant, S., J. Gen. Physiol. *42*:609, 1959). B. Relation between presynaptic depolarization and postsynaptic potentials. Upper recordings are presynaptic; lower ones are postsynaptic. 1 to 3, increasing levels of presynaptic current. Recordings marked A are from normal preparation in which amplitude of psp's increases as response in presynaptic terminals grows, and which gives a spike at 3. Recordings marked B are a series after administration of TTX, showing that synaptic currents are not blocked. Graph shows relation between psp and presynaptic depolarization in both preparations. (From Kusano, K., J. Gen. Physiol. *52*:326–345, 1968.)

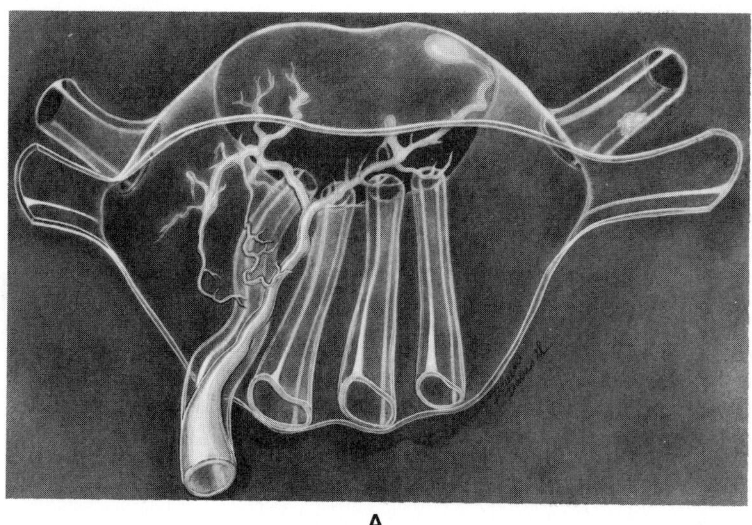

A

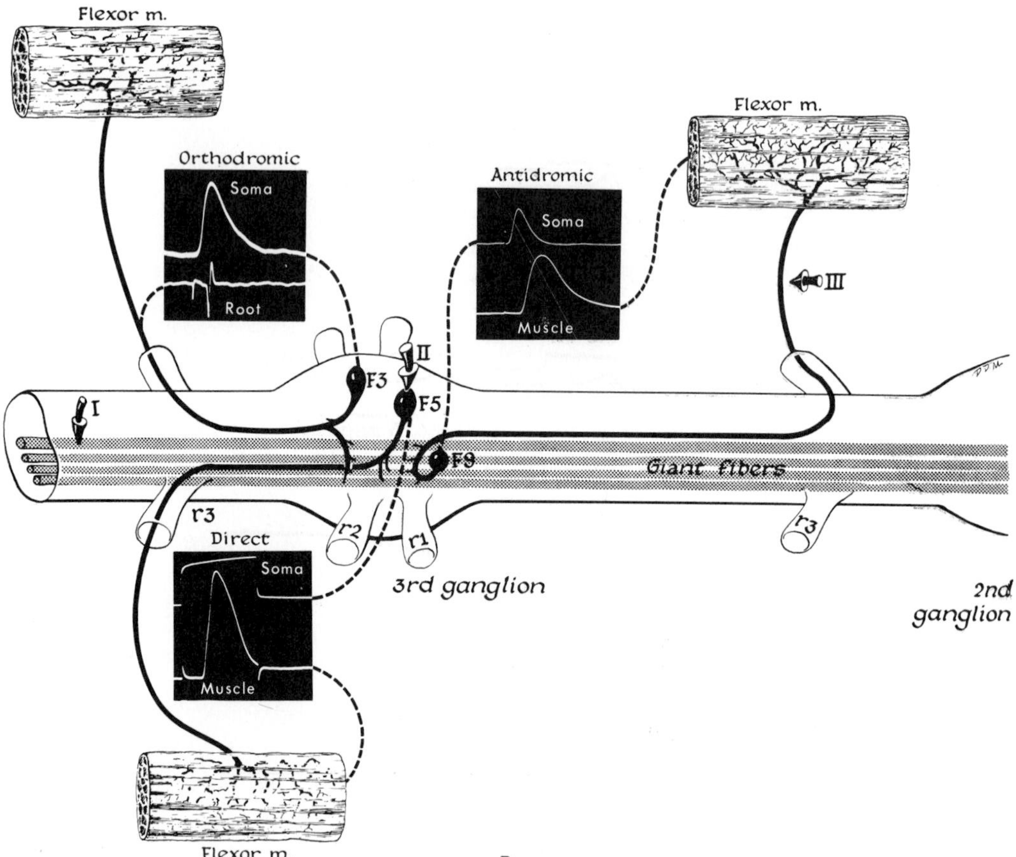

Flexor m.
Orthodromic
Soma
Root
Antidromic
Soma
Muscle
Flexor m.
III
II
I
F3
F5
F8
Giant fibers
r3
r2
r1
r3
Direct
Soma
Muscle
3rd ganglion
2nd ganglion
Flexor m.

B

Figure 15–3. A, Reconstruction of abdominal ganglion of crayfish, showing giant motorneuron (processes based on dye injection) and its relation to one of the four giant fibers. (From Kennedy, D., A. I. Selverston, and M. P. Remler, Science *164*:1488–1496, 1969.) B, Diagram and recordings showing responses of motorneurons and flexor muscles of crayfish to different types of stimulation. Stimulus I in lateral giant fiber gives orthodromic response in neuron soma and in root axons. Stimulus II in soma gives direct soma and muscle junctional responses. Stimulus III in root axon gives antidromic response in soma and junctional response of muscle. (From Kennedy, D., A. I. Selverston, and M. P. Remler, Science *164*:1488–1496, 1969. Copyright 1969 by The American Association for the Advancement of Science.)

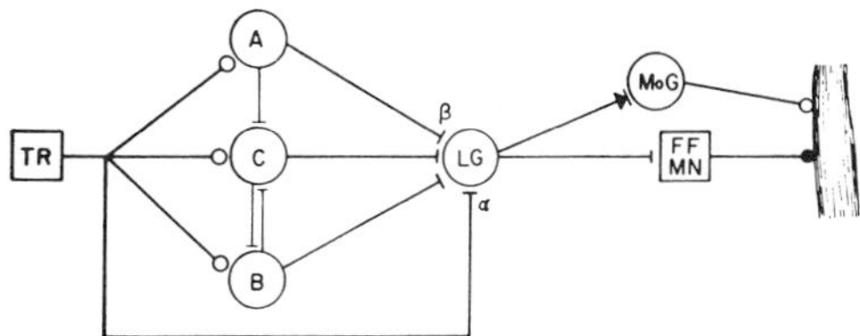

Figure 15–4. Schema of neurons in the escape response circuit in crayfish abdomen. TR, tactile receptors; A, unisegmental tactile interneuron; B and C, multisegmental tactile interneurons; LG, lateral giant fiber; MoG, giant motorneuron; FFMN, fast flexor motorneurons; flexor musculature at right. Synaptic bars indicate electrical synapses; filled circle is a facilitating chemical synapse; open circles are antifacilitating chemical synapses. (From Zucker, R. S., J. Neurophysiol. *35*:599–620, 1972.)

In addition, an impulse in collaterals of a lateral giant leads to recurrent inhibition in the two laterals; this is insensitive to picrotoxin. Recurrent inhibition within the cord limits the number of motor impulses that can be evoked by a single afferent volley.[437] A single giant fiber (median or lateral) impulse results in a massive contraction of flexor abdominal muscles—tail flip—which habituates rapidly. Stimulation of segmental second roots elicits in the lateral giant fibers excitatory junction potentials of two, and sometimes three, components; the early wave is more stable than the late wave, and the late wave represents prolonged activation by interneurons.[302] All of the synapses on processes of the lateral giants are electrical; such transmission may be related to the location of the synapses on ventral dendrites some distance from the site of spike initiation, and to the high threshold of the large lateral giant fiber[617] (Fig. 15–4).

Cockroach

In the last abdominal ganglion of a cockroach are junctions between sensory neurons from anal cerci and large ascending nonsyncytial neurons. Similar "giant" fibers occur in locusts and crickets. These junctions are polarized, normally require spatial or temporal summation, and may show some lability of response according to timing of incoming impulses. Many presynaptic impulses in cercal nerve fibers are needed to trigger a spike in a giant fiber in the sixth abdominal ganglion.[64] The synapses block at 40 to 50/sec in roach, and at 70 to 100/sec in cricket.[448] The giant fiber system participates in the jumping reaction to a puff of air on the anal cerci. The giant fibers do not activate leg motorneurons directly, and the latency of leg response to cercal stimulation is twice as long as it would be if the conduction to leg motorneurons were in the giant fibers. Apparently, the activation of motorneurons is by small fibers, and the giants activate a general alarm system; they also activate antennal movement.[87] They may also provide for continuous sensory input generated by movement during the "escape." Dorsal and ventral giant fibers of *Periplaneta* are continuous from the last abdominal ganglion (a_6) to the subesophageal ganglion (so), but the fibers show segmental narrowing, particularly in the abdomen, and marked tapering in the thorax. There are segmental delays of 0.6 msec, and nicotine can block transmission in each ganglion; these are effects of the segmental narrowing. Conduction is not uniform despite absence of synapses. Conduction velocity variations are due to axon diameter differences.[409, 509]

Mauthner Cells of Fishes

The giant non-syncytial Mauthner neurons of fishes and urodeles have multiple inputs and appear to be capable of more complex input integration than other known neurons. In goldfish, each of the two Mauthner cells (mc) has a large dendrite, which extends caudally and laterally, and two or more ventral dendrites. The axon arises at a prominent axon hillock, which is surrounded by a cap or coil of endings. The axon decussates and then turns caudally; it gives off a branch to the preoptic center, a branch to the contralateral mc, and a collateral branch back to its own soma; and then it passes to motorneurons in the spinal cord[512] (Fig. 15–5). Fibers of the ipsilateral eighth nerve are excitatory, and terminate in club endings on the distal part of the lateral dendrite and in bouton endings on the proximal part. The ipsilateral eighth nerve elicits an early epsp of latency less than 0.1 msec, probably by electrical synapses at club endings which show gap junctions; it also elicits an epsp of latency 0.6 msec, presumably by chemical synapses at bouton endings. The contralateral eighth nerve is inhibitory and has bulb endings, which are chemical synapses, on the proximal part of the lateral dendrite and on the soma. A branch from the axon of the contralateral mc is also inhibitory and acts via one or more interneurons.

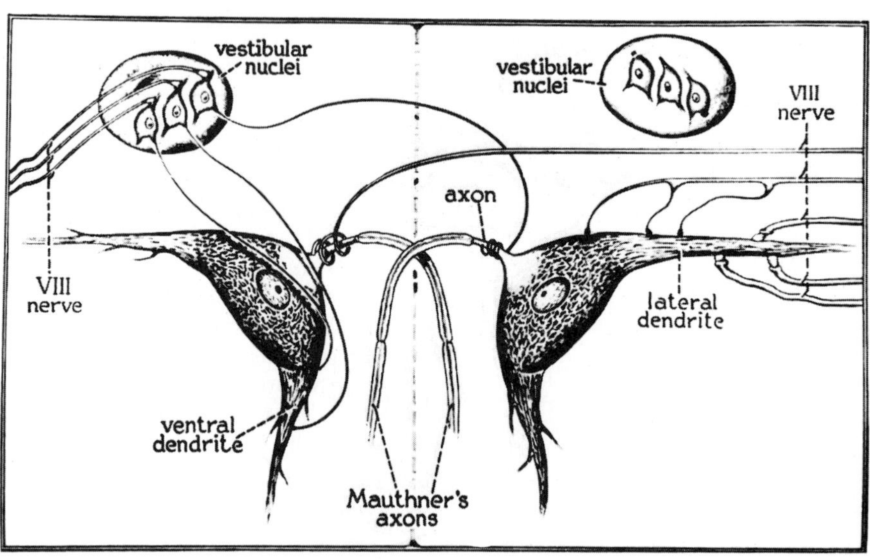

Figure 15-5. Schematic representation of Mauthner cells and connections in goldfish. (From Retzlaff, E., J. Comp. Neurol. *101*:407–445, 1954.)

In addition, a collateral from each axon acts via an interneuron to inhibit recurrently the mc of its origin (as demonstrated by suppression of spread of an antidromic spike by a prior spike), and this inhibition is exerted electrically as an external hyperpolarizing potential (EHP) at the axon hillock. Also, a late collateral recurrent inhibition is mediated by chemical synapses, which give prolonged ipsp's that are due to increased chloride conductance and which are sensitive to hyperpolarization of the neuron soma.[154] The early recurrent inhibition is reflected in a large positivity recorded externally at the axon hillock; this inhibition is not affected by soma hyperpolarization. The external hyperpolarizing potential can block spike initiation if it coincides with an epsp; both mc's show EHP inhibition owing to the collaterals from either axon acting via interneurons (Fig. 15-6). Still another recurrent collateral inhibition from both

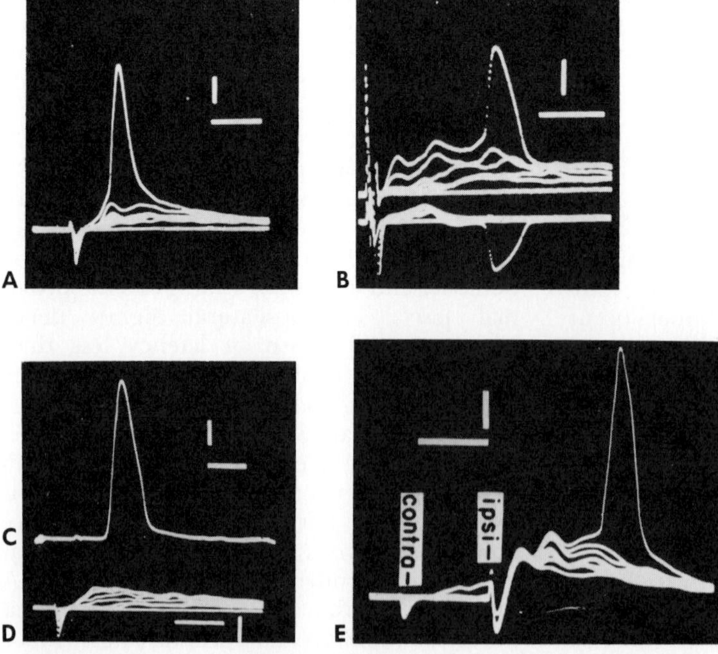

Figure 15-6. Responses of Mauthner cell of goldfish. A and B, Stimulation of ipsilateral eighth nerve; A, recording from soma; B, recording from lateral dendrite. C, Antidromic stimulation. D, Stimulation of contralateral eighth nerve. E, Stimulation of ipsilateral and contralateral eighth nerve. (From Furshpan, E. J., and T. Furukawa, J. Neurophysiol. *25*:732–771, 1962.)

ipsilateral and contralateral Mauthner cells is by reduction in eighth nerve excitation; this action is probably chemical (probably by GABA) and presynaptic via the acoustic nucleus.[156, 157] Gamma-aminobutyric acid in low concentrations reduces epsp responses to ipsilateral eighth nerve stimulation.[103] The mc's receive inputs from other parts of the brain, spinal cord, and especially ventral acoustic nucleus. Recordings from an mc show background epsp's corresponding to random sounds impinging on the fish.[104]

The axon outputs from Mauthner cells go partly to anterior motor centers, where mc impulses modulate movement of jaw, eyeball, and operculum. The main output goes down the spinal cord to activate motorneurons, more on the same side as the axon than on the opposite side. Conduction in the axon is at 50 to 100 m/sec (at 15°C), and the behavioral effect is a strong flexion of the abdomen, fin movement, and flip of the tail, resulting in forward propulsion. It is essential that descending impulses should not arrive at motorneurons simultaneously on the two sides; if they do, the muscle effects are cancelled and the tail does not flip. Part of the control is in the elaborate collateral inhibitory mechanisms listed above. In addition, a reciprocal crossed inhibition at the level of spinal motorneurons (Fig. 15-7) permits a discrimination of 0.15 to 0.2 msec; two mc impulses arriving at less than that interval are without excitatory effect. After a motorneuron response to one mc impulse, the motorneuron cannot respond to a spike in the other mc for about 10 msec.[104]

In summary (Fig. 15-8), the six inputs to one Mauthner cell are as follows: principal excitatory input from ipsilateral eighth nerve via both electrical and chemical synapses; five inhibitory inputs—contralateral eighth nerve on lateral dendrite, contralateral eighth nerve on soma, collateral of mc via interneuron by EHP at axon hillock, collateral by chemical inhibitory synapse, and presynaptic inhibition of acoustic input.[155, 156, 157] The output is partly to the head, and mainly to the spinal motorneurons where a remarkable inhibitory system permits discrimination to within 0.15 msec against activation by both mc axons. The eighth nerve inputs are both vestibular and auditory. The behavioral significance of the Mauthner cells is to mediate rapid startle and escape responses.

To conclude, giant fibers have evolved numerous times, and are of several types, but all of them function for speed and syn-

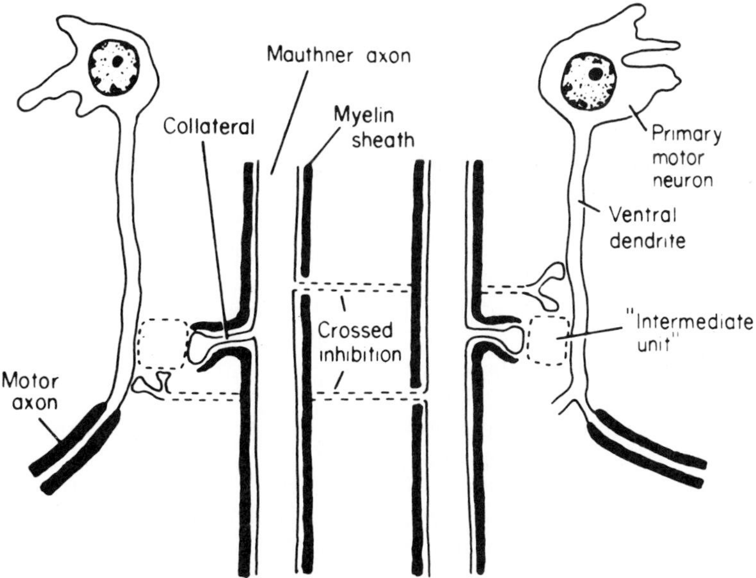

Figure 15-7. Diagram of part of spinal circuitry of Mauthner reflex and its inhibition. (From Diamond, J., pp. 265-346 *in* Fish Physiology, Vol. 5, edited by Hoar, W. S., and D. J. Randall, Academic Press, New York, 1970.)

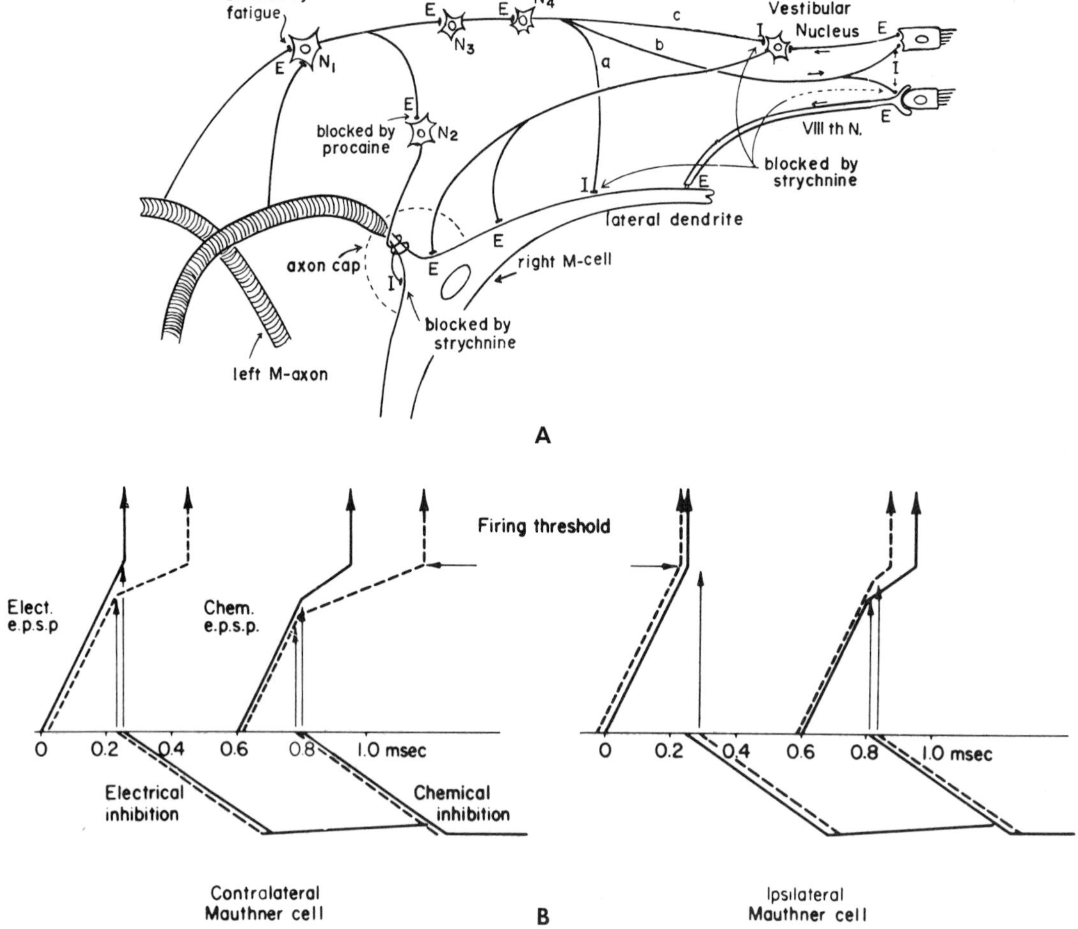

Figure 15–8. A, Synaptic interactions on Mauthner cell (M) of goldfish: E, excitatory synapse; I, inhibitory synapse; N, interneuron. (From Furukawa, T., *in* T. Tokizane and J. P. Schadé, eds., *Correlative Neurosciences,* Progr. Brain Res. *21*A:44–70, 1966. Elsevier, Amsterdam.) B. Diagram showing time relations of responses of goldfish Mauthner cell to stimulation of eighth nerve. Excitatory responses above horizontal line; inhibitory responses below. (From Diamond, J., pp. 265–346 *in* Fish Physiology, Vol. 5, edited by W. S. Hoar and D. J. Randall, Academic Press, New York, 1970.)

chrony. They are distribution systems, complex single cells, or functional syncytia, relatively fixed in response.

NERVE NETS

A nerve net consists of neurons dispersed in a plane or in three dimensions, and so connected as to permit diffuse conduction and to show some integrative properties. Nerve nets comprise the major portion of the nervous system in coelenterates, ctenophores, and enteropneustans; they combine with neuron aggregates in ganglia to form the nervous system of echinoderms, ascidians, and some molluscs; nerve nets may integrate movements in the visceral musculature of animals of many phyla. Nerve nets have been most studied in Cnidaria (coelenterates). Most conduction of signals from one region to another is by nerve net in Scyphozoa and Anthozoa; epithelial conduction is important in Hydrozoa, but nerve nets are also present in this class.[338, 339, 340]

Coelenterates

Early histologists pictured the nerve net of a polyp or medusa as a continuum of anastomosing fibers from multipolar and bipolar neurons, in which axons were not distinct from dendrites. Evidence from polarizing and electron microscopy shows that the processes do not usually fuse; the junctions are regions of apposition between fibers. However, in *Velella* and possibly in *Hydra* there may be fusion in some processes.[337] In the nerve net of the jellyfish *Gonionemus,* typical synapses with dense-core vesicles on one side have been found; these are presumably polarized.[568] In contrast, symmetrical synapses, presumably nonpolarized in transmission, occur in *Aurelia, Phialidium, Cyanea, Cordylophora,* and the siphonophore *Nanomia.*[218, 265, 267] Junctions of *Hydra* are 0.8 μm long, with a 150 Å synaptic cleft and 1200 to 2000 Å vesicles; some junctions are symmetrical, and others have vesicles on only one side.[568] Thus, the ultrastructure indicates junction types ranging from fused contacts through bisymmetrical synapses to polarized synapses. The most common synapses are bisymmetrical, with vesicles on each side.

Conduction in a typical nerve net proceeds freely in all directions; impulses pass around corners, cross tissue bridges, and diverge beyond a narrow neck. The average behavior of a net may not be characteristic of the single neurons, and techniques are not available for examining the properties of single fibers in coelenterate nets. Most coelenterates have two or more nets in parallel;[24] for example, in Scyphomedusae a large-fiber net serves for fast movements of swimming and a diffuse small-fiber net coordinates slow contractions of feeding movements.[411] In a sea anemone, *Calliactis,* one slow system conducts at 4.4 to 14.6 cm/sec, and another at 3 to 5.3 cm/sec, whereas a fast net in the mesenteries conducts at 120 cm/sec;[358] conducting tissues of the first two systems have not been identified.

Rapid action potentials have been recorded from bundles of fibers of the fast net of an anemone.[420] A type of polarity results from different thresholds for net activation in different regions of the animal. The oral disc is 4000 times more sensitive to tactile stimulation than is the column of an anemone.[413] In ctenophores, impulses in the nerve net initiating luminescence go in all directions; those for the comb beat go from the aboral to the oral end.

Several conduction systems exist in hydrozoans, and it is not certain which are epithelial and which are neural. In *Tubularia* the conduction systems are: (1) a system which triggers electrical events in the neck region and conducts at 17 cm/sec; (2) the distal opener system, which initiates opening of tentacles and conducts at 15 cm/sec, and (3) the slow stalk system conducting at 6 cm/sec, with high threshold and not related to any behavior. The first two are blocked by Mg^{++} and probably are conducted by nerves. The third is probably epithelial. The distal opener system inhibits the neck pacemaker.[264–267] *Hydra* shows contraction bursts of potentials, which originate in the hypostome and are conducted at 3 to 5 cm/sec, and which vary in number of spikes per burst according to illumination or feeding state; these potentials are probably in epitheliomuscular cells. *Hydra* also shows rhythmic potentials (1 to 10/min) which are unrelated to contractions, are conducted at 5 cm/sec, and arise near the base of the column and may be in gastrodermal cells.[268, 412, 474]

Coelenterate nerve nets show a wide variety of facilitation requirements. In both anemones and medusae, repeated stimulation is necessary to elicit a maximum contractile response and there may be little mechanical response to a single volley. Experiments with blocking agents, such as magnesium ions, and with regeneration of nerve connections after a transection show much of the facilitation requirement to be neuroneuronal and part of it to be neuromuscular. An apparent decrement in transmission results from decreasing facilitation at junctions as a wave spreads out from its point of origin. In the luminescent *Renilla,* non-polarized non-decremental conduction requires facilitation, which fails at intervals greater than 1.5 sec. In the tentacle of *Anemonia,* facilitation lasts 10 to 30 sec,[91] and in *Tealia* for 2 min.

In an anemone, *Calamactis,* nerve spikes conducted at 0.9 m/sec show some facilitating action at neuromuscular junctions at 30 msec intervals, maximal facilitation at 200 msec, and a small residual effect at 1 to 4 sec.[420] In *Metridium,* a quick contraction is triggered by the net at 1 stimulus/sec; a slow contraction of the same muscle is elicited by 1 stimulus per 5 sec.[444] The parietal and circular muscle sheets give only the slow response, while the retractor and marginal sphincter give both slow and fast responses;[408] the slow response is more dependent on facilitation than is the fast one. The refractory period of the slow net may be 0.3 to 3 sec, and that of the fast net is 60 to 70 msec. The anemone *Stomphia* is caused to swim by chemical stimulus from a starfish; swimming can also be initiated by shocks, but 24 to 40 stimuli are needed, with optimal effect at one per three seconds.[459]

Corals form a series in their dependence on facilitation (Fig. 15–9). In the alcyonarian *Tubipora,* a single wave is conducted throughout the colony like the fast wave in the mesenteries of an actinian. In *Heteroxenia,* the response spreads farther with successive stimuli until the whole colony is involved. In *Acropora* and *Porites,* the polyp response facilitates at 1/sec, but not at 0.25/sec. In *Polythoa,* there is response of the stimulated polyps only, and little spread. Apparently, nonconducting junctions become conducting as successive impulses spread over some coral colonies, and a critical number of active neurons is necessary for a given polyp to contract.[207]

Nerve nets frequently show repeated response, a sort of after-discharge series, which may result from circus or reverberating conduction. A contraction wave, conducted in the nerve net in a doughnut-shaped ring cut from a medusa, may continue to circle for hours or days. In leptomedusae a net which mediates radial conduction is slow, while the net for circular conduction is fast; reciprocal contraction of the circular and radial muscles indicates net interactions, and the radial system can inhibit the circular one but not vice versa.[209] Contractions in rings from the column or pedal disc of *Calliactis* may be stopped by appropriate stimuli to the nerve net.[131]

Another function of some nerve nets is to initiate spontaneous rhythmic activity. The marginal bodies of a jellyfish are neuron aggregates which elicit rhythmic contractions of the bell. Intact anemones show extremely slow (minutes to hours) spontaneous rhythmic movements.[131] In tubularians, the nerve net of the oral disc is a pacemaker, but when this has been removed, spontaneous rhythms originate elsewhere.[264, 266] Any region can be a pacemaker—hydranth neck, tentacles, or hydranth stalk.[265, 267]

Nerve nets mediate some very complex behaviors. Some of these are: rejection of certain foods, acceptance of others, and pointing of the manubrium in hydromedusae; expansion, elongation, defecation, and swaying in anemones. Swimming by *Stomphia* stops when food extract is perceived. *Calliactis* normally lives on snail shells which are inhabited by hermit crabs; when the crab transfers to a new shell, the anemone, by a series of complex maneuvers, moves to the new shell. The hermit crab *Dardanus* assists the anemone to its new shell; the crab *Pagurus* does not.[458]

Except for a few recordings of trains of nerve impulses, cellular function in coelen-

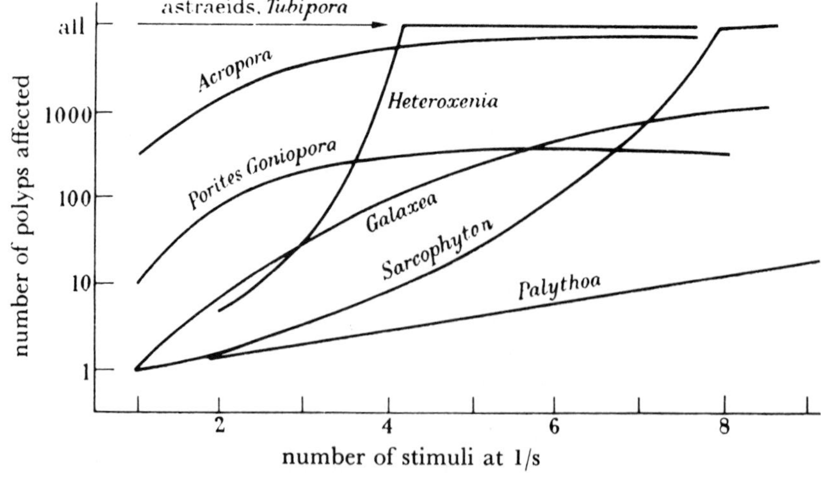

Figure 15–9. Relation between number of stimuli at 1/sec and number of polyps affected for a variety of corals. Whole colony active from first stimulus in *Tubipora* (top line). Accelerating slope in which wave eventually sweeps over whole colony in *Heteroxenia* and *Sarcophyta.* Decreasing slope by which response spreads to limited number of polyps no matter how many stimuli given—*Goniopora* and *Galaxea.* (From Horridge, G. A., Phil. Trans. Roy. Soc. Lond. B *240*:495–529, 1957.)

terate nets is unknown. Chemical transmitters probably occur, as indicated by the presence of synaptic vesicles, but the familiar agents of other complex animals, such as norepinephrine and acetylcholine, are without much effect.[457] Magnesium is commonly used experimentally to block net conduction, and probably it acts at synaptic junctions.

It is not always possible to separate net conduction from intercellular epithelial conduction, particularly in Hydrozoa, and conduction that was formerly attributed to nerve nets now appears to be epithelial.

The cnidarian systems that are clearly nerve nets are sometimes considered primitive, but they show many of the complexities of so-called higher nervous systems. Facilitation is important, and guarantees that an animal does not respond to every stimulus but only to a significant number of stimuli. Facilitation requirement also imposes a sort of polarity. Nerve nets probably range in type from fused neurons through nonpolarized to polarized junctions. They show spontaneity, inhibition, and reciprocal innervation of muscles, and provide for a considerable behavioral repertoire. They may serve as models for complex nervous systems but are in no sense simple.

Transition from Peripheral to Central Integration

Coordination of movements by peripheral networks occurs not only in cnidarians but also in some animals which have central nervous systems. In gastropod and pelecypod molluscs, a subepidermal plexus in the foot and mantle mediates contractions of mantle edge and labial palps. This nerve net shows integrative properties. In earthworms, there is a peripheral neural connection between epidermal receptors and muscles. In balanoglossids, responses of the proboscis may be coordinated by the peripheral plexus.[421] In most animals whose locomotion is centrally controlled, visceral movements may be coordinated or initiated by gastrointestinal plexuses.

The transition between peripheral and central control of locomotion is well shown in echinoderms. The nervous system consists of a central oral ring that gives rise to five radial nerve trunks, which contain nerve cells and connect to the peripheral network in tube feet, spines, pedicellariae, and skin. The peripheral plexus occurs only in the tube feet, in ophiuroids and crinoids. If in a starfish a radial nerve is cut, responses to touch can be transmitted by the peripheral plexus to the ring or to other arms. If all elements of the nerve ring are removed, an isolated arm moves toward its base; but if a piece of the ring remains, the arm moves toward its distal end. If the ring is cut in two places, the starfish may pull itself in two. Coordinated stepping of tube feet requires a radial nerve, but the direction of movement depends on neurons at the junctions of ring and radial nerves.[502] Conduction in the ring is at 11 cm/sec, and in radial nerves at 17 cm/sec.[424]

In an echinoid, each spine has at its base a number of neurons which are part of the general plexus. Conduction in the plexus is at 6 cm/sec and does not require movement of a spine; i.e., it is not by a chain of reflexes.[51] In the echinoid *Diadema,* local shading of impinging light elicits bending of spines toward the shadow, a protective reaction which requires the presence of the radial nerve.[601] Stimulation of a radial element evokes one action potential conducted at 22 cm/sec, which excites spine movement, and also a second action potential at 6 cm/sec which inhibits movement.[375, 376] Righting of a starfish requires radial nerves, and is more readily done if the nerve ring also is present.

In flatworms, nerve trunks (two to eight) run posteriorly from the brain and contain nerve cells not grouped in ganglia. A bit of one of these nerve cords seems necessary for spontaneous coordinated movements in planaria and nemerteans. In a polyclad, *Planocera,* excitation cannot pass from one side of the body to the other via the peripheral network; the "brain" is needed for coordinated locomotion, normal righting, and orientation to food and feeding, but not for swallowing or egg laying.[174] Evidently, in flatworms the peripheral system does not function as an independent nerve net.

Ganglionic Function in Molluscs

Gastropods and Pelecypods

In gastropod and pelecypod molluscs, locomotor and other reflexes are mediated by four paired ganglia—cerebral, pedal, pleural, and visceral—which are interconnected. These ganglia show varying degrees of fusion and symmetry according to the general anatomy of the particular animal. Snails such as *Helix* and nudibranchs such as *Aplysia* have, in addition to many small neurons, a limited number of giant interneurons and motorneurons which are large and arranged in a fixed pattern.[237, 238, 531] In *Helix, Aplysia,* and the tectibranch *Tritonia* the giant neurons have been mapped according to location and function. Each giant interneuron receives input from sensory neurons serving limited regions of the body; the synaptic connections provide for fixed circuits, and the motor output may or may not be symmetrical to the two sides of the nervous system.[73, 74, 106, 107] Most synapses in molluscs are axo-axonic, some are axo-somatic but none are axo-dendritic. Spikes originate some distance from, and independently of, the cell soma.[528]

A subepidermal nerve plexus in excised labial palps of *Anodonta* mediates responses to touch or illumination. Nerve cell bodies are common in the peripheral nerves of many molluscs. In *Aplysia,* a gill disconnected from central ganglia contracts to local tactile stimulation and this response, mediated by peripheral plexus, can become habituated.[414] The degree of interaction between peripheral nerve nets and central ganglia in gastropods and pelecypods remains to be elucidated.

Reflexes of Pelecypods and Gastropods. In general, in bivalve molluscs, contractions (phasic or tonic) and relaxation in the adductor muscles are elicited by different neurons. In *Anodonta,* impulses from the visceral ganglion can elicit either tonic or phasic contractions according to the stimulus to the ganglion. The cerebral ganglion has a center for relaxing adductor muscles.[477, 478] In the surf clam *Spisula,* specific motorneurons in the visceral ganglion receive input from sense cells on each side of the animal, and they trigger bilaterally symmetrical motor responses. Input to the motorneurons is by cerebro-visceral connectives, and by nerves from siphon and mantle; some motorneurons show depolarization followed by hyperpolarizing junctional potentials, while others show only depolarizing responses. Motorneuron output is to the adductor muscles.[362] In a razor clam, *Ensis,* sensory signals from the foot or collar go to one side of the cerebral ganglion, then to the opposite side, and then to the pedal ganglion and out to the foot; a reflex of short latency goes via the pedal ganglion only.[403] Swimming movements in a scallop involve a myotatic reflex, with a motorneuron discharge triggering a contraction which initiates a sensory discharge back to the motorneuron.[363]

There is evidence for *Helix* that the pedal ganglion has both excitatory and relaxing influences on muscles of the foot. One interneuron responds to sensory stimuli on either foot or head. In several bivalves, the reflex pathways and the balance between excitation and inhibition vary with intensity of sensory stimulation. Removal of the cerebral ganglion from *Aplysia* increases locomotor activity; hence, the cerebral ganglion may inhibit the pedal ganglion, which excites foot muscles. Reflex pathways in the central nervous system of the snail *Buccinium* have been mapped; much sensory convergence on central neurons, combined with patterned efferent activity, permits coordinated responses to mechanical and chemical stimulation of receptors.[20] It is concluded that there is some reflex function in all molluscan ganglia, but that the entire nervous system is so interconnected that there is limited localization of function within ganglia.

Function of Identified Neurons and of Clusters of Interneurons. Much progress has been made in correlating specific function with individual identified neurons in gastropods. The abdominal ganglion of *Aplysia* contains 1000 to 2000 neurons, according to size of the animal,[73] and large cells may be identified individually on the basis of position in the ganglion, pigmentation, firing pattern, junctional potentials, responses to orthodromic input, efferent pathways, connections to other neurons, and responses to drugs. Thirty large cells and eight clusters of cells similar to one another have been identified (Fig. 15–10).[150] In addition, there are neurosecretory cells, which include "bag cells" at the exit of nerves, and white and pigmented cells. Some of the pigmented cells respond to illumination by either hyperpolarization or depolarization.[16] All of the cells of a given cluster behave in a similar manner and are called unidentified interneurons. Synchrony between coupled neurons is often by bursts rather than by individual spikes. From tabulations of the properties of identified cells, models of divergent and convergent chains

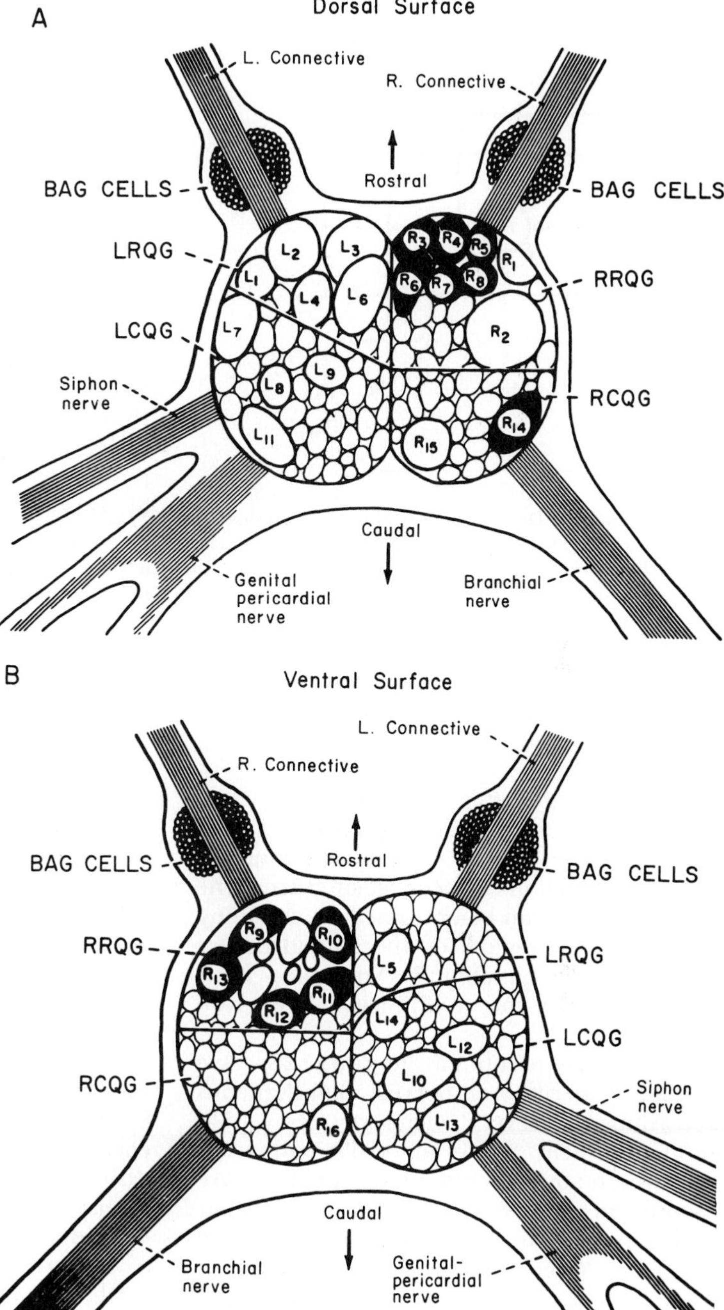

Figure 15–10. Diagrams of (A) dorsal and (B) ventral views of abdominal ganglion of *Aplysia,* showing identified neurons on right (R) and left (L) sides. RRQG, right rostral quarter ganglion; LRQG, left rostral quarter ganglion; RCQG, right caudal quarter ganglion; LCQG, left caudal quarter ganglion. (From Frazier, W. T., E. R. Kandel, I. Kupferman, R. Waziri, and R. E. Coggeshall, J. Neurophysiol. *30*:1288–1351, 1967.)

within the ganglion have been drawn[283] (Fig. 15–11). There is some regional localization of neurons of similar functions; a single multibranched interneuron can exert a multiplicity of actions on follower cells, and different interneurons show a variety of loop interactions.[280] Considerable functional complexity in a system of small neurons results from extreme individuality of single neurons. Coding can be by modulation of endogenous rhythms, by summation of psp's, or by combined inhibitory and excitatory actions.[553]

Of 30 identified neurons in the abdominal ganglion, 27 show endogenous activity, which usually consists of repeated bursts of spikes. Cell L-15 in *Aplysia* is a parabolic burster.[517]

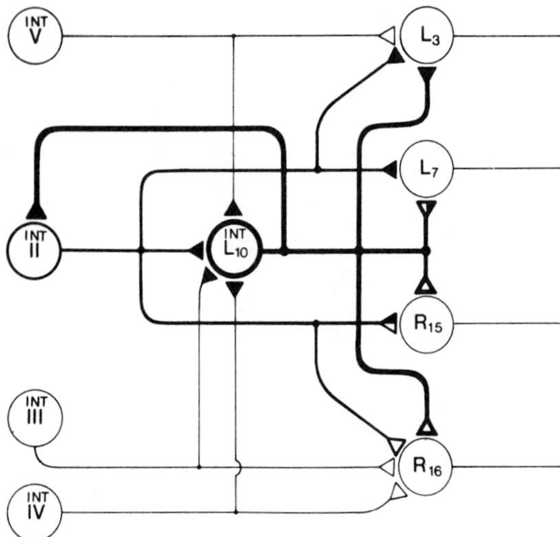

Figure 15–11. Diagram illustrating connections of interneuron L_{10} in abdominal ganglion of *Aplysia*. Feed-forward connections of interneurons II, III, IV, and V and of identified cells L_3, L_7, R_{15}, and R_{16} are shown. Solid endings are inhibitory, and open endings are excitatory. Half-solid endings are both excitatory and inhibitory. (From Kandel, E. R., and H. Wachtel, pp. 17–65 *in* Physiological and Biochemical Aspects of Nervous Integration, edited by F. D. Carlson. Prentice-Hall, 1967.)

The discharge in this cell occurs in diurnal rhythm, with maximum frequency of bursts shortly after dawn; the rhythm is maintained after the ganglion has been isolated from the animal,[518] and can be entrained by changing the light-dark regime of the animal before removing the ganglion.[328] Suggested ionic mechanisms of the rhythmicity were given on p. 737.

Of the 30 identified neurons, 28 have axons in peripheral nerves. Simultaneous recordings by several electrodes show that numerous cells are functionally connected to other neurons. Most of the neurons are modulated, often by inhibition of their endogenous firing, by input in peripheral nerves or from unidentified interneurons. The rhythms of four motor cells in the upper left quadrant are modulated by inhibitory ipsp's, but another cell in the same group shows synchronous epsp's; hence, two branches of one synchronizing interneuron have different actions.[278] Activation in cells L-1 to L-6 is associated with epsp's in R-15 and L-10. One large neuron in the abdominal ganglion, L-10, sends branches to 14 follower cells in three clusters, inhibitory branches to the rostral ganglion, excitatory branches to the right caudal ganglion, and both inhibitory and excitatory branches to the left caudal ganglion.[278, 284]

Five motorneurons in the *Aplysia* abdominal ganglion receive sensory input from siphon and mantle and both excitatory and inhibitory input from spontaneously active interneurons. Change in the balance of these two inputs (numbers of spikes per unit time) can change the motorneuron-mediated response from a simple reflex to a complex withdrawal response.[307] A single large pigmented cell in the visceral ganglion of *Aplysia* can be activated by sensory fibers in each of several peripheral nerves; an impulse in the left visceral nerve which precedes one in the pleural nerve does this more effectively than the same impulses in the reverse order. When groups of three sensory impulses were driven with the interval S_1 to S_3 constant but the time of S_2 in the triplet varied, the interneuron response showed a maximum when S_2 was near S_3 but the effect depended on the mean frequency, i.e., S_1 to S_3; hence, the interneuron recognizes pattern in its input.[492]

Certain interneurons, particularly a group in the upper right quadrant of the abdominal ganglion of *Aplysia,* show long-lasting heterosynaptic facilitation in which one input (a primer) facilitates the response to another input (test). The primer causes no change in conductance in the giant cell, so the effect is presynaptic on the test axons or on an interneuron circuit involving these axons.[281] In another circuit, one input causes presynaptic inhibition which may last as long as one minute; response in an abdominal ganglion cell to stimulation of the right connective is inhibited by impulses in the branchial nerve.[529] Another type of long-lasting electrical change in motorneurons correlates with behavioral habituation and dishabituation (p. 705).

Giant neurons of *Aplysia* and *Helix* show individuality in synaptic responses to common transmitters (p. 475). Acetylcholine excites (depolarizes) R-15 while it inhibits (hyperpolarizes) L-3; two branches of the same interneuron or sensory neuron can excite one cell and inhibit another. In the buccal ganglion of *Aplysia,* two interneurons on each side synaptically (1) hyperpolarize six neurons, (2) depolarize one neuron, and (3) depolarize and later hyperpolarize one other neuron on each side. Applied acetylcholine has similar dual action, and it is concluded that the same cell may have receptors for both Na and Cl conductance responses.[158] The inhibition is due to increased g_{Cl} and the excitation mostly to an increase in g_{Na} (see p. 475). Some cells are excited by ACh and inhibited by dopamine, the inhibition being due partly to increase in g_K. Serotonin (5HT) elicits depolarizing psp's in some cells, and hyperpolarizing psp's in other cells. Thus, individual neurons must have different and sometimes several membrane receptors. It is probable that ACh, 5HT, and dopamine are transmitters (see pp. 489–499).

In snails, similar but less extensive observations than those in *Aplysia* have been made on identifiable neurons. For example, on the ventral surface of the fused cerebral ganglion of *Helix* are two symmetrical cells which send processes to four ipsilateral and four contralateral nerves and to three ipsilateral and three contralateral interganglionic connectives; they have input from two common interneurons.[282] In the buccal ganglion of *Navanax* are 10 cells that are not directly connected synaptically but that activate interneurons which interact and then feed back to the 10 large cells.[326] In *Helix,* as in *Aplysia,* after habituation to one sensory locus, a central neuron can respond to input from another locus.[503]

The nudibranch *Tritonia* has some 50 identifiable neurons in the cerebral mass. An array of electrodes in the brain of an animal, which was suspended so that it was free to move, permitted recording and stimulation of individual interneurons.[582, 583] Stimulation of single neurons elicited specific localized contraction, and stimulation of several interneurons elicited complex behavioral actions (Fig. 15–12). When natural stimuli such as starfish extracts were applied

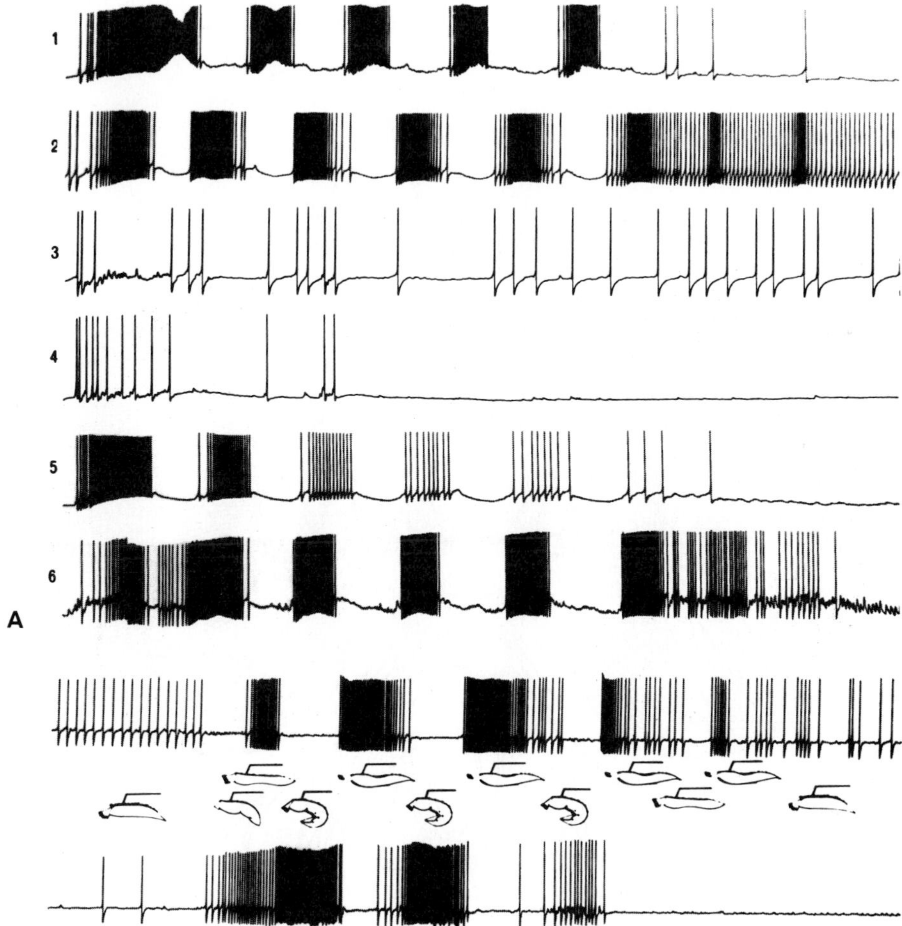

Figure 15–12. Giant brain cells in a mollusc, *Tritonia.* A, Activity in six neurons recorded simultaneously during 55 seconds of swimming. Cell 1 firing correlates with downward flexion, cells 2 and 6 with upward flexion on right and left sides, cells 3 and 4 with cells that trigger escape sequence, and cell 5 with withdrawal of gills on left side of body. B, Simultaneous recordings from two cells that drive escape flexions of body, as illustrated by drawings between two recordings. Upper cell drives dorsal flexion; lower cell drives ventral flexion. (Courtesy of Dr. A. O. D. Willows.)

while recordings were made from identified interneurons, three types of response were found: (1) cells firing in phase with dorsal flexions of the lateral "wings"; (2) cells alternating with (1) and firing in phase with ventral flexions; and (3) cells firing during both swimming movements (general excitor neurons). Apparently there is synchronization of bursts among the cells in one group, and excitation in the two flexor groups is maintained by a positive feedback loop between the flexor neurons and the general excitor neurons. Coordination of alternating movement patterns results from a reciprocal inhibitory connection. In response to a given sensory input, an isolated brain can fire in the same program of bursts as it did in the intact animal.[582a]

In summary, the large neurons of gastropod ganglia show remarkable individuality in synaptic properties and relation to behavior. Approximately 1 to 2 per cent of the neurons in the ganglia are large and can be identified from animal to animal. The functions of the other 98 per cent of the neurons, and of small cells in general, remain to be elucidated. A single central neuron receives from a wide sensory field, and a single cell can elicit a complex motor response.

Cephalopods

In cephalopods, the brain mass represents fusion of several ganglia and permits complex behavior (Figs. 15–13, 15–14). Functions of different brain regions have been analyzed by anatomical observations, effects of lesions, and electrical stimulation. Attempts to record electrical activity from cephalopod brains have been singularly unsuccessful. The role of the brain in cephalopod learning is discussed on p. 700, and

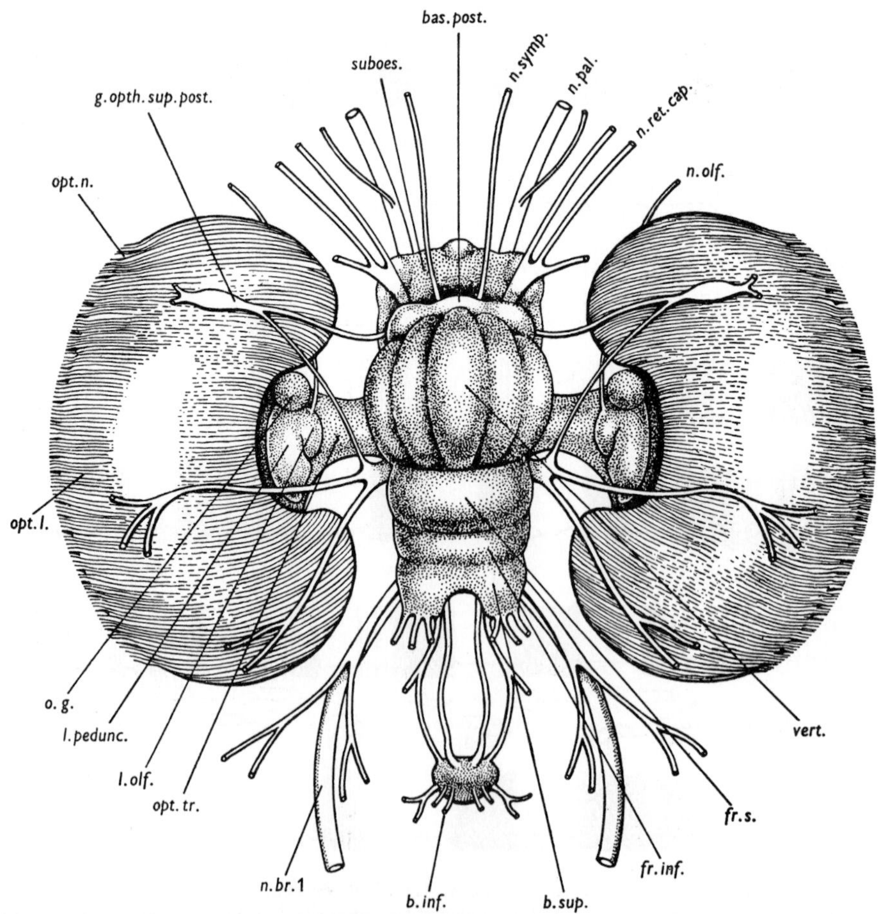

Figure 15–13. Dorsal view of brain and optic lobes of *Octopus vulgaris.* Abbreviations: *b. inf.,* inferior buccal lobe; *bas. post.,* posterior basal lobe; *b. sup.,* superior buccal lobe; *fr. inf.,* inferior frontal lobe; *fr. s.,* superior frontal lobe; *l. olf.,* olfactory lobe; *opt. l.,* optic lobe; *vert.,* vertical lobe. (From Young, J. Z., Proc. Roy. Soc. Lond. B *153*:18–46, 1960.)

the origin of the giant fiber system in the subesophageal magnocellular lobe on p. 638.

In *Sepia,* the supraesophageal portion includes two motor regions—the anterior basal lobes controlling position of head, arms, and eyes, and the posterior basal lobes controlling chromatophores, swimming, breathing, and other actions via six motor centers. Above the basal lobes are the inferior and superior frontal lobes, and dorsally and centrally are the vertical and subvertical lobes. These are association areas, and stimulation of them elicits no movements; they are also connected to the very large optic tract. The subesophageal portion of the brain mass includes the anterior brachial, middle pedal, and posterior palliovisceral lobes. These have discrete motor centers for various regions—mantle, funnel, fins, chromatophores, eye muscles, and so forth—as shown in Table 15-1.[41]

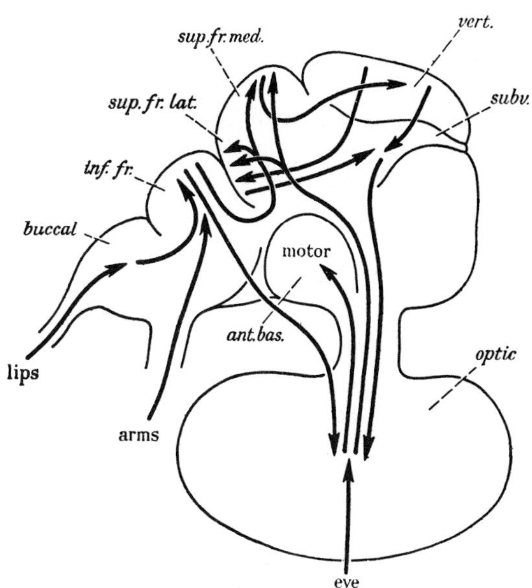

Figure 15-14. Diagram of pathways in visual and tactile learning in *Octopus.* Abbreviations: *inf. fr.,* inferior frontal; *sup. fr. lat.,* superior frontal lateralis; *vert.,* verticalis. (From Young, J. Z., Proc. Roy. Soc. Lond. B *153*:18–46, 1960.)

In *Octopus,* the optic tracts connect with large optic lobes, which serve as reflex and memory centers and which connect with frontal and verticalis complexes. Tactile input from the arms goes directly to the buccal and frontal lobes.[606, 609] In *Octopus,* motor centers comparable to those of *Sepia* have been identified in the subesophageal complex.[561] Removal of the dorsal supraesophageal lobes results in no motor defects but

TABLE 15-1. Motor Connections of Regions of Brain of Cuttlefish (*Sepia*).[41]

Region		*Lobes*	*Parts Responding to Stimulation*
Supraesophageal portion	Vertical complex	Vertical	no motor responses
		Subvertical	
		Superior frontal	
		Dorsal basal	
	Basal	Anterior basal	head, eyes, arms, fins, mantle
		Lateral basal	chromatophores, skin
		Medial basal	head, arms, fins, giant fibers, buccal mass
		Interbasal	tentacles
		Buccal and inferior frontal	mouth and tongue
Subesophageal portion		Anterior pedal	arms, tentacles, chromatophores
		Posterior pedal	head, funnel, fins, tentacles
		Lateral pedal	eye
	Palliovisceral	Central	mantle, collar, funnel, viscera
		Fin	fins
		Chromatophore	chromatophores
		Magnocellular	giant fibers
		Vasomotor	
Optic lobes		Periphery	no motor response
		Center	similar to stimulating basal lobes

seriously interferes with complex learned behavior. Except for the few neurons of the giant fiber system, most cephalopod brain neurons are small (90 per cent of them are less than 5 μm in diameter). The two optic lobes contain some 120×10^6 neurons, the vertical lobe 25×10^6, the superior frontal lobes 1.8×10^6. It is estimated that *Octopus* brain mass contains a total of more than 168×10^6 neurons.[561, 606]

The arms of *Octopus* contain not only afferent and efferent fibers but axial ganglia which serve reflexes of the suckers; interneurons in these ganglia show a variety of patterned responses to stimulation at various sites.[461]

PHYSIOLOGY OF LADDER-TYPE GANGLIONIC NERVOUS SYSTEMS

ANNELIDS

The central nervous system of annelids is a ladder-like chain of ganglia; in some oligochaetes (earthworms) a subepidermal plexus provides for limited connections between epidermal receptors and segmental muscles, but in other annelids (hirudineans and polychaetes) all input-output relations appear to be via the central nerve cord. In an earthworm, several segments of the ventral nerve cord, isolated from body musculature, can conduct a peristaltic wave to the region beyond the ligature. Normally, however, the wave is reinforced by segmental reflexes. A length of twenty to forty intact segments of an earthworm, suspended and mechanically balanced in a saline bath, shows no peristalsis, but a slight stretch or tactile stimulus elicits persistent peristalsis.[79] A tension-induced peristalsis can be inhibited by tactile stimulation. The maximum frequency of rhythmic contractions is about one per two seconds, and rhythmic contractions may persist for minutes after removal of the weight which initiated them. Circular muscles contract when longitudinal muscles relax and vice versa, and correlated reflexes control the setae. The alternating reflex system uses central fibers which conduct slowly at 0.4 to 0.6 m/sec.[441] Peristaltic transmission still occurs after the ventral nerve cord is cut or several segments of it are removed, or even if two pieces of a transected worm are connected by a thread or if a few intervening segments are anesthetized. However, if the nerve cord is removed from more than three segments of an earthworm and these segments are firmly pinned down, no peristaltic wave passes. Each segmental nerve shows a sensory and a motor field extending over three segments. Deganglionated longitudinal muscle shows spontaneous activity (p. 770), and deganglionated segments may contract to tactile or photic stimulation. Hence, the subdermal plexus evidently provides some peripheral connections from sense endings to muscle, as well as providing for convergence of the many sense cells to the few neurons passing in segmental nerves to the central ganglia.

In leeches, transection of the nerve cord disrupts coordinated swimming. Tension, contact of body receptors, or stimuli to the sucker can elicit peristaltic reflexes. Giant interneurons, unlike those of earthworms, can be activated by sense endings at either end of the leech.[318] Each ganglion has 350 neurons; the 14 sensory neurons are in three groups on each side: three respond to light touch, two to maintained pressure, and two others to noxious chemical stimuli (Fig. 15-15). The sensory fields for the tactile neurons overlap; the sensory neurons are connected with one side of a ganglion and with 3 cells in each adjacent ganglion by electrical synapses with double rectification (p. 482). Transmission from these touch cells to motorneurons is also electrical, but with single rectification. Transmission between the sense cells for noxious stimuli and the motorneurons is slow (2 to 4 msec delay), and is abolished by Mg^{++} and enhanced by Ca^{++}; hence, it is by a chemical synapse. Transmission between the pressure sense cells and motorneurons is both chemical and electrical, with one-way rectification as for the touch cells.[392, 393] The motorneurons innervate muscles in overlapping patterns, each motorneuron serving a muscle field in a layer. For each ganglion, there are 14 pairs of excitatory and three pairs of inhibitory motorneurons to the muscles. The motorneurons are electrotonically coupled, but transmission to the muscle is chemical.[520] Connections from one ganglion to the next are not symmetrical; the connection from posterior to anterior ganglion has both inhibitory and excitatory junctions, while that from anterior to posterior has excitatory junctions only. The nervous system of leeches, therefore, shows much neuronal

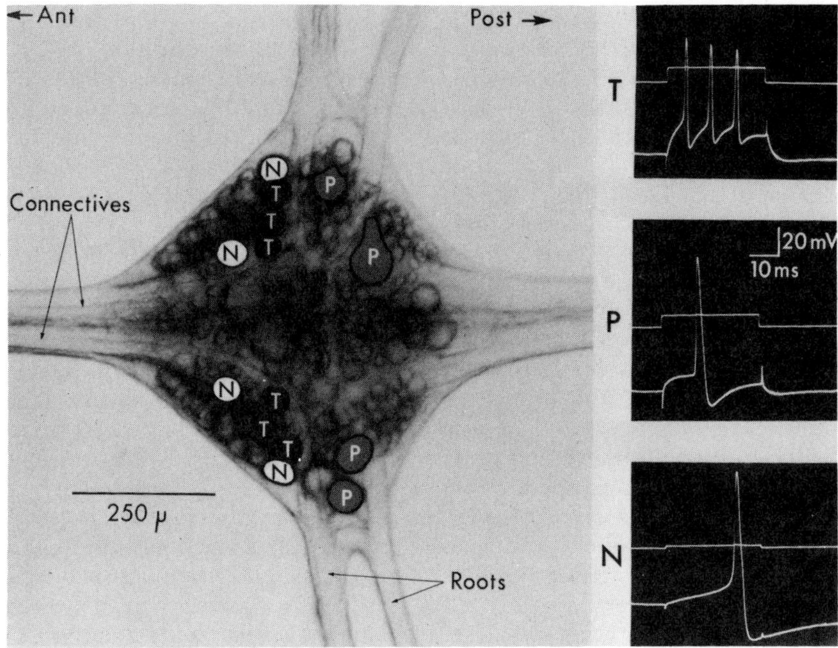

Figure 15–15. Ventral view of ganglion of leech. Sensory interneurons identified by response to stimulation of skin. P, pressure; T, touch; N, noxious stimuli. At right, spike responses to intracellular stimulation. (From Baylor, D. A., and J. G. Nicholls, pp. 3–16 *in* Physiological and Biochemical Aspects of Nervous Integration, edited by F. D. Carlson. Prentice-Hall, 1967.)

specificity and both electrical and chemical synapses.

In most polychaetes, peristalsis is controlled by segmental reflexes and central rhythms. There is no evidence for peripheral connections such as occur in earthworms. During locomotion by *Nereis,* the parapodia move successively in clumps of four to eight. Accompanying the wave of parapodial beat, the longitudinal muscles on the same side contract while those on the opposite side relax; the nereid crawls by a series of S-movements and not by a peristaltic ring wave.[171] In *Harmothoë*, peristalsis can be independent of parapodial stepping; the parapodial wave moves forward at 3 to 10 segments per second, and ipsilateral and contralateral cycles are 180° out of phase. The parapodial wave is driven by the nerve cord and remains after segmental nerves have been cut (leaving the branches to parapodia intact).[319] In tube-dwelling polychaetes, feeding and irrigation cycles are centrally controlled. In *Arenicola marina,* separate but interacting neural pacemakers trigger rhythmic contractions of body wall and tail. In *Arenicola ecaudata,* the isolated extrovert (proboscis and esophagus) shows rhythmic feeding movements as long as it is connected to the supraesophageal ganglion. Some of the centrally controlled cycles of activity are very slow (one every 40 minutes).[559] *Nephthys* makes simpler burrows after decerebration than while its brain is intact; maze learning is retarded by decerebration, but if the brain is removed after conditioning the learned behavior remains.[72, 125, 126]

Ganglionic Function in Arthropods

The nervous systems of arthropods, like those of annelids, are ladder-like. Two symmetrical chains of ganglia joined by connectives run the length of the animal. The pairs of ganglia are joined left and right by transverse commissures. In the thorax there is often fusion of ganglia, both laterally and between segments. The supraesophageal ganglion, the "brain," receives input from important sense organs of the head and has two connectives to the subesophageal ganglion. The number of neurons in an arthropod nervous system is relatively small for the complexity of possible behavior.

Of the "large" neurons in a crayfish, each abdominal ganglion has 500, each thoracic ganglion has somewhat more, the subesophageal ganglion has 6000, the brain has 10,000, and the total is some 29,000; "small" neurons total about 37,000.[574] The second abdominal ganglion of the walking stick *Carausius* has 600 neurons, 1000 glial cells, and 500 perineural cells.[25] The cockroach metathoracic ganglion has 3000 neurons, 230 of them of diameter greater than 20 μm.[76] In addition to the ordinary neurons, many arthropods have giant fiber systems; these may consist of large, long single neurons as in insects and some crustaceans, or they may consist of septate-abutting large axons, such as the lateral giants in decapod crustaceans (p. 640).

Fewer than half of the neurons in arthropods are relatively large (cells 5 μm or larger), and these can be studied as units either after teasing a bundle down to a few fibers or by recording from within a soma or large axon. Most central neurons are unipolar, and activity can spread within and among branching processes without traversing the cell soma (Figs. 15–3, 15–16). Both synaptic sites and points of origin of impulses may be in the neuropil some distance from the soma. Each neuron is independently active in spontaneous firing and response to sensory stimulation; there are no summed slow waves as in "brain waves" or "evoked potentials" of vertebrates, perhaps because of absence of cortical stratification of neurons. Behavior of arthropods, although stereotyped, may be complex. Evidently, individual neurons are rather unique; interneurons function in a hierarchy of control levels, and arthropod behavior may be ascribed to simple neural networks.

Locomotion in arthropods is more complex than in annelids because of jointed appendages, elaborate limb musculature, endogenous rhythmic patterns of central nervous control, and modulation of motor patterns by proprioceptive input and impulses from the brain. Arthropods can be arranged in a series of increasing complexity of locomotion.[345] *Peripatus* presents the basic arthropod pattern in which several gaits are possible. In "low gear" the paired legs move in phase and the legs make a long backstroke, in "middle gear" the strokes are equal, and in "high gear" the back stroke is short and the paired legs alternate. Diplopods have elaborated the low gear pattern and push by their legs; chilopods use the faster gear and tend to pull.

A tarantula has four pairs of walking legs; the walking sequence is inconstant, shifting

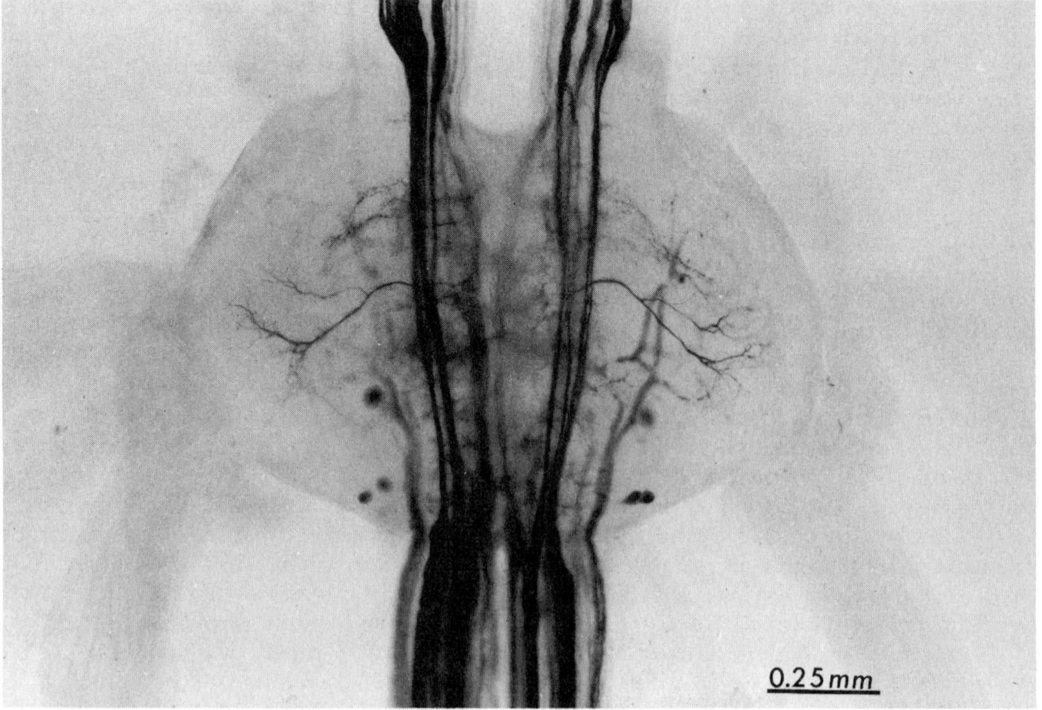

Figure 15–16. Neurons of cockroach ganglia as revealed by iontophoresis of cobalt. A, Ventral view of metathoracic ganglion, showing giant fibers and collaterals. (Courtesy of M. J. Cohen.)

(Figure 15–16 continued on opposite page.)

frequently from alternating to diagonal. Contralateral antagonism is not strong. Each leg and its hemiganglion is an oscillator negatively coupled to the leg on the opposite side and to neighboring legs.[585] Locomotor coordination in insects is described on p. 661.

Crustaceans. SENSORY INTERNEURONS. Primary sensory input to the nervous system of crustaceans has been previously described for several sense organs: stretch and tactile receptors, p. 509; eyes, p. 615; statocysts, p. 542. Axons of interneurons which integrate sensory information have been found in fixed locations in the nerve trunk and optic stalk. Sensory interneurons can be distinguished from primary sensory neurons by greater sensitivity to anoxia, capacity to follow only at low frequencies, and wide (often multiple) peripheral receptive fields.

In one of the connectives between abdominal ganglia III and IV of crayfish, 75 neurons have been identified and mapped. Of 57 *interneurons,* 30 responded to mechanoreceptor input ipsilaterally only, 10 contralaterally only, and 7 bilaterally.[577] The simplest interneurons are unisegmental, whereas others convey sensory information from several ganglia. One interneuron may receive input on several dendrites, and the axons of some may cross and recross in the neural ladder. In the sixth abdominal ganglion, one paired neuron is excited via many sensory roots, and intracellular recordings show unitary epsp's when individual hairs are stimulated; second order interneurons in the same and other ganglia may be triggered by the sensory interneurons.[291]

In a connective between thoracic and abdominal cord of crayfish, 131 distinguishable units were found, of which 22 were primary sensory fibers. The distribution of interneurons, in percentage from different inputs, was as follows:[576]

	ipsilateral	*contralateral*	*bilateral*
from statocysts	2	1	0
from tactile hairs	42	8	5
from tactile hairs & joints	6	2	2
from joints only	13	4	6
from general activity	6	5	7

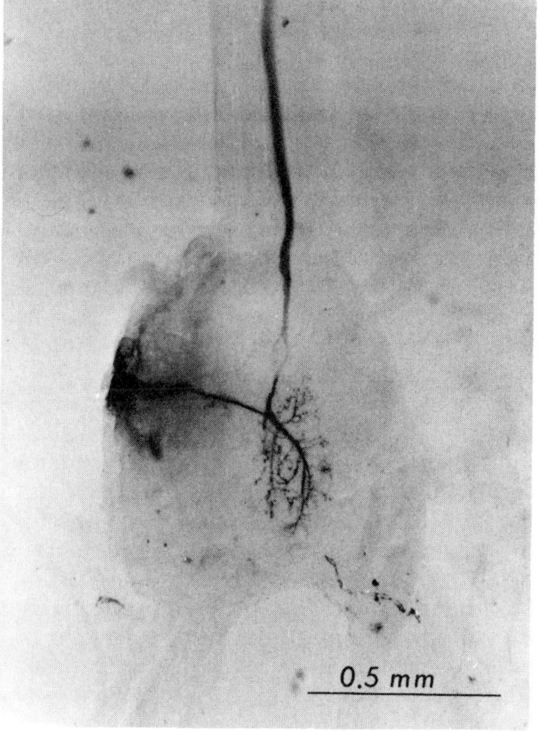

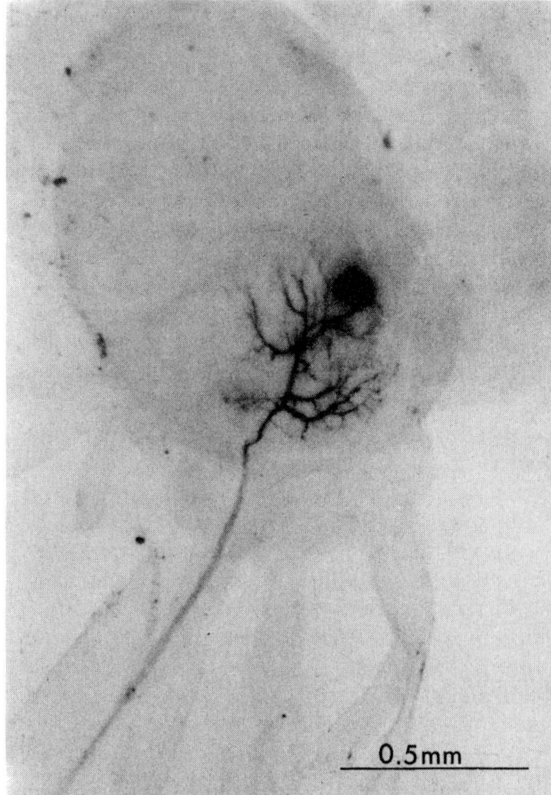

Figure 15–16 *(Continued).* B, Sixth abdominal ganglion, dorsal view, showing giant neuron cell body, dendrites, and part of giant axon. C, Metathoracic ganglion, dorsal view, showing motorneuron (which is the fast extensor to the coxal depressor muscle). (Courtesy of M. J. Cohen.)

Of the interneurons, 31% were descending, 51% ascending and 18% two-way. A few were inhibited by stimulation on the animal's head.[576, 579] Water vibrations on an antennule brought about response of an interneuron of the circumesophageal connective; the response was decreased by simultaneous movement of legs or abdomen, and by visual stimuli, and thus the sensitivity of the interneuron is modulated by various inputs.[532]

The crustacean optic stalk contains three relay stations and is a complex integrating structure, anatomically a part of the brain. In the eyestalk of lobster or crayfish, eleven sensory interneurons have been identified.[13, 575, 580] Some of the fibers code light intensity; some record movement of objects, and respond respectively to fast, moderately fast, to-and-fro, or very slight movement. Other interneurons signal motion of either dark or light objects. There are narrow- and wide-field interneurons, and some which signal unidirectional movement. Multimodal interneurons receive visual input from eyes and tactile input from body hairs. Some sustaining interneurons respond to a small visual field and show surround inhibition with on-off effects at boundaries (p. 615). To-and-fro movement fibers show no directional preference, and space-constant fibers record constancy in location and size of a field when eye position is changed. Sustained-activity fibers, recorded by implanted electrodes, show circadian rhythms with maximum responsiveness at 2200 to 2400 o'clock.[13] Some interneurons in the brain are under statocyst control, but are influenced by visual input and by command motorneurons to optic stalk muscles.[580]

In summary, many second and third order interneurons occur in fixed locations in the crustacean nervous system; these deal with converging and overlapping sensory inputs, they signal specific sensory patterning, and they may respond to more than one modality; some of these interneurons may be influenced in excitability by other interneurons.

MOTOR INTERNEURONS. Interneuron function has been further studied by observing motor responses to stimulation of single cells. A motorneuron is triggered by an interneuron, and a single interneuron may elicit coordinated movements of many muscles by activating many motorneurons. A single motorneuron can be excitatory or inhibitory to whole muscles or to bundles of muscle fibers, and its action differs with frequency of spiking and innervation pattern. A crayfish uropod can move about three axes by using 18 different muscles, which are innervated by at least 55 motorneurons. Tonic muscles of a uropod receive two to three excitors and one inhibitor; phasic muscles receive up to five excitors and either one inhibitor or none.[313]

A single (command) interneuron may activate a group of motorneurons, sometimes via other interneurons and thus elicit a coordinated movement. Command interneurons mediate coordinated output of motorneurons and thus provide for gradation and patterning of contractions. Command neurons may evoke symmetrical movement of the uropods on the two sides of the abdomen. The movements so released are reciprocal: flexors are inhibited both centrally and peripherally during extension, while extensors are inhibited during flexion.[130]

In intact crayfish or lobster, the swimmerets beat rhythmically, with swimmerets in each segment synchronized and those in successive segments sequential. Each swimmeret is moved by 12 muscles, and the power and return strokes are linked between the two sides in a given segment. In one connective between thorax and abdomen are at least five interneurons, stimulation of any one of which evokes bilateral rhythmic movement of all swimmerets of the abdomen. On each side in a segment of the abdominal cord are five command neurons; stimulation of one of these may trigger an oscillator neuron, which in turn causes rhythmic firing of a group of motorneurons. Stimulation at 30/sec elicits a normal beat and rhythmic discharge at 1/sec. Any one of three interneurons can inhibit the rhythmic discharge on both sides.[18, 578] In each of two nerves to a swimmeret are six large excitatory axons and one inhibitory axon, and numerous small motor axons; the twigs from sensory endings may be cut without abolishing the rhythmic firing of motorneurons. Swimmeret motorneurons behave as a group, small ones firing before large ones. It appears, therefore, that the command neurons converge on one or more oscillator neurons (not yet identified), which alternately excite and inhibit groups of motorneurons that act synergistically on the muscles to cause either downstroke or upstroke of a swimmeret.[95, 96, 97]

Fast abdominal flexion and the "escape" response are triggered by tactile hair stimulation; several sets of longitudinal muscles are activated by motorneurons that can be controlled either by small interneurons or by giant fibers. In crayfish, the median or lateral giant fibers act segmentally on a giant motorneuron and several smaller motorneurons to the twitch flexor muscles (Fig. 15-4). The giant motorneuron is stimulated at a rectifying electrical synapse from a branch of the lateral giant axon (p. 483). The motor synapses are on dendrites in which spikes are initiated; these spikes appear to amplify the epsp's to produce enough current to trigger an axon spike beyond a region of low safety factor. Dendritic branching provides for some lability in responses of the giant motorneurons.[617, 618] The repeated tail flips of swimming can be initiated either by giant fibers or by smaller interneurons of the cord.[488] The lateral giant fibers also mediate uropod flaring.[314] Rostral mechanical or visual stimulation activates the median giant, which has its soma in the brain; caudal stimulation activates the lateral giant fibers. Both giant systems have bilateral input and output.[526]

CRUSTACEAN REFLEXES. Reflexes are direct, patterned responses to sensory input, modifiable by convergence of sensory and central signals on interneurons. Movement of leg joints elicits reflex discharge of reciprocal excitation and inhibition. For example, passive extension of the dactyl excites the claw closer and inhibits the opener;[60] stimulation of a specific pair of chordotonal organs excites the claw opener and inhibits the closer.[129]

Two pairs of stretch receptors, tonic and phasic, lie between thorax and abdomen and between abdominal segments. The slowly adapting, low threshold receptor reflexly excites one of the motorneurons to the abdominal extensors on that side; when the extensor contracts, the receptor is unloaded. The tonic stretch receptor thus participates in a "resistance reflex" that tends to stabilize segment length. Its action is mainly unisegmental and ipsilateral, and it functions as an error detector in the servo-control system for position of a segment.[141, 142]

A reflex of remarkable precision in crustaceans is the optokinetic response, the movement of an eyestalk to follow a shifting pattern or small spot of light. A crab eye can follow the sun in a compass reaction, even when landmarks and horizon are obscured. Crustacean eyes can respond to movements as slow as 0.001° of arc per second, and may follow a moving edge through some 15° and then flick back, showing nystagmus. Movement of an image across the retina is converted in a directionally sensitive series of neurons in the brain so that a blinded eye is driven to move by a seeing eye which has been fixed so that it cannot move; no proprioceptive feedback is needed. If an eye fixes the position of a series of stripes, and then the lights are turned off and the stripes are moved and reilluminated, the crab eye moves as if it had seen the displacement in progress. A central memory has correlated ommatidial stimulation before darkness with stimulation of other ommatidia after darkness. Normally, eye tremor enhances perception of edges.[21, 58, 216, 219, 221]

One protective reflex of a crab is withdrawal of the eyestalk, elicited by four motorneurons in the brain. Each motorneuron has two large dendrites which branch extensively and receive input from mechanoreceptors on the body surface. Both ipsp's and epsp's can be recorded, but the large dendrites do not themselves support spikes. Junction potentials in the two ipsilateral motorneurons are synchronous; there is weak electrical coupling between them but their synchrony appears to result mainly from common presynaptic connections. Input from the tegumentary nerve is excitatory, while that from the circumesophageal connective is both excitatory and inhibitory.[481]

The stomatogastric ganglion of a crayfish has 25 nerve cells, and that of a lobster has 33; it innervates muscles of the anterior gut.[312] Sensory impulses in response to mechanical stimulation of the stomach wall pass through the stomatogastric ganglion via the stomatogastric nerve to the thoracic central nervous system.[312] Efferent fibers from the central nervous system end in a rich synaptic neuropil in the stomatogastric ganglion. Most of the neurons are motorneurons and serve the 25 muscles of the stomach (Fig. 15-17). One group (nine) of the motorneurons gives a pattern of discharge which drives the gastric mill; initiation of the discharge seems to be dependent on command neurons in the central nervous system. Another group (14 motorneurons) activates the pyloric filter and related muscles, and their patterned bursts continue in an isolated stomatogastric ganglion. Rhythmicity depends on both the interconnections and the membrane characteristics of specific neurons. Records from identified individual neurons show numerous connections—fast

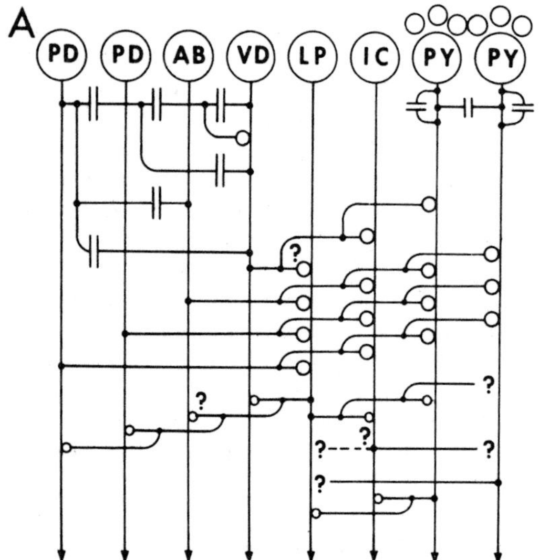

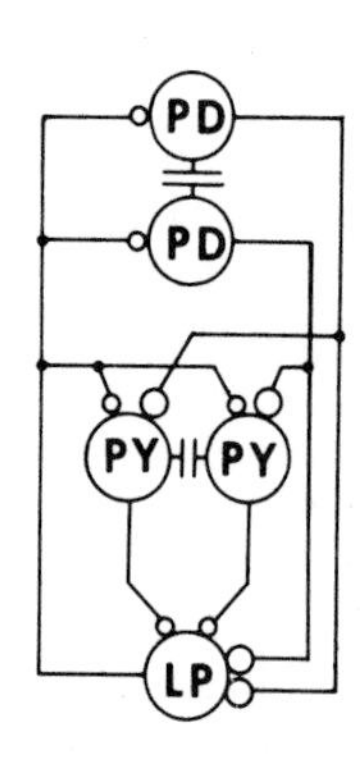

Figure 15–17. A, Wiring diagram of stomatogastric neurons in pyloric cycle of lobster. Parallel lines of electrotonic junctions. Large open circles are slow inhibitory junctions; small open circles are fast inhibitory junctions. PD, pyloric dilator; AB, anterior burster; VD, ventricular dilator; LP, lateral pyloric; IC, inferior cardiac; PY, pyloric neurons. B, Simplified diagram of A. (From Maynard, D. M., Ann. N.Y. Acad. Sci. *193*:59–72, 1972.)

and slow ipsp's, epsp's, both strong and weak electrotonic coupling. Neurons discharging with the same temporal pattern are synaptically connected, and neurons activating the same or synergistic muscles tend to be electrotonically coupled. Most of the chemical synapses are inhibitory. A diagram is given in Figure 15–17.[350, 351]

Another example of programmed interneuron activity in the lobster is in the control of defecation by the sixth abdominal ganglion. Motor impulses to rectal muscles pass in two posterior intestinal nerves, and are initiated by interneurons which can be activated from esophageal connectives or elsewhere in the nerve cord. Interneurons may be unitary in their discharge, either bursting or driving. Coupling between interneurons varies with frequency of cord stimulation; hence, in the circuit, some synapses are more fatigable than others, and synchrony between neurons varies with frequency.[598]

Insects. WALKING AND FLYING. Locomotion by insects apparently results from fixed patterns in thoracic ganglia which can be started by specific stimuli and altered by proprioceptive feedback.[378] Flight is commonly triggered by removal of tarsal contact from a substratum and is maintained by stimulation (by moving air) of tactile receptors on the head. Similarly, giant water bugs' swimming is initiated by lifting the tarsi.[105] Walking can be initiated by any of several converging sensory inputs. Often the six legs of an insect form two alternating tripods and the leg most likely to step is the one bearing the least weight.[236] Normal changes in gait or amputation of a leg alter leg rhythm. The order for quiet walking of a mantis is L (left) 3, R (right) 3, R 2; when the mantis is excited and walking fast the rhythm is L 2, R 2; L 3, R 3; when climbing, L 3, R 1; L 2, R 3; L 1, R 2.[448] In some insects, the order of leg movements depends on kind of surface. In a cockroach, the alternating tripod persists at all speeds and the ratio of leg protraction to retraction increases as speed increases.[98, 587] The alternation of protractors and retractors is under central control, but timing is influenced by feedback from receptors.[136]

For various insects, there may be central oscillators for individual legs that are coupled; sensory input from receptors on the proximal parts of a leg influences the output from a given leg oscillator, and the receptor influence is less in the rapidly moving cockroach than in the slower walking stick (*Carausius*)[567] (Fig. 15–18).

Nymphs of dragonflies, stimulated by prodding the abdomen, eject water so that the animal is propelled forward. The abdominal nerve cord has many units which are spontaneously active, often in regular bursts. Some units are inhibited by sensory stimulation, some are accelerated, and some non-spontaneous neurons are activated by tactile input. An interneuron may be ipsi-, contra-, or bilateral in its sensory field and receives much converging input.[139, 140, 298]

In a locust, the closer muscles of a spiracle are controlled by a complex of motorneurons. Command interneurons in the metathoracic ganglion drive the motorneu-

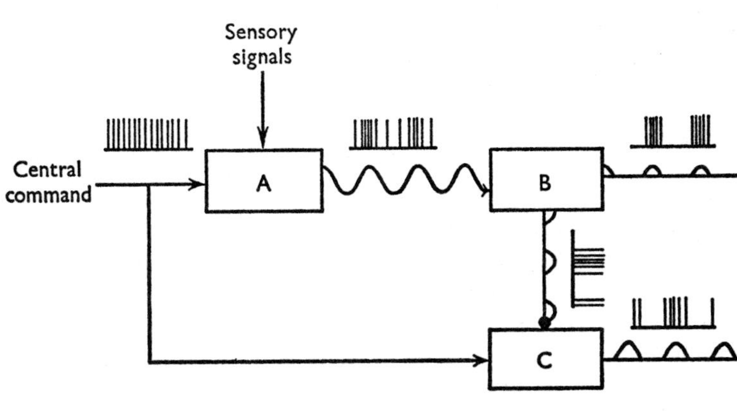

Figure 15–18. Model of control of extension and flexion of leg of cockroach. Activation is by a central command neuron, which sends continuous signals to elements *A* and *C*. *A* converts the train of signals to a sinusoidally varying signal, which is then chopped by the threshold of *B* to produce periodic bursts that excite one set of muscles. Similar but inhibitory bursts from *B* interact with continuing command from *C* to produce another efferent burst pattern, exactly out of phase with *B*, signalling antagonistic muscles. Sensory signals at *A* can modulate level of activity of the system. (From Delcomyn, F., J. Exp. Biol. *54*:453–469, 1971.)

rons, which are themselves rhythmic. The command neurons can drive the motorneurons 1:1 or can facilitate or inhibit their activity; synaptic regions are separate from the pacemaker zones on a motorneuron.[372] In a mantis, mechanical and visual stimuli or CO_2 may cause early ventilation, and movements of the spiracle are linked to locomotion.[374]

Direct flight muscles have a 1:1 correspondence between motorneuron discharge and muscle contraction, for example in moths and locusts (p. 760). A tethered locust begins to open its wings 45 msec after wind stimulation or 30 msec after tarsal release; an untethered locust opens its wings 15 msec after its first jumping movement. During flight the forewing pronation is varied with wind speed so that interneurons activated by wind receptors act in control of lift.[65] Wind from the side causes a yaw correction. Wind detectors can compensate for wing damage, but normally a locust sights the horizon to maintain its orientation.[167, 589, 590, 591] Interneurons signal wind direction and strength and initiate changes in firing patterns of motorneurons, as shown by simultaneous recording from several flight muscles. Motorneurons of one pool may be electrically coupled, but connections between pools are synaptic.[290] Alternation of elevators and depressors of the wing is by endogenous rhythms of two sets of motorneurons. During flight, a burst to the depressor muscles is accompanied by a pause in the discharge to the elevators; that is, the two groups of motorneurons are phasically coupled.[555, 556, 588] Locusts' flight is less plastic than their walking.

Patterned activity in central nervous systems may be released by hormones. For example, discharge in the phallic nerve of a cockroach, associated with copulation, is enhanced by injections of extracts of corpora cardiaca.[368, 369] The behavioral sequence of movements preparatory to emergence of a silk moth from its pupa case is mirrored by a sequence of electrical activity in the nerve cord; the sequence can occur in a deafferented cord that has been treated with eclosion hormone derived from brain and corpora cardiaca.[541]

In locusts, the flight rhythm continues in the absence of brain and subesophageal ganglion; rhythmic efferent discharge and some wing movement continue after sensory nerves from wings are cut, and after wings are clipped and wind receptors covered; thus, the central rhythm is endogenous but capable of some reflex modulation.[584, 591] Most motorneurons fire once per wing stroke, and an increase in frequency reflects increased number of active units and earlier onset of firing so that the wings are braked at top and bottom of stroke.[584, 591] Removal of proprioceptive input reduces stroke frequency; the receptors normally fire once or several times near the top of each wing upstroke and have a tonic effect on the endogenous rhythm. Also, wind stimulation increases the motorneuron discharge. Stretch and wing input are summed over more than one wing cycle, and they control flight maneuvers.[586]

In moths, there may be one or several action potentials per wing stroke. Before flight is possible, warmup occurs, produced by rapid low-amplitude beats, often with two motorneuron spikes per contraction. The warmup pattern is endogenous in prothoracic ganglia. In saturnid moths, partial amputation of a wing decreases the wingbeat period, and cutting sensory nerves from a wing increases the period. At high wing frequency the motorneuron bursts are shortened[276, 277] (Fig. 15–19).

Flight muscles of dipterans and many hymenopterans and coleopterans are oscillatory or myogenically rhythmic, usually indirectly coupled to the wing, and there is no specific phase relation of single motorneuron spikes to wing movement.[385] In these insects, gradation is by a frequency-modulated pulse code so that increased motorneuron frequency increases wing frequency, but not in a regular relation. In dipterans, sensory influences exert tonic modulation on motorneurons, but in locusts there is phasic modulation.[593] The indirect flight muscles of flies beat nearly synchronously at high frequencies, and steering is by nonoscillatory direct muscles; thus, there are two motor systems.[389] In a bumblebee, *Bombus,* synergistic muscle units are synchronous during warmup but they beat sequentially in flight.[386]

It is concluded that insect flight is patterned in the thoracic ganglia, that the discharge is more rhythmic in insects with direct flight muscles than in those with oscillatory muscles, and that in both groups, sensory input has a tonic modulating effect. The neuronal circuitry of the rhythmic centers remains to be elucidated.

NON-LOCOMOTOR REFLEXES. Other examples of behavior which is triggered by a sensory stimulus but then follows a stereotyped preset pattern may be mentioned. The flashing of fireflies is species-specific in duration, intensity, and frequency. In males of species which do not glow continuously, flashing can be induced by brief photic stimulation, and hundreds of individuals of some species flash in synchrony. A flash can be induced by electrical stimulation of the nervous system, with latency of several hundred milliseconds, as compared with 15 to 75 msec for response time to direct stimulation of the luminous organ.[47, 48, 49, 67] The neural connections have not been elucidated.

Reflex control of feeding and emptying of the foregut in blowflies was mentioned previously (p. 556). Taste receptors on the tarsi trigger proboscis extension and labellar sucking; stimulation of stretch receptors in the foregut terminates feeding. After the recurrent nerve in *Phormia* was cut, the fly consumed 37.4 μl sugar solution; after the nerve cord was cut between brain and thorax, 39.4 μl was consumed, but after a sham operation the fly consumed only 15 to 18 μl. Thus, impulses originating in stretch receptors of the foregut pass by the recurrent nerve to inhibit a motor center for drinking.[99, 166]

Detection of sound and the central response to auditory stimulation in noctuid moths were described on p. 523. Binaural primary auditory fibers are distributed ipsilaterally to the three thoracic ganglia; they connect with relay neurons to both ipsilateral and contralateral interneurons. Higher order interneurons in the metathoracic ganglion are: (1) pulse marker neurons, which give one large spike per sound pulse and thus signal pulse repetition rate, (2) train markers, which signal series of pulses and thus duration of trains, (3) binaural interneurons, which fire more when both ears are stimulated than when one is stimulated, and (4) spontaneous neurons,

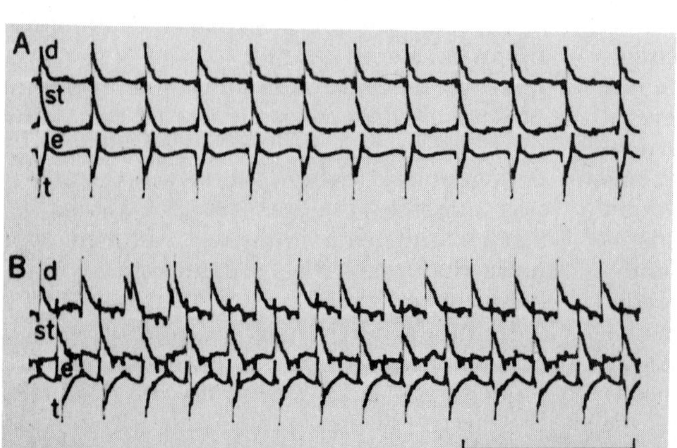

Figure 15–19. Muscle potentials in hawkmoth *Celerio* during shivering (A) and fixed flight (B). Three elevator muscles (*st,* tergosternal; *t,* tergotrochanteral; *e,* unidentified unit) in synchrony with dorsal longitudinal (*d*) during shivering but not during flight. (From Kammer, A. E., Z. vergl. Physiol. *70*:45–56, 1970.)

which stop firing when one ear is stimulated.[447, 448, 449, 616] Other interneurons vary in impulse frequency according to sound intensity, and some auditory units in the brain interrupt firing in response to sound if visual stimuli are presented. The central connections which lead to evasive behavior have not been mapped.

In an acridid grasshopper, two types of large mesothoracic auditory interneurons have been described. (a) One receives both ipsilateral and contralateral auditory input and sends a process to the brain; it habituates to repeated stimulation, requires 25 minutes for recovery, and receives inhibitory input from the brain. (b) The other interneuron signals direction of sound with considerable accuracy.[465] In a tettigonid grasshopper, one large interneuron, T, on each side of a mesothoracic ganglion receives excitatory input from the tympanum; its response is subject to recurrent inhibition from a posterior ganglion and to inhibition from the contralateral tympanum.[521, 522] Another interneuron, S, responds to the species song pattern, and this cell inhibits T via an interneuron, D. Both S and T drive motorneurons.[360]

Sound is produced in crickets and some other orthopterans in species-specific patterns of stridulation. The singing by several males may alternate or synchronize, and different stridulation patterns indicate calling, rivalry, or courtship (two songs). The singing patterns can be modified if sounds are heard by the cricket.[263] The four sound patterns result from different movements of the forewings over the scraper and are due to precise sequences of contraction of a series of tergosternal, basal, and subalar muscles that are evidently well programmed. Intracellular recordings from wing closer and opener motorneurons of a cricket show alternating bursts of inhibitory and excitatory junction potentials. Alternation of firing of the two groups of motorneurons suggests a coupling back to an oscillatory driver and to an interneuron which inhibits the opener and closer motorneurons alternately.[30, 31, 32] Similarly, the sequential movements of courtship are programmed in thoracic ganglia.[123]

A hyperkinetic mutant of *Drosophila* shows fast bursts of impulses from motorneurons in thoracic ganglia. The burst pattern is independent of input from sense organs or higher ganglia; hence, it relates to genetic coding for the motorneurons.[240]

In the locust *Schistocerca*, 26 motorneurons in the metathoracic ganglion have been identified, some of them causing "slow" contractions and others causing "fast" contractions of specific muscles. The motor spikes originate 400 to 500 μm from the motorneuron soma; dendritic regions where epsp's and ipsp's occur are nearer to the soma; the spikes and synaptic potentials spread passively into the somas. Connections between "fast" and "slow" motorneurons are partly by interneurons, and both synergism and timing of motorneurons to different muscle groups are by pre-motor interneurons. Recordings from interneurons show that some give large membrane oscillations which synchronize bursts of spikes in motorneurons. The neuronal patterns for the complex of muscle contractions in walking, jumping, and the termination of these movements may be fully specified in the near future.[225]

In summary, the ganglionic nervous systems of crustaceans and insects have permitted detailed study of the integrative functions of single, identified interneurons. Some of these are connected to sensory neurons; they respond to multiple inputs which represent specific combinations of signals from peripheral sensors. Other interneurons elicit discharge in appropriate sets of motorneurons so as to produce a coordinated movement. "Giant" fibers are special motorneurons of large size that produce phasic responses usually employed in "escape" behavior. Many rhythmic movements are controlled by motorneurons that are activated by oscillatory neurons; these in turn may be activated by a small number of command interneurons. The patterned rhythm may reside in spontaneous single interneurons or in networks. Most integration occurs in neuropil, but the input-output properties of identified interneurons and motorneurons give promise of understanding of the programming of complex movements.

Functions of Arthropod "Brain." The supraesophageal ganglion or brain of an arthropod is inhibitory to the thoracic motor centers; this is shown by release of motor activity after the brain has been destroyed. Complexity of the brain varies considerably in arthropods, but in general there are several regions: the medulla terminalis and accessory lobes are olfactory and optic centers; the middle and anterior regions (proto-, deutero-, and tritocerebrum) are coordinating and include the large dorsal corpora pedunculata, the neuropilar central body and the protocerebral bridge; and the posterior deutero- and tritocerebrum receive tactile input from antennules, contain motor centers, and give rise to circumpharyngeal commissures and to nerves to mouthparts. As in other regions of many invertebrate nervous systems, the central part of the brain is a fibrous neuropil and many of the neurons are unipolar (Fig. 15–20).

Sensory and reflex functions of the brain have been mentioned previously. The proportion of the brain occupied by vision centers is 0.3 to 2.8 per cent in arachnids and myriapods with epithelial photoreceptors and ocelli, 2.9 to 9.9 per cent in lower crustaceans and insects with rudimentary compound eyes, and 33 to 80 per cent in those with highly developed compound eyes.[186] The mushroom bodies or corpora pedunculata consist of an extensive neuropilar stalk with many small unipolar cells in large dorsal lobes. Structures which are similar in appearance occur in some polychaetes, are very large in *Limulus*, vary in extent from 10 to 30% of the brain in decapod crustaceans, and are prominent in most insects. Whether this structure is homologous in all of these annelids and arthropods is uncertain.[52]

Removal of any part of the brain which receives fibers from a sense organ has effects equivalent to removing that sense organ. Many arthropods can compensate for removal of one eye or one antenna, but injury to the brain leaves a more permanent behavior deficiency. If both circumpharyngeal commissures are cut, reflexes of the head

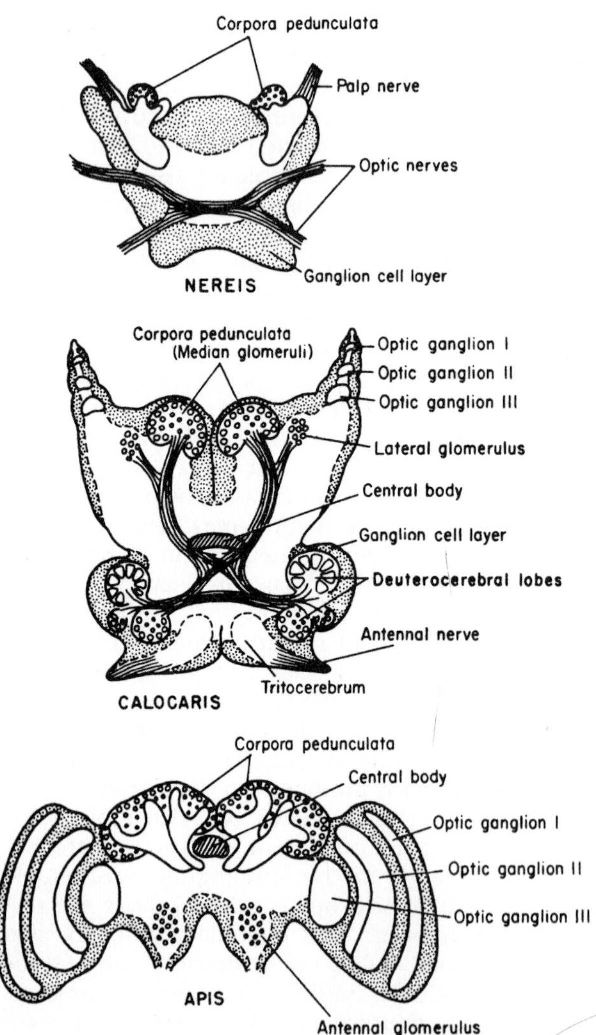

Figure 15–20. Sections of brains (supraesophageal ganglia) of several invertebrates with corpora pedunculata. Polychaete *Nereis;* crustacean *Calocaris;* a worker honeybee *Apis.* (From Hanström, B.: Vergleichende Anatomie des Nervensystems der wirbellosen Tiere. Springer.)

remain, and there may be chewing but not swallowing. Righting and leg reflexes remain, and there may be excessive random activity of the legs. When one commissure is cut or one lateral half of the brain is destroyed, the arthropod circuses toward the intact side. Removing the subesophageal ganglion or cutting behind this ganglion tends to stop spontaneous activity, but local segmental reflexes persist. The subesophageal ganglion is needed for coordinated walking, swimming, or flying, but the other ganglia are capable of reflexes of locomotion and autotomy, and of much spontaneous activity.

After the brain is removed from a dragonfly nymph, there is an increase in breathing frequency, but after the subesophageal ganglion is removed breathing frequency is decreased. Inhibitory action of the brain on ventral ganglia of crabs was shown by cessation of circus movements upon electrical stimulation of a transected circumesophageal commissure. Local destruction of neurons or of the stalk of the corpora pedunculata in a cecropia silkworm results in spinning of an abnormal cocoon.[544] Lesions in the supraesophageal ganglion of a spider cause it to spin small, irregular webs.[598a]

In a cricket, the corpora pedunculata inhibit locomotor activity; the left and right sides of the brain inhibit each other. The central body and subesophageal ganglion excite thoracic motorneurons, and as a complex they are inhibited by the corpora pedunculata.[232, 233, 234, 235] When one side of the protocerebrum of a praying mantis is removed, the legs on the opposite side lose tone while legs on the operated side become more active; circusing is to the normal side. The two sides of the brain give left and right turning signals and also have inhibitory effects on the command center of the subesophageal ganglion, which in turn is excitatory to motorneurons[448] (Fig. 15–21).

Sometimes during mating, after clasping, the female mantis eats the head of the male and copulation continues. If only the brain is removed from the male there is no sexual activity, but if the subesophageal ganglion also is removed, copulation can be completed; thus, the copulatory center in the last abdominal ganglion is normally inhibited by the subesophageal ganglion.[445, 446, 448] After a cockroach is decapitated, leg movements can be coordinated and the phallomeres remain capable of copulation. If a cockroach has been decapitated or a cut made ahead of the terminal ganglion, there is an increase in rhythmic bursts from the last abdominal ganglion. Similar increase occurs on application of an extract from the corpora cardiaca; neuronal information from the brain together with hormone leads to reproductive behavior.[369, 450]

Electrical recording from corpora pedun-

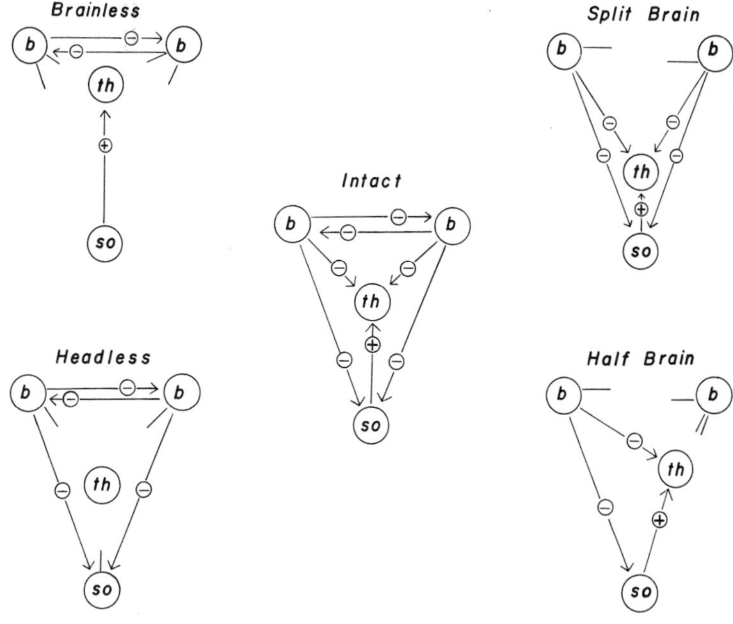

Figure 15–21. Diagram of effects of brain operations on locomotion in praying mantis. Abbreviations: *b*, brain; *so*, subesophageal ganglion; *th*, thoracic motor complex; +, excitatory; –, inhibitory. (From Roeder, K. D., Nerve Cells and Insect Behavior. Harvard University Press, Cambridge, 1967.)

culata of cockroach shows large synchronized spikes in response to antennal nerve stimulation, with 40 to 70 msec latency and fatiguing at stimulation more frequent than 1/sec. In lobster brain, many interneurons are spontaneously active; some neurons, after a brief stimulus to the antennular nerve, give a response persisting for 60 sec.[349, 352]

In crickets, no spontaneous singing occurs if the brain is removed; one corpus pedunculatum is sufficient for singing. Localized stimulation of a mushroom body can release one of the song types; some 28 muscles are involved in singing. Stimulation of the central bodies elicits abnormal singing.[233] Sometimes stimulation can inhibit singing which is in progress. The pattern of sequential muscle activity appears to reside in the corpora pedunculata and to be converted to a command in the central body; muscles seem to be activated by mesothoracic motorneurons.[234] Interference with legs and wings does not alter the pattern of activity of mesothoracic muscles; hence, the song is centrally programmed, and not reflexly defined.[306] Some central neurons show a circadian rhythm which is apparent in motor and singing activity.[19]

The optic lobe of most insects consists of peripheral lamina in the retina, a medulla, and a deeper lobula, with chiasmata connecting all three regions. In sphinx and noctuid moths, many types of sensory interneurons are found: monocular sustaining "on" units; "on-off" units which detect contrast irrespective of movement; binocular directional units responding to visual components of body roll and yaw; units which connect the lobula of one side to the other optic lobe; and a small number of efferent (centrifugal) units.[78] In *Locusta,* a descending contralateral-movement detector sends an axon from the optic lobe to the thoracic ganglia; its response habituates to repeated stimulation at one place on the retina, recovers when the stimulus moves to a new site, and is disinhibited by stimulation of a cervical connective, or by motor activity.[463, 464, 465]

Some interneurons in the brain and subesophageal ganglion of a fleshfly remain active after compound eyes are occluded but are silenced when the ocelli are illuminated; other interneurons respond to antennal movement, and this response may be facilitated by light on the ocelli.[377] In the brain of a locust, some interneurons respond to either visual or tactile input; habituation to one modality leaves the response to the other unaltered.[220]

In summary, the "brain" of arthropods includes supra- and subesophageal parts of the nervous system. Specific sensory interneurons provide for sensory integration. In addition, some regions of the "brain" can inhibit ventral motor centers and contain command interneurons which release complex movements.

BASIC MORPHOLOGY OF VERTEBRATE NERVOUS SYSTEMS

The basic organization of the nervous system is similar in all vertebrates—a tubular neuraxis, dorsal sensory and ventral motor roots, and essentially similar cranial nerves. Details are to be found in many textbooks of comparative anatomy.[452] Relative sizes of brain regions are shown in Fig. 15–22, and a diagrammatic comparison of the brain regions of nonmammals and mammals is given in Fig. 15–23. While the general regions of the nervous system are similar, homologies of subsystems appear much less certain now than they seemed a few years ago. It is relevant to consider functional similarities rather than to postulate homologous subsystems.

The central nervous system of vertebrates is, in general, organized with a greater abundance of neurons for body size than is found in any invertebrates other than cephalopods. There may be redundancy in that some neurons can substitute for others and portions of brain can be removed with little apparent behavioral deficit; however, with refined measurements it appears that the redundancy may be more apparent than real. In a lumbar segment of a dog, one dorsal root carries 12,000 fibers, most of them afferent; a ventral root has 6000 fibers; and there may be in one lumbar segment of the cord 375,000 neurons, many of them small and a few large.[190, 191] The sensory cell bodies are located in spinal ganglia outside the cord, and neurons in the cord are interneurons and motorneurons, each of the latter with some 5500 synaptic knobs. Thus, there is potential for much integration in a single segment of spinal cord.

The midbrain (mesencephalon), consisting of the tectum and tegmentum, is considered to be a principal integrative region in fishes

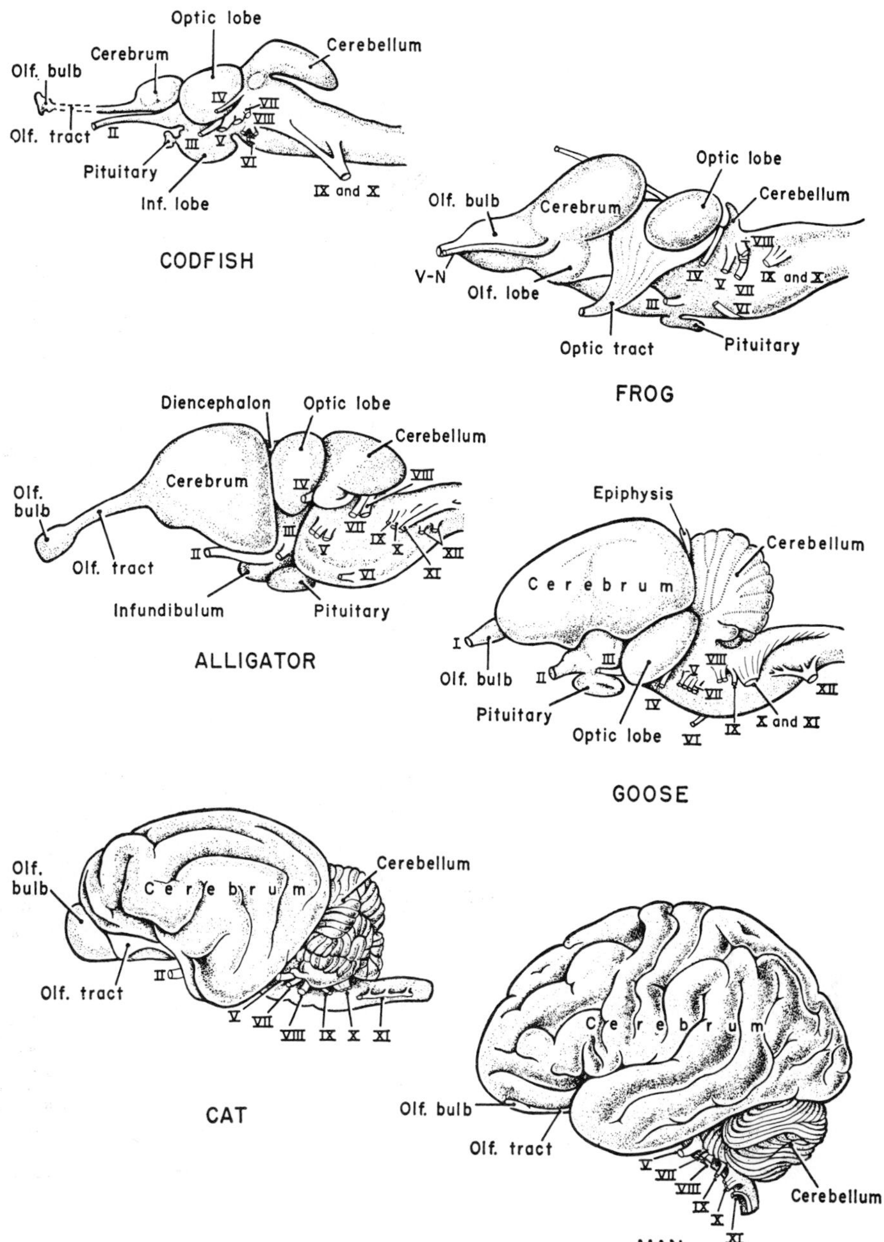

Figure 15–22. Relative sizes of brain regions in several vertebrates. (From Truex, R. C., and M. B. Carpenter, eds., Strong and Elwyn's Human Neuroanatomy, 5th edition. Williams & Wilkins, Baltimore, 1964.)

and amphibians; the diencephalon is increasingly important in reptiles; the basal nuclei of the forebrain are large in birds. The neocortex as a six-layered forebrain cover occurs only in mammals; forebrain structures of similar connections and functions occur in reptiles and birds[102] (Fig. 15–23). It is sometimes said that integrative function moved forward in the evolutionary series; however, it now appears that even in mammals the mesencephalon – superior and inferior colliculi – is important in simple visual identification of objects and reflex reactions to them and in auditory responses. The concept of replacement of function in a "lower" center by that in a "higher" center is less firm than it formerly was. Also, within single classes (e.g., mammals), much parallel

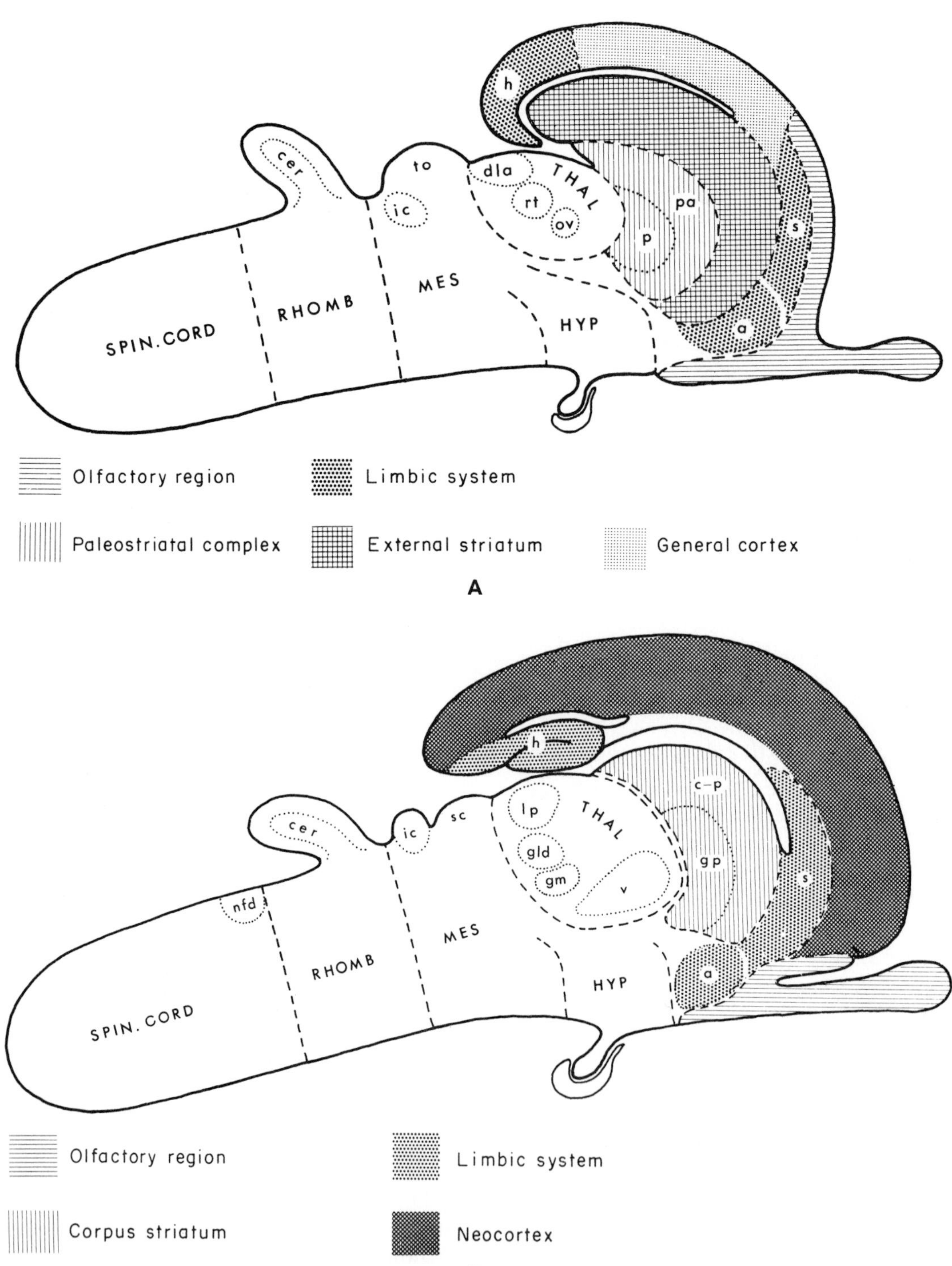

Figure 15–23. A, Schematic representation of nonmammalian vertebrate brain. B, Schematic drawing of the mammalian brain. Abbreviations: *h,* hippocampus; *a,* amygdala; *cer,* cerebellum; *HYP,* hypothalamus; *THAL,* thalamus; *p* and *pa,* paleostriatum; *gm,* medial geniculate; *gld,* lateral geniculate; *c-p,* caudoputamen; *s,* septum; *sc,* superior colliculus; *ic,* inferior colliculus. (From Nauta, W. J. H., and H. J. Karten, pp. 7–26 *in* The Neurosciences: Second Study Program, edited by F. O. Schmitt. Rockefeller University Press, New York, 1970.)

evolution has occurred and generalizations are difficult to arrive at. Examples of development and transfer of functions with increasing cephalization will be given in following sections on both sensory and motor systems.

One part of the brain has remained relatively constant throughout vertebrate evolution, namely the cerebellum. Perhaps the reason this and the spinal cord have retained similar circuitry in all classes is that locomotion is basically similar despite different appendages and modes of movement in different media.

The telencephalon, as seen in cyclostomes and fishes, evolved partly in association with the paired olfactory nerves, and the olfactory bulb is, strictly speaking, a portion of the forebrain. Anatomists are not in agreement concerning the homologies of the parts of the forebrain in various vertebrate classes. The forebrain consists of: olfactory region, limbic system, corpus striatum (very large in birds), and neocortex (large in mammals).[390] The primitive pallium is largely olfactory in function and is considered by some anatomists as homologous with the pyriform lobe of higher vertebrates (Fig. 15–23). However, the forebrain of fishes also receives projections from the diencephalon and gives rise to motor tracts to brainstem and upper spinal cord. The functions of the non-olfactory portions of the forebrain are closely related to those of the diencephalon and are poorly understood in lower vertebrates. In a shark, the forebrain constitutes some 60 per cent of the total brain mass, and in a goldfish it constitutes 12 per cent; in each species, much of it is non-olfactory. In amphibians, the medial pallium has regions of convergence of multiple sensory inputs, and the dorso-lateral pallium represents an infolding which, in reptiles, becomes the dorsal ventricular ridge. The neopallium of reptiles is infolded into the ventricle, forming the dorsal ventricular ridge, the medial part of which receives auditory projections, and the lateral part, visual projections.[396a]

In birds, the infolded pallium is massive; the dorsal ventricular ridge (ectostriatum) receives input from the nucleus rotundus of the diencephalon, the caudal neostriatum receives auditory fibers from the medial geniculate, and the dorso-medial cortex (Wulst) sends motor fibers to the spinal cord. In mammals, the central nuclei pallidus and lentiformis may be homologs of the reptilian and avian paleostriatum. This homology is supported by the occurrence of high concentrations of dopamine and AChEs in bird paleostriatum and mammalian putamen and globus pallidus.[390] The limbic system or non-olfactory part of the archipallium becomes the hippocampus and the amygdaloid complex. In monotremes and marsupials, the neocortex is layered but differently organized than in all other mammals. In higher mammals, the neocortex contains six recognized layers, which may be as much as 100 cells thick in higher primates. The total number of neurons in the human cortex has been estimated at 10^{10}, and many of the connections are within the cortex itself. It is estimated that the total nervous system has 2000 interneurons for every motorneuron.

Regions containing large numbers of small interneurons, such as the central reticulum of the core of the brainstem, are very poorly understood as to cellular organization.

EVOKED POTENTIAL RESPONSES AND BRAIN WAVES

Recordings of spikes from single neurons, made either extracellularly or intracellularly, have provided important information regarding coding in vertebrate nervous systems, as they have for invertebrate ganglia. However, many brain neurons are small, and the central neuropil is a complex mesh of fine processes; hence, methods used with single units are not applicable here. Three types of multiunit recording have been used for the study of synchronized input to central nervous structures and of activity in regions that have a layered or cortical structure: synchronized responses, evoked potentials in cortical regions, and rhythmic electroencephalograms.

In some sensory nuclei, the input in groups of fibers is so well synchronized that extracellular recordings give reproducible waveforms. For example, in the auditory nucleus of the medulla, in inferior colliculus and medial geniculate of the dolphin *Stenella*, compound evoked potentials in response to sound pulses consist of a series of deflections, each corresponding to synchronous firing of a group of units. Changes in amplitude, waveform, and latency are related to sound parameters.[54]

In cerebellum, optic tectum (especially of fishes and amphibians), cerebral cortex, and olfactory bulb, specific neuron types are arranged in strata, often with dendrites oriented perpendicular to the surface and with axons leaving in the direction opposite to the dendrites.[320] The number of layers in one structure varies with the zeal of the histologist and the species observed. Recordings from a dendritic layer give slow negative waves, which represent summed synaptic potentials in response to synchronized inputs. As an electrode moves into the cellular and deeper axonal levels, the polarity reverses so that, typically, a negative sink and positive source sequence is noted (Fig. 15–24). Mapping of isopotential contours in mammalian cerebral cortex[539] or fish optic tectum[525] shows the shape of the sink-source fields (Fig. 15–25). Measurements of current, rather than of voltage, in different positions take account of non-uniformity of electrical resistance in the cortical structure.

Related extracellularly recorded potentials are "brain waves" or EEG's, which represent summed local and synaptic potentials of many neurons synchronized at a given level. Irregular electrical waves occur in the telencephalon of a codfish at 7/sec, and in the optic tecta at 8 to 13/sec.[124] In a goldfish, waves have been recorded from cerebellum at 25 to 35/sec, from optic tecta at 7 to 14/sec, and from forebrain at 4 to 8/sec.[485, 486] In *Caiman* (reptile), forebrain rhythms are at 7 to 12/sec, and optic lobe rhythms are at 6 to 8/sec.[410] In mammals, so-called alpha waves are recorded best (but not exclusively) from occipital regions; they have frequencies of 7 to 8/sec in cat and 10/sec in man. In human infants, slow waves (3/sec) appear at 3 months of age. Puppies show some rhythmic waves at 18 to 20 days, stable alpha waves at 5 weeks, and the dog pattern at 5 months.[297] Electroencephalographers recognize slow delta waves at 1 to 3/sec, theta waves from hippocampus at 4 to 7/sec, alpha waves at 8 to 12/sec, fast beta$_1$ at 13 to 18/sec, beta$_2$ at 19 to 25/sec, and beta$_3$ at 26 to 29/sec.[5] The cortical rhythms are interrupted by sensory stimulation, such as a visual pattern or sound stimulation, or by a change in attention. In sleep, the alpha pattern is replaced by slower waves, sometimes in spindles (p. 692). Recording with multiple electrodes may show conduction of waves over the cortex at less than 1 m/sec. Stimulation of the reticular system in a quiescent brain can initiate an "arousal response" which desynchronizes the alpha and enhances faster waves. Stimulation of medial non-specific

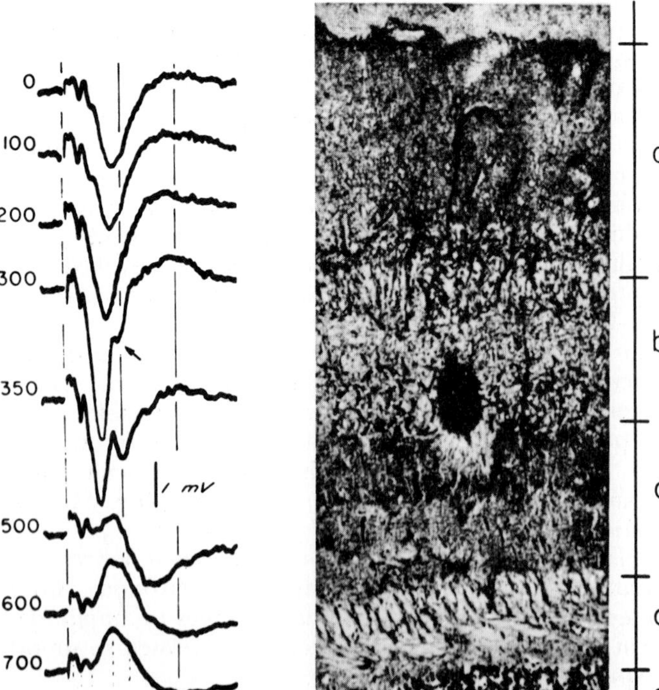

Figure 15–24. A, Evoked potentials from optic nerve of fish, elicited by optic nerve stimulation. Numbers at left indicate depth (in μm) in tectum. B, Photomicrograph showing layers of tectum and dark spot where electrical reversal occurs. (A from Vanegas, H., E. Essayag-Millán, and M. Laufer, Brain Res. *31*:107–118, 1971. B from C. V. Ariens Kappers, G. C. Huber, and E. C. Crosby, The Comparative Anatomy of the Nervous System of Vertebrates, Including Man. Copyright 1965, Macmillan Publishing Co., Inc.)

nuclei of the thalamus brings cortical waves into phase (the recruiting response).[253, 432] Transection of thalamic projections to the cortex interrupts the synchronous waves. Isolated slabs of cortex show irregular spontaneous activity and horizontal conduction. The various "brain waves" result from a combination of neuron spontaneous firing, intracortical spread, and especially summed sequences of epsp's and ipsp's in cortical dendrites.

Slow evoked potentials, as recorded at the surface of the cerebral cortex and optic tectum, have been much used in the study of sensory inputs and their projection. For example, the somatosensory cortex has been mapped for representation of tactile and proprioceptive inputs from various body regions (p. 690). Areas of a retinal field project in a regular pattern on the contralateral optic tectum of the alligator[193] and the goldfish.[431, 489]

Analysis of the waveform of evoked potentials at different depths is commonly done with synchronized sensory inputs or with electrical stimulation of restricted tracts. A few examples may be considered. The olfactory bulb is a nearly spherical structure with primary olfactory fibers synapsing on outwardly directed dendrites of mitral cells which, in addition to giving rise to axons of the olfactory tract, send collaterals to granule cells that return inhibition to the mitral cells (p. 569). Three phases of the evoked response to antidromic input, as well as to orthodromic stimulation recorded at different depths, represent (1) initial negativity at the level of mitral cell synapses and positivity at the surface, (2) positivity at the cell level due to repolarization and a second wave of negativity due to current from granule cells, and (3) final prolonged positivity due to inhibitory psp's.[391, 433]

In the cerebellum, each large Purkinje neuron has a branching dendritic tree; synapses of parallel fibers are mostly on dendritic spines near the periphery of the branches, whereas synapses of climbing fibers occur on the central branches. Stimulation of climbing fibers elicits a negative evoked potential of maximum amplitude at the level of the Purkinje somas, nonreversing with depth, whereas stimulation of parallel fibers in most species (mammals, frog, fishes) elicits an evoked potential which is negative at the surface and positive at the soma. Antidromic stimulation via Pur-

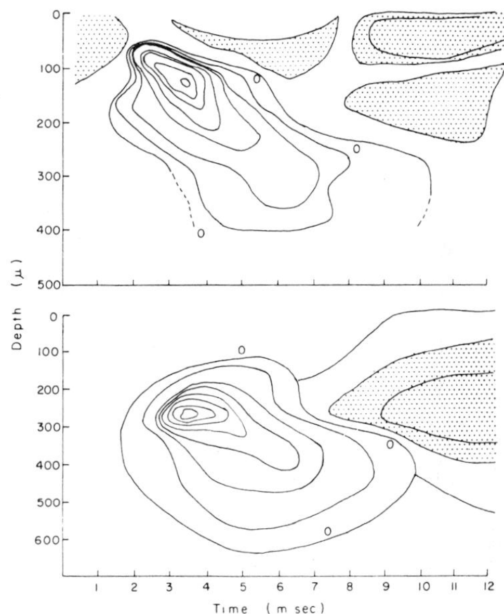

Figure 15-25. Current density profiles of goldfish optic tectum during optic nerve stimulation in two preparations. Contour lines indicate equal currents; stippled areas indicate downward current flow, and clear areas show upward current flow. Current flow is mainly upward during surface-negative field potential, and is of maximum density in the central gray zone of the tectum. The locus of maximum upward current shifts deeper with time in the response. (From Sutterlin, A. M., and C. L. Prosser, J. Neurophysiol. *33*:36–45, 1970.)

kinje axons also elicits reversing responses. Reversing responses are interpreted as passive dendritic depolarizations of sink-source relations. In alligator cerebellum, some responses to parallel fiber stimulation are nonreversing, and these are interpreted as indicating initiation of spikes in the dendrites. In all species, the evoked response to stimulation of climbing fibers does not reverse because their synaptic endings are distributed diffusely to both soma and upper dendritic levels of Purkinje cells.[114, 331, 394]

Stimulation of sensory pathways to the cerebral cortex also elicits localized evoked potentials; responses to stimulation of reticular and some thalamic projections are not so localized. Stimulation of thalamic inputs in cat elicits two types of response—one with initial negative sink in the granular layer and negativity rising to the surface at 20 to 30 msec, and the other with surface negativity and reversing to positivity at depths below 0.3 to 0.5 mm.[505] In rabbit visual cortex, the isopotential point is at 0.8 mm, and single cell recording shows epsp's lasting some 80 msec.[327]

In the hippocampus, the orientation of

pyramidal cells is inverted as compared with the rest of the cortex; i.e., the axons leave superficially. Orthodromic excitatory input elicits surface positive and deep negative evoked potentials.[120] Stimulation of each of several inputs to the hippocampus causes a large positive wave at 0.5 mm depth, which is generated by ipsp's at the level of the pyramidal cell somas and is due to activation of basket cells by collaterals from the pyramidal neurons.[11]

In optic tectum of fish and frog, a photic stimulus to the contralateral eye elicits a slow negative response at the surface, and this reverses in mid-depths; negative and positive waves are not strict mirror images, possibly reflecting the time needed for electrotonic spread from the superficial synaptic layers. Horizontal conduction may also affect the responses. In some fish tecta, a second region of negativity is found at the level of paraventricular or pyriform neurons.[429, 431] By strong electrical stimulation of the optic nerve or with large, high intensity visual fields, multiple waves are noted in the evoked potentials, presumably indicating synaptic responses from fibers of different thresholds and conduction velocities.

Both brain waves and evoked potentials are useful to indicate the state of the brain and to indicate the synchronous activation of large numbers of neurons. The relation between these slow potentials and spikes in single cells has been examined by simultaneous measurement of both types of activity, often by the same electrode connected to short- and long-time-constant amplifiers. In the visual cortex, the highest probability of firing of a single neuron occurs during the time of steepest negative slope of the evoked potential, and the lowest probability occurs at the steepest positive slope; this relation holds at all depths, and hence neuron response is determined by the local electrical field.[547] In both visual and somatosensory cortex of cat, the summing of spikes by an averager gives a pattern corresponding to the evoked potential; also, the probability of spike discharge corresponds to the rhythm of the EEG. However, EEG waves can occur in the absence of pyramidal spikes.[148] A given region of cat cerebellum shows evoked potentials in response to hippocampal stimulation and spike responses to sound clicks; the probability of spike response correlates well with the form and size of the evoked potential.[147] In hippocampus of rabbit, intracellular recordings from pyramidal neurons show spikes that synchronize with rhythmic theta waves, and both are increased by stimulation of the sciatic nerve. By conditioned reflex techniques (reinforcement by milk in the mouth) in cat, it has been possible to enhance either positive or negative phases of an evoked response in cerebral cortex to visual stimulation.[149]

It is concluded that evoked potentials and brain waves are useful indicators of regional activity in clusters or layers of many similar neurons, but that it is difficult to identify the precise origin of the various components. The slow potentials may influence the spike threshold in neurons within the potential field.

MOTOR SYSTEMS OF VERTEBRATES

Spinal Cord and Lower Brainstem

The first detailed studies which established the principles of reflex behavior were done on the mammalian spinal cord.[83, 110, 190, 498] Several afferent inputs enter the cord via the dorsal roots (see p. 513): (1) from stretch receptors, which include primary and secondary fibers from muscle spindles, and tendon organs of Golgi, which respond to tension rather than to muscle length; and (2) from various cutaneous endings, which converge on some of the same interneurons as do the stretch receptors. Some sensory influence is unisegmental, some is multisegmental, and some is relayed to dorsal columns in which impulses ascend to the brainstem. Dorsal column tracts are mainly ipsilateral, and consist of spinothalamic and spinocerebellar tracts. Descending paths—pyramidal, tectospinal, rubrospinal, vestibulospinal, and reticulospinal—converge on motorneurons at all levels of the spinal cord.

Certain rules apply to the actions of the multiple inputs on motorneurons at one spinal segment. Incoming fibers which are excitatory to one motorneuron pool are inhibitory to motorneurons serving the antagonistic muscles; for example, monosynaptic extensor reflexes elicited by extensor spindle afferents are inhibited by afferents from flexors. Conversely, converging inputs onto motorneurons of several pools serving the same muscle action tend to be facilitatory. Inhibition can be postsynaptic, usually of short latency, or it can be presynaptic, requiring some 15 msec to reach a peak. Strychnine blocks postsynaptic but not presynaptic inhibition; the latter is, however, blocked by picrotoxin. Presynaptic inhibition depolarizes afferent endings in spinal cord; postsynaptic inhibition increases chloride conductance and normally hyperpolarizes motorneurons (p. 636).[110, 244] In addition, motorneurons give off collaterals which feed back via an interneuron (Renshaw cell or equivalent) to block continuing discharge of the same and adjacent motorneurons. Antidromic impulses may inhibit or facilitate other motorneurons via collaterals, and may interrupt tonic inhibition by interneurons, thus releasing repetitive firing.[595, 596] Renshaw interneurons give high frequency bursts in response to single antidromic impulses in motorneurons; they act postsynaptically on motorneurons.

The smallest motorneurons (gamma efferents) innervate intrafusal muscle fibers (p. 511); they are commonly spontaneously active and highly excitable electrically and reflexly. Small alpha efferents are activated reflexly to repetitive firing by moderate stretches, and they initiate weak contractions of extrafusal muscle fibers. Large alpha motorneurons elicit maximum tension, have higher thresholds, and are very sensitive to inhibition. Thus, the size of a motorneuron is inversely related to its excitability and directly related to inhibition and to amount of tension elicited.[190]

Control of movement in antagonistic muscles is by a series of feedback interactions (p. 635). The primary muscle spindles (Group Ia afferents) excite α-motorneurons monosynaptically as in extensors of stretch reflex, and they inhibit antagonistic α-motorneurons. Secondary spindle afferents (group II) are in slower fibers; they act via an interneuron, inhibit extensor α-motorneurons, and excite flexor motorneurons. Tendon organs (group Ib afferents) inhibit ipsilateral α-motorneurons, particularly of extensors. The spindles signal muscle length, whereas tendon organs signal tension. Gamma motorneurons elicit contraction of intrafusal muscle fibers and thus increase the frequency of firing of a spindle afferent for a given muscle length.

Interneurons are sometimes spontaneously active and tend to be more repetitive than motorneurons; alpha motorneurons are the least repetitive. Interneurons serve: (1) as distribution paths for convergence and divergence, (2) as amplifiers of sensory input by a series of T-connections, or by reverberating loops of varying numbers of cells, (3) as valves between several inputs as when descending commands and sensory input interact with either facilitation or inhibition, and (4) as signal inverters, as when excitatory input is converted to inhibition.[190]

When the spinal cord of a cat or dog is transected in the upper thoracic region, all motor reflexes are at first depressed by spinal shock. Flexor reflexes in a cat return in a few hours, and extensor reflexes after several days. In frogs and fishes, the duration of spinal shock is very short; i.e., it is inversely related to the animal's degree of cephalization. A "spinal" mammal (one with cord transected) cannot stand, yet when suspended it may make treading motions. Limb reflexes may coordinate many muscles, as in the rhythmic scratch reflex. Visceral functions such as urination and defecation, and vasomotor responses, may be elicited reflexly in a spinal mammal. Sexual reactions may occur. These observations led to postulation of endogenous activity and intrinsic patterns of connections in the spinal cord. However, in decerebrated curarized cats, efferent impulses were recorded in lumbar ventral roots in response to stimulation of sensory fibers in peroneal nerves; motor discharge of rhythmic stepping responses occurred when the nerves on the two sides were stimulated at slightly different frequencies, but not when they were stimulated at the same frequency.[118] This emphasizes the importance of sensory input for timing the pattern of stepping in mammals.

In mammals, a large number of motorneurons in the medulla, which innervate about 20 different muscles, participate in the act of swallowing. A group of neurons situated bilaterally and dorsolateral to the inferior olive is responsible for coordinating the pattern of contraction of swallowing muscles. The coordination of excitatory and inhibitory sequences is not reflexly determined, since it is independent of stimulus used—touch to pharynx, rapid injection of water into the mouth, or electrical stimulation of the superior laryngeal nerve; and disruption of the feedback loops has no significant effect on the patterning.[109]

If the brainstem is transected just behind

the mesencephalon or at the upper pons, a mammal shows exaggerated extensor activity, sometimes called decerebrate or extensor rigidity. Extensor reflexes are then enhanced and spasticity is usual. The effect is due to removal of inhibitory influences originating in the cerebral cortex, the caudate nucleus, and the cerebellum, and to the maintained activity of facilitating pathways, such as the reticulospinal and vestibulospinal tracts.[343] It must be concluded that the precise balance between endogenous patterning and timing of motorneurons by sensory and descending inputs in mammals is far from understood.

In amphibians the input to the spinal cord, as in mammals, is from muscle spindles and tendon, skin, and joint receptors. Intracellular recording from motorneurons in isolated frog spinal cord shows both excitatory and inhibitory junction potentials in response to dorsal root input. The excitatory responses may be monosynaptic (2.2 msec delay), and the inhibitory responses may involve an interneuron (5 msec delay).[305] Extracellular measurements of foci of synaptic potentials on motorneurons indicate that dorsal root impulses act on distal zones of dorsally directed dendrites, whereas descending impulses in lateral column neurons act on the soma and nearby dendrites; motorneuron spikes originate at the initial axon segment, some distance from the synapses.[43] Motor impulses may spread back over the motorneuron soma electrotonically. Motorneurons may also show spontaneous miniature junction potentials.[287] Antidromic stimulation via ventral roots also excites motorneurons, presumably by collaterals, and these give rise to a dorsal root potential. The dorsal root potential can be abolished by cholinergic blocking agents, and it increases in amplitude on warming; presumably it comes from cholinergic synapses. A weak antidromic stimulation in frogs elicits a ventral root potential, and many motorneurons show a short-latency depolarization which may even trigger spikes; the ventral root potential due to brief antidromic volley decreases on warming, and is insensitive to blocking drugs. Apparently there is electrical coupling among the dendrites of many motorneurons.[173, 305]

In amphibians there is evidence for both patterned responses in the spinal cord and behavioral dependence on sensory input. Transection of the brainstem above the medulla does not result in extensor rigidity such as occurs in mammals. A spinal frog can jump in a coordinated fashion if stimulated to do so. Early observations on patterned leg movements indicated that the sensory and motor supply of one segment is necessary for the diagonal pattern of ambulation.[172] The normal pattern of leg movement for walking by a toad, *Bufo marinus*, is: right hind (RH), right front (RF), LH, LF. When all spinal dorsal roots were cut, normal locomotor movements persisted; if three steps were made in normal sequence, the fourth usually followed.[187] The brain and cranial input were intact, but these results indicate that a central nervous program for ambulation is important in amphibians. This evidence for an endogenous rhythm supports observations on reinnervation and development (p. 694). The distinction between intrafusal and extrafusal muscle fibers cannot be made for amphibians and reptiles (see p. 513), and in these classes nothing analogous to the gamma loop control system of mammals has been described. Perhaps this is related to the apparent greater effectiveness of endogenous patterns in frogs than in mammals.

Among fishes there appears to be much species difference in autonomy of the spinal cord. Fishes are said to lack muscle spindles, but the skin adheres so tightly to muscle that skin receptors sense movement.[439] In spinal teleosts and elasmobranchs, fin reflexes can be elicited by localized stimulation of the skin. In an eel, high sectioning of the cord interferes very little with the rhythmic undulatory movements; the rhythmic wave can pass down the cord through a region from which skin and muscles have been removed.[170] A minnow with high spinal section, blinded and free from contact with the substratum, shows spontaneous swimming movements. A dogfish, made spinal by transection behind the head, swims much like a normal fish, and when free from contact shows locomotor movements at 40 waves per minute, which are increased or decreased by tactile stimulation. The rhythmic movement of body and fins of dogfish is abolished by complete deafferentation; however, the movement pattern can pass down the cord through some twelve denervated segments.[201] In goldfish, cord transection ahead of dorsal fin results in paresis—atonia and absence of spontaneous movements—but coordinated fin movements can occur on stimulation; cord regeneration takes place in 30 days.[34]

Electrical recordings from motor nerves of a curarized spinal dogfish show spontaneous bursts of impulses in small motorneurons lasting 1 to 2 hours after cord transection. However, no further motor discharge occurs in the absence of proprioceptive input. Contraction waves in dogfish pass at 0.3 to 0.8 m/sec; conduction of impulses in cord tracts travels at 20 m/sec. Neurons on each side of the cord of a dogfish are organized to fire alternately, but longitudinal coordination of the locomotor wave is disrupted in the absence of phasic sensory excitation.[439] The unpaired fins of a dogfish beat at a rhythm which may be out of phase with that of the body; sensory input initiated by body movement modifies the fin rhythm.[439]

Spinal cyclostomes (hagfishes) can swim normally when stimulated, and a contraction wave may pass through a region of cord transection by local reflexes resulting from movements of segments beyond the cut. A lamprey (*Petromyzon*) has sensory cell bodies, not only in dorsal root ganglia but also within the spinal cord, which give responses to touch, pressure or noxious stimulation of the skin. Stimulation of the sense cells evokes epsp's and ipsp's in motorneurons, probably via interneurons. Habituation can occur at the motor synapse on repeated stimulation of dorsal sense cells or skin receptors.[36, 461] Some of the motorneurons activate units of slow red muscle fibers, other motorneurons serve twitch white muscle, and still others serve ipsilateral muscles which cause fin movements. None of the motorneurons show recurrent inhibition, but contralateral motorneurons are inhibited during a local reflex.[533] The slow motorneurons show tonic after-discharges. The sensory neurons are T-shaped and may initiate both local and long (many-segment) reflexes.

In lampreys, in addition to dorsal sensory neurons, there are in the cord small interneurons and motorneurons, and some uniquely large interneurons. Also, the bulbar region of the medulla contains some eight pairs of Müller (giant) cells with uncrossed axons and two pairs of giant Mauthner neurons with crossed axons. The Müller and Mauthner cells receive input from many sources—optic, trigeminal, vestibular, and posterior lateral line nerves and spinal cord.[460] Much of the input to Müller cells from the spinal cord is inhibitory; one pair of cells responds at voltages below those required for dorsal cell axons, so they are probably excited by medium-sized ascending axons. The Müller axons synapse on motorneurons and interneurons. One pair of Müller axons elicits, in spinal interneurons, excitatory junction potentials having a slow component, which represents chemical transmission (reduced by high Mg, restored by high Ca), and a fast component which is electrically transmitted (Fig. 11–16, p. 485). The epsp's elicited in motorneurons by Müller neurons are small and require high frequency stimulation to reach threshold.[460, 461]

The array of chemical and electrical synapses on Mauthner neurons of fishes was previously described (Fig. 15–8, p. 542). In lampreys, the sources of input may be more extensive than in teleosts. In fishes, each Mauthner axon sends many branches to interneurons of the spinal cord and causes contractions of muscles of trunk and tail, operculum, eyeballs, and lower jaw on the same side as the axon; it also inhibits motorneurons on the opposite side. Pectoral fin adductors of a hatchet fish, *Gasteropelecus*, are innervated by a giant neuron system (Fig. 15–26). In the medulla of this fish, each Mauthner axon activates four contralateral giant fibers by chemical synapses with a 0.4 msec delay and an epsp, which can be reversed by depolarizing the giant fiber. The giant fibers activate some 40 ipsilateral motorneurons in the first spinal segments by electrical synapses with less than 0.05 msec delay. The giant neuron system permits the fish to make a rapid upward dash.[18a]

Nothing analogous to the giant Mauthner neurons is found in terrestrial vertebrates and their derivatives. The medulla and pons contain many functional relay centers of ascending (sensory) and descending (motor) pathways, which are centers for autonomic functions. Reference was made earlier (Ch. 12) to both afferent and efferent centers for auditory, vestibular, and lateral line nerves. Cerebellar connections to midbrain and medulla are numerous in all vertebrates. The red nucleus of reptiles, birds, and mammals is an important focal point for connections from cerebellum, cortex, thalamus, and spinal cord. Here are also found vasomotor and respiratory centers. In mammals the pneumotaxic center of the pons modulates (usually by inhibiting) inspiratory and ex-

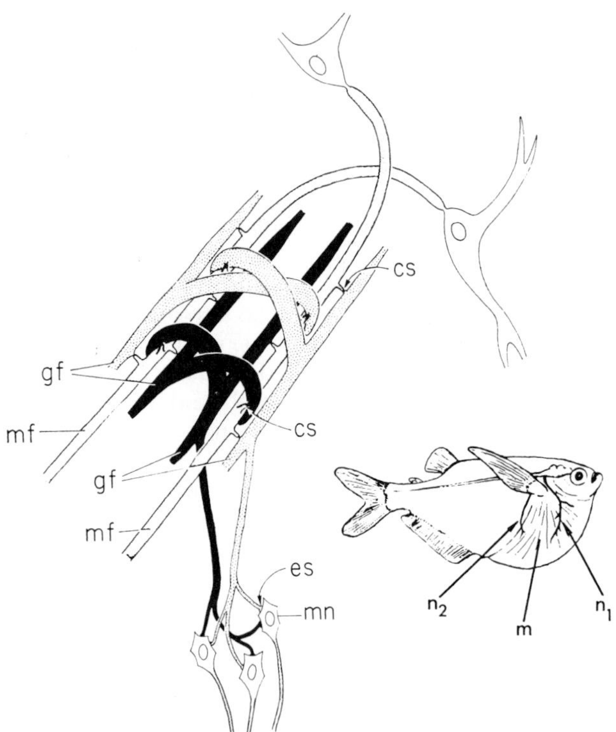

Figure 15–26. Diagram of relations between Mauthner fibers, giant fibers, and pectoral fin adductor motorneurons in hatchet fish (*Gasteropelecus*). Each Mauthner fiber (mf) makes several synapses with ipsilateral giant fibers (gf) and a single synapse with a central giant fiber. Cross branches of giant fibers are paired. Processes of each giant fiber synapse with motorneurons (mn). There are chemical synapses (cs) between Mauthner fiber and giant fiber, and electrical synapses (es) between giant fiber and ipsilateral motorneuron. (From Auerbach, A. A., and M. V. L. Bennett, J. Gen. Physiol. *53*:183–210, 1969.)

piratory neurons. Some respiratory center neurons may be sensitive to CO_2 or may receive input from respiratory receptors in carotid sinus in mammals, and probably in gills or pseudobranchs in fishes (p. 179). The activity of cardiovascular centers can be modified by other sensory inputs, and the output can be conditioned; examples are, in fishes, interruption of breathing or slowing of heart rate.[419]

Locomotor Control by Motor Centers in the Brain

Spinal reflexes and intrinsic rhythms account for much of the automaticity of locomotion in vertebrates, but initiation and control of movement by higher centers is less well understood. Voluntary movement involves a dynamic loop of interactions between neurons of the sensorimotor cortex, basal ganglia, thalamus, cerebellum, and sensory and motor relay nuclei in the brainstem. In monkeys, during voluntary movements of eyes, hands, and legs,[127, 534] neurons in these regions of the brain have been found to change frequencies of firing. Cerebellar Purkinje cells may discharge as movement is stopped, and there is evidence that the cerebellum is primarily inhibitory in its actions on motor systems.[114] In cats conditioned to give an eye blink in response to a click, an evoked potential in the motor cortex had a latency of 13 msec and preceded muscle activity by 7 msec.[599]

Punctate stimulation (electrical or chemical) of the motor cortex of mammals (in the area rostral to the central sulcus) elicits responses of single muscles or of single motor units within a muscle. Gross stimulation elicits discrete movements of legs, neck, trunk, or other regions. A map of the motor cortex matches that of the sensory cortex on the other side of the sulcus—leg representation is dorsal, mouth and face ventral, eyes anterior; all are contralateral. The sizes of areas vary with species according to the extent of limbs; e.g., the tail area in a spider monkey cortex is as large as the leg areas (Fig. 15–27). Stimulation of the somatosensory cortex can lead to movements via connections to the motor cortex. In general, stimulation of a point on the motor cortex can elicit a response on the opposite cortex via transcallosal conduction. In monotremes and marsupials, the sensory and motor areas are superposable and fit onto the same map.[1, 321, 322, 323, 434]

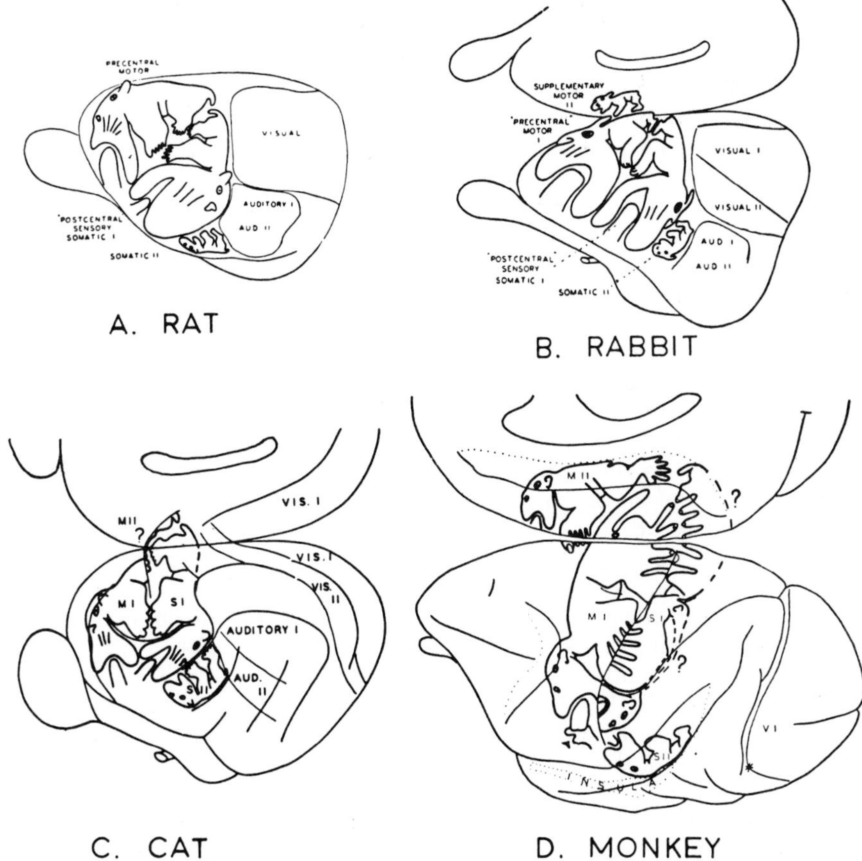

Figure 15–27. Diagrams of cerebral cortex of rat, rabbit, cat, and monkey showing localization of primary and secondary motor (MI and MII) and somatic sensory areas (SI and SII), and also primary and secondary visual and auditory areas. Orientation of body area representation shown as homunculus. Each brain outline corresponds to lateral surface of hemisphere; upper portion of each corresponds to mesial surface. (From Woolsey, C. N., *in* Biological and Biochemical Basis of Behavior, edited by H. F. Harlow and C. N. Woolsey. University of Wisconsin Press.)

The motor outflow from the mammalian cortex is by axons from pyramidal neurons of the motor cortex and by extrapyramidal neurons nearby. There are estimated to be one million pyramidal neurons in man. Volitional skilled movements are initiated via the pyramidal neurons, but reflex maintenance of the motor responses and of posture is controlled largely by extrapyramidal neurons. The pyramidal paths are almost all crossed, and they synapse at various levels in the brainstem and spinal cord; most pyramidal fibers synapse on interneurons, although some synapse directly on motorneurons. All extrapyramidals are relayed at basal ganglia and other subcortical nuclei.

In unanesthesized monkeys with implanted recording electrodes, large pyramidal neurons of highest conduction velocity are silent during motor quiescence and are phasically active during movement; pyramidal neurons with slower velocity are active in the absence of movement and increase or decrease their frequency during activity.[127] After section of the pyramids bilaterally, some species of animals are capable of movements which are adequate for locomotion; other species are incapable of coordinated movements. If the pyramidals are cut on one side, decreased muscle usage and lessened initiative of movement result on the opposite side. There is much species variation in relative importance of pyramidal and extrapyramidal tracts, and monotremes are said to lack a pyramidal tract. In a marsupial possum, *Trichosaurus*, the pyramidal axons run only to the midthorax, where they innervate the forelimbs; extrapyramidals synapse with interneurons which pass down the cord and serve the hindlimbs.[203]

There is no evidence for localized origin of motor tracts in the forebrain of lower vertebrates. Removal of the forebrain from elasmobranch and teleost fishes produces no readily observed disturbance in posture or locomotion.[34] Frogs and toads from which the forebrain has been removed tend to show reduction in spontaneous movements. However, electrical stimulation of the forebrain elicits no specific motor responses.[2]

The midbrain exerts some motor control, and electrical stimulation of the tectum elicits movements, presumably via the tegmentum. For example, in goldfish, tectal stimulation causes movement of the eye on the opposite side. In goldfish, some tectal neurons fire during eye movements; these units do not respond to visual stimuli, but trigger movements which may enhance image fixation.[262] In a toad, tectal stimulation causes turning of the head to right or left, and food snapping and swallowing.[132, 133, 134, 135] A motor representation of the body is found on the tectum of a toad. In a frog, local tectal stimulation elicits orienting and prey-catching movement; pretectal or thalamic stimulation elicits avoidance movement.

In birds, removal of the cerebral cortex causes no motor disturbance. Motor control seems to be localized in the striatum. Electrical stimulation of the striatum elicits movements of appendages; striatum removal causes severe motor and behavioral deficits.[451]

In summary, vertebrate motor systems well illustrate shifts of locus of function with cephalization. In fishes and amphibians, locomotion appears to be mostly under control of endogenous spinal rhythms, whereas in mammals sensory input exerts control via the interaction of cutaneous receptors, muscle afferents of three types, and descending fibers from the brain. Cyclostomes and, to a lesser extent, fishes and urodeles have specialized giant neurons which control large muscle masses. With the evolution of the neocortex of mammals, pyramidal neurons appeared as fast activators of lower motor centers. In all vertebrate classes, some monitoring of locomotion is done by the cerebellum.

CEREBELLUM

The cerebellum is a complex component of all vertebrate nervous systems which is neither sensory, motor, nor integrative in a reflex sense; rather, it stands aside from the main input and output pathways and serves as a monitor and coordinator of neural transactions of other parts of the brain. Anatomical and electrophysiological studies indicate that the basic neuronal organization of the cerebellum is similar in all classes of vertebrates.[112, 315, 330] Details of cerebellar functional circuitry have been elucidated in dogfish,[115] frog,[332, 475] lizard,[296] alligator,[331] cat,[113, 114] and monkey.[534]

In cyclostomes (petromyzonts), the cerebellum is a plate-like structure (archicerebellum) which persists as the flocculo-nodular lobes in birds and mammals. In elasmobranch and teleost fishes, the rostral end of the cerebellar plate folds upward to form the corpus and vulvuli cerebelli (paleocerebellum).[114] In electric fish such as mormyrids, with specialized electroreceptor organs, the cerebellum is enormously developed, occupying nearly half of the brain.[29, 396] In amphibians and reptiles, the cerebellum consists of a small lobe which is homologous with the paleocerebellum in birds and mammals—the anterior lobe or vermis and part of the posterior lobe. In mammals, the cerebellar hemispheres are called the neocerebellum.

In fishes and amphibians, the output from the cerebellar cortex goes both directly to lower motor centers and to deep cerebellar nuclei. In higher vertebrates, most of the output is via basal cerebellar nuclei. In reptiles these nuclei are the fastigial, emboliform, and globosus; in birds and mammals there are these three, plus the dentate nucleus. From the deep cerebellar nuclei, efferent fibers go to lower motor centers.

Two fiber groups in the peduncles conduct the input to the cerebellar cortex, and usually each sense modality uses both paths (Fig. 15–28A, B). These are: (1) climbing fibers, which end directly on Purkinje neurons (PC), and (2) mossy fibers, several of which converge on one granule cell (GC). Axons from the granule cells ascend to the surface layer, where they fan out in T-fashion and run in bundles of parallel fibers (PF) which activate PC's. The dendritic trees of Purkinje cells are isoplanar perpendicular to the parallel fibers (Fig. 15–29). Three layers are identified in the cerebellar cortex: the outer molecular layer consisting of parallel fibers from granule cells and synapses on Purkinje dendrites, the middle or Purkinje cell layer, and the deeper granular layer with GC's. In dogfish,

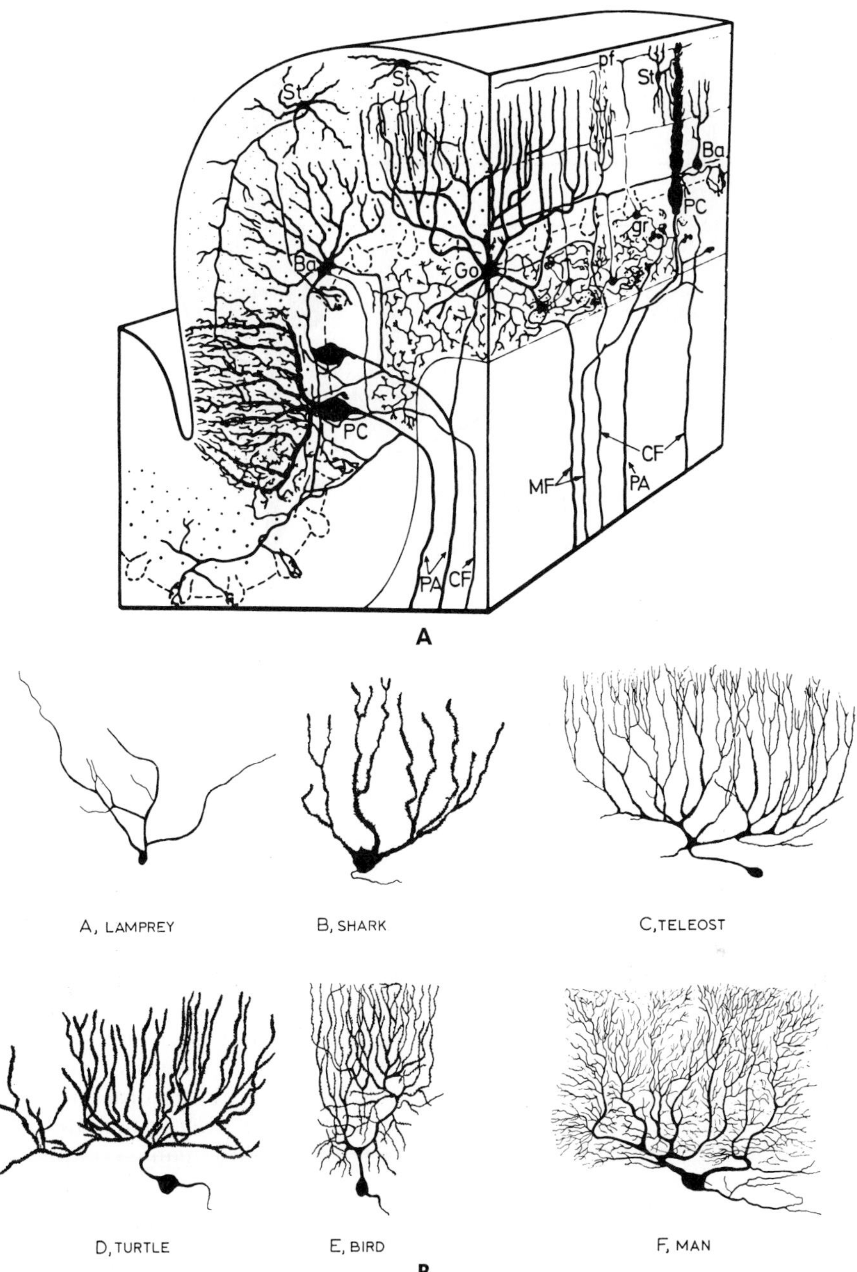

Figure 15–28. A, Stereogram of the five main types of neurons in cerebellum: CF, climbing fiber; MF, mossy fiber; St, stellate cell; Ba, basket fiber; Go, Golgi cell. (From Eccles, J. C., M. Ito, and J. Szentagothai, The Cerebellum as a Neuronal Machine. Springer-Verlag, Berlin, 1967.) B, Purkinje cells of vertebrates. (From Nieuwenhuys, R., *in* C. A. Fox and R. S. Snider, eds., *The Cerebellum,* Progr. Brain Res. *25*:1–93, 1967. Elsevier, Amsterdam.)

the granule cells lie in cords at the sides of the midline and not in a granular layer.

In higher vertebrates, the climbing fiber input arises in relays of the inferior olive, and mossy fiber input comes from many nuclei of ascending tracts. In mammals, the mossy fibers come from the following sources: vestibular system via the vestibular nuclei, proprioceptive and somatic sensory input via dorsal spino-cerebellar and cuneo-cerebellar tracts, sensorimotor cerebral cortex via a relay in the pons, and projections from brainstem reticulum. Information carried in the mossy fiber system is usually modality-specific and exhibits fine gradations with stimulus strength.[388] The climbing fiber

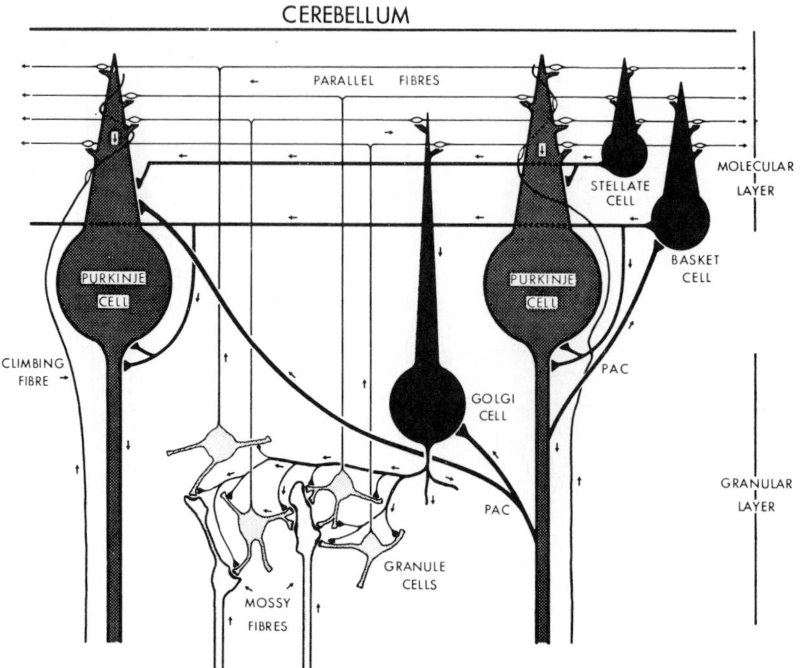

Figure 15–29. Diagram of connections in cerebellum. Climbing fiber input to Purkinje cell; mossy fiber input to cerebellar glomerulus. Golgi, stellate, and basket cells are inhibitory; parallel fibers are excitatory. (From Eccles, J. C., R. Llinas, K. Sasaki, et al., Exp. Brain Res. *1*:161–183, 1966.)

system shows point-for-point projection from the inferior olive to the cerebellar surface. In the cat, stimulation of various leg nerves,[388] rapidly adapting skin receptors,[116] and stimulation of muscle afferents can all activate climbing fibers, via the inferior olive. The fibers from the mossy and climbing fiber paths are arranged at right angles to each other, those from mossy fibers being oriented in the antero-posterior direction.

Purkinje cells are usually active, partly endogenously but mostly by continued stimulation from the mossy fiber system. In cat, the neurons of the inferior olive discharge rhythmically in response to various inputs and may activate Purkinje cells in phase. There has recently been described a third input of small adrenergic fibers coming from nucleus caeruleus, which may be inhibitory to Purkinje cells.[38]

In the frog, stimulation of the vestibular nerve[332] or physiological stimulation of saccule or utricle[330, 332] activates climbing fibers to the cerebellum. Input from the semicircular canals is carried mainly by mossy fibers, and PC's show the following responses: lateral PC's respond to ipsilateral rotation, anterior PC's respond to contralateral rotation, other PC's respond to rotation in either direction, and still other PC's stop spontaneous firing when any rotation occurs.[332] Responses to auditory and visual input vary with class of animal; in bony fishes the visual path is via a relay in the optic tectum.[396, 416] Mammals have extensive input to the cerebellum from the cerebral cortex.

A climbing fiber innervates one Purkinje cell and makes many synaptic contacts on basal dendrites—in cat, some 2000, and in frog, 300. Parallel fibers run in bundles along a folium, perpendicular to the plane of the dendritic espalier. Parallel fibers make synaptic contact near the ends of PC dendrites, and always synapse on spines; in the cat, a single PC has some 200,000 spine synapses, and each PC may receive signals from several thousand parallel fibers.

When a stimulus is applied to a cerebellar peduncle, three excitatory responses in PC's can be distinguished. First there are antidromic spikes due to stimulation of the PC axon. Second, there are short-latency responses to climbing fiber (CF) stimulation; these are recorded as an all-or-none complex of epsp's with superimposed multiple spikes[27]

(Fig. 15–30). The number of spikes varies with the state of excitability of the PC. Third, and after longer latency, is the response via the mossy fiber–granule cell–parallel fiber chain, which consists of a single or multiple epsp with single or multiple spikes and a spike in a PC. The action of mossy fibers on the granule cell is always excitatory and is probably mediated by acetylcholine.[12, 116]

In addition to the excitatory paths, there are three types of inhibitory interneuron. Parallel fibers activate Golgi cells, which are inhibitory to granule cells; i.e., they exert feedback inhibition on the mossy fiber–granule cell–parallel fiber excitation. Also, mossy fiber collaterals excite Golgi cells and thus show feed-forward inhibition on PC's. In mammals, stellate cells and basket cells excited by parallel fibers also inhibit Purkinje cells. Axons of stellate cells end on the dendritic tree, while axons of basket cells end at the axon hillock of a PC. In addition, collaterals of CF's also excite basket and stellate cells, showing feed-forward inhibition. A basket cell may elicit an ipsp in a PC lasting 100 msec or longer. Collaterals of CF's also excite Golgi cells, which disfacilitate the excitatory and inhibitory inputs of the mossy fiber system. After a burst of PC spikes elicited by CF stimulation, activity of the PC is interrupted and its excitability, as tested with antidromic impulses, is reduced because of activation of inhibitory neurons by CF collaterals.[317] This mechanism may provide a "resetting" or "clearing" to erase the ongoing activity or to reset the multiple firing zones of the PC from the excess excitatory or inhibitory inputs.[388]

Stellate cell inhibition has been demonstrated in the dogfish (lasting up to 125 msec as a block of CF response),[111] in the salmon (30 to 150 msec),[554] in frog,[475] and in lizard.[296] In alligator, stellate cells block PC spikes owing to PF input, and Golgi cells block granule cells.[331]

From the observation that parallel fibers are relatively uniform in diameter and that the dendritic trees of Purkinje cells tend to be isoplanar and oriented at right angles to the parallel fibers, it was suggested that sequential activation of PC's along the beam of parallel fibers may function as a timing

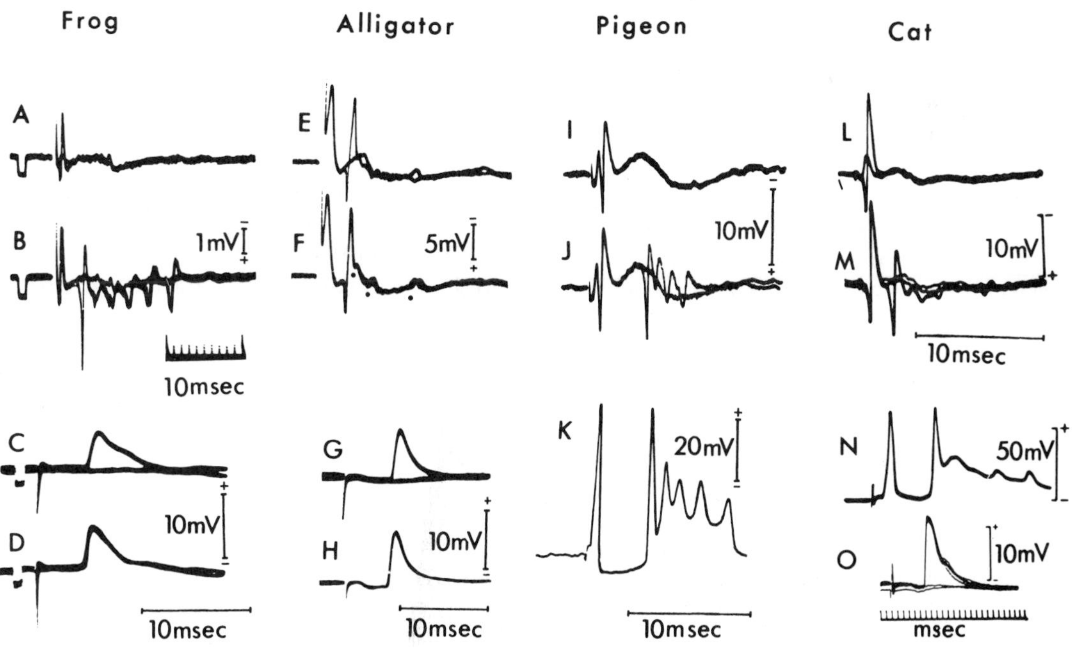

Figure 15–30. Climbing fiber and antidromic activation of Purkinje cells: extracellular recordings in upper two lines, intracellular recordings in lower two lines. Frog: A, antidromic spike; B, antidromic plus climbing fiber (CF) spikes; C and D, epsp's due to climbing fiber activation. Alligator: E and F, antidromic plus CF spikes; G and H, epsp's. Pigeon: I, antidromic; J, antidromic plus CF spikes; K, antidromic plus CF spikes. Cat: L, antidromic; M, antidromic plus CF spikes; N, antidromic plus CF response: O, epsp evoked by CF stimulation. (From Llinas, R. R., et al., *in* W. S. Fields and W. D. Willis, editors. The Cerebellum in Health and Disease. Warren H. Green Publishers, St. Louis, 1969.)

device.[42] In the frog cerebellum, a unidirectional temporal correlation has been found between series of adjacent PC's under a particular beam of parallel fibers.[151] Close apposition between granule cells receiving the same mossy fiber input has been seen in frog[504] and in mormyrid fish;[274] this suggests electrical coupling between granule cells and synchronization of activation of parallel fibers in a beam.

All of the output from the cerebellar cortex is by axons of Purkinje cells. In mammals, the PC's of all areas of the cerebellar cortex (except the flocculus, which projects to the vestibular nuclei) project to deep cerebellar nuclei. The output of the cerebellum is uniformly inhibitory and is exerted on the motor systems via various brainstem nuclei. In lower vertebrates, some PC axons go directly to motor centers in the brain stem and others terminate in subcerebellar (fastigial) nuclei. In the snakefish *Ophiocephalus,* some PC axons synapse with oculomotor neurons, and stimulation of the cerebellum elicits ipsp's in these neurons and blocks their response to excitatory input from vestibular neurons.[294] This system provides for error correction in the rapid and fine adjustments of eye position and movement.[244] In mormyrid electric fishes, several cell types corresponding to deep nuclei are found in the cerebellar cortex.

The transmitter for the inhibitory action of Purkinje cells is probably GABA.[309, 397] There is evidence that the inhibitory actions of the stellate cells[475] and basket cells[86] are also mediated by GABA.

In mammals, feedback and feed-forward loops have been described between the sensorimotor cortex and the cerebellum. Stimulation of some regions of the cerebellum can desynchronize rhythms of the cerebral cortex and induce trains of spikes in the sensorimotor cortex of lower frequency than those that were delivered to the cerebellum.[286] Cerebellar stimulation can suppress movements initiated by cortical stimulation.

Lesions in the cerebellum produce varying degrees of disturbances in posture and movement in different kinds of animals. In a dogfish, cerebellectomy fails to interfere with swimming, although the fish frequently circles toward the side on which the lesion was made. In carp and goldfish, there is some muscle hypotonia on the side of the lesion, side-to-side swaying, and impairment of equilibrium control; the fish cannot be conditioned to light or sound, and previously established conditioned responses may be lost.[286] Frogs and lizards from which the cerebellum has been removed can walk and jump, but there may be reduction of spontaneity of movement and circling or tilting of the body toward the side of the lesion. In birds and non-primate mammals, removal of the cerebellum results in extreme extensor postures, especially of the head and neck. In primates, flexion of limbs results, as well as muscular weakness and decrease in the stretch responses of muscle afferents. Cerebellar lesions bring about difficulties in initiation and coordination of movement, as well as tremor. Small lesions in the flocculonodular lobes in birds and mammals impair posture and equilibrium, whereas lesions in the anterior lobe and cerebellar hemisphere interfere with skilled movement.

Electrical stimulation of the cerebellum may evoke motor activities and postural adjustments, and may either facilitate or suppress motor activities originating elsewhere. Change of stimulus intensity and frequency can convert facilitatory effects into inhibitory ones, and vice versa. In both fishes and mammals, a brief stimulation of the cerebellar cortex evokes a movement or posture which is followed by a long sequence of post-stimulus rebound of other (usually opposite) movements or postures.

The circuitry of the cerebellum suggests that its action may be more subtle than the described motor effects. The basic cerebellar circuit is present in all vertebrate classes, and only inhibitory elements have been refined, and the deep nuclei developed, during vertebrate evolution. There is redundancy of input (via climbing and mossy fibers) in all vertebrate classes, which means convergence of excitation over a considerable time-span. There is inhibition at one, two, or three synaptic levels.[388]

Several functions for the cerebellum have been proposed:[128] (1) it may receive input from muscles which have been activated by the cerebrum and then send out modulating signals to provide refinement of movement; (2) sensory inputs to other brain regions may go by parallel paths to the cerebellum; (3) motor signals originating in the cerebral cortex may go to the cerebellum at the same time as to cord motorneurons, so that the cerebellar output adds to that from the cortex; (4) Purkinje cell discharge may precede and initiate that in the cerebral cortex. Thus, the cerebellum monitors major inputs and outputs. In animals of classes that lack a motor cortex, the cerebellum and centers in the brainstem reticulum may be particularly important in initiating and controlling movements. The usefulness of the cerebellum is indicated by its presence throughout the vertebrates with less change than in any other part of the brain.

CENTRAL SENSORY PATHWAYS IN VERTEBRATES

Previous chapters have discussed peripheral sensory structure and function and have dealt with such parts of the central pathways as are necessary for limited behavior in response to various sensory stimuli.

Central Vision

Properties of that portion of the brain that forms the retina were described previously (Ch. 14), and it was shown that at the levels of bipolar, horizontal, amacrine, and ganglion cells much processing of visual input occurs. Axons of the ganglion cells form the optic nerve.

Fishes and Amphibians. In teleost fishes and amphibians, the optic fibers all cross to the opposite side and are distributed mainly to the tectum (dorsal mesencephalon); in addition, small optic tracts pass to the thalamus and hypothalamus[66, 169] (Fig. 15–31). Lesions in frog tectum reveal the normal presence of output connections to pretectum and thalamus, and to spinal cord, superior olive, reticulum, and optic nerve (efferents).[472] In the optic tectum of fish and frog, many optic fibers synapse in the upper layers, while some pass to deeper layers. Interneurons in the tectum send output to the cerebellum and spinal cord by tecto-cerebellar and tecto-spinal tracts; tectal connections with the forebrain are poorly known. The tectum contains several cell types – pyramidal cells, some horizontal cells, and small pyriform cells in the paraventricular layer, which comprise 90 per cent of the tectal neurons.[320] Especially in frogs and lizards, tectal neurons are in layers, and the number of layers varies with species.[500]

Visual stimulation elicits evoked potentials in the optic tectum, negative just below the surface and positive deeper, at the level of pyramidal neurons. By local stimulation of regions of the retina, point-for-point projections to the tectum are found, with nasal quadrants posterior, temporal quadrants anterior, lower retina medial (dorsal), and upper retina lateral (ventral) in fish[59, 431, 489] (Fig. 15–32), in alligator,[192, 193] in frog[162] (Fig. 15–33), in *Xenopus*,[161] and in goldfish.[160] On the tectum, the center of the retina has a larger representation than does the periphery. In animals with binocular visual fields, such as frogs, with stimulation by light projected on a screen, it is found that the central part of the field is represented via each eye, the binocular field subtending 100° in the horizontal meridian.[159]

When recording spikes with a microelectrode and high-frequency amplifier, it is not always easy to distinguish optic nerve fibers from tectal neurons. In general, the latency for optic fibers is shorter, receptive fields are smaller, the frequency of following of repetitive flashes is higher, and response characteristics are more constant than for tectal cells; these differences are readily evident if the optic nerve is stimulated electrically. Many retinal ganglion cells respond to concentric or side-by-side excitatory and inhibitory fields, so that an "on" cell increases its firing if the center of the field is illuminated, and decreases firing if light is in the surround; the response is reversed for "off" cells. One ganglion cell may respond to a fixed edge, another to change in intensity, and another to movement of a spot in one or several directions (p. 627). The classification of ganglion cell responses varies with species (frog,[178, 348] goldfish[346, 398]). In frogs, optic fibers from edge detectors may terminate more superficially in the tectum than do contrast and dimness fibers.[164, 176] In some species, particularly fishes, some ganglion cells are connected to cones and respond differently to color contrast in the center and in the surround.

In mid-layers of the goldfish tectum, responses of horizontal and pyramidal cells show longer latencies and larger fields than those of ganglion cell axons, and the responses vary according to prior stimulation. Most tectal neurons are spontaneous, and visual stimulation interrupts their firing; a few show accelerated firing or are caused to discharge by visual stimulation.[398, 525, 615] Responses of central tectal neurons habituate

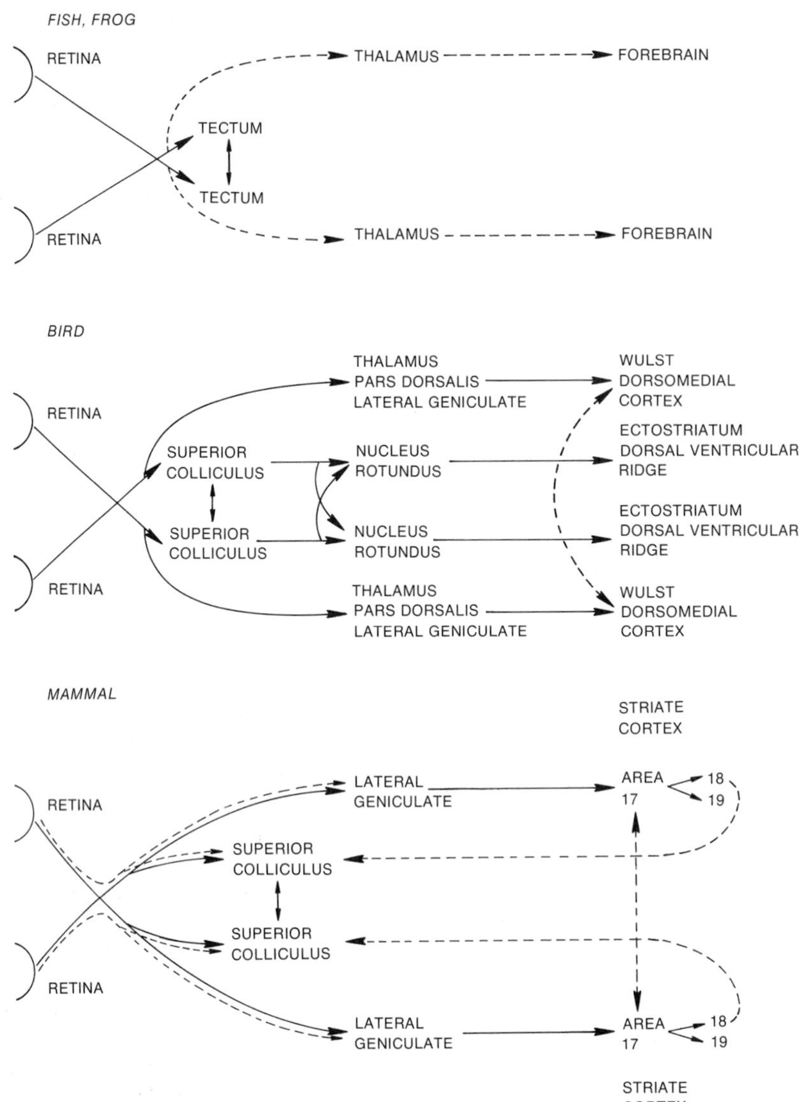

Figure 15–31. Diagrams of principal visual pathways in vertebrates. Many connections, such as those from midbrain to cerebellum, are omitted.

on repetition but can be dishabituated by a change in stimulus, such as color change, light-dark pattern shift, movement, or shock to the fish's flank.[249, 398] In the deep layers of the goldfish tectum, the pyriform neurons are spontaneous in bursts which are interruptable by visual stimulation; their responses are very labile. Output from pyriform cells is not readily traced; their axons are known to cross to the opposite tectum and to make intratectal connections. A neuron in the upper layer of the tectum of the pike *Esox* is sensitive to movement of visual stimuli in one direction; the mid-layer is sensitive to small spots and contrast, and the lower layer is sensitive to changes in intensity.[615] Tectal neurons of frog are of many types. Some respond to edges, and some to moving stimuli (light or dark) or to receding or approaching objects. Tectal neurons habituate on repeated stimulation.[177] In superficial layers of frog tectum, lateral inhibition appears to be exerted by cholinergic connections.[515] In a salamander, staining for AChEs coincides with optic fiber projection, and some tectal neurons respond to either visual or contralateral tactile stimulation.[175]

In tectal neurons, responses have been recorded to vestibular stimulation (frog)[379, 480] and to somatosensory stimulation (urodeles, goldfish). Below the tectum lies the teg-

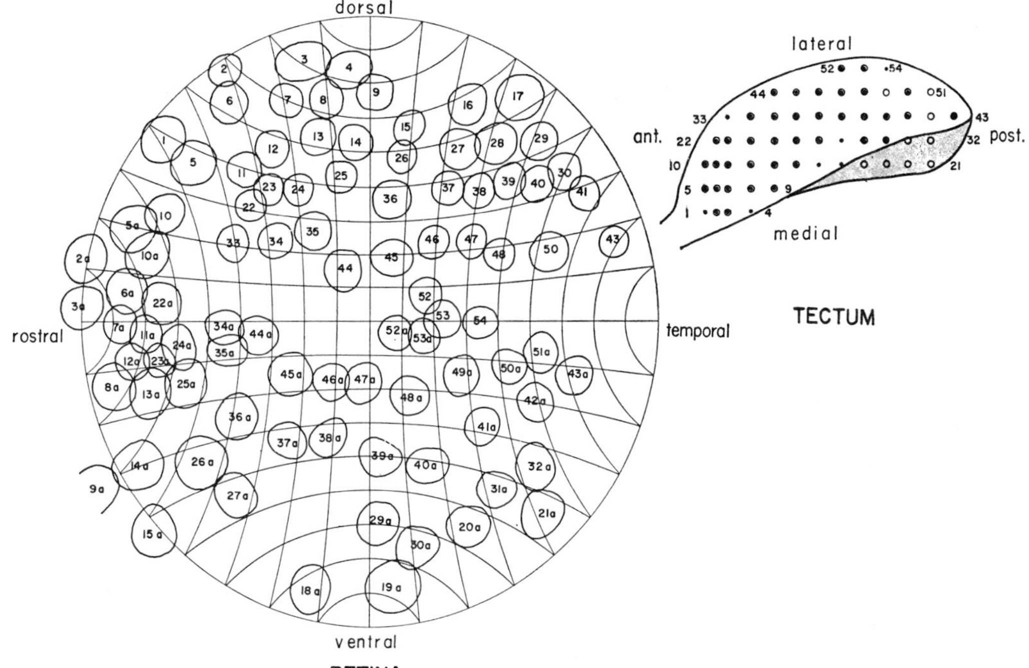

Figure 15-32. Topographic arrangement of visual field in left retina and its projection to corresponding electrode positions on right optic tectum of carp. (From Schwassmann, H. O., and L. Kruger, J. Comp. Neurol. *124*:113-126, 1965.)

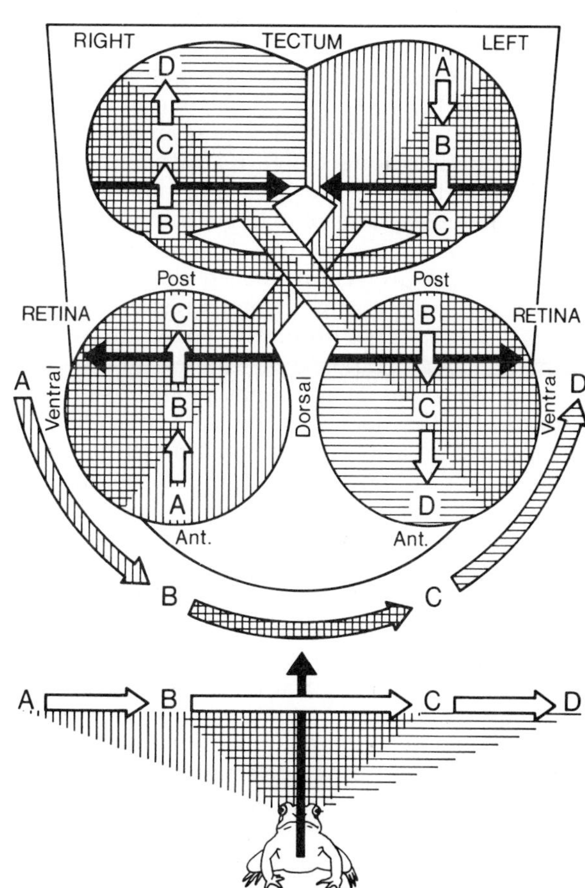

Figure 15-33. Retinotectal projection of frog. Monocular projection crosses in optic chiasma to opposite tectum: A–B from right eye and C–D from left eye. Binocular projection B–C from each eye to tectum of same side. Intertectal commissures connect corresponding points. (From Developmental Neurobiology by M. Jacobson. Copyright © 1970 by Holt, Rinehart and Winston, Inc. Reprinted by permission of Holt, Rinehart and Winston, Inc.)

mentum, which contains motor nuclei. In the tegmentum of goldfish are some neurons that respond to visual and auditory input.[407] Convergence of visual, tactile, and auditory input to subtectal units also occurs in the toad.[135] In frog tectum, many ipsilaterally responding units receive input which comes via the contralateral tectum.[162, 164]

In a turtle, the tectal commissure is essential for optokinetic control.[196] After the superficial tectum has been removed from a goldfish, visual learning is impaired.[346] Local damage to one tectum may lead to rolling and circling movements. Tectal neuron discharges may be associated with saccadic eye movements in goldfish. A local lesion in a fish tectum prevents conditioning of heart rate to visual stimulation in that retinal field projecting to the area of the lesion; hence, retinotectal projection is necessary for localization within the visual field.[489] In a frog, local tectal stimulation elicits patterned movements. The tectum mediates visually guided movements – body turning, eye movement, and optomotor responses. The frog diencephalon inhibits or releases tectal efferents for orienting and feeding behavior; i.e., the thalamus biases tectal output.[242]

In the posterior thalamus of a toad, responses to visual and somatic sensory stimulation can be recorded. Thalamic nuclei project to the primordial hippocampus.[134] A toad hesitates longer before attacking two mealworms than before attacking one, and simultaneous sighting of a large illuminated background inhibits attack; lesions in the thalamus and pretectum abolish the inhibition. Some units in the thalamus respond to visual parameters different from those bringing about response in the tectum. In a frog, the axons of some ganglion cells go directly to the lateral geniculate nucleus (l.g.n.); there are also tectal-l.g.n. projections.[387] Some fibers pass to the forebrain from the thalamus in frogs and fishes, but this projection has not been studied physiologically. Recent evidence suggests that forebrain lesion in a shark interferes with discrimination of visual patterns.[169]

The iguana has two visual nuclei in the thalamus; a dorsal one receives direct retinal projection, and the nucleus rotundus receives from the optic tectum.[61] In a turtle, optic projections to the telencephalon give responses some 40 msec later than those in the tectum; however, some optic fibers bypass the tectum and go via the thalamus to the forebrain, as evidenced by responses obtained there after removal of the tectum.[26, 354, 614] The retina is projected on the tectum in alligator and lizard, and the evoked response consists of one positive and two negative waves, which vary with depth.[192, 193, 344]

Birds. In many birds, the visual system is highly developed. A pigeon is estimated to have 20 million cells in the retina, of which 2.4 million are ganglion cells with processes in the optic nerves. A large fraction of the optic fibers cross to the tectum on the opposite side; output from the tectum goes to the nucleus rotundus of the thalamus, and from there to the ectostriatum of the forebrain. Some optic fibers also run directly from retina to thalamus and then to forebrain.

Each optic tectum of a bird is a large structure; evoked potentials show a broad subsurface negative wave (indicative of two synaptic sinks) and a deeper positive wave. Tectal units usually give only one spike per afferent impulse, and fail to follow at frequencies above about 125/sec. Retinal input can interrupt spontaneous activity, so there may be some inhibition of tectal cells in birds.[200]

In the pigeon, individual tectal cells have visual fields much larger than those of retinal ganglion cells; in a large sample of tectal neurons, 72% were movement selective, 21% were selective for movement in a given direction (excited in one direction and inhibited in the opposite direction), and 7% were sensitive to concentric and contrasting fixed fields.[254, 255] In pigeons, direction and movement are also registered by thalamic cells (corresponding to lateral geniculate); 75% of the cells in the nucleus rotundus are sensitive to movement in the visual field.[84] The importance of the striatum in the integration of vision in birds was shown in pigeons by a deficit in discrimination of brightness and pattern (but not of color) after a lesion was made in the Wulst (hyperstriatum).[427]

Mammals. In mammals, the superior colliculus (homologous with the optic tectum of lower vertebrates) and pretectal nucleus mediate extraocular movements and pupillary reflexes. Units in the pretectal nucleus give both phasic and tonic responses; a greater proportion of them are tonic compared with the colliculus.[499] Units of the superior colliculus of rat and cat show linear projection from the contralateral retina, with greater precision in the colliculus than in the pretectum.[199] The collicular neurons are relatively unresponsive to stationary visual stimuli, but they give sustained responses to moving stimuli. Most of the collicular cells give responses to binocular stimulation, and each cell has an optimum for rate and direction of movement. After removal or cooling of the visual cortex in a cat, the collicular cells become relatively unresponsive to moving stimuli and lose binocular sensitivity; this indicates that collicular responses to visual stimuli are modulated by cortical input.[510, 513, 571] In unanesthetized monkeys, the superficial layer of the superior colliculus showed some units that responded both to smoothly moving and to stationary flashing stimuli, and other units that responded to rapid, jerky movement; the deeper layers showed units that discharged in bursts prior to saccadic eye movements.[487] Neurons in the optic tectum of a squirrel monkey are in layers. Responses to diffuse light occur in upper strata, those to antagonistic fields occur in the middle, and those to moving objects occur in deeper layers; some retinotopic projection occurs.[273]

Motor responses resulting from stimulation of the colliculus in monkey correlate well with retinotopic sensory maps on the colliculus. In rabbit, stimulation of the colliculus elicits deflections of head and eyes toward that part of the visual field corresponding to the stimulation site. A mole has a poorly developed visual system with only 200 fibers in an optic nerve and no apparent projection to the colliculus; the dorsal lateral geniculate appears adequate for light-dark discrimination.[336]

In mammals, most optic fibers pass to the lateral geniculate in the posterior diencephalon, and from there to the visual cortex. It is estimated that the macaque monkey has one million fibers in the optic nerves, two million neurons in the lateral geniculates, and two hundred million neurons in the visual cortex; thus, there is a divergence of optic fibers in the ratio 1:2:100.[423a] This is in contrast to the peripheral retina, where there is considerable convergence from many rods to one ganglion cell.

In fishes and amphibians, all of the optic fibers cross in the chiasma; but in mammals, a certain proportion of the fibers do not cross. In primates, fibers from the nasal half of the retina cross the chiasma, while those from the temporal retina go in the ipsilateral tract. The first relay is in the lateral geniculate nucleus (l.g.n.), a six-layered structure where crossed and same-side optic fibers are distributed in layers. Single geniculate neurons show spike responses, sometimes in bursts; they have a concentric visual field of an "on" excitatory center and an "off" periphery, or vice versa.[226, 227, 230] Most cells in cat l.g.n. have a visual field for each eye; one of the two eyes is dominant, and the non-dominant eye is usually inhibitory.[483] Fibers pass from the lateral geniculate to the occipital lobe of the cerebral cortex, mostly to area 17 (striate cortex), and some then pass to areas 18 and 19.

In the visual cortex of cat and monkey, responses of single neurons have been classified as simple, complex, and hypercomplex.[228] Recent studies with various methods of visual stimulation and several endpoints indicate that these neuron categories are not sharply delimited, but rather that there is a continuum of cell types.[70, 71, 359, 508, 516]

The simplest cells in the visual cortex have receptive visual fields with excitatory and inhibitory areas which are not concentric, as for lateral geniculate cells, but which lie along the two sides of a straight stationary boundary that is inclined at a "preferred" angle. Somewhat less simple are similar cells (with excitatory and inhibitory areas) which respond phasically to movement of the boundary over a small field; some are unimodal or directional in sensitivity, while others are bimodal with respect to movement; all of these are usually quiet when unstimulated. The next level of complexity includes cells for which there are no antagonistic areas, but which respond tonically and at high frequency during movement in a given direction and at a certain angle in a wide field; these cells are highly spontaneous when unstimulated. Still more complex cells (hypercomplex, found in areas 18 and 19) respond to moving oriented boundaries of a particular length, i.e., an excitatory length flanked by regions that elicit responses only when specifically oriented[228] (Fig. 15–34). Many cortical neurons can be stimulated binocularly, and the two visual fields do not coincide; some detect vertical disparity, and others detect horizontal disparity.[418] Of a population of hypercomplex cells in cat, 83% responded binocularly but 68% showed dominance of one eye, with more units responding to the contralateral stimuli than to ipsilateral ones.[228, 229]

If a kitten's eyelid is sewed shut or if an

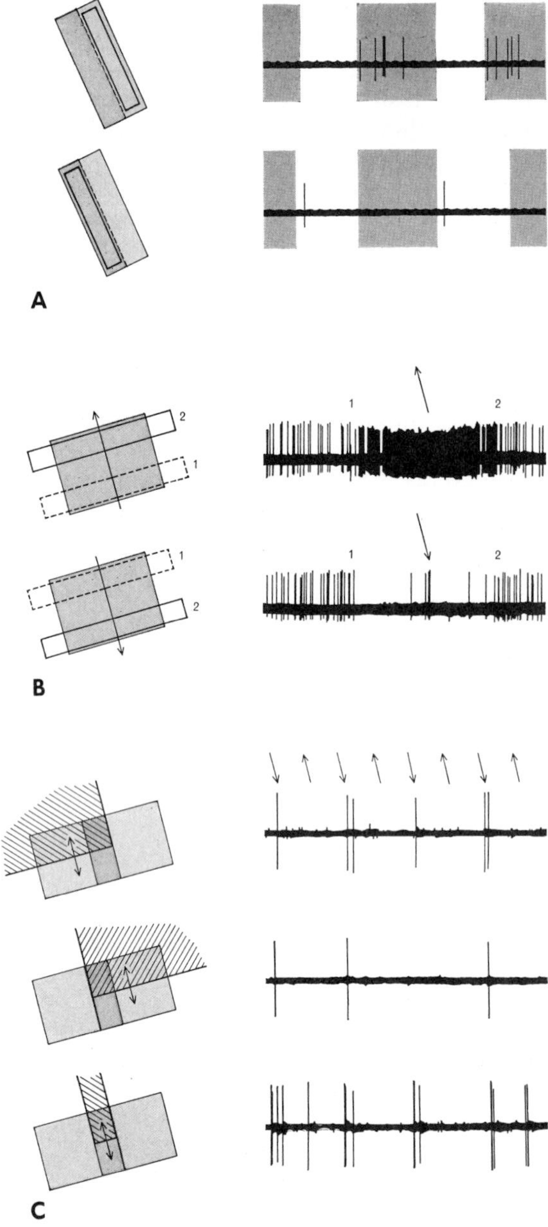

Figure 15–34. Cell types in visual cortex of cat. A, Simple cells. Excitatory (light) and inhibitory (dark) parts of visual field shown at left, and responses to two "on" stimuli (light regions) shown at right. Cell (a) gives "off" responses; cell (b) gives a small "on" response. B, Complex cell is spontaneously active. It is excited by movement of stimulus in one direction, and inhibited by movement in the opposite direction. C, Hypercomplex cell. Central part of field is excitatory; ends of field are inhibitory. Movement of stimulus over central field elicits most spikes. (From Michael, C. R., Retinal Processing of Visual Images, Scientific American, May, 1969, p. 109. Copyright © 1969 by Scientific American, Inc. All rights reserved.)

eye is covered with an opaque shield, and after three months recordings are made, the responses of geniculate cells give normal "on" and "off" areas with concentric fields. However, the binocular dominance in hypercomplex cells has become changed, and cortical cells may fail to respond to stimulation of the eye that was occluded; hence, as a result of lack of usage of a pathway, a defect has occurred between the geniculate and the cortex. A diminution in the number of responding cells can be noted if one eye is kept closed for only three to four days.[231, 581] The most crucial period in kittens is at the 4th and 5th weeks. Little effect is noted if the unilateral deprivation begins after two to three months of age, and no effect is found when deprivation is made with adults. Behavioral indications of the cortical defect resulting from visual deprivation are reductions in visual acuity, pattern discrimination, and paw guidance.

The visual processing observed in cats and monkeys is not typical of all mammals. In rabbits and ground squirrels more integration occurs in the retina. A ground squirrel has an all-cone retina, and half of the ganglion cells respond to contrast (i.e., have center and surround fields); 25% of the ganglion cells are directional (i.e., respond to movement of oriented edges); and 25% are opposite-color receptors.[366] Half of the optic nerve fibers pass to the superior colliculus, and half to the lateral geniculate. Directional units go only to the colliculus, while contrast and color opponent units go to the geniculate. Lateral geniculate neurons of ground squirrel resemble the complex and hypercomplex cells of cortex of cat.[366, 367] A pigeon also has complex ganglion cells.

In summary, cats and monkeys have ganglion cells with concentric center-surround fields; their cortical cells resemble in complexity the ganglion cells of frog and goldfish. Ground squirrels, however, have at least three types of ganglion cells, and complex geniculate cells. Pigeons have complex cell types in the striatum. Frogs and goldfish show diversity of integrating cells in optic tectum.

Auditory Pathways

Sound is an important and complex sensory modality for most vertebrates. Parameters processed in the brain are: temporal patterns of intensity, frequency, and noise; intensity variation at a constant frequency; modulation of frequency, noise, and selection of tone from noise; and position of sound source. The mechanisms of sound reception were discussed previously (Ch. 12), as were central auditory pathways for fishes (p. 530), amphibians (p. 533), reptiles (p. 533), birds (p. 534), and mammals (p. 535). Special features of auditory centers in echolocating animals such as bats were also considered (p. 537).

Every vertebrate has a primary acoustic nucleus in the medulla. In fishes, the next level is the torus semicircularis in the mesencephalon. Some neurons of the acoustic nucleus send axons to the cerebellum. Single-neuron audiograms show narrower ranges in units of torus than in those of cochlear nucleus.[406] Purkinje cells of the fish cerebellum may be either excited or inhibited by auditory input.

In frogs, three nuclei in the torus semicircularis of the tegmentum are activated by sound. Responses are bursts of spikes, and there is no tonotopic organization within the torus.[425] Frogs in their natural habitat are highly responsive to sound patterns of their own species, but the location of the species "template" is unknown (p. 533). In reptiles (e.g., *Caiman*), neurons of cochlear nucleus show wide-band responses, while those of torus have narrower ranges. These reptiles have a region in the forebrain—the striatum—which gives "on" and "off" responses to tones.

In birds, there is much species variation, but in some (e.g., owls) there is binaural interaction in the cochlear nucleus, and a strong response in the cerebellum. Neurons of the lateral mesencephalic nucleus (inferior colliculus) of a dove are spontaneously active and show interruption of spontaneous activity in response to sound stimulation except for a narrow excitatory range. Usually, inhibition occurs above and below a central range. Despite the recognized uniqueness of species and dialect calls, the central representation of these calls is unknown.

In the cochlear nucleus of mammals (e.g., cat), a single cell may show depolarization with accompanying spikes at one tone range and hyperpolarization with depression of spontaneity in another range.[511] Inhibitory effects on cochlear responses caused by stimulation of efferent fibers in the olivocochlear bundle were mentioned on p. 535. The functional meaning of these efferents is not well understood, but stimulation of the efferent tract enhances cochlear microphonics and reduces the initial cochlear nerve response;[300] section of the efferent bundle increases masking of tones by noise.[101a] In goldfish, stimulation in the midbrain can reduce the response in the torus to clicks, and this inhibition may provide a rapid (10 msec) feedback modulation of sound signals.[421a]

The importance of the inferior colliculus in hearing is shown by the ability of animals, including man, to make sound discriminations after the auditory cortex has been removed. Evoked potentials recorded from the inferior colliculus of a porpoise show much variation correlated with sound parameters and an audiogram resembling the range of behavioral response.[54] In a sea lion (pinniped) the auditory system is less specialized than in a porpoise; best collicular responses are in the 4 to 6 kHz range.[53] In bats, collicular neurons respond particularly to frequency-modulated tones.[546]

The auditory cortex consists of primary and secondary areas in the temporal lobe, and in some species a small tertiary area. The two cochleas are represented on each cortex, and there is some tonotopic arrangement; but columnar and spatial representation is less than in somatic sensory and visual cortex.[3, 168] Some cortical units show unstable responses to tones and are either extinguished or facilitated on repeated stimulation; a few units respond not to pure tones but to frequency-modulated stimuli.[545]

For more details of the role of the central nervous system in hearing in vertebrates, see pp. 530–541.

Somatosensory Pathways

Projection of somatosensory inputs has not been much studied in non-mammals. In response to sciatic nerve stimulation, evoked potentials are recorded in frog thalamus and forebrain.[550] In fishes, there may be electrical responses in cerebellum and tectum to fin and body stimulation, but most of the processing seems to be in the spinal cord and medulla. The facial lobes of a goldfish show somatotopic projection from each fin and from skin of head and trunk.[417] In the tiger salamander, the contralateral body surface is projected over the tectum, and endings of ascending axons of the somatosensory system appear to correspond to the distribution of catecholamine fluorescence.[175] Thalamic nuclei in frog and turtle show somatosensory responses.

In a monotreme (the echidna *Tachyglossus*), the cerebral cortex is folded into sulci which have little resemblance to the gyral organization of eutherians; a large area in the posterior half of the cortex receives projections from tactile stimuli in various parts of the body[322] and lies behind the motor area.[365] Similarly, insectivores, such as the hedgehog *Erinaceus,* have an unspecialized forebrain with two dorsal regions of visual projection, a large rostral somatosensory projection, and a postero-ventral auditory region.[272] Marsupials—opossum and wallaby—have in the cortex a posterior visual area, a lateral auditory area, and extensive rostral somatosensory regions; there is completely coincident overlap of sensory and motor representation of body regions.[323, 342]

In mammals of other orders, the somatosensory pathways follow a common pattern. In the spinal cord the fibers from dorsal column nuclei carry messages from touch, pressure, and motion receptors to the medulla, where the trigeminal nerve enters with similar information from face and head. From nuclei in the medulla, tracts go to the thalamus, where the mapped projection is more closely proportional to sensory innervation density than to body geometry. Neurons of the thalamus can distinguish not only place of origin on the body but kind and quality of mechanical stimuli, position of limbs, and serial order of separate stimuli.[382, 383] From the thalamic nuclei there is somatosensory projection to three areas of the cerebral cortex, the first on the postcentral gyrus, the second on the parietal cortex, and the third on the dorsal anterior suprasylvian gyrus[89, 90] (Fig. 15–35). The tactile areas correspond in size to the importance of sensory input; hand and face areas are large in monkey, mouthparts in rabbits, claws and forelimbs in cats, face in dogs, vibrissae in rabbit and rat, mouth in sheep, and forepaw in racoon.[535] In various kinds of monkeys, the size of the area representing the tail corresponds to the prehensile use of the tail.

The cells of the somatosensory cortex are arranged in functionally separated columns of similar sense modality according to region of peripheral receptive field. A few cells in this cortex are non-specific and respond to auditory as well as to mechanosensory stimulation.[90] The cortical representation is a refinement on the good projection in the thalamus and serves to match sensory with motor areas.

Central Olfactory and Gustatory Pathways

In fishes, a large forebrain correlates with a well developed olfactory organ in the eel *Anguilla,* large facial lobes in the medulla with taste in catfish *Ameiurus* and *Nemachilus,* and large midbrain with vision in archer fish *Toxotes.*[23] Olfactory discrimination is used in homing by salmon.[543] The superficial forebrain of fishes shows spontaneous electrical rhythms with superposed responses to olfactory stimulation. Olfactory bulb

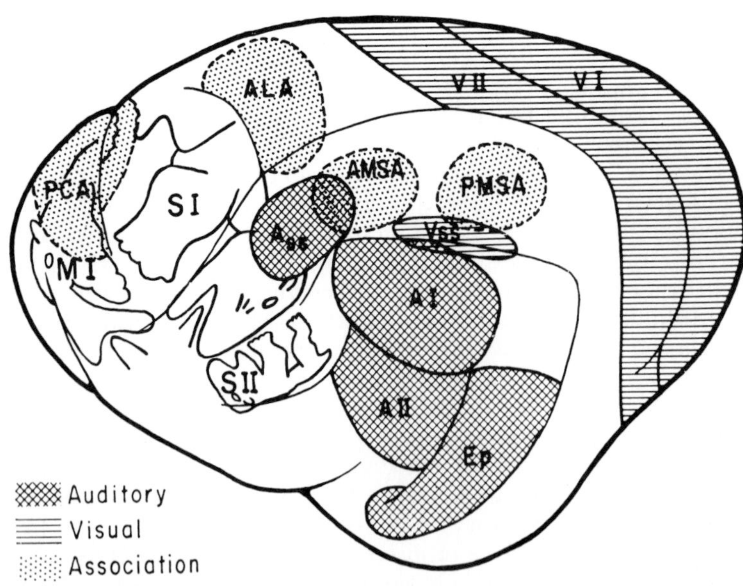

Figure 15–35. Diagram showing primary auditory, somatic sensory, and visual fields and association fields in cat cortex. AI, auditory area I; AII, auditory area II; Ep, ectosylvian auditory area; A_{sg}, auditory field of anterior suprasylvian gyrus; SI, somatic sensory area I; SII, somatic sensory area II; MI, somatic motor area I; VI, visual area I; VII, visual area II; Vss, visual area in suprasylvian sulcus; PMSA, posterior middle suprasylvian association area; AMSA, anterior middle suprasylvian association area; ALA, anterior lateral association area; PCA, pericruciate association area. (From Thompson, R. F., R. H. Johnson, and J. J. Hooper, J. Neurophysiol. *26*:343–364, 1963.)

responses in salmon are modified by contralateral stimulation; hence, efferents pass out from the brain. Ablation of an olfactory bulb in an eel leads to degeneration in the ventrolateral forebrain and posterior and ventromedial portions; there are projections to both pallium and subpallium.[484] Taste input is carried from skin to facial lobes and from mouth and gills to vagal lobes; hence, taste is separate from smell.[17]

In turtles, each olfactory nerve serves both pyriform cortex and central forebrain nuclei and there is some convergence of olfactory, auditory, and visual input onto the forebrain.[404] In an alligator, the olfactory nerve runs to the ventral pyriform cortex.[304]

In most mammals, the olfactory connections are to the olfactory bulb, olfactory tubercle, prepyriform cortex, and cortico-amygdaloid nuclei.[189] Olfaction is said to be the only sensory system without thalamic representation. Taste input is relayed via the ventral posteromedial nucleus of the thalamus to a restricted region within the presylvian sulcus.

EMOTION CENTERS OF FOREBRAIN SERVING SELF-PRESERVATION AND REPRODUCTION

Association functions of the forebrain are observed only by behavior. The non-olfactory part of the ancient forebrain receives sensory input from the diencephalon and has motor connections to the reticulum (Fig. 15–23). It is intimately connected with the mid-diencephalon, particularly the hypothalamus, and may well have functional constancy related to emotions and subtle behavior of self- and species-preservation in all vertebrates.[341] The limbic system consists of hippocampus, septum, and amygdala; it does not show discrete electrical responses to sensory excitation and does not elicit movements when it is stimulated. A useful method of study is to observe behavioral interactions of animals with lesions to, or after stimulation of, this region. In general, the primitive forebrain correlates the biotic environment with behavioral output; i.e., it releases whole patterns of behavior, either stereotyped or not.

When the forebrain has been removed from a goldfish, feeding and spontaneous locomotion are normal, and conditioning to dark or light can occur; however, exploratory behavior is reduced and, after unilateral lesions, the fish tend to swim in the direction of the intact side.[182, 250] In sticklebacks, removal of the non-olfactory portion of the forebrain leaves normal swimming and color and brightness discrimination. However, defects are seen in breeding behavior, schooling, and location of the nest site. The forebrain is responsible for balance of aggression, sex drive, and parental behavior; damage to rostral or lateral regions leads to lowered aggression and reduced periods of sex activity but enhanced parental behavior, while damage to mediocaudal regions leads to heightened aggression and lowered parental action.[182, 490, 491] Similarly, when the forebrain has been removed from wrasse and bream, there is less fighting; on exposure to a rival, the number of bites was reduced from 26 to 8.[138]

Reptiles such as lizards have a complex of ritualized behavior when in groups—display of body regions, general aggressive behavior, and attack movements which lead to a social hierarchy. When the amygdala is removed, lizards become non-aggressive and submissive toward other lizards; they can maintain display postures but do not display in an aggressive context. The anterior part of the amygdala serves assertion display; the posterior part serves attention and response to aggressive action by other lizards.[527] Stimulation via implanted electrodes in the amygdaloid complex of caiman elicited escape responses, and bilateral lesions in this region reduced attack and retreat responses.[288]

When the brain of chickens is stimulated by deeply implanted electrodes, complex behavior can be elicited—clucking, brooding, threatening, attacking, and attempts to escape.[202]

In mammals, the limbic system includes the cortical hippocampus and the subcortical complex of anterior septal nuclei and posterior amygdala nuclei. Stimulation of various parts of the limbic system leads to autonomic responses, both visceral and vascular; stimulation in other limbic regions initiates feeding (chewing, licking, etc.); still other stimuli elicit attack behavior (hissing, claw protrusion) according to species.[295, 443, 542] Stimulation of the septal nucleus in cats is "pleasurable," and animals with electrodes implanted there "enjoy" self-stimulation. Stimulation of the anterior hypothalamus can elicit rage reactions, sex behavior, and, in certain regions, feeding and drinking. In cats, electrical stimulation of the amygdala suppresses attack behavior elicited by hypothalamic stimulation.[119] In a monkey, hippocampal units fire when expectancy of food is high, and prior to pushing a lever to obtain food.[400] Removal of the amygdala often increases food intake (after an initial decrease), and it may lead to hypersexuality. Removal of the amygdala from cat or dog releases rage reactions, but in wild rat or macaque it leads to placidity.[370, 523] Men with hippocampal lesions may lose short-term memory. Removal of the neocortex from cat increases placid behavior. Thus, the behavioral manifestations of damage to the limbic system vary with species. There is a system of checks and balances between neocortex, limbic centers, and hypothalamus.

In summary, the neocortex, especially in mammals, serves to join multiple inputs with the units initiating motor actions. The paleocortex of all vertebrate classes seems to relate elemental or emotional behavior to the biotic environment. Whether the subtle functions responsible for complex behavior require networks of neurons acting *en masse* or whether single neurons have sufficient specificities in coded input requirements to account for complex behavior is being actively debated. In any case, there are in the ancient forebrain of vertebrates endogenous patterns which have survival value.

SLEEP

Near-synchrony in many regions of the brain is found in the states of sleep. In a number of mammals, but best studied in cat, two states of sleep are recognized.[270] In the slow-wave or initial stage of sleep, the brain waves, particularly in frontal lobes, consist of high-amplitude spindles (11 to 16/sec) interrupted by slow (1 to 4/sec) waves; there are no eye movements, but electromyographic recording shows persistent tonic muscular activity, particularly in the neck region. In deep sleep, sometimes called paradoxical or REM (rapid eye movement) sleep, there are low-voltage fast cortical activity, rapid eye movements, and complete absence of muscle tone. In deep sleep, large pyramidal neurons of the motor cortex show bursts of high-frequency spikes, and there are irregular spikes in the lateral geniculate and pontine reticular formation. In the reticulum, from medulla to mesencephalon, are clusters of neurons rich in serotonin (5HT); destruction of this region or blocking of 5HT synthesis by drugs leads to almost permanent wakefulness. Activity in a reticular serotonergic system appears to be responsible for slow wave or spindle sleep.[270] Adrenergic neurons in the pontine tegmentum and nucleus coeruleus appear necessary for the deep or paradoxical sleep, although a cholinergic system is needed for the muscle atony. REM sleep may restore transmitter in catecholamine systems.[380]

Both types of sleep are reported for a mole,[9] but paradoxical sleep is lacking in an echidna.[8] Birds show the slow-wave type of sleep with brief periods of paradoxical sleep. In a tortoise, the slow-wave sleep state has been observed. Recordings from both tectum and forebrain of a frog over 24-hour periods showed both fast and slow rhythms but no sleep pattern, and the frogs were always responsive.[198] Fish show behavioral "sleep," but no electrical correlates like those of mammals have been observed.[270]

Many animals—frogs, beetles, and others—show tonic immobilization, or death feigning, in response to threatening stimuli. The neurophysiology of this capability has not been elucidated.

DEVELOPMENT OF NEURAL FUNCTIONS

Developmental neurobiology is concerned with two general problems: (1) correlation in embryos of spontaneous movements and responses to stimuli with the development of sensory, central nervous, and effector structures, and (2) elucidation of mechanisms of growth and differentiation of neural structures.

Development of Behavior Patterns

Myogenic activity often precedes neural control of movements, as in many lower vertebrates; however, in other animals (e.g., chick), the earliest movements appear to be neurally controlled. Muscles are, however, capable of contraction prior to innervation.

Marked differences are found in various

groups of animals in rates of development and in whether the earliest movements are random or integrated. Several gradients are recognized: the primary one is from the cephalad end posteriorly, and a secondary one is from proximal to distal effectors.

An example of the transition from myogenicity to neurogenicity is given by the earthworm *Eisenia.* Following the ciliated gastrula stage, the embryo shows spontaneous contractions around the stomodeum; then, in response to mechanical stimulation, both local and conducted contractions occur. After ganglionic organization has become visible, the anterior end turns away from a point of stimulation. Peristaltic responses gradually spread backward until the whole animal is responsive as neural development continues.[428]

In embryos of a dogfish, *Scyliorhinus,* rhythmic contractions of myotomes occur on each side independently before neural connections to them are present, and after removal of the neural cord rhythmic contractions continue; thus, the myotomes are capable of initiation and conduction of contraction. After neural control is established, stimulation elicits contractions on both sides.[188] A similar sequence of myogenic flexures, neurogenic movements, and reflexogenic activity was seen in *Salmo.*[569] A bullhead, *Ameiurus,* shows a developmental sequence of control—spinal cord, hindbrain and labyrinth, midbrain and eye.[14]

Embryos of the salamander *Ambystoma* show the following stages: a nonmotile stage when muscles of the somites contract in response to direct stimulation, a stage of simple flexure of the body in response to tactile stimulation, spontaneous bending in an S-stage, and finally locomotor waves of contraction. The first responses are of gross regions—whole limbs and local reflexes are individuated later. Each advance is related to development of more central synapses.[75] In other amphibians, particularly *Xenopus* and *Rana,* local contractile responses occur prior to innervation. When the neural cord is transected during the myogenic stages, a caudal stimulus can elicit a rostral response; but when both the skin and cord are cut, conduction between the two ends ceases. Epithelial cells give electrical responses to tactile or electrical stimulation. Conduction can occur through a sheet of epithelium and can activate myotomes.[438] At slightly later stages (stage 27 or later) conduction requires the nerve cord. Evidence was presented on p. 461 for electrical coupling between cells of *Loligo* embryos. In 3 to 19 somite chick embryos, electrical coupling occurs between cells of the same tissue, and between different tissues (such as between notochord and neural tube cells, and between notochord and mesoderm).[497] However, electrical coupling between epithelial cells and its possible role in conduction during myogenic stages has not been noted in groups other than amphibians.

In chick embryos at the limb-bud stage, random movements occur, first in the neck region, then in the trunk, and finally in limbs and tail. Recordings from the spinal cord during early development show spontaneous electrical activity, which may be correlated with the random movements. Coordinated reflex control does not occur until after 17 days of incubation. Deafferentation was produced by removal from 2-day embryos of the dorsal half of the lumbar cord and transection of the thoracic cord. Periodic leg movements prior to 17 days were similar in deafferented and control embryos. Normally, responses to sensory stimuli begin on the 8th day.[183, 184, 185] Electromyograms show small irregular activity which increases markedly on days 12 to 15, and the adult pattern is established by day 17. Present evidence indicates that in chick embryos, and probably also in mammalian fetuses, the early random movements are caused by motor discharges from the spinal cord in the absence of sensory input.[494] However, diaphragm muscle in tissue culture can show myogenic rhythmicity.

In a sheep fetus (40 days gestation) the first movements are jerky and are mediated by the bulbospinal system. Later, sustained movements require the midbrain. In macaque fetuses, twitch reflexes of forelimbs without crossed or antagonistic inhibition occur before cervical neurons show myelination, but after a few synapses are found on dendrites of motorneurons.[39, 40]

It is concluded that the myogenic stage seen in elasmobranchs and amphibia may be absent from birds and mammals.

The early development of central nervous connections can occur in the absence of nerve impulses. When salamander larvae were anesthetized for many hours, development proceeded and, on emergence, the reflexes were the same as if the animal had been performing the intermediate stages of movement. Explants of rat spinal cord in culture show spontaneous electrical activity; in xylocaine or high concentration of Mg^{++} presumably all electrical activity is blocked, but growth continues. After removal of the anesthetic block, properties of the multisynaptic net are normal. Thus endogenous impulses are not needed for synaptic development.[82] Neurons develop in early embryos, and some regress. The earliest reflexes occur at the time of axo-dendritic closure; formation of neuronal sheaths occurs late in development, possibly permitting neuron-to-neuron recognition in earlier stages.[40]

Development of brain rhythms correlates with degree of maturity at birth or hatching (p. 670). In guinea pig, brain waves are present at birth but reach maximum frequency at seven days; whereas in a rat the first brain rhythms are recorded at four days after birth and the adult pattern appears at 18 days. In chickens, some brain rhythms are seen at 11 to 13 days incubation, good synchrony is seen at 17 days incubation, and adult patterns appear at seven to nine days after hatching. Pigeons, on the other hand, show no brain waves until the fourth day after hatching.[415] Appearance of complex evoked responses in the cortex of kittens is correlated with development of basal dendrites in pyramidal neurons.[432]

Specification of Neuronal Development

Studies of both regeneration and embryonic development show that growth of neurons and of peripheral structures are reciprocally related. In the deafferented chicks mentioned above, motorneurons degenerated after about 17 days; hence, their continuing development requires sensory input.[185] Taste buds in a mammal do not develop in the absence of sensory nerves. When a metathoracic ganglion of a cockroach is transplanted to the coxa of another roach, only previously denervated muscles become innervated; the motorneurons develop a dense ring of RNA during the activation process.[245]

One of the most interesting questions in developmental neurobiology is how the connections of neurons are specified. Some specification is in the genome, and some is induced during development. In general, sensory neurons are specified early in development, and interneurons somewhat later. Examples of the transition from lack of specification to precise specification of function are found in both sensory and motor neurons.

Patches of skin were exchanged between belly and back of tadpoles and frogs; if a graft was made in a late larva or an adult, stimulation of the skin patch elicited reflexes normal to that skin area before grafting, but if the graft was made in an early tadpole the ensuing reflexes were misdirected for the skin; i.e., the specification of sensory neurons had become changed to correspond to the new location[247] (Fig. 15–36). It is concluded that the sensory neurons instruct the central connections and that the sensory neurons become specified according to their body location during late stages of development.

The axons of retinal ganglion cells terminate in the optic tectum of frog and fish in specific patterns of projection, both spatially and in depth. If the optic nerve is cut, the eye inverted, and the optic axons permitted to regenerate in an adult or a late tadpole, the visual reflexes and electrical responses of the tectum show an inversion of the normal projection pattern; i.e., the ganglion cells are specific to certain regions of the tectum. However, if the transplant is made at an earlier stage, the proper projection for the orientation of the eye is achieved; i.e., the ganglion cell axons were not yet specified[248] (Fig. 15–37). In *Xenopus*, surgical rotation of an eye at stage 28 resulted in a normal retinal projection in both axes; after rotation at stage 30, orientation of the projection was inverted in the nasotemporal axis but not in the dorsoventral axis; at stage 31, the projection was inverted for both axes. Thus, specification in one direction occurs earlier than in the other.[246] In *Ambystoma*, the retinal neurons are specified during stages 34 to 36, while in *Xenopus* this occurs at stage 30. The regions of the two tecta that receive similar spatio-temporal patterns of excitation become interconnected in *Xenopus* tadpoles.[163] When, in an adult frog, the optic nerve fibers are displaced at random, the individual fibers grow back to their previous locations in the tectum.[507] Similar sequences of reacquisition of projection patterns have been observed in goldfish.[600]

In chickens, no defects in visual responses occur after the eye has been rotated and reimplanted before 70 hours incubation. In kittens, closure of the eyelids for a few days during the period from 4 to 6 weeks after birth results in loss of pattern discrimination as recorded from cells of the striate cortex.[229] Kittens were reared in cages surrounded by either vertical or horizontal stripes; after several months the kittens were placed in a

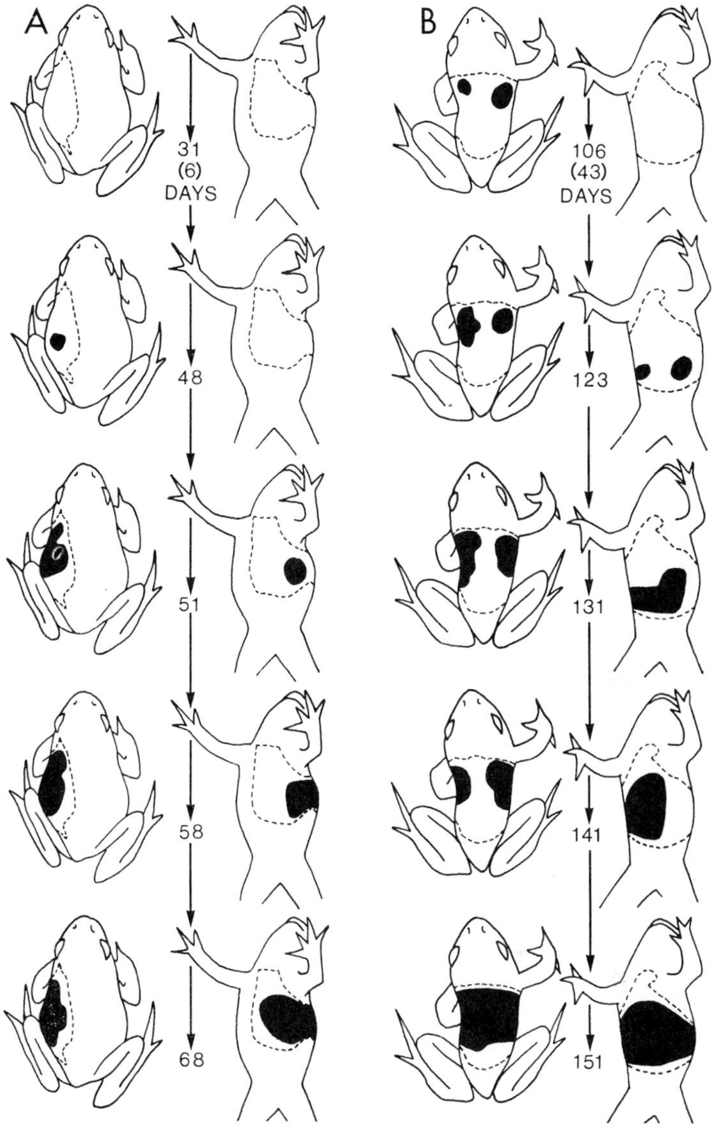

Figure 15–36. Diagrams of frogs with back-to-belly skin grafts, showing areas from which misdirected reflexes are evoked at progressively later times following operation. Misdirected reflexes from black areas, normal from rest of graft. A, Dorsal and ventral views of frog with back-to-belly inverted skin graft made at stage IV and tested at 31, 48, 51, 58, and 68 days later. B, Back-to-belly reversal of trunk skin grafted at stage V and tested 106, 123, 131, 141, and 151 days later. (From Jacobson, M., and R. E. Baker, J. Comp. Neurol. *137*:121–141, 1969.)

complex environment, and they appeared to be blind to stripes perpendicular to the experienced orientation. Binocularly sensitive neurons of their visual cortex showed preference for bars of the experienced orientation.[37, 197] Rats were given varied sensory stimulation daily beginning at birth, and at eight days the number of spines per unit area of pyramidal neurons was significantly greater than in unstimulated rats; no change occurred in cell dimensions.[493] It is concluded that during development sensory neuronal specificity increases, that certain periods may be critical for neuronal connections, and that the nature of specification is in accord with temporal and spatial patterns.

Motor neuron systems can mature without sensory input, but motor organization is normally influenced by limb musculature. More motor column cells develop than are used, and some cell degeneration occurs during central organization; motorneurons degenerate if they are not used, and muscle fibers degenerate if they are not innervated. Within the spinal cord, some topographic representation of motorneurons occurs; distal limb musculature is represented more dorsally in the motor columns than is proximal musculature, and extensor muscles more laterally than flexors.[80] Spinal nerves seek out a transplanted limb for a limited time, and if the transplant is not too far from

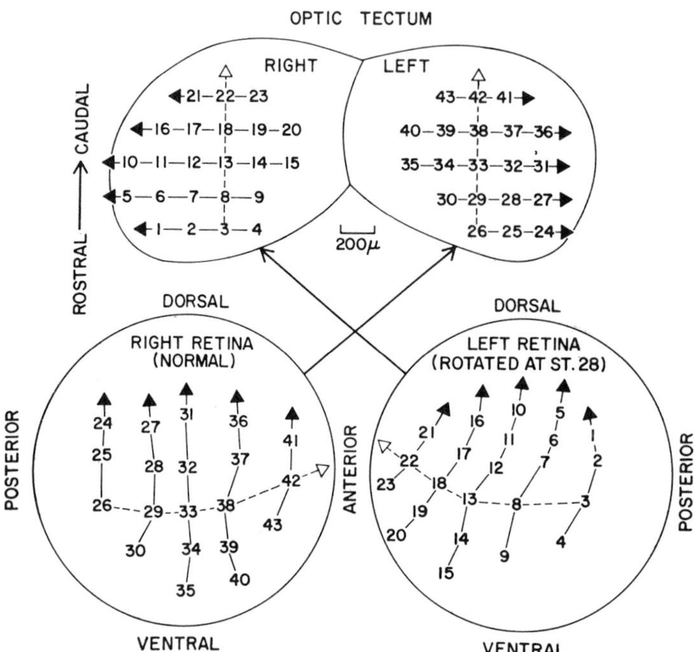

Figure 15-37. Map of retinotectal projection in toad *Xenopus,* to left tectum from normal right eye, and to right tectum from left eye that has been rotated 180 degrees at larval stage 28–29. Projection from inverted eye is normal. Number on tectum represents position where action potential is recorded by microelectrode in response to small spot of light at the position shown by the same number on retina. (From Jacobson, M., Devel. Biol. *17*:202–218, 1968.)

the normal site. For example, when in salamanders a supernumerary foreleg is transplanted in normal orientation to a position near a normal leg, and innervation is established, the corresponding muscles of the two legs contract simultaneously. Each reinnervated muscle of the extra limb contracts synchronously with the corresponding muscle of the normal limb.[558] If a limb is transplanted in a position rotated 180° from normal, the grafted limb moves in temporal coordination but in reverse direction to normal. Similar results have been observed with chickens. It is concluded that there is some myotopic specification of motorneurons.

Central specificity is shown in response to supernumerary appendages in crustaceans. In a crab, an eyestalk was replaced by the outer segments of an antennule. Stimulation of the supernumerary antennule elicited the same reflex responses as did stimulation of the normal antennule. Since new paths were used, the central connections must have been made on the basis of receptor modality.[352] In a cricket, abdominal cerci regenerate and their sensory fibers make normal contacts with giant fibers in the terminal ganglion. If a regenerating cercus is transplanted to a leg stump, the sensory fibers control the giant fiber in the thorax; hence, regenerating sensory fibers recognize central fibers for termination.[117]

What determines the specificity of growing neuron processes is unknown. Hypotheses of thigmotaxis and galvanotaxis appear inadequate, although in tissue culture the growth of nerve processes can be influenced by oriented surfaces and by electrical fields. Long-distance effects of specific chemicals are unlikely.[247] Observations of neurons in tissue culture show that cell processes grow randomly, sometimes apparently testing cells that they pass for appropriateness of contact and then making synapses on specific cells, as if the receptor cells provided membranes of chemical affinity for a given axon. Thus, short-range chemical factors may be very important.

Development of neurons is influenced by many factors. Use and disuse modify the degree of development, and dendrites fail to develop in the absence of the axons that normally terminate on them. There are localized regions on neurons for given types of axonal endings (e.g., dendrites and somata in pyramidal cells). Visual deprivation reduces retinal concentrations of RNA, AchEs, and other critical substances. Nutrition can modify neural development at critical stages, as shown by mental deficiencies in malnourished children, deficiencies that can be related to cortical development and can be reversed up to a certain age. Hormones influence neuronal development; in a tadpole the central connections for a reflex closure of the eye in response to corneal stimulation fail to develop in absence of appropriate levels of thyroxin.[299] Growth of sympathetic neurons and of some sensory neurons is stimulated by a nerve growth factor (NGF), a protein which occurs in some tumors, in salivary glands, and in serum.[324,325] Administration *in vivo* of antibodies to NGF prevents sympathetic maturation.

There is ontogenetic developmental programming of differentiation and degeneration. Mauthner cells in *Xenopus* grow processes and function appropriately in tadpoles; later, during metamorphosis, they atrophy and are absent from adults. Not only do central connections influence the growth of nerve fibers, but it can also be shown that the nature of motorneurons has an influence upon the differentiation of the muscle fibers they innervate. In the young kitten, muscles tend to show intermediate contraction speeds, whereas in the adults they are fast in contraction time and poor in certain mitochondrial enzymes (as, for example, the gastrocnemius muscle) or slow in contraction time and relatively rich in mitochondrial enzymes (e.g., the soleus muscle). Cross-innervation of these muscles carried out early in development results in an altered developmental pattern, so that the soleus muscle develops a faster-than-normal contraction time and the gastrocnemius is slower (p. 752). Motor innervation controls the distribution of acetylcholine-sensitive receptors in vertebrate skeletal muscle; a spread of acetylcholine sensitivity, corresponding to the classically observed "denervation supersensitivity," takes place over the two weeks following section of the motor nerves (see p. 744). Restriction once again occurs when regenerated motor nerves make new connections with the muscle fibers. On the other hand, the muscles are also able to influence the course of differentiation of neurons innervating them. For example, in some crustacean muscles, widely separated terminals from the same axon on the same muscle fiber are very similar, whereas adjacent terminals from the same axon on an immediately neighboring muscle may be very different.

In summary, neurons show a wide range of structural plasticity. Animals in which neuron division persists throughout life are most plastic; in others, the specificity of connections becomes fixed at some developmental stage. All neurons are capable of some changes after damage to an axon occurs; the exact details of reformation processes are uncertain. All developing neurons are subject to induction with regard to connections, which can be by contacts with sense structures, with other neurons, and possibly with muscles. Conversely, growing neurons influence those cells that they contact and can influence protein synthesis in innervated tissues such as muscle.

PERSISTENT CHANGES IN CENTRAL NERVOUS FUNCTION

The events of synaptic transmission, facilitation, and inhibition are transient, lasting at most a few hundred milliseconds. Much of stereotyped behavior can be explained by the switchboard view of the nervous system, yet the most adaptive properties of nervous systems are long-lasting and their analysis is elusive. Among the most important characteristics of central nervous systems are variability and modifiability of responses, alterations which persist for long times—seconds, days, or lifetimes.

Persistent modifications of behavior include effects of use and disuse; short-term habituation and dishabituation; conditioning and learning, which consist of both short-term and long-term memory; and modification of innate behavior patterns. Reviews and textbooks of psychology and ethology present classifications and theoretical models of long-lasting behavioral alterations.[538] The present section considers briefly (1) some examples of behavior modification in diverse groups of animals, (2) evidence for persistent changes in synaptic function, and (3) possible explanations of the first in terms of the second.

CLASSIFICATION OF LASTING CHANGES

Plasticity related to use and disuse includes loss of a skill or of precision in a response after prolonged disuse of a motor system (e.g., decline of a spinal reflex after section of one dorsal root and testing via another); compensatory transfer of function, as from one appendage to another (e.g., change of gait after amputation of one leg in multipedal animals); and changes in specificity of a sensory or motor center during regeneration or development.[495] A related phenomenon is enhanced sensitivity of a muscle to neurotransmitter for a time after denervation (p. 744); this involves induction of synthesis of receptor molecules. Sensitization may be important in the recovery sequence following a lesion of central axons.

Habituation. Habituation has the following characteristics:[537] decrease in magnitude of a response on repetition of a stimulus; spontaneous recovery after the stimulus has been withheld for a time; and decline in responsive state, which may go below zero response. Habituation is greater with frequent and with weak stimuli. Dishabituation is produced by strong stimuli, usually of another sort than those causing habituation. Habituation is distinct from central inhibition, fatigue, and sensory adaptation. Examples of behavioral habituation are found in spinal reflexes of vertebrates, in gill and mantle withdrawal in many molluscs, in orienting and prey-catching in various animals (e.g., spiders and toads), and in mobbing responses to the appearance of a predator (in birds). The components of a complex behavior may habituate at different rates; e.g., in prey-catching spiders the orientation toward prey persists upon repeated presentations longer than do the striking reactions.[426] Many examples of behavioral habituation are given in reference 204.

Conditioning and Extinction. Conditioned behavior is probably of several sorts, involving different mechanisms:

(a) In simple conditioning, a conditioned stimulus (CS) followed by or overlapping the unconditioned stimulus (UCS) leads, on repeated presentation or trials, to a state where the CS alone elicits a response which is virtually the same as that evoked by the unconditioned stimulus. The conditioned response (CR) decays within a short time in the absence of reinforcement. Also, it can be extinguished rapidly by a general or non-specific stimulation.

(b) In Pavlovian conditioning, the CR usually differs from the unconditioned response in having a longer latency, in often being anticipatory, and in persisting for a very long time unless extinguished.

(c) Instrumental conditioning is the basis for maze learning, problem-box solving, and similar tests. A reward is given at the end of solution of the problem, or the responsive action has a reward value in itself. This is similar to trial-and-error learning.

(d) Chains of conditioned reflexes permit shifts from one sensory modality to another. The number of sequential steps that can be added to a chain varies with species.

As a control on conditioning, it is necessary that the UCS be applied alone for an equal number of times as it is with the CS; this may result in sensitization of sensory pathways, sometimes called pseudoconditioning. Similarly, presentation of the UCS prior to the CS must not result in the CR. Much evidence indicates that, at least for vertebrates and octopus, conditioning is by two stages, early and late.[301] The early stage provides for short-term memory, which can be disrupted by certain central lesions or by electroconvulsive shocks to the brain. The sequence of events in most conditioning can be reduced as follows: sensory input (CS), classification or coding at sensory nuclei, consolidation of the classified or coded information, closure or long-term storage, and finally retrieval or readout. Mechanisms for each of these are probably different; hence, to seek a single cellular mechanism for learning approaches absurdity.[251, 252]

Distribution of Capacity for Conditioning

Many Protozoa, especially ciliates, show complex taxic behavior—phototaxes, geotaxes, and galvanotaxes; persistent modifications of taxic behavior have been observed. Paramecia tend to collect around a platinum wire after they have been fed regularly near similar wires.[165]

Planaria have been conditioned in a T-maze with light as CS and shock as UCS to select one arm of the maze. The presence of the cerebral ganglion is necessary for the "learning," but it is claimed that once the habit is learned the worm may be transected and the tail half after regeneration can make the appropriate turn.[258, 356, 357] There have been reports of transfer of some substance from trained to untrained planarians and corresponding transfer of a learned habit, but these observations have not been well confirmed.

Strong objections have been raised by psychologists[256] to characterizing the modified behavior in *Paramecium* and in planaria as true conditioning. A reasoned view by Thorpe is that these changes represent exploratory but not associative learning, that the animals can relate one orthokinetic (directional) movement to another.[538]

In animals with simple nerve nets (coelenterates) and in those with condensed areas of the net (echinoderms), modifications of behavior can occur. Habituation of *Hydra* to mechanical agitation or to illumination is followed by decreased sensitivity for several hours.[473] In sea anemones, some conditioning of the selection of substrate can occur.[458] A starfish can be made to abandon the use of one arm in righting itself if this arm has been restrained a number of times. A starfish, presented with light and food every 48 hours in appropriate sequence, "learns" after a few trials to move downward in an aquarium in response to light alone.[311]

Some annelids show reversal of a response to one sensory stimulus when this is presented with another which normally elicits an opposite response (e.g., reversal of response to light with accompanying tactile stimulation). Nereid worms, which normally respond negatively to illumination or touch, "learn" to come out of their tubes in the light if mussel juice is presented at the same time as the light or touch. *Nereis* in a T-maze continues to learn if all anterior sense organs but one are disconnected from the brain (e.g., if the brain is disconnected from eyes, palps, and antennae but connection remains to tentacular sense organs).[146] Habituation to light can occur in decerebrated nereids, and it thus appears to be the property of all parts of the nervous system.[126] Many investigators have trained earthworms to turn in a specific direction in a T-maze, usually with light or tactile stimulation as a negative UCS. After establishment of the turning, the supraesophageal ganglion can be removed and the conditioned response persists. Earthworms stimulated repeatedly unilaterally by bright light (*negative phototaxis*), dim light (*positive response*), or by electrical or tactile stimulation on one side continued to turn in the "trained" direction in the absence of continued stimulation; this is not due to a change in threshold of positive or negative phototactic response.[303]

Many examples of associative learning, trial-and-error learning, and maze learning have been given for arthropods.[538] Heterogeneous stimuli can release complex patterns of innate behavior. Light has frequently been used as a CS in association with a prod or shock as a UCS. For example, the tail flip of *Limulus* has been conditioned;[501] similar conditioned behavior occurs in crabs, prawns, cockroaches, and other animals. Hymenoptera show complex behavior based on innate stereotyped patterns combined with short-term memory. Bees have been trained to come to specific colors, patterns, or odors associated with feeding sites. The memory persists after a bee has been anesthetized with chloroform or by cold.[436] After training to come for food at a particular time of day, bees were narcotized for four hours; the next day they came to the feeding place both at the trained time and three hours later. When the trained bees were mixed with a guest colony, the visitors became entrained and the first group showed a third peak of visitation.[329, 361] Honeybees inform hive mates concerning distance and whereabouts of food sources by "dance" tempo and direction. They remember aromas.[152] The ant *Formica rufa* can be trained in a maze to follow olfactory cues.

Fish have been conditioned in feeding situations in shuttle boxes and mazes; light of various colors, shape of visual stimuli, and sound have been much used as conditioned stimuli. *Olfactory imprinting* may be important in the homing of migratory fish (p. 563). Goldfish show conditioning of heart or respiration rate to light or tone. If they are conditioned to a light with one eye covered, transfer to the other eye occurs. However, when avoidance behavior is conditioned, no interocular transfer occurs; i.e., both eyes are needed in training for the visual-motor transfer but not for sensory-autonomic transfer.[355] Goldfish from which the forebrain is removed show conditioned heart rate to light as a CS, but they do not learn color discrimination.[33] Also, removal of the goldfish forebrain impairs learning of a shuttle box choice and abolishes the habit previously learned.[181] Conditioned responses are more sensitive to cooling than are swimming reflexes.[430] Goldfish show instrumental learning (e.g., to press a lever for food on a two-minute schedule); after cooling, the timing sense remained unchanged.[417]

Urodele and anuran amphibians have been trained in simple mazes and have been conditioned to visual and auditory conditioned stimuli. Associations are still formed with one side of the forebrain removed, but not with both sides lacking.

Reference is made to extensive accounts elsewhere for examples of conditioned behavior in birds and mammals.[538]

LOCALIZATION OF SITES OF LEARNING

Loss of conditioning following removal of integrative regions of nervous systems has been demonstrated in many animals. It is necessary to exclude sensory and motor losses in the search for location of a "learning center." An excellent example of the use of lesions to analyze complex learning is found in the work of J. Z. Young on the cephalopod mollusc, *Octopus.**

Octopus. An octopus is an intelligent animal that normally makes its home under water in a cave under a pile of rocks, and emerges on feeding forays and to investigate new objects in its territory. Its nervous system is a complicated set of neural centers: the supraesophageal portion contains many sensory relays and several regions of "higher" integrative function, while the subesophageal portion is mainly motor in function.

A normal octopus or cuttlefish, when shown a prawn which then swims out of sight around a corner, will follow the prawn;[563, 565] however, when the verticalis complex of the brain has been removed, the cephalopod attacks a prawn that is in sight but fails to chase one around a corner. In the laboratory, octopi have been conditioned to both visual and tactile stimuli. They readily attack moving objects but can be trained to attack stationary ones. The standard technique is to reward with food an attack on a given object, such as a plastic disc on which a pattern is painted for positive conditioning, or to punish with a shock for negative conditioning. Learning is fast, and by a combination of reward and punishment an octopus can learn to distinguish between two patterns in a few trials. Visual discrimination is better for horizontal than for vertical figures.

The octopus retina contains an array of rhabdomes (visual elements), each with four cells or rhabdomeres (Ch. 14). The axons of the photosensitive cells branch profusely in the outer or plexiform layer of the optic lobe. Each optic lobe has some 6×10^7 neurons. The dendritic fields in the deep plexiform layer of the lobe are oval; some are oriented vertically, and others horizontally. One such dendritic field may have a diameter of more than 500 μm. Deeper neurons have wide conical dendritic fields which receive from the outer cells. Sutherland[524] has evidence that the retinal receptor cells are in an array of horizontal rows and vertical columns. He proposes that the optic lobe codes the ratio of vertical to horizontal elements stimulated by a given pattern, and he relates this coding to the orientation of the dendritic fields in the outer plexiform layer. By testing the ability of the octopus to discriminate different patterns, he finds that such a coding of horizontal to vertical ratios, plus some area perception, can explain form discrimination. Furthermore, when statocysts are removed, the eyes may rotate so that the slit pupil is vertical instead of horizontal; vertical objects are now treated as if they were horizontal. It is proposed that each neuron in the deeper plexiform layer of the optic lobes responds to a particular ratio of horizontal to vertical signals and that this response pattern for one cell is determined during development. The optic lobes send efferent fibers direct to the motor centers and thus elicit attacks.

Tactile learning results from stimulation of sense cells in the arms and is best studied in blinded animals. Tactile discrimination of plastic cylinders with different degrees of roughness or grooving is easily demonstrated by conditioning techniques. There is no discrimination of objects that are similar in size and texture but different in weight. Apparently, tactile discrimination is on the basis of ratio of smooth to rough surfaces as well as by size of object. The sensory neurons from the arms connect to the inferior frontal and subfrontal lobes, removal of which stops discrimination of objects mechanically. By chemical sense an octopus can distinguish between a clam shell containing a live clam and one containing wax.

The superior frontal and the vertical lobes (Fig. 15–38) seem to preserve representations received from either optic (eyes) or subfrontal (arms) lobes. Electrical stimulation of superior and vertical lobes does not elicit any direct motor reactions. Axons from the superior frontal lobe pass to the vertical lobe, which is divided into five lobules. The output of the vertical lobes is via the subvertical lobe to optic and superior and inferior frontal lobes, i.e., back to the regions from which the input came. Thus, for both visual and chemotactile systems, there are two parallel loops.

Removal of the vertical lobes results in no

*See references 603, 604, 605, 607, 608, 610, 612, and 613.

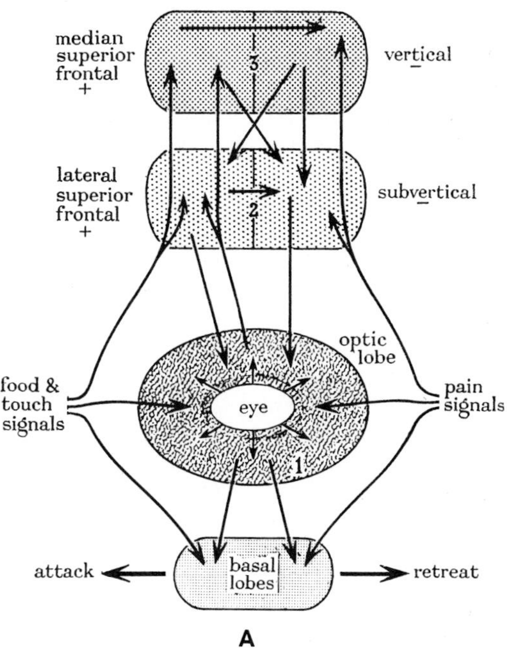

A

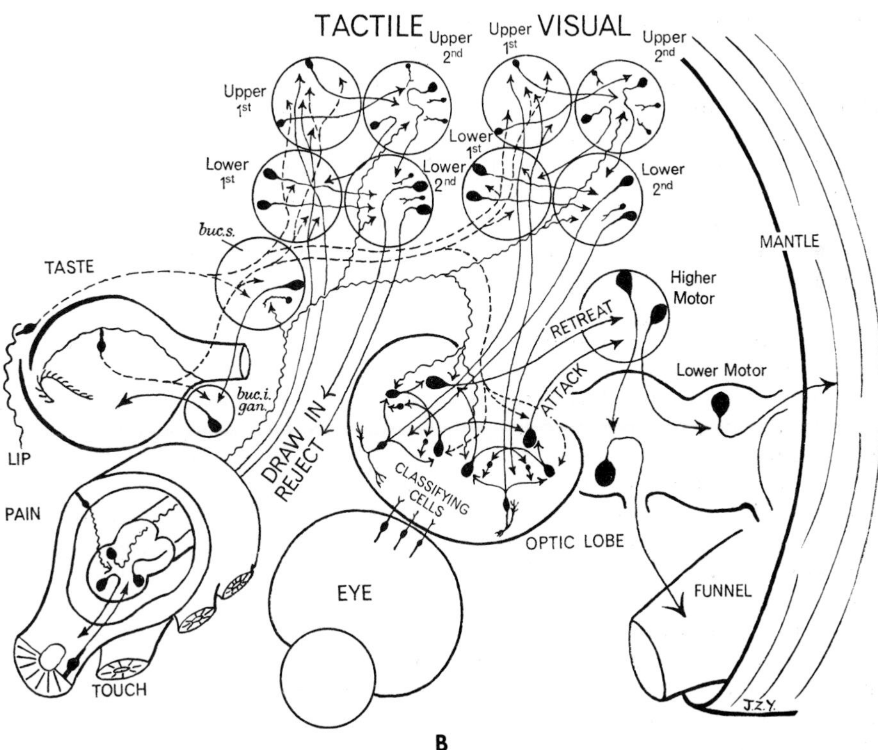

B

Figure 15–38. A, Diagram of centers participating in control of locomotor activity in *Octopus* brain. 1, 2, and 3 are the three levels at which attack and retreat are controlled in response to a visual stimulus. (From Young, J. Z., Proc. Roy. Soc. Lond. B *159*:565–588, 1964.) B, Diagram of two sets of paired centers in visual and tactile memory systems of *Octopus*. (From Young, J. Z., Proc. Roy. Soc. Lond. B *163*:285–320, 1965.)

apparent motor or sensory defects. However, the octopus may make inappropriate responses. If previously trained, there is some conditioned response. If the octopus has been trained to attack an object and then has been subjected to reverse training (i.e., not to attack), removal of the vertical lobe causes reversion to the first pattern (to attack). Interocular transfer of learning discrimination is stopped by verticalis removal even though there are fibers connecting the two optic lobes. After removal of the vertical lobes, learning is possible but it is slower and more mistakes are made than normally. Learning capacity is reduced by 75 per cent.[566] Most important is the fact that the pattern learned without the verticalis lobe persists a short time—only a few hours—whereas normal memory may last for days or weeks. The defect in learning is proportional to the mass of tissue removed from the vertical lobe.

Circuits exist from optic and inferior frontal lobes to superior frontal, thence to vertical, then to subvertical and back to optic and inferior frontal. Interruption of this circuit changes memory capacity. Young's hypothesis is as follows: There is extraction of information and reduction of redundancy in the two plexiform layers according to a two-letter code (vertical and horizontal rows). Probably a similar codification occurs for tactile signals according to a rough-smooth code in inferior frontal lobe. Then, in the superior frontal lobe, a redistribution and mixing of channels occurs; in the vertical lobes there is a great increase in number of parallel channels; from this lobe, transmission to the optic (or subfrontal) occurs by a smaller number of channels which end in these lobes with a two-letter motor code—to attack or not to attack. It is proposed that the initial codification occurs within the optic or subfrontal lobes and that short-term memory can occur there, but that the long-term memory occurs in the verticalis lobe. Codification occurs in an oriented but elaborate dendritic system of cells which are in some way set for particular input/output ratios. Long-term storage occurs in a rich neuropil where the number of channels is very great.[611]

Arthropods. Exclusion of sensory lobes indicates that the neuropil of many arthropods may be capable of "conditioning." Best examples are known for insects.

An ant fails to learn a maze by olfactory cues if antennae are removed or if a cut is made between the antennary lobe and the alpha-lobe of the corpora pedunculata of the brain. The olfactory center is connected to motor regions, but a loop involving the alpha-lobe seems necessary for closure of the conditioned response.[552]

A cockroach normally holds a chemically stimulated antenna with the contralateral foreleg for cleaning; if both forelegs are amputated, the cockroach learns in a week to use the middle legs and a short-delay path is established between the esophageal connective and the motor nerve of the hind leg,[333, 334, 335] with the ganglionic delay dropping from 4 to 2 msec.

A simple preparation for "conditioning" is found in shock avoidance training of an isolated prothoracic ganglion of a cockroach. One tarsus receives a shock each time the leg (P) is extended, while the other leg (R) receives an equal number of shocks at random. After a series of pairings, the first leg (P) is held in a flexed position; this habit may be retained for as much as 24 hours. Tonic excitatory impulses from the ganglion to leg muscles increase in frequency after the conditioning, and the response can be extinguished by ganglionic shocks.[121, 122, 211, 224]

Vertebrates. In fishes the mesencephalon can mediate conditioned responses. When a goldfish had been trained to a visual stimulus with one eye covered and was then tested with the opposite eye covered, transfer from the tectum corresponding to the trained eye to the opposite tectum was demonstrated; cutting the intertectal commissure blocked the interocular transfer and thus demonstrated that the learned engram was in the tectum.[346] Goldfish were conditioned to a visual CS with food as UCS, and then to the olfactory stimulus of amyl acetate as a second order CS with the visual stimulus as the reward. When part of the tectum was removed, the second order learning was upset.[482] Interocular transfer occurred when heart rate was used as CR and light as CS; after tectal lesions were made, the heart rate could not be conditioned.[355] The role of the forebrain in conditioning in fish seems to vary with species. In *Tilapia,* ablation of the forebrain after avoidance training to light resulted in variable slow conditioned responses; ablation of the forebrain before training reduced learning capacity.[285] However, goldfish without forebrain learned a two-way avoidance task as well as did normal goldfish.[101] It is concluded that the tectum (probably with associated tegmentum) is capable of mediating conditioned behavior.

A bird has a very small cerebral cortex lying medial and posterior to the large striatum, which consists of Wulst (hyperstriatum) and neostriatum. Birds can be trained to complex tasks, such as counting and memory of patterns. Removal of the cortex results in no impairment, but damage to any part of the striatum causes marked deficit in learned behavior.[514]

For mammals there have been claims and denials of conditioning of spinal cord reflexes; there is no doubt that the cord can show habituation and dishabituation. Some subcortical structures permit conditioning of autonomic responses to sound or light. After removal of visual cortex from cat, responses to total flux of light, but not to visual patterns, could be conditioned. Recent emphasis has been on the role of the prefrontal cortex, hippocampus, and temporal lobe in relation to memory.[4] There is a reciprocal relation between hippocampus and subthalamus. Rats with subthalamic lesions show a learning defect and fail to show the typical electrical changes in hippocampus waves on approach to a goal.[4] Rats with lesions in the hippocampus or fornix show lower learning ability in a T-maze than normal; if rats are trained and then a lesion is made in the hippocampus, retention and relearning are impaired.[271]

If the corpus callosum that connects the left and right lobes of the cortex is cut, an animal can learn a visual conditioned response. If the learning is by one eye and testing is done with the other, the response is unimpaired in normal animals (cats) but not in those with the corpus callosum cut. Transection after training shows that a memory trace is transferred between the hemispheres only during the learning period.[55] In addition, some subcortical crossing occurs for somatosensory input.[85]

The cerebral cortex is distinguished by some equipotentiality of function. Although there is point-for-point projection of the retina on the visual cortex, deficiencies in learned visual reactions occur in rats in proportion to the amount of cortex removed, largely irrespective of location. Auditory learning is still possible for dogs when any one of the three auditory regions is intact. Such observations led to the concept of "mass action" in the cerebral cortex.[316] Recent observations emphasize the extensive cortico-cortical connections, particularly in the prestriate area, and the fact that some frontal lesions permanently inferfere with associative learning while discrimination remains normal. A cat is unable to learn sound-light sequences after ablation of the two lateral frontal and prefrontal association areas, although it can still learn after lesions of the sensorimotor cortex have been made.[261, 536] Thus, there is localization as well as several areas serving the same function, and some regions of the cortex can substitute for others.

Not only can all sensory inputs be used as conditioning stimuli, but electrical stimulation of the brain is effective. Rats have been conditioned to give foot flexion in response to brain stimulation on the contralateral side.[476] Self-stimulation by inplanted electrodes in the hippocampus and stimulation in the median forebrain bundle can serve as reinforcers much as does food in many tests.[400, 401] Stimuli applied in various parts of the cerebral cortex can serve as CS for cortical conditioned responses.[108]

Autonomic responses which do not ordinarily require cortical pathways but which can be modified by the cortex (e.g., eyelid reactions, pupillary responses, psychogalvanic reflexes, heart rate, and blood pressure) can be conditioned. The degree to which such conditioning occurs in subcortical structures is not clear. Various autonomic responses have been conditioned using stimulation of the medial forebrain bundle as the UCS (reward) and some level of action such as heart rate or blood pressure as the CS.[371] These functions are conditioned as well or better in curarized than in normal rats, hence voluntary movements are not essential for the conditioning.

Persistent Synaptic Events

Post-Tetanic Potentiation (PTP). Postsynaptic responsiveness may be enhanced for some seconds following a repetitive presynaptic stimulation; this enhancement has been noted at a variety of neuromuscular and synaptic junctions. It occurs in curarized muscles or sympathetic ganglia where there may be increased amplitude of junction potentials and quantal liberation of transmitter. In a phrenic nerve-diaphragm preparation, transmitter depletion and increased liberation may give a diphasic curve of response. At a neuromuscular junction in frog sartorius, post-tetanic potentiation depends on the total number of nerve impulses; when external calcium is reduced, the PTP is earlier and briefer.[454] Post-tetanic potentiation in sympathetic ganglia and in monosynaptic spinal reflexes may last from a few seconds to several minutes.[110] In the chicken ciliary ganglion, potentiation occurs for the chemically transmitted synaptic response but not for the electrical transmission.[347] In a crustacean muscle, repetitive stimulation results in increased frequency of miniature epsp's.

In monosynaptic reflexes, the epsp's of motorneurons are enhanced; polysynaptic reflexes have greater amplitude when tested after a tetanus. Stretch reflex amplitude reaches a peak 5 to 30 seconds following an orthodromic tetanus. Tetanization of one dorsal root causes enhancement of polysynaptic reflex discharge elicited by a test volley in the same or nearby dorsal roots; antidromic tetanization has no potentiating effect. Post-tetanic potentiation of inhibition also occurs.[594]

One hypothesis for PTP is that Ca^{++} is accumulated by presynaptic endings during the tetanization and that the effect is proportional to the probability of transmitter release.[454] Another view is that PTP results from hyperpolarization of presynaptic terminals, and hence from an increase of synaptic current.[110, 557]

Habituation and Dishabituation; Heterosynaptic Facilitation. Electrophysiological evidence indicates that, in many nerve centers, repetition of a sensory input leads to a decline in central response. Sensory adaptation is excluded because habituation occurs at low frequency presentation; also, it occurs when sensory nerves are stimulated directly. Similarly, decline in muscle responsiveness can be excluded, as can central fatigue in respect to metabolic failure.

The rate of decline of a response during habituation is a complex function of number of stimuli, frequency of stimulation and intensity of the stimulation. If, after habituation, the sensory stimulation is stopped, a gradual spontaneous recovery of responsiveness occurs, and rapid dishabituation can be initiated by a generalized stimulus or by changing the location or characteristic of stimulus (e.g., to other tactile or visual sites). The rapid recovery in dishabituation shows that the phenomenon of habituation is not one of fatigue.

Habituation in vertebrates. In the spinal cord, reduction in amplitude of both excitatory and inhibitory junction potentials in appropriate interneurons, as well as changes in interneuron spike frequency, accompanies habituation. Motorneuron epsp's in polysynaptic reflexes show a decrease in amplitude during habituation and an increase during dishabituation; monosynaptic reflexes show little or no habituation.[506] In a cat spinal cord, habituation is maximal for one second of afferent stimulation at 50/sec. Bursts of spikes in interneurons show habituation, and habituation of a polysynaptic reflex may result in potentiation of a monosynaptic reflex; hence, the action is not directly on the motorneurons.[570] Recovery from habituation of a reflex to half normal amplitude requires some six minutes.

In a sea lamprey, *Petromyzon,* sensory neurons for receptors to touch, pressure, or noxious stimuli lie dorsally within the spinal cord. Movement of the dorsal fin is elicited reflexly by stimulation of skin, and the response habituates on repetition. Correspondingly, epsp's in motorneurons decline; the locus of habituation and dishabituation for the reflex appears to be at the synapses between sensory and motorneurons.[36]

Habituation of evoked sensory potentials occurs, for example, in tectal responses to visual input. Single neurons of goldfish optic tectum show habituation to repeated stimulation, and dishabituation following a shock to the flanks.[398] Other units habituate to a stimulus moving in one part of the visual field, and then respond to the same stimulus in another part of the field.[525] Evoked potential responses in the cerebral cortex can be reduced on repetition; evoked responses in unanesthetized animals are reduced if the attention of the animal is diverted. This means that response to one sensory modality can be suppressed by stimulation of another.[69, 125] Evoked responses from primary auditory cortex to tones heard at 2-minute intervals showed habituation after 30 presentations.[293] The cortical EEG shows an arousal response of desynchronization following sound or visual stimulation (p. 670). The arousal response can be habituated by repetition.[495]

In higher vertebrates, habituation occurs primarily in polysynaptic systems. Dishabituation is not merely neutralization of habituation, because the response can rise above the control level; it is a superimposed facilitation and represents generalization across several input channels.

HABITUATION IN INVERTEBRATES. Synaptic habituation has been observed in a variety of distributive ganglia. Whether this is identical with behavioral habituation is uncertain, but many of the parameters of the two are similar.

In a squid stellate ganglion, epsp's and spike initiation decline when 0.5 sec trains of presynaptic impulses are repeated at 10 to 15 second intervals, but not at 25 to 40 second intervals. Recovery from the habituation occurs spontaneously after a rest.[206] In a locust brain, single neurons of the tritocerebrum which respond to movement of objects in the contralateral visual field show habituation on repeated stimulation.[463, 464] Habituation occurs even at several minute intervals between stimuli, and stimulation of the neck connectives can cause dishabituation of the response of the tritocerebral neurons.[467, 468]

In a crayfish, the flexion response of the abdomen habituates, and the site of decline is at the sensory interneurons,[597, 618] not at the giant fibers or motorneurons. A crayfish neuromuscular junction potential shows habituation to low-frequency stimulation without decline in muscle membrane conductance; hence, the decline is presynaptic.[44]

Simple preparations for the study of habituation-dishabituation are found in molluscan nervous systems. In gastropods and pelecypods, gill or mantle withdrawal in response to tactile stimulation habituates, as does valve closure in response to photic stimulation.[422] Intracellular recordings from large neurons in *Aplysia* ganglia show habituation of epsp's on repetitive stimulation of appropriate sensory inputs at low frequencies[45, 68] (Fig. 15–39). No change in postsynaptic membrane resistance occurs; hence, the site of habituation is probably presynaptic, a reduction in transmitter release.[307, 308] Habituation is distinct from long-lasting inhibition (ILD) (p. 475). Dishabituation occurs when a different or general sensory input is stimulated after habituation to a restricted input. The dishabituation may be either a form of heterosynaptic facilitation or a post-tetanic potentiation.[279] Habituation can be obtained with an isolated ganglion and attached skin.[308] Also, habituation of an isolated gill can occur in response to repeated touch; the effect lasts for two hours, so the peripheral nerve net, as well as central ganglia, is capable of habituation.[414]

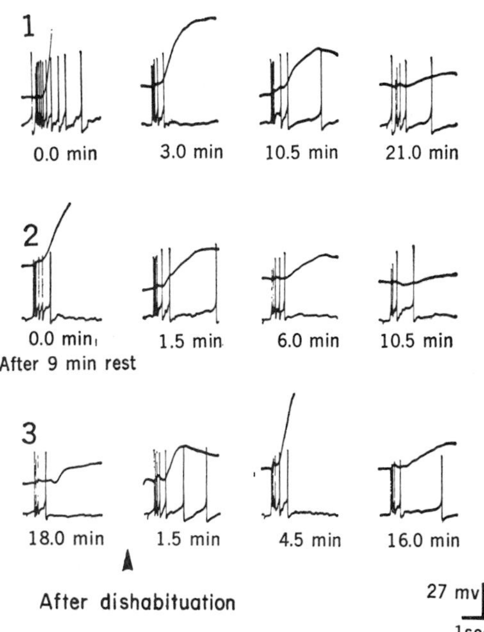

Figure 15–39. Habituation of gill withdrawal (solid line) and corresponding motorneuron spikes during repeated sensory stimulation for number of minutes indicated. Recovery after rest, second line; recovery after dishabituation, bottom line. (From Kupfermann, I., V. Castelucci, H. Pinsker, and E. Kandel, Science *167*:1743–1745, 1970. Copyright 1970 by The American Association for the Advancement of Science.)

HETEROSYNAPTIC FACILITATION. A related phenomenon is heterosynaptic facilitation, known best from work on *Aplysia* ganglia. When, for certain *Aplysia* interneurons, one test input which normally gives a small epsp is paired with a priming input which gives spikes, in such a sequence that the test precedes the primer for 9 or 10 pairings, the response to the test is "conditioned" to give larger epsp's. Similarly, identified giant neurons show enhanced responses to one input when this is paired with tetanic stimuli to various other inputs. The facilitation may last for about 10 minutes, and it appears to be *presynaptic* in location since the effect on epsp's can occur after postsynaptic blocking, and since no change occurs in postsynaptic membrane resistance. The input path may or may not be a specific one according to the facilitated path; the effect is longer-lasting than PTP, it occurs by different input paths, and it is not produced by collateral feedback. Behavioral heterosynaptic facilitation is found for gill withdrawal and other reflexes.[278, 281, 308]

Electrical Concomitants of Conditioning

Enhancement of epsp's in *Aplysia* neurons during heterosynaptic facilitation was mentioned previously, as was conditioning of desynchronization of alpha waves in mammals. Evoked potential responses in the visual cortex of cat showed increased amplitude and decreased latency after visual conditioning.[149] Goldfish conditioned to flashes of light showed new long-latency evoked potential responses in the optic tectum. This conditioned late potential was more sensitive to cooling than was the initial direct response to light, suggesting the establishment of new connections by the conditioning.[431] In unanesthetized rats, acoustic or somesthetic stimuli elicit evoked potentials in cortex and tegmentum, which can be enhanced by conditioning with a shock to cortex or tegmentum as UCS; also evoked responses in cortex and tegmentum habituate.[69] Some cortical neurons show increases, and others show decreases, in frequency of discharge after conditioning.[251, 252] Increments of increase or decrease in frequency of units after conditioning with a tone were found in several parts of the brain (most in hippocampus, and some in dorsal and ventral reticular regions); these were associated with short-term memory.[400, 401] Conditioned changes in spike discharges and behavior can occur without apparent changes in the EEG.

In the motor cortex of the cat, single units showed a decrease in spike responses to flashes on conditioning, whereas random presentations of CS and UCS gave slight increases in frequency.[399] Precentral (sensory) neurons increase in frequency when conditioning is by light, with food as the UCS.[137] Spike frequency in several parts of the brain of rats increased in response to light as CS after conditioning.[56] The percentage of units responding varies with brain region but is, in general, less than half of the neurons tested. Certain thalamic neurons initially gave no response to light, but after pairing with sciatic nerve stimulation as the UCS, responses appeared to the flash. These responses were extinguishable; some units accelerated and others decelerated.[275] Similar frequency modulation of motorneurons occurs in conditioning of cockroach ganglia.[222]

The early stages of visual and auditory conditioning in mammals seem to be due to arousal reactions and alpha-wave desynchronization, probably involving non-specific paths mainly in the reticular system.[259, 260, 381] The alpha brain waves can be reduced in amplitude, desynchronized, and replaced by fast waves; at first, these effects may be in widespread areas, but later they become more localized. The EEG can show conditioning without behavioral correlates.

In addition to changes in evoked potentials and brain waves and in spike frequency of some neurons, direct current or slow potential (SP) changes involving large areas of cerebral cortex have been recorded. These very slow potentials can be either positive or negative and can be modified by cortical state (e.g., in spreading depression induced by locally applied KCl). Changes in the slow potentials of the cortex may be induced by reticular stimulation; pairing of a tone or light stimulus with reticular shock leads to slow changes in response to the tone or light alone.[70, 168, 381, 469, 470] Also, application of direct current polarization to the cortex can modify the responsiveness of cortical neurons.[56] The origin of the very slow potentials in the cortex is unclear, but they presumably represent activity in dendrites.

It is evident that many electrical signs of conditioning have been identified. Whether even the synaptic potential changes are direct signs of memory storage is debatable.

Chemical Changes in Neural Modification

Synthesis of Specific RNA and Protein; Sensitivity to Transmitters. It has been proposed by Hyden[239] that neurons code for particular memory patterns by specific *m*RNA's. Techniques were developed for isolating and analyzing single neurons from critical regions of rat brains. Analyses were made of neurons from Deiter's nucleus of rats that had learned a balancing task, and it was found that the ratio adenine/uracil was increased as compared with controls. Cortical neurons were analyzed for RNA in relation to handedness. When the rats were forced to change handedness, the RNA of cortical neurons increased from 22 to 27 $\mu\mu$g and the ratio G + C/A + U decreased. RNA also increased in concentration in glia. Rats treated with 8-azaguanine, which blocks *m*RNA synthesis, show impaired learning. Goldfish trained to swim in an abnormal posture with an attached float showed changes in the U:C ratio of brain RNA.[496] It was concluded that new RNA is produced during habit acquisition and that this may code for a particular engram.

Evidence for involvement of protein synthesis in conditioning has been obtained by injection of blocking agents. Puromycin, injected into the hippocampus of mice, prevented establishment of short-term memory (less than 9 hours) in a Y-maze, whereas in neocortex it impaired long-term memory consolidation.[22, 143, 144, 145, 442] Acetoxycycloheximide (ACXH) is a more effective blocker of protein synthesis than puromycin, yet it is said to have little effect on memory in rats; in goldfish, however, it is effective. Recent experiments[88] show that injection of ACXH into the hippocampus of rats shortly before Y-maze training causes no impairment at 3 hours after learning, but marked impairment at 1 and 7 days. Injection after the training was ineffective. Thus, it is postulated that short-term memory does not require new protein but that long-term memory depends on it.[88]

In goldfish trained in a shuttle box, electroconvulsive shock (ECS) given 90 minutes after training (or puromycin given at 30 minutes) results in a deficit in learning; cooling prolongs the time of effectiveness of the ECS.[6, 92, 93, 94] Puromycin is effective only if injected within three hours after the training trial; hence, consolidation is thought to occur at that time. In goldfish, ACXH given immediately after training in a shuttle box blocks both short- and long-term memory, but it has no effect when injected six or more hours after training.

In the preceding experiments, secondary effects of the drug injections were, to some extent, controlled by injections in other regions of the body. It is concluded from these and other experiments that RNA and protein synthesis are necessary but not fully sufficient for learning and that the synthesis of these substances is state-dependent.[453]

Other chemical modifications concern chemical transmitters, specifically acetylcholine. Rats, trained in a Y-maze, were injected with anticholinesterases (DFP or physostigmine) at various times relative to training; in general, the injected group performed better than controls for as long as 14 days. Similarly, injection of the ACh-like drug carbachol blocked a habit learned seven days earlier. It is postulated that a postsynaptic membrane becomes more sensitive to acetylcholine during a period of consolidation, and that later the sensitivity declines.[100] Cockroach ganglia may show a decrease in AChEs and glutamate decarboxylase content after conditioning.[402]

All of the preceding chemical changes are interesting accompaniments of conditioning; there is no proof that they represent the storage mechanisms. Their changes in level may reflect effects on arousal, attention, and motor performance. The chemical changes may also underlie alterations in synaptic membrane properties.

Structural Changes

A hypothesis of long standing but without firm support regarding a mechanism of learning is that new interneuronal connections are formed. The destinations of growing neurons are well specified early in development, but less so or not at all according to animal at later stages (see p. 694). When an axon is cut, protein synthesis in the cell soma is stimulated, much new RNA is formed, and some reorganization of processes occurs. In cockroach ganglia, each large neuron has a characteristic pattern of dendrites; when an axon is cut or when colchicine is administered, a perinuclear ring of RNA increases to a maximum and then disappears at 10 days.[77] The neuron soma normally fails to spike on depolarization or antidromic stimulation, but at the time of the RNA ring it gives large spikes. Thus, a close relation exists between regenerative state and excitability.[423]

When mammals are reared in an enriched environment—with patterned illuminated backgrounds, toys, companions, and sounds—they show marked increase in thickness and weight of cerebral cortex (especially in the visual area) as compared with others in a deprived environment. Dendrites of pyramidal neurons have more dendritic spines, synaptic junctions are 50% larger in cross section, and the number of synapses is low; total AChE increases, but its concentration per gram of cortex decreases. Rats reared outdoors in relatively natural environments showed even more cortical development than those in enrichment cages.[28, 455, 456] Effects of visual occlusion were mentioned on page 688, and visual stimulation *per se* is part of environmental enrichment.

Use and disuse result in changes in abundance of synaptic vesicles. Possibly synapses remain structurally modifiable even though a neuron as a whole is relatively fixed.[81] Changes in properties of visual neurons in the cortex of cats according to developmental experience were given on p. 688.

In summary, long-term structural changes can occur in neurons under different conditions of usage. Whether these have a relation to learning is uncertain.

Overview Regarding Development and Persistent Changes

The comparative study of nervous systems emphasizes that generalization from a single hypothesis of neural plasticity to all learning is futile. The following enumeration states some of the persistent neural phenomena that are recognized.

First, neurons are chemically very active cells. Growing neurons are directed by chemical fields to make connections which must be genetically determined. All neurons are actively synthesizing the lipoproteins of their membranes, as well as transmitters, and are transporting complex molecules, especially proteins, by protoplasmic flow from soma to process terminals; all this metabolic activity is under nuclear control. Chemical activity of nerve cells is altered when their processes are cut, or by use or disuse in transmission. Also, certain neurons increase their dendritic branching and synaptic structure when "exercised." Thus growth, various biochemical syntheses, and structural connections may be modified with experience.

On a shorter time scale, synaptic transmission is modifiable. Post-tetanic potentiation may reflect uptake of calcium or presynaptic hyperpolarization, and thus enhanced presynaptic efficacy. Habituation and dishabituation appear to be presynaptic events at interneurons. Heterosynaptic facilitation may result from changes in balance between inhibitory and facilitatory endings. Some conditioning may be by modulation of inherent rhythmicity. These changes occur, to different degrees, in relatively simple nervous systems. Apparently, all nervous systems (nerve nets, ganglia, and cortical structures) are capable of modification by experience.

However, it is doubtful whether the modifications due to growth and degeneration or to synaptic transmission changes in simple monosynaptic or polysynaptic chains can explain learning in complex nervous systems such as those of cephalopods and vertebrates. In such nervous sytems, short-term and long-term conditioning can be distinguished. Also, these nervous systems have hierarchies of parallel paths, possibly with those nearest the sensory inputs for short-term learning and those farther away for long-term memory and command. The analogy of learning to the coding of a sensory input, its storage as on a "tape," and its retrieval on appropriate demand is an oversimplification. It is sharply debated whether the "higher" functions of vertebrate brains are entirely by nerve impulses and synaptic transmission or whether field effects play a role in integration. There is some equipotency of brain regions, much synchronization of electrical responses in different regions, and some evidence for communication among dendrites in cortical structures.

Many concomitants of conditioning have been identified. Some types of conditioning seem to involve protein synthesis, presumably RNA-coded. Changes in sensitivity to transmitters are indicated. Since postsynaptic membranes are heterogeneous, at least for excitation and inhibition, one type of coding may be by variation in the proportion of the patches on the membranes of different postsynaptic sensitivities. A statistical approach seems more hopeful for complex nervous systems than analysis based entirely on unit specificities.

CONCLUSIONS

Major advances have been made in recent years in understanding conduction of nerve impulses and transmission at synapses (Ch. 11). To what extent is neurobiology achieving its goal of providing a description of mechanisms of behavior? Nervous systems and subsystems form a series of levels of integration, a continuum from simple to complex but susceptible to recognition of the different levels.

The simplest nerve centers are for one-to-one distribution, usually with convergence or divergence of input-output. Some such junctions are electrical, others chemical in transmission. Examples are found in giant fiber systems which provide for rapid behavioral responses, often with synchrony of wide muscle fields. One-to-one transmission occurs from sensory neuron to giant fiber, between giant fibers, and from giant fiber to motorneuron. In other nerve centers, electrical coupling between interneurons provides for much synchrony of motor activation (e.g., in the centers controlling discharge of electric organs). At chemical synapses of distributing centers, persistent effects can occur in the form of post-tetanic potentiation. Transmitter release may be triggered by either pulsed impulses or graded potentials, and the postsynaptic action is usually depolarizing (but can be hyperpolarizing).

A simple level of integration, as contrasted to one-to-one transmission, is that mode of central nervous activity in which output is not equal or directly proportional to input, and pulse-coded (all-or-none) input impulses are converted to analog (graded) signals, sorted out, and reconverted to pulse-coded signals. Output may be amplified by facilitation or may be reduced by inhibition; i.e., excitatory inputs may either sum or multiply, and inhibitory inputs either may reduce the action of an excitatory input (presynaptic inhibition) or may reduce the responsiveness of the output neurons (postsynaptic inhibition). In addition, there may be afterdischarge, modulation of spontaneity, and post-inhibitory rebound. Junctions at the facilitating-inhibiting level of integration may be subject to some long-lasting modification such as habituation and dishabituation. Habituation occurs at monsynaptic pathways in some *Aplysia* neurons, where its action is likely due to a decrease of presynaptic efficiency; habituation also occurs in polysynaptic pathways via interneurons in vertebrate spinal cord. This facilitating-inhibiting level of integration is found in the most primitive nervous systems—nerve nets—and persists in subsystems of all central nervous systems. It is well described for spinal reflexes of vertebrates and for locomotor responses of ganglionic chains in many invertebrates. It normally involves chemical synapses.

The next level in integrative control takes the form of switching and selecting between multiple possible inputs and outputs. This may occur by (a) programmed connections within a center, or (b) commands imposed by "higher" or specialized command neurons. The neural programs are determined genetically and developmentally, and once a patterned sequence is started by an appropriate input it is read to termination. Examples of motor programs are the patterns of insect flight or fish swimming. A related sort of coding is in sensory interneurons which respond in a specific way, such as the auditory interneurons of some insects, or in those which respond to specific components of the sensory input, such as visual interneurons in crayfish or interneurons in fish optic tectum or in mammalian lateral geniculate. The motor output of most such programmed systems consists of rhythmic discharges; the timing of the oscillation may reside (1) in a network, as in a respiratory center, or (2) in a single rhythmic neuron, as in the rhythmic control of some crustacean appendages (swimmerets). Command neurons may occur within a programmed center and may be subject to a variety of inputs. How "decisions" are made by these neurons is not clear. However, it is at this level of integration that neurophysiology is closest to a complete description of some behavioral sequences (for example, in initiation, continuation, and termination of swimmeret movements of decapod crustaceans and in flight of locusts and walking of cockroaches). In these processes the role of each neuron can be specified.

The next level of integration is modulation of a reflex center by "higher control" levels. The mechanisms of these controls are poorly known, but they usually take place via neural regions, stimulation of which does not elicit any overt response. For example, regions of the supraesophageal ganglion of arthropods are inhibitory to the thoracic motor centers. The cerebellum of vertebrates monitors incoming messages from sense organs to midbrain and forebrain, as well as outgoing signals to motor regions of brainstem and cord. In the brain mass of *Octopus*, several levels of monitoring are arranged in a hierarchy of control, from simple responses to complex decisions to attack or escape. Related to monitoring functions are neural regions which initiate programs in lower centers. In a cricket, certain regions of the mushroom bodies of the "brain" trigger specific song patterns. Motor cortex in a mammal and motor nuclei in reticular core in some unknown way initiate locomotion.

At what may be the highest level of integration, two sorts of function can be identified, although neither is understood. The first mediates those behaviors which tend to preserve the individual when subjected to insult or stress and to provide for reproduction. In vertebrates, these functions reside largely in the total limbic system—the complex of reticulum, hypothalamus, and nonsensory primitive forebrain, sometimes called "the emotion brain." The second property of the highest level of neural organization is modifiability, association of one input with another as in conditioning, retention of a pattern, and recall on appropriate demand. The association of several inputs to give a long-lasting capacity for response may occur in some ganglionic systems (e.g., in cephalopod molluscs), and conditioning can occur at midbrain and diencephalic levels in vertebrates. There are probably multiple mechanisms of conditioning, but the properties of instrumental learning and of recall of long-term memory traces have not been described in strict neurophysiological terms. Electrical concomitants of conditioning are seen in lowered thresholds for spiking, presynaptic habituation, and changes in endogenous rhythmicity. Chemical and structural concomitants of conditioning are alterations of dendritic growth with experience, and stimulation of RNA and protein turnover in relation to use and disuse (and possibly in relation to conditioning). However, these concomitants do not explain altered patterns or engrams which become established.

In addition to the hierarchical view and the description of behavior in terms of function of individual neurons at intermediate levels of central nervous organization, comparative neurophysiology contributes to the controversy concerning the unitary versus the holistic nature of nervous integration.[52] The information content of identified neurons is surprisingly great. In some gastropod ganglia, each of thirty known neurons is unique in its input-output relations. Some receive sensory input from wide receptor fields; certain interneurons inhibit specific motorneurons and excite others, or have dual action; still other cells can, when stimulated, elicit complex movements involving many muscles. Individuality of function has also been demonstrated for sensory and motor interneurons of arthropods. In vertebrates, as single neurons are intensively explored, particularly in sensory systems, the uniqueness of individual cells becomes evident. This is true irrespective of whether transmission is initiated by all-or-none spikes or by graded potentials. One view is that neural integration could be understood if the information content of each neuron were known. It is not yet possible to extend this concept to regions of small neurons; even in gastropod ganglia, fewer than 1% of the neurons are identifiable at present. In cortical structures of vertebrates—tectum, cerebellum, cortex—synaptic potentials and membrane oscillations are often synchronized over extensive layers. These "field potentials" reflect input to a cortical region; they can be modified, and they relate to the probability of firing of neurons. Thus, a second view is that neural integration in the "higher" subsystems of vertebrates is partly by field potentials. A third view is that integrative properties emerge from the circuitry, from the network of neuronal connections. Much evidence assigns information content to proper connections and indications that the connections formed by a growing neuron are influenced by its input and output fields. Mathematical models show that networks have properties beyond those of their component elements. A reasonable conclusion is that certain cellular mechanisms are common to all neurons, that there is specificity of response by individual cells and by groups of neurons, and that each level of organization builds upon prior levels and may add features unique to its level of integration.

REFERENCES

1. Abbie, A. A., J. Comp. Neurol. *72*:469–488, 1940. Excitable cortex in marsupials.
2. Abbie, A. A., and W. R. Adey, J. Comp. Neurol. *92*:241–292, 1950. Motor mechanisms of anuran brain.
3. Abeles, M., and M. H. Goldstein, J. Neurophysiol. *33*:172–197, 1970. Organization of cat auditory cortex.
4. Adey, W. R., pp. 233–276 *in* Brain Function II, edited by M. Brazier. University of California Press, 1964. Hippocampal mechanisms in processes of memory.
5. Adey, W. R., pp. 224–243 *in* The Neurosciences: Second Study Program, edited by F. O. Schmitt. Rockefeller University Press, New York, 1970. Electrical brain rhythms accompanying learned responses.
6. Agranoff, B. W., R. E. Davis, and J. J. Brink, Proc. Nat. Acad. Sci. *54*:788–793, 1965; *also* Brain Res. *1*:303–309, 1966. Protein synthesis and learning.
7. Akert, K., Helvet. Physiol. Acta 7:112–134, 1949. Motor representation on tectum, fish.
8. Allison, T., and W. R. Groff, Psychophysiology *5*:200–201, 1968. Sleep in echidna.
9. Allison, T., and H. Van Twyver, Exp. Neurol. *27*:564–578, 1970. Sleep in moles.
10. Andersen, P., and J. C. Eccles, Nature *196*:645–647, 1963. Inhibitory phasing of neuronal discharge.
11. Andersen, P., J. C. Eccles, and Y. Loyning, Nature *198*:541–542, 1963. Recurrent inhibition in hippocampus.
12. Andersen, P., J. C. Eccles, and P. E. Voorhoeve, J. Neurophysiol. *27*:1138–1153, 1964. Postsynaptic inhibition of Purkinje cells.
13. Arechiga, H., and C. A. G. Wiersma, J. Neurobiol. *1*:71–85, 1969. Circadian rhythm of crayfish visual units.
14. Armstrong, P. B., and D. C. Higgins, J. Comp. Neurol. *143*:371–384, 1971. Behavior in bullhead embryo.
15. Arvanitaki, A., and N. Chalazonitis, Bull. Inst. Oceanogr. Monaco, No. 1164: 1–83, 1960; No. 1224: 1–15, 1961. Rhythmic synaptic potentials, *Aplysia* neurons.
16. Arvanitaki, A., and N. Chalazonitis, pp. 194–231 *in* Nervous Inhibition, edited by E. Florey. Pergamon Press, New York, 1961. Photostimulation of *Aplysia* neurons.
17. Atema, J., Brain Behav. Evol. *4*:273–294, 1971. Sense of taste in catfish *Ictalurus*.
18. Atwood, H. L., and C. A. G. Wiersma, J. Exp. Biol. *46*:249–261, 1967. Command interneurons in the crayfish central nervous system.

18a. Auerbach, A. A., and M. J. V. Bennett, J. Gen. Physiol. *53*:183–239, 1969. Mauthner cell, giant neuron system in hatchet fish.
19. Azaryan, A. G., and V. P. Tyshchenko, Dokl. Akad. Nauk. S.S.S.R. *186*:464–466, 1969. Cerebro-neuronal regulation of circadian behavior in cricket.
20. Bailey, D. T., and M. S. Laverack, J. Exp. Biol. *44*:131–148, 1966. Neurophysiology of *Buccinium.*
21. Barnes, W. J. P., and G. A. Horridge, J. Exp. Biol. *50*:651–671, 1969. Eyecup movements in crab.
22. Barondes, S. H., and H. D. Cohen, Brain Res. *4*:44–51, 1967. Effects of cycloheximide and puromycin on cerebral protein synthesis and consolidation of memory.
23. Barth, H., Z. wiss. Zool. *167*:238–290, 1962. Comparative anatomy of sensory centers in fishes.
24. Batham, E. J., C. F. Pantin, and E. A. Robson, Quart. J. Micr. Sci. *101*:487–510, 1960. Nerve net of sea anemone *Metridium*, mesenteries and the column.
25. Becker, H. W., Experientia *21*:719, 1965. The number of neurons in an insect ganglion.
26. Belekhova, M. G., and A. A. Kosareva, Brain Behav. Evol. *4*:337–375, 1971. Organization of the turtle thalamus.
27. Bell, C. C., and T. Kawasaki, J. Neurophysiol. *35*:155–169, 1972. Responses of Purkinje cells to climbing fiber input.
28. Bennett, E. L., et al., Science *146*:610–617, 1964. Chemical and anatomical plasticity of brain.
29. Bennett, M. J. V., and A. Steinbach, pp. 207–214 *in* Neurobiology of Cerebellar Evolution and Development, edited by R. Llinas. American Medical Association, Chicago, 1969. Cerebellum of electric fish.
30. Bentley, D. R., Z. vergl. Physiol. *62*:267–283, 1969. Neuronal activity and song patterns, cricket.
31. Bentley, D. R., J. Insect Physiol. *15*:677–699, 1969. Activity of cricket neurons during behavior patterns.
32. Bentley, D. R., and W. Kutsch, J. Exp. Biol. *45*:151–164, 1966. Neuromuscular mechanism of stridulation in crickets.
33. Bernstein, J. J., Exp. Neurol. *3*:1–17, 1961. Loss of hue discrimination in forebrain-ablated fish.
34. Bernstein, J. J., pp. 2–90 *in* Fish Physiology, vol. 4, edited by W. S. Hoar and D. J. Randall. Academic Press, New York, 1970. Review of fish central nervous systems.
35. Binstock, L., and L. Goldman, J. Gen. Physiol. *54*:730–764, 1969. *Myxicola* giant axons.
36. Birnberger, K. R., and C. M. Rovainen, J. Neurophysiol. *34*:983–989, 1971. Habituating fin reflex in sea lamprey.
37. Blakemore, C., and J. D. Pettigrew, Nature *225*:426–429, 1970; *also* Blakemore, C., and G. F. Cooper, Nature *228*:477–478, 1970. Eye dominance in visual cortex.
38. Bloom, T. E., B. J. Hoffer, and G. A. Siggins, Brain Res. *25*:501–521; 523–534; 535–553, 1971. Norepinephrine-containing afferents to Purkinje cells.
39. Bodian, D., Bull. Johns Hopkins Hosp. *119*:129–149, 1966. Embryonic development of spinal cord in monkey.
40. Bodian, D., pp. 129–140 *in* The Neurosciences: Second Study Program, edited by F. O. Schmitt. Rockefeller University Press, New York, 1970. A model of synaptic and behavioral ontogeny.
41. Boycott, B. B., Proc. Roy. Soc. Lond. B *153*:503–534, 1961. Functional organization of brain of cuttlefish.
42. Braitenberg, V., Kybernetik *2*:248–287, 1965. Perception of movement by the frog.
43. Brookhart, J. M., and E. Fadiga, J. Physiol. *150*:633–635, 1960. Potential fields in frog spinal cord.
44. Bruner, J., and D. Kennedy, Science *169*:92–94, 1970. Habituation in crayfish neuromuscular junction.
45. Bruner, J., and L. Tauc, J. Physiol. Paris *57*:230–231, 1965; *also* Nature *210*:37–39, 1966. Habituation at synaptic level, *Aplysia.*
46. Bryant, S. H., J. Gen. Physiol. *41*:473–484, 1958; *42*:609–619, 1959. Transmission in squid giant synapses.
47. Buck, J., and J. F. Case, Biol. Bull. *121*:234–256, 1961. Control of flashing in fireflies.
48. Buck, J., J. F. Case, and J. F. Hanson, Biol. Bull. *125*:251–269, 1963. Peripheral excitation of photic organ.
49. Buck, J., and E. Buck, Science *159*:1319–1327, 1968. Rhythmic synchronous flashing of fireflies.
50. Bullock, T. H., J. Neurophysiol. *8*:55–72, 1945. Function of giant fibers in *Lumbricus.*
51. Bullock, T. H., Amer. Zool. *5*:545–562, 1965. Conduction systems in echinoids and asteroids.
52. Bullock, T. H., and G. A. Horridge, Structure and Function in the Nervous Systems of Invertebrates. W. H. Freeman, San Francisco, 1965. 798 pp.
53. Bullock, T. H., S. H. Ridgway, and N. Suga, Z. vergl. Physiol. *74*: 372–387, 1971. Midbrain responses to sound in sea lion.
54. Bullock, T. H., et al., Z. vergl. Physiol. *59*:117–156, 1968. Central auditory responses in cetaceans.
55. Bures, J., and O. Buresova, pp. 211–238 *in* Neural Control of Behavior, edited by R. E. Whalen et al. Academic Press, New York, 1970. The reunified split brain.
56. Bures, J., and O. Buresova, pp. 363–403 *in* Short-Term Changes in Neural Activity and Behaviour, edited by G. Horn and R. A. Hinde. Cambridge University Press, 1970. Plasticity in single neurons.
57. Burns, B. D., and G. C. Salmoiraghi, J. Neurophysiol. *23*:27–46, 1960. Patterns of rhythmicity in single respiratory neurons of cat.
58. Burrows, M., and G. A. Horridge, J. Exp. Biol. *49*:223–250; 285–297, 1968. Optokinetic movements in *Carcinus.*
59. Buser, P., Arch. Sci. Physiol. *3*:471–487, 1949; *also* C. R. Soc. Biol. (Paris) *143*:817–819, 1949; *also* J. Physiol. (Paris) *43*:673–677, 1951. Responses of optic tectum, catfish.
60. Bush, B. M. H., Comp. Biochem. Physiol. *10*:273–290, 1963; *15*:567–587, 1965. Reflexes elicited by chordotonal organs in crabs.
61. Butler, A. B., and R. G. Northcutt, Brain Res. *26*:1–13, 1971. Retinal projections in *Iguana* and *Anolis.*
62. Cabral, R. J., and J. I. Johnson, J. Comp. Neurol. *141*:17–36, 1971. Mechanoreceptive projections in thalamus of sheep.
63. Callec, J. J., and J. Boistel, C. R. Soc. Biol. (Paris) *160*:1943–1947, 2418–2424, 1966. Rhythmic synaptic potentials, abdominal ganglia, *Periplaneta.*
64. Callec, J. J., et al., J. Exp. Biol. *55*:123–149, 1971. Synaptic transmission in cockroach.
65. Camhi, J. M., J. Exp. Biol. *50*:335–348, 349–362, 363–373, 1969. Locust wind receptors.
66. Campbell, C. B. G., Brain Behav. Evol. *2*:415–430, 1969. Optic systems of teleost.
67. Case, J. F., and J. Buck, Biol. Bull. *125*:234–250, 1963. Central nervous aspects of flashing in fireflies.
68. Castellucci, V., et al., Science *167*:1745–1748, 1970. Synaptic habituation in *Aplysia.*
69. Caviedes, E., and J. Bures, Brain Res. *19*:249–261, 1970. Habituation of evoked responses in rat cortex and tegmentum.
70. Chow, K. L., J. Neurophysiol. *24*:377–390, 1961. Brain electropotentials during visual discrimination learning in monkey.
71. Chow, K. L., et al., Brain Res. *33*:337–357, 1971. Receptive field characteristics of striate cortical neurons in rabbit.
72. Clark, R. B., Soc. Exp. Biol. Symp. *20*:345–379, 1966. Integrative action, worm brain.
73. Coggeshall, R. E., J. Neurophysiol. *30*:1263–1287, 1967. Light and electron microscope study of abdominal ganglion, *Aplysia.*
74. Coggeshall, R. E., et al., J. Cell Biol. *31*:363–368, 1966. Neuronal map, *Aplysia* ganglion.
75. Coghill, G. E., Anatomy and the Problem of Behavior. Macmillan Co., New York, 1929. 113 pp.
76. Cohen, M. J., and J. W. Jacklet, Phil. Trans. Roy. Soc. Lond. B *252*: 561–579, 1967. Organization of motorneurons in insect ganglia.
77. Cohen, M. J., pp. 798–812 *in* The Neurosciences: Second Study Program, edited by F. O. Schmitt. Rockefeller University Press, New York, 1970. A comparison of invertebrate and vertebrate central neurons.
78. Collett, T., J. Neurophysiol. *33*:239–256, 1970. Visual cells, medulla of sphinx and noctuid insects.
79. Collier, H. O., J. Exp. Biol. *16*:286–299, 300–312, 1939. Central reflexes of earthworms.
80. Coulombre, A. J., pp. 108–116 *in* The Neurosciences: Second Study Program, edited by F. O. Schmitt. Rockefeller University Press, New York, 1970. Development of the vertebrate motor system.
81. Cragg, B. G., pp. 1–60 *in* Structure and Function of Nervous Tissue, vol. 4, edited by G. Bourne. Academic Press, New York, 1972. Plasticity of synapses.
82. Crain, S. M., M. B. Bornstein, and E. R. Peterson, Brain Res. *8*:363–372, 1968. Tissue cultures of spinal cord.
83. Creed, R. S., et al., Reflex Activity of the Spinal Cord. Oxford University Press, 1938. 183 pp.
84. Crossland, W., Ph.D. Thesis, University of Illinois, 1971. Visual responses in brain of pigeon.
85. Cuénod, M., pp. 455–506 *in* Structure and Function of Nervous Tissue, vol. 5, edited by G. Bourne. Academic Press, New York, 1972. Split-brain studies.
86. Curtis, D. R., and D. Felix, Brain Res. *34*:301–321, 1971. Synaptic inhibition in cerebral and cerebellar cortex.
87. Dagan, D., and I. Parnas, J. Exp. Biol. *52*:313–324, 1970. Giant fiber and small fiber pathways involved in the evasive response of the cockroach.
88. Daniels, D., J. Comp. Physiol. Psychol. *76*:110–118, 1971. Effects of acetoxycycloheximide on conditioning in rats.
89. Darian-Smith, I., et al., J. Physiol. *182*:671–689, 1966. Somatic sensory cortical projection areas excited by tactile stimulation of the cat.
90. Darian-Smith, I., et al., Science *160*:791–794, 1968. Somatosensory pathways in mammalian brain.
91. Davenport, D., Bull. Inst. Oceanogr. Monaco, No. 1236:1–24, 1962. Responses of tentacles of actinians to electrical stimulation.
92. Davis, R. E., J. Comp. Physiol. Psychol. *65*:72–78, 1968. Environmental control of memory fixation in goldfish.
93. Davis, R. E., and B. W. Agranoff, Proc. Nat. Acad. Sci. *55*:555–589, 1966. Stages of memory formation in goldfish: evidence for an environmental trigger.
94. Davis, R. E., P. J. Bright, and B. W. Agranoff, J. Comp. Physiol. Psychol. *60*:162–166, 1965. Effect of ECS and puromycin on memory in fish.
95. Davis, W. J., J. Exp. Zool. *168*:363–378, 1968. Neuromuscular basis of lobster swimmeret beating.
96. Davis, W. J., J. Exp. Biol. *50*:99–117, 1969; *51*:547–574, 1969. Reflex organization, lobster swimmeret.

97. Davis, W. J., and D. Kennedy, J. Neurophysiol. *35*:1–12, 1972. Command interneurons controlling swimmerets of lobster.
98. Delcomyn, F., J. Exp. Biol. *54*:443–452, 453–469, 1971. Cockroach locomotion.
99. Dethier, V. G., and A. Gelperin, J. Exp. Biol. *47*:191–200, 1967. Hyperphagia in the blowfly.
100. Deutsch, J. A., Science *128*:288–294, 1971. Cholinergic synapse and the site of memory.
101. Dewsbury, D. A., and J. J. Bernstein, Exp. Neurol. *23*:445–456, 1967. Effects of forebrain lesions in goldfish.
101a. Dewson, J. H., J. Neurophysiol. *30*:817–832, 1967; *31*:122–130, 1967. Effects of efferent fibers on auditory responses, monkey.
102. Diamond, I. T., and W. C. Hall, Science *164*:251–262, 1969. Evolution of neocortex.
103. Diamond, J., Nature *199*:773–775, 1963; *also* J. Physiol. *194*:669–723, 1968. Sensitivity to gamma-aminobutyric acid of different regions of Mauthner neuron.
104. Diamond, J., pp. 265–346 *in* Fish Physiology, vol. 5, edited by W. S. Hoar and D. J. Randall. Academic Press, New York, 1970. The Mauthner cell.
105. Dingle, H., Biol. Bull. *121*:117–128, 1961. Flight and swimming reflexes in giant water bugs, *Lethocerus* and *Benacus*.
106. Dorsett, D. A., J. Exp. Biol. *46*:137–151, 1967. Giant neurons and axon pathways in brain of *Tritonia*.
107. Dorsett, D. A., J. Exp. Biol. *48*:127–140, 1968. Pedal neurons of *Aplysia punctata*.
108. Doty, R. W., The Physiologist *5*:270–284, 1962. Behavioral aspects of neurophysiology.
109. Doty, R. W., pp. 1861–1902 *in* Handbook of Physiology, Vol. 4, Sec. 6, edited by C. F. Code. Amer. Physiol. Soc., Washington, D.C., 1968. Neuronal organization of deglutition.
110. Eccles, J. C., The Physiology of Synapses. Springer-Verlag, Berlin, 1964. 316 pp.
111. Eccles, J. C., pp. 408–427 *in* The Neurosciences: A Study Program, edited by G. C. Quarton et al. Rockefeller University Press, New York, 1967. Post-synaptic inhibition.
112. Eccles, J. C., Naturwissenschaften *56*:525–534, 1969. Comparative physiology of cerebellum.
113. Eccles, J. C., Brain Res. *15*:267–271, 1969. Cerebro-cerebellar connections in mammals.
114. Eccles, J. C., M. Ito, and J. Szentagothai, The Cerebellum as a Neuronal Machine. Springer-Verlag, Berlin, 1967. 335 pp.
115. Eccles, J. C., H. Taborekova, and N. Tsukahara, Brain Res. *17*:87–102, 1970. Responses of granule cells of elasmobranch cerebellum.
116. Eccles, J. C., et al., Exp. Brain Res. *13*:15–35, 1971. Mossy fiber input to granule cells.
117. Edwards, J. S., and T. S. Sahota, J. Exp. Zool. *166*:387–396, 1967. Regeneration in sensory system of cricket *Acheta*.
118. Egger, M. D., and R. J. Wyman, J. Physiol. *202*:501–516, 1969. Reflex stepping in cat.
119. Egger, M. D., and J. P. Flynn, Science *136*:43–44, 1962. Effect of amygdaloid on hypothalamus stimulation.
120. Eidelberg, E., J. Neurophysiol. *24*:521–533, 1961. Hippocampal dendritic responses in rabbits.
121. Eisenstein, E. M., and M. J. Cohen, Anim. Behav. *13*:104–108, 1965. Learning in an isolated prothoracic insect ganglion.
122. Eisenstein, E. M., and G. H. Krasilovsky, pp. 329–332 *in* Invertebrate Nervous Systems, edited by C. A. G. Wiersma. University of Chicago Press, Chicago, 1967. Learning in isolated insect ganglia.
123. Elsner, N., and F. Huber, Z. vergl. Physiol. *65*:389–423, 1969. Central nervous basis for song patterns in crickets.
124. Enger, P. S., Acta Physiol. Scand. *39*:55–72, 1957. The electroencephalogram of the codfish.
125. Evans, S. M., Anim. Behav. *11*:172–178, 1963. Learning in polychaetes.
126. Evans, S. M., Biol. Bull. *137*:95–104, 1969. Habituation of withdrawal response in nereid polychaetes.
127. Evarts, E. V., J. Neurophysiol. *31*:14–27, 1968. Motorneuron firing in pyramidal tract correlated with movement.
128. Evarts, E. V., and W. T. Thach, Ann. Rev. Physiol. *31*:451–498, 1969. Motor mechanisms of the CNS: cerebrocerebellar interrelations.
129. Evoy, W. H., and M. J. Cohen, J. Exp. Biol. *51*:151–169, 1969. Sensory and motor interaction in locomotor reflexes of crabs.
130. Evoy, W. H., and D. Kennedy, J. Exp. Zool. *165*:223–238, 1967. Command neurons controlling muscle groups in crayfish.
131. Ewer, D. W., J. Exp. Biol. *37*:812–831, 1960. Inhibition and rhythmic activity of circular muscles of *Calliactis*.
132. Ewert, J. P., Z. vergl. Physiol. *54*:455–480, 1967. Behavior elicited by midbrain stimulation, *Bufo*.
133. Ewert, J. P., Pflüger. Arch. *306*:210–218, 1969; *also* Brain Behav. Evol. *3*:36–56, 1970; *also* Z. vergl. Physiol. *68*:84–110, 1970. Prey catching behavior by *Bufo*.
134. Ewert, J. P., Z. vergl. Physiol. *74*:81–102, 1971. Toad caudal thalamus.
135. Ewert, J. P., and H. W. Borchers, Z. vergl. Physiol. *71*:165–189, 1970. Tectum of toad.
136. Ewing, A. W., and A. Manning, J. Insect Physiol. *12*:1115–1118, 1966. Control of walking in the cockroach.
137. Fetz, E. E., Science *163*:955–958, 1969. Conditioning of cortical units.
138. Fiedler, K., J. Hirnforsch. *9*:480–563, 1967. Forebrain influences on attack behavior in fish.
139. Fielden, A., J. Exp. Biol. *40*:541–552, 1963. Interneurons, abdominal nerve cord of dragonfly nymph.
140. Fielden, A., and G. M. Hughes, J. Exp. Biol. *39*:31–44, 1962. Unit activity in abdominal nerve cord of dragonfly nymph.
141. Fields, H. L., J. Exp. Biol. *44*:455–468, 1966. Proprioceptive control of posture in crayfish abdomen.
142. Fields, H. L., and D. Kennedy, Nature *206*:1232–1237, 1965. Muscle receptor organs in crayfish.
143. Flexner, L. B., J. B. Flexner, and R. B. Roberts, Science *155*:1377–1383, 1967. Memory in mice.
144. Flexner, L. B., and J. B. Flexner, Proc. Nat. Acad. Sci. *55*:369–374, 1966; *also* Science *159*:330–331, 1968. Protein synthesis and memory in mice.
145. Flexner, J. B., et al., Science *141*:57–59, 1963. Effects of puromycin on learning.
146. Flint, P., Anim. Behav. *13*:187–193, 1965. Sensory deprivation and behavior of the polychaete *Nereis* in T-mazes.
147. Fox, S. S., pp. 243–259 *in* The Neurosciences: Second Study Program, edited by F. O. Schmitt. Rockefeller University Press, New York, 1970. Evoked potential, coding, and behavior.
148. Fox, S. S., and J. H. O'Brien, Science *147*:889–890, 1965. Duplication of evoked potential waveform by curve of probability of firing of a single cell.
149. Fox, S. S., and A. P. Rudell, J. Neurophysiol. *33*:548–561, 1970. Cortical potentials and learning.
150. Frazier, W. T., et al., J. Neurophysiol. *30*:1288–1351, 1967. Properties of identified neurons in *Aplysia*.
151. Freeman, J. A., pp. 397–420 *in* Neurobiology of Cerebellar Evolution and Development, edited by R. Llinas. American Medical Association, Chicago, 1969. The cerebellum as a timing device in frog.
152. von Frisch, K., and A. Kratky, Naturwissenschaften *49*:409–417, 1962. Flight distance and dance pattern, bees.
153. von Frisch, K., Kosmos (Stuttgart) *59*:279–283, 1963. Olfactory cues in bees.
154. Fukami, Y., et al., J. Gen. Physiol. *48*:581–600, 1965. Excitability, Mauthner cell; collateral inhibition.
155. Furshpan, E. J., and T. Furukawa, J. Neurophysiol. *25*:732–771, 1962. Synaptic excitation of Mauthner neurons.
156. Furukawa, T., Prog. Brain Res. *21A*:44–70, 1966. Synaptic interaction, Mauthner cell of goldfish.
157. Furukawa, T., et al., J. Neurophysiol. *26*:140–176, 759–774, 1963. Inhibition in Mauthner cell of goldfish.
158. Gardner, D., and E. R. Kandel. Science *176*:675–678, 1972. Synaptic transmission in *Aplysia*.
159. Gaze, R. M., and M. Jacobson, Proc. Roy. Soc. Lond. B *157*:420–448, 1963. Retinotectal projection in frog.
160. Gaze, R. M., and M. Jacobson, J. Physiol. *169*:92*P*–93*P*, 1963. Single-unit responses, different depths in optic tectum, goldfish.
161. Gaze, R. M., M. Jacobson, and G. Szekely, J. Physiol. *165*:484–499, 1963. Retinotectal projections in *Xenopus*.
162. Gaze, R. M., and M. J. Keating, Quart. J. Exp. Physiol. *35*:143–152, 1970. Retinotectal projection in frog.
163. Gaze, R. M., M. J. Keating, G. Szekely, and L. Beazley, Proc. Roy. Soc. Lond. B *175*:107–147, 1970. Binocular interaction in the formation of intertectal neuronal connections, *Xenopus*.
164. Gaze, R. M., et al., Quart. J. Exp. Physiol. *47*:273–280, 1970; *55*:143–152, 1970; *also* J. Physiol. *200*:128*P*–129*P*, 1969. Projection of binocular visual field on optic tecta of frog.
165. Gelver, B., Anim. Behav. *13*(Suppl.):21–29, 1965. Learning behavior, *Paramecium*.
166. Gelperin, A., Z. vergl. Physiol. *72*:17–31, 1971. Negative feedback and feeding behavior of the blowfly.
167. Gettrup, E., and D. M. Wilson, J. Exp. Biol. *41*:183–190, 1964. Lift-control reaction of flying locusts.
168. Goldstein, M. H., et al., J. Neurophysiol. *33*:188–197, 1970. Functional organization, cat primary auditory cortex.
169. Graeber, R. C., and S. O. E. Ebbesson, Comp. Biochem. Physiol. *42A*: 131–139, 1972. Visual learning in normal and tectum-ablated nurse sharks.
170. Gray, J., J. Exp. Biol. *13*:181–191, 200–209, 210–218, 1936. Locomotor reflexes in fishes.
171. Gray, J., J. Exp. Biol. *16*:9–17, 1939. Locomotor reflexes in *Nereis*.
172. Gray, J., and H. Lissmann, J. Exp. Biol. *17*:227–236, 237–251, 1940; *23*:121–132, 133–142, 1946. Locomotor reflexes in amphibians.
173. Grinnell, A. D., J. Physiol. *182*:612–648, 1966. Motorneurons in frog spinal cord.
174. Gruber, S. A., and D. W. Ewer, J. Exp. Biol. *39*:459–477, 1962. Myoneural physiology of polyclad, *Planocera*.
175. Gruberg, E. R., Ph.D. Thesis, University of Illinois, 1969. Organization of tectum of salamander *Ambystoma*.
176. Grüsser-Cornehls, U., Pflüger. Arch. *303*:1–13, 1968. Movement-detecting neurons of frog's retina.
177. Grüsser-Cornehls, U., and O. J. Grüsser, pp. 275–286 *in* Neurophysiologie and Psychophysiologie des visuellen Systems, edited by R. Jung and H. Kornhuber. 1960. Reaction model of neuron in central visual system of fish, rabbits, and cats with monocular and binocular light stimulation.
178. Grüsser-Cornehls, U., O. J. Grüsser, and T. H. Bullock, Science *141*: 820–822, 1963. Visual responses in frog tectum.

179. Guthrie, D. M., and J. R. Banks, J. Exp. Biol. *50*:255–273, 1969. Development of patterned activity by implanted ganglia and their peripheral connections in *Periplaneta*.
180. Hagiwara, S., et al., Comp. Biochem. Physiol. *13*:453–460, 1964. Transmission in giant axons, sabellid worm.
181. Hainsworth, F. R., J. Comp. Physiol. Psychol. *63*:111–116, 1967. Learning in goldfish.
182. Hale, E. B., Physiol. Zool. *29*:107–127, 1956. Effects of forebrain lesions on the aggressive behavior of green sunfish.
183. Hamburger, V., 27th Symp. Soc. Devel. Biol., Suppl. 2, 251–271, 1968. Emergence of nervous coordination.
184. Hamburger, V., pp. 141–151 *in* The Neurosciences: Second Study Program, edited by F. O. Schmitt. Rockefeller University Press, New York, 1970. Embryonic motility in vertebrates.
185. Hamburger, V., E. Wenger, and R. Oppenheim, J. Exp. Zool. *162*: 133–160, 1966. Motility in chick embryo in absence of sensory input.
186. Hanström, B., Vergleichende Anatomie des Nervensystems der wirbellosen Tiere. Springer-Verlag, Berlin, 1928. 628 pp.
187. Harcombe, E. S., and R. J. Wyman, J. Exp. Biol. *53*:255–263, 1970. Diagonal locomotion in de-afferented toads.
188. Harris, J. E., and H. P. Whiting, J. Exp. Biol. *31*:501–524, 1954. Function in the locomotory system of the dogfish embryo.
189. Heimer, L., Ann. N. Y. Acad. Sci. *167*:129–146, 1969. Olfactory connections in vertebrate brain.
190. Henneman, E., pp. 1717–1732 *in* Medical Physiology, edited by V. B. Mountcastle. C. V. Mosby, St. Louis, 1968. Organization of the spinal cord.
191. Henneman, E., et al., J. Neurophysiol. *28*:560–580, 1965. Significance of cell size in spinal motorneurons.
192. Heric, T. M., and L. Kruger, J. Comp. Neurol. *124*:101–112, 1965. Organization of the visual projection upon the optic tectum of reptile (*Alligator*).
193. Heric, T. M., and L. Kruger, Brain Res. *2*:187–199, 1966. Electrical response evoked in reptilian optic tectum by afferent stimulation.
194. Hermann, H. T., Brain Res. *26*:293–304, 1971. Saccade correlated potentials in optic tectum and cerebellum of *Carassius*.
195. Hernandez-Peon, R., Electroenceph. Clin. Neurophysiol. Suppl. *13*:101–114, 1960. Neurophysiological correlates of habituation.
196. Hertzler, D. R., and W. N. Hayes, J. Comp. Physiol. Psychol. *63*:444–447, 1967. Cortical and tectal function, visually guided behavior, turtles.
197. Hirsch, H. B., and D. N. Spinelli, Science *168*:869–871, 1970. Effect of visual experience on visual cortex, kittens.
198. Hobson, J. A., Electroenceph. Clin. Neurophysiol. *22*:113–121, 1967. Frog EEG with reference to sleep.
199. Hoffmann, K.-P., Z. vergl. Physiol. *67*:26–57, 1970. Retinal connection to receptive fields in optic tectum and pretectum of cat.
200. Holden, A. L., J. Physiol. *194*:75–90, 91–104, 1968. Field potentials and unit responses in optic tectum, pigeon.
201. von Holst, E., Z. vergl. Physiol. *20*:582–599, 1934; *26*:481–528, 1939; *also* Pflüger. Arch. *236*:149–158, 1935. Reflexes and locomotor rhythms in fish.
202. von Holst, E., and U. St. Paul, Naturwissenschaften *45*:579, 1960; *47*:409–422, 1960. Brain stimulation, chickens.
203. Hore, J., and R. Porter, Brain Res. *30*:232–234, 1971. Pyramidal neurons in a marsupial.
204. Horn, G., and R. A. Hinde, editors, Short-Term Changes in Neural Activity and Behaviour. Cambridge University Press, 1970. 628 pp.
205. Horn, G., and C. H. F. Rowell, J. Exp. Biol. *49*:143–169, 1968. Long-term changes in behavior of visual neurons in tritocerebrum of locust.
206. Horn, G., and M. J. Wright, J. Exp. Biol. *52*:217–231, 1970. Habituation in squid stellate ganglion.
207. Horridge, G. A., Phil. Trans. Roy. Soc. Lond. B *240*:495–528, 1957. Interneural facilitation, probability basis for transmission.
208. Horridge, G. A., Proc. Roy. Soc. Lond. B *150*:245–262, 1959. Analysis of rapid responses in *Nereis* and *Harmothöe*.
209. Horridge, G. A., J. Exp. Biol. *36*:72–91, 1959. Properties of fast and slow nerve nets of medusae.
210. Horridge, G. A., J. Physiol. *155*:320–336, 1961. Sequence of impulses initiated from ganglia of clam.
211. Horridge, G. A., Proc. Roy. Soc. Lond. B *157*:33–52, 1963. Learning of leg position, ventral nerve cord in insects.
212. Horridge, G. A., Proc. Roy. Soc. Lond. B *157*:199–222, 1963. Proprioceptors, bristle receptors, efferent sensory impulses.
213. Horridge, G. A., J. Exp. Biol. *44*:233–245, 1966. Optokinetic memory in crab.
214. Horridge, G. A., Anim. Behav. Suppl. *1*:163–182, 1966. The electrophysiological approach to learning in isolatable ganglia.
215. Horridge, G. A., Interneurons. W. H. Freeman, San Francisco, 1968. 436 pp.
216. Horridge, G. A., and M. Burrows, J. Exp. Biol. *49*:269–284, 315–324, 1968. Eyecup movement, crab *Carcinus*.
217. Horridge, G. A., D. M. Chapman, and B. MacKay, Nature *193*:889–890, 1962. Electron microscopy of *Aurelia* and *Cyanea*.
218. Horridge, G. A., and G. O. Mackie, Quart. J. Micr. Sci. *103*:531–542, 1962. Electron microscopy of synapses, jellyfish.
219. Horridge, G. A., and D. C. Sandeman, Proc. Roy. Soc. Lond. B *161*: 216–246, 1964. Nervous control of optokinetic responses in crab *Carcinus*.
220. Horridge, G. A., J. H. Scholes, S. Shaw, and J. Tunstall, pp. 165–202 *in* Physiology of the Insect Central Nervous System, edited by J. E. Treherne and J. W. Beament. Academic Press, New York, 1965.
221. Horridge, G. A., and P. R. B. Shepheard, Nature *209*:267–269, 1966. Perception of movement by a crab.
222. Hoyle, G., pp. 203–232 *in* Physiology of the Insect Central Nervous System, edited by J. E. Treherne and J. W. Beament. Academic Press, New York, 1965. Learning in headless insects.
223. Hoyle, G., pp. 349–444 *in* Advances in Insect Physiology, vol. 7, edited by J. W. Beament. Academic Press, New York, 1970.
224. Hoyle, G., J. Exp. Biol. *44*:413–427, 1966. Conditioning of locust ganglion.
225. Hoyle, G., and M. Burrows, personal communication. Neural mechanisms of behavior in *Schistocerca*.
226. Hubel, D. H., J. Physiol. *147*:226–238, 1959; *150*:91–104, 1960. Single unit activity in geniculate and optic cortex, cat.
227. Hubel, D. H., and T. N. Wiesel, J. Physiol. *155*:385–398, 1961. Integrative action in cat's lateral geniculate body.
228. Hubel, D. H., and T. N. Wiesel, J. Physiol. *160*:106–154, 1962; *164*: 559–568, 1963; *also* J. Neurophysiol. *28*:229–289, 1965. Visual fields of cortical neurons.
229. Hubel, D. H., and T. N. Wiesel, J. Neurophysiol. *26*:998–1002, 1965. Receptive fields in striate cortex, young kittens.
230. Hubel, D. H., and T. N. Wiesel, J. Physiol. *195*:215–243, 1968. Receptive fields and functional architecture of monkey striate cortex.
231. Hubel, D. H., and T. N. Wiesel, J. Physiol. *206*:419–436, 1970. The period of susceptibility to physiological effects of unilateral eye closure in kittens.
232. Huber, F., Naturwissenschaften *43*:317–321, 1956. Sound patterns, grasshoppers and crickets.
233. Huber, F., Z. vergl. Physiol. *43*:359–391; *44*:60–132, 1960. Respiration, locomotion, sound production, Orthoptera.
234. Huber, F., pp. 334–406 *in* The Physiology of Insecta, vol. 2, edited by M. Rockstein. Academic Press, New York, 1965. Neural integration (central nervous system).
235. Huber, F., pp. 333–351 *in* Invertebrate Nervous Systems, edited by C. A. G. Wiersma. University of Chicago Press, 1967. Central control of movements and behavior of insects.
236. Hughes, G. M., J. Exp. Biol. *29*:267–284, 1952; *34*:306–333, 1957; *35*:567–583, 1958. Neural coordination of insect movements.
237. Hughes, G. M., J. Exp. Biol. *46*:169–193, 1967. Physiological anatomy of left and right giant cells in *Aplysia*.
238. Hughes, G. M., and L. Tauc, J. Exp. Biol. *39*:45–69, 1962; *40*:469–486, 1963. Electrophysiological evidence for central nervous pathways in *Aplysia*.
239. Hyden, H., and E. Egyhazi, Proc. Nat. Acad. Sci. *49*:618–624, 1963; *52*:1030–1035, 1964. Changes in RNA in cortical neurons of rats in learning.
240. Ikeda, K., and W. D. Kaplan, Proc. Nat. Acad. Sci. *66*:765–772, 1970. Electrical activity in *Drosophila* neurons.
241. Ikeda, K., and C. A. G. Wiersma, Comp. Biochem. Physiol. *12*:107–115, 1964. Rhythmicity in abdominal ganglia of crayfish.
242. Ingle, D., Brain Behav. Evol. *3*:57–71, 1970. Visual-motor functions of frog optic tectum.
243. Ingle, D., Exp. Neurol. *33*:329–342, 1971. Potentials in goldfish optic tectum.
244. Ito, M., et al., J. Physiol. *164*:150–156, 1962; *also* Proc. Roy. Soc. Lond. B *161*:92–108, 132–141, 1964. Anion permeability of inhibitory postsynaptic membrane of cat motorneurons.
245. Jacklet, J. W., and M. J. Cohen, Science *156*:1638–1640, 1967. Synaptic connection of transplanted insect ganglion and muscles of the foot.
246. Jacobson, M., Devel. Biol. *17*:202–218, 1968; *also* Science *163*:543–547, 1969. Neuronal specificity in retinal ganglion cells, *Xenopus*.
247. Jacobson, M., Developmental Neurobiology. Holt, Rinehart and Winston, New York, 1970. 465 pp.
248. Jacobson, M., pp. 116–129 *in* The Neurosciences: Second Study Program, edited by F. O. Schmitt. Rockefeller University Press, New York, 1970. Development, specification, and diversification of neuronal connections.
249. Jacobson, M., and R. M. Gaze, Quart, J. Exp. Physiol. *49*:199–209, 1964. Response of units in optic tectum of goldfish.
250. Janzen, W., Zool. Jahrb. *52*:591–628, 1933. Goldfish forebrain function.
251. Jasper, H. H., and B. Doane, pp. 79–117 *in* Progress in Physiological Psychology, vol. 2, edited by E. Stellar and J. M. Sprague. Academic Press, New York, 1968. Neurophysiological mechanisms in learning.
252. Jasper, H. H., G. Ricci, and B. Doane, Electroenceph. Clin. Neurophysiol. (Suppl.) *13*:137–155, 1960. Cortical cell discharges during avoidance conditioning, monkey.
253. Jasper, H. H., and C. Stefanis, Electroenceph. Clin. Neurophysiol. *18*:541–553, 1965. Oscillatory rhythms in pyramidal neurons, cat.
254. Jassik-Gerschenfeld, D., Brain Res. *24*:407–421, 1970. Pigeon optic tectum.
255. Jassik-Gerschenfeld, D., and J. Guichard, Brain Res. *40*:303–317, 1972. Visual receptive fields of single cells in pigeon's optic tectum.
256. Jensen, D. D., Anim. Behav. *13*(Suppl. I):9–20, 1965. Pseudo-learning in paramecia and planaria.
257. Jha, R. K., and G. O. Mackie, J. Morph. *123*:43–61, 1967. Nerve net structure in *Cordylophora* and siphonophore *Sarsia*.

258. John, E. R., pp. 161–182 *in* Brain Function, vol. 2, edited by M. A. Brazier. University of California Press, 1965. Studies on learning and retention in Planaria.
259. John, E. R., pp. 690–704 *in* The Neurosciences: A Study Course, edited by G. C. Quarton et al. Rockefeller University Press, New York, 1967. Electrophysiological studies of conditioning.
260. John, E. R., D. S. Ruchkin, and J. Villegas, Ann. N.Y. Acad. Sci. *112*: 362–420, 1964. Behavioral correlates of evoked potential configurations in cats.
261. Johnson, R. H., and R. F. Thompson, J. Comp. Physiol. Psychol. *69*: 485–491, 1969. Association cortex in auditory-visual conditional learning in cat.
262. Johnstone, J. R., and R. F. Mark, J. Exp. Biol. *54*:403–414, 1971. Efference copy neuron in goldfish.
263. Jones, M. D. R., J. Exp. Biol. *45*:15–30, 1966. Acoustic behavior of bush cricket.
264. Josephson, R., Comp. Biochem. Physiol. *5*:45–58, 1962. Spontaneous electrical activity in a hydroid polyp.
265. Josephson, R., J. Exp. Biol. *42*:139–152, 1965. Conducting Systems in stalk of hydroid.
266. Josephson, R. K., Cnidarian neurobiology *in* Perspectives in Coelenterate Biology, edited by L. Muscatine and H. M. Lenhoff. Academic Press, New York, 1973.
267. Josephson, R. K., and G. O. Mackie, J. Exp. Biol. *43*:293–332, 1965. Multiple pacemakers in hydroid *Tubularia*.
268. Josephson, R. K., and M. Macklin, J. Gen. Physiol. *53*:638–665, 1969. Electrical properties, body wall of *Hydra*.
269. Josephson, R. K., and J. Uhrich, J. Exp. Biol. *50*:1–14, 1969. Inhibition of pacemaker systems in a hydroid.
270. Jouvet, M., Physiol. Rev. *47*:117–177, 1967; *also* pp. 529–544 *in* The Neurosciences: A Study Program, edited by G. C. Quarton et al., Rockefeller University Press, New York, 1967; *also* et al., Science *159*:112–114, 1968; *163*:32–41, 1969. Physiology of states of sleep.
271. Kaada, B. R., E. W. Rasmussen, and A. Kriem, Exp. Neurol. *3*:333–355, 1961. Hippocampus in memory storage in rodents.
272. Kaas, J., W. C. Hall, and I. T. Diamond, J. Neurophysiol. *33*:595–614, 1970. *Erinaceus* (hedgehog) cortical sensory areas.
273. Kadoya, S., L. R. Wolin, and L. C. Massopust, Jr., J. Comp. Neurol. *142*:495–508, 1971. Optic tectum of squirrel monkey.
274. Kaiserman-Abramof, I. R., and S. L. Palay, pp. 171–205 *in* Neurobiology of Cerebellar Evolution and Development, edited by R. Llinas. American Medical Association, Chicago, 1969. Fine structural studies of the cerebellar cortex in mormyrid fish.
275. Kamikawa, K., J. T. McIlwain, and W. R. Adey, Electroenceph. Clin. Neurophysiol. *17*:485–496, 1964. Response patterns of thalamic neurons during classic conditioning.
276. Kammer, A. E., J. Exp. Biol. *47*:2772–95, 1967. Muscle activity during flight in some large Lepidoptera.
277. Kammer, A. E., J. Exp. Biol. *48*:89–109, 1968. Motor warmup in Lepidoptera.
278. Kandel, E. R., pp. 666–689 *in* The Neurosciences: A Study Program, edited by G. C. Quarton et al., Rockefeller University Press, New York, 1967.
279. Kandel, E., V. Castellucci, H. Pinsker, and I. Kupferman, pp. 281–322 *in* Short-Term Changes in Neural Activity and Behaviour, edited by G. Horn and R. A. Hinde. Cambridge University Press, 1970. Synaptic plasticity in short-term modification of behavior.
280. Kandel, E., W. T. Frazier, and R. E. Coggeshall, Science *155*:346–349, 1967. Synaptic actions mediated by different branches of an identifiable interneuron in *Aplysia*.
281. Kandel, E., and L. Tauc, J. Physiol. *181*:1–27, 28–47, 1965. Heterosynaptic facilitation in neurons of abdominal ganglion of *Aplysia*.
282. Kandel, E., and L. Tauc, J. Physiol. *183*:269–286, 1966. Input organization of two symmetrical giant cells in the snail brain.
283. Kandel, E., and H. Wachtel, pp. 17–65 *in* Physiological and Biochemical Aspects of Nervous Integration, edited by F. D. Carlson. Prentice-Hall, 1967. Functional organization of neural aggregates in *Aplysia*.
284. Kandel, E., et al., J. Neurophysiol. *30*:1352–1376, 1967. Interactions between neurons in *Aplysia*.
285. Kaplan, H., and L. R. Aronson, Anim. Behav. *15*:438–448, 1967. Conditioned behavior in *Tilapia*, fish.
286. Karamian, A. I., V. V. Fanardjian, and A. A. Kosareva, pp. 639–671 *in* Neurobiology of Cerebellar Evolution and Development, edited by R. Llinas. American Medical Association, Chicago, 1969. Functional and morphological evolution of the cerebellum.
287. Katz, B., and R. Miledi, J. Physiol. *168*:389–422, 1963. Spontaneous miniature potentials in spinal motorneurons.
288. Keating, E. G., L. A. Kormann, and J. A. Horel, Physiol. & Behav. *5*:55–59, 1970. Stimulation of amygdala of caiman.
289. Keating, M. J., and R. M. Gaze, Quart. J. Exp. Physiol. *55*:129–142, 1970. Frog retinal ganglion cell units.
290. Kendig, J., J. Exp. Biol. *48*:389–404, 1968. Motorneuron coupling in locust flight.
291. Kennedy, D., The Physiologist *14*:5–30, 1971. Crayfish interneurons.
292. Kennedy, D., and D. Mellon, Comp. Biochem. Physiol. *3*:275–300, 1964. Receptive fields in crayfish interneurons.
293. Key, B. J., Nature *207*:441–442, 1965. Correlation of behavior with changes in amplitude of cortical potentials evoked during habituation.
294. Kidokoro, Y., pp. 257–272 *in* Neurobiology of Cerebellar Evolution and Development, edited by R. Llinas. American Medical Association, Chicago, 1969. Cerebellar and vestibular control of fish oculomotor neurons.
295. King, F. A., and P. M. Meyer, Science *128*:655–656, 1958. Amygdalectomy in cat.
296. Kitai, S., et al., Brain Res. *22*:381–385, 1970; *also* pp. 481–489 *in* Neurobiology of Cerebellar Evolution and Development, edited by R. Llinas, American Medical Association, Chicago, 1969. Cerebellum of reptiles.
297. Klemm, W. R., Animal Electroencephalography. Academic Press, New York, 1969. 292 pp.
298. Knights, A., J. Exp. Biol. *42*:447–461, 1965. Activity of single motor fibers in arthropods.
299. Kollross, J. J., pp. 179–199 *in* Growth of the Nervous System, edited by G. E. W. Wolstenhome and M. O'Connor. Little, Brown Co., Boston, 1968. Endocrine influences in neural development.
300. Konishi, R., and J. Z. Slepian, J. Acoust. Soc. Amer. *49*:1762–1769, 1971. Effect of efferent impulses on cochlear responses, guinea pig.
301. Konorski, J., pp. 115–130 *in* Brain Mechanisms and Learning, edited by J. F. Delafresnaye. Blackwell, Oxford, 1961. The physiological approach to the problem of recent memory.
302. Krasne, F. B., J. Exp. Biol. *50*:29–46, 1969. Excitation and habituation of crayfish escape reflex: depolarizing response in lateral giant fibers.
303. Krivanek, J. O., Physiol. Zool. *29*:241–250, 1956. Habit formation in the earthworm.
304. Kruger, L., and E. C. Berkowitz, J. Comp. Neurol. *115*:125–140, 1960. Afferent connections, reptile telencephalon.
305. Kubota, K., and J. M. Brookhart, Amer. J. Physiol. *204*:660–666, 1963. Inhibitory synaptic potential of frog motorneurons.
306. Kubota, K., and J. M. Brookhart, J. Neurophysiol. *26*:877–893, 1963. Recurrent facilitation of frog muscle.
307. Kulsch, E., and F. Huber, Z. vergl. Physiol. *67*:140–159, 1970. Central vs. peripheral control of cricket singing.
308. Kupfermann, I., et al., Science *164*:847–850, 1969; *167*:1743–1745, 1970. Behavioral habituation, *Aplysia*.
309. Kupfermann, I., et al., Science *174*:1252–1255, 1971. Central and peripheral control of gill movements, *Aplysia*.
310. Kuriyama, K., et al., Proc. Nat. Acad. Sci. *55*:846–852, 1966. GABA as transmitter in cerebellum.
311. Landenberger, D. E., Anim. Behav. *14*:414–418, 1966. Conditioning of starfish.
312. Larimer, J. L., and D. Kennedy, J. Exp. Biol. *44*:345–354, 1966. Afferent signals in crayfish stomatogastric ganglion.
313. Larimer, J. L., and D. Kennedy, J. Exp. Biol. *51*:119–133, 135–150, 1969. Innervation patterns of muscles in responses of crayfish.
314. Larimer, J. L., et al., J. Exp. Biol. *54*:391–402, 1971. Connections of lateral and medial giant fibers of crayfish.
315. Larsell, O., The Comparative Anatomy and Histology of the Cerebellum. University of Minnesota Press, Minneapolis, 1970. 269 pp.
316. Lashley, K. S., Comp. Psychol. Monogr. *11*:3–79, 1935; *also* Quart. Rev. Biol. *24*:28–42, 1949. Neural basis of learning in rats.
317. Latham, A., and D. H. Paul, J. Physiol. *213*:135–156, 1971. Activity of cerebellar Purkinje cells.
318. Laverack, M. S., J. Exp. Biol. *50*:129–140, 1969. Mechanoreceptors, photoreceptors and rapid conduction pathways in leech *Hirudo*.
319. Lawry, J. V., Comp. Biochem. Physiol. *37*:167–179, 1970. Locomotion, polychaete.
320. Leghissa, S., Z. Anat. Entwickl. *118*:427–463, 1955. Structure of optic tectum, fishes.
321. Lende, R. A., J. Comp. Neurol. *121*:395–403, 405–415, 1963; *also* Science *141*:730–732, 1963. Sensory and motor representation, cerebral cortex, opossum.
322. Lende, R. A., J. Neurophysiol. *27*:37–48, 1964. Cerebral cortex of echidna.
323. Lende, R. A., Ann. N. Y. Acad. Sci. *167*:262–275, 1969. Neocortex of monotremes, marsupials, and insectivores.
324. Levi-Montalcini, R., and P. U. Angeletti, Physiol. Rev. *48*:534–569, 1968. Nerve growth factor.
325. Levi-Montalcini, R., R. Angeletti, and P. U. Angeletti, pp. 1–38 *in* Structure and Function of Nervous Tissue, vol. 5, edited by G. H. Bourne. Academic Press, New York, 1972. Nerve growth factor.
326. Levitan, H., L. Tauc, and J. P. Segundo, J. Gen. Physiol. *55*:484–496, 1970. Neuronal interactions in buccal ganglion of mollusc *Navanax*.
327. Li, C., and S. N. Chou, J. Cell. Comp. Physiol. *60*:1–16, 1962. Cortical intracellular synaptic potentials.
328. Lickey, M. E., J. Comp. Physiol. Psychol. *68*:9–17, 1969. Entrainment of circadian rhythm in *Aplysia* neuron.
329. Lindauer, M., and I. Medugorac, Colloq. Internat. Centr. Nat. Recher. Sci. 15–25, 1968. Retention of learned response after narcosis, bees.
330. Llinas, R., and D. E. Hillman, pp. 43–73 *in* Neurobiology of Cerebellar Evolution and Development, edited by R. Llinas. American Medical Association, Chicago, 1969. Comparative morphology of the cerebellum.
331. Llinas, R., and C. Nicholson, J. Neurophysiol. *34*:532–551, 1971; *also* pp. 431–464 *in* Neurobiology of Cerebellar Evolution and Development, edited by R. Llinas, American Medical Association, Chicago, 1969. Properties of dendrites and somata in alligator Purkinje cells.
332. Llinas, R., et al., Brain Res. *6*:371–378, 1967; *also* Exp. Brain Res. *13*:408–431, 1971. Vestibular input to cerebellum, frog.
333. Luco, J. V., pp. 135–159 *in* Brain Function, vol. 2, edited by M. A.

Brazier. University of California Press, 1965. Plasticity of neural function in learning and retention.
334. Luco, J. V., and L. C. Aranda, Nature *201*:1330–1331, 1964. Electrical correlate of learning.
335. Luco, J. V., and L. C. Aranda, Nature *209*:205–206, 1966. Reversibility of electrical correlate to process of learning.
336. Lund, R. D., and J. S. Lund, Exp. Neurol. *13*:302–316, 1965. Vision in mole.
337. Mackie, G. O., Quart. J. Micr. Sci. *101*:119–131, 1960. *Velella*—dual nerve net.
338. Mackie, G. O., Proc. Roy. Soc. Lond. B *159*:366–391, 1964. Locomotion in a siphonophore colony.
339. Mackie, G. O., Amer. Zool. *5*:439–453, 1965. Non-nervous (epithelial) conduction in a cnidarian.
340. Mackie, G. O., and L. M. Passano, J. Gen. Physiol. *52*:600–621, 1968. Epithelial conduction in coelenterates.
341. MacLean, P. D., pp. 336–349 *in* The Neurosciences: Second Study Program, edited by F. O. Schmitt. Rockefeller University Press, New York, 1970. Emotion brain and subjective brain.
342. Magalhaes, C. B., and P. E. S. Saraiva, Brain Res. *34*:291–299, 1971. Sensory and motor representation in cerebral cortex of marsupial *Didelphis*.
343. Magoun, H. W., The Waking Brain. Charles C Thomas, Springfield, Ill., 1958, 188 pp. The reticular system in the brain.
344. Manteifel, Y. B., Dokl. Akad. Nauk. S.S.S.R. *168*:414–417, 1966. Evoked potentials in the tectum of the lizard.
345. Manton, S. M., J. Linn. Soc. Lond. *41*:529–570, 1950; *42*:93–166, 299–368, 1954. Locomotor rhythms in arthropods.
346. Mark, R. F., Exp. Neurol. *16*:215–225, 1966; *21*:92–104, 1968. Intertectal connections, goldfish.
347. Martin, A. R., and G. Pilar, J. Physiol. *168*:443–475, 1963. Synaptic transmission, chick ciliary ganglion.
348. Maturana, H., et al., J. Gen. Physiol. *43*:129–176, 1960. Vision in frog.
349. Maynard, D. M., Symp. Soc. Exp. Biol. *20*:111–150, 1966. Integration in lobster brain.
350. Maynard, D. M., Science *158*:531–532, 1967. Stomatogastric ganglion of mangrove crab.
351. Maynard, D. M., Ann. N. Y. Acad. Sci. *193*:59–72, 1972. Stomatogastric ganglion in crustaceans.
352. Maynard, D. M., and M. J. Cohen, J. Exp. Biol. *43*:55–78, 1965. Heteromorph antennule in lobster.
353. Maynard, D. M., and J. G. Yager, Z. vergl. Physiol. *59*:241–249, 1968. Function of an eyestalk ganglion.
354. Mazurskaya, P. Z., and G. D. Smirnov, Dokl. Akad. Nauk. S.S.S.R. *167*:285–291, 1966. Optic projections of the telencephalon in the turtle.
355. McCleary, R. A., J. Comp. Physiol. Psychol. *53*:311–321, 1960. Interocular transfer in fish.
356. McConnell, J. V., Ann. Rev. Physiol. *28*:107–136, 1966. Learning in invertebrates.
357. McConnell, J. V., et al., J. Comp. Physiol. Psychol. *52*:1–5, 1959; *also* Amer. J. Psychol. *73*:618–622, 1960. Conditioned response in planarian.
358. McFarlane I. D., J. Exp. Biol. *51*:377–385, 1969. Slow conduction systems in sea anemone *Calliactis*.
359. McIlwain, J. T., Ann. Rev. Physiol. *34*:291–314, 1972. Visual cortex and superior colliculus.
360. McKay, J. M., J. Exp. Biol. *53*:137–145, 1970. Cerebral control of an insect sensory interneuron.
361. Medugorac, I., and M. Lindauer, Z. vergl. Physiol. *55*:450–474, 1967. Memory in bees after narcosis.
362. Mellon, D., J. Exp. Biol. *43*:455–472, 1965; *46*:585–597, 1967; *53*: 711–725, 1970. Reflexes in surf clam, *Spisula*.
363. Mellon, D., Z. vergl. Physiol. *62*:318–336, 1969. Reflex control of rhythmic motor output during swimming in scallop.
364. Mendelson, M., Science *171*:1170–1173, 1971. Oscillator neurons in crustacean ganglia.
365. Meulders, M., et al., J. Comp. Neurol. *126*:535–545, 1966. Sensory projections in sloth.
366. Michael, C. R., J. Neurophysiol. *31*:249–282, 1968. Receptive fields of optic nerve fibers in ground squirrel all-cone retina.
367. Michael, C. R., Brain Behav. Evol. *3*:205–209, 1970; *also* Vision Res. Suppl. *3*:299–308, 1971. Superior colliculus of ground squirrel.
368. Milburn, N. S., and K. D. Roeder, Gen. Comp. Endocr. *2*:70–76, 1962. Control of efferent activity in cockroach.
369. Milburn, N. S., E. A. Weiant, and K. D. Roeder, Biol. Bull. *118*:111–119, 1960. Release of efferent nerve activity in *Periplaneta* in extracts of corpus cardiacum.
370. Miller, N. E., Science *148*:328–338, 1965. Chemical coding of behavior in the brain.
371. Miller, N. E., Science *168*:434–445, 1969. Learning of visceral and glandular responses.
372. Miller, P. L., J. Exp. Biol. *46*:349–371, 1967. Derivation of motor command to spiracles in locust.
373. Miller, P. L., pp. 475–498 *in* Short-Term Changes in Neural Activity and Behaviour, edited by G. Horn and R. A. Hinde. Cambridge University Press, 1970. Learning in insect central nervous system.
374. Miller, P. L., J. Exp. Biol. *54*:587–608, 1971. Rhythmic activity in insect nervous system.
375. Millott, N., pp. 465–485 *in* Physiology of Echinodermata, edited by R. Boolootian. Interscience, New York, 1966. Coordination of spine movement in echinoids.
376. Millott, N., and H. Okumura, J. Exp. Biol. *48*:279–287, 1968. Electrical activity of the radial nerve in *Diadema* and certain other echinoids.
377. Mimura, K., et al., Z. vergl. Physiol. *62*:382–394, 1969. Unit recording, brain of fleshfly.
378. Mittelstaedt, H., Ann. Rev. Entomol. *7*:177–198, 1962. Control systems of orientation in insects.
379. Mkrtycheva, L. J., Fed. Proc. Transl. Suppl. *25*(pt. II):T373–T376, 1966. Elements of functional organization of visual system in frog.
380. Morgane, P. J., and W. C. Stern, Ann. N. Y. Acad. Sci. *193*:95–110, 1972. Relation of sleep to neuroanatomical circuits.
381. Morrell, F., Electroenceph. Clin. Neurophysiol. Suppl. *13*:65–79, 1960. Dendritic locus of closure.
382. Mountcastle, V. B., pp. 1345–1371 *in* Medical Physiology, vol. 2, edited by V. B. Mountcastle. C. V. Mosby, St. Louis, 1968. Physiology of sensory receptors: introduction to sensory processes.
383. Mountcastle, V. B., P. W. Davies, and A. L. Bermann, J. Neurophysiol. *20*:374–407, 1957. Responses of somatic sensory cortex, cat.
384. Mulloney, B., Science *168*:994–998, 1970. Structure of the giant fibers of earthworms.
385. Mulloney, B., J. Neurophysiol. *33*:86–95, 1970. Organization of flight motorneurons of flies.
386. Mulloney, B., J. Exp. Biol. *52*:59–78, 1970. Impulse patterns in flight motorneurons of *Bombus*.
387. Muntz, W. R. A., J. Neurophysiol. *25*:699–711, 1962. Diencephalon of frog.
388. Murphy, J. T., and N. H. Sabah, Brain Res. *25*:449–467, 1971. Climbing fiber activation of Purkinje neurons.
389. Nachtigall, W., Z. vergl. Physiol. *61*:1–20, 1968. Flight patterns in flesh flies.
390. Nauta, W. J. H., and H. J. Karten, pp. 7–26 *in* The Neurosciences: Second Study Program, edited by F. O. Schmitt. Rockefeller University Press, New York, 1970. General profile of the vertebrate brain, with sidelights on the ancestry of the cerebral cortex.
391. Nichol, R. F., Exp. Brain Res. *14*:185–197, 1972. Excitation in olfactory bulb.
392. Nicholls, J. G., and D. A. Baylor, J. Neurophysiol. *31*:740–756, 1968. Sensory neurons of leech.
393. Nicholls, J. G., and D. Purves, J. Physiol. *209*:647–668, 1970. Connections between sensory and motor cells in leech.
394. Nicholson, C., and R. Llinas, J. Neurophysiol. *34*:509–531, 1971. Field potentials in alligator cerebellum.
395. Nicol, J. A. C., Quart. J. Micr. Sci. *89*:1–45, 1968. Giant fiber system of *Myxicola*.
396. Nieuwenhuys, R., and C. Nicholson, pp. 107–170 *in* Neurobiology of Cerebellar Evolution and Development, edited by R. Llinas. American Medical Association, Chicago, 1969. Cerebellum of mormyrid fishes.
396a. Northcutt, R. G., Ann. N.Y. Acad. Sci. *167*:180–185, 1969. Homologies in vertebrate brain.
397. Obata, K., et al., Exp. Brain Res. *4*:43–57, 1967. GABA as cerebellar transmitter.
398. O'Benar, J. D., Ph.D. Thesis, University of Illinois, 1971. Electrophysiology of goldfish optic tectum.
399. O'Brien, J. H., and S. S. Fox, J. Neurophysiol. *32*:267–296, 1969. Single cell activity in cat motor cortex.
400. Olds, J., pp. 257–293 *in* Neural Control of Behavior, edited by R. E. Whalen et al. Academic Press, New York, 1970. Behavior of hippocampal neurons during conditioning experiments.
401. Olds, J., and T. Hirano, Electroenceph. Clin. Neurophysiol. *26*:159–166, 1969. Conditioned responses of hippocampal and other neurons.
402. Oliver, G. W. O., et al., Comp. Biochem. Physiol. *38B*:529–535, 1971. Changes in GABA level, GAD, AChE activity in CNS of an insect during learning.
403. Olivo, R. F., Comp. Biochem. Physiol. *35*:787–807, 1970. Reflexes in razor clam.
404. Orrego, F., Arch. Ital. Biol. *99*:425–445, 1961. Olfactory pathways and cortical areas in turtle.
405. Packard, A., Nature *221*:875–877, 1969. Jet propulsion and giant fiber response of *Loligo*.
406. Page, C. H., J. Neurophysiol. *33*:116–128, 1970. Electrophysiology, auditory responses, goldfish brain.
407. Page, C. H., and A. M. Sutterlin, J. Neurophysiol. *33*:129–136, 1970. Visual-auditory unit responses in goldfish tegmentum.
408. Pantin, C. F. A., Amer. Zool. *5*:581–589, 1965. Capabilities of coelenterate behavior.
409. Parnas, I., et al., J. Exp. Biol. *50*:635–649, 1969. Conduction in giant axons of *Periplaneta*.
410. Parsons, L. C., and S. E. Huggins, Proc. Soc. Exp. Biol. Med. *119*: 397–400, 1965. Electrical activity in brain of *Caiman*.
411. Passano, L. M., Amer. Zool. *5*:465–481, 1965. Recordings from Scyphomedusae: giant fiber net for swimming beat and diffuse nerve net.
412. Passano, L. M., and C. B. McCullough, Nature *199*:1174–1175, 1963; *also* J. Exp. Biol. *41*:643–664, 1965; *42*:205–231, 1965. Coordinating system and behavior, *Hydra*.

413. Passano, L. M., and C. F. A. Pantin, Proc. Roy. Soc. Lond. B *143*: 226–238, 1955. Responses to mechanical stimulation in *Calliactis*.
414. Peretz, B., Science *166*:1167–1172, 1969; *169*:379–381, 1970. Control of gill movements in *Aplysia*.
415. Peters, J., A. Vonderahe, and D. Schmid, J. Exp. Zool. *60*:255–262, 1966. Onset of cerebral electrical activity associated with sleep and attention in developing chick.
416. Peterson, R. H., Brain Res. *41*:67–79, 1972. Electrical responses of goldfish cerebellum.
417. Peterson, R. H., Copeia 816–819, 1972. Somatosensory projection to facial lobe in goldfish.
418. Pettigrew, J., T. Nikara, and P. O. Bishop, Exp. Brain Res. *6*:373–390, 1968. Cat striate cortex.
419. Peyraud, C., Thèse, University of Toulouse, 1965. 234 pp. Respiratory movements in carp.
420. Pickens, P. E., J. Exp. Biol. *51*:513–528, 1969. Rapid contractions and associated potentials in a sand dwelling anemone.
421. Pickens, P. E., J. Exp. Biol. *53*:515–528, 1970. Conduction along ventral nerve cord of a hemichordate worm.
421a. Piddington, R. W., J. Exp. Biol. *55*:569–584, 585–610, 1971. Central control of auditory input in goldfish.
422. Pinsker, K., et al., Science *167*:1740–1742, 1970. Habituation and dishabituation, *Aplysia*.
423. Pitman, R. M., C. D. Tweedle, and M. J. Cohen, Science *178*:507–509, 1972. Responses of insect central neurons.
423a. Poggio, G. F., pp. 1592–1625 *in* Medical Physiology, edited by V. B. Mountcastle. C. V. Mosby, St. Louis, 1968. Central visual pathways, mammals.
424. Pople, W., and D. W. Ewer, J. Exp. Biol. *32*:59–69, 1955. Nervous system of holothurian *Cucumaria*.
425. Potter, H. D., J. Neurophysiol. *28*:1132–1152, 1965. Mesencephalic auditory region of the bullfrog.
426. Precht, H., and G. Freytag, Behaviour *13*:143–211, 1958. Habituation of prey capture in spiders.
427. Pritz, M. B., W. R. Mead, and R. G. Northcutt, J. Comp. Neurol. *140*: 81–100, 1970. Effects of Wulst ablation in pigeons.
428. Prosser, C. L., J. Comp. Neurol. *58*:603–630, 1934. Development of behavior in earthworm embryos.
429. Prosser, C. L., Z. vergl. Physiol. *50*:102–118, 1965. Electrical responses, fish optic tectum.
430. Prosser, C. L., and E. Farhi, Z. vergl. Physiol. *50*:91–101, 1965. Temperature effects on conditioned reflexes, goldfish.
431. Prosser, C. L., and T. Nagai, pp. 171–180 *in* The Central Nervous System and Fish Behavior, edited by D. Ingle. University of Chicago Press, 1968. Effect of low temperature on conditioning in goldfish.
432. Purpura, D. P., Int. Rev. Neurobiol. I:47–163, 1959. Nature of electrocortical potentials and synaptic organizations in cerebral and cerebellar cortex.
433. Rall, W., and G. M. Shepherd, J. Neurophysiol. *30*:1138–1167, 1968; *31*:884–915, 1968. Evoked potentials in olfactory bulb.
434. Rees, S., and J. Hore, Brain Res. *20*:439–451, 1970. Motor cortex of brush-tailed possum.
435. Remler, M., A. Selverston, and D. Kennedy, Science *162*:281–283, 1968. Lateral giant fibers of crayfish.
436. Ribbands, C. R., J. Exp. Biol. *27*:302–310, 1950. Memory in bees after anesthesia.
437. Roberts, A., J. Exp. Biol. *48*:545–567, *49*:645–656, 1968. Giant fibers and escape reflexes in crayfish.
438. Roberts, A., Z. vergl. Physiol. *71*:295–310, 1971; *75*:388–401, 1971. Skin impulses in sensory system of young tadpoles.
439. Roberts, B. L., J. Marine Biol. Assoc. U. K. *49*:33–49, 51–75, 1969. Function in spinal cord of dogfish.
440. Roberts, M. B. V., J. Exp. Biol. *39*:219–227, 229–237, 1962. Giant fiber reflexes in earthworm.
441. Roberts, M. B. V., J. Exp. Biol. *45*:141–150, 1966; *46*:571–583, 1967. Fast and slow reflexes, earthworm.
442. Roberts, R. B., and L. B. Flexner, Quart. Rev. Biochem. Biophys. *2*:135–173, 1969. Biochemistry of long-term memory.
443. Roberts, W. W., pp. 175–206 *in* Neural Control of Behavior, edited by R. E. Whalen et al. Academic Press, New York, 1970. Hypothalamic mechanisms for motivational and species-typical behavior (cat, opossum).
444. Robson, E. A., and R. K. Josephson, J. Exp. Biol. *50*:151–168, 1969. Neuromuscular properties of mesenteries from sea anemone.
445. Roeder, K. D., Biol. Bull. *69*:203–220, 1935; *also* J. Exp. Zool. *76*: 353–374, 1937. Nervous control of sexual behavior in mantids.
446. Roeder, K. D., pp. 423–487 *in* Insect Physiology, edited by K. D. Roeder. John Wiley & Sons, New York, 1953. Rhythms in insect nervous systems.
447. Roeder, K. D., Science *154*:1515–1521, 1966. Auditory system of noctuid moths.
448. Roeder, K. D., Nerve Cells and Insect Behavior. Harvard University Press, 1967. 238 pp.
449. Roeder, K. D., J. Insect Physiol. *12*:1227–1244, 1966; *15*:825–838, 1713–1718, 1969. Acoustic interneurons, noctuid moths.
450. Roeder, K. D., L. Tozian, and E. A. Weiant, J. Insect Physiol. *4*:45–62, 1960. Nerve activity and behavior in mantis and cockroach.
451. Rogers, F. T., J. Comp. Neurol. *35*:61–65, 1922; *also* Amer. J. Physiol. *86*:639–650, 1928. Localization of function in pigeon brain.
452. Romer, A. S., The Vertebrate Body, 4th edition. W. B. Saunders Co., Philadelphia, 1970. 601 pp.
453. Rose, S. P. R., pp. 517–551 *in* Short-Term Changes in Neural Activity and Behaviour, edited by G. Horn and R. A. Hinde. Cambridge University Press, 1970. Neurochemical correlates of learning.
454. Rosenthal, J., J. Physiol. *203*:121–133, 1969. PTP at neuromuscular junctions of frog.
455. Rosenzweig, M. R., E. L. Bennett, and M. C. Diamond, Sci. Amer. *226*:22–29, 1972. Brain changes in response to experience.
456. Rosenzweig, M. R., E. L. Bennett, M. C. Diamond, et al., Brain Res. *14*:427–445, 1969. Effect of environmental complexity and visual stimulation on occipital cortex in rat.
457. Ross, D. M., pp. 253–261 *in* Essays on Physiological Evolution, edited by I. Turpayev and J. Pringle. Pergamon Press, New York, 1964. Problems of neuromuscular activity and behavior.
458. Ross, D. M., and L. Sutton, Proc. Roy. Soc. Lond. B *155*:266–281, 1961. Response of sea anemone *Calliactis* to shells of hermit crab.
459. Ross, D. M., and L. Sutton, J. Exp. Biol. *41*:735–749, 1964. Swimming response of sea anemone *Stomphia* to electrical stimulation.
460. Rovainen, C., J. Neurophysiol. *30*:1000–1023, 1024–1042, 1967. Sea lamprey, functions of Müller and Mauthner cells; giant cells of spinal cord.
461. Rovainen, C., and K. L. Birnberger, J. Neurophysiol. *34*:974–982, 1971. Motorneuron function, sea lamprey.
462. Rowell, C. H. F., J. Exp. Biol. *40*:257–270, 1963; *44*:589–601, 1966. Arm of *Octopus*, reflexes.
463. Rowell, C. H. F., J. Exp. Biol. *41*:559–573, 1964. Central control of insect segmental reflex.
464. Rowell, C. H. F., pp. 237–280 *in* Short-Term Changes in Neural Activity and Behaviour, edited by G. Horn and R. A. Hinde. Cambridge University Press, 1970. Incremental and decremental processes in insect central nervous system.
465. Rowell, C. H. F., J. Exp. Biol. *55*:727–747, 749–761, 1971. Visual interneuron in locust.
466. Rowell, C. H. F., and J. M. McKay, J. Exp. Biol. *51*:231–245, 1969. An acridid grasshopper auditory interneuron.
467. Rowell, C. H. F., and G. Horn, J. Exp. Biol. *49*:143–169, 171–184, 1968. Long-term changes in neurons in locust.
468. Rowell, C. H. F., and G. Horn, J. Exp. Biol. *49*:176–183, 1968. Dishabituation and arousal in response of single nerve cell.
469. Rowland, V., pp. 482–495 *in* The Neurosciences: A Study Program, edited by G. C. Quarton. Rockefeller University Press, New York, 1967. Steady phenomena of cortex.
470. Rowland, V., and M. Goldstone, Electroenceph. Clin. Neurophysiol. *15*:474–485, 1963. Conditioned and drive-related baseline shift in cat cortex.
471. Rozin, P. N., and J. Mayer, Science *134*:942–943, 1961. Thermal reinforcement and thermoregulatory behavior in goldfish, *Carassius*.
472. Rubinson, K., Brain Behav. Evol. *1*:529–561, 1968. Projections of optic tectum of frog.
473. Rushforth, N. B., Animal Behavior: Suppl. 1, 30–42, 1965. Habituation in *Hydra*.
474. Rushforth, N. B., Biol. Bull. *140*:255–273; 502–519, 1971. Behavioral and electrophysiological studies of *Hydra*. I. Analysis of contraction pulse patterns.
475. Rushmer, D. S., and D. J. Woodward, Brain Res. *33*:83–90, 315–335, 1971. Responses of Purkinje cell in frog cerebellum.
476. Rutledge, L. J., Science *148*:1246–1248, 1965. Response enhanced by conditional excitation of cerebral cortex.
477. Salánki, J., Symp. Neurobiol. Invert. 493–501, 1967. Cerebral ganglia in regulation of activity in freshwater mussel *Anodonta*.
478. Salánki, J., T. Pécsi, and E. Lábos, Acta Biol. Acad. Sci. Hung. *19*: 391–406, 1968. Ganglionic regulation of tonic muscle, mollusc.
479. Salmoiraghi, G. C., and B. D. Burns, J. Neurophysiol. *23*:2–13, 14–26, 1960. Rhythmicity in respiratory neurons, cat.
480. Samsonova, V. G., Fed. Proc. Transl. Suppl. *25*:T384–T388, 1966. Functional organization of neurons in frog visual center.
481. Sandeman, D. C., J. Exp. Biol. *50*:771–784, 1969; *also* Z. vergl. Physiol. *64*:450–464, 1969; *72*:111–130, 1971. Function of identified motorneurons in crab brain.
482. Sanders, F. K., J. Exp. Biol. *17*:416–434, 1940. Second order learning, tectum of goldfish.
483. Sanderson, K. J., P. O. Bishop, and I. Darian-Smith, Exp. Brain Res. *13*:178–207, 1971. Binocular receptive fields of lateral geniculate neurons.
484. Scalia, F., Brain Behav. Evol. *4*:376–399, 1971. Central projections of olfactory bulb in *Gymnothorax*.
485. Schadé, J. P., Arch. Neerl. Zool. *14*:604–605, 1962. Visual responses of midbrain in fish.
486. Schadé, J. P., and I. J. Beiler, J. Exp. Biol. *36*:435–452, 1959. Electroencephalographic patterns of the goldfish (*Carassius*).
487. Schiller, P. H., and F. Koerner, J. Neurophysiol. *34*:920–936, 1971. Unit activity in superior colliculus of rhesus monkey.
488. Schrameck, J. E., Science *169*:698–700, 1970. Crayfish swimming: motor output and giant fiber activity.
489. Schwassman, H. O., and L. Kruger, J. Comp. Neurol. *124*:113–126, 1965. Visual projection to optic tectum in fish.
490. Seegar, J., Progr. Brain Res. *14*:143–231, 1965. Telencephalon of teleosts in relation to behavior.
491. Seegar, J., and R. Nieuwenhuys, Anim. Behav. *11*:331–344, 1963. Etho-physiological experiments with *Gasterosteus*.
492. Segundo, J. P., G. P. Moore, L. J. Stensaas, and T. H. Bullock, J. Exp. Biol. *40*:643–667, 1963. Sensitivity of neurons in *Aplysia* to input patterns.

493. Shapiro, S., and K. R. Vukovich, Science *167*:292–294, 1970. Developmental effects on cortical dendrites.
494. Sharma, S. C., R. R. Provine, V. Hamburger, and T. T. Sardel, Proc. Nat. Acad. Sci. *66*:40–47, 1970. Unit activity in isolated spinal cord of chick embryo.
495. Sharpless, S. K., Ann. Rev. Physiol. *26*:357–388, 1964. Reorganization of brain in use and disuse.
496. Shashona, V. E., Nature *217*:238–240, 1968. RNA in goldfish brain.
497. Sherman, J. D., J. Cell Biol. *37*:650–659, 1968. Electrical coupling between cells of chick embryo.
498. Sherrington, C. S., The Integrative Action of the Nervous System. Scribner's, New York, 1906.
499. Siminoff, R., H. O. Schwassmann, and L. Kruger, J. Comp. Neurol. *127*:435–444, 1966; *130*:329–342, 1967. Visual projection to superior colliculus, rat.
500. Smirnov, G. D., pp. 263–298 *in* Brain and Behavior, vol. 7, edited by M. A. B. Brazier. Amer. Inst. Biol. Sci., Washington, D.C., 1961. Neurophysiology of vision.
501. Smith, J. C., and H. D. Baker, J. Comp. Physiol. Psychol. *53*:279–281, 1960. Conditioning in *Limulus*.
502. Smith, J. E., pp. 503–511 *in* Physiology of Echinodermata, edited by R. Boolootian. Interscience, New York, 1966.
503. Sokolov, E. N., and V. P. Dulenko, Neurosci. Transl. *5*:592–599, 1968. Neuronal responses of *Helix* to tactile stimulation.
504. Sotelo, C., pp. 327–371 *in* Neurobiology of Cerebellar Evolution and Development, edited by R. Llinas. American Medical Association, Chicago, 1969. Ultrastructural aspects of cerebellar cortex of frog.
505. Spencer, W. A., and J. M. Brookhart, J. Neurophysiol. *24*:26–49, 50–65, 1961. Sensorimotor cortex of cat.
506. Spencer, W. A., R. F. Thompson, and D. R. Neilson, J. Neurophysiol. *29*:221–274, 1969. Conditioning in spinal cord.
507. Sperry, R. W., Proc. Nat. Acad. Sci. *50*:703–709, 1963. Chemoaffinity in orderly growth of nerve fiber patterns and connections.
508. Spinelli, D. N., and T. W. Barrett, Exp. Neurol. *24*:76–98, 1969. Visual cortex, cat.
509. Spira, M. E., I. Parnas, and F. Bergmann, J. Exp. Biol. *50*:615–627, 1969. Organization of giant axons of cockroach *Periplaneta*.
510. Sprague, J. M., Science *153*:1544–1546, 1966. Interaction of cortex and superior colliculus.
511. Starr, A., and R. Britt. J. Neurophysiol. *33*:137–147, 1970. Intracellular recording from cat cochlear nucleus during tone stimulation.
512. Stefanelli, A., Quart. Rev. Biol. *6*:17–34, 1951. Synaptic structure and Mauthner cells.
513. Sterling, P., and B. G. Wickelgren, J. Neurophysiol. *32*:1–16, 1969. Visual responses of superior colliculus, cat.
514. Stettner, L. J., and W. J. Schulz, Science *155*:1689–1692, 1967. Effects of brain lesions on learning in birds.
515. Stevens, R. J., Ph.D. Thesis, University of Illinois, 1969. Inhibition in frog optic tectum.
516. Stewart, D. L., K. L. Chow, and R. H. Masland, Jr., J. Neurophysiol. *34*:139–147, 1971. Lateral geniculate of rabbit.
517. Strumwasser, F., *in* Physiological and Biochemical Aspects of Nervous Integration, edited by F. D. Carlson. Prentice-Hall, 1968. Membrane and intracellular mechanisms of endogenous activity in neurons.
518. Strumwasser, F., pp. 442–462 *in* Circadian Clocks, edited by J. Aschoff, North-Holland Publ. Co., Amsterdam, 1965; *also* pp. 516–528 *in* The Neurosciences: A Study Program, edited by G. C. Quarton et al., Rockefeller University Press, New York, 1967; *also* pp. 291–320 *in* Invertebrate Nervous Systems, edited by C. A. G. Wiersma, University of Chicago Press, 1967. Circadian rhythm in a single neuron.
519. Strumwasser, F., J. Psychiatr. Res. *8*:237–257, 1971. Rhythmic activity in molluscan neurons in culture.
520. Stuart, A. E., Science *165*:817–819, 1969; *also* J. Physiol. *209*:627–646, 1970. Motorneurons in leech.
521. Suga, N., Japan. J. Physiol. *11*:666–677, 1961. Tympanic neurons in noctuid moths.
522. Suga, N., J. Insect Physiol. *12*:1039–1050, 1966. Ultrasonic production and its reception in some neotropical Tettigoniidae.
523. Summers, T. B., and W. N. Kaelber, Amer. J. Physiol. *203*:1117–1119, 1962. Amygdalectomy in cats.
524. Sutherland, N. S., J. Comp. Physiol. Psychol. *54*:43–48, 1961. Discrimination, horizontal and vertical, octopus.
525. Sutterlin, A. M., and C. L. Prosser, J. Neurophysiol. *33*:36–45, 1970. Electrical responses, optic tectum, goldfish.
526. Takeda, K., and D. Kennedy, J. Cell. Comp. Physiol. *64*:165–182, 1964. Modes of activation of crayfish motorneurons.
527. Tarr, R. S., Ph.D. Thesis, University of Illinois, 1971. Neuro-ethological study of ritualized aggressive behavior in the western fence lizard, *Sceloporus*.
528. Tauc, L., J. Gen. Physiol. *45*:1077–1097, 1099–1115, 1962. Active membrane areas in giant neuron of *Aplysia*.
529. Tauc. L., C. R. Acad. Sci. Paris *259*:885–888, 1964. Presynaptic inhibition, *Aplysia* neurons.
530. Tauc, L., pp. 247–257 *in* Mechanisms of Synaptic Transmission, edited by K. Akert and P. G. Waser. Elsevier, Amsterdam, 1969. Polyphasic synaptic activity.
531. Tauc, L., and G. M. Hughes, J. Gen. Physiol. *46*:533–549, 1963. Spikes in mollusc central neurons.
532. Taylor, R. C., Comp. Biochem. Physiol. *33*:911–921, 1970. Control of sensitivity of single crayfish interneuron.
533. Teravainen, H., and C. M. Rovainen, J. Neurophysiol. *34*:990–998, 999–1009, 1971. Motorneurons, sea lamprey.
534. Thach, W. T., J. Neurophysiol. *33*:527–536, 537–547, 1970. Discharge of cerebellar neurons related to two maintained postures and two prompt movements.
535. Thompson, R. F., R. H. Johnson, and J. J. Hooper, J. Neurophysiol. *26*:343–364, 1963. Organization of auditory, somatic sensory, and visual projection to association fields of cerebral cortex in cat.
536. Thompson, R. F., and J. A. Shaw, J. Comp. Physiol. Psychol. *60*:329–339, 1965; *69*:485–491, 1969. Association cortex, cat.
537. Thompson, R. F., and W. A. Spencer, Psychol. Rev. *73*:16–43, 1966. Habituation.
538. Thorpe, W. H., Learning and Instinct in Animals. Methuen, London, 1963. 558 pp.
539. Towe, A. L., Exp. Neurol. *15*:113–139, 1966. On nature of primary evoked response.
540. Treherne, J. E., and J. W. Beament, editors, Physiology of the Insect Central Nervous System. Academic Press, New York, 1965. 277 pp.
541. Truman, J. W., and P. G. Sokolove, Science *175*:1491–1493, 1972. Silk moth eclosion.
542. Turner, B., J. Comp. Physiol. Psychol. *71*:103–113, 1970. Rage syndrome of rat.
543. Ueda, K., T. J. Hara, and A. Gorbman, Comp. Biochem. Physiol. *21*:133–143, 1967. Olfactory discrimination in spawning salmon.
544. Van der Kloot, W. G., and C. M. Williams, Behavior *5*:141–174, 1953; *6*:233–255, 1954. Hormonal and neural control of construction in silkmoths.
545. Vardapetyan, G. A., Neurosci. Transl. *1*:1–11, 1967. Single unit responses in auditory cortex of rats.
546. Vasil'ev, A. G., Dokl. Akad. Nauk. S.S.S.R. *175*:523–526, 1967. Activity in inferior corpora quadrigemina of bats subjected to ultrasonic stimulation.
547. Verzeano, M., pp. 27–54 *in* Neuronal Control of Behavior, edited by R. E. Whalen et al. Academic Press, New York, 1970. Evoked responses and network dynamics.
548. Verzeano, M., R. C. Dill, et al., Experientia *24*:696–698, 1968. Evoked responses, lateral geniculate.
459. Veselkin, N. P., Fed. Proc. Transl. Suppl. *25*:T957–T960, 1966. Electrical reactions in midbrain, medulla and spinal cord of lamprey.
550. Veselkin, N. P., et al., Brain Behav. Evol. *4*:295–306, 1971. Thalamotelencephalic afferent system in frogs.
551. Voronin, L. L., and V. G. Skrebitaku, Fed. Proc. Transl. Suppl. *25*: T574–T576, 1967. Intracellular investigation of cortical neurons in unanesthetized rabbits.
552. Vowles, D. M., J. Comp. Physiol. Psychol. *58*:105–111, 1964. Olfactory learning and brain lesions in wood ant *Formica*.
553. Wachtel, H., and E. R. Kandel, Science *158*:1206–1208, 1967. Synaptic connection mediating both excitation and inhibition.
554. Waks, M. D., and R. D. Westerman, Comp. Biochem. Physiol. *33*: 465–469, 1970. Purkinje cells of *Salmo*.
555. Waldron, J., J. Exp. Biol. *47*:201–212, 1967. Motor output pattern in flying locusts.
556. Waldron, J., Z. vergl. Physiol. *57*:331–347, 1968. Mechanism of coupling of locust flight oscillator to oscillatory inputs.
557. Wall, P. D., and A. R. Johnson, J. Neurophysiol. *21*:148–158, 1958. Post-tetanic potentiation of a monosynaptic reflex.
558. Weiss, P., J. Comp. Neurol. *67*:269–315, 1937; *also* Comp. Psychol. Monogr. *17*:1–96, 1941. Coordination of transplanted amphibian limbs.
559. Wells, G. P., J. Marine Biol. Assoc. U. K. *28*:447–464, 1949; *also* J. Exp. Biol. *28*:41–56, 1951. Feeding rhythms in *Arenicola*.
560. Wells, M. J., J. Exp. Biol. *38*:811–826, 1961. Tactile and visual learning in brain of *Octopus*.
561. Wells, M. J., Brain and Behavior in Cephalopods. Stanford University Press, 1962. 171 pp.
562. Wells, M. J., Ergebn. Biol. *26*:40–84, 1963. Orientation of *Octopus*.
563. Wells, M. J., J. Exp. Biol. *41*:621–642, 1964. Detour experiments with octopuses.
564. Wells, M. J., J. Exp. Biol. *42*:233–255, 1965. Orbital lobe and touch learning in *Octopus*.
565. Wells, M. J., J. Exp. Biol. *47*:393–408, 1967. Short-term learning and interocular transfer in detour experiments with octopuses.
566. Wells, M. J., and J. Z. Young, J. Exp. Biol. *50*:515–526, 1969. Brain lesions and effect of splitting part of brain or removal of median inferior frontal lobe on touch learning in *Octopus*.
567. Wendler, G., Soc. Exp. Biol. Symp. *20*:229–249, 1966. Coordination of walking movements in arthropods.
568. Westfall, J. A., J. Ultrastruct. Res. *32*:237–246, 1970; *also* J. Cell Biol. *51*:318–323, 1971. Synapses in a primitive coelenterate.
569. Whiting, H. P., pp. 85–103 *in* International Neurochemical Symposium, 1st, Biochemistry of the Developing Nervous System: Proceedings, edited by H. Waelsch. Academic Press, New York, 1955. Functional development in the nervous system.
570. Wickelgren, B. G., J. Neurophysiol. *30*:1404–1423, 1967. Habituation of spinal motorneurons.
571. Wickelgren, B. G., and P. Sterling, J. Neurophysiol. *32*:16–23, 1969. Influence of visual cortex on receptive fields in superior colliculus.
572. Wiersma, C. A. G., J. Neurophysiol. *10*:23–38, 1947. Giant nerve fiber system of the crayfish.
573. Wiersma, C. A. G., J. Cell. Comp. Physiol. *40*:399–419, 1952; *also*

Physiol. Rev. *33*:326–355, 1953. Transmission in giant fiber systems, crayfish.

574. Wiersma, C. A. G., pp. 241–279 *in* Physiology of Crustaceans, vol. 2, edited by T. Waterman. Academic Press, New York, 1961. Central nervous systems, crustaceans.
575. Wiersma, C. A. G., pp. 269–284 *in* Invertebrate Nervous Systems, edited by C. A. G. Wiersma. University of Chicago Press, 1967. Visual interneurons, crayfish.
576. Wiersma, C. A. G., and B. M. H. Bush, J. Comp. Neurol. *121*:207–235, 1963. Neuronal connections between thoracic and abdominal cords of crayfish.
577. Wiersma, C. A. G., and G. M. Hughes, J. Comp. Neurol. *116*:209–228, 1961. Functional anatomy of neuronal units in abdominal cord of crayfish.
578. Wiersma, C. A. G., and K. Ikeda, Comp. Biochem. Physiol. *12*:509–525, 1964. Interneurons commanding swimmeret movements in crayfish.
579. Wiersma, C. A. G., and P. S. Mill, J. Comp. Neurol. *125*:67–94, 1965. Descending neuronal units in commissure of crayfish CNS.
580. Wiersma, C. A. G., and T. Yamaguchi, J. Exp. Biol. *47*:409–431, 1967. Integration of visual stimuli by crayfish central nervous system.
581. Wiesel, T. N., and D. H. Hubel, J. Neurophysiol. *26*:1003–1017, 1965; *28*:1060–1072, 1965. Single-cell responses in striate cortex of kittens deprived of vision in one eye.
582. Willows, A. O. D., *in* Physiological and Biochemical Aspects of Nervous Integration, edited by F. D. Carlson, Prentice-Hall, 1967; *also* Science *157*:570–574, 1967. Behavioral acts elicited by stimulation of single identifiable nerve cells, *Tritonia.*

582a. Willows, A. O. D., Sci. Amer. *224*:68–75, 1971. Behavioral function, single neurons, *Tritonia.*

583. Willows, A. O. D., and G. Hoyle, Science *166*:1549–1551, 1969. Neuronal network triggering a fixed action pattern.
584. Wilson, D. M., J. Exp. Biol. *38*:471–490, 1961; *41*:191–205, 1964. Locust flight, nervous control.
585. Wilson, D. M., J. Exp. Biol. *47*:133–151, 1967. Stepping patterns in tarantulas.
586. Wilson, D. M., *in* Neural Theory and Modeling, edited by R. F. Reis. Stanford University Press, 1964. The origin of flight-motor command in grasshoppers.
587. Wilson, D. M., J. Exp. Biol. *43*:397–409, 1965. Proprioceptive leg reflexes in cockroaches.
588. Wilson, D. M., J. Exp. Biol. *39*:669–677, 1962. Bifunctional muscles in thorax of grasshoppers.
589. Wilson, D. M., J. Exp. Biol. *48*:631–641, 1968. Reflex modulation of locust flight.
590. Wilson, D. M., Adv. Insect Physiol. *5*:289–338, 1968. The nervous control of insect flight.
591. Wilson, D. M., and E. Gettrup, J. Exp. Biol. *40*:171–185, 1963. Stretch reflex controlling wingbeat frequency in grasshopper.
592. Wilson, D. M., and T. Weis-Fogh, J. Exp. Biol. *39*:643–667, 1962. Coordinated motor units in flying locusts.
593. Wilson, D. M., and R. J. Wyman, J. Insect Physiol. *9*:857–865, 1963. Phasically unpatterned nervous control of dipteran flight.
594. Wilson, V. J., J. Gen. Physiol. *39*:197–206, 1955; *41*:1005–1018, 1958. Post-tetanic potentiation of polysynaptic reflexes in spinal cord.
595. Wilson, V. J., and P. R. Burgess, J. Neurophysiol. *25*:636–650, 1962. Antidromic conditioning of some motorneurons and interneurons.
596. Wilson, V. J., et al., J. Neurophysiol. *27*:1063–1079, 1964. Inhibitory convergence on Renshaw cells.
597. Wine, J. J., and F. B. Krasne, J. Exp. Biol. *56*:1–18, 1972. Escape behavior in crayfish.
598. Winlow, W., and M. S. Laverack, Marine Behav. Physiol. *1*:1–27, 29–47, 93–121, 1972. Control of hindgut motility in lobster.

598a. Witt, P. N., Amer. Zool. *9*:121–131, 1969. Effects of brain lesions in spider *Araneus.*

599. Woody, C. D., J. Neurophysiol. *33*:838–850, 1970. Conditioned eye blink.
600. Yoon, M., Exp. Neurol. *33*:395–411, 1972. Connections of retina to tectum, goldfish.
601. Yoshida, M., pp. 435–465 *in* Physiology of Echinodermata, edited by R. Boolootian. Interscience, New York, 1966. Photosensitivity of echinoids.
602. Young, J. Z., Quart. J. Micr. Sci. *78*:367–386, 1936; *also* Phil. Trans. Roy. Soc. Lond. B *229*:465–501, 1939. Giant fiber system of cephalopods.
603. Young, J. Z., Proc. Roy. Soc. Lond. B *149*:463–483, 1958. Responses of untrained octopuses to various figures and the effect of removal of the vertical lobe.
604. Young, J. Z., Proc. Roy. Soc. Lond. B *153*:18–46, 1960. Failure of discrimination learning following removal of vertical lobe of *Octopus.*
605. Young, J. Z., Biol. Rev. *36*:32–96, 1961. Learning and discrimination in *Octopus.*
606. Young, J. Z., Phil. Trans. Roy. Soc. Lond. B *245*:1–58, 1962. Retina and optic lobes of cephalopods.
607. Young, J. Z., Quart. J. Exp. Psychol. *14*:193–205, 1962. Reversal of learning in *Octopus* and effect of removal of vertical lobe.
608. Young, J. Z., Nature *198*:626–630, 1963. Essentials of neural memory system.
609. Young, J. Z., Phil. Trans. Roy. Soc. Lond. B *249*:27–44, 45–67, 1965. *Octopus,* buccal nerve system; centers for touch discrimination.
610. Young, J. Z., J. Exp. Biol. *43*:595–603, 1965. Influence of previous preferences on memory of *Octopus.*
611. Young, J. Z., pp. 353–362 *in* Invertebrate Nervous Systems, edited by C. A. G. Wiersma. University of Chicago Press, 1967. Cephalopod brain.
612. Young, J. Z., J. Exp. Biol. *49*:413–419, 1968. Reversal of visual preference in *Octopus* after removal of vertical lobe.
613. Young, J. Z., J. Exp. Biol. *52*:385–393, 1970. Short and long memories in *Octopus* and the influence of the vertical lobe system.
614. Zagorul'ko, T. M., Neurosci. Transl. *6*:659–668, 1969. Evoked responses of general cortex and optic tectum in turtles.
615. Zenkin, Z. M., and I. N. Pigarev, Biophysics *14*:763–772, 1969. Tectal responses in fish *Esox.*
616. Zhantiev, R. D., Dokl. Akad. Nauk. S.S.S.R. *181*:477–479, 1968. Functional properties of neurons in central nervous system of orthopterous insects.
617. Zucker, R. S., J. Neurophysiol. *35*:599–620, 621–637, 1972. Crayfish escape behavior and central synapses.
618. Zucker, R. S., J. Neurophysiol. *35*:638–651, 1972. Excitation of fast flexor motorneurons.

CHAPTER 16

MUSCLES

By C. Ladd Prosser

The speed of locomotion of an animal, and hence its ability to get food and to escape from predators, is limited by the reaction times of its muscles. Muscles perform functions (in addition to locomotion) associated with vocalization, digestion, excretion, and reproduction, and most animals have muscles of several characteristic reaction times. Contractile fibers occur in Protozoa (e.g., myonemes of *Vorticella* and *Stentor*), and contractile myocytes can close the oscula of sponges. The cellular mechanisms of contraction and relaxation reside in certain unique proteins that can develop tension reversibly. Similar contractile machinery can give, at one extreme, the high speed of a mammalian eye movement or the beat of a mosquito's wing and, at the other extreme, the sluggish motion of a vertebrate intestine or of a sea anemone.

No other tissue illustrates so well as muscle the theme of a common mechanism with variations adapted to specific functions. Muscles occur in a spectrum of types with respect to histology, biochemistry, and mechanics. Correlated differences are described for the following properties: (1) morphological arrangement of contractile elements; (2) the nature of contractile proteins; (3) provision of energy for contraction and relaxation; (4) time relations of movement; (5) mode of neuromuscular activation and its dependence on facilitation; (6) coupling of membrane activity to contraction; (7) mechanical (viscoelastic) properties as shown by maintenance of tension, tonus, and effects of quick stretch and release; and (8) spontaneous rhythmicity and its nervous and mechanical control.

Postural muscles generally have origins and insertions on skeletal structures, either endoskeletal or exoskeletal, or on skin. Examples are the muscles which move such appendages as legs, wings, and mouth parts, muscles which protrude or retract proboscises or tentacles, or muscles which close the valves of a mollusc. Such muscles often occur in antagonistic pairs—contraction of one muscle reflexly inhibits contraction of its antagonist. Some other muscles are opposed by an elastic ligament, as are clam adductors. Usually these are relatively fast, or *phasic*, and they form part of a lever system and function by shortening (isotonic contraction) or by developing tension while at a constant length (isometric contraction). No muscle functions purely isotonically or isometrically, but a particular muscle can approach one or the other way of functioning.

Other muscles are arranged around hollow structures, one portion of the muscle inserting into and hence pulling on another portion of itself. Examples are the muscles of the bladder, ureter, uterus, gastrointestinal tract, and heart, as well as the body wall muscles of annelids, holothurians, and coelenterates. In general, muscles of hollow organs are rather slow *(tonic)*, they may occur in pairs (circular and longitudinal), and they may contract against sacs of fluid.

Both the phasic (fast) and tonic (slow) types of muscles are under reflex control, but many of the latter are capable of spontaneous rhythmicity, which may be modulated neurally. Phasic muscles are often organized into motor units, all of the fibers of which are controlled by one motorneuron. Gradation of movement may be by the number of motor units activated, by the frequency of motor impulses, by the balance between fast and slow excitor and inhibitor neurons, or by the mechanical properties of both connective tissue and contractile cells.

HISTOLOGICAL TYPES OF MUSCLES

The basic structural element of muscle is the myofilament of contractile protein; thick filaments (diameter 150 to 250 Å) contain myosin, and thin ones (diameter 45 to 65 Å) contain actin. Filaments are sometimes arranged in fibrils. The basic cellular unit is the muscle fiber, which may be either multinucleate or uninucleate; the fibers may be arranged in bundles. The amount of connective tissue in which muscle fibers are embedded differs greatly in different muscles. The outer layer of the fiber membrane (outer sarcolemma) appears to be in common with connective tissue, while the inner layer is the physiological plasma membrane. Muscles are crudely classified as striated and nonstriated, but many intermediate types occur (e.g., the "spiral" striated muscles of annelids).

Striated Muscles. Striated muscles are characterized by a transverse alignment of thick and thin filaments and by transversely and longitudinally arranged tubules. Striations originated early in the evolution of muscles, and many coelenterate fibers are cross-striated. Cross striations consist of alternating strongly birefringent and weakly birefringent regions, the anisotropic (A) and isotropic (I) bands. The I bands are divided by Z lines (Fig. 16–1) which are transverse boundaries between the structural units, the sarcomeres. The I band consists of thin (actin) filaments, and the A band consists of thick (myosin) filaments. In most of the A band there is overlap of thick and thin filaments; however, in the middle there may be a lighter H region where there are only thick filaments, and in the center of this may be an M line. Cross bridges connect the thick and thin filaments in regions of overlap, and thin filaments are anchored at Z lines.

The tubular system consists of (1) indentations of the sarcolemma which penetrate for varying distances at either the Z line or the A-I junction, called transverse (T) tubules, and (2) a reticulum of longitudinal sarcoplasmic (SR) tubules between myofibrils and vesicles, which forms a bracelet around a sarcomere with bulbs adjacent to the T-tubules. The T-tubules provide for considerable inward extension of the sarcolemmal membrane, and the SR-tubules provide for excitation-contraction coupling.

There are correlations between speed of movement and the following factors: sarcomere length, regularity and density of myofilaments, abundance of mitochondria and fat droplets, and presence or absence of myoglobin in the sarcoplasm. Some fast muscles are sarcoplasm-poor (i.e., they are of "Fibrillen Struktur"), having many filaments arranged uniformly; there are usually six thin filaments around each thick one,

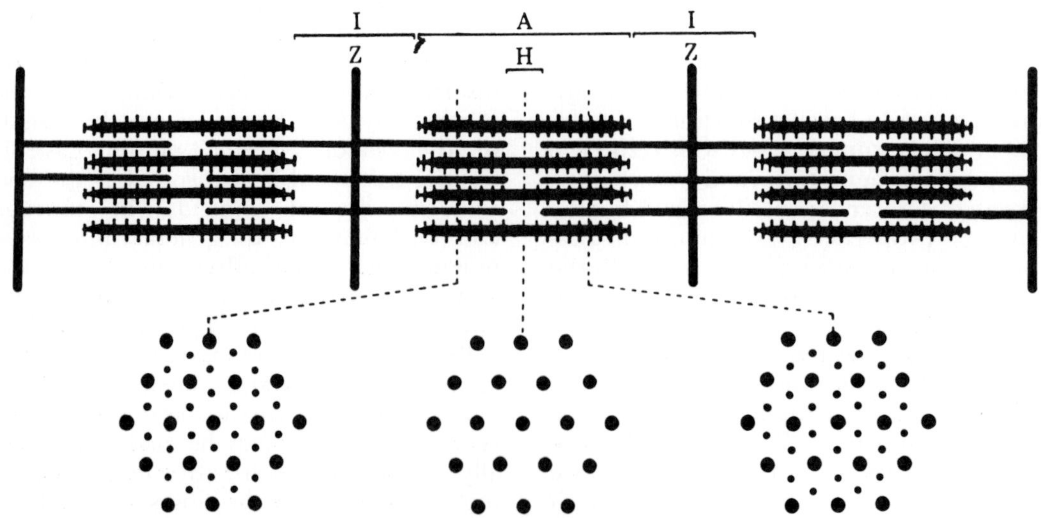

Figure 16–1. Diagram showing structure of striated muscle with overlapping arrays of thick and thin filaments, cross bridges, and Z bands in middle of I bands. Longitudinal section above, and cross sections at different positions in sarcomere below. (From Huxley, H. E., Proc. Roy. Soc. Lond. B *178*:131–149, 1971.)

with the tubular system in a regular pattern and the Z line straight across the fiber (Fig. 16–2). Some slower muscles are sarcoplasm-rich (i.e., they are of "Felder Struktur"), with less regular arrangement of filaments and tubules, more mitochondria, and a zigzag Z line (Fig. 16–3). In general, long sarcomeres are associated with high tension per unit cross section; this is because cross bridge spacings are constant, and whenever there are more bridges in parallel, higher tension is possible. On the other hand, short sarcomeres are associated with fast contraction. Usually the muscles with Fibrillen Struktur have single points of innervation with "en plaque" or single plate nerve endings, while those with Felder Struktur have several "en grappe" or multiple knob nerve endings. Striated muscle fibers are multinucleate and of large diameter (~ 100 μm).

Examples of structure in fast striated muscles are given in Figure 16–4. In frog sartorius, the fibrils or bundles of filaments are small, and T-tubules cross the fiber, forming a flat disc or *cisterna* at the Z level. Adjacent to these are the terminal cisternae of the SR tubular system, forming triads where the two cisternae abut the T-tubules. Terminal tubules connect to longitudinal tubules of the SR system (sarcoplasmic reticulum). The T-tubule area is estimated as seven times the area of the outer surface of the fiber.[276] In "slow" mammalian muscles, the volume of the T system is 0.18 per cent of fiber volume, as compared with 0.39% in "fast" muscles, and the area of diadic contact is five to ten times greater in the fast fibers.

In a garter snake, the twitch fibers have T-tubules at the AI junction (hence there are two per sarcomere), and their communication to the extracellular space is shown by their uptake of particulate ferritin.[286] In fish muscles, the triads may be at the AI junction (as in *Fundulus* extraocular muscle) or at the Z band (as in *Hippocampus* dorsal fin muscle).[32] In the very fast sonic muscle of toadfish swim-bladder, the sarcomeres are only 2 μm long, the filaments are regular (particularly in the core of the fiber), and T-tubules penetrate at the AI junction (and hence there are two per sarcomere).[113, 304] In crayfish leg muscles, there are large sarcolemmal invaginations at many sites per sarcomere; each gives rise to a branching 300 Å T-tubule which terminally contacts an SR-tubule at the AI region, forming a diad[44] (Fig. 16–5). The distance from an SR-tubule to the center of a

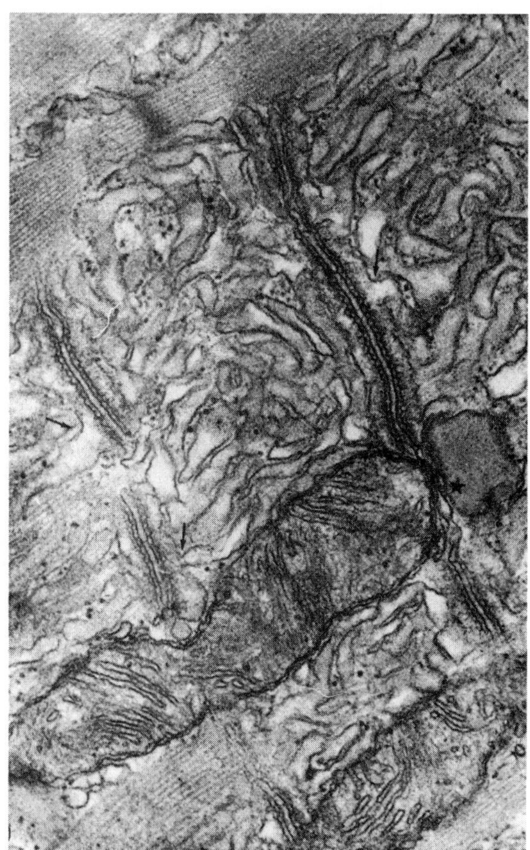

Figure 16–2. Ultrastructure of fast striated muscle, showing T-tubules in relation to fibrils and mitochondria. Electron micrograph of bat cricothyroid muscle. (From Revel, J. P., J. Cell Biol. *12*:571–588, 1962.)

fibril is no more than 0.4 μm.[202] The total membrane area, including T-tubules, is some 10 times that of the external membranes.[303] In crab fibers, similar invaginations or clefts occur, but they give rise to blind tubules at the Z level, whereas at the AI level invaginating T-tubules form diads with S-elements.[275] Fast muscles of a copepod resemble those of crayfish in that tubules penetrate at the Z level and give off branches to the cisternae.[110] Figure 16–6 shows the structure of fast and slow fibers of scorpion muscles.

The ratio of thin filaments to thick filaments varies in different muscles and even in different parts of one fiber.[333, 334] In *Limulus* muscle, which has 8 μm sarcomeres, the thick filaments have a dense core and less dense cortex and there are 12 thin filaments per thick one in the center of the A band; there are more thin filaments at the ends.[197] In some crayfish and cockroach leg muscles the ratio of thin to thick filaments is 6:1, in thoracic muscles of cockroach and butterfly

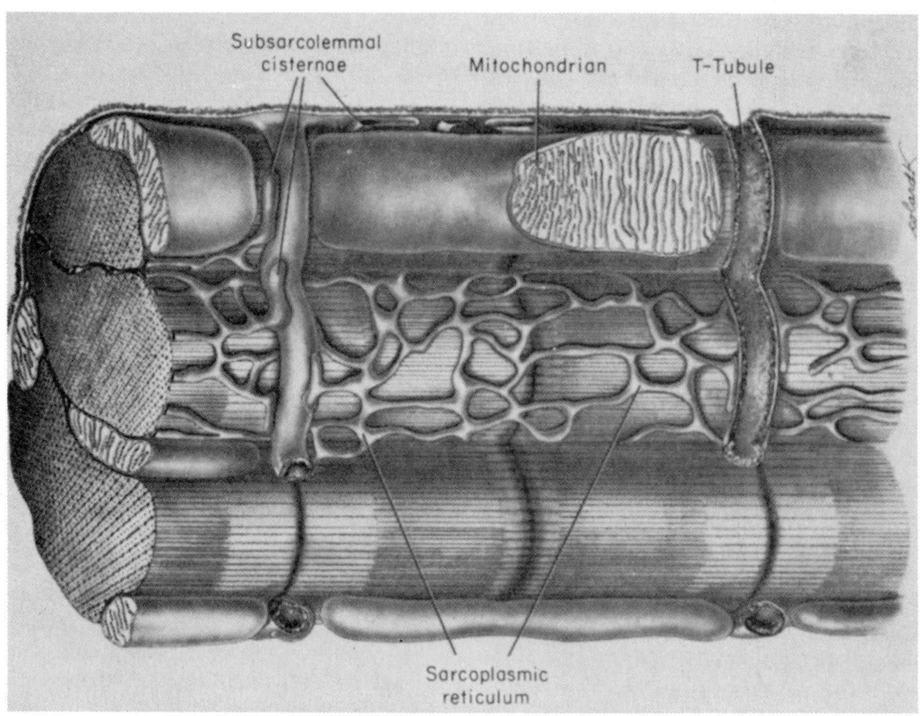

Figure 16-3. Model of ventricular papillary muscle of cat. (From Fawcett, D. W., and N. S. McNutt, J. Cell Biol. *42*:1–45, 1969.)

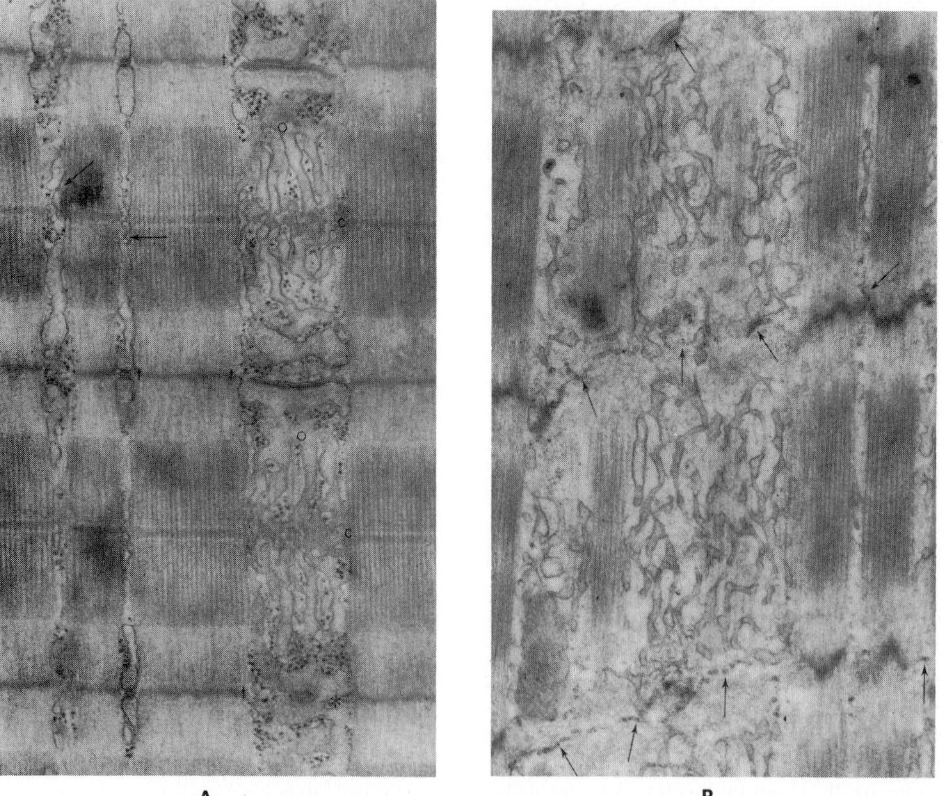

Figure 16-4. Comparison of twitch and tonic muscle fibers of frog. A, Electron micrograph of twitch fiber. B, Longitudinal section of tonic fiber. (From Page, S. G., J. Cell Biol. *26*:477–497, 1965.)

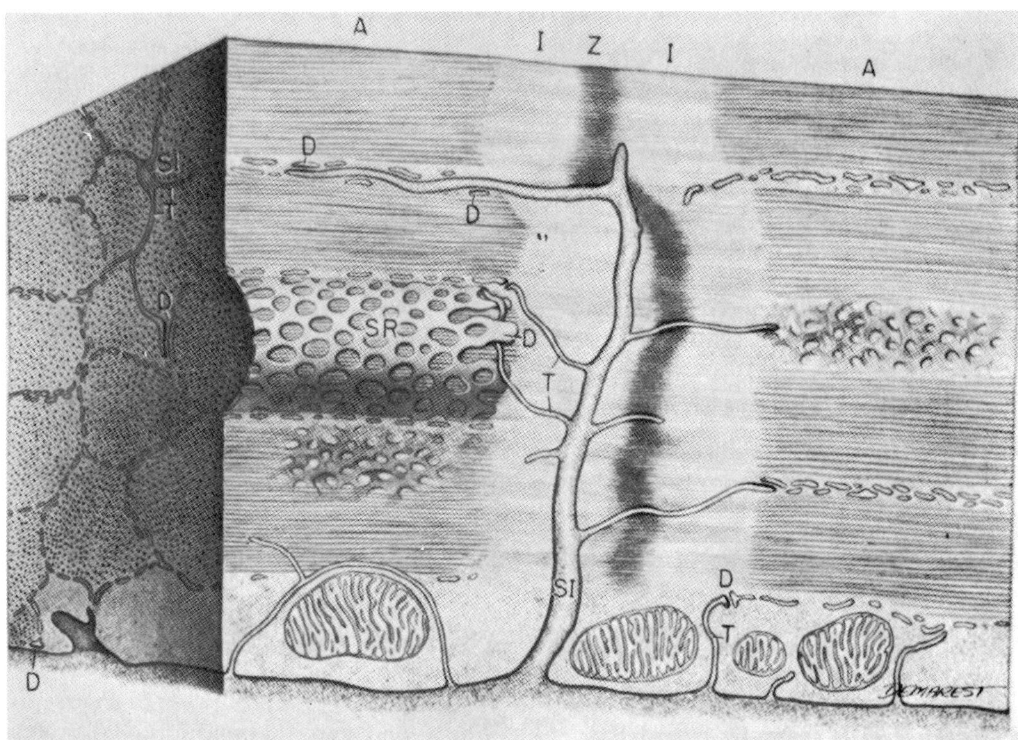

Figure 16–5. Model of crayfish muscle fiber, showing sarcolemmal invagination and T-tubules (SI and T) with connections at diads (D) to S-tubules (SR). (From Reuben, J. P., P. W. Brandt, H. Garcia, and H. Grundfest, Amer. Zool. 7:623–645, 1967.)

4:1, in dragonfly and blowfly flight muscles 3:1, and in most vertebrate striated and cardiac muscle 2:1.[153b] Cross bridges between thick and thin filaments form a left-handed helical pattern, four bridges per 146 Å; these are in a chevron-like array when contracted in rigor.[229] Tracheoles may penetrate inside large muscle fibers of insects, and the mitochondria (sarcosomes) may constitute as much as 40% of the fiber volume.[333]

A direct correlation is found between fiber diameter and abundance of sarcotubules. In papillary fibers of cat heart, 10 to 12 μm cells have large T-tubules and extensive SR-tubules, while in auricle cells that are 5 to 6 μm in diameter, there are few scattered T-tubules and the cisternae of the SR system go to the cell periphery rather than to the T system. In *Branchiostoma* (amphioxus) myotomes the fibers are flat lamellae, 8 μm thick, 100 μm wide, and 600 μm long, and there is little sarcoplasmic reticulum.[274] The large fibers of insects have such a rich tubular system that no myofilaments are farther than 2 μm from a tubule.[333]

In summary, slow striated muscles tend to have filaments in the irregular Felder Struktur, to have irregular Z bands, and to have infrequent or no triads and sparse tubules; they may also have multiple nerve endings with "en grappe" terminals. Slow muscles have long sarcomeres and usually have more mitochondria and fat droplets than phasic muscles. Also, tonic muscles which fatigue slowly often contain myoglobin (which imparts a red color), while rapidly fatiguing phasic muscles are "white." Exceptions to these generalizations will be pointed out later.

Intermediate Types of Muscle. Several muscles have highly modified striated structure. In some invertebrates, especially trematodes and ascidians, the muscle fibers have typical striations along one side of the cell and mitochondria on the opposite half. Other invertebrate muscles show oblique striations. The Z bands can be traced as irregular or staggered in a diagonal fashion.[248] This arrangement is seen in the striated (translucent) muscles of bivalves such as *Pecten,* and in the chromatophore muscles of the squid.[123] In earthworms, the fibers are ribbon-shaped, with bundles of filaments in the cortex and arranged in a helical pattern around the fiber; this helical pattern is particularly evident in polarized light.

The body wall muscles of *Ascaris* and the myotomes of *Branchiostoma* (amphioxus) send processes out from one side of a fiber

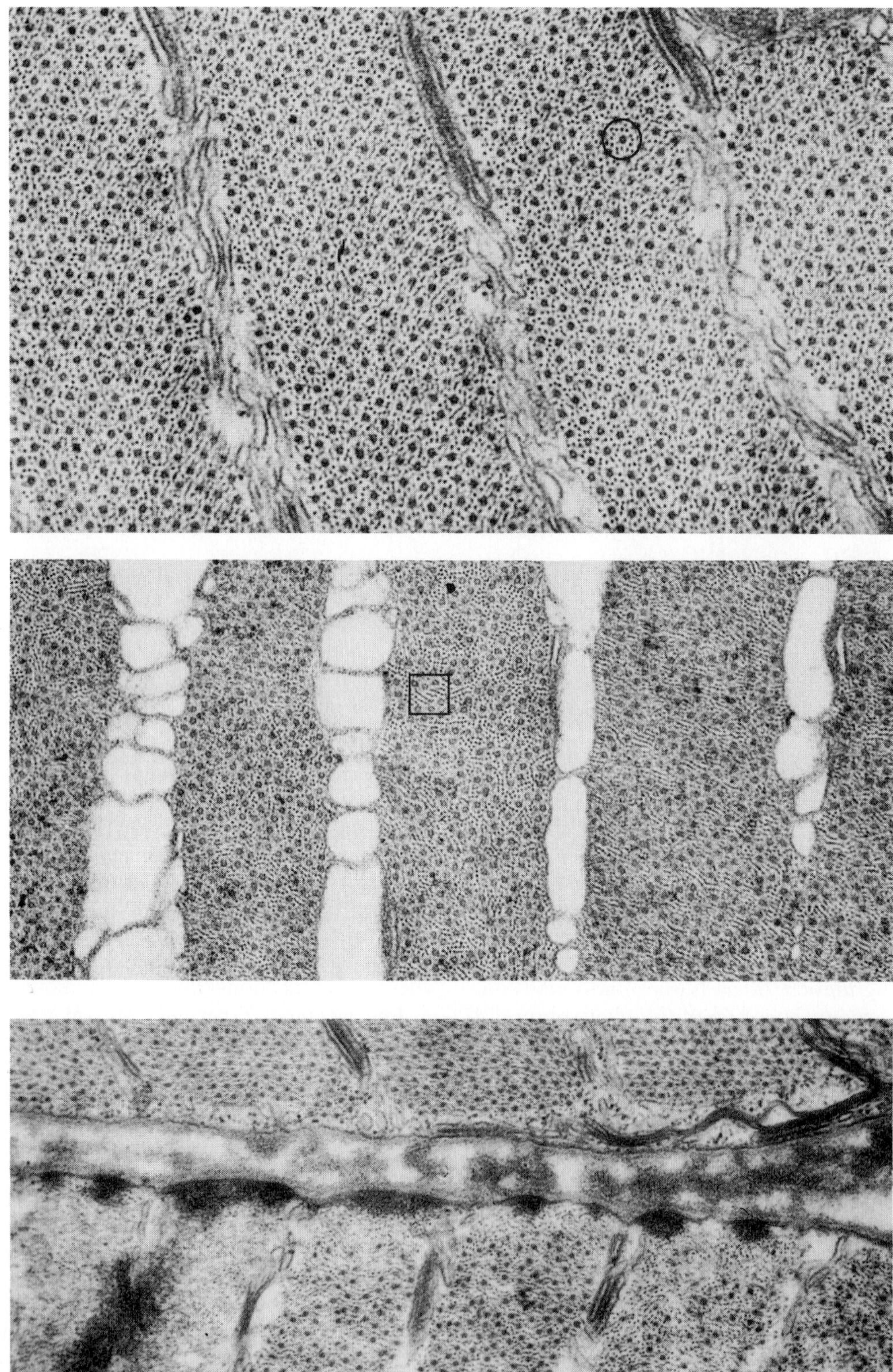

Figure 16–6. A, Cross sections of scorpion muscle fibers. Upper figure from fast fibers, middle figure from slow fibers, and lower figure from intermediate fibers. Note differences in regularity of filament patterns and in T and S tubules.

(Figure 16–6 continued on opposite page.)

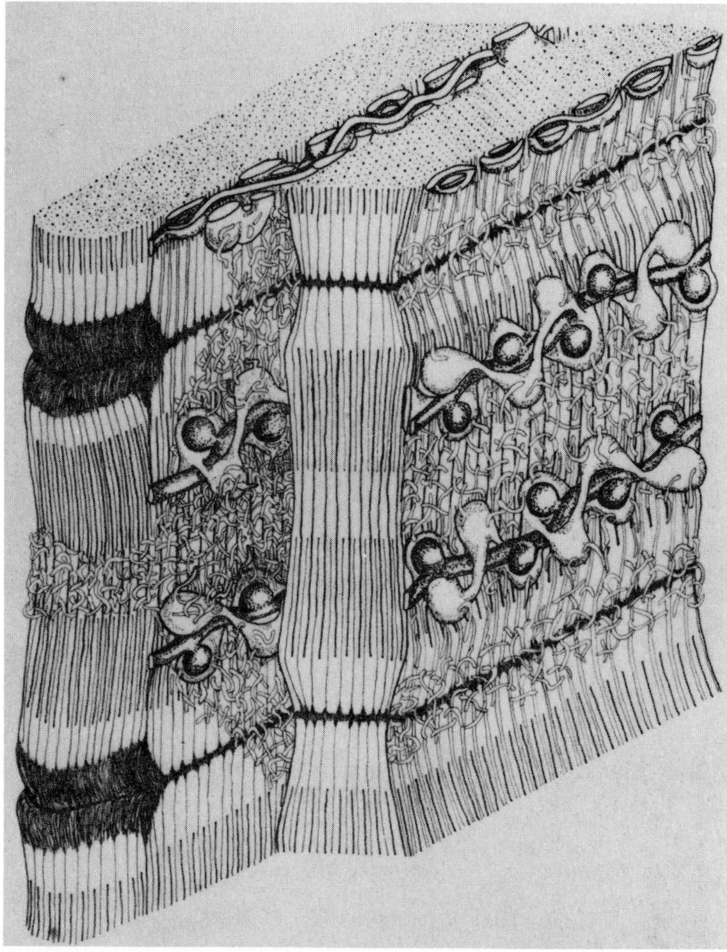

Figure 16–6 (Continued). B. Model of portion of a fiber from scorpion muscle, showing T-tubules and cisternae at A bands and S-tubules between the cisternae. (Courtesy of A. Gilai and I. Parnas.)

toward the nerve cord, where they receive synaptic input. The muscle itself goes to the nervous system.

All intermediate or semi-striated muscles have a limited tubular system, which is less extensive in those fibers that have one very narrow dimension than in thick fibers.

Non-striated Muscles. Non-striated muscles may be long-fibered (up to 2 cm) or short-fibered (100 to 300 μm) and spindle-shaped. All of the fibers are of small diameter, usually 2 to 5 μm in one dimension. They lack T-tubules, and the SR tubular system is either absent or sparse, but some have many pinocytotic vesicles. This observation agrees with the generalization that T-tubules are needed for transfer of signals inward from the membrane and that in the absence of T-tubules, where coupling between membrane and contractile system is direct, fiber diameter is small.

In non-striated muscles there is no transverse alignment of thick and thin filaments. However, many non-striated muscles show thick and thin (myosin and actin) filaments in random array (Fig. 16–7). Scattered among the filaments are "dense bodies" which have been homologized with Z bands. Thick and thin filaments are particularly differentiated in invertebrate non-striated muscles. Even the myocytes of sponges may show two sizes of filaments. In visceral smooth muscle of vertebrates thick filaments are often not seen, yet myosin can be extracted and some contracted smooth muscles (chicken gizzard, taenia coli of guinea pig, and some vascular muscles) show thick as well as thin filaments.[92, 305, 328] It is possible that in the relaxed state the myosin exists as dimers and that it aggregates on contraction. Lateral processes on thin filaments link actin to dense bodies and to membrane patches.[267]

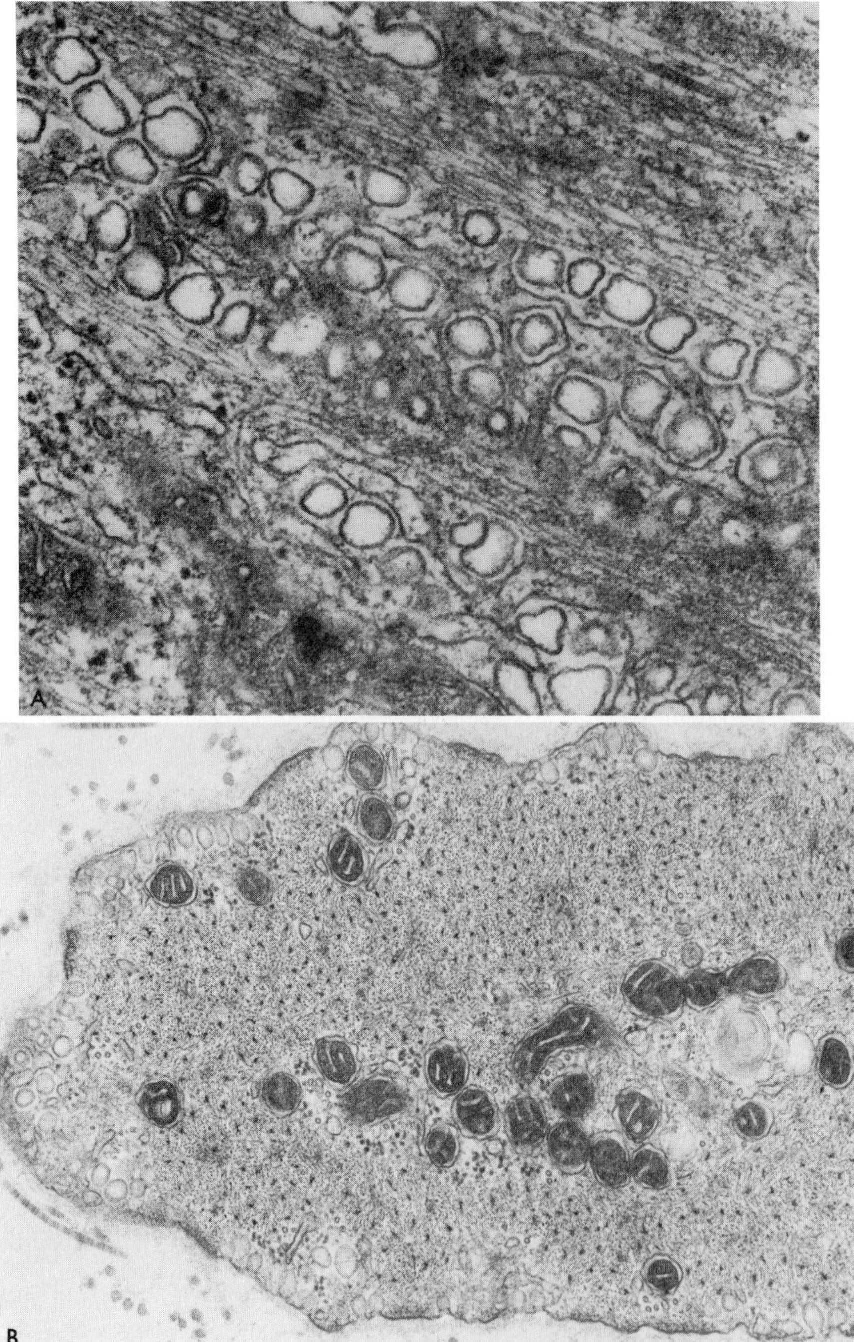

Figure 16–7. A, Longitudinal section of periphery of smooth muscle fiber of portal vein from rabbit, showing filaments; vesicles with S-tubules lie close to them. B, Transverse section of smooth muscle fiber of portal vein. Note thick filaments scattered among many thin filaments, central mitochondria, and peripheral vesicles, some of which connect to S-tubules. (Courtesy of C. Devine and A. Somlyo.)

In some non-striated muscles, both long- and short-fibered, very large filaments (1000 to 1500 Å diameter) are seen, which show clear bands of 145 Å repeated in a period of 760 Å. These very thick filaments contain paramyosin (tropomyosin A), and probably myosin is located on the outside of the large filaments. Paramyosin filaments have been seen in various holding muscles, such as molluscan adductors, *Mytilus* byssus retractors, *Golfingia* proboscis retractors, body wall muscles of gordian worms (*Paragordius*), and the notochord of *Branchiostoma.*

Some non-striated muscles, both long- and short-fibered, have extensive innervation that is often bineuronal. In others, as in vertebrate visceral muscle, nerve fibers are widely scattered; one muscle cell may be 5 to 10 cells removed from a nerve ending. Transmitter presumably diffuses through intercellular spaces and its effect passes through a series of cells, since the functional unit is the group of electrically interconnected cells.[50a] Short-fibered non-striated muscles have been classified as: (1) multiunit—dependent on nerves for activation and conduction; or (2) unitary—myogenically spontaneous, conducting from muscle fiber to fiber through junctions of low electrical resistance and merely modulated by nerves. The unitary muscles are more sensitive to stretch than the multiunitary ones. Most invertebrate postural non-striated muscles are multiunitary—for example, *Golfingia* proboscis retractors, snail radula retractors, and various echinoderm and some coelenterate muscles; multiunitary muscles of vertebrates are found in nictitating membrane, pilomotors, and some blood vessels, whereas unitary muscles are mainly visceral in vertebrates and a few molluscs and echinoderms.

CONTRACTILE PROTEINS AND THEORIES OF CONTRACTION

The general sequence in contraction is as follows: An electrical signal in the fiber membrane is transmitted inward by the T-tubules, and this in some manner activates the SR tubular system. In muscles which lack a tubular system, the electrical events in the cell membrane couple directly to the contractile system. In either case, calcium ions are then released from vesicles of the SR tubules or from cell membranes, and the calcium in the presence of adenosine triphosphate (ATP) triggers an interaction between myosin and actin. ATP is hydrolyzed and conformational changes in the actin-myosin complex occur, the nature of which is obscure; and in striated muscle the thick and thin filaments slide over one another, possibly by a change in angle of the cross bridges (Fig. 16–8). Calcium is then taken up actively by SR membranes and relaxation occurs. Removal of Ca^{++} inhibits actomyosin (AM) ATPase, and the uptake of Ca^{++} by the sarcoplasmic reticulum is associated with an ATPase in the SR vesicles. The basic reaction is

$$\mathrm{AM} \underset{\text{contract}}{\overset{\text{relax}}{\rightleftharpoons}} \mathrm{A} + \mathrm{M}$$

and Ca^{++} favors the aggregation reaction.

It is beyond the scope of this chapter to discuss in detail the physical chemistry of the several myofibrillar proteins that are best known from mammalian striated muscles, but a summary is necessary to provide background for comparative interpretations.

Myosin. Myosin is usually purified from high ionic strength KCl extracts of muscle; it constitutes 50 to 55% of skeletal myofibrillar protein and makes up all or most of the thick filaments. Myosin is a large and complex molecule with a molecular weight of 470,000 to 490,000.[96,97,98] The myosin molecule is large enough to be seen by electron microscopy, and consists of a long, rod-like tail (approximately 1400 Å in length and 20 Å in diameter) and a terminal head region about 200 Å in length and 50 to 70 Å in diameter. The globular head has been shown to be a bilobed structure.[228] From treatment with various structure-disrupting solvents (not breaking covalent bonds) and from physical measurements, it appears that myosin consists of (a) two large (heavy) polypeptide subunits or chains having molecular weights of approximately 200,000 to 212,000 each, which are wound together in a double-stranded α-helical configuration throughout the long tail of the molecule, and (b) two globular halves in the head or globular end. In addition, three or four small polypeptides (light chains or light components) are located in the head portion and constitute 12 to 15 per cent of the mass of the myosin molecule.[84a,228a,325]

The nature of both the light and heavy chains of myosin varies according to the muscle from which they are extracted. The light and heavy chains from white skeletal muscle of mammals differ in number of components from those of red and cardiac muscle.[325] White skeletal muscle of adult rabbits shows four electrophoretic components, while in red and cardiac muscle and in fetal white muscle the two faster components are sparse

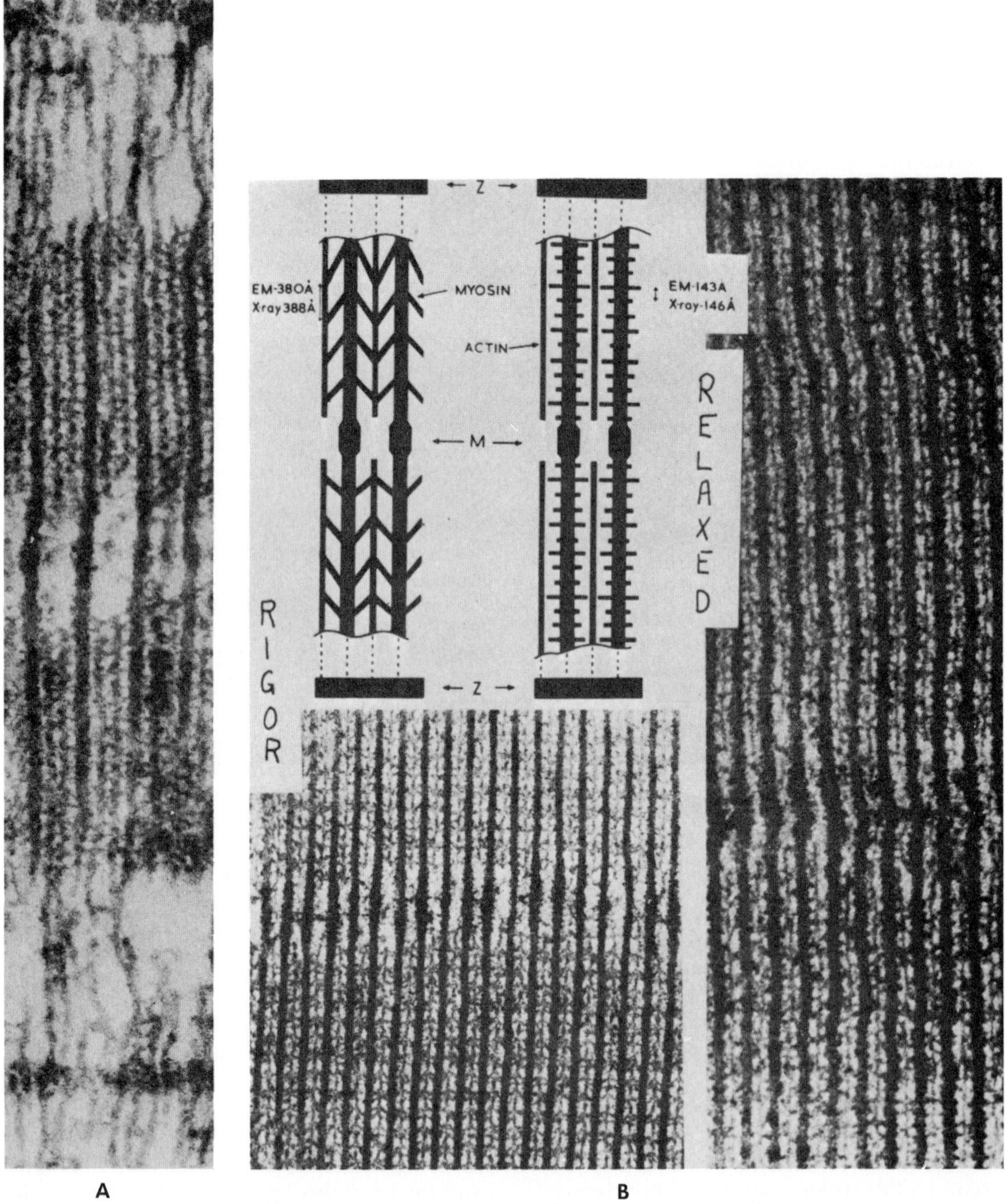

Figure 16–8. A, Longitudinal section of fast skeletal muscle, showing one sarcomere with Z bands at top and bottom, thick filaments in A band, and thin filaments in I bands, overlapping the thick filaments and connected to them by cross bridges. (From Huxley, H. E., J. Molec. Biol. 7:281–308, 1963.) B, Electron micrographs and drawing of insect striated muscle, showing difference in angle of cross bridges in relaxed and rigor states. (From Reedy, M. K., Amer. Zool. 7:465–481, 1967.)

or absent.[279a] Myosin of rabbit white skeletal muscle can be broken into tridecapeptides which contain 3-methylhistidine, but cardiac muscle yields homologous chains in which the histidine is not methylated; differences in amino acid sequences are such as to indicate that cardiac and skeletal myosins are coded by different genes. Cardiac muscle appears to lack the enzyme for methylating histidine.[191a]

The light chains of myosin can be dissociated from the heavy chain by a number of treatments. Removal of all of the light chains results in complete loss of myosin's ability to bind to ATP and loss of all of its ATPase activity.

The myosin molecule may also be broken roughly in half by trypsin hydrolysis into (a) a long portion of the tail called light meromyosin (LMM), of molecular weight 120,000 to 150,000 and length 800 to 900 Å, which corresponds to that portion of myosin that forms the thick filament core (Fig. 16–9); and

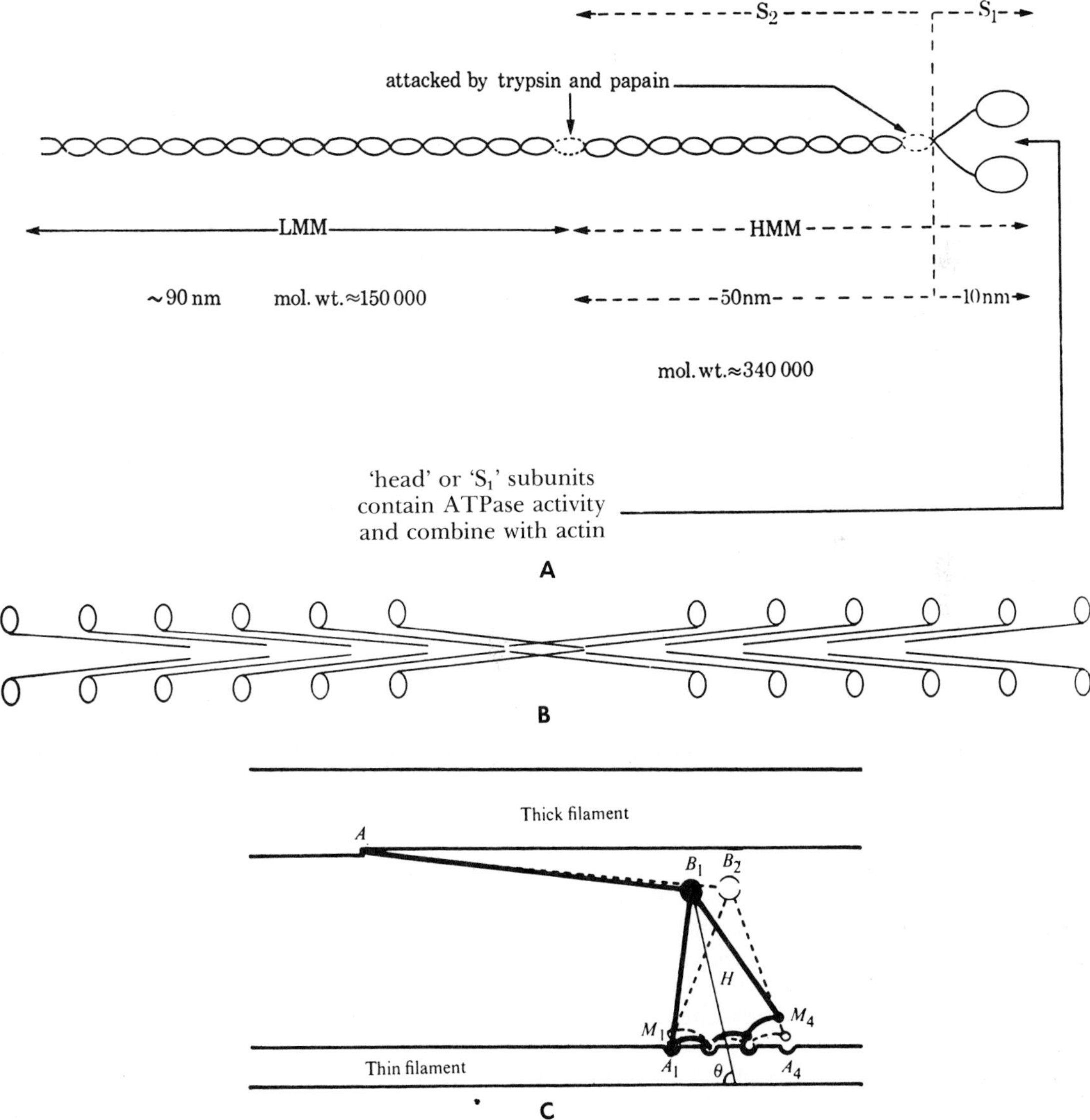

Figure 16–9. A, Diagram representing different portions of a myosin molecule, showing probable arrangement of fragments obtained by fractionation. (From Huxley, H. E., Proc. Roy. Soc. Lond. B *178*:131–149, 1971.) B, Arrangement of myosin molecules in a thick filament. The backbone consists of tails of the molecules, and the molecules form two antiparallel sets. (From Huxley, H. E., *ibid.*) C, Diagram showing a possible mechanical change in cross bridges during contraction. Myosin head H connected to thick filament by an elastic component which can change position from B_2 to B_1, and to thin filament by a portion which can change attachment from M_1 to M_4. (From Huxley, A. F., and R. M. Simmons, Nature *233*:533–538, 1971.)

(b) a head portion with a small part of the rod still attached, termed heavy meromyosin (HMM), of molecular weight about 340,000, which retains the ATPase and actin-binding properties of myosin and forms the cross-bridges. Heavy meromyosin can be digested further by trypsin or papain into three smaller fragments: two particles which correspond to the two globular heads of the myosin molecule, each having a molecular weight of 110,000 and retaining the ATPase and actin-binding properties, and one particle representing the remaining portion of the myosin rod, having a molecular weight of 60,000 and a length of 400 to 500 Å.[136]

From X-ray diffraction and electron microscopy of native myofilaments and reconstituted myosin filaments, together with physical measurements of size and shape, it is concluded that the rod portions of myosin overlap from adjacent molecules in the middle of the thick filaments so that pairs of molecules are oriented with the heads opposite one another[196] (Fig. 16–9). This accounts for the "pseudo" H zone (bare portion of the thick filament) where there are no cross bridges projecting from the thick filaments. In each half of the thick filament, the myosins are packed with the heads oriented in the same direction. The cross bridges project laterally in a staggered pattern to give a 6/2 helix (i.e., the cross bridges project in opposite senses in the two halves of an A band).[193,195] The cross bridges occur in pairs, on opposite sides of the filament; each is separated from the next by 143 Å linearly along the filament axis and is rotated by 120°. Thus, there are three pairs per complete turn, and there is a separation of 429 Å along the filament between cross bridges in any given row.[196] In fast insect muscle, the rotation between adjacent cross bridges is 67.5°, giving a 190° twist for every 380 Å distance along the filament.[299]

Actin. Actin may be isolated as a monomer with a molecular weight of about 46,000, containing only one polypeptide chain; it constitutes 20 to 25% of skeletal myofibrillar protein and is the major constituent of the thin filaments. Actin can transform, in appropriate salt mixtures, from a globular (G) to a fibrous (F) form that resembles the structure of the thin filaments. The polymerized F actin consists of linear aggregates or ovoids of G actin. Actin filaments are of similar diameter (60 to 90Å) in all muscles, and consist of two strands in a right-handed helix with about 13 G actin globules (56.5 Å diameter) per repeating unit of 360 Å (Fig. 16–10A).[158] Identical filaments comprise the actin prepared from rabbit psoas, *Pecten* and *Crassostrea* adductors, *Loligo* mantle, *Lumbricus* longitudinal muscle, and guinea pig taenia coli.[158]

Actin and myosin are less readily separated in frog, fish, *Limulus,* and some insect muscles than in mammalian muscle, and the extracted natural actomyosin (AM) is known as myosin B; in addition to myosin and actin, it contains some other myofibrillar proteins. AM was extracted from non-striated muscles of molluscs[218] and of fish,[77,78] and the product had properties very similar to those of mammalian myosin B. Trout myosin is less heat-stable than mammalian myosin.[53] During development of a tadpole, changes in properties of natural AM indicate chemical transformation as well as an increase in amount.[260] In embryonic myoblasts, myosin can be demonstrated immunologically before striations are apparent microscopically. In a chick embryo, the AM content increases from 1 mg/g at 12 days incubation to 50 mg/g at hatching. In uterine smooth muscle, the actomyosin content increases several-fold under the action of estrogen and in pregnancy. In a solution of 0.6 M KCl, mammalian actomyosin has myosin and F actin in a 3.7/1 weight ratio, or one myosin molecule per 2.7 actin monomers.[105] At low ionic strength, HMM binds actin in molar ratio. It is possible that binding of one head of HMM (or native myosin) to actin may influence the binding of the other head.

Tropomyosin and Troponin. A protein complex, native tropomyosin, extracted from myofibrils is now known to consist of two proteins: tropomyosin B and troponin. Rabbit striated muscle has 54% myosin, 21% actin, and 15% native tropomyosin. Insect muscle yields 55.2% myosin, 21% actin, and 9% native tropomyosin.[62] Tropomyosin B and troponin are found in a proportion of about 1/1.3 in rabbit muscle.[104] The molecular weight of tropomyosin B is 70,000 and its s_{20} is 2.7; the molecular weight of troponin has been estimated at 80,000 to 90,000, and its s_{20} is 3.8. Troponin is now thought to consist of three components; one of these, troponin A (the Ca^{++} binding fraction), has a molecular weight of 18,000 to 20,000, is highly negatively charged, and shows very high affinity for calcium ions – four Ca^{++} ions per molecule or 22 μM Ca^{++} per gram of the

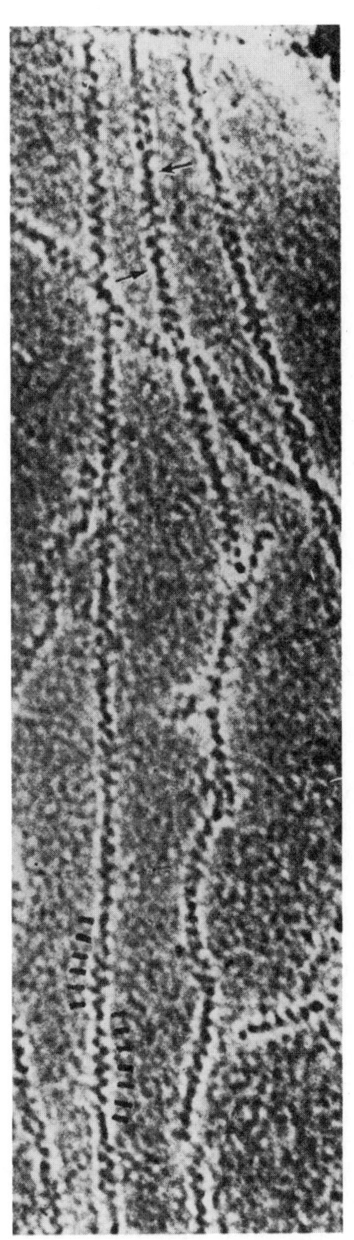

A

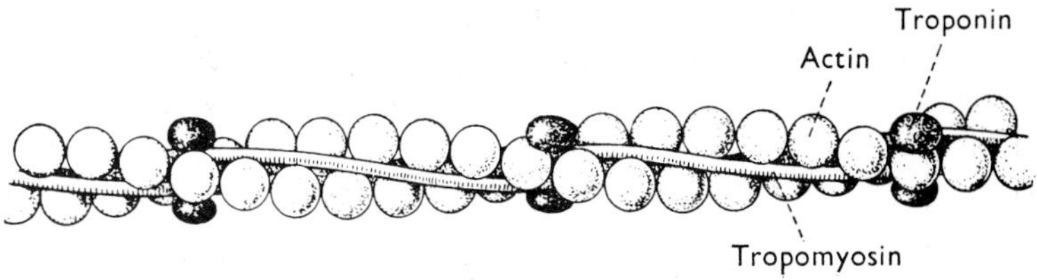

B

Figure 16–10. A. Helical form of F actin molecules. (From Hanson, J., and J. Lowy, J. Molec, Biol. *6*:46–60, 1963.) B. Model of structure of thin filament, showing postulated arrangement of actin and troponin molecules along tropomyosin molecule. (From Ebashi, S., M. Endo, and I. Ohtsuki. Quart. Rev. Biophys. *2*:351–384, 1969.)

protein. Troponin B has a molecular weight of 60,000 and has been dissociated into two components having molecular weights of 23,000 and 37,000.[104]

Tropomyosin B is a rod-shaped molecule, 400 Å in length and 20 to 30 Å wide, and is found along the entire length of the thin filaments (Fig. 16–10B).[105,106] Troponin binds to tropomyosin and occurs at 400 Å intervals along the thin filaments, as shown by immuno-staining methods; it apparently attaches at corresponding points on successive tropomyosin molecules.

One of the essential events in contraction is the making and breaking of cross bridges between myosin and actin; this process has its counterpart *in vitro* in superprecipitation of AM. This event is coupled to the hydrolysis of ATP; *in vivo,* the ATPase of myosin is activated by actin in the presence of Mg^{++}. The energy for muscle contraction is then released by the Mg-ATPase of AM. In the absence of Ca^{++}, troponin apparently depresses the interaction of actin with myosin, and this depressing action is probably due to troponin A and is mediated through tropomyosin B.

Addition of Ca^{++} removes the depressing action of troponin-tropomyosin, and the AM ATPase is then released to hydrolyze Mg-ATP.

Troponin B alone can inhibit Mg-ATPase *in vitro* but is insensitive to Ca^{++}; also, addition of tropomyosin B leads to even greater inhibition of the ATPase. Troponin A has little effect on the Mg-ATPase but avidly binds calcium. Addition of troponin A to troponin-tropomyosin-actomyosin *in vitro* diminishes the inhibition of ATPase in the absence of Ca^{++}, and activates it in the presence of Ca^{++}.[163,164] It must be concluded that, *in vivo,* the troponin-tropomyosin system is a safety catch that, in the absence of Ca^{++}, prevents the interaction of myosin with actin and the activation of Mg-myosin ATPase. When Ca^{++} ions are released from the sarcoplasmic reticulum, these ions become bound to troponin A; this then removes the inhibition by troponin A and tropomyosin, and allows the cyclic interaction of myosin and actin, resulting in contraction. Thus, the role of Ca^{++} is to remove inhibition of AM ATPase rather than to activate the enzyme.

A different method of Ca control is found in molluscs. The thin filament of molluscan muscles contains only actin and tropomyosin and lacks troponin; however, unlike vertebrate myosin, molluscan myosin binds Ca^{++}, the myosin ATPase is dependent on Ca^{++}, and actin does not activate AM ATPase without Ca^{++}. The regulating site which binds Ca^{++} is on a light chain of molluscan myosin. Comparative measurements show the myosin type of regulation in molluscs, brachiopods, nemerteans, and sipunculids. The function of tropomyosin in these phyla is uncertain. Arthropods have a troponin which differs from that of vertebrates in having different electrophoretic bands and in binding less Ca^{++} than vertebrate troponin. Annelids have both myosin and troponin modes of Ca regulation. Echinoderms have a troponin which resembles that of vertebrates. It is postulated that the myosin-linked regulation is more primitive, and that troponin has evolved twice.[214a,227a]

Very little comparative information is available regarding tropomyosin-troponin. The binding constant for Ca^{++} to troponin from white skeletal muscle of rabbit is twice that for troponin from red muscle and three times that for troponin from cardiac muscle.[104] However, the ATPase of cardiac muscle is much more sensitive to strontium than that from white muscle; by various combinations of proteins, it was found that the Sr effect is determined by the troponin, and hence the troponin of cardiac muscle must be different from that of white muscle. White fast muscle contains much more troponin than does red slow muscle; the white portion of rabbit vastus lateralis has three times as much troponin as the red portion of this muscle (expressed as Ca^{++} bound per unit myosin).[130] Mammalian arterial smooth muscle contains much less actomyosin than does skeletal muscle (2 to 10 mg/g compared with 50 to 100 mg/g).[321] However, the tropomyosin content of the smooth muscle is higher—8.2 mg/g in hog carotid arterial muscle compared with 5.8 mg/g in hog striated muscle.[334b] Tropomyosin from several vertebrate and invertebrate muscles forms subunits of 35,000 M.W. which align in the two-chain helix normally associated with thin filaments.[252] Tropomyosins from rabbit, cow, cod, lamprey, squid, clam, and lobster show only minor differences in amino acid composition.[183]

Alpha-actinin. The α-actinin molecule has a molecular weight of about 180,000, consists of two polypeptide chains of equal size (90,000 M.W.), and has an s_{20} of 6.2. α-Actinin accelerates superprecipitation and ATPase activity of actomyosin *in vitro.* α-Actinin exerts this effect through actin, to which it binds in a competitive fashion with tropomyosin. It has been localized in the Z line of muscle structure. Recently it has been purified and found to constitute about 1% of the total myofibrillar protein. Since the Z line constitutes about 6% of the dry mass of the myofibril, it must contain other proteins besides α-actinin.[310]

α-Actinin from fast mammalian (white skeletal) muscle is slightly different in composition from that of slow mammalian (red and cardiac) muscle. Such differences were not found between white, red, cardiac, and smooth muscles of chicken.[310]

Other Myofibrillar Proteins. A number of other proteins have been isolated from muscle extracts, but some are probably contaminants or denaturation products. Two of these proteins appear to have some functional role in the myofibril. β-Actinin has been isolated from crude preparations of actin, and appears to inhibit cross-linking of F actin. M-protein has been described by immuno-chemical methods and probably accounts for at least part of the M line which may link adjacent thick filaments in register.[220a]

Theories of Contraction. There are many

theories of contraction, but most of them are based on the general idea, developed by Huxley, that thick and thin filaments slide over one another. In striated muscle, maximum tension is developed when there is complete overlap of myosin and actin in the region of bridge formation. Tension decreases to zero when the muscle is stretched so that there is no overlap, and at shorter lengths tension declines as the thick filaments pile up on one another.

In addition to bridge formation and ATP hydrolysis, some configurational change in myosin must occur to cause sliding of the thick filaments over the thin ones, and thus cause shortening or the development of tension without change in overall length. One hypothesis is that the angle of the cross bridges changes when ATP is split; this could involve a rotation of the myosin head[195] or a hinging out in the myosin chain.[279] Changes in cross bridge angle have been noted in insect muscle fixed in rigor.[299] Another hypothesis is that of a torsional movement due to sequential cross-bridge interactions.[96, 98] A recent suggestion for tension change without length change is based on the observation that if a muscle is suddenly released while contracted, tension is redeveloped in two steps; it is suggested that each cross bridge consists of a elastic arm which makes contact with the actin filament in one of several energetically stable positions. Redevelopment of tension could be due to (1) elasticity in the base of the bridge and (2) rotation of the head from one position to another, to steps of lower potential energy.[192]

The Role of Calcium. The trigger for contraction in both smooth and striated muscle is Ca^{++}. When various ions were microinjected into striated muscle fibers, only calcium in low concentrations caused contractions. In mammalian striated muscle, the threshold for Ca^{++} is 10^{-6} M and that for maximum activation is about 10^{-5} M. Frog fibers from which the sarcolemma has been skinned away respond at 5×10^{-7} M and give best responses at 10^{-6} M Ca^{++}.[105] In insect glycerinated fibers, the AM ATPase is activated at 10^{-8} M and is maximally activated at 7.5×10^{-7} M Ca^{++};[60] mammalian AM ATPase is active at 10^{-9} M Ca^{++}. The concentration of calcium per milligram of AM is similar in several muscles despite differences on a weight basis:[33]

	AM (mg/g)	Ca (μM/g wet wt.)	Ca (μM/mg AM)
frog sartorius	70	90	1.3
frog heart	38	50	1.3
guinea pig taenia coli	10	13	1.3

The generality of the Ca^{++} requirement for contraction is shown by the fact that myocytes of sponges can contract only in the presence of calcium (or substitutes such as strontium or magnesium).[288]

Most (90%) of the calcium in resting muscle is bound to intracellular membranes (mostly of the SR system). On activation by an action potential, ionized calcium is released. This has been demonstrated by a sudden peak in absorption by the dye murexide-Ca[206] and by emission of light from injected aequorin (a luminescent protein requiring Ca ions)[12] during contraction. The liberation of calcium ions is proportional to the tension developed and corresponds in time to the initial heat liberation.[61]

In vertebrate muscles, the released Ca ions are immediately bound to troponin A, which is then able to remove the inhibition of actin-myosin interaction due to troponin-tropomyosin as described above. In molluscan muscles, the Ca ions bind to a site on a light chain of myosin, and the interaction of other sites of myosin with actin is then permitted.

For relaxation, calcium must again be sequestered. This is accomplished by the active uptake of calcium by membranes of the sarcoplasmic reticulum (SR) system; in homogenized muscle extracts the SR is broken into small vesicles or microsomes which bind calcium and constitute an extractable relaxing system.[370] For active binding of Ca to the SR membranes, energy must be provided, and this comes from ATP and corresponds to the relaxation heat. The ATPase involved is a component of the SR membrane itself; it is Mg^{++}-Ca^{++} activated and has a molecular weight of about 147,000 (102,000 protein and 45,000 lipid). A second, smaller protein of the SR membranes functions in calcium-binding and binds up to 43 moles of Ca^{++} per mole of protein.[231a] Radioautographs of frog muscle labelled with ^{45}Ca show that, immediately after tetanus, the calcium has been pumped back into longitudinal and intermediate tubules of the SR system, and on full recovery the amount in terminal or cisternal tubules increases; excitation releases Ca^{++} from terminal cisternae.[380]

SR particulates (microsome fraction) from several vertebrate and invertebrate muscles have similar calcium-binding and relaxing properties. The rate of uptake by particulates from red muscle is 16% of that by particulates from white muscle of mammals.[336] In lobster muscle, SR vesicles can take up strontium as well as calcium; rat vesicles do not take up strontium. The lobster vesicles take up Ca at the rate of 0.6×10^{-3} mM/mg protein, which is enough to account for relaxation.[364] Agents like NO_3, which prolong contraction, depress the binding of Ca to SR vesicles; other agents, such as Zn, prolong contraction by inhibiting the ATPase of the sarcoplasmic reticulum.[58]

In mammalian smooth muscle, cytoplasmic granules bind calcium as in striated muscle, but the binding of Ca^{++} and the ATPase activity of the particulates from uterine muscle are only one tenth those in heart muscle.[57] Microsomal particles from uterus bind Ca from a concentration of 6×10^{-7} M, but not from lower concentrations.[26] Less calcium is stored in glycerinated smooth muscle than in striated muscle, and the uptake saturates at lower concentrations of Ca.[165,166]

Excitation-Contraction Coupling. Coupling of electrical events to the contractile system requires that the electrical signal from the cell membrane spread inward; in fast fibers, this function is performed by the T-tubules. This was first demonstrated by local stimulation along a sarcomere by an electrode of diameter less than the length of a sarcomere. In frog fibers the threshold for contraction was lowest at the Z-line, and in crab fibers at the AI junction; these points correspond to the location of the T-tubules. With the external membrane voltage-clamped by an external electrode, intracellular recording reveals small spikes which probably come from T- and SR-tubules. Inward conduction at the rate of 7 cm/sec at 20° C has been estimated from sequential activation of fibrils from periphery to center.[145]

The manner by which the electrical signal in the T-tubules triggers the release of Ca^{++} from the SR is uncertain. Myofibrils at the periphery of a fiber can be seen to contract before central ones.[5] After the sarcolemma is removed from a frog striated fiber, local application of Ca causes contractions which are graded according to Ca^{++} concentration. The radial spread is reduced in tetrodotoxin (TTX) or reduced Na; hence, by analogy with surface membrane, it is likely that an action potential is conducted actively along the T-tubules.[81] When muscle fibers are treated with saline made hypertonic with glycerol, the T-tubules swell and may be disrupted; such fibers still conduct normal action potentials (in the outer membrane), but they show no negative after-potential and no contractions.[132] The surface membrane at rest has both K and Cl conductance; the T-tubule of frog muscle shows no Cl-conductance.[133] In crustacean muscle, Cl^- appears to pass out across T membranes during excitation.[44]

Evidently, inward electrical conduction of a signal is from the fiber membrane via the T-tubules; how the signal is then transmitted to the SR-tubules has not been determined.

MECHANICAL PROPERTIES, "ACTIVE STATE," AND HEAT PRODUCTION

Mechanics. Muscles have mechanical properties which affect their contractions, and which can be represented by a model consisting of a contractile element (CE) in series with one or more series elastic elements (SE) and in parallel with another (parallel) elastic element (PE).

For muscles with skeletal attachments, the series elastic element resides mainly in tendon or other connective tissue at the ends of a muscle. In short-fibered muscles, the distinction between series and parallel elastic elements is less clear. The strain of the series elastic element may be measured when an isometrically contracting muscle is released against different loads; it is also observed when the parallel elasticity is taken up by contraction. The compliance* of the series elastic element can be increased by attaching springs at the ends of a striated muscle; the rate of tension development is then reduced, the height of a twitch is lowered, and the tetanus-twitch ratio is raised.

Parallel elasticity is shown by the passive tension-length curve. When an unstimulated muscle is stretched, it develops elastic tension in a non-linear fashion; little tension is developed with initial stretch, and increasingly more tension is developed as stretch approaches the breaking point (Fig. 16–11). Tension in the parallel elastic element increases greatly as length increases beyond the

*Compliance is defined as the displacement per unit of applied force.

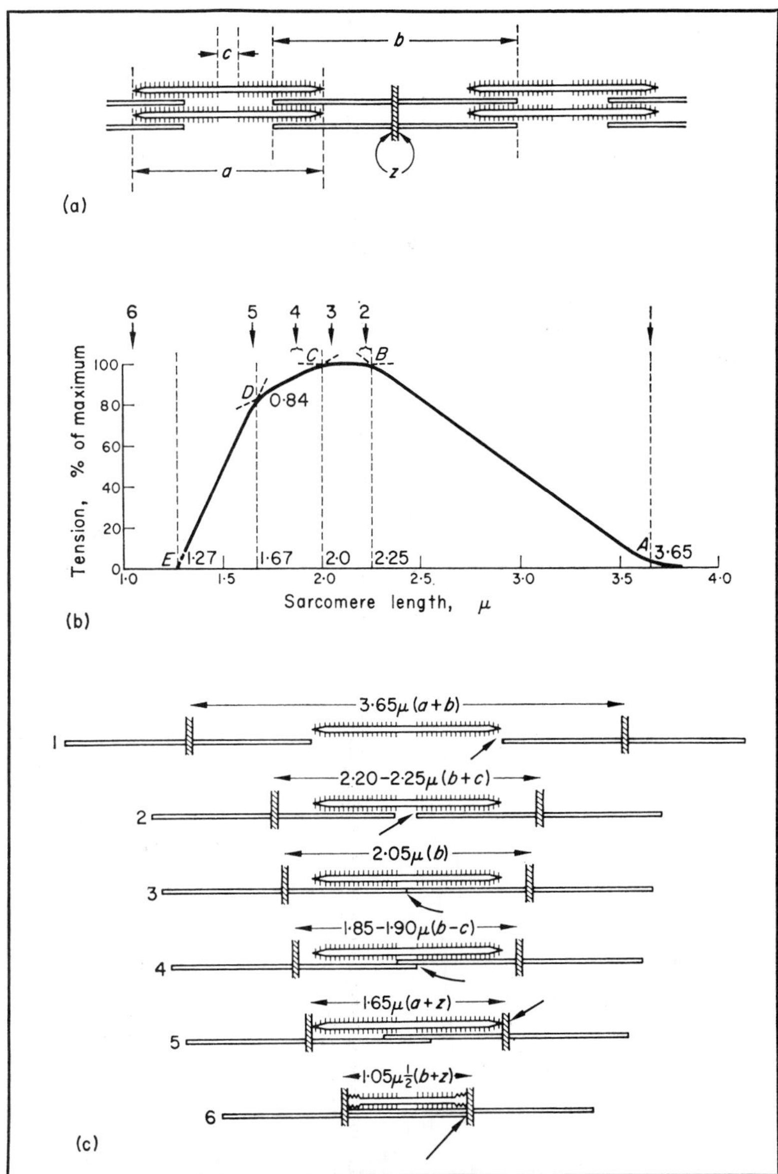

Figure 16–11. Interpretation of tension in percentage of maximum as a function of sarcomere length. Sarcomere lengths and relations between thick and thin filaments in lower half of figure correspond to numbered points on tension-length diagram above. (From Wilkie, D. R., Muscle. St. Martin's Press, Inc., New York, 1968.)

rest length (l_o) (Fig. 16–12A); at less than rest length the mechanical system consists of SE and CE only. The resistance to stretch of the parallel element is in connective tissue between fibers, sarcolemma, and sarcoplasm. Flight muscles of insects, which have little connective tissue, are less extensible than vertebrate muscle. Many short-fibered nonstriated muscles are more extensible than striated muscles. The *Golfingia* proboscis retractor is readily stretched to more than ten times its rest length by unfolding pleats which are transverse to its many fibers (p. 771). For many nonstriated muscles the l_o or the *in vivo* rest length is difficult to determine and may vary with hormonal and other modifying factors.

The contractile element is very compliant in the relaxed state; its stiffness increases when the muscle is activated and may persist after activation is complete. Properties of the contractile element are measured by speed of shortening (isotonic contraction), by effects of stretch and release during contraction, and by maximum tension development (isometric contraction).

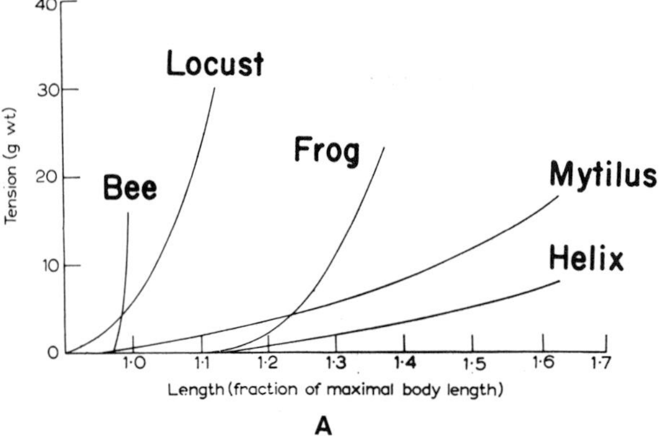

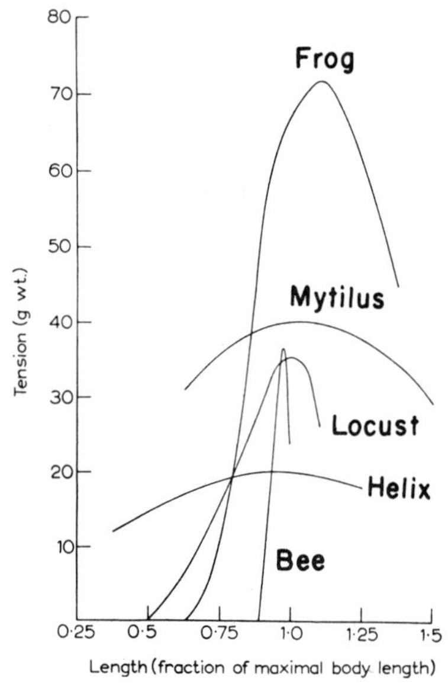

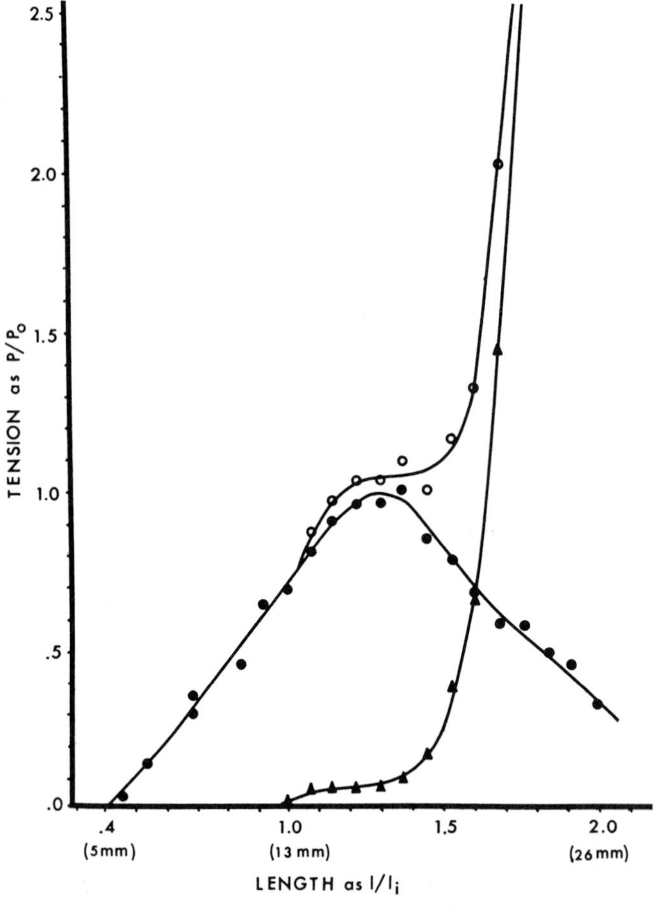

Figure 16–12. Tension-length curves. A, Curves for muscles at rest. Bumblebee *(Bombus)* flight muscle; locust *(Schistocerca)* flight muscle, 11°C; frog sartorius muscle, 0°C; *Mytilus* anterior byssal retractor, 14°C; *Helix* pharynx retractor, 14°C. (From Hanson, J., and J. Lowy, *in* The Structure and Function of Muscle, Vol. I, edited by G. Bourne. Academic Press, New York, 1960.) B, Curves for active muscles in relative units. Bumblebee flight muscle; locust flight muscle; frog sartorius muscle; *Helix* pharynx retractor; *Mytilus* anterior byssal retractor. (From Hanson, J., and J. Lowy, *ibid.*) C, Curve for cat intestinal circular muscle. ▲ rest, ○ active tension, ● difference between active and rest. (From Meiss, R. A., Amer. J. Physiol. *220*:2000–2007, 1971.)

Isotonic contractions are those in which shortening occurs at constant loads, and this kind of contraction provides evidence concerning the kinetics of the shortening process. The velocity of shortening is a function of load (the force-velocity relation, Figure 16–13A) and is given by the equation developed by A. V. Hill:

$$V = \frac{b(P_o - P)}{(P + a)} \text{ or } P = \frac{b(P_o - P)}{V} - a$$

where V is velocity of shortening, P_o is the maximum tension the muscle can develop

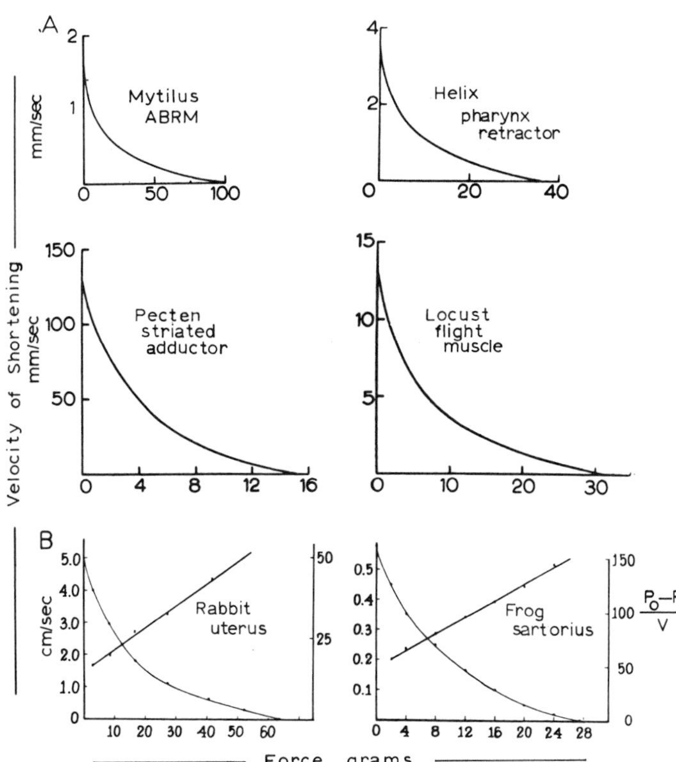

Figure 16–13. Force-velocity curves of different muscles. A, *Mytilus* anterior byssal retractor; *Helix* pharynx retractor; *Pecten*, striated portion of adductor; locust flight muscle. (From Hanson, J., and J. Lowy, *in* The Structure and Function of Muscle, Vol. I, edited by G. Bourne. Academic Press, New York, 1960.) B, Velocity of shortening of frog sartorius and rabbit uterus at different loads; hyperbolic curve gives velocity in cm/sec, and straight line gives the quantity $\frac{P_0 - P}{V}$ against load. (From Csapo, A., *in* The Structure and Function of Muscle, Vol. I, edited by G. Bourne. Academic Press, New York, 1960.)

(i.e., with no shortening, or the maximum tetanic tension), P is the given load, a is a constant with dimensions of force, and b is a constant with dimensions of velocity. V_0 is the maximum velocity, or the extrapolated speed of shortening without load. When the quantity $\left(\frac{P_0 - P}{V}\right)$ is plotted as a function of P, a straight line of slope b and intercept $-a$ results (Fig. 16–13B). As can be seen from the figure, force-velocity curves for several muscles have similar shapes but different constants. V_0, the intrinsic velocity, is expressed in sarcomere lengths or muscle lengths per second. The power output (load times velocity) rises to a maximum at some intermediate force and speed.

Intrinsic speed can be given as V_0/l in reciprocal seconds, where l is muscle length. Thus, an extraocular muscle of a kitten has a V_0/l of 70 sec^{-1}; a frog sartorius (at 0°C) has a value of 1.3 sec^{-1}, and a tortoise muscle has a value of 0.1 sec^{-1}.[177,376] The intrinsic speed can also be given in sarcomere lengths moved per second (or, more accurately, as the rate at which actin filaments move past myosin filaments during shortening). In a small animal (e.g., a mouse), the sarcomeres are not much different in length from those in a large animal (e.g., a horse); however, since muscle lengths are very different, the intrinsic speed (given as V_0/l or as sarcomeres/sec) necessary for movement in space at a given locomotor velocity is much greater for a small animal than for a large one. For example, a hummingbird may move its wings at 100 beats per second, while a pelican need only produce one beat per several seconds to attain the same air speed. The intrinsic speed of a large animal, such as a whale or an elephant, is a small fraction of that of a small porpoise or a jumping mouse traveling at the same locomotor velocity.

The intrinsic speed of a muscle varies as a fractional exponential function of body weight and as a linear function of a limb dimension. Power developed by a muscle is the product of force and velocity, and is usually maximal at about 0.3 V_0 and 0.3 P_0. Mechanical efficiency is given by the work done divided by the energy used, and is maximal at about 0.2 V_0 and 0.5 P_0. Thus, the maximum power and the maximum efficiency do not occur at the same combination of velocity and load. In a delightful essay, A. V. Hill showed[177] that the intrinsic speed of muscles is adapted to body dimensions. It will be shown below that intrinsic speed is determined by specific myosins; hence, it may be that animal myosins are size- and speed-adapted just as hemoglobins of different animals are adapted to oxygen requirements and availability.

Isometric contractions occur at a constant length of muscle, with both ends of the muscle fixed so that no shortening can occur. This kind of contraction can occur over a wide range of lengths and is not a truly isometric response of the contractile element, since internal shortening is allowed by the series elastic component. When a muscle is stimulated at different lengths, the isometric tension reaches a maximum; at extended lengths the total tension approaches the maximum passive tension (Fig. 16–12). The difference between the total tension and the passive tension gives the active tension-length curve. Active tension is maximal at a range of lengths; for most striated muscles it is maximal near l_o, where l_o is the length at which $P = P_o$ (l_o is usually the normal length in the body), while for cardiac and smooth muscles active tension is maximal at somewhat longer lengths. In taenia coli, maximum tension is at 130% to 180% of l_o.

The maximum tension or intrinsic strength is measured in a tetanus, and is given as tension per unit cross sectional area (P/A) in units of g/cm^2. Some representative values of this quantity are given in Table 16–1. In general, muscles with high speeds of shortening develop little tension, while muscles with a holding function develop high tension.

TABLE 16–1. Maximum Force Developed by Muscles.

Muscle	*Maximum Force, kg/cm²*
cat tenuissimus	1.4
rat extensor digitorum longus	3.0
rat gastrocnemius[330]	1.8
rat soleus	0.28, 0.22
rat heart[269]	0.016
sloth gastrocnemius[232]	1.62
sloth diaphragm	2.1
cat papillary[339]	0.8
dog tracheal smooth muscle[337]	0.8
rabbit uterus[339]	0.13, 0.7
guinea pig taenia coli[339]	1.5
rabbit taenia coli[339]	0.89
cat duodenum (circular)[339]	0.42
frog sartorius[339]	2.0
toadfish sonic muscle[266]	0.1
snake costocutaneous[306]	0.7–1.0
lobster slow remotor[244]	2.8
lobster fast remotor	0.025
cockroach coxal muscle	0.8
Lethocerus air guide retractor[367]	1.25
oyster adductor[250]	12 (peak) 0.6 (tonic)

The increase in tension to a maximum as muscle is stretched, followed by a decline at longer lengths, may be explained on the basis of filament overlap. As a striated muscle is stretched, the sarcomere lengths increase. At the maximum length that a muscle can tolerate, there is no overlap between thick and thin filaments. At the length of maximum tension, the overlap in the region of cross bridges is maximal and thin filaments meet in the middle of the A band. At shorter lengths there may be overlap or even crumpling of the thin filaments. Thus, the maximum number of cross bridges can be established at the length corresponding to full overlap.[147] This simplified model applies to cross-striated muscle,[385] and the overlap of thin filaments has recently been well confirmed for oscillatory insect muscles. In non-striated and cardiac muscles (and possibly also in cross-striated ones), alignment of molecules within the filaments also may affect tension development.

Most muscles develop more tension on repetitive stimulation than in single twitches. Stimulation *in vivo* is usually through repetitive nerve impulses, and the higher tetanic tension may be partly the result of the requirement for neuromuscular facilitation and the bringing of more fibers into action with repetitive stimuli. In addition to facilitation of excitation, there may be mechanical summation when maximal stimuli are given at a rate exceeding that permitting complete relaxation. Muscles which give little response to single stimuli and require repetitive activation or facilitation are called *iterative.* Those which give near-maximal responses to single shocks, and hence only summation on repetitive stimulation, are non-iterative. When stimulation is maximal or when stimulation of the muscle fibers is direct, tetanic tension is higher than twitch tension. This is interpreted as being due to the taking up of series elastic compliance in the single contraction; once the series elastic components are extended, the energy of further contractions can go into tension maintenance. Fusion frequency is a measure of the maximum speed with which some relaxation can occur.

The series of events leading from membrane excitation to contraction constitute the "excitation-contraction coupling," and these were formerly lumped together as leading to the "active state." Before an isometric contraction occurs at lengths greater than l_0, there may be initial extension (latency relaxation). The origin of this extension has not been established.

"Active State." The decrease of extensibility (i.e., stiffening) which begins before tension is developed is one measure of the activation of the contractile element. An active state curve is obtained by a combination of mechanical measurements. If a stretch is applied during the latent period of a twitch, the tension which develops may be as great as in a tetanus, because the extension of the elastic elements is done by the stretch. No extra tension is developed when a muscle is stretched at the peak of the twitch; the time of zero tension change at the peak of the twitch lies on the active state curve. If the muscle is suddenly released after the peak of contraction, tension drops briefly and then redevelops to a level slightly less than the corresponding phase of the twitch. A series of quick releases with subsequent tension redevelopment allows the decay of an active state to be plotted. The tension curve resulting from quick stretches before the peak of the twitch and quick releases after the peak gives the time course of the active state of the contractile element.[70] The beginning of the rise of tension in response to the second of two paired stimuli and the time at which tension begins to decline after a tetanus may be used to check the start of the falling phase of the active state. In general, the active state rises rapidly (in 40 msec for frog sartorius at 0°C). It declines to zero at about half the twitch duration. In some non-striated muscles, there is little redevelopment of tension upon release after the peak of contraction; hence, the active state decays rapidly with respect to the tension. The decay of the active state has a larger temperature coefficient than does the speed of shortening. The temperature for peak twitch tension is higher for summer frogs than for winter ones, and seasonal effects have been seen in a number of mechanical properties of frog muscle. The time course of the active state in frog muscle is modified by a variety of agents—nitrate, caffeine, hypertonicity—that are now known to act on specific steps of the coupling sequence.

Some smooth muscles show a residual stiffness long after relaxation of tension has occurred. This indicates a persistent structural change resulting from normal excitation but independent of tension maintenance.[243]

Energetics and Heat Production. Muscles are less than 25% efficient, and a large part of the energy from chemical breakdown goes into heat rather than into mechanical work. In rabbit papillary muscle, only 11.6% of the chemical energy is liberated as mechanical work.[137] The metabolism by a resting cell of a small amount of substrate is accompanied by an equivalent low production of heat. Resting heat in heart muscle increases when the muscle is stretched. When a twitch or tetanus occurs, heat production rises rapidly. Most of the definitive heat studies have been made on frog sartorius at 0°C.

In a brief tetanus of a frog fast striated muscle at 0°C, a rapid burst of heat is liberated, coincident with the rise in tension; this is the *initial heat,* and at l_0 it appears at a rate near 30 mcal/sec/g. After relaxation begins, heat continues to be liberated, but at a slower rate; this is the *delayed heat,* which is coincident with delayed oxidative metabolism and often lasts several minutes. In a prolonged tetanus, a steady heat liberation persists as long as tension is maintained; this is *maintenance heat.* If the muscle is progressively stretched from l_0 and stimulated at different lengths, the isometric heat and tension decrease progressively as filament overlap decreases. Thus, the initial heat is length-dependent and is maximal at l_0. It may be ascribed to the rapid breakdown of phosphocreatine to recharge ATP; a corollary is that contraction *per se* is independent of oxygen. In contrast, cardiac muscle loses contractile strength when oxygen is removed. If a striated muscle is stretched to such a length that the thick and thin filaments no longer overlap, there remains a fraction of the initial heat of some 0.5 to 1.0 mcal/g, which is called *activation heat.* This is the thermochemical sum of all events that constitute coupling between excitation and contraction—movement of Ca^{++}, breakdown of ATP, and other processes.

If the muscle is permitted to shorten and displace a force, it performs work and liberates heat, the *shortening heat,* which is proportional to the change in length multiplied by a coefficient that is a function of the force. The shortening heat appears as extra heat when, during tetanus, a muscle is allowed to shorten. The shortening heat increases and the tension-dependent heat decreases as length decreases below l_0. The work of shortening occurs at a rate $P\,dl/dt$, where P is load and l is distance of shortening. At l_0 the heat components are difficult to separate. The elastic elements absorb heat as tension rises, and this must be released again upon relaxation. Shortening heat and tension-dependent heat are intermingled.

Figure 16–14 is an idealized representation of the energy balance sheet for frog sartorius muscles at 0°C. The shortening heat and tension-dependent heat are reciprocally related, with the former maximal at zero load. In other muscles or at other temperatures, the amount of shortening heat can be altered and the slope of the tension-dependent heat can vary considerably. In skeletal muscle, the magnitude of the activation heat is apparently length-dependent, and in cardiac muscle its magnitude varies with the level of contractility. The work curve is bell-shaped in all muscles, but its magnitude is probably related to the activation heat.

A frog sartorius contains about 24 μM of phosphocreatine (PC) per gram; it liberates some 3 mcal/g in a twitch and 40 mcal/g in a 10-second tetanus.[56] The sum of initial heat and work correlate very well with the breakdown of PC, which liberates 11 kcal/mole.

Phosphagens and Chemical Sources of Energy. It is clear from the preceding account of theories of muscle contraction that the immediate energy for the configurational rearrangements of the contractile proteins comes from adenosine triphosphate (ATP). However, the amount of ATP in a resting muscle (approximately 5×10^{-6} mole/g) is insufficient for more than a few twitches, and maintained contraction requires 10^{-4} to 10^{-3} mole/g/min.[27,376] Thus, ATP must be continually regenerated, and the second level of energy reserve is a phosphagen which is phosphocreatine (PC) in vertebrate muscles and is phosphoarginine (PA) in many invertebrate muscles. The enzyme creatine kinase (CK) catalyzes the transfer of high energy phosphate from PC to ADP, thus forming ATP. When CK is blocked by dinitrofluorobenzene (DNFB), and muscles are rapidly frozen in contraction for analysis, it can be shown that the splitting of ATP is confined to the contraction phase;[254] in a 0.2 sec tetanus at 0°C, 0.34 μM ATP/g is used. ADP is phosphorylated in relaxation or during maintained contraction.[56]

Creatine is rephosphorylated by either aerobic or glycolytic phosphorylation from the breakdown of glycogen. Under aerobic conditions *in vitro*, part of the lactic acid formed glycolytically is oxidized, giving energy for reconversion of the remaining lactate to glycogen. *In vivo*, the lactate leaves the muscle via the blood and is converted to glycogen in the liver; this supplies blood glu-

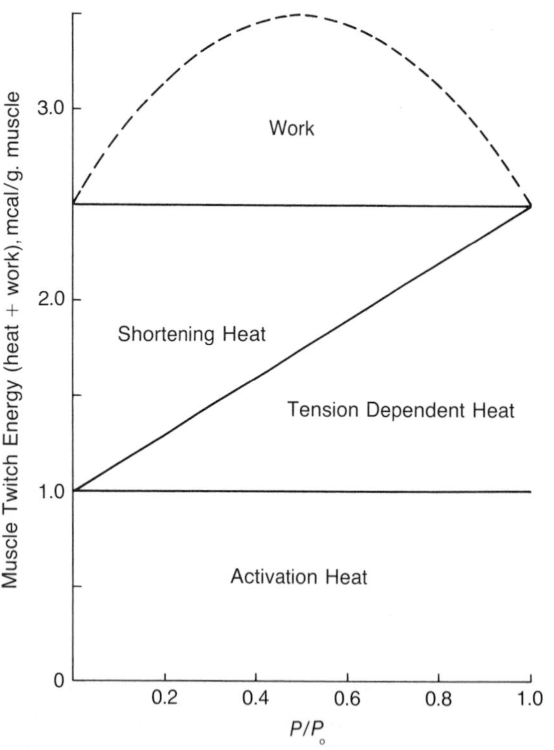

Figure 16–14. Diagram showing relation of energy liberation to load in frog twitch muscle. Shortening heat is maximal at zero load ($P/P_o = 0$) and minimal at maximum load ($P/P_o = 1$). (Courtesy of C. Gibbs.)

cose on demand, which then replenishes muscle glycogen. Payment of an oxygen debt after exercise provides for restoration of muscle glycogen. Under anaerobic conditions, lactate can accumulate and the PC can be recharged so long as glycogen is present in the muscle. In anaerobic muscle poisoned with iodoacetate (IOA) to block phosphoglyceraldehyde dehydrogenase, glycolysis stops at the triose phosphate stage and contraction continues only so long as the store of PC persists for recharging ATP. The amount of PC split is proportional to the number of twitches, and thus to activation heat plus work. An IOA-poisoned frog sartorius in nitrogen splits 0.286 μmole of PC/g muscle/twitch.[56] Cardiac muscle is more dependent on oxidative replenishment of glycogen than is striated muscle, and both cardiac and smooth muscle contain much less PC than does striated muscle. The isozyme of lactate dehydrogenase of heart is better adapted for aerobic metabolism than is the isozyme of the same enzyme of striated muscle (p. 259).

In general, fast muscles are less efficient in maintaining tension than are slow ones. For example, the posterior latissimus dorsi (PLD) of chicken is 15 times faster than the anterior latissimus dorsi (ALD), but the ALD can maintain tension 13 times longer than can the PLD. The ATP content of the ALD is 1 μM/g, and that of the PLD is 3 μM/g; if poisoned with DNFB, the ALD (but not the PLD) can remain active for many minutes. Relative efficiency is shown by the following data:[143,144]

	isometric tension integral in 10^3 g sec/g muscle/μM ATP
hamster biceps brachii	30.8
diaphragm	36.3
extensor digitorum longus	103.
soleus	191.
chicken ALD	540.

Chicken ALD is evidently some 2.5 times more efficient than hamster slow muscle (soleus).

Phosphagens contain various guanidine bases. Many guanidine compounds are formed by transamidination from arginine (see Chapter 7), but only certain of the guanidine bases have the appropriate kinases for phosphorylation and dephosphorylation so that they can serve as phosphagens. For example, octopine is a condensation product of arginine and pyruvate, found in all classes of molluscs, yet the phosphagen of all molluscs is phosphoarginine.

The muscles of all chordates contain PC; muscles of arthropods, molluscs, coelenterates and some lower worms have PA (Table 16–2). The protozoan *Stentor* contains arginine kinase as its only phosphagen kinase.[369] In echinoderms, only arginine phosphagen is found in holothurians and asteroids, and only creatine is found in ophiuroids; some genera of echinoids have both PC and PA, and other genera have only PA. The hemichordates have both PC and PA; the cephalochordates, like the vertebrates, have PC only.[18,352]

The annelids and sipunculids contain a wide variety of guanidine bases; some of these serve as muscle phosphagens, and the appropriate kinases have been found. Table 16–2 lists a few selected examples. In general, free-swimming polychaetes lack PA; sedentary polychaetes have a wide variety of bases but lack phosphoglycocyamine. Nereids use phosphoglycocyamine and phosphocreatine. Most annelid spermatozoids contain PC irrespective of the muscle phosphagen. A leech, *Hirudo*, contains much of a base, hirudine, and *Bonella* contains a different base, bonellidine, yet these bases do not become

TABLE 16–2. Distribution of Phosphagens in Annelids and Related Animals. (Modified from Thoai and Robin.[352])

Polychaeta							
Nereidae		(PC)	PG				
Nephthydae			PG				
Glyceridae		PC					
Eunicidae		PC					
Capitellidae	PA	PC					
Terebellidae	PA	PC					PT
Serpulidae	PA	PC					PT
Sabellidae	PA	PC					PT
Arenicolidae				PH			
Ophelidae	PA				PL	PO	
Oligochaeta							
Lumbricidae					PL		
Phoronida		PC					
Echiurida							
Urechis					PL		
Nemertea	PA						
Sipunculida (except *Sipunculus*)				PH			PT

Key: PA phosphoarginine
PC phosphocreatine
PG phosphoglycocyamine
PH phosphohypotaurocyamine
PL phospholombricine
PO phosphoopheline
PT phosphotaurocyamine

phosphorylated, and the phosphagens for these animals are unknown. Sipunculids (except for genus *Sipunculus*) contain much phosphohypotaurocyamine and some phosphotaurocyamine. The explanation of the extensive biochemical diversity of phosphagens of marine worms is unknown.

NEURAL ACTIVATION OF FAST VERTEBRATE POSTURAL MUSCLES

A vertebrate motor endplate is diagrammed in Figure 16–15. The motor axon in the region near the endplate is without myelin, and then it branches into a number of end-feet; the membrane of each of these together with the underlying sarcolemma folds into many troughs. Neural membrane is separated from muscle membrane by a 500 Å space. At the edges of a trough the Schwann (glial) cytoplasm separates terminal axoplasm from extracellular space. Specific enzyme staining shows a dense layer of acetylcholinesterase (AChE) in the subsynaptic membrane. Mitochondria are abundant in both terminal axoplasm and sarcoplasm. The axon terminals contain many synaptic vesicles 300 to 400 Å in diameter; these appear fewer in number, and empty, after prolonged stimulation.[242] (See p. 488.)

When a motor impulse arrives at a motor endplate, acetylcholine is released during about 2.5 msec (at 20°C), giving rise in about 0.3 msec to an endplate potential (excitatory post-synaptic potential), denoted epp or epsp (Fig. 16–16). This potential is local, and is graded in size according to the amount of transmitter present; it spreads electrotonically for a few millimeters around the endplate, where it initiates the muscle impulse. If curarization blocks initiation of a muscle impulse, the junction potential is seen to persist longer than the refractory period of the nerve so that summation can occur to two or more all-or-none nerve impulses. The epsp is prolonged as much as ten-fold by treatment with the antiesterase drug eserine

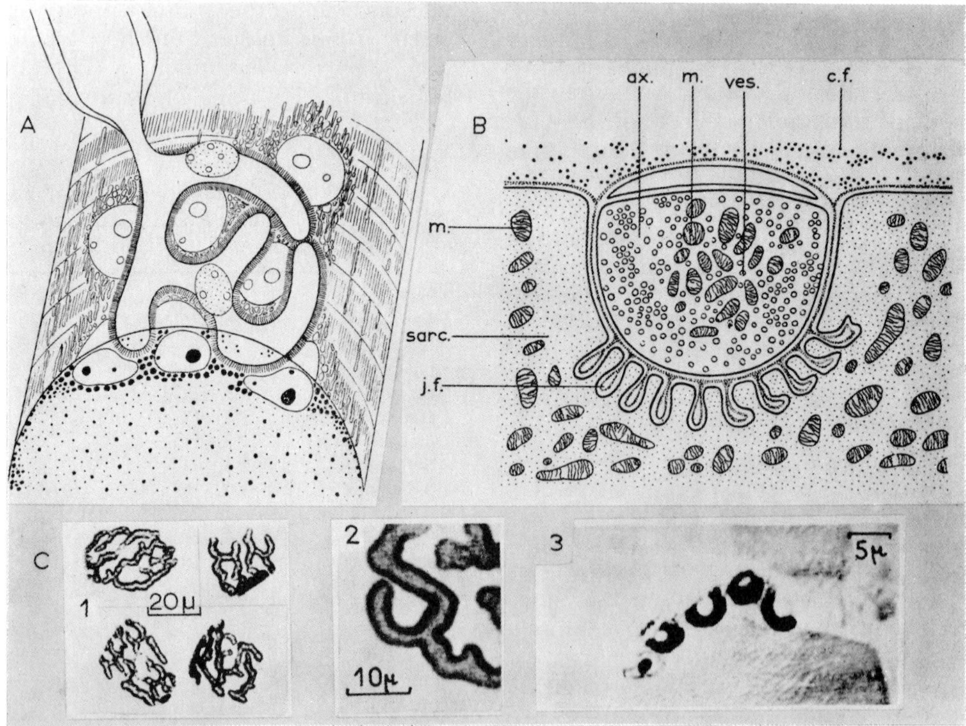

Figure 16–15. A, Diagram of motor endplate on mammalian striated muscle fiber. Myelin sheath of nerve axon lost before ending; nuclei of glial cells outside nerve ending. Rows of "gutters" or invaginations are regions of closest contact between nerve and muscle membranes. B, Detailed cross section of one branch of endplate. *ax.*, axoplasm; *m.*, mitochondria; *ves.*, vesicle; *sarc.*, sarcoplasm; *j.f.*, junctional folds. C, Enlargements of portions of endplates of intercostal muscles of mouse, stained for cholinesterase; *1* and *2*, surface views at low and high magnifications, and *3*, transverse cut through endplate leaflets. (A and C from Couteaux, R., *in* Problèmes de structure, d'ultrastructure et de fonctions cellulaires, edited by J. A. Thomas, Masson et Cie, 1955; B from Couteaux, R., *in* The Structure and Function of Muscle, edited by G. Bourne, Academic Press, New York, 1960.)

(physostigmine). Similar potentials can be elicited by local applications of acetylcholine (ACh) at the endplate region, whereas at other parts of the muscle membrane acetylcholine does not cause a response; ACh is ineffective when injected intracellularly.[211] Tubocurarine blocks neuromuscular transmission and antagonizes the effect of applied ACh by competing with receptor molecules. A snake venom, bungarotoxin, (BuTx), blocks the combining of ACh with receptors; by use of radioactive BuTx, the number of receptor sites per frog sartorius endplate is estimated at 10^9, while for rat diaphragm the value is 4.7×10^7 sites per endplate.[245]

If a nerve impulse is followed immediately by direct stimulation to the muscle fiber, the muscle impulse passes over the endplate without interfering with the junction potential; hence, the same membrane areas are not involved in both electrical events. The subsynaptic membrane is normally excited chemically, not electrically.

Intracellular recording at an endplate shows spontaneous miniature junction potentials (mepsp's) of constant minimal size, each resulting from liberation of a "quantum" of ACh. An epsp represents nearly simultaneous liberation of many quanta of ACh. Release of transmitter to cause endplate depolarization does not require an impulse in the motor axon, since release can be brought about by cathodal depolarization of the nerve endings; frequency (but not size) of mepsp's is increased by imposed depolarization.[212] The release of ACh requires Ca^{++}, which acts on the presynaptic membrane (even in the absence of Na). An epsp increases in amplitude as Ca_o is increased, up to a limit; in frog sartorius four Ca^{++} are required per ACh molecule released, and in rat diaphragm three Ca^{++} are required. Mg^{++} shows competitive inhibition of the Ca^{++} release of ACh, so that the increase in epsp requires more Ca^{++} in the presence of Mg^{++}. Lanthanum and manganese block transmitter release.[132] It is postulated from kinetic analysis in frog that four Ca^{++} ions combine with one receptor site X and that the number of quanta of acetylcholine liberated is proportional to (Ca X).[4] Spontaneous miniature potentials increase in frequency with increasing Ca_o, but they can be produced in Ca-free saline; thus, there is some spontaneous release of ACh in the virtual absence of external Ca^{++}. However, only the Ca complex allows release by nerve impulses.[95,189,296]

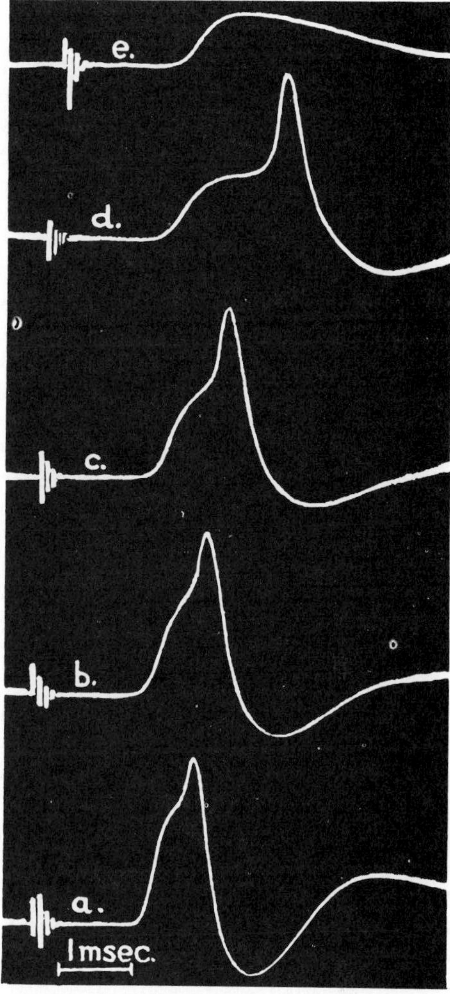

Figure 16–16. Action potentials from region of endplate in frog sartorius muscle: *a*, before curarization; *b* to *d*, increasing degrees of curarization; *e*, curarization complete. Propagated spike arises from junction potential in *a* to *d*. (From Kuffler, S. W., J. Neurophysiol. 5:18–26, 1942.)

Junction current is generated in the subsynaptic membrane in proportion to the amount of ACh liberated. The resistance of the endplate membrane decreases due to simultaneous increase in conductance for Na^+ and K^+. If the resting membrane potential is increased by polarization with a second electrode, the epsp, unlike an action potential, is increased; if the membrane is depolarized, the epsp decreases. The equilibrium potential in frog sartorius (see p. 464) is at −15 to −20 mv. The epsp involves a conductance change which is proportional to acetylcholine and is independent of membrane potential, whereas the action potential involves conductance changes which are voltage dependent. The muscle impulse, like that of a nerve, is caused by increased Na^+ conductance during the rising phase, followed by increased K^+ conductance. The junctional membrane is not affected by the blocking agent tetrodotoxin (TTX), but TTX prevents the Na^+ conduction change of the muscle spike. Disruption of T-tubules by brief treatment in hypertonic glycerol blocks activation of contraction but leaves epsp's and AP's normal.

After denervation the miniature epsp's are lost, and after a few days the entire muscle membrane becomes sensitive to ACh. The action potential in response to direct stimulation continues after denervation or curarization, but becomes resistant to TTX. In rat diaphragm the enhanced sensitivity to ACh becomes maximal at 15 days after denervation. The increased sensitivity can be prevented by agents which block protein synthesis, and the distribution of radioactive bungarotoxin which binds to ACh receptor sites (e.g., at endplates) shows that after denervation new receptors are formed over the surface of the muscle.[164a]

The preceding description of neuromuscular transmission is most complete for the frog sartorius. However, the essential features have been observed in lizards, birds, and mammals. In rat diaphragm, the maximal frequency of mepsp's is at 10 mM Ca^{++}. Decamethonium is a neuromuscular blocking agent which, unlike tubocurarine, blocks initially by depolarization; botulinum toxin poisons the liberation of ACh. Junctions in hen, cat, and man are more sensitive to decamethonium than to tubocurarine; in dog and rabbit they are equally sensitive to both. Tetraethyl ammonium ions (TEA) and procaine reduce outward potassium current and can block postsynaptic responses.

In muscle fibers of myotomes of *Branchiostoma* (amphioxus), when Na spikes are blocked by TTX or by bathing in Na-free saline and when potassium current is blocked by procaine, large spikes are found that result from inward calcium current. Normally, the calcium current is swamped by the inward sodium current, but the muscle is capable of producing either type of spike.[152b]

MUSCLE SPEED

Striated muscles have a broad spectrum of speeds that is highly adaptive for specific functions. Many attempts to classify vertebrate striated muscles as fast or slow, phasic or tonic, have emphasized only a single property. Muscles are adapted in their entirety, and no single property determines speed of movement. A sloth crawls at 50 m/hr and can be forced to 250 m/hr, whereas a cat runs at a peak speed of 40 km/hr and a cheetah at 104 km/hr; these speed differences are clearly related to contractile properties of their limb muscles.[140] Many properties related to speed are modified by trophic effects of motor nerves, which suggests that nerves have chemical influences on muscles which may be unrelated to nerve impulses.

Various measures for speed of contraction and relaxation have been used—isometric contraction time, time for relaxation to half maximum tension, shortening speed in isotonic contraction in sarcomeres or muscle lengths per second, and fusion frequency for tetany; contraction rate is well described by the velocity of shortening as a function of load.

Speed of contraction depends on the amount of series and parallel viscoelastic elements, and on fiber and sarcomere length; it is related to the kinetics or activity of myosin ATPase, to intrinsic properties of the contractile proteins, and to the metabolic patterns of various muscles. Speed is related to whether the muscle is activated by muscle impulses or by junction potentials, as well as to the abundance of the T-tubules. Speed of relaxation is related to the mechanical factors mentioned and to the rate of calcium binding by SR vesicles. Intrinsic speed in sarcomere lengths/sec is related to force-velocity properties, not to active state. Speed is the same for one myosin-actin bridge or for many in parallel so long as load is truly zero and internal friction is negligible. A few fibrils could give as high a velocity as could many.

Table 16–3 gives data on speeds of contraction and of maximum tension; in general, the very fast muscles develop less tension than slow muscles. Rates of relaxation cover a wider range than rates of contraction, and some "holding" muscles can maintain tension for many hours. Fast muscles fatigue sooner on repetition than do slow muscles. Muscles associated with sound production repeat very rapidly—the bat cricothyroid at 200/sec, sonic muscles of swimbladder of toadfish at as high as 300/sec (1 to 1.5 msec total contraction and relaxation time[332]), and the remotor of a lobster at 100 to 300/sec.[244] A hummingbird in hovering or flight beats its wings at 35 to 45/sec, a finch at

22 to 25/sec.[152] The fastest mammalian muscles are the oculomotors. A rat inferior rectus (35°C) contracts in 5 to 6 msec with shortening speed of 65 μm/sec.[69a] Rat anterior gracilis conducts at 3.5 to 5 m/sec at 35°C, and frog sartorius at about 1 m/sec at 20°C.

Innervation, Membrane and Mechanical Properties in Relation to Speed. In general, very few muscles are purely fast or slow; most muscles are mixed—fast and slow fibers. In frog muscles, the motor nerve fibers which elicit fast twitches are 10 to 12 μm in diameter and conduct at 20 to 35 m/sec. In addition, smaller (5 to 8 μm) axons, conducting at 2 to 8 m/sec, innervate many frog muscles. When conduction in the large nerve fibers is blocked or when the small fibers are dissected out and stimulated separately, the muscles give little contraction to single impulses but increasing response on repetitive stimulation, and give much slower contractions than when the muscle is stimulated by the large axons. When the small nerve fibers are stimulated, the muscle shows no all-or-none propagated impulses but only junction potentials which are smaller (7 to 15 mv) than the epsp's of the fast system. The slow epsp's last much longer than the epsp's of fast muscle, and on repetition they facilitate and plateau at 20 to 50 mv depolarization.

The resting potential of slow muscle fibers is lower (about 60 mv) than that of fast muscle fibers in frog (about 90 mv). Slow fibers of frog have apparent membrane capacity three times greater and specific resistance 10 times greater than fast fibers[4] (Table 16–4). Maximum velocity of shortening is linearly related to the reciprocal of contraction time, but twitch/tetanus ratio is independent of contraction time (Fig. 16–17). The slow type muscle fibers have many motor-nerve endings; hence, conduction in the nerve elements is sufficient and muscle impulses are not needed. The slow graded muscle fibers are some one hundred times less fatigable than fast twitch fibers. Acetylcholine or direct current causes a prolonged contraction of the slow fibers, and such muscles as the rectus abdominis may be used to assay for acetylcholine. Only the fast type of response occurs in frog sartorius or adductor longus, while predominantly the slow type occurs in iliofibularis. In frog rectus abdominis, the ventral superficial fibers have low resting potentials, innervation by fibers of high threshold, long latency, small epsp's, and facilitation of epsp's; deeper fibers have high resting potentials, low threshold nerves, short latency, epsp's leading to action potentials, and twitch responses.[125]

In mammals, most striated muscle fibers appear to have single nerve endings, and differences in speed result from properties other than pattern of innervation. However, multiple nerve endings with "en grappe" terminals occur in some fibers of the superior oblique muscle of the mammalian eye.[171] Differences in speed reflect differences in contractile proteins rather than in membrane activation.

Birds have two latissimus dorsi muscles; the posterior one (PLD) is a phasic muscle and the anterior one (ALD) is a tonic muscle. The electrical properties of the membranes

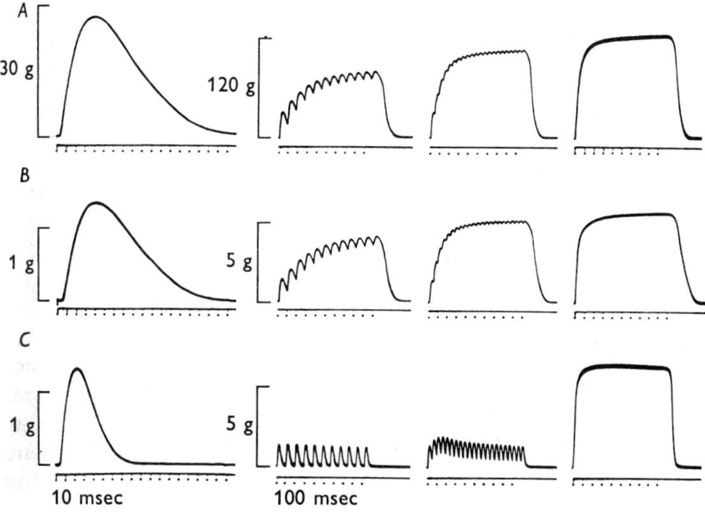

Figure 16–17. Records of isometric twitches and tetanic contractions of rat muscle units: A, slow muscle (soleus); B, intermediate unit (soleus); C, a fast muscle. (From Close, R., J. Physiol. *193*: 45–55, 1967.)

TABLE 16–3. Speed of Movement in Various Muscles.

*Reference**	*Animal and Muscle*	*Contraction time (msec)*	*Time to 1/2 relaxation (msec)*	*Shortening speed (muscle lengths or microns per sec)*	*Fusion frequency* (sec^{-1})	*Tetanus/twitch ratio*
148	Cat EDL†	19	19			
148	Cat soleus	70	109	31 μm/sec		
69	Cat internal rectus	7.5–10		1.3 μm/sec		
69	Cat EDL	19.6				
69	Cat inferior oblique	18			69	
168	Cat gastrocnemius *in vivo*	40				
168	Cat soleus	75–100				4
168	Cat internal rectus	7.5–10				11
179	Cat posterior cricothyroid	22				3.4
47	Cat flexor digitorum longus	18				5
	Kitten anterior tibialis	56				
	Cat anterior tibialis	27				
	Kitten soleus	65				
69	Rat EDL	13	7.5	42.7 μm/sec		5
69	Rat soleus	36	55	18.2 μm/sec		
	Kitten EDL	51	51	22.8 μm/sec		8.3
	Kitten soleus	70	109	12.7 μm/sec		
24	Guinea pig vastus lateralis	19	18.5			
24	Guinea pig medial gastrocnemius	22.2	21.2			
24	Guinea pig flexor digitorum longus	21.2	20.3			
24	Guinea pig flexor hallucis longus	21.2	19.9			
24	Guinea pig soleus	82.3	113.8			
142	Sloth extensor digitorum longus	65–98	169–190			
142	Sloth gastrocnemius	174	111–245			4–5.8
140	*Bradypus* claw flexor	140–300	240–300		15–20	
304	Bat cricothyroid	4			200	
67	Rat newborn EDL	56.5		17.3 μm/sec		2
67	Rat soleus	65		19.3 μm/sec		1.4
154	Rabbit thyroartenoid	6.5				
154	Rabbit cricothyroid	24–30				
154	Rabbit tibialis anticus	24–28				
152	Hummingbird flight muscle	8			250–300	
152	Finch flight muscle	14			100–150	
	Toad semitendinosus, 0°C	800	900			
4	Frog twitch fibers	22 (85% peak)				
4	Frog tonus fibers	46 (85% peak)				
81	Frog fast (peeled)			6–12 μm/sec		
81	Frog slow (peeled)			1 μm/sec		
148	Frog sartorius, 16°C	75	110	6 l/sec 1.3 l/sec	16–20	
168	Snake *Thamnophis* scale muscle	40	30			
	Turtle retractor penis	400	1000	0.36 l/sec		
306	Snake costocutaneous muscle	37–54				
332	Toadfish sonic muscle	5	8	6 l/sec		
134	*Holocentrus* sonic muscle	5	6		200	
266	Toadfish sonic muscle, 29°C	2.5	22.5			
32	*Hippocampus* fin muscle	10–15	13–15		100	

*References not cited are to be found in Table 50, p. 419, in the second edition of this book.
†EDL = extensor digitorum longus; ABRM = anterior byssus retractor muscle.

TABLE 16–3. Speed of Movement in Various Muscles. *(Continued)*

*Reference**	*Animal and Muscle*	*Contraction time (msec)*	*Time to 1/2 relaxation (msec)*	*Shortening speed (muscle lengths or microns per sec)*	*Fusion frequency* (sec^{-1})	*Tetanus/ twitch ratio*
9	*Myxine* fast parietal	150				
9	*Myxine* slow parietal	500				
151	*Branchiostoma* twitch	70–80				
151	*Branchiostoma* tonus	200–250				
186	*Podopthalamus* pink	80	400–2700			
186	*Podopthalamus* white	50	300			
225	*Paralithodes*	70–100	15–200			
242a	*Hemisquilla*					
	median extensor					
	twitch	200				
	tetanus	300	500 (rel. time)			
	lateral flexor					
	twitch	80	100 (rel. time)			
	distal					
	tetanus	50				
	proximal					
	tetanus	500	700 (rel. time)			
244	*Homarus* remotor fast	10				
244	*Homarus* remotor slow	50–60	3000–5000			
269	*Limulus* heart			0.981 l/sec		
208	Katydid sonic muscle	5 (cont. & relax.)				
360	Cockroach coxal muscle	4–10	9–15			
367	*Lethocerus* air guide muscle	20–30	100–200			
	Schistocerca flexor	25	300			
	Limulus abdominal flexor	195	435			
	Helix tentacle retractor	2500	25000	0.31 l/sec		
	Helix pharynx retractor	200–250	1300–1500		5–8	
250	Oyster opaque muscle			1.51 l/sec		
250	*Mytilus* ABRM† (phasic)	1000	4000–17000	0.251 l/sec		
351a,b	*Lumbricus* longitudinal (phasic)	70–80	200			
351b	*Lumbricus* longitudinal (tonic)		0.8–1.5 sec			
351b	*Lumbricus* circular	300–500				
	Pecten striated	46	40			
	Pecten non-striated	2280	5140			
64	*Aurelia*	500–1000	600–1000			
64	*Metridium*	500–30000 (1–3 min)	1–2 min (2–6 min)			
292	*Golfingia* proboscis retractor (phasic)	87	95			
	Golfingia proboscis retractor (tonic)	540	2700			
	Golfingia spindle muscle	1800	3500			
287	*Thyone* lantern retractor	3900	5700			
	Cephalopod chromatophore muscle	110	300			
	Squid mantle	68	1060		35	
339	Tracheal smooth muscle, 37°C	16.6		0.171 l/sec		
	Uterus, 27°C			0.181 l/sec		
	Papillary muscle, 27° C			1.21 l/sec		
3	Taenia coli			0.24 mm/sec		
	Turtle intestine	30000	36000			
	Dogfish mesenteric (phasic)	1800	2500			

*References not cited are to be found in Table 50, p. 419, in the second edition of this book.
†EDL = extensor digitorum longus; ABRM = anterior byssus retractor muscle.

TABLE 16–4. Muscle Membrane Constants.

Muscle	λ (*mm*)	τ (*msec*)	R_m (*ohms cm²*)	C_m (*μF/cm²*)	*Reference*
Frog sartorius				6.1	132
Frog twitch muscle	2.2	15	3140	6.8	4
Sartorius	1.1	18	4000	6	
Extensor digitorum	0.5–0.8	10.8	2080	5.2	
Frog extensor digitorum	1.1	18	4000	4.8	
Sartorius	0.5–0.8	10.8	2080	5.2	
Rat diaphragm	0.42–0.53		360–400	3.4–3.8	
Mouse soleus	0.92	4.4	2170		229a
Mouse EDL	0.55	2.8	708		
Frog tonus muscle			29000	1.6	
Chicken ALD	1.87	35	4388	8.2	116
Chicken PLD	0.68	3.7	561	7.0	
Snake costocutaneous	4.4	37.2	38280	1.04	306
Lizard fast			4000	7	286
Lizard slow			31000	3	286
Lamprey parietal fibers	4	140	36000	4	351c
Lamprey central fibers		5.2			
Carcinus leg 500 μ fibers	0.4	4	88.5	45	
Carcinus leg 210 μ fibers	1.5	15	282	54.6	13
Carcinus leg 105 μ fibers	2.3	41.4	1170	36	
Scorpion claw closer	0.6	3	830	3.6	138
Cockroach rectum	1.5–4	70	2×10^4–1.6×10^5	0.4–3.2	26b
Crab leg	2	5	5000	1.1	
Portunus	0.9	4.6	116	4.2	112
Guinea pig taenia coli			20–30×10^3	3	353
Taenia	1.45	107 (calc.) 6.6 (from foot of spike)	2080	5.2	2
Cat intestine					
Circular muscle	1.0	133 (calc.)	1000	30	220
Longitudinal muscle	1–1.5	52–105 (calc.)	780	12.7	220

are different, the membrane of the PLD having a shorter time constant (τ) and shorter space constant (λ)[116,180] (Table 16–4). The PLD fibers in birds receive single innervation with "en plaque" endings. When explored by local iontophoretic application of acetylcholine, single sensitive points are observed on each fiber, and the threshold to ACh is relatively high. In contrast, the ALD has multiple innervation, shows an ACh-sensitive spot every 740 μm, and has high sensitivity to ACh.

In the garter snake *Thamnophis,* short subdermal muscles run between the scales; their contraction time is 40 msec and $T_{1/2\,relax}$ is 30 msec. The threshold is at depolarization to −40 mv, and maximum contraction is at depolarization to between −10 and 0 mv.[167] Twitch fibers with focal innervation and action potentials are intermingled with tonic fibers having multiterminal innervation and junction potentials.

Fish muscles have both single and multiple innervation. Sonic muscle of the squirrelfish *Holocentrus* contracts at 100/sec in sound production, and can show fused tetanus at 200/sec. Its innervation is polyneural, with nerve endings 100 μm apart along a muscle fiber; both multiple psp's and action potentials are recorded, and hence this muscle has the innervation of a slow type muscle with the membrane properties of a fast one.[134] In

	fast muscle	*slow muscle*
examples	semitendinosus sartorius	rectus abdominus, some fibers iliofibularis
fibrils	1 μm or less, cylindrical Fibrillen Struktur well delineated fibrils	ribbon-like Felder Struktur irregularly distributed fibrils
endings	"en plaque" nerves	"en grappe" nerves
Z band	straight transverse	zigzag
M line	clear M	no M
tubules	extensive, and triads	sparse, no clear triads

snakefish *Ophiocephalus,* both red and white muscles have M bands, triads, and multiple "en grappe" endings; hence, the white fibers resemble frog fast fibers functionally but have endings like frog slow muscle.[259] The red fibers are thin and give graded summing potentials, while white fibers are larger and give epsp's and spikes.[347] The abdominal muscles of the fish *Cottus* are innervated by 10 to 14 μm axons, are fast-contracting, and give spikes; yet they have polyneuronal innervation, and a single fiber may have 8 to 22 terminals, as shown by cholinesterase staining.[190] In the hagfish *Myxine,* the parietal muscle fibers are of two types. The deeper fibers are thick, give relatively fast twitches of 150 msec duration, and have resting potentials of −75 mv and single innervation. More superficial thin fibers show twitch durations longer than 500 msec and have two motor axons with distributed endings and resting potentials of −46 mv.[9, 203] Segmental myotomes of a lamprey have central fibers, with parietal fibers on each side of the units; the central fibers give twitches and overshooting spikes and are electrotonically coupled, whereas the parietal fibers give slow responses and no spikes, have multiple endplates, and give mepsp's.[351c]

In myotomes of *Branchiostoma* the muscle fibers are ribbon-like cells 1 to 2 μm thick and 600 μm long; T-tubules are lacking, sodium and calcium conductance changes can occur in parallel, and calcium can be mobilized in caffeine contractures.[152a] The surface fibers are slow (200 to 250 msec contraction); deep fibers are faster (70 to 80 msec) and give spikes.[151]

Correlations of Muscle Structure with Speed. The ultrastructure of fast and slow muscles has been described on pages 720 to 725. Twitch and tonic postural muscles of frogs may be compared. See table above.

The cat stapedius is a fast muscle, while the tensor tympani is a mixed muscle with some slow fibers which lack M lines and have a jagged Z band.[117]

Correlation of speed with sarcomere length is shown by different fibers in lobster abdominal flexor.[1, 200, 201] See table at bottom of page. It is concluded that fast contractions are correlated with short sarcomeres, regular T-tubules, an extensive SR system, and an orderly pattern of fibrils.

Myoglobin and Metabolic Enzymes in Relation to Speed. Vertebrate striated muscles range from red (myoglobin-rich) to white (lacking myoglobin). In general, red muscles contract more slowly and fatigue less readily than white muscles. For example, the white gastrocnemius of cat is twice as fast as the red soleus, but they develop similar tensions per unit cross section area. Blood flow in red muscles may be three times greater than in white at rest, but in exercise the flow increases more in white than in red muscle. White muscles are better adapted enzymatically for anaerobic glycolysis, while red muscles are adapted for oxidative metabolism. Evidence from histochemistry of enzymes shows, for many muscles, "white" fibers interspersed among "red" ones; there are also fibers of intermediate properties. A general comparison of the two types follows:

red muscles	white muscles
rat soleus	*rat gastrocnemius*
many large mitochondria	few mitochondria
stored lipids	no stored lipids, much glycogen
high in phosphorylase	low in phosphorylase
high in cytochrome oxidase and NADH	high in muscle lactic DH
much succinic DH	little succinic DH
oxidative metabolism	anaerobic glycolysis
fatigue slowly	fatigue rapidly

Heart muscle is very red; it is low in glycogen and high in enzymes of aerobic metabol-

	sarcomere length	*number of thick filaments per* μm^2	*number of diads per mm of fibril*	*contraction duration (msec)*
superficial fibers	6–8 μm	350	600	500–2000
deep fibers	2–4 μm	450	2500	30–50

ism.[99] Exceptions to the above picture are known. A bat cricothyroid and hummingbird breast muscle have many mitochondria and lipid droplets, yet they are fast.

Two forms of lactate dehydrogenase (LDH) have been separated electrophoretically; one is characteristic of heart (H-LDH) and is adapted for high levels of oxygen and oxidative metabolism, while the other is characteristic of skeletal muscle (M-LDH) and is adapted for lower oxygen levels (see p. 259). Other explanations for the two LDH's have been suggested, but the general correlation remains.

Frog twitch fibers have many mitochondria and internal nuclei; frog slow fibers have fewer mitochondria and subsarcolemmal nuclei.[108] Histochemical differences are as shown at the bottom of this page.

The RNA content of chick ALD is greater by 40% (and proteolytic activity is greater by 200%) than in PLD; hence, more turnover of protein is indicated in the slower muscle.[343]

In a dogfish, the radial muscles of the unpaired fins have an outer zone of small red fibers which show only epsp's and are tonic, and an inner zone of large white fibers which have spikes and are twitch-type fibers.[307a]

Vertebrate striated muscles form a series from highly aerobic red muscles to highly glycolytic white muscles, with histochemical intermediates between the two extremes. These enzymatic differences are not necessarily correlated with innervation patterns, but are correlated with relaxation rates and holding power.

Myosin ATPase Correlates with Contraction, Ca-binding with Relaxation. Some properties of myosin ATPase were mentioned on p. 732; whether measured as activated by Mg^{++} actin (as *in vivo*) or by Ca^{++} (*in vitro*), the activity of this enzyme on a protein basis correlates extremely well with rate of contraction in many kinds of animal. The chicken PLD has three-fold greater activity of myosin ATPase than the ALD.[143] Rat red muscle is low and white muscle is high in myosin ATPase.[99] The fast striated portion of scallop adductor has five times the activity of Ca-ATPase of the slow holding scallop muscle.[20] The effect of temperature on myosin ATPase is the same as on shortening.[19] Table 16–5 documents the correlation of myosin ATPase with speed of contraction.[19]

TABLE 16–5. Correlation of Myosin ATPase with Speed of Contraction[19]

	Shortening Speed, Muscle Lengths/sec (at 37-38° C for Mammalian Muscles, 18° C for others)	*Actin Activated ATPase (μM P_i/mg Myosin/min)*
mouse extensor digitorum, longus	24	30.03
rat extensor digitorum, longus	17.2	27.95
mouse soleus	12.8	14.38
rat soleus	7.2	10.74
Pecten striated (mollusc)	3	4.03
frog sartorius	2	2.04
rabbit uterus	0.2	0.23
tortoise iliofibularis	0.1	0.13
Mytilus posterior adductor (mollusc)	0.1	0.15

A series for *Pecten* muscles is:[340]

	Mg ATPase μM P_i/mg/min
striated adductor (fast)	1.75
translucent smooth muscle (slow)	1.21
opaque smooth muscle (holding)	0.36

The mammalian soleus is low in myosin ATPase, and the medial gastrocnemius is high in the enzyme. Muscles of cat have ATPase two to four times more active than that in corresponding muscles of the sloth.[19] In insect glycerinated fibers, tension roughly parallels ATP breakdown as calcium concentration is increased from 10^{-8} M to 10^{-7} M.[60] In general, the rate of contraction is well correlated with myosin ATPase activity.

It was mentioned on page 733 that relaxation occurs when calcium ions are actively re-bound to sarcoplasmic reticular vesicles which contain a Mg-activated ATPase. In general, the rate of *in vitro* uptake of Ca^{++} by isolated vesicles (microsomes) is related to muscle speed, and particularly to speed of relaxation. The Ca-uptake by particles from mammalian smooth muscle is one tenth that by particles of heart.[57] The rate of uptake by microsomes of rabbit white muscle is greater

	diameter	*ATPase*	*DPNH DH*	*phosphorylase*
tonic	small	very low	low	very low
twitch	small	high	high	high
twitch	large	high	low	medium

than by those of rabbit red muscle. Some data on calcium uptake follow:[340]

	Ca uptake, μM/g/10 min at 30°C
locust flight muscle	111
locust jumping muscle	53
mouse gluteus	88
Lumbricus body wall muscle	10
Mytilus anterior byssus retractor	10
Phormia synchronous flight muscle	14

In skinned fibers of the tonus bundle of frog iliofibularis, contraction in response to microperfused Ca spread further in slow fibers than in fast ones, probably because the SR binds Ca more effectively in the fast fibers.[82]

Trophic Effects of Nerves on Muscle Speeds; Correlates with ATPase. Many observations on mammalian muscles show that several enzymatic, membrane, and structural protein changes can occur according to the nature of the muscle innervation, and that muscle speed can change in consequence. Spectacular transformations occur during development, and when nerves from fast muscles are allowed to grow into slow muscles and vice versa. See below.

Force-velocity curves are given in Figure 16–18. The preceding data show that fast muscles can be made slower and slow muscles can be made faster by cross-innervation; specifically, the velocity of shortening is altered, which implies a change in contractile proteins and in duration of active state. Normally, fibers of the soleus are sensitive to acetylcholine over much of their length, and the EDL only at "en plaque" junctions; after innervation by a mixed sciatic nerve, some fibers of the soleus show restricted ACh sensitivity.[247] Similar increase in ACh-sensitive areas in chick PLD and decrease of sensitive areas in ALD occur after cross-innervation.

Myosin ATPase has been isolated from slow (soleus, SOL) and fast (flexor or extensor digitorum longus, EDL) muscles, and from these muscles after mutual cross-innervation. The activities of actin- and Mg-activated ATPase in rat EDL and SOL are proportional to their speeds of shortening. The ATPase from soleus is more heat stable and more alkaline labile, and has different electrophoretic mobility.[135] After cross-innervation, the myosin from the re-innervated fast muscle has reduced ATPase activity and greater alkaline stability, whereas the cross-innervated soleus shows increased ATPase activity and acid lability.[324] The pH profile of the myosin ATPase and the ATP-induced dinitrophenylation showed that the myosin was altered by cross-innervation; dinitrophenylation became reduced in the slow muscle, and the myosin from this muscle became more acid-stable on cross-innervation.[21] Thus, the myosins from fast and from slow muscles are qualitatively different, and the form which is synthesized in a muscle depends upon the innervation. Whether the neural influence is by repeated and patterned impulses or by a "trophic" chemical agent has not been determined.

Evidence for nerve-impulse control of contractile proteins comes from tenotomy and prolonged stimulation of muscles in rabbits. Normally, a soleus receives tonic motor impulses from the spinal cord; fast muscles lack continuous activation. After tenotomy, the contraction rate of soleus accelerates from a contraction time of 62 msec to 20 msec, and if the muscles are stimulated continuously for up to 20 days the speeding of contraction is prevented.[323, 366]

	flexor digitorum longus (FDL) FAST		soleus (SOL) SLOW	
	control (self-innervated)	crossed	control (self-innervated)	crossed
cat contraction time (msec)	25.5	59	75	33.3
contraction time[48]	18–20	60	70–80	35–40
tetanus/twitch ratio[47]	7.1	4.3	5.6	11.7
cat contraction time (msec)	32.8	48.8	69.7	47
rat contraction time (msec)[314]	23.2	32.2	45.8	28.7
	extensor digitorum longus (EDL)		soleus	
rat contraction time (msec)	13	25	34	15
contraction rate μm/sec[68]	45.1	22.5	19.8	33.8

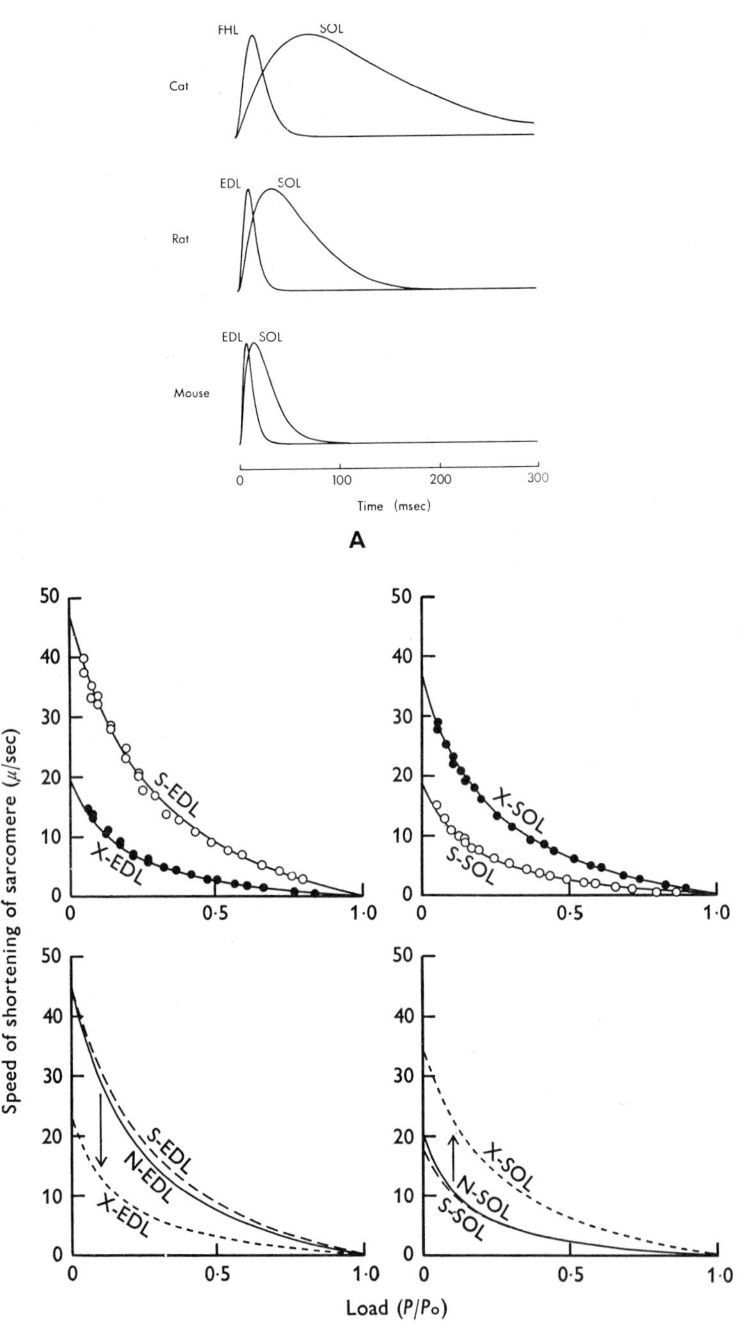

Figure 16–18. A, Isometric twitches of soleus (SOL), flexor hallucis longus (FHL), and extensor digitorum longus (EDL) of three mammals. (From Close, R., Excerpta Med. Internat. Cong. Series, Vol. 147, 1966.) B, Force-velocity curves for normal (N), self-innervated (S), and cross-innervated (X) slow soleus (SOL) and fast extensor digitorum longus (EDL) of rat. (From Close, R., J. Physiol. *204*:331–346, 1969.)

During development of mammals, changes in speed of muscles occur, similar to some of those observed in cross-innervation, as shown by the following data:[70,71]

	Extensor digitorum longus		Soleus	
	kitten	cat	kitten	cat
contraction time (msec)	51	30	70	70–80
$T_{1/2\,relax}$ msec)	51		109	
shortening speed (μm/sec)	22.8	31	12.7	13

Thus, the extensor digitorum longus is speeded during development, but in soleus there is no change. Similarly, in newborn rats the force-velocity curves and contraction speeds of EDL and soleus are similar, but during development the EDL is speeded up some 2.5 times while the soleus is unchanged.[67]

Muscle cell resting membrane potentials increase with age, as shown at the bottom of this page.

The sensitivity to locally applied acetylcholine and the responsiveness to caffeine change with innervation; hence, the SR system and distribution of ACh receptors are under neural control.[247]

In summary, striated muscles adapted for fast contractions differ from those which contract more tonically in many ways: ultrastructure, distribution of nerve endings and ACh sensitivity, oxidative and glycolytic enzymes, mechanical properties relating to myosin, and myosin ATPase. The particular features which are most important in one muscle may not be the same as in another muscle of similar speed, especially in a differrent kind of animal. Many muscles are mixed, i.e., have fibers of two or three types, particularly in regard to distribution of enzymes.

Speed of contraction is correlated with rate of ATP hydrolysis and relaxation speed is correlated with Ca binding. Many of the chemical properties are induced in muscle by influences from their motor nerves. Muscles in turn influence the distribution of endings of nerves growing into them. The effects of innervation are apparent in the synthesis of muscle proteins, and hence must act in part via genetic coding. The relative importance of trophic chemical agents and motor impulses remains to be elucidated.

POLYNEURONAL INNERVATION IN ARTHROPODS, ESPECIALLY CRUSTACEANS

Arthropod muscles show some of the same histological and chemical correlates with speed as do vertebrate fibers. Short sarcomeres, uniform distribution of fibrils, and extensive SR tubules correlate with fast contractions. The range of sarcomere lengths between fast and slow fibers in vertebrates (2.0 to 3.5 μm) is much less, however, than for arthropod fibers (2 to 14 μm). Thin fibers with long sarcomeres and deep sarcolemmal invaginations are slow.[14,15] In the crab *Cancer*, phasic fibers have 4 to 5 μm sarcomeres and one thick to every six thin filaments, whereas tonic fibers have larger fibrils, 12 μm sarcomeres, and one thick to 10 or 12 thin filaments.[110] Speed of contraction is correlated not only with sarcomere length but also with extent of the SR tubule system. In *Hemisquilla* the distal portion of the lateral flexor, which receives a "fast" motor axon, contracts in 50 msec, and the proximal portion of the same muscle, innervated by a "slow" axon, contracts in 500 msec; in each the sarcomere length is 4.4 μm, but the distal part has 24 diads in a given area while the proximal part has twelve.[242a]

Most crustacean muscle fibers are very large; possibly this is related to the peripheral control and sparseness of neurons, which are greater than in vertebrates. Barnacle muscle fibers may be 1 or 2 mm in diameter, and fibers as large as 4 mm occur in leg muscles of king crabs. Barnacle sarcomeres average 9.2 μm, and the muscle can contract to 30% of rest length; in this state, thick filaments can be seen to penetrate the Z disc and overlap in the next sarcomere.[187]

Patterns of Innervation. Nerve endings in arthropods do not occur as discrete endplates, but transmitter may be liberated from axons wherever they make contact with muscle fibers and the Schwann cell is lacking; a dense innervation may envelop muscle fibers (Fig. 16–19).

Whole muscles of many fibers in crustaceans and insects may be innervated by only two (or a few) axons, and each muscle fiber may have polyneuronal innervation, i.e., each fiber may receive branches from two or more axons; as many as five axons have been found to innervate one muscle fiber. Nerve endings occur profusely over the surface of a muscle fiber, and even though most fibers in a muscle are innervated by a given axon, the number of fibers responding may vary with frequency of nerve impulses. This probably reflects differences in neuromuscular facilitation and conduction at axon branch points. Usually one axon is inhibitory; the others are excitatory and may elicit con-

	Extensor digitorum longus		Soleus	
mouse	1 week old	8 weeks	1 week	8 weeks
resting potential (mv)	−57.8	−79.5	−67.4	−77.2

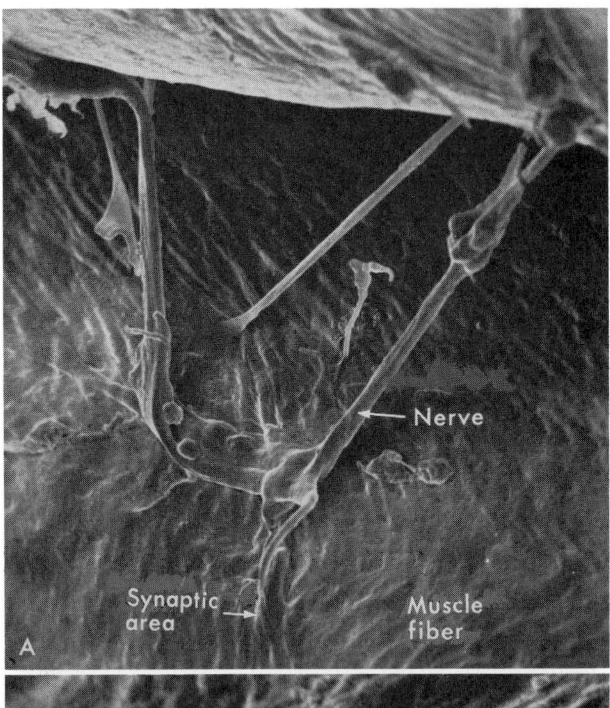

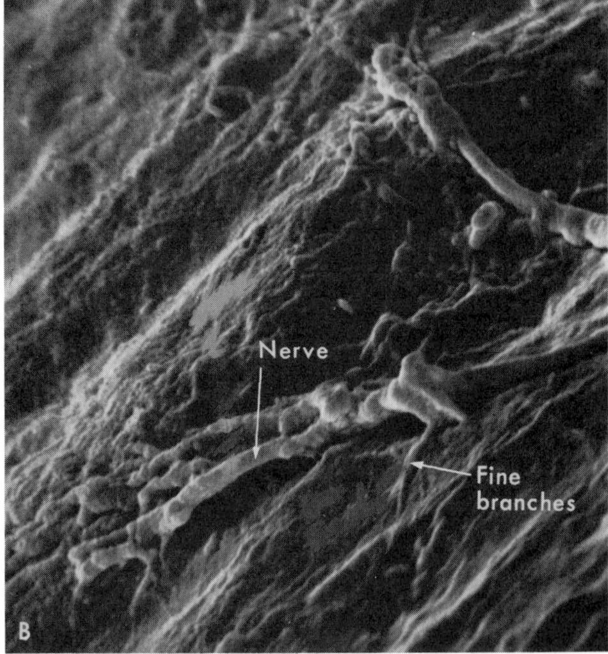

Figure 16–19. A, Scanning electron micrograph of crustacean muscle, showing nerve fiber going from one muscle fiber to another, forming *en passant* junctions and branches, and then returning to the first fiber. B, Scanning electron micrograph showing diffuse nature of nerve endings on a muscle fiber. (A and B courtesy of F. Lang.)

tractions of different speeds and facilitation requirements. Where inhibitor and excitor axons can be identified, it has been found that synaptic vesicles of excitatory axons are spherical and those of inhibitory ones are smaller and oval or irregular.[224a, 357a]

Not all of the fibers within a muscle are innervated by all of the axons that go to the muscle. Sampling of many fibers in the flexor of the claw leg of *Panulirus* showed that 7 per cent received four excitor axons each, 27 per cent received three, 26 per cent received two, and 38 per cent received only one axon.[374, 375] The levator of the eyestalk of a portunid crab, *Podophthalmus,* has an outer portion of fast white fibers and an inner part of tonic pink fibers; the white portion receives two fast axons, while the pink por-

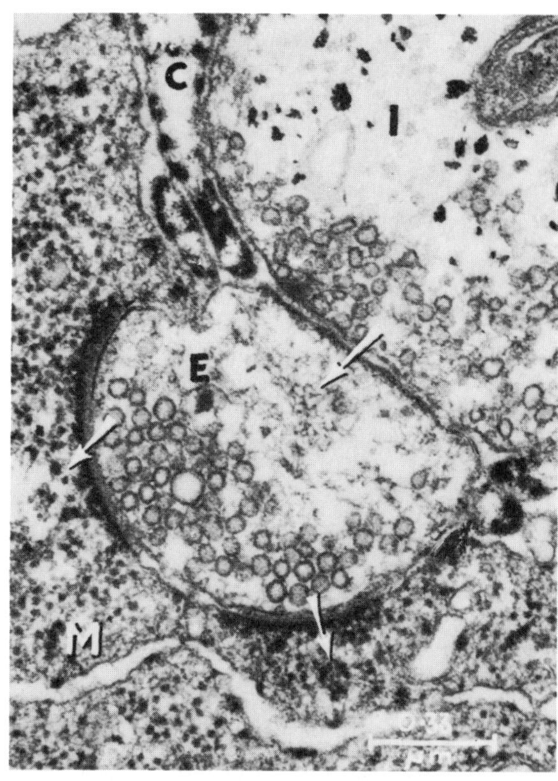

C

Figure 16–19 Continued. C, Neuromuscular and axo-axonal synapses in crayfish opener muscle. Excitatory terminal (E) contains round synaptic vesicles and forms two synaptic membrane areas on the muscle; the inhibitory terminal contains smaller irregular vesicles and forms a presynaptic inhibitory contact on the excitatory nerve terminal. (From Lang, F., H. L. Atwood, and W. A. Morin, Z. Zellforsch., *127*:189–200, 1972.)

tion receives one fast and one slow excitor and one inhibitor axon.[186] In the mantis shrimp, *Hemisquilla,* the median extensor of the merus receives one fast (non-facilitating) excitor and one inhibitor axon; the lateral extensor receives one fast and two slow (facilitating) axons; and the medial flexor receives two slow motor axons.[51a]

There is great variation in crustaceans in innervation pattern from muscle to muscle and for corresponding muscles from animal to animal (Fig. 16–20). Frequently, one axon serves several muscles. In crayfish, the opener and stretcher muscles share a single excitor axon; the closer, bender, and extensor may each have two axons, and the flexor has four.[375] The closer, bender, and extensor of the claw usually have one fast and one slow motor axon each. When one axon, either excitor or inhibitor, serves two muscles, which muscle contracts may be determined by frequency of stimulation.

Peripheral inhibition is of two sorts—presynaptic and postsynaptic. Postsynaptic inhibition hyperpolarizes the muscle membrane; it increases conductance (decreases resistance) to small anions (principally Cl^-) and can be made depolarizing if the membrane is made more negative than normal by intracellular polarization. In crayfish abdominal flexors, all of the inhibition is postsynaptic.[215] Presynaptic inhibition diminishes the release of excitatory transmitter and has no direct effect on the muscle membrane resistance. In the dactyl opener of crayfish, arrival of an inhibitor impulse 1 to 6 msec prior to an excitor impulse reduces the excitatory psp by as much as fourfold.[100, 101] In a number of systems (e.g., *Pachygrapsus* closer) the response to a slow excitor is attenuated presynaptically by the inhibitor, while the response to a fast excitor is less attenuated.[15] Presynaptic inhibition reduces or blocks spontaneous mepsp's. In crayfish claw opener, inhibition elicited at 10 Hz is 90% presynaptic, while at 40 Hz it is 50% presynaptic; hence, the same axon has either action according to frequency. The amount of pre- and postsynaptic inhibition varies from fiber to fiber, and facilitation of inhibition reflects liberation of different amounts of transmitter.[14a]

It is generally assumed that propagated muscle action potentials are not necessary, since the motor axons conduct rapidly and there are many nerve terminations along a

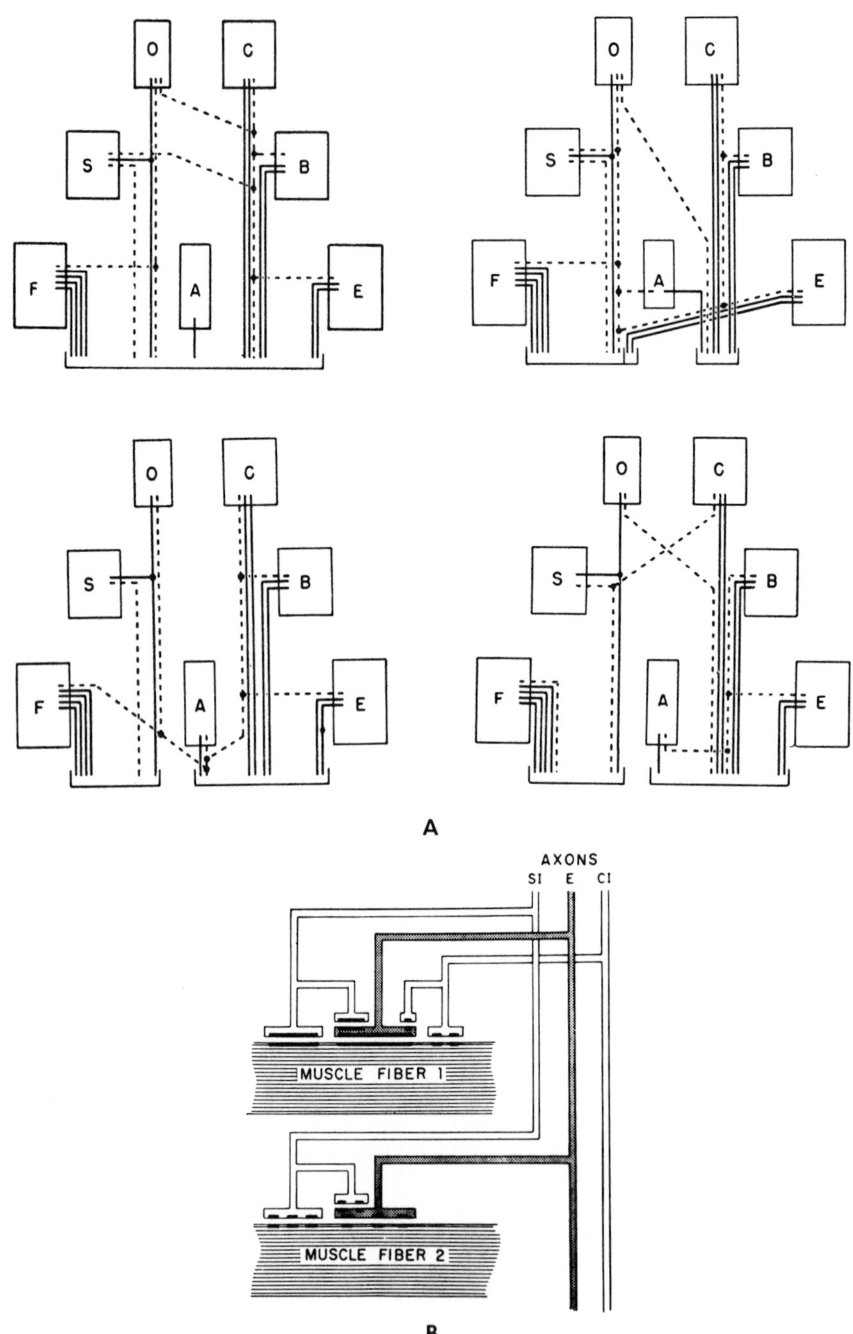

Figure 16–20. A, Distribution of innervation to different muscles in several groups of decapod crustaceans. *O*, opener, and *C*, closer of claw; *S*, stretcher, and *B*, bender of carpopodite; *F*, flexor, *A*, adductor, and *E*, extensor of meropodite. (From Wiersma, C. A. G., *in* Physiology of Crustacea, Vol. II, edited by T. Waterman. Academic Press, New York, 1961.) B, Diagram summarizing synaptic connections of the excitor (E), specific inhibitor (SI), and common inhibitor (CI) in stretcher of crab *Grapsus.* Muscle fiber (1) is long-sarcomere slow fiber; muscle fiber (2) is a short-sarcomere fast fiber. Poorly facilitating synapses have long synaptic membranes, while facilitating synapses have short synaptic areas. (From Atwood, H. L., and G. Bittner, J. Neurophysiol. *34*:157–170, 1971.)

muscle fiber; this results in nearly simultaneous release of transmitter at many points on the muscle fiber. The junction potentials, which are due to increased Na conductance, are adequate to elicit contractions.[264] Junction potentials may be fast, with little or no facilitation, and may reach an amplitude that gives a spike-like depolarization. On the other hand, they may be slow and facilitating; sometimes these psp's reach a depolarization that gives rise to "graded spikes."[14a] In *Hemisquilla,* the median flexor of the merus receives two slow motor axons, and it shows trains of facilitating epsp's with occasional fast spikes breaking through the train.[51a] Direct stimulation by an intracellular electrode can elicit an overshooting action potential, and propagated all-or-nothing spikes can be observed in some fibers in response to stimulation of motor axons.[112]

Spike responses to transmembrane stimulation of crustacean muscles are due to increased conductance for calcium ions. In some crayfish muscles, Sr can substitute for Ca in action potentials but not for contractions.[111] In barnacle fibers, the depolarizing response to transmembrane stimulation is in proportion to the inward gradient of calcium; it is selectively blocked by Mn^{++} or Co^{++} at low concentrations.[153]

Speed of contraction and relaxation, as well as dependence on facilitation, are influenced by the properties of both the muscle fibers—phasic or tonic—and the excitatory motor axons—fast or slow. Some muscles receive a single excitor and a single inhibitor nerve fiber. In some of these, as in the accessory flexor of a crab, the proximal fibers have long sarcomeres and are tonic, while the distal fibers have short sarcomeres and are fast. In the cheliped opener of crayfish, fibers in the central region of the muscle show small, facilitating epsp's, and those in the distal region show larger, poorly facilitating epsp's, all in response to the same excitor axon.[14a] In the stretcher of *Cancer,* phasic fibers have brief but facilitating epsp's and tonic fibers have large and long-lasting epsp's, with both types of response elicited by the same axon.[15] In general, where a muscle with a single motor axon gives both fast and slow contractions, the slow fibers have large, poorly facilitating epsp's and the fast fibers have small, strongly facilitating epsp's, which often elicit a spike at a high frequency of stimulation. Thus, the slow fibers contract at a low frequency of motor axon activity, and the fast fibers contract at a high frequency of stimulation. The facilitation and speed of epsp's differ in various synapses made by the same axon, and relate to the amount of transmitter liberated. Another correlate with contraction speed and amplitude is membrane potential; in several muscles of lobster legs, the rate of tension development increases with the amount of membrane depolarization as applied by an intracellular electrode.[202a]

In muscles having both fast and slow excitors, stimulation of a large motor axon causes a contraction of short latency, rapid shortening, and high tension and twitch nearly as great as tetanus; the muscle potentials are large epsp's that, while not usually overshooting, show little facilitation. Stimulation of the second or intermediate size axon causes a contraction which builds up slowly so that there is an apparent long latency, marked facilitation, and low tension; muscle potentials are initially small epsp's, and on repetition they facilitate greatly. When a whole nerve is stimulated, a transition from fast to slow contraction occurs as the frequency increases. Stimulation of the third, usually the smallest, axon inhibits the response to the excitor fibers. During inhibition, the contraction may be suppressed but facilitation may continue, as shown on release from inhibition.

Thus, in muscles with dual or multiple excitatory innervation, responses may be determined by both type and location of muscle fibers and by the particular axon which is active. In the closer of several crabs, some fibers show much facilitation in response to each excitor, and others give a large epsp to the fast and a small epsp to the slow excitor; differences between responses to the two axons are greater at low than at high frequencies.

The remotor muscle of the lobster second antenna is of two parts, each with multiterminal and polyneuronal innervation. The fast part, containing many sarcoplasmic tubules, gives twitches of less than 10 msec contraction time, is non-facilitating, and develops tension of 25 g/cm^2; the slow part has a contraction time of 50 to 60 msec, requires facilitation of epsp's, relaxes slowly (3 to 5 sec for one-half relaxation), and develops tension of 2800 g/cm^2.[244] In the closer of the claw of *Carcinus,* some fibers give large epsp's, are electrically inexcitable, give twitches, are relatively insensitive to inhibition, have a dense innervation from the fast axon, and have low membrane resistance and a time constant of 5 msec; a second group of fibers shows large, slow epsp's to the slow axon and small or no response to the fast axon, is electrically inexcitable and sensitive to inhibition, and has a membrane constant of 30 to 60 msec. A third group shows both fast and slow epsp's, and both types facilitate.[13] In the crayfish claw opener, distal epsp's are some five times larger than those in central fibers; the probability of transmitter release is greater in the distal fibers.[34] In the king crab, *Paralithodes*, the closer has two excitors; one (the fast excitor) elicits large, weakly facilitating epsp's proximally, small ones distally, and medium size, strongly facilitating ones centrally; the other (slow) excitor gives large, moderately facilitating epsp's distally and small, very strongly facilitating ones centrally[225] (Fig. 16–211).

In the abdomen of crayfish and lobster, the deep or medial extensor is a phasic muscle; some fibers have three excitors and two inhibitors, while others have two excitors and one inhibitor. The superficial or lateral muscle is tonic in contraction and slow to fatigue; it receives five excitors and one inhibitor, and there is much variation in innervation from fiber to fiber.[1, 215] It is concluded that different axons have different effects, especially at low frequencies; that individual muscle fibers differ in their membrane properties; that different fibers in one muscle differ in their innervation; and that all of these differences may be related to location within a muscle.

Still another type of control depends upon the patterning of motor impulses. For example, the claw opener of *Eupagurus* has a single motor (excitor) axon, and stimulation at low frequency gives a slow facilitating contraction, yet interpolated shocks elicit quick, strong contractions; hence, one axon can elicit either slow or fast responses according to how it is stimulated. Some excitor axons discharge repetitively in response to a single brief stimulus; these have long accommodation and short excitation constants. The fast closer of the crayfish claw is not repetitive and has a high threshold and an accommodation constant of 717 msec; the single opener gives repetitive discharges, and has an accommodation constant of 48.2 msec and a shorter excitation time-constant.[384a] In *Uca*, a common excitor axon serves both stretcher and opener muscle, but the inhibitor nerves are separate; the inhibition is coupled reflexly, so that the opener inhibitor fires closely after the excitor and the stretcher inhibitor fires before the excitor. This correlates with the fact that inhibition in the opener is predominantly postsynaptic, while that in the stretcher is presynaptic.[335]

In some arthropods, flexors in the legs are not antagonized by extensors; extension is by hydrostatic means. It has been suggested that inhibition occurs only where antagonists are present. In a scorpion claw, the closing is active and the opening is passive; the flexor has two excitors per fiber, one eliciting psp's that facilitate to spikes and the other eliciting fast, non-facilitating psp's. Each of two or three motor units in the closer has its own pair of axons, but no peripheral inhibition is found.[139] A spider leg flexor muscle receives three motor axons and no inhibitor. A single muscle fiber shows three steps in its response: one axon elicits a fast contraction, another elicits a slow contraction, and the third elicits a fast contraction but has a high threshold.[298] Large-diameter closer muscle fibers of a barnacle also have no antagonists; they have two or three excitor axons and no inhibitor. Responses are non-propagating, but the psp must exceed 18 mv depolarization for a contraction.[187] In *Limulus* the closer muscle receives six excitors, and there may be three axons to one muscle fiber; responses are graded and facilitating. The inhibitor was difficult to isolate and ipsp's were not observed, but mechanical activity was diminished at high stimulus intensities or when GABA was applied.[270]

Transmitters. Drugs that block or potentiate transmission in vertebrate muscles are without effect at reasonable concentrations in crustaceans, and it seems certain that acetylcholine and noradrenaline are not involved in crustacean neuromuscular transmission. The most probable excitatory transmitter is *l*-glutamate; it depolarizes the muscle, causes contraction, acts synergistically with excitatory nerve impulses, and seems to act on the same receptor as does the excitatory transmitter.[349, 350, 351] Excitatory transmission is blocked (in *Pachygrapsus* closer) by high magnesium or low calcium, and synaptic vesicles are abundant in facilitating excitor endings.[15] The abundance of spots locally sensitive to glutamate is proportional to the number of nerve endings.[349]

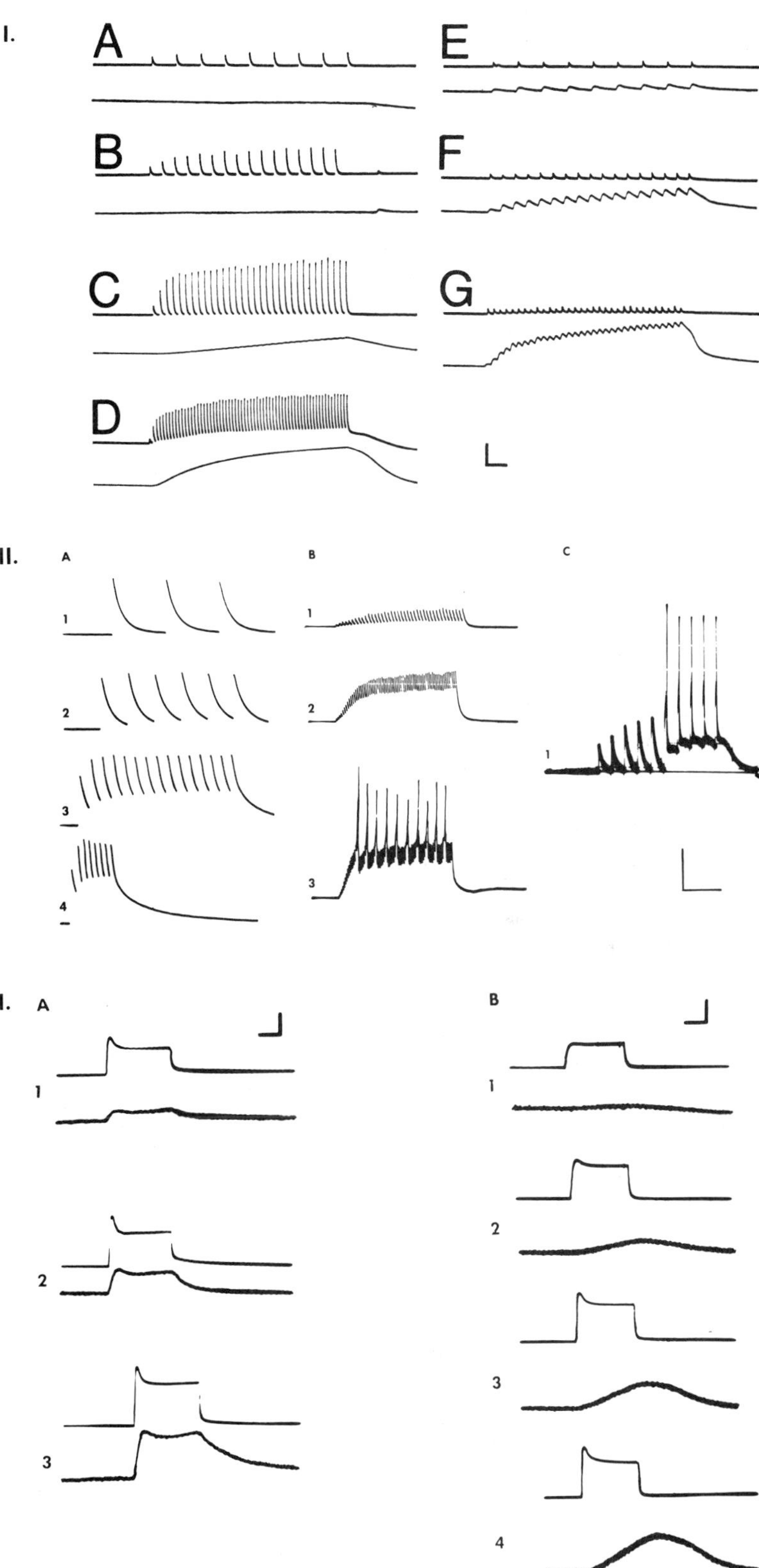

Figure 16–21. I, Electrical (upper) and mechanical (lower) records from fibers of muscle of king crab when stimulated by fast (E – G) and slow (A – D) motor axons. (From Lang, F., A. Sutterlin, and C. L. Prosser, Comp. Biochem. Physiol. *32*:615–628, 1970.) II, Diversity of electrical responses to different frequencies of stimulation of different muscle fibers in stretcher of crab *Grapsus*, a muscle with only one excitatory axon. (A) epsp's are large and poorly facilitating in long-sarcomere slow fibers; (B) epsp's are initially small but facilitate strongly in short-sarcomere fast fibers; (C) epsp's are moderately large and give rise to spike-type responses at low frequency. Calibration: (A), (B) 20 mv and 1 sec; (C) 20 mv and 0.5 sec. (From Atwood, H. L., and G. Bittner, J. Neurophysiol. *34*:157–170, 1971.) III, Tension development in lobster extensor muscle as a function of membrane depolarization. Upper traces, membrane potential responses to injected current; lower traces, tension. (A) In a fast fiber, tension parallels depolarization. (B) In a slow fiber, tension rises slowly in response to a sudden, sustained depolarization. Calibration: 0.2 sec, 10 mv, 20 mg tension. (From Jahromi, S. S., and H. L. Atwood, J. Exp. Zool. *176*: 475–486, 1971.)

Junctional potentials and responses to iontophoretically injected glutamate are abolished when sodium is replaced by Tris or Li.[264] The equilibrium potential for excitatory psp is −20 mv (in crayfish), and during an epsp the membrane resistance is reduced to half its resting value.[303] The crayfish opener (abductor) shows spontaneous mepsp's that are increased by depolarization of the excitor nerve endings; it is calculated that, at 5/sec, each impulse releases on the average 2.2 quanta (one quantum equals the amount for one mepsp) of transmitter. Facilitation increases the probability of release of excitatory quanta.[103] Single terminals at low frequency stimulation release transmitter with greater probability in distal than in central fibers; that is, transmitter release deviates from the Poisson distribution.[35]

Gamma-aminobutyric acid (GABA) mimics the effect of the inhibitory transmitter by hyperpolarizing the membrane potential, by increasing conductance to Cl^-, and by being antagonized by picrotoxin; it inhibits contraction in crayfish at low (10^{-7} M) concentrations. In lobster opener, the reversal potential of the ipsp is −75 mv; in crayfish it is −87 mv,[255] or about 5 to 10 mv negative to the resting potential. GABA is effective in reducing the size of the epsp's in proportion to the chloride concentration. The postsynaptic inhibitory membrane is permeable to small anions (NO_3, SCN, Cl) but not to large ones (isethionate, methyl sulfate), and this anion permeability is increased by GABA.[255] Presynaptic inhibition is also due to increased chloride conductance and is less effective when external chloride is reduced.[101] Picrotoxin reduces ipsp's and hyperpolarization caused by GABA; it has no effect on epsp's. In perfusate from muscles under inhibition, GABA can be isolated in amounts proportional to the inhibitor stimulation; hence, it seems to be established that GABA is the transmitter for both presynaptic and postsynaptic inhibition in crustaceans.

In a scorpion tibial muscle, one axon evokes epsp's and the other elicits spike-like responses; neither GABA nor glutamate has any effect on the muscle, and hence the transmitters in scorpions may be unlike those in crustaceans.[138]

Long-term changes in excitability are observed at the junction of motor giant axons with fast flexors of the abdomen in crayfish. Stimulation of the axon once per minute results in diminished epsp's (habituation), and dishabituation can be caused by sudden increase in frequency.[45] In crayfish and crab muscles, stimulation at frequencies above five per second results in long-term facilitation of transmitter release. Epsp's increase 2 to 5-fold, apparently as a result of sodium accumulation in the terminals.[331a]

In summary, the innervation of crustacean muscles, since it is multiterminal and polyneural, permits much motor integration to occur peripherally. There are several patterns of innervation, and a given axon may have different effects according to frequency of impulses. Some muscle fibers give tonic and others give phasic contractions. The same excitor transmitter—glutamate—may cause fast or slow responses according to the distribution of nerve endings. The inhibitory transmitter—gamma-aminobutyrate—may act either presynaptically or postsynaptically to increase Cl^- conductance. It is evident that crustaceans carry out at neuromuscular junctions much of the function which, in vertebrates, occurs in the spinal cord.

INSECT NON-OSCILLATORY POSTURAL MUSCLES

Motor control in insect leg and abdominal muscles, as well as in muscles of flight in insects with a slow wing-beat, resembles that in other arthropods in having double and sometimes triple innervation. Nerve endings with synaptic vesicles are distributed at intervals along single muscle fibers. One motor axon to a muscle is normally of the fast type, and the other is slow.

Stimulation of the fast axon to an insect leg muscle elicits a large electrical response in a single fiber which may be graded and may have the spike-like property of overshooting zero membrane potential. The fast response does not facilitate; when the preparation is cooled or the magnesium content of the saline is increased, the response is reduced so that a junction potential is separable from the spike. The response is the same at all points along the fiber; thus, the spike is not conducted in the muscle. Transmembrane stimulation elicits graded spikes which do not propagate; hence, they may be called "local spikes." Both fast and slow excitatory junction potentials have been distinguished from the nonpropagating spikes; the junction potentials are augmented by hyperpolarization and decreased by depolarization of the muscle membrane, and spikes may be repetitive on depolarization.[184]

In *Locusta*, stimulation of a slow (6 μm) axon elicits little contraction in leg muscles at frequencies below 10/sec, but with increased frequency the contractions increase to a maximum at 150/sec. Only some 30 to 50 per cent of fibers receive slow innervation; intracellular records show marked differences from fiber to fiber, some giving facilitating psp's and others giving a large potential with an overshoot beyond the 60 mv resting potential. The magnitude of the facilitating response is proportional to the resting membrane potential.[184] The basilar muscles of lamellicorn beetles have single innervation by slow fibers, and the muscle potentials facilitate.

A prothoracic muscle of *Carausius* shows junction potentials (in response to stimulation of its fast nerve fiber) which sum, and spikes which have a refractory period; in response to the slow nerve fiber this muscle shows junction potentials which facilitate on repetition.[381] Not all insect leg muscles have dual innervation. The extensor tibialis of the metathoracic leg of *Locusta* receives a single fast axon, and responses of the trochanter muscle of *Periplaneta* are of the junction-potential type.

In the coxal depressors of a cockroach, the largest motor axons elicit fast contractions, the next largest elicit slow contractions, and the smallest are inhibitory. Some muscles have three inhibitors, while some have one and some have two excitors.[278]

The closer muscle of a locust spiracle has dual innervation. The spiracle closer in a silk moth receives four axons, one being faster than the others. Increased CO_2 opens the spiracle by decreasing the motor output from the nervous system and reducing the closer response; the muscle shows spontaneous pacemaker potentials and action potentials that diminish in high CO_2.[363] In the cockroach *Blaberus*, the left and right spiracles have separate inhibitory nerve fibers but a common excitor.[249]

Some well insulated moths show a period of pre-flight warmup; thoracic muscles contract rhythmically, and the temperature of the thorax (and especially that of the nerve cord) is raised (p. 388). During warmup in saturnids, the wing elevators and depressors may contract synchronously, but as soon as flight starts they alternate.[209]

In the katydid *Neoconocephalus*, stridulation (chirping) is produced by forewing movements due to synchronous contractions as fast as 145 to 212 per second (according to temperature), with one action potential per contraction.[208a] A higher repetition rate is found in stridulating muscle of a cricket, *Orocharis*, which attains 280/sec at 35°C.[367a]

It was formerly considered that insect muscles have two or three excitors but that they lack inhibitor axons; however, inhibitors have now been demonstrated in cockroaches and locusts. The metathoracic posterior coxal levator of the cockroach *Periplaneta* receives four excitor and two inhibitor axons; three of the excitors are small and evoke slow responses, and one is large and elicits a fast response. A common inhibitor serves both synergistic and antagonistic muscles of the leg. Similarly, in *Locusta*, a common inhibitor occurs, possibly to regulate leg position and to facilitate relaxation.[277] For controlling each wing, the locust *Schistocerca* has four muscles—three elevators and one depressor. The depressor receives two fast excitors; one elevator has three fast excitors, and the second and third elevators each receive one fast and one slow excitor and one inhibitor. As in crustaceans, not every fiber in a muscle receives the same innervation.[277] Records from a free-walking locust show that the inhibitor axon to the metathoracic extensor tibiae fires just prior to flexion; hence, sensory input controls central firing, and inhibition speeds relaxation and sets tonic background tension.[362] In a coxal adductor of *Schistocerca* and in muscles of several other insects, there are two excitors plus an inhibitor to each muscle fiber.[280]

In insects, as in crustaceans, excitatory endings are not affected by cholinergic blockers or potentiators, and no marked sensitivity to acetylcholine appears at different stages or in degeneration.[217] L-glutamate was the most effective exciting agent of a wide variety of substances perfused through insect muscles.[361] In cockroach leg, miniature junction potentials and contractions are induced by glutamate at 10^{-6}M.[217] Sensitivity to locally applied glutamate decreases with distance from nerve endings, and after nerve degeneration there is spread of glutamate sensitivity.[359] Contractures are induced in cockroach muscles by glutamate at 10^{-4}M; hemolymph contains glutamate which is normally bound to serum proteins.[361]

Gamma-amino butyric acid (GABA) is the

most likely inhibitory transmitter, since it decreases the effect of neural excitation. Picrotoxin blocks both neural inhibition and action of GABA. Apparently inhibition is due to increased permeability for Cl^-, with a reversal potential in locust muscle of −70 mv.[360] The resting potential of muscle membranes of larvae of the mealworm *Tenebrio* is relatively insensitive to external K concentration, and is normally positive to the K equilibrium potential; in activity, increased K conductance can cause depolarization when the muscle is in a high-K medium.[26b]

INSECT RESONANT FLIGHT AND TYMBAL MUSCLES

The muscles of flight in Hymenoptera and Diptera and the muscles of sound production in Homoptera contract at higher frequencies than are permitted by the recovery rate of excitable membranes. In Odonata, Lepidoptera and Orthoptera, flight muscles are direct; that is, they are attached to thorax or to wing articulations and receive one motor impulse per contraction. The flight muscles of faster insects are indirect, connected with a complex skeletal lever system; they have very large fibrils and a regular arrangement of filaments.

Wing frequencies, measured stroboscopically and by sound frequency in free flight, have been correlated with muscle contractions and action potentials recorded under some restraint. In Diptera, the frequency of beat increases if the wings are clipped or the temperature is raised. Wing frequency is inversely proportional to approximately the cube root of the moment of inertia of the wings. Frequency is also inversely proportional to atmospheric pressure. The maximum wing frequency that has been recorded is 2218 beats/sec in a midge, *Forcips,* maintained at high temperature and with wings clipped. The flight tone of *Chironomus* is 600 to 650/sec; thoracic movements are at a slightly lower frequency when it is restrained. Its indirect flight muscle must be capable of contracting and relaxing within 0.45 msec.[334a] For larger insects, the wing frequency is not affected by clipping or loading.

Numerous analyses of the aerodynamics of insect flight deal with energy requirements and air currents. In *Locusta,* the power output is high—some 5 kcal/kg/hr—and flight speed is influenced by wing frequency, wing inertia, body shape, and environmental factors. Flight speed records reach 150 to 375 cm/sec in some flies.[179a] Lift and thrust are somewhat independent; lift is not much influenced by body angle, but more by wing tilt. In *Locusta,* as in many insects, flight can be initiated by removal of tarsal contact with substrate, by general nonspecific stimulation, and by wind on sense endings on the head; maintenance of flight requires wind against the wings or head.[372]

In insects with direct flight muscles, wing frequencies may reach 20 to 40 per second; there is synchrony between muscle contractions and action potentials, and wing amputation has little effect on frequency. The basic frequency is programmed in the thoracic ganglia. In a moth, *Agrotis,* muscle potentials can repeat at intervals of 20 msec with no facilitation.[311] In flies and hymenopterans, where wing muscle attachments are indirect, there is no synchrony between muscle contractions and action potentials; in intact insects there may be one muscle spike per 5 to 20 wing beats. Amputation of the wings nearly doubles the muscle beat rate without a change in spike frequency. The indirect muscles cannot be driven by nerve stimulation as fast as the normal beat; bumblebee flight muscle tetanizes at 40/sec, and haltere contractions in flies are independent of shock frequencies above about 40/sec. Thus, the myogenic rhythm exceeds the frequency of nerve stimulation.

In some species of cicada (e.g., *Platypleura*), sound is produced at 200 Hz (maximum 5400 to 7000 Hz) by a tymbal organ; the frequency of the muscle action potentials is 50/sec, and one nerve impulse corresponds to one muscle action potential, but the muscle can contract several times for one action potential. One axon supplies the entire muscle. After destruction of the tymbal membrane the relation between muscle potential and contraction is one-to-one; if the tymbal membrane is loaded, the frequency of beat is reduced.[284] In other species (e.g., *Meimura*), the tymbal muscle contractions, as in locust wing muscles, follow the nerve impulses and electrical fusion occurs at about 120/sec. As in flight muscles, when tymbal frequency is 1:1 with action potentials there is little effect of amputation of the tymbal membrane, whereas those tymba in which sound frequency exceeds impulse frequency show an increase of beat on amputation.

The several contractions per nerve impulse and the wing beat speed higher than fusion frequency are possible because of the mechanical properties of the muscles. In Diptera, the indirect flight muscles are connected to skeletal elements that comprise a scutellar lever, which has two stable positions —up and down. When muscle contraction moves one arm of this lever beyond its midpoint the skeletal elements spring in, shorten the muscle slightly, and release tension on it; then the elasticity of skeletal elements (tymba) or paired contractions of an antagonistic muscle (indirect flight muscle) restore the lever to the up position and stretch the muscle (Fig. 16–22). Redevelopment of tension follows this quick stretch, and contraction again pulls the lever down. In the tymbum the muscle pulls the tymbal membrane toward a point of instability; it snaps inward, releasing tension, and then the membrane is restored by its own elasticity and stretches the muscle to give another contraction. This cycle can be repeated as long as the "active state" of the muscle remains, and only enough nerve impulses need be supplied to maintain this state.[284]

In vitro studies reveal oscillatory mechanical behavior not only in intact muscles but also in bundles of fibers extracted with glycerol. If muscle contractions are recorded isometrically or isotonically at tension above critical damping, the tension rises along a typical active tension-length curve (Fig. 16–22B). If, however, the muscle is loaded below a critical damping value, tension lags behind length so that at a critical length, active tension develops slowly, causing slight shortening with subsequent fall in tension; however, continued extension is followed by another rise in tension, and an oscillation starts. Delayed rise of tension on stretch is greater than fall of tension on release. The curve described by the oscillation lies between the resting and active tension-length curves in a position determined by load and intrinsic properties. A glycerinated fiber can be relaxed by EGTA; if it is then immersed in a low ionic strength solution with ATP and Ca^{++} (10^{-8} mEq) and stretched, the fiber starts to oscillate at a frequency which varies with load.[285] At higher Ca^{++} (10^{-7} mEq) the muscle shows more tension and no oscillations. Glycerinated fibers from a belostomid (giant water bug) oscillate at 8×10^{-8} M Ca but contract without oscillation at 10^{-7} M Ca.[23] Glycerinated fibers from a cicada tymbal muscle show oscillatory work when stretched sinusoidally in a medium containing ATP and Ca^{++}.[7] In 5 mM ATP, the sinusoidal tension lags behind length, and oscillation occurs at 10^{-6} M Ca^{++}.[320] Myosin ATPase splits ATP during oscillatory work, but not when the fibers are passively stretched and released.[320] However, stretch increases the activity of Ca-activated myosin ATPase; i.e., stretch acts like an increase in Ca^{++}.[285]

Grasshopper flight muscles show hysteresis in tension-length changes; if they are stretched sinusoidally, the change in tension lags behind the change in length. As frequency declines, the curve of tension decline lies above that for rise in frequency, so that tension at an intermediate frequency is higher during decline than during rise of frequency.[379]

CATCH MUSCLES AND DELAYED RELAXATION

The non-striated, opaque adductor muscles of bivalve molluscs are capable of maintaining tension, once they are contracted, for many hours, as anyone who has tried to open an oyster can attest. An oyster adductor can support 0.560 kg/cm² for 20 to 30 days.[250] Maximum initial tensions developed at minimal or "rest" lengths are as follows, in kg/cm²: *Venus* adductor, 12.8; *Tridacna* adductor, 6.7; *Mytilus* anterior byssus retractor (ABRM), 10 to 12; *Unio* adductor, 3.1; *Pecten* 2 to 8.[226, 250]

In intact scallops (*Pecten*), low-level electrical activity can be recorded during the tonic state, and occasional stimulation of the visceral nerve keeps the muscle contracted. This suggests that the delayed relaxation is in part a tetanic response to low-frequency stimulation. However, if the visceral nerve of *Pecten* is cut while the adductor muscle is contracted, the muscle remains contracted.

The byssus retractor of *Mytilus* consists of thin (5 μm) non-striated fibers[356] which are 1 to 1.5 mm long.[354] When stimulated with direct current (dc) pulses of 1 second or longer, the muscle contracts and may not relax for many minutes. When stimulated with alternating current or a train of brief (5 to 20 msec) dc pulses at 10 to 60/sec for a few seconds, the muscle gives a twitch-like contraction which relaxes rapidly. If a brief series of pulses is delivered while the muscle is in tonic contraction following dc stimula-

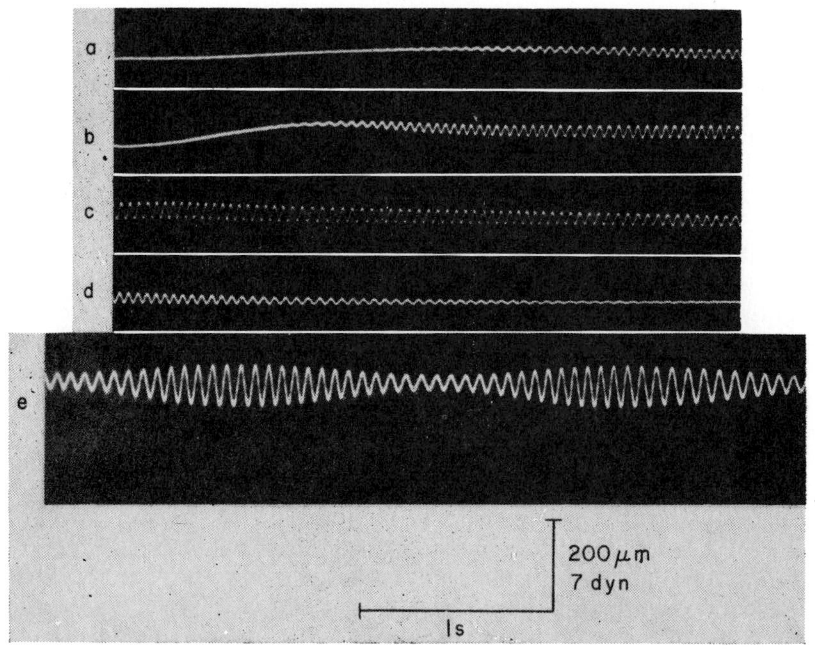

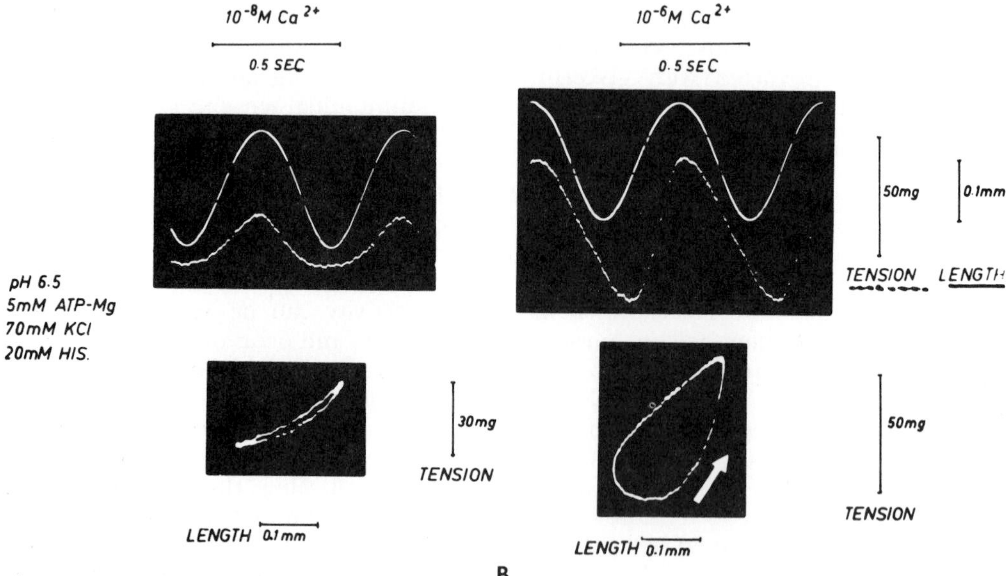

Figure 16–22. A, Oscillatory contractions of glycerinated fiber from insect indirect flight muscle. (From Pringle, J. W. S., Prog. Biophysics *17*:1–60, 1967.) B, Tension-length relations of glycerinated fiber of giant water bug; driven oscillation and loop formed by recording tension as a function of length. (From Ruegg, J. C., Amer. Zool. 7:457–464, 1967.)

tion, relaxation occurs promptly.[207] Nerve fibers which induce contraction, and others which elicit relaxation, have been separated. Tension of the phasic contraction increases with increasing frequency of stimulation up to several impulses per second, and successive potentials increase in amplitude at 0.5/sec and fuse at 5 to 7/sec.[119] Conduction of electrical responses is at 25 to 50 cm/sec. No propagated impulses are observed after treatment with nerve-blocking drugs.

The resting potential of ABRM fibers is

65 mv, and junction potentials up to 25 mv are recorded; these facilitate maximally with nerve stimulation at 2 to 3/sec. Muscle spikes of 50 msec duration and at least 50 mv amplitude precede the phasic contraction.[355]

Applied acetylcholine can cause a contraction which persists long after depolarization. Locally-applied ACh causes junction-type depolarizations with the same reversal potential as that of nerve-induced junction potentials. The ACh contraction is blocked by banthine (=methantheline) which apparently blocks ACh receptors; d-TC, benzoquinonium, atropine, C_6, and C_{10} also block effectively. Serotonin (5-hydroxytryptamine, 5-HT) abolishes the tonic contraction, whether elicited by nerves or by ACh. It is postulated that excitatory nerves are cholinergic and that relaxing nerves (turning off catch) are tryptaminergic. Strength-duration curves for contraction and relaxation in response to stimulation of the visceral nerve show the relaxing fibers to have higher thresholds at all durations than the excitatory fibers.[46] The concentration of 5-HT needed to terminate catch is low—10^{-8} to 10^{-7} M. The pedal ganglion contains significant amounts of 5-HT[322] and of dopamine. Applied 5-HT has no effect on the resting potential, but it does decrease membrane resistance, lower the threshold for spiking, and increase spike amplitude.

The spike heights in *Mytilus* ABRM are proportional to the logarithm of Ca_o (when the muscle is in Na-free medium); they are blocked by Ca antagonists such as La, and they are not affected by tetrodotoxin (TTX). Hence, the muscle spikes are due to increased Ca^{++} conductance. There is evidence that 5-HT favors Ca binding by intracellular particulates; hence, the increase in spike height may be due to reduction of intracellular ionic calcium.[357] Active tension at a given length is reduced in proportion to applied 5-HT.[355]

The catch state (Fig. 16–23) is a change in physical state of the muscle such that it becomes resistant to stretch; a relaxed muscle is plastic, while a muscle in catch is rigid. Stretch resistance is inversely proportional to the concentration of 5-HT; i.e., 5-HT makes the muscle more extensible.[355] However, if a muscle in catch is allowed to shorten, it does not redevelop tension; active state as measured by tension redevelopment lasts only 1 to 7 sec, whereas catch is maintained long after decay of the active state.

Electron micrographs of bivalve muscles capable of catch show filaments 300 to 800 Å in diameter surrounded by thin (55 to 60 Å) filaments (Fig. 16–24). The thick filaments consist of a protein, paramyosin (tropomyosin A), which can be separated from the other proteins by alcohol solubility or by varying ionic strength. The thin filaments are actin, and myosin forms a sheath around the paramyosin filaments.[327,344] The paramyosin filaments (and reconstituted paramyosin) show a characteristic periodicity at 145 Å in repeats of 725 Å bands. Reconstituted paramyosin forms tactoids of 725 Å period, with overlapping molecules (each having a molecular weight of 200,000)[75] (Fig. 16–25). There are 12 or more thin filaments to one thick filament.[318] Among the filaments are scattered dense bodies which have been homologized to Z bands. Removal of thin (actin) filaments reveals an arrowhead bridge structure of thick filaments probably due to myosin surrounding the paramyosin.

One suggestion is that contraction occurs by interaction of actin and myosin, and that this puts the paramyosin into a configuration from which it cannot be readily changed except under the influence of 5-HT. This explanation is supported by the similarity of the dependence of catch in intact muscle and the dependence of change of state in glycerinated fibers on pH and ATP. No catch can be induced above 30°C.[355] Evidence against dependence on a configurational change is that no alteration in x-ray diffraction pattern was detected.[250,251]

Another hypothesis regarding the catch mechanism is that paramyosin stabilizes actin-myosin bridges so that, once formed, they do not readily break. A variant of this idea is that paramyosin may prevent calcium from being taken up by the relaxing system.[261,355] 5-HT could remove Ca from the cross bridges; i.e., 5-HT favors calcium binding. Catch (tension remnant) declines slightly with increased Ca^{++}, whereas peak tension and Ca-ATPase activity increase.[227] Paramyosin may undergo a phase change coupled with cross bridge movement; once a bridge is formed, paramyosin may hinder its detachment.[75]

At one time it was thought that catch is maintained without energy expenditure: Phosphoarginine breaks down during phasic contraction and relaxation, but no measurable increase in inorganic P or loss of PA was

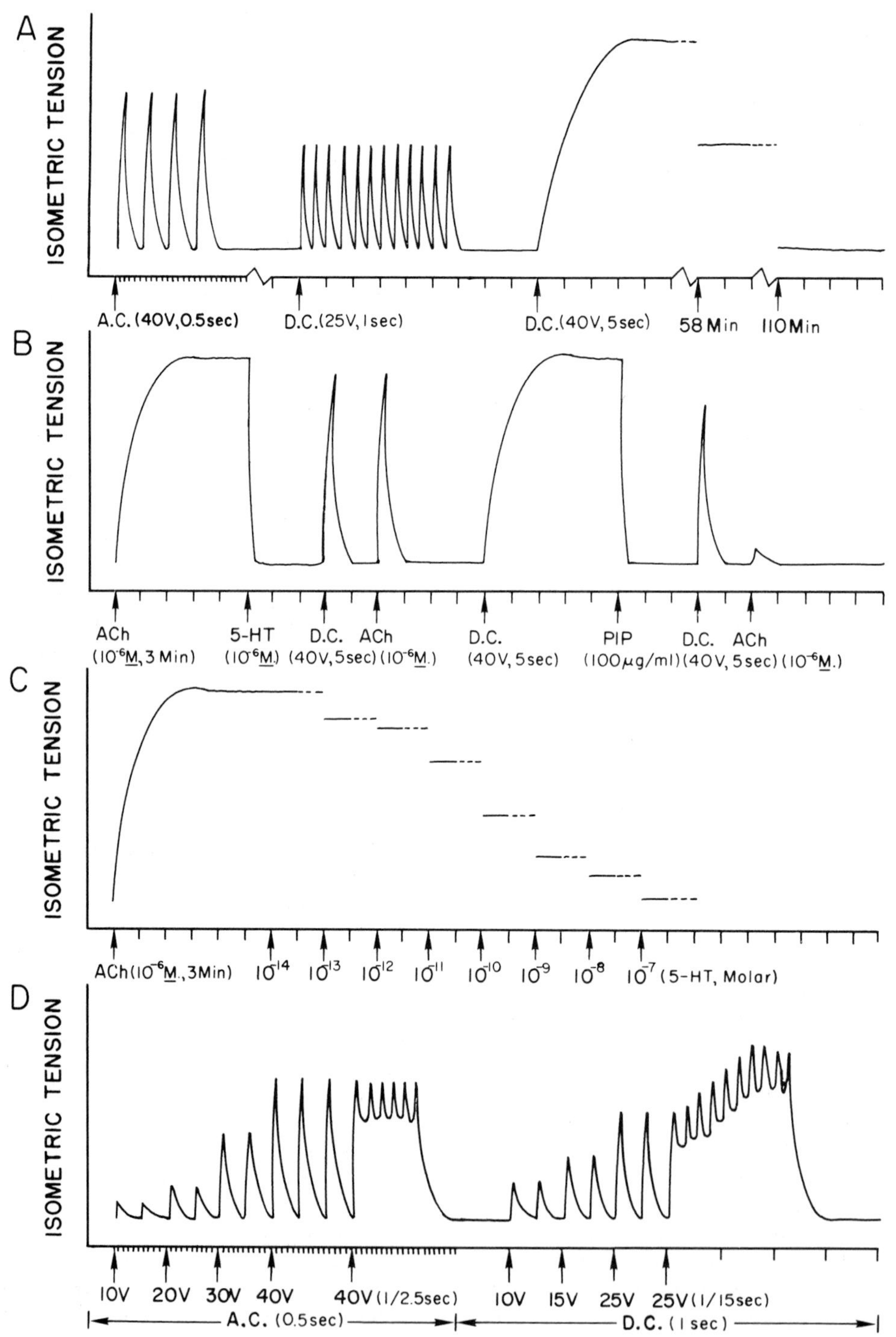

Figure 16–23. Demonstration of catch properties by muscle of worm *Paragordius.* (From Swanson, C. J., Z. vergl. Physiol., *74*:403–410, 1971.)

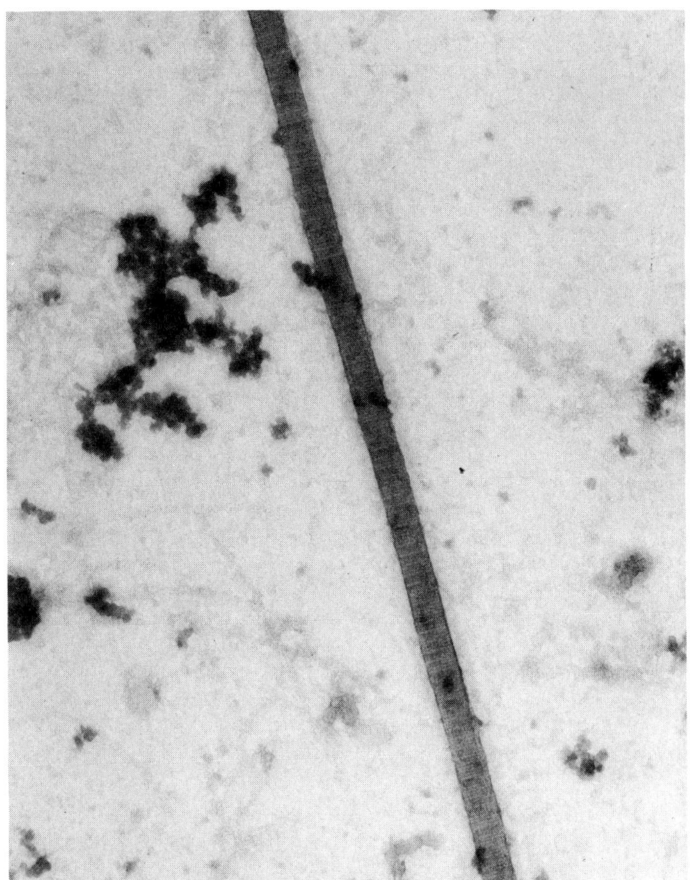

Figure 16–24. Paramyosin-containing filament from proboscis retractor of *Golfingia.* (Courtesy of V. Ernst.)

detected during catch. The amount of phosphoarginine used is the same in phasic as in tonic contraction, and during catch there may actually be reconstitution of phosphoarginine.[261] At rest at 20°C, the ABRM consumes 78 μM O_2/g/min; if the muscle is released to 75% of rest length, the O_2 consumption increases, and if the muscle is stretched the O_2 consumption decreases. Phasic contraction causes an extra O_2 consumption of 83 μM O_2/g/min for a tension of 5 kg/cm², and this extra metabolism recovers by 80% in 20 minutes. The phasic energy expenditure in ABRM is 1/250 that of a frog sartorius. Muscles in catch show elevated oxygen consumption initially, but after 20 minutes the excess metabolism declines to 26 μM O_2/g/min for a tension of 5 kg/cm². If the muscle is released during catch, the O_2 consumption drops 26 to 13 μM O_2/g/min, and if it is then restretched the consumption rises to 27 μM O_2/g/min. 5-HT increases O_2 utilization.

Paramyosin filaments have been observed in muscles in several phyla other than the bivalve molluscs. Thick paramyosin filaments are observed in proboscis retractor of the sipunculid *Golfingia* (cf. Fig. 16–24), in body wall muscles of gordian worms (*Aschelminthes*)[342], and in the notochord of *Branchiostoma.*[122] In *Branchiostoma,* contraction of the notochord serves hydrostatically to give the animal support.[151] The following data for the worm *Paragordius* show a catch much like that in molluscs:[342]

stimulation	*contraction time, sec*	*time for 1/2 relaxation*	*maximum tension, mg*
ac	0.8–1.6	0.3–0.8 sec	450–600
dc, 1 sec	14	18 sec	840
dc, 5 sec	6.4	42 min	6853
ACh	180–240	58 min	8328

Catch in limited amounts may occur widely in non-striated muscles. The role of paramyosin and the mechanism of delayed relaxation in increase in rigidity are not understood. In

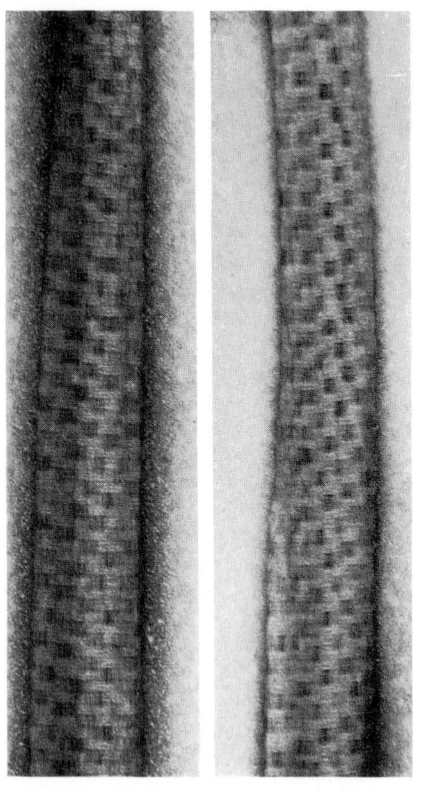

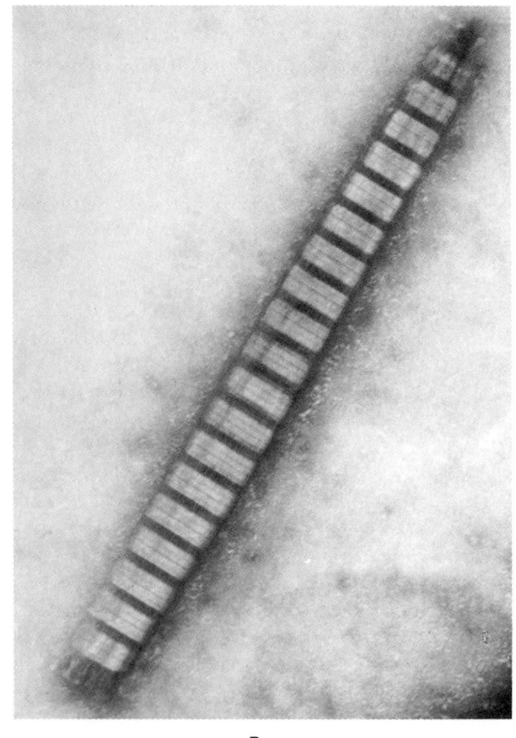

B

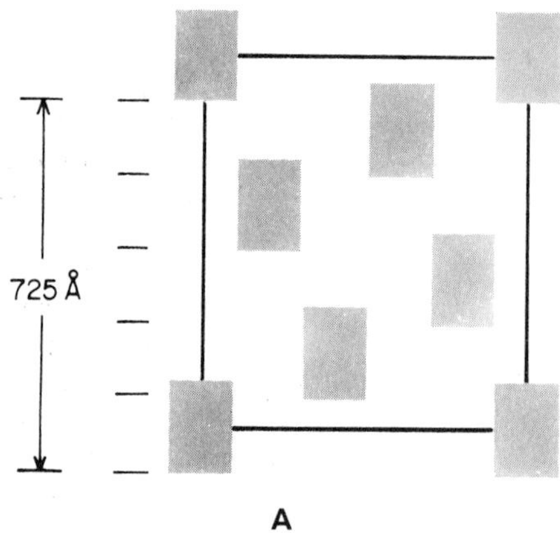

A

Figure 16–25. A, Paramyosin paracrystal formed by precipitation from *Mercenaria* adductor muscle. Periods approximately 725 Å. B, Paramyosin from *Atrina* muscle precipitated with Ba^{+} from thiocyanate solution. (From Cohen, C., A. G. Szent-Györgyi, and J. Kendrick-Jones, J. Molec. Biol. *56*:223–237, 1971.)

molluscan muscles, relaxation is induced by 5-HT, possibly by changes in calcium distribution.

DIAGONALLY STRIATED MUSCLES

***Ascaris* Longitudinal Muscle.** The body wall of the roundworm *Ascaris* contains a complex and unique muscle system of longitudinal fibers. Each muscle fiber is large, and consists of (1) a lateral belly with nucleus, glycogen granules, and mitochondria, (2) a fiber of long arms extending along the hypodermis, and (3) a process that extends toward the nerve cord (Fig. 16–26). Synaptic vesicles are seen in the nerve endings which contact the extended muscle process in the nerve cord. The processes near the nerve cord make tight junctions with those of adjacent

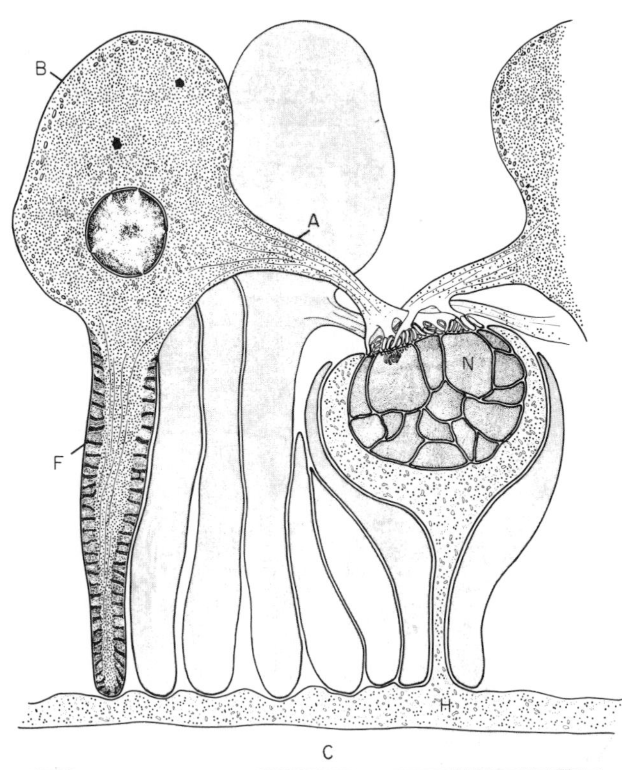

Figure 16–26. Diagram of transverse section of muscle cells of *Ascaris*. *B*, muscle cell with process *A* going to nerve cord *N*. Peripheral myofilaments *F* in body of muscle fiber. (From Rosenbluth, J., J. Cell Biol. *26*:579–591, 1965.)

muscle fibers. The contractile spindles are ribbon-shaped and show oblique striations with thick (230 Å) and thin filaments, with one thick filament for every 10 or 12 thin ones.[315] The nerve cord conducts excitation to the muscle (blocked by tubocurarine), and stimulation of the nerve cord triggers spikes in the muscle.[86] GABA hyperpolarizes the muscle and ACh depolarizes it. The muscle fibers also show spontaneous junction potentials at 1.5 to 7/sec, which give rise to spikes. The spontaneous spiking increases with rise in temperature, and represents activation from the syncytial region. The resting potential of the muscle fibers is unusual in that it decreases when Cl_0 is increased. A hyperpolarizing phase of spikes is due to K-activation. Pulses applied in one fiber spread to adjacent ones, indicating a high degree of electrical coupling.[88, 89, 149]

Another muscle in which the muscle fiber processes go to the nerve cord rather than the reverse is a myotome of amphioxus.[121] Diagonally striated fibers are found in the translucent or fast portion of bivalve adductors. In the scallop (*Pecten*) and oyster (*Crassostrea*), there is clear overlap of thick and thin filaments in a helical arrangement (more transverse in *Pecten*, with 2 to 3 μm sarcomeres).[41] These muscles contract rapidly (twitch duration 100 msec); in *Pecten* the valve hinge is elastic, and by brief contractions and water ejection the scallop swims with a jerking motion.

The muscles of earthworms have thick and thin myofilaments which spiral on the two sides of the ribbon-shaped fibers. Thick filaments are 200 Å in diameter and thin ones are 50 Å, and there is one thick filament to four or more thin ones;[197] T-tubules penetrate from the sarcolemma.[172] Intracellular recordings from the longitudinal muscle show spontaneous psp's, mostly hyperpolarizing (Fig. 16–27). If the cell is depolarized by an intracellular electrode, the inhibitory miniature junction potentials (mipsp's) increase in size, and if the cell is hyperpolarized they decrease. The reversal potential is −58 mv on a resting potential of −35 mv. GABA also hyperpolarizes the cell and decreases membrane resistance; both the mipsp's and the effect of GABA are sensitive to Cl and are antagonized by picrotoxin. Excitatory miniature junction potentials (mepsp's) were also seen, and these could be blocked by tubocurarine. The earthworm longitudinal muscle also shows spontaneous spikes which may overshoot zero membrane

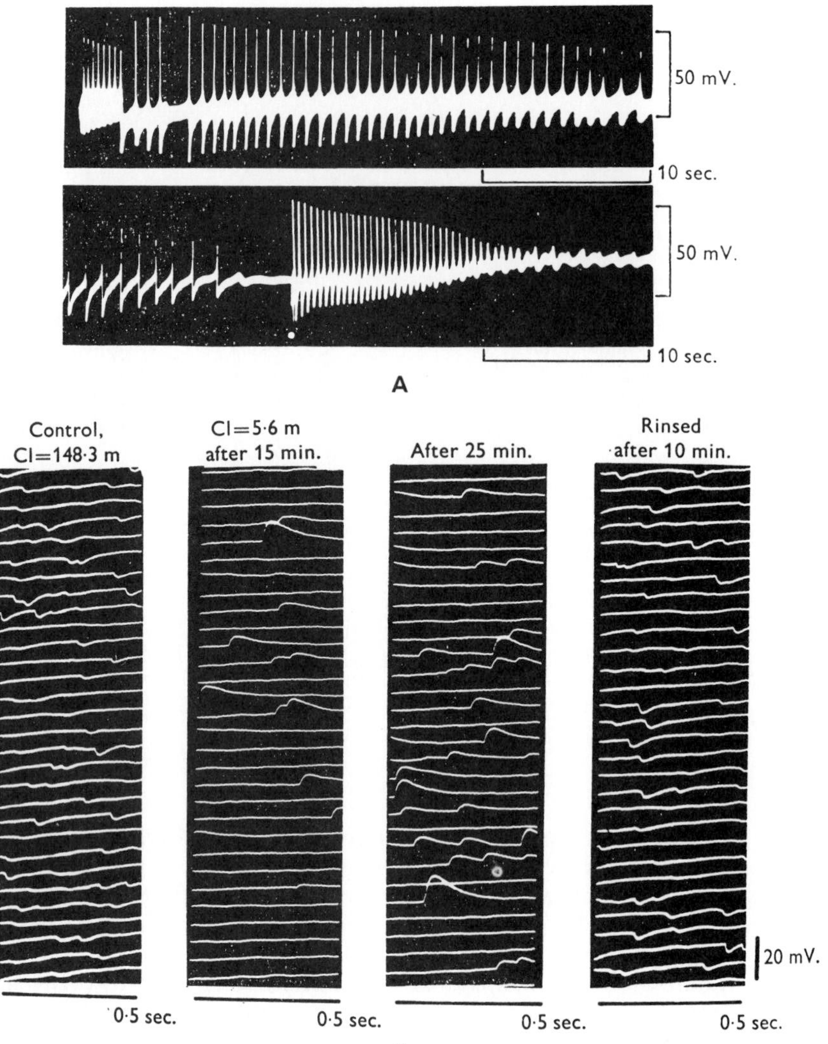

Figure 16–27. Electrical activity in earthworm muscle fibers. A, Trains of spontaneous spikes in longitudinal muscle cells, changing to increased frequency on mechanical stimulation. *B*, Miniature (spontaneous) inhibitory junction potentials. Control followed by replacement of most of chloride by glutamate for 25 minutes, and then returned to control medium. Note reversal of mipsp's in absence of chloride. (From Hidaka, T., Y. Ito, and H. Kuriyama, J. Exp. Biol. *50*:387–403, 1969; *and* Hidaka, T., Y. Ito, H. Kuriyama, and N. Tashiro, J. Exp. Biol. *50*: 417–430, 1969.)

potential by as much as 18 mv, and which are followed by hyperpolarization or undershoot; intracellular depolarization increases the frequency of the spontaneous spikes. Spike amplitude and rate of rise are proportional to external calcium and the spikes are blocked by Mn; hence, they appear to depend on Ca influx.[199] Miniature hyperpolarizing psp's are due to Cl^- conductance, and they may be caused by GABA as an inhibitory transmitter.[198]

Longitudinal muscle of *Lumbricus* shows a smooth tetanus when stimulated (presumably via intrinsic nerves) at 0.4/sec. Facilitatory influence persists for half a second after a single shock and for 30 seconds after a tetanus; it lasts much longer after eserinization.

Ribbon myofibrils are oriented radially in the cortex of a leech fiber; the fibrils have long myofilaments, SR tubules, and some filaments crossing from the sarcolemma to the axial sarcoplasm.[312] Dorsal muscle of leech, like that of earthworm, shows spontaneous spikes followed by undershoot. Electrical coupling between fibers is indicated.[368]

In a nereid worm, the longitudinal muscle

fibers have multineuronal innervation and show Ca-spikes similar to those of the earthworm.[199]

SHORT-FIBERED NON-STRIATED POSTURAL MUSCLES

Neurally controlled non-striated muscles with spindle-shaped fibers some 200 to 500 μm long and with 5 μm maximum diameter, often with thick and thin filaments that lack transverse alignment, constitute the principal postural muscles of many invertebrates.

The entire mantle of a squid or octopus must contract rapidly and synchronously to provide the propulsive jet stream for escape or attack. In *Loligo,* the mantle is innervated from the stellate or mantle ganglion by a set of 10 giant axons on each side, the one to the posterior region being exceptionally large. A single impulse in each giant axon elicits all-or-none contraction of the thousands of muscle fibers in the mantle, and repetitive stimulation does not increase the response. Contraction time is 60 msec and relaxation 200 msec. The mantle also receives many small fibers alongside the giant axons; the former elicit a slow type facilitating response. Electrical recordings indicate a dual (fast and slow) innervation of the muscle fibers in squid and octopus.

Chromatophores of cephalopods are sacs of pigment which are expanded by contraction of radially arranged muscle fibers. Myofilaments are present in the cortex but are absent in the core of these fibers. The muscle fibers receive polyneuronal innervation, and stimulation of nerves elicits epsp's. Acetylcholine increases the frequency of mpsp's, and 5-HT decreases their frequency. No inhibitory nerves were found.[123]

The retractor of the pharynx of the snail *Helix* shows two components in its extracellularly recorded action potential, fast and slow waves. Intracellular recording shows psp's which summate, and also spikes.[326] Contraction of the muscle increases on repetition and shows maximum summation at 20 msec intervals. L-glutamate in low concentrations is excitatory to the muscle.[216] The radula protractor of *Busycon* is a multi-unit muscle; in a mixture of acetylcholine (excitatory) and tryptamine (relaxing) it shows mechanical rhythmicity.[178]

The mantle edge of a bivalve, *Spisula,* provides a muscle strip which shows both phasic and tonic nervous effects. Ganglion cells imbedded in the mantle are spontaneously repetitive and lead to tonic contractions. The tonic activity is decreased by dopamine and 5-HT. Phasic responses to electrical stimulation are blocked by tricaine and enhanced by methysergide.[377] Fibers of the fast adductor of a scallop receive multiple innervation; all junctions are of a fast type.[90]

In echinoderms, the pharyngeal or lantern teeth retractors are strips of short-fibered non-striated muscle. In *Echinus,* the lantern consists of 40 calcareous teeth moved by six sets of muscles. Individual muscle fibers are 8 to 12 μm in diameter and up to 1 mm in length.[74] Conduction in the muscle is at 4 cm/sec, and stimulation elicits both fast and slow electrical and mechanical responses.[72] The pharyngeal retractors of the holothurian *Cucumaria* are innervated from the circumoral ring, and nerve stimulation evokes phasic and tonic contractions with fast and slow action potentials. Fusion is at 2.5/sec and maximum tension at 5/sec. Facilitation of the response is enhanced by eserine, and the response is blocked by tubocurarine. The muscle is very sensitive to acetylcholine. The long body-wall retractors of the holothurian *Thyone* contract with increasing tension as frequency increases up to 1/sec, and they maintain contraction for as long as 10 seconds. The muscles are innervated at intervals of less than 1 mm by branches from radial nerves, and conduction for short distances by these nerves is at 17 cm/sec.[287]

Shortening by Folding of Non-striated Fibers. The proboscis retractors of the sipunculid worm *Golfingia* consist of non-striated fibers 5 μm by 0.5 to 1 mm. These muscles retract the proboscis against high hydrostatic pressures in the body cavity, and they may extend to ten times their contracted length when the proboscis is protracted. The retractors are richly innervated from the cerebral ganglion by axons of two sizes. The muscle gives phasic contractions (87 msec contraction time, 950 msec for ½ relaxation), and tonic contractions (2700 msec for ½ relaxation). External recording reveals two types of action potential – spikes conducted at 1.3 m/sec which fatigue on repetition, and slow waves conducted at 0.3 m/sec which facilitate on repetition. The fast potentials trigger a phasic twitch, and the slow potentials elicit the tonic contraction. The muscle is sensitive to ACh, the slow waves are enhanced by eserine, and tubocurarine blocks both types of response. Nerve impulses can be recorded ahead of fast and slow muscle potentials, and conduction ceases after nerves have been caused to degenerate. Intracellular recording shows that some fibers give only fast, or only slow potentials, and some give both types. Dual innervation is indicated.[292]

When contracted, the *Golfingia* muscle fibers show folds at 10 μm intervals that extend in register across hundreds of fibers (Fig. 16–28). The net effect is a herring-bone appearance which appears in polarized light as birefringent bands. The transverse alignment of the folds indicates a regular pattern of connections between fibers. Active tension-length curves show two peaks.[237] The first is a small increase in tension at lengths where there is no increasing rest tension; the second is at lengths where rest tension is rising. Microscopic observation shows the first peak to be due to extension of the fiber folds, and the second to be due to extension of myofilaments and somewhat irreversible. This muscle appears to use the folding mechanism to permit isotonic contractions over a wide range of lengths.

Another very extensible muscle in *Golfingia* uses mechanical pull from fiber to fiber for conduction. This is the spindle muscle around which the intestine coils. It is stimulated to contract by a slight stretch, and if it is mechanically immobilized in the mid-region, conduction is blocked. There seems to be no nervous conduction such as occurs in the proboscis retractor.

Diverse Contractile Cells of Coelenterates and Sponges. Both striated and non-striated contractile cells may have evolved simultaneously. In some medusoid coelenterates, the exumbrellar surface has non-striated fibers, while the subumbrellar and velum have striated fibers. Tentacles show both striated and non-striated fibers, sometimes in overlapping layers. Thick and thin filaments indicate myosin and actin.[64] The tentacles show both fast and slow contractions, and all regions of the medusa are richly innervated. The subumbrella of the hydrozoan medusa *Obelia* has a middle layer of myocytes with

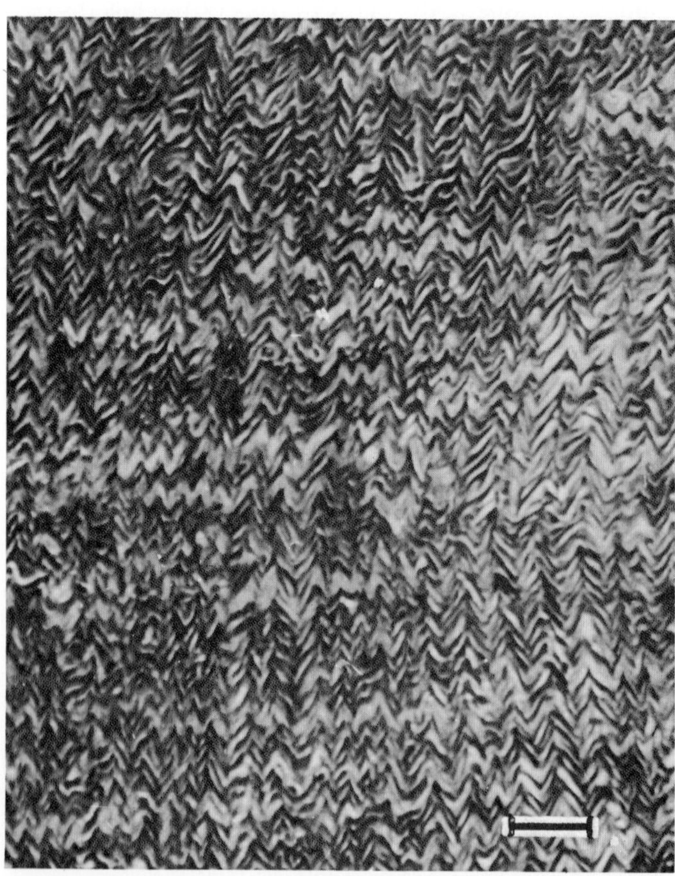

Figure 16–28. Section of proboscis retractor of *Golfingia*, showing folding of fibers in alignment across muscle. (From Prosser, C. L., *in* Invertebrate Nervous Systems, edited by C. A. G. Wiersma. University of Chicago Press, Chicago, 1969.)

striated fibrils in two layers at right angle to each other, sarcomeres 1.4 μm in diameter, and an outer layer of helically banded fibrils; thick and thin filaments occur in the fibrils.[63]

Sea anemones give both fast and slow contractions; sphincters show more fast responses and the muscular column shows more slow responses. The latency of the fast response may be 0.1 sec, while that of the slow response may be 30 or more seconds. Columns of anemones contract spontaneously at intervals of many minutes, presumably under central nervous control.[317] Chemical agents which are active on muscles of other phyla have little effect on coelenterates. Contractile proteins resembling actin and myosin, as well as both Mg- and Ca-activated ATPase have been extracted from sea anemones.[234, 235] It is probable that many coelenterate fibers have a double excitatory innervation.

Sponges have contractile cells, the myocytes, particularly in oscular membranes (where they may control the size of the opening). Myocytes resemble non-striated muscle fibers and may contain both thick and thin myofilaments of proper dimensions for myosin and actin.[16] There are no nerves, but the myocytes respond actively to mechanical stimulation. Contractions could not be elicited by electrical stimulation, no action potentials were recorded, and the mechanically induced contractions continued when Na in the medium had been replaced by potassium (or lithium); hence, typical membrane-contractile protein coupling may be unnecessary. Contractions require calcium (or Mg or Sr), and AM-type ATPase has been detected.[288] It is probable that mechanical stimulation directly activates a calcium coupling to contractile proteins.

VERTEBRATE SMOOTH MUSCLES

Smooth muscles of vertebrates are of two kinds, unitary and multi-unitary. There are some intermediate types. Unitary muscles include those of the viscera—uterus, ureter, gastrointestinal tract; these are capable of spontaneous rhythmicity, and conduction is electrical from muscle fiber to fiber. Unitary muscles can be stimulated by stretch to depolarize and contract; their activity may be modulated by nerves. Multiunitary muscles include nictitating membrane, pilomotor, ciliary, and iris muscles; these are normally activated by nerve impulses or circulating hormones, they contract relatively rapidly, conduction is normally by nerves within the muscle, and many are not stimulated by quick stretch. There may be nexuses (electrical conducting connections) between some cells of multiunitary smooth muscles, but groups of cells are not connected in a functional syncytium as they are in unitary muscles. Some smooth muscles have properties of both categories—vas deferens and large artery walls have more resemblance to multiunitary types in dependence on nerves, while bladder and portal vein more resemble the unitary type in stretch sensitivity and myogenicity.

Nerve-Activated Muscles. The cat nictitating membrane responds to incoming sympathetic impulses by action potentials (externally recorded) which increase in size and complexity as more axons are stimulated. The postganglionic fibers which activate the nictitating membrane are mainly adrenergic; a few may be cholinergic. The innervation of the iris is cholinergic; in some amphibians and fish the iris smooth muscle responds directly to illumination. Tension is developed slowly by such muscles as nictitating membrane, and it increases as the number of excitatory impulses increases up to some limit. Not all the muscle fibers are innervated, and transmitter may be liberated along terminal branches of axons and diffuse among groups of muscle fibers.

The vas deferens of mammals is activated by the hypogastric nerve, and in a guinea pig facilitating epsp's are recorded. Norepinephrine depolarizes the cells, and from graded stimulation it is concluded that many nerve fibers contribute to the response of a single muscle fiber.[129, 182] The epsp's may sum to give rise to a spike.[30, 182] It is estimated that 25% of the muscle fibers in a mouse vas deferens are 0.25 μm from an axon and that 1% of the fibers are within 200 Å (=0.02 μm) from an axon; hence, some of the excitation presumably spreads from muscle fiber to fiber.[50a]

Myogenic Visceral Muscles. PROPERTIES OF VERTEBRATE SMOOTH MUSCLES. Visceral smooth muscles are extremely diverse, particularly with respect to their electrical properties, the effects of neurotransmitters, and modifying drugs; differences exist not only from organ to organ but from species to species for the same organ.

Uterus muscle shows rhythmic spikes, one of which may trigger a contraction; also, in trains the spikes may maintain contractions. In some cells a prepotential or pacemaker potential precedes each spike, while in follower cells only spikes are observed. The spikes resemble those of striated muscle in being proportional to the external concentration of Na.[10, 233] The activity of uterine muscle is enhanced by estrogen, is depressed by progesterone, and varies according to stage in the reproductive cycle.

Taenia coli of guinea pig also shows spontaneous pacemaker potentials, each of which may give rise to one or more spikes. A single contraction follows each spike or burst of spikes. Conduction occurs from muscle fiber to fiber at a rate of about 1 cm/sec.[30] Unlike uterine muscle, the spikes in taenia coli are little affected by external sodium concentration, but rather are caused mainly by inward currents carried by Ca^{++}.[222, 223] As in most visceral smooth muscle, conduction requires many cells in parallel, and the electrical unit is not a single cell. Space constants are many cell lengths (1.5 mm for 100 μm to 200 μm cells which have 50% overlap), and measured time constants are longer (100 to 300 msec) than those calculated from single cell resistance and capacitance (30 msec).[353] Thus, the cable theory which describes conduction in long fibers of neurons or striated muscle cannot be applied to single muscle cells; rather, smooth muscle shows properties of a "smeared out" cable.

Ureter differs from other smooth muscles in showing a spike-like action potential followed by a plateau resembling that in heart muscle (Fig. 16–29B). In some species (guinea pig) oscillations occur on the plateau; various agents, such as procaine, can prolong the plateau. Conduction is slow and corresponds to a single peristaltic wave down the ureter. The amplitude of the spike varies with external calcium, and the plateau is sensitive to sodium; in zero Na or high Ca, repetitive spikes may replace the plateau. The effect of Ca on spike height is greater in the presence of Na than in its absence; hence, it is suggested that there are parallel channels for calcium and sodium.[219]

The electrical activities of the two layers of mammalian small intestine are very different, and there are differences from region to region of the intestine; there are striking differences with species for the same region of the gut. The circular layer of the mammalian small intestine is thicker and does most of the mechanical work; it shows electrical spikes which, *in vitro*, may be spontaneous and random but which, *in vivo*, are driven by the longitudinal layer. Muscle cells of the longitudinal layer show regular, near-sinusoidal, slow electrical waves whose frequency is highest in the duodenum (14 per minute in dog) and decreases caudad (Fig. 16–29A). These slow waves are essentially synchronous around a segment of intestine; they conduct for short distances in the long axis of the fibers, and spread by connecting muscle strands to the circular layer. On the slow waves there may be spikes, at the beginning of each of which a pacemaker potential occurs. Pacemaker potentials may also precede spikes in circular fibers. The slow waves are sensitive to sodium concentration; they are abolished by ouabain and are sensitive to metabolic inhibitors and temperature; and they are associated with sodium efflux.[205a, 227a] They appear to result from a sodium transport system driven by an oxidative rhythm. The spikes and pacemaker potentials are relatively insensitive to sodium, but their amplitude is proportional to calcium up to a limiting concentration. Since calcium is necessary for contraction and since the smooth muscle cells lack a T-tubule and extensive SR tubule system, it is possible that the influx of calcium responsible for the spikes also activates the contractile system.[227b]

The rhythm of the slow waves and the resulting bursts of spikes correspond to the rhythm of segmental intestinal movements. Rhythmic extension and shortening of intestinal segments may be caused by contraction and relaxation of circular muscle with little active movement of the longitudinal layer.[383] Conduction of spikes in sheets of circular muscle is by connections of low electrical resistance between cells, and these may be the regions of gap junctions or nexuses.[93, 94, 220, 258] Treatment with hypertonic solutions (such as sucrose, but not the penetrating molecule urea) causes an increase in resistance between cells, failure of spike conduction, and in some preparations the disappearance of nexuses. However, electrotonic spread can continue and the effect of sucrose may be due partly to cell shrinkage and partly to hyperpolarization. Normally, slow waves are conducted for short distances determined by overlapping pacemakers, and at a given point both conducted and locally generated waves may be recorded, so that a variety of patterns are seen *in vivo*.[289]

Longitudinal and diagonal muscle fiber

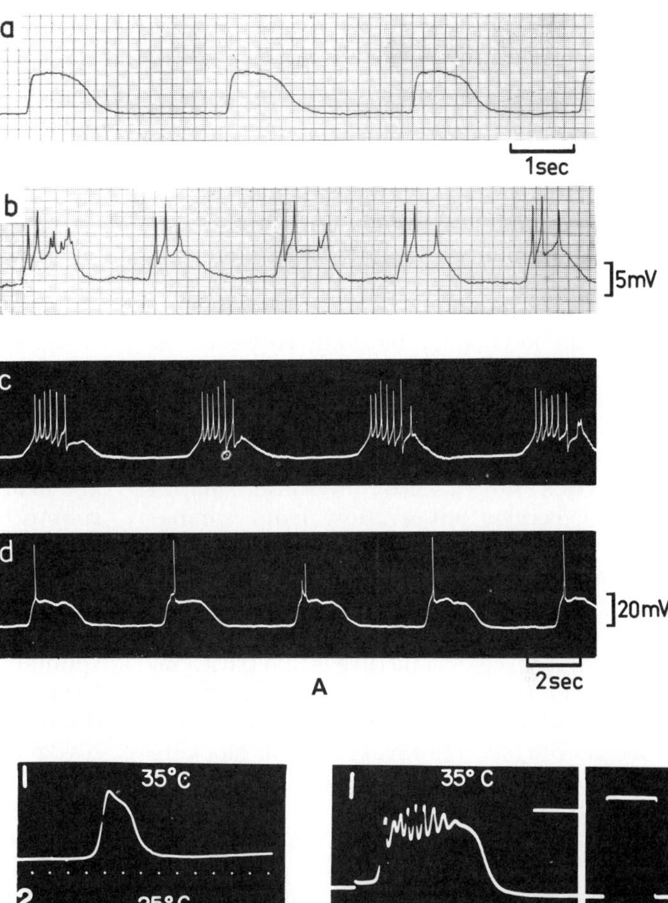

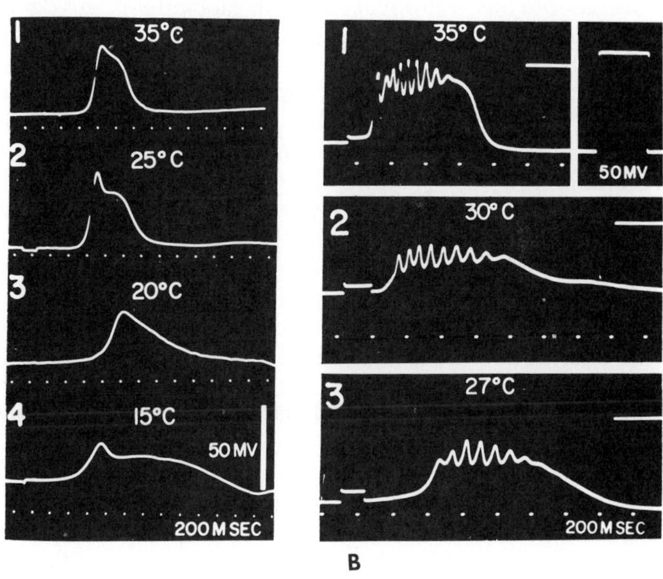

Figure 16–29. A, Recordings from muscle of rabbit ileum. *a* and *b* by double sucrose gap; *c* and *d* by microelectrodes showing slow waves, prepotentials, and spikes. (Courtesy of G. S. Taylor and R. G. Mills.) B, Recordings at different temperatures of intracellular action potentials of smooth muscle of ureter in rat (at left, plateau type potential) and guinea pig (at right, oscillatory type potential). (From Irisawa, H., and M. Kobayashi, Japan. J. Physiol. *13*:421–430, 1963.)

bundles of stomach show much species variation. There appear to be calcium-mediated spikes, as well as two kinds of slow electrical event, one of which is spike-like, can cause contraction, and is dependent on calcium; the other corresponds to the Na slow waves of the small intestine.[83, 268]

In colon, the circular layer is the pacemaker and shows slow waves like the longitudinal layer of the intestine.[65]

Esophagus contains varying amounts of striated and smooth muscle: in man and opossum the upper quarter is striated; in rat, sheep, and beef striated fibers occur along the entire length; in dog and pig the longitudinal layer is striated, and the circular layer is striated except for a region near the cardiac end of the stomach. The esophagus is essentially a multiunit muscle in that it is not spontaneous but is activated by nerves.[66]

Some of the smooth muscle fibers are activated directly, and others via intrinsic ganglia.[230]

Many visceral smooth muscles are excited by appropriate mechanical stimulation, even in the absence of intrinsic reflexes. Increased tension on guinea pig taenia coli lowers membrane potential and increases contractions. A chick amnion, which lacks nerves, is readily stimulated by stretch. Ureter and bladder are stimulated by stretch to contract even after nerves are blocked. However, in these muscles conduction passes a region of mechanical immobilization; hence, it does not depend on successive pull from cell to cell.

Nervous modulation. Visceral smooth muscles receive innervation from sympathetic and parasympathetic nervous systems, and many smooth muscles have intrinsic nerve plexuses which both modulate activity and serve as afferent and efferent relay stations. The classical picture is that parasympathetic fibers liberate acetylcholine as transmitter and that sympathetic fibers liberate norepinephrine. However, use of blocking drugs reveals extensive mixing of cholinergic and adrenergic fibers in the peripheral trunks of each division of the autonomic system.[29, 48a] There are two types of receptors for each agent—muscarinic and nicotinic receptors for acetylcholine, and alpha and beta receptors for norepinephrine. Parasympathetic innervation in mammals is cholinergic and excitatory to stomach and intestine; the effect is blocked by hyoscine (atropine). Excitatory miniature junction potentials are due to ACh liberation (Fig. 16–30). Rebound contraction following inhibition has sometimes been confused with primary excitation.

Much of the confusion regarding inhibitory and excitatory effects of catecholamines arises because different organs, and the same organ in different species, may have alpha receptors (blocked by phenoxybenzamine or phentolamine) or beta receptors (blocked by propranolol); norepinephrine may be excitatory on one smooth muscle system and inhibitory on another. Properties of alpha and beta receptors are given in Table 11–3 (p. 496). The cellular action of catecholamines is given on pp. 494–496.

Transmural stimulation of the gut of mammals activates neurons which are inhibitory to the muscle layers by some transmitter that is not blocked by catecholamine blocking agents. Evidence is accumulating that the non-catecholamine transmitter may be ATP (purinergic nerves); ATP is inhibitory to the gut and excitatory to bladder, and a purinergic system may be more important than catecholamines in inhibiting gut muscles of fishes and amphibians.[48a, 51] Inhibitory (hyperpolarizing) junction potentials in mammalian intestine are not sensitive to adrenergic blockers. Nearly every muscle cell tested in rabbit and guinea pig colon showed ipsp's on transmural stimulation, presumably by purinergic neurons.[128]

Agents such as serotonin (5-HT), prostaglandins, and histamine have been suggested but now seem unlikely as neural transmitters to smooth muscle.

Adrenergic nerve endings in the intestine are mainly on neurons of the myenteric and submucous plexuses, as observed by fluorescence microscopy. Norepinephrine stimulates some plexus neurons which may be inhibitory to spontaneously spiking muscle fibers. Many of the neurons in plexuses show spontaneous activity, and others are stimulated by mechanical distension.[382a] Neuronal discharge shows no regular relation to the myogenic activity of muscle layers; however, blocking the spontaneous ganglionic activity releases intestinal muscle from inhibition, and rhythmicity is enhanced. Mechanoreceptor neurons presumably function in local stretch reflexes.

Study of visceral muscles of lower vertebrates supports the evidence from mammals for diversity of smooth muscles and their autonomic regulation. Autonomic control of vascular smooth muscle is discussed on page 829. In their supply to the gastrointestinal muscles, reptiles resemble mammals in having vagal cholinergic excitatory fibers and vagal preganglionic fibers which activate inhibitory non-adrenergic neurons, as well as sympathetic cholinergic excitatory and adrenergic inhibitory fibers.[50b]

In fishes and amphibians, the cranial parasympathetic and anterior sympathetic outflows run together in the vagosympathetic nerve (Fig. 16–31); hence, identification of the origin of particular fibers is difficult. The stomach of a frog gives rhythmic electrical waves, 1 to 4 per minute and of 10 to 15 sec duration, conducted at 0.3 to 1.4 mm/sec.[365] These waves trigger contractions and appear to be analogous to the "slow spikes" of mammalian stomach. The vagosympathetic nerves of a frog contain cholinergic excitatory fibers of sympathetic origin, and vagal preganglionic fibers which inhibit the stomach via a non-catecholamine

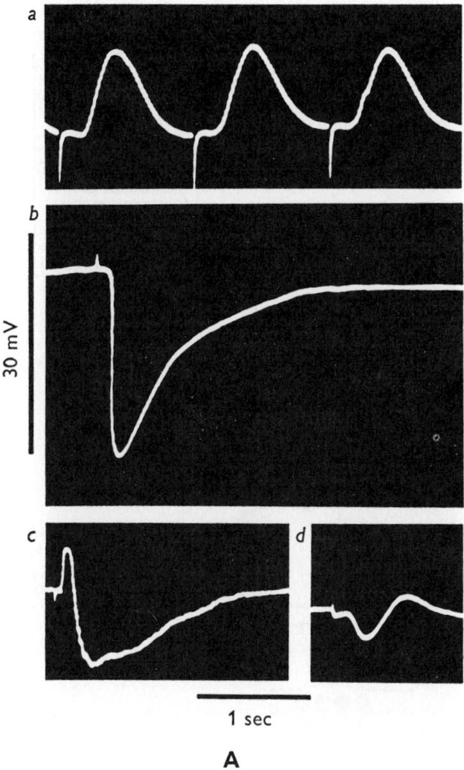

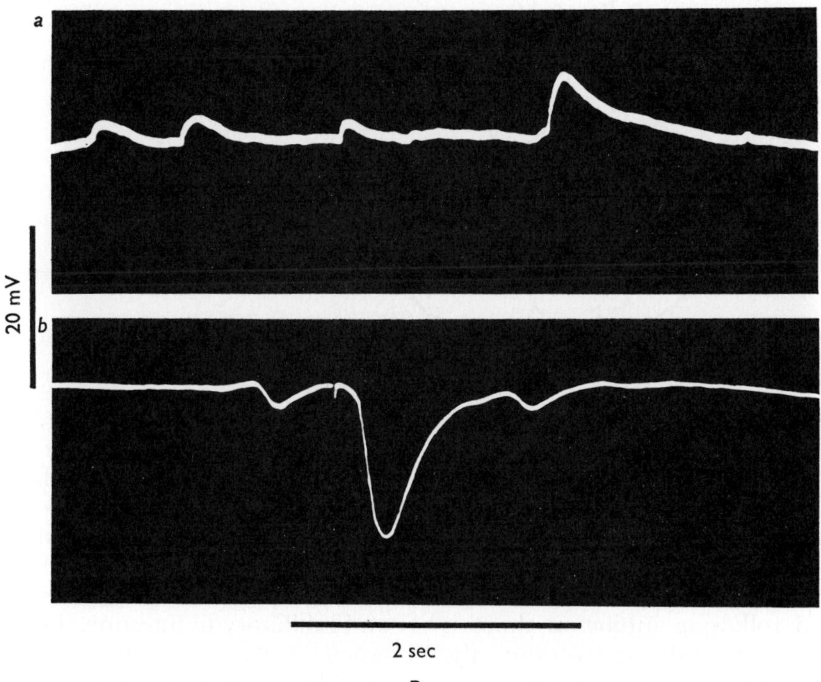

Figure 16–30. Types of junction potentials in smooth muscle cells of guinea pig colon. A, Excitation: *a,* inhibition; *b,* biphasic; *c,* responses to stimulation of intrinsic nerve fibers. B, Spontaneous excitatory miniature junction potentials (*a*) and spontaneous and triggered inhibitory potential (*b*). (From Furness, J. B., J. Physiol. *205*:549–562, 1969.)

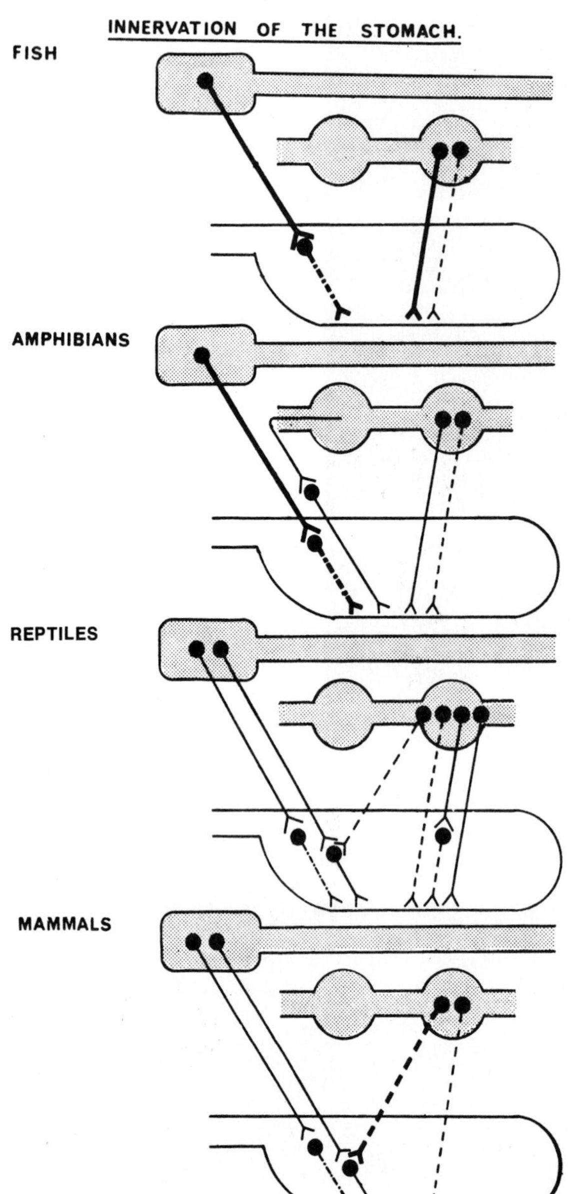

Figure 16–31. Diagrammatic representation of autonomic innervation of muscle of stomach. Vagal outflow is inhibitory in fish and amphibians, and is opposed by excitatory cholinergic sympathetic fibers. In reptiles and mammals, cholinergic excitatory fibers are in the parasympathetic trunk, while adrenergic inhibitory fibers are in the sympathetic trunk. (From Burnstock, G., Pharmacol. Rev. *21*:247–324, 1969.)

neuron.[51] Splanchnic sympathetic nerves carry adrenergic inhibitory fibers to the anuran stomach.[54]

In fishes, the vagal influence does not extend very far posterior to the stomach. In trout, for example, vagal fibers to the stomach are inhibitory (via non-adrenergic neurons), and following inhibition there is post-inhibitory rebound contraction; the vagus also contains a few cholinergic excitatory fibers of sympathetic origin. The stomach also receives excitatory adrenergic fibers in the splanchnic nerve.[48b] In the trout intestine, in contrast to the stomach, epinephrine is inhibitory and high frequency stimulation of the spinal outflow is inhibitory, with post-inhibitory rebound.[54b] Acetylcholine is excitatory to all parts of the fish gut muscle; epinephrine is excitatory to stomach and inhibitory to intestine. The intestine of a tench (*Tinca*) has a coat of striated muscle outside a layer of smooth muscle fibers; hence, when it is stimulated electrically it gives both fast (striated) and slow (smooth) contractions.[263] Many neurons occur in the muscular wall of fish intestine and stomach.

Neural control of stomach muscle in elasmobranchs is unclear. The vagus is apparently inhibitory with rebound excitation; spinal autonomic outflow has both inhibitory and excitatory fibers.[54b] The mesenteric muscle which rotates the spiral valve in dogfish is stimulated by slight stretches; it is unaffected by acetylcholine but is excited by epinephrine, an effect blocked by propranolol. In hagfishes the vagus is inhibitory to the upper gut.

The smooth muscles of lungs are also under autonomic regulation as follows:[48a, 54a]

	excitatory nerves
amphibians	sympathetic cholinergic some adrenergic
reptiles and mammals	parasympathetic cholinergic
	inhibitory nerves
amphibians	vagal non-adrenergic
reptiles	vagal non-adrenergic sympathetic adrenergic
mammals	sympathetic adrenergic

Thus, a change in pattern of innervation of lungs takes place between amphibians and reptiles.

The smooth muscle of the urinary bladder receives cholinergic excitatory innervation in all classes of vertebrates where it occurs. Intramural neurons are numerous in the bladder wall in amphibians, reptiles, and mammals. Adrenergic inhibitory nerves are present in reptiles and mammals. The meaning of epinephrine-containing nerves in frogs is unclear, since this agent contracts the frog bladder.[48c]

The morphological sympathetic system is primarily excitatory cholinergic in elasmobranchs, mixed cholinergic and adrenergic in fishes, amphibians, reptiles, and birds, and predominantly adrenergic inhibitory in mammals. The primitive vagal supply to visceral smooth muscle is primarily non-adrenergic inhibitory with post-inhibitory rebound. The vagus to gastrointestinal muscle seems to have taken on cholinergic excitatory function late in evolution.

INVERTEBRATE VISCERAL MUSCLES

The gut musculature of insects and most crustaceans is striated and quite dissimilar to the visceral muscles of vertebrates. In several insect visceral muscles, each myosin filament is surrounded by 12 thin filaments, rather than by six as in locomotor muscles.[334] Fibers of the striated visceral muscle of the hindgut of the cockroach show spontaneous fluctuations in resting potentials with prepotentials that may trigger spikes, especially when the muscle is stretched; also, nerves from the sixth abdominal ganglion can trigger epsp's which lead to spikes (Fig. 16–32). Innervation of the muscle fibers is multiterminal and polyneural, and miniature spontaneous epsp's are seen.[256, 257] Thus, the hindgut is activated both myogenically and by nerves. In *Locusta,* the foregut is more

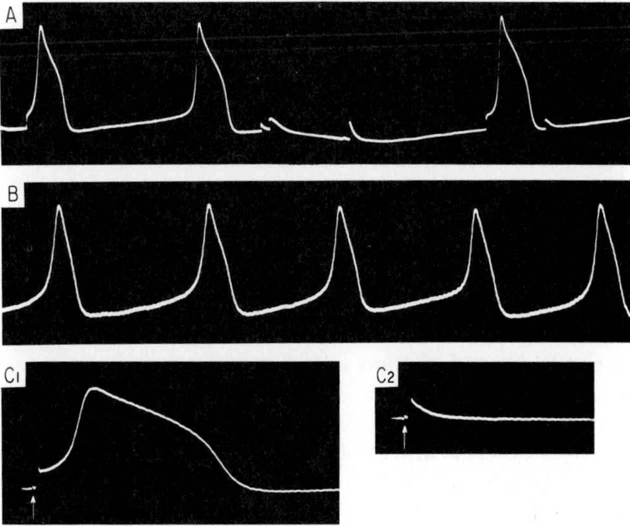

Figure 16–32. Electrical activity recorded by intracellular microelectrode from superior longitudinal muscle of cockroach proctodeum (hindgut). A, Three types of spontaneous responses from intact proctodeum with sixth abdominal ganglion attached: postsynaptic potentials (epsp's), and action potentials with and without epsp's preceding them. B, Spontaneous rhythmic action potentials, with sixth abdominal ganglion removed. C_1: action potential following epsp; C_2: epsp alone. C_1 and C_2 are from same fiber; proctodeal nerve was stimulated at points indicated by arrows. Calibrations: 1 sec for A and B; 0.2 sec for C_1 and C_2; voltage 20 mv. (Courtesy of T. Nagai.)

spontaneous than the hindgut; the foregut is excited by 5-HT at 0.2 μg/ml, while the hindgut requires 20 μg/ml.[126]

Crayfish intestine has a thick inner layer of longitudinal fibers and external circular fibers, all striated and with long sarcomeres. There is a prominent nerve plexus, and contractions appear to be neurogenic.

Molluscs have non-striated muscle fibers in the gut. In the squid stomach, peristalsis is elicited by stretch; spikes are not blocked by TTX, and the peristalsis is inhibited by ACh.[382] Activity resembles that in unitary vertebrate muscles. In the slug *Arion*, patterns of contraction differ in different regions, probably with central nervous control of movement.[307] In a chiton, *Poneroplax*, innervation is dense in the rectum and less so in the posterior intestine, which is spontaneously active with slow waves and spikes. The rectum resembles a vertebrate multiunitary muscle—it is non-spontaneous, there are no slow waves, and nerves elicit psp's; the intestine is spontaneously active, it has slow waves and spikes, and it receives excitatory and inhibitory modulation like vertebrate unitary muscle.[50] In the bivalves *Mercenaria* and *Spisula*, the hindgut shows both phasic and tonic responses to shocks, and there is evidence for dual excitatory and single inhibitory innervation. No spontaneous activity occurs, and ganglionic stimulation elicits both types of contraction; conduction appears to be by nerves.[293] The intestines of the echinoderms *Thyone* and *Arbacia* show conducted contractions in response to stretch; nerve blocking drugs are effective, and hence the muscles resemble vertebrate unitary visceral muscle.

In the tube-dwelling polychaete *Arenicola*, the proboscis consists of buccal mass, pharynx, and esophagus; a stomatogastric nerve plexus extends over the pharynx and esophagus and connects to the circumesophageal nerve ring. The esophageal region is a pacemaker and can be stimulated by epinephrine or acetylcholine. In *Glycera* also, activity of the pharynx and esophagus appears to originate in the stomatogastric plexus. In earthworms, the isolated crop and gizzard are spontaneously active if distended; they are excited by acetylcholine at 10^{-10} (w/v), an effect that is blocked by atropine. Epinephrine inhibition of the rhythm of crop and gizzard can be stopped by a β-blocker; however, epinephrine acts on α receptors in body wall muscle.

SUMMARY

Speed of movement, holding power, tension, and extent of shortening in flexible structures are all adaptive for an animal's life habits and are determined by muscles. The speed of muscles, from the fastest to the slowest, covers a ten-thousand-fold range. The following outline summarizes the general properties of muscles which are related to different speeds of contraction and relaxation. Exceptions to this classification, and intermediate types, have been mentioned previously.

I. Cross-striated; T-tubules and SR tubules; long, straight fibers; skeletal muscles.
   A. One (or two) "en plaque" nerve endings per fiber; unineuronal innervation; "Fibrillen Struktur" and regular T-tubular system; conduction by muscle action potentials.
      1. White phasic; glycolytic; specific myosin: mammal extensor digitorum, chick posterior latissimus dorsi, frog sartorius.
      2. Red tonic; holding muscles; oxidative; weaker myosin ATPase than in (1): mammal soleus, psoas.
   B. Multiple distributed "en grappe" motor endings; graded junction potentials, rarely conducted muscle spikes; "Felder Struktur"; irregular T-system.
      1. Unineuronal innervation; usually tonic or slow: chick anterior latissimus dorsi, frog rectus abdominis, many fish muscles.
      2. Multineuronal innervation; fast or slow muscles; various combinations of excitatory and inhibitory innervation; excitatory may be slow (facilitating) or fast (near all-or-none); most muscle potentials junctional, occasional action potentials.
         a. Synchronous; single activation per motor impulse: crustacean and many insect postural muscles.
         b. Asynchronous; multiple oscillations per motor impulse: indirect flight muscles of Diptera and Hymenoptera.

II. Striated cardiac fibers; often branched; intermediate speeds; oxidative: vertebrate hearts, most arthropod hearts.

III. Intermediate types.
   A. Striations on only one side or in one process of fibers: some trematodes, *Ascaris*, tunicate heart.
   B. Filaments out of register to give spiral pattern or in helix around cortex of fiber: oyster translucent, squid mantle, earthworm body wall.

IV. Non-striated muscles; no T-tubules, weak SR system; small fiber diameters.
   A. Long-fibered, paramyosin-containing; multiple innervation: molluscan "catch" muscles.
   B. Short, spindle-shaped thin fibers.
      1. Multiunitary; nerve activated; nerve conduction: sipunculid proboscis retractors, echinoderm lantern retractors, *Helix* pharynx retractor, mammal pilomotors, nictitating membrane.
      2. Unitary, myogenically spontaneous, conduction from fiber to fiber; often stimulated by stretch; may show modulation by nerves: vertebrate visceral muscles.

The most rapid contractions are in dipteran flight muscles, but in these the frequency of contraction is determined by mechanical resonant properties of the contractile proteins, and there is not a complete activation for each beat. The fastest muscles for complete excitation, contraction, and relaxation are mammalian extraocular muscles and sonic muscles of fishes (5 to 10 msec contraction time). Postural muscles of vertebrates and of many invertebrates show intermediate speed (40 to 100 msec contraction time). Molluscan holding muscles contract quickly and can maintain tension for many hours or days. Vertebrate visceral muscles take seconds for contraction and relaxation, and some sea anemones may contract over periods of minutes.

Speed is correlated with structural and enzymatic properties of muscle. A fast muscle fiber is striated and has short sarcomeres, many fibrils, an extensive sarcotubular system, and one or two nerve endings; it is often richer in glycolytic than in oxidative enzymes. Slow striated muscles usually have much glycogen and fat reserve, are rich in oxidative enzymes, and may have many nerve endings and irregular T-tubules. The slowest muscles are non-striated and thin-fibered, and lack a definite tubular system. Some muscles exhibit properties of both the striated and non-striated types, and have thin fibers, staggered or diagonal sarcomere structure, and irregular tubules. Speed of shortening is described by a force-velocity curve or by muscle lengths moved per second. This measure correlates well with activity of myosin ATPase, and the type of myosin produced in a striated fiber may be determined by the type of innervation. Speed of relaxation correlates well with the rate at which released calcium ions are re-bound by cytoplasmic granules. Holding muscles (invertebrate catch muscles) contain large filaments of paramyosin, a protein which itself does not develop tension but which in some way delays relaxation.

All muscles contain actin, which forms thin filaments and receives bridges of myosin. Myosin is a very large, complex molecule, and there are probably many forms. It consists of several portions—a filamentous tail, a hinge and bridge (to actin), and an active ATPase site. The energy for contraction comes from ATP, the hydrolysis of which, by myosin ATPase *in vivo*, requires magnesium and actin. The interaction between actin and myosin and the resulting ATPase action is initiated by Ca^{++} either by removing tropomyosin-troponin inhibition or by binding to myosin. Release of Ca^{++} from sarcoplasmic vesicles provides for coupling between cell membrane and contractile proteins. Relaxation results when calcium is again sequestered by vesicles. Muscle also contains several actinins of uncertain function—one of these is found in Z bands. ATP is recharged from phosphagens (phosphocreatine or phosphoarginine). The creatine or arginine is rephosphorylated by glycolytic processes; the products of these processes are recycled oxidatively, or their substrates are replaced from other tissues. Only a fraction (about 25%) of the energy from ATP goes into mechanical work, and heat measurements correlate with muscle energetics.

The variety of patterns of activation of muscles is as great as the range of speeds of contraction. Most fast muscles are activated by a single nerve impulse which, by a transmitter (acetylcholine in vertebrates), elicits a junction potential at one (or two) endplates. Liberation of the transmitter requires depolarization of motor nerve endings; it also requires calcium, which appears to combine with some carrier molecule for release of transmitter. The liberated acetylcholine is immediately hydrolyzed by ACh esterase in the subsynaptic membrane.

The junction potential is a gradable event according to the amount of transmitter released; it is due to increased Na and K conductance, and at a critical size it triggers a muscle action potential (sequential Na and K conductance increase), which in turn is conducted inward by the T-tubules. The T-tubule signal in some way activates the SR-tubule system, which releases calcium ions for activation of the actomyosin. The muscle action potential activates fibers in an all-or-none fashion but junctional potentials can trigger graded contractions.

Slow striated muscle fibers (both vertebrate and arthropod) have multiple nerve endings and depend on junction potentials distributed over the muscle membrane for activation. A single motor impulse rarely elicits a maximal response; rather, facilitation is required. Many invertebrate muscles, best known in arthropods, have several excitatory nerve fibers, each apparently using the same transmitter (e.g., glutamate), and each eliciting a response of a certain speed and degree of facilitation. A few muscles of crustaceans receive inhibitory innervation as well as the excitatory type; gamma-amino butyric acid is the transmitter for inhibition, and its action may be presynaptic on motor axon endings or postsynaptic on the muscle membrane. Inhibition here is by an increase in anion conductance.

Non-striated muscles may be activated by motor nerve fibers (e.g., postural muscles of molluscs, sipunculids, coelenterates, and echinoderms, and a few multiunitary muscles of vertebrates). Other non-striated muscles are capable of spontaneous (myogenic) activity, and they may be modulated by regulatory nerves; such muscles are primarily visceral (e.g., in vertebrates, molluscs, and probably echinoderms). Many non-striated muscles, particularly unitary ones, are excited by stretch. In various visceral muscles of vertebrates, acetylcholine and norepinephrine may be transmitters—excitatory or inhibitory according to the muscle. Electrical coupling occurs between many smooth muscle fibers, and in the unitary visceral muscles between bundles of fibers as well. Nervous modulation is either excitatory or inhibitory; it is often coupled with hormonal regulation.

Gradation of movement in vertebrates is largely by variation in the spinal cord of the number of motor units activated. Where slow muscles occur, gradation is by frequency of impulses and by shifting between fast and slow fiber systems, either within the same fiber (in arthropods) or between different fibers in a muscle (in lower vertebrates). In arthropods, further gradation is possible by the involvement of inhibitory axons. All muscles use a similar system of actin-myosin interaction, and in virtually all muscles activation is by electrical signals in the cell membranes; coupling is by calcium ion release. The wide range of speeds of contraction, relaxation, and tension is correlated with differences in gross and fine structure, in kinds of myosins, and especially in patterns of neural activation.

REFERENCES

1. Abbott, B. C., and I. Parnas, J. Gen. Physiol. *48*:919–931, 1965. Electrical and mechanical properties of crustacean muscle.
2. Abe, Y., and T. Tomita, J. Physiol. *196*:89–100, 1968. Cable properties of smooth muscle.
3. Aberg, A. K. G., and J. Axelson, Acta Physiol. Scand. *64*:15–27, 1965. Mechanical properties of intestinal smooth muscle.
4. Adrian, R. H., and L. D. Peachey, J. Physiol. *181*:324–336, 1965. Membrane capacity, frog muscle fibers.
5. Adrian, R. H., et al., J. Physiol. *204*:231–257, 1967. Radial spread of contraction in frog fiber.
6. Ahlquist, R. P., Ann. Rev. Pharmacol. *8*:259–272, 1968. Agents blocking adrenergic β-receptors.
7. Aidley, D. J., and D. C. S. White, J. Physiol. *205*:179–192, 1969. Mechanical properties of tymbal muscles of cicada.
8. Ambache, N., A. S. Dixon, and E. A. Wright, J. Exp. Biol. *21*:46–57, 1945. Properties of intestinal muscles in earthworm.
9. Andersen, P., J. K. S. Jensen, and Y. Loyning, Acta Physiol. Scand. *57*:167–179, 1963. Slow and fast muscle fibers of hagfish.
10. Anderson, N. C., J. Gen. Physiol. *54*:145–165, 1969; *58*:322–339, 1971. Voltage clamp on uterine smooth muscle.
11. Anderson, R., and R. Fänge, Arch. Int. Physiol. Biochem. *75*:461–468, 1967. Pharmacologic receptors in *Lumbricus*.
12. Ashley, C. C., and E. B. Ridgway, Nature *219*:1168–1169, 1968. Calcium transport in barnacle fibers.
13. Atwood, H. L., Comp. Biochem. Physiol. *10*:17–32, 77–81, 1963. Fast and slow contractions in *Carcinus* and *Nephrops*.
14. Atwood, H. L., Comp. Biochem. Physiol. *16*:409–426, 1965. Excitation and inhibition in crab muscle fibers.

14a. Atwood, H. L., and G. Bittner, J. Neurophysiol. *34*:157–170, 1971. Matching of excitatory and inhibitory inputs to crustacean muscle fibers.

15. Atwood, H. L., and H. S. Johnston, J. Exp. Zool. *167*:456–470, 1968; *also* Atwood, H. L., and W. A. Morin, J. Ultrastruc. Res. *32*:351–369, 1970. Neuromuscular synapses in crab and crayfish.

15a. Aubert, X., and J. Lebacq, J. Physiol. *216*:181–200, 1971. Heat of shortening during plateau of tetanic contraction and at end of relaxation.

16. Bagby, R. M., J. Morph. *118*:167–182, 1966. Fine structure, sponge myocytes.
17. Baguet, F., and J. M. Gillis, J. Physiol. *188*:67–82, 1967; *198*:127–143, 1968; *also* Baguet, F., F. J. Gillis, and G. Dainoff, Arch. Int. Physiol. Biochem. *75*:523–527, 1967; *also* Baguet, F., G. Marechal, and X. Aubert, Arch. Int. Physiol. Biochem. *70*:416–417, 1962. Energy cost of tonic contraction in bivalve muscles.
18. Baldwin, E., and W. H. Yudkin, Proc. Roy. Soc. Lond. B *136*:614–631, 1950. Phosphagens in invertebrates.
19. Bárány, M., J. Gen. Physiol. *50* (Suppl.): 197–218, 1967. ATPase activity of myosin in relation to contraction speed.
20. Bárány, M., and K. Bárány, Biochem. Zeitschr. *345*:37–56, 1966. Myosin from adductor of *Pecten*.

21. Bárány, M., and R. I. Close, J. Physiol. *213*:455–474, 1971. Changes of myosin in cross-innervated rat muscles.
22. Bárány, M., et al., Europ. J. Biochem. *2*:156–164, 1967. Myosin and contraction in muscle of cat and sloth.
23. Barber, S. B., and J. W. S. Pringle, Proc. Roy. Soc. Lond. B *164*:21–39, 1966. Flight muscles of belostomatid bug.
24. Barnard, R., et al., Amer. J. Physiol. *220*:410–414, 1971. Histochemistry of red and white muscle fibers.
25. Bastian, J., and H. Esch, Z. vergl. Physiol. *67*:307–324, 1970. Control of flight muscles of honeybee.
26. Batra, S. C., and E. E. Daniel, Comp. Biochem. Physiol. *38A*:369–385, 1971. Calcium uptake by particulates of smooth muscle.
26a. Belton, P., and B. E. Brown, Comp. Biochem. Physiol. *28*:853–863, 1969. Electrical constants, cockroach visceral muscle.
26b. Belton, P., and H. Grundfest, Amer. J. Physiol. *203*:588–594, 1962. K-spikes in muscle of mealworm larvae.
27. Bendall, J. R., Muscles, Molecules and Movement. Heinemann, London, 1969. 219 pp.
28. Bennett, M. R., J. Physiol. *185*:132–147, 1966. Nervous transmission to taenia coli.
29. Bennett, M. R., and G. Burnstock, pp. 1709–1732. *In* Handbook of Physiology, Sec. 6, Vol. IV, edited by C. F. Code. Amer. Physiol. Soc., Washington, D.C., 1968. Effects of extrinsic nerves on smooth muscle.
30. Bennett, M. R., G. Burnstock, and M. E. Holman, J. Physiol. *182*: 527–540, 1966. Neural inhibition of smooth muscle.
31. Bennett, T., Comp. Biochem. Physiol. *32*:669–680, 1970. Excitation and inhibition of bird gizzard.
32. Bergman, R. A., Bull. Johns Hopkins Hosp. *114*:325–343, 344–353, 1964. Dorsal fin muscles of teleosts.
33. Bianchi, C. P., pp. 111–120. *In* Protein Metabolism and Biological Function, edited by C. P. Bianchi and R. Hilf. Rutgers University Press, 1970. Calcium regulation of muscle contraction.
34. Bittner, G. D., J. Gen. Physiol. *51*:731–758, 1968. Differentiation of nerve terminals in crayfish muscle.
35. Bittner, G. D., and J. Harrison, J. Physiol. *206*:1–23, 1970. Crustacean transmitter release.
36. Bolton, T. B., J. Physiol. *216*:403–418, 1971. Intestinal smooth muscle, slow waves, ACh or carbachol.
37. Boltt, R. E., and D. W. Ewer, J. Exp. Biol. *40*:713–726, 1963. Neural transmission to muscles of echinoderms.
38. Bortoff, A., Amer. J. Physiol. *201*:203–208, 1961. Slow potentials of small intestine.
39. Bortoff, A., Amer. J. Physiol. *209*:1254–1260, 1965. Electrical transmission from longitudinal to circular intestinal muscle.
40. Boyd, H., G. Burnstock, and D. Rogers, Brit. J. Pharm. Chemother. *23*:151–163, 1964. Innervation of large intestine of *Bufo*.
41. Bowden, J., Int. Rev. Cytol. 7:295–335, 1958. Structure and innervation of clam muscle.
42. Bozler, E., Amer. J. Physiol. *207*:701–704, 1964. Mechanical properties and excitation of smooth and cardiac muscle.
43. Brady, A. J., The Physiologist *10*:75–86, 1967. Mechanical model of muscle.
44. Brandt, P. W., J. P. Reuben, and H. Grundfest, J. Cell Biol. *38*:115–129, 1968; *also* Brandt, P. W., J. P. Reuben, L. Girardier, and H. Grundfest, J. Cell Biol. *25*:233–260, 1965. Structure and function in single muscle fibers of crayfish.
45. Bruner, J., and D. Kennedy, Science *169*:92–94, 1970. Habituation at a neuromuscular junction.
46. Bullard, B., Comp. Biochem. Physiol. *23*:749–759, 1967. Nervous control of byssus retractor of *Mytilus*.
47. Buller, A. J., and D. M. Lewis, J. Physiol. *178*:343–358, 1965. Cross-innervation of skeletal muscles.
48. Buller, A. J., W. F. H. M. Mommaerts, and K. Seraydarian, J. Physiol. *205*:581–597, 1969. Properties of myosin in fast and slow muscles of cat.
48a. Burnstock, G., Pharmacol. Rev. *21*:247–324, 1969. Autonomic innervation of visceral and vascular muscles of vertebrates.
48b. Burnstock, G., Quart. J. Microscope Sci. *100*:199–219, 1959. The innervation of the gut of the brown trout.
48c. Burnstock, G., and G. Campbell, J. Exp. Biol. *40*:421–437, 1963. Innervation of ringtail possum bladder.
48d. Burnstock, G., Pharmacol. Rev. *24*:509–581, 1972. Purinergic nerves.
49. Burnstock, G., G. Campbell, and M. J. Rand, J. Physiol. *182*:504–526, 1966. Inhibitory innervation of taenia.
50. Burnstock, G., M. J. Greenberg, S. Kirby, and A. G. Willis, Comp. Biochem. Physiol. *23*:407–429, 1967. Pharmacology of visceral muscle of mollusc.
50a. Burnstock, G., and T. Iwayama, Prog. Brain Res. *34*:389–404, 1971. Fine-structural identification of autonomic nerves and their relation to smooth muscle.
50b. Burnstock, G., J. O'Shea, and M. Wood, J. Exp. Biol. *40*:403–420, 1963. Innervation of toad bladder.
50c. Burnstock, G., and M. E. Wood, Comp. Biochem. Physiol. *20*:675–690, 1967; *22*:815–831, 1967. Innervation of lungs of sleepy lizard.
51. Burnstock, G., et al., Brit. J. Pharm. Chemother. *44*:668–688, 1970. ATP as an inhibitory transmitter.
51a. Burrows, M., and G. Hoyle, J. Exp. Zool. *179*:379–394, 1972. Neuromuscular transmission in mantis shrimp *Hemisquilla*.
52. Bush, B. M. H., J. Exp. Biol. *39*:71–88, 1962. Reflex inhibition in claw of *Carcinus*.
53. Buttkus, H., J. Fish. Res. Bd. Canad. *24*:1607–1612, 1967. Amino acid composition of myosin, trout muscle.
54. Campbell, G., Comp. Biochem. *31*:693–706, 1969. Autonomic innervation of stomach of toad.
54a. Campbell, G., Comp. Gen. Pharmacol. *2*:281–286, 1971. Autonomic innervation of lung musculature, toad.
54b. Campbell, G., and G. Burnstock, pp. 2213–2266. *In* Handbook of Physiology, Sec. 6, Vol. IX, edited by C. F. Code. Amer. Physiol. Soc., Washington, D.C., 1968. Comparative physiology of gastrointestinal motility.
55. Carlson, F. D., Prog. Biophys. Molec. Biol. *13*:262–314, 1963. Mechanochemistry of muscular contraction.
56. Carlson, F. D., D. J. Hardy, and D. R. Wilkie, J. Gen. Physiol. *46*:851–882, 1963; J. Physiol. *189*:209–235, 1967. Energetics of muscle contraction.
57. Carsten, M. E., J. Gen. Physiol. *53*:414–426, 1969. Calcium binding in uterine smooth muscle.
58. Carvalho, A. P., J. Gen. Physiol. *52*:622–642, 1968. Calcium binding by SR granules.
59. Cerf, J. A., H. Grundfest, G. Hoyle, and F. V. McCann, J. Gen. Physiol. *43*:377–395, 1959. Dual responses in grasshopper muscle.
60. Chaplain, R. A., Arch. Biochem. Biophys. *115*:450–461, 1966; *121*: 154–168, 1967. Insect actomyosin ATPase.
61. Chaplain, R. A., and E. Pfister, Experientia *26*:505–506, 1970. Heat and calcium transference in frog sartorius.
62. Chaplain, R. A., and R. T. Tregear, J. Molec. Biol. *21*:275–280, 1966. Distribution of myosin and insect fibrillar muscle.
63. Chapman, D. M., J. Marine Biol. Assoc. U. K. *48*:667–688, 1968. Muscle cells in medusa of *Obelia*.
64. Chapman, D. M., C. F. A. Pantin, and E. A. Robson, Rev. Canad. Biol. *21*:267–278, 1962. Muscles in coelenterates.
65. Christensen, J., R. Caprilli, and G. F. Lund, Amer. J. Physiol. *217*: 771–776, 1969. Slow waves in cat colon.
66. Christensen, J., and E. E. Daniel, Amer. J. Physiol. *211*:387–394, 1966. Pharmacology of esophageal smooth muscle.
67. Close, R., J. Physiol. *173*:74–95, 1964. Dynamic properties of fast and slow muscle in developing rats.
68. Close, R., J. Physiol. *204*:331–346, 1969. Fast and slow muscle of rat after cross-innervation.
69. Close, R., pp. 142–149. *In* Exploratory Concepts in Muscular Dystrophy and Related Disorders, Proc. Int. Conf. Harriman, New York, 1966. Dynamic properties of fast and slow skeletal muscle.
69a. Close, R., Physiol. Rev. *52*:129–197, 1972. Dynamic properties of mammalian skeletal muscles.
70. Close, R., and J. F. Y. Hoh, J. Physiol. *192*:815–822, 1967. Force-velocity properties of kitten muscle.
71. Close, R., and J. F. Y. Hoh, J. Physiol. *198*:103–125, 1968. Cross-innervation of fast and slow muscles in toad.
72. Cobb, J. L. S., Comp. Biochem. Physiol. *24*:311–315, 1968. Electrical activity of lantern retractor, *Echinus*.
73. Cobb, J. L. S., and T. Bennett, J. Cell Biol. *41*:287–297, 1969. Nexuses in visceral smooth muscle.
74. Cobb, J. L. S., and M. S. Laverack, Proc. Roy. Soc. Lond. B *164*:624–640, 641–650, 1966. Function and structure in lantern muscle of *Echinus*.
75. Cohen, C., et al., J. Molec. Biol. *56*:223–237, 239–258, 1971. Paramyosin, organization of filaments.
76. Colano, F., and R. Rahamimoff, J. Physiol. *198*:203–218, 1968. Frog neuromuscular junction, Ca and Na effects.
77. Connell, J. J., Biochem. J. *80*:503–509, 1961. Muscle proteins of various animals.
78. Connell, J. J., and I. M. Mackie, Nature *201*:78–79, 1964. Molecular weight of rabbit and cod myosin.
79. Cooke, I. M., and A. D. Grinnell, J. Physiol. *175*:203–210, 1964. Drug effects on denervated skeletal muscle.
80. Cooke, P. H., and R. H. Chase, Exp. Cell. Res. *66*:417–425, 1971. Myosin and actin filaments from vertebrate gizzard.
81. Costantin, L. L., J. Gen. Physiol. *55*:703–715, 1970. Role of sodium in inward spread of contraction in frog muscle.
82. Costantin, L. L., R. J. Podolsky, and L. W. Tice, J. Physiol. *188*:261–271, 1967. Calcium activation of frog slow muscle.
82a. Dancker, P., Pflüger. Arch. *315*:187–197, 1970. Effect of tropomyosin on Ca-activation of actomyosin ATPase.
83. Daniel, E. E., Gastroenterology *49*:403–418, 1965. Properties of stomach muscle.
84. Daniel, E. E., and K. M. Chapman, Amer. J. Digest. Dis. *8*:54–102, 1963. Electrophysiology of gastrointestinal tract.
84a. Daw, J., and A. Stracher, Proc. Nat. Acad. Sci. *68*:1107–1110, 1971. Light chains of myosin in chicken.
85. DeBell, J. T., Quart. Rev. Biol. *40*:233–251, 1965. Neuromuscular junctions in nematodes.
86. DeBell, J. T., J. del Castillo, and V. Sanchez, J. Cell. Comp. Physiol. *62*:159–177, 1963. Electrophysiology of muscle of *Ascaris*.
87. DeBell, J. T., and V. Sanchez, J. Exp. Biol. *48*:405–410, 1968. Effect of temperature on *Ascaris* muscle fibers.
88. del Castillo, J., W. C. de Mello, and T. Morales, J. Exp. Biol. *46*: 263–279, 1967. Action potentials in *Ascaris* muscle.

89. del Castillo, J., W. C. de Mello, and T. Morales, J. Gen. Physiol. *48*:129–140, 1964. Effects of ions on *Ascaris* muscles.
90. de Mellon, F., Science *160*:1018–1020, 1968. Neuromuscular junction in scallop.
91. de Villafranca, G. W., and L. K. Campbell, Comp. Biochem. Physiol. *29*:775–783, 1969; *also* Biochem. J. *104*:263–269, 1967. Activation of actomyosin ATPase, *Limulus*.
92. Devine, C., and A. V. Somlyo, J. Cell Biol. *49*:636–649, 1971. Thick filaments in smooth muscle.
93. Dewey, M. M., and L. Barr, Science *137*:670–672, 1962. Nexuses between smooth muscle fibers.
94. Dewey, M. M., and L. Barr, pp. 1629–1654. *In* Handbook of Physiology, Sec. 6, Vol. IV, edited by C. F. Code. Amer. Physiol. Soc., Washington, D.C., 1968; *also* J. Cell Biol. *23*:553–585, 1964. Ultrastructure of intestinal smooth muscle.
95. Dodge, F. A., and R. Rahamimoff, J. Physiol. *193*:419–432, 1967. Effect of calcium on neuromuscular transmission.
96. Dreizen, P., Ann. Rev. Med. *22*:365–390, 1971. Structure and function of myofibrillar proteins.
97. Dreizen, P., L. C. Gershman, et al., J. Gen. Physiol. *50*:85–110, 1967. Myosin subunits and their interactions.
98. Dreizen, P., and L. C. Gershman, Trans. N.Y. Acad. Sci. Ser. II *32*: 170–203, 1970. Properties of myosin.
99. Dubrowitz, V., and D. L. Neuman. Nature *214*:840–841, 1967. Enzyme changes in cross-innervated muscle.
100. Dudel, J., and S. W. Kuffler, J. Physiol. *155*:514–529, 1961. Neuromuscular transmission in crayfish.
101. Dudel, J., and S. W. Kuffler, pp. 111–113. *In* Nervous Inhibition, edited by E. Florey, Academic Press, New York, 1961; *also* J. Physiol. *155*:543–562, 1961. Presynaptic inhibition in crayfish muscle junctions.
102. Dudel, J., and R. Rüdel, Pflüger. Arch. *301*:16–30, 1968. Electromechanical coupling in crayfish muscle.
103. Ebashi, S., F. Ebashi, and A. Kodama, J. Biochem. (Tokyo) *62*:137–138, 1967. Troponin as calcium receptor protein.
104. Ebashi, S., and M. Endo, Prog. Biophys. Molec. Biol. *18*:125–183, 1969. Muscle proteins, especially troponin.
105. Ebashi, S., M. Endo, and I. Ohtsuki, Quart. Rev. Biophys. *2*:351–384, 1969. Control of muscle contraction.
106. Ebashi, S., and A. Kodama, J. Biochem. *58*:7–12, 13–17, 1965; *59*: 422–424, 1966; *60*:733–734, 1966. Alpha-actinin.
107. Edwards, C., and H. Lorkovic, Amer. Zool. 7:615–622, 1967. Calcium in excitation contraction coupling.
108. Engel, W. K., and R. L. Irwin, Amer. J. Physiol. *213*:511–518, 1967. Histochemistry of muscle enzymes.
109. Faeder, I. R., R. D. O'Brien, and M. M. Saltpeter, J. Exp. Zool. *173*: 187–202, 1970; *also* Faeder, I. R., and R. D. O'Brien, J. Exp. Zool. *173*:203–214, 1970. Pharmacology of cockroach neuromuscular transmission.
110. Fahrenbach, W. H., J. Cell Biol. *17*:629–640, 1963; *35*:69–80, 1967. Fine structure of crustacean muscles.
111. Fatt, P., and B. L. Ginsborg, J. Physiol. *142*:516–543, 1958. Ionic basis for action potentials in crustacean muscle.
112. Fatt, P., and B. Katz, J. Physiol. *120*:171–204, 1953; *121*:374–389, 1953. Properties of crustacean muscle fibers.
113. Fawcett, D. W., Circulation *24*:336–348, 1961. Sarcoplasmic reticulum of skeletal and cardiac muscle.
114. Fawcett, D. W., and N. S. McNutt, J. Cell Biol. *42*:1–45, 46–67, 1969. Ultrastructure of cardiac muscle.
115. Fawcett, D. W., and J. P. Revel, J. Biophys. Biochem. Cytol. (Suppl.) *10*: 89–103, 1961. Sarcoplasmic reticulum in fast muscles of fish.
116. Fedde, M. R., J. Gen. Physiol. *53*:624–637, 1969. Electrical properties of avian muscle.
117. Fernand, V. S. V., and A. Hess, J. Physiol. *200*:547–554, 1969. Slow and fast muscle fibers in muscles of ear.
118. Fleckenstein, A., pp. 71–92. *In* Cellular Functions of Membrane Transport, edited by J. F. Hoffman. Prentice-Hall, New York, 1964. Excitation-contraction coupling.
119. Fletcher, C. M., J. Physiol. *90*:233–253, 415–428, 1937. Responses of *Mytilus* byssus retractor.
120. Flitney, F. W., J. Physiol. *217*:243–257, 1971. T system of slow muscle.
121. Flood, P., J. Comp. Neur. *126*:181–217, 1966. Innervation of muscle in *Amphioxus*.
122. Flood, P., D. M. Guthrie, and J. R. Banks, Nature *222*:87–89, 1969. Paramyosin in notochord of *Amphioxus*.
123. Florey, E., and M. E. Kriebel, Z. vergl. Physiol. *65*:98–130, 1969. Chromatophore muscles of squid.
124. Florey, E., and B. Woodcock, Comp. Biochem. Physiol. *26*:651–661, 1968. Presynaptic actions of glutamate on crab muscles.
125. Forrester, T., and H. Schmidt, J. Physiol. *207*:477–491, 1970. Electrophysiology of slow fibers in frog muscle.
126. Freeman, M. A., Comp. Biochem. Physiol. *17*:755–764, 1966. Pharmacology of alimentary tract of locust.
127. Fuchs, F., and I. N. Briggs, J. Gen. Physiol. *51*:655–676, 1968; *52*: 955–968, 1968. Calcium and troponin in the control of contractions.
128. Furness, J. B., J. Physiol. *205*:549–562, 1969. Innervation of the colon.
129. Furness, J. B., Pflüger. Arch. *314*:1–13, 1970. Neural control of smooth muscle cells.
130. Furukawa, T., and J. B. Peter, Exp. Neurol. *31*:214–222, 1971. Troponin in different fiber types.
131. Gage, P. W., Fed. Proc. *26*:1627–1632, 1969. Excitation-secretion coupling in presynaptic terminals.
132. Gage, P. W., and R. S. Eisenberg, J. Gen. Physiol. *53*:265–278, 298–310, 1969. Electrical properties of transverse tubules in frog sartorius.
133. Gage, P. W., and R. S. Eisenberg, Science *158*:1700–1701, 1702–1703, 1967. Action potentials without contractions.
134. Gainer, H., and J. E. Klancher, Comp. Biochem. Physiol. *15*:159–165, 1965. Neuromuscular junctions in a fast fish muscle.
135. Gergely, J., Ann. Rev. Biochem. *35*:691–722, 1966. Contractile proteins.
136. Gershman, L. C., A. Stracher, and P. Dreizen, J. Biol. Chem. *244*: 2726–2736, 1969. Subunit structure of myosin.
137. Gibbs, C. L., W. F. H. M. Mommaerts, and N. V. Ricchiuti, J. Physiol. *191*:25–46, 1967. Heat production in cardiac muscle.
138. Gilai, A., and I. Parnas, J. Exp. Biol. *52*:325–344, 1970. Neuromuscular transmission in scorpion.
139. Gillespie, C. A., D. R. Simpson, and V. R. Edgerton, J. Histochem. Cytochem. *18*:552–558, 1970. Glycogen content in red and white muscle.
140. Goffart, M., Function and Form in the Sloth. Pergamon Press, New York, 1971. 225 pp.
141. Goffart, M., Electromyography *8*:245–251, 1968. Sloth muscles.
142. Goffart, M., O. Holmes, and Z. M. Bacq, Arch. Int. Physiol. Biochem. *70*:103–106, 1962. Mechanical properties of muscle in the sloth.
143. Goldspink, G., R. E. Larson, and R. E. Davies, Z. vergl. Physiol. *66*: 379–388, 1970. Thermodynamic efficiency of chicken ALD muscle.
144. Goldspink, G., R. E. Larson, and R. E. Davies, Z. vergl. Physiol. *66*: 389–397, 1970. Energetics of contraction in hamster muscles.
145. Gonzalez-Serratos, H., J. Physiol. *212*:777–799, 1971. Inward spread of activation in vertebrate muscle fibers.
146. Goodwin, L. G., and E. M. Vaughan Williams, J. Physiol. *168*:857–871, 1963. Neuromuscular transmission in *Ascaris*.
147. Gordon, G., A. F. Huxley, and F. J. Julian, J. Physiol. *171*:28P–30P, 1964. Length-tension diagram of single vertebrate striated muscle fibers.
148. Gordon, G., and C. G. Phillips, Quart. J. Exp. Physiol. *38*:35–45, 1953. Slow and fast components in a flexor muscle.
149. Grundfest, H., J. Gen. Physiol. *50*:1955–1958, 1967. Anomalous spikes of *Ascaris* esophageal cells.
150. Guth, L., and F. J. Samaha, Exp. Neurol. *25*:138–152, 1969. Actomyosin ATPase of slow and fast muscles.
151. Guthrie, D. M., and J. R. Banks, J. Exp. Biol. *52*:125–138, 1970. Properties of paramyosin in notochord of *Amphioxus*.
152. Hagiwara, S., S. Chichibu, and N. Simpson, Z. vergl. Physiol. *60*: 209–218, 1968. Neuromuscular mechanisms of wingbeat in hummingbirds.
152a. Hagiwara, S., M. P. Henkart, and Y. Kidokoro, J. Physiol. *219*:233–251, 1971. Excitation-contraction coupling in *Amphioxus* muscle.
152b. Hagiwara, S., and Y. Kidokoro, J. Physiol. *219*:217–232, 1971. Ionic components of action potentials in *Amphioxus* muscle.
152c. Hagiwara, S., and K. Takahashi, J. Physiol. *190*:499–518, 1967. Electrical activity in muscles of fish.
153. Hagiwara, S., K. Takahashi, and D. Junge, J. Gen. Physiol. *51*:157–175, 1968. Excitation-contraction coupling in barnacle muscle.
153a. Hagiwara, S., et al., J. Physiol. *219*:233–251, 1971. EC coupling in *Amphioxus* muscle myotomes.
154. Hall-Craggs, E. C. B., J. Anat. *102*:241–255, 1968. Speeds of rabbit laryngeal muscles.
155. Hanson, J., J. Biophys. Biochem. Cytol. *3*:111–122, 1957. Structure of body wall muscle of earthworm.
156. Hanson, J., and J. Lowy, pp. 265–335. *In* The Structure and Function of Muscle, Vol. I, edited by G. Bourne. Academic Press, New York, 1960. Structure and function of invertebrate muscle.
157. Hanson, J., and J. Lowy, Proc. Roy. Soc. Lond. B *154*:173–196, 1961. Structure of muscle fibers in adductor of oyster.
158. Hanson, J., and J. Lowy, J. Molec. Biol. *6*:46–60, 1963. Structure of actin.
159. Hanson, J., and J. Lowy, Proc. Roy. Soc., Lond. B *160*:449–460, 1964. Locations of proteins in smooth muscles.
160. Harris, A. J., and R. Miledi, J. Physiol. *217*:497–515, 1971. Botulinum toxin and neuromuscular transmission block in frogs.
161. Harris, J. B., and A. R. Luff, Comp. Biochem. Physiol. *33*:923–931, 1970. Resting potentials in fast and slow muscles.
162. Harrison, R. G., S. Lowey, and C. Cohen, J. Molec. Biol. *59*:531–535, 1971. Paramyosin in muscle.
163. Hartshorne, D. J., J. Gen. Physiol. *55*:585–601, 1970. Interactions of muscle proteins.
164. Hartshorne, D. J., et al., Biochim. Biophys. Acta *175*:320–330, 1969; *229*:698–711, 1971. Properties of troponin-tropomyosin.
164a. Hartzell, H. C., and D. M. Fambrough, J. Gen. Physiol. *60*:248–262, 1972. Extrajunctional ACh receptors after denervation, rat diaphragm.

165. Hasselbach, W., Ann. N.Y. Acad. Sci. *137*:1041–1048, 1966. Calcium transport by membranes of muscle reticulum.
166. Hasselbach, W., Prog. Biophys. *14*:167–222, 1964. Relaxation of muscle.
167. Heistracher, P., and C. C. Hunt, J. Physiol. *201*:589–611, 1969. Membrane properties in relation to contraction, snake muscle.
168. Henneman, E. C., and B. Olsen, J. Neurophysiol. *28*:581–598, 1965. Structure and function in skeletal muscle.
169. Hess, A., J. Physiol. *157*:221–231, 1961. Fast and slow extrafusal fibers, chickens.
170. Hess, A., J. Cell Biol. *26*:467–476, 1965. Motor terminals of slow and twitch muscle fibers in snake.
171. Hess, A., and G. Pilar, J. Physiol. *169*:780–798, 1964. Slow fibers in extraocular muscles of cat.
172. Heumann, H. G., and E. Zebe, Z. Zellforsch. *78*:131–150, 1967. Fine structure and function of outer muscle layer in earthworm.
173. Hidaka, T., Y. Ito, and H. Kuriyama, J. Exp. Biol. *50*:387–403, 1969. Membrane properties of earthworm muscle.
174. Hidaka, T., Y. Ito, and N. Tashiro, J. Exp. Biol. *50*:405–415, 417–430, 1969. Effects of ions and drugs on excitation and transmission in earthworm muscle.
175. Hidaka, T., H. Kuriyama, and T. Yamamoto, J. Exp. Biol. *50*:431–443, 1969. Mechanical properties of earthworm muscle.
176. Hidaka, T., T. Osa, and B. M. Twarog, J. Physiol. *192*:869–877, 1967. Action of 5-HT on *Mytilus* muscle.
177. Hill, A. V., Sci. Prog. *38*:209–230, 1950. Dimensions of animals and muscle dynamics.
178. Hill, R. B., M. J. Greenberg, H. Irisawa, and H. Nomura, J. Exp. Zool. *174*:331–348, 1970. Electromechanical couplings in muscle of snail.
179. Hirose, H., et al., Ann. Otol. *78*:297–306, 1969. Time constants of laryngeal muscles of cat.
179a. Hocking, B., Trans. Roy. Entomol. Soc. Lond. *104*:223–345, 1953. Intrinsic range and speed of flight of insects.
180. Hoekman, T. H., Ph.D. Thesis, University of Illinois, 1968. Mechanical properties of muscles of chicken.
181. Hoffman-Berling, H., Biochim. Biophys. Acta *27*:247–255, 1958. Contraction properties of *Vorticella*.
182. Holman, M. E., pp. 1665–1708. *In* Handbook of Physiology, Sec. 6, Vol. IV, edited by C. F. Code. Amer. Physiol. Soc., Washington, D.C., 1968. Electrophysiology of visceral muscle.
183. Hoogland, P. L., H. C. Freeman, B. Truscote, and F. E. Waddell, J. Fish. Res. Bd. Canad. *18*:501–512, 1961. Amino acid composition of cod tropomyosin.
184. Hoyle, G., Comparative Physiology of Nervous Control of Muscular Contractions. Cambridge University Press, Cambridge, 1957. 147 pp.
185. Hoyle, G., J. Exp. Zool. *167*:471–486, 1968. Neuromuscular physiology of eyestalk muscle of a crab.
186. Hoyle, G., and P. A. McNeill, J. Exp. Zool. *167*:487–522, 523–550, 1968. Ultrastructure of eyestalk muscles in a crab.
187. Hoyle, G., and T. Smyth, Comp. Biochem. Physiol. *10*:291–314, 1963. Neuromuscular physiology of a barnacle.
188. Hubbard, J. I., S. F. Jones, and E. M. Landau, J. Physiol. *196*:75–86, 1968. End-plate potentials in rat diaphragm.
189. Hubbard, J. I., et al., J. Physiol. *194*:355–380, 1968. Effects of varying calcium and magnesium on spontaneous release of transmitter from motor nerve terminals.
190. Hudson, R. C. L., J. Exp. Biol. *50*:47–67, 1969. Polyneuronal innervation of fast muscles in fish *Cottus*.
191. Huik, P., I. Jirsmanova, L. Vyklicky, and J. Zelena, J. Physiol. *193*: 309–325, 1967. Cross-innervation of chick muscles.
191a. Huszar, G., and M. Elzinga, J. Biol. Chem. *247*:745–753, 1972. Methylated and non-methylated histidines in myosins of skeletal and cardiac muscle.
192. Huxley, A. F., and R. M. Simmons, Nature *233*:533–538, 1971. Quick changes in length, tension redevelopment.
192a. Huxley, A. F., and R. Taylor, J. Physiol. *144*:426–441, 1958. Inward spread of excitation in striated muscle fibers.
193. Huxley, H. E., J. Molec. Biol. *7*:281–308, 1963. Electron microscopy of protein filaments from muscle.
194. Huxley, H. E., Circulation *24*:328–335, 1961. Contractile structure of cardiac and skeletal muscle.
195. Huxley, H. E., Science *164*:1356–1366, 1969. Theory of muscle contraction.
196. Huxley, H. E., Proc. Roy. Soc. Lond. B *178*:131–149, 1971. Structural basis of muscular contraction.
197. Ikemoto, N., and S. Kawaguti, Biol. J. Okayama Univ. *9*:81–126, 1963; *14*:21–33, 1968. Electron microscopy of muscles of earthworm and horseshoe crab.
198. Ito, Y., and N. Tashiro, J. Exp. Biol. *53*:597–609, 1970. Calcium spikes in a polychaete muscle.
199. Ito, Y., H. Kuriyama, and N. Tashiro, J. Exp. Biol. *51*:363–375, 1969; *54*:167–186, 1971. Pharmacology of earthworm muscle.
200. Jahromi, S. S., and H. L. Atwood, Canad. J. Zool. *45*:601–606, 1967. Ultrastructure of crayfish muscle.
201. Jahromi, S. S., and H. L. Atwood, Experientia *25*:1046, 1969. Tension and structure in crayfish muscle.
202. Jahromi, S. S., and H. L. Atwood, J. Exp. Zool. *171*:25–37, 1964. Speeds of various crayfish muscles.
202a. Jahromi, S. S., and H. L. Atwood, J. Exp. Zool. *176*:475–486, 1971. Structure and function of muscle fibers in lobster leg.
203. Jansen, P., J. K. S. Andersen, and Y. Loyning, Acta Physiol. Scand. *57*:167–179, 1963. Fast and slow muscles in hagfish.
204. Jewell, B. R., J. Physiol. *149*:154–177, 1959. Phasic and tonic responses of muscle of *Mytilus*.
205. Jewell, B. R., and J. C. Ruegg, Proc. Roy. Soc. Lond. B *164*:428–459, 1966. Oscillations of insect fibrillar muscles.
205a. Job, D. D., Amer. J. Physiol. *217*:1534–1541, 1969. Ionic fluxes in slow potentials of intestine.
206. Jobsis, I. F., and M. J. O'Connor, Biochem. Biophys. Res. Comm. *25*:246–252, 1966. Calcium release and reabsorption in sartorius muscle.
207. Johnson, W. H., J. S. Kahn, and A. G. Szent-Györgyi, Science *130*: 160–161, 1959. Paramyosin and contraction of catch muscles.
208. Josephson, R. K., and H. Y. Elder, Biol. Bull. *135*:409, 1968. Muscle of sound production in katydid.
208a. Josephson, R. K., and R. C. Halverson, Biol. Bull. *141*:411–433, 1971. High frequency muscles of sound production in katydid.
209. Kammer, A. E., J. Exp. Biol. *48*:89–109, 1968. Motor patterns in Lepidoptera.
210. Katz, B., Biol. Rev. *24*:1–20, 1949. Neuromuscular transmission in invertebrates.
210a. Katz, B., Nerve, Muscle, and Synapse. McGraw-Hill Book Co., New York, 1966. 193 pp.
211. Katz, B., and R. Miledi, J. Physiol. *203*:689–706, 1969; *also* Nature *207*:1096–1097, 1965; *215*:651, 1967. Spontaneous and evoked activity at motor nerve endings.
212. Katz, B., and R. Miledi, Proc. Roy. Soc. Lond. B *161*:453–482, 483–495, 496–503, 1971. Miniature junction potentials.
213. Kendrick-Jones, J., C. Cohen, A. G. Szent-Györgyi, and W. Longley, Science *163*:1196–1198, 1969. Biophysics of paramyosin.
214. Kendrick-Jones, J., W. Lehman, and A. G. Szent-Györgyi, J. Molec. Biol. *54*:313–326, 1971. Ultrastructure of paramyosin.
214a. Kendrick-Jones, J., et al., J. Molec. Biol., 1972 (in press). Myosin-linked regulation of calcium, molluscan muscle.
215. Kennedy, D., and W. H. Evoy, J. Gen. Physiol. *49*:457–468, 1966. Pre- and post-synaptic inhibition in crustaceans.
216. Kerkut, G. A., and L. D. Leake, Comp. Biochem. Physiol. *17*:623–633, 1966. Effect of drugs on snail muscle.
217. Kerkut, G. A., and R. J. Walker, Comp. Biochem. Physiol. *17*:435–454, 1966. Pharmacology of leg muscles of cockroach.
218. Kishimoto, U., Comp. Biochem. Physiol. *2*:81–89, 1961. ATPase activity of myosins.
219. Kobayashi, M., Amer. J. Physiol. *206*:205–210, 1964; *208*:715–719, 1965; *216*:1279–1285, 1969. Electrophysiology of cat ureter.
220. Kobayashi, M., C. L. Prosser, and T. Nagai, Amer. J. Physiol. *213*: 275–286, 1967. Electrical properties of intestinal muscle, measured intracellularly and extracellularly.
220a. Kundrat, E., and F. A. Pepe, J. Cell Biol. *48*:340–347, 1971. M-protein.
221. Kuriyama, H., and F. Mekata, J. Physiol. *212*:667–683, 1971. Longitudinal muscle of guinea pig rectum.
222. Kuriyama, H., T. Osa, and H. Tasaki, J. Gen. Physiol. *55*:48–62, 1970. Electrophysiology of stomach.
223. Kuriyama, H., T. Osa, and N. Toida, J. Physiol. *191*:239–255, 1967. Electrophysiology of intestinal muscle.
224. Kuriyama, H., and T. Tomita, J. Gen. Physiol. *55*:147–162, 1970; J. Physiol. *178*:270–289, 1965. Response of ureter and of taenia coli to applied currents.
224a. Lang, F., H. L. Atwood, and W. A. Morin, Z. Zellforsch., *127*:189–200, 1972. Electron microscopy of nerve supply to crayfish opener muscle.
225. Lang, F., A. Sutterlin, and C. L. Prosser, Comp. Biochem. Physiol. *32*:615–628, 1970. Physiology of leg muscle in *Paralithodes*.
226. Leenders, H. J., Pflüger. Arch. *295*:127–135, 1967. Mechanical properties of *Mytilus* catch muscle.
227. Leenders, H. J., J. Physiol. *192*:681–693, 1967; *also* Comp. Biochem. Physiol. *31*:187–196, 1969. Calcium coupling and ATPase activity in catch muscles.
227a. Lehman, W., J. Kendrick-Jones, and A. G. Szent-Gyorgyi, J. Molec. Biol., 1972 (in press). Comparative studies of Ca regulation in muscles.
227b. Liu, J. C. L. Prosser, and D. D. Job. Amer. J. Physiol. *217*:1542–1547, 1969. Ionic dependence of slow waves and spikes in intestinal muscle.
228. Lowey, S., et al., J. Molec. Biol. *4*:293–308, 1962; *42*:1–29, 1969. Structure of myosin.
228a. Lowey, S., and D. Risby, Nature *234*:81–85, 1971. Light myosin chains from fast and slow muscles.
229. Lowy, J., and B. M. Millman, Phil. Trans. Roy. Soc. *246*:105–148, 1963. Contractile mechanism of *Mytilus* muscle.
229a. Luff, A. R., and H. L. Atwood, Amer. J. Physiol. *222*:1435–1440, 1972. Mouse muscle speeds.
230. Lund, G. F., and I. Christensen, Amer. J. Physiol. *217*:1369–1374, 1969. Electrical stimulation of the esophagus.

231. Lymn, R. W., and E. W. Taylor, Biochemistry *9*:2975–2983, 1970. ATPase activity of myosin.
231a. MacLennan, D. H., et al., J. Biol. Chem. *246*:2702–2710, 1971; *also* Proc. Nat. Acad. Sci. *68*:1231–1235, 1971. Calcium binding and Ca-ATPase of sarcoplasmic reticulum.
232. Marechal, G., M. Goffart, and X. Aubert, Arch. Int. Physiol. Biochem. *71*:236–240, 1963. Muscle speeds in sloths.
233. Marshall, J. M., Amer. J. Physiol. *197*:935–942, 1959; *204*:732–738, 1963. Electrical activity of uterine smooth muscle.
234. Maruyama, K., Biochim. Biophys. Acta *102*:542–548, 1965. Actinin.
235. Maruyama, K., Sci. Papers Coll. Gen. Ed., Univ. of Tokyo *6*:95–111, 1956. Properties of contractile proteins from a sea anemone.
236. Mashima, H., and T. Yoshida, Japan. J. Physiol. *15*:463–477, 1965. Tension-length curves in taenia coli.
237. Matsumoto, Y., and B. C. Abbott, Comp. Biochem. Physiol. *26*:927–936, 1968. Folding muscle fibers of *Golfingia.*
238. Matyushkin, D. P., Fed. Proc. Transl. Suppl. *22*:T728–T731, 1963. Innervation of tonic muscle fibers in oculomotor system.
239. Maynard, D. M., and W. Burke, Comp. Biochem. Physiol. *38A*:339–350, 1971. Tension development by adductor of *Tridacna.*
240. McCann, F. V., and E. G. Boettiger, J. Gen. Physiol. *45*:125–142, 1961. Electrophysiology of fibrillar flight muscles.
241. McCoy, E. J., and R. D. Baker, Proc. Soc. Exp. Biol. Med. *127*:562–563, 1968. Intestinal slow waves.
242. McLennan, H., Synaptic Transmission, 2nd Ed. W. B. Saunders Co., Philadelphia, 1970. 178 pp.
242a. McNeill, P., et al., J. Exp. Zool. *179*:395–416, 1972. Fine structure of muscles of mantis shrimp.
243. Meiss, R. A., Amer. J. Physiol. *220*:2000–2007, 1971. Some mechanical properties of cat intestinal muscle.
244. Mendelson, M., J. Cell Biol. *42*:548–563, 1969. Properties of a very fast lobster muscle.
245. Miledi, R., and L. T. Potter, Nature *233*:599–603, 1971. Acetylcholine receptors in muscle fibers.
246. Miledi, R., and C. R. Slater, Proc. Roy. Soc. Lond. B *174*:253–269, 1969. Ultrastructure of denervated skeletal muscle.
247. Miledi, R., and E. Stefani, Nature *222*:569–571, 1969. Non-selective re-innervation of slow and fast muscle fibers in rat.
248. Mill, P. I., and M. F. Krapp, J. Cell. Sci. 7:233–262, 1970. Fine structure of earthworm muscle.
248a. Miller, M. D., and J. M. Marshall, Amer. J. Physiol. *209*:859–865, 1965. Catecholamine receptors in uterus.
249. Miller, P. L., Nature *221*:171–173, 1969. Inhibitory innervation of insect spiracles.
250. Millman, B. M., J. Physiol. *173*:238–262, 1964; *also* Proc. Roy. Soc. Lond. B *160*:525–536, 1964. Contraction of opaque adductor of oyster.
251. Millman, B. M., and G. F. Elliott, Nature *206*:824–825, 1965. X-ray diffraction of molluscan muscle.
252. Millward, G. R., and E. F. Woods, J. Molec. Biol. *52*:585–588, 1970. Crystals of tropomyosin.
253. Mommaerts, W. F. H. M., Physiol. Rev. *49*:427–508, 1969. Energetics of muscular contraction.
254. Mommaerts, W. F. H. M., and A. Wallner, J. Physiol. *193*:343–357, 1967. Energetics in frog sartorius.
255. Motokizawa, F., J. P. Reuben, and H. Grundfest, J. Gen. Physiol. *54*:437–461, 1969. Ion properties of inhibitory receptors in lobster muscle.
256. Nagai, T., J. Insect Physiol. *16*:437–448, 1970. Properties of insect visceral muscle.
257. Nagai, T., and B. E. Brown, J. Insect Physiol. *15*:2151–2167, 1969. Properties of cockroach visceral muscle.
258. Nagai, T., and C. L. Prosser, Amer. J. Physiol. *204*:915–924, 1963. Electrical parameters of smooth muscles.
259. Nakajima, Y., Tissue and Cell *1*:229–246, 1969. Fine structure of red and white muscle in snake fish.
260. Nass, M. M. K., Develop. Biol. *4*:289–320, 1962. Developmental changes in actomyosin.
261. Nauss, K. M., and R. E. Davies, Biochem. Zeitschr. *345*:173–187, 1966. Energetics of catch muscle of *Mytilus.*
262. Nilsson, S., and R. Fange, Comp. Biochem. Physiol. *30*:691–694, 1969. Neural control of stomach of cod.
263. Ohnesorge, F. K., and R. Rauch, Z. vergl. Physiol. *58*:153–170, 1968. Peristalsis in intestine of *Tinca.*
264. Ozeki, M., and H. Grundfest, Science *155*:478–481, 1967. Ionic mechanisms in crayfish muscle.
265. Page, S. G., J. Cell Biol. *26*:477–497, 1965. Ultrastructure of frog fast and slow fibers.
266. Pak, M. J., and B. C. Abbott, personal communication. Electromechanical properties of toadfish sonic muscle.
267. Panner, B. J., and C. R. Honig, J. Cell Biol. *35*:303–321, 1967. Filament ultrastructure of smooth muscle.
268. Papasova, M. P., T. Nagai, and C. L. Prosser, Amer. J. Physiol. *214*:695–702, 1968. Two component slow waves in cat stomach.
269. Parmley, W. W., L. A. Yeatman, and E. H. Sonnenblick, Amer. J. Physiol. *220*:546–550, 1970. Force-velocity relations, striated muscle.
270. Parnas, I., B. C. Abbott, B. Shapiro, and F. Lang, Comp. Biochem. Physiol. *26*:467–478, 1968. Neuromuscular system of *Limulus* leg muscle.
271. Parnas, I., and H. L. Atwood, Comp. Biochem. Physiol. *18*:701–723, 1966. Neuromuscular system of abdominal muscles of crayfish.
272. Parnas, I., and D. Dagan, Comp. Biochem. Physiol. *28*:359–369, 1969. Abdominal muscles of prawn *Palaemon.*
273. Paul, W. M., E. E. Daniel, C. M. Kay, and G. Monckton, editors, University of Alberta Medical School Symposium: Muscle. Pergamon Press, New York, 1965. 584 pp.
274. Peachey, L. D., Biochem. Biophys. Cytol. Suppl. *10*:159–176, 1961. Ultrastructure of muscle of *Amphioxus.*
275. Peachey, L. D., Amer. Zool. 7:505, 1967. Membrane system in crab muscle.
275a. Peachey, L. D., Ann. Rev. Physiol. *30*:401–440, 1968. Ultrastructure of muscle, correlations with contraction.
276. Peachey, L. D., and A. F. Huxley, J. Cell Biol. *13*:177–180, 1962. Structural identification of fast and slow fibers in frog.
277. Pearson, K. G., and S. J. Bergman, J. Exp. Biol. *50*:445–471, 1969. Neuromuscular inhibition in insects.
278. Pearson, K. G., and J. F. Iles, J. Exp. Biol. *54*:215–232, 1971. Neuromuscular transmission in cockroach.
279. Pepe, F. A., J. Cell Biol. *28*:505–525, 1966; *also* J. Molec. Biol. *27*:203–225, 1967; *also* pp. 75–96 *in* Progress in Biophysics and Molecular Biology, edited by J. A. V. Butler and D. Noble, Pergamon Press, Oxford, 1970; *also* pp. 323–353 *in* Fine Structure of Proteins and Nucleic Acids (Biological Macromolecular Ser. Vol. 4), edited by S. Timasheff and G. Fasman, Marcel Dekker, Inc., New York, 1970. Structural components of the striated muscle fibril.
279a. Perrie, W. T., and S. V. Perry, Biochem. J. *119*:31–38, 1970. Low molecular weight components of myosin.
280. Piek, T., and P. Mantel, Comp. Biochem. Physiol. *34*:935–951, 1970. Electrical activity in insect muscle.
281. Pilar, G., J. Gen. Physiol. *50*:2289–2300, 1967. Electromechanical properties of slow fibers in extraocular muscles.
282. Podolsky, R. D., and L. E. Teickholz, J. Physiol. *211*:19–35, 1970. Ca^{++} and contractions in skinned fibers.
283. Prewitt, M. A., and B. Salafsky, Amer. J. Physiol. *213*:295–300, 1967; *218*:69–74, 1970. Changes in fast and slow muscles after cross-innervation.
284. Pringle, J. W. S., Insect Flight. Cambridge University Press, Cambridge, 1957. 133 pp.
285. Pringle, J. W. S., Prog. Biophys. *17*:1–60, 1967; *also* J. Gen. Physiol. *50*:139–156, 1967. Contractile mechanisms in insect fibrillar muscle.
286. Proske, U., and P. Vaughan, J. Physiol. *199*:495–510, 1968. Histological correlations with functions, lizard skeletal muscle.
287. Prosser, C. L., J. Cell. Comp. Physiol. *44*:247–254, 1954. Activation of muscle of *Thyone.*
288. Prosser, C. L., Comp. Biochem. Physiol. *6*:69–74, 1962; Z. vergl. Physiol. *54*:109–120, 1967. Contractions in sponge.
289. Prosser, C. L., and A. Bortoff, pp. 2025–2050. *In* Handbook of Physiology, Sec. 6, Vol. IV, edited by C. F. Code. Amer. Physiol. Soc., Washington, D.C., 1968.
290. Prosser, C. L., G. Burnstock, and M. Holman, Physiol. Rev. (Suppl. 5) *42*:193–212, 1962. Conduction in non-striated muscles.
291. Prosser, C. L., pp. 133–149. *In* Invertebrate Nervous Systems, edited by C. A. G. Wiersma. University of Chicago Press, Chicago, 1969. Comparative physiology of non-striated muscle.
292. Prosser, C. L., H. J. Curtis, and D. Travis, J. Cell. Comp. Physiol. *38*:299–319, 1951; *also* Prosser, C. L., and C. E. Melton, J. Cell. Comp. Physiol. *44*:255–275, 1954; *also* Prosser, C. L., and N. Sperelakis, J. Cell. Comp. Physiol. *54*:129–133, 1959. Nervous conduction in smooth muscle of Phascolosoma (*Golfingia*).
293. Prosser, C. L., R. Nystrom, and T. Nagai, Comp. Biochem. Physiol. *14*:53–70, 1965. Intestinal muscle of invertebrates.
294. Prosser, C. L., and N. S. Rafferty, Amer. J. Physiol. *187*:546–548, 1956. Electrical activity in chick amnion.
295. Prosser, C. L., C. L. Ralph, and W. W. Steinberger, J. Cell. Comp. Physiol. *54*:135–146, 1959. Responses to stretch in invertebrate muscles.
296. Rahamimoff, R., J. Physiol. *195*:471–480, 1968. Frog neuromuscular junction; ions, glycerol effects.
297. Raj, B. S. D., and M. J. Cohen, Naturwissenschaften *51*:224–225, 1964. Structure-function correlation in crab muscle.
298. Rathmayer, W., Comp. Biochem. Physiol. *14*:673–687, 1965. Neuromuscular transmission in spider.
299. Reedy, M. K., Amer. Zool. 7:465–481, 1967. Ultrastructure of insect flight muscles.
300. Rees, M. K., and M. Young, J. Biol. Chem. *242*:4449–4458, 1967. Properties of actin.
301. Reger, J. F., Biochem. Biophys. Cytol. Suppl. *10*:111–121, 1961. Fine structure of ocular muscles of *Fundulus.*
302. Reis, D. J., G. F. Wooten, and M. Hollenberg, Amer. J. Physiol. *213*:592–596, 1967. Enzymes in red and white skeletal muscle.
303. Reuben, J. P., P. W. Brandt, H. Garcia, and H. Grundfest, Amer. Zool. 7:623–645, 1967. Excitation-contraction coupling in crayfish.
304. Revel, J. P., J. Cell Biol. *12*:571–588, 1962. Sarcoplasmic reticulum of fast muscle in bat.

305. Rice, R. V., et al., Nature *231*:242–246, 1970. Thick filaments in mammalian smooth muscle.
306. Ridge, R. M., J. Physiol. *217*:393–418, 1971. Contraction times of snake muscle fibers.
307. Roach, D. K., Comp. Biochem. Physiol. *24*:865–878, 1968. Rhythmic activity in digestive tract of snail.
307a. Roberts, B. L., J. Marine Biol. Assoc. U.K. *49*:357–378, 1969. Properties of dogfish muscles.
308. Robison, G. A., R. W. Butcher, and E. W. Sutherland, Ann. N.Y. Acad. Sci. *139*:703–723, 1966. Adenyl cyclase as an adrenergic receptor.
309. Robison, G. A., and E. W. Sutherland, Circ. Res. *26* (Suppl.):147–161, 1970. Role of cyclic AMP in adrenergic responses.
310. Robson, R. M., D. E. Goll, N. Arakawa, and M. H. Stromer, Biochim. Biophys. Acta *200*:296–318, 1970. Properties of alpha-actinin.
311. Roeder, K. D., Biol. Bull. *100*:95–106, 1951. Potentials in flight muscles of insects.
312. Rohlich, P., J. Ultrastruct. Res. *7*:399–408, 1962. Fine structure of muscle of leech.
313. Romanul, F. C. A., Arch. Neurol. *11*:355–368, 369–378, 1964. Enzymes in striated muscle.
314. Romanul, F. C. A., and J. P. Van der Meulen, Arch. Neurol. *17*:387–402, 1967. Slow and fast muscles after cross-innervation.
315. Rosenbluth, J., J. Cell Biol. *25*:495–515, 1965; *26*:579–591, 1965. Ultrastructure of muscle of *Ascaris*.
316. Rosenbluth, J., J. Cell Biol. *42*:534–547, 1969. Ultrastructure of a fast crustacean muscle.
317. Ross, D. M., J. Exp. Biol. *37*:732–752, 753–774, 1960. Effects of ions and drugs on muscle of a sea anemone.
318. Ruegg, J. C., Biochem. Biophys. Res. Comm. *6*:24–28, 1961; *also* Proc. Roy. Soc. Lond. B *154*:209–249, 1961. Proteins in tonic muscles of lamellibranchs.
319. Ruegg, J. C., et al., Proc. Roy. Soc. Lond. B *158*:156–176, 177–195, 1963. Contractile properties of tonic molluscan muscles.
320. Ruegg, J. C., Amer. Zool. 7:457–464, 1967; *also* Experientia *24*:529–536, 1968; *also* Ruegg, J. C., and R. T. Tregear, Proc. Roy. Soc. Lond. B *165*:497–512, 1966. Oscillatory flight muscles of insects.
321. Ruegg, J. C., Physiol. Rev. *51*:201–248, 1971. Smooth muscle tone.
322. Salanki, J., and L. Hiripi, Comp. Biochem. Physiol. *32*:629–636, 1970. Transmission in adductor muscles of *Anodonta*.
323. Salmons, S., and G. Vrbova, J. Physiol. *201*:535–549, 1969. Effect of exercise on mammalian fast and slow muscles.
324. Samaha, F. J., L. Guth, and R. W. Albers, Exp. Neurol. *27*:276–282, 1970. Neural regulation of chemical development in muscles.
325. Sarkar, S., F. A. Sreter, and J. Gergely, Proc. Nat. Acad. Sci. *68*:946–950, 1971. White, red, and cardiac myosin of rabbit.
326. Sato, M., M. Tamasige, and M. Ozeki, Japan. J. Physiol. *10*:85–98, 1960. Electrical activity of pharynx of snail.
327. Schmitt, F. O., R. S. Bear, C. E. Hall, and M. A. Jakus, Ann. N.Y. Acad. Sci. *47*:799–812, 1947. Ultrastructure of paramyosin.
328. Schoenberg, C. F., Tissue and Cell *1*:83–96, 1969. Electron microscopy of myosin filaments in chicken gizzard.
329. Scopes, R. K., and I. F. Perry, Biochim. Biophys. Acta *236*:409–415, 1971. Subunits of muscle proteins.
330. Sexton, A. W., and J. W. Gersten, Science *157*:199, 1967. Mechanical properties of red and white muscle.
331. Sherman, R. G., and H. L. Atwood, J. Exp. Zool. *176*:461–474, 1971. Neuromuscular physiology of lobster.
331a. Sherman, R. G., and H. L. Atwood, Science *171*:1248–1250, 1971. Long-term neuromuscular facilitation in crustaceans.
332. Skoglund, C. R., Biochem. Biophys. Cytol. Suppl. *10*:187–200, 1961. Sound producing muscles in toadfish.
333. Smith, D. S., Rev. Canad. Biol. *21*:279–301, 1962. Ultrastructure of insect flight muscle.
334. Smith, D. S., B. L. Gupta, and N. Smith, J. Cell. Sci. *1*:49–57, 1966. Organization of myofilaments in insect visceral muscle.
334a. Sotovalta, O., Biol. Bull. *104*:439–444, 1953. Flight tone and thoracic vibration frequency in midges.
334b. Sparrow, M., personal communication.
335. Spirito, C. P., Z. vergl. Physiol. *68*:211–228, 1970. Control of leg muscles in *Uca*.
336. Sreter, F. A., Arch. Biochem. *134*:25–33, 1969. ATPase in red and white muscle.
337. Stephens, N. L., and B. S. Chiu, Amer. J. Physiol. *219*:1001–1008, 1970. Mechanical properties of tracheal muscle.
338. Stephens, N. L., and U. Kramer, Amer. J. Physiol. *220*:1890-1895, 1971. Series elastic component of tracheal smooth muscle.
339. Stephens, N. L., E. Kroeger, and J. A. Mehta, J. Appl. Physiol. *26*: 685–692, 1969. Force-velocity properties of tracheal smooth muscle.
340. Stossel, W., and E. Zebe, Pflüger. Arch. *302*:38–56, 1968. Control of relaxation in skeletal muscle.
340a. Stromer, M., D. Hartshorne, H. Mueller, and R. V. Rice, J. Cell Biol. *40*:167–178, 1969. Effects of protein fractions on Z- and M-line reconstitution.
341. Sugi, H., and R. Ochi, J. Gen. Physiol. *50*:2145–2176, 1967. Spread of contraction in crayfish fibers.
342. Swanson, C. J., Z. vergl. Physiol. *74*:403–410, 1971; *also* Nature *232*:122, 1971. Paramyosin in Nematomorpha.
343. Syrovy, I., and E. Gutmann, Nature *213*:937–938, 1967. Biochemical differentiation in fast and slow muscles of chick.
344. Szent-Györgyi, A. G., C. Cohen, and J. Kendrick-Jones, J. Molec. Biol. *56*:239–258, 1971. Ultrastructure of paramyosin.
345. Szurszewski, J. H., L. R. Elveback, and C. F. Code, Amer. J. Physiol. *218*:1468–1473, 1970. Electrical slow waves in dog intestine.
346. Takahashi, K., Annot. Zool. Japan. *33*:67–84, 1960. Nervous control in catch muscle of *Mytilus*.
347. Takeuchi, A., J. Cell. Comp. Physiol. *54*:211–220, 1959. Neuromuscular transmission in fish.
348. Takeuchi, N., J. Physiol. *167*:128–155, 1963. Conductance changes in motor endplates.
349. Takeuchi, A., and N. Takeuchi, Fed. Proc. *26*:1633–1638, 1967. Effect of GABA on synapses.
350. Takeuchi, A., and N. Takeuchi, J. Physiol. *170*:296–317, 1964. Effect of glutamate on crayfish muscle.
351. Takeuchi, A., and N. Takeuchi, J. Physiol. *217*:341–358, 1971. Crayfish muscle conductance.
351a. Tashiro, N., J. Exp. Biol. *55*:101–110, 1971. Mechanical properties of earthworm muscles.
351b. Tashiro, N., and T. Yamamoto, J. Exp. Biol. *55*:111–122, 1971. Contraction in earthworm muscle.
351c. Teravainen, H., J. Neurophysiol. *34*:954–973, 1971. Structure and function of lamprey muscles.
352. Thoai, N., and Y. Robin, pp. 163–203. *In* Chemical Zoology, Vol. 4, edited by M. Florkin and B. T. Scheer. Academic Press, New York, 1969. Guanidine compounds and phosphagens.
353. Tomita, T. J., Theor. Biol. *12*:216–227, 1966. Capacity and resistance in mammalian smooth muscle.
354. Twarog, B. M., J. Physiol. *152*:220–235, 236–242, 1960. Innervation and contraction in molluscan smooth muscle.
355. Twarog, B. M., Life Sci. *5*:1201–1213, 1966; *also* J. Physiol. *192*: 847–856, 857–868, 1967. Effects of calcium and 5-HT on molluscan muscles.
356. Twarog, B. M., J. Gen. Physiol. *50*:157–169, 1967. Review of regulation of catch in molluscan muscles.
357. Twarog, B. M., and T. Hidaka, J. Gen. Physiol. *57*:252, 1971. Effect of 5-HT on electrical properties of molluscan muscle fibers.
357a. Uchizono, K., Nature *214*:833–834, 1967. Crayfish vesicles.
358. Usherwood, P. N. R., J. Exp. Biol. *49*:201–222, 1968. Inhibitory innervation of muscle in insects.
359. Usherwood, P. N. R., Nature *223*:411–413, 1969. Glutamate stimulation of insect muscle.
360. Usherwood, P. N. R., and H. Grundfest, Science *143*:817–818, 1964; *also* J. Neurophysiol. *28*:497–518, 1965. Inhibition in grasshopper muscle.
361. Usherwood, P. N. R., and P. Machili, J. Exp. Biol. *49*:341–361, 1968. Pharmacology of excitatory junctions in locust muscle.
362. Usherwood, P. N. R., and H. I. Runion, J. Exp. Biol. *52*:39–58, 1970. Mechanical properties of tibial muscle of locust.
363. Van der Kloot, W. S., Comp. Biochem. Physiol. *9*:317–334, 1963. Control of spiracular muscles in silk moth.
364. Van der Kloot, W. S., Comp. Biochem. Physiol. *15*:547–565, 1965. Uptake of calcium by vesicles of lobster muscles.
365. Van Harn, G. L., Amer. J. Physiol. *215*:1351–1358, 1968. Frog stomach smooth muscle.
366. Vrbova, G., J. Physiol. *169*:513–526, 1963. Activation of striated muscle by motorneurones.
367. Walcott, B., and M. Burrows, J. Insect Physiol. *15*:1855–1872, 1969. Abdominal air-guide retractor muscles, giant bug *Lethocerus*.
367a. Walker, T. J., Ann. Entom. Soc. Amer. *62*:752–762, 1969. Acoustical behavior of cricket *Orocharis*.
368. Washizu, Y., Comp. Biochem. Physiol. *20*:641–646, 1967. Electrical properties of leech muscle.
369. Watts, D. C., B. Moreland, E. C. Tatshell, and L. H. Bannister, Comp. Biochem. Physiol. *25*:553–558, 1968. Phosphoarginine in protozoan *Stentor*.
370. Weber, A., pp. 203–254. *In* Current Topics in Bioenergetics, edited by D. R. Sandi. Academic Press, New York, 1966. Calcium transport and relaxation.
371. Weber, A., R. Herz, and I. Reiss, J. Gen. Physiol. *46*:679–702, 1963. Relaxing factor in sarcoplasmic reticulum.
372. Weis-Fogh, T., J. Exp. Biol. *33*:668–684, 1956; *also* Phil. Trans. Roy. Soc. Lond. B *239*:415–510, 553–584, 1956. Aerodynamics of insect flight.
373. Welsh, J., and M. Moorhead, J. Neurochem. *6*:146–169, 1960. 5-Hydroxytryptamine in molluscan tissues.
374. Wiersma, C. A. G., pp. 143–159. *In* Recent Advances in Invertebrate Physiology, edited by B. Scheer. University of Oregon Press, 1957. Neuromuscular transmission in crustaceans.
375. Wiersma, C. A. G., pp. 191–240. *In* The Physiology of Crustacea, Vol. II, edited by T. Waterman. Academic Press, New York, 1961. Neuromuscular system of crustaceans.
376. Wilkie, D. R., Prog. Biophys. *4*:288–324, 1954; *also* J. Physiol. *134*:

527–530, 1956; *195*:157–183, 1968. Mechanical properties of frog muscle.

376a. Wilkie, D. R., Muscle. St. Martin's Press, Inc., New York, 1968. 63 pp.

377. Wilson, D. F., Comp. Biochem. Physiol. *29*:703–715, 1969. Fast and slow contractions in pelecypod mantle.

378. Wilson, D. M., J. Exp. Biol. *37*:57–72, 1960. Nervous control of muscle in cephalopods.

379. Wilson, D. M., D. O. Smith, and P. Dempster, Amer. J. Physiol. *218*: 916–922, 1970. Length-tension hysteresis in arthropod muscle.

380. Winegrad, S., J. Gen. Physiol. *51*:65–83, 1968; *55*:77–88, 1970. Intracellular localization of calcium in contraction.

381. Wood, D. W., J. Exp. Biol. *35*:850–861, 1958. Fast and slow muscles in stick insect *Carausius*.

382. Wood, J. D., Comp. Biochem. Physiol. *30*:813–824, 1969. Electrical and mechanical activity of stomach of squid.

382a. Wood, J. D., Amer. J. Physiol. *219*:159–169, 1970. Electrical activity from neurons in Auerbach's plexus.

383. Wood, J. D., and W. E. Perkins, Amer. J. Physiol. *218*:762–768, 1970. Interaction between longitudinal and circular axes of the intestine.

384. Wood, M. E., and G. Burnstock, Comp. Biochem. Physiol. *22*:755–766, 1967. Innervation of lung of toad.

384a. Wright, E. B., and P. D. Coleman, J. Cell. Comp. Physiol. *43*:133–164, 1954. Excitation of motor axons of crayfish.

384b. Yasui, V., F. Fuchs, and F. M. Briggs, J. Biol. Chem. *243*:735–742, 1968. Calcium and troponin control of contractions.

385. Zebe, E., W. Meinrenken, and J. C. Ruegg, Z. Zellforsch. *87*:603–621, 1968. Contractions of glycerinated muscle.

386. Zolovick, A. J., R. L. Norman, and M. R. Fedde, Amer. J. Physiol. *219*:654–657, 1970. Membrane constants of rat diaphragm.

387. Zs-Nagy, I., and E. Labos, Ann. Biol. Tihany *36*:123–133, 1969. Ultrastructure of adductor muscles in *Anodonta*.

CHAPTER 17

ELECTRIC ORGANS AND ELECTRORECEPTORS

By C. Ladd Prosser

The first bioelectric potentials observed by man were the discharges from electric fishes; today the large flat cells of electric organs are providing valuable information regarding the nature of biopotentials. The electric catfish *Malapterurus* was figured on Egyptian tombs (ca. 2750 BC),[23] and the Romans named *Torpedo,* an electric ray. "Electric therapy" with electric fish was recommended by Galen and was used by eighteenth century European physicians. The large electric "eel" *Electrophorus* was brought to Europe by South American explorers, and there was speculation as to the nature of its shock. Near the end of the eighteenth century several observers suggested that the shock was similar to lightning or to the electrostatic discharge from a Leyden jar. In 1773, John Walsh showed that the shock is conducted through metals, not through glass or air. The Cambridge physicist Cavendish built a model of *Torpedo,* from which he deduced the distinction between potential difference and quantity of electricity. Faraday used a galvanometer to measure the discharge of *Electrophorus.* He prophetically remarked that if the nature of the electric discharge were understood, one might "reconvert the electric into the nervous force."

Electric organs present a prime example of evolutionary convergence. Muscles, and in two cases nerves, have been modified as electric organs. Electric organs have apparently evolved six times, and functional properties are adapted to particular ecological requirements.[4,8] A few families of fish are strongly electric—marine electric rays or torpedos (and perhaps also stargazers), and freshwater electric "eels" and "catfish"; they discharge at high voltage on reflex demand. Others—marine skates, freshwater mormyrids, gymnarchids, gymnotids, and some others—are weakly electric; with the exception of the skates, they generally discharge more or less continuously, in pulse trains of variable frequency and regularity.

PRODUCTION OF ELECTRIC DISCHARGES

Electric organs are of so many types that a general term "electrocyte" is useful to designate cells, from whatever embryological source, that are specialized for generation of electric discharges. Large, flat electrocytes have been called electroplaques.

The following enumeration, drawn largely from descriptions published by Bennett,[4,8] briefly summarizes the properties of several different types of electric organ (Fig. 17–1).

In the strongly electric *Torpedo* (a marine elasmobranch), each organ consists of some 45 dorso-ventral columns of 700 electrocytes (or electroplaques) per column. In each column the electroplaques are in series electrically, and the columns are in parallel. The discharge from *Torpedo* is usually 20 to 30

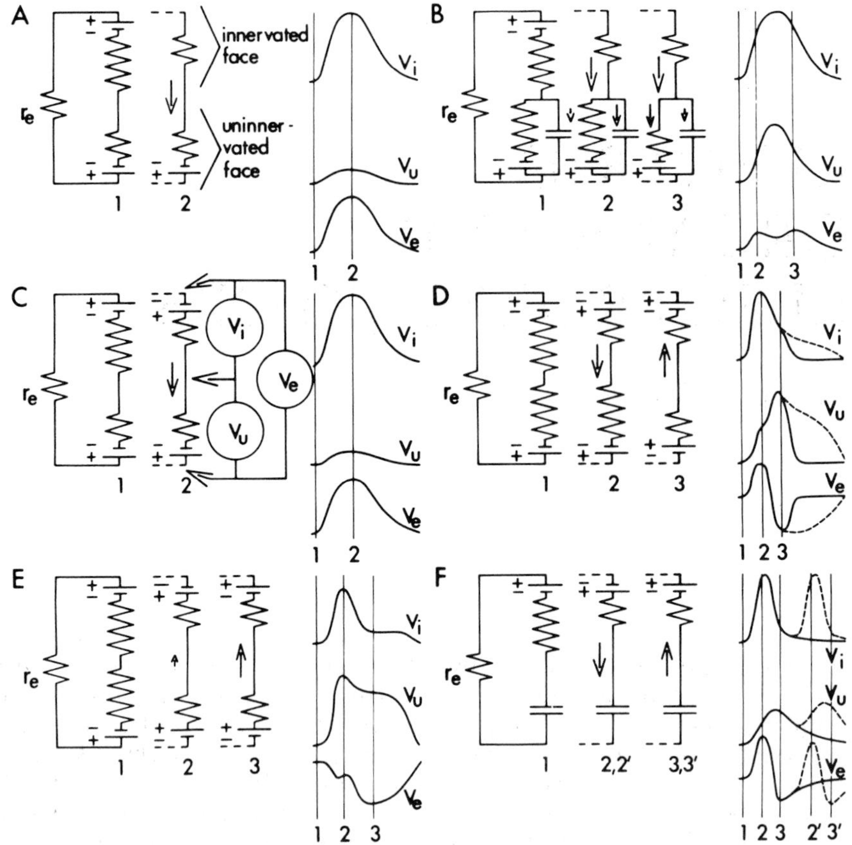

Figure 17-1. Diagrams of equivalent circuits of different kinds of electrocytes and of recordings across innervated surface (V_i), across uninnervated surface (V_u), and across entire cell (V_e). Change from rest to activity at different times in discharge indicated by numbered equivalent circuits and times as numbered on responses. A, a strongly electric marine fish; innervated surface generates an epsp. B, electrocyte of a rajid fish; innervated surface generates an epsp and uninnervated surface shows delayed rectification. C, electrocyte of electric eel and some gymnotids; innervated face generates overshooting spike. D, electrocyte of mormyrids and some gymnotids; both faces generate spike, and V_e is diphasic. E, electrocyte of electric catfish; stalk has high threshold and generates a small spike, and V_e is negative on the non-stalk side. F, electrocyte of *Gymnarchus*; uninnervated face acts as series capacity. (From Bennett, M. V. L., pp. 347–491 *in* Fish Physiology, vol. 5, edited by W. S. Hoar and D. J. Randall, Academic Press, New York, 1971.)

volts, and sometimes 50 volts, which in sea water generates a current of several amperes. *Torpedo occidentalis* has been reported to deliver 1000 watts. All of the plaques in a column are profusely innervated on the ventral face, and the innervated face generates graded, non-propagating postsynaptic potentials (epsp's), with the ventral side negative. The innervated surface membrane is cholinoceptive; the responses are blocked by tubocurarine and prolonged by eserine. The non-innervated membrane has a lower resistance than the innervated one; the skin over the electric organs is also of low resistance, and hence current can flow even though only one cell face is depolarized (Fig. 17–1A).[3, 8, 10]

Another group of marine elasmobranchs, the Rajidae (skates), have electric organs at the base of the tail, with either cup-shaped or disc electrocytes. The innervated face of each electrocyte generates an epsp having a duration of 50 msec or more. In the disc electrocytes, the non-innervated face shows a decrease in resistance (from between 5 and 30 Ω cm² to 1.5 Ω cm²) during the falling phase of the epsp (delayed rectification), thus producing some increase in the external current (Fig. 17–1B).[4]

The marine stargazer *Astroscopus* has extraocular muscles modified as electric organs, which it uses to shock small prey fish. As in the torpedo and skates, the innervated (dorsal) surface of each electrocyte gives a negative epsp (10 to 20 mv) that is a cholinergic response. The non-innervated face has a low electrical resistance.[9]

Thus, in all marine electric fishes the discharge consists of postsynaptic potentials which add arithmetically in electrocytes in series. The electrocyte membranes are not electrically excitable, but the innervated face is stimulated by acetylcholine; the non-innervated face has a low resistance or shows delayed rectification. The external current is high in low-resistance sea water, especially for *Torpedo* and *Astroscopus.*

The strongly electric *Electrophorus* from the Amazon has several organs; the main one has about 1000 plaques per column, oriented antero-posteriorly. Individual pulses can attain some 400 volts in an open circuit, and in short circuit the fish can deliver one ampere. Each discharge is a train, usually consisting of three to five (and sometimes 20 to 30) pulses. A single electroplaque has a resting potential of 90 mv (inside negative) as measured across either surface. Stimulation of the nerve gives a graded epsp, which may elicit a spike on the innervated face. This (posterior) surface reverses its polarity by some 60 mv, but the non-innervated face is of low resistance, is electrically inexcitable, and does not depolarize; hence, the two surfaces in series give a total cell potential of 150 mv (Fig. 17–1C).[1,24] The spike is due to Na activation and can be blocked by TTX. Unlike a nerve or skeletal muscle spike, there is no increase in potassium conductance; rather, the recovery to the resting potential level is due to increased resistance (anomalous rectification) during the sodium current of the rising phase.[8]

The epsp is reduced by tubocurarine; it shows facilitation with repetitive nerve impulses, and this is increased by anticholinesterases. The electric organ is rich in ACh and AChE, and presumably acetylcholine is the transmitter. The spikes are due to an increase in inward sodium current, and they are blocked by TTX. The resistance of the unreactive non-innervated face is less than that of the innervated membrane. Thus, the innervated membrane consists of junctional type sites that are sensitive to ACh and electrically inexcitable, and that are intermingled with spike-producing electrically excitable membrane. ATP is hydrolyzed during discharge and it is regenerated from creatine phosphate (p. 740); the electrocytes contain much Na-K ATPase, which may pump Na out after activity.

Several gymnotid and mormyrid freshwater fishes are weakly electric. Their electrocytes generate spikes, and different species show different properties of the two surfaces so that spikes may be diphasic or triphasic. In *Gymnotus,* both faces of an electrocyte are electrically excitable and each gives a spike response (Fig. 17–1D). However, the lower or innervated face fires first so that the total response is diphasic. In *Hypopomus*, the spike is briefer on the caudal than on the rostral face. In *Sternopygus* and *Eigenmannia,* a series of epsp's and spikes is superimposed on a slow or dc potential; the head of the fish is negative. Measurements of electrocyte resistance show the usual drop in resistance during the rising phase, and then a high membrane resistance at the peak of a spike (anomalous rectification) followed by a decline in resistance (delayed rectification) during the falling phase.[9]

In some mormyrid fishes the electrocyte forms a complex stalk which branches and receives rich innervation. The action potential of the non-innervated face is larger and longer than that of the innervated face, so that the second phase of the spike predominates.[38]

In the strongly electric African catfish *Malapterurus,* a single giant neuron innervates the several million electrocytes on one side of the body. Each electrocyte has a stalk on the innervated face which generates a small brief spike (Fig. 17–1E); this is followed by a large spike, sometimes with two peaks, on the non-innervated face. Thus, the innervated face gives an epsp and a brief spike, which propagates to the non-innervated face where the spike duration is longer.[2,25]

In *Gymnarchus,* another African freshwater fish, the innervated face generates a spike; the non-innervated face is electrically inexcitable and has a high electrical capacity and resistance. Current from the innervated face charges the series capacity (Fig. 17–1F) of the non-innervated membrane, which then discharges passively with a peak later than the current from the innervated face.

In at least two weakly electric fish, the myogenic component has been lost and the electric organs are derived from motor nerves. In *Sternarchus,* spinal nerves form an organ ventral to the vertebral column; many nerve fibers run forward, then loop and pass posteriorly, and end blindly. The nerve fibers expand in diameter during the mid-regions of their forward and backward courses. The discharge in *Sternarchus* is diphasic, with the head initially positive, and results from the synchronous spikes in the nerve fibers conducted forward and back. The widths of the nodal gaps are 1 μm in proximal and central parts of both limbs; in distal parts there are 3 to 5 nodes, 50 μm each in width, and the axon membrane forms irregular processes. Electrical evidence suggests that the regions of 1 μm nodes generate spikes but that the regions of large node surface act as series capacities and conduct electrotonically only. Thus, the initial head-positive phase is generated in the anteriorly running fiber, and the head-negative phase in the posteriorly running fiber.[41] In *Andontosternarchus,* an accessory organ in the chin is derived from sensory neurons.[37]

In general, the non-innervated surface of many electrocytes is highly convoluted, and this increase in membrane area provides for high conductance or capacitance. All strongly electric fish and some of the weak ones emit a direct-current (dc) component. Electrocytes of freshwater fishes have spike generating membranes, while those of marine species do not. Membrane resistances, particularly of the non-innervated face, are low; hence, current spread is favored.[4]

ELECTRORECEPTION

All animals generate low-frequency electrical fields by asymmetric secretory and neuromuscular action. Many nervous systems generate relatively large low-frequency fields, and the spike activity of some neurons in these fields can be altered according to field strength (p. 672). Sense organs for detecting small electric fields are known only for fish and are usually modifications of the lateral line system, if one includes epidermal receptors such as the ampullae of Lorenzini in this system. Electroreceptors occur in considerable variety, but they can be grouped into two major categories[6,8] (Fig. 17–2). (1) Tonic receptors are spontaneously rhythmic and give long-lasting responses to low-frequency or dc electrical stimulation. In each organ, several receptor cells lie in an ampulla at the base of a canal, which opens to the exterior (Fig. 17–3). (2) Phasic receptors adapt rapidly, are sensitive to high-frequency stimulation, and give transient responses to direct current. These organs are tuberous, with epithelial cells between the receptors and the exterior.[39] Often, both types of electroreceptor occur in the same fish. In each type, the receptor cells give sensory or generator potential responses and activate nerve endings at the base of the receptor cell.

Tonic electroreceptors occur in many fishes which do not have electric organs, such as in the lateral line of catfish and in the ampullae of Lorenzini of elasmobranchs.

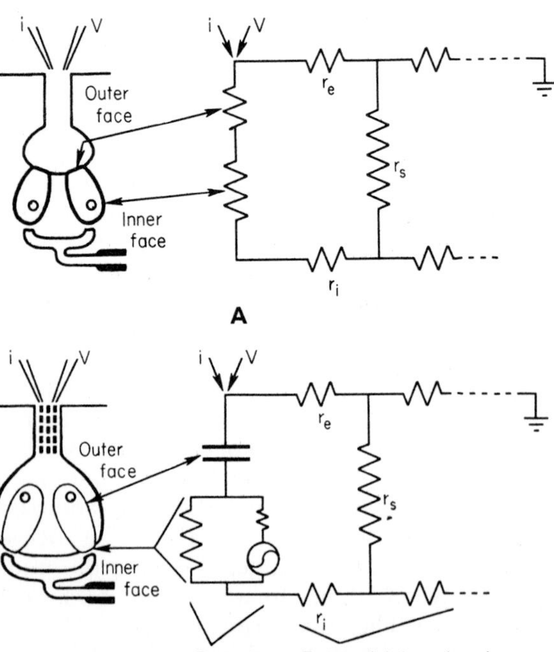

Figure 17–2. Diagrams of structure and equivalent circuits of electroreceptors in freshwater fishes. A, tonic ampullar receptor; B, phasic tuberous receptor. Resistance of external medium, r_e; resistance of skin, r_s; resistance of interior, r_i. Electrodes for stimulating and recording shown at receptor opening. Innervation indicated below receptor cells. (From Bennett, M. V. L., pp. 313–373 *in* Lateral Line Detectors, edited by P. Cahn, Indiana University Press, 1967.)

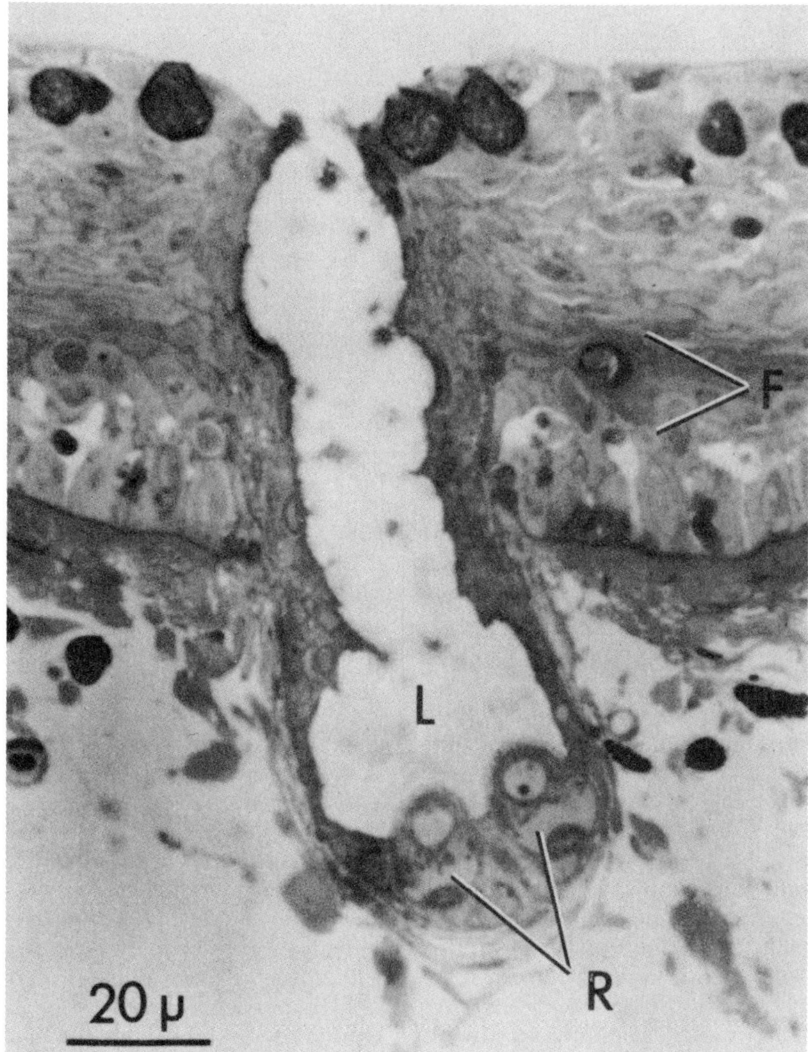

Figure 17–3. Microscopic section of a tonic electroreceptor of the gymnotid *Eigenmannia.* Receptor cavity *L* opens to exterior. Receptor cells *R* show nerve terminal regions as darkening at base of cells. (From Szamier, R. B., and A. W. Wachtel, J. Ultrastruc. Res. *30*:450–471, 1970.)

They detect the action potentials from respiratory and swimming muscles. The frequency of their spontaneous spike activity can be modulated by small applied currents.

The responses of a tonic receptor in *Gymnotus,* as measured *in vitro,* are illustrated in Figure 17–4. The receptor was discharging spontaneously at about 100/sec; an anodal stimulus at the canal opening increased the frequency, and at the end of the pulse a silent period ensued. A cathodal pulse (hyperpolarizing the inner face of the receptor cell) decreased the frequency.[7] A phasic receptor cell shows a receptor potential, which may be oscillatory at a frequency different from that of the sensory spikes. The oscillatory receptor potential may continue steadily or may give damped oscillations after a stimulus ceases.[7]

Transmission to the nerve fibers is mostly by a chemical synapse. This is indicated by presynaptic endings containing many vesicles. The minimum delay for stimulation via the receptor is 1.6 msec; also, the nerve response continues for some time after stimulation stops. The nature of the transmitter is uncertain, but glutamate is indicated in some gymnotids.

Mormyrids have a wide range of sizes of electroreceptors. In large ones, several (1 to

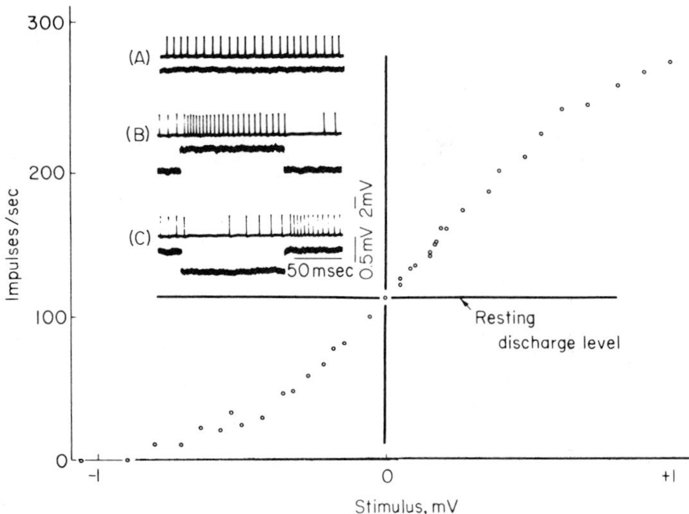

Figure 17–4. Responses of tonic receptor of *Gymnotus.* Inset upper trace gives afferent impulses, and lower trace gives stimulating potential at opening of receptor. A, spontaneous discharge; B, anodal stimulation; C, cathodal stimulation. Graph gives average impulse frequency as a function of stimulating voltage; anodal at right and cathodal at left. (From Bennett, M. V. L., pp. 73–128 *in* Physiological and Biophysical Aspects of Nervous Integration, edited by F. D. Carlson, Prentice-Hall, Englewood Cliffs, N.J., 1968.)

8) cells may be innervated by a single nerve fiber. These have a very brief delay (0.2 msec), and transmission may well be electrical.[20, 21]

The ampullae of Lorenzini are bulbous receptors of elasmobranchs; they are especially abundant on the head. They fire spontaneously and are sensitive to temperature, tactile stimuli, CO_2, and salinity; their primary function, however, seems to be electrodetection.[31] In *Scyliorhinus* and *Raja*, for example, conditioned circulatory responses can be obtained in response to muscle potentials emanating from a flatfish 5 to 16 cm away. The dogfish shows behavioral response to longitudinal gradients as low as 0.02 μv/cm.[17, 22] *Raja* shows cardioacceleration in external fields as low as 0.01 μv/cm, and exclusion of other receptors showed the response to be mediated by ampullae of Lorenzini;[17] lateral line threshold was 3 μv/cm. Ampullae in the skate *Raja* fire spontaneously at about 40/sec; when current passes in the jelly tube, the response is increased by cathodal current and decreased by anodal current.[31] The threshold is calculated to be 5×10^{-11} ampere in the tube, or 10^3 to 10^4 times more sensitive than in the lateral line.[31] Eyelid reflexes and heart rate responses can be obtained in external fields of 0.01 to 0.02 μv/cm.[22]

In intact electric fish (*Gymnotus* or *Eigenmannia*), lateral line nerve fibers give several spikes per electric organ discharge—up to 200 to 750 per second, depending on the individual. If the external field is perturbed, for example by a piece of metal, the rate of sensory discharge changes. Gradation of response is by number of spikes per electric pulse, or by addition or omission of a few spikes in a series.[18, 19]

Impulses in lateral line fibers of electric fishes signal input from electroreceptors in several ways. Some send afferent impulses continuously without reference to the electrical discharge. Others are non-synchronous phasic units that signal the presence of objects of different electrical conductance in the fish's field. Still others discharge synchronously with the fish's frequency, but they either vary in the phase of firing relative to the discharge or code by probability of firing, misfiring, or skipping a discharge.[15]

In electric catfish, pit organs in the skin perceive dc fields of 0.3 mv/cm or 0.005 μA/mm^2, as evidenced by behavior.[35, 36] Electric fish distinguish direction and polarity of field; they also distinguish changes in the shape of the field due to conductors or nonconductors such as grids of wires.

The arrangement of the ampullae is favorable for directing small electric currents to the receptors. The skin of a mormyrid has a resistance of 50 KΩ cm^2, some 100 times that of the skin of a goldfish; the resistance of the skin of *Gymnotus* is 1 to 3 KΩ cm^2. Longitudinally oriented stimuli are most effective on receptors of the head, less so on the tail, and least in the middle of the fish.

Behavioral evidence indicates that many animals can detect magnetic fields whose magnitudes approach that of the earth's magnetic field (slightly less than one gauss). For example, the planarian *Dugesia* deviates from a straight pathway away from a light source and turns to the right when its exit is oriented toward the north or south poles of a magnetic field of less than 5 gauss, and turns to the left when oriented east or west in the same field; in a stronger field (10 gauss), the orientations N-S or E-W are reversed.[12,14] Specimens of a mud snail, *Nassarius*, also deviate in a phototactic path by orienting in reference to a weak magnetic field, and can distinguish north from south poles of a magnet. Responses to weak magnetic fields exhibit periodisms of solar day, lunar day, and semimonthly frequencies.

In honeybees returning from foraging, the deviation in direction of the waggle dance of communication on the vertical comb disappears if the earth's magnetic field is compensated, and the deviation is increased if the magnetic field is enhanced.[27] *Drosophila* shows turning responses on emergence from a dark tube, which vary with the magnetic field.[33] Homing pigeons show conditioned heart rate responses to a magnetic field of 0.8 gauss at 180° to the earth's field.[34]

A number of different species have been shown to detect small electrostatic fields. The mechanisms of detection of small electromagnetic fields are obscure. However, in a skate, the ampullae of Lorenzini have a threshold below the field amplitude caused by sea water moving in the earth's magnetic field; hence, a moving skate could detect magnetic fields by its ampullar system.[22] It seems reasonable that induction of electrical fields around sensitive cells may be one mechanism of magnetic detection.

BEHAVIORAL AND NEURAL CONTROL

Two functions of electric organs are electrolocation and electroparalysis. Most weakly electric fishes are from turbid waters or are nocturnal. They send out steady trains of pulses and detect small changes in the surrounding field; some of them vary the frequency of pulsing as a further aid to probing the environment.[28] Strongly electric fishes give massive discharges which stun prey or predator. Some fish, such as *Electrophorus*, have continuously firing low-voltage organs as well as the major high-voltage organ. In a series of low-voltage Amazon fishes, the low-frequency species have electric organs derived from muscles; high-frequency species have organs derived from nerves.[37] *Sternarchella* discharges at 1200/sec, while *Eigenmannia* discharges at 320 to 750/sec.

The patterns of discharge vary with species and condition. The basal rate in some weakly electric fish, such as mormyrids, is 1 to 15/sec, but when they actively pursue prey the rate may increase up to 130/sec.[28,29,30] *Gymnarchus*' probing frequency may be 250/sec. The stargazer *Astroscopus* lies in the sand with only its eyes exposed; when a small fish approaches, it gives a high-frequency burst before or while opening its mouth, and then after 100 msec it gives a discharge train lasting several seconds.[32]

In *Torpedo*, the electric discharge starts 80 msec after the first movement toward capture of prey, which has been detected via mechanoreceptors; just before "landing" the frequency is 140 to 290/sec, and afterward it is irregular and slow. Prey can be immobilized at a distance of 15 cm.[3]

Central control of strongly electric organs permits near synchrony of thousands of electrocytes. The two organs of a *Torpedo* discharge within a few tenths of a millisecond, and all of the plaques of a six-foot *Electrophorus* are activated within about 1.5 msec. In weakly electric fish, the pattern of discharge is similar for the entire organ.

In the electric catfish, two neurons in the first spinal segment fire the electric organs on their respective sides; the two command neurons are electrotonically coupled, which ensures synchrony. In gymnotids, a cluster of pacemaker cells in the dorsal part of the medulla activates a ventral medullary relay nucleus, which in turn activates spinal neurons; electrotonic coupling occurs at both the medullary and the spinal levels[5,11] (Fig. 17–5). Mormyrids have a bilateral pacemaker nucleus in the midbrain, which activates a second order nucleus in the medulla; this in turn drives spinal neurons. The medullary relay cells fire in doublets, and in the spinal relay there may be triplet impulses.[8,11] Feedback from the electrotonic coupling between neurons at the three levels makes for close integration.

In *Astroscopus*, where the electric organ is modified from eye muscles, the electrocytes are innervated from the oculomotor nucleus. In the electric eel *Electrophorus*, a single command center in the medulla activates both the probing organ of Sachs and the main organ of electric discharge; the main organ

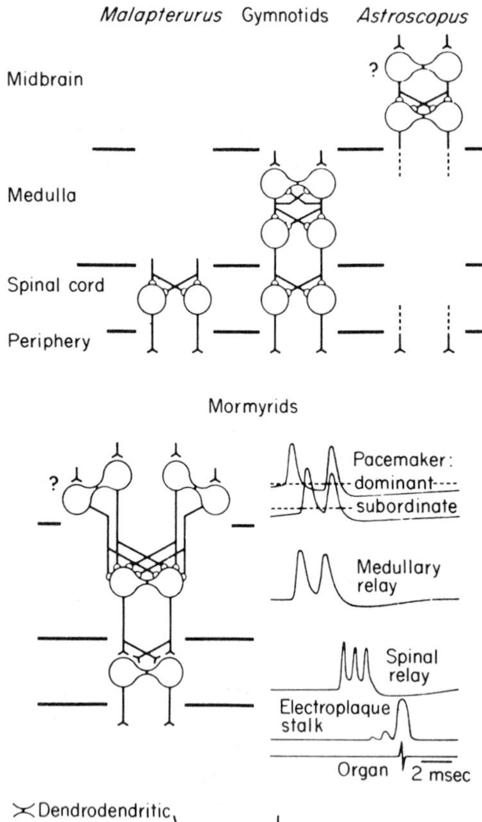

Figure 17–5. Neural circuits of control of electric organs in several electric fish. In *Malapterurus*, gymnotids, and *Astroscopus*, a single command volley at each level precedes each organ discharge, whereas in mormyrids the activity varies in complexity at different levels. (From Bennett, M. V. L., pp. 147–169 *in* The Central Nervous System and Fish Behavior, edited by D. Ingle, University of Chicago Press, Chicago, 1968.)

responds only to facilitating, closely spaced motor impulses, while the organ of Sachs fires in response to single impulses. Synchrony in this long fish results from longer delay at anterior than at posterior spinal relays, and from slower conduction rostrally, so that the electrocytes are activated nearly simultaneously at the two ends of the fish.

In general, a few command and relay neurons activate a larger number of spinal neurons, which in turn activate very large numbers of electrocytes. The command nuclei are subject to feedback control, and relay neurons are often electrotonically coupled. It may be concluded that the combination of electrotonic coupling, the use of a few pacemaker neurons in a command center, feedback interaction, and appropriate conduction rates and synaptic delays make for synchronization of exquisite precision in the electric discharges.

JAMMING AVOIDANCE

A given weakly electric fish responds to applied fields of slightly higher or lower frequency than its own by a jamming avoidance response. When *Eigenmannia* is stimulated repetitively, as by alternating current of various frequencies, there is a reflex change in the discharge rate of its electric organ only if the frequency of stimulation does not vary by more than 10 Hz from the normal firing rate of the organ. At a higher rate of stimulation, the fish decreases its discharge rate, and at a lower rate of stimulation it increases its firing rate as if to avoid jamming by the sensory input. Sensitivities are less than 70 μv/cm along the fish.[40] The threshold in *Hypopomus* is less when the stimuli are given between organ discharges than when they coincide with the discharges.[26]

Eigenmannia shifts its frequency upward for a decrease in stimulus frequency or downward for an increase in stimulus frequency, relative to the fish's discharge frequency; it responds to a frequency modulated sweep better than to a fixed stimulation frequency. Individual fish discharge at slightly different frequencies, and one fish responds to another in such a way as to widen the difference between their rates of discharge; i.e., each maintains its private frequency[15,16] (Fig. 17–6).

The mode of transfer of the input in the brain to the command centers is not well known, but projections have been found in the acoustic lateralis lobe. It is significant that most electric fish have unusually large cerebella. Strongly electric fish are protected against their own discharges by high resistance of the skin, by a fatty insulation around the brain, and by thick myelin sheaths on many nerve fibers; even so, they undoubtedly do "feel" their discharges.[8]

In several species, the behavioral threshold is reported as about 1/100 of the electrophysiological threshold. The reason for this discrepancy is not clear, but such factors as experimental extrapolation from tangential fields at the sides and ends of the fish, possibly poorer condition of fish under recording conditions, failure to be sure of recording from the most sensitive receptors, and central facilitation are cited. After a

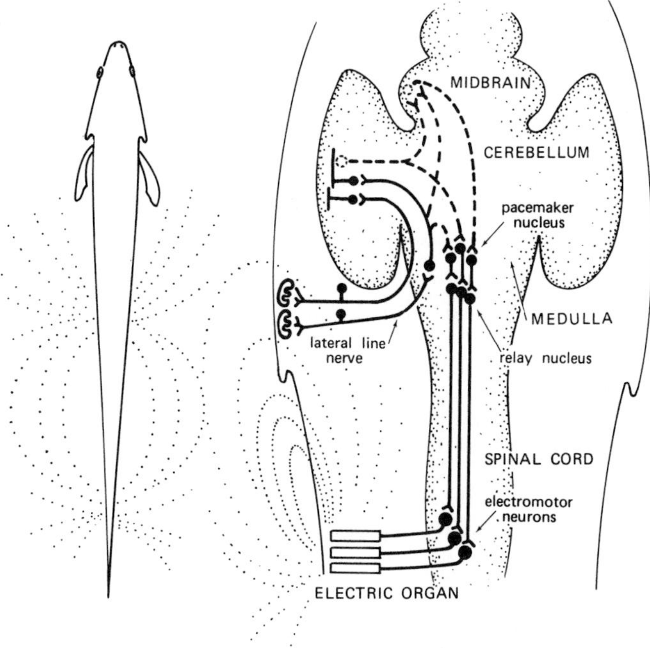

Figure 17–6. Diagram of anatomy of neural circuits of jamming avoidance response. Broken lines are probable but not yet established connections. (From Bullock, T. H., et al., J. Comp. Physiol. 77:1–22, 1972.)

number of corrections, Bennett[4,8] concludes that the actual difference may be only one order of magnitude. In any case, fish with electroreceptors are 10^5 times more sensitive to electric currents than are those fish without such specialized receptors.[4]

SUMMARY

Muscle cells have been modified several times in the evolution of fishes for generation of electric currents; contractile elements have been lost, the fibers have become flattened, and the innervated and non-innervated membranes have taken on different properties. In some fish, an electrogenic stalk attaches to the nerve fibers, and in a few the motor nerve fibers alone constitute an electric organ.

Electrocytes often have asymmetrical potentials, so that the two sides of a cell add electrically; many cells are connected in series electrically. All are under cholinergic activation and are highly synchronized in nervous control.

Patterns of pulse generation include: postsynaptic potentials (epsp's) on one electrocyte face and low resistance on the opposite face, with declining resistance on the inactive face during the pulse; spike generation on the innervated face only; spikes on each face, with conduction from one side to the other; and two sequences of resistance reduction. Thus, electrocytes illustrate many of the known cellular types of electrogenesis.

Weakly electric fishes and some non-electric fishes have electroreceptors formed from a modified lateral line system. These detect longitudinal fields as low as 10^{-8} to 10^{-7} volts/cm. The receptor cells transmit their signals to sensory nerve endings, usually by a chemical synapse, and the sensory coding is mostly by changes in the patterns of spontaneous sequences. Thus, pre-existing structures—muscles and lateral line receptors—have been modified for electric pulse production and detection.

Electric discharges of weakly electric fishes are used for probing the murky environment, while those of strongly electric fishes are used for stunning prey. The discharge of very many (up to millions) of cells is synchronized by a few electrotonically coupled central neurons.

REFERENCES

1. Altamirano, M., et al., J. Gen. Physiol. *38*:319–360, 1955. Direct and indirect excitation of electroplaques of *Electrophorus*.
2. Bauer, R., Z. vergl. Physiol. *59*:371–402, 1968. Electric discharges in *Malapterurus*.

3. Belbenoit, P., Z. vergl. Physiol. *67*:205–216, 1970. Electric discharge in *Torpedo.*
4. Bennett, M. V. L., Ann. N.Y. Acad. Sci. *94*:458–509, 1961; Ann. Rev. Physiol. *32*:471–528, 1970. Modes of operation of electric organs.
5. Bennett, M. V. L., Ann. N.Y. Acad. Sci. *137*:509–539, 1966. Electrotonic junctions.
6. Bennett, M. V. L., pp. 313–393. *In* Lateral Line Detectors, edited by P. Cahn, Indiana University Press, 1967; *also* pp. 147–169 *in* The Central Nervous System and Fish Behavior, edited by D. Ingle, University of Chicago, 1968. Electroreception and neural control.
7. Bennett, M. V. L., pp. 73–128. *In* Physiological and Biophysical Aspects of Nervous Integration, edited by F. D. Carlson. Prentice-Hall, Englewood Cliffs, N.J., 1968. Chemical and electrical synapses.
8. Bennett, M. V. L., pp. 347–491. *In* Fish Physiology, vol. 5, edited by W. S. Hoar and D. J. Randall. Academic Press, New York, 1971. Electric organs and electroreception.
9. Bennett, M. V. L., and H. Grundfest, J. Gen. Physiol. *42*:1067–1104, 1959; *44*:805–843, 1961. Electric organs of *Gymnotus, Narcine,* and *Astroscopus.*
10. Bennett, M. V. L., et al., J. Gen. Physiol. *44*:757–804, 1961. Electric organs of *Torpedo.*
11. Bennett, M. V. L., et al., J. Neurophysiol. *30*:180–300, 1967. Central nervous control of electric organ discharge.
12. Brown, F. A., Biol. Bull. *123*:264–281, 282–294, 1962. Responses of planarian *Dugesia* to very weak horizontal magnetic and electrostatic fields.
13. Brown, F. A., et al., Biol. Bull. *127*:221–231, 1964. Rhythmic modifications of responses of mud snails to geomagnetic fields.
14. Brown, F. A., Ann. N.Y. Acad. Sci. *188*:224–241, 1971. Orientational influences of electromagnetic fields.
15. Bullock, T. H., and S. Chichibu, Proc. Nat. Acad. Sci. *54*:422–429, 1965. Analysis of sensory coding in electroreceptors of electric fish.
16. Bullock, T. H., et al., J. Comp. Physiol. 77:1–22, 23–48, 1972. Jamming avoidance response of high frequency electric fish.
17. Dijkgraaf, S., and A. J. Kalmijn, Z. vergl. Physiol. *53*:187–194, 1966. Function of ampullae of Lorenzini.
18. Hagiwara, S., and H. Morita, J. Neurophysiol. *26*:551–567, 1963. Coding of electroreceptive fibers.
19. Hagiwara, S., T. Szabo, and P. S. Enger, J. Neurophysiol. *28*:775–799, 1965. Properties of electroreceptors, *Electrophorus* and *Sternarchus.*
20. Harder, W., Z. vergl. Physiol. *59*:272–318, 1968. Electroreception in mormyrids.
21. Harder, W., et al., Z. vergl. Physiol. *54*:89–108, 1967. Sensitivity of some weakly electric fish to electrical stimulation.
21a. Hopkins, C. D., Science *176*:1035–1037, 1972. Sex differences in electric signaling in an electric fish.
22. Kalmijn, A. J., J. Exp. Biol. *55*:371–383, 1971. Electric sense of sharks and rays.
23. Kellaway, P., Bull. Hist. Med. *20*:112–137, 1946. Electric fish in history of electrobiology and medicine.
24. Keynes, R. D., and H. Martins-Ferreira, J. Physiol. *119*:315–351, 1953. Membrane potentials in electroplaques of *Electrophorus.*
25. Keynes, R. D., et al., pp. 102–112. *In* Bioelectrogenesis, edited by C. Chagas and A. P. de Carvalho. Elsevier, Amsterdam, 1961. Electrophysiology of electric organ of *Malapterurus.*
26. Larimer, J. L., and J. A. MacDonald, Amer. J. Physiol. *214*:1253–1261, 1968. Sensory feedback from electroreceptors to electropacemakers in gymnotids.
27. Lindauer, M., and H. Martin, Z. vergl. Physiol. *60*:219–243, 1968. Gravity orientation of bees in earth's magnetic field.
28. Lissman, H. W., J. Exp. Biol. *35*:156–191, 1958. Evolution and function of electric organs.
29. Lissman, H. W., and K. E. Machin, J. Exp. Biol. *35*:451–486, 1958. Electrolocation of objects by *Gymnarchus.*
30. Machin, K. E., and H. W. Lissman, J. Exp. Biol. *37*:801–811, 1960. Electroreceptors in *Gymnarchus.*
31. Murray, R. W., J. Physiol. *180*:592–606, 1965. Sensory function of ampullae of Lorenzini.
32. Pickens, P. E., and W. N. MacFarland, Anim. Behav. *12*:362–367, 1964. Electric discharge and behavior in stargazer.
33. Picton, H. D., Nature *211*:303–304, 1966. Orienting response of *Drosophila* to electric fields.
34. Reille, A., J. Physiol. Paris *60*:85–92, 1968. Conditioning of homing pigeons to a magnetic field.
35. Roth, A., Z. vergl. Physiol. *61*:196–202, 1968; *65*:368–388, 1969. Electroreception in catfish *Ictalurus.*
36. Roth, A., and T. Szabo, Z. vergl. Physiol. *62*:395–410, 1969. Receptor function in tuberous organs of mormyrid *Gnathonemus.*
37. Steinbach, A. B., Biol. Bull. *138*:200–210, 1970. Survey of electric fishes in Rio Negro, Brazil.
38. Szabo, T., pp. 20–23 *in* Bioelectrogenesis, edited by C. Chagas and A. P. de Carvalho, Elsevier, Amsterdam, 1961; *also* pp. 295–312 *in* Lateral Line Detectors, edited by P. Cahn, Indiana University Press, 1967. Peripheral and central components in electroreception.
39. Szamier, R. B., and A. W. Wachtel, J. Ultrastruc. Res. *30*:450–471, 1970. Ultrastructure of electroreceptor organs.
40. Watanabe, A., and K. Takeda, J. Exp. Biol. *40*:57–66, 1963. Change in discharge frequency by ac stimulation in weakly electric fish.
41. Waxman, S. G., G. D. Pappas, and M. V. L. Bennett, J. Cell Biol. *53*: 210–224, 1972. Differentiation of nodes along fibers of neurogenic electric organ, *Sternarchus.*

CHAPTER 18

AMOEBOID MOVEMENT

By J. D. Anderson

Motility in biological systems is not restricted to muscular contraction. Since the invention of the microscope, each succeeding generation of biologists has seen the movements of cytoplasm, the gliding of cells, pseudopod formation, protoplasmic streaming, movements of chromosomes during cell division, the beating of cilia and flagella, and numerous other nonmuscular kinds of motion. The importance of such movements has long been appreciated. Phagocytic function of white blood cells, migration of neural crest cells in embryonic development, orderly events in cell division, removal of mucus from the respiratory tract, and sperm motility are but a few of numerous examples of essential physiological processes dependent upon nonmuscular motility.

Protoplasmic streaming occurs in many living cells, possibly in all. Diffusion is too slow a process for the transport of solids from one part of a cell to another, and active cytoplasmic movement provides a rapid transport mechanism. Streaming may proceed in a fixed path and may be fast enough for direct microscopic observation, as the cyclosis around the vacuole of some plant cells, or the transport of food vacuoles and granules about the body of a ciliate protozoan. In other cells, streaming may be slow and may be more of a churning than a fixed current; such cytoplasmic activity is best seen by accelerated motion pictures, as of tissue cultures of fibroblasts or the tips of growing nerve fibers. The effects on streaming of a variety of the environmental factors have been examined, but the exact molecular mechanisms remain unknown.

TYPES OF AMOEBOID CELLS AND PSEUDOPODS

Amoeboid movement has much in common with protoplasmic streaming, and an elucidation of the transformation of chemical energy to mechanical work in one will facilitate an understanding of the other. Amoeboid movement is accompanied by changes in cell shape, by the extension of pseudopods, and often by progressive motion. Amoeboid movement may be directed locomotion as in the rhizopod Protozoa, in the plasmodium of the myxomycetes, in amoeboid leukocytes, and in the amoebocytes or wandering cells of many kinds of animals; or amoeboid movement may consist of extension, flexion, and retraction processes concerned primarily with feeding as in most Foraminifera, Heliozoa, and Radiolaria, vertebrate microphages, and the reticuloendothelial cells. Locomotory amoeboid movement requires attachment to some substrate; nonpolarized amoeboid movement occurs in free pseudopods.

In free-living amoeboid animals, the manner of locomotion differs slightly according to the cell form and the type of the pseudopod.[82] Pseudopods may be lobopods, broad to cylindrical and round at the tip; they may be filopods, slender with pointed tips; they may be reticulopods, thread-like, branching, and anastomosing as in Foraminifera; or they may be axopods, with a central axial rod as in Heliozoa and Radiolaria (Fig. 18–1). Locomotion by lobopods has been much studied; filopods, reticulopods, and axopods may show streaming without accompanying change in length, and axopods may be very contractile.

Amoebae differ greatly in cell form and in rates of cytoplasmic streaming and locomotion.[25] *Amoeba limax* has a single lobose-pseudopod; a stellate form of amoebae has many free pseudopods. *A. proteus* and *A. dubia* are multipodal with several attached pseudopods; streaming occurs in varying amounts in each. Amoebae may have surface ridges as in *A. verrucosa,* or they may have an irregular "tail" or uroid as in *Pelomyxa palustris* (Fig. 18–1). Lymphocytes likewise vary; they sometimes have a broad advancing pseudopod and a smaller tail, or they may be wormlike. A given amoeboid cell may take on different forms under various conditions. For example, *Amoeba proteus* in distilled water becomes stellate or radiate, but in dilute saline it is monopodal. It also is stellate prior to fission. Amoebae have a central fluid endoplasm (plasmasol) and an outer viscous ectoplasm (plasmagel). The ratio of ectoplasm to endoplasm is high in *A. proteus, A. verrucosa, Chaos carolinensis*, and *Pelomyxa palustris*, but it is low in *A. dubia* and in the parasitic *A. blattae.*

The greatest versatility and complexity of pseudopods occur in the Foraminifera and Radiolaria. The filamentous pseudopods often branch and form networks.[5, 6, 67] Protoplasmic streaming in these filamentous pseudopods is bidirectional, outward on one side and inward on the opposite side; it is so different from that found in the typical amoebae that a reclassification of the Sacrodina based on differences in movement has been suggested.[65]

A GENERALIZED PICTURE OF AMOEBOID MOVEMENT

A "typical" amoeba consists of an outer layer, the plasmalemma, which has adhesive properties and is not wetted by water, and which slides freely over the next inner layer, the ectoplasm. Electron micrographs show the plasmalemma to consist of an outer filamentous coat and an inner membrane.[100]

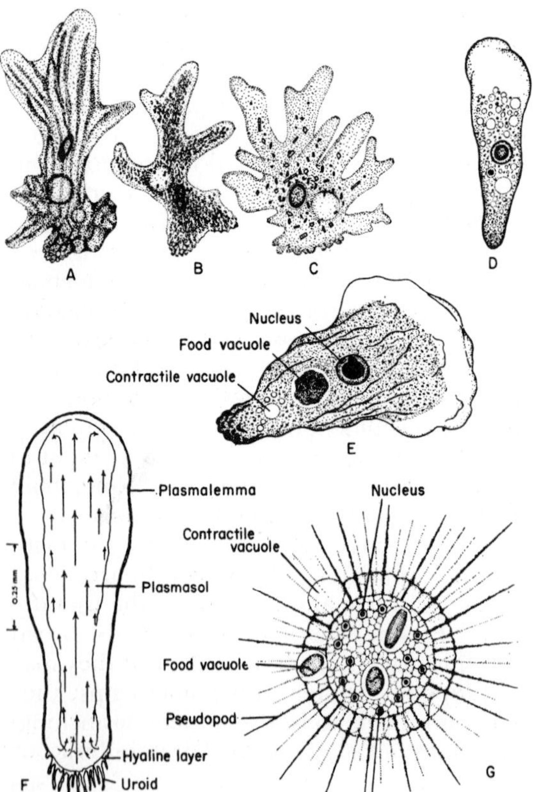

Figure 18–1. Drawings of amoebae with different types of pseudopods. A, *Amoeba proteus;* B, *Amoeba discoides;* C, *Amoeba dubia;* D, *Amoeba limax;* E, *Amoeba verrucosa;* F, *Pelomyxa palustris;* G, heliozoan *Actinosphaerium eichhorni.* (From Schaeffer, A. A., Amoeboid Movement. Princeton University Press; Kuhn, A. R., Morphologie der Tiere in Bildern. Borntraeger; Mast, S. O., Physiol. Zool., vol. 7.)

The filaments are about 80 Å in diameter and extend outward 0.1 to 1.0 μm. The coat contains 35% lipid, 26% protein, and 16% polysaccharide.[98] Beneath the plasmalemma is a hyaline layer, which is fluid as judged by brownian activity when particles enter it. This layer is very thin in the region of attachment to the substrate; it often thickens as an extensive hyaline cap at the front of an advancing pseudopod. Next is the ectoplasm, or the cylinder of plasmagel, which is relatively viscous. In a monopodal amoeba the ectoplasm thickens gradually from the anterior to the posterior end. In many species it extends as a thin plasma gel sheet beneath the anterior hyaline cap, a sheet which frequently ruptures and allows granules to enter the hyaline cap. The frequency of rupture into the hyaline cap may be different in each of the several advancing pseudopods. The endoplasmic core of the amoeba is the plasmasol, in which granules flow freely forward. It stains more for neutral polysaccharides than does the ectoplasm.[112] The nucleus is normally in the plasmasol; in some species it may keep a rather central position even though the streaming may be rapidly taking place around it.[45] Both the endoplasmic core and the ectoplasmic tube contain various granules, food vacuoles, and various types of crystal-like inclusions. The protoplasm is hyperosmotic in fresh water amoebae and nearly iso-osmotic in parasitic and marine amoebae; the ectoplasm imparts a turgidity which results in the nonspherical form.

Locomotion as observed microscopically depends on three basic factors: (1) Attachment to the substrate is facilitated by traces of salts, and particularly by calcium in the medium; Ca^{++}, Mg^{++}, and K^{+} are additive for attachment but not for locomotion. The firmest attachment is at the tips of the pseudopods, and the new pseudopods are more firmly attached than old ones. (2) Ectoplasm is continually being converted to endoplasm at the posterior end or at some fixed region, and the endoplasm is converted to ectoplasm anteriorly or in any extending pseudopod. As the endoplasm flows forward, granules either are deflected laterally to become the ectoplasm or break into the hyaline cap and then gelate as a new cap is formed. (3) As components of the force that causes forward flow of the endoplasm, elastic, contractile, and pressure forces are indicated. There are several theories for the locomotor force of amoeboid movement, but most of them postulate contractile proteins in either "squeezing" or "pulling." Lateral examination of granules in an amoeba or lymphocyte and of particles adhering to the plasmalemma has shown that, in an advancing cell, the granules in the ectoplasm remain fixed until the posterior end reaches them, when they enter into the endoplasm. The particles on the ventral attached plasmalemma are also fixed, but the particles elsewhere on the plasmalemma, both dorsal and ventral, move forward. In *A. verrucosa*, a "rolling" movement of the plasmalemma has been described.[69, 87] In other species, such as *Difflugia*, there may be a "walking" movement, the advancing tip attaching and the remainder being pulled forward. In the reticulopods of the Foraminifera, there is no outer tube of ectoplasm but rather two gel threads, with the movement of granules outward on one thread and back towards the body of the organism on the other thread.[67]

Rates of movement vary greatly, from a barely perceptible rate to 1350 μm/sec for the cytoplasmic streaming in the plasmodium of the acellular slime mold *Physarum polycephalum*.[75] Migrating plasmodia have been observed to advance at 5 to 6 cm/hour.[15] Freely crawling amoebae move at rates of 0.5 to 4.5 μm/sec; most of them move at about 1 μm/sec. Monopodal individuals travel faster (4.6 μm/sec) than multi-podal ones (2.1 μm/sec).[101, 102, 103] Lymphocytes in tissue culture move at an average of 0.55 μm/sec, whereas the non-polarized microphages average 0.004 μm or less per second.[35, 36, 37]

In feeding, those amoebae which travel by small pseudopods (such as *A. proteus*) form food cups consisting of lateral and dorsal pseudopods flowing around the food mass, which in itself may be motile. The pseudopods of the cup meet distally, and the food mass is incorporated into a food vacuole. In amoebae with a single broad pseudopod, as in *A. limax*, there appears to be some adhesion of food to the plasmalemma. Furthermore, the surfaces of reticulopods and axopods appear to be sticky, since food particles adhere to them.[77]

The endoplasm flows in channels, and islands of ectoplasm may be seen in a broad region of the endoplasm. Individual granules only a few microns apart in the slime mold, *Physarum polycephalum*, may move in opposite directions, and the flow generated at a given point reverses with irregular periodicity.[115]

THEORIES OF AMOEBOID MOVEMENT

In contractile systems such as striated muscle, cilia, and flagella, association can be made between structure and movement. Such association becomes difficult in smooth muscle cells and is even more so for amoeba-like cells. The lack of structural criteria makes it difficult to assess whether or not the mechanochemical system responsible for amoeboid movement has indeed been isolated.[129]

In the effort to account for the different interrelated features of amoeboid movement, many theories have been postulated. They are not mutually exclusive, and several of them are simultaneously applicable. It is most unlikely that any one theory would be applicable to all species of amoebae,[6] but it is generally believed that there may be some underlying common ground, possibly at the macromolecular level, in spite of apparent differences.[41]

In the mid-1800's, the cytological studies of protoplasm indicated a reticular structure.[124] It is not surprising that the earliest theories of amoeboid movement postulated contractile strands in various regions, particularly in the advancing pseudopod which pulls the protoplasmic mass ahead.[38] This theory was revived in modified form in the 1960's.[4] As chemistry and physics developed information on surface and interface tensions, theories involving changes in surface tension were postulated as providing the force for amoeboid movement. These have been generally abandoned,[38] although surface forces probably do play some role in shaping cells. Later, studies of colloids and their properties led to gel-sol concepts of amoeboid movement, and models postulating contractility of the ectoplasm (gel) have received wide acceptance.[25]

A wide variety of theories has been published and reviewed.[3, 38, 66, 75, 79, 129] For those organisms possessing a well defined ectoplasm and endoplasm, one can deduce two generalized models which incorporate most of the main aspects of the many theories and models proposed.

*1. **Hydraulic Model.*** Passive or active contraction of the ectoplasm and/or active-shear forces generated in the ectoplasm or at the ectoplasm-endoplasm interface or between microfibrils create pressure on the more fluid endoplasm, which then flows. The ectoplasm solates at specific or general sites, forming endoplasm which then flows from the solation site and gels elsewhere, becoming ectoplasm again.*

*2. **Frontal Contraction Model.*** The contractile force is generated at the front of a pseudopod, and the viscoelastic properties of the endoplasm are sufficient for it to be pulled forward. At the tip of the pseudopod, the endoplasm becomes everted while contracting and converts to ectoplasm. The most anterior portion of the ectoplasmic tube is thought to be recently contracted cytoplasm:[2, 3, 4] when a large *Chaos carolinensis* is broken into a small quartz tube under oil, the isolated fragment shows streaming, first with the axial endoplasm flowing forward and the ectoplasm flowing backward, and then with several channels making U-bends at one end. The velocity of flow toward the bend is faster, and the width of the channel less, than that of the flow away from the bend in the gel. In a normally advancing pseudopod, the cylinder of the endoplasm shortens and thickens in the "fountain" zone. It is postulated that gel strands, attached to the rim of the advancing tube, contract and pull the anterior sol forward; water is removed to the hyaline cap by syneresis and passes back under the plasmalemma to be returned to the posterior zone of recruitment of endoplasm.[7, 8, 9] Flow is chiefly at the boundary or shear zone between the endoplasm and the ectoplasm. It is argued that contraction at a U-bend between the endoplasm and the ectoplasm can equally well explain amoeboid movement and cyclosis. This model was further confirmed by finding a photoelastic effect near the pseudopod tips.[10] Further evidence that frontal contraction rather than hydraulic pressure is responsible for pseudopod formation and extension in *Chaos carolinensis* comes from experiments in which a pseudopod was sucked into a capillary; a pressure reduction of 30 to 35 cm of water placed on the enclosed pseudopod did not prevent extension of other pseudopods exposed to atmospheric pressure.[11]

A great deal of evidence indicates contractility of the ectoplasm. It increases in thickness posteriorly,[88] and electron microscopy reveals filaments in locations where one would expect tension to be developed.[111, 127] Elasticity, combined with contraction, may explain two-way flow from a

*References 42, 43, 45, 46, 48, 63, 64, 65, 66, 68, 80, 86, 87, 92, 99, 107, 108, 114, and 117.

constricted region into a region with weaker elasticity.[63, 92, 114] If microfilaments are part of the contractile apparatus, there is a possibility that a variation of the sliding filament mechanism for muscle contraction (see Chapter 16) may be functioning. Shearing forces have been postulated between sliding filaments in the ectoplasm[97] and in the fine reticulate pseudopods of a foraminiferan, *Allogromia*.[67] These 5 μm thick reticulopods lack a central core of endoplasm and an outer endoplasm layer, but they appear to consist of two semicircular threads of gel, one of which moves forward and the other of which moves backward. It is postulated that one gel filament shears on the other. Multiple contracting sites along the filament have been suggested.[5]

A slightly different characterization of the ectoplasm is based on its colloidal properties. The ectoplasm of amoebae decreases in viscosity under high hydrostatic pressure and at decreasing temperature. Actomyosin behaves similarly. Gels of this kind undergo a volume increase ($+\Delta V$) and absorption of heat (endothermy) on gelation.[80, 86] In slime mold plasmodia, temperature is higher in regions where ectoplasm is being converted into endoplasm than in the areas into which the endoplasm is flowing.[12]

Similarly, when living amoebae are subjected to high hydrostatic pressure, the "viscosity" falls, the ectoplasm solates, and locomotion stops. At 2,000 lbs/in^2, the pseudopods are long and cylindrical; at about 6,000 lbs/in^2, no new pseudopods are formed; and at about 6,500 lbs/in^2, terminal spheres appear on pseudopods and the organisms become balls of fluid. A series of functions—amoeboid movement, chromatophore expansion, cyclosis in *Elodea*, cleavage in *Arbacia* eggs—are similarly stopped as the protoplasmic rigidity decreases under high pressures.[30, 53, 81, 85] These are in contrast to bioluminescence, ciliary movement, muscle contraction, and nerve conduction, which may be enhanced by pressure, probably because of membrane effects.[29] The critical pressure to cause collapse of pseudopods and to give a spherical form is less at low than at high temperatures, and the gel strength as measured by centrifugation decreases more with high pressure at 15°C than 25°C.[85] When ATP is added to the medium, the pressure required to make the amoebae spherical is greater; this agrees with the initial increase in viscosity of the gel protein on treatment with ATP and the initial apparent contraction of live amoebae on ATP injection.[46]

CONTRACTILE PROTEINS

The first demonstration that nonmuscular systems possessed contractile proteins was reported by Loewy in 1952. From the plasmodium of the acellular slime mold, *Physarum polycephalum*, he extracted a protein which had many of the properties of actomyosin: it was soluble in 1 M KCl at pH 8.2; it formed a viscous solution and showed decreased viscosity when ATP was added; it appeared to dephosphorylate ATP to AMP with an increase in viscosity.[83] This protein, called myxomyosin, was further purified and its physical properties were determined by Ts'o and his colleagues. Its molecular weight is about 6×10^6; it is rod-shaped with a length of 4000 to 5000 Å and a diameter of 70 Å.[120, 121, 122] Supplemental evidence for the existence of a contractile system came when it was found that ATP increased streaming in slime molds,[75] and that ATP initiated streaming movements in the isolated cytoplasm of *Amoeba proteus*.[131] Later, Nakajima demonstrated that purified myxomyosin extracted from slime mold has many of the properties of muscle myosin B.[95] The ATP sensitivity of plasmodial myosin B was found to be as high as that of rabbit myosin B, and Mg^{++} increases the rate of precipitation while Ca^{++} inhibits it.[54, 58]

An actin-like protein isolated from slime molds, when added to myosin B from rabbit, forms an actomyosin complex. Conversely, slime mold myosin B reacts with mammalian actin.[56, 57, 59] Actin has also been isolated from sea urchin eggs[55] and from amoebae.[123] There are many similarities between the actins isolated from these various sources and the actin isolated from muscle: (1) there is one mole of 3-methylhistidine per mole of protein; (2) the molecular weights are very similar; (3) they are precipitated by myosin at low ionic strength; (4) the increase in viscosity upon addition of myosin at high ionic strength is reversed by ATP; and (5) they possess the ability to undergo a reversible G-F transition.[59, 123] The similarities of the actins suggest a common evolutionary background for motility. Various forms of cell motility have quite different structural organization visible from the gross to the electron microscope level. However, the similarities of the contractile proteins suggest there may be a fundamental functional requirement for transforming chemical energy into mechanical work.[41]

FIBRILS, MICROFILAMENTS, AND MICROTUBULES

Wohlfarth-Botterman and his colleagues first described microfilaments of approximately 50 Å in diameter in both *Amoeba proteus* and the slime mold, *Physarum polycephalum*.[111, 125, 126, 127] The number was few compared to the total amount of ground substance. However, the microfilaments were found in areas where one would postulate tensile activity, that is, around vacuoles, in pseudopods, and in the ectoplasm of slime mold channels. A contractile function was postulated for these filaments, but the small number of filaments and their lack of attachment to membranes or other structures raised doubts as to the significance of their role in cytoplasmic movement.

The existence of microfilaments has been amply verified.[89, 90, 94] Aside from their possible role in cytoplasmic movements, it has been postulated that they restrain organelles during cytoplasmic flow[60] and that they are associated with vacuole contraction.[106]

Wolpert and his colleagues were the first to demonstrate movement and contraction of isolated cytoplasm of *Amoeba proteus* upon the addition of ATP in the presence of EDTA. Electron micrographs of the contracted gel showed numerous microfilaments of 70 Å diameter.[130, 131] Two phases of contraction and movement have been reported in such preparations.[104] At first there is an apparent increase in viscosity and the appearance of 50 to 70 Å diameter filaments, which is followed by the appearance of 160 Å diameter filaments that align into fiber-like structures of 0.1 to 1.0 μm diameter. Binding of heavy meromyosin by such microfilaments has been reported.[105] Formation of these filaments is inhibited by a detergent, Salyrgan,[1] but not by colchicine.[93]

Microfilaments are present in glycerinated preparations of both amoebae and slime molds, and contraction follows the application of ATP.[61, 78] Actomyosin threads prepared from slime molds and amoebae shorten and thicken when treated with ATP.[16, 17, 18]

Microtubules and microfilaments are abundant in heliozoan axopods,[109] but are not found in the slime mold plasmodia or in the large amoebae under normal conditions. They have been reported in *Amoeba proteus* that had been starved for 16 days[33] and in the small amoeba *Hartmannella* sp.[60] The microtubules are irreversibly degraded by cupric and nickelous ions[110] and are reversibly degraded by dilute urea treatment.[113] The movement of cultured microphages treated with colchicine or vinblastine changes from the gliding form of locomotion to amoeboid movements. No microtubules are found in the treated cells, but microfilaments are present.[21] Colchicine treatment transforms cultured fibroblast cells from their normal shape to an epithelial-like form, and cell locomotion is inhibited.[47]

There is some question as to whether microfilaments and microtubules have the same chemical composition and subunit structure. Colchicine degrades formed microtubules and inhibits the formation of new ones,[47] but does not inhibit formation of microfilament bundles.[93] Microfilaments and microtubules from squid axon differ in molecular weight, amino acid composition, immunological responses, and electrophoretic mobility.[34, 62]

The birefringence seen in amoebae,[10] slime molds,[96] and axopods[5] suggests an orderly array of macromolecular structures. The finding of microfilaments and microtubules strengthens this assumption. It is speculated that filaments and microtubules play some role in movement, either by processes similar to those postulated in the sliding filament model for muscle contraction[107] or by the development of shear forces.[117] The finding of Ca^{++} accumulating vesicles in slime molds further suggests the similarity to muscle contraction.[26] However, experimental evidence is lacking for an explanation of the rapid, and sometimes cyclic, changes in birefringence that are sometimes seen in amoebae and slime molds, or for the periodicity in the shuttle-type streaming in slime mold plasmodia.

THE ROLE OF THE NUCLEUS

Amoeba proteus has been used extensively in nuclear-cytoplasmic interaction studies.[27, 28] The nucleus can be transferred from one individual to another, or the nucleus and practically all of the cytoplasm can be removed from one amoeba and then the cytoplasm from a second amoeba and the nucleus from a third inserted into the first. The reconstituted amoeba survives.[73] Only by removing the nucleus was its role in amoeboid movement discovered. Enucleated amoebae soon lose the organized progressive cytoplasmic flow that results in movement. There is some cyclosis, but pseudopod formation is absent or erratic. The ability to attach to substrate and to feed is lost. Insertion of a nucleus quickly brings back normal activities and is called reactivation. The nucleus is necessary for the formation and maintenance of contractile vacuoles and Golgi bodies.[39, 40] Nuclei from one strain inserted into the cytoplasm of another strain creates a new strain with characteristics different from either of the original strains.[84] Hereditary function is lost in nuclei kept in salt solution for 20 minutes, but these nuclei still retain their ability to reactivate the cell.[70] It would appear that protoplasmic streaming in *Physarum polycephalum* is dependent on an intact interphase nucleus, since the rapid shuttle-like flow ceases during nuclear division.[15, 50]

The nuclear-cytoplasmic interaction(s) that impart(s) some control and regulation to cytoplasmic streaming and amoeboid movement is unknown, but the discovery of the synthesis of an actin-like protein in the nuclei of slime mold may lead to more experiments which will elucidate the role of the nucleus in amoeboid movement.[74]

THE ROLE OF THE MEMBRANE

It is obvious that total surface area must greatly increase as an amoeboid cell changes from a nearly spherical shape to a multipodal form; it then decreases as pseudopods are retracted. The theories on how these changes occur fall into three groups: (1) The membrane is a fluid or plastic surface that slides freely over the ectoplasm. (2) New membrane is rapidly formed as needed, and resorbed when the surface area decreases. (3) The total amount of membrane material is relatively constant and the membrane folds or unfolds.

Several studies of the movement of attached particles on moving amoebae favor the fluid or plastic concept. Particles move forward past a fixed point and proceed outward to the tip of the pseudopod,[49] and may round the tip and flow back, giving the impression that the plasmalemma is rolling.[69, 87]

It has also been observed that particles attached to the posterior part of the cell move backward and that the membrane is very wrinkled at the rear. These observations led to the postulation that the membrane is absorbed at the rear and that the cell surface is renewed each time the cell passes through its own length.[44] The wrinkling of the membrane at the uroid of *Amoeba proteus* follows a cycle. During rapid pseudopod formation, the wrinkling disappears, and then increases at the end of the forward movement.[128]

Wolpert and O'Neill labelled the surface of *Amoeba proteus* with fluorescent-labelled antibodies and found that the half-life of the label was about 5 hours, indicating a turnover rate of 0.2% per minute or 100 times less than it would be if the membrane were completely renewed for one cell's length of locomotion.[129, 130, 131] However, the antibody binds only to the surface coat, and if this coat is mechanically held back during pseudopod extension, the membrane covering the new pseudopod is not labelled,[72] so the question of the membrane turnover rate still remains obscure.

Some electron microscope and cinemagraphic studies support the folding and unfolding concept and indicate that the amount of membrane remains relatively constant.[32, 51, 116]

Although there is no evidence that the membrane generates forces of the magnitude required for locomotion, its mechanical and physiological properties may influence the cell shape and control of movement. At points of attachment there is less space between the plasmalemma and the ectoplasm than elsewhere.[20] It has been suggested that contact between the plasmalemma and the ectoplasm in *Amoeba proteus* initiates contraction.[44, 45]

Amoeboid cells respond to an electrical field and show biopotentials. In an electrical field, *Amoeba proteus* shows solation on the cathodal side, and pseudopodia advance in that direction.[52] Migration of the plasmodium of *Physarum* toward the cathode results from inhibition on the anodal side, and there is 15 to 20 per cent less K^+ in the posterior part of the plasmodium.[13, 14]

In slime molds, mechanical stimulus elicits an electrical response of a few seconds duration,[118] and it has been reported that "action" potentials can be recorded following electrical stimulation;[31] but since these are not propagated, it is questionable whether they should be called action potentials.

A slime mold can be kept in an hour-glass form in a chamber of two compartments connected by a channel. Hydrostatic pressure applied on one side can stop the flow in that direction, and the pressure necessary to prevent flow fluctuates periodically; i.e., the balancing pressure is ±5 to 10 inches of H_2O.[75] The potential between the two compartments also fluctuates periodically, but the change in potential lags behind the change in balancing pressure. Therefore, the potential changes do not initiate protoplasmic streaming.

Transmembrane potentials of 40 to 70 mv can be recorded in amoebae and slime molds.[22, 23, 24, 91, 119] In an amoeba, the extending pseudopod tip is negative with respect to the internal rear and is positive with respect to the outside;[24] chemotactic agents which cause pseudopod formation reduce the potential across the membrane.[19, 71] Recordings of potential changes in amoebae[119] and slime molds[91] show rapid potential changes of 6 to 10 mv, which cannot be correlated with direction of streaming. It is assumed these are the result of membrane changes, but the nature of these changes is unknown. Therefore, although the plasmalemma has some electrical properties similiar to those of other cell membranes, their role in amoeboid movement is obscure.

REFERENCES

1. Achterrath, M., Cytobiologie *1*:159–168, 1969. The inhibition of formation of fibrillar pattern in slime mold plasmodia by Salyrgan.
2. Allen, R. D., J. Biophys. Biochem. Cytol. *8*:379–397, 1960. The consistency of ameba cytoplasm and its bearing on the mechanism of ameboid movement.
3. Allen, R. D., pp. 135–216. *In* The Cell, Vol. II, edited by J. Brachet and A. E. Mirsky, Academic Press, New York, 1961. Ameboid movement.
4. Allen, R. D., Exp. Cell Res., Suppl. *8*:17–31, 1961. A new theory of amoeboid movement and protoplasmic streaming.
5. Allen, R. D., pp. 407–432. *In* Primitive Motile Systems in Cell Biology, edited by R. D. Allen and N. Kamiya, Academic Press, New York, 1964. Cytoplasmic streaming and locomotion in marine Foraminifera.
6. Allen, R. D., Symp. Soc. Exp. Biol. *22*:151–168, 1968. Differences of a fundamental nature among several types of amoeboid movement.
7. Allen, R. D., J. W. Cooledge, and P. J. Hall, Nature *187*:896–899, 1960. Streaming in cytoplasm dissociated from the giant amoeba, *Chaos chaos*.
8. Allen, R. D., R. R. Cowden, and P. J. Hall, J. Cell Biol. *12*:185–189, 1962. Syneresis in ameboid movement.
9. Allen, R. D., and D. W. Francis, Symp. Soc. Exp. Biol. *19*:259–271, 1965. Cytoplasmic contraction and the distribution of water in the amoeba.
10. Allen, R. D., D. W. Francis, and H. Nakajima, Proc. Nat. Acad. Sci. *54*:1153–1161, 1965. Cyclic birefringence changes in pseudopods of *Chaos carolinensis* revealing the localization of the motive force in pseudopod extension.
11. Allen, R. D., D. Francis, and R. Zeh, Science *174*:1237–1240, 1971. Direct test of the positive pressure gradient theory of pseudopod extension and retraction in amoebae.
12. Allen, R. D., W. R. Pitts, Jr., D. Speir, and J. Brault, Science *142*: 1485–1487, 1963. Shuttle-streaming: synchronization with heat production in slime mold.
13. Anderson, J. D., J. Gen. Physiol. *35*:1–16, 1951. Galvanotaxis of slime mold.
14. Anderson, J. D., J. Gen. Physiol. *45*:567–574, 1962. Potassium loss during galvanotaxis of slime mold.
15. Anderson, J. D., pp. 125–136. *In* Primitive Motile Systems in Cell Biology, edited by R. D. Allen and N. Kamiya, Academic Press, New York, 1964. Regional differences in ion concentration in migrating plasmodia.
16. Beck, R., H. Hinssen, H. Komnick, W. Stockem, and K. E. Wohlfarth-Botterman, Cytobiologie *2*:259–274, 1970. Extensive fibrillar protoplasmic differentiations and their significance for protoplasmic streaming. V. Contraction, ATPase-activity and fine structure of actomyosin-threads from *Physarum polycephalum*.
17. Beck, R., H. Komnick, W. Stockem, and K. E. Wohlfarth-Botterman, Cytobiologie *1*:99–114, 1969. Extensive fibrillar protoplasmic differentiations and their significance for protoplasmic streaming. IV. Comparative studies on actomyosin-threads and glycerinated cells.
18. Beck, R., H. Komnick, W. Stockem, and K. E. Wohlfarth-Botterman, Cytobiologie *2*:413–428, 1970. Extensive fibrillar protoplasmic differentiations and their significance for protoplasmic streaming. VI. Comparative studies on actomyosin-threads isolated from obliquely striated and smooth muscles.
19. Bell, L. G. E., J. Theoret. Biol. *1*:104–106, 1961. Surface extension as the mechanism of cellular movement and cell division.
20. Bell, L. G. E., and K. W. Jeon, Nature *198*:675–676, 1963. Locomotion of *Amoeba proteus*.
21. Bhisey, A. N., and J. J. Freed, Exp. Cell Res. *64*:419–429, 1971. Ameboid movement induced in cultured macrophages by colchicine or vinblastine.
22. Bingley, M. S., J. Exp. Biol. *45*:251–267, 1966. Membrane potentials in *Amoeba proteus*.
23. Bingley, M. S., L. G. E. Bell, and K. W. Jeon, Exp. Cell Res. *28*:208–209, 1962. Pseudopod initiation and membrane depolarization in *Amoeba proteus*.
24. Bingley, M. S., and C. M. Thompson, J. Theoret. Biol. *2*:16–32, 1962. Bioelectric potentials in relation to movement in amoebae.
25. Bovee, E. C., pp. 189–219. *In* Primitive Motile Systems in Cell Biology, edited by R. D. Allen and N. Kamiya, Academic Press, New York, 1964. Morphological differences among pseudopodia of various small amebae and their functional significance.
26. Braatz, R., and H. Komnick, Cytobiologie *2*:457–463, 1970. Histochemical demonstration of a calcium accumulating system in myxomycete plasmodia.
27. Brachet, J., Ann. N.Y. Acad. Sci. *78*:688–695, 1959. Cytoplasmic dependence in amoeba.
28. Brachet, J., pp. 771–841. *In* The Cell, Vol. II, edited by J. Brachet and A. E. Mirsky, Academic Press, New York, 1961. Nucleocytoplasmic interactions in unicellular organisms.
29. Brown, D. E. S., J. Cell. Comp. Physiol. *8*:141–157, 1936. The effects of rapid compression upon events in the isometric contraction of skeletal muscle.
30. Brown, D. E. S., and D. A. Marsland, J. Cell. Comp. Physiol. *8*:159–165, 1936. The viscosity of amoeba at high hydrostatic pressure.
31. Burr, H. S., J. Exp. Zool. *129*:327–341, 1955. Certain electrical properties of the slime mold.
32. Czarska, L., and A. Grebecki, Acta Protozool. *4*:201–239, 1966. Membrane folding and plasma-membrane ratio in the movement and shape transformation in *Amoeba proteus*.
33. Daniels, E. W., and E. P. Breyer, Z. Zellforsch. *91*:159–169, 1968. Starvation effects on the ultrastructure of amoeba mitochondria.
34. Davison, P. F., and F. C. Huneeus, J. Molec. Res. *52*:429–439, 1970. Fibrillar proteins from squid axons: microtubule protein.
35. DeBruyn, P. P. H., Anat. Rec. *89*:43–63, 1944. Locomotion of blood cells in tissue cultures.
36. DeBruyn, P. P. H., Anat. Rec. *93*:295–315, 1945. The motion of the migrating cells in tissue cultures of lymph nodes.
37. DeBruyn, P. P. H., Anat. Rec. *95*:177–191, 1946. The amoeboid movement of the mammalian leukocyte in tissue culture.
38. DeBruyn, P. P. H., Quart. Rev. Biol. *22*:1–24, 1947. Theories of amoeboid movement.
39. Flickinger, C. J., J. Cell Biol. *49*:221–226, 1971. Decreased formation of Golgi bodies in amebae in the presence of RNA and protein synthesis inhibitors.
40. Flickinger, C. J., and R. A. Coss, Exp. Cell Res. *62*:326–330, 1970. The role of the nucleus in the formation and maintenance of the contractile vacuole in *Amoeba proteus*.
41. Gibbons, I. R., Ann. Rev. Biochem. *37*:521–546, 1968. The biochemistry of motility.
42. Goldacre, R. J., Rev. Cytol. *1*:135–164, 1952. The folding and unfolding of protein molecules as a basis of osmotic work.
43. Goldacre, R. J., Symp. Soc. Exp. Biol. *6*:128–144, 1952. The action of general anaesthetics on amoebae and the mechanism of the responses to touch.
44. Goldacre, R. J., Exp. Cell Res. Suppl. *8*:1–16, 1961. The role of the cell membrane in the locomotion of amoebae and the source of the motive force and its control by feedback.
45. Goldacre, R. J., pp. 237–255. *In* Primitive Motile Systems in Cell Biology, edited by R. D. Allen and N. Kamiya, Academic Press, New York, 1964. On the mechanism and control of amoeboid movement.
46. Goldacre, R. J., and I. J. Lorch, Nature *166*:497–500, 1950. Folding and unfolding of protein molecules in relation to cytoplasmic streaming, amoeboid movement and osmotic work.

47. Goldman, R. D., J. Cell Biol. *51*:752–762, 1971. The role of three cytoplasmic fibers in BHK-21 cell motility: microtubules and the effects of colchicine.
48. Griffin, J. L., pp. 303–327. *In* Primitive Motile Systems in Cell Biology, edited by R. D. Allen and N. Kamiya, Academic Press, New York, 1964. The comparative physiology of movement in the giant, multinucleate amebae.
49. Griffin, J. L., and R. D. Allen, Exp. Cell Res. *20*:619–622, 1960. The movement of particles attached to the surface of amebae in relation to current theories of ameboid movement.
50. Guttes, E., and S. Guttes, Exp. Cell Res. *30*:242–244, 1963. Arrest of plasmodial motility during mitosis in *Physarum polycephalum.*
51. Haberey, M., K. E. Wohlfarth-Botterman, and W. Stockem, Cytobiologie *1*:70–84, 1969. Pinocytosis and locomotion of amoebae. VI. Kinematographic studies on the behavior of the cell surface of moving *Amoeba proteus.*
52. Hahnert, W. F., Physiol. Zool. *5*:491–526, 1932. A quantitative study of reactions to electricity in *Amoeba proteus.*
53. Harvey, E. N., and D. A. Marsland, J. Cell. Comp. Physiol. *2*:75–97, 1932. The tension at the surface of *Amoeba dubia* with direct observations on the movement of cytoplasmic particles at high centrifugal speeds.
54. Hatano, S., Exp. Cell Res. *61*:199–203, 1970. Specific effect of Ca^{++} on movement of plasmodial fragment obtained by caffeine treatment.
55. Hatano, S., H. Kondo, and T. Miki-Noumura, Exp. Cell Res. *55*:275–277, 1969. Purification of sea urchin egg actin.
56. Hatano, S., and F. Oosawa, J. Cell Physiol. *68*:197–202, 1966. Extraction of an actin-like protein from the plasmodium of a myxomycete and its interaction with myosin A from rabbit striated muscle.
57. Hatano, S., and F. Oosawa, Biochim. Biophys. Acta *127*:488–498, 1966. Isolation and characterization of plasmodium actin.
58. Hatano, S., and M. Tazawa, Biochim. Biophys. Acta *154*:507–519, 1968. Isolation, purification and characterization of myosin B from myxomycete plasmodium.
59. Hatano, S., T. Totsuka, and F. Oosawa, Biochim. Biophys. Acta *140*:109–122, 1967. Polymerization of plasmodium actin.
60. Holberton, D., Nature *222*:680–681, 1969. Microtubules in the cytoplasm of an amoeba.
61. Holberton, D. V., and T. M. Preston, Exp. Cell Res. *62*:473–477, 1970. Arrays of thick filaments in ATP-activated *Amoeba* model cells.
62. Huneeus, F. C., and P. F. Davison, J. Molec. Biol. *52*:415–428, 1970. Fibrillar proteins from squid axons: neurofilament protein.
63. Jahn, T. L., Biorheology *2*:133–152, 1964. Protoplasmic flow in the mycetozoan, *Physarum*: The mechanism of flow; a re-evaluation of the contraction-hydraulic theory and of the diffusion drag hypothesis.
64. Jahn, T. L., pp. 279–302. *In* Primitive Motile Systems in Cell Biology, edited by R. D. Allen and N. Kamiya, Academic Press, New York, 1964. Relative motion in *Amoeba proteus.*
65. Jahn, T. L., and E. C. Bovee, Amer. Midl. Nat. *73*:30–40, 1965. Mechanisms of movement in taxonomy of Sarcodina as a basis for a new major dichotomy into two classes, Autotractea and Hydraulea.
66. Jahn, T. L., and E. C. Bovee, Physiol. Rev. *49*:793–862, 1969. Protoplasmic movements within cells.
67. Jahn, T. L., and R. A. Rinaldi, Biol. Bull. *117*:100–118, 1959. Protoplasmic movement in the foraminiferan, Allogromia, *Lati collaris*; and a theory of its mechanism.
68. Jahn, T. L., R. A. Rinaldi, and M. Brown, Biorheology *2*:123–131, 1964. Protoplasmic flow in the mycetozoan, *Physarum*: Geometry of the plasmodium and the observable facts of flow.
69. Jennings, H. S., Carnegie Inst. Wash. Publ. No. 16, 129–234, 1904. The movements and reactions of amoeba.
70. Jeon, K. W., Exp. Cell Res. *50*:467–471, 1968. Nuclear control of cell movement in amoebae: nuclear transplantation study.
71. Jeon, K. W., and L. G. E. Bell, Exp. Cell Res. *27*:350–352, 1962. Pseudopod and foodcup formation in *Amoeba proteus.*
72. Jeon, K. W., and L. G. E. Bell, Exp. Cell Res. *33*:531–539, 1964. Behavior of cell membrane in relation to locomotion in *Amoeba proteus.*
73. Jeon, K. W., I. J. Lorch, and J. F. Danielli, Science *167*:1626–1627, 1970. Reassembly of living cells from dissociated components.
74. Jockusch, B. M., D. F. Brown, and H. P. Rusch, J. Bact. *108*:705–714, 1971. Synthesis and some properties of an actin-like nuclear protein in the slime mold *Physarum polycephalum.*
75. Kamiya, N., Protoplasmatologia *8*:1–199, 1959. Protoplasmic streaming.
76. Kamiya, N., pp. 257–277. *In* Primitive Motile Systems in Cell Biology, edited by R. D. Allen and N. Kamiya, Academic Press, New York, 1964. The motive force of endoplasmic streaming in the ameba.
77. Kitching, J. A., pp. 445–455. *In* Primitive Motile Systems in Cell Biology, edited by R. D. Allen and N. Kamiya, Academic Press, New York, 1964. The axopods of the sun animalcule *Actinophrys sol* (Heliozoa).
78. Komnick, H., W. Stockem, and K. E. Wohlfarth-Botterman, Zellforsch. *109*:420–430, 1970. Extensive fibrillar protoplasmic differentiations and their significance for protoplasmic streaming. VII. Experimental induction, contraction and extraction of the protoplasmic fibrils of *Physarum polycephalum.*
79. Komnick, H., K. E. Wohlfarth-Botterman, Fortschr. Zool. *17*:1–154, 1966. Morphology of cytoplasms.
80. Landau, J. V., Ann. N.Y. Acad. Sci. *78*:487–500, 1959. Gel-sol transformations in amoebae.
81. Landau, J. V., A. M. Zimmerman, and D. A. Marsland, J. Cell. Comp. Physiol. *44*:211–232, 1954. Temperature-pressure experiments on *Amoeba proteus;* plasmagel structure in relation to form and movement.
82. Leidy, J., Fresh-Water Rhizopods of North America. Government Printing Office, Washington, D.C., 1879, 324 pp.
83. Loewy, A. G., J. Gen. Physiol. *40*:127–156, 1952. An actomyosin-like substance from the plasmodium of a myxomycete.
84. Lorch, I. J., and K. W. Jeon, Exp. Cell Res. *57*:223–229, 1969. Character changes induced by heterologous nuclei in amoeba heterokaryons.
85. Marsland, D. A., pp. 127–161. *In* A Symposium on the Structure of Protoplasm; A Monograph of the American Society of Plant Physiologists, edited by W. Seifriz, Iowa State College Press, Ames, Iowa, 1942. Protoplasmic streaming in relation to gel structure in the cytoplasm.
86. Marsland, D. A., Int. Rev. Cytol. *5*:199–227, 1956. Protoplasmic contractility in relation to gel structure: Temperature-pressure experiments on cytokinesis and amoeboid movement.
87. Mast, S. O., J. Morphol. *41*:347–425, 1926. Structure, movement, locomotion, and stimulation in *Amoeba.*
88. Mast, S. O., and C. L. Prosser, J. Cell. Comp. Physiol. *1*:333–354, 1932. Effects of temperature, salts and hydrogen-ion concentration on rupture of the plasmagel sheet, rate of locomotion, and gel/sol ratio in *Amoeba proteus.*
89. McManus, S. M. A., Amer. J. Bot. *52*:15–25, 1965. Ultrastructure of myxomycete plasmodia of various types.
90. McManus, S. M. A., and L. E. Roth, J. Cell Biol. *25*:305–318, 1965. Fibrillar differentiation in myxomycete plasmodia.
91. Miller, D. M., J. D. Anderson, and B. C. Abbott, Comp. Biochem. Physiol. *27*:633–646, 1968. Potentials and ionic exchange in slime mold plasmodia.
92. Miller, D. M., and X. B. Reed, Acta Protozool. *8*:119–128, 1970. The distributed solation hypothesis: an explanation of protoplasmic movement in slime mold plasmodia.
93. Morgan, J., Exp. Cell Res. *65*:7–16, 1971. Microfilaments from *Amoeba* proteins.
94. Nagai, R., and N. Kamiya, Proc. Japan Acad. *42*:934–939, 1966. Movement of the myxomycete plasmodium: electron microscopic studies on fibrillar structures in the plasmodium.
95. Nakajima, H., pp. 111–123. *In* Primitive Motile Systems in Cell Biology, edited by R. D. Allen and N. Kamiya, Academic Press, New York, 1964. The mechanochemical system behind streaming in *Physarum.*
96. Nakajima, H., and R. D. Allen, J. Cell Biol. *25*:361–374, 1965. The changing pattern of birefringence in plasmodia of the slime mold, *Physarum polycephalum.*
97. Noland, L. E., J. Protozool. *4*:1–6, 1957. Protoplasmic streaming: a perennial puzzle.
98. O'Neill, C. H., Exp. Cell Res. *35*:477–496, 1964. Isolation and properties of the cell surface membrane of *Amoeba proteus.*
99. Pantin, C. F. A., J. Marine Biol. Assoc. U.K. *13*:24–69, 1923. On the physiology of amoeboid movement.
100. Pappas, G. D., Ann. N.Y. Acad. Sci. *78*:448–473, 1959. Electron microscope studies on amoeba.
101. Pitts, R. F., and S. O. Mast, J. Cell. Comp. Physiol. *3*:449–462, 1933. The relation between inorganic salt concentration, hydrogen ion concentration and physiological processes in *Amoeba proteus.* I. Rate of locomotion, gel/sol ratio, and hydrogen ion concentration in balanced salt solutions.
102. Pitts, R. F., and S. O. Mast, J. Cell. Comp. Physiol. *4*:237–256, 1934. The relation between inorganic salt concentration, hydrogen ion concentration and physiological processes in *Amoeba proteus.* II. Rate of locomotion, gel/sol ratio, and hydrogen ion concentration in solutions of single salts.
103. Pitts, R. F., and S. O. Mast, J. Cell. Comp. Physiol. *4*:435–455, 1934. The relation between inorganic salt concentration, hydrogen ion concentration and physiological process in *Amoeba proteus.* III. The interaction between salts (antagonism) in relation to hydrogen ion concentration and salt concentration.
104. Pollard, T. D., and S. Ito, J. Cell Biol. *46*:267–289, 1970. Cytoplasmic filaments of *Amoeba proteus*: the role of filaments in consistency of changes and movement.
105. Pollard, T. D., and E. D. Korn, J. Cell Biol. *48*:216–219, 1971. Filaments of *Amoeba proteus*: binding of heavy meromyosin by thin filaments in motile cytoplasmic extracts.
106. Prusch, R. D., and P. B. Dunham, J. Cell Biol. *46*:431–434, 1970. Contraction of isolated contractile vacuoles from *Amoeba proteus.*
107. Rinaldi, R. A., and W. R. Baker, J. Theoret. Biol. *23*:463–474, 1969. A sliding filament model of amoeboid motion.
108. Rinaldi, R. A., and T. L. Jahn, J. Protozool. *10*:344–357, 1963. On the mechanism of ameboid movement.
109. Roth, L. E., D. J. Pihlaja, and Y. Shigenaka, J. Ultrastruct. Res. *30*: 7–37, 1970. Microtubules in the heliozoan axopodium. I. The gradion hypothesis of allosterism in structural proteins.
110. Roth, L. E., and Y. Shigenaka, J. Ultrastruct. Res. *31*:356–374, 1970. Microtubules in the heliozoan axopodium. II. Rapid degradation by cupric and nickelous ions.
111. Schneider, L., and K. E. Wohlfarth-Botterman, Protoplasma *51*:377–389, 1959. Unicellular animal studies. IX. Electron microscopic investigation of *Amoeba,* with particular consideration of the fine structure of cytoplasms.
112. Sheen, S. J., F. B. Gailey, D. M. Miller, J. D. Anderson, T. J. Bargman,

and D. A. Carter, Bioscience *19*:1003–1005, 1969. Sol-gel differences in plasmodia of the acellular slime mold, *Physarum polycephalum*.
113. Shigenaka, Y., L. E. Roth, and D. J. Pihlaja, J. Cell Sci. *8*:127–151, 1971. Microtubules in the heliozoan axopodium. III. Degradation and reformation after dilute urea treatment.
114. Stewart, P. A., pp. 69–78. *In* Primitive Motile Systems in Cell Biology, edited by R. D. Allen and N. Kamiya, Academic Press, New York, 1964. The organization of movement in slime mold plasmodia.
115. Stewart, P. A., and B. T. Stewart, Exp. Cell. Res. *17*:44–58, 1959. Protoplasmic movement in slime mold plasmodia: the diffusion drag force hypothesis.
116. Stockem, W., K. E. Wohlfarth-Botterman, and M. Haberey, Cytobiologie *1*:37–57, 1969. Pinocytosis and locomotion of amoebae. V. Outline-changes and folding-degree of the cell surface of *Amoeba proteus*.
117. Subirana, J. A., J. Theor. Biol. *28*:111–120, 1970. Hydrodynamic model of amoeboid movement.
118. Tasaki, I., and N. Kamiya, Protoplasma *39*:333–343, 1950. Electrical response of a slime mold to mechanical and electrical stimuli.
119. Tasaki, I., and N. Kamiya, J. Cell. Comp. Physiol. *63*:365–380, 1964. A study on electrophysiological properties of carnivorous amoeba.
120. Ts'o, P. O. P., J. Bonner, L. Eggman, and J. Vinograd, J. Gen. Physiol. *39*:325–347, 1956. Observations on an ATP-sensitive protein system from the plasmodia of a myxomycete.
121. Ts'o, P. O. P., L. Eggman, and J. Vinograd, J. Gen. Physiol. *39*:801–812, 1956. The isolation of myxomyosin, an ATP-sensitive protein from the plasmodium of a myxomycete.
122. Ts'o, P. O. P., L. Eggman, and J. Vinograd, Biochim. Biophys. Acta *25*:532–542, 1957. Physical and chemical studies of myxomyosin, an ATP-sensitive protein in cytoplasm.
123. Weihing, R. R., and E. D. Korn, Biochem. Biophys. Res. Comm. *35*: 906–912, 1969. Ameba actin: the presence of 3-methylhistidine.
124. Wilson, E. B., The Cell in Development and Heredity, 3rd Ed., MacMillan Co., New York, 1934, pp. 63–65.
125. Wohlfarth-Botterman, K. E., Protoplasma *54*:514–539, 1962. Extensive fibrillar protoplasmic differentiations and their significance for protoplasmic streaming. I. Electron microscopic evidence and fine structure.
126. Wohlfarth-Botterman, K. E., Protoplasma *57*:747–761, 1963. Extensive fibrillar protoplasmic differentiations and their significance for protoplasmic streaming. II. Light microscopic description.
127. Wohlfarth-Botterman, K. E., pp. 79–109. *In* Primitive Motile Systems in Cell Biology, edited by R. D. Allen and N. Kamiya, Academic Press, New York, 1964. Differentiations of the ground cytoplasm and their significance for the generation of the motive force of ameboid movement.
128. Wohlfarth-Botterman, K. E., and W. Stockem, Z. Zellforsch. *73*:444–474, 1966. Pinocytosis and locomotion of amoebae. II. Information on permanent and induced pinocytosis by *Amoeba proteus*.
129. Wolpert, L., Symp. Soc. Gen. Microbiol. *15*:270–293, 1965. Cytoplasmic streaming and amoeboid movement.
130. Wolpert, L., and C. H. O'Neill, Nature *196*:1261–1266, 1962. Dynamics of the membrane of *Amoeba proteus* studied with labelled specific antibody.
131. Wolpert, L., C. M. Thompson, and C. H. O'Neill, pp. 143–168. *In* Primitive Motile Systems in Cell Biology, edited by R. D. Allen and N. Kamiya, Academic Press, New York, 1964. Studies on the isolated membrane and cytoplasm of *Amoeba proteus* in relation to ameboid movement.

CHAPTER 19

CILIA

By F. A. Brown, Jr.

Cilia are organelles of highly characteristic internal structure differentiated at the cell surface.[19, 20, 38, 60, 64, 69, 76] Their primary function is the propulsion of fluids over surfaces by lashing movements. They are often classified into two types: (1) flagella, which are relatively larger organelles, usually occurring singly or in small numbers upon cells, and (2) cilia proper, which are relatively much smaller and occur characteristically in large numbers upon each cell. Typical flagella characterize the Mastigophora of the phylum Protozoa. Also possessing flagella are the choanocytes of the Porifera, the gastroderm of many coelenterates, the flame end-bulbs of certain rotifers, the solenocytes of annelids, and the sperm cells of most groups throughout the animal kingdom. Cilia proper characterize the ciliated protozoans, and are found commonly over the body surface of coelenterates, Turbellaria, and Nemertea. In all other phyla of animals except the Nematoda and the Arthropoda, cilia are usually found at specific locations in or on the body.

Ciliary activity is restricted to an aqueous medium, and hence is found only on surfaces which are submerged, or at least covered by an aqueous film. Typical ciliary movement produces either, or both, of two results, depending on the inertia of the ciliated surface. If the inertia is small, then a movement of the ciliated surface through the medium, or locomotion, is effected; if, on the other hand, it is large, or the ciliated structure is not free to move, the external medium is moved over the ciliated surface. It is therefore obvious that ciliated surfaces can become most effective for rapid locomotion only in such small organisms as protozoa, rotifers, and ciliated larvae. In these forms, acceleration to maximum speed is very rapid, and the organisms stop very quickly on cessation of their ciliary activity.

Organisms more than about a millimeter in diameter may use cilia for locomotion. For these, however, movement is sluggish. Free-swimming organisms such as ctenophores must possess a density very close to that of the surrounding medium. The comb-jelly, *Pleurobrachia*, has a water content of 94.73 per cent and a density of 1.02741, very close indeed to that of the surrounding sea water.

The many small worms and snails, e.g., *Nassarius*, using cilia for locomotion are restricted to creeping over surfaces in water or on damp surfaces by ciliary movement in a layer of secreted mucus.

It is perhaps significant that, whereas activity of most ciliated surfaces is continuous throughout the lifetime of the animal, cilia whose primary function is locomotion are always under the control of the coordinative mechanism of the organism. This enables the animal adaptively to alter its rate and direction of movement in response to stimuli. This control of locomotor ciliation appears, however, to be superimposed upon a fundamental ciliary automaticity.[28]

In those instances in which the surrounding medium, rather than the ciliated cells, is caused to move, cilia are concerned in one or more of numerous functions, such as feeding, circulation, cleansing, respiration, and the movements of materials within ducts. The action of cilia establishes feeding currents for many ciliate protozoans and for sessile or sluggish species in a wide variety of animal groups. The feeding currents of rotifers are easily observed. The ciliary activity of the tentacles and oral regions of numerous sea anemones and corals serves both feeding and cleansing; it sweeps non-nutritious particles away from the mouth and off the tips of the tentacles.[7,87] In response to food, the tentacles may be tipped toward the oral aperture and thereby may pass these materials into the gastrovascular cavity. In *Metridium marginatum,* among other species, it has been reported that the direction of the effective beat of the cilia in the stomodeum is reversed under the stimulus of food.[61] The unstimulated cilia normally beat outward, but in the presence of food, such as crab flesh, they beat inward. KCl produces an action similar to that produced by food. In *Actinoloba* there are longitudinal grooves in the stomodeum, the cilia on the ridges beating outward and those in the grooves beating inward. The degree of muscular contraction and folding of this organ would therefore be expected to influence strongly the direction of the dominant currents.

Probably nowhere in the animal kingdom are ciliary mechanisms more intricately developed than in the ciliary filter-feeders, principally the lamellibranchs, certain gastropods, and protochordates.[5,65] Here the systems are organized to set up feeding currents, filter out suspended particles, collect them into specific ciliary tracts, and convey them to the oral opening of the digestive tract. Along the route, special ciliary mechanisms for sorting out the particles on the basis of size, discarding the larger ones, are not uncommon.

Materials are transported within the digestive systems of numerous invertebrates by ciliary action. Definite courses of circulation are present in the coelenteron canal systems of *Aurelia* and *Pleurobrachia.* Ciliary movement contributes importantly to the transport of food in the digestive system of some echinoderms, numerous molluscs,[11] and many other organisms.

A cleansing role of cilia is common for sessile or sluggish animals. In many starfish, cilia beat from mouth toward anus over the entire surface, sweeping away debris. The action of the cilia of the epithelium of the frog's mouth and that of the cilia of the respiratory epithelium of mammals appears also to play this role.[35]

Ciliary movement either on the general body surface of aquatic skin breathers or on specially differentiated respiratory surfaces facilitates respiratory gaseous exchange by circulation of the ambient medium.

Cilia are essential to the normal circulation of the body fluids in some annelids such as the pelagic polychaete, *Tomopteris*, and the sipunculoids and echiuroids,[49,50] where a true blood-vascular system is vestigial or absent. In these worms a special system of ciliary tracts causes the coelomic fluids to circulate in a very effective manner. The coelomic fluids of starfishes are circulated in a similar fashion, the action of cilia providing effective circulation even within the dermal papillae. The cerebrospinal fluid of the frog has been reported to be circulated by ciliated epithelia lining the ventricles.[14]

Very commonly among animals, cilia are responsible for facilitating the passage of the materials normally conducted within tubules, particularly those of nephridia and kidneys and the ducts of the genital system.

The foregoing account of cilia suggests the abundant and widespread occurrence and varied functional roles played by these organelles. Although they are by no means able to produce movements of the power and conspicuousness of those effected by muscle contraction, there are numerous instances where cilia normally are indispensable functional constituents of the organism. In many of these instances they perform their assigned functions far more efficiently than would be the case for any conceivable typical muscular mechanism.

THE STRUCTURE OF CILIA

Cilia show a great uniformity of structure.[2,3,13,30,32,33,85] The cilium appears, optically, to be quite homogeneous.

The flagellum of some flagellates, such as *Euglena,* has attached to the sheath a series of diagonally oriented rodlets or mastigonemes, which give the flagellum a feathery appearance.[7,23,62] Such flagella are variously called "feather-type," "ciliary," or "stichonematic," in contrast to the simpler type. The proximal end of the axial filament is invariably associated with a basal granule, which is believed to be derived from the cell centrosome. Many cytologists have described systems of fibrils proceeding from the basal granules to the vicinity of the cell nucleus.

The ultrastructure has been disclosed with the aid of electron micrographs for the cilia of numerous unicellular and metazoan organisms.

The diameter of cilia and flagella usually lies within the limits of 0.20 to 0.25 μm. Cilia typically are 10 to 20 μm in length, while flagella may be somewhat longer, ranging from about 20 μm up to a few millimeters.[76] Some of the longest known flagella are sperm tails.

A cilium characteristically comprises a bundle of longitudinal fibrils,[1,20,27] the axoneme (Fig. 19–1A), enclosed within a bounding membrane which is continuous with the surface membrane of the cell. The axoneme is composed of an outer cylinder of nine double fibrils surrounding two central fibrils. All these fibrils are microtubules.[1,22,27] The central ones have a diameter of about 20 nm with walls about 4 nm thick. Both the two central and the nine double peripheral microtubules have walls made up of about 13 rows of a globular protein, tubulin, whose amino acid composition is quite similar to that of the protein, actin, of muscle fibers.[25,66,68] Each member of the pairs of the nine peripheral microtubules shares with its mate a common wall of about three rows of the tubulin, and hence the pairs each have a total of about 23 rows.

The axoneme becomes altered in structure as it enters the cell. The central fibrils terminate at the surface of the basal body of the cilium, and the peripheral doublets become triplets by the addition of a third microtubule. To the basal body, which is about 0.5 μm in length, are attached fibrous roots whose chief function appears to be to anchor the cilium securely at the cell surface. The peripheral fibrils of the axoneme bear arms (Fig. 19–1B) except at their base and tip. One of the fibrils of each doublet bears two rows of these arms, which are directed toward the armless member of the next adjacent pair in a clockwise direction as one views a cross-section of a cilium from its base. The arms, about 15 nm long and at about 17 nm intervals along the fibril, consist of molecules of a protein termed dynein, which acts as an enzyme in the presence of suitable concentrations of the divalent cations, cal-

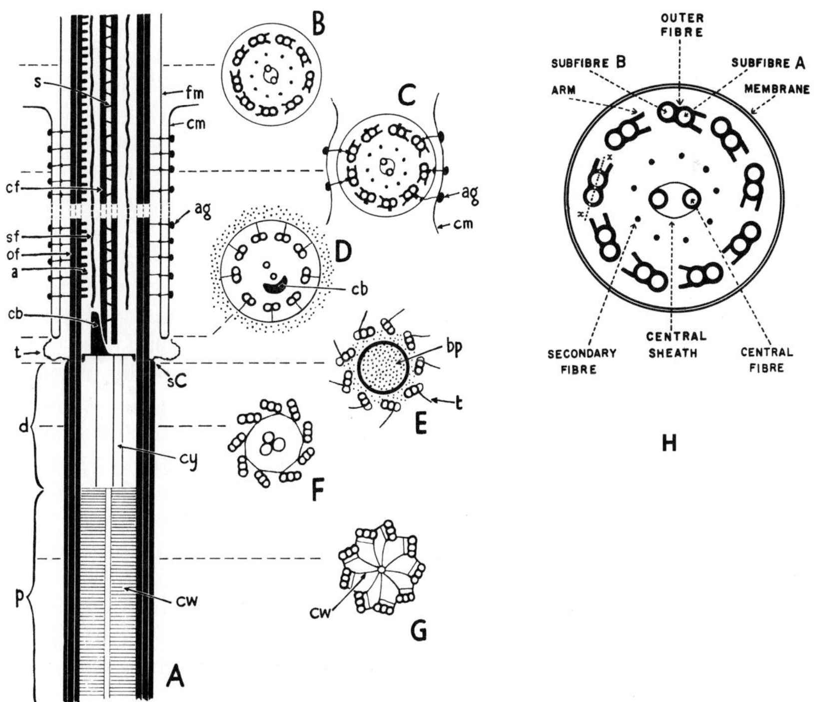

Figure 19–1. A, Diagrammatic reconstruction of a flagellum and basal body in the anterior body region of *Pseudotrichonympha.* A combines features seen in median and tangential longitudinal sections. B to G show transverse sections at the levels indicated. *a,* arms; *ag,* anchor granule; *bp,* basal plate; *cf,* central fiber; *cb,* crescentic body; *cm,* cell membrane, *cw,* cartwheel structure; *cy,* cylinders; *d,* distal region of basal body; *fm,* flagellar membrane; *of,* outer fiber; *p,* proximal region of basal body; *s,* central sheath; *sc,* distal end of subfiber *c*; *sf,* secondary fiber; *t,* transitional fiber. H, Diagrammatic transverse section of a flagellum. The line *xx* indicates the axial plane of the outer fiber. (From Gibbons, I. R., and A. V. Grimstone, J. Biophys. Biochem. Cytol. 7:697–716, 1960.)

cium and magnesium, to split ATP to release its contained energy.

Other structures have been described within the axoneme. Rods extending radially inward from the peripheral doublets and terminating in small knobs about half way to the central fibrils have been variously interpreted as a second ring of smaller longitudinal fibrils lying within the outer ring, or as a fibrillar system interconnecting the peripheral and central fibrils. A sheath has also been described surrounding the two central fibrils, but its nature, whether a spiral filament or a membrane, remains in doubt.

While the structure of the cilium just described is the rule, modifications of this organization are known. Simplifications, such as the loss of the two central fibrils, commonly occur in flagella and in cilia specially modified for roles other than motor action. Complications include the superimposition of other components peripheral to the typical nine microtubules, especially seen in some flagella such as many sperm tails.

Cilia often unite to form compound organelles. The cirri of *Euplotes* and the large cilia of *Nephthys* gills may be regarded as composite bodies comprising numerous cilia arising from a number of basal granules in a circular or oval field. The ciliary elements adhere closely to one another in a viscous matrix. The cilia of such a cirrus beat in unison. Membranelles of the adoral region of many peritrichs are small platelets of fused cilia beating synchronously. The undulating membranes of ciliates such as *Blepharisma* are long rows of cilia, each adhering to its neighbors. The compound character of the membrane may often be demonstrated by its fragmentation by appropriate manipulation with a microdissecting needle. The component cilia, when separated from one another, beat quite independently; when they have reunited, the characteristic coordinated activity giving rise to the undulatory movement is restored.

CHARACTERISTICS OF CILIARY MOVEMENT

Cilia usually beat at between 10 and 40 cycles per second. The activity of any cilium may be resolved into one or another, or some combination, of four fundamentally different types of movement.[29] One is pendular movement, in which the cilium bends back and forth, flexing only at its base.[37] This type is seen in frog pharynx or hypotrich cirri. The effective and the recovery phases of its stroke differ only in their rapidity. A second type is flexural. Bending begins first at the tip of the cilium and passes toward its base; in recovery, the cilium progressively straightens from base to tip. Such hooklike bending occurs in the laterofrontal cilia of lamellibranch gills. A third type of ciliary activity is undulatory movement. This type is characteristic of flagella. Waves usually pass along the flagellum from base to tip; more rarely they pass in the reverse direction[31] (see Figure 13–4).

Ciliary activity involving a combination of pendular and flexural activities is perhaps the most common. In the frontal cilia of the gill of *Mytilus* or on the gill of *Sabellaria*, the effective stroke is a rapid, stiffly sweeping, pendular movement (Fig. 19–2). The recovery stroke involves a bending back again beginning at the base of the cilium,[75, 78] and rapidly progressing to the tip. The recovery stroke is initiated before the end of the effective sweep.

In ciliated epithelia in which the direction of beat is constant, this occurs at right angles to an axis through the two central fibrils.[20, 26, 82] In guinea pig sperm the beat is, however, 20° to 30° off the perpendicular.[21]

Ciliary beat may occur with both effective and recovery stroke, or undulations on the same plane, or there may be three-dimensional movement.[57, 70] The last is seen for the locomotory cilia of protozoans like *Paramecium* and *Colpoda*, in which during the recovery stroke the cilial base appears to rotate counterclockwise while the effective stroke occurs in a single plane. Undulatory movements of flagella may be helical rather than planar. The vibratile organs of the mammalian epididymis and in a number of flagellates as, for example, in *Trypanosoma*, move in three dimensions. In such organisms as the latter, the undulatory and pendular movements need not occur in a single plane, nor need the two occur simultaneously in the same plane. Thus, the tip of the flagellum in the course of its beat may trace out an elliptical orbit, a figure 8, or a more complicated figure.

The single vibratile element of the flagellate *Monas* has been described as possessing the capacity to carry out numerous complex activities.[40] Forward movement is accom-

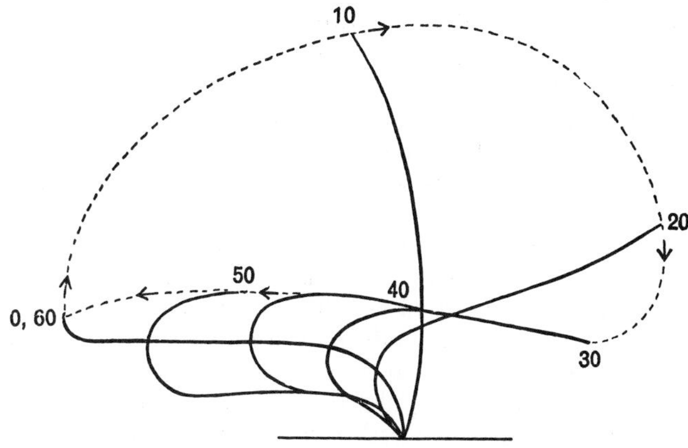

Figure 19–2. Stages in beat cycle of a polychaete gill cilium. Time (msec) to each stage from the initiation of the effective stroke is indicated. (From Sleigh, M. A., Endeavour *30*:11–17, 1971.)

plished by a rapid pendular movement of the flagellum from a position directed forward to one at about a right angle to the direction of progression (Fig. 19–3A). The recovery stroke is typical flexural recovery. The effective stroke may, instead, be initiated near the base and rapidly pass toward the tip. In slower forward movement, the total sweep of the flagellum may be reduced to about half the amplitude (Fig. 19–3B), but with the typical form of beat seen in rapid progression. Backward progression is brought about by undulatory activity, with waves passing from base to tip of the organ (Fig. 19–3C). Lateral movements result from undulatory movements with the flagellum flexed at about 90 degrees (Fig. 19–3D), and forward movement by directing the flagellum backward. The undulatory activity may involve only the tip of the flagellum, or practically its whole length.

The activity of the flagellum has been investigated with the aid of high-speed cinema photomicrography.[43] The technique of instantaneous fixation, in which cilia are halted abruptly and fixed at whatever stage of their cycle they happen to be in, has been especially productive in resolving the successive form changes during the beat cycles. The undulatory activity most commonly passes from base to tip around the flagellum, as well as along it. Such movement imparts rotational and gyrational components, along with an outward thrust. The resulting rotation and gyration of the body of the flagellate around an axis which constitutes the direction of locomotion is considered to provide an important, if not the chief, force propelling the organism forward. The body proper acts under these circumstances as a screw propeller.[42]

From the preceding paragraphs, we see that ciliary movement is highly variable even in a single cilium; activity seems to be under the control of the general response mechanism of the body.

Cilia are able to propel an organism through the water or to propel the surrounding medium past a stationary ciliated cell as a result of a directed thrust upon the medium. More work must be done upon the medium during the effective phase of a stroke than during recovery. In those cases where the effective stroke is pendular and recovery is flexural, the mechanism is obvious. Simple

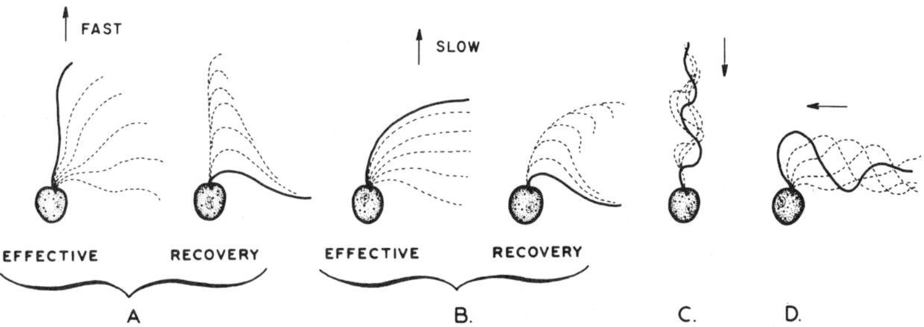

Figure 19–3. Some activities of the flagellum of *Monas* during locomotion. Arrows indicate direction of movement of the organism. (Redrawn from Krijgsman, B. J.: Arch. Protist., vol. 52).

pendular ciliary activity would be expected to be quite an inefficient type for directed movement.

Although cilia appear, on observation, to move through the medium at a very high rate, they are actually moving very slowly, as a simple calculation will show. The angular velocity of cilia is high, as is seen from the fact that a cilium may show many cycles per second, but the tip of a cilium is actually moving through the medium at a maximum rate of only a few feet per hour, a rate quite slow in terms of propulsive instruments of the type of which we ordinarily think. The rate of the effective stroke is usually two to ten times that of the recovery.

The reversal of the direction of effective beat of cilia is common. Among the ciliated protozoans, small turbellarians such as *Stenostomum*, and ciliated larvae, reversal of direction of locomotion reflects this reversal of activity. Some early reported cases of ciliary reversal among certain metazoans appear, however, to have an alternative explanation. The apparent reversal on the labial palps of *Ostrea*, and probably also in the gullet of the coelenterate *Actinoloba*, appears to be explicable in terms of opposite directions of beat of cilia in longitudinal grooves and on ridges, together with muscular movements emphasizing action of one or the other of the two tracts.

The cilia of the lips and gullet of the sea anemone, *Metridium*, and also of other species of anthozoans, whose tentacles are too short for successful transport of food to the mouth, appear capable of reversal of beat in response to direct stimulation by chemical agents. In the normal unstimulated anemone, the cilia beat outward. When, however, a piece of crab flesh is brought into contact with the ciliated epithelium, those cilia in the immediate region of the food reverse their beat, now carrying the food into the gullet. Immediately following passage of the particle, the cilia resume their outward beat.

The effect of crab flesh in inducing ciliary reversal in the *Metridium* can be imitated through the application of 2½ per cent KCl in sea water or the addition of glycogen. After such treatment the direction of both the effective stroke and the metachronic wave is reversed. Such treatment does not reverse the direction of beat of the cilia of the tentacles or siphonoglyphs.

In *Paramecium*, ciliary reversal is effected by treatment with KCl and salts of other monovalent cations.[8,56] When the animal is subjected to a field of electric current flow, there is a reversal of beat of cilia on the end of the animal nearer the cathode. Reversal of the direction of spiraling also occurs but may be due to other factors, such as medium viscosity.[74]

Ciliary reversal occurs for the ectodermal cilia of amphibian larvae;[84] local ciliary reversal is observed in response to mechanical stimulation. In higher metazoans in general, it appears impossible to effect ciliary reversal.

CILIARY CONTROL AND COORDINATION

Cilia appear to show a well developed automaticity. The cilia of bits of isolated epithelial tissue continue to beat after removal from the organism. In protozoans, the cilia of small enucleated fragments of an individual continue to beat for some time. Sperm cells with the head removed continue to show active locomotion. The automaticity of ciliary action therefore resides entirely in the cilium.

Ciliated surfaces typically exhibit metachronism, indicating a general ciliary coordination of action. Metachronism is a term applied to the orderly succession of initiation of beat of cilia located in a spatial sequence in a ciliated surface (Fig. 19–4). Each cilium along one axis is, in consequence, very slightly out of phase with its neighbor in front of and behind it in the direction of the metachronic wave propagation. At right angles to this wave, cilia beat isochronically, or in phase with one another, giving an optical picture of waves passing over the epithelium. The crests of the waves include cilia at the peak of their effective stroke; the troughs include those cilia at the ends of their effective strokes and about to commence their recovery.

The direction of the metachronic wave of a ciliated surface often appears to be as fundamental a property of a surface as is the direction of ciliary beat itself. Even a small isolated portion of a ciliated surface may continue to show its own inherent direction of metachronism. This is not disturbed by removing and then replacing, after rotating through 180 degrees, a portion of the ciliated epithelium of the roof of a frog's mouth. In these circumstances the transplanted portion

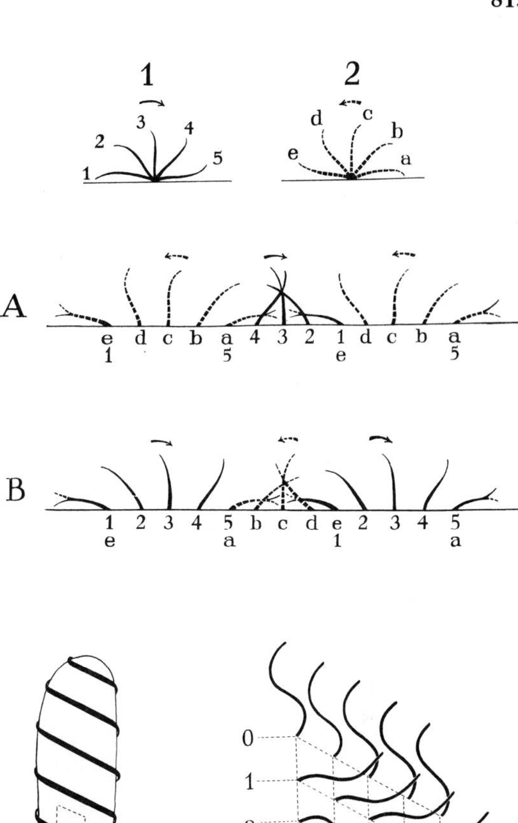

Figure 19–4. A, Schematic representation of the two types of metachronism. 1, individual stages of the effective stroke to the right; and 2, stages of the recovery stroke to the left. A and B, Metachronal ciliary rows in which the subsequent stages follow each other from right to left (symplectic metachronism, A) or from left to right (antiplectic metachronism, B). C, *Paramecium multimicronucleatum.* Scheme of the metachronal waves on the animal. The arrow shows the direction of swimming. D, *Paramecium multimicronucleatum.* Scheme of a single beating cycle corresponding to the dotted area in B. The spacing of the cilia is increased in both directions for the sake of clarity. (From Parducz, B., Int. Rev. Cytol. *21*:91–128, 1967.)

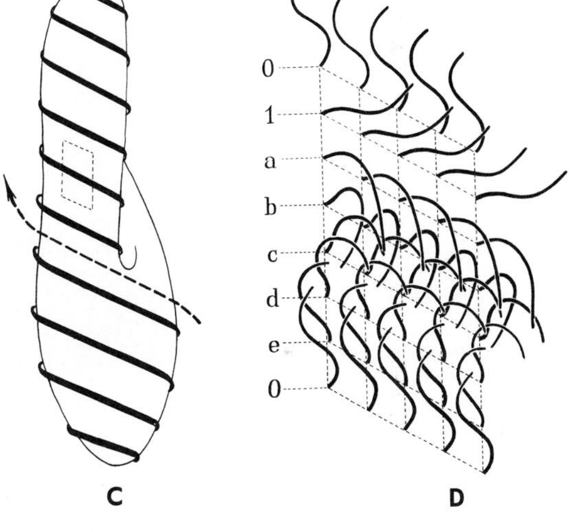

has its metachronic wave proceeding in a direction opposite that of the surrounding tissue.

Among ciliated tissues from various sources, the direction of the metachronic wave bears various relationships to the direction of the effective stroke. A general terminology to describe the major relationships has been proposed (Fig. 19–5), symplectic if in the same direction, antiplectic if about 180°, and laeoplectic or dexioplectic if about 90° to left or right respectively.[39] There appears to be a correlation between type of wave and function, with particle and slime transport associated with the orthoplectic types and water pumping with the diaplectic.

In the frog's mouth, the metachronic wave and ciliary effective strokes are in the same direction. In the rows of ciliated combs of ctenophores the two are in opposite directions.[15] In the cilia of the gills of the annelid *Nephthys*[16] and in the lateral ciliated epithelium of *Mytilus* gills[29, 37] the metachronic wave is always at right angles to the effective stroke. In the ciliates *Paramecium* and *Colpoda*, in which rapid metachronic waves pass from posterior to anterior tips, the effective stroke may be oblique.[57] In *Opalina*, the direction of the metachronal wave changes freely with the direction of swimming, passing backward, to the right, or to the left[75] (Fig. 19–6). It has been suggested that under

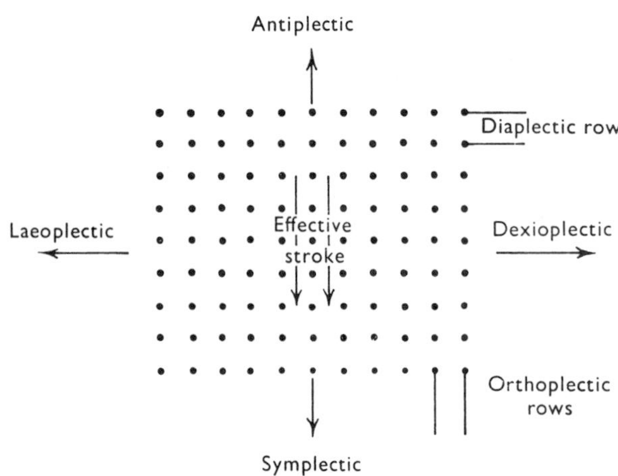

Figure 19–5. The nomenclature of ciliary patterns and metachronal patterns proposed by Knight-Jones (1954) (arrows indicate the movement of metachronal waves). (From Sleigh, M. A., Symp. Soc. Exp. Biol. *20*:11–31, 1966.)

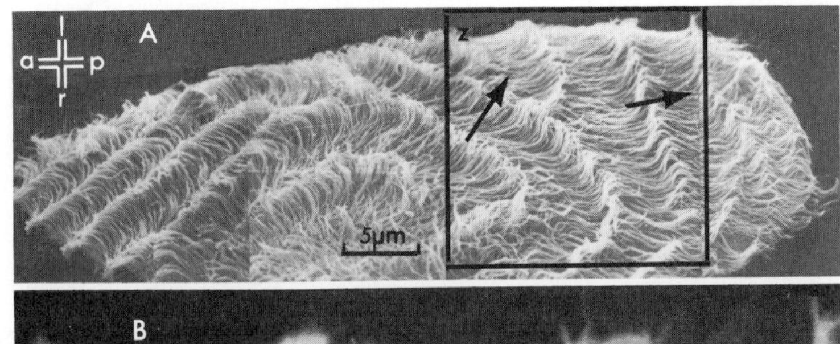

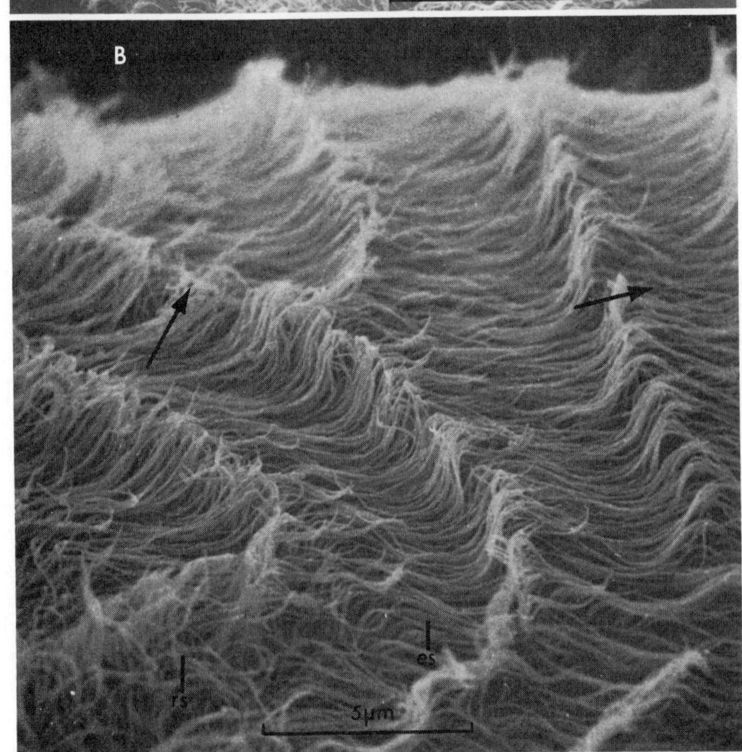

Figure 19–6. A, Instantaneously fixed metachronal waves travelling in different directions on different regions of an *Opalina*. The organism was fixed while turning in Ringer solution. B, Enlargement of region *z* in A, showing cilia in effective (*es*) and recovery strokes (*rs*). The waves are viewed from behind and almost end-on. *a*, anterior; *p*, posterior; *r*, right; *l*, left. (From Tamm, S. L., and G. A. Horridge, Proc. Roy. Soc. Lond. B *175*:219–333, 1970.)

some circumstances such as high medium viscosity, the wave as a whole is the propulsive agent, in contrast with to effects of the individual cilia.

The activity of flagella, when several occur close together, whether on a single cell or on closely packed, contiguous, independent ones such as among sperm tails, tend to display synchronous undulations of nearly the same wavelength and frequency.[7] The functional significance of such synchrony is evident. Although a smoothly continuous propulsive force can be generated by a single flagellum, when that one occurs together with others their efficiency is maximized by reduction to a minimum of the interferences among them due to interactions by way of the viscous forces of the medium. For the cilia whose propulsive force is periodic, the metachronism would smooth out the propulsive action, and would also raise the efficiency of their action by the contributions of near synchrony of the immediately adjacent cilia to the reduction of mutual viscous interference. The generation of the metachronism is a consequence of the eccentric ciliary beat, with the temporal differences corresponding to times of maximum hydrodynamic interference for the effective and recovery strokes.[79] Laeoplectic and dexioplectic metachronism are most common in cases where narrow bands of cilia occur, and the plane of ciliary beat is oblique to the band direction.

Theories of the mechanism of ciliary coordination[38, 58, 60, 76, 77] have included (a) a neuroid transmission of an excitatory wave along the cell surface, (b) the triggering of the beat of the next adjacent cilium by either mechanical contact of the cilia themselves or by some type of transmission from one cilium base to the next, and (c) simply the operation of the viscous interaction through the medium. The last explanation presently seems adequate to account for most observations and experimental results.

Although ciliary beat appears in general to be the result of an automaticity of the ciliary mechanism, the cilia in all investigated cases are controlled in their activities by more basic response mechanisms within the organism. This control ranges from mere inhibition, through regulation of rate of beating, to reversal of beat,[4] and even to further regulation of beat direction in some protozoans such as *Opalina* and *Paramecium* to regulate swimming direction.[59, 60]

The hypotrichs *Uronychia* and *Euplotes* show several types of locomotory patterns. A role of the neuromotor system in control of locomotion in *Euplotes* is suggested from experimental destruction of parts of this system. In a number of metazoans, both direct and indirect evidence has been advanced pointing to nervous modification of ciliary activity.[28, 44]

The rows of ciliary combs of ctenophores are controlled by impulses passing out over the paddle rows from the aboral sense organs.[5] The ciliary systems of the turbellarian *Stenostomum* and of the ciliated larvae of many molluscs and annelids are known to be nerve controlled. The cilia may be active or motionless or, in the case of *Stenostomum*, reversed. A freshly isolated piece of ciliated lip of the snail *Physa* shows no ciliary activity unless a nerve fiber innervating this region is stimulated. On the other hand, pedal ciliary activity in *Alectrion* appears to be inhibited by nervous action. If the foot is excised, there is temporary cessation of ciliary beat, followed by a resumption, after which the beat continues unabated for the remainder of the life of the ciliated epithelium.

The pharyngeal cilia of the frog are accelerated greatly and the amplitude of the beat increased by electrical stimulation of certain cranial nerves.[45, 73] This response, which has a latent period of about a second, persists for as long as 10 minutes following stimulus cessation. It is not known whether the fibers are sympathetic or parasympathetic, though the bulk of the evidence suggests that they belong to the facialis group. Acetylcholine has been reported to increase ciliary activity.[9, 10, 48, 72] In further support of normal nervous control of the pharyngeal cilia of the frog is the observation that these cilia are normally quiescent;[46] they commence to beat in response to stimuli such as the addition of foreign particles and cease beating after these particles are swept away.

The mechanisms of control of cilia are largely unknown. There is reason to believe that it is more complex among protozoans, in which a number of complex locomotory patterns are exhibited by a single individual and the same cilia may perform a variety of other functions, than in the ciliated epithelia of the larger metazoan invertebrates and vertebrates. In these higher animals the pattern of coordination seems for the most part to be invariable.

Studies on the progressive anesthetization of the ciliary mechanism have shown that in protozoans the first thing to be lost is the power of ciliary reversal. At a later stage metachronic coordination is lost, leaving the cilia beating independently. Last of all, the cilia entirely cease their activity. This observation has led to the hypothesis that ciliary beat, metachronism, and reversal are controlled by three separate mechanisms. Reversal must be governed by a mechanism of relatively rapid transmission affecting all cilia of the organism nearly simultaneously and passing in all directions over the animal. Transverse cuts passing deeply into those protozoans investigated in this regard do not measurably interfere with the transmission of excitation in response to stimuli which induce ciliary reversal. Changes in ciliary beat are accompanied in ciliate protozoans by changes in membrane potential which are associated with the movements of ions, particularly calcium.[18, 18a, 34, 50a, 51, 52, 53, 55]

THEORIES OF CILIARY MOVEMENT

Historically, theories have fallen into two general categories: (1) those that assumed the moving force to occur in the cell body proper, with the cilia acting only passively, and (2) those that assumed that the moving force occurred in the cilium itself, the cilia therefore moving as a result of intrinsic contractile capacity. Schafer in 1891 proposed that cilia were passive bodies comparable to hollow elastic structures with differences in the degree of elasticity of various portions of their walls. Ciliary beat was held to be a result of rhythmic surging and ebbing of fluid into and out of these cellular processes.

Much more compatible with modern theories of the mechanism of action, Heidenhain in 1911 advanced the view that the cilium itself contains actively contractile elements responsible for the movements. Supporting this view are numerous observations. The waves passing out along a flagellum often show no reduction in amplitude as they pass from base to tip. Indeed, increases in amplitude are commonly observed. A reduction would be expected were kinetic energy generated only in the cell body proper at the flagellar base. In species such as *Peranema*, the undulatory activity may be restricted to the tip region of the flagellum, the rest of the flagellum remaining relatively rigid. Also, the flagellar wave passes from tip to base in trypanosomes.

Current theories of the mechanism of activity of cilia are based upon the presumption that the complex, detailed structural organization of a cilium, so nearly universally similar among all organisms, is intimately associated with the process. Within this ciliary structure are sought the principal functional properties that are displayed in ciliary movement. One is the ability to bend, and to bend in two or more directions. Another is the capacity to propagate bending activity along the shaft, and a third is a means to retain at appropriate times a high degree of stiffness. Nearly all portions of the ciliary shaft may display all of these properties at one time or another within a single beat cycle. Clearly, these are properties of an intrinsically active, in contrast with a passive, cell organelle. It is readily evident that the system of longitudinal fibrils could account for the stiffness, especially if cross-linkages were developed among the fibrils. The properties of bending and its propagation are less obviously explained.

Two kinds of hypotheses have been advanced for the bending. One is that the fibrils possess the ability to contract and that the bending is accountable by differential contraction of the fibrils of the peripheral ring. Bending could thus occur in any plane, propagated directly along the shaft, or spirally along it to produce circular or helical bending. Evidence for such postulated shortening of the fibrils is still wanting. The other kind of hypothesis is that, as has been demonstrated for muscle fibers, the peripheral fibrils do not contract but rather that they slide actively relative to one another, to the central fibrils, or to the ciliary matrix, and produce bending as a consequence of all the fibrils being firmly rooted at the base of the cilium. A result from such a mechanism for fibrils of unchanging length would be that the peripheral doublet fibrils on the concave side of the bent cilium, whatever its direction of bending, would extend further into the ciliary tip than the doublets on the convex side. Evidence has been presented that this is the case.[71]

While it is conceivable that the force effecting fibril sliding is exerted between a peripheral doublet and its immediately surrounding matrix, or even the two central fibrils, the orientation of the ATP-splitting dynein arms toward an adjacent peripheral doublet suggests that the sliding force operates between these doublets. It has been postulated, therefore, that the mechanism involves dynein linkages between a tubulin molecule of one doublet and one of the next doublet within the ring.

It has been postulated that within a cilium, erect at the initiation of the effective stroke, all fibris terminate at the same level in the ciliary tip.[80] To commence the stroke, linkages between adjacent doublets are postulated to occur all along the fibrils except their base and tip, where arms are lacking, and to generate a sliding force which induces bending at the less stiff, linkage-less ciliary base of the otherwise relatively stiff shaft. The shaft is held stiffened by an intrinsic stiffness of the fibrils augmented by the linkages. The required rate of sliding, varying with rate of ciliary beat, would be only a few μ/sec. As the effective stroke draws toward its conclusion, the base commences to return to its erect state by reversal of sliding direction in the bend near the base. The localized region of reversed sliding is thereupon propagated up the shaft toward the tip until the whole is returned to the initial state, preparatory for another effective stroke.

The rate of propagation of the bending during the recovery stroke appears to be related to the rate of fibrillar sliding, suggesting that the rate of propagation itself could depend upon the rate at which linkages were broken during sliding and freed for the formation of new ones.[6]

The same general kind of hypothesis may be extended to the undulatory movement of flagella by the simple further postulation that localized regions of sliding and bending in essentially opposite directions are propagated along the shaft.

ENERGY FOR CILIARY ACTIVITY

The force of ciliary beat in *Paramecium* appears to remain constant; the velocity of the ciliary effective stroke is inversely proportional to the viscosity of the medium in solutions of gum arabic.[63]

Ciliary activity normally involves a utilization of oxygen, the rate of consumption varying with ciliary rate.[29] O_2 consumption increases and beat ceases upon treatment with 2,4-dinitrophenol, probably by uncoupling energy-using steps.[86] However, like muscle, cilia appear capable of acting for a time anaerobically. After administration of 0.1 per cent NaCN to a fragment of *Mytilus* gill, O_2 consumption drops off abruptly, but ciliary activity diminishes much more slowly. Fragments of *Mytilus* gill placed in weak hemoglobin solutions in chambers perfused with damp hydrogen show ciliary beat continuing for some time after deoxygenation of the hemoglobin. After ciliary beat has become very slow under the anaerobic conditions, rapid beat is restored quickly on readmission of oxygen and reoxygenation of the hemoglobin. The cilia of the gill of *Pecten*, on the other hand, are reported to cease beating immediately on removal of oxygen, as indicated by the reduction of dyestuffs such as janus green and neutral red, but before nile blue is reduced. The cilia of *Paramecium* cease activity within a few seconds in an O_2-free medium.[24] It is known, nevertheless, that some ciliates are able to carry on normal ciliary movement in natural environments which are practically oxygen-free. One cannot yet conclude whether the need for O_2 differs qualitatively or only quantitatively among different species of animals.

The work performed by a cilium or flagellum is chiefly for overcoming the viscous resistance of the surrounding medium together with the elastic resistance to bending of the shaft.[67, 81] Inertial forces of the minute structures are insignificant. The force needed during the effective stroke of a cilium to overcome the viscous resistance is proportional to the angular velocity, the cilium length and diameter, and viscosity of the medium. For a 32 μ cilium at 20°C, for example, the work done in the effective stroke was calculated to be about 4×10^{-9} ergs, about four times that done during the recovery. For such a cilium it has been calculated that, if in a single cycle each dynein molecule were to split one ATP, the released energy would be about 4×10^{-7} ergs/sec. Work rate of a single flagellum has been found to be of the order of 1 to 4×10^{-7} ergs/sec.

INFLUENCE OF ENVIRONMENTAL FACTORS ON CILIARY ACTIVITY

Temperature and hydrogen ion concentration strongly influence ciliary activity. The maximum rate of ciliary activity occurs at about 34°C for *Mytilus* gill and 35°C for frog epithelium.[29, 45, 61] The Q_{10} varies with the temperature range. For *Mytilus* gill cilia it ranges from near 3 at the lower temperatures to slightly less than 2 at the upper end of the physiological range. Above a critical temperature there is a rapid drop in rate; this critical temperature would be expected to vary with mean habitat temperature.

Hydrogen ion concentration, which varies considerably in nature, has a greater influence on the rate of ciliary activity than does concentration of any other ion.[12, 54, 83] Acids, like carbonic, which penetrate cells more rapidly have a greater effect than those which penetrate less rapidly. High hydrogen ion concentration can completely inactivate cilia. A biological significance of this response of cilia to increased acidity is evident for the bivalve mollusc. After two or three hours of valve closure there is adequate concentration of CO_2 to inhibit ciliary activity. Inasmuch as O_2 consumption of ciliated tissue is a direct function of the rate of ciliary activity, this inhibitory influence of CO_2 is obviously adaptive. The minimum hydrogen ion concentration effecting inhibition of gill cilia varies with species. For example, it is lower in *Mya*, which customarily dwells in well aerated water, than in *Ostrea* or *Mytilus*.

A very striking instance of adaptive variation in the minimal hydrogen ion concentration essential for ciliary inhibition has been reported for the cilia of the different portions of the digestive tract of *Mya*. There is a definite correlation between the normal pH of each region of the digestive tract and the pH at which the cilia of that region are brought to rest.

Other cations, and also anions, influence ciliary activity.[29] Potassium produces a transient augmentation of ciliary beat in *Mytilus*.[36] In general, these ions are relatively constant in concentration in any given natural environment. The effects of altering their concentrations and ratios appear to be basic ones, quite comparable to their action on all cells. As with certain other cellular phenomena, cations of the medium have more influence on ciliary beat than have anions.

REFERENCES

1. Afzelius, B. A., J. Biophys. Biochem. Cytol. *5*:269–278, 1959. Ultrastructure of cilium.
2. Afzelius, B. A., J. Biophys. Biochem. Cytol. *9*:383–394, 1961. Fine structure of ctenophore ciliary plates.
3. Allen, R. D., J. Cell Biol. *37*:825–831, 1968. Electron micrograph cross sections of *Tetrahymena* cilia.
4. Andrivon, C., L'Annee Biol. Ser. 4, *8*:99–114, 1969. The phenomenon of ciliary reversal in protozoans.
5. Atkins, D., Quart. J. Micr. Sci. *80*:321–435, 1938. Ciliary mechanisms, lamellibranchs.
6. Brokaw, C. J., Symp. Soc. Exp. Biol. *22*:101–116, 1965. ATP in ciliary movement.
7. Brown, H. P., Ohio J. Sci. *45*:247–301, 1945. Structure of protozoan flagellum.
8. Brutkowska, M., Acta Protozool. *4*:353–364, 1967. Ciliary reversal.
9. Bulbring, E., J. H. Burn, and H. J. Shelley, Proc. Roy. Soc. Lond. B *141*: 445–466, 1953. Acetylcholine and ciliary movement.
10. Burn, J. H., Pharmacol. Rev. *6*:107–112, 1954. Acetylcholine and ciliary movement.
11. Carriker, M. R., Biol. Bull. *91*:88–111, 1946. Ciliation in gut of *Lymnaea*.
12. Chase, A. M., and O. Glaser, J. Gen. Physiol. *13*:627–636, 1930. H^+ ion concentration and ciliary activity.
13. Chasey, D., J. Cell. Sci. *5*:453–458, 1969. Fine structure of *Tetrahymena* cilia.
14. Chu, Hsiang-Yao, Amer. J. Physiol. *136*:223–228, 1942. Ciliary circulation of anuran cerebrospinal fluid.
15. Coonfield, B. R., Biol. Bull. *66*:10–21, 1934. Coordination of swimming combs of ctenophores.
16. Coonfield, B. R., Biol. Bull. *67*:399–409, 1934. Ciliary movement on *Nephthys* gills.
17. Drai, A. D. G., Neth. J. Sea Res. *3*:391–422, 1967. Activity and roles of latero-frontal cilia of *Mytilus*.
18. Dryl, S., Acta Protozool. 7:325–333, 1970. Excitation in ciliary control.
18a. Eckert, R., Science *176*:473–481, 1972. Control of ciliary action.
19. Fauré-Fremiet, E., Biol. Rev. *36*:464–536, 1961. General review ciliary movement.
20. Fawcett, D. W., pp. 217–297. *In* The Cell, Vol. 2, edited by J. Brachet and A. A. Mirsky, Academic Press, New York, 1961. General treatment of cilia and flagella.
21. Fawcett, D. W., J. Cell. Sci. *3*:187–198, 1968. Beat direction in relation to central tubules.
22. Fawcett, D. W., and K. R. Porter, J. Morphol. *94*:221–281, 1954. Fine structure of cilia.
23. Foster, E., M. B. Baylor, N. A. Meinkoth, and G. I. Clark, Biol. Bull. *93*:114–121, 1947. Structure of protozoan flagellum.
24. Gersch, M., Protoplasma *27*:412–441, 1937. O_2 and ciliary activity in *Paramecium*.
25. Gibbons, I. R., Symp. Internat. Soc. Cell. Biol. *6*:99–113, 1968. Chemical composition of cilia.
26. Gibbons, I. R., J. Biophys. Biochem. Cytol. *11*:179–205, 1961. Relation of fine structure to beat direction in lamellibranch.
27. Gibbons, I. R., and A. V. Grimstone, J. Biophys. Biochem. Cytol. *7*: 697–716, 1960. Ciliary ultrastructure.
28. Gothlin, G. L., Skand. Arch. Physiol. *58*:11–32, 1930. Influence of nerves on ciliary activity.
29. Gray, J., Ciliary Movement. Cambridge University Press, 1928.
30. Grim, J. N., J. Protozool. *14*:625–634, 1967. Ultrastructure of *Euplotes* ciliary structure.
31. Holiwell, M. E. J., Physiol. Rev. *46*:696–785, 1966. General review of cilia.
32. Horridge, G. A., Proc. Roy. Soc. Lond. B *162*:351–364, 1965. Macrocilia of ctenophore (*Beroe*) lips.
33. Horridge, G. A., and S. L. Tamm, Science *163*:817–818, 1969. Scanning electron micrographic studies of *Opalina* cilia.
34. John, T. L., J. Cell. Physiol. *70*:79–90, 1967. Mechanism of ciliary movement.
35. Kilburn, K. H., Ed. Suppl. Amer. Rev. Resp. Dis. *93*, 1966. Respiratory cilia.
36. Kinosita, H., Annot. Zool. Japan. *25*:8–14, 1952. Effect of KCl on cilia.
37. Kinosita, H., and T. Kamada, Japan. J. Zool. *8*:291–310, 1939. Ciliary movement in *Mytilus*.
38. Kinosita, H., and A. Murakami, Physiol. Rev. *47*:53–82, 1967. General review of ciliary movement.
39. Knight-Jones, E. W., Quart. J. Micr. Sci. *95*:503–521, 1954. Ciliary metachronism.
40. Krijgsman, B. J., Arch. Protist. *52*:478–488, 1925. Flagellar movement in *Monas*.
41. Lowndes, A. G., Nature *150*:579–580, 1942. Ciliary movement and density, *Pleurobrachia*.
42. Lowndes, A. G., Proc. Zool. Soc. Lond. *114*:325–338, 1944. Flagellar activity, *Monas* and *Peranema*.
43. Lowndes, A. G., School Sci. Rev. *100*:319–332, 1945. Locomotion in *Euglena*.
44. Lucas, A. M., J. Morphol. *53*:243–263, 1932. Influence of nerves in ciliary activity.
45. Lucas, A. M., J. Morphol. *53*:265–276, 1932. Influence of temperature on ciliary beat.
46. Lucas, A. M., Amer. J. Physiol. *112*:468–476, 1935. Control of cilia of frog mouth.
47. Machin, K. E., Proc. Roy. Soc. Lond. B *158*:88–104, 1963. Ciliary coordination.
48. Maroney, S. P. J., and R. R. Ronkin, Biol. Bull. *105*:378, 1953. Cholinesterase and ciliary movement.
49. Meyer, A., Z. wiss. Zool. *135*:495–538, 1929. Coelomic ciliation of annelids.
50. Meyer, A., Zool. Jahrb. Abt. und Ont. Tiere *64*:371–436, 1938. Ciliated coelom in *Tomopteris*.
50a. Murakami, A., and R. Eckert, Science *175*:1375–1377, 1972. Ca ions in ciliary regulation.
51. Naitoh, Y., Science *154*:660–662, 1966. Ciliary reversal in *Paramecium*.

52. Naitoh, Y., J. Gen. Physiol. *51*:85–103, 1968. Mechanism of ciliary reversal, "a calcium hypothesis."
53. Naitoh, Y., J. Gen. Physiol. *53*:517–529, 1969. Regulation of cilia in *Paramecium*.
54. Nomura, S., Protoplasma *20*:85–89, 1933. H^+ ions and ciliary movement; O_2 consumption of cilia.
55. Okaljima, A., Japan. J. Zool. *11*:87–100, 1953. Membrane potential and beat direction.
56. Oliphant, J. F., Physiol. Zool. *15*:443–452, 1942. Influence of chemicals on ciliary beat in *Paramecium*.
57. Parducz, B., Acta Biol. Acad. Sci. Hungaricae *5*:169–212, 1954. Ciliary activity in infusoria.
58. Parducz, B., J. Protozool. *6*:Suppl. 27, 1962. Ciliary coordination.
59. Parducz, B., Acta Protozool. *2*:367–374, 1964. Environmental regulation of ciliary movement.
60. Parducz, B., Int. Rev. Cytol. *21*:91–128, 1967. General review of ciliary movement.
61. Parker, G. H., and A. P. Marks, J. Exp. Zool. *52*:1–7, 1928. Ciliary reversal, *Metridium*.
62. Peterson, J. B., Bot. Tidskr. *40*:373–389, 1929. Structure of feathered flagella.
63. Pigon, A., and H. Szarski, Bull. Acad. Polonaise Sci. *3*:99–102, 1955. Velocity and force of ciliary beat.
64. Pitelka, D. R., and F. M. Child, pp. 131–198. *In* Biochemistry and Physiology of Protozoa, Vol. 3, edited by S. H. Hutner. Academic Press, New York, 1964. Review of ciliary structure and function.
65. Purchon, R. D., Proc. Zool. Soc. Lond. *124*:859–911, 1955. Ciliation of Pholadidae.
66. Renaud, F. L., A. J. Rowe, and I. R. Gibbons, J. Cell. Biol. *36*:79–90, 1968. Proteins of cilia.
67. Rikmenspoel, R., and M. A. Sleigh, J. Theoret. Biol. *28*:81–100, 1970. Ciliary work.
68. Ringo, D. L., J. Ultrastruct. Res. *17*:266–277, 1967. Ciliary fine structure.
69. Rivera, J. A., Cilia, Ciliated Epithelium and Ciliary Activity. Macmillan Co., New York, 1962. General review of cilia.
70. Satir, P., J. Cell. Biol. *18*:345–365, 1963. Form of ciliary movement.
71. Satir, P., J. Cell. Biol. *39*:77–94, 1968. Mechanism of ciliary bending.
72. Seaman, J. R., Biol. Bull. *99*:347, 1950. Acetylcholine and ciliary movement.
73. Seo, A., Japan. J. Med. Sci. III Biophysics *2*:47–75, 1931. Nervous control of frog mouth.
74. Seravin, L. N., Acta Protozool. *7*:313–323, 1970. Regulation of protozoan spiralling.
75. Sleigh, M. A., J. Exp. Biol. *37*:1–10, 1960. Ciliary beat, *Opalina* and *Stentor*.
76. Sleigh, M. A., The Biology of Cilia and Flagella. Pergamon Press, Oxford, 1962. General account of cilia.
77. Sleigh, M. A., Symp. Soc. Exp. Biol. *20*:11–31, 1966. General review of ciliary coordination.
78. Sleigh, M. A., Symp. Soc. Exp. Biol. *22*:131–150, 1968. Form of ciliary beat.
79. Sleigh, M. A., Int. Rev. Cytol. *25*:31–54, 1969. Mechanism of metachronal coordination.
80. Sleigh, M. A., Endeavour *30*:11–17, 1971. Review of ciliary movement.
81. Sleigh, M. A., and M. E. J. Holiwell, J. Exp. Biol. *50*:733–743, 1969. Ciliary work.
82. Tamm, S. L., and G. A. Horridge, Proc. Roy. Soc. Lond. B *175*:219–333, 1970. Central fibrils and ciliary beat in *Opalina*.
83. Tomita, G., J. Shanghai Sci. Inst. *1*:Sec. IV, 69–76, 77–84, 1934. H^+ ions and ciliary movement.
84. Twitty, V. E., J. Exp. Zool. *50*:319–344, 1928. Ciliary reversal in Amphibia.
85. Warner, F. D., J. Cell. Biol. *47*:159–182, 1970. Flagella fine structure.
86. Weller, H., and R. R. Ronkin, Proc. Soc. Exp. Biol. Med. *81*:65–66, 1952. 2,4-Dinitrophenol on cilia.
87. Yonge, C. M., Great Barrier Reef Expedition Sci. Rep. *1*:13–58, 1930. Ciliary feeding in corals.

CHAPTER 20

CIRCULATION OF BODY FLUIDS

By C. Ladd Prosser

All animals have some mechanisms in addition to diffusion for the transport of essential gases, nutrients, and products of metabolism from one part of the body to another. All animals, including unicellular ones, have multiple fluid compartments.

CLASSIFICATION OF CIRCULATORY MECHANISMS

Several kinds of circulatory mechanisms serve the transport function (Fig. 20–1):

Intracellular Transport. In protozoans there is usually protoplasmic movement, which supplements diffusion. Protoplasmic streaming may follow a definite course, as in circulation of food vacuoles in ciliates. In metazoans some protoplasmic streaming occurs in most cells, if not in all; it can be observed well by time-lapse motion photography.

Movement of External Medium. In some animals, particularly sponges and coelenterates, the water in which they live provides the medium of transport. The medium passes through definite body channels and may be propelled by ciliary, flagellar, or muscular activity.

Movement of Body Fluids by Somatic Muscles. Body fluid is circulated in a pseudocoelom in nematodes, bryozoans, and rotifers. Some transport is associated with body movement in the mesodermally lined coelom of echinoderms, annelids, sipunculids, and chordates. The coelom is reduced, in some molluscs, to the pericardial cavity and the lumen of gonads and kidneys, and in crustaceans to the latter. In molluscs and arthropods with "open" hemocoel, much propulsion of fluid is by movement of somatic muscles. In vertebrates the coelom persists as peritoneal, pericardial, and pleural cavities.

Movement of Hemolymph in an Open System. Most arthropods, many molluscs, and ascidians have a circulatory system in which a heart pumps hemolymph through vessels and a hemocoel derived from the primary body cavity (blastocoel). Hemolymph passes from the vessels into tissue spaces and returns via sinuses to the heart. The extent of the vessels varies greatly with animal class. In echinoderms the coelomic fluid, haemal fluid, and intercellular space are in communication to different degrees; ambulacral fluid is separate and may in fact be sea water.

Movement of Blood in a Closed Vascular System. A closed system of tubes, often with one or more pumps, is found in oligochaetes, many polychaetes, leeches, phoronids, nemerteans, cephalopod molluscs, holothurian echinoderms, and vertebrates. The blood comes into intimate association with the tissues by capillaries, with or without sinuses. In some vertebrates the blood makes a single cycle from heart through respiratory organs to tissues; in others it makes a double transit through the heart.

Lymph Channels. In vertebrates, the intercellular space (primary body cavity) is connected with the blood vascular system through lymph channels. These converge on veins and form a lined network which may be as extensive as the capillary bed. In some animals there are lymph hearts.

A circulatory system, whether open or closed, may be considered as consisting of two parts: the peripheral channels and the pump, the heart.

FLUID COMPARTMENTS

Fluid compartments of several typical kinds of animal may be compared as follows:

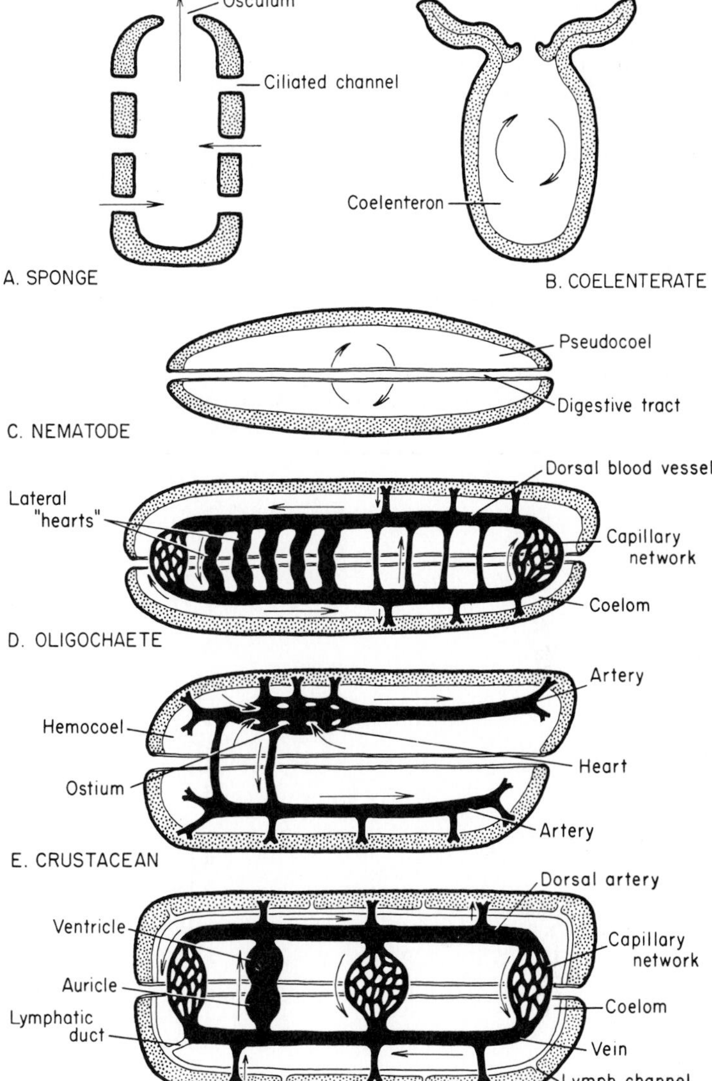

Figure 20–1. Diagram of systems of internal transport. A, "sponge" type, with flagellated channels; B, "coelenterate" type; C, "nematode" type, with pseudocoel primary body cavity; D, "oligochaete" type, with coelom and closed blood system; E, "crustacean-molluscan" type, with blood system open to hemocoel which is derived from primary cavity or blastocoel; F, "vertebrate" type, with coelom, closed vascular system, and lymphatic channels.

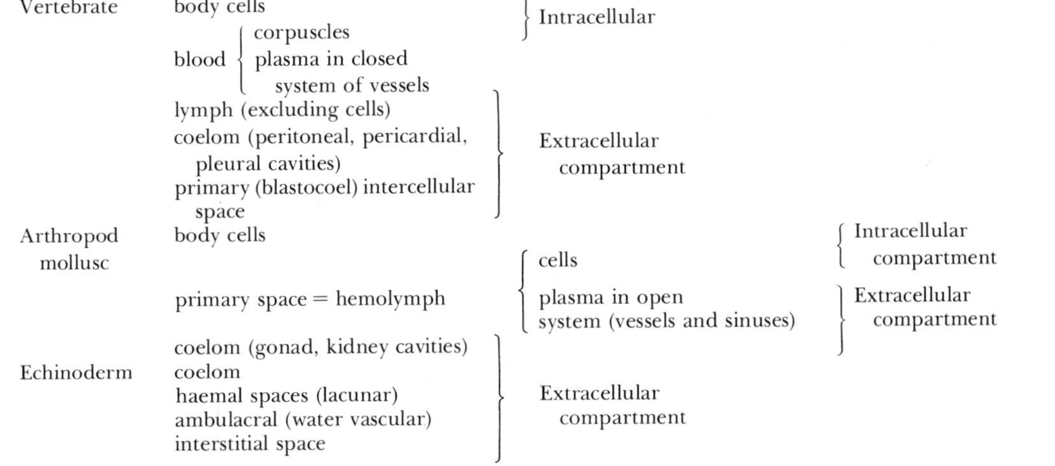

Vertebrate	body cells		Intracellular
	blood: corpuscles		
	blood: plasma in closed system of vessels		Extracellular compartment
	lymph (excluding cells)		
	coelom (peritoneal, pericardial, pleural cavities)		
	primary (blastocoel) intercellular space		
Arthropod mollusc	body cells		Intracellular compartment
	primary space = hemolymph	cells	
		plasma in open system (vessels and sinuses)	Extracellular compartment
	coelom (gonad, kidney cavities)	Extracellular compartment	
Echinoderm	coelom		
	haemal spaces (lacunar)		
	ambulacral (water vascular)		
	interstitial space		

Total body water can be estimated in several ways. Desiccation is the most accurate; in intact animals, D_2O or H_3O can be used as tracer. Water content of an animal varies from 60 to 80% (see Ch. 1). The turnover of water varies with osmotic habits; half-time for body water turnover is: for laboratory mouse, 1.1 day; man, 9.5 days; *Dipodomys,* 12 days.[149]

Estimation of extracellular space is made by measuring the dilution of some substance which fails to penetrate tissue cells and which is not rapidly excreted or phagocytized. Time for complete distribution must be determined. No single material is useful for all animals or for all organs, and different values of extracellular space are obtained with different solutes. Those which are commonly used are inulin, thiocyanate ion, radioactive sucrose and, for some tissues, chloride and sodium. For precision it is preferable to speak of the space occupied by the tracer substance

TABLE 20–1. Fluid volume as percentage of body weight (based on specific gravity of 1.0). Data from cited references, or from second edition of this book, or from the Handbook of Biological Data.

	Plasma Volume % body wt.	*Blood Volume (method) % body wt.*	*Extracellular Space % body wt.*
Mammals			
man male	4.3–4.8	7.1–7.8	17.5
female	4.1–4.6	6.5–7.1	
at 14,900′ altitude		8.3	
dog	5	8.6 (dye)	
goat		7.3 (dye)	
		6.1 (Cr^{51})	
cow	3.9	5.2–5.7 (tagged cells)	
horse	4.2–6.2	7.6, 10.3	
cat	4.1	5.6 (Cr^{51})	
lab rat[137]	4.0	8.5 (dye, P^{32})	
		7.5	
guinea pig		7.5 (I^{131})	
bat *Myotis*	6.5	13.0 (dye)	
Marmota	5.1	10.0 (dye)	
hedgehog summer[45]		8 (dye)	
hibernating		2–3	
opossum		5.8 (dye)	
		5.5 (P^{32} cells)	
rabbit		5.7 (dye)	
Birds			
pigeon		7.8	
chicken male		9	
female		7	
duck	6.5	10.2 (I^{131})	
pigeon[19a]	4.4	9.2	
Reptiles and Amphibians			
Pseudemys[77]		12 for 400 g turtle	
		4 for 1600 g turtle	
turtle[184]	7.4	9.1 (dye)	
alligator[184]	6.0	7.3 (I^{131})	
alligator	5.8	13.2	
crocodile[184]	3.5	15.4	
iguana	4.2	16.8	
Rana pipiens[183]	8.0	9.5	22
Necturus[183]			24
Cryptobranchus[183]	3.5		22
Fishes			
Ictalurus	1.2	1.8 (dye)	
Salmo	1.6	2.8 (I^{131})	
Anguilla		2.9	
Anguilla[123a] F.W.		12.2 (inulin)	
		13.9	
several teleosts[182]	1.8	2.2–3.6 (dye)	11–16.6 (sucrose)
several holosteans[182]	2.1	3.1	16 (sucrose)
chondrosteans[182]	2.5	3.5	18.4 (sucrose)

(Table 20–1 continued on opposite page.)

rather than of the extracellular space (ECS). Plasma volume is measured by injecting a tracer agent which does not pass into intercellular spaces in tissues; blood volume is then determined by adding to this the volume of the blood cells. Plasma volume is commonly measured by dilution of a dye such as Evans blue or of radioactive iodinated albumin; blood cell volume is measured by hematocrit or by ^{32}P-labelled red blood cells.[58] Blood cell concentration of blood removed from large vessels may be significantly smaller than that from total vessels which include storage capillaries. Both extracellular space and blood volume are usually expressed as percentage of body weight on the assumption of a specific gravity of 1.0; deviation of specific gravity may be sufficient that blood volume is better given as percentage of body volume rather than of weight.

Representative values of plasma volume, blood volume, and ECS are given in Table

TABLE 20–1. Fluid volume as percentage of body weight (based on specific gravity of 1.0). Data from cited references, or from second edition of this book, or from the Handbook of Biological Data. (*Continued.*)

	Plasma Volume % body wt.	*Blood Volume (method) % body wt.*	*Extracellular Space % body wt.*
Salmo[39]		3.5 (dye)	
F.W. shark[181]	5.1	6.1	19.7 (sucrose)
marine shark[181]	5.4	7.0	21.2
dogfish[181]	5.5	6.8	21.2
sockeye salmon[170]	3.3	4.8 (dye)	
pink salmon[170]	7.8	9.8	
		6.2	
		4.5 (blood donor)	
Petromyzon[181]	5.5	8.5	23.9
Lampetra adult	5.5	8.5	
larva	4.8	7.9	
Carcharhinus F.W.	5.1	6.8	19.7
Negaprion[181] S.W.	5.4	7.0	21.2
Chaenocephalus[70]		7.6	
Worms			
Sipunculus			50
Dendrostomum			47
Arthropods			
Limulus[151]			51.6 (inulin)
Carcinus[151]			32.6 (inulin)
Maja (freshly molted)			29
Homarus			17 (dye)
Eriocheir			33 (SCN)
Cambarus			25.1 (dye)
			25.6 (SCN)
Locusta[13]			18.1
Apis			25–30
Periplaneta			15.7–17.5
Schistocerca[113]			14.4 (dye)
Carausius[189]			10.4–15.2
scorpion			34 (dye)
tarantula[175] male			19.6
female			18.1
Molluscs			
Acmaea[202]			25.9
Strombus[117a]			64.6
Viviparus[117a] F.W.			23 (inulin)
20% S.W.			33 (inulin)
Margaritifera			49
Mytilus			51 (inulin)
Aplysia			79 (inulin)
Octopus		5.9 (dye)	28 (inulin)
Sepia[151]		8.5 (inulin)	14.5 (sucrose)
Eledone		14.7	21.8
Bullia			61–64 (dye)

20–1. The blood volume in mammals and birds normally comprises 7 to 10 per cent of body weight; extracellular space corresponds to 18 to 25 per cent of body weight. The blood volume in man is slightly less in females than in males. The actual volume is relatively constant in adults; hence, on a percentage of weight basis, it often decreases with age. Blood volume is relatively lower in species of large body size than in small ones (see Table 20–1). Blood volume decreases as much as 4-fold in hibernation (see p. 417). Values for amphibians are similar to those for mammals: *Rana catesbiana* is 79% water and has ECS of 22%; a toad, *Bufo*, is 74% water with ECS of 25%.[183] In elasmobranch fishes the blood volume is less (about 5% of B.W.) and teleosts have the least blood of all vertebrates (usually 1.5 to 3% of B.W.). Blood volume in a salmon estimated by means of a dye was 6.2% B.W., while that measured by means of labelled blood from a donor fish was a more accurate 4.5%.[169] The octopus (a mollusc with closed circulation) has blood volume of 5.8% and ECS of 28%.[122]

In animals with open circulation, blood volume is essentially the same as total extracellular space (Table 20–1). Values for several crustaceans are from 25 to 30% of body volume; on a weight basis the blood volume may be 30% just after molting and 10% after the shell hardens, and hence the weight basis is inaccurate. The blood volume of many adult insects is less than that of larval insects. The value for larvae of the fly *Celerio* is 18.6%, while that for the adult is 7.2 to 7.8% of B.W. In the fly *Sarcophaga*, the larval hemolymph is 35 to 42% and pupal hemolymph 24 to 33% of body weight.

The range of values of hemocoel volume in gastropods and pelecypods is greater than in any other animal group. In *Chiton* the hemolymph volume may be 90% of body volume, and in the land snail *Achatina* it is 40%.[122]

Blood volume may vary under different stress conditions. At high altitudes the plasma volume of mammals decreases, red cell volume increases, and blood volume may increase slightly (8.3% B.W. at 14,900 feet compared to 7% at sea level).[23] The blood volume helps to maintain normal blood pressure, and after severe hemorrhage cells and fluid are poured into the circulation from body stores such as the spleen. Mammalian blood volume is partly regulated by reflexes originating in visceral receptors (p. 513). Loss of more than 39% of blood volume by rapid hemorrhage is fatal to man. In contrast, a beetle can survive loss of 50% of its hemolymph. A small circulating volume, such as is found in teleost fish, is efficient in that the same blood is reused more frequently in transport. However, a large volume provides a reserve of fluid and a store of essential solutes; in some animals it also serves as a hydrostatic skeleton (p. 833).

PRESSURE AND FLOW IN VERTEBRATE CIRCULATORY SYSTEMS

Mammals. In any closed tubular system containing a pump, a head of hydrostatic pressure is developed at the pump and the pressure declines with frictional loss in the tubes; the loss of pressure is less if the tubes are distensible. The principles of blood flow hemodynamics have been extensively studied.[29] The total energy in a unit volume of blood is given by

$$E = P + \rho\, gh + \frac{1}{2}\rho V^2$$

TABLE 20–2. Blood pressures of selected animals in mm Hg. Data from cited references, or from Second edition of this book, or from the Handbook of Biological Data.

Animal	*Site of Measurement*	*Systolic/Diastolic Pressure or Mean*
man, 20 years	radial artery	120/75
40 years		138/84
60 years		140/85
man	pulmonary artery	25/10
horse		171/103
dog		112/56
rat		130/91
guinea pig		77/47
mouse		113/81

(Table 20–2 continued on opposite page.)

TABLE 20–2. Blood pressures of selected animals in mm Hg. Data from cited references, or from second edition of this book, or from the Handbook of Biological Data. (*Continued.*)

Animal	*Site of Measurement*	*Systolic/Diastolic Pressure or Mean*
giraffe (anesthetized)[53]		240–315/185–240
giraffe (standing, not anesthetized)[37]		120/75
sheep (not anesthetized)[120]		135/112
elephant seal (not anesthetized)[36]		120/90
Tursiops[172]	carotid	142/111 to 160/130
	pulmonary	55/35 to 76/50
duck (not anesthetized)		100–175
chicken (14 mo.)		149/43
pigeon (local anesthesia)		135/105
lizard *Tiliqua*[26b]		14/10
Sauromalus[180]		80/60
Pseudemys		42/32
Xenopus[161]		38/34
Rana temporaria[161]		32/19
Rana pipiens[161]		30/20
Bufo[161]		32/19
Salamandra[161]		22/12
Amphiuma[161]		30/25
Bufo marinus[26b]		25/16
Salmo (not anesthetized)[174]	dorsal aorta	29/25
	ventral aorta	40/32
Oncorhynchus (not anesthetized)[20a]	dorsal aorta	44/38
Gadus (not anesthetized)	bulbus arteriosus	29/18
Mustelus	ventral aorta	26/19
Squalus[155]	dorsal aorta	17/16
	ventral aorta	30/24
Raja	ventral aorta	16/14
Myxine	branchial artery[33]	17/7
	dorsal aorta[87]	5/3
	ventral aorta[87]	9
Octopus[90]	cephalic aorta	44/22
	afferent branchial artery	18–37/11
	efferent branchial artery	7.5–18/3.7–11
	vena cava	0–12.5
Glossoscolex[93]	dorsal vessel	17.6/10.3
	ventral vessel	48/34.5
Locusta[13]	dorsal aorta	6.3/2.3
Homarus	ventricle	
	resting	13/1
	active	27/13
Maja	heart	4/3.3
	thoracic sinus	1.8/2.5
crayfish	leg sinus	7.4
Mytilus[171]		1.9/0
Pecten[171]		0.5/0
Crassostrea[171]		1.4/0
Macoma[171]		0.8/0
Mya[171]		0.6/0
Helix[95]	heart	24/5
Arenicola	body cavity	
	resting	0.73
	active	11
Neanthes	body cavity	
	resting	1.1–7.2
	active	17.6
Lumbricus	body cavity	
	resting	1.5
	active	10
Golfingia	body cavity	
	resting	2.2–2.9
	active	14–79

where P is pressure (dynes/cm^2), ρ is density (g/cm^3), g is gravitational acceleration (980 cm/sec^2), h is height (cm), and V is velocity (cm/sec). Representative values of pressure in a large artery (usually aorta or carotid) are given in Table 20–2. In young adult men the pressure at the time of heart contraction (systole) is about 120 mm Hg, and at heart relaxation (diastole) it is 80 mm Hg; this is designated as 120/80. The difference, or pulse pressure, decreases as the blood passes through smaller vessels. Pulmonary arterial pressure is about 25/10 mm Hg.[172] In a porpoise, aortic pressures range from 142/111 to 160/130, and in the pulmonary artery the pressures are 55/35 to 76/50 mm Hg.[172] Velocity of flow is slow in the vast capillary bed; in the veins, velocity increases slightly although pressure continues to decrease. In mammals the cross-section of total capillaries is some 800 times that of the aorta. When two capillary beds are in series, pressure in the second is very low; in the hepatic artery, pressure may be 120 mm Hg, while in the hepatic portal vein which brings blood from intestinal capillaries the pressure is only 5 mm Hg. In a llama the mean pulmonary arterial pressure at sea level was 14 mm Hg, while at 3400 meters altitude it was 23 mm Hg because of increased resistance to flow.[10]

Blood pressure is normally determined by (1) the peripheral resistance, (2) the head of pressure built up by the heart, and (3) the volume of blood. Peripheral resistance is varied by constriction and dilatation of arterioles and capillaries; passive changes in size are due to the properties of elastin and collagen, while active changes are brought about by vascular smooth muscle and capillary endothelial cells. In birds and mammals, the basal pressure is usually higher in males than in females, and it rises with age. In general, the resting pressure is higher in large mammals than in small ones; carotid pressure in a horse may be 190 mm Hg. During the first three weeks after birth, the arterial pressure of a rat rises from 20 to 75 mm Hg.[117] The effect of gravity may be sufficient to increase the pressure in the foot of an erect 6-foot-tall man by 60 mm Hg and to reduce the cerebral pressure by a similar amount. Compensation for gravity is provided by valves in veins, by pressure developed by somatic muscle, and by vasomotor reflexes. The effect of vertical posture is compensated better in animals which are normally vertical than in those normally horizontal. A giraffe has a carotid artery nearly 2 meters long; the carotid pressure when lying prone was 260/160 and that when standing was 120/75 mm Hg, and carotid flow was 50 ml/sec when prone, 35 ml/sec when standing. Thus, the cerebral pressure is less when standing than when lying down.[36]

Exercise usually increases both pressure and cardiac output. In man, isometric exercise such as sustained hand-grip causes a marked rise in blood pressure and a moderate rise of heart rate, whereas dynamic exercise, such as treadmill walking, elicits a large rise in heart rate and little change in blood pressure.[116] Work done by the heart is given by $W = P \times SV$, where P is systolic pressure and SV is stroke volume. Increase in cardiac output results from both increased heart rate and stroke volume; in mild exercise the heart rate is dominant, but in trained athletes the increase in stroke volume may be relatively more important for increasing cardiac output. Venous pressure and flow, particularly in the limbs, are increased by contraction of muscles. Wing veins of a bat contract in a 10 second cycle, and the contractile wave travels at 0.33 to 3.5 mm/sec.[40]

The control of size of vessels and heart beat involves a series of direct reflex and hormonal mechanisms. Local responses of capillaries to irritation may be by direct action (as by histamine) or by axon reflexes (reflexes by efferent branches of sensory neurons). Pressure receptors are located in the walls of the carotid sinus, the aortic arch, the pulmonary artery, and the heart (particularly auricles) and also in vessels of the adrenal glands.

Sensory endings in the carotid sinus fire in proportion to the arterial pulse; atrial receptors of cats fire at higher rates than aortic receptors as recorded in vagal afferent fibers.[74] Frequency of sensory impulses in the nerve from adrenal glands is proportional to systemic blood pressure.[135] Sensory impulses go to the vasomotor center in the medulla of the brain, from which vasomotor messages leave via autonomic nerves. In the sympathetic supply to blood vessels in mammals, cholinergic fibers serve vasodilatation in some muscle beds and in the skin of face and neck; cholinergic fibers also occur in parasympathetic nerves, both cranial and sacral—the well known vasodilators of genitalia. Vasoconstriction is mediated by sympathetic fibers which are adrenergic; here the receptors are of the alpha type, while the fibers may be inhibitory (dilator) when receptors are of the beta type (p. 496).

Norepinephrine (liberated from sympathetic nerves or as 20 to 50% of the catecholamine from the adrenal medulla) is a vasoconstrictor and causes a rise in blood pressure. Epinephrine (from the adrenal glands) is a constrictor of skin and kidney vessels but is a dilator of muscle, liver, and coronary vessels; at high concentrations epinephrine is a constrictor of vessels in postural muscles, possibly by indirect metabolic action.

Acetylcholine generally causes dilation and a fall in blood pressure; histamine is also a dilator. Serotonin (5-hydroxytryptamine) in equivalent doses raises blood pressure in the dog and lowers it in the cat. Posterior pituitary extracts raise blood pressure in mammals by action of vasopressin; oxytocin has little effect. In birds, however, a depressor action of oxytocin is dominant. *In vitro,* pulmonary arteries of dog contract in response to acetylcholine, while the renal, mesenteric, and cerebral arteries are relatively insensitive; in response to epinephrine, the aorta, mesenteric, and renal arteries contract, and the pulmonary and cerebral arteries are insensitive.[19] Mammalian vessels also differ markedly in the spontaneous rhythmicity of their smooth muscle; for example, for rat vessels the series is portal vein > umbilical artery > aorta > cerebral vessels.[18] The isolated femoral artery of a rabbit is very sensitive to cooling; responses to norepinephrine disappeared on cooling to 13° C, whereas responses of the artery to the ear were maximal at 24° C and did not disappear until 7° C. Hence, the smooth muscle of these two arteries differs in temperature sensitivity.[52a]

When renal arterial pressure falls, renal baroreceptors change their activity and renin is released from juxtaglomerular cells; sympathetic impulses can also stimulate renin liberation. Renin then activates angiotensin in circulating plasma, and this regulates blood pressure by its vasoconstrictor action. Renin release also occurs when the sodium load to the kidney is low.[195] Renin is produced in kidneys of fish, more in freshwater than in marine species; whether it regulates vascular pressure in addition to its action in ionic regulation in fish is unknown (p. 70).

By *vasomotor autoregulation* is meant peripheral control of blood flow by constriction or dilatation of resistance vessels according to metabolic need. For example, coronary vessels respond by vasodilation to local hypoxia; hence, the oxygen supply to the heart muscle is kept constant under wide variations in demands. Similarly, flow of blood to the kidney increases linearly with increasing pressure up to about 100 mm Hg, above which flow remains relatively constant; i.e., the peripheral vessels of the kidney autoregulate. In muscle, as in an arm or leg, after occlusion or exercise, a marked increase in flow occurs during payment of an oxygen debt. In addition to neural and hormonal control, there may be local actions of metabolic products such as lactic acid. Not only are there differences in smooth muscle of different vessels, but the same vessels respond differently according to animal species.

An important effect of oxygen supply and demand is the variable opening of capillaries and change in use of arteriolar-venule shunts. In a rat at $P_{a_{O_2}}$ of 100 mm Hg, the intercapillary distance was 20 μm and only half of the capillaries in a muscle were open; in hypoxia of $P_{a_{O_2}} = 32.2$ mm Hg all of the capillaries opened and the average separation was 18 μm.[123] When a capillary tube is inserted in contact with interstitial fluid, subcutaneously or intramuscularly, a negative pressure is recorded, ranging from -0.7 to -6.8 cm H_2O. Interstitial pressure in a series of terrestrial animals (mammals and reptiles) averaged -5.6 cm H_2O, and in two frogs it was -1.0 cm H_2O. Interstitial pressure becomes more negative during dehydration and is caused by a combination of capillary absorption pressure and plasma protein colloid osmotic pressure. Its net effect is to restrict loss of fluid from vessels.[155b]

In summary, regulation of blood flow is precise homeostasis; flow must be adequate for varying conditions, yet not excessive. Blood pressure and flow are regulated largely by peripheral dilatation, constriction, and shunting between arteries and veins. The smooth muscles of peripheral vessels make response locally to chemicals, temperature, and stretch, and make response reflexly to pressure and volume sensors and central nervous influences. Different circulatory beds—coronary, pulmonary, renal, and muscle—may respond differently to the same stimuli. The interrelations between stimuli and responses in different parts of the circulatory system have been modelled cybernetically so that effects of changes in variables are predictable.

Lower Vertebrates. In amphibians and reptiles, blood makes a double cycle to the heart, even though the two sides of the ventricle may not be separated. A two-channel heart not only separates oxygenated from oxygen-depleted blood but also permits higher pressure in systemic arteries than is possible when the blood has first passed through respiratory capillaries. In a frog, the difference between systemic and pulmocutaneous artery pressures decreases during systole; as pulmocutaneous resistance is relatively greater than peripheral resistance, pressure drops more than does the systemic pressure[161] (Fig. 20–2). In frogs and other amphibians, the adrenergic receptors are of the alpha type, and vasoconstriction occurs in some vascular beds. In a toad, the vagus nerve increases pulmonary vascular pressure and sympathetic fibers cause a slight fall. Blood vessels of reptiles receive both cholinergic and adrenergic innervation. In turtle heart, epinephrine constricts and acetylcholine dilates the coronaries (in contrast to mammals, in which epinephrine dilates the coronaries).[26a]

In fish and cyclostomes, the heart is a gill heart and must drive the blood through gill vessels before it is distributed via the dorsal aorta to the periphery. In rainbow trout, pressure in the ventral aorta was 40/32 mm Hg at rest, and in the dorsal aorta it was 29/25 mm Hg. In exercise, the venous pressure was 9 mm Hg; pressure in the ventral aorta increased by 40% and that in the dorsal aorta increased by 16%.[174] In a shark, the ventral aorta has three times more elastin than does the dorsal aorta; elastin has a much lower elastic modulus than does collagen, and hence the ventral aorta is relatively rigid and, as pressure increases, the rush of blood through the gills is minimized.[155a] Swimming in trout is accompanied by cardioacceleration, and return of blood is favored by the pumping action of the caudal muscles.[177] In a shark, pericardial pressure goes from −2 cm H_2O to −5.6 cm H_2O in systole; suction helps to fill the heart in elasmobranchs and cyclostomes, but not in teleosts.[155a]

Cyclostomes have accessory hearts to boost flow in a low pressure system in which there are quasi-open sinuses. In hagfish, for example, blood passes from the branchial (systemic) heart under pressure of 14–16/5–8 mm Hg to the gills, which are muscular; then the blood passes by the aorta at a pressure of about 5 mm Hg to the sinuses, from which part of the blood goes through caudal accessory hearts and part goes through the gut and portal accessory heart with a pressure of 3 to 6 mm Hg, and ultimately back to the systemic heart.[33, 87]

The heart of fishes receives vagal innervation, and the peripheral vessels receive sympathetic innervation. If water flow over the gills of a skate or shark is stopped, the

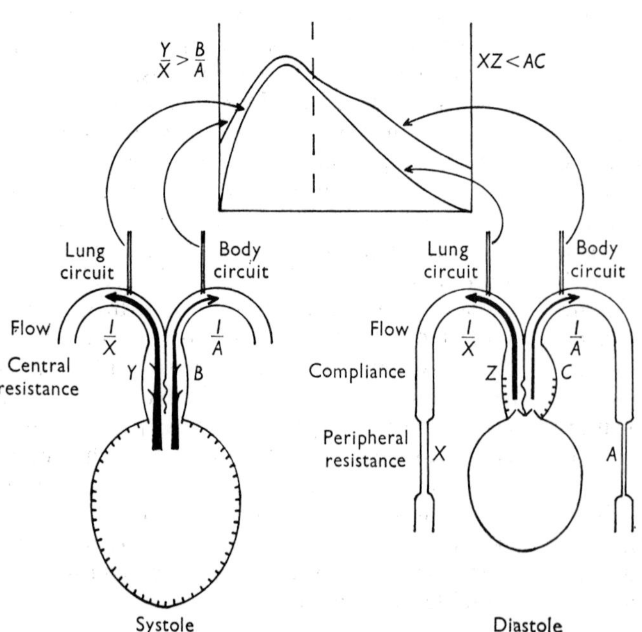

Figure 20–2. Above, pulse curves for systemic (body) and pulmocutaneous (respiratory) circulations in frog. Below, conditions in cardiac systole and diastole which would account for decreasing pressure differences between the two circuits during systole and increasing differences during diastole. (From Shelton, G., and D. R. Jones, J. Exp. Biol. *42*:339–357, 1965.)

heart stops and blood pressure falls; if the fish has been atropinized, cessation of water flow does not affect heart rate and blood pressure. Increase in pressure in perfused branchial arteries sets up sensory impulses in branchial nerves, and a sensory discharge occurs at each heart beat.[83] The pseudobranch vessels are homologous with the carotid sinus and carotid bodies of mammal and bird. Gill receptors can be stimulated by brief touch or by maintained pressure. The pseudobranch of some teleosts has baroreceptors that send afferent fibers in the glossopharyngeal nerves. Injection of epinephrine has a pressor action in dogfish; the branchial vessels dilate in response to epinephrine, as do mammalian coronaries, in contrast to visceral vessels, which constrict to epinephrine.[32] In eel, acetylcholine constricts branchial vessels; eipinephrine dilates them but constricts systemic arterioles. In an eel gill, blood can traverse the lamellae or it can pass through a shunt in the filament core and bypass the respiratory surface. Lamellar channels are opened and flow is increased by norepinephrine when oxygen demand is high; the shunt is opened by acetylcholine and is used when low flow is adequate.[100]

In teleost fish, control of peripheral vessels is both cholinergic and adrenergic. Acetylcholine constricts and epinephrine dilates gill vessels of teleosts[138] and lung vessels of amphibians.[96a] Comparison of responses of large arteries of fish, amphibians, and reptiles to catecholamines led to the view that there is a progressive increase in adrenergic control in the evolutionary series.[26b] In contrast to mammals, there appears to be no catecholamine dilator action on systemic vessels in fish, amphibians, or reptiles.[26a] In mammals, many vessels have adrenergic control only. Few veins have adrenergic innervation in fishes and amphibians, while in reptiles, birds, and mammals many veins are so supplied. Norepinephrine as an inhibitor (dilator) of large vessels may be present only in birds and mammals.[26a, 49a] In fish in general, acetylcholine constricts most arteries and raises blood pressure, whereas in mammals it is a vasodilator and lowers pressure. Epinephrine constricts most systemic vessels in both mammals and fish, and dilates coronaries in mammals and branchials in fish.

In turtles, warming the anterior hypothalamus by as little as 1° C increases blood pressure by 10 to 20%, and cooling it by 1° C lowers the pressure so long as the body temperature is above 20° C.[67]

In summary, control of pressure and flow in fishes, where blood traverses the heart only once per cycle, is brought about by regulation at the two capillary beds in series—in the gills and the periphery. Shunts are evidently much used, and blood volume is low. In amphibians and reptiles, where the blood passes twice per cycle through an incompletely divided heart, flow on the right side is nevertheless well separated from flow on the left.

Diving Animals; Responses to Hypoxia. A universal circulatory response to diving by air-breathers is marked bradycardia. In an elephant seal (*Mirounga*) the heart rate at rest was 60/min, and when the animal was immersed the rate was 4/min. At rest, carotid flow was 1200 ml/min and iliac artery flow was 1600 ml/min; in a dive, flow in the iliac artery stopped while that in the carotids remained constant, as did carotid pressure.[36] After a dive, tachycardia and vasodilation were observed. In a seal during a dive, there is general arterial constriction in all organs except heart and brain; blood flow in viscera is virtually zero.[21] Likewise, in a diving duck, blood flow decreased: to gizzard by 4%, to skin by 29%, and to gastrocnemius by 11%; flow to the head increased, and flow to the myocardium increased 4-fold.[90] In a diving duck, carotid arterial pressure remained constant, whereas central venous pressure and the diastolic pressure of the left ventricle increased.[90] In good divers, such as porpoises and whales, a rete separates the systemic and cerebral circulations, and vasoconstriction in the systemic circulation provides for continued flow to the brain.[132]

Reptiles and amphibians show reflex slowing of the heart on submergence. In an iguana, heart rate may drop to 1% of normal with marked lengthening of the interval between T and P waves; atropine abolished the response, and hence the reflex is vagal.[208] A turtle, *Pseudemys*, shows a bradycardia to 41% of the normal rate, a slowing which can be prevented by atropinization; this turtle has sufficient oxygen stores for 2 to 3 hours of submergence.[14] A marine turtle, *Chelonia*, shows accentuated bradycardia when hydrostatic pressure is increased during submergence; at 19 atmospheres the heart rate was slowed to 9 minutes between beats.[15] In a frog, *Rana*, immersion causes an early slowing of the heart by a vagal reflex, whereas asphyxiation (in nitrogen) causes a much delayed bradycardia that is not vagal in origin.[119] Systemic pressure in *Rana* at the surface was 32/21 mm Hg, while submerged it was 23/16 mm Hg; but pulmocutaneous pressure at the surface was 31/16, and submerged it was 26/15. Thus, the systolic pressure in each arch declines in submergence, but the diastolic pressure is relatively constant. Heart rate and stroke volume decrease in submergence.[161]

Fish removed from water show a bradycardia like that of diving air-breathers. An intertidal clingfish, *Sicyases,* had a normal heart rate of 70 to 80/min in water and 40/min in air.[54] A carp has a constant heart rate from air saturation down to 2.5 ml O_2/liter; then heart rate accelerates slightly, and at below 1.8 ml O_2/liter it slows rapidly. The hypoxic bradycardia is delayed by vagotomy.[158] Similarly in a trout, *Salmo gairdneri,* heart rate is constant at 65/min at Po_2 from 155 to 60 mm Hg, but drops to 22/min at Po_2 of 40 mm Hg; no such bradycardia occurs if the fish is atropinized. The O_2 receptors are in gill or buccal membrane.[145] In general, hypoxia in fishes causes a reflex bradycardia, increase in stroke volume, increase in resistance of branchial circulation and of peripheral vascular beds, and increased use of glycolytic metabolism.[155a] Exercise of fish leads to a large net decrease in peripheral vascular resistance due to dilation of gill and muscle vessels, accompanied by constriction of visceral vessels.[155a]

PERIPHERAL CIRCULATION IN INVERTEBRATES

Closed Circulatory Systems. Cephalopod molluscs have closed circulatory systems with large systemic hearts which send blood to the periphery, and accessory hearts in the branchial vessels which pump the blood through the gill capillaries. In a large octopus, the systemic pressure may be 60/30 mm Hg and the pressure in the branchial vessels may be 10–25/7–15 cm H_2O.[91] As in other molluscs, filtration of protein-free plasma occurs through the wall of the heart into the pericardium, and fluid then passes to the kidneys.

In oligochaete and hirudinean annelids, the blood is in a closed system. In the giant earthworm *Glossoscolex,* pressure in the dorsal vessel is 24/14 cm H_2O and the vessel contracts at 6 to 8/min; at 18° C a contraction wave passes at a velocity of 12 segments/sec. The dorsal vessel fills the lateral hearts, which beat at 20/sec, and pressure in the ventral vessel (filled from the hearts) may exceed 100 cm H_2O at the time of cardiac systole.[93]

Open Circulatory Systems. The distinction between closed and open circulations may not be as sharp as formerly thought; some crustaceans have vessels which approach the dimensions of vertebrate capillaries before ending in tissue sinuses. However, pressure in open systems in general is low and extremely variable, and hemolymph volume is large. In large crustaceans, such as lobster, systolic pressure at rest may be 12 to 17 cm H_2O and diastolic pressure may be less than 1 cm H_2O.[26] During leg movement, the ventricular pressure in a lobster may increase to 37 cm H_2O at systole and 17 cm H_2O at diastole (Fig. 20–3). Thus, resting differences between systole and diastole, as well as variations with motor activity, are greater in open than in closed circulatory systems. Simultaneous measurements in several regions of a crayfish show that the pressure increases are not equal throughout the system and may be greater in an active appendage than in an inactive one; the pressure at the distal end of a leg may be temporarily higher than at the proximal end. The flow in sinuses may be controlled more by activity in somatic muscles than by the heart and vessels. Septa separate afferent and efferent channels in gills of crayfish and lobster and in legs of dragonfly nymphs.

In bivalve molluscs, systolic pressures at the heart are only 7 to 26 mm H_2O; the pressure is zero at diastole. In *Anodonta* a negative pressure is recorded in the pericardium at systole; hence, cardiac suction favors auricular filling. Muscular contraction which adducts the valves may increase pressure in the heart to 170 mm H_2O. Urine is ejected under positive pressure.[69]

In the insect *Locusta,* aortal pressures are 8.6/3.2 cm H_2O for a heart rate of 79/min. Body and leg movements increase the pressure in the hemocoel, and in some sinuses negative pressures of -2.4 to -3.6 cm H_2O were recorded.[13] Stroke volume was calculated as 0.17 μl.

In molluscs (except cephalopods) and crustaceans the blood volume is comparable to the total extracellular volume of a vertebrate (Table 20–1), yet the blood pressure of a 20-

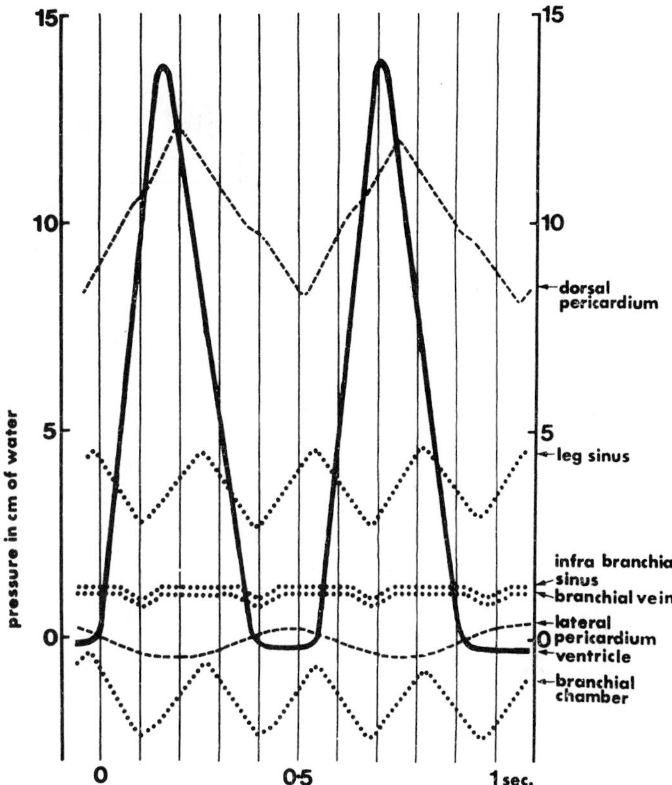

Figure 20–3. Pressures in various regions of the circulatory system of the crab *Carcinus* during two ventricular systoles. (From Blatchford, J. G., Comp. Biochem. Physiol. *39*A:193–202, 1971.)

gram crayfish is only about 30% of that of a 20-gram mouse. The velocity of circulation in the crayfish is therefore very low; a lobster circulates its blood volume in 3 to 8 minutes[26] as compared to 21 seconds in man. The sluggishness of blood flow may be a limiting factor in size and activity of animals with an open circulation. Insects are active animals yet they have an open circulation; their respiratory exchange is by tracheae rather than by blood.

Hydrostatic Skeletons. In many animals in which the pressure in either hemocoel or coelom is related to motor activity, the fluid-filled system provides rigidity which has a skeletal function. Hydrostatic skeletons are most important in soft-bodied animals, particularly those which burrow. However, hydrostatic pressures in sinuses may be important for movement of appendages of some animals with open circulation. In spiders, for example, the legs are flexed by muscles, but extension is due to pressure on fluid in the legs; if blood volume is reduced the legs are flexed, and if fluid volume is increased the legs remain extended.[46]

Hydraulic mechanisms are important for locomotion in echinoderms which move by means of tube feet. Extension of the tube feet of a starfish results from contraction of the ampullae with flow of fluid into the tube feet; withdrawal results from contraction of longitudinal muscles of the tube feet.[168] The tube feet of a starfish can develop a suction pressure of 3000 grams, and by periodic pull they can open a bivalve mollusc.[111] In holothurians, the anterior ambulacral system consists of two portions separated by a valve; these are the tentacular ducts which supply the tentacles with fluid, and the reservoir or Polian complex of radial ducts, ambulacral ring, and Polian vesicles. In the sea cucumber *Thyone,* pressure is normally higher in the tentacular duct (10 to 25 cm H_2O) than in the Polian complex (3 to 3.5 cm H_2O); in the body cavity, pressure is 0 to 2.5 cm H_2O when *Thyone* is relaxed, and is up to 35 cm H_2O when it is contracted.[211] In *Holothuria,* the coelomic pressure at rest was 1.7 cm H_2O, and in activity it was 19 cm H_2O.

Burrowing by bivalve molluscs makes use of the hydrostatic skeleton in the foot.

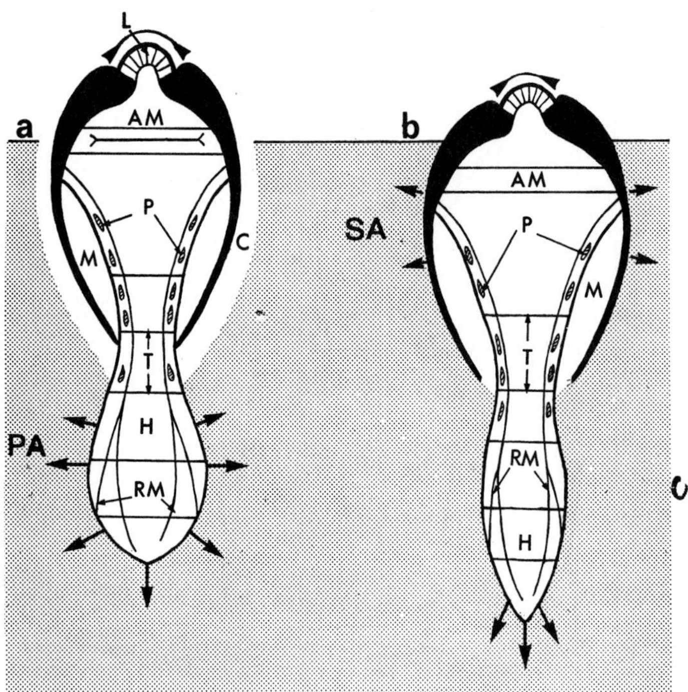

Figure 20–4. Diagram of stages of burrowing of a bivalve mollusc; pedal anchorage (PA) and shell anchorage (SA) are shown by arrows. (a) Valves adducted by muscle (AM), producing pedal dilation and a cavity (C) in sand around valves. (b) Valve reopened by elasticity of ligament, holding shell fast while contraction of protractor (P) and transverse muscles (T) cause pedal protraction. H, pedal hemocoel; M, mantle cavity. (From Trueman, E. R., Biol. Bull. *131*:369–377, 1966.)

Teredo in its burrow is turgid and has an intramantle pressure of 5 to 17 mm H_2O; when it is outside the burrow the pressure is 2 to 5 mm H_2O. Pressures up to 50 cm H_2O have been recorded in burrowing *Ensis,* and 40 cm H_2O has been recorded in *Anodonta*[193] (Fig. 20–4). Water ejected from the mantle cavity loosens the sand, and then the foot is protruded under pressure; the end of the foot is dilated to provide a pedal anchor, and the siphons close. The valves then partly adduct or close, and retractor muscles of the foot pull the shell downward; the shell then abducts, forming a shell anchor, and the foot may be retracted. The opening of the shell (abduction of valves) is by pressure of foot and siphon in some species, and by elastic ligament in others. *Mya* can alternately contract anterior and posterior adductor muscles.[8, 193] In some bivalves the pedal sinuses can be separated from the rest of the hemocoel.

Burrowing worms depend on changes in pressure in coelomic fluid to provide turgidity or rigidity. *Arenicola* shows rhythmic pressure changes as it burrows—dilation of the proboscis by increasing coelomic pressure, and flanging and eversion of the proboscis as an anchor so that the body can be pulled downward. The coelomic pressure in *Arenicola* at rest was recorded as 2 cm H_2O, and that at the peak of burrowing as 110 cm H_2O. Pressure waves are developed by the circular muscle of the body every 5 to 7 seconds, and circular tension was calculated to be 3 kg/cm^2 (Fig. 20–5B).[190, 191] Removal of coelomic fluid from *Arenicola* delays burrowing. In free *Glycera* and *Nereis* the pressure increased from 1 to 2 cm H_2O at rest to 15 to 18 cm H_2O in activity. Resting pressure in an earthworm was 0.26 cm H_2O; longitudinal and circular contractions gave active pressures of 7 and 12 cm H_2O[160] (Fig. 20–5A). For a body diameter of 0.6 cm, a thrust of 8.5 grams is provided by the increase in pressure from 2 to 13.5 cm H_2O. In burrowing, the anterior end is inserted into soil and inflated, and then peristalsis pulls the body down; setae help to anchor it. Pressure may differ in the two ends, and flexible septa transmit pressure changes throughout the system.

In sipunculids and echiuroids, high turgidity is essential for burrowing. In the body cavity of large *Sipunculus,* pressures as high as 600 cm H_2O have been recorded, and in small *Golfingia,* pressure may rise from 2.8 cm when relaxed to 108 cm H_2O when contracted.[211] *Sipunculus* is unable to burrow if the posterior part of the body is paralyzed

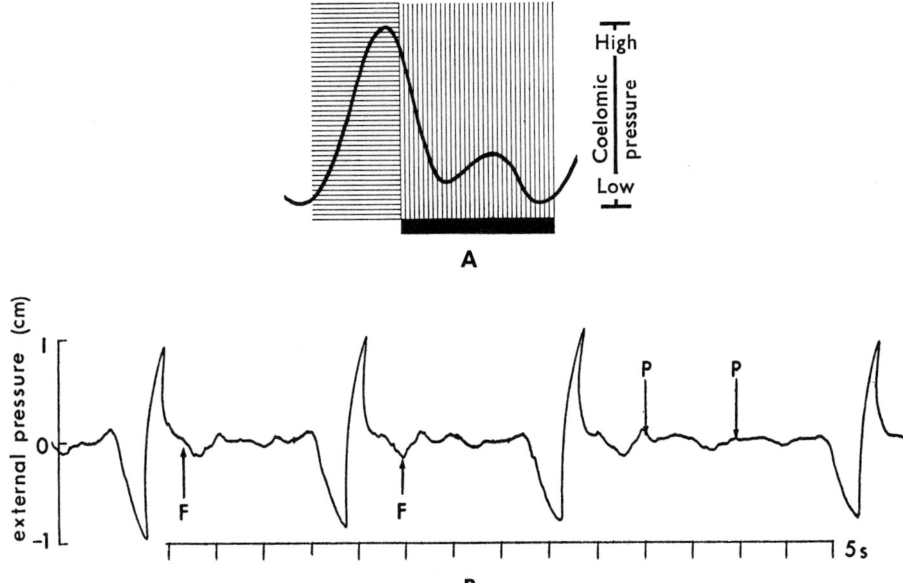

Figure 20–5. A, Pressure fluctuations in coelom of earthworm *Lumbricus* during passage of a peristaltic wave. Horizontal hatching represents thinning and elongation by contraction of circular muscle; vertical hatching indicates thickening and shortening of segment. Black bar is setal protrusion. (From Seymour, M. K., J. Exp. Biol. *51*:47–58, 1969.) B, Pressures produced in soil by mudworm *Arenicola* when burrowing; F, flanging; P, proboscis eversion. (From Trueman, E. R., Biol. Bull. *131*:369–377, 1966.)

by deganglionation, but if a ligature has been placed ahead of the paralyzed portion, the anterior part can disappear rapidly into the sand. In tube-dwelling worms, water is pumped partly by ciliary currents, but more by peristaltic contractions. *Urechis* shows peristaltic waves at 25 second intervals; a pacemaker is located in the circumesophageal nerve ring, and the wave is reflex in nature. One worm can pump 29 liters of H_2O/24 hours.[110] Each pumping wave moves 20 to 30 ml H_2O, and pumping pressures are 3 to 4 cm H_2O compared to 100 cm H_2O for burrowing.[34]

The pressure in the coelenteron of a sea anemone is small (1 to 2 cm H_2O) at rest but may rise to 10 to 15 mm H_2O on contraction of the body.

Hydrostatic skeletons which provide rigidity on demand by body wall contractions furnish less independence of parts than a segmented exoskeleton or endoskeleton, and they can be maintained only by muscular work. However, the hydrostatic skeleton can be made to appear or to diminish by variations in internal pressure according to the animal's need, so that the body can go from flaccidity to rigidity. In many animals, then, the fluid system functions not only for circulation but to provide a hydrostatic skeleton.

TYPES OF HEARTS

Any system for circulating a mass of fluid requires a repeating pump. To assure that fluid goes in a constant direction, the pump either must be equipped with suitable valves to prevent back flow or must compress its contained fluid in a progressive wave. Morphologically, hearts may be classified as (1) chambered hearts, (2) tubular hearts, (3) pulsating vessels and (4) ampullar accessory hearts.

Chambered Hearts. Chambered hearts are well known in vertebrates and molluscs. In vertebrate hearts the sequence of contractions is essentially similar whether the heart is two-sided (four- or three-chambered) or one-sided (two-chambered). During ventricular contraction, while the aortal (semilunar) valves are open, the aorta fills and the heart empties; ventricular systole (contraction) stops when ventricular pressure falls below that of the aorta and the aortal valves close (Fig. 20–6). During ventricular contraction the auricular pressure is below that in the ventricle, and the auriculoventricular (AV) valves remain closed; as the auricles fill, their pressure gradually rises while the ventricular pressure is falling and the AV valves open, blood enters the ventricles, and the auricles

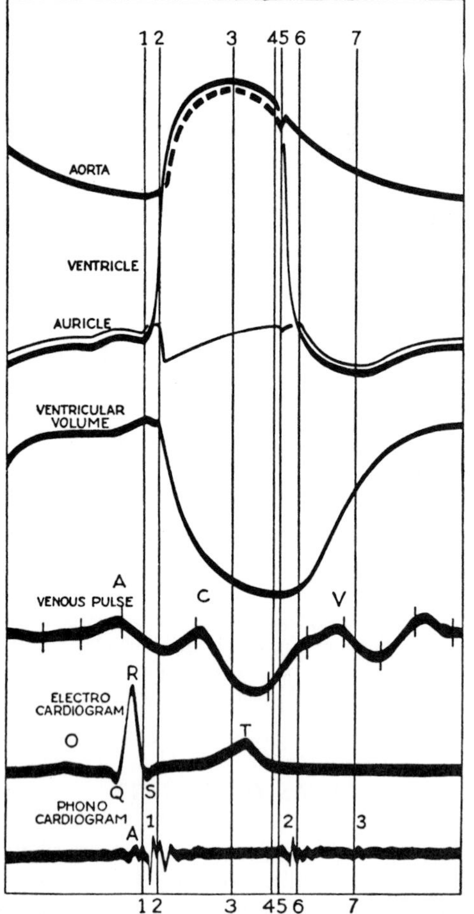

Figure 20-6. Correlated events of the cardiac cycle. Venous pulse, electrocardiogram, and heart sounds in man; aortic pressure, auricular pressure, ventricular pressure, and ventricular volume adapted from the dog. *1*, closure of the AV valves and beginning of ventricular contraction; *2*, opening of the aortic valves; *2-3*, maximal ejection phase; *3-4*, reduced ejection phase; *4-5*, protodiastolic phase with closure of aortic valves at *5*; *5-6*, isometric relaxation, opening of AV valves at *6*; *6-7*, rapid filling; *7* to auricular systole is phase of diastasis. Time marks at 0.1 sec on venous pulse curve apply to lower three curves. (From Hoff, H. E., *in* Howell's Textbook of Physiology, 1946.)

contract until the high pressure in the ventricles closes the AV valves. In man the ventricles first contract with the AV valves closed (isometric contraction) for 0.05 sec, and then during ejection of blood (isotonic) for 0.22 sec; ventricular relaxation lasts 0.53 sec. The auricles contract for 0.11 sec and are in diastole for 0.69 sec.

In fishes, with a single auricle and ventricle, the blood passes through gill vessels before entering the aorta; hence, the heart must provide pressure for traversing two capillary beds. The evolution of the two-sided heart began with air-breathing fish. In modern dipnoans the return of blood from the air bladder is separate from return from the rest of the fish. Amphibians, with lung and skin breathing, have a divided auricle, and hence a two-sided circulation; the pulmonary vein is separate from the vena cava, but a pulmocutaneous artery runs from the aortic arch. Some reptiles (crocodiles) have a complete ventricular septum; in lizards, turtles, and snakes the septum is incomplete. By radiography, by dyes, and by oxygen determinations it has been shown that in ventricles of lizard, caiman, and turtle there is relatively complete separation of streams of aerated and deaerated blood; also, variable shunting of blood from the right to the left side occurs when need for passage through the lungs is reduced. In a giant toad, the blood in the pulmonary vein is 96% saturated, that in the left side of the ventricle is 87.6% saturated, and that in the right side is 8.9% saturated; hence, there is very little mixing between the two sides.[92] In *Amphiuma,* shunts of blood in the heart are labile according to conditions.[92] In an alligator, bradycardia (slowing of the heart) occurs during diving (p. 831) and blood is shunted from the right to the left side of the ventricle. Thus, oxygen supply to the brain is favored, an adaptation not possible in birds and mammals.[207,208] In *Iguana,* when the animal is heated, a shunting of 20 to 25% of the blood occurs, which favors more blood going to the skin and less going to the lungs.[9]

The chambered hearts of molluscs consist of one or two auricles (four in tetrabranch cephalopods) and one ventricle. In cephalopod molluscs, blood returns from the body by veins to branchial hearts at the base of the gills, and passes from the gills by veins to the systemic auricles. In other molluscs, the blood leaves arterioles for tissue sinuses, from which it passes via gills and kidneys in variable proportions to venous channels for return to the auricles.

Tubular Hearts. The systemic hearts of most arthropods consist of contractile tubes. There may be a thin receiving chamber (atrium) surrounding part of the heart, as in *Limulus,* or the heart may be free in a large pericardial sinus. The heart is usually anchored at several corners and receives blood through paired valved ostia. In crustaceans, blood leaves the heart by arteries, and passes from small arterioles into tissue sinuses which ultimately carry it to the gills; branchial veins return the blood to the pericardium.

In many insects the heart is suspended by variously arranged alary muscles which maintain tension on the heart. The entire tubular heart may contract nearly simultaneously, or it may show a wave of contraction.

The heart of tunicates is a delicate nonvalved tube lying in the pericardium. The heart pumps for a time in one direction and then reverses, so that blood does not flow in the same circuit continuously.

Pulsating Vessels. Blood vessels which contract with peristaltic waves are widely distributed. In annelids, many blood vessels show rhythmic peristalsis. In the earthworm, contraction waves in the dorsal vessel pass from the posterior to the anterior end; blood then passes through several pairs of lateral "hearts" to the contractile ventral vessel. Leeches have two lateral vessels which contract alternately. In *Arenicola,* blood flows forward in the dorsal vessel, and then by a gastric plexus to lateral "hearts" and to the ventral vessel; the dorsal, lateral, and esophageal vessels and some nephridial vessels are contractile.

In some holothurians, pulsating vessels of the haemal system contract peristaltically. The distinction between tubular hearts and contractile vessels is not sharp.

In the prochordate *Branchiostoma* (amphioxus) many vessels are contractile and the heart is little more than a "sinus venosus" and a "conus arteriosus." In higher vertebrates pulsating veins, as in bat wings, help to propel blood back toward the heart.

Ampullar Hearts. Accessory hearts are boosters which propel blood through peripheral channels. The branchial hearts of cephalopods consist of a spongy musculo-epithelial tissue surrounding many small vessels. Accessory hearts are common in insects, particularly at the base of antennae and at the attachment of wings and in legs.

In the cyclostome *Myxine,* three accessory hearts occur in the portal, cardinal, and caudal veins; striated muscle also forms accessory hearts in the gills which propel blood forward.[87]

Fishes, amphibians, and reptiles have lymph hearts which are contractile enlargements of lymph vessels, and which tend to force lymph into the veins. Lymph enters veins at many points and not through thoracic ducts as in birds and mammals. The lymph hearts are composed of striated anastomosing fibers and have valves which prevent backflow.

HEART RATE AND OUTPUT

Measurement of heart rate is probably the commonest measure in comparative physiology, yet in a given animal heart rate is one of the most variable of characteristics. Table 20–3 gives a few selected data, measured mostly on unanesthetized and often unrestrained animals. Telemetered values from freely moving animals in a natural environment yield lower and more stable heart rates. A few trends may be mentioned.

In general, heart rate is faster in small animals of a given kind than in large ones. The heart rate in an elephant or horse is 25 to 40/min, in a dog 80/min, in a cat 125/min, in a rabbit 200/min and in mice 300 to 500/min. In hummingbirds rates of 500 to 600/min are common, in sparrows 400 to 500/min, and in domestic fowl 150 to 300/min. Among crustaceans, *Daphnia* heart beats at 250 to 450/min (20° C), *Asellus* 180 to 200/min, and crayfish 30 to 60/min. In a series of crabs, the heart rate declined exponentially with increasing body size but with a slope slightly less than the decline in metabolism with increasing size.

Heart rate varies greatly with exercise, rest, or locomotor activity and with nervous excitement. In general, heart rate is lower in sluggish animals than in related active ones. In clams, the heart rate ranges from 0.2 to 22/min; in squid and octopus it is 40 to 80/min. Heart rate in fast fish exceeds that in slow-moving ones. The heart rate of a sphinx moth is 40 to 45/min when at rest, and 110 to 140/min when in flight. Heart rate varies with internal pressure, particularly in animals with open circulation; in clams, foot retraction is accompanied by rise in hemolymph pressure and heart rate. In an awake but resting bat, the heart rate was 250 to 450/min, when excited 880/min, in diurnal lethargy 120–180/min, and in hibernation 18/min.[105]

Heart rate is very sensitive to temperature and to oxygen need and availability. In most poikilotherms the rate rises two to three times per 10° C rise in temperature in the physiological range. Similar increases occur in homeotherms, but the basal rate at the same temperature is higher than in poikilotherms. Diving mammals, birds, and turtles show a bradycardia (slowing of heart rate) during a dive (p. 831; also p. 201, Chapter 5). Many fish show a similar slowing of heart rate on removal from water which is greater

TABLE 20–3. Resting heart rate, stroke volume, cardiac output, and weight specific cardiac output. Data from cited references or from Handbook of Biological Data.

Animal	*Resting Heart Rate, beats/min*	*Stroke Volume*	*Cardiac Output*	*Weight Specific Cardiac Output*
Mammals				
man[59]	75	77 ml	5.5–6 li/min	80–90 ml/kg/min 2.1–4.9 li/m²/min
Tursiops[172]	84–140		12.2 li/min	47–105 ml/kg/min
Phoca			5.8 li/min	
Phoca[131] surface, 10° C			5.2 li/min predive	
diving			0.62 li/min diving 8.3 li/min postdive	
elephant seal[36, 37]				
surface	60			
diving	4			
giraffe[53]	90	100 ml	20–40 li/min	
elephant	30–40			
cow	45–60		45.8 li/min	113 ml/kg/min
pig			4.5 li/min	146 ml/kg/min
horse	35–40		18–24 li/min	
llama (at sea level)				118 ml/kg/min
dog	90–100		2–3 li/min	
rat[141]	350			286 ml/kg/min
mouse	498			
Citellus[141] awake	276		69 li/min	313 ml/kg/min
hibernating	2–7		1.04 li/min	4.6 ml/kg/min
bat[105]				
quiet awake	250–450			
excited	880			
daily torpor	120–180			
deep hibernation	18			

(Table 20–3 continued on opposite page.)

than the rate reduction in hypoxic water, but they do show bradycardia in water which is very low in oxygen. Both diving air-breathers and fish out of water show a tachycardia (speeding of heart rate) on return to normal respiratory conditions. In a mollusc such as *Mytilus* the heart rate is lower in hypoxic water or in air than in well-aerated water.

Some changes in heart rate result from direct environmental action on the heart; most of them are by complex reflexes which are only partially understood.

Cardiac output (C.O.) is a measure of the volume of blood discharged by each ventricle of a two-sided heart or by the single ventricle of a one-sided heart. One method of measurement is to divide the rate of O_2 utilization by an animal by the difference between arterial O_2 and the O_2 in blood going to the respiratory surface (Fick method):

$$\text{C. O. (liters/min)} = \frac{\text{ml } O_2 \text{ used/min}}{\text{arterial } O_2 - \text{venous (or pulm. art.) } O_2}$$

Other methods involve similar measurements with foreign gases or with dyes. The cardiac index expresses output on the basis of surface area or $W^{2/3}$. Most comparative data is in weight-specific cardiac output in liters/min/kg. Cardiac output can be varied in two ways—by varying stroke volume and by varying heart rate. Stroke volume (ml per beat) is the cardiac output divided by the heart rate.

Table 20–3 gives some selected examples of cardiac output for a variety of animals. In man, cardiac output (C.O.) rises from birth to a maximum at 3 to 5 years and then declines slightly thereafter throughout life. Average values for a 45-year-old man are: 5.2 liters/min output; 3 liters/min/m² surface area at 68 beats/min and stroke volume of 74 ml/beat.[59] Cardiac output increases (up to 6 times) with exercise, mainly because of in-

TABLE 20-3. Resting heart rate, stroke volume, cardiac output, and weight specific cardiac output. Data from cited references or from Handbook of Biological Data. *(Continued)*

Animal	*Resting Heart Rate, beats/min*	*Stroke Volume*	*Cardiac Output*	*Weight Specific Cardiac Output*
Birds				
duck[176] male	175			287 ml/kg/min
female	185			253 ml/kg/min
hen	178–460		400 ml/min	262 ml/kg/min
sparrow	460			
Troglodytes	450			
Reptiles				
Iguana[194], 30° C	38.2	1.5 ml/kg		58 ml/kg/min
Iguana[207]				60 ml/kg/min
Iguana,[9] normal temp.				44 ml/kg/min
heated				80 ml/kg/min
Tiliqua,[26b] winter	26			
summer	38			
Varanus		1.0 ml		
turtle[208] surface	40	3.7 ml	148 ml/min	50–60 ml/kg/min
diving	2	3.3 ml	6.6 ml/min	
Amphibians				
Amphiuma[89]		0.67 ml		30 ml/kg/min
Rana pipiens[161] in air		35 μl	20–30 ml/min	
		10–15 μl	5–7 ml/min	
Fishes				
Gadus[88]	30			9.3 ml/kg/min
Cyprinus carpio[51]	16			18.3 ml/kg/min
	2.1			7.6 ml/kg/min
Ictalurus[63]	21.6			11.3 ml/kg/min
Amia[63]	13.9			6.2 ml/kg/min
sucker	20			4.5 ml/kg/min
Electrophorus[88]	65			40–70 ml/kg/min
Salmo[84]	46			16 ml/kg/min
Anguilla[84]	32			18.8 ml/kg/min
Oplegnathus[84]				36.6 ml/kg/min
Scyliorhinus[12]	39–48			22 ml/kg/min
Squalus				25 ml/kg/min
Raja[155]				21 ml/kg/min
Myxine[87]	30–40			
Invertebrates				
Octopus[91]	16–23	6.3–12.1 ml	57–81 ml/min	6.6–20.5 ml/kg/min
Mytilus[69] in water	24–26			
in air	10			
Crassostrea[171]	20–30			
Ciona[102]	20	0.045 ml		66 ml/kg/min
Panulirus[146]	60–100	0.6–1 ml		80 ml/kg/min
Locusta[13]	80		0.015 ml/min	
Glossoscolex[93]				
lateral vessel	20			
dorsal vessel	6–8			

creased heart rate; C.O. decreases on standing and increases with rise in body temperature. C.O. in mammals decreases with body size but is nearly constant with respect to body surface area; in a series of mammals, C.O. = $W^{0.776}$. For rats, C.O. = 3.83 $W^{0.582}$ where surface area = 0.09 $W^{0.667}$ (reference 115); for a series of mammals, C.O. = 0.76 $W^{0.776}$ (reference 209). In a series of mammals (rat to horse), cardiac output increases with the end-diastolic volume (EDV) according to the relation C.O. = 106 $EDV^{0.77}$ (Fig. 20-7).[73] In an active ground squirrel, C.O. is 69 ml/min (313 ml/kg/min), while in hibernation C.O. is 1.04 ml/min (4.6 ml/kg/min); the decrease is mainly due to decreased heart rate rather than to change in stroke volume.[141]

In a frog, *Rana pipiens*, the effects of submergence are as follows:[161]

	in air	submerged
heart rate	40	25–32/min
stroke volume	35 μl	10–15 μl
cardiac output	20–23	5–7 ml/min

Similarly, a turtle shows decreased cardiac

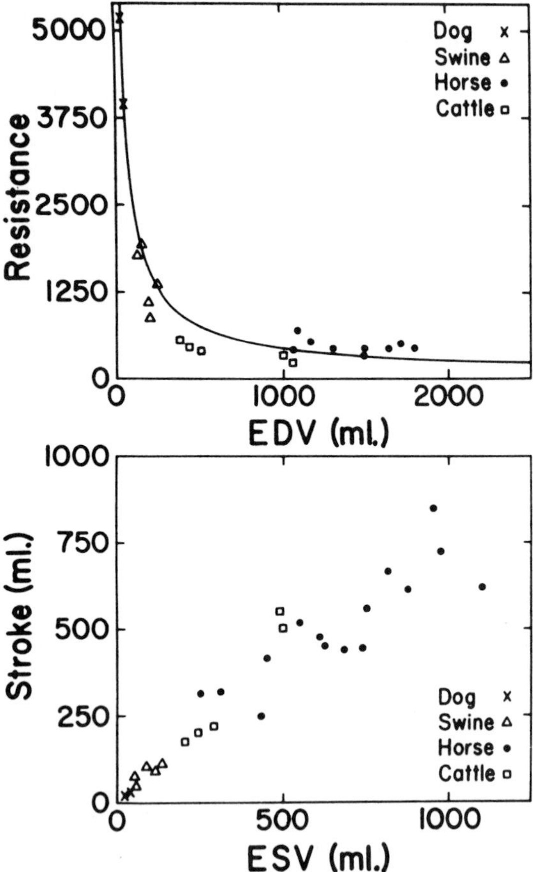

Figure 20–7. Relation between end-diastolic volume (EDV) and peripheral resistance, and between end-systolic volume (ESV) and stroke volume, for mammals of different sizes. (From Holt, J. P., E. A. Rhode, S. A. Peoples, and H. Kines, Circ. Res. *10*:798–806, 1962.)

output during a dive because of marked bradycardia.

Early measurements with immobilized fish tended to show a higher C.O. for elasmobranchs than for teleosts. With catheterized unanesthetized carp, a C.O. of 18.3 ml/min/kg was obtained,[51] and in dogfish it was 22 ml/kg/min.[12]

The work done by a heart is given by: Work = $P_{ventricle} \times$ vol/second. For man, the work done by the left ventricle (pressure 100 mm Hg) is 1.1×10^7 ergs/sec; work by the right ventricle (P = 20 mm Hg) is 0.2×10^7 ergs/sec.[29]

VERTEBRATE SYSTEMIC HEARTS

Initiation and Conduction of Excitation. The beat of the heart of a vertebrate, tunicate, or mollusc is myogenic. In frog or fish, the beat normally originates in the sinus venosus; in bird or mammal, it originates in the sinoauricular (SA) node. The fibers of embryonic heart muscle in tissue culture contract rhythmically. In a chick embryo, coordinated contractions start in the aortal end of the ventricle after only 29 hours of incubation, and with development the pacemaker moves toward the sinus where recovery between beats is fastest. If the sinus venosus is removed from an amphibian heart, the normal rhythm stops and other regions take over initiation of regular but slower contractions. The rate in sinoatrial fragments of embryonic rat hearts is twice as fast as in ventriculobulbar fragments.[62] In adult mammalian hearts *in situ,* local cooling or warming is most effective in altering rate when applied in the sinoauricular region. Upon local warming of fish hearts, the sinus, the auricular floor, or the AV junction can be made the pacemaker.

Cells of the sinus pacemaker region of a frog heart show graded, slowly rising potentials which, at about 15 mv depolarization, give rise to 70 mv spikes which are conducted over the auricles.[76] The pacemaker potential in frog or turtle sinus is diminished by impulses in the vagus nerve, and the cells may be driven from resting potential of −55 to −65 mv by vagal impulses.[76] Similarly, in rabbit SA node, the vagus hyperpolarizes and diminishes the rate of rise of the pacemaker potential.[186] Pacemaker activity in both trout heart and the aneural heart of a hagfish shows up to a 75% increase in frequency with rise in internal pressure.[85]

The wave of excitation from the pacemaker is conducted over the atrial myocardium (in mammals at 1 m/sec) and then, after an AV delay, spreads over ventricular muscle (in poikilotherms) or at 5 to 6/sec over specialized conducting Purkinje tissue (in birds and mammals). The electrical wave front passing over the heart is of sufficient magnitude that it can be recorded as an electrocardiogram (ECG) at some distance from the heart. Typically the ECG in all vertebrate hearts, as recorded extracellularly, consists of a series of slow waves, upward (negative) deflections called P, R, and T, and downward (positive) deflections, Q and S. The P wave corresponds to conduction in the auricles, the PQ interval represents delay at the auriculoventricular junction, and the QRS complex corresponds in time to conduction in the ventricles. The T wave represents repolarization of the ventricular surface, an upright T indicating earlier repolarization of the left side and an inverted T indicating earlier repolarization of the right side of the heart. In amphibians and fishes, a V wave originating in the sinus venosus precedes the P wave; in elasmobranchs, a B wave due to depolarization of the conus arteriosus occurs between the S and T waves. Records obtained with intracellular microelectrodes show that each fiber depolarizes rapidly; then, after a partial repolarization, it remains partly depolarized in a plateau which may last for 0.1 to 1.5 sec according to the kind of heart (Fig. 20–8A). Auricular fibers show less plateau than ventricular ones. The excitation wave travels down the inner conducting bundles and then turns outward over the myocardium; the resulting electrical dipoles are directed oppositely in the left and right ventricles, or in right and left sides of a single ventricle. An algebraic summation of ventricular action potentials as recorded from the two sides of the body shows that the apparent "isoelectric" S-T segment results from neutralization of the two plateaus before repolarization and that the T wave results from the earlier repolarization of the left ventricular muscle than of the right.[29]

Histologically, a vertebrate heart consists of branched striated fibers. The bundles of Purkinje fibers of mammalian and bird ventricles are muscle modified for rapid conduction (2 m/sec). Frequently, cardiac muscle fibers are joined by transverse intercalated

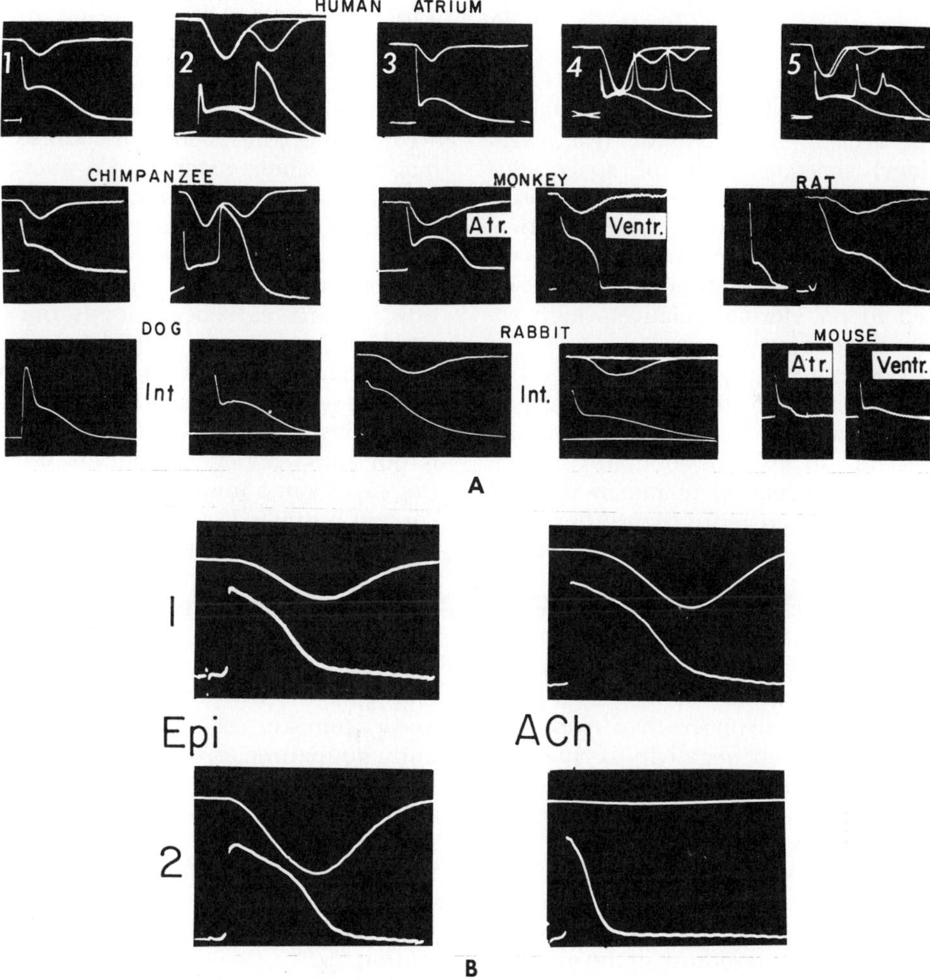

Figure 20–8. *A*, Intracellular action potentials and contraction records from hearts of various mammals. Atrium sometimes gives double spikes (human record 2, chimpanzee record 2). Rat atrium shows two recordings at different sweep speeds. In dog and rabbit atria, second records are 10 and 5 seconds, respectively, after first. (From Sleator, W., and T. de Gubareff, Amer. J. Physiol. *206*:1000–1014, 1964.) *B*, Effects of epinephrine and acetylcholine on contractions (upper traces) and intracellular action potentials (lower traces) in guinea pig atrium. 1, controls; 2, effects of 1.2×10^{-6}M epinephrine and 4.4×10^{-8}M acetylcholine. (Courtesy of W. Sleator.)

discs which include desmosomal regions, probably for mechanical coupling, and nexuses or gap membrane junctions which appear to be paths for intercellular electrical conduction (p. 459).[11] The space constant (length for an applied potential to fall to $1/e$ of its applied value) in rat atria is 130 μm parallel to the fiber axis and 65 μm perpendicular to it; interfiber junctional resistance is calculated to be about 1 ohm/cm^2.[203,204,210] In Purkinje fibers (in sheep) the space constant is two to four times longer than in striated muscle, and the intercalated disc junctions appear to impose virtually no resistance to electrical conduction.[76,188] The space constant of single fibers of right ventricle of sheep or calf is 880 μm, and the time constant is 4.4 msec for R_m of 9100 Ω cm^2 and C_m of 0.81 mF/cm^2. The apparent capacitance of Purkinje fibers is much greater—10 μF/cm^2 as measured on bundles of electrically coupled cells; allowance for their surfaces yields smaller capacitances per unit area.[204] Acetylcholine (the vagal transmitter) shortens both the space constant and time constant of cardiac muscle and reduces the duration of the electrical plateau. There is evidence that, in tissue cultures of heart muscle, conduction can go from cell to cell despite relatively high resistance connections.[173]

The time occupied by the P-R interval in the ECG is about 30 to 50 per cent of the total duration, whether the heart rate is 36/minute (crocodile, horse) or over 600/min (mouse). However, in hibernating mammals the P-R interval is prolonged more than other components; i.e., there is longer delay at the auriculoventricular junction.

When mammalian heart tissue is stored in Tyrode solution in a refrigerator, it gains sodium and loses potassium; on rewarming, the resting potential may be increased from the normal 70 mv to more than twice that value. This increased negativity is due to a transiently overactive sodium extrusion, and can be prevented by poisoning the sodium pump with ouabain.[178] Purkinje fibers of a dog heart have a resting potential of 90 mv and an action potential of 120 mv which may last 300 to 500 msecs. The potential overshoot is reduced when sodium in the medium is lowered, and conduction stops when 70% of Na_o is replaced. Lithium can replace sodium for the spike in cat Purkinje fibers, but no spikes occur with choline as replacement.[31] Similarly, in frog ventricle the overshoot is proportional to external sodium concentration.[20] Membrane resistance drops to a low value during the rising phase of the spike; then it increases to a higher value than the resting resistance during the plateau and falls again during repolarization. The plateau is prolonged by barium; it is shortened and membrane resistance is increased by external potassium and by acetylcholine. From voltage-clamping of cardiac fibers, combined with ion replacements, it appears that normally the initial spike is due to increased sodium conductance, followed by a decrease in K conductance during the plateau; then there is slow recovery of potassium conductance during the repolarization following the plateau.[48,136] Pacemaker cells have high resting inward sodium current; this, combined with time- and voltage-dependent K conductance, permits rhythmic activity (p. 470).[66,188] Epinephrine activates the late K current at a lower membrane potential than normally; hence, pacemaker frequency is increased. In frog ventricle, calcium uptake accompanies contractions; also, calcium influx increases when external Ca increases or sodium decreases.[134] In rabbit auricle, high external Ca increases the rate of rise of pacemaker potential.[157] Also, in Purkinje fibers clamped to a partially depolarized state, a slow inward current depends on Ca_o; inward calcium current is small in normal Na_o, and increases if external sodium concentration is low.[147] It appears that in mammalian heart muscle the inward current of the spike is normally carried by sodium, but calcium can carry some inward current; more of the current is due to Ca^{++} in frog heart. In addition, calcium modifies permeability to sodium and potassium. In guinea pig ventricle the plateau is prolonged in zero Mg and Ca; high Ca shortens the plateau, thus hastening repolarization by outward potassium current.[52] Acetylcholine shortens the plateau in atrial and pacemaker cells (Fig. 20–8B). In human and chimpanzee atrium at reduced temperature, the voltage level of the plateau is lowered and plateau duration is lengthened. A single stimulus may elicit two spikes, the second one slower than the first and also depending on sodium conductance, which is determined by external calcium concentration.[166,167]

Contraction and Regulation. By virtue of the electrical spread over the entire cardiac muscle, a vertebrate heartbeat is all-or-none. When tetanized, the heart shows spasmodic uncoordinated contractions, the condition of "delirium cordis." The heart is absolutely refractory to electrical stimulation during most of systole; it can be stimulated (is relatively refractory) at the end of systole and during diastole, but an extra contraction elicited during this time is of submaximal height. An extra contraction superimposed in a series of normal beats may be followed by a compensatory pause longer than the normal diastolic interval, depending on the placement of the extra beat.

The all-or-none nature of cardiac contraction applies for a given state of stretch or cardiac volume. However, when strips of cardiac muscle are extended, the developed tension rises to a maximum, and if a ventricle is distended under pressure the force of each contraction increases; i.e., the heart compensates mechanically for load. Above a limiting diastolic volume (radius of curvature), the pressure developed for a given wall tension actually decreases and stroke volume diminishes; this is the phase of decompensation.[29] In exercise, or in sympathetic nerve stimulation, the stroke work may increase for a given pressure; i.e., ventricular contractility increases. In different species the velocity of shortening of cardiac muscle is adjusted to the normal heart rate; in a dog with beat duration of 161 msec, left ventricle strips shorten at 3.1 muscle lengths per sec, while in rat with 50 msec duration the shortening is 6.5 lengths per sec.[71]

Reflex regulation of heartbeat is correlated with peripheral regulation, in that heart rate rises in compensation for a fall in blood pressure and the heart slows when blood pressure rises. In exercise, both rate and amplitude increase but the rate change is more important. The vagus nerves slow the heart and reduce its amplitude of contraction; the right vagus is more effective than the left in slowing. The primary vagal fibers terminate in ganglia located in the SA node and in the walls of the atria, not in the ventricles. Acetylcholine (ACh) is released at vagal endings and has negative inotropic and chronotropic action. Atropine blocks the inhibitory action of the vagi and of applied ACh, and physostigmine (eserine) enhances the inhibitions by blocking acetylcholinesterase (AChEs). Atropine may act at the muscle receptor sites, whereas nicotine may block the ganglia.

Sympathetic nerve fibers from the stellate ganglion innervate the SA node, atrium, and ventricle; in addition, epinephrine-containing cells are found in the heart wall. The secondary (postganglionic) sympathetic fibers cause cardiostimulation by liberated norepinephrine (NE). Sympathetic outflow may also elicit liberation of epinephrine (E) from the adrenal glands. Stimulation is both on rate (on pacemakers) and on amplitude of contraction. Action of adrenergic fibers is on β-receptors, as evidenced by the sensitivity series: isoproterenol > epinephrine > norepinephrine, and as indicated by blocking by β-blocking agents (see p. 496).

Cardioregulation is essentially similar in mammals, birds, reptiles, and amphibians. In all these animals, vagotomy results in increase in heart rate and rise in blood pressure; tonic action of the vagi probably keeps the heart slightly depressed normally. Stimulation of the vagi causes bradycardia, and stimulation of the central end of one vagus causes cardioinhibition and a fall in blood pressure. Afferent fibers occur in the vagus and in carotid sinus nerves.[89] In diving animals, atropinization or section of the vagi prevents the bradycardia characteristic of submergence (p. 831). Vagal stimulation, as well as applied ACh, shortens the plateau and reduces overshoot in cardiac action potentials. During entry and recovery from torpor, the heart rate of mice (*Peromyscus*) drops faster and recovers faster than body temperature; hence, the vagus regulates entrance and sympathetics regulate arousal.[121]

In amphibians and fishes, the sympathetics to the heart are often in the same trunk as the vagus. In anurans, all sympathetic innervation is via the vagus, while in urodeles both vagosympathetic and direct sympathetic innervation are found.

Preganglionic vagal and parasympathetic nerves may end on endocardiac ganglia in frog heart.[187] Individual parasympathetic neurons can be visualized in the interauricular septum; each presynaptic axon forms some 25 synaptic boutons, and the synaptic potentials (excitatory junction potential, epsp) show persistence of transmitter for 40 msec. The reversal potential at the synapse is 0 to −12 mv, and calcium ions increase the frequency of miniature synaptic potentials (mepsp). Acetylcholine sensitivity is normally restricted to the synaptic spots, and after denervation the whole cell becomes sensitive.[104]

The heart of a jawed fish is slowed by acetylcholine, and has inhibitory cholinergic vagal innervation to the sinus venosus and atrium (but not to the ventricle, which is insensitive to ACh). Vagal inhibition of the heart of the elasmobranch *Scyllium* can be elicited reflexly by afferents in the vagus, hypobranchial, and lateral line nerves, and in the ganoid *Acipenser* stimulation of the skin, especially on the head, fins, and barbels, results in reflex standstill of the heart, an effect antagonized by atropine. In *Tinca*, hypoxia initially stimulates enhanced breathing; later it decreases heart rate.[144] In a trout, electrical stimulation of the vagus is inhibitory; but when this action is blocked by atropine, further stimulation causes acceleration, an effect blocked by guanethidine and pronethalol. Hence, the fish vagus contains adrenergic excitatory fibers as well as cholinergic inhibitory ones.[50] Fluorescence microscopy of trout heart reveals many nerve endings containing catecholamines with very few intrinsic adrenergic ganglion cells; epinephrine and norepinephrine have positive inotropic and chronotropic actions.[50]

Three modes of adrenergic excitation are recognized: (1) Sympathetic nerve fibers pass in the vagus, in trout to the whole heart and in tench to the sinus venosus only, or they pass directly from the sympathetic chain to the heart (e.g., to the sinus in the shark *Heterodontus*). (2) Chromaffin cells occur along veins (in trout) leading to the heart or as chromaffin masses in the posterior carotid sinus (in *Heterodontus*). (3) Chromaffin cells occur in the walls of ventricle and atrium (in skate and lungfish).[50]

Among cyclostomes, hagfishes differ from lampreys. In hagfishes, no nerves are found going to the heart, and the heart is insensitive to acetylcholine as high as 10^{-2} g/ml. However, some chromaffin cells are present, and these can be depleted by reserpine.[87] In lampreys, there are no adrenergic nerves but cholinergic fibers of the vagus (or applied acetylcholine) accelerate heart rate and diminish amplitude of beat; the heart has nicotinic cholinergic receptors (blocked by tubocurarine). Chromaffin cells are present and epinephrine stimulates the heart.[85]

It may be concluded that (1) in cyclostomes innervation is lacking, or is excitatory and

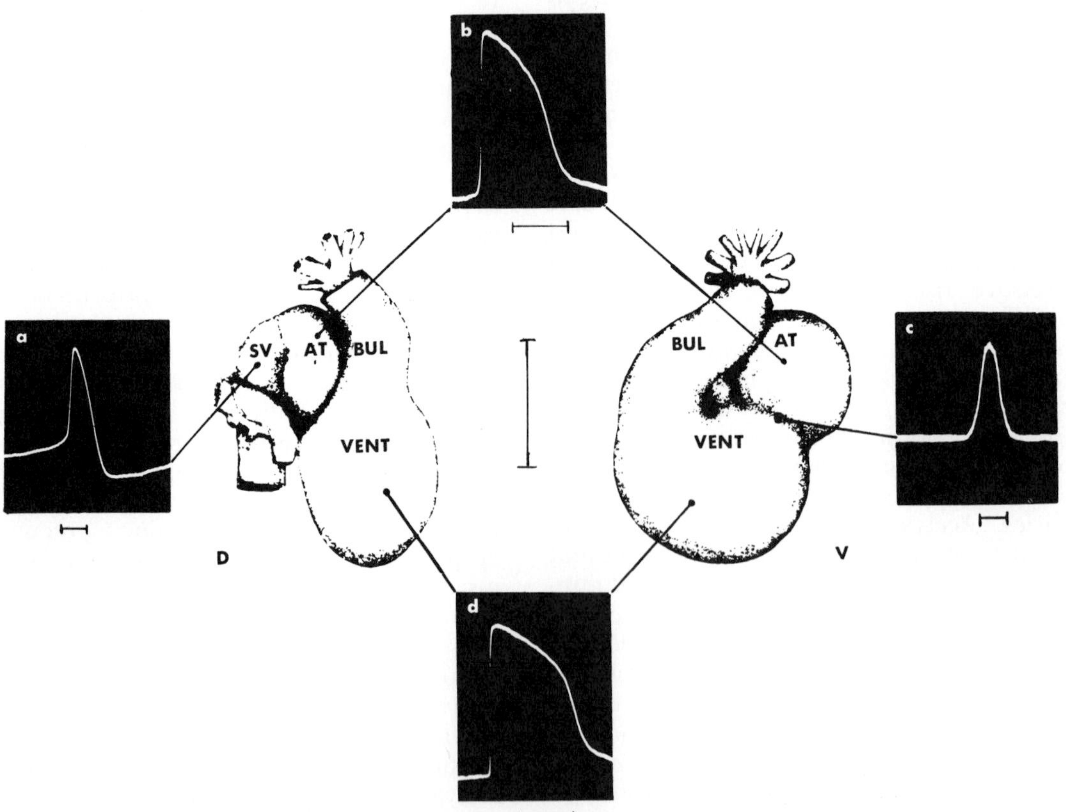

Figure 20–9. Intracellular action potentials from different regions of heart of chick embryo, as indicated in (D) dorsal and (V) ventral views. (From Lieberman, M., and A. P. de Carvalho, J. Gen. Physiol. *49*:351–363, 1965.)

cholinergic; (2) all other vertebrates have inhibitory vagal cholinergic innervation that is muscarinic (blocked by atropine); and (3) excitatory adrenergic fibers are mixed with the vagus in some fishes and amphibians, but are in separate sympathetic trunks in reptiles, birds, and mammals. Chromaffin cells containing catecholamines are found in all vertebrate hearts, but their function is not clear.

The hearts of chick embryos show rhythmic action potentials at the 9 somite stage and spikes with overshoots at 20 somites.[101] Embryonic heart muscle shows plateau-type potentials similar to adult heart[114] (Fig. 20–9). Embryonic hearts prior to vagal innervation either are insensitive to ACh, as in the fish *Fundulus,* or are inhibited by high concentrations of ACh, as in chick or rat. AChEs increases during development and reaches a maximum at the time of innervation.

Lymph hearts of amphibians and a few fish (eel) are normally under control of the spinal cord. After connections to the spinal cord are cut, the lymph hearts stop but may later show irregular beats. Anterior and posterior homolateral hearts are normally synchronized, but each heart is driven by a local portion of the spinal cord. Local cooling or warming of the cord slows or speeds the lymph heart. A single spinal neuron may fire 5 to 7 times per beat of the lymph heart. Lymph hearts transplanted to the lymph sac of the tongue beat strongly although spasmodically. A toad lymph heart shows an oscillatory action potential, apparently with motor impulses preceding the muscle potentials.[42] In intact frogs, ACh may increase contraction amplitude and tone; high concentrations of ACh may stop the heart.[41]

In summary, the vertebrate systemic heart maintains constancy of activity, yet it adapts to changing demands for blood. Generalized cardiac muscle is myogenically active or spontaneous, and conducts as a functional syncytium. It has a plateau action potential, i.e., depolarization prolonged for one heart beat. It contracts with increasing tension as it is stretched. Some cardiac muscles have become specialized as pacemakers, some as conducting strands, and some as tension-developing muscle. Regulation of cardiac muscle is partly by direct responses of the muscle fibers to distension and to local chemicals (ions, catecholamines), partly by hormones and partly by autonomic reflexes.

MOLLUSCAN HEARTS

Molluscan hearts are myogenic; nerve cells occur in or near the heart in cephalopods, but these are secondary neurons of regulating nerves. In the heart of clam, mussel, or snail, the beat can originate at any point and the contraction can be local or complete. The heart of a bivalve has two fragile auricles and a muscular ventricle, through which the intestine passes. Cephalopods have systemic hearts, each with two auricles and a ventricle, and they also have branchial hearts.

The hearts of molluscs are very sensitive to pressure; they fail to give maximum beats unless distended, and the beats are stronger and more frequent when internal pressure is high. In *Octopus* no beat was recorded at pressures below 2 cm H_2O; the frequency was 69/min at 8.5 cm H_2O, and 52/min at 4.5 cm H_2O.[49] Beat in the heart of a snail, *Rapana,* occurs when pressure reaches 40 cm H_2O; in *Aplysia* a beat occurs at 2 cm H_2O.[72]

The electrocardiograms of molluscan hearts consist of several slow waves, often a diphasic faster component at the beginning of contraction and a slower one during contraction. In oyster heart, intracellular microelectrodes record pacemaker potentials in some fibers before spikes appear. In snail hearts, pacemaker potentials, spikes, and plateaus are recorded.[72] The spike may be followed by a plateau, much as in the vertebrate ventricle. The spike amplitude in oyster heart reaches 53 mv on a 57 mv resting potential.[78, 80] When a bridge of muscle connects two portions of an oyster heart, electrical spread from one half to the other occurs.[43, 44] Both Na and Ca are necessary for normal spikes; at a ratio of Ca/Na^2 of 4.2×10^{-5} the spike amplitude is proportional to Na_o, whereas in Na-free medium the spike is proportional to Ca_o with a 22 mv slope in *Mytilus,* and 19 mv in oyster (per log unit of Ca_o^{++}); Mn reduces these spikes as it does Ca spikes in other systems.[82] Anodal polarization of fibers of the oyster heart abolishes the plateau; hyperpolarized fibers (large E_m) give large spikes which are due to Na^+ influx, whereas at low resting potentials small spikes are due to Ca^{++}.[78, 81]

Cardiac fibers of oyster contain thick and thin myofilaments; there are no T-tubules, but nexuses join fibers and conduction appears to be electrical as in vertebrate cardiac muscle.[78, 81]

Most molluscan hearts receive both inhibitory and accelerator innervation. In pelecypods and gastropods both types of fiber originate in the visceral ganglion. In some (e.g., *Mya, Anodonta,* and *Mercenaria*), inhibition is dominant; in others (e.g., chitons, *Aplysia, Haliotis,* and *Ariolimax*), acceleration is more evident on visceral ganglion stimulation. In *Helix, Limax* and a nudibranch, *Triopha,* inhibitor and accelerator fibers arise in the pleural ganglion. In *Mercenaria,* nervous inhibition is often followed by acceleration and, when inhibitory endings are blocked by Mytolon (benzoquinonium), acceleration results from nerve stimulation.[205] In the gastropod *Dolabella,* the posterior part of the visceral ganglion inhibits and the anterior part accelerates the heart. Visceral nerves of *Octopus* slow the heart; they also contain accelerator fibers of higher threshold than the inhibitory fibers. Beat of cephalopod branchial hearts is regulated from the visceral ganglion. In the cockle *Clinocardium,* the cerebral ganglion is primarily excitatory and the visceroparietal is mainly inhibitory. Excitation of the heart precedes foot extension, and inhibition of the heart precedes foot retraction.[165] Stimulation of a visceral nerve to the heart of *Aplysia* can give depolarizing (excitatory) junction potentials of 35 msec rising phase, or hyperpolarizing (inhibitory) junction potentials of 27 msec rising phase.[106]

Acetylcholine inhibits the heart of all classes of molluscs. Eserine enhances neural inhibition and the action of ACh in molluscs such as *Mercenaria* and *Dolabella.* Benzoquinonium antagonizes both ACh and nervous inhibition in *Mercenaria* (but not in *Helix*). Perfusion during neural inhibition in *Mercenaria* and *Dolabella* yields an inhibitory agent which is probably ACh. Perfusates from heart of *Octopus* and *Eledone* during visceral nerve stimulation did not contain ACh.[49] In *Helix,* visceral nerve stimulation liberates an agent active in inhibiting other snail hearts, but it is apparently not ACh.[96] Some bivalve hearts are inhibited by extremely small amounts of ACh (10^{-12} to 10^{-10}M in *Mercenaria*); this inhibition may be blocked by tetraethylammonium and Mytolon but not by atropine. Thus the ACh-receptor molecules are different from those of vertebrate heart. Also, in the snail *Lymnaea,* the heart is inhibited by 10^{-12}M ACh and is blocked by benzoquinonium.[152, 153] In most species with low threshold for inhibition of rate and amplitude, there may be tonic excitation at high concentrations; also, a quiescent heart may be stimulated by ACh to beat. In other species of bivalve, such as *Mytilus, Gryphaea,* and *Modiolus,* ACh is ineffective at low concentrations, and at 10^{-5} to 10^{-4}M it may be excitatory with respect to amplitude (Fig. 20–10). A survey of 40 species with respect to inhibition or excitation showed no taxonomic or ecological correlation.[56] In the heart of an oyster, ACh reduces frequency and amplitude and hyperpolarizes muscle fibers, a response which is absent in Cl-free medium; hence, the inhibitory action of ACh is an increase in Cl conductance with a reversal potential of 68 mv.[164] In *Mytilus,* ACh increases frequency and depolarizes the muscle fibers, a response which is lost in Na-free medium; hence, the excitatory action is to increase Na conductance.[164] The membranes of the hearts which are inhibited must differ in their molecular organization from those which are accelerated by ACh.

The excitatory transmitter in visceral nerve fibers may be a catecholamine. In *Mercenaria,* 5HT is excitatory at 10^{-7}M and BOL (methyl-d-lysergic acid) blocks both 5-HT and nerve excitation.[55] Depletion of the

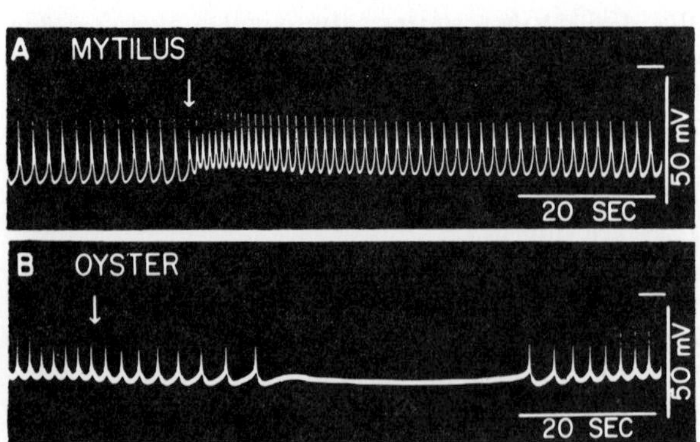

Figure 20–10. Effects of acetylcholine, applied as indicated by arrows, on electrical activity of hearts of *Mytilus* and oyster. (From Shigeto, N., Amer. J. Physiol. *218*:1773–1779, 1970.)

nerves of 5HT by treatment with reserpine stops nerve excitation.[118] Similarly, in the snail *Lymnaea*, 5HT is excitatory at 10^{-10} M, an effect blocked by BOL; dopamine at 10^{-10} M and epinephrine at 10^{-9} M increase both rate and amplitude.[153] Perfusion experiments indicate 5HT to be the excitatory transmitter to the heart of *Helix*.[152]

Both the systemic and branchial hearts of cephalopods are richly innervated,[5] but neurotransmitters have not been identified.

TUNICATE HEARTS

The hearts of tunicates are fragile tubes without valves; they beat for a time in one direction, then in the reverse direction, and thus there is a pacemaker at each end. Frequency is maximal when internal pressure is 5 mm H_2O.[102,103] Reversal occurs in isolated hearts where no pressure changes occur, and if a heart is ligated in the middle the two ends beat independently; hence, reversal is not the result of build-up of pressure. Electrical stimulation can drive the heart at intervals of about 2.6 sec; the threshold is lowest at the ends. Apparently groups of cells at each end, coupled together electrically, show periodically fluctuating excitability levels.[7] In *Ciona* the normal rate is 50/min; the threshold to elicit contractions is highest in the mid-region and is lowest at the pacemaker ends.[103] Intracellular recordings show spikes of 75 mv on a resting potential of −71 mv with duration of 1.2 sec[102] (Fig. 20–11). The muscle fibers of the single-layered wall are 5 × 100 μm and are oriented at 70° to the long axis of the heart. The wave front of conduction is parallel to the cell axis, having a velocity of 7.7 cm/sec at 10° C; in addition, there is spread from a point of stimulation at 0.87 cm/sec perpendicular to the cell axis.[103] Spread of pulses as recorded intracellularly shows electrical coupling of 25 to 50% between adjacent fibers, with a cell-to-cell resistivity of 0.2 ohm cm². Conduction is faster at the two ends than in the middle of the heart.[103]

Removal of the central ganglion does not affect the activity of the heart, and reflex regulation appears to be absent. According to one report[156] acetylcholine at 10^{-6}M and eserine are without effect on the heart of *Ciona*. Another report[102] indicates that an opened heart can be stopped for a minute by ACh at 10^{-8}M, an effect prevented by atropine. Epinephrine at 10^{-5}M accelerates

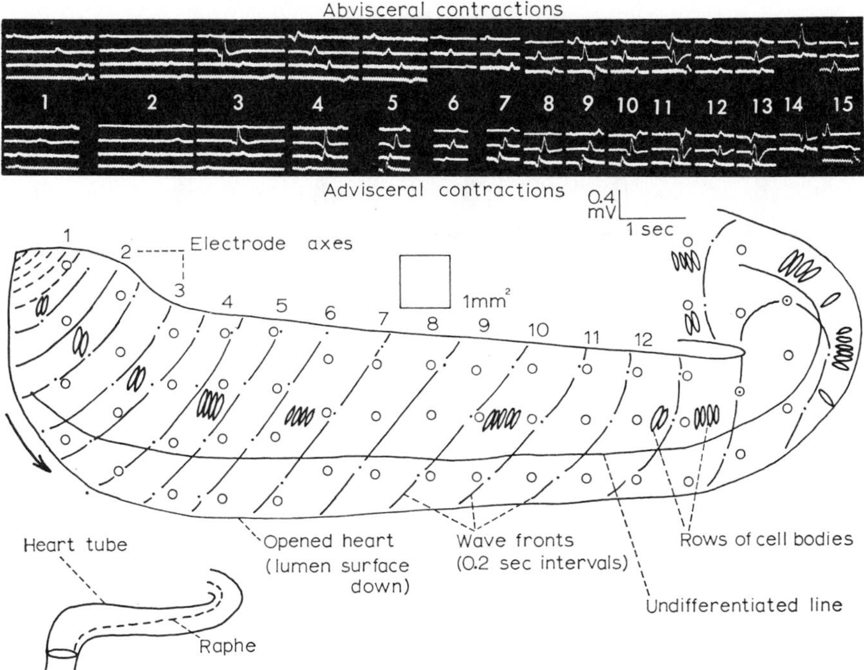

Figure 20–11. Drawing of heart of ascidian *Ciona*, showing wave fronts of contraction. Above, successive action potentials recorded from 15 points along the heart as indicated by drawing. (From Kriebel, M. E., Amer. J. Physiol. *218*:1194–1200, 1970.)

and then stops the heart. No ganglion stimulation experiments have been reported.

HEARTS OF CRUSTACEANS AND LIMULUS

The hearts of most crustaceans and of the horseshoe crab *Limulus* have nerve cells in ganglia on the dorsal surface which originate the excitation wave for the heartbeat. The cardiac ganglion of *Ligia* has six neurons, *Astacus* has 15, and *Squilla* has 16. A number of macrurans, anomurans, and brachyurans have nine neurons.[4] The lobsters *Panulirus* and *Homarus* have five large and four small neurons in the ganglion. The dorsal ganglion of *Limulus* heart has many small multipolar neurons intermingled with large unipolar neurons.

The *Limulus* heart has eight pairs of ostia which divide it into nine segments; blood enters through the valved ostia from the surrounding pericardium. The anterior half of the heart has five pairs of arteries plus one anteromedian artery. The pacemaker median ganglion is connected by side nerves to a lateral nerve on each side of the heart. Local warming accelerates the heart rate most when applied to the fourth and fifth segments. The heart stops beating after the pacemaker ganglion is removed, but if tension is then applied by inflation, peristaltic waves may appear; if the heart is placed in Ca-free solution, there may be local contractions. Both distension and Ca-free beats are

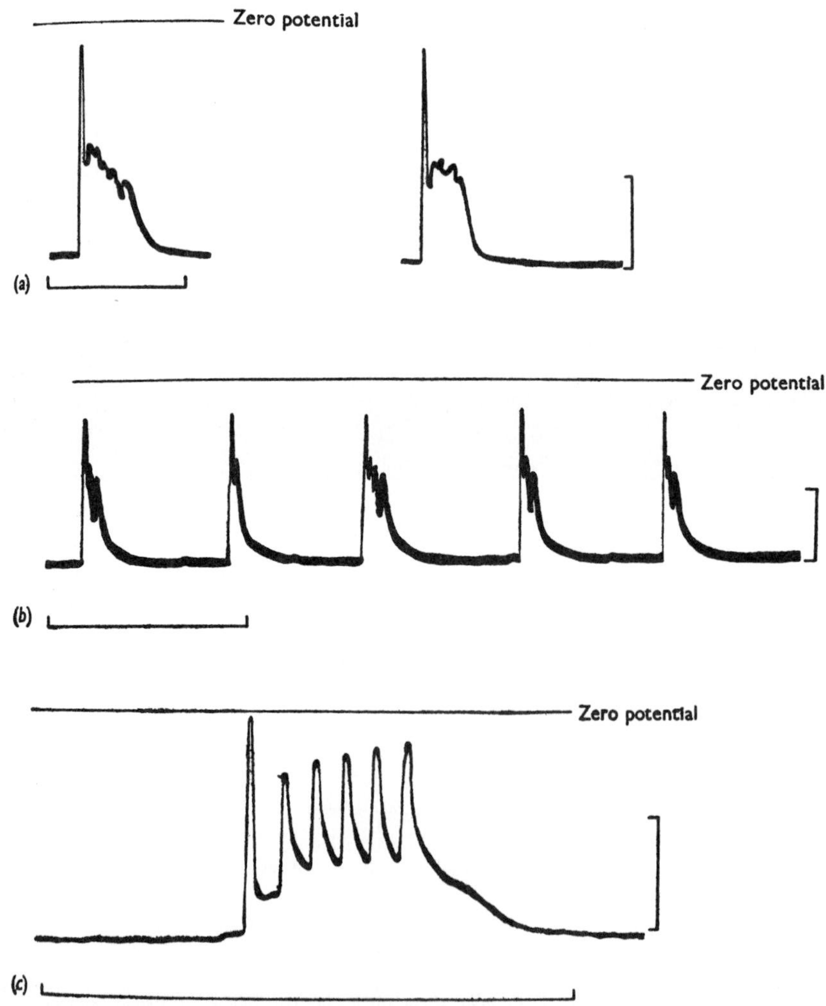

Figure 20–12. Intracellular action potentials from hearts of three crustaceans: (a) *Maja*, (b) *Carcinus*, and (c) *Palaemon*.

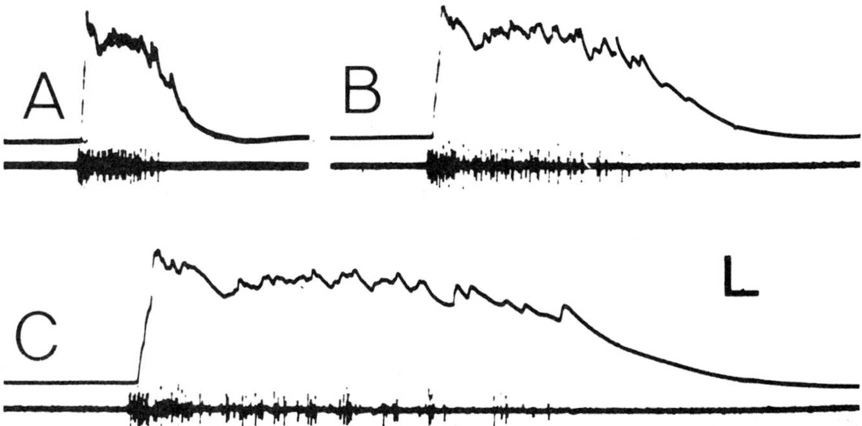

Figure 20–13. Electrocardiograms from *Limulus* for three heart beats at different speeds of recording. Upper records are intracellular muscle potentials, and lower records are extracellular ganglion discharges. Rising phase of initial muscle depolarization shows steps corresponding to junction potentials. Calibration: upper traces 10 mv and lower traces 40 μv; A, 500 msec; B, 200 msec; C, 100 msec. (From Lang, F., Biol. Bull. *141*:269–277, 1971.)

unlike the normal synchronized beat.[107] In *Limulus* embryos, a myogenic peristaltic beat begins at 22 days, but no nerves are present until the 28th day.[30]

Electrocardiograms from crustacean and *Limulus* heart consist of an initial fast depolarization followed by variable number of oscillations; this pattern is characterized as a tetanus (Fig. 20–12). Each single muscle fiber shows a series of fused junction potentials, each corresponding to one nerve impulse (Fig. 20–13). In a hermit crab heart two junction potentials (psp's) may fuse to give an overshooting spike. The cardiac muscle potentials are caused by increased Ca conductance.[197] In *Squilla* cardiac muscle, the graded psp's facilitate when closer together than 630 msec; amplitudes of psp's are enhanced by hyperpolarization, and reduced by depolarization.[24] In *Limulus,* the psp's may lead to a prolonged depolarization; stimulation of side nerves shows that one muscle fiber may be innervated by as many as six different axons, some originating several segments away.[1] The psp's are graded, and in *Limulus* the maximum depolarization may be 35 mv on a −45 mv resting potential.[150] Hyperpolarization of a muscle fiber increases the amplitude of psp's and depolarization decreases their amplitude.[179] With multiple innervation of muscle fibers, near-synchrony of heartbeat is provided by conduction in the cardiac neurons.

The cardiac ganglion of a lobster has five large anterior and four smaller posterior neurons (Fig. 20–14). External recording shows that each cell discharges many impulses in each cardiac burst (Fig. 20–15). The four smaller neurons are the pacemaker or driver cells and the larger ones are motor or follower cells, but when separated from the drivers the anterior cells may show some spontaneity. Intracellular recordings to date are from the large follower cells; these show small graded waves which are partly synaptic potentials (psp's) initiated from driver spikes. These waves may give rise to spikes which originate at the base of the axon, and which spread back to the soma and are conducted out by the axon to the muscle. The nerve cell somas apparently do not spike, but double microelectrode recording shows that the follower cells are weakly electrically coupled with attenuation of 2- to 4-fold and that slow potentials (but not spikes) may spread from cell to cell.[61] Recordings from the large cells indicate that spikes may occur in axon branches without spreading to the entire neuron.[64, 65] Since the large cells can be spontaneous in the absence of driver cells, some of the slow waves recorded in them may be pacemaker potentials rather than synaptic potentials.[38] The origin of pacemaker waves may be in the neuropil which connects cell processes. Muscle fibers of intact heart of *Homarus* show spontaneous mepsp's, and stimulation of intracardiac neurons elicits facilitating psp's.[7a]

The *Squilla* cardiac ganglion contains 16 neurons; pacemaker and follower cells are not distinguished. Each neuron fires 4 to 6 times per heartbeat. Cells at the ends of the ganglion are more frequently spontaneous than those in the middle; conduction is at 1.5

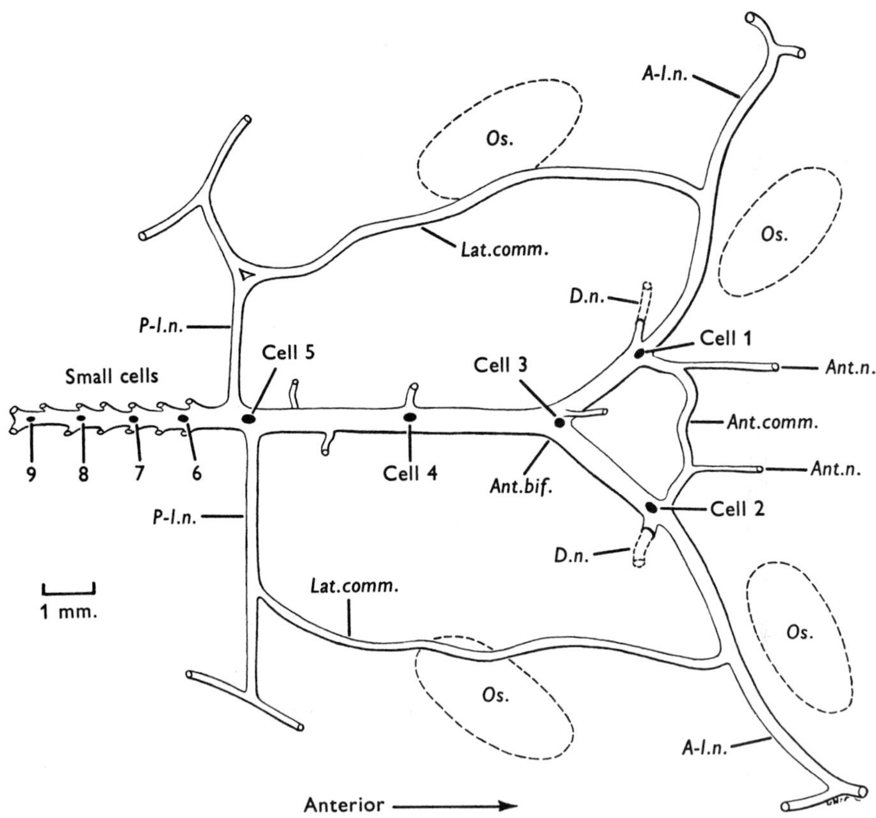

Figure 20–14. Diagram of nerves and nerve cell locations in cardiac ganglion of lobster. (From Hartline, D. K., J. Exp. Biol. *47*:327–340, 1967.)

m/sec.[24] Pacemaker cells show pacemaker potentials and spikes. The bipolar neurons may show a spike component from each axon plus a soma potential; i.e., the soma may be invaded from two directions. The neurons are electrically coupled.[201] Spikes arise at a distance from the neuron soma, farther away in unipolar than in bipolar neurons.[3]

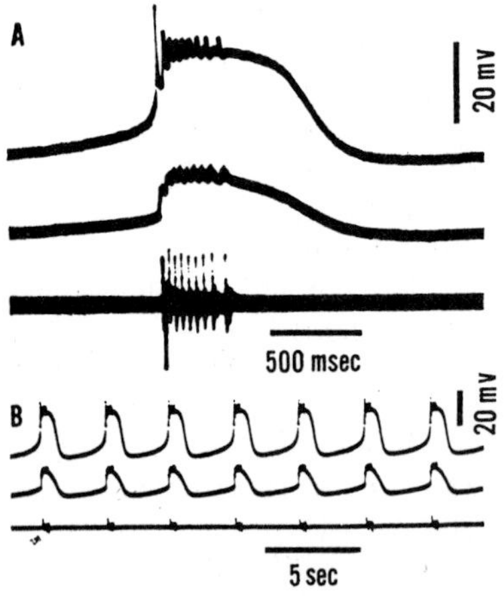

Figure 20–15. Electrocardiograms from nerve cells of cardiac ganglion of crustacean *Squilla.* A, Top record from neuron 5, middle record from neuron 6, lower record from extracellular recording. B, Records from same source at lower sweep speed. (From Watenabe, A., S. Obara, and T. Akiyama, J. Gen. Physiol. *50*:813–838, 1967.)

In *Limulus,* recordings from the somas of large neurons show a train of spikes which corresponds to the total ganglionic discharge. Occasionally spikes occur between bursts, and sometimes a slow potential, probably an epsp, precedes the first spike (Fig. 20–16). The spikes in a burst vary in amplitude and presumably originate in various axon branches. Some records show two patterns of spikes from the same soma, indicating two independent activation sites. Rare recordings from single small neurons show pacemaker potentials giving rise to one or two large spikes. Apparently, one or two pacemaker spikes trigger trains of spikes in follower cells.[107]

The mechanical properties of heart muscle from crustaceans and molluscs differ from those of vertebrate hearts; graded contractions occur and the hearts can be tetanized. Deganglionated *Limulus* heart can beat spontaneously with large Na-dependent spikes when in a Ca-free medium, and when

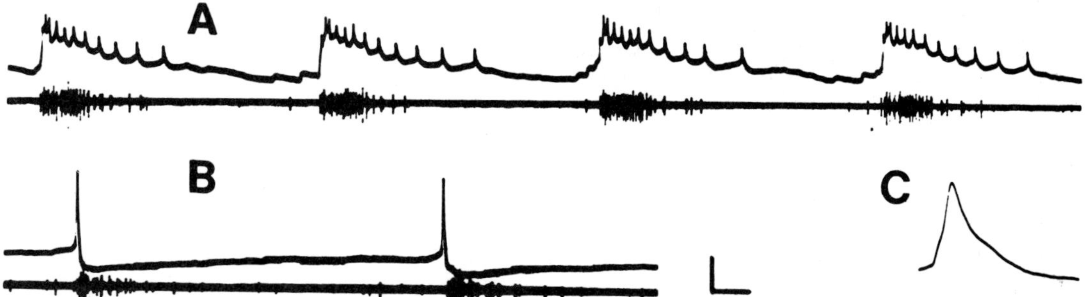

Figure 20–16. Records from neurons in cardiac ganglion of *Limulus.* In A and B, upper records are intracellular and lower records are extracellular. A, Bursts from large follower neurons. B, Single spikes from small pacemaker neuron. Each follower neuron gives multiple spikes on a sustained depolarization in response to one spike in a driver (pacemaker) cell. C, Expanded record of pacemaker cell, to show pacemaker potential preceding spike. Calibration: A, top 10 mv, bottom 20 μv, 500 msec. B, top 20 mv, bottom 20 μv, 500 msec. C, 20 mv, 40 msec. (Courtesy of F. Lang.)

distended to two times normal diameter. The *Limulus* heart then shows peristalsis, with fast spikes. The heart of a *Limulus* embryo beats locally before ganglion cells are differentiated. Thus, the muscle is capable of myogenic activity but normally is driven by the ganglionic pacemaker.[107]

The hearts of crustaceans and of *Limulus* receive from the central nervous system several regulating nerves which are distributed to the pacemaker ganglion and to the muscle. The heart of *Callinectes* receives two pairs of accelerator and one pair of inhibitory nerves. In *Limulus,* inhibitory nerves arise from the posterior part of the brain; in Crustacea both inhibitory and accelerator nerves arise from the subesophageal ganglion, the inhibitors lying anterior to the accelerators.

In *Squilla,* inhibitory impulses hyperpolarize cardiac neurons; the threshold for a spike is normally 7 mv depolarization, but under inhibition it is 15 mv. Accelerator nerves act on pacemaker membranes and increase the rate of rise of pacemaker potentials.[201] In *Limulus,* nerves 7 and 8 from the central ring are mostly inhibitory and nerves 9 to 13 are excitatory.[27] Stimulation of the accelerators is most effective at 5/sec and the latency may be 30 to 40 secs. The inhibitor is most effective when stimulated at 40/sec.[27, 140]

Neither excitor nor inhibitor transmitter is known with certainty in the crustacean or *Limulus* heart. Acetylcholine is excitatory to the ganglion, but relatively high concentrations are required and the effects of blocking and potentiating drugs are not indicative of a cholinergic innervation. Glutamate, which may be an excitatory transmitter at some arthropod junctions, contracts deganglionated muscle of *Limulus* heart at 10^{-6}M; it stimulates a ganglionated heart at 10^{-4}M.[1] Gamma aminobutyric acid (GABA), an inhibitor in crustacean neuromuscular junctions, reduces the duration and frequency of ganglionic bursts at 10^{-4}M.[1] In *Squilla,* 5HT is excitatory to the ganglion.[24] In several brachyuran crabs, a cluster of neurosecretory cells constituting the pericardial gland is excitatory to the heart by virtue of a substance that resembles but is not 5HT.[6] A substance extracted from the heart of *Carcinus* accelerates other hearts and is probably a tryptamine derivative, possibly 6HT.[98]

Acetylcholine is without effect on the heart of the cladoceran *Daphnia;* also, the electrocardiogram is not oscillatory.

The cardiac ganglia of crustaceans and of *Limulus* have provided models for nervous integration in that a very few neurons interact precisely, yet with some variation in timing. They may be expected to yield important information concerning the nature of spontaneous rhythms in neurons. The cardiac muscle in these animals, although it may be capable of myogenic activity, comes under nervous control and resembles somatic striated muscle in most of its performance.

HEARTS OF SPIDERS, SCORPIONS, AND INSECTS

The heart of the tarantula *Eurypelma* has a ganglion with unipolar, bipolar, and multipolar neurons. The electrocardiogram is oscillatory, much as in *Limulus;* hence, the heart is probably neurogenic. Its ganglion is excited by ACh and epinephrine and norepinephrine at 10^{-6}M, and by glutamate and 5HT at 10^{-4}M; it is inhibited by GABA at

10^{-5} M.[162,163] The heart of the scorpion *Urodacus* has a ganglionic pacemaker, the removal of which stops the heart and which gives rhythmic volleys in isolation. The heart receives both acceleratory and inhibitory innervation from the central nervous system; acetylcholine fails to accelerate the heart and at 10^{-4} M slows it. However, experiments with various drugs show that it is unlikely that ACh is the inhibitory transmitter.[212]

Since the hearts of insects are not needed for respiratory transport and since the development of insects is of several types, it is not surprising that their hearts show considerable diversity. Some are neurogenic and others are myogenic. Intracellular recordings from heart muscle of the fly *Sarcophaga* show summed oscillations like those of *Limulus*.[25] Oscillatory ECG's occur in *Melanoplus, Galleria,* and *Dytiscus.* In a cicada, ganglionic pacemakers are located in segments 2 to 7 of the heart.[79] The hearts of cockroach, grasshopper, cricket, and honeybee are accelerated by acetylcholine, and some of these have been shown to be very sensitive to ethyl ether, as are crustacean neurogenic hearts. The hearts of many insects are normally kept under tension by the dorsal suspensory ligaments and alary muscles; transection of these muscles and ligaments stops the heart in some species, but not in others. In the cicada *Cryptotympana* the ECG is simplified but still oscillatory after the alary muscles are cut.[79]

Absence of neurons in the heart has been reported for numerous insects, especially in larvae. Hearts of larval *Anax* and adult *Belostoma* lack nerve cells, and they beat only so long as the dorsal suspensory ligaments are intact. Hearts of larval *Galleria* and *Anopheles* are not affected by ACh, and probably not by ether.

The heart of an adult cockroach, *Periplaneta,* has a cardiac ganglion of six neurons which discharges in coordination with the heartbeat. However, rhythmic heartbeats persist after the ganglion is removed. It is suggested that the cardiac ganglion is stimulated by distension of the heart and that in turn it enhances the myogenic excitability; ACh is excitatory, possibly by stimulating the ganglionic control.[129] Peristaltic beating continues after ganglionic activity is stopped by TTX. An isolated cardiac ganglion shows no bursts of spikes, but in an attached ganglion, spiking is enhanced by inflation of the heart.[171a] The muscle fibers receive multiterminal and polyneuronal innervation.[130] Nerve fibers from the central nervous system appear to modulate both the muscle and the cardiac ganglion.

The heart of the moth *Hyalophora cecropia* can beat after the alary muscles are themselves stimulated by stretch to contract. The heart lacks nerve cells and has no localized pacemaker region. Its beat is not affected by acetylcholine, epinephrine, or ether; hence it is myogenic. The intracellularly recorded ECG is an overshooting spike with delayed repolarization, somewhat like the plateau of vertebrate heart (Fig. 20–17). The blood of this moth, as is that of many insects (p. 90), is rich in potassium and magnesium and low in sodium. The resting potential of the heart muscle is −47 mv, whereas the equilibrium potential for potassium is −11.6 mv; thus, other ions than potassium determine the resting potential. The large spike-like action potentials continue in a sodium-free medium, but if calcium is omitted they disappear and the heart stops. The myogenicity depends on calcium, and the spikes reflect changes in Ca^{++} conductance.[124,125,126,127,128]

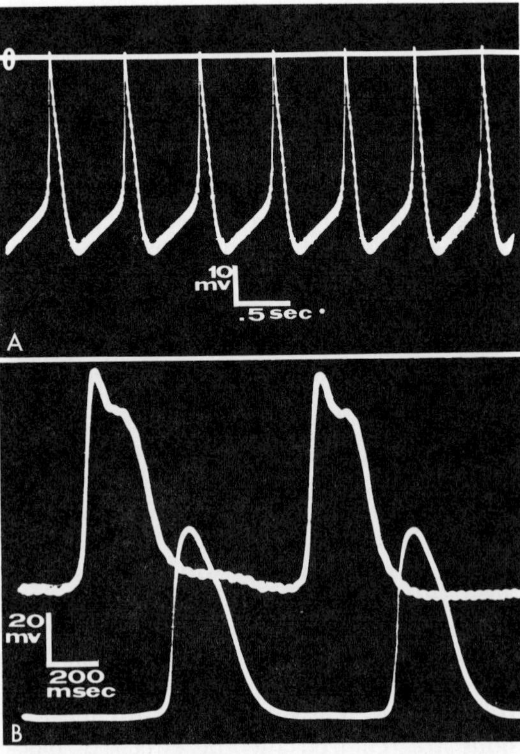

Figure 20–17. Intracellular electrocardiograms from moth *Hyalophora.* A, Cell showing pacemaker potentials and spikes. B, Follower cells (upper record) from posterior end and (lower record) from anterior end. (Courtesy of F. McCann.)

Regulation of insect hearts by the central nervous system has not been much examined. In a sphinx moth the heart (in the abdomen) changes its rate and amplitude of beat according to the temperature of the thorax, presumably by excitatory and inhibitory nerves.[68]

It may be concluded that hearts of scorpions, spiders, and insects show great diversity in origin of beat—they may be myogenic, neurogenic, or controlled by local reflexes. The pacemaker ganglia, where they occur, may not respond to drugs as do the pacemaker ganglia of other arthropods.

ANNELID HEARTS

Many blood vessels of annelids are spontaneously active; in the giant earthworm *Glossoscolex,* the dorsal and ventral vessels are independent in frequency, and the ventral one synchronizes with the lateral "hearts." Nerve cells have been described in the hearts of *Arenicola* and *Lumbricus*; these hearts are accelerated by acetylcholine,[142] and by epinephrine and norepinephrine.[154] Distension by blood is important in establishment of contraction. The hearts of a leech lack nerve cells and are accelerated by epinephrine and slowed by muscarine. Anterior roots of segmental nerves are excitatory and posterior roots are inhibitory to leech hearts.[196]

CONCLUSIONS

To be of adaptive value, a circulatory system must respond to stress; blood must be available where it is needed, in adequate amounts, when required. Open systems—coelom, pseudocoel, or hemocoel—without hearts are adequate where respiratory and nutritional demands are not great. With increase in body size and activity, hearts came into use in some open systems, as in crustaceans and molluscs. These hearts are incapable of developing high pressures, however, and circulatory flow is accomplished in part by contraction of somatic muscles. The blood volume is large and transport of oxygen from gills to muscles is relatively slow. Annelids have closed blood vessels, particularly for oxygen transfer, but the coelom remains the main circulatory mechanism for excretion and nutrition. The open system is used in insects, in which the hemocoel performs minor functions only, and transport of air is in tubes directly to the muscles. In some crustaceans, the arterial vessels branch extensively before opening to sinuses. In cephalopods and vertebrates, most of the circulatory system is in closed vessels, with the result that systemic pressure is high, velocity of blood flow is high, and blood volume is low. The smooth muscle of blood vessels in different organs and species varies adaptively in kind of response to regulatory substances such as acetylcholine and catecholamines. In lower vertebrates, the lymph system with accessory lymph hearts is more important than in higher vertebrates. Among vertebrates there are complex reflex and hormonal systems, as well as local responses of blood vessels, for maintenance of blood pressure; the nature of pressure-regulating systems in invertebrates is unknown. Frequently fluid-filled cavities may be less important for circulation than for providing a hydrostatic skeleton; such skeletons become rigid or flaccid on demand, are useful for burrowing, but lack the local control and protection afforded by segmented and external skeletons.

Heart muscle in all animal groups has some of the properties of visceral and some of somatic muscle. The vertebrate heart tends to contract in an all-or-none fashion, molluscan and arthropod hearts less so. Myogenicity, which is more primitive, is retained in adult vertebrates and molluscs but is replaced by neurogenicity in many adult arthropods. The small number of neurons in crustacean pacemaker ganglia constitutes a simple neural integrating system. Cardiac muscles generally have prolonged action potentials, modifications as pacemakers, fast conducting cells, and muscles with varying durations of depolarization; there is a variety of membrane types, most having Na conductance and others, Ca.

In most animals, nervous regulation of heartbeat occurs, often with inhibitory fibers and sometimes with both inhibitory and excitatory fibers. Most embryonic hearts beat before they are innervated, and in some animals—*Artemia,* many insects such as *Anopheles* larvae, and myxinoid fish—the hearts remain myogenic and noninnervated. Activity in cardiac nerves is associated with liberation of chemical mediators. The inhibitory mediator is acetylcholine in vertebrates and some molluscs; the accelerator is epinephrine-norepinephrine in vertebrates and serotonin in some molluscs. Neural transmitters can alter the time course of membrane conductance changes; they may also have intracellular actions.

The evolution of circulatory mechanisms shows many parallel developments, each tending to make a given volume of fluid serve more efficiently the transport needs of the animal. The varied patterns of peripheral circulation, hemodynamics, and cardiac physiology and pharmacology provide systems for cybernetic or model studies in homeostasis.

REFERENCES

1. Abbott, B. C., F. Lang, and I. Parnas, Comp. Biochem. Physiol. *28*: 149–158, 1969; *also* pp. 232–243 *in* Comparative Physiology of the Heart, edited by F. McCann, Birkhaüser, Basel, 1969. Muscle and ganglion of heart of *Limulus.*
2. Ai, N., Sci. Rep. Tokyo Kyoiku Daigaku B *12*:131–149, 1966. Pacemaker of heart of *Ligia.*
3. Akiyama, T., Comp. Biochem. Physiol. 27:197–211, 1968. Pacemaker ganglion of *Squilla.*
4. Alexandrowicz, J. S., J. Marine Biol. Assoc. U. K. *31*:85–96, 1952; *33*:709–719, 1954; *34*:47–53, 1955; *also* Quart. J. Micr. Sci. 75:291–309, 1932. Innervation of crustacean hearts.
5. Alexandrowicz, J. S., Acta Zool. *41*:65–100, 1960. Innervation of heart of *Sepia.*
6. Alexandrowicz, J. S., and D. B. Carlisle, J. Marine Biol. Assoc. U. K. *32*:175–192, 1953. Pericardial organs in crustacea.
7. Anderson, M., J. Exp. Biol. *49*:363–385, 1968. Initiation and reversal of heart beat in *Ciona.*

7a. Anderson, M., and I. M. Cooke, J. Exp. Biol. *55*:449–468, 1971. Activation of heart muscle in lobster.

8. Ansell, A. D., and E. A. Trueman, J. Exp. Biol. *46*:105–115, 1967. Burrowing by *Mercenaria.*
9. Baker, L. A., and F. N. White, Comp. Biochem. Physiol. *35*:253–262, 1970. Cardiac output in *Iguana.*
10. Banchero, N., R. F. Grover, and J. A. Will, Amer. J. Physiol. *220*:422–427, 1971. Pulmonary hypertension in llama at high altitude.
11. Barr, L., and W. Berger, Pflüger. Arch. *279*:192–194, 1964; *also* Barr, L., M. M. Dewey, and W. Berger, J. Gen. Physiol. *48*:797–823, 1965. Conduction and structure in cardiac muscle.
12. Baumgarten-Schulmann, D., and J. Piiper, Respir. Physiol. *5*:317–325, 1968. Cardiac output in dogfish.
13. Bayer, R., Z. vergl. Physiol. *58*:76–135, 1968. Circulation in *Locusta.*
14. Belkin, D. A., Copeia 321–330, 1964. Changes in heart rate during diving in *Pseudemys.*
15. Berkson, H., Comp. Biochem. Physiol. *21*:507–524, 1967. Physiology of diving in *Chelonia.*
16. Blatchford, J. G., Comp. Biochem. Physiol. *39*A:193–202, 1971. Hemodynamics of *Carcinus.*
17. Bloom, G., et al., Acta Physiol. Scand. *53* (Suppl. 185):1–38, 1961. Catecholamines in hearts of cyclostomes.
18. Bohr, D. F., pp. 342–355. *In* Electrolytes and Cardiovascular Diseases, edited by E. Basjusz, Karger, Basel, 1965. Differences among vascular smooth muscles.
19. Bohr, D. F., et al., Angiology *12*:478–485, 1961. Mechanical properties of resistance vessels.

19a. Bond, C. F., and P. W. Gilbert, Amer. J. Physiol. *194*:519–521, 1958. Blood volume of pigeon.

20. Brady, A. J., and J. W. Woodbury, J. Physiol. *154*:385–407, 1960. Bioelectrics of frog ventricle.

20a. Brett, J. R., and J. C. Davis, J. Fish. Res. Bd. Canad. *24*:1775–1790, 1967. Cardiovascular dynamics of salmon.

21. Bron, K. M., et al., Science *152*:540–542, 1966. Vasomotor responses in diving mammals.
22. Brown, A. C., J. Exp. Biol. *41*:837–854, 1964. Blood distribution in sinuses of snail *Bullia.*
23. Brown, E., J. Hoffer, and R. Wennesland, Ann. Rev. Physiol. *19*:231–254, 1957. Blood volume and its regulation.
24. Brown, H. F., J. Exp. Biol. *41*:689–700, 701–734, 1964. Electrophysiology of heart of *Squilla.*
25. Bruen, J. P., and R. C. Ballard, Comp. Biochem. Physiol. *32*:227–236, 1970. Action potentials in heart of fly *Sarcophaga.*

25a. Bull, J. M., and R. Morris, J. Exp. Biol. *471*:485–494, 1967. Blood volume in lamprey.

26. Burger, J. W., and C. M. Smythe, J. Cell. Comp. Physiol. *42*:369–383, 1953. Pattern of circulation in *Homarus.*

26a. Burnstock, G., Pharmacol. Rev. *21*:247–313, 1969. Evolution of autonomic innervation of visceral and cardiovascular systems in vertebrates.

26b. Burnstock, G., and S. Kirby, J. Pharm. Pharmacol. *20*:404–406, 1968. Absence of inhibitory effects of catecholamines on lower vertebrate arterial strip preparations.

27. Bursey, C. R., and R. A. Pax, Comp. Biochem. Physiol. *35*:41–48, 1970. Cardioregulation in *Limulus.*
28. Bursey, C. R., and R. G. Sherman, Comp. Gen. Pharm. *1*:160–170, 1970. Spider cardiac physiology.
29. Burton, A. C., Physiology and Biophysics of Circulation. Yearbook Medical Publishers, Chicago, 1968. 217 pp.

29a. Campbell, G., Comp. Gen. Pharm. *2*:287–294, 1971. Pulmonary vascular bed in toad *Bufo marinus.*

30. Carlson, A. J., Amer. J. Physiol. *12*:55–66, 67–74, 1904; *14*:16–53, 1905; *15*:207–234, 1906; *17*:179–210, 1906. Invertebrate hearts.
31. Carmeliet, E. E., J. Gen. Physiol. *47*:501–530, 1964. Effects of ions on cardiac muscle.
32. Carvalho, O., et al., Amer. J. Physiol. *196*:483–488, 1959. Electrical properties of rabbit atrium.
33. Chapman, C. B., D. Jensen, and K. Wildenthal, Circ. Res. *12*:427–440, 1963. Circulatory control of hagfish.
34. Chapman, G., J. Exp. Biol. *49*:657–667, 1968. Hydraulic system of *Urechis.*
35. Chapman, G., and G. E. Newell, Proc. Roy. Soc. Lond. B *145*:564–580, 1956. Hydraulics in molluscs.
36. van Citters, R. L., et al., Comp. Biochem. Physiol. *16*:267–276, 1965. Cardiovascular adaptations to diving in elephant seal.
37. van Citters, R. L., W. S. Kemper, and D. L. Franklin, Comp. Biochem. Physiol. *24*:1035–1042, 1968; Science *152*:384–386, 1966. Blood pressure in giraffe.
38. Connor, J. A., J. Exp. Biol. *50*:275–295, 1969. Pacemaker ganglion of lobster heart.
39. Conte, F. P., H. H. Wagner, and T. O. Harris, Amer. J. Physiol. *205*:533–540, 1963. Blood volume in *Salmo.*
40. D'Agrossa, L. S., Amer. J. Physiol. *218*:530–535, 1970. Venous vasomotion in bat wing.
41. Day, J. B., R. H. Rech, and J. S. Robb, J. Cell Comp. Physiol. *62*:33–42, 1963. Electrophysiology of frog lymph heart.
42. Del Castillo, J., and V. Sanchez, J. Cell. Comp. Physiol. *57*:29–46, 1961. Bioelectrics of amphibian lymph heart.
43. Ebara, A., Japan J. Physiol. *16*:371–379, 1966. Plateau potential of oyster heart.
44. Ebara, A., Sci. Rep. Tokyo Kyoiku Daigaku B *13*:129–137, 1967. Conduction in heart of oyster.
45. Eliassen, E., Nature *192*:1047–1049, 1961. Blood volume in hibernating hedgehog.
46. Ellis, C. H., Biol. Bull. *86*:41–50, 1944. Hydraulics of leg of spider.
47. Folkow, B., et al., Acta Physiol. Scand. *70*:347–361, 1967. Cardiac output of duck.
48. Fozzard, H. O., and W. Sleator, Amer. J. Physiol. *212*:945–952, 1967. Ionic basis of conductance in guinea pig atrial muscle.
49. Fredericq, H., and Z. Bacq, Arch. Int. Physiol. Biochem. *49*:490–496, 1939; *50*:169–184, 1940. Effects of drugs on cephalopod hearts.

49a. Furness, J. B., and J. Moore, Z. Zellforsch. *108*:150–176, 1970. Adrenergic innervation of cardiovascular system in lizard *Trachysaurus.*

50. Gannon, B. J., and G. Burnstock, Comp. Biochem. Physiol. *29*:765–773, 1969. Excitatory innervation of fish heart.
51. Garey, W., Comp. Biochem. Physiol. *33*:181–189, 1970. Blood volume and cardiac output of fishes.
52. Garnier, D., E. Coraboeuf, and A. Paille, Compt. Rend. Soc. Biol. *155*:2430, 1961; *155*:1349–1357, 1962. Effects of ions on potentials in heart of guinea pig and rat.

52a. Glover, W. E., et al., J. Physiol. *194*:78–79*P*, 1967. Cooling of rabbit arteries.

53. Goetz, R. H., et al., Circ. Res. *8*:1049–1058, 1960. Circulation of giraffe.
54. Gordon, M. S., S. Fischer, and E. Tarifeno, J. Exp. Biol. *53*:559–572, 1970. Physiology of heart of amphibious fish.
55. Greenberg, M. J., Brit. J. Pharmacol. Chemother. *15*:365–374, 375–388, 1960. Pharmacology of heart of *Venus.*
56. Greenberg, M. J., Comp. Biochem. Physiol. *14*:513–539, 1965. Responses of bivalve hearts to acetylcholine.
57. Greenberg, M. J., Comp. Biochem. Physiol. *33*:259–295, 1970. Effect of acetylcholine on hearts of various molluscs.
58. Gregersen, M. I., and C. Shu, pp. 244–261. *In* Medical Physiology, edited by B. B. Mountcastle. C. V. Mosby, St. Louis, 1968. Blood volume.
59. Guyton, A. C., Cardiac Physiology. W. B. Saunders Co., Philadelphia, 1963. 468 pp. Cardiac output and its regulation.
60. Hagiwara, S., Ergebn. Biol. *24*:287–311, 1961. Electrophysiology of heart in crustaceans.
61. Hagiwara, S., A. Watanabe, and N. Saito, J. Neurophysiol. *22*:554–572, 1959. Cardiac ganglion of lobster.

62. Hall, E. K., Anat. Rec. *111*:381–400, 1951; *118*:175–184, 1954; *also* J. Cell. Comp. Physiol. *49*:187–200, 1957. Properties of embryonic rat heart.
63. Hart, J. S., Canad. J. Res. D *21*:77–84, 1943. Cardiac output of fish.
64. Hartline, D. K., J. Exp. Biol. *47*:327–340, 1967. Mapping of cardiac ganglion of lobster.
65. Hartline, D. K., and I. M. Cooke, Science *164*:1080–1082, 1969. Control of cardiac ganglion in lobster.
66. Hauswirth, O., D. Noble, and R. W. Tsien, Science *162*:916–917, 1968. Action of epinephrine on cardiac muscle.
67. Heath, J. E., E. Gasdorf, and R. G. Northcutt, Comp. Biochem. Physiol. *26*:509–518, 1968. Blood pressure of turtle.
68. Heinrich, B., Science *169*:606–607, 1970. Nervous control of heart of sphinx moth.
69. Helm, M. M., and E. R. Trueman, Comp. Biochem. Physiol. *21*:171–177, 1967. Heart rate in *Mytilus.*
70. Hemmingsen, E. A., and E. L. Douglas, Comp. Biochem. Physiol. *33*:733–744, 1970. Blood volume in Antarctic fish.
71. Henderson, A. H., et al., Proc. Soc. Exp. Biol. Med. *134*:930–932, 1970. Species differences in cardiac contractility.
72. Hill, R. B., and H. Irisawa, Life Sci. *6*:1691–1696, 1967. Pressure relations in heart of marine gastropod.
73. Holt, J. P., E. A. Rhode, S. A. Peoples, and H. Kines, Circ. Res. *10*: 798–806, 1962. Cardiac output in animals of different sizes.
74. Homma, S., and S. Suzuki, Japan. J. Physiol. *16*:31–41, 1966. Sensory discharge from aortic and atrial receptors.
75. Huggins, S. E., and R. A. Percoco, Proc. Soc. Exp. Biol. Med. *119*: 678–682, 1965. Blood volume in alligators.
76. Hutter, O. F., and W. Trautwein, J. Gen. Physiol. *39*:715–733, 1956. Bioelectrics of the sinus venosus.
77. Hutton, K. E., Amer. J. Physiol. *200*:1004–1006, 1961. Blood volume in turtles.
78. Irisawa, H., and M. Kobayashi, Japan. J. Physiol. *13*:421–430, 1963. Molluscan heart.
79. Irisawa, H., et al., Japan. J. Physiol. *6*:150–161, 1956. Electrocardiogram of cicada.
80. Irisawa, H., et al., Japan. J. Physiol. *11*:162–168, 1961. Action potentials of oyster heart.
81. Irisawa, H., et al., pp. 176–191. *In* Comparative Physiology of the Heart, edited by F. McCann. Birkhaüser, Basel, 1969. Ion effects on bivalve heart.
82. Irisawa, H., et al., Comp. Biochem. Physiol. *23*:199–212, 1969; Japan. J. Physiol. *18*:157–168, 1968. Effects of calcium and sodium on oyster and *Mytilus* heart.
83. Irving, L., et al., J. Physiol. *84*:187–190, 1935. Branchial pressure receptors in dogfish.
84. Itazawa, Y., Bull. Japan. Soc. Sci. Fish. *36*:926–931, 1970. Cardiac output of fish.
85. Jensen, D., Comp. Biochem. Physiol. *2*:181–201, 1961. Cardiac regulation in aneural heart of hagfish.
86. Jensen, D., Comp. Biochem. Physiol. *30*:685–690, 1969. Cardiac regulation in lamprey and trout.
87. Johansen, K., Biol. Bull. *118*:289–295, 1960. Circulation in hagfish *Myxine.*
88. Johansen, K., Comp. Biochem. Physiol. *7*:169–174, 1962. Cardiac output and aortic flow in *Gadus.*
89. Johansen, K., Acta Physiol. Scand. *60* (Suppl. 217):1–82, 1963. Cardiovascular dynamics in *Amphiuma.*
90. Johansen, K., Acta Physiol. Scand. *62*:1–17, 1964; *also* Johansen, K., and T. Aarhus, Amer. J. Physiol. *205*:1167–1171, 1963. Blood flow during submersion in duck.
91. Johansen, K., J. Exp. Biol. *42*:475–480, 1965; *also* Comp. Biochem. Physiol. *5*:161–176, 1962. Cardiac output in *Octopus.*
92. Johansen, K., and A. S. F. Ditadi, Physiol. Zool. *39*:140–150, 1966. Double circulation in giant toad.
93. Johansen, K., and A. W. Martin, J. Exp. Biol. *43*:337–347, 1965. Circulation of giant earthworm *Glossoscolex.*
94. Johansen, K., and O. B. Reite, Comp. Biochem. Physiol. *12*:479–487, 1964. Cardioregulation in birds.
95. Jones, H. D., Comp. Biochem. Physiol. *39*A:289–295, 1971. Circulatory pressures in *Helix.*
96. Jullien, A., and J. Ripplinger, C. R. Soc. Biol. *144*:544–545, 1950; *146*:1326–1329, 1952. Cardiac regulation in snails and fishes.
96a. Kadatz, R., Pflüger. Arch. *252*:1–16, 1949. Effects of acetylcholine and noradrenaline on lung vessels in amphibians.
97. Kawakami, Y., B. H. Natelson, and A. B. DuBois, J. Appl. Physiol. *23*:964–970, 1967. Cardiovascular reflexes of diving in man.
98. Kerkut, G. A., and M. A. Price, Comp. Biochem. Physiol. *11*:45–52, 1964. Cardioaccelerators of *Carcinus.*
99. Keyes, A., and B. Bateman, Biol. Bull. *63*:327–336, 1932. Branchial responses to epinephrine in eel.
100. Kirschner, L. B., Amer. J. Physiol. *217*:596–604, 1969. Circulation in gills of eel.
101. Kreski, V., and W. W. Sleator, Life Sci. *5*:1441–1446, 1966. Development of electrical activity in chick embryo hearts.
102. Kriebel, M. E., J. Gen. Physiol. *50*:2097–2107, 1967; *52*:46–59, 1968; *also* Biol. Bull. *134*:434–455, 1968; *also* Life Sci. 7:181–186, 1968; *also* Amer. J. Physiol. *218*:1194–1200, 1970. Action potentials of pacemakers and conduction in tunicate heart.
103. Kriebel, M. E., Biol. Bull. *135*:166–173, 1968. Pacemaker properties of tunicate heart.
104. Kuffler, S. W., M. J. Dennis, and A. J. Harris, Proc. Roy. Soc. Lond. B *177*:488–508, 509–539, 541–553, 555–563, 1971. Properties of synapses in vagal secondary neurons of frog heart.
105. Kulzer, E., Z. vergl. Physiol. *56*:63–94, 1967. Heart beat in hibernating bats.
106. Kuwasawa, K., Sci. Rep. Tokyo Kyoiku Daigaku B *13*:111–128, 1967. Nervous regulation in molluscan heart.
107. Lang, F., J. Exp. Biol. *54*:815–826, 1971. Intracellular studies on pacemaker and follower neurons in *Limulus* cardiac ganglion.
107a. Lang, F., Biol. Bull. *141*:269–277, 1971. Myogenic activity in *Limulus* heart.
108. Laplaud, J., et al., C. R. Soc. Biol. *155*:1990–1993, 1961. Electrocardiogram of *Carcinus.*
109. Larimer, J. L., and J. T. Tindel, Animal Behav. *14*:239–245, 1966. Reflex regulation of heart in crayfish.
110. Laury, J. V., J. Exp. Biol. *45*:343–356, 357–368, 1966. Hydraulics of burrowing *Urechis.*
111. Lavoie, M. E., Biol. Bull. *111*:114–122, 1956. Hydrostatic movement in starfish.
112. Leblanc, J., Ann. N. Y. Acad. Sci. *134*:721–732, 1966. Cardiovascular adaptations to cold in man.
113. Lee, R. M., J. Insect Physiol. *6*:36–51, 1961. Blood volume in locust.
114. Lieberman, M., and A. P. de Carvalho, J. Gen. Physiol. *49*:351–363, 1965. Electrocardiogram of chick embryo.
115. Lin, Y., C. A. Dawson, and S. M. Horvath, Comp. Biochem. Physiol. *33*:901–909, 1970. Cardiac output in rat.
116. Lind, A. R., Circulation *41*:173–176, 1970. Cardiovascular response to exercise in man.
117. Litchfield, J. B., Physiol. Zool. *31*:1–6, 1958. Blood pressure in infant rats.
117a. Little, C., J. Exp. Biol. *46*:459–474, 1967. Hemolymph volume in snail *Strombus.*
118. Loveland, R. E., Comp. Biochem. Physiol. *9*:95–104, 1963. Serotonin as excitatory transmitter to heart of *Mercenaria.*
119. Lund, G. F., and H. Dingle, J. Exp. Biol. *48*:265–277, 1968. Vagal control of diving bradycardia in frog.
120. Magarey, F. R., and W. E. Stebbens, J. Exp. Biol. Med. Sci. *35*:347–351, 1957. Blood pressure in sheep.
121. Markardt, J. E., Comp. Biochem. Physiol. *33*:441–457, 1970. Regulation of heart rate in *Peromyscus.*
122. Martin, A. W., J. Exp. Biol. *35*:260–279, 1958. Blood volume in molluscs.
123. Martini, J., and C. R. Honig, Microvasc. Res. *1*:244–256, 1969. Capillary distribution in rat heart.
123a. Mayer, N., and N. Nibelle, Comp. Biochem. Physiol. *31*:589–597, 1969. Blood volume in *Anguilla.*
124. McCann, F. V., J. Gen. Physiol. *46*:803–821, 1963. Electrophysiology of insect heart.
125. McCann, F. V., Ann. N. Y. Acad. Sci. *127*:84–99, 1965; *also* Comp. Biochem. Physiol. *17*:599–608, 1966. Electrical properties of moth myocardium.
126. McCann, F. V., Comp. Biochem. Physiol. *40*A:353–357, 1971. Ionic basis of moth heart potentials.
127. McCann, F. V., and C. R. Wira, Comp. Biochem. Physiol. *22*:611–615, 1967. Ionic gradients in lepidopteran hearts.
128. McCann, F. V., ed., Comparative Physiology of the Heart. Birkhaüser, Basel, 1969.
129. Miller, T., J. Insect Physiol. *14*:1265–1275, 1968. Neuronal control of cockroach heart.
130. Miller, T., and P. N. R. Usherwood, J. Exp. Biol. *54*:329–348, 1971. Cardioregulation in cockroach.
131. Murdaugh, H. V., et al., Amer. J. Physiol. *210*:176–180, 1966. Cardiac output in seal.
132. Nagel, E. L., et al., Science *161*:898–900, 1968. Cerebral circulation in dolphin.
133. Nayler, W. G., and D. Chipperfield, Amer. J. Physiol. *217*:609–614, 1969. Calcium binding by cardiac sarcoplasmic reticulum.
134. Niedergerke, R., J. Physiol. *143*:486–503, 1958; *167*:515–550, 551–586, 1963. Movements of calcium ions in frog heart.
135. Nijima, A., and D. L. Winter, Science *159*:434–435, 1968. Baroreceptors in adrenal gland.
136. Noble, P., and R. W. Tsien, J. Physiol. *195*:185–214, 1968. Membrane properties of cardiac muscle.
137. Ormond, A. P., and J. M. Rivera-Velez, Proc. Soc. Exp. Biol. Med. *118*:600–602, 1965. Blood volume in rat.
138. Ostlund, E., and R. Fänge, Comp. Biochem. Physiol. *5*:307–309, 1962. Circulation in gill of fish.
139. Parer, J. T., and J. Metcalfe, Resp. Physiol. *3*:151–159, 1967. Cardiac output, monotremes.
140. Pax, R. A., Comp. Biochem. Physiol. *28*:293–305, 1969. Cardioregulation in *Limulus.*
141. Popovic, V., Amer. J. Physiol. *207*:1345–1348, 1964. Cardiac output of hibernating ground squirrels.
142. Prosser, C. L., J. Cell. Comp. Physiol. *21*:295–305, 1943; *also* Biol. Bull. *98*:254–257, 1950. Pacemaker activity in *Limulus* and *Arenicola.*
143. Prosser, C. L., and S. J. Weinstein, Physiol. Zool. *23*:113–124, 1950. Blood volume of catfish and crayfish.

144. Randall, D. J., and G. Shelton, Comp. Biochem. Physiol. *9*:229–239, 1963. Respiratory and cardiovascular reflexes in fish.
145. Randall, D. J., and J. C. Smith, Physiol. Zool. *40*:104–113, 1967. Cardiac regulation in hypoxia, fish.
146. Redmond, J. R., J. Cell. Comp. Physiol. *46*:209–247, 1955. Cardiac output in crustaceans.
147. Reuter, H., J. Physiol. *192*:479–492, 1967; *197*:233–253, 1968. Membrane properties of Purkinje fibers.
148. Rhodin, J., A. G. P. Missier, and L. C. Reid, Circulation *24*:349–367, 1961. Conduction in beef heart.
149. Richmond, C. R., W. H. Langham, and T. T. Trujillo, J. Cell. Comp. Physiol. *59*:45–53, 1962. Body water in mammals.
150. Robb, J. S., and R. H. Rech, J. Cell. Comp. Physiol. *63*:299–307, 1964. Electrocardiogram of *Limulus.*
151. Robertson, J. D., Biol. Bull. *138*:157–183, 1970. Inulin space in arthropods.
152. Rozsa, K., and I. Z. Nagy, Comp. Biochem. Physiol. *23*:373–382, 1967. Neuroendocrines of heart of snail.
153. Rozsa, K., and L. Perenyi, Comp. Biochem. Physiol. *19*:105–113, 1966. Serotonin as excitor in *Helix* heart.
154. Rozsa, K., and I. Szoke, Ann. Biol. Tihany *37*:85–97, 1970. Pharmacology of heart of *Limulus.*
155. Satchell, G. H., Fed. Proc. *29*:1120–1123, 1970; *also* Satchell, G. H., D. Hanson, and K. Johansen, J. Exp. Biol. *52*:721–726, 1970. Regulation of peripheral circulation in fish.
155a. Satchell, G. H., Circulation in Fishes. Cambridge University Press, 1971, 131 pp.
155b. Scholander, P. F., et al., Science *161*:321–328, 1968; *also* Stromme, S. B., et al., J. Appl. Physiol. *27*:123–126, 1969. Interstitial fluid pressures in vertebrates.
156. Scudder, C. L., T. K. Akers, and A. G. Karczmar, Comp. Biochem. Physiol. *9*:307–312, 1963. Electrocardiogram of tunicate.
157. Seifen, E., H. Schaer, and J. M. Marshall, Nature *202*:1223–1224, 1964. Membrane properties of rabbit atrium.
158. Serfaty, A., R. Labat, and A. Berant, Rev. Canad. Biol. *24*:1–5, 1965. Cardiac reaction to anoxia in carp.
159. Seyama, I., Amer. J. Physiol. *216*:687–692, 1969; *also* Seyama, I., and H. Irisawa, J. Gen. Physiol. *50*:505–518, 1967. Effect of sodium and calcium on heart of skate.
160. Seymour, M. K., J. Exp. Biol. *51*:47–58, 1969. Hydrodynamics of *Lumbricus.*
161. Shelton, G., and D. R. Jones, J. Exp. Biol. *42*:339–357, 1965; *43*:479–488, 1965; *49*:631–643, 1968. Blood pressure, volume regulation, and cardiac output in lower vertebrates.
162. Sherman, R. D., and R. A. Pax, Comp. Biochem. Physiol. *26*:529–536, 1968; *also* Comp. Gen. Pharm. *1*:171–184, 185–195, 1970. Spider cardiac physiology.
163. Sherman, R. G., and R. A. Pax, Comp. Biochem. Physiol. *28*:487–489, 1969. Electrical activity in spider hearts.
164. Shigeto, N., Amer. J. Physiol. *218*:1773–1779, 1970. Acetylcholine excitation and inhibition in molluscan heart.
165. Silvey, G. E., Comp. Biochem. Physiol. *25*:257–269, 1968. Nervous regulation of heart in cockle.
166. Sleator, W., and T. de Gubareff, Amer. J. Physiol. *206*:1000–1014, 1964. Electrical activity of human atrium.
167. Sleator, W., and T. de Gubareff, pp. 107–119. *In* Paired Pulse Stimulation of the Heart, edited by P. F. Cranefield and B. F. Hoffman. Rockefeller University Press, New York, 1969. Action potentials of the human heart.
168. Smith, J. E., Phil. Trans. Roy. Soc. B *232*:279–310, 1946. Hydraulics of starfish tube feet.
169. Smith, L. S., J. Fish. Res. Bd. Canad. *23*:1439–1446, 1966; *also* Smith, L. S., and G. Bell, J. Fish. Res. Bd. Canad. *21*:711–717, 1964. Blood volume in salmon.
170. Smith, L. S., J. R. Brett, and J. C. Davis, J. Fish. Res. Bd. Canad. *24*:1775–1790, 1967. Cardiovascular dynamics in salmon.
171. Smith, L. S., and J. C. Davis, J. Exp. Biol. *43*:171–180, 1965. Hemodynamics in molluscs.
171a. Smith, N. A., Experientia *15*:200–205, 1969. Rhythmicity in cardiac ganglion of cockroach.
172. Sommer, L. S., et al., Amer. J. Physiol. *215*:1498–1505, 1968. Hemodynamics and coronary flow in dolphin *Tursiops.*
173. Sperelakis, N., and D. Lehmkuhl, J. Gen. Physiol. *47*:895–927, 1964. Action potentials in cultured heart cells.
174. Stevens, E. D., and D. J. Randall, J. Exp. Biol. *46*:307–315, 1967. Cardiovascular changes in swimming in trout.
175. Stewart, D. M., and A. W. Martin, Z. vergl. Physiol. *70*:223–246, 1970. Blood volume in tarantula.
176. Sturkie, P. D., Proc. Soc. Exp. Biol. Med. *123*:487–488, 1966. Cardiac output in ducks.
177. Sutterlin, A. M., Physiol. Zool. *42*:36–52, 1969. Effect of exercise on heart of teleosts.
178. Tamai, T., and S. Kagiyama, Circ. Res. *22*:423–433, 1968. Effect of hypothermia on cardiac muscle.
179. Tanaka, I., Y. Sasaki, and H. Shin-mura, Japan. J. Physiol. *16*:142–153, 1966. Electrical activity of heart of horseshoe crab.
180. Templeton, J. R., Physiol. Zool. *37*:300–306, 1964. Cardiovascular response in lizard.
181. Thorson, T. B., Science *138*:99–100, 688–690, 1959. Partitioning of body fluids in sea lamprey, marine sharks, and freshwater shark.
182. Thorson, T. B., Biol. Bull. *120*:234–254, 1961. Partitioning of body water in fishes.
183. Thorson, T. B., Physiol. Zool. *37*:395–399, 1964. Partitioning of body water in amphibians.
184. Thorson, T. B., Copeia 592–601, 1968. Body fluid partitioning in reptiles.
185. Tille, J., J. Gen. Physiol. *50*:189–202, 1966. Electrotonic interaction between muscle fibers in ventricle.
186. Toda, N., and T. C. West, Nature *205*:808–809, 1965. Effect of vagal stimulation on rabbit atrium.
187. Tokchieva, E. P., Fed. Proc. Transl. Suppl. *25*:T739–T742, 1966. Responses of neurons in frog heart.
187a. Trautwein, W., Pharmacol. Rev. *15*:277–332, 1963. Ionic mechanisms in heart conduction.
188. Trautwein, W., and D. G. Kassebaum, J. Gen. Physiol. *45*:317–330, 1961. Pacemaker activity of heart.
189. Treherne, J. E., J. Exp. Biol. *42*:7–27, 1965. Blood volume in stick insect, *Carausius.*
190. Trueman, E. R., Biol. Bull. *131*:369–377, 1966. Hydrodynamics of burrowing in *Arenicola.*
191. Trueman, E. R., J. Exp. Biol. *44*:93–118, 1966. Burrowing reactions in *Arenicola.*
192. Trueman, E. R., Science *152*:523–525, 1966. Fluid dynamics of burrowing.
193. Trueman, E. R., A. R. Brand, and P. Davis, J. Exp. Biol. *44*:469–492, 1966. Burrowing in bivalves.
194. Tucker, V. A., J. Exp. Biol. *44*:77–92, 1966. Circulation in *Iguana.*
195. Vander, A. J., Physiol. Rev. *47*:359–382, 1967. Control of renin release.
196. Van der Kloot, W. G., Fed. Proc. *26*:975–980, 1967. Regulation of heart of leech.
197. Van der Kloot, W. G., J. Exp. Zool. *174*:367–380, 1970. Electrophysiology of cardiac muscle of crustaceans.
198. Van der Kloot, W. G., and B. Dane, Science *146*:74–75, 1964. Conduction in frog ventricle.
199. Vogel, J. A., and P. D. Sturkie, Science *140*:1404–1406, 1963. Cardiac output in hen.
200. Watanabe, A., Japan. J. Physiol. *8*:305–318, 1958. Interaction between neurons and lobster cardiac ganglion.
201. Watanabe, A., S. Obara, and T. Akiyama, J. Gen. Physiol. *50*:813–838, 839–862, 1967; *52*:908, 1968; *54*:212–231, 1969. Cardiac ganglion of *Squilla.*
202. Webber, H. H., and P. A. Dehnel, J. Exp. Zool. *168*:327–336, 1968. Water compartments in snail *Acmaea.*
203. Weidmann, S., J. Physiol. *118*:348–360, 1952. Electrical constants of Purkinje fibers.
204. Weidmann, S., J. Physiol. *210*:1041–1054, 1970. Electrical constants of mammalian heart.
205. Welsh, J. H., and A. C. McCoy, Science *125*:348, 1957. Chemical transmitters of molluscan heart.
206. Whartun, D., M. L. Whartun, and J. Lola, J. Insect Physiol. *11*:391–404, 1965. Blood volume in *Periplaneta.*
207. White, F. N., Fed. Proc. *29*:1149–1153, 1970. Vascular shunt in reptiles.
208. White, F. N., and G. Ross, Amer. J. Physiol. *211*:15–18, 1966. Circulatory changes in diving in turtle.
209. White, L., et al., Comp. Biochem. Physiol. *27*:559–566, 1968. Cardiac output of various mammals.
210. Woodbury, J. W., and W. E. Crill, pp. 124–135. *In* Nervous Inhibition, Proceedings of the International Symposium, edited by E. Florey. Pergamon Press, New York, 1961. Conduction in the atrium.
211. Zuckerkandl, E., Biol. Bull. *98*:161–173, 1950. Pressure in holothurians and sipunculids.
212. Zwicky, K. T., Nature *207*:778–779, 1965; Comp. Biochem. Physiol. *24*:799–808, 1968. Pharmacology of scorpion heart.

CHAPTER 21

ENDOCRINE MECHANISMS

By R. R. Novales, L. I. Gilbert, and F. A. Brown, Jr.

The production and dispersal within the organism of chemical substances which subserve definite integrating and coordinating roles, and thereby supplement the activity of the nervous elements, are characteristic of all living things. Such substances may be referred to descriptively as chemical coordinators. In the broadest sense, every substance which enters the body fluids from the external environment or from the constituent cells of a higher organism and thus contributes to the normal composition of the internal medium is a chemical coordinator. O_2 and CO_2, for example, operate importantly in the coordination of organismic activities. Other coordinatory substances, such as the D vitamins, may enter the body or may under certain circumstances be synthesized within certain cells of the organism and thence be liberated into the blood. Chemical substances may be more restricted in their region of origin within the body and adaptively participate in a specialized activity within the organism, e.g., secretion of pancreozymin (cholecystokinin) and secretin by the duodenal mucosa, which stimulates the liberation of pancreatic juice in response to the presence of food in the duodenum. Many groups of higher organisms have specialized glandular cells, tissues, or organs which elaborate coordinatory substances for the organism as a whole. This latter development seems to have paralleled the specialization and restriction within the organism of numerous other organs and organ systems; these developments could come about only as soon as an effective mechanism for internal transport, or a circulatory system, was provided. Thus, specialized endocrine organs are found chiefly among the annelids, molluscs, and crustaceans,[78, 201] as well as the insects,[166] other arthropods, and vertebrates.[409]

The terms hormones or endocrines are applied to special chemical coordinators which are produced at some more or less restricted region or regions within the organism and which possess specific physiological action, usually elsewhere in the body. They are usually, but not always, produced in well-defined glandular organs. Endocrinologists disagree on how broadly one should interpret the definition of a hormone. Obviously, in view of the complete intergradation of all types of chemical coordinators with one another it is impossible to draw a sharp line between the hormones, on the one hand, and other chemical coordinators, on the other. Ascorbic acid may fulfill the definition of a hormone for the rat, for example, and yet appear as a vitamin for man.

Considerable progress has been made in the elucidation of the mechanism of action of the hormones. Some (e.g., epinephrine, glucagon, parathyroid hormone, antidiuretic hormone, thyroid-stimulating hormone, corticotropin, luteinizing hormone, hypothalamic releasing hormones, and melanocyte-stimulating hormone) appear to act by stimulating the membrane-bound enzyme adenyl cyclase in their target cells, resulting in an increase in cellular content of adenosine 3′,5′-monophosphate (cyclic AMP).

Cyclic AMP then goes on to produce the hormonal effect, acting as a "second messenger" according to the concept of Sutherland and his coworkers[346, 347] (Fig. 21–1). Much effort is currently being invested in establishing precisely how cyclic AMP produces these hormonal effects.

Others (e.g., thyroid hormones, estrogens, androgens, progesterone, cortical steroids, ecdysone, and growth hormone) appear to stimulate the synthesis of specific proteins as a result of an action on RNA or protein synthesis. The site of action depends upon the type of hormone and its target cell.[405] Thus, it may be on the nucleus, as in the case of estradiol action on the uterus,[175] or it may be cytoplasmic, as in the case of growth hormone action on polysomes in liver.[240] A general concept in endocrinology is that hormones combine with receptors in their target cells. This concept has been particularly well applied to the action of estradiol on the uterus, where it appears that the estradiol combines with a specific cytoplasmic receptor protein and then enters the nucleus bound to the receptor, which itself has become modified.[175] The action on RNA synthesis then takes place. Still other hormones (e.g., insulin) appear to act mainly to regulate the permeability of the plasma membrane to specific metabolites.[344] The possibility of combinations of the three above mechanisms exists, and such combinations are known to occur in the case of some hormonal effects.

A physiological definition of the term "hormone" is difficult, but one should be given that is at least useful as a basis for discussion: "A hormone is a specific chemical substance, secreted by specific cells in a particular part of the body (an endocrine gland), which passes into the blood stream and then exerts specific effects on certain distant target cells or target organs, resulting in overall coordination of the organism." The present chapter will be concerned primarily with substances which conform to the features of this definition.

The important point to emphasize for comparative physiological purposes is that in a number of phyla and classes of animals special chemical substances are produced which are essential to normal development and functional integration of the body. The points of origin within the organism, the specific chemical nature of the hormones, and the methods of transport are secondary in importance. The nature of the effects which are produced depends as much on the nature of the reacting tissues as on the chemical properties of the circulating hormone. Hormones spreading randomly through the body in the body fluids are obviously powerless to produce tissue and organ differentiation, or induce any directed or organized activities, in the absence of an underlying gene-determined differentiation. The activities of hormones are in a sense, therefore, superimposed on the basic pattern of the organism, and serve to bring into full functional development and activity the numerous and complex latent differentiations. In a study of endocrine mechanisms, the phylogenetic and ontogenetic development of ability to respond to an endocrine is therefore as important as the appearance of the endocrine itself.

The two major integrative systems of the body, the nervous and endocrine systems, are intimately interrelated functionally.[276, 367] Nerve cells generally play the dual role of conduction of excitation and the secretion of such neuroendocrine substances as the neurotransmitters, acetylcholine,[280] norepinephrine,[423] and 5-hydroxytryptamine,[260, 310, 432] which

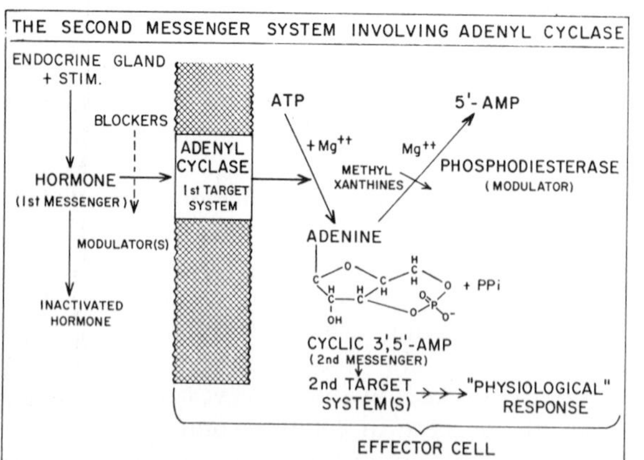

Figure 21–1. The two-messenger system of hormone action involving adenyl cyclase. (From Sutherland, E. W., Øye, I., and Butcher, R. W., Rec. Progr. Horm. Res., *21*:623–646, 1965.)

have important roles at interneuronal and neuromotor synapses. In addition, certain groups of nerve cells have become specialized as important sources of neurosecretory substances which, transported by the blood, function in manners quite comparable to the true hormones produced by gland cells unrelated to nerves. Examples of such neurohormones are oxytocin and vasopressin, produced in cells of the vertebrate hypothalamus; active principles (urotensins) liberated by a cell cluster in the caudal region of the spinal cord of fishes, the urophysis;[150] and insect hormones produced in nervous elements in the brain and nerve cord. The adrenal medulla is itself embryologically of neural crest origin, arising from centrifugally migrating cells which are homologues of the sympathetic postganglionic nerve cells.

Neurosecretion is one of the most important principles of endocrinology.[158] Strictly speaking, it could represent the secretion of any substance by a neuron, e.g., a neurotransmitter. However, this term is usually used to designate the secretion of true hormones by nerves which contain granules visible with the light microscope, usually after specific staining techniques. The hormones are carried down the axons in the neurosecretory granules and then released at their endings, which may form neurohemal organs specialized for their release. There is also evidence that the hormones of the vertebrate neurohypophysis are bound to a carrier protein, neurophysin.

The relationship between the nervous system and the endocrine system extends even further. The secretion of the trophic hormones of the anterior lobe of the pituitary is regulated by the action of other hormones (hypothalamic releasing or inhibiting hormones) originating in the hypothalamus and transported to the anterior lobe by way of a highly specialized vascularization known as the hypothalamo-hypophyseal portion system (Fig. 21–2). The excitation of the adrenal medulla, and perhaps of the pars intermedia of some species, is directly nervous, and the remaining glands may be indirectly subjected to nervous modification. In brief, it is apparent that the endocrine system is to a great extent subservient to, and in part evolved from, the nervous system. This intimate, hierarchical organization of the two systems leads to highly efficient coordinatory biological mechanisms.

The three animal groups in which the endocrine system has been most extensively investigated, the vertebrates, insects, and crustaceans, while displaying among them remarkable similarities of function and superficial organizational plans, show no recognizable true homologies. The hormones, which are more or less similar in chemical nature among the animals of any one of these main groups, appear usually to be different from

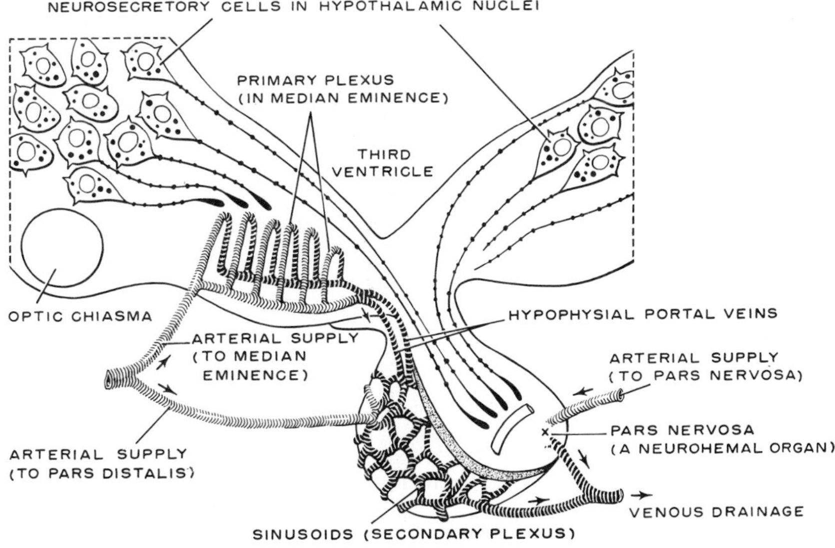

Figure 21–2. Diagram of the anatomic connections between the mammalian hypothalamus and the pituitary gland. The origin of fibers to the median eminence is not as certain as designated. (From Turner, C. D., and Bagnara, J. T., General Endocrinology, 5th ed. W. B. Saunders Co., Philadelphia, 1971.)

those in the other groups. Consequently, it has been decided to deal with comparative aspects within each group separately.

VERTEBRATES

Endocrine Glands

Thyroid. This gland, phylogenetically related to the endostyle of the protochordates,[20] arises in the mammal as an unpaired entodermal evagination of the pharyngeal floor. It loses its connection with the gut, separating as a bilobed gland. The gland becomes organized in the form of numerous follicles bounded by secretory epithelium and containing an iodine-rich proteinaceous colloid. The height of the epithelial cells varies directly with the level of glandular activity. Secretion normally occurs as a result of the resorption of the thyroid hormone from the colloid by the follicular epithelium, followed by its secretion into the blood. The colloid includes a protein of high molecular weight (680,000), thyroglobulin.

The thyroid takes up inorganic iodides derived from the diet. The thyroid epithelium has a greater capacity to concentrate iodide than any other tissue in the body. After concentration, the iodide is then enzymatically oxidized to iodine or IO^-. The iodine thus formed iodinates the tyrosine residues in the follicular proteins. Combination of the iodinated tyrosine results in the formation of a series of thyronines: diiodothyronine, triiodothyronine (T_3), and tetraiodothyronine or thyroxine (T_4) (see below). Thyroid hormone secretion then requires the liberation of the T_3 and T_4 as a result of the proteolysis of thyroglobulin by a thyroidal protease. The hormones circulate in the blood, mainly bound to certain α-globulins, the thyroid binding proteins. The liver and kidneys carry out the major catabolism of the thyroid hormones. T_3 and T_4 are differentially active on various processes, but T_4 is usually regarded as the thyroid hormone.

The most characteristic manifestation of thyroxine action in homoiotherms is the calorigenic action that results from an enhanced oxygen consumption by most of the body tissues. This effect has not been found to any degree in poikilothermic vertebrates. In fishes, such studies have been complicated by technical considerations (e.g., the need to prevent excessive movements after thyroid hormone administration). In a study wherein excessive movement was prevented, it was found that although thyroxine had a variable effect on the oxygen consumption of dogfish (*Squalus suckleyi*) pups, a thyroxine analogue (triiodothyroacetic acid) was able to increase oxygen consumption above control levels.[335] The low temperature at which poikilotherms sometimes live may be relevant to the lack of effect. It was found that a significant effect of T_4 or T_3 could be obtained on oxidative metabolism by maintaining lizards (*Anolis carolinensis*) at temperatures significantly higher than "room temperature."[273] However, such studies do not change the basic finding that cold-blooded vertebrates are essentially unresponsive to the calorigenic action of thyroid hormones.[172]

Although the molecular basis of the calorigenic action is still unknown, there is now good evidence for the concept that stimulation of nuclear RNA synthesis occurs during the early stages of the effect. Actinomycin D, a potent inhibitor of DNA-dependent RNA synthesis, blocks the calorigenic effect of thyroxine in thyroidectomized rats.[403] It has become apparent that there is a sequence of events leading to the calorigenic effect.[404] First, incorporation of RNA precursors is enhanced. Then, the level of the Mg^{++}-activated RNA polymerase in nuclei rises, synthesis of ribosomal RNA is increased, and finally, the incorporation of amino acids into proteins is enhanced. Since it had earlier been reported that the synthesis of oxidative enzymes increases after thyroxine administration, it can be supposed that the synthesis of these enzymes accounts for the calorigenic effect.

Thyroid hormone is essential to normal growth and differentiation, particularly of the hard or cornified derivatives such as bone or hair, and exerts a regulatory action on

---

Thyroxine (Tetraiodothyronine)

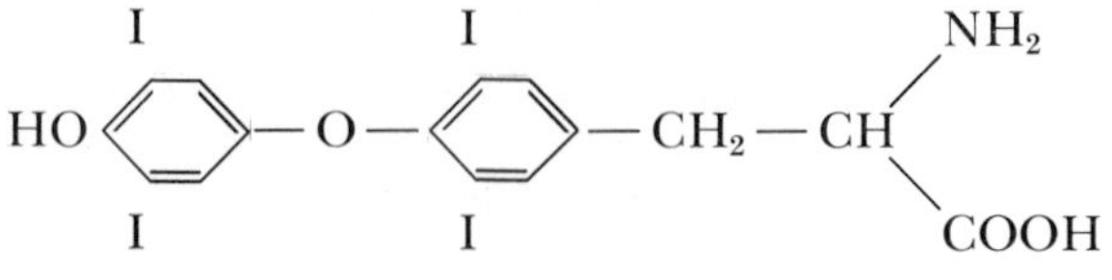

$$\mathrm{HO}-\underset{\mathrm{I}}{\overset{\mathrm{I}}{\bigcirc}}-\mathrm{O}-\underset{\mathrm{I}}{\overset{\mathrm{I}}{\bigcirc}}-\mathrm{CH_2}-\mathrm{CH}(\mathrm{NH_2})\mathrm{COOH}$$

amounts of intercellular substances of tissues. It facilitates utilization of carbohydrate. Thyroid hormones maintain normal excitability of such elements as central nervous centers and cardiac muscle.

The thyroid appears also to exercise influence upon other endocrine glands. It stimulates the adrenal cortex. Normal gonadal activity depends upon thyroid function. Thyroxine seems in some manner to cooperate with antidiuretic hormone (ADH) in regulation of water. And participating in its own regulation, thyroid hormone depresses the secretion of thyrotropic hormone by the anterior lobe and, possibly also, directly, its own production of hormone.

Control of the thyroid is predominantly by way of thyroid-stimulating hormone (TSH). All phases of synthesis and release of thyroid hormone are facilitated by TSH.

TSH secretion is under the control of a hypothalamic hormone, the thyrotropic-hormone-releasing-hormone (TRH), whose structure has now been elucidated.[48] It is (Pyro)Glu-His-Pro(NH_2), and it stimulates the release of TSH from rat anterior pituitary glands *in vitro* as well as *in vivo*. Earlier evidence had indicated that TSH secretion requires an intact hypothalamus.[100] The chief environmental factor which activates thyroidal secretion is lower temperature,[58] which favors increased TSH and thus thyroidal secretion in most species. However, this may not occur in lizards, because it was found that a temperature range from 21° to 38° C had no significant effect on thyroid secretion in *Sceloporus occidentalis*.[84] Thus, the role of temperature in thyroid activation in this species may be only a permissive one.

TSH is now known to act on the thyroid by increasing the formation of cyclic AMP, as a result of stimulation of thyroidal adenyl cyclase.[371] Cyclic AMP levels increase in the thyroid in response to TSH, and there is strong evidence that cyclic AMP brings about the following effects in the thyroid: stimulation of iodide trapping, iodide organification, iodothyronine synthesis, iodine secretion, and phospholipid turnover.[371]

Research is presently being devoted to the mechanism of adenyl cyclase stimulation by TSH and to the mode of action of cyclic AMP in producing these effects.

It was discovered many years ago[179] that the feeding of thyroid gland material to amphibians hastens metamorphosis and, conversely, that removal of the rudiment of the thyroid from larval forms prevents metamorphosis. Metamorphosis can be obtained in thyroidectomized individuals by administration of thyroglobulin or thyroxine. Less effective is diiodotyrosine, and still less effective are inorganic salts of iodine or elemental iodine, which, however, will induce metamorphosis if they are present in sufficient quantities.[6] This response to thyroxine is quite distinct from the increase in the general rate of metabolism. Dinitrophenol, a powerful stimulant of metabolism, has no influence on metamorphosis. Acetylated thyroxine, on the other hand, retains its normal capacity to induce metamorphosis although it has lost entirely its metabolism-stimulating action.

Various amphibian species differ greatly among themselves in their responsiveness to thyroxine. During their early development, none of them shows ability to respond to this hormone, but most acquire the capacity at some particular stage in their development. Some species, like the Mexican axolotl, never develop any considerable degree of reactivity to this metamorphosis-inducing factor, and hence normally do not metamorphose. Axolotls can be made to metamorphose by large doses of thyroxine; other species, such as *Necturus*, appear never to develop the reactivity and never metamorphose. That the failure of metamorphosis in *Necturus* is the result of failure of tissue response rather than of the absence of an appropriate thyroid principle is shown by the fact that *Necturus* thyroid will accelerate metamorphosis in *Rana clamitans*.[398]

In amphibian development, including metamorphosis, there is an orderly sequence of changes. This sequence appears due, at least in part, to gradually increasing thyroid activity and developing responsiveness of the various tissues of the organism to the hormone.[25, 124, 238, 272, 320]

The capacity of amphibians to metamorphose is lost after hypophysectomy.[387] This activity has been traced to the anterior lobe; the principle involved is ineffective in inducing metamorphosis after complete thyroidectomy. The thyroids are also known not to accumulate any colloid in the absence of the hypophysis. This demonstrates that, in amphibians, the production and liberation of the thyroid hormone, and hence metamorphosis, are under the control of the hypophysis. Prolactin may also be important for the control of metamorphosis.[125]

Among teleost fishes it is known that the thyroid gland is concerned in metamorphosis and that changes in the amount of thyroxine available will, in other species, result in alterations in growth rates and in body form.[12, 203, 205] Furthermore, as in higher vertebrates, the pituitary of lower vertebrates yields a thyrotropic principle (TSH) on whose activity the development and secretion of the thyroid depends.[169]

The developmental effects of thyroxine are also mediated by an effect on RNA and protein synthesis. Actinomycin D injection inhibits the tail atrophy that normally occurs during metamorphosis of *Xenopus* larvae.[427] Furthermore, the increase in cathepsin which normally occurs during tail atrophy is much reduced, indicating that thyroxine probably normally stimulates the synthesis of cathepsin, rather than its release. A number of the other developmental changes caused by thyroxine are also associated with a stimulation of RNA synthesis,[151] just as in the case of the effect on oxygen consumption.

It is interesting that, despite the influences of the thyroid on teleost growth and development, there is still no decisive demonstration that in fishes the thyroid influences the general metabolic rate.[126] In fact, in poikilotherms in general, reports from work with various species indicate a variety of kinds of results.

Experiments involving thyroid administration to fishes appear to indicate that this hormone has no influence on O_2 consumption in guppies and goldfish.[126] Treatment with thiourea does not significantly alter O_2 consumption in *Fundulus*. Injection of extracts of thyroid glands of Bermuda parrot fish increased O_2 consumption in white grunts, but only when the latter fish weighed more than about 15 g.[385] As in most fish, administration of thyroid to *Rana pipiens* tadpoles prior to changes of metamorphosis resulted in no alteration in O_2 consumption. In adult *Rana pipiens*, on the other hand, administration of thyroid increased O_2 consumption substantially,[425] and if the animals were kept at temperatures higher than about 13°C the thyroid extract caused reduction in body weight. Among the reptiles and birds, the effect of thyroid administration on the basal metabolic rate appears quite similar to that observed in mammals. Pigeons shows a marked reduction in basal heat production after complete thyroidectomy. Young pythons, during long-continued thyroid feeding, show greatly increased excitability and weight reduction.

Other thyroid roles have been reported for fishes. It is concerned in the regulation of liver glycogen and with osmoregulation.[386] In euryhaline fishes, greater thyroid activity is associated with more saline media.[199] In the stickleback *Gasterosteus*, it is increase in thyroid activity which is responsible for the migration from salt to fresh water, a selection related to its breeding habit. This change is subject to photoperiodic regulation.[15]

Thyroxine has neurophysiological effects in the goldfish brain. Thus T_4 affects the electrical responses obtained in the olfactory tract after infusion of the olfactory sac with weak salt solutions. There was a local facilitatory action of T_4 in the olfactory bulb, but an inhibition of the telencephalic centrifugal action upon the NaCl-evoked response.[307] Further study showed that there are three types of discharge patterns of single cells in the various layers of the olfactory bulbs. T_4 produced an increase in the "spontaneous" firing rate of one type (Type I).[308] In addition, T_4 is important for the production of a stage in the development of the salmon. This involves the conversion of the parr to the smolt (smoltification), with deposition of guanine in the dermis, among other changes.[172]

Thyroid response in acclimatization to temperature differs among fishes. Whereas most fishes show higher thyroid activity at higher temperatures, the trout *Salmo* and the minnow *Umbra* show higher activity at lower temperatures as does the homoiotherm.[171]

The thyroid gland is an organ highly specialized organically for binding iodine. This capacity is not only widespread among the endostyles of amphioxus[19] and Urochordates such as *Ciona*,[18] but is observed in the dermal glands of hemichordates. Iodine binding appears to occur not uncommonly with the exoskeletal scleroproteins of many invertebrates.[30, 170, 173]

Parathyroid. Parathyroid glands are lacking in fishes, but are present in amphibians and higher vertebrates. In mammals, the four parathyroid glands are buried in the thyroids and derive from the entoderm of the third and fourth pharyngeal pouches. There are two cell types: the chief cells, source of the parathyroid hormone (PTH), and the mitochondria-rich oxyphile cells, of unknown function. Bovine PTH is a single polypeptide chain of 83 amino acids, with a molecular weight of 9500.[333]

The hormone appears concerned almost wholly with the regulation of calcium and phosphate metabolism. The hormone effects a fall in blood phosphate and its elimination by the kidneys together with a rise in blood calcium and its loss by a similar route. The hormone regulates only the ionized calcium and phosphate, not that which is bound to protein or in the body fluids in un-ionized form.

Control of secretion is effected by the serum level of calcium. Drop in calcium stimulates the gland. A rise in serum phosphate exerts a similar action but only indirectly through the correlated change in calcium.

An action of PTH on adenyl cyclase explains its action at both renal and osseous sites. Physiological amounts of PTH increase the urinary excretion of cyclic AMP,[79] and PTH stimulates the adenyl cyclase of both skeleton and kidney *in vitro*.[81] Furthermore, the dibutyryl derivative of cyclic AMP has a PTH-like action on bone explants in tissue culture.[412] Present studies are focussing on the mechanism by which cyclic AMP brings about the movement of calcium from bone to blood. Since calcium has numerous essential physiological roles, among them the excitability of nerve and muscle, the coupling of excitation to contraction in muscle, and the permeability of membranes to various ions, its regulation is of great importance. Furthermore, the role of phosphate in buffering, in carbohydrate metabolism, and as a constituent of nucleic acids also makes its regulation important.

Ultimobranchial Glands. The hormone calcitonin is secreted by cells derived from the ultimobranchial bodies of the embryo, which form discrete glands in adult non-mammalian vertebrates. They become incorporated into the thyroid gland in mammals.[91] Calcitonin is active in lowering the concentration of calcium in the blood by inhibiting the resorption of calcium from bone.[292] In mammals, the cells appear in the thyroid as the C-cells (calcitonin-cells), also referred to as parafollicular cells because of their frequent position between the follicles. Hog and salmon calcitonin both have a sequence of 32 amino acids, but they differ in structure. The salmon hormone is more active than the hog hormone in the standard mammalian assay system.[91] Furthermore, it is active longer than the hog hormone, because it is degraded less rapidly.[181] The signal for the release of calcitonin is a hypercalcemia, and this fact proved important in the discovery of the hormone.[292] There have been difficulties proving that the calcitonin is secreted normally in response to a change in serum calcium; thus, its true status as a hormone is perhaps somewhat questionable.[292]

Pancreas. In the pancreas, of endodermal origin by budding from the ducts, are the islets of Langerhans. These contain three kinds of cells. Alpha cells with acidophilic granules, more abundant in the peripheral than central regions of the islets, are believed to be the source of glucagon.[208] Beta cells, comprising the majority, are the source of insulin. In addition, a third cell type is the delta, without granular contents and of unknown function. However, they may secrete gastrin. More recently, Epple[123] has described amphiphil cells in the islets of lower vertebrates, where they possess the staining characteristics of both beta cells and delta cells. They are not present in birds or mammals. In the toad (*Bufo bufo*), these cells are only present during the spring. In the shark (*Scyliorhinus canicula*), other cells are present which exhibit features intermediate between alpha and beta cells.

Both insulin and glucagon are proteins. Insulin (M.W. 6000; isoelectric point, pH 5.4) contains fifty-one amino acids whose arrangement has been fully worked out. In structure the hormone comprises two peptide chains, one acidic and the other basic; the two chains are connected by two disulfide linkages. Species difference has been found in the sequence of three amino acids at specific loci.[191] Insulin is formed from proinsulin.[391] This is a single protein, consisting of the two chains of insulin connected by a separating peptide. Enzymatic removal of this connecting peptide yields insulin, its two chains connected by disulfide bridges. While there is a general immunological similarity among insulins from different species, this is not total, as is evident from antibody production.[288] Glucagon is a straight polypeptide chain of twenty-nine amino acids.[27, 143]

Insulin plays an important role in regulation of carbohydrate metabolism. It acts to increase the conversion of glucose into glycogen in both muscle and liver. It facilitates the production of fat from carbohydrate and stimulates protein synthesis. It also acts to accelerate the oxidation of glucose in muscle and elsewhere. The modes of action of insulin are not yet known with surety. The one for which the most evidence has been amassed is that insulin facilitates the movement of glucose and amino acids through cell membranes.[344]

When the insulin level falls below minimally adequate levels, blood glucose increases because of depression of the mechanisms for its removal and use, and it may be lost in the urine. Fat becomes oxidized in far larger than normal proportions and acidic ketone bodies such as acetone, acetoacetic acid, and hydroxybutyric acid accumulate. These are excreted as sodium and potassium salts, reducing the body's alkali reserve and lowering blood pH. Protein is catabolized and the nitrogen is excreted.

The second islet hormone, glucagon, antagonizes insulin in that it stimulates glycogenolysis in the liver by a "cascade" type of effect. Glucagon stimulates the enzyme adenyl cyclase, resulting in the formation of more cyclic AMP from ATP.[346] The cyclic AMP then goes on to increase the liver content of active phosphorylase, which catalyzes the formation of glucose-1-phosphate from glycogen. After conversion to glucose-6-phosphate, the glucose is liberated into the blood by the action of glucose-6-phosphate phosphatase, resulting in the hyperglycemia produced by glucagon.

The secretion of the two islet hormones appears chiefly regulated directly by the blood sugar level, a rise stimulating insulin liberation, and a fall, glucagon. Glucagon at high levels stimulates insulin secretion and vice versa. Growth hormone from the adenohypophysis stimulates insulin, and probably also glucagon, liberation. Thus, it is hyperglycemic in its effect.

There is considerable variability in relative importance among vertebrates of the two pancreatic hormones. Among mammals there is high sensitivity to insulin lack in dogs, cats, and rats, and generally less sensitivity among herbivores. Reptiles and birds are generally less sensitive to insulin;[396] the effect of insulin lack is transitory in birds.[284] Tortoises exhibit hyperglycemia following pancreatectomy.[144] On the other hand, reptiles and birds are highly responsive to glucagon, with reptiles being especially well provided with α-cells. Amphibians contrast with the sauropsidans in being generally very sensitive to insulin, or its lack, and parallelly often having an absence of α-cells.[286, 287, 456]

Adrenal Medulla. The medullary tissue is of ectodermal origin, arising from the same embryonic tissue, the neural crests, that generates the sympathetic ganglia. The secretory cells are, in effect, postganglionic elements of the sympathetic system and remain synaptically associated with cholinergic preganglionic fibers. The gland is especially richly vascularized. The mammalian adrenal medulla is known to secrete two closely related hormones, one the primary amine, epinephrine, and the other the secondary amine, norepinephrine.[423] The levorotatory isomers are the natural, active ones. In structure, these hormones are related as shown below.

Glands differ in the relative amounts of these two hormones which they contain. The proportion of the two which is liberated varies with the origin of nervous stimulation. Although not fully established, evidence suggests that the two hormones are secreted by two types of medullary cells. Within the medullary cells the hormones are contained inside vesicles.[36]

Epinephrine and norepinephrine effect in general a qualitatively similar action on heart, blood vessels, and smooth muscle and also upon carbohydrate metabolism. However, there are important quantitative differences. In the vascular system, both increase the excitability and contractility of heart muscle with a consequent rise in blood pressure. While both hormones constrict visceral vessels and those of skin and mucous membranes, and both elevate pulmonary blood pressure, epinephrine effects increased blood flow through the muscles, heart, liver and brain, causing thereby an over-all drop in peripheral resistance, while norepinephrine constricts to varying degrees all peripheral vessels except those of somatic muscle and liver, yielding a slight rise in peripheral resistance. Following norepinephrine administration, heart output is reflexly held reduced by way of pressure receptors.

Epinephrine produces a rapid increase in

Transmethylation (from methionine)

Norepinephrine → Epinephrine

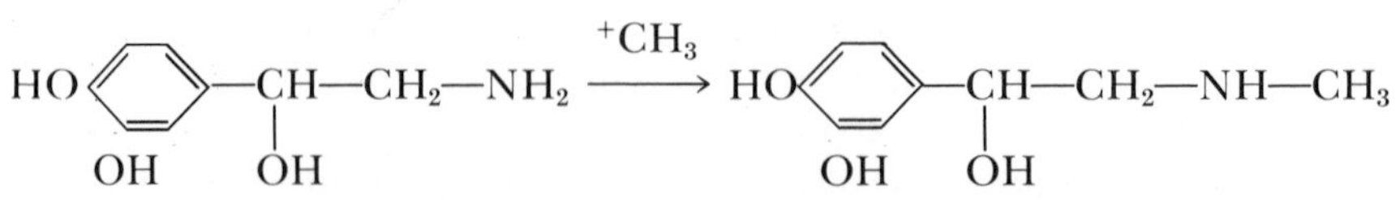

$$\mathrm{HO(OH)C_6H_3{-}CH(OH){-}CH_2{-}NH_2} \xrightarrow{^{+}CH_3} \mathrm{HO(OH)C_6H_3{-}CH(OH){-}CH_2{-}NH{-}CH_3}$$

blood sugar; norepinephrine is only about 25 per cent as effective in this role. Sugar is released from glycogen in the liver, as a result of the action of glucose-6-phosphatase on the glucose-6-phosphate formed, as with glucagon. Cyclic AMP was discovered by Sutherland and Rall in biological material as a result of the study of the mode of action of epinephrine and glucagon on hepatic glycogenolysis. A heat-stable adenine ribonucleotide was formed from cellular particles containing plasma membranes in response to epinephrine or glucagon,[397] and this turned out to be cyclic AMP. Epinephrine and glucagon act on different receptors which regulate adenyl cyclase, because epinephrine action can be specifically blocked by ergotamine. However, the net effect of the action of both hormones is to stimulate adenyl cyclase and then phosphorylase content. In muscle, which lacks the appropriate phosphatase, the formation of the glucose-6-phosphate is followed by its oxidation to lactic acid, which then appears in the blood. The lactic acid is largely converted into glycogen by the liver. There are also some differences in the phosphorylase.

Adrenal medullary hormones act on two different types of receptors, designated as alpha and beta by Ahlquist.[3] Although the exact determination of which is involved in a given action can often be difficult, three criteria are used: (1) The relative potency of a series of catecholamines;[3] (2) the action of blocking agents;[4,300] (3) the nature of the cyclic AMP changes in the tissues.[411] For example, the following three responses have been designated as mediated by alpha receptors: arteriolar constriction, spleen contraction, and gastrointestinal sphincter contraction.[299] On the other hand, the following three responses are mediated by beta receptors: arteriolar dilation, stomach relaxation, and relaxation of the non-pregnant cat uterus.[299] Many other examples could be given to show that the study of the adrenergic receptor involved is an important aspect of the elucidation of any adrenal medullary response.

Epinephrine has a much stronger action than norepinephrine in relaxing the pulmonary bronchial muscles. This difference, together with the substantially different cardiovascular action of the two hormones, probably accounts in large measure for the apprehensive and other nervous responses to increased epinephrine but not to norepinephrine.

Numerous other responses to epinephrine occur. These include pupillary dilatation, pilomotor response, sweat-gland inhibition, relaxation of gut and bladder muscles (except for the sphincters, which are stimulated), contraction of spleen capsule, secretion of saliva, and facilitated blood coagulation.

The adrenal medulla is controlled exclusively by the sympathetic nervous system. The degree of its tonic activity is believed to be related to total afferent nervous activity and perhaps to fluctuate with it. The gland appears dispensable, its absence capable of being fully compensated by the norepinephrine production by adrenergic fibers. However, normally, any rapid drop in blood sugar, or stress, stimulates production of epinephrine, this hormone contributing to a rapid compensatory response. The normal stimuli for specific liberation of norepinephrine are less clear.

Adrenal Cortex. Four glands secrete steroid hormones: the adrenal cortex, testis, ovary, and placenta. Steroids all contain the cyclopentanoperhydrophenanthrene nucleus in their structures, and a knowledge of their chemistry is important to an understanding of their biological activities.[409] The relationship of the cortical to the medullary elements varies in the different vertebrates.[172]

The situation of a cortex surrounding a medulla is absent in almost all non-mammalian forms. In cases where the cortical tissue is imbedded in the kidneys (teleosts), it is called interrenal tissue. The adrenal cortex is derived from mesoderm to the right and left near the base of the dorsal mesentery at the level of the primary urogenital differentiation. In the mammal, the adrenal cortex, as the term indicates, wholly encompasses the medullary elements. The mammalian cortex is relatively thick, is yellow in color, and appears stratified as three regions: a thin outermost zona glomerulosa; a deeper, middle zone fasciculata; and an innermost zona reticularis.

Extracts of the gland have revealed almost fifty different steroids, though relatively few of these have ever been demonstrated to reach the blood stream. Some of them are precursors in hormone synthesis.

The cortical hormones fall into three categories: glucocorticoids, mineralocorticoids, and sex hormones. Although there are still some questions about their exact sites of origin, there is general agreement that glucocorticoids are secreted by the fasciculata-reticularis area and mineralocorticoids by the zona glomerulosa. The sex hormones from the cortex include estrogens, androgens, and progesterone. Androgens appear to be the chief type secreted in the mammal. In both genetic males and females, adrenal androgens appear to contribute to regulation of normal muscular and skeletal development, differentiation of external genitalia and hair-growth patterns, and sexual behavior and drive.

Glucocorticoids effect profound regulatory actions upon metabolic transformations in the body, which in turn underlie behavioral changes. This group of steroids is therefore of great importance for normal metabolism of the adult mammal. Glucocorticoids by themselves will permit survival following adrenalectomy. The best known of these steroids are:

Cortisone
(Compound E)

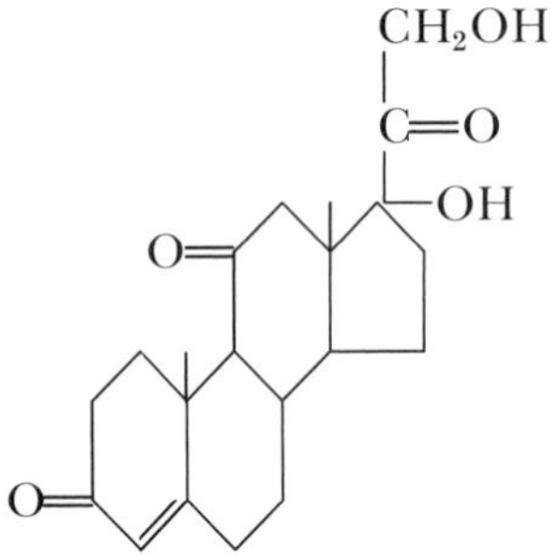

CH_2OH | C=O | —OH; O= at position 11; O= at position 3

Cortisol
(Compound F)

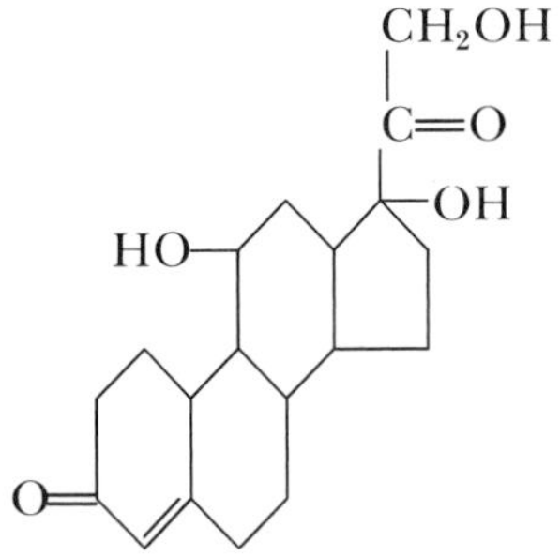

CH_2OH | C=O | —OH; HO— at position 11; O= at position 3

Corticosterone

CH_2OH | C=O; HO— at position 11; O= at position 3

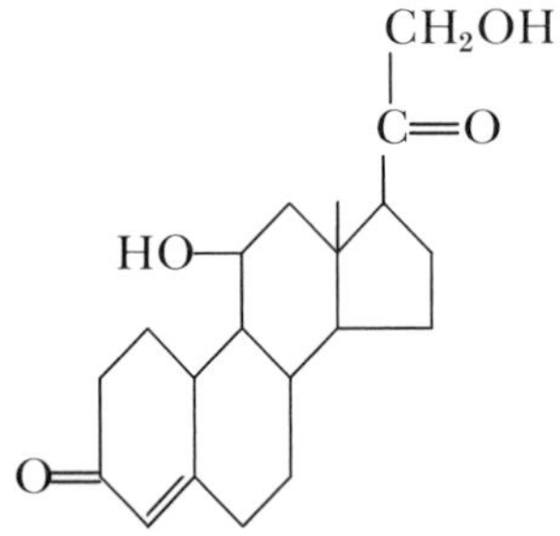

Extensive studies have disclosed the following rather uniform general distribution of corticosteroid synthesis among the various vertebrates.[357] The cyclostomes synthesize cortisol and corticosterone; the cartilaginous fishes, 1-alpha-hydrocortisone,[213] cortisol and corticosterone; the bony fishes, cortisol, cortisone, and 11-deoxycortisol; the amphibians, reptiles, and birds, corticosterone, aldosterone, and 18-hydrocorticosterone; and most mammals, cortisol, cortisone, corticosterone, 18-hydroxycorticosterone, and aldosterone. Thus, either corticosterone, cortisol, or both are secreted by members of all the vertebrate classes, indicating the ancient origin and persistence of the glucocorticoids. However, aldosterone, a potent mineralocorticoid, is not secreted by fishes and first appears in the tetrapods. Furthermore, cortisone is only rarely secreted.

The glucocorticoids and mineralocorticoids are characterized by having ketones at positions 3 and 20, and a hydroxyl group at 21. Glucocorticoids bear an oxygen at position 11, while mineralocorticoids are usually without it.

The actions of glucocorticoids on protein metabolism are extensive and large. They direct activities toward increasing carbohydrate by increasing gluconeogenic processes. Liver glycogen increases. Glycosuria results from action of the hormone to reduce tubular resorption of glucose in the kidneys. On the other hand, glucocorticoid-influenced metabolism favors protein catabolism, and consequent increased nitrogen excretion. This is superficially evident in such histological changes as a reduction in connective and lymphatic tissues and muscle, alterations in bone structure, and inhibition of various phenomena involved normally in the body's reaction to irritants, injury, and allergenic agents. Glucocorticoids other than cortisol exert other actions, too. They participate in the distribution and regulation of water and electrolytes, in a manner qualitatively similar to that of mineralocorticoids, and encourage rapid regulation by the body in response to excess water. However, they are less active in this regard. They also maintain the normal state of low permeability of the membranes, including the synovial ones. They stimulate gastric secretion and hence are ulcer producing. And as would be expected from their basic metabolic functions, derangements of mental function are related to abnormal amounts of these hormones.

The best known mineralocorticoids are 11-deoxycorticosterone and aldosterone:

11-Deoxycorticosterone

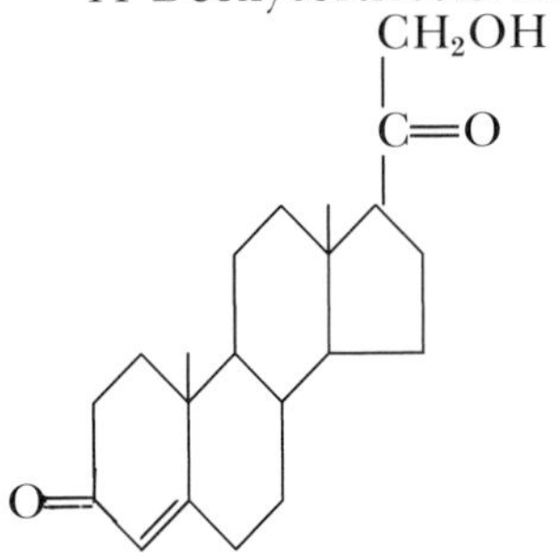

CH_2OH

C=O

O

Aldosterone

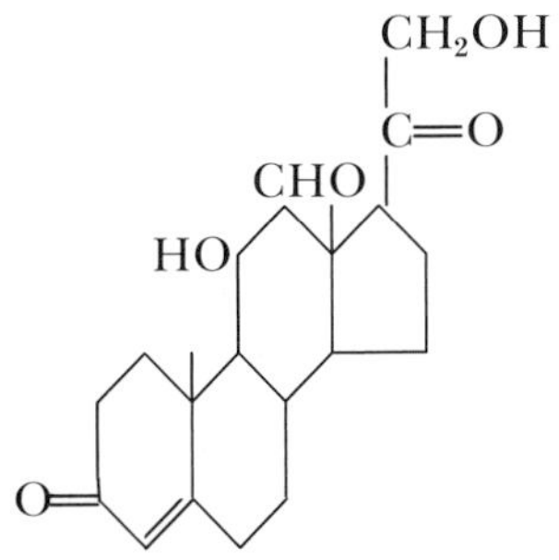

CH_2OH

C=O

CHO

HO

O

As would be expected perhaps from its hydroxyl group at position 11, aldosterone also possesses glucocorticoid properties, if adequate amounts are present.

Mineralocorticoid action is chiefly upon regulation of body electrolytes and water. 11-Deoxycorticosterone, which lacks entirely glucocorticoid properties, effects increased urinary excretion of potassium and phosphate and reduced excretion of sodium, chloride, and water. It is believed to exert its action chiefly upon the renal tubules, regulating in the ascending limb of Henle's loop and the distal tubule the resorption of sodium in exchange for hydrogen and potassium ions which are excreted.[289]

The adrenal cortex exhibits altering rates in its secretion of hormones in response to physiological demands as a consequence of extrinsic controls. Most prominent by far is the hormone corticotropin (ACTH) from the adenohypophysis, which seems to stimulate the glucocorticoids.

The response of the body to any stress involves secretion of ACTH by the adenohypophysis. The consequent production of adrenocortical hormones assists the body to meet the increased demands upon it. There are three theories as to the means by which stress effects ACTH secretion. One view is that depletion of the cortical hormones through use constitutes the chief stimulus. A second theory is that epinephrine liberated in response to stress stimulates the hypophysis. A third theory proposes that a resting level of ACTH production is maintained under hypothalamic direction and that stress elevates this level. There is some evidence that the level of blood potassium itself regulates directly the cortical production of mineralocorticoids. Another substance that may be important in bringing about the release of aldosterone is angiotensin II. This peptide is formed in the blood by the action of renal renin upon angiotensin I. It can act on the zona glomerulosa cells to release aldosterone.

Adrenocorticotropic hormone secretion is mediated by the hypothalamus by means of a corticotropin-releasing hormone (CRH), since the interruption of blood and nerve connections between the hypothalamus and the pituitary abolished the response to stress.

There are adequate reasons to postulate that there is a common pattern of regulation of water and electrolyte and of carbohydrate metabolism in all vertebrates. Interestingly enough, cortisol, which is generally regarded as a glucocorticoid, has important osmoregulatory effects in teleosts. This osmoregulatory action of cortisol in fishes, in contrast to its largely metabolic role in mammals, is an example of the change of use to which a hormone is put as a result of evolution. Thus, it may not have been the hormones themselves which have evolved, but rather, the uses to which they have been put (their functions). Another example of this is the thyroid hormone, already considered.

In frogs, an interesting seasonal difference in response to total adrenalectomy has been noted. The frogs tolerate this condition in winter but not in summer. In summer, unless they are kept in isotonic saline, the operated frogs accumulate H_2O and K^+, lose Na^+ and succumb in a day or two.[219, 220]

Corticosteroids appear to facilitate ovulation in both mammals[400] and amphibians,[72, 451, 453] even *in vitro*.[457] In addition, aldosterone has the ability to stimulate sodium transport by toad bladder, presumably as a result of enhancing the synthesis of a protein, which then stimulates the sodium transport.[329]

Ovary. This gland secretes two kinds of hormones, both steroids. The first type is the estrogens. These are secreted before ovulation by the theca interna of the developing follicle and later, after ovulation, by the granulosa and theca-lutein cells. Estrogens are also secreted by the placenta. Although still others are known, two of the more common estrogens are:

Estradiol-17β

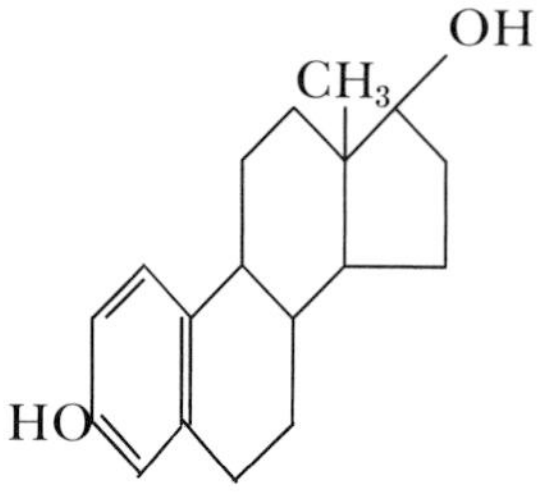

Estrone

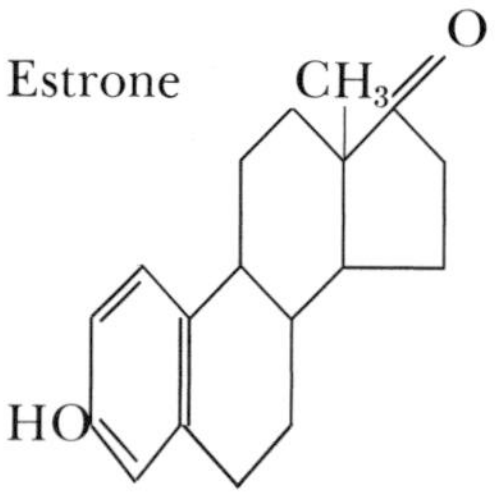

Extensive information on the secretion and actions of estrogens, androgens, and progesterone in sub-mammalian vertebrates[417] shows that most lower vertebrates secrete estradiol-17β or estrone. Thus, there is far less difference in secretion among animal types than was present even with the adrenocortical steroids.

The estrogens are transported in the blood largely bound to protein. Estradiol is the more potent in action. These hormones are essential to the normal sexual development of the accessory and secondary characters of the female mammal. They stimulate growth of the uterine endometrium, its glands and vascularization, and increase activity of the myometrium. They induce cornification of the vaginal epithelium and complete the development of the oviducts. In low titers, estrogens normally contribute to follicle differentiation, but in higher titers, depress this by reducing FSH secretion. They contribute to the normal growth and differentiation of the mammary glands.

Estrogens contribute to the feminine form and stature by influence on bone growth. They depress growth of long bones and stimulate closure of epiphyses. They cause increased protein synthesis and retention of sodium, calcium, phosphate, and water. They reduce activity of the sebaceous glands.

A second hormone of the ovary is progesterone:

Progesterone

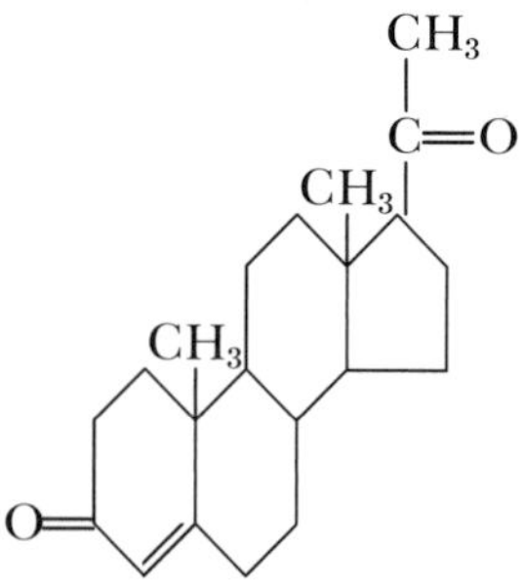

In addition, the following other hormones are known to occur naturally in mammals and to have progestational effects:

20β-Hydroxypregn-4-en-3-one

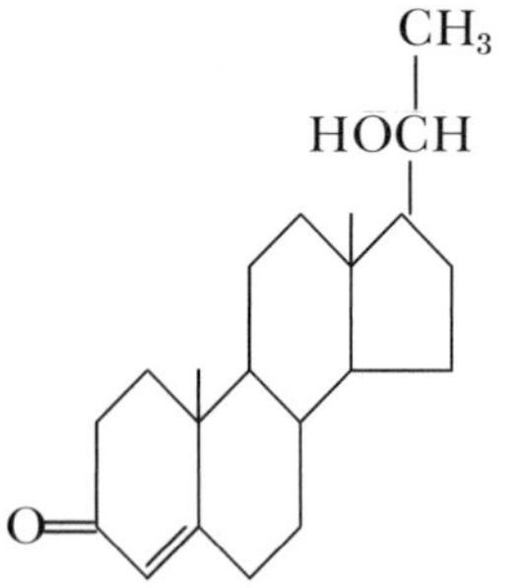

20α-Hydroxypregn-4-en-3-one

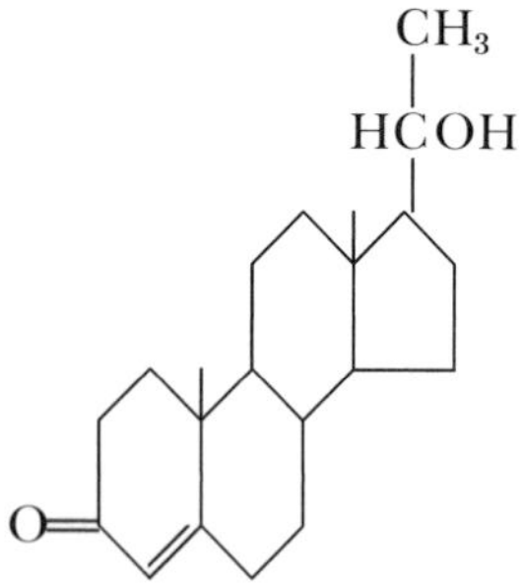

These are produced by the granulosa-lutein and theca-lutein cells of the corpus luteum.

Progesterone is also secreted by the placenta and the adrenal cortex. This hormone is responsible for the activation of the secretory phase of the uterine endometrium, thereby supplementing the estrogenic activity. Contrary to the action of estrogens, progesterone depresses myometrial activity. It effects mucification of the vaginal mucosa. In cooperation with estrogen, low titers of progesterone stimulate LH production; higher titers inhibit it. Progesterone supplements estrogen in inducing mammary gland development.

Ovarian secretion[17] is controlled by three hormones of adenohypophyseal origin (Fig. 21–3). FSH stimulates follicle development and estrogen secretion. In this it is assisted by LH, which also produces ovulation and corpus luteum development. Prolactin, the third hormone, or LH itself, stimulates secretion of progesterone by the corpus luteum, depending upon the species.

Estrogens can depress FSH secretion and induce LH and prolactin production. Progesterone inhibits LH secretion and encourages FSH production by the adenohypophysis. The actions and interactions of the ovary and hypophysis and these five hormones are responsible for the regulation of the menstrual and estrus cycles of mammals.

During pregnancy the placenta assumes the role of an endocrine organ, producing hormones assuring more or less its own maintenance until parturition. The inner cytotrophoblast of the placenta secretes chorionic gonadotropin, a hormone resembling LH or prolactin in its physiological action. Human placental lactogen has recently been purified and its structure determined.[380] It is remarkably similar to human STH, being identical in 85% of corresponding positions. The outer syncytial trophoblast secretes estrogens and progesterone, both essentially the same hormones which are of normal ovarian origin.

Although the mammary glands are initiated in their development by estrogens, and this action is supplemented by action of progesterone, the relative roles of the two hormones vary from species to species. Prolactin contributes in its role in stimulating progesterone secretion, but also contributes directly to the mammary glandular development and its final secretion of milk. Also essential to milk production are corticosteroids from the adrenal. Oxytocin, from the neurohypophysis, causes milk ejection by stimulating the contractile elements of the glandular ducts.

Testis.[267] Androgens are produced by the testis. Most important and potent of them is testosterone, though others may be secreted.

Testosterone

OH
CH_3
CH_3
O

In addition, there is another important circulating androgen:

Androstenedione

O
O

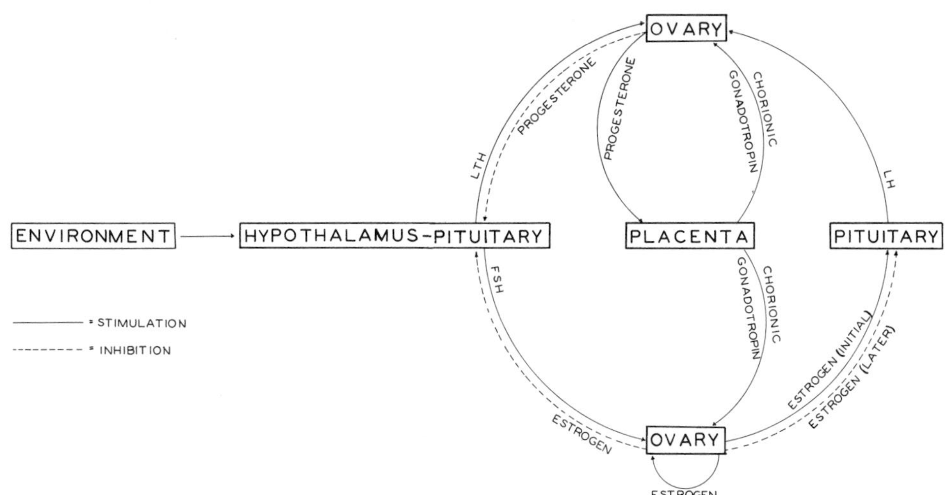

Figure 21–3. Diagram representing the major endocrine influences operating during the normal reproductive cycle of the adult female mammal.

Furthermore, evidence has accumulated that dihydrotesterone, formed in androgen-responsive tissues, may be the active form of the hormone *in vivo*.[409]

Androgens are produced in the interstitial cells of Leydig, and possibly elsewhere. The roles of androgens are numerous. One important group of actions is on the male accessory and secondary sexual characters and sexual behavior, being essential to normal development of these. In general metabolic action, androgens supplement STH in accelerating tissue growth, including protein synthesis. It promotes bone growth and final closure of the epiphyses. It stimulates red cell production and blood flow through the tissues. It stimulates the glands of the skin and the production of melanin.

Control of the testis is exercised primarily by the adenohypophysis, though its activity is modified by both the thyroid and the adrenal cortex (Fig. 21-4). FSH and ICSH (= LH) from the hypophysis are principally involved, themselves controlled by a hypothalamic releasing hormone. The former stimulates the germinal epithelium and initiates spermatogenesis, and the latter stimulates testosterone production and thereby controls development of the sex characters. The testis in turn influences the adenohypophysis. After castration the beta-cells of the latter organ become enlarged and FSH production is increased. This appears due to the absence of testosterone, and possibly estrogens which normally act to depress FSH secretion.

Low titers of androgens appear essential for spermatogenesis, though higher titers have deleterious action by suppressing FSH production or even by direct action on the germinal epithelium.

Adenohypophysis. This gland, of ectodermal origin, arises as an outpocketing of the roof of the pharynx. This outpocketing, Rathke's pouch, comes into close association with a comparable outpocketing of the infundibulum which will become the neurohypophysis. Rathke's pouch gives rise to the three major divisions of the adenohypophysis as follows: The posterior wall remains thin, becoming the pars intermedia (in species in which one forms), the anterior wall becomes greatly thickened to form the pars distalis, and a dorsal portion of the pouch extends toward the infundibulum forming the pars tuberalis. Seven hormones have been demonstrated in the adenohypophysis. Since at least five of these exert important regulatory actions on other endocrine glands, this gland is in large measure a master gland of the endocrine system.

GROWTH OR SOMATOTROPIC HORMONES. These are proteins with a molecular weight of an average of 22,000, comprising either large branched or small unbranched amino acid chains.[263, 264, 439] Each is most active in the species in which it is secreted, but there is a suggestion that they all possess a common core structure. Evidence suggests their source to be the orange-staining acidophils of the pars distalis.

Growth hormone (STH) normally stimulates growth in young animals, acting to accelerate protein synthesis and depress catabolism of amino acids. This results in nitrogen retention, and an increase in water and salts as in normal growth, and a reduction in fats. The hormone also promotes increased bone growth, both periosteal ossification and normal cartilage proliferation and bone synthesis at the diaphyso-epiphyseal region to produce bone lengthening.

In adult carnivores the growth hormone is

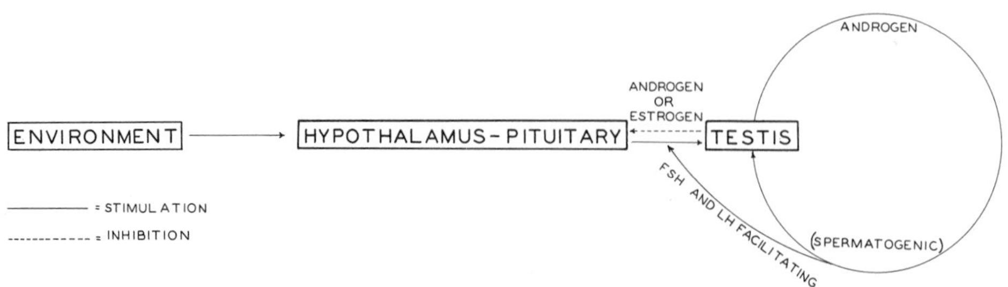

Figure 21-4. Diagram representing the major endocrine influences operating during the normal reproductive cycle of the adult male animal.

diabetogenic, neutralizing the action of insulin. The islet beta-cells are first stimulated by growth hormone and later atrophy, resulting in permanent diabetes.

In fishes, STH is essential to growth in length.[324, 325, 439]

Secretion of growth hormone is regulated by a growth hormone releasing hormone from the hypothalamus.

THYROID-STIMULATING HORMONE. This is a glycoprotein, with a molecular weight of about 30,000, produced by the basophils. The specific chemical structure of the hormone is known.[328] The principal role of TSH is to stimulate thyroid activity in all its aspects. It increases the quality of intracellular colloid and induces hormone liberation. It increases the height of the follicle cells, their rate of iodine uptake, and hormone synthesis. There are species differences in the structure of TSH.[147]

There is a reciprocal regulating action of thyroid hormone and TSH. The former, by blocking the pituitary response to TRH, depresses TSH secretion. Though this relationship is the primary regulating mechanism for TSH production, epinephrine and the corticosteroids also reduce TSH production, thereby accounting for reduced thyroid activity in initial response to stress other than cold stress, which has the opposite action. Stress, therefore, normally stimulates ACTH liberation while inhibiting TSH, except for cold stress which stimulates production of both.

ADRENOCORTICOTROPIC HORMONES (ACTH). These hormones are polypeptides; their site of origin in the gland is the basophils. There appear to be three corticotropic hormones. Their molecular weight is about 4500. They possess up to thirty-nine amino acids, of which a certain sequence of twenty-four is essential. Of these, a series of thirteen is in common with the melanocyte-stimulating hormone, α-MSH.[209, 264, 265, 379] ACTH is, correspondingly, reported to melanize goldfish.[83]

ACTH stimulates the adrenal cortex, inducing hypertrophy particularly of the zona fasciculata, and causing secretion of glucocorticoids. Mineralocorticoid secretion appears unaffected, in the main.

Hypophyseal secretion of ACTH is normally regulated by the hypothalamus through a corticotropin-releasing hormone (CRH) which, in turn, is probably influenced by blood titer of the corticosteroid. In stress, accelerated ACTH secretion is in part neurally effected by way of the hypothalamus, and in part by an exciting action of epinephrine on the hypothalamic center as well as the gland itself. The midregion of the median eminence appears normally involved in the regulation,[159, 160] though a basal level of secretion is maintained without such regulation.

FOLLICLE-STIMULATING HORMONE (FSH). This is a glycoprotein believed to be produced by the pale basophils of the pars distalis. Its role is to stimulate growth of ovarian follicles in the female and promote spermatogenesis in the male. It has a molecular weight of about 30,000.

LUTEINIZING HORMONE (LH) OR INTERSTITIAL CELL-STIMULATING HORMONE (ICSH). This hormone is a glycoprotein, of molecular weight 32,000.[266, 327] It is probably produced by the pale basophils of the pars distalis.

Luteinizing hormone stimulates in the female the ovarian secretion of both estrogen and progesterone, and in the male, the testicular secretion of androgens. The secretion of the gonadotropins is regulated by the hypothalamus.[338] There is now evidence that a releasing hormone for both FSH and LH (LH-RH/FSH-RH) is secreted by the hypothalamus. Its structure has been elucidated.[14] The posterior portion of the median eminence appears to be involved. The hypothalamus is, in turn, regulated by the level of circulating gonadal hormones as well as by numerous other factors such as light and temperature, presence of mates, and genital stimulation.

PROLACTIN OR LUTEOTROPIC HORMONE (LTH). This is a protein generally considered to be produced by the red-staining acidophils of the pars distalis. It has a molecular weight of about 25,000 but may show species differences, as for example between ox and sheep.[162, 263]

The roles of this hormone are known in the mammal for the female, in which it is responsible for maintenance of the corpus luteum and continued production of progesterone. There are species differences with regard to this function, however. It supplements the actions of the gonadal hormones in effecting mammary gland development, thereafter being essential for normal lactation, in collaboration with thyroid hormone and adrenocorticoids.[298] A third possible action of prolactin is to encourage maternal instincts and behavior. There is also evidence indicating that prolactin influences prostate development in the male,[82] synergizing with sex hormones and gonadotropins.

The secretion of prolactin is inhibited by the hypothalamus, even though a basal level of production seems maintained without regulation.[300] During lactation its continued production depends upon nipple stimulation and activation of afferent pathways to the brain which, in turn, stimulates the adenohypophyseal activity as well as oxytocin liberation from the neurohypophysis for milk ejection. Prolactin is regulated mainly by hypothalamic prolactin-inhibiting factor, but there may also be an activating factor.

Among vertebrates other than mammals, in addition to its well known action on the crop gland of pigeons in stimulating "milk" secretion, prolactin cooperates with melanocyte-stimulating hormone in increasing integumentary pigmentation in fishes[326] and is responsible for the water-drive of the red eft stage of the newt *Triturus.*[176] It also has many other actions,[32] e.g., sodium ion retention in fish.

MELANOCYTE-STIMULATING HORMONE (MSH).[303] This hormone is secreted by the pars intermedia and is polypeptide in nature. Two types have been isolated, α-MSH, with 13 amino acids[257] and a variety of β-MSH's with from 18 to 22 amino acids.[165, 190] A seven amino acid sequence (methionyl-glutamyl-histidyl-phenylalanyl-arginyl-tryptophyl-glycyl), known as the heptapeptide core, is common to the MSH's and the ACTH's, accounting for the intrinsic melanin-dispersing activity of ACTH.

Although it is established that MSH functions to bring about the darkening of certain lower vertebrates by melanin dispersion in melanophores (Chapter 23), iridophore contraction, melanin synthesis and purine degradation, its function in birds and mammals is less certain.[303] Pure MSH stimulates melanin synthesis and hair or skin darkening in mice[163] and men;[258] it also has extrapigmentary effects of uncertain physiological significance. Thus, MSH is thyrotrophic, lipolytic,[13] anti-gonadal,[69] hypocalcemic,[152] and cardiostimulating,[241] but the functional role these effects might play is not known. Another possible role of MSH that derives from its function in background adaptation may be a general role in adaptation. MSH or ACTH are required for the learning and retention of certain conditioned avoidance responses in rats,[105] and thus MSH functions to facilitate adaptations. In addition, MSH or ACTH can induce yawning and stretching behavior in dogs[133] and may be important in the etiology of this behavior in the intact animal.

MSH secretion is mainly under inhibitory control by the hypothalamus, although reports of a stimulatory control exist. Transplantation of the salamander pituitary is followed by excessive MSH secretion.[40] Frog hypothalamic extracts contain a substance that inhibits MSH secretion,[340] but there is also an adrenergic nerve plexus that could be responsible for the inhibitory control. Recently, it has been reported that the bovine hypothalamic MSH-release-inhibiting hormone (MRIH) has the structure prolyl-leucyl-glycine amide.[294]

Neurohypophysis. This organ is not an endocrine gland in the same sense as the others we have discussed. Not only does it arise as an outpocketing of the floor of the brain, but it retains its connection throughout life with the latter organ. Its hormones now are clearly established to be formed in the supraoptic and paraventricular nuclei of the hypothalamus and later transported within nerve fibers to the neurohypophysis, there to be liberated into capillaries upon appropriate stimulation. Thus, it is a neurosecretory gland. Electron microscopy has shown the secretory material within the cells to be contained within vesicles. The organ in mammals liberates two hormones, the antidiuretic hormone (ADH), or arginine or lysine vasopressin, and oxytocin, each polypeptide comprising eight amino acids, of which six are common to the two.[419, 420, 421]

ANTIDIURETIC HORMONE. The chief action in the mammal is to increase the absorption of water in the distal kidney tubules. Correlations have been found between nerve cell content of ADH and water retention.[449] ADH acts directly upon the tubules, particularly the loop of Henle, the distal convoluted tubule, and the collecting tubule. A second, less important action is upon the vascular system, in which it causes principally constriction of the arteriovenous capillaries. However, the physiological role of this effect has not been demonstrated.

OXYTOCIN. This hormone induces contraction of smooth muscle, especially that of the uterine wall, and to a lesser extent of the

Oxytocin (Pitocin)

glutamine
isoleucine
asparagine
tyrosine
cystine
proline
leucine
glycinamide

urinary bladder, gallbladder, and intestine. Its uterine effect is enhanced by estrogen and counteracted by progesterone. Oxytocin plays a special role in lactation to bring about the "letting down" of the milk and its ejection, through an action on the myoepithelial cells surrounding the mammary alveoli. It also increases the water permeability of frog skin[358] and bladder,[359] although vasotocin is the natural hormone for these effects.

The neurohypophysis is dually controlled through the stalk upon which it depends for inflow of its hormones[67, 315] and for the stimulus for release of its hormones into the blood.[332] The means for differential control of liberation of the two hormones is still not known. The secretion of ADH is continuous, but its rate is regulated by osmotic pressure of the blood, lowered osmotic pressure inhibiting hormone release.[311] The receptor for osmotic pressure is thought to be the cells of the supraoptic nucleus itself. Many other factors also influence ADH secretion. Increased secretion is effected by stress, reduction in blood fluid volume, and drugs such as morphine and nicotine. Reduced secretion is brought about by increased blood fluid volume and by cold exposure.

Oxytocin is secreted at the time of mammary gland activity in response to stimulation of the nipple. Afferent nervous impulses pass to, and activate, the hypothalamic center. Other stimuli believed to initiate oxytocin liberation by afferent nervous activity are uterine distention late in pregnancy and, in some species, also mating. Epinephrine, on the other hand, inhibits secretion.

Among mammals there are two vasopressins which differ by a single amino acid. Lysine vasopressin is found in the pig and

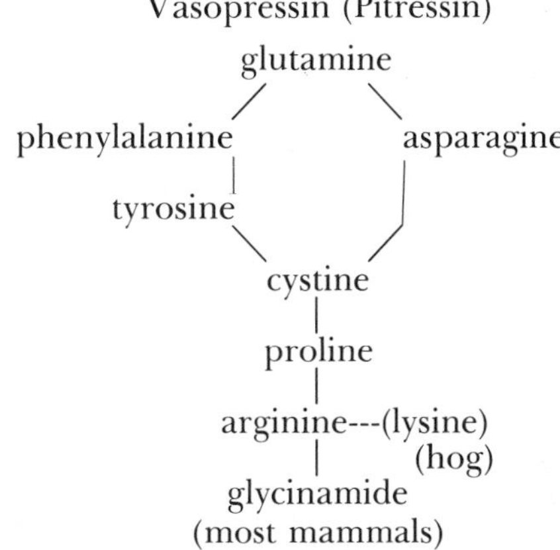

hippopotamus and arginine vasopressin in all other mammals investigated. The actions of the two may differ substantially. On the dog, for example, the arginine form possesses about six times the potency of the lysine form in antidiuretic action.[360, 415, 416]

In birds and poikilothermic vertebrates, extensive differences exist in the nature and actions of the neurohypophyeal peptides. For example, ADH has a *depressor* action on the blood pressure in birds, and oxytocin can bring about the spawning reaction in the killifish, *Fundulus heteroclitus*.[440] Birds, amphibians, and fishes secrete another hormone, arginine vasotocin.

Arginine Vasotocin

glutamine
isoleucine
asparagine
tyrosine
cystine
proline
arginine
glycinamide
(birds, teleosts,
amphibians,
cyclostomes)

In addition, three others are known: mesotocin, found in primitive bony fishes, amphibians, and reptiles; isotocin, found in holostean and teleostean fishes; and glumitocin, found in rays. This brings the total of neurohypophyseal peptides known to at least seven, permitting interesting speculations about their evolution.

ADH appears to be another hormone which acts to stimulate the enzyme adenyl cyclase, resulting in the formation of cyclic AMP. ADH is able to increase the cyclic AMP content of isolated toad bladder, along with its action on water permeability.[185] Cyclic AMP also imitates the action of ADH on toad bladder.[306] It is probable that cyclic AMP is involved in the action of ADH on the kidney, because an ADH-sensitive adenyl cyclase has been found in the renal medulla, where the distal convoluted tubules lie.[80]

Urophysis. The urophysis is a neurosecretory endocrine organ present in the caudal region of the spinal cord of teleost fishes.[150] Neurosecretory cells in the spinal cord (Dahlgren cells) lead by a tract to a neurohemal organ, the urophysis. Although a distinct urophysis is present in teleosts, it is lacking in elasmobranchs, which may represent the primitive condition. The most frequently postulated function has been osmoregulation, and effects on both sodium movements and water retention have been obtained. The teleost urophysis also contains a substance which is active in promoting the contraction of the isolated bladder of the rainbow trout.[251,252] In addition, there are hypertensive and hypotensive principles.

Hormones and Reproduction

The gonadotropins, FSH, LH, and prolactin, occur in virtually all classes of vertebrates.[164] In the absence of gonadotropins, such as normally accompanies hypophysectomy, the ovary shows an arrest of development in young mammals and a reduction in size in adults. In the latter instances existing follicles become atretic. FSH in excess, on the other hand, will stimulate the simultaneous maturation of an excessive number of follicles. In some mammals, such as the rabbit and ferret, ovulation from mature follicles occurs only after mating or after some other effective stimulus. Such stimulation is known to induce the liberation of LH, which causes the ovulation.[127] In other mammals, including the human, no special stimulation of the accessory complex is essential to induce ovulation. In all mammals LH is normally responsible for the formation of corpora lutea in the ruptured follicles.

Estrous cycles cease upon removal of the hypophysis. These may be restored, however, by hypophyseal implants, but only when the implants are made under the median eminence of the hypothalamus.[301] The gonadotropins are not sex specific. They regulate the activity of the testis as well as of the ovary. Removal of the pituitary in the male results in an arrest of testicular function and reduction in size of the testes. Activity of the testes can be maintained in hypophysectomized individuals by pituitary implants or extract injection. Many mammals which breed only at certain sharply delimited periods of the year may have their testes rendered unseasonably active by treatment with gonadotropins. This has been demonstrated for such species as the ground squirrel *Citellus* and the alpine marmot *Marmota.* After the mating season in such species the testes normally recede into the abdomen and spermatogenesis ceases. Administration of pituitary gonadotropin, or even chorionic gonadotropin, found normally only in pregnant females, will reverse these processes at any season. This is a result of its LH-like properties.

An enlargement of the ovaries of the horned lizard *Phrynosoma* has been obtained by FSH and LH from hog pituitary and by serum from pregnant mares.[282] Hypophysectomy in the turtle *Emys* has shown the gonad and secondary sex characters to be dependent on the pituitary.[89]

Female frogs and toads may be induced to ovulate at any time of the year, except immediately after an egg-laying period, by injection of macerated anterior lobe of the pituitary.[351,352,353] Preparations of pituitaries from females are about twice as effective as those from males. The ovulation, which will occur between the second and fourth days, includes rupture of the follicles and expulsion of the eggs by contraction of smooth muscle fibers in the follicle wall. It can be made to occur *in vitro.*[349] Injection of pituitary extract into male frogs induces amplexus. These sex reactions in amphibians may also be induced by pituitary extracts from other amphibians, fishes,[448] or mammals,[458] or may even occur in response to chorionic gonadotropin from pregnancy urine. Reciprocal injection experiments have indicated substantial differences in the active factor, LH, among various vertebrates.[453]

Premature spawning has been induced in

several species of fishes after injection of fresh pituitary glands from other fishes. The ovoviviparous fish *Cnasterodon* was caused to spawn more than 2 weeks before the normal time after treatment with pituitaries from *Micropagon* and *Luciopimelodus.* The Brazilian species, *Pimelodus* and *Prochilodus,* which normally spawn after the heavy rains that in Brazil follow a long period of drought, were induced to produce eggs and sperm during the period of drought within 1 to 3 days after an intramuscular injection of pituitaries from these same species. Rainbow and brown trout, *Salmo gairdneri* and *S. fario,* in Wisconsin, were caused to produce mature eggs and sperm 6 to 7 weeks before the normal spawning season by intraperitoneal injections of fresh or acetone-dried pituitary glands from the carp.[194] FSH from sheep and serum from pregnant mares were without such effect. Increase in size of the ovocytes and ovaries of the lamprey *Petromyzon* and precocious sexual maturity were observed after treatment with human pregnancy urine.[107] Hypophysectomy in the killifish *Fundulus* is followed by regressive changes in the ovaries and testes, compared with controls. When the gland is removed in the autumn the gonads fail to undergo their normal spring enlargement. Furthermore, implantation of adult pituitaries at 3-day intervals into immature *Fundulus* induces within 4 weeks a considerable degree of gonadal activation in both sexes, together with production of secondary sexual pigmentary changes in the males characteristic of the breeding season. Spawning in *Fundulus* appears to be under direct pituitary control.[324]

Androgens are found in all vertebrates. The primary function of androgens is to stimulate the development and activity of the male accessory reproductive organs. They are also responsible in part for differences in the conformation of the male body, the lower pitch of the human male voice, the characteristic distribution of hair, male pugnacious assertiveness, and many special male secondary sex characteristics such as the swollen clasping digits of the male frog, the comb of the cock, the crest of the newt, and the nuptial coloration and gonapodal appendages of certain viviparous fishes. Many of these are characters whose presence is largely restricted to the breeding season.

The various androgens differ from one another in their over-all potency and their relative effects on the numerous individual characters within the body, and these differences in turn may vary from species to species. This complicates greatly the problem of the roles of androgens within the body.

The testis comprises the principal source of androgens. This organ, furthermore, shows the capacity to extract less potent androgens from the circulating blood and convert them to substances of increased potency. Numerous observations have indicated the interstitial glandular cells to be the specific source. Reduction or disappearance of the seminal epithelium following x-radiation or ligation of the vasa efferentia or in cryptorchidism has been reported to cause no atrophy of the interstitial glandular tissue and no loss of androgenic potency. In the newt *Triton cristatus* there is no interstitial glandular tissue in the testes until the approach of the breeding season. At this time certain portions of the seminiferous epithelial tissue become transformed to produce a glandular body, and this transformation is temporally associated with the differentiation of the characteristic seasonal secondary sexual adornments in the male. Destruction of this glandular tissue is the equivalent of total castration in preventing development of these characters. In the stickleback *Gasterosteus* there is very close correlation between the annual cycle of differentiation of interstitial glandular cells and the cycle of secondary sex characters and behavior, such as skin coloration, secretion of nest-building mucus, and mating behavior. The latter characteristics in this species, on the other hand, show no relationship to the amount of spermatogenic activity. That the seminiferous epithelium is not a significant source of androgen has also been demonstrated by removal of the greater part of the testes of cocks and assay of these organs after regeneration. The regenerated organs are principally seminiferous epithelial tissue with a great paucity of interstitial tissue, and correspondingly, there is evidence of only little androgen production by such organs.

These and numerous other experiments suggest that the interstitial glandular cells are the points of origin of testicular androgens.

Androgens are also produced by organs other than the testes. Male accessory reproductive structures have been maintained in full functional activity in castrated specimens by grafted ovarian tissue. Among fowl, the combs of both the cock and the hen are stimulated by androgens. The ovaries of the hen normally produce enough androgen to maintain a certain degree of development of the comb; after ovariectomy there is atrophy of the comb. The seasonal yellowing of the bills of both male and female starlings and blackening of the bills of male and female English sparrows are responses to androgens, estrogens being ineffective in this respect. Black-crowned night herons show, during the breeding season, certain changes in plumage coloration common to both sexes. Many observations such as the foregoing establish decisively the ovaries as a site of androgen production.

Another source of androgens is the adrenal cortex. Implantation of adrenal tissue in young cocks results in precocious development of male sex characters and behavior. Castration of young rats does not result in atrophy of accessory genital structures for many days unless there is simultaneously complete adrenalectomy. Androgens are still excreted in the urine of animals after castration. Extracts of adrenal tissues have been found to yield the androgenic substance, adrenosterone, and the mineralocorticoid, deoxycorticosterone, the latter having an activity about the equivalent of that of androsterone, and the former having activity of about one-fifth of this value.

There is little evidence of any significant storage of androgens in the body. They are usually rapidly destroyed in the liver, and therefore must be constantly produced if their influence is to be maintained. Certain poeciliid fishes can, however, be treated with androgens for only a few hours, but the activity of the androgens continues for many days. The pigmented fat-body of the hibernating woodchuck, the so-called hibernating gland, has a remarkably high androgen content. The androgens are inactivated principally in the liver. The androgenic activity of blood leaving the liver is much less than that of blood entering this organ. Products of the inactivation are both inactive substances and androgens of somewhat lower activity, such as androsterone. Estrogens are also commonly produced in the process. These products of metabolic conversion are excreted by the kidney.

The relation of androgens to the development of the characteristic male suspensor organs, gonopodia, and to the smaller body size in male poeciliid fishes has been carefully analyzed.[410] Castration of males during the metamorphosis of the anal fin into a gonopodium results in immediate cessation of the process. Application of androgenic hormones to castrated males or to normal females can induce full development of the male characters which are typical of the species. It has been shown that progressively larger concentrations of hormones are required for consecutive steps in the development of the gonopodium, with a concentration of one part of hormone in 4.2×10^{10} parts of water being sufficient to induce the first step. The gonopodium normally develops under the influence of a gradually increasing concentration in the blood of androgen from the fish's own testes. A certain optimum concentration is required for normal development of each step, higher concentrations inhibiting growth and inducing precocious differentiation. In toads the clasp reflex is lost following castration.[59]

Estrogen is a term applied to any substance which will produce the characteristics of normal estrus, including cornification in the vagina, in an adult mouse. Estrogens are produced in ovaries of all vertebrates.

The placenta, both the maternal and the embryonic portion, contains abundant estrogen of much the same character as ovarian estrogen. Furthermore, oophorectomy in certain pregnant mammals does not result in significant change in the term of pregnancy or in any permanent reduction in the amount of estrogen excreted in the urine. In such cases the pubic symphyses also become separated normally, a change ordinarily conditioned by joint action of estrogen and progesterone.

Estrogens are also produced by the testes of males. In fact, the tissue with the highest known estrogen yield is the stallion testis, in which estrone is the substance present. The testicular source appears to be, as for the androgens, the interstitial glandular cells. Estrogens have also been extracted from the adrenals. Androgens are necessary precursors for estrogen biosynthesis.

Estrogens are very rapidly inactivated in the body by formation of glucuronides or sulfate esters. Studies on the relation between site of transplantation of ovarian tissue and the physiological actions of the implants indicate the liver to be the principal site of destruction. Estrogen-containing blood passing through the liver reappears largely free of active concentrations of the hormone. Destruction of hepatic cells by carbon tetrachloride permits the hormone titer in the blood to rise with expected effects on estrus of the animal. Liver slices inactivate estradiol *in vitro* through the action of some CN-sensitive mechanism. Progesterone, on the other hand, diminishes the rate of estrogen inactivation and thus permits a higher titer to be maintained in the blood, from which much is then able to escape into the urine. The kidneys and liver are the principal organs of estrogen excretion.

The organ and tissue changes induced by estrogens are characterized by large variations in degree of responsiveness of the different cells and tissues to the hormone, thereby giving rise to gradients of response. Also characterizing the responses is a very high degree of reversibility of the changes when the concentration of the hormone declines. Such reversibility of the changes is most evident in many of the secondary and accessory sex character changes paralleling the reproductive cycles in many species.

As with the androgens, the specific threshold and character of the responses are determined by the inherent cellular characters, as can be seen in the unchanged character of the response as the individual tissues are transplanted to novel sites within the body. The relation of the response to the genetic constitution of the species is seen in reciprocal transplantations of feather-bearing skin between the sexes in fowl, resulting in the characteristic feather types for the host sex.

A large amount of information has become available bearing on the mechanism of action of estrogens on the rat uterus. Meuller[290] pointed out the sequence of events that occurs after the administration of estrogen. During the first 4 hours, there is a hyperemia and imbibition of water. In the period from 6 to 24 hours there is an accumulation of RNA followed by protein. This is followed by DNA synthesis 24 to 72 hours after estrogenization. Most recently, work has centered on the elucidation of the mechanism by which estrogen enhances RNA synthesis, which seems to be a common factor in the action of many steroid hormones.[384] Although the possibility exists that cyclic AMP *may* play a role in estrogen action,[399] most explanations have centered on the binding to cytoplasmic estrogen receptors, followed by movement of the estrogen into the nucleus. There is evidence for the concept that estradiol-17β binds to a cytoplasmic protein in the uterus sedimenting at 9.5S (Svedberg units). It is then transported into the nucleus bound to this 9.5S protein, where the protein dissociates into 5S subunits with the estradiol bound to them. Presumably, the estradiol-protein complex then triggers the messenger RNA synthesis involved in the hormonal effects.[175] Steroid hormones have also been localized by autoradiography of the tritiated derivatives.[395]

Progesterone. This substance is concerned primarily with those changes in the mammal associated with pregnancy and parturition. It is also an obligatory intermediate in steroidogenesis in, for example, the adrenal cortex. The chief sources of progesterone are the corpora lutea of the ovary.

Progesterone is also produced in the adrenals. Not only has progesterone been demonstrated in this organ, but deoxycorticosterone also in part simulates the action of progesterone. Extracts of adrenals can produce typical progestational alterations in the uterus in rabbits if, as a preliminary, they are treated with estrogen.

The placenta is also an important natural source of progesterone. In the mammal the placenta is believed normally to take over, after a time, the major share of production of the normal progesterone of pregnancy. Even after the total removal of the ovaries of pregnant rats, sufficient progesterone is liberated from the placenta to carry the animals to normal terms. In humans, this changeover of control from corpora lutea to placenta is considered to occur between the seventy-ninth and the ninetieth day of pregnancy.

Progesterone is not stored in the body; it disappears very rapidly from the blood. Injection of progesterone, or the normal production of this hormone, is followed by the appearance of corresponding amounts of a relatively inert alcohol, pregnanediol, in the urine. The preponderance of evidence points to the progestational endometrium of the uterus as the principal organ concerned in this conversion, which is principally in evidence in the progestational phase of the sexual cycle and usually ceases on hysterectomy. Furthermore, it is greatly reduced by injections of estradiol, which is known to suppress the progestational condition. That the progestational endometrium is not the sole site of the conversion is seen in the excretion of pregnanediol in males, and occasionally in females after hysterectomy. In addition, the corpuscles of Stannius, in the kidney of the rainbow trout (*Salmo gairdneri*) are able to convert progesterone to 11-deoxycorticosterone[88] and thus may be important for steroid metabolism.

The corpora lutea which are formed from the follicles, either without ovulation, as in the case of atretic follicles, or after ovulation, vary considerably in the extent of their development. Their greatest development accompanies gestation; lesser development is associated with lactation and pseudopregnancy, and the typical corpora lutea of ovulation are still smaller. The corpora lutea are dependent for their development on a supply of LH from the pituitary. Mammals appear to vary considerably in their need for external stimuli to encourage the necessary production of LH. An adequate stimulus has been shown to be copulation in the case of the rabbit. Suckling and lactation have been shown to stimulate its production in the rat, and psychic stimuli, such as the mere presence of a second individual, have been found effective in such animals as the pigeon and the rabbit. That the last stimulus is visual, at least for the pigeon, is demonstrated by the fact that a mirror image is often sufficient to produce the effect.

While all investigators are agreed that LH is essential for the initial development of the corpora lutea, and LTH or LH to activate their secretion, there is much controversy as to other endocrine factors participating in their maintenance. Gonadal hormones assist in their maintenance, probably through a suppression of FSH liberation by the pituitary. Comparison of experiments in which the embryos have been removed, leaving the placentae *in situ,* with those in which the placentae as well are excised, clearly demonstrates that the latter organs yield hormonal material which can operate to maintain the corpora lutea. These results also appear capable of interpretation in terms of suppression of FSH production by placentally derived gonadal hormones. That still other factors may contribute to the total explanation is seen in the observation that the corpora lutea may be maintained even after complete removal of the fetuses and placentae, provided the uteri are kept distended by such inert bodies as pellets of paraffin wax.

Corpora lutea have been described even for the poikilothermic vertebrates, but their highest development appears to be associated with viviparity in the mammal. They are said to be absent in birds and oviparous reptiles, but present in certain pregnant viviparous snakes,[7,49] as *Crotalus* and *Bothrops,* the viviparous lizard *Zootoca,*[312] and the viviparous anuran, *Nectophrynoides.*[245–248] Outside of the mammal there is no clear evidence that they are essential to the maintenance of pregnancy.

It has been postulated that corpora lutea arose evolutionally as a clean-up process following ovulation, with progesterone being one simple end-product.[202] In the primitive mammals, the progesterone came to function in holding eggs to time of laying. In fact, in marsupials the period of gestation is only the length of the luteal phase of the estrous cycle. In rats and mice, where gestation is still longer, a chorionic LTH arose to maintain the corpora lutea. In horses and primates, where gestation is still longer, the luteal phase was first extended by chorionic LTH and later a placental autonomy developed, with the placenta providing not only progesterone but also many additional essential factors, such as estrogens and prolactins.

Relaxin. Relaxin, a polypeptide having a molecular weight of about 9000, is found within the corpora lutea and can be isolated from the blood of pregnant rabbits. It is capable of markedly relaxing the pelvic girdle of a guinea pig in 6 to 12 hours after a single injection. Progesterone and estradiol induce the secretion of relaxin in female rabbits unless they are castrated and hysterectomized. The latter observations indicate the ovaries and uterus to be essential to this response to the gonadal hormones. Relaxin may also increase the dilation of the cervix in pregnant women very near or at term.

Prolactin (Luteotrophin). The production of milk is an adaptation of all mammals, and such birds as the pigeon, for postpartum nutrition of their young. Hence it is not surprising to find that there is a considerable influence of gonadal hormones on milk-producing glands of the organism. The mammary glands of the mammalian fetus become enlarged under the influence of the hormonal complex of the maternal blood supply, but regress on parturition and typically remain so until sexual maturity, when the animal's own hormonal supply becomes adjusted for their development. Prolactin plays the important role of stimulating the formation of milk by the mammary acini.

Control of Reproductive Cycles and Behavior. The pituitary gland, through the production of gonadotropins, is the principal endocrine organ through which reproductive activities are governed. We, therefore, with reason, look to a study of factors controlling this organ to supply us with fundamental information as to the mechanism of control of reproductive rhythms and adaptive reproductive responses of the organism to its external environment. The releasing hormones are of great importance in this regard.

The pituitaries of young mammals are bipotential organs which normally develop into either a male or female type, under the influence of the particular gonad present. The gland is typically a larger organ in the adult female than in the male. Both female and male pituitaries normally produce FSH and LH in relatively large amounts. The primary difference between male and female pituitaries is one of cycle differences. Ovarian implants, but not testis implants, suppress FSH production by the adult pituitary in castrated females; in castrated males only testis implants will suppress FSH production. Thus, the adult pituitary has developed into differently responding male and female types, in contrast with its sexually indifferent condition in newborn mammals.

During the growth of a young mammal there is a gradual increase in the production of gonadotropins up to the time of sexual maturity, when the gonads are consequently brought to their complete functional state and their estrogens and androgens contribute to the full differentiation of the secondary and accessory sexual characteristics. The animal is now capable of normal reproductive activity. This is, in the vertebrates, typically a cyclical phenomenon, with the periodicity often correlated with the annual solar cycle in such an adaptive fashion as to assure the young of a favorable time of year for birth and early postnatal development. In some species the reproductive periodicity bears much less or no relationship to external environmental stimuli and appears to result from a wholly inherent rhythmical mechanism, as in the cases of the rat, mouse, and man.

A vast amount of research has been done on the contributions of the interactions of endocrine sources to the maintenance of the normal cycles of mammals. The endocrine actions and interactions of the major organs participating in the normal reproductive cycles are indicated for the female mammal in Figure 21–3, and for the male in Figure 21–4.

In the female the pituitary, through FSH production, stimulates the development of the ovarian follicles and interstitial tissues, with a consequent increase in estrogen production. The rising concentration of estrogen in the blood mimics FSH and furthers the ovarian development. At the same time, the estrogen gradually suppresses FSH production by the pituitary while at first strongly encouraging LH liberation by the pituitary. Over a longer period, estrogen suppresses LH liberation. LH liberated from the pituitary stimulates ovulation in many mammals, and in all it stimulates the differentiation of corpora lutea with the subsequent production by them of progesterone. Although LH induces differentiation of the corpora lutea, an additional pituitary principle, luteotrophin, is needed for stimulating the actual secretion of progesterone *in some species.* Luteotrophin is believed to be identical with prolactin. Progesterone cooperates with estrogen in the further suppression of the production of FSH by the pituitary. In case of pregnancy, the placentae of the developing embryos are activated in the presence of progesterone and proceed to liberate substantial amounts of chorionic gonadotropin, the activity of which very closely resembles that of pituitary LH. This gonadotropin contributes to the suppression of production of both pituitary gonadotropins through its stimulation of greater gonadal hormone production.

The corresponding endocrine interactions participating in the male reproductive cycle are simpler. Gonadotropin from the pituitary activates the interstitial and spermatogenic tissues of the testes, with the resultant increase in testosterone production in the former. This androgen simultaneously stimulates further development of the spermatogenic epithelium and facilitates the influences of both FSH and LH. As the blood titer of androgen gradually rises, this hormone suppresses pituitary gonadotropin production. The decline of androgen production after pituitary suppression eventually leaves the latter free to initiate a new cycle of reproductive activity.

In a few species there tends to be an annual reproductive period at a particular season of the year, without any regard to any particular factor of the external environment. Such an operation of an inherent rhythm has been clearly demonstrated for certain organisms which continue to breed indefinitely during the same calendar months after transfer from the northern to the southern hemisphere, or vice versa. Other species on similar transfer more or less rapidly undergo a readjustment of their breeding period to the corresponding season for their new locality.[275] The readiness with which the organism makes such an adjustment appears to be a function of the relative degree to which the breeding cycle depends on external factors.

Of all the external stimuli known to exert an influence on reproductive cycles, one of the most effective so far demonstrated is light and photoperiodism.[21, 128, 129, 350, 455] Among certain north temperate zone birds it has been shown that the testis of the male normally reaches a peak of its annual cyclical activity in the late spring. Experiments on male juncos, sparrows, and starlings have clearly shown that gradual increase in length of the daily period of illumination through supplementary artificial lighting will bring the activity of the testis to a maximum at a time not typical for the species. Properly controlled experiments have shown that it is the light itself rather than the resulting longer daily periods of activity or feeding which is the actual determining factor. Immediately following one period of breeding activity, the bird is refractory to response to photoperiod again.[285] A similar subjection to increased periods of illumination over a period of about 2 months has been found to bring female ferrets into estrus in winter, a time at which in their normal reproduction cycle they are in anestrus. Similar observations have been made on mice (*Peromyscus*). Such reptiles as *Anolis, Xantusia,* and *Lacerta* respond to photoperiod.[23] Among the fishes, the reproductive cycle of trout, sunfishes, and several species of minnows has also been shown to be importantly influenced by photoperiodism.

In the field mouse, *Microtus,* the gonads have been shown to diminish in size during treatment of the animal with gradually decreasing periods of illumination.

Such experiments as these appear to indicate that the gradual changes in day length in the annual solar cycle constitute a very effective stimulus in determining the normal annual reproductive cycles of many species of vertebrates.[454]

In the immature duck whose gonads are normally stimulated by increased illumination, it has been shown that pituitaries from illuminated specimens show increased gonadotropic activity over those of untreated controls when implanted into immature mice. Red and orange lights are more effective than other colors in inducing the effect. The eyes are not essential to the reaction; direct illumination of the pituitary resulting from illumination of the orbit after enucleation effectively activates the gland. When light is conducted directly to the pituitary through a quartz rod, blue light, which otherwise shows only small influence, becomes quite as effective as red and orange. Reproductive cycles continue in the duck in either constant light or dark, but the period ceases to be annual.[29]

A similar effect of light on the reproductive cycle through other routes than the eyes has been shown for sparrows. In the ferret, on the other hand, division of the optic nerves completely inhibits the action of light, indicating the retina of the eyes to be the effective receptor in this species.

Nervous connection of the hypothalamus with other parts of the nervous system is usually essential to normal sexual functioning. Nervous connection between the hypothalamus and the pituitary is essential for certain aspects of control of the reproductive cycle,[274] as for example ovulation in the rabbit, but appears to be nonessential in numerous other instances, the conduction in these latter cases apparently being humoral. Secretory granules accumulate in nerve loops of the median eminence in white-crowned sparrows; held on a 8-hour light regimen they disappear, and the testes recrudesce when the light periods are increased to 20 hours.[305]

The temperature of the external environment has been shown to play an important role in determining the time of gonadal development and appearance of secondary sex characters in the stickleback (*Gasterosteus*), a sufficient rise in temperature rapidly inducing these changes. Low temperature and darkness such as normally obtain during its annual period of hibernation will activate the gonads in the ground squirrel *Citellus* at any season. Both light and temperature influence the reptilian sexual cycle.[22]

Factors other than light and temperature are often important. Periods of rainfall with the correlated increase in green vegetation serve as the stimulus to reproduction in the equatorial weaver finch *Quelea.*[106]

The Pineal Gland and Reproduction. The pineal gland is known to secrete melatonin (N-acetyl-5-methoxytryptamine), which has a direct melanin-aggregating action on amphibian melanophores.[302, 304] However, the pineal has also been implicated in the regulation of reproduction. There is a correlation between precocious puberty and the presence of non-parenchymal pineal tumors in men.[226] Such tumors destroy the secretory parenchymal tissue, and the idea was thus developed that the pineal may be exerting an inhibitory action on gonadal development. Pineal extracts and melatonin do have significant antigonadal actions in some species.[343] Furthermore, light deprivation enhances the antigonadal capacity of the pineal, presumably as a result of an enhancement of the activity of the enzyme responsible for melatonin synthesis, hydroxyindole-O-methyl transferase.[459] In addition, pineal removal accelerates growth in immature animals and can bring about a short-term enlargement of the reproductive organs in adults. There is some question of the direct nature of these effects, because melatonin has been reported to be able to affect the pituitary secretion of tropic hormones. In addition, it is still not possible to state that the pineal is *the* mediator for the light-controlled regulation of reproduction in any one species.

Hormones and Gastrointestinal Coordination[293]

In the mammal the integration of the activities of various parts of the digestive system in dealing with ingested food is in part endocrine in character. The major hormones and their general sites of formation and regions of actions are shown diagrammatically in Figure 21–5.

The gastric pyloric mucosa, on appropriate stimulation, secretes into the blood a hormone, gastrin.[178] Noninnervated, transplanted gastric pouches are caused to secrete in response to the presence of food in the stomach of the animal. For a long time gastrin was believed to be identical with histamine, but it is now known that histamine-free extracts of the pyloric mucosa will stimulate secretion of a highly acid gastric juice with very little peptic activity and simultaneously stimulate pancreatic secretion. However, there is some reason to believe that histamine also plays here the role of a hormone.[368] By fractional precipitation, the gastric stimulant, gastrin, has been separated from the activator of the pancreas; the latter resembles very closely the secretin obtained from the duodenal mucosa. Two pure poly-

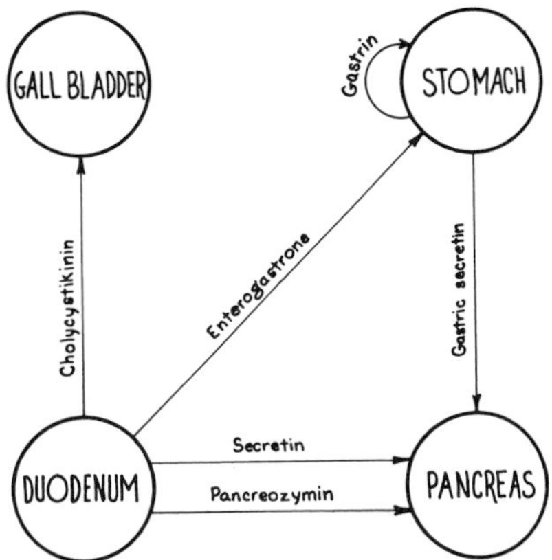

Figure 21–5. Diagrammatic representation of the sources and points of action of the principal hormones operating in gastrointestinal coordination.

peptides, gastrins I and II, have been isolated from porcine gastric mucosa and synthesized. In addition to stimulating the stomach, they also have the capacity to stimulate the pancreas.

The duodenal mucosa quantitatively stimulates the pancreas to liberate its digestive juice through the action of a 27-amino-acid polypeptide hormone, secretin.[177] It was formerly thought that secretin stimulated mainly the secretion of water and bicarbonate by the pancreas and that the production of enzymes was nervously controlled, but no consistent data were obtained to support this hypothesis. The variable concentrations of enzyme in pancreatic juice that have been reported to be produced in response to secretin injections appear to find their explanation in terms of the presence of a second hormone in the crude extract. One fraction stimulates the secretion chiefly of water and inorganic salts from a denervated pancreas. A second fraction stimulates pancreatic enzyme secretion. The active factor of this latter fraction has been termed pancreozymin.

Another active principle from the duodenum, and one which has been separated from crude secretin extracts, is cholecystokinin, a powerful agent in producing gallbladder contraction. This is the same as pancreozymin.

A fourth factor, enterogastrone, has been isolated from the duodenal mucosa. This factor inhibits gastric secretion of HCl and gastric motility. The activity of enterogastrone appears to be principally antisecretory, and since in the presence of enterogastrone the gastric juice produced under the influence of gastric stimulants such as histamine is low in acid content and rich in pepsin it would appear that enterogastrone preferentially inhibits the acid-secreting parietal cells of the stomach.

Other gastrointestinal chemical coordinators have been proposed on the basis of brief experiments, but these appear to have a more questionable existence than the aforementioned ones.

Coloration and Seasonal Color Changes of Birds and Mammals

Among the numerous species of birds which show differentiation of hen and cock plumages a wide variety of mechanisms is involved.[99, 108] In the English sparrow the control of plumage type is exclusively genetic. Among pheasants, the plumage type is determined by simultaneous action of genes and hormones. In perhaps the majority of birds, however, the plumage typical of one sex is neutral; that of the other sex is determined by blood-borne hormones. In the common domestic fowl the neutral type appears to be the cock plumage, with the hen type the result of action of estrogens. In African weaver finches, the neutral type is the hen plumage; the cock plumage is the result of the indirect action of hypophyseal hormones.[102, 339, 376, 452] Only in the herring gull, among the birds thus far studied, does the cock type of plumage depend on action of androgens.[45]

In the majority of common birds, the adults undergo a rather complete postnuptial molt immediately after the breeding period; the new regenerating plumage becomes the winter or eclipse plumage. A second, usually much less extensive molt, the prenuptial, occurs in the spring. The regenerating plumage constitutes the breeding or nuptial plumage. This latter molt is most extensive in those species which exhibit conspicuous breeding coloration. The times at which birds assume the nuptial and winter plumages are governed by a number of factors. In some species, such as the African weaver finch *Pyromelanox* and the starling, the changes appear to be due to an inherent cyclically rhythmic hypophyseal activity,[450] although the occurrence of the rhythm may depend in part on length of daily light periods.[56] Some other species, such as mallard ducks, the white-throated sparrow, and the bobwhite, can be induced to molt and undergo a plumage-type change at a time other than their regular season by subjection to artificially increasing or decreasing light periods. Therefore, the annual plumage changes of most birds, like migrating and breeding activity, appear to be controlled in good measure by the annual cycle of day lengths.

Certain northern birds and mammals show a seasonal color change from brown in summer to white in winter. The times of these changes in such forms as the ptarmigan *Lagopus*,[207] the ermine *Mustela*,[35] and the varying hare *Lepus*[271] have been shown to be determined by the seasonal changes in day lengths; the animals could be caused to whiten out of season by appropriate experimental alteration of the daily lighting. Experiments involving masking of the varying hare[271] indicate that the eyes are the normal receptors. Both in the natural cycle of color change and in color changes induced artificially by modification of illumination, the varying hare is physiologically brown when large amounts of gonadotropic hormones are present in the blood and is physiologically white when these hormones are low in concentration. Molting in a physiologically brown animal is followed by production of brown hair; in a physiologically white one, by production of white hair. Extracts of whole pituitary containing gonadotropic hormones will convert physiologically white animals into physiologically brown ones and simultaneously induced shedding of the white hair. Hypophysectomy in ferrets abolishes the cyclic molting.[34] No endocrine gland other than the pituitary appears to be involved in these color changes. Castrated and thyroidectomized hares undergo the normal seasonal color changes.[271]

Additional proof of the role of the pituitary in seasonal color changes is the finding that hypophysectomy of the short-tailed weasel *(Mustela erminea)* results in the regrowth of white hair after molting, rather than the expected brown hair.[354] Injection of MSH or ACTH into hypophysectomized weasels caused the regrowth of brown hair.[355] Pituitary transplants favored molting and the growth of brown hair in hypophysectomized weasels, perhaps because of the MSH secreted by the transplants.

Prostaglandins[206]

These are a group of hormone-like substances, first discovered because of their ability to lower arterial blood pressure and stimulate the contraction of intestinal and uterine smooth muscle. They have been chemically identified in at least semen, lung, iris, brain, thymus, and renal medulla, and are probably present in other tissues. They are all derivatives of prostanoic acid and are classified in four series, E, F, A, and B. Since they are biosynthetic derivatives of an essential fatty acid, the symptoms of essential fatty acid deficiency may be related to their absence, but this is speculative. Among the most significant of their effects clinically is their potent abortifacient activity. In addition, they have been implicated with cyclic AMP in their mode of action. In most cases (e.g., adrenal cortex, thyroid, and islets) they tend to increase the cyclic AMP content of the tissue. However, in some (e.g., adipose tissue) they inhibit the increase in cyclic AMP that normally takes place after hormonal stimulation.

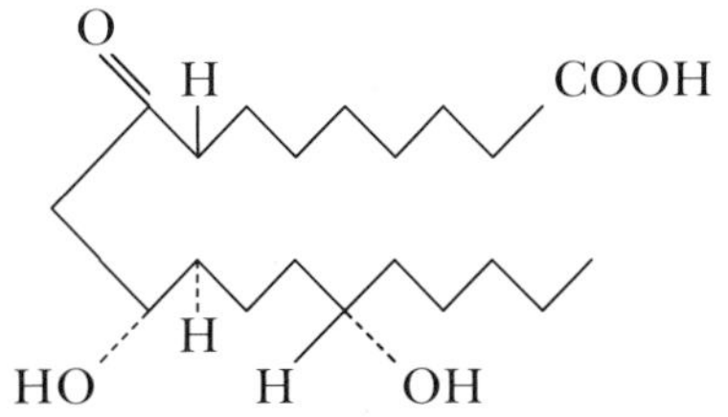

Prostanoic acid

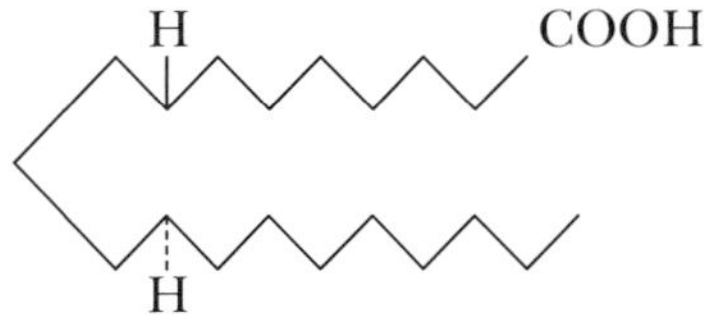

H COOH

H

Prostaglandin E_1 (PGE_1)

O H COOH

H

HO H OH

INSECTS

The insects possess an endocrine system organized, in its functional plan, remarkably like that of the vertebrates. Dominating the system are the central nervous organs, themselves in part neuroendocrine. Clusters of neurosecretory cells in the brain, comprising the pars intercerebralis, secrete hormonal material liberated into the hemolymph by way of paired organs posterior to the brain known as the corpora cardiaca. Such a secretory complex is highly comparable to the hypothalamic-neurohypophyseal system. This complex regulates the activity of an endocrine gland (sometimes paired) of the anterior thoracic region of the body, the prothoracic or ecdysial gland. Closely associated with the corpora cardiaca in most insects are other glands, the corpora allata, which also have their activity partially regulated by the brain. Other parts of the central nervous system, particularly the subesophageal ganglia, are also sites of neuroendocrine cells. Hence, again as in the vertebrates, the neural and endocrine systems are very intimately interrelated in a unitary coordinatory system.

Molt, Pupation, and Metamorphosis in Insects. The postembryonic development of insects comprises an orderly series of stages, in the course of which the insect becomes transformed from a larva to an adult or imago. The process involves growth by means of a series of molts and a metamorphosis, in which the larval characters are lost and imaginal ones are differentiated. In the hemimetabolous insects, such as Orthoptera (e.g., grasshoppers), Dictyoptera (e.g., cockroaches, walking sticks), Hemiptera (e.g., bedbugs and *Rhodnius*), and Odonata, the transition from larval to adult condition is characteristically a gradual one in which the developing young, typically known as nymphs, in succeeding molts come gradually to resemble more and more nearly the adult. In such hemimetabolous insects, however, the last nymphal molt is usually associated with by far the most striking transformation and hence is the one which is commonly considered the metamorphic molt.

In holometabolous insects, such as the Lepidoptera (e.g., moths and butterflies), Diptera (e.g., *Drosophila* and blowflies), and Coleoptera (beetles), very little change of form in the direction of the adult occurs during the larval series of molts, these being primarily growth changes. These rapidly growing larvae are usually given such terms as maggots and caterpillars. The last larval molt is associated with the formation of a pupa. The pupal stage, whose duration often depends on external factors such as temperature and light,[254, 316] often includes an extended period of quiescence or diapause. Within the pupal exoskeleton the organism becomes completely transformed into an adult, a spectacular metamorphosis. There is no fundamental difference between the two foregoing types of development, the observed dissimilarities being primarily ones in the times at which the processes of imaginal differentiation occur. The latter are extended to a considerable extent over the whole of the larval period in the Hemimetabola, and are largely confined to the pupal (pharate adult) stage in the Holometabola.

The number of larval or nymphal stages in development differs from species to species, but is usually a constant for each species. The bug *Rhodnius,* for example, has five nymphal stages or instars, the orthopteran *Carausius* has six, and the dipteran *Drosophila* has three. It is now well established that the sequence of events in the growth and differentiation in the postembryonic insect is in good measure under the integrating action of hormones derived from the animal's endocrine system (Fig. 21–6).[42, 373, 438]

A common plan of the endocrine mechanism regulating growth and differentiation appears to extend throughout the insects.* Its essence is as follows: The molt-initiating stimulus, which is usually environmental and varies from species to species, induces secretion of a brain hormone which is liberated into the hemolymph from the corpora cardiaca after being transported within nerve axons to that point.[10, 200, 365, 366, 408] Recent studies suggest that the brain hormone is a polypeptide or a small protein,[161, 214, 462] and this would agree with what we know of vertebrate neurosecretory substances. The brain hormone activates the prothoracic glands, which are believed to produce the molting hormone. The molting hormone (α-ecdysone) is a unique water-soluble steroid having the following structure:[211]

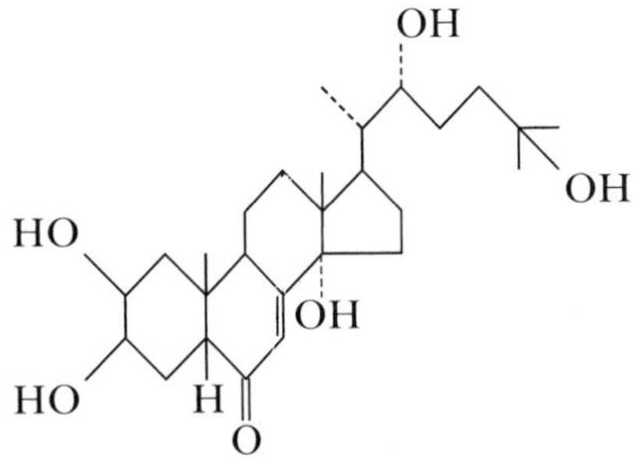

In the original purification of the molting hormone, a second component was noted to have hormone activity, and it was denoted β-ecdysone, the principal component having been termed α-ecdysone.[62] β-Ecdysone is 20-hydroxyecdysone, and has also been named ecdysterone, crustecdysone, and iso-inokosterone.[167] Recent studies show that α-ecdysone is rapidly converted to β-ecdysone by arthropods, and β-ecdysone may in fact be the true insect molting hormone.[174] It is of interest to note that insects cannot synthesize sterols from simple precursors as can mammals; they demand an exogenous supply of sterol in their diet, such as cholesterol for omnivorous insects or β-sitosterol for phytophagous insects.[168] Thus, dietary sterols are the probable precursors for the insect molting hormone. In this regard, one of the most intriguing findings in recent years is that more than 20 compounds (including α- and β-ecdysone) with insect molting hormone activity have been extracted from plants. All are sterols with ecdysone-like structures. Some of these compounds are in much greater concentration in plant tissues than in insect tissues, and have even greater hormonal activity than the natural α- or β-ecdysone. The role of these sterols in the physiology of plants is not known, although it has been suggested that they may have evolved in order to deter insects from eating the plants (ferns and yews have particularly large concentrations of these "phyto-ecdysones" and are relatively resistant to insect attack).

Ecdysone stimulates the molting processes, effecting immediately such transformations in the epidermis as increases in protein and ribonucleic acid, and increases in mitochondria and the endoplasmic reticulum,[378] irrespective of whether the molt leads to another larval stage, a pupal stage, or the imago. Studies with the giant polytene chromosomes of certain flies suggest that ecdysone may elicit the synthesis of mRNA at specific gene loci,[87, 261] and that this is the initial event in the molting process. Recent experiments indicate that the effect is indirect via changes in the permeability of the nuclear envelope that causes changes in the ionic milieu of the chromosomes.[243, 262] In addition, ecdysone appears to stimulate adenyl cyclase activity in epidermal tissues, suggesting a role for cyclic AMP in molting hormone action.[9]

For larval or nymphal molts a second hormone, whose secretion is also partially regulated through the brain,[119, 122, 216, 270, 365] is liberated. This one, termed the juvenile hormone, arises from the corpora allata and suppresses imaginal differentiation by promoting the synthesis of larval structures. It is secreted in smaller amounts at the pupal molt and is absent at the pupal-adult molt. The juvenile hormone identified from extracts of

*See references 85, 212, 218, 330, 337, 373, 437, 438, 445, and 446.

the adult male *Hyalophora cecropia* moth has a terpene-derived configuration:[348]

$COOCH_3$

O

This molecule duplicates all of the morphogenetic effects of implanted corpora allata, including induction of egg development (see p. 888). The structure of the *H. cecropia* juvenile hormone has recently been confirmed, and a lower homolog of that structure has been identified that also possesses juvenile hormone activity.[283] Since the application of juvenile hormone to an immature insect prevents adult development, this material may have potential in insect control. More than 500 analogs with juvenile hormone activity in various insect species have been prepared.[383] It is suspected that the juvenile hormone may act at the molecular level to antagonize the effects of ecdysone.[90, 261, 318]

An additional hormone participating in insect development is a diapause hormone secreted by the subesophageal ganglion.[154, 192] The presence or absence of this hormone in the female moth determines whether the embryos produced will enter diapause. Active extracts of this hormone have been prepared.[192] It has been demonstrated that there is a hormone in the hemolymph of newly emerged flies which is necessary for tanning of the cuticle of the adult fly; it has been termed bursicon.[149, 422] The chemical structure of this hormone is not known with certainty, but it is probably released from neurosecretory cells in the central nervous system.

Underlying the action of endocrine-regulating factors in insect growth and differentiation is a gradually changing pattern of responsiveness of the target tissues to the hormones. These changing competencies contribute quite as importantly as do the hormones themselves to the orderly character of insect differentiation.[41–44, 389]

Experiments on the tropical bug *Rhodnius prolixus* were the first to provide a clear demonstration of the action of brain hormone and juvenile hormone.[434–436] This insect, at each nymphal instar, molts a definite number of days after a meal of blood. The distension of the abdomen serves as the adequate stimulus. Following decapitation these insects

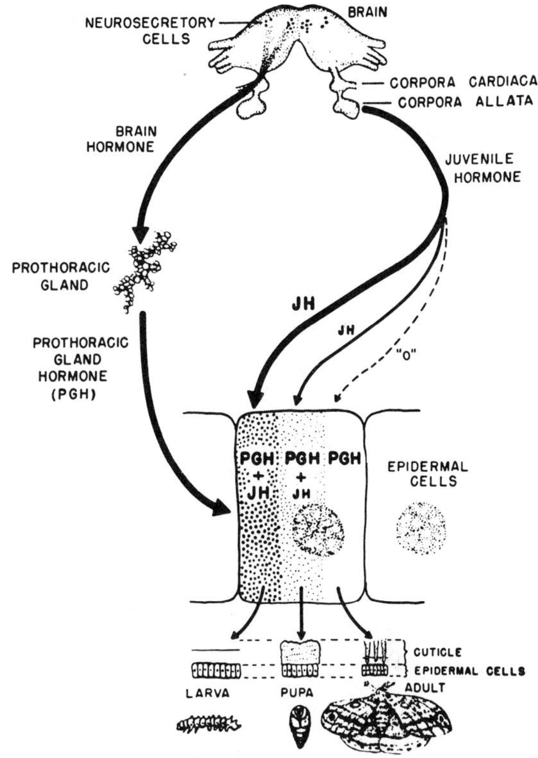

Figure 21–6. Diagrammatic representation of the insect endocrine system involved in growth and differentiation. (From Schneiderman, H. A., and Gilbert, L. I., *in* Cell, Organism and Milieu. Ronald Press Co.)

will survive for 6 to 10 months. It was shown by decapitation at different intervals after feeding that there was a "critical period" occurring a few days after the feeding. Nymphs decapitated before this period never underwent molt; those decapitated afterward did. That the critical period constituted a time when a blood-borne molt-initiating hormone was being liberated from an organ in the head was proved by experimental telobiosis (Fig. 21–7*A*) of two decapitated individuals. Under such conditions, for example, molting in a fourth stage nymph decapitated before the critical period was induced by a fourth stage nymph decapitated after the critical period. Both thereby became fifth stage nymphs.

The fifth nymphal stage, upon molt, typically metamorphosed at that time into an adult. When a fifth stage nymph was decapitated after its critical period it was shown by telobiotic union with a first or second stage nymph, decapitated before the critical period, that the fifth stage nymph possessed

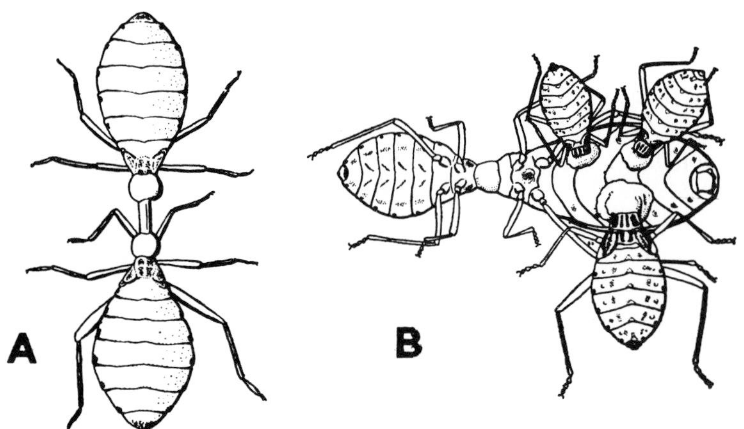

Figure 21–7. Experimental telobiosis and parabiosis in the bug, *Rhodnius*. *A*, telobiosis involving only nymphal instars. *B*, fourth and fifth instar nymphs united telobiotically and parabiotically with an imago. (From Wigglesworth, V. B., Quart. J. Micr. Sci., vol. 77.)

within its hemolymph a hormonal factor which elicited the molt. The attached first or second stage nymph became in this instance a diminutive adult since it had not yet secreted juvenile hormone. When a fourth stage nymph which would normally molt to become a fifth stage nymph was decapitated after the critical period and united with a fifth stage nymph decapitated before the critical period, the latter molted again and became a giant, supernumerary sixth stage nymph instead of an imago. Furthermore, although the adult normally does not molt again, it was made to do so by telobiotic and parabiotic union to two or three fifth stage nymphs decapitated after the critical period. When fourth stage instead of fifth stage nymphs were used, the adult at molt showed a partial return to the nymphal condition (Fig. 21–7*B*).

From such experiments it was readily seen that the hemolymph of a nymph after the critical period contains factors that determine molting either without, or with, an accompanying extensive metamorphosis. A careful study of the results of decapitation of third and fourth instar nymphs at different times during the critical period itself showed that, of those molting after decapitation, those decapitated early in the period showed a significant tendency toward premature metamorphosis, while those decapitated later did not do so. This observation indicated that during the critical period of the first four instars a molting hormone was first liberated and that quickly thereafter a metamorphosis-inhibiting principle (juvenile hormone) was also secreted. The fifth instar alone in *Rhodnius* fails to produce the latter principle.

The source of the metamorphosis-inhibiting or juvenile hormone was found to be the corpus allatum, a median unpaired gland in *Rhodnius* located in the posterior region of the head. The source of the hormone responsible for initiating molting and differentiation is the neurosecretory elements of the pars intercerebralis of the dorsal midregion of the brain. The elongated form of the head of *Rhodnius* rendered it very simple to cut the head transversely in such a manner as to retain the corpus allatum while the brain was removed. Utilizing this technique in conjunction with telobiotic experiments, as well as by implanting corpora allata and pars intercerebralis regions of brains into nymphs decapitated before the critical period, one could obtain at will either a nymphal molt or a metamorphic one in *Rhodnius*.

Studies on lepidopterans first demonstrated the role of the prothoracic glands and clarified the relationship of these to the brain hormone. From the time of the original experiments of Kopec[239] on lepidopteran pupation and metamorphosis it was known that at a certain "critical period" prior to pupation a hormone was liberated from the anterior portion of the larva and that this was essential for pupation and metamorphosis. In the silkworm moth *Bombyx* this hormone (or another factor requisite for hormone synthesis at another site) is produced in the prothoracic glands located in the prothoracic segment.[153] If the prothoracic segment was cut off by ligation from the more posterior regions of the body immediately after pupation, the posterior portion failed to develop further. If the constriction was made 12 to 18 hours after pupation, the posterior portion metamorphosed normally. If a prothoracic gland was implanted into the posterior portion of a pupa ligated before the 12- to 18-hour "critical period," metamorphosis of this part also proceeded normally; or an abdomen cut off by constriction was induced to metamorphose by connection of its hemocoel with that of a normally metamorphosing specimen, even when the connection was made by way of a glass capillary tube.

In the giant silkworm moth *Hyalophora cecropia*, after pupation, a dormant period of diapause exists for some 5 to 6 months. When the brain was removed from such a diapausing pupa, the insect never metamorphosed although the animal usually survived for about a year.[441] If a pupa was chilled by exposure to a temperature of 3 to 5°C for 1½ months, metamorphosis then occurred in a little over 1 month after the pupa was restored to the higher temperature. When a chilled pupa was united parabiotically with a unchilled one, both metamorphosed synchronously in about 1½ months (Fig. 21–8*A* and *B*). Implants of brain from a chilled pupa induced metamorphosis in an unchilled pupa.

A reconciliation of the observed influences on metamorphosis of the prothoracic gland, on the one hand, and of the brain, on the other, was effected. In *Hyalophora cecropia* the active fraction from the brain arises in the inner mass of each cerebral lobe of the brain, within which are found two groups of neurosecretory cells, a median and a lateral group. The median group of cells corresponds with the pars intercerebralis cells of the hemipterans.[443] To induce metamorphosis of a brainless pupa a portion of a brain containing both groups of neurosecretory cells had to be implanted, suggesting that these two groups collaborate in the production of the material involved.

When pupae of *H. cecropia* were transected just anterior to the sixth abdominal segment and then a plastex coverslip sealed over the cut end of each portion, preparations were obtained which were viable for 8 months or more and development could be readily observed in them. Chilled anterior portions, or brainless anterior portions which had received an implant of a chilled brain, underwent normal metamorphosis. The posterior portion did not undergo metamorphosis even when a number of chilled brains were implanted into a single abdomen. Such abdomens metamorphosed, however, when they were grafted to metamorphosing anterior portions. An endocrine factor in addition to

Figure 21–8. *A*, brainless, diapausing pupa (right) of *Telea polyphemus*, grafted to a chilled pupa of *Platysamia cecropia* (left). *B*, same animals after metamorphosis. (From Williams, C. M.: Biol. Bull., vol. 90.) *C*, posterior portion of a diapausing pupa. *D*, metamorphosis of the portion illustrated in *C* occurring after implantation of both chilled brain and prothoracic glands from a diapausing pupa. (From Williams, C. M., Biol. Bull., vol. 93.)

that from the brain was obviously essential. That the source of the second factor was the prothoracic gland was shown by the induction of pupation of isolated abdomens by the implantation of both chilled brains and prothoracic glands (Fig. 21–8*C* and *D*). The latter could come from either chilled or unchilled pupae. Prothoracic glands by themselves were not adequate. It was clear, therefore, that metamorphosis normally depends on the production of at least two factors. The first, produced from the brain, activated the prothoracic glands. The brain, molting hormone, and juvenile hormone are not species or even order specific.

Among the orthopterans and dictyopterans, *Carausius*,[323] *Leucophaea*,[120, 363, 365] and *Melanoplus*,[321] which have been investigated at some length and which show hemimetabolous development, the site of origin of the molt-initiating hormone is the brain, and the brain hormone activates the prothoracic glands. The corpora allata, in all, in addition, produce a juvenile hormone. In *Carausius*, which normally has six nymphal stages, removal of the corpora allata in the third nymphal stage results in a premature metamorphosis, although two nymphal molts usually intervene between the operation and the metamorphosis. This latter observation is interpreted to mean that the hormone from the gland is stored somewhere in the body over this period. Transplantation of corpora allata from third or fourth instar nymphs into sixth instar nymphs of *Carausius* may produce giant imagos which have undergone as many as three or four additional nymphal molts. In *Leucophaea*, which normally possess eight nymphal instars, allatectomy of a seventh instar before the critical period results in an adult-like stage following the molt. Such an individual has been termed an "adultoid" and differs from adults in having a smaller size and shorter wings. Allatectomy in fifth or sixth instars results in molt in "preadultoids" which require one additional molt to produce "adultoids."

In the holometabolous dipterans the ring gland of the larva, located dorsal to the brain and between the brain lobes, is the source of a hormone inducing pupation.[182] This gland is a composite including the functional corpora allata, corpora cardiaca, and prothoracic glands. The role of this gland is readily demonstrated by ligating the larva to constrict it into two portions, one containing the brain, and the other without it.[148] If this operation is performed before a certain "critical period," only that portion containing the ring gland pupates; if constriction is produced after the critical period, both portions pupate. Transplantation of a ring gland from a last larval instar into a first larval instar produces premature pupation in the latter. Ring-gland implants will also induce pupation in a portion of a larva cut off by ligation from its own ring gland before the critical period. This hormone is the effective one operating in metamorphosis as well as in pupation.[26] Imaginal discs implanted into the hemocoel of adult flies will not differentiate unless ring glands of late larvae are also implanted.[182] Furthermore, the molting hormone of Lepidoptera will induce pupation in dipterans,[26] and the juvenile hormone of lepidopterans is also effective in preventing metamorphosis in dipterans.[389]

The corpora allata of the Lepidoptera, like those of the hemipterans and orthopterans, are the source of a juvenile hormone. Allatectomy in caterpillars of younger instars is followed by a premature pupation, and, conversely, implantation of corpora allata from early instars into caterpillars ready to pupate will significantly delay the pupation.[46]

Reproduction. In insects, despite extensive observations on the effects of parasitic castration, surgical castration, and gonad implantation, there is only one species in which gonadal or other blood-borne hormones have been shown to influence significantly the differentiation of secondary sex characters. In *Lampyris noctiluca*, ablation of the corpora allata and corpora cardiaca of larval males near the third molt (prior to gonadal differentiation) results in the feminization of these animals.[295] It is suggested that the neurosecretory cells of the male larva are responsible for the development of testicular apical tissue, which is the source of an androgenic (male) hormone.[296] If this apical tissue is transplanted into a genetic female larva, the latter will develop into a masculinized adult with both the primary and secondary sexual characteristics altered. These experiments represent our only knowledge regarding sex hormones in insects.

The majority of species of insects so far investigated show a hormonal relationship between the corpora allata and the ovaries. Allatectomy in late larval stages or young adults is accompanied by a failure of the eggs in the ovary to undergo their normal growth and development. This has been demonstrated for the hemipteran, *Rhodnius*;[435] the dipterans *Calliphora*,[406] *Lucilia*,[103] and *Sarcophaga*;[103] the orthopteran *Leucophaea*,[364] and many other insects.[121] Implantation of corpora allata into allatectomized individuals restores the ability to produce normal eggs. On the other hand, no such relationship appears to exist in the orthopteran *Carausius*[323] or in non-feeding adult Lepidoptera.[47] That a brain secretion plays a role in reproductive regulation is suggested by the correlation between observed increased neurosecretion in the adult and reproductive activity.[16, 111, 297] Evidence for the humoral nature of this relationship was clearly demonstrated by telobiotic experiments in *Rhodnius*, in which the factor concerned was shown to be bloodborne, which indicated that the corpora allata exert their typical action irrespective of their new location.[435] The influence of the active corpus allatum principle appears to operate through its influence on the deposition of yolk within the eggs rather than on the earlier development of the oocytes. The corpora allata are apparently essential only throughout the period of oocyte growth and yolk deposition of each successive reproductive cycle. It has been shown that allatectomy is followed by profound alterations in the general metabolism of *Melanoplus*.[322] This has led some investigators to suggest that the reproductive functions of the juvenile hormone are secondary to more basic metabolic ones. There is good evidence, however, that the juvenile hormone triggers the synthesis and release of specific "female" proteins in the fat body of insects requiring the juvenile hormone for oogenesis. These proteins are incorporated into the yolk of the developing oocyte and are requisite for normal egg development.[121]

Allatectomy also depresses the growth and activity of certain female accessory organs in *Calliphora, Melanoplus*, and *Leucophaea*, this influence being independent of the presence of the gonads and hence not operating through them. There is considerable evidence that the ovary also exerts an action on the corpora allata in female insects. Ovariectomy in *Melanoplus, Calliphora*, and *Lucilia* leads to hypertrophy of the allata, and perhaps leads to a functional alteration in still other insects, such as *Sarcophaga* and *Leucophaea*.

Allatectomy leads to less distinct results in male insects than in females. The operation in no manner interferes with the production of sperm cells; in fact, allatectomized males of *Leucophaea* show ability to mate with, and effectively fertilize the eggs of, normal females. There are reports, however, that the male accessory glands of *Rhodnius* and of *Calliphora* fail to show normal development after allatectomy. Most of the data accumulated concern the colleterial glands of cockroaches. The colleterial glands of *Blattella germanica* respond to juvenile hormone by increased rates of synthesis of RNA and protein.[463] It appears that the juvenile hormone is important in controlling the accumulation of protocatechuic acid glucoside, a component of the secretion of these accessory glands.[377, 447] Castration of male *Leucophaea* and *Lucilia* has led to no observable modifications in the corpora allata.

Hormone Action. Several recent reviews[106a, 168a, 460] have described the many morphological and biochemical effects noted after injection of ecdysone or juvenile hormone into a variety of insects. One can summarize by noting that ecdysone initiates molting (apolysis and secretion of a new cuticle) and that all the biochemical alterations associated with molting and development (e.g., stimulation of protein, RNA, and DNA synthesis) are at least indirectly attributed to ecdysone. Juvenile hormone and a host of mimetic substances, on the other hand, have been shown to deter adult development in a large number of insect species and to promote those morphological and biochemical changes indicative of a more juvenile developmental stage. Juvenile hormone does not act alone (except in its role as a gonadotropic hormone in many female adult insects), but appears to moderate the effects of ecdysone. For example, ecdysone stimulates the epidermal cells to secrete a new cuticle, but it is the concentration of juvenile hormone that determines whether the cuticle will be that characteristic of the larva, pupa, or adult. However, the great majority of reported studies shed little light on the means by which the molting hormone and juvenile hormone elicit the observed effects. To convince the reader that progress in this direction is being made, we will discuss several recent studies that may form a substratum upon which future experiments can be designed so that definitive answers on hormone action can be attained.

In order to be effective, the hormones must reach their target sites. This means that ecdysone must leave the prothoracic glands and be carried in the open circulatory system of the insect and that juvenile hormone must be similarly transported from the corpora allata. (In our discussion we will assume that α-ecdysone is a pro-hormone synthesized in the prothoracic glands and that it is converted to β-ecdysone (the true molting hormone) by the target tissues.[225a]) In the case of α-ecdysone, the possibility exists that it is carried in the circulatory system by being bound to specific hemolymph proteins,[117a, 117b] although other data suggest that this relatively water-soluble steroid may simply be carried free in the hemolymph.[83b] The juvenile hormone, on the other hand, is a classic lipid and cannot be carried free in the aqueous circulation. Recent studies reveal that, indeed, juvenile hormone is bound to a specific hemolymph lipoprotein and is probably transported from the corpora allata to its target tissue in the form of a lipoprotein-juvenile hormone complex.[433] Whether the entire complex enters the target cell or whether cleavage occurs at the cell surface, resulting in only juvenile hormone entrance, is not known at this time.

Since available evidence suggests that ecdysone makes its way into all cells and tissues studied,[174] why do some cells respond (target cells) while others appear not to respond? A great deal of work in mammalian endocrinology has demonstrated the existence of specific proteins (receptors) that possess the capability of binding certain hormones (e.g., estradiol, progesterone) or active hormone metabolites (5α-dihydrotestosterone), and it may be these hormone-receptor complexes that are responsible for action at the nuclear level (see page 877). With the recent availability of [^{3}H] α-ecdysone of high specific activity, two reports have appeared suggesting the presence of receptors in arthropod target tissues. Emmerich[117b] showed that when [^{3}H]α-ecdysone was injected into *Drosophila* larvae, he could recover two proteins (3.6S and 2.0S) from the salivary glands that appeared to bind the labelled α-ecdysone. It is suggested that the hormone then enters the nucleus, where action is exerted, as a component of the receptor complex. When similar studies were conducted utilizing the crustacean hepatopancreas as a model system, it was shown that two proteins (11.3S and 6.35S) from the cytosol exhibited radioactivity.[173a] Further study indicated that the smaller receptor protein was a subunit of the larger. Finally, it was demonstrated that the label associated with these receptor proteins was no longer due to [^{3}H] from the [^{3}H]α-ecdysone, but a previously undescribed ecdysone metabolite which has yet to be fully characterized. This may be analogous to the situation in the mammal where testosterone is converted to the active metabolite 5α-dihydrostestosterone. The above, although preliminary, suggests the possibility of hormone receptors in arthropod target tissues, and it will be of great interest to conduct similar experiments with as yet unavailable high specific activity [^{3}H]β-ecdysone.

Once the hormone finds its way to the target cell and is presumably in an active form, how does it elicit the wide range of effects noted at the more macroscopic level? The answer is certainly not known, but one can examine some effects that appear to be close to the site of action. β-Ecdysone, for example, stimulates RNA synthesis in epidermal cells *in vitro*[460a] as it does *in vivo*, and several studies suggest that juvenile hormone inhibits this stimulation of RNA synthesis.[90, 318] Preliminary data indicate that different types of RNA are synthesized by isolated fat body nuclei, depending on the hormonal regimen to which the nuclei are exposed.[90a] Some elegant *in vitro* experiments with *Drosophila* imaginal discs[83a, 152a, 337a] demonstrated a rapid stimulation of ribosomal (r) RNA by β-ecdysone which could be inhibited by addition of juvenile hormone. The stimulation of rRNA synthesis may be important in the transport of new messenger (m) RNA from the nucleus to the cytoplasm. These studies suggest that β-ecdysone and juvenile hormone act at the transcriptional level. However, equally cogent studies at the molecular level suggest that juvenile hormone, at least, acts to control translational events.[213a] There is no firm answer at this time, and it is possible that both molting hormone and juvenile hormone act at several levels, since there are some twelve possible points of control between transcription and translation. For example, as noted above, the data of Ilan and his colleagues suggest that juvenile hormone acts at the translational level. Other studies on the effects of juvenile hormone revealed that it could induce the *de novo* synthesis of the female specific protein needed for normal egg development in adult female cockroaches[121a] and could induce the synthesis of specific hemolymph proteins in certain lepidopteran pupae.[433a]

In both cases, administration of actinomycin D abolishes the protein induction by juvenile hormone, indicating that the hormone acts at the transcriptional level.

*CRUSTACEANS**

Like that of the vertebrate and the insects, the crustacean endocrine system, a complex one, is functionally very closely related to the central nervous system.[51] Neurosecretory cells in ganglia of the nervous system form hormones which are transported within their axons to places of their liberation into the blood. Conspicuous elements of this system are the X-organs associated with ganglia in the eyestalks of most stalk-eyed crustaceans and in the heads of others. Best known of these are the medulla terminalis X-organ (MTGX) and the sensory papilla X-organ (SPX)[237] (Fig. 21–9). The sinus gland of the eyestalk (or head) is generally believed to be simply an organ comprising chiefly the hormone-charged terminations of the X-organ neurosecretory cells.[37, 112, 118, 186, 277, 317] The secretory materials in the cells appear to be contained in membrane-bounded vesicles.[204, 236, 331]

A second important crustacean endocrine complex comprises secretory nervous cells in the brain, and perhaps also esophageal connective ganglia, with hormone-conducting axons terminating in the post-commissural organs (Fig. 21–10) where the substances are released into the blood.[52, 135, 235, 279] Following the same organizational pattern are cells in the ventral ganglionic chain liberating their hormone some distance away in the pericardial organs at the openings of the large veins into the pericardial cavity.[5]

It has been postulated that crustacean neurosecretory hormones, like those found in the vertebrate neurohypophysis, are polypeptides. Consistent with this view are the recent findings that the neurosecretory hormones such as the chromatophorotropic hormones,[131, 231] retinal pigment hormones,[229] hyperglycemic hormones,[223] molt-inhibiting hormone,[341] and cardio-accelerator hormones[28, 31] all appear to be polypeptides.

But not all the crustacean endocrine glands are based directly upon nervous secretions. Quite distinct is a pair of glands in the anterior thoracic region in the antennary or maxillary segment. These glands are the Y-organs[115, 155] and are apparently without any direct innervation. They are regulated by secretions from the eyestalk-gland complex. In addition, the ovary serves in an endocrine capacity.[73] There is also an important organ, the androgenic gland, morphologically associated with the sperm duct.[73, 75] In some isopods they are found close to the testes.

Molting in Crustaceans. The decapod crustaceans in their development typically undergo a number of molts, passing through a series of characteristic larval stages, and after having achieved the body form of the adult they continue to grow through periodic molting of the exoskeleton.

The molting process is a complex one. The crustacean is, in fact, at all times during its growing lifetime either preparing for the next molt, or completing the past one. The molting cycle may be divided into four periods: (1) premolt, or proecdysis, a period of active preparation for molt, including gradual thinning of the cuticle, storage in the gastroliths or hepatopancreas of inorganic constituents for a new exoskeleton; acceleration of any regenerating tissues, glycogen deposition in epidermal tissues, and numerous other changes; (2) molt, or ecdysis, the splitting and shedding of the old, partially resorbed cuticle, and an abrupt size increase due immediately to absorption of water; (3) postmolt, or postecdysis, a period of rapid redeposition of chitin and inorganic salts to produce a new cuticle, and of tissue growth; and (4) intermolt, or "interecdysis," a period of relative quiescence during which physiological processes normally associated with the active molting process are largely absent, but there is, however, a storing of reserves in the hepatopancreas and elsewhere in anticipation of the next molt. Some crustaceans such as *Maja* and *Carcinus* have terminal growth, passing into a permanent intermolt termed anecdysis.

The molting cycle has been accorded precise means of staging.[75, 109, 110, 180, 366, 369] In the freshwater crayfish *Cambarus* there is no true larval stage; the individual hatches as a diminutive adult. During its first year of life it molts at intervals of about 12 to 13 days, probably without intervention of any significant intermolt period. After the first growth season there are usually two molts a year, one occurring in the spring, in late April or May, and the other in the summer, in July or August. In these older, mature

*With contributions by M. Fingerman.

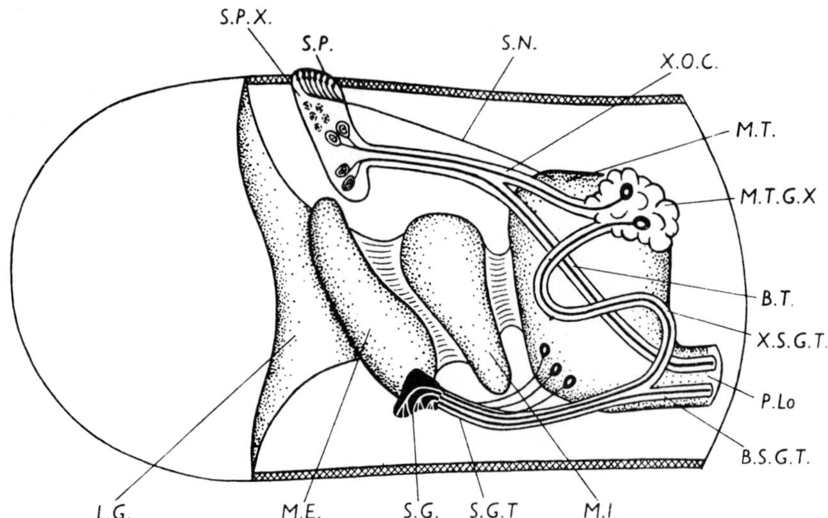

Figure 21–9. Diagram of the eyestalk neuroendocrine complex of the natantian, *Lysmata. B.S.*, brain-sinus gland tract; *B.T.*, brain-X-organ tract; *L.G.*, lamina ganglionaris; *M.E.*, medulla externa; *M.I.*, medulla interna; *M.T.* medulla terminalis; *M.T.G.X.*, medulla terminalis ganglionic X-organ; *P.Lo.*, peduncle of optic lobe; *S.G.*, sinus gland; *S.G.T.*, combined sinus gland tract; *S.N.*, sensory nerve; *S.P.*, sensory pore; *S.P.X.*, sensory papilla X-organ; *X.O.C.*, X-organ connective; *X.S.G.T.*, X-organ sinus gland tract. (From Carlisle, D. B., and Knowles, F. G. W., Endocrine Control in Crustaceans. Cambridge University Press.)

crayfishes the premolt period is 3 to 5 weeks. During this time there is gradual resorption of the exoskeleton and a deposition of calcium salts in the form of gastroliths in the anterolateral walls of the cardiac stomach. There is also a gradual increase in the rate of oxygen consumption and of water content for a week or so prior to molt, reaching a peak at the time of molt. The period of postmolt is one in which these changes proceed

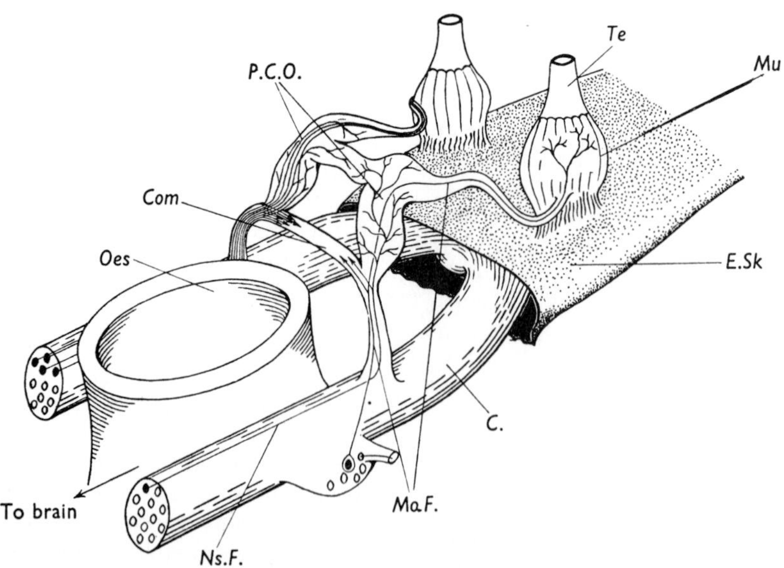

Figure 21–10. Diagram of neurosecretory-postcommissural organ complex of the prawn, *Leander. C.*, circumenteric connective; *Com.*, commissure; *E.Sk.*, endophragmal skeleton; *Mo.F.*, motor fiber; *Mu.*, muscle; *NsF.*, neurosecretory fiber; *Oes.*, Esophagus; *P.C.O.*, postcommissural organ. *Te.*, tendon. (From Carlisle, D. B., and Knowles, F. G. W., Endocrine Control in Crustaceans. Cambridge University Press.)

in the opposite direction and require approximately the same time as the corresponding processes of premolt. Postmolt is followed by intermolt, which is of longer duration after the summer than after the spring postmolt.

Observations on changes associated with the molting cycle in other crustaceans show that the hepatopancreas is a site of storage of calcium, phosphates, glycogen, and lipid. The stored salts are not sufficient to account for the total hardening of the exoskeleton, thus making it necessary that the postmolt period be one of rapid absorption of calcium, both directly from the external medium and from ingested food. Gastroliths appear to be an adaptation of such forms as crayfish and *Gecarcinus* to relative inaccessibility of calcium; these are typically absent in marine crustaceans. A study of apparent respiratory quotients shows that although animals in intermolt have a quotient of about 0.8, freshly molted crayfishes show values as low as 0.1 to 0.2 during the first few postmolt hours because of CO_2 fixation during carapace hardening; this value gradually increases to 0.7 to 0.8 during the first postmolt week. This indicates that calcium is avidly taken up from the surrounding medium immediately after molt, and that the rate declines rapidly during the first few days.

It is known that the removal of the eyestalks from *Astacus*,[373] *Uca*,[233] *Eriocheir*,[187] *Palaemonetes*,[50] *Cambarus (Orconectes)*,[53] or *Cambarus (Procambarus)*[388] results in a more rapid onset of the following and succeeding molts. In young *Cambarus* (*Procambarus*), in their first year of life, removal of the eyestalks results in a shortening of the period between molts at 20° to 22°C from about 12 days to about 8 days.[388] That this influence is not the result of general operative injury is seen in that other operative injuries, such as destruction of the retinas, which are at least as severe, do not result in such acceleration; if any influence is seen there is a retardation. The remaining two possibilities, that the results are due to (1) the destruction of important nerve centers, or (2) the removal of endocrine organs important in molt regulation, have been resolved in favor of the latter. If both eyestalks are removed from mature crayfishes molt will occur a significant time in advance of that of unoperated controls. If, however, sinus glands from other crayfishes are implanted into the abdomen of such eyestalkless specimens, molting will be delayed beyond the time of that of the controls.[50, 53]

A molt-inhibiting role of the crustacean X-organ–sinus gland complex has received confirmation in studies of the control of gastrolith formation in crayfishes (Fig. 21–11). These concretions, normally produced only during preecdysis, may be caused to form at other times by excision of both eyestalks or by surgical extirpation of the eyestalk-gland complex.[375] After eyestalk removal during a nonmolting period, such as between September and March, gastroliths of crayfishes commence to form in less than 24 hours at about 20°C and then increase in size slowly during 8 to 10 days, thereafter accelerating rapidly to the time of molt, which usually occurs between the fifteenth and twentieth days. Those individuals which survive the molt immediately proceed to form a new set of gastroliths. If, however, one implants a sinus gland into the abdominal region of eyestalkless animals at 3- to 4-day intervals gastrolith formation is suppressed. If the sinus gland implantations are discontinued, gastroliths begin to form about a week after the last implant, indicating that the implanted glands are no longer effective.

Histological changes in the sinus glands of crustaceans have been shown to be correlated with the molt cycle.[336] Just prior to molt, acidophilic secretory granules appear to be the predominant ones; after molt completion, basophilic ones are more prevalent.

The molting process which is set into operation by eyestalk removal resembles that observed in normal molting animals,[375] and all of the molt-correlated changes ensuing upon eyestalk removal are prevented by implantation of sinus glands into newly de-

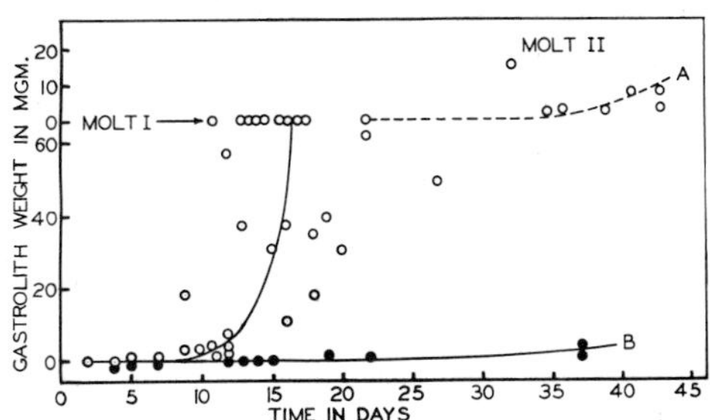

Figure 21–11. *A*, increase in gastrolith size and molting in crayfishes after removal of the eyestalks with their included sinus glands. Note that an animal once molted proceeds almost immediately to prepare for another molt. *B*, eyestalkless crayfishes into which sinus glands are implanted at 3- or 4-day intervals show no gastrolith production. (From Scudamore, H. H., Tr. Illinois Acad. Sci., vol. 34.)

stalked animals. It appears probable, therefore, that all are under the influence, directly or indirectly, of a single molt-inhibiting hormone from the sinus gland. Molting in the crustacean is prevented if the Y-organs are extirpated, and the capacity is restored when Y-organs are reimplanted.[66,116,156] The Y-organs also exhibit histological cycles correlated with the molting cycle and become greatly reduced in size in crabs which are in anecdysis. The relationship in the regulation of molting between the X-organ–sinus gland complex and the Y-organ has been extensively elucidated. The MTGX-organ produces a hormone which inhibits the Y-organ. This inhibitory hormone is continuously produced during postmolt and intermolts. Cessation of its secretion liberates the Y-organ to secrete molting hormone. Once premolt is fully initiated, the eyestalk inhibitory factor is ineffective in stalling the molting sequence. The molting hormone may not merely serve as a trigger for the initiation of premolt but may also be required for the successful completion of the later stages of premolt.

One of the main molting hormones of insects, 20-hydroxyecdysone (= β-ecdysone, ecdysterone, crustecdysone) also occurs in the marine crayfish *Jasus*.[184] The blue crab *Callinectes* contains, in addition to 20-hydroxyecdysone, two other steroids which are probably inokosterone and makisterone. Injection of 20-hydroxyecdysone and related steroids into the crayfish *Procambarus* and the fiddler crab *Uca* resulted in the precocious induction of premolt events such as apolysis, gastrolith formation, acceleration of regenerative limb growth, secretion of a new exoskeleton, and shortening of the intermolt period.[130,242,268,342] Definitive identification of any of the ecdysones as true crustacean molting hormones will be possible when they are tested on X-organless individuals.

The control of this molt-triggering mechanism is clearly complex and variable in crustaceans, since numerous factors influence the initiation of molt. These include nutrition,[94,345] temperature, light and photoperiod,[37,393] parasitism, injury, and state of reproductive activity. The molt-inhibiting hormone from the sinus glands appears to be responsible for the failure of egg-bearing female *Crangon* to molt until the young have hatched, and for the fact that at the annual spring molt of *Cambarus* (*Orconectes*) the egg-bearing females molt several weeks later than the males and only after the young have left the maternal pleopods. Sinus gland removal is just as effective in inducing molt in egg-bearing female crayfishes as is a similar operation in males.[375]

There is evidence suggesting further hormonal regulation of crustacean molting. A molt-accelerating factor has been reported from the MTGX- and SPX-organs of the eyestalk.[63] There is also reason to suspect that at least one additional factor, arising in some region of the body other than the eyestalks, cooperates in the control of molt.[64,375] A careful study of the deposition of calcium salts in the gastroliths shows that this process is rhythmical, with rapid deposition during the night and little deposition during the day. Suggestive in this regard is the observation that injection of extract of the brain tissue or strong electrical stimulation of the cut ends of the optic nerves of eyestalkless animals will cause a transitory acceleration in rate of oxygen consumption.

Recent studies on the control of molting in brachyuran larvae revealed that the regulatory mechanisms may vary from species to species. Removal of both eyestalks from the zoea of the mud crab *Rhithropanopeus* resulted in one or two supernumerary zoeal stages, but acceleration of larval molts was not observed.[97] Similarly, the duration of the megalops stage of *Rhithropanopeus* was not reduced by eyestalk ablation at any period of development. The production of supernumerary zoeal stages in eyestalkless larvae supports the hypothesis that a factor within the eyestalks may control the rate of morphological development and may be partly responsible for the initiation of physiological processes culminating in metamorphosis. The absence of accelerated molting during larval development suggests that the molt-inhibiting X-organ–sinus gland complex of the adult is not functional during the larval development of *Rhithropanopeus* and that larval molting is controlled by some other mechanism. However, the molt-inhibiting control becomes apparent during the megalops stage of development in the crabs *Callinectes*[95] and *Sesarma*.[96] Ablation of the eyestalks at a specific time during the megalops stage of the latter two species leads to the acceleration of molting.

Reproduction. Experiments have indicated that oogenesis, particularly vitellogenesis, in female shrimp of the genus *Leander* (*Palaemon*), is under the control of a hormone liberated in the sinus glands.[313, 314, 401, 461] This hormone appears to be formed in the MTGX-organ.[65, 418] *Leander* (*Palaemon*) reaches the end of its breeding season late in the summer, and its ovaries become tremendously reduced in size and activity and normally remain so during the fall, winter, and early spring. Removal of the eyestalks, or only the sinus glands from the eyestalks, in such a nonbreeding season as September or October results in a very rapid increase in weight of the ovaries, these organs increasing about seventyfold in 45 days (Fig. 21–12). Normal eggs may be laid at the end of this period. Unoperated controls show almost no increase during the same period. Implantation of sinus glands into the abdomens of eyestalkless animals will inhibit ovarian development, depressing it even more than is observed in unoperated controls. A similar sort of hormonal relationship between the sinus gland and the ovary, with the sinus gland acting as inhibitor for ovarian maturation, has been demonstrated for the fiddler crab *Uca*.[55] A similar situation appears to obtain for the crayfish. It is presumed to be through this route that light may affect the reproductive cycle.[394]

A further reproductive function of the crustacean sinus gland is observed in the fact that female crayfishes bearing eggs on their pleopods normally postpone their spring molt beyond the time of molting of males and until the young have become free. This adaptive response is apparently the result of activity of a sinus gland hormone, inasmuch as egg-bearing females after sinus gland extirpation molt as readily as do males. The molt-inhibiting and ovary-inhibiting factors are not identical.[92, 104]

An additional endocrine dependence of the ovary has been shown for *Carcinus*. The gonads fail to develop in the absence of the Y-organ. This organ, however, is not essential to the continued maintenance and function of these organs.[11]

There have been numerous observations of the influence of parasitic castration of male decapod crustaceans upon such secondary sex characters as forms of the pleopods, chelipeds, and abdomen. The partial to complete castration which normally results from parasitization by rhizocephalans (parasitic Cirripedia) or bopyrids (parasitic isopods) is commonly accompanied by the failure of these portions of the body to assume their

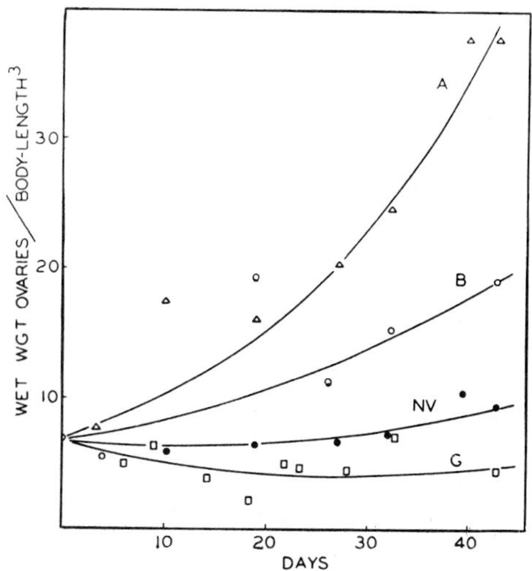

Figure 21–12. Rate of increase in ovarian weight in the shrimp, *Leander*, after: *A*, eyestalk amputation; *B*, sinus gland extirpation; *G*, eyestalk amputation followed by sinus gland implantation. *NV*, normal control. (Redrawn from Panouse, J., C. R. Acad. Sci., vol. 218.)

typical masculine form, the specimens approaching the female form in their secondary sexual differentiation.

The interpretations of these results of parasitic castration of male crustaceans in general fall into either one of two categories: (1) those which assume that the presence of the parasite has influenced the expression of the genetic mechanism of sex determination, and (2) those which assume that the testes or some other endogenic endocrine source has been destroyed. In support of the first of these two types of explanations is the observation that certain decapod crustaceans normally possess hermaphroditic gonads or are readily induced to develop them under the influence of parasitization. Some species, such as *Leander* (*Palaemon*), which show no secondary sexual character changes on castration are thought to possess a relatively stable mechanism of sex determination, not easily influenced by the parasite, whereas others like *Upogebia* are considered to possess very labile mechanisms and consequently show considerable feminization. A number of investigators have noticed that the higher fat content of the blood and liver characteristic of normal females is also often observed in parasitized males. It is postulated by these investigators that the parasite imposes much the same metabolic demands on the host as are normally made by the developing eggs of the ovary, and that associated with the increased fat metabolism is the production of a "sexual formative" stuff which parallelly influences the development of both the gonads and the secondary sex characters.

The second type of explanation of the observed influences of parasitic castration in males assumes the operation of a masculinizing hormone. According to this hypothesis the animal after parasitization does not become feminized but assumes a neutral form, which chances to resemble more clearly the female than the male. Suggestive support for this view is the influences of three species of parasites on the crab *Munida sarsi.* Two smaller parasitic crustaceans, *Triangulus munidae* and *Lernaeodiscus ingolfi,* totally or partially castrate the crab and produce striking modifications in the male secondary sex characters; a much larger parasite, *Triangulus boschmai,* leaves the gonads functional and does not modify the sex characters. Such observations appear to exclude in these instances a direct influence of the metabolic demands of the parasite in inducing the observed changes. Other investigators, however, have failed to find a correlation between the degree of gonadal atrophy and the extent of suppression of the male characters and have suggested that a tissue other than the testis produces the hormone in question.

It seems probable that the answer lies in the presence of androgenic glands first discovered in *Orchestia* (Fig. 21–13) but since found in virtually all crustaceans in which they have been sought.[74–77] Extirpation of these glands results in feminization, an effect reversed by implantation of the glands. Immature females implanted with androgenic glands become masculinized in nearly all sex characters, primary, accessory, and secondary. The ovaries become transformed into testes. Even mating behavior of the masculinized females is male-like. In the protandrous crustacean *Lysmata* the male to female transition is correlated with degeneration of the androgenic glands.[77]

There is a seasonal cycle of changes in the copulatory appendages of the crayfish. These appendages assume a sexually functional form (form I) at the time of the late summer molt. This is a time when the testes are large and active. At the time of the spring molt, correlated with a period of low gonadal activity, they revert to a nonfunctional condition (form II). Experimental induction of molt during the winter months when testis activity is similarly low always produces form

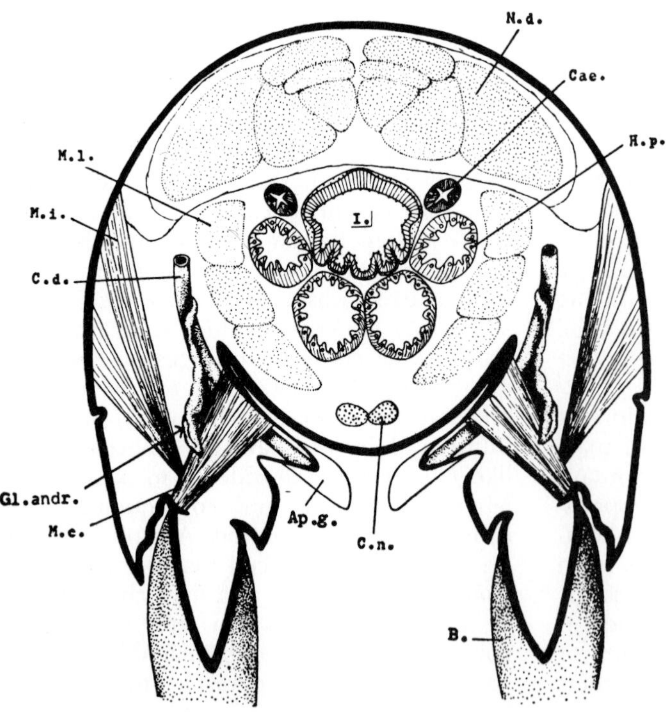

Figure 21–13. Cross-section of the amphipod, *Orchestia,* indicating the location of the androgenic glands (*Gl. andr.*). *Ap.g.,* genital papilla; *B,* seventh periopod; *Cae,* intestinal caecum; *C.d.,* vas deferens; *C.n.,* nerve cord; *H.p.,* hepatopancreas; *M.d., M.l., M.i., M.c.,* muscles. (From Charniaux-Cotton, H., Bull. Soc. Zool. France, vol. 83.)

II. These changes are endocrine regulated in all probability by the same endocrine complex regulating male reproductive activity.[95]

Parasitic castration of female crustaceans is in general accompanied by little or no change in the general form of the body and appendages. It has been observed, however, that the brood pouch of *Asellus* and that of *Daphnia* fail to develop after injury to the ovaries by irradiation. In the amphipod *Gammarus pulex,* suppression of the ovaries by a parasitic worm, *Polymorphus minutus,* or by irradiation has been observed to be associated with failure of the typical marginal bristles of the oostegites to develop. The ability to develop the marginal bristles was restored parallelly with oogenesis after cessation of the irradiation treatments. Female shrimp, *Leander,* castrated by bopyrids or by x-ray irradiation, showed absence of development of the abdominal incubatory chamber and the special guanophores associated with the corresponding abdominal segments. These observations and others strongly support the hypothesis that in these crustaceans the ovaries produce a gynecogenic hormone normally determining certain feminine modifications concerned with provision for the developing young.[73] There is suggestive evidence that still other hormones are concerned in the regulating normal sexual development,[16, 93, 392] including one from the eyestalk. Both sexes appear to have gonad-inhibiting and gonad-stimulating hormones of neurosecretory origin. Those in the female operate directly on the ovary, whereas those in the male exert their influence on the testis indirectly by inhibiting or activating the androgenic glands.[2, 309]

Retinal Pigment Migration. The movements of pigments within the eyes of many animals such as vertebrates, insects, and crustaceans contribute importantly to the mechanical adaptation of these organs to changes in light intensity. Only among the crustaceans, however, has clear evidence been presented that hormones are involved in the control of these movements.

The compound eye of crustaceans is made up of a number of units, the ommatidia (Fig. 21–14). Each ommatidium possesses three functionally distinct groups of pigments. The distal retinal pigment, either ommochrome or melanin, is located in cells which surround the distally placed dioptric apparatus of the ommatidium. In the dark-adapted eye this pigment occupies only a

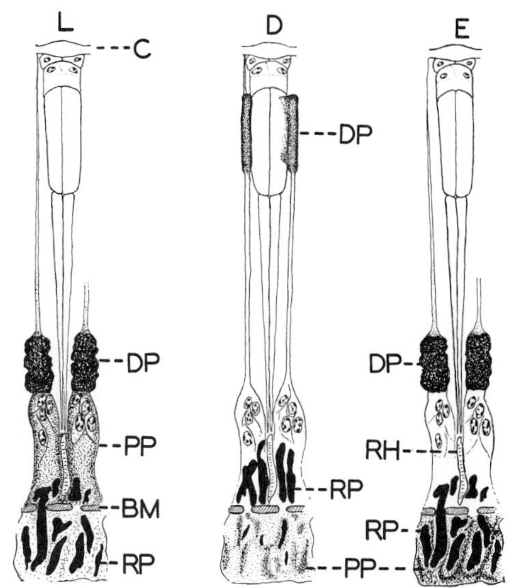

Figure 21–14. Position of the retinal pigments in the ommatidium of the eye of the shrimp, *Palaemonetes, L,* in the light-adapted state; *D,* in the dark-adapted state, and *E,* after injection of eyestalk extract into a dark-adapted specimen kept in darkness. *C,* cornea; *DP,* distal retinal pigment; *PP,* proximal retinal pigment; *RP,* reflecting pigment; *RH,* rhabdome; and *BM,* basement membrane. (From Kleinholz, L. H., Biol. Bull., vol. 99.)

distal position; the pigment disperses proximally in daylight to envelop the whole of the ommatidium as far as the retinula elements. The proximal pigment, chemically like the distal, is located in the retinula cells and migrates to a position proximal to the basement membrane in darkness, and distally to meet the distal retinal pigment in light. In the light-adapted state the whole ommatidium is therefore enclosed in a light-absorbing sleeve of pigment. A third pigment, the reflecting white, comprises probably purines and pteridines.[230] In darkness this pigment occupies a position surrounding the retinula elements, thus constituting a functional tapetum. It migrates to a position proximal to the basement membrane in light.

The distal pigment, also commonly termed the iris pigment, like the vertebrate iris, exhibits a graded response to light over a wide intensity range.[137, 356] Investigations to determine the extent to which the pigment cells of the right and left eyes of an individual are capable of independent responses to illumination have led to various results. The more recent experiments of this type have led to the conclusion that there is at least a partial interdependence between the two eyes, a darkened eye becoming more or less light-adapted when the contralateral eye is subjected to illumination. Numerous observations have also indicated that one or more of the retinal pigments of numerous species of crustaceans may undergo diurnally rhythmic alterations in their position in animals kept in constant conditions in respect to illumination, especially in constant darkness.[139, 426] *Leander* (*Palaemon*) kept on an illuminated black background has been observed to show a dorsoventral differentiation in position of retinal pigment, apparently the result of the considerably lesser illumination of the ventral than of the dorsal elements of the eyes. These various responses of the retinal pigments suggest that the control of the retinal pigments is not a simple one, but probably involves a direct reaction of the retinal pigment cells to illumination and, in addition, endocrine and possibly also nervous activities.

The eyestalk-gland complex possesses a hormone which influences the position of the retinal pigment,[227, 431] as does also the subesophageal ganglion in *Palaemonetes* and *Cambarellus*.[137] This has been called retinal pigment hormone or RPH. Injection of extracts of the eyestalks of light-adapted *Palaemonetes* into dark-adapted specimens kept in darkness induces a movement of the distal and reflecting pigments to the position characteristic of the light-adapted state. The eyestalks of all crustaceans which have been examined show the presence of RPH in larger or smaller quantities. That this principle is normally concerned with the retinal pigment movements is indicated by the fact that the RPH in the supraesophageal ganglia of *Palaemonetes* kept in darkness for 14 days showed a 4-fold increase.[142] However, it has been reported that the diurnal variation in retinal pigments of certain grapsoid crabs persists even following removal of the sinus glands. A hormone from the eyestalks contributes at least in part to the control of the distal retinal pigment of *Leander* (*Palaemon*) and *Uca*.[134, 136]

The retinal pigments of crayfish show different thresholds of response to RPH. In low concentrations of the hormone only the distal pigment is influenced; with higher concentration both distal and proximal retinal pigments move to the light-adapted state.[228]

The origin of RPH is the MTGX-sinus gland complex of the eyestalk. The sinus gland when extracted alone is able to induce a strong retinal pigment response.

The retinal pigment hormone of the sinus gland will withstand boiling as in the case of the eyestalk chromatophorotropins. RPH from eyestalks of the prawn *Pandalus* is an octadecapeptide (molecular weight about 2000) containing 12 different amino acids.[131, 134] There is ample reason to believe that it is not identical with any of the principal chromatophorotropins, inasmuch as the pigmentary system of the integument ordinarily undergoes its complete gamut of activities in color changes in response to illuminated backgrounds while the eye remains continuously light-adapted. This latter is true despite the fact that the threshold of response of the retinal pigments to eyestalk extract is substantially higher than the threshold of response of the body chromatophores. Such a situation obviously could not obtain were RPH identical with one of the major chromatophorotropins. However, since RPH has so far not been separated from the chromatophoric pigment-dispersing material, the suggestion has been made that distal retinal pigment light adaptation and chromatophoric pigment dispersion are both caused by either the same molecule or closely related different molecules.[232]

A search for a possible comparable endocrine influence on the state of the retinal pigments of the insect *Ephestia* disclosed no evidence of such an endocrine activity. Injection of extracts of the heads of light-adapted moths into either dark- or light-adapted specimens produced no modification in the state of the pigments. Extracts of crustacean sinus glands also showed no activity on the retinal pigments of the moth.

An additional hormone of neurosecretory source exercises a dark-adapting influence on the distal retinal pigment.[54, 137] It was shown possible to induce differential secretion of the two by light stimuli of regulated duration and intensity.[57] There is also evidence that the two hormones are differentially secreted as a daily rhythmic phenomenon.[57, 426] The two retinal-pigment hormones have been separated by electrophoretic means[140, 141] and by gel filtration chromatography.[138] Both hormones evoke graded responses directly related to the injected dosage.

Other Phenomena. Heart rate in crustaceans is influenced by endocrine factors.

An eyestalk-arising principle accelerates the beat,[430] and in *Palaemonetes* its secretion is correlated with that of certain color-change hormones.[374] There are reasons to believe that the heart and chromatophore activators are not identical.[189, 237] A more specific heart regulatory principle is secreted by a neurosecretory system with secreting areas in the pericardial cavity. These have been termed pericardial organs (Fig. 21–15). Extracts of these organs increase amplitude of heart beat and exercise other influences, apparently in manners differing from species to species. The hormone released from the pericardial organs appears to be free polypeptide, not bound to a carrier protein.[31] Removal of the eyestalks or of the sinus glands of these stalks of *Cambarus (Orconectes)* and *Gecarcinus* is promptly followed by an elevation in basal metabolism, as evidenced by the observed increase in O_2 consumption.[37, 375] This can be reduced by injection of aqueous extracts of the glands. Injection of extracts of eyestalks of *Uca,* the fiddler crab, into *Callinectes,* the blue crab, results in a rapid rise in the blood sugar from about 20 mg/100 cc to more than 80.[229] The maximum is reached in 1 hour, and then there is a slow decline to normal. This latter action has been attributed to the presence of a diabetogenic factor in the crustacean sinus gland, and recent work has confirmed this conclusion for the spider crab *Libinia.* This hormone is a heat-labile molecule, apparently a polypeptide.[224, 231] That from the crayfish *Orconectes* has a molecular weight of about 10,000. It shows some species specificity of action. The hormone from crabs has very little effect on the crayfish and vice versa.[225] There is also evidence in crustaceans for diuretic and antidiuretic hormones controlling salt and water balance.[38, 222, 223]

Miscellaneous. Many other invertebrates also show annual or other reproductive rhythms, with periods of sexual activity alternating with periods of inactivity. In most instances there is as yet no knowledge of the pathways through which the gonads are activated or inhibited.

Many species of invertebrate organisms representing many phyla possess a sexual dimorphism which has long been suspected to owe its origin, in some measure at least, to hormones comparable to the gonadal hormones of vertebrates.[187] Even to the present, however, no incontrovertible evidence has been advanced to prove this is so except for crustaceans. Nevertheless, there are numerous reports in the literature suggesting hormones to be playing roles in this regard. These come from observations on (1) parallel effects of parasites on gonads and sexual differentiation; (2) parallel cyclical changes in the degree of gonadal activity and certain secondary sexual characters; (3) results of irradiation of the gonads or of the whole

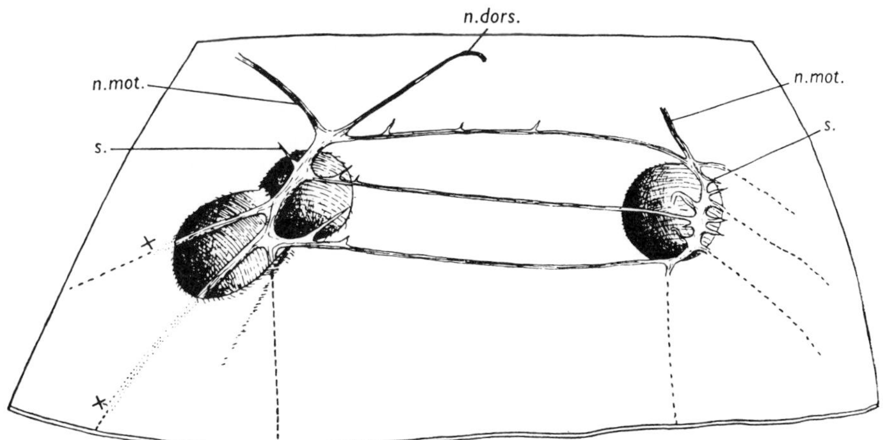

Figure 21–15. The pericardial organs of the crab *Maia* and their relationship to openings into the pericardial cavity of the branchiocardiac veins. Broken lines depict nerves from the central nervous organs. *n.mot.,* nerves innervating muscles; *n.dors.,* dorsal cardiac nerve; *S.,* strands suspending the trunks. (From Alexandrowicz, J. S., J. Mar. Biol. Assn. U. K., vol. 31.)

organism with x-rays or radium rays; and (4) results of surgical extirpation of gonads.

Among arthropods other than insects and crustaceans, physiological evidence for endocrine mechanisms is relatively sparse. In *Xiphosura (Limulus)*, a correlation between density of neurosecretory cells and chromatophorotrophic activity of extracts has been reported.[362] Myriapods[334, 370] and some arachnids[157, 198] appear to have neurosecretory cells in their brains associated with neurohemal structures resembling those of insects (corpora cardiaca) and crustaceans (sinus glands). Also, glandular organs whose histological changes with molt suggest similarities with the prothoracic glands of insects have been described.[117, 217, 253]

Neurosecretory cells have been described for the hypostomal region of *Hydra*. Since hypostomal extracts stimulate growth and formation of new heads, this comprises circumstantial evidence that the nerve cells are sources of a growth hormone. In sexual forms, the growth factor is greatly reduced or absent in hypostomal extracts; it has been suggested, therefore, that presence of the growth factor induces somatic growth by the interstitial cells, and absence of it induces the production of sex cells.[60, 61, 256]

Planarian regeneration appears to be in part regulated by chemical mediators released at cut or injured surfaces. Also, the brain possesses neurosecretory cells, and in multi-eyed *Polycelis* the presence of the brain is necessary for normal eye regeneration.[255] The possibility has been suggested that neurosecretory agents derived from the brain may be a mediating agent for regeneration and also for fissioning.[33]

In the nematode *Phocanema*, neurosecretory cells have been described for the dorsal and ventral ganglia of the anterior nerve ring, and evidence suggests that one of their products may be concerned with ecdysis.[101] Nemerteans possess a unique structure composed of neuroglandular elements and termed the cerebral organ because of its usual close association with the cerebral ganglion. Its function has not been discovered.[361]

The radial nerves of starfish provide an extract that induced shedding of eggs and sperm when injected into a brachial cavity.[70, 71, 381] This shedding hormone appears normally to be released directly into the environment, to be taken up by the tube feet of both the individual producing it and neighboring starfish. The hormone is a polypeptide with a molecular weight of about 2600. The maturation divisions in starfish are believed to be induced by a hormone arising in the ovaries. Evidence suggests also that the radial nerves provide a hormone which inhibits shedding, with the shedding hormone able to perform its role only when the titre of inhibiting hormone drops.

A moderate amount of evidence has now related to endocrine regulation of growth and regeneration of polychaete annelids as well as their reproduction.[24] Brain-originating hormones in nereids control growth in ovocytes, as well as epitokal metamorphosis. These are inhibited by brain hormonal action, but their normal development appears to depend on an orderly, progressive decrease in the titre of the inhibiting hormone.[86, 113, 195] Sperm production is also inhibited in *Arenicola;* the brain stimulates reproductive activity in the female and is essential for spawning,[210] and the brain is also required for normal spermatogenesis.

In the oligochaete *Eisenia,* gonadal development and the maintenance of the clitellum require the presence of the brain; neurosecretory cell changes have been correlated with stages in development of reproductive activity.[197]

Correlation between testicular activity and histological evidences of activity in neurosecretory cells in *Hirudo,* as well as influences of injection of extracts of leech brains, have given evidence of a gonad-stimulating action by a brain hormone in these organisms.[183]

Among gastropod molluscs there is suggestive evidence for endocrine control of egg production and ovulation by dorsal bodies of the cerebral ganglia of *Lymnaea*[221] and by ganglia of the optic tentacles of *Helix* and *Arion* and cerebral ganglia of *Arion*.[249, 250, 319, 382] Neuroendocrine control of egg laying has been described in *Aplysia*.[244] Neurosecretory cells in the right pleural ganglion appear to produce a hormone involved in water regulation.[259] Evidence for endocrine roles in lamellibranchs is rather weak. Correlations between reproductive activities and histological evidences of neurosecretory changes have been described for the mussels *Mytilus*[269] and *Dreissena*.[8]

The cephalopods possess two endocrine glands, the branchial and the optic. In the octopus, the former are large, widely vascularized organs whose extirpation leads to effects suggestively similar to those seen following removal of the adrenal cortex of the mammal, effects which may be partially and briefly compensated by injections of extracts of the glands.[402] The optic glands regulate development of the reproductive activities and appear in turn to be regulated by the central nervous system.[429]

SUMMARY

In any comparative survey of endocrine mechanisms one is impressed by the fact that, in all those animal groups in which endocrine systems have differentiated, the same general types of functions are being subserved by them. In the two very widely separated animal groups, the Arthropoda and the Vertebrata, which are considered by most zoologists to possess no common ancestry short of relatively primitive forms of life, we see the same general distribution of integrative functions between nervous and endocrine systems and coordinated activity of the two. In both the vertebrate and invertebrate, growth, differentiation, reproduction, metabolism, and pigment cells are importantly regulated by hormones. In brief, the endocrine system seems to dominate primarily (a) those functions for which the time required to induce a response is long (as in those activities involving growth and cell differentiation), (b) those processes for which the controlling influence needs to be maintained over long periods (as in the control of various aspects of basic maintenance and metabolism), and (c) situations in which effector organs are to be maintained in one or another condition for extended periods (as with chromatophores).

Within even relatively large natural groups of animals there is commonly a lack of specificity of the hormonal substances. Among the vertebrates the same general functional types of hormones seem rather widely distributed, typically producing within any species a type of response characteristic for that species. It is chiefly the proteinaceous ones which show evidences of differences in detailed chemical character from species to species. Similarly, among the insects, the juvenile hormone, brain hormone, and ecdysone appear interchangeable among the various species and often even among orders. The same appears to hold true for the chromatophorotropins, retinal pigment hormones, and molt-inhibiting hormone among various species of the Crustacea.

Some similarities have been noted among active principles among the three major groups with endocrine systems. The corpora cardiaca of insects yield a principle highly active on the chromatophores of crustaceans. A rather extensive literature has developed regarding the influence of invertebrate hormones on vertebrates and especially of vertebrate hormones on invertebrates. Much of the work is confusing and contradictory and gives us little reason for believing that the results have anything other than interesting pharmacological value. The vertebrate chromatophorotropin, melanocyte-stimulating hormone (MSH), and the crustacean hormone, UDH, appear qualitatively to resemble one another in their chromatophorotropic action and in a number of their physicochemical properties, but there is adequate reason for believing that they are not identical.[1,51]

The nervous systems of a wide variety of animals show histological and, in many cases, excellent physiological evidence for the differentiation of endocrine elements or neurosecretory cells.[158,188,366] Such cells have been described particularly for certain portions of the nervous systems of *Hydra,* flatworms, roundworms, annelids, starfishes, molluscs, arthropods, and vertebrates. In the last group, they are located in the nucleus preopticus of fishes and amphibians, and in the homologous nuclei supraopticus and paraventricularis of reptiles and mammals. These form part of the intimate hypothalamic-hypophyseal complex demonstrated to be a key portion in coordinating neurons and endocrine activities.

In a fully comparable manner, in the insect the pars intercerebralis–corpora cardiaca complex interrelates the nervous and endocrine systems. In the crustaceans, X-organ–sinus gland complexes and brain-secretory cell-postcommissural organ associations form the fully analogous systems. Evidence that still another large group, the Arachnoidea, possesses the same kind of organization is seen in spiders. Also, there has been found in the nervous system of *Limulus* a quantitative distribution of chromactivating factors for crustaceans closely paralleling the frequency of neurosecretory cells known to occur there.[53,362] The existence of histological and physiological evidence for neurosecretion in such a gamut of kinds of creatures argues for high probability that a neuroendocrine complex will be found to be a general metazoan characteristic.

Certain large ganglion cells in annelids give a chromaffin-staining reaction characteristic of adrenalin-producing tissues, and extraction and assay of such nervous tissue gives positive physiological tests for adrenaline. Among the vertebrates the medulla of the adrenal gland is derived embryologically from nervous tissue.

In brief, there appears to have been an evolution of certain essential endocrine sources from tissues possessing a simple nervous or conductile function and located within the central nervous system, through an intermediate stage where the cells exhibited their secretory function while still retaining the special conductile ability of nerve cells, to a condition in which the cells have become specialized for endocrine activity alone and form glandular tissues or organs apart from the nervous system. And in those instances where the secreting cell bodies are still retained within nervous organs, special intracellular conductile means are utilized to convey the secreted product beyond the bounds of the highly specialized "brain-blood barriers" and into proximity to their own vascular beds. Secondarily, other endocrine organs appear to have developed morphologically, but never physiologically, independent of either direct or indirect regulation by the nervous system.

It is interesting to note that, in the vertebrates, those hormones from sources of ectodermal or endodermal origin are proteins or polypeptides (e.g., STH, LH, ADH, ACTH, MSH, TSH) or at least contain nitrogen (epinephrine); hormones arising from tissues of mesodermal origin are characteristically steroids (sex hormones, cortical principles).

Interesting similarities also exist between vertebrate and invertebrate hormones with regard to their mechanisms of action. The action of serotonin on the muscular tissue of *Fasciola hepatica* is mediated by cyclic AMP. There is also some evidence for the involvement of cyclic AMP in some actions of insect hormones and of crustacean chromatophorotropins. In addition, RNA and protein synthesis are important to the action of insect and crustacean hormones. The studies on the effects of insect hormones on these processes have provided information of great importance to the mode of action of hormones in general.

There has been some speculation in the literature as to which is probably the more primitive integrating mechanism within animals—nervous or endocrine. Obviously both mechanisms, when broadly interpreted, extend to all forms of living organisms, both unicellular and multicellular. The phenomena of excitation and conduction and chemical intercellular transmission of information, the basic underlying activities in the physiology of the nervous system, are common to all cells. As was pointed out in the introduction to this chapter, some organizing and differentiating forces obviously had to precede both of these coordinating mechanisms in both phylogeny and ontogeny. In development, gene-induced differentiation must precede organizer activity. There is no good reason to postulate, therefore, that either the excitatory or the chemical coordination factor is phylogenetically the more primitive. Both types of coordinatory mechanisms probably evolved simultaneously and entirely parallelly. In short, in response to the functional needs of larger size, multicellular character, and division of labor within organisms, highly specialized, cooperating nervous and endocrine systems became differentiated.

REFERENCES

1. Abramowitz, A. A., Physiol. Zool. *11*:299–311, 1938. Similarity of crustacean sinus gland hormone to intermedin.
2. Adiyodi, K. G., and R. G. Adiyodi, Biol. Rev. *45*:121–165, 1970. Endocrine control of reproduction in decapod Crustacea.
3. Ahlquist, R. P., Amer. J. Physiol. *153*:586–600, 1948. Original elucidation of the adrenergic receptors.
4. Ahlquist, R. P., Ann. Rev. Pharmacol. *8*:259–272, 1968. Agents which block adrenergic beta-receptors.
5. Alexandrowicz, J. S., J. Marine Biol. Assoc. U. K. *31*:563–580, 1953. Pericardial organs of crustaceans.
6. Allen, B. M., Biol. Rev. *13*:1–19, 1938. Hormones and amphibian metamorphosis, review.
7. Amoroso, E. C., Ann. Endocrinol. *16*:435–447, 1955. Reptile corpus luteum.
8. Antheunisse, L. J., Arch. Neerl. Zool. *16*:237–314, 1963. Neurosecretory activities in mussel *Dreissena*.
9. Applebaum, S. W., and L. I. Gilbert, Devel. Biol. *27*:165–175, 1972. Ecdysone stimulation of cyclic AMP.
10. Arvy, L., J. B. Bounhiol, and M. Gabe, C. R. Acad. Sci. *236*:627–629, 1953. Neurosecretion by corpora cardiaca; increased brain secretion in adult *Bombyx*.
11. Arvy, L., G. Échalier, and M. Gabe, Ann. Sci. Nat. Zool. Biol. Animale *18*:263–268, 1956. Gonad regulation by Y-organ in *Carcinus*.
12. Arvy, L., M. Fontaine, and M. Gabe, J. Physiol. *49*:685–697, 1957. Hypothalamic control of ACTH and TSH.
13. Astwood, E. B., pp. 529–532 *in* Handbook of Physiology, Sec. 5, Adipose Tissue, edited by A. E. Renold and G. F. Cahill, Jr. Amer. Physiol. Soc., Washington, D. C., 1965. The pituitary gland and the mobilization of fat.
14. Baba, Y., H. Matsuo, and A. V. Schally, Biochem. Biophys. Res. Comm. *44*:459–463, 1971. Structure of the porcine LH- and FSH-releasing hormone.
15. Baggerman, B., Arch. Neerl. Zool. *12*:105–318, 1957. Photoperiod and sticklebacks; thyroid increases induction of F. W. selection.
16. Balesdent-Marquet, M. L., C. R. Acad. Sci. *236*:1086–1088, 1953. Hormone other than ovarian on female sex characters.
17. Barr, W. A., pp. 164–238 *in* Perspectives in Endocrinology, Hormones in the Lives of Lower Vertebrates, edited by E. J. W. Barrington and

C. Barker Jørgenson. Academic Press, New York, 1968. Patterns of ovarian activity.
18. Barrington, E. J. W., J. Marine Biol. Assoc. U. K. *36*:1–15, 1957. Protein binding of I in *Ciona* endostyle.
19. Barrington, E. J. W., J. Marine Biol. Assoc. U. K. *37*:117–125, 1958. *Amphioxus* endostyle cell binding of I.
20. Barrington, E. J. W., Hormones and Evolution. English Universities Press, London, 1964. 154 pp.
21. Bartholomew, G. A., Bull. Mus. Comp. Zool. Harvard *101*:433–476, 1949. Photoperiodism and English sparrow.
22. Bartholomew, G. A., Anat. Rec. *106*:49–60, 1950. Light, temperature and reptile sex cycles.
23. Bartholomew, G. A., pp. 669–676 *in* Photoperiodism and Related Phenomena in Plants and Animals, edited by R. B. Withrow. Amer. Assoc. Adv. Sci., Washington, D. C., 1959. Photoperiodism and reptiles.
24. Baskin, D. G., and D. W. Golding, Biol. Bull. *139*:461–475, 1970. Endocrine control of reproduction in a viviparous polychaete.
25. Beaudoin, A. R., Anat. Rec. *125*:247–259, 1956. Differential CNS growth with thyroid.
26. Becker, E., and E. Plagge, Biol. Zentralbl. *59*:326–341, 1939. Hormonal control of dipteran metamorphosis.
27. Behrens, D. K., A. Staub, M. A. Root, and W. W. Bromer, Ciba Colloq. Endocrinol. *9*:167, 1956. Glucagon, a polypeptide, crystallized.
28. Belamarich, F. A., and R. C. Terwilliger, Amer. Zool. *6*:101–106, 1966. Cardio-excitor hormone from pericardial organs of *Cancer*.
29. Benoit, J., I. Assenmacher, and E. Brard, C. R. Acad. Sci. *242*:3113–3115, 1956. Cycles in constant light or dark.
30. Berg, O., A. Gorbman, and H. Kobayashi, pp. 302–319 *in* Comparative Physiology, edited by A. Gorbman. John Wiley & Sons, New York, 1959. Variability in thyroid activity and hormones.
31. Berlind, A., and I. M. Cooke, J. Exp. Biol. *53*:679–686, 1970. Release of a neurosecretory hormone as peptide by electrical stimulation of crab pericardial organs.
32. Bern, H. A., and C. S. Nicoll, Rec. Prog. Hormone Res. *24*:681–720, 1968. Comparative endocrinology of prolactin.
33. Best, J. B., A. B. Goodman, and A. Pigon, Science *164*:565–566, 1969. Brain control of planarian fissioning.
34. Bissonnette, T. H., Endocrinology *22*:92–103, 1938. Influence of the hypophysis upon molting in the ferret.
35. Bissonnette, T. H., and E. E. Bailey, Ann. N. Y. Acad. Sci. *45*:221–260, 1944. Control of seasonal changes in coat color of the ermine, *Mustela*.
36. Blaschko, H., P. Hagen, and A. D. Welch, J. Physiol. *129*:27–49, 1955. Membranes of vesicles.
37. Bliss, D. E., Biol. Bull. *104*:275–296, 1953. X-organ source of sinus gland hormones; hormonal regulation of crustacean metabolism; inhibition of growth and regeneration of crab by light.
38. Bliss, D. E., S. M. E. Wang, and E. A. Martinez, Amer. Zool. *6*:197–212, 1966. Water balance in the land crab, *Gecarcinus lateralis*, during the intermolt cycle.
39. Bliss, D. E., and J. H. Welsh, Biol. Bull. *103*:157–169, 1952. X-organ sinus gland in brachyurans.
40. Blount, R. F., J. Exp. Zool. *63*:113–141, 1932. Pituitary transplantation and embryonic pigmentation.
41. Bodenstein, D., J. Exp. Zool. *123*:189–232, 1953. Hormones in development of *Periplaneta*.
42. Bodenstein, D., J. Exp. Zool. *123*:413–434, 1953. Molting in cockroaches.
43. Bodenstein, D., J. Exp. Zool. *124*:105–116, 1953. Molting in cockroaches.
44. Bodenstein, D., pp. 197–211 *in* Recent Advances in Invertebrate Physiology, edited by B. T. Scheer. University of Oregon Publications, 1957. Importance of development of competence to respond to hormones.
45. Boss, W. R., J. Exp. Zool. *94*:181–209, 1943. Hormones and plumage type in gulls.
46. Bounhiol, J. J., C. R. Soc. Biol. *126*:1189–1191, 1937. Corpora allata and Lepidopteran development.
47. Bounhiol, J. J., Z. Zool. Exp. Gen. *81*:54–64, 1939. Hormones in Lepidopteran reproduction.
48. Bowers, C. Y., A. V. Schally, F. Enzmann, J. Boler, and K. Folkers, Endocrinology *86*:1143–1153, 1970. Porcine thyrotropin releasing hormone.
49. Bragdon, D. E., E. A. Lazo-Wasem, M. X. Zarrow, and F. L. Hisaw, Proc. Soc. Exp. Biol. Med. *86*:477–480, 1954. Progesterone in blood of pregnant snakes.
50. Brown, F. A., Jr., Anat. Rec. *75* (Suppl.):129, 1939. Sinus gland and molting in *Palaemonetes*.
51. Brown, F. A., Jr., Quart. Rev. Biol. *19*:32–46, 118–143, 1944. Endocrines in crustaceans, review.
52. Brown, F. A., Jr., Physiol. Zool. *19*:215–223, 1946. Postcommissural gland.
53. Brown, F. A., Jr., and O. Cunningham, Biol. Bull. 77:104–114, 1939; *81*:80–95, 1941. Sinus gland hormone and molting in Crustacea; *Limulus* neurosecretion.
54. Brown, F. A., Jr., M. N. Hines, and M. Fingerman, Biol. Bull. *102*:212–225, 1952. Hormonal control of *Palaemonetes* retinal pigment.
55. Brown, F. A., Jr., and G. M. Jones, Biol. Bull. *91*:228–232, 1949. Eyestalk-hormone inhibition of ovaries in *Uca*.
56. Brown, F. A., Jr., and M. Rollo, Auk *57*:485–498, 1940. Influence of light periods upon plumage type in birds.
57. Brown, F. A., Jr., H. M. Webb, and M. I. Sandeen, J. Cell. Comp. Physiol. *41*:123–144, 1953. Control of secretion of two retinal hormones.
58. Brown-Grant, K., C. von Euler, G. W. Harris, and S. Reichlin, J. Physiol. *126*:1–28, 1954. Thyroid response to cold.
59. Burgos, M. H., Rev. Soc. Argent. Biol. *26*:359–371, 1950. Loss of clasp reflex of toads on castration.
60. Burnett, A. L., and N. A. Diehl, J. Exp. Zool. *157*:217–226, 1964. Neurosecretion in *Hydra* reproduction.
61. Burnett, A. L., N. A. Diehl, and F. Diehl, J. Exp. Zool. *157*:227–236, 1964. Neurosecretion in *Hydra* growth and regeneration.
62. Butenandt, A., and P. Karlson, Z. Naturforsch. *9B*:389–391, 1954. Crystallization of insect molting hormone.
63. Carlisle, D. B., Pubbl. Staz. Zool. Napoli *24*:279–285, 1953. Molt accelerating principle of MTGX-organ.
64. Carlisle, D. B., Pubbl. Staz. Zool. Napoli *24*:285–292, 1953. Thoracic ganglia accelerate molt in Crustacea.
65. Carlisle, D. B., Pubbl. Starz. Zool. Napoli *24*:(Suppl.) 79–80, 1954. Intra-axonal transport.
66. Carlisle, D. B., J. Marine Biol. Assoc. U. K. *36*:291–307, 1957. Y-organ activity induces molt.
67. Carlisle, D. B., pp. 18–19. 2. Internat. Symp. Neurosekr. Springer-Verlag, Berlin, 1958. Intra-axonal transport.
68. Carpenter, M. B., and R. de Roos, Gen. Comp. Endocrinol. *15*:143–157, 1970. Seasonal changes in androgenic gland of crayfish *Orconectes*.
69. Cehovic, G., C. R. Acad. Sci. *261*:1405–1408, 1965. Antigonadal action of MSH.
70. Chaet, A. B., Amer. Zool. *6*:263–271, 1966. Gamete-shedding factors in starfishes.
71. Chaet, A. B., Symp. Zool. Soc. Lond. *20*:13–24, 1967. Gamete-shedding factor of starfish.
72. Chang, C. Y., and E. Witschi, Endocrinology *61*:514–519, 1957. Facilitation of ovulation by cortisone.
73. Charniaux-Cotton, H., C. R. Acad. Sci. *239*:780–782, 1954. Ovarian hormone in crustacean; androgenic gland of Crustacea.
74. Charniaux-Cotton, H., Ann. Biol. *32*:371–398, 1956. Androgenic gland.
75. Charniaux-Cotton, H., Ann. Sci. Nat. Zool. Biol. Animale *19*:411–559, 1957. Molting in *Orchestia;* androgenic gland of Crustacea.
76. Charniaux-Cotton, H., Bull. Soc. Zool. France *83*:314–336, 1958. Crustacean sex differentiation.
77. Charniaux-Cotton, H., C. R. Acad. Sci. *246*:2817–2819, 1958. Androgenic gland of crustaceans.
78. Charniaux-Cotton, H., and L. H. Kleinholz, pp. 135–198 *in* The Hormones, Vol. 4, edited by G. Pincus, K. V. Thimann, and E. B. Astwood. Academic Press, New York, 1964. Hormones in invertebrates other than insects.
79. Chase, L. R., and G. D. Aurbach, Proc. Nat. Acad. Sci. *58*:518–525, 1967. Effect of PTH on cyclic AMP excretion.
80. Chase, L. R., and G. D. Aurbach, Science *159*:545–547, 1968. Renal adenyl cyclases responsive to PTH and ADH.
81. Chase, L. R., S. A. Fedak, and G. D. Aurbach, Endocrinology *84*:761–768, 1969. *In vitro* activation of adenyl cyclase by PTH.
82. Chase, M. D., I. I. Geschwind, and H. A. Bern, Proc. Soc. Exp. Biol. Med. *94*:680–683, 1957. Prolactin synergism with sex hormones and gonadotropins. Direct influence of LTH on prostate.
83. Chavin, W., J. Exp. Zool. *133*:1–46, 1956. Goldfish melanization by ACTH.
83a. Chihara, C. J., W. H. Petri, J. W. Fristrom, and D. S. King, J. Insect Physiol. *18*:1115–1123, 1972. Ecdysone and juvenile hormone on imaginal discs *in vitro*.
83b. Chino, H., L. I. Gilbert, J. B. Siddall, and W. Hafferl, J. Insect Physiol. *16*:2033–2040, 1970. Ecdysone transport in lepidopteran hemolymph.
84. Chiu, K. W., W. G. Lynn, and J. P. Leichner, Biol. Bull. *139*:107–114, 1970. Effect of temperature on thyroid in lizard.
85. Church, N. S., Canad. J. Zool. *33*:339–369, 1955. Hormones and insect molt.
86. Clark, R. B., Oceanography and Marine Biology, an Annual Review, Vol. 3, edited by H. Barnes, 1965. Endocrine activity in polychaetes.
87. Clever, U., and P. Karlson, Exp. Cell. Res. *20*:623–626, 1960. Ecdysone activation of gene activity.
88. Colombo, L., H. A. Bern, and J. Pieprzyk, Gen. Comp. Endocrinol. *16*:74–84, 1971. Steroid transformations by salmon corpuscle of Stannius.
89. Combescot, C., Bull. Soc. Hist. Nat. Afrique Nord *45*:366–377, 1955. Hypophysectomy and gonad and secondary sex character dependence in reptiles.
90. Congote, L. F., C. E. Sekeris, and P. Karlson, Exp. Cell. Res. *56*:338–346, 1969. Juvenile hormone and ecdysone on RNA synthesis.
90a. Congote, L. F., C. E. Sekeris, and P. Karlson, Z. Naturforsch. *25b*:279–284, 1970. Specific RNA synthesis in isolated fat body nuclei after ecdysone and juvenile hormone stimulation.
91. Copp, D. H., pp. 377–398 *in* Fish Physiology, Vol. 2, edited by W. S.

Hoar and D. J. Randall. Academic Press, New York, 1969. Ultimobranchial glands.
92. Cornubert, G., C. R. Acad. Sci. *238*:952–953, 1954. Distinction between ovary-inhibiting hormone and molting one.
93. Cornubert, G., N. Demeusy, and A. Veillet, C. R. Acad. Sci. *234*: 1405–1407, 1952. Acceleration of sexual development by eyestalk removal.
94. Costlow, J. D., Jr., and C. G. Bookhout, Biol. Bull. *113*:224–232, 1957. Nutrition and molting in barnacles.
95. Costlow, J. D., Jr., Gen. Comp. Endocrinol. *3*:120–130, 1963. Eyestalk extirpation and metamorphosis of blue crab megalops.
96. Costlow, J. D., Jr., pp. 209–224 *in* Some Contemporary Studies in Marine Science, edited by H. Barnes. George Allen and Unwin, London, 1966. Eyestalk extirpation and larval development of the crab *Sesarma*.
97. Costlow, J. D., Jr., Gen. Comp. Endocrinol. 7:255–274, 1966. Eyestalk extirpation and larval development of mud crab *Rhithropanopeus*.
98. Courrier, R., and G. Cehovic, C. R. Acad. Sci. *251*:832–834, 1960. Thyrotropic action of MSH.
99. Danforth, C. H., Biol. Symp. *9*:67–80, 1942. Hormones and plumage changes in birds.
100. D'Angelo, S. A., and R. E. Traum, Endocrinology *59*:593–596, 1956. Hypothalamic damage disturbs TSH secretion.
101. Davey, K. G., and S. P. Kan. Canad. J. Zool. *46*:893–897, 1968. Neurosecretory function in nematodes.
102. Davis, D. E., Science *126*:253, 1957. Effects of castration on starling male singing and aggressiveness.
103. Day, M. F., Biol. Bull. *84*:127–140, 1943. Corpus allatum hormone and dipteran reproduction.
104. Demeusy, N., and R. Lenel, C. R. Soc. Biol. *148*:156–158, 1954. Ovary-inhibiting factor and molting hormone.
105. De Wied, D., pp. 97–140 *in* Frontiers in Neuroendocrinology, edited by W. F. Ganong and L. Martini. Oxford University Press, New York, 1969. Effects of peptid hormones on behavior.
106. Disney, H. J., and A. J. Marshall, Proc. Zool. Soc. Lond. *127*:379–387, 1956. Cycles in equatorial birds.
106a. Doane, W. W., *in* Developmental Systems: Insects, edited by S. J. Counce. Academic Press, New York, 1972. Review of insect endocrinology.
107. Dodd, J. M., P. J. Evennett, and C. K. Goddard, Symp. Zool. Soc. Lond. *1*:77–103, 1960. Reproductive endocrinology of cyclostomes and elasmobranchs.
108. Domm, L. V., pp. 227–327 *in* Sex and Internal Secretions, edited by W. C. Young. Williams & Wilkins, Baltimore, 1939. Hormones and plumage changes in birds.
109. Drach, P., Ann. Inst. Oceanogr. Paris N. S. *19*:103–391, 1939. Crustacean molt-cycle stages, Brachyura.
110. Drach, P., Bull. Biol. France Belg. *78*:40–62, 1944. Crustacean molt-cycle stages, Natantia.
111. Dupont-Raabe, M., Bull. Soc. Zool. France *76*:386–397, 1952. Increased brain secretion in adult phasmids.
112. Durand, J. B., Biol. Bull. *111*:62–76, 1956. X-organ in brachyuran.
113. Durchon, M., J. Montreuil, and Y. Boilly-Marer, C. R. Acad. Sci. *257*:1807–1808, 1963. Hormonal activities in polychaetes.
114. Eales, J. G., Canad. J. Zool. *41*:811–824, 1963. Seasonal thyroid changes in the salmon.
115. Echalier, G., C. R. Acad. Sci. *238*:523–525, 1954. Y-organ and crustacean molt.
116. Echalier, G., C. R. Acad. Sci. *242*:2179–2180, 1956. Y-organ and crustacean molt.
117. Eckert, M., Acta Soc. Zool. Biochem. *32*:34–38, 1968. Hormonal control of arachnid molt.
117a. Emmerich, H., J. Insect Physiol. *16*:725–747, 1970. Binding of ecdysone to hemolymph proteins.
117b. Emmerich, H., Z. vergl. Physiol. *68*:385–402, 1970. Binding proteins for ecdysone in *Drosophila*.
118. Enami, M., Biol. Bull. *101*:241–258, 1951. Secretory droplets in X-organ–sinus gland complex; cyclic changes.
119. Engelmann, F., J. Insect Physiol. *1*:257–278, 1957. Ovarian stimulation of brain; inhibition of corpus allatum by brain.
120. Engelmann, F., Z. vergl. Physiol. *41*:456–470, 1959. Prothoracic glands in adult *Leucophaea*.
121. Engelmann, F., The Physiology of Insect Reproduction. Pergamon Press, Oxford, 1970. Review of insect reproduction.
121a. Engelmann, F., Arch. Biochem. Biophys. *145*:439–447, 1971. Juvenile hormone induction of female egg protein.
122. Engelmann, F., and M. Luscher, Verhandl. Deutsch. Zool. Ges. Hamburg 215–220, 1956. Nerve inhibition of corpora allata.
123. Epple, A., Gen. Comp. Endocrinol. *9*:137–142, 1967. Amphiphil cells in pancreatic islets of lower vertebrates.
124. Etkin, W. N., *in* Analysis of Development, edited by B. H. Willier, P. A. Weiss, and V. Hamburger, W. B. Saunders Co., Philadelphia, 1955. Review of thyroid and amphibian development.
125. Etkin, W., and A. G. Gona, J. Exp. Zool. *165*:249–258, 1967. Prolactin-thyroid antagonism in amphibian development.
126. Etkin, W. N., R. W. Root, and B. P. Mofskin, Physiol. Zool. *13*:415–429, 1940. Thyroid and oxygen consumption in fishes.
127. Everett, J. W., Endocrinology *58*:786–796, 1956. LH stimulation of ovulation.
128. Farner, D. S., pp. 198–237 *in* Recent Studies in Avian Biology, edited by A. Wolfson. University of Illinois Press, 1955. Annual stimulus for bird migration; independent of gonads or perhaps all hormones.
129. Farner, D. S., pp. 717–750 *in* Photoperiodism and Related Phenomena in Plants and Animals, edited by R. B. Withrow. Amer. Assoc. Adv. Sci., Washington, D.C., 1959. Photoperiodic control of annual gonadal cycles in birds.
130. Faux, A., D. H. S. Horn, E. J. Middleton, H. M. Fales, and M. E. Lowe, Chem. Comm. 175–176, 1969. Moulting hormones of crab during ecdysis.
131. Fernlund, P., Biochim. Biophys. Acta *237*:519–529, 1971. Chromactivating hormones of *Pandalus borealis;* light-adapting hormone.
132. Fernlund, P., and L. Josefsson, Biochim. Biophys. Acta *158*:262–273, 1968. Chromactivating hormones of *Pandalus borealis*; red-concentrating hormone.
133. Ferrari, W., G. L. Gessa, and L. Vargui, Ann. N. Y. Acad. Sci. *104*: 330–343, 1963. Behavioral effects of ACTH and MSH.
134. Fielder, D. R., K. R. Rao, and M. Fingerman, Marine Biol. *9*:219–223, 1971. Control of distal retinal pigment migration in *Uca*.
135. Fingerman, M., Amer. Zool. *6*:169–179, 1966. Neurosecretory control of pigmentary effectors in crustaceans.
136. Fingerman, M., J. Interdiscipl. Cycle Res. *1*:115–121, 1970. Circadian rhythm of distal retinal pigment migration in *Uca*.
137. Fingerman, M., J. Cell. Comp. Physiol. *50*:357–370, 1957. Evidence for light- and dark-adapting hormones for retinal pigments of crayfish.
138. Fingerman, M., R. A. Krasnow, and S. W. Fingerman, Physiol. Zool. *44*:119–128, 1971. Retinal pigment light-adapting and dark-adapting hormones in *Palaemonetes*.
139. Fingerman, M., and M. E. Lowe, J. Cell. Comp. Physiol. *50*:371–380, 1957. Daily rhythms of crayfish retinal pigment.
140. Fingerman, M., M. E. Lowe, and B. I. Sundararaj, Amer. Midl. Nat. *62*:167–173, 1959. Electrophoretic separation of dark- and light-adapting retinal pigment hormones.
141. Fingerman, M., M. E. Lowe, and B. I. Sundararaj. Biol. Bull. *116*:30–36, 1959. Dark- and light-adapting hormones of *Palaemonetes* retinal pigment.
142. Fingerman, M., and W. C. Mobberly, Jr., Biol. Bull. *118*:393–406, 1960. Hormones controlling the distal retinal pigment of *Palaemonetes*.
143. Foa, P. P., G. Galansino, and G. Pozza, Rec. Prog. Hormone Res. *13*: 473–510, 1957. Properties and role of glucagon.
144. Foglia, V. G., E. M. Wagner, M. de Barros, and M. Marques, C. R. Soc. Biol. *149*:1660–1661, 1955. Pancreatectomy and hyperglycemia and glucosuria in tortoises.
145. Fontaine, M., Mem. Soc. Endocrinol. *5*:69–81, 1956. Thyroxin in osmoregulation in poikilotherms.
146. Fontaine, M., M. M. Baraduc, and J. Hately, C. R. Soc. Biol. *147*: 214–216, 1953. Thyroid in liver glycogen control in poikilotherms.
147. Fontaine, M., and Y. A. Fontaine, J. Physiol. Paris *49*:169–173, 1957. Species differences in TSH.
148. Fraenkel, G., Proc. Roy. Soc. Lond. B *118*:1–12, 1935. Hormonal control of dipteran development.
149. Fraenkel, G., and C. Hsiao, J. Insect Physiol. *11*:516–556, 1965. Bursicon in flies.
150. Fridberg, G., and H. A. Bern. Biol. Rev. *43*:175–199, 1968. The urophysis of fishes.
151. Frieden, E., and J. J. Just, pp. 1–52 *in* Biochemical Actions of Hormones, Vol. 1, edited by G. Litwack. Academic Press, New York, 1970. Hormonal responses in amphibian metamorphosis.
152. Friesen, H., Endocrinology *75*:692–697, 1964. Hypocalcemic effect of MSH.
152a. Fristrom, J. W., R. Raikow, W. Petri, and D. Stewart, *in* Problems in Biology: RNA in Development, edited by E. W. Hanly. University of Utah Press, Salt Lake City, 1969. Ecdysone and juvenile hormone on imaginal discs.
153. Fukuda, S., Ann. Zool. Japan *20*:9–13, 1941. Prothoracic glands and lepidopteran metamorphosis.
154. Fukuda, S., Proc. Japan. Acad. *27*:272–677, 1951. Diapause hormone from subesophageal ganglion.
155. Gabe, M., C. R. Acad. Sci. *237*:1111–1113, 1953. Y-organ and molt in crustaceans.
156. Gabe, M., Ann. Sci. Nat. Zool. Biol. Animale *18*:145–152, 1956. Y-organ and molt in crustaceans.
157. Gabe, M., Arch. Anat. Micr. *44*:351–383, 1955. Neurosecretion in arachnids.
158. Gabe, M., Neurosecretion. Pergamon Press, New York, 1966.
159. Ganong, W. F., N. I. Gold, and D. M. Hume, Fed. Proc. *14*:54, 1955. Absence of corticoids in stress after lesions in median eminence.
160. Ganong, W. F., and D. M. Hume, Endocrinology *55*:474–483, 1954. Basal level of ACTH secretion without hypothalamus.
161. Gersch, M., and J. Stürzebecher, J. Insect Physiol. *14*:87–96, 1968. Protein nature of brain hormone.
162. Geschwind, I. I., pp. 421–443 *in* Comparative Endocrinology, edited by A. Gorbman. John Wiley & Sons, New York, 1959. Prolactin differences between ox and sheep.
163. Geschwind, I. I., Endocrinology *79*:1165–1167, 1966. Action of MSH on mouse hair color.
164. Geschwind, I. I., pp. 180–189 *in* Progress in Comparative Endocrinology (Gen. Comp. Endocrinol. Suppl. 2), edited by M. R. N. Prasad, 1969. Comparative biochemistry of gonadotropins.
165. Geschwind, I. I., C. H. Li, and L. Barnafi, J. Amer. Chem. Soc. *78*:

4494–4495, 1956. Structure of β-MSH. Chemistry of MSH.
166. Gilbert, L. I., pp. 67–134 *in* The Hormones, Vol. 4, edited by G. Pincus, K. V. Thimann, and E. B. Astwood. Academic Press, New York, 1964. Hormones regulating insect growth.
167. Gilbert, L. I., Proc. Third Int. Cong. Endocrinol., pp. 340–346. Chemistry of ecdysones.
168. Gilbert, L. I., Adv. Insect Physiol. *4*:69–211, 1967. Lipids and sterols in insects.
168a. Gilbert, L. I., and D. S. King, *in* Physiology of Insecta, 2nd edition, edited by M. Rockstein. Academic Press, New York, 1973. Review of insect endocrinology.
169. Gorbman, A., Proc. Soc. Exp. Biol. Med. *45*:772–773, 1940. Thyrotropic principle in fish pituitary.
170. Gorbman, A., Physiol. Rev. *35*:336–346, 1955. Bound I in dermal glands of Hemichorda; bound I with exoskeletal scleroproteins.
171. Gorbman, A., pp. 266–282 *in* Comparative Endocrinology, edited by A. Gorbman. John Wiley & Sons, New York, 1959. Problems in comparative morphology and physiology of vertebrate thyroid gland.
172. Gorbman, A., and H. A. Bern, Textbook of Comparative Endocrinology. John Wiley & Sons, New York, 1962. 468 pp.
173. Gorbman, A., M. Clements, and R. O'Brien, J. Exp. Zool. *127*:75–89, 1954. Protein binding of I in stolonic canals, *Botryllus.*
173a. Gorell, T. A., L. I. Gilbert, and J. B. Siddall, Proc. Nat. Acad. Sci. *69*:812–815, 1972. Binding proteins in a crustacean for an ecdysone metabolite.
174. Gorell, T. A., L. I. Gilbert, and J. Tash, Insect Biochem. *2*:94–106, 1972. Uptake of ecdysone by insect tissues.
175. Gorski, J., D. Toft, G. Shyamala, D. Smith, and A. Notides, Rec. Prog. Hormone Res. *24*:45–72, 1969. Estrogen interaction with uterine receptors.
176. Grant, W. C., and J. A. Grant, Biol. Bull. *114*:1–9, 1958. Prolactin in amphibians; responsibility for water drive in red eft stage of newt *Triturus.*
177. Grossman, M. I., Vitam. Horm. *16*:179–263, 1958. Secretin.
178. Grossman, M. I., Gastrin. University of California Press, Berkeley, 1966. 337 pp.
179. Gudernatsch, J. F., Arch. Entwick-Mech. *35*:457–483, 1912. Thyroid and amphibian metamorphosis.
180. Guyselman, J. B., Biol. Bull. *104*:115–137, 1953. Molting in *Uca.*
181. Habener, J. F., F. R. Singer, L. J. Deftos, R. M. Neer, and J. T. Potts, Jr., Nature *232*:91–92, 1971. Explanation for the high potency of salmon calcitonin.
182. Hadorn, E., and J. Neel. Arch. Entwick-Mech. *138*:281–304, 1938. Larval ring gland and dipteran development.
183. Hagadorn, I. R., Amer. Zool. *6*:251–261, 1966. Neurosecretion in leech reproduction.
184. Hampshire, F., and D. H. S. Horn, Chem. Comm. 37–38, 1966. Structure of crustecdysone, a crustacean moulting hormone.
185. Handler, J. S., R. W. Butcher, E. W. Sutherland, and J. Orloff, J. Biol. Chem. *240*:4524–4526, 1965. ADH increases toad bladder/cyclic AMP.
186. Hänstrom, B., Kungl. svensk. Vetensk. Handl. *16*:1–99, 1937. X-organs and sinus glands.
187. Hänstrom, B., Hormones in Invertebrates. Oxford University Press, 1939. 198 pp.
188. Hänstrom, B., Colston Papers, Butterworths Sci. Publ., London *8*:23–37, 1956. Comparative neurosecretion.
189. Hara, J., Ann. Zool. Japan. *25*:411–414, 1952. Hormonal regulation of crustacean heart.
190. Harris, J. I., and P. Roos, Nature *178*:90, 1956. Amino acid sequence in pig β-MSH. Chemistry and structure of MSH.
191. Harris, J. I., F. Sanger, and M. A. Naughton, Arch. Biochem. *65*:427–438, 1956. Structure of insulin.
192. Hasegawa, K., Proc. Japan. Acad. *27*:667–671, 1951. Diapause hormone from subesophageal ganglion.
193. Hasegawa, K., Nature *179*:1300–1301, 1957. Extract of diapause hormone.
194. Hasler, A. D., R. K. Meyer, and H. M. Field, Endocrinology *25*:978–983, 1939. Hypophyseal gonadotropins in fishes.
195. Hauenschild, C., Zool. Anz. Suppl. *27*:111–120, 1964. Neuroendocrine roles in polychaetes.
196. Henry, S., and H. B. van Dyke, J. Endocrinol. *16*:310–325, 1958. Immunological differences in ICSH.
197. Herlant-Meewis, H., *in* Advances in Morphogenesis, Vol. 4, edited by M. Abercrombie and J. Brachet. Academic Press, New York, 1964. Neurosecretory roles in oligochaetes.
198. Hertault, J., C. R. Acad. Sci. *272*:1981–1983, 1971. Neuroendocrine system of pseudoscorpions.
199. Hickman, C. P., Jr., Canad. J. Zool. *37*:997–1060, 1959. Greater thyroid activity in euryhaline fish in more saline medium.
200. Highnam, K. C., Quart. J. Micr. Sci. *99*:73–88, 1958. Neurosecretion by corpora cardiaca.
201. Highnam, K. C., and L. Hill, The Comparative Endocrinology of the Invertebrates. American Elsevier Publ. Co., New York, 1969. 270 pp.
202. Hisaw, F. L., *in* Comparative Endocrinology, edited by A. Gorbman. John Wiley & Sons, New York, 1959. Endocrine adaptation of the mammalian estrous cycle and gestation.
203. Hoar, W. S., *in* The Physiology of Fishes, edited by M. Brown. Academic Press, New York, 1957. Thyroid and fish development.
204. Hodge, M. G., and G. B. Chapman, J. Biophys. Biochem. Cytol. *4*:571–574, 1958. Vesicles in sinus gland.
205. Hopper, A. F., J. Exp. Zool. *119*:105–109, 1952. Low thyroid prolongation of sex immaturity; thyroxin treatment and early sex development—*Lebistes.*
206. Horton, E. W., Prostaglandins. Monographs on Endocrinology, No. 7. Springer-Verlag, New York, 1972. 197 pp. Up-to-date survey of prostaglandins.
207. Höst, P., Auk *59*:388–403, 1942. Light periods and seasonal plumage changes in the ptarmigan.
208. Houssay, B. A., *in* Comparative Endocrinology, edited by A. Gorbman. John Wiley & Sons, New York, 1959. Comparative physiology of the endocrine pancreas.
209. Howard, K. S., R. G. Shepherd, E. A. Eigner, D. S. Davies, and P. H. Bell, J. Amer. Chem. Soc. *77*:3419, 1955. Three corticotropins.
210. Howie, D. I. D., Gen. Comp. Endocrinol. *6*:347–361, 1966. Endocrine factors in *Arenicola* reproduction.
211. Huber, R., and W. Hoppe, Chem. Ber. *98*:2403–2424, 1965. Structure of ecdysone.
212. Ichikawa, M., and J. Nishutsutsuji, Ann. Zool. Japan. *24*:205–211, 1951. Brain and imaginal development in Lepidoptera.
213. Idler, D. R., and B. Truscott, Steroids *9*:457–478, 1967. Discovery of 1-α-hydrocorticosterone in ray interrenal.
213a. Ilan, J., J. Ilan, and N. Patel, J. Biol. Chem. *245*:1275–1281, 1970. Juvenile hormone action at the translational level.
214. Ishizaki, I., and M. Ichikawa, Biol. Bull. *133*:355–368, 1967. Protein nature of brain hormone.
215. Iwanoff, P. P., and K. A. Mestscherskoja, Zool. Jahrb. Abt. allg. Zool. Physiol. *55*:281–348, 1935. Gonadal hormone in insect reproduction.
216. Johannsen, A. S., Nature *181*:198–199, 1958. Inhibition of corpus allatum by brain.
217. Joly, R., Bull. Biol. Lance Belg. *100*:379–480, 1966. Molt regulation in myriapods.
218. Jones, B. M. J. Exp. Biol. *33*:174–185, 1956. Brain and prothoracic gland in embryonic molts.
219. Jones, I. C., The Adrenal Cortex. Cambridge University Press, 1957. Role of adrenal cortex; annual rhythm.
220. Jones, I. C., *in* The Neurohypophysis, edited by H. Heller. Butterworth & Co., London, 1966. Adrenalectomy in winter and summer frogs.
221. Joose, J., Arch. Neerl. Zool. *16*:1–103, 1964. Neurosecretory activities in *Lymnaea.*
222. Kamemoto, F. I., K. N. Kato, and L. E. Tucker, Amer. Zool. *6*:213–219, 1966. Neurosecretion and salt and water balance in the Annelida and Crustacea.
223. Kato, K. N., and F. I. Kamemoto, Comp. Biochem. Physiol. *28*:665–674, 1969. Neuroendocrines in osmoregulation in grapsid crab *Metopograpsus.*
224. Keller, R., Verh, Deutsch. Ges. Innsbruck 628–835, 1968. Diabetogenic hormone of *Orconectes.*
225. Keller, R., Z. vergl. Physiol. *63*:137–145, 1969. Species specificity of a crustacean hormone.
225a. King, D. S., Gen. Comp. Endocrinol. Suppl. *3*:221–227, 1972. Metabolism of ecdysone.
226. Kitay, J. I., and M. D. Altschule, The Pineal Gland. Harvard University Press, Cambridge, 1954. 280 pp.
227. Kleinholz, L. H., Biol. Bull. *70*:159–184, 1936. Eyestalk hormone and retinal pigment migration in crustaceans.
228. Kleinholz, L. H., Biol. Bull. *75*:510–532, 1938. Hormonal control of crustacean retinal pigment movements.
229. Kleinholz, L. H., Biol. Bull. *99*:454–468, 1950. Hormonal regulation of blood sugar in crustaceans.
230. Kleinholz, L. H., Biol. Bull. *109*:362, 1955. Reflecting pigment of crustacean eye.
231. Kleinholz, L. H., Amer. Zool. *6*:161–167, 1966. Separation and purification of crustacean eyestalk hormones.
232. Kleinholz, L. H., Gen. Comp. Endocrinol. *15*:578–588, 1970. Separation and purification of crustacean neurosecretory pigmentary-effector hormones.
233. Kleinholz, L. H., and E. Bourquin, Proc. Nat. Acad. Sci. *27*:145–149, 1941. Eyestalks and molting in crabs.
234. Kleinholz, L. H., F. Kimball, and M. McGarvey, Gen. Comp. Endocrinol. *8*:75–81, 1967. Separation of diabetogenic hormone from the crustacean eyestalk.
235. Knowles, F. G. W., Proc. Roy. Soc. Lond. B *141*:248–267, 1953. Postcommissural organ.
236. Knowles, F. G. W., *in* Comparative Endocrinology, edited by A. Gorbman. John Wiley & Sons, New York, 1959. Secretory granules.
237. Knowles, F. G. W., and D. B. Carlisle, Biol. Rev. *31*:396–473, 1956. Terms MTGX-, SPX-organ. Crustacean chromatophorotropin polypeptides. Crustacean endocrinology, review.
238. Kollros, J. J., pp. 340–350 *in* Comparative Endocrinology, edited by A. Gorbman. John Wiley & Sons, New York, 1959. Complex reacting system of amphibian development to thyroid activity.
239. Kopec, S., Biol. Bull. *42*:324–342, 1922. Endocrine control of insect development.
240. Korner, A., Proc. Roy. Soc. Lond. B *176*:287–290, 1970. Hormonal control of protein synthesis.

241. Krayer, O., E. B. Astwood, D. R. Waud, and M. H. Alper, Proc. Nat. Acad. Sci. *47*:1227–1236, 1961. Effect of ACTH and MSH on heart rate.
242. Krishnakumaran, A., and H. A. Schneiderman, Biol. Bull. *139*:520–538, 1970. Control of molting in mandibulate and chelicerate arthropods by ecdysones.
243. Kroeger, H., *in* Metamorphosis: A Problem in Developmental Biology, edited by W. Etkin and L. I. Gilbert. Appleton-Century-Crofts, New York, 1968. Control of gene activity by ions.
244. Kupferman, I., J. Neurophysiol. *33*:877–881, 1970. Neuroendocrine regulation of egg-laying in *Aplysia*.
245. Lamotte, M., and P. Prum, C. R. Soc. Biol. *151*:1187–1191, 1957. Endocrine physiology of a viviparous amphibian.
246. Lamotte, M., and P. Rey, C. R. Acad. Sci. *238*:393–395, 1954. Corpora lutea in amphibian.
247. Lamotte, M., and P. Rey, C. R. Soc. Biol. *151*:1191–1194, 1957. Corpora lutea in amphibian.
248. Lamotte, M., P. Rey, and V. Vilter, C. R. Soc. Biol. *150*:393–396, 1956. Endocrine physiology of viviparous amphibian.
249. Lane, N. J., Quart. J. Micr. Sci. *103*:211–223, 1962. Neurosecretory cells in snail optic tentacles.
250. Lane, N. J., Quart. J. Micr. Sci. *105*:31–34, 1964. Neurosecretory cells in *Helix*.
251. Lederis, K., Gen. Comp. Endocrinol. *14*:417–426, 1970. Bioassay of the teleost urophysis.
252. Lederis, K., Gen. Comp. Endocrinol. *14*:427–437, 1970. Characterization of the teleost urophysis principle.
253. Legendre, R., Ann. Sci. Nat. Zool. *1*:339–474, 1959. Neurosecretory cells in arachnids.
254. Lees, A. D., pp. 585–600 *in* Photoperiodism and Related Phenomena in Plants and Animals, edited by R. B. Withrow. Amer. Assoc. Adv. Sci., Washington, D.C., 1953. Photoperiodism in insects.
255. Lender, T., and N. Klein, C. R. Acad. Sci. *253*:331–333, 1961. Neurosecretion in *Polycelis* regeneration.
256. Lentz, T. L., Science *150*:633–635, 1965. Neurosecretion and *Hydra* growth.
257. Lerner, A. B., and T. H. Lee, J. Amer. Chem. Soc. *77*:1066–1067, 1955. Structure of MSH.
258. Lerner, A. B., and J. S. McGuire, Nature *189*:176–179, 1961. Effect of MSH on human skin color.
259. Lever, J., J. Jansen, and T. A. De Vlieger, Proc. K. ned. Akad. Wet. c. *64*:532–542, 1961. Neurosecretion in *Lymnaea* water regulation.
260. Lewis, G. P., ed., 5-Hydroxytryptamine. Pergamon Press, London, 1958. Serotonin (5-hydroxytryptamine) as a neurohumoral agent.
261. Lezzi, M., and L. I. Gilbert, Proc. Nat. Acad. Sci. *64*:498–503, 1969. Gene activation by ecdysone and juvenile hormone.
262. Lezzi, M., and L. I. Gilbert, J. Cell. Sci. *6*:615–628, 1970. Effects of ions on gene activity in isolated chromosomes.
263. Li, C. H., *in* Symposium on Protein Structure, edited by A. Neuberger. John Wiley & Sons, New York, 1958. Three classes of growth hormone; differences in growth hormones and in sheep and ox prolactin.
264. Li, C. H., Persp. Biol. Med. *11*:498–521, 1968. Current concepts of pituitary hormone biochemistry.
265. Li, C. H., et al., Nature *176*:687–689, 1955. Three corticotropins.
266. Liao, T. H., and J. G. Pierce, J. Biol. Chem. *245*:3275, 1970. Structure of LH.
267. Lofts, B., pp. 239–304 *in* Perspectives in Endocrinology, Hormones in the Lives of Lower Vertebrates, edited by E. J. W. Barrington and C. Barker Jørgenson. Academic Press, New York, 1968. Patterns of testicular activity.
268. Lowe, M. E., O. H. S. Horn, and M. Galbraith, Experientia *24*:518–519, 1968. The role of crustecdysone in molting crayfish.
269. Lubet, P., Ann. Sci. Nat. Zool. *18*:175–183, 1956. Neurosecretion and reproduction in *Mytilus*.
270. Luscher, M., and F. Engelmann, Rev. Suisse Zool. *62*:649–657, 1955. Corpora allata function in *Leucophaea*.
271. Lyman, C. P., Bull. Mus. Comp. Zool. *93*:391–461, 1943. Control of seasonal changes in coat color of the varying hare, *Lepus*.
272. Lynn, W. G., and H. E. Wachowski, Quart. Rev. Biol. *26*:123–168, 1951. Growth and differentiation effects, thyroid.
273. Maher, M. J., and B. H. Levedahl, J. Exp. Zool. *140*:169–189, 1959. Thyroid effect on lizard oxidative metabolism.
274. Marshall, A. J., Mem. Soc. Endocrinol. *4*:75–93, 1955. Hypothalamic control of pituitary in bird.
275. Marshall, F. H. A., Biol. Rev. *17*:68–90, 1942. Factors in sex periodicity.
276. Martini, L., and W. F. Ganong, eds., Neuroendocrinology. Academic Press, New York, 1967. 1551 pp.
277. Matsumoto, K., Biol. J. Okayama Univ. *4*:103–176, 1958. X-organ in crabs.
278. Matthews, L. H., Mem. Soc. Endocrinol. *4*:129–148, 1955. Reptile corpus luteum.
279. Maynard, D. M., Biol. Bull. *121*:316–326, 1961. Thoracic neurosecretory structures in Brachyura.
280. McLennan, H., Synaptic Transmission, 2nd edition. W. B. Saunders Co., Philadelphia, 1970. 134 pp.
281. Megusar, F., Arch. Entwick-Mech. *33*:462–665, 1912. Eyestalks and molting in Crustacea.
282. Mellish, C. H., and R. K. Meyer, Anat. Rec. *69*:179–189, 1937. Gonadotropic activity in reptiles.
283. Meyer, A. S., E. Hanzmann, H. A. Schneiderman, L. I. Gilbert, and M. Boyette, Arch. Biochem. Biophys. *137*:190–213, 1970. Structure of the two juvenile hormones.
284. Mialhe, P., J. Physiol. Paris *47*:248–250, 1955. Ducks and pancreas.
285. Miller, A. H., Condor *56*:13–20, 1954. Refractory period in birds.
286. Miller, M. R., and D. H. Wurster, *in* Comparative Endocrinology, edited by A. Gorbman. John Wiley & Sons, New York, 1959. Sensitivity of amphibians to insulin; insensitivity to glucagon; absence of α-cells.
287. Miller, M. R., and D. H. Wurster, Endocrinology *63*:191–200, 1958. Sensitivity of reptiles to glucagon and abundance of α-cells.
288. Molony, P. J., and L. Goldsmith, Canad. J. Biochem. Physiol. *35*:79–92, 1957. Antibodies to administered insulin.
289. Morel, F., Aldosterone. J. & A. Churchill, London, 1958.
290. Mueller, G. C., Cancer Res. *17*:490–506, 1957. Discussion of steroid hormone mechanism of action.
291. Muhlbach, O., and L. M. Boot, Ann. Endocrinol. *17*:338–343, 1956. Series of pseudopregnancies in mice with extra pituitary subcutaneous-LTH production.
292. Munson, P. I., et al., Rec. Progr. Hormone Res. *24*:589–637, 1968. Thyrocalcitonin.
293. Mutt, V., and J. E. Jorpes, Rec. Progr. Hormone Res. *23*:483–503, 1967. Gastrointestinal hormone biochemistry.
294. Nair, R. M. G., A. J. Kastin, and A. V. Schally, Biochem. Biophys. Res. Comm. *43*:1376–1381, 1971. MSH release-inhibiting hormone.
295. Naisse, J., Arch. Biol. 77:139–201, 1966. Sex hormone in an insect.
296. Naisse, J., Gen. Comp. Endocrinol. *7*:105–110, 1966. Sex hormones in an insect.
297. Nayar, K. K., Curr. Sci. *22*:149, 1953. Correlation of insect brain secretion with oviposition.
298. Nelson, W. O., Mécanisme Physiologique de la Secretion Lactée. Paris, Publ. de CNRS, 1951. LTH affect mammary tissue.
299. Nickerson, M., Ann. N. Y. Acad. Sci. *139*:571–579, 1967. New developments in adrenergic blocking drugs.
300. Nickerson, M., and L. S. Goodman, Fed. Proc. *7*:397–409, 1948. Adrenergic blockage by dibenamine.
301. Nikitovitch-Winer, M., and J. W. Everett, Endocrinology *63*:916–930, 1958. Cycles in rats only when pituitary implants under median eminence.
302. Novales, R. R., Ann. N. Y. Acad. Sci. *100*:1035–1047, 1963. Melanophore responses to MSH, melatonin, and epinephrine.
303. Novales, R. R., *in* Handbook of Physiology, Sec. 7, Endocrinology: The Hypothalamus, edited by E. Knobil. Amer. Physiol. Soc., Washington, D.C., 1971. Physiological actions of MSH.
304. Novales, R. R., and B. J. Novales, Progr. Brain. Res. *10*:507–519, 1965. Melatonin antagonisms on melanophores.
305. Oksche, A., D. Laws, F. I. Kamemoto, and D. S. Farner, Z. Zellforsch. *51*:1–42, 1959. Secretory granules in nerve loops of median eminence of white-crowned sparrows under 8 hours' light.
306. Orloff, J., and J. A. Handler, Biochem. Biophys. Res. Comm. *5*:63–66, 1961. ADH-like effects of cyclic AMP on toad bladder.
307. Oshima, K., and A. Gorbman, Gen. Comp. Endocrinol. *7*:398–409, 1966. Effect of T_4 on electrical activity in goldfish brain.
308. Oshima, K., and A. Gorbman, Gen. Comp. Endocrinol. *7*:482–491, 1966. Further information on thyroxine and goldfish brain.
309. Otsu, T., Embryologia *8*:1–20, 1963. Bihormonal control of sexual cycle in *Potamon*.
310. Page, I. H., Physiol. Rev. *38*:277–335, 1958. Serotonin as a neurohumoral agent.
311. Palay, S. L., Anat. Rec. *121*:348, 1955. Vesicles in nerve fibers: the secretion.
312. Panigel, M., Ann. Sci. Nat. Zool. *18*:569–668, 1956. Reptile corpus luteum; presence in all reptiles, whether oviparous or viviparous.
313. Panouse, J., C. R. Acad. Sci. *218*:293–294, 1944. Sinus glands and ovarian growth in crustaceans.
314. Panouse, J., Ann. Inst. Oceanogr. *23*:65–147, 1946. Sinus glands and ovarian growth in crustaceans.
315. Pardoe, A. Y., and M. Weatherall, J. Physiol. *127*:201–212, 1955. Vasopressin and oxytocin from granules of neurohypophysis.
316. Paris, O. H., Jr., and C. E. Jenner, pp. 601–624 *in* Photoperiodism and Related Phenomena in Plants and Animals, edited by R. B. Withrow. Amer. Assoc. Adv. Sci., Washington, D.C., 1959. Photoperiodic control of diapause in insects.
317. Passano, L. M., Physiol. Comp. Oecol. *3*:155–189, 1953. X-organ-sinus gland complex.
318. Patel, N., and K. Madhaven, J. Insect Physiol. *15*:2141–2150, 1969. Juvenile hormone and ecdysone on RNA synthesis.
319. Pelluet, D., and N. J. Lane, Canad. J. Zool. *39*:789–805, 1961. Neurosecretory roles in slugs.
320. Pesetsky, I., and J. J. Kollros, Exp. Cell. Res. *11*:477–482, 1956. Differential CNS growth with thyroid.
321. Pfeiffer, I. W., J. Exp. Zool. *82*:439–461, 1939. Corpora allata and reproduction in Orthoptera. Hormonal control of development.
322. Pfeiffer, I. W., J. Exp. Zool. *99*:183–233, 1945. Corpus allatum and general metabolism in insects.
323. Pflugfelder, O., Z. wiss. Zool. *153*:108–135, 1940. Hormones in *Dixippus* development and reproduction.

324. Pickford, G. E., Bull. Bingham Oceanogr. Coll. *14*:46–68, 1953. Control of spawning in *Fundulus*; fish response to beef growth hormone.
325. Pickford, G. E., Endocrinology *55*:274–287, 1954. Fish growth to fish growth hormone.
326. Pickford, G. E., pp. 404–420 *in* Comparative Endocrinology, edited by A. Gorbman. John Wiley & Sons, New York, 1959. Prolactin cooperation with MSH in fish pigmentation.
327. Pickford, G. E., and J. W. Atz, The Physiology of the Pituitary Gland of Fishes. New York Zoological Society, 1957.
328. Pierce, J. G., Endocrinology *89*:1331–1344, 1971. Structure of TSH.
329. Porter, G. A., R. Bogoroch, and I. S. Edelman, Proc. Nat. Acad. Sci. *52*:1326–1333, 1964. Aldosterone action on sodium transport.
330. Possompès, B., Arch. Zool. Exp. Gen. *89*:203–364, 1953. Hormones and *Calliphora* metamorphosis.
331. Potter, D. D., pp. 113–118 *in* 2. Int. Symp. Neurosekr. Springer-Verlag, Berlin, 1958. Secretory granules.
332. Potter, D. D., and W. R. Lowenstein, Amer. J. Physiol. *183*:652, 1955. Conduction of impulses by axons of pituitary stalk.
333. Potts, J. T., Jr., and G. D. Aurbach, pp. 53–67 *in* The Parathyroid Glands, edited by P. J. Gaillard, R. V. Talmage, and A. M. Budy. University of Chicago Press, Chicago, 1965. Chemistry of PTH.
334. Prabhu, V. K. K., Z. Zellforsch. *54*:717–733, 1961. Neurosecretory organs of myriapods.
335. Pritchard, A. W., and A. Gorbman, Biol. Bull. *119*:109–119, 1960. Thyroid hormone and dogfish oxygen use.
336. Pyle, R. W., Biol. Bull. *85*:87–102, 1943. Cycles in histological picture in sinus gland and molting in crustaceans.
337. Rahm, U. H., Rev. Suisse Zool. *59*:173–237, 1952. Hormones and *Sialis* development.
337a. Raikow, R. B., and J. W. Fristrom, J. Insect Physiol. *17*:1599–1614, 1971. Effect of ecdysone on RNA metabolism in imaginal discs.
338. Ralph, C. L., and R. M. Frapps, Anat. Rec. *130*:360–361, 1958. Gonadal control via hypothalamus.
339. Ralph, C. L., D. L. Grinwich, P. F. Hall, J. Exp. Zool. *166*:289–294, 1967. Hormones and weaver bird pigmentation.
340. Ralph, C. L., and S. Sampath, Gen. Comp. Endocrinol. 7:370–374, 1966. Inhibition of MSH release.
341. Rao, K. R., Experientia *21*:593–594, 1965. Moult-inhibiting hormone of the crustacean eyestalk.
342. Rao, K. R., M. Fingerman, and C. Hays, Amer. Zool. *11*:644, 1971. α-Ecdysone and 20-hydroxyecdysone on regeneration and molting in *Uca*.
343. Reiter, R. J., and S. Sorrentino, Jr., Amer. Zool. *10*:247–258, 1970. Pineal gland and reproduction.
344. Riggs, T. R., pp. 157–208 *In* Biochemical Actions of Hormones, Vol. 1, edited by G. Litwack. Academic Press, New York, 1970. Hormones and transport across cell membranes.
345. Roberts, J. L., Physiol. Zool. *30*:232–242, 1957. Nutrition and crab molting.
346. Robison, G. A., R. W. Butcher, and E. W. Sutherland, Ann. Rev. Biochem. *37*:149–174, 1968. Cyclic AMP.
347. Robison, G. A., R. W. Butcher, and E. W. Sutherland, Cyclic AMP. Academic Press, New York, 1971, 531 pp. Thorough treatment of all aspects.
348. Röller, H., K. H. Dahm, C. C. Sweeley, and B. M. Trost, Angew. Chem. *79*:190–191, 1967. Structure of juvenile hormone.
349. Rondell, P. A., and P. A. Wright, Physiol. Zool. *31*:236–243, 1958. Metabolic studies of frog ovulation.
350. Rowan, W., Biol. Rev. *13*:374–402, 1938. Light and reproductive cycles.
351. Rugh, R., Biol. Bull. *66*:22–29, 1934. Gonadotropic activity of hypophysis in amphibians.
352. Rugh, R., J. Exp. Zool. *71*:149–162, 1935. Gonadotropic activity of hypophysis in amphibians.
353. Rugh, R., J. Exp. Zool. *71*:163–193, 1935. Gonadotropic activity of hypophysis in amphibians.
354. Rust, C. C., Gen. Comp. Endocrinol. 5:222–231, 1965. Hormonal control of weasel pelage.
355. Rust, C. C., and R. K. Meyer, Gen. Comp. Endocrinol. *11*:548–551, 1968. Pituitary grafts and weasel hair color.
356. Sandeen, M. I., and F. A. Brown, Jr., Physiol. Zool. *25*:233–230, 1952. Retinal pigment response of *Palaemonetes* to illumination.
357. Sandor, T., Gen. Comp. Endocrinol. Suppl. *2*:284–298, 1969. Survey of steroids throughout the vertebrates.
358. Sawyer, W. H., Amer. J. Physiol. *164*:44–48, 1951. Posterior pituitary and frog skin water permeability.
359. Sawyer, W. H., Endocrinology *66*:112–120, 1960. Neurohypophyseal hormones and water regulation in *Rana*; neurohypophyseal hormones and frog bladder permeability.
360. Sawyer, W. H., R. A. Munsick, and H. B. van Dyke, Circulation *21*: 1027–1037, 1960. Distribution of posterior lobe hormones.
361. Scharrer, B., J. Comp. Neurol. *74*:109–130, 1941. The nermertean cerebral organ.
362. Scharrer, B., Biol. Bull. *18*:96–104, 1941. Neurosecretory cells in *Limulus*.
363. Scharrer, B., Endocrinology *38*:35–45, 1946. Hormonal control of orthopteran development.
364. Scharrer, B., Endocrinology *38*:46–55, 1946. Corpus allatum hormone and ovarian growth in Orthoptera.
365. Scharrer, B., Biol. Bull. *102*:261–272, 1952. Intra-axonal transport in insects; role of corpus cardiacum; nerve inhibition of corpus allatum.
366. Scharrer, E., and B. Scharrer, Rec. Progr. Hormone Res. *10*:183–240, 1954. Neurosecretion; neuroendocrinology.
367. Scharrer, E., and B. Scharrer, Neuroendocrinology. Columbia University Press, New York, 1963. 287 pp.
368. Schayer, R. W., and A. C. Ivy, Amer. J. Physiol. *189*:369–372, 1957. Histamine as a hormone.
369. Scheer, B. T., and M. A. R. Scheer, Pubbl. Staz. Zool. Napoli *25*:419–426, 1954. Crustacean molt-cycle stages, Natantia.
370. Scheffel, H., Zool. Jahrb. *71*:359–370, 1965. Hormone in myriapod molting.
371. Schell-Frederick, E., and J. E. Dumont, pp. 415–463 *in* Biochemical Actions of Hormones, Vol. 1, edited by G. Litwack. Academic Press, New York, 1970. Mechanism of action of TSH.
372. Schneiderman, H. A., and L. I. Gilbert, Biol. Bull. *115*:530–535, 1958. Widespread distribution among animals of juvenile hormone activity.
373. Schneiderman, H. A., and L. I. Gilbert, pp. 157–187 *in* Cell, Organism and Milieu, edited by D. Rudnick. Ronald Press Co., New York, 1959. Substances with juvenile hormone activity among animals.
374. Scudamore, H. H., Trans. Illinois Acad. Sci. *34*:238–240, 1941. Hormonal regulation of crustacean heart rate.
375. Scudamore, H. H., Physiol. Zool. *20*:187–208, 1947. Sinus gland and oxygen consumption in crayfishes; hormonal control of gastrolith formation in crayfishes.
376. Segal, S. J., Science *126*:1242–1243, 1957. Gonadotropic control of nuptial plumage in African finches.
377. Shaaya, E., and D. Bodenstein, Proc. Nat. Acad. Sci. *59*:1223–1230, 1968. Juvenile hormone on collateral glands.
378. Shappirio, D. G., and C. M. Williams, Proc. Roy. Soc. Lond. B *147*: 233–246, 1957. Ecdysone induction of increase in mitochondria and endoplasmic reticulum of epithelial cells.
379. Shepherd, R. G., et al., J. Amer. Chem. Soc. *78*:5067–5076, 1956. Three corticosteroids.
380. Sherwood, L. M., L. M. Handwerger, W. D. McLaurin, and M. Lanner, Nature New Biol. *233*:59–61, 1971. Structure of human placental lactogen.
381. Shirai, H., H. Kanatanui, and S. Taguchi, Science *175*:1366–1368, 1972. Action of the starfish gonad-stimulating factor.
382. Simpson, L., H. A. Bern, and R. S. Nishioka, Amer. Zool. *6*:123–138, 1966. Neurosecretion in gastropods.
383. Slâma, K., Ann. Rev. Biochem. *40*:1079–1102, 1971. Review of juvenile hormone analogues.
384. Smellie, R. M. S., ed., Biochemical Society Symposium 32, 1971. The Biochemical Society, London, 1971, 178 pp. The biochemistry of steroid hormone action.
385. Smith, D. C., and S. A. Matthews, Amer. J. Physiol. *153*:215–221, 1948. Thyroid and oxygen consumption in fishes.
386. Smith, D. C. W., Mem. Soc. Endocrinol. *5*:83–98, 1956. Thyroid in poikilotherm osmoregulation.
387. Smith, P. E., Anat. Rec. *11*:57–64, 1916. Hypophysis in amphibian metamorphosis.
388. Smith, R. I., Biol. Bull. *79*:145–152, 1940. Eyestalks and molting in young crayfishes.
389. Srivastava, U. S., and L. I. Gilbert, Science *161*:61–62, 1968. Juvenile hormone effects on flies.
390. Sroka, P., and L. I. Gilbert, J. Insect Physiol. *17*:2409–2420, 1971. Juvenile hormone and post-emergence egg maturation.
391. Steiner, D. F., et al., Rec. Progr. Hormone Res. *25*:207–272, 1969. Proinsulin.
392. Stephens, G. C., Biol. Bull. *103*:242–258, 1952. Eyestalk hormone and cement glands.
393. Stephens, G. C., Biol. Bull. *108*:235–241, 1955. Molt induction in crayfish by photoperiod.
394. Stephens, G. J., Physiol. Zool. *25*:70–84, 1952. Light influence upon cyclic ovarian activity in crayfishes.
395. Stumpf, W. E., Amer. Zool. *11*:725–739, 1971. Autoradiographic localization of estrogen.
396. Sturkie, P. D., Avian Physiology. Comstock Publishing Associates, Ithaca, N. Y., 1954. Insulin-resistant reptiles and birds.
397. Sutherland, E. W., and T. W. Rall, J. Amer. Chem. Soc. *79*:3608, 1957. Isolation of cyclic AMP.
398. Swingle, W. W., J. Exp. Zool. *36*:397–421, 1922. Thyroid and amphibian metamorphosis.
399. Szego, C. M., and J. S. Davis, Proc. Nat. Acad. Sci. *58*:1711–1718, 1967. Effect of estrogen on uterine cyclic AMP.
400. Takasugi, N., J. Fac. Sci. Univ. Tokyo, Sec. IV 7:605–623, 1956. Rat ovulation by stress.
401. Takewaki, K., and Y. Yamamoto, Ann. Zool. Japan. *23*:187–190, 1950. Hormonal control of *Paratya* ovary.
402. Taki, I., J. Lac. Fish. Anim. Husb. Hiroshima *5*:345–417, 1964. Branchial gland in cephalopods.
403. Tata, J. R., Nature *197*:1167–1168, 1963. Actinomycin inhibits thyroid hormone action.
404. Tata, J. R., Progr. Nucleic Acid Res. Mol. Biol. *5*:191–250, 1966. Hormones and nucleic acids.
405. Tata, J. R., pp. 89–133 *in* Biochemical Actions of Hormones, Vol. 1, edited by G. Litwack. Academic Press, New York, 1970. Regulation of protein synthesis by hormones.
406. Thomsen, E., Vidensk. Meddel. Dansk. Naturhist. Foren. *106*:320–405, 1942. Corpus allatum hormone and dipteran reproduction.

407. Thomsen, E., J. Exp. Biol. *29*:137–172, 1952. Neurosecretory brain cells and corpus cardiacum in *Calliphora*.
408. Thomsen, E., J. Exp. Biol. *31*:322–330, 1954. Intra-axonal transport in *Calliphora*.
409. Turner, C. D., and J. T. Bagnara, General Endocrinology, 5th edition. W. B. Saunders Co., Philadelphia, 1971, 659 pp.
410. Turner, C. L., Physiol. Zool. *15*:263–280, 1942. Gonadal hormones and secondary sex characters of fish.
411. Turtle, J. R., and D. M. Kipnis, Biochem. Biophys. Res. Comm. *28*:797–802, 1967. Adrenergic receptors and cyclic AMP.
412. Vaes, G., Nature *219*:939–940, 1968. PTH-like action of cyclic AMP.
413. Van der Kloot, W. G., Biol. Bull. *109*:276–294, 1955. Controls of diapause in *H. cecropia*.
414. Van der Kloot, W. G., and C. M. Williams, Behavior *5*:157–174, 1953. Hormonal control of cocoon formation.
415. van Dyke, H. B., K. Adamsons, and S. L. Engel, *in* The Neurohypophysis, edited by H. Heller. Academic Press, New York, 1957. Differing activities of various vasopressins.
416. van Dyke, H. B., S. L. Engel, and K. Adamsons, Proc. Soc. Exp. Biol. Med. *91*:484–486, 1956. Differing activities of vasopressins.
417. van Tienhoven, A., Reproductive Physiology of Vertebrates. W. B. Saunders Co., Philadelphia, 1968, 498 pp.
418. Veillet, A., G. Cornubert, and N. Demeusy, C. R. Soc. Biol. *147*:1264–1265. 1953. MTGX hormone inhibiting ovaries.
419. du Vigneaud, V., Science *123*:967–974, 1956. Differences in pig and beef vasopressin.
420. du Vigneaud, V., H. C. Lawler, and E. A. Popenoe, J. Amer. Chem. Soc. *75*:4880–4881, 1953. Octapeptides oxytocin and vasopressin.
421. du Vigneaud, V., H. C. Lawler, E. A. Popenoe, and S. Trippett, J. Biol. Chem. *205*:949–957, 1953. Octapeptides oxytocin and vasopressin.
422. Vincent, J. F. V., J. Insect Physiol. *17*:625–636, 1971. Bursicon in locusts.
423. von Euler, U. S., Noradrenaline. Charles C Thomas, Springfield, Ill., 1956. Noradrenaline at nerve endings; catecholamines.
424. von Hagen, F., Zool. Jahrb. Abt. Anat. *61*:467–538, 1936. Thyroid and metamorphosis in fish.
425. Warren, M. R., J. Exp. Zool. *83*:127–156, 1940. Thyroxin and oxygen consumption in frogs.
426. Webb, H. M., and F. A. Brown, Jr., J. Cell. Comp. Physiol. *41*:103–122, 1953. Daily rhythm in *Palaemonetes* retinal pigment.
427. Weber, R., Experientia *221*:665–666, 1965. Actinomycin inhibits tail atrophy.
428. Wells, M. J., and J. Wells. J. Exp. Biol. *36*:1–33, 1959. Hormonal control of sexual maturity in *Octopus*.
429. Wells, M. J., and J. Wells, Nature *222*:293–294, 1969. Pituitary analogue in *Octopus*.
430. Welsh, J. H., Proc. Nat. Acad. Sci. *23*:458–460, 1937. Hormonal influence on crustacean heart beat.
431. Welsh, J. H., J. Exp. Zool. *86*:35–49, 1941. Sinus glands and retinal-pigment migration in crayfishes.
432. Welsh, J. H., Ann. N. Y. Acad. Sci. *66*:618–630, 1957. Serotonin as a neurohumoral agent.
433. Whitmore, E., and L. I. Gilbert, J. Insect Physiol. *18*:1153–1168, 1972. Lipoprotein transport of juvenile hormone.
433a. Whitmore, D., Jr., E. Whitmore, and L. I. Gilbert, Proc. Nat. Acad. Sci. *69*:1592–1595, 1972. Juvenile hormone; induction of enzymes.
434. Wigglesworth, V. B., Quart. J. Micr. Sci. 77:191–222, 1934. Hormones in insect molt and metamorphosis: Hemiptera.
435. Wigglesworth, V. B., Quart. J. Micr. Sci. *79*:91–121, 1936. Hormones in molt and metamorphosis and reproduction: Hemiptera.
436. Wigglesworth, V. B., J. Exp. Biol. *17*:201–222, 1940. Hormones in insect molt and metamorphosis: Hemiptera.
437. Wigglesworth, V. B., J. Exp. Biol. *29*:561–570, 1952. Hormones and insect molt.
438. Wigglesworth, V. B., The Physiology of Insect Metamorphosis. Cambridge University Press, 1954.
439. Wilhelmi, A., p. 59 *in* The Hypophyseal Growth Hormone, Nature and Actions, edited by R. W. Smith, Jr., O. H. Gaebler, and C. N. H. Long. Blakiston Co., New York, 1955. Crystalline fish growth hormone.
440. Wilhelmi, A., G. E. Pickford, and W. H. Sawyer, Endocrinology *57*: 243–252, 1955. Spawning in *Fundulus* by neurohypophysis. Oxytocin and vasopressin as sex controllers; mammalian posterior lobe hormones give spawning in *Fundulus*.
441. Williams, C. M., Biol. Bull. *90*:234–243, 1946. Brain hormones in lepidopteran metamorphosis.
442. Williams, C. M., Biol. Bull. *93*:89–98, 1947. Roles of brain and prothoracic gland principles in lepidopteran metamorphosis.
443. Williams, C. M., Anat. Rec. *99*:671, 1947. Brain and termination of pupal diapause.
444. Williams, C. M., Biol. Bull. *94*:60–65, 1948. Prothoracic glands and insect development.
445. Williams, C. M., Biol. Bull. *103*:120–138, 1952. Brain and prothoracic gland in insect metamorphosis.
446. Williams, C. M., Harvey Lect. *47*:126–155, 1952. Brain and prothoracic gland in larval-pupal molt.
447. Willis, J. H., and P. C. J. Brunet, J. Exp. Biol. *44*:363–378, 1966. Juvenile hormone on colleterial glands.
448. Wills, I. A., G. M. Riley, and E. M. Stubbs, Proc. Soc. Exp. Biol. Med. *30*:411–412, 748–786, 1933. Hypophysis and sex behavior in Amphibia.
449. Wingstrand, K. G., Arkiv. Zool. Soc. *26*:41–67, 1953. Low neurosecretory material, ADH, and water retention in fetuses and young.
450. Witschi, E., Wilson Bull. *47*:177–188, 1935. Hypophyseal control of plumage in weaver finches.
451. Witschi, E., J. Clin. Endocrinol. *13*:316–329, 1953. Estrogens give adrenal hyperplasia.
452. Witschi, E., Mem. Soc. Endocrinol. *4*:149, 1955. Control of nuptial plumage in African finches.
453. Witschi, E., *in* Comparative Endocrinology, edited by A. Gorbman. John Wiley & Sons, New York, 1959. Taxonomic specificity of LH; endocrine basis of reproductive adaptations in birds.
454. Wolfson, A., pp. 38–70 *in* Comparative Endocrinology, edited by A. Gorbman. John Wiley & Sons, New York, 1959. Photoperiodism and bird physiology.
455. Wolfson, A., *in* Photoperiodism and Related Phenomena in Plants and Animals, edited by R. B. Withrow. Amer. Assoc. Adv. Sci., Washington, D.C., 1959. Photoperiodism and annual cycles in birds.
456. Wright, P. A., Endocrinology *64*:551–558, 1959. Blood sugar studies in the bullfrog.
457. Wright, P. A., Biol. Bull. *119*:351, 1960. *Rana* ovulation *in vitro* by steroids.
458. Wright, P. A., and F. L. Hisaw, Endocrinology *39*:247–255, 1946. Ovulation in frogs with mammalian FSH and LH.
459. Wurtman, R. J., J. Axelrod, and L. S. Phillips, Science *142*:1071–1072, 1963. Light controls melatonin synthesis.
460. Wyatt, G. R., *in* Biochemical Actions of Hormones, Vol. 2, edited by G. Litwack. Academic Press, New York, 1972. Review of insect endocrinology.
460a. Wyatt, S. S., and G. R. Wyatt, Gen. Comp. Endocrinol. *16*:369–374, 1971. *In vitro* stimulation of RNA synthesis in epidermis by ecdysone.
461. Yamamoto, Y., Ann. Zool. Japan. *28*:92–99, 1955. Ovariectomy influence on sinus gland.
462. Yamazaki, M., and M. Kobayashi, J. Insect Physiol. *15*:1981–1990, 1969. Purification of brain hormone.
463. Zalokar, M., J. Insect Physiol. *14*:1177–1184, 1968. RNA synthesis in colleterial glands.

CHAPTER 22

NEMATOCYSTS AND NEMATOCYST ANALOGUES

By F. A. Brown, Jr.

Members of the phylum Cnidaria and certain protozoans contain explosive bodies which discharge an everting thread. These bodies are not cells, but are secretory products of cells. In cnidarians these bodies, which are formed within cnidoblasts or nematocytes and are termed nematocysts, share certain common characteristics not observed for a variety of comparably explosive bodies which occur in many protozoans. These latter nematocyst analogues include trichocysts of many ciliates and some flagellates, cnidocysts of certain dinoflagellates, and polar capsules of cnidosporans.

NEMATOCYSTS

Nematocysts are extremely small intracellular structures of spherical, ovoid, or spindle form. They consist of a capsule within which a hollow thread, continuous with one end of the capsule, is introverted and coiled.[17,37,48] The nematocysts appear largely comprised of protein. A wide variety of mono- and diphenols, amino acids, and enzymes have been identified from them.[28,34] Particularly prominent in the contents appear to be glutamic and aspartic acid. The capsule itself is probably collagen-like.[4,13,14,20,27,31,34] In electron microscope studies of isorhizas of *Corynactis,* the undischarged thread has been found to have the form of a tapering tube pleated deeply to form a triple screw. Upon eversion the thread smooths, expanding to about three times its undischarged length, but holding close to the same diameter and surface area. The discharged thread of the large holotrichous isorhiza (Fig. 22–1) is about 2½ times as large in diameter at its base as at its tip. It shows a cross-banding of 190 Å and a longitudinal banding of about 50 Å. The discharge of the nematocyst involves a rapid expulsion of the thread-like process by eversion. There is usually a cap or operculum covering the region of the undischarged nematocysts through which the thread is ejected. In the undischarged thread the points of the spines are tightly packed. As the thread discharges the spines are first directed forward, quickly flicking backward as the discharge continues.

The nematocysts are produced within interstitial cells either at their final site or at some distance from this site, to which they move by amoeboid movement or by passive transportation as, for example, in the course of normal centrifugal growth of hydra tentacles.[7] The portion of the cnidoblast containing the nematocyst eventually comes to occupy a superficial position and usually develops at its outer end a bristle-like projection, the cnidocil. The latter is imbedded in a small crater-like elevation on the cell. The cnidoblast also often differentiates trichite-like supporting rods in its peripheral regions and often, too, a fibrillar network associated

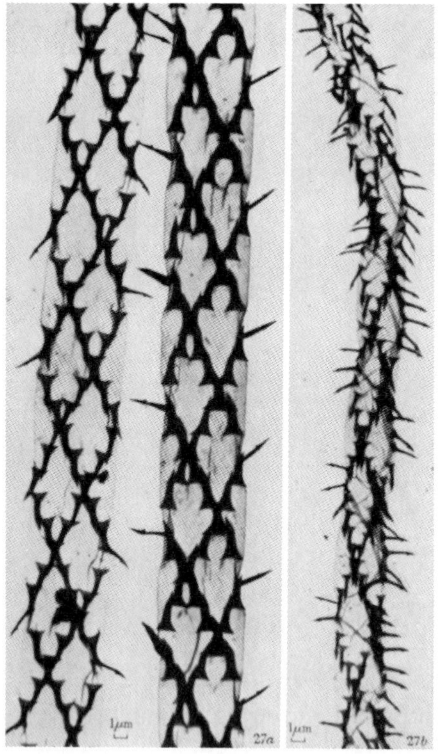

Figure 22–1. Three discharged threads from holotrichous isorhizas, flattened to different degrees to show barb-whorls and the staggering of whorls to form three barb rows. (From Skaer, R. J., and Picken, L. E. R., Phil. Trans. Roy. Soc. Lond. B *250*:131–164, 1965.)

with the capsule of the nematocyst and extending proximally in the cell from it. In *Diadumene,* it is held that the fully developed nematocyst is extracellular.[51,54]

The nematocysts of Cnidaria are divisible into major groups, the spirocysts of Zoantharia (which are often, but not always, acid-staining bodies possessing peculiar adhesive threads), and the basic-staining nematocysts proper. The latter are widely distributed through the whole phylum. Nematocysts show a tremendous variety of detailed forms of their discharged threads, even enabling them to serve to some degree as criteria for cnidarian evolution and systematics.[45,46] In desmonemes, the thread forms a tightly coiled filament on discharge, wrapping itself about bristles or fibers of organisms in whose presence they are discharged. Most other types have open tips and are believed to penetrate prey, injecting into them a toxic substance. The threads of these latter types (e.g., isorhizas and stenoteles) commonly possess an armature in the form of three spiralling rows of spines which serve effectively to anchor the thread into tissues which have been penetrated. The penetrating capacity of at least some types is so great that even the chitinous cuticles of small organisms can be punctured.

The sting of numerous species gives rise to severe itching and other skin disorders in man. The toxicity is very great in a few cnidarians such as *Physalia,* and some larger scyphozoan jellyfish such as *Dactylometra,* even enough to render them highly dangerous to man. Their sting may in some cases produce serious illness or even death. The toxin associated with nematocysts has been investigated for the anemones *Metridium*[32] and *Aiptasia,*[3] for a cubomedusan,[11] and especially for *Physalia.*[15,25,26] The effective agents are probably peptides. Assayed on crustaceans, frogs and toads, and mammals, the action appears to be predominantly on nerve and muscle excitation with the most prominent influences upon cardiovascular function.

The cnidoblasts respond predominantly as independent effectors in the discharge of their nematocysts. They are discharged by many chemical agents, including both strong acids and bases.[53] In their normal responses, and in response to highly localized electrical stimuli, there is complete restriction of the discharge to the specific region excited.[29,30] Intense mechanical stimulation by inert objects will induce only a weak discharge, whereas mild mechanical stimulation by natural foods is sufficient to evoke a strong response (Fig. 22–2). Submersion of cnidoblasts in a weak extract of a normal food which will typically not itself induce discharge will greatly lower the threshold of these effectors in *Anemonia* to purely mechanical stimuli.[29] The specific food factors that are involved in this sensitization are lipoidal substances adsorbed upon proteins. The adsorptive forces are so strong that the factors cannot be removed by ether extraction, but they can be removed with alcohol. The active chemical substances appear to have properties resembling sterols and phospholipins and to be highly surface-active. In view of the rapid and thorough discharge of nematocysts even to dried foods, it would appear that the normal reaction must be "contact-chemical" in character, chemical sensitization to mechanical stimulation occurring almost instantaneously.

These highly specialized characteristics of excitation of nematocysts in *Anemonia* are peculiarly adapted to the normal functional

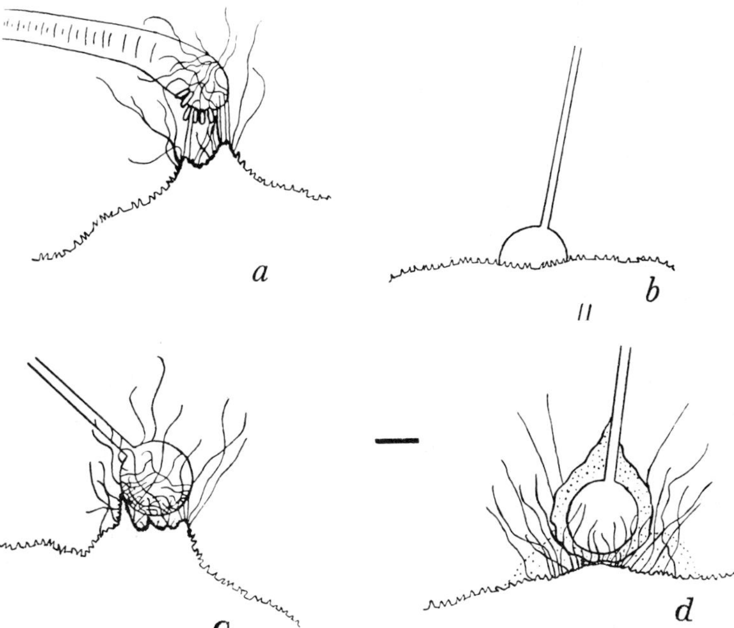

Figure 22–2. Discharge of nematocysts of *Anemonia* to various stimuli. *a*, Touch by a human hair. *b*, Touch by a clear, blunt glass rod fails to cause discharge. *c*, Response to clean glass rod after sensitization of the cnidoblasts with dilute saliva extract. *d*, Response to glass rod smeared with alcohol extract of *Pecten* gill. (From Pantin, C. F. A.: J. Exp. Biol., vol. 19.)

roles of these effectors. The value of having the nematocysts withhold their discharge until the instant of mechanical contact is readily seen in comparing the results of stimulating discharge by a piece of cotton soaked in bile salts with that by a similar piece soaked in an extract of normal food material. In the former case the nematocysts are induced to discharge before contact by the diffusion outward of the bile salts, which are highly potent cnidoblast excitors, and therefore before the discharge can result in any attachment of the discharging filaments to the cotton fibers, as occurs in the case of normal food substances.

The cnidocils, when present, are probably concerned with the excitation process. That they are nonessential, however, is indicated by the fact that they are absent in many of the anthozoans. It would appear that in their absence the cnidoblast surface comprises the natural receptive area for the response.[27] Nematocysts may be extruded, unexploded, in response to K^+ or NH_4^+ or in absence of ions, in an active process capable of suppression by Mg^{++}, Cl^-, Br^-, and anesthetics.[33,51] Isolated nematocysts may retain for a time their capacity to discharge, longest in Ml glycerol solution and shorter in isotonic $CaCl_2$, NaCl, KCl, or sea water.[49,52] Discharge has been postulated to result from weakening of an operculum or plug.[50]

While the cnidoblasts are independent effectors, there is evidence that there is some regulation of their responsiveness. This may change with degree of satiation of the organism,[8] with season,[5] with concurrent state of locomotor activity,[39] or in the case of the commensal, *Calliactis parasitica,* with whether the anemone is on, or not yet on, the whelk that it normally inhabits.[38,39] Also, whether a discharged nematocyst is extruded or held by the organism may be similarly regulated within the organism. Suggestive in this regulation are reported instances of neurite endings on cnidoblasts.[28]

The mechanism of discharge of nematocysts is not well understood. Two types of theories in particular have found favor among students in this field; both types have in common the postulation that an increased pressure within the capsule forces out the filament. One theory considers the pressure increase following stimulation to be the result of a passage of water into the capsule. Some have believed the nematocyst operates as a simple osmometer. Supporting this belief is the finding of freezing-point depressions of the capsular contents of the order of –6°C, becoming substantially less upon discharge.[34] Others think the water inflow is associated with hydration and swelling of colloids within the nematocyst in response to change in pH, and a consequent swelling of colloidal material contained therein. A second type of theory attributes the increase in pressure to a contraction of the capsule, either of itself or by a fibrillar network associated with the capsule. It is possible that mechanisms fitting both theories are to be found among the numerous nematocyst types of the various cnidarian species. The nematocysts of *Metridium* appear to discharge as a result of inflow of water; those of *Physalia* appear to require the activity of contractile fibrils of the nematoblast.

Functional nematocysts have been found in acoelous flatworms, including *Microstomum,* in some polyclads, and in the sea slug, *Aeolis.*[21, 22, 23] It has been conclusively demonstrated that these nematocysts have been derived from the cnidarians on which these animals feed. It is interesting that, in the appropriation of the nematocysts, these flatworms and molluscs selectively utilize only certain types. For example, *Microstomum* digests the volvents and utilizes the penetrants. The sea slugs, feeding on *Pennaria,* utilize only the highly effective type known as the microbasic mastigophores, to the exclusion of other types. The appropriate nematocysts are quite concentrated in the bodies of their new carriers and would appear to serve as effective defensive weapons. This phenomenon involves, obviously, loss of responsiveness to their new host. A comparable loss is seen between the commensally related hermit crab and sea anemone.[2, 38]

NEMATOCYST ANALOGUES

Numerous species of protozoans contain explosive bodies known as trichocysts in their ectoplasm.[1, 34] In ciliates these are usually fusiform bodies oriented obliquely or at right angles to the surface; they may be uniformly distributed over the surface, as in *Paramecium,* or restricted to particular regions as in the proboscis of *Dileptus.* They may be projected outward on the ends of long tentacles in feeding specimens of *Actinobolina.* In the dinoflagellate, *Polykrikos,* the cnidocysts lie deep within the cytoplasm.

Electron micrographic studies have disclosed many details of the origin, development, and structure of the nematocyst analogues as well as the regenerative capacities of organisms for them.[6, 9, 10, 24, 36, 55, 56] They appear to originate within or close to the macronucleus of *Frontonia,* making their way from there to their definitive position in the ectoplasm, completing their differentiation as they migrate. The formation of trichocysts may be effectively prevented by strong ultraviolet light.[12] Like the nematocysts, the trichocysts usually possess the form of a capsule containing a tube attached at one end, and similarly, upon discharge the thread emerges in a manner resembling eversion of a glove-finger.[34] In *Paramecium,* the undischarged trichocysts possess an ovoid body proper, about 2 to 3 μm long by 2/3 μm in diameter, and a cap-covered tip of slightly smaller diameter. On discharge, the tip with its covering cap is separated from the body proper through the elongation of a shaft.[6]

The trichocysts can readily be induced to discharge their filaments through chemical (acid or base), mechanical (pressure), or electrical (condenser discharge or induction shock) stimulation. Extract of geranium appears to be a highly effective chemical agent for inducing discharge in *Paramecium,* probably due to contained tannic acid.[42] With electrical stimulation there is an increase in the number discharged as the strength of the stimulus is increased.[41] The total discharge is very rapid, occurring in a matter of a few milliseconds. The discharged trichocyst is needle-like in general form, being ten or more times as long as the undischarged body. The discharged trichocysts of *Paramecium* may be as much as 40 μ long. The trichocysts of *Paramecium* show birefringence, indicating an orientation of elongated submicroscopic particulates in the long axis of the organelle. A study of the extended threads of discharged trichocysts of *Paramecium* with the electron microscope reveals them to be without bounding membrane and to display a periodic transverse banding at intervals of 600 to 650 Å (Fig. 22–3). The characteristics of the shaft indicate it to be composed of elongated cross-striated protein fibrils showing a periodic structure somewhat resembling collagen fibers. The overall striation of the organelle appears to be a consequence of the alignment of these fibers in phase with one another. At the tip of the shaft there is a relatively opaque, thornlike body indicative of the presence of a dense proteinaceous structure or of elements of high atomic mass.[18, 19, 35]

Little or nothing is known of the mechanism of trichocyst discharge or to what extent trichocysts are activated other than in direct

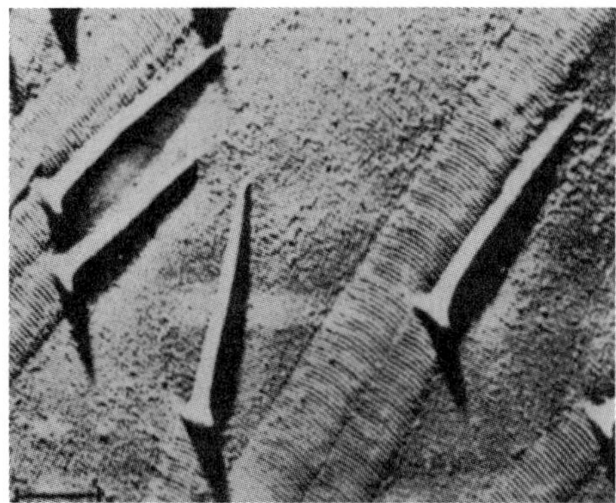

Figure 22-3. Electron micrograph of discharged *Paramecium* trichocysts, shadow-cast with chromium. (From Jakus, M. A., and Hall, C. E.: Biol. Bull., vol. 91.)

response to environmental stimuli. The trichocysts appear to be in contact with the silver-line system, which some consider to have a nerve-net type of conductile function. The mechanism of discharge probably involves either a hydration of some protein within the trichocyst or a rapid osmotic inflow of water on excitation of these organelles.

The role of trichocysts in protozoans is still not definitely established. They have been considered by many to possess a protective function. This appears doubtful in such species as *Paramecium,* which are readily ingested by *Didinium* even after the *Paramecium* has exploded many of its trichocysts. More probably the trichocysts function here as organelles of attachment.[19] In ciliates such as *Dileptus* and *Actinobolina,* prey coming in contact with a region of the body bearing trichocysts appears to be paralyzed instantly, as if a toxin were associated with the discharged trichocysts. Among the dinoflagellates, *Polykrikos* and *Nematodinium,* there are trichocysts that bear close superficial resemblance structurally to the nematocysts of cnidarians and which are termed cnidocysts. Trichocysts in various stages of differentiation have been described within these dinoflagellates.[16] In *Noctiluca* the trichocysts appear to be of mitochondrial origin.[44]

Another type of effector organelle very closely resembling the cnidarian nematocyst is the polar capsule of sporozoans of the groups Myxosporidia and Actinomyxidia. Within the digestive tract of host organisms, the action of the digestive juices induces these capsules to discharge their spirally coiled hollow thread after the fashion of a nematocyst. The discharged filament serves as a means of temporary anchorage of the parasite to the host tissues. For most of the Microsporidians the total spore appears usually to behave as a polar capsule, with the whole spore contents being ejected into the everting tube.

REFERENCES

1. Anderson, E., J. Protozool. *9*:380–395, 1962. *Chilomonas* trichocysts.
2. Berner, L., Bull. Soc. Zool. France *78*:221–226, 1953. Nematocyst response in commensal relationship.
3. Blanquet, R., Comp. Biochem. Physiol. *25*:893–902, 1968. Toxicity of sea anemone nematocyst.
4. Boisseau, J. P., Bull. Soc. Zool. France *77*:151–169, 1952. Chemical contents of nematocysts.
5. Bouchet, C., C. R. Acad. Sci. (Paris)*252*:327–328, 1961. Seasonal variation in cnidoblast responsiveness.
6. Bouck, G. B., and B. M. Sweeny, Protoplasma *61*:205–223, 1966. Fine structure of dinoflagellate trichocysts.
7. Burnett, A. L., and T. Lentz, Ann. Soc. Zool. Malacol. Belg. *90*:281–293, 1960. Nematocyst migration in *Hydra.*
8. Burnett, A. L., T. Lentz, and M. Warren, Ann. Soc. Zool. Malacol. Belg. *90*:247–267, 1960. Nutrition and nematocyst discharge.
9. Doroszewski, M., and K. Golinska, Acta Protozool. *4*:343–350, 1967. Trichocyst regeneration in *Dileptus.*
10. Dragesco, J., G. Auderset, and M. Baumann, Protistologica *1*:81–90, 1965. Structure and genesis of trichocysts of *Dileptus.*
11. Endean, R. C., Duchemin, D. McColm, and E. H. Fraser, Toxicon *6*: 179–204, 1969. Nematocysts in a cubomedusan.
12. Haller, G. de, and B. ten Heggeler, Protistologica *5*:115–120, 1969. Inhibition of trichocyst formation by ultraviolet light.
13. Hamon, M., Nature *176*:357, 1955. Chemical nature of nematocysts.
14. Hamon, M., Bull. Soc. d'Hist. Nat. l'Afrique du Nord *46*:169–179, 1955. Chemistry of nematocysts.
15. Hastings, S. G., J. B. Larsen, and C. E. Lane, Proc. Soc. Exp. Biol. Med. *125*:41–45, 1967. Effects of *Physalia* nematocyst toxin on dog cardiovascular system.
16. Hovasse, R., Arch. Zool. Gen. *102*:189–198, 1963. Trichocysts of *Polykrikos.*
17. Hyman, L. H., The Invertebrates. McGraw-Hill Book Co., New York, 1940. Trichocysts and nematocysts.
18. Jakus, M. A., J. Exp. Zool. *100*:457–485, 1945. Trichocysts; structure and properties.
19. Jakus, M. A., and C. E. Hall, Biol. Bull. *91*:141–144, 1946. Electron micrographs of trichocysts.
20. Johnson, F. B., and H. M. Lenhoff, J. Histochem. Cytochem. *6*:394, 1958. Nature of *Hydra* nematocyst capsule.
21. Karling, T. G., Acta Zool. Fennica *116*:3–28, 1966. Nematocysts in turbellarians.
22. Kepner, W. A., J. Morphol. *73*:297–311, 1943. Nematocysts in nudibranchs.
23. Kepner, W. A., W. C. Gregory, and R. J. Porter, Zool. Anz. *121*:114–124, 1938. Nematocysts of *Microstomum.*
24. Kudo, R. R., and E. W. Daniels, J. Protozool. *10*:112–120, 1963. Fine structure of microsporidian spore.
25. Larsen, J. B., and C. E. Lane, Toxicon *4*:199, 1966. Effect of nematocyst toxin on land crabs, *Cardisoma,* and myogenic rat heart.
26. Larsen, J. B., and C. E. Lane, Toxicon *8*:21–23, 1970. Action of *Physalia* toxin on frog nerve and muscle.
27. Lenhoff, H. M., E. S. Kline, and R. Hurley, Biochem. Biophys. Acta *26*:204–205, 1957. Chemistry of *Hydra* nematocyst.
28. Lentz, T. L., and J. G. Wood, J. Histochem. Cytochem. *12*:37, 1964. Neurite endings on cnidoblasts in hydra and sea anemones.
29. Pantin, C. F. A., J. Exp. Biol. *19*:294–310, 1942. Control of nematocyst discharge.
30. Parker, G. H., and M. A. Van Alstyne, J. Exp. Zool. *63*:329–344, 1932. Control and mechanism of nematocyst discharge.
31. Phillips, J. H., Nature *178*:932, 1956. Chemical nature of *Metridium* nematocysts.
32. Phillips, J. H., and D. P. Abbott, Biol. Bull. *113*:296–301, 1957. Toxin of *Metridium* nematocyst.
33. Picken, L. E. R., Quart. J. Micr. Sci. *94*:203–227, 1953. *Corynactis* nematocysts.
34. Picken, L. E. R., and R. J. Skaer, Symp. Zool. Soc. Lond. *16*:19–50, 1966. Review of researches on nematocysts.

35. Potts, B., Biochem. Biophys. Acta *16*:464–470, 1955. Electron microscopic study of trichocysts.
36. Puytorac, P. de, Acta Protozool. *2*:147–152, 1964. Ultrastructure of ciliate trichocyst.
37. Robson, E. A., Quart. J. Micr. Sci. *94*:229–235, 1953. Electron microscopic study of nematocysts.
38. Ross, D. M., Proc. Zool. Soc. Lond. *134*:43–57, 1960. Nematocyst response in commensal relationships.
39. Ross, D. M., and L. Sutton, J. Exp. Biol. *41*:751–757, 1964. Locomotion and nematocyst discharge.
40. Saunders, J. T., Proc. Cambridge Phil. Soc. (Biol. Sci.) *1*:249–269, 1925. Function of trichocysts of *Paramecium.*
41. Schmitt, F. O., C. E. Hall, and M. A. Jakus, Biol. Symp. *10*:261–276, 1943. Trichocysts.
42. Schuster, F. L., B. Prazak, and C. F. Ehret, J. Protozool. *14*:483–485, 1967. Trichocyst discharge in *Paramecium.*
43. Skaer, R. H., and L. E. R. Picken, Phil. Trans. Roy Soc. Lond. B *250*: 131–164, 1965. Structure and discharge of nematocyst of *Corynactis.*
44. Soyer, M.-O., Protistologica *5*:327–334, 1969. Trichocysts of *Noctiluca* (dinoflagellate).
45. Weill, R., Trav. Station Zool. Wimereux 10, 11, I, II, 1934. Nematocyst classification.
46. Werner, B., Helg. wiss. Meeresunters. *12*:1–39, 1965. Cnidarian nematocysts, and systematics and evolution.
47. Wood, J. G., and T. L. Lentz, Nature *201*:88–90, 1964. Amines in sea anemone and hydra nematocysts.
48. Yanagita, T. M., J. Fac. Sci. Tokyo Univ. *6*:97–108, 1943. Nematocyst discharge.
49. Yanagita, T. M., Nat. Sci. Report Ochanomizu Univ. *2*:117–123, 1951. Physiology of isolated nematocysts.
50. Yanagita, T. M., Japan. J. Zool. *12*:363–375, 1959. Responses of isolated nematocysts.
51. Yanagita, T. M., J. Exp. Biol. *36*:478–494, 1959. Extrusion of nematocysts.
52. Yanagita, T. M., J. Fac. Sci. Tokyo Univ. *8*:381–400, 1959. Response of isolated nematocysts.
53. Yanagita, T. M., and T. Wada, Nat. Sci. Report Ochanomizu Univ. *4*:112–118, 1953. Nematocyst discharge by acids and bases.
54. Yanagita, T. M., and T. Wada, Cytologia *24*:81–97, 1959. Physiology of acontia and nematocysts.
55. Yusa, A., J. Protozool. *10*:253–262, 1963. Fine structure of regenerating *Paramecium* trichocysts.
56. Yusa, A., J. Protozool. *12*:51–60, 1965. Fine structure of *Frontonia* trichocysts.

CHAPTER 23

CHROMATOPHORES AND COLOR CHANGE

*By F. A. Brown, Jr.**

INTRODUCTION

The ability to change color through movements of pigments within certain integumentary cells or organs is widely distributed among animals. It has been observed for numerous cyclostomes, fishes, amphibians, and reptiles among the vertebrates; among the invertebrates it is exhibited by many higher crustaceans, cephalopods, and leeches and a few insects, echinoderms, and polychaetes. A comparable activity has been described for a euglenoid protozoan. The spectacular color changes of the chameleon between black and green and the rapid color changes of the octopus were described as early as the fourth century B.C. by Aristotle, and those of fishes were described somewhat later by Pliny, who observed the changes of the dying mullet. The first changes recorded in amphibians were in the frog, and those in crustaceans were in the prawn, *Hippolyte*.[179] The relatively rapid color changes in the cephalopods were early demonstrated to be due principally to the activity of special organs in the skin, to which the name cromofora was given. Later, the movements of pigments in special integumentary organs were shown clearly to account for color changes in the chameleon,[202] the frog,[11] and crustaceans.[261] These special organs have come to be known as chromatophores.

*With contributions by M. Fingerman.

Brücke[53] made studies on the physiology of color change in the chameleon; Pouchet,[242] in crustaceans and fishes; and Gamble and Keeble,[159,160] in crustaceans. All these early investigators concluded that the chromatophore systems were under the control of the nervous system or the chromatophores responded directly to the action of environmental stimuli. The possibility of a role of hormones in color changes was suggested first by the discovery that frogs are blanched by injection of adrenaline.[188] The early work on chromatophores has been thoroughly reviewed in the extensive accounts of van Rynberk[257] and Fuchs.[118] Later general summaries include those of Hogben,[141] Parker,[214,230] and of Fingerman,[89,90,92] and ones covering separate groups, the vertebrates,[240,312] insects,[262] and crustaceans.[35,58,172] The pigments of animals have been described in other summaries.[109,128]

CHROMATOPHORES: STRUCTURE AND METHODS OF ACTION

Chromatophores are special pigment cells located in the skin or often even in certain deeper tissues of the body of an animal. Chromatophores possess the ability to bring about redistributions of their pigment in such a manner as to influence the general coloration of the animal. A pigment that is concentrated into a small ball (punctate) contributes little or nothing to the gross coloration of the individual, whereas its dispersion to cover a larger surface (reticulate) results in its imparting its tint to the animal. The foregoing mechanism of action (concentration and dispersion of pigment) is referred to as physiological color change. The chromatophores may also influence the coloration of an animal through their accumulation or production of pigment or their loss or destruction of it. This latter mode of action is termed morphological color change.

Physiological Color Changes. Chromatophores are of two major types. One type is seen among the cephalopod molluscs. It is a complex organ with a pigment-containing cell with numerous radially arranged smooth muscle fibers associated with it (Fig. 23–1). The second, more common, type is found in most color-changing species. It comprises a single cell or small syncytium, usually of highly branched outline, and within which pigment distribution is altered by streaming movements.

The chromatophore of cephalopods comprises a central uninucleate cell filled with pigment. The pigment is actually contained within an intracellular sac, the cytoelastic sacculus. The elastic properties of the chromatophore are due to this sac and not the cell membrane. The latter becomes highly folded when the sac retracts.[61] Radiating out from the central cell in the plane of the skin are from 6 to 20 or more uninucleate smooth-muscle fibers. All the fibers of a chromatophore usually contract simultaneously, stretching out the small, spherical, central, pigmented cell into a disc having a diameter fifteen to twenty times that of the original sphere. The spherical form is restored by the elasticity of the cytoelastic sacculus of the central pigmented cell after relaxation of the radiating fibers. A single nerve fiber is said to supply each muscle fiber;[140] its terminal arborizations disperse broadly over the surface of the muscle. No motor end-plates are present, and eserine is reported to be ineffective in blocking nervous activation. The nature of the transmitter substance is unknown.[106] The muscle fibers show a rapidity of contraction to electrical stimuli approaching that of striated muscle fibers (Chapter 16).

Little is known about the chemical nature of the pigments of the cephalopods. Their chromatophore pigments appear to be ommochromes.[269] *Octopus* possesses two kinds of chromatophores, one containing a reddish-brown pigment, and the other a yellow. The squid, *Loligo,* has three types: brown, red, and yellow. Underlying these chromatophores is an immobile layer of light-reflecting pigment. By means of their special type of chromatophores the cephalopods are able to show more rapid color changes than other animals.

In animals other than the cephalopods, the chromatophores are single cells (e.g., in most vertebrates) or closely associated groups of cells or syncytia (e.g., in many crustaceans). These were once thought to be ameboid cells, contraction of whose processes resulted in a concentration of the pigment mass into a small sphere, and whose extensive pseudopodial production resulted in a broad dispersal of the pigments. Now it is generally believed that the chromatophore has a permanent arborized form, and that the pigment granules either become concentrated into the

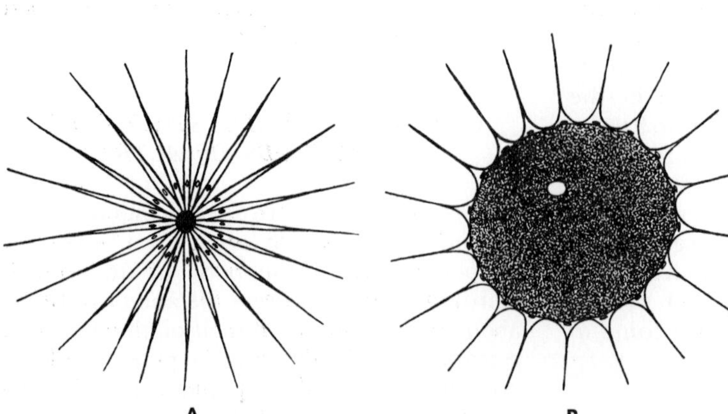

Figure 23–1. The cephalopod chromatophore. *A*, pigment concentrated; *B*, pigment dispersed. (From Bozler, E., Z. vergl. Physiol., vol. 8.)

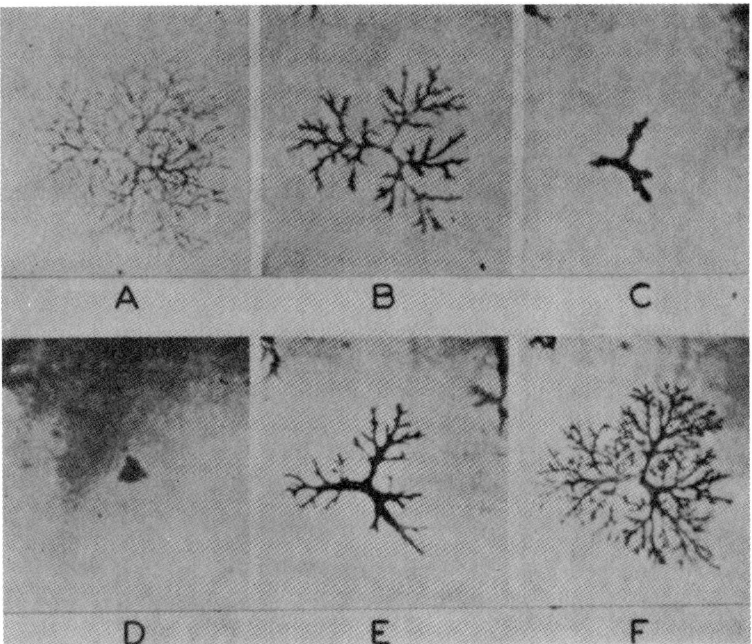

Figure 23–2. A series of photographs of a single white chromatophore of *Palaemonetes* as its pigment fully concentrates in response to a black background, and then redisperses on return of the animal to a white background. (From Brown, F. A., Jr., J. Exp. Zool., vol. 71.)

chromatophore center to form a punctate mass or become dispersed to varying degrees through the intricately branching structure[194] (Fig. 23–2) to impart color to the macroscopic appearance of the animal. Matthews and others using the light microscope have reported observing branches of chromatophores whose pigment was in the punctate condition. This observation has been confirmed by electron microscopy.[21] Also supporting this view has been the demonstration of the striking similarity, even to the minute terminals of a chromatophore, after pigment dispersal on two different occasions.[27, 237, 288]

In typical details of form, pigments, and reactions, however, each animal, species, or group has its own chromatophoral peculiarities. Chromatophores are known as monochromatic, dichromatic, or polychromatic, depending on whether they possess one, two, or more kinds of pigment. The crustaceans commonly possess dichromatic or polychromatic chromatophores, with each pigment typically dispersing out into its own processes and, when concentrated into the chromatophore center, possessing its own distinct individuality. In fact, in the responses of the crustacean chromatophore system to colored backgrounds the several pigments within a single chromatophore may show a considerable degree of independence of one another.[29, 176]

In some insects all of the epidermal cells themselves serve as functional chromatophores. During darkening of the skin the dark brown-black pigment within these cells migrates from small concentrated masses below an evenly dispersed yellow and green

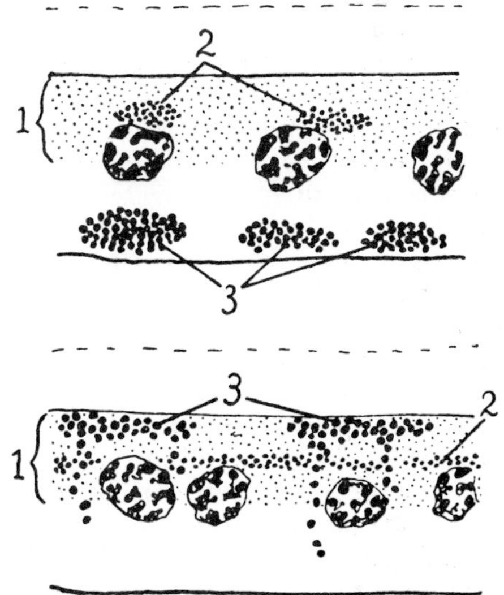

Figure 23–3. Diagrammatic sections through the epidermis of light-(*upper*) and dark-(*lower*) adapted *Carausius.* The coarsest stippling (*3*) indicates brown-black pigment; the intermediate stippling (*2*), red pigment; and the finest (*1*), yellow-green pigment. (From Giersberg, H., Z. vergl. Physiol., vol. 7.)

pigmented layer to a more superficial position, where it disperses (Fig. 23–3). The red pigment during skin darkening disperses from small spherical masses to form a continuous sheet of pigment. Thus the forces which operate in pigment concentration and dispersal in chromatophores in general are apparently of such a nature as can operate within the limits of conventionally shaped cells.

Migration of a red hematochrome pigment in *Euglena rubra*, from a deeper position in the body internal to the chloroplasts to a dispersed, superficial position, results in a green to red color change in response to elevated light intensity.[154]

In and associated with the chromatophores of crustaceans are a number of different colors of pigments. The kinds which are present vary with the species and even occasionally among individuals within the same species having different histories. Yellow and red pigments appear to be of quite general occurrence within the group. These are carotenoids of various kinds, some often conjugated with proteins.[183] Also of quite general presence is a reflecting white pigment. The majority of the macruran decapods possess a transparent blue pigment, which is a carotenoid conjugated with a protein.[309] The application of heat or alcohol to integument containing the blue pigment results in its rapid transformation from a water-soluble blue to a fat-soluble red pigment. In some crustaceans a black or brownish-black pigment is present. Such black pigment, found in the natantian *Crago (Crangon)* and in isopods such as *Ligia*, appears to be an ommochrome; in some of the true crabs the pigment is melanin.[128]

The chromatophores of the vertebrates resemble those of crustaceans. Unlike in the latter, however, they usually comprise single cells and are for the most part monochromatic. The predominant pigment is melanin, and it is the activity of the melanophores which is principally responsible for the conspicuous color changes in this group. In many vertebrates, reflecting white chromatophores, termed leucophores, iridoleucophores, or antaugophores, are also found.[116] Chromatophores known as lipophores, containing fat-soluble red pigment (erythrophores) or yellow pigment (xanthophores), are often present. These pigments are carotenoids.[108] In addition to these more conventional types of pigment cells, there are sometimes glistening bluish-green bodies, the iridophores, whose color and color changes are structural ones dependent on the form, arrangement, and movements of fine, plate-like crystals.

Supplementing the contribution of the chromatophores themselves to the coloration of the vertebrates there is often, as in the cephalopods, an immobile layer of whitish or yellowish pigment. This pigment is responsible entirely for the tint of the animal when the active chromatophores contribute little or nothing, or it cooperates with the chromatophores in producing the normal coloration. In species such as the lizard, *Anolis*, the central bodies of the melanophores lie beneath such a passive layer of pigment. As the animal darkens in response to appropriate stimulation the melanin streams within melanophore processes to a position superficial to the inert light-colored layer, thereby concealing the latter[165] (Fig. 23–4).

The rate of physiological color change is limited by the rates of mechanical response of the effector organ and of its controlling mechanism. There is, however, great variation in such rates among animals. The squid is able to carry out maximum color change in a matter of seconds, as is also the squirrel fish, *Holocentrus*. A few minutes suffice for maximum color change in the minnows, *Fundulus* and *Lebistes*. From one to several hours are needed by many crustaceans, insects, and the catfish *Ameiurus*, and days are required for comparable maximum changes in flatfishes, the eel, *Anguilla*, elasmobranchs, and amphibians.

A number of methods have been utilized in the measurements of physiological color changes. These have been critically reviewed by Parker.[229] None of the methods permits complete differentiation between influences which are in part the result of morphological color changes and those which are purely physiological. Instead, the measurements are generally based on time intervals of sufficient brevity to assure that morphological changes would not have influenced the results significantly. One group of methods employs simply the gross changes in color of the animal as an index of the extent of dispersion of the dark pigments. This may involve a visual determination in which the animal is merely described as being light, dark, or intermediate, or in which subjective grades of variation between known extremes are estimated and expressed numerically[33] in four or five grades. Some of the subjective aspects

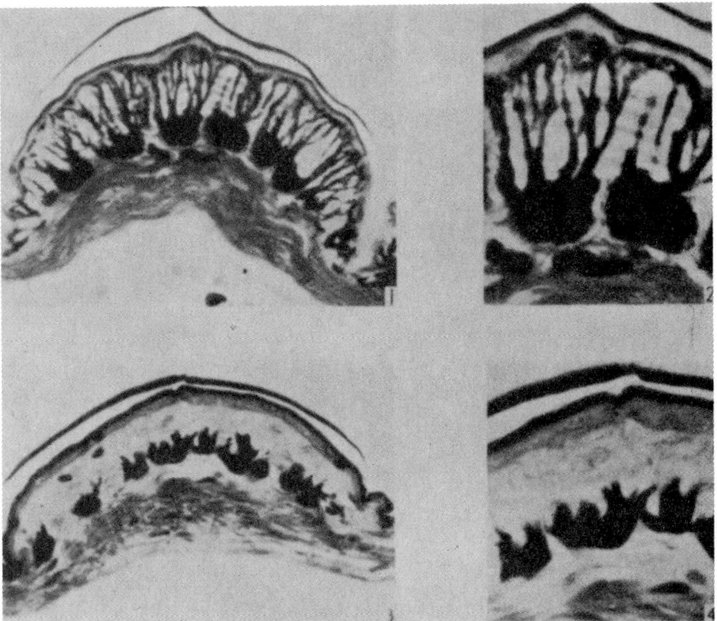

Figure 23-4. Sections through scales of *Anolis* showing the condition of the chromatophores, 1,2, in the brown state and 3,4, in the green state. (From Kleinholz, L. H., J. Exp. Biol. *15*:474-499, 1938.)

have been removed by a method employing photometric determination of the fraction of the incident light reflected from a unit area of skin surface[207,316] or the relative amounts of light transmitted by isolated fish scales.[283]

A second group of methods has been based, not on the gross light absorptive changes in the skin, but rather on the changes in chromatophores themselves. One of these methods is measurement of the actual diameter changes in the individual chromatophores. This method has been employed for fish and crustacean chromatophores.[27,30,289] Another method, and one rather extensively adopted, describes the chromatophore state numerically as follows[192,280] (Fig. 23-5):

1 = Punctate; 2 = Punctostellate; 3 = Stellate; 4 = Reticulostellate; 5 = Reticulate. This system has the advantage that quick inspection can, after a little practice, yield numerical values adequate for many comparative purposes.

The preceding methods have sometimes been supplemented with photomicrography[217] or the rapid (often heat) fixation of the animal to provide temporary or permanent skin preparations.[30,317]

Morphological Color Changes. Morphological color changes involve actual changes in quantity of pigment within the animal or its integument. In the normal adaptive color changes of animals, physiological and morphological color changes proceed simultaneously. The morphological changes may result in both an increase in the amount of pigment within each pigment cell and an increase in the number of functional chromatophores per unit area of skin.[28,209] Quantitative studies of morphological color changes have correspondingly involved two types of techniques: (1) the determination of changes in the number of functional chromatophores per unit area of surface, and (2) the determination of changes in the total pig-

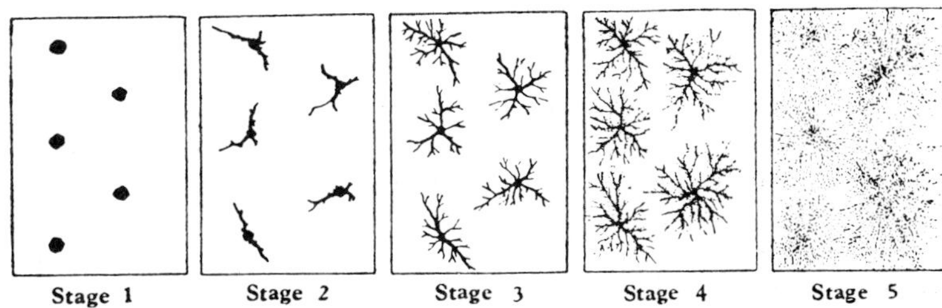

Figure 23-5. Melanophores showing various degrees of pigment dispersion. *1*, punctate; *2*, punctostellate; *3*, stellate; *4*, reticulostellate; *5*, reticulate. (From Matsumoto, K., Biol. J. Okayama Univ., vol. 1.)

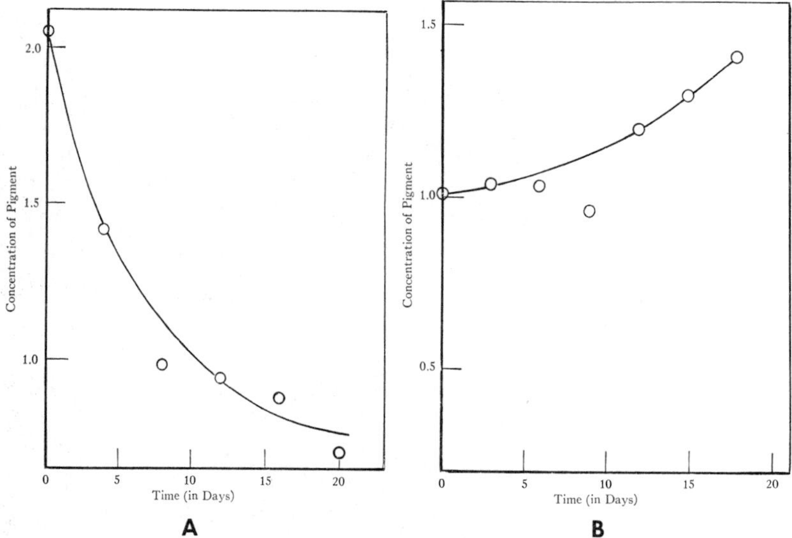

Figure 23–6. *A*, graph showing rate of loss of red pigment from the bodies of *Palaemonetes* kept on a white background. *B*, graph showing increase in quantity of red pigment in the bodies of animals kept on a black background. (From Brown, F. A., Jr., Biol. Bull., vol. 57.)

ment content of the animals by chemical extraction and colorimetric determinations of pigment quantity in the extracts (Fig. 23–6). This subject has been ably reviewed for vertebrates by Sumner,[296] and a quite similar picture seems to obtain for invertebrates.

There appears to be a close functional relationship between physiological and morphological color changes. A correlation between the two has been noted by many investigators.[15, 160] Maintained concentration of a pigment within a chromatophore seems usually to be correlated with a reduction in quantity of that pigment, and, conversely, pigment dispersion appears associated with the pigment production. This relationship has been termed Bábàk's law.[15] It would appear either (1) that pigment formation or destruction results from the state of dispersion of the pigment, or (2) that both physiological and morphological color changes are effected in a parallel manner by the same controlling mechanisms. The latter view seems to prevail, supported especially by the fact that the process may even involve the production of new chromatophores.

COLOR CHANGES IN ANIMALS

The chromatophore system of animals is influenced by a number of factors.

Temperature. Temperature appears to influence importantly the coloration of many animals. Low temperatures usually induce darkening in *Hyla*[71] and *Phrynosoma*.[224] Elevation of temperature results in concentration of the dark pigment with consequent lightening of the animal.[224] Among the invertebrates, on the other hand, the situation is not as uniform. Both *Callinectes* and *Palaemonetes* lighten with increasing temperature.[81] The shrimp *Macrobrachium* darkens at both high and low temperatures,[281] as does also the isopod *Idothea*.[83, 198] *Hippolyte* and *Uca*, on the other hand, tend to blanch at both high and low temperatures.[43] The walking-stick, *Carausius*, is black at 15°C and green at 25°C.

Humidity. *Carausius* is influenced by a change in humidity. High humidity induces darkening.[123] *Rana* darkens in a damp environment and lightens in a dry one.[256]

Tactile Stimulation. In general, tactile stimulation seems to have only little influence on the chromatophore system. It was once believed that the color changes of the tree frog were to a great extent response to the texture of the background to which it was attached, the frog becoming dark on a rough background and green on a smooth one, but it is now generally agreed that this is not the case. Tactile stimulation of the suctorial discs of certain cephalopods has been reported to influence the chromatophores reflexly.[291] Schlieper[266] has reported that *Hyperia galba*, a crustacean parasite on jellyfishes, becomes pale when normally attached to its host but darkens when swimming freely. Attachment to any surface, whether black or white, is said to induce in this parasite the paling response which is obviously adaptive, since it usually becomes attached only to its highly transparent normal host.

Psychical Stimuli. The chromatophores of some animals appear to be influenced by psychic states. An excited squid or cuttlefish shows extraordinary plays of color. Such color plays may be caused by the presence of a predator, such as a large crab. The changes often take the form of waves of change passing smoothly and rapidly over the surface of the body. Color plays also frequently appear to contribute in some manner to mating behavior in these animals.

Reptiles and some fishes show also characteristic color changes when excited. The horned toad *Phrynosoma* on strong excitement exhibits a blanching known as "excitement pallor." The frog *Xenopus* darkens during excitement; adrenin administration produces the same response.[56,150,161] *Anolis* when going into combat with another or when manipulated roughly shows a peculiar change of coloration to a mottled condition.[165]

Light. The most important single environmental factor influencing the state of the chromatophore system of animals is light, and in the great majority of animals the method of action of the light involves principally the eyes, central nervous system, and various types of efferent pathways, nervous, hormonal, or both. The importance of the eyes is clearly manifested in the immediate cessation, or great change in character, of color changes on the blinding of an animal. Color changes which are controlled by way of the eyes are known as secondary responses, in contrast with primary responses, which are those proceeding under the influence of light through routes other than the eyes. The latter may involve either a direct action of light on the chromatophores or an influence of light operating reflexly on the chromatophores through extraocular receptor mechanisms.

Secondary color changes dominate the situation in most adult animals. Through a wide range of light intensities these changes are determined by the values of the ratios of the amount of light directly striking the eye from above to the amount of light reflected from the background on which the organism resides. On an illuminated black background where the ratio is large, the animal becomes dark, and on an illuminated white background where the ratio is small, the animal becomes pale, irrespective of the total illumination. There is often a good direct correlation between the value of the ratio of incident light to reflected light, the albedo, reaching the eye of the animal and the degree of black pigment dispersion[30,300] (Fig. 23–7*B*), or the amount of melanin formed in melanophores.[296] Furthermore, either of these

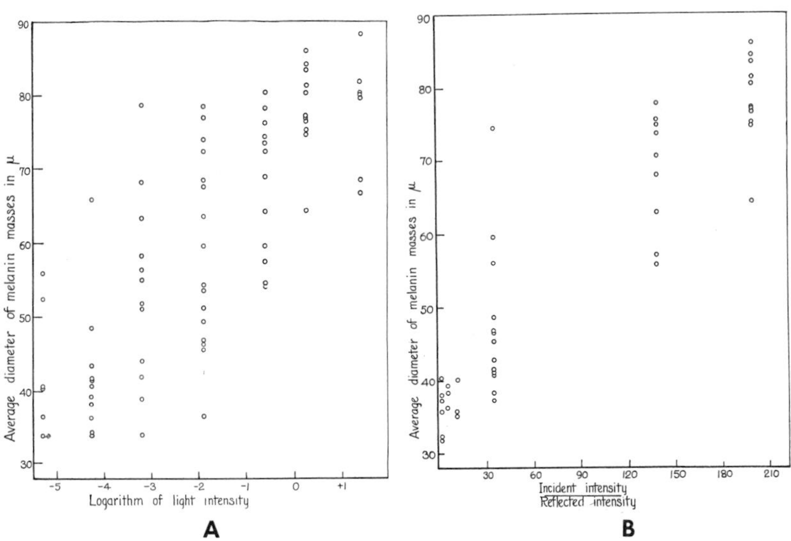

Figure 23–7. *A*, relationship between the log of the incident light intensity and the average diameter of the melanophores of the minnow, *Ericymba*. *B*, relationship between the ratio of *incident intensity* to *reflected intensity* striking the eye and the average melanophore diameter. (From Brown, F. A., Jr., Biol. Bull., vol. 70.)

melanophore responses varies significantly with variation in intensity within a wide range of total illumination.

Adaptive morphological color changes in response to background have been reported for the chromatophores of crustaceans[28] and for melanophores,[297] lipophores,[3] and iridoleucophores[298] of a number of fishes. Functional melanophores have been induced to form on the normally unpigmented ventral surface of flounders either by placing normal fish in black tanks illuminated from below, or by blinding the animals and illuminating them from below.[212]

The influence of a black background may be simulated in many animals by opaquing the lower half of the eye. This has been demonstrated for the walking-stick *Carausius*,[13, 244] the shrimps *Palemonetes*[132] and *Leander* (*Palaemon*),[133] and teleost fishes.[57]

The character of the influence of light in the secondary color responses of animals is obviously one especially adapted to provide the animal with a certain degree of protective or obliterative coloration with respect to its background. As might be anticipated, the function appears in general better developed in animals which are bottom dwellers, or which spend much of their time quietly attached to objects, than in forms which are more freely mobile.

The adaptation to color of background is, in many instances, not restricted to simple blanching and darkening of the skin on white and black backgrounds, respectively. Adaptations of the prawn *Hippolyte* to many colors and tints of background were described many years ago.[203] More recently it has been shown that *Crago* (*Crangon*) will change its coloration to match red and yellow backgrounds, in addition to black and white.[176] *Palaemonetes* has been shown to change its color within a few days to accord with black, white, red, yellow, blue, or green backgrounds.[27] The crab *Portunus ordwayi*[4] adapts to black, white, red, and yellow, as does also *Planes*.[139] The cephalopods *Sepia* and *Octopus*, possessing black, orange, and yellow chromatophores, in addition to iridophores, show striking color responses to background.[181]

Some of the most striking examples of color adaptation in vertebrates are seen in flatfishes. The changes in the flounder *Paralichthys albiguttus* on red, green, yellow, and blue backgrounds have been described.[184, 191] This fish, moreover, is able to simulate rather strikingly the color pattern of the background, thus rendering its protective coloration even more effective. Many other teleosts and amphibians are able to show yellow tints in response to yellowish backgrounds.

These adaptations to color of background are the result of appropriate differential movements of the various pigments within the chromatophores, supplemented by morphological color changes to reinforce these physiological ones and by immobile integumentary pigments. In these activities the animal may show the capacity to induce changes in the distributions or amounts of its various pigment types more or less independently of one another, thus indicating complex mechanisms of chromatophore control. Species may be limited in the colors of background to which they can adapt themselves, through lack of suitably colored pigments. For example, *Crago* (*Crangon*) lacks blue pigment and hence is unable to adapt to either blue or green. *Palaemonetes*, possessing red, yellow, and blue pigments, can, by appropriate pigment manipulation, become almost any color, including black. On the other hand, some species possess pigments of particular colors yet appear to show no ability to adapt to backgrounds of those colors. An example of this is seen in *Carausius*, which is apparently unable to become reddish, despite the possession of a red pigment in its integument.

For the color-adaptive ability, the eyes function not only in their capacity to differentiate incident and reflected light but also in their capacity to distinguish lights of different wavelengths by that portion of the retina stimulated by the reflected light.

The primary chromatophore responses of most adult higher animals are subordinated to the secondary. In the typical adult organism they are best made evident through blinding the animal. Many animals are able to show considerable dispersion and concentration of their pigments in light and darkness respectively, after blinding, but usually there is no longer any response to the background. Examples of chromatophore response in blinded animals have been described for all the major animal groups which exhibit color changes.

A striking exception to the rule of loss of background-color matching is reported for a number of orthopterans.[74, 75] The color is also apparently independent of food. It is believed to involve the epidermal cells acting as independent effectors.[162]

The degree of influence of the primary responses, relative to the secondary responses, appears to vary from species to species or even among the various pigment types within a single species. The chromatophores of zoea larvae of *Crangon* appear to exhibit only a primary response to illumination,[235] whereas the chromatophores in the zoeae of the crabs *Ocypode*[247] and *Rhithropanopeus*[187] exhibit both primary and secondary responses to light. The black and red chromatophores of the megalopa of *Ocypode* exhibit both primary and secondary responses to light, whereas the white chromatophores exhibit only the primary response.[248] Furthermore, the ability of the crab *Ocypode* to change color varies with the age of the individual, occurring more rapidly in juveniles than in older individuals. Long-term adaptation of crayfishes and *Ocypode* to a black or a white background resulted in a marked decrease in their ability to alter their color after a background change.[94, 97, 249]

The most general mechanism of primary response is one of a direct influence of light on the chromatophores. In this response the chromatophore acts as an independent effector. Primary responses of this character have been observed in (1) vertebrate chromatophores after complete denervation of the pigment cells by nerve transection and degeneration, (2) localized light responses in species whose chromatophores are normally not innervated, and (3) in young specimens whose chromatophores have not yet come under the control of a typical secondary mechanism.

Color changes in response to light may be reflexly induced in the absence of eyes. It has been shown for *Phoxinus* that the midbrain is a receptive mechanism for this response.[263] The pineal body has been reported to be a receptor organ in the pigmentary response of *Lampetra* larvae,[321] with the eyes dominating this response in the adult.

The majority of adult animals showing color changes have their coloration correlated within wide ranges of illumination with background color rather than with general intensity of illumination. In the total absence of light there is typically a blanching of the animal, but often not to the extent observed in response to an illuminated white background. Some species, such as *Crago* (*Crangon*) and *Xenopus*, become intermediate in shade through partial dispersion of their dark pigments. Note: *Crangon* does not have melanin. Its dark pigments are ommochromes. It has been shown for the minnow, *Ericymba*, that upon a black background there is no influence of amount of illumination on the coloration as long as the illumination is higher than 1.75 foot candles; at illuminations below this value the average diameter of the melanophores is a linear function of the logarithm of the incident light, down to 0.00053 foot candles, which has the same influence as complete darkness[30] (Fig. 23–7 *A*).

Diurnal Rhythms. Another important factor operating in the control of chromatophores in many animals is a persisting daily rhythm.[293] Many species of animals continue to show their characteristic night-day color changes even when kept under constant conditions as regards temperature and illumination, especially in constant darkness. Such rhythms may persist for a considerable time. The rhythm in *Idothea* kept in total darkness persists at least 8 or 9 weeks.[198] In *Uca*, the diurnal rhythm whereby the animals pale by night and darken by day[9,47] (Fig. 23–8) is so strong that only slight color change can be induced by variation in background or illumination.[43] Some interesting characteristics of the rhythmical mechanism controlling *Uca* chromatophores have been reported.

The times of darkening and lightening in the daily cycles persisting in constant conditions in the laboratory may be reset by a few daily cycles of appropriately timed light or temperature changes[48, 292, 313] to occur at any arbitrarily desired time of day, and then these new temporal relations to solar-day and night will persist in the 24-hour rhythm until they are again reset. Furthermore, even detailed changes in the character of the daily patterns of color changes, and ranges of daily change, may be altered experimentally, to persist later under constant laboratory conditions.[40, 45] The timing mechanism responsible for measing the recurring 24-hour cycles is independent of temperature over at least a 20°C range,[47] and its accuracy during the measurement of one cycle is not affected even when crabs are concurrently being flown by airplane west through 50° of longitude,[49] yielding about a 27½-hour artificial day of all local-time related geophysical fluctuations.

The discovery that, under some conditions, a rhythm of color change, with events reset to occur at a different time of day, would gradually drift back to its initial times, required the postulation that a serial depend-

Figure 23–8. The fiddler crab *Uca pugnax* in dark daytime and pale nighttime phases.

ence of at least two timing centers was present.[48] To effect a permanent rhythm shift, both centers needed to be reset.

Evidence has been obtained that the resetting of the phases of the color change by light and temperature depend upon a diurnal rhythm of responsiveness of the crabs,[38, 292, 313] the crabs responding to increased illumination or temperature only during the nighttime phase of their normal rhythm by a resetting of the color-change time. This has been postulated to be the mechanism by which the crabs attune their color-change rhythm in nature adaptively to the day-night cycles.[36]

Superimposed upon the persistent diurnal rhythm of color change is a lower amplitude tidal rhythm, paralleling the 12.4-hour ocean tides,[39, 79, 80, 86] which modulates the daily one to yield persistent semimonthly patterns of fluctuation in the day by day color-change cycles. The tidal rhythms of color change, with period related to the earth's rotation relative to the moon, also appear to persist indefinitely in constant conditions in the laboratory and to display the same kind of extraordinary properties as those mentioned earlier for the diurnal ones.[50] There is every evidence that the tidal rhythms of color change are adjusted to tidal times of local beaches by the time of uncovering of the crab burrows by the receding water.[80, 85, 99]

To the extent they have been analyzed, the properties of the persistent rhythms of color change in other animals appear similar to those described for the crabs. Persistent rhythms have been found for color changes not only for many other crustaceans but also for vertebrates such as *Lampetra*,[321] *Phoxinus*, salamander larvae, frogs, *Phrynosoma*,[253] and the chameleon,[323] and in the phasmid, *Carausius*, and the echinoderm, *Diadema*.[200]

In some instances where the diurnally rhythmic changes are normally masked by secondary responses to illuminated backgrounds, an underlying influence of the rhythm may be evident in an increased rapidity of those responses which tend to support, and the sluggishness of those which tend to antagonize, the particular phase of the underlying persistent rhythm at that moment. Thus many animals more readily adapt to an illuminated white background during the night phase and to an illuminated black one during the day phase, or show corresponding differing susceptibilities to injection of color-change hormones between the night and day phases.

FUNCTIONAL ORGANIZATION OF CHROMATOPHORE SYSTEMS

Annelids. Certain annelids become pale in darkness and dark when illuminated. This has been observed for the polychaete *Nereis* (*Platynereis*) *dumerilii*,[136] and leeches *Piscicola geometra*,[152, 153] *Protoclepsis tessellata*,[294] *Hemiclepsis marginata*,[152, 153] *Glossiphonia complanata*,[152, 153, 314] and *Placobdella parasitica* and *P. rugosa*.[287] In no case has any response to color of the background been demonstrated.

The rhynchobdellid leech *Placobdella parasitica*, a common parasite on turtles of the central United States, possesses three types of pigment cells. These pigments contribute to the mottled brownish and white coloration shown by these animals. The amount of white varies considerably from specimen to specimen, ranging from the greater part of the dorsal surface to a few minute lateral papillae and a short median line, both restricted to the anterior half of the animal. One type of cell containing a pale yellowish, granular pigment occupies the characteristic longitudinal mid-dorsal stripe, the numerous light papillae, and segmental blocks along the margin. These cells show no physiological responses. Distributed over all the darker areas of the body are relatively large chromatophores containing dark greenish pigment. This latter pigment participates in physiological color changes. Another pigment, reddish brown, is also located within functional chromatophores. These last chromatophores are much smaller than the greenish ones and are located more superficially. The green pigment is alcohol soluble; the reddish-brown pigment is alcohol insoluble. The European leech *Protoclepsis tessellata* possesses three pigments: brown, green, and yellow. The Mediterranean polychaete *Nereis (Platynereis) dumerilii* possesses iridophores and brick red and brownish violet chromatophores. There is a circadian rhythm of migration of these pigments. They are more dispersed by day than at night. This rhythm appears to have an endogenous component.[104, 255]

The leech *Placobdella parasitica,* like other annelids showing physiological color changes, blanches in darkness and darkens when illuminated. The time required for the greenish pigments, which are predominant in these changes, to complete their concentration in darkness or dispersion in light is approximately 1 hour.

There are several types of evidence all pointing to nervous control of the green chromatophores. Decapitation or any other transection or injury to the nervous system results in a darkening of pale animals kept in darkness. This darkening may persist for many hours. If a uniformly pale leech in darkness is stimulated faradically at either the anterior or posterior end, the whole animal darkens. If, however, the experiment is repeated with a specimen whose nerve cord has been transected in the middle, only that half of the animal receiving the stimulus darkens, the body beyond the point of transection remaining light. There is little evidence for direct action of light on the chromatophores.

The two eyespots at the anterior end appear to play an active but by no means an exclusive role in the responses. Decapitated specimens show the characteristic changes even though responding more sluggishly. When specimens are brought from darkness into light and one-half of the body is immediately covered with opaque paper, the covered portion remains largely in the dark-adapted phase while the uncovered portion becomes completely light adapted. The results are more striking when the posterior end is covered than when the anterior end is. These experiments suggest a role of generally distributed photoreceptors operating through segmental reflexes.

The reddish-brown pigment of *Placobdella parasitica* responds independently of the green and shows no predictable responses to background or to light intensity. Its condition is more or less variable in normal specimens. When dispersed it can usually be made to concentrate by intense stimulation of the animal. In animals with transected cords, the response to electrical stimulation, when it occurs, passes as for the green cells, only to the point of transection. Thus it would appear that the nervous system controls the reddish-brown chromatophores directly, but that excitation induces concentration rather than dispersion as for the green.

Echinoderms. The sea urchins, *Arbacia pustulosa, Centrostephanus longispinus, Diadema antillarum,* and *Diadema setosum,* become lighter in color on transfer from light to darkness.[166, 199, 200, 307, 319] Illuminated *Arbacia pustulosa* are blackish in color whether on a white or on a black background, but in darkness become brown in color. *Centrostephanus longispinus,* which are dark purple in light, change in darkness to gray. The physiological color changes require about 1 or 2 hours for their completion. Microscopic examination of tube feet removed from light-adapted and dark-adapted individuals shows numerous reddish-brown chromatophores with their pigment dispersed in the light-adapted and concentrated in the darkness-adapted individuals. The color changes in these urchins are due to the movements of pigments within definite chromatophores, but morphological color change also occurs. The chromatophores of isolated tube feet which have been mounted on a microscope slide respond to illumination and darkness in the same manner as when they are present in the intact animal, indicating that the chromatophores are responding to the light either directly or by way of local reflex pathways comparable to those known to function in the locomotor movements of the tube feet. Studies with highly restricted points of illumination suggest the chromatophores to respond directly to light.

The American species of *Arbacia, Arbacia punctulata,* appears to show no color changes comparable to those just described.[215]

Cephalopods. Many cephalopods show remarkably rapid color changes as a result of the activity of their peculiar type of chromatophores. These changes may result from many different types of stimuli, but light is one of the most important.

The chromatophores are primarily controlled by the nervous system. Cutting a nerve innervating a particular region of the body results in an immediate cessation of all color changes in that region. It was shown long ago that after a mantle connective is cut[110] there is a paralysis of the chromatophores of the corresponding half of the body and a persisting blanched coloration. Since the connectives between mantle ganglion and the chromatophores are still intact, it is evident that the ganglion contributes little to the control by itself and that the normal control over the responses resides in higher centers of the nervous system. The cerebral ganglion of the brain appears to possess an inhibitory center for the chromatophore system;[239, 270] after its destruction or inactivation there is tonic expansion of the chromatophores. The inhibitory center is believed to operate through control of the color center[271] located in the central ganglia, which in turn operates through motor centers found in the subesophageal ganglia. The motor centers each control the chromatophores of the corresponding halves of the body.

The eyes are the chief sense organs influencing the central nervous centers. Bilateral blinding does not eliminate changes of color, but the changes which then occur are in no sense adaptive ones. If only one eye is blinded, the responses of the chromatophores on the corresponding side of the body are diminished.

Another significant source of influence on the color-control centers is the suckers on the arms. If these are all extirpated there is a considerable loss of tone in the chromatophores and, hence, skin lightening occurs.[291] Removal of both eyes and suckers, however, does not entirely eliminate the chromatophore responses. After such an operation vigorous stimuli will still result in color changes, probably as a result of stimulation of organs such as tactile ones and organs of equilibrium.

The chromatophores on the side of the animal that is lowermost when the animal is in contact with the substrate always are more contracted than those of the remainder of the body.[271] This is not a direct influence of illumination as one might first suspect, for it cannot be reversed by illumination from below instead of above. Rather, it appears to be part of a postural response involving stimulation of tactile receptors reflexly through the central nervous system, resulting in the localized chromatophore contraction.

The substances tyramine and betaine are known to be present in the blood of cephalopods. The former, like adrenin, increases the tonus of the motor centers, resulting in a darker coloration.[272] Betaine, on the other hand, like pilocarpine or acetylcholine, appears to decrease the chromatophore tone by stimulation of the inhibitory center. If one transfuses blood from a characteristically darker species, such as *Eledone* or *Octopus macropus,* into a lighter species, such as *O. vulgaris,* the latter darkens.[273] Interconnection of the circulatory system of the two will yield comparable results. Tyramine is known to be more concentrated in the blood of these darker species than in that of the lighter ones. Surgical removal of the posterior salivary gland, known to be the important source of tyramine in the blood of these animals, results in an increase in paleness and a complete loss of tone of the chromatophores.[273] Darkening may be induced again by injection of tyramine solutions. The results of Sereni suggest that tyramine, and probably betaine as well, function as humoral agents operating to modify through the central nervous centers the general tone of the chromatophores. Furthermore, studies of denervated chromatophores[274, 275] suggest that both tyramine and betaine also exert a tone-increasing action on the chromatophores themselves, thus functioning directly as well as indirectly on the chromatophores. However, Sereni's results have been criticized by Taki, who found that removal of the posterior salivary gland had no effect on the color changes of the cephalopods he was using.[302] The entire matter of endocrine control of color changes in these animals needs to be investigated further.

Superimposed upon the possible, slower humoral influence is the more conspicuous nervous mechanism responsible for the rapid color changes. Earlier observations have led to the hypothesis that the chromatophores have a double innervation. If an isolated piece of the integument of *Loligo* containing chromatophores is allowed to stand for some time, the chromatophores first contract and later reach a condition of maintained partial expansion. Chromatophore contraction results from electrical stimulation of these latter. Here, therefore, electrical stimulation appears to act to inhibit the tonus of the chromatophore musculature.[24] On the other hand, single electrical shocks to the chromatophore nerve, or to the chromatophores themselves, give single twitches of the chromatophore muscle. Repeated shocks give tetanus and, consequently, chromatophore expansion. It has therefore been concluded that *Loligo* chromatophore muscles receive double innervation, both an excitatory and an inhibitory element. However, these inhibitions could have been due simply to Wedensky inhibition and not to an inhibitory axon. Furthermore, recent electrophysiological investigations revealed no evidence for an inhibitory innervation. The action of the motor neurons can account for all of the behavior of the chromatophore muscle cells.[105, 106]

Insects. A number of insects can change their coloration in response to external stimuli, usually as the result of morphological color changes only. Among factors effecting the changes are temperature, humidity, general activity, and illumination. In only a few species are there the relatively rapid physiological color changes.

Many butterfly pupae are darker or lighter in coloration, depending on whether they are reared at lower or higher temperatures, respectively. The effect of temperature operates in *Vanessa* through the head, the pupa taking on a coloration which is determined by the temperature of that portion of the body when the head and body are maintained at different temperatures.[124] The prepupal color changes of caterpillars have been found by ligation experiments to depend normally on hormones. Clearly involved are ecdysone from the prothoracic glands and probably also the juvenile hormone.[55, 137] A similar influence of temperature on the degree of development of the dark pigments has been observed for the wasp, *Habrobracon*,[157, 268] and the bug, *Perillus*.[168]

The colors of *Pieris brassicae* are due to melanin in the cuticle, white pigment in the epidermal cells, and green pigment in the deeper tissues. The coloration of these pupae is also influenced by the background. On a black or red background the pupae are grayish white, whereas on green or orange backgrounds they are clear green, the latter background suppressing formation of the black and white pigments.[67] Exposing the pupae to colored lights gives the same results as the colored backgrounds. The eyes or some other head structure are essential for the response, for it ceases and the animals behave as in darkness when the eyes are extirpated or the animal is decapitated.[25]

The migratory locust, *Locusta migratoria*, shows a limited ability to adapt its coloration to backgrounds.[77] Through variations in the quantities of yellowish and black pigments the locusts may become yellowish white, brown, or black. The quantity of black pigment formed seems to depend on the ratio of incident to reflected light striking the eyes; the amount of the yellowish pigment appears to depend on the predominant wavelengths present, being formed more rapidly at longer wavelengths (550 to 660 nm) and less rapidly at shorter ones (450 to 500 nm). When this species enters its swarming migratory phase, a skin coloration darker than in the solitary phase is produced. There is some evidence that the darker coloration is a result of the more intense metabolic activity. When migrating locusts are returned to solitary conditions they will regain their lighter color phase, but this color change is delayed if the isolated locust is kept in a constantly excited state.[77]

A few insects show color change which can be traced to redistribution of pigments within pigment cells or chromatophores. A larva of *Corethra (Chaoborus)* shows a rapid physiological color change of its air sacs.[64, 190] This last is due to the presence of special pigment cells. On a black background the pigment becomes dispersed, and the pigment cells are scattered uniformly over the sacs. On a white background the pigment concentrates, and the cells appear to wander to one side of the sacs. The eyes are involved in these reactions, as is also the brain. A hormonal factor present in the brain, and to a lesser extent in the corpora allata, disperses the dark pigment.[64, 265]

The color changes of the phasmid *Carausius morosus* have been investigated rather extensively. The hypodermal cells of this species contain four pigments, brown (melanin), orange-red and yellow (lipochromes), and green.[123, 265] The brown and orange-red pigments show active concentration and dispersion within the cells in response to external stimuli. The green and yellow pigments show no such activity. Therefore, the green varieties found in nature show no physiological color changes while the brown ones do. Brown specimens are usually dark by night and pale by day, as a result of dispersion and concentration, respectively, of the brown and orange-red pigments. A partial independence of a direct influence of light in these changes is indicated by the persistence of typical day-night cycles of color change in animals kept in constant darkness.[123, 265] It is possible to reverse the rhythm by keeping the animal in illumination by night and in darkness by day, whereupon the newly established rhythm will continue in constant darkness.

Utilizing the fact that high humidity also produces body-darkening, Giersberg[123] ingeniously proved that the effect of this stimulus on the chromatophores is indirect, operating by way of afferent nervous pathways, the brain, an endocrine source, and finally a blood-borne agent. When the posterior half of a pale *Carausius* is inserted into a small moist chamber the whole animal darkens in 30 to 60 minutes (Fig. 23–9). The darkening commences at the anterior end, which lies outside of the chamber in dry air, and gradually spreads backward over the whole body. Returned to dry air, the animal lightens again in 1 to 2 hours. If a ligature is then drawn tightly around the anterior thoracic region and the experiment repeated, the darkening spreads back only as far as the constricted region. If the ventral nerve cord is carefully transected between the subesophageal and the first thoracic ganglia, or between the subesophageal and the supraesophageal ganglia of an otherwise normal pale animal and the experiment repeated, there is no darkening whatsoever, but if the animal is turned about and the head inserted into the moist chamber, the head darkens first and the darkening then spreads out over the body in just the same fashion as seen in unoperated specimens. In the ligatured animal the nerve pathways are apparently still able to conduct anteriorly the nerve impulses arriving from the nerve endings of the abdominal region which are stimulated by the moisture. A blood-borne hormone, liberated at the anterior end, is unable to diffuse posteriorly past the ligature. After nerve-cord transection the nerve impulses are prevented from reaching the brain and stimulating the liberation of a hormonal substance. Hormone production can still be effected by stimulation of nerve endings anterior to the operated region, and then the active principle is free to pass posteriorly in the body fluids.

Further evidence that the physiological color changes in *Carausius* are predominantly controlled by hormonal material has come from transplantation of portions of the skin of one animal to another.[151] The transplanted tissue begins to show color changes entirely paralleling those of the host in 2 or 3 days. It is unlikely that the transplant tissue would have received any innervation from the host nervous system in such a short period. In fact, normal epidermal tissue shows no indications of innervation.

The eyes of *Carausius* are essential to the normal responses to light.[13] Section of the optic tracts or blackening of the eye surface stops the responses.

Carausius also shows responses to background when humidity and illumination are kept constant. It turns dark colored on black and red surfaces and light colored on white and yellowish ones. The background responses are determined by the ratio of light striking the dorsal and the ventral halves of the eyes. Painting the lower half of the eyes black brings about darkening as on a black background.

The brain appears to be the chief source of a color-change hormone. After brain extirpation the animals become pale gray. Injection of brain extract darkens the insects.[65, 66] An active factor, termed the C-substance, has been isolated electrophoretically.[173]

Morphological color changes in response to

Figure 23–9. Diagram illustrating the use of a moist chamber as a stimulus for producing darkening in *Carausius.* The darkening in this instance is in response to abdominal stimulation and commences at the head and passes posteriorly only as far as an anterior thoracic ligature. (From Giersberg, H., Z. vergl. Physiol., vol. 9.)

illumination, background, and humidity also occur in *Carausius* and appear to involve the same mechanism of control as does the physiological change. Such change, however, requires stimulation over a much longer period.[13,124] It has been suggested that the substances normally influencing physiological changes so modify the general nutrition and metabolism of the insects as to result in further changes in color through pigment formation and destruction.

Certain mantids also exhibit diurnal color changes[117] which correspond very closely with the diurnal movements of the retinal pigments in the lateral regions of the compound eye.

Crustaceans. The crustaceans exhibit some of the most remarkable instances of color adaptation to be found in the animal kingdom. Most crustaceans possess within their chromatophores white, red, yellow, and often also black, brown, and blue pigments. By appropriate rearrangements of the individual pigments within the chromatophores, many crustaceans are able to approximate rather closely the colors of the backgrounds upon which they come to lie.

Although it was believed by all the early investigators that the chromatophores were controlled by nerves, there was never any satisfactory demonstration of nerve terminations at the chromatophores, nor did nerve transection ever appear to interfere directly with the responses of the chromatophores within the animal. Koller,[175,176] working with *Crago vulgaris* (*Crangon crangon*), provided the first clear evidence that a blood-borne agent was active in controlling the chromatophores. He found that transfusion of blood from a specimen darkened on a black background into a light animal kept on a white background would cause darkening of the light animal. No evidence for a lightening factor was obtained, however, by the reciprocal transfusion. He also observed that blood from a yellow-adapted specimen would render yellow a white-adapted specimen.

Perkins[237] discovered that, although denervation of an area of the body of *Palaemonetes* in no way interfered with the responses of the region when the animal was placed on a black or white background, occlusion of the blood supply to any region resulted in an immediate cessation in the responses of that region. Upon readmission of blood to the region, that region changed its color at once to harmonize with the color of the remainder of the body. These results were interpreted to indicate that factors for dark-pigment dispersion and concentration were conveyed to the individual chromatophore by way of the blood. Extraction and injection of various parts of the body showed that the eyestalks contain a potent factor for concentrating the predominant red pigment and dispersing the white[238] in *Palaemonetes* and hence for blanching the animal. Removal of the eyestalks results in a permanently darkened condition of the animal. These results were quickly confirmed by Koller,[177] working with *Crago* (*Crangon*), *Leander* (*Palaemon*), and *Processa*. The hormonal substance involved was shown by reciprocal injection experiments to be neither species nor genus specific. Since these pioneering efforts, numerous investigators have shown that either

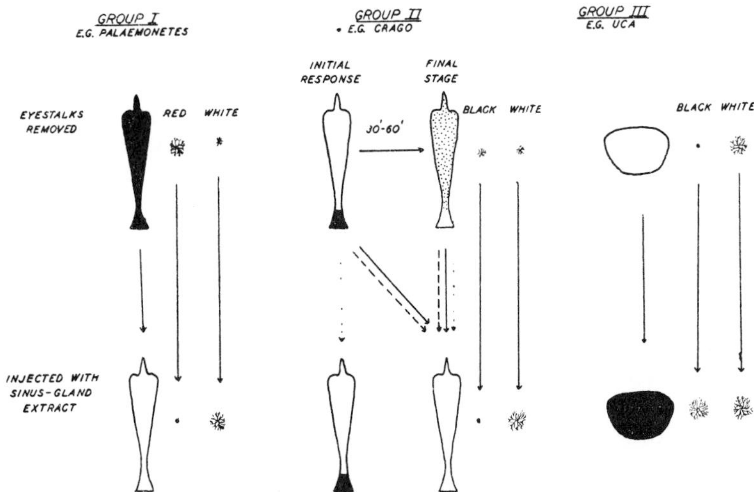

Figure 23–10. Schematic representation of the coloration of eyestalkless crustaceans and state of the dominant chromatophores for each of the three differently responding groups, and of the responses of these to injection of eyestalk or sinus gland extract. Solid arrows indicate extract of total water-soluble material, dashed arrows indicate an alcohol-insoluble fraction, and dotted arrows indicate an alcohol-soluble fraction. (From Brown, F. A., Jr., Action of Hormones in Plants and Invertebrates. Academic Press.)

the eyestalks or, in a few species, the anterior thoracic region contains the source of a material influencing the state of the chromatophores.

Decapod crustaceans which have been investigated extensively with respect to their eyestalk hormonal activities in color changes appear to fall, with a rare exception, into three groups with respect to roles of eyestalks in their chromatics (Fig. 23–10). Group I contains such genera as *Palaemonetes, Peneus, Hippolyte, Leander* (*Palaemon*), *Orconectes,* and *Cambarellus.* Their chromatophore systems usually contain red, yellow, blue, and white pigments. Group II includes only the genus *Crago* (*Crangon*), which has a complex pigmentary system with no less than eight differently responding chromatophore types, enabling the shrimp to show not only general shade and tint changes, but also a certain degree of change in color pattern. The chromatophores contain black, brown, red, yellow, and white pigments. Group III includes all true crabs (brachyurans) which have been investigated. These include *Eriocheir, Hemigrapsus, Callinectes, Uca,* and *Sesarma.*[10, 100] The best known of these is the fiddler crab *Uca,* which commonly contains black, red, yellow, and white pigments in the chromatophore system.

After removal of the eyestalks from a member of any one of the three groups, the characteristic type of response for that particular group is observed. In Group I the animals darken rapidly through complete dispersion of their red and yellow pigments and become quite dark (although usually not as dark as in normal response to a black background) in an hour or two. They remain in this condition indefinitely. The white pigment usually undergoes a transitory concentration and thereafter exhibits a variable state. *Crago* (*Crangon*), of Group II, most commonly shows a more complex change after eyestalk removal. First there is a transitory darkening of the telson and uropods and a blanching of the remainder of the body, which lasts from ½ to 1 hour. This is followed by a complete blanching of the telson and uropods and darkening of the body to an intermediate and mottled coloration. The white pigment on the body initially concentrates but then assumes an intermediate condition. The crab *Uca,* of Group III, blanches rather quickly after removal of its eyestalks, its black chromatophores becoming for the most part punctate, and its white ones commonly broadly reticulate. This condition is maintained without significant change indefinitely. Thus, we see eyestalk removal from various crustaceans resulting in three types of conditions: body darkening, adoption of an intermediate coloration, and body blanching. In all three, the animals lose practically all of their responses to changes in background or illumination.

Injection of eyestalk extract into Group I animals results in rapid blanching.[84, 87, 102, 264] In Group II there is complete blanching of both the body and telson and uropods. In Group III, on the other hand, there is a blackening of the whole body.[23, 72, 192] These strikingly different results observed for the animals of Groups I, II, and III are explained chiefly in terms of differences in the responses of the chromatophore systems to the eyestalk hormones, since reciprocal injections demonstrate that extracts of eyestalks from specimens of other groups produce in the specimens of any given group a response qualitatively the same as that produced by the specimen's own eyestalks. For example, eyestalk extracts prepared from animals of Group I darken eyestalkless specimens of Group III.

The place of normal liberation of the chromatophorotropins from the eyestalk is the sinus gland (Fig. 21–9). This was postulated and strongly supported by Hanström.[132] This gland had been described earlier,[131, 278] and at that time was called the "blood gland," since it was first believed to be homologous with a blood-forming gland in the eyestalks of *Crago* (*Crangon*), which had erroneously been considered to be the source of the active material.[178] This gland occurs in practically all of the numerous crustaceans in which it has been sought. Hanström, using eyestalkless specimens of *Uca, Palaemonetes,* or *Penaeus* as animals for bioassay in his numerous experiments, found that eyestalks of animals whose sinus glands were located in the head near the brain (*Gebia* and *Hippa* [*Emerita*]) showed no chromatophorotropic activity. When the eyestalks of other species were sectioned in various ways, the sections possessing the sinus gland always showed activity; other parts were relatively inactive. By utilizing the species differences in the anatomical arrangement of eyestalk organs, Hanström was able to get, one by one, all the remaining organs of the eyestalk into portions of the stalk showing little or no activity. Furthermore, no other structure in the stalk gave histological evidence of having secretory activity except a glandular organ called the X-organ. Sections containing the X-organ, but not the sinus gland, were inactive; and removal of the X-organ from eyestalks did not diminish their chromatophorotropic activity. It was thus concluded that the sinus gland was the only eyestalk source of hormones influencing the red or black pigments in the test animals employed.

Hanström's conclusions were confirmed by Brown,[32] who found that the sinus glands could be removed and extracted by themselves, and that such extracts possessed essentially all the activity of total stalks despite the fact that their volume was only about 1 per cent of that of the stalk tissue. Such extracts elicited about 80 per cent of the response from whole stalks for both *Palaemonetes* red and *Uca* black chromatophores. Thus, for these two widely different chromatophore types, the sinus glands are the chief eyestalk reservoir of hormonal material. Furthermore, implantation of a sinus gland into the ventral abdominal sinus of *Palaemonetes* results in a blanching of the animal which lasts about 100 times as long as the effects of an injection of extract that is the equivalent of approximately one gland.

It is generally conceded that the sinus glands are not the site of synthesis of the chromatophorotropins. Rather, these are believed to be produced within neurosecretory cell bodies, clustered into groups termed X-organs. (See Chapter 21.) The sinus gland, a composite of enlarged terminations of these neurosecretory cells, is only the organ of storage and liberation of the active hormones. This functional relationship between X-organs and sinus glands, initially established from studies on hormonal control of molt, from morphological interrelationships, and from observations of the actual passage of secretory granules to the sinus gland within the X-organ–cell axons, was given direct experimental support by the finding that X-organ extirpation in *Leader (Palaemon)* terminates color changes, the prawns remaining with dispersed dark pigment.[234]

The X-organ–sinus gland complex of the eyestalk provides more than one chromatophorotropin. This was demonstrated first by means of a comparative study of the relative influences of extracts of the eyestalks of seven genera of crustaceans in concentrating *Palaemonetes* red pigment and dispersing *Uca* black pigment.[44] The ratio of the effect of extracts on two chromatophore types, *Uca* black and *Palaemonetes* red, was called the U/P ratio. The U/P ratios obtained for eyestalks or sinus glands from different sources varied from one genus to another. For example, *Crago (Crangon)* sinus glands showed a relatively high value, whereas those of *Palaemonetes* showed a relatively low value. *Uca* yielded an intermediate value. The order for the seven species investigated showed no correlation with either the sizes of the animals or the relative potencies as assayed on *Uca*. The hypothesis proposed to explain these data—namely, that sinus glands differed from one another in the proportions of two principles, (1) a factor predominantly darkening *Uca* (UDH), and (2) a factor predominantly lightening *Palaemonetes* (PLH)—was given strong support by the discovery that the sinus glands of each species yielded two active fractions, one alcohol soluble, and the other alcohol insoluble. The former gave a very low U/P ratio, as if possessing a larger proportion of PLH, and the latter gave a high U/P ratio, suggesting a larger proportion of UDH. It was possible to restore the initial U/P ratio for the glands of a species simply by recombining the two fractions. These results indicate that species representing Groups I, II, and III possess two eyestalk chromatophorotropins, but in differing proportions.

The sinus glands of the crustaceans from Group III which have been so far examined lack an activity shown by glands of species of Groups I and II. This has been established through the discovery that, whereas extracts of the eyestalks or sinus glands of species of Group I or II will lighten the telson and uropods of *Crago* (*Crangon*) within 3 or 4 minutes after injection, sinus glands of species of Group III fail to do so.[37, 52] This is so despite the fact that Group III eyestalks are apparently as effective in lightening the bodies of *Crago* (*Crangon*) as are eyestalks from other groups. This principle, which is present in Groups I and II but absent in Group III, has been called *Crago* (*Crangon*) "tail"-lightening hormone (CTLH). Though at least three chromatophorotropins occur, therefore, in crustacean sinus glands, two possibilities exist: (1) species in Groups I and II possess three principles and those of Group III possess two principles, or (2) all possess only two principles, with one of the two differing in physiological properties between Groups I and II and Group III.

The eyestalks are not the sole sites of production of chromatophorotropins in crustaceans. It is well known that undisturbed eyestalkless specimens of *Crago* (*Crangon*) not uncommonly exhibit random color changes. Eyestalkless crustaceans may be induced to undergo color changes through the action of blood-borne factors by electrical or other stimulation of the cut ends of the optic nerves. Crustaceans of Group I are induced to blanch, but whereas blanching under the influence of sinus gland principles includes white-pigment dispersion, by this means the white usually concentrates. Members of Group III darken, but here, too, the white concentrates as the black disperses, unlike under sinus gland influence. The responses of *Crago* (*Crangon*) of Group II appear more complex, and its reactions to electrical stimulation of its optic nerve stubs provided a clue which led to the discovery of an additional source of the chromatophorotropins in crustaceans. If stimulation is mild, the whole animal blackens; if stimulation is intense only the telson and uropods darken, the remainder of the body blanching. Some chromatophorotropins arising outside of the eyestalks could obviously antagonize the activity of sinus gland factors in some responses and supplement them in others.

Extracts of the central nervous organs of certain species of Group I, on injection, lighten eyestalkless specimens and concentrate white pigment[26, 51, 84, 88, 149, 169] (Fig. 23–11). Similarly, extracts of the *Uca* nervous system disperse red and simultaneously concentrate white pigment. Extracts of the nervous system of *Crago* (*Crangon*), of Group II, lighten the body and darken the telson and uropods while concentrating white pigment. Therefore, in all three groups, injection of extracts of central nervous system organs produces the same major results as strong stimulation of the eyestalks of eyestalkless animals.

The major portion of the telson- and uropod-darkening action, as well as body-lightening activity of *Crago* (*Crangon*), resides in an organ of the head region which has been termed the postcommissural organ.[33, 170] (See Chapter 21.) The remainder of the central nervous system, however, possesses substantial body-lightening activity. This latter fact, together with the discovery that an alcohol-soluble fraction of the tritocerebral commissures possesses very strong body-lightening and no "tail"-darkening activity, while the alcohol-insoluble fraction causes strong "tail"-darkening and simultaneously strong body-darkening,[41] establishes the presence of two chromatophorotropins in *Crago* (*Crangon*) nervous systems (Fig. 23–12). One lightens the body but not the "tail" of *Crago* (*Crangon*), and hence has been called *Crago* (*Crangon*) body-lightening hormone (CBLH); the other, in the presence of CBLH, darkens only the "tail," but in the absence of CBLH darkens the whole body and is hence called *Crago* (*Crangon*) darkening hormone (CDH). The striking difference between strong and weak stimulation of the eyestubs in this species appears to be a consequence of the liberation of both principles with strong stimulation, and exclusively CDH with weak stimulation.

A survey of the influence on *Crago* (*Crangon*)

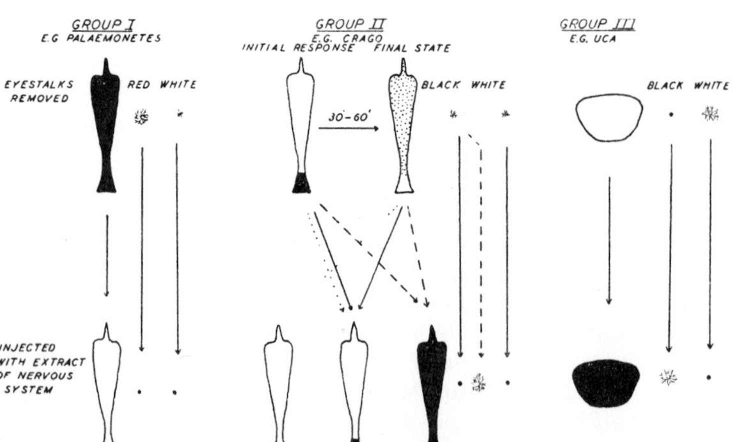

Figure 23–11. Schematic representation of the coloration of eyestalkless crustaceans and state of their dominant chromatophores for each of the three differently responding groups, and of the responses of these to injection of extracts of the nervous system. Solid arrows indicate extract of total water-soluble material, dashed arrows indicate an alcohol-insoluble fraction, and dotted arrows indicate an alcohol-soluble fraction. (From Brown, F. A., Jr., Action of Hormones in Plants and Invertebrates. Academic Press.)

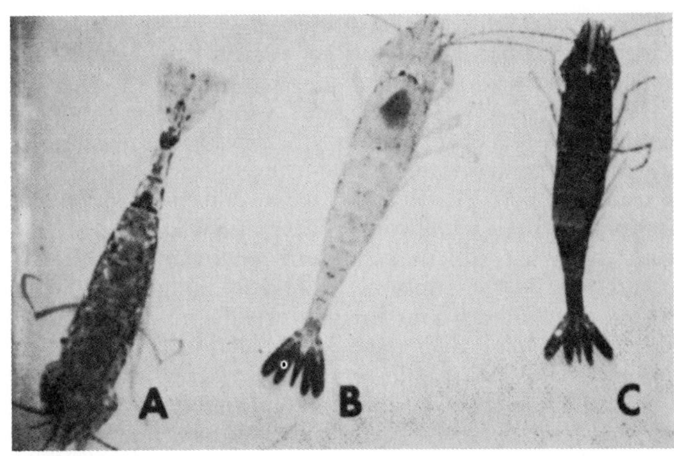

Figure 23–12. Appearance of three *Crago* (*Crangon*) closely matched in coloration initially, in response to injection of various solutions: *A*, Sea water (control, exhibiting no change); *B*, total water-soluble fraction of tritocerebral commissures; *C*, an alcohol-insoluble residue of the commissures. (From Brown, F. A., Jr., and Klotz, I. M., Proc. Soc. Exp. Biol. Med., vol. 64.)

of extracts of the nervous systems of various species of Group I led to the conclusion that all possess both CDH and CBLH, although in none of these other species is it as localized as in *Crago* (*Crangon*).[42] These hormones are usually widely distributed along the nervous system (*Palaemonetes, Cambarus, Homarus*) or restricted to regions other than the tritocerebral commissures (e.g., to the posterior portion of the thoracic cord in *Pagurus* and other anomurans). In some nervous systems, and more especially in some portions of the nervous systems, the ratio of CDH activity to CBLH activity is so large that the extracts blacken eyestalkless *Crago* (*Crangon*) at least as effectively as the alcohol-insoluble fraction of the *Crago* (*Crangon*) tritocerebral commissures (e.g., lobster or crayfish abdominal cords). The high CDH activity of lobster abdominal cords (Fig. 23–13) appears to be associated chiefly with the sheath rather than in ganglionic cells and nerve fibers. In contrast to crustaceans of Group I, when extracts are prepared of freshly dissected eyestalks or central nervous organs from the body proper of Group III animals, no CDH activity is apparent. However, the presence of a *Crangon*-darkening substance in brachyurans can be demonstrated when their eyestalks are subjected to gel filtration chromatography, thereby separating this substance from the body-lightener.[96] By this technique CDH has also been found in the eyestalks of *Crangon*.[279] CBLH is present in the nervous systems of all crustaceans examined.

It had been established earlier that *Uca* of Group III possesses at least two chromatophorotropins in its nervous system,[260] a black dispersing and a white concentrating factor. This evidence comes from comparison of the relative influences of extracts of the brain, circumesophageal connectives, and thoracic cord on eyestalkless *Uca*. Extracts of the brain and thoracic cord show strong activity in black-pigment dispersal and simultaneously leave white pigment dispersed or induce its dispersal. Extracts of connectives, on the other hand, produce only a weak black-dispersing action but cause concentration of

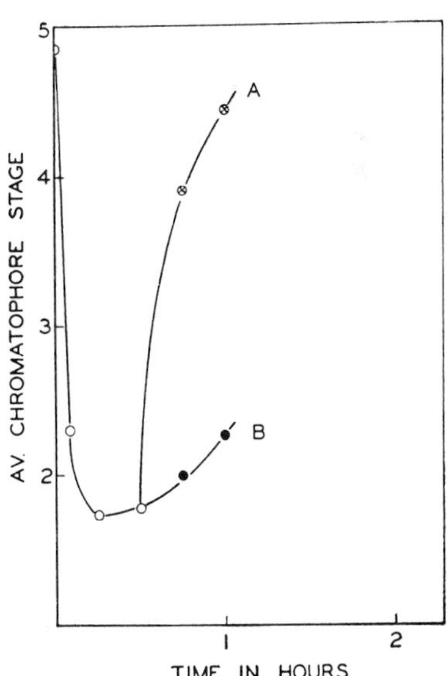

Figure 23–13. Demonstration of dual control of red chromatophores of *Palaemonetes*. Response of red chromatophore to injection into eyestalkless animals of an extract of postcommissural extract, followed, at the end of 30 minutes, by injection of extract of abdominal nerve cord in half the animals *(A)* and of sea water, as control, into the remaining half *(B)*. (From Brown, F. A., Jr., Webb, H. M., and Sandeen, M. I., J. Exp. Zool., vol. 120.)

the white pigment. The differing results cannot be duplicated simply by varying the concentration of nervous system extract. Although these observations established the presence of two chromatophorotropins, they did not permit one to decide between two possibilities: (1) all portions have a black-dispersing and white-concentrating action, with the brain and thoracic cord having in addition an antagonistic white-dispersing agent; or (2) all portions have a white-dispersing and a black-dispersing action, but the connectives alone contain a white-concentrating principle. More recent experiments showed that the first suggestion is the correct one.[252] The presence of a black-concentrating factor has been demonstrated for *Uca*; this pigment, therefore, is under dual control by two antagonistic hormones. The red-pigment cells are also under dual control by pigment-dispersing and pigment-concentrating substances.[34] The red and white pigment-dispersing activities in the eyestalks of *Uca* have been separated from the pigment-concentrating activities, but it is not yet clear whether the same substance can disperse (or concentrate) more than one of these pigments.[95] The red-pigment- and white-pigment-concentrating hormones in the eyestalks of *Crangon* appear to be different substances.[279]

Numerous reciprocal-injection experiments among *Crago* (*Crangon*), *Uca*, and *Palaemonetes*, in which the comparative distributions of *Crago* (*Crangon*) darkening, *Crago* (*Crangon*) body-lightening, *Uca* black-dispersing, and *Uca* white-concentrating activities within the various nervous systems are compared, show rather clearly that CDH and UDH cannot be the same, or like CBLH or UWCH. However, there is still a possibility that CBLH and UWCH are identical, since there is a qualitative parallel in their distributions within the three nervous systems studied. Certain quantitative differences of activity, however, cast doubt on this identity.

Until the active principles are available fully free of one another, great difficulty will be experienced in determining their individual roles. In the earlier attempts to separate the crustacean chromatophorotropins, the technique of paper electrophoresis was employed.[88,93,173,174] By this technique it was shown that the red pigment dispersing and concentrating hormones from the brain and circumesophageal connectives of the crayfish *Cambarellus* have at pH 7.4 to 7.5 opposite charges, − and + respectively.[88,93] In recent years, attempts have been made to isolate and purify the chromatophorotropins in crustaceans using the techniques of gel filtration, ion exchange chromatography, and disc electrophoresis.[78,167,250]

The red pigment concentrating hormone isolated from the eyestalks of *Pandulus borealis* is a peptide containing eight amino acids: aspartic acid, glutamic acid, glycine, leucine, phenylalanine, proline, serine, and tryptophan.[78] The hormone that evokes pigment dispersion in the melanophores of *Uca* also appears to be a peptide because it is inactivated by proteolytic enzymes.[19,236] It is also inactivated by oxidizing agents such as hydrogen peroxide and performic acid, but the activity of the oxidized hormone can be fully restored by treating it with the reducing agent thioglycollate.[251]

In extended periods of adaptation to background there appears to be an accumulation in crayfishes of that color-change hormone, darkening or lightening, which is not being used under the particular conditions. The amounts of pigment-dispersing and -concentrating hormones in the blood are altered when the background is changed. For example, when a crayfish is on a black background it has more red pigment-dispersing hormone in its blood than when it is on a white background.[94,97,98]

In summary: Decapod crustacean nervous systems possess at least three or four chromatophorotropins. Their roles in the total chromatic responses of the whole organism have not been worked out for any one species. It seems likely, however, that the nervous-system chromatophorotropins, together with those from the sinus glands, will be shown to go far toward accounting for the intricate control of the crustacean pigmentary systems.

A study of the time relations of melanophore changes in the isopod *Ligia oceanica* during light to dark and background changes led to the hypothesis of a dual endocrine control of these pigment cells, with operation of both a darkening or B-substance, and a lightening or W-substance.[285] The chromatophore systems of isopods show considerable variability in their capacity to respond to background or illumination changes. It has been established that extracts of their heads concentrate the dark pigment in a variety of species of isopods including *Sphaeroma*, *Idothea*, and *Liga*.[59,205,211] In some other species of *Ligia*, extract of heads disperses the dark pigment.[83] In *Trachelipus* there is evidence suggesting the presence of both concentrating and dispersing hormones.[196] In brief, the evidences from assay of head extracts appears to support the hypothesis that isopod dark chromatophores, like most other crustacean chromatophores, are regulated by a pair of antagonistic hormones. Since the pigment remains fully dispersed after extirpation of the organ of Bellonci, an organ closely associated with the optic lobes and sinus glands of *Sphaeroma* and believed to be homologous with the decapod X-organs, it has been suggested that this is a source of a pigment-concentrating factor.[240]

In the stomatopod *Squilla*,[101,171] extracts of the sinus glands, postcommissural organs and central nervous organs disperse dark pigment. The brownish-black chromatophoric pigment of *Squilla*, which has the properties of an ommochrome, is maximally concentrated in eyestalkless individuals.

Experiments designed to elucidate the mechanism of action of crustacean chromatophorotropins at the molecular level revealed that the red pigment-concentrating hormone of *Palaemonetes* caused hyperpolarization of the transmembrane potential of the erythrophore as the pigment concentrated.[111] The primary action of this hormone is most likely stimulation of a pump which exchanges sodium ions from inside the chromatophore for potassium ions from the outside, resulting in a high internal K^+:Na^+ ratio which in some manner is responsible for the triggering of the red pigment-concentrating mechanism.[91]

Vertebrates. All the vertebrates possessing chromatophores show, in general, a fundamental similarity in their functional organization of this system. It seems profitable to develop the evolutionary trends separately within each of the major divisions of the vertebrates in which functional chromatophores are found.

Amphibians. The amphibians may show in their early development a period in which only primary color responses occur, the animals darkening in light and becoming pale in darkness. This was originally described for very young axolotl by Bábàk[14] and has since been observed in very young *Rana pipiens*[148] and *Amblystoma*.[185] These changes do not involve the eyes. It has been suggested that this is a period during which the eyes are still nonfunctional.[311] Other amphibians appear to show the secondary types of response involving stimulation through the eye immediately on hatching. Such species are *Bombinator* and *Hyla*,[14] and *Xenopus*.[311] In the secondary phase, primary responses are still operating but are dominated by the secondary ones.

Investigation of the physiology of color changes in amphibians has provided no evidence of any direct nervous control of the integumentary melanophores. The striking discovery was made very early that hypophysectomized tadpoles remain pale indefinitely.[286] This strongly suggested that the pituitary might be the normal source of a melanin-dispersing hormone in amphibians. The relation of the hypophysis to color changes in *Rana* was carefully investigated by Hogben and Winton.[146,147] Upon hypophysectomy the animals were rendered pale and refractory to further color changes. Removal of the anterior lobe by itself, however, showed no significant interference with the background responses. Extracts of the posterior lobe showed a tremendous capacity to darken pale frogs. All attempts to produce specific chromatophore responses through nerve stimulation or nerve transection failed. Therefore, these workers concluded that the color changes in *Rana* were accountable in terms of the activity of a single hormone arising in the posterior lobe of the pituitary, whose concentration in the blood was controlled by environmental stimuli operating through the eyes. This conclusion has been confirmed through transfusion of blood from a dark to a light *Rana*.[233]

More recently much work has been done on *Xenopus*,[144,145] in which, also, direct innervation of the melanophores seems to be lacking. *Xenopus*, too, lightens after hypophysectomy and darkens on injection of extracts of posterior lobe of the pituitary. A critical and detailed examination of the characteristics of change of the melanophores of *Xenopus* after transfer from a white to a black background and the reverse, from black to darkness and reverse, and from white to darkness and reverse (Fig. 23–14) failed to permit an adequate explanation in terms of a single principle and led to the postulation that the hypophysis produces two principles, one with melanin-dispersing action, referred to as the B-substance (MSH, melanophore-stimulating hormone = intermedin), and the other with melanin-concentrating action, called the W-substance (MCH). All the data obtained appeared readily interpretable in terms of an excitation of secretion of the W-

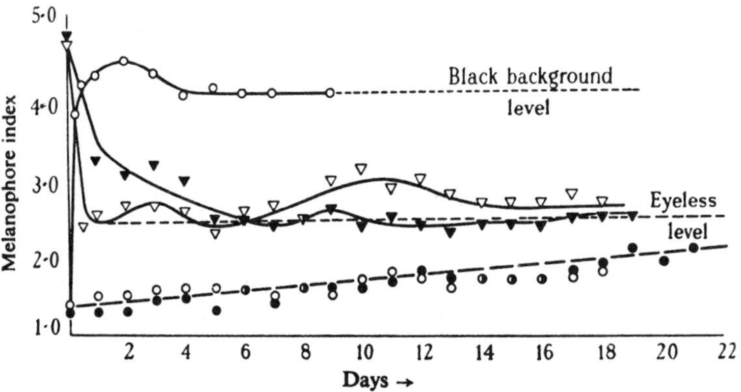

Figure 23–14. Changes in melanophore state in *Xenopus* after transfer from white to black backgrounds and vice versa, and from black background to darkness. (From Hogben, L., and Slome, D., Proc. Roy. Soc. Lond. B, vol. 120.)

substance by stimulation of peripheral retinal elements and excitation of secretion of the B-substance by stimulation of basal retinal elements (Fig. 23–15). Furthermore, the data require the additional assumption that the W-substance is added to the blood more slowly than the B-substance and also disappears from the blood more slowly. The B-substance and the W-substance are antagonistic. Thus, according to this hypothesis, in an illuminated environment B-substance is always secreted, but its secretion is reduced in

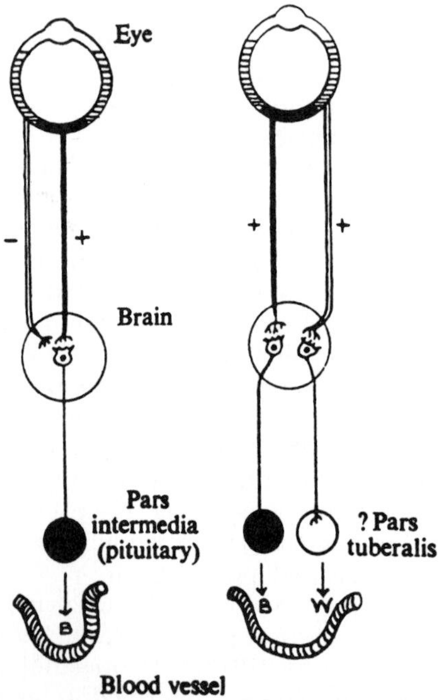

Eye
Brain
Pars intermedia (pituitary)
? Pars tuberalis
Blood vessel

Figure 23–15. Schematic representation of two hypotheses of hormonal control of melanophores in *Xenopus*, one involving a single hormone, and the other, two. (From Hogben, L., and Slome, D., Proc. Roy. Soc. Lond. B, vol. 120.)

low illumination and in darkness. The response to black and white illuminated backgrounds involve changes in the amount of the W-substance and consequently changes in the B/W ratio. The very slow responses observed in the change from an illuminated white background to darkness and the reverse appear explicable in terms of concomitant decreases or increases, respectively, of the two antagonistic substances. In the change from darkness to an illuminated black background the melanophores pass through a supernormal phase (more dispersed than typical for the background), a fact explainable in terms of a more rapid secretion of B-substance than of W-substance. As the latter increases to its full quantity, the melanin reaches its slightly less dispersed final state. Similarly, in passing from an illuminated black background to darkness the melanophore passes through a transitory stage of greater concentration than the ultimate state, apparently due to a more rapid reduction in B-substance than in W-substance.

Direct evidence for the existence of the W-substance in *Xenopus* has come from observations on the responses of the animals to environmental stimuli after various types of operative procedures.[145] When the anterior lobe of the hypophysis is removed the animal responds quite as it does normally. When the anterior lobe and the pars tuberalis are extirpated, the melanophores become maximally dispersed and show no background response. When the posterior lobe is removed, the animals are maximally pale. The source of the darkening hormone appears definitely to reside in the intermediate lobe.[12,145] Removal of the whole gland leaves the pigment slightly dispersed and nonresponsive. All these facts fit the hypothesis of the existence of two factors, with perhaps the pars tuberalis

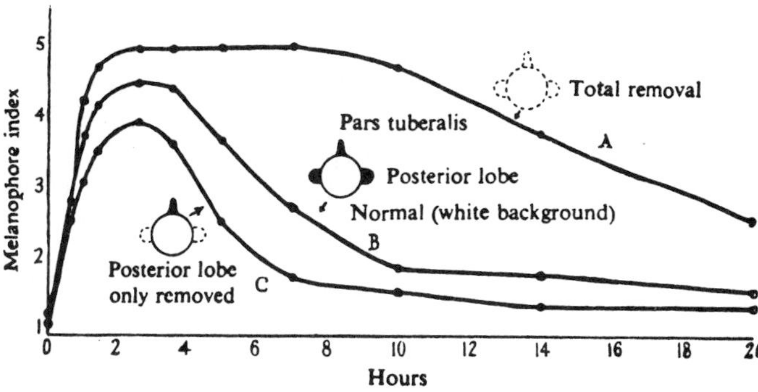

Figure 23–16. Responses of the melanophores of *Xenopus* on a white background to injection of equivalent doses of pituitary extracts into totally hypophysectomized specimens, specimens with only posterior lobe removed, and normal specimens. (From Hogben, L., and Slome, D., Proc. Roy. Soc. Lond. B, vol. 120.)

responsible for the W-substance and the posterior or the intermediate lobe for the B-substance. Support for the presence of two factors also comes from the relative effects of injection of B-substance into completely hypophysectomized *Xenopus*[145] as compared with its effects in normal animals or animals with only the posterior lobes removed (Fig. 23–16). As would be expected, a larger dose is required in the last two types of recipients to bring about a given response, while in those with complete hypophysectomy a smaller dose has an equivalent action. The latter observation finds a most logical explanation in terms of the resulting absence of an antagonist to MSH. However, the existence of a hypophysial blanching hormone in *Xenopus* has been questioned because Hogben and Slome may not have completely removed the intermediate lobe of the pituitary gland as they believed. When this is done the melanin becomes virtually fully concentrated and is not submaximally concentrated as Hogben and Slome reported.[155] Release of MSH is probably regulated in *Xenopus* by inhibitory and stimulatory nerves.[156] A melanocyte-concentrating factor has been separated from MSH-containing extracts by differential solubility in alcohol;[70, 73] the MCH was believed to be from the hypothalamus.

MSH darkens *Rana pipiens, R. clamitans,* and *Hyla,* and adrenalin blanches them,[204, 316] these reactions even proceeding in isolated frog skin *in vitro.*[316] Sodium ions are specifically essential to the MSH response,[207] and the MSH response is also inhibited by iodoacetate, which suppresses glycolysis.[316] Thyroxin darkens *Xenopus,* apparently by indirect stimulation of MSH secretion.[60] ACTH disperses the melanin in *Hyla.*[70] There is suggestive evidence that frog skin itself possesses a melanocyte-concentrating factor.[277] Tadpoles of *Xenopus* exhibit a body-blanching reaction when placed in darkness. Pinealectomy abolishes this response. There is strong evidence that melatonin is the hormone produced by the pineal gland of amphibians that induces the body-blanching reaction.[16, 17]

Adenosine 3′-5′ monophosphate (cyclic AMP) mimics the action of MSH on *Rana* skin.[22] It appears normally to mediate the action of MSH. MSH stimulates cyclic AMP synthesis in frog skin.[2]

FISHES. The fishes, which have probably been more intensively investigated than any other group, with respect to their chromatophore system and color changes, have several types of chromatophores. The most common and conspicuous type is the melanophore. Other common types contain yellow pigment (xanthophores), red pigment (erythrophores), and white pigment (iridophores). In addition, many fishes possess chromatophores containing small clusters of glistening platelike crystals that impart a bluish green structural coloration (iridosomes).

The activities of the melanophores are the principal ones involved in the conspicuous responses to light and darkness and to black and white backgrounds in fishes. The mechanism of response in fishes, as with amphibians, commonly shows a change from primary to secondary color responses during their early development. Young *Perca* and *Salmo*[68, 206] *Macropodus,*[303] and *Hoplias*[197] show a transition in response mechanism. On the other hand, a number of other species appear not to pass through a phase of primary responses but to have initially the secondary type. The latter species include *Fundulus,*[126, 318] *Lebistes, Xiphophorus, Gambusia,*[303] *Scyllium,*[311] and

Mustelus.[219] A number of fishes normally showing secondary color responses will revert to primary responses after blinding.[303]

The secondary color responses of fishes are dependent typically upon the eyes. They involve nervous pathways to the central nervous system, thence either to endocrine glands affecting the chromatophores through blood-borne hormones, or by way of efferent nervous pathways directly to the chromatophores where chemical mediators are liberated by the nerve terminations. Both hormonal and nervous mechanisms may cooperate in many cases. There is considerable variation among fishes as to the normal mechanism of control. Parker,[228] the leading investigator in the field of animal color changes, divided the fishes into three groups on the basis of the degree to which direct innervation of the melanophores is found. Dineuronic chromatophores possess double innervation with separate dispersing and concentrating fibers. Mononeuronic chromatophores possess single innervation in which the activity is always pigment concentrating. Aneuronic chromatophores possess no innervation, their secondary responses being regulated solely by activity of blood- and lymph-borne chemical factors.

The great majority of the teleost fishes thus far carefully investigated appear to possess dineuronic melanophores. One type of innervating nerve fiber, a pigment-concentrating one, is readily demonstrated by electrical stimulation of appropriate loci in the central nervous system, of central nerve tracts, or of peripheral nerves. The extent of the area of the skin blanching under this stimulation parallels the area of the skin innervated by these nervous elements. Furthermore, after denervation of any area of the skin the denervated area no longer exhibits blanching responses to stimulation of the nerves central to the point of denervation, even though responses may continue in the adjacent areas. Therefore, the blanching produced by nerve stimulation in the animals is a result of localized responses at the region of the nerve terminations and is not due to freely diffusible substances in the blood. The action of the concentrating fibers is believed[225] to be mediated through adrenin-like material which diffuses away from the nerve terminations.

The presence of a second set of fibers, pigment-dispersing ones, has been demonstrated for many teleosts through experimentation with melanophores of the tail fins. If, in a teleost such as *Fundulus*, kept on an illuminated white background, a group of the radiating caudal nerves is cut by a transverse incision, the band of fin innervated by these fibers darkens quickly and then over the course of a few days fades again[216] (Fig. 23–17). If now a second cut is made parallel to the first cut, and somewhat distal to it, the faded band will repeat its transient darkening. This latter behavior has led to the hypothesis that the melanophore response observed is due to a restimulation of the dispersing nerve fibers that have been transected and stimulated by the first incision. It cannot be explained simply in terms of transection of the concentrating fibers, as possibly the response to the first cut could be. Interpolation of a cold block between the point of nerve transection and the melanophores abolishes the response.[221] Such redarkening of faded bands after a second incision has been observed by a number of investigators for a number of species of teleost fishes. These include *Holocentrus*,[222] *Parasilurus*,[193] *Pterophyllum*,[305] *Ameiurus*,[226] and *Gobius*.[114] An activation of melanin-dispersing fibers in the catfish tail has also been produced by electrical stimulation.[232]

Other lines of evidence, both morphological and physiological, have given further support to the hypothesis of a dual innervation of teleost melanophores. Ballowitz[18] many years ago clearly demonstrated that the melanophores of the perch receive nerve terminations from more than one fiber, thus providing an anatomical basis for the conclusions reached by more recent physiological experimentation. The electron microscope also reveals two types of presynaptic terminals for the melanophores of the teleost *Lebistes*, one containing vesicles 500 Å in diameter and the other having larger vesicles of 1000 Å

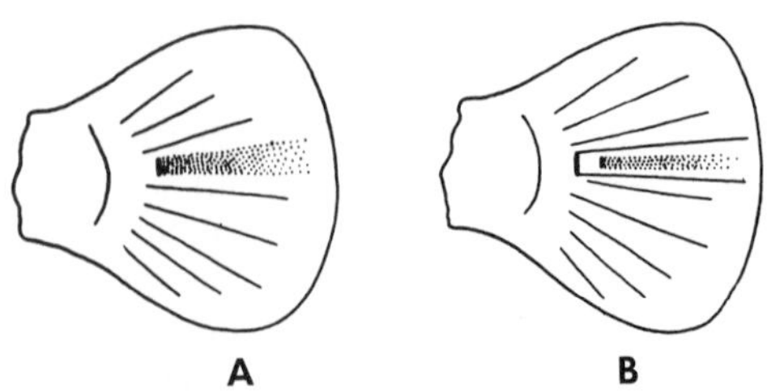

Figure 23–17. *A*, dark band in caudal fin of white-adapted *Fundulus* produced by severance of radial nerves. *B*, Redarkening of a faded band following a second, more distal, cut. (From Parker, G. H., Proc. Nat. Acad. Sci., vol. 20.)

diameter.[120] Furthermore, a critical examination has been made of the responses of chromatophores at the edges of denervated caudal bands of *Fundulus*,[201] and of those near the regenerating front of nerve fibers in the course of reinnervation of denervated bands,[6] as the animals darken on black backgrounds and lighten on white ones. These observations provide suggestive evidence that many of the melanophores located in these regenerating fronts possess only one type of fiber, either concentrating or dispersing, but not both as under normal circumstances. Some of these melanophores show rapid pigment concentration and very slow dispersion; others show the reverse. Studies of the influence of drugs on chromatophores of *Phoxinus*[125] and *Fundulus*[282] also support the concept of a dual innervation. Cyclic AMP has a melanin-dispersing action in *Fundulus*.[208] The nerve transmitters (pigment-dispersing and pigment-concentrating) may act by controlling the level of cyclic AMP in these melanophores.

The dispersing fibers appear to exert their action on the melanophores through the mediation of acetylcholine. Acetylcholine is known to cause dispersion of the melanin of fishes when the latter are eserinized to prevent rapid destruction of the material. In fact, a bioassay of the acetylcholine content of the skin of a dark-adapted catfish *Ameiurus*[225] or snakefish *Ophiocephalus* showed its presence in a concentration of about 0.078 gamma per gram of skin. This is approximately the concentration of acetylcholine which, when injected into the body fluids of eserinized fish, was in general nontoxic and at the same time quite effective upon melanophores. However, it should be noted that some investigators have claimed that acetylcholine evokes no[2] or at best only a weak melanin-dispersing response in teleosts.[135]

Of the fishes thus far investigated, the dogfishes *Mustelus* and *Squalus* appear to possess mononeuronic melanophores. If a transverse cut is made in the pectoral fin of a dogfish of intermediate tint, a light band is produced distal to the point of the cut. Such light bands may be revived after they have redarkened[220] or may be produced by electrical excitation[218] of the integumentary nerves. These bands follow the distribution of the cut nerves and not necessarily of the blood vessels. Furthermore, light bands may be produced by a similar transverse cut in a fin from which the blood supply has been cut off. All these facts point to a nerve supply to the melanophore whose function it is to induce pigment concentration. There is no indication whatsoever that pigment-dispersing nerve fibers are present in this animal.

Parker[227] postulated that the concentrating fibers influence the melanophores through the production of a chemical mediator which he called selachine. If skin from a pale animal is extracted with ether or olive oil, but not water, the material may be extracted. Injections of an olive-oil extract of this substance into a dark-adapted animal will produce temporary lightening, which spreads very slowly from the point of injection. Olive oil by itself produces no such effect.

The other elasmobranch fishes which have been investigated, the skate *Raja*,[223] the dogfish *Scyllium*,[320] and the lamprey *Lampetra*,[321] show no evidence of direct innervation of their melanophores, and hence their melanophores are believed to be aneuronic.

Blood-borne agents typically supplement the nervous system in the secondary responses of the melanophores of fishes to light stimuli. In the more primitive fishes, such as the cyclostomes and the elasmobranchs which possess aneuronic color cells, hormones alone are the agents involved. Among these hormones an important substance is a pigment-dispersing principle from the pituitary, the B-substance of Hogben, or MSH.[322] This substance is secreted from the posterior lobe of the pituitary. Lundstrom and Bard[189] observed that hypophysectomized *Mustelus* become and remain pale. The fish may be darkened again by injection of extract of the posterior lobe. A similar role of a posterior-lobe principle has since been demonstrated for other elasmobranchs, *Raja* and *Scyllium*,[142] and the cyclostome *Lampetra*.[321] Implantation of the pineal complex into the ammocoete larva of *Geotria* causes localized pallor. Intraperitoneal injection of melatonin into this lamprey also causes blanching.[69]

There is some evidence that in *Scyllium* and *Raja* a second neurohumor from the pituitary acts as a pigment-concentrating agent.[142] Evidence for a role of such a body-blanching principle has been derived from studies of the characteristics of the melanophore responses to background and light-intensity changes and to the influence of hypophysectomy on the state of the pigments, and of pigmentary responses to injections of posterior-lobe extract. The general methods of experimentation and logic involved in these experiments are the same as those developed in the studies of amphibian melanophore control. According to the bihumoral concept of Hogben and his associates, the state of the melanophores in these fishes is determined by the ratio of MSH to MCH present in the blood at any given instant, and this ratio is in turn controlled by visual stimuli. The visual stimuli, through differential, dorsoventral, retinal stimulation, result in different rates of secretion of the two principles by their respective sources.

The chromatic pituitary hormone involved in melanin dispersion seems also to be present and active in normal color change to a greater or lesser degree in most teleost fishes. The eel *Anguilla* shows sluggish color changes requiring days for completion.[206] In this fish, despite the apparent presence of both concentrating and dispersing nerve fibers, color changes seem predominantly determined by hormones. In its activity in the eel, MSH is believed to be assisted by an MCH.[310] That direct innervation does play some role in color changes in this fish is seen in the limited background response after hypophysectomy.

An MSH is found to be slightly less important in normal color changes in the catfish *Ameiurus.* Hypophysectomized catfishes continue to show color changes in response to black and white backgrounds but show only an intermediate degree of darkening on black.[5] Injection of posterior lobe extract will, however, completely blacken these fish. Here we must assume that blood-borne MSH supplements the action of dispersing nerve fibers in the normal responses to black backgrounds; there is as yet no evidence for the operation of an MCH in this species. The pencil fish *Nannostomus beckfordi anomalus* has separate day and night markings. Melatonin induces the night coloration by causing concentration of the pigment in some melanophores and dispersion in others.[254]

The killifish *Fundulus,* on which a vast amount of research has been done, is a species in which the dominant mechanism of melanophore control is nervous. Color changes are very rapid, only a minute or two being required for nearly maximal color change. These changes continue to occur in hypophysectomized specimens.[195] Furthermore, injection of extracts of posterior lobe into pale fish on a white background does not produce significant darkening. Since such extracts will induce darkening in denervated areas of the skin, we must conclude that the chromatophores are normally influenced to some extent by MSH which is shown to be present in their pituitaries.[8, 164] However, its normal influence is probably seen only in the production of extreme conditions of dark adaptation maintained over relatively long periods. In *Macropodus,* studies of hypophyseal influences after interference with nerve supply to the melanophores indicated that a pituitary MSH was not significantly involved.[308]

All teleost fishes thus far investigated appear to fall into a series in the relative influences of direct innervation and blood-borne hormones. A reasonable hypothesis has been advanced that the humoral control is phylogenetically the older, and that direct nervous control has been superimposed upon it in those fishes of more recent evolutionary origin.[311] Direct nerve control is associated with more rapid response to background, which has become possible through a simultaneous increase in the speed of the melanophore change itself. The typical teleost controlling mechanism for melanophores is diagramed in Figure 23–18.

A survey of the characteristics of the response of melanophores of fishes to war gases suggests the presence of two differently responding types, with catfishes possessing one type and scaly fishes the other.[245] Sodium ions appear to be peculiarly effective in inducing melanin dispersion in *Fundulus,*[119] recalling its role in contributing to MSH action in frogs.

Very much less is known about the control of the erythrophores, xanthophores, iridophores, and iridosomes than about the control of the melanophores. The erythrophores of the squirrelfish *Holocentrus* are rapidly responding effector organs.[222] Through their activity the fish can change from red to white in about 5 seconds, and make the reverse change in about 20 seconds. These responses may be induced by change from black to white background and vice versa. Transection of nerve tracts in a fish on a white background results in dispersion of red pigment. The areas blanch again in a short while and may be darkened again by a second more distal cut, indicating the presence of dispersing nerve fibers. The presence of concentrating nerve fibers can be demonstrated by electrical stimulation of the medulla, result-

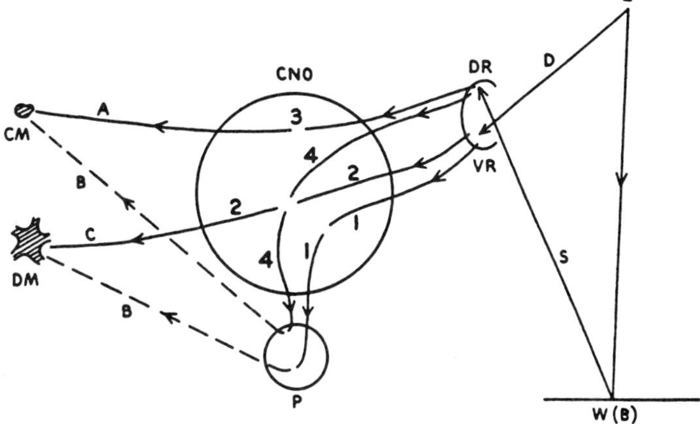

Figure 23-18. Diagram of the controlling mechanism of melanophores in the eel *Anguilla*. *L*, incident light; *W(B)*, white or black background; *DR*, dorsal retina; *VR*, ventral retina; *CNO*, central nervous system; *P*, pituitary; *A*, adrenergic fibers; *C*, cholinergic fibers; *B*, blood and lymph; *DM*, dispersed melanin; *CM*, concentrated melanin. (From Parker, G. H., Quart. Rev. Biol., vol. 18.)

ing in rapid concentration of pigment of all innervated erythrophores. Experimentally denervated cells fails to give this response. Adrenin concentrates the pigment. Pituitary extracts from other squirrelfish produce no effect when injected into normal light-adapted or dark-adapted specimens. It thus appears that the erythrophores of *Holocentrus* normally are exclusively under nervous control.[284] The erythrophores of *Phoxinus*, on the other hand, have been shown to be influenced by a principle from the hypophysis.[125, 134]

The xanthophores of *Fundulus* appear to possess double innervation, comprising concentrating and dispersing fibers.[113] A concentrating hormone, probably adrenin, also appears to be responsible for the concentration of the pigment which results from handling of the fish. This latter concentration occurs as rapidly in denervated xanthophores as in innervated ones. On the other hand, intraperitoneal implantation of *Fundulus* pituitaries into hypophysectomized specimens induces pigment dispersion in denervated xanthophores, regardless of color of background. Such implants also impede the typical pigment concentration in innervated cells in response to blue or white backgrounds.[115] It therefore appears that several factors normally influence the state of the xanthophores in this species. Melanophores and xanthophores of *Fundulus* react independently in background responses.[7] Hypophysectomy does not affect the melanophores of the mudsucker *Gillichthys*, but results in concentration of the pigment in its xanthophores. Prolactin dispersed this yellow pigment.[258] The release of prolactin is probably normally controlled by an inhibitory factor from the brain. MSH release in the catfish *Ictalurus* appears to be under inhibitory control also.[163]

A number of fishes, including *Fundulus*, possess reflecting-white chromatophores known as guanophores or iridoleucophores.[112] These may show physiological changes in the adaptive responses of the fish to background. They continue to respond to background after hypophysectomy and after sufficient dosage of the fish with ergotamine to prevent any response of the accompanying melanophores. They disperse their pigment under the influence of adrenin.[210] In *Bathygobius*, experiments involving nerve transections and hypophysectomy led to the conclusion that the iridoleucophores are regulated importantly by local innervation, though they are also hormonally influenced.[116]

Iridosomes play only a passive role in adaptive color changes, becoming more or less obscured through activity of the other chromatophore types. Normally these bodies are green or blue. They are highly responsive to certain environmental stimuli, changing reversibly through the spectral colors to red on excitation and in the opposite direction on recovery. This response is direct and does not involve coordinating mechanisms within the animal, either nervous or humoral.[107]

Reptiles. The regulatory system of the melanophores of reptiles shows, as in the fishes, a great diversity. It appears to involve, to differing degrees in different reptiles, the activities of hormones and nerves.

The melanophore responses of the iguanid *Anolis* have been investigated extensively by Kleinholz.[165] These lizards show a color change ranging from bright green to dark brown. They typically assume the former color in an illuminated white container and the latter in an illuminated black one. The response to change from a white to black background is usually completed in 5 to 10 minutes. The reverse change normally requires 20 to 30 minutes. These background-induced responses depend on the eyes; they cease after bilateral blinding. However, such blinded specimens still are capable of color change; they darken in light and become pale in darkness, through primary responses.

For many years after Brücke's[53] classic studies of color change in the chameleon, in which he demonstrated nervous control of the melanophores by the sympathetic system, it was considered that reptilian chromatophores generally were thus controlled. Studies on *Phrynosoma*,[253] however, provided a basis for a strong suspicion that nerves were not the sole method of control. The lizard *Hemidactylus* becomes pale in color after hypophysectomy. Hypophysectomized *Anolis* remain permanently bright green[165] (Fig. 23–19). They no longer darken in response to a black background or bright light. They can be darkened readily by injection of extracts of whole pituitary of fishes or the intermediate lobes of frogs or reptiles. The melanophores of the rattlesnake appear also to be normally dispersed by a principle from the pituitary.[246]

Transection of nerves, such as section of the spinal cord at various levels, or cutting the sciatic nerve in *Anolis*, in no way interferes with the normal color responses. In fact, attempts at histologic demonstration of nerve terminations at the melanophores have been uniformly unsuccessful. Skin grafts very soon show color changes which are synchronous with those of the host. Exclusion of the blood supply from any region, on the other hand, results in a paling of the region in about 15 minutes.

The roles of the animal's own hypophysis and adrenals can be shown by electrical stimulation with one electrode placed in the cloaca and the other in the mouth. Stimulated pale animals kept in darkness become uniformly dark brown. Denervated areas respond just as do the innervated ones. Similar stimulation of hypophysectomized specimens gives, on the other hand, a characteristic mottling of the body. This last is not obtained after both

Figure 23–19. A normal dark and a hypophysectomized light *Anolis*. (From Kleinholz, L. H., J. Exp. Biol., vol. 15.)

adrenalectomy and hypophysectomy. Furthermore, injection of adrenin or of extracts of the animal's own adrenals in Ringer solution produces the typical mottling. Adrenalectomized animals lighten in response to a white background. All of these observations, and others, point strongly to the conclusion that the melanin in *Anolis* normally disperses in response to MSH from the intermediate lobe, and that its gradual disappearance from the circulation suffices to account for lightening. Rapid blanching which normally follows electrical stimulation or excitement may perhaps be accounted for by integumentary vasoconstrictor activity, and under some circumstances by the production of adrenin. Under conditions of stress, stimulation of alpha adrenergic receptors by catecholamines leads to excitement pallor, followed by excitement darkening (mottling) resulting from stimulation of beta adrenergic receptors.[130] Melanin dispersion in *Anolis* also may be mediated by cyclic AMP.[127] MSH has a specific calcium requirement for its action in *Anolis*.[306]

Light appears to have no significant influence on the melanophore state in intact *Anolis*, other than through the eyes.

The available evidence indicates roles of both nerves and hormones in the melanophore responses of the iguanid, *Phrynosoma*. The early work of Redfield[253] on this form has been largely confirmed and considerably extended by Parker.[224] These animals are normally gray with characteristic black patches. The latter patches show no color changes, whereas the intervening area varies with appropriate stimulation from dark gray to pale grayish white. These changes are due primarily to melanophore activities. Darkening of the animal occurs in about 15 minutes, and lightening in approximately twice that time.

Phrynosoma darkens on an illuminated black background, at low temperatures, and in response to very strong illumination. It lightens on an illuminated white background, at high temperatures, and in darkness.

The melanophores are normally under the influence of pigment-concentrating nerve fibers. Stimulation of a sciatic nerve will induce lightening in the corresponding hind leg. Electrical stimulation of the roof of the mouth or of the cloaca results in a paling of the whole animal, which is quickly reversible. Following denervation of a region of the body, leg, or lateral trunk, similar stimulation results in a lightening of all regions except the denervated one, despite the fact that the melanophores in the denervated area still show pigment concentration in response to an injection of adrenin. No indication of pigment-dispersing fibers has been uncovered in this species.

Phrynosoma, like *Anolis*, after hypophysectomy becomes pale and remains so indefinitely. Injection of extract of *Phrynosoma* pituitary induces strong darkening either in normal pale or in hypophysectomized individuals. Furthermore, injection of defibrinated blood from a dark specimen into the leg of a pale one produces darkening in the latter. These results provide strong evidence that a pituitary MSH is normally concerned in body darkening. An action of a pigment-concentrating hormone is also seen in that adrenalin or extract of *Phrynosoma* adrenals strongly blanches dark specimens. The presence of a similarly acting agent in the blood of white-adapted animals is seen in that their defibrinated blood will, on injection into a leg of a dark specimen, lighten the latter. It has been known for a long time that animals held on their back or otherwise stimulated to struggle vigorously will lighten rapidly. By comparison of the influences of denervation and occlusion of blood supply on the production of this type of lightening it can be clearly demonstrated that a blood-borne agent is involved. These observations lead to the obvious conclusion that the paling is brought about by two methods, nervous and hormonal, either one alone capable of producing the response.

Both the dispersing hormone from the pituitary and the concentrating hormone from the adrenals operate directly on the melanophores; each is capable of exhibiting its complete action after nerve transection and degeneration of the nerve fibers.

The influence of temperature and of light and darkness on the melanophores of *Phrynosoma* is a direct one, in which the melanophores act as independent effectors. The responses may be obtained locally by application of the stimulus to the specific region in completely denervated portions of the body. Since even degeneration of the innervating fibers does not result in termination of the response, axon reflexes cannot be responsible.

The coordination of the melanophores of chameleons *Chamaeleo* or *Lophosaura*, unlike that in the iguanids, appears to be exclusively nervous.[143] Nerve transection is followed by a darkening of the area normally innervated. These nerves are of the autonomic nervous system. The melanophores are readily caused to concentrate their pigment by electrical stimulation of the nerves. The results of nerve transection have been interpreted to be the result of the absence of tonic impulses reaching the pigment cells.[259, 323] There is no clear evidence as yet that the pigment is actively dispersed by a second set of nerve fibers, as appears to be true of many teleosts.

There is little or nothing known for the most part as to possible roles of hormones in the color changes in the chameleon, but the fact that color patterns can be produced on the body by alternate light and shadow[53] argues against any substantial importance of such factors.

The melanophores of intact animals respond to light and darkness by pigment dispersion and concentration, respectively. Dark regions produced by denervation show no such responses. These results were interpreted by Zoond and Eyre[323] to prove that the responses of the melanophores can occur only by way of reflexes involving the central nervous system. But Parker[224] has pointed out the possibility of an alternative interpretation, that the pigment cells in these forms may show direct response when their state is not determined by a dominating mechanism.

In summary: We appear to find among the reptiles, as with fishes, a spectrum of mechanisms of coordination of the chromatophores, ranging from systems involving probably primitive hormonal control, through those in which both nerves and hormones cooperate, to those largely dominated by direct innervation.

FUNCTIONAL SIGNIFICANCE OF CHROMATOPHORES AND COLOR CHANGE

Since the responses of the chromatophores are predominantly responses to color or shade of background, one is led to the hypothesis that the color changes contribute significantly to the obliterative coloration of the animal for protection or aggression and hence increase its chances for survival. Among the few experimental demonstrations that animal color changes do actually increase chances for survival, a view often questioned, is that of Sumner,[295] in which he found that fishes given time to change their coloration were seized in smaller percentages by a predatory bird than were unadapted ones. Similarly, in insects, *Acrida* and *Oedipoda* larvae were found to be captured to smaller extent by birds when the insects were on backgrounds of their adaptation coloration.[76] It has also been clearly demonstrated that fishes which are black adapted tend to select black backgrounds when given choice of black or white, more commonly than do white-adapted individuals[46, 191] (Fig. 23–20), and that the rate of change of choice with change of background in *Ericymba* is approximately the same as the change in skin coloration (Fig. 23–21). Fishes with more rapid and striking color changes seem to have their choice more strongly modified along with background adaptation than do ones with less effective changes. The crayfish *Cambarus* (*Orconectes*) also appears to possess an adaptive background selection.[31]

Chromatophores appear also to serve in the protection of animals from bright illumination which may be deleterious. In some animals, e.g., certain leeches and sea urchins, the chromatophore pigment disperses only in re-

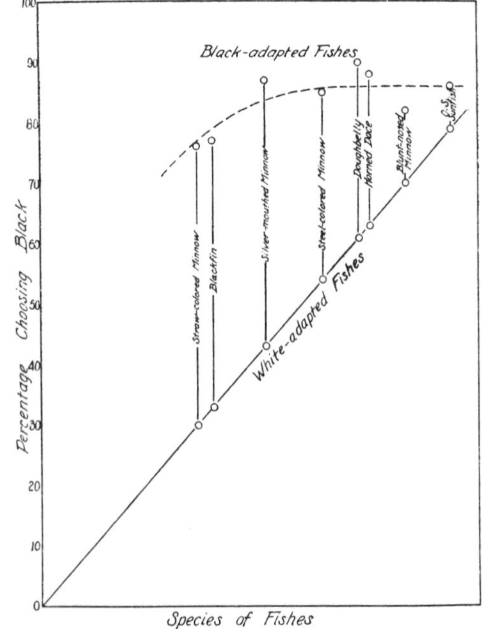

Figure 23–20. Percentage choice of a black over a white background for each of eight species of fishes adapted to black and white backgrounds. (From Brown, F. A., Jr., and Thompson, D. H., Copeia, vol. 3.)

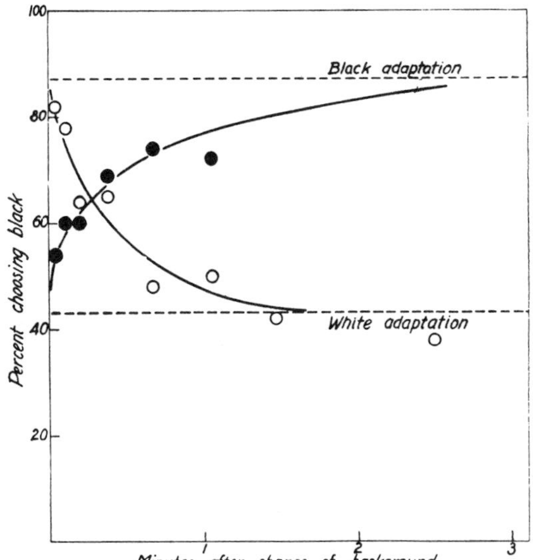

Figure 23–21. The rate of change in percentage choice of a black background following transfer of *Ericymba* from a black to a white background and vice versa. (From Brown, F. A., Jr., and Thompson, D. H., Copeia, vol. 3.)

sponse to bright illumination and with no regard to a possible protective coloration in relation to the background. The black and white pigments in the crab *Uca* are normally dispersed in light during the daytime and concentrated in darkness at night. This is due partly to an inherent rhythm and partly to a response to total illumination. The melanophores of *Uca* are most sensitive to near-ultraviolet (300 to 400 μm) light. This light is more than 200 times as effective in causing direct melanin dispersion than is visible light.[62] It is also noteworthy in this regard that the white chromatophores of such brilliant-light–inhabiting Sargasso-weed crustaceans as *Latreutes fucorum, Leander tenuicornis,* and *Hippolyte acuminata* are very abundant and richly charged with reflecting white pigment. In bright illumination this pigment disperses broadly, providing a continuous layer of a very effective diffusing reflector. Since the primary color responses of animals are those of pigment dispersal in light and pigment concentration in darkness, it appears reasonable to postulate that this light-protective function of chromatophores is a primitive one, with a role in obliterative coloration appearing later in evolution of chromatophore systems. The formation of melanin in human skin in response to bright sunlight, is, in effect, a morphological color change of this same character.

Another function which has been attributed to chromatophores is thermoregulation.[20, 213] The desert lizard *Phrynosoma* is light at night and during midday, and dark during the early morning and late afternoon.[180] These and similar observations have led to the hypothesis that the chromatophores function in thermoregulation in this species. The animal apparently is adaptively controlling heat absorption and radiation at the various times of day by chromatophore activities. Strongly supporting this view are the observations of numerous investigators of reptilian color change that elevation of the body temperature to approximately 40°C leads to melanin concentration and that lowering of the temperature to about 5°C leads to dispersion.[224] The black pigment of the crab *Uca* also tends to concentrate as the body temperature is elevated to about 25 to 30°C.[43] These conditions obviously result in control of the amount of light and heat absorbed by the black pigment in a manner beneficial to the animal. Experimental evidence has been obtained in support of this conclusion. In only five minutes after having been exposed to sunlight, the body temperature of dark *Uca* whose melanin was maximally dispersed when the crabs were first exposed to the sunlight was 2°C higher than that of pale crabs having maximally concentrated melanin at the outset.[315] The differential reflection of light from frog skin, and from the dorsal surface of the cephalothorax of *Uca,* favoring reflection of the longer, heating wavelengths, together with reflectance changes correlated with the color changes, also supports a thermoregulatory role.[63, 315]

One additional role of color changes in animals is suggested by the color displays that sometimes accompany mating behavior in such species as *Anolis.*[129] During pairing the males show a striking change from green to brown. Color displays are also associated with the breeding season in certain teleosts and cephalopods. The importance of these mating color displays is unknown. Special nuptial morphological color changes are not infrequently associated with the breeding season.[231]

SUMMARY

Color changes involving movements of pigments, and increases or decreases of pigments, within special bodies in the integument, the chromatophores, have been observed among reptiles, amphibians, teleosts, elasmobranchs, cyclostomes, crustaceans, insects, cephalopods, annelids, and echinoderms. These color changes are most characteristically ones involving adaptive adjustment of the animal to the color of its background through a functional relationship between the eyes and the chromatophores, although total illumination, temperature, humidity, tactile stimuli, endogenous rhythms, and numerous other factors may also influence the color changes.

The reflex pathways, initiated at the eyes, which are responsible for the control of the color changes in all crustaceans, insects, amphibians, and many fishes and reptiles, involve hormones. Both blood-borne hormones and direct innervation operate importantly in most other species, with the possible exception of the chameleon among the reptiles and such fishes as *Fundulus* and *Macropodus* among the teleosts, and possibly also in leeches where nervous reflexes possess dominant or exclusive control. Responses of the chromatophore system to light continue in most animals in the absence of the eyes, either through the responses of the chromatophores acting as independent effectors or reflexly through other light receptors, e.g., the pineal complex. These latter are responses only to intensity of illumination and bear no relation to color of background, and typically operate to render the animal darker in higher illumination and lighter in lower.

Several functions have been ascribed to the chromatophore system, among them being: (1) protective and aggressive coloration; (2) thermoregulation; (3) protection of the body tissues from intense illumination; and (4) mating color displays.

REFERENCES

1. Abbott, F. S., Can. J. Zool. *46*:1149–1161, 1968. Drugs and biogenic substances on *Fundulus* melanophores.
2. Abe, K., R. W. Butcher, W. E. Nicholson, C. E. Baird, R. A. Liddle, and G. W. Liddle, Endocrinology *84*:362–368, 1969. Cyclic AMP as mediator for frog melanocytes.
3. Abramowitz, A. A., Proc. Nat. Acad. Sci. *21*:132–137, 1935. Morphological color change, *Fundulus* xanthophores.
4. Abramowitz, A. A., Proc. Nat. Acad. Sci. *21*:677–691, 1935. Adaptation to colored backgrounds, Crustacea.
5. Abramowitz, A. A., Biol. Bull. *71*:259–281, 1936. Hypophysis and color change in *Ameiurus*.
6. Abramowitz, A. A., Proc. Nat. Acad. Sci. *22*:233–238, 1936. Dual innervation of teleost melanophores.
7. Abramowitz, A. A., Amer. Nat. *70*:372–378, 1936. Independence of action of melanophores and xanthophores, *Fundulus*.
8. Abramowitz, A. A., Biol. Bull. *73*:134–142, 1937. Role of hypophysis in *Fundulus* color change.
9. Abramowitz, A. A., and R. K. Abramowitz, Biol. Bull. *74*:278–296, 1938. Diurnal rhythm and chromatophore state, Crustacea.
10. Aoto, T., Annot. Zool. Japan. *34*:193–196, 1961. Melanophore regulation in *Sesarma*.
11. Ascherson, Arch. Anat. Physiol. wiss. Med. 7:15–23, 1840. Color change, frog.
12. Atwell, W. J., and E. Holley, J. Exp. Zool. *73*:23–42, 1936. Pituitary in amphibian color change.
13. Atzler, M., Z. vergl. Physiol. *13*:505–533, 1930. Relation of eyes to *Carausius* color change.
14. Bábàk, E., Arch. ges. Physiol. *131*:78–118, 1910. Color change in young Amphibia.
15. Bábàk, E., Arch. ges. Physiol. *149*:462–470, 1913. Relationship of morphological to physiological color change, vertebrates.
16. Bagnara, J. T., Science *132*:1481–1483, 1960. Pineal in amphibian color change.
17. Bagnara, J. T., Gen. Comp. Endocrinol. *3*:86–100, 1963. Pineal in amphibian color change.
18. Ballowitz, E., Z. wiss. Zool. *56*:673–706, 1893. Innervation of melanophores of teleosts.
19. Bartell, C. K., K. R. Rao, and M. Fingerman, Comp. Biochem. Physiol. *38A*:17–36, 1971. Melanin-dispersing factors in *Uca*.
20. Bauer, V., Z. allg. Physiol. *16*:191–212, 1914. Melanophores in heat regulation.
21. Bikle, D., L. G. Tilney, and K. R. Porter, Protoplasma *61*:322–345, 1966. Microtubules in fish melanophores.
22. Bitensky, M. W., and S. R. Burstein, Nature *208*:1282–1284, 1965. Cyclic AMP and frog melanophores.
23. Bowman, T. E., Biol. Bull. *96*:238–245, 1949. Chromatophore hormones in *Hemigrapsus*.
24. Bozler, E., Z. vergl. Physiol. *8*:371–390, 1929. Cephalopod chromatophores and their innervation.
25. Brecher, L., Arch. Mikr. Anat. *102*:501–548, 1924. Color change in insect pupae.
26. Brown, F. A., Jr., Proc. Nat. Acad. Sci. *19*:327–329, 1933. Chromatophorotropins from the crustacean central nervous system.
27. Brown, F. A., Jr., J. Morphol. *57*:317–333, 1935. Adaptation to colored backgrounds. Crustacea. Behavior of chromatophores in color change in *Palaemonetes*.
28. Brown, F. A., Jr., Biol. Bull. *57*:365–380, 1934. Morphological color change; *Palaemonetes*.
29. Brown, F. A., Jr., J. Exp. Zool. *71*:1–15, 1935. Functional independence of pigments, crustaceans.
30. Brown, F. A., Jr., Biol. Bull. *70*:8–15, 1936. Total illumination and chromatophore state, fishes; ratio of incident to reflected light and color change; measurement of chromatophore activity.
31. Brown, F. A., Jr., Ecology *20*:507–516, 1939. Adaptive background selection by crayfishes.
32. Brown, F. A., Jr., Physiol. Zool. *13*:343–355, 1940. Crustacean sinus gland and color change.
33. Brown, F. A., Jr., Physiol. Zool. *19*:215–233, 1946. Central nervous sources of crustacean chromatophorotropins; measurement of chromatophore activity.
34. Brown, F. A., Jr., Biol. Bull. *98*:218–226, 1950. Responses of *Uca* red chromatophores.
35. Brown, F. A., Jr., Action of Hormones in Plants and Invertebrates. New York, Academic Press, 1952. Review of crustacean color change.
36. Brown, F. A., Jr., Science *130*:1535–1544, 1959. Timing of biological rhythms.
37. Brown, F. A., Jr., and H. E. Ederstrom, J. Exp. Zool. *85*:53–69, 1940. Chromatophorotropins from the crustacean central nervous system.
38. Brown, F. A., Jr., M. Fingerman, and M. N. Hines, Biol. Bull. *106*: 308–317, 1954. Shifting phases of color change rhythm.
39. Brown, F. A., Jr., M. Fingerman, M. I. Sandeen, and H. M. Webb, J. Exp. Zool. *123*:29–60, 1953. Tidal and monthly rhythms of color change in *Uca*.
40. Brown, F. A., Jr., and M. N. Hines, Physiol. Zool. *25*:56–70, 1952. Persistent modifications of daily color change cycles.
41. Brown, F. A., Jr., and I. M. Klotz, Proc. Soc. Exp. Biol. Med. *64*:310–313, 1947. Chemical separation of chromatophorotropins from Crustacean central nervous system.
42. Brown, F. A., Jr., and L. M. Saigh, Biol. Bull. *91*:170–180, 1946. Central nervous sources of chromatophorotropins, Crustacea.
43. Brown, F. A., Jr., and M. I. Sandeen, Physiol. Zool. *21*:361–371, 1948. *Uca* color response to light and temperature; light intensity and *Uca* color change.
44. Brown, F. A., Jr., and H. H. Scudamore, J. Cell. Comp. Physiol. *15*: 103–119, 1940. Two chromatophorotropins from crustacean sinus glands.
45. Brown, F. A., Jr., and G. C. Stephens, Biol. Bull. *101*:71–83, 1951. Persistent modification of daily color-change cycles by photoperiod.
46. Brown, F. A., Jr., and D. H. Thompson, Copeia *3*:172–181, 1937. Adaptive background selection by fishes.
47. Brown, F. A., Jr., and H. M. Webb, Physiol. Zool. *21*:371–381, 1948. Diurnal rhythm in *Uca* color change.
48. Brown, F. A., Jr., and H. M. Webb, Physiol. Zool. *22*:136–148, 1949. Shifting by light of phases of color-change rhythm.
49. Brown, F. A., Jr., H. M. Webb, and M. F. Bennett, Proc. Nat. Acad. Sci. *41*:93–100, 1955. Color change rhythms during change of longitude.

50. Brown, F. A., Jr., H. M. Webb, M. F. Bennett, and M. I. Sandeen, Physiol. Zool. *27*:345–349, 1954. Temperature independence, tidal rhythm of color change.
51. Brown, F. A., Jr., H. M. Webb, and M. I. Sandeen, J. Exp. Zool. *120*: 391–420, 1952. Dual control of *Palaemonetes* red chromatophores.
52. Brown, F. A., Jr., and V. J. Wulff, J. Cell. Comp. Physiol. *18*:339–353, 1941. Chromatophorotropins from the crustacean central nervous system.
53. Brücke, E., Denkschr. Akad. Wiss. Wien. *4*:179–210, 1852. Color changes in the Chameleon.
54. Buchholz, R., Arch. Anat. Physiol. wiss. Med., pp. 71–81, 1863. Color changes, fishes.
55. Bückmann, D., J. Insect Physiol. *3*:159–189, 1959. Control of prepupal color change in *Cerura* caterpillars.
56. Buergers, A. C. J., T. Boschman, and J. C. van der Kamer, Acta Endocrinol. *14*:72–82, 1953. *Xenopus* darkening by adrenin and in excitement.
57. Butcher, E. O., J. Exp. Zool. *79*:275–297, 1938. Relation of portions of eyes to color changes, fishes.
58. Carlisle, D. B., and F. G. W. Knowles, Endocrine Control in Crustaceans. Cambridge University Press, 1959, 120 pp.
59. Carstam, S. P., and S. Suneson, Kungl. Fysiogr. Sällskap. Lund. Förhandl. *19*:1–5, 1949. Isopod chromatophorotropins.
60. Chang, C. Y., Science *126*:121–122, 1957. Thyroxine and *Xenopus* melanophores.
61. Cloney, R. A., and E. Florey, Z. Zellforsch. *89*:250–280, 1968. Ultrastructure of cephalopod chromatophore organs.
62. Coohill, T. P., C. K. Bartell, and M. Fingerman, Physiol. Zool. *43*:232–239, 1970. Ultraviolet and visible light effects directly on *Uca* melanophores.
63. Deanin, G. G., and F. R. Steggerda, Proc. Soc. Exp. Biol. Med. *67*:101–104, 1948. Light reflection from frog skin.
64. Dupont-Raabe, M., Arch. Zool. Exp. Gen. *86*:32–39, 1949. Regulation of color change in *Corethra*.
65. Dupont-Raabe, M., C. R. Acad. Sci. *228*:130–132, 1949. Chromatophorotropins of insects.
66. Dupont-Raabe, M., Ann. Biol. *32*:247–282, 1956. Mechanism of color change in insects.
67. Dürken, B., Arch. Mikr. Anat. *99*:222–389, 1923. Color change in insect pupae.
68. Duspiva, F., S.-B. Akad. Wiss. Wien. Math.-Nat. Kl. *140*:553–596, 1931. Color change in young fishes.
69. Eddy, J. M. P., and R. Strahan, Gen. Comp. Endocrinol. *11*:528–534, 1968. Pineal influences upon chromatophores of lampreys.
70. Edgren, R. A., Proc. Soc. Exp. Biol. Med. *85*:229–230, 1954. *Hyla* darkening by ACTH.
71. Edgren, R. A., Proc. Soc. Exp. Biol. Med. *87*:20–23, 1954. *Hyla* color change and temperature.
72. Enami, M., Biol. Bull. *100*:28–43, 1951. Neurosecretion and color change in *Eriocheir*.
73. Enami, M., Science *121*:36–37, 1955. A melanocyte-concentrating factor.
74. Ergene, S., Z. vergl. Physiol. *32*:530–551, 1950; *34*:69–74, 159–165, 1952; *35*:36–41, 1953; *37*:221–229, 1955; *38*:311–316, 1956. Adaptation to colored backgrounds in *Orthoptera*.
75. Ergene, S., Zool. Anzeiger. *153*:110–113, 1954. Color change in adult *Oedaleus* without molting.
76. Ergene, S., Mitt. Zool. Mus. Berlin *29*:127–133, 1953. Selective value of background color adaptation.
77. Faure, J. C., Bull. Ent. Res. *23*:293–405, 1932. Color change in locusts.
78. Fernlund, P., and L. Josefsson, Biochim. Biophys. Acta *158*:262–273, 1968. Chromactivating hormones of *Pandalus borealis*.
79. Fingerman, M., Biol. Bull. *109*:255–264, 1955. Tidal rhythms of color change in *Callinectes*.
80. Fingerman, M., Biol. Bull. *110*:274–290, 1956. Tidal rhythms of color change in *Uca*; phasing of color-change tidal rhythm by tides.
81. Fingerman, M., J. Exp. Zool. *133*:87–106, 1956. Response of *Callinectes* melanophores to background, light and temperature.
82. Fingerman, M., Science *123*:585–586, 1956. *Uca* pigment-concentrating factor.
83. Fingerman, M., Tulane Stud. Zool. *3*:139–148, 1956. Hormones and *Ligia* chromatophores.
84. Fingerman, M., Tulane Stud. Zool. *5*:137–148, 1957. Dual control of *Cambarellus* red chromatophores; background responses in *Cambarellus*.
85. Fingerman, M., Biol. Bull. *112*:7–20, 1957. Phasing of tidal rhythm of color change in nature.
86. Fingerman, M., Amer. Nat. *91*:167–178, 1957. Review of crustacean tidal rhythms of color change.
87. Fingerman, M., Amer. Midl. Nat. *60*:71–83, 1958. Chromatophore system of *Orconectes* (*Faxonella*).
88. Fingerman, M., Tulane Stud. Zool. *7*:21–30, 1959. Electrophoretic separation of crayfish chromatophorotropins.
89. Fingerman, M., Internat. Rev. Cytol. *8*:175–210, 1959. Review: physiology of chromatophores.
90. Fingerman, M., The Control of Chromatophores. New York, Pergamon Press, 1963, 184 pp.
91. Fingerman, M., Amer. Zool. *9*:443–452, 1969. Cellular aspects of the control of crustacean color changes.
92. Fingerman, M., Ann. Rev. Physiol. *32*:345–372, 1970. Comparative physiology: chromatophores.
93. Fingerman, M., and T. Aoto, J. Exp. Zool. *138*:25–50, 1958. Electrophoretic separation of color-change hormones of crayfish.
94. Fingerman, M., and T. Aoto, Physiol. Zool. *31*:193–208, 1958. Crayfish hormones and long-term color adaptation.
95. Fingerman, M., C. K. Bartell, and R. A. Krasnow, Biol. Bull. *140*:376–388, 1971. Chromatophorotropins from *Limulus* and *Uca*.
96. Fingerman, M., and S. W. Fingerman, Amer. Zool. *11*:646, 1971. A *Crangon*-darkening substance in brachyurans.
97. Fingerman, M., and M. E. Lowe, Physiol. Zool. *30*:216–231, 1957. Crayfish color-change ability facilitated by use, sluggish with disuse.
98. Fingerman, M., and M. E. Lowe, Tulane Stud. Zool. *5*:149–171, 1957. Rate of hormone secretion following background change in *Cambarellus*.
99. Fingerman, M., M. E. Lowe, and W. C. Mobberly, Limnol. Oceanog. *3*:271–282, 1958. Phasing of tidal rhythm of color change.
100. Fingerman, M., R. Nagabhushanam, and L. Philpott, Biol. Bull. *120*: 337–347, 1961. Physiology of *Sesarma* melanophores.
101. Fingerman, M., and K. R. Rao, Physiol. Zool. *42*:138–147, 1969. Physiology of brown-black chromatophores of *Squilla*.
102. Fingerman, M., M. I. Sandeen, and M. E. Lowe, Physiol. Zool. *32*:128–149, 1959. Color-change hormones of *Palaemonetes*.
103. Fingerman, M., and D. W. Tinkle, Biol. Bull. *110*:144–152, 1956. Response of white chromatophores of *Palaemonetes* to light, temperature, and background.
104. Fischer, A., Z. Zellforsch. *65*:290–312, 1965. Chromatophores and color-change in *Platynereis*.
105. Florey, E., Comp. Biochem. Physiol. *18*:305–324, 1966. Nervous control of cephalopod chromatophores.
106. Florey, E., and M. E. Kriebel, Z. vergl. Physiol. *65*:98–130, 1969. Electrical and mechanical responses of *Loligo* chromatophore muscle.
107. Foster, K. W., Proc. Nat. Acad. Sci. *19*:535–540, 1933. Control of iridosome changes in *Fundulus*.
108. Fox, D. L., Ann. Rev. Biochem. *16*:443–470, 1947. Chemical nature of pigments, review.
109. Fox, D. L., Animal Biochromes and Structural Colors. Cambridge University Press, 1953.
110. Fredericq, L., Arch. Zool. Expér. Gén. *7*:535–583, 1878. Control of color change, Cephalopods.
111. Freeman, A. R., P. M. Connell, and M. Fingerman, Comp. Biochem. Physiol. *26*:1015–1029, 1968. Electrophysiological study of *Palaemonetes* red chromatophores.
112. Fries, E. F. B., Proc. Nat. Acad. Sci. *28*:396–401, 1942. Control of *Fundulus* leucophores.
113. Fries, E. F. B., Biol. Bull. *82*:261–272, 1942. Control of *Fundulus* xanthophores.
114. Fries, E. F. B., Biol. Bull. *82*:273–283, 1942. Dual innervation of teleost melanophores.
115. Fries, E. F. B., Physiol. Zool. *16*:199–212, 1943. Control of *Fundulus xanthophores*.
116. Fries, E. F. B., J. Morphol. *103*:203–254, 1958. Physiology of *Fundulus* reflecting chromatophores.
117. Friza, F., Z. vergl. Physiol. *8*:289–336, 1928. Diurnal color changes in mantids.
118. Fuchs, R. F., Winterstein, Handb. vergl. Physiol. *4*, pt. 1:1189–1656, 1914. Color change, review.
119. Fugh, R., J. Fac. Sci., Tokyo, Sec. IV, *8*:371–380, 1959. Melanin dispersion in *Fundulus* by sodium ions.
120. Fujii, R., and R. R. Novales, Amer. Zool. *9*:453–463, 1969. Cellular aspects of control of fish color changes.
121. Gabritschevsky, E., J. Exp. Zool. *47*:251–267, 1927. Color changes in spider, *Misumena*.
122. Gamble, F. W., and F. W. Keeble, Quart. J. Micr. Sci. *43*:589–698, 1900. Color changes, crustaceans; influence of temperature.
123. Giersberg, H., Z. vergl. Physiol. *7*:657–695, 1928. Color change in *Carausius*; influence of humidity.
124. Giersberg, H., Z. vergl. Physiol. *9*:523–552, 1929. Temperature and morphological color changes, insects.
125. Giersberg, H., Z. vergl. Physiol. *13*:258–279, 1930. Control of erythrophores of *Phoxinus*; influence of drugs.
126. Gilson, A. S., Jr., J. Exp. Zool. *45*:415–455, 1926. Color changes in young fishes.
127. Goldman, J. M., and M. E. Hadley, Brit. J. Pharmacol. *39*:160–166, 1970. Melanophore stimulating hormone and cyclic AMP in melanophore responses.
128. Goodwin, T. W., pp. 101–140 *in* The Physiology of the Crustacea, Vol. 1, edited by T. H. Waterman. New York, Academic Press, 1960.
129. Hadley, C. E., Bull. Mus. Comp. Zool. *69*:108–114, 1929. Color changes during mating of lizards.
130. Hadley, M. E., and J. M. Goldman, Amer. Zool. *9*:489–504, 1969. Physiological color changes in reptiles.
131. Hanström, B., Zool. Jahrb. Abt. Anat. Ontog. Tiere. *56*:387–520, 1933. The crustacean sinus gland.
132. Hanström, B., Kungl. Svenska Vetenskap. Handl. *16*:1–99, 1937. Relation of portions of eyes to color changes, Crustacea. Crustacean sinus gland and color change.
133. Hanström, B., Kungl. Fysiogr. Sällsk. Handl. N. F. *49*:1–10, 1938. Relation of portions of eyes to color changes, Crustacea.

134. Healey, E. G., J. Exp. Biol. *31*:473–490, 1954. Color change control in *Phoxinus*.
135. Healey, E. G., and D. M. Ross, Comp. Biochem. Physiol. *19*:545–580, 1966. Drug effects on the background color response of *Phoxinus*.
136. Hempelmann, F., Z. wiss. Zool. *152*:353–383, 1939. Color change, polychaetes.
137. Hidaka, T., Annot. Zool. Japan. *29*:69–74, 1956. Control of pupal color in Lepidoptera.
138. Hill, A. V., J. L. Parkinson, and D. Y. Solandt, J. Exp. Biol. *12*:397–399, 1935. Measurement of chromatophore activity.
139. Hitchcock, H. B., Biol. Bull. *80*:26–30, 1941. Adaptation to colored backgrounds, Crustacea.
140. Hofmann, F. G., Arch. Mikr. Anat. *70*:361–413, 1907. Innervation, cephalopod chromatophore.
141. Hogben, L. T., The Pigmentary Effector System. Edinburgh, Oliver and Boyd, 1924. 152 pp.
142. Hogben, L. T., Proc. Roy. Soc. Lond. B *120*:142–158, 1936. Hypophysis in elasmobranch melanophore control.
143. Hogben, L. T., and L. Mirvish, Brit. J. Exp. Biol. *5*:295–308, 1928. Control of color changes in chameleons.
144. Hogben, L., and D. Slome, Proc. Roy. Soc. Lond. B *108*:10–53, 1931. Mechanism of chromatophore control, *Xenopus*.
145. Hogben, L., and D. Slome, Proc. Roy. Soc. Lond. B *120*:158–173, 1936. Mechanism of chromatophore control, *Xenopus*.
146. Hogben, L., and F. R. Winton, Proc. Roy. Soc. Lond. B *94*:151–162, 1922. Hypophysis and amphibian color change.
147. Hogben, L., and F. R. Winton, Proc. Roy. Soc. Lond. B *95*:15–30, 1923. Hypophysis and amphibian color change.
148. Hooker, D., Amer. J. Anat. *16*:237–250, 1914. Color change in young *Rana*.
149. Hosoi, T., J. Fac. Sci. Imp. Univ. Tokyo *3*:265–270, 1934. Chromatophorotropins from the crustacean central nervous system.
150. Hudson, B., and G. A. Bentley, Lancet *1*:775, 1955. Adrenin darkening in *Xenopus*.
151. Janda, V., Zool. Anz. *115*:177–185, 1936. Color change in skin transplants in *Dixippus*.
152. Janzen, R., Z. Morph. Okol. Tiere *24*:327–341, 1932. Color change, leeches.
153. Janzen, R., Zool. Anz. *101*:35–40, 1932. Color changes, leeches; *Hemiclepsis* and *Glossiphonia*.
154. Johnson, L. P., and T. Jahn, Physiol. Zool. *15*:89–94, 1942. Color changes, *Euglena*.
155. Jørgenson, C. B., Gen. Comp. Endocrinol. *2*:610, 1962. Hypophysectomy influences on melanophores of *Xenopus*.
156. Jørgenson, C. B., and L. O. Larsen, Nature *186*:641–642, 1960. Control of colour change in amphibians.
157. Kaestner, H., Arch. Entw.-Mech. Org. *124*:1–16, 1931. Temperature and morphological color change, insects.
158. Kalmus, H., Z. vergl. Physiol. *25*:494–508, 1938. Diurnal rhythm in *Carausius* color change.
159. Keeble, F. W., and F. W. Gamble, Phil. Trans. Roy. Soc. Lond. B *196*:295–388, 1904. Relationship of morphological to physiological color change, crustaceans.
160. Keeble, F. W., and F. W. Gamble, Phil. Trans. Roy. Soc. Lond. B *198*: 1–16, 1905. Color changes, crustaceans.
161. Ketterer, B., and E. Remilton, J. Endocrinol. *11*:7–18, 1954. *Xenopus* darkening by adrenin and in excitement.
162. Key, K. H. L., and M. R. Day, Australian J. Zool. *2*:309–363, 1954. Primary color change in grasshopper, *Kosciuscola*.
163. Khokhar, R., Experientia *27*:340–341, 1971. Inhibitory control of pars intermedia function in the teleost *Ictalurus*.
164. Kleinholz, L. H., Biol. Bull. *69*:379–390, 1935. Role of hypophysis in *Fundulus* color changes.
165. Kleinholz, L. H., J. Exp. Biol. *15*:474–499, 1938. Control of color changes in *Anolis*.
166. Kleinholz, L. H., Pubbl. Staz. Zool. Napoli *17*:53–57, 1938. Color change, echinoderm.
167. Kleinholz, L. H., Gen. Comp. Endocrinol. *14*:578–588, 1970. Isolation of crustacean chromatophorotropins.
168. Knight, H. H., Ann. Ent. Soc. Amer. *17*:258–272, 1924. Temperature and morphological color change, insects.
169. Knowles, F. G. W., Physiol. Comp. Oecol. *2*:284–296, 1952. Color changes after sinus gland removal.
170. Knowles, F. G. W., Proc. Roy. Soc. Lond B *141*:248–267, 1953. Post-commissural organs and color change.
171. Knowles, F. G. W., Pubbl. Staz. Zool. Napoli *24* (suppl.):74–78, 1954. Color change control in *Squilla*.
172. Knowles, F. G. W., and D. B. Carlisle, Biol. Rev. *31*:396–473, 1956. Review of crustacean color change.
173. Knowles, F. G. W., D. B. Carlisle, and M. Dupont-Raabe, J. Marine Biol. Assoc. U. K. *34*:611–635, 1955. Chemical properties of *Carausius* chromatophorotropin.
174. Knowles, F. G. W., D. B. Carlisle, and M. Dupont-Raabe, C. R. Acad. Sci. *242*:825, 1956. Chemical properties of crustacean chromatophorotropin.
175. Koller, G., Verh. deutsch. zool. Gesell. *30*:128–132, 1925. Mechanism of chromatophore control, Crustacea.
176. Koller, G., Z. vergl. Physiol. *5*:191–246, 1927. Functional independence of pigments, crustaceans; color adaptations; mechanism of chromatophore control.
177. Koller, G., Z. vergl. Physiol. *8*:601–612, 1928. Mechanism of chromatophore control, Crustacea.
178. Koller, G., Z. vergl. Physiol. *12*:632–667, 1930. Hormones in crustacean color change.
179. Kröyer, H., Kong. Dansk. Videnskap. Selskabet *9*:209–361, 1842. Color changes, *Hippolyte*.
180. Krüger, P., and H. Kern, Arch. ges. Physiol. *202*:119–138, 1924. Melanophores and heat regulation.
181. Kühn, A., Z. vergl. Physiol. *32*:572–598, 1950. Background adaptation in *Sepia* and *Octopus*.
182. Kühn, A., and R. F. Heberdey, Zool. Anz. *231*:suppl. 4, 1929. Adaptation to colored backgrounds, Cephalopoda.
183. Kühn, A., and E. Lederer, Ber. deutsch. chem. Ges. *66*:488–495, 1953. Chemical nature pigments, crustaceans.
184. Kuntz, A., Bull. U. S. Bur. Fish. *35*:1–29, 1916. Adaptation to colored backgrounds, fishes.
185. Laurens, H., J. Exp. Zool. *16*:195–210, 1914. Color change in young *Amblystoma*.
186. Laurens, H., J. Exp. Zool. *18*:577–638, 1915. Color change in young *Amblystoma*.
187. Lawinski, L., and F. Pautsch, Acta Biol. Med. Soc. Gedan. *9*:5–14, 1965. Chromatophores and their behaviour in the crab, *Rhithropanopeus*.
188. Lieben, S., Centralbl. Physiol. *20*:108–117, 1906. Adrenaline and color change, frog.
189. Lundstrom, H. M., and P. Bard, Biol. Bull. *62*:1–9, 1932. Hypophysis in elasmobranch melanophore control.
190. Martini, E., and I. Achundow, Zool. Anz. *81*:25–44, 1929. Color changes in *Corethra*.
191. Mast, S. O., Bull. U. S. Bur. Fish. *34*:173–238, 1916. Adaptation to colored backgrounds, fishes; adaptive background selection.
192. Matsumoto, K., Biol. J. Okayama Univ. *1*:234–248, 1954. Neurosecretion and color change in crustaceans; color-change hormones of *Eriocheir*.
193. Matsushita, K., Sci. Rys. Imp. Univ. Sendai, 4, Biol. *13*:171–200, 1938. Dual innervation of teleost melanophores.
194. Matthews, S. A., J. Exp. Zool. *58*:471–486, 1931. Mechanism of pigment migration in chromatophores, fish.
195. Matthews, S. A., Biol. Bull. *64*:315–320, 1933. Color changes in hypophysectomized *Fundulus*.
196. McWhinnie, M. A., and H. M. Sweeney, Biol. Bull. *108*:160–174, 1955. Color change in *Trachelipus*.
197. Mendes, E. G., Bol. Gac. Filos. Ciên. Letr. Univers. São Paulo, 15. Zool. *6*:285–299, 1942. Color change in young fishes.
198. Menke, E., Arch. ges. Physiol. *140*:37–91, 1911. Diurnal rhythm and chromatophore state, Crustacea; influence of temperature on chromatophores.
199. Millott, N., Nature *170*:325–326, 1952. Color change daily rhythm in *Diadema*.
200. Millott, N., Experientia *9*:98–99, 1953. Color changes in *Diadema* in response to illumination.
201. Mills, S. M., J. Exp. Zool. *64*:231–244, 1932. Dual innervation of teleost melanophores.
202. Milne-Edwards, H., Ann. Sci. Nat., Sec. 2, Zool. *1*:46–54, 1834. Color change, Chameleon.
203. Minkiewicz, R., Bull. Acad. Sci. Cracovie. 918–929 (November), 1908. Adaptation to colored backgrounds, Crustacea.
204. Mussbichler, A., and K. Umrath, Z. vergl. Physiol. *32*:311–318, 1950. Effect of MSH and adrenin on *Hyla* melanophores.
205. Nagano, T., Sci. Rep. Tohoku Univ. ser. 4, *18*:167–175, 1949. Color changes in isopods.
206. Neill, R. M., J. Exp. Biol. *17*:74–94, 1940. Color change in young fishes; color change in *Anguilla*.
207. Novales, R. R., Physiol. Zool. *32*:15–28, 1959. MSH action on isolated frog skin.
208. Novales, R. R., and R. Fujii, J. Cell. Physiol. *75*:133–136, 1970. Melanin-dispersing effect of cyclic AMP in *Fundulus*.
209. Odiorne, J. M., Proc. Nat. Acad. Sci. *19*:329–332, 1933. Morphological color change, fishes.
210. Odiorne, J. M., Proc. Nat. Acad. Sci. *19*:750–754, 1933. Adrenaline and teleost leucophores.
211. Okay, S., C. R. Ann. Arch. Soc. Turqu. Sci. Phys. Nat. *12*:101, 1946. Color change hormones in isopods.
212. Osborn, C. M., Proc. Nat. Acad. Sci. *26*:155–161, 1940. Induced pigmentation on ventral surface of flounder.
213. Parker, G. H., J. Exp. Zool. *3*:401–414, 1906. Melanophores and heat regulation.
214. Parker, G. H., Biol. Rev. *5*:59–90, 1930. Chromatophores, review.
215. Parker, G. H., Proc. Nat. Acad. Sci. *17*:594–596, 1931. Color change, echinoderm.
216. Parker, G. H., Proc. Nat. Acad. Sci. *20*:306–310, 1934. Dual innervation of *Fundulus* melanophores.
217. Parker, G. H., Proc. Amer. Philosoph. Soc. *75*:1–10, 1935. Measurement of chromatophore activity.
218. Parker, G. H., Biol. Bull. *68*:1–3, 1935. Innervation of *Mustelus* melanophores.

219. Parker, G. H., Biol. Bull. *70*:1–7, 1936. Color change in young elasmobranchs.
220. Parker, G. H., Biol. Bull. *71*:255–258, 1936. Single innervation of *Mustelus* melanophores.
221. Parker, G. H., Cold Spring Harbor Symp. Quant. Biol. *4*:358–370, 1936. Cold-block of chromatophore nerves.
222. Parker, G. H., Proc. Nat. Acad. Sci. *23*:206–211, 1937. Control of erythrophores of *Holocentrus*; dual innervation of melanophores.
223. Parker, G. H., Proc. Amer. Philosoph. Soc. 77:223–247, 1937. Control of elasmobranch melanophores.
224. Parker, G. H., J. Exp. Biol. *15*:48–73, 1938. Control of color changes in *Phrynosoma*; influence of temperature.
225. Parker, G. H., Proc. Amer. Philosoph. Soc. *83*:379–409, 1940. AcCh and fish color change.
226. Parker, G. H., Proc. Amer. Philosoph. Soc. *85*:18–24, 1941. Dual innervation of teleost melanophores.
227. Parker, G. H., J. Exp. Zool. *89*:451–473, 1942. Chemical mediator in *Mustelus* melanophore control.
228. Parker, G. H., Quart. Rev. Biol. *18*:205–227, 1943. Vertebrate color changes, review.
229. Parker, G. H., Biol. Bull. *84*:273–284, 1943. Methods of measurement of color changes.
230. Parker, G. H., Animal Colour Changes and Their Neurohumors. New York, Cambridge University Press, 1948, 377 pp.
231. Parker, G. H., and H. P. Brower, Biol. Bull. *68*:4–6, 1935. Seasonal development of nuptial secondary sex coloration, *Fundulus*.
232. Parker, G. H., and A. Rosenblueth, Proc. Nat. Acad. Sci. *27*:198–204, 1941. Electrical stimulation of melanophore nerves.
233. Parker, G. H., and L. E. Scatterty, J. Cell. Comp. Physiol. *9*:297–314, 1937. Hormonal control *Rana* color change.
234. Pasteur, C., C. R. Acad. Sci. *246*:320–322, 1958. Effect of X-organ extirpation on *Leander* (*Palaemon*) color change.
235. Pautsch, F., Bull. Internat. Acad. Polon. Sci. Classe. Sci. Math. Nat. ser. B 7:511–523, 1951. Primary response in *Crangon* zoea.
236. Pérez-González, M. D., Biol. Bull. *113*:426–441, 1957. Crustacean chromactivator destroyed by chymotrypsin.
237. Perkins, E. B., J. Exp. Zool. *50*:71–105, 1928. Pigment migration in chromatophores, crustacean. Mechanism of chromatophore control.
238. Perkins, E. B., and T. Snook, J. Exp. Zool. *61*:115–128, 1932. Mechanism of chromatophore control, Crustacea.
239. Phisalix, C., Arch. Physiol. Norm. Path., ser. 5, *6*:92–100, 1894. Control of color change, cephalopods.
240. Pickford, G. E., and J. W. Atz, The Physiology of the Pituitary Gland of Fishes. New York Zoological Society, 1957.
241. Pigeault, N., C. R. Acad. Sci. *246*:487–489, 1958. Organ of Bellonci and isopod color change.
242. Pouchet, G., J. Anat. Physiol. *12*:1–90, 113–165, 1876. Color changes, crustaceans, fishes.
243. Poulton, E. B., The Colors of Animals. New York, D. Appleton and Co., 1890, 360 pp.
244. Priebatsch, I., Z. vergl. Physiol. *19*:453–485, 1933. Relations of portions of eyes to color changes, insects.
245. Prosser, C. L., B. von Limbach, and G. W. Bennett, Physiol. Zool. *20*:349–354, 1947. Reactions of fish chromatophore to war gases.
246. Rahn, H., Biol. Bull. *80*:228–237, 1941. Hypophysis and color change in rattlesnake.
247. Rao, K. R., Experientia *23*:231, 1967. Crustacean larval chromatophore responses to light and endocrines.
248. Rao, K. R., Zool. Jahrb. Physiol. *74*:247–291, 1968. Chromatophorotropin and color change variations during the life of the crab, *Ocypode*.
249. Rao, K. R., Gen. Comp. Endocrinol. *12*:547–585, 1969. Influence of prolonged adaptation to a background upon responses of crab chromatophores.
250. Rao, K. R., D. R. Fielder, and M. Fingerman, Amer. Zool. *10*:495, 1970. Melanin-dispersing substances from the eyestalks of *Uca*.
251. Rao, K. R., D. R. Fielder, and M. Fingerman, Amer. Zool. *10*:495–496, 1970. Effects of oxidation and reduction on melanin-dispersing material from *Uca* eyestalks.
252. Rao, K. R., M. Fingerman, and C. K. Bartell, Biol. Bull. *133*:606–617, 1967. Physiology of the white chromatophores in *Uca*.
253. Redfield, A. C., J. Exp. Zool. *26*:275–333, 1918. Control of color change in *Phrynosoma*; diurnal rhythm of color change.
254. Reed, B. L., Life Sci., Part II:961–973, 1968. Melatonin in circadian color changes in pencil fish.
255. Röseler, I., Z. vergl. Physiol. *70*:144–174, 1970. Die Rhythmik der Chromatophoren des Polychaeten *Platynereis dumerilii*. [Chromatophore rhythms in *Platynereis*.]
256. Rowlands, A., J. Exp. Biol. *29*:127–136, 1952. Humidity and *Rana* color change.
257. van Rynberk, G., Ergebn. Physiol. *5*:347–571, 1906. Color change, review.
258. Sage, M., J. Exp. Zool. *173*:121–128, 1970. Prolactin role in color change in the teleost, *Gillichthys*.
259. Sand, A., Biol. Rev. *10*:361–382, 1935. Control of color changes in chameleons.
260. Sandeen, M. I., Physiol. Zool. *23*:337–352, 1950. Chromatophorotropins of *Uca*.
261. Sars, G., Histoire naturelle des Crustacés d'eau douce de Norvège. Christiania, 1867, 145 pp. Color changes, Crustaceans.
262. Scharrer, B., Action of Hormones in Plants and Invertebrates. New York, Academic Press, 1952. Review of insect color change.
263. Scharrer, E., Z. vergl. Physiol. 7:1–38, 1928. Pineal body and color changes, fishes.
264. Scheer, B. T., and M. A. R. Scheer, Pubbl. Staz. Zool. Napoli *25*:397–418, 1954. Color changes in *Leander* (*Palaemon*).
265. Schliep, W., Zool. Jahrb. Physiol. *30*:45–132, 1910. Color change in *Carausius*; diurnal rhythm in *Carausius* color change.
266. Schlieper, C., Z. vergl. Physiol. *3*:547–557, 1926. Color change in *Hyperia*.
267. Schlottke, E., Z. vergl. Physiol. *3*:692–736, 1926. Temperature and morphological color change, insects.
268. Schlottke, E., Z. vergl. Physiol. *20*:370–379, 1934. Temperature and morphological color change, insects.
269. Schwinck, I., Naturwissenschaften *40*:365, 1953. Über den Nachweis eines Redox-Pigmentes (Ommochrom) in der Haut von *Sepia officinalis*. [Pigments in *Sepia* skin.]
270. Sereni, E., Boll. Soc. Ital. Biol. Sper. *2*:377–381, 1927. Control color change, cephalopods.
271. Sereni, E., Z. vergl. Physiol. *8*:488–600, 1928. Inhibitory color-change center, cephalopods. Postural chromatophore reflexes, cephalopods.
272. Sereni, E., Boll. Soc. Ital. Biol. Sper. *3*:707–711, 1928. Betaine and tyramine in cephalopod color change.
273. Sereni, E., Boll. Soc. Ital. Biol. Sper. *4*:749–753, 1929. Salivary gland (tyramine) and cephalopod color change.
274. Sereni, E., Z. vergl. Physiol. *12*:329–503, 1930. Color change mechanism, cephalopods.
275. Sereni, E., Biol. Bull. *59*:247–268, 1930. Direct action of betaine and tyramine on cephalopod chromatophores.
276. von Siebold, K. T. E., Die Susswasserfische von Mittleuropa. Leipzig, 1863, 431 pp. Color change, fishes.
277. Sieglitz, G., Z. vergl. Physiol. *33*:99–124, 1951. Melanin-concentrating factor in frog skin.
278. Sjögren, S., Zool. Jahrb. Abt. Anat. Ontog. Tiere *58*:145–170, 1934. The crustacean sinus gland.
279. Skorkowski, E. F., Marine Biol. *8*:220–223, 1971. Chromatophorotropic hormones from the eyestalk of *Crangon*.
280. Slome, D., and L. T. Hogben, South African J. Sci. *25*:329–335, 1928. Measurement of chromatophore activity.
281. Smith, D. C., Biol. Bull. *58*:193–202, 1930. Temperature and color change, Crustacea.
282. Smith, D. C., J. Exp. Zool. *58*:423–453, 1931. Autonomic drugs and teleost color change.
283. Smith, D. C., J. Cell Comp. Physiol. *8*:83–87, 1936. Measurement of chromatophore activity.
284. Smith, D. C., and M. T. Smith, Biol. Bull. *67*:45–58, 1934. Control of erythrophores of *Holocentrus*.
285. Smith, H. G., Proc. Roy. Soc. Lond. B *125*:250–263, 1938. Dual control of isopod melanophores.
286. Smith, P. E., Anat. Rec. *11*:57–64, 1916. Hypophysis and amphibian color change.
287. Smith, R. I., Physiol. Zool. *15*:410–417, 1942. Mechanism of color change, leeches.
288. Spaeth, R. A., Anat. Anz. *44*:520–524, 1913. Pigment migration in chromatophore, fish.
289. Spaeth, R. A., Amer. J. Physiol. *41*:597–602, 1916. Measurement of chromatophore activity.
290. Steggerda, F. R., and A. L. Soderwall, J. Cell. Comp. Physiol. *13*:31–37, 1939. Dual humoral control of melanophores, *Rana pipiens*.
291. Steinach, E., Pflügers Arch. Physiol. *87*:1–37, 1901. Tactile stimuli and color change, cephalopods.
292. Stephens, G. C., Physiol. Zool. *30*:55–59, 1957. Phase shifting of color-change rhythm by temperature cycles; sensitivity rhythm.
293. Stephens, G. C., Amer. Nat. *91*:135–152, 1957. Review of crustacean daily rhythms of color change.
294. Steschegolew, G. G., Rev. Zool. Russe 7:149–166, 1927. Color change, leeches.
295. Sumner, F. B., Amer. Nat. *69*:245–266, 1935. Protective value of fish color change.
296. Sumner, F. B., Biol. Rev. *15*:351–375, 1940. Morphological color change in fishes and amphibians, review.
297. Sumner, F. B., Biol. Bull. *84*:195–205, 1943. Morphological color changes, *Girella, Fundulus*.
298. Sumner, F. B., Proc. Nat. Acad. Sci. *30*:285–294, 1944. Morphological color change, guanine.
299. Sumner, F. B., and P. Doudoroff, Biol. Bull. *84*:187–194, 1943. Assay of melanin in fishes.
300. Sumner, F. B., and A. B. Keys, Physiol. Zool. *2*:495–504, 1929. Ratio of incident to reflected light and color change.
301. Suneson, S., Kungl. Fysiogr. Sallsk. Handl. N.F. *58*:5, 1947. Color change system of *Idothea*.
302. Taki, I., J. Fac. Fish. Anim. Husb. Hiroshima Univ. *5*:345–417, 1964. Physiology of branchial gland in Cephalopoda.
303. Tomita, G., J. Shanghai Sci. Inst. IV 2:237–264, 1936. Color change in young fishes; color change in blinded teleosts.
304. Tomita, G., J. Shanghai Sci. Inst. IV *4*:1–8, 1938. Dual innervation of teleost melanophores.
305. Tomita, G., J. Shanghai Sci. Inst. IV *5*:151–178, 1940. Dual innervation of teleost melanophores.

306. Vesely, D. L., and M. E. Hadley, Science *173*:923–925, 1971. Calcium requirement for melanophore-stimulating hormone action on melanophores.
307. von Uexküll, J., Z. Biol. *34*:319–339, 1896. Color change, echinoderm.
308. Umrath, K., and H. Walcher, Z. vergl. Physiol. *33*:129–141, 1951. Control of *Macropodus* melanophores.
309. Verne, J., Arch. Morph. Gén. Exp. 1923, 168 pp. Chemical nature of pigments, crustaceans.
310. Waring, H., Proc. Roy. Soc. Lond. B *128*:343–353, 1940. Dual hormonal control of *Anguilla* melanophores.
311. Waring, H., Biol. Rev. *17*:120–150, 1942. Vertebrate color changes, review.
312. Waring, H., and F.W. Landgrebe, *in* The Hormones, edited by G. Pincus and K. V. Thimann, New York, Academic Press, 1950. Vertebrate color change, review.
313. Webb, H. M., Physiol. Zool. *23*:316–336, 1950. Shifting by light, phases of color-change rhythm.
314. Wells, G. P., Nature *129*:686–687, 1932. Color change, leeches.
315. Wilkens, J. L., and M. Fingerman, Biol. Bull. *128*:133–141, 1965. Temperature relationships and *Uca* body coloration.
316. Wright, P. A., Physiol. Zool. *28*:204–218, 1955. Isolated *Rana* melanophore response to MSH and adrenin.
317. Wykes, U., J. Exp. Biol. *14*:79–86, 1937. Measurement of chromatophore activity.
318. Wyman, L. C., J. Exp. Zool. *40*:161–180, 1924. Color change in young fishes.
319. Yoshida, M., J. Exp. Biol. *33*:119–123, 1956. Color change in *Diadema*, a direct chromatophore response.
320. Young, J. Z., Quart. J. Micr. Sci. *75*:571–624, 1933. Control of elasmobranch melanophores.
321. Young, J. Z., J. Exp. Biol. *12*:254–270, 1935. Pineal body in color change, *Lampetra*; control of *Lampetra* melanophores.
322. Zondek, B., and H. Krohn, Klin. Wchnschr. *11*:405–408, 1932. Intermedin in melanophore responses.
323. Zoond, Z., and J. Eyre, Phil. Trans. Roy. Soc. Lond. B *223*:27–55, 1934. Control of color change in chameleons; diurnal rhythm of color changes.

CHAPTER 24

BIOLUMINESCENCE

*By F. A. Brown, Jr.**

The ability to produce light is very widely distributed among bacteria, fungi, algae, and animals.[38, 43–45] It seems to have arisen independently numerous times during evolutionary history. Among animals, luminous representatives have been described for all the major phyla from Protozoa through Chordata. Minor phyla for which reports of luminous species are lacking are the Mesozoa, Entoprocta, Phoronidea, Echiuroidea, Sipunculoidea, Brachiopoda, Onychophora, Linguatula, and Tardigrada. There appears to be little or no general pattern in the distribution of the capacity among or within the animal groups; its occurrence is quite sporadic. It may occur in one species of a genus and be absent in another. Practically all of the known luminous species are marine or terrestrial. Luminescent marine species are found among the abyssal, littoral, and planktonic faunas. The only reported luminescent animals from fresh water are a glowworm and a snail. The only described luminescent cavernicolous animal is a dipteran larva.

Luminescence in animals is the result of chemiluminescent reactions in which a substrate is oxidized. The chemiluminescent reaction is sometimes associated with the presence of special granules (possibly organelles) in the cytoplasm of the luminous tissue. In some organisms the light-producing reactants are expelled to the exterior where the actual reactions in production of light occur. This type of light production is known as *extracellular* luminescence, in contrast with *intracellular* luminescence, in which the light-yielding reaction proceeds within cells. In animals with extracellular luminescence the light-producing organs take the form of unicellular or multicellular glands which secrete to the exterior. Sometimes a differentiation of two types of secretory cells (enzyme and substrate) is observed in luminous glands, e.g., in *Cypridina.* Both contribute to light production.

In higher animals with intracellular luminescence there is a tendency toward an evolution of specialized photogenic organs. In the protozoans, the luminescing granules are dispersed throughout the cytoplasm. In higher animals, the photogenic cells or organs exhibit characteristic patterns of distribution. In animals such as cephalopods, crustaceans, insects, and fishes, the light-producing cells may form only a portion of organs which possess, in addition to these cells, light-absorbing and light-reflecting layers, light filters, refractive bodies, and nerve supply. Such organs superficially resemble photoreceptors.

It is by no means always an easy matter to determine whether any particular luminescent animal possesses of itself the ability to generate light. Luminescence may result instead from the presence of luminescent bacteria in or on the organism in question. One criterion for distinguishing between systems involving bacterial symbionts and those which are truly self-luminous is that in the former case the light is continuous, whereas in the latter it is usually produced intermittently, and commonly only in response to external stimuli. There are a number of exceptions to this, however. Many truly photogenic cells of animal origin exhibit a continuous glow. On the other hand, luminescence of bacterial origin can be made to give the semblance of

*With contributions by J. W. Hastings.

intermittency as, for example, in the fish *Photoblepharon*, where the symbiotic luminous bacteria are in a light-producing organ with a movable lid capable of screening the light.

Occurrence of Bioluminescence Among Animals. Among the Protozoa there are numerous luminescent marine radiolarians and dinoflagellates. The best known examples of the latter group are *Noctiluca* and *Gonyaulax*.[55] The light-producing granules are located throughout the cytoplasm, which is confined to the peripheral regions.

The dinoflagellate, *Noctiluca*, normally flashes in response to mechanical or electrical stimulation. Such stimulation evokes a slow, graded, generator-like electrical potential which can give rise to all-or-none, flash-triggering action potentials. Gross electric stimulation elicits a synchronous flash over the whole cell with very brief latency. The normally observed longer cell flash appears to be due to the conduction time of the exciting action potential over the cell, with sequential firing of the microsomes which comprise the bioluminescent elements.[32]

Gonyaulax, *Pyrodinium*, and *Pyrocystis* display both a spontaneous and a stimulable bioluminescence, the relative strengths of which vary with time of day.[4, 53]

Numerous coelenterates are known to be luminous. These include a number of hydroid polyps,[74–76] jellyfishes,[23, 93] siphonophores, sea pens, and sea pansies.[84, 85, 89] Luminescence occurs only in response to stimuli. The natural stimulus is probably mechanical. The region immediately stimulated is first to respond, followed by a wave of luminescence proceeding out from that point. Electrical changes correlated with luminescence in *Obelia* have been described.[73] In the jellyfish, *Pelagia*, the extent of spread of the luminescence is a direct function of the strength of the stimulus. Light production in some coelenterates appears to be extracellular in part, since luminous mucus can usually be readily rubbed from the surface of the organisms, but is also intracellular in sea pens, sea pansies, and hydromedusae. Light cells occur on the marginal canal of hydromedusae.

Probably all Ctenophora are luminous.[18, 47] Here the light-producing cells are located along the meridional gastrovascular canals in the vicinity of the germ cells. These organisms luminesce in response to stimuli, but the response is inhibited by light.[58] After subjection to daylight, the organisms regain full responsiveness after a short time in darkness. A photoprotein has been isolated from the ctenophore, *Mnemiopsis*, which unlike any other photoprotein is light-sensitive; it is inactivated by light.[50]

Only a single luminous nemertean, *Emplectonema kandai*, has been described. Photogenic cells, producing light intracellularly, are distributed over the whole worm. A local response to localized tactile stimulation is seen. A generalized response follows stretching of the animal.

Among the Annelida, luminescence is restricted to species of terrestrial oligochaetes and marine polychaetes. Many earthworms, e.g., *Eisenia submontana*, on irritation, eject a luminous slime. This may come from the oral or anal opening or from the dorsal pores. The polychaete *Chaetopterus*[78–80, 82] exhibits a striking luminosity, much of the surface becoming luminous whenever the worm is disturbed (Fig. 24–1). The photogenic cells are located in the hypodermis along with mucus cells, both of which secrete their products to the exterior (Fig. 24–2*C*). There is nerve control of light production; stimulation of the anterior end of the worm results in a wave of light production passing posteriorly. In the transparent pelagic *Tomopteris* the photogenic organs are specialized nephridial funnels. In luminous scale worms the light originates in certain cells in the dorsal overlapping scales.[81, 83] Stimulation of any portion of the body results in a wave of light production passing up and down the body from the point of excitation, moving posteriorly with more facility than anteriorly and thus indicating a role of a partially polarized nervous system.

Luminous species of arthropods are numerous. They are largely crustaceans[26] and insects, rarely myriapods[25] and arachnids. In fact, it is from a luminous ostracod, *Cypridina*, and the firefly that we have learned much of the fundamental chemistry of bioluminescent systems. The light producing organ of *Cypridina* is a large gland located near the mouth. There are two kinds of secretory cells, one producing substrate (luciferin) and the other producing enzyme (luciferase), both in the form of small granules. The granules are ejected by muscle contraction, and the luminescence is extracellular.

In those copepods showing photogenic capacity, the sources of the active agents which are expelled into the sea water are small groups of greenish secretory cells on various parts of the body.

Many euphausids and shrimp possess

rather highly differentiated organs[3] having a reflecting layer and lens associated with the light-producing cells (Fig. 24–2*A*). These organs are distributed widely over the surface of the body. The numerous organs appear to be coordinated through the nervous system, inasmuch as the sequence of their activity may, as in *Sergestes*, for example, follow an anteroposterior progression. The deep-sea shrimp, *Acanthephyra purpurea*, has, in addition to typical photophores (luminous organs) of the general type just described, glands near the mouth from which luminescent substances may be forcefully ejected so as to permit the shrimp to escape from predators, leaving a luminous cloud behind.

Among the insects, luminescence is restricted to members of very few orders. A few species of Collembola have been described as glowing continuously, although varying in intensity with the state of excitation of the individual. The larvae of the fungus gnat, *Ceratoplanus*, and of the tipulid fly, *Bolitophila*, have been described as luminescent, and in the latter species the adults are also luminescent. The light appears to arise in the malpighian tubules.

Perhaps the most familiar and striking examples of light production in organisms are to be found among the Coleoptera, specifically the lampyrids and elaterids.[8] In these "fireflies" or "lightning bugs" the photogenic organ is typically located ventrally in the posterior abdominal region. The organ is composed of two cellular layers, a ventral layer of light-producing cells and a dorsal or internal layer of "reflecting" cells. The reflecting cells contain minute particles of a purine base, probably "urates." This layer is said to serve as a diffusing reflector for the photogenic cells, while simultaneously being a dorsal shield. The photogenic organ is richly provided with tracheal vessels which terminate in numerous tracheal end-cells. Nerves do not innervate photogenic cells. The detailed organization of the organs has been worked out.[10, 42, 102]

Flashing of the firefly is controlled by the nervous system. One remarkable aspect of the activity of these organs, which indicates this, is the synchronous flashing of some tropical fireflies.[7] Synchrony among males of the Thai species *Pteroptyx malaccae*, which has an extremely regular flashing rhythm of 560 ± 6 msec at 28°C, has been shown to be within ± 20 msec of exact coincidence.[11] Since the minimum eye-to-light-organ latency is 60 to 80 msec, the synchrony must be

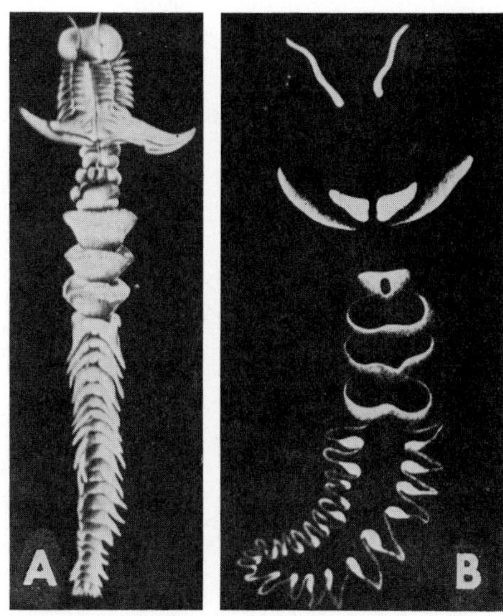

Figure 24–1. *Chaetopterus. A*, dorsal view (after Trojan). *B*, a luminescent individual in darkness (after Panceri). (From Harvey, E. N., Living Light. Princeton University Press.)

anticipatory: i.e., the inter-individual cues which regulate each mass flash must operate in the preceding cycle.

The elaterid "cucujo" beetle, *Pyrophorus*, of the West Indies has a pair of greenish luminescent organs dorsolaterally on the prothorax and an orange-yellow organ on the ventral surface of the first abdominal segment. The South American "railway worm," *Phrixothrix*, possesses a reddish luminescence on the head and greenish luminous spots segmentally arranged along the body. A North American species of the related genus, *Phengodes* (Fig. 24–3), in its photogenic organs resembles somewhat the South American species, except that it lacks the red head organ and glows continuously, unlike *Phrixothrix*, which glows only upon excitation.

Several species of myriapods secrete a luminous slime in much the same manner as the luminous earthworms. A luminous pycnogonid has been described.

The rock-boring clam *Pholas* has photogenic glands which secrete a protein luciferin.[87] Some nudibranchs have distributed over the body luminescent cells which flash when the animal is appropriately stimulated. Of the molluscs, however, the cephalopods as a group show the highest development of this capacity. A large number show luminescence. In many, such as *Loligo*, the luminous organs are structures which communicate with the outside, in which luminous bacteria are cultured. In others, probably the majority,

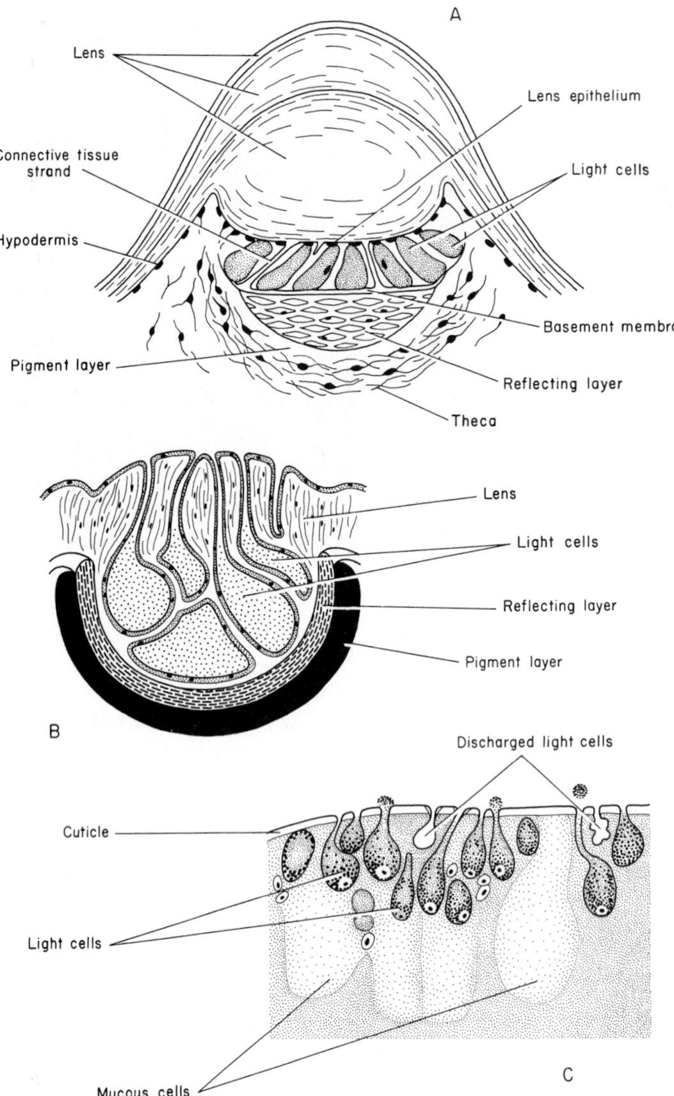

Figure 24–2. *A,* section through a light organ of the deep-sea shrimp *Sergestes. B,* section through light organ of the cephalopod, *Rondeletia. C,* section of a luminous area of *Chaetopterus* epidermis (after Dahlgren). (*A* and *C* from Harvey, E. N., The Nature of Animal Light. J. B. Lippincott Co.)

light is produced by cells of the animal itself (Fig. 24–2). In one squid, *Heteroteuthis,* there is an unpaired luminous organ which opens into the mantle cavity and expels a luminous cloud through the siphon when the animal is disturbed. Some species of squids, such as the firefly squid *Watsenia,* have complex patterns of luminous organs of the intracellular type over the body. *Watsenia* has three luminous organs at the tip of each of its arms. The organs contain small rod-like structures measuring about 2 μm by 4 μm which were once thought to be bacteria; they have now been shown not to be bacteria.[88]

A still higher state of differentiation is observed in such deep-sea forms as *Lycoteuthis diadema.* In the latter, as many as four different colors are produced by the various luminescent organs.

Only the Ophiuroidea among the echinoderms contain luminous species. A number of these have unicellular photogenic organs scattered over the body, and luminesce in response to any disturbance of the animal.

Light production among the chordates is restricted to the protochordates and fishes. A luminous slime is produced by some species of *Balanoglossus.* Among the tunicates, the luminescence of *Pyrosoma,* a colonial species, is the best known. This organism luminesces on stimulation, and a wave of photogenic activity spreads out over the whole colonial organism. Since activity may be induced in such a colony by using a light flash as a stimulating agent, light perhaps operates as a stimulus for the natural spread of activity in the colony from one luminescing individual to adjacent ones.

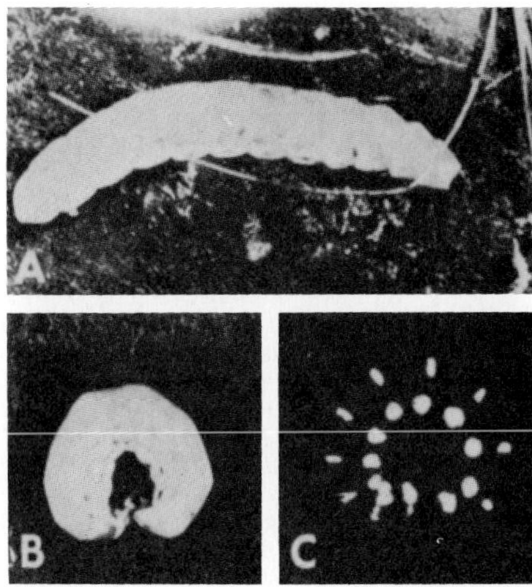

Figure 24-3. Female larva of the beetle *Phengodes*. *A*, dorsal view. *B*, lateral view. *C*, luminous specimen in darkness. (From Harvey, E. N., Living Light. Princeton University Press.)

The elasmobranch and especially the teleost fishes have numerous luminescent representatives, particularly among those inhabiting the depths of the seas. In some, like *Photoblepharon, Anomalops, Physiculus, Leiognathus,* and *Monocentris,* the light is due to the presence of symbiotic bacteria in special organs which are found in several different specific locations. In *Malacocephalus,* the luminescent bacteria are expelled from ventral sacs onto the ventral surface of the body when the fish is excited. *Astronesthes* and *Stomias* possess well differentiated headlight organs with lenses and pigmented cups. In some fishes, the appearance of light on stimulation is slow. In *Porichthys,* a latent period of 8 to 10 seconds is observed upon sympathetic neural stimulation at the level of the spinal cord.[37] Repetitive electrical excitation immediately adjacent to a photophore induces photogeny in it with a minimum latency of 1.1 sec. The possibility that an adrenaline-like compound is the transmitter is supported by the facts that latencies of less than 5 seconds are observed for local injection of noradrenaline and the adrenergic releasing agent, amphetamine, and that histochemical tests for catecholamines are positive in the neuro-photocyte layer of the photophore.[15]

Physical Aspects of Bioluminescence. Although heat is commonly associated with light, bioluminescence occurs without substantial heat production—it is a cold light. Physically we can distinguish two types of light emission: *incandescence,* in which the emission is a function of temperature and occurs due to thermal excitation of electrons, and *luminescence,* in which there is a *specific* non-thermal mode for the excitation of electrons. In bioluminescence, the energy required for this specific excitation is obtained from a chemical reaction.

In the case of incandescence—the fire or the light bulb—the color of the emission depends on temperature: at higher temperatures there is a greater proportion of higher energy photons. In luminescence (TV screen phosphor, fluorescent bulbs) the color is determined instead by the characteristic emitting molecule which specifically absorbs the energy and loses it via a specific electronic transition, when a photon is produced. The energy of the electronic transition determines the color. A variety of different colors of bioluminescence are produced by different organisms, all the way from blue through red (Fig. 24-4 and Table 24-1).

In some organisms, notably *Phrixothrix,* the South American railway worm, different organs on the same animal have different colors. It is possible that these color differences could be due to the presence of colored membranes acting as filters, but no such cases have been shown. Rather, the differences appear attributable to the bioluminescent reaction itself, and in particular to differences in the specific luciferase.[90, 91]

The exact color of bioluminescence in each specific case is presumed to have adaptive significance in connection with the specific function of the light emission. Blue light is commonly emitted by marine forms, matching well the color of light transmitted at

TABLE 24-1. Wavelength of maximum intensity of several different bioluminescent organisms. In all cases examined except the coelenterates, the color of the light in the *in vitro* reaction is identical or very similar to that emitted by the living organism. In the coelenterates there is an additional fluorescent compound *in vivo* which serves as the emitter.

	in vivo	*in vitro*
Cypridina (crustacean)	–	465
Gonyaulax (dinoflagellate)	470	470
Photobacterium, Achromobaite (bacteria)	480–490	490
Aequorea, Obelia, Renilla (coelenterates)	508	460–485
Omphalia (fungus)	525	520
Photinus and other species (fireflies)	550–580	550–580
Phrixothrix	red (~ 600)	

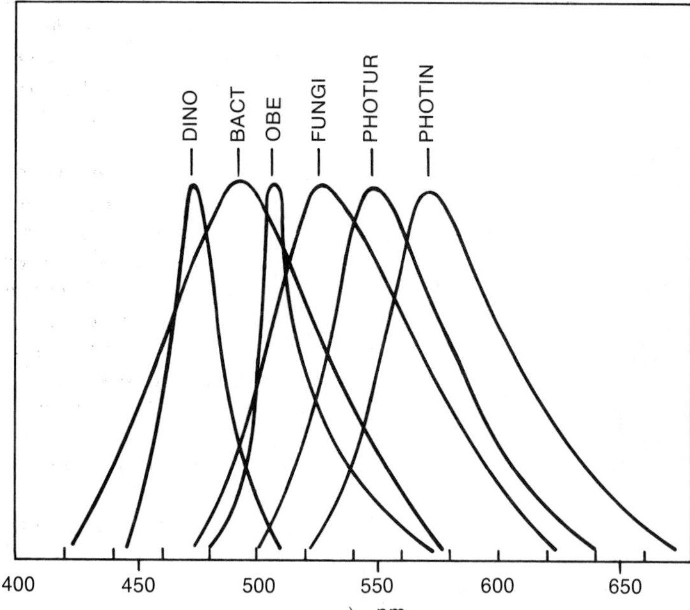

Figure 24–4. Emission spectra for the *in vivo* bioluminescence of a number of different organisms, illustrating the color differences and the band width differences. It also shows an example of the rather large difference in color between two species of fireflies.
DINO = *Gonyaulax polyedra*
BACT = *Photobacterium fischeri*
OBE = *Obelia geniculata*
FUNGI = *Omphalia flavida*
PHOTUR = *Photuris pennsylvanica*
PHOTIN = *Photinus scintillans*

depths of 30 m or more, and matching as well the visual sensitivity of many marine fish.[19,27,28,107] The intensity of light produced by photogenic organs is relatively low, but readily visible to the dark-adapted eye. The intensity can best be expressed in terms of photons (quanta), which relate directly to the number of excited molecular species (see below). A single isolated light organ from the midshipman fish (*Porichthys*) emits at an intensity of about 10^9 q sec^{-1}, lasting for almost one second; there result 10^9 quanta altogether in a flash.[2] The midshipman fish has hundreds of button-like (hence the name) organs distributed along its ventral and lateral surfaces. A firefly organ is much brighter; between 10^{14} and 10^{15} photons are emitted in a single flash of a firefly, which lasts perhaps half a second, while a single dinoflagellate (*Gonyaulax* or *Noctiluca*) emits between 10^8 and 10^{10} photons in a flash of similar duration. Bright strains of bacteria may emit as many as 10^4 quanta per second; integrated over 10^4 sec ($\sim$3 hours), this amounts to a substantial production for a cell only a few microns in size.

The efficiency of bioluminescence may be expressed in several ways. The energetic efficiency is calculated as the percentage of the energy theoretically available from the chemical reaction which actually appears as light. Values quoted for this are no more than estimates, but they are generally high—between 10% and 50%.[92] Experimentally, it is more straightforward to measure the chemiluminescent quantum yield, i.e., the number of photons produced for each molecule of substrate oxidized. The value for the reaction in the firefly system is about 0.86 ±; that is, nearly 100% of the luciferin molecules that react result in photons. In other bioluminescent systems, high values have been reported, such as 0.27 in *Aequorea* and 0.3 in the bacteria. Finally, the efficiency may be described in terms of the detector sensitivity, that is, how well the spectrum of the emitted light matches the sensitivity of the detecting element. This is referred to as luminous efficiency. For an eye with the spectral sensitivity of the human, the light of the firefly *Photuris* is about 92% efficient; for the blue light of *Cypridina,* the luminous efficiency is much less, about 20%. We must bear in mind, however, that the value of concern in this regard is the ecologically relevant system, and while good comparative data have not yet been summarized on this question, it is to be expected that the color of bioluminescent light will be found to match closely the visual sensitivity of the organisms which interact with the luminous species.

The Chemistry of Light Production. Bioluminescence is an enzymatically catalyzed chemiluminescence. In chemiluminescence, the reaction releases energy; but instead of being lost as heat or coupled to some synthetic reaction, the energy is used for the specific excitation of a molecule capable of releasing the energy as a photon.

The energy of the photon is not fixed, but depends upon the color (i.e., the frequency or wavelength) of the light. The energy, E, is given by the fundamental equation $E = h\nu$, where h is Planck's constant and ν is the frequency. For bioluminescence, where the light is in the frequency range corresponding to wavelengths between 450 and 600 nm (Fig. 24–4), the energy involved ranges from about 65 to 45 kcal per mole of photons (an einstein). This is a large amount of energy. By comparison, "high energy" ATP hydrolysis yields about 7 kcal per mole, almost an order of magnitude less. Knowledge of the biochemistry and enzymology of such reactions may be important to our understanding of physiological processes, especially in connection with energy metabolism.

It was shown almost a century ago by the French chemist Dubois[31] that the chemical substance responsible for light emission will function in the test tube. One must simply preserve the compounds involved, for example, by rapid freezing or drying. In his studies of the luminous clam, *Pholas*, Dubois showed that the reaction involves the oxidation by molecular oxygen of a substrate (for which he coined the term luciferin, i.e., light bearing) catalyzed by an enzyme, luciferase. In the equation the first product is designated as L*, signifying an electronically excited state

$$\text{Luciferin} + O_2 \xrightarrow{\text{luciferase (E)}} \text{E–L*}$$

$$\text{E–L*} \longrightarrow \text{E} + \text{L} + h\nu$$

which falls to the ground state with the emission of a photon.

The lifetime of such excited states is known to be very short, between 10^{-9} and 10^{-8} sec. On the assumption that the excited state is produced enzymatically, it is reasonable to assume that it emits before it diffuses away. Thus, the excited state is designated as an enzyme-bound excited species, E–L*, L being the product of luciferin oxidation.

Luciferin and luciferase are now used as generic terms to refer to the substrate and enzyme, respectively, involved in *any* bioluminescent reaction. Although different specific molecular species are involved in each different class or group of organisms, the general reaction mechanisms have certain close similarities, and can thus be represented by the generalized equations above.

The biochemistry of the reaction in several different organisms is now known in some detail. In the crustacean *Cypridina,* where enzyme and substrate are expelled together into the sea water, the luciferin structure was deduced in 1966.[61–64] The structure of the active compound (Fig. 24–5) comprises tryptamine, arginine, and L-isoleucine moieties. Synthetic luciferin was judged identical with natural luciferin by paper chromatography, specific absorption, and measurements of luminescent rates. *Cypridina* luciferase has been purified and partially characterized. It has a molecular weight of about 50,000, and requires calcium in order to function.[108] In the bioluminescent reaction, it is known[103] that one molecule of oxygen is taken up and one mole of CO_2 is evolved for each luciferin molecule that reacts, with a chemiluminescence quantum yield of about 0.3.[59] It is hypothesized that oxygen reacts at the position indicated, and that (via several intermediate steps) CO_2 and the excited product are formed. The product molecule is unstable and suffers cleavage as shown, yielding α-methylbutyric acid and etioluciferin. Since the products are not recoverable by the organism, the question of reutilization and/or turnover does not arise.

The firefly reaction is intracellular, controlled by a nerve impulse. If a firefly light organ is ground up in water, the light from the soluble fraction will continue for several minutes. The light can be restored by the addition of adenosine triphosphate (ATP). It is now known that the firefly contains a luciferin (Fig. 24–6) which is "activated" by reaction with ATP to form luciferyl adenylate —a reaction like that which occurs between amino acids and ATP as the first step in protein synthesis. Like that in *Cypridina,* the reaction then involves the uptake of one molecule of oxygen and the production of one CO_2 molecule per luciferin molecule oxidized. Here the chemiluminescent quantum yield is remarkably high, 0.89. The intermediate steps proposed (Fig. 24–6)[29] are based on isotope experiments.

Firefly luciferase was crystallized in 1956,[36] and much is now known about its chemical structure. No metal or prosthetic groups appear to be obligatorily associated with the activity. Its molecular weight is in the vicinity of 100,000, with two (possibly

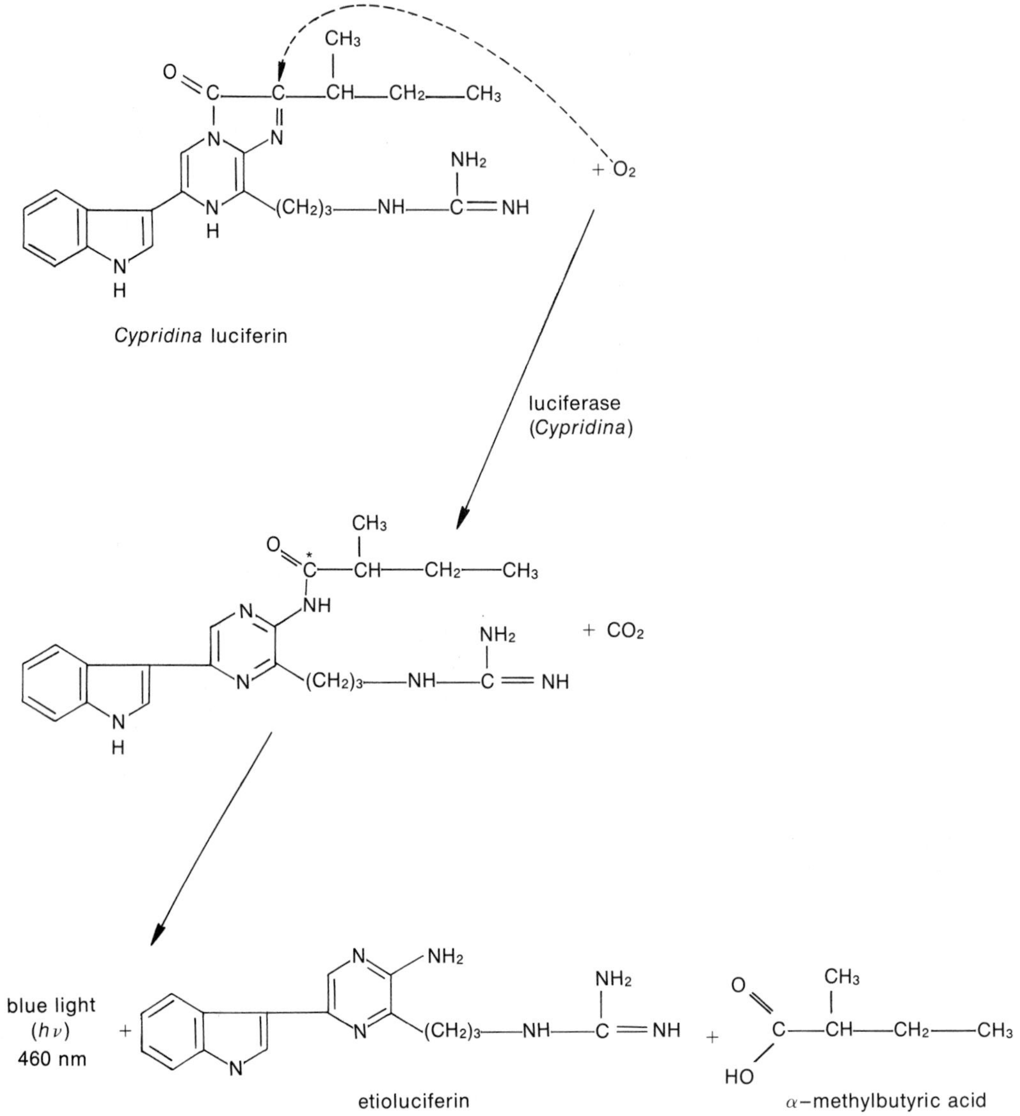

Figure 24–5. Postulated series of reactions in production of light by *Cypridina.*

different) subunits and one active center per molecule. As mentioned above, the color differences of firefly luminescence from different species have received considerable attention; enzyme structure or conformation, or both, are believed to have a substantial effect upon the emission spectrum of the luminescence.[90, 92] The possibility that the luciferin is responsible appears to have been eliminated by the use of synthetic luciferin and by the demonstration that luciferins from different fireflies are identical, even though the color of the luminescence may be different. However, exactly how luciferases differ in these cases is not yet known.

The molecular basis for the control of the *in vivo* firefly flash, which characteristically has a duration of about 0.5 sec, is not known. One of the necessary requirements for the "dark" condition is that reactants somehow be separated, compartmentalized, or otherwise prevented from reaction. Photogenic cells undoubtedly contain all chemical components required for flashing; knowledge of their cellular distributions in the intact cell would be of interest and importance in a solution of this problem.

The bioluminescent reaction in bacteria provides a third, distinctly different case, both in terms of the chemistry of the reaction and insofar as its cellular organization and control are concerned. The reaction is intracellular and continuous, biochemically derived from the oxidation of a reduced flavin cofactor, riboflavin 5′ phosphate, also referred to as flavin mononucleotide (FMN). Connected directly with cofactors involved in the oxidation of substrates, this system provides a pathway whereby electrons are shunted directly to oxygen, the energy being used for light production instead of the ATP synthesis which otherwise occurs during oxidative phosphorylation.

The precise chemistry of the reaction is not completely known, but it is believed to be a reaction analogous to a mixed function

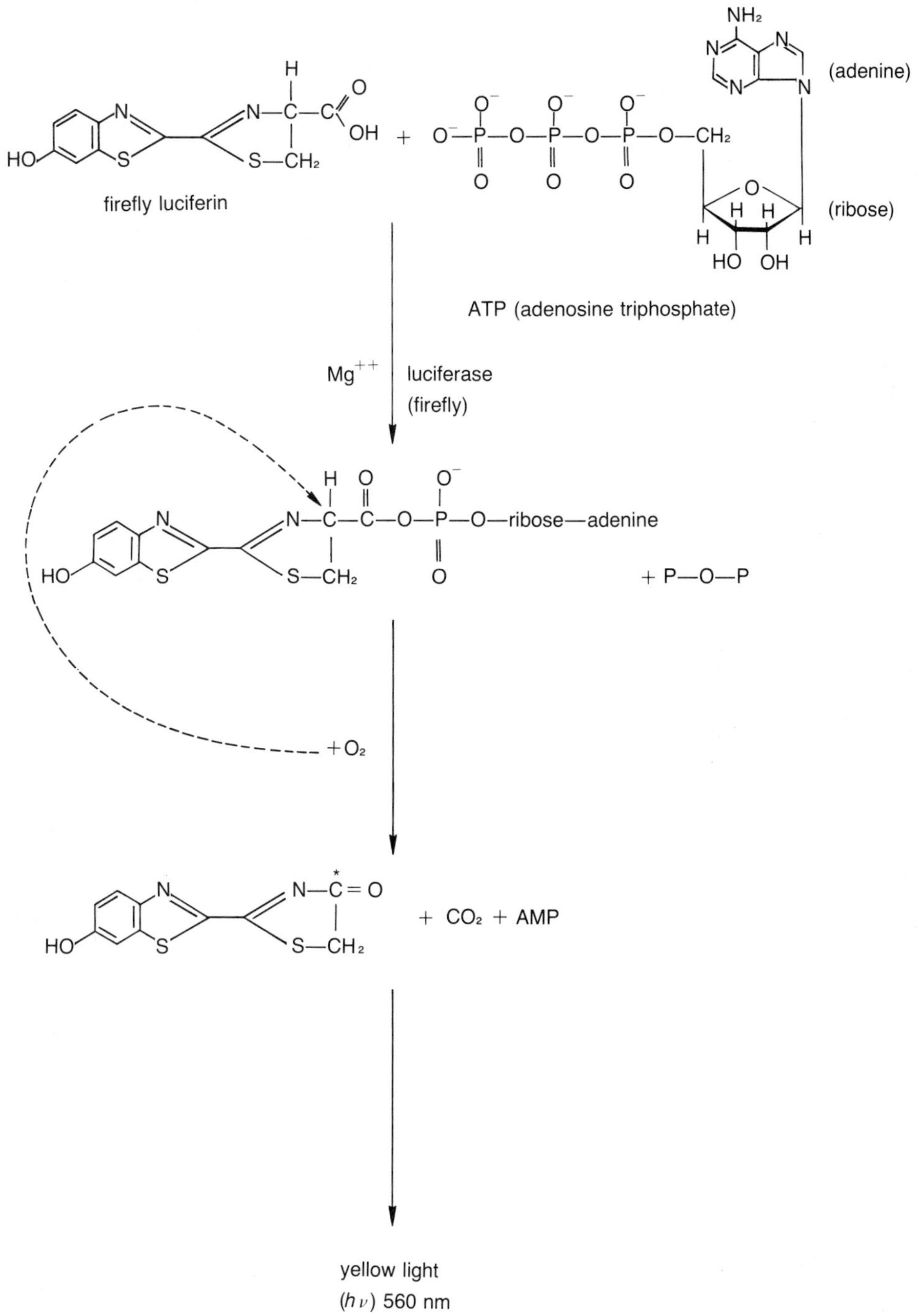

Figure 24–6. Postulated reaction of firefly luciferin with ATP and subsequent oxidation of intermediate.

oxidase, for a long chain aldehyde (> 6 carbons) is also required for light emission (see below). Again, the electronically excited state is designated as an enzyme-flavin complex (E–FMN*); in this case there is additional reason to so represent it, since FMN fluorescence has a green color, peaking at 530 nm, while bacterial bioluminescence is blue-shifted, at about 490 nm. Interaction with the protein, or with a transient enzyme-bound species, might be the cause for this shift.

Pure bacterial luciferase has been prepared in the crystalline form and well characterized. It is a simple protein (M.W. 79,000) comprised of two non-identical subunits: alpha, 42,000; beta, 37,000.[34, 56] The catalytic site has been shown to be located on the alpha subunit; the beta function is uncertain, but it may be important for solubilizing the system.[70, 71]

One of the more interesting physiological phenomena associated with the bacterial luminescent system is the fact that the control of luciferase synthesis appears to be rather independent of the synthesis of other cellular proteins. Thus, in a freshly inoculated culture in the laboratory no luciferase synthesis occurs during the first several hours; when a certain cell density has been reached, synthesis begins and proceeds at a rate several times faster than cell growth, and then abruptly stops.[77] This quasi-independent control of the luciferase gene may be related to the symbiotic involvement of bacteria, in which case their luminescence is developed within a specialized host photogenic organ.[49]

A second and even more extreme restriction of the luciferase gene occurs in spontaneous dark mutants. In these the gene activity is completely repressed; no luciferase synthesis occurs. The gene is not lost or irreversibly changed, however, since revertants can be readily isolated.[60] This represents a second type of control which might be of importance to an organism which may alternate between two states: the symbiotic, where luminescence is relevant and required; and the free living, where it is not.

The biochemistry of luminescence in several of the coelenterates has also been extensively studied. From the anthozoan *Renilla*, Cormier and his associates[20, 21] have isolated a substrate, *Renilla* luciferin, and enzymes, demonstrating two steps which they represent as follows:

$$\mathrm{DPA} + \mathrm{LH_2\text{-}x} \xrightarrow[\mathrm{Enz\ 1}]{\mathrm{Ca^{++}}} \mathrm{LH_2} + \mathrm{PAPS}$$

$$\mathrm{LH_2} + \mathrm{O_2} \xrightarrow[\mathrm{Enz\ 2}]{} \text{light}$$

In the first step, the activation of luciferin (LH_2) involves the removal of a group (x), identified as sulfate, stimulated by the cofactor 3′,5′ diphosphoadenosine (DPA), thereby forming activated luciferin and PAPS (phosphoadenosylphosphorylsulfate). Activated luciferin appears to be nearly identical in structure to *Cypridina* luciferin, and its oxidation—which also may be similar to that of *Cypridina* chemically—yields light.

From other coelenterates, first the hydrozoans *Aequorea* and *Halistaurea*[93] and later still other coelenterates,[51, 75] an unusual type of protein was isolated—one which emits light when it is simply mixed with calcium ions.

$$\text{Photoprotein} \xrightarrow{\mathrm{Ca^{++}}} \text{light}$$

This *in vitro* emission is rapid,[52] occurring as a flash lasting about 1 sec at 20°C, and the quantum yield is high. Aside from the protein and calcium, no other factors, including oxygen, are required. Hastings and Morin[51] postulated a unitary scheme for coelenterate bioluminescence, in which the photoprotein was postulated to be a relatively stable enzyme intermediate of a pathway similar to that described in *Renilla*. The photoprotein is past the stage of oxygen entry, requiring calcium to activate a terminal intramolecular energy-yielding reaction. Physiologically, this is of special interest, for calcium is mobilized in other systems, also as a consequence of an action potential, to control specific reactions—notably muscle contraction. The fact that a biochemical intermediate that is triggered by calcium can be isolated is very suggestive.

Moreover, the bioluminescent system in coelenterates appears in most instances studied[76] to be localized in cellular granules or organelles, in which the intermediates are

$$\mathrm{NADH^+\ (or\ NADPH)} + \mathrm{FMN} + \mathrm{H^+} \xrightarrow{\text{flavin reducatase}} \mathrm{NAD} + \mathrm{FMNH_2}$$

$$\underset{\text{(reduced flavin mononucleotide)}}{\mathrm{FMNH_2}} + \underset{\text{(aldehyde)}}{\mathrm{RCHO}} + \mathrm{O_2} \xrightarrow{\text{luciferase}} \underset{\text{(riboflavin 5′ phosphate)}}{\mathrm{E\text{–}FMN^*}} + \mathrm{RCOOH} + \mathrm{H_2O}$$

presumably synthesized and stored, and whose membranes presumably function in the flashing phenomena.

The studies of coelenterates have led to still another unique observation, namely that of energy transfer.[76] In other luminescent organisms for which spectral data is available, the color of the light produced by the living organism is the same as that of the isolated biochemical reaction, corresponding to the transition in reaction (2) below:

(1) $A + B \longrightarrow C^* + D$
(2) $C^* \longrightarrow C + h\nu$ (λ_1, blue)
(3) $C^* + E \longrightarrow E^* + C$
(4) $E^* \longrightarrow E + h\nu$ (λ_2, green)

In several of the luminescent coelenterates, the color of the light emission of the living organism is green, whereas the emission of the isolated *in vitro* reaction is blue (Table 24–1). This indicates that there is energy transfer from the excited product molecule (C^*) to a second species (E), with emission from its excited state. It was further shown that the components involved are localized within some kind of subcellular particle, which can be isolated with the active components of the *in vivo* luminescent system intact.

Still another system which involves some type of particle or organelle is that of the dinoflagellates, unicellular marine forms which emit a brief (0.1 sec) flash upon stimulation. Hastings and associates[30, 55] have found that a large, rapidly sedimenting subcellular particle (sedimentation coefficient = ~ 15,000s) can be caused to emit a flash of light (which closely resembles the *in vivo* flash) by simply lowering the pH from 8 to 5.7. It is hypothesized that this particle exists in a similar form *in vivo,* and that it is responsible for the luminescent flash of the cell via some membrane-associated event which causes the permeability of the particle to protons to be transiently increased.[30] Both luciferin and luciferase may be isolated in the supernatant, and these are presumed to be components of the active scintillon.[33] A "discharged" scintillon can be caused to flash a second time if it is incubated with fresh substrate (luciferin) at pH 8, and then shifted to the lower pH.

The biochemistry of several other light emitting systems has been studied in at least a preliminary way.

Fungal bioluminescence has been demonstrated to involve at least two steps: (1) Reduction of dehydro- or oxyluciferin by $NADH_2$ ($NADPH_2$) with a soluble enzyme, and (2) light-emitting oxidation of the reduced product by O_2 with a particulate enzyme.[1]

A freshwater limpet from New Zealand, *Latia,* is the only luminous animal that has been discovered to date whose whole life cycle is spent in fresh water. Its luciferin has been isolated and purified and found to have the empirical formula $C_{15}H_{24}O_2$. *Latia* luciferase has a molecular weight of 173,000. The system also requires the presence of a cofactor, which has been called the "purple protein," with a molecular weight of 39,000. The quantum yields (i.e., the photons emitted/molecules reacted) at 8°C have been determined as 0.63 for luciferase and 0.0068 for luciferin.[95, 98, 99]

The luminous system of the annelid *Chaetopterus* involves a photoprotein which has been obtained in crystalline form and found to have a molecular weight of 184,000. For luminescence there are also required two cofactors (one nucleoprotein-like and the other lipid-like), a peroxide (H_2O_2 or an organic one), O_2, and highly specifically ferrous iron.[94, 97]

The luminescent system of the euphausid, *Meganyctiphanes,* appears to be a two-component one involving a photoprotein and a still unidentified fluorescent substance.[96, 100]

Among the fishes which have been investigated, the luminescence appears to be dependent upon symbiotic bacteria in the genera *Acropoma, Cleidopus, Gazza, Leiognathus, Paratrachychthys* and *Siphamia,* and possibly *Photoblepharon* and *Anomalops.*[40] No bacteria have been found in the luminous organs of *Apogon, Parapriacanthus,* and *Pempheris.* The cross-reactions of the luciferin-luciferase systems of *Apogon* and *Parapriacanthus* with the *Cypridina* one suggest the possibility that the fish luciferin might be obtained from the feeding of the fish on *Cypridina.*[39] The luciferases, however, are dissimilar for *Apogon* and *Cypridina,* as is evidenced from differences in reaction kinetics and immunological tests; even though they are alike superficially, they hence have endogenous origins in each group.[106]

A comparative study of the structure of luminous organs has disclosed that in some organs, such as those of *Pholas* and *Latia,* euphausids, and some fishes, there are two types of glandular cells. In other organisms, such as the polynoid worms that possess only a single cell type, electron micrographic examination suggests that separation of reactants for the luminescent process may be accomplished by means of walls of microtubules.[3]

The Control of Bioluminescence. Among animals the production of light is typically not a continuous process. Light is usually produced intermittently in response to external stimuli.[9,86] This fact is of such generality that, when an animal appears to have continuous luminescence, it can usually be assumed that this is the result of the presence of symbiotic, pathogenic, or transient luminescent bacteria within or upon that animal; a possible exception is *Phengodes.*

The means of control of light production in animals are fundamentally of three types. The first two are observed in extracellular luminescence. In those instances where luminescent materials are expelled from a photogenic sac into the surrounding sea water on stimulation (e.g., the cephalopod *Heteroteuthis,* the ostracod *Cypridina,* and the shrimp *Acanthephyra*) the control is indirect and operates by means of typical neuromuscular mechanisms. On the other hand, a slightly different situation obtains in those organisms which can secrete a luminescent slime over the surface of the body. Such animals include the luminescent earthworms, *Chaetopterus,* myriapods, some coelenterates, and the clam *Pholas.* In these there is a control of a secretory process by direct nervous or, possibly also in some cases, endocrine excitation. The third general type of control applies to those numerous animals where the luminescence is an intracellular phenomenon. This is obviously the situation in such well known animals as *Noctiluca,* jellyfishes, ctenophores, insects, and certain fishes. Here we have to do with some mechanism whereby excitation of the luminescent cell results in the contained photogenic substances reacting to produce light. This may come about (a) through rapidly making small quantities of one of the essential organic photogenic agents available for reaction within the cell, (b) through temporarily rendering the intracellular medium favorable for the luminescent reactions such as through control of oxygen, water, or hydrogen ions, or (c) through altering the direction of paths of chemical change to favor chemiluminescent ones.

We can at present do little more than speculate on the means by which the animal can, in response to stimuli, give very brief and intense flashes of light which vie in rate of development and decline of intensity with the best of incandescent lamps. In multicellular organisms the flashing is typically associated reflexly with tactile or photic receptors. The reflex pathways may involve the nervous system alone, or both nerves and endocrines. In the coelenterate *Renilla,* the spread of luminescence over a stimulated animal proceeds in all directions at a rate characteristic of nerve-net transmission. Excessive stimulation can give rise to a state of excitation causing luminescence to persist for some time. In the ctenophore *Mnemiopsis,* the tactile receptors involved in reflex light production lie along the conducting pathways underlying the rows of ciliated combs. In the response the neuroeffector mechanism displays the phenomenon of facilitation quite like that seen in neuromuscular junctions.

Most studies of the control of luminescence have concerned fireflies.[13,14,16,17] Among other stimuli, visual cues normally excite a responsive flash in these insects. The photogenic organ (Figs. 24–7 and 24–8) is innervated, but the nerve supply goes to the tracheal endings within the light organ. The nerves appear to innervate some mechanism located in the tracheal end-cell.[41,42,102] This system is adrenergic.[101]

Two general types of theories for the control of flashing in the firefly have been advanced. The first type presumes that the flash depends on a rapid admission of oxygen to anoxic photogenic cells. This is considered to be brought about either by direct nervous control of admission of O_2 in the tracheal end-cells or by means of a stimulated increase in metabolites in the photogenic tissue and a consequent osmotic withdrawal of water from the terminal portion of the tracheal tubules, with the result that oxygen would be brought directly to the glandular elements. The presence of oxygen would then permit an oxidation of the metabolites, again reducing the osmotic pressure of the cell contents, and permit restoration of water to the tubules and re-exclusion of the oxygen. In support of an oxygen-control mechanism is the fact that microscopic observations of a luminescing gland show the brightest light to come from the immediate vicinity of the ends of the tracheal tubules. An end-cell valve mechanism is not essential, however, since some insects, especially larvae of fireflies, which show at least some degree of intermittency in light production, even though not true flashing, do not have differentiated tracheal end-cells. And, of course, rapid flashing can occur in many organisms having no tracheal system.

The second general type of postulated

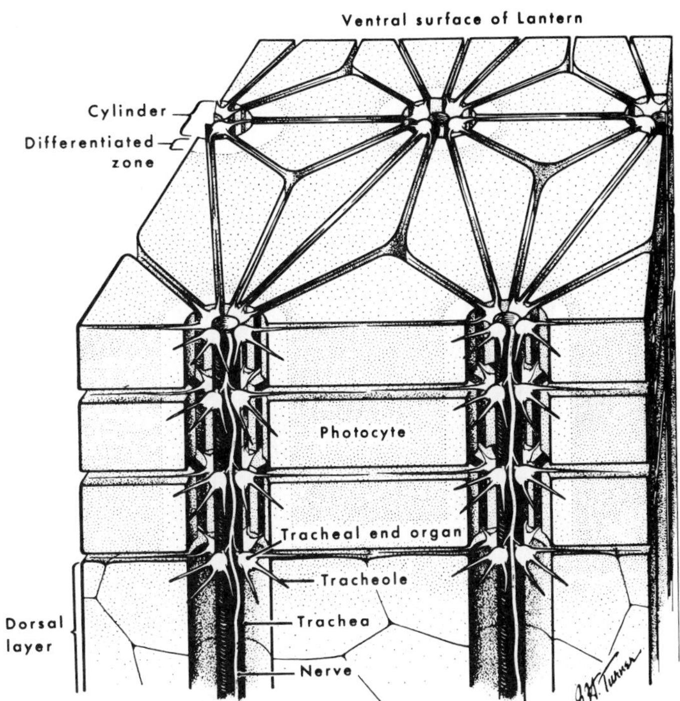

Figure 24–7. A three-dimensional diagrammatic reconstruction of the firefly light organ, showing the cylinders with nerves and tracheal elements and the pattern of arrangement of 12 photocytes as a rosette around each cylinder. From John B. Buck, "Unit Activity in Firefly Lanterns," in Frank H. Johnson & Yata Haneda, Bioluminescence in Progress (copyright © 1966 by Princeton University Press) Fig. 1, p. 462. Reprinted by Permission of Princeton University Press.

mechanism is a direct nervous excitation of the light-producing cells through photogenic nerves (Fig. 24–9). Such a postulated direct nervous excitation has potentially a much wider application, including control in animals without tracheal systems.

In the fishes *Porichthys* and *Echiostoma*, there is evidence that the normal reflex pathways of control of flashing in the system of photogenic organs include sympathetic nervous innervation. A slow response is also seen in response to epinephrine injection.

There have been a number of observations indicating that bright light inhibits the production of light by animals, although in most cases we know little or nothing of the mechanism of this inhibition, whether directly upon the photogenic reaction system or on mechanisms concerned normally with reflex excitation. It may result, as some evidence indicates for ctenophores, from a direct destruction of photogenic material within the light-producing cells. In these animals, extracts made from specimens exposed to sunlight show practically no capacity to luminesce. Furthermore, after exposure to bright light, animals such as ctenophores and the cnidarian *Renilla* have to be kept in darkness often for several minutes before they will again luminesce.

There have been a number of reported observations of a persistent daily rhythmicity in the capacity of certain species of animals to luminesce. Such a rhythm has been described as persisting for several days in constant darkness in the firefly *Photinus*,[5] in the balanoglossid *Ptychodera*,[22] in the dinoflagellate *Gonyaulax*,[54,104] and probably also in the jellyfish *Pelagia*.[72]

The luminescent responses of ctenophores and hydromedusae have been shown to be readily fatigable and to exhibit recovery in much the same fashion as other typical sensory-neuroeffector mechanisms.

Functional Significance of Bioluminescence. The significance of light production in the lives of many of the organisms possessing this capacity is far from obvious. That some survival value is commonly associated with this function is, however, likely, inasmuch as most higher organisms which luminesce have evolved specialized light cells or organs for the purpose. Moreover, the flashing itself may become incorporated into the general response mechanism of the organism. A striking instance of this is the aggressive mimicking by females of some species of *Photuris* of the female codes of some species of *Photinus*. The males thereby attracted are eaten.[65, 67]

Three general types of functions have been ascribed to animal luminescence. One of these is the luring of food.[65] A second function ascribed to the organs is that they serve as protective devices. They may operate through warning or frightening predators; by

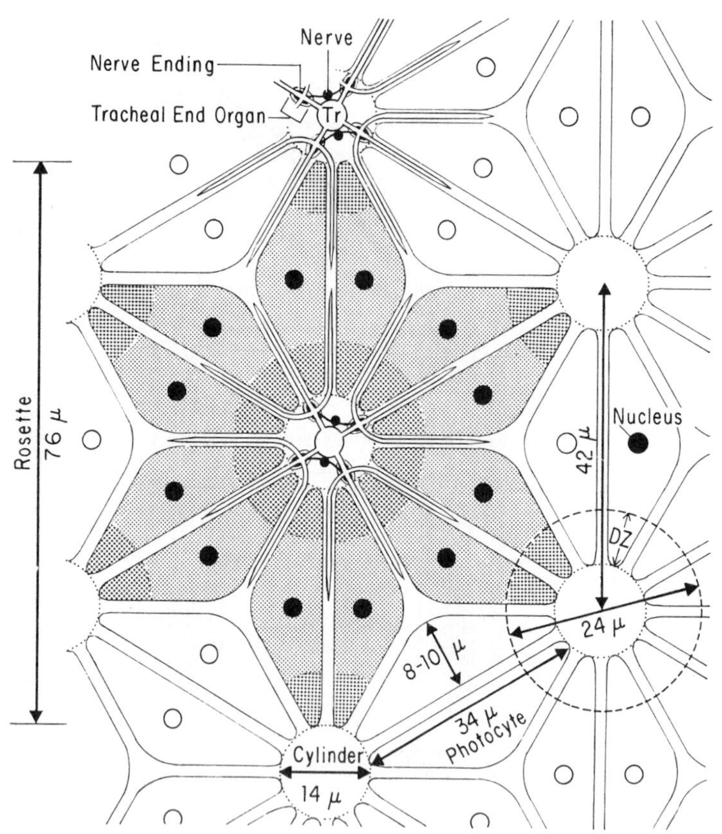

Figure 24–8. A diagrammatic tangential section of the light organ of *Photuris*, illustrating its histological organization. The average dimensions are indicated in the figure. The cylinders include tracheae and end organs, and from the cylinders radiate the photophores extending from one cylinder to another. (From Hanson, F. E., J. Miller, and G. T. Reynolds, Biol. Bull. *137*:447–464, 1969.)

concealing the light-emitting animals, as in the ejection of a luminous cloud by the deep-sea shrimp *Acanthephyra* and the squid *Heteroteuthis*; or through distracting predators. The pattern of nervous control of luminescence in such an annelid as the scale worm *Acholoe* appears to suggest this last role. When the worm is transected, as by an attacking predator, the posterior portion alone luminesces brightly, while luminescence is inhibited in the anterior portion. This behavior conceivably results in a greater chance for the more viable anterior part to escape. Autotomized scales of this worm also flash rhythmically. Suggestive also of such an emergency protective role of luminescence is the fact that adrenin supplements the nervous system in inducing bright and extensive luminescence in certain fishes. An interesting variation of the protective function of bioluminescence in organisms has been advanced, namely, that stimulation to luminescence of one individual in a group might serve as a warning to its fellows and permit their escape. An interesting role of bioluminescence has been described for the pony fish. The underside of this fish, luminescing with a color resembling sky light, would tend to camouflage the fish with respect to deeper lying predators.[19, 48]

The function of luminescent organs having best observational and experimental support up to the present is that of serving as signals for bringing together the two sexes in mating. The fireworm of Bermuda, *Odontosyllis*, provides a well-established illustration of this role.[35] The females, which exhibit a marked lunar periodicity, appear during mating periods at the surface of the sea where they swim in small circles, luminescing brilliantly. Males from deeper water swim directly toward the luminescing females and join in the mating "dance"; both sexes then liberate their gametes into the water together. If, perchance, no male joins a luminescing female in a short time, her light gradually fades, but after a brief period of rest her photogenic organs again become active and the luminescent "dance" is repeated. Males moving directly toward a luminescing female but failing to reach her before the end of a luminous period have been observed to cease their directed movement and wander aimlessly until the female again luminesces.

Other well-established cases of the use of light production as mating signals have been described for various species of fireflies.[6] In many species the females, which may even be wingless, remain in the grass while the males fly about. The males fly toward females which signal in response to their flashes. The various species differ in characteristics of their flashing, such as in the frequency, total number of flashes, color, intensity, and duration of each flash. The female responds in a characteristic manner to flashing of the male of the same species. The signal of males of *Photinus greeni* is two flashes 1.1 to 1.7 seconds apart at 27°C. Frequency of occurrence of the pairs, or lengths of the individual flashes, are not important. The female response occurs 0.8 to 1.1 seconds after the second flash of the male.[12] The male continues to signal to the female until the two sexes have met. The attracting light response of the female can be imitated by use of a flashlight which is operated to flash with the response delay of a typical responding female for a species. Numerous other suggestive examples of a role of luminescence in the mating reactions of animals have been described, but most of them require more convincing descriptions, or experimental study, before they can be accepted as such.

However, inasmuch as it is difficult to imagine any functional significance of bioluminescence in bacteria or fungi, we probably can assume that bioluminescence has arisen as a fortuitous correlate of the cellular oxidative mechanism, persisting in many animals, especially lower ones, despite no obvious survival value.[46] Secondarily, the phenomenon became utilized in special manners in higher animals, as potentially useful roles of bioluminescence came into existence.

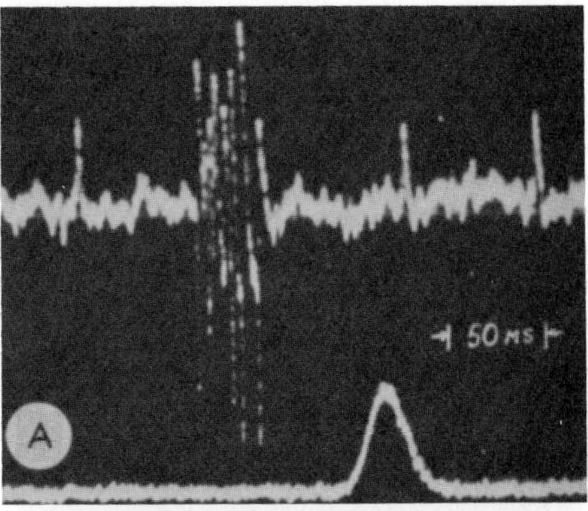

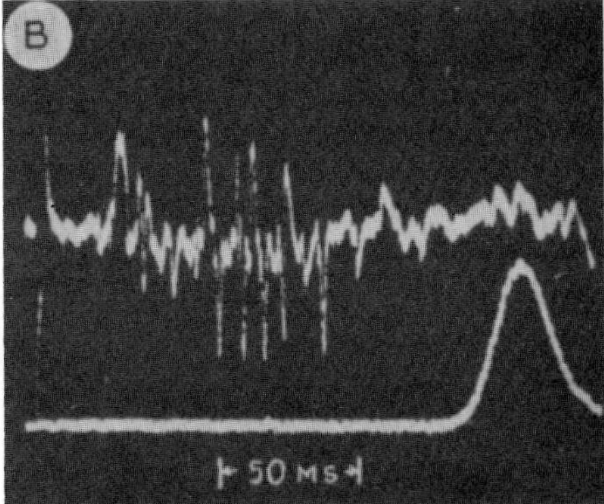

Figure 24-9. Spontaneous and electrically driven flashes in the light organ of *Photuris pennsylvanica* together with associated neural action potentials recorded from the ventral surface of the light organ. *A*, spontaneous. *B*, electrical stimulus to brain (stimulus artifact at extreme left) followed by neural volley and flash. (Courtesy of Case, J., and Buck, J. B., Biol. Bull. *125*: 234–250, 1963.)

REFERENCES

1. Airth, R. L., G. E. Foerster, and P. Q. Behrens, pp. 203–223 *in* Bioluminescence in Progress, edited by F. H. Johnson and Y. Haneda. Princeton University Press, 1966. Fungal bioluminescent system.
2. Baguet, F., and J. Case, Biol. Bull. *140*:15–27, 1971. *Porichthys* photophore excitation.
3. Bassot, J.-M., pp. 557–610 *in* Bioluminescence in Progress, edited by F. H. Johnson and Y. Haneda. Princeton University Press, 1966. Fine structure of luminous organs.
4. Biggley, W. H., E. Swift, R. J. Buchanan, and H. H. Seliger, J. Gen. Physiol. *54*:96–122, 1969. Bioluminescence in marine dinoflagellate.
5. Buck, J. B., Physiol. Zool. *10*:45–58, 1937. Diurnal rhythm in *Photinus* luminescent mechanism.
6. Buck, J. B., Physiol. Zool. *10*:412–419, 1937. Luminescence in firefly mating.
7. Buck, J. B., Quart. Rev. Biol. *13*:301–314, 1938. Sychronous flashing of fireflies.
8. Buck, J. B., Ann. N. Y. Acad. Sci. *49*:397–482, 1947. Luminescence in fireflies.
9. Buck, J. B., pp. 323–332 *in* The Luminescence of Biological Systems, edited by F. H. Johnson. Amer. Assoc. Adv. Sci., Washington, D.C., 1955. Control of bioluminescence.
10. Buck, J. B., pp. 459–474 *in* Bioluminescence in Progress, edited by F. H. Johnson and Y. Haneda. Princeton University Press, 1966. The firefly lantern.
11. Buck, J. B., and E. Buck, Science *159*:1319–1327, 1968. Mechanism of rhythmic synchronous firefly flashing.
12. Buck, J. B., and E. Buck, Biol. Bull. *142*:195–205, 1972. Bioluminescent signals in courtship of *Photuris greeni.*
13. Buck, J. B., and J. Case, Biol. Bull. *121*:234–256, 1961. Control of firefly flashing.
14. Buck, J. B., J. Case, and F. E. Hansen, Jr., Biol. Bull. *125*:251–269, 1963. Control of firefly flashing.
15. Case, J., F. Baguet, and J. Burns, Amer. Zool. *10*:504, 1970. Photophore control in *Porichthys*.
16. Case, J., and J. B. Buck, Biol. Bull. *125*:234–250, 1963. Electrical activity in firefly flash.
17. Case, J., and M. S. Trinkle, Biol. Bull. *135*:476–485, 1968. Light inhibition of *Photuris* flashing.
18. Chang, J. J., J. Cell. Comp. Physiol. *44*:365–394, 1954. Luminescence of *Mnemiopsis*.
19. Clarke, W. D., Nature *198*:1244–1246, 1963. Functional significance of bioluminescence.
20. Cormier, M. J., K. Hori, and P. Kreiss, pp. 349–362 *in* Bioluminescence in Progress, edited by F. H. Johnson and Y. Haneda. Princeton University Press, 1966. Light production in sea pansy, *Renilla.*
21. Cormier, M. J., Y. D. Karkhanes, and K. Hori, Biochem. Biophys. Res. Comm. *38*:962–964, 1970. Bioluminescent system of *Renilla.*
22. Crozier, W. J., Anat. Rec. *20*:186–187, 1920. Diurnal rhythm in *Ptychodera* luminescence.
23. Davenport, D., and J. A. C. Nicol, Proc. Roy. Soc. Lond. B *144*:399–411, 1955. Hydromedusoid luminescence.
24. Davenport, D., and J. A. C. Nicol, Proc. Roy. Soc. Lond. B *144*:480–496, 1955. Luminescence of sea pens.
25. Davenport, D., O. M. Wootton, and J. E. Cushing, Biol. Bull. *102*:100–110, 1952. Luminous millipede.

26. Dennell, R., J. Linn. Soc. *42*:393–406, 1955. Luminescence in Crustacea, Decapoda.
27. Denton, E., Phil. Trans. Roy. Soc. Lond. B *258*:285–313, 1970. Nature of reflectors in bioluminescence.
28. Denton, E., Sci. Amer. *224*:64–72, 1971. Biological significance of color of luminescence.
29. De Luca, M., and M. E. Dempsey, Biochem. Biophys. Res. Comm. *40*: 117, 1970. Firefly bioluminescent reaction.
30. DeSa, R., and J. W. Hastings, J. Gen. Physiol. *51*:105–122, 1968. Luminescent system in *Gonyaulax*.
31. Dubois, R., C. R. Soc. Biol. *39*:564–566, 1887. Luciferin and luciferase in *Pyrophorus* and *Pholas*.
32. Eckert, R., pp. 269–300 *in* Bioluminescence in Progress, edited by F. H. Johnson and Y. Haneda. Princeton University Press, 1966. Electrophysiology of luminescence in *Noctiluca*.
33. Fogel, M., and J. W. Hastings, Arch. Biochem. Biophys. *142*:310–321, 1971. *Gonyaulax* luminescent system.
34. Friedland, J. M., and J. W. Hastings, Proc. Nat. Acad. Sci. *58*:2336, 1967. Bacterial luciferase.
35. Galloway, T. W., and P. S. Welch, Trans. Amer. Micr. Soc. *30*:13–39, 1911. Luminescence of *Odontosyllis* during mating.
36. Green, A. A., and W. D. McElroy, Biochem. Biophys. Acta *20*:170–176, 1956. Crystal luciferase.
37. Greene, C. W., and H. H. Greene, Amer. J. Physiol. *70*:500–507, 1924. Luminescence in *Porichthys*; action of adrenaline.
38. Haneda, Y., pp. 335–385 *in* The Luminescence of Biological Systems, edited by F. H. Johnson. Amer. Assoc. Adv. Sci., Washington, D.C., 1955. Luminous organisms of Far East.
39. Haneda, Y., F. H. Johnson, and O. Shimomura, pp. 533–545 *in* Bioluminescence in Progress, edited by F. H. Johnson and Y. Haneda. Princeton University Press, 1966. Origin of luciferin in fishes.
40. Haneda, Y., and F. I. Tsugi, Science *173*:143–145, 1971. Light production in the luminous fishes *Photoblepharon* and *Anomalops*.
41. Hanson, F. E., J. Insect Physiol. *8*:105–111, 1962. Control of firefly flashing.
42. Hanson, F. E., J. Miller, and G. T. Reynolds, Biol. Bull. *137*:447–464, 1969. Subunit coordination in fireflies.
43. Harvey, E. N., The Nature of Animal Light. J. B. Lippincott, Philadelphia, 1920, 182 pp.
44. Harvey, E. N., Living Light. Princeton University Press, 1940, 328 pp.
45. Harvey, E. N., Bioluminescence. Academic Press, New York, 1952, 649 pp.
46. Harvey, E. N., Fed. Proc. *12*:597–606, 1953. Evolution of bioluminescence.
47. Harvey, E. N., and J. J. Chang, Science *119*:581, 1954. Luminous response of *Mnemiopsis*.
48. Hastings, J. W., Science *173*:1016–1017, 1971. Bioluminescence as camouflage in pony fish.
49. Hastings, J. W., and G. Mitchell, Biol. Bull. *141*:261–268, 1971. Photogenic organ for symbiotic bacteria.
50. Hastings, J. W., and J. G. Morin, Biol. Bull. *135*:422, 1968. Calcium activated bioluminescent proteins of *Mnemiopsis* and *Obelia*.
51. Hastings, J. W., and J. G. Morin, Biochem. Biophys. Res. Comm. *37*: 493–498, 1969. *Renilla* bioluminescence.
52. Hastings, J. W., G. Mitchell, P. H. Mattingly, J. R. Blinks, and M. Van Leeuwen, Nature *222*:1047–1050, 1969. Aequorin and calcium response.
53. Hastings, J. W., and B. M. Sweeney, J. Cell. Comp. Physiol. *49*:209–226, 1957. Diurnal rhythms in light reactants.
54. Hastings, J. W., and B. M. Sweeney, Biol. Bull. *115*:440–458, 1958. Diurnal rhythm of *Gonyaulax* light.
55. Hastings, J. W., M. Vergin, and R. DeSa, pp. 301–329 *in* Bioluminescence in Progress, edited by F. H. Johnson and Y. Haneda. Princeton University Press, 1966. Particulate bioluminescent system in *Gonyaulax*: "scintillon."
56. Hastings, J. W., K. Weber, J. Friedland, A. Eberhard, G. W. Mitchell, and A. Gunsalus, Biochemistry *8*:4681, 1969.
57. Heymans, C., and A. R. Moore, J. Gen. Physiol. *6*:273–280, 1923. Diurnal rhythm in *Pelagia* luminescence.
58. Heymans, C., and A. R. Moore, J. Gen. Physiol. *7*:345–348, 1925. Inhibition of ctenophore luminescence by light.
59. Johnson, F. H., O. Shimomura, Y. Saiga, L. Gershman, G. T. Reynolds, and J. R. Waters, J. Cell. Comp. Physiol. *60*:85–104, 1962. Quantum efficiency in *Cypridina*.
60. Keynan, A., C. Veeder, and J. W. Hastings, Biol. Bull. *125*:382, 1963. Studies on the survival of dark and bright mutants of luminescent bacteria in seawater.
61. Kishi, Y., T. Goto, Y. Hirata, O. Shimomura, and F. H. Johnson, pp. 83–113 *in* Bioluminescence in Progress, edited by F. H. Johnson and Y. Haneda. Princeton University Press, 1966. *Cypridina* luciferin.
62. Kishi, Y., T. Goto, Y. Hirata, O. Shimomura, and F. H. Johnson, Tetrahedron Letters *29*:3427, 1966. *Cypridina* luciferin.
63. Kishi, Y., T. Goto, S. Inoue, S. Sugiura, and H. Kishimoto, Tetrahedron Letters *29*:3445, 1966. *Cypridina* luciferin.
64. Kishi, Y., T. Goto, S. Eguchi, Y. Hirata, E. Watanabe, and T. Aoyama, Tetrahedron Letters *29*:3434, 1966. *Cypridina* luciferin.
65. Lloyd, J. E., Science *149*:653–654, 1965. Aggressive mimicry in *Photuris*.
66. Lloyd, J. E., Misc. Publ. University of Michigan, No. 130, 1966, 93 pp. Studies on the flash communication system in *Photinus* fireflies.
67. Lloyd, J. E., Ann. Rev. Entomol. *16*:97–122, 1971. Bioluminescent communication in insects.
68. McElroy, W. D., pp. 463–505 *in* The Physiology of Insecta, Vol. 1, edited by M. Rockstein. Academic Press, New York, 1964. General review.
69. McElroy, W. D., M. De Luca, and J. Travis, Science *157*:150, 1967. Conformational changes in bioluminescent enzymes.
70. Meighen, E. A., M. Z. Nicoli, and J. W. Hastings, Biochemistry *10*:4062–4068, 1971. Bacterial luciferase.
71. Meighen, E. A., M. Z. Nicoli, and J. W. Hastings, Biochemistry *10*: 4069–4073, 1971. Chemically modified bacterial luciferase.
72. Moore, A. R., J. Gen. Physiol. *9*:375–381, 1926. Diurnal rhythms in *Pelagia* luminescence.
73. Morin, J. G., and I. M. Cooke, J. Exp. Biol. *54*:707–721, 1971. Electrical activity in *Obelia* bioluminescence.
74. Morin, J. G., and I. M. Cooke, J. Exp. Biol. *54*:723–735, 1971. Bioluminescent system of *Obelia*.
75. Morin, J. G., and J. W. Hastings, J. Cell. Physiol. *77*:305–311, 1971. Coelenterate and ctenophore photoprotein and Ca^{++}.
76. Morin, J. G., and J. W. Hastings, J. Cell. Physiol. *77*:313–338, 1971. Coelenterate and ctenophore photoprotein and calcium.
77. Nealson, K., T. Platt, and J. W. Hastings, J. Bacteriol. *104*:313–322, 1970. Induction and control of bacterial luciferase synthesis.
78. Nicol, J. A. C., J. Marine Biol. Assoc. U. K. *30*:417–431, 1952. *Chaetopterus* light glands.
79. Nicol, J. A. C., J. Marine Biol. Assoc. U. K. *30*:433–452, 1952. Nerve control of light production in *Chaetopterus*.
80. Nicol, J. A. C., J. Marine Biol. Assoc. U. K. *31*:113–144, 1952. Nerve control of light production in *Chaetopterus*.
81. Nicol, J. A. C., J. Marine Biol. Assoc. U. K. *32*:65–84, 1953. Luminescence in polynoid worms.
82. Nicol, J. A. C., J. Marine Biol. Assoc. U. K. *33*:173–175, 1954. *Chaetopterus* luminescence.
83. Nicol, J. A. C., J. Marine Biol. Assoc. U. K. *33*:225–255, 1954. Light control in polynoid worms.
84. Nicol, J. A. C., J. Exp. Biol. *32*:299–320, 1955. Luminescence in *Renilla*.
85. Nicol, J. A. C., J. Exp. Biol. *32*:619–635, 1955. Nerve control of luminescence in *Renilla*.
86. Nicol, J. A. C., pp. 299–319 *in* The Luminescence of Biological Systems, edited by F. H. Johnson. Amer. Assoc. Adv. Sci., Washington, D.C., 1955. Control of luminescence.
87. Henry, J. P., and A. M. Michelson, Biochim. Biophys. Acta *205*:451–458, 1970. *Pholas* luciferin, a protein.
88. Okada, Y. K., pp. 611–625 *in* Bioluminescence in Progress, edited by F. H. Johnson and Y. Haneda. Princeton University Press, 1966. Fine structure of *Watsenia* luminous systems.
89. Parker, G. H., J. Exp. Zool. *31*:475–513, 1920. Spread of luminescence in *Renilla*.
90. Seliger, H. H., and W. D. McElroy, Proc. Nat. Acad. Sci. *52*:75–81, 1964. Enzyme structure and conformation in bioluminescence.
91. Seliger, H. H., and W. D. McElroy, J. Marine Res. *26*:244–255, 1968. Diurnal rhythms in bioluminescence.
92. Seliger, H. H., and R. A. Morton, pp. 253–314 *in* Photophysiology, Vol. 4, edited by A. C. Giese. Academic Press, New York, 1968. Specificity of luciferase; efficiency of photosynthesis.
93. Shimomura, O., F. H. Johnson, and Y. Saiga, Science *140*:1339–1340, 1963. Luminescence in *Aequorea*.
94. Shimomura, O., and F. H. Johnson, pp. 495–522 *in* Bioluminescence in Progress, edited by F. H. Johnson and Y. Haneda. Princeton University Press, 1966. Luminous system of *Chaetopterus*.
95. Shimomura, O., F. H. Johnson, and Y. Haneda, pp. 391–404 *in* Bioluminescence in Progress, edited by F. H. Johnson and Y. Haneda. Princeton University Press, 1966. Isolation of *Latia* luciferin.
96. Shimomura, O., and F. H. Johnson, Biochemistry *6*:2293–2306, 1967. Bioluminescent system of euphausid.
97. Shimomura, O., and F. H. Johnson, Science *159*:1239–1240, 1968. Isolation of *Chaetopterus* photoprotein.
98. Shimomura, O., and F. H. Johnson, Biochemistry *7*:2574–2580, 1968. Luciferase and protein cofactor in *Latia*.
99. Shimomura, O., and F. H. Johnson, Biochemistry *7*:1734–1738, 1968. Luciferin of *Latia*.
100. Shimomura, O., and F. H. Johnson, Proc. Nat. Acad. Sci. *59*:475–477, 1968. Photoprotein system in euphausid.
101. Smalley, K. N., Comp. Biochem. Physiol. *16*:467–477, 1965. Adrenergic transmission in *Photinus* light organ.
102. Smith, D. S., J. Cell Biol. *16*:323–359, 1963. Fine structure of firefly luminescent organ.
103. Stone, H., Biochem. Biophys. Res. Comm. *31*:386, 1968. *Cypridina* luminescent reaction.
104. Sweeney, B. M., and J. W. Hastings, J. Cell. Comp. Physiol. *49*:115–128, 1957. Diurnal rhythm in *Gonyaulax* light.
105. Swift, E., and W. R. Taylor, J. Physiol. *3*:77–81, 1967. Luminescence in *Pyrocystis*.
106. Tsuji, F. I., and Y. Haneda, pp. 137–149 *in* Bioluminescence in Progress, edited by F. H. Johnson and Y. Haneda. Princeton University Press, 1966. Luciferases of *Cypridina* and *Apogon*.
107. Tyler, J. E., Bull. Scripps Inst. Oceanogr. *7*:363–411, 1960. Color in bioluminescence and depth in ocean.
108. Lynch, R. V., F. I. Tdaji, and D. H. Donald, Biochem. Biophys. Res. Comm. *46*:1544–1550, 1972. Calcium requirement for *cypridina* bioluminescence.

INDEX

INDEX